W9-AZY-050

ENVIRONMENT AND HEALTH

HEALTH NOTES

Biology

The Web of Life

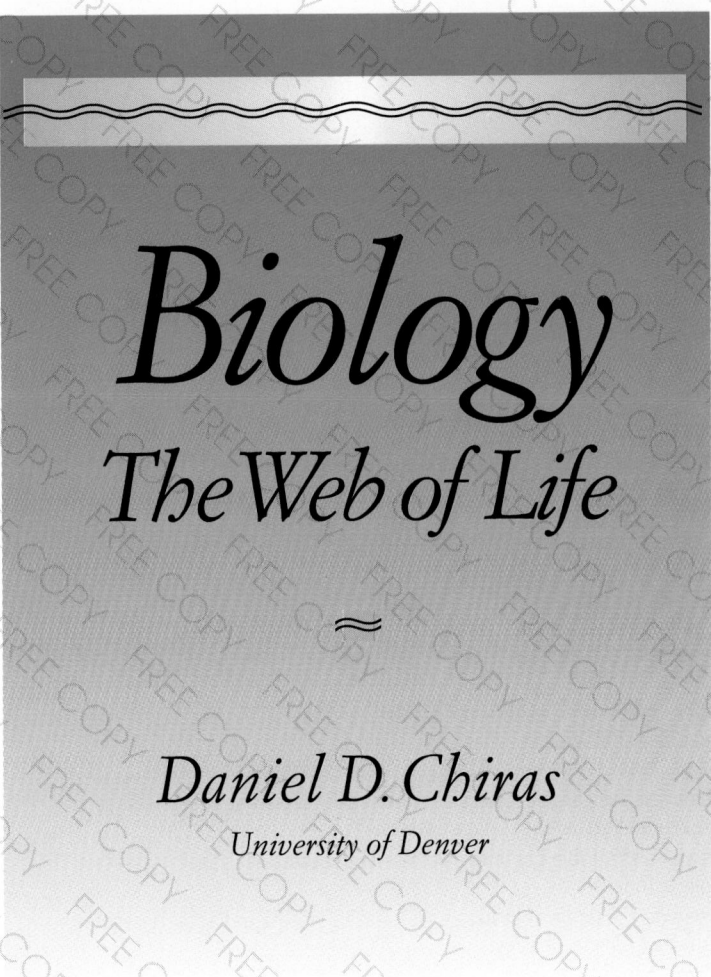

Biology
The Web of Life

≈

Daniel D. Chiras
University of Denver

CONTRIBUTORS

ANIMAL BIOLOGY AND BEHAVIOR
David Armstrong
University of Colorado at Boulder

PLANT BIOLOGY
Mathew Nadakavukaren
Illinois State University

ANIMAL BEHAVIOR
Stephen Freedman
Loyola University of Chicago

West Publishing Company
Minneapolis/St. Paul
New York
Los Angeles
San Francisco

To my wife Kathleen,
and my delightful sons, Skyler and Forrest,
with love and affection

Interior and cover design Diane Beasley
Copyediting Bill Waller
Artwork Carlyn Iverson, Elizabeth Morales-Denney, Publications Services, Pat Rossi, Cyndie Wooley, J/B Woolsey & Associates (Individual art credits follow Index)
Photo research John Cunningham/Visuals Unlimited (Photo credits follow Index)
Composition Graphic World
Page layout Diane Beasley
Indexing Sandi Schroeder
Cover image "Chipping Sparrow, Great Spruce Head Island, Maine." Copyright Amon Carter Museum, Eliot Porter Collection.
Production, prepress, printing and binding by West Publishing Company

00 99 98 97 96 95 94 93   8 7 6 5 4 3 2 1 0

Library of Congress Cataloging-in-Publication Data

Chiras, Daniel D.
    Biology, the web of life / Daniel Chiras.
        p.   cm.
    Includes bibliographical references (p.   ) and index.
    ISBN 0-314-01251-6 (hardcover).   —ISBN 0-314-01343-1 (paperback).
    —ISBN 0-314-01344-X (Volume I).   —ISBN 0-314-01345-8 (Volume II).
    —ISBN 0-314-01346-6 (Volume III).
    1. Biology.   I. Title.
QH308.2.C45   1993
574—dc20   92-34101
                                                    CIP

West's Commitment to the Environment

In 1906, West Publishing Company began to recycle gold shavings, sheepskin scraps, and paper trimmings left over from the production of books. This began a tradition of efficient and responsible use of resources that has continued longer than most printers and publishers have been in existence.

Today, West Publishing's operation is environmentally conscious in almost every aspect of its operation. West was one of the first publishers to use recycled paper in books. Today, up to 95% of our legal books and 65 to 70% of our college texts are printed on recycled, acid-free stock. In addition, West recycles virtually all scrap paper left over from production. In the production plant, over 22 million pounds of scrap paper is recycled annually—the equivalent of 182,000 trees.

Being a responsible publisher and printer goes beyond paper. Since the 1960s, West has also recycled steel, sheet metal, and other metals. We have devised ways to capture and recycle waste inks, solvents, oils, and vapors created in the printing process and have eliminated the use of styrofoam book packaging. Today, we also recycle plastics of all kinds, wood, glass, corrugated cardboard and batteries.

Employees at West's home office and production plant are provided with containers for recycling aluminum, glass, and paper. Employees are encouraged not only to recycle, but also to suggest additional ways to use our resources more efficiently. It is employee initiative and involvement that has contributed to West's long-standing tradition of environmental responsibility.

When you order a West book, you get more than a token commitment to a clean environment. We thought you would appreciate knowing this.

About the Author

Daniel D. Chiras received his Ph.D. in reproductive biology in 1976 from the University of Kansas Medical School, where his research on ovarian physiology earned him the Latimer Award. In September 1976, Dr. Chiras joined the Biology Department at the University of Colorado in Denver in a teaching and research position. Since 1976, he has taught numerous undergraduate and graduate courses, including general biology, cell biology, histology, endocrinology, and reproductive biology. Dr. Chiras also has a strong interest in environmental issues and has taught a variety of courses on the environment. Currently an adjunct professor at the University of Denver and at the University of Colorado in Denver, he has also been a visiting professor at the University of Washington.

Dr. Chiras is the author of numerous technical publications on ovarian physiology, critical thinking, sustainability, and models for environmental education, which have appeared in the *American Biology Teacher, Biology of Reproduction,* the *American Journal of Anatomy,* and other journals. He has also written numerous articles for newspapers and magazines on environmental issues, and is the author of the environment section for World Book Encyclopedia's *Science Year 1993.* Dr. Chiras has also published five college and high school textbooks, including *Human Biology: Health, Homeostasis, and the Environment* and *Environmental Science: Action for a Sustainable Future.* He is the coauthor of *Natural Resource Conservation: An Ecological Approach* (with Oliver S. Owen). He currently serves as an editor of *Environmental Carcinogenesis and Ecotoxicology Reviews.*

Dr. Chiras has also written books for a general audience, including *Beyond the Fray: Reshaping America's Environmental Response*, and, most recently, *Lessons from Nature: Learning to Live Sustainably on the Earth.* Besides his active scientific pursuits, he is an avid kayaker, skier, bicyclist, and organic gardener.

Brief Contents

PART IV

Homeostasis, Integration, and Control: Organs and Organ Systems of Animals 225

PART V

The Continuation of Life: Reproduction and Development 537

PART VI

The Evolutionary Legacy: Unity and Diversity of Living Things 609

PART VII

Interactions: Behavior, Ecology, and the Environment 773

Contents

PART I

Principles of Biology: Evolution, Homeostasis, and the Chemistry of Life 1

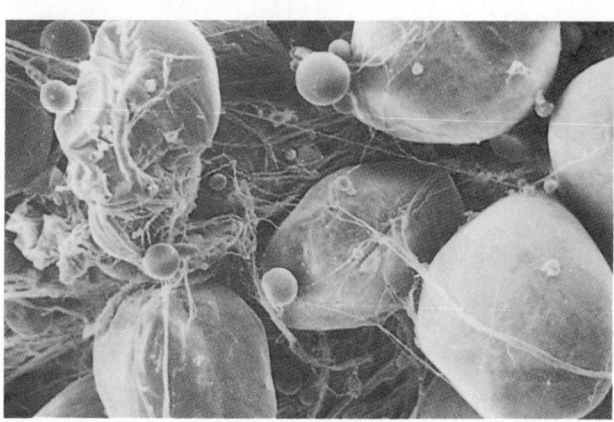

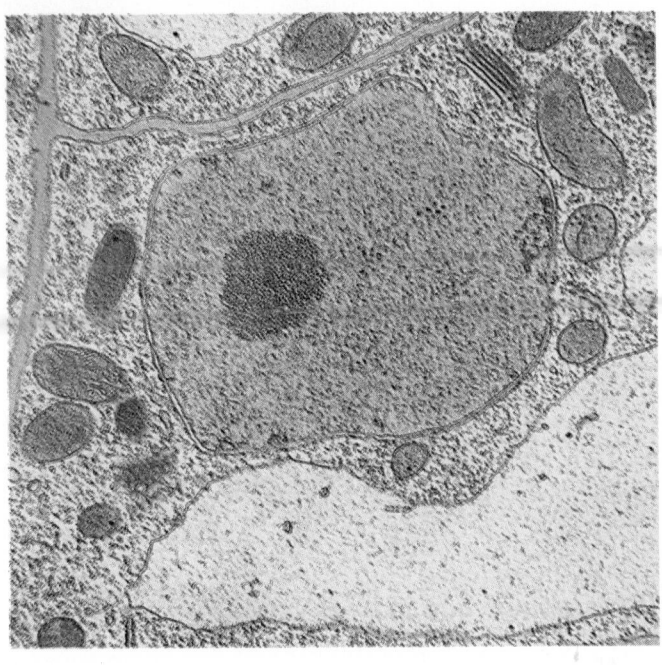

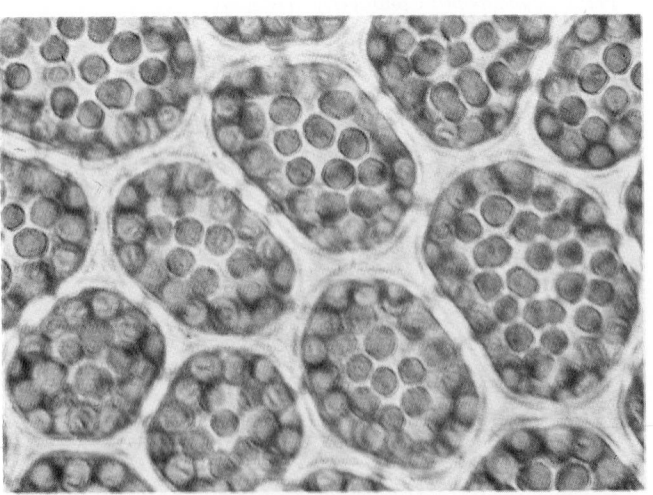

PART III

Chromosomes, Cell Division, and Heredity 137

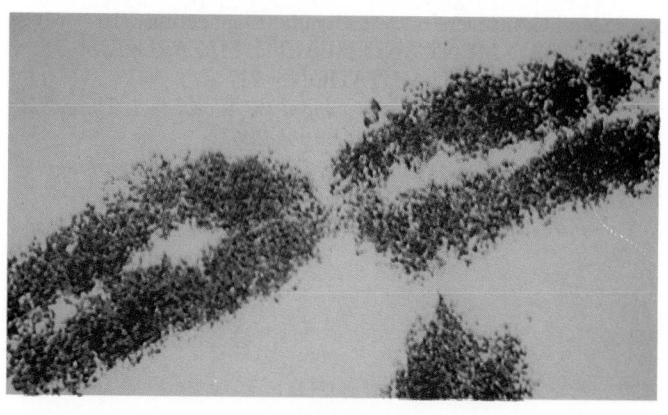

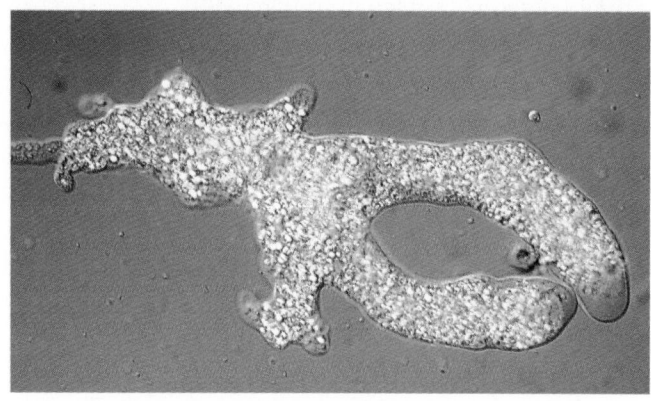

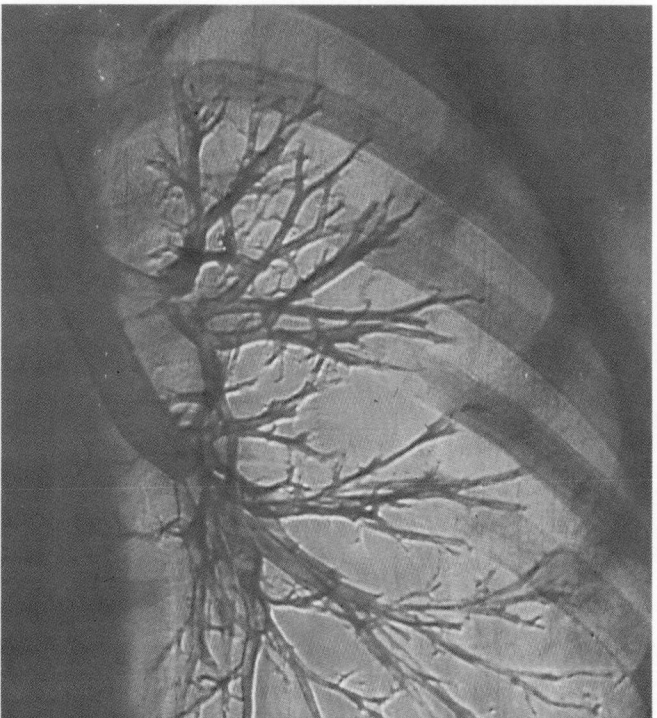

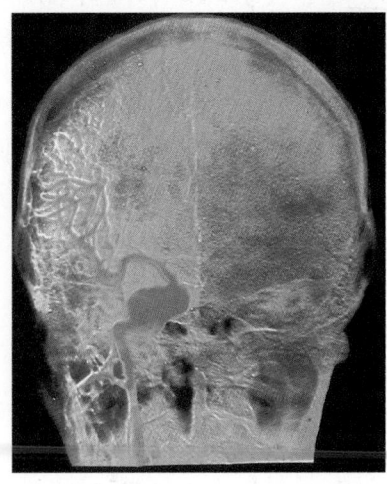

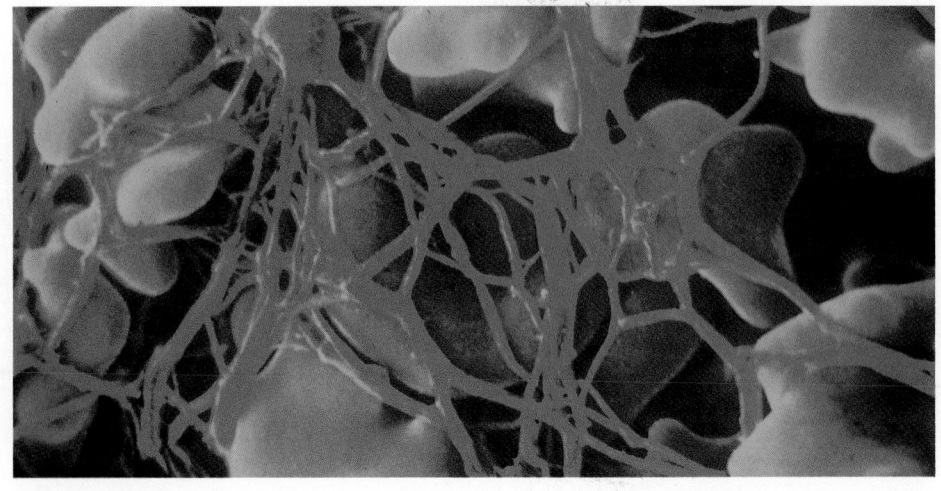

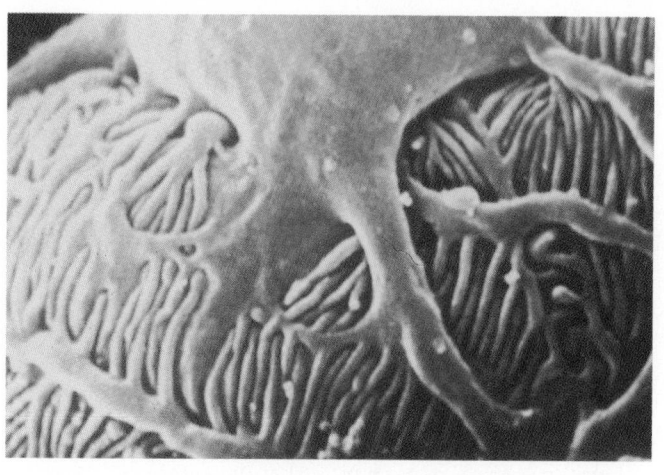

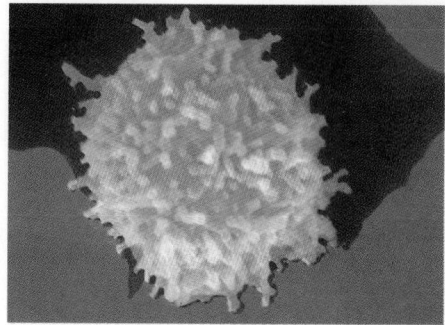

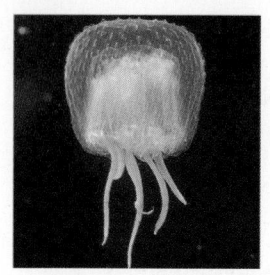

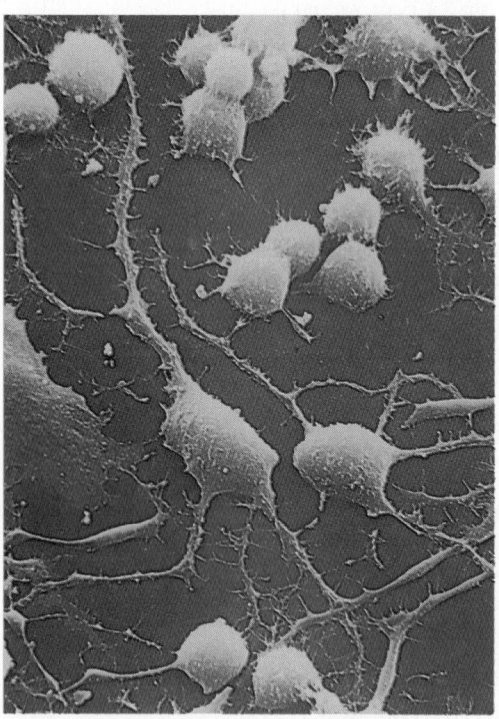

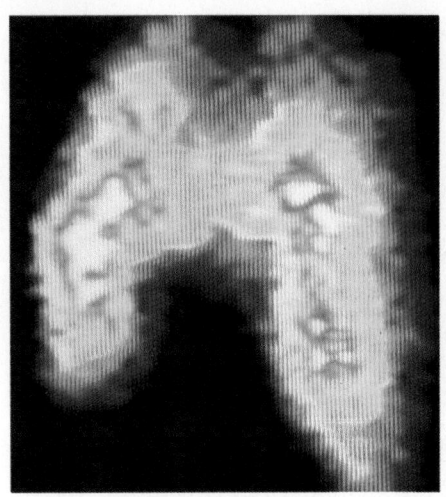

PART V

The Continuation of Life: Reproduction and Development 537

PART VI

The Evolutionary Legacy: Unity and Diversity of Living Things 609

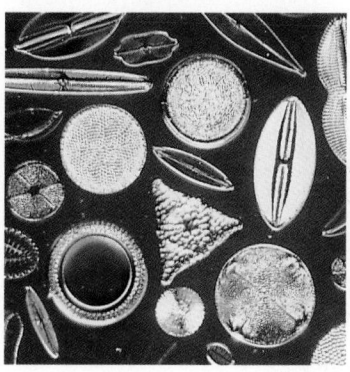

Preface

Biology: The Web of Life, written for the introductory course in biology, explores the living world, showing how organisms function and how they interact. It also highlights ways in which organisms evolved to meet different needs and offers an overview of the vast and diverse array of species on the Earth. It explains our part in the web of life and our dependence on other life forms.

ORGANIZATION

Biology: The Web of Life is divided into seven parts. Part I provides background information vital to students' understanding of biology. The two chapters in this part describe biological and chemical principles that apply to all living things and to our everyday lives. They also show how humans are both similar to and different from other organisms that share this planet with us. Part I closes with an important discussion of science and the scientific method and introduces critical thinking skills that are applied throughout the book.

Part II lays additional groundwork for understanding life as it journeys through the world of the cell. The four chapters in this part describe the structure and function of cells and show how they acquire energy.

Part III delves into the fascinating topic of heredity. It describes cell division, the structure of chromosomes, and the many ways in which cells control genes. Students will learn how gametes are made and acquire the basic principles of heredity.

Part IV outlines the structure and function of animals. The 11 chapters in this section provide an overview of the evolution of the major organ systems and show how vertebrate organ systems operate, relying principally on the human model. Homeostasis and evolution are emphasized in these chapters as unifying principles of biology.

Part V discusses reproduction and development. These two chapters outline the evolution, structure, and function of reproductive systems in animals and portray the dramatic events that lead to the formation of new individuals.

Part VI gives an overview of the diversity of life on Earth. It explains how life-forms evolved and how organ-

isms evolve to meet changing conditions of their environment. The six chapters in this part also survey the five major groups (kingdoms) of organisms.

Finally, Part VII focuses on interactions. It examines animal behavior, ecology, the study of ecosystems. The final chapter surveys the problems that modern society has created in the natural world and offers solutions for redirecting human society onto a sustainable course vital to protecting the web of life.

SPECIAL FEATURES

Biology was written with the student in mind. Presented in a friendly, clear style, this book describes basic principles of biology and offers many examples relevant to the student's life. Interesting analogies are interspersed throughout the text to help clarify the more difficult concepts. I have attempted to hold complex terminology down wherever possible and have provided pronunciations for the more tongue-twisting terms that students will encounter.

Study Skills

Immediately following the preface is a brief list of study skills—tips on how students can improve their memory, get more out of what they read, and, in general, study more efficiently. The study skills are designed to help in this and virtually every other course that students will take. Many of the tips will be useful later in life as well.

Scientific Discoveries That Changed the World

Special essays highlight important scientific discoveries, such as the discovery of the structure of DNA, and feature the work of some of the world's most important scientists. These essays also illustrate how scientific discoveries have changed our view of the world and help further students' understanding of the scientific method. Finally, these case studies illustrate the fact that scientific advances usually require the work of many scientists, sometimes working in seemingly unrelated areas.

Spotlight on Evolution

To emphasize and elaborate on points made in the text about evolution, I have included numerous essays entitled Spotlight on Evolution. These essays show how plants and animals have adapted to various evolutionary challenges, and they provide a contrast to human systems. They also help stimulate student interest in evolution and weave a thread of evolution through the book.

Health Notes

Health Notes are included in many chapters to present some of the more exciting developments in health and medicine—for example, new procedures or other discoveries that could revolutionize medical care. Health Notes will help students learn to reduce stress in their lives and avoid unhealthy foods. Some health notes show how the health of other species is affected by human activities.

Point/Counterpoints

Many discoveries in biology have profound impacts on our lives. However, the application of these discoveries often results in new and controversial techniques, such as genetic engineering. This book presents a number of modern-day controversies in Point/Counterpoint sections. Some address social and political issues that require a good biological background, and others focus on scientific debates.

Each Point/Counterpoint consists of two brief essays written by distinguished writers and thinkers. These lively essays present opposing views on such important issues as genetic engineering, fetal cell transplantation, cancer, and global warming. Point/Counterpoints also offer students a chance to practice critical thinking skills.

Critical Thinking Skills

As noted earlier, Chapter 1 presents students with a number of guidelines for improving their critical thinking skills. These will help students become more discerning thinkers, an ability that will prove useful in this and many other college courses—not to mention later in life.

Additional emphasis is placed on critical thinking throughout the text. Each chapter, for example, contains a section on Exercising Your Critical Thinking Skills, which calls on students to use their critical thinking skills. These exercises include case studies, hypothetical scenarios, or summaries of news or scientific reports. They ask students to analyze situations or reports and give alternative interpretations or find flaws in reasoning. Each exercise emphasizes one or two critical thinking rules presented in the first chapter. Critical thinking questions are also included after each Point/Counterpoint.

Environment and Health

The health of the Earth's organisms and the health of the environment in which they live are closely connected. To help illustrate these connections, each chapter ends with an Environment and Health section. These brief sections illustrate some of the ways in which the physical and chemical environments affect our health and the health of other species.

In-Text Summaries

Chapter section heads are written as summary statements that capture key concepts or facts presented in the material that follows. These in-text summaries provide students a way to review major concepts in preparation for exams.

End-of-Chapter Summaries

Unlike most other biology textbooks, this one contains a fairly detailed summary at the end of each chapter. These summaries cover key concepts and ideas. Students can use them for a quick review after reading the chapter and as they prepare for exams. The in-text summaries, detailed end-of-chapter summaries, and extensive questions (discussed below) provide an excellent study guide.

Summary Tables

To help students summarize key concepts, processes, and systems, I have included summary tables in many chapters. Students can use these tables to prepare for exams or to review material after reading the chapter.

Test of Terms

To help students review the key terminology in each chapter, a Test of Terms has been included at the end of each chapter. These tests contain fill-in-the-blank questions and can be used by students to assess their grasp of the main terms and concepts presented in the chapter. They may also prove useful when preparing for exams. Students can fill in the blanks immediately after reading the chapter or after they have spent some time studying the material.

Test of Concepts

Each chapter also contains a number of brief essay questions that enable students to assess their understanding of the material.

Art Program

This book contains a remarkable collection of drawings and photographs. These colorful illustrations supplement the text, helping make the more complex concepts and processes understandable.

≋ SUPPLEMENTARY ITEMS FOR INSTRUCTORS

To help you teach this course, an extensive ancillary package has been developed. It includes an instructor's guide, test bank (also available in a computerized test-generation program), study guide, transparency acetates, slides, laboratory manual, videotapes, and videodisk, among other items.

Instructor's Guide

An instructor's guide with a 2000-question test bank is available from West Publishing Company. For each chapter, the instructor's guide contains lecture outlines with tips and suggested enrichment topics, page-referenced key terms found in the chapter, food-for-thought questions that can be used to provide students with something to think about as they read the next assignment from the book, and a list of film and video sources that might prove valuable in lectures.

Computerized Testing

The test bank is also available in a form that allows tests to be computer generated. Contact your West sales representative for details.

Study Guide

A study guide with helpful review items is available to supplement the text's built-in study guide material. The study guide is also available in electronic form on disk for learning laboratory usage.

Transparency Acetates and Slides

A set of transparency masters of important diagrams and acetates of key full-color art pieces will be available from the publisher for adopters to use in classes. Many combine photos and art. A slide set with other important pieces of art and photographs from the text will also be available.

Laboratory Manual

A laboratory manual with 33 class-tested lab exercises for introductory biology classes is available.

Videotapes

West offers a video library of films from which adopters of the text may choose. Contact your local West sales representative for further information.

Videodisk

A videodisk is available from West with animations, some of which are art from the book, still images (photos and art), and film clips on relevant topics.

≋ ACKNOWLEDGEMENTS

A project of this magnitude is the fruit of a great many people. I wish to thank the thousands of scientists and teachers who have contributed to our understanding of the web of life. A special thanks to the extraordinary teachers who have made tremendous contributions to my education, especially the late Dr. Weldon Spross, the late Dr. H. T. Gier, Dr. Gilbert Greenwald, Dr. Floyd Foltz, and Dr. Douglas Poorman.

I am also deeply indebted to many people for their assistance during the writing of this book. A special thanks to my editors, Jerry Westby and Theresa O'Dell, for their exceptional patience, guidance, perserverance, and inspiration. Thanks to my outstanding production editor, Tom Hilt, for his calmness in the midst of turmoil, cordiality, attention to detail, and diligence. My appreciation goes to our copyeditor, Bill Waller, for a skillful job of editing. A word of thanks also to my colleague Dr. John Cunningham of Visuals Unlimited who supplied the excellent photographs. Thanks also go to my team of artists: Carlyn Iverson, Elizabeth Morales-Denney, Publications Services, Pat Rossi, Cyndie Wooley, and J/B Woolsey & Associates. Thanks also to Richard Shippee for his excellent review of the artwork. I greatly appreciate West's marketing team, Ann Hillstrom and Amelia Jacobson, and the many sales representatives who have helped make this book a success. It has been a pleasure and an honor to have worked with such a fine and talented group of people.

I would like to thank Drs. Dave Armstrong, Mathew Nadukavukaren, and Steve Freedman for their excellent chapters on the animal kingdom, behavior, and plants. Their timeliness, willingness to go the extra mile to revise their early drafts and help with the art program have been a godsend. Thanks also to the many authors who contributed the Point/Counterpoints in this book. Your work will make this a more exciting journey, and will help students see the practical applications of the study of biology.

Throughout this extraordinarily difficult time, my wife, Kathleen, and our two sons, Skyler and Forrest, have offered considerable support and a much-needed counterbalance to the stresses and strains of a project of this scope. A very special thanks to Kathleen for acquiring the Point/Counterpoints in this volume and making all of the copyediting changes on disk. Thanks to all of you for seeing me through this project and helping me along.

Finally, a special thanks to all the reviewers who offered many useful comments throughout this project. Their insights and attention to detail have been greatly appreciated. Following is a list of those who reviewed the manuscript.

Reviewers

D. DARYL ADAMS
Mankato State University

JAMES S. BACKER
St. Vincent College

WILLIAM E. BARSTOW
University of Georgia

LAWRENCE J. BELLIPANNI
University of Southern Mississippi

CLYDE E. BOTTRELL
Tarrant County Junior College

KATHLEEN BURT-UTLEY
University of New Orleans

ROY B. CLARKSON
West Virginia University

BARBARA CRANDALL-STOTLER
Southern Illinois University

KENNETH J. CURRY
University of Southern Mississippi

PETER DALBY
Clarion University

GARY E. DOLPH
Indiana University at Kokomo

JAMIN EISENBACH
Eastern Michigan University

SALLY K. FROST-MASON
University of Kansas

BERNARD L. FRYE
University of Texas-Arlington

ROBERTA GIBSON
Glendale Community College

MADELINE M. HALL
Cleveland State University

LASZLO HANZELY
Northern Illinois University

JOHN P. HARLEY
Eastern Kentucky University

ANN T. HARMER
Orange Coast College

JAMES G. HARRIS
Utah Valley Community College

WILLIAM R. HAWKINS
Mt. San Antonio College

RONALD K. HODGSON
Central Michigan University

ROBERT R. HOLLENBECK
Metropolitan State University

HARRY R. HOLLOWAY
University of North Dakota-Main Campus

GEORGE A. HUDOCK
Indiana University, Bloomington

ARTHUR B. JANTZ
Western Oklahoma State University

TOM KANTZ
California State-Sacramento

ARNOLD J. KARPOFF
University of Louisville

GEORGE KLEE
Kent State University

MARK LEVINTHAL
Purdue University

JAMES LUKEN
Northern Kentucky University

ANN S. LUMSDEN
Florida State University

KATHY MARTIN
Central Connecticut State University

PATRICIA MATTHEWS
Grand Valley State University

PRISCILLA MATTSON
University of Lowell

HEATHER McKEAN
Eastern Washington University

MICHAEL P. McKINLEY
Glendale Community College

JAN MERCER
Tarrant County Junior College

JAMES E. MICKLE
North Carolina State University

GLENDON R. MILLER
Wichita State University

STEVEN N. MURRAY
California State University—Fullerton

DEBRA K. PEARCE
Northern Kentucky University

CHRIS E. PETERSEN
College of Dupage

BARBARA Y. PLEASANTS
Iowa State University

DAVID J. PRIOR
Northern Arizona University

RODNEY A. ROGERS
Drake University

EDWIN DANIEL SCHREIBER
Mesa Community College

JAY TEMPLIN
Widener University

ROBERTA WILLIAMS
University of Nevada

STEPHEN WILLIAMS
Glendale Community College

LARRY D. WILSON
Miami Dade Community College

DANIEL D. CHIRAS
Evergreen, Colorado

Study Skills

College is a demanding time in the lives of many students. Term papers, tests, reading assignments, and classes require a new level of commitment and intellectual activity. The work load can become overwhelming and frustrating. Fortunately, there are ways to lighten the load, to manage your time efficiently, to increase your chances of getting good grades, and to improve your knowledge and understanding.

This section offers some suggestions to help you manage your studies and improve your mastery of the subjects you study. If you already are adept and efficient at studying and taking tests, you may still benefit from reading this section. Every suggestion you can use to your advantage will help.

To begin, read over the suggestions listed below. Pick a few that seem right for you under each category, then put them into action. Applying these ideas could pay huge dividends—not just in college, but throughout your entire life. Learning is a lifetime endeavor, and those that learn fastest seem to get the most out of life.

Mastering basic study skills will require some work at first and may require that you break some bad habits. In the long run, the additional time investment could save you lots of time, help you get better grades, and become a more efficient learner. Most important, it could help you improve your knowledge and understanding.

≈ GENERAL STUDY SKILLS

- Study in a quiet, well-lighted space.
- Work at a desk or table. Don't lie on a couch or bed.
- Establish a specific time each day to study and stick to your schedule.
- Study when you are most alert. Many people find they retain more if they study in the evening a few hours before bedtime.
- Let your friends and family know when you study and ask them to respect that time.
- Take frequent breaks—one every hour or so. Exercise or move around during your study breaks to help you stay alert.
- Reward yourself after a study session with an ice cream cone or a mental pat on the back.

- Study each subject every day to avoid cramming for tests. Some courses may require more hours than others, so adjust your schedule accordingly.
- Look up new terms or words whose meaning is unclear to you in the glossaries in your textbooks or in a dictionary.

≈ IMPROVING YOUR MEMORY

You can improve your memory by following the PMC method. The PMC method involves three simple learning steps: (1) paying attention, (2) making information memorable, and (3) correlating new information with facts you already know.

Step 1. Paying attention means taking an active role in your education—taking your mind out of neutral. Eliminate distractions when you study. Review what you already know and formulate questions about what you are going to learn *before* a lecture or *before* you read a chapter in the text. Reviewing and questioning help prime the mind.

Step 2. Making information memorable means finding ways to help you retain information in your memory. Repetition, mnemonics, and rhymes are three examples.

- Repetition can help you remember things. The more you hear or read something, the more likely you are to remember it. Scribble notes while you read or study.
- You can also use learning tools, mnemonics, to help remember lists. For example, *k*eep *p*iling *c*hocolate *o*n *f*or *g*oodness *s*akes helps you remember the taxonomic classification scheme: kingdom, phylum, class, order, family, genus, and species.
- Rhymes and sayings are also helpful. If you are having trouble remembering a list of facts, try making up a rhyme.
- If you're having trouble remembering key terms, look up their roots in the dictionary. This helps them stick in your memory. Use the list of prefixes, suffixes, and roots on the back endsheets of this book.
- Draw pictures and diagrams of processes.

Step 3. Correlating with things you know means tying facts together or making sense of the bits and pieces of information you are learning and have learned previously.

■ Instead of filling your mind with disjointed facts and figures, try to see how they relate to previous information you have learned. Stop and scan your memory for similar facts. Correlating facts with previous knowledge enables you to comprehend the big picture. The end-of-chapter questions will assist you in this function.

■ After studying your notes or reading a chapter in your textbook, determine the main points. How does the new information you have learned fit into your view of life or the general subject under discussion? How can you use the information?

≈ BECOMING A BETTER NOTE TAKER

■ Spend 5–10 minutes before each lecture reviewing the material you learned in the previous lecture. This is extremely important for learning.

■ Know the topic of each lecture *before* you enter the class. Spend a few minutes *before* each class reflecting on facts you already know about the subject that is to be discussed.

■ If possible, read the text *before* each lecture. If not, at least look over the main headings in the chapter, read the topic sentences, and look over the figures.

■ Develop a shorthand system of your own. Symbols such as = (equals), w/o (without), w (with), > (greater than), < (less than), ↑ (increase), and ↓ (decrease) can save you time.

■ Develop special abbreviations. For example, if you find yourself writing the word human over and over again in your notes, abbreviate it to H. Muscle could be abbreviated as m or mm. Species is sometimes abbreviated sp.

■ Omit vowels and abbreviate words to decrease writing time (for example: omt vwls shrten wrtng tme). This takes some practice.

■ Don't take down every word your professor says, but be sure your notes contain the main points, supporting information, and important terms.

■ Watch for signals from your professor indicating important material ("This is an extremely important point. . . ").

■ If possible, sit near the front of the class to avoid distractions.

■ Review your notes soon after lecture while they're still fresh in your mind. Be sure to leave room in your notes during class to add material you missed. Recopy your notes if you have the time.

■ Compare your notes with those of your classmates to be sure you understood everything and did not miss anything important.

■ Attend lecture regularly.

■ Use a tape recorder, if necessary and if it's acceptable to your professor, if you have trouble catching all the points.

■ If your professor talks too quickly, politely ask him or her to slow down.

■ If you are unclear about a point, ask during class. Chances are other students are confused as well. If you are too shy, go up after lecture and ask, or visit your professor during his or her office hours.

≈ HOW TO GET THE MOST OUT OF WHAT YOU READ

■ Before you read a chapter or other assigned readings, preview the material by reading the main headings or chapter outline to see how the material is organized.

■ Pause over each heading and ask a question or two about each main heading.

■ Next, read the first sentence of each paragraph. When you have finished, turn back to the beginning of the chapter and read it thoroughly.

■ Take notes in the margin or on a separate sheet of paper. Underline or highlight key points.

■ Don't skip terms that are confusing to you. Look them up in the glossary in the back of your textbook or in a dictionary. Make sure you understand each term before you move on.

■ Use the study aids in your textbook, including end-of-chapter questions and summaries. Don't just look over the questions and say, "Yeah, I know that." Write out the answer to each question as if you were turning it in for a grade and save your answers for later study. Look up answers to questions that confuse you. This book has questions that test your understanding of the terms, concepts, and processes. Critical thinking questions are also included to help you sharpen your critical thinking skills.

≈ PREPARING FOR TESTS

■ Don't fall behind on your reading assignments and review lecture notes as frequently as possible.

■ If you have the time, you may want to outline your notes and your assigned readings. Try to prepare the outline with your book and notes closed. Determine weak areas, then go back to your text or class notes to study those areas.

■ Space your study to avoid cramming. One week before your exam, go over all of your notes. Study for two nights, then take a day off. Study again for a couple of days. Take another day off, then make one final push before the exam, being sure to study not only the facts and concepts, but also how the facts are related. Unlike cramming, which puts a lot of information into your brain for a one-time event, spacing will help you retain information for the test and for the rest of your life.

- Be certain you can define all terms and give examples of how they are used.
- Draw key structures over and over until they stick in your memory.
- You may find it useful to write flash cards to review terms and concepts.
- After you have studied your notes and learned the material, look at the big picture—the importance of the knowledge and how the various parts fit together.
- You may want to form a study group to discuss what you are learning and to test one another.
- Attend review sessions offered by your instructor or by your teaching assistant. Study before the review session and go to the session with questions.
- See your professor or class teaching assistant with questions as they arise.
- Take advantage of free or low-cost tutoring offered by your school or, if necessary, hire a private tutor to help you through difficult material. Get help quickly, though. Don't wait until you are way behind. Remember that learning is a two-way street. A tutor won't help unless you are putting in the time.
- If you are stuck on a concept, it may be that you may have missed some important previous material. Look back over your notes or ask your tutor or professor what facts might be missing and causing you to be confused.
- If you have time, write and take your own tests, including all types of questions.

- Study tests from previous years, if they are available legally.
- Determine how much of a test will come from notes and how much will come from the textbook.

≈ TAKING TESTS

- Eat well and get plenty of exercise and sleep before tests.
- Remain calm during the test by deep breathing.
- Arrive at the exam early or on time.
- If you have questions about the wording of a test question, ask your professor. Don't be shy.
- Look over the entire test first so you can budget your time.
- Skip questions you can't answer right away and come back to them at the end of the session if you have time.
- Read each question carefully and be sure you understand its full meaning before you answer it.
- For essay questions and definitions, organize your ideas on a piece of scrap paper or the back of the test *before* you start writing.

Now take a few moments to go back over the list. Check off those things you already do. Then, mark the new ideas you want to incorporate into your study habits. Make a separate list, if necessary, and post it by your desk or on the wall and keep track of your progress. Good luck!

CHAPTER

1

An Introduction to Biology

Monarch butterfly on a poinsettia flower. Insects like this feed off the flower's nectar and pollinate flowers.

Some people think of humans as the crowning achievement of nature. They see humankind as separate from nature and immune to the laws that govern life on Earth. Nonetheless, although we humans have unparalleled skills and talents and differ in significant ways from other creatures, we are still a part of the environment—one component in the complex web of life. Although unique, *Homo sapiens* is just one of millions of species living on the Earth (▷ Figure 1–1).

Diverse and astonishingly beautiful, the web of life we belong to flourishes within a closed system. A **closed system** is one in which all of the materials necessary for life are present in a relatively fixed amount. For life to exist in such a system, these materials must be used over and over, or recycled.

In many ways, then, the Earth is like a terrarium. Sealed from the outside, a properly constructed terrarium containing soil, water, air, plants, microorganisms, and some forms of animal life, such as snails, can persist indefinitely with only one contribution from outside the system: energy from the sun. Sunlight nourishes the plants of the terrarium, and these, in turn, feed the snails. Microorganisms feed on waste materials. The animals and microorganisms in a terrarium utilize foodstuffs made by plants and release carbon dioxide, which plants reuse to make additional food molecules. But the plants also produce oxygen, essential for all other organisms in the closed system. Other nutrients are recycled as well, as you will see in Part VII of this book.

In the terrarium called Earth, humans participate in the global recycling of nutrients. Each day, we consume millions of tons of foodstuffs derived from plants. We also consume oxygen and release carbon dioxide, helping the cycle to continue. From a biological perspective, then, humans are clearly not apart from nature but, rather, are a thread in a much larger, interdependent community of living things called the **biosphere**. In many ways, the biosphere, which consists of plants, animals, and a wide range of microorganisms, is like the living skin of the planet.

This book will take you on a journey through the community of life and will show how we humans fit into the scheme of things. During this journey, you will learn about our coinhabitants, the plants, animals, and microorganisms that share this Earth with us. You will see how they evolved, survive, and reproduce.

Some scientists believe that as many as 30 million to 50 million species live on Earth. A **species** is a distinct group of organisms capable of interbreeding and producing viable offspring. Don't worry, though—in this book we won't be looking at all of the species that grace this planet. Only a relative handful of species will be mentioned here. From these carefully chosen representatives of the living world, however, we can draw many generalizations, or biological principles, that apply to other species, even ourselves. Some of these principles of biology have a direct bearing on modern society, offering guidance on ways to live sustainably on a finite planet.

▷ **FIGURE 1–1 Humans and Nature** Despite what many think, humans are an integral part of nature, dependent on the natural environment for a host of products and for many free services.

On this journey, you will also learn a great deal about yourself—how you got here, how you inherited certain characteristics from your parents, and how your body functions. You will learn some of the basics of nutrition. You will see how broken bones mend and discover how your immune and nervous systems, among others, operate. You will learn about many common diseases and—perhaps more important—ways to prevent them. The information you learn here may help you decide whether you should support an environmental candidate in the next election. Or it may help potential attorneys understand the details of complex environmental litigation or a personal injury case.

In this day and age, it is clearly important to be biologically literate, but lest we forget, the study of life is also an aesthetic experience. As you proceed through this course, be sure to take some time to marvel at the wonders of the living world—the intricate details of the cell, the remarkable structure of the body, and the brightly colored animals and plants that live in splendid coral reefs and rich tropical forests. Take some time to appreciate the interconnections of living things—how the body's systems work together and how organisms depend on one another. In college courses, it is easy to get bogged down with memorization for tests. Although studying hard for tests is vital to your success, don't let it keep you from experiencing the wonder of the phenomenon we call life. As Professor James Backer of St. Vincent College tells his students, each detail you learn is a dot in a bigger picture. Don't forget to connect the dots.

≋ ORGANIZATION AND APPROACH

The biological journey on which you are about to embark begins at the submicroscopic level in Part I. This Part gives you a quick overview of the chemicals and chemical reactions that are the basis of all life. Part II moves to the next higher level of organization, the cell, the fundamental unit of all organisms. Part II examines the structure and function of cells and shows how cells acquire and use energy. Part III examines chromosomes, cell division, and heredity. Part IV widens the focus, discussing the structure and function of animals, with special emphasis on humans. Part V discusses reproduction and development in animals, once again using humans as the main example. Part VI describes the evolution of life and the five major groups of organisms (kingdoms) that have evolved since the Earth's beginning. Finally, Part VII describes behavior, principles of ecology, and major environmental problems facing humankind. It seeks to show how we fit into the web of life and how we need to function in order to preserve it.

≋ LIFE IN THE BALANCE: HOMEOSTASIS

Biology classes, like all other science courses, offer a number of facts and principles. The principles of biology are basic concepts that apply to all life forms or to large groupings. They provide an overview that makes the study of life more understandable.

One of the key principles of biology, central to your understanding of life, is homeostasis. Simply stated, the **principle of homeostasis** says that all organisms have evolved mechanisms that ensure internal constancy in an ever-changing environment. These mechanisms are vital to survival and reproduction. In order to understand what this principle means, let's consider this subject in a little more detail.

Homeostasis Is a State of Internal Constancy

The term **homeostasis** is derived from two Greek roots, *homeo*, which means "the same," and *stasis*, which means "standing." Literally translated, it means "staying the same." Some people like to think of homeostasis as a state of internal constancy. But this view is not entirely accurate. In reality, homeostasis is a kind of **dynamic equilibrium**. Let me explain what this means by looking at body temperature.

Your body temperature is fairly constant—about 37°C, or 98.6°F. If you took your temperature at regular intervals during the day over long periods, however, you would find that it actually fluctuates slightly. It falls at night when you sleep and rises in the day when you are active. Although body temperature changes over time, body systems prevent wide swings. Thus, body temperature stays more or less the same over long periods. This is what is meant by a dynamic equilibrium. In this book, I'll use the term *internal constancy* for the sake of tradition, recognizing that homeostasis really means a dynamic equilibrium.

During the Course of Evolution, Numerous Systems of Checks and Balances Have Emerged.[1] Organisms achieve internal constancy through a variety of automatic mechanisms. As you will see in Chapter 10, these mechanisms involve some means of sensing internal and external change. The organism's sensors then elicit a response, which offsets the change and maintains constancy. Shivering, for example, is a rhythmic contraction of muscles that generates body heat when the outside temperature drops. Shivering is one of several homeostatic mechanisms in warm-blooded animals that are elicited by falling ambient (outside) temperature and help maintain body temperature.[2] One scientist discovered recently that some small (nonmigrating) birds shiver through much of the winter to stay warm (▷ Figure 1–2).

Many homeostatic mechanisms help control the levels of nutrients in the blood. In humans and other vertebrates, for example, blood sugar levels are regulated by two hor-

---

[1] An ecosystem is a biological system consisting of organisms and their environment.

[2] Warm-blooded animals such as horses, humans, and dogs maintain a fairly constant body temperature. They are also known as homeotherms.

▷ **FIGURE 1–2 Rosy Finch** Beneath its blanket of feathers, this tiny bird stays warm in winter by shivering most of the time.

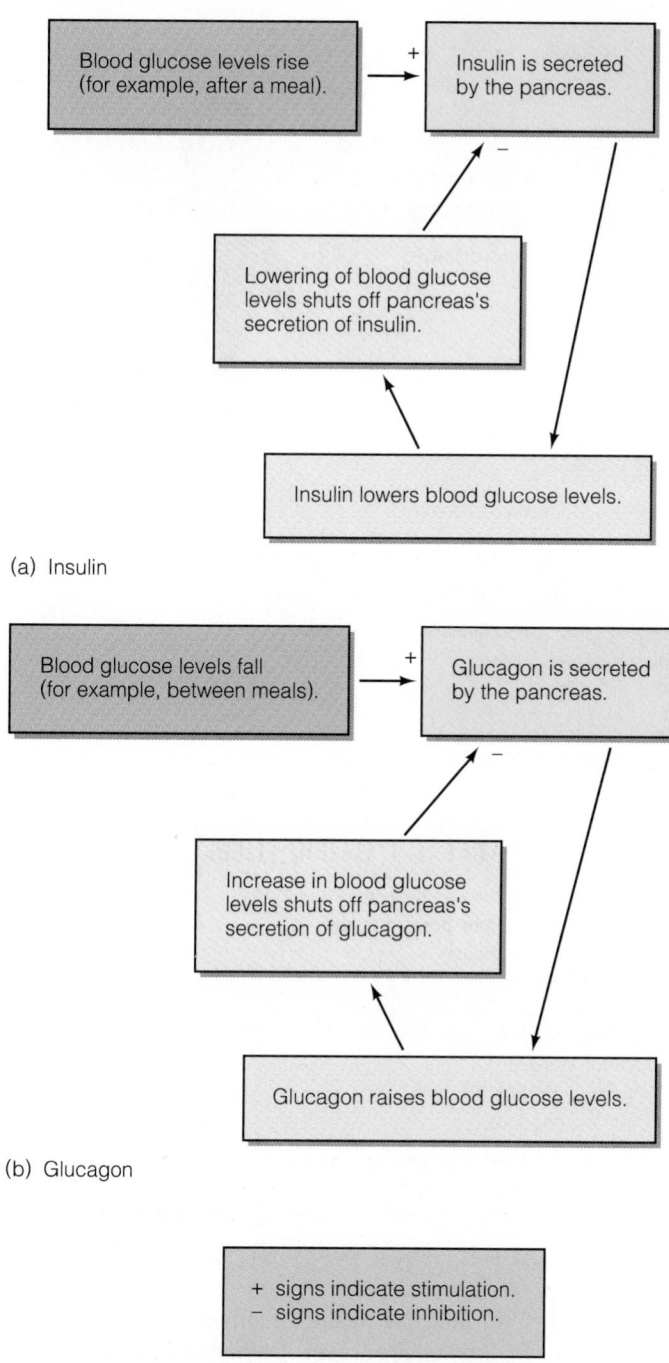

(a) Insulin

(b) Glucagon

+ signs indicate stimulation.
- signs indicate inhibition.

▷ **FIGURE 1–3 Actions of Insulin and Glucagon** (a) Insulin helps lower blood glucose after a meal. (b) Glucagon raises blood glucose levels between meals. Both hormones help maintain constant blood glucose levels, necessary for normal body function.

mones, **insulin and glucagon** (▷ Figure 1–3). (A hormone is a substance that is produced in one part of the body and travels to another part, where it elicits a certain effect.) In this system, insulin is released by the pancreas when blood glucose levels rise above a certain level, usually after a meal. Insulin causes glucose levels to drop, for reasons explained later. In between meals, when glucose levels fall too low, glucagon is released by the pancreas. It causes levels to rise. Through the interaction of these hor-

mones—and several others—blood glucose levels remain fairly constant.

Homeostatic mechanisms also exist in ecosystems. A highly simplified example illustrates the point. In the grasslands of Kansas, rodent populations generally remain about the same size year after year.[3] The relative constancy of natural systems (ecosystem homeostasis) results from dozens of factors, one of the most important being predators (animals that hunt and kill other organisms). Coyotes, hawks, and other predators feed on rodents, in the process helping hold rodent populations in check. Predators are therefore a crucial element in the maintenance of ecological balance. Weather, food supplies, and a host of other factors also help regulate rodent populations and maintain ecosystem balance.

In this book, the term *homeostasis* will be used to refer to the balance that occurs at all levels of biological organization—from cells to organisms to entire ecosystems. Maintaining this balance is essential to life. Imbalances in these systems, in fact, often upset the workings of cells, organisms, and ecosystems. If they cannot be corrected by natural mechanisms, they may result in severe repercussions.

Health Is a State of Physical and Mental Well-being

For many years, human **health** was defined as the absence of disease (▷ Figure 1–4a). As long as a person had no obvious symptoms of a disease, he or she was considered healthy. Recently, many experts have recognized the limitations of this definition and have sought to revise our thinking on the subject. In the process, they have created a much broader and more useful definition of the word.

Today, human health is viewed as a state of physical and emotional well-being characterized by few **risk factors**— that is, early signs of disease that precede discernible illness. High blood pressure, for example, is a risk factor for heart disease. A person with high blood pressure may feel fine but is much more likely to suffer a heart attack than a person with normal blood pressure. All other things being equal, a person with high blood pressure is less healthy than one with normal blood pressure.

Health experts today also encourage us to include mental and physical fitness in our assessment of health. Mental and physical fitness are measures of our psychological and physical abilities to meet the demands of everyday life. In general, fit people are able to cope with day-to-day psychological stresses and are able to walk up a flight of stairs without becoming short of breath. (If you can't, it may be time to seek counseling and start an exercise program.)

Figure 1–4b illustrates the new concept of health. It shows that human health is measured on a continuum. The healthiest people have no obvious illness, have no risk factors, and are mentally and physically fit. At the other

[3]A population is a group of organisms of the same species living within a prescribed region.

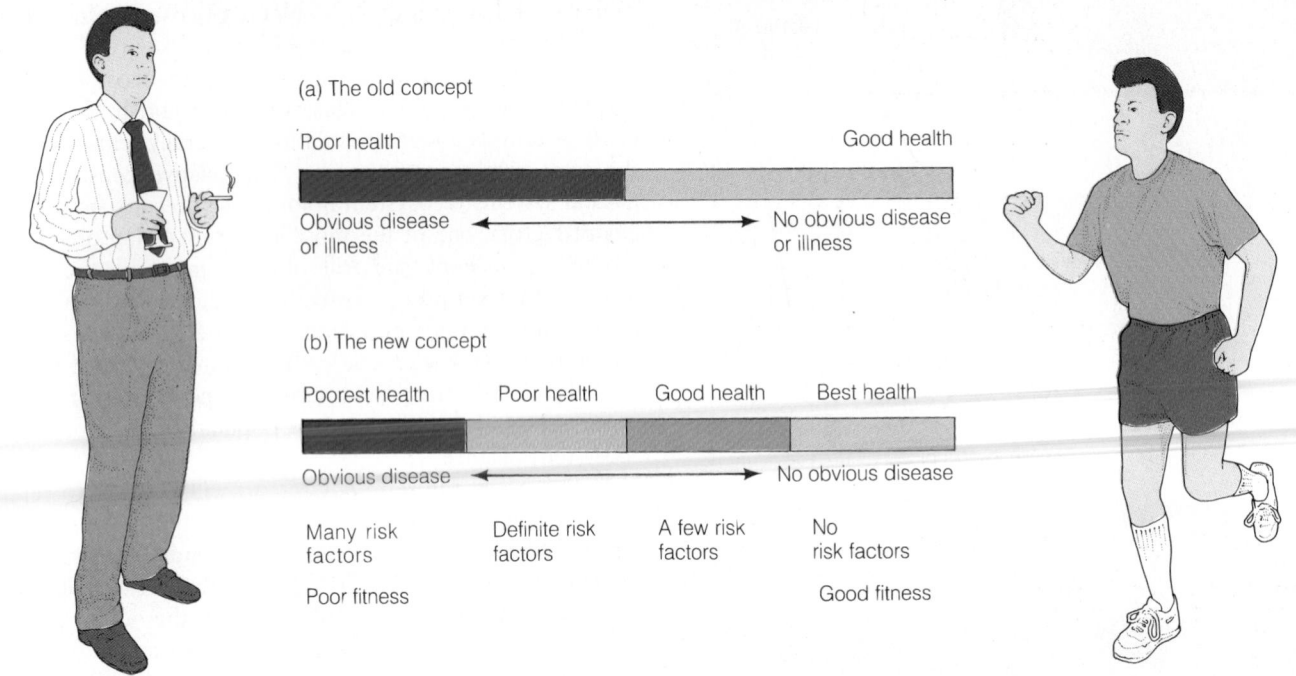

FIGURE 1–4 Old and New Concepts of Health

end of the continuum are those people in the poorest of health. They have obvious symptoms of disease and are in poor physical condition. In between these poles on the health continuum lies a universe of possibilities.

Maintaining good health is a lifetime job. Table 1–1 lists healthy habits. By following its advice, you can generally live a long, healthy life.

Health Is Dependent on Well-functioning Homeostatic Mechanisms and a Healthy Environment

More and more, it is becoming clear that our physical well-being depends on well-functioning homeostatic mechanisms—that is, regulatory controls in the body that help maintain homeostasis; when these controls are not functioning properly or when they break down completely, illness may result. This is not to say that all diseases result from homeostatic imbalance. Some are produced by genetic defects; others are caused by bacteria or viruses. Both genetic defects and microorganisms, however, can disturb the regulatory mechanisms of the body, severely disrupting homeostasis. A good example is acquired immune deficiency syndrome, or AIDS, which is caused by a virus that attacks certain cells of the immune system. The AIDS virus causes a breakdown of immune protection, which is vital to homeostasis. In other diseases, temporary upsets in homeostasis (for example, stress) may make us more susceptible to infectious agents. According to a recent study, people under stress are twice as likely to suffer from colds and the flu as those who are not. (For a discussion of ways to reduce stress, see Health Note 1–1.)

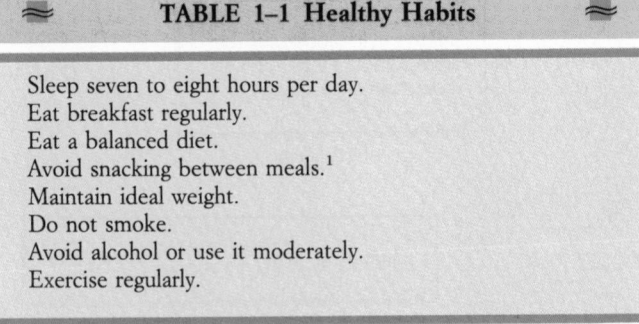

≈ TABLE 1–1 Healthy Habits ≈
Sleep seven to eight hours per day.
Eat breakfast regularly.
Eat a balanced diet.
Avoid snacking between meals.[1]
Maintain ideal weight.
Do not smoke.
Avoid alcohol or use it moderately.
Exercise regularly.

[1]Some nutritionists favor eating numerous smaller meals throughout the day, contending that the body can better metabolize food ingested in this way.

Human health, and the health of other species, is dependent on a healthy environment. Polluted air, water, and soils take a toll on humans and other species, as pointed out in the Environment and Health features that appear in some chapters of this book. But the health of organisms requires more than a clean environment. It requires conditions conducive to mental health. In Colorado, for example, construction activity and road traffic involved in building a dam southwest of Denver caused stress in a bighorn sheep population nearby that wiped out half of the herd.

From a strictly human perspective, it should become clear that planet care is the ultimate form of self-care. Living responsibly reduces our need for resources and minimizes our damage to the environment. It not only helps humankind but also assists the millions of species that depend on the water, air, and land in their struggle for survival.

REDUCING STRESS IN YOUR LIFE

Stress is a normal occurrence in everyday life. In some instances, a little stress may help improve your performance. As you are crossing a busy street, for example, the sound of screeching brakes may start your heart racing and increase your rate of breathing. It also results in the release of hormones that elevate your blood levels of glucose, providing additional energy for muscles. All these changes enable you to perform at an extraordinary level, darting out of the way of the oncoming car.

Other external threats stimulate the stress response as well, but for most of us stress comes from within. That is, we create stress through our thoughts. We worry about our performance on exams, and we worry about embarrassing ourselves at parties. The physical reaction from psychological stress is the same as the reaction that occurs when we are threatened by an outside factor.

Studies show that severe stress, if prolonged, can cause great physical harm. It may interfere with performance at home, at work, and at school. It may lead to cardiovascular disease, including heart attacks. It increases the risk of developing ulcers and can even weaken the immune system.

Reducing stress requires conscious effort, but it can be done. One of the most successful ways is through exercise. A single workout at the gym, a ride on your bicycle, a vigorous hike, or a day of cross-country skiing reduces tension for two to five hours. And regular exercise may reduce the overall stress in your life. An individual who is easily stressed may find that stress levels decline after several weeks of exercise. Exercise may also relieve depression in some people.

Exercise can be supplemented by relaxation training to reduce stress. As you prepare for a difficult test or get ready for a date that you are nervous about, tension often builds in your muscles. Periodically stopping to release that tension helps you reduce physical stress. For some people, getting up and stretching or taking a walk can help reduce the tension. Others find it useful to tighten their muscles forcefully, then let them relax. Still others may need more concentrated efforts. Cassette tapes can teach you relaxation methods. If that's not for you, consider signing up for stress-reduction classes through a trained therapist. The more you practice relaxation, the better you will get at relieving tension.

Stress from within is often blown out of proportion. By dealing with the thoughts that exaggerate your stress, you can help eliminate stress before it begins. Start paying attention to the thoughts that provoke anxiety in your life. Are they exaggerated? If so, why? For example, are you nervous before exams? Why? Do you always study adequately? Could you prepare better? Would better preparation help reduce your anxiety? Do people asking you questions about a test right before class make you nervous? If so, can you isolate yourself from them? Finding the source of your anxiety and taking positive action to alleviate it are helpful ways of reducing stress.

Stress reduction is not always so easy. Test anxiety, for example, may be deeply rooted in feelings of insecurity and inadequacy. You can struggle with low self-esteem your entire life or seek professional help. A trained psychologist can help you find the roots of your problem and help you learn to feel better about yourself. Psychological help is as important as medical help these days. Given the complexity and pace of our society, there is no shame in seeking counseling. In fact, just asking for help is the first-step to recovery. It is a positive step toward healing yourself.

Biofeedback is another form of stress relief. A trained health worker hooks you up to a machine that monitors heart rate, breathing, muscle tension, or some other physiological indicator of stress (see ▷ Figure 1). During a biofeedback session, your trainer will help you relax, then perhaps discuss a stressful situation. When one of the indicators shows that you are suffering from stress, a signal is given off. Your goal is to con-

sciously reduce the frequency of the signal. For example, if your heart started beating faster when you thought about taking an exam, a clicking sound might be heard. By deep breathing and relaxing, you consciously slow down your heart rate, and the clicking sound slows down as well, then disappears. Learning to recognize the symptoms of stress and to counter them is the goal of biofeedback. Eventually, you should be able to do it without the aid of a machine.

You can also reduce tension by managing your time efficiently and managing your work load. Numerous books on the subject can help you learn to budget your time more effectively. This alone can help you keep from feeling stressed.

It is important to challenge yourself in college, but be realistic in what you expect of yourself. If you must work, for example, sign up for a class load that you can handle. Even if you are not working, take a reasonable class load, and be sure to exercise regularly and get plenty of sleep.

Relieving stress in our lives helps us reduce the risk of cardiovascular disease. It helps us relax and enjoy life. Lest we forget, it also makes us more pleasant to be around. All in all, it is best to start learning early in life how to cope with stress. Lessons learned now will be useful for years to come.

▷ **FIGURE 1 Biofeedback** Student in a biofeedback session.

≋ EVOLUTION: THE UNITY AND DIVERSITY OF LIFE

Another common thread woven through this book is evolution. The **theory of evolution** says that all life forms developed from earlier forms. As described more fully in later chapters, the process of evolution results from modifications in existing species that leave them better suited to their environment—that is, better able to survive and reproduce. Changes in the coloration of a species of butterfly during the course of evolution, for example, might make it better able to avoid being eaten. Evolution also results in the formation of brand new species. For example, forty million years ago, a small, shrewlike creature is believed to have given rise to the first primates, the prosimians, or premonkeys (▷ Figure 1–5).

According to modern evolutionary theory, all species that are alive today evolved from the cells that formed over 3.7 billion years ago. As you will see in Chapter 23, these cells are thought to have evolved from molecules in the Earth's primitive environment. The common ancestry of living things has resulted in a commonality among the Earth's organisms—for example, biochemical similarities among diverse species. Bacteria and humans, for instance, use the same type of genetic material, and they make protein in similar ways. They even have many of the same enzymes (specialized protein molecules that speed up chemical reactions) for extracting energy from sugar molecules. The biochemical similarity of all species illustrates an important principle of evolution—notably, that successful characteristics were maintained and then were often modified to meet the particular demands of different species. A comparison of certain anatomical features, like the bones in a person's arm and the wings of birds, also reveals remarkable similarities. This line of evidence supports the notion that species evolved from a common ancestry and that common characteristics were modified over time by evolutionary processes to accommodate the particular needs of different species.

Despite a common origin, organisms have evolved to form a remarkably diverse array of species (▷ Figure 1–6). A quick comparison of the giraffe, the lion fish, the crayfish, the frog, the saquaro, and the human being illustrates the point well, especially if you remember that this is only a tiny fraction of the diverse living world. With this in mind, let's look at some of the commonalities.

Organisms Share Seven Characteristics

All organisms share seven features—commonly referred to as the characteristics of life. Keep these features in mind as you proceed through your study of biology.

First, all organisms are made of cells. Cells, in turn, consist of molecules, nonliving particles composed of smaller units called atoms (▷ Figure 1–7). Glucose molecules, for instance, are composed of carbon, hydrogen, and oxygen atoms. Molecules combine to form the parts of the cell.

Organisms fit into two broad categories: unicellular and multicellular. Unicellular organisms, like the paramecia and bacteria you may have studied in high school biology or the free-floating algae in a nearby pond consist of one cell. These unicellular creatures live independently, often floating or swimming about in water.

Multicellular organisms consist of aggregations of cells. In some species of algae, for example, individual cells, although capable of living independently, combine with others of like kind to form colonies. Colonies are aggregations of similar cells and are the simplest type of multicellular organism.

More complex multicellular forms exist in a wide range of shapes and sizes, but they are all characterized by one common feature: the presence of cells that become structurally and functionally specialized during development—that is, structurally altered to perform specific functions. In these organisms, specialized cells combine to form distinct tissues, and tissues unite to form organs that perform specific functions. Thus, these organisms are charac-

▷ **FIGURE 1–5 The Evolution of a New Species** (*a*) A creature very similar to the tree shrew may have given rise to the first primates, (*b*) the prosimians.

(a)

(b)

(a)

(b)

(c)

(d)

(e)

(f)

▷ **FIGURE 1–6 The Diversity of Life** Evolution has resulted in an amazing variety of species. A few are shown here to illustrate that remarkable diversity: (*a*) giraffe, (*b*) lion fish, (*c*) crayfish, (*d*) poison arrow frog, (*e*) saquaro, and (*f*) human being.

terized by a division of labor not unlike that found in an automobile factory or even your own family. The division of labor evolved over many millions of years and undoubtedly gives them decided advantages over the simpler colonial forms of life.

The second characteristic of life is that all organisms grow and maintain their complex organization by taking in molecules and energy from their surroundings. As you will see in subsequent chapters, living things must expend considerable amounts of energy to maintain their rather complex internal organization. In fact, 70% to 80% of the energy that adult humans need is required just to maintain our bodies—to transport molecules in and out of cells, to build protein, and to perform other basic body functions. The rest is used for activity—walking, running, talking, and so on. Where does the energy that organisms require come from?

As you will see in later chapters, all energy in living systems comes from the sun. Plants capture the energy they need from sunlight, but virtually all other organisms must acquire energy from the food they eat—plants or other animals; even the energy of a meat-eating animal (carnivore) ultimately comes from sunlight through plants. For example, a mountain lion that feeds exclusively on deer and rabbits gets its energy from the molecules of its prey, which, in turn, derived theirs from plant matter.

The third characteristic of life is metabolism. **Metabolism** refers to the chemical reactions occurring in cells and organisms. These reactions consist of two types—those in which food substances are built up into living tissue (anabolic reactions) and those in which food is broken down into simpler substances, often releasing energy (catabolic reactions). In human body cells, thousands of reactions occur each second just to maintain life.

The fourth characteristic of life is homeostasis, the maintenance of internal constancy, described earlier. Vital in the constantly shifting world of many organisms, maintaining a constant internal environment is a never-ending task.

A fifth feature common to all forms of life is irritability. **Irritability** is the ability to perceive and respond to stimuli. A young girl demonstrates this vital property when she withdraws her hand from a hot stove or closes her eyes in response to bright light. A plant demonstrates irritability by bending toward the light.

The ability to perceive and respond to stimuli allows all other organisms, including single-celled ones, to respond to their environment and is, therefore, an important survival tool. Not surprisingly, evolution has resulted in some rather remarkable ways by which organisms both detect and respond to stimuli. The Spotlight on Evolution in Chapter 18, for example, describes special electrical sensors in the bill of the duck-billed platypus that detect tiny electrical

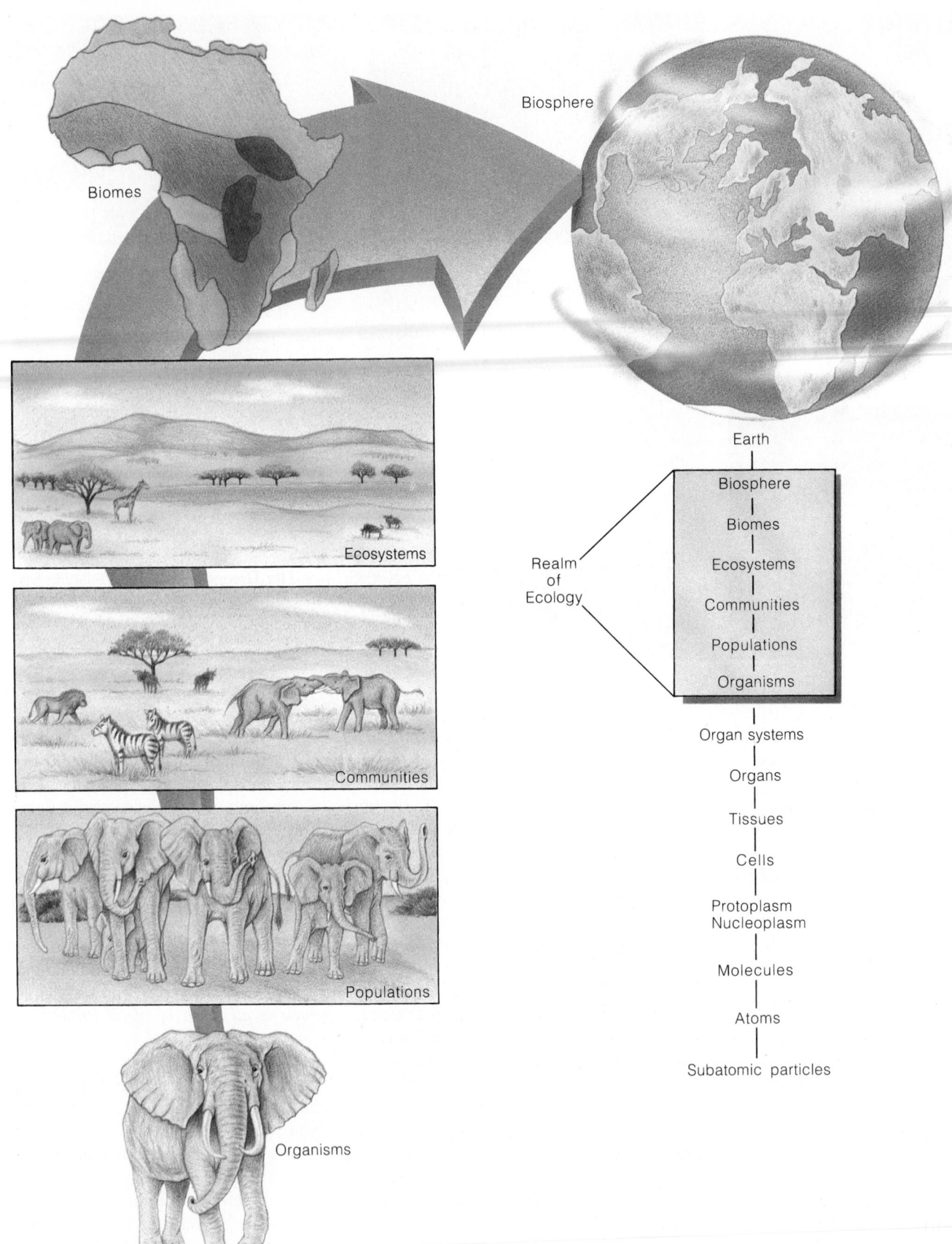

Biosphere

Biomes

Ecosystems

Communities

Populations

Organisms

Earth

Realm
of
Ecology

Biosphere

Biomes

Ecosystems

Communities

Populations

Organisms

Organ systems

Organs

Tissues

Cells

Protoplasm
Nucleoplasm

Molecules

Atoms

Subatomic particles

▷ **FIGURE 1–7 The Organization of Life**

signals from prey while this odd-looking animal swims under water with its eyes closed.

The sixth characteristic of life is reproduction and growth. All organisms are capable of reproduction and growth. Two types of reproduction are known to exist: sexual and asexual. **Sexual reproduction** occurs in organisms that produce offspring by combining sex cells. In vertebrates such as birds and mammals, these cells come from male and female individuals. Attesting to its evolutionary importance, most groups of organisms have sexual reproduction of some sort. A rudimentary form can even be found in bacteria.

Asexual reproduction is common in single-celled organisms such as the amoeba (▷ Figure 1–8a). In these and other single-celled organisms, reproduction generally occurs by simple cell division—that is, when a parent cell divides and forms two identical offspring. As noted above, however, even microorganisms may exhibit crude forms of sexual reproduction.

Some multicellular organisms reproduce asexually by budding. Budding is a process in which new individuals are produced from small outgrowths called buds. Tiny buds enlarge and soon develop into a fully formed organism. Figure 1–8b shows how an aquatic organism known as a hydra buds. Like cell division in unicellular organisms, budding produces generation after generation of identical offspring.

The seventh, and final, characteristic of life is that organisms are subject to evolution.. Evolution, as mentioned earlier, is a process that leads to structural, functional, and behavioral changes in species, known as **adaptations.** Favorable adaptations increase an organism's chances of survival and reproduction. (You might want to take a few moments to read Spotlight on Evolution 1–1 for a discussion of adaptations in butterflies.)

Evolutionary change occurs through the interplay of genetic variation and environment. **Genetic variation** refers to naturally occurring genetic differences in various members of a population, groups of genetically similar organisms. These differences result in variations in structure, function, and behavior of organisms—some that enhance survival and reproduction and some that don't. Those that give members of a population an advantage over others in survival and reproduction tend to persist. Charles Darwin, the nineteenth century British naturalist who spent much of his life studying evolution, proposed a **theory of natural selection.** He described natural selection as a process in which slight variations, if useful, are preserved. Thus, natural selection is a process by which organisms become better adapted to their environment. Because organisms endowed with beneficial variations are more likely to survive and reproduce, they pass on the favorable genetic material. Over time, the genetic composition of the species may change. Individuals of a species may become better camouflaged or faster, and thus are better able to escape being eaten or to capture prey.

The details of evolution are discussed in Chapter 23. It is important to point out here, however, that individuals do not evolve, only populations do. (As noted, a population is a group of organisms of the same species.) Put another way, although evolution results from the emergence of adaptations in individuals, it occurs only when there is a shift in the genetic composition of a population. Profound changes in the genetic composition of a population can result in the formation of entirely new species.

Organisms Can Be Grouped Into Five Kingdoms

The common characteristics of the Earth's organisms permit scientists to group them into groups, the largest of which is the **kingdom.** Each kingdom includes all organisms that exhibit one or perhaps a few common features. For example, the plant kingdom includes all photosynthetic organisms whose cells have cell walls. In this book you will study five kingdoms: monerans, protistans, fungi, plants, and animals.

Within a kingdom, biologists create subcategories, based on anatomical, behavioral, and physiological similarities

▷ **FIGURE 1–8 Asexual Reproduction** (*a*) Asexual reproduction takes place when an amoeba divides, producing identical offspring. (*b*) Small cellular buds in some organisms like this hydra enlarge to produce genetically identical offspring, which eventually break free from the parent.

(a)

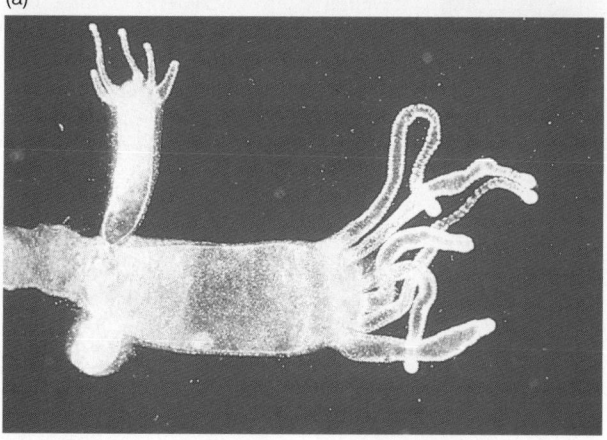

(b)

BUTTERFLIES: DELICATE CREATURES ADAPTED TO A HARSH WORLD

Few creatures inspire as much awe as butterflies. On gossamer wings, these delicate creatures flutter precariously in a world filled with danger. But pity them not. As a group, they display many adaptations that permit survival in this apparently harsh world. Collectively, in fact, butterflies' arsenal of adaptations is as impressive as that of any other group of animals on the planet, writes the science editor Mike Toner. Consider coloration first.

Butterflies come in a wide range of colors and color patterns that provide camouflage from hungry predators. The wings of the brush-footed butterfly, for example, have dull undersides. When the insect folds its wings, it resembles a dead leaf, protecting it from hungry birds. In other butterflies, plain-colored wings or dark, mottled colors help the butterfly blend in with the bark of trees.

Camouflage is not restricted to adults, either. In the chrysalis, or cocoon, stage, which occurs between the larval (caterpillar) and adult stages, many species take on the appearance of dead leaves and thus avoid predators who pass them by in search of food.

The viceroy exhibits one of the more interesting camouflages. After their first molting, the caterpillars of viceroy butterflies turn a dull brown color. A white spot on the insect makes it look a lot like a bird dropping, an adaptation that no doubt fails to catch the eye of its predators.

Butterflies also rely on evolved deception. The prominent eyespots on the wings of some butterflies may resemble the eyes of snakes and thus startle predatory birds (see ▷ Figure 1).

In some species coloration varies depending on habitat. In the subalpine meadows of the Cascade Mountains, for example, one species of caterpillar appears a dull gray-brown, a color that blends into the environment. The same species in the lowland rain forests of the Pacific coast, however, has conspicuous black and yellow spots, which, scientists believe, mimic the warning colors of a species of inedible millipede, thus protecting it from attack.

One of the most interesting butterfly adaptations occurs in the Ross's metalmark. This species has entered into a symbiotic relationship with carpenter ants in its home in southeastern Mexico that is beneficial to both organisms. The symbiotic relationship is dependent on the secretion of sugary fluid. Soon after the caterpillar hatches, it begins to produce its sweet secretion, which attracts the nocturnal carpenter ants. The ants, in turn, feed on the "honeydew" and protect their source by digging a small hole at the base of plants.

Each morning, the ants coax the caterpillar down the stem of the plant into its new home. Once the caterpillar is inside, the ants cover the hole, but they open it up each night. Before releasing the caterpillar, though, the ants search out and destroy or drive away potential predators such as red ants. Carpenter ants have a chemical fumigant, formic acid, which they spray on intruders. Once the coast is clear, the ants usher the caterpillar back up onto the plant for a night of uninterrupted grazing. When dawn comes, though, they herd it back into its protective hole. Scientists believe that without this relationship, the caterpillar would probably have little chance of survival. When the caterpillar spins its cocoon (forms a chrysalis), the ants remain behind to protect it, leaving only just before the butterfly emerges.

Butterflies are remarkable flying machines, too. Some butterflies migrate hundreds of miles, even traversing mountain ranges, to reach their winter homes. The master of long-distance migration has to be the monarch, which migrates 2000 miles each year in its annual migration from the eastern United States to Mexico. Most others migrate, but for smaller distances.

Monarch butterflies are relatively slow fliers, capable of maximum speeds of only 12 miles per hour. The trip to Mexico would take many days and would probably exhaust the insects. How do they migrate, then?

Studies show that, like some birds, butterflies hitch rides on south-heading storm fronts, which whisk them along at rapid rates. When they reach the southern United States they climb thermals, warm air rising from Earth's surface. Once aloft, the insects generally head southwest for Mexico, soaring as much as they can. Pilots of ultralight planes and gliders have seen butterflies cruising at 1500 to 7000 feet.

As they get closer to their winter home, the monarchs do less soaring and more purposeful flying, staying closer to the ground. And each year, approximately 200 million monarch butterflies, traveling alone or in large groups, descend unerringly on the same hillside in Mexico. Why, no one knows, but it is likely that this delicate creature has some innate guidance system to bring it to its warm winter ground, where it waits until the next spring, when it repeats the journey in reverse.

▷ **FIGURE 1 Predators Beware**
These false eyespots frighten many would-be predators, thus protecting this otherwise vulnerable organism.

SOURCE Adapted from Mike Toner, "The Hidden Strength of Gossamer Wings," *National Wildlife* (1988) 26(5): 4–11.

TABLE 1–2 Taxonomic Classification of Humans

Kingdom	Animalia
Phylum	Chordata
Class	Mammalia
Order	Primates
Family	Hominidae
Genus	Homo
Species	Sapiens

and differences. The broadest grouping in a kingdom is the **phylum** (Table 1–2).[4] In the kingdom Animalia, for example, are many phyla. Phyla are divided into subgroups called **classes.** Classes are broken into **orders,** and orders are broken into **families.** Families consist of **genera** (singular, **genus**), and each genus contains one or more **species.**[5] Table 1–2 shows the classification of humans.

Although most of us refer to animals, plants, and other organisms by their common name, scientists usually use scientific names. The scientific name of a species is unique and often descriptive. It consists of two parts. The robin, for example, goes by the scientific name *Turdus migratorius* (the migratory thrush). The grizzly bear goes by the name *Ursus horribilis* (the horrible bear). The first word in the scientific name is the genus to which the organism belongs. The second is a descriptive term that distinguishes it from other members of the same genus. You will note that the scientific name is always underlined or in italics because these words are Latin or Latin derivatives, and foreign words are italicized or underlined when they appear in English. As you will note, the genus is always capitalized and the second word of the scientific name begins with a lower case letter.

Humans Are One Form of Life on Earth and Have Many Unique Characteristics

Humans share many characteristics with the millions of other species on Earth, yet we are also a unique form of life. One of the key differences between humans and other animals is our ability to acquire language. Another is our culture. Culture has been defined in many ways. One humorist, for example, remarked that culture is anything we do that monkeys don't. A bit more technically, **culture** can be defined as the ideas, values, customs, skills, and arts of a given people in a given time. Culture clearly varies from place to place and changes over time. American culture, for example, differs markedly from the culture of the Japanese, the Chinese, and the British and has changed dramatically from the early days of American settlement.

[4] Botanists use the term *division* instead of phylum.
[5] Here's a good opportunity to use a mnemonic (learning aid). Try this one: *keep putting chocolate on for goodness sake.*

Humans differ in other important ways as well. One noteworthy difference is our ability to plan for the future. Although a few other animals seem to share this ability, most of what appears to be planning is probably the result of instinct. A bird's nest-building activities, for example, are programmed by its genetic material and are probably not the product of conscious planning. In contrast, building skyscrapers and launching rockets are distinctly human activities that require an extraordinary amount of forethought.

Another unique characteristic of humans is our unrivaled ability to reshape the environment. The ability of our minds to shape images and the ability of our hands to translate those images into reality have given us a power almost beyond our control. Despite the benefits of our remarkable technologies, attempts to control nature sometimes backfire, creating larger problems. Efforts to control upstream flooding on the Mississippi River, for instance, have led to more frequent floods and more widespread damage downstream (▷ Figure 1–9). By building levees along the river, upstream communities prevent water from spilling over their banks, but this protection delivers a larger slug of water to downstream sections when the rains fall.

Another control measure that has backfired is the attempt to keep the Mississippi's channel free of silt as the river courses through its delta. Levees in the delta reduce the deposition of sediment. At one time, sediment deposited along the river's banks in the delta offset the natural subsidence (sinking). Today, over 1 million acres have been reclaimed by the sea. Much of New Orleans is now below sea level, protected from the ocean's waters only by extensive levees.

The marvelous human achievements over the past 200 years have led many to think of our species as the crowning accomplishment of nature. This view, however, can be dangerously misleading. It makes us think of ourselves as apart from nature and immune to its laws, which we are not. It leads many to believe that nature (the environment) is at our disposal and that we can do with it as we like.

To live on the planet without destroying the natural world around us, disrupting evolution, and ruining the prospects of the human race requires a new way of thinking and acting, one that fosters a cooperative relationship between humans and nature. Barry Commoner, a biologist well known for his strong stand on environmental issues, put it best when he said that "nothing can survive on the planet unless it is a cooperative part of a larger, global life." It is our challenge to find a way to build such a way of life—and this book outlines some of the steps that are needed.

≋ SCIENCE, CRITICAL THINKING, AND SOCIAL RESPONSIBILITY

Over the years, human societies have accumulated enormous amounts of information about the universe through

(a)

(b)

▷ **FIGURE 1–9 Human Control of Nature** (*a*) Earthen walls, or levees, along the Mississippi River have grown considerably in the past hundred years in an effort to stop flooding. (*b*) Levees and other measures to control the river have led to more frequent flooding, as explained in the text.

systematic study. The systematic study of the universe and its many parts today falls into the realm of science.

Science Is Both a Method of Accumulating Knowledge and the Body of Knowledge

The term *science* comes from the Latin word *scientia,* which means "to know" or "to discern." Today, **science** is defined as systematized knowledge derived from observation, study, and experimentation. It is also a method of accumulating knowledge. In other words, it involves ways to learn facts. Unfortunately, many people view science as dull and uninteresting, considering it an endeavor best left to a select few. If the truth be known, science is one of the most exciting endeavors of modern society. It teaches us about the workings of the universe, a source of great fascination to many. As paleontologist Robert Bakker once noted, science is "fun for the mind." But science is also a practical endeavor, as pointed out earlier in the chapter. It provides information that has improved our lives in many ways.

Science helps explain the fascinating phenomena around us—for example, the weather, heredity, disease, and nutrition. A knowledge of science makes us better voters—better able to discern fact from fiction in a political debate. An understanding of science can help us decide on the safest forms of energy to meet future demands. Finally, it can help make us make smarter buying decisions or select harmless ways to control insects in our gardens.

The Scientific Method Is a Process of Inquiry

The methods of science are not as foreign as you might think. In fact, most of us use the scientific method in hundreds of ways nearly every day of our lives. When your car fails to start, for example, or when your computer acts

up, you probably engage in the type of thinking that scientists use to gather information and test their ideas. This process, called the **scientific method,** is summarized in ▷ Figure 1–10. As illustrated, the scientific method generally begins with observations and measurements, which often stem from carefully conducted experiments.

To see how the scientific method works, let us take a familiar example. Suppose you sat down at a computer, turned on the switch, and nothing happened. You might

▷ **FIGURE 1–10 Scientific Method** Scientific study begins with observation and measurement. These activities lead to hypotheses that can be tested by experiments. New and revised hypotheses are derived from experimentation.

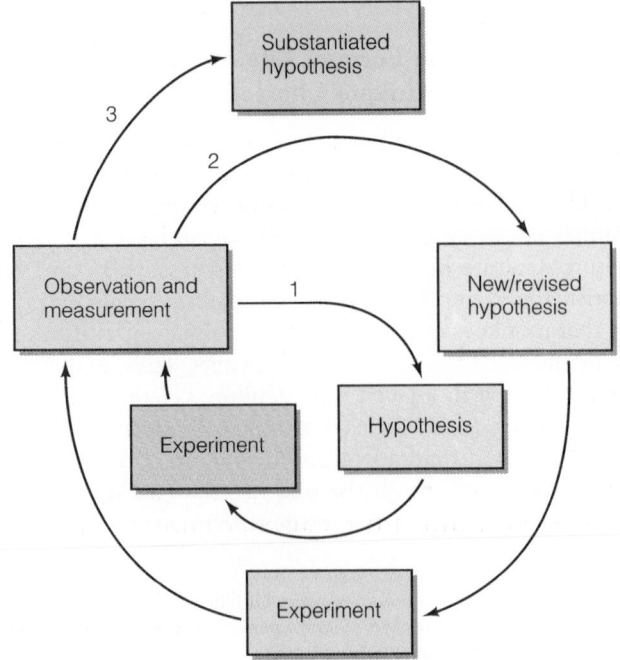

also have noticed that the lights in the room hadn't come on. These two observations would lead to a **hypothesis,** a tentative explanation of the phenomenon. From your observations, you might hypothesize that the electricity in your house was off.

Your next step is to test your hypothesis by performing an **experiment.** In this case, you wouldn't need a $500,000 research grant, all you would need to do to test your hypothesis would be to traipse into the kitchen or the bathroom and try out the light switch.

If the lights in the kitchen and bathroom worked, you would reject your original hypothesis and form a new one. Perhaps, you hypothesize, the circuit breaker to your study had been tripped. To test this idea, you would "perform" another experiment, locating the circuit breaker to see if it was turned off. If the circuit breaker was off, you would conclude that your second hypothesis was valid. To substantiate your conclusion, you would throw the switch, and then test your computer and the room light.

This simple process involving observations and measurements, hypothesis, and experimentation is the basis of all science. Although the subject may be complicated, the process itself is really quite simple.

The body of scientific knowledge is more than just a grab bag of facts and figures. It contains numerous **theories**—broad generalizations about the universe and the way it works.[6] Theories are supported by numerous facts gathered through observation, measurement, and experimentation. It is important to point out, however, that theories generally cannot be tested by single experiments because they encompass many pieces of information. Atomic theory, for instance, explains the structure of the atom and fits observations made in different ways over decades.

A theory commands respect in science because it has stood the test of time—and survived. This does not mean, however, that theories are always correct. As history bears out, theories may be modified or even discarded as new scientific evidence is gathered. Even widely held theories that have persisted for many decades can be overturned. In A.D. 140, for example, the Greek astronomer Ptolemy proposed a theory that placed Earth at the center of our solar system. It was called the **geocentric theory**. For nearly 1500 years, the geocentric view held sway. Many fellow astronomers vigorously defended this position while ignoring observations that did not fit the theory. In 1580, the astronomer Nicolaus Copernicus dared to expound a new theory—the heliocentric view—which places the sun at the center of the solar system. His work created considerable debate, but it eventually prevailed because it better fit the observations.

Because theories may require modifications or rejection, scientists must be open-minded and must be willing to analyze new evidence that throws into question their most cherished beliefs. For the most part, though, theories are talked about as if they were fact. Some people even object

[6]Unfortunately, the word *theory* is often used incorrectly in everyday conversation and writing to mean hypothesis.

to calling a theory a "theory" for fear that it makes it sound tentative.

Science Helps Shape Our Lives and Our Values

Besides forming the backbone of technology, science also bears heavily on political decisions. What we do or do not do about major environmental or medical problems depends, in part, on the scientific facts.

Most decisions in the public-policy arena, however, are not made on the basis of scientific facts. Many are made after considering human values—what we view as right or wrong—and economic needs. When humans values are framed in the absence of scientific knowledge, however, they can lead to a lopsided way of viewing the world. The political and economic decisions that emerge may be fundamentally flawed.

Few people realize it, but science can enrich our value system. Scientific principles, especially those derived from the discipline of ecology, can teach us a great deal about living on the planet. As I will point out in the final chapter of the book, these principles could help reshape human ethics with profound impacts on our actions. Today, scientific knowledge of the factors that sustain natural ecosystems is changing thinking about the way society conducts its affairs.

Science also broadens the mind. It helps us understand the interconnections among living things. It helps uncover vital relationships that are not obvious to most people, such as the role of bacteria in recycling nutrients. In widening our understanding of the relationships among living organisms and the environment, science helps us understand our dependence on other species and act more thoughtfully and compassionately. It therefore helps us widen our ethical boundaries, assisting us in learning to temper our actions.

Critical Thinking Helps Us Analyze Problems, Issues, and Information More Clearly

Science provides us with methods to learn facts. Another hidden benefit of your study of science is that it will help you learn to think more critically. **Critical thinking** is the capacity to distinguish between beliefs (what we believe to be true) and knowledge (facts that are well supported by research). In other words, critical thinking is a process by which we separate judgment from facts. It is our most ordered kind of thinking. In the strictest sense, it is objective analysis. It is not being critical, and it is not just thinking deeply about a subject, although that is necessary. Critical thinking is subjecting facts and conclusions to careful analysis, looking for weaknesses in logic and other errors of reasoning. Critical-thinking skills, therefore, are essential to analyzing a wide range of problems, issues, and information.

Table 1–3 summarizes eight critical-thinking rules to remember as you read the newspaper, watch the news,

DEBUNKING THE THEORY OF SPONTANEOUS GENERATION
Featuring the work of Aristotle, Redi, and Pasteur

The Greek philosopher and scientist Aristotle, who lived from 384 to 322 B.C., proposed a theory to explain the origin of living things. It was called the theory of **spontaneous generation** and asserted that living things arose spontaneously from innate matter. Mice, for instance, arose from a pile of hay and rags placed in the corner of a barn. Flies could be produced by first killing a bull, then burying it with its horns protruding from the ground. Flies emerged from one of the horns sawed off several days after burial.

As absurd as these recipes for spontaneous generation sound to us today, the view remained compelling to many respected scientists well into the 19th century. Debunking the theory of spontaneous generation, in fact, engaged some of the best scientific minds of the century and, as is often the case in science, resulted in findings that laid the foundation for other important discoveries, notably the germ theory of disease.

Although many scientists were involved in shooting down the theory of spontaneous generation, I will highlight the work of two prominent scientists, Francesco Redi and Louis Pasteur.

Redi, an Italian naturalist and physician, was one of the first scientists to refute spontaneous generation through experimentation. Around 1665, Redi performed a simple but compelling experiment to determine whether ordinary houseflies were spontaneously generated in rotting meat. Redi began by placing three small pieces of meat in three separate glass containers. The first one was covered with paper. The second was left open, and the third was covered with gauze. Left at room temperature, the meat quickly began to rot and attract swarms of flies.

Soon, the meat in the open container began to seethe with maggots—larvae hatched from fly eggs. The paper-covered container showed no evidence of maggots, nor did the meat in the gauze-covered container, although maggots did appear in the gauze itself. Redi's conclusion from this simple experiment was as follows: maggots (which give rise to flies) do not come from the meat itself but from the eggs deposited by flies.

Redi's experiments convinced many people that flies and other organisms did not arise by spontaneous generation, but it did not put the debate to rest.

Soon after his now-famous experiment, in fact, a Dutch linen merchant by the name of Anton Leeuwenhoek discovered bacteria using simple microscopes he had built. This discovery revived arguments for spontaneous generation on the microscopic level. Many scientists asserted that although flies and other organisms did not arise spontaneously, microorganisms probably did. As evidence, they cited studies showing that microorganisms could arise from boiled extracts of hay or meat. Nonliving plant and animal matter was thought to possess a vital, or life-generating, force that could give rise to these microorganisms.

The debate over spontaneous generation persisted for the next two hundred years. A number of studies showed that a sterilized medium sealed from the outside air remained free of bacteria, disproving spontaneous generation. But supporters of spontaneous generation argued that these experiments were flawed, for the experimenters had eliminated some vital force needed to give rise to new life.

In 1861, Louis Pasteur published the results of an experiment that helped put this debate to rest. He placed sterilized

---

listen to speeches, and study new subjects in school. Here is a brief description of each rule, with some examples to make them more meaningful.

The first rule of critical thinking is to take an active part in acquiring information. Critical thinking requires facts. Don't sit back and accept everything you read and hear. If you do, you may at times run the risk of learning only what the news media or biased special-interest groups want you to think. Question what you learn, and seek other sources of information. Gathering lots of information keeps you from falling into the trap of mistaking ignorance for perspective. By keeping an open mind to alternative views and new facts, you can develop an enlightened viewpoint yourself.

Second, critical thinking requires a clear understanding of all terms. As you study biology, you will encounter many terms. Learning new terms, although sometimes tedious, is essential. Without a clear understanding of terms, you will not be able to master this or any other subject.

Understanding terms and making sure that others define them in discussions also help bring clarity to a variety of issues. The Greek philosopher Socrates, in fact, destroyed many an argument in his time by insisting on clear, concise definitions of terms. As you analyze any information or issue, always be certain that you understand the terms—and make sure that others define their terms as well.

Third, critical thinking requires individuals to question the methods by which facts are derived. Were they derived from experimentation, or were they derived from casual, unscientific observations?

Proper experimentation in biology usually requires two groups: experimental and control. The **experimental group** is tested or manipulated in some way. The control group is not. Both groups are identical except for one variable, known as the **experimental variable**. In order to test the effect of excess vitamin ingestion on laboratory

broth in a sterilized swan-necked flask, a flask with a long curved neck (see ▷ Figure 1). The design of the flask permitted air to enter, thus eliminating criticism that he had destroyed any vital forces necessary for spontaneous generation, but it blocked airborne bacteria from entering. (Bacteria were probably deposited on the tube leading to the flask and thus were prevented from entering the broth.) Through his work, Pasteur clearly showed that bacteria could not arise spontaneously.

Besides helping to put an end to the debate about the origin of living organisms, Pasteur's work helped lay the foundation for the modern understanding of infectious disease. Before his work, some scientists argued that the diseases we now call infectious were caused by a malfunction in one or more body parts. This malfunction, they claimed, resulted in the production of poisons that caused the illness. The microbes present in the diseased individuals, they went on to say, arose spontaneously as a result of the malfunction and were not the cause of the disease per se.

This brief history points out two important lessons. First, scientific discovery is usually the result of many scientists, each working on different parts of the puzzle. Often, though, one scientist is credited for an important discovery that is, in reality, built on a foundation of scientific research by many others. Second, this brief discussion also shows how discoveries open up new ways of thinking.

▷ **FIGURE 1 Pasteur's Experiment** Pasteur's simple but elegant experiment helped debunk the theory of spontaneous generation. (*a*) These specially designed swan flasks allowed air to enter, but prevented bacteria from entering the broth. (*b*) The broth was boiled to kill microorganisms. (*c*) Microorganisms appeared only if the flasks' necks were removed. (*d*) No microbes grew if the neck remained intact.

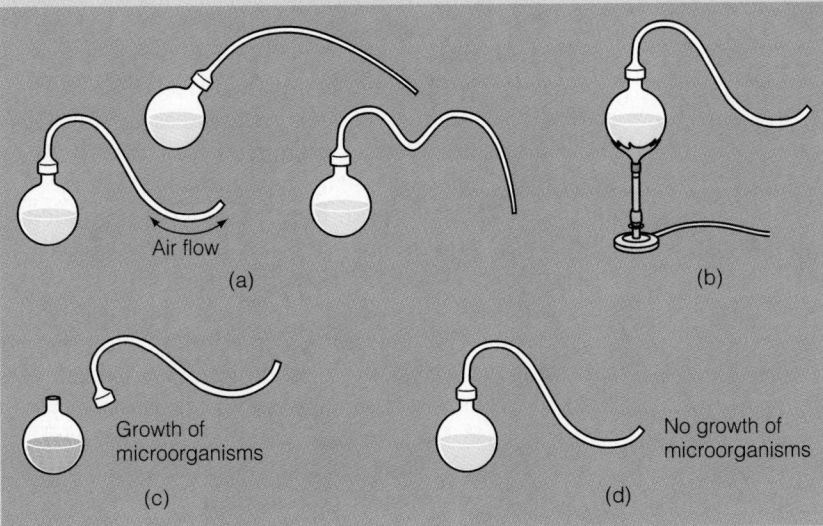

mice, for example, you would start with two groups of mice of the same age, sex, weight, genetic composition, and so on. Both groups would be treated the same throughout the experiment, receiving the same diet and being housed in the same type of cage at the same temperature. In a well-run experiment, the only difference in the two groups should be the vitamin supplements given to the experimental group. Consequently, any observed differences between the groups could be attributed to the treatment.

In studies of human health, especially many that will be cited in this book, choosing a control group and experimental group is not always easy. Unlike laboratory rats, we vary considerably in our genetic makeup, and no two of us have experienced identical conditions in our lives. For this reason, studies of humans are sometimes less reliable than those of other animals.

Besides having an experimental group and a control group, good experimentation requires an adequate number of subjects to be sure that any observed differences are real. Individual variation is natural, and, as a rule, the smaller the number of animals in each group, the larger the variation. In laboratory experiments, at least 10 test animals are generally required for reliable statistical analysis. Groups larger than 10 are even better. For human health studies, much larger groups are generally used.

When you analyze new information, then, first check to see how it was derived. If the results were obtained from experimentation, were the experiments well planned and executed? Did the experimenters use an adequate number of subjects? Did the experiment have a control group? Were the control and experimental groups treated identically except for the experimental variable? Did the results support the conclusions? (The critical-thinking exercise at the end of this chapter presents an experiment in which one young researcher failed to use a control group and reached an erroneous conclusion.)

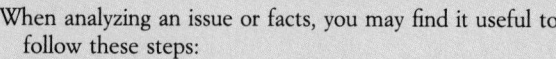

≋ TABLE 1-3 Critical-Thinking Skills ≋

When analyzing an issue or facts, you may find it useful to follow these steps:

1. Take an active part in gathering new information on the subject.
2. Understand and define all terms.
3. Question the methods by which data and information were derived:

- Were the facts derived from experiments?
- Were the experiments well executed?
- Did the experiment include a control group and an experimental group?
- Did the experiment include a sufficient number of subjects?
- Has the experiment been repeated?

4. Question the source of the information:

- Is the source reliable?
- Is the source an expert or supposed expert?

5. Question the conclusions:

- Are the conclusions appropriate?
- Was there enough information on which to base the conclusions?

6. Uncover assumptions and biases:

- Was the experimental design biased?
- Are there underlying assumptions that affect the conclusions?

7. Tolerate ambiguity:

- Don't expect all of the facts to be established.
- Expect controversy.

8. Examine the big picture:

- Look for multiple causes and effects.
- Look for hidden effects.
- Look for hidden relationships.

In science, one experiment is rarely adequate to allow firm conclusions. Careful scrutiny, for example, may show small but significant design flaws: perhaps the mice being tested were resistant to the drug under study or were hypersensitive to it. In some cases, a second researcher may repeat the experiment and find different results. This requires a third test. As a result, caution is often advisable when interpreting the results of new experiments.

Nowhere are caution and a large measure of healthy skepticism more necessary than in the announcements of scientific discoveries made on the television news, and in magazine and newspaper articles. Ever eager to showcase new scientific studies before they have been verified by further experimentation, the media sometimes fail to publicize follow-up studies that contradict early results. The general public is left with a false impression of reality.

Fourth, critical thinking requires us to question the source of the facts—that is, who is reporting them. Beware of "experts" who have a hidden agenda or have such a narrow

focus that they miss the big picture. So-called experts from industry who swear under oath to the safety of their product are likely to be biased or may be downright deceitful.

Also beware of people who may not know as much as you think they should. Many physicians, for instance, have little training in nutrition. Although we think of physicians as experts on human health, most of them received little or no training in nutrition in medical school. In the 1970s, one study reported that many hospital patients suffered severe malnutrition when under a doctor's care for prolonged periods. Although this research resulted in a dramatic re-analysis of nutritional instruction in medical schools, a study of U.S. medical schools in the mid-1980s showed that training in nutrition was still inadequate in many schools. Many medical students graduate without a full understanding of the role of nutrition in preventing disease and promoting good health.

Fifth, critical thinking requires us to question the conclusions derived from facts. Two questions you should ask when analyzing a study are: (1) Do the facts support the conclusions? and (2) Are there alternative conclusions? Consider an example. One of the earliest studies on lung cancer showed that people who consumed large quantities of table sugar had a higher incidence of the disease than those who ate moderate amounts. The researchers concluded that lung cancer was somehow caused by sugar. Did the facts support this conclusion? Was there an alternative explanation? Possibly. What about smoking or occupation?

A reexamination of the patients showed that the group with lung cancer had a noticeably higher percentage of smokers. Thus, the link between sugar and lung cancer was probably false.

This example illustrates a key principle of biological and medical research: *a correlation does not necessarily mean causation.* In other words, two factors that appear to be related may, in fact, not be linked at all. In 1989, for example, doctors found that many patients taking the amino acid L-tryptophan to help ease insomnia and symptoms of premenstrual syndrome (tension, irritability, and bloating that may occur in many women before menstruation) were becoming extremely ill. Physicians reported over 1200 cases of L-tryptophan "poisoning" to the Centers for Disease Control that year and as of February 1990, 12 people had died. Many others were left nearly paralyzed or severely impaired, and L-tryptophan was quickly taken off the market. Headlines in newspapers proclaimed its dangers. Further research, however, showed that the culprit was not L-tryptophan, as first thought, but rather a chemical contaminant in the pills.

The sixth rule of critical thinking is to look for assumptions and biases that may underlie conclusions. Popular health magazines, for example, sometimes promote taking megadoses of vitamins based on the belief that if a little vitamin is good for you, a lot is even better. Nutritionists point out that much of the work on the benefits of vitamin megadoses is derived from studies on rats and mice, which cannot always be extrapolated to humans. They also point

(a)

(b)

▷ **FIGURE 1–11 The Changing Face of Agriculture**
(*a*) For years farmers rotated crops to ensure soil fertility and to control insects. (*b*) Today many farmers plant acre after acre of the same crop, which is susceptible to disease and insects. When planted on the same field year after year, these crops deplete the soil of important nutrients that are not replenished by synthetic fertilizers.

out that large doses of many vitamins cause significant harm (Chapter 11).

Seventh, critical thinking sometimes requires a tolerance for uncertainty. Although this rule may seem contradictory at first, and maybe a little absurd in a section that seeks to crystallize your thinking, remember that hard and fast answers are not always available in science. We must be patient as scientists work out the details and must therefore be comfortable with uncertainty.

Scientists believe, for example, that the temperature of the Earth is warming as a result of carbon dioxide and other gaseous pollutants produced by modern society, which trap heat in the atmosphere. Many scientists stake their reputation on global warming and cite an impressive body of information in support of their view. (For more on this topic, see the Point/Counterpoint in Chapter 33.) Not all scientists agree, however. Some think that the conclusions of their colleagues, which are based on computer projections, may be in error. Because too many uncertainties exist to allow a firm conclusion, global warming may be an issue on which critical thinkers would reserve opinion, which leads us to the next rule.

Eighth, critical thinking requires us to examine the big picture. Global warming, if it occurs, could cause a dramatic shift in world climate that could turn productive farmland into desert and could melt glaciers and the Antarctic ice cap, flooding 20% of the world's land mass. The possibility of a drastic change in the climate of the planet therefore poses such a serious threat that steps to mitigate the problem may be merited, even without 100% certainty on the probability of the event.

Reducing the threat of global warming can be brought about by marked improvements in energy efficiency in automobiles, factories, office buildings, and homes (Chapter 33). These changes, say proponents of swift action, are a kind of insurance policy against a possible global disaster. They would also benefit the Earth and human society by reducing other forms of harmful pollution, stretching our

limited supplies of coal and oil, protecting wildlife habitat, and saving substantial amounts of money.

Consider another slightly different example of the benefits of examining the big picture. In 1988, researchers at the Monsanto Company announced that they had discovered a way to alter the genetic material of wheat to make it resistant to a harmful fungus that causes enormous crop damage. To control the fungus now, farmers sometimes rotate wheat from year to year with crops that do not support the pest (▷ Figure 1–11).

With the new genetically altered strain, the researchers say, farmers will no longer have to rotate their crops. In other words, they can plant their fields with wheat year after year and can even plant larger crops without worrying about fungus.

Although this practice may sound like a good idea, a closer, more critical analysis shows that it could be an invitation to disaster. First, crop rotation helps build soil fertility. Rotating beans, clover, alfalfa and other legumes with wheat, for example, adds nitrogen to the soil and helps maintain soil fertility. Not rotating crops often drains a soil of its nutrients, reducing its productivity over time. Second, crop rotation also helps reduce insect pest populations. By planting a new crop in a field every year, farmers reduce food sources for insects that tend to prefer one crop over another. Because the food supply is not constant from one year to the next, pest populations remain low and manageable. Eliminating crop rotation, therefore, is very likely to result in an outbreak of harmful wheat-eating insects and to require costly and potentially dangerous insecticides. Third, resistant crop varieties could eventually lead to the evolution of genetically resistant fungus populations—that is, populations that can overcome the plant's resistance. Thus, the new fungus-resistant species may not solve the fungus problem in the long term but only make it better for a while.

In the long run, it is possible that Monsanto scientists may be creating more problems than they are solving.

Elliot M. Katz

Elliot M. Katz, DVM (here with companion, Manco) is a graduate of the Cornell University School of Veterinary Medicine. He is president and founder of In Defense of Animals, a national animal rights organization.

Vivisection is the purposeful burning, drugging, blinding, infecting, irradiating, poisoning, shocking, addicting, shooting, freezing, and mutilating of perfectly healthy animals. In psychological studies, baby monkeys are separated from their mothers and driven insane; in smoking research, some dogs have tobacco smoke forced into their lungs through holes cut in their throats, while others have electrodes implanted in their penises to see how nicotine affects erection; in addiction research, chimpanzees are addicted to cocaine, heroin, alcohol, and amphetamines, then forced to go through convulsions and withdrawal symptoms; in vision studies, kittens are blinded and electrodes placed in their brains; in the classroom, mice and rats are thrown into convulsions and die in repetitive teaching demonstrations; in spinal cord and bone studies, the backs and legs of kittens and dogs are broken.

Started at a time when the scientific community did not believe animals felt pain, vivisection has left a legacy of animal suffering of unimaginable proportions. Descartes, the father of vivisection, said that the cries of a laboratory animal have no more meaning than the metallic squeak of an overwound clock spring. Though many members of the research community call vivisection a "necessary evil," a rapidly growing portion of the scientific, veterinary and medical communities see vivisection as simply *evil*.

As a veterinarian, I was taught and, like most of my colleagues, believed that vivisection was essential to medical progress, that all of it was carried out in a responsible manner, and that untold numbers of my friends and family would suffer and die if vivisection were stopped. In 1982, my eyes were opened to the reality of vivisection by a series of unfolding events on the University of California, Berkeley campus. Media reports described a long standing history of negligence, cruelty, and animal abuse. I investigated the situation by meeting with faculty and campus veterinarians.

I discovered an appalling reality. Animals were dying by the thousands from suffocation, brain infections, viral infections, heatstroke, and gangrene because vivisectors and administrators were unwilling to follow the recommendations of the campus veterinarians. It took a lawsuit filed by In Defense of Animals against the United States Department of Agriculture to end this egregious negligence and irresponsibility. Throughout this entire time, spokespersons for the biomedical community attempted to discredit the efforts of those of us who were working to end these tragedies. They implied that existing laws were adequate—that the federal Animal Welfare Act, Public Health Service and National Institutes of Health regulations would never allow such waste, cruelty and negligence to occur, and that vivisectors would never permit such horrors either. Their assertions were wrong then, and they are wrong today.

I also discovered that assertions by the biomedical community that vivisection is an indispensable part of protecting the public's health are incorrect. It has become quite apparent that in this day and age vivisection can be eliminated. The advent of sophisticated scanning technology, including computerized tomography (CT), positron emission tomography (PET) and magnetic resonance imaging (MRI), has given scientists the ability to examine people and animals non-invasively. The technology has isolated abnormalities in the brains of patients with Alzheimer's disease, epilepsy and autism, revolutionizing diagnosis and treatment of these diseases. Tissue and cell cultures can be, and are being, used to screen cancer and AIDS drugs. Progress in AIDS has come from areas entirely unrelated to animal experimentation. Clinical research and epidemiological studies revealed the modes of transmission of AIDS. Human skin cell cultures are used to test new products and drugs for toxicity and irritancy.

Why then is vivisection so entrenched and so defended with an almost religious fervor? Dr. Murry Cohen, a physician, summed it up when he stated: "change is difficult for most people, but it is particularly painful for scientific and medical bureaucracies, which fight to maintain the status quo, especially if required change might imply admission of previously-held incorrect ideas or flawed axioms." Vivisection continues today because of vested interests, habit, economics, and legal considerations, not for the real advancement of science or medicine.

Members of the general public when presented with the facts, almost unanimously express their desire to see an end to vivisection. Thousands of professionals like myself have re-evaluated the sense, efficacy, and worth of vivisection and have formed or joined organizations to demand an end to this outdated research methodology. The movement away from vivisection will lead to truly effective research, advance the cause of public health, and restore to medicine and science much needed compassion and caring.

ANIMAL RESEARCH IS ESSENTIAL TO HUMAN HEALTH

Frankie L. Trull

Frankie L. Trull is president of the foundation for Biomedical Research, a nonprofit educational organization dedicated to informing the public about the importance of humane animal research.

Virtually every major medical advance of the last 100 years—from chemotherapy to bypass surgery, from organ transplantation to joint replacement—has depended on research with animals. Animal studies have provided the scientific knowledge that allows health-care providers to improve the quality of life for humans and animals by preventing and treating diseases and disorders and by easing pain and suffering.

Some people question animal research on the ground that data from animals cannot be extrapolated to humans. But physicians and scientists agree that the many similarities that exist provide the best insights into the complex systems of both humans and animals. Knowledge gained from animal research has contributed to a dramatically increased human life span, which has increased from 47 years in 1900 to more than 75 years in 1988. Part of this increase can be attributed to improved sanitation and better hygiene; however, most of this increased longevity is a result of health and medical advances made possible in part through animal research.

Research on animals has also led to countless treatments, techniques, and medical technologies. Animal research was indispensable to the development of immunization against many diseases, including polio, mumps, measles, diphtheria, rubella, and hepatitis. One million insulin-dependent diabetics survive today because of the discovery of insulin and the study of diabetes using dogs, rabbits, rats, and mice. Organ transplantation, considered a dubious proposition just a few decades ago, has become commonplace because of research on mice, rats, rabbits, and dogs.

Animal research has contributed immeasurably to our understanding of tumors and led to the discoveries of most cancer treatments and therapies. Virtually all cardiovascular advances, including the heart-lung machine, the cardiac pacemaker, and the coronary bypass, could not have been possible without the use of animals. Other discoveries made possible through animal research include an understanding of DNA; X-rays; radiation therapy; hypertension; artificial hips, joints, and limbs; monoclonal antibodies; surgical dressings; ultrasound; the artificial heart; and the CT scan.

Animal research will be essential to medical progress in the future. With the use of animals, researchers are gaining understanding into the cause of—and treatments for—AIDS, Alzheimer's disease, cystic fibrosis, sudden infant death syndrome, and cancer in the hopes that these problems can be eliminated.

Researchers care about the welfare of laboratory animals. Like everybody else, scientists don't want to see animals suffer or die. It is not a controversial position; there is no constituency for inhumane treatment. In fact, treating animals humanely is good science; animals that are in poor health or under stress will provide inaccurate data.

Many people are under the false impression that laboratory animals are not protected by laws and regulations. In fact, many safeguards are in place to guarantee the welfare of animals used in research. A federal law, entitled the Animal Welfare Act, stipulates standards for care and treatment of all laboratory animals except rats, mice, and birds. Too, the U.S. Public Health Service (PHS), the country's major source of funding for biomedical research, sets forth requirements with which research institutions must comply in order to qualify for PHS grants for any biomedical research involving *any* kind of animal.

Both the Animal Welfare Act and the PHS animal welfare policy mandate review of all research by an animal-care committee set up in each institution to ensure that laboratory animals are being used responsibly and cared for humanely. The committee, which must include one veterinarian and one person unaffiliated with the institution, has the power to reject any research proposal and stop projects if it believes proper standards are not being met.

Those who work in medical science and see firsthand the effects of disease feel no ambivalence about the value of animal research. Although animal research opponents portray the medical community as deeply divided over the merits of animal experimentation, the percentage of physicians opposed to animal research remains very small. A 1989 survey by the American Medical Association found that 99% of the more than 500,000 active physicians who were asked believed animal research had contributed to medical progress, and 97% supported the continued use of animals for basic and clinical research.

The general public, when presented with the facts, has also been supportive of animal research. This support must not be allowed to erode through apathy or misconceptions put forth by those opposed to animal research. Should animal research be lost to the scientific community, the victims would not be scientists. The victims would be all people: our families, our neighbors, our fellow man.

≋ SHARPENING YOUR CRITICAL THINKING SKILLS

1. Summarize the main points of each author.
2. Can you spot flaws in either essay?
3. Do these authors use data or ethical arguments to make their cases?

Thus, looking more carefully at the big picture throws into question the wisdom of the action and shows how important critical thinking is to our society.

Examining the big picture also requires that we look for multiple causes and effects. The self-proclaimed "thinking man," *Homo sapiens,* exhibits remarkable intelligence. But as intelligent as our species is, we are also rather narrow-minded at times. That is, we frequently fall back on simplistic thinking when analyzing problems.

In the 1970s, for example, one noted ecologist argued vigorously that the world's environmental problems stemmed from overpopulation—too many people for the available resources. Another equally notable scientist argued that the problems were due to technology and its by-product, pollution.

A more critical analysis of the environmental crisis shows that our problems are the result of many factors. Overpopulation and technology are two of the many. Inadequate laws and education must be factored into the equation. So must various psychological and cultural factors—for instance, our view of the world as something to overcome. Many more could be added to the list.

Critical thinking demands a broader view of cause and effect. Look at a variety of parameters when you assess problems. In other words, you can avoid simplistic thinking by considering all of the contributing factors.

As you read this text, you will be presented with examples to help you sharpen your critical-thinking skills. The Point/Counterpoints will help you put these important tools into practice. You may also want to use your skills as you study the material. But whatever you do, enjoy this journey. It may be a once-in-a-lifetime opportunity.

SUMMARY

1. The Earth is a closed system, in which all of the materials necessary for life are recycled. The only outside contribution is sunlight, the source of energy for virtually all life.
2. Humans are an integral part of the biosphere, participating in a major way in the global recycling network that ensures life.

LIFE IN THE BALANCE: HOMEOSTASIS

3. One of the central concepts of biology is the principle of homeostasis. It states that all organisms have evolved mechanisms that ensure internal constancy in an ever-changing environment. These homeostatic mechanisms are vital to survival and reproduction in an ever-changing world.
4. Numerous mechanisms exist in our cells and our bodies to maintain the dynamic equilibrium commonly referred to as internal constancy. Similar mechanisms are also found in ecosystems.
5. The health of all species, even the biosphere itself, is dependent on homeostasis. When homeostatic systems break down, illnesses often result.
6. Human health has traditionally been defined as the absence of disease, but a broader definition of health is now emerging. Under this definition, good health is a state characterized by an absence of risk factors that lead to disease and a high level of mental and physical fitness.
7. Human health and the health of many species depends on a well-functioning, healthy ecosystem. Thus, alterations of the environment can have severe repercussions for species.

EVOLUTION: THE UNITY AND DIVERSITY OF LIFE

8. Another major principle of biology is evolution. The theory of evolution says that all life evolved from earlier forms. The process of evolution results in improvements in existing species, modifications that make a species better suited to its environment, and the formation of new species.
9. Evolution is responsible for the great variety of life forms, but since the Earth's organisms evolved from early cells that arose over 3.7 billion years ago, organisms share many common characteristics, listed in points 10–16 below.

10. All living things are made up of cells.
11. All organisms grow and maintain their complex organization by taking in chemicals and energy from the surroundings.
12. All living things have metabolism.
13. All organisms possess homeostatic mechanisms.
14. All organisms exhibit irritability—the capacity to perceive and respond to stimuli.
15. All organisms are capable of reproduction and growth.
16. All organisms are the product of evolutionary development and are subject to evolutionary change.
17. Although humans are similar to many other organisms, they also possess many unique abilities. One of the key differences is culture. In addition, humans can plan for the future and possess enormous abilities to reshape the Earth through ingenuity and technology. This, however, does not imply that humans are in any way superior to other forms or are separate from nature—just unique.

SCIENCE, CRITICAL THINKING, AND SOCIAL RESPONSIBILITY

18. Science is both a systematic method of discovery and a body of information about the world around us.
19. Scientists gather information and test ideas through the scientific method. The scientific method begins with observation and measurements, often made during experiments. Observations and measurements often lead to hypotheses, testable explanations of natural phenomena. Hypotheses are tested in experiments. The results of experiments help scientists support or refute their hypotheses.
20. The body of scientific knowledge also contains theories, or broad generalizations about the way the world works. Theories can change over time as new information becomes available.
21. Scientific discovery can influence ethics. New knowledge about our place in the biosphere, for example, may help temper the prevalent notion of human dominance and help humans the world over build a more sustainable relationship with nature.

22. Critical thinking is an integral part of science and is best defined as careful analysis that helps us distinguish knowledge from beliefs or judgments.

23. Critical thinking provides a way to analyze issues and information. It requires that you first define an issue, then study the evidence.

24. Table 1–3 summarizes the critical-thinking rules.

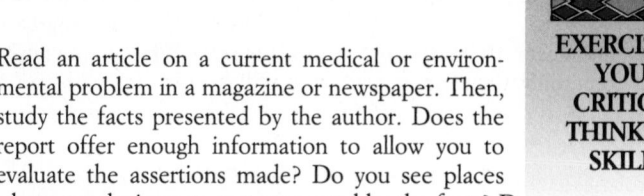

EXERCISING YOUR CRITICAL THINKING SKILLS

1. Read an article on a current medical or environmental problem in a magazine or newspaper. Then, study the facts presented by the author. Does the report offer enough information to allow you to evaluate the assertions made? Do you see places where conclusions are not supported by the facts? Do you see places where bias has entered into the author's writing? Do you see authorities cited to support issues who may not be as knowledgeable as you would like or who may be biased? Do you see faulty experiments?

2. Red-sided garter snakes breed in a mad frenzy in the spring in western Canada. In order to see what stimulates their sex drive, a young postdoctoral student, who will remain nameless, placed six red-sided garter snakes in cold storage in the fall. After weeks of hibernation, he warmed the reptiles to awaken them. He then implanted tiny capsules of testosterone under the snakes' skin. When the males were placed in cages with female snakes, each one displayed courtship behavior. After removing the implants, the researcher found that the males lost interest in sex. From this work, he suggested that high levels of the hormone testosterone found in the blood during the spring stimulated the male garter snake's sex drive.

Before reading on, make a list of flaws in the experimental design. You may want to review the material on experimentation on pages 14–15 and review the critical thinking rules in Table 1–3 before you perform this exercise.

What did you find? If you said he used a small group and failed to include a control group, you are right on both counts. In fact, had he included a control group with untreated males, he would have found that these males also showed intense interest in mating *without* testosterone implants. Had he studied testosterone levels in the controls, he would also have found that they were low in the spring during mating.

A more detailed study of the snake by the same researcher, now a professor at a prominent university, shows that testosterone levels in the snake's blood peak in the late summer, more than eight months before the breeding season. This suggests that the hormonal signal responsible for the male sex drive in red-sided garter snakes has a delayed effect. That is, it starts into motion certain nervous system changes that culminate in male courtship eight months later.

This is an unusual occurrence in the field of biology, for most hormones act fairly quickly—in a matter of hours or a few days. In fact, this is such an oddity among animals that it clearly requires more careful examination. Could there be another type of signal that corresponds more closely with the spring breeding season?

TEST OF TERMS

1. The _____ is the living skin of the planet, consisting of the plants, animals, and microorganisms.

2. _____ is the term that describes the internal constancy maintained by automatic mechanisms in plants and animals.

3. The pancreas produces a substance called _____ that lowers blood sugar after a meal to help maintain constancy.

4. _____ reproduction occurs when an organism forms offspring by cell division.

5. The chemical reactions occurring in an organism are collectively referred to as _____ .

6. _____ is the ability to perceive and respond to stimuli and is one of the chief characteristics of all life forms.

7. Changing environmental conditions may weed out certain members of a population of organisms, leaving only those that are capable of surviving and reproducing under the new conditions. This weeding-out process is called _____ _____ .

8. The process by which scientists gather and test knowledge about the world around us is called the _____ _____ .

9. _____ are tentative, testable explanations of observations.

10. A _____ is a principle of science supported by considerable scientific research.

Answers to the Test of Terms are given in Appendix B.

TEST OF CONCEPTS

1. How would you define life?

2. In what ways are a rock and a living organism (for example, a bird) similar, and in what ways are they different?

3. In what ways is planet care a form of self-care? Give as many examples as you can.

4. Describe the concept of homeostasis. How does it apply to humans? How does it apply to ecosystems? Give examples from your experiences.

5. In what ways are humans different

from other animals? In what ways are they similar?

6. Describe the scientific method, and give some examples of how you have used it recently in your own life.

7. How do a hypothesis and a theory differ?

8. How can a group of animals (including humans) serve as an experimental group and a control group in the same experiment?

9. Find examples of important issues in your daily newspaper on which scientific knowledge can help you make decisions.

10. List and discuss the eight critical thinking skills presented in this chapter. Which skills seem to be most important for the kind of thinking you normally do?

11. A graduate student injects 10 mice with a chemical commonly found in the environment and finds that all of his animals die within a few days. Eager to publish his results, the student comes to you, his adviser. What would you advise the student to do before publishing his results?

12. "It is meaningless to compare species . . . to say one is superior to another." Do you agree or disagree with this statement? Why or why not?

C H A P T E R

2

The Chemistry of Life

≈

Light micrograph of crystals of ATP molecules, which serve as energy carriers in all organisms

Surveys of public knowledge are almost always embarrassing. In a recent poll of world geography, for example, Americans (ages 18 to 24) scored tenth among industrial nations. Many Americans couldn't locate Egypt or Germany on a world map. Fourteen percent of those polled couldn't even pinpoint the United States. Like other similar surveys, this poll illustrates how uninformed many people are about the world around us.

Nowhere is public lack of knowledge more blatant than in the field of chemistry. Considering the importance of chemistry to our lives, it is surprising how little most people know about this fascinating field of study. Why is chemistry so important to us?

Chemistry is important, in part, because it offers so many useful applications. Today, tens of thousands of drugs developed in chemistry laboratories help humankind combat a wide range of often deadly or debilitating diseases (▷ Figure 2–1). Synthetic antibiotics have helped us make dramatic improvements in human health and have greatly reduced infant mortality.

Advances in chemistry have also led to many improvements in household products that make our lives safer. House paints are a case in point. At one time, most interior and exterior paints contained lead-based pigments. Because lead is toxic to the nervous system, chips of paint flaking off walls and ceilings pose a health hazard to young children, who routinely mouth foreign objects as they explore their new world. Modern paints lacking lead-based pigments are a definite improvement.

A little knowledge of chemistry is also useful to us in many other ways. It can, for example, help us understand environmental problems, such as ozone depletion, urban air pollution, and acid rain. Closer to home, it can help us determine which solvent to use to remove dirt or stains

from carpets. On a personal level, a knowledge of chemistry is important to those of us who want to eat well and live long, healthy lives. But why study chemistry at the beginning of a biology course?

The answer is simple. First, cells and organisms are elaborate combinations of **molecules**, structures composed of two or more atoms. Many molecules undergo chemical reactions that are the basis of all life. In order to understand the structure and function of cells and organisms, then, you must understand at least some of the basics of chemistry. Without a grounding in the basic principles of chemistry, the study of biology wouldn't be much different from studying medicine without a basic understanding of the structure and function of the human body.

This chapter introduces you to several key principles of chemistry, laying the foundation for the study of biology. It begins with a discussion of the structure of atoms and molecules and then looks at water, acids, bases, and buffers—providing information useful to you throughout your life. It concludes with an overview of the four major biological molecules encountered in the study of biology.

≈ ATOMS AND SUBATOMIC PARTICLES

The physical world in which we live is composed of a wide variety of matter—for example, wood, water, plastic, metal, air, and food. **Matter** is defined as anything that has mass and occupies space.[1] Technically, mass is a measure of the amount of matter in an object. Thus, a water balloon has a greater mass than an air-filled balloon, as anyone who has ever been hit by the former will tell you.

Atoms Are the Fundamental Unit of All Matter

Matter is composed of tiny particles called atoms. The word *atom* comes from the Greek *atomos,* which means "incapable of division." The ancient Greek philosophers who proposed the existence of atoms believed that all matter consisted of small particles that lacked any internal structure. Over time, though, evidence from a variety of experiments has shown that atoms consist of smaller particles, the **subatomic particles.** Subatomic particles can be obtained from an atom by physical methods—for instance, by smashing an atom with a particle traveling at high speed. So far, more than 100 subatomic particles have been identified, and the search for more continues. The role of most subatomic particles in the atom is not well understood.

The current atomic theory includes three well-known subatomic particles: electrons, protons, and neutrons (▷ Figure 2–2). **Protons** are the heaviest of the subatomic particles and are located in the center of the atom, the **nucleus** (Figure 2–2). Each proton has a positive charge.

▷ **FIGURE 2–1 Better Living through Chemistry** Store shelves are packed with drugs to cure everything from warts to ugly facial hair. We are a chemical society, dependent on chemistry's benefits and subject to its hazards.

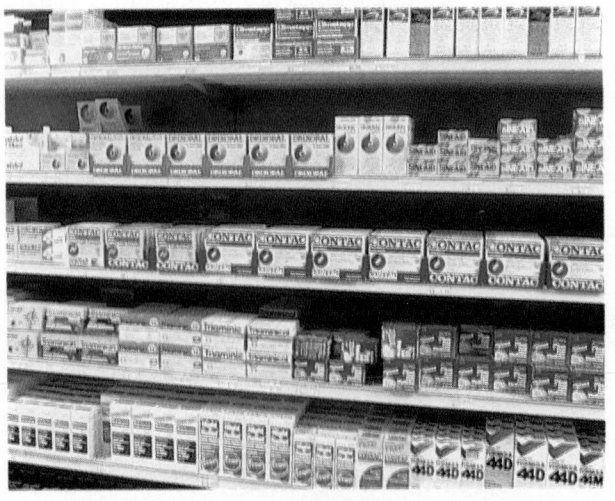

---

[1] Mass is a measure of the amount of matter in an object. Two balls of equal size, one filled with air, the other filled with lead, have markedly different masses.

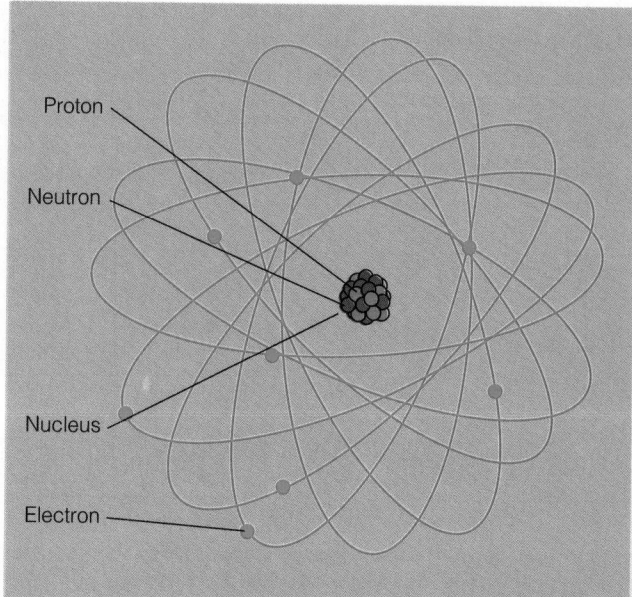

FIGURE 2–2 The Atom The atom consists of two regions. The central nucleus contains protons and neutrons and makes up 99.9% of the mass. Surrounding the nucleus is the electron cloud, where the electrons spin furiously around the nucleus. (Figure not drawn to scale.)

▷ FIGURE 2–3 The Elements Elements are the purest form of matter. They consist of tiny particles called atoms.

The Elements Are the Purest Form of Matter

Early scientists accounted for the diversity of matter by assuming that there must exist a number of pure substances and that they could be combined into an infinite variety of different forms. They called these pure substances **elements.** Today, modern chemists define an element as a pure substance (like gold or lead) that contains only one type of atom. Elements are substances that cannot be separated into different substances by ordinary chemical means. Some common examples are aluminum, hydrogen, tin, silver, and carbon.

To date, 92 naturally occurring elements have been discovered (▷ Figure 2–3). Another dozen or so can be made in the laboratory. Of the 92 naturally occurring elements, only about 20 are found in organisms. Four elements in this group: carbon, oxygen, hydrogen, and nitrogen (remember: COHN) comprise 98% of the atoms of all living things.

Elements are listed on a chart called the **periodic table of elements** (Table 2–1; Appendix A). The periodic table is a handy summary of vital statistics on each element. As illustrated in Table 2–1, the elements are represented on the table by a one- or two-letter symbol. (There are a few three-letter symbols.) The element hydrogen, for example, is indicated by H. Oxygen is designated by O, and carbon is represented by C. Helium is signified by He.

Above each symbol on the periodic table is its atomic number. The **atomic number** of an element is the number of protons found in the nuclei of the element's atoms. Because all atoms are electrically neutral, the atomic number also equals the number of electrons in the atoms. The elements are listed on the periodic table in ascending numerical order in rows, with the lowest numbers on the left side. Below the chemical symbol for each element is the mass number. The **mass number** is the average mass of the atoms of each element. Mass number (also called atomic mass) is measured in **atomic mass units** and is approximately equal to the number of protons and neutrons in a given atom. Protons and neutrons each have an atomic weight of 1 atomic mass unit. (Recall that neutrons

Also found in the nuclei of virtually all atoms are **neutrons,** uncharged particles slightly lighter than protons.

Most of the mass of an atom comes from the protons and neutrons, which are so tightly packed in the nucleus and are so heavy that a single cubic centimeter (about 1/3 teaspoon) of protons and neutrons would weigh 100 million tons.

Because protons are positively charged and neutrons have no charge, the nucleus of an atom carries a positive charge. Surrounding the positively charged nucleus is a region called the **electron cloud.** It consists of **electrons,** tiny, negatively charged particles. Electrons move rapidly through a large volume of space surrounding the nucleus and are held in orbit, in large part, by the electrostatic attraction between them and the positively charged protons of the nucleus. The more energetic electrons tend to orbit the nucleus in the outer regions of the electron cloud, and the less energetic ones remain close to the nucleus.

Carl Sagan, one of America's leading astronomers, once remarked that atoms are mostly space, or "electron fluff," referring to the fact that the electron cloud accounts for nearly all the volume of an atom but virtually none of its mass. In fact, if you could enlarge an atom to the size of Mount Everest (29,028 feet), the nucleus would be about the size of a football. The electrons would spin around the nucleus in the vast electron cloud, a space 29,000 feet in diameter.

Despite the fact that they contain charged particles, atoms are electrically neutral. Because they contain the same number of protons and electrons, the positive charges cancel out the negative ones.

I	II	III	IV	V	VI	VII	VIII
1 H Hydrogen 1.0	Atomic number Atomic symbol Mass number						2 He Helium 4.0
3 Li Lithium 7.0	4 Be Beryllium 9.0	5 B Boron 11.0	6 C Carbon 12.0	7 N Nitrogen 14.0	8 O Oxygen 16.0	9 F Fluorine 19.0	10 Ne Neon 20.2
11 Na Sodium 23.0	12 Mg Magnesium 24.3	13 Al Aluminum 27.0	14 Si Silicon 28.1	15 P Phosphorus 31.0	16 S Sulfur 32.1	17 Cl Chlorine 35.5	18 Ar Argon 40.0
19 K Potassium 39.1	30 Ca Calcium 40.1						

Note that each element is represented by a one- or two-letter symbol. Elements are listed according to their atomic number (the number of protons). Their atomic mass is also shown.

are actually slightly lighter than protons.) Because electrons have virtually no mass, they contribute very little to the mass of an atom.

Before moving on, let's take a few moments to study the periodic table in Table 2–1. We'll start with hydrogen, the first entry. As indicated, the atomic number of hydrogen is 1. This means that all hydrogen atoms have one proton. Hydrogen atoms also have one electron each, but because electrons are so light, the mass number of most hydrogen atoms is only 1.

Helium is the next element on the chart. Its atomic number is 2. How many protons does helium have? How many electrons? If you said it has two protons and two electrons, you are right. Now look at the mass number. The table lists the mass number as 4, which means that helium must contain two protons and two neutrons.

Let's move to the element carbon. As indicated, the atomic number of carbon is 6, but the mass number is 12. This means that carbon atoms contain six protons and six neutrons. Interestingly, a small fraction of the carbon atoms have seven neutrons, and another subgroup contains eight (▷ Figure 2–4). These alternative forms of the carbon atom are known as **isotopes.**

Isotopes Are Alternative Forms of Atoms, Differing in the Number of Neutrons They Contain

Most elements are made up of mixtures of two or more isotopes. Hydrogen, for instance, has three forms. The most prevalent hydrogen atom contains one proton and one electron—and no neutrons. It is represented by the symbol $_1^1H$. The upper 1 is the mass number, and the lower

▷ **FIGURE 2–4 Isotopes of Carbon** (a) Carbon-12 is the main isotope of carbon; it constitutes 99.89% of the naturally occurring carbon in the world. (b) Carbon-13 forms 1.1% of the world's carbon. (c) Carbon-14 is found only in trace amounts. All three isotopes have six electrons; carbon-14 is radioactive, but

carbon-12 and carbon-13 are not. These atoms are drawn according to the Bohr model, with the nucleus in the center surrounded by concentric circles. Each circle represents an energy level, and all of the electrons within a given energy level are drawn on the circle.

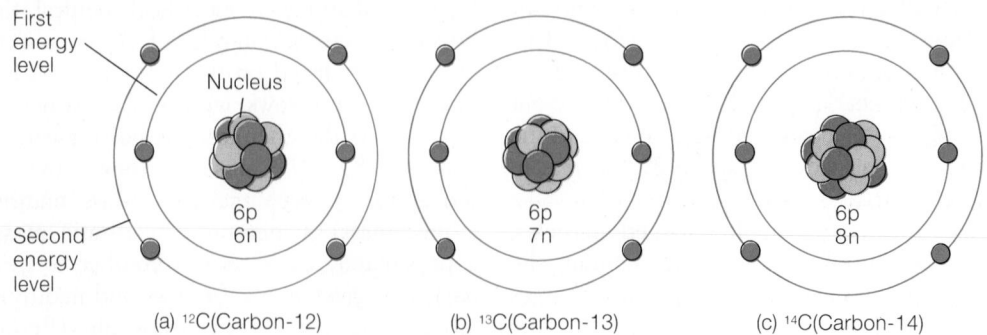

(a) ^{12}C(Carbon-12)   (b) ^{13}C(Carbon-13)   (c) ^{14}C(Carbon-14)

RADIOACTIVE MARKERS
Featuring the Work of Schoenheimer

Biochemist Rudolf Schoenheimer (1898–1941) and his colleagues offered biological science a tool that would yield important new information and result in significant scientific and medical advances. The tool is radioactively labeled molecules—that is, common biological molecules to which are attached radioactive isotopes, such as two isotopes of hydrogen, hydrogen-2 or hydrogen-3.

A radioactive atom incorporated in a biological molecule is akin to a radiotransmitter collar on a grizzly bear. The atom emits tiny bursts of radioactivity, which permit scientists to track molecules on their course through the body, in much the same way that the collar emits radio signals that allow wildlife biologists to electronically track the bear through its forbidding terrain.

Marking common molecules like amino acids with radioactive atoms permits scientists not just to track molecules but also to follow what happens to them along the way and, of course, to determine their fate. Before the emergence of this revolutionary new technique, substances such as fatty acids or amino acids administered to an animal were lost almost immediately, becausethey were indistinguishable from molecules in the body. As a result, there was no way of determining their fate. For example, if substance A were administered to an animal and an excess of substance B soon showed up, one could not be sure that the original substance had been converted into B. It might, for instance, have merely stimulated B's production.

Radioactive markers therefore permit scientists to distinguish labeled molecules from "native" ones and to follow these marked molecules as they progress through the body's maze of biochemical reactions. Radioactive markers also let scientists determine excretion and storage sites, as well as rates at which various processes, such as excretion, occur. In fact, many of the physiological processes you will read about in this book—and particularly those in Part III—were determined by using radioactively labeled molecules.

Clearly, radioactive markers are a godsend to modern science. Before Schoenheimer's discovery, in fact, the inability to track molecules resulted in a huge gulf in our knowledge about metabolism and other body functions. Futile attempts to "tag" molecules with phenyl groups or other molecular markers generally failed, because the altered molecules differed so much from their natural substances that they were treated differently by the body.

Through experimentation Schoenheimer showed that radioactively labeled molecules were so similar to the unlabeled molecule as to be indistinguishable, thus making them ideal markers and opening the door to many additional discoveries.

From his brilliantly conceived experiments, a new picture of living things emerged. For example, he and his colleagues showed that the "permanent" structural elements of the body, such as DNA, which had previously been considered stable, were actually in a continuous state of flux. Even apparently dormant fat deposits in the body were not unchanging. In addition, studies showed that amino acids were rapidly broken down and rebuilt into new ones.

Since their introduction in the 1930s, radioactive markers have been widely used and have yielded some important insights into biology and other sciences.

Radioactively labeled molecules are also used in medicine to determine the physiological status of various organs. Radioactive iodide ions, for instance, are used to assess the function of the thyroid gland, a hormone-producing gland in the throat. Radioactively labeled glucose molecules can be used to determine the relative activity of different parts of the brain, permitting scientists to more accurately map brain functions. These are but a few examples of the tremendous contribution of radioactively labeled molecules to medicine and science.

---

1 is the atomic number. The next most common form has one neutron and is represented by $_1^2H$. The 2 refers to the number of protons and neutrons, and the 1 indicates the atomic number (number of protons). The least common hydrogen atom has two neutrons and is indicated by $_1^3H$.

The three isotopes of hydrogen can also be indicated by the elemental names followed by the mass number. In conversation, then, a chemist might talk about hydrogen-1, hydrogen-2, or hydrogen-3. Three isotopes of carbon are shown in Figure 2–4.

Additional neutrons often make isotopes of various elements unstable. In order to achieve a more stable state, many isotopes carrying an extra neutron or two release **radiation**—small bursts of energy, or tiny energetic particles—from their nuclei. Radiation carries away energy and

mass from an atom's nucleus and helps atoms achieve more stable states.

Radiation has provided numerous benefits to humankind, many of which were discovered by scientists at the turn of the century. Since then, scientists have found a wide variety of additional uses for radiation in medicine and research, including X-rays and radioactive tracers used in medical tests. One of the most important to the advancement of biological knowledge is the use of radioactive isotopes as markers in biological experiments, a topic discussed in Scientific Discoveries 2–1. Another extremely important application is the use of naturally occurring radioactive isotopes to date rocks and fossils. Information from this technique, described more fully in Chapter 23, has helped scientists pinpoint important evolutionary

FOOD IRRADIATION: TOO MANY QUESTIONS

Donald B. Louria

Donald B. Louria teaches at the New Jersey Medical School, where he is the chairman of the Department of Preventive Medicine and Community Health.

The debate about the safety of irradiating foods with large doses of cobalt-60 or cesium-137 raises five salient issues:

1. **The safety issue.** Irradiated food does not become radioactive, but the radiation causes chemical changes. The major concerns are over possible deleterious effects on the elderly and on malnourished persons. Some animal experiments raise concerns about the former. A highly controversial but flawed study on small numbers of children in India suggested that consumption of freshly irradiated wheat could produce chromosomal abnormalities. However, a study of a much larger group of healthy adults showed no such abnormalities. The proponents of food irradiation have derided the first Indian study and pointed to the larger study that showed no problems. But that does not answer the concerns. If food irradiation becomes an accepted technique, millions of malnourished people will eat irradiated foods. What is needed is a careful study with adequate numbers of malnourished children and adults. I am prepared to believe that consumption of irradiated foods is unlikely to produce significant harm, but such a belief must be supported by proper data.

2. **The nutrition issue.** Irradiated food loses some of its nutritional value; the extent of loss of vitamin content depends on the type of food and the dose of radiation—the higher the radiation dose, the greater the loss. Furthermore, some evidence suggests that when irradiated foods are processed (frozen, thawed, heated), there is an accelerated loss of vitamins. The irradiation proponents suggest that the effects of irradiation on vitamin content are not different from the effects of processing foods in conventional ways (heating, freezing, and the like). They also maintain that the diet in the United States contains redundant vitamins, so some loss through irradiation would not be of concern. Of course, this would not be true for millions of people in other countries, people over 60 in the United States, and people with various diseases. Arguing that our diet provides adequate vitamins and that destroying the vitamin content of foods by irradiation is of no importance is not likely to be viewed with equanimity by a large proportion of the American public.

3. **The necessity of this technology.** Proponents say that food irradiation will reduce diarrheal illness from infected poultry. However, proper cooking of the poultry would be just as effective. Proponents say irradiating meats will reduce the dangers of trichinosis, a parasitic disease, but trichinosis occurs infrequently. Will irradiating foods prolong their shelf lives and thus feed the world? Shelf lives will be increased: that is an important benefit for the less-developed world, but the proponents have provided no adequate data on the extent of the benefit. It is likely that the foods will be sold primarily in affluent countries where the shelf life issue is of less concern.

4. **The issue of safer competing technologies.** Food irradiation could reduce the use of toxic postharvest fumigants, chemicals applied to foods in storage. During the next decades, scientists will develop food crops that resist pests, grains that do not require fumigants, and foods that have longer shelf lives. Advances in biotechnology are likely to give us much safer alternatives to food irradiation, if we will only have a little patience.

5. **The environmental pollution issue.** It seems a bit incongruous that as our society is becoming increasingly concerned about our inability to get rid of nuclear wastes, we appear to be endorsing food irradiation, a technology that will result in numerous food irradiation plants in the United States. The few irradiation plants operating in the United States have contaminated their workers and the environment. Imagine the potential for contamination if there were hundreds of such plants using radioactive materials.

Food irradiation does have potential benefits: it also raises substantial concerns. Whether we should adopt the technology is obviously a matter for continuing debate. Certainly the technology should not be adopted until the issues have been resolved.

---

events and given us important insights into the history of life on Earth.

As in many other cases, the exploitation of radiation has not occurred without a price. For example, radiation exposure damages molecules in body cells and causes changes (mutations) in the genetic material that can result in birth defects and cancer. Individuals whose acne was treated with radiation in the 1960s, for example, are more likely to contract skin cancer.

Despite these and other problems, radiation is still widely used today. It is even being used to sterilize foods, thus eliminating refrigeration (▷ Figure 2–5). The Point/Counterpoint examines the pros and cons of this process.

FOOD IRRADIATION: SAFE AND SOUND

George G. Giddings

George G. Giddings has been involved in food irradiation research and development since 1963 and has written and spoken extensively on the subject.

Irradiated foods are safe and wholesome. The ionizing radiation process applied to foods offers certain proven public health and economic benefits without significant public health risks when carried out according to well-established principles and procedures. Decades of worldwide research and testing by competent, knowledgeable, objective, and responsible scientists led to this conclusion. This conclusion is also supported by some 30 years of experience in the sterilization of medical devices and other health care products to prevent infections, plus a growing list of other industrial and consumer products, including foods and their raw materials, ingredients, and packaging materials.

There is organized political opposition to food irradiation by a network of special-interest activists serving various political agendas, notably the antinuclear/antiradiation one. This network includes a handful of scientists and medical professionals from other fields who act as "expert witnesses" against food irradiation to serve their hidden agendas. Despite some short-term political successes, their campaign is doomed to failure in the longer term in the face of the unshakable facts, including a growing appreciation for public health and other proven benefits of food irradiation, and its growing worldwide regulatory approval, industrial usage, and public acceptance.

Ionizing-radiation processing is undoubtedly the most versatile physical process yet applied to food materials in terms of the range and variety of objectives it can accomplish. Radiation can:

- inhibit the sprouting of foods, such as potatoes and onions, and delay spoilage
- rid fruits, vegetables, and grains of insect pests
- prevent parasites from infecting consumers of fish and meats
- delay microbial spoilage of a wide variety of animal and plant products by reducing microbe levels
- sterilize packaging materials, eliminating microorganisms that would otherwise contaminate products
- sterilize a wide variety of prepared or cooked foods such as meat, poultry, and fishery products, which have already been used to feed astronauts and immune-compromised patients

All of these beneficial effects and more can be readily accomplished by the application of ionizing (gamma, electron, and X-ray) radiation according to well-established principles and procedures. Further, all of these benefits are already being realized in one country after another as food irradiation gradually increases worldwide. Nevertheless, irradiation must compete with a number of other new technologies. As a result, it is not likely to be applied to all, or even a high percentage, of the national and world food supply. It will therefore be used in cases in which it is clearly the best all-around choice.

≈ SHARPENING YOUR CRITICAL THINKING SKILLS

1. What is food irradiation? Why is it used?
2. List and summarize the key points of each author.
3. Using your critical thinking skills, analyze each author's position. Are you inclined to agree with either one on all issues? Why or why not?
4. In your opinion, what is the best course for the development and implementation of this technology?
5. What facts would you need to know to form a personal conclusion about this issue?

≈ CHEMICAL BONDS AND MOLECULES

The diversity of matter in the world around us is, as noted above, a result of the presence of a wide range of elements, each with its own unique properties. It also results from the fact that atoms can combine to form new substances.

The forces that unite atoms to create these "atomic partnerships" are called **bonds** and will be the central focus of this section. Before we proceed, however, we must take a closer look at the atom to study the arrangement of electrons, because electrons play a central role in chemical bonding.

▷ **FIGURE 2–5 Food Irradiation** Passing radiation through food kills bacteria and other microorganisms, thus eliminating decay. These products tend to keep longer because radiation has killed bacteria that would cause them to rot.

Electrons Are Found in Regions Called Orbitals

Electrons whiz around the nucleus in specific regions, or **shells,** within the electron cloud. These are not hard, encased enclosures, as the name implies, but rather regions of the amorphous electron cloud where an electron is likely to be found.

Each electron shell can hold a certain number of electrons, but no more. The first shell lying close to the nucleus, for example, holds up to two electrons. The electron in an atom such as hydrogen is located in the first shell. Helium, the next element on the periodic table, has two electrons, both of which are located in the first shell.

As noted in the far-right column of Table 2–2, the second shell can hold up to 8 electrons. In lithium, which has 3 electrons, 2 electrons are located in the first shell and the third is found in the second shell. Beryllium, the 4th element on the table, has 4 electrons; 2 are in the first shell and 2 in the second. Neon, the 10th element on the periodic table, has 10 electrons—2 in the first shell and 8 in the second shell.

Electrons within a shell actually occupy specific regions, referred to as **atomic orbitals** or **subshells.** Atomic orbitals are not like the orbits of the planets, which are rather fixed trajectories in space. An orbital is really a subregion within a shell occupied by one or more electrons.

Orbitals have very definite shapes, as shown in ▷ Figure 2–6. As illustrated, the *s* orbitals are spherical, the *p* orbitals are dumbbell-shaped, and the *d* orbitals are shaped like three-dimensional four-leaf clovers.

To understand how shells and orbitals are related, take a look at Table 2–2. It notes that the first shell has only one orbital, an *s* orbital, which holds 2 electrons. The electrons in the *s* orbital move around inside the volume in paths that cannot be determined. The second shell has two orbitals, an *s* orbital and a *p* orbital. The *s* orbital can hold up to 2 electrons, and the *p* orbital contains up to 6. Thus, neon atoms, which contain 10 electrons, each have two shells; the first shell contains one *s* orbital, and the second shell contains one *s* and one *p* orbital.

Additional shells are also present. Table 2–2 shows the shells and orbitals in atoms with up to 60 electrons. As you can see, the number of orbitals in a shell corresponds to the shell number. That is, the first shell has one orbital, the second shell has two, and so on.

Atoms Bond to Form More Stable Configurations

The arrangement of electrons in shells and orbitals is important because it helps us understand the chemical reactivity of the elements—that is, how atoms bond to one another. In general, atoms tend to react with others to

SHELL NUMBER (N)	NUMBER OF SUBSHELLS IN SHELL	ORBITAL DESIGNATION	MAXIMUM NUMBER OF ELECTRONS IN EACH ORBITAL	MAXIMUM NUMBER OF ELECTRONS IN SHELL
		TABLE 2–2 Relationships between Shells, Subshells, Orbitals, and Electrons		
1	1	1s	2	2
2	2	2s	2	
		2p	6	8
3	3	3s	2	
		3p	6	
		3d	10	18
4	4	4s	2	
		4p	6	
		4d	10	
		4f	14	32

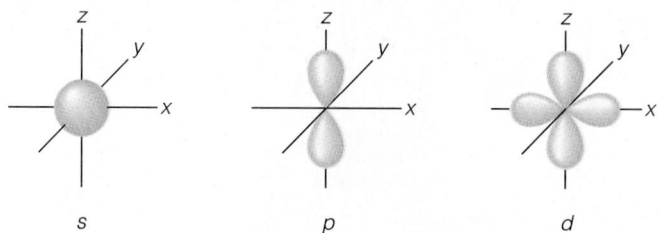

▷ **FIGURE 2–6 Orbitals** Each orbital has a definite shape. The electrons can be found within orbitals.

form complete outer shells—which are more stable atomic configurations.

During chemical reactions, electrons are either passed from one atom to another or are shared between two atoms. The result is that both participants end up with a complete outer shell. This rule is commonly called the **octet rule.**[2]

Ionic Bonds Are Electrostatic Attractions Between Two Oppositely Charged Particles

During some chemical reactions, the octet rule is satisfied when electrons are transferred from one atom to another. Transferring an outer-shell electron from one atom to another results in the production of charged atoms, called **ions.** An electrostatic attraction is set up between the two charged particles. This weak attractive force holds them together and is called an **ionic bond.**

▷ Figure 2–7 illustrates how an ionic bond forms between a chlorine atom and a sodium atom. As illustrated, the chlorine atom contains 17 electrons, 2 in the first shell, 8 in the second, and 7 in the third, or outermost, shell. Sodium contains 11 electrons—2 in the first shell, 8 in the second, and 1 in the third. Because the outer-shell elec-

[2]The name refers to the fact that the outer shells of many atoms are full when they have eight electrons.

trons are the only ones involved in chemical reactions, we will focus our attention on them.

As illustrated, sodium "donates" an electron to the chlorine atom, essentially completing the outer shell of the chlorine atom and forming a more stable configuration. In the process, the sodium atom empties its outermost shell. The second shell of sodium, however, is full, and this configuration is also stable.

As a result of the electron transfer, the sodium atom becomes positively charged and is now represented by the symbol Na^+. Chlorine, in turn, gains an electron and is converted to a negatively charged ion, the **chloride ion,** represented by the symbol Cl^-. The sodium and chloride ions are drawn toward each other by an electrostatic force. That force is the ionic bond.

Single molecules of sodium chloride do not exist in nature. Instead, sodium and chloride ions combine to form crystals containing numerous sodium and chloride ions. Sodium chloride is therefore known as an **ionic compound,** a substance composed of two or more different ions held in place by ionic bonds.

▷ Figure 2–8a shows the arrangement of sodium and chloride ions in sodium chloride crystals. As illustrated, each sodium ion is surrounded by six chloride ions. In addition, each chloride ion is surrounded by six sodium ions. The ions in crystals arrange themselves in such a way that attractive forces between unlike charges are maximized and repulsive forces between like charges are minimized.

In water, sodium chloride crystals break apart, releasing sodium and chloride ions into solution (Figure 2–8b). In solution, the sodium and chloride ions exist independently.

Ionic bonds form between atoms with markedly different affinities for electrons. In the sodium chloride example, sodium atoms have a low affinity for electrons, whereas chlorine atoms have a high affinity. Thus, sodium gives up its electrons to chlorine atoms. As a general rule, the fewer the number of electrons in the outer shell of an atom, the lower its affinity for electrons. The greater the number

▷ **FIGURE 2–7 Ions and Ionic Bonds** Sodium (Na) and chlorine (Cl) atoms differ in their electron affinity. Sodium has a weaker affinity and tends to give up an electron to atoms like chlorine. As a result, sodium becomes a positively charged ion, and chlorine becomes a negatively charged ion. The oppositely charged ions attract each other, forming an ionic bond.

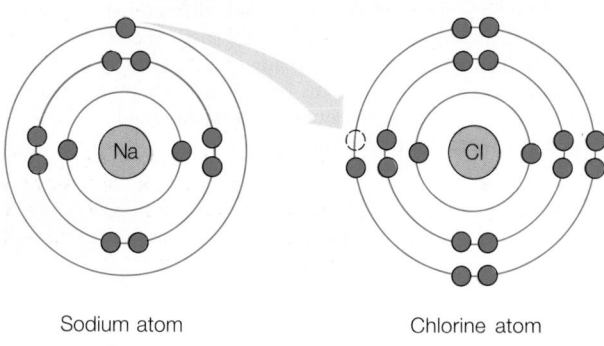

Sodium atom   Chlorine atom

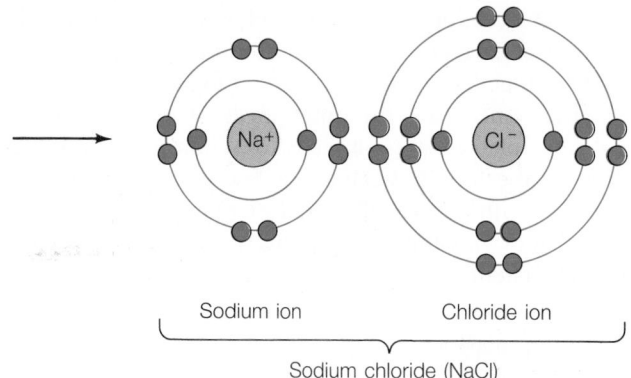

Sodium ion   Chloride ion

Sodium chloride (NaCl)

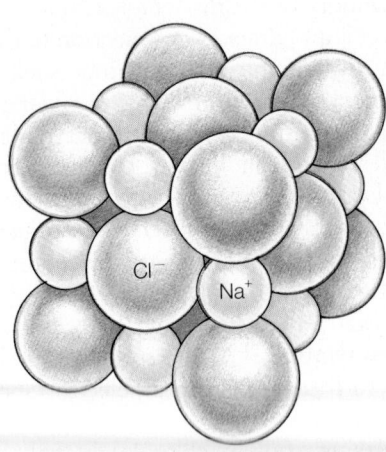

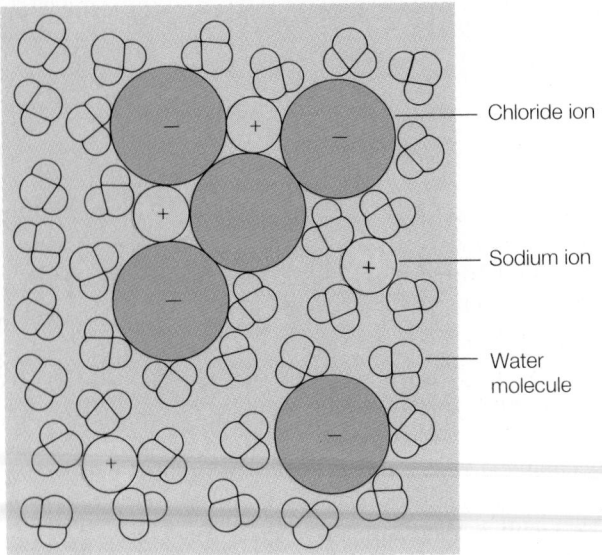

Chloride ion

Sodium ion

Water molecule

(a) Arrangement of the Na⁺ and Cl⁻ in a sodium chloride crystal.

(b) Breakup of salt crystal by water molecules.

▷ **FIGURE 2–8 A Sodium Chloride Crystal and Its Dissolution** (*a*) Sodium chloride crystal. Each sodium ion is surrounded by six chloride ions, and each chloride ion is surrounded by six sodium ions. (*b*) Dissolution of sodium chloride. Water molecules are polar and are attracted to both positively and negatively charged ions. When the sodium chloride crystal is dropped into water, the water molecules pull ions from the salt and tend to push themselves between the positively and negatively charged ions of the crystal, breaking it apart. When this occurs, sodium chloride is said to be dissolved.

(that is, the closer a shell is to being full), the greater the affinity. On the periodic table, elements on the left side tend to be electron donors, and elements on the right side tend to be electron acceptors.[3] (Take a moment to note where sodium and chlorine are located.)

Covalent Bonds are Formed by the Sharing of Electrons Between Atoms

Ionic bonds form because one atom "gives up" its outer shell electron(s) to another. In contrast, **covalent bonds** occur when two atoms "share" outer-shell electrons so that both atoms fill their outer shells. This "sharing" of electrons satisfies the octet rule and, incidentally, produces a much stronger bond.

Unlike ionic bonds, covalent bonds form between two atoms of approximately equal electron affinity. Consider the formation of hydrogen gas, H_2, from two hydrogen atoms.

Two hydrogen atoms, each with one electron, unite to form a molecule of hydrogen gas, H_2. In this molecule, the electrons of each atom now whiz around both nuclei, holding them together as illustrated in ▷ Figure 2–9. Because this arrangement fills the outer shells of both atoms—at least most of the time—it leads to a more stable configuration of the atoms. Consider a few common examples, beginning with a molecule called methane.

Methane is a flammable gas given off by rotting vegetation. The chemical formula of methane is CH_4. This indicates that the molecule contains four hydrogen atoms bonded to one carbon atom. To understand how the molecule forms, we begin with carbon, its central atom.

Carbon atoms have six electrons, two in the inner shell and four in the outer shell. A full outer shell would contain eight electrons. In order for a carbon atom to fill its outer shell (and reach a more stable state), it must acquire four more electrons. It can do this by covalently bonding to four hydrogen atoms.

Covalent bonding can be simplified by representing the outer-shell electrons of an atom by dots and x's, as shown in ▷ Figure 2–10. I'll use this technique to show the electron sharing occurring in a molecule of methane.

In this example, carbon is represented by the letter C surrounded by four outer shell electrons, indicated by dots. Each hydrogen atom is represented by H, and the outer

▷ **FIGURE 2–9 Covalent Bond** Like a good marriage or friendship, nonpolar covalent bonds are characterized by a more or less equal sharing of electrons. (*a*) An artist's rendition of two separate hydrogen atoms. The shaded regions represent the atoms' electron clouds. (*b*) Two atoms bound together by a covalent bond. The electrons spin around both nuclei, holding them together to create a molecule.

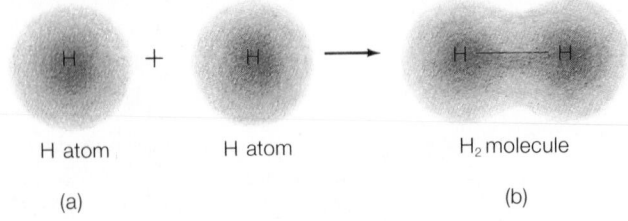

H atom   H atom H_2 molecule

(a)   (b)

[3] There is one exception to this rule. That is, the far-right column is a stable group of gases with complete outer shells that do not accept electrons.

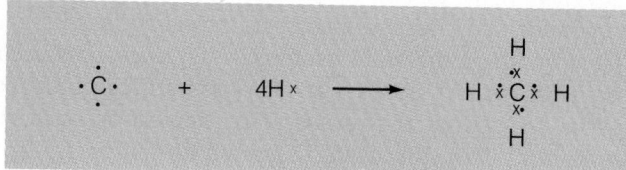

▷ **FIGURE 2–10 Covalent Bonding Simplified** You can keep track of the electrons being shared in covalent bonds by representing the electrons as dots or x's. See the text for explanation.

shell electron is indicated by an x. In the methane molecule, illustrated in Figure 2–10, each hydrogen atom shares its electron with the carbon atom, giving the carbon atom four additional electrons and thus creating a full outer shell for carbon.

In this nifty arrangement, each of the hydrogen atoms also ends up with a filled outer shell because the carbon atom shares its four outer-shell electrons with the four hydrogen atoms. The eight electrons from carbon and hydrogen now circulate around all five atoms, holding them together in a single molecule.

Consider another example, ammonia (NH_3), a pungent gas also released from decaying matter. Ammonia consists of a central atom of nitrogen and three atoms of hydrogen. Each nitrogen atom has seven electrons, two in the inner shell and five in the outer shell. Filling its outer shell requires only three electrons. By entering into covalent bonds with three hydrogen atoms, nitrogen fills its outer shell. Of course, hydrogen atoms also "gain" from this arrangement.

Our last example is another important molecule, water, or H_2O. As the formula shows, a water molecule consists of a single atom of oxygen bonded to two atoms of hydrogen. Oxygen atoms contain eight electrons, two in the inner shell and six in the outer shell. By combining with two hydrogen atoms, oxygen fills its outer shell.

When two atoms share a pair of electrons—one coming from each of the participants—the attractive force that holds them together is referred to as a **single covalent bond.** In chemical nomenclature, it is often indicated by a single line joining two atoms, as in H—H.

Some atoms can share two pairs of electrons with another atom, thus forming a **double covalent bond.** Oxygen, for instance, can share two of its outer shell electrons with a carbon atom, forming a double covalent bond (C=O), as shown in ▷ Figure 2–11. In this arrangement, two electrons once privately held by the oxygen atom now circulate around both carbon and oxygen, helping fill carbon's outer shell. In addition, two electrons once located in carbon's outer shell also circulate around the oxygen atom, helping fill its outer shell. The four shared electrons—or two electron pairs—bind the carbon and oxygen atoms together and constitute a double covalent bond. Other atoms can share three pairs of electrons, forming **triple covalent bonds** (Figure 2–11).

Polar Covalent Bonds Occur Anytime There Is an Unequal Sharing of Electrons by Two Atoms

Although elements with low electron affinity combine to form covalent molecules, the electrons shared between two such atoms are not always shared equally. Unequal sharing occurs if one of the two atoms has a slightly higher affinity for electrons than the other. For example, oxygen and hydrogen join to form a single covalent bond, but oxygen has a slightly higher affinity for electrons than hydrogen. Consequently, the electron of the hydrogen atom tends to spend more time around the oxygen atom than around the hydrogen atom. Because of this, the oxygen atom has a slightly negative charge. The hydrogen atom is visited less frequently by the electron and has a slightly positive charge (▷ Figure 2–12a). The result is a **polar covalent bond.** Technically speaking, a polar covalent bond is the result of electron sharing between two atoms of slightly different

▷ **FIGURE 2–11 Double and Triple Covalent Bonds** (*a*) When atoms share two pairs of electrons, a double covalent bond is formed. Chemists have devised several ways to draw the bonds. In the formula on the left, each line represents a pair of shared electrons. On the right, electrons are indicated by x's and dots. (*b*) On rare occasions, atoms share three pairs of electrons, creating triple covalent bonds.

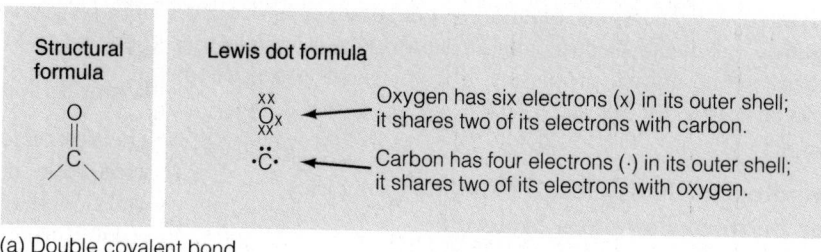

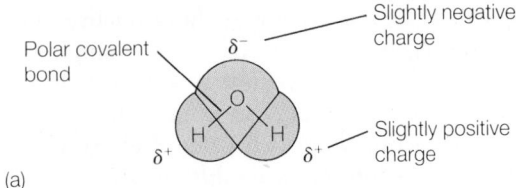

Polar covalent bond

Slightly negative charge

Slightly positive charge

(a)

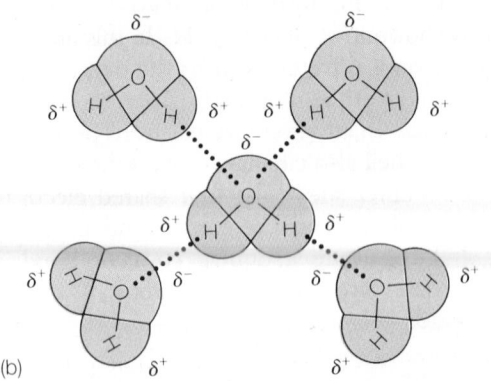

(b)

▷ **FIGURE 2–12 Hydrogen Bond** (*a*) Slightly unequal sharing of electrons in the water molecule creates a polar molecule. Electrons tend to spend more time around the oxygen nucleus, making it slightly negative. The hydrogen atoms are, therefore, slightly positive, as indicated by the symbols δ^+. (*b*) Because of the polarity, hydrogen atoms of one molecule are attracted to oxygen atoms of another. The attraction is called a hydrogen bond.

electron affinities. The greater the difference in electron affinity, the more polar the bond.

The presence of polar covalent bonds in molecules is important because it can make entire molecules polar. One of the most familiar examples is water. As shown in Figure 2–12a, water contains two polar covalent bonds and three slightly charged atoms.

The slightly charged atoms of a polar molecule are often attracted to oppositely charged atoms on neighboring molecules, as illustrated in Figure 2–12b. In the case of water, the electrostatic attraction between the positively charged hydrogen atoms of one water molecule and the negatively charged oxygen atoms of another constitute a **hydrogen bond.** This weak electrostatic attraction is responsible for many of the unique properties of water and is essential to life on Earth (as discussed later).

Hydrogen Bonds Occur Between Molecules or Between Parts of a Molecule

Hydrogen bonding occurs in gases, liquids, and solids that contain polar molecules. Polar molecules are formed when hydrogen atoms covalently bond to atoms such as oxygen, nitrogen, and fluorine, which have a slightly higher affinity for electrons. These bonding pairs result in polar covalent bonds and slightly positive and slightly negative atoms that in turn create polar molecules that are attracted to one another.

An important distinction, not often made is that ionic and covalent bonds form between atoms, resulting in ionic

compounds and covalent molecules, respectively. The hydrogen bond, on the other hand, generally occurs between individual polar molecules, such as water molecules, or between atoms on parts of very large molecules, such as protein and DNA molecules. (More on this later.)

≈ WATER, ACIDS, BASES, AND BUFFERS

All molecules fall into two broad groups: organic and inorganic. **Organic molecules** are compounds made primarily of carbon atoms. In these molecules, many of which can be quite large, the carbon atoms are joined by covalent bonds. Attached to the carbon "backbone" of these compounds are hydrogen, oxygen, and other atoms.

Inorganic molecules are defined by exclusion. That is, they are compounds that are not organic. As a rule, inorganic molecules are generally smaller than organic molecules and frequently consist of atoms joined by ionic bonds. Sodium chloride and magnesium chloride are examples. However, not all inorganic compounds contain ionic bonds. Water, for example, is an inorganic compound whose atoms are joined by covalent bonds.

The important difference between organic and inorganic molecules, then, is not the kind of bonds they contain but the presence of carbon as a major component. If carbon is the main atomic constituent, the molecule is organic. If not, it is inorganic.

Organic chemistry deals with compounds composed primarily of one element, carbon, and inorganic chemistry takes as its domain compounds formed by the more than 100 remaining elements. Although that may sound a bit lopsided, it isn't. In fact, organic compounds are far more numerous than inorganic compounds. One estimate puts the total number of organic compounds at 6 million, with thousands of new ones being synthesized or isolated each year. In contrast, the number of inorganic compounds is a comparatively paltry 250,000.

The next section looks at several important inorganic and organic molecules that are essential to life, beginning with water.

Water Is Vital to Life

The scientific historian and naturalist Loren Eiseley once wrote that if there is magic on the planet, it has to be water. This remarkable substance produces billowy clouds, pointed icicles, and elegant waterfalls. But as aesthetically pleasing as water is, it is also of great practical importance to living things.

Water is a major component of all cells and organisms. Nearly two-thirds of the human body is water. If you weigh 100 pounds, nearly 70 pounds of your body weight is water. In plants, water is stored inside cells, helping create rigidity, as anyone who has neglected his or her houseplants or garden for a few days will attest.

Water is an important biological solvent. A **solvent** is a fluid that dissolves other chemical substances. Blood, for

▷ **FIGURE 2-13 Perspiration Cools the Body**
Perspiration is an automatic homeostatic reaction that helps rid the body of heat.

example, is about 55% water. In blood, water dissolves and transports nutrients, hormones, and wastes throughout the bodies of vertebrates (animals with backbones).

Water also participates in many chemical reactions—for example, the breakdown of protein (described later). Water also functions as a lubricant. Saliva, which is largely water, helps lubricate food in the mouth and esophagus of vertebrates, easing its passage to the stomach. A watery fluid in the joints called **synovial fluid** helps enable bones to slide over one another, thus facilitating body motion.

Finally, water helps us regulate body temperature. Perspiration, for example, rids our bodies of heat, for when it evaporates, water draws off heat (▷ Figure 2-13).

Water Molecules Dissociate Into Hydrogen and Hydroxide Ions

Water molecules are fairly stable; nonetheless, many molecules dissociate, or break apart, forming two oppositely charged ions as shown in the following reaction:

$$H_2O \rightleftarrows H^+ + OH^-$$

water   hydrogen ion   hydroxide ion

The hydrogen and hydroxide ions formed in this reaction can react to reform water molecules, hence the double arrow. Although water dissociates freely, the ratio of water molecules to H^+ and OH^- in the human body is about 500 million:1. Nevertheless, even the slightest change in the hydrogen-ion concentration can alter cells and organ-

isms, shutting down biochemical pathways and sometimes killing organisms.

Acids Are Substances That Add Hydrogen Ions to Solution, Bases Remove Hydrogen Ions

In pure water, the concentrations of H^+ and OH^- ions are equal, and a solution containing an equal number of these ions is said to be chemically neutral. Chemical neutrality can be upset by adding or removing H^+ or OH^-. Hydrochloric acid (HCl), for instance, dissociates into hydrogen ions (H^+) and chloride ions (Cl^-) when added to water, increasing the hydrogen-ion concentration.

A substance that adds hydrogen ions to a solution is known as an **acid,** and solutions with proportionately more H^+ than OH^- are said to be acidic. In contrast, **bases** are substances that remove H^+ from solution. Sodium hydroxide, for example, dissociates when added to aqueous solutions, forming Na^+ and OH^-. The hydroxide ions react with hydrogen ions already in the solution, forming water molecules. As more and more sodium hydroxide is added to a solution of pure water, the hydrogen-ion concentration declines, and the hydroxide ion concentration increases. A solution with a greater concentration of OH^- than H^+ ions is said to be basic.

Acidity is measured on the **pH scale** (▷ Figure 2-14).[4] On this scale, neutral substances are assigned a pH of 7. Basic substances have a pH greater than 7, and acidic substances have a pH less than 7.

As shown in Figure 2-14, the pH scale ranges from 0 to 14 and is designed to condense a wide range of values into a relatively small scale. Don't be fooled by a small change in pH. A difference of only one pH unit actually represents a tenfold change in acidity. Thus, a solution with a pH of 3 is 10 times more acidic than one with a pH of 4 and 100 times more acidic than one with a pH of 5.

Most biochemical reactions occur at pH values between 6 and 8. Human blood, for example, has a pH of 7.4. A slight shift in pH, say, to 7.8, for even a short period can be fatal.

One of the only places in the human body where high acidity (pH 1.5–3.5) is tolerated and beneficial is the stomach. The stomach's high acidity results from hydrochloric acid secreted by cells in its lining. Stomach acid coagulates ingested proteins, making them more susceptible to enzymatic digestion, as described later in the book. It also activates an enzyme in the stomach that begins the breakdown of all proteins.

Changes in the pH of lakes and streams can have profound effects on organisms. In the eastern United States, sulfuric acid draining from abandoned coal mines has destroyed nearly 2000 miles of streams, killing fish and other aquatic organisms (▷ Figure 2-15). Many abandoned mines in the western United States, once the source of gold and silver, also release acid into nearby streams.

[4]pH $= -\log [H^+]$ where $[H^+] =$ hydrogen ion concentration.

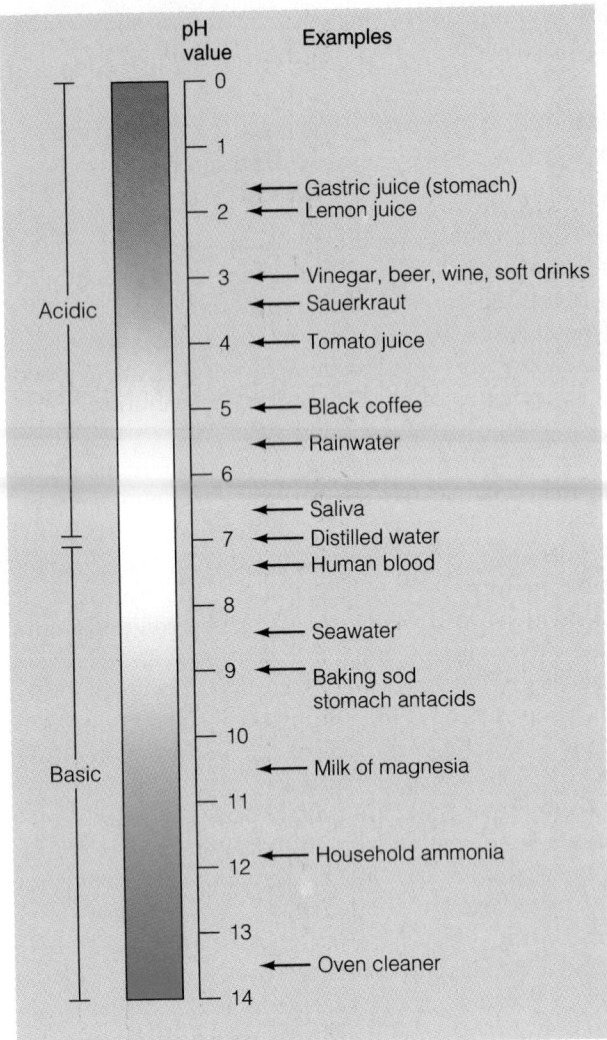

FIGURE 2–14 The pH Scale

FIGURE 2–15 Acid Mine Drainage More than 7000 miles of U.S. streams, mostly in Appalachia, are contaminated by abandoned coal mines that leak sulfuric acid.

Acids are also produced in the atmosphere from gaseous pollutants (sulfur and nitrogen dioxide) given off by automobiles, power plants, and factories. These acids fall in rain and snow and are called **acid deposition.** Acid deposition is a worldwide problem that has increased the acidity of many lakes and streams, causing widespread biological impoverishment, for when the pH level of a lake or stream falls below 4.0–4.5, virtually all life disappears. (Chapter 33 describes this problem in more detail.)

Buffers Are Molecules That Help Regulate pH Within a Fairly Narrow Range, Protecting Life.

Biological systems operate within a narrow pH range that is maintained by a simple homeostatic mechanism created by chemical buffers. **Buffers** are chemical substances that protect against drastic shifts in pH. One professor I know calls them "hydrogen ion sponges," for they help maintain a constant pH by removing hydrogen ions from solution when levels increase. Buffers give back the hydrogen ions when levels fall. A good example of a buffer is **carbonic acid.** Found in the blood of animals, carbonic acid is

formed from water and carbon dioxide, a waste product of cellular metabolism.[5] In blood, carbonic acid dissociates into bicarbonate and hydrogen ions:

$$H_2CO_3 \quad \rightleftharpoons \quad H^+ \quad + \quad HCO_3^-$$

carbonic acid   hydrogen ion   bicarbonate
(weak acid)   (weak base)

This reaction shifts back and forth in response to changing levels of hydrogen ions. Thus, when hydrogen ions are added to the blood, they combine with bicarbonate, driving the reaction to the left. When hydrogen-ion concentrations fall, the reaction is driven to the right. In either case, the pH of the blood remains constant.

Many lakes and streams in the United States contain buffers that protect them from acid rain and snow. Others, such as those in high mountain regions of the Sierra Nevada, Cascades, Rockies, and Adirondacks have little buffering capacity because they are situated on rock or soil that contains few buffers. Lakes in these regions are, therefore, highly vulnerable to acids from pollution.

≋ MORE BIOLOGICALLY IMPORTANT MOLECULES

Organic molecules form a rather large and diverse group. This diversity results chiefly from the carbon atom. As

[5] As you will see in later chapters, carbonic acid is produced inside red blood cells in the blood.

noted earlier, a carbon atom has four electrons in its outermost shell and can share these electrons with a variety of atoms, including oxygen, hydrogen, nitrogen, and even other carbon atoms. The carbon atom can also enter into single, double, and triple covalent bonds (▷ Figure 2–16).

Further adding to molecular diversity, many smaller organic molecules, such as amino acids and sugars, can combine to form large organic molecules, generically referred to as polymers. A **polymer** is a very large molecule consisting of many smaller molecules. The individual molecules of a polymer are called **subunits** or **monomers.** Starch, for example, is a polymer composed of hundreds of glucose molecules. Protein is a polymer of amino acid molecules.

The biologically important organic molecules found in plants, animals, and microorganisms fall into four classes: (1) carbohydrates, (2) lipids, (3) amino acids, peptides, and proteins, and (4) nucleic acids. The following sections discuss these important molecules, giving you information useful throughout the rest of this book. We begin with carbohydrates.

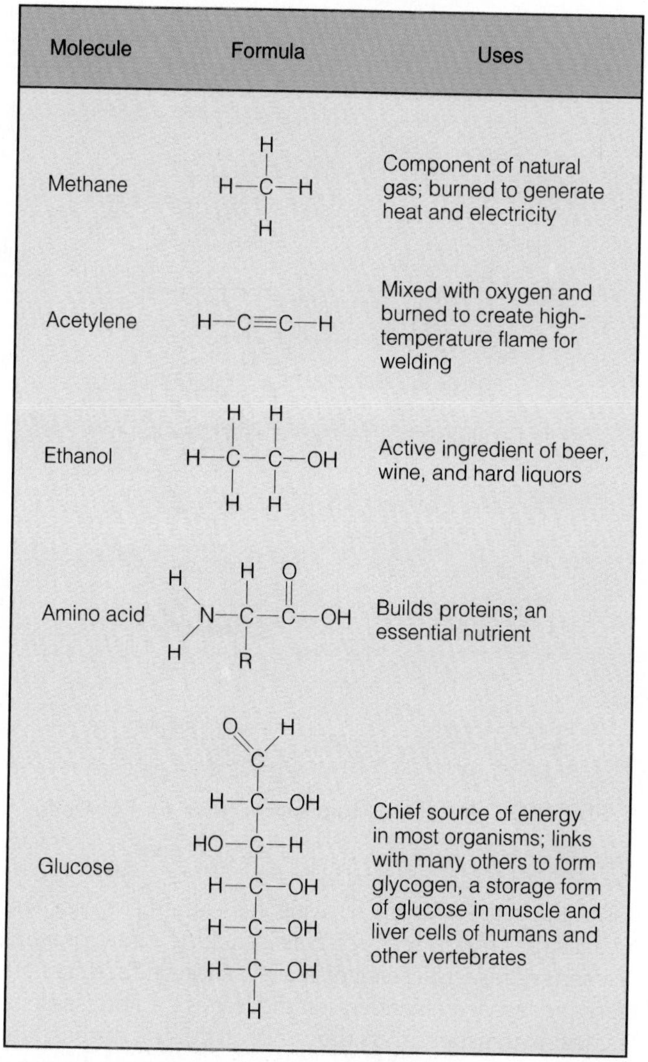

Carbohydrates Are a Diverse Group of Molecules That Serve as a Source of Energy for Many Organisms and Have Many Other Functions

Carbohydrates comprise a familiar group of organic compounds, including glucose, starch, glycogen, and cellulose. These molecules, consisting chiefly of carbon, oxygen, and hydrogen, have a general formula $(CH_2O)_n$. This formula tells us that for every carbon atom in the molecule, there are two atoms of hydrogen and one atom of oxygen. The general formula, in fact, gives this structurally diverse group of molecules its name—"hydrates of carbon," or, more simply, carbohydrates.[6]

Carbohydrates fall into three broad groups: monosaccharides, disaccharides, and polysaccharides.

Monosaccharides Are the Smallest Carbohydrates and Often Combine to Form Polymers. Monosaccharides are often called simple sugars. These water-soluble molecules contain three to seven carbons, but many of them form interesting ring structures when dissolved in water, as shown in ▷ Figure 2–17.

One of the most common and most important monosaccharides is glucose. **Glucose** is a six-carbon sugar, the structure of which is shown in Figures 2–17 and ▷ 2-18. Made by plants during photosynthesis, glucose molecules can be joined to form long, branching molecules known as **starch.** Starch is stored in plant roots (as in potatoes) or seeds (as in wheat and rice) and is an important nutrient for plant-eating animals. In the digestive tract of vertebrates, starch is broken down into glucose molecules, which enter the bloodstream. These molecules are then distributed to body cells, where they may either be stored for later use or broken down to release energy.

Molecule	Formula	Uses
Methane		Component of natural gas; burned to generate heat and electricity
Acetylene		Mixed with oxygen and burned to create high-temperature flame for welding
Ethanol		Active ingredient of beer, wine, and hard liquors
Amino acid		Builds proteins; an essential nutrient
Glucose		Chief source of energy in most organisms; links with many others to form glycogen, a storage form of glucose in muscle and liver cells of humans and other vertebrates

▷ **FIGURE 2–16 Carbon's Bonding Options Create Molecular Diversity** Carbon can covalently bond to many other atoms, creating a diverse array of small organic molecules. Some of these molecules, in turn, can bond to form large chains, or polymers (not shown here).

Figure 2–17 illustrates three other important monosaccharides that you will encounter in your study of biology. Ribose and deoxyribose are components of the genetic material. Fructose is a component of table sugar, a disaccharide known as sucrose. Take a few moments to familiarize yourself with their structures.

Disaccharides Consist of Two Monosaccharides. Disaccharides are carbohydrates that consist of two monosaccharides covalently bonded to each other. As Figure 2–18 illustrates, the disaccharide **sucrose** consists of two monosaccharides, glucose and fructose. Sucrose is produced in the leaves of plants, during photosynthesis, from fructose and glucose, which are also synthesized there.[7]

Sucrose produced in the leaves of a tree is transported

---

[6]The word *hydrates* refers to the water in the formula.

[7]Some plants also store fructose in their fruits.

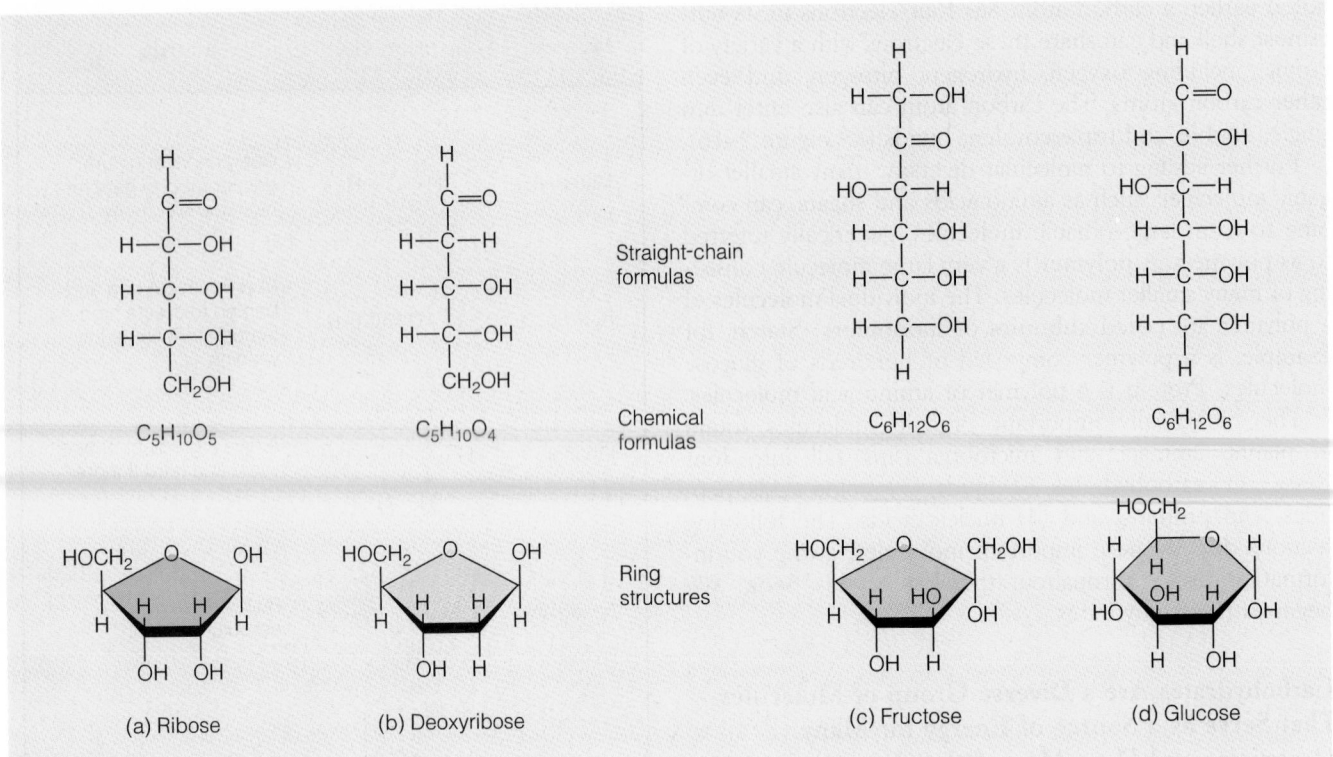

Straight-chain formulas

Chemical formulas

Ring structures

$C_5H_{10}O_5$ $C_5H_{10}O_4$ $C_6H_{12}O_6$ $C_6H_{12}O_6$

(a) Ribose   (b) Deoxyribose   (c) Fructose   (d) Glucose

▷ **FIGURE 2–17 Four Monosaccharides** (*a*) Ribose. (*b*) Deoxyribose. (*c*) Fructose. (*d*) Glucose.

down the trunk in vessels lying beneath the bark. This disaccharide nourishes root cells. Lacking photosynthetic machinery, these cells cannot make their own food and are therefore heavily dependent on the leaves for nourishment. Sucrose is transported to the roots of the tree throughout the growing season. In the fall in the temperate zone, however, trees transport extraordinarily large amounts of sucrose to their roots for use over the long winter. (The roots stay active in the winter because they are buried underground in warm subsoil.) In the early spring, the sugar-rich sap stored in the roots runs upward, providing nourishment to growing buds. New Englanders have long tapped the sugary fluid of maples to obtain the raw material for maple syrup (▷ Figure 2–19).

Polymers of Monosaccharides Are Called Polysaccharides. In animals and plants, many monosaccharides react to form long polymers, known as **polysaccharides.** The most common building block of polysaccharides is glucose.

Plants use glucose to synthesize two extremely important polysaccharides: (1) **cellulose,** one of the molecules found in the walls around plant cells, which gives wood its rigidity and protects the cells of leaves and other tissues; and (2) starch, a molecule that serves, as mentioned earlier, as a storage depot for glucose molecules.

In animals, glucose combines to form **glycogen,** a polysaccharide that is often called animal starch because it serves a similar purpose and has a structure similar to

starch found in plants. In humans and many other animals, glycogen is composed of thousands of glucose molecules. Glucose molecules derived from the food we eat may be used immediately, but most are stored in muscle and liver cells as glycogen (Figure 2-18c). When blood glucose levels fall between meals, glycogen in the liver is broken down into glucose molecules. Muscle glycogen is broken down to provide energy for active muscles.

Animals derive glucose—from which they make glycogen—from two major sources: plant starch and animal glycogen (found in the muscle of other animals they eat). A few animals such as cattle, sheep, and other grazers are able to acquire glucose from cellulose in the plants they consume.

Cellulose is also prevalent in the diet of other animals, but it cannot be digested because most animals lack the enzymes needed to break the bonds joining its glucose subunits. (Enzymes are protein molecules that speed up the rate of many chemical reactions occurring in cells and organisms.) Consequently, cellulose passes through the digestive system of most animals relatively unchanged.

Even though it is not digested, cellulose is important to normal body function. In the human diet, for example, nondigestible cellulose is a principal form of dietary fiber. For reasons explained in Chapter 10, cellulose facilitates the passage of feces through the large intestine and also reduces the incidence of colon cancer.

Those species that can digest cellulose rely on bacteria and single-celled organisms known as protists within their

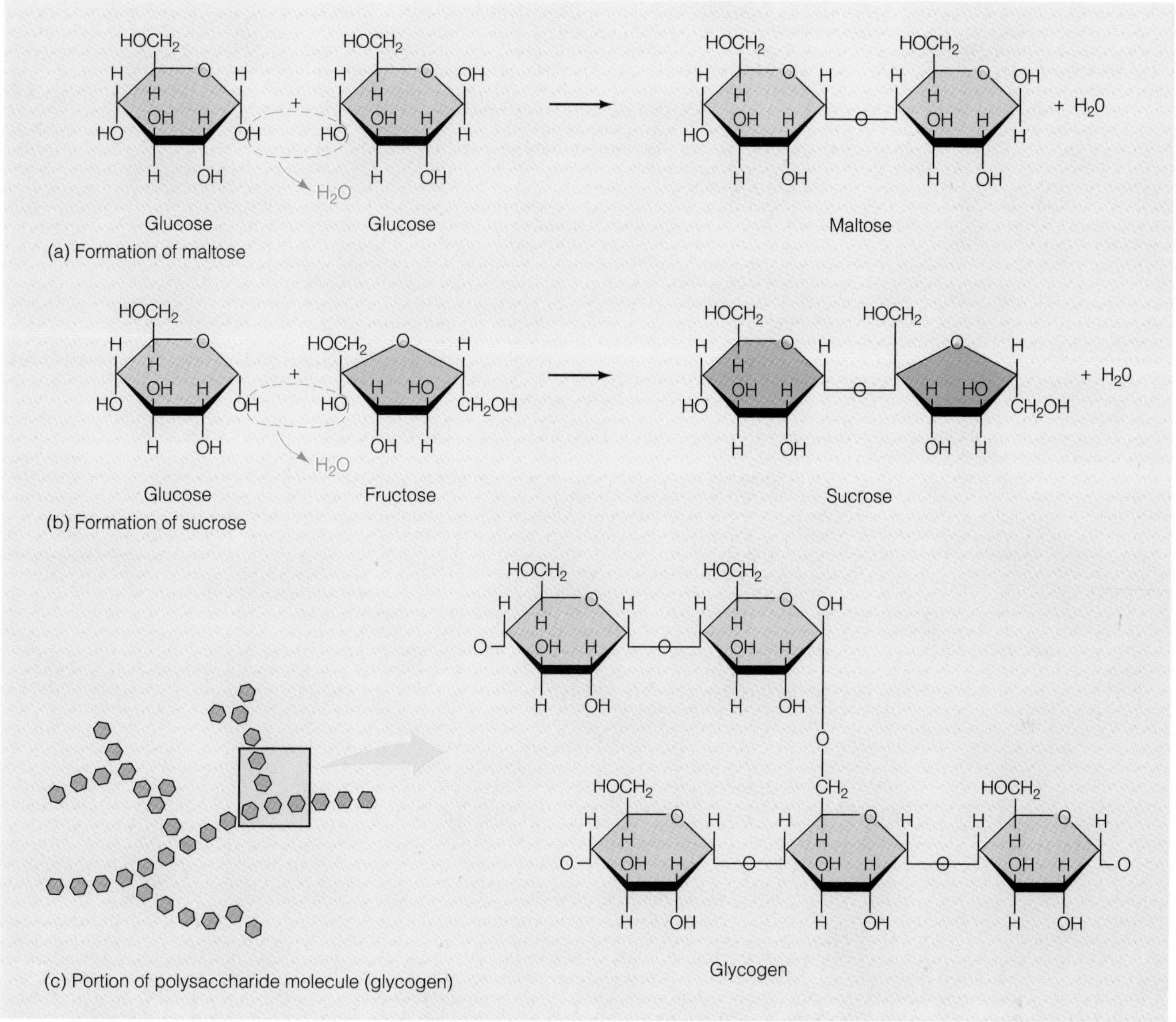

Figures (carbohydrate structures)

(a) Formation of maltose

Glucose + Glucose → Maltose + H_2O

(b) Formation of sucrose

Glucose + Fructose → Sucrose + H_2O

(c) Portion of polysaccharide molecule (glycogen)

Glycogen

▷ **FIGURE 2–18 Disaccharides and Polysaccharides** Simple sugars like glucose can combine to form disaccharides, such as (a) maltose and (b) sucrose, and polysaccharides, like (c) glycogen.

digestive system that contain the enzymes required to break the bonds linking the glucose molecules in cellulose.

Lipids are Fat-soluble Molecules That Serve as a Source of Energy and Are Also Important Structural Components of Cells

Lipids are a structurally diverse group of organic molecules. Unlike carbohydrates, which are defined by their common structure, lipids are defined in terms of one common property—solubility. As a rule, then, **lipids** are biological molecules that are insoluble in water and soluble in nonpolar solvents such as ethyl alcohol and benzene.

Lipids are the waxy, greasy, or oily compounds found in plants and animals. These compounds repel water, a highly useful characteristic for the protective wax coatings of some plants. Fats and oils are also energy-rich molecules

▷ **FIGURE 2–19 The Early Spring Harvest of Sugar** New Englanders tap the sugary sap of a sugar maple and boil it to make maple syrup. Tubes driven into the tree draw off the sap without harming the trees.

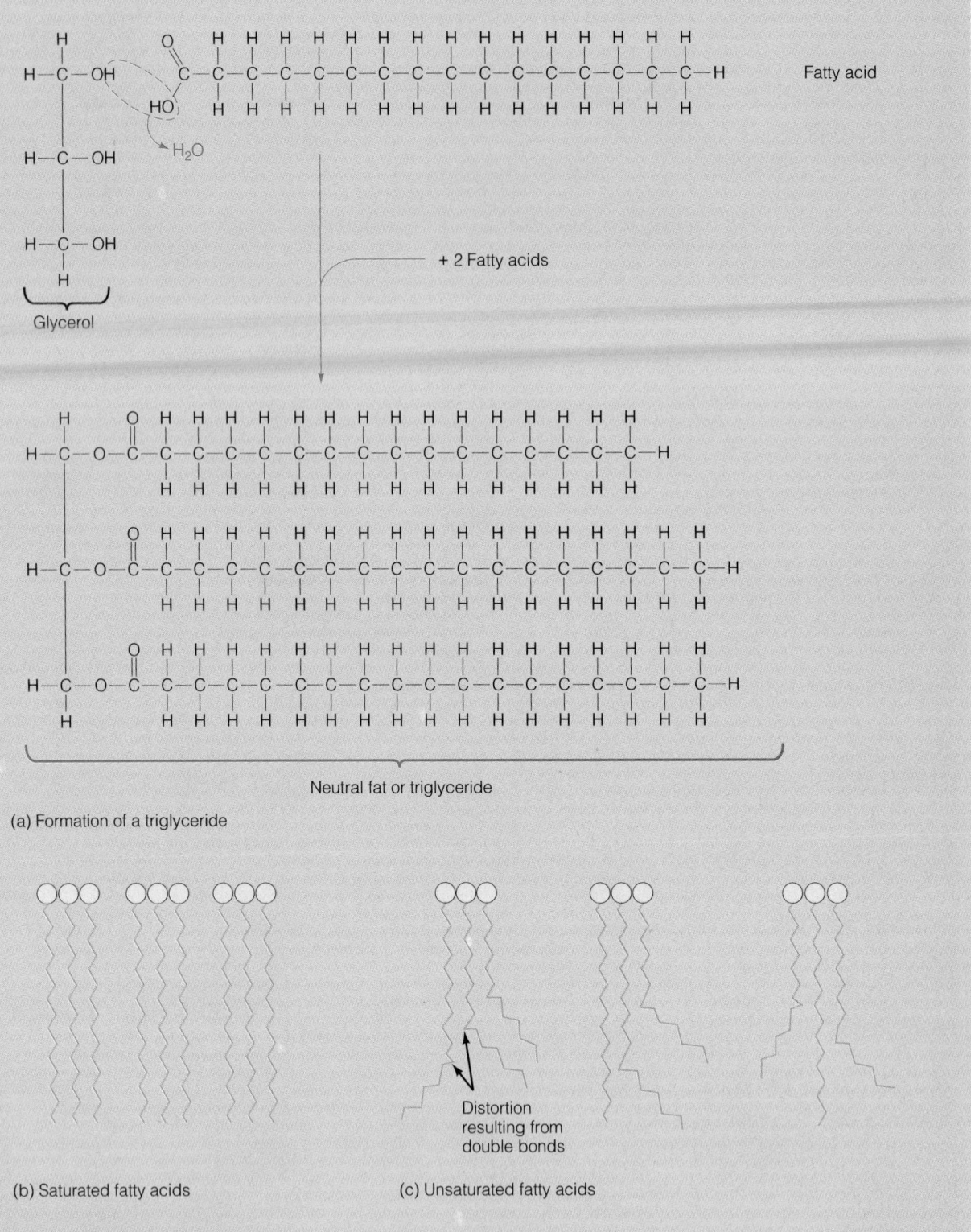

(a) Formation of a triglyceride

Neutral fat or triglyceride

Fatty acid

Glycerol

+ 2 Fatty acids

(b) Saturated fatty acids

(c) Unsaturated fatty acids

Distortion resulting from double bonds

▷ **FIGURE 2–20 Triglycerides** The triglycerides are the fats and oils. (*a*) Triglycerides consist of glycerol and three fatty acids, covalently bonded as shown. (*b*) Saturated fatty acids are principally derived from animal fats. The side chains are relatively straight and allow the molecules to pack tightly together which explains why fats are solid at room temperature. (*c*) Double bonds in unsaturated fatty acids in oils cause the fatty acid chains to bend and prohibit tight packing. Oils, derived chiefly from plants, are therefore liquid at room temperature.

and therefore serve as an important storage form of energy in both plants and animals. Still other lipids are structural components of cells.

The biologically important lipids fall into four main categories: (1) triglycerides, (2) waxes, (3) phospholipids, and (4) steroids.

Triglycerides. Triglycerides are known to most of us as fats and oils. Cooking oil, for example, is a triglyceride, as is the fat on a steak or the butter on a piece of bread.

Triglycerides are composed of four parts: one molecule of glycerol and three fatty acid molecules (▷ Figure 2–20a). **Glycerol** is a three-carbon molecule shown in the top portion of Figure 2–20a. A **fatty acid** is a long molecule with many carbons and hydrogens and a COOH, or carboxyl group, on one end. The most common fatty acids usually have an even number of carbons, ranging from 10 to 20. Each fatty acid attaches to one of the oxygen atoms of glycerol. Take a few moments to study Figure 2–20a to see how the glycerol and fatty acids combine.

Triglycerides contain many covalent bonds, each of which stores a small amount of energy. When cells break down (catabolize) triglycerides, these bonds are broken and energy is released. Energy liberated during the catabolic breakdown of triglycerides is used to drive a variety of cellular processes. Triglycerides yield more than twice as much energy per gram as carbohydrates.

In humans and many other mammals, triglycerides are stored in fat cells under the skin and in other locations (▷ Figure 2–21). In adults, triglycerides provide about half of the cellular energy consumed at rest, with carbohydrates providing most of the remainder. During moderate (aerobic) exercise, triglycerides provide a larger proportion of the body's energy demand, explaining why aerobic exercise helps people lose weight (▷ Figure 2–22). (This topic is discussed in more detail in Chapter 11.)

With a few exceptions, fats (from most animals) are solid at room temperature, and oils (from plants and fish) are liquid (▷ Figure 2–23). The reason for this difference lies in their chemical structures. In fats, for example, the carbon atoms of the fatty acids are joined by single covalent bonds (see Figure 2–20b). The remaining bond sites on the carbon atoms are taken up by hydrogens, and the fatty acids are said to be **saturated** with hydrogens. When

▷ **FIGURE 2–21 Fat Cells** (a) A light micrograph of fat tissue. The clear areas are regions where the fat has dissolved during tissue preparation. Notice that the cytoplasm is reduced to a narrow region just beneath the plasma membrane. (b) A scanning electron micrograph of fat cells.

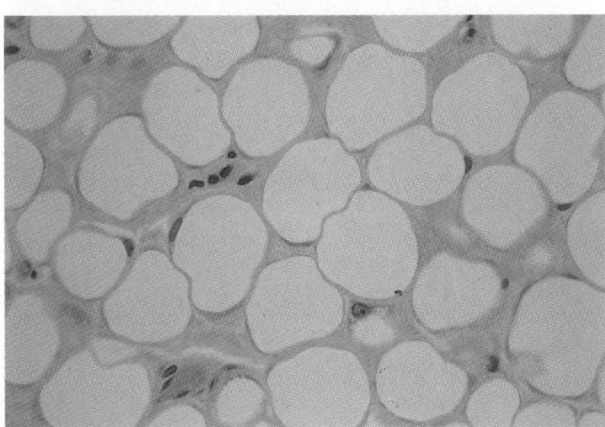
(a)

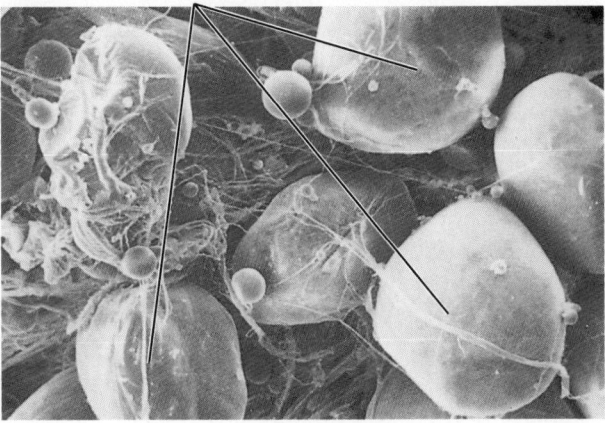
Fat cells
(b)

▷ **FIGURE 2–22 Aerobic Exercise** Besides being good for your heart, aerobic exercise can burn off calories and help you lose weight.

▷ **FIGURE 2–23 Fats and Oils** At room temperature, fats are solid, whereas oils are liquid. The reason for the difference is explained in the text.

the carbon backbone of a fatty acid consists of single covalent bonds, the fatty acid zigzags but remains fairly linear, as shown in Figure 2–20b. This allows the triglyceride molecules of a fat to pack together, forming a solid at room temperature.

In contrast, the fatty acids in oils contain a number of double covalent bonds; these molecules are therefore said to be **unsaturated** (see Figure 2–20c). Double covalent bonds cause a bend in the fatty acid molecules that is believed to prevent tight packing. This somewhat "looser" arrangement of triglyceride molecules produces a liquid at room temperature.

When numerous double bonds exist in a fatty acid molecule, the molecule is said to be **polyunsaturated.** Studies show that polyunsaturated fats reduce one's risk of developing **atherosclerosis,** a disease that results from a buildup of cholesterol deposits, or **atherosclerotic plaque,** on the walls of arteries (▷ Figure 2–24). Plaque restricts blood flow to the heart and other vital organs and can cause heart attacks and strokes.

Because saturated fatty acids increase cholesterol production by the liver, diets rich in saturated fat (animal fats) increase the chances of developing atherosclerosis. Atherosclerosis is also more common in sedentary people and smokers. Some individuals are genetically predisposed to develop atherosclerosis. Their livers simply produce abnormally high levels of cholesterol.

Lowering the level of saturated fat in the diet is therefore only one step in reducing the concentration of cholesterol in the blood. This step can be accomplished by switching from whole milk to low-fat milk, reducing the consumption of red meat, trimming fat from chicken and other meats, and cooking with unsaturated vegetable oil

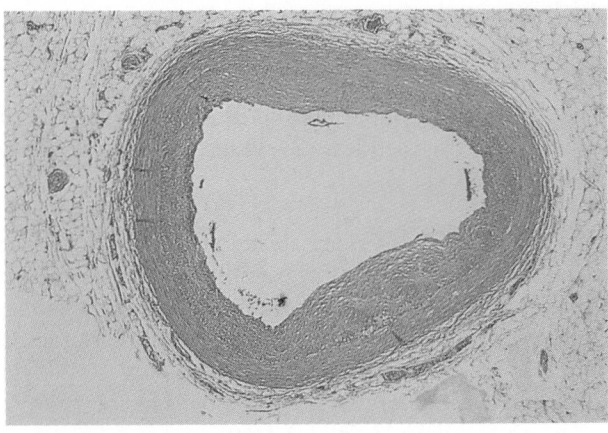

(a)

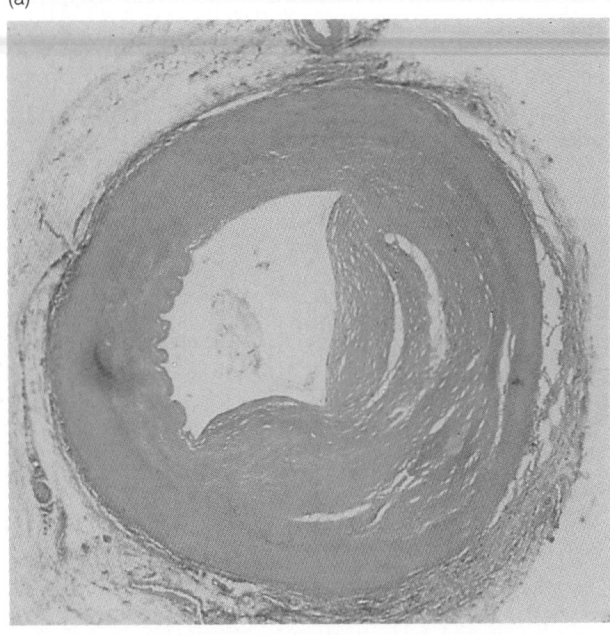

(b)

▷ **FIGURE 2–24 Atherosclerosis** These cross sections of (*a*) a normal artery and (*b*) a diseased artery show how atherosclerotic plaque can obstruct blood flow.

instead of animal fat (lard). (For more on cholesterol, see the discussion of steroids below and Health Note 11–1.)

Waxes. **Waxes** are lipids composed of long-chain fatty acids joined to long-chain alcohols. The structure of beeswax, shown in ▷ Figure 2–25, gives you an idea of their general structure. Waxes are water-insoluble and resistant, and therefore they often occur in nature as protective coatings on fur, skin, feathers, leaves, and fruits. The oil produced by tiny glands (sebaceous glands) in the skin, in fact, contains many waxes that help keep the skin soft and protect us from dehydration. Waxes are also commonly used to make candles, cosmetics, and polishes.

Phospholipids. Phospholipids are a class of lipids that contain phosphate, PO_4. Shown in ▷ Figure 2–26a, the most common type of phospholipid is **phosphoglyceride,** a

▷ **FIGURE 2–25 Structure of Beeswax** As shown here, beeswax is composed of a long-chain fatty acid and a long-chain alcohol.

▷ **FIGURE 2–26 Phosphoglyceride** Phosphoglycerides are the predominant phospholipid in the body (a) Each phosphoglyceride consists of a glcyerol backbone, two fatty acids, a phosphate, and a variable group generally designated by the letter R. In this case the R group is choline, which is polar. (b) Because of the R group, the molecule has a polar head. The nonpolar tail region is formed by the two fatty acid chains.

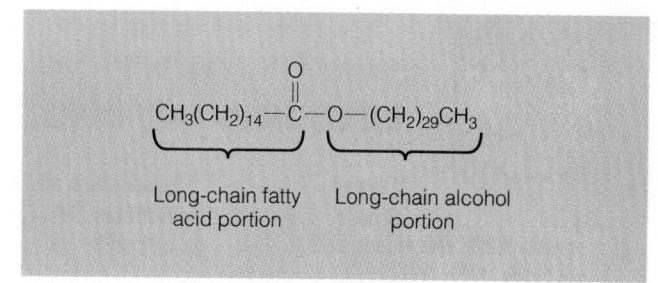

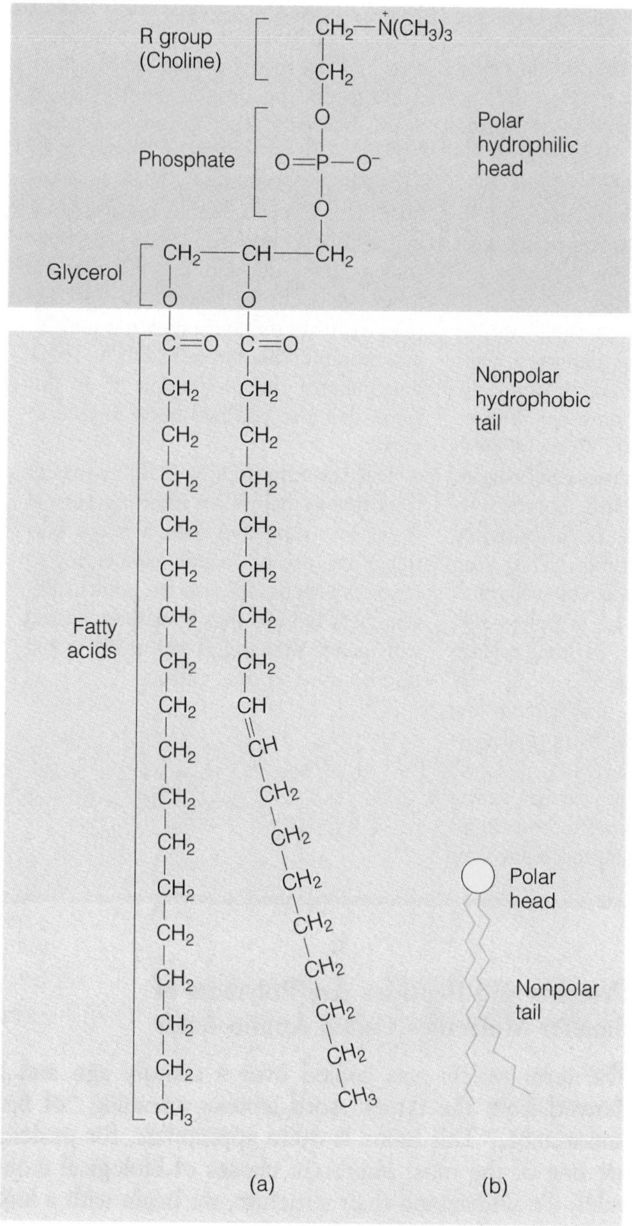

Long-chain fatty
acid portion   Long-chain alcohol
                    portion

oms. As shown in Figure 2–26a, a small organic molecule, choline, is attached to the phosphate. Choline is just one of a half dozen or more polar, or charged, organic molecules found in phospholipids. These molecules create a polar region (the phosphate head) on the molecule and are generally designated by the letter R, a shorthand chemists use to indicate any variable portion of a molecule (Figure 2–26b). The long hydrocarbon chains of the fatty acids form a large nonpolar region.

In water, phosphoglycerides form tiny globules called **micelles,** shown in ▷ Figure 2-27. As illustrated, the polar region of phosphoglyceride molecules orient outward toward the polar water molecules. The nonpolar ends orient inward, away from the polar water molecules.

Phosphoglycerides in water may also form microspheres, more commonly known as liposomes, each with a watery core (Figure 2–27b). Health Note 2–1 describes how liposomes can be used to treat cancer.

Steroids. Steroids are lipids that are similar to phosphoglycerides and triglycerides in only one regard—their solu-

▷ **FIGURE 2–27 Micelles and Microspheres** (a) In a solution of water, phosphoglyceride molecules form tiny globules called micelles. Notice how the nonpolar tails of the molecules bunch together while the polar heads stick out. (b) Phosphoglycerides may also form microspheres, tiny spheres containing an aqueous core. Again, notice how the polar and nonpolar ends of the molecule seek a chemically similar environment.

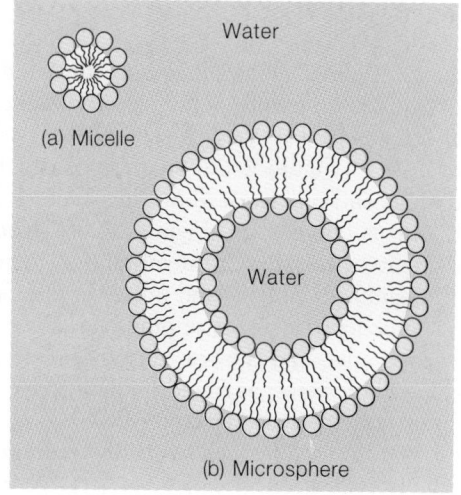

major component of the various cellular membranes. Like triglycerides, phosphoglycerides consist of a molecule of glycerol. However, only two fatty acids attach to the glycerol molecule. The third site is occupied by a phosphate group containing a phosphorus atom and four oxygen at-

LIPOSOMES AND OTHER TREATMENTS FOR CANCER

Cancer is a deadly disease that results from the uncontrolled division of body cells, producing a rapidly growing mass of cells known as a tumor. Cancer cells often migrate from the site of origin, invading other body tissues and organs, where they form secondary tumors.

One of every three of us will contract cancer in our lifetime, and one of every four of us will die from it. For most patients, cancerous tumors are removed surgically. This procedure is usually followed by radiation treatments or chemotherapy. Radiation kills cancer cells not removed during surgery and can also kill cancer cells in secondary tumors. Radiation can also be used in place of surgery to destroy tumors in hard-to-reach locations, as can chemotherapy, a regiment of chemical agents that kills cancer cells. Unfortunately, radiation and chemotherapy also kill normal body cells. Especially vulnerable are the rapidly dividing cells of hair follicles and the lining of the intestine. As a result, hair loss, nausea, and diarrhea are common side effects of these treatments. Patients may feel ill for three days to a week or more after a treatment.

Thanks to a relatively new and simple technique, physicians may have a way to eliminate such treatments. The potential new treatment uses tiny spheres of lipid, called liposomes, packed with cancer-killing drugs.

Liposomes are made from synthetic lipid molecules or natural products, such as egg yolks and soybeans, that contain lipids. Each tiny, fat-soluble sphere contains a watery interior that traps water-soluble drugs. Fat-soluble drugs can be trapped in the fatty exterior. The microspheres act as a drug reservoir that releases its cancer-killing chemicals as a liposome breaks up or after it is ingested by a cell.

Medical researchers are optimistic about the potential use of liposomes for cancer therapy. By packaging chemotherapeutic drugs in the liposomes, physicians may be able to protect noncancerous body cells from harm. Experiments with an anticancer drug, doxorubicin, have yielded some promising results. This highly toxic drug damages heart muscle when administered intravenously. Injection in liposomes, however, reduces its harmful effects on the heart by 80%, because much smaller doses can be used.

Researchers have also successfully packaged cytochalasin B, a potential cancer drug, in liposomes. This drug halts the spread of malignant cells but, to be effective, must be present at all times. Liposomes act as time-release capsules, administering the drug very slowly and therefore maintaining low levels that are nontoxic to normal cells but fatal to tumor cells.

Researchers are now experimenting with ways to get liposomes to hone in on specific targets. One approach involves the use of antibodies—that is, proteins that react with foreign substances or foreign cells, such as cancer cells. By coating drug-containing liposomes with antibodies, medical researchers hope, they can produce site-specific "bullets" that bind specifically to cancer cells. Liposomes engulfed by tumor cells would release their contents into the cancerous cells and kill them.

Target-specific liposomes may also be useful in treating genetic diseases. Scientists at the University of Tennessee in Knoxville, for example, have injected mice with pieces of DNA encapsulated in antibody-coated liposomes. The liposomes attach to and are ingested by mouse cells. Inside the cells, the DNA is released from the liposomes and may be incorporated into the cell's DNA. There it stimulates the production of an enzyme that the cell had been unable to make.

This startling finding could prove extraordinarily helpful in inserting normal genes into defective cells, a major barrier in the use of genetic engineering to reverse genetic disease in adults. Researchers believe that liposomes loaded with genes that trigger cell death could also be used to treat cancer.

---

bility in nonpolar solvents. As a group, the steroids are structurally quite different from the triglycerides and phospholipids.

As shown in ▷ Figure 2–28, steroids consist of three six-carbon rings and one five-carbon ring. One of the best known steroids is **cholesterol** (Figure 2–28a). Cholesterol is a component of the plasma membrane in animal cells (but not in plants). In humans and other vertebrates, it is a raw material needed to synthesize other steroids, including vitamin D, bile salts, and the sex hormones estrogen and testosterone. Cholesterol, as noted earlier, is also a major component of atherosclerotic plaque.

In humans, cholesterol comes primarily from the liver. A lesser amount comes from the diet.

Proteins and Peptides Are Polymers of Smaller Molecules Called Amino Acids

The term *protein* was coined over a century ago and is derived from the Greek word *proteios,* meaning "of first importance." This name is quite appropriate, for proteins are one of the most important classes of biological molecules. To understand their structure, we begin with a look at amino acids.

Amino Acids. **Amino acids** are the building blocks of proteins. As shown in ▷ Figure 2–29a, each amino acid contains a central carbon attached to four functional groups: an **amino group** (NH_2), a **carboxyl group** (COOH), a hydrogen, and a variable group, indicated by

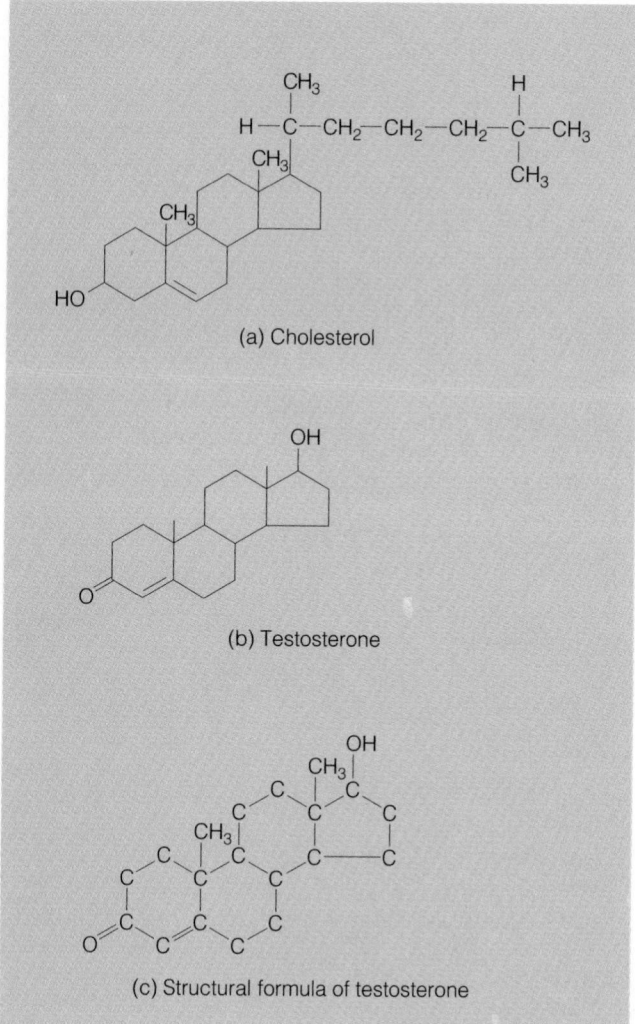

▷ **FIGURE 2–28 Steroids** Steroids, like (a) cholesterol and (b) testosterone, consist of four rings joined together. The drawings in (a) and (b) are a shorthand way of drawing (c) the structural formula.

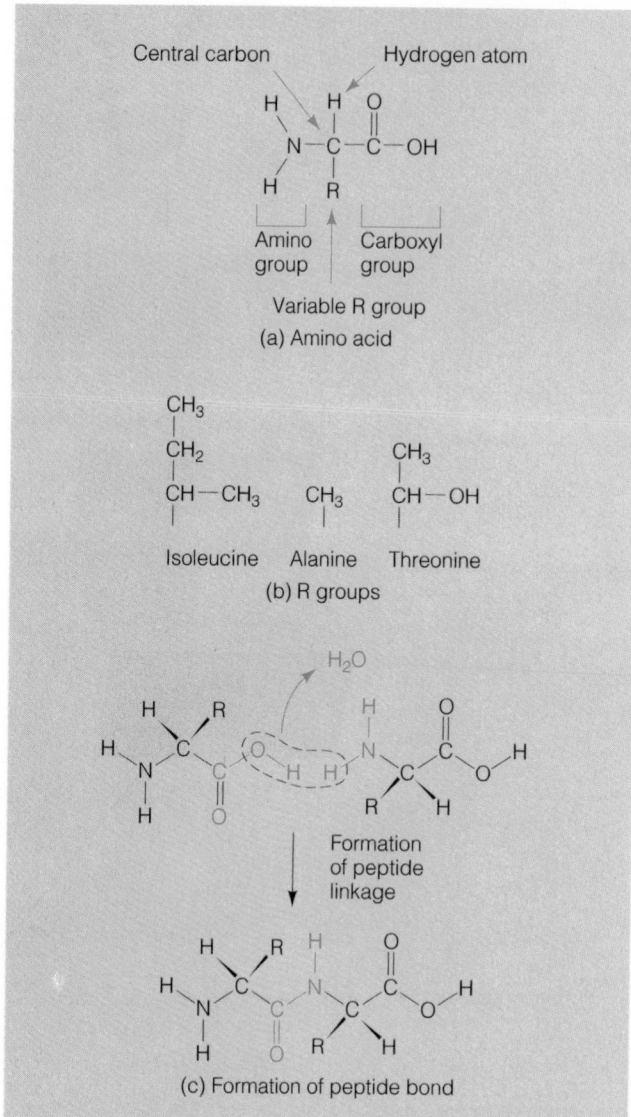

▷ **FIGURE 2–29 Structure of Amino Acids and Formation of Peptide Bonds** (a) The amino acid is a small organic molecule with four groups, one of which is variable and is designated by the letter R. (b) Some representative R groups. (c) The carboxyl and amino groups react in such a way that a covalent bond is formed between the carbon of the carboxyl group of one amino acid and the nitrogen of the amino group of another. This bond is called a peptide bond. Water is released during this reaction—hence the name dehydration synthesis.

the letter R.[8] To date, approximately 250 amino acids have been discovered in plants and animals or synthesized in the laboratory, but only 20 of them are commonly found in natural proteins. From this seemingly limited pool, cells build thousands of different proteins in much the same way that composers and songwriters create a diversity of unique music from a relatively small number of notes.

Proteins are polymers of amino acids and serve many functions in the human body. One important class of proteins, the **enzymes,** accelerates chemical reactions in the body. Without enzymes, few reactions would occur. Other proteins play a structural role. **Keratin,** for example, is a protein that makes up human hair and nails. **Collagen,** the most abundant protein in the human body, forms fibers that help hold tissues together and are the major component of bones. If the calcium is dissolved from a bone, a rubbery collagenous replica of the bone remains.

[8] The R groups found in amino acids are different from those found in phospholipids.

Proteins and Peptides. Proteins are synthesized in the cells, one amino acid at a time. During this process, amino acids join by covalent bonds that are given a special name, **peptide bonds.** The formation of these bonds is shown in Figure 2–29b. As illustrated, when two amino acids link, a hydrogen and a hydroxyl group are lost. These ions combine to form water and a reaction of this sort is therefore called a **dehydration synthesis.**

Dehydration syntheses are quite common in biology. In fact, if you look back at the synthesis of triglycerides in Figure 2–20a, you will see that water is also given off when

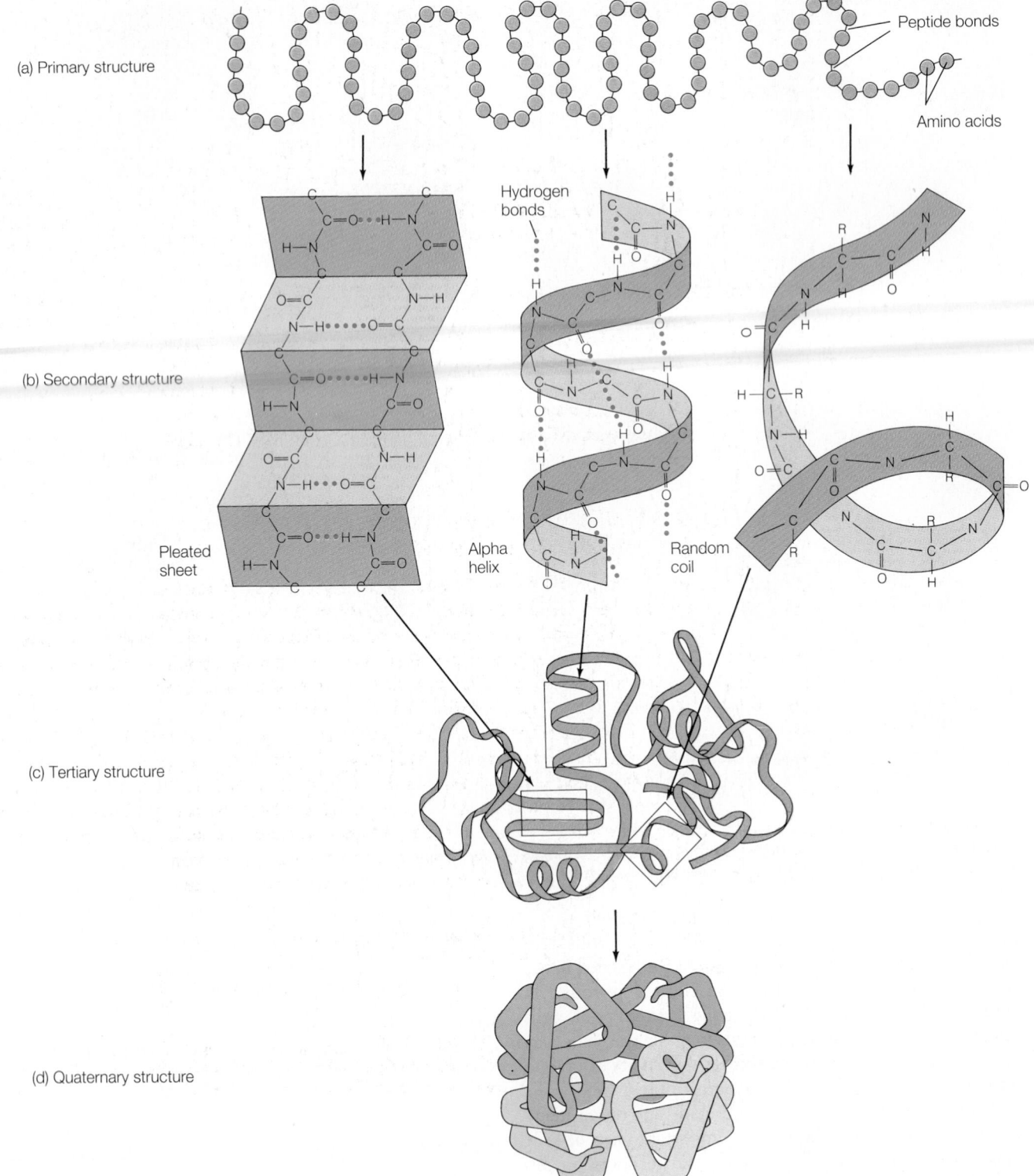

(a) Primary structure

Peptide bonds

Amino acids

Hydrogen bonds

(b) Secondary structure

Pleated sheet

Alpha helix

Random coil

(c) Tertiary structure

(d) Quaternary structure

▷ **FIGURE 2–30 Protein Structure** (*a*) Primary structure is the sequence of amino acids. It determines to a large extent the complex three-dimensional shape a protein assumes. (*b*) Secondary structure results from a bending or coiling of the primary structure. Three general types are found: pleated sheet, alpha helix, and random coil. (*c*) Tertiary structure results when the secondary structure of a protein is compacted to form the three-dimensional shape on which its function is dependent. (*d*) Quarternary structure results when two or more globular proteins unite to form a "superprotein."

fatty acids attach to glycerol. Dehydration syntheses are also encountered in the formation of nucleic acids (discussed below) and carbohydrates.

Protein synthesis occurs with remarkable speed and protein molecules with hundreds, even thousands, of subunits can be manufactured in a matter of minutes.

In nature, the length of amino acid chains varies considerably from the simplest **dipeptides** (a molecule made of two amino acids) to the largest proteins, containing 3000 amino acids. In general, a string of amino acids is referred to as a **polypeptide.** Polypeptides with fewer than 20 amino acids are called **oligopeptides,** or simply **peptides,** and those polypeptides with more than 20 amino acids are called *proteins.*

Protein Structure. Thousands of different types of protein are found in the cells of organisms. Each of these proteins is structurally and functionally unique. This uniqueness results primarily from the sequence of amino acids in the protein, which is often known as the protein's **primary structure** (▷ Figure 2–30a).

As proteins are produced in cells, the chain of amino acids begins to bend and fold, creating the **secondary structure.** The secondary structure may consist of a spiral (alpha helix), a random coil, or, in some cases, a pleated sheet (Figure 2–30b). The bending and coiling of the primary structure is brought about by hydrogen bonds that form between parts of the polypeptide chain. Although some proteins have all three types of secondary structure in different parts of the molecule, most contain only alpha helix and random coil, because the pleated sheet form is rather inflexible and cannot be compacted. Random coils and alpha helices, however, can compact further, thus converting the polypeptide chain into a globular protein.

The three-dimensional shape of a protein, resulting from the compaction of the secondary structure, is known as the protein's **tertiary structure** (Figure 2–30c). It results from a variety of interactions—too numerous and complicated to discuss here—between R groups of the amino acid chain. These include hydrogen bonds, ionic interactions, and covalent bonds. The tertiary structure also results from interactions between R groups and the predominant molecules in the protein's environment, usually water or lipid.

Some proteins consist of two or more globular subunits, which are held together by a variety of noncovalent forces, including ionic bonds and hydrogen bonds (Figure 2–30d). The result is a **quaternary structure.** One of the best-studied examples is hemoglobin, the oxygen-carrying molecule of red blood cells of vertebrates. Hemoglobin contains four polypeptide subunits, each in a globular form (tertiary structure).

As you will learn in Chapter 8, the sequence of amino acids in a protein is determined by the genes. Thus, **mutations**—minute changes in the genes caused by radiation, chemical substances, or other factors—can alter the three-dimensional structure of a protein. This, in turn, may result in the formation of a defective molecule, one that functions improperly or not at all. A good example of a mutation occurs in **sickle-cell anemia,** a disease that primarily afflicts African Americans, black Africans, and several other groups[9] (Chapter 8).

In individuals with sickle cell anemia, the genetic defect results in one wrong amino acid being substituted for the correct one during protein synthesis. When abnormal hemoglobin molecules are exposed to low oxygen levels—for example, as the blood flows through body tissues—the hemoglobin molecules unfold, forming odd-shaped dysfunctional molecules. The misshapen protein, in turn, changes the shape of the red blood cell, making it sickle shaped and inflexible (as shown in ▷ Figure 2–31). These cells clog tiny blood vessels in the internal organs of victims, (especially the brain and heart), and can cause severe blockage and death.

---

[9] People from parts of Italy, Greece, Arabia, and India and their descendants.

▷ **FIGURE 2–31 Sickle-Cell Anemia** Scanning electron micrographs of (*a*) normal blood cell and (*b*) blood cell taken from patient with sickle-cell anemia. Because a single amino acid is incorrect, the structure of the entire hemoglobin molecule and the red blood cell changes when the cell encounters low-oxygen environments, creating inflexible, sickle-shaped cells that clog in tiny blood vessels.

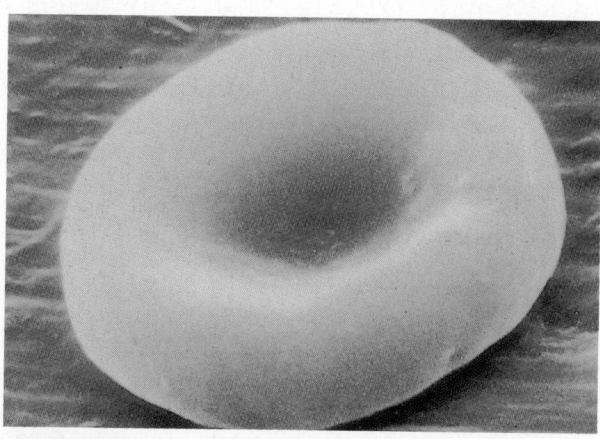

(a)

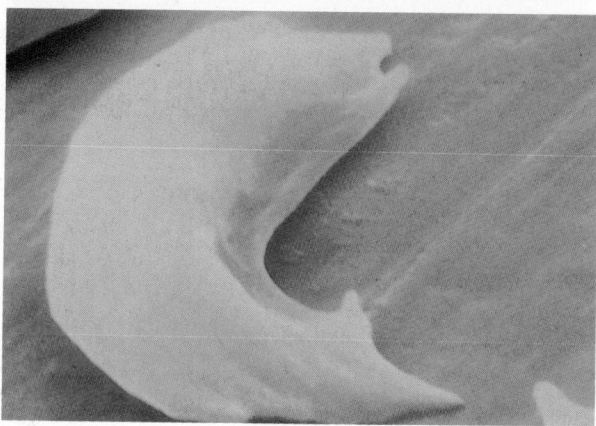

(b)

Changes in the physical and chemical environment can also disrupt the three-dimensional structure of a protein, a process called **denaturation.** Because heat denatures protein, dropping an egg in a skillet causes it to solidify; heat causes the protein molecules to unfold, entangling with chains of other molecules.

An increase in acidity can also cause a protein to unfold. If you were to crack an egg and drop it into an acidic solution, such as vinegar, it would solidify just as if you had boiled it. Hydrogen ions in the solution interfere with the normal relationships between side groups, which are responsible for the tertiary structure of the protein, and cause the proteins to unravel.

In humans and other animals, for example, acids in the stomach denature proteins in food. This process facilitates enzymatic digestion that occurs in the small intestine, a topic discussed in a bit more detail in Chapter 11.

Nucleic Acids Are Linear Polymers of Nucleotides

Nucleic acids are an amazing and extraordinarily important group of organic molecules with two renowned members: ribonucleic acid (RNA) and deoxyribonucleic acid (DNA). Both molecules are long-chained polymers consisting of many smaller organic molecules, known as nucleotides. We begin our discussion of nucleic acids with nucleotides.

Nucleotides, shown in ▷ Figure 2–32, consist of a phosphate, a sugar molecule, and a nitrogen-containing organic base. Nucleotides join by covalent bonds to form RNA and DNA. (More on this in Chapter 7.)

A molecule of **DNA,** or **deoxyribonucleic acid,** actually consists of two polynucleotide chains, which combine to form a double-stranded **helix,** akin to a spiral staircase (▷ Figure 2–33a). As illustrated, the two strands of the DNA molecule are held together by hydrogen bonds.

▷ **FIGURE 2–32 Nucleotides** (a) Nucleotides are the building blocks of RNA and DNA. Nucleotides consist of three molecules: (b) a phosphate, (c) a simple sugar, and (d) a nitrogen-containing base. RNA and DNA nucleotides contain different sugars and bases. Uracil replaces thymine in RNA.

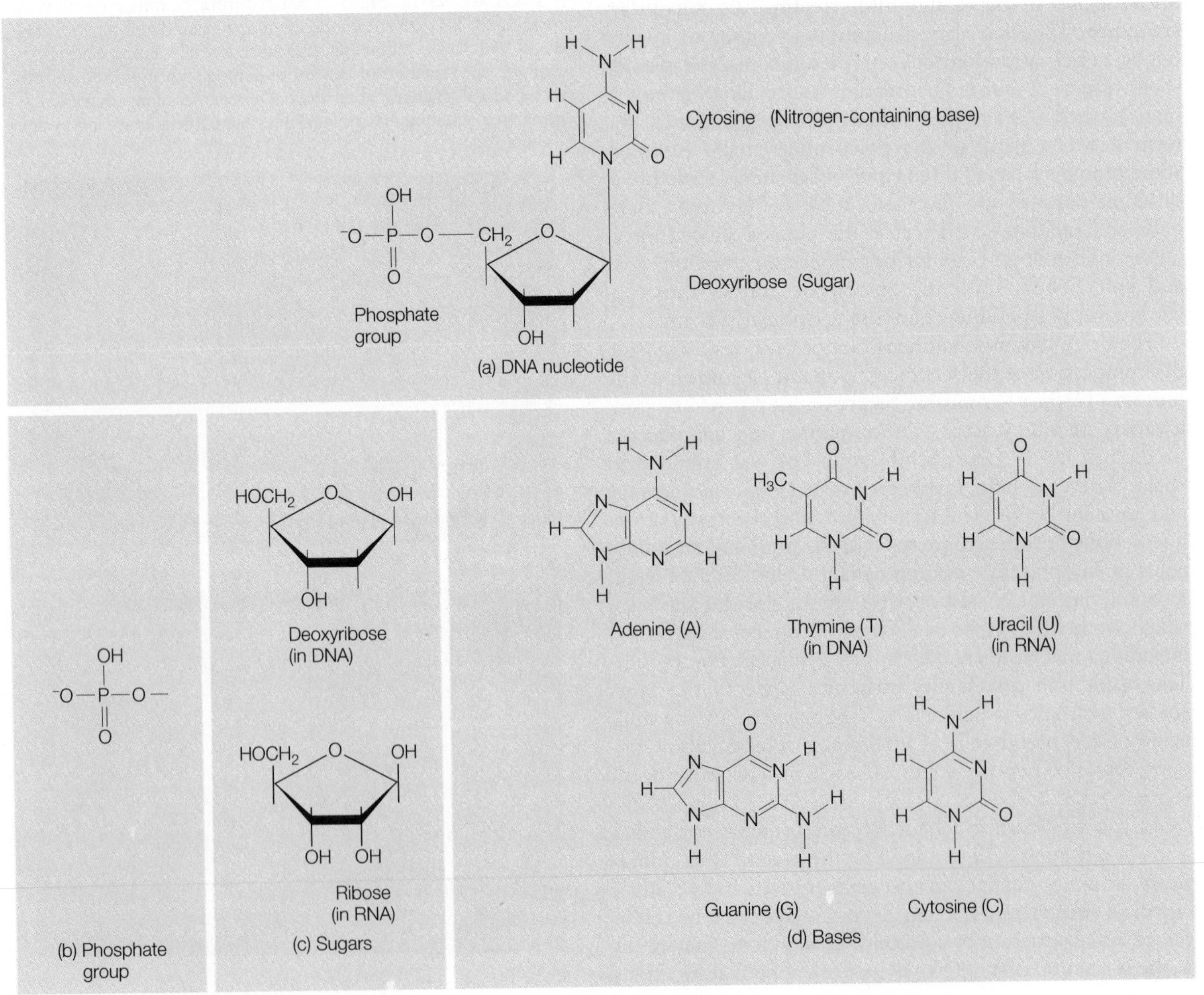

Cytosine (Nitrogen-containing base)

Deoxyribose (Sugar)

Phosphate group

(a) DNA nucleotide

Deoxyribose (in DNA)

Adenine (A)

Thymine (T) (in DNA)

Uracil (U) (in RNA)

Ribose (in RNA)

Guanine (G)

Cytosine (C)

(b) Phosphate group

(c) Sugars

(d) Bases

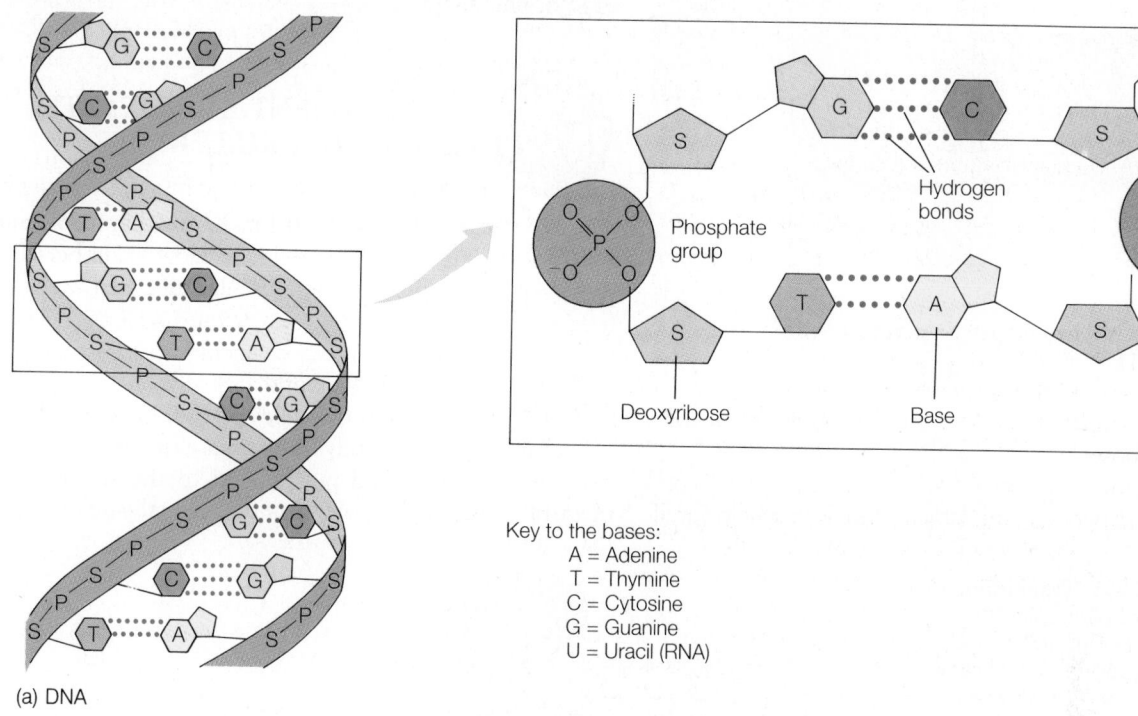

▷ **FIGURE 2–33 DNA and RNA** (*a*) DNA is a double-stranded molecule containing two polynucleotide chains joined by hydrogen bonds between the bases. (*b*) RNA is a single-stranded polynucleotide chain. It is sometimes twisted into unusual shapes.

Hydrogen bonds

Phosphate group

Deoxyribose

Base

Key to the bases:
A = Adenine
T = Thymine
C = Cytosine
G = Guanine
U = Uracil (RNA)

(a) DNA

Ribose

OH

Ribose

OH

Ribose

OH

Ribose

OH

(b) RNA

DNA forms the genes of all cells and carries the instructions required for proper development and normal function. The nucleotides in the DNA molecule contain the sugar **deoxyribose,** from which this nucleic acid gets its name.

RNA, or **ribonucleic acids,** consists of a single strand of nucleotides containing the sugar molecule ribose—hence its name (Figure 2–33b). RNA is a key player in protein synthesis (Chapter 8).

ATP. Some nucleotides do not bond with others. A good example is a nucleotide molecule with the somewhat forbidding name of **adenosine triphosphate,** or simply **ATP** (▷ Figure 2–34). ATP shuttles energy from chemical reactions in the body that release energy to chemical reactions that require it. This nucleotide consists of a nitrogen-containing organic base, adenine, a five-carbon sugar, ribose, and three phosphate groups (Figure 2–34). ATP is formed as follows:

$$ADP \quad + \quad P_i \quad + \text{ Energy } \leftrightarrows \quad ATP$$

adenosine diphosphate   inorganic phosphate   adenosine triphosphate

In order for cells to form ATP, energy must be present. This energy is used to covalently bond the inorganic phosphate molecule (P_i) to ADP and is "stored" in the bond for later use.

The reverse reaction—the breakdown of ATP into ADP and inorganic phosphate—releases energy stored in the

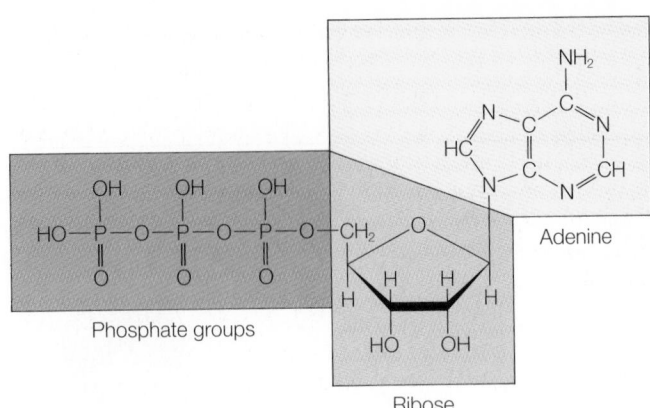

Phosphate groups

Adenine

Ribose

▷ **FIGURE 2–34 Molecular Structure of the Nucleotide ATP**

covalent bond. That energy can be used by the cell for a variety of purposes.

This concludes our brief look at chemistry. It won't make you an expert on the subject, but it will help famil-

iarize you with some biologically important molecules, which will be essential to your studies. The chapter concludes with a case study that involved an intimate knowledge of chemistry.

ENVIRONMENT AND HEALTH
TRACKING A KILLER

For 2000 years, the people in Lin Xian, China, 250 miles south of Beijing, have been dying in record numbers from cancer of the esophagus, the muscular tube that transports food to the stomach. One of every four persons succumbs to this mysterious killer, whose incidence is higher there than anywhere else in the world.

In 1959, scientists in the valley of 70,000 people around Lin Xian began to study the disease in an attempt to discover its origins and put an end to the scourge. The mysterious deaths led scientists on a lengthy search. They

▷ **FIGURE 2–35 Tracking a Deadly Killer** In Lin Xian, China, the link between esophageal cancer and the soil was uncoverd by the diligent work of medical researchers.

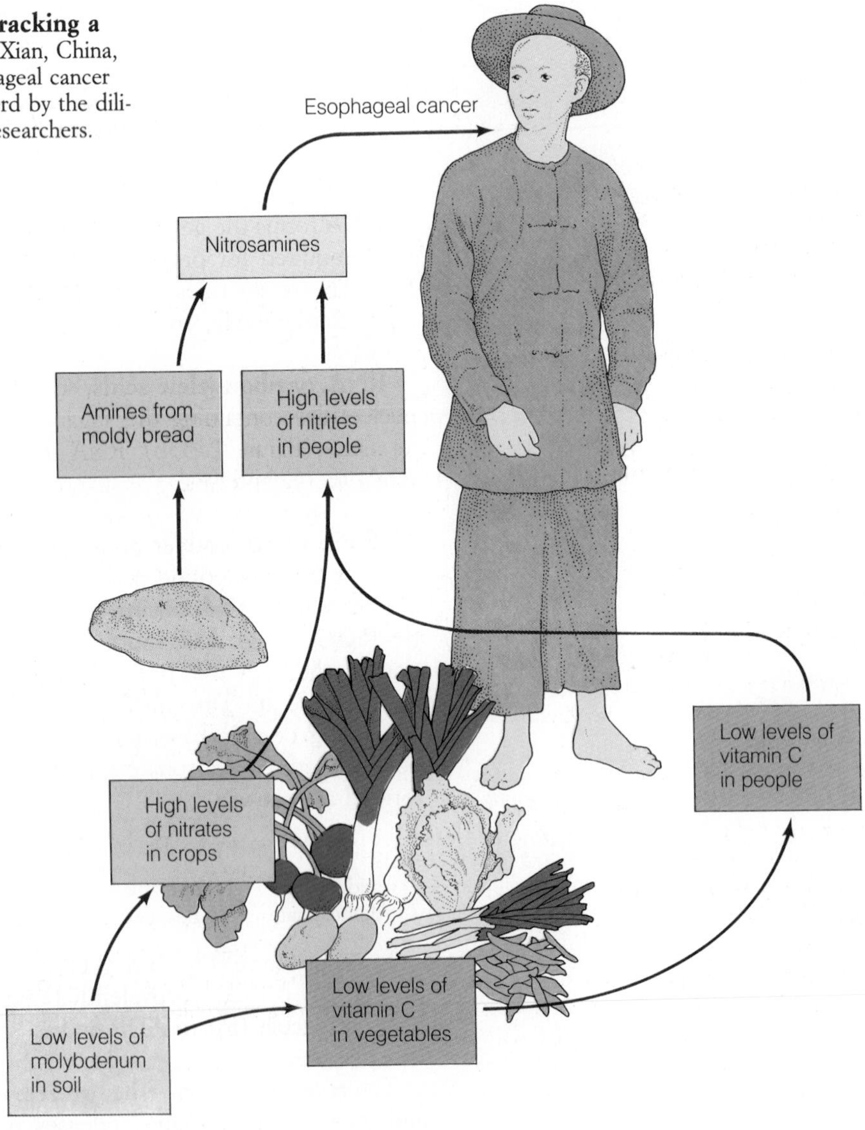

Esophageal cancer

Nitrosamines

Amines from moldy bread

High levels of nitrites in people

Low levels of vitamin C in people

High levels of nitrates in crops

Low levels of vitamin C in vegetables

Low levels of molybdenum in soil

first found that esophageal cancer in Lin Xian was the result of a group of chemicals called nitrosamines, which were being produced in the stomachs of the residents from two other chemicals: nitrites and amines. Research showed that the nitrites came from the vegetables that residents ate and that the amines were present in moldy bread, which is considered a delicacy in the region. The complex set of circumstances that led to cancer is outlined in ▷ Figure 2–35.

Studies of the area's residents showed that their bodies contained high levels of nitrites. Studies also showed that the vitamin C levels in their diets were low. When the residents were given vitamin C tablets, their nitrite levels dropped. The researchers probed further to find out what was causing the elevation in nitrites and the deficiency of vitamin C.

A careful study showed that the amount of nitrates in vegetables grown in the region was much higher than expected. In the stomach of the residents, nitrates were apparently being converted to nitrites. Vitamin C levels were lower than normal. That explained the concentrations found in the residents, but why were the levels in plants different from areas a few hundred miles away?

The answer, scientists found, lay in the soil. Chemical analyses showed that the topsoil around Lin Xian was deficient in molybdenum, an element required by plants in minute quantities. This deficiency caused the plants to concentrate nitrites and reduce their vitamin C production. This put the local residents who ate moldy bread rich in amines at risk for esophageal cancer.

This intriguing example shows one of the many links between the quality of the environment and the health of people. Low levels of molybdenum in the soil, caused by long-term farming or resulting from natural deficiencies, resulted in changes in the plants with a profound impact on the residents of the valley.

To prevent esophageal cancer, the villagers now coat wheat and corn seeds with molybdenum. As a result, nitrite levels in vegetables have dropped 40%, and vitamin C levels have increased 25%. Although it is still too early to tell whether restoring the soil's nutrient health will put a halt to this deadly killer, scientists are optimistic.

SUMMARY

1. Chemistry has many practical applications. An understanding of chemistry is essential to understanding biology, because all cells and organisms are composed of molecules, and many life processes are chemical reactions.

ATOMS AND SUBATOMIC PARTICLES

2. The physical world we live in is composed of matter. Matter is anything that has mass and occupies space.
3. All matter is composed of tiny particles called atoms.
4. Atoms contain many subatomic particles, three of which are important to a basic understanding of chemistry: electrons, protons, and neutrons.
5. Electrons, the smallest of these three subatomic particles, orbit in the electron cloud around the dense, central nucleus.
6. Protons are positively charged particles and are found in the nucleus with neutrons, which are noncharged. Neutrons and protons are the most massive subatomic particles.
7. Atoms are electrically neutral, because the number of electrons always equals the number of protons.
8. A pure substance (like gold or lead) that contains only one type of atom is called an element. The elements are listed on the periodic table by atomic number, the number of protons in the nucleus.
9. Most elements are made up of mixtures of two or more isotopes, or atoms with slightly different atomic mass resulting from additional neutrons.
10. Additional neutrons often make isotopes of various elements unstable. In order to achieve a more stable state, many isotopes release radiation—small bursts of energy, or tiny energetic particles—from their nuclei.
11. Radiation causes much damage in cells and organisms, but it has also provided many benefits to humankind.

CHEMICAL BONDS AND MOLECULES

12. Atoms bond to form molecules. Chemical reactivity results, in large part, from electrons.
13. Electrons orbit around the nucleus in specific regions, or shells, within the electron cloud. Each electron shell can hold a set number of electrons.
14. Electrons within a shell actually occupy specific regions called atomic orbitals, subregions within shells that are occupied by one or more speeding electrons. Orbitals have very definite shapes.
15. The arrangement of electrons in shells and orbitals helps us understand the chemical reactivity of the elements—that is, how atoms bond. In very general terms, atoms tend to react with others to form complete outer shells, creating more stable atomic configurations.
16. During some atomic reactions, electrons are transferred from one atom to another. Transferring an outer shell electron from one atom to another creates two oppositely charged ions. An electrostatic attraction forms between them and is referred to as an ionic bond.
17. Covalent bonds form between atoms with similar electron affinity that share one or more pairs of electrons. This sharing of electrons holds the atoms together.
18. Unequal sharing of electrons results in the formation of a polar covalent bond, which is often sufficient to result in the formation of polar molecules. Polar molecules are frequently attracted to one another by hydrogen bonds, weak electrostatic attractions between the oppositely charged atoms of neighboring molecules.

WATER, ACIDS, BASES, AND BUFFERS

19. Two types of covalent molecules exist: organic and inorganic.

20. Organic molecules are compounds made up primarily of carbon and hydrogen. The atoms of these molecules are held together by covalent bonds.

21. Inorganic molecules are frequently much smaller molecules whose atoms are often joined by ionic bonds.

22. Water is an important inorganic molecule and a major component of all body cells. It serves as a solvent, which transports many substances in the blood, body tissues, and cells. Water also participates in many chemical reactions, serves as a lubricant, and helps regulate body temperature.

23. Although fairly stable, water molecules can dissociate, forming hydrogen and hydroxide ions.

24. In pure water, the concentration of these ions is equal and water has a pH of 7. Adding acidic substances increases the concentration of hydrogen ions, causing the pH to fall.

25. Substances that remove hydrogen ions from solution cause the pH to climb and are called bases.

26. A solution with a pH less than 7 is acidic; a solution with a pH greater than 7 is basic.

27. Biological systems contain buffers, chemicals that offset changes in the concentration of hydrogen and hydroxide ions.

MORE BIOLOGICALLY IMPORTANT MOLECULES

28. The diversity of organic matter is largely the result of the carbon atom. Carbon atoms can enter into single, double, and triple covalent bonds and can react with a number of other atoms, forming a wide assortment of molecules.

29. Many smaller organic molecules, such as amino acids and sugars, combine to form large molecules called polymers.

30. The organic molecules found in organisms can be divided into four classes: (a) carbohydrates, (b) lipids, (c) amino acids, peptides, and proteins, and (d) nucleic acids.

31. Carbohydrates are a group of organic compounds with the general formula $(CH_2O)_n$.

32. Lipids are a diverse group of organic chemicals that are characterized by their lack of water solubility. The biologically important lipids serve many functions and fall into four main categories: (a) triglycerides, (b) waxes, (c) phospholipids, and (d) steroids.

33. Amino acids are small organic molecules that join by peptide bonds, forming a variety of peptides (containing fewer than 20 amino acids) and proteins (containing more than 20 amino acids).

34. Thousands of different types of proteins are found in human cells, and each type is structurally and functionally unique. This uniqueness results from the sequence of amino acids in a protein. The sequence of amino acids determines the ultimate shape and function of the protein. A change of even one amino acid can severely disrupt the shape and function of a protein.

35. The nucleic acids RNA and DNA are polymers of smaller organic molecules known as nucleotides. A nucleotide consists of a phosphate, a sugar molecule, and a nitrogen-containing base.

36. DNA, or deoxyribonucleic acid, forms the genes of our cells and carries all of the genetic information required for cell structure and function.

37. RNA, or ribonucleic acid, plays a key role in protein synthesis.

38. Some nucleotides do not bond with others. A good example is an adenosine triphosphate, or ATP. ATP shuttles energy from chemical reactions in the body that give off energy to chemical reactions that require energy. This nucleotide consists of a nitrogen-containing organic base, adenine; a five-carbon sugar, ribose; and three phosphate groups.

EXERCISING YOUR CRITICAL THINKING SKILLS

In 1971, Joseph Chatt and his colleagues at the University of Sussex in England synthesized two compounds with the same chemical formula. Interestingly, however, he found that the crystals formed by one compound were sapphire-green, whereas the crystals formed by the other were bright green. Chatt hypothesized that he had created two isomers, compounds that have the same molecular weight and atomic composition but differ in chemical or physical properties. Typically, isomers occur when the atoms are linked in different molecular geometries. However, when Chatt subjected his isomers to X-ray analysis, which helps determine the structure of molecules, he found that the atoms of his compounds were in the same relative positions. The only difference appeared to be in the molybdenum-oxygen bond, which was longer in the bright green compound than in the sapphire-green one.

Based on the assumption that impurities are rare in single, well-formed crystals, Chatt's group concluded that the color difference in the two compounds resulted from differences in the length of a single bond, a phenomenon not previously reported.

At first, Chatt's findings were largely discounted, because chemists believe that molecular bonds have characteristic lengths. Most scientists concluded that his experiments were flawed.

(Can you find any reason(s) in the above information that might have led Chatt to an incorrect conclusion?)

In 1985, however, chemists at the Ruhr University in Germany reported the synthesis of another pair of isomers like Chatt's, but from tungsten-containing compounds. According to the researchers, their isomers were chemically and structurally identical except for the length of one metal-oxygen bond.

Several other laboratories also reported similar findings, creating a great deal of excitement among scientists, for it appeared that a long-held belief (the constancy of bond length) was about to topple. Who knows what new insights might have developed?

Chemists were by now intrigued by this new isomer, and theorists began looking for an explanation for the phenomenon. In 1988, the Nobel Prize–winning chemist Roald Hoffman and his colleagues published an explanation. In 1991, this new type of isomerism was discussed in a new inorganic chemistry textbook.

As chemists began to repeat the experiments to learn more about the isomers, however, enthusiasm began to fade. In one replication of Chatt's experiment, scientists found that in some preparations chlorine atoms substituted for oxygen atoms in the same type of molecules that Chatt had studied. When this occurred, the pure blue compound mixed with the yellowish impurity, producing green crystals. The impurity, therefore, explained the presence of crystals with two different colors. This explanation has been accepted by Chatt and his colleagues. And after the German experiment was repeated, it was shown that those results had also been the result of contamination.

If you said earlier that underlying assumptions affected the conclusions, you were right. "Bond-stretch isomerism," as it was called, nearly became accepted theory based on the widely-accepted, but faulty, assumption that impurities were rare in single, well-formed crystals. The lesson here is that underlying assumptions may lead to erroneous conclusions.

The textbook author who included the findings in his book learned the hard way to beware of preliminary findings and wait for validation.

TEST OF TERMS

1. Any substance that occupies space and has mass is called _____ .
2. _____ are the purest forms of matter and cannot be separated into other substances by chemical means.
3. Atoms consist of three subatomic particles: protons, _____ , and _____ .
4. Protons are found in the _____ of the atom and have a positive charge.
5. Elements are listed on the periodic table of elements by ascending _____ _____ .
6. An isotope is an alternative form of an atom containing one or more additional _____ .
7. _____ released from isotopes helps stabilize their nuclei.
8. An _____ is formed when an atom either gains or loses electrons.
9. A _____ bond forms when two atoms share a pair of electrons.
10. A _____ bond forms between the charged ends of two water molecules.

11. Organic molecules are principally made of the elements _____ and _____ and contain _____ bonds.
12. The _____ molecule is a storage form for glucose in human liver and muscle cells.
13. Water dissociates into two ions, _____ and _____ .
14. An _____ is any chemical substance that adds hydrogen ions to solution, and a _____ is any chemical substance that removes them.
15. On the pH scale substances with pH values lower than 7 are _____ . The lower the pH reading, the _____ the concentration of hydrogen ions.
16. A substance that helps maintain a constant pH is called a _____ .
17. Simple sugars are also called _____ and are used to synthesize long-chained molecules, such as starch and glycogen, which are

in a group of carbohydrates called _____ .
18. A triglyceride consists of three _____ _____ molecules and one molecule of _____ .
19. The principal lipid in the cell membrane is _____ .
20. _____ is the lipid that deposits in the walls of arteries, causing atherosclerosis.
21. Proteins are composed of amino acids linked by _____ bonds.
22. _____ are a class of proteins that speed up chemical reactions.
23. The sequence of amino acids in a protein is called its _____ structure. The final three-dimensional shape of a protein is its _____ structure.
24. DNA and RNA are polymers of molecules called _____ .
25. ATP is formed from ADP, _____ , and _____ .

Answers to the Test of Terms are found in Appendix B.

TEST OF CONCEPTS

1. Why is the matter on Earth so varied in its appearance?
2. Describe the structure of an atom, using a diagram to illustrate your explanation.
3. Define the following terms: mass number, atomic number, and isotope.
4. How do ionic and covalent bonds form? Describe what holds the atoms together in both cases. In what ways are the bonds different?
5. Describe how polar covalent bonds can result in the formation of hydrogen bonds.
6. Temperature is a measure of the speed of molecules: the higher the

temperature, the higher the speed. With this knowledge and your knowledge of water and the hydrogen bond, do you think water has a higher boiling point than a nonpolar liquid like alcohol? Why or why not?
7. Describe the biological significance of water.
8. How does the body regulate H^+ levels in the blood? Describe the process.
9. List the four major kinds of biological molecules described in the chap-

ter, and explain why each one is important.
10. Define the following terms: monosaccharide, disaccharide, and polysaccharide. Give an example of each.
11. Describe the biological importance of each of the following lipids: triglycerides, steroids, and phosphoglycerides.
12. List some of the functions that proteins serve in cells and organisms.
13. Why do changes in the primary structure of a protein alter its tertiary structure and function?
14. How are DNA and RNA similar? How are they different?

The Life of the Cell

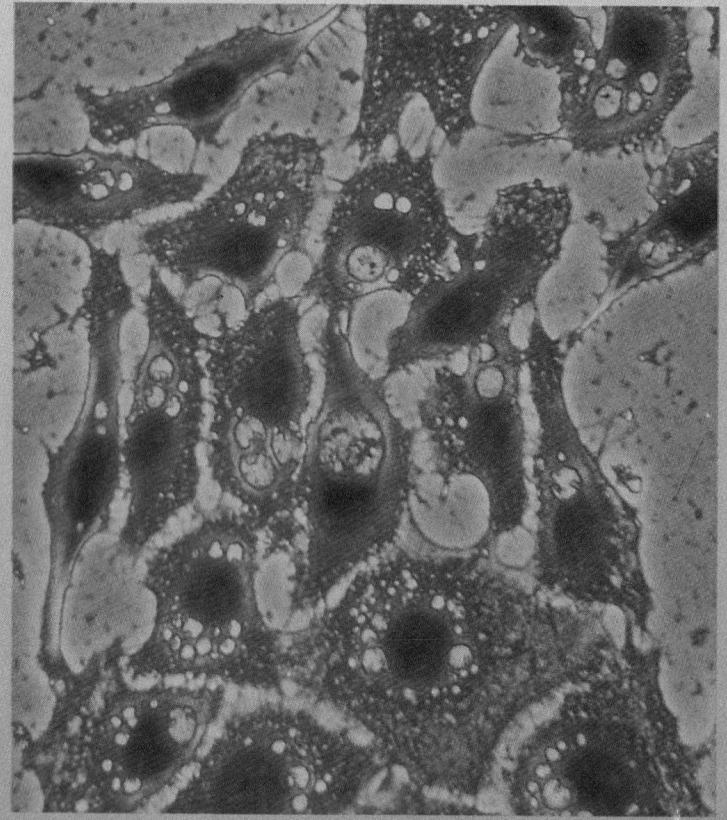

Mammalian cells in a culture showing nuclei and cytoplasm.

I n Sweden and Mexico, medical researchers are experimenting with a bold and controversial procedure called fetal cell transplantation. By transplanting healthy cells from human fetuses into adults, medical scientists hope to be able to cure a number of chronic, debilitating diseases such as diabetes. The general goal of this research is to replace defective cells in organs and tissues with normal-functioning fetal cells. In time, the procedure could reduce the pain and suffering of tens of thousands of people worldwide, people who had little hope of a cure.

Although promising, fetal cell transplantation poses a number of technical challenges, many of which are discussed in Health Note 3–1. It also poses several significant moral and political dilemmas, which are discussed in the Point/Counterpoint in this chapter.

This chapter focuses on the cells of plants and animals. The information presented here, like that on chemistry in the previous chapter, is part of the foundation on which a good understanding of biology is built. Thus, in order to understand the function of multicellular organisms like ourselves, we must understand the workings of the cells.

≈ CELLULAR EVOLUTION AND HOMEOSTASIS

One of the central tenets of modern biology is the cell theory. The **cell theory** holds that all organisms are composed of cells and that all cells come from preexisting cells. There is, of course, one exception, the very first cells to form during evolution. These cells are believed to have been formed from organic molecules present in the early Earth's environment. All succeeding cells, however, were derived from these first cells, which appeared on Earth about 3.7 billion years ago. Resembling modern-day bacteria, these cells lived primarily in the oceans and were the dominant form of life for over 2 billion years.

The earliest cells and modern-day bacteria are known as **prokaryotes** (*pro* = before; *karyon* = nucleus). Prokary-

otic cells contain a single circular strand of DNA but otherwise show very little internal differentiation (▷Figure 3–1a). Biologists place bacteria and their prokaryotic predecessors in a large group, or kingdom, of organisms known as the Monera.

Structurally more complex cells arose from the prokaryotes about 2.5 billion years later and are referred to as eukaryotes. **Eukaryotes** (*eu* = true) contain DNA within a membrane-bound structure, the nucleus (Figure 3–1b). The earliest eukaryotic cells lived independently, as do their modern descendants, aquatic amoebae, paramecia, and the like. The single-celled eukaryotes belong to the kingdom Protista.

The Formation of Complex, Multicellular Organisms Resulted from Evolution

During evolution, protists gave rise to three additional kingdoms: plants, fungi, and animals (▷Figure 3–2). All five kingdoms remain today.

The evolution of the three new kingdoms was characterized by two major developments: (1) multicellularity, the development of organisms composed of many cells, and (2) cellular specialization, the emergence of specialized cells that perform specific tasks.

Multicellularity and cellular specialization led to an increase in the overall complexity of life on Earth. The emergence of cells that perform specialized functions resulted in many developments, such as the complex motor and sensory systems of numerous animal species, which enable them to move about in their environment and to respond to a wide range of external stimuli.

Growing complexity required homeostatic control mechanisms. Distinct body systems of multicellular animals, such as the nervous and endocrine systems, evolved to control a wide variety of functions. These and other body systems maintain homeostasis, which is necessary for the proper functioning of the cells and, thus, the body as a whole.

▷ **FIGURE 3–1 Prokaryotes and Eukaryotes** (*a*) Prokaryotes are relatively simple organisms like this bacterium. They contain DNA, but the DNA is not contained within a nucleus.

(*b*) Eukaryotes evolved from the prokaryotes. Their DNA is membrane-bound, and the cells contain numerous cellular organelles.

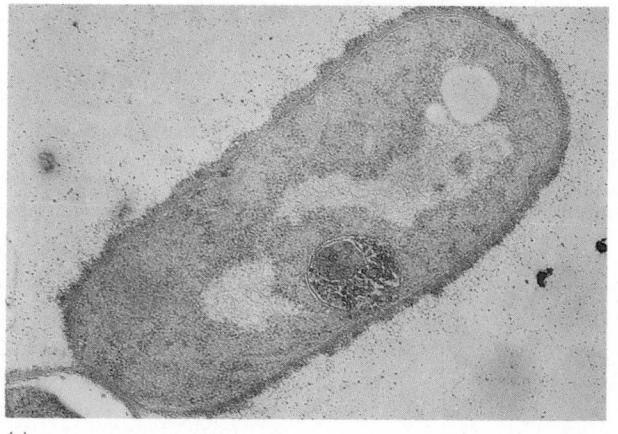

(a)

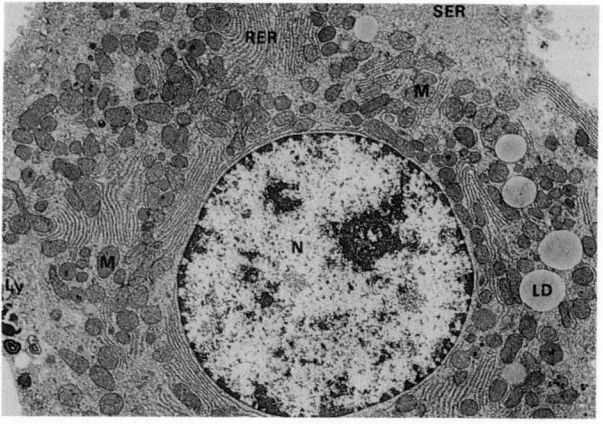

(b)

FETAL CELL TRANSPLANTS: THE ALLURE OF A NEW CURE

Treating and curing diseases are two different things. Consider diabetes. This disorder afflicts an estimated 12 million Americans and is characterized by the body's inability to control blood glucose levels. Physicians recognize two types of diabetes, one of which is generally treated by diet and the other by daily injections of insulin, a hormone lacking in many victims of the disease. (For more on diabetes, see Chapter 20.)

Despite the widespread use of insulin, a diabetic's life is never easy. Too much exercise or too much insulin can cause blood glucose levels to plummet. Victims are left feeling dizzy and weak. Excess blood sugar, caused by overeating or a failure to take an insulin shot on time, can cause irritability, depression, fatigue, and even death. Short-term fluctuations in blood sugar are mostly an annoyance, but diabetics are also plagued by several serious long-term complications due to periodic bouts of hyperglycemia (elevated blood sugar), which damages blood vessels and nerves. Blindness and kidney damage are two common results at the cellular level.

Today, thanks to developments in fetal cell transplantation, a cure may be possible for some victims of this disease. Fetal cell transplantation can be used to insert healthy fetal pancreatic cells into diabetics who cannot produce insulin hormone on their own. Here, research suggests, the fetal cells will establish a permanent residence and begin secreting insulin, ending a patient's dependency on insulin injections. As noted in the chapter, fetal cells are not rejected, because they have not yet developed their cellular ID.

Fetal cell transplantation offers promise in other areas as well. Medical researchers are hoping that fetal brain cells transplanted into patients with Parkinson's disease will cure this disorder. In 1988, U.S. researchers announced the successful transplantation of human fetal brain cells into monkey brains. Seventy days after transplantation, the grafted cells had formed a dense tangle of maturing nerve cells. Two of three grafts showed signs of producing dopamine, a chemical lacking in the brains of Parkinson's victims. Researchers hope that these cells will restore normal brain levels of dopamine, which could eliminate tremors and rigidity.

Since their announcement, American and Swedish scientists have transplanted human fetal brain tissue into human subjects. Early results have been disappointing, but new studies show more promise, raising hopes that the technique can help millions of people worldwide (▷Figure 1).

On another front, researchers are looking to fetal cell transplantation to provide a way to reverse the neurological effects of long-term alcohol abuse in humans. Rats treated with alcohol suffer memory loss similar to that of chronic alcoholics. Recently, medical researchers announced the successful transplantation of fetal brain tissue into the brains of alcohol-treated rats. About nine weeks after receiving such transplants, alcohol-treated and non-alcohol-treated rats were given performance tests that measured their memory. Twelve of fourteen alcohol-treated rats performed remarkably well, almost as well as control rats that had received no alcohol. Researchers do not know why the brain transplants work, but they note that similar transplants could improve the lives of many recovering alcoholics whose brains are severely damaged after years of alcohol abuse.

Fetal cell tissue transplantation offers great hope to humankind. It could provide physicians a way to cure diseases long thought to be incurable. But this

▷ **FIGURE 1 Fetal Cell Transplant Patient** Once suffering from Parkinson's disease, Donald Nelson now leads a fairly normal life thanks to fetal cell transplantation.

technique has stirred considerable controversy in the United States. (See the Point/Counterpoint.)

In most states, the use of fetal tissue for experimentation is legal, and a federal panel convened to debate the issue concluded that clinical trials should be conducted. The possibility of relieving suffering and saving life, the panel said, "cannot be a matter of moral indifference to those who shape and guide public policy." Unfortunately, say some, the U.S. Department of Health and Human Services, which funds such research, has decided to withhold research money, forcing researchers to seek private funding. In 1992, the U.S. Congress passed a law to authorize funding of research on fetal cell transplantation but it was vetoed by President Bush.

To avoid the controversy, some researchers are developing lines of human fetal cells that could be grown in culture and used instead of tissues taken from aborted fetuses. Unfortunately, many scientists believe that tissue-culture techniques suitable for fetal cells are a decade away from perfection.

Despite these and other dramatic evolutionary changes, the cell's basic features have endured, reinforcing the basic principle of evolution that successful patterns persist (Chapter 1). This chapter focuses on general cell characteristics, providing knowledge applicable to a wide range of organisms. Before we peer into the internal architecture of the cell, however, let's take a brief look at the microscope, an important tool for understanding the structure and function of cells.

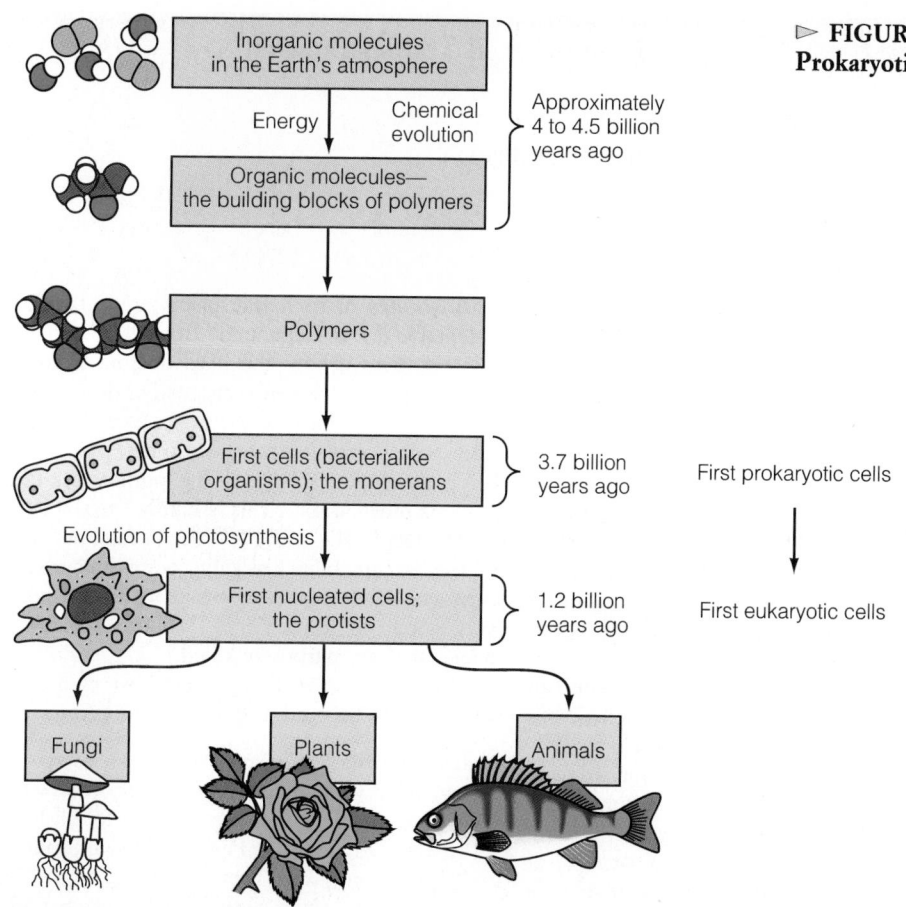

Inorganic molecules in the Earth's atmosphere

Energy · Chemical evolution

Approximately 4 to 4.5 billion years ago

Organic molecules—the building blocks of polymers

Polymers

First cells (bacterialike organisms); the monerans — 3.7 billion years ago — First prokaryotic cells

Evolution of photosynthesis

First nucleated cells; the protists — 1.2 billion years ago — First eukaryotic cells

Fungi

Plants

Animals

≈ MICROSCOPES: ILLUMINATING THE STRUCTURE AND FUNCTION OF CELLS

Cells are exceedingly small and, with few exceptions, cannot be seen without the aid of a **microscope**, a special instrument consisting of a lens or combination of lenses that enlarges tiny objects. (For a discussion of the discovery of cells, see Scientific Discoveries 3–1.)

Microscopes fall into two broad categories: (1) **light microscopes**, which use ordinary visible light to illuminate the specimen under study, and (2) **electron microscopes**, which use a beam of electrons to create a visual image of the specimen.

▷ Figure 3–3a shows the design and operation of a light microscope. As illustrated, light from below passes through

▷ **FIGURE 3–3 The Light Microscope** (*a*) In the ordinary light microscope, the image is illuminated by light that is cast from a mirror or (in this case) light bulb located below the object. Two lenses (one in the eyepiece and one just above the object) magnify the object. (*b*) Light micrograph of a section through a paramecium. (*c*) Photomicrograph of paramecium taken through a dissecting microscope.

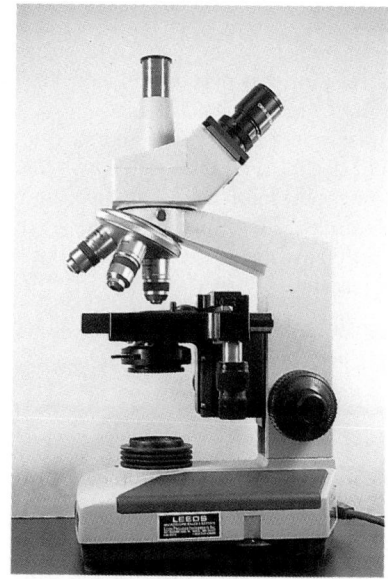

(a)

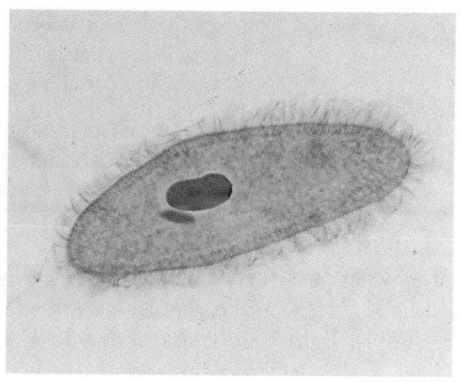

(b)

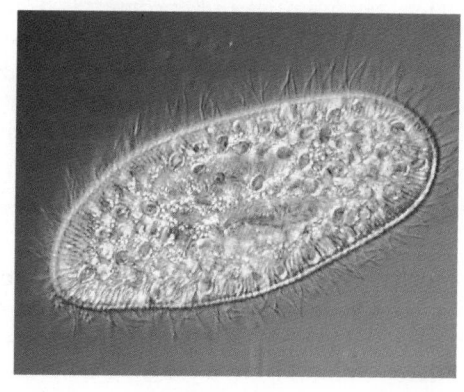

(c)

FETAL TISSUE TRANSPLANTS: AUSCHWITZ REVISITED

Thomas J. Longua

Thomas J. Longua has been an educator for 30 years and is currently instructor of anatomy and physiology at the Denver Academy of Court Reporting. He has been active in the pro-life movement for 17 years and is vice-president of the Colorado Right to Life Committee.

Opposition to the practice of using tissue from aborted human babies for medical purposes starts with opposition to abortion itself.

In the mid-1960s, Western society decided that it needed an "acceptable" way to rid itself of the many unwanted babies it was creating in its newfound sexual permissiveness. Therefore, abortion was legalized, even though it was recognized—as it had been for over 2000 years—as an act of killing an innocent human life.

Pro-life citizens predicted that legalization would lead to other evils, including infanticide and euthanasia. What most of us did not foresee was the use of aborted human fetuses as medical "spare parts." Even when this practice began to be discussed, we could not believe that a "civilized" society could take it seriously. After all, Auschwitz was still a recent memory.

After World War II, the Nuremburg Tribunal was convened for the Nazi war criminals, including the physicians who had experimented on concentration camp victims. In his opening statement, the prosecutor of the Doctors Trials said, "The defendants in this case are charged with . . . atrocities committed in the name of medical science. . . . The wrongs which we seek to condemn and punish have been so calculated, so malignant, and so devastating that civilization cannot tolerate their being ignored, because it cannot survive their being repeated."

Yet with fetal tissue transplantation, we are again committing the same "malignant wrongs" as the Nazi doctors committed. Even the arguments are the same, including the pretense that the victims are "subhuman." (It is ironic that abortion proponents who once denied that the fetus was human now endorse fetal transplants precisely because the tissue *is* human.)

Compare a Nazi doctor to a modern-day advocate:

- Wouldn't it be ridiculous to send the bodies to the crematory oven without giving them an opportunity to contribute to the progress of society?" (Nazi Doctor August Hirt, 1942)
- We are simply using something which is destined for the

incinerator to benefit mankind." (Transplant Doctor Lawrence Lawn, 1970)

Advocates of such transplants argue that the procedure is acceptable if it is "separated" from the abortion itself. But even this argument echoes the Nazi doctors' defense: "We caused no deaths. They were all consigned to death by legal authorities; with . . . professional correctness we tried to salvage some good from their plight." Furthermore, new abortion techniques are being developed solely for the purpose of ensuring that the tissue is more usable. This is hardly "separation."

The fact is, the entire practice of fetal tissue transplants rests on the "acceptability" of killing some humans in the first place. (Certainly there is nothing wrong with using tissue from ectopic pregnancies or miscarriages, but the debate revolves around babies that are purposely killed.)

Besides the Nazi-like perverseness of the practice itself, it will certainly have other detrimental effects:

- It will further legitimize—and even *sanctify*—the original killing. What woman contemplating abortion will not be swayed by the argument that the decision to kill her baby will help others? Dr. James J. Parks, a Denver abortionist, bragged, "[My patients] say, 'Thank God, some good is going to come out of this.' " (*New York Times,* November 19, 1989)
- It will certainly increase the number of abortions. Some medical journals claim that fetal tissue could be used to treat a vast array of diseases. In a guest editorial in the November 7, 1988, *Wall Street Journal,* Dr. Emanuel D. Thorne estimated that the current 1.6 million abortions in America every year will not be enough to keep up with the demand for such tissue!
- It will increase the pressure to legalize euthanasia. If harvesting unwanted humans becomes standard practice, who can doubt that there will be further demand for organs from older humans whose lives, like those of aborted babies, are deemed "meaningless?"
- It will lead to trafficking in human "spare parts"—even to pregnancy planned specifically for that purpose. Indeed, this has already happened in both America and Europe.
- It will decrease research that would oterhwise be pursued to find alternative methods of dealing with disease, such as the use of autografts, transgene cell lines, animal donors, and synthetic and neurotropic drugs. These methods hold great promise, yet acceptance of fetal tissue as the "easy way out" is already curtailing such research.

Advocates of fetal tissue transplants depict themselves as "humanitarians." But then, so did the doctors at Nuremberg.

---

two lenses that magnify the object. Photographic images can be produced by a camera mounted on a light microscope and are called **photomicrographs** or, simply, **light micrographs** (Figure 3–3b).

Biological specimens to be studied with a light microscope are often treated with chemicals that harden them

and preserve the molecules. Specimens are then cut into thin, light-penetrable sections. The sections are mounted on glass slides and stained with special dyes that help biologists distinguish their internal architecture. Living cells and tissues can also be studied with specially designed light microscopes and may or may not be stained (Figure 3–3c).

HUMAN FETAL TISSUE SHOULD BE USED TO TREAT HUMAN DISEASE

Curt R. Freed

Curt R. Freed, M.D., is a professor of medicine and pharmacology at the University of Colorado School of Medicine in Denver. He has written some 50 articles on medical topics as well as numerous abstracts, chapters, and reviews.

Despite its legalization, abortion remains a controversial issue and will continue to stir debate in the future. In the United States, the debate centers on whether a woman has the right to control her reproduction. The future developments of this political debate are uncertain, but recent elections suggest that pro-choice candidates have been victorious when elections are based on the abortion issue.

Currently, over 1 million legal abortions are performed in the United States each year; most abortions are performed in the first trimester. For nearly all women, having an abortion is an anguishing choice filled with regret and ambivalence. Nonetheless, the difficult personal decision to terminate a pregnancy is made. After the abortion, fetal tissue is usually discarded.

As an alternative to throwing this tissue away, research has shown, fetal tissue may be useful for treating patients with disabling diseases. For over 50 years, research in animals has demonstrated that fetal tissue has a unique capacity to replace certain cellular deficiencies and so may be useful for treating some chronic diseases of humans. These diseases include Parkinson's disease, diabetes, and some immune system disorders.

Cadaver fetal tissue offers the promise of helping large numbers of Americans with crippling diseases. Parkinson's disease, for example, affects hundreds of thousands of Americans. By reducing the ability to move, the disease can end careers and turn people into invalids. The disease is caused by the death of a small number of critically important nerve cells that produce a chemical called dopamine. Experiments in animals and early experiments in humans indicate that fetal dopamine cells transplanted into the brains of these patients may restore a patient's capacity to move and may eliminate the disease. Patients whose minds work perfectly well and whose bodies are otherwise

normal may become healthy and productive citizens once again.

Using cadaver tissue to treat humans has been debated for nearly 40 years. As kidney, cornea, and other organ transplants were developed in the 1950s, many objected to recovering organs from cadavers. In the intervening decades, opinion has changed so that the practice of recovering these organs from cadavers has gone from a provocative and controversial practice to an accepted policy endorsed by most states. In fact, in most states a check-off box on the back of driver's licenses is used to give permission for organ donation in the event of the death of the driver. Because abortions are induced, some argue that fetal tissue should be regarded differently than other cadaver tissue. Given the facts that abortion is legal and that fetal tissue is ordinarily discarded, there should be no moral dilemma in using fetal tissue for therapeutic purposes. As with the use of all human tissue for transplant, specific informed consent by the woman donating the tissue must be obtained.

Some have proposed that using fetal tissue for therapeutic purposes will increase the number of abortions. This is preposterous. It strains the imagination to think that a woman would get pregnant and have an abortion simply on the chance that the aborted fetal tissue might be used to treat a patient unknown to her. An unwanted pregnancy is an intimate and deeply personal crisis; it is inconceivable that the pregnancy would be seen primarily as a philanthropic opportunity.

Politics and medicine have frequently mixed in the past and will continue to do so in the future. As a physician, I think it is important to try to improve the health of patients with serious diseases. Legally acquired fetal cadaver tissue that would otherwise be discarded should be used to treat humans with disabling diseases.

☰ SHARPENING YOUR CRITICAL THINKING SKILLS

1. Summarize the positions of each author.
2. Using your critical thinking skills, analyze the view of each author. Is each stand well substantiated? Do the author's biases play a role in each argument?
3. Do you see this debate as scientific or ethical? Explain your answer.
4. Each position is based on at least one key argument. Can you pinpoint them?
5. Which viewpoint do you agree with? Why? What factors (biases) affect your decision?

The light microscope has provided a wealth of information about organisms for hundreds of years. This information has been considerably refined with the introduction of the electron microscope, which uses a beam of electrons in a vacuum to study a specimen rather than a beam of light (▷Figure 3–4a). Because electron beams cannot penetrate

the thick sections used in light microscopy, tissues must be cut into much thinner sections. Two types of electron microscopes are in common use today: *transmission* and *scanning*.

Figure 3–4a shows a **transmission electron microscope (TEM)**, so named because it transmits an electron

THE DISCOVERY OF CELLS

Featuring the work of Hooke, Leeuwenhoek, Brown, Schleiden, Schwann, and Virchow

The study of biology offers a variety of delightfully interesting facts about a wide array of living things. But stepping back from the maze of facts often yields important generalizations, or principles, that apply to all living things and thus make the study of biology more meaningful and manageable.

One of the fundamental principles of biology is known as the cell theory, discussed briefly in the chapter. The cell theory consists of three parts: (1) all organisms consist of one or more cells, (2) the cell is the basic unit of structure of all organisms, and (3) all cells arise from preexisting cells. Although this may seem rather elementary, it was not so obvious to early scientists, who labored with relatively crude instruments in a time when facts that many of us take for granted were unknown.

One of those pioneers who opened our eyes to the world of cells was Robert Hooke, a 17th-century scientist who, besides discovering cells, invented the vacuum pump and the forerunner of the balance spring used in many watches today. Equipped with a relatively crude microscope, Hooke made observations on just about everything he could lay his hands on—from tree bark to fabric—which he described in his book *Micrographia*, published in 1665.

One especially useful description was that made on a thin slice of cork (▷Figure 1). Peering through his microscope, Hooke beheld a network of tiny, boxlike compartments that reminded him of a honeycomb. He called these compartments cellulae, meaning "little rooms." Today, we know them as cells.

Hooke did not really see cells, but rather cell walls. The cytoplasm and cellular organelles had disappeared.

Hooke's work was complemented a few years later by Antony van Leeuwenhoek, a Dutch shopkeeper who spent much of his free time designing simple microscopes. Like Hooke, Leeuwenhoek examined just about everything he could catch and wrote extensively on his observations in the publication of the Royal Society. Wayne Becker, a cell biologist at the University of Wisconsin writes, "His detailed reports attest to both the high quality of his lenses and his keen powers of observation." Becker continues "They also reveal an active imagination, since at one point he reported seeing a 'homunculus' (little man) in the nucleus of a human sperm cell."

Leeuwenhoek discovered bacteria (monerans) and protozoans (single-celled protists) in his lifetime. Although his microscopes were superior to any others around, they were still crude by modern standards. This fact and the general focus of the scientific community on observing and describing life forms restricted further growth in our understanding of cells for some time. In fact, more than a century passed before cell biology moved significantly forward.

Aided by improved lenses, the 18th-century English botanist Robert Brown noted that every plant cell he studied contained a nucleus. A German colleague, Matthias Schleiden, concluded that all plant tissues consisted of cells. One year later, Theodor Schwann, a German scientist, arrived at a similar conclusion regarding animal cells. This discovery laid to rest an earlier hypothesis that plants and animals were structurally different.

Based on his and earlier work, Schwann proposed the first two parts of the cell theory—that all organisms consist of one or more cells and that the cell is the basic unit of structure of all organisms. Less than 20 years after the discovery of cell division, the German physiologist Rudolf Virchow added the third tenet of the cell theory, that all cells arise from preexisting cells.

Like other discoveries, the cell theory is the work of many people over many years. Although a few people receive credit for this important discovery, the credit really belongs to an entire line of scientists who, through observation and experimentation, have changed our view of the world.

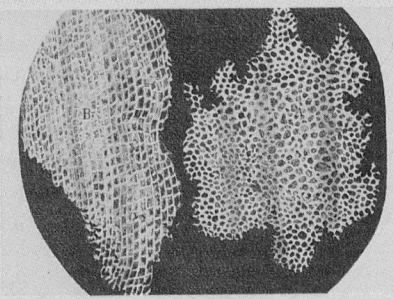

▷ **FIGURE 1 Thin Slice of Cork** Hooke discovered tiny, boxlike compartments that reminded him of a honeycomb. He called them cellulae, meaning "little rooms."

---

beam through the specimen under study. The TEM gives a detailed black-and-white view of the internal architecture of cells. The photographic images it produces are known as **transmission electron micrographs** (Figure 3–4b).[1]

The **scanning electron microscope (SEM)** yields three-dimensional images of specimens (Figure 3–4c). The black and white photographic images of the SEM are referred to as **scanning electron micrographs**. Tissues prepared for a SEM are often frozen, then fractured with a special instrument. The surfaces of the specimen are customarily coated with gold, which is impermeable to electrons. A finely focused electron beam then scans the surface of the specimen, and the scattered electrons are

[1] In this book, transmission and scanning electron micrographs have been colorized to enhance their visual appeal. A few have been color-enhanced, a computer procedure that highlights subtle differences in the density of images.

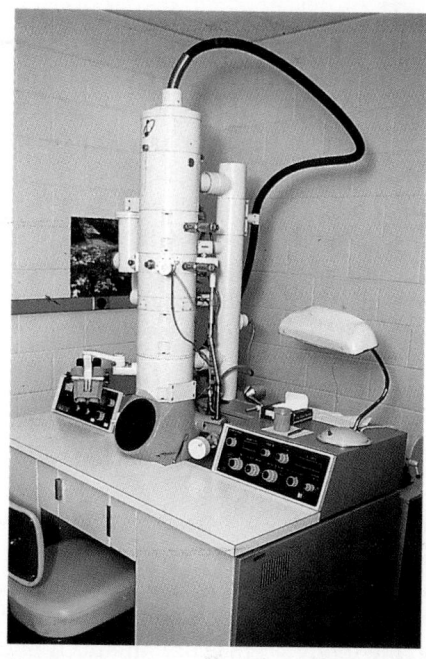

(a)

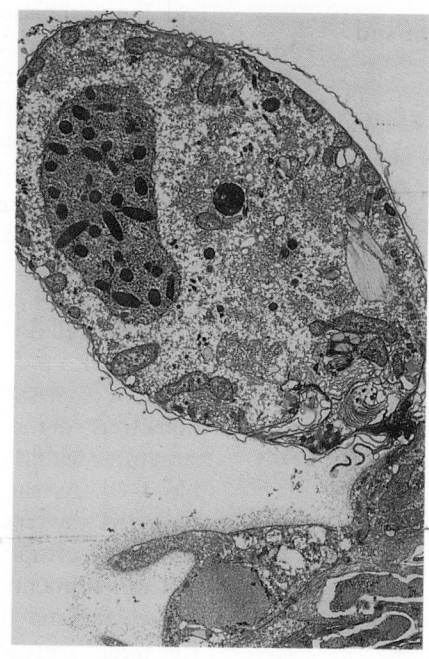

(b)

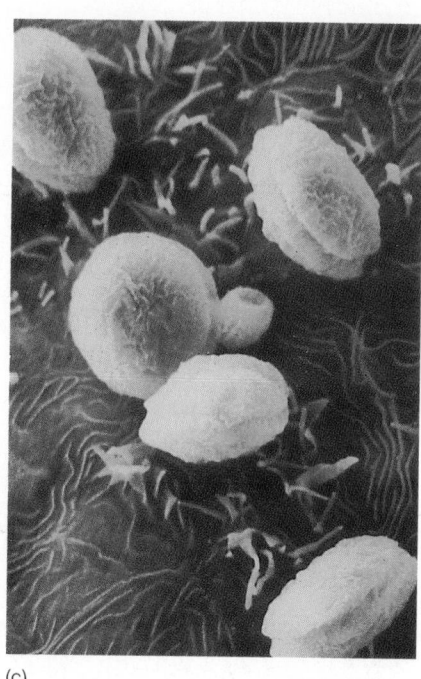

(c)

▷ **FIGURE 3–4 The Electron Microscope** (*a*) Transmission electron microscope. (*b*) Transmission electron micrograph of dinoflagellate. (*c*) Scanning electron micrograph of dinoflagellate.

collected by a detector, producing a three-dimensional image of the surface that is not only instructive but also often exquisitely beautiful (Figure 3–4c).

All microscopes provide two basic services: magnification and resolution. Magnification is simply the enlargement of an object, making it visible to the naked eye. As a general rule, light microscopes magnify objects from 100 to 1400 times their original size, whereas electron microscopes magnify 100,000 times.

▷ **FIGURE 3–5 Tunneling Electron Micrograph** This is a protein molecule as seen with the tunneling electron micrograph.

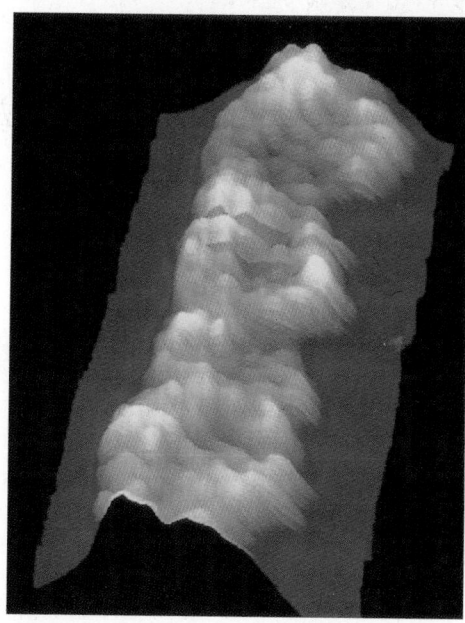

The second service provided by a microscope, **resolution,** is the ability to see detail. The higher the resolving power of a microscope, the greater one's ability to distinguish fine details. Resolution is a function of the energy source used to "illuminate" an object under study. Transmission electron microscopes, for example, use a more energetic source than light microscopes and therefore provide far greater resolution (400 times greater). Thus, electron microscopes permit scientists to examine the minute details of a cell, an inquiry that often yields valuable information.

A new class of electron microscopes may prove extremely useful in unraveling the chemical construction of cells. Called tunneling electron microscopes, they provide even greater resolution and magnification than standard electron microscopes, so that they enable scientists to view individual atoms and molecules (▷Figure 3–5).

≈ **PLANT AND ANIMAL CELLS: SIMILARITIES AND DIFFERENCES**

With this background information in mind, we begin our study of the cell. We will focus our attention primarily on eukaryotic cells found in plants and animals. A word of warning, however: the cells we will examine are composites that contain all of the structures found in the various types of eukaryotic cells. In a sense, they are fictional entities that exist only on the pages of biology textbooks. In real life, most cells exhibit a high degree of structural and functional specialization and rarely have all of the features illustrated here.

Despite the wide range of structural diversity, plant and animal cells share many characteristics, and it is these fea-

TABLE 3-1 Comparison of Plant and Animal Cells

FEATURE	PLANT CELL	ANIMAL CELL
Plasma membrane	Yes	Yes
Cell wall	Yes	No
Nucleus	Yes	Yes
Mitochondria	Yes (generally few)	Yes
Chloroplast	Yes	No
Endoplasmic reticulum	Yes	Yes
Golgi complex	Yes	Yes
Ribosomes	Yes	Yes
Lysosomes	Yes (rare)	Yes
Central vacuole	Yes	No
Cytoskeleton	Yes	Yes
Centrioles	No	Yes

tures that will prove useful to you in your study of biology (Table 3–1).

Plant and Animal Cells Consist of Two Main Compartments

Plant and animal cells are elaborate compartmentalized units. As illustrated in ▷Figures 3–6a and 3–6b, both types of cells consists of two major compartments, the nuclear and the cytoplasmic. The nuclear compartment, or **nucleus,** is the control center of the cell. Delimited by a double membrane, the nucleus contains the genetic information that controls the structure and function of the cell.

The cytoplasmic compartment lies between the nucleus and the plasma membrane, the outermost structure of the cell. As you might surmise, it contains a substance known as **cytoplasm.** Cytoplasm consists of a wide range of molecules, including water, protein, ions, nutrients, vitamins, dissolved gases, and waste products. It forms a pool from which the cell draws the chemicals it needs for metabolism, and it is also a dumping ground for wastes.

The cytoplasm of plant and animal cells also contains numerous organelles. **Organelles** ("little organs") are structures within cells that carry out specific functions (Table 3–2). As shown in Figure 3–6, many organelles are bounded by membranes and form subcompartments within the cytoplasm.

Compartmentalization is mentioned here because it permits plant and animal cells to segregate many of their functions, increasing the body's efficiency in much the same way as a modern factory does, with different tasks performed at different locations. Non-membrane-bound organelles such as ribosomes and microtubules, which perform many important functions, are also present in the cytoplasm and will be described shortly.

Giving shape to the cell is a network of protein tubules and filaments found in the cytoplasm and known as the

▷ **FIGURE 3–6 Structure of the General Eukaryotic Cell** (*a*) Animal cell. (*b* — opposite page) Plant cell.

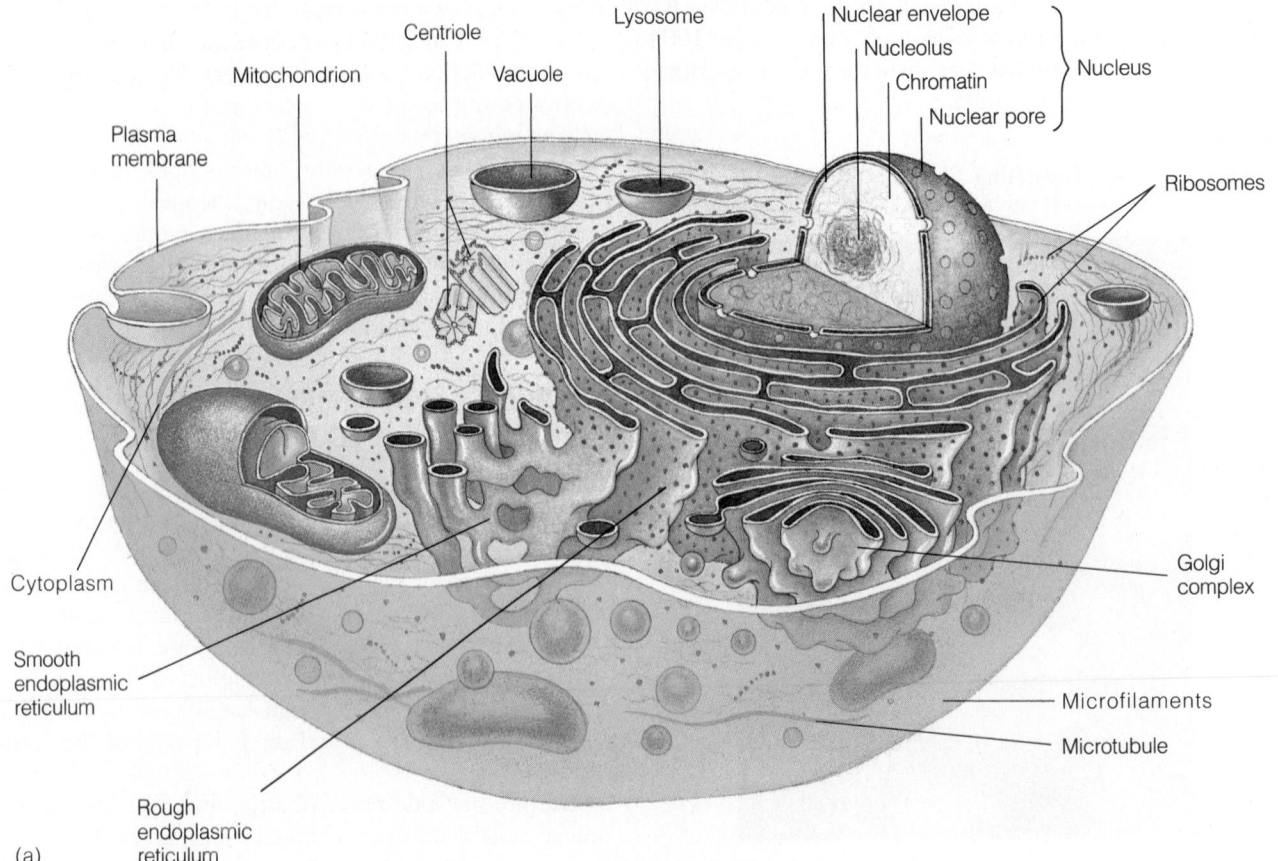

cytoskeleton (▷Figure 3–7). Like many other organelles, the cytoskeleton helps organize the cell's activities, increasing cellular efficiency. It does this by binding with enzymes. As noted in the previous chapter, enzymes are proteins that dramatically increase the rate of chemical reactions in cells. Many cellular chemical reactions occur in series, with one reaction giving rise to a substance used in the next. A series of linked reactions is called a **metabolic pathway** and is the cellular equivalent of the assembly line. As on an assembly line, molecules enter a metabolic pathway and are modified along their course by the enzymes (▷Figure 3–8). Each reaction in a metabolic pathway requires a specific enzyme. Like the workers in a factory, enzymes alter the molecule as it proceeds down a pathway. The cytoskeleton increases metabolic efficiency by binding the enzymes of certain metabolic pathways in their proper order. In that way the product of one reaction is conveniently situated for the next reaction.

Although Plant and Animal Cells Share Many Characteristics, There Are Important Differences

Plants and animals evolved from the same stock, and for this reason, they have many common features. However, plant and animal cells are adapted to quite different conditions. Animals are mobile and are unable to produce their own food. Plants are generally immobile and are capable of producing many of their own nutrients.

As noted in Tables 3–1 and 3–2 and Figure 3–6b, most plant cells contain a rigid, outer (nonliving) layer known as the **cell wall,** which helps give them their shape. Most plants cells also contain **chloroplasts,** small organelles in which light is captured and food production (photosynthesis) takes place. In addition, many plant cells contain large vacuoles that store water and other molecules. Finally, cells of higher plants (more recently evolved species like the flowering plants) lack **centrioles,** small cylindrical organelles made of microtubules that, in animals, play a role in cell division, as discussed in Chapter 7. Despite the absence of centrioles, plant cells remain capable of division. With these important differences in mind, we turn our attention to the plasma membrane.

≋ THE STRUCTURE AND FUNCTION OF THE PLASMA MEMBRANE

The **plasma membrane** is a thin layer of lipid and protein that controls what goes in and out of the cell and thus helps control the cell's internal chemical environment—a function essential to cellular homeostasis and survival. Let's begin with a look at the structure of the plasma membrane.

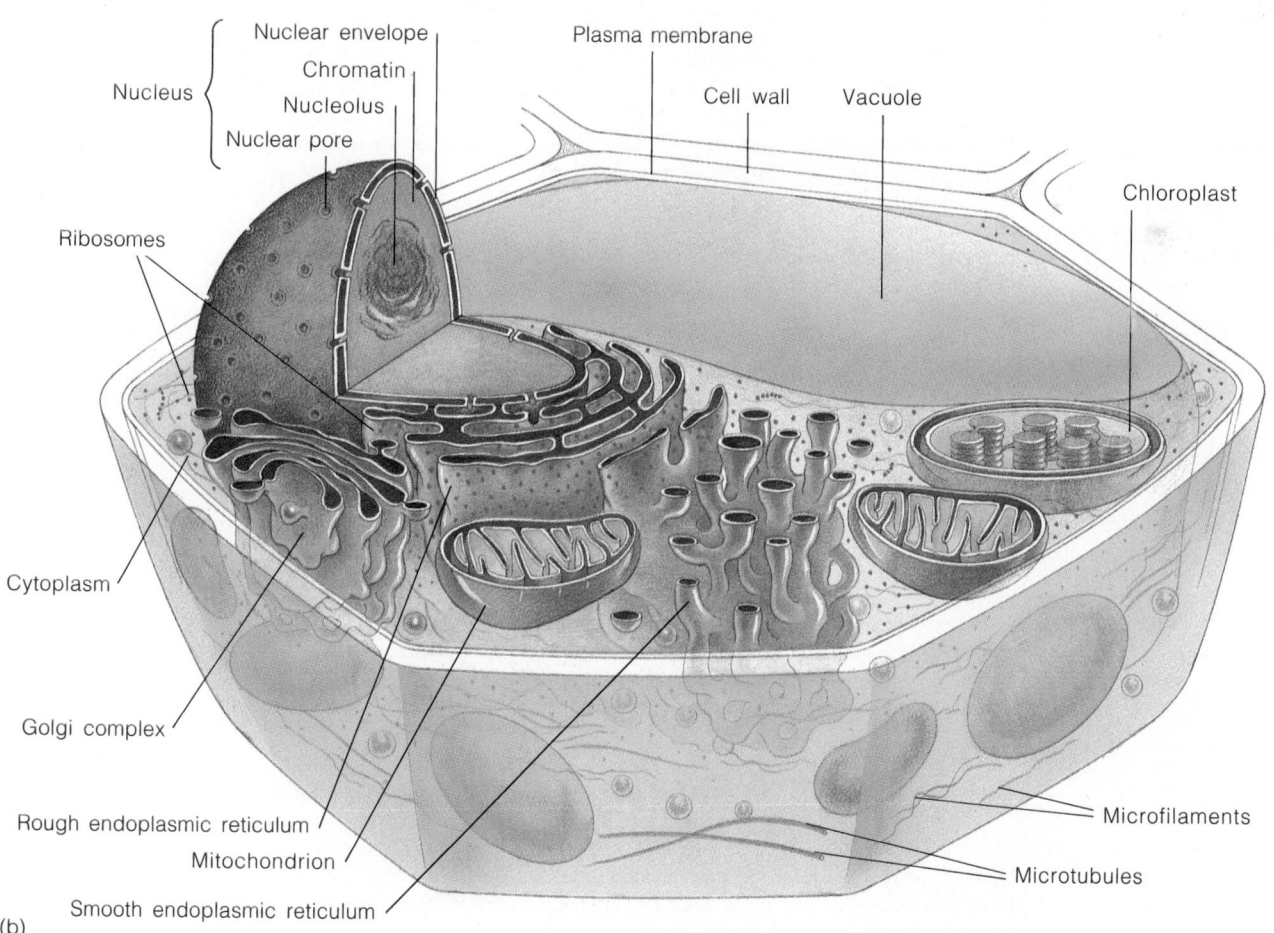

(b)

TABLE 3–2 Overview of Cell Organelles

ORGANELLE	STRUCTURE	FUNCTION
Nucleus	Round or oval body; surrounded by nuclear envelope	Contains the genetic information necessary for control of cell structure and function. DNA contains hereditary information.
Nucleolus	Round or oval body in the nucleus consisting of DNA and RNA	Produces ribosomal RNA
Endoplasmic reticulum	Network of membranous tubules in the cytoplasm of the cell. Smooth endoplasmic reticulum contains no ribosomes. Rough endoplasmic reticulum is studded with ribosomes.	Smooth endoplasmic reticulum (SER) is involved in the production of phospholipids and has many different functions in different cells; rough endoplasmic reticulum (RER) is the site of the synthesis of lysosomal enzymes and proteins for extracellular use.
Ribosomes	Small particles found in the cytoplasm; made of RNA and protein	Aid in the production of proteins on the RER and polysomes
Golgi complex	Series of flattened sacs usually located near the nucleus	Sorts, chemically modifies, and packages proteins produced on the RER
Secretory vesicles	Membrane-bound vesicles containing proteins produced by the RER and repackaged by the Golgi complex; contain protein hormones or enzymes	Store protein hormones or enzymes in the cytoplasm awaiting a signal for release
Food vacuole	Membrane-bound vesicle containing material engulfed by the cell	Stores ingested material and combines with lysosome
Lysosome	Round, membrane-bound structure containing digestive enzymes	Combines with food vacuoles and digests materials engulfed by cells
Mitochondria	Round, oval, or elongated structures with a double membrane. The inner membrane is thrown into folds.	Complete the breakdown of glucose, producing NADH and ATP
Cytoskeleton	Network of microtubules and microfilaments in the cell	Gives the cell internal support, helps transport molecules and some organelles inside the cell, and binds to enzymes of metabolic pathways
Cilia	Small projections of the cell membrane containing microtubules	Propel materials along the surface of a cell
Flagella	Large projections of the cell membrane containing microtubules	Provide motive force for cells
Cell wall	Layer of cellulose fibers lying outside the plasma membrane of plant cells	Provides support and protection
Chloroplast	Ovoid or disk-shaped organelle in plant cells; delimited by two membranes. Its inner membrane is not infolded like inner membrane of mitochondrion. It contains numerous stacks of thylakoid disks, called grana.	Captures solar energy and produces ATP and carbohydrates
Central vacuole	Large membrane-bound cavity in plant cells	Stores water, ions, toxic materials, pigments, protein, and starch
Centrioles	Small cylindrical bodies composed of microtubules arranged in nine sets of triplets; found in animal cells, not plants	Help organize spindle apparatus necessary for cell division

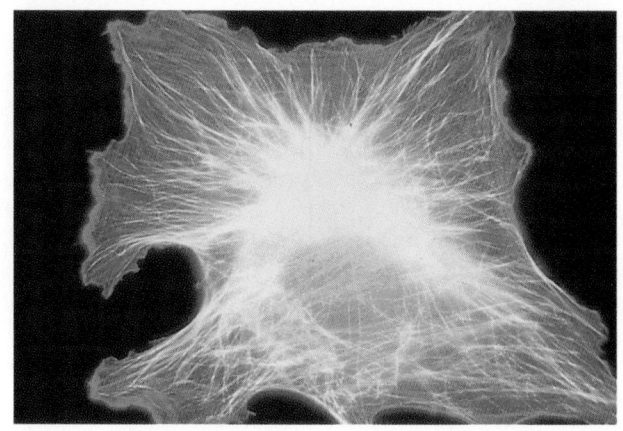

(a)

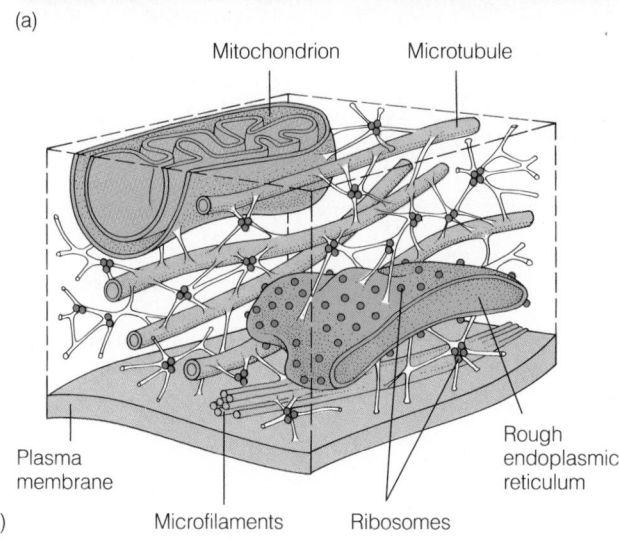

Mitochondrion   Microtubule

Plasma
membrane

(b)   Microfilaments   Ribosomes

Rough
endoplasmic
reticulum

▷ **FIGURE 3–7 The Cytoskeleton** (*a*) Photomicrograph of the cytoskeleton of a human fibroblast (connective tissue cell). Microtubules are yellow, and microfilaments are red. (*b*) Artist's rendition of the cytoskeleton, showing its two major components: microtubules (tubes made of the protein tubulin) and smaller microfilaments (solid fibers made of actin and myosin proteins).

The Lipids of the Plasma Membrane Form a Double Layer in Which Many of the Proteins Float Freely

The plasma membrane and all of the other internal membranes of the cell consist of lipids and proteins, as mentioned above, and a small amount of carbohydrate (▷Figure 3–9). Most of the lipid molecules in the plasma membrane are phosphoglycerides, described in Chapter 2. Experimental studies suggest that the phosphoglyceride molecules are arranged in a double layer. Figure 3–9 shows the most widely accepted model (theory) of the plasma membrane's structure, the **fluid mosaic model.** As shown, the polar heads of the outer layer of phosphoglyceride molecules face the aqueous extracellular fluid, and the polar heads of the inner layer of molecules are directed inward toward the aqueous cytoplasm.

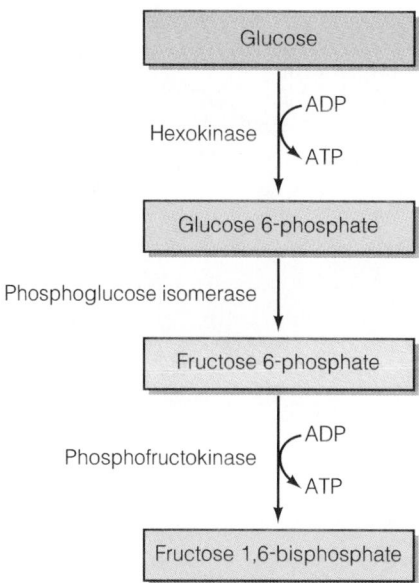

▷ **FIGURE 3–8 Metabolic Pathway** Reactions in the cell occur as part of larger pathways where the product of one reaction becomes the reactant of another, as in this biochemical pathway. This illustrates the early steps in glycolysis, the breakdown of glucose in the cytoplasm of eukaryotic cells. Note that each reaction has its own enzyme.

Interspersed in the lipid bilayer of plant and animal cells are large protein molecules known as integral proteins. The **integral proteins** are globular proteins that float freely in their sea of lipid like giant icebergs. As illustrated, some of them extend entirely through the plasma membrane; others penetrate only part way. Several integral proteins may join to form pores that permit the movement of molecules in and out the cell. Others may attach to the underlying cytoskeleton, helping anchor the plasma membrane in place.

Another group of proteins, known as **peripheral proteins,** is also found in the plasma membrane. On the cytoplasmic side of the membrane, the peripheral proteins often attach to the exposed surfaces of integral proteins and may also bind to the underlying cytoskeleton.

Plasma membranes contain small but significant amounts of carbohydrate. Most carbohydrate consists of oligosaccharides (short chains of monosaccharides), which are attached to integral proteins that protrude into the extracellular fluid of cells. A protein combined with carbohydrate is referred to as a glycoprotein. The glycoproteins of the plasma membrane are part of a cellular identification system described below.

Cholesterol molecules are also found in the plasma membranes of animal cells, wedged in between the phosphoglyceride molecules. Their importance is not well understood.

On an evolutionary note, the composition of the plasma membranes of many organisms changes in response to temperature. This adaptation, found in bacteria, yeast, plants, and cold-blooded animals, allows organisms to

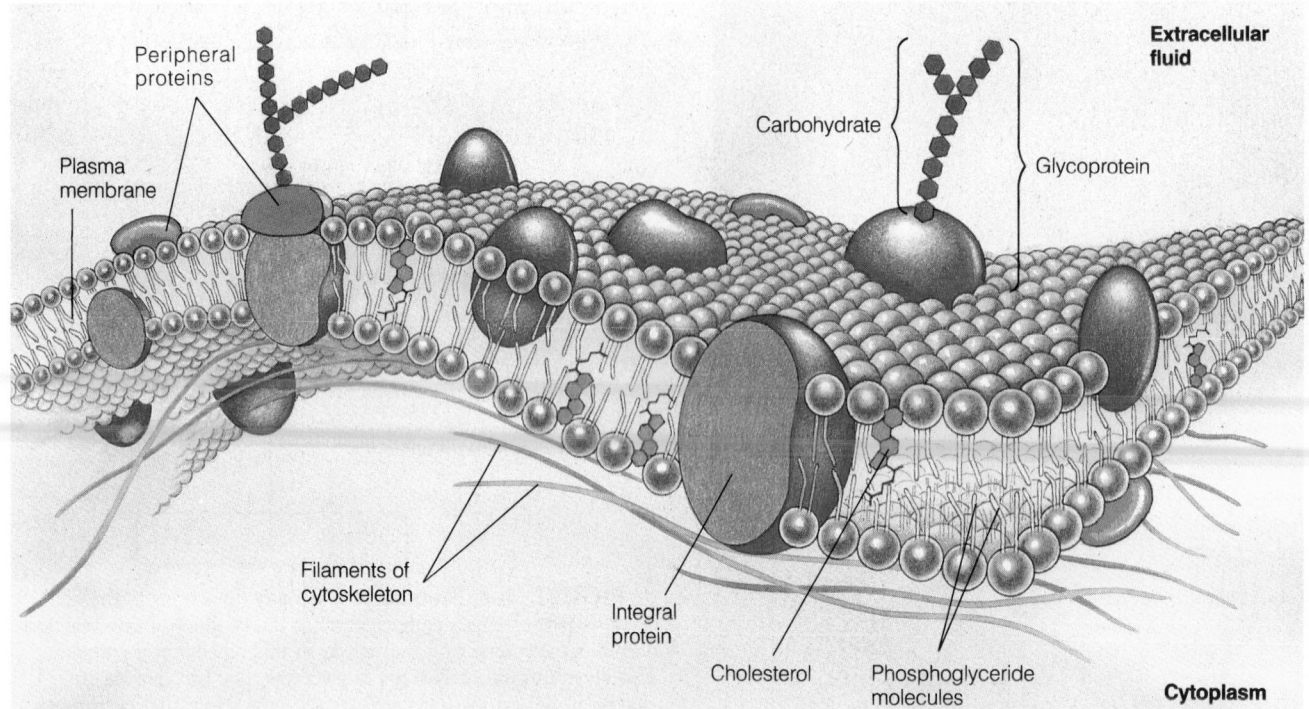

Peripheral proteins

Plasma membrane

Carbohydrate

Glycoprotein

Extracellular fluid

Filaments of cytoskeleton

Integral protein

Cholesterol

Phosphoglyceride molecules

Cytoplasm

▷ **FIGURE 3–9 Fluid Mosaic Model of the Plasma Membrane of Animal Cells** The fluid mosaic model is the most widely accepted theory of the plasma membrane. Phospho-glycerides are the chief lipid component. They are arranged in a bilayer. Integral proteins float like icebergs in a sea of lipid.

maintain a fairly constant membrane fluidity despite decreases in environmental temperature. Membrane fluidity is a measure of the freedom of movement of integral proteins and is essential for normal cellular function. Organisms unable to maintain body temperature ensure a constant membrane fluidity over a range of temperatures by replacing membrane phosphoglycerides with ones whose side chains are of appropriate length and degree of unsaturation. This subtle change maintains the proper level of fluidity and keeps cells alive.

The Plasma Membrane Serves Many Functions

Table 3–3 summarizes the functions of the plasma membrane. As noted, the plasma membrane separates the cytoplasm from the cell's external environment and protects the cell's structural integrity. It also helps maintain cellular homeostasis by regulating the flow of molecules and ions into and out of the cell. Because the plasma membrane regulates the molecular traffic, it is said to be **selectively permeable**. In other words, the plasma membrane "selects" or "controls" what enters and leaves the cell. It also controls the rate at which molecules and ions cross it. This homeostatic function helps maintain the precise chemical concentrations inside and outside the cell so essential for optimal cellular function.

The functions of the plasma membrane do not stop here. As you will see in later chapters, the plasma mem-

brane also plays a vital role in cellular communication.[2] Some hormones, for example, attach to specific integral proteins (membrane receptors) in the plasma membranes of cells. For reasons explained in Chapter 20, the attachment of a molecule to the membrane receptor sends a message to the cell's interior, triggering physiologically important changes in the cell's structure and function. Thus, hormones are said to "communicate" with the cell through its membrane.

The plasma membrane is also part of a cellular identification system, as mentioned above. The cells in each of us have a unique glycoprotein "cellular fingerprint." Because the glycoprotein composition of the plasma membrane of the cells of each individual is unique, a person's immune system can recognize its own cells from foreign cells—such as bacteria and single-celled fungi—that occasionally invade. The immune system also identifies cancer cells, whose cellular fingerprints have become altered. It then mounts an attack on the tumor cells or microbial invaders. Without the ability to recognize foreign invaders, a multicellular animal would quickly perish from infections or tumors.

As an interesting sidelight, the immune system, so beneficial to animals, also poses a major challenge to surgeons who perform tissue and organ transplants. Failing kidneys

[2] *Communication* is defined very broadly here to include the transmission or receipt of any kind of message.

TABLE 3-3 Overview of Plasma Membrane Functions

Ensures the cell's structural integrity

Regulates the flow of molecules and ions into and out of the cell

Maintains the chemical composition of cytoplasm and extracellular fluid

Participates in cellular communication

Forms a cellular identification system

or hearts, for instance, can be surgically replaced by healthy organs from donors. Unless the organ comes from an identical twin, however, the immune system of the recipient will recognize the cells of a transplanted organ as foreign and attempt to destroy them.[3]

In order to prevent rejection, physicians administer drugs that suppress the immune system. Unfortunately, as you might guess, this treatment makes a patient much more vulnerable to bacterial and viral infections. Therefore, although they were heralded as a major medical breakthrough, early heart transplants all failed because the patients succumbed to massive infections. Happily, new drugs are available that suppress the immune system's rejection of organ and tissue grafts without completely eliminating the body's protection from bacteria and viruses.

Continuing along this line, fetal cell transplantation, discussed at the opening of the chapter and in Health Note

[3] The only people with identical cells are identical twins.

3–1, is successful because it avoids the problem of tissue rejection altogether. Because fetal cells have not developed their glycoprotein fingerprint, they can be transplanted without triggering an immune reaction. The immunologic immaturity of fetal cells is probably an evolutionary adaptation that protects fetuses from attack by the mother's immune system.

Molecules Move Through the Plasma Membrane in Five Major Ways

The cytoplasm and the interstitial fluid outside the plasma membrane contain a variety of materials dissolved or suspended in water. Some of these can pass through the membrane with ease; others cannot. Ultimately, the membrane itself "determines" which molecules pass through and how quickly they pass. How does the plasma membrane "control" the traffic? Control over plasma membrane traffic is achieved in five ways, as summarized in Table 3–4.

Lipid-soluble Substances Pass Directly Through the Membrane via Diffusion. In chemistry class, you probably learned that "likes dissolve likes." In other words, chemically similar substances mix readily. A polar compound, for example, will readily dissolve in water. A nonpolar compound like oil will not.

Because the plasma membrane is principally lipid, lipid-soluble materials, such as steroid hormones, oxygen, and carbon dioxide, pass directly through the lipid bilayer of the plasma membrane with ease (Figure 3–10a). Moving from areas of high concentration to low concentration, these substances are said to diffuse through the membrane. **Diffusion** refers to any movement of molecules or ions down a concentration gradient—that is, from high to low

TABLE 3-4 Overview of Plasma Membrane Transport

PROCESS	DESCRIPTION
Simple diffusion	Flow of ions and molecules from high concentrations to low. Water-soluble ions and molecules probably pass through pores; water-insoluble molecules pass directly through the lipid layer.
Facilitated diffusion	Flow of ions and molecules form high concentrations to low concentrations with the aid of protein carrier molecules in the membrane
Active transport	Transport of molecules from regions of low concentration to regions of high concentration with the aid of transport proteins in the cell membrane and ATP
Endocytosis	Active incorporation of liquid and solid materials outside the cell by the plasma membrane. Materials are engulfed by the cell and become surrounded in a membrane.
Exocytosis	Release of materials packaged in secretory vesicles
Osmosis	Diffusion of water molecules from regions of high water (low solute) concentration to regions of low water (high solute) concentrations

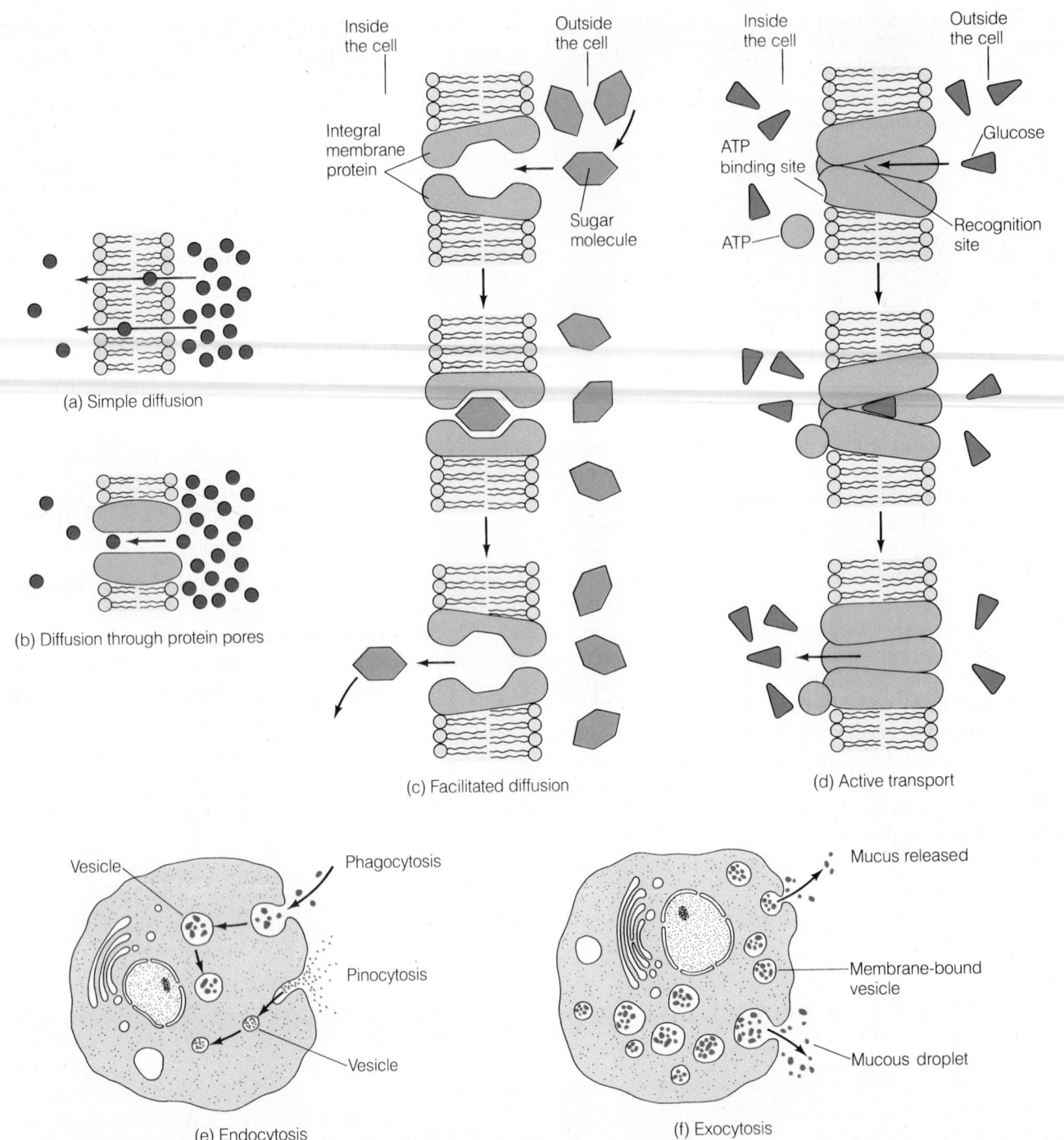

Inside the cell | Outside the cell

Integral membrane protein

Sugar molecule

(a) Simple diffusion

(b) Diffusion through protein pores

(c) Facilitated diffusion

Inside the cell | Outside the cell

ATP binding site

Glucose

ATP

Recognition site

(d) Active transport

Vesicle

Phagocytosis

Pinocytosis

Vesicle

(e) Endocytosis

Mucus released

Membrane-bound vesicle

Mucous droplet

(f) Exocytosis

▷ **FIGURE 3–10 Membrane Transport** Molecules move through the plasma membrane primarily in five ways. (*a*) Lipid-soluble substances pass throug the membrane directly via simple diffusion. (*b*) Water-soluble molecules may diffuse passively through pores formed by protein molecules. (*c*) Water-soluble molecules can also diffuse through membranes with the assistance of proteins in facilitated diffusion. (*d*) Other proteins use energy from ATP to move against concentration gradients in a process called active transport. (*e*) Finally, cells can engulf large particles, cell fragments, and even entire cells via endocytosis. (*f*) Exocytosis, the reverse process, rids the cell of large particles.

concentration. Two types of diffusion exist, simple and facilitated. Movement through the membrane without assistance, as described above, is **simple diffusion.**

Water-soluble Molecules May Diffuse Passively Through Pores. Water-soluble materials cannot pass through the lipid bilayer of the plasma membrane and must travel via other routes. It is widely believed that water and many small water-soluble molecules and ions must pass through **pores** in the plasma membrane and that these pores are formed by integral proteins (Figure 3–10b). Movement through pores is another form of simple diffusion. The evidence for membrane pores is fairly circumstantial. No one, in fact, has ever even seen a pore in electron micrographs of cell membranes, and their existence is based entirely on physiological research. As a result of this work, researchers believe that some protein pores remain open all the time, thus permitting small molecules and ions to cross the plasma membrane.

Water-soluble Molecules Can Also Diffuse Through Membranes with the Assistance of Protein Carrier Molecules. Carrier proteins are molecules that transport small water-soluble molecules and ions across the membrane in ways not yet completely understood. Carrier proteins "shuttle" molecules across membranes from regions of high concentration to low concentration, a movement often said to be "down" or "with" the concentration gradient. This process is called **facilitated diffusion** to distinguish it from simple diffusion occurring through pores (Figure 3–10c).

Molecules Are Also Actively Transported Across the Membrane. The fourth transport mechanism found in plasma membranes is known as active transport. **Active transport** is the movement of molecules across membranes with the aid of protein carrier molecules in the plasma membrane and energy supplied by ATP (Figure 3–10d). In this process, molecules and ions are transported from regions of low to high concentration—movement up or against the concentration gradient.

Active transport occurs in cells that must concentrate chemical substances to function properly. For example, thyroid cells require large quantities of the iodide ion (I^-) to manufacture the hormone thyroxine. Thyroid cells actively transport iodide ions from the bloodstream, where they are present in fairly low concentrations, into the cytoplasm, where the iodide concentration is many times higher. If the thyroid cell had to rely on diffusion, it probably would not be able to produce enough hormone to meet the body's needs. It is important to note that cells also use active transport to move materials out of their cytoplasm against concentration gradients.

Active transport proteins operate in several ways. The simplest active transport proteins have two binding sites, as illustrated in Figure 3–10d. One of them attaches to the molecule that is to be moved across the membrane; the other binds to ATP. Biologists hypothesize that the breakdown of the ATP molecule releases energy that causes the protein molecule to change its shape; this, in turn, causes the protein to propel the molecule attached to its other binding site across the plasma membrane, moving it into or out of the cell.

Most active transport molecules move two substances across the membrane at the same time, often in opposite directions. In nerve cells, for example, an active transport protein pumps sodium ions out of the cytoplasm while pumping potassium ions in. This pump ensures the proper functioning of the nerve cell.

The Fifth Mechanism That Enables Cells to Ingest Large Molecules or Even Other Cells Is Endocytosis. Endocytosis (literally "into the cell") is illustrated in Figure 3–10e. This process requires ATP and consists of two related activities: phagocytosis and pinocytosis. **Phagocytosis** (cell eating) occurs when cells engulf larger particles, such as bacteria and viruses. In humans, phagocytosis is limited to a relatively few types of cells: those involved in protecting the body against foreign invaders (Chapter 15). **Pinocytosis** (cell drinking) occurs when cells engulf extracellular fluids and dissolved materials. Most, if not all, cells are capable of pinocytosis.

Cells release large molecules, such as hormones, by a process called **exocytosis** (out of the cell), the reverse of endocytosis (Figure 3–10f). In the cells of the pituitary gland, for instance, protein hormones are packaged internally into tiny membrane-bound vesicles called **secretory vesicles** (discussed later). Secretory vesicles migrate to the plasma membranes and fuse with them. At the point of fusion, the membranes break down, and the protein hormone is released into the extracellular fluid. This process is exocytosis.

The Diffusion of Water Across the Plasma Membrane Is Known as Osmosis

Like any other small molecule, water moves from one side of a plasma membrane to the other by diffusion. The diffusion of water across a selectively permeable membrane, however, is given a special name, **osmosis** (which comes from a Greek word meaning to push). To understand osmosis, consider a simplified model.

First of all, suppose we have a large bag made of a selectively permeable material that permits water but not sucrose (table sugar) to flow through it. The bag is filled with water and sucrose and is then submerged in a large beaker of distilled water (▷Figure 3–11). As soon as the bag is immersed in the water, it begins to swell. This swelling results from the inward movement of water molecules. Water moves from a region of high water concentration (distilled water) to a region of low water concentration (inside the bag).[4]

The concentration difference across the membrane "drives" the water across a selectively permeable membrane. Therefore, whenever two fluids with different concentrations of solute (dissolved substance) are separated by a selectively permeable membrane, water will flow from one to the other, moving down the concentration gradient, and the driving force is called **osmotic pressure.** The greater the difference in water concentration, the greater the osmotic pressure, and the more quickly water moves.

[4] Water concentration is measured by the number of water molecules per milliliter. Pure water has a higher water concentration than salt water; that is, there are more water molecules per unit volume in pure water than in salt water.

▷ FIGURE 3–11 Osmosis

Osmosis is the diffusion of water molecules from a region of high water concentration to one of low water concentration across a semipermeable membrane. (*a*) To demonstrate the process, immerse a bag of sugar water in a solution of pure water. (*b*) Water diffuses into the bag, causing it to swell.

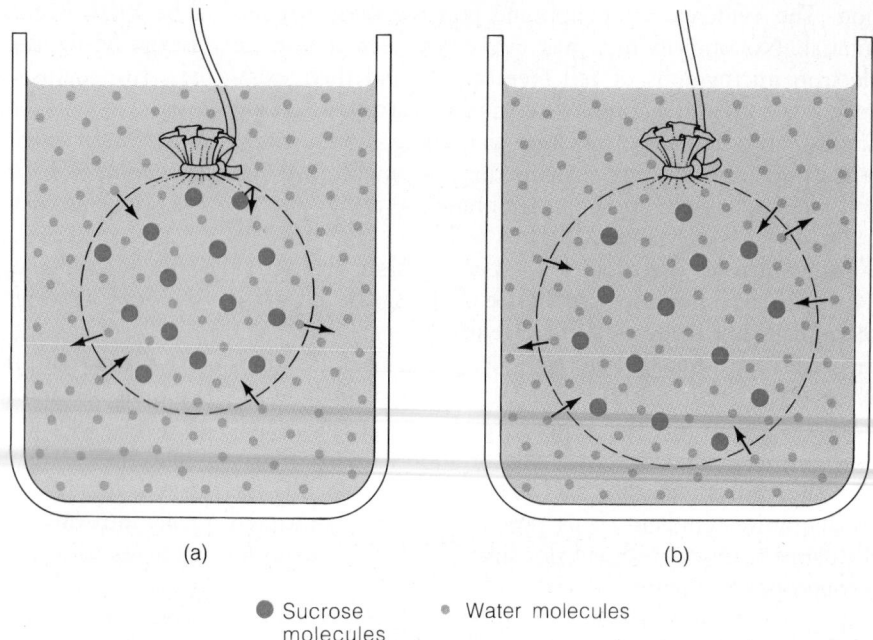

(a)   (b)

● Sucrose molecules   · Water molecules

In plants and animals, osmotic pressure is responsible for the movement of water across many membranes. Combined with blood pressure and other forces, osmotic pressure plays an important role in the filtering of blood in the kidney (discussed in more detail in Chapter 16).

Because osmosis helps equalize water concentrations on opposite sides of membranes, it is an important homeostatic mechanism in cells. That is to say, osmosis helps regulate the concentration of the fluid surrounding the cells of multicellular organisms (called **extracellular fluid**), keeping it the same as the concentration of the cytoplasm of the cells. In such cases the extracellular fluid is said to be **isotonic** (having the same strength).

If the fluid surrounding a cell is not isotonic, serious problems can be expected. For example, if blood cells are immersed in a solution that is more concentrated than their cytoplasm, water will move out of the cells, and the cells will shrink. A solution with a higher solute concentration than the cell's cytoplasm is said to be **hypertonic** (having a greater strength). A solution with a solute concentration lower than the cell's cytoplasm is said to be **hypotonic** (having a lesser strength). If blood cells are placed in such a solution, water will rush in, causing the cells to swell and possibly burst. Early studies on the structure of the plasma membrane used this knowledge to produce cell "ghosts," isolated membranes whose composition could be studied.

Plant cells respond similarly. When placed in a hypertonic solution, a plant cell shrinks. When placed in a hypotonic solution, it expands. Much of the water that flows into a plant cell is stored in the central vacuole; as the vacuole enlarges, the plant cell expands, pushing outward against the **cell wall,** a rather permeable but generally inflexible layer that surrounds the cell (described next). The cell wall prevents the plant cell from bursting.

The Cell Wall Lies Outside the Plasma Membrane of Plant Cells, Providing Support and Protection

Plants are immobile structures lacking the internal skeletons of many animals, which permit our kind to upright stance and movement. The lack of a skeleton, however, does not limit plants to being slimy masses of green protoplasm that hug the ground. Many species stand upright, and some like the redwood and Douglas fir achieve extraordinary heights. This rigidity is made possible by the cell wall.

All plant cells have a primary cell wall composed of cellulose molecules (▷Figure 3–12). Cellulose molecules form thin, threadlike strands, which are embedded in

▷ FIGURE 3–12 The Cell Wall

Composed primarily of cellulose, the primary cell wall (CW1) protects the plant cell and gives support. The secondary cell wall (CW2) forms between the plasma membrane and the primary cell wall. ER is a label for endoplasmic reticulum, and T is a label for tonoplast (the membrane around the central vacuole).

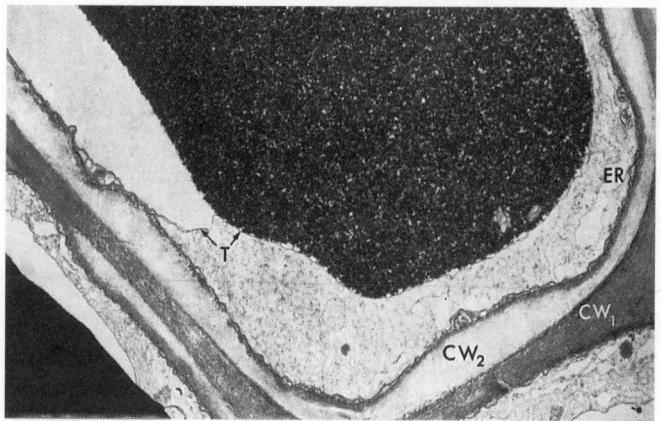

polysaccharides and small amounts of protein. The primary cell wall is formed by the plant cell itself. A thin layer of sticky substance on the cell wall, known as pectin, holds adjacent plant cells together.[5]

When first laid down, the cell wall is flexible and thus allows the plant cell to grow. When the cell matures and stops growing, however, it produces chemical substances that harden the cell wall, making it rigid.

Some cells produce a secondary cell wall that lies between the primary cell wall and the plasma membrane. The secondary wall is much more rigid and provides even better protection and support for cells. Wood, in fact, is composed largely of secondary cell walls. Wood chips are used to make paper, and thus much of the paper we use is nothing more than processed cell walls of plant cells.

≈ CELLULAR COMPARTMENTALIZATION: ORGANELLES

As noted earlier, the cytoplasm of eukaryotic cells is partitioned into distinct compartments, or organelles, each with a specific function (see Table 3–2). Cellular organelles have evolved and persisted over time because they offer many advantages, efficiency being perhaps the most impor-

---

[5] Pectin is a sticky polysaccharide that is used to make jams and jellies.

tant. This section takes you on a tour through the interior of the cell, showing each compartment and discussing its function. We begin with the nucleus.

The Nucleus Is the Cell's Command Center

The nucleus is one of the cell's most conspicuous and most important organelles (▷Figure 3–13a).[6] It is often likened to a cellular command center. The nucleus of the cell consists of (1) the nuclear envelope, (2) the chromatin, (3) the nucleoplasm, and (4) the nucleolus.

The **nuclear envelope** is a double membrane that isolates the nuclear material from the cytoplasm. Minute channels in the envelope, the **nuclear pores,** allow materials to pass into and out of the nucleus (Figure 3–13b).

The bulk of the nucleus contains long, threadlike **chromosomes**, fibers of DNA and protein. These DNA-protein fibers, also known as **chromatin,** appear as fine granules in transmission electron micrographs of nuclei (Figure 3–13a). Proteins, water, and other small molecules and ions are also found in the nucleus, forming a semifluid material known as the **nucleoplasm.** Just before cell division, the chromosomes coil to form short, compact bodies (▷Figure 3–14). This compaction makes the chromo-

---

[6] In plants, the vacuole is usually the largest organelle.

▷ **FIGURE 3–13 The Nucleus**
(a) The nucleus houses the genetic information that controls the structure and function of the cell. The nuclear envelope, made of lipid and protein like the plasma membrane, actually consists of two membranes, separated by a space. Pores in the membrane allow the movement of molecules into the nucleus, providing raw materials for the synthesis of DNA and RNA. They also allow RNA molecules to travel into the cytoplasm, where they participate in the production of protein. Chromatin may be densely packed or loosely arranged in the nucleus. (b) Colorized scanning electron micrograph of the nuclear membrane, showing numerous pores.

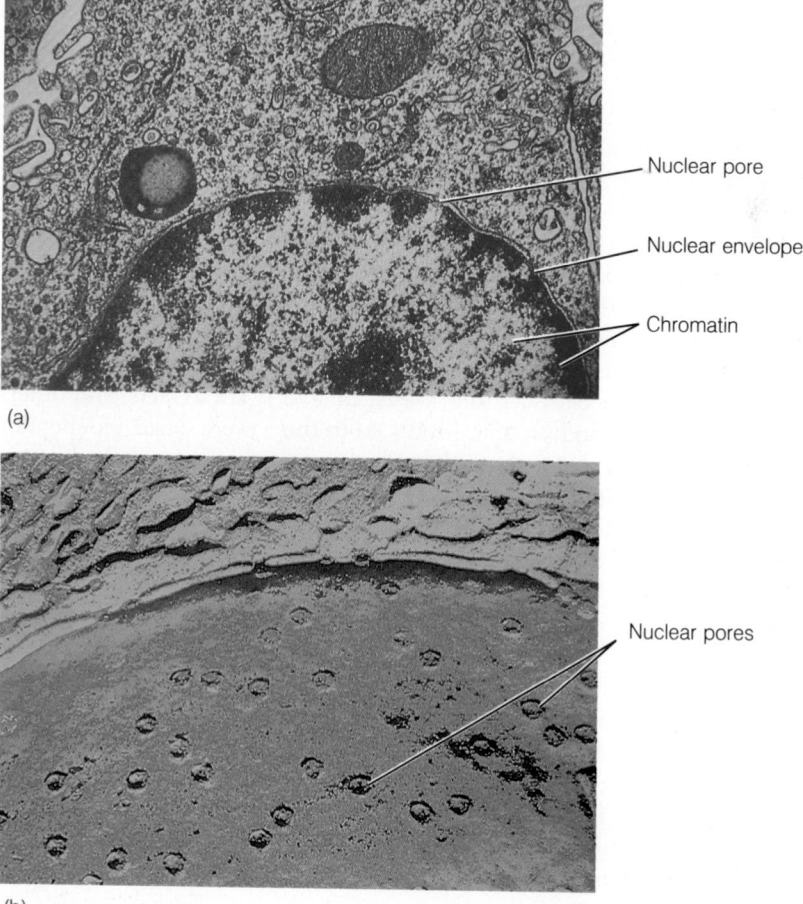

Nuclear pore

Nuclear envelope

Chromatin

(a)

Nuclear pores

(b)

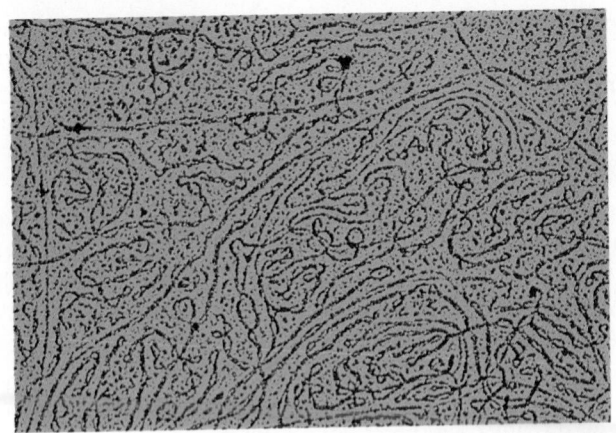

▷ **FIGURE 3–14 Chromosomes** The threadlike chromosomes, made of chromatin fibers consisting of protein and DNA, must condense before the nucleus can divide.

somes easier to separate when the nucleus and cell divide.

The DNA in the nucleus is the genetic material, and it houses all of the information a cell needs for proper development and functioning—hence the description of this organelle as a central command center. Interestingly, each cell in a multicellular organism contains the same genetic information. As an example, muscle cells have the same DNA as brain cells, even though they perform markedly different functions and look very different from each other. The structural and functional differences between these cells result from a selective repression of part of each cell's genetic material.

Nucleoli ("little nuclei") are temporary structures found in the nuclei of cells between cell divisions (see Figure 3–13a). They appear as small, clear, oval structures in light micrographs and as dense bodies in transmission electron micrographs. Nucleoli are regions of the DNA actively engaged in the production of **ribosomal RNA (rRNA)**, so named because it combines with certain proteins to form ribosomes. **Ribosomes** are nonmembranous organelles that appear as small, dark granules in electron micrographs. These tiny organelles actually consist of two subunits, each made of RNA and protein. Produced in the nucleus, the subunits enter the cytoplasm through the nuclear pores. In the cytoplasm they combine to play an important part in protein synthesis, described in Chapter 9.

The Mitochondrion Is the Site of Cellular Energy Production in Animal Cells

During the evolution of eukaryotic cells, an organelle known as the **mitochondrion** arose (▷Figure 3–15). Specialized to liberate energy from organic molecules, mitochondria (plural) are active cellular factories that crank out enormous quantities of ATP. Cells use this energy for many purposes: to synthesize chemical substances, to transport molecules across membranes, to divide, to contract, and to move about.

Mitochondria vary considerably in form and number from cell to cell. Nevertheless, they all have several common characteristics. All of them, for instance, are membrane-bound. In addition, all mitochondria contain an inner membrane, which is usually thrown into folds known as **cristae** (Figure 3–15b). The inner membrane divides the organelle into two distinct compartments, the inner compartment and the outer compartment, each with its own function.

The breakdown of glucose, the principal source of energy, is by various estimates 30% to 60% efficient; that is, one-third to nearly two-thirds of the energy released during the chemical reactions is captured by the cell to produce ATP. The rest of the energy contained in the glucose molecule is given off as heat. This "waste" product creates body heat, which is necessary for maintaining normal enzymatic function in warm-blooded animals like you and me. Thus, mitochondria play a key role in homeostasis by liberating energy and by maintaining body temperature. Their function is vital for individual cells as well as the entire organism.

Most biologists believe that mitochondria were derived from free-living organisms similar to bacteria. During cellular evolution, these free-living organisms probably took up residence inside other cells. There they remained, living off the food acquired by their host and providing heat and ATP in return. Most biologists believe that this relationship persisted over many millions of years and that the once free-living organisms became a permanent fixture of the new cells, the eukaryotes. The relationship persisted because it gave the first eukaryotes an advantage over others, for reasons explained later in the book.

The seemingly outlandish claim that mitochondria were once free-living, bacterialike organisms is supported by a variety of observations. First, mitochondria resemble bacteria in many respects. Like bacteria, they are capable of dividing. In addition, they contain circular DNA not unlike that found in bacteria. Mitochondria are also capable of producing some of their own proteins, and they contain ribosomes that closely resemble those found in bacteria. These facts, and others, strongly suggest that mitochondria were once independent organisms. Reflecting this belief, biologist and philosopher Lewis Thomas calls the mitochondria "essentially foreign creatures" within our cells.

Plastids Are Membranous Organelles Found in Plant Cells

Plants and animals both contain mitochondria, but only plant cells contain **plastids,** membranous organelles that perform a variety of functions, depending on the cell in which they are located. The plastids in the cells of the fruits, flowers, and leaves of plants, for example, contain pigments that give these structures their color. Cells in roots and other locations contain colorless plastids that merely act as storage depots for fat, starch, and protein.

The best known and most abundant plastid is the **chlo-**

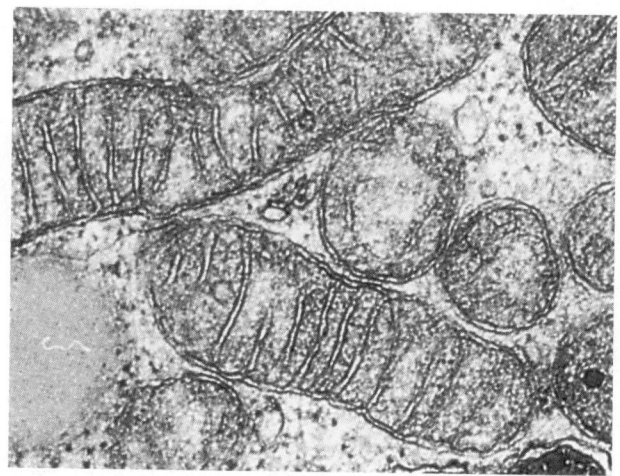

(a)

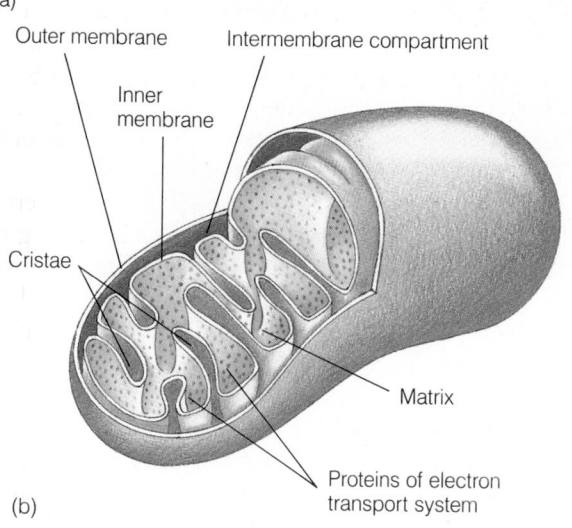

Outer membrane   Intermembrane compartment

Inner membrane

Cristae

Matrix

(b)

Proteins of electron transport system

▷ **FIGURE 3–15 The Mitochondrion** (*a*) Like the nucleus, the mitochondrion is delimited by a double membrane made of protein and lipid. The inner membrane, however, is infolded, forming cristae. (*b*) The infolding creates two distinct compartments: an inner compartment filled with a material called the matrix, and an outer compartment lying between the two membranes. Compartmentalization is essential for ATP synthesis.

roplast (▷Figure 3–16). Chloroplasts contain the green pigment **chlorophyll** and the enzymes and other molecules necessary to produce ATP and carbohydrates from sunlight, water, and carbon dioxide, a process known as **photosynthesis** and described in detail in Chapter 5. The overall reaction for photosynthesis can be summarized as follows:

sunlight + carbon dioxide + water →

carbohydrate + oxygen

Found in the photosynthetic cells of the higher plants and in certain unicellular eukaryotes (the photosynthetic protists), the chloroplast is generally an ovoid or disk-shaped organelle. As illustrated in Figure 3–16, the chloroplast contains an elaborate system of flattened membra-

nous sacs piled on top of one another. Each sac is called a **thylakoid disk,** and many disks combine to form a **granum** (plural, *grana*), resembling a stack of poker chips. Chlorophyll and other light-absorbing molecules needed for photosynthesis are embedded in the membrane of the thylakoid disks.

Lying outside the grana is a substance generally referred to as the **stroma.** It is here that the chloroplast manufactures glucose (and other carbohydrates) from carbon dioxide and water using energy supplied by ATP. Glucose molecules produced in the stroma often combine to form starch, which may be stored in the chloroplast as small granules.

Like mitochondria, chloroplasts also contain their own DNA, divide, and can synthesize some of their own proteins, leading evolutionary biologists to hypothesize that they, too, were once free-living organisms.

Three Organelles Are Involved in Manufacturing Protein and Other Cellular Products

Cells are miniature factories that synthesize a variety of molecules vital for growth, reproduction, and day-to-day maintenance. Many of these molecules are used within the cell; others, such as hormones, are released from the cell, then travel to distant cells through the bloodstream, eliciting a response. This section examines three organelles principally involved in cellular synthesis: the endoplasmic reticulum, ribosomes, and the Golgi complex.

Endoplasmic Reticulum. Coursing throughout the cytoplasm of many cells is a branched network of membranous channels known as the **endoplasmic reticulum** (meaning intracellular network) (▷Figure 3–17a). These membranous channels are derived chiefly from the nuclear envelope and may be coated with ribosomes (Figure 3–17b). Ribosome-studded endoplasmic reticulum is referred to as the **rough endoplasmic reticulum (RER).**

The RER produces a variety of proteins. In humans, for example, digestive enzymes are produced on the RER of cells in the pancreas. These enzymes are released from the cell by exocytosis and travel in ducts to the small intestine, where digestion takes place. The RER also synthesizes protein hormones, which are released from the cells that produce them. In addition, the RER produces proteins destined for inclusion in the plasma membrane as well as enzymes bound for inclusion in a cellular organelle known as the **lysosome.** Described more fully shortly, lysosomes digest materials engulfed (phagocytized) by the cell.

All of these proteins are produced on the outside surface of the RER but are quickly transferred into the interior of the organelle. Known as the **cisterna** (plural, *cisternae*), the interior of the RER is another compartment containing 30 to 40 enzymes specialized to convert proteins into glycoproteins. After chemical modification, the proteins and glycoproteins are transferred to the Golgi complex, another cellular organelle, for final packaging.

▷ **FIGURE 3–16 The Chloroplast** (*a*) Transmission electron micrograph of the chloroplast; (*b*) three-dimensional drawing. Like the mitochondrion, the chloroplast has two membranes, but the inner membrane is not thrown into folds. Note the presence of the extensive system of thylakoid disks.

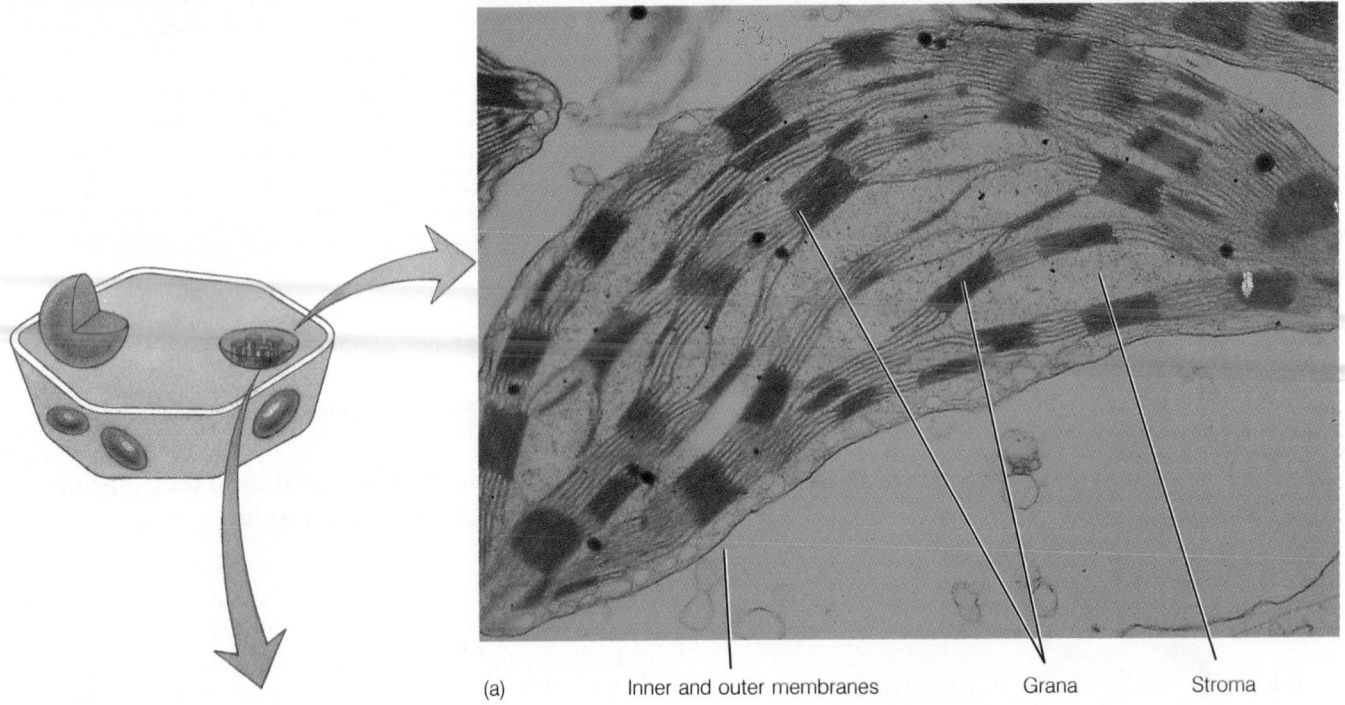

(a) Inner and outer membranes   Grana   Stroma

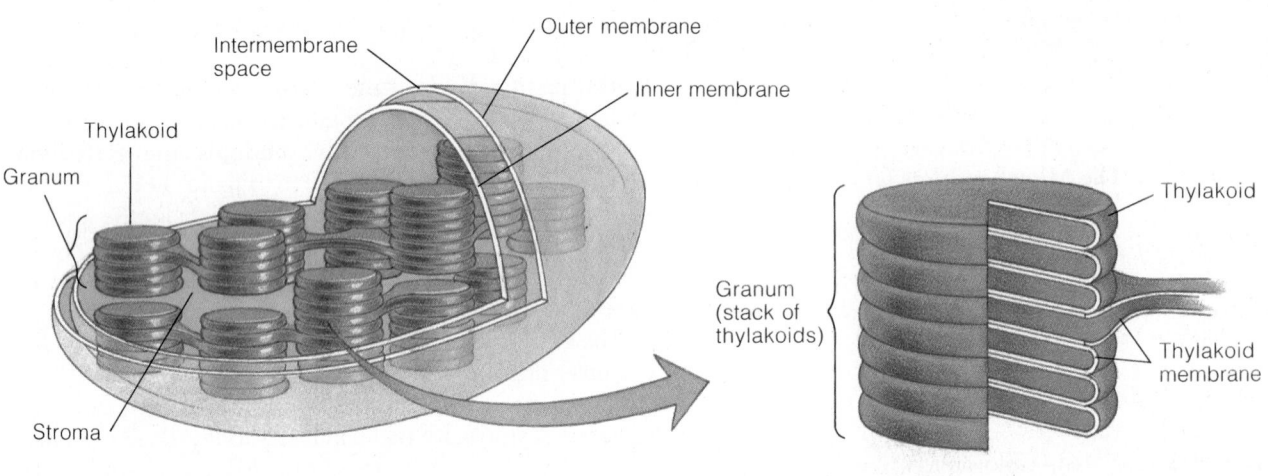

(b)

Cells also contain endoplasmic reticulum lacking ribosomes (Figure 3–17c). Known as the **smooth endoplasmic reticulum (SER),** this organelle specializes in the production of phosphoglycerides needed to replenish the plasma membrane. The SER performs a variety of functions in different cells. In the human liver, for instance, it detoxifies certain drugs, such as barbiturates, one type of sedative. In the stomach of humans and other mammals, the SER of certain cells produces hydrochloric acid, necessary for protein digestion. In the adrenal glands, the SER produces steroid hormones.

Ribosomes. As noted earlier, ribosomes are tiny particles made of protein and ribosomal RNA (rRNA). In the cytoplasm, ribosomes attach to one class of RNA molecules known as **messenger RNA (mRNA).** These polynucleotides produced in the nucleus of the cell contain all of the information required to synthesize protein.

In the cytoplasm several ribosomes attach to a single strand of mRNA, forming a structure known to as a **polyribosome** or **polysome.** Together, mRNA and its attached ribosomes begin to produce protein. If it is a protein destined for inclusion in lysosomes or the plasma membrane

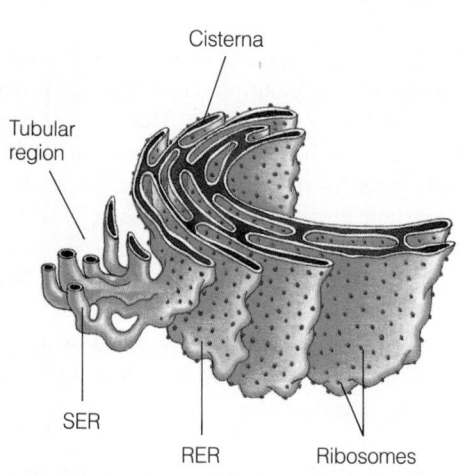

Cisterna

Tubular region

SER

RER   Ribosomes

(a)

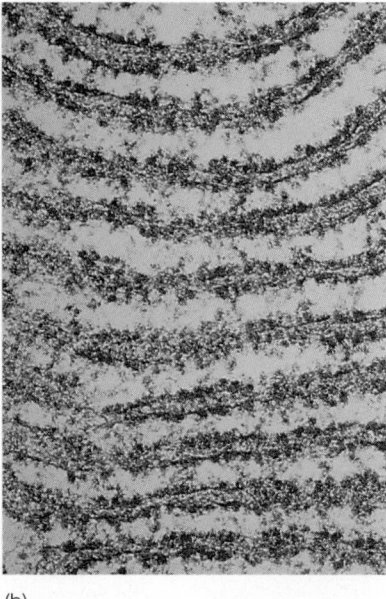

(b)

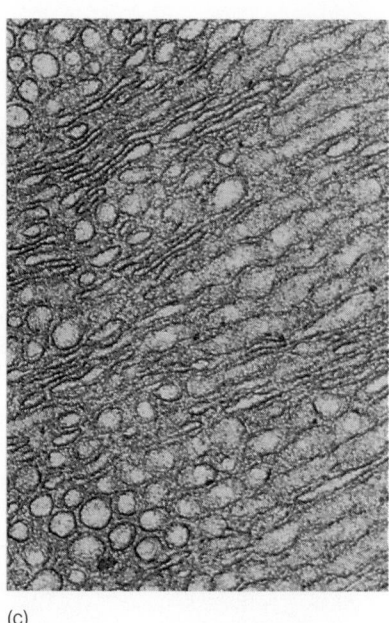

(c)

▷ **FIGURE 3–17 The Endoplasmic Reticulum** (*a*) Created by a network of membranes in the cytoplasm, the endoplasmic reticulum is often studded with small particles, the ribosomes, forming rough endoplasmic reticulum. This is the site of the synthesis of protein for lysosomes and extracellular use (digestive enzymes and hormones). (*b*) Electron micrograph of RER; (*c*) one of SER.

or is intended for extracellular use (hormones and digestive enzymes), the protein-ribosome-mRNA complex attaches to the RER, and the protein is transferred into the cisterna. If it is a protein for intracellular use, like those of the cytoskeleton or cytoplasmic enzymes, the polysome remains free within the cytoplasm.

The Golgi Complex. The RER transports the proteins it produces to the terminal ends of the cisternae (▷Figure 3–18). As the protein builds up inside the RER, the ends of the cisternae bulge, eventually pinching off as tiny membrane-bound vesicles. These vesicles migrate to the Golgi complex.

The **Golgi complex** consists of a series of membranous sacs, lying on top of one another like a stack of pancakes (▷Figure 3–19). It performs three functions. First, it sorts the molecules it receives, in much the same way that postal workers sort the outgoing mail. As a result, lysosomal enzymes are segregated from hormones and plasma membrane proteins. Second, like the RER, the Golgi complex chemically modifies many proteins, often adding carbohydrates or other small molecules. Finally, the Golgi complex packages proteins into lysosomes and **secretory vesicles** (also called **secretory granules**) (▷Figure 3–20b).

Secretory vesicles are membranous organelles that act as temporary storage sites for hormones and digestive enzymes bound for distant sites. Stored in the cytoplasm until the signal for release comes, secretory vesicles migrate to the plasma membrane. The membrane of the vesicle then fuses with the plasma membrane and the fused membranes dissolve away, releasing the contents of the secretory vesicle into the extracellular space.

▷ **FIGURE 3–18 Protein Synthesis and Secretion** Protein packed in lysosomes and secretory granules (for later export) is synthesized on the RER and transferred in tiny transport vesicles to the Golgi complex. Protein is sometimes chemically modified in the cisternae of both the RER and the Golgi complex. The Golgi complex separates protein by destination and repackages it into secretory granules, which remain in the cytoplasm until secreted by exocytosis.

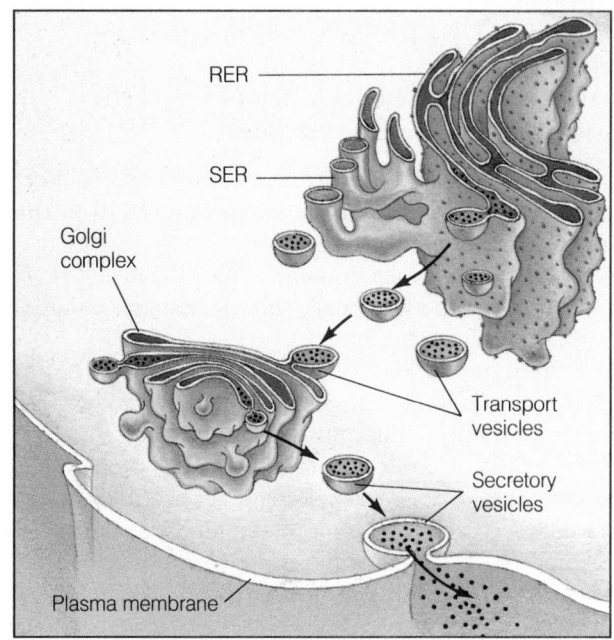

RER

SER

Golgi complex

Transport vesicles

Secretory vesicles

Plasma membrane

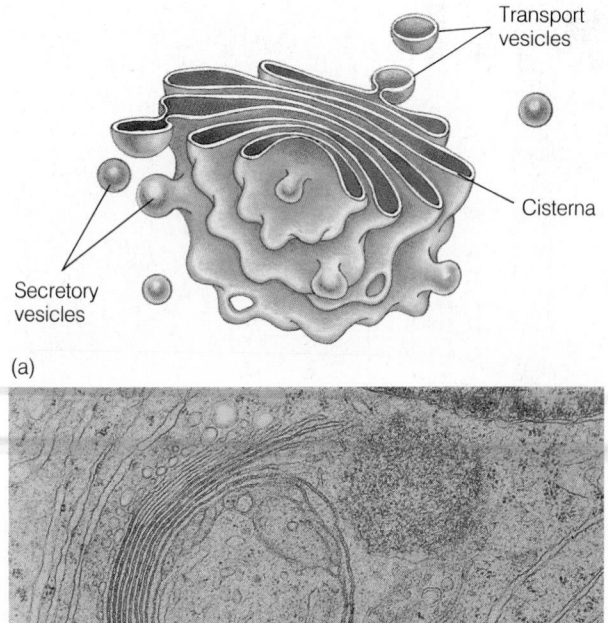

Transport vesicles

Cisterna

Secretory vesicles

(a)

(b)

▷ **FIGURE 3–19 Golgi Complex** (*a*) Artist's rendition of the three-dimensional structure of the Golgi complex. Transport vesicles carry protein produced by the RER to the Golgi complex and fuse to it, releasing the protein into its cisternae. Here the protein is sometimes chemically modified and repackaged. (*b*) Electron micrograph of the Golgi complex.

Lysosomes Are Membrane-bound Organelles That Contain Digestive Enzymes

Lysosomes (literally, digestive bodies) are tiny time bombs inside cells, that, fortunately, rarely go off. Containing numerous digestive enzymes produced by the RER and packaged by the Golgi complex, these organelles—when functioning properly—break down materials that are phagocytized by cells.

As illustrated in Figure 3–20a, material engulfed by the cell is enclosed in a membrane derived from the plasma membrane during endocytosis. The resultant structure is called a **food vacuole.** When functioning properly, lysosomes attach to food vacuoles and release their enzymes into them. The molecules inside the food vacuole are then broken down into smaller molecules that diffuse out of the vacuole into the cytoplasm, where they are often reused by the cell. The undigested material left behind is expelled from the cell by exocytosis.

Besides helping provide cellular nutrients, lysosomes are cellular custodians, destroying aged or defective cellular organelles, such as mitochondria. Lysosomes may also digest protein and other cellular molecules that have become

defective. This process helps cells maintain their structure and function. How cells recognize defective organelles remains a mystery.

Lysosomes also play an important role in embryonic development. In humans, for example, the fingers of an early embryo are webbed (see page 576). The cells of the webbing, however, are genetically programmed to self-destruct. At a certain point in development, the lysosomes of these cells release their enzymes, causing the cells to be destroyed and thus eliminating the webbing. A programmed release of lysosomal enzymes also causes the reabsorption of a tadpole's tail as it develops into a frog.

Finally, lysosomes help rid the body of injured or aged cells. In such cases, lysosomes break open, releasing their enzymes, which quickly destroy the cell from the inside out. Interestingly, lysosomal enzymes are released from damaged cells and are the basis of a useful diagnostic tool. During a heart attack, for example, heart muscle cells that die release very specific enzymes into the blood, which can be measured to confirm a physician's diagnosis. Several other diseases can also be detected by blood enzyme levels.

Most cells in the human body contain only a few lysosomes, used primarily to recycle outdated organelles or perhaps to put an end to the cell when its useful lifetime is up. Certain specialized cells, however, contain hundreds of lysosomes. A prime example is the neutrophil, a type of blood cell. Neutrophils scavenge the blood and body tissues looking for bacteria and viruses, which they phagocytize and digest internally with the aid of lysosomes. Neutrophils therefore help reduce the spread of infection and are one of the body's protective mechanisms.

Plant Cells Contain Large Central Vacuoles That Store Water

As the preceding material has shown, plant and animal cells contain membranous chambers of varying size. As you may have surmised by now, small ones are generally referred to as **vesicles**—for example, the secretory vesicle. Larger ones are called **vacuoles**.

The largest of all vacuoles occurs in plant cells and is called the **central vacuole.** Occupying up to 80% of the volume of the cell, it leaves only a thin rim of cytoplasm between itself and the plasma membrane (▷Figure 3–21). The central vacuole of plants is delimited by a membrane and serves many functions. Most vacuoles hold water and dissolved ions needed by plant cells for cellular metabolism. In the storage cells of seeds, however, the central vacuole stores proteins and starch. Vacuoles in the cells of certain flowers contain red and blue pigments, which attract pollinating insects. Some vacuoles contain poisons or unpalatable chemicals that help protect plants against hungry predators. The plant vacuole also contains digestive enzymes that break down stored polymers, such as starch.[7]

---

[7] Plant cells normally lack lysosomes.

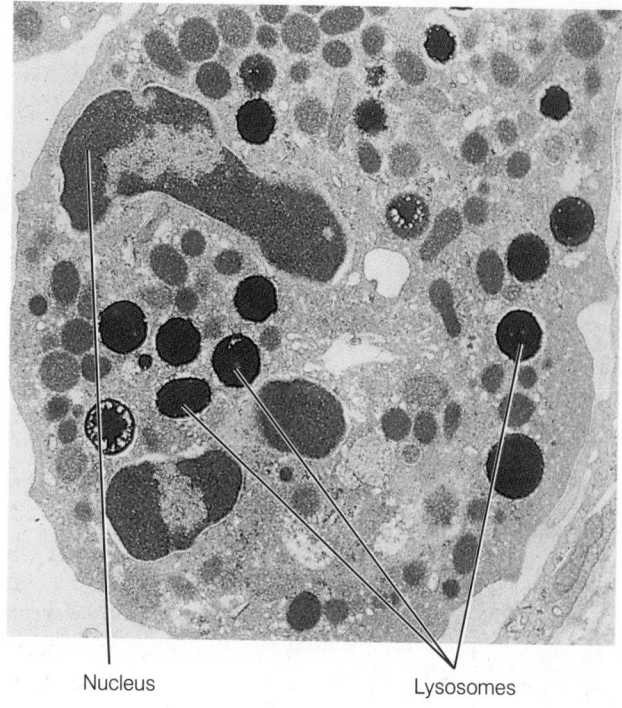

Nucleus   Lysosomes

(a)

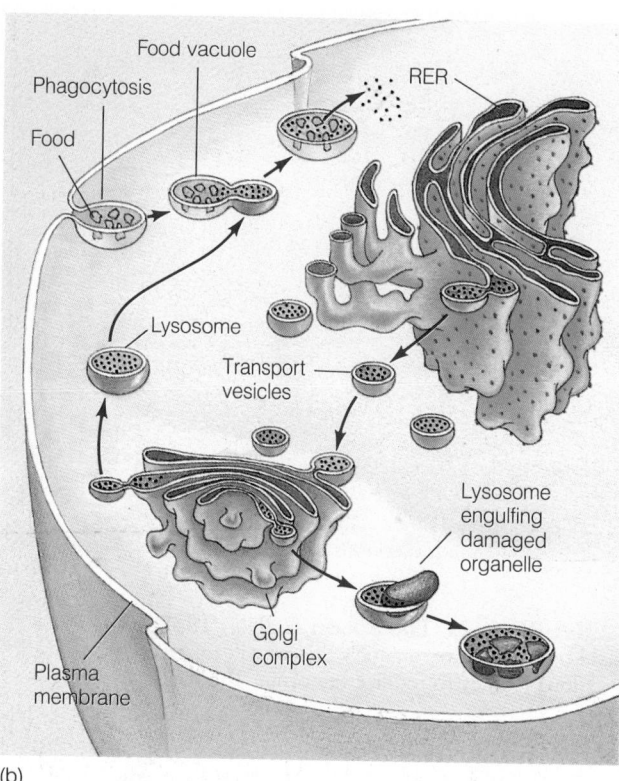

(b)

▷ **FIGURE 3–20 The Lysosome** (*a*) Electron micrograph of a neutrophil, a type of white blood cell that phagocytizes bacteria and cellular debris in wounds. The dark-staining bodies are lysosomes, filled with digestive enzymes that are used by these cells to break down ingested material. (*b*) The digestive enzymes of the lysosome are produced on the RER and transported to the Golgi complex for repackaging. Lysosomes produced by the golgi fuse with food vacuoles, which allows their enzymes to mix with the contents of the food vacuole. The enzymes digest the contents, which diffuse through the membrane into the cytoplasm, where they are used. The membrane surrounding the lysosome helps protect the cell from digestive enzymes.

Cilia and Flagella Are Organelles That Permit Cellular Motility

Mobility is a key to success in the animal world and, not surprisingly, many animal cells are motile (capable of movement). So are many protists, single-celled eukaryotic organisms. Two cellular organelles participate in movement in animals and protists, the flagellum and the cilium.

The mammalian sperm cell moves with the aid of a **flagellum** (plural, *flagella*, Latin for whip), a long, whiplike extension of the plasma membrane containing numerous small tubules, known as **microtubules,** that produce and transmit the motive force (▷Figure 3–22). Flagella are also found in single-celled protists like the Euglena (see Figure 24–13) and help these organisms move about in the aqueous environments they inhabit.

As ▷Figure 3–23a illustrates, the nine pairs of microtubules of the flagellum are arranged in a central pair. Biologists typically refer to this as the **"9 + 2"** arrangement. Flagella have additional fibers outside the central 9 + 2 array, which give the organelle extra strength.

At the base of each flagellum is an anchoring structure, the **basal body** (Figure 3–23b). The basal body gives rise to the flagellum during development and contains nine sets of microtubules (with three microtubules each) but no central pair.

Paramecia, single-celled aquatic eukaryotes, contain another organelle involved in motility, the **cilium.** Cilia (plural) are smaller extensions of the cell membrane, that, like flagella, have a central core of microtubules in a 9 + 2 arrangement (Figure 3–23c). Hundreds, even thousands, of cilia can be found on a single cell.

▷ **FIGURE 3–21 The Central Vacuole** Found only in plant cells, the central vacuole serves many functions.

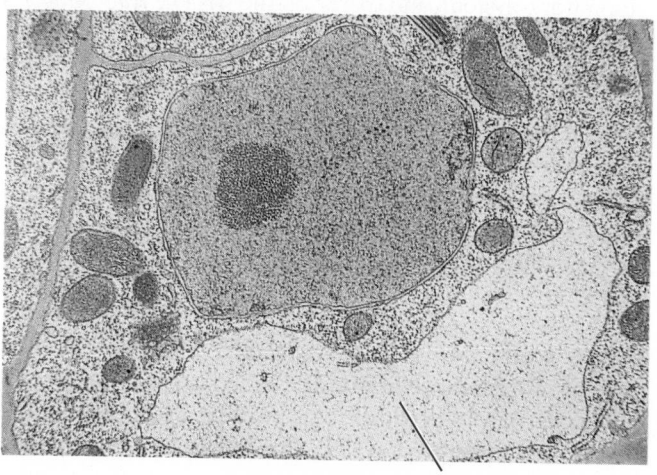

Central vacuole

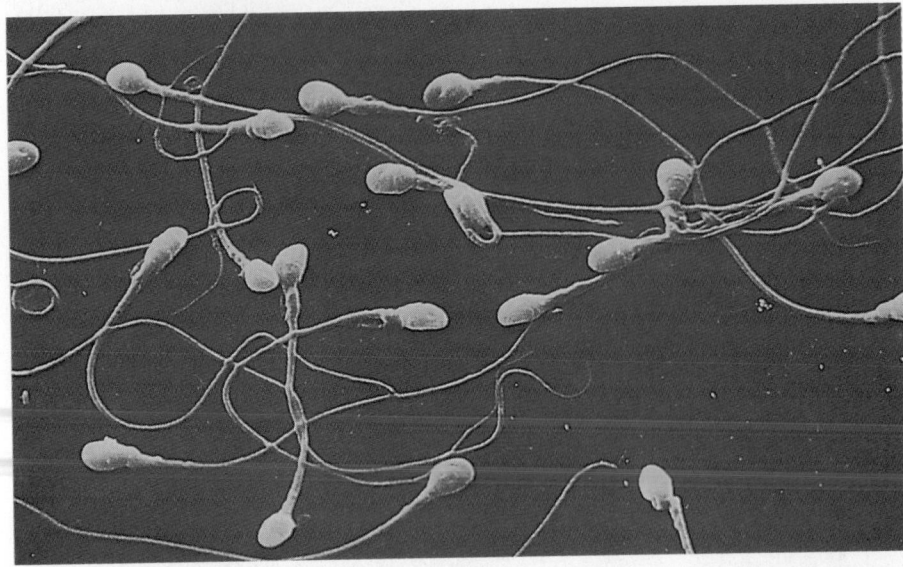

(a)

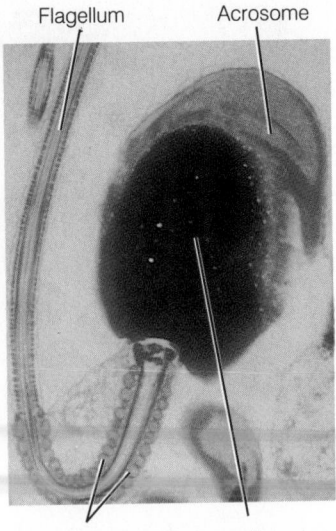

Flagellum   Acrosome

Mitochondria   Sperm cell nucleus

(b)

▷ **FIGURE 3–22 The Sperm Cell** (*a*) The sperm cell is a marvel of architecture, uniquely "designed" to streamline the cell for its long journey to fertilize the ovum. The nuclear material is compacted into the sperm head. A flagellum propels the sperm through the female reproductive tract. (*b*) Transmission electron micrograph of sperm cells.

Cilia are also found in multicellular organisms. In humans, for instance, hundreds of thousands of cilia protrude from the cells lining the trachea (Figure 3–23d). These cilia beat toward the mouth and transport mucus produced by other cells in the lining to the oral cavity, where it can be either swallowed or expectorated (spit out). Since the mucus traps dust particles in the air we breath, the cilia are part of an automatic cleansing mechanism, vital for maintaining our health.

At the base of each cilium is a basal body, consisting of nine sets of microtubules (with three microtubules each) but no central pair. The basal body gives rise to the cilium during development and anchors the organelle in place.

In sum, cilia and flagella are structurally similar, but flagella are much longer than cilia and far less numerous. Normal human sperm cells, for example, have one flagellum each, whereas each of the ciliated cells lining the respiratory system and parts of the female reproductive system (notably the Fallopian tubes) contains thousands of cilia. Flagella and cilia also beat differently. Cilia perform a stiff rowing motion with a flexible return stroke, whereas flagella beat more like whips, undulating in a continuous motion (▷ Figure 3–24).

Because cilia and flagella often beat continuously, they require a more or less constant supply of energy, provided by ATP. As noted earlier, most ATP is produced by mitochondria, which are located nearby.

Some Cells Move by Amoeboid Motion. Another type of movement seen in single-celled organisms and certain cells of multicellular organisms is called **amoeboid motion**. As illustrated in ▷ Figure 3–25, during amoeboid movement, the cell sends out many small cytoplasmic projections, known as **pseudopodia** (false feet). Pseudopodia attach to solid surfaces "in front" of the cell. Cytoplasm flows into the pseudopodia, moving the cell forward.

Phagocytic neutrophils, described earlier, are capable of amoeboid movement. These cells can escape the tiny **capillaries,** thin-walled blood vessels in body tissues, and migrate through tissues, gobbling up bacteria and damaged cells. Amoeboid movement also occurs in a group of single-celled organisms, appropriately called the amoeboids (Figure 24–18, page 661).

ENVIRONMENT AND HEALTH: PARKINSON'S AND POLLUTION?

In 1983, Dr. J. W. Langston, a California neurologist, was confronted with a novel medical puzzle unlike any he had ever seen. Several relatively young men and women were admitted to the hospital in a catatonic stupor. Confined to their beds, the patients lay immobile day after day. They could neither talk nor feed themselves nor move their limbs. It was as if they had been frozen.

Langston began an intensive study of the victims, all drug addicts. He found that each of them had taken a synthetic, heroinlike drug that had been contaminated with a paralyzing chemical known as MPTP. Made in the basement of one of California's small-time drug pushers, the heroin-like drug is one of the latest in a long list of "designer" drugs now on the market. Sloppy chemistry resulted in the contamination.

MPTP attacks nerve cells in a part of the brain called the substantia nigra. Degeneration of the brain cells in this

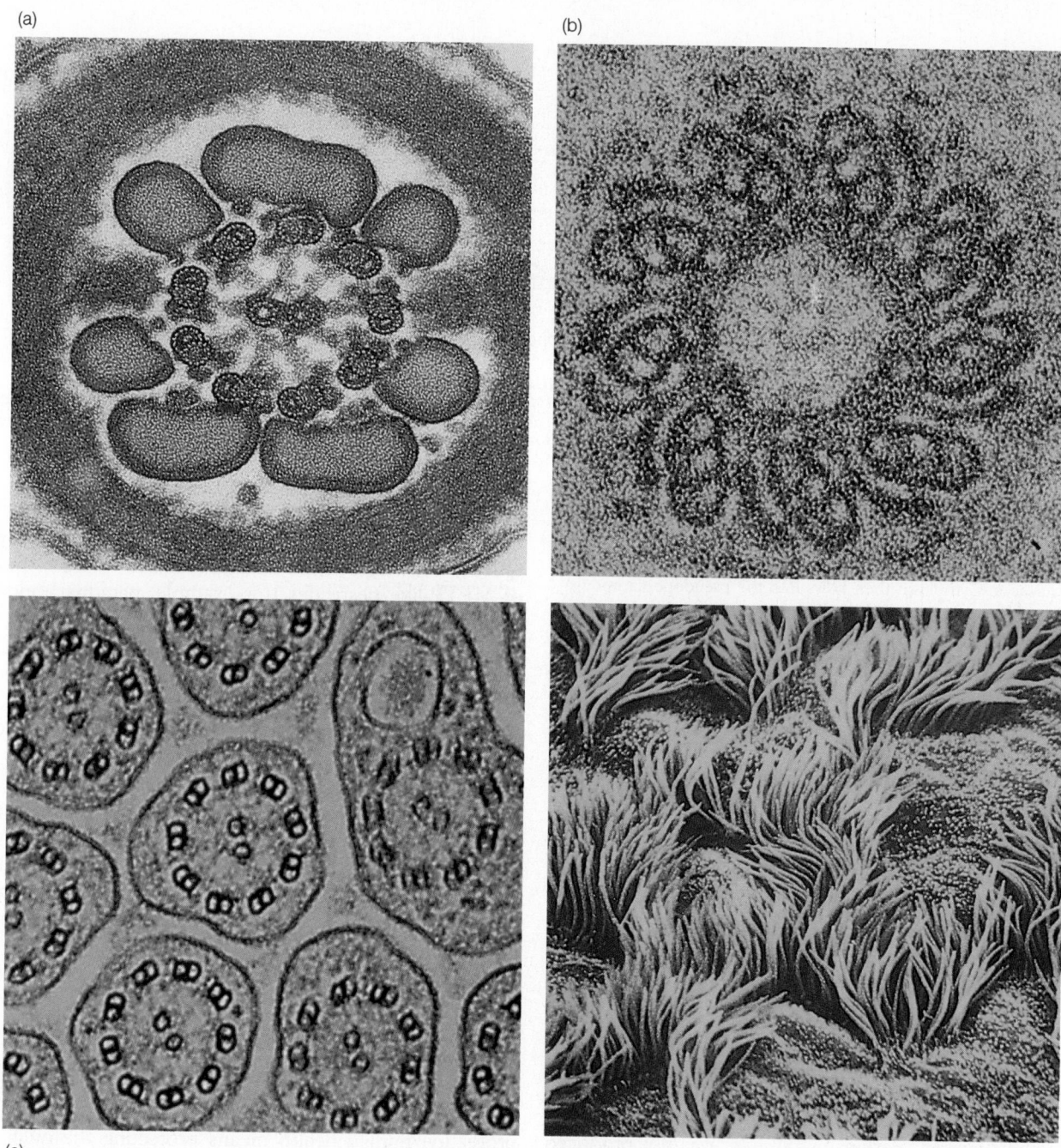

(a)   (b)

(c)   (d)

▷ **FIGURE 3–23 Celia and Flagella** (*a*) A sperm flagellum, showing 9 + 2 arrangement and additional fibers thought to provide support and strength to the vigorously beating tail. (*b*) A basal body is found at the base of each cilium. It consists of nine sets of triplets that give rise to the microtubule doublets of the cilium. (*c*) Numerous cilia in cross section, showing the 9 + 2 arrangement of microtubules. (*d*) Cilia on cells lining the trachea of the human respiratory tract.

region has long been thought to cause Parkinson's disease, a disorder that afflicts older individuals and is characterized by tremors, or involuntary rhythmic shaking, of the hands or head. If the disease worsens, patients have difficulty talking and writing. Walking becomes awkward and clumsy. In the later stages, memory and thinking can be severely impaired.

Individuals exposed to even minute amounts of MPTP develop symptoms similar to those seen in Parkinson's patients. A Swiss researcher who had worked briefly with the chemical was hospitalized because his body muscles, like those of the addicts whom Langston had encountered, "froze up" on him. Puzzled physicians put him in a mental hospital, assuming that he was suffering from a psychological disorder called catatonic schizophrenia.

MPTP is similar to several common industrial chemicals and several popular pesticides. Some scientists now speculate that MPTP-like chemicals in the air, water, and soil may be the underlying cause of the nerve cell damage in Parkinson's disease. Opponents of this hypothesis, how-

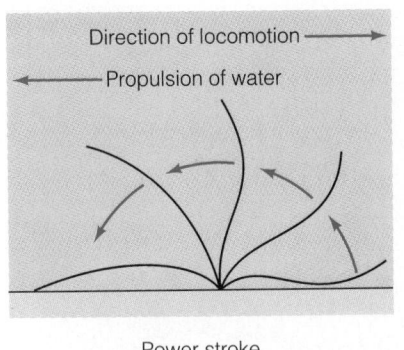

Direction of locomotion →
← Propulsion of water

Power stroke   Return stroke

(a) Cilium

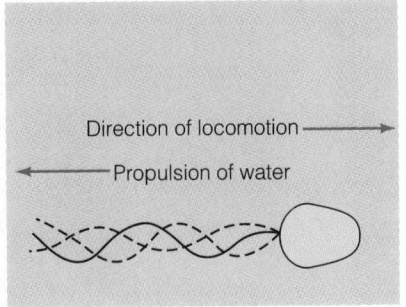

Direction of locomotion →
← Propulsion of water

Continuous propulsion
(b) Flagellum

▷ **FIGURE 3–24 Beating of Cilia and Flagella** (*a*) Cilia beat with a stiff power stroke—something of a rowing motion. During the return stroke, the cilium is flexible. (*b*) The flagellum beats like a whip, over and over, propelling the sperm cell forward.

▷ **FIGURE 3–25 Amoeboid Movement** Some cells in the human body move by amoeboid movement. The cell sends out minute extensions called pseudopodia (false feet), which attach to solid surfaces. Cytoplasm then flows into the pseudopodia, moving the cell forward.

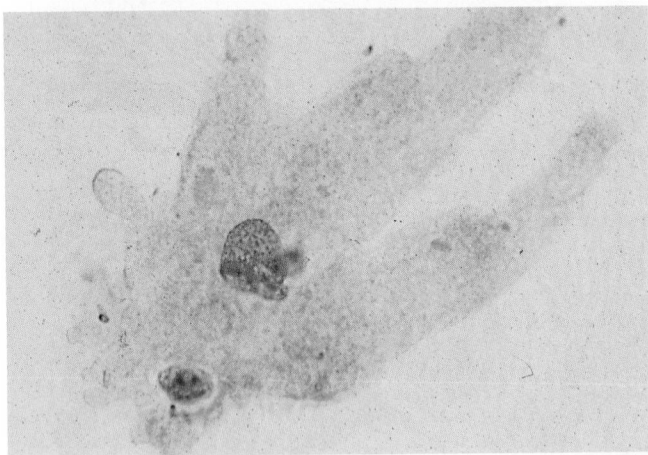

ever, point out that several studies have failed to show a link between Parkinson's disease and industrial pollution. At a 1985 meeting at the National Institutes of Health, researchers agreed: Parkinson's disease was not produced by environmental contamination.

No sooner had the researchers arrived home than a Canadian scientist announced the results of a study in Quebec showing a remarkable correlation between the use of MPTP-like pesticides (for example, paraquat) and the incidence of Parkinson's disease. He argued, however, that pesticides were probably not the only chemical villain. Many neurotoxins could have been present in the area as discharges from nearby chemical factories.

In support of the hypothesis that Parkinson's disease is linked to industrial pollution and pesticide use, researchers note that this disease was unheard of before the Industrial Revolution. As industrialization and environmental pollution spread, the incidence of the disease rose sharply, reaching a plateau in the early 1900s. Although no increase has been seen since 1940, some researchers warn that an increase in the use of paraquat and other MPTP-like pesticides and an increase in industrial pollution could substantially raise the incidence of the disease in years to come.

This example shows how certain nerve cells can be affected by a chemical contaminant. It also shows how alterations in cellular function level can have dramatic effects on the whole body and underscores the fact that protecting human health means protecting the quality of the environment.

SUMMARY

1. The fundamental unit of all living organisms is the cell. The cell theory states that all organisms are composed of cells and that all existing cells arise from preexisting cells.

2. Two types of cells are found: prokaryotes and eukaryotes. Prokaryotes lack nuclei and include the bacteria. Eukaryotes have true nuclei.

CELLULAR EVOLUTION AND HOMEOSTASIS

3. During cellular evolution, prokaryotic organisms emerged first, about 3.7 billion years ago. These early cells and their modern descendants, the bacteria, are members of the kingdom Monera.

4. Eukaryotes evolved about 2.5 billion years later. The first eukaryotic cells were single-celled organisms. These early organisms and their descendants, such as amoebae, belong to the kingdom Protista.

5. As evolution proceeded, organisms became multicellular and began to acquire specialized cells. Over time, eukaryotes gave rise to three additional kingdoms: plants, animals, and fungi.

6. The increasing complexity of organisms was paralleled

by the development of complex systems that help ensure homeostasis.

MICROSCOPES: ILLUMINATING THE STRUCTURE AND FUNCTION OF CELLS

7. The microscope is one of many scientific instruments used to study cells. Basically consisting of a lens or combination of lenses, a microscope provides two services: magnification (enlargement) and resolution (the ability to see detail).

8. Microscopes fall into two broad categories: (1) light microscopes, which use ordinary visible light to illuminate the specimen, and (2) electron microscopes, which use beams of electrons to visualize an image.

9. Two types of electron microscopes are commonly used. The transmission electron microscope provides a view of the interior of cells, and the scanning electron microscope produces three-dimensional images of organelles, cells, tissues, and even small organisms.

PLANT AND ANIMAL CELLS: SIMILARITIES AND DIFFERENCES

10. Plant and animal cells share many features. Both types consist of two major compartments, the nuclear and the cytoplasmic.

11. The nuclear compartment, or nucleus, contains the genetic information that controls the structure and function of all eukaryotic cells.

12. The cytoplasmic compartment contains cytoplasm, which, in turn, contains numerous cellular organelles, specialized structures that perform specific functions.

13. Plant and animal cells also contain cytoskeletons, networks of tubules and filaments that help give cells their three-dimensional shape and also bind to enzymes in metabolic pathways.

14. Although plant and animal cells share many characteristics, they have important differences. All plant cells contain a rigid outer layer, the cell wall. Plant cells also contain chloroplasts, organelles in which photosynthesis takes place, and many also contain large vacuoles used to store water and other chemical substances. Finally, plant cells lack centrioles.

THE STRUCTURE AND FUNCTIONS OF THE PLASMA MEMBRANE

15. The outermost boundary of the cell is the plasma membrane, which is made of lipid, protein, and carbohydrate.

16. The plasma membrane ensures the structural integrity of the cell, regulates the flow of materials into and out of the cell, and participates in cellular communication and cellular identification.

17. Because it is mostly lipid, the membrane is a natural barrier to water-soluble molecules. Lipid-soluble materials pass through with ease.

18. Water-soluble molecules and ions are believed to pass through pores in the membrane by diffusion, the flow of a substance from an area of higher concentration to an area of lower concentration.

19. Carrier proteins may also help transport water-soluble molecules across the membrane down concentration gradients. This process is called facilitated diffusion.

20. Some molecules are actively transported across the membrane. Active transport mechanisms use energy to pump materials into or out of the cell and require a special class of

membrane proteins, the transport molecules. Active transport permits cells to transport substances from regions of low concentration to regions of high concentration.

21. Cells engulf liquids and their dissolved materials in a process called pinocytosis. They may engulf large molecules and cells by phagocytosis. Pinocytosis and phagocytosis are collectively referred to as endocytosis.

22. During endocytosis, the plasma membrane indents and surrounds the material to be ingested. The material engulfed by the cell is incorporated in a membranous vesicle, which is taken into the cell.

23. Water is believed to move through tiny pores in plasma membranes. The diffusion of water molecules through a selectively permeable membrane from a region of higher water concentration to a region of lower water concentration is known as osmosis.

CELLULAR COMPARTMENTALIZATION: ORGANELLES

24. The nucleus is one of the most conspicuous cellular organelles. It houses the DNA, which contains the genetic information that determines the structure and function of the cell.

25. The nucleus is bounded by a double membrane, the nuclear envelope, which contains numerous pores that allow many materials (but not the DNA) to pass freely between the nucleus and the cytoplasm.

26. In the cell, energy is produced principally by the breakdown of glucose. Glucose breakdown takes place in the cytoplasm and in the mitochondrion. Most of the ATP produced during this process is produced in the mitochondrion.

27. All mitochondria have two membranes, an outer one and an inner one. The inner membrane is thrown into folds called cristae.

28. Cellular energy production captures 30% to 60% of the energy contained in glucose; the rest is liberated as heat.

29. Plant cells contain plastids, membranous organelles that serve a variety of functions. Some plastids contain pigments; others are colorless and merely act as storage sites for organic molecules produced by plant cells.

30. The best known and most abundant plastid is the chloroplast. This disk-shaped or ovoid organelle contains numerous stacks of flattened membranous sacs (thylakoid disks) containing chlorophyll and other light-absorbing molecules needed for photosynthesis.

31. The endoplasmic reticulum, a network of membranous channels in the cytoplasm of the cell, is a major site of cellular protein synthesis.

32. Endoplasmic reticulum may be smooth or rough. Smooth endoplasmic reticulum has no ribosomes. Rough endoplasmic reticulum contains many ribosomes on its outer surface.

33. Smooth endoplasmic reticulum produces phosphoglycerides, needed to make more plasma membrane, and has many additional functions in different cells. The rough endoplasmic reticulum is involved in synthesizing protein for extracellular use (for example, hormones) and enzymes of the lysosome.

34. Proteins produced on the surface of the RER enter the cavity, where they may be chemically modified. Proteins are transferred from the RER to the Golgi complex, where they are sorted, chemically modified, and repackaged into secretory vesicles or lysosomes.

35. Secretory vesicles accumulate in the cytoplasm and are re-

leased when needed. Lysosomes remain in the cell, where they bind to food vacuoles (material phagocytized or pinocytized by the cell) or destroy malfunctioning cellular organelles.

36. Plant cells contain large central vacuoles that serve many functions, including the storage of water, ions, pigments, and macromolecules.

37. Most cells in the body are fixed. Many of the mobile cells, however, are transported passively in the blood. Still other cells can move about in tissues via amoeboid movement.

38. The human sperm cell utilizes another kind of motive force, the flagellum, a long, whiplike structure.

39. In the human, some cells contain numerous smaller extensions, called cilia, which propel fluids and materials along their surfaces. Some single-celled organisms are covered by cilia.

40. Both cilia and flagella contain nine pairs of microtubules arranged in a circle around a central pair.

ENVIRONMENT AND HEALTH: PARKINSON'S AND POLLUTION

41. MPTP is a chemical contaminant in a synthetic, heroinlike drug. MPTP attacks certain cells in the brain, the same cells, in fact, long thought to be asociated with Parkinson's disease, a disorder that generally afflicts older individuals and is characterized by involuntary rhythmic shaking of the hands or head.

42. Individuals exposed to even minute amounts of MPTP develop symptoms similar to those seen in Parkinson's patients.

43. MPTP is similar to several common industrial chemicals and several popular pesticides. Some scientists now speculate that MPTP-like chemicals in the air, water, and soil may be the underlying cause of Parkinson's disease.

44. This example shows how alterations in the level of cellular function can have dramatic effects on the whole body and, if the link between Parkinson's and industrial chemicals proves valid, underscores the importance of protecting human health by protecting the quality of our environment.

EXERCISING YOUR CRITICAL THINKING SKILLS

Garbage from modern society contains a great deal of organic matter: grass clippings, leaves, kitchen wastes, paper, cardboard, and so on. This waste is dumped in landfills, where it is covered with dirt and supposedly decomposed by bacteria over time. Recently, however, U.S. newspapers and magazines have featured stories about researchers who recovered newspapers and food buried in an Arizona landfill 35 years ago and still not decomposed. This story has been often repeated, and some people are now under the impression that our garbage is not decomposing in landfills. One person, in fact, recently suggested that it does not matter whether plastic cups or paper cups are used, because neither decompose in a landfill.

This is an issue where critical thinking skills come in handy. Critical thinking requires us to be an active participant: to take some time to study matters more carefully. In keeping with this goal, you may want to read a section on landfills in an environmental science textbook. Can you find any useful information to make you think that the conclusion that our trash is not decomposing is false?

Reread the first paragraph above carefully. Can you find any clues that would suggest why the conclusions from the study on garbage decomposition might not apply to landfills in Massachusetts or Michigan?

TEST OF TERMS

1. The first cells probably resembled modern-day _____ .

2. The _____ contains the cell's hereditary information contained in molecules of _____ .

3. The network of protein tubules in the cytoplasm of a cell is called the _____ . It binds to organelles and to _____ , protein molecules that increase the rate of chemical reactions.

4. A series of chemical reactions in the cell is called a(n) _____ .

5. The plasma membrane consists of a _____ layer of lipid, mostly _____ . Proteins in the mem-

brane can be embedded in the lipid layer of the membrane; these are called _____ proteins.

6. A membrane that is permeable to some substances and impermeable to others is called a _____ membrane.

7. A protein that has carbohydrate units attached is called a _____ . Carbohydrate units are added to protein in the cell in two organelles, the _____ and the _____ .

8. The movement of a molecule (other than water) across the plasma membrane from a region of high concen-

tration to one of low concentration is called _____ .

9. Facilitated diffusion requires a(n) _____ protein in the membrane to shuttle molecules and ions from one side to the other down a concentration gradient.

10. The process of engulfing another cell, such as a bacterium, is called _____ .

11. Water moves across membranes through pores. It is driven by concentration differences. The driving force is called _____ pressure.

12. When the concentration of a fluid is the same as that of a cell, the fluid is said to be _____ . When

the fluid has a solute concentration lower than the cell's cytoplasm, the fluid is said to be _____ . Dropping cells into this type of solution will cause them to _____ .

13. The double membrane surrounding the nucleus is called the nuclear _____ . It is pierced by small openings called nuclear _____ .

14. Strands of DNA with associated protein are called _____ . They condense before cell division to produce dark-staining bodies called _____ .

15. _____ is synthesized at the nucleoli. It combines with _____ to form ribosomes.

16. The _____ is the site of the bulk of the energy production occurring in a cell.

17. The _____ is the site of photosynthesis in plants and is a member of a general group of organelles called _____ .

18. Protein for extracellular use and digestive enzymes found in lysosomes are produced on the _____ . The instructions for these proteins come from the nucleus in the form of _____ molecules.

19. Secretory vesicles are packages of protein and other cellular products. Secretory granules are produced by the _____ .

20. Digestive enzymes are held in a membrane-bound organelle called the _____ . It fuses with engulfed material contained in a(n) _____ _____ .

21. The long, whiplike organelle on human sperm cells is called a _____ . The motive force is created by microtubules arranged as _____ pairs surrounding a central doublet. This is the _____ arrangement.

22. Small motile projections on the cells lining much of the respiratory tract are called _____ .

Answers to Test of Terms are located in Appendix B.

TEST OF CONCEPTS

1. Describe cellular and organismic homeostasis. How are they related?
2. Cellular compartmentalization allows for greater efficiency. Why? Give an example to support your argument.
3. Draw a diagram of the plasma membrane, and describe the five routes by which molecules pass through the membrane.
4. In what ways does the plasma membrane participate in cellular communication?
5. Define the terms *osmosis, isotonic, hypotonic*, and *hypertonic*. In what way(s) is osmosis similar to diffusion, and in what way(s) is it different?
6. Describe the similarities and differences between plant and animal cells.
7. Describe the organelles involved in the synthesis, storage, and release of protein for extracellular use, starting with the basic instructions needed to make the protein.
8. In what ways are cilia and flagella similar, and in what ways are they different?

Energy and Metabolism

≈

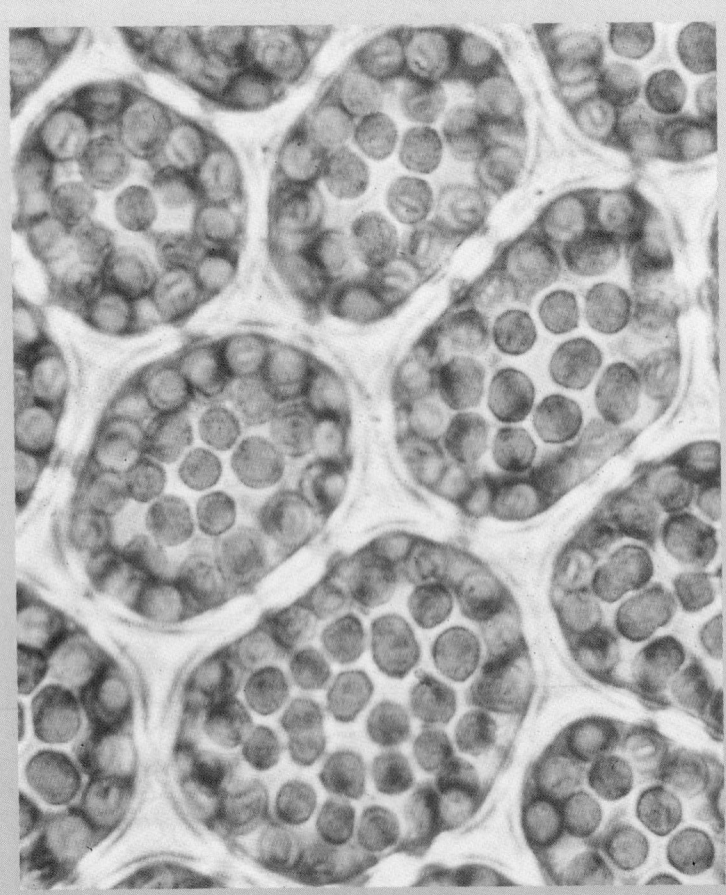

Plant cells and mitochondria, which trap sunlight energy and produce organic food molecules.

I n the woods of Canada, a rabbit races to escape a hungry wolf. Overhead, a Steller's jay, startled from a snooze, sounds a warning. Squirrels dart for shelter, and a hungry raven flaps its wings on the branches of a pine. A tired owl, exhausted from a night of hunting, blinks its eyes open and shut, then settles back into sleep. A camper turns her head to hear the commotion.

These activities are united by one common feature: they all require energy. All organisms need a fairly constant supply of energy to live. The cells of organisms, for example, use energy to synthesize protein, to transport molecules across membranes, to reproduce, to secrete hormones, and to do much more. Much of the energy is used to drive chemical reactions.

This chapter describes the general nature of chemical reactions, paying special attention to energy and the role it plays in the chemical reactions occurring in the cells of all organisms. In this discussion, you will discover the importance of enzymes and how cells regulate the thousands of chemical reactions that make life possible.

≈ THE NATURE OF CHEMICAL REACTIONS

Many different molecules form the structures of our bodies and the bodies of other organisms. Arranged in a fascinating variety of ways, these molecules form cells, which in turn form the tissues and organs of multicellular organisms. The molecules found in the bodies of organisms participate in thousands of chemical reactions, which are essential for growth, reproduction, and development. Chemical reactions also play a key role in homeostasis. All living things are really complex chemical machines.

As noted in Chapter 1, the chemical reactions occurring in an organism are known collectively as metabolism. You may also recall that metabolic reactions fall into two major categories: (1) synthetic reactions, or **anabolic reactions,** in which small molecules combine with others to form larger molecules; and (2) breakdown reactions, or **catabolic reactions,** in which molecules split apart, or dissociate. In anabolic reactions, new chemical bonds are formed; in catabolic reactions, existing chemical bonds are broken. Anabolic reactions require energy to run; catabolic reactions release it. What is energy?

Physicists define **energy** as the capacity to do work—to move an object some distance against the force of gravity or friction. Energy is, therefore, an agent of change. Unlike matter, which occupies space and has mass, energy is intangible. It takes up no space. It has no mass.

In the study of biology, we concern ourselves largely with a type of energy called potential energy. **Potential energy** is stored in several different forms. A rock perched on a ledge, for instance, is said to contain potential energy. When released, this stored energy causes work to be done—for example, squashing a picnic table in a park at the base of the ridge.

In biology, potential energy is stored in the chemical bonds between the atoms of molecules. As Chapter 5 notes, the energy stored in these bonds originally comes from sunlight captured by plants. Plants use the energy to synthesize carbohydrates, such as glucose, from atmospheric carbon dioxide and water (▷Figure 4–1). Most organisms break down these carbohydrates in an elaborate series of chemical reactions, releasing the stored energy for use in their cells.

Another important form of energy of somewhat less concern in the study of biology is **kinetic energy,** the energy of a body that results from its motion. A moving car is said to have kinetic energy. Moving molecules possess kinetic energy. When they collide, the energy may be sufficient to cause them to react.

Anabolic Reactions Are Synthetic Reactions and Require Energy

Ben Johnson, Canada's extraordinary track star (▷ Figure 4–2), was stripped of a gold medal in the 1988 Summer Olympics because he had allegedly taken anabolic steroids—hormones that help build muscle protein, boosting athletic performance (Chapter 19). Protein synthesis is one of many anabolic reactions encountered in the study of biology.

Anabolic reactions require energy and, as noted earlier, in animals most of this energy comes from the breakdown of carbohydrates. The day before running marathons, for example, many athletes consume large quantities of carbohydrate-rich food, a process often referred to as "carbo-loading." Foods including potatoes, rice, and pasta, contain a lot of starch, which is broken down into glucose in the intestinal tract and then absorbed into bloodstream. Glucose molecules are stored in muscle cells or the liver. During the marathon, glucose molecules in the muscle are broken down in a series of chemical reactions that release large quantities of energy needed by the runner.

Anabolic reactions in plants, also require energy. Plants acquire energy for these reactions from glucose they synthesize with ATP that is generated during photosynthesis, a topic described in more detail in Chapter 5.

Catabolic Reactions Liberate Stored Energy

Potential energy stored in covalent bonds is released by the cells as they catabolize carbohydrates and other energy-rich molecules like triglycerides. In a precisely controlled series of reactions, covalent bonds are broken apart, and the energy stored in the bonds is released.

Energy Released from Catabolic Reactions Drives Anabolic Reactions

In the cells of your body, energy released from catabolic reactions is used to power anabolic, or energy-requiring, reactions. As a result, anabolic and catabolic reactions are

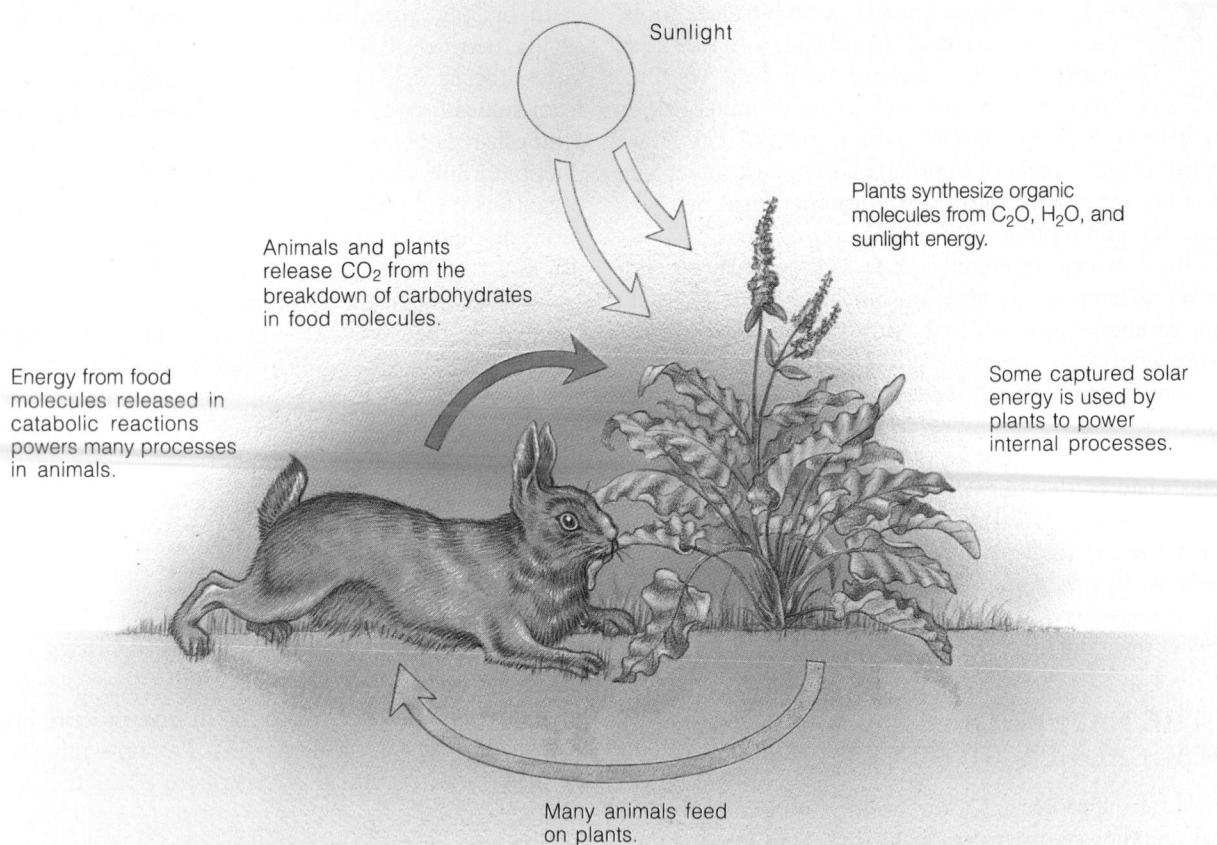

Sunlight

Plants synthesize organic molecules from C_2O, H_2O, and sunlight energy.

Animals and plants release CO_2 from the breakdown of carbohydrates in food molecules.

Energy from food molecules released in catabolic reactions powers many processes in animals.

Some captured solar energy is used by plants to power internal processes.

Many animals feed on plants.

▷ **FIGURE 4–1 All Energy Comes from Sunlight** Plants capture the sun's energy and store it in the covalent bonds of the carbohydrates they produce. In both plant cells and animal cells, this energy is released by the catabolic breakdown of glucose.

said to be **coupled.** ▷ Figure 4–3b illustrates one of the most important coupled reactions. As shown, the energy released by the breakdown of glucose in muscle cells is used to drive the synthesis of **adenosine triphosphate (ATP).** As you may recall, ATP is produced from ADP and inorganic phosphate. Energy needed to add the phosphate usually comes from catabolic breakdown of glucose. ATP, in turn, stores the energy it "acquires" for later use— for example, to power muscle contraction or synthesize protein, or to transport molecules across the plasma membrane of cells. (ATP is discussed in more detail later in this chapter and also in Chapters 5 and 6.)

During coupled reactions, the transfer of energy is never 100% efficient. Some energy is lost as heat.

Chemical Reactions Are Reversible and Reach a Point of Chemical Equilibrium

Most chemical reactions are a two-way street; that is, traffic proceeds in both directions. In the language of a chemist, chemical reactions are said to be reversible. In the blood, for instance, carbon dioxide released from cells combines with water to form a weak acid, known as carbonic acid:

$$H_2O \ + \ CO_2 \ \leftrightarrows \ H_2CO_3$$

water   carbon dioxide   carbonic acid

The double arrows indicate that this reaction proceeds in

both directions. By convention, the molecules on the left are called **reactants.** The molecule on the right, which is formed in the reaction, is called the **product.**

Chemical reactions such as this reach a point at which the speed of the forward reaction equals that of the reverse reaction: **chemical equilibrium.** Because the forward and reverse reactions occur at the same rate once equilibrium has been achieved, the concentrations of chemicals on both sides of the reaction remain constant. It is important to note, however, that chemical equilibrium does not mean that the concentrations of the products and the reactants are the same, only that the rates of the forward and reverse reactions are equal and that the amounts of reactant and product are constant.

Chemical equilibria can be upset by changes in the concentration of molecules on either side of the equation. For example, additional carbon dioxide entering the bloodstream from actively contracting muscle cells results in the formation of additional carbonic acid, thus shifting the equilibrium temporarily. Other factors, such as temperature, may also affect chemical equilibria.

⇌ ENZYMES AND CHEMICAL REACTIONS

Each chemical reaction in the cells of organisms requires a specific enzyme. In humans, each cell contains approxi-

▷ FIGURE 4–2 Anabolic Athlete? Officials at the 1988 Olympics in Seoul, South Korea, found traces of an anabolic steroid in the blood of Ben Johnson (center). Anabolic steroids stimulate protein synthesis in muscle tissue. Combined with exercise and training, they can result in an extraordinary physique and astounding athletic performance. Steroids may have adverse psychological and physiological side effects and are banned for use by Olympic athletes.

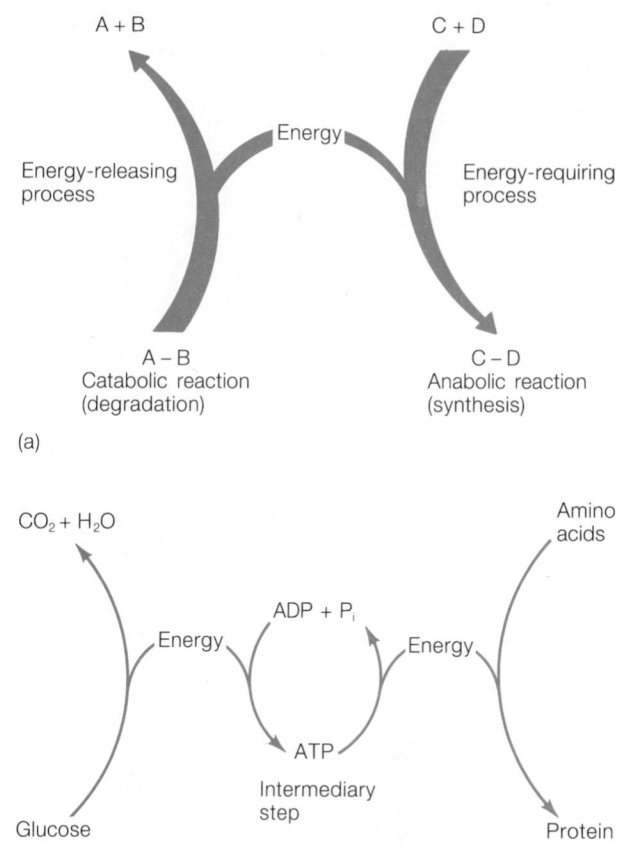

▷ FIGURE 4–3 Coupled Reactions (*a*) In cells, energy-releasing reactions are coupled to energy-requiring reactions. (*b*) Glucose breakdown, for example, is coupled to the synthesis of ATP from ADP and inorganic phosphate. ATP synthesis absorbs energy given off from the catabolic breakdown of glucose. The breakdown of ATP is coupled to the synthesis of protein from amino acids. However, the transfer of energy is never 100% efficient.

mately 2000 different enzymes. All told, there are an estimated 10,000 *different* enzymes in the human body.

Enzymes are essential to cellular metabolism. A missing enzyme in a metabolic pathway, for example, will shut down a process—in much the same way that a missing worker on an assembly line interrupts production. (For a discussion of the possible implications of such a defect, see Health Note 4–1.)

Active Sites of Enzymes Provide Specificity and Help Cells Regulate Their Reactions

Each enzyme is a large, globular protein with a small region known as the active site (▷ Figure 4–4a). The **active site** is an indentation, or pocket, where the chemical reaction occurs. Its shape corresponds to that of the **substrate(s),** the molecule(s) undergoing reaction. Substrates fit into the active site in much the same way that a hand fits into a glove. Because each enzyme has its own uniquely shaped active site capable of binding to one or a few chemical substrates, enzymes are said to be **specific.**

Specificity helps the cells regulate chemical reactions and maintain homeostasis. Imagine the biochemical disorder that would result if one enzyme regulated all of the thousands of chemical reactions occurring in the cell. By producing many different enzymes, each specific for one or—at the very most—a few reactions, the cell can control metabolism with an extraordinary degree of precision. Because of this, the cell produces only what it needs when it needs it, an evolutionary adaptation that conserves energy and cytoplasmic supplies of important biochemicals.

Enzymes Are Unchanged in Chemical Reactions and Can Be Used Over and Over

Enzymes belong to a class of compounds called **catalysts,** substances that speed up chemical reactions. Although they take an active role in the process, catalysts are unchanged by the reaction. Because they are unchanged, enzymes can be used over and over.

One familiar nonbiological catalyst is the catalytic converter in an automobile. The catalytic converter contains an inorganic catalyst that converts the poisonous carbon

DAIRY PRODUCTS LINKED TO OVARIAN CANCER

In recent years, scientists have shown that diet is a leading cause of cancer in humans. Naturally occurring carcinogens in the food we eat have become increasingly suspect. The latest chemical is galactose, a monosaccharide generated by the breakdown of lactose in the small intestine. Lactose is a disaccharide found in milk. Galactose itself is also found in a variety of milk products, including yogurt and cottage cheese, where it is produced from lactose by a bacterium.

Galactose is thought to cause ovarian cancer in certain women (see ▷Figure 1). A study of about 500 women (half of whom had ovarian cancer) suggests that eating yogurt or cottage cheese just once a month may predispose certain women to ovarian cancer. This study supports the results of animal research.

Ovarian cancer strikes approximately 20,000 women each year, but physicians warn that for most women, there is no reason to stop eating yogurt and other

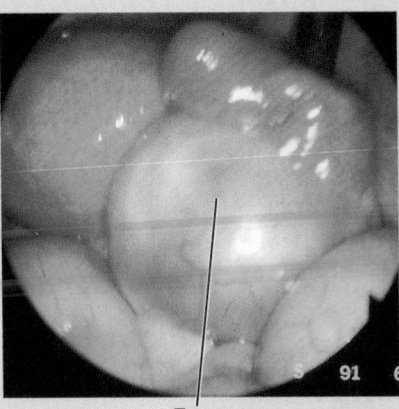

Tumor

▷ **FIGURE 1 Ovarian Tumor**

milk products. The study suggests that the only women affected by galactose are those with an abnormal galactose-catabolizing enzyme (called galactose-1-phosphate uridyltransferase). The defective enzyme results in higher-than-normal galactose levels, which may cause ovarian cancer.

How galactose could cause cancer remains a mystery. Galactose is a component of glycoproteins in the plasma membrane, which may function in contact inhibition—a process in which cells stop dividing when they contact other cells. When higher-than-normal levels of galactose are present, researchers hypothesize, glycoprotein synthesis may be impaired, thus decreasing contact inhibition.

The research linking ovarian cancer with a defective enzyme and consumption of certain milk products is the first of its kind. Critical thinking principles therefore suggest that we view these results with caution until additional studies are performed to determine whether this connection is real.

If further research supports these findings, a simple enzymatic test to assess galactose metabolism may become a routine medical procedure for women, saving thousands of lives each year.

---

monoxide in the car's exhaust into less harmful carbon dioxide.[1]

Enzymes Are Often Controlled by End Products

Unlike the catalytic converter in an automobile, biological catalysts can be regulated—that is, switched on or off. The most common regulators of enzyme activity are the end products of chemical reactions. These molecules often bind to specific regions on one of the enzymes in a metabolic pathway. The binding site is known as an **allosteric site.** Allosteric sites are molecular switches. When an end product binds to one, it turns the enzyme off, shutting down the entire metabolic pathway (Figure 4–4b). How does it act?

The binding of a chemical to an enzyme's allosteric site causes changes in the protein's structure that alter the shape of the active site. This change, in turn, prevents the

substrate from binding to the enzyme and is appropriately referred to as **end-product inhibition.**

In some metabolic pathways, end products may bind directly to the active site of an enzyme. This binding also blocks substrates from entering the site, inhibiting the enzyme and shutting down the metabolic pathway.

Control processes such as these are called feedback mechanisms. A **feedback mechanism** is any process that is regulated by its own end product(s). A familiar example is the furnace in a house. A furnace is controlled by a temperature sensor, the thermostat. When the temperature in the house drops below the desired setting, the thermostat sends a signal that turns on the furnace. Heat produced by the furnace raises the room temperature, and when the temperature reaches the appropriate level, the thermostat switches the furnace off. Heat, the product of the furnace, is therefore said to "feed back" on the system.

Enzymes' Active Sites Help Break Chemical Bonds

How do enzymes speed up catabolic reactions? As shown in ▷ Figure 4–5, substrates fit tightly into the active site of enzymes. The insertion of a molecule into an active site

---

[1] Even though current levels of carbon dioxide gas in the environment are not harmful to human health, the buildup of carbon dioxide in the atmosphere may be partly responsible for global warming, a gradual increase in global temperature, which is discussed in detail in Chapter 33.

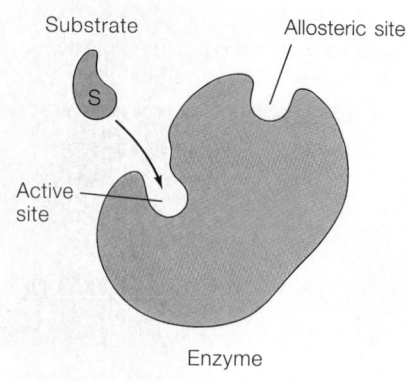

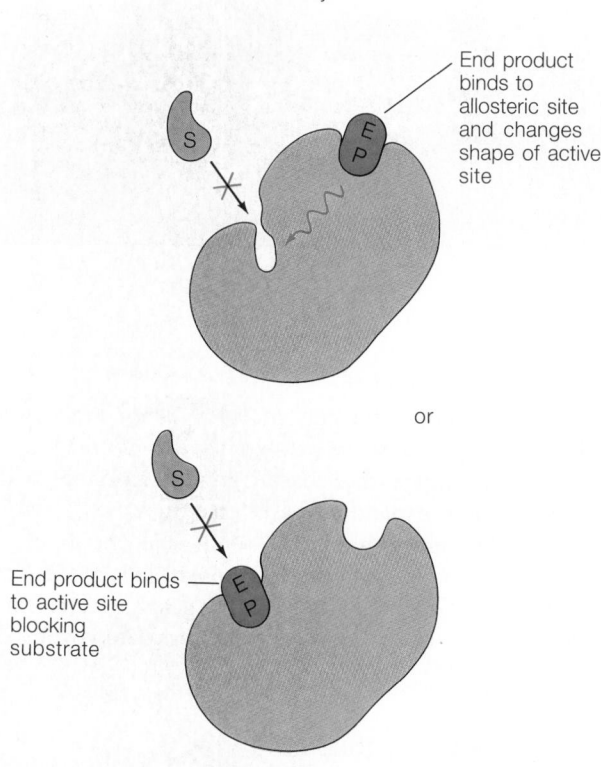

▷ **FIGURE 4–4 Enzyme Structure and Function**
(*a*) Three-dimensional drawing of an enzyme. The active site conforms to the shape of the reacting molecule(s). The allosteric site regulates the enzymes. When products bind to the allosteric site, they can turn the active site on or off, depending on the enzyme. (*b*) End products of a reaction can inhibit the enzyme by binding to the active site or by binding to the allosteric site.

causes the site to distort, in much the same way that a hand distorts a glove when it is inserted into it. The change in the shape of the active site strains the covalent bonds of the substrate molecule. If the strain is great enough, a bond may break, and the molecule will split apart.

Bond breakage in the absence of enzymes requires a considerable amount of energy, which might come from molecular collisions. In general, though, very little bond breakage occurs in organisms without enzymes. Enzymes, therefore, offer a site where bonds can be broken more readily with considerably less energy.

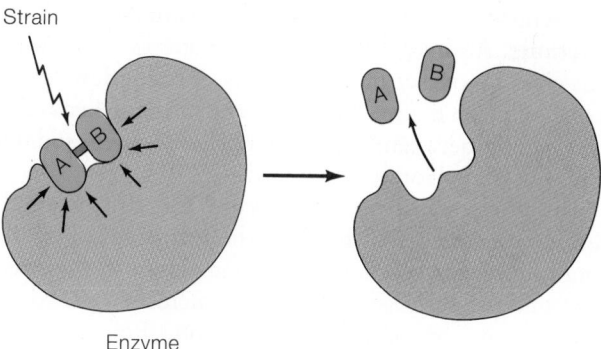

▷ **FIGURE 4–5 Enzyme-Regulated Cleavage** As the substrate molecule fits into the active site, it distorts the site slightly, which in turn puts a strain on the molecule, enabling bonds to break.

Enzymes Provide a Place for Molecules to React

When two molecules react to form a new molecule in the absence of enzymes, the molecules must first be energized. In the chemistry laboratory, anabolic reactions are generally initiated by heating reactants in a test tube. Heat accelerates the molecules, increasing the rate and force of their collisions. Forceful collisions bring the atoms of the molecules close enough to react.

Organisms cannot generate the heat required to increase collisions and force molecules to react. Instead, they must "rely" on enzymes, the active sites of which provide a place for molecules to come into close proximity and overcome repulsive forces that prevent them from reacting. Thus, enzymes eliminate the need for high-speed collisions and greatly enhance the rate of anabolic reaction. Active sites may also orient the reacting molecules in such a way as to increase the likelihood of chemical reaction.

Enzyme-catalyzed reactions occur many times faster than those in which enzymes are absent. Remarkably, calculations show that a single enzyme molecule can catalyze as many as 1000 reactions in a single second! Enzyme-catalyzed reactions occur 100 million to 10,000 million times faster than uncatalyzed reactions. Without them, most of the chemical reactions in the human body and the bodies of other organisms would occur much more slowly or, more likely, would not occur at all.

≈ OXIDATION AND REDUCTION REACTIONS

Many chemical reactions involve the transfer of electrons from one atom to another. In Chapter 2, for example, you learned that sodium and chlorine atoms react to form an ionic bond. As you may recall, a sodium atom gives up an outer-shell electron to a chlorine atom. In this reaction, the sodium atom becomes a positively charged sodium ion, and the chlorine atom becomes a negatively charged chloride ion. The electrostatic attraction between the two ions holds them together.

Scientists refer to such reactions as **oxidation-reduction reactions.** An **oxidation reaction** is one in which a reactant gives up one or more electrons. A **reduction reaction** is one in which the reactant gains one or more electrons. In the reaction between sodium and chloride, then, sodium is oxidized, and chlorine is reduced.

The terms *oxidation* and *reduction* can also be used to describe reactions that result in the formation of covalent bonds. In these reactions, oxidation refers to the loss of electrons or hydrogen atoms. (Remember, a hydrogen atom contains an electron and a proton.) Reduction refers to the gain of electrons or hydrogen atoms.

One well-known reduction reaction is the conversion of unsaturated fats to saturated fats by the addition of hydrogen atoms. Another important reduction reaction occurs in photosynthesis (Chapter 5). In this reaction, carbon dioxide molecules are joined to produce carbohydrates, such as glucose. Hydrogen atoms are added.

≈ ENERGY, ENZYMES, AND CELLULAR METABOLISM: A CLOSER LOOK

The previous material has shown how enzymes work—that is, how enzymes facilitate chemical reactions. However, it doesn't explain why reactions occur. To understand that, we must look a little more closely at energy and energy conversions, key considerations in whether a chemical reaction will occur.

Energy Can Be Converted from One Form to Another But Cannot Be Created or Destroyed

Energy exists in many substances. For example, wood contains a great deal of potential energy locked up in chemical bonds. When burned, wood releases heat and light, both forms of energy (▷ Figure 4–6). This process illustrates a key principle of energy—that is, that energy can be converted from one form to another. The study of energy conversions is called **thermodynamics.** Energy conversions follow two important laws.

The **first law of thermodynamics,** often called the **law of conservation of energy,** states that energy can be converted from one form to another but cannot be created or destroyed. The energy contained in a gallon of gasoline burned in an automobile is not destroyed but is simply converted into an equivalent amount of mechanical energy (movement) and heat (thermal energy). Whenever one form of energy is converted to another, some energy is always lost as heat. Thus, when a cell breaks down glucose to tap the potential energy in its chemical bonds, some heat is given off. In many systems, heat is generally diffuse and is therefore not very useful when it comes to performing work. The **second law of thermodynamics** is concerned with the useful work generated during an energy transformation. It is more subtle than the first law and can be stated in several ways. Let's consider one that will be helpful in understanding photosynthesis and cellular energy production.

▷ **FIGURE 4–6 The First Law of Thermodynamics**
Combustion releases the stored energy in the wood in this fire. The stored energy is converted to light and heat.

One way of stating the second law is that no energy conversion is 100% efficient. *In other words, never is the amount of energy in a system fully converted to work.* In a car engine, for example, the combustion of gasoline in the cylinders creates expanding gases that move the pistons, propelling the vehicle along the road. The energy of the moving automobile, however, is much lower than the chemical energy originally found in the gasoline. The "missing" energy has been "lost" as heat dissipated from the engine.

Energy Plays a Key Role in Chemical Reactions in the Cell

Cells are the scene of thousands of chemical reactions. As noted earlier, anabolic reactions require energy and are also called **endergonic reactions**. Because outside energy drives endergonic reactions, the products of the reaction have more energy than the reactants, as shown in ▷ Figure 4–7.

Catabolic reactions give off energy and are also known as **exergonic reactions.** In exergonic reactions, the reactants contain more potential energy than the products, as shown in ▷ Figure 4–8. The energy released during the reaction can be determined by subtracting the potential energy of the products from the potential energy of the reactants.

Although exergonic reactions give off energy, they require a little bit of energy to initiate them. For example, the combustion of gasoline is an exergonic reaction. During this reaction, hydrocarbons in the gasoline react with oxygen, forming carbon dioxide and water. For gasoline to burn, though, it must first be ignited. That is, a tiny amount of energy (supplied by a spark) is required to start the

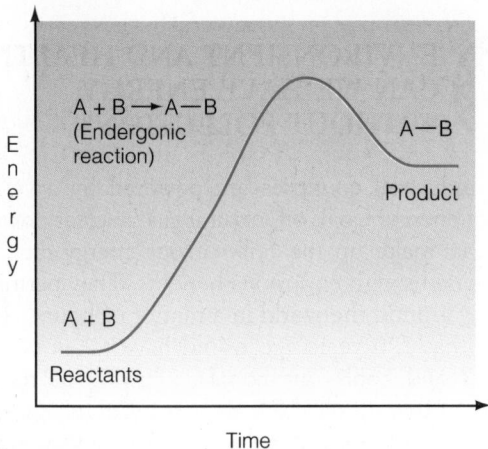

▷ FIGURE 4–7 Energy Profile of an Endergonic Reaction In an endergonic reaction, the energy of the products is greater than the energy of the reactants. This is possible because energy is added to the system.

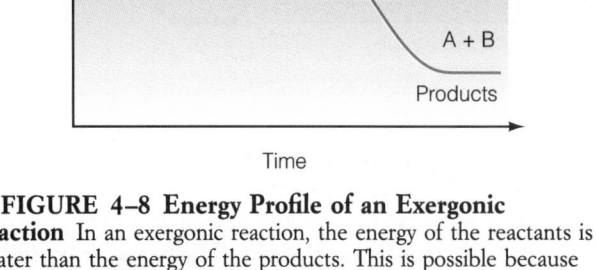

▷ FIGURE 4–8 Energy Profile of an Exergonic Reaction In an exergonic reaction, the energy of the reactants is greater than the energy of the products. This is possible because energy is given off, or released.

chemical reaction between oxygen and the hydrocarbons of the gasoline. Once the reaction has commenced, energy released from the reaction maintains it.

Figure 4–8 depicts the energy content of the molecule before, during, and after an exergonic reaction. As noted above, the initial reactants have more potential energy than the products. The energy that must be invested to make the reaction go is known as the **activation energy.** The energy released by the reaction is the **net energy yield.**

Endergonic reactions also require activation energy. Why? As pointed out in Chapter 2, atoms consist of positively charged nuclei and negatively charged electron clouds. For two atoms of different molecules to react and form a new covalent bond, their electron clouds must be forced together. That is, the atoms must overcome strong electrical repulsion. Activation energy overcomes this repulsive force and permits the endergonic reaction to occur.

To understand why energy is required to break internal chemical bonds in molecules, allowing new bonds to form, consider a hypothetical molecule, A—B. For this molecule to react, the existing bond must break. The activation energy destabilizes the chemical bond in A—B, causing it to break. The unstable condition of the molecules is called the **transition state.**

In the chemistry lab, the usual source of activation energy is the kinetic energy of movement. Only molecules moving with sufficient speed can collide with enough force to cause their electron clouds to combine or cause bonds to break. Because molecules move faster at higher temperatures, most reactions occur more readily when heat is added. In organisms, additional heat cannot be added to drive anabolic and catabolic reactions. Cells sidestep the activation energy requirement through the use of enzymes, which greatly lower the activation energy needed for anabolic and catabolic reactions, as shown in ▷ Figure 4–9.

ATP, an Energy Carrier Molecule, Links Many Exergonic and Endergonic Reactions in Cells

Earlier in the chapter, you learned that catabolic (exergonic) and anabolic (endergonic) reactions are coupled. Energy given off by exergonic reactions is used to drive endergonic reactions. But the energy is not transferred directly. It has to be shuttled from one set of reactions to the other by an energy-carrier molecule, adenosine triphosphate (Chapter 2). ATP consists of three subunits: a nitrogen-containing base, adenine; a monosaccharide, ribose; and three phosphate groups (▷ Figure 4–10).

The bonds joining the last two phosphate groups of ATP are special covalent bonds that house extraordinary amounts of potential energy. As a result, these bonds are

▷ FIGURE 4–9 Enzymes and Activation Energy Enzymes reduce the activation energy required for endergonic (shown here) and exergonic chemical reactions occurring in the body.

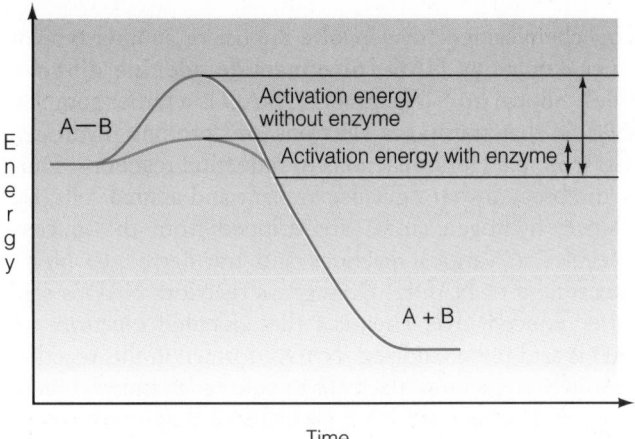

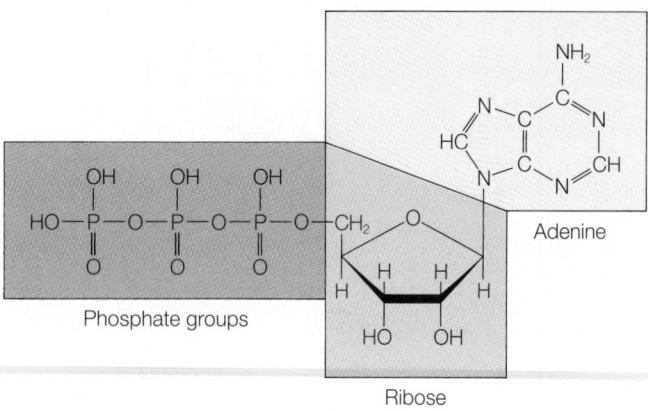

NH₂

Adenine

Phosphate groups

Ribose

▷ FIGURE 4-10 Structure of ATP

referred to as high-energy bonds. High-energy bonds require a lot of energy to create but are also fairly unstable—that is, easily broken. Under most circumstances, only the last high-energy bond in ATP is broken. When this bond is cleaved, ATP is converted to ADP and inorganic phosphate. The breakdown is accompanied by the release of stored energy, which is used by the cell to drive anabolic (endergonic) reactions.

In the cells of plants, animals, fungi, single-celled protists, and many bacteria, ATP is produced during the catabolic breakdown of glucose. ATP then transports this energy to many endergonic processes. One familiar endergonic process is the active transport of molecules across cell membranes. Protein synthesis on the RER is another.

ADP and P_i are recycled by the cell—that is, used to regenerate ATP. In the bacterium *E. coli,* found in the digestive tracts of mammals, 2.5 million ATP molecules are broken every second. Because each bacterium has only about 10,000 to 100,000 ATP molecules, each ATP must be recycled about 25 to 250 times per second! For this reason ATP serves as a short-term form of energy storage. More stable molecules such as glucose, glycogen, starch, or fat are used by cells to store energy for longer periods.

Oxidation and Reduction Reactions Are Coupled by NAD

Many chemical reactions involve the use of another type of carrier, known as **NAD** (**nicotinamide adenine dinucleotide**). Shown in ▷ Figure 4-11, NAD is a rather complex molecule that transports electrons and protons (hydrogen ions) from oxidation reactions to reduction reactions. During the breakdown of glucose in plant and animal cells, for instance, hydrogen atoms are stripped from the glucose molecules in various reactions and transferred to NAD, converting it to NADH. During this reaction, NAD is said to be reduced. The chemical that donated electrons to NAD is said to be oxidized. NADH formed in this reaction eventually gives up its electrons to another chemical, regenerating NAD. You will learn more about these reactions in Chapters 5 and 6.

ENVIRONMENT AND HEALTH: CAN WE HAVE ENERGY WITHOUT POLLUTION?

Modern industrial countries are powered by a variety of sources of energy. Coal, oil, natural gas, nuclear power, and hydropower make up the bulk of our energy diet. These fuels provide us with enormous benefits. They permit us to jet halfway around the world in a matter of hours, covering distances that 120 years ago would have taken years. Energy heats and cools our homes, allowing us to live in deserts and other inhospitable climates. Energy lets us melt iron ore and produce steel to fashion skyscrapers and bridges. It gives us the power to explore the far reaches of space.

Unfortunately, our use of energy is something of a Faustian bargain, for the extraction and use of energy threaten the health of the planet. The combustion of fossil fuels by industrial and nonindustrial nations, for example, releases billions of tons of carbon dioxide into the air each year. Carbon dioxide accumulates in the atmosphere, trapping heat like the glass in a greenhouse. As you will see in the next chapter and in Chapter 33, carbon dioxide (and other pollutants from human society) may change the climate of Earth—and the prospects for all organisms.

Fossil-fuel combustion produces two additional harmful pollutants—sulfur dioxide and nitrogen dioxide. Released into the atmosphere, these gases react with water vapor and sunlight energy, producing sulfuric acid and nitric acid, respectively. These acids rain down on the land, causing enormous damage to buildings, crops, streams, rivers, forests, fish, and wildlife.

Even nuclear energy, touted by some as a clean form of energy, has its problems. The extraction and processing of nuclear fuels, for instance, expose workers to dangerous levels of radiation. Today, millions of tons of radioactive waste from mining and milling operations litter the West, washing into streams and rivers or being blown by the wind. Radiation released from nuclear power plants is small under normal operating conditions, but accidents can release large amounts of harmful materials. The accident at the Chernobyl nuclear reactor in 1986, for example, killed over 200 people, some immediately, and others within a few years. Hundreds of thousands have been exposed to radiation released from the plant; over the next 50 years, some estimates suggest, 13,000 people will die of cancer caused by radiation from the accident.

Clearly, energy comes at a cost—not just to humans but to all species that inhabit the Earth. The fate of the planet's millions of species depends more than ever on finding alternative sources of energy that do not emit dangerous radiation or harmful gases. Fortunately, there are many clean alternatives. Perhaps the best is conservation—that is, using energy we have much more efficiently.

For some, calling energy savings a "source" of energy seems unjustified. However, for the same reason that a

The figure at the top of the page shows the structure of NAD.

$$Substrate \text{ (Reduced)} \longrightarrow Substrate \text{ (Oxidized)}$$

$2e^- + 2H^+$

NAD$^+$

dehydro-genase

H$^+$

NADH

▷ **FIGURE 4–11 Structure of NAD** NAD shuttles electrons and protons (hydrogen atoms) from oxidation to reduction reactions. When a hydrogen is released from a substrate its proton (H$^+$) and electron (e$^-$) are transferred to NAD. See the text for an explanation.

penny saved is the same as a penny earned, every kilowatt of energy we save is the same as a kilowatt of new energy discovered. Consider one example. The high-efficiency light bulbs in my house produce the same amount of light as typical 75-watt incandescent bulbs, but they require only 18 watts of energy. In essence, each bulb is a tiny 57-watt power plant. That's how much energy it conserves, freeing the energy for later use.

In the long run, however, an environmentally compatible energy supply will require new sources, like solar energy, wind energy, and biomass. As you will see in Chapter 33, the potential of these energy sources is enormous. In addition, these sources offer us a way to live sustainably on the Earth, coexisting with the millions of other species that evolved long before we did.

EXERCISING YOUR CRITICAL THINKING SKILLS

Water is an essential ingredient of life for all organisms. It is a major component of our bodies and supplies habitat for many species. Water conservation programs are therefore important to humans, because they help ensure an adequate supply in times of drought. They are also important because they ensure adequate supplies for future generations. Human water conservation programs are vital means of protecting the homes of the species that share the planet with us.

Members of a family living near Lake Superior are interested in helping protect the environment, but when someone suggests that they could use water more efficiently, they disagree. They maintain that they do not need to practice water conservation. Their argument is this: the lake provides plenty of water and is not in danger of being depleted, so why should they use water efficiently?

Using your critical thinking skills, analyze this issue. Are they right? Stop and think about this for a while before reading further.

Now that you have thought about the issue, what have you concluded? Ask yourself the following questions: Has this family failed to examine the bigger picture—for example, the other benefits of water conservation. What are some of those benefits?

THE NATURE OF CHEMICAL REACTIONS

1. Molecules form the structures of the cells and bodies of all organisms. Many molecules also participate in reactions, which are essential for growth, reproduction, development, and homeostasis.

2. The chemical reactions in the body constitute metabolism. Those metabolic reactions in which chemicals combine with others to form new molecules are synthetic, or anabolic, reactions, and those reactions in which chemicals are split apart, or dissociate, are called catabolic reactions.

3. Most of the energy required by body cells comes from the catabolic breakdown of carbohydrates and fats. The energy released from these chemicals ultimately comes from the sun via plants, which capture sunlight energy and use it to synthesize organic molecules, such as glucose.

4. In the cells of the body, energy released from catabolic reactions is used to drive anabolic reactions. Such reactions are said to be coupled.

5. ATP (adenosine triphosphate) shuttles energy between energy-yielding reactions and energy-requiring reactions.

6. Most chemical reactions are reversible. When the speed of the forward reaction equals the speed of the reverse reaction, the reaction is said to have reached chemical equilibrium.

ENZYMES AND CHEMICAL REACTIONS

7. Many chemical reactions occur as part of elaborate metabolic pathways. A metabolic pathway is a series of reactions, each catalyzed by its own enzyme.

8. Enzymes are protein molecules that speed up the rate of anabolic and catabolic reactions.

9. Each enzyme is specific for one or a few reactions. Specificity is conferred by the shape of the active site, a region of the enzyme that binds to the reactants.

10. Enzyme activity is often regulated by levels of the by-products of metabolic pathways. The active sites can be physically blocked by the products of chemical reactions. Enzymes also contain allosteric sites that bind to products. Binding may activate or deactivate the enzyme.

OXIDATION AND REDUCTION REACTIONS

11. Many chemical reactions involve the loss or gain of electrons or hydrogen atoms. A reaction in which electrons or hydrogen atoms are lost is called an oxidation reaction. A reaction in which a molecule gains electrons or hydrogen atoms is called a reduction reaction.

ENERGY, ENZYMES, AND CHEMICAL REACTIONS: A CLOSER LOOK

12. Energy is defined as the capacity to do work. Energy is essential to living things. It is used to synthesize protein, to move molecules across cell membranes, to contract muscles, and to do much more.

13. In biology, we concern ourselves principally with potential energy, or stored energy. Energy is stored in chemical bonds and can be released by cells to perform many functions.

14. Energy exists in many forms. The first law of thermodynamics, the law of conservation of energy, states that energy is neither created nor destroyed but is merely transformed from one form to another.

15. The second law of thermodynamics states that when energy is converted from one form to another, the conversion is never 100% efficient. Heat is a by-product of all energy conversions.

16. Two types of chemical reactions occur in organisms: endergonic reactions, in which energy is absorbed, and exergonic reactions, in which energy is given off.

17. In endergonic reactions, the products of the reaction have more energy than the reactants; outside energy is used to drive these reactions.

18. In exergonic reactions, the reactants give off energy; the products, therefore, have less energy than the reactants.

19. Both endergonic and exergonic reactions require a small additional amount of energy to make them go—activation energy.

20. Exergonic and endergonic reactions in cells are coupled, so that the energy given off by one can be used to drive the other.

21. Energy is shuttled from exergonic to endergonic reactions by energy-carrier molecules—most notably, ATP.

22. ATP is synthesized from ADP and inorganic phosphate, using energy given off from exergonic reactions.

23. Energy is stored in the terminal phosphate bond of ATP and is released when the bond is broken down. The energy released by the breakdown of ATP is used to drive endergonic processes.

24. Oxidation and reduction reactions are often linked by a molecule called NAD. It shuttles electrons and protons (hydrogen atoms) from one reaction to another.

ENVIRONMENT AND HEALTH: CAN WE HAVE ENERGY WITHOUT POLLUTION?

25. Modern society depends on energy, but many of the sources of energy we rely on threaten the health and well-being of the planet.

26. Carbon dioxide from fossil fuels such as coal, oil, gasoline, and natural gas appears to be warming the planet's atmosphere. Heating may occur so quickly that it destroys many species.

27. Sulfur dioxide and nitrogen dioxide emitted from the combustion of coal and other fossil fuels are also troublesome pollutants. In the atmosphere, they are converted into sulfuric acid and nitric acid, which fall in rain and snow. Acid deposition accelerates the deterioration of buildings and statues and damages forests, crops, rivers, lakes, fish, and wildlife.

28. Nuclear power releases radioactive waste into the environment.

29. To protect the planet and the many species that share it with us will require more environmentally benign sources of energy, including solar energy, wind, and biomass.

1. Chemical reactions in organisms consist of two types, those in which chemicals are broken down, known as _____ reactions, and those in which chemicals are synthesized, known as _____ reactions.
2. Exergonic reactions _____ energy, and endergonic reactions _____ energy. These reactions are linked, or _____ , so that energy given off by one can be used by the other.
3. The chemicals on the left-hand side of a chemical equation are called _____ .
4. The speed of the forward and reverse reactions are equal at a point called chemical _____ .
5. Enzymes speed up the rate of chemical reactions. They belong to a group of chemicals known generally as _____ .

The depression, or pocket, in an enzyme in which a chemical reaction occurs is called the _____ .
6. Each enzyme is uniquely shaped, fitting only one or a few similar substrates. This property, called _____ , helps cells regulate chemical reactions in a precise way.
7. A molecule produced by an enzymatically controlled reaction sometimes binds to a special site on the enzyme, called the _____ site, thus helping regulate enzyme activity.
8. The production of carbohydrates from carbon dioxide and water using sunlight energy is called photosynthesis and is an example of a(n) _____ reaction.

9. The addition of hydrogen atoms to a molecule is called _____ .
10. Energy is defined as the capacity to do _____ .
11. The first law of thermodynamics is also known as the law of _____ of _____ . It states that energy can be neither _____ nor destroyed but can be _____ from one form to another.
12. A small amount of energy is required to initiate an endergonic or exergonic reaction. Enzymes accelerate these reactions because they _____ the _____ energy.

Answers to the Test of Terms are found in Appendix B.

1. List some of the cellular processes that require energy.
2. Define the terms *anabolic* and *catabolic*. Give some examples of each.
3. What does it mean when it is said that two reactions are coupled?
4. Define the term *chemical equilibrium*.
5. Carbon dioxide reacts in water to form carbonic acid, which dissociates to form bicarbonate and hydrogen ions. Describe what happens after carbon dioxide is dissolved in water. When is an equilibrium set up? What factor(s) can upset the equilibrium? Are the concentrations of products and reactants on both sides of the

equation the same at the equilibrium point?
6. Describe the structure and function of enzymes. Be sure to include the following terms: specificity, active site, allosteric site, end-product inhibition, and catalyst.
7. How do enzymes lower the activation energy required for catabolic and anabolic reactions?
8. Define the terms *oxidation* and *reduction*.
9. What is the first law of thermodynamics? Give some examples. Why

is this law relevant to biological systems?
10. Describe the second law of thermodynamics. What implications does it have for biological systems?
11. Draw energy graphs for an endergonic reaction and an exergonic reaction. Label the activation energy and net energy yield. Explain the need for activation energy. How do enzymes affect activation energy?
12. Describe the role of ATP as an energy-carrier molecule. Why is ATP considered a short-term storage form of energy?
13. What chemical reactions require NAD? Why?

Energy from Sunlight: Photosynthesis

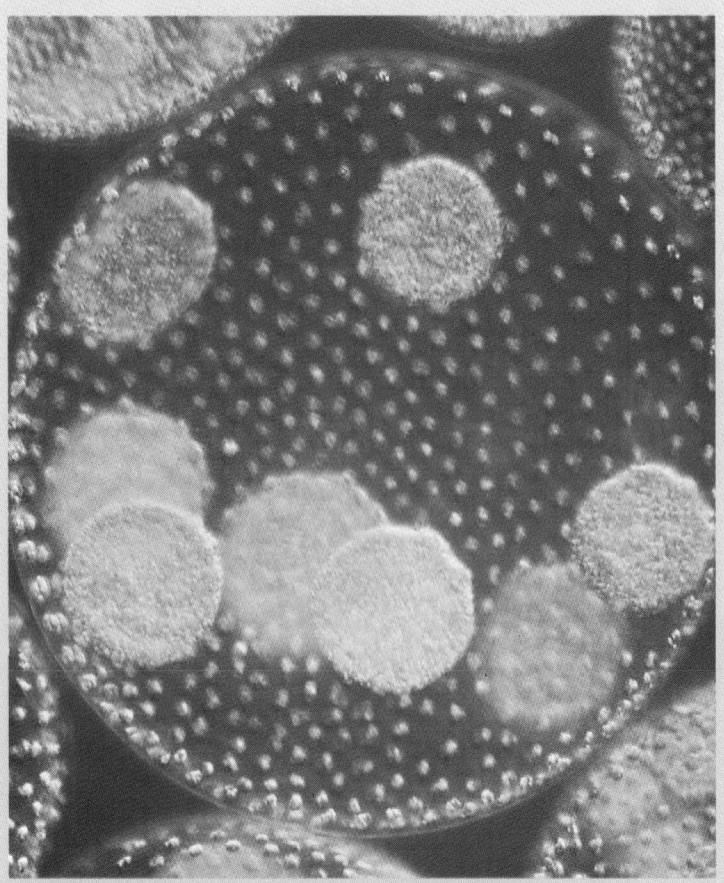

Photomicrograph of a colony of algal cells, known as volvox. The sphere consists of many smaller photosynthetic cells.

U nbeknownst to many, the sun is the source of virtually all energy on Earth. Located 93 million miles from the Earth, the sun supplies energy to the smallest microorganisms, the tallest trees in the tropical rain forest, and everything in between. Without sunlight, life on Earth could probably not exist.

Like the vast majority of other species, humans depend heavily on the sun's radiant energy. In fact, every ounce of energy we burn in our bodies came from sunlight energy beamed to Earth. Surprisingly, even most of the energy we use in modern society originates in the sun ($\triangleright$ Figure 5–1).

With few exceptions, sunlight energy must first be intercepted and converted to a form useful to life. The interceptors are members of the plant kingdom and certain photosynthetic monerans (cyanobacteria) and protists, a vast and diverse array of organisms specially adapted to trap sunlight energy.

$\triangleright$ Figure 5–2 traces the path that energy takes from the sun to its end use. In each case, the intermediary between the sun and the end use is a plant or some other photosynthetic organism. As you will soon see, plants capture sunlight energy and use it to produce organic food molecules. These molecules are then combined to produce complex carbohydrates, such as starch and cellulose, and other organic molecules. Animals acquire organic nutrients by eating plants or other animals, which often eat plants.

The energy that plants absorb from the sun goes into making covalent bonds and remains trapped in these bonds until it is liberated by enzymatic reactions in cells or by combustion—for example, of coal. Mined in the United States and many other countries, coal was formed from plants that lived 250 to 325 million years ago. During this period, coal-producing regions supported extraordinary plant growth. Leaves and dead plants fell into inland swamps and other shallow waters, accumulated in thick layers on the bottom, and eventually became covered with sediment. Over time, heat and pressure converted the dense mat of plant matter into peat, then into coal. Despite these dramatic changes, the solar energy trapped within the molecules of the original plants remained. Today, when coal is burned in a power plant, the sunlight energy locked up for many millions of years is finally released as light and heat. The heat is used to drive turbines that produce electricity, which powers refrigerators, computers, and electric motors. The light illuminating this page from a nearby lamp is very probably ancient solar energy that struck the Earth millions of years ago.

This chapter examines the process of photosynthesis, showing how plants capture sunlight and also how they use it to synthesize a wide range of biologically important organic molecules.

≈ AN OVERVIEW OF PHOTOSYNTHESIS

Photosynthesis is one of the most fascinating and important processes encountered in biology. Why is it so important?

First, photosynthesis provides plants with a ready supply of energy-rich carbohydrates, the most important being glucose. Glucose molecules are broken down in plant mitochondria to produce ATP.[1]

Second, photosynthesis produces organic building blocks of the larger molecules, such as starch and cellulose, that serve as structural components of plant cells. Fats, proteins, nucleic acids, and other molecules in plants also owe their origin to photosynthesis.

Third, photosynthesis produces oxygen, as shown in the following chemical reaction:

$$6\,CO_2 + 6\,H_2O + \text{sunlight energy} \rightarrow C_6H_{12}O_6 + 6\,O_2$$

In the closed system of the biosphere, in which the only input of significance is the sun, all materials are recycled. Oxygen is no exception. Within the mitochondria of plants and animals, oxygen produced during photosynthesis is used in the breakdown of glucose. The metabolic pathways that catabolize glucose, in turn, liberate the energy captured during photosynthesis. This energy is stored briefly in ATP molecules before being tapped for cellular metabolism, transport, and a variety of other functions.[2]

Finally, photosynthesis is important because it consumes carbon dioxide, a waste product of energy catabolism in plants, animals, fungi, and microorganisms.[3]

$\triangleright$ **FIGURE 5–1 Sunlight Powers the World** Virtually all energy is derived from the sun. The energy contained in cereals, grain, coal, meat, and gasoline was once sunlight energy.

[1] Photosynthesis also produces ATP directly.
[2] Energy released from the mitochondria is important to all plant cells, especially those in the roots, which are unable to photosynthesize.
[3] A few organisms rely on a process called chemosynthesis. Instead of acquiring energy from sunlight, these microorganisms acquire energy from the breakdown of certain molecules, such as hydrogen sulfide.

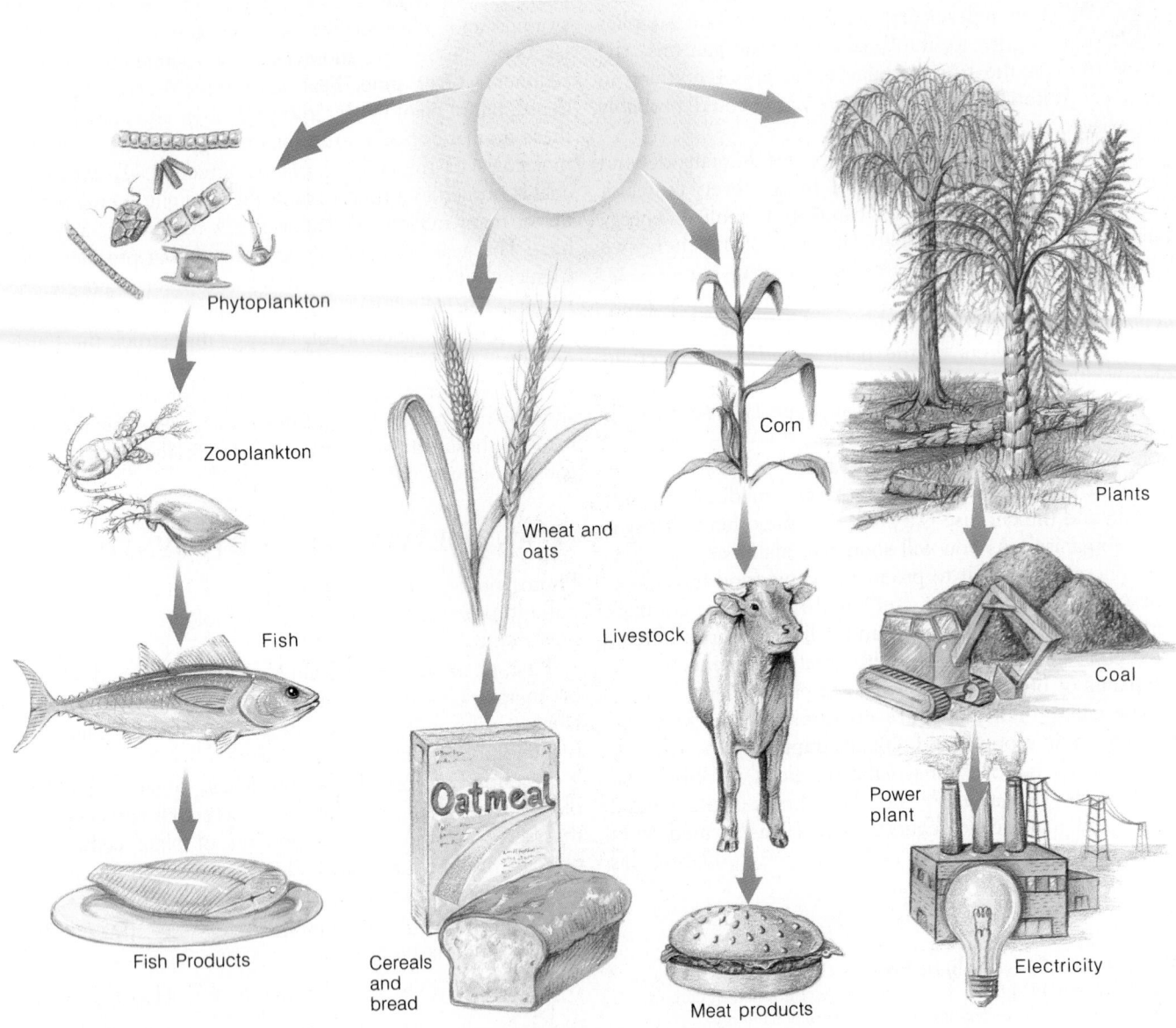

FIGURE 5–2 How Sunlight Energy makes Its Way to Us Plants capture sunlight energy and trap it in organic matter. It then makes its way to humans and other organisms through plant material, animal matter, or coal and oil.

Photosynthesis Takes Place in the Chloroplast in Plants and Protists

The **chloroplast** is a cellular organelle found in single-celled protists (for example, single-celled algae) and in certain cells of plants—for example, the cells in leaves and stems (▷ Figure 5–3). As noted in Chapter 3, chloroplasts contain pigments that capture sunlight energy and enzymes used to synthesize organic molecules from atmospheric carbon dioxide and water.

The chloroplast has a double outer membrane, which encloses a protein-rich, gel-like material, the **stroma** (Figure 5-3). Within the stroma are numerous flattened membranous sacs, the **thylakoid disks,** which are stacked on

one another like poker chips. Each stack is called a **granum** (plural, *grana*).

Photosynthesis Consists of Two Separate But Interdependent Sets of Reactions

Photosynthesis consists of two interdependent parts. The first part is a series of reactions that captures sunlight energy. These reactions are referred to as the **light-dependent reactions,** or, simply, the **light reactions** (▷ Figure 5–4). The second part of photosynthesis consists of a series of chemical reactions responsible for the synthesis of energy-rich carbohydrate molecules, known as

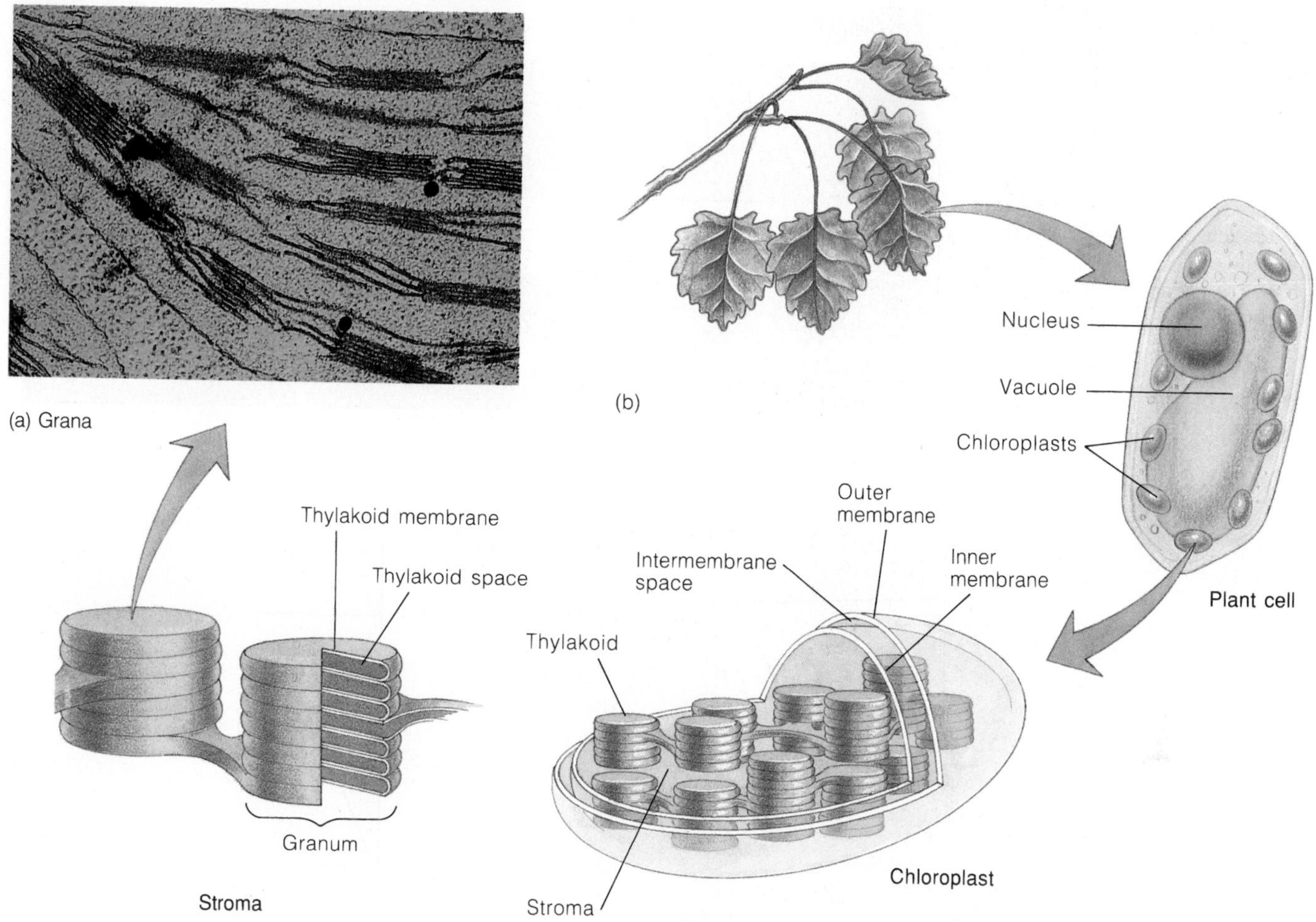

(a) Grana

(b)

Nucleus

Vacuole

Chloroplasts

Plant cell

Thylakoid membrane

Thylakoid space

Outer membrane

Intermembrane space

Inner membrane

Thylakoid

Granum

Stroma

Stroma

Chloroplast

▷ **FIGURE 5–3 The Chloroplast** (*a*) Transmission electron micrograph of the chloroplast. (*b*) Artist's rendition showing the grana and thylakoid disks.

the **light-independent reactions,** or **dark reactions** (Figure 5–4), so named because they are not directly dependent on sunlight. In other words, they can occur in either light or dark conditions as long as carbon dioxide and ATP are available. It is during the dark reactions that chloroplasts produce glucose and other carbohydrates. We begin our study of photosynthesis by examining sunlight and looking at the light-dependent reactions.

≋ THE NATURE OF SUNLIGHT

Sunlight is the driving force behind photosynthesis. As shown in   Figure 5–5a, sunlight is a mixture of many wavelengths of electromagnetic radiation, covering a wide spectrum from radio waves to visible light to gamma rays. All forms of electromagnetic radiation travel in waves (Figure 5–5b). The distance between two wave peaks constitutes the **wavelength.** The closer the peaks are to one another, the shorter the wavelength (Figure 5–5c). The shorter the wavelength, the higher the energy of any given form of electromagnetic radiation.

As shown in Figure 5–5a, only a small portion of the sun's emission spectrum is visible to animals, and it is

appropriately called **visible light.** High-energy, shorter-wavelength radiation, such as gamma rays, X-rays, and ultraviolet light, is invisible to most organisms, as is low-

▷ **FIGURE 5–4 Summary of Photosynthesis**

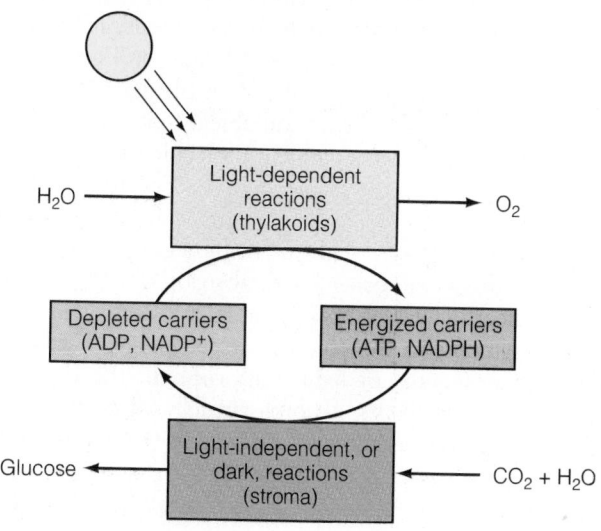

CHAPTER 5   *Energy from Sunlight: Photosynthesis*   **103**

▷ **FIGURE 5–5**
Electromagnetic Radiation
(*a*) Electromagnetic radiation from the sun consists of many wavelengths, ranging from radio waves to gamma rays. (*b*) All electromagnetic radiation travels in waves. Wavelength is the distance from one peak to the next. (*c*) The longer the wavelength, the lower the energy.

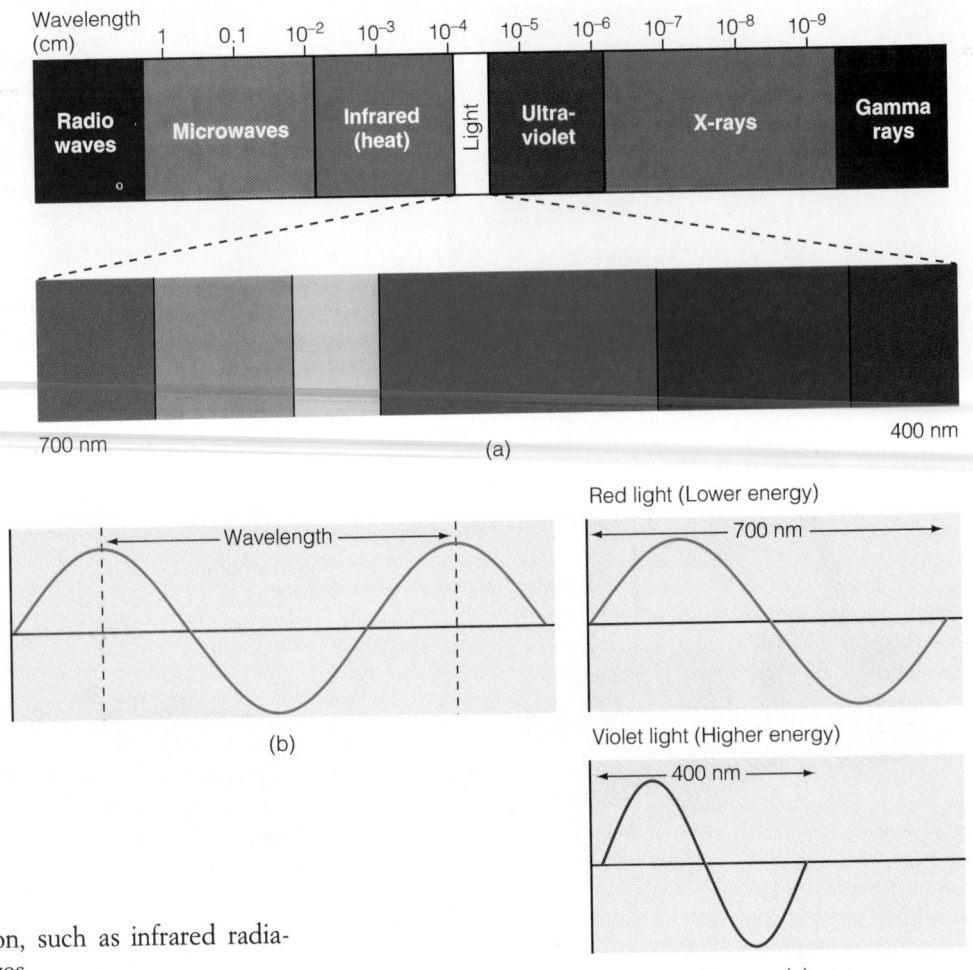

energy, long-wavelength radiation, such as infrared radiation, microwaves, and radio waves.

Visible light from the sun appears more or less colorless, but when it passes through a prism, it reveals itself to be a rainbow of colors (▷ Figure 5–6). As you can see from the spectrum, the shortest wavelength of visible light seen by humans is violet light; the longest is red.

≈ THE LIGHT-DEPENDENT REACTIONS: CAPTURING SUNLIGHT ENERGY

Sunlight energy striking the leaves of plants is absorbed by pigment molecules embedded in the membranes of the thylakoid disks of the chloroplast. Objects on Earth absorb and reflect different wavelengths of visible light, depending on the pigments they contain. An object that appears red to us, for example, contains pigment molecules that absorb all of the wavelengths of visible light except red, which it reflects (▷ Figure 5–7). The pigments in a blue object absorb all of the wavelengths of visible light except blue. A black object, on the other hand, absorbs all light. A white object reflects all wavelengths; that is, it reflects white light.

The pigments in the chloroplasts of plants absorb very specific wavelengths of light. **Chlorophyll,** the dominant pigment in most plants, absorbs virtually all of the visible light spectrum except green and yellow (▷ Figure 5–8). The abundance of chlorophyll in a leaf is therefore responsible for the lush green color of tree leaves and most plants.

Another group of pigments, found in smaller amounts, is the **carotenoids.** These pigments reflect either yellow or red light, depending on their chemical structure, but are frequently masked by the more abundant chlorophyll. During the fall, the carotenoids make a brief but splendid

▷ **FIGURE 5–6 The Rainbow** Prisms break visible light into a rainbow of constituent colors.

> **FIGURE 5-7 Pigments and Color** Pigments absorb and reflect visible light. The paint on this shiny car absorbs all colors except red, which it reflects.

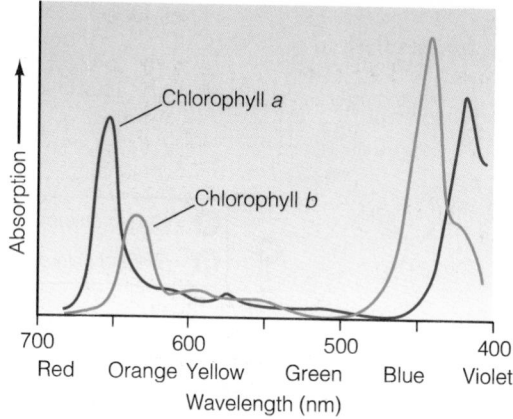

> **FIGURE 5-8 Absorption Spectrum for Chlorophyll** Note that chlorophyll a and b absorb virtually all light except for green and yellow, which they reflect. The height of the peaks indicates the amount of absorption.

appearance. The flourish of autumn colors is the result of a decline in chlorophyll production caused by decreasing daylight. As chlorophyll production drops off, the chlorophyll content of cells in the leaves falls, and the brilliant carotenoids become visible (▷ Figure 5–9).

Besides being the dominant pigment, chlorophyll is also the dominant player in photosynthesis. As illustrated in ▷ Figure 5–10, each chlorophyll molecule consists of a large ring structure, the **porphyrin ring,** containing magnesium. The porphyrin ring is a light-absorbing part of the molecule and is attached to a long hydrocarbon chain.

Two types of chlorophyll are found in plants, chlorophyll a and chlorophyll b, differing only very slightly in their structure (Figure 5–10). This difference, in turn, accounts for a noticeable disparity in their absorption spectra—that is, a difference in the wavelengths of the visible spectrum they absorb. Chlorophyll a participates directly in the light-absorbing reactions of photosynthesis, which convert solar energy into chemical energy. Chlorophyll b does not sit idly by, however. Lying alongside chlorophyll a molecules in the membranes of the thylakoid disks, it captures visible light within its absorption spectrum and transmits the energy to chlorophyll a molecules.

Also located in the membranes of the thylakoid disks, the carotenoids absorb still other wavelengths outside the "grasp" of the chlorophyll molecules, and they therefore serve to broaden the spectrum of colors that can power photosynthesis.

As a side note, photosynthesis also occurs in cyanobacteria. These monerans lack chloroplasts but contain many of the same pigments, which are embedded in layers of internal membranes within the bacterial cytoplasm.

The Photosystems Are Aggregations of Pigments and Other Molecules That Capture Sunlight Energy

The chloroplast is often likened to a solar collector. Sunlight collection in the chloroplast, as noted above, is made possible by the pigments in the membranes of the thylakoid disks. These membranes contain aggregations of three different molecules: chlorophylls, carotenoids, and special proteins, described below. These aggregations are called photosystems.

Photosystems are light-gathering, or light-harvesting, complexes, akin to the solar cells that generate electricity from sunlight (▷ Figure 5–11a). Sunlight strikes the pigments of the photosystems and is absorbed by them. The absorbed energy is then passed from one molecule to the other in the aggregation, eventually being "funneled" to a region of the photosystem called the **reaction center** (Figure 5–11b). In the reaction center is a molecule of chlorophyll a, which absorbs the energy it receives from outlying molecules and responds by emitting an electron.

To understand why chlorophyll reacts to energy input by ejecting an electron, take a brief look at the atom. As

> **FIGURE 5-9 Autumn Colors** These bright aspen leaves have recently turned from green. As day length decreases, chlorophyll production declines, unmasking the carotenoid pigments also found in the leaf.

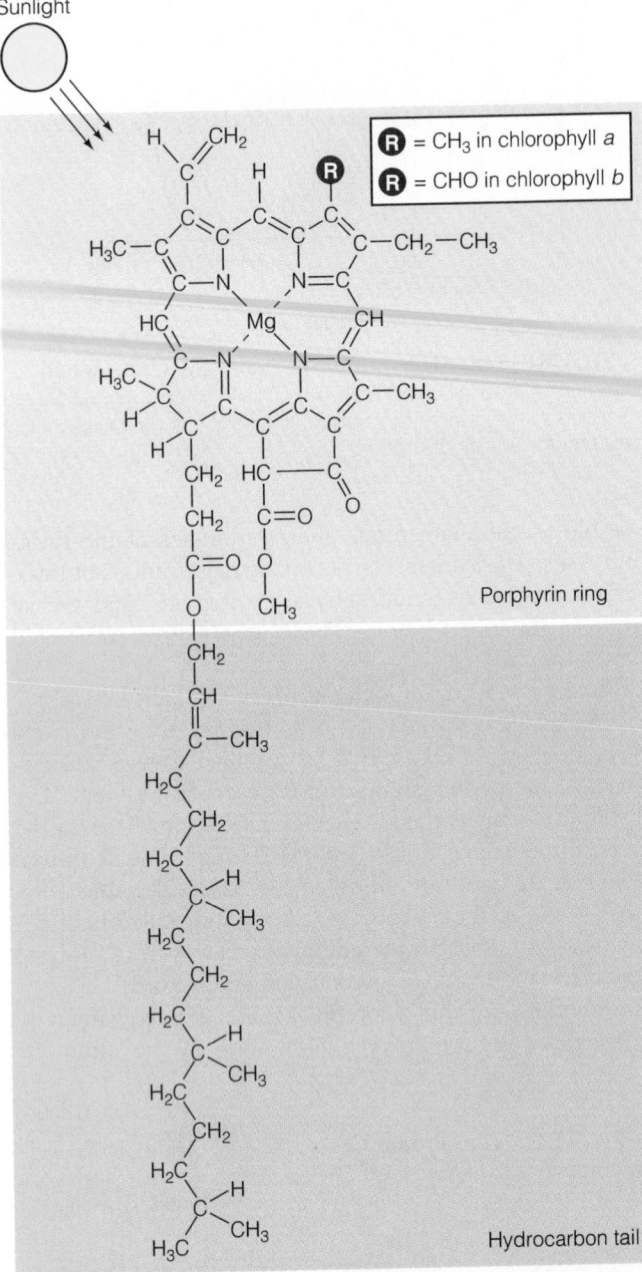

Sunlight

R = CH₃ in chlorophyll *a*
R = CHO in chlorophyll *b*

Porphyrin ring

Hydrocarbon tail

▷ **FIGURE 5–10 Molecular Structure of Chlorophyll a and b.**

Chapter 2 noted, all atoms are orbited by energetic particles called electrons. These electrons contain a certain amount of energy and occupy a given energy level (or shell) within the atom. When excited by an outside source of energy, electrons can be boosted out of their shell to a higher energy level. A good example of this phenomenon is the incandescent light bulb.

Electricity running through the tungsten filament of an incandescent light bulb imparts energy to the electrons of the tungsten atoms. This energy boosts the electrons of the tungsten atoms into higher energy shells. These electrons soon fall back into their previous shells, releasing the energy they absorbed from the electrical current. The released energy is emitted from the bulb as light and heat.

In the chloroplast, sunlight energy absorbed by chlorophyll a in the reaction center is so great that the electron is actually ejected from its orbital. This high-energy electron is then picked up by a protein molecule in the photosystem, known as the **primary acceptor molecule.** The primary acceptor molecule passes the electron to a molecule of $NADP^+$, which is converted to NADPH. NADPH, as you will soon see, is used to synthesize carbohydrates in the light-independent reactions.

Two Photosystems Are Involved in the Light Reactions

As Figure 5–11 shows, two types of photosystem are found in plant cells, **photosystem I** and **photosystem II,** named in the order of their discovery. Both systems are connected by an intervening series of protein molecules, also embedded in the membrane of the thylakoid disk. These proteins comprise the **photoelectron transport system,** or, more commonly, the **electron transport system (ETS).** To understand how this entire system operates, we begin with photosystem II.

Light penetrates the leaves of plants in packets, or units, of energy known as **photons** (▷ Figure 5–12). Light striking the pigments of photosystem II causes electrons to be emitted from the reaction center. Boosted out of their shells, the electrons are immediately given over to a nearby protein molecule, known as **PQ (plastoquinone),** shown in Figure 5–12. Plastoquinone molecules transfer the high-energy electrons to the electron transport chain. Electrons, therefore, soon find themselves traveling along a series of connected proteins bridging the two photosystems. As the electrons travel down the electron transport chain, they lose energy, which the chloroplast uses to generate ATP.

Because light streams into the plant cell continuously during daylight hours, photosystem I is also illuminated, and electrons are emitted concurrently from reaction centers in both photosystems. The electrons given off by photosystem I, however, are passed to an iron-containing protein molecule known as **ferredoxin.** Also embedded in the membrane near photosystem I, ferredoxin rapidly transfers the electrons to other carrier proteins and eventually to $NADP^+$, forming NADPH. NADPH is a phosphorylated form of NADH.[4]

NADPH is a reducing agent. It diffuses from the thylakoid disks into the stroma of the chloroplast, where it provides hydrogen atoms needed to reduce atmospheric carbon dioxide in the light-independent, or synthetic, reactions of photosynthesis. After relinquishing its electron and hydrogen ion, $NADP^+$ is available for reuse in photosystem I.

[4] NADPH production requires the addition of two electrons and two H^+ ions (protons), which—as you will soon see—comes from the splitting of water in photosystem II.

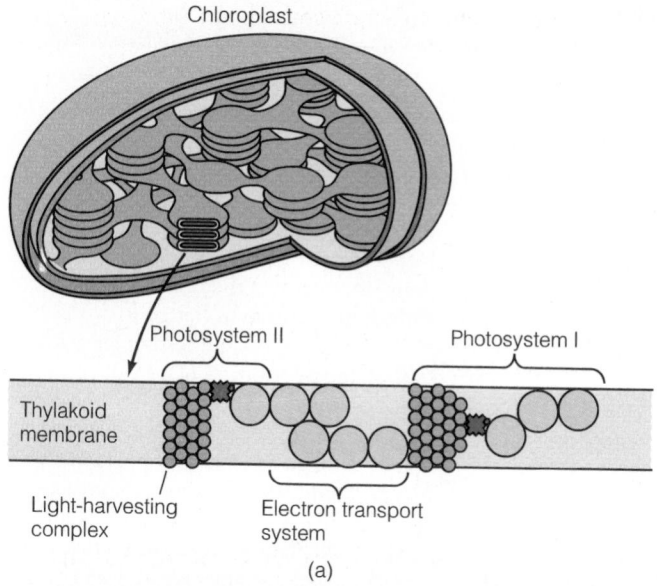

Chloroplast

Photosystem II

Photosystem I

Thylakoid membrane

Light-harvesting complex

Electron transport system

(a)

Sunlight

Reaction center (chlorophyll a molecule)

2e⁻

Electrons ejected from reaction center

Energy passed from molecule to molecule

Chlorophyll and accessory pigment molecules of photosystem

(b)

▷ **FIGURE 5–11 The Photosystems** (*a*) Light energy is absorbed by pigments of the two photosystems embedded in the membrane of the thylakoid disk. (*b*) Light energy is transmitted from pigment molecule to pigment molecule until it reaches the reaction center. Chlorophyll a in the reaction center absorbs the energy, emitting two electrons in the process.

The photosystems and the intervening electron transport system form an elaborate apparatus to capture sunlight and create a stream of electrons that flows to NADP⁺ molecules. This system runs perfectly well as long as sunlight is available and as long as there is some way to fill the electron deficiencies created in the reaction centers.

As illustrated in Figure 5–12, water molecules provide a ready source of electrons needed to fill electron holes in photosystem II. Enzymes within the chloroplast break down water molecules in the following reaction:

$$H_2O \rightarrow 1/2\ O_2 + 2\ H^+ + 2e^-$$

The electrons given off in this reaction are donated to chlorophyll a of the reaction center of photosystem II, and as long as water and sunlight are available, the light reactions can continue.

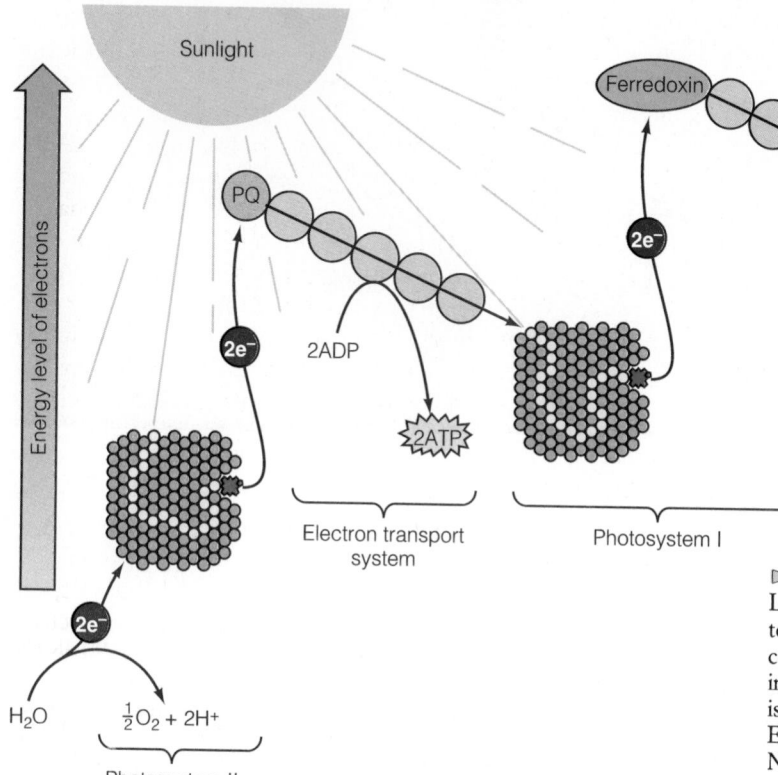

Sunlight

Energy level of electrons

PQ

2ADP

2ATP

2e⁻

Ferredoxin

NADPH

2e⁻

NADP⁺

2e⁻

Electron transport system

Photosystem I

H₂O

$\frac{1}{2}O_2 + 2H^+$

2e⁻

Photosystem II

▷ **FIGURE 5–12 Function of the Photosystems**
Light absorbed by both photosystems causes electrons to be ejected from chlorophyll a molecules in the reaction centers. Electrons from photosystem II fill electron holes in photosystem I. The electron hole in photosystem II is filled by the enzymatic splitting of water molecules. Electrons ejected from photosystem I are passed to NADP⁺. ATP is produced by the electron transport system.

THE REMARKABLE LEAF

Leaves of plants are miniature photosynthetic laboratories containing chloroplasts, the organelles that trap sunlight and use it to produce the organic molecules that, in turn, are the foundation of all life on Earth. These remarkable structures come in a variety of shapes and sizes, illustrating a host of intriguing adaptations that have emerged during evolution by natural selection. These adaptations not only improve photosynthetic efficiency but also protect plants from hungry grazers.

The prickly margins of thistle leaves, for example, protect these plants from grazers. Many leaves contain poisons that ward off hungry browsers, a topic discussed in Spotlight on Evolution 26–1. The royal water lily offers a lethal delivery system. The leaves of this plant may be up to 8 feet in diameter and are equipped with numerous sharp spines that protect the plant from fish and the Amazon sea cow, an aquatic mammal related to the manatee. Some leaves, like those of the nettle, have hollow, needlelike hairs that penetrate the skin of animals. The pressure of touch forces poison from a bulb at the base of each hair into the skin.

The leaves of many tropical trees have numerous fine "drip tips," which prevent water from pooling on the leaves during heavy rainfall. (Tropical rain forests receive 140 to 400 inches of rain a year.) This anatomical adaptation protects the leaves from being colonized by lichen, moss, and algae. Combined with a thick waxy coating, it also prevents leaching of leaf nutrients by rain.

In some plants, leaves act as funnels, diverting water toward the roots. The leaves of zucchini plants, for instance, collect water and send much of it inward toward the roots.

Leaf coloration is another adaptation of great interest. Green and white leaves of some species, for example, are an adaptation that permits plants to grow in very sunny locations by reducing the overall amount of sunlight they capture. One of the most striking leaves belongs to a shade-tolerant species of fern that grows in the tropics. This plant has a deep iridescent blue color, which studies have shown allows the fern to make use of high-energy red wavelengths of light, thus gathering more energy from the scant amount of sunlight that strikes the forest floor. Another adaptation to low-level lighting is large leaves. Most tropical species that live on the forest floor, for example, have large, thin leaves to intercept the available sunlight.

Leaves of aquatic plants may contain air-filled sacs at the base of each leaf stalk (petiole), adaptations that permit the leaves to float on the surface, where they gather sunlight. Other aquatic plants have evolved a buoyant, spongy tissue that permits them to float on the surface like a piece of flat cork.

One of the most interesting leaves is that of the neotropical plant called *Mimosa sensitivum*. These leaves actually fold up when touched, as shown in Figure 1. This adaptation presumably makes them less apparent to grazing animals.

These adaptations have one feature in common: they enhance survival and reproduction of the plant. By capturing additional sunlight or warding off predators, the leaves ensure levels of photosynthesis adequate for plant growth and reproduction.

Adapted from G. T. Prance, "In Praise of Leaves," *International Wildlife* (1986), 16(5): 12–17.

(a)

(b)

▷ **FIGURE 1 Sensitive Plants** (*a*) Open leaves, (*b*) closed leaves after being touched.

In photosystem I, electron holes are filled by electrons arriving from photosystem II via the electron transport system. Electrons emitted from photosystem II, therefore, serve two purposes: (1) they emit energy (gleaned from the sun), which is used to drive ATP synthesis in the electron transport system, and (2) they fill the electron holes in chlorophyll a of photosystem I.

ATP Production Is Coupled to the Electron Transport System

One of the mysteries of biology that baffled scientists for some time was how plants produce ATP. In 1961, a British biochemist, Peter Mitchell, proposed a mechanism—known as **chemiosmosis**—that explained how the electron transport system of plants made ATP. Mitchell's research,

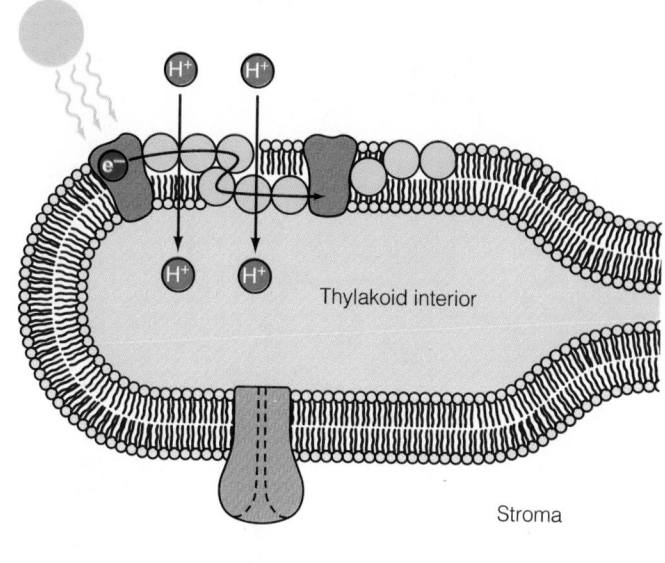

(a)

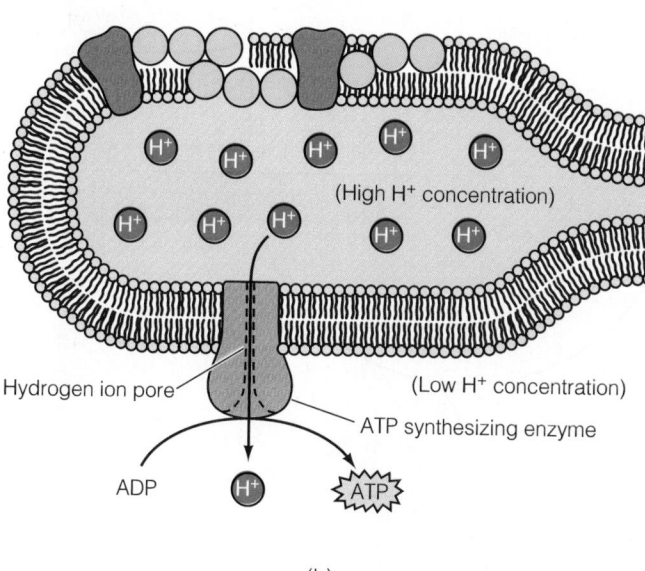

Hydrogen ion pore

(Low H⁺ concentration)

ATP synthesizing enzyme

ADP

(b)

▷ **FIGURE 5–13 Chemiosmosis** (*a*) The electron transport proteins in the thylakoid membrane are dual-purpose molecules. They transport electrons from photosystem II to photosystem I. Energy lost along the way is used to pump hydrogen ions into the interior of the thylakoid disk. (*b*) These ions leak out of the thylakoid disk into the stroma. The movement of hydrogen ions through the pores is believed to drive the synthesis of ATP.

for which he later won the Nobel Prize, suggests that the proteins (electron transport molecules) of the electron transport system are really dual-purpose molecules. That is, they transfer electrons from photosystem II to photosystem I but also serve as hydrogen ion pumps. These molecules transport hydrogen ions produced by the breakdown of water (as a part of photosystem II) into the interior of the thylakoid disk, as shown ▷ Figure 5–13a. Energy released

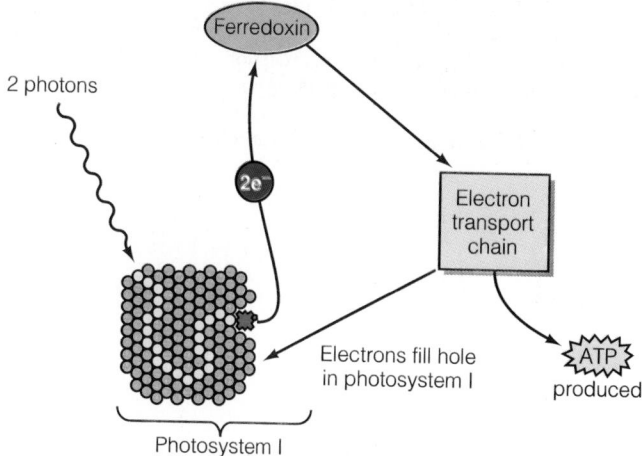

▷ **FIGURE 5–14 Cyclic Photophosphorylation** When carbon dioxide levels fall, NADPH levels increase. Electrons are cycled back to chlorophyll a in the reaction center of photosystem I.

by the electrons as they zip along the electron transport system drives the pumps.

Pumping hydrogen ions into the thylakoid disks results in an accumulation of these ions in the interior of the disks. Mitchell's own measurements, in fact, show that the concentration of hydrogen ions inside the thylakoid disk is approximately 1000 times greater than the levels in the outlying stroma. Mitchell hypothesizes that once inside, the hydrogen ions begin to leak back into the stroma through special protein pores in the membrane (Figure 5–13b). These pores are also enzymes that catalyze the formation of ATP from ADP. As the hydrogen ions diffuse down their concentration gradient, they drive the synthesis of ATP in much the same way that the flow of electrons through a wire powers a light bulb or runs a small electric motor.

In plants and photosynthetic protists, electrons generally flow from photosystem II to photosystem I; in the process, ATP and NADPH are formed. The production of ATP in this manner is called **noncyclic photophosphorylation.**[5]

Cyclic Photophosphorylation Produces ATP But Not NADPH

Electrons emitted from photosystem I in plants and algae can be shunted from ferredoxin to the electron transport system, then to chlorophyll a, bypassing $NADP^+$ altogether, as shown in ▷ Figure 5–14. On their cyclic trip back to chlorophyll a, the electrons travel along the electron transport system and generate at least one ATP. Thus, although no NADPH is produced in this process, ATP synthesis continues unabated. This process, known as **cyclic photophosphorylation,** helps ensure the proper ratio of ATP to NADPH required for the dark reactions as carbon dioxide levels fluctuate.

[5]Photophosphorylation refers to the phosphorylation of ADP in the presence of sunlight.

The C₃ cycle takes one carbon dioxide molecule at a time, adding them to ribulose biphosphate. This reaction produces an unstable six-carbon compound that splits, forming two three-carbon compounds (PGA). PGA is eventually converted into PGAL, which is used to produce glucose and other carbohydrates.

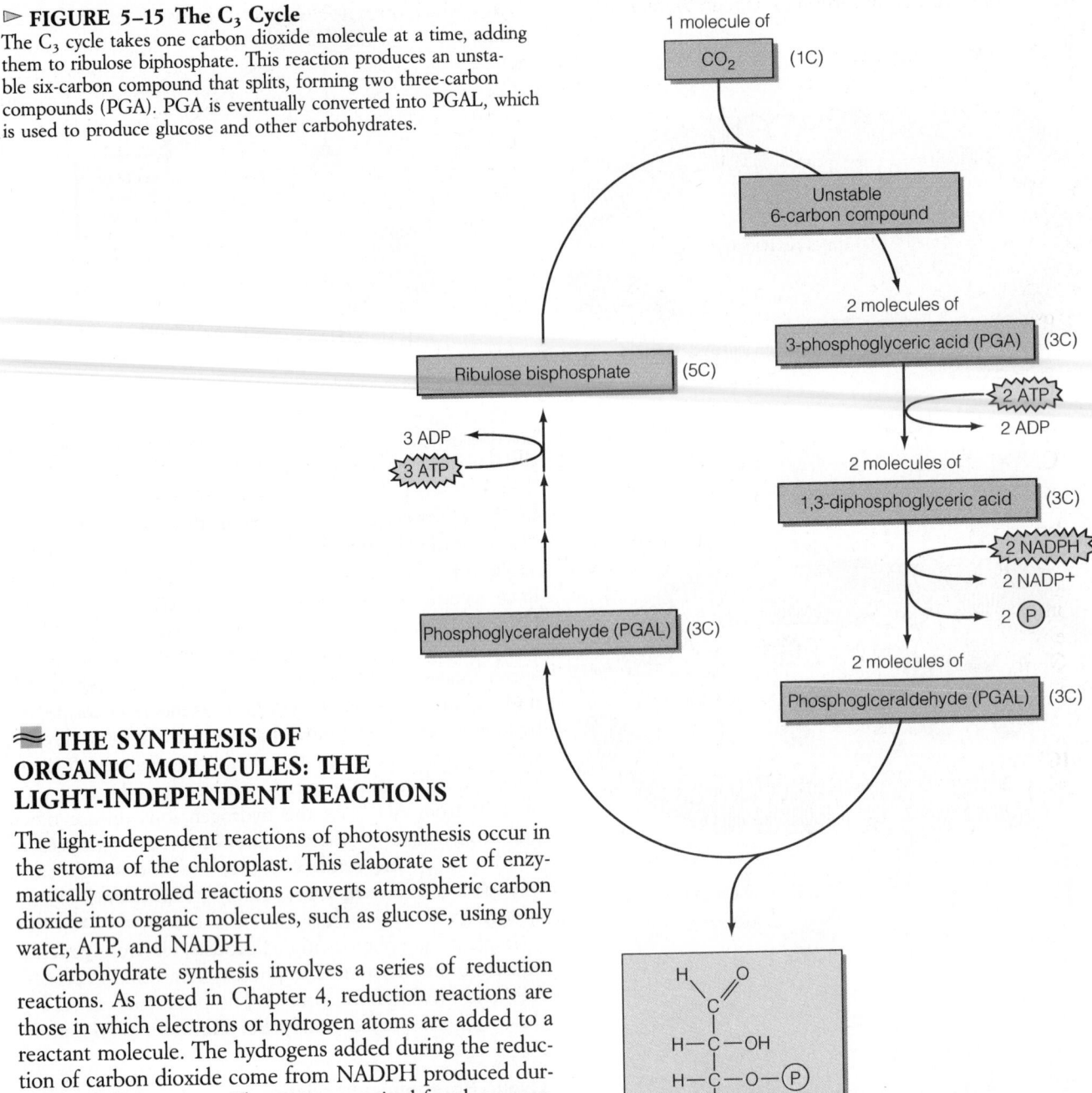

≋ THE SYNTHESIS OF ORGANIC MOLECULES: THE LIGHT-INDEPENDENT REACTIONS

The light-independent reactions of photosynthesis occur in the stroma of the chloroplast. This elaborate set of enzymatically controlled reactions converts atmospheric carbon dioxide into organic molecules, such as glucose, using only water, ATP, and NADPH.

Carbohydrate synthesis involves a series of reduction reactions. As noted in Chapter 4, reduction reactions are those in which electrons or hydrogen atoms are added to a reactant molecule. The hydrogens added during the reduction of carbon dioxide come from NADPH produced during the light reactions. The energy required for these reactions comes from ATP generated by chemiosmosis (see Figure 5–4).

The Calvin-Benson Cycle, or C₃ Cycle, Produces Carbohydrates in all Plants

All photosynthetic organisms produce carbohydrates via the **Calvin-Benson cycle,** a series of chemical reactions each catalyzed by its own enzyme. Named after the scientists who elucidated the steps of the process, the Calvin-Benson cycle first produces a three-carbon compound called glyceraldehyde phosphate—hence its alternative name, the C₃ cycle (⊳ Figure 5–15). Although the details of this process are beyond the scope of this book, a few observations are merited.

As illustrated in Figure 5–15, the C₃ cycle begins with the incorporation of a single molecule of carbon dioxide into a five-carbon molecule, **ribulose bisphosphate (RuBP).** This step is called **carbon fixation.** As illustrated, the product of the first chemical reaction in the C₃ cycle is an unstable six-carbon compound. Being unstable, it breaks down almost immediately, forming two three-carbon molecules of **3-phosphoglyceric acid (PGA).** One of the carbons came from carbon dioxide in the first step of the cycle.

In the next reaction, two molecules of ATP from the light-dependent reactions are used to convert both molecules of PGA into 1,3-diphosphoglyceric acid. These molecules are then reduced by phosphoglyceraldehyde (PGAL), producing two molecules of **glyceraldehyde phosphate.** Glyceraldehyde phosphate molecules serve two purposes: some combine to form glucose and others combine to regenerate ribulose bisphosphate, thus permitting the cycle to continue.

Chloroplasts contain millions of molecules of ribulose bisphosphate and are, therefore, capable of fixing thousands of carbon dioxide molecules every second. As a general rule, 2 of every 12 molecules of PGAL produced in the chloroplast go to make glucose. The remaining 10 molecules are used to regenerate ribulose bisphosphate.

The C_4 Cycle Is an Alternative Way to Fix Carbon in Certain Plants

All plants produce glucose via the C_3 cycle. However, the first step in the C_3 cycle, the reaction of carbon dioxide and ribulose bisphosphate, is extremely sensitive to levels of carbon dioxide and oxygen. When carbon dioxide levels inside the leaf fall, the rate of glucose synthesis drops.

CO_2 levels inside the plant are controlled by adjustable pores in the leaves of plants called **stomata** (singular, *stoma*), which open and close in response to environmental conditions (▷ Figure 5-16). On hot days, the stomata close down, reducing water loss. Although this mechanism protects the plant from desiccation, it also reduces the inflow of CO_2 from the atmosphere. The result is a decrease in the internal carbon dioxide levels. The closing of the stomata results in a concomitant increase in oxygen, which, you may recall, is a waste product of the light reactions. When oxygen levels increase, the enzyme that catalyzes the very first reaction of the C_3 begins fixing oxygen, rather than carbon dioxide, to ribulose bisphosphate. Because oxygen and carbon dioxide compete for the active site on the first enzyme in the C_3 cycle, the higher the level of oxygen, the more oxygen is fixed. This, in turn, impairs glucose production via the C_3 cycle.

Certain plants have evolved an alternative biochemical pathway that lets them continue producing glucose even when the internal concentration of carbon dioxide is low. It is known as the **C_4 cycle,** or the **Hatch-Slack cycle** (after the scientists who discovered it), and operates in conjunction with the C_3 cycle. In C_4 plants, two cells are involved in photosynthesis, the mesophyll cells and the bundle sheath cells (Figure 5–16). The **mesophyll cells** are located just beneath the epidermal cells of the leaf and have ready access to atmospheric carbon dioxide. (The term *mesophyll* means "middle of the leaf"). In these cells, atmospheric carbon dioxide combines with a three-carbon

▷ **FIGURE 5–16 The Anatomy of a Leaf** (*a*) Cross section of a C_3 leaf showing stomata and various cell layers. Stomata open and close in response to water levels. (*b*) C_4 leaf.

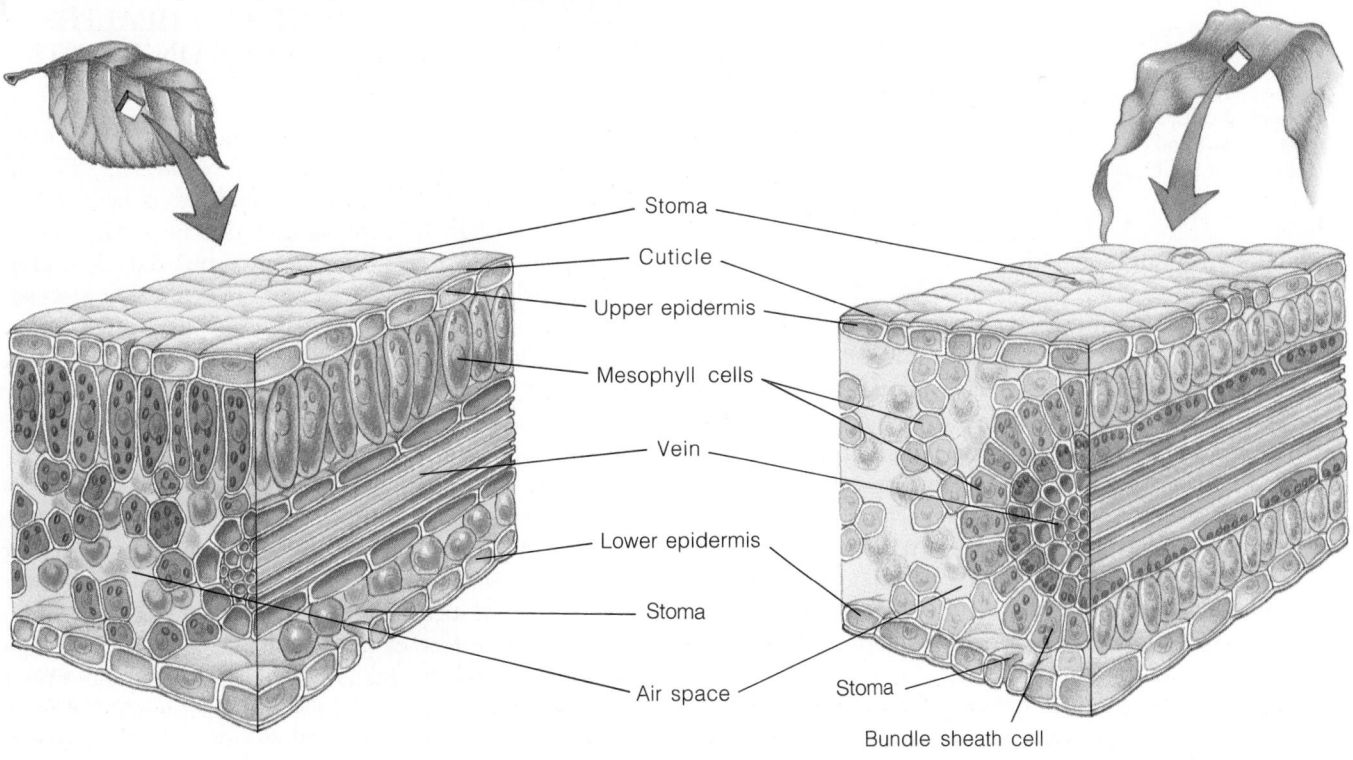

(a) C_3 leaf

(b) C_4 leaf

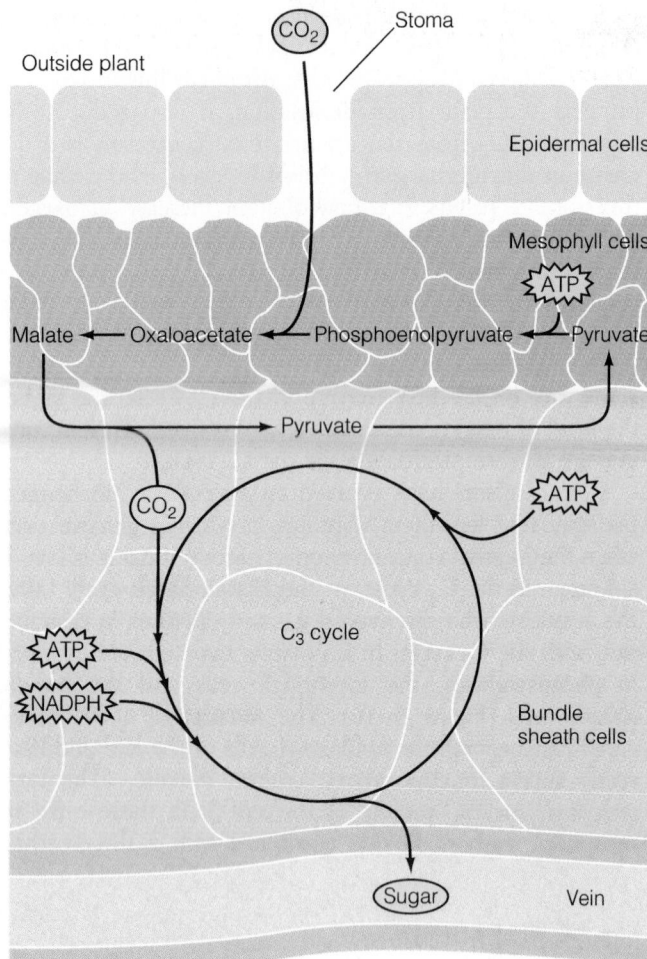

Outside plant

CO₂ — Stoma

Epidermal cells

Mesophyll cells

ATP

Malate ← Oxaloacetate ← Phosphoenolpyruvate ← Pyruvate

Pyruvate

CO₂

ATP

ATP
NADPH

C₃ cycle

Bundle sheath cells

Sugar   Vein

▷ **FIGURE 5–17 The C₄ Cycle** Present only in select plants, the C₄ cycle allows them to fix atmospheric carbon dioxide when internal levels are low. Carbon dioxide is fixed in the C₄ cycle in mesophyll cells. Malate produced in the mesophyll cells diffuses into nearby bundle sheath cells, where carbon dioxide is liberated and used in the C₃ cycle.

compound, phosphoenolpyruvate. The enzyme that catalyzes this reaction has high affinity for carbon dioxide and can operate efficiently at low levels.

The resulting four-carbon product is **oxaloacetate** (▷ Figure 5–17). Oxaloacetate is next converted into malic acid, a four-carbon sugar (hence the name C₄ cycle). As shown in Figure 5–17, malic acid is transported out of the mesophyll cells and into neighboring cells, the **bundle sheath cells,** which surround the veins in the leaf. Enzymes in the bundle sheath cells break down malic acid, releasing carbon dioxide and a three-carbon compound, pyruvate. Carbon dioxide builds up inside the bundle sheath cells in concentrations sufficient for the first enzyme of the C₃ cycle to function.

The outer mesophyll cells essentially "collect" carbon dioxide from the air and transport it as four-carbon compounds to the bundle sheath cells, where it enters the C₃ cycle. The cell biologist Wayne Becker likens the C₄ pathway to a carbon dioxide pump, which maintains carbon

dioxide levels in the bundle sheath cells sufficient for fixation.

Most C₄ plants are found in deserts and arid tropical regions (▷ Figure 5–18). (Remember that not all tropical regions are lush, humid rain forests.) Corn, sorghum, sugarcane, crabgrass, and portulaca (moss rose) are some of the more common members of this small but important group of plants (Figure 5–18).[6]

One well-known C₄ plant in the temperate zone is crabgrass; it grows well in open farm fields and suburban lawns because it can fix atmospheric carbon dioxide even when sunlight and heat are intense. Kentucky bluegrass, favored by many homeowners, can be overrun by crabgrass because it is a C₃ plant whose stomata close down on hot, dry days, reducing the rate of carbon fixation and organic synthesis. C₄ plants like corn and sugarcane often have yields two to three times greater than C₃ plants, such as wheat and other cereal grains.

The C₄ cycle is not a substitute for the C₃ cycle. It is, rather, a carbon-fixing sequence that precedes the C₃ cycle in certain plants. The C₄ cycle, however, uses considerably more ATP than the C₃ cycle to regenerate starting materials. Therefore, C₄ plants are favored in deserts and in midsummer in temperate climates, where sunlight is plentiful and ATP production occurs at a high rate. C₃ plants, on the other hand, tend to have the advantage in cool, wet, and cloudy climates, where sunlight energy is limited (Figure 5–18c).

ENVIRONMENT AND HEALTH: OUR DEPENDENCE ON PLANTS

Plants and other photosynthetic organisms make many contributions to human health and well-being. Besides feeding billions of people, aquatic and terrestrial plants fix about 170 billion tons of carbon dioxide each year, or about 20% of the total atmospheric carbon dioxide. Before the Industrial Revolution, which began nearly 200 years ago in Europe and about 100 years ago in the United States, carbon dioxide fixation was in balance with carbon dioxide production by animals and other organisms. Today, this balance has been dangerously tipped by the combustion of coal and other fossil fuels, which release carbon dioxide trapped by prehistoric plants.

As you will recall, carbon dioxide in the atmosphere acts like the glass of a greenhouse, reflecting heat back to Earth. Many scientists believe that its buildup in the atmosphere may drastically shift global climate in the coming decades. The effect of global climate shift is potentially profound, not just for humans but also for all other life on Earth. A rising temperature, for example, might kill many species of plants and animals that are unable to adapt to the rapid climb.

[6]It is important to note that this is not a taxonomic group. C4 plants are found in several different families.

(a)

(b)

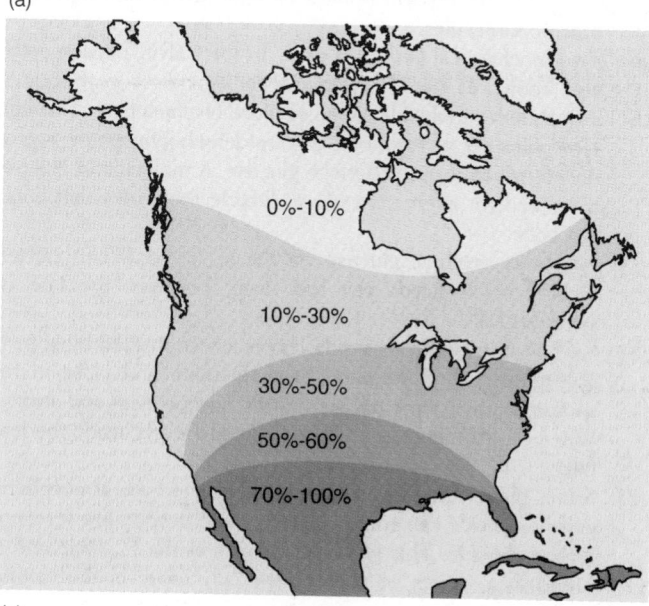

(c)

▷ **FIGURE 5–18 C₄ Plants** (*a*) Corn and (*b*) portulaca (moss rose) are both C_4 plants. (*c*) Distribution of C_4 grass species as a percentage of all grass species.

Because trees and other plants absorb carbon dioxide, rapid deforestation occurring throughout the world also contributes to the rise in atmospheric carbon dioxide. Approximately 13% of the annual increase in carbon dioxide can be attributed to deforestation.

Plants also contribute directly to human health. Today, approximately half of all prescription and nonprescription drugs originate in plants. Naturally occurring chemicals discovered in tropical forest plants and others help cure leukemia and a host of other diseases. These drugs are worth an estimated $20 billion a year to the U.S. economy and about $40 billion a year to the global economy.

Wild plants also provide genes that are used to improve cultivated plants. Genes from wild plants may help re searchers create drought- and insect-resistant crop plants or plants that resist disease. According to estimates from the U.S. Department of Agriculture, genetic infusions are worth about $1 billion a year in increased crop production in the United States alone.

Clearly, plants are essential to human health and well-being and to the health and well-being of virtually all other organisms. Unfortunately, the plant kingdom is under direct assault. Global warming, ozone depletion, deforestation, pollution, and human development are the five principal assaults.

Global warming, for example, could cause a massive die-off of plants in many regions of the world. Ozone depletion from the upper atmosphere, now being addressed internationally, may result in an increase in ultraviolet light reaching Earth. The ozone layer, high above our heads, normally filters out most of the UV light. With its gradual decline, plants might suffer, because UV light damages plants and impedes photosynthesis.

Deforestation, if it continues, could worsen global warming and alter global climate in profound ways. Approximately 40% to 50% of the tropical forests of the world are already gone. Most of the remaining ones could be destroyed within your lifetime if current trends continue.

Certain pollutants also affect plants. One notable example is ground-level ozone produced in the lower atmosphere from automobile exhaust and other industrial pollutants. Ground-level ozone adversely affects plant reproduction and may destroy foliage and root tissues. One of the most significant effects, however, is its interference with photosynthesis and cellular respiration in plant cells. At intermediate levels, plants may be subtly harmed, but at high levels, ozone can cause acute disease and mortality.

Development is also eroding our rich biological legacy. In the United States, about 7000 acres of farmland and grassland are destroyed each day by the expanding human population.

Each trend is dangerous by itself. Taken together, they suggest reason for rapid action—significant steps that protect the photosynthetic organisms so vital to human survival and the future ecological health of the planet. Destroying plant life is like tearing down the foundation of a home. Without plants, the walls of our world are sure to crumble.

AN OVERVIEW OF PHOTOSYNTHESIS

1. Solar energy, captured by photosynthetic organisms, is the source of virtually all energy used by living things. Plants and photosynthetic protists collect solar energy and use that energy to make simple organic molecules from carbon dioxide and water.

2. In plants, photosynthesis generates oxygen, needed by animal cells to catabolize glucose efficiently.

3. Photosynthesis occurs in the chloroplasts in the leaves and other photosynthetic tissues of green plants and in photosynthetic protists.

4. Photosynthesis consists of two sets of reactions, the light-dependent and the light-independent reactions, both of which take place in the chloroplast.

5. The light-dependent reactions are those by which the cell captures sunlight energy. The light-dependent reactions consist of two photosystems. They produce ATP and NADPH, which are necessary for the light-independent, or synthetic, reactions.

6. The light-independent reactions produce organic molecules from carbon dioxide and water using ATP and NADPH produced in the light reactions.

THE NATURE OF SUNLIGHT

7. The sun emits electromagnetic radiation of many wavelengths. Visible light is that part of the spectrum detectable by animals.

8. Part of the visible portion of the spectrum also stimulates photosynthesis.

THE LIGHT-DEPENDENT REACTIONS: CAPTURING SUNLIGHT ENERGY

9. Plants and photosynthetic protists capture sunlight energy via pigments, such as chlorophyll and carotenoids, found in the thylakoid membrane of the chloroplast.

10. The pigments absorb sunlight energy and transmit that energy to a special molecule of chlorophyll a in the reaction center of the photosystems. Chlorophyll a responds to the energy input by emitting electrons.

11. The light-dependent reactions consist of three parts: photosystem I, photosystem II, and a photoelectron transport system—all found in the membrane of the thylakoid disk.

12. Light strikes both photosystems simultaneously, creating a steady flow of electrons from photosystem II to photosystem I to NADP.

13. Electrons emitted from the reaction center of both photosystems are picked up by molecules called primary acceptors. The primary acceptor of photosystem I transfers electrons to an iron-containing protein called ferredoxin. Ferredoxin relinquishes its electrons to $NADP^+$, thus forming NADPH.

14. Electrons ejected from photosystem I create an "electron hole," which is filled by electrons coming from photosystem II. These electrons are transferred to photosystem I via the electron transport system (ETS).

15. Along the ETS, energy lost by the electrons is captured by the cell to produce ATP from ADP and inorganic phosphate.

16. To fill the hole created in photosystem II, water is enzymatically split. The splitting of water produces oxygen.

17. The electron transport system consists of a series of carrier proteins. The carrier proteins are dual-purpose molecules. They not only transfer electrons from photosystem II to photosystem I but also use some of the energy lost by the electrons to pump hydrogen ions into the interior of the thylakoid disk, creating a concentration gradient.

18. Also found in the thylakoid membrane are enzymes that catalyze the formation of ATP. In the center of the ATPase molecules are pores, through which hydrogen ions diffuse back into the stroma. This passive diffusion drives the synthesis of ATP.

THE SYNTHESIS OF ORGANIC MOLECULES: THE LIGHT-INDEPENDENT REACTIONS

19. The light-independent reactions are elaborate chemical reactions that convert atmospheric carbon dioxide and water into organic molecules.

20. Two biochemical pathways exist. In the Calvin-Benson cycle, also known as the C_3 cycle, carbon dioxide is incorporated into a parent molecule called ribulose bisphosphate. Through a complex set of reactions, glyceraldehyde phosphate is produced, which is used to make glucose. A molecule of glucose is formed for every six carbon dioxide molecules that enter the cycle.

21. All plants produce glucose via the C_3 cycle. When carbon dioxide levels inside the leaf drop, however, the rate of synthesis falls.

22. Carbon dioxide levels inside leaves are controlled by adjustable pores called stomata. Stomata open and close in response to environmental conditions, such as heat and aridity. When the stomata close, carbon dioxide levels inside the leaf fall.

23. Some plants have evolved a system by which atmospheric carbon dioxide can continue to be fixed despite low levels of carbon dioxide. This process is known as the C_4 cycle and is found in a variety of plants, especially those in the tropics and deserts.

24. During the C_4 cycle, carbon dioxide reacts with a three-carbon compound, phosphoenolpyruvate, producing a four-carbon compound, oxaloacetate. Oxaloacetate is converted to malic acid, which leaves the cells and is transported to the bundle sheath cells surrounding veins in the leaves. Enzymes in these cells break down malic acid, giving off carbon dioxide, which enters the C_3 cycle. The C_4 system, thus, helps maintain high levels of carbon dioxide in the bundle sheath cells so that the C_3 cycle can operate.

ENVIRONMENT AND HEALTH: OUR DEPENDENCE ON PLANTS

25. Plants make many vital contributions to human health and well being. Besides feeding billions of people each year, plants absorb carbon dioxide, helping offset global warming.

26. Plants also contribute directly to human health. Today, approximately half of all prescription and nonprescription drugs owe their origin to plants.

27. Wild plants also provide genes that are used to improve cultivated plants. Genes from wild plants may help researchers create drought- and insect-resistant crop plants or plants that resist disease.

28. Plants, which renew atmospheric oxygen levels, are essential to the health and well-being of animals.

29. Unfortunately, the plant kingdom is under direct assault. Global warming, ozone depletion, deforestation, pollution, and human development are the five principal means of assault.

30. Concerted action is needed to protect this group of endangered organisms so vital to human survival and the future ecological health of the planet.

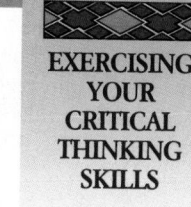

EXERCISING YOUR CRITICAL THINKING SKILLS

1. Write a list of the critical thinking skills you learned in Chapter 1. As you talk with friends, relatives, and family members and as you listen to the news or television documentaries or read letters from congressional representatives, use your list to keep track of "violations" of these rules. Write brief notes on each incident. How many "violations" did you notice in a week? Did you notice any rules that were violated more often than others?

2. A species of cordgrass, known to scientists as *Spartina anglica*, has established itself in the last 100 years in the mud flats at the mouths of rivers along the coast of Great Britain. Interestingly, these intertidal marsh zones lacked higher plant life before the invasion of this hardy species.

   Ecologists are interested in this species because it can shed some light on invasive species. Of special interest are the reasons why species such as this are so successful.

   A closer look at *S. anglica* has revealed several important adaptations, including its ability to trap sediment from tidal water, which then settles to the bottom creating a substrate on which the plant grows well, and its tolerance of long periods of tidal flooding. These and other characteristics, which give this species an adaptive advantage over others, are the result of its unique genetic makeup. *S. anglica,* in fact, is a hybrid plant, produced by two different species of cordgrass. The new combination of genes presumably gives it its advantages.

   Photosynthesis is another characteristic that determines the success of an invasive species. With your knowledge of photosynthesis, which pathway would you expect to see in *S. anglica?* Which photosynthetic pathway is the most efficient? Which photosynthetic pathway is more common in cooler climates?

If you answered the first question by saying a C_4 plant, you are right; C_4 plants are more efficient. If you answered the second by saying a C_3 plant, you are right; C_3 plants are more common in temperate zones.

It turns out that *S. anglica* is a C_4 plant. As you have learned in this chapter, C_4 plants operate under lower concentrations of carbon dioxide than C_3 plants, and are able to continue making carbohydrates and other organic molecules when carbon dioxide levels are low. Photosynthesis in this species is also more efficient because the chloroplasts of its leaves have twice the cross-sectional area of those typically found in C_3 species, which increases the plant's ability to absorb sunlight.

As a rule, C_4 species are very productive, but their high productivity is usually attained only at leaf temperatures of 30° C and above, temperatures found in climates warmer than that of the coastal waters of Great Britain. Could the C_4 pathway contribute to *S. anglica's* success?

It turns out that *S. anglica* is an anomaly among C_4 plants. That is, when it comes to temperature, it behaves more like a C_3 plant than a C_4 plant. Put another way, unlike other C_4 plants, it can tolerate cool temperatures.

Although you haven't had to do much critical thinking here, this exercise helps illustrate one of the rules presented in Chapter 1. What critical thinking rule does this example illustrate?

If you look over the list on page 18, you'll find that it illustrates the rule to uncover hidden assumptions. Note also that it illustrates a rule that I haven't included earlier: in biology there are sometimes exceptions to the rules. Be willing to question assumptions.

TEST OF TERMS

1. Photosynthesis is the process by which plants and other organisms use sunlight energy to produce carbohydrates from _____ _____. This process occurs in the _____ of plants.

2. Photosynthesis consists of two separate but interdependent parts: the _____-_____ reactions, which capture sunlight, and the _____-_____ reactions, which use the captured energy to synthesize carbohydrates.

3. The light-absorbing pigment _____ gives many plants their green color and is the dominant pigment found in leaves.

4. Electromagnetic radiation from the sun consists of many _____, but only a small portion of the spectrum can be detected by the human eye. This light is called _____ light.

5. The light-harvesting complexes in photosynthetic plant cells are aggregations of chlorophyll molecules, carotenoids, and special proteins; these complexes, or aggregations, are called _____.

6. When light strikes these aggregations of chlorophyll, the energy is funneled to the _____ center, which contains a special molecule of chlorophyll called chlorophyll a. Chlorophyll a responds by emitting _____.

7. Photosystems I and II are connected by a series of proteins, the _____ _____ _____, located in the thylakoid membrane.

8. The light-harvesting reactions produce three molecular products: _____, _____, and _____, which are used in the synthesis of carbohydrates in the _____-_____ reactions.

9. In photosystem II, _____ is split, and electrons released during

the process are used to fill the electron hole in the systems.

10. Peter Mitchell proposed a model for the production of _____ during the _____ _____ _____ in photosynthesis. In his theory, Mitchell states that energy lost from electrons is used to pump _____ ions into the interior of the thylakoid disk. These ions flow back out through _____ pores in the membrane, and the outward flow of these ions creates energy that drives the synthesis of _____ .

11. When _____ _____ levels fall inside photosynthetic plant cells, carbohydrate synthesis plummets. Electrons from photosystem I are recycled back to chlorophyll a in a process called _____ _____ .

12. All plants contain the enzymes needed to convert carbon dioxide into glyceraldehyde phosphate, which is then used to make glucose. These enzymes are part of a cyclic biochemical path called the _____ cycle.

13. Plants adapted to hot, dry conditions have a supplementary biochemical mechanism for fixing atmospheric carbon dioxide. This is called the _____ cycle.

14. The leaves of plants contain small pores, the _____ , which open and close in response to water levels.

Answers to the Test of Terms are found in Appendix B.

Answers to the Test of Terms are found in Appendix B.

TEST OF CONCEPTS

1. Define photosynthesis in one or two sentences. Is it an endergonic or an exergonic process?
2. Describe how sunlight energy is captured during the light-dependent reactions. Be sure to include the following terms in your definition: photosystems I and II, excited electron, chlorophyll, reaction center, primary acceptor molecule, electron transport system.
3. Compare and contrast photosystems I and II. How are they similar, and how are they different?
4. What role does water play in the light-dependent reactions of photosynthesis? Where does the oxygen come from in photosynthesis?
5. Define the term *light-independent reactions*. What inputs do they require, and where do these inputs come from?

6. What are the main products of the C_3 cycle? Is it found in all plants?
7. Describe the C_4 cycle. How does it differ from the C_3 cycle? How is it related to the C_3 cycle? Is it found in all plants?
8. Why is a leaf green throughout the spring and summer? Why does it turn color in the fall?
9. List the environmental conditions that reduce the rate of photosynthesis, and explain why they do.

CHAPTER 6

The Energy of Life
How Cells Produce Energy

≈

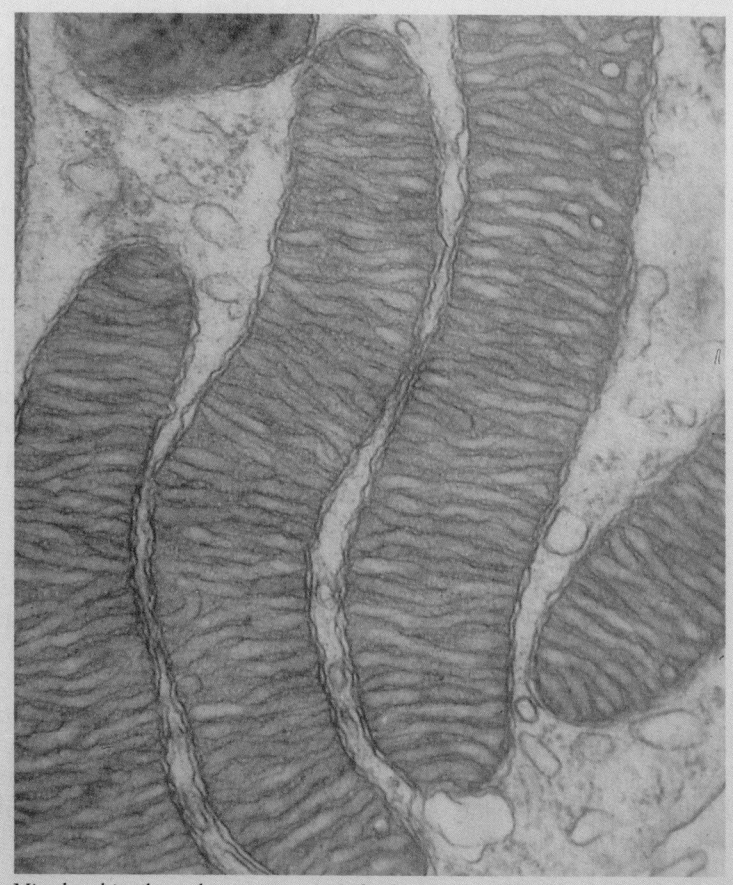

Mitochondria, shown here, strip organic food molecules of their energy and convert it to ATP, a form cells can use directly.

Sunlight streams onto the surface of the planet. But plants capture only a small portion (1% to 2%) of the sun's incoming visible light. That small fraction, though, powers the entire living world—providing energy for a wide range of microorganisms, among them fungi, plants, and animals.

As noted in the last chapter, the energy of the sun is stored in chemical bonds in carbohydrates and other compounds. Carbohydrates (principally glucose molecules) are broken down in the cells of organisms, a process that releases stored energy. This chapter examines the metabolic processes responsible for the breakdown of glucose, focusing primarily on eukaryotic cells. It shows how these cells liberate energy stored in the covalent bonds of glucose and use it to produce ATP.

≈ CELLULAR RESPIRATION: AN OVERVIEW

In eukaryotic cells (protists, fungi, plants, and animals), glucose breakdown occurs in a series of chemical reactions that begins in the cytoplasm and is completed in an organelle known as the mitochondrion. During these reactions, glucose containing six carbon atoms is broken down into six molecules of carbon dioxide and six of water. The overall reaction is:

$$\underset{\text{glucose}}{C_6H_{12}O_6} + \underset{\text{oxygen}}{6\,O_2} \rightarrow \underset{\text{carbon dioxide}}{6\,CO_2} + \underset{\text{water}}{6\,H_2O} \text{ (catabolic reaction)}$$

$$\underset{\substack{\text{adenosine} \\ \text{diphosphate}}}{38\,ADP} + \underset{\substack{\text{inorganic} \\ \text{phosphate}}}{38\,P_i} \rightarrow \underset{\substack{\text{adenosine} \\ \text{triphosphate}}}{38\,ATP} \text{ (anabolic reaction)}$$

This reaction is highly exergonic, releasing lots of energy. The cell captures a sizable portion of this energy and stores it in ATP molecules for later use. As shown in the equation above, the complete breakdown of glucose generates 38 molecules of ATP.

Also shown in the reaction, the breakdown of glucose requires oxygen, which comes from photosynthesis in plants and photosynthetic protists and Monerans. Without oxygen, glucose can be only partially broken down, and only a fraction of its potential energy can be captured. (You'll see why later.)

The carbon dioxide released from glucose catabolism in eukaryotic cells is released into the atmosphere. Photosynthetic plants and algae use the carbon dioxide to produce more carbohydrate in an endless cycle of life.

The Complete Breakdown of Glucose in Plants and Animal Cells Is Called Cellular Respiration

The term *respiration* is commonly used to denote the act of breathing. During respiration, many animals inhale oxygen-rich air and exhale air relatively high in carbon dioxide. As you have just seen, cells also take in oxygen and give off carbon dioxide during the breakdown of glucose. As a result, glucose catabolism is often referred to as **cellular respiration.**

In evolution, cellular respiration actually developed well before pulmonary respiration, which arose as an adaptation in multicellular organisms. Unlike the early single-celled organisms, which acquire oxygen and rid themselves of carbon dioxide by simple diffusion, most of the cells of multicellular organisms are separated from the surrounding environment, rendering diffusion as a means of gaseous exchange wholly inadequate. Multicellular organisms clearly required a more elaborate means of getting the oxygen its cells needed and getting rid of carbon dioxide wastes. Over time, organisms evolved gills, respirable skin, and lungs and circulatory systems to transport oxygen and carbon dioxide to and from their cells.

Cellular respiration in eukaryotes consists of four interconnected parts: (1) glycolysis, (2) the transition reaction, (3) the citric acid cycle, and (4) the electron transport system (▷ Figure 6–1). Table 6–1 summarizes the stages of cellular respiration and lists their products. Take a few moments to familiarize yourself with each stage. It may help to refer to this table as you read the rest of this chapter.

Glycolysis Is the Breakdown of Glucose into Two Pyruvate Molecules. The first phase of cellular respiration in eukaryotic cells is known as glycolysis. (We'll explore energy production in prokaryotes later.) **Glycolysis** (sugar breakdown) is a metabolic pathway found in the cytoplasm of all eukaryotic cells. As ▷ Figure 6–2 shows, during glycolysis, glucose is split in half, forming two three-carbon molecules of **pyruvic acid,** or **pyruvate.** The energy released during the catabolic breakdown nets the cell two molecules of ATP and two molecules of NADH. All in all, it is not a very impressive gain.

The Transition Reaction Is an Intermediate Step in Which One Carbon Atom Is Cleaved Off Each Pyruvate. The remaining stages of cellular respiration occur in the mitochondrion. As you may recall from Chapter 3, mitochondria have two membranes, forming two distinct compartments: an inner compartment enclosed by the inner membrane and an intermembrane, or outer, compartment lying between the two membranes. The second and third phases of cellular respiration occur within the inner compartment, or within the matrix.

As Figure 6–1 illustrates, pyruvate generated in glycolysis diffuses into the inner compartment of the mitochondrion and there reacts with a molecule called coenzyme A. This reaction is known as the **transition reaction.** During the transition reaction, one carbon atom is cleaved off each pyruvate molecule, forming a two-carbon compound (acetyl group) that binds to Coenzyme A. The product of this reaction is a molecule known as **acetyl CoA.**

The Citric Acid Cycle Completes the Breakdown of Glucose. Acetyl CoA enters the third stage of cellular

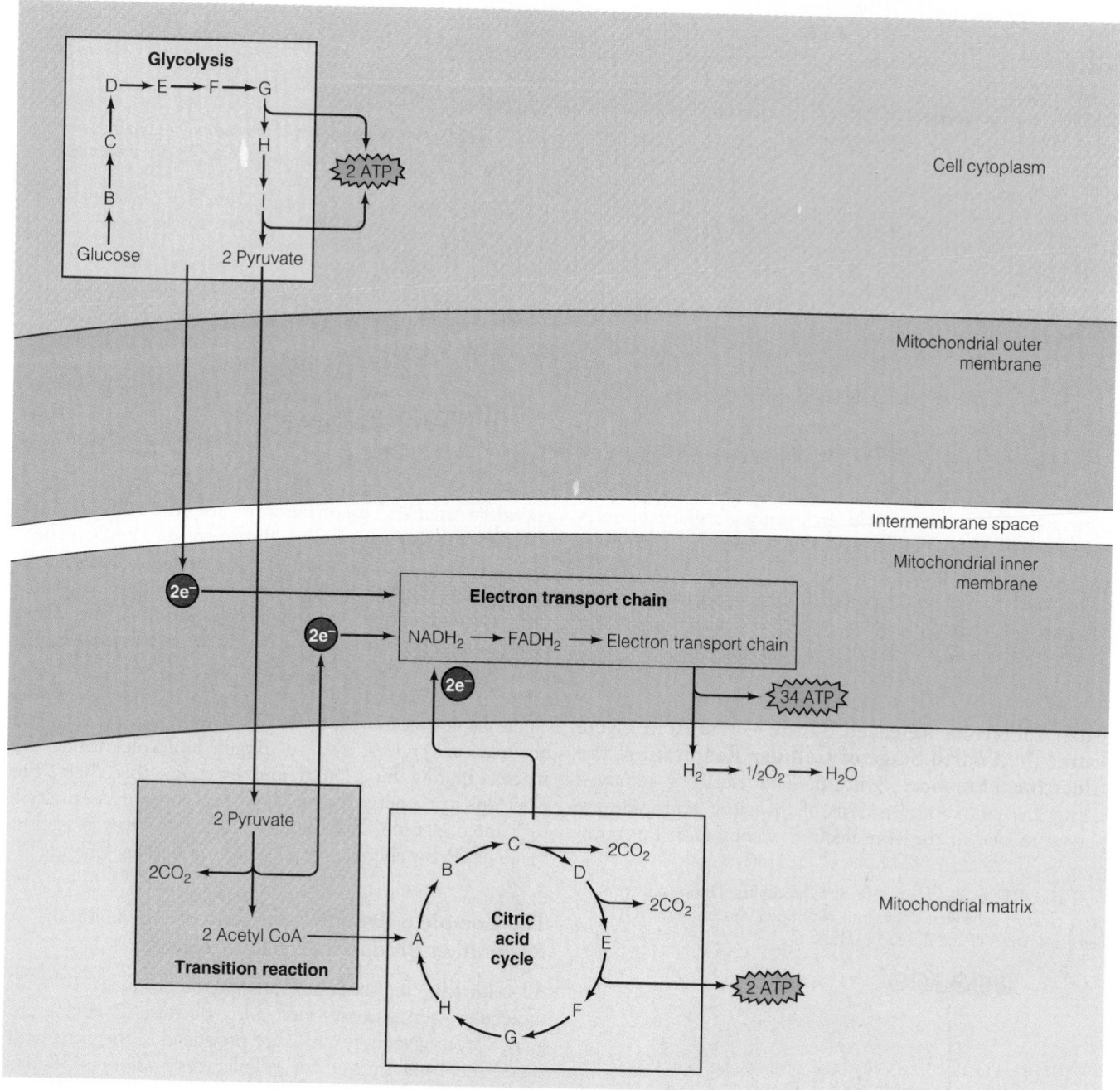

▷ **FIGURE 6–1 Cellular Respiration** This process involves four steps: glycolysis, the transition reaction, the citric acid cycle, and the electron transport system. Glycolysis occurs in the cytoplasm; the remaining steps take place in the mitochondrion. Most of the ATP is produced in the electron transport system from the energy stripped from electrons given off by the previous three stages.

respiration, a cyclic metabolic pathway located inside the inner compartment of the mitochondrion. This pathway is called the **Krebs cycle,** after the scientist who worked out many of its details, or the **citric acid cycle,** after the very first product formed in the reaction sequence (▷ Figure 6–3).

During the cycle, the two-carbon compound produced in the transition reaction binds to a four-carbon compound known as **oxaloacetate.** The result is a six-carbon product, **citric acid.** Citric acid proceeds through a series of chem-

ical reactions, each catalyzed by its own enzyme. Along the way, the molecule is modified many times. During the complex molecular rearrangements, two molecules of carbon dioxide are removed from the original six-carbon chain, and oxaloacetate is regenerated at the end, thus ensuring the continuation of the cycle.

As shown in Figure 6–3, one of the chemical reactions of the citric acid cycle yields an ATP molecule. That is to say, one exergonic reaction is coupled to the production of ATP (an endergonic reaction). Because two molecules of

TABLE 6-1 Overview of Cellular Energy Production

REACTION	LOCATION	DESCRIPTION AND PRODUCTS
Glycolysis	Cytoplasm	Breaks glucose into two three-carbon compounds, pyruvate; nets two ATPs; nets two NADH molecules
Transition reaction	Mitochondrion	Removes one carbon dioxide from each pyruvate, producing two acetyl CoA molecules; produces two NADH molecules
Citric acid cycle	Inner compartment of mitochondrion	Completes the breakdown of acetyl cycle CoA; produces two ATPs per glucose; produces numerous NADH and FADH molecules
Electron transport	Inner membrane of mitochondrion	Accepts electrons from NADH and FADH, generated by previous system reactions; produces 34 ATPs

pyruvate enter the citric acid cycle for each glucose molecule broken down by glycolysis, the citric acid cycle nets the cell two ATP molecules.

Most of the energy acquired from glucose comes from the oxidation reactions of the cycle. As you will soon see, these reactions yield high-energy electrons. The cell soon captures this energy to make additional ATP.

Most Electrons Released by the Citric Acid Cycle Enter the Fourth Stage of Cellular Respiration, the Electron Transport System. The electrons removed during the oxidation reactions of the citric acid cycle are passed to one of the two electron acceptors in the mitochondrial matrix: **nicotinamide adenine dinucleotide (NAD)** and **flavine adenine dinucleotide (FAD)**. These molecules, in turn, transfer the electrons to a series of protein carrier molecules embedded on the inner surface of the inner membrane of the mitochondrion.[1] The carrier proteins constitute the **electron transport system (ETS)**. The electron transport system accepts electrons from the reduced forms of FAD and NAD (Figure 6–1). These electrons are then passed from one protein to another and are eventually given over to oxygen molecules inside the matrix. During their rapid journey along this chain, the electrons lose energy, much like a hot potato passed along by a line of people. The energy lost along the way is used to make ATP by chemiosmosis, as described in Chapter 5.

The Complete Breakdown of Glucose in Cellular Respiration Produces 38 ATP Molecules

All told, the electron transport system produces 34 ATP molecules per glucose molecule. Because 2 ATPs are gained from glycolysis and 2 are produced in the citric acid cycle, the total output for cellular respiration is 38 per glucose molecule.

Metabolic Water is Produced from Electrons, Hydrogen Ions, and Oxygen

As illustrated in Figure 6–1, oxygen molecules in the mitochondria of plants and animals combine with electrons from the electron transport system and hydrogen ions to form water, called **metabolic water**. Metabolic water supplements the water you and I (and many other animals) must drink to stay alive. For animals like the kangaroo rat that live in arid environments, however, metabolic water

▷ **FIGURE 6–2 Overview of Glycolysis** During this process, glucose is broken down into two molecules of pyruvate. The cell nets two ATPs and two NADHs.

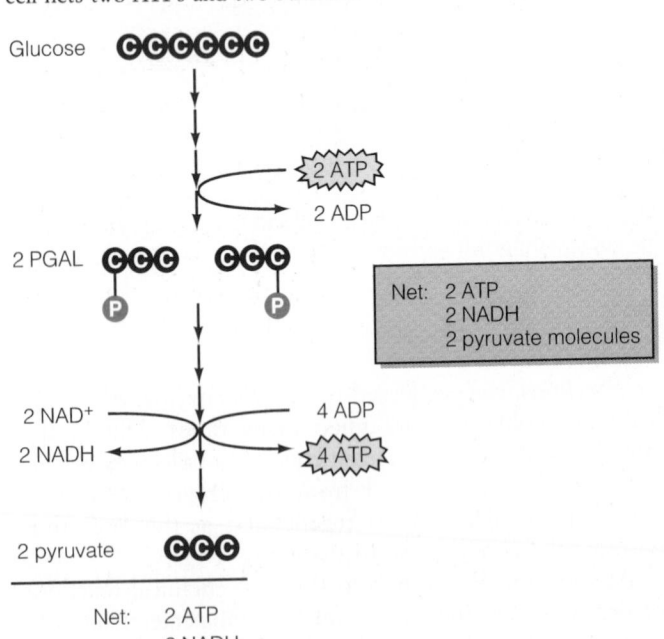

Glucose

2 ATP
2 ADP

2 PGAL

Net: 2 ATP
2 NADH
2 pyruvate molecules

2 NAD⁺   4 ADP
2 NADH   4 ATP

2 pyruvate

Net: 2 ATP
2 NADH
2 pyruvate molecules

[1] Four types of carrier molecule are found here: cytochromes, the predominant one; flavoproteins; a quinone called Coenzyme Q; and several iron-sulfur proteins.

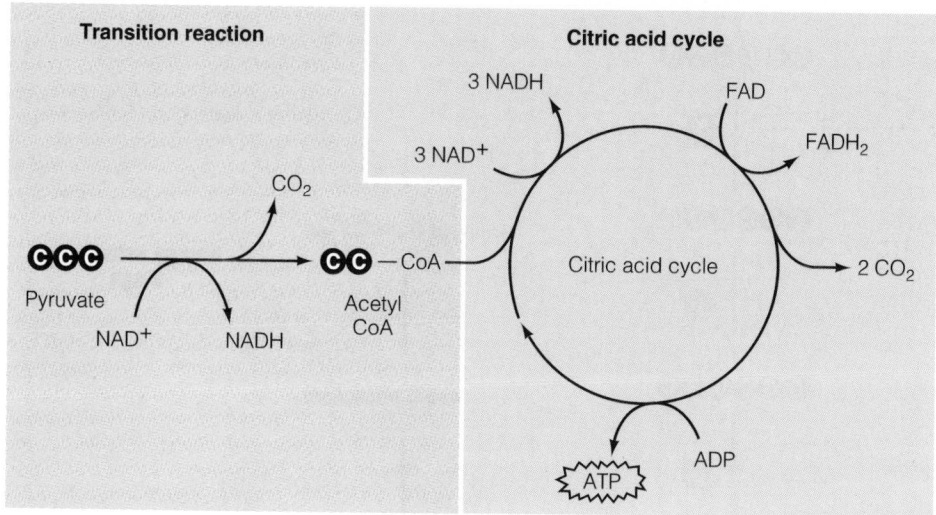

Transition reaction

Pyruvate

NAD^+ → NADH

CO_2

Acetyl CoA — CoA

Citric acid cycle

3 NAD^+ → 3 NADH

Citric acid cycle

FAD → $FADH_2$

2 CO_2

ATP ← ADP

▷ **FIGURE 6–3 Overview of the Citric Acid Cycle**
Occurring in the matrix of the mitochondrion, the citric acid liberates two carbon dioxides and produces one ATP per pyruvate molecule. Its main products, however, are NADH and FADH, bearing high-energy electrons that are transferred to the electron transport system.

provides nearly all of their body needs (see Spotlight on Evolution 16–1).

≈ CELLULAR RESPIRATION: A DETAILED LOOK

Cellular respiration is a finely tuned process. To understand it more fully, we will take a closer look at the reactions and the logic behind them.

During Glycolysis, Glucose Is First Activated, Then Energy Is Harvested From the Molecule

The details of the chemical reactions occurring during glycolysis are shown in ▷ Figure 6–4. At first glance, glycolysis may appear to be a perplexing series of molecular rearrangements generating very little energy output and lots of student confusion. (Recall that glycolysis nets the cell only two ATPs and two NADHs.)

A closer look, however, reveals a two-part strategy to this process. During the early reactions, glucose is activated by the addition of ATP (Figure 6–4). The investment of ATP is a little like giving a boost to get a skier to the top of a hill, so gravity can propel him or her downhill. After activation, the cell is able to harvest some of the energy stored in the glucose molecule. Let's examine these steps in more detail.

As illustrated in Figure 6–4, activation occurs during the first and third chemical reactions of glycolysis. During the first reaction, glucose is converted to glucose-6-phosphate. The formation of glucose-6-phosphate is an endergonic reaction, which is coupled to the breakdown of ATP, an exergonic reaction. The addition of phosphate makes the molecule less stable and easier to break apart. In reaction 2, the glucose-6-phosphate is rearranged and converted into fructose-6-phosphate. In reaction 3, fructose-6-phosphate is energized further by the addition of one more phosphate from ATP. The result is a highly reactive molecule, fructose-1,6-diphosphate (Figure 6–4). So far, two

molecules of ATP have been invested to activate glucose. (The skier has been pushed to the top of the hill.)

In the energy harvest steps, fructose-1,6-diphosphate splits into two three-carbon molecules called **phosphoglyceraldehyde (PGAL)**. In the next reaction, a phosphate group is added to each molecule of PGAL, but the addition of these phosphates does not require ATP. This turns out to be a very important step. As shown in Figure 6–4, each molecule of PGAL ends up with two phosphates, but the cell has only had to use two ATPs to get to this stage. In reactions 7 and 9, the cell strips the phosphates off, using them to produce ATP. Thus, two molecules of ATP are used to activate glucose (reactions 1 and 3), and four ATP molecules are produced during the energy harvest steps (reactions 7 and 9). The cell consequently produces two more ATPs than it uses. In other words, for every molecule of glucose that enters the glycolytic pathway, the cell nets two ATPs.

Besides generating a small amount of ATP, glycolysis produces two molecules of NADH. NADH, introduced in Chapter 4, is an electron carrier. It transfers highly energetic electrons from one reaction and deposits them in another. NADH is produced in reaction 6 during the conversion of glyceraldehyde-3-phosphate into 1,3-diphosphoglycerate. Because two molecules of PGAL are formed for every molecule of glucose and because each of these molecules proceeds through the second half of the glycolytic pathway, the cell generates two NADHs for each glucose molecule.

The NADH molecules produced in reaction 6 transport the high-energy electrons to the electron transport system where the energy of the electrons is used to produce ATP from ADP and inorganic phosphate (explained later).

Overall, glycolysis is a rather inefficient process. That is to say, for all of the chemical activity that occurs during these reactions, the cell strips very little potential energy from the glucose molecule. Glycolysis, however, has served an important purpose: it has chemically modified glucose in preparation for subsequent reactions.

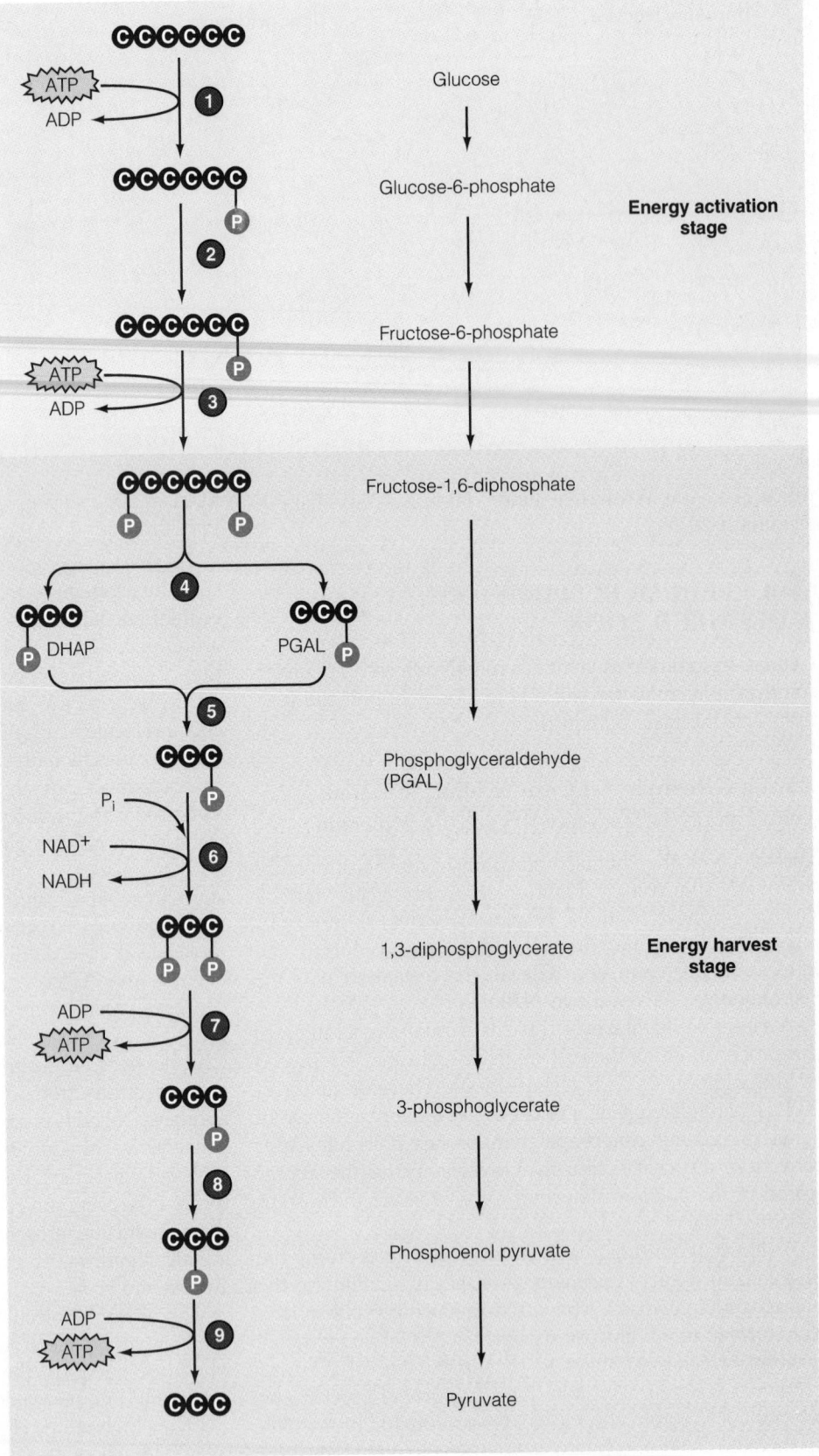

The Transition Reaction Links Glycolysis and the Citric Acid Cycle

In eukaryotes, the two pyruvate molecules formed from glucose during glycolysis pass into the mitochondrial matrix, traversing both mitochondrial membranes. Within the semifluid matrix, each pyruvate molecule reacts with a molecule of Coenzyme A, producing acetyl CoA (▷ Figure 6–5). During this reaction, a carbon dioxide molecule is stripped from each pyruvate, and one molecule of NADH is produced. Because two pyruvates enter the reaction, the cell nets two molecules of carbon dioxide and two molecules of NADH.

The Citric Acid Cycle Consists of Three Stages

The citric acid cycle, the next stage in energy metabolism, can be divided into three general stages: a modest activation stage, an energy harvest stage, and a regeneration stage (▷ Figure 6–6).

In the activation series, acetyl CoA reacts with the four-carbon compound oxaloacetate, forming citric acid, or citrate. Coenzyme A is liberated during this reaction and can be reused.

Citric acid is a six-carbon compound. It is quickly converted to isocitric acid, a slightly energized molecule that is now chemically prepared to proceed through the energy harvest stage of the cycle.

As shown in Figure 6–6, during the energy harvest stage, two carbon dioxide molecules are stripped away for every molecule of pyruvate entering from glycolysis.[2] A single molecule of ATP is also generated for each pyruvate. Because two pyruvates are produced for each molecule of glucose that enters glycolysis, the citric acid cycle yields four carbon dioxide molecules and two ATPs.

[2]Thus, four carbon dioxide molecules are released from each glucose molecule.

The organic substrates passing through the energy harvest reactions of the citric acid cycle release most of their

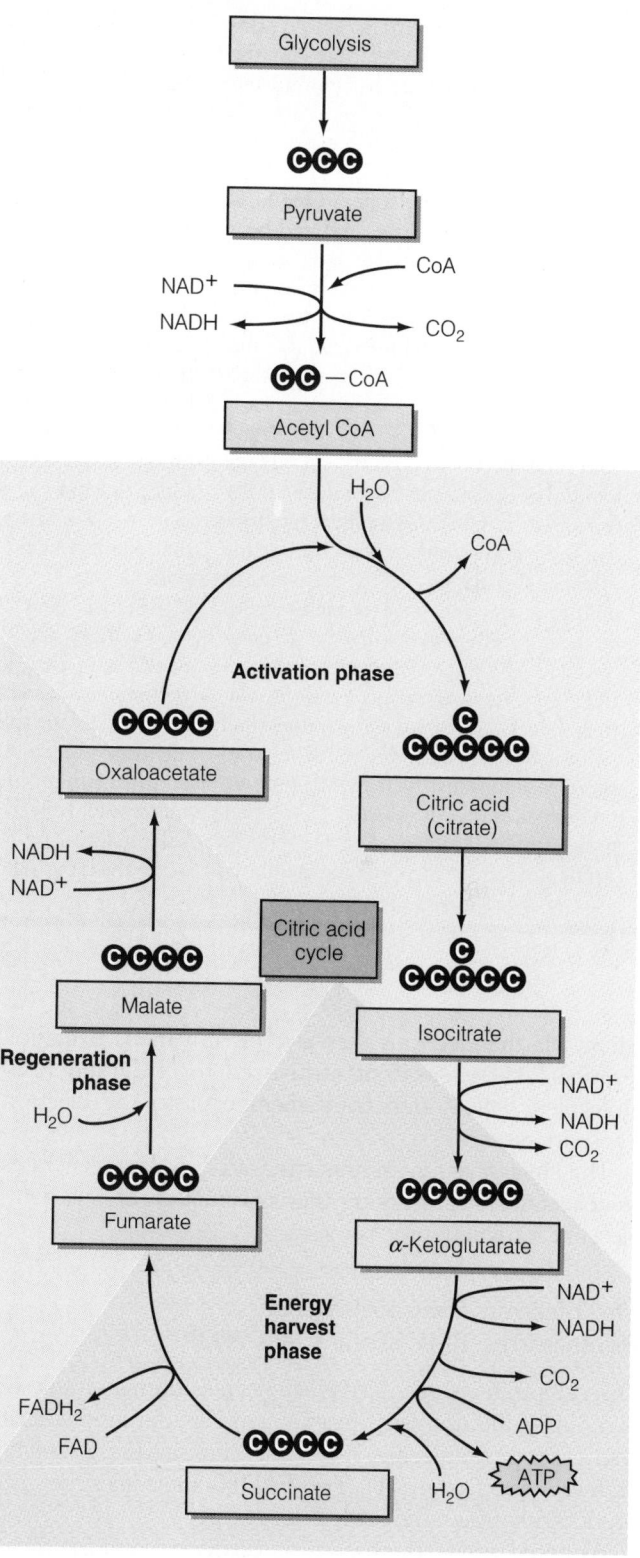

▷ **FIGURE 6–6 A Detailed Look at the Citric Acid Cycle**

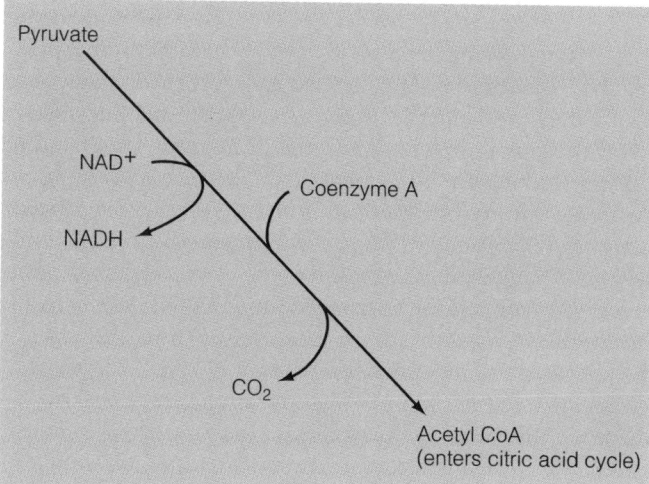

▷ **FIGURE 6–5 The Transition Reaction**

DENIZENS OF THE DEEP FREEZE: ENERGY CONSERVATION IS THE NAME OF THE GAME

It is a cold winter day in the Arctic. The temperature has plummeted to -30°F, and the wind whips across the land with a vengeance. Far out on the ice, three Arctic foxes circle a polar bear dining on a seal it has killed. One hungry fox comes too close, and the bear turns, swiping its massive paw at the tiny, cat-sized fox. The fox turns and scurries across the slippery ice. Its sharp claws and fur-covered feet give the animal traction and permit it to escape unharmed.

Remarkably adapted to the frigid cold, the Arctic fox is one of only a handful of species that lives in the polar regions year round. Getting energy from the food it eats, this animal, like many others, also relies on a number of unique adaptations that conserve cellular heat production in this harsh environment. Its fur-covered feet and its thick coat, for example, provide insulation. In fact, not until temperatures fall below -40°F does the Arctic fox need to produce additional heat by stepping up the breakdown of glucose. Scientists have estimated that even at −95°F, the Arctic fox must increase its metabolic rate (the rate of catabolism generally measured by determining oxygen consumption) by only 37% to stay warm.

Besides helping conserve energy, the fox's thick, white coat provides perfect camouflage in the winter months. During the summer, the coat turns a brownish gray, letting the fox blend with its environment.

Another denizen of the deep freeze is the polar bear. Its thick, white coat, like the fox's, provides extraordinary protection against the cold, minimizing energy loss. But evolution has also "provided" this animal with several additional adaptations conducive to life in the Arctic. First, the hair of a polar bear's coat is akin to optic fibers—clear glass threads that conduct light. The clear hairs of the polar bear pelt, which appear white from a distance, transmit sunlight to the bear's black skin, not visible under its thick coat. This remarkable adaptation permits light to penetrate the insulating fur coat to the black skin, which, like any dark object, absorbs the visible rays and turns them into heat. A truly white fur would reflect sunlight, eliminating solar gain.

At the South Pole lives the emperor penguin (▷Figure 1). In the autumn, when all other Antarctic species are flying or swimming northward to warmer climes, emperor penguins remain, breeding in the depths of winter. In a place where the temperature regularly reaches −50°F and where there is no vegetation for nesting material or protection from bitterly cold winds, emperor penguins rely on three anatomical adaptations to conserve energy and keep warm.

The first is a thick, waterproof layer of feathers. The outer segments of the bird's feathers are short and stiff, and these portions fit over one another like roof tiles, creating a rather impervious layer. The base of each shaft is covered with fluffy strands called down, which form a 1.5-inch layer of insulation beneath the outer layer. The second adaptation is a 1- to 2-inch layer of fat covering the body. This also helps to hold in the heat and keep out the cold. The third adaptation is the animal's compact, round shape, which scientists believe helps it conserve body heat by minimizing surface area.

Life on the windswept ice and snow of Antarctica is also made a little more

---

energy via the electrons they give off in various oxidation reactions. These electrons are passed to NAD and FAD molecules, which soon hand them over to the electron transport system.

The final reaction of the citric acid cycle regenerates oxaloacetate, thus ensuring the continuation of this complex but intriguing process.

The Electron Transport System Produces the Bulk of the Cell's ATP

The cell produces two NADH molecules during glycolysis, two during the transition reaction, and six during the citric acid cycle per molecule of glucose entering the glycolytic pathway (see Table 6–1). Two $FADH_2$ molecules are also produced by the citric acid cycle.

As noted above, these electron carriers siphon the high-energy electrons from oxidation reactions and donate them to the electron transport system, embedded in the inner membrane of the mitochondrion (▷ Figure 6–7). It is here that the energy contained in the electrons is finally released and used to produce most of the cell's ATP.

The energetic electrons move down the electron transport chain, passing from one protein carrier to the next in rapid succession. As the electrons proceed along the chain, they lose energy. According to the chemiosmotic theory, described in Chapter 5, the proteins of the electron transport chain are dual-purpose molecules: they convey electrons from one to another but also serve as hydrogen ion pumps, moving protons (H^+) from the matrix into the outer compartment (Figure 6–7). The energy required to pump hydrogen ions across the membrane comes from the electrons traveling along the electron transport chain, as described in the previous chapter.

As illustrated in Figure 6–7, the hydrogen ions that accumulate in the outer compartment set up an electrochemical gradient (that is, an electrical and concentration difference). Hydrogen ions flow back into the matrix, and as they do, they create a tiny electrical current that drives the production of ATP from inorganic phosphate and ADP.

bearable by a behavioral adaptation. Emperor penguins huddle together in large groups containing several hundred birds. The outer birds shield those in the middle from the wind. But the birds on the outer fringes constantly migrate inward, changing positions and sharing the duty.

Surviving reproduction is no easy task for the emperor penguin. In fact, emperor penguins go without food for up to three months as they huddle together on the ice—that's three months of nearly complete darkness with wind-chill temperatures reaching -150°F. Despite the close quarters and the absence of food, there is very little aggression in emperor penguin rookeries, which some scientists believe is another behavioral adaptation that minimizes energy loss.

When the eggs are laid, the emperor penguin males hold them on their feet and cover them with a flap of abdominal skin, protecting the eggs from the cold from above and below. The females head north for open water while the males take over the two-month incubation. When the eggs hatch, the females begin to reappear, and the males quickly transfer the babies to their mother, then head for open water to feed and recover body weight lost during the previous two months.

Soon the ice begins to break up, and both parents share in feeding squid and fish to their young. By the time spring arrives, the young have developed a thick plumage and a layer of fat necessary to survive the rigors of the sea. In fact, scientists believe that it is this requirement for adequate insulation that necessitates a midwinter breeding and incubation.

▷ **FIGURE 1 Emperor penguin** This magnificent creature endures some of the coldest temperatures on Earth.

Oxygen Is Essential for Energy Production

After having given up most of their energy, electrons combine with protons (hydrogen ions) and oxygen to form water (Figure 6–7). Oxygen is therefore the final electron acceptor in the ETS. In a sense, it clears the electron transport system, leaving it free to accept additional electrons from the citric acid cycle. Without oxygen, the flow of electrons through the ETS would halt. Protons would no longer be pumped into the outer compartment and ATP synthesis would stop. NADH and $FADH_2$, which accept electrons from the citric acid cycle, would have nowhere to dump their load, and the citric acid cycle would grind to a halt. At this point, the cell would have to rely on glycolysis for energy. Thus, when oxygen levels inside a cell fall, energy production becomes much less efficient—so inefficient that a prolonged interference with the citric acid cycle and the electron transport system may result in death.

The Citric Acid Cycle Provides a Pathway for Energy Production From Fats and Proteins

The citric acid cycle is much more than a pathway for glucose catabolism. It also provides a means by which cells can generate ATP from fats and proteins (▷ Figure 6–8).

At rest, the body generates energy from two principal sources: carbohydrates and fats. The most common form of fat, triglycerides, is broken down to form glycerol and fatty acids. Fatty acids are transported into the mitochondria, where they are catabolized in a series of reactions that remove two carbons at a time. The two-carbon groups react with Coenzyme A, forming acetyl CoA, which then enters the citric acid cycle (Figure 6–8). Glycerol, the other major product of triglyceride breakdown, enters the glycolytic pathway. Gram for gram, fats produce 2.5 times more

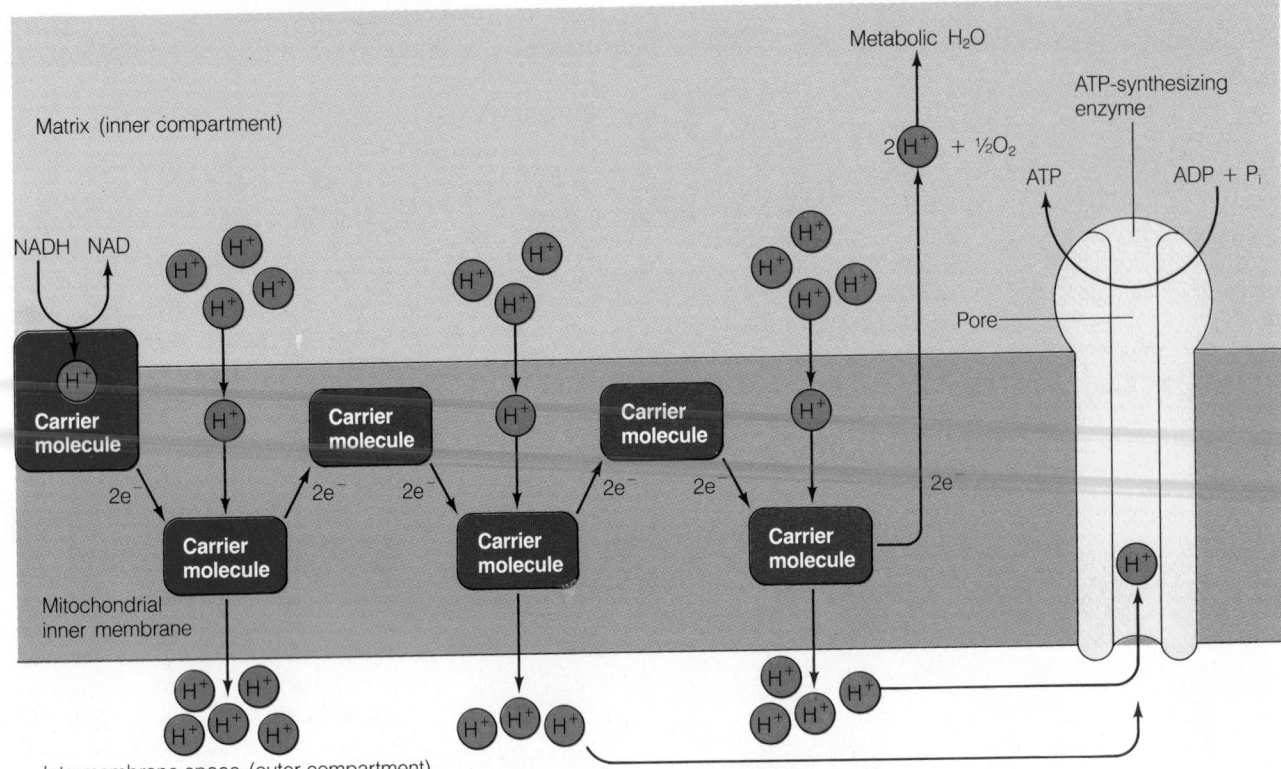

▷ **FIGURE 6–7 The Electron Transport System** Located on the inner surface of the mitochondrial inner membrane, the electron transport system produces ATP via chemiosmosis, described in the text.

ATP than carbohydrates, and they are therefore an important source of energy.

In individuals severely deprived of carbohydrates and fats or in those feeding primarily on protein, amino acids become a major source of energy. As shown in Figure 6–8, some amino acids are converted to pyruvate. Others are converted to acetyl CoA. Still others are converted to substrates of the citric acid cycle. These molecules then proceed through the remaining stages of cellular respiration. Gram for gram, protein produces as much energy as carbohydrates.

The Citric Acid Cycle Provides Chemical Building Blocks for Making Amino Acids and Fatty Acids

Further adding to their usefulness, the metabolic pathways of cellular respiration also produce chemical building blocks needed to synthesize certain amino acids and fatty acids, as illustrated in ▷ Figure 6–9.

Amino acids are produced from pyruvate and two substrates of the citric acid cycle. Although much more complicated than shown here, the reactions provide about a dozen amino acids.

Figure 6–9 shows that fatty acids can be produced from acetyl CoA. This explains why eating too many sweets can make a person fat. As shown in the figure, glucose from table sugar is broken down during glycolysis and converted into acetyl CoA. If the cell needs energy, acetyl CoA enters the citric acid cycle, where it is broken down. If the cell has sufficient energy, however, acetyl CoA is used to synthesize fatty acids. Fatty acids, in turn, combine with glycerol to produce triglycerides. In humans, fatty acids and triglycerides are synthesized in the liver. The liver stores some of the triglycerides in its cells, but most are stored in specialized fat cells distributed throughout the body but often concentrated in certain areas like the waist and hips.

In summary, the metabolic pathways of cellular respiration provide an extraordinary measure of flexibility for cells—producing energy (ATP) when it is needed from carbohydrates or other sources and providing molecular building blocks for amino acids and fats in times of surplus.

Overall, Cellular Respiration Is an Exergonic Reaction

As you have just seen, cellular respiration is a series of enzyme catalyzed reactions that gradually releases the potential energy stored in the carbon-carbon bonds of glucose in a controlled "cellular burn." As ▷ Figure 6–10 shows, except for the initial stages of the reaction series (during glycolysis) and one reaction in the citric acid cycle (the conversion of citric acid to isocitrate), potential energy is lost at each and every reaction. This energy is carried away from the reactions by electrons donated to NAD and FAD and also by ATP molecules.

Because no energy conversion is 100% efficient, some of

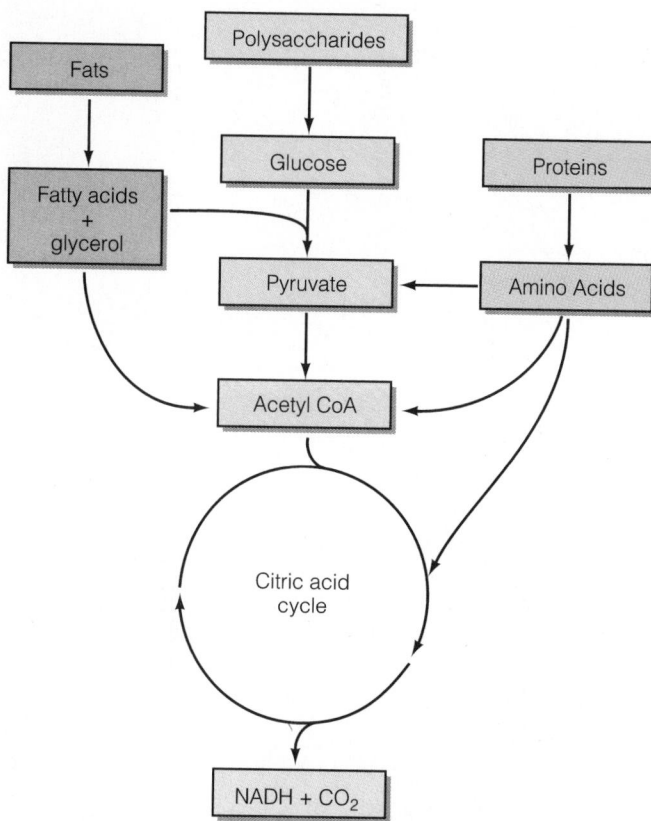

▷ **FIGURE 6–8 Source of Cellular Energy** The metabolic pathways of cellular respiration accept substrates from three sources: carbohydrates (polysaccharides), fats, and proteins. All three sources, therefore, can provide cellular energy. Although not shown here, all of these reactions are reversible.

the energy is lost as heat. Heat given off by these chemical reactions is responsible for body heat. Body heat keeps warm-blooded animals warm and is essential for many enzymatic reactions.

≋ THE CONTROL OF CELLULAR RESPIRATION

Oxygen levels profoundly influence cellular respiration. As noted earlier, when oxygen levels fall below critical levels the citric acid cycle shuts down. But oxygen is not the primary controlling mechanism in cellular energy production under most circumstances. The most important regulators are two products of cellular respiration, ATP and citric acid, which form a feedback loop with an enzyme in glycolysis known as **phosphofructokinase (PFK)** ▷ Figure 6–11). PFK catalyzes the conversion of fructose-6-phosphate to fructose-1,6-diphosphate (reaction 3 of glycolysis). Phosphofructokinase has an allosteric site to which ATP and citric acid bind, shutting off the enzyme. (Enzyme inhibition by end products was discussed in Chapter 4.) In the same way that a dam across a river halts downstream flow, inhibition of PFK terminates all "downstream" processes, including the citric acid cycle and the electron transport system. When this happens, cellular energy production comes to a standstill.

Enzyme inhibition by citric acid and ATP occurs when a cell's demand for energy falls. When energy demands are low, ATP and citric acid levels increase. These molecules bind to the allosteric site of phosphofructokinase, shutting down the enzyme.

▷ **FIGURE 6–9 Fat and Protein Synthesis** Fats and proteins can be synthesized from substrates of cellular respiration.

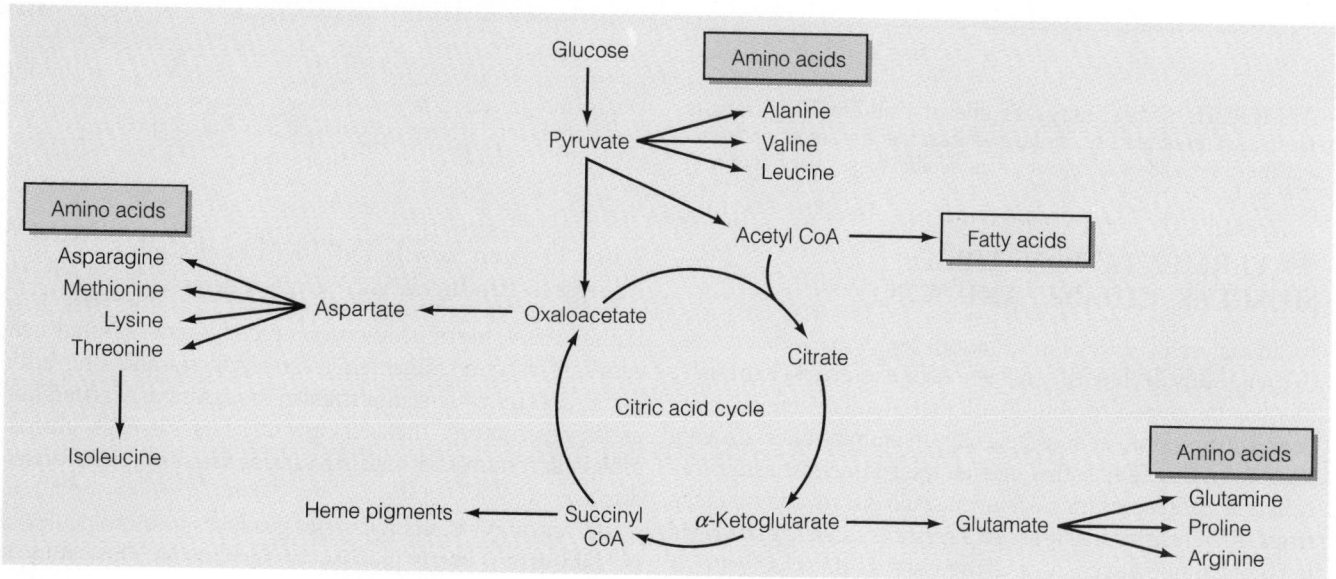

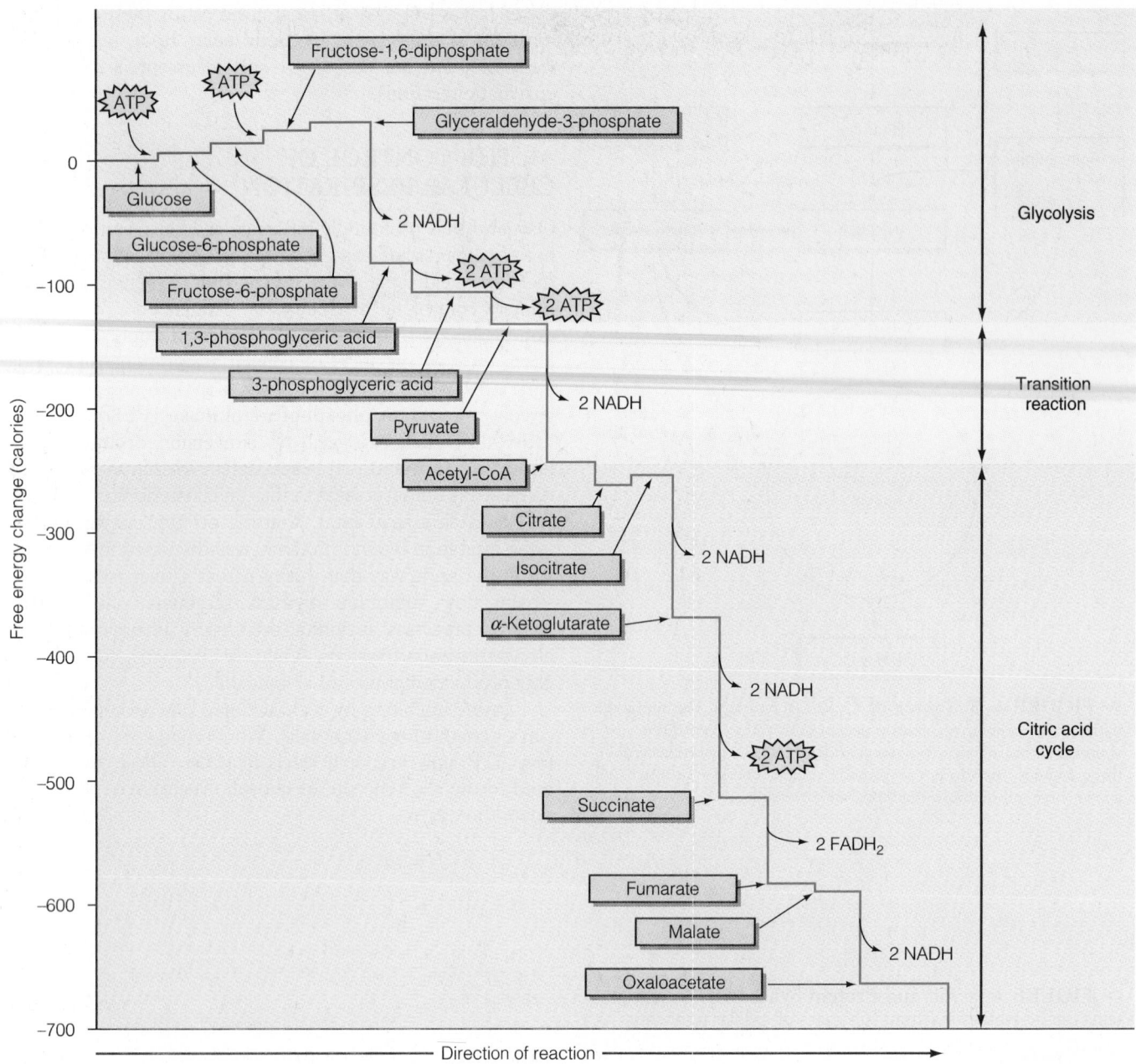

FIGURE 6–10 Energy Profile of Cellular Respiration
The oxidation of glucose during glycolosis, the transition reaction, and the citric acid cycle releases considerable amounts of energy. During this process, however, there are two activation stages, one in glycolysis and one brief one during the citric acid cycle.

≋ FERMENTATION: TIRED MUSCLES, CHEESE, AND WINE

In animal cells, cellular respiration comes to a halt when oxygen levels decline. If you have ever exercised vigorously, this has happened to you. During strenuous exercise, oxygen consumption in muscle cells of animals may exceed replenishment. When this occurs, oxygen levels inside the cells fall. With oxygen no longer available to accept electrons, the electron transport system shuts down. If the ETS becomes inoperative, so does the citric acid cycle.

When Oxygen Levels Fall in Animal Cells, Energy Is Produced via Fermentation

Rather than shutting down completely, most animal cells continue to generate energy via glycolysis and one additional reaction. This alternative, however, is exceedingly inefficient, netting the cell only 2 ATPs for each glucose molecule, compared with 38 during the complete breakdown.

As you may recall, glycolysis produces two molecules of NADH from every glucose molecule. As illustrated in

▷ Figure 6–12a, when the electron transport system is no longer able to accept electrons, NADH produced during glycolysis finds an alternative donor, pyruvate (the end product of glycolysis). Animal cells convert pyruvate to lactic acid in a reduction reaction. This reaction liberates NAD, thus keeping the glycolytic pathway running. The production of lactic acid from pyruvate is called **lactic acid fermentation.**

Intense Exercise Depletes Muscle Oxygen, Causing the Buildup of Lactic Acid

In humans, heavy exercise, such as weightlifting and running, depletes muscle oxygen and results in the buildup of lactic acid in muscle cells. This causes muscle fatigue.[3]

[3]At one time scientists thought that muscle soreness resulted from the buildup of lactic acid, but today it is believed to be caused by physical damage to muscle fibers caused by intense exercise.

▷ **FIGURE 6–11 Feedback Control of Cellular Respiration** Citric acid (citrate) and ATP levels regulate cellular respiration through a feedback mechanism with phosphofructokinase (PFK), an enzyme in the glycolytic pathway.

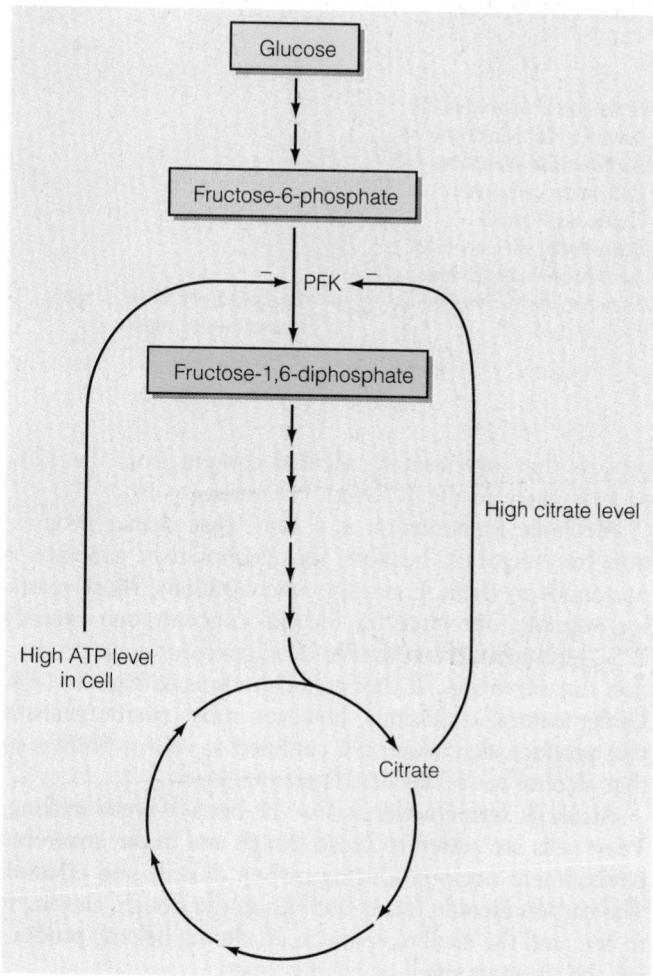

Lactic acid, however, diffuses out of the muscle within hours and is carried by the blood to the liver. There it is converted to pyruvate, then to glucose in a series of chemical reactions that is essentially the reverse of glycolysis ▷ Figure 6–13). Some of the glucose is used to synthesize glycogen, and the rest is released into the blood, where it is redistributed to body cells and used to generate energy.

Fermentation Is the Chief Source of Energy for Bacteria and Other Single-celled Organisms

Organisms can be classified according to their need for oxygen as an electron acceptor. Many organisms require oxygen and are referred to as **obligate aerobes.** Humans are a good example. Without oxygen we die.[4] Other organisms (some bacteria) cannot tolerate oxygen at all and are known as **obligate anaerobes.** Not surprisingly, they live in oxygen-free environments, such as in the mud beneath a pond or in soil. Still others can function in either mode and are known as **facultative anaerobes.** Many bacteria and fungi are facultative anaerobes.

Obligate anaerobes—those that live in oxygen-free environments—meet their energy needs by fermentation, as do facultative aerobes that are subjected to oxygen deprivation. In these organisms, the fermentation pathway involves the reactions of glycolysis plus one of several anaerobic options. The most common fermentation option is lactic acid fermentation, discussed above. It is the major energy-yielding pathway in many bacteria and in animal cells operating under anaerobic or relatively anaerobic conditions.

Besides providing energy, lactic acid fermentation is also an important source of food for humans. Table 6–2 shows some of the more popular foods produced by fermentation, including cheeses and yogurts, and shows the starting material and the bacteria and fungi involved.

Most cheeses, for instance, are produced from milk and selected bacteria. To make cheese, lactic acid bacteria are added to milk. These bacteria catabolize milk sugar (lactose), releasing glucose molecules in the process. Glucose molecules are then broken down in the glycolytic and fermentation pathways, producing lactic acid. Proteolytic enzymes are then added to cause the liquid to curdle. The thin watery part (whey) is drawn off, leaving behind the unripened cheese (for example, cottage cheese). Additional bacteria and fungi may be added to break down milk fat and protein, yielding specific ripened cheeses, such as the ever-popular blue cheese.

Some Organisms Produce Ethanol and Carbon Dioxide During Fermentation

Some microorganisms undergo **alcoholic fermentation,** a process that produces ethanol and carbon dioxide from

[4]Some cells of aerobic organisms, like muscle cells, can function temporarily under anaerobic conditions if necessary.

TABLE 6–2 Foods Produced by Fermentation

| FOOD OR PRODUCT | RAW STARTING MATERIAL | FERMENTING AND FLAVOR-CONTRIBUTING MICROORGANISMS | |
		Bacteria	Fungi
Breads			
Cakes, rolls	Wheat flours		Saccharomyces cerevisiae
San Francisco sourdough bread	Wheat flours	Lactobacillus sanfrancisco	Saccharomyces exiguu
Dairy Products			
Acidophilus milk	Milk	Lactobacillus acidophilus	
Cheeses (ripened)	Milk curd	Brevibacterium spp.	
		Lactic acid bacteria	
		Micrococcus caseolyticus	
		Propionibacteria	
		Streptococcus cremoris	
		Streptococcus lactis	
		Streptococcus thermophilus	
Cultured buttermilk	Milk	Leuconostoc cremoris	Geotrichum spp.,
		Streptococcus cremoris	Penicillium camemberti,
		Streptococcus lactis	Penicillium roqueforti
Yogurt	Milk and milk solids	Lactobacillus bulgaricus	
		Streptococcus thermophilus	
Meat Products			
Country cured hams	Pork hams		Aspergillus spp.
			Penicillium spp.
Dry sausages	Pork, beef	Pedicoccus cerevisiae	
Nonbeverage plant products			
Olives	Green olives	Leuconostoc mesenteroides	
		Lactobacillus plantarum	
Pickles	Cucumbers	Lactobacillus plantarum	
		Pedicoccus cerevisiae	
Poi	Taro roots	Lactic acid bacilli	
Sauerkraut	Cabbage	Leuconostoc mesenteroides	
		Lactobacillus plantarum	
Soy sauce (shoyu)	Soybeans, wheat, rice	Lactobacillus delbrueckii	Aspergillus oryzae or A. soyae, Saccharomyces rouxii

pyruvate (Figure 6–12b).[5] As in lactic acid fermentation, NADH produced during the glycolytic pathway is used to reduce pyruvate.

Alcoholic fermentation is the basis of the production of beer, wine, and hard liquor. Beer is the product of a yeast that catabolizes carbohydrates in barley and other grains. The carbon dioxide released during the process makes beer carbonated.

Wine is made by fermenting fruit juices and extracts of many plants, such as dandelions. Hard liquors come from a variety of carbohydrate sources. Whiskey, for example, comes from rye, and bourbon comes from corn. The higher alcohol content in hard liquors results from distillation, a process that increases the alcohol content from the 12%-to-18% range to the 50%-to-100% range.

Alcoholic fermentation is a somewhat chancy proposition for microbes, because the alcohol they generate is poisonous to them at certain concentrations. Most yeasts, for example, die when the alcohol concentration exceeds 12%, although certain strains developed for wine production can survive in alcohol concentrations as high as 18%. Under natural conditions, however, most microorganisms that produce alcohol are not confined to vats or bottles, so that alcohol never reaches dangerous levels.

Alcoholic fermentation is also the basis of bread making. Yeast cells are added to bread dough and in the anaerobic environment begin producing carbon dioxide and ethanol. The carbon dioxide forms bubbles in the dough, causing it to rise, and the alcohol is driven off during baking, producing the pleasant smell of baking bread.

---

[5] This process actually involves two chemical reactions, the first of which produces carbon dioxide and acetaldehyde, a two-carbon compound. In the second, acetaldehyde is converted into ethanol.

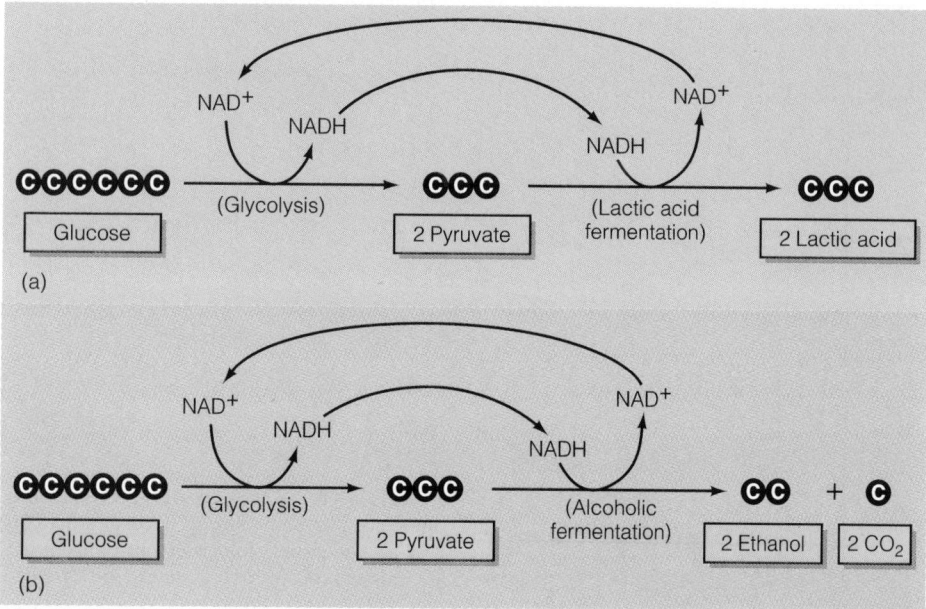

▷ **FIGURE 6–12 Fermentation** (*a*) Pyruvate or pyruvic acid, from glycolysis is converted into lactic acid in animal cells when oxygen levels fall. (*b*) Fermentation can produce a variety of products in bacteria and yeast cells. In certain microorganisms, pyruvate is converted into ethanol and carbon dioxide.

 **ENVIRONMENT AND HEALTH: COMPOUND 1080, COYOTES, AND COMMON SENSE**

Humankind has made a science out of killing species that compete with our needs. Mountain lions, eagles, bears, various insects, and a host of other organisms have, over the years, been targeted for extermination. The chief by-product of the science of pest control has been an arsenal of chemical biocides. For weeds in farm fields, scientists have devised a wide range of herbicides. For insects, an equally diverse group of insecticides has emerged. For rodents, there is an assortment of toxic rodenticides.

In the Midwest and West, many ranchers and farmers see the coyote as an enemy (▷ Figure 6–14). This shy predator and scavenger is a relative of the dog and flourishes today on an ever-widening range, despite intense efforts to "control" its numbers. Where wolves once howled, the coyote makes its home, feeding on insects, rodents, small birds, carrion of all sorts, and even an occasional sheep or a house cat that wanders too far from home.

Called by the Navajo Indians "God's dog," the coyote receives high praise for its cunning and resistance to human control efforts. But to ranchers, the coyote is a loathsome animal. They maintain that coyotes kill thousands of sheep annually, costing the industry $100 million a year. Environmentalists, however, believe that this estimate is a gross exaggeration meant to stir public dislike for the animal.

Their studies put the economic damage caused by the coyote, at about $10 million a year. They assert that the economic costs of control far exceed the benefits and argue that society can reduce coyote damage to herds for a fraction of the current cost.

Over the years, humans have attempted to keep coyote populations under control by shooting, poisoning, trapping, and even burning or gassing the animals in their dens. One of the most popular weapons in the battle against the coyote was a poison called Compound 1080, or simply 1080.

Compound 1080 is sodium monofluoroacetate, one of the most toxic biocides known to humankind. It works by binding to coenzyme A in the transition reaction of the citric acid cycle, bringing the process to an abrupt halt. Cells end up producing fluoroacetyl coenzyme A, which reacts with oxaloacetate, forming fluorocitrate. But that is as far as it goes, for fluorocitrate inhibits the next enzyme in the pathway, shutting down the citric acid cycle and causing death.

So toxic is 1080 that a single teaspoon can kill 100 adults. A healthy adult who handles meat soaked with this deadly compound without gloves is very likely to become ill from poison absorbed through the skin. A child who licks his or her fingers after handling poisoned meat may die.

Ranchers and government predator control agents once treated meat and animal carcasses with 1080, spreading them around the land. Although the poison was intended for coyotes that preyed on sheep, many "nontarget" species such as badgers, eagles, dogs, foxes, and raccoons also fed on the bait and died.

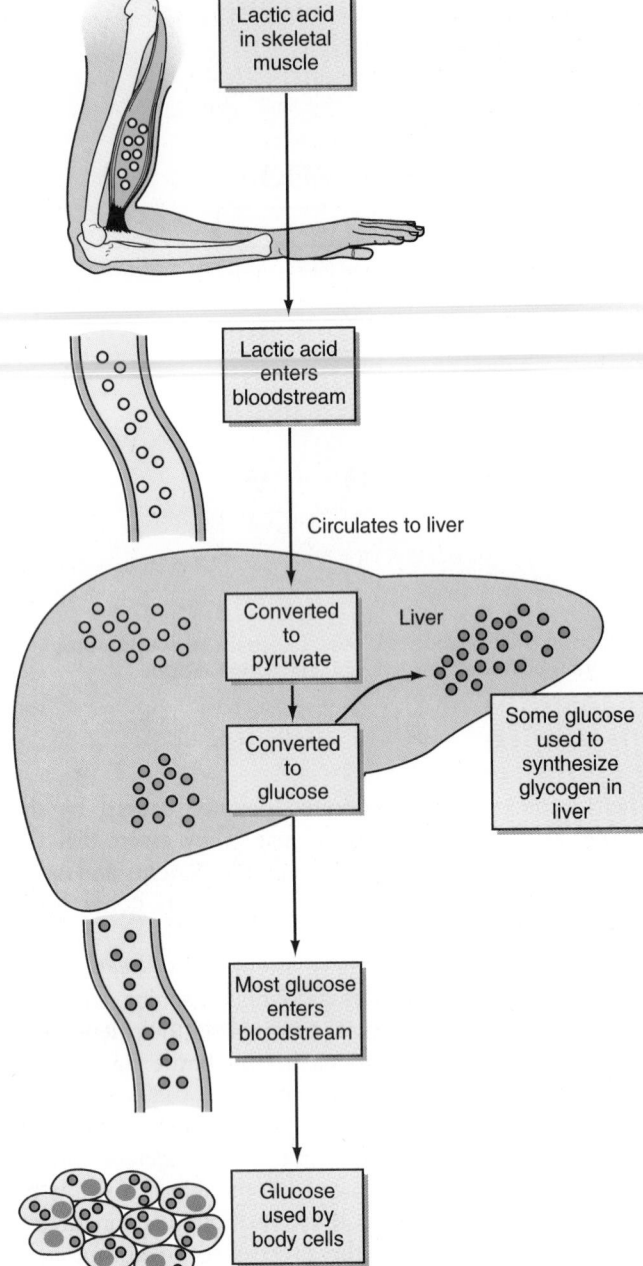

▷ **FIGURE 6–14 Coyote** Opinions on this animal differ. Some see it as a menace in need of strict control. Others recognize it as an important ecological control agent itself, helping to regulate populations of rabbits, mice, and even insects.

▷ **FIGURE 6–13 Fate of Lactic Acid in Humans** In humans, lactic acid builds up in body muscles that have depleted their supply of oxygen during vigorous exercise. Lactic acid leaves the muscle and circulates in the bloodstream to the liver, where it is converted into glucose.

In 1972, President Richard Nixon banned the use of 1080 for predator control on federal lands and in programs sponsored by the federal government. That same year, after determining that 1080 was responsible for 13 human fatalities and six nonfatal poisonings, the EPA banned the chemical for commercial sale, putting an end to its use on private property.

In February 1982, under heavy pressure from the livestock industry, President Ronald Reagan rescinded the EPA ban. This action left it up to the EPA to determine the fate of 1080. Despite pressure from environmentalists, the EPA reauthorized the use of 1080 in sheep collars, beginning in 1986.

Sheep collars contain small packets of liquid 1080 and are worn around the sheep's neck. When a coyote bites into the collar, it receives a lethal dose of 1080. This technique, say supporters, is far more selective than other measures and should eliminate only sheep-killing coyotes and reduce incidental poisoning of nontarget species, such as eagles and bears. However, biologists point out that scavengers such as ravens and vultures will very probably feed on the coyotes killed by 1080. Many nontarget species will die as a result. Another concern of wildlife biologists and environmentalists is that unscrupulous ranchers will extract 1080 from the collars, use it to treat meat cubes, which they spread around their property. This, in turn, could kill many animals besides coyotes.

Opponents of 1080 argue that there are much safer ways to control troublesome coyotes. In Kansas, for example, only coyotes that raid sheep herds can be shot. Thus, the state has a large population of coyotes and a healthy sheep population; costs of control are minimal. In other states, like Wyoming, hundreds of thousands of dollars are spent to kill coyotes.

In many western states, ranchers release their flocks into pastures or mountain meadows, where the sheep are left unattended throughout much of the grazing season. Coyotes with young to feed find the helpless sheep easy prey. Even in fenced-in pastures or pens, lambs fall victim to coyotes. In these situations, guard dogs can provide another economical and effective means of controlling coyote predation (▷ Figure 6–15).

Guard dogs have been used successfully in Europe for hundreds of years and, where they have been tried in the United States, have greatly reduced coyote predation. Liv-

ing with the flock, guard dogs watch over their sheep with great care. The dogs' protective instincts are so powerful, in fact, that they need virtually no training. Pups are simply placed in a pen containing a flock of sheep and kept there day and night. Exposed to sheep early in life, the dog becomes part of the flock and is soon transformed into a fierce protector.

Controlling coyotes by guard dogs is just one of many environmentally sound ways of protecting human interests without damaging the environment. It symbolizes a strategy of cooperation that replaces outmoded ways of domination that have wrought so much destruction to the web of life.

▷ **FIGURE 6–15 Guard Dogs**

SUMMARY

1. Sunlight powers the entire living world—not just plants but also animals and virtually all microorganisms.
2. Solar energy trapped by plants is used to produce carbohydrates and is stored in their covalent bonds.
3. This energy can be released in a controlled series of reactions occurring in the cytoplasm and mitochondria of plant and animal cells.

CELLULAR RESPIRATION[6]

4. The complete breakdown of glucose is called cellular respiration.
5. Cellular respiration consists of four interdependent parts: glycolysis, the transition reaction, the citric acid cycle, and the electron transport system.

GLYCOLYSIS

6. Glycolysis is a series of reactions occurring in the cytoplasm of plant and animal cells and in microorganisms. During glycolysis, glucose is split in two. Each molecule of glucose yields two molecules of pyruvic acid, or pyruvate.
7. Reduced to its essentials, glycolysis consists of two major processes, glucose activation steps and energy harvest steps. During glucose activation, the cell invests some energy in the form of ATP to activate the molecule. In the energy harvest steps, the cell splits the glucose molecule in two and then systematically strips off the phosphates added in the first part of the reaction sequence.
8. Glycolysis nets the cell two molecules of ATP. ATP is used to drive endergonic reactions.
9. Glycolysis also nets the cell two NADH molecules containing high-energy electrons. The electrons are passed to the electron transport chain.

THE TRANSITION REACTION

10. Pyruvate produced during glycolysis diffuses into the matrix of the mitochondrion and reacts with Coenzyme A. During this reaction, one carbon dioxide molecule is released. Two NADH molecules are also produced for every molecule of glucose entering glycolysis above.

[6]In this summary, I have combined the overview and detailed view of cellular respiration.

11. Acetyl CoA, the product of the transition reaction, next enters the citric acid cycle, located in the matrix of the inner compartment of the mitochondrion.

THE CITRIC ACID CYCLE

12. Acetyl CoA reacts with oxaloacetate to produce citric acid. Citric acid then proceeds through a series of reactions. During these reactions, two carbon atoms are removed, regenerating oxaloacetate.
13. The citric acid cycle produces two ATPs and numerous molecules of NADH and $FADH_2$. These electron carriers accept high-energy electrons from various reactions in the citric acid cycle and transport them to the electron transport system.

THE ELECTRON TRANSPORT SYSTEM

14. The electron transport system consists of a series of protein molecules embedded in the inner surface of the inner membrane of the mitochondrion.
15. Electrons from NADH and $FADH_2$ are passed down the chain and are eventually given to oxygen.
16. Electrons moving along the transport system cause these dual-purpose molecules to pump protons (H^+) out of the matrix into the intermembrane compartment. This creates an electrochemical gradient.
17. The electrochemical imbalance drives hydrogen ions inward through special pores. As the hydrogen ions flow through the pores in the inner membrane, they create a tiny but measurable electrical current. This is used by the cell to drive the endergonic production of ATP from inorganic phosphate and ADP.
18. Electrons moving down the electron transport chain ultimately combine with oxygen and two hydrogen ions from the matrix to form metabolic water.
19. Without oxygen, electrons accumulate and shut down the electron transport system. Hydrogen ions are no longer pumped into the outer compartment, and ATP synthesis stops. NADH and $FADH_2$ recycling stops, halting the citric acid cycle. At this point, cells must rely on glycolysis to generate energy.
20. Cells use the metabolic pathways of cellular respiration

to harvest energy from fats and proteins under certain conditions.

21. Substrates of the citric acid cycle can be used to synthesize various amino acids and fatty acids.

THE CONTROL OF CELLULAR RESPIRATION

22. Cellular respiration is primarily controlled by the levels of ATP and citric acid in a negative feedback mechanism with the enzyme phosphofructokinase. Phosphofructokinase catalyzes the third reaction of glycolysis.

23. Both ATP and citric acid bind to the allosteric site on the enzyme, shutting it off. This stops the breakdown of glucose and halts energy production when the cell's energy demands fall.

FERMENTATION: SORE MUSCLES, CHEESE AND WINE

24. When oxygen levels fall, animal cells fall back on fermentation, a chemical reaction in which pyruvate is converted into lactic acid or some other product. In skeletal muscle cells, lactic acid produces muscle fatigue.

25. Alcoholic fermentation produces ethanol from pyruvate and occurs in certain microorganisms.

26. Wines, beer, and hard liquor are all produced by combining plant by-products (grains or fruits) with microorganisms that are alcoholic fermenters.

27. Lactic acid fermentation is also an important source of food for humans, including cheeses and yogurts, which are made from milk and selected bacteria.

ENVIRONMENT AND HEALTH: COMPOUND 1080, COYOTES, AND COMMON SENSE

28. Humans rely on an arsenal of biocides to control pests, but most of these chemicals are also harmful to people and to other species.

29. In the Midwest and West, one of the chief enemies of ranchers has been the coyote. Over the years, humans have attempted to keep coyote populations under control by a variety of means.

30. One of the most popular has been Compound 1080, or sodium monofluoroacetate. It works by inhibiting the citric acid cycle.

31. Although the poison is intended for coyotes that prey on sheep, many "nontarget" species such as badgers, eagles, dogs, foxes, and raccoons have also fed on bait and died.

32. Although 1080 was banned in 1972, its use in sheep collars has since been reinstated. This technique, say supporters, is far more selective than other measures, should eliminate only sheep-killing coyotes, and should reduce incidental poisoning of nontarget species, such as eagles and bears.

33. However, biologists point out that scavengers such as ravens and vultures will feed on the coyotes killed by 1080. Despite assurances from the ranching community, many nontarget species will die.

34. Another concern of wildlife biologists and environmentalists is that unscrupulous ranchers may extract 1080 from the collars, use it to treat meat cubes, which they spread around their property, killing many additional species.

35. Opponents of 1080 argue that there are much safer ways to control troublesome coyotes. One of them is the guard dog.

36. Guard dogs greatly cut down on coyote predation and have been used successfully in Europe for hundreds of years. They do the job with minimum impact on the environment, promoting a healthy sheep population and healthy populations of bears, badgers, eagles, and others.

EXERCISING YOUR CRITICAL THINKING SKILLS

You own a guest ranch in northern Wyoming. Besides horseback riding, your guests are treated to abundant opportunities to view wildlife. They have spotted bear, deer, elk, eagles, hawks, and many other species. You also have a large garden that provides fresh vegetables for your guests and helps hold down costs.

Because the summer season does not net enough income to sustain your operations, you also raise sheep for wool and meat. One day a neighbor approaches you and asks for help in controlling coyotes, which he says are killing off his lambs and ewes (pronounced, yous). In fact, he found a dead ewe the other day on a ride through his property. The animal had been eaten by coyotes, or so he said.

Your neighbor wants you to join him in spreading Compound 1080 on your land. It's not legal, but he can extract the poison from sheep collars, which are. Your neighbor asks for permission to spread bait on your property. "We've gotta kill them coyotes," he says, leaving you to ponder the proposition.

After rereading the Environment and Health section, what concerns would you have about joining a 1080 control program?

Make a list of questions you will need to answer before deciding whether to help your neighbor.

This exercise encourages you to examine the big picture—all the ways in which this effort might affect your operations.

Here's a list of questions to compare with yours: Will this effort affect your wildlife-watching opportunities and your future income? Will it affect the number of rabbits on your range and the amount of available grasses for your sheep? Will it affect the safety of your guests? Will it affect your garden by increasing pests? Do you really need to kill all of the coyotes to make the sheep perfectly safe? Are there other ways to control coyote predation on sheep?

Another important line of questioning goes to the basic premise of your neighbor—that is, that coyotes are killing lots of his sheep. Is this true? Is he exaggerating? What is the true extent of the damage? Does it merit the economic and environmental costs of the drastic control measure your neighbor is proposing? How would you find answers to the latter questions?

1. _____ _____ is the complete breakdown of glucose into carbon dioxide and water.
2. Glycolysis is the breakdown of glucose into two molecules of _____ . During glycolysis, the cell nets _____ ATPs.
3. Glycolysis occurs in the _____ of plant and animal cells.
4. The product of the transition reaction is called _____ _____ .
5. The citric acid cycle occurs in the _____ of the mitochondrion. The cycle begins and ends with a four-carbon compound called _____ .
6. During some of the chemical reactions of the citric acid cycle, electrons are given off and accepted by two electron-carrier molecules, _____ and _____ .
7. The bulk of the ATP produced during the complete breakdown of glucose in animal cells is produced in the _____ _____ _____ via a process called _____ .
8. The citric acid cycle also provides a way to harvest energy from _____ and _____ .
9. The breakdown of glucose is regulated by one key enzyme found in glycolysis. That enzyme is _____ and is controlled by levels of _____ and _____ .
10. _____ is the process by which the product of glycolysis, _____ , is converted to ethanol or _____ .
11. The breakdown of glucose in glycolysis yields _____ ATPs.
12. The complete breakdown of glucose in cellular respiration (including glycolysis and the electron transport system) yields _____ ATP molecules.

Answers to the Test of Terms are found in Appendix B.

1. Explain the statement: "All life depends on sunlight energy."
2. Draw a diagram that illustrates the four stages of cellular respiration. Note the major products of each stage.
3. Describe the process of glycolysis in detail. What are its major products? Is it a very energy-efficient process?
4. Glycolysis is said to consist of two basic parts—a glucose activation stage and an energy harvest stage. Explain how glucose is activated and how the energy is harvested. Explain why the cell uses only two ATPs to activate glucose during glycolysis yet reaps four ATPs in the energy-harvest stage.
5. Describe the citric acid cycle. What are the major products? What are the major reactants?
6. Describe the role of NADH and FADH$_2$ produced during the Krebs cycle.
7. Explain how ATP is formed in the electron transport system.
8. What is metabolic water? How is it formed?
9. Describe how the following molecules control cellular respiration: oxygen, ATP, and citric acid.
10. What is fermentation? What are the chief fermentation processes described in this chapter?
11. How does fermentation keep the glycolytic pathway functioning?
12. Explain what happens in a muscle cell when oxygen levels fall.
13. A camel stores fat in its hump, which is gradually used up as it wanders the desert. Besides providing energy, what is the purpose of this fat? (Hint: take a look at Figures 6–7 and 6–8.)
14. Trained muscle cells are more efficient at cellular respiration than untrained muscles. That is, they can do more work before lactic acid levels build up. Using your knowledge of cellular respiration and fermentation, what features of these cells would you think are well developed to allow such high energy output.

Chromosomes, Cell Division, and Cancer

≈

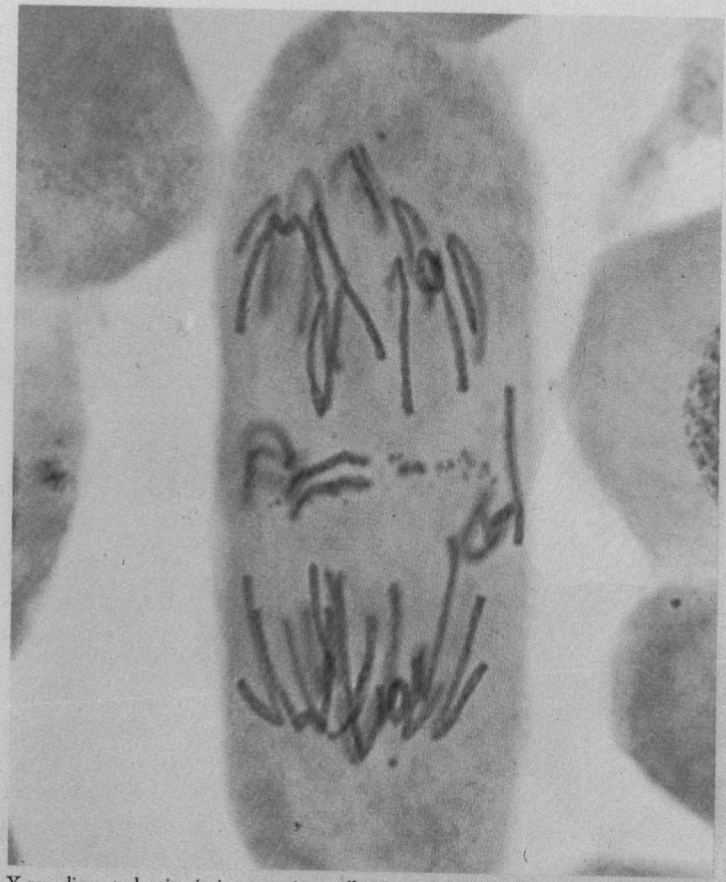

X-ray disrupted mitosis in an onion cell. Note that the cell's chromosomes have
failed to separate completely.

E lectricity flowing through wires produces a magnetic field. Since the discovery of this phenomenon, scientists have been interested in it mostly as a matter of curiosity. In 1979, two researchers from the University of Colorado reported a potentially disturbing finding on the health effects of extremely low frequency (ELF) magnetic waves generated by high-voltage power lines. Their studies, which sparked an immediate controversy, suggest that ELF magnetic waves may increase the incidence of leukemia (a cancer of the white blood cells) in children living near high-voltage power lines. In the children they studied, cancer death rates were twice what would have been expected in the general public.

In November 1986, researchers from the University of North Carolina published reports showing a fivefold increase in childhood cancer (particularly leukemia) in families living 25 to 50 feet from wires transporting electricity from power substations to neighborhood transformers (▷ Figure 7–1).

Studies of human health, such as these, are part of the science of **epidemiology** (ep-uh-deem-ee-ol-uh-je), literally, "the study of epidemics." Epidemiological studies depend primarily on statistical analysis to show correlations between cause and effect. In isolation, they rarely prove a point. But when several studies uncover the same correlation, as in this case, the results are more reliable.

Experimental work on laboratory animals also lends support to epidemiological studies. In the ELF controversy, additional credibility was lent to the hypothesis by a Texas researcher, who found that ELF fields increased the growth rate of cancer cells in tissue culture and rendered the cells 60% to 70% more resistant to the immune system's naturally occurring killer cells, which attack cancer cells.

Fortunately, few Americans live very close to high-voltage wires and power substations. Nonetheless, these findings are troubling, because common household items, such as water bed heaters and electric blankets, produce small magnetic fields. Research suggests that these devices may increase the likelihood of miscarriage (spontaneous abortion of an embryo or fetus). Some researchers even suggest that ELF fields may also cause birth defects. (For a debate on the controversy now raging over the alleged effects of ELF fields, see the accompanying Point/Counterpoint.)

Cancer—like that thought to result from ELF fields—is a disease caused by the uncontrolled replication of cells. It is induced by alterations, or mutations, of the genetic information (DNA) contained in the chromosomes. Multiple sequential mutations accumulate over time and may, if unrepaired, free cells from the strict controls that govern cellular replication.

In short, cancer is a malady afflicting the genetic information (DNA) of an organism, which affects cell division, the principal focus of this chapter. Knowledge of these topics is therefore important to your understanding of this all-too-common disease. It is also important because it provides insights into key biological processes like inheritance and development.

≋ EUKARYOTIC CELL DIVISION AND THE CELL CYCLE

As noted in Chapter 1, all unicellular organisms (bacteria, amoebae, and the like) and all of the cells in multicellular organisms (plants, animals, and fungi) arise from existing cells through division. Cell division therefore contributes to two basic characteristics of life: growth and reproduction. In multicellular organisms, cell division also plays an important role in the repair of damaged tissues and organs.

Cell division is a remarkably well-ordered process that is essential to the lives of most cells and the survival of organisms. As illustrated in ▷ Figure 7–2, the life cycle of a cell, commonly referred to as the **cell cycle,** is divided into two distinct phases: cell division and interphase. During **cell division,** the nucleus and the cytoplasm divide, splitting the contents of the cell more or less equally and forming two new cells. **Interphase** is the period between cell divisions.

Interphase Is Divided Into Three Parts

Once thought to be a resting period, interphase is now known to be a time of intense metabolic activity. It is in interphase, for example, that cells replicate their DNA and many of their cytoplasmic components, including organelles. This process prepares a cell for division. When a cell divides, its offspring, or **daughter cells,** receive an adequate supply of the organelles and molecules they need to function on their own.

Interphase is divided into three parts: G_1, S, and G_2 (Figure 7–2 and Table 7–1). G_1 (gap 1) begins immediately

▷ **FIGURE 7–1 Too Close for Comfort?** Some scientific evidence suggests that extremely low frequency radiation from electrical substations and power lines, such as these, may increase the incidence of cancer, especially childhood leukemia, in nearby residents.

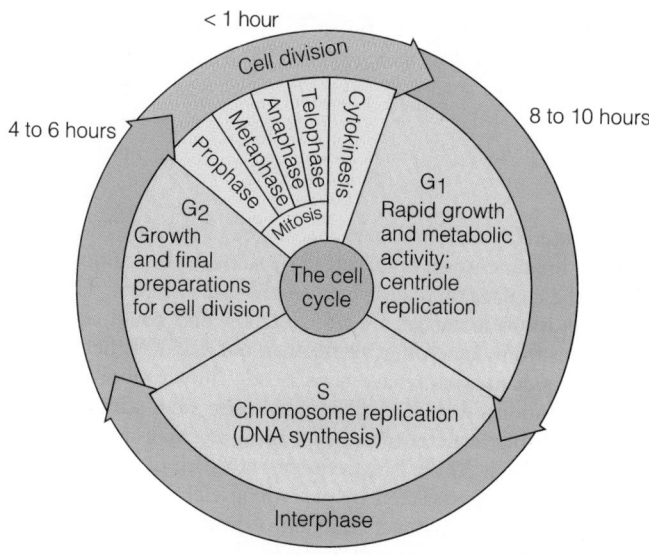

FIGURE 7–2 The Cell Cycle The cell cycle consists of two parts: interphase and cell division. Interphase is divided into three phases. The times given here represent the approximate time that mammalian cells spend in each phase when grown in the laboratory.

after a cell divides. During G_1, the chromosomes in the nucleus consist of a molecule of DNA (double helix) and associated protein; the DNA-protein complex is called **chromatin** (Table 7–2). Human cells at this stage in development contain 46 unreplicated chromosomes, carrying an estimated 100,000 genes. A **gene** is a segment of the DNA that helps control cell structure and functions in ways described in Chapter 9.

During G_1, the nuclear DNA orchestrates the cell's synthesis of RNA, proteins, and other molecules vital for proper cellular function and growth. The cell also stockpiles molecules in preparation for division.

The length of G_1 varies considerably from one cell to another. Some cells in an organism spend only a few minutes or hours in G_1; others spend weeks or even months. Cells such as nerves that never divide remain locked in G_1 for life. As a general rule, mammalian cells in culture generally spend 8 to 10 hours in G_1 (Figure 7–2).

In the next phase of the cell cycle—the **S**, or **synthesis, phase**—DNA replicates. As briefly noted in Chapter 2, the DNA molecules of the chromosomes unwind, and complementary strands form on the unwound portions, or templates. After replication, the chromosomes consist of two chromatin threads—sometimes called **chromatids**—which are attached at a region called the **centromere** ▷ Figure 7–3, page 144).

As a general rule, cultured mammalian cells spend six to eight hours in the S phase. Immediately following this phase is a period known as G_2, which lasts four to six hours in cultured mammalian cells. During this phase, mitochondria divide and the precursors of the spindle fibers (microtubular structures essential to nuclear division) form. The chromosomes begin to condense at this time.

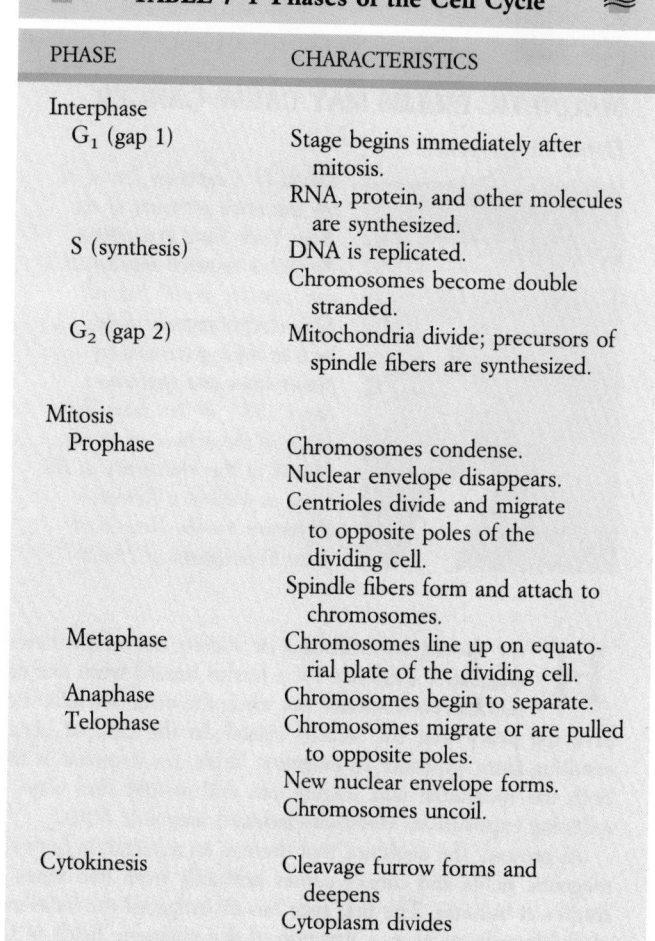

TABLE 7–1 Phases of the Cell Cycle

PHASE	CHARACTERISTICS
Interphase	
G_1 (gap 1)	Stage begins immediately after mitosis.
	RNA, protein, and other molecules are synthesized.
S (synthesis)	DNA is replicated.
	Chromosomes become double stranded.
G_2 (gap 2)	Mitochondria divide; precursors of spindle fibers are synthesized.
Mitosis	
Prophase	Chromosomes condense.
	Nuclear envelope disappears.
	Centrioles divide and migrate to opposite poles of the dividing cell.
	Spindle fibers form and attach to chromosomes.
Metaphase	Chromosomes line up on equatorial plate of the dividing cell.
Anaphase	Chromosomes begin to separate.
Telophase	Chromosomes migrate or are pulled to opposite poles.
	New nuclear envelope forms.
	Chromosomes uncoil.
Cytokinesis	Cleavage furrow forms and deepens
	Cytoplasm divides

TABLE 7–2 Review of Terminology

Chromatin	General term referring to a strand of DNA and associated histone protein
Chromatin fiber	Strand of chromatin
Chromosome	Structure consisting of one or two chromatin fibers
Chromatid	Generally used to refer to one of the chromatin fibers of a replicated chromosome
Centromere	Region of each chromatid to which a sister chromatid attaches

Nuclear and Cytoplasmic Division Occur Separately

Interphase is followed by cell division, a rapid event by cell standards, which lasts only about one hour in cultured mammalian cells. During cell division, the nucleus and cytoplasm both divide.

Nuclear division, commonly known as **mitosis,** precedes cytoplasmic division and involves a series of rather dramatic (and fascinating!) morphological changes. In order for the nucleus to divide, however, the replicated chromo-

MAGNETIC FIELDS MAY CAUSE CANCER

David O. Carpenter

David O. Carpenter served as the executive secretary of the New York State Powerlines Project, a research investigation into possible health hazards from electromagnetic fields such as those generated by power lines and appliances. Since 1985, he has been the Dean of the School of Public Health at the University at Albany, as well as a Research Physician for the New York State Department of Health.

How should an individual or society act when there is suggestive evidence for a health hazard from use of a modern convenience but when the evidence falls short of being proof that the hazard exists? In the case of cancer resulting from exposure to magnetic fields, my response is that both the individual and society can and should find ways of reducing exposure to electricity-induced magnetic fields.

At present, the evidence that there is an association between magnetic fields and cancer comes primarily from two types of studies in humans. The first type has investigated the incidence of childhood cancer as a function of the magnetic fields in the home. These studies have consistently shown a small increase in cancers, especially leukemia and brain cancer, in children living in homes where there is an elevated magnetic field caused by the neighborhood power line. For reasons not presently understood, the relation is stronger with indirect than direct measures of the magnetic field. In general, these studies show a two- or three-fold increase in the approximately 20% of homes with higher magnetic fields from the neighborhood power line.

In addition, over 30 studies have shown that individuals who work in "electric" occupations tend to develop and die from some specific cancers—notably, leukemia and brain tumors. More recently, three studies indicate that male breast cancer, a rare disease that is not different from female breast cancer, is strikingly increased in individuals exposed to elevated magnetic fields at work. This observation is important because female breast cancer was, until recently, the most common cancer in women (exceeded only by lung cancer secondary to smoking).

I take this issue very seriously because there is a consistency in the results, indicating a relatively low increase in risk, and there is a number of reasons why one would expect that these studies considerably underestimate the degree of hazard. There are also legitimate reasons, emphasized by some, to be cautious in drawing the conclusion that magnetic fields cause cancer. The reported elevations in cancer are relatively low and could conceivably result from some other factor than the magnetic fields. Also, the indirect measures have consistently shown more significant associations than the direct measures, and scientists have not been able to determine a biological mechanism.

Magnetic fields at 60 Hz (the frequency at which our electric system operates) are produced by everything electric, and the strength of the field is a direct function of how much current flows through a wire. The fields fall off with distance away from a current source, but they cannot be easily shielded. Magnetic fields have a direction, and one easy way of reducing them is to pass the same current, but of opposite direction, in a second wire placed close to the first. Sixty-Hz fields are a part of the electromagnetic spectrum, which includes visible light, radio and TV transmission frequencies, microwaves and X-rays. Microwaves and the transmission frequencies are of lower frequencies and energy, but they are still much more energetic than the 60-Hz fields. Since visible light, microwaves, and transmission frequencies are not known to be associated with significant hazards, it is surprising that at least some evidence suggests that 60-Hz magnetic fields are.

Magnetic fields are induced by many things besides power lines. In fact, it is likely that for most of us at least half of our total exposure to magnetic fields comes from appliances and the normal electrical wiring of our homes, offices, and schools. Some appliances are particularly "bad," like the electric blanket used in close proximity to the body for relatively long periods. The hair dryer is another appliance that produces very high magnetic fields, although for a short period. However, every appliance generates magnetic fields, and the fields are largest if one is close to it. The obvious way to reduce exposure is to keep a reasonable distance from appliances whenever possible. If magnetic fields cause cancer (as I strongly suspect), it is not going to make any difference whether the exposure comes from a power line or a hair dryer.

What should we do? Without question there is a great need for more research, both epidemiological studies, where one looks to see what cancers people get as a function of some aspect of their lifestyle, as well as basic research into mechanisms to determine how magnetic fields affect cells. It probably is too early to "regulate," but it is not, in my judgment, too early to inform people of the wisdom of avoiding unnecessary exposure. Each of us has some choice in whether and how we use electricity, and in my judgment it should be used with caution.

MAGNETIC FIELDS DO NOT CAUSE CANCER

Eleanor R. Adair

Eleanor R. Adair is an environmental physiologist who studies the biological consequences of exposure to radio frequency and microwave fields.

The public perception that magnetic fields may cause certain cancers (such as leukemia in children) has arisen principally from exploitation in the media of the results of more than 30 weak epidemiological studies. These studies, conducted since the late 1970s, attempted to characterize the magnetic-field environments in which people who died of cancer had lived or worked and to compare them with the environments of control cases (without cancer). Unfortunately, in nearly all of the studies, no measurements of magnetic field strengths were made; instead, the proximity of the home to power lines, power substations, or transformers was noted, or the size of the wires carrying electric current to the home was determined. In the work environment, job descriptions such as electrician and cable-splicer were loosely interpreted to indicate degree of exposure to magnetic fields on the job; in many cases, the "exposed" had far less exposure to electric or magnetic fields than the "controls," and all may have been confounded by other environmental variables such as chemical carcinogens. The bottom line is this: of all the epidemiological studies reported, about half suggest a positive relationship between purported electric and magnetic field (EMF) exposure and certain cancers, and the other half do not. Some studies even indicate a protective effect of increased EMF exposure!

Prominent epidemiologists have pointed out that studies of this kind are highly susceptible to error. Epidemiologists use a measure called relative risk to measure the impact of an environmental agent on a disease such as cancer. If the relative risk for a factor like EMF is 2 for a disease like leukemia, this means that EMF exposure doubles the probability of getting leukemia. A relative risk of 1 means there is no effect. An eminent epidemiologist, Dr. Philip Cole of the University of Alabama, says that "in the context of epidemiology, increases of two-fold or less in risk are not generally considered highly credible." Nearly all of the epidemiological results that are held as evidence for an EMF-cancer link are statistically consistent with

risk factors less than 2. In contrast, the relative risk for lung cancer from cigarette smoking is about 10 for a few cigarettes per day and rises to about 60 for three-pack-a-day smokers.

There are many serious problems with the epidemiological attempts to demonstrate an EMF-cancer link, some of which have been mentioned above. Failure to measure any electric or magnetic fields in the test environment is unforgivable; but even spot measurements, if attempted, can shed little light on the ever-changing fields to which people may be exposed over many, many years. Then there is the failure to account for other variables that may influence the tendency to develop cancer. Smoking, drug use, diet, genetic predisposition, X-ray exposure, and chemical pollutants in air and water are only a few of the many confounders in the test environment that are usually ignored. Indeed, one recent study pinpointed traffic density outside the home, passive smoking, or whether the child was breast-fed as variables of equal significance to EMF with respect to childhood leukemia. Finally, no valid dose-response relationship between the strength of the electric or magnetic field and the severity of a biological effect or the rapidity of its induction has been reported. This is as true for laboratory experiments as for epidemiological studies and casts grave doubt on the validity of any EMF-cancer connection.

Overall, the strongest evidence that electric or magnetic fields from power lines and electrical appliances do not induce leukemia and other cancers comes from cancer statistics compiled over the last 50 years. Epidemiologist Philip Cole has observed that if childhood leukemia derives mainly from exposures to the low-frequency EMF from our electrical environment, then leukemia should be rare in populations exposed to very little EMF. He tested this hypothesis by comparing the past and present incidence of leukemia in Connecticut (which has the oldest cancer registry in the United States) with per capita electric power use. This comparison shows that as electric power consumption has increased by a factor of 12 over the last half-century, the reported leukemia rate has remained nearly constant, particularly from 1960 to 1980 (when power use increased by a factor of 2.5). These data compel the conclusion that the use of electricity in modern society has no significant bearing on cancer incidence and that the poor-quality studies that show otherwise are simply wrong.

⇌ SHARPENING YOUR CRITICAL THINKING SKILLS

1. After you read the first essay, were you convinced that magnetic fields were a health problem? What statistics and points made by the author affected your view on this issue?
2. After completing the second essay, did your opinion change? If so, why?
3. Given the disagreement, what course of action would you recommend?

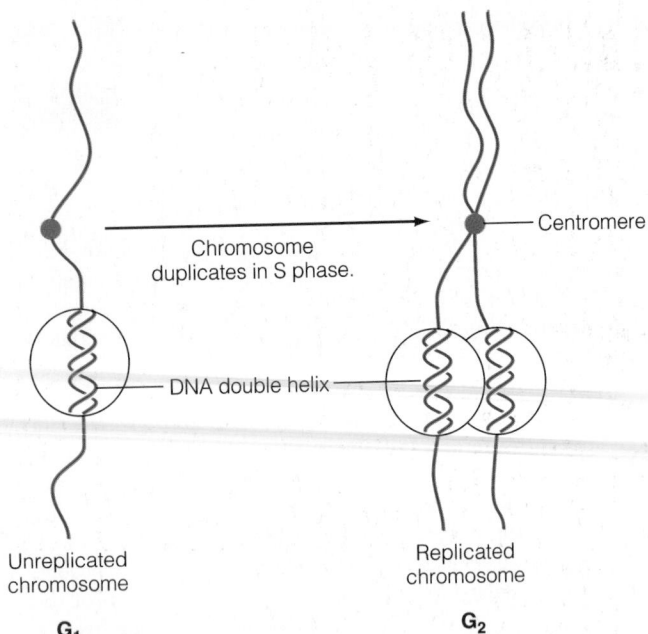

Centromere

Chromosome
duplicates in S phase.

DNA double helix

Unreplicated
chromosome

Replicated
chromosome

G_1

G_2

▷ **FIGURE 7–3 DNA Replication** During the S phase of the cell cycle, the DNA molecule unwinds and duplicates, forming two strands, each containing a DNA double helix.

somes must first condense, becoming compact structures clearly visible in ordinary light microscopic preparations ▷ Figure 7–4). Condensation facilitates chromosomal segregation. Without it, cell division would be a nightmare. Dividing up the 46 unravelled chromosomes in humans would very probably result in a tangled mass of broken chromosomes. Cell division without condensation might be likened to the task of separating a plate of spaghetti into two equal piles without breaking a single strand.

▷ **FIGURE 7–4 Metaphase Chromosomes** Taken from a human cell during metaphase, these chromosomes are clearly visible under a light microscope.

In order for mitosis to occur smoothly, the chromosomes must also line up in the center of the cell. From this position, the two chromatids of each chromosome can easily be drawn apart and then divvied up, with half going to each new daughter cell.

Cytoplasmic division, or **cytokinesis** (literally, "cell movement"), is an independent process that occurs toward the end of mitosis. Before we study cell division in more detail, though, let's look more closely at one of the chief actors in this remarkable drama, the chromosome.

≋ THE CHROMOSOME

As you have learned, chromosomes contain the genetic information of cells, DNA, which directs most cellular events. (For a discussion of the scientific work that led to the discovery of DNA, see Scientific Discoveries 7–1.) All organisms have a set number of chromosomes. The cell nuclei in your body, except the gametes (sex cells—the sperm and egg), have 46 chromosomes. The cells of the fruit fly you find swarming around overripe fruit in your kitchen have 8 chromosomes. The cells of the common peas growing in your garden have 14.

Each type of chromosome contains very specific genetic information and is different from every other type. Chromosomes also differ in several other key respects, such as size, location of the centromere, and banding patterns ▷ Figure 7–5). A careful examination of a photomicrograph of the 46 human chromosomes shows that they can be matched in pairs. Humans therefore are said to have 23 pairs of chromosomes.

The chromosomes of each pair are the same length and have the same banding pattern and centromere position.

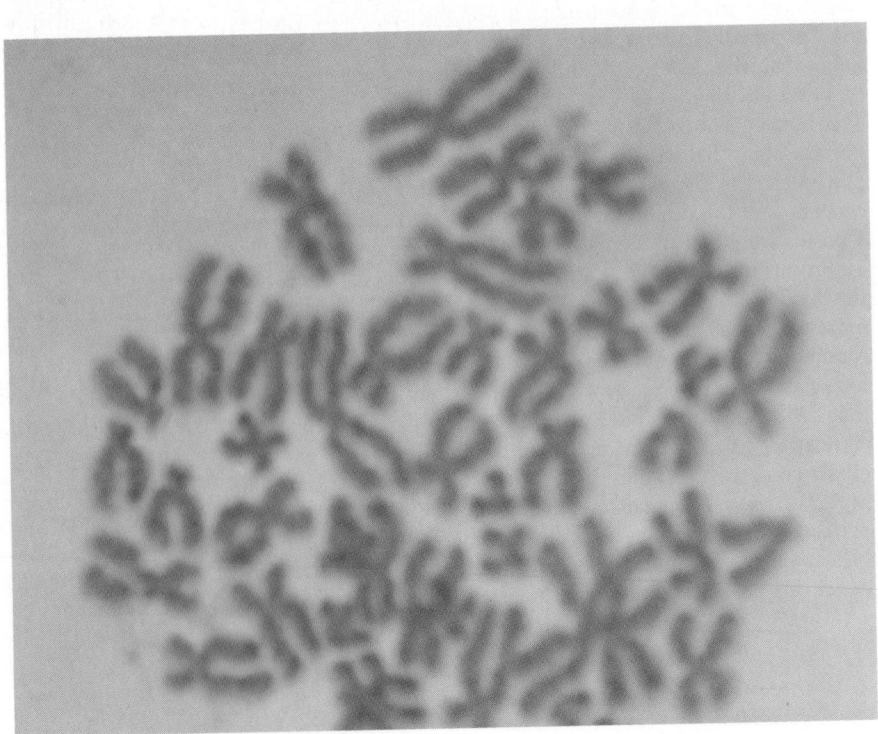

These chromosomes are described as **homologous.** That is, they contain genes that control the same inherited traits. For example, if a gene for hair color is located on one chromosome of a pair, it will also be found at the same location (or **locus**) on the other member of the duo.

The presence of homologous chromosome pairs results from sexual reproduction, with one member of the pair coming from each parent. In humans, then, 23 of the chromosomes come from the mother, and 23 come from the father.

Cells that contain the full complement of chromosomes are said to be **diploid** (diploid = twofold). In humans and other animals, all body cells are diploid. These cells are often called **somatic cells** to distinguish them from reproductive, or **germ, cells,** the sperm and ovum, which carry paternal and maternal chromosomes, respectively. Germ cells are all **haploid;** that is, they contain half the chromosomes of somatic cells. Germ cells are produced by a special kind of cell division known as meiosis, which occurs in the gonads (ovaries and testes).

Although Somatic Cells Contain a Set Number of Chromosomes, the Number of Chromatids Varies, Depending on the Stage of the Cell Cycle

As noted above, during G_1, each chromosome consists of a single DNA molecule (a double helix) and associated protein. The 46 unreplicated chromosomes in the human cell are packed loosely in the nucleus in an apparently random fashion. During the S phase, the DNA strands replicate (see Figure 7–3). Each chromosome now consists of two identical chromatids held together at the centromere.

Chromosomes Condense After Replication, Forming Tightly Coiled Structures in the Nucleus

Soon after replication, the chromosomes of the cell begin to coil, compacting in much the same way that a stretched phone cord shortens and compacts when the tension is removed. Chromosomes in the condensed state are metabolically inactive—that is, unable to produce either RNA or DNA.

The structure of a condensed chromosome is shown in ▷ Figure 7–6 (page 148). As illustrated, each chromatid consists of a strand of DNA and associated globular proteins called histones. Also present when chromosomes are unwound, the **histones** are globular proteins, which are thought to play a role in regulating the DNA's activities (Chapter 9). As illustrated in Figure 7–6, loops of the double helix of DNA encircle the histones, forming small clusters that compact to form hollow tubules.

Besides facilitating nuclear division, chromosomal condensation offers physicians and geneticists a great opportunity to study the chromosomal composition of cells. This is particularly useful to expectant parents who are concerned about possible genetic defects in their baby.

In order to acquire fetal cells, physicians extract fluid from the liquid-filled cavity surrounding the growing fetus through a long needle inserted through the mother's abdomen (▷ Figure 7–7, page 149). This process is called **amniocentesis.** Fetal cells naturally present in the fluid are separated, then grown in culture dishes, where they divide by mitosis. After the number of cells has increased, the tissue culture is treated with a chemical substance that arrests cell division. (More on amniocentesis and another method to

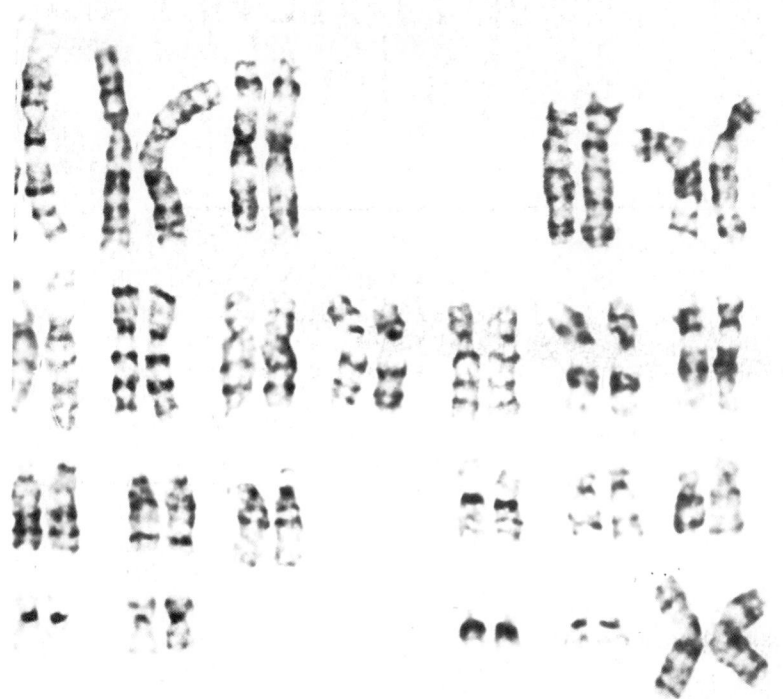

▷ **FIGURE 7–5 Karyotype** Chromosomes from a normal human female arranged in order of decreasing size. Note that there are 46 chromosomes—or 23 pairs—in somatic cells. Each pair is similar in size, banding pattern, and location of centromere.

UNRAVELING THE STRUCTURE AND FUNCTION OF DNA
Featuring the Work of Miescher, Griffith, Avery, Watson, Crick, Wilkins, and Franklin

When most people think about the structure of DNA, two names come to mind, James Watson and Francis Crick. In 1953, these two scientists published a brief paper proposing the double helix model for DNA structure (▷ Figure 1). Nine years later, they were awarded the Nobel Prize along with Maurice Wilkins, who played a key role in the discovery.

Like many other scientific advances that changed our understanding of the world and, in some cases, our way of life, the discovery of the structure of DNA resulted from the efforts of many scientists over many years. In fact, work on DNA began in the 1860s, when Frederick Miescher launched an inquiry into the chemical composition of white blood cells. In 1868, Miescher separated the nuclei of white blood cells from their cytoplasm, then began studying the material he had extracted from these nuclei. Research by a number of other scientists revealed the presence of two types of molecules, RNA and DNA, within the nucleus.

Although evidence accumulated over the next 80 years that DNA played a role in heredity, most scientists believed that proteins were the hereditary molecules. This notion persisted until about 1950 because many scientists believed that nucleic acids, which contain only four different nucleotide building blocks, lacked the complexity necessary to carry and transmit the vast amount of genetic information present in many species. Proteins, on the other hand, contain as many as 20 different amino acid subunits, offering a level of complexity scientists thought was needed to transmit thousands of traits. This is an excellent example of the way in which biases can thwart the advancement of scientific knowledge.

Our understanding of the structure and function of DNA was also aided by the work of Frederick Griffith. In 1920, Griffith discovered a chemical substance in bacteria he called "transforming factor." When transmitted from one bacterium to another, this substance caused specific structural changes in the recipient. In 1934, Oswald Avery and his colleagues began a 10-year study that identified the transforming factor as DNA. Avery and his co-workers showed that DNA extracted from one strain of bacteria produced an altered capsule in a recipient strain and that the trait was then passed to all of the offspring of the recipient.

To confirm the hypothesis that DNA was the transforming agent, Avery and his co-workers treated the extracts from the first bacteria with enzymes that destroy protein and RNA. Removing residual protein and RNA did not affect transformation. The researchers also treated the preparation with an enzyme that destroys DNA, and, as expected, they found that transformation did not take place.

This work resulted in two important conclusions vital to the modern understanding of genetics. First, DNA is the genetic material. Second, DNA con-

▷ **FIGURE 1 The Double Helix Model for DNA** Watson and Crick and their Double Helix Model.

---

examine fetal chromosome structure and number are found in Chapter 8.)

Cells are removed from the culture, then placed on a glass slide and flattened, forcing the chromosomes to spread out. The chromosomes are photographed through the microscope, cut out, and arranged in homologous pairs with the largest first, as shown in Figure 7-5. The resulting display is called a **karyotype.**

Thanks to modern technology, computers can also be used to produce a karyotype without scissors and glue. A camera mounted on the microscope projects the image on a computer screen. Technicians isolate the chromosomes on the screen and assign each one a number; then a computer creates a karyotype.

Karyotypes help physicians find gross morphological defects—for example, extra chromosomes or missing segments. These defects may have serious consequences and are described in Chapter 8. In many cases, chromosome analysis assures parents that their child will show no obvious genetic defects, alleviating worry during a period of heightened anxiety. Chromosome analysis also permits physicians to determine the sex of the fetus.

≈ CELL DIVISION IN ANIMALS: MITOSIS AND CYTOKINESIS

One of the most fabulous of all cellular events is cell division, a process during which the cell divides its chromosomes equally and distributes its cytoplasm and organelles more or less evenly between the two daughter cells. Pre-

trols the synthesis of specific cellular products.

Unraveling the actual structure of DNA came in the years from the mid-1940s through 1953. In the early 1950s, Watson and Crick began to develop a number of models for the structure of DNA, based on available chemical and physical information from other researchers. One vital piece came from Rosalind Franklin, who worked in the laboratory of British researcher Maurice Wilkins. Franklin produced X-ray photographs of crystals of highly purified DNA. Her remarkable photographs suggested that the DNA molecule was helical (▷ Figure 2).

Another vital piece came from the laboratory of Erwin Chargaff, who had spent years, with the help of his colleagues, studying the chemical nature of DNA. Chargaff's lab showed that the amount of purine in DNA always equals the amount of pyrimidine and that the ratios between adenine and thymine and between cytosine and guanine are always 1:1.

Armed with this information, Watson and Crick began to work out a model of DNA structure. After a series of setbacks, they finally devised an ingenious model that fit the chemical and physical data. They published their results in 1953 in the journal *Nature*, and two months later they published a second paper proposing that the complementary strands of their model could explain the process of replication occurring in genetic material. In addition, they hypothesized that genetic information could be stored in the sequence of bases and that mutations might result from a change in that sequence.

In the ensuing years, a number of scientists provided evidence that supported the hypothetical model proposed by Watson and Crick. And in 1962, Watson, Crick, and Wilkins were awarded the Nobel Prize for this monumental discovery. Although much of the X-ray data on which the model was based had been provided by Franklin, she was not included in the celebrations. She couldn't be. Rosalind Franklin died of cancer in 1958, and Nobel Prizes are awarded only to living scientists. Had she been alive, it is likely that she would have shared in this coveted prize.

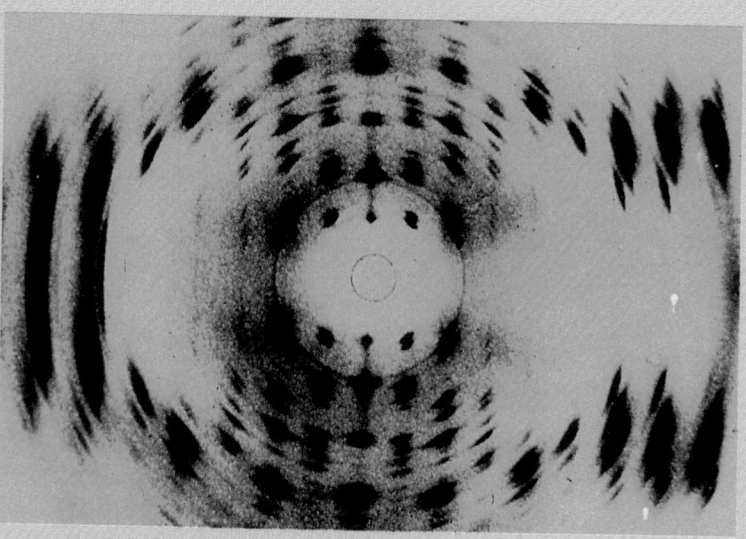

▷ **FIGURE 2 Discovering DNA's Helical Shape** X-ray diffraction photograph of DNA by Rosalind Franklin.

cisely choreographed to ensure a smooth apportionment of cellular contents, cell division occurs in a series of events that takes place over many hours and involves virtually every cellular organelle and other structure. Let's take a look at nuclear division in animals first.

The Division of the Cell's Nucleus Is Called Mitosis

Mitosis is divided into four stages, occurring in the following order: prophase, metaphase, anaphase, and telophase (see Figure 7–2). Although we are concentrating on the nucleus, we will also consider some of the accompanying cytoplasmic changes—events essential to proper nuclear division.

Prophase. Prophase begins immediately after interphase. During prophase, the replicated chromosomes shorten and thicken considerably, forming compact structures from the threadlike structures characteristic of the interphase nucleus (▷ Figures 7–8a and 7–8b, page 150). During prophase, the nucleoli, regions of active rRNA synthesis, gradually disappear from view.

In the cytoplasm, two events of particular importance occur during prophase. The first is the division of the cell's centrioles. Shown in ▷ Figure 7–9 (page 151), each centriole consists of two small, cylindrical structures identical in architecture to the basal bodies (described in Chapter 3). Like other organelles, centrioles replicate during interphase. Then, during prophase, the centrioles separate, migrating to opposite ends of the nucleus.

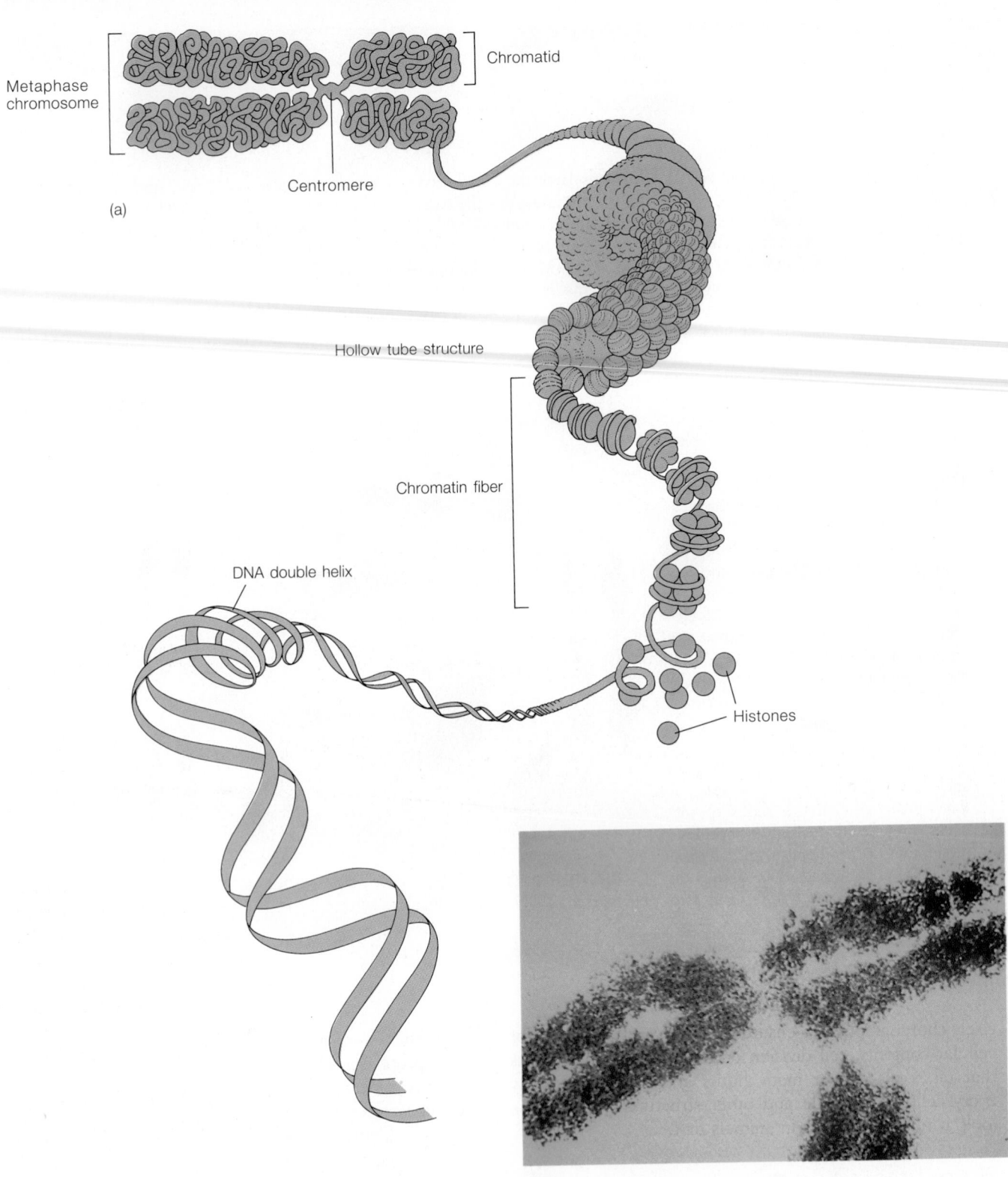

Metaphase chromosome

Chromatid

Centromere

(a)

Hollow tube structure

Chromatin fiber

DNA double helix

Histones

(b)

▷ **FIGURE 7–6 Chromosome Structure** (*a*) The DNA helix wraps around histone molecules forming a chromatin fiber. During prophase, the chromatin fiber coils tightly forming the densely packed chromosome, (*b*) Transmission electron micrograph of metaphase chromosome.

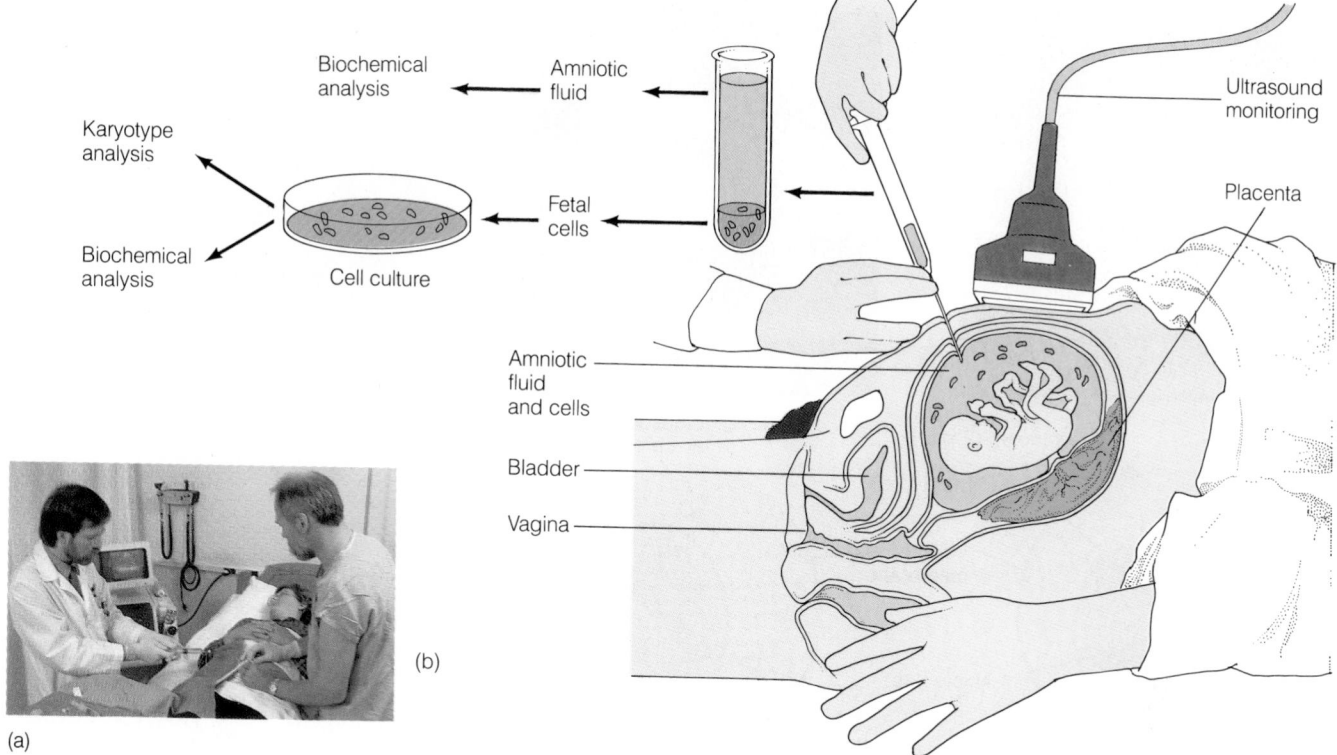

▷ **FIGURE 7–7 Amniocentesis** (*a*) A patient undergoing amniocentesis. (*b*) A needle is inserted into the fluid-filled space surrounding the fetus to withdraw fluid containing fetal cells that can be subjected to chromosomal and biochemical analyses. Ultrasound monitors are used to direct the needle to prevent injury to the fetus.

The second cytoplasmic event of great importance to the success of mitosis is the formation of the mitotic spindle. The **mitotic spindle** is an elaborate array of microtubules that forms during prophase and is responsible for subsequent movement of the chromosomes. As Biologist Normal Wessels writes, "Just as a dancing puppet is operated by moving strings, the dance of the chromosomes is choreographed and controlled by the action of the spindle fibers."

As shown in ▷ Figure 7–10, two types of fibers are found in the mitotic spindle: **polar fibers,** which extend from the spindle pole inward toward the center of the cell, and **chromosomal fibers,** which extend from the poles inward to attach to individual chromosomes via their centromeres. Each fiber is actually a bundle of microtubules composed of tubulin molecules made available by the disassembly of the cell's cytoskeleton.

Also associated with the centrioles in animal cells is a star burst of microtubules called the **aster.** Its role in cell division remains unclear, although it may help anchor the spindle to the plasma membrane.

Late in prophase, the nuclear envelope begins to disintegrate. By then, chromosomal fibers of the mitotic spindle have attached to the chromosomes, and the stage is set for this most dramatic cellular event.

Metaphase. Before chromosomes can be divided equally, they are lined up in the center of the cell an equal distance from the two poles of the mitotic spindle with their centromeres located along the equatorial plane (see Figure 7–8c). This alignment occurs during metaphase and greatly facilitates the separation of chromatids.

Anaphase. When the chromatids of each chromosome begin to separate, the cell enters anaphase, the shortest part of mitosis, lasting only a few minutes (see Figure 7–8d). During anaphase, the chromatids of the homologous chromosomes are drawn toward opposite poles of the mitotic spindle and therefore toward opposite ends of the cell. As shown in Figure 7–8d, the chromosomes are dragged centromere-first, with their arms trailing behind. Movies of mitosis show that all of the chromosomes begin to separate simultaneously.

What causes the movement of chromosomes during anaphase no one knows for sure, but two mechanisms are clearly operating in the cell: elongation of the cell and shortening of the polar fibers, which draws the chromosomes to their respective poles.

Time-lapse photographs of mitosis show that during anaphase, the distance between the poles increases. The cell elongates as if being pulled apart at the poles. Experimental evidence suggests that the polar fibers play an important role in cellular elongation. The exact mechanism, however, is unknown. Some researchers believe that the polar fibers slide along one another and in so doing push the poles farther apart and "drag" the chromatids away

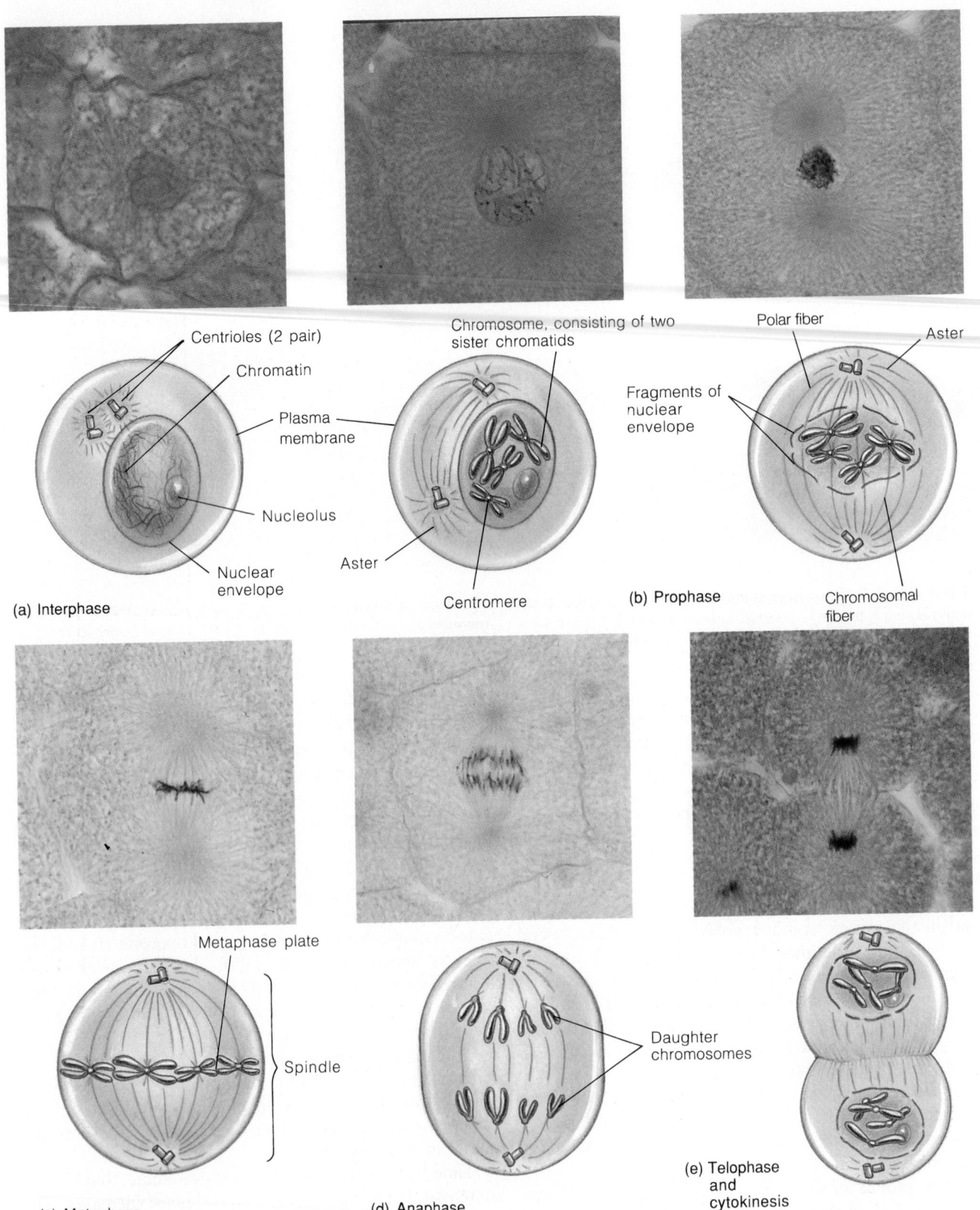

(a) Interphase

Centrioles (2 pair)

Chromatin

Plasma membrane

Nucleolus

Nuclear envelope

Chromosome, consisting of two sister chromatids

Aster

Centromere

(b) Prophase

Polar fiber

Aster

Fragments of nuclear envelope

Chromosomal fiber

(c) Metaphase

Metaphase plate

Spindle

(d) Anaphase

Daughter chromosomes

(e) Telophase and cytokinesis

▷ FIGURE 7–8 Mitosis

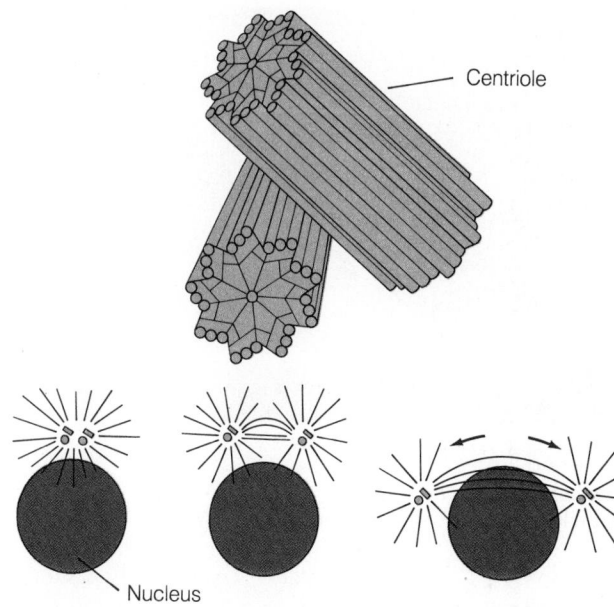

▷ **FIGURE 7–9 Centriole** The centriole, like the basal body, consists of nine sets of microtubules with three in each set. Centrioles are found in the cytoplasm and replicate during interphase. They migrate to opposite poles of the nucleus during mitosis.

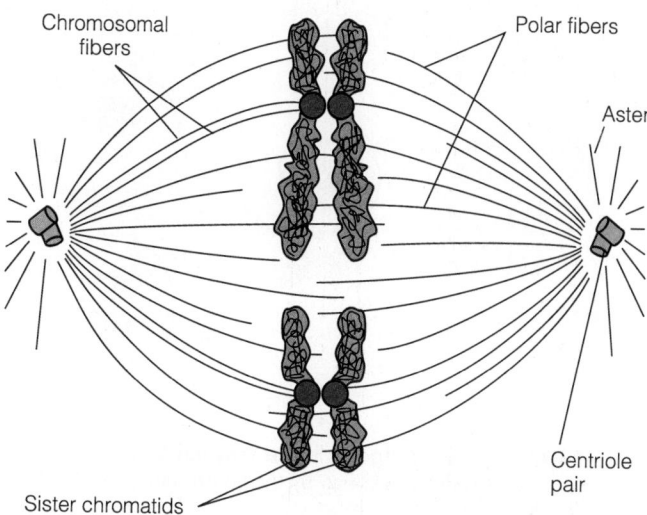

▷ **FIGURE 7–10 Mitotic Spindle** The mitotic spindle, made of microtubules, forms during prophase in the cytoplasm of the cell. Chromosomal fibers connect to the chromosomes, and polar fibers extend from a pole to the equatorial region of the spindle. In ways not clearly understood, the mitotic spindle helps separate each double-stranded chromosome during mitosis.

from the center of the cell. Another hypothesis holds that the polar fibers elongate and that as they lengthen, the cell elongates, and the chromatids pull apart.

Although these hypotheses explain why the chromosomes separate initially, they do not explain what causes the chromosomes to migrate the rest of the way—that is, to complete their journey to the poles. One hypothesis holds that after initial separation of the chromatids, the chromosomal fibers somehow shorten, drawing the chromosomes to the poles.

Telophase. The final stage of nuclear division is telophase (see Figure 7–8e). Telophase begins when the chromosomes complete their migration to the poles, and it involves a series of changes in the nuclei that is essentially the reverse of prophase.

During telophase, nuclear envelopes form around the chromosomes of each daughter cell. The new nuclear envelopes are produced by the endoplasmic reticulum. During telophase, the spindle fibers disappear, and the chromosomes uncoil, regaining their threadlike appearance. The nucleoli also form. Telophase ends when the daughter cell nuclei appear to be in interphase.

Cytokinesis Is the Division of the Cytoplasm

Nuclear and cytoplasmic division do not always occur in tandem. In some species of fungi and algae, for example, cells undergo many nuclear divisions that are unaccompanied by cytoplasmic division, or cytokinesis. These divisions result in the formation of large multinucleated cells. In most species, though, nuclear and cytoplasmic division are

linked. Cytokinesis usually begins late in anaphase or early in telophase.

In animal cells, cytokinesis is brought about by a dense network of contractile fibers, or **microfilaments,** that lies beneath the plasma membrane. These microfilaments, which are part of the cytoskeleton of the cell, are composed of two contractile proteins, **actin** and **myosin,** the same proteins found in muscle cells.

When the time comes for the cytoplasm to divide, the microfilaments along the midline of the cell contract and pull the membrane inward (▷ Figure 7–11). The membrane furrows as if it were being constricted by a thread that had been tied around the cell, then tightened. As cytokinesis proceeds, the furrow deepens, eventually pinching the cell in two and producing two daughter cells with more or less equal amounts of cytoplasm. With this equal allocation of cytoplasm to the daughter cells comes a sharing of the cell's organelles.

≋ CELL DIVISION IN PLANTS

Cell division in plants differs from that in animals in two main respects. The first difference is the obvious lack of centrioles and asters during plant mitosis (Figure 7–12). Despite their absence, a fully functional spindle forms, although it appears a bit less organized.

The second major difference between plants and animals occurs during cytokinesis. Unlike cytokinesis in animal cells, which involves a furrowing of the plasma membrane caused by the contraction of underlying microfilaments, cytoplasmic division in plants involves the assembly of a new cell wall between the two daughter nuclei.

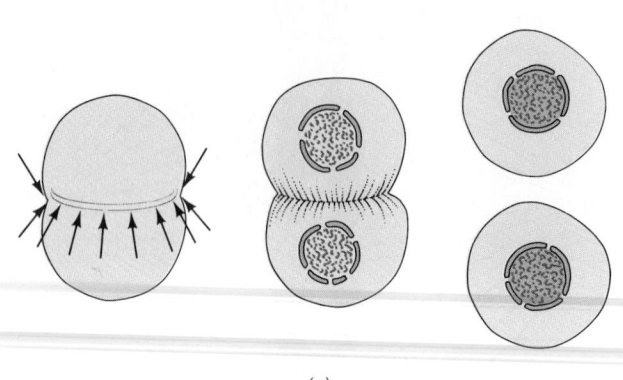

(a)

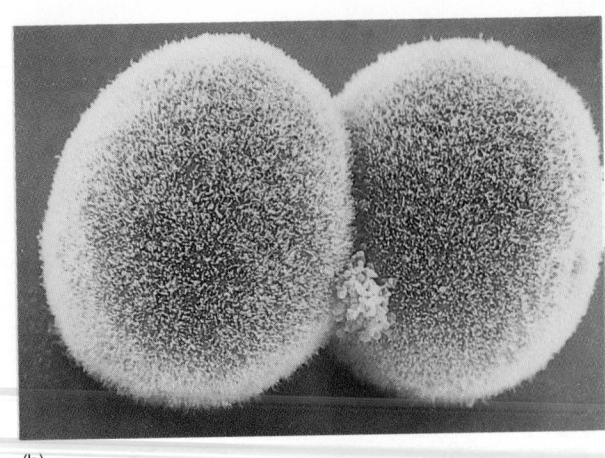

(b)

▷ **FIGURE 7–11 Cytokinesis in an Animal Cell** (*a*) The cytoplasm of a cell is divided in two by the contraction of a ring of contractile microfilaments. Cytokinesis begins late in anaphase or early in telophase. (*b*) Cell undergoing cytokinesis.

As illustrated in ▷ Figure 7–13, the new cell wall is formed from small membranous vesicles containing cell-wall precursors. Produced by the Golgi complex, the vesicles migrate to the center of the cell along microtubules. Along the midline between the two daughter nuclei, the vesicles fuse and form a large flattened sac, known as the **early cell plate.** The cell wall precursors form the primary cell wall, which expands outward as additional vesicles fuse along the lateral edges of the plate. Eventually, the forming primary cell wall makes contact with the original cell wall, and the two daughter cells are completely separated. Cellulose fibers are then deposited on the primary cell wall.

Despite these differences, cell division in plants and animals bears a remarkable similarity, a reflection of their common evolutionary ancestry.

≈ CONTROL OF THE CELL CYCLE

What stimulates the replication of DNA and the division of organelles? What causes a cell to divide? Despite years of intensive research, our understanding of the controls of the cell cycle is still incomplete. Some experiments suggest that chemical messages from the cytoplasm play a key role in choreographing the events of the cell cycle. In one experiment, in fact, cell nuclei in G_1 were transplanted to the cytoplasm of actively dividing cells (cells in mitosis). In their new location, the nuclei quickly entered the S phase. Soon after the chromosomes replicated, the nuclei entered G_2 and began dividing. Researchers hypothesize that this and similar experiments suggest the presence of a chemical within the cytoplasm that controls the cell cycle.

If cytoplasmic signals control the nucleus, the next logical question is, What controls the cytoplasm? No one knows for sure.

Research shows that certain external signals may exert some control over cytoplasmic events. Studies of **cell cultures,** cells grown in the laboratory under precise conditions, have found that normal cells migrate and divide until they spread evenly across the bottom surface of culture dishes (▷ Figure 7–14). At this point, the cells stop dividing, even though nutrients are plentiful. The contact of cells with one another apparently inhibits further growth; this process is called **contact inhibition.** Contact inhibition also occurs in the tissues of the body.

One of the reasons cancer cells grow uncontrollably is that they lose contact inhibition. In a culture dish, cancer cells proliferate wildly, growing on top of one another and quickly utilizing nutrients in the culture dish. In the body, cancer cells grow in a similar way, forming large growths, called **tumors,** that may eventually destroy vital organs and cause death.

Hormones affect the cell cycle as well. During each menstrual cycle, estrogen released from a woman's ovaries stimulates the growth of breast tissue and the uterine lining in preparation for pregnancy. Several other hormones also stimulate the division of body cells.

Growth-promoting and growth-inhibiting factors likewise play a role in controlling the cell cycle. Skin cells, for example, produce a growth-inhibiting factor that prevents cell division. Damage to the skin—say, a cut or burn—reduces the number of skin cells and presumably reduces the amount of inhibiting factor at the site of injury. Cells released from inhibition proliferate and repair the damaged area. The rate of cell division returns to normal when the tissue is fully repaired.

Obviously, there is no single answer to the question of what controls the cell cycle. Many factors play a role, depending on the type of cell. In time other factors are sure to be discovered. These factors may act directly on the cytoplasm, but it is more likely that they induce changes in the genes in the nucleus that regulate the cell cycle. Activation of these genes may, for example, lead to the production of chemicals in the cytoplasm that, in turn, feed back on the nucleus, choreographing the nuclear changes found in the cell cycle.

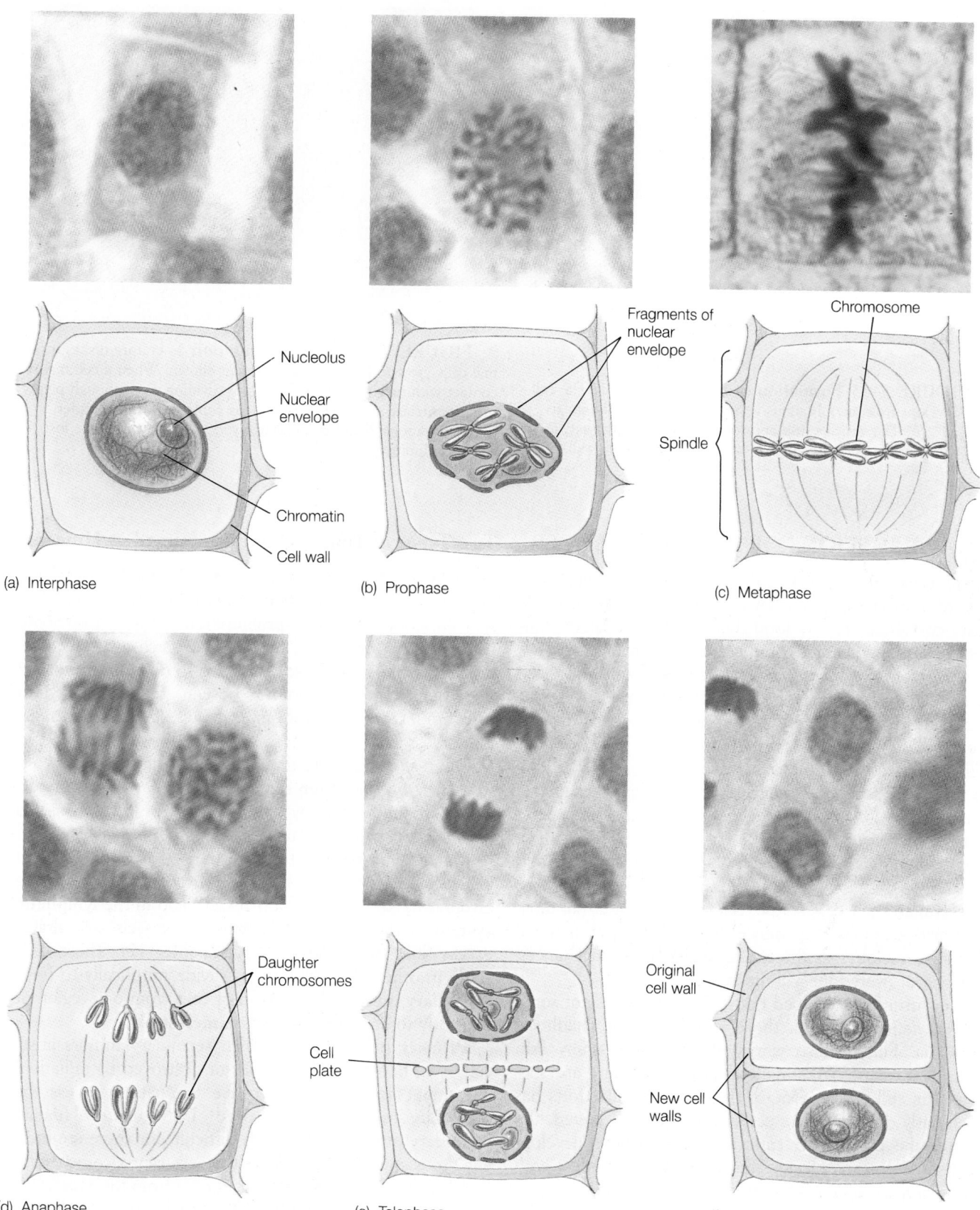

(a) Interphase

Nucleolus

Nuclear envelope

Chromatin

Cell wall

(b) Prophase

Fragments of nuclear envelope

(c) Metaphase

Chromosome

Spindle

(d) Anaphase

Daughter chromosomes

(e) Telophase

Cell plate

(f) Daughter cells

Original cell wall

New cell walls

▷ **FIGURE 7–12 Cell Division in Plants** Note the absence of centrioles and astral fibers and the formation of the cell plate between the two nuclei of the daughter cells.

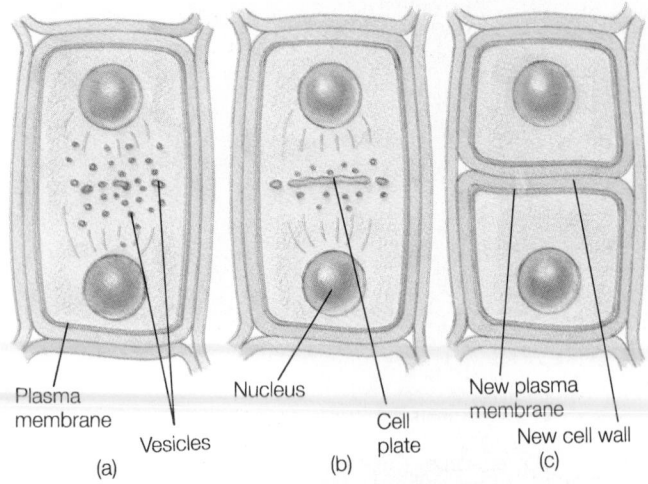

Plasma membrane
Vesicles
(a)

Nucleus
Cell plate
(b)

New plasma membrane
New cell wall
(c)

▷ **FIGURE 7–13 Formation of the Cell Wall** (*a*) The cell wall develops from membranous vesicles produced by the golgi complex. (*b*) The vesicles fuse and form a large flattened sac. (*c*) Cell wall precursors in the vesicles form the primary cell wall.

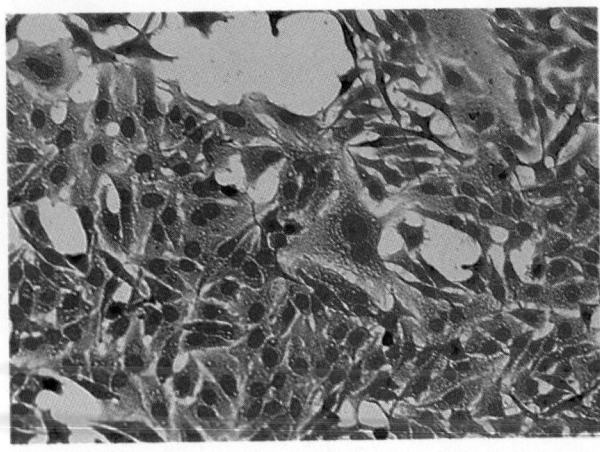

▷ **FIGURE 7–14 Contact Inhibition** In culture dishes, normal cells grow to form a single layer, as shown. When a cell makes contact with neighboring cells, it stops dividing, even though plenty of nutrients are available. Cancer cells continue to grow under these conditions, forming multiple layers, because they have lost contact inhibition.

≈ APPLICATIONS OF BIOLOGY: CANCER

In the early 1970s, Americans were shocked by reports of hazardous wastes seeping into their drinking water, pesticides contaminating the food they ate, and pollution fouling the air they breathed. Some experts estimated that 90% of all cancers were caused by environmental and workplace pollution and by other agents such as X-rays, ultraviolet light, and viruses. That conjecture caused a storm of controversy and a flurry of research into cancer, a disease that kills nearly 500,000 Americans each year.

More recent studies suggest that the early estimates were probably in error. Today, in fact, many health experts believe that only 20% to 40% of all cancers arise from environmental and workplace pollutants. Most of the rest are caused by smoking, diet, and natural causes.

No matter what the cause, cancer is a disease that most of us will have close experience with. According to national statistics, one of every three Americans will contract cancer, and one of every four will die from it.

New drugs have helped boost the survival rate of some cancer victims, such as those suffering from childhood leukemia. Surgery also works effectively for many skin cancers. Radiation treatment and chemotherapy are successful in still others. One of the most helpful efforts has been early detection: the earlier a cancer is detected, the greater the chances of survival. Despite these developments, the war on cancer seems to be a losing battle—with only marginal gains in survival rates.

We often think of cancer as a strictly human disease, but in actuality cancer occurs in a wide range of animals and plants (▷ Figure 7–15). This section takes a closer look at cancer, concentrating on humans. It will familiarize you with terms and concepts that you will encounter throughout your lifetime.

Two Types of Tumors Are Encountered in Humans

As mentioned, cancer is a disease in which cells divide uncontrollably, often invading other parts of the body. But not all abnormal cellular proliferation is cancerous. Some cells, in fact, form small masses that reach a certain size, then stop growing. These are called **benign tumors.** As a rule, benign growths do not pose a significant medical problem, except when they put pressure on nerves or block blood vessels.

Of great concern are the **malignant tumors,** which continue to grow and often spread to other parts of the body, where they may destroy vital organs. The term *cancer* is reserved for malignant tumors.

Malignant tumors spread by the release of individual cells or clusters of cells that travel through the body in the **circulatory system,** or blood vessels, and in the **lymphatic system,** a supplementary network of vessels that drains excess fluid from body tissues (Chapter 12). These cells settle in other sites, where they divide mitotically to form **secondary tumors.** The spread of cells from one region of the body to another is known as **metastasis.**

Primary tumors, from which secondary tumors arise, can often be removed surgically or destroyed by radiation. Treatment generally fails to save the patient unless the secondary tumors are destroyed. Unfortunately, finding secondary tumors is extremely difficult, for there are literally thousands of places where the cells can become established and begin to divide. Cancer cells from the breast, for example, may lodge in the brain or lungs or in **lymph nodes,** organs interspersed along the lymphatic system to filter the fluid flowing through the vessels. Doctors can take tissue samples, or **biopsies,** of lymph nodes to assess the spread of the cancer. Cancerous nodes can be removed surgically. When a tumor is difficult to remove, radiation

(a)

(b)

▷ **FIGURE 7–15 Cancer in Plants** Cancers occur in a number of different species, not just humans. (*a*) A tumor on a tomato plant. (*b*) A cancerous tumor on a tree.

can be used. Patients can also be treated with **chemotherapeutic agents,** highly toxic drugs that attack rapidly dividing cells of secondary tumors. This shotgun approach has many side effects, including nausea, hair loss, and general sickness, but it may be a patient's only hope. Some new developments in cancer treatment are described in Health Note 7–1.

No Question Interests Science More Than What Causes a Cell to Begin Dividing Uncontrollably

The conversion of a normal cell to a cancerous one is quite complex and is presented here in a simplified form. In many cases, the production of a malignant tumor (cancer) involves two steps: (1) conversion and (2) development and progression (▷ Figure 7–16).

The conversion of a normal cell to a cancerous one often begins with a **mutation,** a change in the DNA caused by a chemical, physical, or biological agent. Those substances or agents that cause the initial transformation are technically known as **co-carcinogens.** Mutations occurring in the genes that control the replication of cells may partially or completely release a cell from the normal growth controls. (These genes are called proto-oncogenes and are discussed in Chapter 9.)

To date, over 50 growth-control genes (proto-oncogenes) have been found to have transforming capabilities; that is, when mutated, they transform normal cells into cancer cells. These genes code for proteins that play a role in the regulation of cell division.

A mutated cell that is fully released from growth controls and thus able to divide uncontrollably is often held in check by neighboring cells or tissue factors from nearby cells. These homeostatic controls keep the cancer cell from dividing and thus provide an important measure of protection. This protection, however, can be overridden by inter-

▷ **FIGURE 7–16 Stages Leading to Cancer**

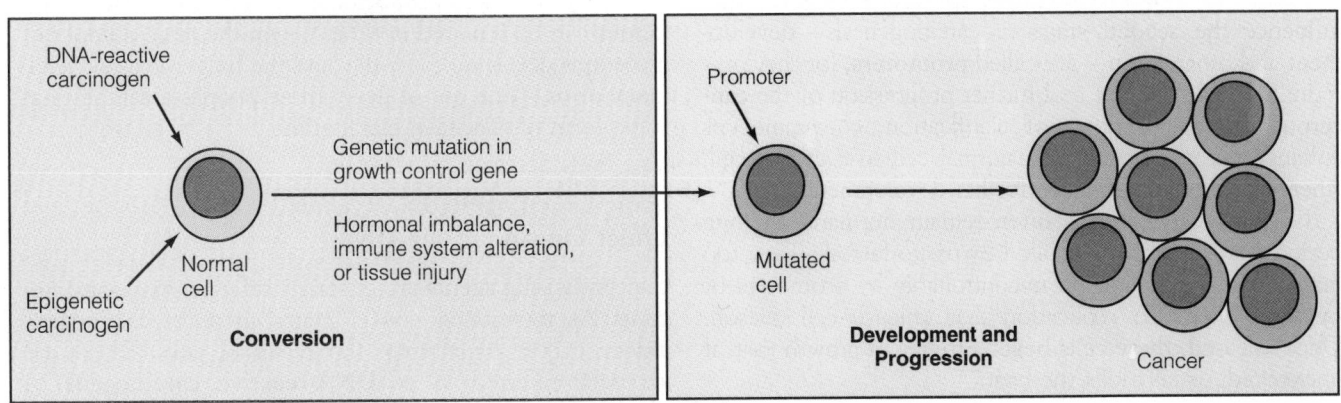

HEALTH NOTE 7–1

CANCER: ULTIMATELY, YOU MAY HOLD THE KEY TO A CURE

Hardly a year goes by without headlines announcing several promising new treatments for cancer.

Anticancer Vaccine

In 1988, for example, researchers reported preliminary results of a clinical study using a "vaccine" against lung cancer, which annually kills 150,000 Americans. The new vaccine, said proponents, could double survival rates in patients whose lung cancers were diagnosed early.

The lung cancer vaccine is really not a vaccine in the traditional sense, because it does not prevent the disease. This treatment is actually an immune system boost, which helps fight off an existing disease. In the study, 34 patients with early-stage lung cancer were treated with a plasma membrane protein extracted from lung cancer cells. When given to patients in early stages of lung cancer, the protein appears to stimulate the patient's immune system. The immune system, in turn, mounts an assault on the foreign protein floating freely in the blood. But because the protein is also attached to the plasma membrane of lung cancer cells, the immune system attacks and kills the cancer cells, especially those that have spread to other sites.

In clinical studies, this experimental therapy greatly increased the subjects' five-year survival rates. Normally, only 1 patient in 10 who is diagnosed with lung cancer is alive five years after diagnosis and conventional treatment (surgery followed by chemotherapy). Five years after the immune system boost, 5 of 10 patients were still alive.

Although this work is encouraging, medical scientists caution that it must be repeated before they can be sure of the effect. If the treatment lives up to its promise, researchers believe, other cancers might be treatable by the same technique.

Microspheres and Monoclonal Antibodies

In recent years, pharmaceutical manufacturers have become excited over the cancer-fighting potential of lipid microspheres containing cancer-killing chemicals, or **oncotoxins.** As described in Health Note 2–1, liposomes target cancer cells, delivering small doses of the lethal drugs to cancer cells and sparing the rest of the body. Targeting cancer cells is the job of antibodies, special proteins incorporated in the membrane of the liposome. These antibodies bind to proteins in the plasma membrane of cancer cells, allowing the "toxic bullet" to find its target.

Oncotoxins may also be attached to monoclonal antibodies. **Monoclonal antibodies** are synthesized by the immune system of animals that have been injected with human cancer cells. These antibodies can be chemically bonded to oncotoxins, then administered to patients. Inside the body of the patient, the antibody-oncotoxin complex binds to the cell surface of cancer cells and destroys them.

Prevention

Despite the promising new developments in cancer treatment, many experts stress the importance of prevention. By quitting smoking or not taking it up in the first place, by limiting exposure to X-rays and ultraviolet light, and by choosing a safe occupation and dwelling, you can greatly reduce the likelihood of contracting cancer.

As you will see in Chapter 11, proper diet may also help prevent cancer. Dietary fiber in grains and vegetables, for example, may help reduce your chances of contracting colon cancer, a leading killer of men ($\triangleright$ Figure 1). Other substances in food may also help reduce the risk of cancer. Recent research suggests that a chemical substance called ellagic acid, which is found in strawberries and Brazil nuts, may destroy carcinogenic

---

nal conditions, such as hormones. In other words, certain hormones produced by the body can stimulate the growth of cancerous cells. These cancers are hormone-dependent.

Chemical substances can also stimulate the growth of a mutated cell. Collectively, those chemical substances that influence the second stage of carcinogenesis—development and progression—are called **promoters,** for they promote the development and further progression of the cancerous cell. As a point of clarification, co-carcinogens enhance the conversion of a normal cell to a cancer cell, whereas promoters enhance further development.

Partially released cells often remain dormant for long periods, presumably controlled by tissue factors. They too may be spurred to divide uncontrollably by promoters or by errors in DNA replication that unleash cell divison. Once released, these cells begin a frenzy of growth that, if unchecked, usually kills the host.

Mutations occur with remarkable frequency. Fortunately, cells have evolved an intricate mechanism to repair genetic damage. If a mutation that transforms a cell into a cancerous one can be repaired before the cell divides, the cancer is eliminated. Unfortunately, not all cancerous mutations can be repaired in time. Given the large number of mutations occurring every day and the longer lives we lead, many of us—one out of every three people—will develop some form of cancer in our lifetime.

Some Carcinogens Exert Their Effect Outside of the DNA

Cancer-causing agents are generally referred to as **carcinogens.** As mentioned above, many chemical carcinogens induce cancer by altering the DNA of cells. These are generally referred to as **DNA-reactive carcinogens.** A

molecules in the body, reducing the likelihood of developing cancer.

Scientists studying carcinogenic agents found in tobacco smoke, auto exhaust, and foods noted that ellagic acid reduced DNA damage caused by one prevalent carcinogen by 45% to 70%. Using cultures of mouse and human lung tissue, the researchers also found that ellagic acid inhibited the formation of cancer caused by nitrosamines.

Ellagic acid has to be added before or during exposure to a carcinogen, and researchers have not yet identified the mechanism by which this substance works in people. They suspect that it competes for binding sites on the DNA molecule to which carcinogens attach.

Emotions and Cancer

Recent studies indicate that there may be a link between one's emotional state and the likelihood of developing cancer. Researchers at Johns Hopkins University studied medical students subjected to personality tests between 1948 and 1964 and compared the results of the tests with subsequent health records. They found that over a 30-year period students characterized as "loners"—who suppressed emotions beneath a bland exterior—were 16 times more likely to develop leukemia and several other cancers than a group that gave vent to emotions and took active measures to relieve anger and frustration. Why this might be so is anyone's guess.

Corroborating these results, a 17-year study of nearly 7000 people in Alameda County, California, showed that two types of social isolation (having few close friends and feeling alone even in the presence of friends) elevated the risk of certain cancers in women. A study in Yugoslavia also showed that repression and denial of emotions on a regular basis and in response to stress were related to an increase in cancer and heart disease.

Despite these gains from years of research and billions of dollars spent studying cancer, modern science may actually be losing the war against many types of cancer. For example, long-term survival rates in cancer patients over the past decade or so have increased only 4.2% compared with an overall increase of 5.1% in survival rates for all other diseases. Because of this statistic, many agree that the cheapest and most effective cure is prevention.

☞ FIGURE 1 Reducing Cancer by Eating Right Fruits, vegetables, and grains provide fiber that helps reduce the incidence of colon cancer.

growing body of evidence, however, indicates that many chemical substances act outside the DNA; that is, they do not react with DNA. These carcinogens, known as **epigenetic carcinogens,** operate by eliciting other biological effects. For example, some may cause a hormonal imbalance that leads to cellular proliferation. Others may alter the function of the immune system, which in turn results in cellular proliferation. Still others may cause chronic (persistent) tissue injury. Evidence suggests that some epigenetic carcinogens may stimulate changes in the DNA indirectly and may also facilitate tumor development of already-converted cells.

Interestingly, seemingly innocuous chemicals can turn into dangerous substances inside the body. A good example are the nitrites. Although fairly harmless, nitrites are converted to carcinogenic nitrosamines inside the human body. (See the Environment and Health section in Chapter 2.) The enzymes responsible for this conversion reside in the smooth endoplasmic reticulum (SER) of liver cells and possibly others. Although the liver's SER is usually an ally in detoxifying chemical substances, in this instance it converts a relatively harmless substance into a carcinogen with lethal consequences.

Cancers Develop Many Years After the Initial Exposure to a Carcinogen

As a rule, mutations do not manifest themselves immediately in cancer. In fact, many tumors form 20 or 30 years after the mutation. A few cancers such as leukemia occur within 5 years. The period between exposure and the emergence of a cancerous tumor is called the **latent period.** Because the latent period is long, it is often extremely difficult for medical researchers to determine the exact causes of many cancers, a fact that has hindered public health research considerably.

Several Physical and Biological Agents Are Also Responsible for Producing Cancer

Chemicals are not the only cause of cancer. X-rays that a physician uses to diagnose disease, ultraviolet radiation from the sun or a tanning salon, and radon gas given off by soils in some parts of the country can lead to mutations that result in cancer. Even extremely low frequency radiation given off by high-voltage electric lines, mentioned in the opening paragraphs of this chapter, may contribute.

Biological agents also contribute to cancer. In 1911, Peyton Rous discovered that a certain virus caused cancer in chickens. Since that time, researchers have discovered many viruses that cause cancer in a wide variety of animals, including humans. Unlike physical and chemical carcinogens, which alter DNA structure, carcinogenic viruses contain genes that, when inserted into the genetic material of the cells, may cause uncontrollable cell division.

ENVIRONMENT AND HEALTH: DEPLETING THE OZONE LAYER

Encircling the Earth, 12 to 30 miles above your head, is a layer of air rich in ozone (O_3) (▷ Figure 7–17). Of what importance is the **ozone layer** to a discussion of chromosomes, cell division, and cancer? Plenty.

The ozone layer forms an invisible shield that filters out most (99%) of the ultraviolet radiation emitted from the sun. Consequently, it protects all living things—from plants to humans—from damaging UV light, which can cause mutations in cells that may lead to cancer and death. The formation of the ozone layer over 1 billion years ago, in fact, was probably necessary for the evolution of life on land (Chapter 23). Without it, terrestrial life could probably not exist or would have evolved thick, protective shields.

Today the ozone layer is slowly disappearing. In 1974, two chemists warned that chemical substances used as spray-can propellants and refrigerants, known as the **chlorofluorocarbons (CFCs)**, were eroding the ozone layer. Their conclusions stunned the scientific world, because CFCs had previously been thought to be chemically inert (unreactive). Subsequent research showed that these compounds, although fairly stable, migrate from ground level into the upper atmosphere, where they are broken down by sunlight. Chlorine atoms released in the breakdown of these molecules react with ozone molecules in the upper atmosphere, destroying them and slowly depleting the ozone layer.

In response to the threat, the United States and Sweden soon banned the use of CFC spray-can propellants, but not refrigerants. Much of the rest of the industrial world remained skeptical and continued to use CFCs for spray cans, refrigerants, blowing agents (to make styrofoam), and cleaning agents. Over the next decade and a half, numerous estimates of ozone depletion were presented, but because of conflicting results and because the ozone concentrations in the ozone layer fluctuate naturally from year to year, atmospheric scientists could not tell for certain whether the projections were accurate.

Then, in 1988 a group of 100 scientists, working under the auspices of the National Aeronautics and Space Administration (NASA), gathered to review the data of nearly 20 years of atmospheric monitoring. They agreed that the ozone layer was indeed declining and that levels over the

▷ **FIGURE 7–17 The Ozone Layer** High above our heads lies the ozone layer of the atmosphere. The ozone layer shields the Earth from most of the ultraviolet radiation in sunlight.

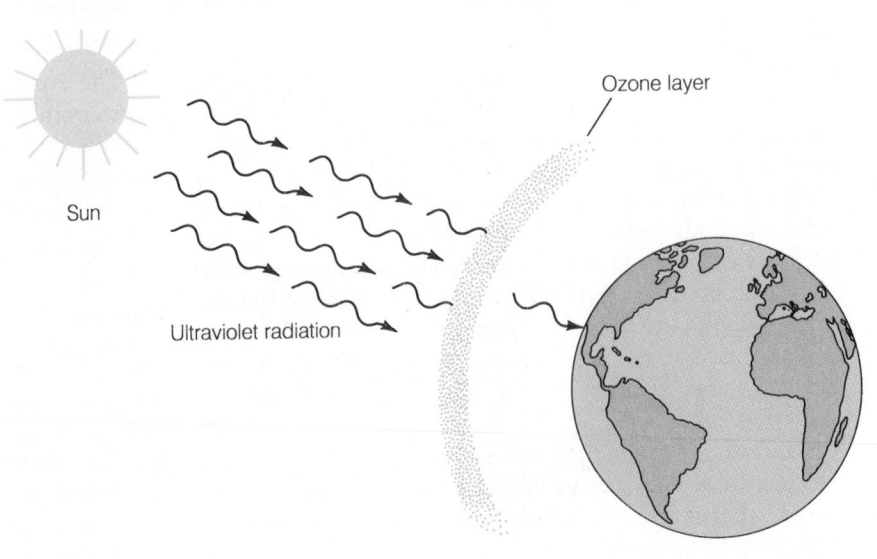

poles had fallen by as much as 10% since 1969. The decline over the heavily populated regions of North America and Europe over the same period was 3%.

Scientists predict that each 1% decrease in the ozone layer will increase the incidence of skin cancer by at least 2%.[1] If this prediction proves correct, the 3% decline over North America could increase the skin cancer rate by 6%. This would result in an additional 20,000 to 60,000 cases of skin cancer each year in the United States alone. Given that 4% of all skin cancers are fatal, an additional 800 to 2400 Americans could die each year from skin cancer.

Increased ultraviolet light may also harm other animals and plants. Plants are particularly vulnerable. High ultraviolet radiation interrupts photosynthesis and can result in death.

In September 1987, about a year before reports of ozone depletion were endorsed by the atmospheric scientists, a group of 24 nations met in Montreal to work out a treaty to restrict the use of most CFCs and a few related chemicals also thought to destroy the ozone layer. The nations agreed to cut their use to one-half of the 1986 levels by 1999. Given the accumulation of CFCs in the upper atmosphere since the 1950s, many critics warned

[1] More recent estimates suggest that each 1% decline in the ozone layer will result in a 4% increase in skin cancer.

that this reduction would not save the ozone layer. Estimates showed that an eventual 10% reduction of the ozone layer was probable under the Montreal agreement.

Much to the surprise of the critics, Du Pont, a world leader in CFC production, agreed in 1988 that further cuts were necessary and called for a total worldwide ban on CFC production. The company argued that stronger regulations could eliminate CFC production by the end of the century. Two years after the Montreal Treaty, the signatories met and agreed to a complete plase-out of CFCs by 1999. Manufacturers are now actively pursing ways to reduce CFC emissions by using materials more conservatively and by finding safer alternatives.

Shifting to alternative chemicals, the so-called HCFCs, will cost the U.S. economy an estimated $5.5 billion between 1990 and 2010, according to manufacturers. The Environmental Protection Agency projects that the cost of complying with the Montreal agreement will equal $27 billion through the year 2075. This price, however, is insignificant compared with the savings from reducing skin cancer deaths alone. In dollars and cents, the cutbacks called for by the Montreal accord would save $6.5 trillion in health costs by 2075. Add to that the incalculable reduction in human pain and suffering, and it becomes clear that good health and a healthy planet are economic as well as social and scientific concerns.

SUMMARY

1. Researchers have found a disturbing correlation between extremely low frequency (ELF) magnetic radiation and leukemia in children. At least one study shows that ELF radiation accelerates the growth of cancer cells and increases their resistance to the body's natural killer cells.
2. ELF magnetic fields are also generated by water beds and electric blankets, and some scientists think they may increase miscarriages or spontaneous abortions.

THE CELL CYCLE

3. The life cycle of a cell is called the cell cycle and consists of two parts: interphase and cell division.
4. Interphase is a period of active cellular synthesis and is divided into three phases: G_1, S, and G_2.
5. During G_1, the cell produces RNA, proteins, and other molecules.
6. During the S phase, the DNA replicates. After replication, each chromosome in the nucleus of the cell contains two chromatin fibers, or chromatids.
7. G_2 is a much shorter period and is relatively inactive.
8. During interphase, the cell replicates its organelles and molecules needed by the two daughter cells.
9. Cell division follows interphase. It requires two separate but related processes: mitosis, or nuclear division, and cytokinesis, or cytoplasmic division.

THE CHROMOSOME

10. Each organism has a set number of chromosomes. Somatic cells, containing a full complement, are described as diploid.
11. Gametes contain half the number of chromosomes of somatic cells and are referred to as haploid cells.
12. Gametes are produced by a special type of cell division known as meiosis, which occurs in the gonads (ovaries and testes).
13. The chromosomes are loosely arranged in the nucleus during interphase, but condense during prophase. Condensation facilitates chromosome separation.
14. Chromosomal condensation also allows physicians to study the morphology of chromosomes. Cells taken from fetuses are placed on a microscope slide, then flattened, which causes the chromosomes of cells in metaphase to spread out. Photographs of the chromosomes are then taken, and the chromosomes are cut out and arranged according to size, forming a karyotype. This allows technicians to count the chromosomes and to determine if there are any abnormalities.

CELL DIVISION IN ANIMALS: MITOSIS AND CYTOKINESIS

15. During cellular division, the cell divides its chromosomes equally and distributes its cytoplasm and organelles more or less equally between the two daughter cells.
16. In mature organisms, cell division helps repair tissue damage and replaces cells lost through normal wear and tear. In growing organisms, it is a key element of growth but also helps repair damaged tissue.

17. Mitosis is divided into four stages: prophase, metaphase, anaphase, and telophase. Table 7–1 summarizes the major changes in each stage.

18. Successful mitosis depends on the presence of the mitotic spindle, an array of microtubules that forms during prophase. Some of the spindle fibers attach to the chromosomes and help pull them apart during anaphase.

19. Several mechanisms have been proposed to explain the movement of chromosomes. According to one theory, the polar fibers slide along one another and push the poles of the cell farther apart, dragging the chromatids away from the metaphase plate.

20. Elongation of the polar fibers may also draw the chromosomes apart.

21. The final journey to the poles of the spindle may be accomplished by a shortening of the chromosomal fibers.

22. Cytokinesis, the division of the cytoplasm, begins in late anaphase or early telophase. In animal cells, cytokinesis results from the contraction of microfilaments lying beneath the plasma membrane.

CELL DIVISION IN PLANTS

23. Plant cell division differs from that of animals in two main respects: (a) centrioles and asters are absent, and (b) cytokinesis does not involve furrowing but, rather, the formation of a membranous cell plate between the two daughter cells.

CONTROL OF THE CELL CYCLE

24. Research suggests that the cell cycle is controlled in part by substances produced in the cytoplasm. These chemicals cause changes in the nucleus of the cell.

25. External controls, such as hormones, growth regulators, cell contact, and others, are also imposed on the cell.

CANCER: UNDERSTANDING A DEADLY DISEASE

26. Cancer is a disease characterized by uncontrolled cell division. Current estimates suggest that 20% to 40% of all cancers arise from environmental and workplace pollutants. The remaining cases are caused by smoking, diet, and natural causes.

27. No matter what the cause, cancer is a disease that most of us will have close experience with. One out of three Americans will contract the disease, and one out of four will die from it.

28. Cells that divide uncontrollably form tumors. Tumors that fail to grow and spread are benign and rarely cause medical problems.

29. Malignant tumors are those that continue to grow, often spreading to other parts of the body and invading vital organs. The term *cancer* is reserved for malignant tumors.

30. Malignant tumor cells spread through the body in the circulatory and lymphatic systems, often becoming established in distant sites, where they form secondary tumors.

31. Tumors can be removed surgically or can be destroyed with radiation, but treatment generally succeeds only if all of the secondary tumors are located and destroyed.

32. Patients can also be treated with chemotherapeutic agents, highly toxic chemical substances that attack rapidly dividing cells.

33. The transformation of a normal cell into a cancerous one often results from a mutation, a change in the DNA resulting from a chemical, biological, or physical agent. Not all mutations result in cancer.

34. Cancer-causing agents are called carcinogens. Chemical carcinogens bind to the DNA and alter its structure.

ENVIRONMENT AND HEALTH: DEPLETING THE OZONE LAYER

35. The ozone layer is found in the upper atmosphere. It filters out most of the incoming ultraviolet light, protecting many life forms from harmful burns and damaging mutations that can lead to cancer.

36. Unfortunately, certain chemicals called chlorofluorocarbons (CFCs) are destroying the ozone layer. In industrialized nations, CFCs are used as spray-can propellants, refrigerants, coolants, cleaning agents, and blowing agents for foam.

37. According to a recent study, CFCs have caused a 3% decline in ozone levels in the atmosphere over North America.

38. Many nations have agreed to phase out CFC production and release by 2000, but efforts are now under way to phase them out more rapidly.

EXERCISING YOUR CRITICAL THINKING SKILLS

1. In 1987, the industrial nations met in Montreal to negotiate a worldwide treaty to limit ozone-depleting chemical production and use. A few years later, faced with ever-worsening news on the depletion of the ozone layer, they met again and agreed to a complete ban on many ozone-depleting chemicals by the year 2000.

On February 12, 1992, the *New York Times* announced that President George Bush, concerned about the decline in the ozone layer, had "directed American manufacturers to end virtually all production of chemicals that destroy ozone" by the end of 1995. For this move, the president received high praise from supporters.

However, there was a loophole, which the newspaper glossed over—that is, that Bush had decided to permit limited production and use of these chemicals. From the above quote, can you detect a hint of the loophole?

The key phrase in the newspaper was "virtually all production." An analysis of Bush's plan shows that it would actually permit production of ozone-destroying chemicals at 15% of the 1987 levels after the December 1995 deadline. It would not eliminate the production of ozone-destroying chemicals by that time as many people thought from reading the news report. Furthermore, it would accelerate the complete U.S. phaseout called for by the 1987 Montreal agreement by only one year.

Take a moment to reflect on this issue. What is the critical thinking lesson?

If you said, "read very carefully and seek information," you were right. Beware of television and newspaper reports as well, for they often gloss over important details.

2. Professor Arthur Kronberg has spent a lifetime exploring the secrets of cells, and with great success. In 1956, for example, he discovered DNA polymerase, the enzyme cells require to

synthesize DNA. In 1959, he was awarded the Nobel Prize for his contributions. In 1967, Kronberg synthesized an entire viral DNA chain, which successfully replicated inside host cells.

Kronberg has focused much of his research on cellular enzymes. In his work, he grinds up cells and extracts the liquid contents. He then manipulates the extract to perform the specific process in which he is interested. After that, he begins removing enzymes from the extract until the process stops. When it stops, he knows he has isolated the enzyme responsible for catalyzing the function he is studying. Kronberg used this technique to find out what stops a cell from dividing—in other words, what factor or factors are responsible for turning off DNA replication in nondividing cells. Kronberg and his colleagues began this search for this "off switch" by extracting enzymes from the cell extracts. Through this work, they discovered several enzymes and proteins that stopped replication. Unfortunately, the researchers found that the proteins worked by destroying the DNA, obviously not the way nature works.

Although you're not a trained cell biologist, can you see anything wrong with the way Kronberg approached this problem? One of Kronberg's postdoctoral students did.

In 1988, Deog Su Hwang, who was working in Kronberg's lab, began to study the same problem but from a different angle. He had observed that two intertwined strands of DNA separate slightly just before replication. This separation takes place at a site along the DNA strands known as the "origin." Many enzymes involved in DNA replication attach to the origin. This separation forms a kind of bubble that promptly causes the strands to separate further. The nucleotides necessary for making complementary strands are then assembled by DNA polymerase.

Hwang thought that this separation site was a likely place for the off switch and concentrated his work on this region. In other words, he located a probable location for the switch, then began the search for the protein that binds to that site, hoping that it would somehow block DNA replication. His approach focused more narrowly on the problem and very probably increased his chance of success.

Hwang used enzymes to slice off portions of the DNA chain and associated protein around the origin. These fragments were then isolated and studied. In late 1989, he succeeded in isolating a pure sample of the protein that inhibited DNA replication when added before initial separation of the DNA helix.

How did the experimental procedures differ? Kronberg's enzyme purification procedure searched for an enzyme that catalyzed a specific cellular process. In a sense, he was searching for a needle in a haystack. Hwang's method involved searching for the protein that fit a specific site and then testing it to determine its function. The two procedures are obviously quite different.

What is to be learned from this exercise? With all due respect to Arthur Kronberg, question the experts and their assumptions. At times, even Nobel-prize winning scientists can embark on paths of scientific inquiry that yield little information. In this case, Kronberg's line of inquiry was based on techniques that worked in other situations but weren't applicable to the problem at hand.

TEST OF TERMS

1. The cell cycle consists of two major phases, _____ in which the cell divides, and _____ , the period of growth and preparation between divisions.

2. The cell replicates its DNA during the _____ phase.

3. The segment of the DNA that controls a trait is called a(n) _____ .

4. Division of the nucleus during the cell cycle is called _____ , and division of the cytoplasm is called _____ .

5. Gametes contain half the number of chromosomes of somatic cells; gametes are, therefore, referred to as _____ cells.

6. The unlimited growth of cancer cells results from a loss of _____ _____ , a signal moderated by the plasma membrane, which stops cell division.

7. During G_2, the chromosomes consist of two chromatids, each consisting of a molecule of DNA and associated protein. The DNA/protein complex is called _____ .

8. Chromatids of the replicated chromosomes are attached at the _____ .

9. A photograph of chromosomes aligned in pairs is called a(n) _____ .

10. The _____ appears in prophase and is composed of microtubules that attach to the chromosomes.

11. Chromosomes line up in the middle of the cell during _____ .

12. A(n) _____ _____ lying just beneath the plasma membrane contracts and draws the plasma membrane in, creating a furrow that eventually divides the cytoplasm in two.

13. In plant cells, membranous vesicles produced by the Golgi complex fuse in the center of dividing cells, forming the _____ _____ .

14. The spread of cancer cells through the lymphatic and circulatory systems is called _____ . This process results in the establishment of _____ tumors at different sites.

15. The conversion of a normal cell to a cancerous one is called _____ .

16. A chemical or physical agent that causes cancer is a(n) _____ .

17. The chemical substances believed responsible for destroying the _____ layer, a protective region in the atmosphere that blocks ultraviolet radiation, belong to a group of compounds called _____ .

Answers to the Test of Terms are found in Appendix B

1. Describe the cell cycle, and explain what happens during each part.
2. Discuss the factors that control the cell cycle. Is it likely that many factors play a role in determining when and how often a cell divides?
3. Draw a diagram of the cell cycle, and note how many chromosomes are found in the cell at each stage and whether they are single- or double-stranded.
4. List the stages of mitosis, and describe the major nuclear changes occurring in each stage. When does cytokinesis occur? What are the major cytoplasmic changes?
5. Discuss the similarities and differences between plant and animal cell division.
6. In what ways are cancer cells different from normal cells? What events lead to a cancer? Why don't more cells become cancerous? What is the difference between a benign and a malignant tumor?
7. The Environment and Health section argues that chemical substances produced by modern society—notably, the chlorofluorocarbons—may be making our environment hazardous to human health. Describe how.
8. Cancer is often considered a disease of aging. Given the fact that cancer is caused by DNA mutation, why is this assertion correct? Can you think of some reasons why the incidence of some cancers has risen in the past 20 years?

8

An Introduction to Heredity
Meiosis, Mendelian Genetics, and Human Genetics

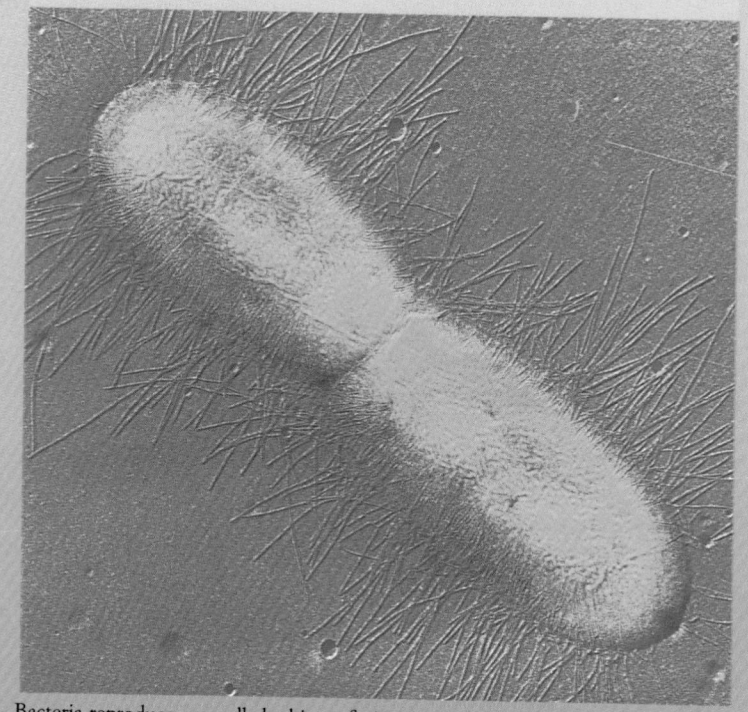

Bacteria reproduce asexually by binary fission, a simple splitting in two.

For most of us, life is a bit of a roller coaster. Our emotions tend to rise and fall with changing circumstances. Mild emotional peaks and troughs are quite normal. For some, however, the roller coaster ride is more intense. Their feelings swing widely. One minute they are in a state of elated overactivity (called mania), the next they fall into a state of depressed inactivity. In this condition, known as manic-depression, cyclic mood swings are unrelated to external events.

Manic-depression varies considerably from one person to the next. In some, the cyclic mood swings last only a short time, but in others the bouts of depression and elation continue for years (▷ Figure 8–1). When in a state of deep depression, victims sometimes threaten suicide, but their lack of energy often prevents them from following through. In some cases, the desire may persist in the period following depression, resulting in suicide. During mania, the animated and highly energetic state, manic-depressives often act irrationally or obnoxiously. Such behavior can ruin relationships and can lead to financial disaster, especially if it interferes with decision making.

Manic-depression occurs in about 3% of the U.S. population and tends to run in families, a fact that suggests a genetic link. Researchers, in fact, recently located a gene near the tip of chromosome 11 that may predispose its bearers to manic-depression and possibly severe depression.

Chapter 7 described the genes and chromosomes, noting how they affect the anatomy, physiology, and behavior of all organisms from the simplest single-celled protist to the complex, multicellular mammal. This chapter examines another aspect of genetics, the inheritance of traits. We begin with an overview of meiosis, a special kind of cell division that is found only in gamete formation in sexually reproducing organisms. We then focus our attention on some basic principles of heredity.

≈ MEIOSIS AND GAMETE FORMATION

For successful reproduction to occur, sexually reproducing organisms must have some mechanism to reduce by half the number of chromosomes in their germ cells. Thus, when the male and female gametes of a flower or a horse unite, the offspring will have the normal (diploid) number of chromosomes. This process of reduction is called **meiosis.** Defined more precisely, meiosis is a type of nuclear division that occurs in germ cells and yields haploid gametes. Germ cells are formed in the reproductive structures of sexually reproducing diploid plants and animals.

As noted in the last chapter, diploid organisms have two sets of chromosomes, one maternal in origin, the other paternal. In the human, for instance, there are 46 chromosomes, 23 from the mother and 23 from the father. In diploid organisms, chromosomes occur in pairs, which are called homologous pairs because they are the same length, have the same appearance, and contain the same genes.

▷ **FIGURE 8–1 Manic-Depressive** Manic-depression results in periodic outbursts of mania (hyperactivity and elation) followed by periods of depression. Scientists think that this disease may be genetically based.

During meiosis, each gamete ends up with one member of each homologous pair.

Meiosis Involves Two Cellular Divisions

You will better understand the process of meiosis if you keep in mind a few key facts. First, remember that meiosis occurs during the formation of germ cells. Second, note that meiosis involves two nuclear divisions. Third, during the first division, the chromosome number is halved; that is, a diploid cell produces two haploid cells. The second division is virtually identical to a mitotic division.

Meiosis I Is a Reduction Division

The two nuclear divisions are referred to as meiosis I and II. **Meiosis** I is often referred to as a reduction division, for it is during this process that the diploid cell divides, splitting its homologous pairs evenly. ▷ Figure 8–2 illustrates the process. As shown, the chromosomes condense, and the homologous chromosomes pair up during prophase I. During metaphase I, the homologous pairs line up along the equatorial plate. During anaphase I, the pairs are separated. During telophase, each daughter cell ends up with half the number of chromosomes of the parent cell from which it formed. Note, however, that each chromosome contains two chromatids.

Meiosis II Is Akin to Mitotic Division

The second division resembles mitosis, which you studied in the last chapter, except for one simple difference: the cells are haploid. As illustrated in Figure 8–2, the chromosomes of the haploid cells condense during prophase II,

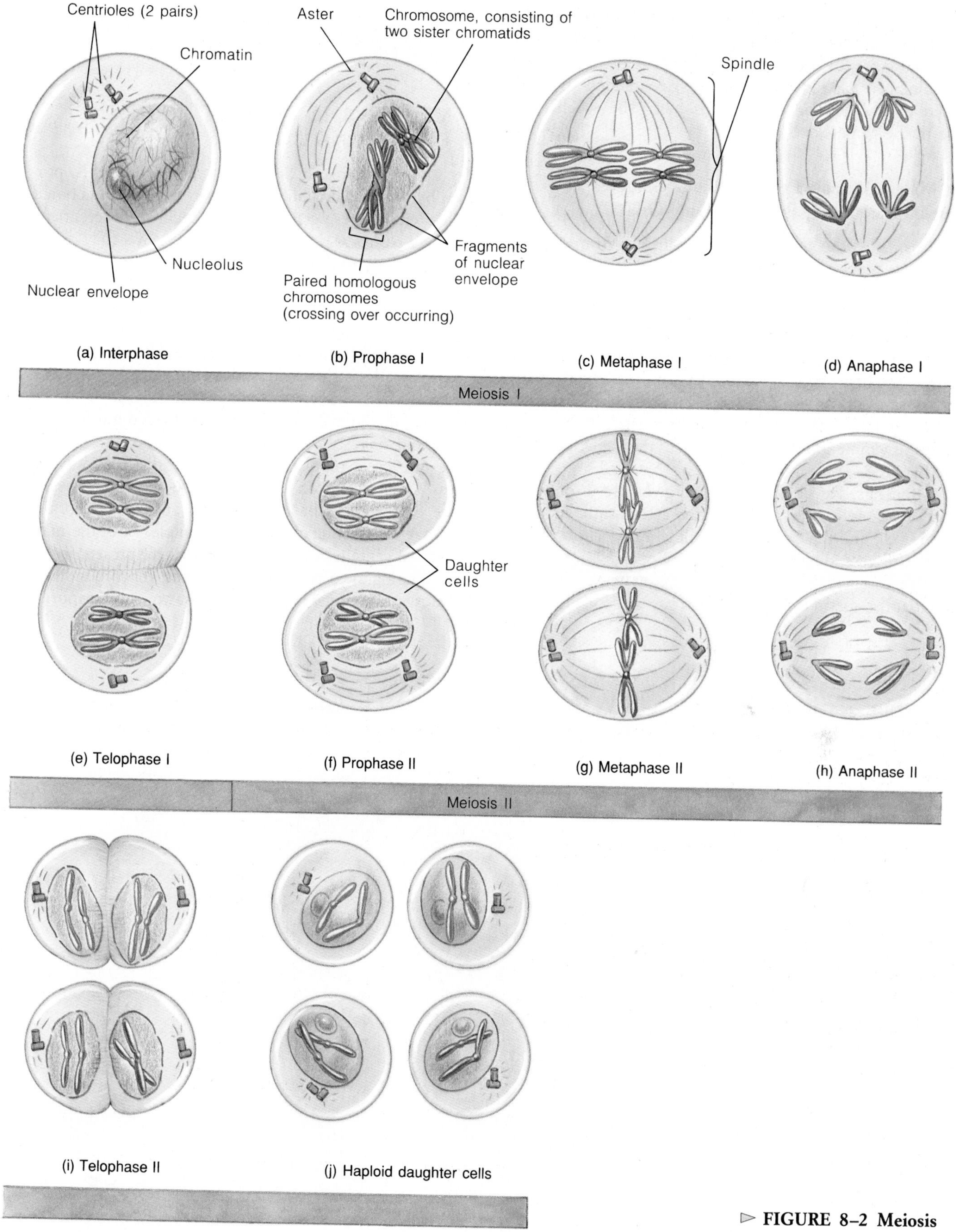

Centrioles (2 pairs)

Chromatin

Aster

Chromosome, consisting of two sister chromatids

Spindle

Nucleolus

Nuclear envelope

Paired homologous chromosomes (crossing over occurring)

Fragments of nuclear envelope

(a) Interphase

(b) Prophase I

(c) Metaphase I

(d) Anaphase I

Meiosis I

Daughter cells

(e) Telophase I

(f) Prophase II

(g) Metaphase II

(h) Anaphase II

Meiosis II

(i) Telophase II

(j) Haploid daughter cells

▷ **FIGURE 8–2 Meiosis**

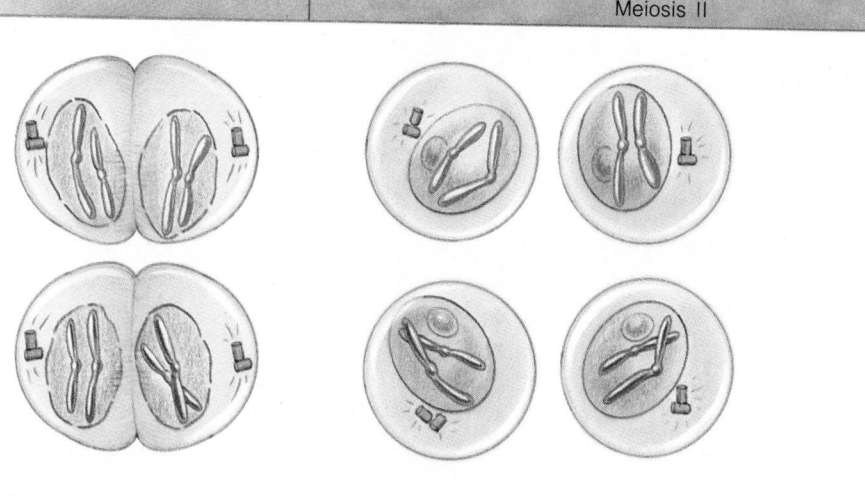

then line up in single file along the equatorial plate during metaphase II. During anaphase II, the chromatids dissociate, one going to each pole. The result is two cells, each containing a haploid number of single-stranded (unreplicated or one-chromatid) chromosomes. These cells give rise to the gametes.

In Males, Meiosis Produces Four Gametes, But in Females, It Produces Only One

Meiosis differs between males and females. In males, the process is much as I have described above. Thus, a single diploid cell in the gamete-producing structure like the human testis gives rise to four functional sperm cells. Consequently, the cytoplasm of the original cells that give rise to the sperm is divided equally among the four germ cells. During the final stages of sperm development, much of the cytoplasm is discarded to produce streamlined sperm capable of swimming up the female reproductive tract, or in plants, swimming in water to reach the sedentary egg (▷ Figure 8–3).

In females, the single diploid cell gives rise to only one egg (▷ Figure 8–4). Thus, all of the cytoplasm is retained in one cell, and the extra chromosomal material is systematically discarded during the two meiotic divisions. This difference reflects the fact that the egg is fairly sedentary and that the sperm must make their way to it for successful fertilization. This division of labor also ensures that adequate cellular organelles (supplied by the female) are available for the zygote.

≈ MEIOSIS AND THE LIFE CYCLES OF PLANTS AND ANIMALS

After a zygote forms, it divides by mitosis. The cells produced by the first division divide again and again. As noted in Chapter 10, these cells will soon begin to differentiate, producing tissues and organs that are part of a fully developed organism. One of those organs is the reproductive structure, which produces gametes, capable of fertilizing or being fertilized and thus capable of producing offspring.

This greatly simplified series of events is called the life cycle. In sexually reproducing organisms, the life cycle can be condensed to the following events: meiosis, gamete formation, fertilization, and mitosis (▷ Figure 8–5). As illustrated, the animal life cycle can be divided into haploid and diploid stages.

The life cycle of plants can also be divided into two stages. As illustrated in ▷ Figure 8–6, however, several unique events occur between meiosis and gamete formation. Take a flowering plant as an example. Both the male and female gamete-producing structures are present in the flower. During meiosis in these structures, germ cells divide by meiosis to produce haploid cells known as **meiospores.**

▷ **FIGURE 8–3 Meiosis in Sperm Formation** Note that four haploid cells are produced during sperm formation. The cytoplasm is discarded in the final conversion to sperm.

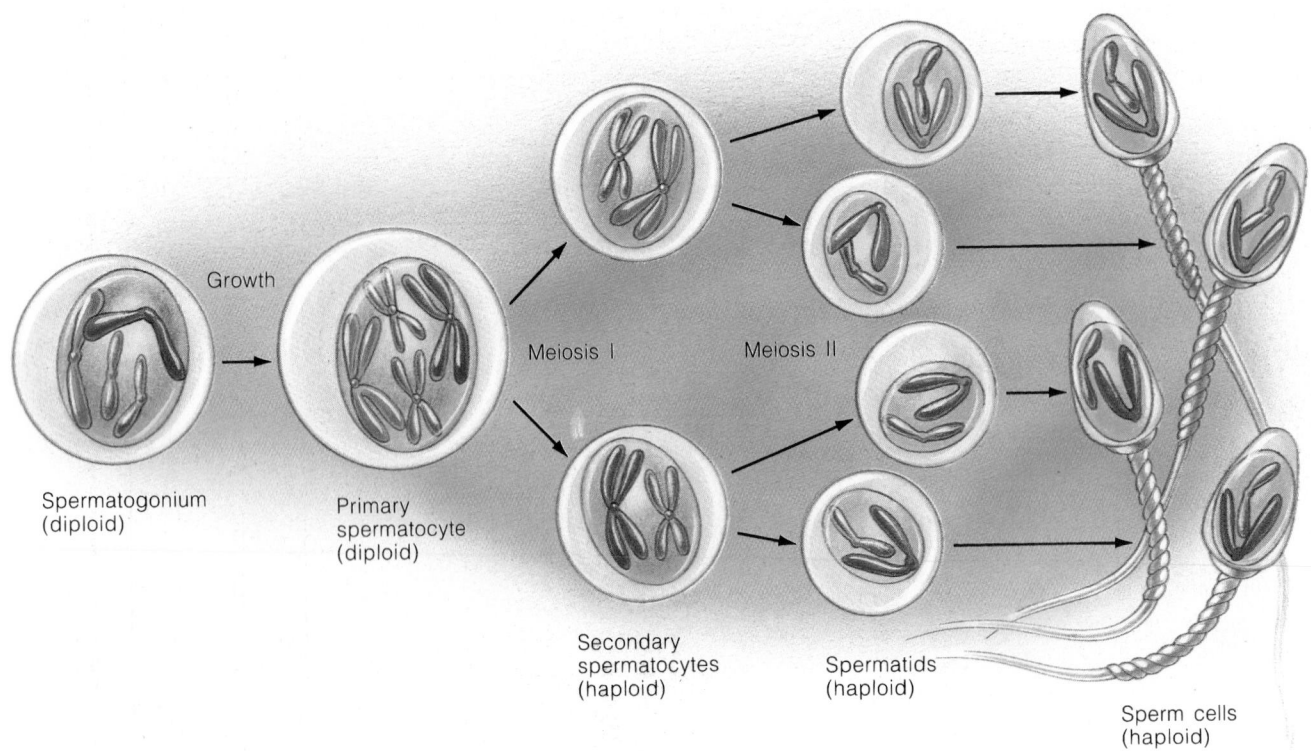

Spermatogonium (diploid)

Growth

Primary spermatocyte (diploid)

Meiosis I

Meiosis II

Secondary spermatocytes (haploid)

Spermatids (haploid)

Sperm cells (haploid)

Both male and female meiospores are produced in the same flower. The meiospores are not sperm and eggs, however. Rather, these cells divide by mitosis to produce multicellular haploid structures called **gametophytes** (gamete-producing bodies). The cells of the male and female gametophytes divide by mitosis to produce eggs and sperm. The male gametophyte in flowering plants is a pollen grain. In females, the gametophyte is a structure that gives rise to an egg.

The life cycle is completed when an egg and a sperm unite. This union produces a diploid embryo that is enclosed in a protective seed and covered by a fruit. (That is why the flowers of plants often become fruit.) If the seed falls to the ground and receives the right amount of moisture, it can grow into a full-fledged plant, known as a **sporophyte,** thus starting the cycle all over again.

≈ PRINCIPLES OF HEREDITY: MENDELIAN GENETICS

With this basic overview of gamete formation in mind, we turn our attention to heredity, the transmission of genetic traits from one generation to the next. Heredity is the subject of **genetics,** defined as the study of the structure and function of genes and the transmission of genes from parents to offspring. The formal study of genetics began with the work of a 19th-century monk, Gregor Mendel (▷ Figure 8–7, page 170). Born in Eastern Europe in 1821, Mendel entered an Augustinian monastery in Brno, in what is now Czechoslovakia, at the age of 21.[1] After completing his studies, Mendel enrolled at the University of Vienna, where he studied botany, mathematics, and other sciences. Returning to the monastery, he began a series of experiments on garden peas—studies that would reveal several key principles of genetics still valid today.

During Mendel's time, many scientists believed that the traits of a child's parents were somehow blended in the offspring, producing a child with intermediate characteristics. In addition, because the ova of sexually reproducing organisms are much larger than the sperm, some scientists believed that the female had a greater influence on the characteristics of the offspring than the male.

Mendel's studies were designed to answer two questions: First, are the physical characteristics of an offspring the result of a blending of parental traits? Second, if blending does occur, do both parents contribute equally to the offspring?

Mendel was a careful scientist. He spent two years selecting an organism to study. He eventually settled on one species of garden pea. Over a 10-year period, he studied inheritance in approximately 28,000 pea plants, thus avoiding problems that might arise from small sample size, a requirement of careful experimentation, as noted in the critical thinking section of Chapter 1. Mendel also repeated

[1] Brno is now the city of Brünn.

▷ **FIGURE 8–4 Meiosis in Egg Formation** Note that only one haploid cell is formed, containing all of the cytoplasm. Nuclear material is discarded along the way.

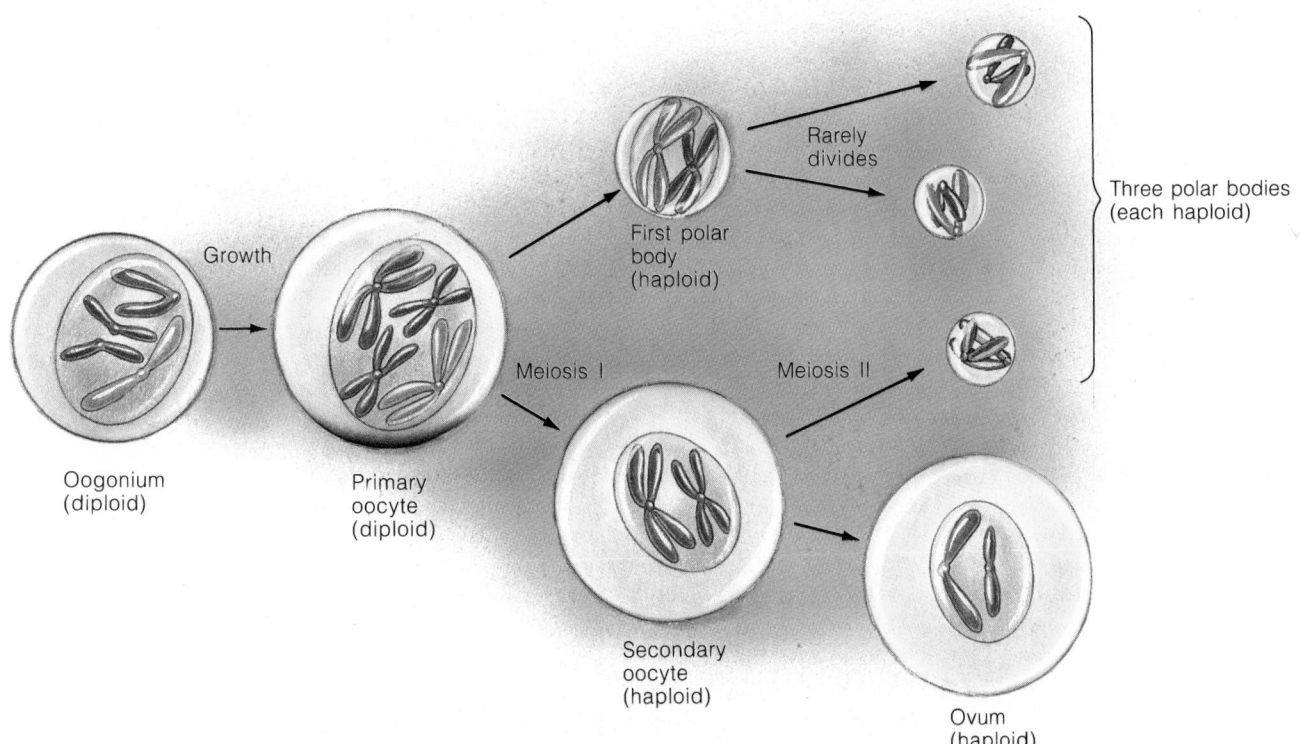

Oogonium (diploid)

Growth

Primary oocyte (diploid)

Meiosis I

First polar body (haploid)

Rarely divides

Meiosis II

Three polar bodies (each haploid)

Secondary oocyte (haploid)

Ovum (haploid)

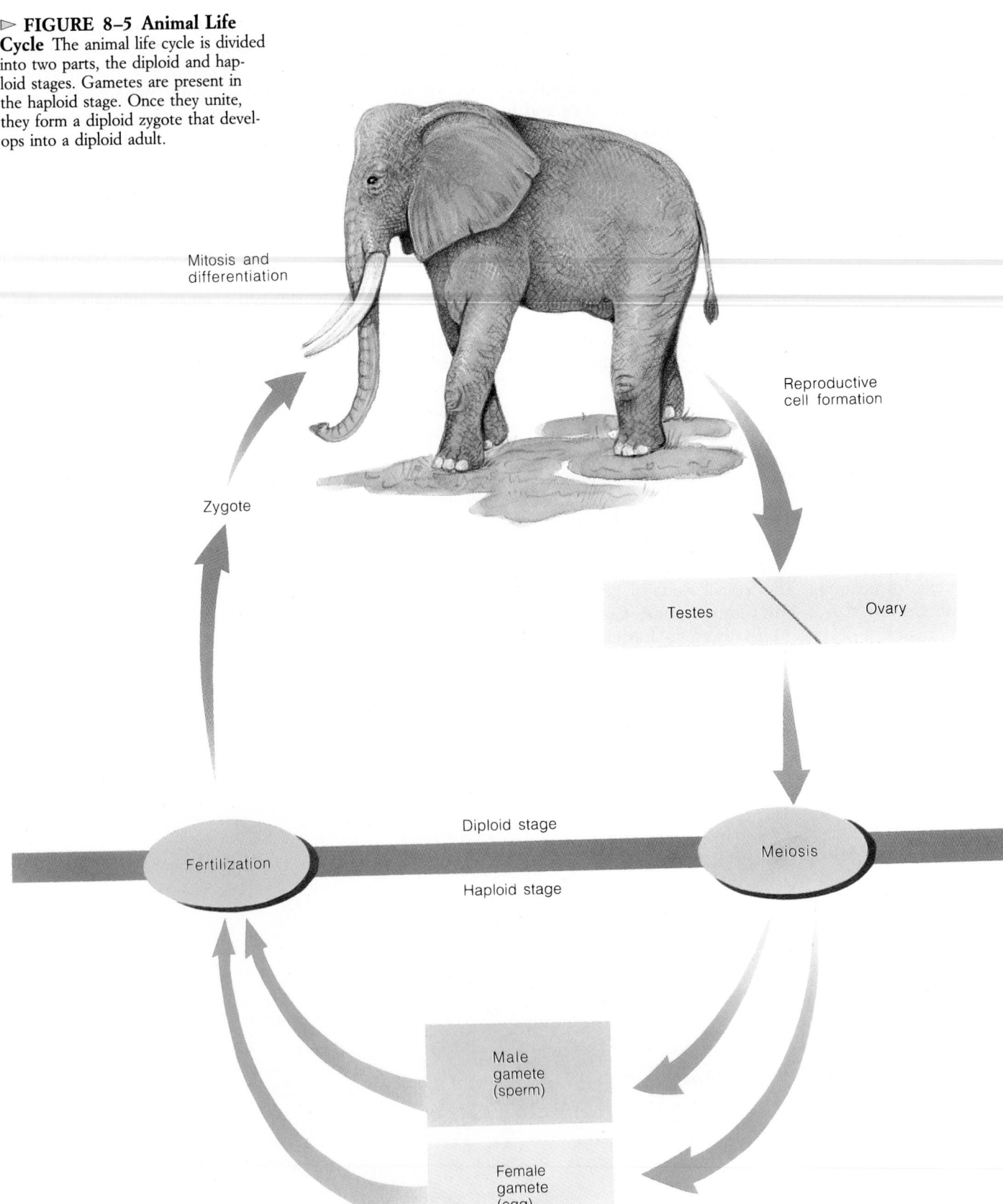

▷ **FIGURE 8–5 Animal Life Cycle** The animal life cycle is divided into two parts, the diploid and haploid stages. Gametes are present in the haploid stage. Once they unite, they form a diploid zygote that develops into a diploid adult.

Mitosis and differentiation

Reproductive cell formation

Zygote

Testes   Ovary

Diploid stage

Fertilization   Meiosis

Haploid stage

Male gamete (sperm)

Female gamete (egg)

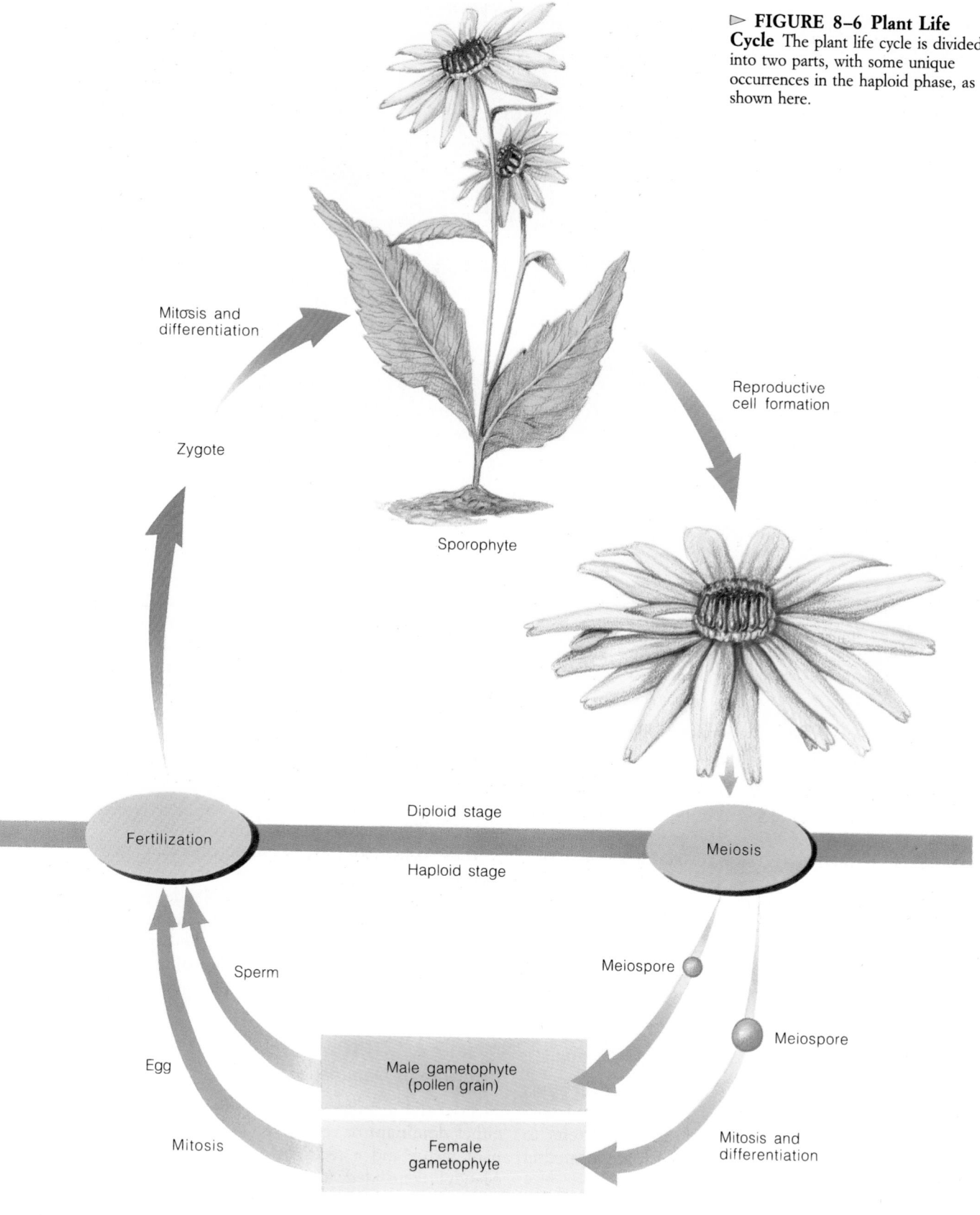

▷ **FIGURE 8–6 Plant Life Cycle** The plant life cycle is divided into two parts, with some unique occurrences in the haploid phase, as shown here.

Mitosis and differentiation

Zygote

Reproductive cell formation

Sporophyte

Diploid stage

Fertilization

Haploid stage

Meiosis

Sperm

Meiospore

Egg

Meiospore

Mitosis

Male gametophyte (pollen grain)

Female gametophyte

Mitosis and differentiation

(a) (b)

▷ **FIGURE 8–7 Mendel and His Garden** (*a*) Gregor Mendel worked out some of the basic rules of inheritance in the mid-1800s in his research on garden peas. (*b*) Mendel's garden at the monastery in Czechoslovakia as it appears today.

his experiments, ensuring that his results were reproducible. Finally, he analyzed his results according to principles of probability. In fact, he was probably the first scientist to use this kind of analysis in biological research.

Mendel Discovered That the Traits He Was Studying Did Not Blend

Mendel's first discovery in his analysis of inheritance in peas was that the physical characteristics of the parents were not blended to produce offspring with intermediate traits. When he bred a pea with a white flower to a pea with a purple flower, he did not produce pea plants with intermediate pink flowers but, rather, plants with purple flowers, for reasons that will be discussed later. Experiments that Mendel performed on the inheritance of other traits showed similar results.

Mendel Discovered That the Parents Contributed Equally to the Characteristics of Their Offspring

Mendel's research led him to conclude that adult plants contain pairs of hereditary factors and that each pair governed the inheritance of a single trait. Today, we refer to Mendel's hereditary factors as genes, which are located in the DNA of the chromosomes.

Mendel reasoned that because a male gamete and a female gamete combine to form a new organism and an adult contains only two hereditary factors for any given trait, each gamete must contain only one hereditary factor (gene)

for each trait. Mendel's studies therefore suggested to him that the paired hereditary factors of the mother and father must separate during gamete formation so that each gamete contributes one and only one hereditary factor to the zygote. The separation of hereditary factors during gamete formation is known as the **principle of segregation.**

Because the gametes of the parents combine to produce an offspring and each gamete contains one hereditary factor for each trait, Mendel concluded, the contributions of the parents must be equal. That is, each parent contributes one hereditary factor, or gene, for each trait.

These conclusions may seem unremarkable to you now, especially with your knowledge of meiosis and gamete formation. However, remember that Mendel performed his experiments in the 1850s and 1860s, long before chromosomes had been discovered! In fact, he hypothesized the presence and behavior of genes long before science knew even the most basic facts of cell biology and inheritance.

Mendel Also Discovered the Principle of Dominance

Mendel also postulated that hereditary factors (genes) are either **dominant** or **recessive** (Table 8–1). When a dominant factor and a recessive factor are present in a pea, Mendel concluded, the dominant factor is always expressed. A recessive factor is expressed only when the dominant factor is missing. An alternative form of the same gene is called an **allele.** The dominant and recessive factors are known today as dominant and recessive alleles. Domi-

TABLE 8–1 Genetic Traits Studied by Mendel		
STRUCTURE STUDIED	DOMINANT TRAIT	RECESSIVE TRAIT
Seeds	Smooth Yellow	Wrinkled Green
Pods	Full Green	Constricted Yellow
Flowers	Axial (along stems) Purple	Terminal (top of stems) White
Stems	Long	Short

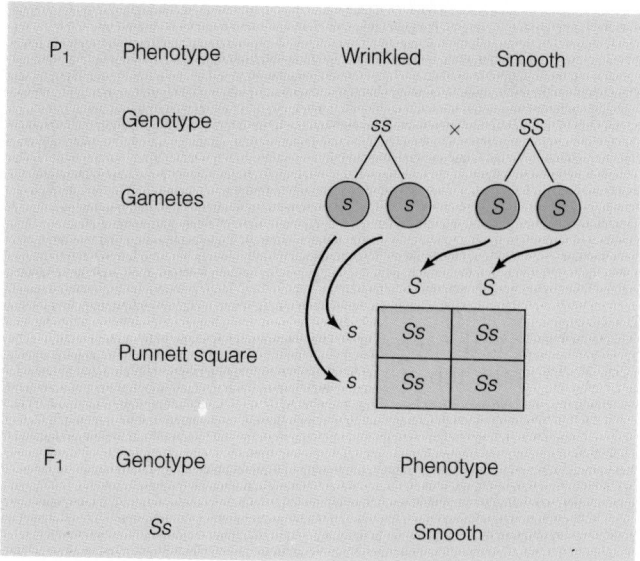

▷ **FIGURE 8–8 Mendel's Early Experiments** When studying the inheritance of seed conformation, Mendel first crossed homozygous recessive (*ss*) and homozygous dominant (*SS*) plants. The offspring were all heterozygotes (*Ss*).

nant alleles are designated by capital letters; recessive alleles are signified by lowercase letters.

In human somatic cells, genes exist in pairs, with one member of the pair (allele) on each homologous chromosome. In Mendel's peas, for instance, the gene for flower color is the *P* gene. The dominant form is *P* (purple flowers), and the recessive form is *p* (white flowers). The *P* and *p* genes are both alleles of the flower-color gene. For genes with two alleles, such as these, three combinations are possible. The first consists of two dominant genes—for example, *PP*. This individual is said to be **homozygous dominant** for that particular trait. The second consists of two recessive alleles—in this example, *pp*—and the individual is said to be **homozygous recessive** for that trait. The third occurs when dominant and recessive alleles are present (*Pp*), and the individual is **heterozygous**.

The Genotype of an Organism Is Its Genetic Makeup; the Phenotype Is Its Appearance

Mendel determined that blending did not occur in the inheritance of traits, such as flower color. Thus, a pea plant with purple flowers (*PP*) bred with a pea plant with white flowers (*pp*) produced offspring with purple flowers and not pink flowers, which would occur if blending took place. Mendel proposed that the purple-flowered offspring of this cross were heterozygous—*Pp*. In such cases, he argued, the recessive hereditary factor (genes) are *masked* by the dominant hereditary factors (genes). Because of dominance, Mendel noted, the outward appearance of a plant—its **phenotype**—does not always reflect its genetic makeup, or **genotype**. In this example, two different genotypes can lead to the same phenotype (*PP* = purple and *Pp* = purple). As a general rule, a homozygous dominant individual and a heterozygous individual are indistinguishable on the basis of phenotype.

Genotypes and Phenotypes Can Be Determined by the Punnett Square

To track the genotypes and phenotypes of breeding experiments, such as those that Mendel performed, geneticists use a relatively simple tool known as the Punnett square (▷ Figure 8–8). To illustrate the process, consider one of the traits that Mendel studied, seed coat texture. The gene for seed coat texture has two alleles: one that codes for smooth seeds (*S*) and one that codes for wrinkled seeds (*s*). As indicated by the capital *S*, the smooth-seed allele is dominant.

Suppose that you bred a homozygous-dominant plant (*SS*) with a homozygous-recessive plant (*ss*). Figure 8–8 lists the genotypes of each parent and lists the possible genotypes of the gametes. Determining the genotype is a rather simple affair. If a diploid parent is *SS*, its haploid gametes are all *S*. If a parent is *ss*, the gametes are all *s*.

To determine the outcome of crossing these two plants, all of the possible gametes produced from one plant are listed along the top of the Punnett square, and all of the possible gametes from the other are listed along the side. The gametes are then combined as shown, producing all of the possible genotypes present in the offspring.

As this example illustrates, a cross between a homozygous-dominant individual and a homozygous-recessive individual produces only heterozygous offspring. The offspring are phenotypically uniform, and all of them resemble the homozygous-dominant parent; that is, they have a smooth seed coat.

In the language of genetics, when one organism is bred to another to study the transmission of a single trait, the procedure is called a **monohybrid cross.** In a monohybrid cross, the organisms that are bred initially constitute the P_1 generation (*P* for parents). The first set of offspring constitutes the F_1 generation (first filial generation). In this example, the F_1 offspring could be bred to produce additional offspring, the F_2 generation (second filial generation).

To determine the genotypes and phenotypes of such a cross, the Punnett square can be used once again. As ▷ Figure 8–9 shows, the gametes of the F_1 generation are *S* and *s*. All of the possible gametes from the father are placed along the top of the Punnett square, and all of the possible gametes of the mother are placed along the side. The gametes are combined as before in the appropriate boxes. As shown in Figure 8–9, this cross produces three different genotypes: *SS, Ss,* and *ss*. The ratio of genotypes, or the genotypic ratio of the F_2 generation is one *SS* to two *Ss* to one *ss*, or 1:2:1. Thus, if 100 offspring are produced, you would expect 25 of them to be *SS* (homozygous dominant), 50 to be *Ss* (heterozygous), and 25 to be *ss* (homozygous recessive). Because of dominance, though, this cross yields only two phenotypes: plants with smooth seeds (*SS* and *Ss*) and plants with wrinkled seeds (*ss*). The ratio of phenotypes, or the phenotypic ratio, is three smooth to one wrinkled, or 3:1.

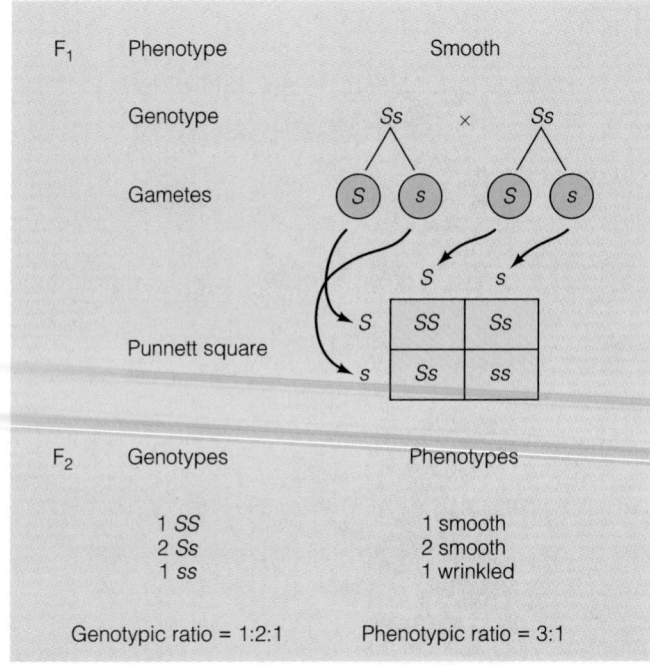

▷ **FIGURE 8–9 Monohybrid Cross** In Mendel's early work, he crossed offspring from his F_1 generation (*Ss*) and found a variety of genotypes and phenotypes.

Genes on Different Chromosomes Segregate Independently of One Another During Gamete Formation

In his later experiments, Mendel tracked two traits at the same time, a procedure geneticists refer to as a **dihybrid cross** (▷ Figure 8–10). This work led to the **principle of independent assortment**—the idea that genes located on different chromosomes segregate independently during meiosis. To understand what this principle means and how Mendel arrived at it, consider another example that examines seed coat texture and seed color.

As noted above, peas contain a gene for seed coat texture, the *S* gene. The dominant form is *S* (smooth), and the recessive form is *s* (wrinkled). Peas also contain a gene for seed color; the dominant form (*Y*) yields yellow seeds, and the recessive form (*y*) yields green seeds.

To study the inheritance of these two genes at the same time, Mendel crossed homozygous dominant plants with smooth, yellow seeds (*SSYY*) with homozygous recessive plants (*ssyy*). As Figure 8–10 shows, the genotype of the F_1 generation was *SsYy*. Consequently, the offspring were phenotypically identical, and all displayed the dominant characteristics (smooth, yellow seeds).

When Mendel crossed two of the F_1 offspring (genotype = *SsYy*), however, he got a mixture of phenotypes (Figure 8–10). To produce this combination of offspring, Mendel concluded, the hereditary factors *S* and *s* must have segregated independently of *Y* and *y* during gamete formation. Translated into modern terms, this means that the *Y* and *S* genes were on different chromosomes and separated independently of each other during gamete formation (meiosis). Because of independent assortment, the gametes of the F_1 generation contain all combinations of the alleles in equal proportions: *SY, Sy, sY,* and *sy*. Fertilization occurred at random, giving rise to 16 possible combinations (Figure 8–10).

As noted above, this research led Mendel to propose the principle of independent assortment, which states that the segregation of the alleles of one gene on one chromosome is independent of the segregation of the alleles of another gene on a second chromosome during gamete formation.

Mendel may have been one of the luckiest scientists in the world. He chose to study seven traits, listed in Table 8–1, each of which scientists now believe is on a different chromosome (This assumption has not yet been proved.) If two genes under study are located on the same chromosome, they are said to be *linked,* and they tend to be inherited together. Therefore, independent assortment holds only for genes located on different chromosomes.[2]

---

[2]Widely separated genes on the same chromosome often assort independently if the distance between them is great *and* if crossing over occurs with great frequency.

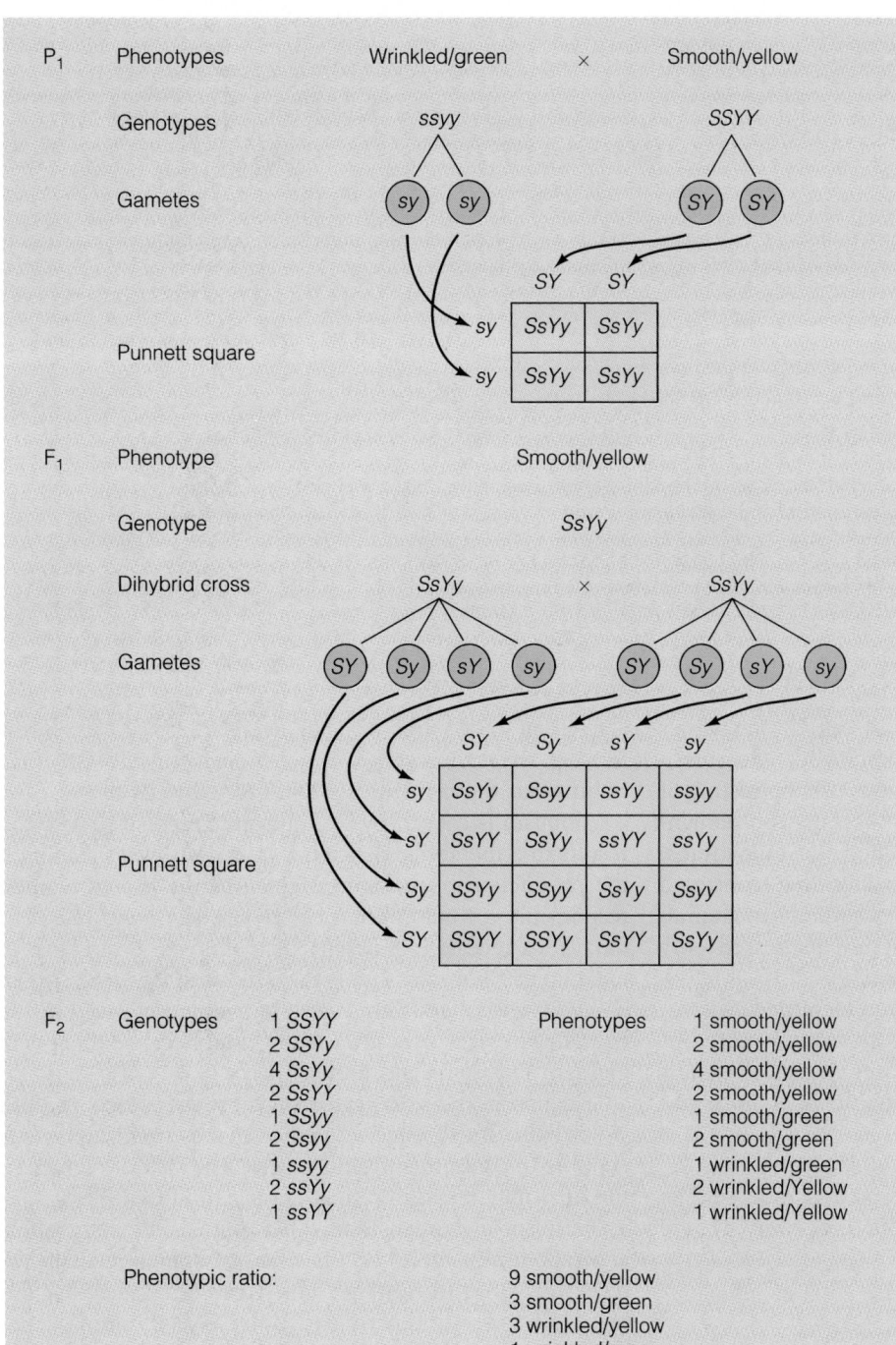

P₁ Phenotypes   Wrinkled/green   ×   Smooth/yellow

Genotypes   *ssyy*   *SSYY*

Gametes   sy   sy   SY   SY

Punnett square

	SY	SY
sy	SsYy	SsYy
sy	SsYy	SsYy

F₁ Phenotype   Smooth/yellow

Genotype   *SsYy*

Dihybrid cross   *SsYy*   ×   *SsYy*

Gametes   SY Sy sY sy   SY Sy sY sy

Punnett square

	SY	Sy	sY	sy
sy	SsYy	Ssyy	ssYy	ssyy
sY	SsYY	SsYy	ssYY	ssYy
Sy	SSYy	SSyy	SsYy	Ssyy
SY	SSYY	SSYy	SsYY	SsYy

F₂ Genotypes   1 *SSYY*   Phenotypes   1 smooth/yellow
                   2 *SSYy*   2 smooth/yellow
                   4 *SsYy*   4 smooth/yellow
                   2 *SsYY*   2 smooth/yellow
                   1 *SSyy*   1 smooth/green
                   2 *Ssyy*   2 smooth/green
                   1 *ssyy*   1 wrinkled/green
                   2 *ssYy*   2 wrinkled/Yellow
                   1 *ssYY*   1 wrinkled/Yellow

Phenotypic ratio: 9 smooth/yellow
                                     3 smooth/green
                                     3 wrinkled/yellow
                                     1 wrinkled/green

≈ MENDELIAN GENETICS IN HUMANS

Mendelian genetics applies to many human traits. Table 8–2 lists a number of human traits and diseases whose inheritance follows basic Mendelian principles. This section describes the inheritance of those traits, using several common diseases as examples.

Autosomal-Recessive Traits Are Expressed Only When Both Alleles Are Recessive

As noted in Chapter 7, human cells contain 23 pairs of chromosomes, consisting of two types: the sex chromosomes and the autosomes. The **sex chromosomes** are involved in sex determination (and a few other functions). As

TABLE8-2 Traits and Diseases Carried on Human Chromosomes

Autosomal recessive	
Albinism	Lack of pigment in eyes, skin, and hair
Cystic fibrosis	Pancreatic failure, mucus buildup in lungs
Sickle-cell anemia	Abnormal hemoglobin leading to sickle-shaped red blood cells that obstruct vital capillaries
Tay-Sachs disease	Improper metabolism of a class of chemicals called gangliosides in nerve cells, resulting in early death
Phenylketonuria	Accumulation of phenylalanine in blood; results in mental retardation
Attached earlobe	Earlobe attached to skin
Hyperextendable thumb	Thumb bends past 45° angle
Autosomal dominant	
Achondroplasia	Dwarfism resulting from a defect in epiphyseal plates of forming long bones
Marfan's syndrome	Defect manifest in connective tissue, resulting in excessive growth, aortic rupture
Widow's peak	Hairline coming to a point on forehead
Huntington's disease	Progressive deterioration of the nervous system beginning in late twenties or early thirties; results in mental deterioration and early death
Brachydactyly	Disfiguration of hands, shortened fingers
Freckles	Permanent aggregations of melanin in the skin

noted in earlier chapters, two types of sex chromosomes exist, X and Y. Females have two homologous X chromosomes, and males have a nonhomologous pair, consisting of one X chromosome and one Y chromosome. The remaining 22 pairs of chromosomes are called the **autosomes.** The autosomes contain numerous genes that control a variety of traits. This section examines several **autosomal-recessive traits**—that is, traits that are expressed *only* when both recessive alleles are present.

Over 600 traits in humans have been identified as autosomal recessive, and another 800 are strongly suspected of being autosomal recessive. Some autosomal-recessive traits cause serious medical problems, among them albinism and cystic fibrosis.

Albinism. **Albinism** is one of the most common genetic defects known to science. It occurs in 1 of every 38,000

▷ **FIGURE 8-11 Albinism** Albinism is an autosomal-recessive trait. Albino man with his wife in India.

Caucasian births but in 1 of every 22,000 black births. Among the Hopi and Navajo Indians of the American Southwest, the incidence is 1 in every 200.

Albinism is caused by an autosomal recessive allele that results in a metabolic deficiency in the pathways leading to the production of melanin. Melanin is the brown pigment responsible for coloration of the eyes, skin, and hair. Homozygous recessive individuals for albinism often have no melanin at all or may simply have reduced levels (▷ Figure 8–11). Consequently, the skin of an albino is pale, and the hair is white. The eyes are usually pink, because there is no pigment in the iris and retina.

Melanin in the skin protects against the effects of ultraviolet light. Its absence in albinos makes them highly susceptible to sunburn and skin cancer. The lack of pigment in the eyes may result in retinal damage and blindness.

Albinism in humans and other animals as well (rabbits and rats, for example) occurs when an animal receives two recessive genes from its parents. Consider the albino rabbit. Coat color is determined by several different genes. Two genes of particular importance are the *B* and *C* genes. For example, the *B* allele is dominant and when present in homozygous-dominant (*BB*) or heterozygous (*Bb*) animals produces a black coat. The *b* allele is recessive and yields a brown coat when the individual is homozygous recessive (*bb*). The *C* gene, however, has a profound influence on the actual phenotype, for it controls the synthesis of tyrosinase, the first enzyme in the biochemical pathway leading to melanin production. If a rabbit (or a human, for that matter) inherits two recessive genes, no tyrosinase is made. Nor is melanin, and the result is an albino. Gene pairs like this one that mask the expression of other genes are called **epistatic genes.**

Cystic Fibrosis. **Cystic fibrosis** is an autosomal-recessive disease that leads to early death. The presence of two recessive alleles for the disease alters the function of sweat glands, mucous glands, and the pancreas. Defective sweat glands, for example, release excess amounts of salt, a marker that helps physicians diagnose the disease. The most significant symptoms, however, occur in the pancreas and lungs. In patients with cystic fibrosis, the ducts that drain digestive enzymes from the pancreas into the small intestine become clogged. This not only impairs digestion but also causes a buildup of enzymes in the pancreas that results in the formation of cysts in the organ. Over time, the pancreas begins to degenerate, and fibrous tissue replaces glandular tissue—hence, the name cystic fibrosis (▷ Figure 8–12).

Despite adequate nutritional intake, victims of cystic fibrosis often show signs of malnutrition. To enhance digestion, patients are often given powdered or granular extracts of animal pancreases containing digestive enzymes. Massive doses of vitamins and nutrients must also be given.

The respiratory systems of most victims of cystic fibrosis produce copious amounts of mucus. Mucus blocks the respiratory passages, making breathing difficult. Victims must be treated several times a day to remove the mucus (▷ Figure 8–13). Despite antibiotic therapy and other treatments, most cystic fibrosis patients live only into their late teens or early 20s.

Cystic fibrosis is one of the most common genetic diseases known to medical science. Surprisingly, 1 of every 22 Caucasians carries a gene for this disease, and approximately 1 of every 2000 Caucasians born in the United States suffers from it. In the black population, only about 1 in 100,000 to 150,000 individuals is a carrier.

Autosomal-Dominant Traits Are Expressed in Heterozygotes and Homozygous-Dominant Individuals

Many human traits are **autosomal dominant;** that is, they are carried on the autosomes and are expressed in heterozygotes and homozygous-dominant individuals. To date, nearly 1200 human traits have been identified as autosomal dominant, and 1000 others are suspected. The absence of dermal ridges (which give rise to fingerprints), short fingers and toes, freckles, cleft chin, and drooping eyelids are all autosomal-dominant traits. This section discusses three other examples: widow's peak, achondroplasia, and Marfan's syndrome.

Widow's Peak. Take a moment to examine the hairlines of your friends and classmates. You will notice that in some individuals the hairline runs straight across the forehead. In others, it juts forward in the center, forming a "widow's peak" (▷ Figure 8–14). Widow's peak results from an autosomal dominant gene, indicated by W. Because the W allele is dominant, this phenotype is expressed in homozygous dominant individuals (WW) and also heterozygotes

▷ **FIGURE 8–13 Cystic Fibrosis** Inhalants, antibiotics, and special physical therapy techniques are used to treat patients with cystic fibrosis. To remove mucus from the lungs, parents or physical therapists must treat the patient two or three times a day. Pounding on the rib cage with a cupped hand (clopping) loosens mucus.

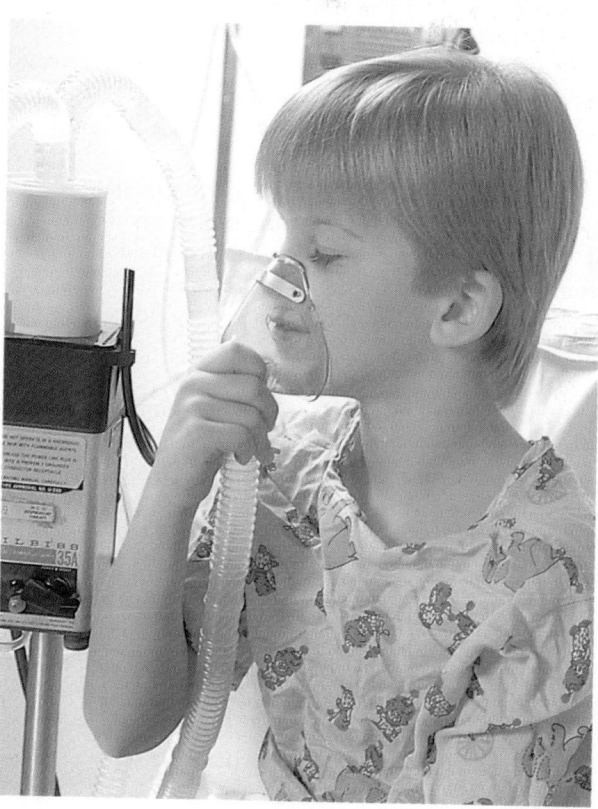

▷ **FIGURE 8–12 Pancreas from Patient with Cystic Fibrosis** (a) Normal pancreas. Cystic fibrosis is a disease caused by an autosomal-recessive gene. (b) It results in a blockage of the ducts draining the pancreas, leading to cysts in the pancreatic tissue. The tissue degenerates and is replaced by fibrous connective tissue.

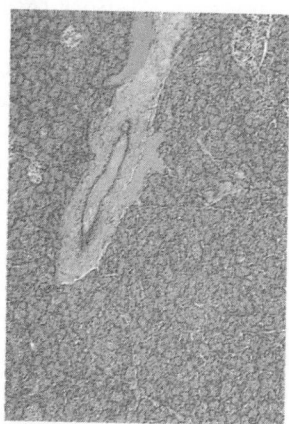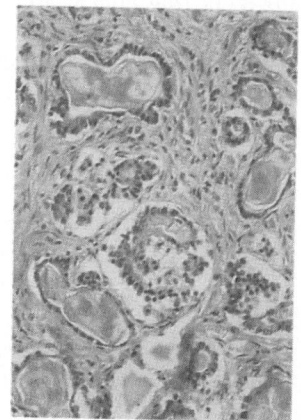

(a)   (b)

(a)

(b)

▷ **FIGURE 8–14 Widow's Peak** (*a*) Widow's peak is a dominant trait carried on one of the autosomes. (*b*) A straight hairline is a recessive trait. Simple Mendelian genetics can be used to determine the genotype of the offsping.

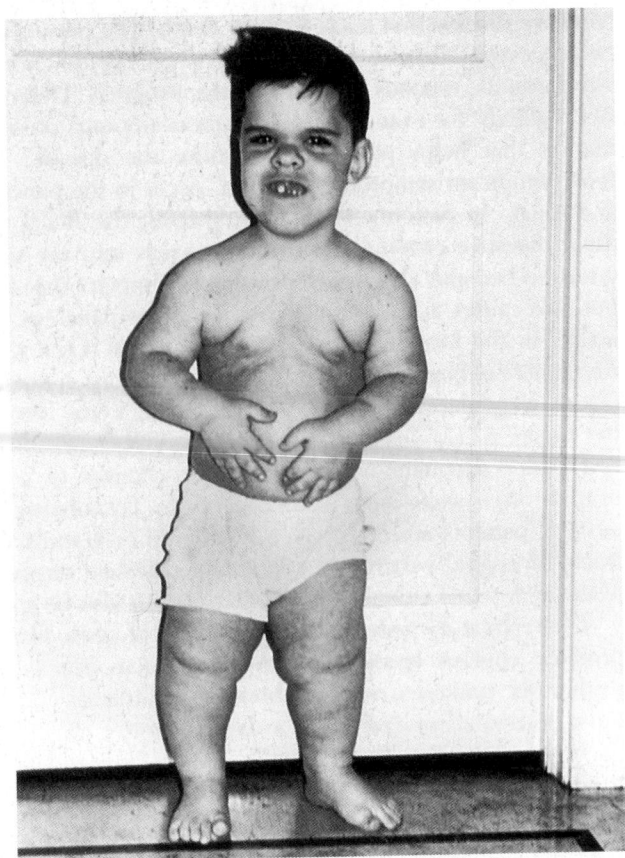

▷ **FIGURE 8–15 Achondroplasia** Achondroplasia, one form of dwarfism, is an autosomal-dominant disorder.

(*Ww*). Individuals with the genotype *ww* have a straight hairline.

Achondroplasia. The boy shown in ▷ Figure 8–15 suffers from a genetic disease called **achondroplasia,** one form of dwarfism. Victims of the disease have short, stubby legs and arms but a relatively normal-sized trunk. As a rule, they do not grow taller than 4 feet. Interestingly, surgeons recently developed and successfully tested a technique to lengthen the arms and legs of people with this disorder. The leg and arm bones are fractured (under anesthesia), and the patient is fitted with a traction device. Traction is applied to the broken bones, causing them to separate. New bone is formed in the gap, and the bones slowly elongate.

Achondroplasia afflicts about 1 of every 10,000 children born in the United States and results from an autosomal dominant gene. Most cases are believed to arise from spontaneous mutations, because the majority of children with the condition are born to phenotypically normal parents.

Marfan's Syndrome. More than a hundred years after his death, President Abraham Lincoln has become the subject of scientific controversy. The debate among scientists is not over who killed him or why but, rather, over the possibility that he suffered from a genetic disorder called **Marfan's syndrome.**

Marfan's syndrome is an extremely rare autosomal dominant disorder that affects the skeletal system, the eye, and the cardiovascular system. It is difficult to diagnose and is usually identified only after its victims die. Flo Hyman, the star of the 1984 U.S. women's Olympic volleyball team and at one time believed to be the best female volleyball player in the world, had Marfan's syndrome (▷ Figure 8–16). In 1986, Hyman was taken out of a game because of ill health. Shortly afterward, she collapsed on the floor and died. Upon autopsy, pathologists found that her aorta had ruptured.

Marfan's victims are characterized by exceedingly long arms and legs, making them excellent candidates for volleyball and basketball. In normal individuals, the arm span (the distance from fingertip to fingertip) is equal to their height. In Marfan's syndrome, however, arm span exceeds height.

Nearsightedness and lens defects are also common among Marfan's patients. The most serious problem, however, is enlargement and weakening of the aortic arch, caused by a degeneration of the connective tissue in the wall of the aorta. This defect causes the aorta to gradually enlarge; if untreated, it will burst.

As for Lincoln, no one really knows whether he had the disorder. Some evidence suggests that he did. Lincoln was tall and lanky, for example, and he wore glasses. His great-great-grandfather had the disease, making it possible that Abe had received the Marfan's gene. The critical thinking skills section, however, reminds us of the importance of a

▷ **FIGURE 8–16 Marfan's Syndrome** Flo Hyman, volleyball player extraordinaire, suffered from Marfan's syndrome, an autosomal-dominant trait.

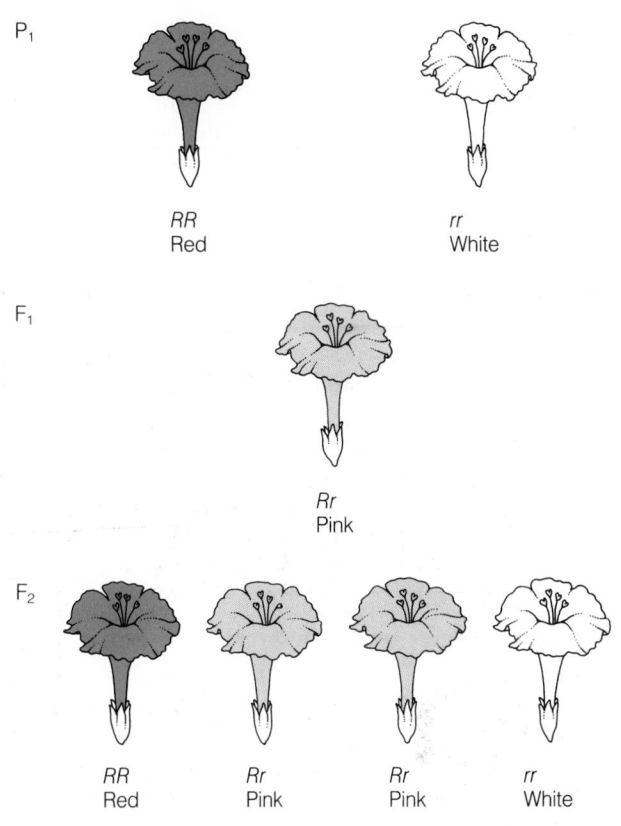

▷ **FIGURE 8–17 Incomplete Dominance** Incomplete dominance involves two alleles, neither of which is dominant over the other. The result is an intermediate phenotype as shown here is the flower of the plant *Mirabilis.*

close examination of the facts. First of all, Lincoln's glasses were to correct farsightedness, not myopia, a characteristic of victims of the disease. Secondly, he also had shown no sign of cardiovascular disease, and his limbs were within the normal dimensions of tall people, suggesting that he did not have this genetic disorder.

≈ VARIATIONS IN MENDELIAN GENETICS

Mendel presented his work at meetings of the Natural Science Society in Czechoslovakia in 1865 and published his results the following year. Like many ideas ahead of their time, Mendel's conclusions went largely unnoticed. It was not until 1900, 16 years after he died, that his work received the attention it deserved. At that time, the publication of three other studies confirmed Mendel's findings. More studies followed, and excitement began to grow. The scientific community came to realize that Mendel's principles pertained to a great many organisms. Since that time, additional research has uncovered new modes of inheritance and gene expression that are more common than the simple dominance he proposed.

Incomplete Dominance Results in Intermediate Traits—That Is, a Kind of Blending of Traits

One example is a phenomenon called incomplete, or partial, dominance. **Incomplete dominance** occurs when heterozygous offspring exhibit intermediate phenotypes. In other words, incomplete dominance produces F_1 offspring with phenotypes intermediate to the parental phenotypes. (This explains some observed cases of apparent blending of traits.) Incomplete dominance occurs in a plant called *Mirabilis.* ▷ Figure 8–17 shows a cross between two plants, one with red flowers *(RR)* and one with white flowers *(rr)*. This cross produces offspring with pink flowers, an intermediate phenotype (Figure 8–17). If the *R* gene were completely dominant, you would expect all of the F_1 offspring to be red. In this case, however, the gene does not exert complete dominance. Incomplete dominance also occurs in a number of human disorders, including sickle-cell anemia.

Sickle-Cell Anemia. Sickle-cell anemia is a disorder that affects the blood and is chiefly found in African Americans and in Caucasians of Mediterranean descent. Sickle-cell anemia is caused by a genetic defect that leads to abnormal hemoglobin formation. Hemoglobin is a protein that is found in red blood cells, which carry oxygen to body cells. Sickle-cell anemia occurs in individuals who are homozygous recessive. The abnormal hemoglobin causes red blood cells to convert to sickle-shaped cells when they encounter low oxygen levels in the blood—for example, when blood cells flow through capillaries in metabolically active tissues. Sickle-shaped red blood cells clog capillaries and reduce oxygen flow to brain cells, heart cells, and other organs. In homozygous-recessive individuals, sickle-cell anemia is usually lethal. In fact, most individuals with the disease die by their late 20s.

Individuals who are heterozygous for the trait are referred to as **carriers,** because they can pass the gene on to their children. Carriers generally lead relatively normal lives, but they are subject to occasional problems. Although their red blood cells appear normal, they actually contain 50% normal hemoglobin and 50% abnormal. Moderate sickling occurs when these cells are exposed to low oxygen.

Approximately 1 in every 500 African Americans born in the United States is homozygous recessive, and about 1 of every 12 is heterozygous (a carrier) for the sickle-cell trait. Why is this allele so prevalent? In tropical climates, from which African Americans come, the sickle-cell allele protects carriers and homozygous-recessive individuals from malaria, a deadly disease prevalent in humid, tropical regions. Although homozygous-recessive individuals die earlier, the selective advantage that carriers enjoy has caused the allele to increase in frequency in the population. Malaria is caused by a microscopic parasite called *Plasmodium,* which is transmitted from one person to the next by the *Anopheles* mosquito. Inside the body, the parasites invade and colonize the liver, where they multiply rapidly. New parasites leave the liver and enter the bloodstream, where they attack and destroy red blood cells. Many of the parasites remain in the liver, however, continuing to reproduce and periodically releasing new offspring. Unless treated, a victim of malaria suffers from repeated attacks of chills followed by fever. Each attack corresponds with the release of a new batch of *Plasmodia* from the liver.

How does the sickle-cell allele protect people? The hemoglobin in the red blood cells of carriers and individuals who are homozygous recessive is altered by the sickle-cell allele. For reasons not entirely clear, the presence of altered hemoglobin changes the plasma membrane of red blood cells in ways that prevent the parasite from entering. As a result, both homozygous-recessive individuals and carriers (heterozygotes) are relatively immune to the parasite. Because the sickle-cell allele confers a selective advantage on those who have it and because heterozygotes suffer relatively few problems, in East Africa, where the malaria parasite is quite common, 45% of all blacks are carriers of the trait (heterozygotes). In the United States, where malaria is virtually nonexistent, the frequency of the sickle-cell allele has decreased considerably.

Some Genes have Multiple Alleles

Mendel studied seven characteristics of peas. Each characteristic is determined by a gene with two alleles—that is, two alternative forms. In human and other vertebrates, however, research shows that some genes can have more than two alleles; such a gene is said to have **multiple alleles.**

Consider blood types. One gene that controls blood type is called the *I* **gene** (for isoagglutinin). This gene can exist in one of three biochemically distinct forms. The three alleles are I^A, I^B, and I^O. The *I* gene is located at a particular site (or locus) on one pair of chromosomes. Note, however, that even though there are three possible alleles in human beings, an individual can have only two of the alleles in his or her genome, one on each homologous chromosome.

Four possible blood types exist: A, B, AB, and O. Table 8–3 lists the four blood types and the six different genotypes that give rise to them. The *I* gene codes for glycoproteins (proteins with carbohydrate attached) that project from the surface of the red blood cell. The A and B alleles of the *I* gene produce A and B glycoproteins respectively, as illustrated in ▷ Figure 8–18. The O allele produces neither type of cell-surface glycoprotein. Therefore, in AO and AA individuals, the red blood cells contain only type A glycoproteins (Figure 8–18). In BO and BB individuals, the type B glycoproteins are present in the plasma membrane. When the A and B alleles are both present, the result is a cell surface with both types of glycoprotein. Both A and B are dominant genes, and the O allele is recessive. Interestingly, therefore, the I^A and I^B genes are said to be codominant. **Codominant genes** are expressed fully and equally.

Codominance and incomplete dominance are two exceptions to Mendel's principles. These examples do not negate Mendel's discoveries. They are simply additional modes of gene expression. Mendelian principles, in fact, can even be used to predict the genotypic ratios and patterns of inheritance in instances of codominance and incomplete dominance.

TABLE 8–3 Phenotype and Genotype in ABO system	
PHENOTYPE (BLOOD TYPE)	GENOTYPES
Type A	$I^A I^A$, $I^A I^O$
Type B	$I^B I^B$, $I^B I^O$
Type AB	$I^A I^B$
Type O	$I^O I^O$

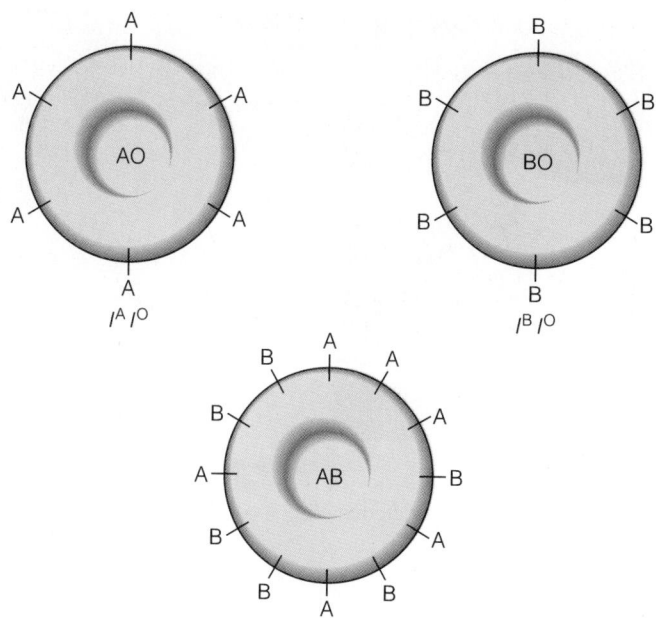

GENOTYPE	PHENOTYPE	NUMBER OF RECESSIVE GENES
AABB	Black	0
AABb	Dark	1
AaBB	Dark	1
AaBb	Mulatto	2
AAbb	Mulatto	2
aaBB	Mulatto	2
Aabb	Light	3
aaBb	Light	3
aabb	White	4

TABLE 8–4 Possible Skin-Color Genotypes and Phenotypes with Two Skin-Color Genes*

*Skin color probably involves many more genes.

▷ **FIGURE 8–18 Codominance** Codominant genes are expressed fully when present in the same cell. In blood type AB, both *A* and *B* genes produce their characteristic glycoproteins. Type A (genotype AO) cells have only type A glycoprotein, and type B (genotype BO) cells have only type B glycoprotein. Type O cells have neither. Note that type A blood may also be in people with the AA genotype and that type B blood may be in people with the BB genotype.

Some Traits are Determined by More Than One Gene Pair

A rule of thumb presented in Chapter 7 is that each gene controls a single trait. In humans and other animals, however, many traits are controlled by not one but by a number of genes—from a few to perhaps hundreds. Skin color, for example, is controlled by as many as eight genes. This type of inheritance is called **polygenic inheritance.** Polygenic inheritance results in incredible phenotypic variation.

To see how polygenic inheritance leads to such wide phenotypic variation, consider an example simply using two genes for skin color, designated *A* and *B*. In this example, we will say that the genotype of an African American is *AABB* and the genotype of a Caucasian is *aabb*. Table 8–4 lists all of the possible genotypes in this example with possible phenotypes (▷ Figure 8–19). With eight genes, the variation would be even greater.

Polygenic inheritance is also responsible for height, weight, intelligence, and a number of behavioral traits. As a rule of thumb, the genotype establishes the range in which a phenotype will fall, but environmental factors determine exactly how much of the potential will be realized for many genes. For example, one's height is partly determined by one's diet. A healthy diet throughout infancy, childhood,

▷ **FIGURE 8–19 Polygenic Inheritance** Skin color is probably determined by at least two genes, resulting in a wide range of phenotypes: (*a*) black, (*b*) dark, (*c*) mulatto, (*d*) light, and (*e*) white.

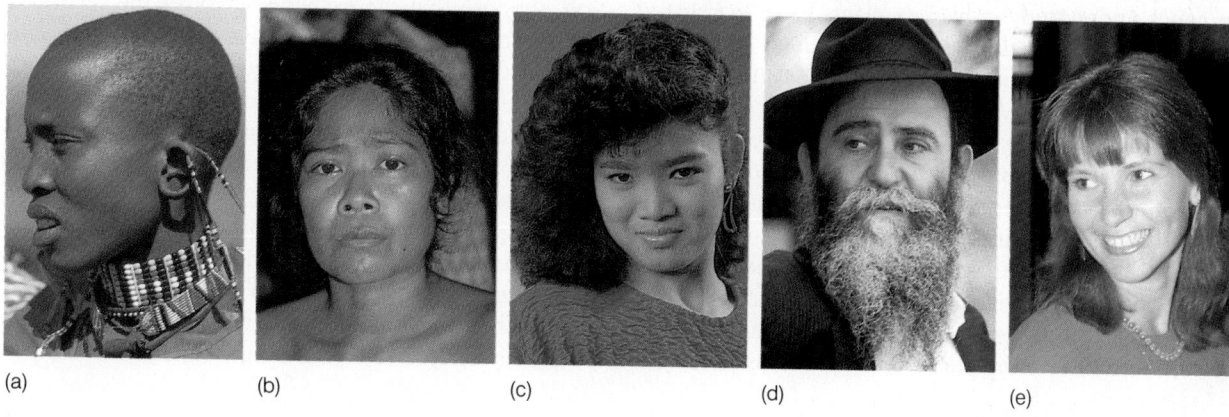

(a)   (b)   (c)   (d)   (e)

(a) Walnut   (b) Rose   (c) Pea   (d) Single

▷ **FIGURE 8–20 Polygenic Inheritance in Chickens** The shape of a cock's comb is determined by two different genes.

and adolescence, for instance, contributes to body growth.

Another example of polygenic inheritance in vertebrates is the determination of the shape of a cock's comb in chickens (▷ Figure 8–20). Research suggests that two genes are involved, the *R* and the *P*. The most common phenotype, known as the single comb, is *rrpp*. Various other combinations are also possible, as shown in Figure 8–20.

A Single Gene Can Also Exert Effects on More Than One Trait

The opposite of polygenic inheritance is **pleiotrophy** (pronounced ple-ah-trophy)—a fancy name that refers to a single gene affecting more than one trait. This phenomenon may occur because the gene influences a biochemical pathway common to more than one trait. In the fruit fly, for example, a single gene affects eye color, body size, wing size, vigor, growth rate, fertility, the arrangement of veins in the wing, and several others. Even Mendel noticed that one of his hereditary factors simultaneously affected the color of flowers, seeds, and parts of the leaves.

Genes Located on the Same Chromosome Are Said to be Linked

Mendel found that during gamete formation (meiosis), the seven genes under study in his pea plants were segregated independently. As noted earlier, independent assortment generally occurs only when the genes under study are on different chromosomes. As a rule, if two genes are on the same chromosome, they do not segregate independently.

Humans have an estimated 100,000 genes on their 46 chromosomes. Those genes found on the same chromosome tend to be inherited together and are said to be **linked**. To illustrate the concept, imagine that you are studying two traits. Designate the dominant form of the first trait *A* and the recessive allele *a*. The dominant form of the second trait is designated *B* and the recessive form is

b. For comparison, we will first cross a homozygous-dominant individual *(AABB)* with a homozygous-recessive mate *(aabb)* (▷ Figure 8–21a). If the *A* and *B* genes are not linked, this cross will produce an F$_1$ generation that is entirely heterozygous *(AaBb)*. As shown, four different gametes will be produced by the F$_1$ offspring. Thus, if an F$_1$ heterozygote mates with another F$_1$ heterozygote, the result is an F$_2$ generation with nine distinct genotypes and four different phenotypes.

If the two genes are on the same chromosome, the outcome changes dramatically. As Figure 8–21b shows, the F$_1$ generation is heterozygous, as in the previous case. However, because the *A* and *B* genes are linked—that is, on the same chromosome—the F$_1$ generation produces only two types of gametes, *AB* and *ab*. Consequently, only two phenotypes are produced in the F$_2$ generation.

As a rule, independent assortment does not occur when genes are linked. There is one exception—a phenomenon called **crossing over.** Crossing over occurs during meiosis. When the homologous chromosomes come together, or "pair up," in prophase I, many of them exchange strands of chromatin (▷ Figure 8–22a). During this process, a segment of one chromosome is "traded" with the corresponding section of the homologous chromosome.

Figure 8–22b presents a hypothetical example to illustrate this important phenomenon. For the sake of illustration, suppose that the exchange involves the section of the chromosome bearing the *A* and *B* genes. Suppose the break occurs between the two genes during the production of female gametes. As illustrated in Figure 8–22b, because of crossing over, four distinct gametes can be produced, instead of the two that would be expected when dealing with linked genes (Figure 8-22c).

The upshot is that crossing over increases genetic variation in gametes and this, in turn, leads to genetic variation in offspring. (Other factors also contribute to genetic variation.) The more variation, the more genotypes in a population. As Chapter 23 points out, variation is essential to

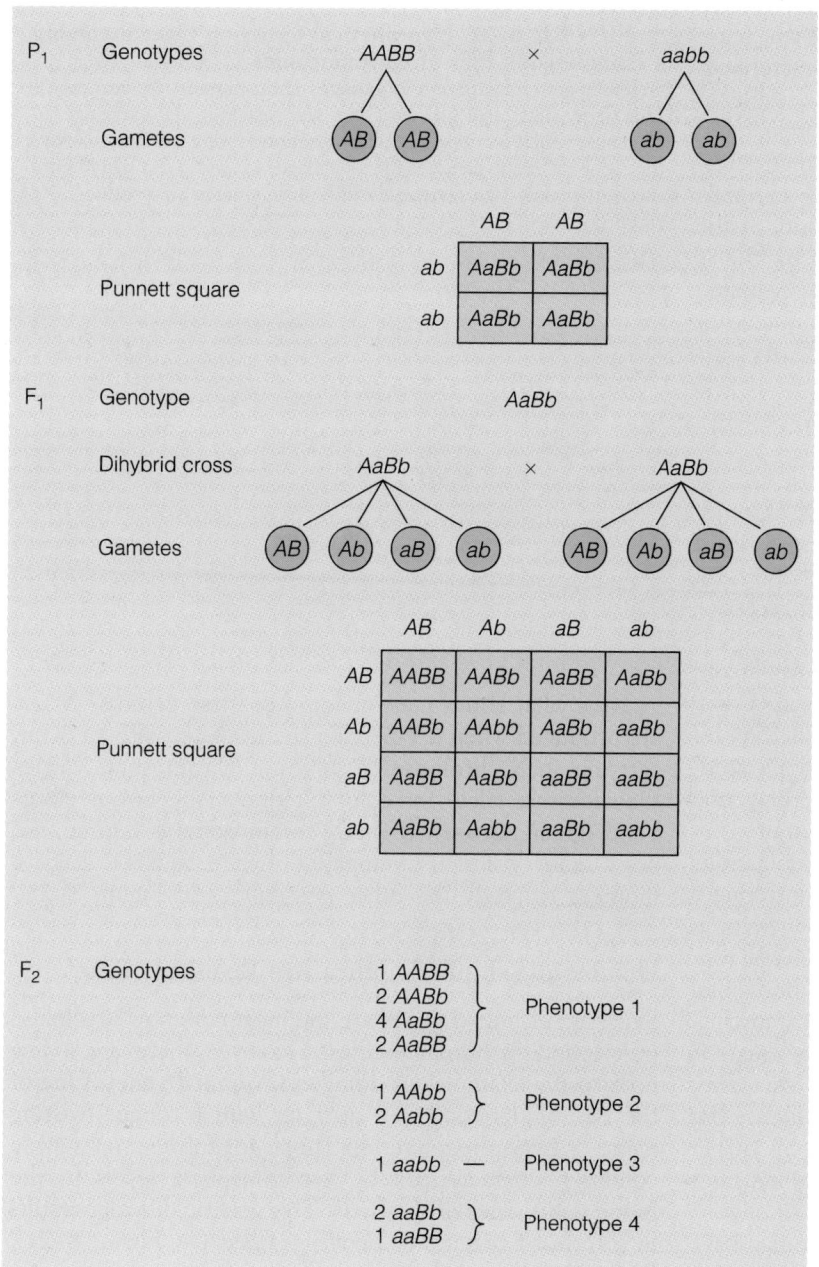

(a)

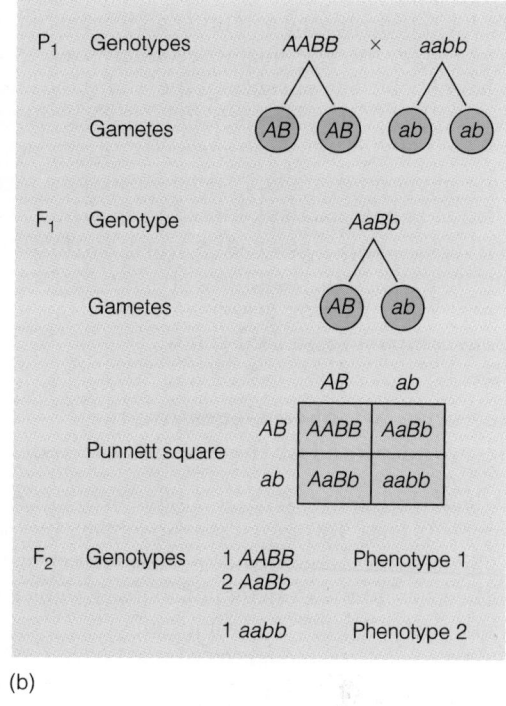

(b)

▷ **FIGURE 8–21 Linkage**
(a) Hypothetical dihybrid cross with no linkage. (b) Hypothetical dihybrid cross with linkage. Linkage reduces the number of genotypes in the offspring.

evolution. Genetic variants may have characteristics that give one organism an advantage over another.

Crossing over can occur anywhere along the length of a chromosome. However, the greater the distance between two genes on the same chromosome, the more likely it is that a crossover will occur between them. This fact is helping scientists map the human genome, a monumental project now under way. Supported by the U.S. government and private agencies, the Human Genome Project hopes to pinpoint the location of all of the genes but also to determine the precise order of the bases of the DNA of all chromosomes. This project could take a decade or more to complete and cost $3 billion.

SEX-LINKED GENES

Sex is determined by the Y chromosome. If the Y chromosome is present, an individual becomes a male. The Y chromosome exerts its effect during embryonic development. Early in the embryo's development, the gonads, which develop in the abdominal cavity, are structurally identical. When the Y chromosome is present, however, the embryonic gonad becomes a testis thanks to the presence of the t gene (testis-determining gene).[3] The testis, in turn,

[3] Sex determination may require one or more genes.

produces testosterone and other androgens, all of which are responsible for the male secondary sex characteristics (Chapter 21). The absence of the Y chromosome results in the development of an ovary and a female phenotype.

The X and Y chromosomes also carry genes that determine many other traits. A gene located on a sex chromosome is known as a **sex-linked** gene. The majority of the sex-linked genes known to science are located on the X chromosome and, therefore, are also known as X-linked genes. The following sections describe the inheritance of some common sex-linked genes.

Recessive X-Linked Genes Are the Best Understood of the Sex-linked Genes

At least 124 genes have been assigned to the X chromosome. At least 160 more are thought to be located on it. Color blindness, discussed in more detail in Chapter 18, and certain forms of hemophilia are examples of recessive traits carried on the X chromosome.

In order for a female to display a recessive sex-linked trait, each of her X chromosomes must carry the recessive gene. For males, however, only one recessive allele is required. That is because the Y chromosome is not genetically equivalent to the X chromosome.[4] Thus, only one recessive gene is needed for men to exhibit the trait.

▷ Figure 8–23 shows four possible genetic combinations leading to color blindness. This illustration also introduces you to a genetic tracking system used to follow traits in families, known as a **pedigree.** In a pedigree, squares represent men and circles represent women. The horizontal line linking a square to a circle (□—○) indicates a mating. Offspring are shown below their parents. When a square is lightly shaded (pink in this figure), the individual is a carrier of the gene. He or she does not suffer from the disease but carries the recessive gene and can pass it on to his or her offspring. A darkly shaded square or circle (green) indicates the person has the disease.

In Figure 8–23a, for example, a man and a woman have four children, two boys and two girls. The woman (pink circle) is a carrier of color blindness. Her cells contain one X chromosome with a recessive allele for color blindness and another X chromosome with the normal allele. Consequently, half of her ova will contain the recessive allele, and the other half will contain the normal allele. As shown, the woman's husband (white square) is not color blind. He produces sperm containing either an X or a Y chromosome. When one of his X-bearing sperm unites with an ovum carrying an X chromosome with the recessive allele for color blindness, the result is a daughter who is a carrier, indicated by the pink circle. When one of his X-bearing sperm unites with an ovum carrying a normal X chromosome, the result is a daughter who is neither a carrier nor a victim (white circle).

[4]The X and Y chromosomes are believed to share few genes, and only part of the Y chromosome is homologous with the X.

▷ **FIGURE 8–22 Hypothetical Crossing Over involving a Homologous Pair of Chromosomes** (a) Crossing over increases genetic variation in gametes and offspring. (b, next page) This figure illustrates how crossing over results in the exchange of genes, which increases genetic variation among gametes. Note that four different gametes are produced. (c, next page) This figure shows that only two genetically different gametes are produced when no crossing over occurs.

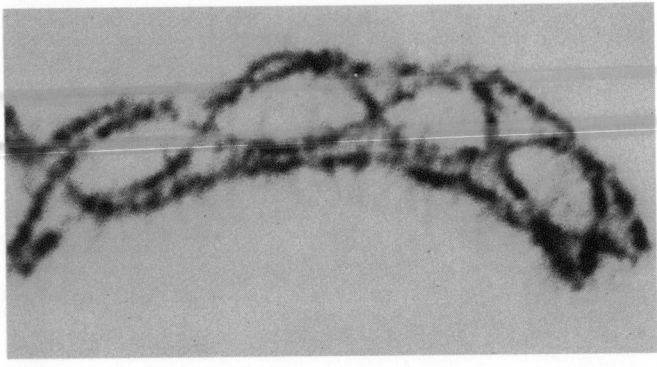

(a)

Now what about male children of these parents? Males are produced when a Y-bearing sperm unites with an ovum carrying an X chromosome. If the X chromosome carries the recessive allele for color blindness, the boy is color-blind (green square). If the X chromosome is normal, the boy's color vision is unimpaired (white square). Take a moment to study the other possibilities in Figure 8–23.

Dominant X-Linked Genes Are Relatively Rare

Recessive X-linked genes are the most common sex-linked trait found in human beings. However, there are a few noteworthy examples of dominant X-linked genes. One of the best understood is a disorder with a tongue-twisting name of **hypophosphatemia** (hi-po-fos-fuh-teem-ee-uh), a condition characterized by low phosphate levels in the blood and tissues of the body. This genetic disorder results in a form of rickets, or bowleggedness (▷ Figure 8–24). Rickets usually results from a dietary deficiency of vitamin D or insufficient exposure to sunlight (Chapter 11). Alleviating the dietary deficiency usually solves the problem. In this genetic disease, however, vitamin D cannot reverse the symptoms.

Hypophosphatemia occurs when either the male (XY) or the female (XX) has one X chromosome bearing the dominant gene (X'). ▷ Figure 8–25 illustrates the pattern of inheritance when a woman who is heterozygous for the trait mates with a man who does not carry the trait. Study the figure and see if you can determine any rules that apply to the inheritance of X-linked dominant genes.

Y-Linked Genes Are Those Found Only on the Y Chromosome

As noted above, the genes that affect gonadal differentiation, spermatogenesis, and other male secondary sex char-

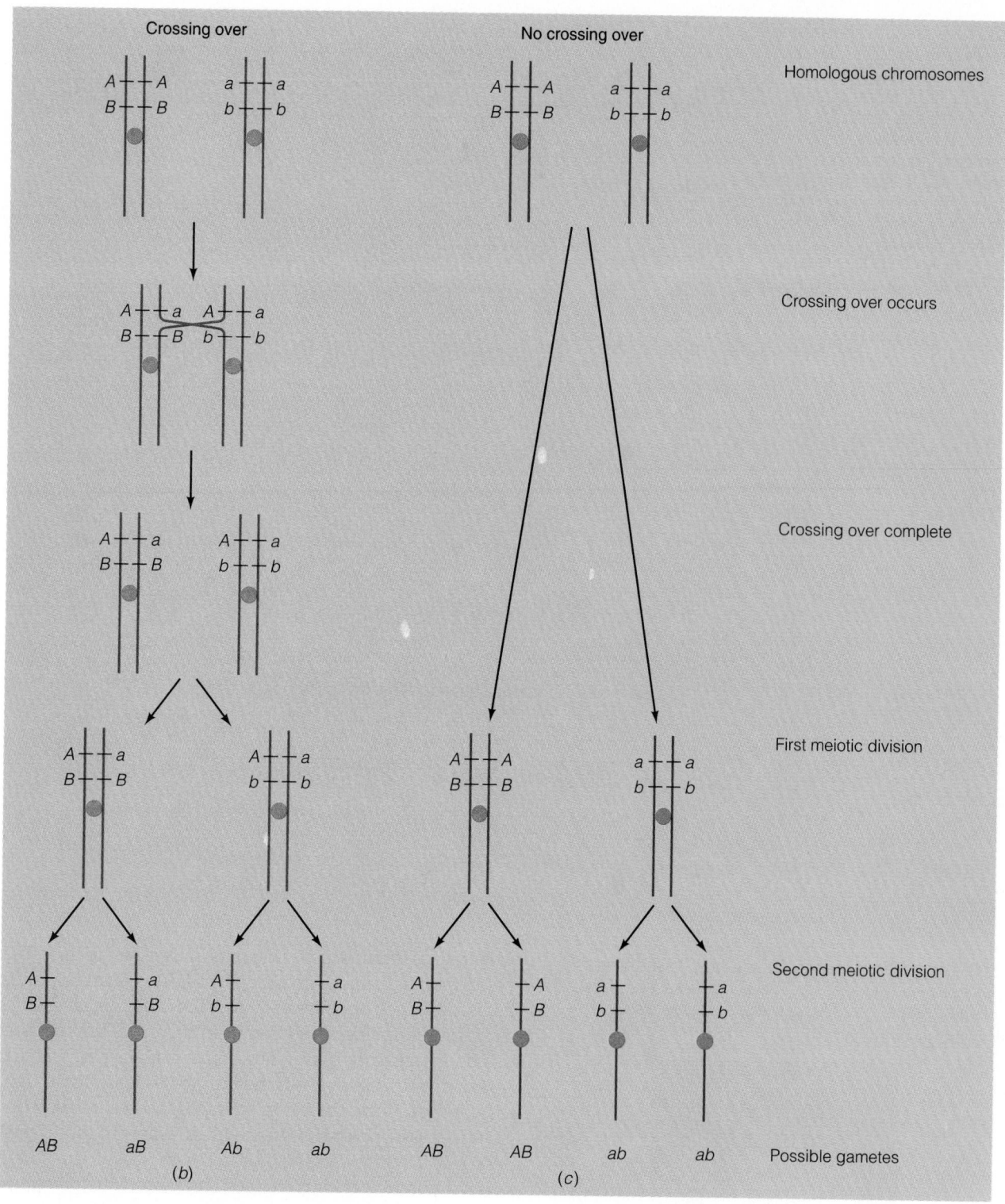

Crossing over | No crossing over

Homologous chromosomes

Crossing over occurs

Crossing over complete

First meiotic division

Second meiotic division

AB   aB   Ab   ab   AB   AB   ab   ab   Possible gametes

(b)   (c)

acteristics are thought to be located on the Y chromosome. Y-linked genes have a simple but fairly distinct pattern of inheritance. Because only males have Y chromosomes, Y-linked traits only appear in males, and Y-linked genes are transmitted only from fathers to sons. Because the X and Y chromosomes in men are not homologous, each gene on the Y chromosome has only one allele, which is always expressed.

Sex-Influenced Genes Act Differently in the Two Sexes

Certain autosomal genes behave differently in the two sexes. In one sex, for example, an allele will be dominant; in the other sex, it will be recessive. These genes are known as **sex-influenced genes.** The best-known example is the gene for pattern baldness. Pattern baldness is the loss of

▷ **FIGURE 8–23 Inheritance of Color Blindness** Four possible genetic ways a sex-linked recessive gene like color blindness can be passed to offspring. Males are indicated by □ and females by ○. In this scheme, green boxes and circles represent men and women with color blindness. Pink boxes represent men and women who are carriers. Blue boxes and circles represent men and women without the color blindness gene. X^c indicates X chromosome carrying gene for color blindness.

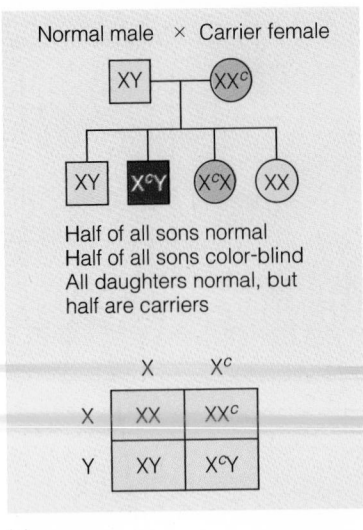

Normal male × Carrier female

Half of all sons normal
Half of all sons color-blind
All daughters normal, but half are carriers

(a)

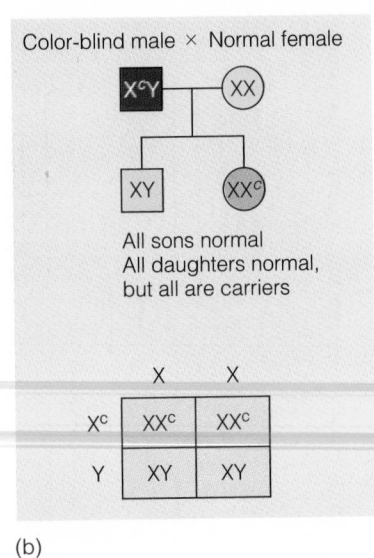

Color-blind male × Normal female

All sons normal
All daughters normal, but all are carriers

(b)

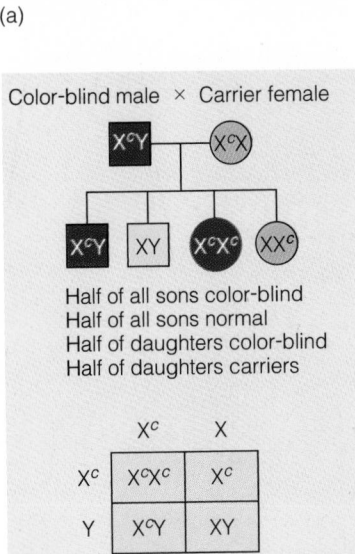

Color-blind male × Carrier female

Half of all sons color-blind
Half of all sons normal
Half of daughters color-blind
Half of daughters carriers

(a)

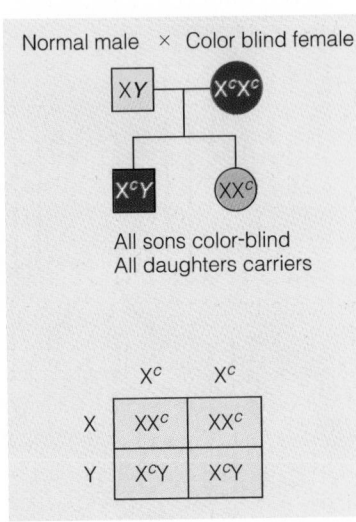

Normal male × Color blind female

All sons color-blind
All daughters carriers

(b)

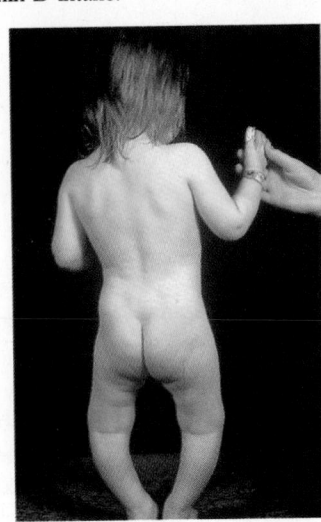

▷ **FIGURE 8–24 Hypophosphatemia** People with hypophosphatemia, a dominant X-linked genetic disorder, resemble this patient with rickets, which usually results from inadequate vitamin D intake.

hair that often begins in a man's 20s (▷ Figure 8–26). Affected individuals do not go completely bald but retain a rim of hair on the temples and back of the head. The gene for pattern baldness is present in men and women, but in men the gene acts as an autosomal dominant and is therefore expressed in both heterozygous and homozygous-dominant individuals. In women, the allele acts as an autosomal-recessive. Only women who are homozygous recessive for the trait exhibit baldness. What accounts for the different behavior of these genes in men and women? Geneticists believe that the genes are influenced by testosterone, a sex steroid hormone found in far greater concentration in the blood of men than women. In this example, then, the genotype and hormonal environment interact in determining the expression of the pattern-baldness gene.

≈ CHROMOSOMAL ABNORMALITIES AND GENETIC COUNSELING

A young couple wait in their doctor's office for the results

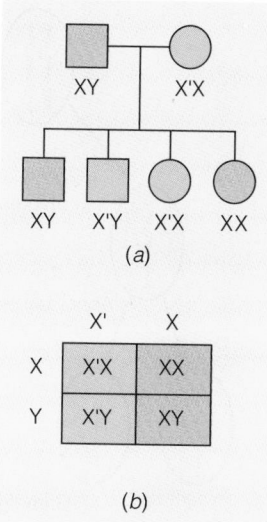

(a)

	X'	X
X	X'X	XX
Y	X'Y	XY

(b)

▷ **FIGURE 8–25 Inheritance of a Sex-Linked Dominant Gene** (*a*) pedigree, (*b*) Corresponding Punnett square.

▷ **FIGURE 8–26 Pattern Baldness** The autosomal gene responsible for pattern baldness is dominant in men and recessive in women, making pattern baldness much more common among men.

of a genetic test on cells of their unborn baby, which were drawn via amniocentesis, described briefly in Chapter 7. They will soon learn that the studies have revealed an abnormal number of fetal chromosomes. Abnormal chromosome numbers are surprisingly common in humans, but this is only one of several genetic defects with which physicians and parents must contend. Chemical changes may also occur in the DNA. As explained in Chapter 7, such mutations may be beneficial. Others can lead to debilitating diseases or death. Chromosomes can also be torn apart in meiosis, resulting in missing segments. A segment from one chromosome may attach to another, which can lead to serious medical problems. This section examines some chromosome abnormalities—how they arise and the problems they create in the physiological mechanisms of the body.

Abnormal Chromosome Numbers Generally Result From a Failure of Chromosomes to Separate During Gamete Formation (Meiosis)

During meiosis I, homologous chromosomes pair, then separate. One double-stranded (replicated) chromosome migrates to each pole, and the other chromosome migrates to the other pole (▷ Figure 8–27a). If a homologous pair fails to separate during meiosis, however, one of the new cells will end up with an extra chromosome (Figure 8–27b). The other cell will be short one chromosome. The failure of homologous chromosomes to separate is called **nondisjunction.**

Nondisjunction can also occur in the second meiotic division (Figure 8–27c). In this division, you may recall, the 23 double-stranded (replicated) chromosomes split apart, with one chromatid going to each daughter cell. If a chromosome fails to separate into its two chromatids, the result is the same as nondisjunction in meiosis I—a daughter cell with an extra chromosome and another daughter cell missing one chromosome.

When a gamete with an extra chromosome unites with a normal gamete, the resulting zygote will contain 47 chromosomes. The zygote may be able to divide successfully by mitosis, producing an embryo all of the cells of which have an additional chromosome. Thus, instead of the normal 23 chromosome pairs, each cell in the embryo contains 22 pairs and one triplet. This condition is called **trisomy** (literally, "three bodies").

Gametes with a missing chromosome can unite with normal gametes, producing individuals with 45 chromosomes—22 chromosome pairs and a chromosome singlet. This condition is called **monosomy.**

Monosomy and trisomy are collectively referred to as **aneuploidy.** Aneuploidy (literally, "not a true number") has profound effects on human reproduction and development. Surprisingly, one of every two conceptions is aneu-

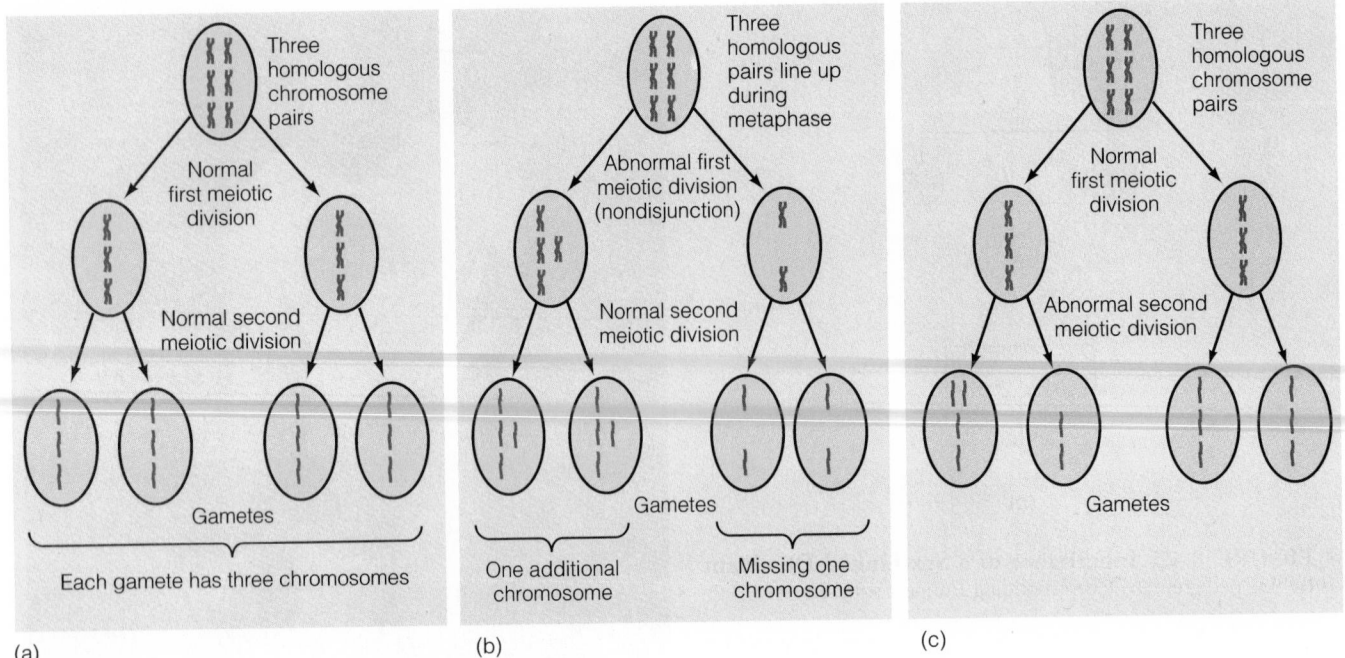

(a)   (b)   (c)

▷ **FIGURE 8–27 Meiosis and Abnormal Chromosome Numbers** (*a*) A simplified version of meiosis. During the first meiotic division, the homologous pairs line up, then separate, producing daughter cells with one-half the number of chromosomes. In the second meiotic division, the chromosomes line up single file and separate, with one chromatid going to each daughter cell.

(*b*) Nondisjunction in the first meiotic division. A chromosome pair may fail to separate during meiosis I, resulting in abnormal gametes. Half are missing a chromosome, and the other half have an extra chromosome. (*c*) Nondisjunction in second meiotic division, resulting in two normal gametes, one gamete with two chromosomes, and one gamete with four.

ploid. Most aneuploid embryos and fetuses die *in utero*, and aneuploidy is believed to be responsible for 70% of all early embryonic deaths and 30% of all fetal deaths. Aneuploidy is also associated with an increased miscarriage rate in older mothers.

Down Syndrome Is Trisomy 21. One of the most common trisomies is **Down syndrome,** or **trisomy 21.**[5] Approximately 1 of every 700 babies born in the United States has Down syndrome. Down syndrome children are typically short and have round, moonlike faces (▷ Figure 8–28). Their tongues protrude forward, forcing their mouths open. The eyes of Down syndrome children slant upward at the corners, and these children are mentally retarded, with IQs rarely over 70. A significant number of Down syndrome babies die in the first year of infancy from heart defects and respiratory infections. Modern medical care, especially antibiotics, has helped reduce early death, and many individuals with Down syndrome live to age 20 or beyond.

The incidence of Down syndrome (and many other aneuploidies) increases with maternal age. As ▷ Figure 8–29 shows, a woman's chances of having a Down syndrome baby increase dramatically after age 35. For this reason, many pregnant women over 35 chose to have amniocente-

[5] Contemporary geneticists generally refer to this as Down syndrome, rather than Down's syndrome.

sis. If the test shows that the fetus has Down syndrome, a couple may decide to abort the fetus. Amniocentesis also helps parents for whom abortion is not an option to prepare psychologically and to receive the education to help them care for their child.

Why does the incidence of Down syndrome increase with age? Some researchers believe that the rise may be the result of exposure to radiation or potentially harmful chemicals—from natural and human sources. How do these agents contribute to Down syndrome? The reasoning is as follows: All of the primary oocytes that a woman will have during her lifetime are present in her ovaries at birth. Each menstrual cycle, a small number (about 12) begin to develop, but only one of these typically ovulates. Thus, the older a woman is, the older her oocytes are and the more likely it is that these cells have been exposed to some potentially harmful agent. The more exposure she has had to such agents, some biologists hypothesize, the more likely nondisjunction is to occur. When nondisjunction involves chromosome 21, the result is a Down syndrome baby. A new study discussed in the Exercising Your Critical Thinking Skills section in Chapter 22, suggests that the real reason for the rise in Down syndrome children with maternal age may be related to the mother's ability to carry such a fetus to term. This study suggests that the number of trisomy 21 conceptions is the same in all women regardless of age. The embryos are simply less likely to be naturally aborted in older women. Why? No one knows.

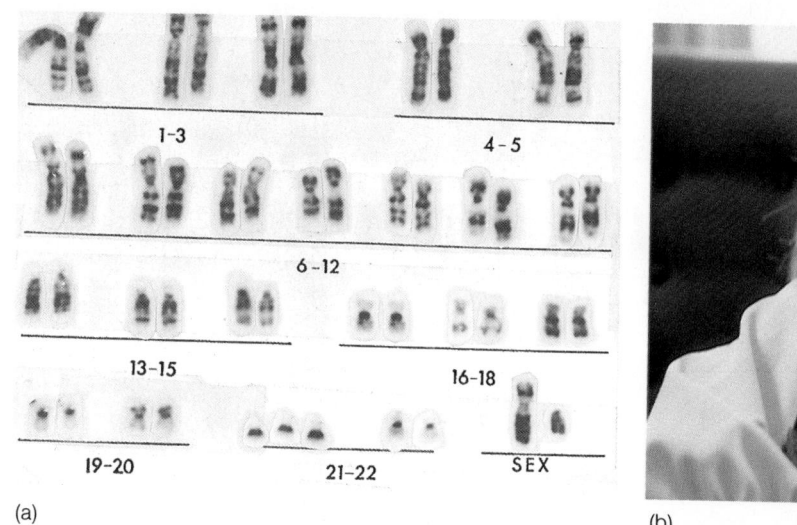

(a)

(b)

▷ **FIGURE 8–28 Down Syndrome** (*a*) Karyotype of Down syndrome girl. Note trisomy of chromosome 21. (*b*) Notice the distinguishing characteristics described in the text.

Nondisjunction of the Sex Chromosomes. Nondisjunction of the sex chromosomes can lead to a variety of nonlethal genetic disorders. If an ovum with two X chromosomes is fertilized by a Y-bearing sperm, for example, the result is a XXY genotype. This condition, known as **Klinefelter syndrome,** occurs in about 1 of every 700 to 1000 males born.[6] Although Klinefelter syndrome victims are males, masculinization is incomplete. The victims' ex-ternal genitalia and testes are unusually small, and about half of the victims develop breasts (▷ Figure 8–30). Spermatogenesis is abnormal, and Klinefelter patients are generally sterile.

Another common disorder involving the sex chromosomes is **Turner syndrome,** a monosomy. Turner syndrome results when an ovum lacking its X chromosome is fertilized by a X-bearing sperm. It may also result when a genetically normal ovum is fertilized by a sperm lacking an X or Y chromosome. In either case, the result is an offspring with 22 pairs of autosomes and a single, unmatched X chromosome (XO).

[6]Nondisjunction can also occur in sperm development, resulting in an XY sperm. The XY sperm can fertilize a normal ovum, resulting in an XXY individual.

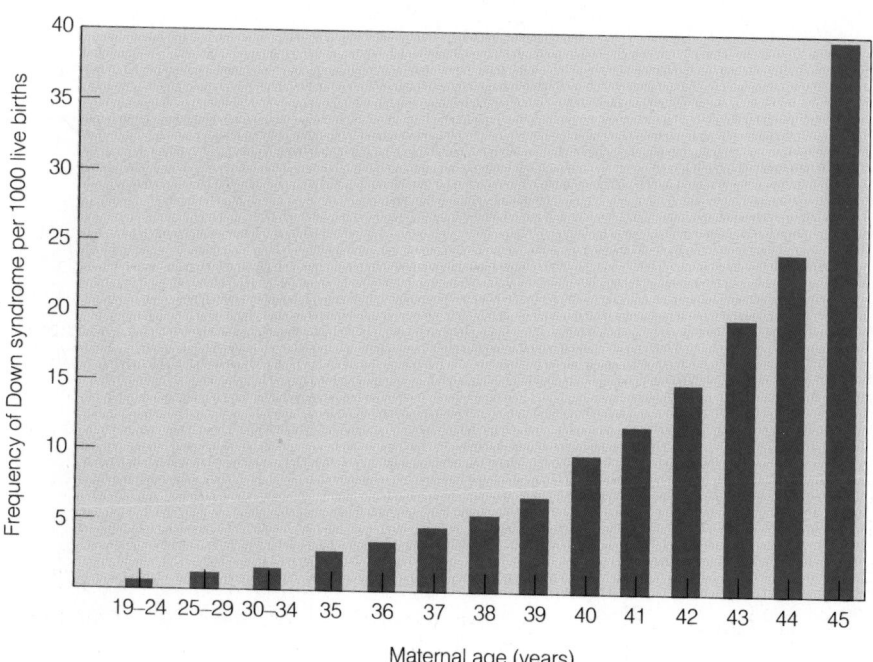

▷ **FIGURE 8–29 Incidence of Down Syndrome Babies at Various Maternal Ages** The indicence of Down syndrome rises quickly after maternal age 35.

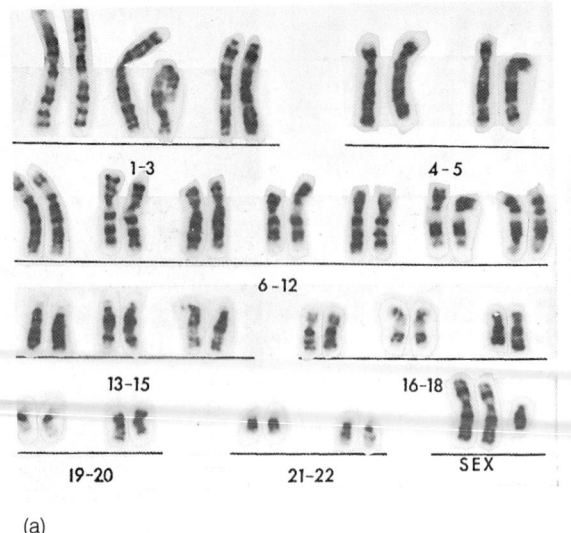

▷ **FIGURE 8–30 Klinefelter Syndrome** (*a*) Karyotype of Klinefelter. Notice that the sex chromosomes include two x and one y. (*b*) Breast development in a male with Klinefelter syndrome.

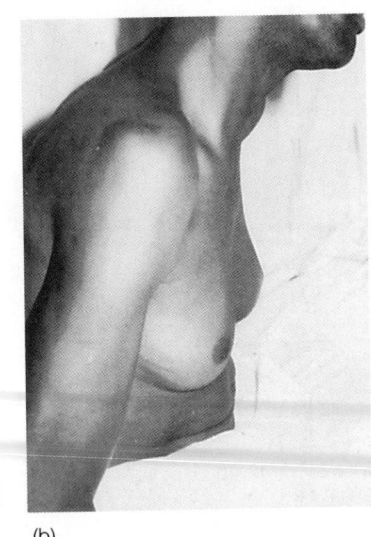

1-3   4 - 5

6 - 12

13-15   16-18

19-20   21-22   SEX

(a)   (b)

Turner syndrome patients are phenotypically female and are characteristically short with wide chests and a prominent fold of skin on their necks (▷ Figure 8–31). Because their ovaries fail to develop at puberty, Turner syndrome patients are sterile, have low levels of estrogen, and have small breasts.[7] For the most part, they lead fairly normal lives. Mental retardation is not associated with the disorder, although some studies suggest that Turner patients are not as capable at numerical skills and spatial perception as genetically normal children. Turner syndrome occurs in 1 of every 10,000 female births. The rarity of this condition, compared with Klinefelter's syndrome, is due to the fact that the XO embryo is more likely to be spontaneously aborted.

[7] Some estrogen is released from the adrenal cortex, but not enough to permit breast development.

Polyploidy. Occasionally, zygotes are formed with a complete extra set of chromosomes. Instead of having the normal 46 chromosomes, the cells have 69—that is, 23 triplets rather than 23 pairs. This condition, known as **triploidy,** results when a normal (haploid) gamete combines with a gamete that has twice the normal number of chromosomes (a diploid gamete). A diploid gamete can be produced during meiosis by a complete nondisjunction— a complete failure of chromosome separation during meiosis I or meiosis II.

Remarkably, triploidy occurs in 1 of every 100 conceptions, but over 99% of all triploid embryos die and are aborted or resorbed. Thus, the incidence of triploidy at birth is fairly small—1 in 10,000. Most of the triploid offspring that are born die shortly after birth.

Triploidy is a form of **polyploidy**—any abnormality of chromosome number resulting from complete non-

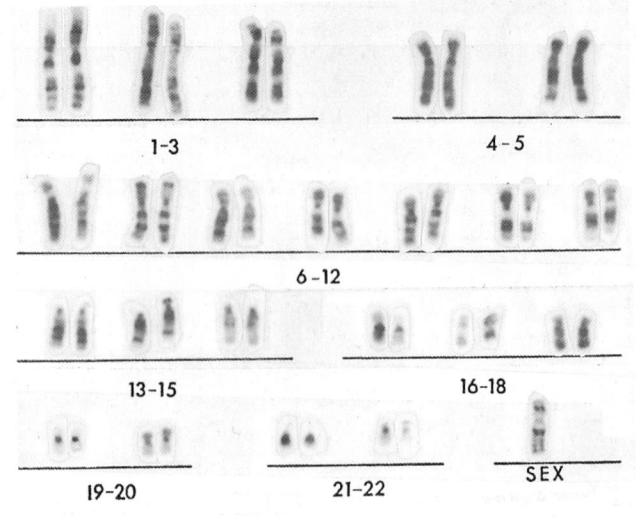

▷ **FIGURE 8–31 Turner Syndrome** (*a*) Karyotype of Turner Syndrome. Notice the single X chromosome. (*b*) Characteristic physical features of a Turner syndrome girl.

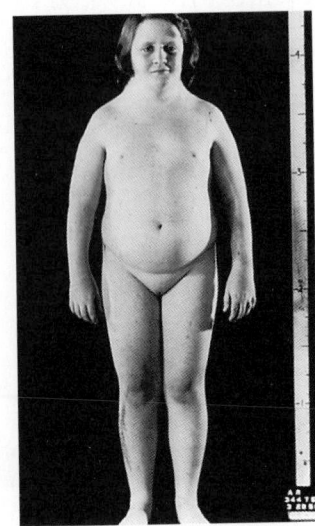

1-3   4 - 5

6 - 12

13-15   16-18

19-20   21-22   SEX

(a)   (b)

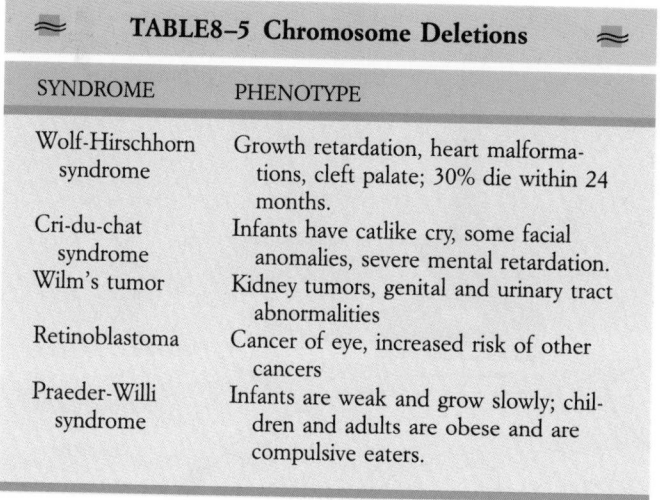

TABLE 8–5 Chromosome Deletions	
SYNDROME	PHENOTYPE
Wolf-Hirschhorn syndrome	Growth retardation, heart malformations, cleft palate; 30% die within 24 months.
Cri-du-chat syndrome	Infants have catlike cry, some facial anomalies, severe mental retardation.
Wilm's tumor	Kidney tumors, genital and urinary tract abnormalities
Retinoblastoma	Cancer of eye, increased risk of other cancers
Praeder-Willi syndrome	Infants are weak and grow slowly; children and adults are obese and are compulsive eaters.

disjunction. An even rarer polyploidy results when a diploid sperm and ovum unite forming a tetraploid. Tetraploidy is usually lethal and therefore rare in newborns.

Genetic Disorders May Also Result From Variations in Chromosome Structure

Chromosome structure may be altered by numerous events. Two important ones are (1) **deletions,** the loss of a piece of chromosome, and (2) **translocations,** breakage followed by reattachment elsewhere.

Deletions. Most deletions are fairly deleterious, and embryos with them are usually eliminated fairly early in pregnancy. Nevertheless, some embryos whose cells contain deletions do survive. Table 8–5 lists some of the more prevalent disorders caused by deletions. One of the more striking is called **Praeder-Willi syndrome.** This condition is caused by the loss of one of the arms of chromosome 15 during gamete formation. Praeder-Willi syndrome is characterized by slow infant growth, compulsive eating, and obesity. Babies born with the syndrome are weak. They have a poor suckling reflex and do not feed well. By age 5 or 6, however, these children become compulsive eaters. Parents must lock their cupboards and refrigerators. Neighbors must be warned to discourage begging and must keep their garbage cans under lock and key. The urge to eat results in obesity, which often leads to diabetes. If food intake is not restricted, victims can literally eat themselves to death. Interestingly, researchers believe that the eating disorder may result from an endocrine imbalance caused by the deletion. Praeder-Willi syndrome occurs in an estimated 1 in 10,000 to 25,000 births.

Translocations. Translocations occur when a segment of a chromosome breaks off but reattaches to another site on the same chromosome or to another chromosome. The movement of a segment of a chromosome to another site

can upset the delicate balance of gene expression and alter homeostasis. Translocations, for example, may be the cause of certain forms of leukemia.

Genetic Screening Allows Parents to Determine If They Will Have a Genetically Normal Baby

Thanks to advances in modern medicine, parents can now find out the sex of their child and the presence of certain genetic defects well before birth. One of the procedures for studying the genome of an unborn child, amniocentesis, was discussed in Chapter 7. Studying the fetus's chromosomal makeup and characteristic banding patterns in cells removed via amniocentesis allows doctors to identify numerous chromosomal and biochemical disorders before birth, although only a dozen or so are routinely screened.

Amniocentesis increases the risk of spontaneous abortion by about 1%. It also slightly increases the risk of maternal uterine infection. Therefore, this procedure is usually recommended only if (1) a woman is over 35, (2) she has already delivered a baby with a genetic defect, (3) she is a carrier of an X-linked biochemical disorder, or (4) she or the father has a known chromosomal or genetic abnormality.

As a rule, amniocentesis is usually not performed until the 16th week of pregnancy.[8] Before this time, there is not enough fluid in the amnion surrounding the fetus, and the needle inserted into the amnion could damage the fetus. Analysis of the fetal cells withdrawn from the amnion requires a further 10 to 15 days. If a serious defect is observed, a couple may elect to have an abortion. However, the risk of an abortion to the mother is slightly greater at this time than it is in the 12- to 16-week period, and some state laws do not permit abortion after 16 weeks.

To permit earlier detection of genetic defects, a new procedure, known as **chorionic villus biopsy,** has been developed. Placental or chorionic villi form from embryonic tissue early in development. These structures form part of the placenta, which nourishes the growing fetus. Physicians insert a catheter into the uterus through the vagina and remove a small sample of a villus (▷ Figure 8–32). The cells of the villus are then examined for chromosomal abnormalities. If a defect is found, an abortion can be performed earlier—at 8 to 12 weeks of gestation, which is safer than an abortion at 16 to 20 weeks. Although chorionic villus biopsy allows for earlier detection, biopsies pose a slightly higher risk to the mother and her fetus.

≈ DNA ABNORMALITIES IN MITOCHONDRIA

Mitochondria contain about 0.3% of a cell's DNA. As noted in Chapter 3, the DNA inside a mitochondrion may

---

[8]Improvements in the technique allow physicians to perform the procedure as early as the 14th week, although risks of spontaneous abortion increase somewhat.

▷ **FIGURE 8–32 Chorionic (Placental) Villus Biopsy** This procedure allows for early detection of fetal chromosomal abnormalities.

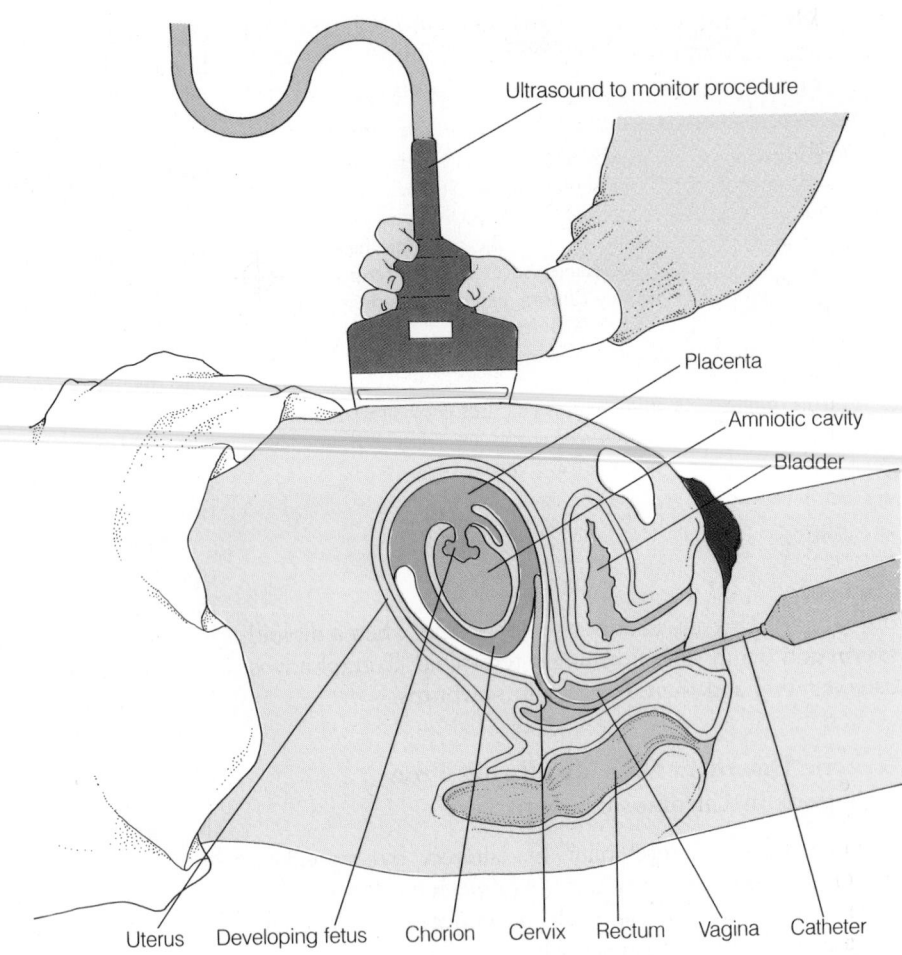

Ultrasound to monitor procedure

Placenta

Amniotic cavity

Bladder

Uterus   Developing fetus   Chorion   Cervix   Rectum   Vagina   Catheter

be an evolutionary remnant from the time the mitochondrion was a free-living organism. Supporting this hypothesis is the fact that the genes of the mitochondrial DNA actually code for some of the enzymes that mitochondria need to produce ATP. Researchers recently discovered a rare form of blindness in humans linked to a defect in the mitochondrial DNA. This finding confirms suspicions that defective mitochondrial genes do indeed produce genetic defects and also suggests an additional mechanism for inheriting genetic diseases.

The defective gene in the mitochondria that leads to blindness codes for a protein required in the first step of ATP production. The absence of this protein in the neurons of the optic nerve results in their death, which, in turn, leads to blindness by age 20. Because mitochondria are passed on by the mother, all of the children of a woman with the defective gene will inherit it. Only a small fraction of the children who inherit the defective gene actually go blind, so it is thought that the mutation only predisposes people to blindness. Other factors may contribute to blindness.

Several other rare genetic diseases may also result from mitochondrial DNA defects. Researchers suggest that even some of the more common diseases may be caused by genetic defects in mitochondrial DNA. Some cases of heart, kidney, and central nervous system failure, whose causes are now unknown, may one day be linked to defective mitochondrial DNA.

 ## ENVIRONMENT AND HEALTH: NATURE VERSUS NURTURE

Throughout this book, a central theme has emerged repeatedly—notably, the idea that humans are profoundly influenced by their environment. Our social, psychological, and physical environments—even our diet—have a profound impact on our internal environment and, therefore, our health.

In this section, we examine a controversy that has raged in science for decades over the connection between our genes and our environment. We will look at two basic questions: How do our genes contribute to our personalities? What is the role of our environment in determining our personality and behavior?

At one time, psychologists viewed the baby as a blank slate. A child's personality, they said, develops through interaction with its environment—its parents, friends, teachers, and so on. Extensive research suggests, however, that our personalities are also influenced by our genes. Thus, our genes and our environment probably operate together in determining personality.

Michael Lewis, a researcher at the Robert Wood Johnson Medical School, studies infant response to stress. His research shows that newborn babies differ markedly in how they respond to the stress of a blood test performed well before their environment could have affected their personality. Lewis found that some children wail when poked with a needle during routine blood tests in the first few days of life; others hardly seem to notice. Of the newborns who cry, some quickly dampen their response. Others seem to go on forever.

Lewis notes that a child's reaction to stress at this time is likely to be repeated three months later when the child receives an inoculation. He believes that the difference in response to stress is genetically based and that the inherent differences will persist.

Lewis has also found that babies differ in how they react to frustration. He performed a series of experiments to test infant response to a frustrating situation. Most children responded with anger. Some, however, showed no response at all, and others displayed sadness. These innate differences in behavior, occurring too early to stem from differences in upbringing, probably result from genetic differences, Lewis maintains. They may help account for the profound differences in responses to stress seen in adults.

The psychologist Nathan Fox has performed important studies on shy and extroverted children. His studies show that shy children cling to their mothers when a clown suddenly appears; extroverted children eagerly engage the clown in play. Fox has found that shy children show greater electrical activity in the right part of the brain; extroverted children show a higher level of activity in the left side. Because the children's personalities differ at birth and because the personality differences are reflected in sharp differences in brain activity, Fox argues, genetics may be playing a powerful role in personality development.

Allison Rosenberg, a researcher at the National Institutes of Health, has also studied shy and outgoing children. Her work shows marked differences in heart rate and cortisol secretion between these two groups, further supporting the notion that there are inherent physiological differences in children from the outset. These are related to differences in early personality and, very possibly, to differences in genetic makeup.

Infant monkeys display a wide range of personality traits early in life. Research shows that timid animals differ physiologically from their braver counterparts. Stephen Soumi, a researcher at the National Institutes of Health, believes

▷ **FIGURE 8–33 Type T Behavior** Thrill seeking, or type T behavior, is thought to have a genetic basis.

that the individual differences seen in personality and in underlying physiology in response to stress are so pronounced that they must be genetically based. Furthermore, he and his colleagues have found that shyness and extroversion tend to run in families, further suggesting a link between early behavior and genetics.

Soumi notes, however, that environmental effects, especially events very early in life, can modify genetically programmed behavior. If a shy baby monkey is paired with an exceptionally nurturant mother, he finds, the shy animal develops rapidly and actively explores its environment.

Another example of behavioral modification by environment comes from research on thrill seeking. Psychologists believe that some individuals are naturally born thrill seekers, or type T people (▷ Figure 8–33). Some researchers believe that thrill seeking may be genetically based. One theory is that risk takers, the type T or "big T" individuals, may be hard to excite and, therefore, may attempt to seek out arousal. At the opposite end of the spectrum are "small t" people, risk avoiders, who are easily aroused. They seek to avoid stimulation.

Some researchers believe that type T individuals have an imbalance of a neurotransmitter known as monoamine oxidase (MO) in the brain. Thrill seeking supposedly increases the MO levels in the brain, creating a feeling of exhilaration.

Type T behavior can be modified by an individual's upbringing and turned in a negative or positive direction, some psychologists say. A positive direction might lead an individual to play for the Los Angeles Rams football team. A negative direction might lead that same individual, under different environmental conditions, into gang fighting and crime in the streets.

Peers, teachers, relatives, ministers, parents, and others make up the environment. Their influence may turn the type T child to healthy, constructive opportunities or un-healthy, destructive ends. A healthy psychological and so-cial environment proves itself a positive asset.

SUMMARY

MEIOSIS AND GAMETE FORMATION

1. Sexually reproducing organisms halve the num-ber of chromosomes in germ cells through meiosis, a type of nuclear division found only in germ-cell production in the reproductive structures of sexually reproducing diploid plants and animals.

2. Meiosis involves two nuclear divisions. During the first divi-sion, meiosis I, the chromosome number is halved. Thus, a diploid cell produces two haploid cells. The second meiotic division is virtually identical to mitosis, except for the fact that the cells are haploid.

3. When the haploid cells divide in meiosis II, they produce two new cells, each containing a haploid number of single-stranded (unreplicated, or one-chromatid) chromosomes.

4. In males, meiosis produces four gametes, but in females it produces only one.

MEIOSIS AND THE LIFE CYCLES OF PLANTS AND ANIMALS

5. The life cycles of plants and animals are divided into two stages: haploid and diploid.

6. In plants, several unique events occur between meiosis and gamete formation. During meiosis in the male and female gamete-producing structures of a flower, germ cells divide by meiosis to produce haploid meiospores. The meiospores di-vide by mitosis to produce a multicellular haploid gameto-phyte, which produces eggs and sperm. The male gameto-phyte in flowering plants is the pollen grain. In females, the gametophyte gives rise to the egg.

7. In plants, the life cycle is completed when an egg and a sperm unite. This union produces a diploid embryo that is enclosed in a protective seed and covered by a fruit.

PRINCIPLES OF HEREDITY: MENDELIAN GENETICS

8. Gregor Mendel, a 19th-century monk, derived several impor-tant principles of inheritance from his work on garden peas.

9. Mendel determined that, in peas, traits did not blend as was commonly thought at the time. He also postulated that each adult had two hereditary factors for a given trait and that these factors (genes) separated during gamete formation. He called this notion the principle of segregation.

10. Mendel also postulated that a hereditary factor might be either dominant or recessive. A dominant factor masks a recessive factor. A recessive factor is expressed only when the dominant factor is missing. The dominant and recessive genes are alternative forms of the gene, or alleles.

11. Three genetic combinations are possible for a given trait: heterozygous, homozygous-dominant, and homozygous-recessive.

12. The genetic makeup of an organism is called its genotype. The physical appearance, which is determined by the geno-type and the environment, is the phenotype.

13. From his studies, Mendel concluded that the hereditary fac-tors were separated independently of one another during gamete formation. This is the principle of in-dependent assortment and holds true only for nonlinked genes.

MENDELIAN GENETICS IN HUMANS

14. Human cells contain 23 pairs of chromosomes: 22 pairs of autosomes and 1 pair of sex chromosomes. Chromosomes carry dominant and recessive traits, and inheritance of these traits is consistent with Mendel's principles of inheritance, although additional mechanisms are at work in humans and other organisms.

15. Sickle-cell anemia, cystic fibrosis, and albinism are autosomal-recessive traits and are expressed only in homozygous-recessive individuals.

16. Widow's peak, achondroplasia, and Marfan's syndrome are autosomal-dominant traits or diseases and are expressed in heterozygous and homozygous-dominant genotypes.

VARIATIONS IN MENDELIAN GENETICS

17. Genetic research since Mendel's time has turned up several additional modes of inheritance. One mode is incomplete dominance. Incomplete dominance occurs when an allele exerts only partial dominance, producing intermediate phenotypes.

18. Some genes have more than two possible alleles. Multiple alleles result in more genotypes and phenotypes in a popula-tion. Because chromosomes exist in pairs, individuals can have only two of the possible alleles.

19. Codominance occurs in multiple-allele genes. Codominant genes are expressed fully and equally.

20. Some traits are controlled by many genes. This phenomenon is referred to as polygenic inheritance.

21. Genes that are found on the same chromosome are said to be linked. If crossing over does not occur, these genes are inherited together.

SEX-LINKED GENES

22. The X and Y chromosomes are commonly referred to as the sex chromosomes. However, studies suggest that the real determinant of sex is the Y chromosome. If it is absent, the individual (XX) is a female. If it is present, the individual (XY) is a male.

23. The sex chromosomes carry genes that determine physical traits. A trait determined by a gene on a sex chromosome is a sex-linked trait. Most sex-linked traits occur on the X chromosome. Both dominant and recessive sex-linked traits are present.

CHROMOSOMAL ABNORMALITIES AND GENETIC COUNSELING

24. Abnormalities in the human genome arise from mutations (changes in DNA structure), abnormalities in chromosome

number (aneuploidy and polyploidy), and alterations in chromosome structure (deletions and translocations).

25. Alterations in the number of chromosomes result chiefly from errors in gamete formation when chromosomes fail to separate during meiosis. This process is called nondisjunction. If a sperm or ovum with an additional chromosome fertilizes or is fertilized, the chromosomes will be passed on to the offspring, resulting in a trisomy.

26. A sperm or ovum missing a chromosome that unites with a normal gamete produces a monosomy, an individual with 45 chromosomes.

27. Nondisjunction of the sex chromosomes during meiosis can result in extra or missing sex chromosomes in offspring.

28. Variations in chromosome structure result from two occurrences: deletions, or the loss of a piece of chromosome, and translocations, or breakage followed by reattachment elsewhere.

29. Embryos with abnormal chromosome numbers or abnormal chromosome structure are likely to die and to be aborted spontaneously.

ENVIRONMENT AND HEALTH: NATURE VERSUS NURTURE

30. At one time, psychologists thought that a child's personality developed principally through interaction with the environment—parents, friends, and teachers. New research suggests, however, that our personalities are also influenced by our genes.

EXERCISING YOUR CRITICAL THINKING SKILLS

The sex of a child is determined by the presence or absence of the Y chromosome. As you learned in Chapter 7, the sex chromosomes in males are XY; in females, they are XX. Thus, the presence of a Y chromosome yields a male; its absence produces a female.

Since 1959, geneticists have been searching for the specific gene on the Y chromosome responsible for male characteristics. In 1983, however, a group of scientists was studying a very rare condition—males with XX chromosomes, an anomaly occurring in approximately 1 of every 20,000 males. When they examined the X chromosomes of these rare males, the researchers discovered that the X chromosomes (presumably from the father) carried a small fragment from a Y chromosome. Maleness, they concluded, comes from the presence of a gene on the segment exchanged with the X chromosome.

The magazine article in which I read this report noted that the researchers had studied several dozen XX males with these fragments. After examining the fragments, they isolated a specific gene that they identified as the possible sex determiner. This astounding finding was reported with much fanfare by the news media in 1987.

When I went to the original research, I found that the researchers had not reported how many XX males were included in the study. Was it all XX males or just a few? (The editors should have insisted on full disclosure.) Given the conclusions of the study and my faith in scientific publications, I surmised (uneasily) that all of the XX males studied had the Y fragment—that is, that this phenomenon occurred in 100% of the XX males.

A few years later, though, a second report hit the press. In this study, researchers found that three XX males in their study group did not have the much-heralded gene, suggesting that in the earlier research the same phenomenon might have occurred.

This study clearly illustrates the importance of good follow-up research, which allows scientists to corroborate the results of earlier work and find weaknesses in reporting. As I pointed out in the critical thinking section in Chapter 1, follow-up research is essential to good science. Critical thinkers must expect it, perhaps even demand it.

Continuing the saga, a more recent study uncovered another gene that researchers think may be the sex determiner. Their studies suggest that this gene may be something of a master gene; that is, it may control other genes involved in sexual development. The sex-determining gene controls sexual development by inducing testicular development. Subsequent male sexual differentiation is a consequence of testosterone, a hormonal product of the testis. Interestingly, the sex-determining gene also appears to work just prior to the development of male sex organs in the embryo. If this gene is indeed responsible for sex determination, it would be expected to be found in all mammals. Research shows that an almost identical gene is present in many mammals, including mice, rabbits, chimpanzees, horses, and tigers.

The research team that discovered the "master gene" is hesitant to announce that it has found the sex determiner gene even with all this evidence. Because any announcement would be met with at least some skepticism, the evidence must be concrete before definite conclusions can be made. In other words, more studies are needed.

Using your knowledge of genetics, can you think of any way to test whether this gene is the sex determiner? Pause for a few minutes, and jot your ideas down on a piece of scrap paper.

Here's what the research team is doing: It is using a procedure that allows it to insert the gene into mouse embryos with two X chromosomes. These embryos would normally develop into females. If the gene is the sex determiner, the researchers assert, the embryos should develop into males.

1. Many of the basic principles of inheritance arose from the work of the Austrian monk _____ _____ .

2. During his research, Mendel hypothesized that each adult plant carried a pair of _____ factors, now known as genes. These factors separate during gamete formation, a concept called the principle of _____ .

3. A gene that can be expressed only in the homozygous condition is called a(n) _____ gene.

4. An alternative form of a gene is called a(n) _____ .

5. The genotype *Aa* is described as _____ .

6. The outward appearance of an organism, or its _____ , is determined by its genotype.

7. A genetic cross in which a single pair of genes is under study is called a(n) _____ cross.

8. The principle of _____ _____ holds true for genes that are not linked.

9. Human cells contain 23 pairs of chromosomes. A female possesses a pair of sex chromosomes, _____ , and 22 pairs of _____ .

10. Cystic fibrosis and albinism are _____ - _____ genetic disorders. A heterozygote is said to be a _____ of these diseases.

11. Widow's peak, achondroplasia, and Marfan's syndrome are _____ _____ traits or diseases. They occur in the _____ - dominant and _____ genotypes.

12. The *H* gene determines whether hair is curly or straight, but neither allele is dominant. An individual with curly hair mates with an individual with straight hair, and the result is an intermediate form, wavy hair. This is an example of _____ .

13. Blood types are determined by the *I* gene. It has three possible alleles. This phenomenon is called _____ alleles.

14. _____ genes are expressed fully and equally.

15. Skin color and height are probably controlled by two or more genes. They are examples of _____ inheritance.

16. Two genes located on the same chromosome are said to be _____ .

17. A trait carried on a sex chromosome is called a _____ trait.

18. Color blindness results from a _____ gene located on the _____ chromosome.

19. Certain autosomal genes are influenced by sex hormones and are called _____ - _____ genes.

20. Monosomy and trisomy are caused by _____ during gamete formation.

21. Down syndrome is also known as _____ .

22. Klinefelter and Turner syndromes result from _____ of sex chromosomes.

23. Tetraploidy and triploidy are conditions that are collectively called _____ .

24. Variations in chromosome structure result from _____ , the loss of a piece of chromosome, and _____ , breakage followed by reattachment elsewhere.

25. To permit earlier detection of embryonic genetic defects, medical scientists have developed a procedure called _____ _____ biopsy.

1. Explain the process of meiosis in general terms. Where does it occur? What does it accomplish?

2. Draw a diagram showing the various stages of meiosis I and meiosis II. Make a note of the number of chromosomes at each stage and their condition—that is, whether they have one chromatid or two. Which division is the reduction division?

3. How is mitosis different from meiosis? How is it similar?

4. Mendel's research was designed to answer two basic questions. What were the questions, and what were his findings?

5. Define the following terms: principle of segregation, principle of independent assortment, allele, phenotype, genotype, heterozygous, homozygous, monohybrid cross, and dihybrid cross.

6. Freckles are an autosomal-dominant trait. A woman with freckles *(Ff)* marries and has a baby by a man without freckles *(ff)*. What are the chances that their children will have freckles?

7. Two freckled adults marry and have children. The first baby has no freckles. What are the genotypes of the parents?

8. Attached earlobes *(A)* are an autosomal dominant trait. The *(A)* allele is dominant over the *(a)* allele, which produces unattached earlobes in homozygous-recessive individuals. A woman with freckles and attached earlobes *(FfAa)* marries a man who has freckles and attached earlobes *(FfAa)*. Draw a Punnett square showing the various gametes as well as the genotypes of the offspring. List all possible phenotypes and the genotypes that correspond to them.

9. An albino guinea pig with the genotype *ccBB* is mated to a brown guinea pig with the genotype *CCbb*. Using a Punnett square, determine the genotype and phenotype of the F_1 generation.

10. What is sickle-cell anemia? What causes it? Why can a person be a carrier of the disease but not display outward symptoms?

11. How do incomplete dominance and codominance differ? Give examples of each.

12. Assuming that two genes control height, *(A* and *B),* list all of the possible genotypes, and indicate the phenotype associated with each.

13. Describe how crossing over works. What impact does it have on the genotype of a person's gametes?

14. Color blindness is a recessive, X-linked gene. A color-blind man and his wife have four children, two boys and two girls. One boy and one girl are both color-blind, and the other two are normal. What is the genotype of the woman?

15. What is a sex-influenced gene? Give some examples. What criteria would you use to assess whether a trait was sex-influenced?

16. Explain how each of the following genetic defects could arise: trisomy 18, monosomy 10, triploidy, and tetraploidy.

9
Molecular Genetics
How Genes Work,
How Genes Are Controlled,
and Genetic Engineering

≈

Tunneling electron micrograph of the DNA molecule, which is responsible for heredity.

In 1988, a U.S. military court sentenced a serviceman in Korea to 45 years in prison for rape and attempted murder. Ten days later, a Florida court convicted a man on two counts of first-degree murder. What makes these two convictions important is that prosecutors relied heavily on the results of a new technique known as "DNA fingerprinting." In this procedure criminologists analyze the composition of DNA in samples of hair, semen, or blood found at the scene of the crime. Then they compare the DNA in these samples with the DNA of the accused. Without this procedure, prosecutors believe, these two convictions would not have been won.

Many people believe that DNA fingerprinting could revolutionize criminal investigations. King County, Washington, in fact, has already begun taking DNA samples from all convicted sex offenders. Like fingerprints, the results of DNA analysis will be kept on file, readily available for future cases. The FBI is also actively developing the technique to aid in its work.

DNA fingerprinting holds great promise, say supporters, because there is only 1 chance of a mistaken identity in 4 or 5 trillion. In contrast, the best conventional methods, such as blood typing and blood-enzyme analysis, run the risk of error in about 1 of every 1000 cases.

Although DNA fingerprinting sounds promising, not all geneticists think that it is as reliable as its supporters would have us believe. In December 1991, in fact, two prominent geneticists published an article in the prestigious journal *Science* in which they assert that the probabilities (noted above) are based on improper assumptions (violating one of the critical thinking rules) and comparatively little scientific data on the genetic composition of populations. Clearly, further work is needed to refine the technique.

DNA fingerprinting is only one of several exciting and controversial developments in biology. Another promising advance is genetic engineering. **Genetic engineering** emerged in 1973 as the result of research by Stanley Cohen at Stanford University and Herbert Boyer at the University of California at San Francisco. These scientists cut segments of DNA from one bacterium and spliced them into the genetic material of another bacterium. The cutting and splicing of DNA is now a routine procedure and goes by one of several names: genetic engineering, gene splicing, and recombinant DNA technology.

Today, genetic engineering techniques find many uses. One of the most important is in the production of hormones for medical uses. Researchers begin by transferring human genes that control the production of insulin and growth hormone into bacterial cells. The bacteria, in turn, are grown in culture. These miniature hormone factories produce large quantities of human insulin and growth hormone, greatly increasing the supply of the hormones for medical use. (These hormones were previously extracted from animal tissues, a costly and low-yielding process.)

Scientists are also tinkering with ways in which genetic engineering can be used to insert normal human genes into genetically defective body cells, curing several serious genetic diseases. Advances in genetic engineering could also help farmers produce faster growing and more disease-resistant crops and livestock. Disease-resistant crops, in turn, might help reduce the use of pesticides worldwide, providing enormous environmental and health benefits.

Although genetic engineering offers great promise, critics warn that we must proceed with caution. Understanding genetic engineering, of course, requires an understanding of genes—how they work and how they are controlled. These processes are the primary focus of this chapter.

≈ BUILDING TWO MACROMOLECULES: DNA AND RNA

In 1953, two biologists, James Watson, an American scientist, and Francis Crick, a British scientist, proposed a model for the structure of the DNA molecule. This model was based on research by Rosalind Franklin, Maurice Wilkins, and many other scientists, as outlined in Scientific Discoveries 7–1.

In 1962, Watson, Crick, and Wilkins, a British biophysicist, received the Nobel Prize in Physiology and Medicine for their model—a discovery that opened the doors to a new and infinitely fascinating field of study known as **molecular genetics.** Molecular genetics concerns itself with the study of the structure and function of RNA and DNA in cells. We begin with a review of the structure and the synthesis of these molecules.

DNA Is the Molecular Basis of the Gene and Consists of a Double Helix Held Together by Hydrogen Bonds

DNA is an intriguing molecule containing millions of nucleotides, each of which consists of three parts: a nitrogenous base, a phosphate group, and a monosaccharide, deoxyribose (▷ Figure 9–1). The nucleotides join by covalent bonds, forming long, polynucleotide chains. As shown in ▷ Figure 9–2, covalent bonds between nucleotides form between the sugar molecules and phosphates.

The two polynucleotide chains of DNA intertwine, forming an elaborate structure resembling a spiral staircase, known as the **double helix.** As shown in this figure, the bases of the nucleotides of each chain project inward, lying inside the helix. Hydrogen bonds form between the bases of the two strands of the double helix—as indicated by the dotted lines in Figure 9–2—and hold the two polynucleotide chains together. These bonds, which are much weaker than covalent bonds, can be easily broken so the molecule can unwind for replication.

As illustrated in Figure 9–1, two types of bases are found in the two strands of the DNA molecule: purines and pyrimidines. The **purines** are the more complex of the two, consisting of two fused rings. The **pyrimidines** consist

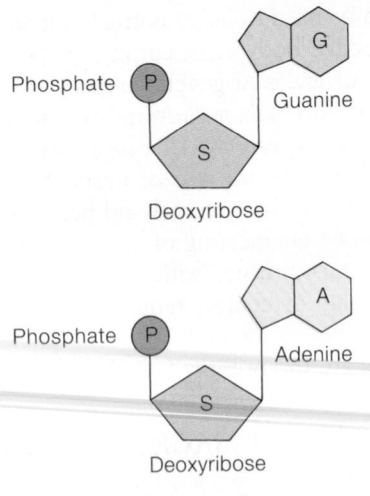

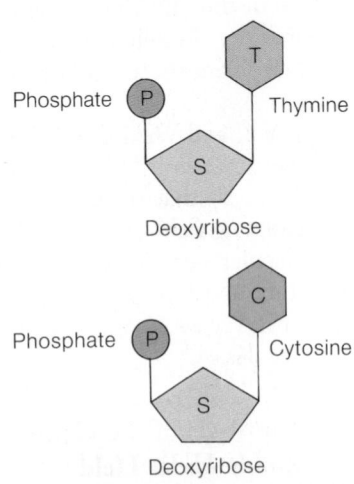

(a) DNA nucleotides containing purine bases

(b) DNA nucleotides containing pyrimidine base

▷ **FIGURE 9–1 Nucleotides Containing Purine and Pyrimidine Bases** All nucleotides consist of three parts: a phosphate group; an organic, nitrogen-containing base; and a five-carbon sugar. DNA nucleotides contain the sugar deoxyribose. Two types of bases are found: purines and pyrimidines. (*a*) The purines are adenine and guanine. (*b*) The pyrimidines are thymine and cytosine.

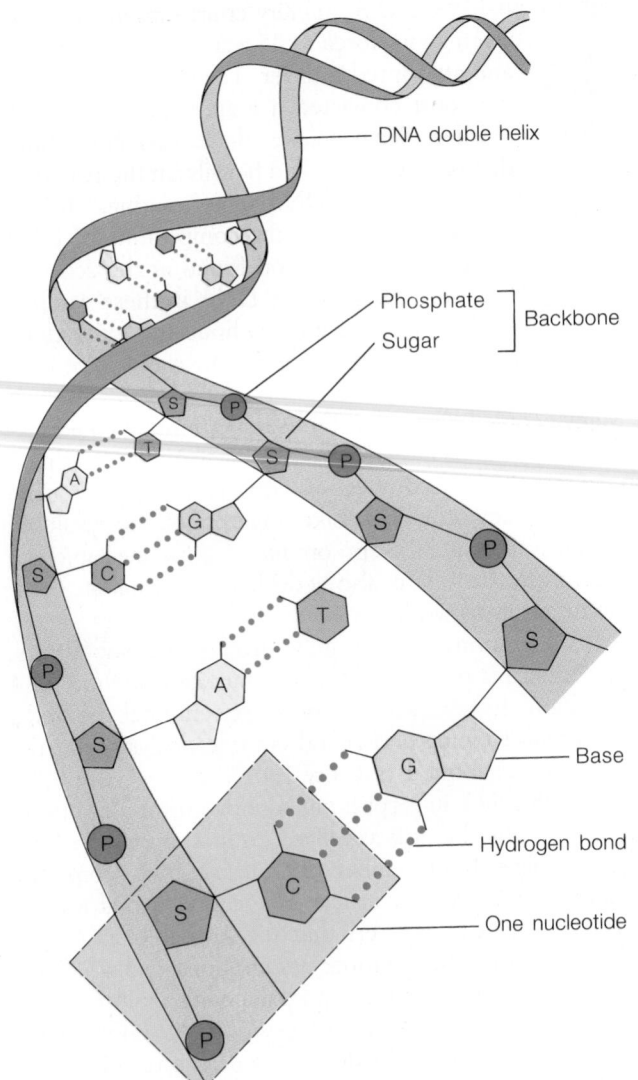

▷ **FIGURE 9–2 DNA** DNA consists of two polynucleotide chains wrapped around each other to form a double helix. The sugars and phosphates form the backbone of each chain with the bases projecting inward. The bases on opposite strands are connected by hydrogen bonds.

of only one ring. The purines in DNA are adenine (A) and guanine (G).[1] The pyrimidines are cytosine (C) and thymine (T).

As shown in Figure 9–2, purines on one strand bind (via hydrogen bonds) to pyrimidines on the opposite strand. But the relationship between the two is even more specific. If you look at ▷ Figure 9–3, you will see that the adenine (a purine) binds only to the pyrimidine thymine. Guanine (a purine) binds only to the pyrimidine cytosine. Adenine

and thymine are therefore said to be complementary bases, as are guanine and cytosine. As you will soon see, this unalterable coupling, called **complementary base pairing,** ensures the accurate replication of DNA and the accurate transmission of the genetic information from a parent cell to its daughter cells during cell division.

DNA Replication Takes Place on the Individual Polynucleotide Strands After the Double Helix Unwinds

As you learned in Chapter 7, before a cell can divide, it must first make an exact copy of all of its DNA. This

[1] To remember this, you can use the mnemonic "pure Ag" offered by a professor who taught my genetics course. He was from the agricultural college at my university.

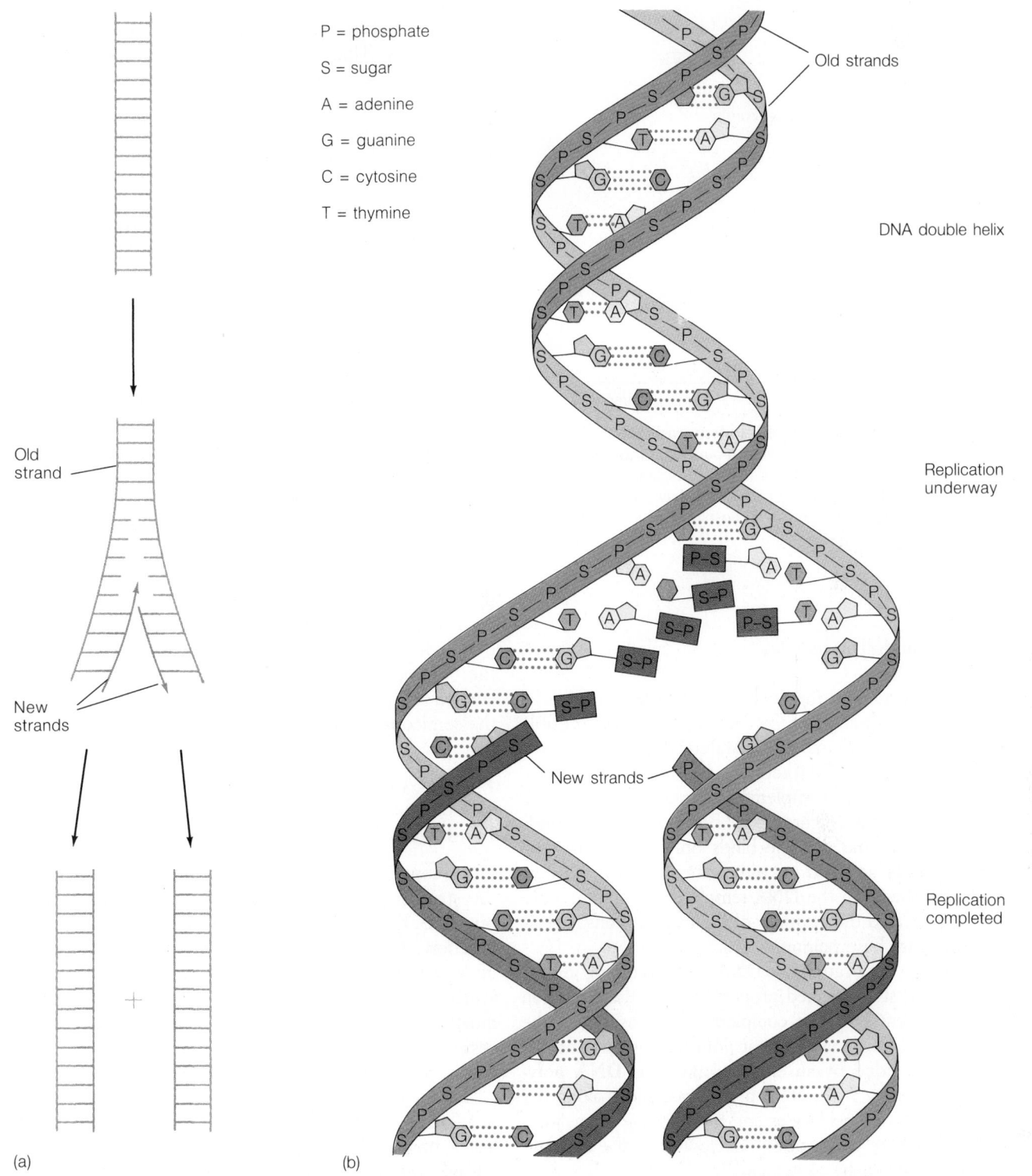

P = phosphate

S = sugar

A = adenine

G = guanine

C = cytosine

T = thymine

Old strands

DNA double helix

Replication underway

New strands

Replication completed

Old strand

New strands

(a)

(b)

▷ **FIGURE 9–3 Semiconservative Replication** DNA replication is semiconservative. (*a*) Each double helix unwinds, and each half of the helix serves as a template for the production of a new strand of DNA. When replication is complete, each new helix contains one old and one new strand. (*b*) Nucleotides attach to the template one at a time and are joined to others with the aid of enzymes.

ensures that a cell about to divide has two sets of genetic information, one for each daughter cell. DNA replication occurs during the S phase of interphase. During this period, cells replicate their DNA by first splitting each DNA double helix along the hydrogen bonds that unite the complementary bases. Each polynucleotide strand then serves as a template (described below) on which a new strand is produced (Figure 9–3). The new strand is called a **comple-**

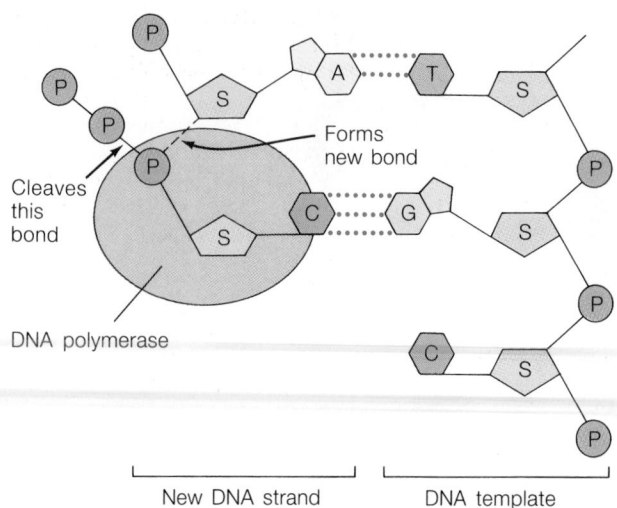

Cleaves this bond

Forms new bond

DNA polymerase

New DNA strand

DNA template

▷ **FIGURE 9–4 Role of DNA Polymerase** DNA polymerase binds loosely to the DNA template and to the nucleotide, aligning it for bonding to the previous nucleotide. The enzyme catalyzes the formation of the bond between the phosphate and sugar and cleaves off two phosphates of each nucleotide.

mentary strand. The template is often called the original strand.

DNA replication during the S phase begins when special enzymes start to pull apart, or "unzip," the DNA double helix. As the two strands are separated, the bases on each strand are exposed and therefore become free to form hydrogen bonds with complementary nucleotides floating in the nucleoplasm (Figure 9–3). Because adenine binds only to thymine and guanine binds only to cytosine, the original strands are said to direct the synthesis of new strands. Synthesis on the DNA templates occurs one base pair at a time, and the accuracy of replication is ensured by complementary base pairing.

During the replication of DNA, the incoming nucleotides must first be aligned properly so that the hydrogen bonds can form between complementary bases and the sugar and phosphate groups can join. Alignment and attachment are aided by an enzyme known as **DNA polymerase** (▷ Figure 9–4). DNA polymerase slides along the template, aligning nucleotides and then linking them, joining the phosphate group of one nucleotide to the deoxyribose molecule (sugar) of its neighbor.

When DNA synthesis is finished, two new DNA molecules exist, each containing one strand from the original double helix and one new strand. The chromosome, once a single molecule of DNA and protein, now consists of two molecules with associated proteins. Each strand of DNA and associated protein is called a chromatid. The two chromatids of each chromosome are joined at their centromeres after synthesis, but they separate during anaphase of mitosis.

DNA replication is often referred to as a semiconservative process. That is simply another way of saying that during replication, each double helix unwinds and that each polynucleotide chain of the helix serves as a template for the production of a new strand of DNA. When replication is complete, each new helix contains one old and one new strand, as shown in Figure 9–3a. The classic experiment that demonstrated this important biological principle is shown in ▷Figure 9–5. In this experiment, Mathew Meselson and Franklin Stahl cultured the bacterium *E. Coli* in a medium containing ^{15}N-containing ammonium ions, which were incorporated into their nucleic acids. The bacteria were then transferred to a second culture dish containing only ^{14}N, a lighter isotope. During this incubation, the bacterial DNA replicated, and the cells reproduced. Cells were then collected, and DNA was analyzed using a technique called density gradient ultracentrifugation. In this technique, DNA is extracted and centrifuged. The DNA settles in the centrifuge tubes at different levels, depending on the weight of its molecules. Slightly heavier DNA molecules (double helices) containing lots of ^{15}N settle closer to the bottom.

The results of the study suggested that both strands of the DNA were labeled by ^{15}N in the first culture. After replication in the ^{14}N-medium, only one strand of each double helix was labeled. Permitting another duplication in the ^{14}N-medium resulted in a mixture of labeled and unlabeled DNA molecules. These results are consistent with the semiconservative nature of replication.

Three Types of RNA Exist, and Each Consists of a Single Strand of RNA Nucleotides and Is Involved in Protein Synthesis

DNA contains the genetic information, which determines the structure and controls most of the functions of cells, but DNA does not exert its influence on the structure and function of the cell directly. Instead, it must rely on two intermediaries: RNA, which carries the instructions for the synthesis of proteins into the cytoplasm, and proteins, mostly enzymes, which regulate virtually all cellular processes. In order to understand exactly how this process works, we begin with a look at RNA.

Three types of RNA exist: ribosomal RNA (rRNA), messenger RNA (mRNA), and transfer RNA (tRNA). Each has a unique function in protein synthesis (Table 9–1). Despite the differences in these molecules, all three RNA molecules share some features. In humans, for example, all RNAs are single-stranded molecules, and all are polynucleotides (▷ Figure 9–6a). RNA nucleotides consist of three molecules: a sugar, a nitrogenous base, and a phosphate group. Unlike the DNA nucleotides, however, the RNA nucleotides contain the sugar ribose instead of deoxyribose and the pyrimidine uracil instead of thymine. See Table 9–2 for a summary of the key differences between RNA and DNA.

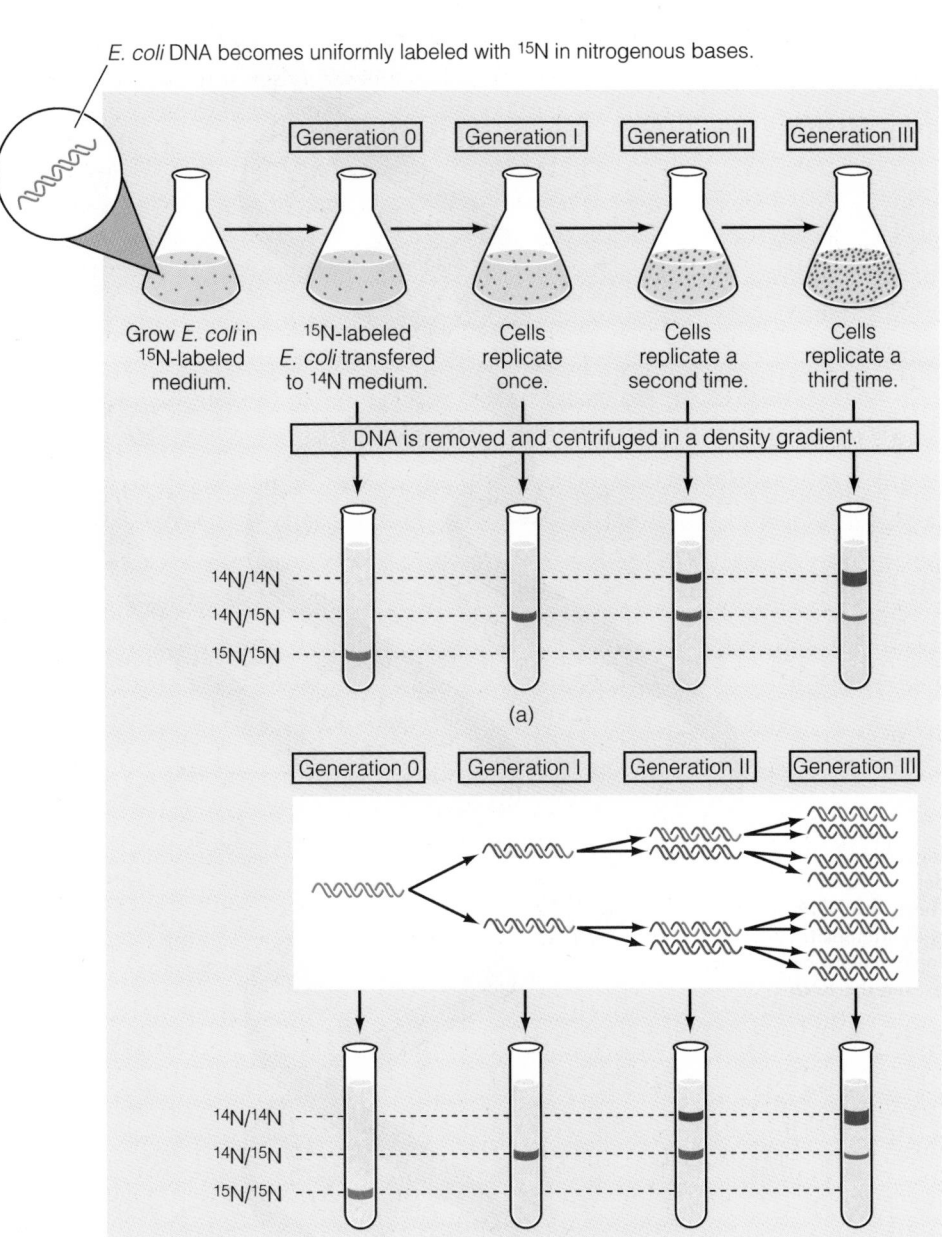

E. coli DNA becomes uniformly labeled with ^{15}N in nitrogenous bases.

| Generation 0 | Generation I | Generation II | Generation III |

Grow E. coli in ^{15}N-labeled medium.

^{15}N-labeled E. coli transfered to ^{14}N medium.

Cells replicate once.

Cells replicate a second time.

Cells replicate a third time.

DNA is removed and centrifuged in a density gradient.

^{14}N/^{14}N
^{14}N/^{15}N
^{15}N/^{15}N

(a)

| Generation 0 | Generation I | Generation II | Generation III |

^{14}N/^{14}N
^{14}N/^{15}N
^{15}N/^{15}N

(b)

▷ **FIGURE 9–5 The Meseleson-Stahl Experiment**
(a) By growing the bacterium E. coli in culture with ^{15}N labeled medium and following subsequent generations, which were grown in culture with ^{14}N labeled medium, the researchers were able to produce DNA strands with and without heavy nitrogen. Extraction of the DNA and centrifugation yielded various combinations of light and heavy DNA, from which the researchers were able to determine that DNA replication is semiconservative, (b) This drawing compares the experimental results (bottom) with the moleculare details.

RNA Synthesis Is Called Transcription and Takes Place on a DNA Template in the Nucleus of the Cell

RNA is synthesized on a DNA template. The information coded in the DNA is therefore transferred from the DNA molecule to RNA. Because RNA is free to leave the nucleus and DNA is not, RNA serves as a kind of shuttle service that transfers the genetic information necessary for protein synthesis into the cytoplasm, where proteins are made.

RNA synthesis on a DNA template is called **transcription**. Like note taking, it is merely a transfer of information from one medium (speech) to another (written form).

≋	**TABLE 9–1 Role of RNA Molecules**	≋

MOLECULE	ROLE
Messenger RNA (mRNA)	Carries the information needed to make proteins from the nucleus to the cytoplasm
Transfer RNA (tRNA)	Binds to specific amino acids, transports them to the mRNA, and inserts them in the correct location on the mRNA
Ribosomal RNA (rRNA)	Component of the ribosome

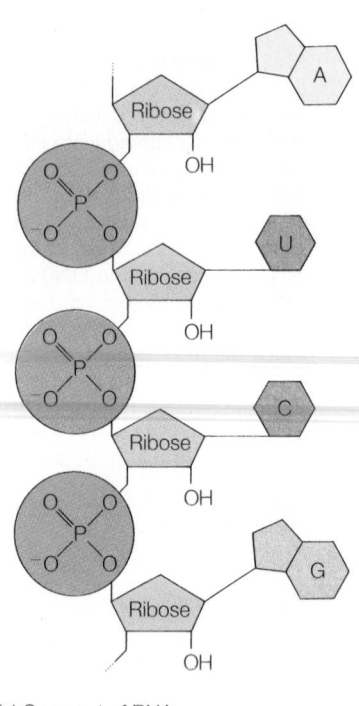

(a) Segment of RNA

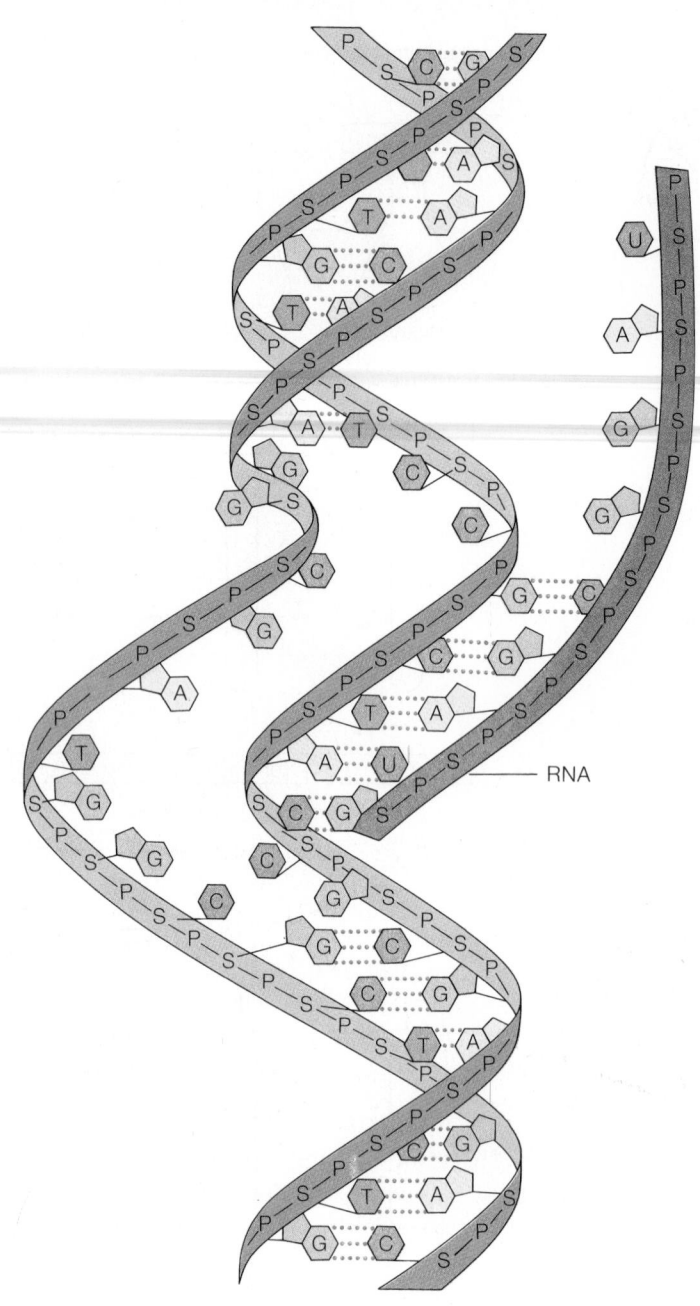

RNA

▷ **FIGURE 9–6 RNA** (*a*) RNA is a single-stranded molecule consisting of many RNA nucleotides. (*b*) RNA is synthesized on a DNA template. As shown here, the DNA double helix unwinds, and one strand serves as a template for the production of a strand of RNA. Complementary base pairing determines the exact sequence of bases in the RNA molecule.

TABLE 9–2 Differences between RNA and DNA in Prokaryotic and Eukaryotic Cells

DNA	RNA
Double-stranded	Single-stranded
Contains the sugar deoxy-ribose	Contains the sugar ribose
Contains adenine, guanine, cytosine, and thymine	Contains adenine, guanine, cytosine, and uracil
Functions primarily in the nucleus	Functions primarily in the cytoplasm

Transcription occurs during all three phases of the cell cycle at more or less the same rate. During transcription, small sections of the DNA helix unwind temporarily with the aid of special enzymes. This creates a DNA template on which RNA can be made (Figure 9–6b). Interestingly, RNA is produced on only one of two DNA strands, and only a small portion (less than 1%) of a cell's DNA is used to make RNA. (This explains why most mutations have no effect on cell function.)

During RNA synthesis, an enzyme called **RNA polymerase** helps align the RNA nucleotides on the DNA template in much the same way that DNA polymerase functions during DNA synthesis. RNA polymerase also

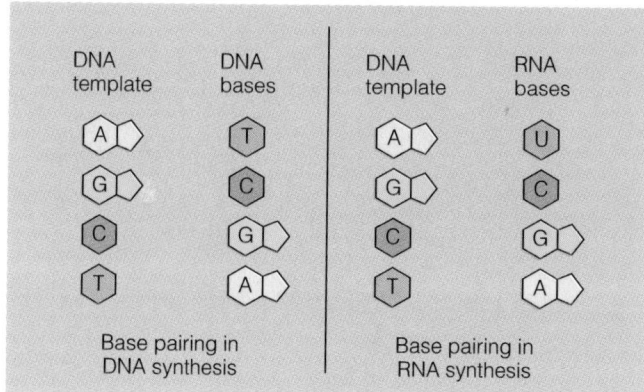

▷ **FIGURE 9–7 DNA and RNA Base Pairing** The sequence of bases on the DNA template determines the sequence of bases on the complementary strand of DNA and RNA.

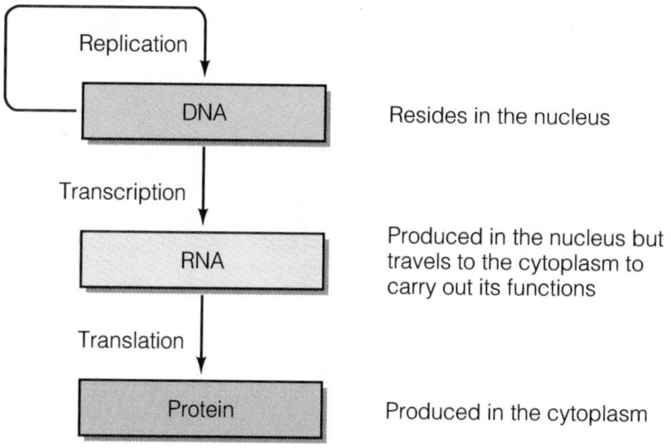

Replication → **DNA** — Resides in the nucleus

Transcription

RNA — Produced in the nucleus but travels to the cytoplasm to carry out its functions

Translation

Protein — Produced in the cytoplasm

▷ **FIGURE 9–8 The Central Dogma of Molecular Genetics** The DNA controls the cell through protein synthesis. RNA serves as an intermediary, carrying genetic information to the cytoplasm, where protein is synthesized.

catalyzes the formation of covalent bonds between ribose and phosphate groups, forming a polynucleotide chain. When the synthesis is complete, the RNA molecule is released from the DNA template, and the two strands of the DNA molecule reunite, re-forming the double helix.

During the synthesis of RNA, base pairing ensures the proper sequence of nucleotides on the RNA strand. ▷ Figure 9–7 shows which bases pair up during RNA synthesis. The only difference between RNA and DNA synthesis, is that adenine on the DNA template pairs with uracil.

In summary, the DNA helix is a dual-purpose molecule, serving as a template for the synthesis of both DNA and RNA, which require different enzymes. By accurately replicating itself, DNA ensures the precise transmission of genetic information from cell to cell and, ultimately, from generation to generation. By making RNA, DNA provides instructions for cells to make protein in their cytoplasm.

≈ HOW GENES WORK: PROTEIN SYNTHESIS

The information required to synthesize protein is transported out of the nucleus on one special type of RNA, known as messenger RNA (mRNA). In the cytoplasm, mRNA serves as a template for protein synthesis, which may occur either on the endoplasmic reticulum or free within the cytoplasm, as noted in Chapter 3.

The synthesis of protein on a mRNA template is called **translation** (▷ Figure 9–8). In effect, the RNA message (the genetic code) is translated into a new "molecular language"—that of the protein.

Protein synthesis requires two additional "players," transfer RNA (tRNA) and ribosomes. **Transfer RNA** molecules are relatively small strands of RNA that bind to specific amino acids in the cytoplasm and transport them to the mRNA, where the amino acids are incorporated into

the protein (▷ Figure 9–9). A tRNA molecule bound to an amino acid is generally written like this: tRNA-AA.

Ribosomes, mentioned briefly in Chapter 3, are composed of ribosomal RNA (rRNA) and protein and consist of two subunits, a small one and a large one. The subunits are often detached in the cytoplasm, joining for protein synthesis on the mRNA template.

Protein synthesis consists of three stages: (1) chain initiation, (2) chain elongation, (2) and chain termination. During chain initiation, the small subunit of the ribosome attaches to the mRNA at a specific site called the initiator codon. The **initiator codon** is a sequence of three bases on the mRNA molecule that marks where protein synthesis should begin. The large subunit of the ribosomes quickly attaches, and the process can start.

▷ **FIGURE 9–9 Messenger RNA** Messenger RNA, either free within the cytoplasm or bound to the endoplasmic reticulum, serves as a template for protein synthesis. The codons, each consisting of three bases, determine the sequence of amino acids by binding to anticodons on the tRNA molecules. Each tRNA, in turn, delivers a specific amino acid to the mRNA template. The ribosome provides binding sites, catalyzes the formation of the peptide bonds, and slides down the mRNA template to permit chain elongation.

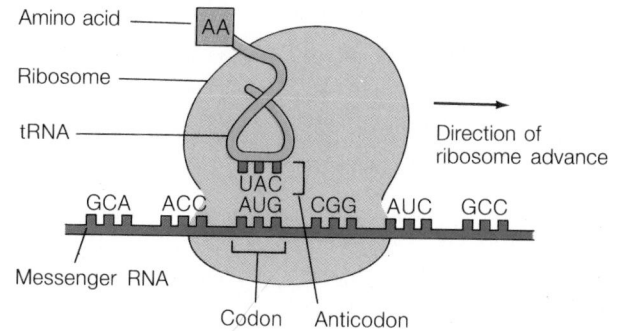

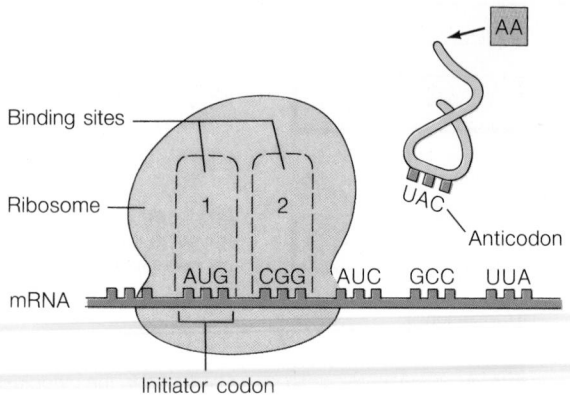

(a) The ribosome binds to the mRNA template. The tRNA binds to a specific amino acid in the cytoplasm.

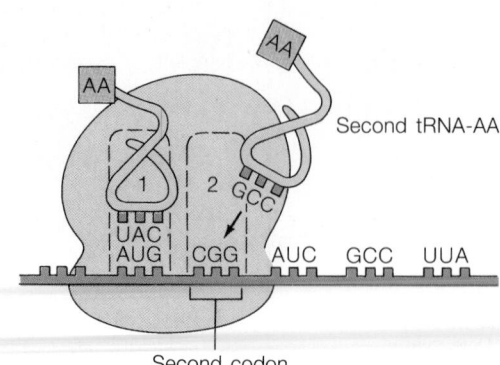

(b) The tRNA–amino acid (tRNA-AA) complex binds to the first codon and is held in place by the first binding site. A second tRNA-AA complex binds to the second codon.

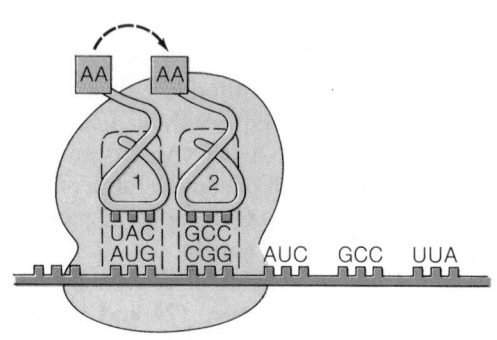

(c) The ribosome contains an enzyme that catalyzes the formation of a peptide bond between the two amino acids. The dipeptide is then attached to the second tRNA. This frees up the first tRNA, which vacates the first binding site.

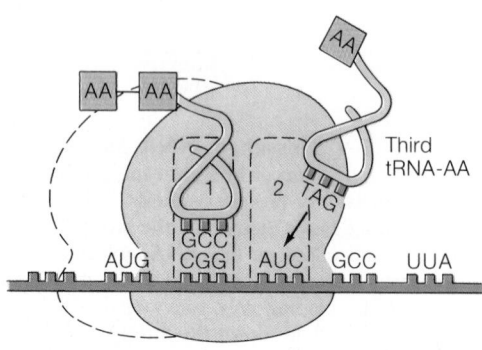

(d) The ribosome next slides down the mRNA, transferring the tRNA-dipeptide to the first binding site and opening up the second binding site to a third amino acid.

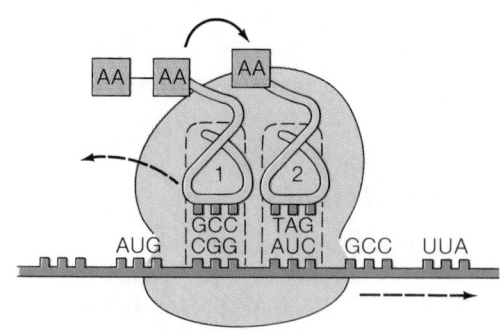

(e) The dipeptide is then linked by a peptide bond to the third amino acid, forming a tripeptide. This frees what was the second tRNA. The ribosome next slides down one more codon, exposing the second binding site and freeing it up for the addition of another tRNA-AA. This process repeats itself until the terminator codon is reached.

▷ **FIGURE 9–10 Protein Synthesis**

TABLE 9-3 Codons on Messenger RNA and Their Corresponding Amino Acids

CODON	AMINO ACID	CODON	AMINO ACID	CODON	AMINO ACID	CODON	AMINO ACID
AAU AAC	Asparagine	CAU CAC	Histidine	GAU GAC	Aspartic acid	UAU UAC	Tyrosine
AAA AAG	Lysine	CAA CAG	Glutamine	GAA GAG	Glutamic acid	UAA UAG	Terminator
ACU ACC ACA ACG	Threonine	CCU CCC CCA CCG	Proline	GCU GCC GCA GCG	Alanine	UCU UCC UCA UCG	Serine
AGU AGC	Serine	CGU CGC		GGU GGC		UGU UGC	Cysteine
AGA AGG	Argenine	CGA CGG	Arginine	GGA GGG	Glycine	UGA UGG	Terminator Tryptophan
AUU AUC	Isoleucine	CUU CUC		GUU GUC		UUU UUC	Phenylalanine
AUA AUG	Methionine	CUA CUG	Leucine	GUA GUG	Valine	UUA UUG	Leucine

*Terminator codons signal the end of the formation of a polypeptide chain.

As shown in ▷ Figure 9–10a, the ribosome contains two binding sites for tRNA-AA. Soon after the ribosome attaches to the mRNA, a tRNA bearing a specific amino acid enters the first binding site. At the base of the tRNA molecule is a sequence of three bases called an **anticodon.** The anticodon contains complementary bases that pair with the bases of the initiator codon of the mRNA template. Soon thereafter, a second tRNA-AA enters the scene, inserting itself into the second binding site on the ribosome.

The second binding site is located above the next codon, the next three bases on the mRNA template (Table 9–3). In the example shown in Figure 9–10, the codon CGG binds to the tRNA with the anticodon GCC. This tRNA, in turn, binds to only one amino acid, arginine. The proper sequence of amino acids therefore results from two factors: (1) the specificity of tRNA molecules for amino acids and (2) complementary base pairing between codons (mRNA) and anticodons (tRNA).[2]

Once the two amino acids are in place, the next step is to link them via a peptide bond. This occurs with the assistance of an enzyme in the ribosome (Figure 9–10c). After the peptide bond is formed, the first tRNA (minus its amino acid) leaves the binding site. It is then free to retrieve another amino acid for use later on. As illustrated in Figure 9–10c, the dipeptide formed during the first reaction is now attached to the second tRNA still located in the second binding site.

In order for the chain to grow, the ribosome must move down the mRNA strand. This is accomplished with the aid of a contractile protein in the ribosome. Thanks to this protein, the ribosome literally slides along mRNA one codon at a time. As illustrated in Figure 9–10d, as the ribosome moves down the mRNA, the dipeptide is shifted to the first binding site. This frees up the second site, permitting it to accept another tRNA-AA. A new tRNA-AA enters the vacant site and is joined to the dipeptide, thus forming a tripeptide.

Chain elongation takes place by the addition of one amino acid at a time, but this does not mean that protein synthesis is a slow process. Quite the contrary, protein synthesis occurs with remarkable speed. In bacteria, many proteins are synthesized in 15 to 30 seconds. In humans, the large alpha and beta chains of the very large hemoglobin molecule are synthesized in three minutes.

As the peptide chain is formed, hydrogen bonds between amino acids on different parts of the chain cause it to bend and twist, forming the secondary structure of the protein or peptide, described in Chapter 2 (▷ Figure 9–11). When the ribosome reaches the **terminator codon,** the sequence of bases that signals the end of the protein chain, the protein is released. If the protein is produced on mRNA on the surface of the rough endoplasmic reticulum, the peptide chain is transferred into its interior. The protein may then be chemically modified and packaged into transport vesicles, which pinch off and transport their cargo to the Golgi complex. If this process occurs on mRNA molecules in the cytoplasm, the protein is released into the cytoplasm.

[2]This process is a little more complicated than presented here, but it has been simplified for ease of comprehension. In reality, there are 64 codons that can be read from the mRNA, but only 40 distinct types of tRNA molecules. The reason is that some tRNA anticodons can pair with two or three different mRNA codons that specify the same amino acid.

UNRAVELING THE MECHANISM OF GENE CONTROL
Featuring the Work of Jacob and Monod

One of the questions that intrigued geneticists for many years—and still does—is how the genetic information contained in a cell's DNA is controlled. Many experiments have been performed over the years in an attempt to answer this question, but none so important as those of two French molecular geneticists, François Jacob and Jacques Monod, who performed a series of experiments on the common intestinal bacterium *E. coli* (▷Figure 1). Today, much of what is known about gene regulation in bacteria and most of the terms that describe the process come from their studies.

Working at the Pasteur Institute in the late 1950s, Jacob and Monod focused much of their attention on the bacterial uptake and catabolism of lactose, a disaccharide also known as milk sugar. As noted in the text, *E. coli* absorb lactose and catabolize it with the aid of an enzyme known as galactosidase. The catabolic breakdown of lactose results in the release of energy and carbon atoms useful in bacterial biosynthesis.

E. coli living in a lactose-free medium, however, contain very little lactose. It is only when lactose is added that the galactosidase is synthesized.

The introduction of lactose also increases the amount of a carrier molecule (galactoside permease) that transports lactose across the membrane of the bacterium, and it increases the intracellular concentration of another enzyme, which may play a role in lactose catabolism but is not essential to the process. The carrier protein and two enzymes are part of an inducible enzyme system; that is, their production can be brought on (induced) by the presence of lactose.

Mapping studies have shown that three adjacent genes on the bacterial DNA are responsible for the production of these proteins. These studies led Jacob and Monod to hypothesize that the three genes all belong to a single unit, which they called an operon. They defined an operon as a cluster of genes with related functions that is regulated in such a way that all the genes in the group are activated and inactivated simultaneously.

In a paper published in 1961, Jacob

and Monod proposed a mechanism by which these genes might be controlled, creating the model you have studied in this chapter. This intriguing model is "one of the truly important conceptual advances in biology,"said University of Wisconsin cell biologist Wayne Becker.

The bulk of the evidence in support of the model came from genetic analysis of mutant bacteria. These bacteria fit into one of two categories: those that produced abnormal amounts of all three proteins and those that responded abnormally to the addition or removal of the inducer, lactose. Mapping studies showed that the first group of bacteria resulted from mutation of the structural genes, which transcribed RNA responsible for enzyme production. The second group resulted from mutations of the regulatory components of the system.

Extensive studies of the mutants and how they responded to lactose helped Jacob and Monod piece together a sensible picture of gene regulation in *E. coli*. In these studies of the inducible *lac* operon, the researchers found that structural gene mutations resulted in an alteration of one specific protein. If the

▷ **FIGURE 9–11 Protein Synthesis** As the protein is synthesized on the mRNA, it begins to coil and bend, forming its secondary structure. Several ribosomes may "work" a single strand of mRNA simultaneously.

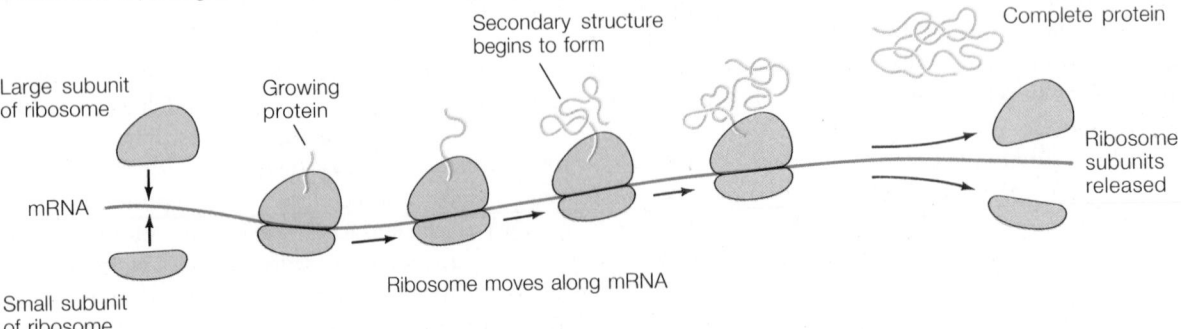

bacterium had a mutation in the *y* gene, one of the structural genes, the permease molecule would not be producedwhen inducer was added to the system. Galactosidase was formed, but very little lactose could enter the cell. If it had a mutation in the *z* gene, another structural gene, galactosidase would not be produced but permease would be. The bacterial cell could absorb lactose, it just could not do much with it.

In contrast, mutations in the regulatory regions affected nearly *all* of the structural genes and proteins they produced. A defective operator site, which binds the repressor, for example, resulted in the production of galactosidase and permease whether or not the inducer, lactose, was present. (This response was due to the fact that RNA polymerase was free to move down the *lac* operon.)

Bacteria with a mutation in the regulator gene, the segment that coded for the production of mRNA that gave rise to the repressor protein, also produced galactosidase and permease. They did so because this mutation blocked all repressor protein production, letting RNA polymerase wander freely down the *lac* operon, transcribing the genes.

Data from these and other mutants helped the scientists piece together a model. As a testimony to the thoroughness of their work, the original model has undergone very little change over 30 years.

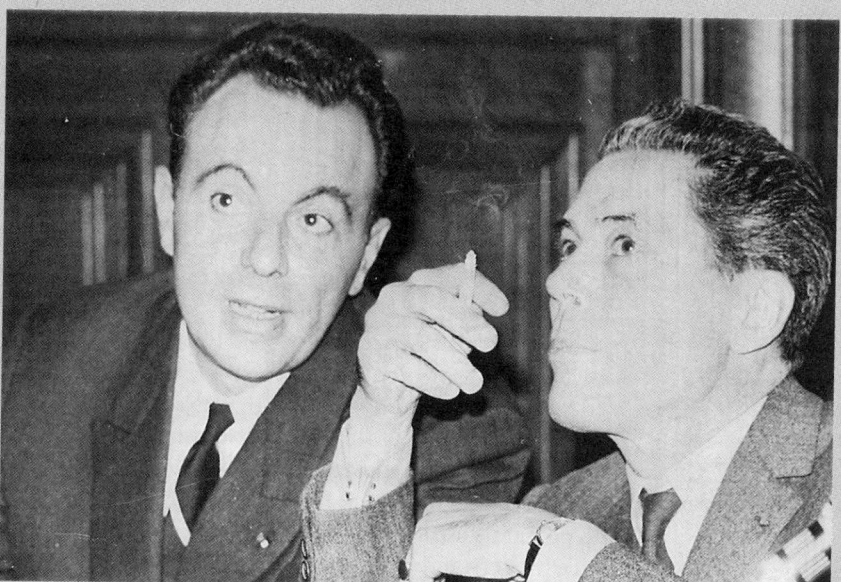

▷ **FIGURE 1 Jacob and Monod** These two French scientists proposed the operon hypothesis.

≋ CONTROLLING GENE EXPRESSION

In science it is one thing to describe a process but quite another to understand how the process is controlled. This is certainly true of genetic expression—that is, the control of the activity of genes. After decades of research, however, molecular geneticists (scientists who study genes) have pieced together a fairly good understanding of the mechanisms behind gene regulation, although the process is still not completely understood. Understanding the control mechanisms is important because it could help scientists discover new ways to treat, or even cure, diseases, such as cancer, that originate in the genes of an organism. The following section describes gene regulation in bacteria, which provides considerable insight into human gene regulation.

Bacterial Genes Consist of Three Interdependent Segments of DNA: the Regulator Gene, Control Regions, and Structural Genes

As explained in Scientific Discoveries 9–1, two French researchers, François Jacob and Jacques Monod, worked out the details of bacterial gene control using a common intestinal bacterium with a tongue-twisting name, *Escherichia coli,* or *E. coli* for short (▷ Figure 9–12). *E. coli* lives in the large intestine of humans and other mammals, living off undigested and unabsorbed food matter.

After extensive studies on *E. coli,* the two French geneticists proposed a model of bacterial gene control. It states that bacterial genes consist of three distinct parts: structural genes, regulator genes, and control regions. **Structural genes,** shown in light blue in ▷ Figure 9–13, code

Nucleoid

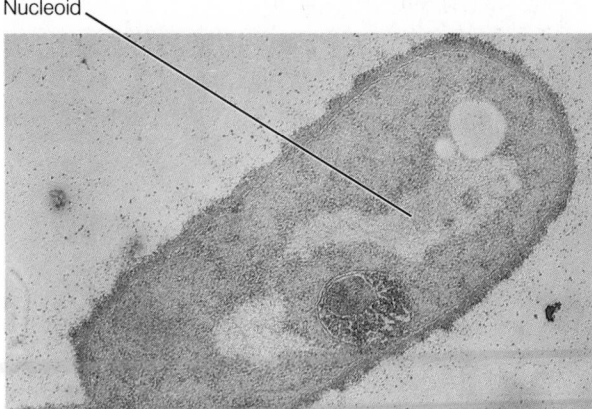

(a)

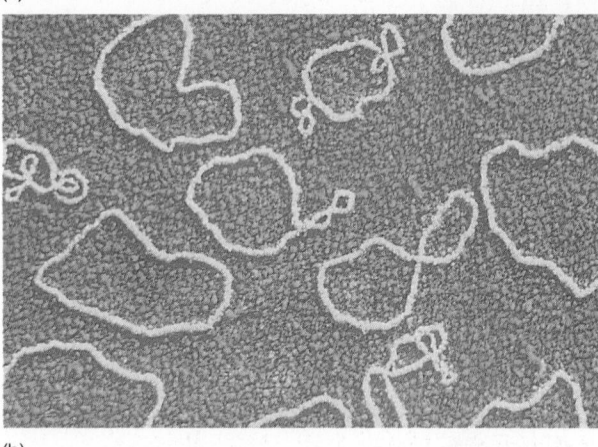

(b)

▷ **FIGURE 9–12 E. coli** (*a*) A transmission electron micrograph of the bacterium *E. coli*, a common inhabitant of the large intestine of mammals. Note that the region of the bacterium containing the circular DNA is not enclosed by a membrane and is therefore called a nucleoid. (*b*) A circular molecule of DNA from *E. coli*.

for the production of enzymes and other proteins. **Regulator genes,** shown in dark blue, and **control regions,** shown in yellow and green, regulate the production of mRNA on the structural genes, as described below.

Jacob and Monod hypothesized that the control regions and structural genes of a metabolic pathway were located side by side on the chromosome, as shown in Figure 9–13. Together, the structural genes and the control regions form a functional unit known as an **operon**. The regulator genes are located some distance from the operon. Based on the results of their experiments, Jacob and Monod proposed two types of operons: inducible and repressible.

Inducible Operons Are Switched on When Enzymes Are Needed. *E. coli* is capable of producing many enzymes needed to digest leftover foodstuffs, but it produces these enzymes only on demand by activating **inducible operons**— a set of genes that remains inactive until required. Jacob and Monod worked out the details on an inducible operon that is responsible for the production

of three different proteins associated with the uptake and metabolism of lactose, or milk sugar, in *E. coli* (Figure 9–13). They called it the *lac* **operon.**

When lactose is absent, the *lac* operon is inactive. (Like a light bulb in a room, it is kept off until needed.) When lactose is present in a human digestive system, small quantities diffuse into the bacteria. These molecules activate the *lac* operon. Lactose is therefore said to be an **inducer,** for it activates the genes responsible for the production of the proteins involved in its transport and metabolism.

To understand how lactose "turns on" the structural genes, we look first at the operon in the absence of lactose, illustrated in Figure 9–13a. This figure shows the location of the structural genes, control regions (operator and promoter), and the regulator gene. As illustrated, the regulator gene produces a protein known as a **repressor.** In the absence of lactose, the repressor binds to the **operator site.** Just "upstream" is the **promoter site,** where the enzyme RNA polymerase attaches. RNA polymerase catalyzes the formation of mRNA from the structural genes needed for the production of the proteins responsible for the transport and metabolism of lactose. When the repressor protein is present, however, it physically blocks RNA polymerase from binding to the strand of DNA and transcribing the structural genes. In the absence of lactose, then, the repressor protein acts something like a molecular switch, turning off the operon. As a result, the operon cannot produce proteins needed to transport lactose into the cell and break it down.

To understand how the operon is switched on, refer to Figure 9–13b. As illustrated, the lactose molecules (shown in purple) bind to the repressor protein, causing it to change shape. This, in turn, causes the protein to detach from the operator site. When the operator site is vacant, RNA polymerase is no longer blocked from the structural genes. The cell then produces the mRNAs it needs, which are used to produce the enzymes and membrane-transport molecules necessary for the uptake and breakdown of lactose.

This complex, but elegant, system provides the bacterium with an on-off switch for many of its genes. This mechanism no doubt evolved to save the cell from having to produce a lot of unnecessary proteins. It is a conservation mechanism aimed at saving ATP and amino acids. The importance of this evolutionary adaptation is even more evident when one considers what the cell's demand for ATP and molecular building blocks might be if all of the genes were on all of the time.

Repressible Operons Provide a Steady Supply of Enzymes and Other Proteins and Remain On Until Switched Off. During evolution, another strategy to control genes evolved, the **repressible operon.** Unlike the inducible operon, this one remains active unless turned *off.* This strategy provides a constant supply of enzymes and other proteins that bacteria require to produce the molecules, such as amino acids, necessary for routine cellular metabolism.

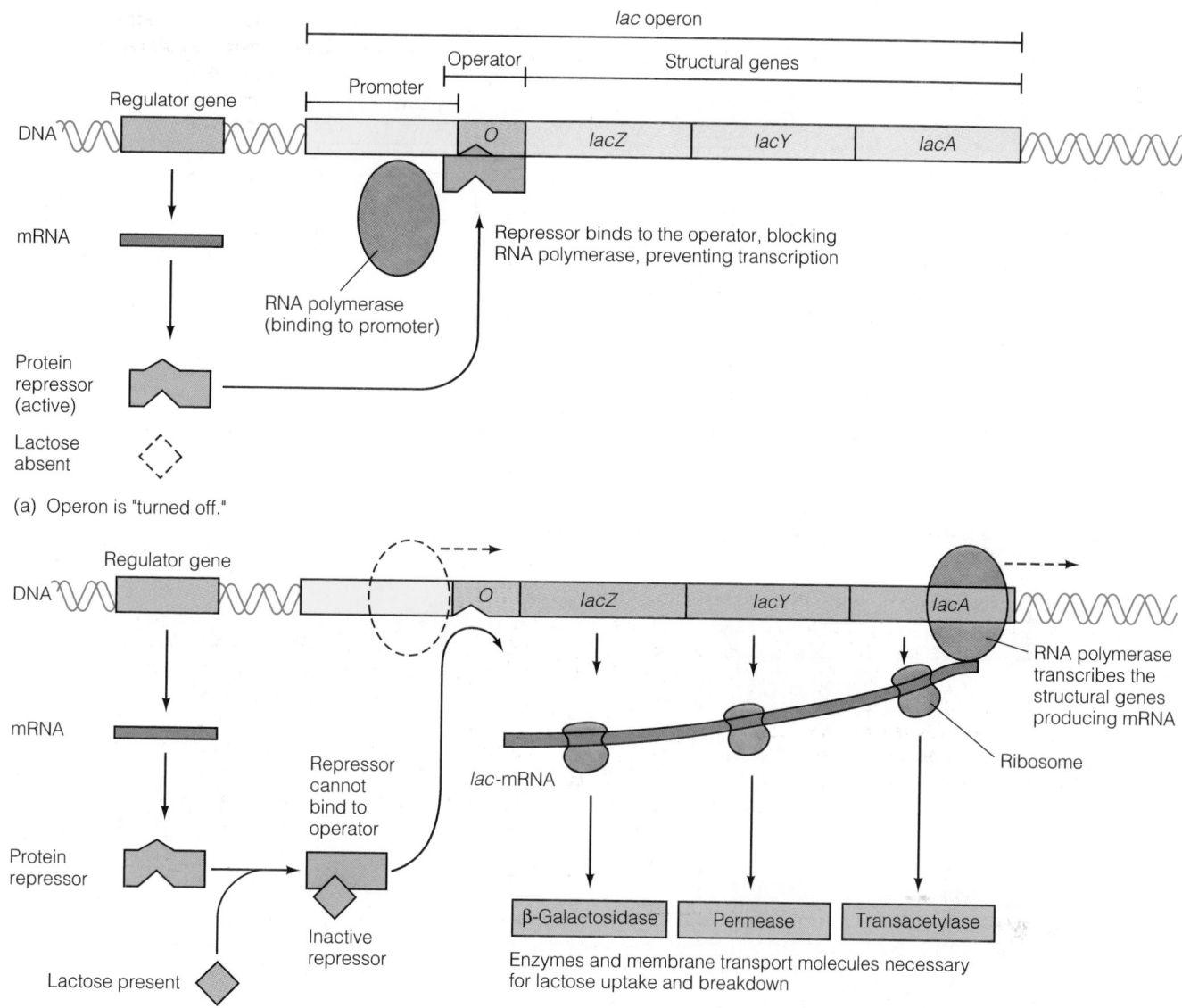

(a) Operon is "turned off."

(b) Operon is "turned on."

▷ **FIGURE 9–13 Inducible Operon** (*a*) The lactose, or *lac,* operon of bacteria is an inducible operon. It is shut off until activated. Note that the regulator gene of the DNA produces mRNA that codes for a repressor protein. It binds to the operator site of the operon, blocking RNA polymerase from transcribing the structural genes. (*b*) When lactose is present, it binds to the repressor protein, releasing it from the operator site. This allows RNA polymerase to transcribe the structural genes.

▷ Figure 9–14 illustrates how repressible operons work. This particular operon controls the production of enzymes needed to synthesize the amino acid tryptophan. As shown in this diagram, the repressible operon is similar in many ways to the inducible operon. It contains structural genes, a regulator gene, and two binding sites, the operator and promoter. The regulator gene of the repressible operon also produces a repressor protein, but this is where the similarities end.

As shown in the top panel, in this operon the repressor protein binds to the operator site only when another molecule, the **corepressor,** is present. In this example, tryptophan is the corepressor. Thus, when tryptophan levels are high, the operon is switched off (Figure 9–13b). As shown, the repressor-corepressor complex binds to the operator site, blocking RNA polymerase and shutting down the production of mRNA from the structural genes, in much the same way that an assembly line in a factory might be shut down if it were producing parts faster than they could be used elsewhere in the plant. When tryptophan levels are low, the repressible operon (**trp operon**) is switched on, and the cell produces a continuous supply of the enzymes it needs for its synthesis. As long as the cell is using tryptophan and levels are low, this operon functions.

In repressible operons, corepressors are usually molecules produced by the enzymes generated by the operon

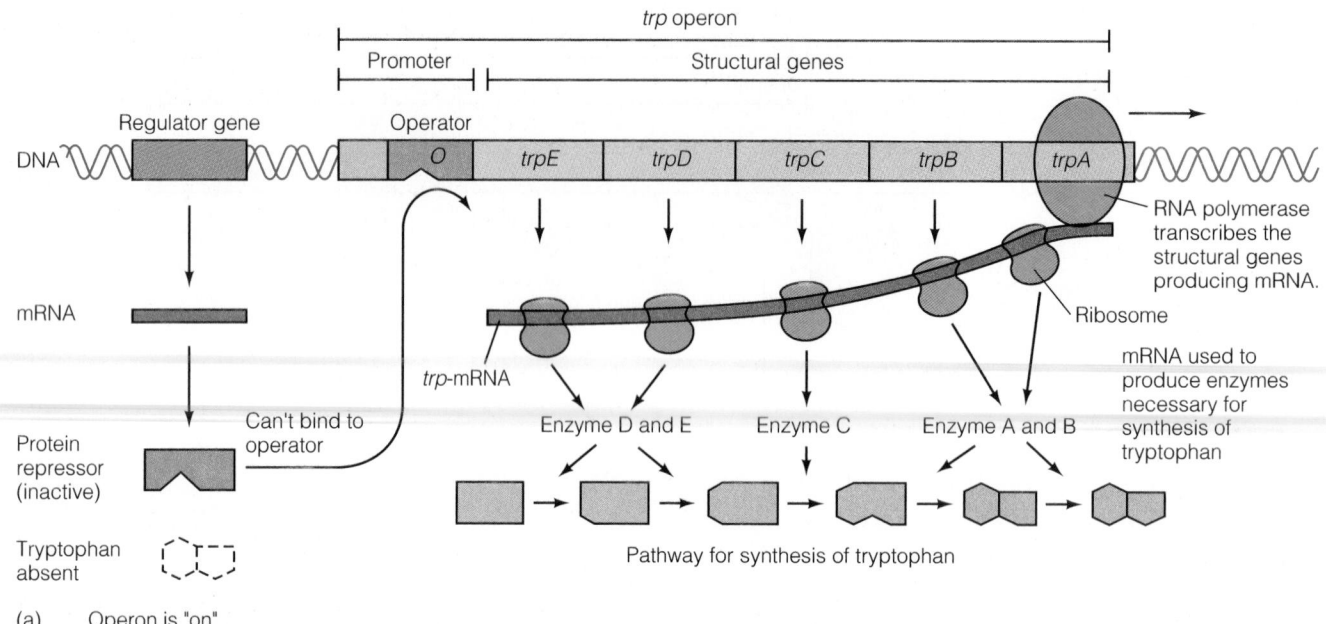

(a)   Operon is "on"

(b)   Operon is "turned off"

▷ **FIGURE 9–14 Repressible Operon** (*a*) Structurally similar to the inducible operon, this group of genes is customarily turned on. The regulator gene codes for a repressor protein that cannot bind to the operator site unless a corepresssor is present. In the absence of a corepressor, RNA polymerase is free to transcribe the structural genes. (*b*) In this example, when tryptophan, the corepressor, is present in excess, it binds to the repressor protein. This allows the repressor protein to bind to the operator site, blocking RNA polymerase and turning off the operon.

through the mRNA it produces. When a cell is producing excess product, the corepressor binds to the repressor molecule, and the operon is temporarily switched off, as illustrated in Figure 9–14b.

In Humans, Genetic Expression Is Controlled at Four Levels

Inducible and repressible systems are present in human cells, but geneticists believe that most human genes are inducible. The control of human genes, however, also involves several other mechanisms. For simplicity, these can

be divided into four major categories: (1) control at the chromosome level, (2) control at transcription, (3) control after transcription before translation, and (4) control at translation (▷ Figure 9–15).[3]

Control at the Chromosome: Access to the Genes Is Controlled by Coiling and Uncoiling of the Chromosomes During Interphase. Chapter 7, you may recall, noted that chromatin fibers in the nucleus condense and become inactive during prophase of mitosis.

[3]Bacteria also control gene expression at these levels.

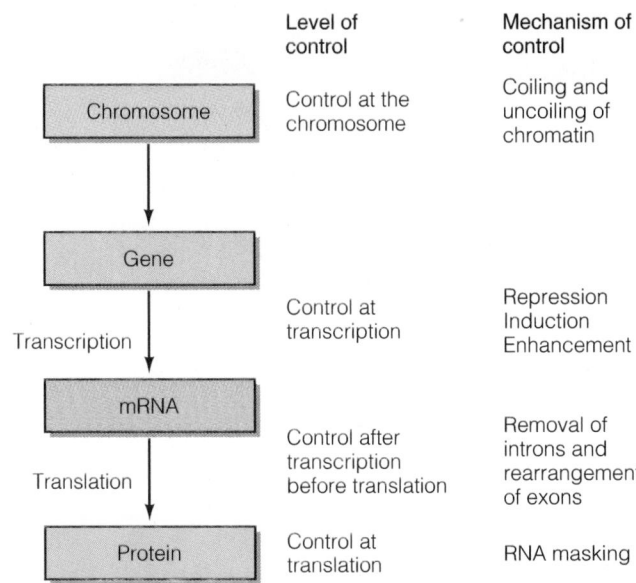

Level of control	Mechanism of control
Control at the chromosome	Coiling and uncoiling of chromatin
Control at transcription	Repression Induction Enhancement
Control after transcription before translation	Removal of introns and rearrangement of exons
Control at translation	RNA masking

▷ **FIGURE 9–15 Gene Expression** In humans, gene expression is regulated at four levels.

In the condensed state, they cannot produce RNA or new DNA. The molecule is essentially shut down in preparation for nuclear division.

During interphase, the cell also selectively inactivates some of its chromatin. Inactivation is an evolutionary adaptation essential for cellular specialization in complex multicellular organisms. In humans, for example, each cell in the body contains all of the genetic information that is needed for all other cells. Therefore, a muscle cell contains the same genes as a liver cell or a brain cell. However, the muscle cell does not express all of the genes that a liver cell or a brain cell expresses, and vice versa. Instead of getting rid of the genes it does not need, the muscle cell simply inactivates them.

Inactive chromatin in the interphase nucleus is tightly coiled, or compacted, and appears as dark clumps called **heterochromatin.** Some heterochromatin remains in this state throughout the cell cycle, except during replication (S phase). The remaining chromatin in the interphase nucleus is diffusely organized—consisting of fine threads unraveled in the nucleus—and is called **euchromatin** ("true chromatin"). Euchromatin is metabolically active.

Three Control Mechanisms Operate During Transcription. In eukaryotes, control of the gene at transcription (RNA production) involves repression and activation of operons, similar to that which occurs in bacteria. Human cells also modulate gene activity via a third mechanism, known as enhancement. **Enhancement** is a process in which already active genes increase their production of mRNA as a result of the action of nearby control segments

of the DNA called **enhancers.** Research suggests that certain protein molecules bind to enhance regions of the chromosome; these proteins somehow enhance the attachment of RNA polymerase to a DNA strand, increasing RNA production by the structural gene. Enhancers do not turn nearby structural genes on and off but simply amplify their activity. An enhancer is therefore a little like the accelerator in your car—it doesn't turn the engine on, it simply speeds it up.

Control After Transcription, But Before Translation: mRNA Molecules Are Altered in Predictable Ways, Producing Specific Proteins They Require. Genetic studies show that eukaryotic DNA contains two types of genetic material: (1) sections that are used to produce mRNA and (2) sections that are not. As ▷ Figure 9–16 shows, the noncoding segments of DNA are called **introns** and are interspersed within the functional segments of DNA (functional genes), which are referred to as **exons.** Where do these interesting names come from? The name *intron* signifies that these are *in*tervening segments of noncoding DNA. Exons are *ex*pressed segments—that is, those sections of the DNA that can be used to produce mRNA that is used to make protein.

Both the introns and exons are transcribed during the cell cycle, producing a **RNA transcript** that is a complete "readout" of the DNA. Eukaryotic cells, however, process the transcript, cutting out the introns then joining the exons together in order to produce a functional mRNA molecule. This process is necessary because in eukaryotes, the structural genes that produce the enzymes of a metabolic pathway are not lined up next to one another as they are in the operons of prokaryotes.

The exons from a single gene can be spliced together in different ways. This, in turn, produces slightly different mRNAs that code for the synthesis of slightly different types of protein. In other words, splicing can produce new proteins and therefore represents another mechanism for controlling the expression of genes. This mechanism alters the DNA blueprint in predictable ways to meet the needs of the cell. One place this comes in handy is in the production of antibody molecules, which play a role in immune protection. Splicing permits cells to produce a wide variety of antibody molecules to stave off a wide variety of foreign invaders.

Control at Translation: Some Cells Can Mask mRNA, Rendering It Inactive, Until It Is Required for Protein Synthesis. Eukaryotic cells can also control gene expression at the level of translation (protein synthesis). Messenger RNA, for example, which is produced in the nucleus, may be transferred to the cytoplasm in an inactive state. The mRNA is said to be **masked.** Masking allows the cell to build up large supplies of mRNA in preparation for a sudden burst of protein synthesis. When needed, the mRNA is activated and begins producing large

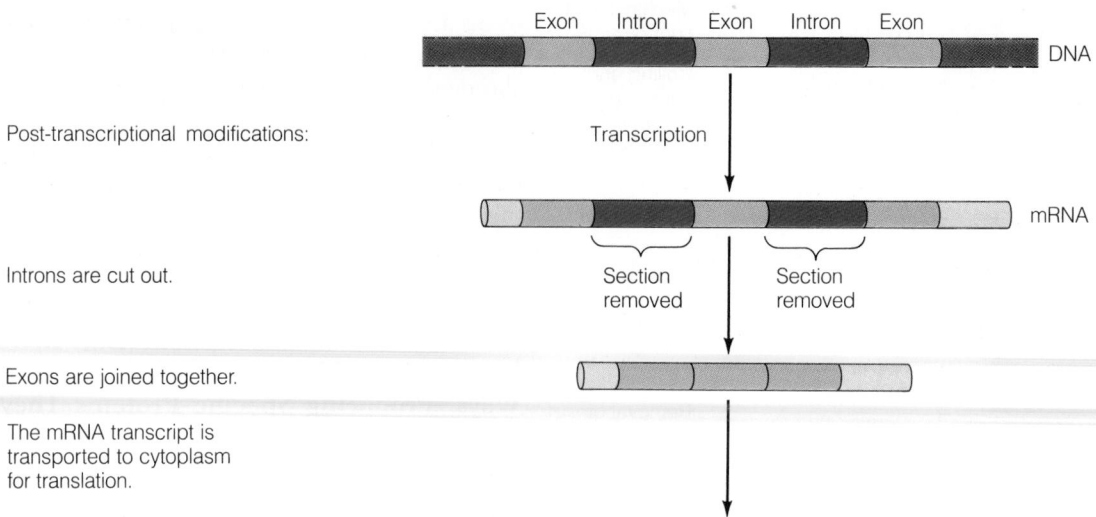

Exon Intron Exon Intron Exon

DNA

Post-transcriptional modifications:

Transcription

mRNA

Introns are cut out.

Section removed

Section removed

Exons are joined together.

The mRNA transcript is transported to cytoplasm for translation.

▷ **FIGURE 9–16 Posttranscriptional Control of Human Genes**   One key mechanism of control occurs after transcription. DNA contains noncoded segments (introns) and useful segments (exons). An mRNA transcript is produced from the DNA, and the segments produced by introns are removed. The final product is a strand of mRNA containing only RNA copies of the exons. The exon copies can be linked differently, producing slightly different products.

quantities of protein. This phenomenon occurs in the human ovum, the female sex cell produced in the ovaries. The ovum produces enormous amount of masked mRNA as it awaits fertilization. When the sperm arrives and fertilizes the ovum, the mRNA that has accumulated in the cytoplasm is unmasked, and there is an explosion of protein synthesis necessary for subsequent cell divisions.

Gene expression in humans and other eukaryotes is a complex phenomenon. Like the cell cycle, it is no doubt under the influence of many controlling factors.

≈ APPLICATIONS OF BIOLOGY: ONCOGENES—THE SEEDS OF CANCER WITHIN US

Chapter 7 described cancer and noted a causal link between this disease and numerous biological, chemical, and physical agents, known as carcinogens. It also noted that most cancers arise from **mutations,** alterations of parts of the genetic material. Broadly defined, mutations include (1) alterations of the DNA, such as the deletion or addition of a purine or pyrimidine base; (2) alterations of the chromosome, such as missing segments; and (3) alterations in the chromosome number—that is, too many or too few chromosomes. (These were explained in Chapter 8.)

Mutations may occur in germ cells, the sperm and ovum, or in body cells. This section concerns itself with body cell, or **somatic, mutations.** In most instances, somatic mutations in DNA are rapidly repaired by enzymes in the nucleus. Some somatic mutations, however, escape repair. Among those that escape repair are three basic types: (1) mutations that can improve cellular function, (2) mutations that are neutral, having neither a positive nor a negative effect, and (3) mutations that affect vital sections

of the genome, killing the cell outright or releasing the cell on the rampage of growth we call cancer. It is the mutations that lead to cancer that will concern us here.

Within the last 10 years, researchers have made a startling discovery about somatic cell mutations that may explain the underlying cause of most , if not all, cancers. They have found that humans and other organisms actually conitan the seeds of cancer in their own genes. Each of us contains a group of genes that, when mutated, can lead to cancerous growth. These genes are called **proto-oncogenes.**

Proto-oncogenes are genes that normally code for the production of important cellular structures or functions—for example, cell adhesion and the production of the plasma membrane receptors that bind to growth factors or hormones. Collectively, the proto-oncogenes play a key role in cellular growth. When mutated by radiation, ultraviolet light, and chemical carcinogens, these genes produce the uncontrolled cellular proliferation we call cancer. (For a discussion of methods used to assess carcinogenicity, see the Point/Counterpoint feature on page 214.)

Viruses may also cause cancer. Cancer-causing viruses fall into two groups. Some viruses stimulate cancer by activating proto-oncogenes in body cells. Others contain cancer-causing genes. These genes, known as true **oncogenes,** are carried into viral-infected cells, where they migrate to the nucleus. Here, viral oncogenes may become incorporated in the cell's own DNA, turning on the cell's genes that control cellular division.

The discovery of proto-oncogenes and viral oncogenes has resulted in a quantum leap in our understanding of cancer. Some scientists believe that the key to treating cancer may ultimately lie in techniques that turn off activated proto-oncogenes. Switching off these genes may end

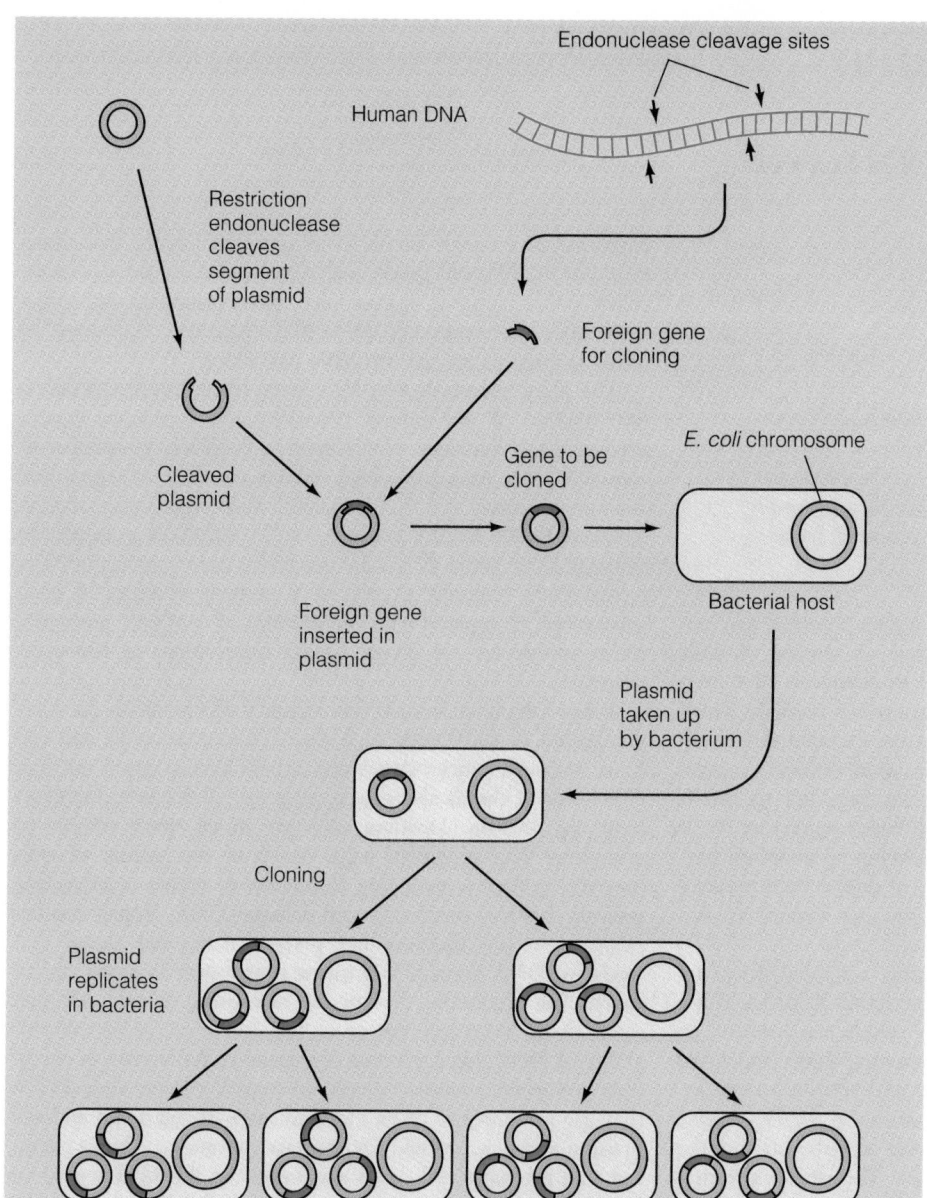

▷ FIGURE 9–17 Genetic Engineering Genes can be snipped from chromosomes and inserted in plasmids, then transferred to bacteria or other microorganisms, where they are replicated in large quantity, a process known as cloning.

Endonuclease cleavage sites

Human DNA

Restriction endonuclease cleaves segment of plasmid

Foreign gene for cloning

Cleaved plasmid

Gene to be cloned

E. coli chromosome

Foreign gene inserted in plasmid

Bacterial host

Plasmid taken up by bacterium

Cloning

Plasmid replicates in bacteria

the cancerous proliferation of cells. Who knows, maybe someday you or someone you know will benefit directly from this now-only-theorectical idea. (For more on cancer treatment, see Health Notes 4–1 and 7–1.)

≈ A PRIMER ON GENETIC ENGINEERING

Genetic engineering is a popular name for something geneticists prefer to call **recombinant DNA technology.** Recombinant DNA technology is a procedure by which scientists remove segments of DNA from one organism (a human, for example) and insert them into the DNA of another organism (for instance, a bacterium). The technique evolved from the discovery of a group of enzymes that snips off segments of the double-stranded DNA molecule. Known as **restriction endonucleases,**

these enzymes cut through both polynucleotide strands of the DNA and produce numerous fragments, as illustrated in ▷ Figure 9–17.

To transplant DNA fragments into bacterial cells, the fragments must first be inserted into small circular strands of DNA isolated from bacteria. These strands, known as **plasmids,** lie outside a bacterium's circular chromosome and contain important bacterial genes—for example, the genes that provide some bacteria with resistance to antibiotics.

As shown in Figure 9–17, plasmids are also treated with restriction endonucleases to open up the circular DNA. Foreign genes are then inserted into the opening. Another enzyme is used to catalyze the splicing of the foreign gene into the plasmid. In order to insert the plasmid with its foreign gene into a cell, the modified plasmids are added to

ANIMAL TESTING FOR CANCER IS FLAWED

Philip H. Abelson

Philip Abelson is the Deputy Editor of Science. *This article is copyrighted (© 1990) by the American Association for the Advancement of Science.*

The principal method of determining potential carcinogenicity of substances is based on studies of daily administration of huge doses of chemicals to inbred rodents for a lifetime. Then by questionable models, which include large safety factors, the results are extrapolated to effects of minuscule doses in humans. Resultant stringent regulations and attendant frightening publicity have led to public anxiety and chemophobia. If current ill-based regulatory levels continue to be imposed, the cost of cleaning up phantom hazards will be in the hundreds of billions of dollars with minimal benefit to human health. In the meantime, real hazards are not receiving adequate attention.

The current procedures for gauging carcinogenicity are coming under increasing scrutiny and criticism. A leader in the examination is Bruce Ames, who with others has amassed an impressive body of evidence and arguments. Ames and Gold summarized some of their recent data and conclusions in *Science* (31 August 1990, p. 970). Three articles in the *Proceedings of the National Academy of Sciences* provide an elaboration of the information with extensive bibliographies. The articles also provide data about other pathologic effects of natural chemicals.

A limited number of chemicals tested, both natural and synthetic, react with DNA to cause mutations. Most chemicals are not mutagens, but when the maximum tolerated dose (MTD) is administered daily to rodents over a lifetime, about half of the chemicals give rise to excess cancer, usually late in the normal life span of the animals. Experiments in which synthetic industrial chemicals were administered in the MTD to both rats and mice resulted in 212 of 350 chemicals being labeled as carcinogens. Similar experiments with chemicals naturally present in food resulted in 27 of 52 tested being designated as carcinogens. These 27 rodent carcinogens have been found in 57 different foods, including apples, bananas, carrots, celery, coffee, lettuce, orange juice, peas, potatoes, and tomatoes. They are commonly present in quantities thousands of times as great as are the synthetic pesticides.

The plant chemicals that have been tested represent only a tiny fraction of the natural pesticides. As a defense against predators and parasites, plants have evolved a large number of chemicals that have pathologic effects on their attackers and consumers. Ames and Gold estimate that plant foods contain 5,000 to 10,000 natural pesticides and breakdown products. In cabbage alone some 49 natural pesticides have been found. The typical plant contains a total of a percent or more of such substances. Compared with the amount of synthetic pesticides we consume, we eat about 10,000 times more of the plant pesticides.

It has long been known that virtually all chemicals are toxic if ingested in sufficiently high doses. Common table salt can cause stomach cancer. Ames and others have pointed out that high levels of chemicals cause large-scale cell death and replacement by division. Dividing cells are much more subject to mutations than quiescent cells. Much of the activity of cells involves oxidation, including formation of highly reactive free radicals that can react with and damage DNA. Repair mechanisms exist, but they are not perfect. Ames has stated that oxidative DNA damage is a major contributor to aging and to cancer. He points out that any agent causing chronic cell division can be indirectly mutagenic because it increases the proability of DNA damage being converted to mutations. If chemicals are administered at doses substantially lower than MTD, they are not likely to cause elevated rates of cell death and cell division and hence would not increase mutations. Thus a chemical that produces cell death and cancer at the MTD could be harmless at lower dose levels.

Diets rich in fruits and vegetables tend to reduce human cancer. The rodent MTD test that labels plant chemicals as cancer-causing in humans is misleading. The test is likewise of limited value for synthetic chemicals. The standard carcinogen tests that use rodents are an obsolescent relic of the ignorance of past decades. At that time, extreme caution made sense. But now tremendous improvements of analytical and other procedures make possible a new toxicology and far more realistic evaluation of the dose levels at which pathological effects occur.

CURRENT METHODS OF TESTING CANCER ARE VALID

Devra Davis

Devra Davis is a Scholar in Residence for the National Research Council/National Academy of Sciences.

The vast majority of the scientific community endorses the conduct of experimental studies in order to try to identify those materials that could cause disease and prevent harmful exposures. In the typical toxicologic study of 50 rodents, each animal is a stand-in for 50,000 people. Because rodents live about two years on average, they are usually exposed to amounts of the suspect agent that approximate what a human would encounter in an average lifetime of 70 years.

Some have argued that the use of the maximum tolerated dose (MTD) in these studies produces tissue damage and cell proliferation, which in turn lead to cancer. This is biological nonsense that ignores a fundamental characteristic of cancer biology that has been known for more than 50 years: cancer is a multistage disease, with multiple causes, which arises by a stepwise evolution that involves progressive genetic changes, cell proliferation, and clonal expansion. Thus, swamping of tissues with high doses alone may well kill an animal or damage its tissues, but high doses alone are not sufficient to cause cancer.

A two-year study at the National Institute of Environmental Health Sciences (NIEHS) by David Hoel and colleagues provides good evidence on this point. They looked for signs of damage in tissues taken from rats and mice used in cancer studies. Cancers occurred in organs that did not show apparent damage, and some damaged organs were completely free of tumors.

In addition, most compounds tested do not cause cancer only in the highest dose group tested but typically produce a dose-response relationship, where the amount of cancer developed is proportional to the dose administered. Some chemicals cause toxicity only, others cause only cancer. Not all those that cause cancer do so through organ toxicity. In fact, almost 90% of the substances shown to cause cancer in the National Toxicology Program do so without producing any increased cellular toxicity, and they also cause cancer at both lower and higher doses.

Animal studies are evolving and being further refined, as is our understanding of differences between species that need to be taken into account in conducting these studies. Every compound known to cause cancer in humans also produces cancer in animals, when adequately tested. For 8 of the 54 known human cancer-causing agents, evidence of carcinogenicity was first obtained in laboratory animals; in many cases, the same target organs and doses have been involved in producing cancer in both animals and humans.

The NIEHS has carried out nearly 400 long-term rodent studies, which have been published following public peer review by specialists in the field. Chemicals nominated for testing usually represent a sample of potentially "problematic" materials. About 40% of the "suspect" pesticides evaluated to date have been found to cause cancer. As to the role of so-called natural pesticides, rodent diets are also loaded with many of these materials. Nevertheless, a number of test compounds added to these diets markedly increase the amount of tumors produced. Thus, animals are more sensitive to certain synthetic compounds than to the background level of natural materials. Humans are also likely to have acquired some resistance to these persisting, natural materials throughout our evolution.

To date, only about 20% of all synthetic organic chemicals have been adequately tested for their potential human toxicity. Those who must set public policy on the use of chemicals need a rational basis on which to stake their actions. Continued advances in animal testing provide an important contribution to environmental health sciences and to public health efforts to predict, and then prevent, the development of environmentally caused disease.

In summary, the current system should be used until it can be replaced by a demonstrably better one. There is no scientific basis for rejecting the MTD as capable of inducing cancer.

SHARPENING YOUR CRITICAL THINKING SKILLS

1. Summarize each author's main points and supporting data. Do you see any inconsistencies or examples of faulty logic in either essay? Where?
2. Why is it that two scientists can disagree on an issue such as this?
3. Given the disagreement, what course of action would you recommend?

a culture dish containing bacteria or yeast cells, In culture, the plasmids are rapidly taken up by the new hosts.

Bacteria or yeast cells with their newly acquired plasmids can be grown in the laboratory. As the bacteria duplicated, so do the plasmids, producing many copies of the gene. The gene is then said to be **cloned.**

The Applications of Genetic Engineering Are Many

Genetic engineering has potential applications in three areas: (1) the mass production of commercially important plants or animals in order to increase disease resistance, stimulate growth, or introduce some other desirable traits; and (3) gene therapy, the treatment of genetic diseases in humans. Critical thinking suggests, however, that these applications may not be problem-free. As you read this section, you may want to think of some of the problems that might arise.

Mass Production of Gene Products Is Already Bearing Fruit.
Historically, the first practical application of genetic engineering was the production of human insulin and growth hormone. The genes responsible for the production of these hormones are now routinely inserted into bacteria, which in turn are grown in massive quantities, creating large amounts of these proteins for commercial use. Before the advent of genetic engineering, insulin for diabetics was extracted from the pancreases of animals, usually pigs or cattle. This task was costly and time-consuming and produced a hormone that was not chemically identical to human insulin. Diabetics treated with pig or cow insulin eventually developed problems, because their immune systems reacted to the foreign substances (Chapter 15). Although the immune response was generally low-keyed, it did require physicians to switch their patients to other insulin preparations from time to time.

Genetic-engineering firms today also routinely produce growth hormone, a protein needed to treat children whose pituitary glands produce insufficient amounts of this substance. Before the advent of this technique, growth hormone was difficult to procure and prohibitively expensive.

Gene Transfer in Plants and Animals Is Becoming Commonplace But Is Highly Controversial.
For years, livestock and plant breeders have been selectively breeding hardy animals to produce genetically superior livestock and crops. Dogs and cats and a host of other animals were selectively bred over thousands of years. Thus, people have been performing a kind of genetic engineering—albeit a slow one—for millennia. Genetic engineering offers a potentially quicker way of achieving the same goals.

It is important to point out, however, that although genetic engineering and selective breeding achieve the same end points in some cases, the two techniques are quite different. Selective breeding alters gene frequencies. Genetic engineering often seeks to introduce new genes into species, producing new alleles (that is, new genetic combinations). Scientists, for example, recently implanted the gene that codes for human growth hormone into cattle embryos in hopes of increasing their levels of the hormone and producing marketable cattle more quickly. Faster growing cattle, they hope, might produce more meat per pound of grass, reducing the impact of cattle on grazing land. In dairy cattle, growth hormone increases milk production. The Critical Thinking section later in the chapter will help you explore the advantages and disadvantages of this idea.

Genetic scientists are working to improve crops as well. Already, they have introduced genes that allow oats to grow in salty soil, combating a problem in irrigated farmland throughout the world. Transplanting these same genes into other commercial crop species could allow farmers to use the vast acreage now idle because of the buildup of salts caused by the irrigation of poorly drained soils with water high in various salts.

Researchers are also working on ways to provide crop plants resistance to **herbicides,** chemicals applied to crops to control weeds. Although herbicides generally act only on weed species, they sometimes impair the growth of crop species. Proponents hope that herbicide-resistant crops might help farmers increase yields. One problem with this development, however, is that making plants resistant to herbicides might increase a farmer's dependency on costly chemicals that could have harmful effects on other species like birds and fish. The logic is that if herbicide-resistant crops are used, farmers may apply larger amounts of herbicide to their fields. These chemicals, in turn, may be washed into nearby waterways or may poison soils, killing important bacteria and other microorganisms essential for the long-term health of the soil.

Genetic engineering is also being used to protect plants from insects—a development that may reduce the use of potentially harmful insecticides (toxic chemicals that kill insects). For example, genes found in one species of bacteria code for the production of a protein toxic to many insects. Scientists have successfully transplanted these genes to another bacterium that grows on the roots of crop plants. Soil-dwelling insects that eat the roots therefore also ingest the genetically altered bacteria. Inside the stomach of insects, the genetically engineered bacteria release their toxic protein, killing the insects and protecting the plant against further damage.

Genetic researchers have also developed a bacterium that retards the formation of frost on plants. The genetic variant is found naturally in the environment, but not in sufficient quantities to protect crops. Thus, researchers have cloned the bacterium and have begun testing its efficacy in the field. Should it prove successful and safe, the bacterium could be sprayed on crops late in the growing season when frosts could be devastating, potentially saving farmers millions of dollars a year.

Geneticists are also working on ways to genetically alter plants themselves to increase their resistance to pests. By transferring the genes that protect wild species from insects into crop species, scientists may be able to produce many varieties that require little, if any, chemical pesticide.

Genetic Diseases May Be Cured by Gene Therapy.

Approximately 1 of every 100 children born in the United States suffers from a serious genetic defect, such as sickle-cell anemia or hemophilia. Thanks to advances in genetic engineering, medical science may soon find ways of replacing defective human genes with normal genes, curing some previously incurable diseases that cost our society millions of dollars and account for enormous human suffering. The largest obstacle, however, is getting the genes into body cells where they are missing.

Bone marrow transplants may provide one avenue. Consider an example. Krabbe's disease is a rare genetic disorder resulting from a deficiency in one enzyme in human cells. The absence of this enzyme causes fat to accumulate in the nervous system, causing nerve cells to degenerate. In humans, seizures and visual problems occur early in life, and most victims die within the first two years of life. Scientists recently injected a strain of mice that suffers from a similar condition with bone marrow cells containing the gene that codes for the missing enzyme. These cells became established in the lungs and liver and restored enzymatic activity. The bone marrow cells even found their way into the brain, which is an important breakthrough, because the brain cells of patients with Krabbe's disease are severely affected by this enzyme deficiency. In order for this disease to be cured, the enzyme must become available in human brain cells.

Another somewhat promising technique involves the use of microspheres—tiny lipid spheres that can be coated with antibodies that allow them to deliver their contents—in this case, a replacement gene or two—to body cells (see Health Note 2–1). Researchers have also developed a novel transplantation technique that could help surgeons introduce genetically engineered cells into specific organs in the human body. In a recent experiment, scientists injected genetically altered liver cells into a foamlike material. The foam was then transplanted into rats after being impregnated with a hormone that stimulates the growth of blood vessels from nearby larger vessels. Within a week after implantation, the partially artificial tissue was riddled with a network of blood vessels.

This technique could be used to introduce genetically altered cells (or even fetal cells) into patients with genetic disorders. It could even provide a way to replace insulin-deficient pancreas cells in diabetics or dopamine-producing cells in patients with Parkinson's disease.

Despite the Promises, Genetic Engineering Is Fraught With Controversy

Although genetic engineering has spawned a great deal of enthusiasm, the procedure has its critics. Many safety questions still remain unanswered. Of greatest concern is the possibility of unleashing genetically altered bacteria or viruses. Critics fear that a new strain might spread through the environment, wreaking havoc on natural ecosystems and, possibly, human populations. Once unleashed, it would be impossible to retrieve. The genetically altered bacterium that retards frost formation, some worry, could enter the atmosphere on dust particles, reducing cloud formation and altering global climate. No one knows for sure how serious this threat really is.

Other critics object to genetic tinkering, especially the transfer of genes from one species to another. Do humans have the right, they wonder, to interfere with the course of evolution? Unfortunately, human experience with genetic engineering is too limited to answer the concerns of critics. Preliminary work suggests that the dangers have been exaggerated and that genetically engineered bacteria are not a threat to ecosystem stability. Still, further research is needed to be certain that in our zeal to make life better, we don't do ourselves in.

ENVIRONMENT AND HEALTH: HAZARDOUS WASTES AND MUTATIONS

Thousands of mutations occur in each cell of your body every day, many of which could lead to serious problems, including cancer. Fortunately for you, the cells of your body repair much of the damage. This repair is brought about by special enzymes in the nucleus that snip off damaged sections of the DNA, then rebuild the molecule. In a sense, evolution has provided us with our own internal medical care system.

Remaining healthy, though, requires that we not stretch that resiliency to the breaking point. And yet some individuals fear that modern industrial society may be doing just that. Today, 60,000 chemical substances are in commercial use. The National Academy of Sciences recently noted that few of these substances have been adequately tested for their ability to cause mutations, cancer, and birth defects. Of the pesticides in use, for example, only about 10% have been adequately tested for their propensity to cause cancer and birth defects.

The extensive use of chemicals and widespread publicity over the ill effects of some have created a chemical paranoia—sometimes referred to as chemophobia—in our society. New research suggests that many people are overreacting to the threat. Numerous studies have found that naturally occurring chemicals in our food, like one type of mold that grows on peanut butter, probably cause more cancer than pesticide residues. (For a debate over chemical contamination and the relative importance of pesticides and other industrial chemicals on human health, see the Point/Counterpoint on pages 218-219.)

Regardless, some critics argue that the debate over human health is misleading, for it ignores the potentially

THE MYTHS OF THE CANCER EPIDEMIC

David L. Eaton

David L. Eaton is Associate Professor of Environmental Health and Environmental Studies, and Director of Toxicology at the University of Washington. He has an active research program on the mechanisms by which chemicals cause cancer.

There is no debate that cancer is a devastating and deadly disease. One in three people living in the US today will contract some form of cancer in his or her lifetime, and one in four will die from it, if current rates continue.

Are cancer rates increasing in epidemic proportions? The total number of people and the fraction of all deaths caused by cancer have increased dramatically in the past 50 years. However, cancer is largely a disease of old age, and thus it is necessary to adjust such statistics for changes in the age distribution of our population. A 1988 report from the National Cancer Institute states that "the age adjusted mortality rates for all [types of] cancers combined, except lung cancer, have been declining since 1950 for all age groups except 85 and above." Statistics from the American Cancer Society yield the same conclusion. The incidence of some childhood leukemias and brain tumors has indeed increased significantly in the past decade. How much of this increase is a result of better diagnosis and reporting, rather than a "true" increase, remains controversial.

We *are* in an "epidemic" of lung cancer. For most of the first half of this century, lung cancer mortality was not even in the "top five" types of cancer-related deaths. Lung cancer is now the leading cause of cancer-related deaths in both men and women. About 85–90% of all lung cancers in men, and perhaps 70% in women, is directly attributable to smoking. *Per capita* consumption of cigarettes increased 5-fold in men from 1900 to 1960, and with it a concomitant increase in lung cancer. The risks of several other types of common cancers are also increased by smoking (e.g., cancers of the bladder and esophagus). Approximately one-third of *all* cancer deaths could be eliminated by eliminating smoking from our society.

Of the variety of environmental factors other than smoking, dietary factors are now generally thought to represent the largest source of cancer risk, perhaps related to 30%–40% of all cancers. Although synthetic chemicals such as industrial pollutants and pesticides present in trace amounts in our food supply may contribute to dietary risk, recent studies have suggested that this contribution is trivial relative to other "non-pollutant" factors. For example, the risk of breast cancer in women (second only to lung cancer in incidence and mortality) is significantly increased by high fat diets, and the amount of fiber in the diet substantially influences the risk of colon cancer, a major site of cancer in both men and women.

The largest source of exposure to cancer-causing chemicals may not be industrial pollution, but chemicals that occur naturally in our diet. All plants produce toxic chemicals to protect against insects, fungi, and animal predators. It has been estimated that we ingest in our diet about 10,000 times more of "nature's pesticides" than man-made chemical residues. Many of these chemicals are potent mutagens and carcinogens, and are frequently present at levels thousands of times higher than the trace levels of synthetic pesticide residues and industrial chemicals sometimes found foods. Taken together, the dietary risk factors from natural sources often present in relatively high amounts, are far more important than the pesticide residues and industrial chemicals that can often be detected at exceedingly small concentrations in our diets. Unfortunately, because of the relatively high exposure to carcinogens from natural sources, the complete elimination of synthetic industrial chemicals from our diet, if it were possible, would not likely have any significant beneficial effect on cancer incidence and mortality.

Finally, recent advances in the understanding of the biology of cancer suggest that "spontaneous" or "background" alterations in DNA may explain much of the cause of cancer. The use of modern techniques in molecular biology has revealed that DNA is inherently unstable and can be altered by normal errors in DNA replication. DNA is subject to extensive damage from processes associated with normal cellular metabolism. Within our life span, our cells undergo about 10 million billion cell divisions. Spontaneous errors in this process, which lead to mutations and cancer, accumulate with age. It is not surprising then that cancer seems to be a frequent outcome of old age.

The view that we are in cancer epidemic largely due to industrial chemicals is not supported by the majority of cancer researchers throughout the world. Unfortunately, it will take time for the political arena to come to grips with the fact that further reduction in public exposure to synthetic chemicals will not have much impact on cancer incidence, and that such results will come only at great social and economic expense. The U.S. is currently spending about $80 billion per year on pollution reduction, about nine times the total budget for all basic scientific research. I believe that much of this is justified to enhance the quality of our environment, and ensure the habitability of our planet for humans and other species. However, there is also little question that huge sums of money are spent each year to reduce what is in all likelihood a trivial cancer risk, with few other environmental benefits. If our society is truly concerned about reducing the human tragedy from cancer, more efforts should be focused on eliminating smoking and alcohol abuse, better research and education on dietary risk factors, more research into the biochemical and molecular events that lead to cancer (which in turn will lead to more effective preventive and curative measures), and continued identification and reduction in occupational exposures to those chemicals which pose a significant cancer risk.

THE POISONING OF A NATION: AMERICA'S EPIDEMIC OF CANCER

Lewis G. Regenstein

Lewis G. Regenstein, an Atlanta writer and conservationist, is the author of How to Survive in America the Poisoned *and director of the Interfaith Council for the Protection of Animals and Nature, an affiliate of the Humane Society of the United States.*

America is in the throes of an unprecedented cancer epidemic, caused in large part by the pervasive presence in our environment and food chain of deadly, cancer-causing pesticides and industrial chemicals.

Today, significant levels of hundreds of toxic chemicals known to cause cancer, miscarriages, birth defects, and other health effects, are found regularly in our food, our air, our water—and our own bodies. Accompanying this widespread pollution has been a dramatic and alarming rise in the cancer rate in recent decades.

Each year, over a million Americans (an estimated 1,130,000 for 1992, not including 450,000 skin cancers) are diagnosed as having cancer—over 3000 people a day! The disease now strikes almost one American in three, and kills over a thousand of us *every day!* This means that of the Americans now alive, some 70 to 80 million people can expect to contract cancer in their lifetimes. More Americans die of cancer *every year* (an estimated 520,000 in 1992) than were killed in combat in World War II, Korea, and Vietnam *combined.*

And cancer has now become a common disease of the young as well as the old, with incidence of childhood cancers, especially leukemia and brain tumors, mounting sharply in recent years.

Man-made chemicals are also depleting the Earth's protective ozone layer, which makes life on the planet possible by shielding us from most of the sun's ultraviolet rays. The US Environmental Protection Agency (EPA) has projected that the increase in radiation hitting the Earth will cause Americans to suffer 40 million cases of skin cancer, 800,000 deaths in the next 88 years, and 12 million incidences of eye cataracts.

In 1978, the President's Council on Environmental Quality (CEQ) reported unequivocally that "most researchers agree that 70% to 90% of all cancers are caused by environmental influences and are hence theoretically preventable."

Evidence continues to mount demonstrating that toxic chemicals are heavily contributing to the cancer epidemic. In general, the most polluted areas of the country have the highest cancer rates, with heavily industrialized New Jersey having the greatest concentration of chemical and petroleum facilities. In July, 1982, the University of Medicine and Dentistry of New Jersey released a study showing a correlation between the presence of toxic waste dumps and elevated cancer death rates (up to 50% above average) in areas of the state.

In February, 1984, a report by researchers at the Harvard School of Public Health demonstrated a link between consumption of chemically contaminated well water near Woburn, Massachusetts and the extraordinary incidence of childhood leukemia, stillbirths, birth defects, and disorders of kidneys, lungs, and skin among local residents.

Dr. Samuel Epstein of the University of Illinois Medical Center, perhaps the foremost authority on the subject, points out that apart from AIDS, "cancer is the only major killing disease which is on the increase," with incidence rising by at least 2% a year, and death rates rising at 1% annually over the last decade. He concludes that "the facts show very clearly that we are in a cancer epidemic now," in large part because of "the carcinogenizing of our environment, the increasing contamination of our air and our water and our food and our workplace."

Today, every American is exposed to a variety of health-destroying chemicals. Dozens of pesticides used on our food are known or thought to cause cancer in animals. By the time restrictions were placed on some of the deadliest chemicals, such as DDT, dieldrin, BHC, and PCBs, these carcinogenic poisons were being found in the flesh tissues of literally 99% of all Americans tested, as well as in the food chain and even mother's milk. In fact, breast milk is heavily contaminated with high levels of banned, cancer-causing chemicals. And virtually all Americans carry in their bodies traces of dioxin (TCDD), the most deadly synthetic chemical known.

The response of the US government has been largely weak or nonexistent enforcement of the nation's health and environmental protection laws. For example, with few exceptions, the EPA has refused to carry out its legal duty to ban or restrict pesticides known to cause cancer. Nor has the government adequately implemented or enforced the laws regulating hazardous waste, which is being generated at a rate of up to 275 million metric tons a year—over a ton for every man, woman, and child in the nation. Much of this is disposed of in a manner that will ultimately threaten the health of nearby residents.

Thus we are even now sowing the seeds for cancer epidemics of the future. Only time will tell what will be the effect on this generation, and future ones, of Americans—the chemical industry's ultimate guinea pigs.

SHARPENING YOUR CRITICAL THINKING SKILLS

1. Regenstein asserts that America is in the midst of a cancer epidemic caused in large part by pesticides and industrial chemicals. Eaton argues the opposite. He says that the death rate for all types of cancer except lung cancer is actually declining. Summarize the supporting data of each author.
2. Using the critical thinking skills described in Chapter 1, analyze each argument.

▷ **FIGURE 9–18 Hazardous Wastes** The careless dumping of hazardous materials in rivers and lakes, in abandoned fields, along highways, and in abandoned warehouses creates a health hazard to humans and many other species and results in costly cleanup efforts. Thankfully, new laws have put tighter controls on hazardous waste disposal, although some industries and waste disposal companies still violate the law.

▷ **FIGURE 9–19 Victim of Love Canal**

widespread effects of toxic chemicals on plants, animals, and microorganisms that share this planet with us. They say that modern society is heading blindly into the future with little concern for the long-term impacts of our actions. Nowhere has the impact been more noticeable than around toxic waste dumps, where for decades some chemical companies have disposed of highly dangerous substances, often in cardboard containers or steel barrels that rust within a few years of burial (▷ Figure 9–18).

Love Canal in the city of Niagara Falls, New York, was the scene of one of the nation's worst toxic nightmares. Here, in an abandoned canal, Hooker Chemical Company dumped over 20,000 metric tons of highly toxic and carcinogenic wastes from 1947 to 1952.[4] In 1952, the city of Niagara Falls began condemnation proceedings, not to shut down the operations but to legally seize the land to build a school and housing development. Hooker sold the land to the city for $1 in exchange for a release from future liability.

A few years later, trouble began. Bulldozers preparing the site for construction of the school removed the clay lid that Hooker had placed over the dump site to protect it. In the late 1950s, after the construction was complete, rusty barrels began to work their way up through the ground. Toxic chemicals oozed to the surface, killing trees and gardens and causing chemical burns in children who played in the ooze. Some children even died (▷ Figure 9–19).

The problem came to a head in the 1970s. After a period of heavy rainfall, toxic wastes began to leak into basements of local residents, and the chemical stench be-

came unbearable. Over 80 different chemical substances turned up in studies of water, air, and soil. Many of the chemicals were known or suspected carcinogens. A New York State Health Department study showed that one of every three pregnant women who lived in the area had miscarried. The miscarriages could have been caused by chemically induced mutations in the early embryo or in the germ cells of the parents. Birth defects were present in one of every five children, far in excess of the expected rate. Birth defects may also result from chemically induced mutations. The chemical substances released from the old dump site irritated lungs, gave residents headaches, and brought on convulsions in some people. Genetic studies showed mutations in chromosomes of residents. Nearly 1000 families were evacuated from the site over the years, and now the homes sit idle, boarded up, in solemn tribute to human carelessness.

New laws and concerted efforts by industries have cut back on the reckless disposal of hazardous wastes, but the practice still occurs. Some companies, in fact, have even attempted to ship their hazardous wastes to cash-poor Third World countries where disposal is far cheaper and not regulated, spreading the problem throughout the globe.

Cleaning up past mistakes like Love Canal has proved costly. Over $40 million has been spent so far to clean up the Love Canal area, and millions more may have to be spent. What is striking, though, is that some experts estimate that there are nearly 10,000 hazardous waste sites, some worse than Love Canal, in the United States in need

---

[4]A metric ton is 2240 pounds.

of cleanup. Government estimates put the cost of cleaning up the mess at $100 billion. Critics believe that the actual cost will be far greater. Adding to the problem, however, are the more than 250 million metric tons of hazardous materials that American industry generates annually—over a ton of hazardous waste for every man, woman, and child each year!

Improper hazardous-waste disposal now pollutes groundwater in many areas, and that concerns public health officials because half the people in the United States get their drinking water from wells. By one estimate, more than 10 million Americans now use tap water contaminated with chemical pollutants in excess of EPA standards. Because of years of careless waste disposal, groundwater contamination is expected to grow worse. Even if we stop polluting now, the problem will linger for many years.

A healthy population requires clean water. But how do we get it? Many changes are needed in American society. Especially important are ways to reduce hazardous waste production. By redesigning chemical processes, for example, manufacturers can make significant reductions in their waste output. They can also reuse and recycle hazardous wastes. A purified waste product from one process may actually become the raw material for another process. These steps are preventive—they help avoid the problems of disposal altogether.

Individuals can also help reduce hazardous waste by becoming conscientious consumers and reducing unnecessary consumption. Buy environmentally safe cleaning products, for example, and reduce your overall purchasing (▷ Figure 9–20). These actions can have a profound effect on the production of toxic waste and the health of the environment. Using energy and other resources more frugally and recycling household products, paper, aluminum, glass, and plastics can also make substantial inroads into hazardous waste production.

[5] Most garbage incinerators in the United States are not designed to eliminate harmful emissions in their entirety. Japan's incinerators have been so designed, but they are expensive to build and operate.

Wastes can be burned in high-temperature furnaces, specially designed to eliminate harmful emissions.[5] Wastes can be chemically neutralized or treated in other ways to render them less harmful. After reductions and chemical modifications come new and improved disposal techniques. However, all of these ideas should be a last resort. Even new landfills lined with thick clay bottoms or synthetic liners may eventually leak, releasing toxic chemicals into the ground. Prevention is the best cure. The health of the planet depends on it.

▷ **FIGURE 9–20 Environmentally Friendly Cleaning Products** Gentle Earth products are packed in recycled plastic, which can be recycled again once they are emptied. Products are produced in concentrated form and diluted for use, thus reducing energy required for shipping and reducing packaging. These products are made from citrus oils and other natural products as well.

SUMMARY

BUILDING TWO MACROMOLECULES: DNA AND RNA

1. The DNA molecule houses the genetic information. DNA is a double helix, consisting of two polynucleotide strands joined by hydrogen bonds between purine and pyrimidine bases on the opposite strands.

2. Each nucleotide in the DNA molecule consists of a purine or pyrimidine base, the sugar deoxyribose, and a phosphate group. The nucleotides are joined by covalent bonds between phosphate groups and deoxyribose molecules.

3. Complementary base pairing is an unalterable coupling in which adenine on one strand of the DNA molecule always binds to thymine on the other, and guanine always binds to cytosine.

4. Complementary base pairing ensures accurate replication of the DNA and accurate transmission of genetic information from one cell to another and from one generation to another.

5. To replicate, DNA must first unwind with the aid of a special enzyme. After unwinding, the strands provide templates

for the production of complementary DNA strands, a process aided by the enzyme DNA polymerase.

6. The cell contains three types of RNA: transfer RNA, ribosomal RNA, and messenger RNA. All play a role in the synthesis of protein in the cytoplasm.

7. All three RNA molecules are single-stranded polynucleotide chains. RNA nucleotides contain the sugar ribose instead of deoxyribose, which is found in DNA. RNA nucleotides contain four bases: adenine, guanine, cytosine, and uracil (instead of thymine).

8. RNA is synthesized in the nucleus on a template of DNA. The synthesis of RNA is called transcription.

HOW GENES WORK: PROTEIN SYNTHESIS

9. The genetic information contained in the DNA molecule is transferred to messenger RNA. Messenger RNA molecules carry this information to the cytoplasm, where proteins are synthesized.

10. Messenger RNA serves as a template for protein synthesis. Ribosomes are required to produce proteins on the mRNA template.

11. Transfer RNA molecules deliver amino acid molecules to the mRNA and insert them in the growing chain. Each tRNA binds to a specific amino acid and delivers it to a specific codon, a sequence of three bases on the mRNA. Thus, the sequence of codons determines the sequence of amino acids in the protein.

12. Proteins are synthesized by adding one amino acid at a time.

13. During protein synthesis, the ribosome first attaches to the mRNA at the initiator codon. Soon after, the large subunit attaches. A specific tRNA bound to an amino acid binds to the initiator codon and the first binding site of the ribosome. A second tRNA-amino acid then enters the second site.

14. An enzyme in the ribosome catalyzes the formation of a peptide bond between the two amino acids. After the bond is formed, the first tRNA (minus its amino acid) leaves the first binding site.

15. The ribosome moves down the mRNA, shifting the tRNA bound to its two amino acids to the first binding site and opening the second site for another tRNA-amino acid. This process repeats itself many times in rapid succession.

16. As the peptide chain is formed, hydrogen bonds begin to form between the amino acids, and the chain begins to bend and twist, forming the secondary structure of the protein or peptide. When the ribosome reaches the terminator codon, the peptide chain is released.

CONTROLLING GENE EXPRESSION

17. Bacterial DNA contains functional units called operons, consisting of three parts: structural genes, regulator genes, and control regions.

18. Structural genes code for the production of enzymes and other proteins and are controlled by the regulator genes and the control regions.

19. Two types of operons are present: inducible and repressible.

20. Inducible operons are switched off until needed. In an inducible operon, the regulator gene produces a repressor protein. The repressor protein binds to the operator site next to the structural genes and prevents the transcription of the structural genes by physically blocking RNA polymerase. When an inducer substance is present, it binds to the repressor, causing it to release from the operator site and permitting the transcription of the DNA.

21. Repressible operons are generally continuously activated but can be switched off by the presence of a chemical substance called a corepressor, usually an end product of a metabolic pathway.

22. The corepressor binds to the repressor protein, thus permitting it to bind to the operator site and block RNA polymerase.

23. Gene control in humans occurs at four levels: at the chromosome, at transcription, after transcription before translation, and at translation.

24. At the chromosome: Condensation, or coiling, of the chromatin fibers, for instance, inactivates genes. How the cell condenses a segment of the chromatin is not clearly understood.

25. At transcription: research shows that inducible and repressible systems are present in human cells.

26. At transcription: Research also shows that certain segments of the DNA, called enhancers, can greatly increase the activity of nearby genes. Geneticists think that protein molecules bind to enhancer regions of the chromosome and facilitate gene activity.

27. After the production of RNA before translation: Human DNA contains far more genetic material than it needs. Those segments of the DNA used to produce RNA that will be used to make protein are called exons. The intervening segments of noncoding DNA are called introns.

28. Both the introns and exons are transcribed during the cell cycle. The cell, however, removes the RNA from the introns and joins the RNA fragments produced by exons to produce a functional mRNA molecule.

29. The exon-produced RNA can be spliced together in different ways; the resulting mRNAs produce different proteins.

30. At translation: Messenger RNA may be transferred to the cytoplasm in an inactive, or masked, state. Masking permits the cell to build up large supplies of mRNA in preparation for a sudden burst of protein synthesis.

31. Most cancers arise from mutations. Mutations include three basic changes: (a) alterations of the DNA itself, (b) alterations of the chromosome, and (c) alterations in the chromosome number.

APPLICATIONS OF BIOLOGY: ONCOGENES— THE SEEDS OF CANCER WITHIN US

32. Humans and other organisms contain specific genes, called proto-oncogenes, that when mutated lead to cancerous growth. Proto-oncogenes are normal genes that code for cellular structures and functions, such as cell adhesion, mitotic proteins, and the production of plasma-membrane receptors for growth factors or hormones.

33. Radiation, ultraviolet light, and chemical carcinogens can mutate these genes, resulting in uncontrolled cellular proliferation (cancer).

34. Viruses can also cause cancer. Some viruses, for example, possess oncogenes, which enter human body cells and are incorporated in the cell's genome, stimulating uncontrolled cellular division. Other viruses carry genes that stimulate cancer by activating human proto-oncogenes.

A PRIMER ON GENETIC ENGINEERING

35. Genetic engineering, or recombinant DNA technology, is a procedure by which scientists remove segments of DNA from one organism and insert them into the DNA of another organism.

36. To transplant DNA fragments into bacterial cells, the fragments must first be inserted into plasmids, circular strands of DNA that lie outside a bacterium's chromosome and contain important bacterial genes.

37. The genetically modified plasmids are added to a culture dish containing bacteria or yeast cells and are taken up by the hosts.

38. Bacteria or yeast cells with their newly acquired plasmids can be grown in the laboratory. As the bacteria duplicate, so do the plasmids, producing many copies of the gene. The gene is then said to be cloned.

39. Genetic engineering has potential applications in three areas: (a) the mass production of commercially important chemicals, such as hormones; (b) the transfer of genes to plants or animals in order to increase disease resistance, stimulate growth, or introduce some other desirable traits; and (c) gene therapy, the treatment of genetic diseases in humans.

40. Although genetic engineering has stimulated a great deal of enthusiasm, many safety questions remain unanswered. Of greatest concern is the possibility of unleashing genetically altered bacteria or viruses.

ENVIRONMENT AND HEALTH: HAZARDOUS WASTES AND MUTATIONS

41. Numerous chemicals are released into the environment from factories, farms, automobiles, and other sources. Many chemicals now in common use have not been tested to determine their ability to cause cancer and birth defects.

42. New research suggests that people may be overreacting to the threat of low-level chemical pollutants. Naturally occurring chemicals in our food, in fact, may cause more cancer than pesticide residues.

43. In some locations, such as toxic waste dumps, chemicals are present in extremely high concentrations. Decades of improper waste disposal have left a legacy of pollution, ill health, and contaminated groundwater.

44. Reducing the contamination of our environment by hazardous wastes will require concerted efforts on the part of governments, businesses, and individuals. Especially important are ways to reduce the production of toxic waste by redesigning processes, recycling and reusing wastes, and finding substitutes.

EXERCISING YOUR CRITICAL THINKING SKILLS

One of America's leading drug companies is spending about $300 million to construct a facility that will produce synthetic growth hormone to be sold to dairy farmers. The hormone will be produced by genetically engineered bacteria that contain growth-hormone genes isolated and transplanted from dairy cattle. Given to cows, the hormone dramatically increases milk production. Business economists believe that an increase in milk production will reduce the cost of milk to the consumer. Based on this information, does introducing synthetic growth hormone seem like a good idea?

After you have thought about it for a while, consider some additional facts. The American dairy industry already produces an excess of milk. The federal government buys the surplus, dehydrates some of it, and makes cheese out of the rest. The government stockpiles the dehydrated milk and cheese at considerable cost to taxpayers. Now what do you think?

Before you draw a firm conclusion, however, remember that cheese and dehydrated milk are given to the needy. The food is not going to waste. Will increasing the surplus produce more food for the poor? Do we need more cheese and milk for indigent Americans?

Before you draw any conclusions, consider another factor—the effects on small dairy operations of increasing the surplus of milk. Many owners of small dairy herds believe that the large producers who use the hormone will increase their milk production. As noted above, this increase will probably drive down the cost of milk and milk products, and it could put many small dairy farmers out of business. This loss could adversely affect the economies of many rural regions.

Some consumer groups are concerned about the potential health effects of using synthetic growth hormone. A trace of growth hormone is present in milk produced normally. Will the use of synthetic growth hormone increase the concentration in milk? What effect will that have on your health or the health of children?

Make a list of the pros and cons of using synthetic growth hormone. Make a list of questions, those that you can answer and those that you cannot answer. Do you need more information to decide?

If you are able to make up your mind, what factors swayed your opinion? Did the concerns of one group outweigh the concerns of another? What critical thinking rules did you use in this exercise?

TEST OF TERMS

1. DNA consists of two polynucleotide chains held together by _____ bonds, which form a spiral staircase, or _____ .

2. A purine molecule is joined to a sugar molecule and a phosphate group, forming a(n) _____ .

3. In the DNA molecule, adenine forms hydrogen bonds only with _____ on the opposite strand, and _____ binds only with cytosine.

4. The production of RNA on a DNA template is called _____, whereas the production of protein on messenger RNA is called _____ .

5. A series of three bases on the messenger RNA molecule codes for a specific amino acid and is called a(n) _____ .

6. _____ molecules bind to amino acids in the cytoplasm and deliver them to the peptide chain being synthesized on the mRNA.

7. _____ _____ pairing is responsible for the accurate production of protein on mRNA and the accurate replication of a strand of DNA.

8. The _____ genes and the control region form a unit called the _____ .

9. A(n) _____ operan is a set of genes that remains inactive until needed. In order for the structural genes to produce mRNA, RNA polymerase must be able to transcribe them. It cannot bind if the _____ protein is bound to the _____ site.

10. Densely packed chromatin is metabolically inactive and is called _____ .

11. _____ are segments of the DNA that increase the activity of nearby genes in humans.

12. The human genome contains segments of DNA that are not coded, called _____ , and segments that code for protein, called _____ .

13. Viruses carry cancer-causing genes called _____ .

14. Human cells also have genes that, when mutated, can result in cancer. These are called _____ .

Answers to the Test of Terms are found in Appendix B.

TEST OF CONCEPTS

1. What is the genetic code of DNA? How is the genetic code, housed in the DNA, translated into instructions the cell can understand?

2. Describe how DNA is synthesized. How does the cell ensure the accurate replication of DNA?

3. List the three types of RNA, and briefly describe their function in protein synthesis.

4. In what ways are RNA and DNA similar? In what ways are they different?

5. Describe, in detail, the production of protein on mRNA. Be sure to note the enzymes involved and the role of the ribosome.

6. Discuss how bacterial cells control their genes. In what way does an end product of a metabolic pathway control genes? In what way does a starting material control the genes?

7. Discuss the various levels at which human genes are controlled. Give an example of each.

8. What is a proto-oncogene? How is it affected by a mutagen, a chemical or physical agent that causes mutation?

9. What is genetic engineering? Describe some of its applications, and note some of the potential problems.

Principles of Structure and Function

≈

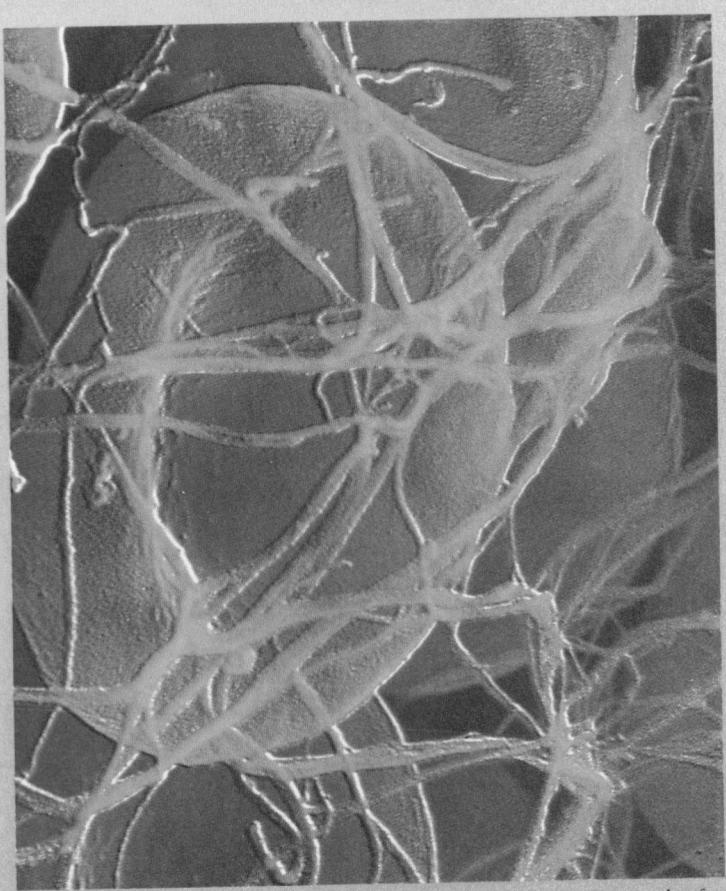

Colorized electron micrograph of red blood cells trapped in fibrin network of a blood clot.

A human being, a horse, and a bald eagle are strikingly different organisms. One walks on two feet, the second on four hooves, and the third depends on powerful wings to soar above the landscape. Despite their marked differences, these three organisms share several features. First, they are all members of the animal kingdom, a large and diverse group of multicellular organisms. Second, the cells of each of these uniquely adapted organisms are arranged in tissues, which, in turn, combine to form various organs, such as the heart, stomach, and brain. Third, each organism contains a number of homeostatic mechanisms that maintain internal constancy, permitting survival in an ever-changing world.

This chapter focuses on the second feature, the organizational pattern common to many multicellular animals, and the third feature, homeostasis. Throughout this chapter, and in the remaining chapters of Part IV, we will focus on organ systems and their role in homeostasis, studying representative members of each major group of animals.

≈ FROM CELLS TO ORGAN SYSTEMS

To begin our study of the structure and function of animals, we look at the beginning of life in sexually reproducing vertebrate (backboned) animals to put things in context. Life begins at fertilization. The fertilized ovum contains all of the information needed to develop into a fully functional adult and all of the information needed to control complex life functions, such as growth, reproduction, and homeostasis.

Cells Unite to Form Tissues, and Tissues Combine to Form Organs

During embryonic development, the fertilized ovum divides many times, first producing a ball of cells called the morula (▷ Figure 10–1). Soon thereafter, the cells undergo a process called **differentiation,** during which they become structurally and functionally specialized. The first sign of differentiation is the emergence of three distinct cell types: ectoderm, mesoderm, and endoderm. **Ectoderm** lies on the outside of the embryo and gives rise to the skin, the eyes, and the nervous system. **Mesoderm** lies in the middle and forms muscle, bone, and cartilage. **Endoderm** is the innermost layer and forms the lining of the intestinal tract and several digestive glands.

During embryonic development in most animals, ectodermal, mesodermal, and endodermal cells give rise to a variety of highly differentiated cell types, each of which carries out very specific functions. These specialized cells are often bound together by extracellular fibers and other extracellular materials, forming **tissues** (from the Latin "to weave"). Extracellular materials may be liquid (as in blood), semisolid (as in cartilage), or solid (as in bone). Tissues, in turn, combine to form **organs,** discrete structures in the body that carry out specific functions that benefit the entire organism. Organs, like their counterparts in cells, the organelles, provide for a division of labor.

Plants undergo a similar process during development, forming specialized cells and tissues. In addition, the cells and tissues of a plant give rise to highly specialized structures, such as leaves and roots, that carry out specific functions much as the organs of animals do. (The structure and function of plants are described in Chapter 26.)

Cells Combine to Form Four Primary Tissues

Four major tissue types are found in vertebrate animals: epithelial, connective, muscle, and nervous. These are called **primary tissues.** Table 10–1 lists the primary tissues and shows that each has two or three subtypes. Muscle tissue, for instance, comes in three varieties: cardiac, skeletal, and smooth. Cardiac muscle cells are found exclu-

▷ **FIGURE 10–1 Early Embryonic Development in Humans** The fertilized ovum (*a*) divides mitotically, eventually forming a solid ball of cells, the morula (*b*), which later becomes a hollow blastocyst (*c*). The inner cell mass shown here becomes the embryo, differentiating into ectoderm, endoderm, and mesoderm in later stages of development (*d*).

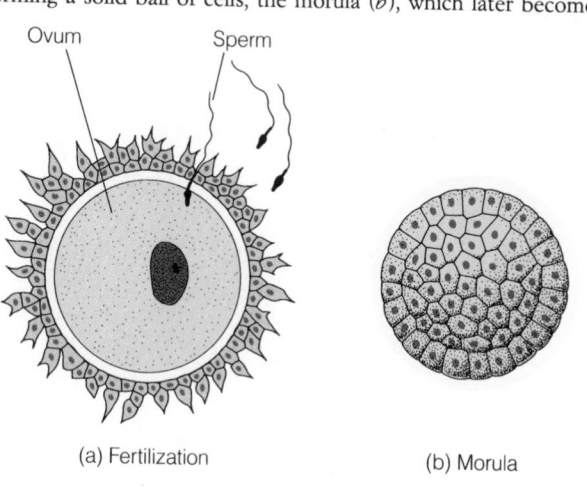

Ovum Sperm

(a) Fertilization

(b) Morula

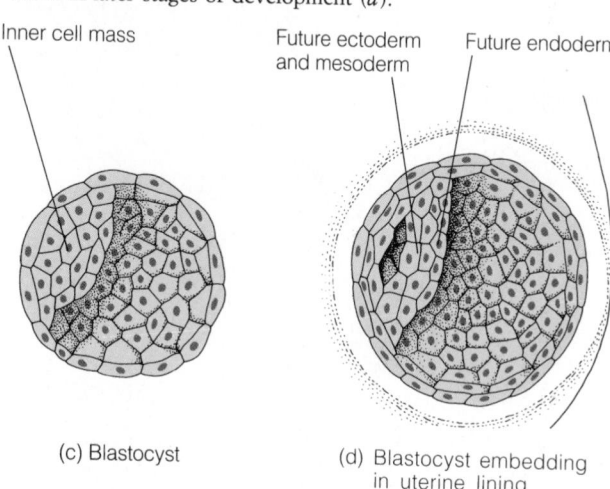

Inner cell mass

Future ectoderm and mesoderm Future endoderm

(c) Blastocyst

(d) Blastocyst embedding in uterine lining

sively in the heart. Skeletal muscle cells are located chiefly in body muscles. Smooth muscle cells are found in the walls of the stomach, intestinal tract, and blood vessels.

The primary tissues exist in all organs in varying amounts. The lining of the stomach, for example, consists of a single layer of epithelial cells called the surface epithelium (▷ Figure 10–2). Just beneath the lining is a layer of connective tissue. A thick sheet of smooth muscle cells forms the bulk of the stomach wall. Smooth muscle cells are also found in blood vessels supplying the tissues of the stomach. Nerves enter with the blood vessels and control the flow of blood.

Epithelium Forms the Lining or External Covering of Organs and Also Forms Glands

Epithelial tissue exists in two basic forms: glandular and membranous. The **membranous epithelia** consist of sheets of cells tightly packed together, forming the external coverings or linings of organs. ▷ Figure 10–3 shows some of the remarkable variety in epithelial membranes and illustrates the presence of two basic types of membranous epithelium: **simple epithelia,** consisting of a single layer of cells, and **stratified epithelia,** consisting of many layers of cells.

Glandular epithelia are clumps of cells that form many of the glands of the body. Epithelial glands arise during embryonic development from tiny invaginations of surface epithelia, as illustrated in ▷ Figure 10–4. Some glands remain connected to the epithelium by hollow ducts and

are called **exocrine glands** (glands of external secretion); products of the exocrine glands flow through ducts into some other body part. In humans, sweat glands in the skin are exocrine glands that produce a clear, watery fluid that is released onto the surface of the skin by small ducts. This fluid evaporates from the skin, helping cool the body. Salivary glands are another type of exocrine gland. Located around the oral cavity, the salivary glands produce saliva, a fluid that is released into the mouth via small ducts.

Some glandular epithelial cells break off completely from their embryonic source, as shown in Figure 10–4, to form **endocrine glands** (glands of internal secretion). The endocrine glands produce hormones that are released from

TABLE 10–1 The Primary Tissues and Their Subtypes	

Epithelial tissue
  Membranous
  Glandular
	Connective tissue
	  Connective tissue proper
	
	  Specialized connective tissue
	    Blood
	    Bone
	    Cartilage

Muscle tissue
  Cardiac
  Skeletal
  Smooth	

Nervous tissue
  Conductive
  Supportive	

▷ **FIGURE 10–2 Human Stomach** (*a*) The stomach, like all organs, contains all four primary tissues. (*b*) These are shown here in cross section.

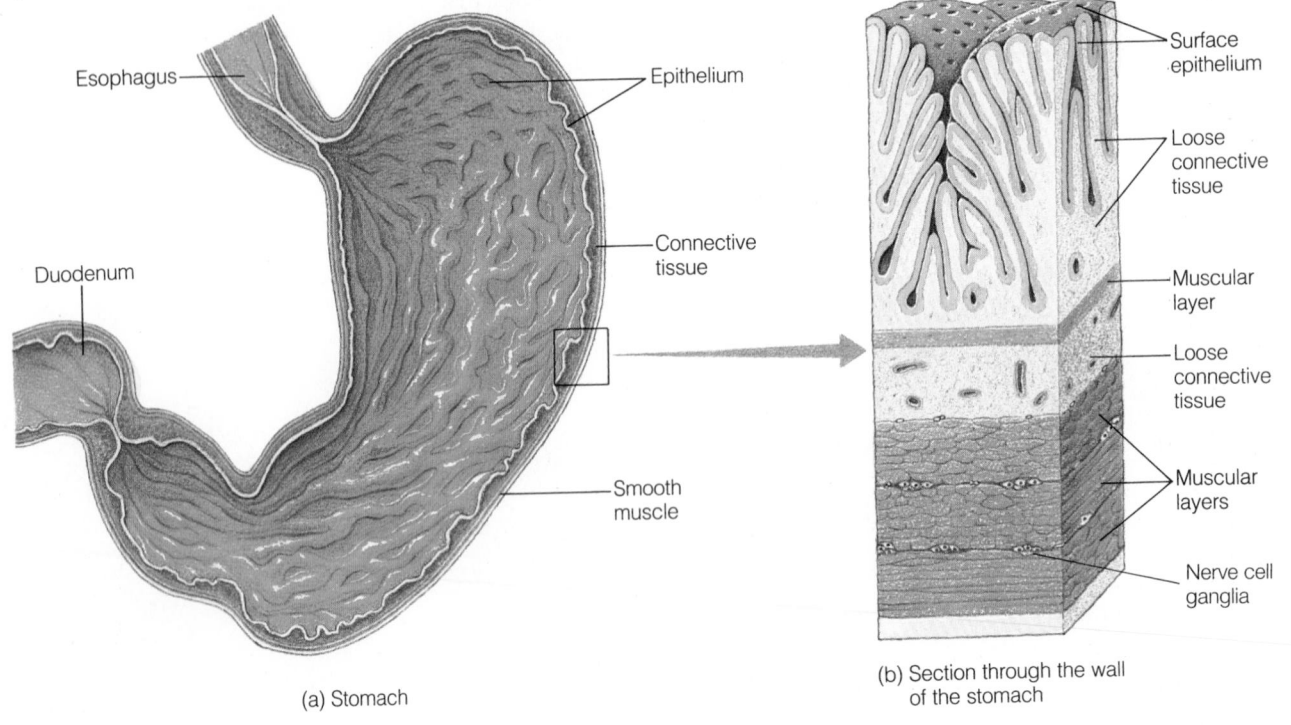

(a) Stomach

(b) Section through the wall of the stomach

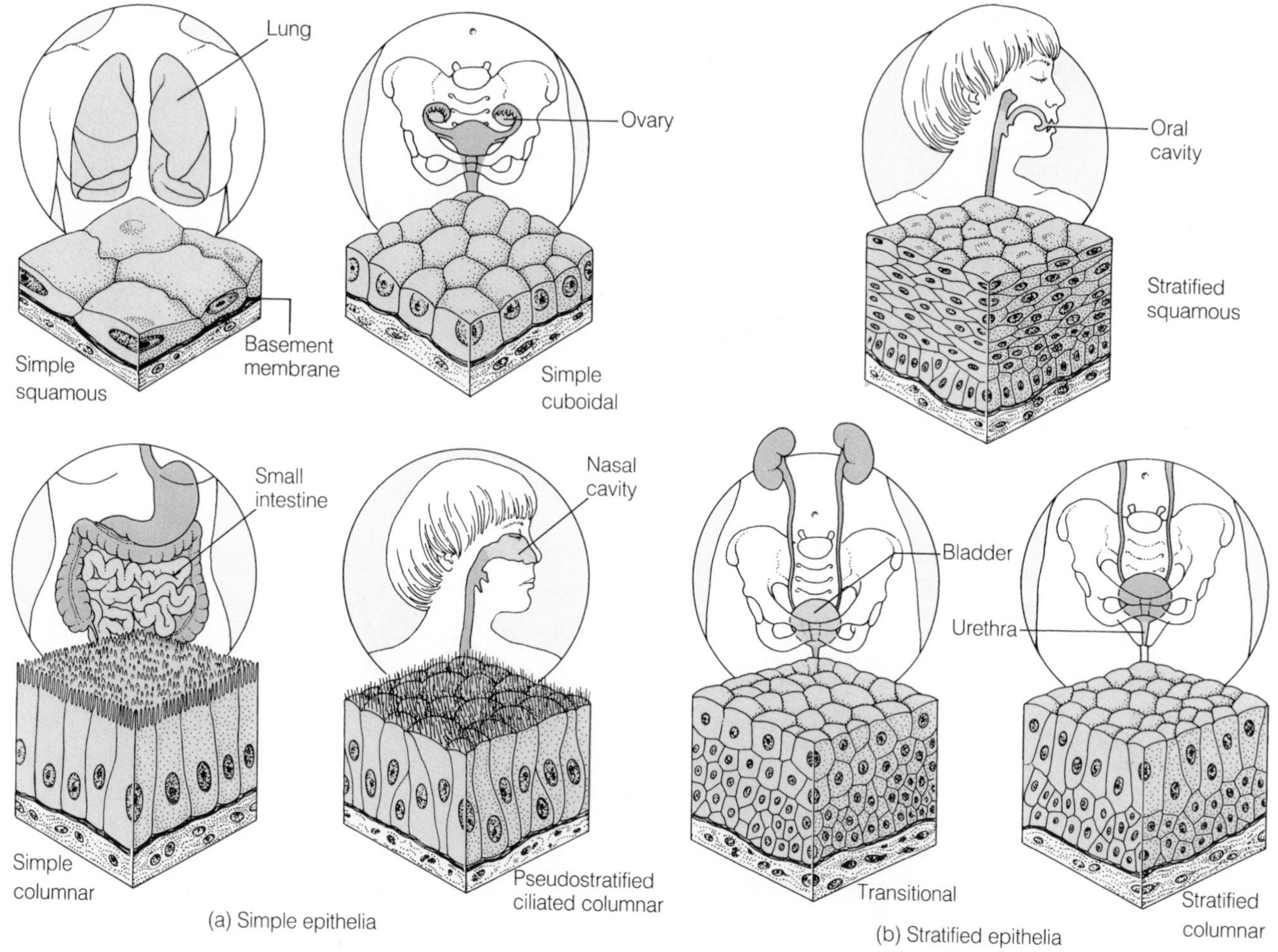

FIGURE 10–3 Membranous Epithelia

Simple squamous

Lung

Basement membrane

Simple cuboidal

Ovary

Stratified squamous

Oral cavity

Simple columnar

Small intestine

Pseudostratified ciliated columnar

Nasal cavity

(a) Simple epithelia

Transitional

Bladder

Urethra

Stratified columnar

(b) Stratified epithelia

▷ **FIGURE 10–3 Membranous Epithelia** Single-celled (simple) epithelia (*a*) and stratified epithelia (*b*) exist in different parts of the body.

the cell and diffuse into the bloodstream, where they travel to other parts of the body (Chapter 20).

Epithelial Tissues Illustrate the Basic Biological Principle That Structure Correlates with Function.
One of the basic rules of architecture is that form (the structure of a building) often follows function—in other words, architectural design reflects underlying function. Animals exhibit a similar relationship achieved through evolution.

The membranous epithelia provide many examples of the correlation between structure and function. A good example is the outer layer of the human skin, the **epidermis,** which, among other things, protects us humans from water loss. Shown in ▷ Figure 10–5, the epidermis con-

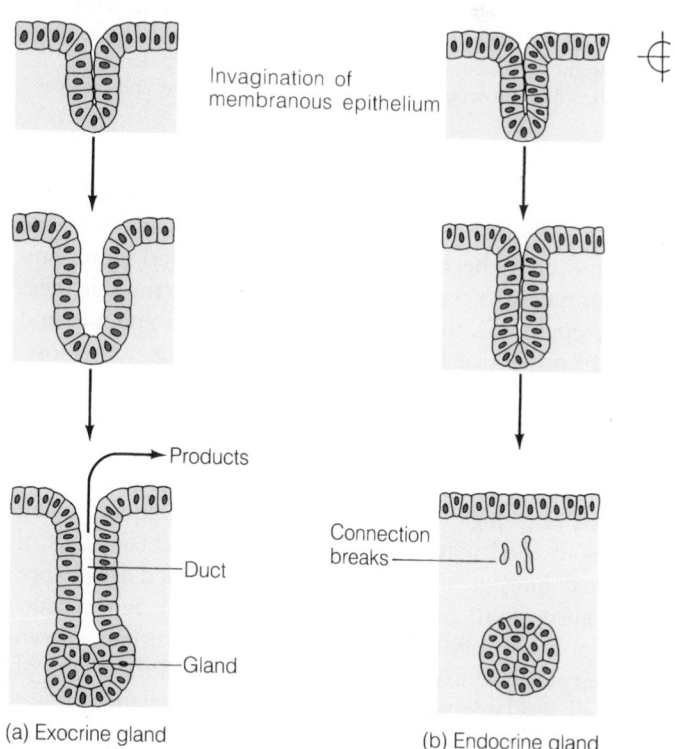

Invagination of membranous epithelium

Products

Duct

Gland

(a) Exocrine gland

Connection breaks

(b) Endocrine gland

▷ **FIGURE 10–4 Formation of Endocrine and Exocrine Glands** (*a*) Exocrine glands arise from invaginations of membranous epithelia that retain their connection. (*b*) Endocrine glands lose this connection and must secrete their products into the bloodstream.

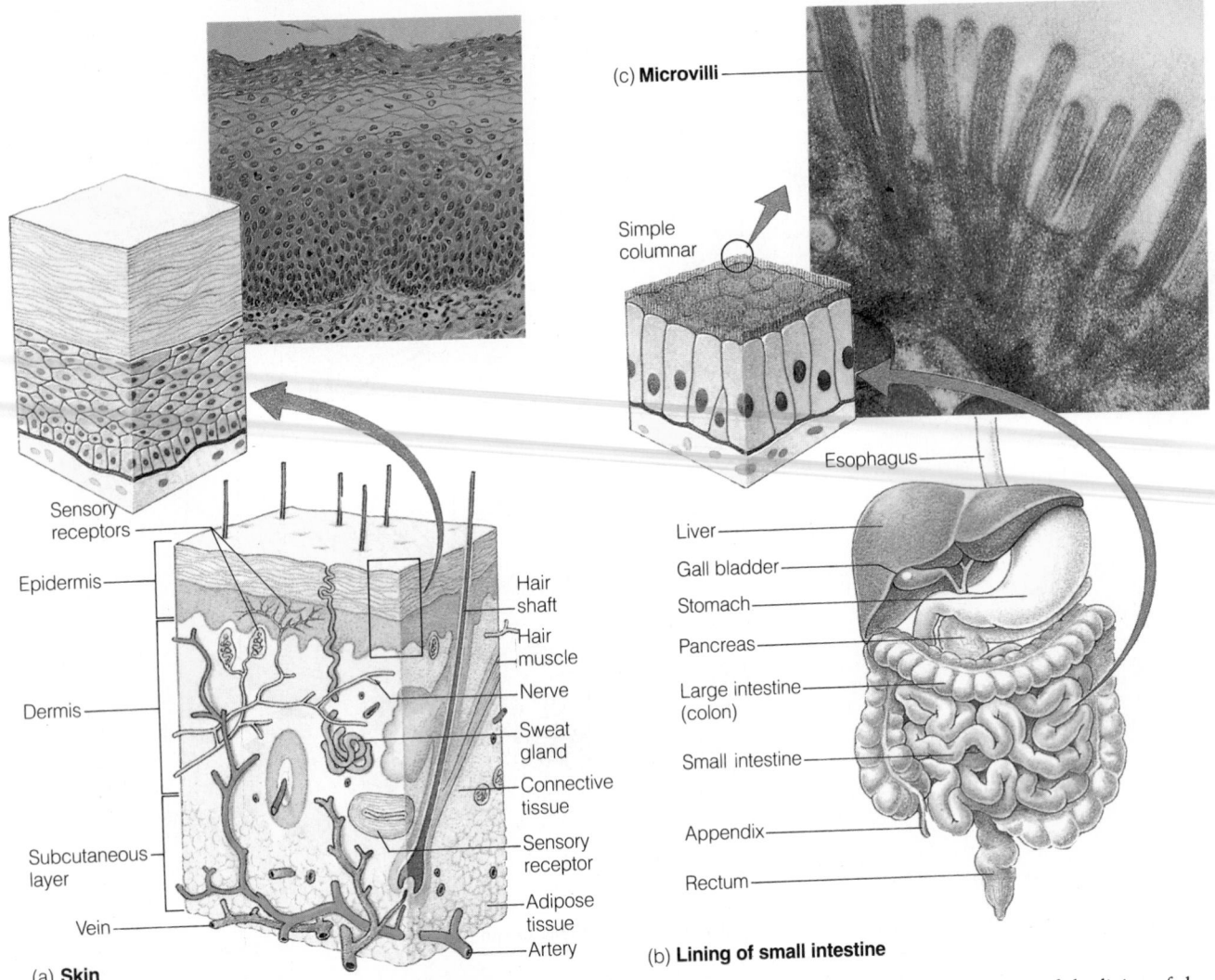

(c) **Microvilli**

Simple columnar

Sensory receptors

Epidermis

Dermis

Subcutaneous layer

Vein

Hair shaft

Hair muscle

Nerve

Sweat gland

Connective tissue

Sensory receptor

Adipose tissue

Artery

(a) **Skin**

Esophagus

Liver

Gall bladder

Stomach

Pancreas

Large intestine (colon)

Small intestine

Appendix

Rectum

(b) **Lining of small intestine**

▷ **FIGURE 10–5 Comparison of Two Epithelia with Markedly Different Functions** (*a*) A cross section of the skin showing the stratified squamous epithelium of the epidermis (above), which protects underlying skin from sunlight and dessica- tion, and (*b*) the simple columnar epithelium of the lining of the small intestine, which is specialized for absorption. (*c*) The plasma membranes of the cells lining the intestine are thrown into folds (microvilli) that greatly increase the surface area for absorption.

sists of numerous cell layers. The cells flatten toward the surface and are tightly joined by special connections. The outermost cells become isolated from the blood supply and die, forming a dry protective layer. Together, the thickness of the epidermis, the adhesion of one cell to another, and the dry protective layer of dead cells impede water loss. They also present a formidable barrier to microorganisms.

Another example of the relationship between structure and function is the epithelium of the small intestine. As shown in Figure 10–5b, the lining of the small intestine consists of a single layer of columnar cells, uniquely suited to absorb food materials. The columnar epithelial cells of the small intestine are also structurally modified to enhance food absorption. As illustrated in Figure 10–5c, the sur- faces of these cells have numerous tiny protrusions known as **microvilli,** which markedly increase the surface area of the cell available for absorption. The larger the surface area, the more efficient is food absorption.

Connective Tissue Binds the Cells and Organs of the Body Together

As the name implies, **connective tissue** is the body's glue. Connective tissues bind cells and other tissues together and are present in all organs in varying amounts.

The body contains several types of connective tissue, each with specific functions (▷ Figure 10–6). Despite the differences, all connective tissues consist of two basic components: cells and varying amounts of extracellular ma- terial. Two types of connective tissue will be discussed here: connective tissue proper and the specialized connec- tive tissues—bone, cartilage, and blood.

Connective Tissue Proper Consists of Two Sub- types, Determined by the Relative Proportion of Fibers and Cells. Connective tissue proper is an im- portant structural component of the vertebrate body and is

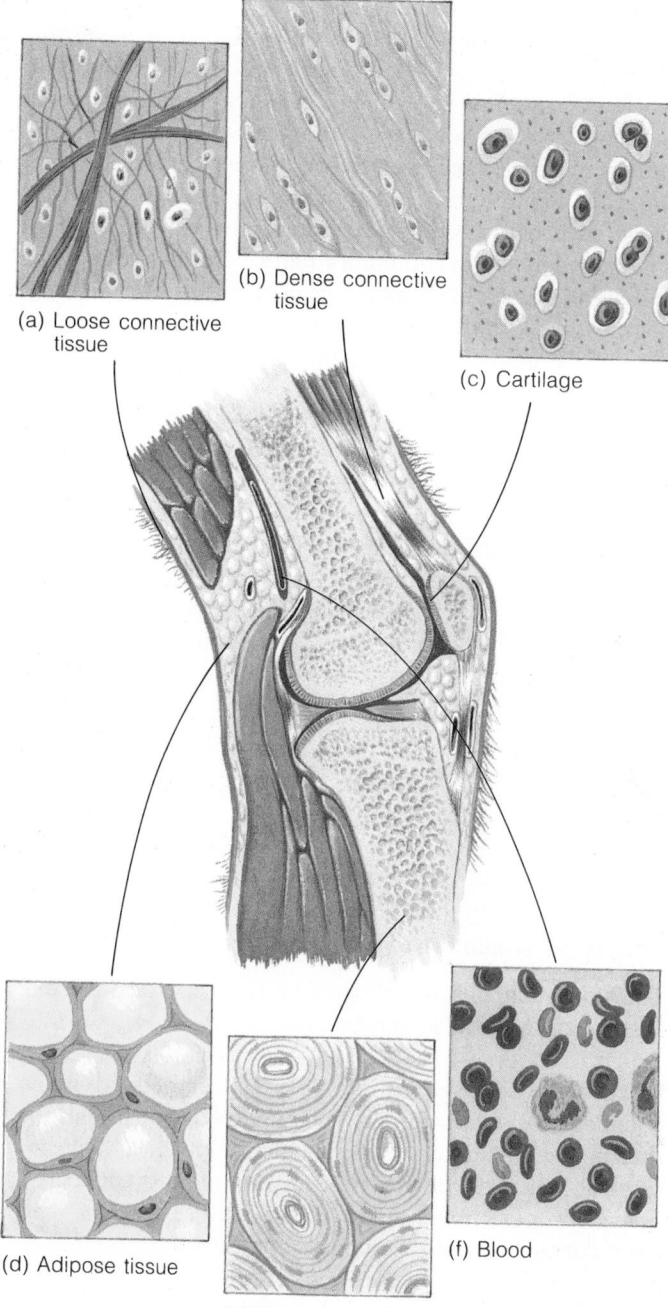

(a) Loose connective tissue

(b) Dense connective tissue

(c) Cartilage

(d) Adipose tissue

(e) Bone

(f) Blood

▷ **FIGURE 10–6 Connective Tissue** Connective tissue consists of many diverse subtypes.

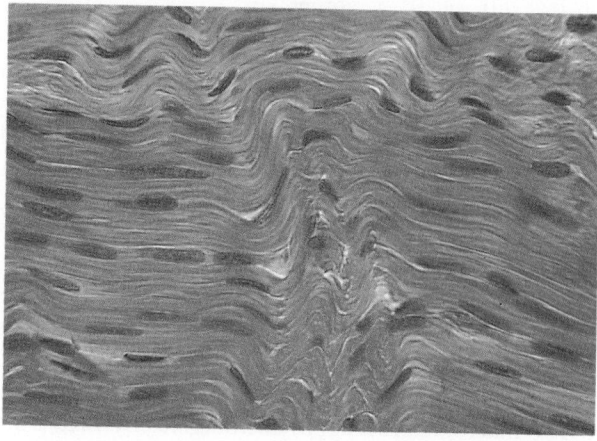

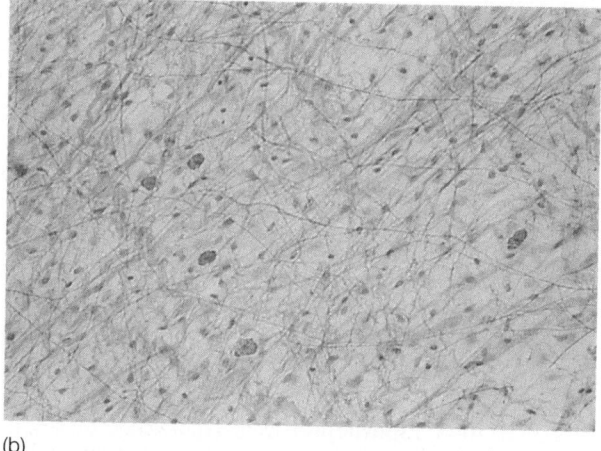

(b)

▷ **FIGURE 10–7 Light Micrographs of Connective Tissue** (*a*) Dense connective tissue. (*b*) Loose connective tissue.

composed of dense connective tissue and loose connective tissue. **Dense connective tissue (DCT)** consists primarily of densely packed fibers produced by cells interspersed between them. DCT is found in ligaments and tendons (▷ Figure 10–7a). Ligaments join bones to bones at joints and provide support for joints; tendons join muscle to bone and aid in body movement.[1] The layer of the skin underlying the epidermis, the **dermis,** is also dense connective

tissue, although the fibers are less regularly arranged than those in ligaments and tendons (see Figure 10–5a). The dermis binds the epidermis to underlying muscle and bone.

Loose connective tissue (LCT) is the body's packing material. As shown in Figure 10–7b, LCT contains cells in a loose network of collagen and elastic fibers. Both of these fibers are made of protein. Loose connective tissue forms around blood vessels in the body and in skeletal muscles, where it binds the muscle cells together. Loose connective tissue also lies beneath epithelial linings of the intestines and trachea, anchoring them to underlying structures.

The chief difference between dense and loose connective tissues lies in the ratio of cells to extracellular fibers. As you can see in Figure 10–7, dense connective tissue has far more fibers than loose connective tissue.

The extracellular fibers found in dense and loose connective tissue are produced by a connective tissue cell known as the **fibroblast.** In addition to producing the extracellular fibers that hold many tissues together, fibroblasts repair damage created by cuts or tears in body tissues. When the skin is cut, for example, fibroblasts in the dermis migrate into the injured area, where they begin producing large quantities of collagen. The collagen fibers

---

[1] I use the mnemonic *LBJ* to remember that ligaments connect bones at joints.

fill the wound, closing it off. The epidermis soon begins to grow over the damaged area, helping repair the damage and restoring the integrity of the skin. When the cut is small, the epidermis covers the damaged area completely, leaving no scar, but in larger wounds, the epidermal cells are often unable to grow over the entire wound, leaving some of the underlying collagen exposed and producing a scar.

Besides helping to hold the body together, loose connective tissue gives refuge to several cell types that protect us against bacterial and viral infections. One of the most important of the protective cells is the **macrophage** ("big eater"). Containing numerous lysosomes to digest foreign material, macrophages engulf microorganisms that penetrate the skin and underlying loose connective tissues as a result of injury. This helps prevent bacteria from spreading to other parts of the body and producing a systemic infection—that is, one that affects the entire organism. Macrophages also play a role in immune protection, which will be discussed in Chapter 15.

Loose connective tissues also contain lymphocytes and neutrophils, two types of blood cells that play an important role in protecting the body from invaders described in later chapters.

Some loose connective tissues contain large, conspicuous fat cells. The **fat cell** is one of the most distinctive of

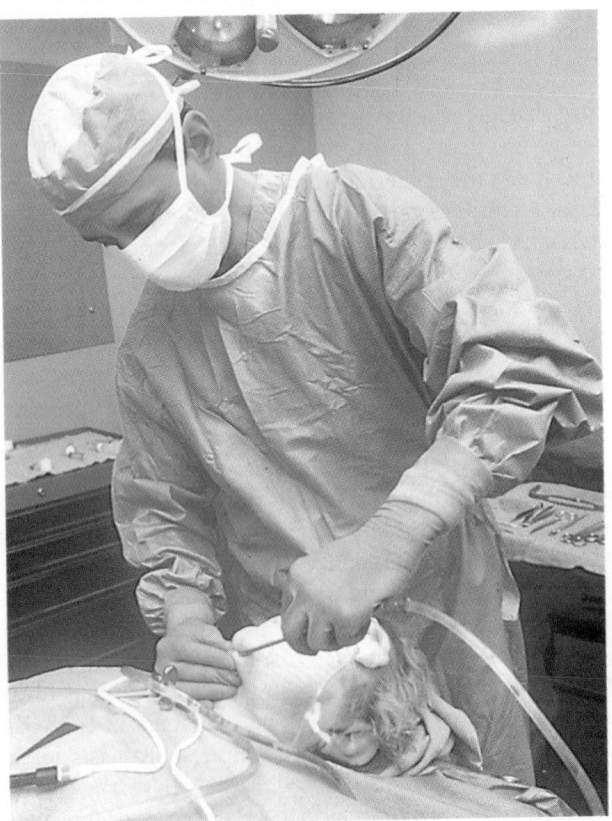

▷ **FIGURE 10–8 Liposuction** During liposuction surgery, the physician aspirates fat from deposits lying beneath the skin, helping to reduce unsightly accumulations.

all body cells. When fully formed, it contains huge fat globules that occupy virtually the entire cell, pressing the cytoplasm and the nucleus to the periphery. Fat cells occur singly or in groups of varying size. Large numbers of fat cells in a given region form a modified type of loose connective tissue known as **adipose tissue,** or, less glamorously, fat. Adipose tissue is an important storage depot for lipids, particularly triglycerides, which are used as an energy source under certain conditions. Fatty deposits also provide a degree of heat-conserving insulation for humans and many other mammals and some birds, especially those that live in aquatic environments where heat loss can be substantial.

For humans, fatty deposits often become unsightly. To rid the body of these deposits, individuals can begin exercise programs and reduce their intake of food. The combination of the two can prove quite effective. Others opt for a surgical measure called **liposuction** (▷ Figure 10–8). In this procedure, a small incision is made in the skin through which surgeons insert a suction device to aspirate fat deposits under the skin in various locations, such as the thighs, buttocks, and abdomen. The fat cells extracted from one region can even be transferred to other regions, such as the breast, to resculpt the human body. Liposuction is a relatively safe technique, but it is not free from risk.

The Specialized Connective Tissues Are Structurally and Functionally Modified to Perform Specific Functions Essential to Homeostasis

The body contains three types of specialized connective tissue: cartilage, bone, and blood.

Cartilage. Cartilage consists of cells embedded in an abundant and rather impervious extracellular material, the **matrix** (▷ Figure 10–9). Surrounding virtually all types of cartilage is a layer of dense, irregularly packed connective tissue, the **perichondrium** ("around the cartilage"). This layer contains the blood vessels that supply nutrients to cartilage cells through diffusion. No blood vessels penetrate the cartilage itself. Because cartilage cells are nourished by diffusion from perichondral capillaries, damaged cartilage heals very slowly. For this reason, joint injuries that involve the cartilage often take years to repair or may not heal at all.

Three types of cartilage are found in vertebrates: hyaline, elastic, and fibrous. Each serves a special purpose, which is reflected in its underlying structure. The most prevalent type of cartilage is **hyaline cartilage** (Figure 10–9a). Hyaline cartilage contains numerous collagen fibers, which appear white to the naked eye. Found on the ends of many bones, hyaline cartilage greatly reduces friction, so bones move over one another with ease. It also makes up the bulk of the nose and is found in the larynx (voice box) and the rings of the trachea, which you can feel below the larynx. The ends of the ribs that join to the

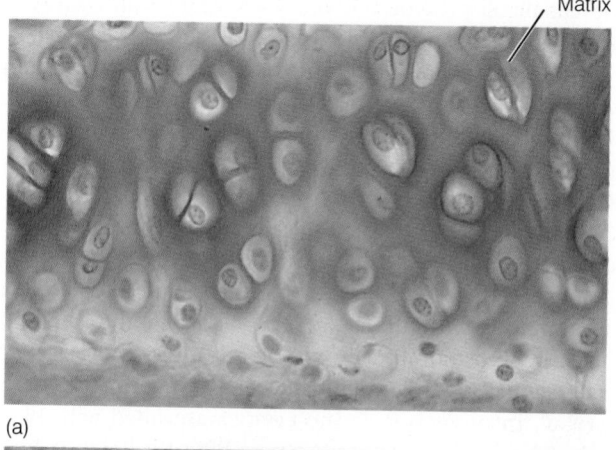

Matrix

(a)

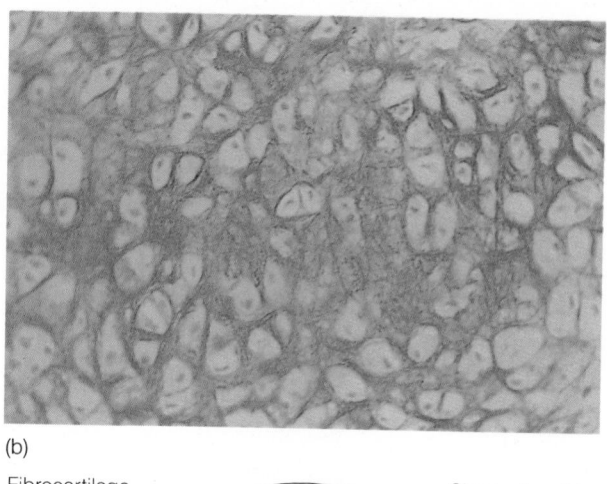

(b)

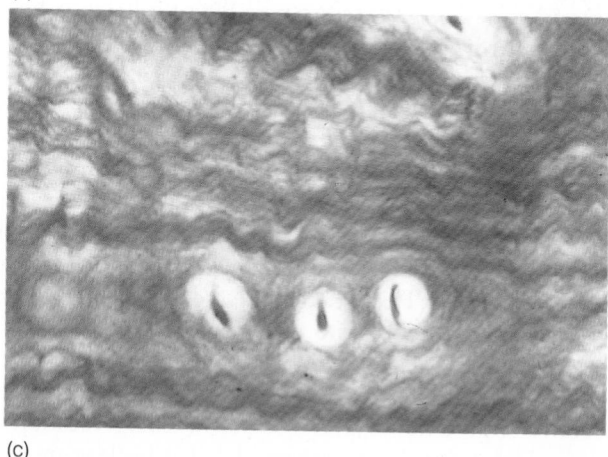

(c)

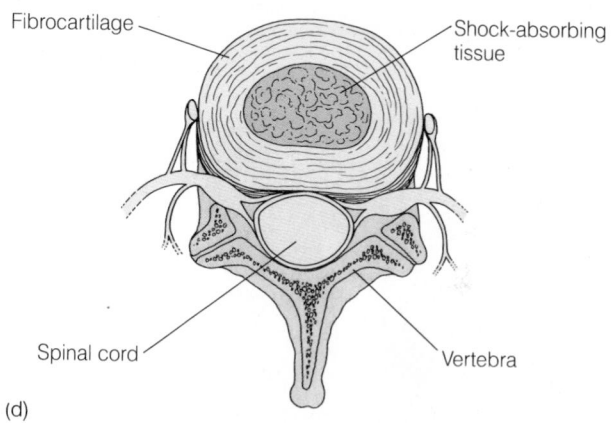

Fibrocartilage

Shock-absorbing tissue

Spinal cord

Vertebra

(d)

▷ **FIGURE 10–9 Light Micrographs of Cartilage** (*a*) Hyaline. (*b*) Elastic. (*c*) Fibrous. (*d*) Intervertebral disk showing arrangement of fibrocartilage in a protective ring around the soft, spongy part of the disk that absorbs shock.

sternum (breast bone) are composed of hyaline cartilage. In embryonic development, the first skeleton is hyaline cartilage, much of which is later converted to bone.

Elastic cartilage is similar to hyaline cartilage but contains many wavy elastic fibers, which give it much greater flexibility (Figure 10–9b). Elastic cartilage is found in regions where support and flexibility are required—for example, the external ears and eustachian tubes, cartilaginous ducts that help equalize pressure in the inner ear (Chapter 18).

Fibrocartilage is the rarest of all cartilage. Like hyaline cartilage, it consists of an extracellular matrix containing numerous bundles of collagen fibers. As shown in Figure 10–9c, however, fibrocartilage contains far fewer cells than either hyaline or elastic cartilage.

Fibrocartilage is found in the outer layer of **intervertebral disks,** the shock-absorbing tissue between the vertebrae of the spine (Figure 10–9d). An intervertebral disk consists of a soft, cushiony central region that absorbs shock. Fibrocartilage forms a ring around the central portion of the disk, holding it in place. Over time, the fibrocartilage ring may weaken or tear, so that the central part of

the disk bulges outward (herniates). Referred to as a slipped or herniated disk, this condition may result in a significant amount of pain in the neck, back, or one or both legs, depending on the location of the damaged disk. Pain is generated when the disk presses against nearby nerves. Surgeons can correct the problem by removing the herniated portion of the disk. You can reduce your chances of "slipping" a disk in the first place by watching your weight, sitting upright (not slouching) in a chair, and lifting heavy objects carefully (▷ Figure 10–10).

Bone. Bone is another form of specialized connective tissue. Contrary to what many might think, bone is a dynamic living tissue. Besides providing internal support and protecting internal organs such as the brain, heart, and lungs, bone plays an important role in maintaining optimal blood calcium levels and is therefore a homeostatic organ. Calcium is required for many body functions: muscle contraction, normal nerve functioning, and even blood clotting.

Like all connective tissues, bone consists of cells embedded in an abundant extracellular matrix (▷ Figure

Correct

Incorrect

▷ **FIGURE 10–10 Protecting Your Back** When lifting heavy objects, bend your legs, and grasp the object. With your back straight, stand up, lifting with your legs.

10–11a). Bone matrix consists primarily of collagen fibers, which give bone its strength and resiliency, interspersed with numerous needlelike salt crystals containing calcium, phosphate, and hydroxide ions, which give bone its hardness. Calcium in bone can be dissolved by weak acids, leaving behind a collagen replica that can be turned into a thick paste, called **demineralized bone matter (DBM)**. DBM is rather remarkable stuff, and it is being used to repair severe bone damage, as described in Health Note 10–1.

Two types of bone tissue are found in the body: compact bone and spongy bone (Figure 10–11a). Compact bone is dense and hard. As illustrated in the photomicrograph in Figure 10–11b, the cells in compact bone (**osteocytes**) are located in concentric rings of calcified matrix. These surround a **central canal** through which the blood vessels and nerves pass. Each osteocyte is endowed with numerous processes that course through tiny canals in the bony matrix, known as **canaliculi** (literally, "little canals").

The canaliculi provide a route for nutrients and wastes to flow to and from the osteocytes and the central canal.

Inside most bones of the body is a tissue known as spongy bone. **Spongy bone** consists of an irregular network of calcified collagen spicules (Figure 10–11b). As illustrated in Figure 10–11c, on the surface of the spicules are numerous **osteoblasts**, bone cells that produce collagen, which later becomes calcified. Once these cells are surrounded by calcified matrix, they are referred to as osteocytes.

Between the spicules are numerous cavities, which often communicate with much larger cavities in the center of bones. In most of the bones of an adult, the large and small cavities are filled with fat cells and form the **yellow marrow**. In other bones, the cavities are filled with blood cells and cells that give rise to new blood cells, thus forming **red marrow**, described in later chapters.

On the surfaces of many bony spicules of spongy bone are large, multinucleated cells called **osteoclasts** ("bone breakers") (Figure 10–11d). These cells are part of a homeostatic system that ensures proper blood calcium levels. When calcium levels in the blood fall, osteoclasts are activated by the parathyroid hormone produced by the thyroid gland. So activated, the osteoclasts digest small portions of the spongy bone. This in turn releases calcium into the bloodstream and helps restore proper levels.

Spongy bone is also remodeled when bones are subjected to new stresses. For example, the leg bones of a desk-bound executive from Atlanta who goes on a skiing vacation in Jackson Hole, Wyoming, are remodeled as his legs are subjected to the rigors of skiing. This adjustment accommodates for the new stresses and strains. During this process, osteoclasts tear down some of the spongy bone, while osteoblasts rebuild new bone in other areas to meet the new stresses. By the end of the two-week ski trip, the executive's bones have been considerably refashioned and are much stronger than when he left home. When he is back at his desk, his bones revert to their previous, weaker state.

Blood. Blood is also a specialized form of connective tissue and consists of two components: (1) the formed elements (cells and platelets) and (2) a large amount of extracellular material, a fluid called **plasma** (▷ Figure 10–12).

The formed elements of blood consist of the red and white blood cells and the platelets. **Red blood cells**, or **erythrocytes**, transport oxygen and small amounts of carbon dioxide to and from the lungs and body tissues. **White blood cells**, or **leukocytes**, are involved in fighting infections. **Platelets** are fragments of large cells (megakaryocytes) that are located in the red bone marrow, the principal site of blood cell formation. Platelets play a key role in blood clotting.

Muscle Tissue Consists of Specialized Cells That Contract When Stimulated

Muscle, the third in our series of primary tissues, is found in virtually every organ in the body. Muscle gets its name

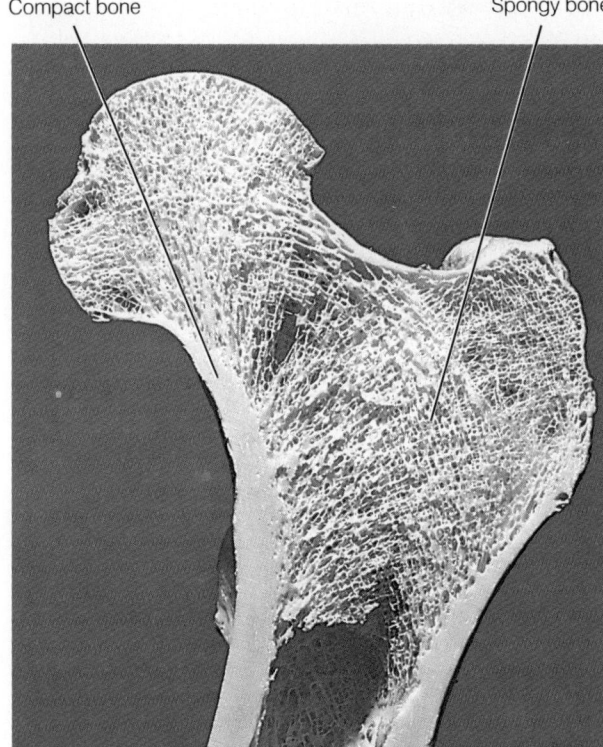

Compact bone   Spongy bone

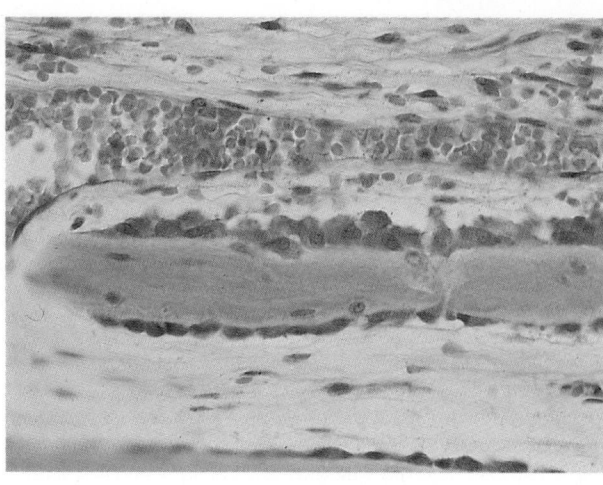

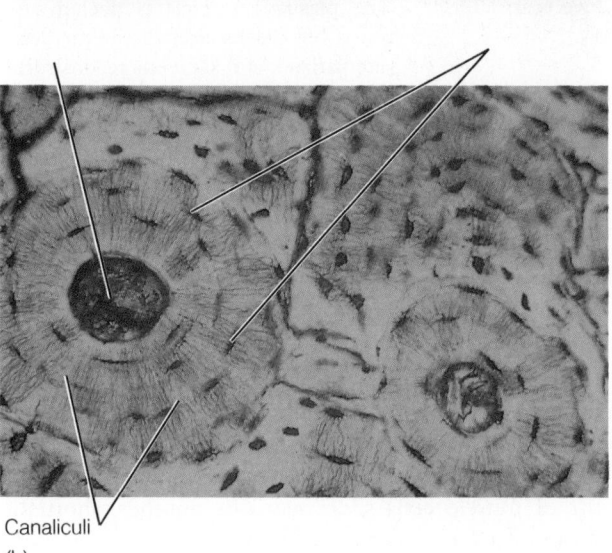

Canaliculi

(b)

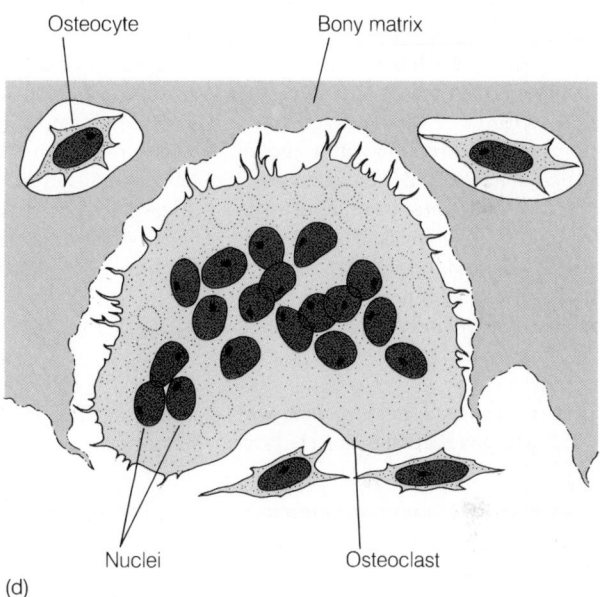

Osteocyte   Bony matrix

Nuclei   Osteoclast

(d)

▷ **FIGURE 10–11 Bone** (*a*) Compact and spongy bone, shown in a section of the humerus. (*b*) Light micrograph of compact bone. (*c*) Photomicrograph of spongy bone, showing osteoblasts. (*d*) Osteoclast digesting surface of bony spicule.

from the Latin word for "mouse" (*mus*). Early observers likened the contracting muscle of the biceps to a mouse moving under a carpet.

Like nervous tissue, muscle demonstrates a characteristic of life described in the first chapter, irritability—that is, the ability to perceive and respond to stimuli. Muscle is therefore an excitable tissue that, when stimulated, is capable of contracting, producing mechanical force. Muscle cells working in large numbers can create enormous forces. Muscles of the jaw, for instance, create a pressure of 200 pounds per square inch, forceful enough to snap off a finger. (Don't try this at home.) Muscle also moves body

parts, propels food along the digestive tract, and expels the fetus from the uterus during birth. Muscle of the heart contracts and pumps blood through the 50,000 miles of blood vessels in the human body. Acting in smaller numbers, muscle cells may be responsible for more intricate movements, such as those required to play the piano or move the eyes.

As mentioned earlier, three types of muscle are found in humans: skeletal, cardiac, and smooth. The cells in each type of muscle contain two types of contractile protein filaments, **actin** and **myosin.** These very same fibers were first encountered in your study of biology in the microfila-

HEALTH NOTE 10-1

REMAKING THE HUMAN BODY

David Eastland was driving home from work late one evening when he fell asleep at the wheel. His car swerved off the road, striking a tree. The impact of the accident crushed Eastland's right leg, destroying a large segment of his femur (thigh bone).

In earlier times, this accident would have cost Eastland his leg, because surgeons would have had no way of repairing such extensive bone damage. Even bone grafts were insufficient in such instances. Today, however, thanks to advances in medical science, surgeons can literally remake the human skeleton using a material called demineralized bone matter (DBM). Demineralized bone matter consists primarily of a protein known as collagen. DBM is produced by immersing the bones from human cadavers and other animals in a weak acid solution. As noted in the text, this treatment dissolves the mineral matter of the bone, leaving behind a thick, rubbery paste, which can be used to repair severe bone damage.

After replacing the missing bone with DBM, surgeons actually inject bone cells into it. These cells, usually derived from small fragments taken from a patient's hip bone, grow and divide in the DBM, eventually converting the implant into a functional bone, practically indistinguishable from normal bone.

Demineralized bone matter can also be used to repair birth defects and to replace segments of bone that have been surgically removed because of bone cysts, tumors, or severe infections. One of the most remarkable success stories is that of a boy born with a disease known as cloverleaf syndrome. In this disease, growth in the bones of the skull fails to keep pace with the brain's growth, eventually killing its victims. Dr. John Mulliken, a surgeon at Boston's Children's Hospital, removed the skull of a 6-year-old boy suffering from the disease. He then fashioned a new and larger skull from DBM. Five years later, the boy was alive and well.

Artificial Skin

Another promising development is artificial skin. Researchers at the Massachusetts Institute of Technology have developed an artificial skin that can be applied to burned areas, helping to reduce fluid loss, prevent infection, and promote faster recovery.

Artificial skin consists of two layers that resemble the body's natural skin. The lower layer, the artificial dermis, is made of collagen fibers extracted from cowhide and a substance extracted from the cartilaginous skeleton of sharks. The upper layer, the artificial epidermis, is made of plastic.

In severely burned patients, physicians clean away the burned flesh, then sew the artificial skin in place. Within a short time, fibroblasts and blood vessels from the surrounding region invade the artificial dermis. Over a period of several months, these cells produce new collagen fibers, replacing the fibers of the implant.

The plastic epidermis is peeled off within a few weeks after the dermis has revascularized, and surgeons reconstruct a new cellular epidermis using some of the patient's own epidermal cells, which are grafted in thin sheets onto the underlying dermis.

Like DBM, artificial skin is relatively

▷ **FIGURE 10–12 Blood** Blood is about 55% liquid (plasma) and 45% formed elements: red blood cells, white blood cells, and platelets.

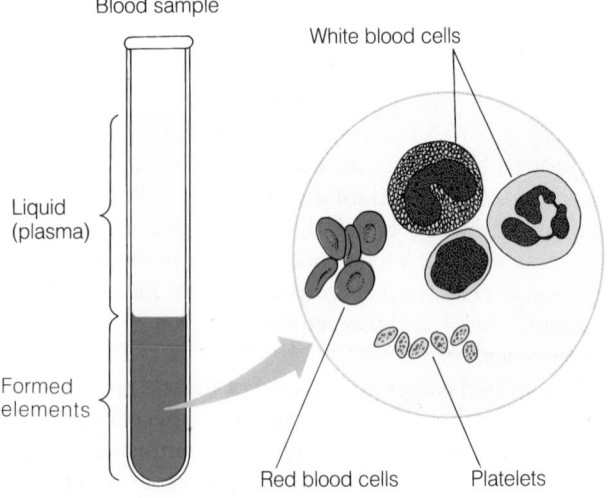

Blood sample

White blood cells

Liquid (plasma)

Formed elements

Red blood cells

Platelets

mentous network lying beneath the plasma membrane of animal cells that is responsible for cytokinesis, the division of the cytoplasm. When stimulated, actin and myosin filaments of muscle cells slide over one another, shortening them and causing contraction.

Skeletal Muscle. The majority of the body's muscle is called **skeletal muscle**—so named because it is frequently attached to the skeleton. When skeletal muscle contracts, it causes body parts (arms and legs, for instance) to move. Most skeletal muscle in the body is under voluntary, or conscious, control. Signals from the brain brought about by thoughts cause the muscle to contract. A notable exception is the skeletal muscle of the upper esophagus, which contracts automatically during swallowing.

Skeletal muscle cells are long cylinders formed during embryonic development by the fusion of many embryonic muscle cells. Because of this, skeletal muscle cells are usually referred to as **muscle fibers.** Each muscle fiber contains many nuclei (▷ Figure 10–13a). Because of the

easy to make and transplant. New skin grows in without the severe scarring that accompanies more conventional burn treatment.

Rebuilding Ailing Hearts

Each year, 400,000 people in the United States develop end-stage heart failure, aserious weakening of the heart. This disease results from the death of cardiac muscle cells caused by excess stress on a heart whose arteries are clogged with cholesterol. Nearly all of these people die within a year.

Many patients with end-stage heart failure could benefit from heart transplants, but because of a lack of donors, only 1400 heart transplants were performed in 1987. Artificial hearts may provide some hope, but they are costly and plagued with problems, not the least of which is that they strap patients to a machine for the rest of their lives.

One promising technique that could benefit thousands of patients each year is known as the skeletal muscle wrap. As ▷ Figure 1 illustrates, in this procedure a segment of the latissimus dorsi, a large muscle of the back, is cut loose, then wrapped around the diseased heart. The muscle is stapled in place, forming a contractile basket. Then it is attached to a pacemaker, which senses the heart's natural electrical pulses and delivers bursts of electricity to the skeletal muscle, causing it to contract in step with the heart. This technique leaves the blood supply and nerve supply of the muscle intact, so there is little worry over graft survival. Skeletal muscle wraps improve cardiac function in patients by about 20%, a change that means the difference between living as an invalid and leading a fairly normal life.

Using a patient's own skeletal muscle has many benefits. First, it is readily available, and in most cases, patients are more than willing to "donate" tissue for their own survival. Second, because the graft is taken from a person's own body, surgeons do not have to worry about the immune system rejecting the graft. Promising as this technique is, it is no substitute for prevention of heart disease.

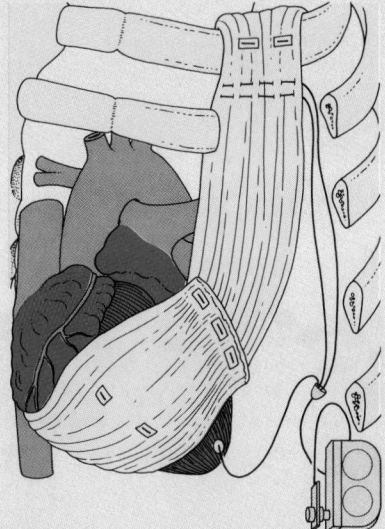

▷ **FIGURE 1 Skeletal Muscle Wrap** Patients with ailing hearts may be aided by skeletal muscle wrapped around the failing organ. Pacemakers can stimulate the muscle to beat in step with the heart muscle, thus increasing cardiac output.

dense array of contractile fibers in the cytoplasm of the muscle fiber, the nuclei are generally pressed against the plasma membrane. As you might suspect, this highly specialized cell cannot divide, and damaged muscle cells cannot be replaced.

Skeletal muscle fibers appear banded, or **striated,** when viewed under the light microscope, as illustrated in Figure 10–13a. The striations result from the unique arrangement of actin and myosin filaments inside muscle cells, a topic discussed in Chapter 19.

Cardiac Muscle. Like skeletal muscle, **cardiac muscle** is striated (Figure 10–13b). Unlike skeletal muscle, cardiac muscle is involuntary; that is, it contracts without conscious control. Found only in the walls of the heart, cardiac muscle cells contain a single nucleus. They also branch and interconnect freely, but individual cells are tightly connected to one another, an adaptation that helps maintain the structural integrity of the heart, an organ subject to incredible strain as it pumps blood through the body day and night for years on end. The points of connection also provide pathways for electrical impulses to travel from cell to cell, allowing the heart muscle to contract uniformly when stimulated.

Smooth Muscle. The third and final type of muscle is smooth muscle. **Smooth muscle,** so named because it lacks visible striations, is involuntary. Actin and myosin filaments are present but are not organized like those found in other types (Figure 10–13c). Smooth muscle cells may occur singly or in small groups. Small rings of smooth muscle cells, for example, surround tiny blood vessels. When these cells contract, they shut off or reduce the supply of blood to tissues. Smooth muscle cells are most often arranged in sheets in the walls of organs, such as the stomach, uterus, and intestines (see Figure 10–2). Smooth muscle cells in the wall of the stomach churn the food, mixing the stomach contents, and force tiny spurts of liquified food into the small intestine. Smooth muscle contractions also propel the food along the intestinal tract.

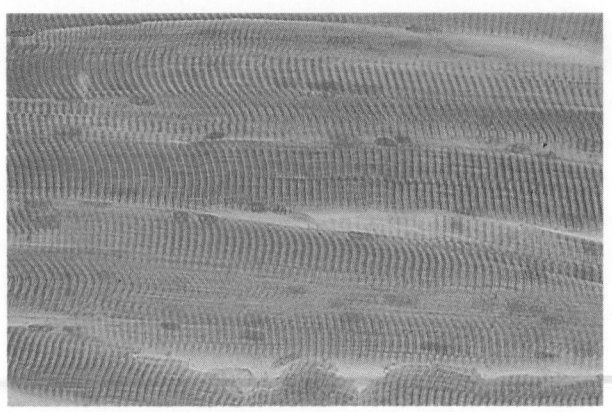

(a)

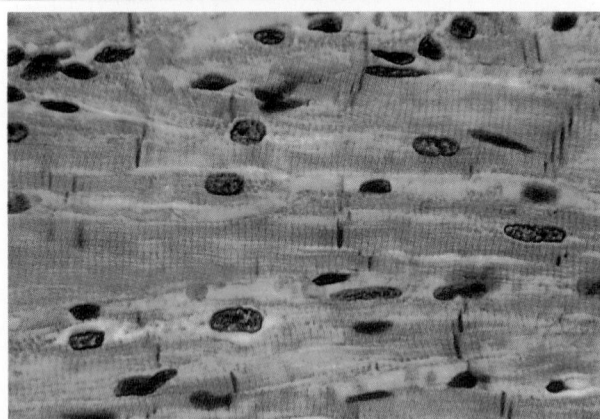

(b)

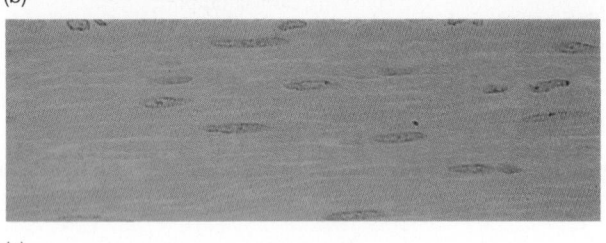

(c)

▷ **FIGURE 10–13 Light Micrograph of the Three Types of Muscle** (*a*) Skeletal. (*b*) Cardiac. (*c*) Smooth.

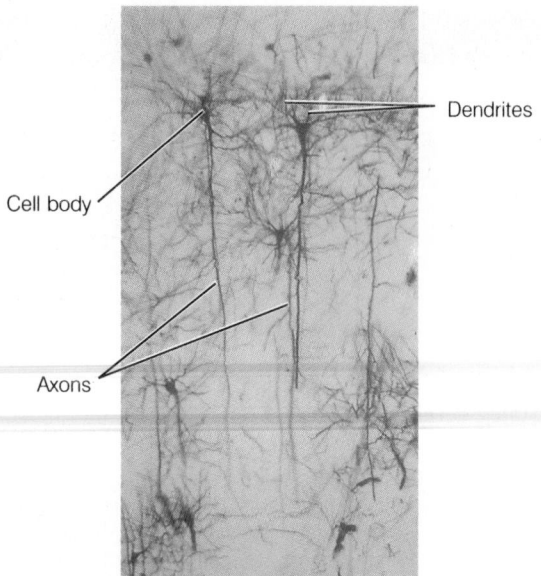

▷ **FIGURE 10–14 Multipolar Neuron** Attached to the cell body of the multipolar neuron are many highly branched dendrites, which deliver impulses to the cell body. Multipolar neurons have one long, unbranching fiber called the axon, which transmits impulses away from the cell body.

Nervous Tissue Contains Specialized Cells Characterized by Irritability and Conductivity

Last but not least of the primary tissues is nervous tissue. **Nervous tissue** consists of two types of cells: conducting cells and supportive cells. Many of the conducting cells, known as **neurons,** are modified to respond to specific stimuli, like pain or temperature. Stimulation results in bioelectric impulses, which the neuron transmits from one region of the body to another. As noted above, the ability to respond to stimuli is a characteristic of all living things and is called irritability. The ability to transmit an impulse is called **conductivity.** The properties of irritability and conductivity in neurons allow animals to be aware of their environment and to respond to a variety of internal and external stimuli. The evolution and refinement of the nervous system has been vitally important to the emergence of the more complex organisms in the animal kingdom.

The supportive cells of the nervous system are a kind of nervous system connective tissue. These cells are incapable of conducting impulses, but they help transport nutrients from blood vessels to neurons and help guard against toxins by creating a barrier to many potentially harmful substances. As you will see in later chapters, the supportive cells also help increase the rate of conduction in neurons with which they are associated. Together, the neurons and their supporting cells combine to form the brain, spinal cord, and nerves of the nervous system, which are described in more detail in Chapters 17 and 18.

Although Three Types of Neurons Can Be Found in the Body, They Share Many Characteristics. At least three distinct types of neurons are found in the body. Despite their differences, they have several common features. We will study these similarities by looking at one of the most common nerve cells, the **multipolar neuron,** shown in ▷ Figure 10–14. The multipolar neuron contains a prominent, multiangular **cell body** to which are attached several short, highly branched processes, known as **dendrites.** (It is the presence of these processes that give this neuron its name, multipolar.) The dendrites receive impulses from receptors or other neurons and transmit them to the cell body. Also attached to the cell body of the multipolar neuron is a large, fairly thick process, the **axon,** which transports bioelectric impulses away from the cell body.

Like muscle cells, nerve cells are highly differentiated and cannot divide. When a cell is destroyed, it degenerates and cannot be replaced by cell division. A cut nerve axon, however, may partially regenerate. A new axon may grow

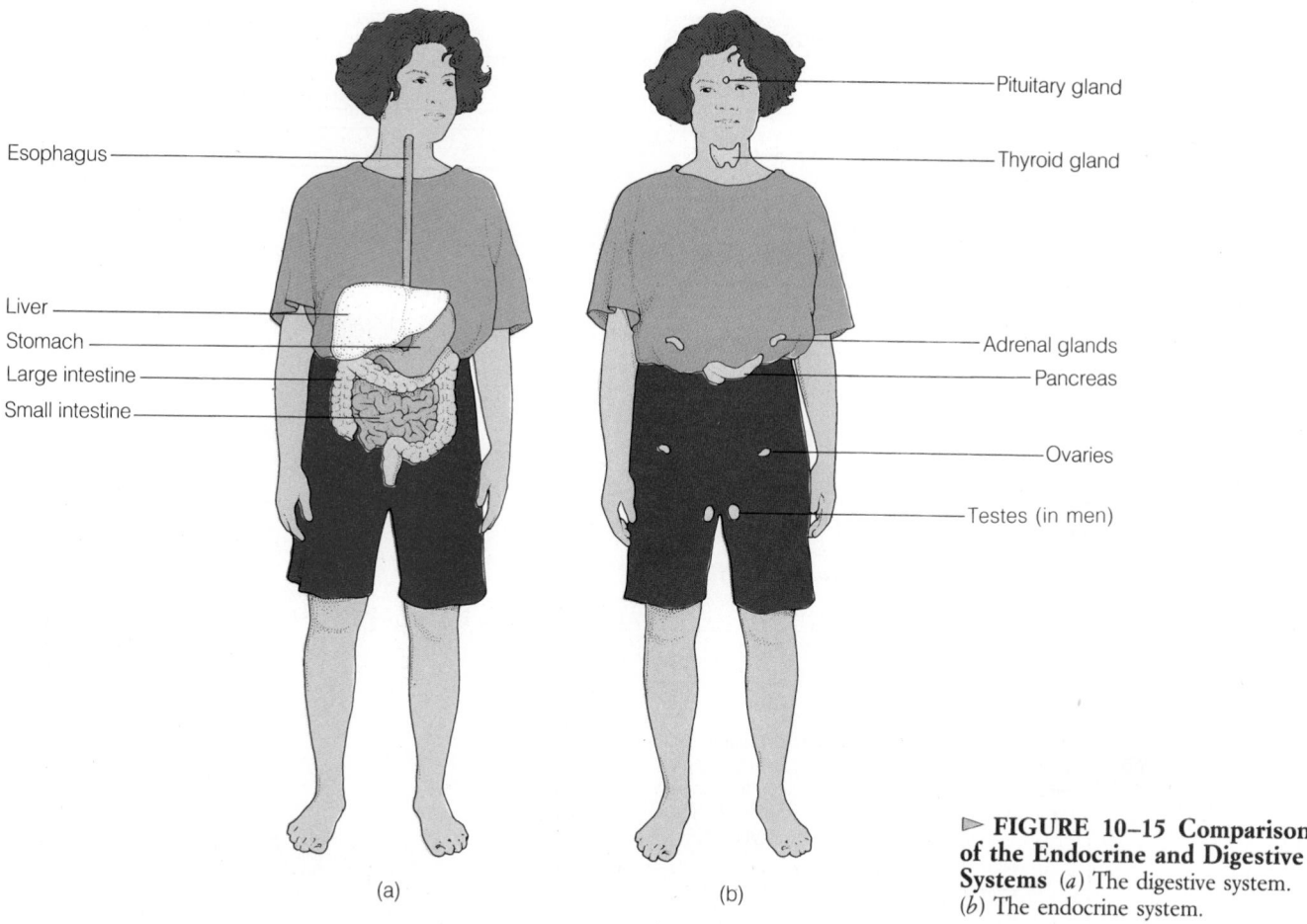

Esophagus

Liver
Stomach
Large intestine
Small intestine

Pituitary gland

Thyroid gland

Adrenal glands
Pancreas

Ovaries

Testes (in men)

(a)   (b)

▷ **FIGURE 10–15 Comparison of the Endocrine and Digestive Systems** (*a*) The digestive system. (*b*) The endocrine system.

from the damaged end, reestablishing previous connections and restoring some sensation or control over muscles. New research may someday provide ways to stimulate regeneration of nerve cells artificially, thus helping physicians more fully restore nerve function to victims of accidents.

Organs Often Function in Groups Called Organ Systems

The cell contains organelles ("little organs") that carry out many of its functions in isolation from the biochemically active cytoplasm. Compartmentalization is an evolutionary adaptation that also occurs at the organismic level in organs. Discrete structures called **organs** evolved to perform specific functions, such as digestion, enzyme production, and hormone synthesis. Most organs, however, do not function alone. Instead, they are part of a group of cooperative organs, called an **organ system.** The brain, spinal cord, and nerves are part of the organ system known as the nervous system.

As you will see in the upcoming chapters, components of an organ system are sometimes connected—as in the digestive system—and are sometimes dispersed throughout the body—as in the endocrine system (▷ Figure 10–15). Some organs belong to more than one system. For

example, the pancreas produces digestive enzymes that are secreted into the small intestine, where they break down food materials. The pancreas is therefore part of the digestive system. The pancreas also contains cells that produce insulin and glucagon, two hormones that help control blood glucose levels. The pancreas therefore also belongs to the endocrine system.

The following 10 chapters describe the major organ systems and the functions they perform, paying special attention to their role in homeostasis and highlighting important aspects of their evolution. ▷ Figure 10–16 summarizes the functions of each organ system and lists the role each plays in the overall economy of the human organism. You may want to take a moment to read the descriptions in the light blue boxes before proceeding.

As Figure 10–16 shows, organ systems are the functional machinery of the body. In sickness and in health, the 10 major organ systems remain interconnected and mutually dependent.[2] This interdependence is often apparent when one system "breaks down." Such an event can have catastrophic effects on other organ systems and the organism

[2]You will note that this figure includes the integumentary system as an organ system. I prefer to discuss the skin in Chapter 9 and others, rather than devote a full chapter to this important system.

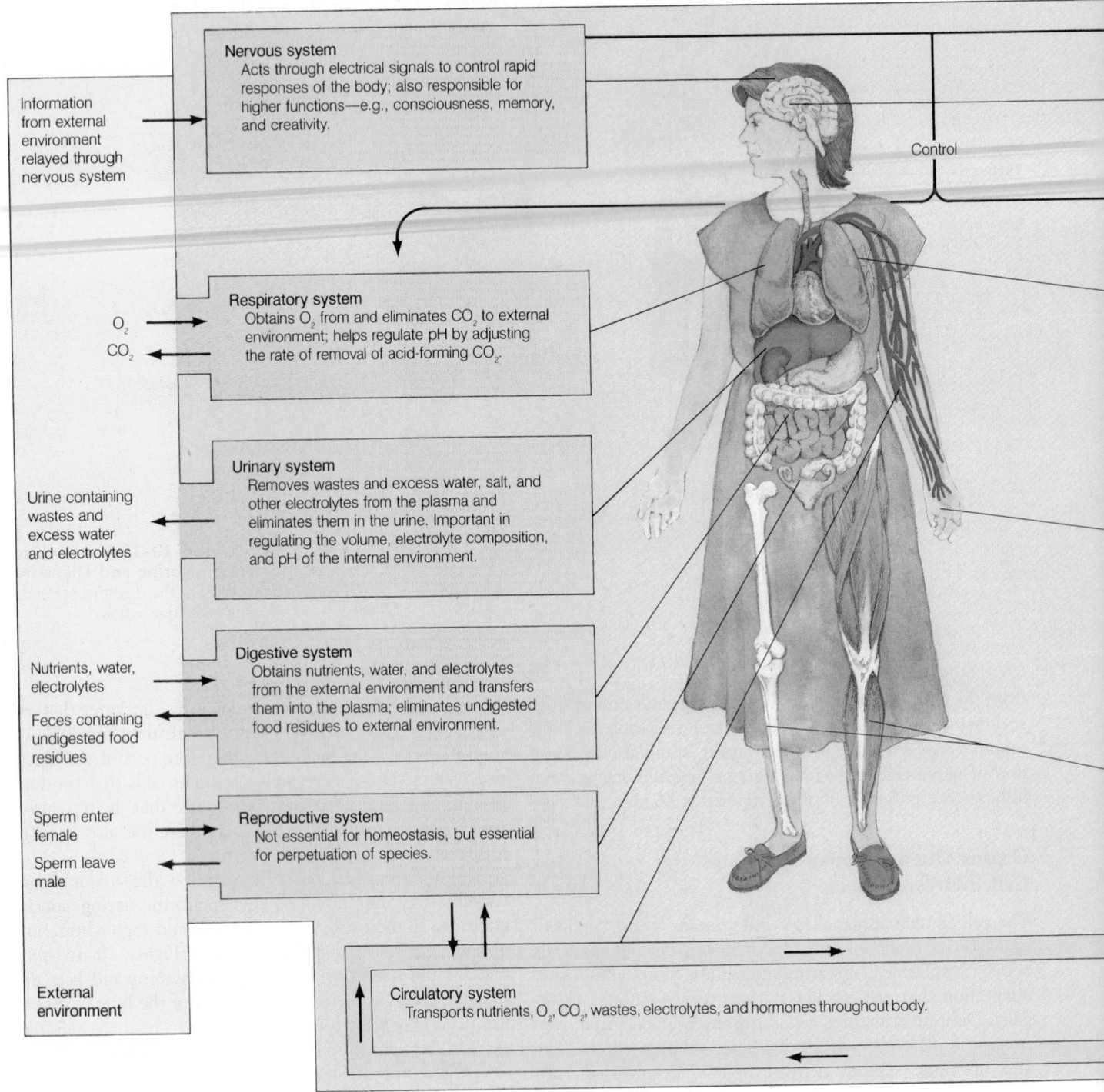

BODY SYSTEMS
Made up of cells organized according to specialization to maintain homeostasis.

Nervous system
Acts through electrical signals to control rapid responses of the body; also responsible for higher functions—e.g., consciousness, memory, and creativity.

Information from external environment relayed through nervous system

Control

Respiratory system
Obtains O_2 from and eliminates CO_2 to external environment; helps regulate pH by adjusting the rate of removal of acid-forming CO_2.

O_2
CO_2

Urinary system
Removes wastes and excess water, salt, and other electrolytes from the plasma and eliminates them in the urine. Important in regulating the volume, electrolyte composition, and pH of the internal environment.

Urine containing wastes and excess water and electrolytes

Digestive system
Obtains nutrients, water, and electrolytes from the external environment and transfers them into the plasma; eliminates undigested food residues to external environment.

Nutrients, water, electrolytes

Feces containing undigested food residues

Reproductive system
Not essential for homeostasis, but essential for perpetuation of species.

Sperm enter female

Sperm leave male

External environment

Circulatory system
Transports nutrients, O_2, CO_2, wastes, electrolytes, and hormones throughout body.

▷ **FIGURE 10–16 Role of the Body Systems in Maintaining Homeostasis**

as a whole. When defective kidneys fail to remove toxins from the blood, for instance, other cells and other organ systems become poisoned. If the disease is untreated, death may occur in a matter of days.

≈ **PRINCIPLES OF HOMEOSTASIS**

Homeostasis is one of the primary themes of this book. Defined in Chapter 1 as the internal constancy of the body, homeostasis occurs on a variety of levels—in cells, tissues,

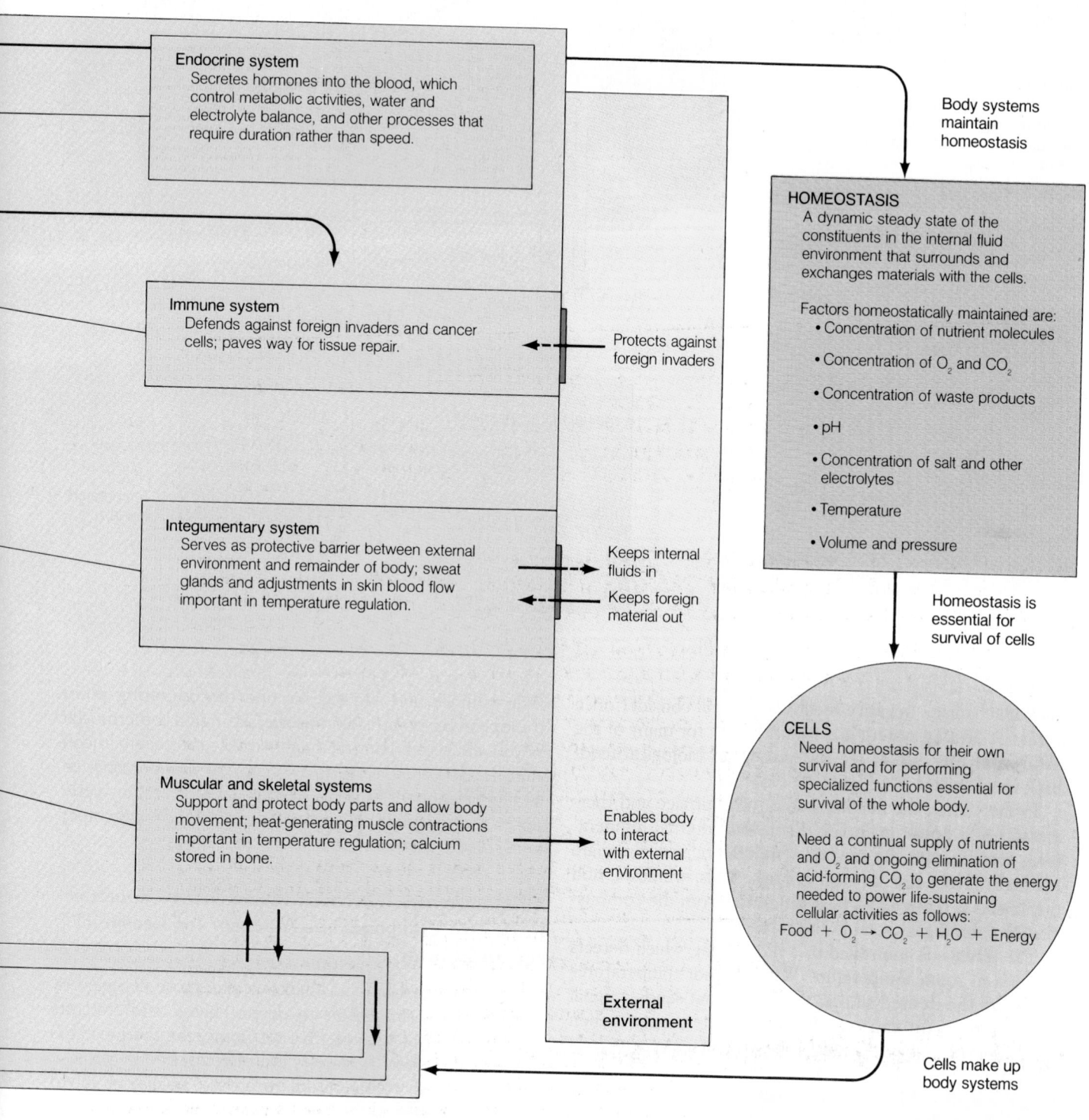

Endocrine system
Secretes hormones into the blood, which control metabolic activities, water and electrolyte balance, and other processes that require duration rather than speed.

Immune system
Defends against foreign invaders and cancer cells; paves way for tissue repair.

Protects against foreign invaders

Integumentary system
Serves as protective barrier between external environment and remainder of body; sweat glands and adjustments in skin blood flow important in temperature regulation.

Keeps internal fluids in

Keeps foreign material out

Muscular and skeletal systems
Support and protect body parts and allow body movement; heat-generating muscle contractions important in temperature regulation; calcium stored in bone.

Enables body to interact with external environment

External environment

Body systems maintain homeostasis

HOMEOSTASIS
A dynamic steady state of the constituents in the internal fluid environment that surrounds and exchanges materials with the cells.

Factors homeostatically maintained are:
• Concentration of nutrient molecules

• Concentration of O_2 and CO_2

• Concentration of waste products

• pH

• Concentration of salt and other electrolytes

• Temperature

• Volume and pressure

Homeostasis is essential for survival of cells

CELLS
Need homeostasis for their own survival and for performing specialized functions essential for survival of the whole body.

Need a continual supply of nutrients and O_2 and ongoing elimination of acid-forming CO_2 to generate the energy needed to power life-sustaining cellular activities as follows:
Food + $O_2 \rightarrow CO_2 + H_2O$ + Energy

Cells make up body systems

organs, organ systems, organisms, and even the environment. Homeostatic systems at all levels of biological organization have several common features worth noting before we begin our study of organ systems.

Homeostatic Systems Maintain Constancy Chiefly Through Negative Feedback Mechanisms

Feedback mechanisms were briefly described in Chapter 4. To expand your understanding of the most common feed-

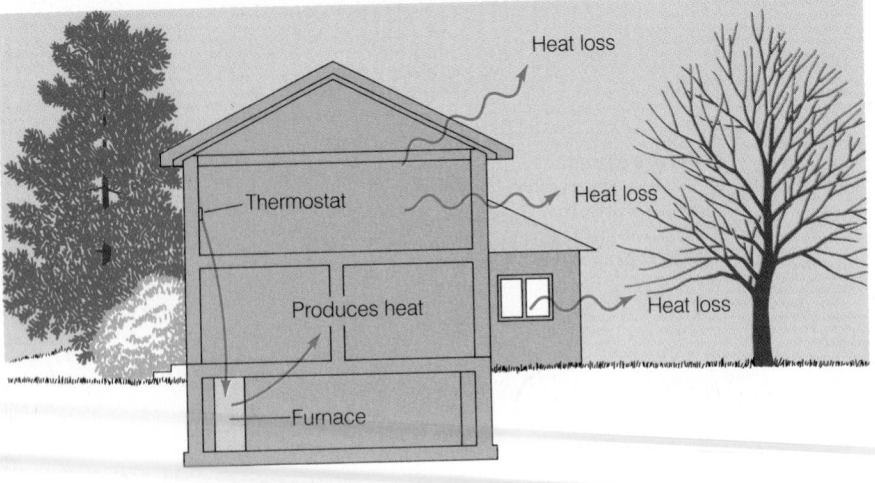

FIGURE 10-17 Homeostasis and the House (a) Heat is maintained in a house by a furnace, which produces heat to balance heat loss. The thermostat monitors the internal temperature and switches the furnace on and off in response to temperature changes. (b) A hypothetical temperature graph showing temperature fluctuation around the operating point.

(a)

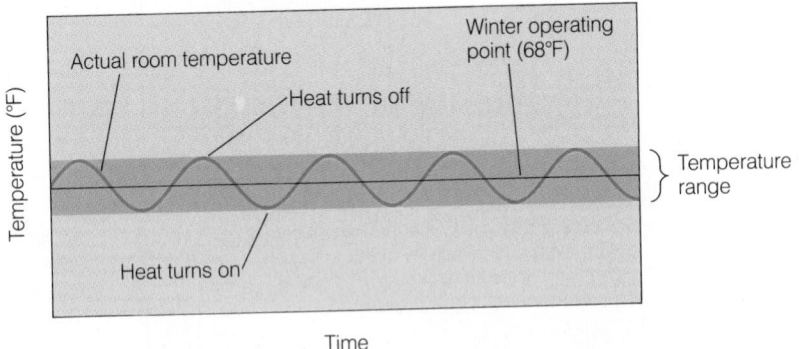

(b)

back mechanism, negative feedback, we will consider once again the heating system in a typical house, for many of the features of this system are also present in biological feedback mechanisms (▷ Figure 10-17a).

In the winter, the heating system (the furnace and thermostat) of a house maintains a constant internal temperature, even though the outside temperature may fluctuate markedly. Heat lost through ceilings, walls, windows, and tiny cracks is replaced by heat generated from the combustion of natural gas or oil in the furnace.

The furnace is controlled by a thermostat, which detects changes in room temperature. When indoor temperatures fall below the desired setting, the thermostat sends a signal to the furnace, turning it on. Heat from the furnace is then distributed through the house, raising the room temperature. When the room temperature reaches the desired setting, the thermostat shuts the furnace off. Like all negative feedback mechanisms, the product of the system (heat) "feeds back" on the process, shutting it down.

A graph of a hypothetical house temperature is shown in Figure 10–17b and illustrates another important principle of homeostasis: *homeostatic systems do not maintain absolute constancy*. Rather, they maintain conditions (such as body temperature) within a given range.

All homeostatic mechanisms in vertebrates operate in a similar fashion, maintaining conditions by negative feedback within a narrow range around the **operating point.** The operating point is akin to the setting on a thermostat. As you will see in later chapters, vertebrates maintain constant levels of a great many chemical components, including hormones, nutrients, wastes, and ions. Many vertebrates also maintain physical parameters like body temperature, blood pressure, blood flow, and others.

All Homeostatic Feedback Mechanisms Contain at Least Two Components—A Sensor (or Receptor) and an Effector

Biological homeostatic mechanisms, like those designed by engineers, contain sensors to detect change and effectors to correct for the change. In your home, the thermostat is the sensor, and the furnace is the effector.

In your body, temperature receptors, special modified nerve cell endings in the skin, are one type of sensor. They detect changes in the ambient (outside) temperature and send signals to the brain, alerting it when the temperature rises or falls. When the ambient temperature falls, the brain sends signals to the body to reduce heat loss and increase heat output. The options for generating more heat are many. Put another way, the body contains several different types of effectors.

One of the main sources of heat is the catabolism of

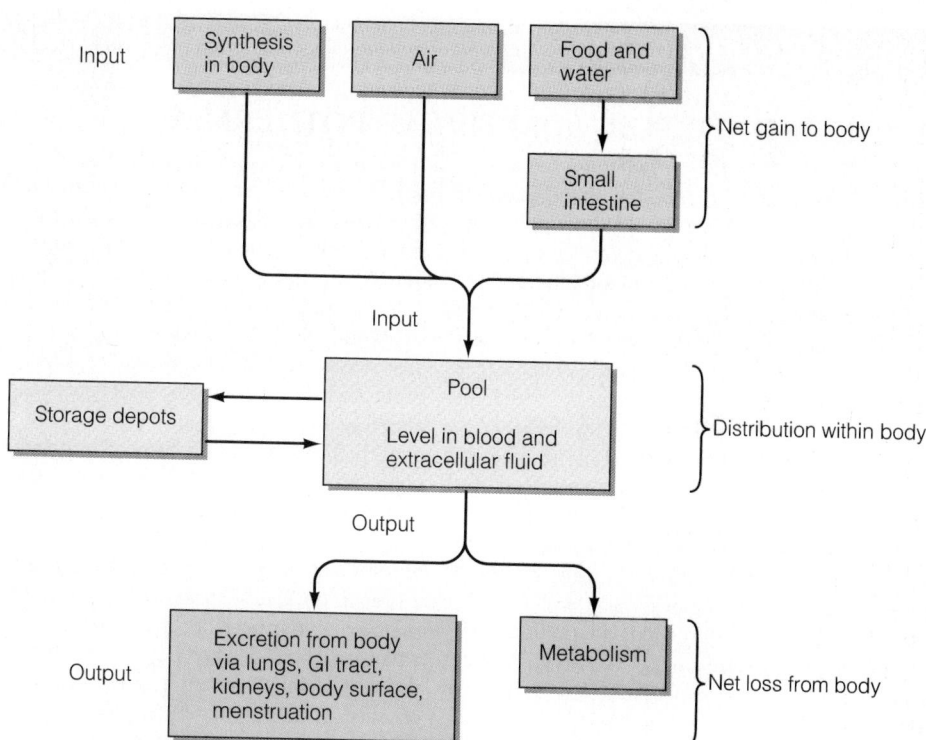

Inputs and outputs are balanced to maintain more or less constant levels of chemical and physical parameters.

glucose and other molecules. In earlier chapters, you learned that the cells of the body break down glucose to make ATP. During this process, heat is given off. Each cell, then, is a tiny furnace whose heat radiates outward and is distributed throughout the body by the blood. Unlike the furnace in your home, the cellular "furnaces" cannot be turned up very quickly. They respond much more slowly than a furnace and are part of a delayed response to low temperature. As winter progresses, energy catabolism increases, and the body produces more heat.

In order to respond to sudden changes in outside temperature, the body must rely on other, more rapid mechanisms. If you walked outdoors on a cold winter night dressed only in a light sweater and blue jeans, for example, receptors in your skin would sense the cold and send signals to the brain. The brain in turn would send signals to blood vessels in the skin, causing them to constrict, reducing the flow of blood in the skin. The restriction of blood flow through the skin reduces heat loss. If it is cold enough outside, the brain may also send signals to the muscles, causing them to undergo rhythmic contractions, commonly known as shivering. Shivering burns additional glucose, releasing extra heat. (It's not unlike adding additional logs to a fire. The more fuel that's burned, the more heat you get.) For a discussion on how some animals respond to seasonal changes, see Spotlight on Evolution 10–1.

Many voluntary actions may also be "ordered" by the brain to help reduce heat loss or generate more heat. For example, you might put on a hat or turn around and go back inside. In humans, conscious acts are often crucial components of homeostasis.

Homeostasis is Maintained by Balancing Inputs and Outputs

The generalized diagram in ▷ Figure 10–18 illustrates the many ways the body regulates the level of various chemicals. Let's consider the input side first. As illustrated, ions (for example, calcium) and other chemical substances (for example, glucose) are ingested in food or water. Other essential chemicals like oxygen enter our bodies in the air we breathe. Lastly, some substances like certain amino acids are produced in the body. All three routes tend to raise internal concentrations.

Excretion and metabolism tend to lower concentrations. **Excretion** is the loss of materials through a number of specialized body systems. Carbon dioxide, for example, is excreted by the lungs, as is water. Water is also excreted by the kidneys. Metabolism, chemical reactions occurring in the body cells, also removes certain substances from the body. Blood glucose levels, for instance, are decreased by cellular catabolism.

Internal concentrations are kept in balance, in large part by input and output. However, internal storage depots (regions where chemicals are stored) also participate in homeostatic balance. For example, as you learned in previous chapters, glucose is stored in the liver, and blood glucose levels are therefore also determined by the input and output occurring in this important organ.

The relative importance of the various homeostatic pathways depends on the substance in question. Water, for example, enters the body primarily through ingestion (the food and liquids we consume) and is removed by the

HIBERNATION AND TORPOR: ADAPTIVE HYPOTHERMIA

Stress is a daily occurrence in the lives of animals. Predators, cold spells, hot spells, and shortages of food, among other factors, all contribute to stress. In many regions, stressful conditions may be seasonal; for example, periods of plenty may alternate with periods of great scarcity. Such cycles are conspicuous in the Arctic, with its long summer days. During the Arctic summer, plants grow rapidly, and many species make their home there, feeding off the abundance of life. But Arctic summers are followed by prolonged winters in which bitter cold brings production to a halt and sends many species to warmer grounds. Surprisingly to many, some tropical rain forests also exhibit annual cycles with distinct rainy (monsoon) seasons punctuated by hot, dry seasons.

Animals respond to seasonal stress in several ways, including hibernation (adaptive hypothermia), aestivation, torpor, and migration.

Some mammals, like chipmunks, ground squirrels, woodchucks (groundhogs), skunks, and bears, for instance, enter a period of greatly reduced activity called winter hibernation. Thus, during the coldest months of the year when food is scarce, these animals hole up in their dens or burrows and enter into a state of suspended animation. Metabolism, breathing, and heart rate all decrease. During hibernation, for instance, a black bear's heart and breathing rates may slow to as little as 10% of the active rates. That would be like your pulse dropping to 7 beats per minute and your respiration rate to one or two breaths per minute. The bear's body temperature drops by about 5°C, and metabolism drops by about 50%—an adaptation that results in a dramatic energy savings.

Mammals prepare for hibernation in at least two ways—by storing large amounts of body fat, which they live off during the long months of hibernation, and by growing thick winter pelts, which reduce heat loss. As days become shorter, internal biological clocks trigger changes in feeding behavior, which in turn result in increases in food consumption and body fat content. The internal clock also triggers growth of the animal's fur coat. These adaptive changes, found in animals that remain active in the winter as well, serve all species during lean days of winter.

Another interesting adaptation that helps conserve energy is known as torpor.

▷ **FIGURE 1 The Hummingbird** This lovely bird saves energy in temperate climates by reducing metabolic activity during periods of inactivity.

Torpor is a temporary reduction of the metabolic set point and is found in small animals, such as bats and hummingbirds ▷Figure 1). These species enter a state of decreased metabolism every day during periods of inactivity. The reason is that animals with a small body size in relation to body surface area lose heat faster than it can be generated through metabolism. Thus, at night or during periods of rest, a hummingbird's body temperature drops relative to that of its environment, saving enormous amounts of energy. Resting on a branch on a cool spring day, the hummingbird's temperature plummets, but when it is time to move again, the bird flexes its wings, shivers a little, and increases its metabolic heat production, raising its body temperature to about 40°C, a temperature at which it can perform.

Bats exhibit both hibernation and torpor. During the daylight hours, bats repair to caves or dark places, where they remain inactive, falling into a state of torpor. During the winter months, they enter hibernation for many weeks.

Some animals undergo a summer dormancy, known as aestivation, which permits them to cope with temperature extremes or seasonal shortages of food and water. Like hibernation, aestivation is characterized by slow metabolism and inactivity, which permits animals to survive long periods of elevated temperatures and water shortages. Still others migrate long distances, moving from warm spring and summer habitat to warm wintering grounds, finding in each location ample food and suitable temperature for survival.

All of these adaptations save energy and promote survival and reproduction. They are but a handful of the evolutionary strategies for survival.

---

kidneys, lungs, and skin. For iron, the rate of absorption by the intestinal tract is a key determinant of its levels in the blood. When iron levels decrease—say, because of a decrease in iron intake—the body reestablishes the balance by increasing the rate of absorption in the small intestine.

Homeostasis Can Be Upset by Changes in the Input, Output, or Storage Occurring in the Body's Complex Homeostatic Network

Consider output first. On a hot day, water escapes our

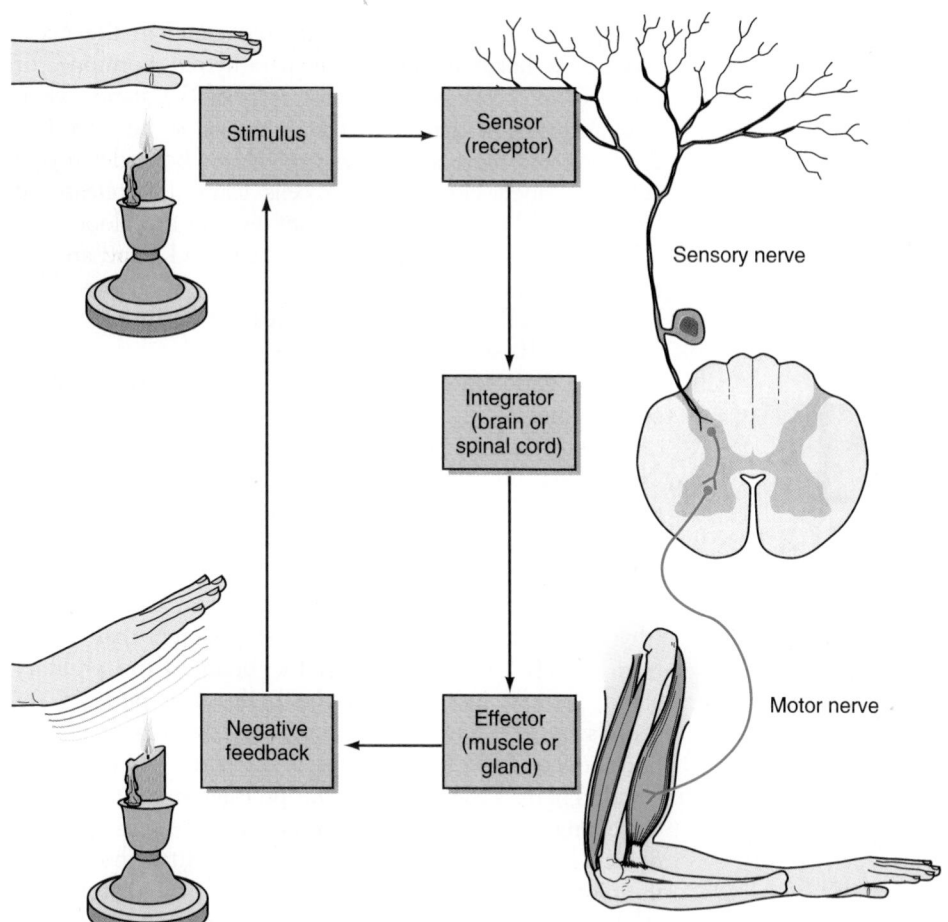

▷ **FIGURE 10–19 Nervous System Reflex** Reflexes involve some kind of stimulus, a sensor, an integrator, and an effector. The sensor detects a change and sends a signal to the integrator, the brain or spinal cord, which elicits a response in the effector organs. Negative feedback from the effector eliminates the stimulus.

bodies very rapidly as a result of perspiration, an adaptation that helps cool us down. Heavy perspiration can result in a severe decrease in water volume (dehydration) that may upset homeostasis—so much so, in fact, that death may occur. Severe diarrhea—a drastic increase in water output—also results in a dangerous depletion of body fluids that can be so severe that it results in death. In fact, dehydration resulting from severe diarrhea kills millions of Third World children each year.

Just as these changes in the output of a substance drastically alter homeostasis, so do changes in input. For example, going without water for a prolonged period (approximately three days) can kill a human being. Excess input can also prove dangerous. Excess salt intake, for example, can result in hypertension (high blood pressure) in some people.

Whatever the cause, imbalances in homeostasis can have dramatic impacts on human health.

Homeostatic Control Requires the Action of Nerves, Hormones, and Various Chemicals that Operate Over Short Distances

Homeostatic mechanisms are **reflexes**—that is, automatic physiological responses triggered by certain stimuli. Reflexes occur without conscious control. Homeostatic re-

flexes in the body involve two main mechanisms: nervous and chemical. Chapter 31 describes homeostatic mechanisms that operate in the environment.

The Nervous System Reflex. The line between cause and effect in a homeostatic system that involves the nervous system is fairly easy to trace. As shown in ▷ Figure 10–19, changes are detected by a sensor, or **receptor,** a nerve cell ending or a special structure that responds to various stimuli in the internal or external environment. In vertebrates, the sensor sends nerve impulses to the brain or spinal cord by a nerve. The brain and spinal cord, in turn, direct an appropriate response to counterbalance the change, maintaining or restoring balance.

At any one time, the brain and spinal cord receive thousands of signals from receptors in the body. These signals alert the nervous system to a variety of internal and external conditions. The brain and spinal cord are centers of integration able to make sense of information and generate a meaningful response. They control the activity of many **effectors,** generally muscles and glands. On a cold winter night, for example, nerve impulses travel from temperature sensors in your skin to your brain. The brain, in turn, sends signals to the blood vessels in the skin, causing them to constrict and reduce blood flow through the skin, thus conserving heat. As noted earlier, if it is cold enough, the

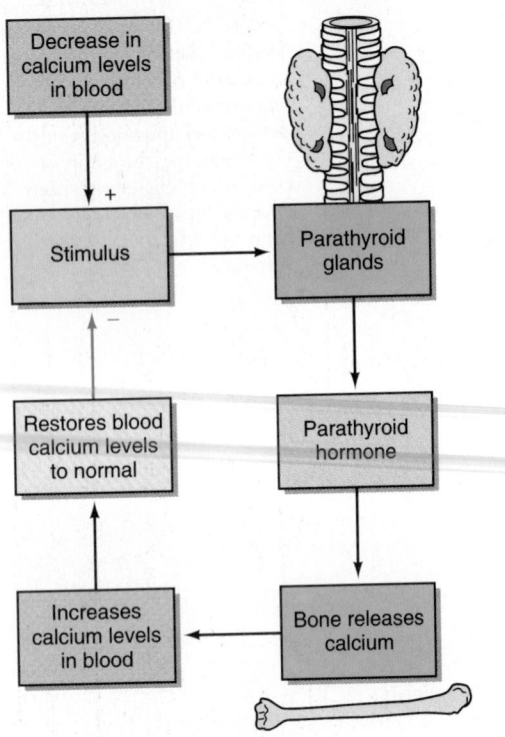

▷ FIGURE 10–20 Endocrine Reflex Not Involving the Nervous System This reflex operates through the bloodstream. A decrease in calcium in the blood stimulates a series of reactions that helps restore normal blood calcium levels. A plus sign indicates that the stimulus increases parathyroid gland activity; a minus sign indicates the opposite effect.

brain may also send impulses to the muscles, causing shivering. The muscles surrounding the blood vessels of the skin and those involved in shivering are effectors in this reflex.

Chemical Control. Hormones also participate in reflexes. Consider an example. The parathyroid glands, four

tiny glands embedded in the thyroid gland in the neck, produce a hormone known as parathyroid hormone, or PTH. PTH is released from the cells of the glands when calcium levels in the blood fall. It travels in the blood to the bone and there stimulates osteoclasts, bone-destroying cells described earlier. These cells cause the release of calcium and help raise the calcium level of the blood.

In this reflex, the cells of the parathyroid gland are the receptors, which detect the drop in blood calcium. The bone is the effector, which produces the desired response (▷ Figure 10–20).

Some endocrine reflexes operate through the nervous system, making the lines of cause and effect a bit more difficult to follow. A good example is the thyroid gland. The thyroid produces a hormone called thyroxine (▷ Figure 10–21). This hormone increases the metabolic rate of cells and thus increases heat production. Thyroxine levels are monitored by cells in the brain (the receptors). When blood levels fall, these cells release a chemical substance (thyroid-stimulating hormone releasing factor) that travels from its site of production in the brain to the pituitary gland, located just beneath the brain. The pituitary responds by releasing another hormone, thyroid-stimulating hormone, or TSH, which travels in the blood to the thyroid gland. There, TSH steps up the production of thyroxine, correcting the deficiency. This reflex has one sensor, which detects levels of thyroxine, but three effectors: the cells of the pituitary that release TSH, the cells of the thyroid that release thyroxine, and body cells that respond to thyroxine by increasing their rate of metabolism.

Local Chemical Control. Nervous and endocrine mechanisms generally occur over considerable distances. In the nervous system, for example, nerves carry impulses from the brain and spinal cord to distant parts of the body, often several feet away. In the endocrine system, the bloodstream carries the messages from endocrine glands to

▷ FIGURE 10–21 A Neuroendocrine Reflex This reflex involves both the endocrine and nervous systems. Sensors are located in the brain.

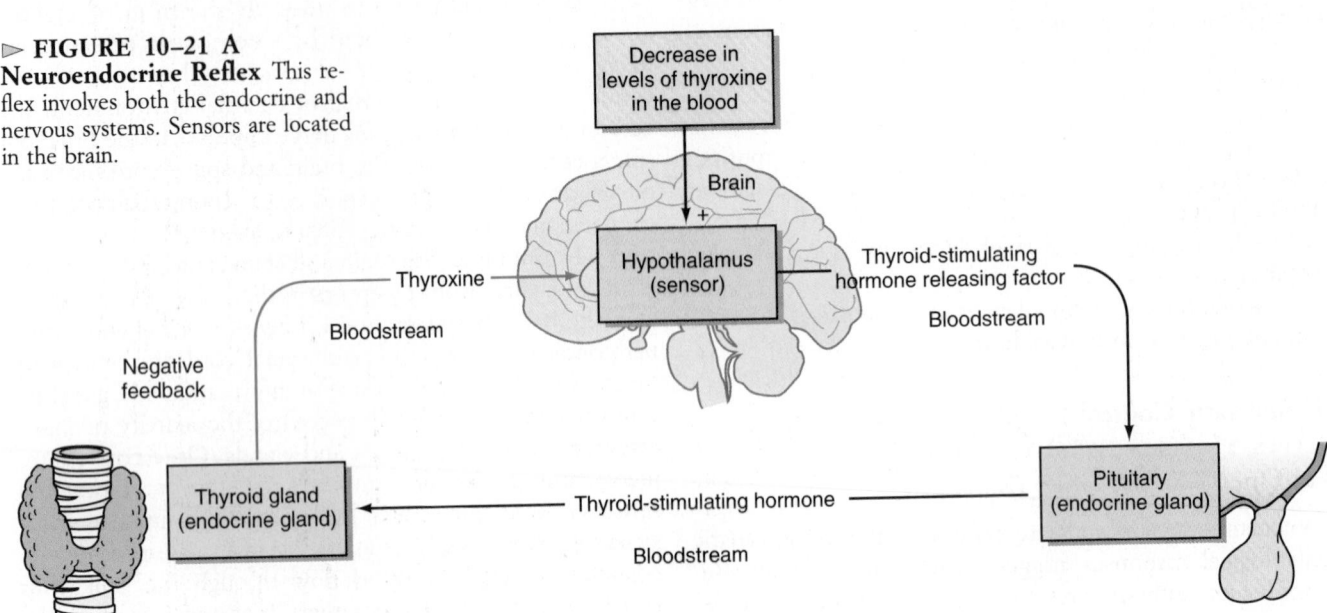

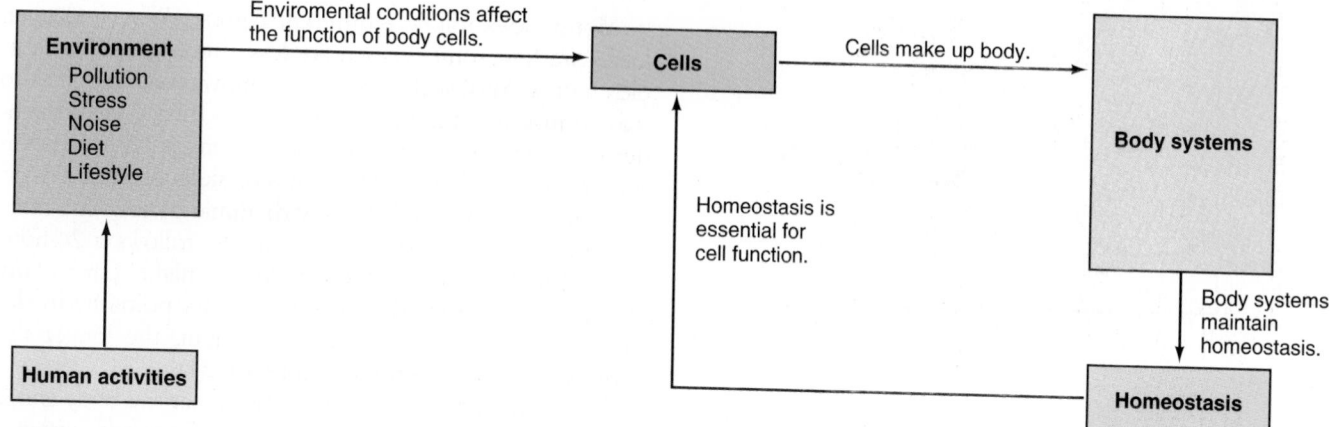

► FIGURE 10–22 Health, Homeostasis, and the Environment Human health is dependent on maintaining homeostasis. Homeostasis, however, is affected by the condition of our environment. Stress, pollution, noise, and other environmental factors upset the function of cells and body systems, thus upsetting homeostasis and human health. As a result, human health is dependent on a healthy environment.

distant effectors. But not all systems require messages that are transported over long distances. In some cases, chemical control is exerted through the extracellular fluid only a cell or two away. Chemicals that elicit local effects are called **paracrines.** Produced by individual cells, paracrines diffuse to neighboring cells via the extracellular fluid. Epidermal growth factor produced by skin cells is one example. These molecules stimulate cell division when skin cells are damaged or lost, providing a degree of local control over cell growth.

One of the best-known paracrines is a group of chemical substances known as **prostaglandins.** Prostaglandins comprise a rather large group of molecules with diverse functions. Some help stimulate blood clotting; others stimulate smooth muscle contraction.

An interesting evolutionary adaptation akin to paracrines are the **autocrines,** molecules produced by a cell that affect the function of that cell. Autocrines are part of the simplest chemical reflexes in the body. Some prostaglandins, for example, function as autocrines as well as paracrines.

Human Health Depends on Homeostasis, Which in Turn Requires a Healthy Environment

This book emphasizes a theme that is important to all of us—notably, that human health is dependent on homeostasis. Homeostasis, in turn, requires a healthy, clean environment. The health of the environment, of course, is also dependent in part on well-functioning homeostatic systems in nature that help maintain conditions conducive to life.

The relationship between homeostasis and health is shown in ► Figure 10–22. Take a moment to study this diagram. Notice that the body systems help maintain overall homeostasis, which is essential for proper cell function. Cells also contain homeostatic mechanisms, that are vital to their own function and to the overall economy of the organism. Notice too that environmental factors, such as pollution, affect the function of body systems and cells in ways that can upset the internal balance, thus altering human mental and physical health.

Just as human health is affected by the condition of the environment, so too is the health of many other species. Numerous examples of this connection will be pointed out in this book.

≈ BIOLOGICAL RHYTHMS

The previous discussion may have given the impression that homeostasis establishes an unwavering condition of stability that remains more or less the same, day after day, year after year. In truth, many physiological processes undergo rhythmic change.

Not All Physiological Processes Remain Constant Over Time but Those That Fluctuate Do So in Predictable Ways

Body temperature varies during a 24-hour period by as much as 0.5° C. Blood pressure may change by as much as 20%, and the number of white blood cells, which fight infection, can vary by 50% during the day. Alertness also varies considerably. About 1:00 P.M. each day, for instance, most people go through a slump. For most of us, activity and alertness peak early in the evening, making this an excellent time to study. Daily cycles, such as these, are called circadian rhythms ("about a day"). **Circadian rhythms** are natural body rhythms linked to the 24-hour day-night cycle.

Not all cycles occur over 24 hours, however. Some can be much longer. The **menstrual cycle,** for instance, is a recurring series of events in the reproductive functions of women that lasts, on average, 28 days. During the menstrual cycle, levels of the female sex hormone estrogen

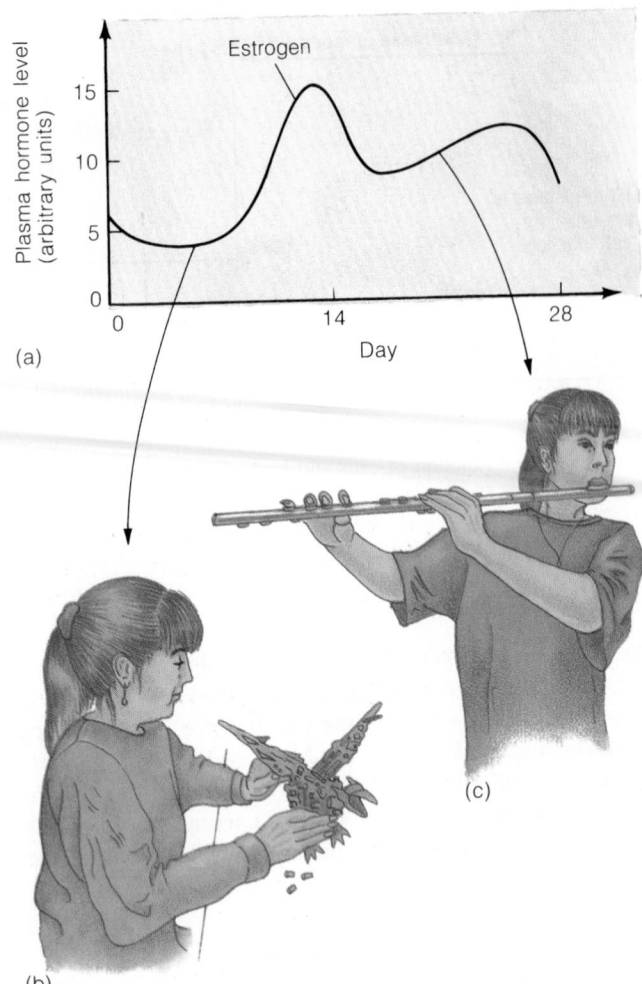

(a)

(b)

(c)

▷ **FIGURE 10–23 Estrogen Levels during the Menstrual Cycle** (a) Note that over a 28-day period, estrogen levels vary considerably. The variation is part of a finely tuned reproductive cycle that prepares the woman for pregnancy. (b) During the early part of the cycle when estrogen levels are low, studies show that women excel at tasks involving spatial relationships, but they do less well at tasks involving fine motor skills. (c) During the later half of the cycle when estrogen levels are high, women excel at tasks involving motor skills, like playing a musical instrument, but do less well at tasks involving spatial relationships.

undergo dramatic shifts. As illustrated in ▷ Figure 10–23a, estrogen concentrations in the blood are low at the beginning of each cycle and peak on day 14, when ovulation normally occurs. Throughout the remaining 14 days, estrogen levels are rather high. Then they drop off again when a new cycle begins.

Estrogen follows this cycle month after month in women of reproductive age. Interestingly, one recent study suggests that estrogen levels may exert profound influences on neural functions in women. Early in the cycle when estrogen levels are low, the study showed, women excel at tasks involving spatial relations—for example, solving three-dimensional puzzles. When estrogen levels are elevated in the second half of the cycle, many women may find such tasks more difficult (Figure 10–23b).

Motor skills follow a different pattern. When estrogen levels are low, some women are less able to perform complex motor skills, such as the finger movements required to play a musical instrument, than they are during the remainder of the cycle (Figure 10–23c). When estrogen concentrations in the blood are high, motor skills become easier.

Many hormones follow daily rhythmic cycles. The male sex hormone testosterone, for example, follows a 24-hour cycle. The highest levels occur in the night, particularly during dream sleep. Dream sleep occurs primarily in the early morning hours—the later the hour, the longer the periods of dream sleep (▷ Figure 10–24).

The important point of all of this is that the body is not static. *Although many chemical substances are held within a fairly narrow range by homeostatic mechanisms, others fluctuate widely in normal and quite essential cycles.* Just as the seasons change, altering the face of the landscape, so do the internal body seasons. But over the long run, these changes are predictable. They also occur within prescribed physiological limits; they do not run out of control. They are part of the body's dynamic balance, just as the weather changes throughout the year are part of the dynamic balance of the planet's climate.

In Humans, Internal Biological Rhythms Are Apparently Controlled by the Brain

Just how the body controls its many internal rhythms remains a question. Research suggests that the brain controls many biological cycles. One region in particular, the **suprachiasmatic nucleus (SCN),** is thought to play a major role in coordinating several key rhythms. The SCN, a clump of nerve cells in the base of the brain in a region called the hypothalamus, may also regulate other control centers. As a result, the suprachiasmatic nucleus is often referred to as the "master clock."

Like a clock, the SCN ticks off the minutes, faithfully imposing its control on the body, turning body functions on and off like the master control in an automated factory. Even in complete darkness, the master clock ticks on, directing many circadian rhythms.

In humans, research suggests, the master clock operates on a 25-hour cycle. That is, if isolated in a dark room, many of us would fall into a 25-hour sleep-wake cycle. But as with most other biological phenomena, there is considerable variability. Some individuals, for example, operate on 28-hour sleep-wake cycles. Some are shorter than 24 hours.

Sleep researchers believe that in the real world the natural 25-hour clock is modulated by the 24-hour day-night cycle. In other words, the environment alters the clock and maintains many biological rhythms on a 24-hour cycle. Control of the clock is thought to reside in a gland in the brain known as the **pineal gland.** It secretes a hormone thought to keep the suprachiasmatic nucleus in sync with the day-night cycle.

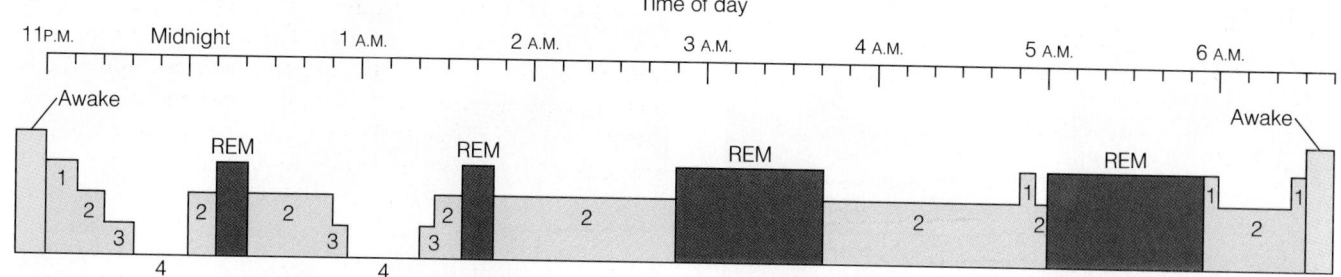

Time of day

| 11 P.M. | Midnight | 1 A.M. | 2 A.M. | 3 A.M. | 4 A.M. | 5 A.M. | 6 A.M. |

▷ **FIGURE 10–24 Stages of Sleep** Numbers indicate the stages of sleep: the higher the number, the deeper the sleep. Note that around midnight sleep is deepest. As morning approaches, a person's sleep is lighter and REM sleep, or dream sleep, occurs in longer increments.

The study of biological rhythms is a fascinating field that has yielded some important information and insights. One practical application is a better understanding of jet lag, that drowsy, uncomfortable feeling people get from the disruption of sleeping patterns caused by long-distance airplane travel. Studies suggest that jet lag occurs when the body's biological clock ticks out of synchrony with the day-night cycle of a traveler's new surroundings. A businesswoman who travels from Los Angeles to New York, for instance, may be weary and irritable the first few days in the new time zone. At 10:00 P.M. New York time, when New Yorkers are heading to bed, she is wide awake because her body is still on Los Angeles time—three hours earlier. When the alarm goes off at 6:00 A.M. New York time, our weary traveler crawls out of bed exhausted, because as far as her body is concerned, it is 3:00 A.M. Los Angeles time. To avoid the ill feelings, some sleep researchers suggest that you abide by your internal clock, following "home time" when in a new time zone, or, barring that, try to reset your biological clock before you get there. You can do this by going to bed an hour or two earlier when you are traveling from west to east. Do just the opposite for east-to-west travel.

Research on body rhythms has also shown that people respond to drugs differently at different times of the day and night. By administering drugs at the body's most receptive times, physicians may be able to reduce the doses, lowering toxic side effects and fighting diseases more effectively.

ENVIRONMENT AND HEALTH: TINKERING WITH OUR BIOLOGICAL CLOCKS

Modern life with its stress, noise, pollution, hectic pace, and weird work schedules can deal a blow to our internal biological rhythms, with serious consequences. Dr. Richard Restak, a neurologist and author, notes that the "usual rhythms of wakefulness and sleep . . . seem to exert a stabilizing effect on our physical and psychological health." The greatest disrupter of our natural circadian rhythms, he says, is the variable work schedule, surprisingly common in the United States and other industrialized countries. Today, one out of every four working men and one out of every six working women is on a variable work schedule, shifting frequently between day and night work. In many industries, workers are at the job day and night to make optimal use of equipment and buildings. As a result, more restaurants and stores are open 24 hours a day, and more health-care workers must be on duty at night to care for accident victims.

What American business has forgotten is that for millions of years, humans have slept during the night and been awake during the day. Turn that around, and you're asking for trouble. Make a person work at night when he or she normally sleeps, say sleep experts, and you can expect more accidents and lower productivity. Consider an example.

At 4:00 o'clock in the morning in the control room of the Three Mile Island nuclear reactor in Pennsylvania, three operators failed to notice warning lights and to note that a valve in the system had remained open. When the morning-shift operators entered the control room, they quickly discovered the problem, but it was too late. Pipes in the system had burst, sending radioactive steam and water into the air and into two buildings. John Gofman and Arthur Tamplin, two radiation health experts, estimate that the radiation released from the accident, the worst nuclear accident in U.S. history, will result in at least 300 and possibly as many as 900 additional fatal cases of cancer in the residents living near the troubled reactor, although other experts (especially in the nuclear industry) contest these projections, saying that the accident will not have a noticeable effect. Whatever the outcome of this debate, one thing is certain: the 1979 accident at Three Mile Island will cost several billion dollars to clean up.

Late in April 1986, another nuclear power plant ran amok. This crisis, in the former Soviet Union, was far more severe (▷ Figure 10–25). In the wee hours of the morning, two engineers were testing the reactor. They deactivated key safety systems in violation of standard operational pro-

▷ **FIGURE 10–25 Disaster at Chernobyl Nuclear Power Plant** This costly accident was caused primarily by operator error. What price do we pay to have industries running 24 hours per day to keep production levels high?

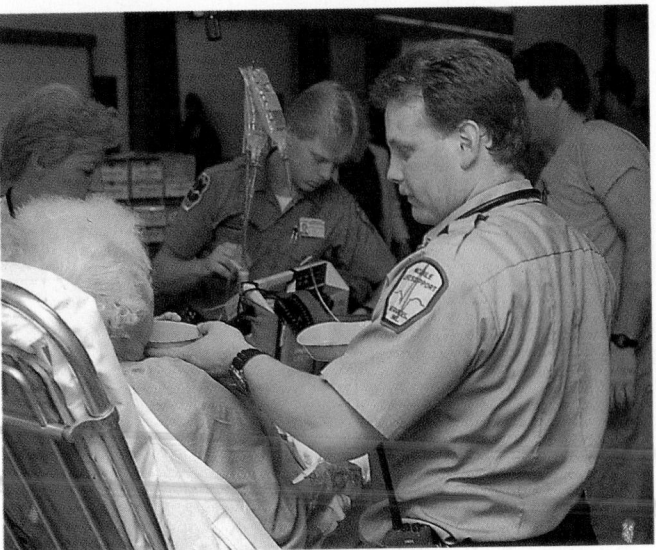

▷ **FIGURE 10–26 Late Night in the Emergency room.** Increasing numbers of men and women working the night shift may increase the number of judgment error accidents on the job. This causes an increase in the need for medical care. However, late-night medical care workers may also suffer from errors in judgment.

tocol. This single error in judgment (possibly due to fatigue) led to the largest and most costly nuclear accident in history. Steam built up inside the reactor and blew the roof off the containment building. A thick cloud of radiation rose skyward and then dissipated throughout Europe and the world. Workers battled for days to cover the molten radioactive core, which spewed radiation into the sky while the world watched in horror.

Although no one will ever know for sure, the Chernobyl disaster, like the accident at Three Mile Island, may have been the result of workers operating at a time unsuitable for clear thinking. One has to wonder how many plane crashes, auto accidents, and acts of medical malpractice can be traced to judgment errors resulting from our insistence on working against the natural body rhythms.

Making matters worse, many companies that maintain shifts round the clock spread the burden evenly among employees. One week, workers are on the day shift; the next week, they are switched to the "graveyard shift" (midnight to 8:00 A.M.). The next week, they are put on the night shift (4:00 P.M. to midnight). Many workers subject

to such disruptive changes report that they often feel run down and have trouble staying awake at the job. Their work performance suffers. When employees who have been on the graveyard shift arrive home, they are physically exhausted but cannot sleep, because they are trying to sleep at a time when the body is trying to wake them up. What is more, weekly changes in schedule never permit workers' internal clocks to adjust. Studies show that most people require 4 to 14 days to adjust to a new schedule.

Not surprisingly, workers on alternating shifts suffer more ulcers, insomnia, irritability, depression, and tension than workers on unchanging shifts. Many suffer from impaired judgment on the job, and in some circumstances they may pose a threat to society (Figure 10–26).

Thanks to studies of biological rhythms, researchers are finding ways to reset the body's clock. These efforts could help lessen the misery and suffering of shift workers and could improve the performance of those on the graveyard shift. For instance, one simple measure is to place all shift workers on three-week cycles instead of weekly cycles. This gives their biological clocks time to adjust. And instead of shifting workers from daytime to a graveyard shift, transfer them forward, rather than backward (for example, from a daytime to an evening shift and from an evening shift to the graveyard shift). Studies suggest that moving forward is far easier for workers than shifting backward.

For reasons not well understood, special treatment with bright lights helps reset the biological clock. Patients who are suffering from insomnia, because their biological clocks are out of phase, for example, receive daily doses of bright light, which somehow reset their internal clocks. Similar treatments are being tested for shift workers. They are a small price to pay for a healthy work force and a safer society.

FROM CELLS TO ORGAN SYSTEMS

1. The basic structural unit of animals (and all other organisms) is the cell. Cells and extracellular material form tissues, and tissues, in turn, combine to form organs.

2. Four primary tissues are found in the bodies of most multicellular animals such as humans: epithelial tissue, connective tissue, muscle tissue, and nervous tissue. Most of these tissues have two or more subtypes.

3. Organs contain all four primary tissues in varying proportions.

4. Epithelial tissues consist of two types: membranous epithelia, which form coverings or linings of organs, and glandular epithelia, which form exocrine and endocrine glands.

5. Connective tissues bind other tissues together, provide protection, and support body structures. All connective tissues consist of two basic components: cells and extracellular fibers.

6. Two types of connective tissue are found in the body: connective tissue proper (tendons, ligaments, and loose connective tissue) and specialized connective tissue (bone, cartilage, and blood).

7. Cartilage consists of specialized cells embedded in a matrix of extracellular fibers and other extracellular material. The cells are supplied with nutrients from blood vessels in the periphery of the cartilage, which is why damaged cartilage is repaired slowly.

8. Bone is a dynamic tissue that provides internal support, protects organs such as the brain, and helps regulate blood calcium levels.

9. Bone consists of bone cells (osteocytes) and a calcified cartilage matrix. Two types of bone tissue exist: spongy and compact.

10. The osteoclast, one type of bone cell, plays a major role in reshaping bone to meet the changing demands of the body and in releasing calcium to help maintain blood calcium levels.

11. Blood is another form of specialized connective tissue. It consists of numerous blood cells and platelets and an extracellular fluid, called plasma.

12. Muscle is an excitable tissue that contracts when stimulated. Three types of muscle tissue are found in the human body: cardiac, skeletal, and smooth muscle.

13. Cardiac muscle is found in the heart and is involuntary. Skeletal muscle is under voluntary control, for the most part, and forms the muscles that attach to bones. Smooth muscle is involuntary and forms sheets of varying thickness in the walls of organs and blood vessels.

14. Nervous tissue is the fourth primary tissue. It consists of two types of cells: conducting and nonconducting (supportive). The conducting cells, called neurons, transmit impulses from one region of the body to another. The nonconducting cells are a type of nervous system connective tissue.

15. Tissues combine to form organs in which specialized functions are carried out. Most organs are part of an organ system, a group of organs that cooperate to carry out some complex function.

PRINCIPLES OF HOMEOSTASIS

16. The nervous and endocrine systems coordinate the functions of organ systems, helping the body achieve homeostasis, or internal constancy. Homeostasis also occurs at the level of the cell.

17. All homeostatic systems maintain constancy by balancing inputs and outputs as well as movement to and from storage depots. Homeostatic systems do not maintain absolute constancy and can be upset by alterations in input or output or by changes in movements in or out of storage depots. Imbalances can have serious effects on an individual's health.

18. Homeostatic mechanisms are reflexes, occurring without conscious control.

19. Homeostatic reflexes occur at all levels of biological organization and involve two main mechanisms: nervous and hormonal.

20. Nervous and endocrine mechanisms generally occur over considerable distances. In some cases, control is exerted locally through the extracellular fluid.

BIOLOGICAL RHYTHMS

21. Many physiological processes undergo definite rhythmic changes. These natural rhythms may take place over a 24-hour period or over much longer or shorter periods.

22. The brain controls many biological cycles. A clump of nerve cells (the suprachiasmatic nucleus) located in the hypothalamus is thought to play a major role in coordinating several key functions and several other control centers. It is therefore often referred to as the "master clock."

ENVIRONMENT AND HEALTH: TINKERING WITH OUR BIOLOGICAL CLOCKS

23. The hectic pace of modern life, the shifting work schedules that many people follow, and the stressful environments in which we live can upset internal rhythms, with disastrous consequences.

24. The greatest disrupter of our natural circadian rhythms is the variable work schedule, which is surprisingly common in industrialized nations.

25. Workers on alternating shifts suffer from a higher incidence of ulcers, insomnia, irritability, depression, and tension than workers on regular shifts. Making matters worse, tired, irritable workers whose judgment is impaired by fatigue pose a threat not only to themselves but also to society.

26. Researchers are finding ways to reset the biological clock, which could help lessen the misery and suffering of shift workers and could increase the performance of those on the graveyard shift.

1. In the Environment and Health section, I noted that workers on night shifts are more accident-prone and suffer from lower productivity. I then cited two possible examples: the Three Mile Island and Chernobyl nuclear reactor accidents, both of which occurred at night. One of the academic reviewers of this text wrote, "I am certain I can cite terrible worker error in daytime circumstances." The examples, she contended, are "biased and used for effect."

    Looking at this issue more carefully, and using your critical thinking skills, how can you determine whether my statement that night shift workers are more accident-prone is valid? Is there a way to determine whether the accidents at the nuclear power plants were the result of the late hour? If so, how? If not, why not? Is there a way to determine if accidents at nuclear plants are more frequent at night or during the day? Is the reviewer's statement "I am certain I can cite terrible worker error in daytime circumstances" necessarily a valid refutation of the points I made? What is wrong with this line of reasoning?

2. Health care costs are skyrocketing, and many critics are arguing that our health care dollars are being misused. In particular, they argue that expensive procedures, such as organ transplants for indigent people costing taxpayers $100,000 or more each, are draining dollars that could be invested in preventive medicine, such as prenatal care. Prenatal care is medical care and advice on nutrition and other matters that are given to pregnant women and are crucial to the development of a healthy fetus. By various estimates, the cost of one organ transplant could supply medical care for 1000 pregnant women, assistance that could prevent many of the problems that result in ailments in newborns that necessitate costly transplant procedures.

    What is your opinion on this matter? Should Americans spend more on prevention and less on dramatic life-saving measures, such as organ transplants? Why? Can you think of ways to provide both types of health care without raising taxes?

    After you have given your opinion on this matter, put yourself in the position of a parent whose child needs a kidney transplant to survive. How does this perspective affect your position?

1. The first three cell types to emerge during embryonic development are _____, _____, and _____ .

2. All tissues are made of cells and a variable amount of _____ _____ .

3. The four primary tissues are _____, _____, _____, and _____ .

4. Glands that secrete their products into ducts that, in turn, carry the products to some other part of the body are called _____ glands.

5. The outer layer of the skin is called the _____ ; it is a type of _____ tissue.

6. Tendons are a form of _____ tissue.

7. The cell that produces collagen in tendons and other similar primary tissues is the _____ .

8. The soft tissue forming most of the nose is made of _____ _____ .

9. The intervertebral disk contains a ring of _____ that holds the soft, cushiony center of the disk in place.

10. The blood vessels of bone are found in tunnels called _____ canals. Nutrients flow out of these canals through even smaller tunnels in the bony matrix called _____ , providing the bone cells, or _____ , with nourishment.

11. The dense outer layer of most bones is called _____ bone.

12. The fluid component of blood is known as _____ .

13. The blood contains two components that do not have nuclei or organelles; they are the _____ _____ _____ and the _____ .

14. The involuntary muscle found in the lining of the small intestine and stomach is called _____ muscle. It contains two contractile proteins, _____ and _____ , which are also present in other muscle types.

15. A conducting nerve cell, or _____ , is a specialized cell that generates and transmits bioelectric impulses from one part of the body to another.

16. A group of organs that cooperate to perform some key function is called a(n) _____ _____ .

17. All homeostatic systems maintain chemical and physical parameters within a narrow range, called the _____ _____ .

18. Endocrine and neural _____ are a series of events that begins with a stimulus and ends with a response and is crucial for maintaining homeostasis.

19. The process of making sense of various chemical and nervous inputs is called _____ .

20. Organs and glands that carry out the instructions of the brain and spinal cord are called _____ .

21. A chemical released into the bloodstream that travels to distant sites where it elicits some response is called a(n) _____ ; a chemical that acts on cells close to its point of origin is called a(n) _____ .

22. Body temperature and some hormone levels vary over a 24-hour period; these and other similar cycles are called _____ rhythms.

23. A clump of nerve cells called the _____ nucleus in the hypothalamus of the brain is thought to be the master clock, controlling several key internal body rhythms and several other clocks.

24. Travelers suffer from _____ _____ because their biological clocks are out of synchrony with the 24-hour day-night cycle.

Answers to the Test of Terms are found in Appendix B.

1. Define the following terms: tissues, extracellular material, organs, and organ systems.
2. List the four primary tissues and their subtypes.
3. Describe the similarities and differences in the embryonic origin of endocrine and exocrine glands. Explain why the two secrete differently.
4. From your study of biology, discuss some examples in which structure reflects function.
5. Describe the two types of connective tissue and how they function.
6. Why do cartilage injuries repair so slowly? Bone is repaired much more easily than cartilage. Why do you think this is true?
7. In what way is bone part of a homeostatic system?
8. Describe the chief differences among cardiac, skeletal, and smooth muscle.
9. What is an organ system? List some examples.
10. Define homeostasis, and describe the major principles of homeostasis presented in this chapter. Use an example to illustrate your points.
11. How do storage depots enter into the input-output concept of homeostasis?
12. Homeostatic mechanisms are largely reflexes, involving chemical or nervous impulses. Describe each type of reflex, giving an example. You may find it helpful to include drawings of the systems to help explain these reflexes.
13. Are biological rhythms an exception to the principle of homeostasis?
14. Describe the biological clock, and explain how it is synchronized with the 24-hour day-night cycle.
15. How does shift work upset the biological clock? How can these problems be mitigated?

Nutrition and Digestion

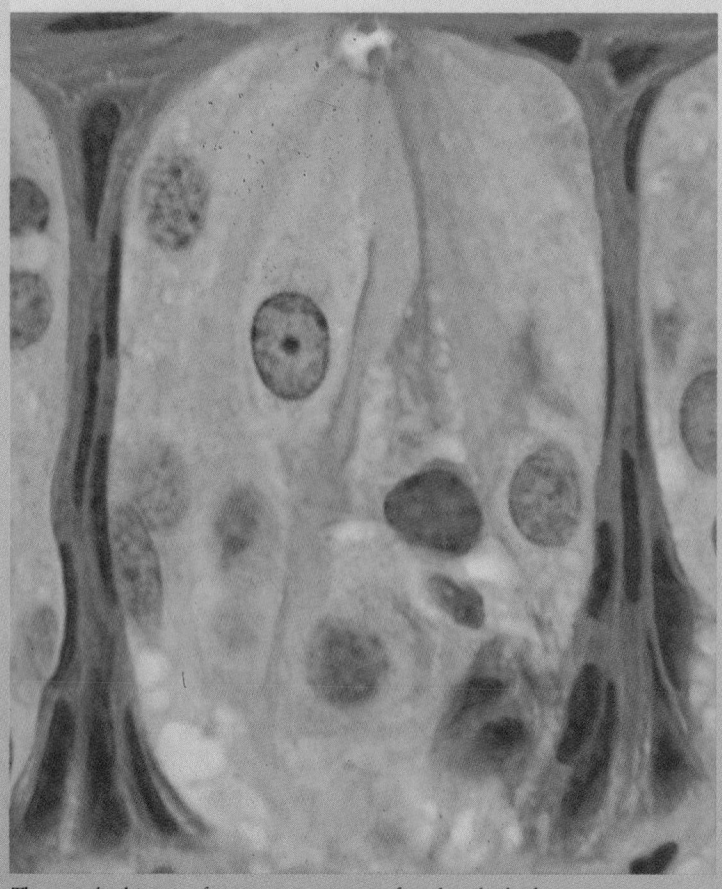

The taste bud is one of many sensory organs found in the body.

I t must be a law of human nature. Ask almost any couple, and they will tell you: she lies shivering under the covers on a cold winter night while he bakes. Out on a hike in winter, he stays warm in a light jacket while she bundles up in stocking cap, down coat, and gloves. What causes this difference between many men and women?

Part of the answer may lie in iron—not pumping iron, but dietary iron. Quite simply, many American women do not consume enough iron to offset losses that occur during **menstruation,** the monthly discharge of blood and tissue from the lining of the uterus. Iron deficiencies in women may reduce internal heat production.[1]

John Beard, a researcher at Pennsylvania State University, recently published a study that supports this conclusion. In his research, Beard compared two groups of women, one with low levels of iron in the blood and another with normal levels. Beard found that body temperature dropped more quickly in iron-deficient women exposed to cold than in those with normal iron levels. He also found that iron-deficient women generated 13% less body heat.

Adding credence to the hypothesis that iron deficiency reduces body heat and makes women colder, Beard gave the iron-deficient women iron supplements for 12 weeks, after which they responded normally to cold.

At least two hypotheses can explain these results. The first is that iron deficiencies reduce the amount of oxygen carried in the blood. Iron is a vital component of the hemoglobin molecule, a protein in the red blood cells, which binds to oxygen. Because red blood cells transport oxygen from the lungs to body tissues, a decrease in iron in the hemoglobin molecule may reduce the amount of oxygen available for cellular respiration. Because cellular respiration produces ATP and heat, a shortage of oxygen may lower body heat.

The second hypothesis is based on the role of iron in energy production. Iron is a vital component of the electron transport proteins found in the mitochondria. As noted in Chapter 6, the electron transport system produces most of the ATP in animal cells. If iron levels are low, energy production could be impaired. Because ATP production also releases waste heat, iron deficiencies could reduce internal heat production.

The research on iron may help explain why many women are colder than their male partners and also why many women who take iron supplements during pregnancy report a marked improvement in their response to cold. These findings, and others presented in this chapter, illustrate how important proper nutrition is to normal physiological function.[2]

To further your understanding of nutrition, this chapter begins with a discussion of the major dietary requirements of humans, then describes the process of digestion.

A PRIMER ON NUTRITION

Some people eat to live, but many others seem to live to eat. No matter what your orientation, food probably occupies a central part of your life. You plan your daytime activities around meals. You spend a good part of your life shopping for, preparing, and eating meals. And, depending on your income, 10% to 20% of the money you earn goes to buy food. Thus, one and a half to three hours of each workday goes to providing money for food. (The rest goes to income tax!)

If you live to be 65, you will consume over 70,000 meals. Because foods affect your body in many ways, what you eat will determine how you feel in your later years. That's how important nutrition is to your health. Despite the increased emphasis on nutrition today, studies suggest, most Americans still pay little attention to their diet. To perform our very best, though, we must eat well, acquiring energy and nutrients required by our cells, tissues, and organs.

A **balanced diet** provides proper levels of both nutrients and energy and contains a variety of foods. In 1992, the U.S. Department of Agriculture released a food pyramid to supplant the long-standing four food groups. The food pyramid places foods in six major groups and prescribes allotments from each group necessary for proper nutrition (▷ Figure 11–1).

Table 11-1 lists the basic nutrients required by humans and includes some of the foods and beverages that provide them. As indicated in the table, nutrients are divided into two groups: **macronutrients** and **micronutrients.** Take a

---

[1] Differences in body temperature between men and women may also result from other factors, such as body mass and surface area. Low blood pressure or poor circulation would also make the extremities feel cold.

[2] Because iron can be toxic if ingested in excess, people considering supplementing their diets with iron tablets should consult a physician.

▷ **FIGURE 11–1 Food Guide Pyramid** Replacing the four food groups, the food pyramid divides food into six groups. Recommended daily intake is shown for each group.

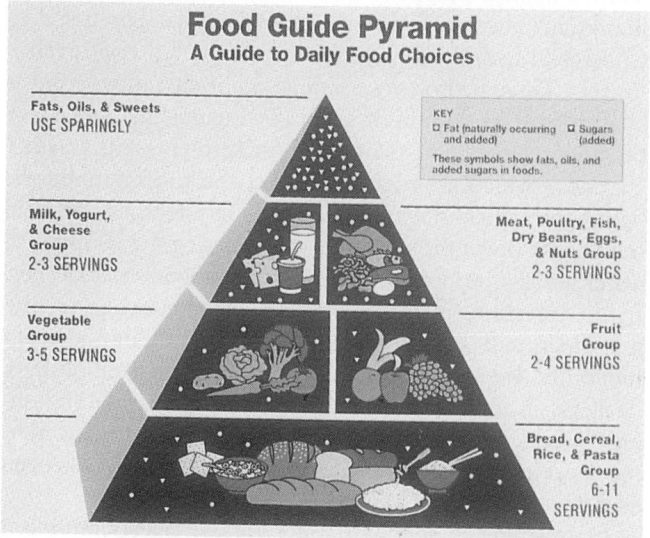

TABLE 11–1 Macronutrients and Micronutrients

NUTRIENTS	FOODS CONTAINING THEM
Macronutrients	
Water	All drinks and many foods
Amino acids and proteins	Milk and milk products, meats, eggs
Lipids	Milk and milk products, meats, eggs, nuts, oils
Carbohydrates	Breads, pastas, cereals, sweets
Micronutrients	
Vitamins	Many vegetables, meats, and fruits
Minerals	Many vegetables, meats, fruits, nuts, and seeds

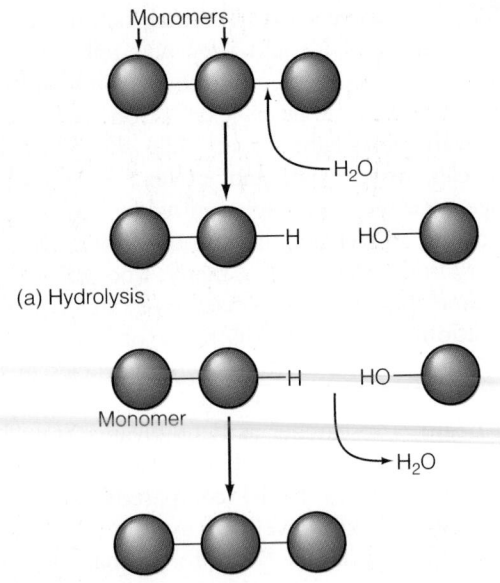

▷ **FIGURE 11–2 Hydrolysis and Dehydration Synthesis** (*a*) Hydrolysis is a reaction in which water is added across a covalent bond, causing the bond to split. (*b*) Dehydration synthesis is the opposite reaction, in which water is lost from two reacting molecules.

few minutes to study the table. If you would like to assess your eating habits, take the test on page 266.

The following section describes the macronutrients and micronutrients more fully and explains why they are important to the proper functioning of the human body and the maintenance of health.

Macronutrients Are Required in Relatively Large Quantities and Include Four Substances: Water, Carbohydrates, Lipids, and Proteins

Water. Water is one of the most important of all the substances we ingest. Without it, a person can survive only about three days. Despite its importance, water rarely shows up on nutrition charts. In part, that is because it is supplied in so many different ways. For example, water constitutes the bulk of the liquids we drink and is present in virtually all of the solid foods we eat; water is even produced internally during cellular metabolism (Chapter 6).

Maintaining the proper level of water in the body is important for several reasons. First, water participates in many chemical reactions in the body. One of the most important reactions is known as **hydrolysis.** Discussed in Chapter 2, hydrolysis means to break up (*lyse*) with water (*hydro*). Thus, hydrolytic reactions occur anytime a bond is broken by the addition of water (▷ Figure 11–2). Most of the food we eat consists of polymers (proteins and polysaccharides), which are too large to be absorbed by the cells lining the small intestine. Enzymes inside the intestinal tract break down the polymers by catalyzing the addition of water across the bonds that hold the monomers together. In this reaction, a hydrogen from water attaches to one monomer, and a hydroxyl group attaches to the adjacent monomer. Because of water's importance in metabolism, a decrease in its level can impair metabolism, including energy production. According to some studies, athletic performance may drop significantly when water level falls even slightly.

Maintaining an adequate water volume is also important

because it helps stabilize body temperature. A decline in the amount of water in your body, for example, decreases the volume of your blood and your extracellular fluid. These declines cause your body temperature to rise, because the heat normally produced by body cells is being absorbed by a smaller volume of water. A rise in body temperature can impair cellular function and can, if high enough, lead to death. One reason physicians tell you to drink plenty of water when you are suffering from a fever is to help control the increase in your body temperature.

Maintaining proper water levels also helps individuals maintain normal concentrations of nutrients and toxic waste products in the blood and extracellular fluid. If you don't drink enough liquid, your urine will become more concentrated; the rise in the concentration of chemicals in the blood and urine increases your chances of developing a kidney stone, a deposit of calcium and other materials that can block the flow of urine, causing extensive damage to this organ.

Maintaining proper water levels is so important that animals have evolved a variety of homeostatic mechanisms to help carry out this function, described in Chapter 16. As the previous discussion suggests, however, water homeostasis also plays a role in maintaining homeostasis in other areas—for example, energy production, body temperature, and blood levels of various ions and molecules.

Carbohydrates. Carbohydrates serve many purposes in the biological world. One of the most important is providing a source of energy.

All active organisms, humans included, need a continuous

TABLE 11–2 Energy Sources in the Average American Diet	
SOURCE	PERCENTAGE
Fat	42
Sucrose	24
Starch	22
Protein	12

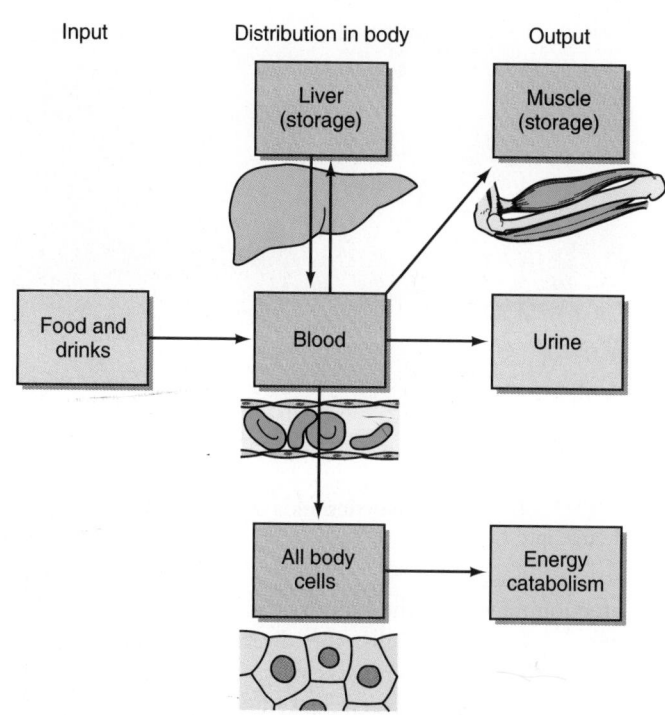

▷ **FIGURE 11–3 Glucose Balance** Glucose levels in the blood result from a balance of input and output.

supply of energy. Cells, for instance, require energy to carry out thousands of functions needed to maintain homeostasis, to grow, to divide, and to transport molecules across their membranes. Surprisingly, 70% to 80% of the total energy required by fairly sedentary humans (individuals who get little exercise) goes to perform basic functions—metabolism, food digestion, absorption, and so on. The remaining energy is used to power body movements, such as walking, talking, and turning on the television. All waste energy (heat), of course, helps maintain body temperature. For more active people, these percentages shift considerably. A basketball player, for example, uses proportionately more energy for vigorous muscular activity than a sedentary office worker.

At rest, the body relies on nearly equal amounts of carbohydrate (mostly glucose) and fat (triglycerides) to supply basic body needs. As you may recall from Chapter 6, glucose is catabolized during cellular respiration.

Glucose comes from a variety of sources. The most common source is the molecule known as starch. Produced by plants, starch is present in grains and their by-products (pasta and bread). It is also provided by many vegetables. Starch contributes on average about 22% of the total energy requirements of most Americans (Table 11–2). Glucose also comes from glycogen, a polysaccharide found in small amounts in meat.

In the digestive system of humans and other vertebrates, glycogen and starch are broken down into glucose molecules, which are absorbed into the bloodstream and distributed to body cells. The cells break down some glucose immediately to produce energy. Most of the rest is converted into glycogen in muscles and the liver and stored for later use (▷ Figure 11–3). Because most of us eat only a few times a day, the liver plays a crucial role in maintaining homeostasis. Between meals and during exercise, the liver breaks down glycogen to form glucose, which is released into the bloodstream. Glycogen in muscle is catabolized to provide energy for muscular contraction but is not released into the bloodstream.

Another important carbohydrate, not involved in energy production, is fiber. Dietary **fiber** is nondigestible polysaccharide (such as cellulose) found in fruits, vegetables, and grains. As noted in Chapter 2, humans cannot digest cellulose, because the body lacks the enzymes needed to break the covalent bonds joining the monosaccharide units (glu-

cose) in the molecule. Consequently, fiber passes through the intestine largely unaffected by stomach acidity or digestive enzymes.

Fiber exists in two basic forms: water-soluble and water-insoluble. Water-soluble fibers are gummy polysaccharides (for example, pectins) in fruits, vegetables, and some grains—including apples, bananas, carrots, barley, and oats. Several studies suggest that water-soluble fiber helps lower blood cholesterol by acting as a sponge that absorbs dietary cholesterol inside the digestive tract, thus preventing it from being absorbed into the bloodstream. Some water-soluble fiber may change the pH of the intestine, making cholesterol insoluble and more difficult to absorb. (The chemical action of water-soluble fibers is discussed in more detail in the section of this chapter on the liver.)

In contrast, water-insoluble fibers are rigid cellulose molecules in celery, cereals, wheat products, and brown rice. Some foods, such as green beans and green peas, contain a mixture of both types. Water-insoluble fiber increases the water content of the **feces,** the semisolid waste produced by the large intestine. This makes the feces softer and facilitates their transport through the large intestine. Increasing the water content also reduces constipation and helps prevent pressure from building up in the large intestine. In some cases, pressure from constipation causes small pouches, or diverticulae, to form in the wall of the large intestine, resulting in a condition known as **diverticulosis** (▷ Figure 11–4). The pouches in the intestinal wall can become inflamed and infected with bacteria, producing a condition known as **diverticulitis.** Bacteria from the local

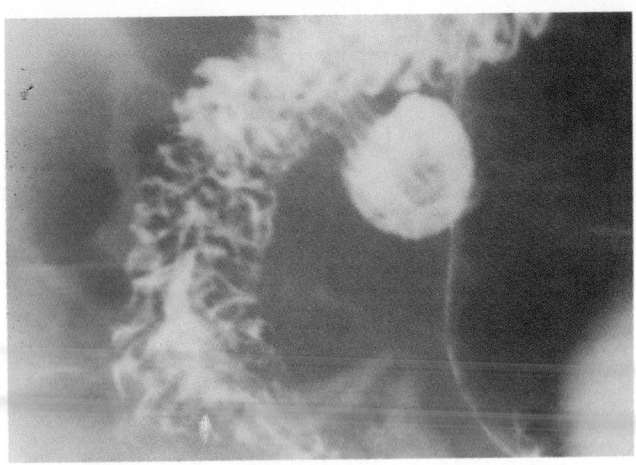

▷ **FIGURE 11–4 Diverticulosis** X-ray showing outpocketing of large intestine.

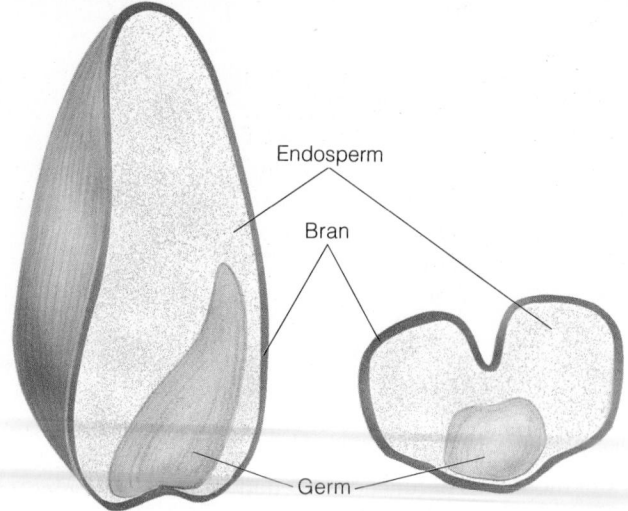

(a)

(b)

▷ **FIGURE 11–5 The Wheat Seed** (*a*) Parts of a wheat grain and the nutrients and minerals they supply. (*b*) White flour made by removing the bran and the germ, and whole-wheat flour made from the entire wheat grain.

infection may enter the bloodstream, causing a dangerously high fever. Occasionally, the diverticulae burst, releasing feces into the abdominal cavity. Because feces contain billions of bacteria, their release into the abdominal cavity results in a massive infection that is difficult to treat and is sometimes fatal.

The incidences of diverticulosis and diverticulitis, although not great, have increased substantially in the United States since the introduction of white flour made from wheat. Why? The wheat grain consists of three parts: (1) the germ, the vitamin- and mineral-rich portion that becomes a new plant; (2) the endosperm, the starchy portion that provides nutrients for the growing plant; and (3) the bran, or shell, which contains most of the fiber (▷ Figure 11–5a). When whole-wheat grain is ground, it produces whole-wheat flour, a brown powder containing the entire wheat grain (Figure 11–5b). In contrast, white flour is produced from grain whose bran and germ have been removed (Figure 11–5b). Removing the bran eliminates most, if not all, of the water-insoluble fiber (cellulose). Removing the germ eliminates most of the vitamins and minerals. White flour, therefore, consists mostly of starch.

The widespread use of white flour in the United States and elsewhere has increased the incidence of diverticulosis and diverticulitis because of the elimination of water-insoluble fiber from our diets. Not surprisingly, these conditions are rare in areas, such as Africa, where fiber intake is higher.

Research suggests that water-insoluble fiber, such as that present in the bran, also reduces the incidence of colon cancer, which afflicts about 3% of the U.S. population. How does fiber function? A recent study showed that some bacteria in the large intestine produce a potent chemical mutagen that may cause cancer. Researchers believe that by accelerating the transport of wastes through the intestine, water-insoluble fiber either reduces the formation of the mutagen or reduces the time in which the intestinal cells are exposed to it. Interestingly, in rural Africa, colon

cancer is practically unheard of. Although other factors may also be responsible for the difference, researchers recommend an increase in the amount of fiber we eat.

Lipids. When most of us think about fats, or lipids, in our diets, we think about their harmful effects. In truth, lipids are absolutely essential to proper nutrition and, as noted earlier, provide about half of the body energy during rest. Fatty acids are broken down in a series of cellular chemical reactions whose products enter the biochemical pathways of cellular respiration, as noted in Chapter 6. The complete breakdown of fat yields slightly more than twice as much ATP as the complete breakdown of carbohydrates (about 9 calories per gram).

Although lipids provide cellular energy, they serve other functions as well. The layer of fat beneath the skin in humans, for instance, helps insulate the body from heat loss and helps maintain homeostasis. Fatty deposits around certain organs also cushion them from damage. Horseback riders, motorcyclists, and runners, for instance, can engage in their sports for hours without damage to internal organs in part because of nature's natural cushions.

Certain lipids are also needed by body cells to synthesize steroid hormones, and lipids are a principal component of the plasma membrane of all body cells. In the intestine, lipids increase the uptake of the fat-soluble vitamins (A, D, E, and K).

Most Americans consume lipids in excess. On average, fats provide about 42% of the dietary caloric intake (see Table 11-2), but to lower the risk of heart attack, fat intake should only be about 30%, perhaps even lower, and animal fat should be eaten in small quantities. (For more on lipids and their effects on heart disease, see Health Note 11–1.)

Amino Acids and Protein. Although many people commonly think of proteins as a source of energy, they are only used for this purpose under two conditions: either when dietary intake of carbohydrates and fats is severely restricted or when protein intake far exceeds demand. Dietary deficiencies occur in millions of children throughout the world. The arms and legs of children deprived of protein are often emaciated because their muscle cells break down protein in an effort to provide energy (▷ Figure 11–6).

On the other end of the spectrum are the overfed populations of the world. Many Americans, for example, eat far more food than they need. In fact, the average American consumes twice the daily requirement for protein. Amino acids derived from the surplus dietary protein are broken down in the body to produce energy and fat, as explained in Chapter 6.

In healthy, well-nourished individuals, dietary protein is used to synthesize enzymes, hormones, and structural proteins such as collagen. Dietary proteins, however, cannot be absorbed by the small intestine; they must first be broken down into amino acids by hydrolytic enzymes in the digestive tract. Amino acids liberated during this process are absorbed into the bloodstream, distributed throughout the body, and used to synthesize protein.

Proteins in the human body contain 20 different amino acids—all of which can be provided from the diet. The body, however, is capable of synthesizing 11 of the amino acids it needs from nitrogen and smaller molecules derived from carbohydrates and fats. Thus, if they are not present in the diet, they can be made. The remaining 9 amino acids cannot be synthesized by body cells; that is, they must be provided by the diet. These amino acids are appropriately called **essential amino acids.** A deficiency of even one of the essential amino acids can cause severe physiological problems. As a result, nutritionists recommend a diet containing many different protein sources. In this way, individuals can receive all of the amino acids they need.

Proteins are divided into two major groups by nutritionists: complete and incomplete. **Complete proteins** contain ample amounts of all of the essential amino acids. Complete proteins are found in milk, eggs, meat, fish, poultry, and cheese. **Incomplete proteins** lack one or

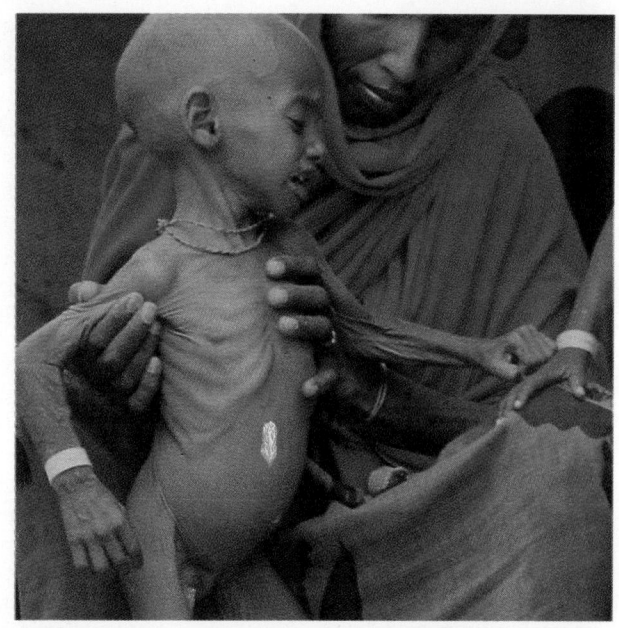

▷ **FIGURE 11–6 A Protein-Deficient Child** This child suffers from severe protein deficiency (kwashiorkor). His arms and legs are emaciated because muscle protein has been broken down to supply energy. His belly is swollen because of a buildup of fluid in the abdomen.

more essential amino acids and include those found in many plant products: nuts, seeds, grains, most legumes (peas and beans), and vegetables. Vegetarians who avoid all animal products, including milk and eggs, must acquire the essential amino acids they need by combining two incomplete protein sources, as shown in ▷ Figure 11–7 (page 262).

Micronutrients are Substances Needed in Small Quantities and Include Two Broad Groups: Vitamins and Minerals

Vitamins. Vitamins are a diverse group of organic compounds present in very small amounts in many of the foods we eat. These molecules are absorbed by the lining of the digestive tract without being broken down. Some vitamins occur in foods in a form known as precursors, or **provitamins.** Inside the body, these precursor molecules are chemically altered to one or more active forms.

The 13 known vitamins play an important role in many metabolic reactions. Because vitamins are recycled many times during metabolic reactions, they are needed only in very small amounts. In fact, 1 gram of vitamin B-12, about 1 teaspoon, would supply over 300,000 people for a day.

Most vitamins are not synthesized in the cells of the body or, if they can be made, are not produced in sufficient amounts to satisfy cellular demands, making dietary intake essential. Vitamin D, for instance, is manufactured by the skin when exposed to sunlight. However, most Americans spend so much time indoors that dietary input is absolutely essential to good health. Table 11–3 (page 263) lists the vitamins and their functions.

LOWERING YOUR CHOLESTEROL

Let there be no doubt about it: diseases of the heart and arteries are leading causes of death in the United States.

Atherosclerosis, the accumulation of plaque on artery walls, and the problems it creates are responsible for nearly two of every five deaths in the United States each year. Thanks to improvements in medical care and diet, however, the death rate from atherosclerosis has been falling steadily in recent years, but it is still a major concern. New research, in fact, shows that atherosclerotic plaque is present even in children.

Researchers believe that atherosclerotic plaques begin to form after minor injuries to the lining of blood vessels. High blood pressure, they think, may damage the lining, causing platelets and cholesterol in the blood to adhere to the injured site. The blood vessel responds by producing cells that grow over the fatty deposit. This thickens the wall of the artery, reducing blood flow. Additional cholesterol is then deposited in the thickened wall, forming a larger and larger obstruction.

Cholesterol deposits impair the flow of blood in the heart and other organs, cutting off oxygen to vital tissues. Blood clots may form in the restricted sections of arteries, further reducing blood flow. When the oxygen supply to the heart is disrupted, cardiac muscle cells can die, resulting in heart attacks and death. Blood clots originating in other parts of the body may also lodge in diseased vessels, obstructing blood flow. Oxygen deprivation can weaken the heart, impairing its ability to pump blood. When the oxygen supply to the heart is restricted, the result is a type of heart attack known as a **myocardial infarc-**

tion. Victims of a myocardial infarction feel pain in the center of the chest and down the left arm. If the oxygen-deprived area is extensive, the heart may cease functioning altogether.

Atherosclerotic plaque also impairs the flow of blood to the brain. Blood clots catch in the restricted areas and block the flow of blood to vital regions of the brain. Victims may lose the ability to speak or to move limbs. If the damage is severe enough, they may die. Thanks to advances in medical treatment, however, many victims can be saved. And over time, they can recover lost functions as other parts of the brain take over for the damaged regions. Atherosclerosis also affects other organs, such as the kidneys.

Atherosclerosis and cardiovascular disease result from nearly 40 factors, some more important than others. Several of these risk factors, such as old age and sex (being male), cannot be changed. Other factors are controllable. These include high blood pressure, high blood cholesterol, stress, smoking, inactivity, and excessive food intake. Of all the risk factors, three emerge as the primary contributors to cardiovascular disease: elevated blood cholesterol, smoking, and high blood pressure.

Consider cholesterol. Cholesterol is essential to normal body function. It is part of the plasma membrane and is needed to synthesize certain hormones. Interestingly, in most people, the majority of the cholesterol in your blood is produced by the liver. The liver synthesizes and releases about 700 milligrams of cholesterol per day. Only about 225 milligrams of cholesterol are derived from the food we eat each day. Normally, the concentration of cholesterol in the blood is constant. If dietary input

falls, the liver increases its output. If the amount of cholesterol in the diet rises, the liver reduces its production. So what's all the fuss about cholesterol in a person's diet?

Even though the liver regulates cholesterol levels, it cannot work fast enough. That is, it may simply be unable to absorb, use, and dispose of cholesterol quickly enough. Consequently, excess cholesterol circulates in the blood after a meal and is deposited in arteries.

Cholesterol is carried in the bloodstream bound to protein. These complexes of protein and lipid fall into two groups: **high-density lipoproteins (HDLs)** and **low-density lipoproteins (LDLs).** HDLs and LDLs function very differently. LDLs, for example, transport cholesterol from the liver to body tissues. In contrast, HDLs are scavengers, picking up excess cholesterol and transporting it to the liver, where it is removed from the blood and excreted in the bile. Research shows that the ratio of HDL to LDL is an accurate predictor of cardiovascular disease. The higher the ratio, the lower the risk of cardiovascular disease.

A high cholesterol level (or hypercholesterolemia) tends to run in families. Thus, if a parent has died of a heart attack or suffers from this genetic disease, his or her offspring are more likely to have high cholesterol levels.

For many years, physicians have advised patients to cut back on high-cholesterol foods, especially eggs, to reduce blood levels of cholesterol. Interestingly, though, a reduction in dietary cholesterol in one individual may result in very little decline in total blood cholesterol, but in another a reduction may result in a much larger drop. The difference in response can be attributed to

Because vitamins are needed in almost all cells of the body, a dietary deficiency in just one vitamin can cause wide-ranging effects. Vitamins also interact with other nutrients. Vitamin C, for example, increases the absorption of iron in the small intestine. Large doses of vitamin C, however, decrease copper utilization by the cells. Consequently,

maintaining good health requires the proper balance of vitamins and other nutrients.

Vitamins fall into two broad categories: water-soluble and fat-soluble. **Water-soluble vitamins** include vitamin C and eight different forms of vitamin B. Water-soluble vitamins are transported in the blood plasma unassisted and

▷ **FIGURE 1** (*a*) Fatty foods like these increase cholesterol levels. (*b*) Foods like these will help you reduce cholesterol levels.

exercise, genetics, initial cholesterol levels, and age.

To reduce your chances of atherosclerosis, the American Heart Association recommends (1) limiting dietary fat to less than 30% of the total caloric intake, (2) limiting dietary cholesterol to 300 milligrams per day, and (3) acquiring 50% or more of one's calories from carbohydrates, especially polysaccharides (notably, starches found in potatoes, rice, and other foods). Reductions in saturated fats (animal fats) can also help lower cholesterol levels, for reasons explained in Chapter 2. You can cut back on saturated fat by reducing your consumption of red meat and trimming the fat off all meats before cooking. You can also increase your consumption of fruits, vegetables, and grains, letting these low-fat foods displace some of the fatty foods you might otherwise have eaten (▷ Figure 1).

New and still controversial research also indicates that a diet rich in fish oils can help reduce blood cholesterol. Fish oils contain polyunsaturated fatty acids called omega-3 fatty acids. These fatty acids stimulate the release of prostaglandins, which increase the flexibility of the red blood cells and reduce their stickiness, which is essential for blood clotting. Research on mice shows that a diet rich in omega-3 fatty acids doubles their life span. One of the conclusions of the study, though, is that omega-3 fatty acids, which are extremely susceptible to oxidation, are effective only if oxidation can be prevented. Mincing fish prior to cooking increases oxidation and lowers the level of omega-3 fatty acids.

One's cholesterol level can be lowered with drugs, diet, and exercise. Research spanning several decades shows that a lower cholesterol level translates into a decline in cardiovascular disease. Unfortunately, experts disagree on several key issues. One is exactly who will benefit from a reduction in cholesterol. Some researchers say that only high-risk people with a cholesterol level over 250 milligrams per 100 milliliters of blood should take steps to cut back. Because two-thirds of the American adult population has a blood cholesterol level over 200 milligrams, some experts believe that the entire adult population should take steps to reduce cholesterol.

High cholesterol is also surprisingly common in children, leading many health experts to believe that steps should be taken to prevent problems later in life. In children under the age of 2, however, diets should not be restricted. A diet that is too restrictive may actually impair physical growth and development. What children need is a well-balanced diet, low in fats, especially animal fat, with sufficient calories from other sources. If nothing else, it could help create the eating habits necessary for good health throughout adult life.

are eliminated by the kidneys. Because they are water-soluble, most are not stored in the body in any appreciable amount.

Water-soluble vitamins generally work in conjunction with enzymes, promoting the cellular reactions that supply energy or synthesize cellular materials. Contrary to common myth, the vitamins themselves do not provide energy.

Many proponents of vitamin use believe that megadoses of water-soluble vitamins are harmless because these vitamins are excreted in the urine and do not accumulate in the body. That is, many people think that we can ingest unlimited amounts of water-soluble vitamins with impu-

▷ **FIGURE 11–7**
Complementary Protein Sources
By combining protein sources, a vegetarian who consumes no animal by-products can be assured of getting all of the amino acids needed. Legumes can be combined with foods made from grains or nuts and other seeds.

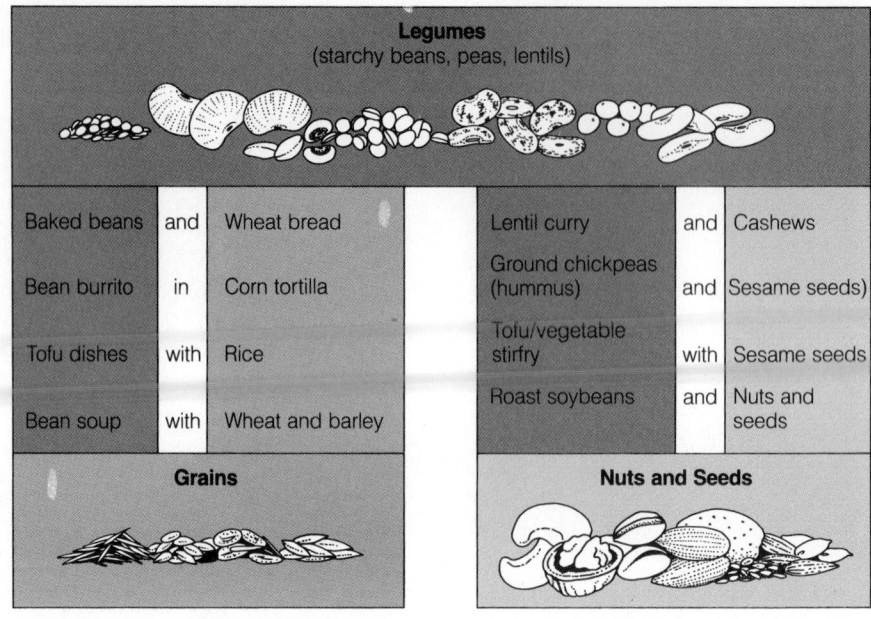

Legumes (starchy beans, peas, lentils)					
Baked beans	and	Wheat bread	Lentil curry	and	Cashews
Bean burrito	in	Corn tortilla	Ground chickpeas (hummus)	and	Sesame seeds)
Tofu dishes	with	Rice	Tofu/vegetable stirfry	with	Sesame seeds
Bean soup	with	Wheat and barley	Roast soybeans	and	Nuts and seeds
Grains			**Nuts and Seeds**		

nity. New research, however, shows that this is not entirely true. Some water-soluble vitamins, such as vitamin C, when ingested in excess, can be toxic. (See Table 11–3 for some examples.)

The **fat-soluble vitamins** are vitamins A, D, E, and K. They perform many different functions. Vitamin A, for example, is converted to light-sensitive pigments in receptor cells of the retina, the light-sensitive layer of the eye. These pigments play an important role in vision. Another member of the vitamin A group removes harmful chemicals (oxidants) from the body. Unlike water-soluble vitamins, the fat-soluble vitamins are stored in body fat and accumulate in the fat reserves. The accumulation of fat-soluble vitamins can have many adverse effects (Table 11–3). An excess of vitamin D, for example, can cause weight loss, nausea, irritability, kidney stones, weakness, and other symptoms. Large doses of vitamin D taken during pregnancy can cause birth defects.

Vitamin excess is encountered largely in the developed countries and usually occurs only in people taking vitamin supplements. Each year, in fact, approximately 4000 Americans are treated for vitamin supplement poisoning. To avoid problems from excess vitamins, nutritionists recommend eating a balanced diet that provides all of the vitamins the body needs, rather than taking vitamin pills. Megadoses, they say, should be avoided.

Vitamin deficiencies, like dietary excesses, can lead to serious problems. A deficiency of vitamin D, for example, can produce rickets, a disease that results in bone deformities. Vitamin K deficiencies can result in severe bleeding on injury and internal hemorrhaging. Vitamin C deficiency can result in delayed wound healing and reduced immunity, making people more susceptible to infectious disease.

Most people afflicted by vitamin deficiencies are those who fail to get enough to eat, although even apparently well-fed individuals may suffer from a vitamin deficiency if they are not eating a well-rounded diet. Sadly, about one of every six people living in the nonindustrialized nations of the world, or about 700 million to 800 million people, go to bed hungry; most of these people suffer from multiple vitamin deficiencies and exhibit many symptoms. People suffering from vitamin deficiency typically complain of weakness and fatigue. Children with insufficient vitamin intake fail to grow. All told, about 10 million children under the age of 5 suffer from extreme malnutrition in the less-developed nations of the world. Another 90 million under the age of 5 are moderately malnourished.

Deficiencies of vitamin A afflict over 100,000 children worldwide each year. If not corrected, vitamin A deficiency causes the eyes to dry. Ulcers may form on the eyeball and can rupture, causing blindness (▷ Figure 11–8).

Vitamin deficiencies are also surprisingly common in the industrialized world, where poverty is a way of life for many millions of people. In the United States, one of every five children is born into poverty and is a candidate for vitamin deficiency.

Minerals. On average, an adult contains about 5 pounds of **minerals,** naturally occurring inorganic substances vital to many life processes. Humans require about two dozen minerals, such as calcium, sodium, iron, and potassium, to carry out normal body functions.

Minerals, like vitamins, are micronutrients and are derived from the food we eat and the beverages we drink. Minerals are divided into two groups: the **major minerals** and the **trace minerals** (Table 11–4). The major minerals are present in the body in amounts larger than 5 grams; the trace minerals are found in lesser quantities.

Calcium and phosphorus, for example, are major minerals, and make up three-fourths of the total mineral content

TABLE 11–3 Important Information on Vitamins

VITAMIN	MAJOR DIETARY SOURCES	MAJOR FUNCTIONS	SIGNS OF SEVERE, PROLONGED DEFICIENCY	SIGNS OF EXTREME EXCESS
Fat-soluble				
A	Fat-containing and fortified dairy products; liver; provitamin carotene in orange and deep green produce	Component of visual pigments in eye; still under intense study	Keratinization of epithelial tissues including the cornea of the eye (xerophthalmia); night blindness; dry, scaling skin; poor immune response	From preformed vitamin A: damage to liver, kidney, bone; headache, irritability, vomiting, hair loss, blurred vision. From carotene: yellowed skin.
D	Fortified and full-fat dairy products; egg yolk	Promotes absorption and use of calcium and phosphorus	Rickets (bone deformities) in children; osteomalacia (bone softening) in adults	Gastrointestinal upset; cerebral, cardiovascular, kidney damage; lethargy
E	Vegetable oils and their products; nuts, seeds; present at low levels in other foods	Antioxidant that prevents plasma membrane damage; still under intense study	Possible anemia	Debatable; perhaps fatal in premature infants given intravenous infusion
K	Green vegetables; tea, meats	Aids in formation of certain proteins, especially those for blood clotting	Severe bleeding on injury: internal hemorrhage	Liver damage and anemia from high doses of the synthetic form menadione
Water-soluble				
Thiamin (B-1)	Pork, legumes, peanuts, enriched or whole-grain products	Coenzyme used in energy metabolism	Beriberi (nerve changes, sometimes edema (swelling), heart failure)	?
Riboflavin (B-2)	Dairy products, meats, eggs, enriched grain products, green leafy vegetables	Coenzyme used in energy metabolism	Skin lesions	?
Niacin	Nuts, meats; provitamin tryptophan in most proteins	Coenzyme used in energy metabolism	Pellagra (which may be multiple vitamin deficiencies)	Flushing of face, neck, hands; liver damage
B-6	High protein foods in general, bananas, some vegetables	Coenzyme used in amino acid metabolism	Nervous and muscular disorders	Unstable gait, numb feet, poor hand coordination, abnormal brain function
Folacin	Green vegetables, orange juice, nuts, legumes, grain products	Coenzyme used in DNA and RNA metabolism; single carbon utilization	Megaloblastic anemia (large, immature red blood cells); gastrointestinal disturbances	Masks vitamin B-12 deficiency
B-12	Animal products	Coenzyme used in DNA and RNA metabolism; single carbon utilization	Megaloblastic anemia; pernicious anemia when due to inadequate intrinsic factor; nervous system damage	?
Pantothenic acid	Widely distributed in foods	Coenzyme used in energy metabolism	Fatigue, sleep disturbances, nausea, poor coordination	?
Biotin	Widely distributed in foods	Coenzyme used in energy metabolism	Dermatitis, depression, muscular pain	?
C	Fruits and vegetables, especially broccoli, cabbage, cantaloupe, cauliflower, citrus fruits, green pepper, strawberries	Maintains collagen; is an antioxidant; aids in detoxification; still under intense study	Scurvy (skin spots; bleeding gums, weakness); delayed wound healing; impaired immune response	Gastrointestinal upsets, confounds certain lab tests, poorer immune response

SOURCE: Adapted from J. L. Christian and L. L. Greger, *Nutrition for Living*, 2d ed., copyright © 1988 by the Benjamin/Cummings Publishing Company. Used with permission.

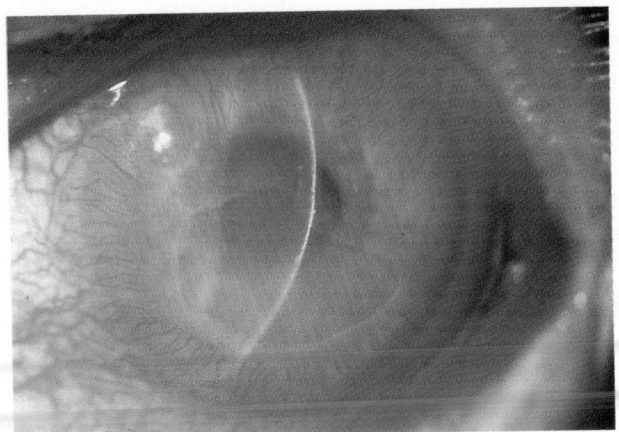

▷ **FIGURE 11–8 Vitamin A Deficiency** Vitamin A deficiency causes the cornea of the eye to dry and become irritated. If the deficiency is not corrected, ulcers may form and break, causing blindness.

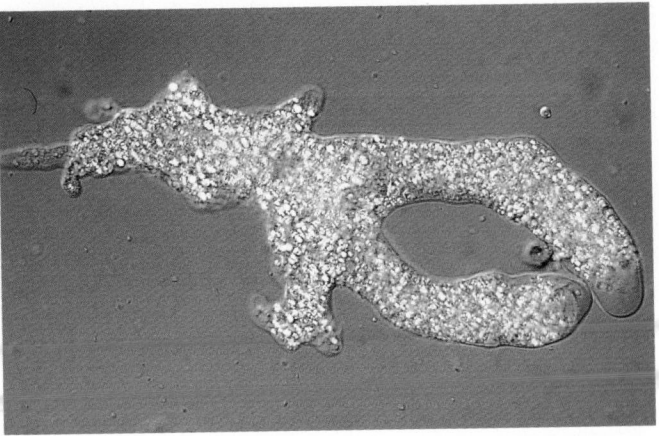

▷ **FIGURE 11–9 The Simplest Form of Digestion** Amoebae and other protists phagocytize materials, then digest them intracellulary within food vacuoles.

of the human body. These two minerals form part of the dense extracellular matrix of bone and are required in a much greater quantity than zinc or copper, two trace minerals that are components of some enzymes.

The distinction between major and trace minerals is not meant to imply that one group is more important than the other. In fact, a daily deficiency of a few micrograms (a microgram is 1/1000 of a gram, and a gram is 1/454 of a pound) of iodine is just as serious as a deficiency of several hundred milligrams of calcium.

Table 11–4 lists some of the minerals, their function, and problems that arise when they are ingested in excess or deficient amounts.

If you'd like to assess your eating habits and learn how to eat a more healthy diet, complete the survey in Table 11–5.

≈ THE DIGESTIVE SYSTEM

The human digestive system is a product of millions of years of evolution. During the course of evolution, strategies for digesting food have generally increased in complexity. In single-celled protists, such as *amoebae,* food is phagocytized, then digested intracellularly in food vacuoles (▷ Figure 11–9). This is the simplest form of digestion and, no doubt, the earliest to evolve. Its inefficiency probably accounts for its relatively rare occurrence within the animal world.

Much more efficient modes of digestion are provided by various types of digestive tracts, tubular canals that run through organisms. Digestive tracts provide a site for extracellular digestion in a wide number of vertebrates and invertebrates alike.

The simplest digestive tract is found in flatworms and a group of animals known as the **cnidarians** (ny-dar-ee-uns), which includes jellyfish, hydras, sea anemones, and coral (▷ Figure 11–10). In these creatures, the gut consists of a saclike or branched tube with only a single opening.

Food enters this opening and is digested by enzymes secreted into the gut. Waste is ejected through the same opening.

Most other animals have a complete gut, consisting of a tube opened at both ends. Food enters the mouth and is digested by enzymes secreted into the gut. Waste is excreted at the opposite end, the **anus** (▷ Figure 11–11,

▷ **FIGURE 11–10 The Incomplete Gut** Food enters the gut, or coelenteron, of the jellyfish and other cnidarians through an opening that also serves to rid these organisms of undigested waste. Enzymes released by some of the cells lining the sac digest the food inside the gut. (*a*) Schematic and (*b, next page*) photograph.

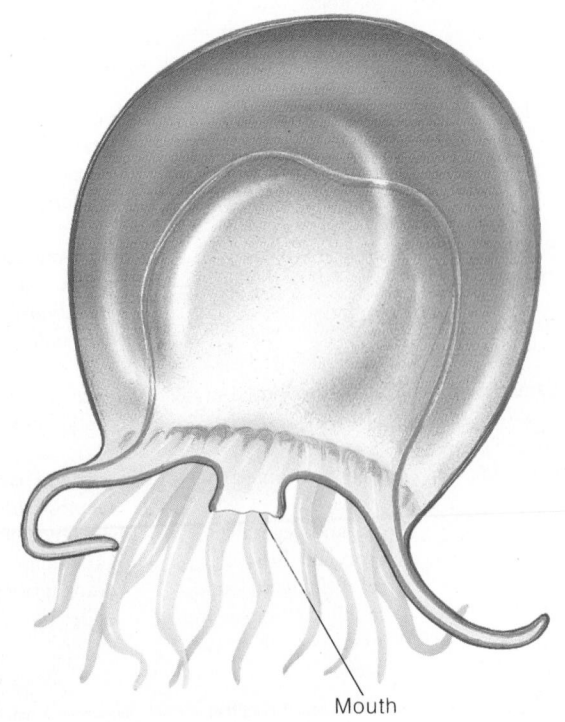

(a)

Mouth

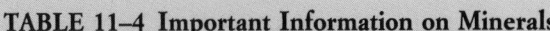

TABLE 11–4 Important Information on Minerals

MINERAL	MAJOR DIETARY SOURCES	MAJOR FUNCTIONS	SIGNS OF SEVERE, PROLONGED DEFICIENCY	SIGNS OF EXTREME EXCESS
Major minerals				
Calcium	Milk, cheese, dark green vegetables, legumes	Bone and tooth formation; blood clotting; nerve transmission	Stunted growth; maybe bone loss	Depressed absorption of some other minerals
Phosphorous	Milk, cheese, meat, poultry, whole grains	Bone and tooth formation; acid-base balance; component of coenzymes	Weakness; demineralization of bone	Depressed absorption of some minerals
Magnesium	Whole grains, green leafy vegetables	Component of enzymes	Neurological disturbances	Neurological disturbances
Sodium	Salt, soy sauce, cured meats, pickles, canned soups, processed cheese	Body water balance; nerve function	Muscle cramps; reduced appetite	High blood pressure in genetically predisposed individuals
Potassium	Meats, milk, many fruits and vegetables, whole grains	Body water balance; nerve function	Muscular weakness; paralysis	Muscular weakness; cardiac arrest
Chloride	Salt, many processed foods (as for sodium)	Plays a role in acid-base balance; formation of gastric juice	Muscle cramps; reduced appetite; poor growth	Vomiting
Trace minerals				
Iron	Meats, eggs, legumes, whole grains, green leafy vegetables	Component of hemoglobin and enzymes	Iron-deficiency anemia, weakness, impaired immune function	Acute: shock, death; chronic: liver damage, cardiac failure
Iodine	Marine fish and shellfish; dairy products; iodized salt; some breads	Component of thyroid hormones	Goiter (enlarged thyroid)	Iodide goiter
Fluoride	Drinking water, tea, seafood	Maintenance of tooth (and maybe bone) structure	Higher frequency of tooth decay	Mottling of teeth; skeletal deformation

SOURCE: Adapted from J. L. Christian and L. L. Greger, *Nutrition for Living*, 2d ed., copyright © 1988 by the Benjamin/Cumming Publishing Company. Used with permission.

(b)

page 268). In the course of evolution, the various parts of the digestive tract have evolved to perform specific functions, as shown in the figure. Some regions may store food; others grind and mix the food. Still others play a role in digestion and absorption. The earthworm digestive tract is shown in ▷ Figure 11–12 (page 268), along with the functions of each region.

In humans and most animals, food digestion requires physical and chemical processes that take place in the digestive tract (▷ Figure 11–13, page 269). In the mouth, for example, food is sliced, crushed, and torn by the teeth into smaller particles, greatly increasing the surface area presented to the digestive enzymes in the stomach and small intestine and therefore increasing the efficiency of digestion. Enzymes participate in a chemical breakdown of food. In the small intestine, amino acids, monosaccharides, and other small molecules produced by enzymatic digestion are absorbed into the bloodstream for distribution to body cells.

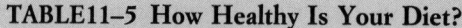 **TABLE11-5 How Healthy Is Your Diet?**

This quiz enables you to assess your eating habits. The more points you get, the better your nutritional health is likely to be.

PART I

1. I usually limit my meat, fish, poultry, or egg servings to once or twice a day. — Yes/No
2. I eat red meats (beef, ham, lamb, or pork) not more than about three times a week. — Yes/No
3. I remove fat or ask that fat be trimmed from meat before cooking. — Yes/No
4. I eat about three or four eggs per week, including those cooked with other foods. — Yes/No
5. I sometimes have meatless days and eat such protein-rich foods as legumes and nuts. — Yes/No
6. I usually broil, boil, bake, or roast meat, fish, or poultry; I usually don't fry it. — Yes/No

Total "YES" answers: _____

PART II

7. I have two or more cups of milk or the equivalent in milk products every day. — Yes/No
8. I drink low-fat or nonfat milk (2% or less butterfat) rather than whole milk. — Yes/No
9. I eat ice cream or ice milk only twice a week or less. — Yes/No
10. I seldom have more than about 3 tsp of margarine or butter per day. — Yes/No

Total "YES" answers: _____

PART III

11. I usually have one serving (½ c) of citrus fruit or juice (oranges, grapefruit, etc.) each day. — Yes/No
12. I have at least one serving of dark green or deep orange vegetables each day. — Yes/No
13. I eat fresh fruits and vegetables when I can get them. — Yes/No
14. I cook vegetables without fat (if I use margarine, it's measured and added after cooking). — Yes/No
15. I eat fresh fruit for dessert more often than pastries. — Yes/No

Total "YES" answers: _____

PART IV

16. I generally eat whole-grain breads. — Yes/No
17. Most of the cereals I use are whole-grain and good sources of fiber. — Yes/No
18. The cereals I use have little or no sugar added. — Yes/No
19. I use brown rice in preference to white rice. — Yes/No
20. I generally have at least four servings of bread or cereal grain products each day. — Yes/No

Total "YES" answers: _____

PART V

21. I am usually within 5 to 10 lb of the weight considered appropriate for my height. — Yes/No
22. I drink no more than 1½ oz of alcohol (one to two drinks) per day. — Yes/No
23. I do not add salt to food after preparation and prefer foods salted lightly or not salted at all. — Yes/No
24. I try to avoid foods high in refined sugar and use sugar sparingly. — Yes/No
25. I always eat a breakfast of at least cereal and milk, egg and toast, or other protein-carbohydrate combination with fruit or fruit juice. — Yes/No

Total "YES" answers: _____

For each "yes" answer, give yourself 1 point. How are your points distributed among the various areas of nutrition?

		Excellent	Good	Fair	Poor	Your score
Part I	Meat and meat alternate choices	5–6	4	3	0–2	_____
Part II	Dairy choices	4	3	2	0–1	_____
Part III	Fruit and vegetable choices	5	4	3	0–2	_____
Part IV	Grain choices	5	4	3	0–2	_____
Part V	Weight control, other choices	5	4	3	0–2	_____

How is your over-all nutrition score? **Total points earned:** _____

(24–25 is excellent; 19–23 is good; 14–18 is fair; 13 or lower is poor.)

SOURCE: Adapted with permission from Roger Sargent, *Have a Good Life Series,* Greenville, S.C.: Liberty Life.

INSECT ADAPTATIONS FOR FEEDING

Insects are one of the most successful groups of animals that inhabit the Earth. In fact, there are more insect species than all other forms of life combined! What is more, some scientists believe that hundreds of thousands—perhaps millions—of additional species are yet to be discovered. Most of these will probably be discovered in tropical rain forests, where one recent study showed that as many as 20,000 species may be found on a single tree!

Adapted to a wide range of habitats, insects have been on the planet for about 400 million years. They owe their great success in large part to the evolution of flying. This ability helps them escape predators, find food, locate mates, and move to new habitats more quickly than most other animals.

Insects occupy nearly every terrestrial and aquatic habitat known to humankind, and they are rare only in the seas. With so many species occupying so many habitats, this remarkable class of animals displays a wide range of adaptations. Here we focus our attention on a few adaptations related to the ingestion and digestion of food.

As anyone who has ever walked through the woods or tended a garden knows, many insects feed on plant or animal juices. These species come equipped with a tiny, needlelike proboscis (pro-bos-is). These hollow structures can penetrate flowers, stems, leaves, or even skin (▷ Figure 1) and can be used to suck nutrient-rich fluids from plants and animals.

Perhaps the most familiar example is the mosquito. The female mosquito requires blood for successful egg production. Besides having a sharp proboscis, which she inserts into the skin, the female mosquito has tiny glands that release a chemical anticoagulant. This substance prevents blood from clotting, thus ensuring a steady flow while she dines on an animal's blood.

Some insects have evolved strong mandibles composed of chitin (a type of polysaccharide). The mandibles are used for biting into wood, leaves, or prey. Leaf-cutting ants, for instance, have extremely strong chitinous mandibles, which aid in the chewing of leaves, which they deposit in their underground tunnels. Here the leaves support a healthy colony of fungi, which are eaten by the ants.

A number of insects have auxiliary food storage capacity—that is, sites where they can store ingested foodstuffs for later use. This permits many insects to feed intermittently. The mosquito comes to mind as a good example. It contains a small, thin-walled sac, that can store blood for up to a week before it passes into the digestive tract. Honeybees also have a storage site for nectar (sugar secreted by the base of a flower's petals). Nectar is stored in a special segment of the gut where it is converted to honey. When the bee returns to its hive, it regurgitates the honey into a waxy honeycomb.

The grasshopper's digestive system—like that of so many other insects—contains three segments, the foregut, the midgut, and the hindgut. The foregut contains a region known as the crop, which functions primarily in food storage. In some insects, the crop has a roughened surface that helps grind food, which then passes on to the midgut, where digestion and absorption occur.

Most insects digest their food within the digestive tract, but the common housefly has a unique mode of digestion. As a fly walks across the icing of cake, it tastes the sweet substance with its feet. (In fact, many insects have their taste receptors in their feet.) As it wanders around, the fly disgorges enzymes, which it then stirs into the food with its mouth. These enzymes digest food molecules. But don't worry, says the biologist Robert Wallace. When it's done, the housefly sucks any mess it has made with its tubelike mouthparts, leaving the rest for you.

▷ **FIGURE 1 Proboscis of the mosquito.**

The Mouth Is the Site of the Physical Breakdown of Food

The mouth is a complex structure in which food is broken down mechanically and, to a much lesser degree, chemically. Food taken into the mouth is sliced into smaller pieces by the sharp teeth in front; it is then ground into a pulpy mass by the flatter teeth toward the back of the mouth. As the food is pulverized in the mouth, it is liquefied by **saliva,** a watery secretion released by the **salivary glands,** three sets of exocrine glands located around the oral cavity (▷ Figure 11–14). The release of saliva is triggered by the smell, feel, taste, and sometimes even the thought of food.

Saliva (1) liquifies the food, making it easier to swallow, (2) kills or neutralizes some bacteria via the enzymes and antibodies it contains, (3) dissolves substances so they can

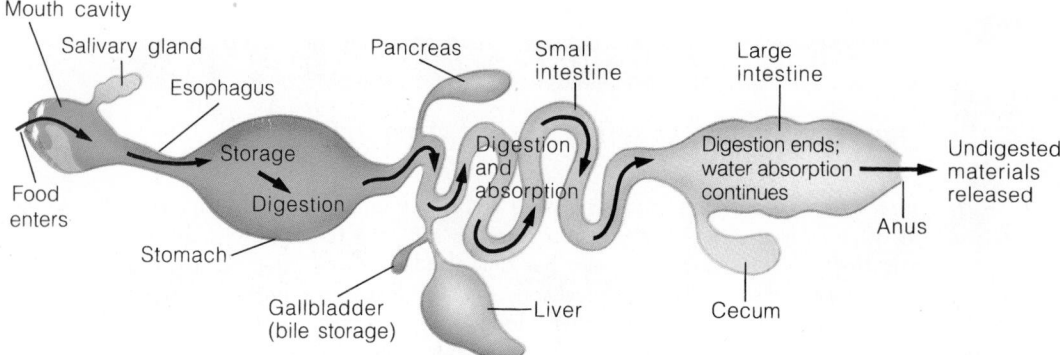

▷ **FIGURE 11–11 The Complete Gut** This diagram illustrates the general anatomy of the tubular digestive tract present in many animals. Food enters the mouth and is excreted at the opposite end through the anus. Along the way, it is chemically and physically broken down, then transported into the bloodstream.

be tasted, and (4) begins to break down starch molecules with the aid of the enzyme **amylase.** Saliva also cleanses the teeth, washing away bacteria and food particles. Because the release of saliva is greatly reduced during sleep, bacteria tend to accumulate on the surface of the teeth, where they break down microscopic food particles, producing some foul-smelling chemicals that give us "dragon breath," or "morning mouth."

Controlling the bacteria that live on the teeth by regular brushing is important, because these organisms secrete a rather sticky material called **plaque.** Plaque adheres to the surface of the teeth, trapping bacteria. Entombed in their own secretions, these bacteria release small amounts of a weak acid that dissolves the hard outer coating of our teeth, the **enamel.** This acid forms small pits in the enamel, commonly referred to as **cavities.** Continued acid secretion may cause the cavities to deepen into the softer layer beneath the enamel. If the cavity is left untreated, an entire tooth can be lost to decay.

Brushing helps remove plaque and helps reduce the incidence of cavities. Most toothpastes also contain small amounts of fluoride, which hardens the enamel and helps reduce cavities. In most cities and towns, small amounts of fluoride are added to drinking water as a preventive measure. However, one recent study suggests that excess fluoride may be carcinogenic, and health officials are reexamining this practice.

A recent study showed that chewing sugarless gum increases the flow of saliva and cleanses the teeth. By chewing sugarless gum within 5 minutes after a meal, and for at least 15 minutes, you can reduce the incidence of cavities.

After food is chewed, it must be swallowed. The tongue plays a key role in swallowing by pushing food to the back of the oral cavity into the **pharynx,** a funnel-shaped chamber that connects the oral cavity with the **esophagus,** a long muscular tube that leads to the stomach (see Figure 11–13).

The tongue, which also aids in speech, contains taste receptors, the **taste buds,** on its upper surface. Taste buds are stimulated by four basic flavors: sweet, sour, salty, and bitter (▷ Figure 11–15). Various combinations of these flavors (combined with odors we smell) give us a rich assortment of tastes.

Food propelled from the pharynx into the esophagus is prevented from entering the trachea, or windpipe, which carries air to the lung and lies in front of the esophagus, by the epiglottis (▷ Figure 11–16). The **epiglottis** is a flap of tissue that acts like a trap door, closing off the trachea during swallowing. (Details of epiglottis function are found in Chapter 14.)

Swallowing begins with a voluntary act—the tongue pushing food into the back of the oral cavity. Once food enters the pharynx, however, the process becomes automatic. Food in the pharynx stimulates stretch receptors in the wall of this organ, which, in turn, trigger the **swallowing reflex,** an involuntary contraction of the muscles in the wall of the pharynx. This forces the food into the esophagus.

▷ **FIGURE 11–12 The Digestive System of the Earthworm** This valuable organism has a complete gut, which consists of a single tubular structure with a mouth and anus.

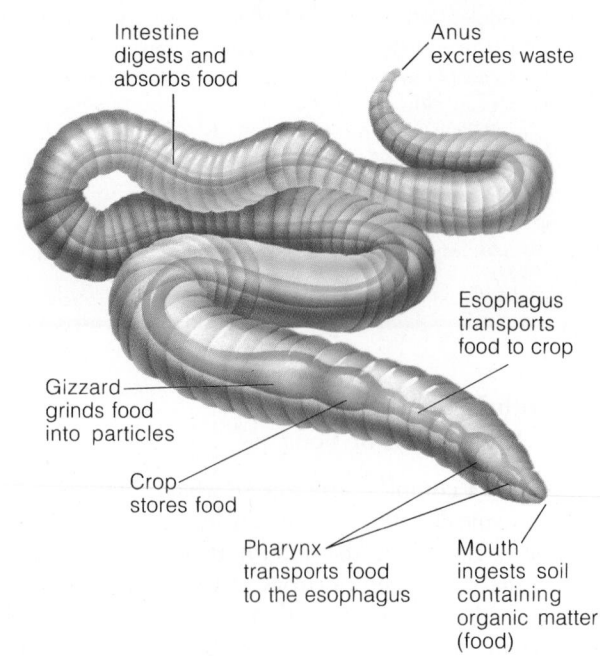

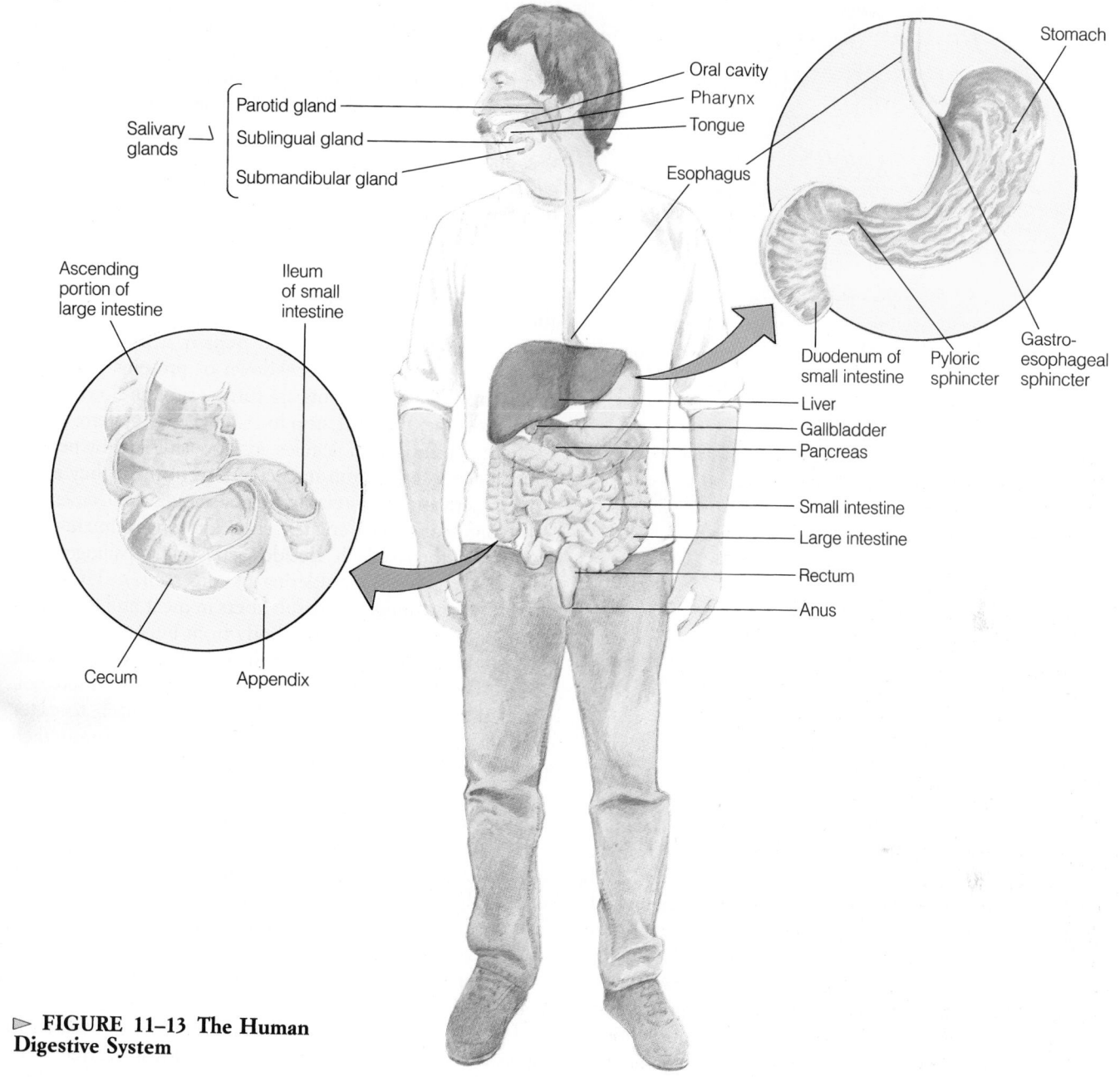

Salivary glands {
- Parotid gland
- Sublingual gland
- Submandibular gland
}

Oral cavity
Pharynx
Tongue
Esophagus

Stomach
Duodenum of small intestine
Pyloric sphincter
Gastro-esophageal sphincter

Ascending portion of large intestine
Ileum of small intestine

Liver
Gallbladder
Pancreas
Small intestine
Large intestine
Rectum
Anus

Cecum
Appendix

▷ **FIGURE 11–13 The Human Digestive System**

The Esophagus Transports Food to the Stomach via Peristalsis

Involuntary contractions of the muscular wall of the esophagus propel the food to the stomach. The muscles of the esophagus contract above the swallowed food mass, squeezing it along (▷ Figure 11–17a, page 272). This involuntary muscular action is called **peristalsis.** It is so powerful that you can swallow when hanging upside down. Peristalsis also propels food (and waste) along the rest of the digestive tract.

Scientists once thought that esophageal peristalsis could proceed in the opposite direction under certain conditions and called this process **reverse peristalsis,** or, more commonly, vomiting. On closer examination, however, they found that the stomach and esophagus are both relaxed during vomiting. Food is expelled from the stomach as a result of contractions in the muscles of the abdomen and the diaphragm, which separates the thoracic and abdominal cavities and plays an important role in breathing.

Vomiting is a reflex action that occurs when irritants are present in the stomach. It is, therefore, a protective adaptation that allows the body to rid the stomach of bad food and harmful viruses and bacteria, so important to vertebrates. Vomiting may also be caused by (1) emotional factors, such as stress, (2) rotation or acceleration of the head, as in motion sickness, (3) stimulation of the back of the throat, and (4) elevated pressure inside the brain caused by injury.

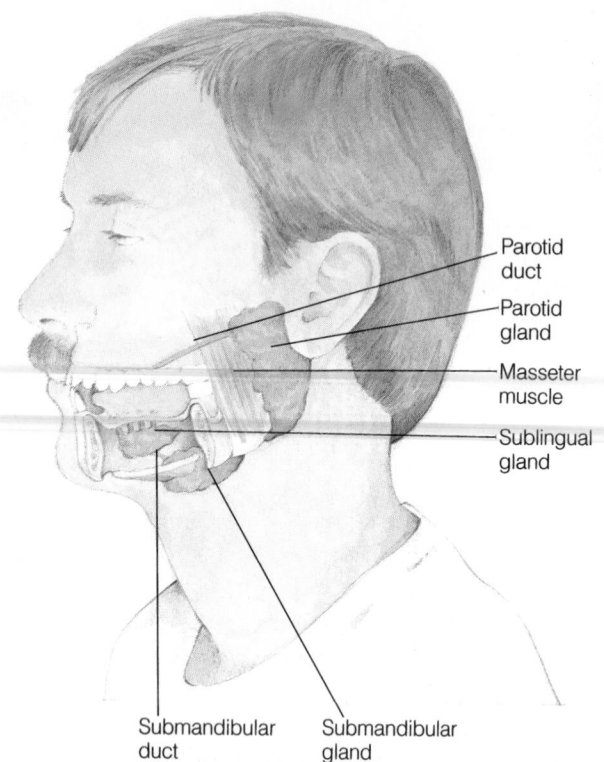

Parotid
duct

Parotid
gland

Masseter
muscle

Sublingual
gland

Submandibular
duct

Submandibular
gland

▷ **FIGURE 11–14 Salivary Glands** Three salivary glands (parotid, submandibular, and sublingual) are located in and around the oral cavity and empty into the mouth via small ducts.

The Stomach Stores Food, Releasing It into the Small Intestine in Spurts

In humans, the stomach, as shown in Figure 11–13, lies on the left side of the abdominal cavity, partly under the protection of the rib cage. Food enters the stomach via the esophagus. The opening to the stomach, however, is closed off by the **gastroesophageal sphincter,** a thickened layer of smooth muscle at the juncture of the esophagus and stomach (▷ Figure 11–18). As food enters the lower esophagus, the gastroesophageal sphincter opens, allowing the food to enter the stomach. The sphincter then promptly closes, preventing food and acid from percolating upward. If the sphincter fails to close, acid rising in the esophagus causes irritation, a condition commonly known as "heartburn" (see Figure 11–17b).

Inside the stomach, food is liquified by acidic secretions of tiny glands in the wall of the stomach, the **gastric glands.** These glands produce a watery secretion called **gastric juice.** It contains hydrochloric acid (HCl), pepsin, and a proteolytic (protein-digesting) enzyme precursor known as **pepsinogen** (discussed in more detail below). These substances come from two distinctly different cell types in the gastric gland.

Inside the stomach, food is churned by peristalsis and mixed with gastric juice. The churning action of the stomach's muscular walls may help break down large pieces of food. Combined with the liquid from salivary glands and

the gastric glands, the food becomes a rather thin, watery paste known as **chyme.** The stomach can hold 2 to 4 liters (2 quarts to 1 gallon) of chyme, which is gradually released into the small intestine at a rate suitable for proper digestion and absorption. The periodic release maximizes the efficiency of digestion.

Contrary to what many think, very little enzymatic digestion occurs in the stomach. The stomach's role is largely to prepare most food for enzymatic digestion that will occur in the small intestine. There are some exceptions to this rule, however. Protein is one of them.

Proteins are denatured by hydrochloric acid (HCl) in the stomach. Hydrochloric acid also acts on pepsinogen, converting it to the active form **pepsin,** a proteolytic enzyme that catalyzes the breakdown of proteins into large fragments. These fragments are further broken down in the small intestine, the next stop in the digestive system. Interestingly, pepsinogen molecules are also activated by pepsin. Thus, once a few pepsin molecules are formed, they assist HCl in activating the remaining pepsinogen molecules.

Hydrochloric acid creates an acidic environment (pH 1.5–3.5) in the stomach. Besides activating pepsinogen and denaturing protein (rendering it digestible), HCl also breaks down connective tissue fibers in meat and kills most bacteria, helping protect the body from infection.

Also, contrary to popular belief, the stomach does not absorb foodstuffs. Only a few substances, such as alcohol, can actually penetrate the lining of the stomach to enter the bloodstream. Alcohol consumed on an empty stomach passes quickly into the bloodstream, often producing a rather immediate dizzying effect. The presence of food in the stomach, however, retards alcohol absorption.

The stomach lining is protected from destruction by an alkaline secretion known as **mucus,** produced by certain cells in the lining. Mucus coats the epithelium, protecting it from acid and pepsin. The tissues beneath the epithelium are protected from acid leakage by the cells of the epithelium, which are tightly joined to one another, forming a leak-proof barrier.

Unfortunately, the stomach's protective mechanisms can break down. A number of factors—stress, coffee (caffeine), excess aspirin, nicotine, and alcohol or combinations of them—can increase acid levels in the stomach, overwhelming the mucous layer. When this happens, hydrochloric acid and pepsin come in contact with the epithelial cells and may digest parts of the wall of the stomach, forming painful **ulcers** (▷ Figure 11–19).[3] When detected early, most ulcers can be treated by reducing stress and changing one's diet—for example, eating smaller amounts of food and reducing coffee, aspirin, and alcohol. Stress reduction and dietary changes, such as these, reduce acid

[3] Ulcers also occur in the esophagus and, more commonly, in the small intestine. Excessive acid percolating into the esophagus and the release of excess acid into the small intestine can overwhelm the protective mucous layer found in both organs.

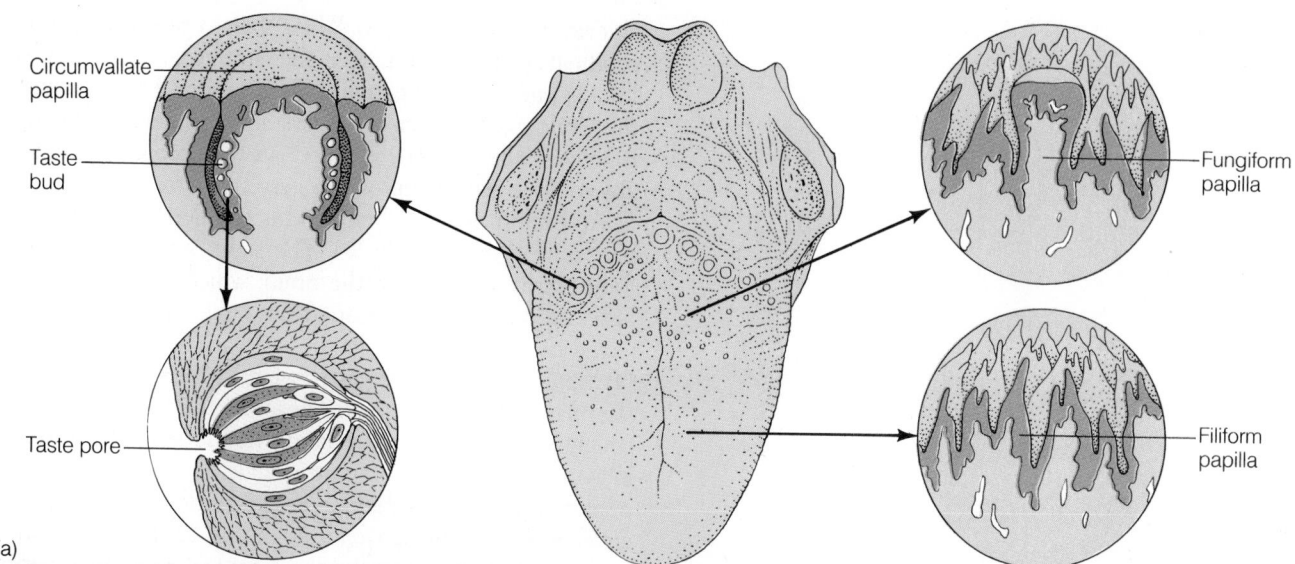

(a)

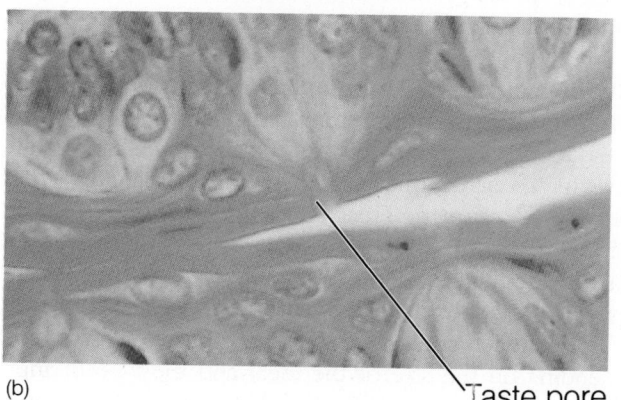

(b)

Taste pore

▷ **FIGURE 11–15 The Tongue and Taste Buds** (*a*) The tongue is a muscular organ that aids in swallowing and phonation (producing sounds). Its upper surface is dotted with protrusions called papillae. Three types are present: the fungiform, filiform, and circumvallate. Taste buds are located on the fungiform and circumvallate papillae. Four types of taste buds are found: those that detect salt, bitter, sweet, and sour flavors. Each type is found on a specific region on the tongue. Food molecules dissolved in saliva enter the taste pore and stimulate these cells, which in turn trigger nerve impulses to the brain. (*b*) A photomicrograph of taste buds; the arrow indicates a taste pore.

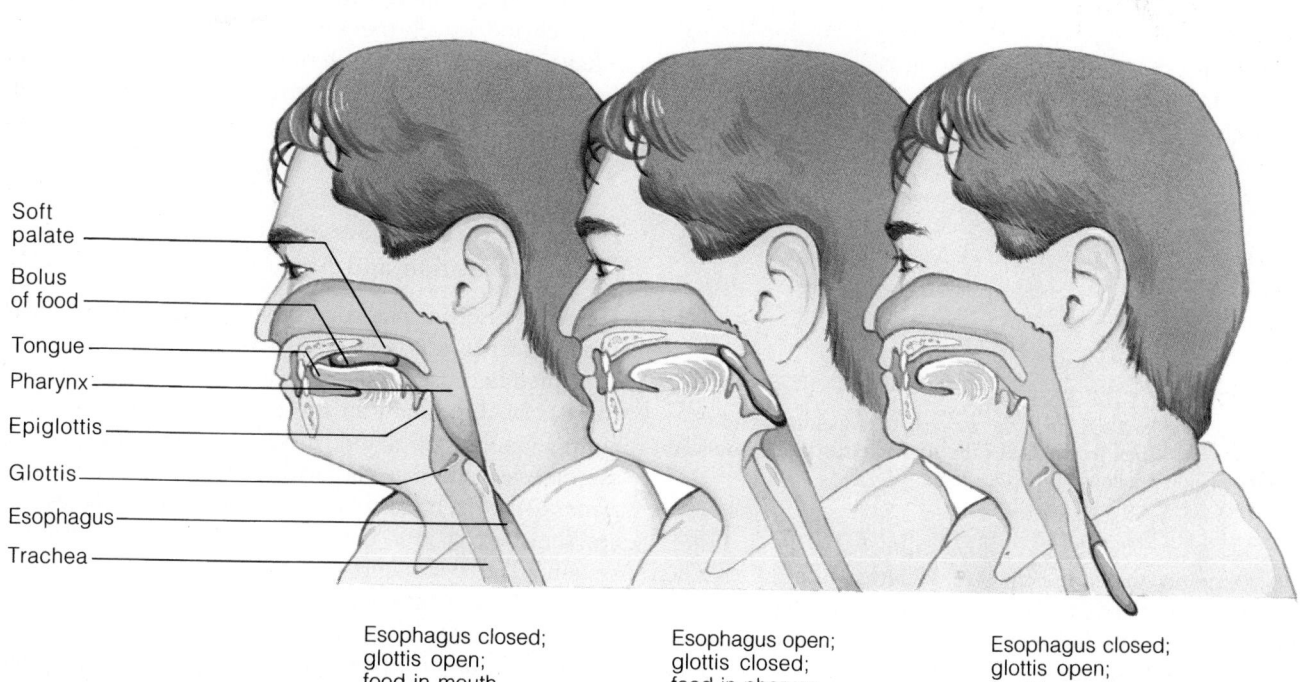

Esophagus closed; glottis open; food in mouth.

Esophagus open; glottis closed; food in pharynx.

Esophagus closed; glottis open; food in esophagus.

▷ **FIGURE 11–16 The Epiglottis** This trapdoor prevents food from entering the trachea during swallowing. As illustrated, the trachea is lifted during swallowing, pushing against the epiglottis, which bends downward.

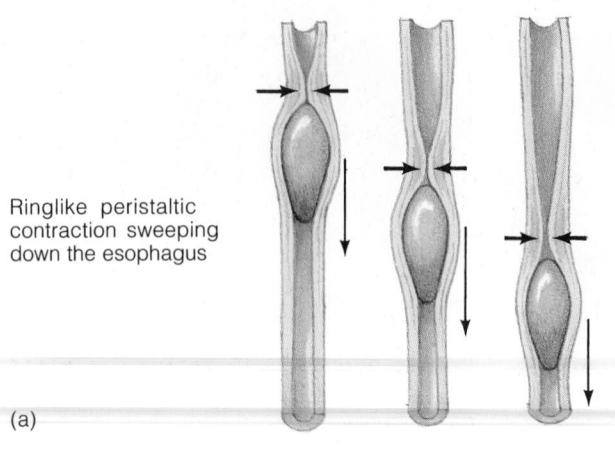

Ringlike peristaltic contraction sweeping down the esophagus

(a)

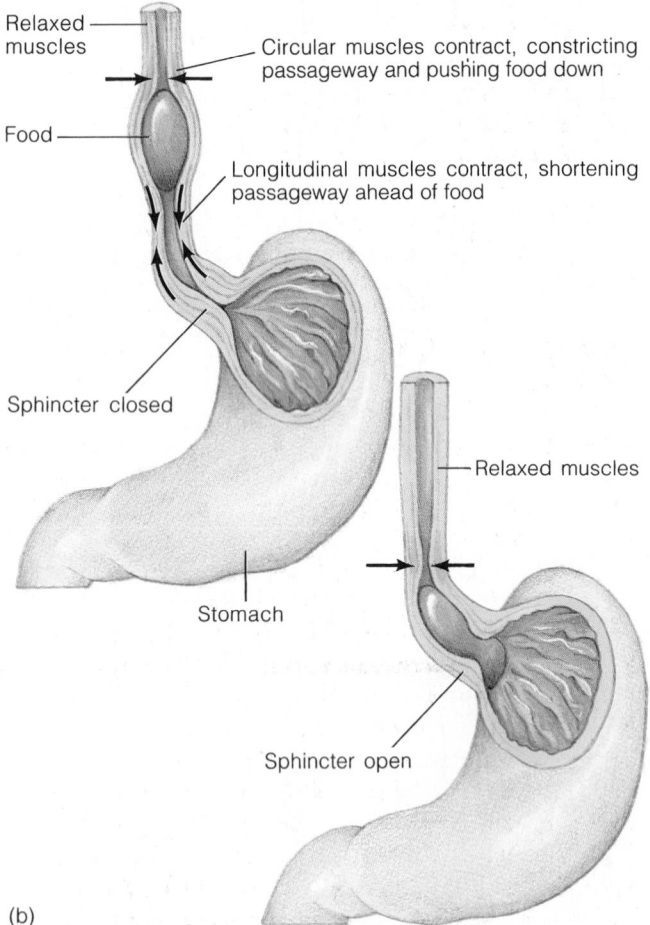

Relaxed muscles

Circular muscles contract, constricting passageway and pushing food down

Food

Longitudinal muscles contract, shortening passageway ahead of food

Sphincter closed

Stomach

Relaxed muscles

Sphincter open

(b)

▷ **FIGURE 11–17 Peristalsis**
(*a*) Peristaltic contractions in the esophagus propel food into the stomach. (*b*) When food reaches the stomach, the gastroesophageal sphincter opens, allowing food to enter.

chloric acid continuously. Were that to occur, continuous production might endanger the stomach lining. Instead, HCl is produced on demand. Its secretion is controlled by the endocrine and nervous systems, sometimes working together. These highly evolved systems provide a remarkable degree of control over a very complex process.

Activation begins when food is smelled, tasted, or simply seen (▷ Figure 11–20). The sight, smell, and taste of food activate centers of the brain, which, in turn, transmit nerve impulses to the stomach via the **vagus nerve.** The vagus nerve terminates in the stomach wall and activates HCl production by the gastric glands. Nerve impulses also stimulate the synthesis and release of the hormone gastrin, which is produced by the lining of the stomach. **Gastrin** increases the output of gastric juice, containing HCl and pepsinogen. Acid production is also stimulated by the presence of proteins and peptides in the stomach and small intestine. In fact, protein in the stomach is the most potent stimulus of gastric gland secretion.

Chyme Leaves the Stomach and Enters the Small Intestine. Chyme in the stomach is ejected into the small intestine by peristaltic muscle contractions. A peristaltic wave travels across the stomach every 20 seconds. When the wave of contraction reaches the far end of the stomach, the **pyloric sphincter** (a ring of smooth muscle at the juncture of the small intestine and stomach) opens, and chyme squirts into the small intestine.

The stomach contents are emptied in two to six hours, depending on the size of the meal and the type of food. The larger the meal, the longer it takes to empty. Solid foods (meat) empty faster than liquid foods (a milkshake). Peristaltic contractions continue after the stomach is empty and are felt as hunger pangs.

As chyme leaves the stomach, the major stimulus for enhanced gastric-gland secretion, the presence of protein in the stomach, is removed. For this and other reasons, acid and pepsinogen release decline, shutting down the stomach's HCl production until the next meal.

The Small Intestine Serves as a Site of Food Digestion and Absorption

The small intestine is a coiled tube in the abdominal cavity about 6 meters (20 feet) long in adults (see Figure 11–13). So named because of its small diameter, the small intestine consists of three parts in the following order: the duodenum, jejunum, and ileum. Inside the small intestine, macromolecules are digested with the aid of numerous enzymes—that is, broken down into smaller molecules—that are transported into the bloodstream and the lymphatic system. The lymphatic system, discussed more fully in Chapter 12, is a network of vessels that carries extracellular fluid from the tissues of the body to the circulatory system. In addition, tiny lymphatic vessels in the wall of the small intestine, known as **lacteals,** absorb fats and transport them to the bloodstream.

secretion in the stomach (see Health Note 1-1). When ulcers are not detected early and when damage is severe, parts of the stomach may have to be removed surgically.

The Stomach's Function Is Regulated by Nerves and Hormones. The stomach does not produce hydro-

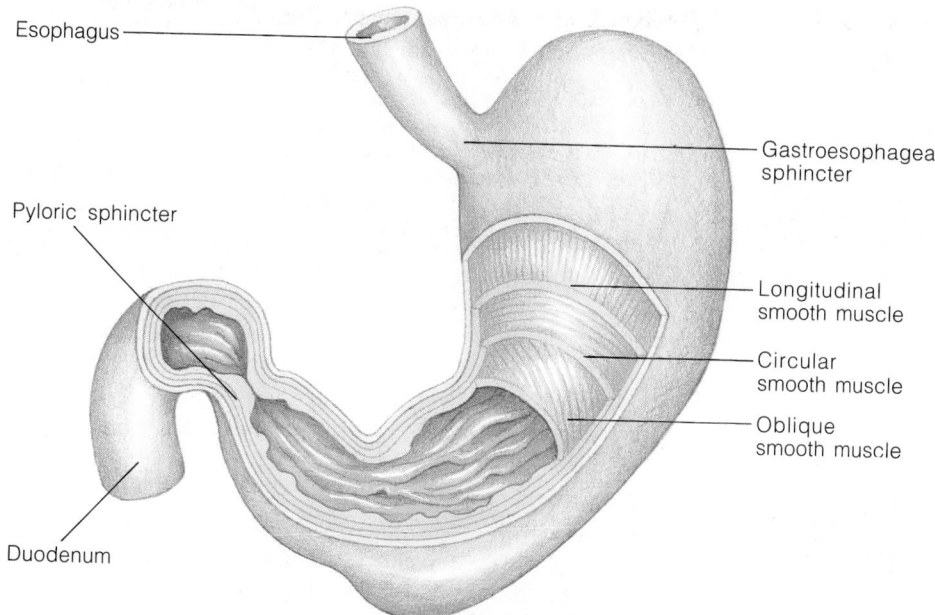

Esophagus

Gastroesophageal sphincter

Pyloric sphincter

Longitudinal smooth muscle

Circular smooth muscle

Oblique smooth muscle

Duodenum

▷ **FIGURE 11–18 The Stomach** The stomach lies in the abdominal cavity. In its wall are three layers of smooth muscle that help mix the food and force it into the small intestine, where most digestion occurs. The gastroesophageal and pyloric sphincters control the inflow and outflow of food, respectively.

Digestion in the Small Intestine Requires Enzymes from Two Major Sources.

The digestion of food molecules inside the small intestine requires enzymes produced from two distinctly different sources: the pancreas, an organ that lies beneath the stomach, and the lining of the small intestine itself.

The **pancreas** is nestled in a loop formed by the first portion of the small intestine, the duodenum (▷ Figure 11–21). The pancreas is a dual-purpose organ; it has endocrine and exocrine functions. As an exocrine gland, it produces enzymes and sodium bicarbonate essential for the digestion of foodstuffs in the small intestine. Its endocrine function is fulfilled by special cells that produce hormones that help regulate blood glucose levels and thus help maintain homeostasis.

The digestive enzymes of the pancreas are produced in small glandular units. These glands empty into many small ducts that converge to form the large pancreatic duct. The pancreatic duct joins with a duct draining the gallbladder and liver and then empties into the duodenum, the first segment of the small intestine. Each day approximately

▷ **FIGURE 11–20 Pathway Leading to the Release of HCl by the Stomach**

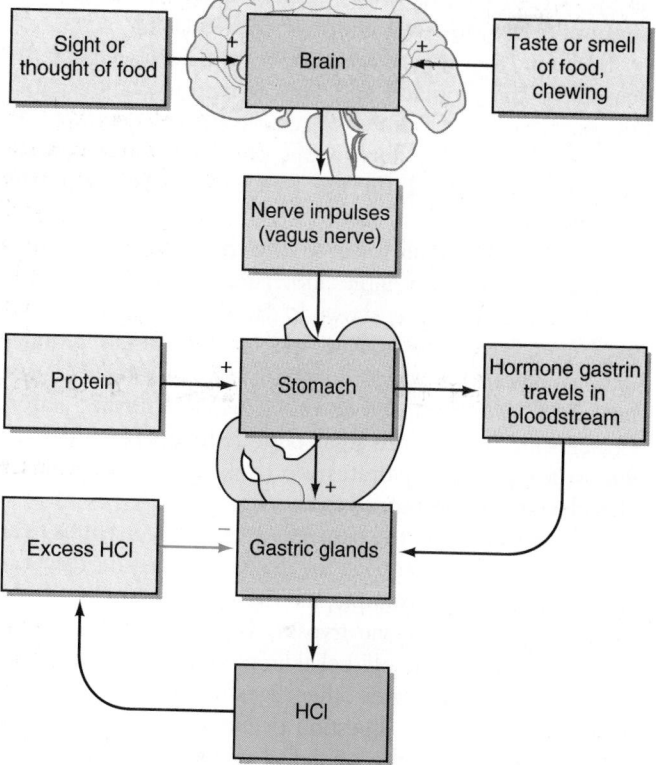

Sight or thought of food + Brain + Taste or smell of food, chewing

Nerve impulses (vagus nerve)

Protein + Stomach → Hormone gastrin travels in bloodstream

Excess HCl − Gastric glands

HCl

▷ **FIGURE 11–19 Ulcer**

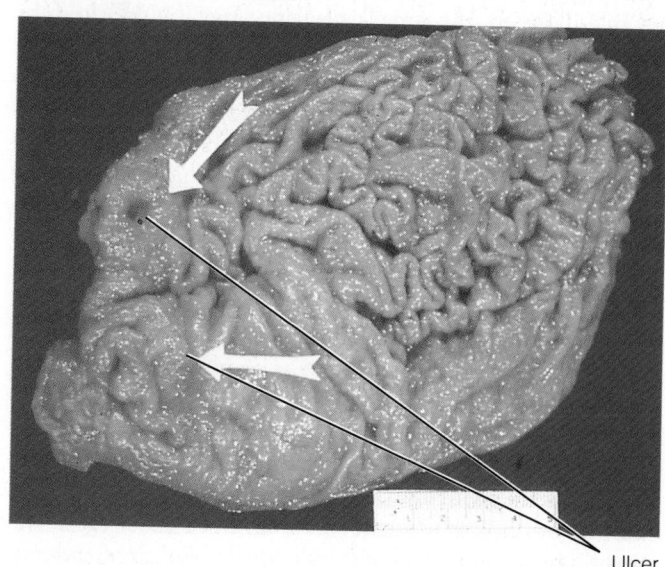

Ulcer

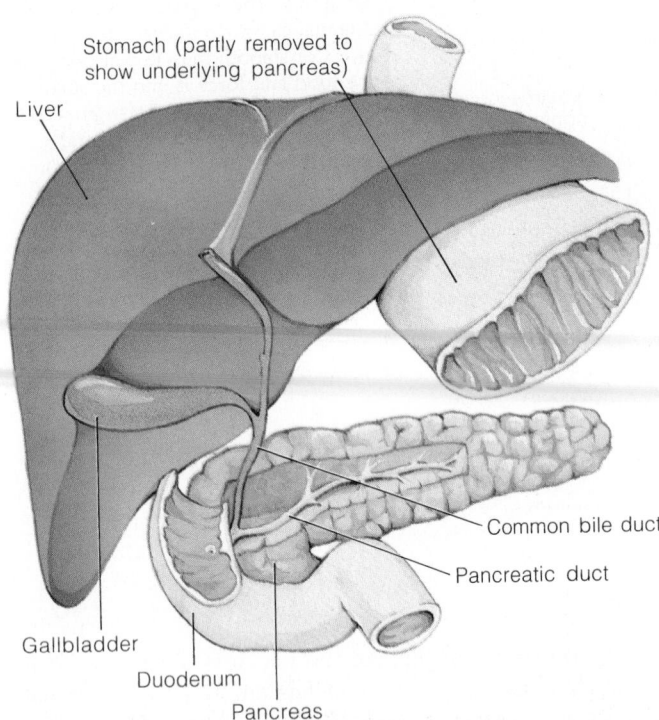

Stomach (partly removed to show underlying pancreas)

Liver

Common bile duct

Pancreatic duct

Gallbladder

Duodenum

Pancreas

▷ **FIGURE 11–21 Organs of Digestion** The liver, gallbladder, and pancreas all play key roles in digestion and empty by the common bile duct into the small intestine, in which digestion takes place.

1200 to 1500 milliliters (1.0 to 1.5 quarts) of pancreatic juice is produced and released into the small intestine. This liquid is composed of water, sodium bicarbonate, and several important digestive enzymes (Table 11–6).

Sodium bicarbonate neutralizes the acidic chyme released by the stomach and thus helps protect the small intestine from the harmful effects of stomach acid. It also gives the pancreatic juice a pH of about 8, creating an environment optimal for the function of the pancreatic digestive enzymes.

Pancreatic enzymes released into the lumen of the small intestine act on the large molecules in food (proteins, starches, and so on), as shown in Table 11–6. As a result of pancreatic enzymatic activity, fats are completely reduced to monoglycerides (glycerol attached to one fatty acid) and fatty acids, which can be absorbed without further action. Proteins are broken into small peptide fragments and some amino acids. Carbohydrates are broken down into disaccharides and some monosaccharides.

Like the stomach, the pancreas secretes its enzymes in an inactive form. This protects the gland from self-destruction. **Trypsinogen,** for example, is the inactive form of the protein-digesting enzyme trypsin. Trypsinogen is activated by a substance on the epithelial lining of the small intestine. Trypsin, in turn, activates other digestive enzymes.

The final stage of digestion occurs with the aid of enzymes produced by the epithelial cells of the small intestine. These enzymes are embedded in the membranes of the epithelial cells (Table 11–6). As a result, the final phase of digestion occurs just before the nutrient is absorbed into the cell.

The Liver Produces an Emulsifying Agent, Bile, Which Plays a Key Role in the Digestion of Fat.

The **liver** is one of the largest and most versatile organs in the body, performing many functions essential to homeostasis. By various estimates, the liver performs as many as 500 different functions. Situated in the upper-right quadrant of the abdominal cavity, under the protection of the rib cage, the liver is one of the body's storage depots for glucose (Figure 11–13). It also stores fats, iron, copper, and many vitamins. By storing glucose and lipids and releasing them as they are needed, the liver helps ensure a constant supply of energy-rich molecules to body cells. The liver also synthesizes some key blood proteins involved in clotting and is an efficient detoxifier of potentially harmful chemicals, such as nicotine, barbiturates, and alcohol. These functions contribute significantly to homeostasis.

The liver also plays a key role in the digestion of fats through the production of a fluid called **bile,** which contains water, ions, and molecules such as cholesterol, fatty acids, and bile salts. **Bile salts** are steroids that emulsify fats—that is, break fat globules into smaller ones. This process is essential for lipid digestion, because lipid-digesting enzymes in the small intestine do not work well on large fat globules. Unless fat globules are broken into smaller ones, fat digestion will be impaired.

Produced by the cells of the liver, bile is first transported to the **gallbladder,** a sac attached to the underside of the liver (see Figure 11–13). The gallbladder concentrates the bile by removing water from it. (This action also helps conserve water.) Bile is stored in the gallbladder until needed. When chyme is present in the small intestine, the gallbladder contracts, causing bile to flow out through the duct system and into the small intestine (Figure 11–21).

Bile flow to the small intestine may be blocked by **gallstones,** deposits of cholesterol and other materials that form in the gallbladder of some individuals. Gallstones may lodge in the ducts draining the organ, thus reducing—even completely blocking—the flow of bile to the small intestine. The lack of bile salts greatly reduces lipid digestion. Because lipid digestion is reduced, fat is not absorbed. Passed on with the undigested food mass into the large intestine of a person with gallstones, these fats are then decomposed by bacteria in the large intestine, but they are not absorbed. The decomposition of these fats often gives the feces a foul odor. The higher percentage of fat also makes the feces quite buoyant and difficult to flush.

Approximately 1 of every 10 American adults has gallstones, although many (30% to 50%) of these people exhibit no symptoms whatsoever. Gallstones occur more frequently in older, overweight individuals, and the incidence

TABLE 11–6 Digestive Enzymes

SITE OF PRODUCTION	ENZYME	ACTION
Salivary glands	Amylase	Digests polysaccharides in oral cavity
Stomach	Pepsin	Breaks proteins into peptides
Pancreas	Trypsin	Cleaves peptide bonds of polypeptides and proteins
	Chymotrypsin	Same as trypsin
	Carboxypeptidase	Cleaves peptide bonds on carboxy end of polypeptides
	Amylase	Breaks starches into smaller units (maltose)
	Phospholipase	Cleaves fatty acids from phosphoglycerides to form monoglycerides
	Lipase	Cleaves two fatty acids from triglycerides
	Ribonuclease	Breaks RNA into smaller nucleotide chains
	Deoxyribonuclease	Breaks DNA into smaller nucleotide chains
Epithelium of small intestine	Maltase	Breaks maltose into glucose subunits
	Sucrase	Breaks sucrose into glucose and fructose subunits
	Lactase	Breaks lactose into glucose and galactose subunits
	Aminopeptidase	Breaks down peptides into amino acids

in the elderly is about 1 in every 5. When they cause problems, gallstones are usually removed surgically. This procedure requires that the entire gallbladder be removed. Interestingly, bile continues to be produced in these patients but is stored in the common bile duct, which becomes distended to accommodate the liquid bile. Scientists are now testing a new drug that dissolves gallstones in many patients, hoping that it may someday eliminate the need for surgery in many instances.

The Intestinal Epithelium Is Specially Modified for Absorption. Virtually all food digestion occurs in the duodenum and jejunum. Once food molecules are digested, they must be absorbed—that is, transported across the epithelial lining of the first two portions of small intestine into the bloodstream or lymphatic system. In these regions, absorption is facilitated by three structural modifications of the small intestine. ▷ Figure 11–22a shows, for instance, that the lining of the small intestine is thrown into circular folds, which increase the overall surface area. As shown in Figure 11–22b, on the surfaces of the circular folds are many fingerlike projections known as **villi**, which also increase the surface area available for the absorption of food molecules. The intestinal surface area is further increased by **microvilli**, tiny protrusions of the plasma membranes of the epithelial cells lining the villi, which are located on the surface facing the lumen, or cavity (Figure 11–22d). Each epithelial cell that functions in absorption contains an estimated 3000 microvilli. Two hundred million microvilli occupy a single square millimeter of intestinal lining. Together, the circular folds, villi, and microvilli result in a dramatic increase in the surface area of the lining of the small intestine, making it 600 times greater than if it were merely a flat layer, lined by epithelial cells.

Numerous mechanisms are involved in absorption—too numerous to be discussed in this book. Three of the most common are diffusion, osmosis, and active transport, described in Chapter 3. Many nutrients enter the epithelial cells via active transport, then diffuse out of the intestinal cells into the bloodstream or lymphatic vessels. As illustrated in ▷ Figure 11–23, each villus of the small intestine is endowed with a rich supply of blood and lymph capillaries. These tiny vessels absorb nutrients that have passed through the lining of the small intestine, then transport these nutrients elsewhere.

Although most nutrients diffuse from the epithelial cells into the blood capillaries, fatty acids and monoglycerides produced by the enzymatic digestion of triglycerides follow a different route. As shown in ▷ Figure 11–24 (page 278), these molecules diffuse into the cells lining the villi and inside the cells they combine to reform triglycerides. The triglycerides, in turn, combine to form fat droplets within the cytoplasm of the epithelial cells of the small intestine. These tiny fat droplets are then coated with a layer of lipoprotein (lipid bound to protein), which is produced by the endoplasmic reticulum. This treatment renders the fat droplets, or **chylomicrons**, water-soluble. The coated fat droplets are then released by the epithelial cells by exocytosis into the interstitial fluid (Figure 11–24). Because blood capillaries are relatively impermeable to chylomicrons, most of the lipid globules enter the more porous lymph capillaries in the villi. You will note in Figure 11–24 that small or medium chain fatty acids not incorporated in triglycerides pass directly into blood capillaries in the villi.

The Large Intestine Is the Site of Water Resorption

The **large intestine** is about 1.5 meters (5 feet) long and consists of four basic regions, the cecum, appendix, colon, and rectum (▷ Figure 11–25, page 279). The **cecum** is simply a pouch that forms below the juncture of the large and small intestine. A small, wormlike structure, the **appendix**, attaches to the bottom of the cecum. Most of the large

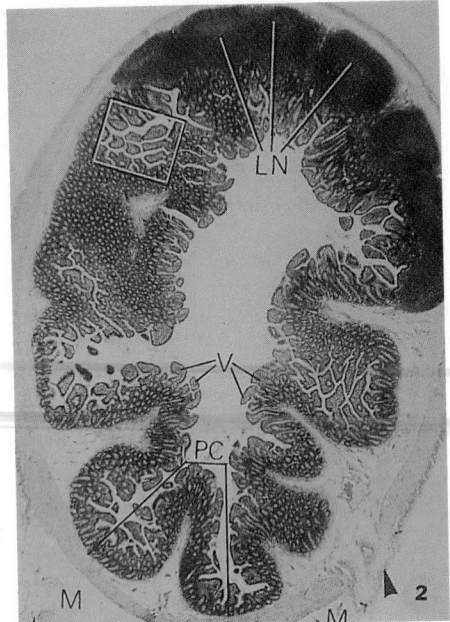

(a)

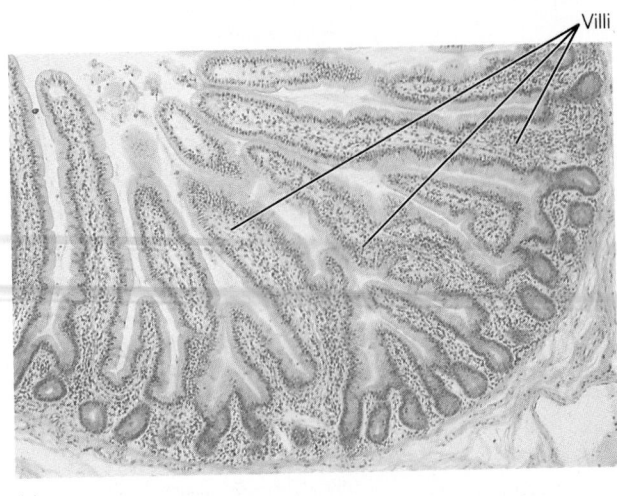

Villi

(c)

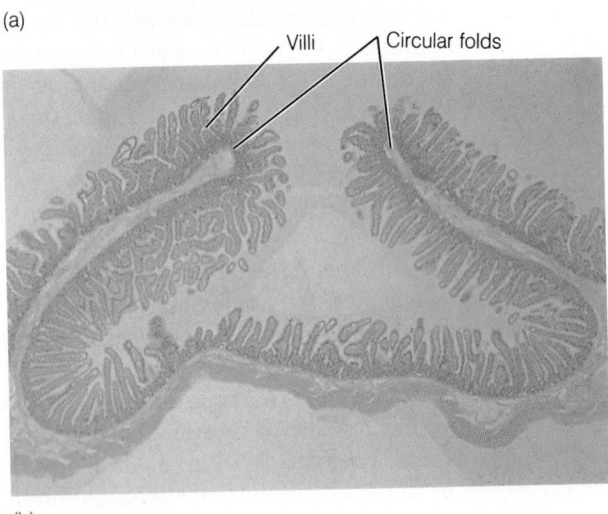

Villi   Circular folds

(b)

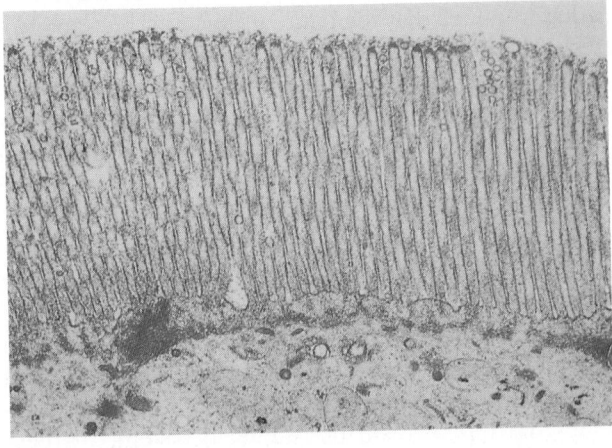

(d)

▷ **FIGURE 11–22 The Small Intestine** The small intestine is uniquely "designed" to increase absorption. (*a*) A cross section showing the folds. LN means lymph nodules (aggregation of lymphocytes); V means villi; PC means plica circulares (arculan fold). (*b*) A light micrograph of folds and villi. (*c*) Higher magnification of villi. (*d*) An electron micrograph of the apical surface of the absorptive cells showing the microvilli.

intestine consists of the **colon.** Unlike the small intestine, which is coiled and packed in a small volume, the colon consists of three relatively straight portions, the **ascending colon,** the **transverse colon,** and the **descending colon.** The colon empties into the **rectum.**

So named because of its large diameter, the large intestine receives materials from the small intestine. The material entering the large intestine consists of a mixture of water, undigested or unabsorbed food molecules, and undigestible food residues, such as cellulose. It also contains sodium and potassium ions.

The colon absorbs approximately 90% of the water and sodium and potassium ions that enter it. The undigested or unabsorbed nutrients feed a rather large population of

bacteria in the large intestine. These bacteria synthesize several key vitamins: B-12, thiamine, riboflavin, and, most importantly, vitamin K, which is often deficient in the human diet. These vitamins are absorbed by the large intestine.

The contents of the large intestine (after water and salt have been removed) are known as the **feces.** The feces consist primarily of undigested food, indigestible materials, and bacteria. Bacteria, in fact, account for about one-third of the dry weight of the feces.[4]

[4]The feces are about 30% dead bacteria, 10% to 20% fat, 10% to 20% inorganic matter, 2% to 3% protein, and about 30% undigested roughage (cellulose).

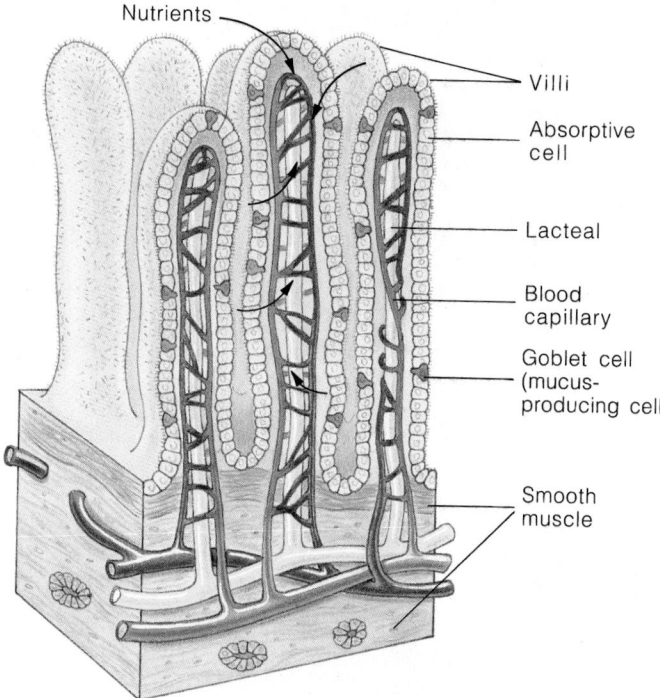

Nutrients

Villi

Absorptive cell

Lacteal

Blood capillary

Goblet cell (mucus-producing cell)

Smooth muscle

▷ **FIGURE 11–23 The Villus** Each villus contains a loose core of connective tissue, a lacteal, or lymph capillary, and a network of blood capillaries. Nutrients pass from the lumen through the epithelium and into the interior of the villi, where they are picked up by the lymph and blood capillaries.

The feces are propelled by peristaltic contractions along the colon until they reach the rectum, the last segment of the large intestine (Figure 11–25). As fecal matter accumulates in the rectum, it distends the organ. This action, in turn, stimulates stretch receptors in the wall of the rectum. These receptors stimulate the **defecation reflex.** In this reflex, nerve impulses from the stretch receptors in the rectum travel via sensory nerves to the spinal cord. In the spinal cord, the nerve impulses stimulate nerve cells that supply the smooth muscle in the wall of the rectum, causing them to contract and expel the feces. Other nerve impulses from the spinal cord travel to the **internal anal sphincter,** a ring of smooth muscle that holds the feces back. The internal anal sphincter normally keeps fecal matter from entering the anal canal. In the defecation reflex, however, the sphincter relaxes and the fecal matter can escape. Defecation does not occur until the **external anal sphincter** relaxes. This sphincter is composed of skeletal muscle and is under conscious control. If the time and place are appropriate, the external anal sphincter is relaxed, and defecation can occur. If circumstances are inappropriate, voluntary tightening of the external anal sphincter prevents defecation, despite prior activation of the reflex. When defecation is delayed, the muscle in the wall of the rectum eventually relaxes, and the urge to defecate subsides—at least until the next movement of feces into the rectum occurs. Defecation is usually assisted by volun-

tary contractions of the abdominal muscles and a forcible exhalation, which increase intra-abdominal pressure.

If defecation is delayed too long, a person may become constipated. Constipation results from excess water absorption, which makes the feces hard and dry. Besides being uncomfortable, constipation may result in a dull headache, loss of appetite, and depression.

Constipation is generally caused by (1) ignoring the urge to defecate; (2) decreases in colonic contractions, which occur with age, emotional stress, or low-bulk diets; (3) tumors or colonic spasms that obstruct the movement of feces; and (4) nerve injury, which impairs the defecation reflex.

Constipation is not only uncomfortable, it can create serious medical problems. Hardened fecal material, for instance, may become lodged in the appendix, mentioned above. This, in turn, may lead to inflammation of the organ, a condition called **appendicitis.** When this occurs, the appendix becomes swollen and filled with pus and must be removed surgically to prevent the organ from bursting and spilling its contents into the abdominal cavity. Fecal matter leaking into the abdominal cavity as mentioned earlier, introduces billions of bacteria and can result in a deadly infection.

≈ CONTROLLING DIGESTION

Digestion is a complex and varied process that is controlled largely by nerves and hormones. This section discusses some of the key events involved in the control of digestion, building on material you have learned earlier.

Salivation Is Stimulated by a Nervous Reflex

Digestion begins in the oral cavity. As noted earlier, the sight, smell, taste, and sometimes even the thought of food stimulate the release of saliva. Chewing has a similar effect. The secretion of saliva is therefore controlled by the nervous system and is largely a reflex (▷ Figure 11–26). For a discussion of the discovery of this phenomenon see the Scientific Discoveries 11–1.

Gastric-Gland Secretion Is Controlled by a Variety of Different Mechanisms

Besides activating salivary production, these stimuli also cause the brain to send nerve impulses along the vagus nerve to the stomach (Figure 11–20). As described earlier, these nerve impulses initiate the secretion of hydrochloric acid and pepsinogen from cells of the gastric glands. Nerve impulses also stimulate the secretion of the hormone gastrin. Gastrin is released into the blood and acts on the gastric glands, stimulating additional HCl and pepsinogen secretion.

The most potent stimulus for gastric secretion is the presence of protein in the stomach. Protein stimulates chemical receptors inside the stomach that activate net-

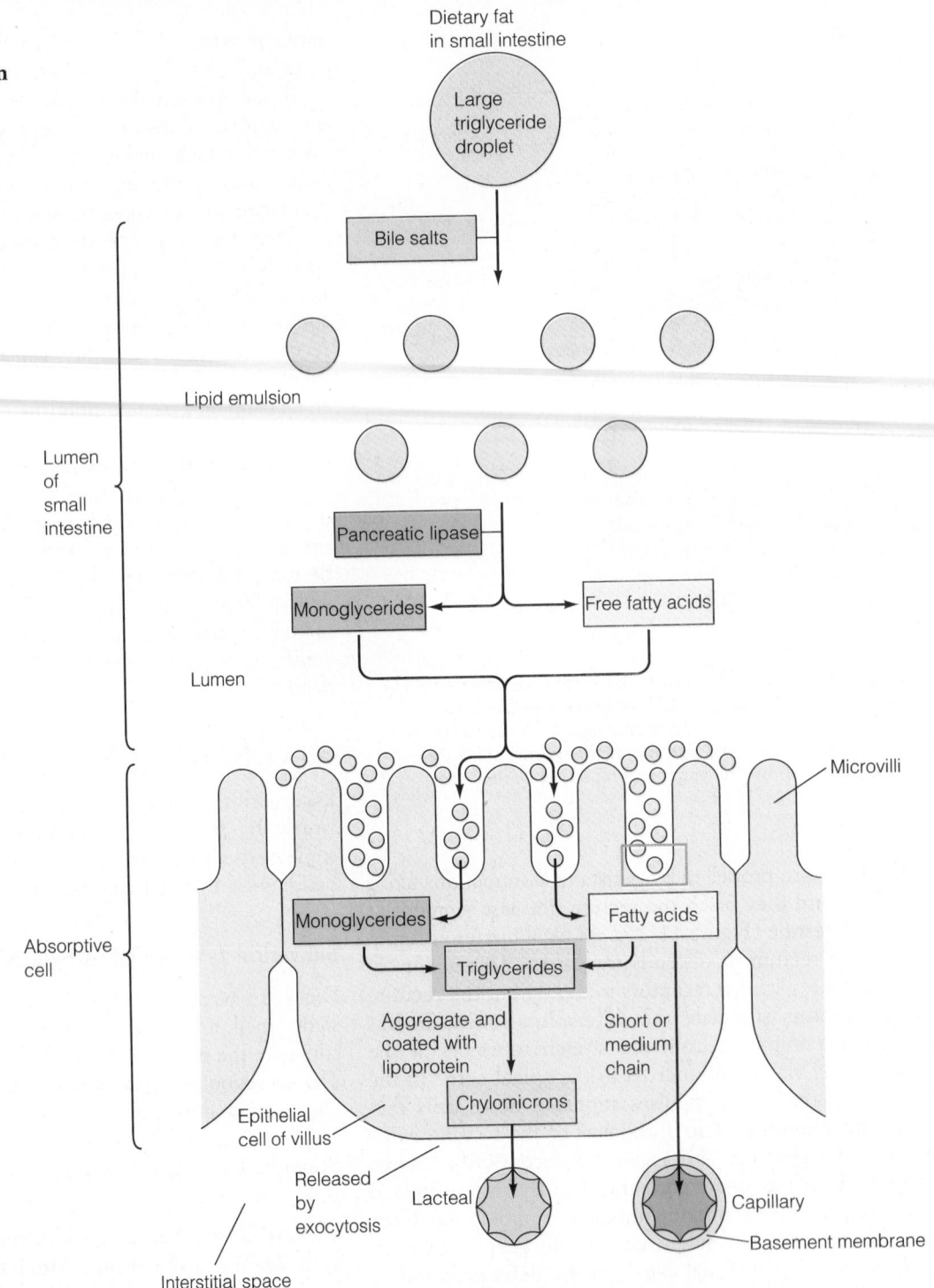

Dietary fat
in small intestine

Large triglyceride droplet

Bile salts

Lipid emulsion

Lumen of small intestine

Pancreatic lipase

Monoglycerides → ← Free fatty acids

Lumen

Microvilli

Monoglycerides

Fatty acids

Absorptive cell

Triglycerides

Aggregate and coated with lipoprotein

Short or medium chain

Chylomicrons

Epithelial cell of villus

Released by exocytosis

Lacteal

Capillary

Basement membrane

Interstitial space

works of nerves in the stomach wall. These nerves, in turn, stimulate the gastric glands to secrete HCl and pepsinogen. Protein also stimulates gastrin secretion directly, in ways beyond the scope of this book.

The concentration of HCl in the stomach is also regulated by a negative feedback mechanism. When the acid content rises too high, it inhibits gastrin secretion, thus shutting off HCl production (see Figure 11–20).

Pancreatic Secretions Released into the Small Intestine Are Stimulated by Two Intestinal Hormones

Acidic chyme leaves the stomach and enters the small intestine, where it stimulates the release of a hormone known as secretin. **Secretin** is produced by the cells of the duodenum and travels in the bloodstream to the pancreas. Here it stimulates the release of sodium bicarbonate ▷ Figure 11–27). Sodium bicarbonate, in turn, is excreted

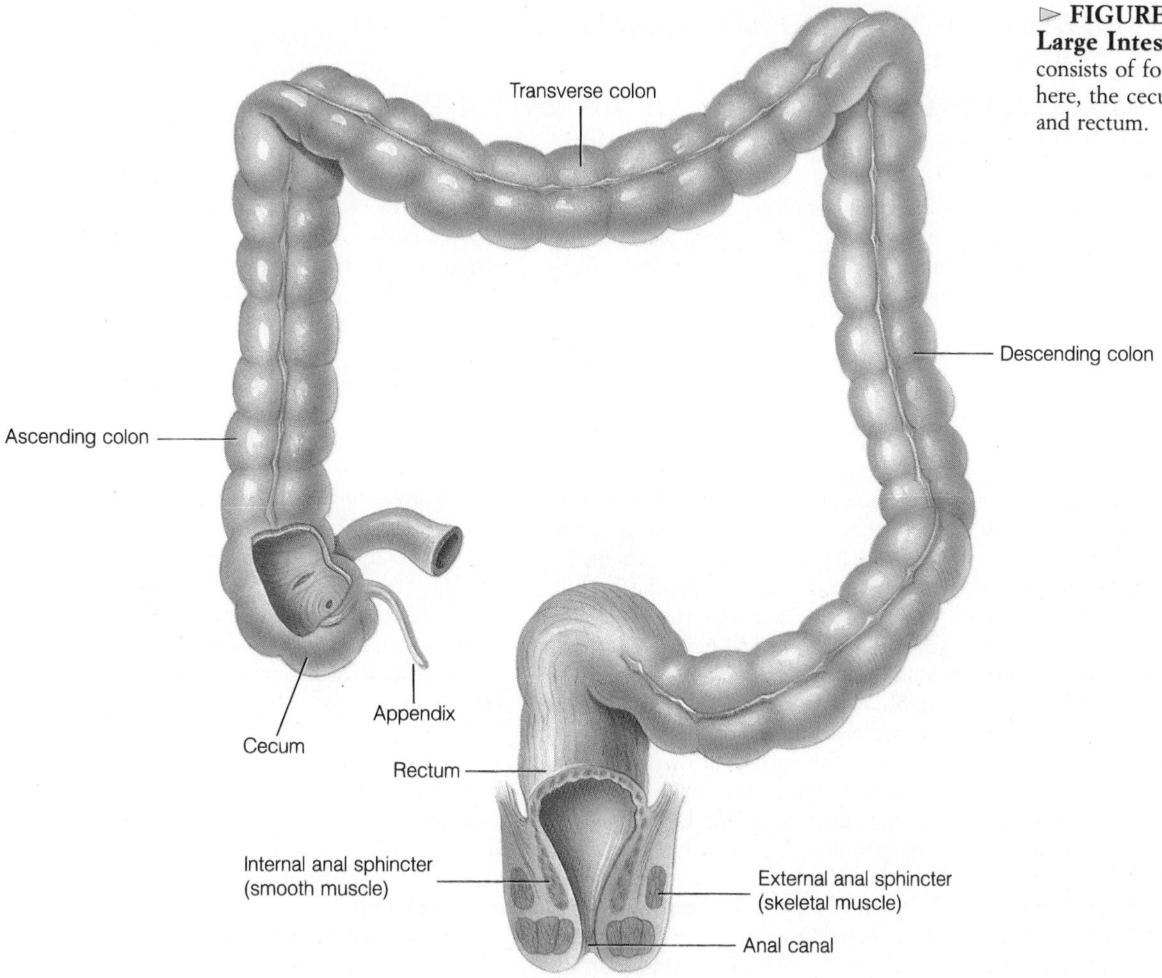

Transverse colon
Descending colon
Ascending colon
Appendix
Cecum
Rectum
Internal anal sphincter
(smooth muscle)
External anal sphincter
(skeletal muscle)
Anal canal

FIGURE 11–25 The Large Intestine This organ consists of four basic parts, shown here, the cecum, appendix, colon, and rectum.

FIGURE 11–26 Pathway Leading to the Release of Saliva

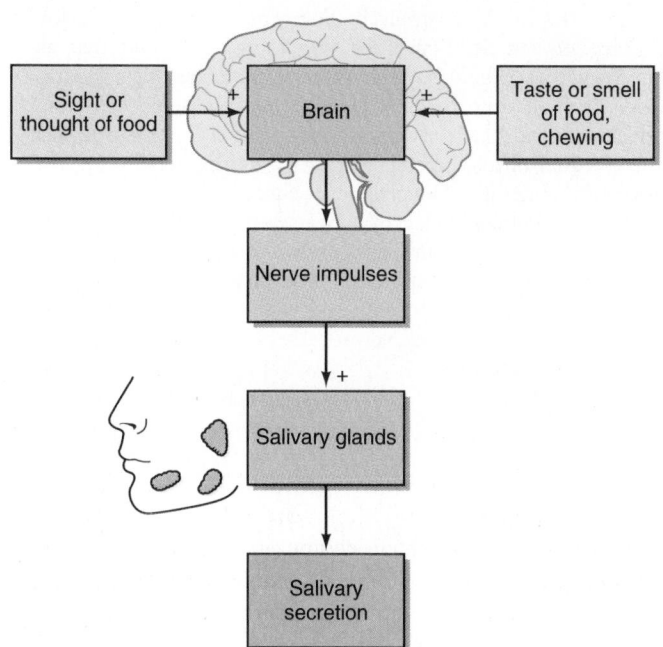

Sight or thought of food
+
Brain
+
Taste or smell of food, chewing
Nerve impulses
+
Salivary glands
Salivary secretion

into the small intestine, where, as noted earlier, it neutralizes the acidic chyme and creates an environment optimal for pancreatic enzymes.

The release of pancreatic enzymes is triggered by another intestinal hormone, **cholecystokinin (CCK),** produced by cells of the duodenum in the presence of chyme (Figure 11–27). CCK also stimulates the gallbladder to contract, releasing bile into the small intestine.

Interestingly, recent evidence links abnormally low secretion of CCK to **bulimia,** an eating disorder characterized by recurrent binge eating followed by vomiting. CCK has been found in the brain's hypothalamus with other hormones and may be involved in a range of behaviors, including bulimia. Approximately 4% of America's young women, and a far smaller fraction of men, suffer from this disorder. Bulimia is thought to have both biological and psychological roots, but researchers had failed to identify a biochemical cause until recently. Although no single chemical is likely to control a complex behavior like appetite, it appears that CCK plays an important role.

DISCOVERING THE NATURE OF DIGESTION
Featuring the work of van Helmont and Beaumont

The foundation for the modern understanding of digestion was laid by a number of pioneering scientists. Two noteworthy contributors were a 17th-century Flemish physician, Jan Baptista van Helmont, and a 19th-century American, William Beaumont.

Van Helmont was a man of great originality and ingenuity. In his time, most people thought that the digestion of food in the human was akin to cooking—that is, that food inside the stomach was broken down by body heat. Van Helmont, however, dismissed this theory with the simple observation that digestion occurs in cold-blooded animals such as fish. The lack of heat in such animals suggested to him that an alternative process was at work. But what was it?

The road to discovery began in a rather bizarre way. One day, a tame sparrow that used to visit the scientist attempted to bite van Helmont's tongue. The physician detected a slight acidity in the bird's throat and hypothesized that acid must digest food.

To test this hypothesis, van Helmont tried to digest meat with a vinegar solution, but he could not. Consequently, he hypothesized that the body must contain "ferments," chemical substances in the stomach and small intestine that are specific for different types of food. Today, we call these "ferments" the digestive enzymes produced by the stomach lining, pancreas, and small intestine.

Little insight into digestion was forthcoming in the period between van Helmont's work and the studies of the American scientist William Beaumont. Beaumont was the adventurous son of a Connecticut farmer who left his home in 1806. Soon thereafter, he became a schoolmaster in New York, where he studied medicine and associated sciences in his spare time. Several years later, he became an apprentice to a physician in Vermont and two years later he received a license to practice medicine.

Soon after becoming a licensed physician, Beaumont joined the army, where he would make his important discoveries about digestion. He owes his great work to an accident. In 1822, a French-Canadian porter and servant in the army was wounded in the abdomen by a musket that had accidentally discharged at close range. The servant, Alexis St. Martin, who was only 18, was brought to Beaumont for treatment. The doctor found that the bullet had ripped a hole in St. Martin's stomach, out of which poured food the man had eaten.

As was the custom of the time, Beaumont bled the victim, removing a substantial amount of his blood. Despite this treatment, the patient survived, but healing became a protracted process. For over eight months, Beaumont tried in vain to close the hole in the stomach. During this time, it dawned on the young physician that he had stumbled across a golden opportunity to study digestion.

Beaumont's experiments consisted of feeding the patient various types of food, then studying what took place inside the poor man's stomach. In the next nine years, Beaumont studied St. Martin's stomach contents with a wide assortment of foods, looking at the rate and temperature of digestion and the chemical conditions that favored different stages of the process. During his work, he noticed that the stomach lining could be injured by drinking alcohol.

Beaumont's studies helped settle a controversy over the nature of digestion. Some of his contemporaries contended that gastric juice was a kind of chemical solvent. Others believed that it merely liquified food and that digestion occurred as a result of a vital force present in living organisms. By showing that some digestion could take place outside the stomach in the presence of gastric juice, Beaumont demonstrated that gastric secretions did more than moisten food and that no "vital force" was at work. Digestion, Beaumont asserted, was a purely chemical phenomenon.

Beaumont did not address one question that could have been answered with existing techniques—that is, what caused gastric juice to be secreted? Was it the presence of food in the stomach?

The answer to that question came in 1889 from the work of Ivan Pavlov, another notable scientist, whose experiments on behavior are discussed in Scientific Discoveries 29–1. In studies on dogs, Pavlov showed that the secretion of gastric juice was mediated by the nervous system. How did he make this determination?

In one of many experiments, the Russian scientist surgically connected a fold of a dog's stomach to an opening in the animal's side, which permitted him to examine the production of gastric secretions. He next cut and tied off the esophagus of the dog, so that food the animal swallowed could not enter the stomach. Following the surgery, Pavlov gave the dog food and found that as soon as food entered the dog's mouth, gastric juice began to be secreted. These secretions continued as long as food was present, thus suggesting nervous system involvement. Since that time, a great deal has been learned about digestion and its control. For example, as you learned in this chapter, hormones are also involved in the control of many digestive processes.

The small intestine also produces the hormone, **gastric inhibitory peptide (GIP).** This hormone is released in response to fatty acids and sugars in chyme. GIP inhibits HCl production and peristalsis in the stomach, slowing down the rate at which chyme is discharged from the stomach and therefore enhancing digestion and absorption of food already in the small intestine.

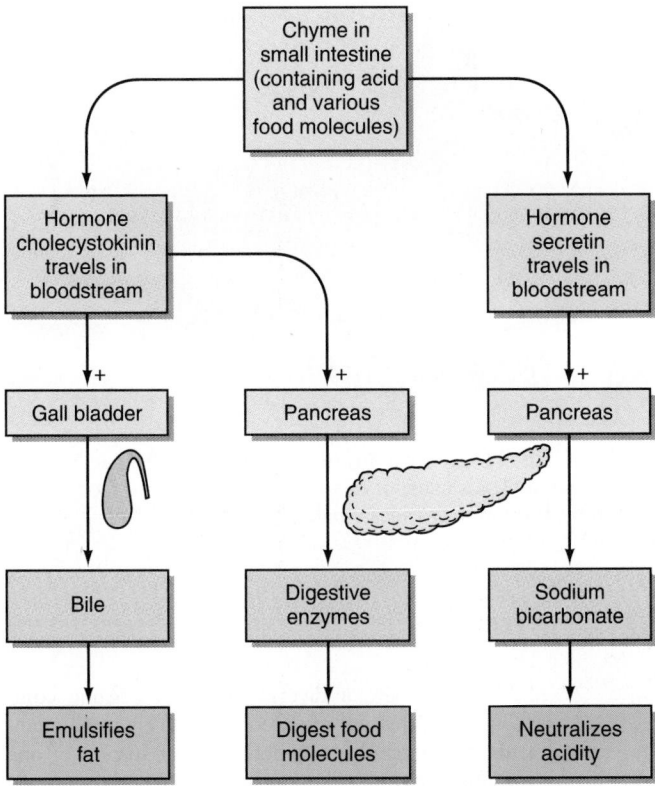

▷ **FIGURE 11–27 Control of the Pancreas and Gallbladder**

ENVIRONMENT AND HEALTH: EATING RIGHT/LIVING RIGHT

In few places is the relationship between homeostasis and human health as evident as in human nutrition. Because homeostasis requires an adequate supply of nutrients, nutritional deficiencies can have noticeable impacts on body functions and human health. Supporting this idea, many studies show that an unhealthy diet can increase the risk of contracting certain diseases, including cancer, heart disease, hypertension, and others.

Consider a few examples. Magnesium, one of the major minerals, is routinely ingested in insufficient amounts. New research suggests that such deficiencies may underlie a number of medical conditions, including diabetes, high blood pressure, pregnancy problems, and cardiovascular disease.

Research shows that adding magnesium to the drinking water of rats with hypertension can eliminate high blood pressure. Studies in rabbits show that magnesium reduces lipid levels in the blood and also reduces plaque formation in blood vessels. Rabbits on a high-cholesterol, low-magnesium diet, for example, have 80% to 90% more atherosclerotic plaque than rabbits on a high-cholesterol, high-magnesium diet.

Researchers have found that in humans, magnesium de-

ficiencies during pregnancy result in migraine headaches, high blood pressure, miscarriages, stillbirths, and babies with low birth weight. Research suggests that magnesium deficiency causes spasms in the blood vessels of the placenta, which reduce blood flow to the fetus. This can retard fetal growth and may even kill a fetus in utero. Magnesium supplements greatly reduce the incidence of these problems.

Researchers believe that 80% to 90% of the American public may be magnesium deficient. One reason the American diet may be deficient in magnesium is that phosphates in many carbonated soft drinks bind magnesium in the intestine, preventing it from being absorbed into the blood. Magnesium deficiencies can be reversed by eating more green leafy vegetables, seafood, and whole grain cereals. Mineral supplements could help as well, but they should be used with extreme caution, because excess magnesium can cause neurological disorders.

Zinc is a trace mineral that has also been implicated in a wide range of health problems. Rats fed diets severely deficient in zinc, for example, have more birth defects, are often stunted, and reach sexual maturation later than their counterparts fed normal diets. Concerned that less severe zinc deficiencies may cause problems in humans, researchers followed 10 monkeys from birth through adolescence. One group was fed a diet low in zinc. The other received far more than it required. Monkeys fed the zinc-deficient diet showed several curious symptoms. Their immune function was suppressed 20% to 30%, making them more susceptible to disease. Significant learning impairments were also observed. The monkey studies substantiate studies in rodents and suggest concern for people in less developed countries, who often subsist on low-zinc diets consisting primarily of cereals.[5]

Over the years, numerous dietary recommendations have been issued to help Americans live healthier lives and reduce the risk of cancer and heart attack. Nutritionists recommend that we daily consume (1) fruits and vegetables, especially cabbage and greens, (2) high-fiber foods, such as whole-wheat bread and celery, and (3) foods high in vitamins A and C. A healthy diet also minimizes the consumption of animal fat; red meat; and salt-cured, nitrate-cured, smoked, or pickled foods, including bacon and lunch meat.

As I noted earlier in the chapter, the American public has not taken these recommendations to heart. A survey of the eating habits of nearly 12,000 subjects conducted from 1976 to 1980, in fact, showed that the diets of both black and white Americans were typically deficient in the very foods that nutritionists recommend, such as fruits and vegetables.[6] When asked to recall everything they had

---

[5] To determine if you are receiving an adequate supply of micronutrients, you can undergo a blood test or an analysis of your diet at a nutritional clinic. If there are problems, a trained nutritionist will be able to make recommendations to correct the problem.

[6] Critical thinking suggests caution in interpreting these results. Studies of the dietary habits of people in 1990 would be more informative.

eaten in the previous 24-hour period, fewer than one in five people in the study reported having eaten any of these foods. In sharp contrast, many of the people surveyed had eaten red meat, bacon, and lunch meat.

A healthy diet results largely from habit and circumstance—that is, our social environment. How does our environment affect our nutrition and health? In the hustle and bustle of modern society, many of us ignore proper nutrition, grabbing fat-rich foods when we are hungry because we haven't the time to sit down to a nutritionally balanced meal (▷ Figure 11–28). Our fast-paced world places a high premium on saving time, often ignoring the importance of eating right. Fast food may allow us to hurry on to our next appointment, but unless it is nutritionally sound, it probably decreases the quality of your lives in the long run.

▷ **FIGURE 11–28 Fast Food Restaurants** The hectic pace of modern life leads many of us to eat at places whose food is rich in fat and rarely provides a balanced meal.

SUMMARY

1. Studies suggest that iron levels in the body may affect heat production in women, a relationship that, if substantiated by additional research, helps illustrate the importance of diet for physiological processes.

A PRIMER ON NUTRITION

2. Humans acquire energy and nutrients from the food they eat. There are two types of nutrients: macronutrients, substances needed in large quantity, and micronutrients, substances required in much lower quantities.

3. The four major macronutrients are water, carbohydrates, lipids, and proteins.

4. Water is in the liquids we drink and the foods we eat. Maintaining adequate water intake is important, because water is involved in many chemical reactions in the body. It also helps maintain body temperature and a constant level of nutrients and wastes in body fluids, so vital to homeostasis.

5. Carbohydrates and lipids are major sources of cellular energy; 70% to 80% of all energy required by the body goes for basic functions.

6. Contrary to popular myth, protein does not supply much energy, except when lipid and carbohydrate intake is low or when protein intake exceeds daily requirements. Dietary protein is chiefly a source of amino acids for building proteins.

7. Some amino acids can be synthesized in the body. These are known as nonessential amino acids. Others cannot be synthesized in the body and must be supplied in the food we eat. These are known as essential amino acids.

8. To ensure an adequate supply of all amino acids, individuals should eat complete proteins, such as those found in milk or eggs, or combine lower quality protein sources.

9. Lipids provide energy during rest and aerobic activity. They serve many other functions in the body, such as insulation.

10. Besides providing energy, carbohydrates serve other important functions. Dietary fiber, for example, increases the liquid content of the feces, reducing constipation, the incidence of diverticulosis, and the risk of colon cancer.

11. Micronutrients are needed in much smaller quantities and include two groups: vitamins and minerals.

12. Vitamins are a diverse group of organic compounds that are required in relatively small quantities for normal metabolism. A deficiency or surplus of one or more vitamins may alter homeostasis, with serious effects on human health.

13. Human vitamins fit into two categories: water-soluble and fat-soluble. The water-soluble vitamins include vitamin C and the B-complex vitamins. The fat-soluble vitamins include vitamins A, D, E, and K.

14. Minerals fit into one of two groups: trace minerals, those required in very small quantity, and major minerals, those required in greater quantity. Deficiencies and excesses of both types of minerals can lead to serious health problems.

THE DIGESTIVE SYSTEM

15. Food is physically and chemically broken down in the digestive system. Small molecules produced during digestion are absorbed by the intestinal tract into the bloodstream or lymphatic system and circulated throughout the body for use by the cells.

16. Food digestion begins in the mouth. The teeth mechanically break down the food. Saliva liquifies it, making it easier to swallow. Salivary amylase begins to digest starch molecules.

17. Food is pushed by the tongue to the pharynx, where it triggers the swallowing reflex. Peristaltic contractions propel the food down the esophagus to the stomach.

18. The stomach is an expandable organ that stores and further liquifies the food. The churning action of the stomach, brought about by peristaltic contractions, mixes the food, turning it into a paste referred to as chyme. The stomach releases food into the small intestine in timed pulses, ensuring efficient digestion and absorption. Very limited chemical digestion and absorption occur in the stomach.

19. The stomach produces hydrochloric acid, which denatures protein, allowing it to be acted on by enzymes. The stomach also produces a proteolytic enzyme called pepsin, which breaks proteins into peptides. The lining of the stomach is protected from acid by mucus. When the mucous protection fails, however, the lining may be eroded by acids, creating an ulcer.

20. The functions of the stomach are regulated by neural and hormonal mechanisms.

21. The small intestine is a long, coiled tubule in which most of the enzymatic digestion and absorption of food take place. Enzymes for digestion come from the pancreas and the lining of the intestine itself.

22. Pancreatic enzymes break macromolecules into smaller fragments. The intestinal enzymes break these molecules into even smaller fragments that can be absorbed by the epithelial lining of the small intestine.

23. The pancreas also releases sodium bicarbonate, which neutralizes the acid entering the small intestine with the chyme and creates an environment suitable for pancreatic enzyme function.

24. The liver also plays an important role in digestion. It produces a liquid called bile that contains, among other substances, bile salts. Bile is stored in the gallbladder and released into the small intestine when food is present. Bile salts emulsify fats, breaking them into small globules that can be acted on by pancreatic lipase.

25. Undigested food molecules pass from the small intestine into the large intestine, which absorbs water, sodium, potassium, and vitamins that are produced by intestinal bacteria. It also transports the waste, or feces, to the outside of the body.

CONTROLLING DIGESTION

26. Digestive processes are largely controlled by the nervous and endocrine systems. The release of saliva is stimulated by the sight, smell, taste, and sometimes even the thought of food. These stimuli also cause the brain to send nerve impulses to the gastric glands of the stomach, initiating the secretion of HCl and gastrin, a hormone that also stimulates HCl secretion.

27. Chyme entering the small intestine stimulates the release of two hormones, secretin and cholecystokinin. Secretin travels in the bloodstream to the pancreas, where it stimulates the release of sodium bicarbonate. Cholecystokinin stimulates the release of pancreatic enzymes and also stimulates the gallbladder to contract, releasing bile into the small intestine.

ENVIRONMENT AND HEALTH: EATING RIGHT/LIVING RIGHT

28. Human health is dependent on good nutrition. Numerous studies suggest that a healthy, balanced diet can decrease the risk of cancer, heart disease, hypertension, and other diseases.

29. Nutritionists recommend the daily consumption of (a) fruits and vegetables, especially cabbage and greens, (b) high-fiber foods, such as whole-wheat bread and celery, and (c) foods high in vitamins A and C.

30. In addition, they recommend reducing consumption of animal fat; red meat; and salt-cured, nitrate-cured, smoked, or pickled foods, including bacon and lunch meat.

31. Studies suggest, however, that Americans have not taken these recommendations to heart. Many of us ignore proper nutrition, because we don't take the time to sit down to a nutritionally balanced meal. Living fast-paced lives, we often ignore the importance of eating right.

EXERCISING YOUR CRITICAL THINKING SKILLS

1. *Science News* recently reported the results of a study on premenstrual syndrome, a condition that results in irritability, tension, bloating, and discomfort, among other symptoms, prior to menstruation. This study showed that high doses of calcium successfully reduce mood swings and physical discomfort in women before and during menstruation.

Researchers at the U.S. Agricultural Research Service in Grand Forks, North Dakota, studied 10 healthy women who experienced mild behavioral and physical symptoms before and during menstruation. The women were assigned to one of two groups. The first group received a high daily dose of calcium (1300 milligrams); the second got a lower dose (600 milligrams) in liquid form added to their food. Halfway through the six-month study, the women switched doses.

According to the report, 9 of the 10 women reported a reduction in crying, irritability, and depression while on the high dose. The high dose also seemed to reduce physical discomfort. In fact, 7 out of the 10 women reported a reduction in cramps and backaches while on the high-calcium diet. (The daily requirement for women 25 and older is 800 milligrams per day.)

Unfortunately, the *Science News* report failed to mention whether the low-dose group benefited from the treatment. It did, however, note that the project leader suggested that women boost their intake of calcium by eating more calcium-rich foods, such as skim milk and nonfat yogurt.

Knowing what you do about proper experimentation, how would you critique this study? What are its flaws? What further work is needed to be certain of these results? Why?

2. Corn is a staple for 200 million people, including nearly half of the world's chronically malnourished people. Because corn is such an important source of calories and protein throughout the world, researchers have developed a new strain of corn called quality-protein maize (QPM).

QPM has about the same amount of protein as common maize but has twice the usable amount because its protein is more complete; that is, it supplies more of the essential amino acids. Normal maize has about 40% of the biological value of milk protein. QPM approaches that of milk, a common standard of nutritional excellence.

In countries where maize is a staple, QPM could double the efficiency of subsistence farming. A study by the National Research Council reports that QPM could allow families to combat malnutrition without outside help. QPM could prove helpful in Mexico, Central America, and Africa, where hunger and starvation are common.

Using your critical thinking skills, examine this optimistic finding. What questions come to mind about the conclusion that QPM could help reduce malnutrition? You may want to refer to an environmental science book (see the Suggested Readings in Chapter 33) to find out more on world hunger—that is, to get a view of the big picture.

Prepare a list of questions on this issue, then consult the list below. How does this list compare with your list?
  a. Will QPM be affordable to peasants in the Third World?
  b. Will it be more susceptible to insect pests than currently used native strains?
  c. Will it require costly pesticides?
  d. Will it require costly irrigation?
  e. Will it deplete the soil?
  f. How have efforts to improve crop yield worked in the past?
  g. How does QPM compare economically with other measures?
  h. What are other solutions to the problem?
  i. Should QPM be supplemented by other strategies?

TEST OF TERMS

1. Water, carbohydrates, lipids, and proteins are all members of a class of nutrients called _____ .
2. The two main sources of energy in a well-balanced diet are _____ and _____ .
3. Excess protein in the diet can be used to generate energy or be used to make _____ .
4. The amino acids the body cannot produce are called _____ amino acids. Proteins that supply all of these amino acids are said to be _____ .
5. Indigestible carbohydrates like the cellulose in vegetables and grains are called water-insoluble _____ and are thought to reduce the incidence of _____ cancer.
6. _____ are a diverse group of organic compounds that are found in minute amounts in our diet and are essential for metabolism. They fit into two broad groups: _____ and _____ .
7. Minerals required in minute quantities, such as zinc and copper, are called _____ _____ .
8. The salivary glands produce the enzyme _____ .
9. The taste receptors on the tongue are called _____ _____ .
10. The _____ connects the oral cavity with the esophagus.
11. The involuntary muscle contractions that propel food along the digestive tract are called _____ .
12. At the juncture of the esophagus and stomach is a ring of muscle called the _____ sphincter.
13. In the stomach, food is converted to a liquified mass called _____ .
14. The proteolytic enzyme _____ , produced by the stomach, is released in an inactive form, _____ .
15. The ring of muscle at the junction of the stomach and small intestine that controls the passage of food into the small intestine is called the _____ sphincter.
16. The _____ produces bile, a fluid that is stored in the gallbladder and later released into the small intestine where its chief chemical component, _____ _____ emulsifies fats.
17. The pancreas produces two major exocrine products: _____ _____ and _____ .
18. Three major structural modifications increase the surface area for absorption in the small intestine; they are _____ _____ , _____ , and _____ .
19. Fats absorbed by the small intestine are carried away by _____ capillaries.
20. The hormone _____ , produced by the small intestine, stimulates the contraction of the gallbladder.

Answers to the Test of Terms are found in Appendix B.

TEST OF CONCEPTS

1. The body requires proper nutrient input to maintain homeostasis. Give an example, and explain how the nutrient affects homeostasis.
2. Describe the conditions during which protein provides cellular energy.
3. If you were considering becoming a strict vegetarian, eating no animal products, even milk and eggs, how would you be assured of getting all of the amino acids your body needs?
4. Describe how the different types of dietary fiber help protect human health.
5. What are vitamins, and why are they needed in such small quantities?
6. A dietary deficiency of one vitamin can cause wide-ranging effects. Why?
7. What organs physically break food down, and what organs participate in the chemical breakdown of food?
8. Describe the process of swallowing.
9. Describe the function of hydrochloric acid and pepsin in the stomach. How does the stomach protect itself from these substances?
10. How do ulcers form, and how can they be treated?
11. Describe the endocrine and nervous system control of the stomach function.
12. The small intestine is the chief site of digestion and absorption. Where do the enzymes needed for this process come from, and how is the release of these enzymes stimulated? What other molecules are needed for proper digestion?
13. Describe the functions of the large intestine.

The Circulatory System

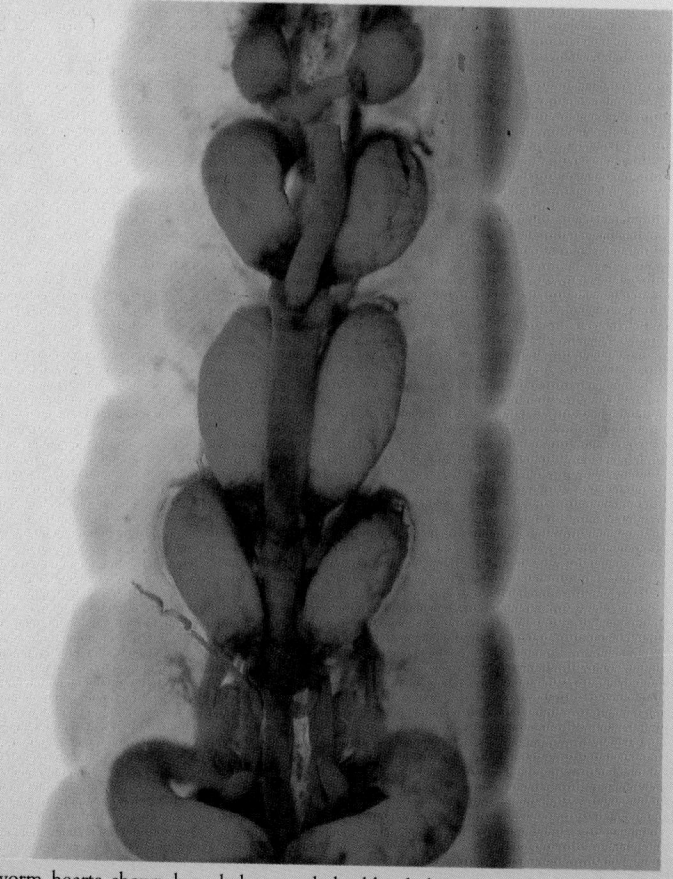

Earthworm hearts shown here help propel the blood through a closed system of blood vessels.

D avid McMahon, a 48-year-old New York attorney, collapsed in his office one morning. Dazed and in great pain, he was rushed to a nearby hospital. There a team of cardiologists discovered a blood clot lodged in a narrowed section of his right coronary artery, which supplies the heart. The physicians began immediate action to prevent further damage to the oxygen-starved heart muscle. Through a small incision in McMahon's groin, they inserted a tiny plastic catheter into the large artery that carries blood to the leg. They then threaded the catheter up through the arterial system to his heart (▷ Figure 12–1). Once they reached the heart, physicians guided the catheter into the clogged coronary artery, where they injected an enzyme, called streptokinase, through the catheter. Streptokinase dissolved the blood clot, restoring blood flow to McMahon's heart muscle.

David McMahon survived. Unfortunately, many others who suffer similar attacks are not so lucky. They either arrive at the hospital too late to be treated or, if they do make it to the hospital, do not receive blood-clot-dissolving agents or other treatments in time to avoid extensive damage to their heart muscle.

Heart attacks strike thousands of Americans each year. They are one of a handful of diseases of the circulatory system caused by the stressful conditions of modern life, poor eating habits, and a host of other factors. This chapter examines the circulatory system of animals. This chapter also takes a look at the lymphatic system, which is involved in both circulation and immune protection.

▷ **FIGURE 12–1 The Circulatory System** (*a*) The circulatory system consists of a series of vessels that transport blood to and from the heart, the pump. (*b*) The circulatory system has two major circuits, the pulmonary circuit, which delivers blood to the lungs, and the systemic circuit, which delivers blood to the rest of the body.

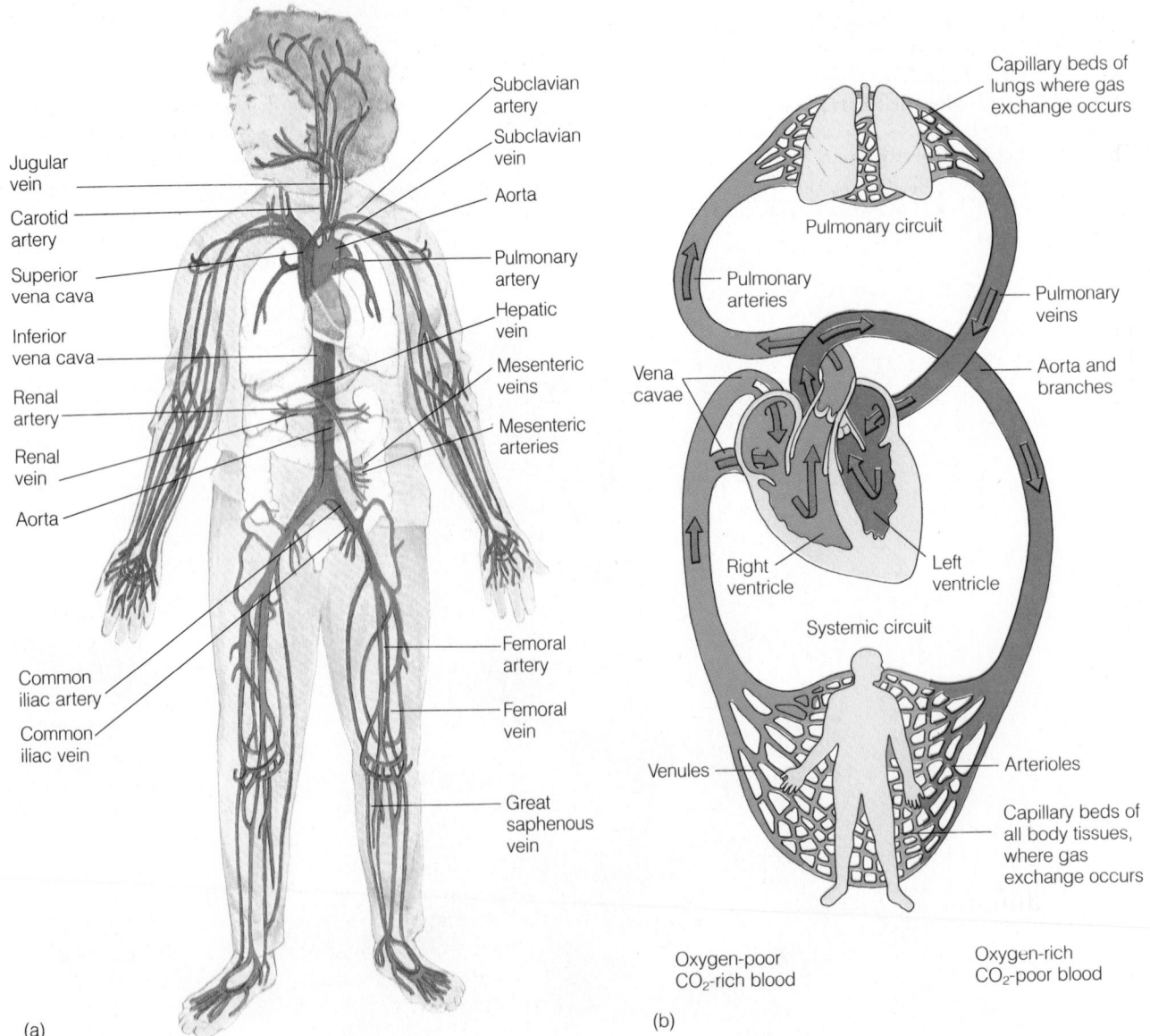

(a)

(b)

≈

(a)

Jugular vein

Carotid artery

Superior vena cava

Inferior vena cava

Renal artery

Renal vein

Aorta

Common iliac artery

Common iliac vein

Subclavian artery

Subclavian vein

Aorta

Pulmonary artery

Hepatic vein

Mesenteric veins

Mesenteric arteries

Femoral artery

Femoral vein

Great saphenous vein

(b)

Capillary beds of lungs where gas exchange occurs

Pulmonary circuit

Pulmonary arteries

Pulmonary veins

Vena cavae

Aorta and branches

Right ventricle

Left ventricle

Systemic circuit

Venules

Arterioles

Capillary beds of all body tissues, where gas exchange occurs

Oxygen-poor CO₂-rich blood

Oxygen-rich CO₂-poor blood

≈ CIRCULATORY SYSTEMS: AN EVOLUTIONARY IMPERATIVE

In the early days of evolution, life was a rather simple affair. For the first 2.5 billion years, in fact, all that existed were single-celled monerans (bacteria) living in the shallow coastal waters where life presumably arose. The monerans were later joined by protists, single-celled eukaryotes, and together these two life forms dominated the scene for many more millions of years.

Monerans and protists acquired nutrients from the seawater. Seawater also carried wastes away. In a sense, then, the watery environment bathing the cells served as a huge external circulatory system. But with the advent of multicellularity, organisms began to evolve elaborate ways to circulate nutrients to and remove wastes from cells that had become increasingly distant from the external environment.

The earliest form of internal circulation is seen in the sponge, one of the earliest animals (▷ Figure 12–2). Seawater enters through tiny pores in the sponge's body and then flows through a network of internal channels. This series of channels forms a primitive, but effective, means of transporting fluids to cells and ensuring proper nutrition and waste disposal. As in the protists and monerans, nutrients and wastes move in and out of the cell by diffusion.

This basic type of "circulatory system" is found in a wide assortment of the Earth's earliest multicellular organisms, such as jellyfish, hydras, and anemones. Far more elaborate circulatory systems would soon evolve.

▷ **FIGURE 12–2 Primitive Internal Circulation** (*a*) Sponge. (*b*) Cross section showing how water enters the body cavity of the sponge to nourish the cells and wash away metabolic wastes.

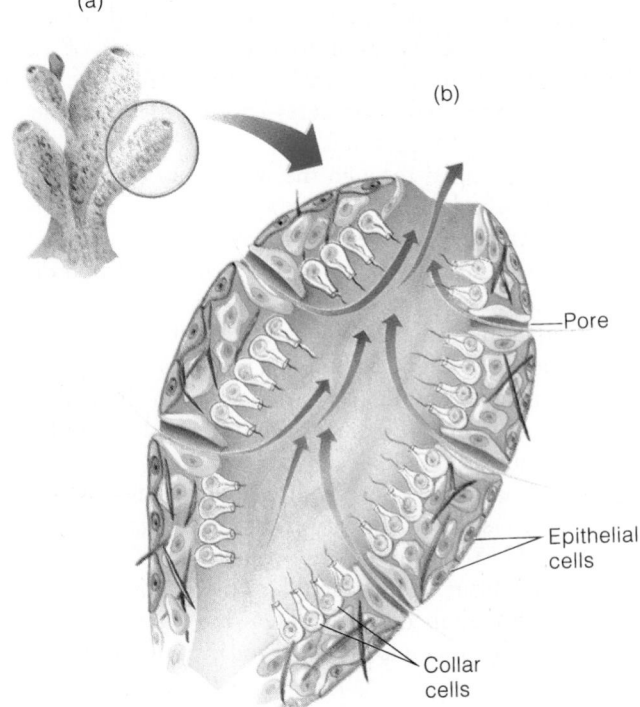

(a)

(b)

Pore

Epithelial cells

Collar cells

These more complex systems fall into two major categories: open and closed. An **open circulatory system** is one in which the blood vessels empty into the body cavity, where the organs are located. A good example is the grasshopper (▷ Figure 12–3). The grasshopper's organs receive ample nutrition and are adequately rid of wastes by this crude but effective system. Blood is removed from the body cavity through tiny openings (ostia) in the dorsal blood vessel. The walls of the dorsal vessel are thickened and serve as pumps. When the muscles are relaxed, the holes in the wall open, permitting blood to enter. When the muscles contract, the holes close, and blood is propelled forward in the dorsal vessel.

Open circulatory systems are rather inefficient and are generally found in organisms that do not exhibit high rates of metabolism: arthropods (insects, spiders, and crustaceans) and molluscs (clams and snails). **Closed circulatory systems** transport blood through a closed system of vessels, delivering it more directly to cells through thin-walled capillaries.

One of the first closed circulatory systems appears in the segmented worms (annelids) such as the earthworm, shown in ▷ Figure 12–4. The earthworm feeds on organic matter as it burrows through soil. The food it ingests is digested in the primitive digestive system, and nutrients diffuse through the wall of the intestine into blood vessels, which transport them throughout the body with the aid of five pumps that connect the dorsal and ventral vessels.

Closed circulatory systems permit a more efficient transport of nutrients than open systems and are an important adaptation in warm-blooded species, such as birds and mammals, which require a sufficient supply of nutrients to maintain body temperature and the high metabolic demand of cells. The human circulatory system, described in this chapter, is an excellent example of a closed system and a testament to the remarkable improvements in animal physiology resulting from evolution.

≈ THE CIRCULATORY SYSTEM'S FUNCTION IN HOMEOSTASIS

The circulatory system of humans, like that of other vertebrates, consists of a muscular pump and a network of vessels to transport blood to and from the heart (see Figure 12–1). The functions of the human circulatory system are listed in Table 12–1. The blood coursing through the system transports oxygen from the lungs to the body cells, where it is used in cellular energy production, as noted in Chapter 6. The circulatory system also helps distribute nutrients from the digestive tract as well as hormones produced by the endocrine glands to body tissues and cells. Waste produced by cells is swept away by the blood and carried in the blood vessels to excretory organs, such as the kidneys and lungs, which rid the body of potentially harmful chemical substances. By pumping blood throughout the body, the circulatory system also helps distribute body heat in warm-blooded animals. Each of these functions—and a great

▷ FIGURE 12–3 Open Circulatory System

In circulatory systems such as those in insects, the heart (actually a series of hearts) pumps blood into a series of vessels that empty into a body cavity, or hemocoel. Each organ of the body is directly bathed with blood in the hemocoel. When the hearts relax, blood is sucked back into them through openings guarded by one-way valves. When the hearts contract, the valves are pressed shut, forcing the blood to travel through the vessels.

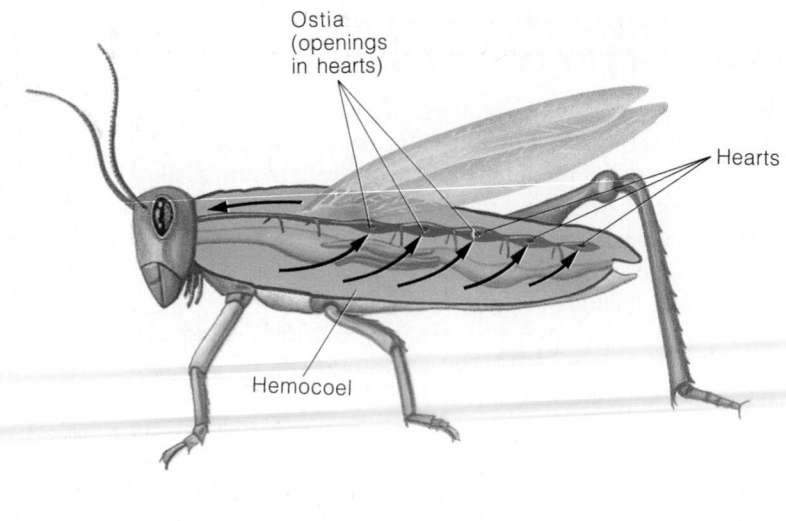

Ostia (openings in hearts)

Hearts

Hemocoel

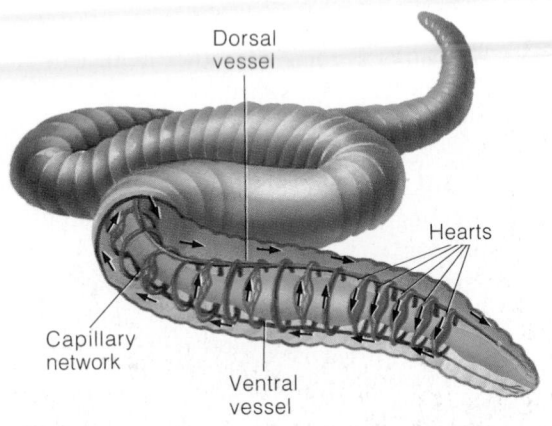

Dorsal vessel

Hearts

Capillary network

Ventral vessel

▷ FIGURE 12–4 Circulatory System of the Earthworm

In this closed circulatory system, blood remains within the vessels and hearts. Nutrients must be transported out of the vessels to reach body cells.

TABLE 12-1 Functions of the Circulatory System

Transports oxygen to body cells
Transports nutrients from the digestive system to body cells
Transports hormones to body cells
Transports wastes from body cells to excretory organs
Distributes body heat

---

many more—serves a higher purpose, the maintenance of homeostasis; that is, it helps maintain the relatively constant internal conditions necessary for cellular function.

≈ THE HEART

The heart is a muscular pump in the thoracic (chest) cavity. Sometimes referred to as the workhorse of the cardiovascular system, the heart propels blood through the 50,000 miles of blood vessels in the human body. Each day, the heart beats approximately 100,000 times, adjusting its rate to meet the changing needs of the body. If you had a dollar for every heartbeat, you would be a millionaire in 10 days. Over a 70-year lifetime, you would collect $2.5 billion for your heart's work.

The heart, shown in ▷ Figure 12–5, is a fist-sized organ whose walls are composed of three layers, the pericardium, the myocardium, and the endocardium. The **pericardium** forms a thin, closed sac surrounding the heart and the bases of large vessels that enter and leave the heart. The pericardial sac is filled with a clear, slippery fluid that reduces friction produced by the heart's repeated contraction. The middle layer, the **myocardium,** is the thickest part of the wall and is composed chiefly of cardiac muscle cells, briefly described in Chapter 10. The inner layer, the **endocardium,** is the endothelial layer, which forms the lining of the heart chambers.

The Circulatory System Has Two Distinct Circuits through Which Blood Flows

The circulatory system, shown in simplified form in Figure 12–1b, consists of two distinct circuits, the **pulmonary circuit,** which carries blood to and from the lungs, and the **systemic circuit,** which transports blood to and from the rest of the body. The central driving force in both of these circuits is the heart. As shown in Figure 12–1b, the heart consists of four hollow chambers—two on the right side of the heart and two on the left.

Blood is pumped through the pulmonary circuit by the right side of the heart—that is, the right atrium and right ventricle. As noted above, the pulmonary circuit delivers blood to the lungs, where it loses most of its carbon dioxide and replenishes its supply of oxygen. The oxygenated blood is then returned to the heart and distributed to body tissues via the systemic circulation, which is served by the left atrium and left ventricle. In the tissues of the body, the blood releases much of the oxygen it gained in the lungs and takes up carbon dioxide released by cells during cellular respiration. The blood is then returned to the heart, where it enters the pulmonary circuit once again.

Figure 12–5 illustrates the course that blood takes through the heart. Drawn in blue, blood low in oxygen (and rich in carbon dioxide) enters the right side of the heart from the systemic circuit through the **superior** and

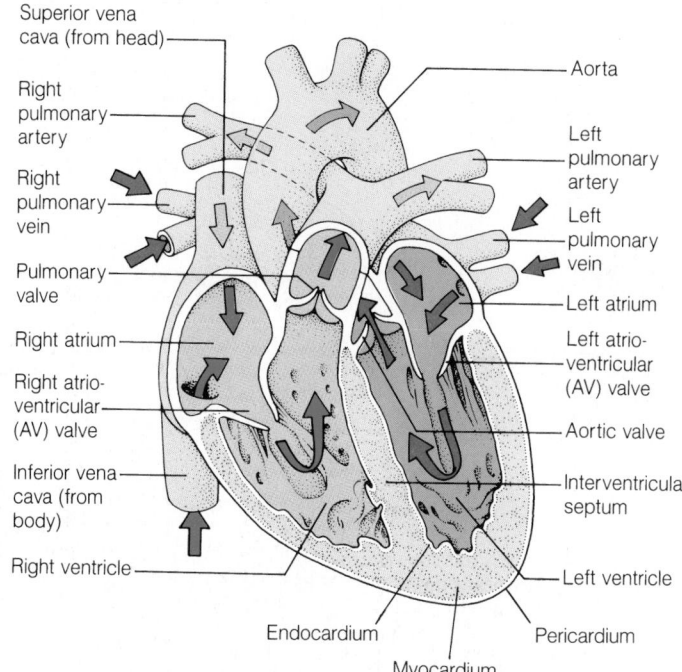

Superior vena cava (from head)

Right pulmonary artery

Right pulmonary vein

Pulmonary valve

Right atrium

Right atrio-ventricular (AV) valve

Inferior vena cava (from body)

Right ventricle

Aorta

Left pulmonary artery

Left pulmonary vein

Left atrium

Left atrio-ventricular (AV) valve

Aortic valve

Interventricular septum

Left ventricle

Endocardium

Myocardium

Pericardium

▷ **FIGURE 12–5 Blood Flow through the Heart**
Deoxygenated (carbon-dioxide-enriched) blood (blue) flows into the right atrium from the systemic circulation, then is pumped into the right ventricle. The right ventricle, in turn, pumps the blood into the pulmonary artery, which delivers it to the lungs, where carbon dioxide is released and oxygen is picked up. Reoxygenated blood (red) is returned to the left atrium, then flows into the left ventricle, which pumps it to the rest of the body through the systemic circuit.

inferior vena cavae. These large veins empty directly into the **right atrium,** the uppermost chamber on the right side of the heart. The blood is pumped from here into the **right ventricle,** the lower chamber on the right side. When the right ventricle is full, the muscles in its wall contract, forcing blood into the pulmonary arteries, which lead to the lungs.

Blood whose oxygen supply has been restored, drawn in red in Figure 12–5, flows back to the heart from the lungs in the **pulmonary veins.** The pulmonary veins, in turn, empty directly into the **left atrium,** the upper chamber on the left side of the heart. From here, the oxygen-rich blood is pumped to the **left ventricle.** When the left ventricle is full, its thick, muscular walls contract and propel the blood into the aorta. The **aorta** is the largest artery in the body. It carries the oxygenated blood away from the heart, delivering it to the cells and tissues of the body via many smaller branches (discussed below).

The flow of blood I have just described presents a slightly misleading view of the way the heart really works. As shown in ▷ Figure 12–6, both atria actually fill and contract simultaneously, delivering blood to their respective ventricles. The right and left ventricles also fill simultaneously, and when both ventricles are full, they too contract in unison, pumping the blood into the systemic and pulmo-

nary circulations. The coordinated contraction of heart muscle is brought about by an internal timing device, or pacemaker (described later).

Heart Valves are Located Between the Atria and Ventricles and Between the Ventricles and the Large Vessels into Which They Empty

In the evolution of circulatory systems, one of the most important features was the emergence of valves to help regulate the flow of blood through the heart. The human heart contains four valves that control the direction of blood flow, ensuring a steady flow from atria to ventricles to the large vessels that are part of the systemic and pulmonary circuits (▷ Figure 12–7a).

The valves between the atria and ventricles are known as **atrioventricular valves.** Each valve consists of two or three flaps of tissue anchored to the inner walls of the ventricles by slender tendinous cords, the **chordae tendineae,** resembling the strings of a parachute (Figure 12–7a). The right atrioventricular valve, between the right atrium and right ventricle, is called the **tricuspid valve,** because it contains three flaps. The left atrioventricular valve is the **bicuspid valve.**[1] To remember the valves, imagine you are wearing a jersey with the number 32 on the front. This reminds you that the tricuspid valve is on the right side and the bicuspid valve is on the left.

Between the right and left ventricles and the arteries into which they pump blood (pulmonary artery and aorta, respectively) are the **semilunar valves** (Figures 12–7a and 12–7b). The semilunar valves (literally, "half moon") consist of three semicircular flaps of tissue (Figures 12–7b and 12–7c).

The atrioventricular and semilunar valves are one-way valves, opening when blood pressure builds on one side and closing when it increases on the other, much like the purge valves in scuba diving masks, which allow divers to force water out of their masks, or the ball valves in snorkels, which operate similarly. When the ventricles contract, blood forces the semilunar valves open. Blood flows out of the ventricles into the large arteries. The backflow of blood causes the valve to close, preventing blood from draining back into the ventricles. The atrioventricular valves function in like fashion.

Heart Sounds Result from the Closing of Various Heart Valves

When physicians listen to your heart, they are actually listening to sounds of the heart valves closing. The noises they hear are called the **heart sounds** and are often described as "LUB-dupp." The first heart sound (LUB) results from the closure of the atrioventricular valves. It is longer and louder than the second heart sound (dupp),

[1] The bicuspid valve is also called the mitral valve, because it resembles a miter, a hat worn by the Pope and Catholic bishops.

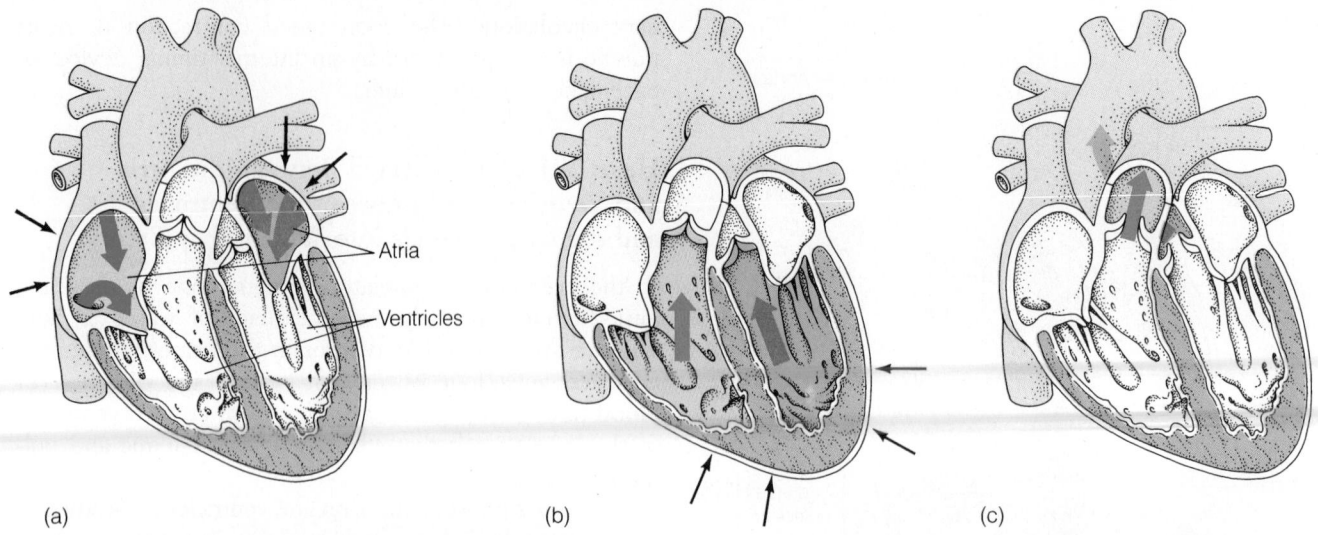

(a)   (b)   (c)

▷ **FIGURE 12–6 Blood Flow through the Heart**
(a) Blood enters both atria simultaneously from the systemic and pulmonary circuits. When full, the atria pump their blood into the ventricles. (b) When the ventricles are full, they contract simultaneously, (c) delivering the blood to the pulmonary and systemic circuits.

produced when the semilunar valves snap shut. Interestingly, the right and left atrioventricular valves do not close at precisely the same time. Nor do the semilunar valves. Thus, by careful placement of the stethoscope, a physician can listen to each valve individually to determine if it is functioning well.

For most of us, our heart valves function flawlessly throughout life. However, some diseases can alter the function of the valves, greatly affecting the efficiency of the heart and the circulation of blood. Rheumatic fever, for example, is caused by a bacterial infection and affects many parts of the body, including the heart. Although relatively rare in developed countries, rheumatic fever is still a significant problem in the Third World. Rheumatic fever begins as a sore throat caused by certain types of **streptococcus** bacteria. The sore throat—known as strep throat—is usually followed by general illness. During this infection, antibodies (proteins made by cells of the immune system) to the streptococcus bacteria, which circulate in the blood, damage the heart valves, preventing them from closing completely. This damage causes blood to leak back into the atria and ventricles after contraction and results in a distinct "sloshing" sound, commonly called a **heart murmur.** This condition, generally referred to as **valvular incompetence,** reduces the efficiency of the heart and causes the organ to work harder than usual to make up for the inefficient pumping. Increased activity, in turn, causes the walls of the heart to enlarge. In severe cases, valvular incompetence can result in heart failure. To prevent heart failure, damaged valves can be replaced by artificial implants.

Tumors (benign and malignant) and scar tissue have an opposite effect; that is, they can obstruct the valve, reducing blood flow through the heart. This condition is known as **valvular stenosis** (from the Greek word *steno,* meaning "narrow"). Valvular stenosis prevents the ventricles from filling completely. As in valvular incompetence, the heart must beat faster to ensure an adequate supply of blood to the body's tissues. This acceleration also puts additional stress on the organ.

Heart Rate Is Controlled by an Internal Pacemaker

The activity level of multicellular animals is highly variable. For this reason, they need a way to control the rate at which the heart beats and, consequently, the rate at which blood flows to body tissues. That is to say, they need a way to accelerate heart rate and blood flow when tissues demand more oxygen and decelerate it when demand falls.

During the evolution of multicellular animals, several mechanisms evolved to control heart rate. One of them is an internal pacemaker, known as the **sinoatrial (SA) node** (▷ Figure 12–8). Located in the wall of the right atrium, the SA node is composed of a clump of specialized cardiac muscle cells. The cells of the SA node contract spontaneously and rhythmically. Each contraction produces a bioelectric impulse, akin to those produced by nerve cells. The impulse given off by the SA node spreads rapidly from muscle cell to muscle cell in both atria. Because cardiac muscle cells are tightly joined and because the impulse travels quickly, the two atria contract simultaneously and uniformly.

Left to their own devices, cardiac muscle cells would contract independently and in a disorderly way, creating an ineffective pumping action. The SA node, therefore imposes a single rhythm on all of the atrial heart muscle cells.

The electrical impulse next passes from the atria to the ventricles; however, its passage is briefly slowed by a barrier of unexcitable tissue that separates the atria from the ventricles. The impulse is delayed approximately 1/10 of a second. After this brief holdover, the impulse is channeled

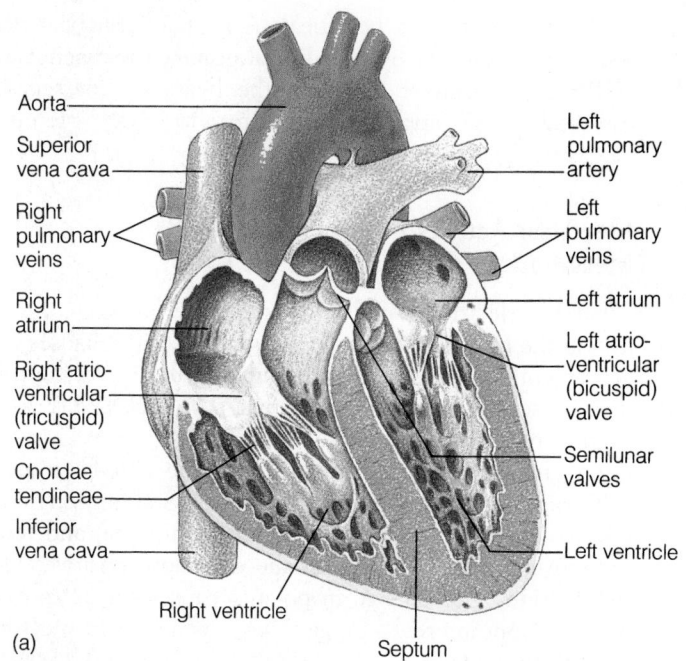

(a)

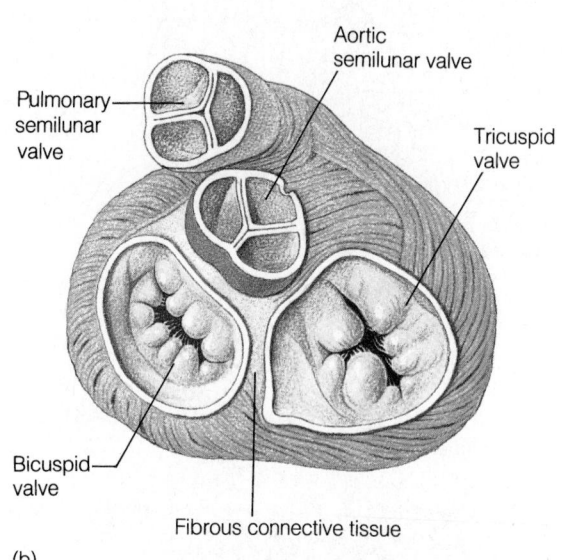

(b)

▷ **FIGURE 12–7 Heart Valves** (*a*) A cross section of the heart showing the four chambers and the location of the major vessels and valves. (*b*) A view of the heart from above, with the major vessels removed to show the valves. (*c*) Pulmonary semilunar valve.

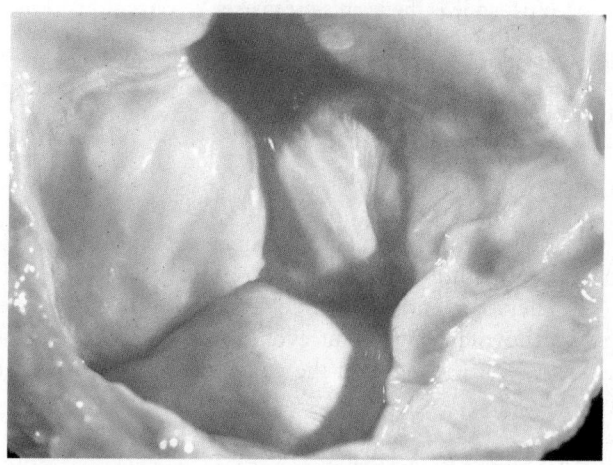

(c)

through a second mass of specialized muscle cells, the **atrioventricular (AV) node,** shown in Figure 12–8. From the AV node, the impulse travels along a tract of specialized cardiac muscle cells, known as the **atrioventricular bundle.** As shown in Figure 12–8, the atrioventricular bundle divides into two branches (called the bundle branches) that travel on either side of the wall separating the two ventricles. The bundle branches give off smaller branches that terminate on **Purkinje fibers,** specialized muscle cells that come in contact with cardiac muscle cells of the ventricles.

The slight delay created when the impulse travels from the atria to the ventricles gives the blood-filled atria ample time to contract and empty into the ventricles. It also provides the ventricles plenty of time to fill before they are stimulated to contract. Unlike the muscle cells of the atria, the cardiac muscle cells of the two ventricles do not contract in unison, in large part, because the impulse is not transmitted as quickly and as uniformly through the ventricles as it is through the atria. Contraction begins at the bottom of the heart and proceeds upward, squeezing the blood out of the ventricles into the aorta and pulmonary arteries.

The SA node of the human heart produces a steady rhythm of about 100 beats per minute when isolated from outside influences, but this is much too fast for most human activities. To bring the heart rate in line with body demand, the SA node must be dampened. This dampening is brought about by nerves that supply the heart with impulses from a control center in the brain. At rest or during nonstrenuous activity, these impulses slow the heart to about 70 beats per minute, thus aligning heart rate with body demands. (They dampen much as the brakes reduce a car's speed.) During exercise or stress, when the heart rate must increase to meet body demands, the decelerating impulses from the brain are reduced. (In other words, the body lets up on the brakes, permitting the heart rate to increase.) Other nerves also influence heart rate. These nerves carry impulses that accelerate the heart rate even further, allowing the heart to attain rates of 180 beats or more when body demand for oxygen is great. (Continuing with the car analogy, these nerves are akin to the accelerator pedal, which increases the RPMs of the engine.)

Several hormones also play a role in controlling heart rate. One of these hormones is **epinephrine,** commonly known as **adrenalin.** During stress or exercise, the adrenal glands, located on top of the kidneys, secrete this hor-

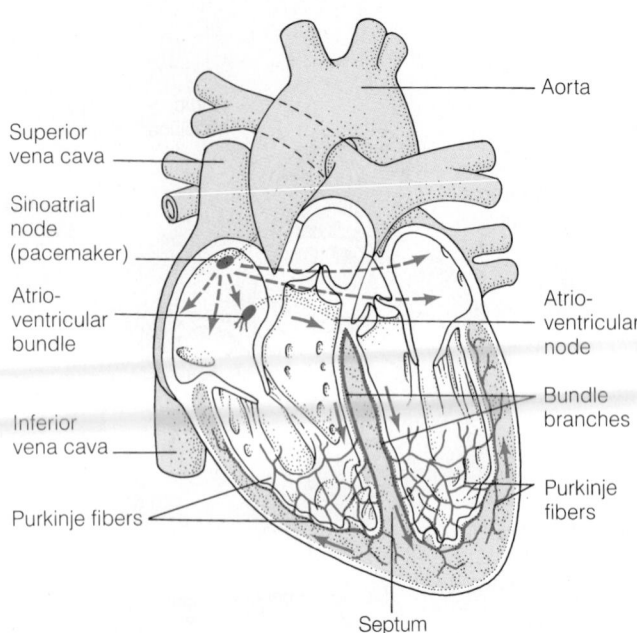

Aorta

Superior
vena cava

Sinoatrial
node
(pacemaker)

Atrio-
ventricular
bundle

Atrio-
ventricular
node

Bundle
branches

Inferior
vena cava

Purkinje
fibers

Purkinje fibers

Septum

▷ **FIGURE 12–8 Conduction of Impulses in the Heart**
The sinoatrial node is the heart's pacemaker. Located in the right atrium, it sends timed impulses into the atrial heart muscles, co-ordinating muscle contraction. The impulse travels from cell to cell in the atria, then passes to the atrioventricular node and into the ventricles via the atrioventricular bundle and its two branches, which terminate on the Purkinje fibers.

mone. Epinephrine stimulates the heart rate, increasing the flow of blood through the heart.

The nervous and endocrine system mechanisms described above are important evolutionary adaptations that help vertebrates cope with the changing demands of their lives.

Heart Muscle Cells Unleashed from Their Control Beat Independently, Greatly Reducing the Heart's Effectiveness

A patient is wheeled into the emergency room of a major hospital, where the doctors find that his heart is in a kind of cardiac anarchy known as **fibrillation.** Individual cells are beating on their own accord, and little blood is actually moving in and out of the heart chambers.

This patient, like thousands of others each year, is suffering one kind of heart attack. Having lost control from the SA node, the patient's heart has been converted into an ineffective, quivering mass of muscle tissue with individual cardiac muscle cells contracting at their own rate. During fibrillation, the heart pumps very ineffectively and may cease pumping altogether, a potentially fatal condition known as **cardiac arrest.**

Physicians treat fibrillation by applying a strong electrical current to the chest, a procedure appropriately known as **defibrillation.** The electrical current passes through the wall of the chest and is often sufficient to restore normal

electrical activity and heartbeat. A normal heartbeat can also be restored by **cardiopulmonary resuscitation (CPR),** a procedure in which the heart is "massaged" externally by applying pressure to the sternum (breastbone).

Electrical Activity in the Heart Can Be Measured on the Surface of the Chest

When the electrical impulse that stimulates muscle contraction in the heart reaches a cardiac muscle cell, it causes the cell to contract. Normally, the outside surface of the cardiac muscle cell is slightly positive. The inside surface is slightly negative. When the impulse arrives, it causes a rapid change in the permeability of the cardiac muscle cell's plasma membrane to sodium ions. Sodium ions flow inward, changing the polarity of the membrane, and temporarily making the inside of the cell more positive than the outside. This change in polarity causes the release of calcium from internal storage depots, which in turn causes the cell to contract.

The shift in cardiac muscle cell polarity, or **depolarization,** can be detected by surface electrodes, small metal plates connected to wires and a voltage meter (▷ Figure 12–9a). The electrodes are placed on a person's chest. The resulting reading on a voltage meter is called an **electrocardiogram (ECG),** or sometimes also **EKG** (Figure 12–9b).

For a normal person, the tracing produced on the voltage meter has three distinct waves (Figure 12–9b). The first wave, the **P wave,** represents the electrical changes occurring in the atria of the heart. The second wave, the **QRS wave,** is a record of the electrical activity taking place during ventricular contraction, and the third wave, the **T wave,** is a recording of electrical activity occurring as the ventricles relax.

Diseases of the heart may disrupt one or more waves of the ECG. As a result, an ECG is often a valuable diagnostic tool for cardiologists. Bear in mind, though, that the ECG detects only those diseases that alter the heart's electrical activity.

Cardiac Output Varies from One Person to the Next, Depending on Activity and Conditioning

The total amount of blood pumped by the ventricles each minute is called the **cardiac output.** Cardiac output is a function of two variables: **heart rate,** or the number of contractions the heart undergoes per minute, and **stroke volume,** or the amount of blood pumped by each ventricle during each contraction. At rest, the heart beats approximately 70 times per minute, and the stroke volume is about 70 milliliters. This produces a cardiac output of 5000 milliliters, or 5 liters per minute.

Cardiac output varies among individuals, depending on their physical condition and their level of activity. The heart of a trained athlete, for example, can pump 35 liters of blood per minute (seven times the cardiac output at rest).

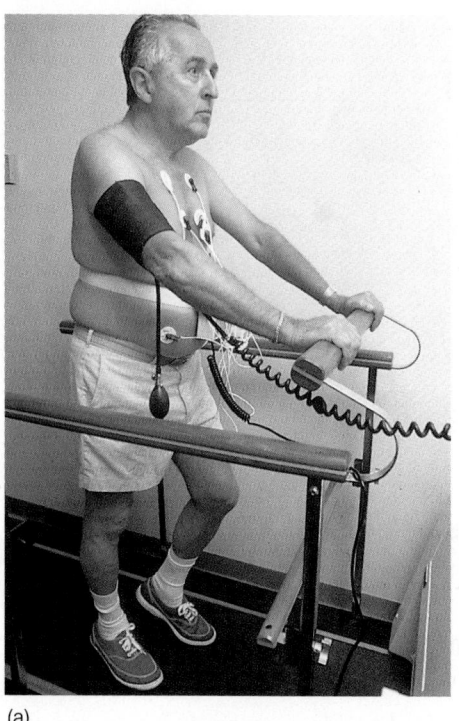

(a)

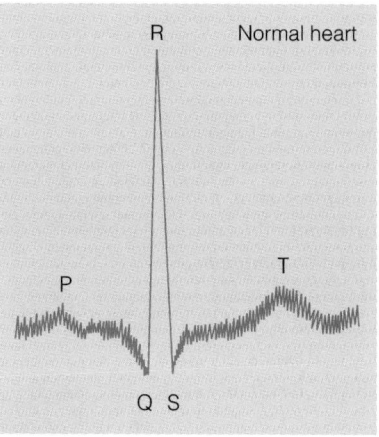

Normal heart

R

P   T

Q S

P = atrial depolarization, which triggers
    atrial contraction.

QRS = depolarization of AV node and
      conduction of electrical impulse
      through ventricles. Ventricular
      contraction begins at R.

T = repolarization of ventricles.

P to R interval = time required for
                  impulses to travel from
                  SA node to ventricles.

(b)

▷ **FIGURE 12–9 The
Electrocardiogram** (*a*) This patient taking a treadmill test to check his heart's
performance is wired to a meter that detects electrical activity produced by the
heart. (*b*) An electrocardiogram.

Most nonathletes, however, can increase the cardiac output
to only about 20 liters per minute.

≈ HEART ATTACKS: CAUSES, CURES, AND TREATMENT

Heart attacks come in several varieties. The most common
heart attack is called a **myocardial infarction** (mentioned
in Chapter 11 and described below) and is caused by a
thrombosis, a clot blockage of one or more of the **coronary arteries,** small vessels that supply blood to the heart.
A blood clot lodged in a coronary artery restricts the flow
of blood to the heart muscle, cutting off the supply of
oxygen and nutrients. Depriving the heart muscle of oxygen can damage and even kill the cells. The region damaged by a thrombosis is called an **infarct,** hence the name
myocardial infarction.

Myocardial Infarctions Usually Occur When Blood Clots Lodge in Arteries Narrowed by Atherosclerosis

As noted in Health Note 11–1, the formation of atherosclerotic plaque results from a combination of stress, poor
diet, lack of exercise, smoking, heredity, and several other
factors. Narrowing of a coronary artery by plaque does not
necessarily result in a heart attack, however, unless the
narrowing is quite severe. Nonetheless, less severe narrowing does makes the vessel more susceptible to blood clots.

When a clot forms in the vessel at the site of narrowing or
when a clot originating elsewhere in the body lodges in the
narrowed vessel, trouble begins.

The outcome of heart attacks varies. If the size of the
infarct (damaged area) is small and if the change in electrical activity of the heart is minor and transient, a heart
attack is usually not fatal. If the damage is great or electrical activity is severely disrupted, however, myocardial in-
farctions can prove fatal.

Heart attacks can occur quite suddenly, without warning, or may be proceeded by several weeks of **angina,** pain
that is felt when the supply of oxygen to the myocardium is
reduced. Anginal pain appears in the center of the chest
and can spread to a person's throat, upper jaw, back, and
arms (usually just the left one). Angina is a dull, heavy,
constricting pain, that appears when an individual is active,
then disappears when he or she ceases the activity.

Angina attacks may also be caused by stress and exposure to carbon monoxide, a pollutant that reduces the
oxygen-carrying capacity of the blood. Angina begins to
show up in men at age 30 and is nearly always caused by
coronary artery disease. In women, angina tends to occur at
a much later age. Interestingly, about 90% of all "chest
pain" turns out to be unrelated to the heart. Patients go to
their doctors thinking they are suffering from angina, when,
in fact, what they are feeling is tension and pain in the wall
of the chest, usually in the muscles between the ribs. Deep
breathing, relaxation, and stress reduction are effective in
reducing, even eliminating, this pain.

Prevention is the Best Cure, but in Cases Where Damage has Already Occurred, Medical Science has a Great Deal to Offer

Proper diet, exercise, and stress management can help reduce the risk of heart problems later in life, as noted in Health Notes 1–1 and 11–1). Research also shows that a daily dose of aspirin (half a tablet a day is sufficient) taken over long periods can help reduce an individual's chances of having a heart attack. A recent study of 22,000 male physicians in the United States, in fact, showed that aspirin in small doses reduced the risk of first heart attacks by nearly half.[2] Studies suggest that aspirin reduces heart attacks by impairing blood clotting.

Prevention should be the first line of attack against heart disease. It could save Americans hundreds of millions of dollars each year in medical bills, lost work time, and decreased productivity. But given human nature, the fast pace of modern life, and our inattentiveness to exercise and proper diet, heart disease will probably be around for a long time. To reduce the death rate, physicians therefore also look for ways to treat patients after they have had a heart attack.

One promising development is the use of blood-clot-dissolving agents like streptokinase, mentioned at the beginning of the chapter. When administered within a few hours of the onset of a heart attack, streptokinase can greatly reduce the damage to heart muscle and accelerate a patient's recovery.

Ironically, streptokinase is an enzyme derived from the bacterium that causes rheumatic fever. Because streptokinase is a foreign substance, it evokes an immune reaction. In some people, the reaction is quite severe and may even cause death. The immune reaction to this drug has inspired a search for similar chemicals without the dangerous side effects. One promising chemical is urokinase, a clot-busting enzyme produced by human cells that does not trigger an immune reaction.

Scientists are also testing another naturally occurring clot dissolver, called TPA (tissue plasminogen activator). Tests in humans suggest that TPA may also be free of the dangerous side effects of streptokinase. Consequently, TPA has been approved for use in humans since 1987. Nevertheless, its use is not without its problems. Two of the most significant are (1) the recurrence of blood clots in many patients and (2) the high cost of the drug.

In cases where the coronary arteries are completely blocked by atherosclerotic plaque, it is necessary to reestablish full blood flow to the heart muscle. To restore blood flow, physicians often perform **coronary bypass surgery,** in which they transplant segments of intact vein from the leg into the heart ($\triangleright$ Figure 12–10). These venous bypasses transport blood around the clogged coronary arteries, restoring blood flow. Once hoped to be a long-term

[2] You should consult your physician if you are thinking about taking aspirin as a preventive measure.

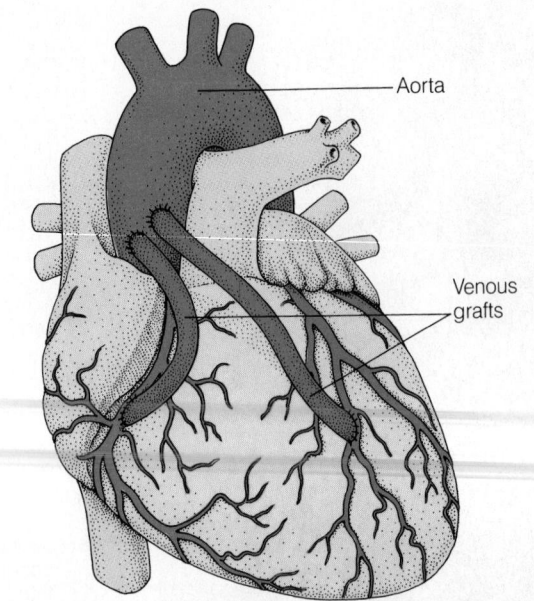

Aorta

Venous grafts

$\triangleright$ **FIGURE 12–10 Coronary Bypass Surgery** Venous grafts bypass coronary arteries occluded by atherosclerotic plaque.

solution to a widespread problem, coronary bypass surgery may provide only temporary relief with little, if any, long-term benefits. Research studies show that bypass patients have a significantly higher rate of survival in the five years following surgery than patients who just receive drugs. In the next seven years, however, these studies show, long-term survival from coronary bypass surgery is about the same as that of patients treated with diet and medications. In the long run, say researchers, bypass surgery is only slightly more effective than nonsurgical medical treatments.

The problem with coronary bypasses is that the venous grafts often fill fairly quickly with cholesterol plaque, and the heart becomes starved for oxygen once again, especially when forced to work harder (for example, during exercise). The recurrence of plaque in venous grafts has led researchers to turn to marginally important arteries, such as the internal mammary artery, for coronary bypass surgery. Researchers believe that these arteries will prove more resistant to plaque buildup than veins. Physicians can also clean clogged blood vessels by using a small catheter with a tiny balloon attached to its tip in conjunction with the clot-dissolving agents. After the chemical clot dissolvers are administered to a patient, the balloon is inflated. This procedure, called **balloon angioplasty,** forces the artery open and apparently loosens the plaque from the wall, allowing it to be washed away by the blood.

Scientists are also experimenting with lasers to burn away plaque in artery walls. Catheters containing fine glass fibers can be inserted into the blocked arteries during open-heart surgery. They can also be inserted through the artery in the thigh and snaked through the arterial system until they reach the clot. Laser beams transmitted through the glass fibers burn away the plaque. Unfortunately, as in other techniques, cholesterol builds up again in the walls of arteries within a few months.

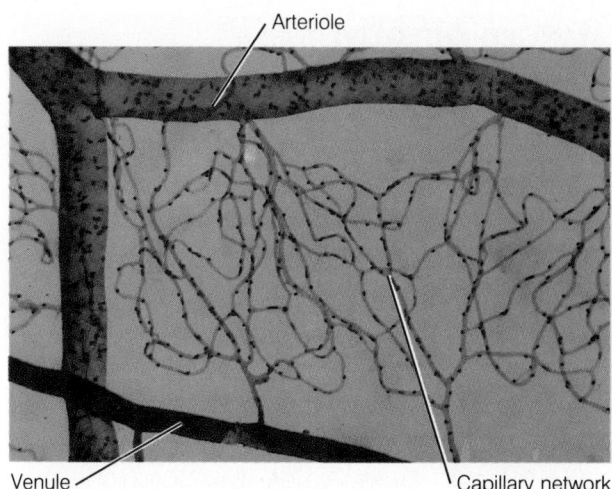

> **FIGURE 12–11 Capillary Network** A network of capillaries between the arteriole and the venule delivers blood to the cells of body tissues, not shown.

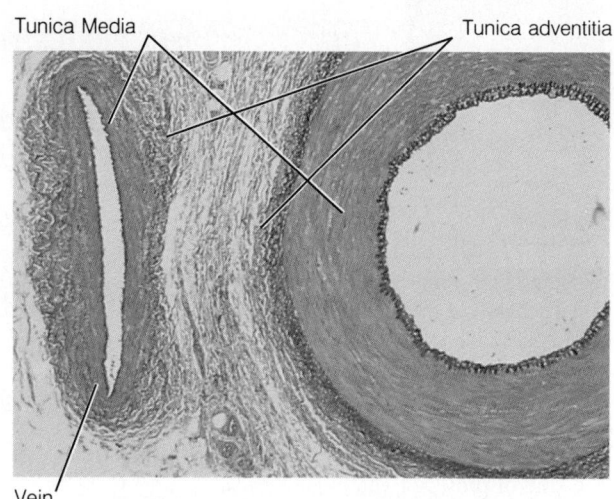

> **FIGURE 12–12 Artery and Vein** A cross section through a vein shows that the muscular layer, the tunica media, is much thinner than it is in an artery. Veins typically lie alongside arteries and in histological sections such as these usually have irregular lumens.

≈ THE BLOOD VESSELS

The circulatory system can be divided into four functional parts. The first is the heart, which pumps blood throughout the body. The second is the arteries, which form a delivery system that carries blood from the heart to the body tissues. The third is an exchange system, consisting of networks of tiny vessels known as capillaries, found in body tissues. The fourth is the return system, consisting of veins that carry oxygen-depleted and waste-enriched blood back to the heart from the body tissues. (For a discussion of the discoveries that led to our understanding of circulation, see the Scientific Discoveries 12–1.)

Arteries, which transport blood away from the heart, branch many times, forming smaller and smaller vessels. The smallest of all arteries is the **arteriole.** As illustrated in ▷ Figure 12–11 arterioles empty into **capillaries,** tiny, thin-walled vessels that permit wastes and nutrients to pass through with relative ease. Capillaries form extensive, branching networks in body tissues, referred to as **capillary beds,** which provide an avenue for exchange between the blood and the tissue fluid surrounding the cells of the body.

Blood flows out of the capillaries into the smallest of all veins, the **venules.** Venules, in turn, converge to form small veins, which unite with other small veins, in much the same way that small streams unite to form a river.

▷ Figure 12–12 shows a cross section of an artery and a vein. As illustrated, these two vessels are structurally different. Veins, for example, tend to be smaller and to have thinner walls. Despite their obvious differences, arteries and veins have a common architecture. For example, both consist of three layers: (1) an external layer of connective tissue, the **tunica adventitia,** which binds the vessel to surrounding tissues; (2) a middle layer, the **tunica media,** which is primarily made of smooth muscle; and (3) an internal layer, the **tunica intima,** which is composed of a layer of flattened cells, the **endothelium,** and a thin, nearly indiscernible layer of connective tissue, which binds the endothelium to the tunica media (▷ Figure 12–13).

> **FIGURE 12–13 General Structure of the Blood Vessel** The artery shown here consists of three major layers, the tunica intima, tunica media, and tunica adventitia.

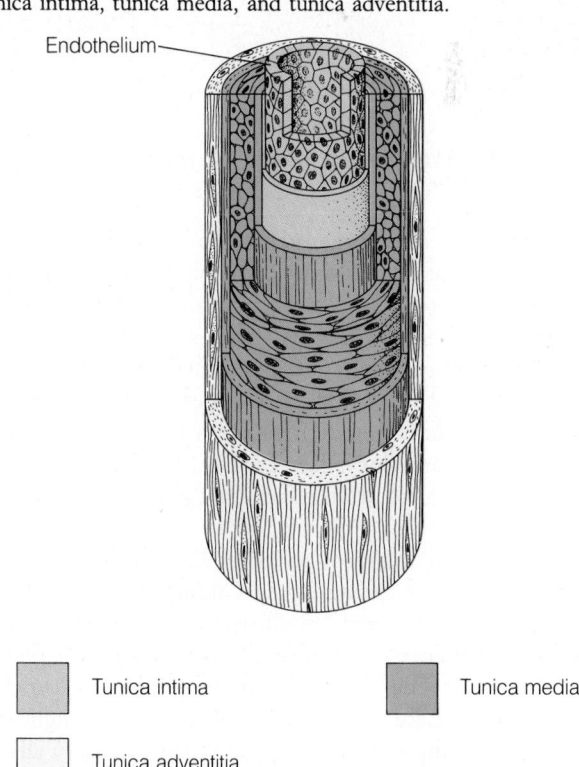

- Tunica intima
- Tunica media
- Tunica adventitia

THE CIRCULATION OF BLOOD IN ANIMALS
Featuring the Work of William Harvey and Stephan Hales

The 17th-century British physician William Harvey is generally credited with the discovery of the circulation of blood in animals. Harvey is known as a scientist with a short temper who wore a dagger in the fashion of the day, which he reportedly brandished at the slightest provocation (▷ Figure 1). He was probably not the kind of professor you might "argue" with over grades.

Temperament aside, Harvey is generally honored for his work on the role of the heart and the flow of blood in animals and is often praised as a pioneer of scientific methodology. His application of quantitative procedures to biology, some say, ushered in the modern age of this science.

In Harvey's medical school days, anatomists thought the intestines produced a substance called chyle, a kind of lymph that passed from the intestines to the liver. Chyle was, they thought, derived from the food people ate. The liver converted the chyle to venous blood and then distributed the blood through arteries and veins. As a medical student, Harvey learned that blood oozing through arteries and veins supplied organs and tissues with nourishment. He also learned that there was no real circulation, that blood merely ebbed back once in a while to the heart and lungs, where impurities were removed.

▷ **FIGURE 1 William Harvey**
This important scientific figure greatly advanced our knowledge on circulation in animals.

These ideas were proposed by the Greek physician Galen 14 centuries earlier and had persisted nearly without challenge until Harvey's time.

As a teacher in the Royal College of Physicians in 1616, Harvey began to describe the circulation of blood, based on the results of experiments and observations on animals. Apparently rebuked for his ideas by some of his colleagues, Harvey engaged in many years of research to provide supporting evidence. In these studies, he described the muscular character of the heart and the origin of the heartbeat in the right atrium. He also demonstrated that the pulse felt in arteries resulted from the impact of blood pumped by the heart. Furthermore, he described the pulmonary and systemic circuits and proposed that blood flowed to the tissues and organs of the body via the arteries and returned via the veins.

A brief examination of just one of his experiments illustrates that even though Harvey is a key figure in the history of biological science and played a key role in promoting quantitative study, some of

Arteries and Arterioles Deliver Oxygen-Rich Blood to Body Tissues and Organs

The largest of all arteries is the aorta. As noted earlier, this massive vessel carries oxygenated blood from the left ventricle of the heart to the rest of the body. The aorta loops over the back of the heart, then descends through the chest and abdomen, giving off large branches along its way. These branches carry blood to the head, the extremities (arms and legs), and major organs, such as the stomach, the intestines, and the kidneys (see Figure 12–1). The very first branches of the aorta are the coronary arteries, which supply the heart muscle.

The aorta and many of its chief branches are **elastic arteries,** so named because they contain numerous wavy elastic fibers interspersed among the smooth muscle cells of the tunica media (▷ Figure 12–14a). As blood pulses out of the heart, the elastic arteries expand to accommodate the blood. Like a stretched rubber band, though, the elastic fibers in the tunica media cause the arterial walls to recoil. This provides an ancillary pump that helps push the blood along the arterial tree and also helps maintain an even flow of blood through the capillaries.

The elastic arteries branch to form smaller vessels, the **muscular arteries** (Figure 12–14b). Muscular arteries contain fewer elastic fibers but can still expand and contract with the flow of blood. You can feel this expansion and contraction in the arteries lying near the skin's surface in your wrist and neck. It's the pulse that health-care workers use to measure your heart rate.

The smooth muscle of the tunica media of muscular arteries responds to a number of stimuli, including nerve impulses, hormones, carbon dioxide, and lactic acid. These stimuli cause the blood vessels to open or close to varying degrees. This, in turn, allows the body to adjust blood flow through its tissues to meet increased demands for nutrients and oxygen. Arterioles in muscles, for instance, dilate when an animal is threatened by danger. This opening increases blood flow to the muscle, allowing the animal to flee or to meet the danger head on. At the same time, vessels in the digestive system constrict, temporarily reducing the digestive process and increasing the amount of blood available

his work was less than exceptional, based as it was on some poor assumptions and inaccurate observations. As an example, consider the work he used to rebut Galen's hypothesis that the blood was produced by the food people ate.

Harvey first approximated the amount of blood the heart ejected with each heartbeat (stroke volume), then determined the pulse rate. He called on earlier observations of a heart from a human cadaver to determine stroke volume. At that time, he had noted that the left ventricle contained more than two ounces of blood and then, for reasons not entirely clear to historians of science, had hypothesized that the ventricle ejected "a fourth, a fifth, a sixth or only an eighth" of its contents. (Today, studies indicate that the heart ejects nearly all of its contents.) Harvey estimated that the stroke volume was about 3.9 grams of blood per beat. Modern estimates put it at 89 grams per beat. Harvey also made a grave error in determining pulse. His value of 33 beats per minute is about half of the actual rate in humans. No one knows how he could

have been so wrong. Armed with two erroneous measurements, Harvey derived a figure for the amount of blood that circulated through the body that was 1/36 of the lowest value accepted today.

Regardless, Harvey "proved" his point that each half-hour the blood pumped by the heart far exceeds the total weight of blood in the body. From this he concluded that blood must be circulated. It is not, as Galen proposed, produced by the food we eat. The amount of food one eats could not produce blood in such volume.

Harvey debunked another falsehood perpetrated through the centuries—the Galenic myth that blood flowed into the extremities in both arteries and veins. Harvey first wrapped a bandage around an extremity. This obstructed the flow of blood through the veins but not the arteries. He noted that the veins swelled because, as he conjectured, blood was being pumped into them via underlying arteries and there was nowhere for the blood to go. Tightening the bandage further cut the blood flow in the arteries

as well and thus prevented the veins from swelling. From these observations, Harvey correctly surmised that the arteries deliver blood to the extremities, and the veins return it to the heart.

Harvey's work laid the foundation of modern cardiovascular physiology, but it left many questions unanswered. Many of these questions were addressed by the highly industrious English biologist Stephan Hales, who was born a century after Harvey. In a long series of rather gruesome but scientifically rigorous experiments on horses, dogs, and frogs, Hales explored many aspects of the cardiovascular system. Benefiting from more modern methods of study, he charted blood pathways and examined blood flow and blood pressure in different parts of the circulatory system. After settling many of the unanswered questions left by Harvey, Hales went on to study plant physiology and is perhaps best known for his work on the circulation of sap in plants, discussed in Scientific Discoveries 26–1.

to the muscles. Regulating the flow of blood to body tissues is required to control body temperature as well, as noted in earlier chapters.

Blood Pressure. The force that blood applies to the walls of a blood vessel is known as the **blood pressure.** Like many other physical conditions in the human body, blood pressure varies from time to time. For example, it changes in relation to one's activity and stress levels. In a given artery, it rises and falls with each heart beat. Blood pressure also varies throughout the cardiovascular system, being the highest in the aorta and dropping considerably as the arteries branch. When blood reaches the capillaries, the flow of blood and blood pressure are greatly reduced, enhancing the rate of exchange between the blood and the tissues.

Blood pressure is measured by using an inflatable device with the tongue-twisting name of **sphygmomanometer** (sfig-mo-ma-nom-a-ter), or, more commonly, blood pressure cuff (▷ Figure 12–15a). The blood pressure cuff is first wrapped around the upper arm. A stethoscope is po-

sitioned over the artery just below the cuff. Air is pumped into the cuff until the pressure stops the flow of blood through the artery (Figure 12–15b). The pressure in the cuff is then gradually reduced as air is released. When the blood pressure in the artery exceeds the external pressure of the cuff, the blood starts flowing through the vessel once again. This point represents the **systolic pressure,** the peak pressure at the moment the ventricles contract. Systolic pressure is detected by a thump heard through the stethoscope and is the higher of the two numbers in a blood pressure reading (120/70).[3] The pressure at the moment the heart relaxes to let the ventricles fill again is the **diastolic pressure** and is the lower of the two readings. It is determined by continuing to release air from the cuff until no arterial pulsation is audible. At this point, blood is flowing continuously through the artery.

A typical reading for a young, healthy adult is about 120/70, although readings vary considerably from one individual to the next. Thus, what is normal for one person

[3] Blood pressure is measured in millimeters of mercury (mm Hg).

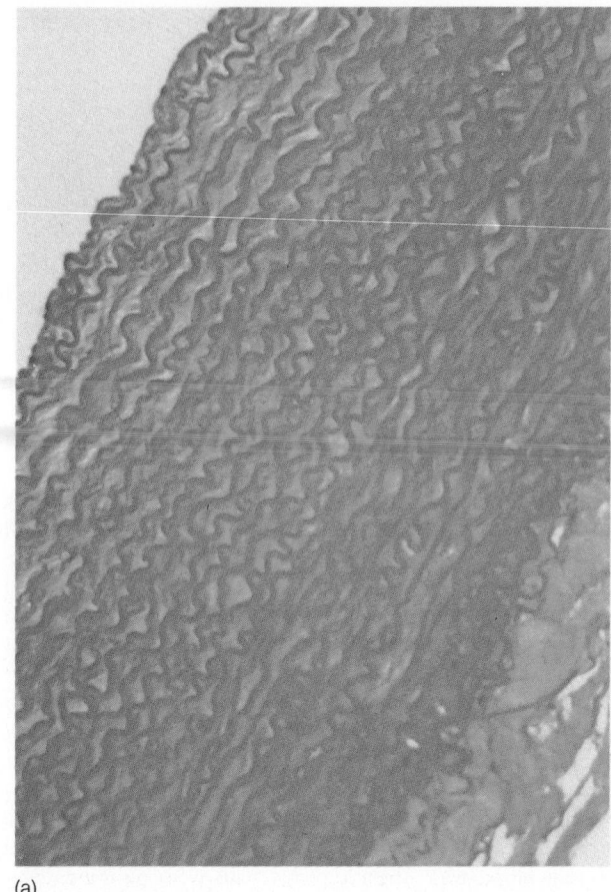

(a)

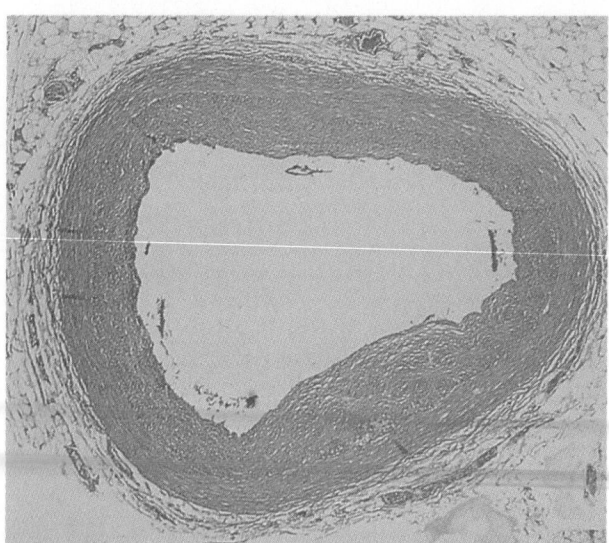

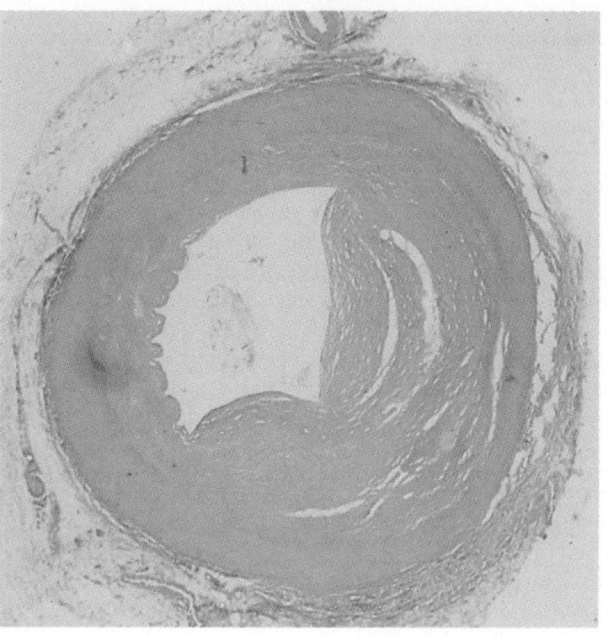

(c)

▷ **FIGURE 12–14 Arteries** (*a*) A cross section of the wall of the aorta showing numerous wavy elastic fibers common in all elastic arteries. (*b*) The muscular artery has a thick tunica media as well but fewer elastic fibers. (*c*) Muscular artery partly occluded by atherosclerotic plaque.

may be abnormal for another. As a person ages, blood pressure tends to rise. Therefore, a healthy 65-year-old might have a blood pressure reading of 140/90. In a 30-year-old, this reading might be cause for alarm.

Hypertension is a prolonged elevation in blood pressure. It has many causes, including kidney disease, high salt intake, and obesity. Nearly always a symptomless disease, hypertension is often characterized by a gradual increase in blood pressure over time. A person may feel fine and display no physical problems whatsoever for years. Symptoms, such as headaches, palpitations (rapid, forceful beating of the heart), and a general feeling of ill health, usually occur only when blood pressure is dangerously high. Consequently, early detection and treatment are essential to prevent serious problems, including heart attacks. (For more on hypertension, see Health Note 12–1.)

Capillaries Permit the Exchange of Nutrients and Wastes Between Blood and Body Cells

As described above, the heart, arteries, and veins form an elaborate system that propels and transports blood to and

from the capillaries. As also noted earlier, capillaries form branching networks, known as capillary beds, among the cells of body tissues (▷ Figure 12–16a). It is in these extensive networks of vessels that wastes and nutrients are exchanged between the cells of the body and the blood.

As illustrated in Figure 12–16b, the walls of the capillaries consist of flattened endothelial cells; they permit dissolved substances to pass through them with great ease—and provide another illustration of the correlation between structure and function in the body. In some places, such as the kidneys and small intestine, small windows, or **fenestrae,** are present in the cells of capillary walls (Figure 12–16c). Fenestrae permit even greater movement of molecules to and from the capillary.

If you could remove all of the capillaries from the body and line them up end to end, they would extend over

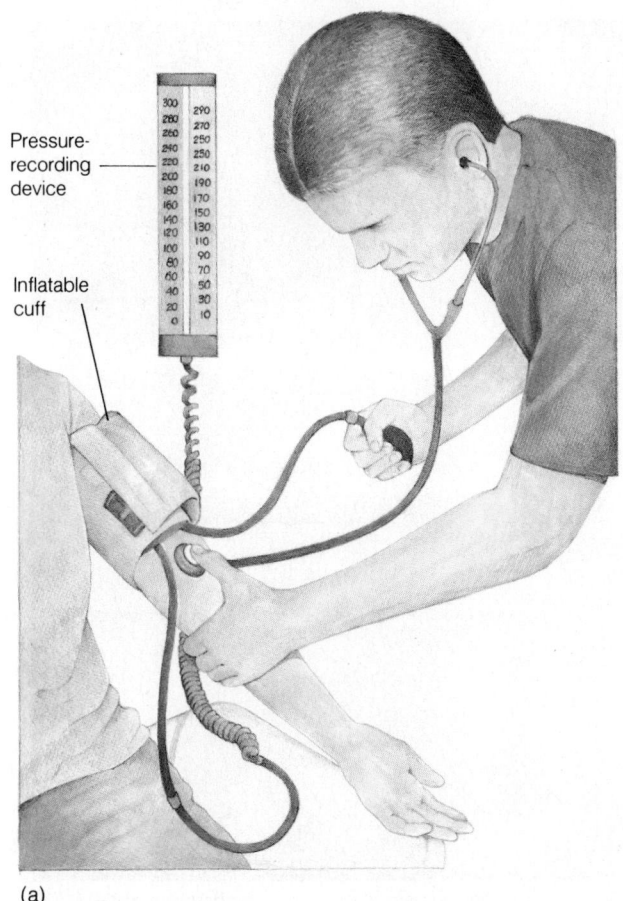

Pressure-recording device

Inflatable cuff

(a)

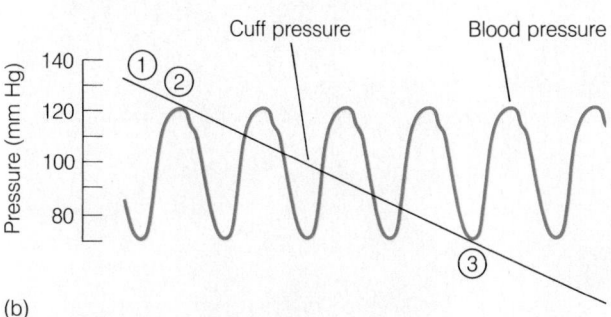

Cuff pressure

Blood pressure

(b)

▷ **FIGURE 12–15 Blood Pressure Reading** (*a*) A sphygmomanometer (blood pressure cuff) is used to determine blood pressure, indicated by the red line. As shown, the blood pressure in a given region rises and falls with each contraction of the heart. (*b*) When the pressure in the cuff exceeds the arterial peak pressure, blood flow stops ①. No sound is heard. Cuff pressure is gradually released. When pressure in the cuff falls below the arterial pressure, blood starts flowing through the artery once again. This is the systolic pressure ②. The first sound will be heard. Cuff pressure continues to drop. When cuff pressure is equal to the lowest pressure in the artery, the artery is fully open and no sound is heard ③. This is the diastolic pressure.

80,000 kilometers (50,000 miles)—enough to circle the globe at the equator two times. The extensive branching of capillaries brings them in close proximity to body cells but also slows the rate of blood flow through capillary networks

and decreases pressure, both of which increase the efficiency of capillary exchange.

A "typical" capillary network is shown in ▷ Figure 12–17a. As illustrated, most capillary beds contain **thoroughfare channels,** or **metarterioles,** circulatory "shortcuts" that connect the arterioles with the venules. The metarterioles give off smaller branches, the **true capillaries,** the site of the exchange of nutrients and wastes.

The flow of blood through the capillary bed, and therefore the supply of nutrients to tissues, is largely controlled by the constriction and dilation of the metarteriole. When the metarteriole is open, blood flows into the capillary network, providing ample nutrients. When it constricts, blood passes by, traveling on to other tissues in need of nutrients.

The constriction and dilation of the metarteriole also help regulate body temperature. On a cold winter day, for example, the metarterioles of the capillary networks in the skin close down, restricting blood flow and conserving body heat. Just the reverse happens on a warm day. The flow of blood through the skin increases, releasing body heat and often creating a pink flush.

Blood flow through capillary networks is also controlled at another level. As Figure 12–17a shows, tiny rings of muscle surround the capillaries as they branch from the metarterioles. These muscle rings are called **precapillary sphincters.** They open and close like floodgates in response to local chemical signals (such as carbon dioxide levels) from nearby tissues. The relaxation of the precapillary sphincters causes blood to rush into the capillary bed. The precapillary sphincters therefore provide a means of delivering blood on demand to cells that need it or reducing the flow to cells that do not.

As blood flows *into* a capillary bed, nutrients, gases, water, and hormones immediately begin to diffuse out of the tiny vessels. As the blood flows *through* the capillary bed, water-dissolved wastes, such as carbon dioxide, begin to flow into the capillaries by diffusion. Wastes and nutrients can travel (1) between the endothelial cells of the capillary, (2) through the endothelial cells by diffusion, (3) through fenestrae, and (4) through the endothelial cells in minute pinocytotic vesicles (Chapter 3). The forces that control movement across the capillary wall are explained in ▷ Figure 12–18 (page 302).

Veins and Venules Transport the Oxygen-Poor and Waste-Laden Blood Back to the Heart

Blood leaves the capillary beds stripped of its nutrients and loaded with cellular wastes. Blood draining from the capillaries first enters the smallest of all veins, the venules. The venules converge with others, forming small veins. **Veins** carry blood toward the heart. Unlike the arteries, the veins start off small and converge with other veins, forming larger and larger vessels. Eventually, all blood returning to the heart from the systemic circulation enters either the superior or inferior vena cavae, the two main veins that

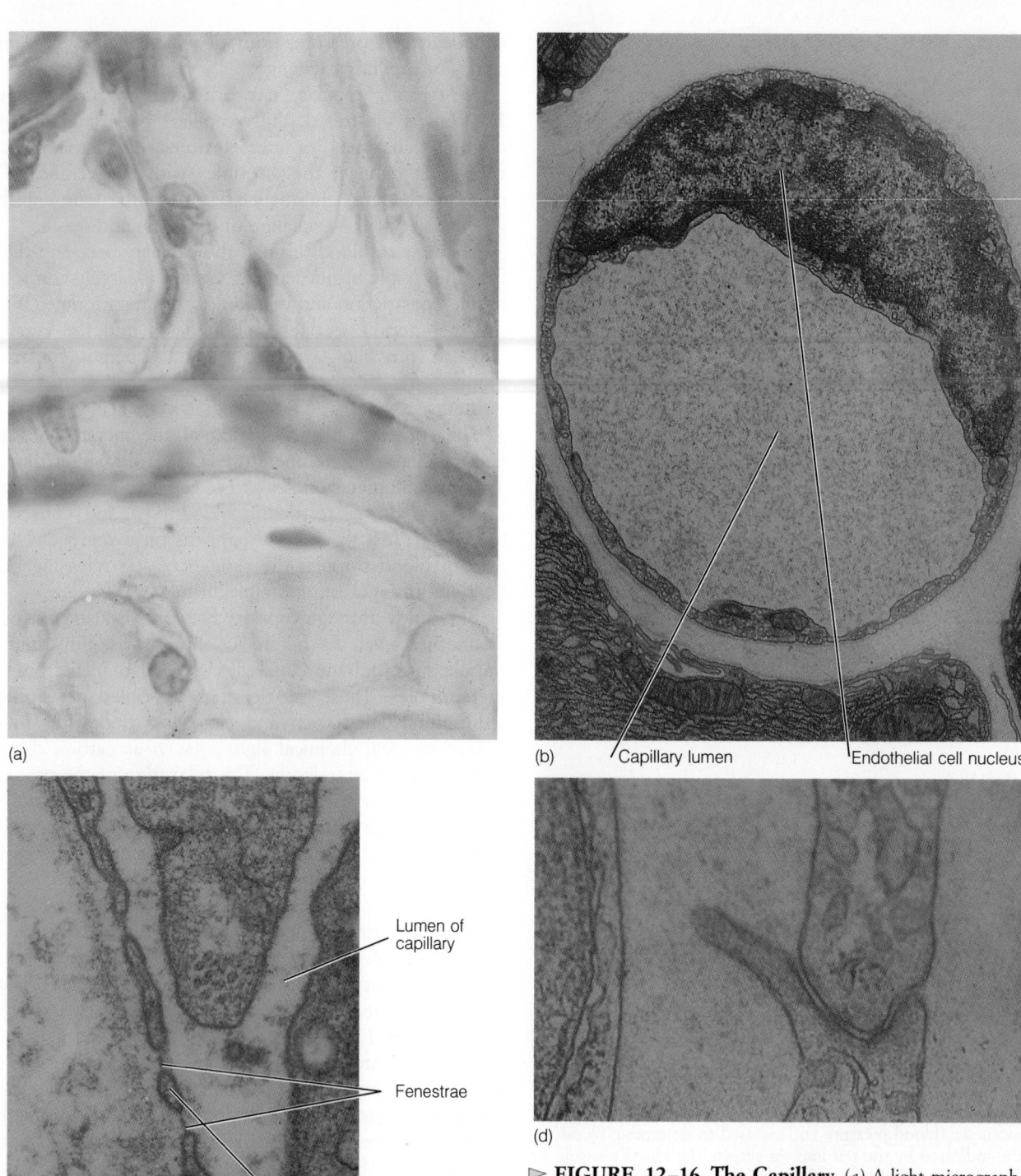

(a)

(b)   Capillary lumen   Endothelial cell nucleus

(c)

Lumen of capillary

Fenestrae

Tissue fluid

Endothelial cell of capillary

(d)

▷ **FIGURE 12–16 The Capillary** (*a*) A light micrograph of a capillary showing the endothelial cells that make up the wall of this vessel. (*b*) A cross section of a capillary showing the nucleus of an endothelial cell and capillary lumen. (*c*) A section through the wall of a highly porous capillary showing the fenestrae. Note that each window is spanned by a thin membrane that permits rapid movement of molecules into and out of the capillary. (*d*) A cross section through the wall of a capillary showing how materials flow through via endocytosis.

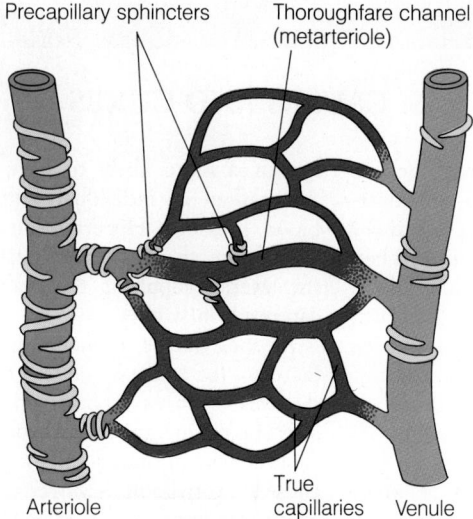

Precapillary sphincters   Thoroughfare channel
                           (metarteriole)

Arteriole   True   Venule
                   capillaries

Sphincters open

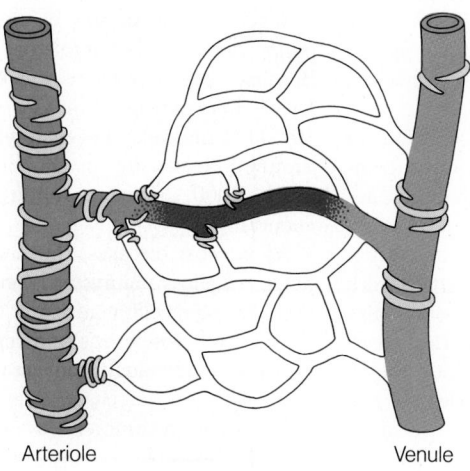

Arteriole   Venule

Sphincters closed

(a)

▷ **FIGURE 12–17 Control of Blood Flow through Capillaries** (*a*) The flow of blood into a capillary bed is controlled by the metarterioles and the precapillary sphincters. When the metarterioles are open, they allow blood to flow into the capillary network. Precapillary sphincters must open to permit further influx. When precapillary sphincters close, they reduce blood flow through the capillary bed. (*b*) A scanning electron micrograph of an arteriole showing the smooth muscle cells that contract and relax, controlling flow through these vessels.

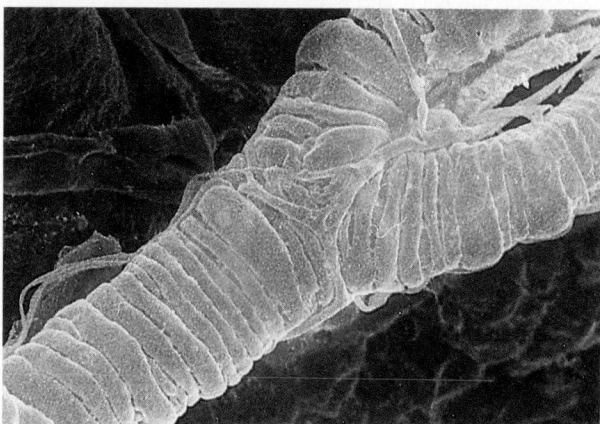

Smooth muscle cells

(b)

drain into the right atrium of the heart (see Figure 12–5). These vessels drain the upper and lower parts of the body, respectively.

Veins and arteries generally run side by side throughout the body much like opposing lanes on a freeway. The arteries take blood away from the heart and toward body tissues, and the veins return blood to the heart.

Blood pressure in the veins is low, and veins have relatively thin walls with fewer smooth muscle cells (see Figure 12–12). Because the veins' walls are so thin, obstructions can cause them to balloon, in much the same way that a tree down across a small stream can cause water to pool upstream. Blood pools in the obstructed veins, forming rather unsightly bluish bulges called **varicose veins** (▷ Figure 12–19, page 304).

Some people inherit a tendency to develop varicose veins, but most cases can be attributed to factors that reduce the flow of blood back to the heart. Abdominal tumors, pregnancy, obesity, and even sedentary lifestyles

often result in varicose veins in the lower extremities. The restriction of blood flow may result in muscle cramps and the buildup of fluid, **edema** (swelling), in the ankles and legs.

Varicose veins may also form in the wall of the anal canal. The veins in this region are known as the internal hemorrhoidal veins. A swelling of the internal hemorrhoidal veins results in a condition known as **hemorrhoids.** Because the internal hemorrhoidal veins are supplied by numerous pain fibers, this condition can be quite painful.

How Do the Veins Work? ▷ Figure 12–20 (page 304) shows that blood pressure declines fairly rapidly as blood travels through the arteries and that it continues to decrease, but at a slower rate, as it flows through the capillaries and veins. The lowest blood pressure readings occur in the superior and inferior vena cavae. With such low pressure, how do the veins return blood to the heart?

For blood in veins above the heart, gravity is the chief means of propulsion. But for veins below the heart, which have very little pressure to force the blood along, return flow depends on the movement of body parts, which "squeezes" the blood upward. As you walk to class, for example, the contraction of muscles in your legs pumps the blood in the veins, like the hands of a person milking a cow. This forces the blood upward, slowly and surely causing it to move against the force of gravity. Even the nervous muscle contractions that occur during study help move the blood back to the heart.

HYPERTENSION AND ANEURYSMS: CAUSES AND CURES

The stresses of modern life and the unhealthy diets of many people take their toll on the heart and blood vessels. One of the most common problems of modern times is atherosclerosis, the buildup of cholesterol plaque in the walls of arteries, discussed in previous chapters. Arteries clogged with cholesterol force the heart to work harder, putting strain on this organ. Perhaps the most significant problem arises from blood clots that lodge in narrowed coronary arteries, reducing the flow of blood to the heart muscle, often with devastating consequences.

Hypertension, or high blood pressure, is another altogether-too-common problem of our times. In some individuals, hypertension is hereditary, passed from parent to offspring. Hypertension is also thought to result from stress and diet, especially high salt intake in some individuals. In other instances, it is caused by disorders such as kidney failure or hormonal imbalances. Pregnancy can lead to hypertension, and so can use of oral contraceptives. Low levels of cadmium in food, air, and water may also contribute to this disease, as noted in this chapter.

A few facts about hypertension are clear, though. People who are overweight when they are young are more likely to suffer from hypertension when they reach adulthood. An adult who is hypertensive and overweight, however, can often control the disease by losing weight and by reducing salt intake.

If this disease is untreated, blood pressure rises steadily over the years. As noted in the chapter, hypertension is a nearly symptomless disease for many people. Individuals feel fine for many years, despite their gradually rising blood pressure, and may not exhibit any signs of the disease until it has progressed to the dangerous stage. Thus, it is important for people over the age of 40 to have their blood pressure checked each year.

Hypertension is more common in men than in women and is more common in African Americans than in Caucasians. The disease is dangerous because the increased pressure in the circulatory system forces the heart to work harder. Elevated blood pressure may also damage the lining of arteries, creating a site at which atherosclerotic plaque forms. Atherosclerosis increases the risk of heart attack: a hypertensive person is six times more likely to have a heart attack than an individual with normal blood pressure. Hypertension also increases the chances for an occlusion in the arteries supplying the brain, which can result in strokes.

Arteries can be weakened by various factors—for example, certain infectious diseases (such as syphilis) and atherosclerotic plaque as well as hypertension. A weakening of the wall of an artery may cause it to balloon, a condition known as an **aneurysm** (▷ Figure 1). Like a worn spot on a tire, an aneurysm can give way when pressure builds inside or when the wall becomes too thin.

When an aneurysm breaks, blood pours out of the circulatory system. Because they happen so quickly, most aneurysms lead to death. An estimated 30,000 Americans die each year from ruptured aneurysms in the brain, and nearly 3000 die from ruptured aortic aneurysms.

As in most diseases, the first line of defense against aneurysms is prevention. By reducing or eliminating the two main causes—atherosclerosis and high blood pressure—individuals can greatly lower their risk.

Physicians recommend a number of

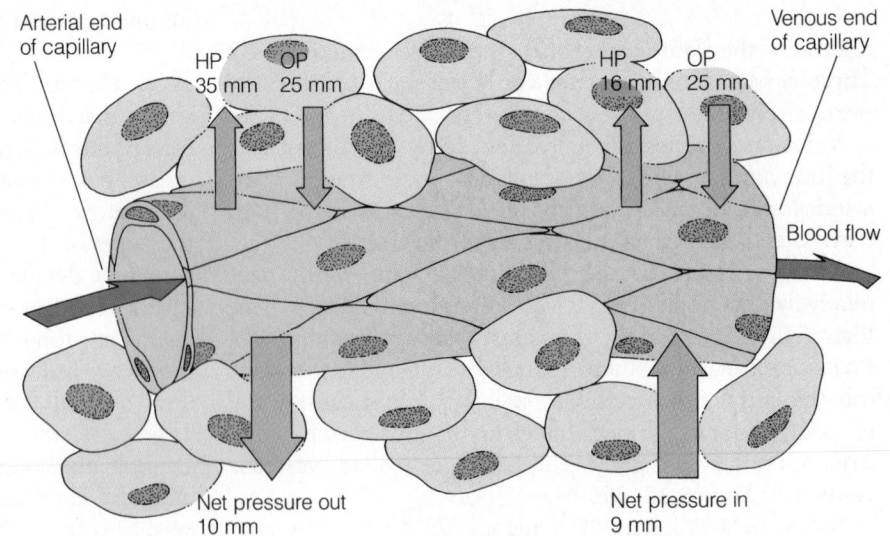

▷ **FIGURE 12–18 Capillary Exchange** Fluid begins to flow out of the capillary as soon as it enters the vessel from the arteriole. A slight hydrostatic pressure (HP), resulting from blood pressure, creates a slight outward force of 35 mm Hg. Osmotic pressure (OP), the tendency for water to flow inward because of a higher internal concentration of dissolved solute, is only 25 mm Hg. As a result, there is a net outward force on the arterial side of the capillary network of 10 mm Hg. The hydrostatic pressure declines as the blood flows through the capillary bed, and therefore the osmotic pressure forcing water inward exceeds the outward pressure. This drives water back into the capillary. Notice that the net movement inward does not equal the net movement out. As a result, excess fluid accumulates in the tissue and must be drained by the lymphatic system.

Arterial end of capillary

HP 35 mm OP 25 mm

HP 16 mm OP 25 mm

Venous end of capillary

Blood flow

Net pressure out 10 mm

Net pressure in 9 mm

steps to reduce your chances of developing atherosclerosis and hypertension. If you smoke, stop. If you are overweight, exercise and lose weight. If you ingest excess salt, gradually cut back on your intake. If you are stressed at work and at home, find ways to reduce stress levels. If you drink alcohol, drink in moderation, or quit. If you eat foods rich in fats and cholesterol, cut down on them and consume more water-soluble fiber like that found in apples, bananas, citrus fruits, carrots, barley, and oats, which reduces cholesterol uptake by the intestine, as explained in Chapter 11.

The second line of defense against cardiovascular disease is early detection and treatment. Hypertension can be detected by regular blood pressure readings. Atherosclerosis can be discovered by blood tests to measure cholesterol. Aneurysms can be detected by X-ray. Pain also alerts patients and physicians that something is wrong. Once any of these diseases is detected, physicians have many options.

In the event of an aneurysm, surgeons can remove the weakened section of the artery and replace it with a section of a vein. In other instances, where venous grafts are more difficult (for example, in the brain), surgeons can clamp or tie off the artery just before the bulge, preventing blood flow through the damaged section. This works only when other arteries provide adequate blood flow to the area served by the damaged artery.

In larger arteries, pieces of Dacron or other synthetic materials can be sewn into the wall of the artery, protecting it from breaking. Researchers are also experimenting with an alloy of nickel and titanium called nitinol. Nitinol is a "metal with a memory." When a fine nitinol wire is heated and wrapped around a cylinder, it forms a tightly coiled spring. When the spring is cooled, it reverts to a straight wire. When reheated, the metal returns to a coil.

In experiments with dogs, scientists create a wire coil that corresponds to the internal diameter of an artery. Next they cool the wire, causing it to revert to the straight form, and then push it through a catheter inserted in the artery. As the wire emerges from the cooled catheter inside the artery, body heat causes it to coil again. In place, the coil adds strength to the wall of the artery, preventing rupture. Experiments with dogs show that the endothelial cells of the tunica intima soon grow over the implant, making it a permanent part of the artery's wall. If successful in humans, this procedure could help save hundreds, perhaps thousands, of lives each year. It is, however, no substitute for a healthy diet and a healthy environment.

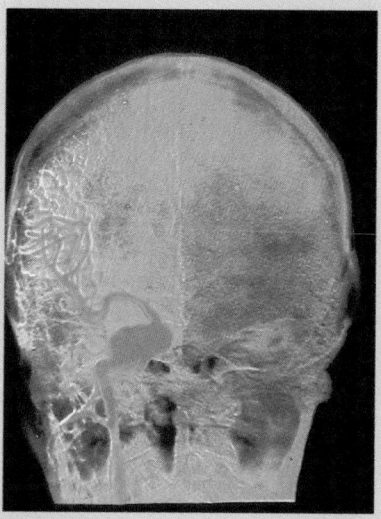

▷ **FIGURE 1 Aneurysm** This X-ray shows a ballooning of one of the arteries in the brain. If untreated, an aneurysm can break, causing a stroke.

Important as muscles are, they are not enough to ensure return flow. Evolution has "provided" an additional factor: valves. **Valves** are flaps of tissue that span the veins and prevent the backflow of blood. The structure of the valves is shown in Figure 12–19a. As illustrated, the semilunar flaps of the veins resemble those found in the heart. Just as in the valves of the heart, blood pressure, however slight, pushes the flaps open (Figure 12–19a). This allows the blood to move forward. As the blood fills the segment of the vein in front of the valve, it pushes back on the valve flaps and forces them shut. You can locate the valves in the superficial veins on your forearm by pressing gently on a vein, then running your finger toward your wrist. You will see that the vein will collapse behind your finger until it crosses a valve.

≈ THE LYMPHATIC SYSTEM

The lymphatic system is functionally related to two systems: the circulatory system and the immune system.

This section examines the role of the **lymphatic system,** an extensive network of vessels and glands, in circulation (▷ Figure 12–21).

You may recall from Chapter 10 that the cells of the body are bathed in a liquid called interstitial fluid. Interstitial fluid is in equilibrium with the blood plasma and provides a medium through which nutrients, gases, and wastes can diffuse between the capillaries and the cells.

Tissue fluid is replenished by water that diffuses out of the capillaries. The flow of water out of the capillaries, however, normally exceeds the return flow by about 3 liters per day. The "excess" water is picked up by small **lymph capillaries** in tissues. Like the capillaries of the circulatory system, these vessels have thin, highly permeable walls through which water and other substances pass with ease. As illustrated in ▷ Figure 12–22, the cells in the walls of the lymph capillaries overlap, creating a number of one-way valves, which resemble swinging doors. The accumulation of interstitial fluid in the tissues forces the "doors" to open, causing the fluid to flow into

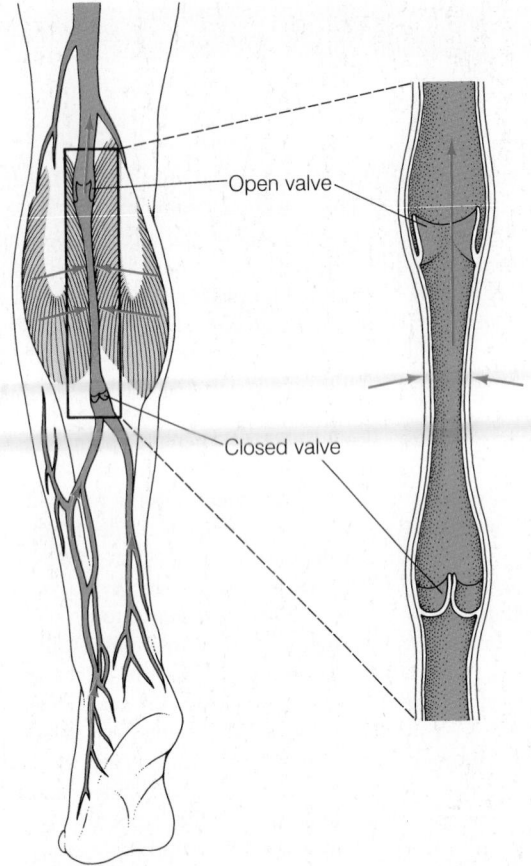

Open valve

Closed valve

(a)

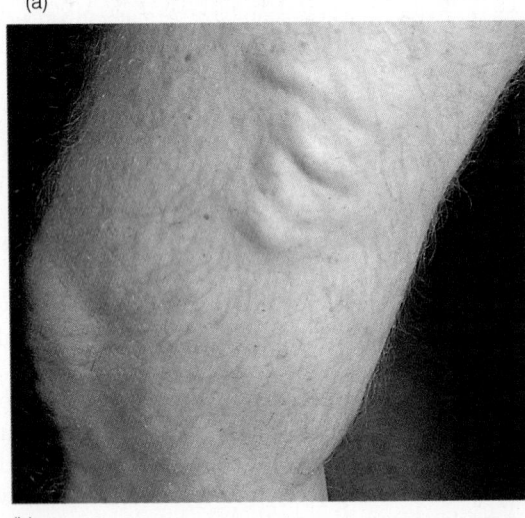

(b)

▷ **FIGURE 12–19 Valves in Veins** (*a*) The slight hydrostatic pressure in the veins and the contraction of skeletal muscles propel the blood along the veins back toward the heart. The one-way valves stop the blood from flowing backward. (*b*) Any restriction of venous blood flow to the heart causes veins to balloon out, creating bulges commonly known as varicose veins.

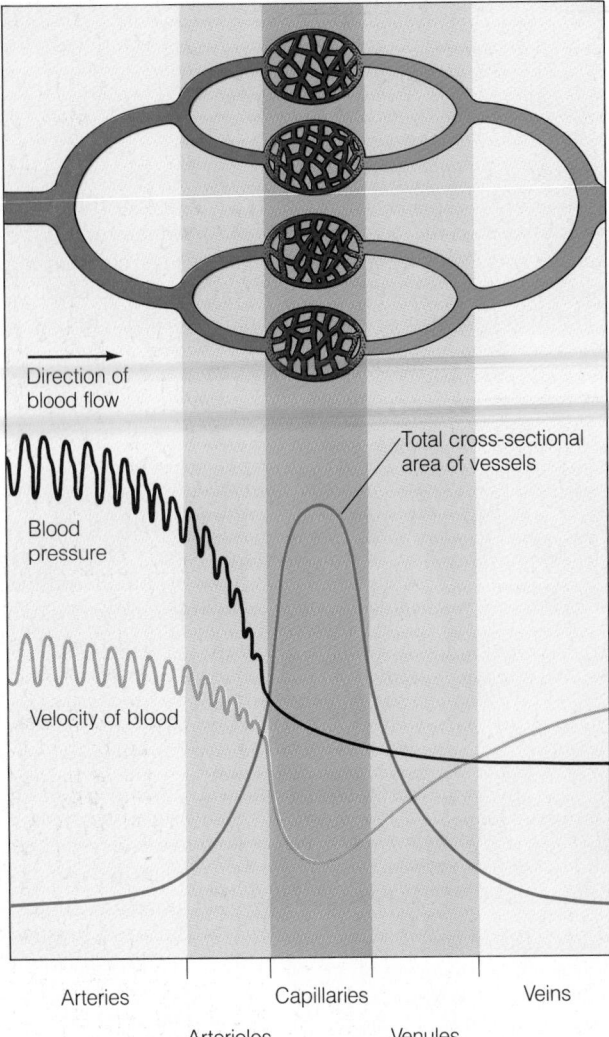

Direction of blood flow

Total cross-sectional area of vessels

Blood pressure

Velocity of blood

Arteries   Capillaries   Veins

Arterioles   Venules

▷ **FIGURE 12–20 Blood Pressure in the Circulatory System** Blood pressure declines in the circulatory system as the vessels branch. Arterial pressure pulses because of the heartbeat, but pulsation is lost by the time the blood reaches the capillary networks, creating an even flow through body tissues. Blood pressure continues to decline in the venous side of the circulatory system.

the lymphatic capillaries. Once inside the fluid is called **lymph.**

Lymph drains from the capillaries into larger ducts, the collecting vessels. These vessels, in turn, merge with others, creating larger and larger ducts, eventually forming the **thoracic duct** and the **right lymphatic duct,** which empty into the large veins at the base of the neck (Figure 12–22).

Lymph moves through the vessels of the lymphatic system in much the same way that blood is transported in veins. In the upper parts of the body, lymph flows by gravity. In regions below the heart, lymph is propelled largely by muscle contraction. Breathing and walking, for example, pump the lymph out of the extremities. Lymphatic flow is also assisted by valves akin to those found in the veins.

The lymphatic system also consists of several lymphatic organs: the lymph nodes, the spleen, the thymus, and the tonsils. The lymphatic organs function primarily in immune protection and are discussed in more detail in Chapter 15.

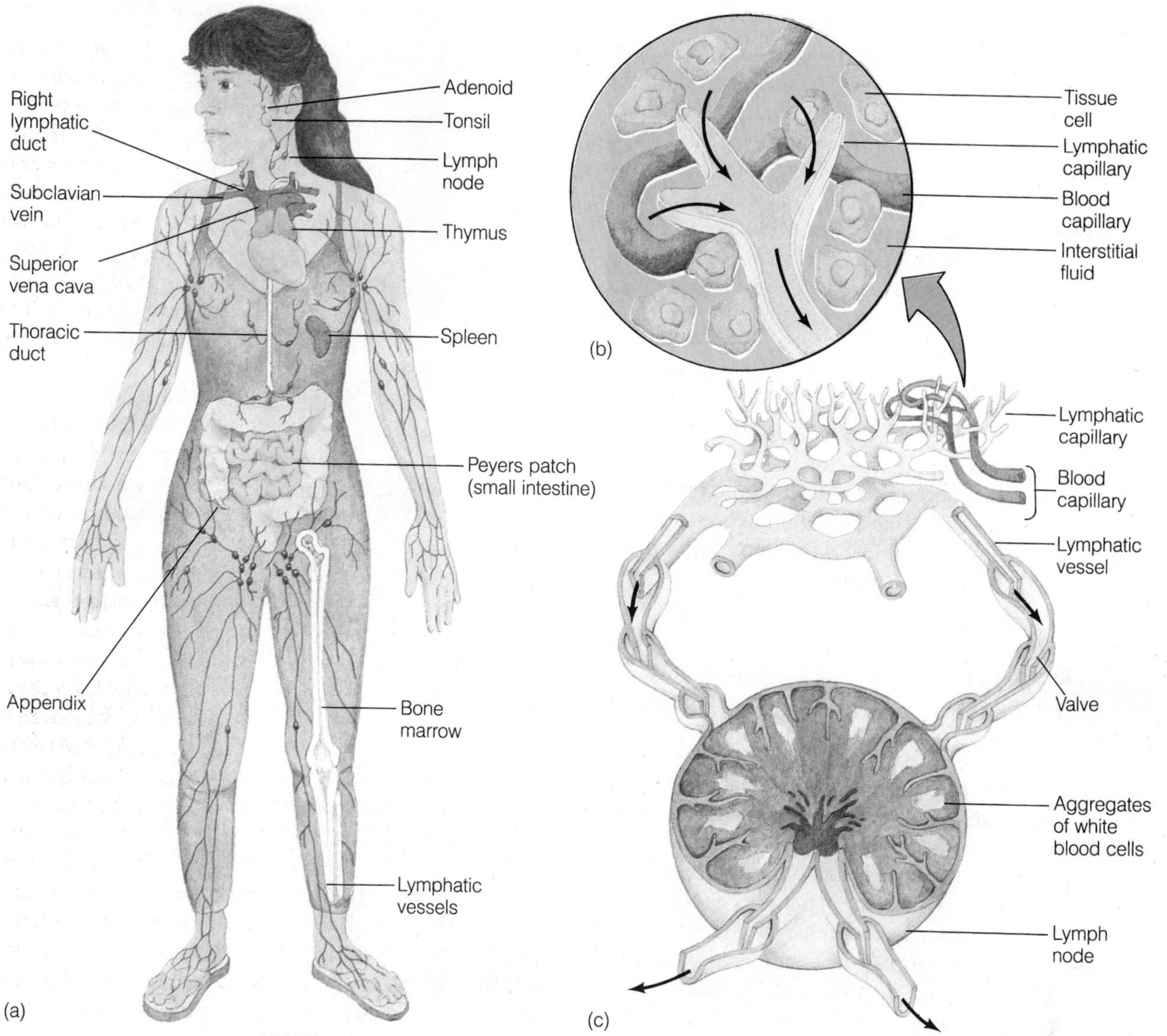

Figure 12-21 (a)

Right lymphatic duct

Adenoid

Tonsil

Lymph node

Subclavian vein

Superior vena cava

Thymus

Thoracic duct

Spleen

Peyers patch (small intestine)

Appendix

Bone marrow

Lymphatic vessels

(a)

Figure 12-21 (b)

Tissue cell

Lymphatic capillary

Blood capillary

Interstitial fluid

(b)

Figure 12-21 (c)

Lymphatic capillary

Blood capillary

Lymphatic vessel

Valve

Aggregates of white blood cells

Lymph node

(c)

▷ **FIGURE 12–21 The Lymphatic System** (*a*) The lymphatic system consists of vessels that transport lymph, excess tissue fluid, back to the circulatory system. (*b*) Like the veins, the lym phatic vessels contain valves that prohibit backflow. Lymph nodes are interspersed along the vessels and serve to filter the lymph.

Lymph nodes, however, deserve attention here (see Figure 12–21a). Varying in size and shape, the lymph nodes are found in association with lymphatic vessels in small clusters in the armpits, groin, neck, and numerous other locations. A lymph node consists of a network of fibers and irregular channels that slow down the flow of lymph and filter out bacteria, viruses, cellular debris, and other particulate matter transported in the lymph (Figure 12–21b). Lining the channels are numerous **macrophages,** which phagocytize microorganisms and other materials in the lymph.

▷ **FIGURE 12–22 Lymphatic Capillaries** These thin-walled vessels absorb excess tissue fluid. The cells of the wall push inward to allow the fluid to enter the vessel.

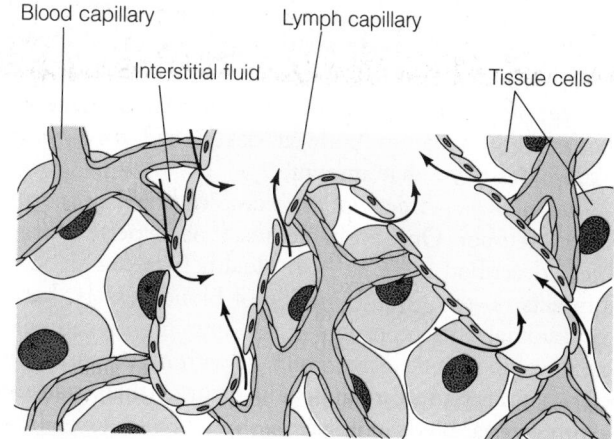

Blood capillary

Interstitial fluid

Lymph capillary

Tissue cells

Under normal circumstances, lymph is removed from tissues at a rate equal to its production. In some instances, however, lymph production exceeds the capacity of the system. A burn, for example, may cause extensive damage to blood capillaries, increasing their leakiness and overwhelming the lymphatic capillaries. This "flood" results in a buildup of fluid in tissues, referred to as edema.

Lymphatic vessels may also become blocked. One of the most common causes of blockage is an infection by tiny parasitic worms that are transmitted by mosquitoes in the tropics. The worm larvae (an immature form) enter lymph vessels and take up residence in the lymph nodes. An inflammatory reaction causes the buildup of scar tissue in the nodes, which blocks the flow of lymph. After several years, the lymphatic drainage of certain parts of the body may become almost completely obstructed. A leg may swell so much, in fact, that it weighs as much as the rest of the body (▷ Figure 12–23). This condition, known as **elephantiasis,** (el-uh-fun-tie-a-sis) often affects the scrotum, the sac of skin that holds the testes.

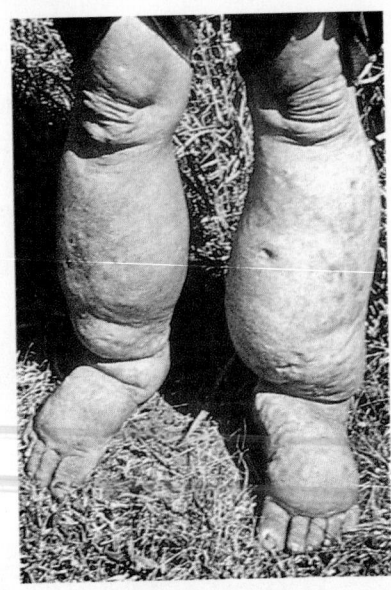

▷ **FIGURE 12–23 Elephantiasis** A parasitic worm that invades the body blocks the flow of lymph through the lymph nodes, causing tissue fluid to build up. This condition is known as elephantiasis.

ENVIRONMENT AND HEALTH: CADMIUM AND HYPERTENSION

Cadmium is a heavy metal once used to treat syphilis and malaria, a remedy abandoned when physicians learned of its toxic effects. Today, cadmium is used for many commercial purposes. One of the most common uses is as a chemical stabilizer added to polyvinyl chloride, a compound used to make vinyl (found in children's toys such as beach balls). Cadmium is also combined with gold and other metals to make jewelry. It is even used to make camera batteries, engine parts, and radio and television sets.

Cadmium is released into the air and water during the manufacture of many products. Incinerators that burn municipal garbage also release small amounts of cadmium found in rubber tires, plastic bottles, furniture, and other items (▷ Figure 12–24). Recycling facilities that melt down radiators and scrap steel also emit cadmium. Together, these facilities emit over 4 million pounds of cadmium into the air each year in the United States alone.

Cadmium, which ends up in food, air, and water, is toxic to practically all the body systems of humans and other animals. It is absorbed into the body, and its levels increase as we get older. The average American has 30 milligrams of cadmium in his or her body at death, and the average Japanese has twice that amount.

What are the effects of low-level cadmium exposure over a lifetime? One probable result is hypertension, a subject described more fully in Health Note 12–1. Cadmium acts on the smooth muscle of blood vessels. Laboratory animals that consume small amounts of cadmium soon develop hypertension. Humans with hypertension secrete 40 times more cadmium in their urine than individuals with normal blood pressure.

No one knows how many people are suffering from hypertension caused by cadmium, but the number is probably fairly large. Given the rise in cadmium use and the increases in recycling and incineration of trash, exposure can only get worse unless concerted efforts are made to tighten controls on facilities that release this toxic element and, perhaps more importantly, to reduce the use of cadmium in the first place.

▷ **FIGURE 12–24 Municipal Waste Incinerator** To deal with growing mountains of garbage, many cities and towns are installing incinerators to burn their garbage. But these incinerators release toxic chemicals, including cadmium, in small amounts, which add to emissions from other sources.

CIRCULATORY SYSTEMS: AN EVOLUTIONARY IMPERATIVE

1. Primitive circulatory systems appear in the earliest animals—the sponges, jellyfish, hydras, and anemones.
2. More complex systems are found in higher animals. These systems fall into two major categories: open and closed.
3. In open systems, hearts propel blood through vessels which empty their contents into body cavities to bathe organs.
4. In closed systems, blood remains in the blood vessels. The exchange of nutrients and wastes occurs through thin-walled vessels—capillaries.

THE CIRCULATORY SYSTEM'S FUNCTION IN HOMEOSTASIS

5. The circulatory system is one of the body's chief homeostatic systems. It helps maintain constant levels of nutrients and wastes, helps regulate body temperature, distributes body heat, protects against microorganisms, and, through clotting, protects against blood loss.

THE HEART

6. The vertebrate circulatory system consists of a pump, the heart, and two circuits, the pulmonary circuit, which transports blood to and from the lungs, and the systemic circuit, which delivers blood to the body and returns it to the heart.
7. The human heart consists of four chambers: two atria and two ventricles. The right atrium receives blood from the superior and inferior vena cavae. This blood, returning from body tissues, is low in oxygen and rich in carbon dioxide.
8. From the right atrium, blood is pumped into the right ventricle, then to the lungs, where it is resupplied with oxygen and stripped of most of its carbon dioxide. Blood returns to the heart via the pulmonary veins, which empty into the left atrium.
9. Blood is pumped from the left atrium to the left ventricle, then out the aorta to the body, where it supplies cells of tissues and organs with oxygen and picks up cellular wastes.
10. Heart valves help control the direction of blood flow. The atrioventricular valves permit blood to flow from the atria to the ventricles and then prevent it from flowing in the reverse direction when the ventricles contract. The semilunar valves in the base of the aorta and pulmonary artery prevent blood from flowing back into the ventricles.
11. The closing of the valves produces distinct heart sounds, which can be detected through the chest using a stethoscope. Irregularities in heart sounds indicate the presence of diseased or damaged valves.
12. Cardiac muscle cells contract rhythmically and independently; contraction is coordinated by the heart's pacemaker, the sinoatrial node located in the upper wall of the right atrium.
13. The cells of the SA node discharge periodically, sending impulses to all atrial muscle cells, which causes them to contract in unison.
14. The impulse next travels to the ventricles, but its passage is delayed, providing time for the ventricles to fill before contracting.
15. The impulse travels from the atrioventricular node down the atrioventricular bundle into the myocardium of the ventricles.
16. Contraction begins at the tip of the heart and proceeds upward, squeezing the blood out of the ventricles.
17. Left on its own, the SA node would produce 100 contractions per minute. Nerve impulses, however, reduce the heart rate to about 70 beats per minute when a person is inactive. During exercise or stress, the heart rate increases to meet body demands.
18. Depolarization of the heart muscle produces weak electrical currents that can be detected by surface electrodes. The change in electrical activity is detected by a voltage meter.
19. The tracing produced on the voltage meter is called an electrocardiogram (ECG) and has three distinct waves. Diseases of the heart may disrupt one or more of the waves, making the ECG a valuable diagnostic tool for many heart diseases.
20. The amount of blood pumped by the ventricles each minute, or cardiac output, is about 5 liters (about 5 quarts) per minute in a person at rest. Cardiac output is a function of stroke volume (the amount of blood pumped with each contraction of the left ventricle) and heart rate.

HEART ATTACKS: CAUSES, CURES AND TREATMENT

21. The most common type of heart attack is a myocardial infarction, caused by a blood clot that either forms in an already narrowed coronary artery or arises elsewhere and breaks loose, only to become lodged in a coronary artery. The obstruction decreases the flow of blood and oxygen to heart muscle, sometimes killing the cells.
22. Heart attacks can occur quite suddenly without warning, or they may be preceded by several weeks of angina, pain felt when blood flow to heart muscle is reduced. Angina appears when an individual is active, then disappears when he or she rests.
23. The risk of heart attack can be reduced by proper diet, exercise, and an aspirin taken daily. Numerous treatments are available to treat patients who have had a heart attack.

THE BLOOD VESSELS

24. The heart pumps blood into the arteries, which distribute the blood to the body tissues, where nutrient and waste exchange occurs in capillary beds. Blood is returned to the heart in the veins.
25. The largest of all arteries is the aorta. It carries oxygenated blood from the left ventricle of the heart. The aorta descends through the chest and abdomen, giving off large branches that carry blood to the head, arms, legs, and major organs. The first branches are the coronary arteries to the heart.
26. The aorta and many of its chief branches are elastic arteries. As blood pulses out of the heart, the elastic arteries expand to accommodate the blood, then contract, helping to pump the blood and ensuring a steady flow through the capillaries.
27. As they course through the body, the elastic arteries branch to form the muscular arteries, which also expand and contract with the flow of blood.
28. The smooth muscle in the walls of muscular arteries responds to a variety of stimuli. These stimuli cause the blood vessels to open or close to varying degrees, controlling the flow of blood through body tissues.

29. Blood pressure and flow rate are highest in the aorta and drop considerably as the arteries branch. By the time the blood reaches the capillaries, its flow and pressure are greatly reduced. The declines enhance the rate of exchange between the blood and the tissues.

30. Capillaries are thin-walled vessels that form branching networks, or capillary beds, among the cells of body tissues. Cellular wastes and nutrients are exchanged between the cells of body tissues and the blood in capillary networks.

31. Blood flow through a capillary network is regulated by constriction and relaxation of the metarterioles, circulatory shortcuts that connect the arterioles on one side with the venules on the other. Blood flow is also controlled by the precapillary sphincters, rings of smooth muscle that surround the capillaries that branch from the metarterioles.

32. Blood draining from capillary beds enters venules, which join to form veins. Veins return blood to the heart and generally run alongside the arteries.

33. Because the walls of veins have very little smooth muscle, they are easily affected by obstructions, which cause the walls to balloon, forming bluish bulges called varicose veins.

34. In the veins above the heart, blood drains by gravity. Veins below the heart, however, rely on the movement of body parts to squeeze the blood upward and on valves, flaps of tissue that span the veins and prevent the backflow of blood.

THE LYMPHATIC SYSTEM

35. The lymphatic system is a network of vessels that drains interstitial fluid from body tissues and transports it to the blood.

36. The lymphatic system also consists of several lymphatic organs, such as the lymph nodes, the spleen, the thymus, and the tonsils, which function primarily in immune protection.

37. Along the system of lymphatic vessels are small nodular organs called lymph nodes, which filter the lymph.

38. Under normal circumstances, lymph is removed from tissues at a rate equal to its production, keeping tissues from swelling. In some cases, however, lymph production exceeds the capacity of the lymphatic capillaries, and swelling results.

ENVIRONMENT AND HEALTH: CADMIUM AND HYPERTENSION

39. Cadmium is a toxic heavy metal used for many commercial purposes, including the manufacturing of vinyl, batteries, engine parts, and televisions.

40. Cadmium is released into the air and water during the manufacture of many products, the incineration of municipal garbage, and the recycling of steel and other metals.

41. Cadmium ends up in food, air, and water and increases in humans and other organisms over time. Cadmium acts on the smooth muscle of blood vessels, causing hypertension.

EXERCISING YOUR CRITICAL THINKING SKILLS

In June 1988 an EPA scientist, Joel Schwartz, published data showing a strong relationship between levels of lead in the blood of men and hypertension. In March 1989 Schwartz published another study showing that even relatively low lead levels in men's blood increased their risk of developing heart disease. That same month, a professor at the University of Rochester, in New York, published the results of studies suggesting that elderly men have a high and previously unrecognized sensitivity to lead.

Schwartz and two other scientists noted that bone loss (osteoporosis) in women who have passed menopause releases large quantities of lead stored in bone. In the blood, this lead may harm sensitive organs such as the brain, liver, and kidneys. Lead is also a neurotoxin, which, even in low doses, has been shown to affect mental development and coordination.

In 1989, in response to these and other studies, the EPA announced new guidelines for lead levels in drinking water. The rules allow lead levels no higher than 5 parts per million, compared with an old standard of 50 parts per million. Some newspapers praised the action as a major accomplishment.

Would it affect your view of this announcement if you knew that the measurements for the old standard were taken at people's faucets and that the measurements for the new standard are to be taken at water treatment plants, before the water is released into the pipes that distribute it to customers? What do you need to know before you can answer that question?

Would it affect your view of the announcement if you knew that most of the lead in a water system came from lead pipes in old homes and lead solder used in copper pipes? What questions need to be answered before you can determine the importance of the EPA's announcement? What does this exercise suggest about analyzing news reports?

1. The vessels that supply blood to the lungs are part of the _____ circuit. The vessels that supply blood to the rest of the body are part of the _____ circuit.

2. The superior and inferior vena cavae empty into the _____ _____ of the heart.

3. The large artery carrying blood from the left ventricle of the heart to the body is called the _____ . It is an _____ artery.

4. The _____ valves are found in the larger arteries leading from the heart.

5. The _____ valves are anchored to the inside wall of the heart via the chordae tendineae.

6. The heart's internal pacemaker is called the _____ node and is located in the wall of the _____ _____ .

7. The impulse generated by the pacemaker travels down the septum between the ventricles along the _____ _____ .

8. Arteries and veins contain three layers; starting from the inside, they are the _____ _____ , _____ _____ , and _____ _____ .

9. _____ _____ give the large arteries the ability to recoil when blood is pumped into them.

10. The pressure created by the contraction of the ventricles of the heart is called _____ pressure.

11. The exchange of nutrients and wastes occurs in the _____ , thin-walled vessels that form extensive networks in body tissues. These vessels empty into _____ , the smallest veins.

12. _____ _____ result when the wall of a vein balloons because blood flow is impaired.

13. A swelling resulting from the buildup of tissue fluid is called _____ .

14. The backward flow of blood in veins is prevented by _____ .

15. The _____ _____ pick up excess _____ fluid in body tissues, preventing tissues from swelling.

16. _____ _____ filter the lymph and remove bacteria, viruses, and other particulate matter.

Answers to the Test of Terms are found in Appendix B.

1. List and describe several ways in which the circulatory system functions in homeostasis.

2. Describe how the pacemaker, the sinoatrial node, coordinates muscular contractions of the heart.

3. Define the pulmonary and systemic circuits.

4. Based on what you know about the heart, what would happen if the septum separating the two atria failed to form completely during embryonic development?

5. Describe how the valves help control the direction of blood flow in the heart. What are valvular incompetence and valvular stenosis? What causes these conditions, and how do they affect the heart?

6. How does atherosclerosis of the coronary arteries affect the heart?

7. Describe the general structure of the arteries and veins. How are they different? How are they similar?

8. How do the elastic fibers in the major arteries help ensure a continuous blood flow through the capillaries?

9. Capillaries illustrate the remarkable correlation between structure and function. Do you agree with this statement? Why or why not?

10. Describe how the body controls the movement of blood through capillary beds. Why is this homeostatic mechanism so important? Give some examples.

11. Explain how the blood is returned via the veins.

12. Explain the role of the lymphatic system in circulation.

CHAPTER

13

The Blood

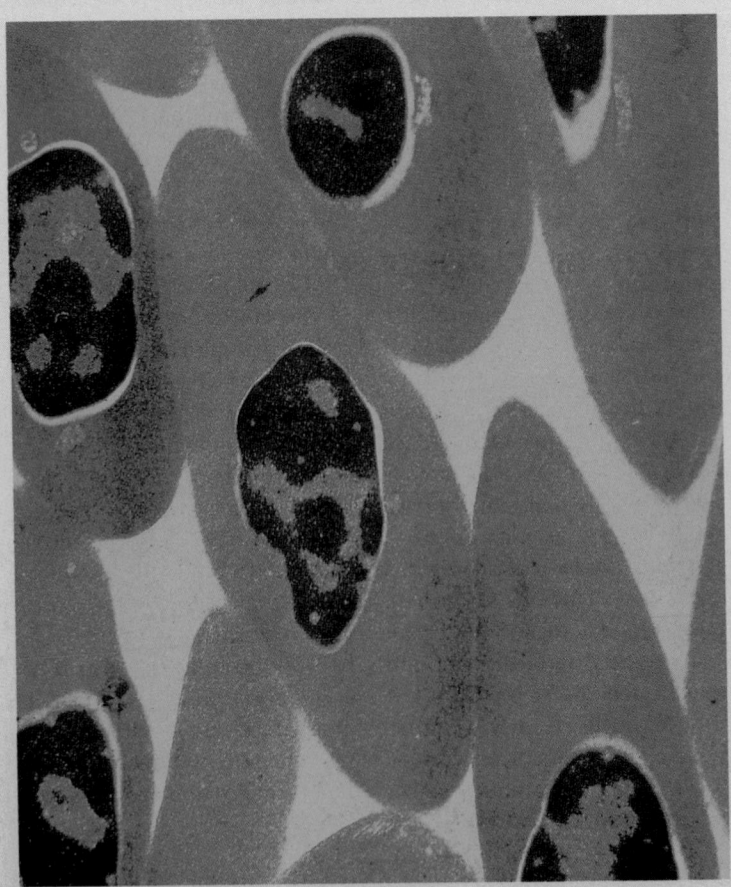

Nucleated red blood cells from a sparrow.

In 1966, Dr. Leland C. Clark of the University of Cincinnati's College of Medicine immersed a live laboratory mouse in a clear fluorocarbon solution saturated with oxygen. To the amazement of the audience, the mouse did just fine, "breathing in" the oxygen-rich liquid as if it were air. Some time later, Clark extracted the mouse from the solution, and after a moment, the rodent began to move about, apparently unharmed by the ordeal.

The point of this demonstration was not to show that an animal could "breathe" this fluid into its lungs and survive but, rather, to show that the solution held enormous amounts of oxygen, so much so that it could possibly be used as a substitute for blood. Clark and his colleagues hoped that their "artificial blood" would someday be a boon to medical science, helping paramedics keep accident victims who have lost substantial amounts of blood alive while they were being transported to the hospital.[1] In rural America, where a trip to the hospital takes considerable time, artificial blood could save thousands of lives a year.

≈ BLOOD: ITS COMPOSITION AND FUNCTIONS

Animal blood is a far cry from Clark's artificial substitute. The **blood** in the circulatory system of vertebrates, in-

---

[1] "Artificial blood," of course, is a misnomer. The fluorocarbon solution takes over the role of the plasma and the red blood cells, which transport oxygen and carbon dioxide, but not the role of the white blood cells, which play a role in immune protection.

---

cluding humans, for example, is a water-based fluid, not a fluorocarbon, and consists of two basic components: plasma and formed elements. **Plasma** is the liquid portion of the blood and is about 90% water. It contains many dissolved substances. Three types of formed elements are suspended in the plasma: (1) white blood cells (or leukocytes), (2) red blood cells (or erythrocytes), and (3) platelets (or thrombocytes).

In a man weighing 70 kilograms (150 pounds), the cardiovascular system contains 5 to 6 liters of blood, or 1.3 to 1.5 gallons. On average, women have about a liter less. Blood accounts for about 8% of our total body weight.

▷ Figure 13–1 shows that the blood plasma makes up about 55% of the blood volume and formed elements about 45%. Technically speaking, the volume of the blood occupied by blood cells is referred to as the **hematocrit**. To determine the hematocrit, blood is withdrawn from a patient and placed in a test tube, then inserted into a centrifuge, a device that spins the samples at high speeds (Figure 13–1). Centrifugation causes the formed elements to separate from the plasma and "settle" to the bottom of the test tube. The white blood cells and platelets settle out on top of the RBCs. They make up slightly less than 1% of the total blood volume. Therefore, the hematocrit is largely determined by the RBCs.

The hematocrit varies in individuals inhabiting different altitudes. In people living in the thin air of the Mile High City of Denver, for example, hematocrits are typically about 5% higher than in people living at sea level. The slight increase in the hematocrit (RBC concentration) in Denver residents compensates for the slightly lower level of

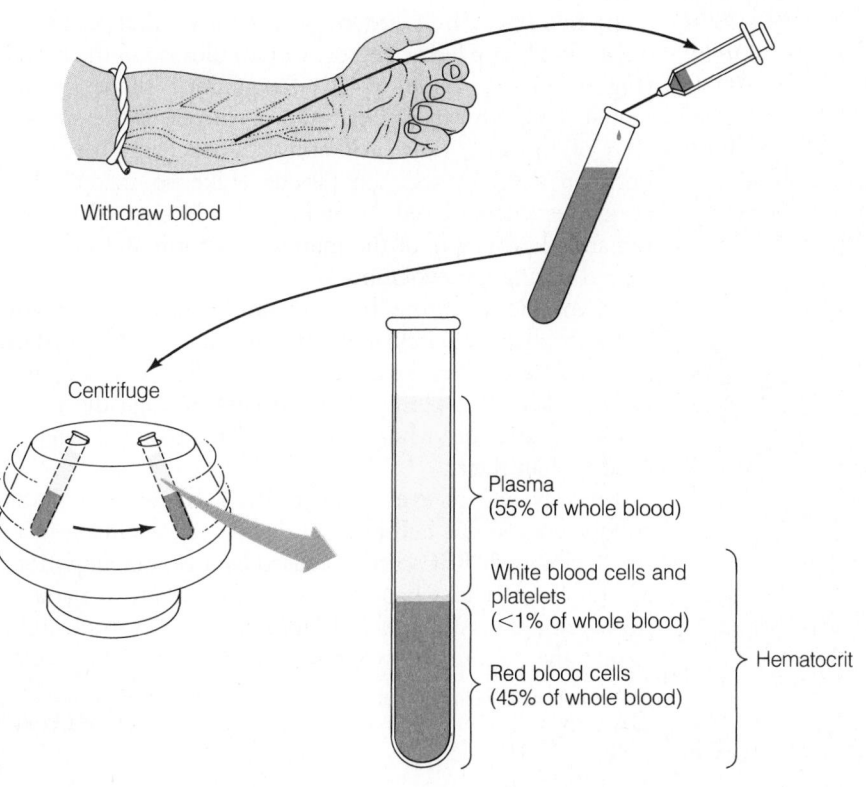

Withdraw blood

Centrifuge

Plasma
(55% of whole blood)

White blood cells and platelets
(<1% of whole blood)

Red blood cells
(45% of whole blood)

} Hematocrit

▷ **FIGURE 13–1 Blood Composition** Blood removed from a person can be centrifuged to separate plasma from the cellular component. Red blood cells constitute about 45% of the blood volume, except at higher altitudes where they make up about 50% of the volume to compensate for the lower oxygen levels.

oxygen in the atmosphere at the higher altitude and is the result of a homeostatic mechanism that helps maintain normal oxygen delivery to body cells. The effect of elevation on hematocrit is one reason why the U.S. summer Olympic team has its headquarters in Colorado Springs. Officials think that the higher hematocrit may help many athletes outperform their lowland competitors.

The Blood Plasma Is a Watery Transport Medium

The plasma is a light yellow (straw-colored) fluid. Dissolved in the plasma are: (1) gases, such as nitrogen, carbon dioxide, and oxygen; (2) ions, such as sodium, chloride, and calcium; (3) nutrients, such as glucose and amino acids; (4) hormones; (5) proteins; and (6) various wastes. Lipid molecules are also found in the plasma, but they are either suspended in tiny globules or bound to certain plasma proteins, which transport them through the bloodstream.

Plasma proteins are the most abundant of all dissolved substances in the blood plasma. Three major types of protein are found in the blood plasma: (1) albumins, (2) globulins, and (3) fibrinogen (Table 13–1).

All of these proteins contribute to osmotic pressure (described briefly in Chapter 3 in the section on membrane transport), which is essential to capillary exchange and, therefore, the proper distribution of wastes and nutrients, a function vital to homeostasis.[2] Blood proteins also augment the effects of bicarbonate, a buffer described in Chapter 2, and therefore help regulate the pH of the blood. They do it by binding to hydrogen ions, preventing the hydrogen ion concentration in the blood from rising.

In addition to these common functions, many plasma proteins serve very specific functions. Albumins and two types of globulins, for example, bind to hormones, ions, and fatty acids, helping transport these molecules through the bloodstream. These carrier proteins are large, water-soluble molecules. Their binding to much smaller lipid molecules renders the latter water-soluble, and facilitates

[2] Osmotic pressure is the force responsible for the movement of water across a selectively permeable membrane and is created by the difference in solute concentrations on either side of the membrane. The greater the concentration difference, the greater the osmotic pressure.

their transport through the largely aqueous bloodstream. Carrier proteins also protect smaller molecules from destruction by the liver.

Another group of globulins, known as the gamma globulins, are **antibodies,** proteins that "neutralize" viruses and bacteria or target them for destruction by phagocytic cells in body tissues known as macrophages, a topic discussed in more detail in Chapter 15.

Still another important blood protein is **fibrinogen.** This unique protein is converted into **fibrin,** which forms blood clots when blood vessels are injured. Blood clots prevent blood loss and thus also help maintain homeostasis. (More on blood clotting later in the chapter.)

Red Blood Cells Are Flexible, Highly Specialized Cells that Transport Oxygen and Small Amounts of Carbon Dioxide in the Blood

The **red blood cell (RBC),** or **erythrocyte,** is the most abundant cell in the blood of vertebrates. In fact, in humans, 20 drops of blood equal 1 milliliter and contain approximately 5 billion RBCs.[3] If RBCs were people, a single milliliter of blood would contain the entire world population.

RBCs are highly specialized "cells." In humans and other mammals, in fact, RBCs lose their nuclei and organelles during cellular differentiation (▷ Figure 13–2a). RBCs found in all other vertebrates, however, retain their nuclei (Figure 13–2b).

In terrestrial (land-dwelling) vertebrates, the bone marrow continuously produces new RBCs to replace the billions that die each day. In fish, RBCs are made in the kidney.

In humans, RBCs are biconcave disks that transport oxygen and, to a lesser degree, carbon dioxide in the blood (Figure 13–2c; Table 13–2). The unique shape of the human RBC, shown in Figure 13–2, increases the surface area of the cell, thus facilitating the exchange of gases between the cell and the plasma. Like so many other structures encountered in biology, the human RBC is a remarkable example of the marriage of form and function forged during the evolution of life on Earth.

Swept along in the bloodstream, RBCs travel many times through the circulatory system each day. When they reach the capillaries, the highly flexible cells often bend and twist in ways that permit them to pass through the many miles of channels whose internal diameters are slightly smaller than theirs.

Roughly one of every 500 to 1000 African Americans suffers from **sickle-cell anemia.** Sickle-cell anemia affects the flexibility of RBCs and is caused by a genetic mutation, as you learned in Chapter 8. In places where malaria is prevalent, like Africa, this mutation provides protection against the disease. However, protection exacts a high cost.

[3] Many texts note the concentration per cubic millimeter, which is about 5 million RBCs. There are, of course, 1000 cubic millimeters in 1 cubic centimeter, or 1 milliliter.

≋	TABLE 13–1 Summary of Plasma Proteins	≋

PROTEIN	FUNCTION
Albumins	Maintain osmotic pressure and transport smaller molecules, such as hormones and ions
Globulins	Alpha and beta globulins transport hormones and fat-soluble vitamins; gamma globulins (antibodies) bind to foreign substances.
Fibrinogen	Converted into fibrin network that help form blood clots

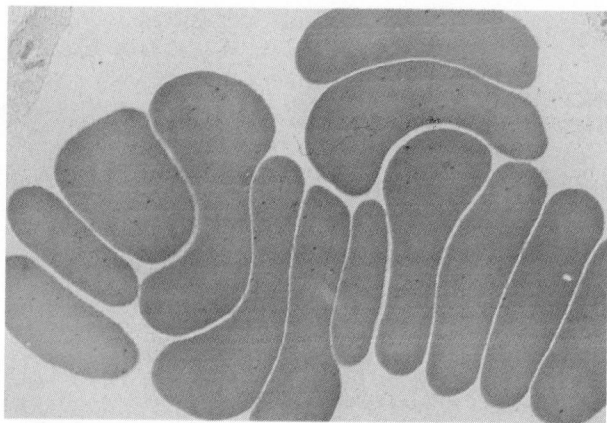

(a) Transmission electron micrograph of human RBC

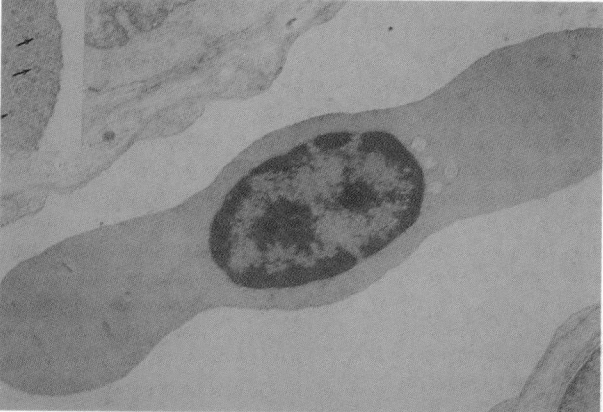

(b) Nucleated RBC from a bird

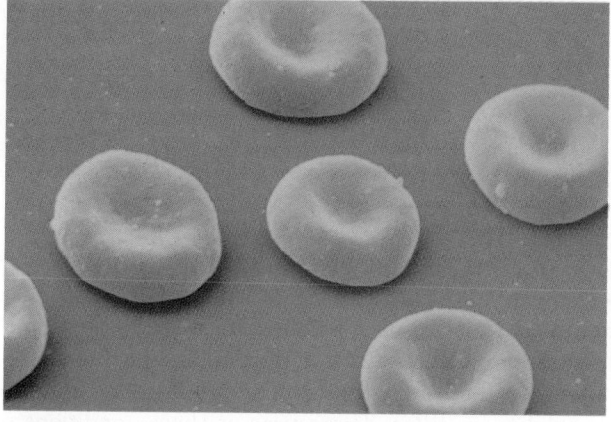

(c) Scanning electron micrograph of human RBC

▷ **FIGURE 13–2 Red Blood Cells**

Why? The RBC hemoglobin of those who carry the gene for sickle-cell anemia contains one incorrect amino acid. This defect alters the tertiary structure of the hemoglobin molecule, causing RBCs to transform from biconcave disks into sickle-shaped cells when they encounter low levels of oxygen in capillaries (▷ Figure 13–3). The sickle-shaped cells are considerably less flexible and unable to bend and twist. As a result, the RBCs collect at branching points in capillary beds much like logs in a logjam. Here, they block capillary blood flow, disrupting the influx of nutrients and oxygen to tissues. This reduces oxygen levels in body tis-

sues and results in a condition known as **anoxia.** Anoxia causes considerable pain and can kill body cells. Blockages in the lungs, heart, and brain can even be life-threatening and often lead to heart attacks and brain damage. Many people who have sickle-cell anemia die in their late 20s and 30s; some die even earlier.

On average, RBCs live about 120 days before the liver and spleen remove them from circulation. The iron contained in the hemoglobin, however, is recycled by these organs—as many nutrients are—and is used to produce new RBCs in the red bone marrow. The recycling of iron is not 100% efficient, however, so small amounts of iron must be ingested each day in the diet. Loss of blood from an injury or, in women, during menstruation increases the body's demand for dietary iron. Without adequate intake, oxygen transport may become impaired.

In the red bone marrow, RBCs are produced by **stem cells,** undifferentiated cells that trace back to embryonic development. These cells give rise to 2 million RBCs per second! (For a discussion of some exciting efforts to manipulate stem cells to fight cancer and other diseases, see Health Note 13–1.)

In infants and children, almost all of the bone marrow is dedicated to the production of RBCs. As growth slows, though, the red marrow of many bones becomes inactive and gradually fills with fat cells, becoming **yellow marrow,** a fat storage depot. By the time an individual reaches adulthood, only a few bones, such as the hip bones, sternum (breastbone), ribs, and vertebrae are engaged in RBC production. In severe, prolonged anemia, however, yellow marrow can be converted back into red marrow to produce RBCs.

The number of RBCs in the blood remains more or less constant over long periods. Maintaining a constant concentration of RBCs is essential to homeostasis and is controlled by the hormone erythropoietin. The kidney secretes **erythropoietin** when blood oxygen levels decline—for example, when a person moves to high altitudes or loses a significant amount of blood. Erythropoietin is transported throughout the bloodstream. In the red bone marrow, it stimulates the stem cells to multiply, increasing RBC production. As the RBC concentration increases, oxygen supplies increase. When oxygen levels return to normal, erythropoietin levels fall, reducing the rate of RBC formation in a classical negative feedback mechanism so common among homeostatic mechanisms.

Hemoglobin Is an Oxygen-Transporting Protein Found in RBCs. Hemoglobin is a large protein molecule composed of four protein subunits. Found exclusively in the RBCs, it accounts for about a third of the RBC's weight. As shown in ▷ Figure 13–4, each hemoglobin subunit contains a **heme group,** which consists of a large, organic ring structure, called a **porphyrin ring.** In the center of the ring is an iron ion.[4]

[4]The iron is Fe^{++}, the ferrous ion. To be effective, iron supplements should contain ferrous iron.

TABLE 13-2 Summary of Blood Cells

NAME	LIGHT MICROGRAPH	DESCRIPTION	CONCENTRATION (NUMBER CELLS/mm³)	LIFE SPAN	FUNCTION
Red blood cells (RBCs)		Biconcave disk; no nucleus	4 to 6 million	120 days	Transports oxygen and carbon dioxide
White blood cells					
Neutrophil		Approximately twice the size of RBCs; multi-lobed nucleus; clear-staining cytoplasm	3000 to 7000	6 hours to a few days	Phagocytizes bacteria
Eosinophil		Approximately same size as neutrophil; large pink-staining granules; bilobed nucleus	100 to 400	8 to 12 days	Phagocytizes antigen-antibody complex; attacks parasites
Basophil		Slightly smaller than neutrophil; contains large, purple cytoplasmic granules; bi-lobed nucleus	20 to 50	Few hours to a few days	Releases histamine during inflammation
Monocyte		Larger than neutrophil; cytoplasm grayish-blue; no cytoplasmic granules; U- or kidney-shaped nucleus	100 to 700	Lasts many months	Phagocytizes bacteria, dead cells, and cellular debris
Lymphocyte		Slightly smaller than neutrophil; large, relatively round nucleus that fills the cell	1500 to 3000	Can persist many years	Involved in immune protection, either attacking cells directly or producing antibodies
Platelets		Fragments of mega-karyocytes; appear as small darkstaining granules	250,000	5 to 10 days	Play several key roles in blood clotting

When blood from the pulmonary arteries flows through the capillary beds of the lungs, oxygen diffuses into the blood, then into the RBCs. Inside the RBC, oxygen binds to the iron in hemoglobin molecules for transport through the circulatory system. The importance of this process is underscored by the fact that 98% of the oxygen in the blood is actually transported bound to iron in hemoglobin. The remaining 2% is dissolved in the blood plasma.

Carbon dioxide also binds to hemoglobin, but to a much lesser degree. Most carbon dioxide molecules react with water in the RBC to form bicarbonate ions (HCO_3^-). This reaction is catalyzed by carbonic anhydrase inside the RBCs. Most of the bicarbonate ions then diffuse out of the RBCs and are transported in the plasma.

Disease of RBCs Can Alter Homeostasis. Homeostasis requires the normal operation of the heart and blood vessels. It also requires that the blood absorb a sufficient amount of oxygen as it passes through the lungs. Unfortunately, the oxygen-carrying capacity of the blood may decrease for one reason or another. A reduction in the oxygen-carrying capacity of the blood is generally known as **anemia** and may result from (1) a decrease in the number of circulating RBCs, (2) a reduction in the hemoglobin content of the RBCs, or (3) the presence of abnormal hemoglobin in RBCs. The causes of these conditions are many. Consider just a few. First, the number of RBCs in the blood may decline because of excessive bleeding or because of the presence of a tumor in the red bone marrow that reduces RBC production. Several infectious diseases (such as malaria) also decrease the RBC concentration in the blood. A reduction in the amount of hemoglobin in RBCs may be caused by iron deficiency or a deficiency in vitamin B-12, protein, and copper in the diet. Abnormal hemoglobin is produced in sickle-cell anemia and other genetic disorders.

Anemia generally results in weakness and fatigue. Individuals are often pale and tend to faint or become short of breath easily. People suffering from anemia often have an increased heart rate, because the heart beats faster to com-

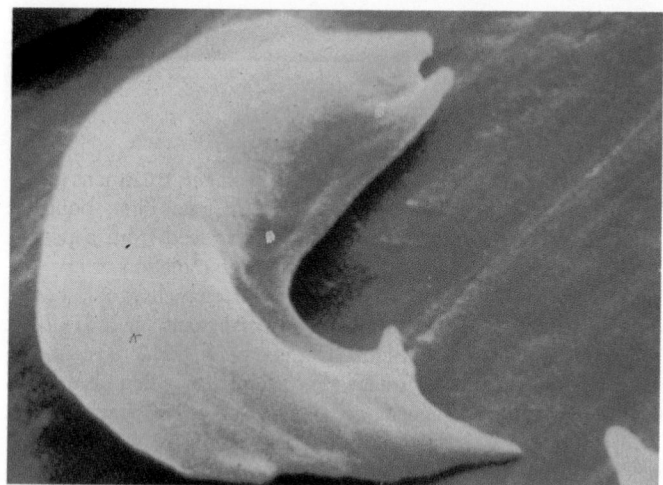

▷ **FIGURE 13–3 Sickle-Cell Anemia** Scanning electron micrograph of a sickle cell.

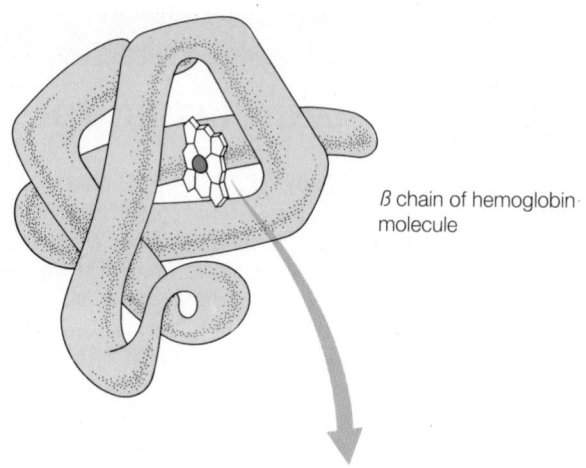

β chain of hemoglobin molecule

pensate for a reduction in the oxygen-carrying capacity of the blood.

As a rule, anemia is not a life-threatening condition. However, it does weaken one's resistance to other diseases or injuries and also limits a person's productivity and energy level. Therefore, no matter what the cause, anemia should be treated quickly.

White Blood Cells Are a Diverse Group That Helps Protect the Body from Infections

White blood cells (WBCs), or **leukocytes,** are nucleated cells that are part of the body's protective mechanism, which combats harmful microorganisms, such as bacteria and viruses. White blood cells are produced in the bone marrow and circulate in the bloodstream, but as noted earlier, they constitute less than 1% of the blood volume. White cells do most of their work outside of the bloodstream, in the tissues. The bloodstream, therefore, serves as a transport vehicle, delivering WBCs to sites of infection. When WBCs arrive at the scene, they escape through the walls of the capillaries by squeezing between the endothelial cells. This process is known as **diapedesis** (from the Greek *dia,* meaning "through" and *pedesis,* meaning "leaping") (▷ Figure 13–5).

Table 13–2 lists and describes the five types of WBCs found in the blood. The three most numerous, which are discussed here, are neutrophils, monocytes, and lymphocytes.

Neutrophils are the most abundant WBC. Approximately twice the size of the RBC, these cells are distinguished by their multilobed nuclei. So named because their cytoplasm has a low affinity for stains, neutrophils circulate in the blood like a cellular police force awaiting microbial invasion. Attracted by chemicals released from infected tissue, neutrophils escape from the bloodstream, then migrate to the site of infection by amoeboid movement.

▷ **FIGURE 13–4 Porphyrin Ring** The porphyrin ring of the hemoglobin molecule contains an iron ion that binds to oxygen and carbon monoxide.

Neutrophils are usually the first WBCs to arrive on the scene. When they arrive, they immediately begin to phagocytize (engulf) microorganisms, helping prevent the spread of bacteria and other organisms from the site of invasion.

▷ **FIGURE 13–5 Diapedesis** White blood cells (leukocytes) escape from capillaries by squeezing between endothelial cells.

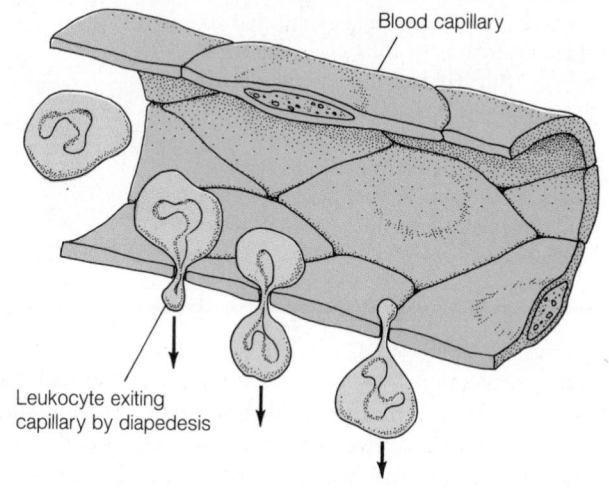

Blood capillary

Leukocyte exiting capillary by diapedesis

MANIPULATING THE STEM CELL

A baby is born into the world. After the nurse-midwife cuts the umbilical cord, which supplies oxygen and nutrient-rich blood to the fetus, technicians whisk the cord off to the lab, where they carefully extract the remaining blood. From the blood sample they extract stem cells, bone marrow cells that routinely escape into the blood (▷ Figure 1). These cells are frozen and placed in a storage bank for future use. If in the ensuing years the donor develops cancer or any of a handful of diseases affecting the blood cells or immune system, medical workers will thaw the stem cells they have put into cold storage and place them in a tissue culture, where they will be treated in one of a handful of ways to produce normal, healthy body cells that can be transfused into the patient.

Although this scenario is science fiction, if researchers at a number of universities throughout the world succeed, it may become commonplace in the coming years.

Stem cells in bone marrow transplants are currently used to restore the blood-producing cells and immune systems of people exposed to high levels of radiation (like those exposed at the Chernobyl nuclear reactor in the former Soviet Union). They are also used to restore these systems in patients treated with chemotherapy, a systemic drug treatment for cancer that kills cancer cells but also wipes out all actively dividing cells, among them the cells of the bone marrow. Bone marrow transplants are also performed in patients subjected to radiation treatment for cancer and individuals with leukemia, a type of cancer that results from the proliferation of lymphocytes in the bone marrow, which, in turn, crowds out other cells, greatly reducing the production of red blood cells, platelets, and other cells vital to homeostasis.

The stem cell's desirability for medical treatment results from its ability to form a whole host of useful cells. As noted in this chapter, the bone marrow stem cell gives rise to five different white blood cells, red blood cells, and platelets. If medical researchers could find a way to manipulate the cell to replace diseased cells, they could cure an array of diseases.

One example is sickle-cell anemia, the genetic disorder described in the chapter. If immunologically compatible stem cells (that is, cells that would not be rejected by the recipient's immune system) could be transplanted into a person with sickle-cell anemia, researchers say, they could repopulate the bone marrow, replacing the person's own stem cells, which give rise to sickle cells, with ones that give rise to normal RBCs.

Chronic enzyme defects in white blood cells, which are genetic in origin, could also be treated, as could osteopetrosis, a bone disorder resulting in overly dense bone. Osteopetrosis is caused by a metabolic abnormality in macrophages. (You will recall that macrophages are derived from monocytes, which are derived from bone marrow stem cells.)

Some researchers believe that stem cells could be used in the treatment of cancers such as leukemia. First, bone marrow would be removed from a patient about to undergo radiation or chemotherapy. Normal stem cells would be isolated from the malignant ones. (This procedure has not been perfected.) After treatment, presumably when all malignant cells in the body had been destroyed, normal stem cells would be reinjected into the patient. Setting up residence in the marrow, the cells would begin to proliferate and would eventually restore blood formation and immune functions. This procedure would use the person's own cells and, if successful, bypass immune reactions that result when bone marrow is transplanted from another person.

Researchers see all kinds of intriguing possibilities for stem cell manipulation. With further research, they hope that stem cells could someday be intentionally altered in the laboratory in ways that cause them to produce lymphocytes specifically targeted to fight cancer or viral infections, such as AIDS. Although a great deal of basic research remains to be done, many scientists believe that they are on the road to a radically new form of treatment for many heretofore untreatable diseases.

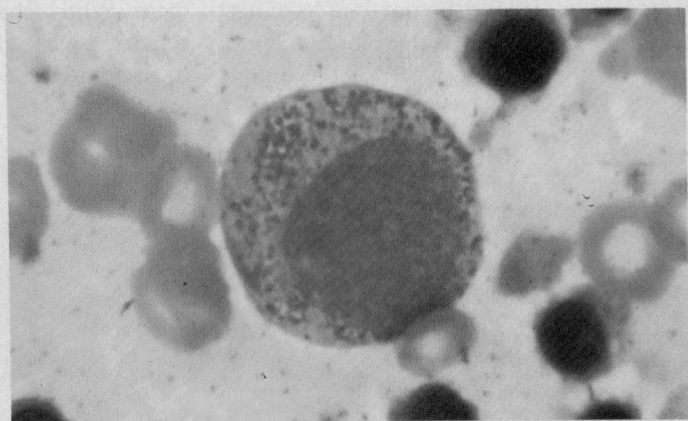

▷ **FIGURE 1 Stem Cell from Human Bone Marrow**

When a neutrophil's lysosomes are used up, the cell dies and becomes part of the yellowish liquid, or **pus,** that exudes from wounds. Pus is a mixture of dead neutrophils, cellular debris, and bacteria, both living and dead.

Monocytes are also phagocytic cells. Slightly larger than neutrophils, monocytes contain distinctive U-shaped or kidney-shaped nuclei. Like neutrophils, monocytes leave the bloodstream to do their "work." They migrate through body tissues via amoeboid motion. Once on the scene, they begin phagocytizing microorganisms, dead cells, and dead neutrophils. Thus, neutrophils are the "first-line" troops, and monocytes are something of a mop-up crew.

Monocytes also take up residence in connective tissues of the body, where they are referred to as **macrophages.** These cells remain more or less stationary, like watchful soldiers ever ready to attack and phagocytize invaders.

The second most abundant WBC is the **lymphocyte.** Most lymphocytes exist outside the circulatory system in **lymphoid organs,** such as the spleen, thymus, and lymph nodes, and **lymphoid tissue,** aggregations of lymphocytes beneath the lining of the intestinal and respiratory tracts. It is in these locations that they attack microbial intruders (▷ Figure 13–6).

Two types of lymphocytes are found in the body, both of which play a vital role in immune protection (Chapter 15). The first type, the **T lymphocyte,** or **T cell,** attacks foreign cells, such as fungi, parasites, and tumor cells.[5] T lymphocytes are thus said to provide "cellular immunity." The second type is called the **B lymphocyte,** or **B cell.** When activated, it actually transforms into another kind of cell, known as the **plasma cell.** Plasma cells, in turn, synthesize and release antibodies. **Antibodies** are proteins that circulate in the blood and bind to foreign substances, neutralizing them or just "marking" microorganisms and tumor cells for destruction by macrophages. (More on this in Chapter 15.)

The WBCs, like the other formed elements of blood, are involved in homeostasis. Their numbers can increase or decrease greatly during a microbial infection and other

[5] T lymphocytes usually attack large eukaryotic cells, such as fungi and parasites. Most bacteria are controlled by antibodies. Only the few bacteria that are intracellular parasites, such as *M. tuberculosis,* are attacked by T cells, but even then, the lymphocyte attacks the host cell, not the bacterium directly.

diseases. In fighting off a disease, they help return the body to normal function once again.

An increase in the number of WBCs, called **leukocytosis,** is a normal homeostatic response to intruders. It ends when the microbial invaders have been destroyed. Increases and decreases in various types of blood cells can be used to diagnose many medical disorders. For example, an increase in eosinophils, not discussed here, is sometimes an indication of an allergy. A dramatic increase in lymphocytes and lower abdominal pain are usually signs of appendicitis (discussed in Chapter 11). Because variations in the WBC count accompany many diseases, a blood test is a standard procedure for patients undergoing diagnostic testing.

Diseases Involving White Blood Cells Also Affect the Body's Internal Balance. By protecting the body, WBCs are part of one of the most important homeostatic mechanisms. WBCs, however, can run amok. For example, some white blood cells can become cancerous, dividing uncontrollably in the bone marrow, then entering the bloodstream. A cancer of WBCs is called **leukemia** (literally, "white blood"). The most serious type of leukemia is **acute leukemia,** so named because it kills victims quickly. Children are the primary victims of this disease.

In acute leukemia, WBCs fill the bone marrow and crowd out the cells that produce RBCs and platelets. A decline in the production of RBCs leads to anemia. A reduction in platelet production reduces clotting, increasing internal bleeding. Making matters worse, the cancerous WBCs produced in leukemia are often incapable of fighting infection. Because of this defect and because the production of platelets is reduced, victims of acute leukemia typically succumb to infections and internal bleeding.

Leukemia can be treated by irradiating the bone marrow and by administering a drug (vincristine) that stops mitosis. Twenty years ago, only one of every four children with leukemia survived. Today, thanks to vincristine, the odds have changed dramatically: three of every four children

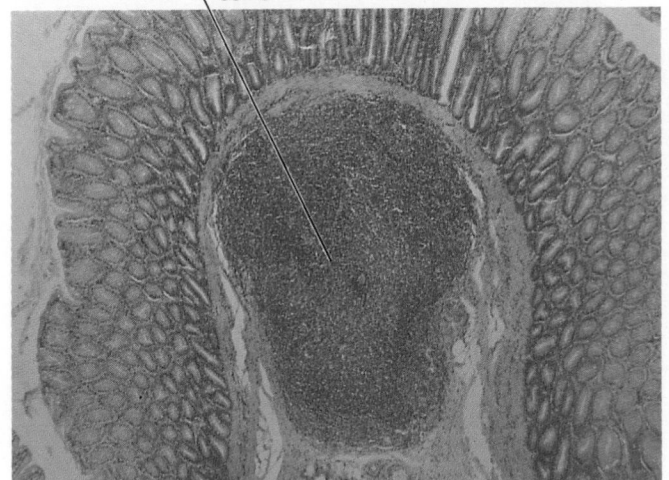

Aggregation of Lymphocytes

▷ **FIGURE 13–6 Lymphoid Tissue** The loose connective tissue beneath the lining of the large intestine and other sites is often packed with lymphocytes that have proliferated in response to invading bacteria.

▷ **FIGURE 13–7 Rosy Periwinkle** Many tropical plants contain chemicals that provide extraordinary medical benefits. One substance from the rosy periwinkle has helped physicians treat leukemia. Unfortunately, the tropical rain forests are being cut at an alarming rate, reducing our chances of finding other cures.

with the disease survive! Interestingly, vincristine was derived from a plant known as the rosy periwinkle, which is found in tropical rain forests (▷ Figure 13–7). Thousands of other drugs have a similar origin, underscoring the importance of protecting the rain forests from decimation (▷ Figure 13–8 a and b).

Another common disorder of the WBCs is **infectious mononucleosis**, commonly called "mono" or "kissing disease." Mono is caused by a virus transmitted through saliva and may be spread by kissing; by sharing silverware, plates, and drinking glasses; and possibly even through drinking fountains. The virus spreads through the body and affects many organs.

Although the virus infects only lymphocytes in the bloodstream, the number of monocytes and lymphocytes in the blood increases rapidly during an infection. Individuals suffering from mono complain of fatigue, aches, sore throats, and low-grade fever. Physicians recommend that victims get plenty of rest and drink lots of liquids while the immune system eliminates the virus. Within a few weeks, symptoms generally disappear, although weakness may persist for two more weeks.

Platelets Are Fragments of Cells Found in the Bone Marrow and Are a Vital Component of the Blood-Clotting Mechanism

The circulatory system is a delicate structure. Even minor bumps and scrapes can cause it to leak. Leakage is normally prevented by **blood clotting,** surely one of the most intricate homeostatic systems to have evolved in the living world. A simplified version is discussed here.

Among the most important agents of blood clotting are the platelets, tiny formed elements produced in the bone marrow by fragmentation a huge cell known as the **megakaryocyte** (▷ Figure 13–9). Platelets lack nuclei and organelles and therefore are not true cells. Like RBCs, they cannot divide. Platelets are carried passively in the bloodstream. Coated by a layer of a sticky material, platelets tend to adhere to irregular surfaces, such as tears in blood vessels or cholesterol-containing plaque.

The process of blood clotting and the role of platelets are summarized in ▷ Figure 13–10. As illustrated, the process begins when cells in the damaged tissue release a substance into the bloodstream called **thromboplastin.** Thromboplastin is a lipoprotein that converts an inactive plasma enzyme, **prothrombin** (produced by the liver) into its active form, **thrombin.** Thrombin, in turn, acts on another blood protein, **fibrinogen,** also produced by the liver. When activated, fibrinogen is converted into **fibrin,** creating long, branching fibers that produce a weblike network in the wall of the damaged blood vessel (▷ Figure 13–11). The fibrin web traps RBCs and platelets, forming a plug that cuts off the flow of blood to the tissue. Platelets captured by the fibrin web release additional thromboplastin, known as platelet thromboplastin, which causes more

▷ **FIGURE 13–8 Forest Destruction** (*a*) Rampant deforestation leaves the tropics denuded but also robs humankind of many potential cures for disease. (*b*) The loss of forests and other habitats in the industrial countries is also of grave importance. The bark of this tree, the Pacific yew, contains a chemical that may prove effective in fighting breast cancer. However, each year thousands of acres of old-growth forest, where the tree lives, are being cut down.

(a)

(b)

Megakaryocyte Platelets

▷ **FIGURE 13–9 Megakaryocyte** A light micrograph of a megakaryocyte, a large, multinucleated cell found in bone marrow, which fragments, giving rise to platelets.

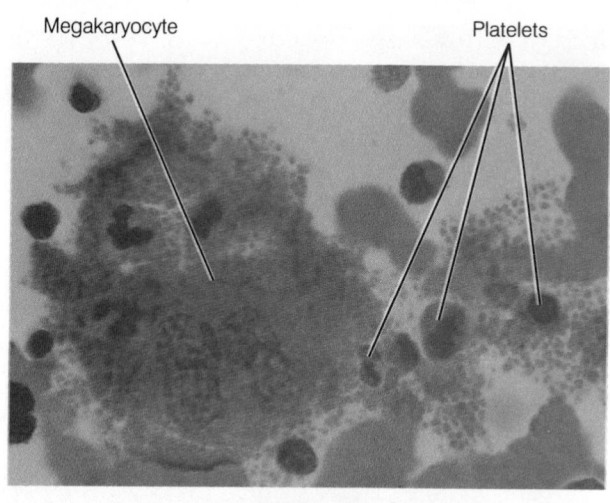

Megakaryocyte   Platelets

fibrin to be laid down and therefore serves to reinforce the fibrin network.

Blood clotting occurs fairly quickly. In most cases, a damaged blood vessel is sealed by a clot within 3 to 6 minutes of an injury; 30 to 60 minutes later, platelets in the clot begin to draw the clot inward, stitching the wound together. How? Platelets contain contractile proteins like those in muscle cells. Contraction of the protein fibers draws the fibrin network inward, pulling the edges of the

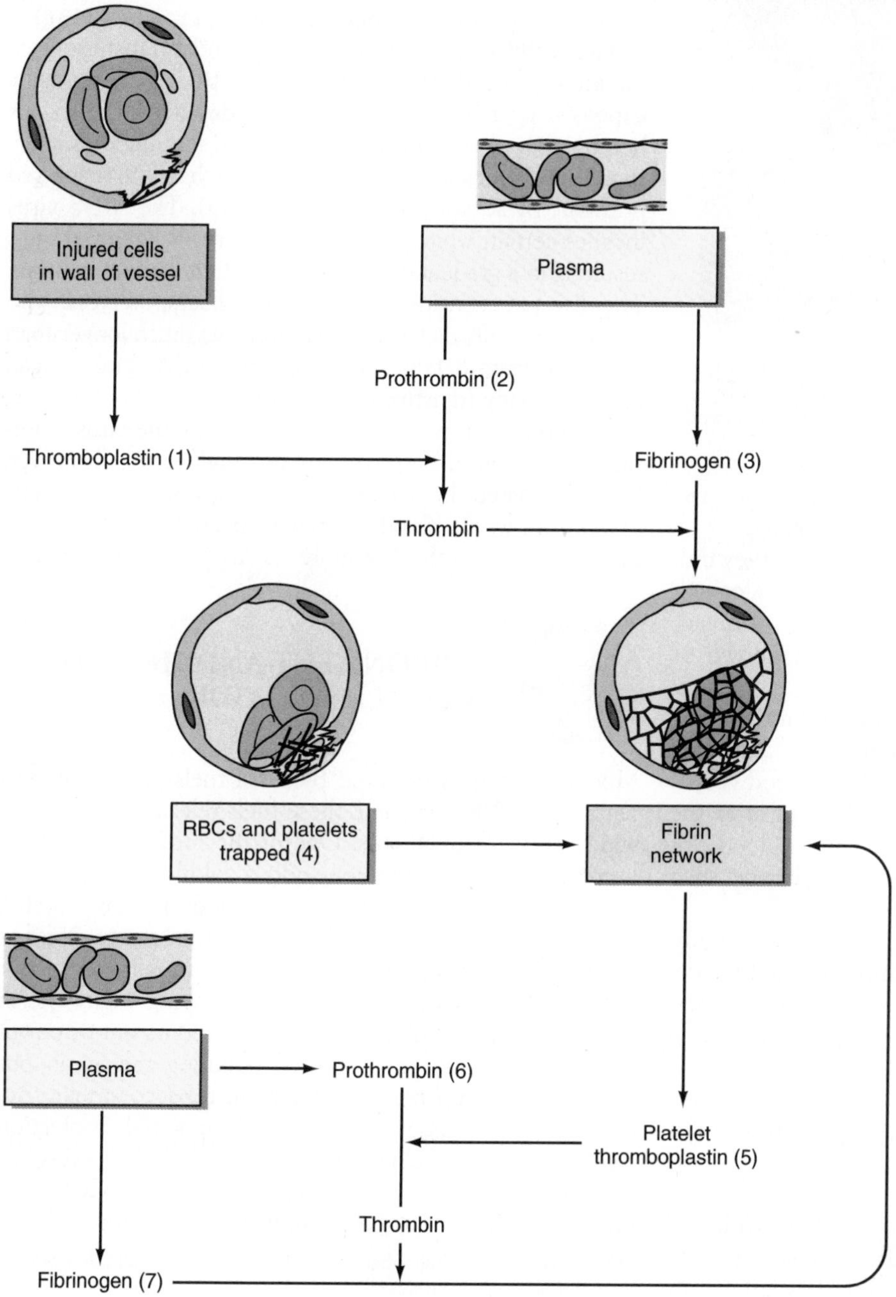

▷ **FIGURE 13–10 Blood Clotting Simplified** Injured cells in the wall of blood vessels release the chemical thromboplastin (1). Thromboplastin stimulates the conversion of prothrombin, found in the plasma, into thrombin (2). Thrombin, in turn, stimulates the conversion of the plasma protein fibrinogen into fibrin (3). The fibrin network captures RBCs and platelets (4). Platelets in the blood clot release platelet thromboplastin (5), which converts additional plasma prothrombin into thrombin (6). Thrombin, in turn, stimulates the production of additional fibrin (7).

Injured cells
in wall of vessel

Plasma

Thromboplastin (1) ⟶

Prothrombin (2)

Fibrinogen (3)

Thrombin ⟶

RBCs and platelets
trapped (4)

Fibrin
network

Plasma ⟶ Prothrombin (6)

Platelet
thromboplastin (5)

Thrombin

Fibrinogen (7) ⟶

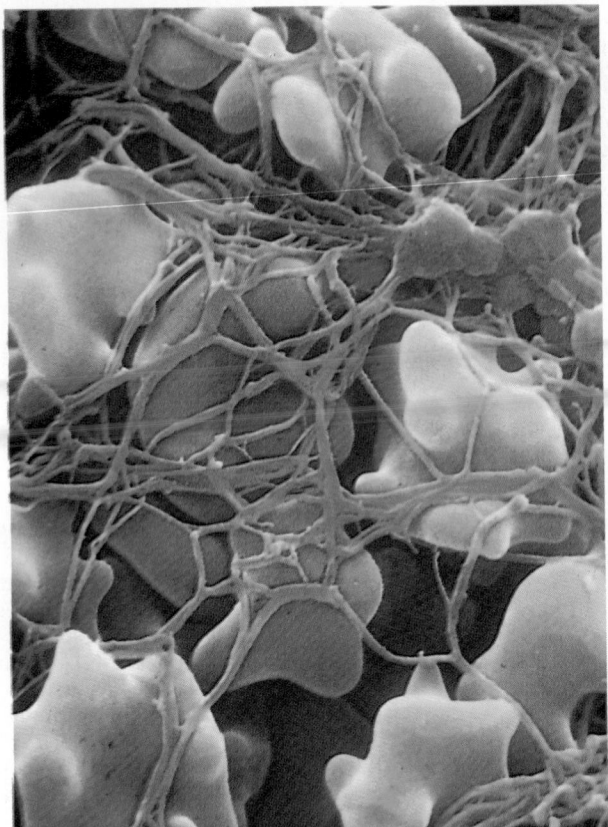

▷ **FIGURE 13–11 Blood Clot** A scanning electron micrograph of a fibrin clot that has already trapped platelets and RBCs, helping plug up a leak in a vessel. RBCs are yellow. Fibrin network is red.

cut or damaged blood vessel together. This closes the wound like a surgical stitch and accelerates healing.

Blood clots do not stay in place indefinitely. If they did, the circulatory system would eventually become clogged, and blood flow would come to a complete halt. Instead, clots are dissolved by a blood-borne enzyme known as **plasmin,** which is produced from an inactive form, **plasminogen.** Plasminogen is incorporated in the clot as it forms. It is then gradually converted to plasmin by an activating factor secreted by the endothelial cells of the blood vessel. This ensures that the plasmin dissolves the clot after the blood vessel damage has been repaired. TPA, a blood clot buster discussed briefly in the last chapter, is one such activator.

As important as blood clots are in protecting the body, normal clotting can also cause problems. As noted in previous chapters, blood clots can break loose from their site of formation, circulate in the blood, and become lodged in arteries narrowed by atherosclerosis, further restricting blood flow and causing considerable damage. Blood clots typically lodge in the narrow vessels serving the heart muscle, brain, and other vital organs.

Clotting Disorders Upset the Homeostatic Balance and Can Be Life Threatening. Tiny breaks occur in blood vessels with surprising regularity, but they are usually repaired without our ever knowing it. In some individuals, blood clotting is impaired because of: (1) an insufficient number of platelets; (2) liver damage, which hinders the production of clotting factors; or (3) genetic disorders that result in a lack of clotting factors.

A reduced platelet count may result from leukemia, as noted earlier, or from an exposure to excess radiation, which damages bone marrow where the megakaryocytes reside. Liver damage, which impairs the production of blood-clotting factors, can be caused by hepatitis, liver cancer, or excessive alcohol consumption. The most common genetic defect is called **hemophilia.** In people with the disease, the liver fails to produce necessary clotting factors. Problems begin early in life, and even tiny cuts or bruises can bleed uncontrollably, threatening the person's life. Because of repeated bleeding into the joints, victims suffer great pain and often become disabled; they often die at a young age.

Hemophiliacs can be treated by periodic transfusions of the missing blood-clotting factors. This therapy, however, is expensive and is required every few days. It has also put hemophiliacs at risk for **AIDS,** acquired immunodeficiency syndrome, which is caused by a virus that is transmitted primarily by sexual contact (Chapter 15). The AIDS virus invades certain white blood cells, known as helper T cells, resulting in a gradual deterioration of the immune function. The disease is considered 100% fatal; that is, everyone who gets it eventually dies, although some live much longer than others. Unfortunately, testing for the AIDS virus began late, and many transfusions of whole blood, blood plasma, and clotting factors were contaminated by the virus. Clotting factors produced by genetic engineering, however, have eliminated the need for transfusions of clotting agents taken from whole blood, reducing the risk to hemophiliacs of contracting AIDS. (For more on AIDS, see Chapter 15.)

 **ENVIRONMENT AND HEALTH: CARBON MONOXIDE**

Modern society is powered by fossil fuels—coal, oil, and natural gas. When burned, these fuels release energy and a wide variety of pollutants. One of the most dangerous to human health is carbon monoxide, a colorless, odorless gas that results from incomplete combustion of organic fuels.[6]

Carbon monoxide (CO) emanates from stoves and furnaces in our homes and spews out of the tailpipes of our automobiles, causing increased levels along highways, in parking garages, and in tunnels. It also pours out of power plants and factory smokestacks, polluting the air in our cities. It is even a major pollutant in tobacco smoke. According to estimates by the Environmental Protection Agency, over 40 million Americans, or one of every six people in this country, are exposed to levels of CO in the outside air thought to be harmful to their health.

[6]When gasoline, coal, and other organic fuels burn completely, they produce carbon dioxide and water.

What makes CO so dangerous? Carbon monoxide, like oxygen, binds to hemoglobin. However, hemoglobin has a much greater affinity for CO than for oxygen (about 200 times greater). Consequently, CO "outcompetes" oxygen for the binding sites on the hemoglobin molecules and thus reduces the blood's ability to carry oxygen. For healthy people, CO does not create much of a problem, as long as levels are low. The body simply produces more RBCs or increases the heart rate to augment the flow of oxygen to tissues. CO does create a problem, however, when levels are so high that the body cannot compensate. At very high levels, CO becomes a deadly killer.

New research suggests that CO levels currently deemed acceptable by federal standards can, in some people, trigger chest pain (angina), which results from a lack of oxygen in the heart muscle. Levels such as those once thought safe in many workplaces or levels in the blood encountered after one hour in heavy traffic cause angina and abnormal ECGs in moderately active adults with coronary artery disease, according to a study published in the *New England Journal of Medicine*.

In levels commonly found in and around cities, CO is especially troublesome for the elderly or for individuals suffering from coronary artery and lung diseases. Carbon monoxide places additional strain on their heart, making the already weakened organ work harder. Thus, the elderly and infirm are advised to stay inside on high-pollution days.

Carbon monoxide is just one of many pollutants we breathe. No one knows what, if any, long-term effects it will have on our health. In the short term, however, it is clear that CO upsets oxygen delivery, which is so essential to homeostasis in the elderly and infirm. It has become another risk factor in an increasingly risky society, but it is not an insolvable problem. Much more efficient automobiles, widespread use of mass transit, overall reductions in driving, and alternative automobile fuels (such as hydrogen) can all reduce the level of CO in and around the cities where most Americans live and work (▷ Figure 13–12). More efficient factories, home furnaces, and power plants can also help reduce CO levels in our cities. Passive-solar homes and improvements in insulation can also cut our demand for fossil fuel energy (▷ Figure 13–13). Bans on smoking indoors could go a long way, too, in ridding our air of this transparent killer.

▷ **FIGURE 13–12 Air Saver** Commuter lines like this one in Portland, Oregon, significantly reduce air pollution, because light rail systems are far more efficient—per passenger mile traveled—than cars.

▷ **FIGURE 13–13 Passive-Solar Heating** The sun heats the author's home, located at 8000 feet above sea level in the Colorado Rockies. In the fall and winter, sunlight streams through south-facing windows, heating the interior of the house. Superinsulated walls and ceilings keep the heat from escaping. This house costs only about $100 per year to heat even though winter temperatures often remain well below zero for long stretches. The low use of heating fuel greatly reduces the output of pollution.

SUMMARY

BLOOD: ITS COMPOSITION AND FUNCTIONS

1. Blood is a watery tissue containing two basic parts: the plasma—which contains dissolved nutrients, proteins, gases, and wastes—and the formed elements—red blood cells, white blood cells, and platelets, which are suspended in the plasma.
2. The functions of the formed elements are summarized in Table 13–2.
3. The plasma constitutes about 55% of the volume of a person's blood, and the formed elements make up the remainder. The volume occupied by the blood cells and platelets is called the hematocrit.
4. Red blood cells (RBCs) are highly specialized cells that lack nuclei and organelles and are produced by the red bone marrow. The RBCs transport oxygen in the blood.
5. The concentration of RBCs in the blood is maintained by the hormone erythropoietin, produced by the kidneys in response to falling oxygen levels.
6. White blood cells (WBCs) are nucleated cells and are part of the body's protective mechanism to combat microorganisms.

WBCs are produced in the bone marrow and circulate in the bloodstream, but they do most of their work outside it, in the body tissues.

7. The most abundant WBCs are the neutrophils, which are attracted by chemicals released from infected tissue. Neutrophils leave the bloodstream and migrate to the site of infection by amoeboid movement.

8. Neutrophils are the first WBCs to arrive at an infection, where they phagocytize microorganisms, helping prevent the spread of bacteria and other organisms.

9. The second group of cells to arrive are the monocytes, which phagocytize microorganisms, dead cells, cellular debris, and dead neutrophils.

10. Lymphocytes are the second most numerous WBCs and play a vital role in immune protection.

11. Platelets are fragments of large bone marrow cells and are involved in blood clotting.

12. Platelets are coated by a layer of a sticky material, which causes them to adhere to irregular surfaces, such as tears in blood vessels. The process of blood clotting is summarized in Figure 13–10.

ENVIRONMENT AND HEALTH: CARBON MONOXIDE

13. Carbon monoxide is produced by the incomplete combustion of organic fuels in our homes, automobiles, factories, and power plants.

14. Carbon monoxide binds to hemoglobin and reduces the oxygen-carrying capacity of the blood.

15. At high concentrations, carbon monoxide can be lethal. At lower concentrations, it is harmful to people with cardiovascular disease and lung disease. For individuals with heart disease, it puts additional strain on the heart.

EXERCISING YOUR CRITICAL THINKING SKILLS

The stem cell in bone marrow is the subject of intense research, for scientists believe that by manipulating the cell they can find ways to better treat—perhaps even cure—a number of diseases, such as cancer, sickle-cell anemia, and AIDS. In order to manipulate the cell, however, scientists recognize that they need to know more about cell replication and differentiation. Through basic research, they hope to learn what, if anything, causes the cell to replicate and differentiate. Two schools of thought exist on the subject. The first says that the stem cell is subject to external influences, such as hormones, which direct its differentiation. The second school of thought contends that replication and differentiation occur randomly; that is, there are no outside controls.

At this point, can you see any problems in the way the debate is framed? You may want to review the critical thinking rules on page 18.

If after reviewing the debate and the critical thinking rules, you said the problem might be that researchers were over simplifying matters, you have pinpointed the problem. Put another way, researchers seem to have fallen into the dualistic thinking trap and may have set up a false dichotomy. It's quite possible that outside influences and random replication and differentiation both occur.

In an effort to settle this debate, Makio Ogawa of the Medical University of South Carolina, in Charleston, devised a method for growing human stem cells in a semisolid gel. He removed single cells from the stem cell colonies, then used these cells to start new colonies, which he could study. He found that the cells in the secondary colonies developed in a variety of cell types. That is, they appeared to differentiate randomly along several lines.

Think about the experiment for a few moments. Can you find any weaknesses with the design?

If you said that the researcher had isolated the cells from outside influences, such as hormones, that might be present in bone marrow and might influence differentiation, you're right. Cell culture studies—that is, studies of cells in culture dishes—although extremely useful, are often criticized because they do not expose cells to potentially important influences, such as hormones.

Can you think of any ways to solve this problem—that is, to study cell differentiation while preserving potential hormonal influences?

TEST OF TERMS

1. The liquid portion of blood is called _____.

2. The proteins in the blood that destroy or inactivate viruses and bacteria are _____ .

3. The _____ is a biconcave disk that contains the protein _____, which binds to oxygen.

4. Blood cells are produced in _____ _____ marrow.

5. The hormone _____ , produced by the kidney, stimulates the synthesis of _____ when oxygen levels in the blood fall.

6. The heme group consists of a _____ ring and an atom of iron.

7. _____ is a reduction in the oxygen-carrying capacity of the blood.

8. The two white blood cells that phagocytize bacteria and viruses are _____ and _____ .

9. The _____ is the white blood cell involved in immune protection.

10. _____ is a cancer of the white blood cells.

11. The _____ is a cell fragment produced from megakaryocytes.

12. _____ , an inactive protein in the blood, is activated by thrombin and converted into long, branching fibers that form a weblike network in the wall of blood vessels that have been damaged.

13. _____ is an enzyme in the blood that dissolves blood clots.

14. _____ is a genetic defect that results in a lack of clotting factors and leads to uncontrollable bleeding.

15. _____ _____ is a colorless, odorless gas that binds to the hemoglobin molecule, reducing the oxygen-carrying capacity of the blood.

Answers to the Test of Terms are found in Appendix B.

TEST OF CONCEPTS

1. Describe the structure and function of each of the following: red blood cells, platelets, lymphocytes, monocytes, and neutrophils.

2. Define each of the following terms: leukemia, anemia, and infectious mononucleosis.

3. Explain how a blood clot forms and how it helps prevent bleeding.

CHAPTER

14

The Vital Exchange: Respiration

≈

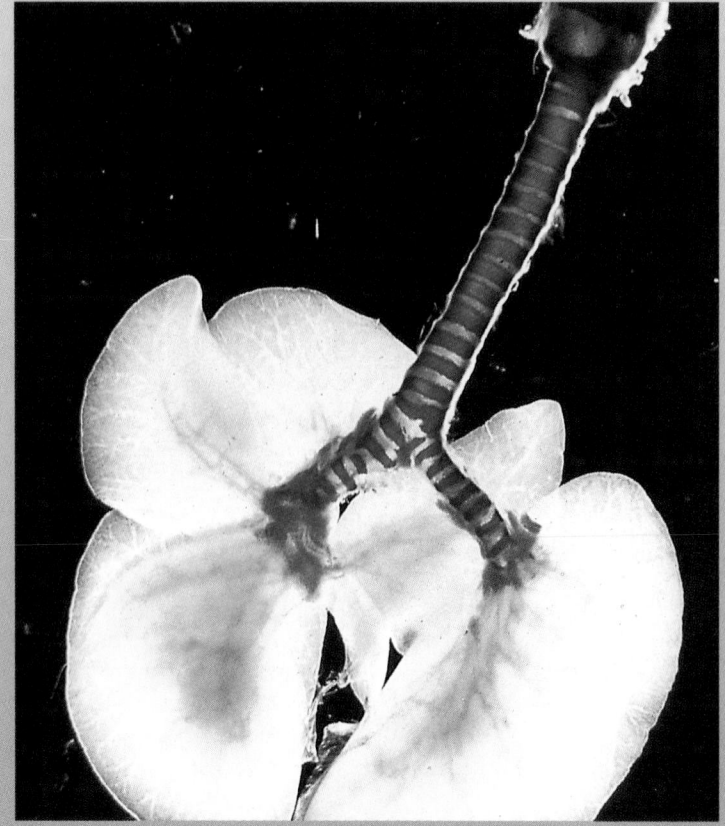

Respiratory system of a mammal showing larynx, trachea, bronchi, and lungs.

George F. Eaton was a robust and handsome Irishman who grew up in eastern Massachusetts, married, and raised three children. To support his family, Eaton worked as a fire fighter for 30 years. Fire fighting is dangerous work, in part because it exposes men and women to smoke containing numerous potentially harmful and sometimes lethal air pollutants.

Upon retirement, Eaton moved to Cape Cod, but his retirement years were cut short by **emphysema,** a debilitating respiratory disease resulting from the breakdown of the air sacs, or **alveoli,** in the lungs, where oxygen and carbon dioxide are exchanged between the air and the blood. As the walls of the alveoli degenerate, the surface area for the diffusion of oxygen and carbon dioxide greatly decreases. Emphysema worsens by the month and is an irreversible and incurable disease. Patients complain of shortness of breath; eventually, even mild exertion becomes trying, prompting one patient to describe the disease as a kind of "living hell."

The degeneration of the lungs in patients with emphysema creates a domino effect. One of the dominoes is the heart. Because oxygen absorption in the lungs declines substantially, the heart of an emphysemic patient must work harder. This effort puts additional strain on this already hard working organ and can lead to heart failure.

George Eaton died a slow and painful death as his lungs grew increasingly more inefficient. Relatives mourned his death and blamed it on the pollution to which he was exposed while working for the fire department. Fire fighting, however, was probably only part of the cause, for Eaton had smoked most of his adult life. Cigarette smoking is the leading cause of emphysema.

This chapter describes the respiratory system—its evolution, its function, and the diseases that affect it, like emphysema.

≈ THE EVOLUTION OF RESPIRATORY SYSTEMS: AN OVERVIEW

What do the gills of a fish, the skin of a frog, and the lungs of mammals have in common? The answer is simple: all three are evolutionary adaptations to the same problem. That is, each one represents "evolution's answer" to an important biological challenge: getting oxygen into the body of a multicellular animal and getting rid of carbon dioxide.

Why is oxygen so important? If you think back on class discussions and your readings on cellular energy production, you may recall that all eukaryotic cells need oxygen for cellular respiration, the energy-yielding breakdown of glucose. That process produces abundant amounts of carbon dioxide that must be disposed of.

In the early history of life, single-celled eukaryotic organisms relied on diffusion to acquire oxygen and get rid of carbon dioxide. With the advent of multicellular animal life, though, diffusion was replaced by more efficient systems. Thus arose the respirable skin of the amphibian, the gill, the lung, and others. These specialized structures accommodated the needs of large, multicellular animals for oxygen.

Gills are structures found in aquatic vertebrates, mostly fishes, that increase the surface area for gas exchange. As shown in ▷ Figure 14–1, the gills of a fish consist of arched structures appropriately called **gill arches.** Each arch consists of many **gill filaments,** featherlike projections heavily endowed with blood vessels. Water containing oxygen enters a fish's mouth and is pumped backward over the gill filaments. Oxygen in the water is then absorbed across the relatively thin membrane of the gill filaments, moving into the bloodstream by diffusion. Carbon dioxide diffuses in the opposite direction.

If you have ever watched an aquarium fish floating idly in a tank, you may have noticed that it opens and closes its mouth, as if gulping water. This gulping action is vital to gaseous exchange, for it produces suction that draws water into the oral cavity. The water is then pushed out of the mouth and over the gills by movements of the gill covers, special plates that protect the gills from injury. When a fish is swimming, it ventilates its gills by opening its mouth and gill covers. This permits water to flow through the mouth and over the gill filaments. Some fish, such as mackerel, lack a pumping mechanism, and must therefore swim continuously with their mouth and gill covers open or die for lack of oxygen.

In terrestrial animals, several kinds of respiratory strategies have evolved. A few of them are presented here to highlight the diversity of ways in which animals meet the same requirement for gaseous exchange.

In the earthworm (a segmented worm, or annelid), gases are exchanged across the body wall. If you've ever picked up an earthworm, you know that its skin is slimy. That slime, or mucus, helps keep the earthworm's exterior moist so that gases can move in and out more easily.

Insects and other land-dwelling arthropods have tubular gas-exchange systems, as shown in ▷ Figure 14–2. Air diffuses into this system through tiny openings in the body wall called **spiracles** (spy-ruh-kuls). The spiracles empty into relatively large tubes, called **tracheae** (tray-key-eye). These tubes divide and subdivide, forming smaller and smaller tubules that course throughout the insect body. Gas exchange occurs in very tiny, dead-end tubules that terminate in body tissues. Moisture in these tubules, as on an earthworm's skin, is vital for the diffusion of oxygen and carbon dioxide.

In larger insects, the tracheal system is supplemented by air sacs, shown in Figure 14–2. These sacs are located near large muscles. As the insect moves, the muscles pump air out of the sacs into the tracheae. Air is drawn back into the sacs when the muscles relax. This interesting adaptation operates similarly to our lungs, but because the movement of air in and out of the sacs is incidental to movement, no additional energy is required for respiration.

Amphibians (frogs and salamanders) live on land and in

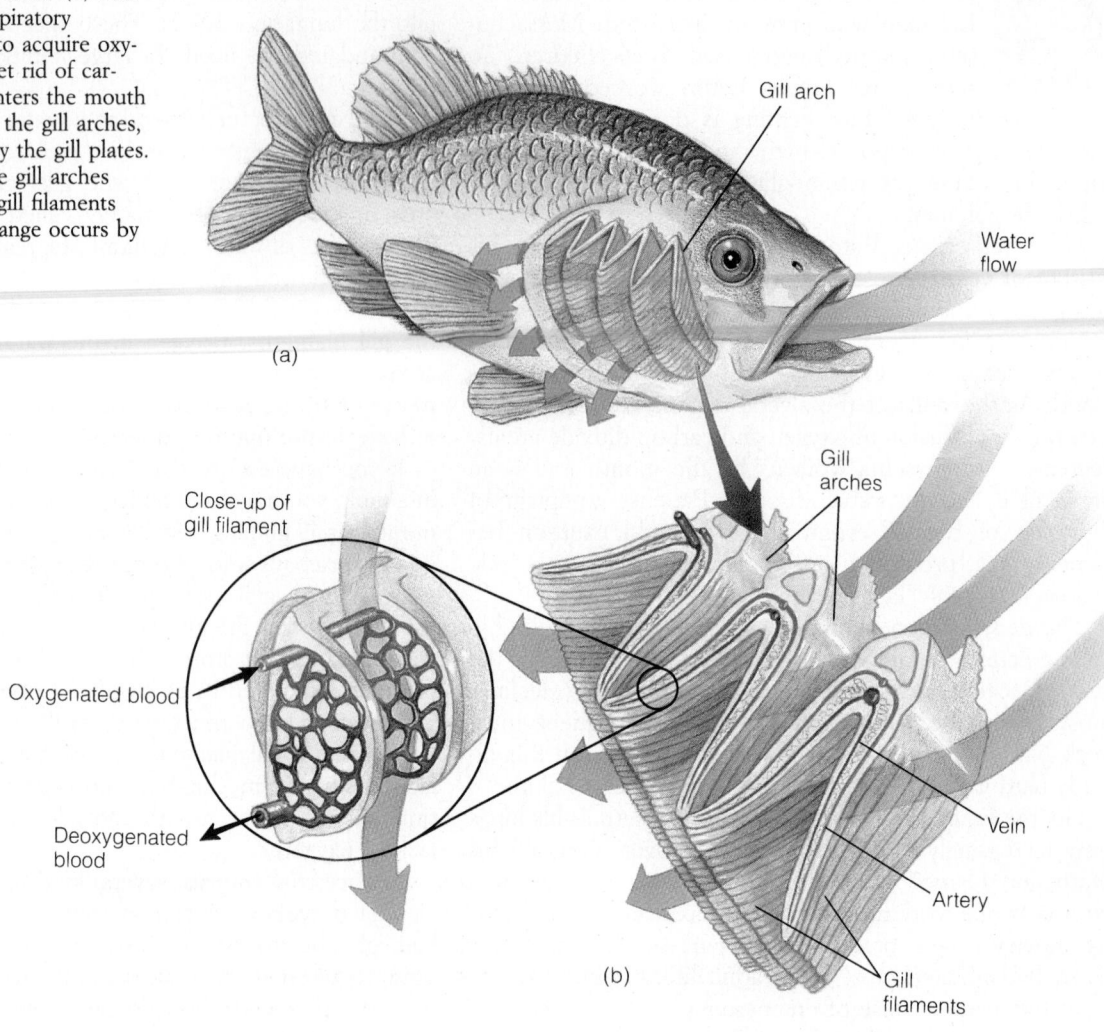

▷ **FIGURE 14–1 Gills** (*a*) Fish rely on specialized respiratory structures called gills to acquire oxygen from water and get rid of carbon dioxide. Water enters the mouth and is pumped across the gill arches, which are protected by the gill plates. (*b*) As shown here, the gill arches consist of featherlike gill filaments across which gas exchange occurs by diffusion.

Gill arch

Water flow

(a)

Gill arches

Close-up of gill filament

Oxygenated blood

Deoxygenated blood

Vein

Artery

Gill filaments

(b)

water. (The word amphibian comes from the Greek word *amphibios,* "living a double life.") Frogs have primitive, saclike structures called **lungs,** shown in ▷ Figure 14–3. As illustrated, the frog draws air into its mouth, closes it, then pushes the air into its lungs. However, lung respiration is less important than gas exchange through the animal's moist skin and across the moist surfaces of its mouth.

Mammals live on land, in the air, and in the oceans, lakes, and rivers, often moving from one medium to another in the course of their day. As a group, the mammals exhibit a higher rate of metabolism than reptiles, amphibians, and insects. This is due in part to the fact that mammals rely on internally generated heat to stay warm. That heat, of course, comes from glucose catabolism.

Endothermy, or internal heat production, no doubt led to the evolution of more efficient respiratory systems—systems that could acquire greater amounts of oxygen. Thus arose the mammalian lung and associated structures. We turn to the human respiratory system to study the mammalian lung.

≈ STRUCTURE OF THE HUMAN RESPIRATORY SYSTEM

The human respiratory system, like those of other mammals, functions automatically, drawing air into the lungs, then letting it out, in a cycle that repeats itself about 16 times per minute at rest, or about 23,000 times per day. The respiratory system supplies oxygen to the body and gets rid of carbon dioxide. Oxygen is needed for cellular respiration; carbon dioxide is a waste product of this process.

In its role as a provider of oxygen and a disposer of carbon dioxide waste, the respiratory system helps maintain a constant internal environment necessary for normal cellular metabolism. Thus, like many other systems, it plays an important role in homeostasis.

▷ Figure 14–4 shows the structure of the human respiratory system. As illustrated, the respiratory system consists of two basic parts: an air-conducting portion and a gas-exchange portion (Table 14–1). The air-conducting portion is an elaborate set of passageways that transports air to and

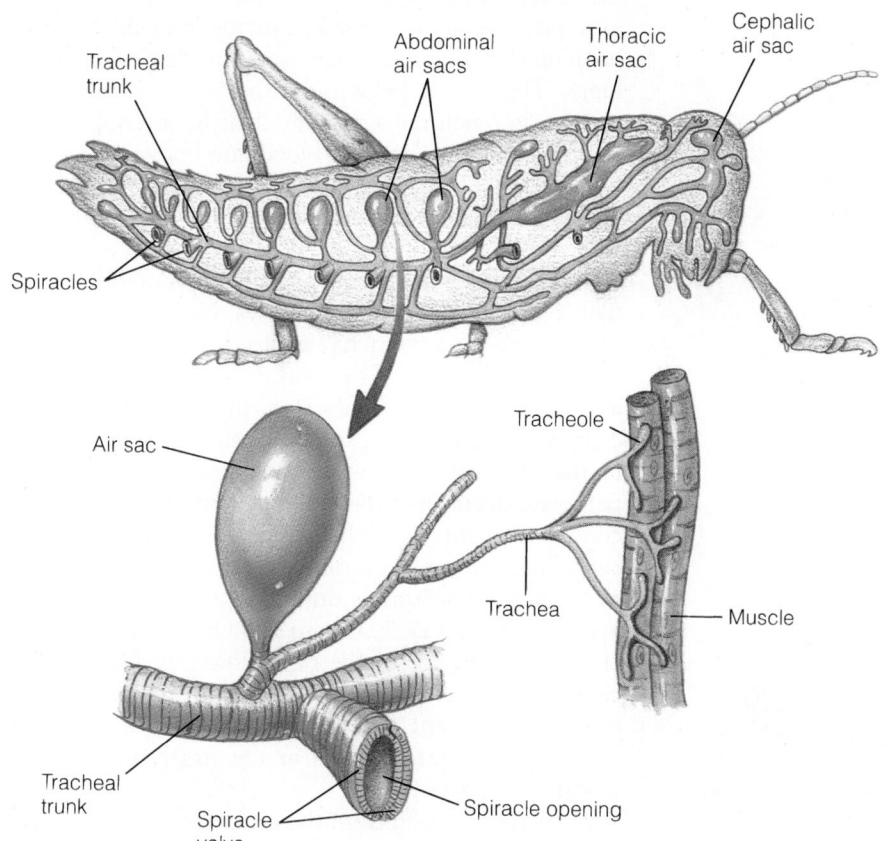

▷ **FIGURE 14–2 Tracheal System of an Insect** As shown here, insects contain an internal system of channels to transport oxygen to body cells. Air enters and leaves the system via tiny openings called spiracles. Air sacs are present in some insects and form reservoirs that store air.

Labels on figure:
Tracheal trunk
Abdominal air sacs
Thoracic air sac
Cephalic air sac
Spiracles
Air sac
Tracheole
Trachea
Muscle
Tracheal trunk
Spiracle valve
Spiracle opening

from the lungs, two large, saclike organs in the thoracic cavity (Figure 14–4). Like the arteries of the body, these passageways start out large, then become progressively smaller and more numerous after entering the lung tissue, where the exchange of oxygen and carbon dioxide between the air and the blood takes place.

The lungs are the gas-exchange portion of the respiratory system. Each lung has millions of tiny, thin-walled alveoli (Figure 14–4). The walls of the alveoli contain numerous blood capillaries that absorb oxygen from the inhaled air and release carbon dioxide (▷ Figure 14–5, page 330).

The Conducting Portion of the Respiratory System Moves Air In and Out of the Body but also Filters the Air and Moistens the Incoming Air

Air enters the respiratory system through the nose and mouth, then is drawn backward into the pharynx (▷ Figure 14–6, page 330). The **pharynx** opens into the nose and mouth in the front and joins below with the larynx. The **larynx,** or voice box, is a rigid but hollow structure that houses the vocal cords (Figure 14–6). To feel it, gently put your fingers alongside your throat, then swallow. As shown in Figure 14–4, the larynx opens into the **trachea** below. You can feel the trachea below your Adam's apple, the protrusion of the laryngeal cartilage on your neck.

As explained in Chapter 11, food is prevented from entering the larynx by the epiglottis, a flap of tissue that closes off the opening to the larynx during swallowing. Occasionally, however, food goes the wrong way, accidentally entering the larynx and trachea. It may lead to violent coughing, a reflex that helps eject the food from the trachea. If the food cannot be dislodged by coughing, steps must be taken to remove it—and fast—or a person will suffocate. Health Note 14–1 (page 331) explains what to do when a person chokes.

As Figure 14–4 shows, the trachea is a short, wide duct. Starting in the neck below the larynx, it enters the thoracic cavity where it divides into two large branches, the right and left bronchi. The **bronchi** (singular, *bronchus*) enter the lungs alongside the arteries and veins. Inside the lungs, the bronchi branch extensively, forming progressively smaller tubes that carry air to the alveoli, mentioned earlier.

The trachea and bronchi are reinforced by hyaline cartilage in their walls (▷ Figure 14–7, page 332). The C-shaped cartilage "rings" of the trachea prevent the organ from collapsing during breathing, ensuring a steady flow of air in and out of the lungs. Plates of cartilage in the walls of the bronchi have the same effect. You can feel the cartilage rings in the trachea by gently rubbing your trachea just below the larynx.

The smallest bronchi in the lungs branch to form **bronchioles,** which lead to the alveoli. Like the arterioles of the

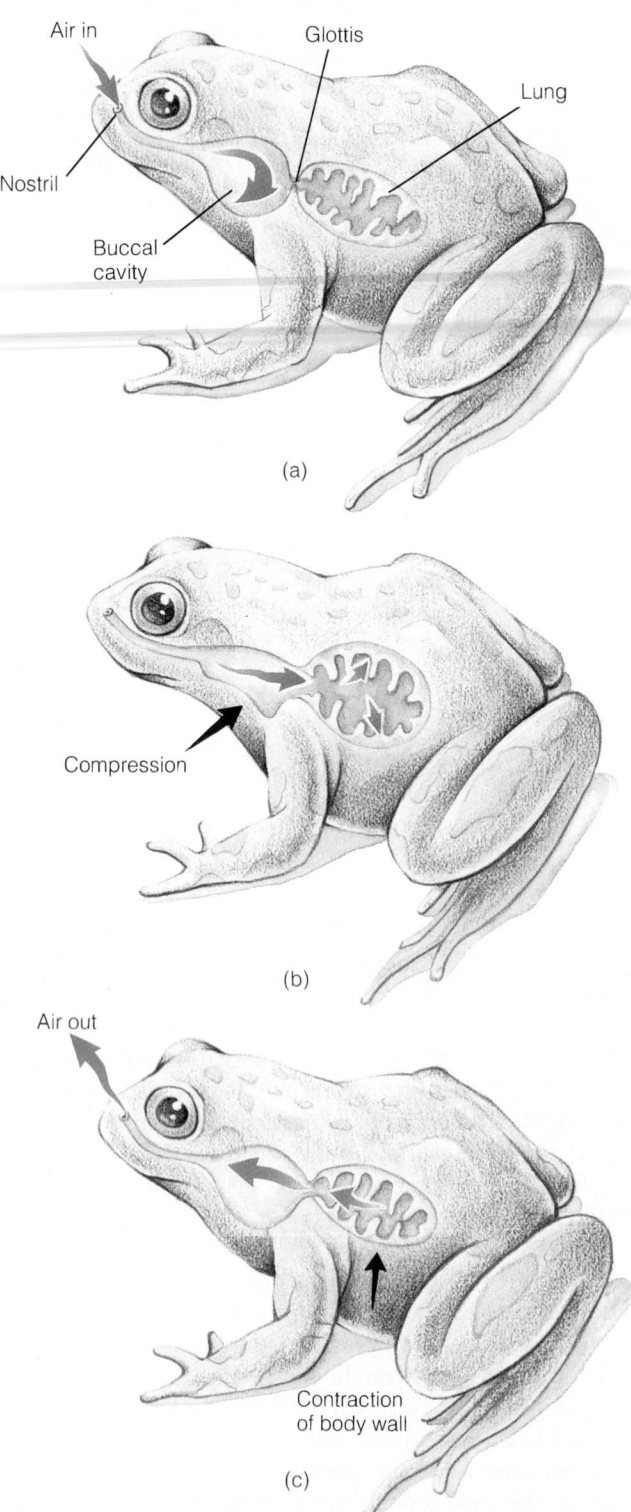

Air in
Glottis
Lung
Nostril
Buccal
cavity

(a)

Compression

(b)

Air out

Contraction
of body wall

(c)

▷ **FIGURE 14–3 The Lungs of the Frog** Frogs rely principally on their skin for gas exchange but also have lungs. As illustrated here, air is drawn into the mouth cavity, then pushed back into the lungs.

circulatory system, the walls of the bronchioles consist largely of smooth muscle. As a result, the bronchioles can open and close, providing a means of controlling air flow in

the lung. During exercise or during times of stress, for example, the smooth muscle in the bronchioles relaxes, opening the tubes and increasing the flow of air into the lungs. This action helps meet the body's need for more oxygen, in much the same way that the arterioles of capillary beds dilate to let more blood into body tissues in times of need.

The respiratory system is constantly in direct contact with the external environment and is therefore exposed to many infectious organisms as well as a number of potentially hazardous particles and gases. Not surprisingly, then, the respiratory system has evolved a number of mechanisms for protection. The conducting portion, for example, also filters many impurities from the air we breathe, especially airborne particles, such as dust and bacteria. Particles in the air exist in many sizes. Some are small and can penetrate deeply into the lung. Some of these particles may contain toxic metals, such as mercury and nickel, which can cause lung cancer. As a rule, larger particles are deposited as the inhaled air travels through the maze of passageways leading to the lungs. Many are also filtered in the nose. The convoluted interior of the nose slows the flow of air and causes the larger particles to drop out, in much the same way that sediment falls to the bottom in a slow-moving section of a stream. Hairs in the nasal cavity may also physically trap particles.

Particles removed from the air in the nose, trachea, and bronchi are trapped in a layer of **mucus,** a thick, slimy secretion deposited on the inside of much of the respiratory tract (▷ Figure 14–8, page 332). Mucus is produced by cells in the epithelium, called **mucous cells.**[1] The epithelium of the respiratory tract also contains numerous ciliated cells, an additional adaptation vital to keeping the respiratory system healthy. The cilia of these cells beat upward toward the mouth, slowly transporting mucus and its cargo of bacteria and dust particles. Operating day and night, they sweep the mucus toward the oral cavity, where it can be swallowed or expectorated (spit out). This process protects the respiratory tract and lungs from bacteria and potentially harmful particulates.

Like all homeostatic mechanisms, however, the respiratory mucous trap is not invulnerable. Bacteria and viruses do occasionally penetrate the lining, where they may proliferate, causing respiratory infections. Making matters worse, sulfur dioxide, a pollutant in cigarette smoke and urban air pollution, temporarily paralyzes, and may even destroy, cilia. Sulfur dioxide gas in the smoke of a single cigarette, for instance, can paralyze the cilia for an hour or more, permitting bacteria and toxic particulates to be deposited on the lining of the respiratory tract, and in some cases, even enter the lungs. Ironically, the cilia of a smoker are paralyzed when they are needed the most!

Because smoking impairs this natural protective mechanism, it should come as no surprise that smokers suffer

---

[1]You will notice that the adjective form is spelled *mucous* (as in mucous cell) and the noun form is spelled *mucus*.

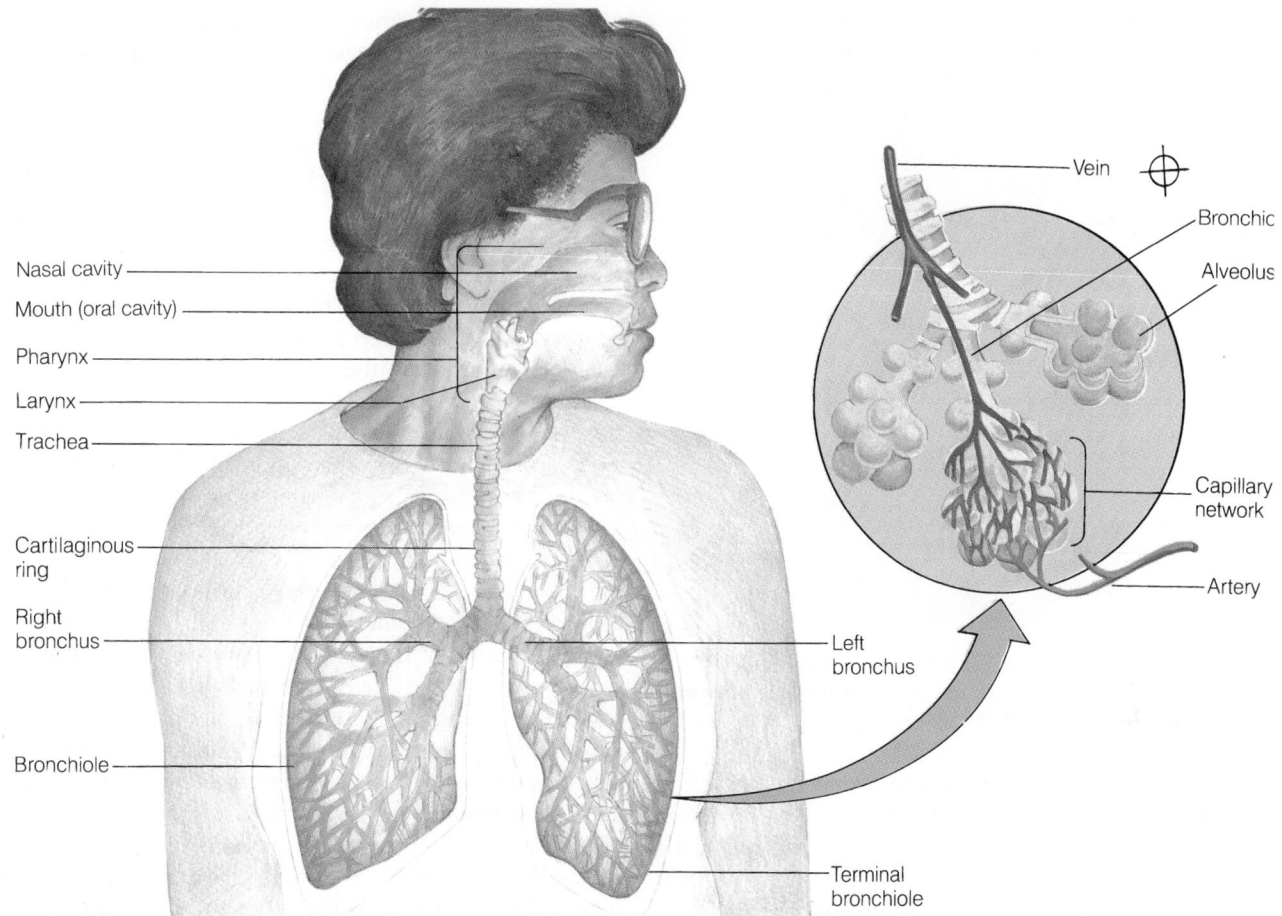

Nasal cavity
Mouth (oral cavity)
Pharynx
Larynx
Trachea

Cartilaginous ring
Right bronchus

Bronchiole

Vein
Bronchic
Alveolus

Capillary network

Artery

Left bronchus

Terminal bronchiole

▷ **FIGURE 14–4 The Human Respiratory System** A cutaway showing the air-conducting portion and the gas-exchange portion of the human respiratory system. The insert shows a higher magnification of the alveoli, where oxygen and carbon dioxide exchange occurs.

more frequent respiratory infections than nonsmokers. Interestingly, alcohol also paralyzes the cilia of the respiratory system. As a result, alcoholics are much more prone to certain types of respiratory infections.

The conducting portion of the respiratory system also moistens and warms the incoming air. Beneath the epithelium of the respiratory tract is a rich network of capillaries that releases moisture and heat. Moisture protects the lungs from drying out, and heat protects them from cold temperatures. Except in extremely cold weather, by the time inhaled air reaches the lungs, it is nearly saturated with water and is warmed to body temperature.

≋	TABLE14–1 Summary of the Respiratory System	≋

ORGAN	FUNCTION
Air conducting	
Nasal cavity	Filters, warms, and moistens air; also transports air to pharynx
Oral cavity	Transports air to pharynx; warms and moistens air; helps produce sounds
Pharynx	Transports air to larynx
Epiglottis	Covers the opening to the trachea during swallowing
Larynx	Produces sounds; transports air to trachea; helps filter incoming air; warms and moistens incoming air
Trachea and bronchi	Warm and moisten air; transport air to lungs; filter incoming air
Bronchioles	Control air flow in the lungs; transport air to alveoli
Gas exchange	
Alveoli	Provide area for exchange of oxygen and carbon dioxide

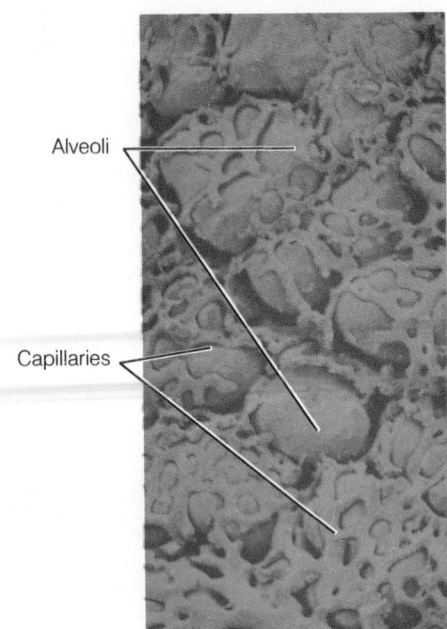

FIGURE 14–5 The Alveoli and Their Capillaries
A scanning electron micrograph of the alveoli of the lung, showing the rich capillary network surrounding them.

On the way out, much of the water that was added to the air condenses on the lining of the nasal cavity which was cooled by the evaporation of water as the air was being drawn in. This mechanism is an adaptation that conserves water and also accounts for the reason our noses tend to drip in cold weather.

The Alveoli are the Site of Gaseous Exchange

The air we breathe consists principally of nitrogen and oxygen with small amounts of carbon dioxide and other gases (Table 14–2). Oxygen in the atmosphere is generated by the photosynthetic activity of plants and photosynthetic protists and is, of course, vital to humans and virtually all other living organisms. A constant supply must be delivered to human body cells in order for us to maintain cellular energy production.

As noted earlier, the conducting portion of the respiratory system delivers warmed, moistened, filtered air that is rich in oxygen to the bronchioles, which, in turn, deliver the air to the alveoli. Each lung contains an estimated 150 million alveoli, giving the lung the appearance of an angel food cake (▷ Figure 14–9a and 14–9b). Each alveolus is surrounded by an extensive capillary bed. The alveoli provide a surface area for absorption of oxygen in the lungs measuring 60 to 80 square meters, an area approximately the size of a tennis court.

As shown in Figures 14–9c and ▷ 14–10a (page 334), the alveoli are lined by a single layer of flattened cells, called **Type I alveolar cells.** These cells permit gases to move in and out the alveoli with ease. Thus, the large surface area of the lungs created by the alveoli and the relatively thin barrier between the blood and the alveolar air combine to produce a rather rapid and efficient diffusion of gases across the alveolar wall. The alveoli provide another example of the marriage of form and function produced by the process of evolution.

Another important cell in the alveoli is the **alveolar macrophage,** sometimes known as the **dust cell.** Alveolar macrophages remove dust and other particulates that reach the lungs (Figure 14–10a). Dust cells wander freely around and through the alveoli, engulfing foreign material that has escaped filtration. Once filled with particulates, the macrophages accumulate in the connective tissue surrounding the

FIGURE 14–6 Uppermost Portion of the Respiratory System
Bony protrusions into the nasal cavity create turbulence that causes dust particles to settle out on the mucous coating. Notice that air passing from the pharynx enters the larynx. Food is kept from entering the respiratory system by the epiglottis, which covers the laryngeal opening during swallowing.

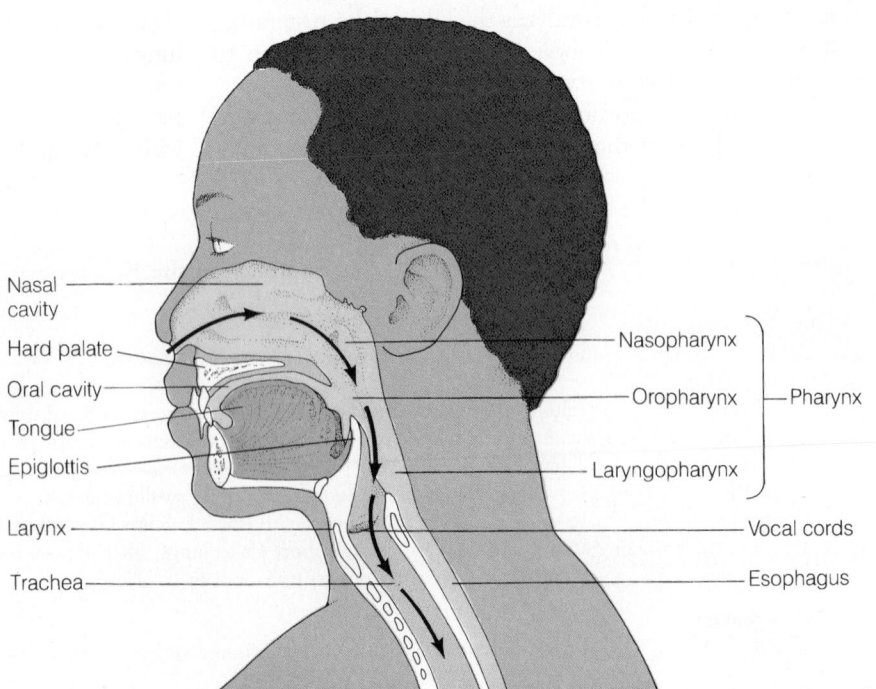

FIRST AID THAT MAY SAVE SOMEONE'S LIFE: THE HEIMLICH MANEUVER

During swallowing, food is normally excluded from the trachea by the epiglottis, as explained in the chapter. Each year, however, approximately 3000 Americans—on average, about 8 people a day—will choke to death on food that becomes lodged in the larynx or trachea. Many of these deaths could be prevented. Here's what should be done if you encounter a person who is choking on food. First, stand behind the victim, who may be either standing or sitting. Position yourself slightly to one side, as shown in ▷ Figure 1a. Place an arm across the victim's chest for support, and lean the person forward. With the heel of your other hand, give four hard thumps between the shoulder blades. This should clear the trachea. If it doesn't, don't give up. Try the Heimlich maneuver.

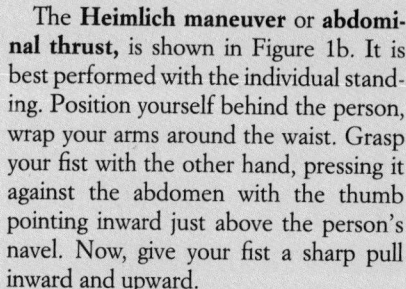

▷ **FIGURE 1 First Aid for Choking Victims** (*a*) As a first measure, stand behind and slightly to the side of the victim, and let him lean over one arm. Hit the victim's back between the shoulder blades with the heel of your hand. (*b*) If this does not dislodge the food, try the Heimlich maneuver. While standing behind the victim, wrap your arms around his waist. Press your fist against his waist with your thumb inward. Grasp your fist with your other hand, and pull sharply in and up. (*c*) A similar procedure can be performed on young children and babies in the positions shown here.

The **Heimlich maneuver** or **abdominal thrust,** is shown in Figure 1b. It is best performed with the individual standing. Position yourself behind the person, wrap your arms around the waist. Grasp your fist with the other hand, pressing it against the abdomen with the thumb pointing inward just above the person's navel. Now, give your fist a sharp pull inward and upward.

This should dislodge the food. If it doesn't, try it three more times. If that doesn't work, repeat the back blows. That should dislodge the food. If that doesn't work, call an ambulance.

Children are often victims of choking, and many parents make the mistake of trying to dislodge food by sticking their fingers down their child's throat to extract the food. Unfortunately, this may force the food deeper into the larynx or trachea. Figure 1c shows how to treat babies or young children who are choking on food or other objects. For a toddler, sit down and put the child across your knees with the head down. Give several thumps on the back between the shoulder blades with the heel of your hand (but softer than you would for an adult). Babies can be held face down with one arm supporting the body. Several light raps on the back should dislodge the food. Babies can also be held by the ankles while you rap on the back.

It is a good idea to practice these techniques on friends and family members, so that when the time comes you will be prepared. When you practice, however, be gentle, and be sure your arms are around the person's waist, not the chest.

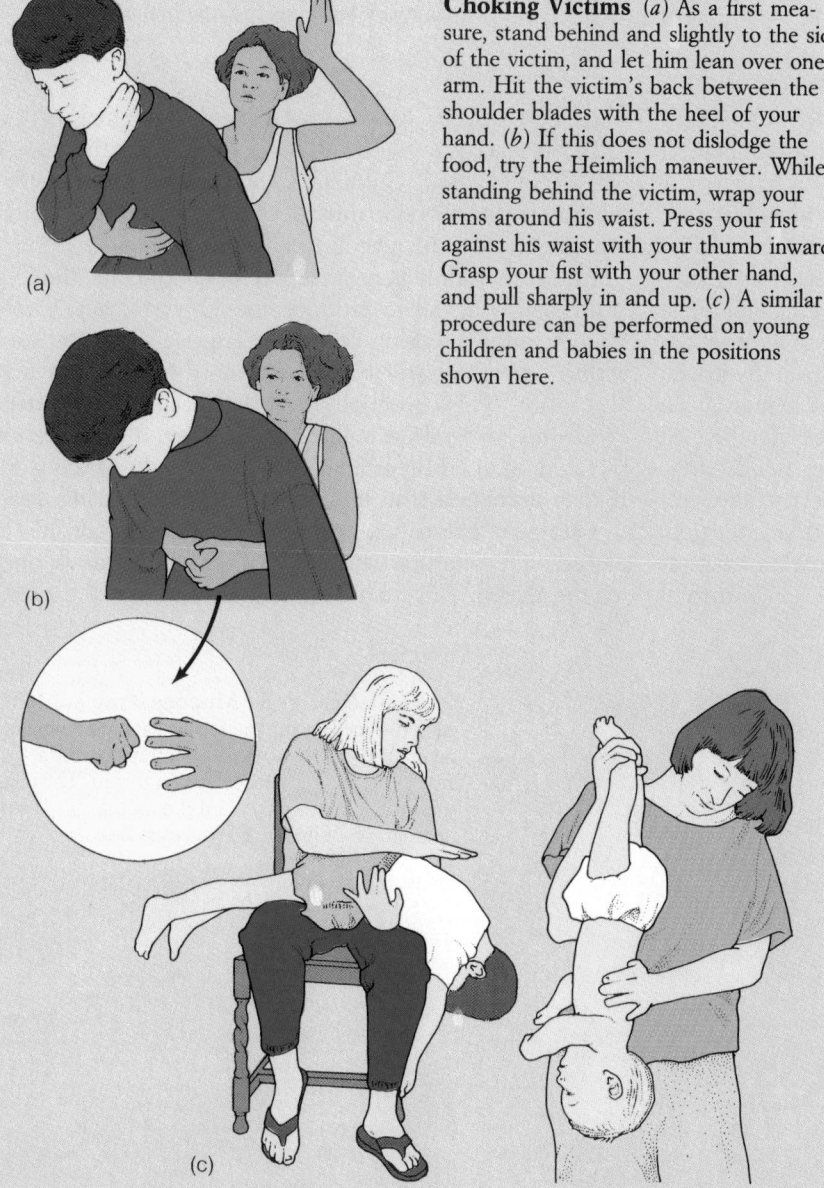

(a)

(b)

(c)

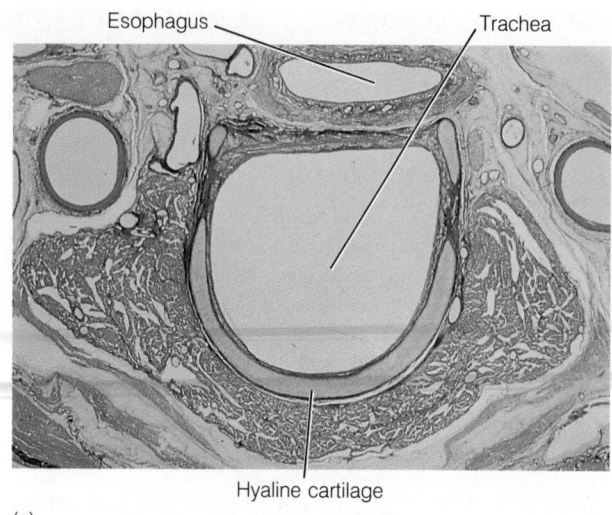

Esophagus   Trachea

Hyaline cartilage

(a)

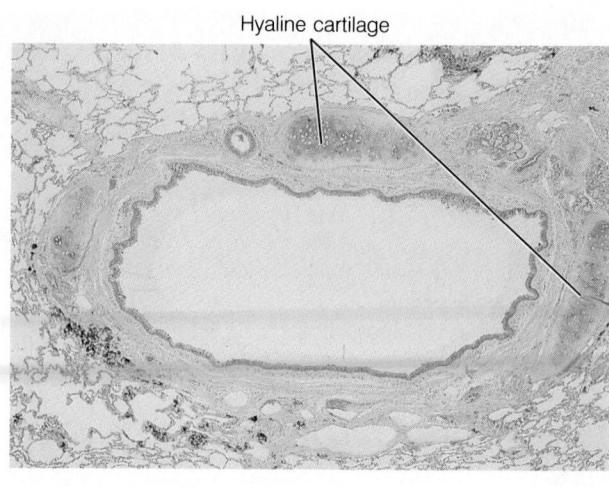

Hyaline cartilage

(b)

▷ **FIGURE 14–7 Cross Sections of the Trachea and Bronchus** (*a*) The trachea contains hyaline cartilage ribs,

C-shaped segments of cartilage that give the organ internal support. (*b*) An intrapulmonary bronchus showing hyaline cartilage.

alveoli. At death, a smoker's lungs or the lungs of an urban resident are therefore often blackened by the accumulation of smoke and dust particles.

Another important cell is the **Type II alveolar cell,** shown in Figure 14–10a. Type II alveolar cells are large, round cells that produce **surfactant,** a detergentlike substance that dissolves in the thin layer of water lining the alveoli. The water covering the alveolar lining produces surface tension. **Surface tension** results from hydrogen bonds that form between water molecules (Chapter 2). In water, hydrogen bonds draw water molecules together. At the surface of a watery fluid, the hydrogen bonds draw water molecules together more tightly than elsewhere, creating a slightly denser region referred to as surface tension. Surface tension on a pond permits some insects, such as water striders, to walk on water and is the reason a drop of

water beads up on your car windshield (▷ Figure 14–11).

In the alveoli, surface tension tends to draw the walls of the alveoli inward. Surfactant, however, reduces surface tension in the alveoli, thus decreasing forces that might otherwise cause the alveoli to collapse.

At birth, in some very small, premature infants, the Type II alveolar cells fail to produce enough surfactant. Consequently, surface tension causes the larger alveoli to collapse within a few hours of birth, reducing the surface area for absorption. This condition, known as **respiratory distress syndrome,** or **hyaline membrane disease,** is characterized by rapid, labored breathing, which may lead to exhaustion. If it is untreated, the lungs may collapse, causing death.

Currently, babies with respiratory distress syndrome are placed on respirators, which force air into their lungs, opening the alveoli. This treatment must continue until the lungs

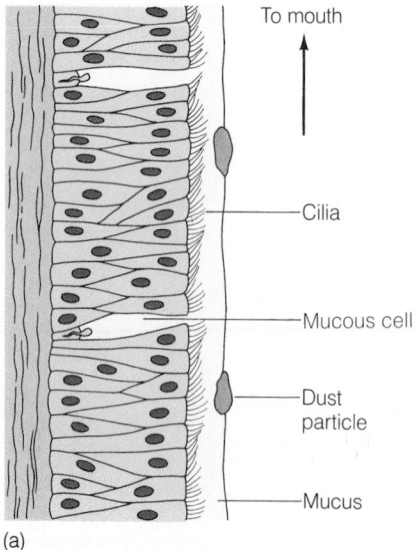

To mouth

Cilia

Mucous cell

Dust particle

Mucus

(a)

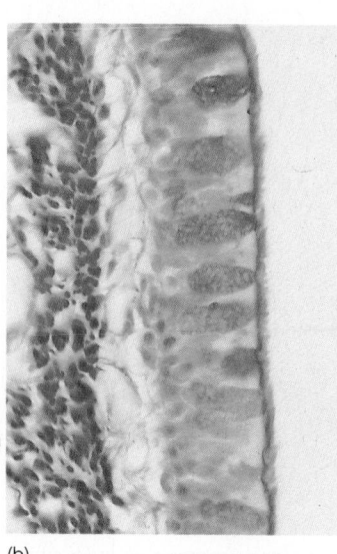

(b)

▷ **FIGURE 14–8 Mucous Trap** (*a*) Mucus produced by the mucous cells in the lining of much of the respiratory system traps bacteria, viruses, and other particulates in the air. The cilia transport the mucus toward the mouth. (*b*) Light micrograph of lining. (*b*) Mucous cell.

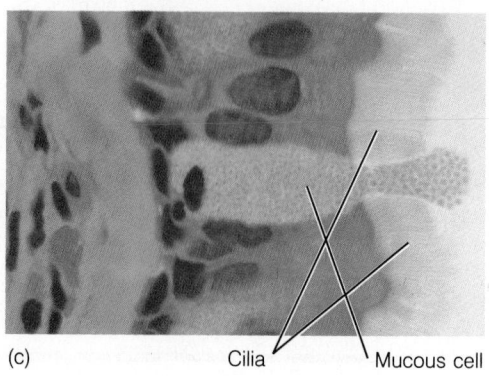

(c)   Cilia   Mucous cell

TABLE 14–2 Composition of Air	
GAS	PERCENTAGE COMPOSITION
Nitrogen (N_2)	78
Oxygen (O_2)	21
Argon (Ar)	0.9
Carbon dioxide (CO_2)	0.03
Water vapor (H_2O)	Variable (0–4)
Pollutants	Variable

begin to produce surfactant on their own. One experimental treatment that may help in the future involves the use of a surfactantlike substance. Mixed with the air the baby breathes, this substance penetrates the lung, where it dissolves in the water lining the alveoli. This reduces surface tension and prevents the alveoli from collapsing. Treatment continues until a baby's own surfactant-producing cells mature and begin producing adequate amounts of surfactant. Although fairly successful, this procedure has not been approved by the Food and Drug Administration because of

some unanswered ethical questions—notably, whether treating these babies, which often have multiple and often severe birth defects, is a prudent goal.

≈ FUNCTIONS OF THE RESPIRATORY SYSTEM

The chief functions of the respiratory system are to (1) replenish the blood's oxygen supply, and (2) rid the blood of excess carbon dioxide. The respiratory system serves other functions as well. The vocal cords, described below, produce sounds that allow people and other mammals to communicate. The respiratory system houses the **olfactory membrane,** a specialized patch of epithelium in the roof of the nasal cavity that allows humans and other vertebrates to perceive odors. The respiratory system helps maintain pH balance by its influence on carbon dioxide levels.

In Humans, Sound Is Produced by the Vocal Cords and Is Influenced by the Tongue and Oral Cavity

Phonation, the production of sounds, is critical to many members of the animal kingdom. The eerie cry of the coyote, for example, signals to the pack a member's whereabouts and helps the members of the pack stay in contact (▷ Figure 14–12). The coyote's growl may signal to an intruder its intention to defend itself.

Humans exhibit a wider range of sounds for communication than other animals. These sounds are produced by two **vocal cords,** elastic ligaments inside the larynx. The vocal cords vibrate as air is expelled from the lungs (▷ Figure 14–13, page 336). The sounds generated by the vocal cords are further modified by changing the position of the tongue and the shape of the oral cavity.

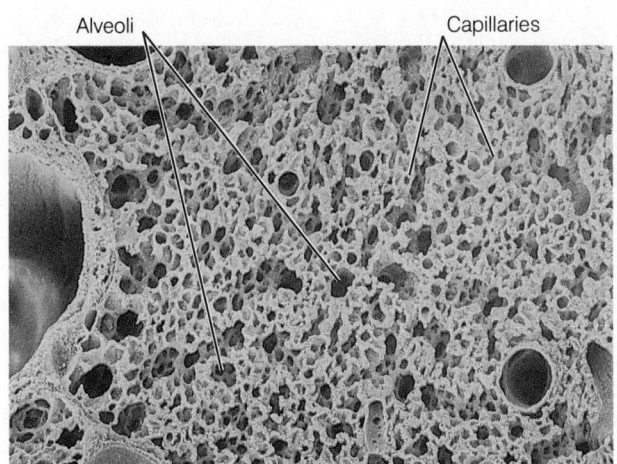

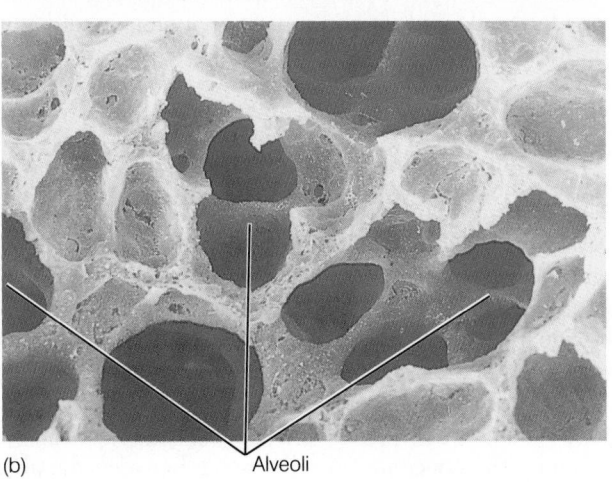

▷ **FIGURE 14–9 The Alveoli** (*a*) A scanning electron micrograph of the lung showing many alveoli. The smallest openings are capillaries in the alveolar walls. (*b*) A scanning electron micrograph of lung tissue showing alveoli. (*c*) An electron micrograph showing several alveoli and the close relationship of the capillaries.

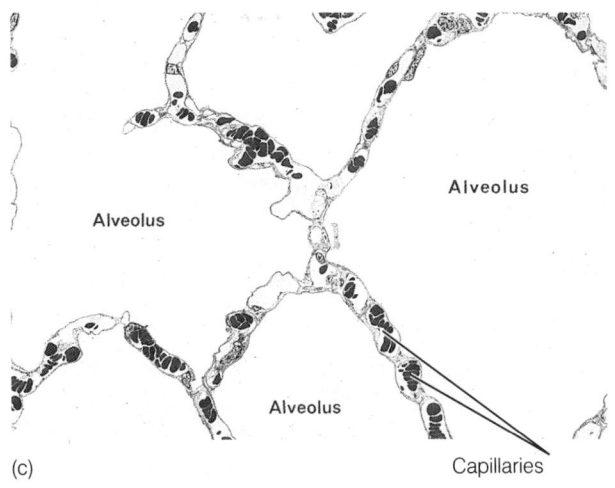

▷ **FIGURE 14–10 The Alveolar Macrophage, or Dust Cell** (*a*) Drawing of the alveolus showing Type I and Type II alveolar cells and dust cells. (*b*) An electron micrograph of a dust cell from the lung of a nonsmoker. (*c*) Compare this with a dust cell from the lung of a smoker, which is filled with carbon particles that have been engulfed by the cell.

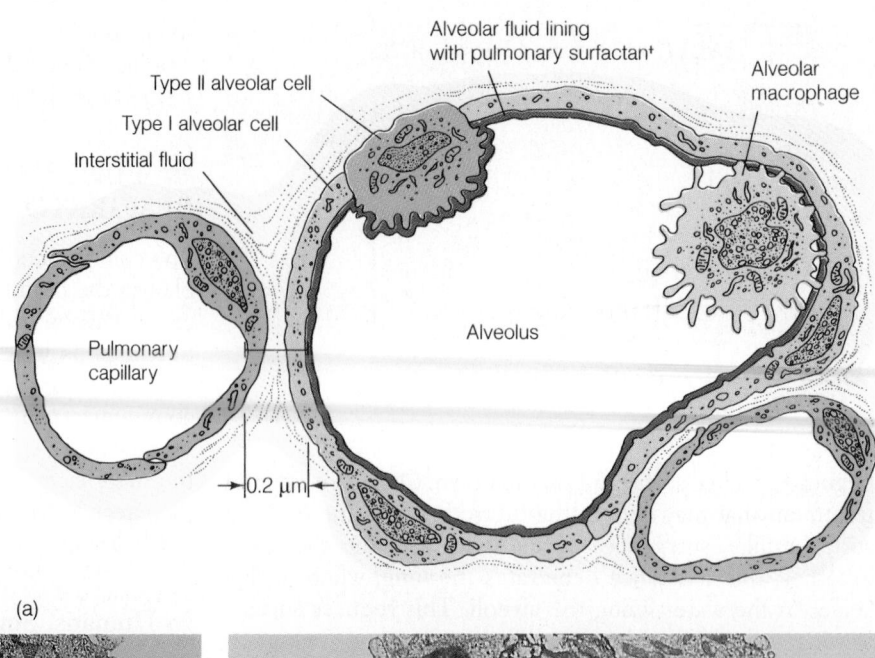

(a)

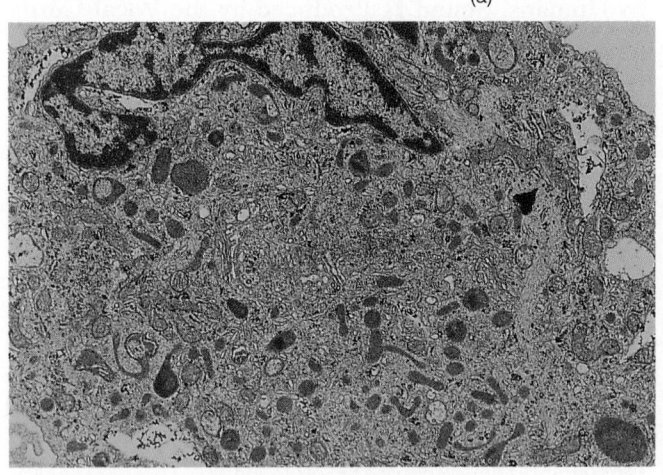

(b)

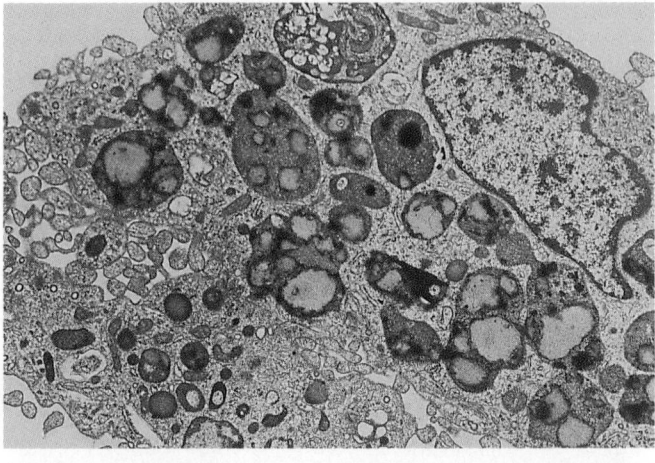

(c)

The vocal cords vary in length and thickness from one person to the next. They also vary between men and women. Most men, for example, have longer, thicker vocal cords than women and, therefore, have deeper voices. Testosterone, a male sex hormone produced by the testes, is responsible for the growth of the larynx and the characteristic length and thickness of the male vocal cords.

In some respects, the vocal cords are like the strings of a guitar or piano. That is, they can be tightened or loosened, producing sounds of different pitch. The tighter the string on a guitar, the higher the note. In humans, muscles in the larynx that attach to the vocal cords make this adjustment. Relaxing the muscles lowers the tension on the cords, dropping the tone. Tightening the vocal cords has the opposite effect.

Bacterial and viral infections of the larynx result in a condition known as laryngitis. **Laryngitis** is characterized by inflammation of the lining of the larynx and the vocal cords. Inflammation thickens the cords, causing a person's

voice to lower. Laryngitis may also be caused by irritation of the larynx from tobacco smoke, alcohol, excessive talking, shouting, coughing, or singing. In young children, inflammation results in a swelling of the lining that may impede the flow of air and impair breathing, resulting in a condition known as the croup.

Oxygen and Carbon Dioxide Move Rapidly Across the Alveolar and Capillary Walls, Flowing Down Concentration Gradients

Deoxygenated blood entering the lungs arrives from the right ventricle of the heart via the pulmonary arteries. This blood, you may recall from Chapter 12, laden with carbon dioxide picked up as the blood circulates through body tissues giving off oxygen. In the capillary beds of the lungs, carbon dioxide is released, and oxygen is added. These gases readily diffuse across the capillary and alveolar walls, driven by the concentration difference between the alveoli

(a)

(b)

▷ **FIGURE 14–11 Surface Tension** (*a*) Some insects can walk on water because of surface tension, the tight packing of water molecules along the surface of a pond. (*b*) Water forms droplets because hydrogen bonds hold the molecules together and because of surface tension along the droplet's surface.

and capillaries.[2] ▷ Figure 14–14 illustrates the direction in which carbon dioxide and oxygen flow.

Oxygen in the alveolar air first diffuses through the alveolar epithelium, then into the extracellular fluid surrounding the capillaries. It then diffuses through the capillary wall and into the blood plasma (▷ Figure 14–15). From here, oxygen molecules cross the plasma membrane of the red blood cells (RBCs) and bind to hemoglobin molecules in their cytoplasm. About 98% of the oxygen in the blood is carried in the RBCs bound to hemoglobin; the rest is dissolved in the plasma and the cytoplasm of the RBCs.

In order to understand the details of carbon dioxide diffusion in the lung, we must go back to the body cells, where CO_2 is formed. In body tissues, carbon dioxide

---

[2] Physiologists actually speak of differences in partial pressure. The partial pressure of a gas is caused by the collision of moving gas molecules with a surface. The partial pressure of oxygen is proportional to the impact of all the oxygen molecules striking the alveolar wall. Thus, the total pressure is directly proportional to the concentration of the gas molecules.

▷ **FIGURE 14–12 A Coyote Howls**

diffuses out of the cells and enters the blood plasma. Much of it then diffuses into the RBCs, where it is chemically converted to carbonic acid, H_2CO_3 (▷ Figure 14–16). This reaction is catalyzed by the enzyme **carbonic anhydrase** inside all RBCs. Carbonic acid molecules readily dissociate, however, forming bicarbonate ions and hydrogen ions. Many of the bicarbonate ions diffuse out of the RBCs into the plasma, where they are carried along with the blood. A small percentage (15% to 25%) of the carbon dioxide given off by body cells binds to hemoglobin (but not at the oxygen binding site), and an even smaller percentage (7%) dissolves in the plasma.

When blood rich in carbon dioxide reaches the lungs, bicarbonate ions in the plasma reenter the RBCs, where they combine with hydrogen ions to form carbonic acid. Carbonic acid, in turn, reforms carbon dioxide, as illustrated in ▷ Figure 14–17. Carbon dioxide then diffuses out of the RBC into the blood and finally into the alveoli down its concentration gradient. It is then expelled from the lungs during exhalation.

The uptake of oxygen and discharge of carbon dioxide occurring in the lungs "replenish" the blood in the alveolar capillaries. The oxygenated blood then flows back to the left atrium of the heart via the pulmonary veins. From here it empties into the left ventricle and is pumped to the body tissues via the aorta and its multitude of branches.

≈ DISEASES OF THE RESPIRATORY SYSTEM

Given that the air we breathe is laden with bacteria, viruses, and other potentially dangerous pathogens and that the respiratory organs are in direct contact with the outside environment, the respiratory system is one of the main routes for infectious agents to enter the body. Protected by adaptations such as the mucous layer, the cilia along much

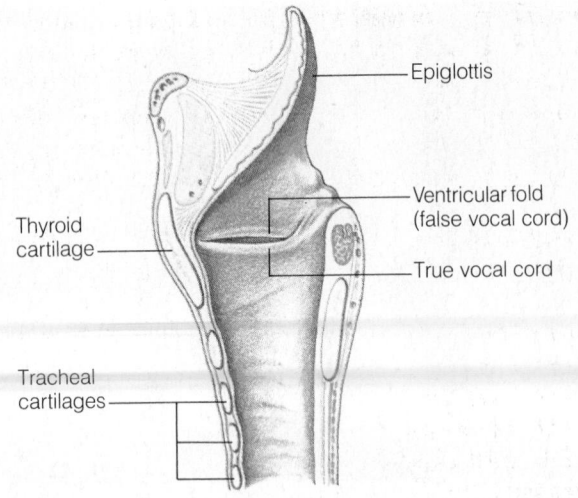

Epiglottis

Ventricular fold
(false vocal cord)

Thyroid
cartilage

True vocal cord

Tracheal
cartilages

(a)

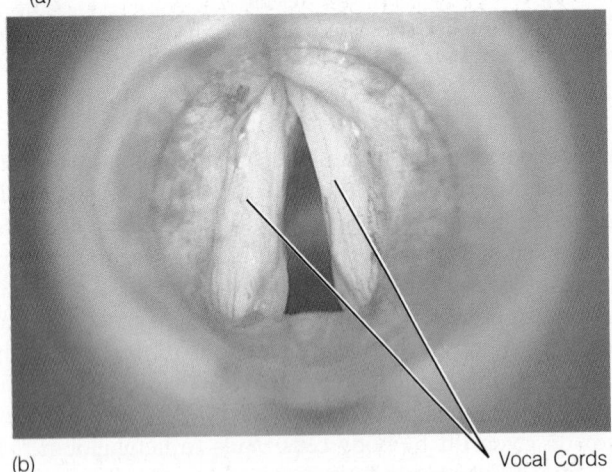

(b)

Vocal Cords

▷ **FIGURE 14–13 Vocal Cords** (*a*) Longitudinal section of
the larynx showing the location of the vocal cords. Note the
presence of two false vocal cords. They do not function in phonation. (*b*) View into the larynx of a patient also showing the vocal
cords.

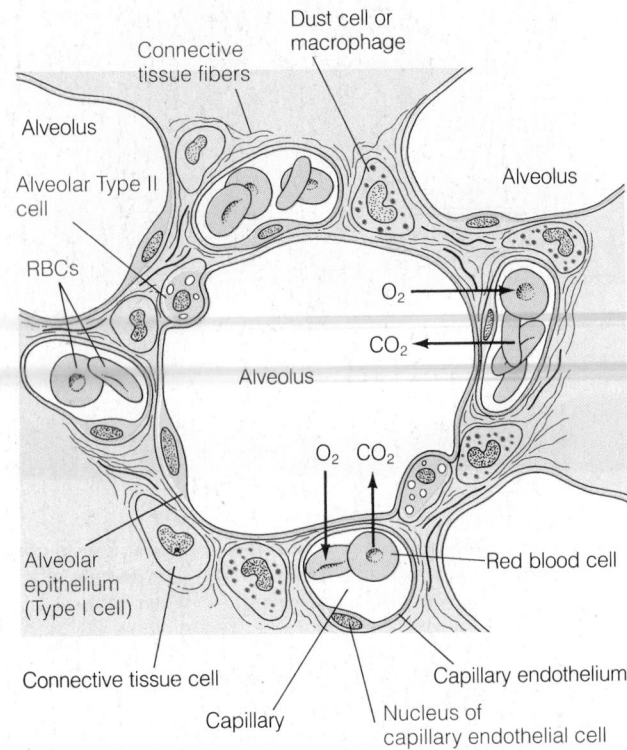

Connective
tissue fibers

Dust cell or
macrophage

Alveolus

Alveolar Type II
cell

Alveolus

RBCs

O_2

CO_2

Alveolus

O_2 CO_2

Alveolar
epithelium
(Type I cell)

Red blood cell

Connective tissue cell

Capillary endothelium

Capillary

Nucleus of
capillary endothelial cell

▷ **FIGURE 14–14 Close-Up of the Alveolus** Oxygen diffuses out of the alveolus into the capillary. Carbon dioxide diffuses
in the opposite direction, entering the alveolar air that is expelled
during exhalation.

▷ **FIGURE 14–15 Oxygen Diffusion** Oxygen travels from
the alveoli into the blood plasma, then into the RBCs, where much
of it binds to hemoglobin. When the oxygenated blood reaches
the tissues, oxygen is released from the RBCs and diffuses into the
plasma, then interstitial fluid and into body cells.

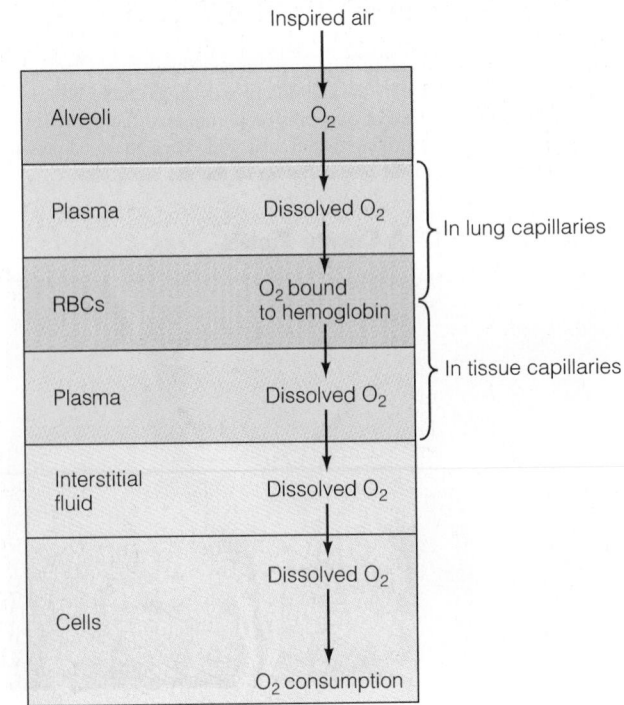

Inspired air

Alveoli — O_2

Plasma — Dissolved O_2

RBCs — O_2 bound
to hemoglobin

In lung capillaries

Plasma — Dissolved O_2

In tissue capillaries

Interstitial
fluid — Dissolved O_2

Dissolved O_2

Cells

O_2 consumption

of the epithelium, the phagocytic cells in the lung, and
others, the respiratory system nevertheless still falls victim
to bacterial and viral invasion. These infectious agents may
penetrate the epithelium, entering the underlying tissues
where they proliferate madly. Some viruses, such as the
influenza virus, multiply in the epithelial cells lining the
respiratory tract.

Bacterial and viral infections of the respiratory tract can
cause considerable discomfort, and some can be fatal. Infections may settle in many different locations in the respiratory system and are named by their site of residence. An
infection in the bronchi is therefore known as **bronchitis.**
An infection of the sinuses is known as **sinusitis.** (Bacterial
and viral infections are discussed in more detail in the next
chapter, but a few of the more common ones are listed in
Table 14–3, on page 340.)

Once inside the respiratory tract, bacteria, viruses, and
other microorganisms can spread to other organ systems. A
good example is **meningitis,** a bacterial or viral infection of

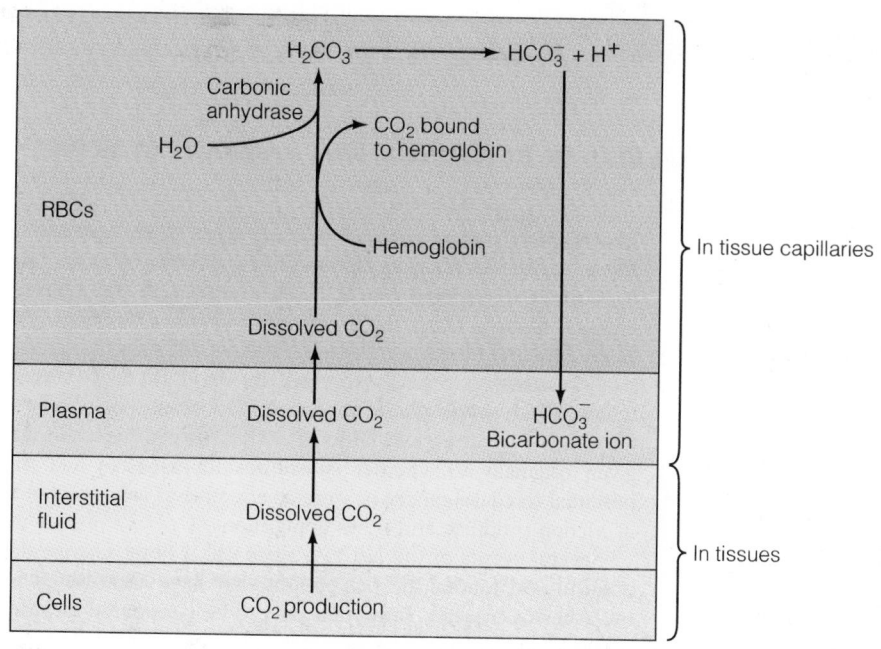

▷ **FIGURE 14–16 Bicarbonate Ion Production** Carbon dioxide (CO_2) diffuses out of body cells where it is produced and into the tissue fluid, then into the plasma. Although some carbon dioxide binds to hemoglobin and some is dissolved in the plasma, most is converted to carbonic acid (H_2CO_3) in the RBCs. Carbonic acid dissociates and forms hydrogen ions and bicarbonate, most of which is transported in the plasma.

the meninges, the fibrous layers surrounding the brain and spinal cord. This potentially fatal disease usually starts out as an infection of the sinuses or the lungs.

The lungs are also susceptible to airborne materials, among them asbestos fibers, which can cause two types of lung cancer and a debilitating disease known as **asbestosis**. Asbestosis is a build up of scar tissue that reduces the lung capacity. Because it is believed to be dangerous, many uses of asbestos have been banned in the United States, and asbestos used for insulation and decoration is being removed from buildings. For a discussion of the controversy over removal, see the Point/Counterpoint in this chapter.

Another common disease of the respiratory system worth noting is asthma. Unlike sinusitis, colds, and other respiratory diseases, **asthma** is a chronic disorder—a disease that persists for many years. Marked by periodic episodes of wheezing and difficult breathing, asthma is not an infectious disease. Most cases of asthma, in fact, are caused by allergic reactions, or abnormal immune reactions, to common stimulants such as dust, pollen, and skin cells (dander) from pets. In some individuals, certain foods, such as eggs, milk, chocolate, and food preservatives, trigger asthma attacks. Still other cases are triggered by drugs, vigorous exercise, and physiological stress.

In asthmatics, irritants, such as pollen and dander, cause a rapid increase in the production of mucus by the bronchi and bronchioles, making breathing considerably more difficult. Irritants also stimulate the contraction of the smooth muscle cells in the walls of the bronchioles, constricting the

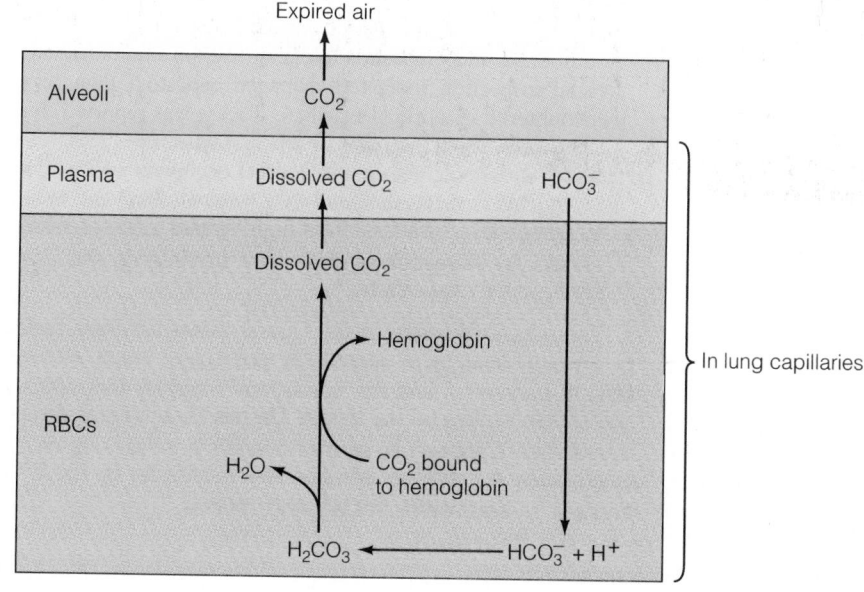

▷ **FIGURE 14–17 Carbon Dioxide Production from Bicarbonate** When the carbon-dioxide-laden blood reaches the lungs, bicarbonate ions combine with hydrogen ions in RBCs form carbonic acid, which dissociates, forming carbon dioxide gas. CO_2 diffuses out of the RBCs into the plasma, then into the alveoli.

MAINTENANCE OF ASBESTOS-CONTAINING MATERIAL IN BUILDINGS, NOT RAMPANT REMOVAL

Brooke T. Mossman

Brooke T. Mossman, Ph.D., is an associate professor of pathology at the University of Vermont College of Medicine. Her research, which is supported by the National Institutes of Health and the Environmental Protection Agency, concerns lung disease caused by asbestos fibers and implications for public policy on asbestos.

Despite the fact that scientific panels have concluded that the health risks of environmental (as opposed to worker) exposures to asbestos are low, the United States has been caught up in an asbestos removal fiasco, with projected expenditures of as much as $150 billion in the next 10 years. There is evidence that poorly performed asbestos removal may even increase air levels of fibers for extended periods and may increase the risk of disease in untrained individuals removing asbestos. The expenses incurred in removing asbestos over the past several years are approximately equivalent to the overall annual budgets of the National Institutes of Health. Thus, we are spending funds on asbestos removal desperately needed to remedy other significant health problems such as the AIDS epidemic, smoking, and drugs and alcohol.

Asbestos refers to a diverse group of naturally occurring fibers that are ubiquitous in the environment and mined for their use in insulation, friction materials, floor and ceiling tiles, and the like. Classic studies in the 1960s established that "blue" (crocidolite) asbestos, mined in parts of Africa and western Australia, caused a unique tumor called "mesothelioma" in asbestos miners and their family members. U.S. studies on insulation workers exposed to both amosite ("brown") asbestos and chrysotile ("white") asbestos and on shipyard workers (exposed largely to blue asbestos) demonstrated increases in numbers of mesotheliomas and lung cancers. Lung cancers were almost exclusively reported in heavy smokers, but occupational exposure to asbestos increased the risks of tumors in a multiplicative fashion.

Although a report by the National Academy of Sciences in 1984 suggested that environmental health risks of asbestos were low in comparison with cancer risks in workers, Congress in 1986 passed the Asbestos Hazard Emergency Response Act (AHERA), calling for inspection of asbestos-containing material (ACM) in private and public schools. If ACM is found, schools are required to notify parents and develop management plans for its control. The passage of AHERA, misconceptions about environmental cancer risks due to asbestos, and public panic encouraged the rampant removal of asbestos from schools and other public buildings. Costs of removal decimated school systems and engendered widespread litigation. More importantly, many poorly performed removals resulted in higher airborne concentrations of fibers than those observed before removal.

Inadequately trained removal workers were often exposed to levels of asbestos fibers as high as those encountered in the past by workers developing cancer. Concurrently with the asbestos removal fiascoes occurring in schools and other buildings, the Environmental Protection Agency (EPA) in 1989 also initiated a ban on asbestos, recently reversed by the Fifth U.S. Circuit Court, which would phase out various industrial uses of asbestos over a 10-year period. Unfortunately, little consideration was given originally to whether substitutes for asbestos had the potential to cause cancer or were as effective as asbestos fibers in friction products and other applications.

Several events of the last two years will, I hope, change the political and journalistic perceptions that have contributed to the asbestos hysteria. Important data to be considered include:

1. Types of asbestos differ both chemically and physically. The types of asbestos (crocidolite, amosite) associated in the workplace with the development of mesothelioma are not the types most often encountered in buildings. Ninety percent to 95% of the bulk ACM in buildings is chrysotile (white) asbestos.
2. Asbestos fibers must be airborne and of a certain respirable size range to cause cancers in humans. Longer fibers (>5 microns), which are associated with the development of tumors in animals, are rarely found in building air samples.
3. In many situations, asbestos fibers are in a matrix or behind walls and are thus inaccessible. Unnecessary removal causes disruption of surfaces and may result in increased levels of airborne fibers and increased exposures to building occupants and removal workers.
4. The incidence of mesothelioma, a tumor associated with occupational exposure to asbestos, in U.S. women and men under 65 has been stable over the past several years, indicating that environmental exposure to asbestos is not resulting in increases in this disease.
5. Air monitoring of asbestos fibers in buildings, homes, and schools has indicated that levels are comparable to levels in outdoor air and thousands-fold less (even in situations where visual inspection indicated damaged asbestos) than levels encountered in workplaces where disease was reported. Risk estimations show that risks of asbestos-associated premature deaths in schoolchildren and general occupants of buildings are much less than calculated risks from smoking and radon.
6. A compendium of information suggests that asbestos-related cancers are dosage-dependent, thus invalidating the "one-fiber-can-kill" hypothesis.

The asbestos hysteria in the United States has been fueled by asbestos abatement companies, politicians, union activists, plaintiff lawyers . . . and the 'courtroom' scientists they employ, and sensationalism by the media. Despite these counterforces, sound science appears to be instrumental in influencing recent environmental policy on asbestos and statements by the EPA urging a "management, not removal" policy.

APPROPRIATE BALANCE BETWEEN REMOVAL AND IN-PLACE MANAGEMENT IS NEEDED

John M. Dement, Ph.D.

John M. Dement, Ph.D., is Director of Disease Prevention Research at the National Institute of Environmental Health Sciences. His research includes epidemiological studies of human populations exposed to asbestos and other fibers.

Extensive use of asbestos in building products in the United States has resulted in millions of tons of "in-place" material, much of which is deteriorating. The magnitude of asbestos exposure and the resulting risks to building occupants have been the subject of considerable debate. Although many would agree that some asbestos is being removed from schools and buildings that would pose less human health risk if properly managed in place, some authors go much further to imply that asbestos removal is *seldom* warranted and that "chrysotile asbestos, the type of fiber found predominantly in U.S. schools and buildings, is not a health risk in the nonoccupational environment." This conclusion is not supported by published literature concerning the carcinogenicity of chrysotile asbestos. Furthermore, most risk estimates for building occupants fail to consider exposure levels of airborne asbestos and exposure in buildings containing friable asbestos as a result of normal cleaning and maintenance operations.

The term *asbestos* refers to several fibrous mineral species and includes both serpentines (chrysotile) and amphiboles (amosite, crocidolite, tremolite, anthophyllite). Although these minerals have different crystal structures and chemical compositions, all produce respirable fibers that are capable of penetrating deeply into the lungs. Epidemiological studies of human populations exposed to asbestos have clearly demonstrated excess risks for a number of diseases, including asbestosis, lung cancer, mesothelioma, gastrointestinal cancers, and cancers at several other sites. Extensive reviews of the epidemiological literature clearly demonstrate the ability of both chrysotile and amphiboles to cause all of these diseases. The lung cancer exposure-response slope has been shown to vary widely; however, the steepest exposure-response relationship for lung cancer in any asbestos-exposed study has been observed among workers processing chrysotile. Furthermore, animal bioassay data clearly establish the carcinogenicity of chrysotile. The ability of chrysotile to produce asbestosis, lung tumors, and mesotheliomas in rats by inhalation has been amply demonstrated.

Although the incidence of all asbestos-related diseases increases with increasing dose, mesotheliomas have been observed among individuals with very brief periods of exposure. A recent study of mesothelioma cases in Wisconsin identified nine cases where the only identifiable source of asbestos exposure was living or working in a building containing asbestos. These data are consistent with a detailed case report of mesothelioma due to building materials containing amosite asbestos.

Assessing airborne exposures in buildings for purposes of risk assessment is problematic due to the variable nature of asbestos fiber release. Although *average* exposures are very low, *peak* exposures can occur during normal building maintenance and cleaning. Significantly elevated exposure levels have been demonstrated during routine cleaning operations such as sweeping and dusting in asbestos-contaminated buildings. Elevated levels of airborne fibers have been clearly demonstrated in schools containing damaged, friable asbestos. The significance of these peak exposures becomes more clear upon review of recent data that demonstrate asbestos-related diseases among school maintenance and custodial personnel. In these studies, the prevalence of asbestos-related changes in chest X-rays ranged from 11% to 27% for workers whose only known asbestos exposure had occurred in schools.

The public health implications for asbestos-containing materials in place are not insignificant. Risk assessments must consider risks to cleaning and maintenance personnel as well as to building occupants where damaged materials create elevated exposure for general occupants. Although the individual risk to building occupants may be small, the impact on the overall population is very significant given the number of people exposed. EPA regulations promulgated under the Asbestos Hazard Emergency Response Act (AHERA) *do not* require that asbestos be removed but require assessments and management plans. The solution is appropriate triage to ensure that the least-risk alternative is chosen. When asbestos is removed, it must be done using established methods that will minimize exposures to building occupants and removal workers. Sweeping conclusions concerning the lack of a health risk for chrysotile asbestos in the nonoccupational environment are not supported by the available literature and may lead to inappropriate asbestos-control strategies.

⮑ SHARPENING YOUR CRITICAL THINKING SKILLS

1. Summarize the key points and supporting arguments of each author.
2. Do you see any errors in the reasoning of either author? If so, where? What critical thinking rules were violated?
3. Given the extremely technical nature of this debate, can you decide whether asbestos in public schools should be removed?

TABLE14-3 Common Respiratory Diseases

DISEASE	SYMPTOMS	CAUSE	TREATMENT
Emphysema	Breakdown of alveoli; shortness of breath	Smoking and air pollution	Administer oxygen to relieve symptoms; quit smoking; avoid polluted air. No known cure.
Chronic bronchitis	Coughing, shortness of breath	Smoking and air pollution	Quit smoking; move out of polluted area; if possible, move to warmer, drier climate.
Acute bronchitis	Inflammation of the bronchi; yellowy mucus coughed up; shortness of breath	Many viruses and bacteria	If bacterial, take antibiotics, cough medicine; use vaporizer.
Sinusitis	Inflammation of the sinuses; mucus discharge; blockage of nasal passageways; headache	Many viruses and bacteria	If bacterial, take antibiotics and decongestant tablets; use vaporizer.
Laryngitis	Inflammation of larynx and vocal cords; sore throat; hoarseness; mucus buildup and cough	Many viruses and bacteria	If bacterial, take antibiotics, cough medicine; avoid irritants, like smoke; avoid talking.
Pneumonia	Inflammation of the lungs ranging from mild to severe; cough and fever; shortness of breath at rest; chills; sweating; chest pains; blood in mucus	Bacteria, viruses, or inhalation of irritating gases	Consult physician immediately; go to bed; take antibiotics, cough medicine; stay warm.
Asthma	Constriction of bronchioles; mucus buildup in bronchioles; periodic wheezing; difficulty breathing	Allergy to pollen, some foods, food additives; dandruff from dogs and cats; exercise	Use inhalants to open passageways; avoid irritants.

openings and making it even more difficult for asthmatics to move air in and out of their lungs.

Asthma is fairly common in school children but often disappears as they grow older. As a result, about 2% of the adult population suffers from asthma. Nevertheless, asthma is a serious disease. Periodic attacks can be quite disabling; some may even lead to death. By one estimate, several thousand Americans die each year from severe asthma attacks. Victims are generally elderly individuals who are suffering from other diseases.

Although asthma is incurable, the severity of an attack can be greatly lessened by proper medical treatment. One of the most common treatments is an oral spray (inhalant) containing the hormone epinephrine, which stimulates the bronchioles to open (▷ Figure 14–18). Screening tests can help a patient find out what substances trigger an asthmatic attack so they can be avoided.

≈ BREATHING AND THE CONTROL OF RESPIRATION

Air moves in and out of the lungs in much the same way that it moves in and out of the bellows that blacksmiths use to fan their fires. Breathing, however, is largely an involuntary action, controlled by the nervous system.

Air Is Moved In and Out of the Lungs by Changes in the Intrapulmonary Pressure

During breathing, air must first be drawn into the lungs. This process is known as **inspiration,** or **inhalation** (Table 14–4). Following inspiration, air must be expelled. This is known as **expiration,** or **exhalation.** For a discussion of an

▷ **FIGURE 14–18 Asthma Relief** Constriction of the bronchioles can be released by epinephrine inhalant spray.

interesting evolutionary variation on this theme, see Spotlight on Evolution 14-1.

Inhalation is an active process that is controlled by the brain. Nerve impulses traveling from the brain stimulate the **diaphragm,** a dome-shaped muscle (unique to mammals) that separates the abdominal and thoracic cavities (▷ Figure 14–19a). These impulses cause the diaphragm to contract. When it contracts, the diaphragm flattens and lowers. Much in the same way that pulling the plunger of a syringe draws in air, the contraction of the diaphragm draws air into the lungs.

Inhalation also requires the **intercostal muscles,** short, powerful muscles that lie between the ribs. (They are the meat on barbecued ribs.) Nerve impulses traveling to these muscles cause them to contract as the diaphragm is lowered. When the intercostal muscles contract, the rib cage lifts up and out (Figure 14–19a). Together, the contractions of the intercostal muscles and the diaphragm increase the volume of the thoracic cavity (Figure 14–19b). This, in turn, decreases the **intrapulmonary pressure,** the pressure in the alveoli, which draws air in through the mouth or nose downward into the trachea, bronchi, and lungs. At rest, each breath delivers about 500 milliliters of air to the lung. This is known as the **tidal volume,** the amount of air inhaled or exhaled with each breath when a person is at rest.

As you have just seen, inhalation is an active process that requires muscle contraction to occur. In a person at rest, exhalation is a decidedly passive process (not requiring muscle contraction) beginning after the lungs fill with air. At this point in the cycle, the diaphragm and intercostal muscles relax. The relaxed diaphragm rises and resumes its domed shape, and the chest wall falls slightly inward. These changes reduce the volume of the thoracic cavity, raising the pressure and forcing the air out. The lungs also participate in passive exhalation. Containing numerous elastic connective tissue fibers, they fill like balloons during inspiration. When inhalation ceases, the lungs simply recoil (shrink), forcing air out.

Although exhalation is a passive process in an individual at rest, it can be made active by enlisting the aid of the muscles of the wall of the chest and abdomen. The forceful expulsion of air is known as **forced exhalation.** Try this by taking in a breath of air, then actively forcing it out. During forced exhalation, contraction of the abdominal muscles increases the intra-abdominal pressure, forcing the abdominal organs upward against the diaphragm. Contraction of the muscles in the wall of the chest also reduces the volume of the chest, helping force the air out of your lungs.

Inhalation can also be consciously augmented by a forceful contraction of the muscles of inspiration. (You can test this by taking a deep breath.) Forced inhalation increases the amount of air entering your lungs. Athletes often actively inhale and exhale just before an event to increase oxygen levels in their blood. A competitive swimmer, for example, may take several deep breaths before diving into the pool for a race. Deep breathing, while effective, can be dangerous, for reasons explained shortly.

TABLE14–4 Summary of Inhalation and Exhalation

Inhalation

Nerve impulses from the breathing center stimulate the muscles of inspiration—the diaphragm and intercostal muscles.

Contraction of the intercostal muscles causes the rib cage to move up and out.

Contraction of the diaphragm causes it to flatten.

Volume of the thoracic cavity increases.

Intrapulmonary pressure decreases.

Air flows into the lungs through the nose and mouth.

Exhalation

Nerve impulses from the breathing center feed back on it, shutting off stimuli to muscles of inspiration.

The intercostal muscles relax and the rib cage falls.

The diaphragm relaxes and rises.

The lungs recoil.

Air is pushed out of the lungs.

The Health of a Person's Lungs Can Be Assessed by Measuring Air Flow In and Out of Them Under Various Conditions

Studies show that children who are exposed to tobacco smoke at home suffer a decrease in lung capacity—that is, a decrease in their ability to move air in and out of their lungs. Several measurements of lung capacity are routinely used by physicians to determine the "health" of a person's lungs.

Measurements of lung function are taken under controlled conditions. As shown in ▷ Figure 14–20a (page 344), patients breathe into a machine that measures the amount of air moving in and out of the lung at various times.

Figure 14–20b shows a graph of some of the common measurements. The first is the tidal volume, which, as noted earlier, is the amount of air that moves in and out during passive breathing. Another important measurement is the **inspiratory reserve volume**—the amount of air that can be drawn into the lungs during active inspiration. Deep inhalation pulls in four to six times more air than the tidal volume, or 2000 to 3000 milliliters, depending on the size of the individual.

After exhalation, under resting conditions, the lungs still contain a considerable amount of air—about 2400 milliliters. Forced exhalation will expel about half of that air. The amount that can be exhaled after a normal exhalation is called the **expiratory reserve volume.** The remaining 1200 milliliters is known as the **residual volume.**

Lung disease often results in changes in the amount of air that can be moved in and out of the lung or changes in the residual volume. Asthma, for example, reduces the inspiratory reserve volume—the amount of air that can be drawn into the lungs by forced inspiration—because the constricted bronchioles limit air flow.

AVIAN ADAPTATIONS TO HIGH-ALTITUDE LIVING

Radar observers scan the night skies monitoring airline traffic. Time and again they track flocks of birds flying as high as 6500 meters (21,000 feet) above sea level, where oxygen is low and the air bitterly cold. But birds are capable of surviving at even higher altitudes. One species, in fact, is routinely found on the slopes of Mount Everest, 8200 meters (27,000 feet) above sea level. Even more remarkably, bar-headed geese pass directly over the summit of the Himalayas at altitudes of 9200 meters (30,000 feet) during the annual migration. How do birds manage to breathe at such altitudes?

The answer lies in the bird's unique respiratory system, unmatched among vertebrates. Shown in ▷ Figure 1a, the respiratory system of the bird consists of a trachea, eight or nine supplemental air sacs, and two lungs. The air sacs hold nine times more air than the bird's lungs, but are poorly vascularized and therefore are not involved in gas exchange.

Unlike the human respiratory system, in which air flows into the lungs, then back out, retracing its steps like the tides, the bird's respiratory system is characterized by a unidirectional air flow, an evolutionary adaptation that greatly increases efficiency. In order for air to pass through the respiratory system of the bird, however, two cycles of inhalation and exhalation are required. During the first inhalation, the volume of the thoracic cavity increases, drawing air into the trachea. As illustrated in Figure 1b, however, most of the air bypasses the lungs and enters the posterior air sacs. During exhalation, the volume of the thoracic cavity decreases, and the air in the posterior air sacs passes into the lungs.

During the next inhalation, while new air is being drawn into the posterior air sacs, the air in the lungs is pushed into the anterior air sacs. During the second exhalation, the air in the anterior air sacs is expelled, while the air in the posterior air sacs is pushed into the lungs. As a result, the lung always contains a relatively fresh supply of air.

A peek inside the bird's lungs reveals another interesting adaptation that improves the efficiency of respiration. Unlike mammalian lungs, which contain a series of branching tubules that end in dead-end sacs (alveoli) where oxygen exchange occurs, bird lungs contain channels through which air flows continuously in one direction. Tiny blind-ended tubules branch off from these channels, coming in intimate contact with blood capillaries that draw off the oxygen and dump carbon dioxide. In the bird lung, moreover, air and blood flow in opposite directions, which is more efficient than the mammalian system. Combined with the one-way flow of air, this highly efficient system helps birds acquire the oxygen they need at very high altitudes.

Besides providing birds an efficient means of acquiring oxygen and ridding the body of carbon dioxide, this unique arrangement of lungs and air sacs helps lighten the bird's body, reducing energy use during flight. Hollow, air-filled bones found in many species may also assist the bird in flight. Ornithologists (biologists who study birds) believe that the air sacs probably also serve as a repository for heat generated by metabolism in the bird's flight muscles, thus helping the bird stabilize its body temperature and maintain homeostasis.

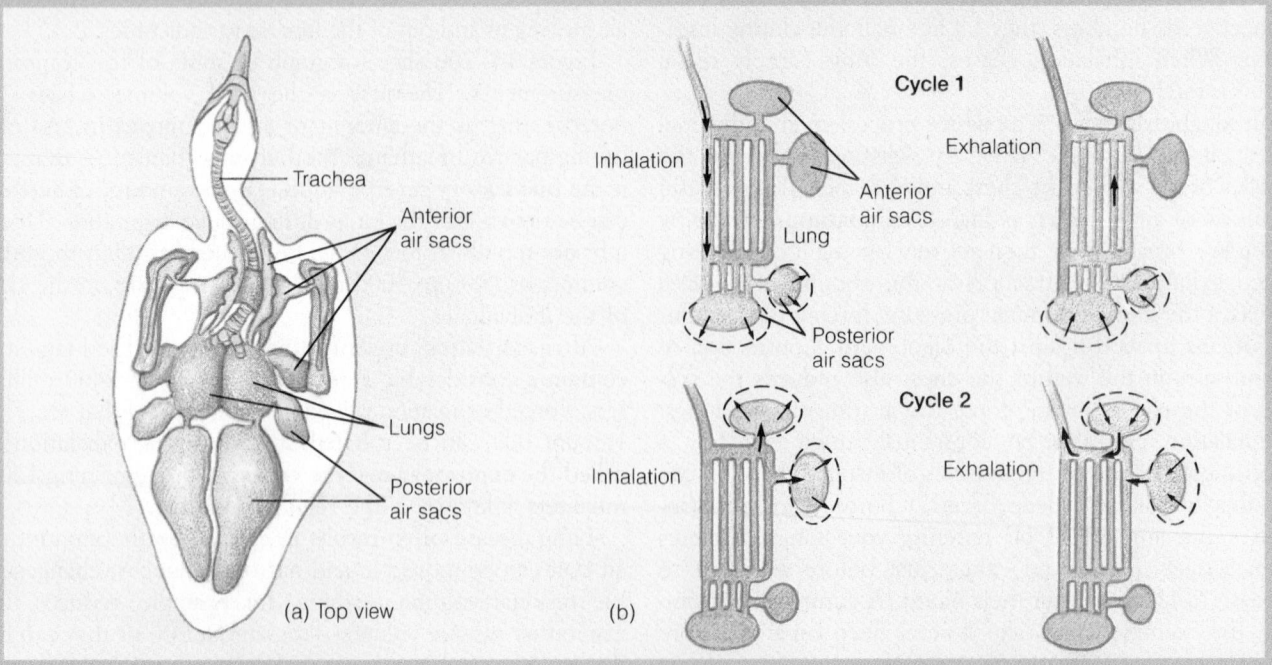

▷ **FIGURE 1 Respiratory System of the Bird** (*a*) Note the air sacs on both sides of the lungs. (*b*) Schematic illustrating flow of air in the respiratory system of the bird.

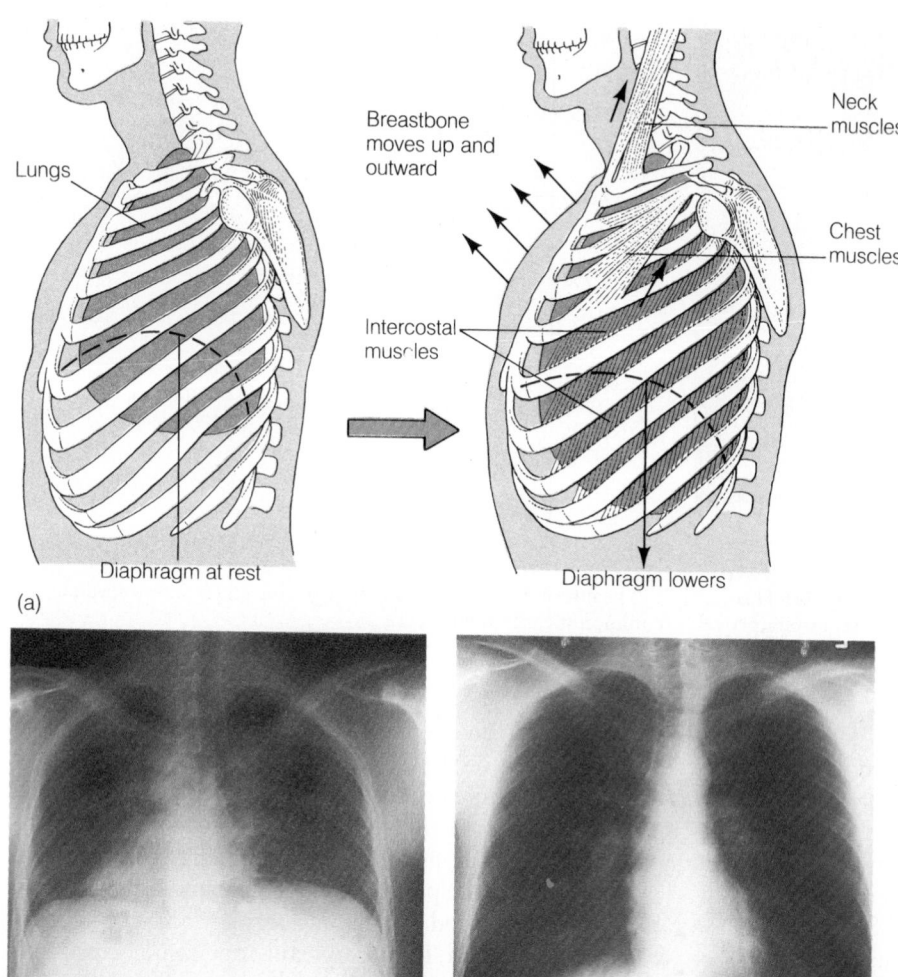

(a)

(b)

▷ **FIGURE 14–19 The Bellows Effect** (*a*) The rising and falling of the chest wall caused by contraction of the intercostal muscles is shown in this diagram, illustrating the bellows effect. Inspiration is assisted by the diaphragm, which lowers, like the plunger on a syringe, drawing air into the lungs. (*b*) X-rays showing the size of the lungs in full exhalation (*left*) and full inspiration (*right*).

Breathing Is Controlled Principally by the Breathing Center in the Brain

The **breathing control center** is located in a region of the brain called the brain stem (or medulla). In some ways similar to the sinoatrial node of the heart, the breathing center contains nerve cells that give off periodic impulses that stimulate contraction of the intercostal muscles and the diaphragm, resulting in inhalation. When the lungs fill, the nerve impulses cease and, as noted earlier, the muscles relax. This, in turn, decreases the volume of the thoracic cavity and forces air out of the lungs.

Several mechanisms are responsible for the termination of these impulses. The first is a negative feedback loop, shown on the right side of ▷ Figure 14–21. Here's how it works. The breathing center sends nerve impulses to the diaphragm and intercostal muscles; it also sends impulses to a nearby region of the brain stem, which is a kind of "relay center" that transmits nerve impulses back to the breathing center. When these impulses arrive, they inhibit the neurons

in the breathing center, which shuts off the signals to the muscles of inspiration, terminating the inspiration.

Changes in the depth and rate (frequency) of breathing are thought to result from neural input arising from chemical receptors in the brain and certain arteries. These receptors detect the concentration of carbon dioxide, hydrogen ions, and oxygen in the body. Since they are key components of cellular metabolism, it is therefore no surprise that they are involved in the control of breathing. Perhaps the most important chemical involved in controlling respiration is carbon dioxide. Carbon dioxide levels are monitored by receptors in the aorta, the large artery that carries oxygenated blood out of the heart, and the carotid arteries, which carry oxygenated blood to the brain from the aorta. When carbon dioxide levels rise, these receptors transmit impulses to the breathing center, as shown in Figures 14–21 and ▷ 14–22 and described above. This increases the rate of breathing. A decline in carbon dioxide levels has the opposite effect.

(a)

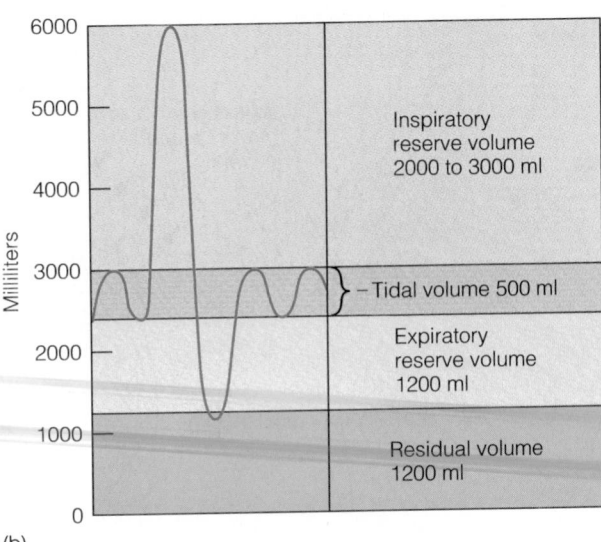

(b)

▷ **FIGURE 14–20 Measuring Lung Capacity** (*a*) This machine allows physicians to determine tidal volume, inspiratory reserve volume, and other lung-capacity measurements to determine the health of an individual's lung. (*b*) This graph shows several common measurements.

In the brain, carbon dioxide diffuses into the **cerebrospinal fluid (CSF),** a clear liquid found within cavities in the brain known as the **ventricles.** In the CSF, carbon dioxide is converted into carbonic acid, which then dissociates to form bicarbonate and hydrogen ions. A rise in carbon dioxide in the blood, therefore, results in an increase in the H^+ concentration of the CSF. The increase in H^+, in turn, is detected by chemical receptors, or **chemoreceptors,** in the brain. These receptors send impulses to the breathing center, triggering an increase in the rate and depth of breathing (Figure 14–22).

The chemoreceptors in the brain and arteries allow the body to align respiration with cellular demands. During exercise, for example, cellular energy production increases to meet body demands. As noted in Chapter 6, energy production requires oxygen and generates carbon dioxide waste. The carbon dioxide produced during exercise increases the depth and rate of breathing. This has two effects. First, it helps lower the concentration of carbon dioxide in the blood. (Breathing slows when levels return to normal.) Second, increased ventilation also makes more oxygen available for energy production.

The body also contains a set of oxygen receptors. These receptors are not as sensitive as the H^+ receptors so oxygen levels must fall considerably before the oxygen receptors begin generating impulses. This fact can have pro-

▷ **FIGURE 14–21 Breathing Center** The breathing center controls respiration. It sends periodic impulses along the nerves to the muscles of inspiration, causing them to contract. The center also sends impulses along another route to a relay center in the brain stem. Impulses from here feed back to the breathing center, shutting off the impulses that stimulate inspiration. Chemical receptors in the brain and certain arteries and stretch receptors in the lung also alter the activity of the breathing center.

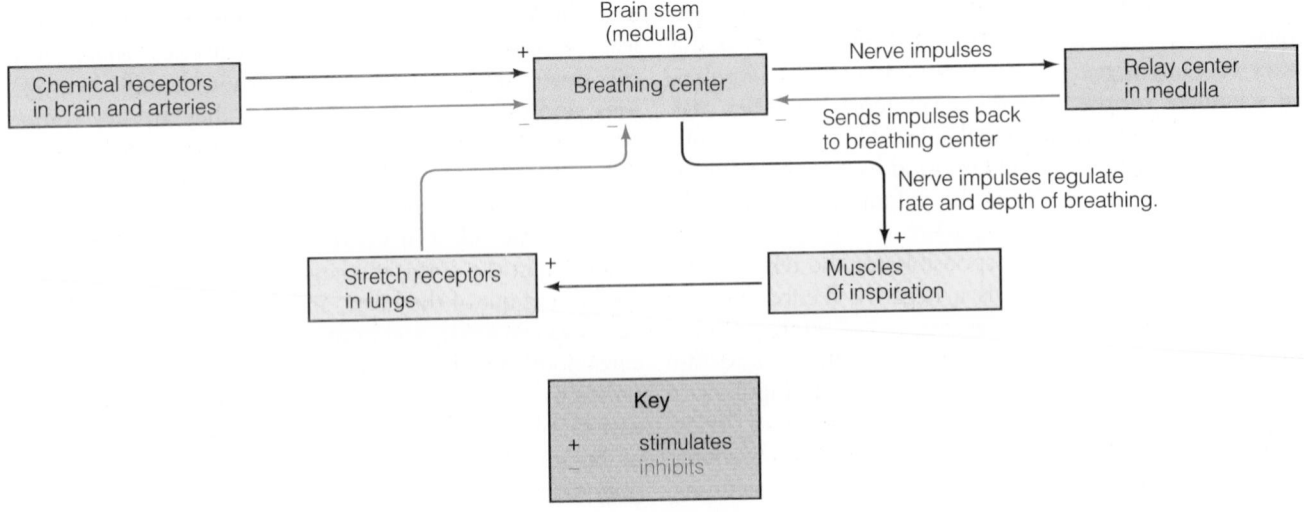

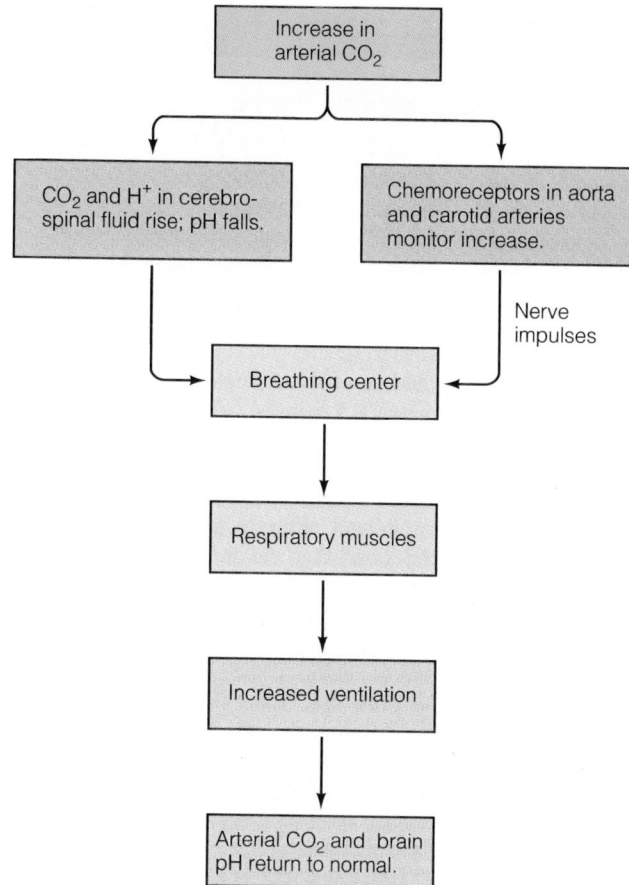

▷ FIGURE 14–22 Chemical Control of Breathing
CO_2 and hydrogen ions are the chief chemical controls of breathing.

found consequences for divers and swimmers. Repeated deep and rapid breathing, or **hyperventilation,** for example, makes it possible for divers to hold their breath underwater longer. Hyperventilation decreases carbon dioxide levels in the blood and H^+ concentrations in the cerebrospinal fluid, reducing the urge to breathe. When the diver enters the water, oxygen levels in the blood may fall so low that the brain is deprived of oxygen, causing the individual to lose consciousness. Ironically, the decrease in blood oxygen levels is not enough to stimulate breathing; the diver blacks out well before the H^+ concentration in the CSF reaches the level needed to stimulate breathing.

The third control mechanism consists of sensory nerve fibers known as stretch receptors, which are found in the lungs. When the lung is full, nerve impulses from the stretch receptors are transmitted to the breathing center. These impulses turn off the breathing center. Stretch receptors, however, probably function only during exercise, when large volumes of air are moved in and out of the lungs.

If the breathing center or the nerves that convey impulses to these muscles are destroyed—for example, by the polio virus or a head injury—breathing ceases.

ENVIRONMENT AND HEALTH: AIR POLLUTION

The air of the industrialized world contains a multitude of potentially harmful chemicals, or pollutants, that are damaging to human health. Air pollutants generated in our homes, factories, and cities in fact now claim the lives of thousands of Americans each year.

Air pollution upsets the normal homeostatic balance of our bodies, affecting millions of us in often subtle, but sometimes pronounced, ways. Unfortunately, most people are unaware of the dangers of air pollution because the line between cause and effect is not always clear. Consider, for instance, the headache you experienced in traffic going home from school or work. Was it caused by tension, or could it have been caused by carbon monoxide emissions from cars, buses, and trucks? And what about the runny nose and sinus condition you experienced last winter? Were they caused by a virus or bacterium, or by pollution?

One classic study of air pollution on the East Coast illustrates the relationship between air pollution and upper respiratory problems. The researchers found that the level of sulfur dioxide, a pollutant produced by automobiles, power plants, and factories, increased during the winter months in New York City. In one episode, certain weather conditions trapped the pollutants, raising ground-level sulfur dioxide concentrations. During this episode, upper respiratory illness in New York residents skyrocketed (▷ Figure 14–23). Colds, coughs, nasal irritation, and other symptoms increased fivefold in a few days. Soon after the pollution levels returned to normal, the symptoms subsided.

Air pollution is partly responsible for other long-term diseases. One of these is **chronic bronchitis,** a persistent irritation of the bronchi characterized by excess mucus production, coughing, and difficulty breathing. Today, one out of every five American men between the ages of 40 and 60 suffers from chronic bronchitis.

Although the leading cause of chronic bronchitis is cigarette smoking, studies show that urban air pollution also contributes to this disease. Three air pollutants have been linked to chronic bronchitis: sulfur dioxide, nitrogen oxides, and ozone. Each of these irritates the lung and bronchial passages and arises (directly or indirectly) from the combustion of fossil fuel by cars, buses, power plants, factories, and homes.

A far more troublesome disease is emphysema. Emphysema, discussed earlier in the chapter, is one of the fastest growing causes of death in the United States. Resulting principally from smoking and air pollution, emphysema afflicts over 1.5 million Americans and is more common in men than women. As noted earlier, it is a progressive, incurable disease. As the disease worsens, lung function deteriorates, and victims eventually require supplemental oxygen to perform even routine functions, such as walking or speaking.

SMOKING AND HEALTH: THE DEADLY CONNECTION

Urban air pollution worries many Americans, and for good reason. However, the city air that many of us breathe is benign compared with the "air" that 65 million Americans voluntarily inhale from cigarettes. Loaded with dangerous pollutants in concentrations far greater than those of our cities, cigarette smoke takes a huge toll on citizens of the world. In the United States, for example, an estimated 390,000 people—about 1000 every day—die from the many adverse health effects of tobacco smoke, including heart attacks, lung cancer, and emphysema.

Smoking also costs society a great deal in medical bills and lost productivity. According to the Worldwatch Institute, every pack of cigarettes sold in the United States costs our society $1.25 to $3.17 in medical costs, lost wages, and reduced productivity—that's about $125 billion to $400 billion a year!

Smoking is a principal cause of lung cancer, claiming the lives of an estimated 130,000 men and women in the United States each year. Depending on how many packs they smoke each day, smokers are 11 to 25 times more likely to develop lung cancer than nonsmokers.

Unfortunately, nonsmokers are also affected by the smoke of others. Nonsmokers inhale tobacco smoke in meetings, in restaurants, at work, and at home. For years, these "passive smokers" have had little to say about their exposure to other people's smoke. Today, however, as a result of a growing awareness of the dangers, new regulations are banning smoking in many public places and in the workplace, except in designated areas.

This trend has been spurred, in part, by research showing that passive smokers are more likely to develop lung cancer than people who manage to steer clear of smokers. In a study of Japanese women married to men who smoked, researchers found that the wives were as likely to develop lung cancer as people who smoked half a pack of cigarettes a day! A recent report by the U.S. Environmental Protection Agency estimates that passive smoking causes 500 to 5000 cases of lung cancer a year in the United States. Passive smokers who are exposed to tobacco smoke for long periods also suffer from impaired lung function equal to that seen in light smokers (people who smoke under a pack a day).

Cigarette smoke in closed quarters may also cause angina (chest pains) in individuals who are suffering from atherosclerosis of the coronary arteries. Carbon monoxide in cigarette smoke is responsible for angina attacks. Research also shows that smokers are more susceptible to colds and other respiratory infections. And smoking affects children. One study showed that children from families in which both parents smoked suffered twice as many upper respiratory infections as children from nonsmoking families. Recently, researchers from the Harvard Medical School reported finding a 7% decrease in lung capacity in children raised by mothers who smoked. The researchers believe that this change may lead to other pulmonary problems later in life.

Smoking has been shown to affect fertility in women as well. Studies show, for example, that women who smoke over a pack of cigarettes a day are half as fertile as nonsmokers. Smoking may also affect the outcome of pregnancy. According to the 1985 U.S. Surgeon General's Report, women who smoke several packs a day during pregnancy are much more likely to miscarry and are also more likely to give birth to smaller children. On average, the children of these women are 200 grams (nearly 0.5 pounds) lighter than children born to nonsmoking mothers.[1] Finally, children of women who smoke heavily during pregnancy generally score lower on mental aptitude tests during early childhood than children whose mothers do not smoke.

Tobacco smoke contains numerous hazardous substances that damage the delicate lining of the respiratory system. Nicotine and sulfur dioxide, for example, paralyze the cilia lining the respiratory tract. As noted in the chapter, one cigarette can knock the cilia out of action for an hour or more, eliminating the natural cleansing mechanism.

Tobacco smoke is also laden with microscopic carbon particles. Many toxic chemicals adhere to these particles and are transported into the lungs. A dozen or so of these chemicals are known to cause cancer. Carbon particles penetrate deeply into the lungs, where they accumulate in the alveoli and alveolar walls, turning healthy tissue into a blackened mass that often becomes cancerous (▷ Figure 1). Tobacco smoke may also paralyze the alveolar macrophages, making a bad situation even worse.

Toxin-carrying particles adhere to the lungs, larynx, trachea, and bronchi. Virtually any place they stick, they can cause cancer. That is why smokers are five times more likely than nonsmokers to develop laryngeal cancer and four

---

The leading cause of emphysema is smoking, a habit of 65 million Americans. But emphysema is also caused by urban air pollution. Not surprisingly, smokers who live in polluted urban settings have the highest incidence of the disease. Like chronic bronchitis, emphysema is caused by ozone, sulfur dioxide, and nitrogen oxides.

Researchers believe that 85% of the nearly 150,000 cases of lung cancer in the United States are caused by smoking (for more on the effects of smoking and the controversy surrounding tobacco advertising, see Health Note 14–2). The remaining 15,000 or so cases are thought to be caused by a variety of other factors, including urban and workplace air pollutants as well as natural pollutants, such as radioactive radon gas. Radon is emitted from radium, which occurs naturally in the soil in many parts of the country.

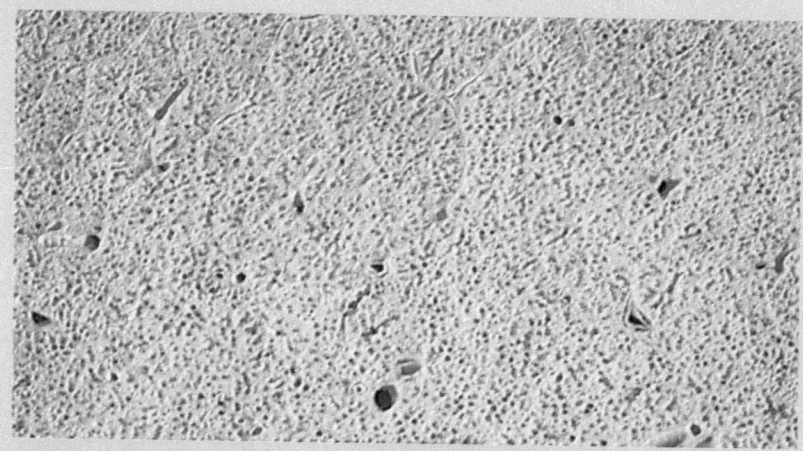

(a)

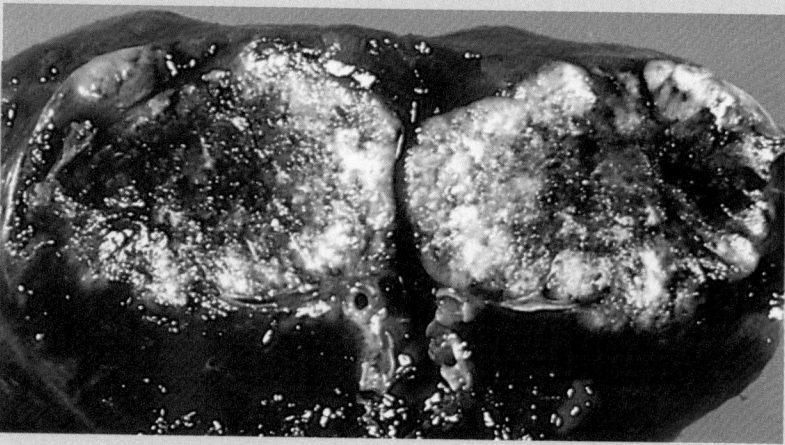

(b)

▷ **FIGURE 1 The Normal and Cancerous Lung** (*a*) The normal lung appears spongy. (*b*) The cancerous lung from a smoker is filled with particulates and tumor tissue.

discontinue generous government subsidies to tobacco growers? Surely, an air pollutant or food contaminant that killed hundreds of thousands of Americans each year would be banned.

Part of the answer lies in the fact that tobacco use has a long history in the United States, and many people think that because smoking is a voluntary act, individuals should have the right to make their own decision. Furthermore, smoking supports a $30-billion-a-year tobacco industry that employs about 2 million people, including tobacco farmers, advertising people, retailers, and so on. The tobacco industry lobbies diligently to protect the rights of smokers.

The dangers of smoking are becoming well known. As a result of widespread publicity and public pressure, smoking has dropped substantially in the United States. In 1990, for example, only 26% of American adults smoked, down from 34% in 1985 and down substantially from the 1950s and 1960s, when well over half of all men and over one-third of all women engaged in this potentially lethal habit.

Despite the downturn in smoking, an estimated 65 million Americans still smoke. Literally millions of people around the world will continue to die from smoking over the coming decade. And while cigarette smoking is on a decline in the United States, tobacco manufacturers have set their sights on the Third World, hoping to capture a rapidly growing market.

[1] The Surgeon General's office publishes an annual report that summarizes new findings on the effects of smoking on reproduction and health.

times more likely to develop cancer of the oral cavity.

Nitrogen dioxide and sulfur dioxide in tobacco smoke penetrate deep into the lungs, where they dissolve in the watery layer inside the alveoli. Nitrogen dioxide is converted to nitric acid; sulfur dioxide is converted to sulfuric acid. Both acids erode the alveolar walls, sometimes leading to emphysema.

If tobacco smoke is so dangerous, you may ask, why don't we ban smoking or

No one knows the exact toll of urban air pollution. It is not something that can be determined easily, if at all, because people are exposed to so many different pollutants over their lifetimes. However, a recent report issued by the federal government estimates that approximately 51,000 Americans die each year from lung disease caused by urban air pollution. The authors of the study predict that, by the year 2000, the number of victims will climb to nearly 60,000 per year, illustrating once again that human health is clearly dependent on a clean environment. This statistic also illustrates that the respiratory system of humans is not well adapted to the physical environment we have made for ourselves. Cultural evolution has progressed in ways that overwhelm past biological evolution.

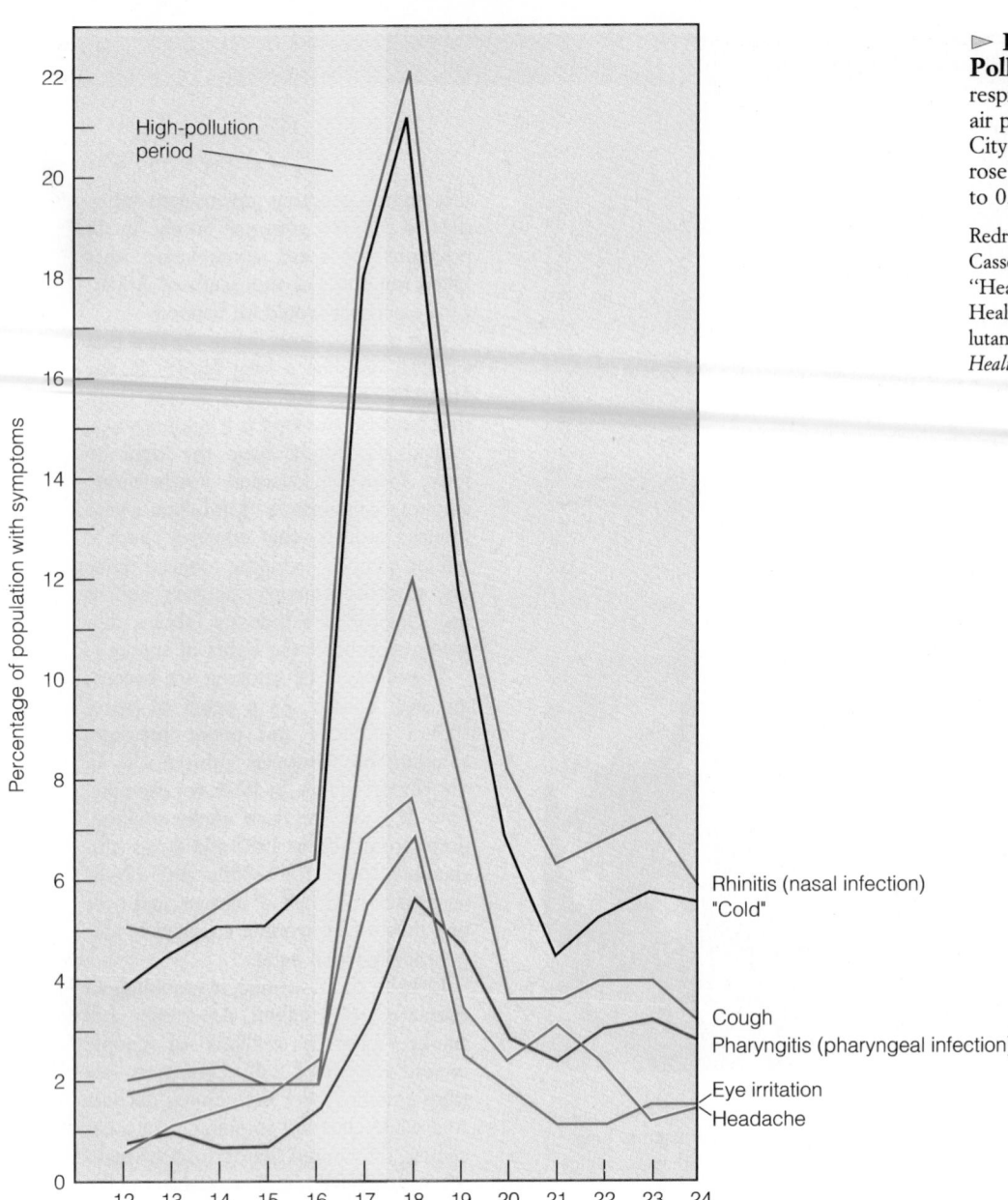

Rhinitis (nasal infection) "Cold"

Cough
Pharyngitis (pharyngeal infection)

Eye irritation
Headache

High-pollution period

Percentage of population with symptoms

Date

▷ **FIGURE 14–23 Air Pollution and Health** Graph of respiratory illnesses associated with an air pollution episode in New York City in 1962. Sulfur dioxide levels rose on days 16 and 17 from 0.2 ppm to 0.8 to 0.9 ppm.

Redrawn from J. R. McCarroll, E. J. Cassell, W. T. Ingram, and D. Wolter, "Health and the Urban Environment: Health Profiles versus Environmental Pollutants," in *American Journal of Public Health* 56(1966): 266–75.

SUMMARY

EVOLUTION OF RESPIRATORY SYSTEMS: AN OVERVIEW

1. Animals have evolved a wide array of respiratory systems to acquire oxygen and get rid of carbon dioxide.
2. Single-celled eukaryotic organisms relied on diffusion to acquire oxygen and get rid of carbon dioxide.
3. With the advent of multicellular animal life, though, more efficient life-supporting systems evolved, among them the respirable skin of the amphibian, the gill, and the lung.

STRUCTURE OF THE HUMAN RESPIRATORY SYSTEM

4. The respiratory system consists of an air-conducting portion and a gas-exchange portion.

5. The air-conducting portion transports air from outside the body to the alveoli in the lungs, the site of gaseous exchange.
6. The alveoli are tiny, thin-walled sacs formed by a single layer of flattened epithelial cells that facilitate diffusion. Surrounding the alveoli are capillary beds that pick up oxygen and expel carbon dioxide.
7. The lining of the alveoli is kept moist by water. Surfactant, a phospholipid produced in the lung, reduces the surface tension inside the alveoli and prevents their collapse.

FUNCTIONS OF THE RESPIRATORY SYSTEM

8. The respiratory system conducts air to and from the lungs,

exchanges gases, and helps produce sounds. Sound is generated as air rushes past the vocal cords, causing them to vibrate. The sounds are modified by movements of the tongue and changes in the shape of the oral cavity.

9. Oxygen and carbon dioxide gases diffuse across the alveolar wall, which is driven by concentration differences between the blood and alveolar air. In the lungs, oxygen diffuses from the alveoli into the blood plasma, then into the RBCs, where most of it binds to hemoglobin. Carbon dioxide diffuses in the opposite direction.

10. Carbon dioxide, a waste product of cellular energy production, is picked up by the blood flowing through capillaries. Some carbon dioxide is dissolved in the blood. Most, however, enters the RBCs in the bloodstream where it combines with hemoglobin or is converted into carbonic acid. Carbonic acid dissociates, forming hydrogen and bicarbonate ions, which diffuse out of the RBCs and are transported primarily in the blood plasma.

BREATHING AND THE CONTROL OF RESPIRATION

11. Breathing is an involuntary action with a conscious override. It is controlled by the breathing center in the brain stem. Nerve cells in the breathing center send impulses to the diaphragm and intercostal muscles, which contract. This increases the volume of the chest cavity, which draws air into the lungs through the nose or mouth.

12. When the impulses stop, inspiration ends. Air is then expelled passively as the chest wall returns to the normal position, and the diaphragm rises. The recoil of the lungs also assists in expelling the air.

13. The breathing center is regulated by a negative feedback loop that it generates itself. It is also regulated by outside influences—notably, levels of carbon dioxide in the blood.

14. Expiration can be augmented by enlisting the aid of abdominal and chest muscles, as can inspiration.

15. The rate of respiration can be increased by rising blood carbon dioxide levels or falling oxygen levels and by an increase in physical exercise.

ENVIRONMENT AND HEALTH: AIR POLLUTION

16. The proper functioning of the respiratory system is essential for health. Respiratory function can be dramatically upset by microorganisms as well as by pollution from factories, automobiles, power plants, and even our own homes.

17. Chronic bronchitis, a persistent irritation of the bronchi, is caused by sulfur dioxide, ozone, and nitrogen oxides, lung irritants sometimes found in dangerous levels in urban air.

18. Emphysema, a breakdown of the alveoli that gradually destroys the lung's ability to absorb oxygen, is similarly induced. Despite the role of air pollution in causing emphysema and chronic bronchitis, smoking remains the number one cause of these diseases.

EXERCISING YOUR CRITICAL THINKING SKILLS

An inventor has devised a pollution-control device that removes particulates from the smokestacks of factories. He claims that the device removes 80% of all particulates and will bring factories into compliance with federal law, thereby protecting nearby residents from harmful pollutants.

What is your reaction? What questions might you ask? Would you approve installing the device, given this information? Would your decision be affected by data showing that although the device does reduce total particulates by 80%, it does not capture finer particulates? These are particles that can be inhaled deeply into the lung because they do not precipitate out of the respiratory tract. Would your decision be affected if you found that the small (respirable) particles in the smokestack contained toxic metals, such as mercury and cadmium? What rule(s) of critical thinking does this example illustrate?

TEST OF TERMS

1. _____ is a disease of the lung characterized by the progressive breakdown of the alveoli.

2. The _____ is that part of the respiratory system that conducts air from the nose and mouth to the larynx.

3. Air travels from the larynx to the, _____ a ribbed duct that leads to the lungs, then splits into right and left _____ , which penetrate the lungs.

4. The _____ , the muscular ducts that conduct air to the alveoli, can contract and relax like arterioles, thus providing a way to control the flow of air inside the lung.

5. Much of the lining of the conducting portion of the respiratory system contains _____ cells, which produce a secretion that traps dust particles and bacteria.

6. A chemical, _____ , is produced by certain cells in the alveoli and helps reduce surface tension, preventing the alveoli from collapsing.

7. The phagocytic cell that wanders in and out of the alveoli is called a _____ cell.

8. Inside the larynx are two elastic cords that vibrate when air breezes past. These are the _____ _____ _____ .

9. The patch of epithelium in the roof of the nasal cavity that perceives smell is called the _____ _____ .

10. Most oxygen is carried in the blood bound to _____ in the RBCs. Most carbon dioxide is transported as _____ .

11. The region of the _____ that controls inspiration is called the _____ _____ .

12. The active process in which air is drawn into the lungs is called _____ . It is caused by an enlargement of the chest cavity, which results from the contraction of the _____ and the _____ muscles.

13. The passive expulsion of air is called _____ .

14. Persistent irritation of the bronchi, leading to buildup of mucus and coughing, is called _____ _____ and is often caused by pollution in cigarette smoke and urban air.

15. The _____ volume is the amount of air moved in and out of the lungs at rest. The _____ _____ volume is the amount of air that can be expelled from the lungs after a normal, passive exhalation.

Answers to the Test of Terms are found in Appendix B.

TEST OF CONCEPTS

1. Trace the flow of air from the mouth and nose to the alveoli, and describe what happens to the air as it travels along the various passageways.

2. Draw an alveolus, including all cell types found there. Be sure to show the relationship of the surrounding capillaries. Show the path that oxygen and carbon dioxide must take.

3. Trace the movement of oxygen from alveolar air to the blood in alveolar capillaries. Describe the forces that cause oxygen to move in this direction. Do the same for the reverse flow of carbon dioxide.

4. Why would a breakdown of alveoli in emphysemic patients make it more difficult for them to receive adequate oxygen?

5. Describe how sounds are generated and refined.

6. A baby is born prematurely and is having difficulty breathing. As the attending physician, explain to the parents what the problem is and how it could be corrected.

7. Smoking irritates the trachea and bronchi, causing mucus to build up and paralyze cilia. How do these changes affect the lung?

8. Describe inspiration and expiration, being sure to include discussions of what triggers them and the role of muscles in bringing about these actions.

9. How does the breathing center regulate itself to control the frequency of breathing?

10. Exercise increases the rate of breathing. How?

11. Turn to Figure 14–20. Explain each of the terms.

12. What is asthma? What are its symptoms? How does it affect one's inspiratory reserve volume?

13. Debate this statement: urban air pollution has very little overall impact on human health.

14. Do you agree or disagree that the single most effective way of reducing deaths in the United States would be to ban smoking. Explain.

C H A P T E R

15

The Immune System

≈

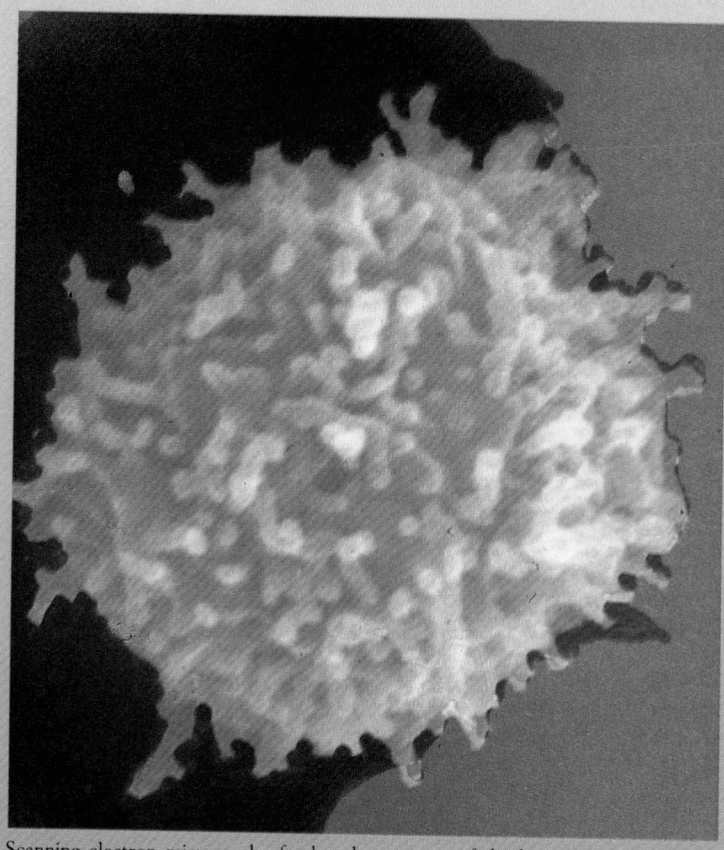

Scanning electron micrograph of a lymphocyte, one of the key players in immune protection.

ulticellular plants and animals evolved in a world teeming with viruses and single-celled organisms—bacteria and protistas. These organisms were the evolutionary ancestors of multicellular life. The vast majority of these microorganisms do not interact with multicellular animals directly. Therefore, even though they may be part of important environmental functions, essential for the web of life, most of them live their lives seemingly apart from the animals all around them. Anyone who has suffered from a cold or the flu will attest, however, that some microorganisms do indeed cross our paths, with sometimes dismal consequences. Fortunately, multicellular animals have evolved a series of highly effective defenses against bacteria, viruses, and other infectious agents.

This chapter examines the human body's defense mechanisms against disease-causing, or pathogenic, microorganisms. It looks briefly at inherent defenses against cancer and also discusses the devastating disease known as AIDS. The discussion of AIDS examines the causes of the disease, how it can be prevented, and some of the more recent efforts to find a cure. We begin with a brief overview of infectious agents.

≈ VIRUSES AND BACTERIA: AN INTRODUCTION

Two of the most important infectious agents are viruses and bacteria. A brief discussion of these agents will help you understand the information that follows on the immune system. (Chapter 24 provides a much more detailed coverage of bacteria and viruses.)

A **virus** is a submicroscopic intracellular parasite. It consists of a nucleic acid core, consisting of either DNA or RNA, and an outer protein coat, the **capsid** (▷ Figure 15–1). Some viruses have an outer envelope that lies outside the capsid. The envelope is structurally similar to the plasma membranes of eukaryotic cells.

▷ **FIGURE 15–1 General Structure of a Virus** (*a*) The virus consists of a nucleic acid core of either RNA or DNA. Surrounding the viral core is a layer of protein known as the capsid. Each protein molecule in the capsid is known as a capsomere. (*b*) Some viruses have an additional protective coat known as the envelope.

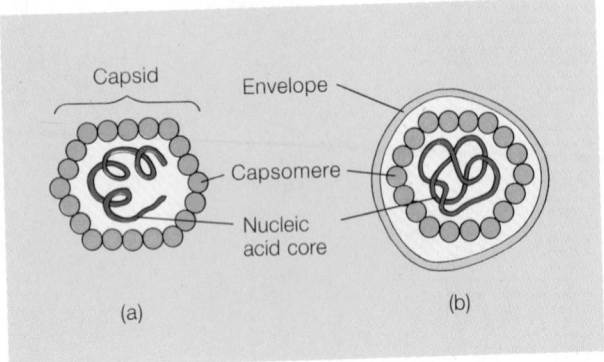

Viruses invade cells and commandeer their metabolic machinery. In the process, they convert host cells into miniature virus factories, often killing their hosts.

Viruses most often enter the body through the respiratory and digestive systems and spread from cell to cell in the bloodstream and lymphatic system. However, other avenues of entry are also possible—for example, sexual contact.

The immune system kills many viruses, but some may take refuge in cells, reemerging under stress or some other influence. One example is the virus responsible for genital herpes, tiny sores that periodically emerge on the genitals, thighs, and buttocks of men and women.

Bacteria (singular, **bacterium**) are microorganisms that do not require host cells to replicate. Bacteria are prokaryotic cells containing cytoplasm enveloped by a plasma membrane (▷ Figure 15–2). Outside the plasma membrane is a thick, rigid cell wall. Bacteria contain a single circular "strand" of DNA. Many bacteria also contain tiny circular pieces of extrachromosomal DNA, known as **plasmids** (described in Chapter 24).

Although they are best known for their role in causing sickness and death, most bacteria perform useful functions. Soil bacteria, for example, help recycle nutrients, thus helping ensure the continuation of life. This chapter concerns itself with potentially harmful bacteria that invade the human body.

≈ THE FIRST AND SECOND LINES OF DEFENSE

Thousands of different bacteria and viruses and other microorganisms (for example, single-celled protistas) are found in the air we breathe and the food and water we consume. The protective mechanisms that evolved to protect multicellular organisms like humans from these hordes of potentially harmful microorganisms can be divided into three groups.

In Humans, the First Line of Defense Consists of the Skin, Epithelial Linings of the Respiratory, Digestive, and Urinary Systems, and Body Secretions that Destroy Harmful Microorganisms

Human skin forms an outer protective layer, which repels many potentially harmful microorganisms. As noted in Chapter 10, the skin consists of a relatively thick layer of epidermal cells overlying the rich vascular layer known as the dermis. Epidermal cells are produced by cell division in the base of the epidermis. As the basal cells proliferate, they move outward, become flattened, and die. The dead cells contain a protein called **keratin,** which forms a fairly waterproof protective layer that not only reduces moisture loss but also protects underlying tissues from microorganisms. The cells of the epidermis are tightly joined by special structures referred to as tight junctions. These structures also help impede water loss and microbial penetration.

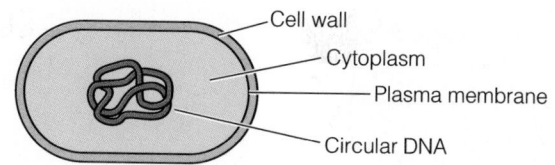

Cell wall
Cytoplasm
Plasma membrane
Circular DNA

▷ **FIGURE 15–2 General Structure of a Bacterium**
Bacteria come in many shapes and sizes, but all have a circular strand of DNA, cytoplasm, and a plasma membrane. Surrounding the membrane of many bacteria is a cell wall.

The epithelial linings of the respiratory, digestive, and urinary systems are also part of the body's wall of protection. They keep potentially harmful microorganisms from invading the underlying tissues. A break in these linings, like an opening in a fortress wall, permits microorganisms to enter. Chapter 11, for example, pointed out how a break in the lining of the digestive tract, caused by diverticulosis, can spawn a dangerous bacterial infection that spreads through the body.

The first line of defense also includes several protective chemicals. The skin, for example, produces slightly acidic secretions that impair bacterial growth. The stomach lining produces hydrochloric acid, which destroys many ingested bacteria. Tears and saliva contain an enzyme called **lysozyme,** which dissolves the cell wall of bacteria, killing them. Cells in the lining of the trachea and bronchi produce mucus, which not only traps bacteria but also has antimicrobial properties. Cilia in the lining of the respiratory system sweep the mucus toward the mouth, where it can be expectorated or swallowed (Chapter 14).

The protective mechanisms described above are all nonspecific; that is, they operate indiscriminately against all microorganisms.

The Second Line of Defense Combats Infectious Agents that Penetrate the First Line and Consists of Cellular and Chemical Responses

The first line of defense is not impenetrable. Even tiny breaks in the skin or in the lining of the respiratory, digestive, and urinary tracts permit viruses, bacteria, and other microorganisms to enter the body. Fortunately, a second line of defense exists. It involves a whole host of chemicals and cellular agents that work together to combat the invaders. Like the first line of defense, these mechanisms are nonspecific.

The Inflammatory Response Is a Major Part of the Second Line of Defense. Damage to body tissues triggers a series of reactions, that is part of the **inflammatory response.** The word *inflammatory* comes from the Latin word *inflammare* meaning "in flame" and refers to the heat given off by a wound. A protective measure, the inflamma-

tory response is also characterized by redness, swelling, and pain—symptoms not unknown to anyone who has ever been cut or had a splinter.

The inflammatory response is a kind of chemical and biological warfare waged against bacteria, viruses, and other microorganisms. It begins with the release of a variety of chemical substances by injured tissue. Some chemicals attract resident macrophages that reside in body tissues and neutrophils found in the blood (Chapter 13). These cells quickly begin phagocytizing bacteria that enter the wound. Soon after these cells begin to work, a yellowish fluid begins to exude from the wound. Called pus, it contains dead white blood cells (mostly neutrophils), microorganisms, and cellular debris, which accumulate at the site of inflammation.

Tissue injury also stimulates the release of chemical substances that cause blood vessels to dilate and leak (▷ Figure 15–3). One such substance is **histamine.** Histamine stimulates the arterioles to dilate, causing the capillary networks to swell with blood.[1] The increase in the flow of blood through an injured tissue is responsible for the heat and redness around a cut or abrasion. Heat increases the metabolic rate of cells in the injured area and therefore accelerates the healing process.

Still other substances released by injured tissues increase the permeability of capillaries, augmenting the flow of plasma into a wounded region. Plasma carries with it oxygen and nutrients that facilitate healing. It also carries the molecules necessary for blood clotting. As described in Chapter 13, the clotting mechanism walls off injured vessels and helps reduce blood loss.

Plasma leaking into injured tissues causes swelling, which stimulates pain receptors in the area. Pain receptors send nerve impulses to the brain. Pain also results from chemical toxins released by bacteria and from chemicals released by injured cells themselves. One important pain-causing chemical is prostaglandin. Aspirin and other mild painkillers inhibit the synthesis and release of prostaglandins, reducing pain.

Although it evokes pain, the flow of fluid into body tissues is helpful in certain circumstances. Injury to joints, for example, results in local swelling that helps immobilize joints. Swelling is nature's way of protecting joints and allowing tissues to mend.

Inflammation occurs in virtually all tissues invaded by bacteria and viruses. It even comes equipped with its own cleanup crew—late-arriving monocytes—that mops up after the battle, phagocytizing dead cells, cell fragments, dead bacteria, and viruses.

Three Additional Chemicals Are Part of the Second Line of Defense. The second line of defense includes three additional substances: pyrogens, interferons, and complement.

[1] Histamine is produced by mast cells, platelets, and basophils, a type of white blood cell. Mast cells are a connective tissue cell described later in the chapter.

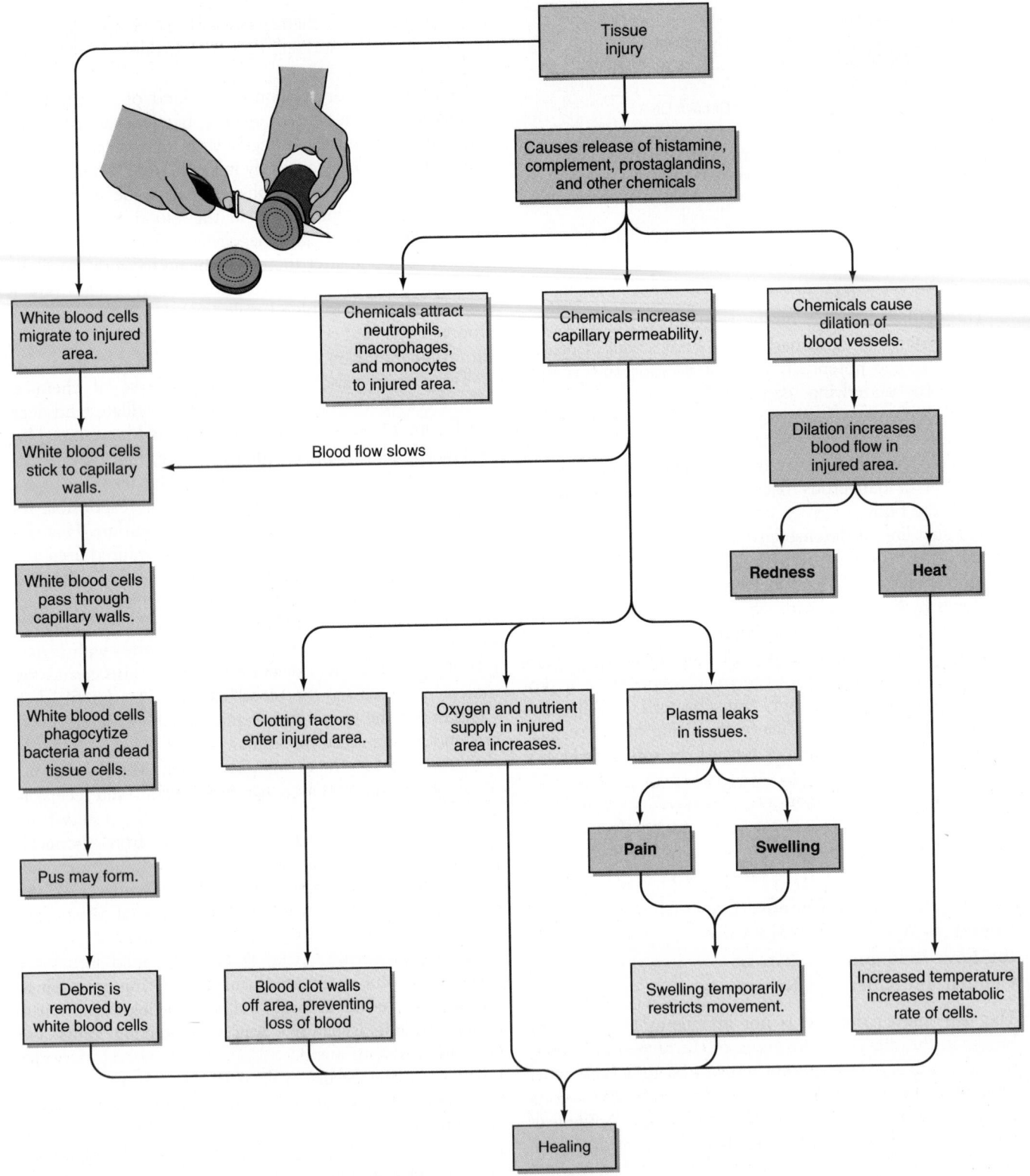

▷ **FIGURE 15–3 The Inflammatory Response**

Pyrogens are molecules released primarily from macrophages that have been exposed to bacteria and other foreign substances. Pyrogens travel to a region of the brain called the hypothalamus. In the hypothalamus is a group of nerve cells that controls the body's temperature, in much the same way that a thermostat regulates the temperature of a room. Pyrogens turn the thermostat up, increasing body temperature and causing a fever.

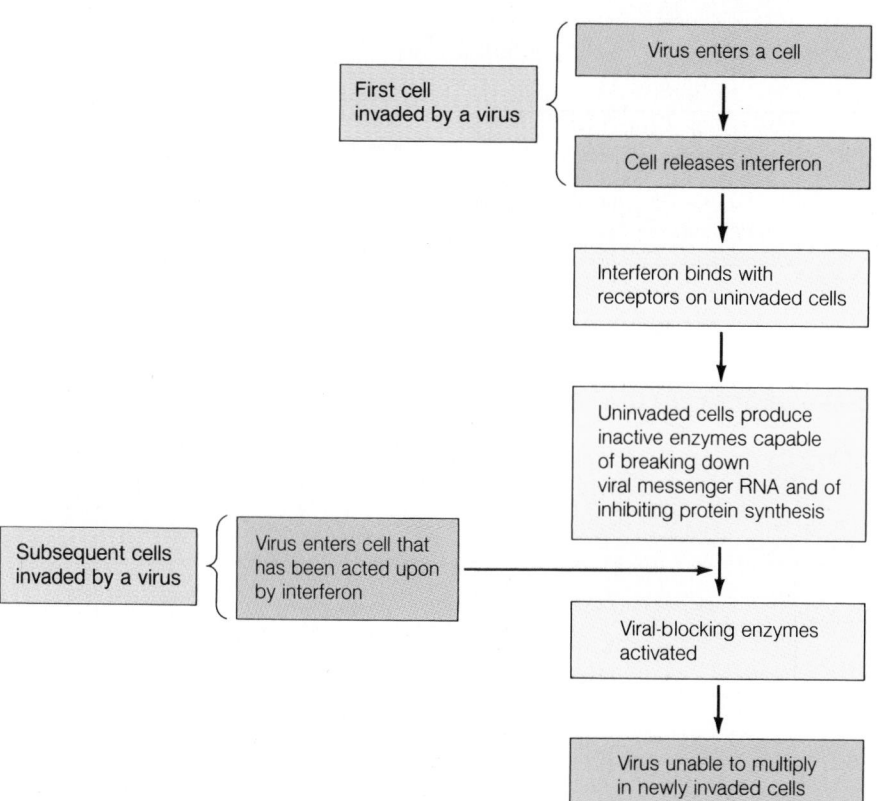

Fever, usually poorly regarded because of the discomfort it causes us, is actually an adaptation that helps combat bacterial infections. For example, mild fevers cause the spleen and liver to remove additional iron from the blood. Many pathogenic bacteria require iron to reproduce. Fever therefore reduces the replication of bacteria and helps the body battle them. Fever also increases metabolism, which facilitates healing and accelerates cellular defense mechanisms, such as phagocytosis. Important as it is, fever can also be debilitating, and a severe fever (over 105°F) is potentially life-threatening because it begins to denature vital body proteins, especially enzymes needed for the biochemical reactions occurring in body cells.

Another chemical safeguard in the second line of defense is a group of small proteins known as the interferons. **Interferons** are released from cells infected by viruses. Research suggests that each type of cell produces a slightly different form of interferon. Interferons released by infected cells bind to receptors on the plasma membranes of noninfected body cells (▷ Figure 15–4). The binding of interferon to these cells triggers the synthesis of cellular enzymes capable of breaking down viral mRNA and blocking viral protein synthesis. These enzymes, however, remain inactive until a virus enters the cell.

Interferons do not protect cells already infected by a virus, they simply stop the spread of viruses from one cell to another. In essence, the production and release of interferon are the dying cell's last acts to protect other cells of the body. Interferons are a remarkable chemical adaptation that stops the spread of viruses while the immune response

attacks and destroys the viruses outside the cells. Additional effects of the interferons are listed in Table 15–1.

Another group of chemical agents used to fight infection are the **complement proteins.** These blood proteins form the **complement system,** so named because it complements the action of antibodies, briefly described in Chapter 13 and elsewhere.

The details of the complement system are far too complex for this book, but a few generalizations will demonstrate how this system works. Complement proteins circulate in the blood in an inactive state, much like the proteins involved in blood clotting (fibrinogen and prothrombin). When foreign cells, such as bacteria, invade the body, the complement system is activated. This triggers a cascade of reactions in which one complement protein activates the next in the series, much as the chain of command is awakened when a nation's surveillance system detects an invasion (▷ Figure 15–5).

Five proteins in the complement system join to form a

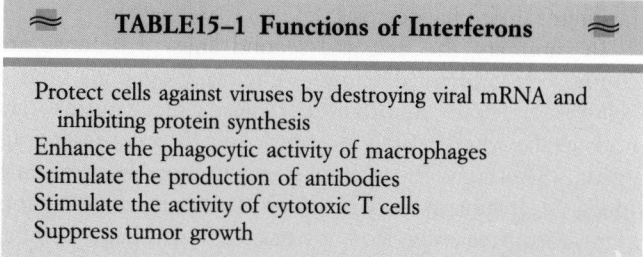

≈ TABLE15–1 Functions of Interferons ≈
Protect cells against viruses by destroying viral mRNA and inhibiting protein synthesis
Enhance the phagocytic activity of macrophages
Stimulate the production of antibodies
Stimulate the activity of cytotoxic T cells
Suppress tumor growth

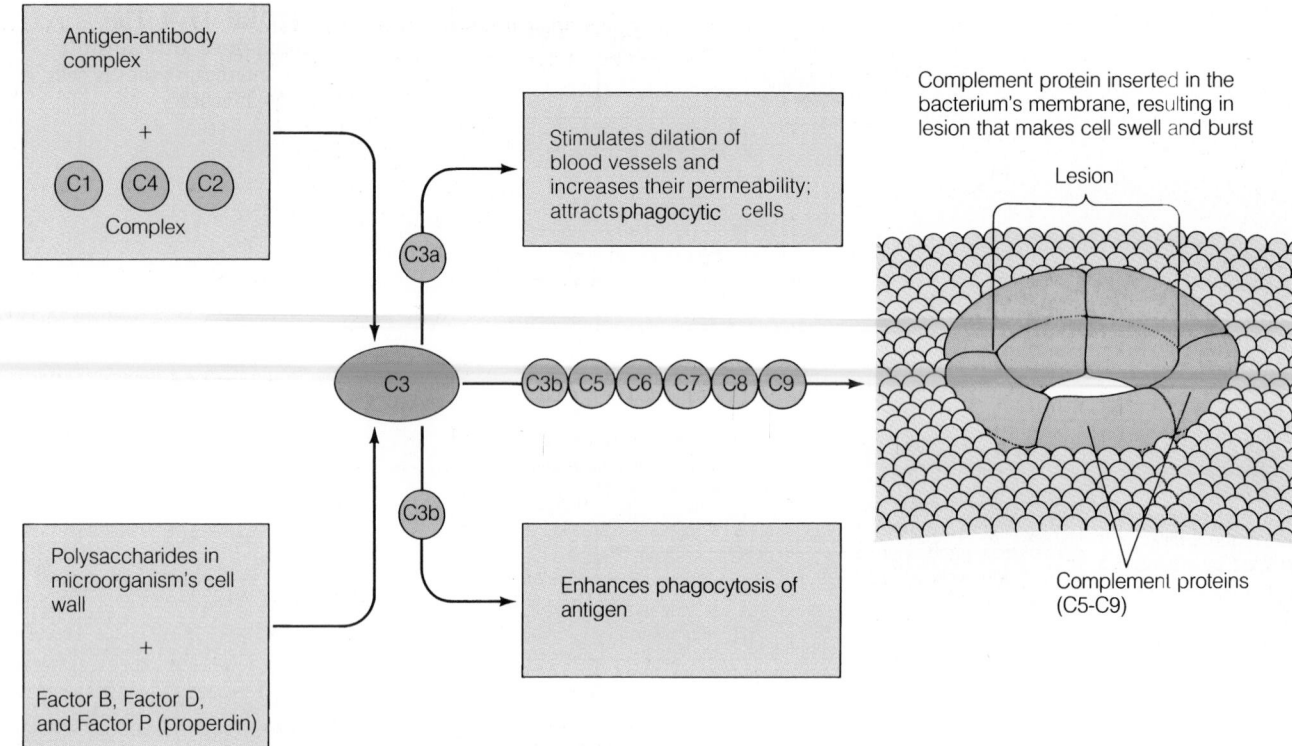

Complement protein inserted in the bacterium's membrane, resulting in lesion that makes cell swell and burst

Lesion

Complement proteins (C5–C9)

▷ **FIGURE 15–5 The Complement System** The complement system is activated by the presence of antibodies bound to their antigens and by the polysaccharides in the cell wall of some bacteria and fungi. These triggers activate C3 protein, which splits, forming two fragments. The C3a fragment stimulates the inflammatory response. The C3b fragment binds to bacteria, making them easier to phagocytize. C3b also activates additional complement proteins, which are inserted into the membrane of bacteria. They create an opening, or lesion, in the membrane, which, if large enough, can kill the cell.

large protein complex, known as the **membrane-attack complex** (▷ Figure 15–6). The membrane-attack complex embeds in the plasma membrane of bacteria, creating an opening into which water flows. The influx of water causes bacterial cells to swell, burst, and die.

Several of the activated complement proteins also function on their own and are part of the inflammatory response. Some of them, for example, stimulate the dilation of blood vessels in an infected area, described earlier. Others increase the permeability of the blood vessels, allowing white blood cells and nutrient-rich plasma to pass more readily into an infected zone. Certain complement proteins may also act as chemical attractants, drawing macrophages, monocytes, and neutrophils to the site of infection, where they phagocytize foreign cells. Another complement protein (C3b) binds to microorganisms, forming a rough coat on the intruders that facilitates their phagocytosis.

In summary, the first and second lines of defense are composed of physical barriers, chemical weapons, and a cellular defense mechanism. These mechanisms are nonspecific; that is, they do not target specific infectious agents. The skin, for example, repels most bacteria and fungi. Macrophages and neutrophils devour whatever foreign substances enter the body tissues. Fever helps combat dividing bacteria by reducing iron levels. In so doing, the first and second lines of defense lighten the work load of the third and final line of defense, the immune system.

≈ THE THIRD LINE OF DEFENSE: THE IMMUNE SYSTEM

The **immune system** is not a distinct organ system like the digestive or respiratory system. Rather, it is a functional system consisting of many millions of lymphocytes, a type of white blood cell. Humans, for instance, contain an estimated 2 trillion lymphocytes—that's 2000 billion. These cells circulate in the blood and lymph but also take up residence in the lymphoid organs, such as the spleen, thymus, lymph nodes, and tonsils, as well as other body tissues (Figure 12-21). The cells of the immune system selectively target foreign substances and foreign organisms. As a result, the immune system is said to be specific.

The immune system is an important homeostatic mechanism that eliminates foreign organisms—including bacteria, viruses, single-celled protists, and even many parasites—that penetrate the outer defenses of the body. It also helps reduce cellular dissent from within—that is, the emergence of cancer. Thus, in a world filled with infectious agents and natural mutagens (agents that cause mutation, some of

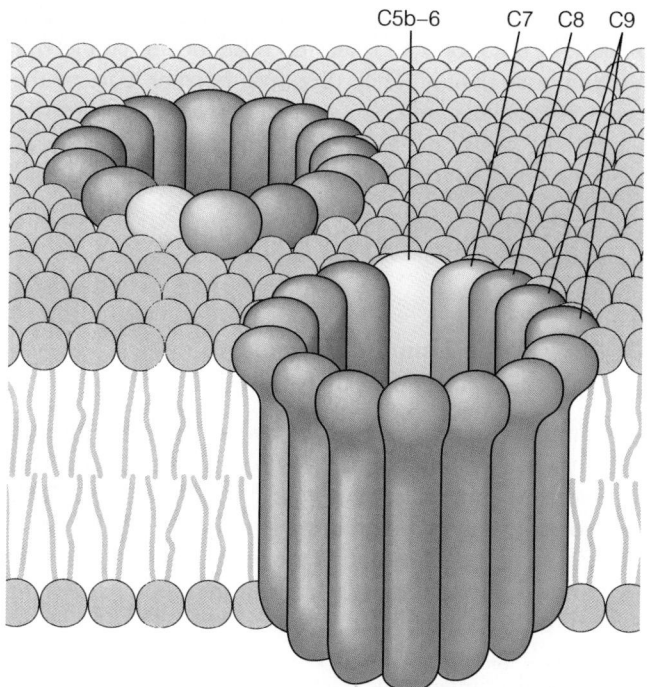

C5b–6   C7   C8   C9

▷ **FIGURE 15–6 The Membrane-Attack Complex**
Complement proteins embed in a cell's membrane, causing it to
leak, swell, and burst. By Dana Burns from John Ding-E Young and
Zanvil A. Kohn, "How Killer Cells Kill." Copyright © January 1988 by Scientific American, Inc. All rights reserved.

which might lead to cancer), the immune system is an important evolutionary advance.

Lymphocytes Detect Foreign Substances in the Body and Mount an Attack on Them

One of the chief functions of the immune system is to
identify what belongs in the body and what does not. The
ability to recognize foreign materials, while essential to
survival, is a rather difficult task. It is a little like having to
sort through the thousands of items in your family's home,
including the contents of every drawer in the kitchen and
every jar of nails in the garage, each and every day of the
year to determine if someone has brought something as
tiny as a pin into the house.

Once a foreign substance has been detected, the immune system mounts an attack to eliminate it. Therefore,
like all homeostatic systems, the immune system requires
receptors to detect a change and effectors to bring about a
response. In the immune system, the lymphocytes serve
both of these functions.

Foreign Substances that Trigger an Immune Response are Proteins and Polysaccharides with Large Molecular Weights

The immune response is triggered by large foreign molecules, notably proteins and polysaccharides. These molecules are called **antigens** (*anti*body-*gen*erating substances).
The larger the molecule is, the greater its antigenicity.

As a rule, small molecules generally do not elicit an
immune reaction. In some individuals, however, small,
nonantigenic molecules like formaldehyde, penicillin, and
the poison ivy toxin bind to naturally occurring proteins in
the body, forming complexes. The large complexes so
formed are unique compounds foreign to the body and
capable of eliciting an immune response.

The immune system responds to viruses, bacteria, and
single-celled fungi in the body. It also responds to parasites, such as the protozoan that causes malaria. Viruses,
bacteria, fungi, and parasites elicit a response because they
are enclosed by a membrane, or coat, that contains
large-molecular-weight proteins or polysaccharides—that
is, antigens.

Cells transplanted from one person to another also elicit
an immune response, because cells from another individual
contain a unique "cellular fingerprint," resulting from the
unique array of plasma membrane glycoproteins, as noted
in Chapter 3. The immune system is activated by these
antigens on the foreign cells. Cancer cells also present a
slightly different chemical fingerprint, making them essentially foreign cells within our own bodies to which the
immune system responds. Although cancer cells evoke an
immune response, it is often not sufficient to stop the
disease.

Antigens stimulate the proliferation and differentiation
of two types of lymphocytes: the **T lymphocytes,** commonly called **T cells,** and the **B lymphocytes,** also known
as **B cells** (Chapter 13). The **immune reaction,** therefore,
is the response of B and T cells. As you will see in later
sections, B and T cells react differently and respond to
different types of antigens. As a rule of thumb, B cells
recognize and react to microorganisms such as bacteria.
They also respond to a few viruses and bacterial toxins,
chemical substances released by bacteria. When activated,
B cells produce antibodies to these antigens. In contrast, T
cells recognize and respond to body cells that have gone
awry, such as cancer cells, or cells that have been invaded
by viruses. T cells also respond to transplanted tissue cells
and larger disease-causing agents, such as single-celled
fungi and parasites. Unlike B cells, T cells attack their
targets directly.

Immature B and T Cells Are Incapable of Responding to Antigens But Soon Mature and Gain that Ability

Lymphocytes are produced in the red bone marrow and
released into the bloodstream. These immature cells circulate through the blood and lymph (▷ Figure 15–7).

T Cells. Cells destined to become T lymphocytes take up
temporary residence in the thymus, a lymphoid organ located above the heart. Inside the thymus, these lymphocytes mature in two to three days. During this time, lym-

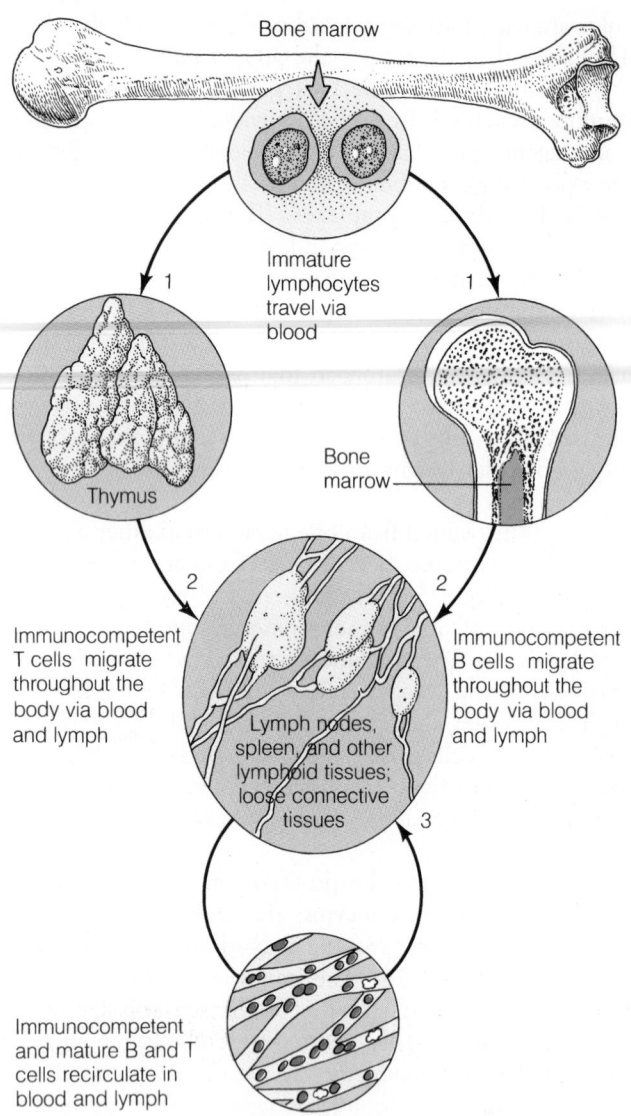

marrow.[2] Afterwards, immunologically competent B cells circulate in the blood and take up residence in connective and lymphoid tissues. They therefore become part of the body's vast cellular reserve, stationed at distant outposts, awaiting the arrival of the microbial invaders.

By various estimates, several million immunologically distinct B and T cells are produced in the body early in life. Over a lifetime, only a small fraction of these cells will be called into duty.

B Cells Provide Humoral Immunity Through the Production of Antibodies

The immune response consists of two separate but related reactions: humoral immunity, provided by the B cells, and cell-mediated immunity, involving T cells (Table 15–2). Let's consider humoral immunity and the B cells first.

When an antigen first enters the body, it binds to B cells programmed during their residence in the bone marrow to respond to that particular antigen (▷ Figure 15–8).[3] These cells soon begin to divide, producing a population of immunologically similar cells, also known as a **clone.** As the clone expands, some of the B cells begin to differentiate, forming another kind of cell, the plasma cell. **Plasma cells** are differentiated B cells and contain a prominent rough endoplasmic reticulum on which they produce prodigious amounts of antibody. Antibodies released from plasma cells circulate in the blood and lymph, where they bind to the antigens that triggered the response. Because the blood and lymph were once referred to as body "humors," this arm of the protective immune response is called **humoral immunity.**

[2]B cells are so named because they develop immunocompetence in a part of the chicken's digestive system known as the bursa. Humans lack this organ.

[3]As you will soon see, this process is a bit more complex and involves the macrophage.

▷ **FIGURE 15–7 B- and T-Cell Immunocompetence**
Immunocompetence, the ability to respond to specific antigens, is conferred in the bone marrow, in the case of B cells, or the thymus, in the case of T cells. The cells then migrate in the blood and lymph to lymphoid organs, such as the lymph nodes, spleen, and loose connective tissue underlying many epithelia.

phocytes are said to develop **immunocompetence,** because they gain the capacity to respond to specific antigens. During this process, each differentiated T cell (so named because it becomes immunocompetent in the thymus) produces a unique type of membrane receptor that will bind to one—and only one—type of antigen. Over an individual's lifetime, millions of antigens will be encountered. Thanks to immunocompetence developed during fetal development, each of us is equipped with millions of different programmed T cells to respond to the onslaught of antigens long before we encounter them.

B Cells. B cells mature and differentiate in the bone

TABLE 15-2 Comparison of Humoral and Cell-Mediated Immunity

HUMORAL	CELL-MEDIATED
Principal cellular agent is the B cell.	Principal cellular agent is the T cell.
B cell responds to bacteria, bacterial toxins, and some viruses.	T cells responds to cancer cells, virus-infected cells, single-celled fungi, parasites, and foreign cells in an organ transplant.
When activated, B cells form memory cells and plasma cells, which produce antibodies to these antigens.	When activated, T cells differentiate into memory cells, cytotoxic cells, suppressor cells, and helper cells; cytotoxic T cells attack the antigen directly.

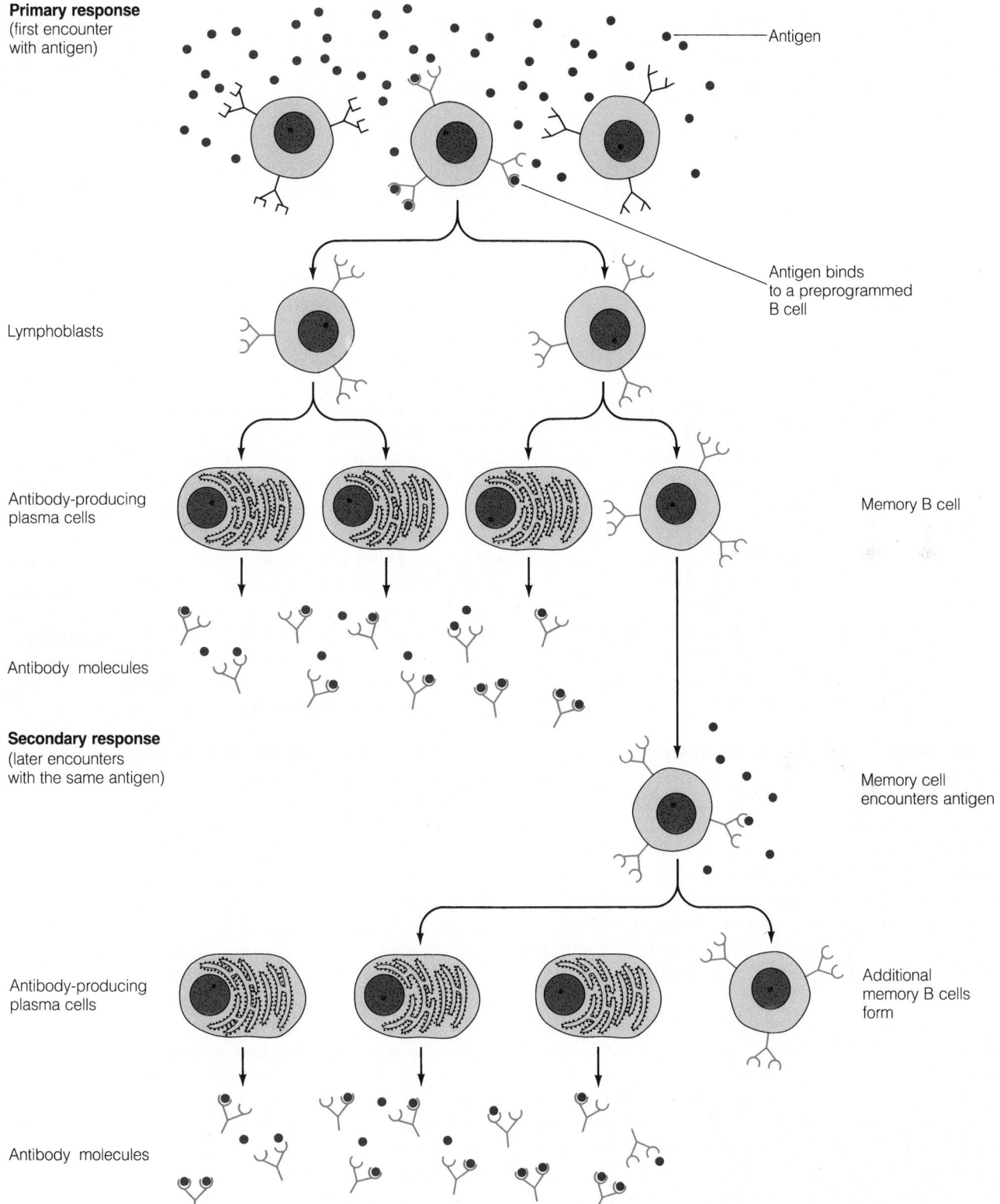

Primary response
(first encounter with antigen)

Antigen

Antigen binds to a preprogrammed B cell

Lymphoblasts

Antibody-producing plasma cells

Memory B cell

Antibody molecules

Secondary response
(later encounters with the same antigen)

Memory cell encounters antigen

Antibody-producing plasma cells

Additional memory B cells form

Antibody molecules

▷ **FIGURE 15–8 B-Cell Activation** Immunocompetent B cells are stimulated by the presence of an antigen, producing an intermediate cell, the lymphoblast. The lymphoblasts divide, producing plasma cells and some memory cells. Memory cells respond to subsequent antigen encroachment, yielding a rapid, secondary response.

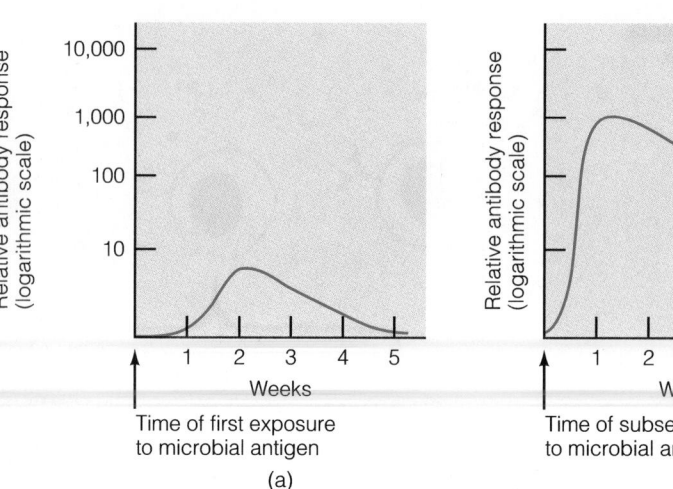

▷ **FIGURE 15–9 Primary and Secondary Responses** (*a*) The primary (initial) immune response is slow. It takes about 10 days for antibody levels to peak. Almost no antibody is produced during the first week as plasma cells are being formed. (*b*) The secondary response is much more rapid. Antibody levels rise almost immediately after the antigen invades. T cells show a similar response pattern.

The Initial Reaction to an Antigen Is Slower and Weaker Than Subsequent Responses. The first time an antigen enters the body, it elicits an immune response, but the initial reaction—or **primary response**—is relatively slow and of small magnitude (▷ Figure 15–9). During the primary response, antibody levels in the blood do not begin to rise until approximately the beginning of the second week *after* the intruder was detected. This delay occurs because it takes time for B cells to multiply and form a sufficient number of plasma cells. Antibody levels usually peak about the end of the second week, then decline over the next three weeks, partly explaining why it takes most people about a week to 10 days to combat a cold or the flu.

If the same antigen enters the body at a later date, however, the immune system acts much more quickly and more forcefully (Figure 15–9). This greatly fortified reaction constitutes the **secondary response**. As Figure 15–9 illustrates, during a secondary response antibody levels increase rather quickly, only a few days after the antigen has entered the body. The amount of antibody produced also greatly exceeds quantities generated during the primary response. Consequently, the antigen is quickly destroyed, and a recurrence of the illness is prevented.

The Rapidity of the Secondary Response Is the Result of the Production of Memory Cells During the Primary Response. As shown in Figure 15–8, during the primary response, some lymphocytes divide to produce memory cells. **Memory cells** are immunologically competent B cells that do not transform into plasma cells. Instead, they remain in the body awaiting the antigen's reentry. These cells therefore create a relatively large reserve force of antigen-specific B-cells. When the antigen reappears, memory cells proliferate rapidly, producing numerous plasma cells that quickly crank out antibodies to combat the foreign invaders. As illustrated in Figure 15–8, during the secondary response, the memory cells also generate additional memory cells that remain in the body in case the antigen should reappear at some later date.

Immune protection afforded by memory cells can last 20 years or longer and explains why once a person has had a childhood disease, such as the mumps or chicken pox, it is unlikely that he or she will contract it again. Resistance to disease that is provided by the immune system is known as **immunity.** From an evolutionary standpoint, this adaptation serves us extremely well, greatly reducing the incidence of infectious disease. Without it, humans would probably not be able to survive.

Antibodies Act in Four Ways to Destroy Antigens. Antibodies belong to a class of blood proteins called the globulins, introduced in Chapter 13. Antibodies are specifically called **immunoglobulins.** Each antibody consists of four peptide chains (▷ Figure 15–10). The chains are joined by disulfide bonds—that is, covalent bonds that form between sulfur atoms of different amino acid side groups. Two small chains intertwine with two larger chains, forming T-shaped molecules. It is the arms of the T that bind to antigens and confer specificity, in much the same way that the active sites of enzymes result in enzyme specificity. Immunologists have discovered five classes of antibodies, each with a slightly different role (Table 15–3).

Antibodies destroy foreign organisms and antigens through four mechanisms: (1) neutralization, (2) agglutination, (3) precipitation, and (4) complement activation. We will consider each one very briefly.

Neutralization. During **neutralization,** antibodies bind to viruses, forming a complete coating around them. This action prevents viruses from binding to plasma membrane receptors of body cells. If a virus cannot bind to a plasma membrane receptor, it cannot get inside most cells (▷ Figure 15–11, far right).

Neutralization also helps destroy bacterial toxins. A toxic protein, for instance, may be so heavily coated with antibody that it is rendered ineffective. Toxins and viruses neutralized by their antibody coating are eventually engulfed by macrophages and other phagocytic cells, as shown in Figure 15–11.

Agglutination. Antibodies deactivate foreign cells (bacteria and red blood cells transfused into another person) by **agglutination**—a clumping of the foreign cells (Figure

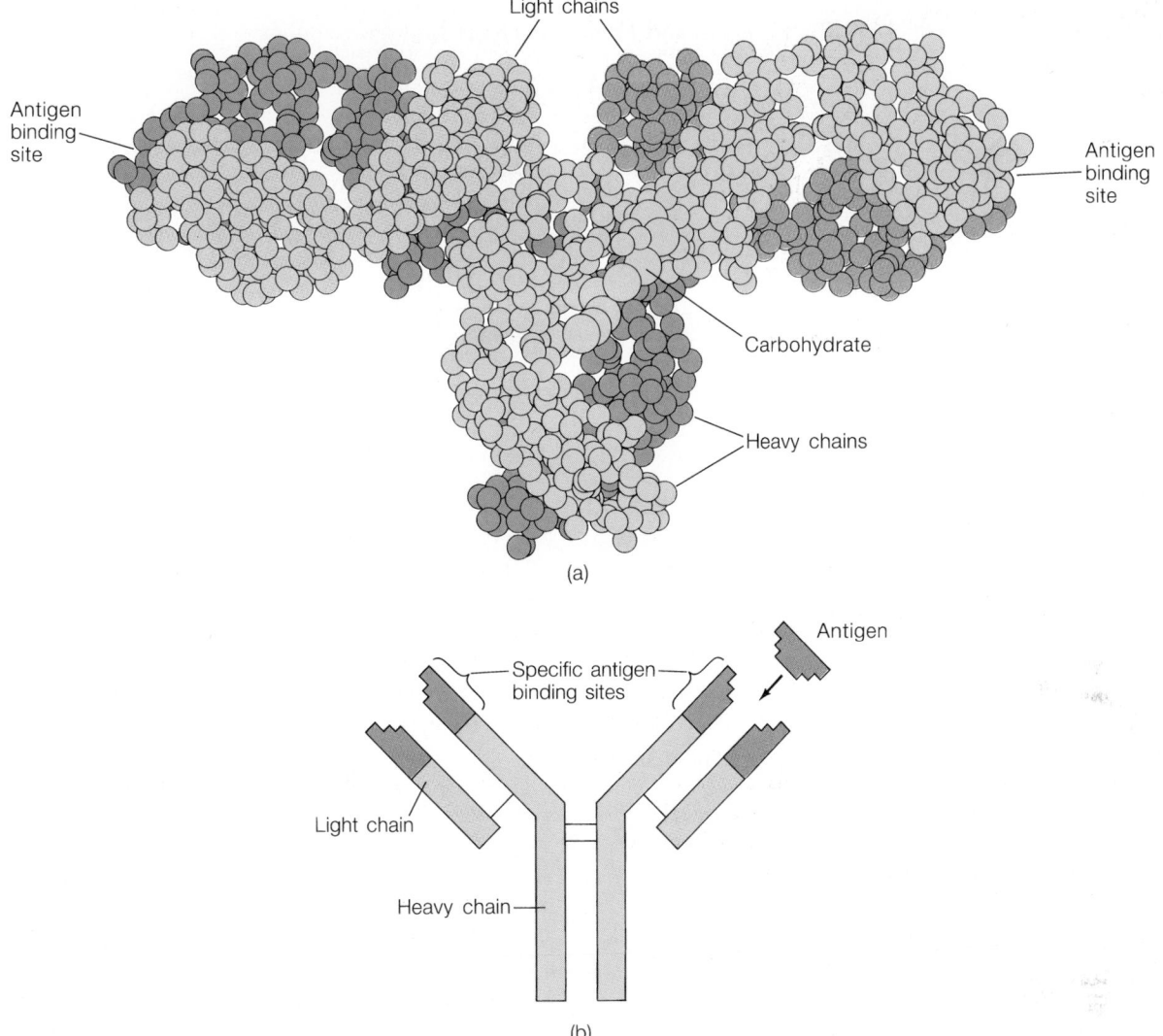

(a)

(b)

FIGURE 15–10 Antibody Structure (*a*) A three-dimensional model of an antibody showing the four chains. The molecule is T shaped before binding to an antigen. After it binds, it becomes Y shaped, as shown in (*b*), a diagrammatic represen

tation of the structure of an antibody molecule. It shows the four protein chains, two large (heavy chains) and two small (light chains). Note that the antigens bind to the arms of the molecule.

15–11). During agglutination, a single antibody may bind to several antigens, causing them to clump together. These antigen-antibody complexes are then phagocytized and removed from blood and body fluids by macrophages and other phagocytic cells.

Precipitation. Antibodies also bind to soluble antigens (for example, a protein), forming much larger, water-insoluble complexes that precipitate (fall out) out of solution, where they are engulfed by phagocytic cells.

Activation of the complement system. The final mechanism by which antibodies help rid the body of bacteria is through the activation of the complement system. As noted earlier in this chapter, the complement system is a family of blood-borne proteins that is part of the nonspecific immune response to antigens. The complement system is activated by the presence of antigen-antibody complexes (antibodies bound to antigens). When activated, the complement sys-

tem produces a membrane-attack complex that embeds in the plasma membrane of bacterial cells, causing them to leak, swell, and burst. Some proteins of this system stimulate the inflammatory process, and some coat microorganisms, facilitating their phagocytosis by macrophages.

Macrophages in the Body's Tissues Play a Key Role in Activating B Cells. Macrophages are phagocytic cells found in connective tissue, lymphoid tissue, and the organs of the lymphatic system (for example, lymph nodes); as you may recall from Chapter 13, these cells arise from monocytes, which escape from the bloodstream and set up residence in body tissues.

Macrophages play several important roles in the immune response. First, they phagocytize bacteria and other antigens at the site of infection. They also phagocytize antigen-antibody complexes, dead cells, and dead microorganisms,

TABLE15–3 Types and Functions of the Immunoglobulins

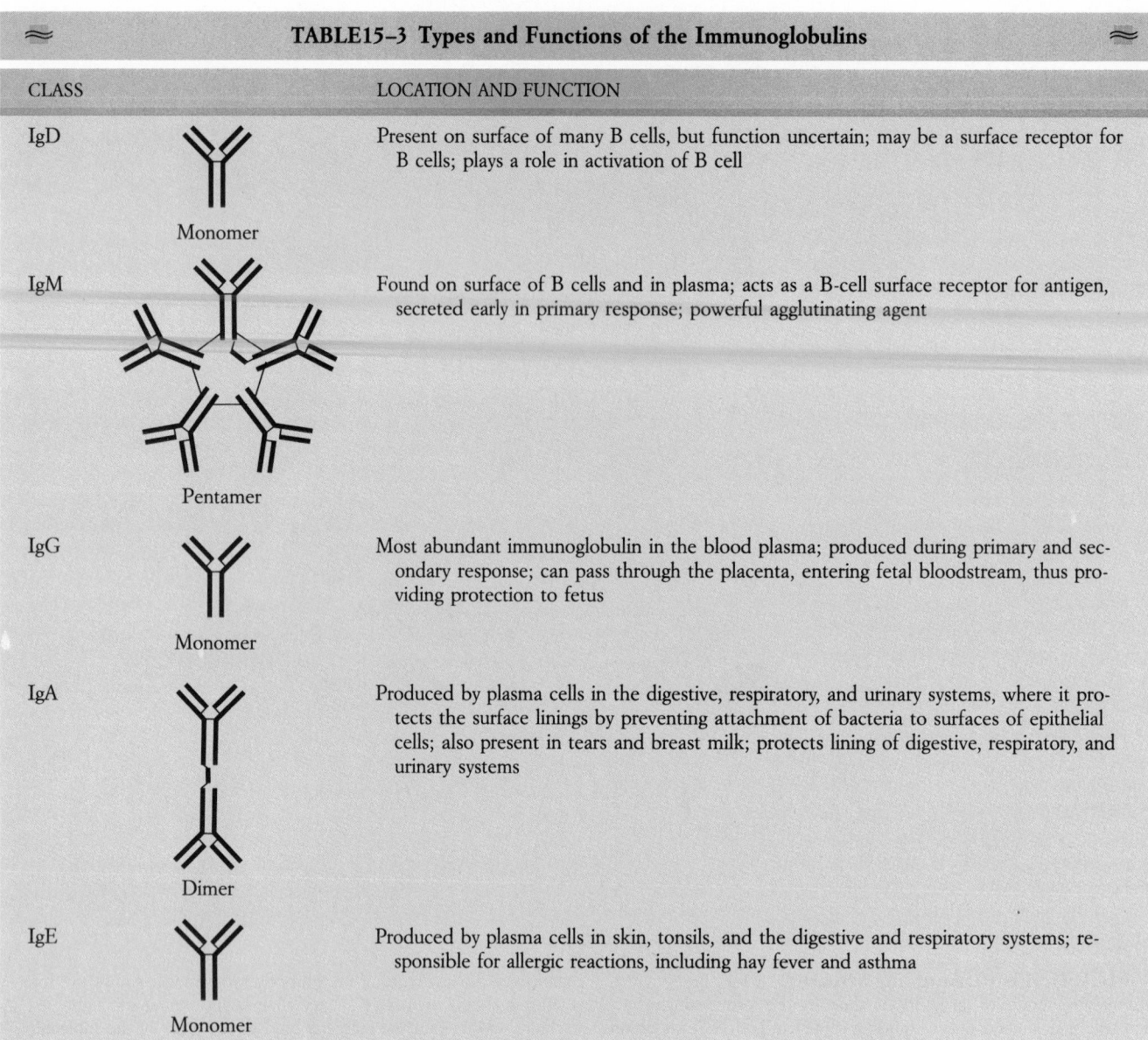

CLASS		LOCATION AND FUNCTION
IgD	Monomer	Present on surface of many B cells, but function uncertain; may be a surface receptor for B cells; plays a role in activation of B cell
IgM	Pentamer	Found on surface of B cells and in plasma; acts as a B-cell surface receptor for antigen, secreted early in primary response; powerful agglutinating agent
IgG	Monomer	Most abundant immunoglobulin in the blood plasma; produced during primary and secondary response; can pass through the placenta, entering fetal bloodstream, thus providing protection to fetus
IgA	Dimer	Produced by plasma cells in the digestive, respiratory, and urinary systems, where it protects the surface linings by preventing attachment of bacteria to surfaces of epithelial cells; also present in tears and breast milk; protects lining of digestive, respiratory, and urinary systems
IgE	Monomer	Produced by plasma cells in skin, tonsils, and the digestive and respiratory systems; responsible for allergic reactions, including hay fever and asthma

helping mop up the debris of the battle. Third, macrophages also help to activate T- and B-cell differentiation. B cells, in fact, cannot differentiate into plasma cells and produce antibodies without macrophages.

▷ Figure 15–12 is a simplified illustration showing the role of the macrophage in B-cell activation. As illustrated, macrophages first engulf invading bacteria. The macrophages then transfer antigens from the surface of the bacterium to their own plasma membrane. The macrophages then cluster around B cells, "presenting" the bacterial antigen to them. B cells that are programmed to respond to that antigen are activated when they encounter the concentrated bacterial antigen on the macrophage plasma membrane. As a result, the B cells begin to divide and differentiate, forming antibody-producing plasma cells and memory cells. Macrophages also secrete a chemical called **interleu-**

kin 1, which enhances the proliferation and differentiation of activated B cells.

Macrophages also present antigen to certain T cells, called helper T cells (described in more detail shortly and shown in Figure 15–12). When activated, the helper T cells produce a chemical substance known as **B-cell growth factor.** B-cell growth factor enhances the proliferation and differentiation of B cells much as does interleukin 1. It also enhances antibody production by the plasma cells (Figure 15–12).

T Cells Differentiate Into At Least Four Cell Types, Each with a Separate Function in Cell-Mediated Immunity

T cells provide a much more complex form of protection

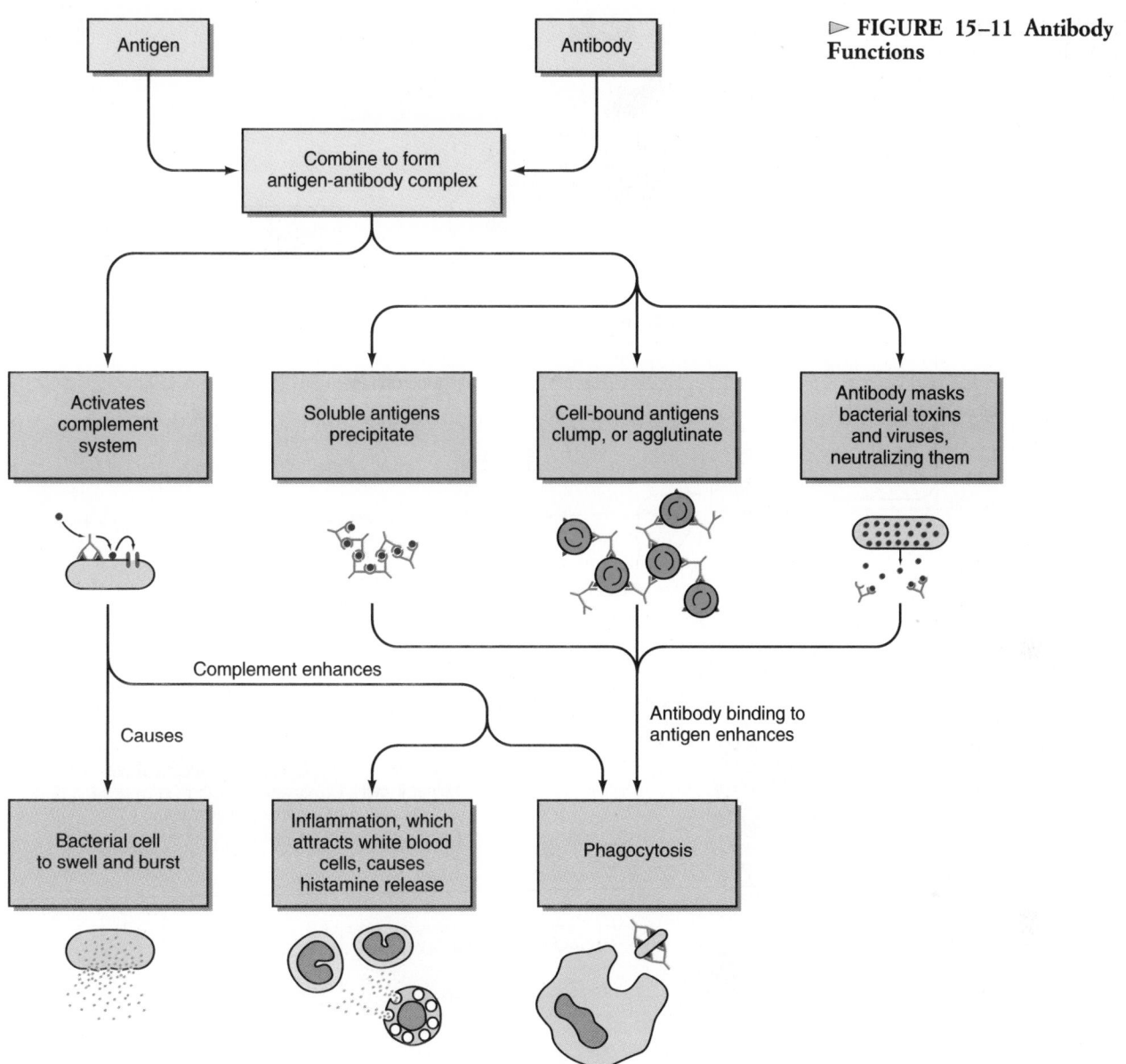

than B cells. Like B cells, they respond to the presence of antigens by undergoing rapid proliferation. T cells, however, differentiate into at least four cell types: (1) memory T cells, (2) cytotoxic T cells, (3) helper T cells, and (4) suppressor T cells (Table 15–4).

Memory T cells form a cellular reserve force that plays a crucial role in the secondary response. **Cytotoxic T cells** perform two essential roles (Table 15–4). Some cytotoxic T cells attack and kill body cells that have been infected by viruses. When a virus infects a cell, antigenic proteins in the virus's envelope become incorporated in the plasma membrane of the host cell. Cytotoxic T cells bind to that antigen and destroy the host cell. They also attack and kill bacteria, parasites, single-celled fungi, cancer cells, and foreign cells introduced during blood transfusions or tissue or organ transplants. Cytotoxic T cells bind to antigenic mol-

ecules in the membranes of these cells and release a chemical called **perforin-1** (▷ Figure 15–13). Perforin-1 molecules become embedded in the plasma membrane of the target cell. These molecules join to form pores, similar to those produced by the membrane-attack complex of the complement system. These pores cause the plasma membrane to leak, destroying the target cell within a few hours. After it has delivered its lethal payload, the cytotoxic cell detaches and is free to hunt down other antigens.

Helper T cells enhance the immune response and are activated by the presence of certain antigens. As noted earlier, helper T cells produce a B-cell growth factor, enhancing humoral immunity. They also enhance cell-mediated immunity by releasing a molecule known as **interleukin 2.** This substance increases the activity of cytotoxic T, suppressor T, and even helper T cells.

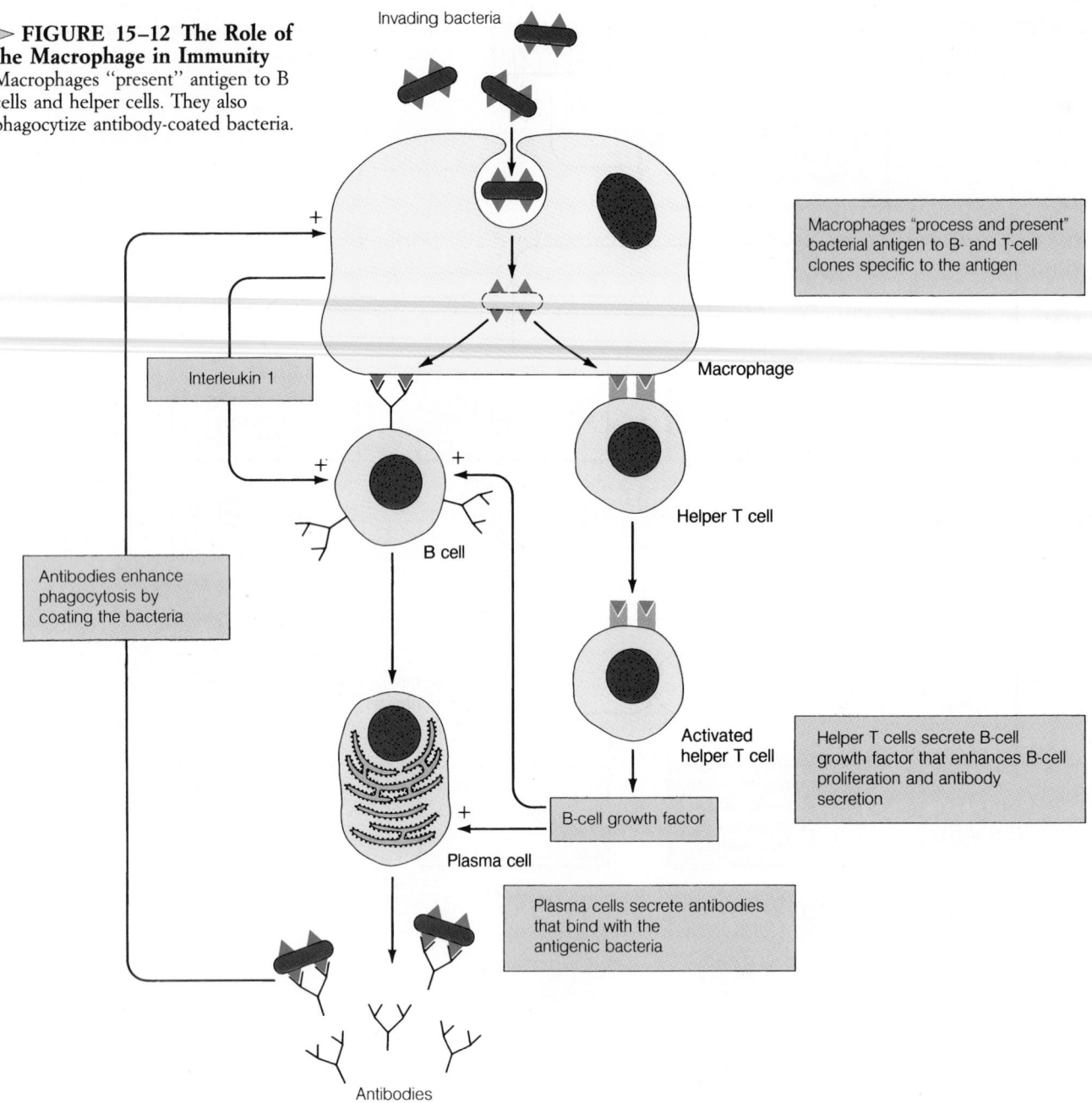

▷ **FIGURE 15–12 The Role of the Macrophage in Immunity** Macrophages "present" antigen to B cells and helper cells. They also phagocytize antibody-coated bacteria.

Invading bacteria

Macrophages "process and present" bacterial antigen to B- and T-cell clones specific to the antigen

Interleukin 1

Macrophage

Helper T cell

Antibodies enhance phagocytosis by coating the bacteria

B cell

Activated helper T cell

Helper T cells secrete B-cell growth factor that enhances B-cell proliferation and antibody secretion

B-cell growth factor

Plasma cell

Plasma cells secrete antibodies that bind with the antigenic bacteria

Antibodies

Helper T cells are the most abundant of all the T cells (making up 60% to 70% of the circulating T cells). They have been likened to the immune system's master switch. Without them, antibody production and T-cell activity would be greatly reduced. In fact, the immune response without helper T cells would be almost nonexistent. In their absence, antigens would stimulate a few B and T cells, then the process would come to a halt. Because the AIDS virus preferentially infects helper T cells, people suffering from the disease are unable to mount an effective immune response and eventually die from bacterial infections or cancer.

The role of **suppressor T cells** is less well understood. Research suggests that they "turn off" the immune reac-

tion as the antigen begins to disappear—that is, as the antigen is phagocytized. The activity of suppressor T cells, therefore, increases as the immune system finishes its job. Suppressor cells release chemicals that reduce B- and T-cell division.

Two Types of Immunity Are Possible: Active and Passive

One of the major medical advances of the last century was the discovery of **vaccines,** which help prevent bacterial and viral infections. Vaccines contain inactivated or weakened viruses, bacteria, or bacterial toxins. When injected into the body, the antigens in vaccines elicit an immune response.

TABLE 15-4 Summary of T Cells

CELL TYPE	ACTION
Cytotoxic T cells	Destroy body cells infected by viruses, and attack and kill bacteria, fungi, parasites, and cancer cells
Helper T cells	Produce a growth factor that stimulates B-cell proliferation and differentiation and also stimulates antibody production by plasma cells; enhance activity of cytotoxic T cells
Suppressor T cells	May inhibit immune reaction by decreasing B- and T-cell activity and B- and T-cell division
Memory T cells	Remain in body awaiting reintroduction of antigen, at which time they proliferate and differentiate into cytotoxic T cells, helper T cells, suppressor T cells, and additional memory cells

Many vaccines provide immunity or protection from microorganisms for long periods, sometimes for life. Others, however, give only short-term protection.

Vaccines stimulate the immune reaction because the weakened or deactivated organisms (or toxins) they contain still possess the antigenic proteins or carbohydrates that trigger B- and T-cell activation. Because their antigens have been seriously weakened or deactivated, however, vaccines usually do not cause disease.

Vaccination provides a form of protection that immunologists call **active immunity**—so named because the body actively produces memory T and B cells that protect a person against future infections. Viral or bacterial infections also produce active immunity.

Vaccinations are important in helping us avoid deadly diseases such as polio, typhus, and smallpox that often kill their victims before the immune system can mount an effective response. In fact, in the wealthier nations of the world, like the United States, vaccines have nearly eliminated many infectious diseases like small pox. Today, many

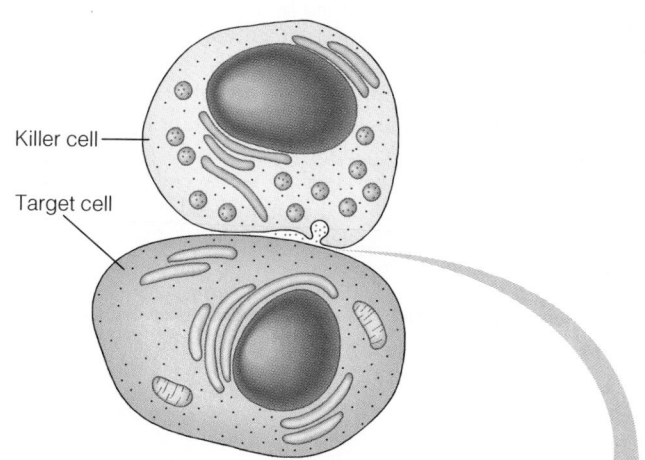

Killer cell

Target cell

▷ **FIGURE 15–13 How Cytotoxic T Cells Work**
Cytotoxic T cells, containing perforin-1 granules, bind to their target and release perforin-1, then detach in search of other invaders. Perforin-1 molecules congregate in the target plasma membrane, forming a pore that disrupts the plasma membrane, causing the cell to die. By Dana Burns from John Ding-E Young and Zanvil A. Kohn, "How Killer Cells Kill." Copyright © January 1988 by Scientific American, Inc. All rights reserved.

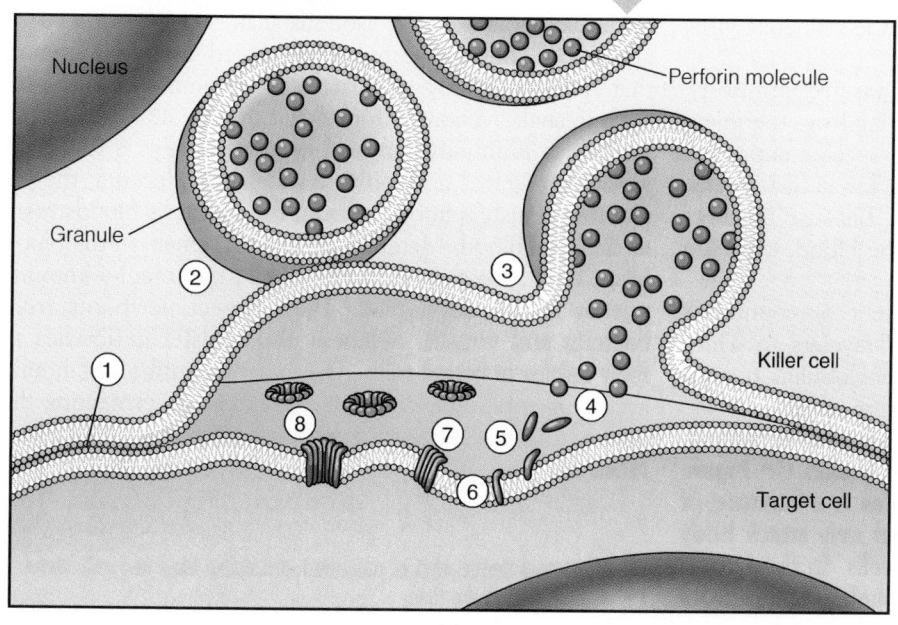

Nucleus

Perforin molecule

Granule

Killer cell

Target cell

(a)

BRINGING BABY UP RIGHT: THE IMMUNOLOGICAL AND NUTRITIONAL BENEFITS OF BREAST MILK

A baby is born into a dangerous world in which bacteria and viruses abound. Complicating matters, the immune system of a newborn child is poorly developed. Fortunately, newborns are protected by passive immunity—antibodies that have traveled from their mothers' blood. Antibodies also travel to the infant in breast milk (▷ Figure 1).

Several immunoglobulins are present in breast milk. One of these is called **secretory IgA.** It is present in very high quantities in **colostrum,** a thick fluid produced by the breast immediately after delivery—before the breast begins full-scale milk production. Colostrum is so important, in fact, that some hospitals give "colostrum cocktails" to newborns who are not going to be breast fed by their mothers. Nurses remove the colostrum from the mother's breast with a breast pump and feed it to the baby in a bottle.

Colostrum, says Sarah McCamman, a nutritionist at the University of Kansas-Medical School, coats the lining of the

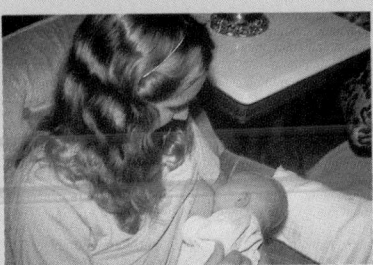

▷ **FIGURE 1 Breast Feeding**
Mother breast feeding her newborn infant.

intestines. The IgA antibodies in colostrum prevent bacteria ingested by the infant from adhering to the epithelium and gaining entrance. Breast milk also contains lysozyme, an enzyme that breaks down the cell walls of bacteria, destroying them.

Unfortunately, not all medical personnel agree on the benefits of breast milk. One problem, they say, is that breast milk has unusually low levels of iron. This fact has led many physicians to recommend iron supplements for newborns. A more careful analysis, however,

shows that breast-fed infants generally do not suffer from iron deficiency because the percentage of iron absorbed from breast milk is extraordinarily high. Thus, low levels of iron in breast milk are offset by the high absorption.

The wisdom of iron supplements has also been questioned on other grounds. Researchers, for example, have found that iron supplements increase the incidence of harmful bacterial infection in newborns. As noted in the chapter, iron is a limiting factor in many pathogenic bacteria. Low levels of iron in breast milk, therefore, may reduce bacterial replication in an infant's intestinal tract, aiding in protecting the newborn.

In general, breast-fed babies are healthier than bottle-fed babies. The incidence of gastroenteritis (inflammation of the intestine), otitis (ear infections), and upper respiratory infections is lower in breast-fed babies. Studies also show that children breast fed for at least six months contract fewer childhood cancers than their bottle-fed counterparts. The incidence of childhood lymphoma, a cancer of the lymph glands, in bottle-

---

physicians recommend vaccines against the influenza virus for young children and the elderly.

The second type of immunity, called **passive immunity,** is a temporary form of protection, resulting from the injection of immunoglobulins (antibodies to specific antigens). Immunoglobulins remain in the blood for a few weeks, protecting an individual from infection. Because the liver slowly removes these molecules from the blood, a person gradually loses protection.

Immunoglobulins are given to prevent or counteract certain infections already under way. Travelers to Third World countries are often given immunoglobulins to viral hepatitis (liver infection) as a preventive measure.

In addition, immunoglobulins are used to treat individuals who have been bitten by poisonous snakes (▷ Figure 15-14). The venom in poisonous snakes is a mixture of proteins, enzymes, and polypeptides that may attack body cells, especially nerve cells and cardiac cells. In the United States, the most common poisonous snakebite comes from

rattlesnakes.[4] Bites of poisonous snakes can be treated by antivenom, immunoglobulins produced in other animals that quickly destroy or deactivate the immunogenic proteins in snake venom before they can have adverse effects.

Passive immunity can also occur naturally. A fetus, for instance, receives antibodies through the placenta, the organ that transfers nutrients from the mother's bloodstream to the fetal blood. Maternal antibodies remain in the blood of an infant for several months while the infant's immune system is still developing. They protect newborns from bacteria and viruses. Mothers also transfer antibodies to their babies in breast milk. The maternal antibodies in milk attack bacteria and viruses in the intestine, protecting the infant from infection. (For more on this topic, see Health Note 15–1.)

---

[4] In the United States, 15% of untreated rattlesnake bites and only about 1% of treated bites are fatal.

fed babies is nearly double the rate in breast-fed children for reasons not yet understood.

New research also suggests that certain proteins in breast milk may stimulate the development of a newborn's immune system. In laboratory experiments, the still-unidentified proteins speed up the maturation of B cells and prime them for antibody production. These soluble proteins may also activate macrophages, which play a key role in the immune system.

Breast milk is also more digestible and more easily absorbed by infants than formula. Formula is a mixture of cow's milk, proteins, vegetable oils, and carbohydrates. It is only an approximation of mother's milk and is not broken down and absorbed as completely as breast milk.

Because of a growing awareness of the benefits of breast feeding, virtually every national and international organization involved with maternal and child health supports breast feeding, says McCamman.

Breast feeding can have a major health impact in this country. Unfortunately, the benefits are not as widely known as many people would like, even among health-care professionals. Fortunately, more and more health workers, including physicians, are beginning to understand the benefits of breast feeding and are promoting this option. As a result, many middle-class American women are now choosing to breast-feed.

Unfortunately, says McCamman, there is "a huge population of low-income, poorly educated . . . women who choose not to nurse." The federal government may be playing an unwitting role in their decision. A national program aimed at improving child nutrition provides free formula to needy women, perhaps discouraging mothers from breast feeding.

Another reason is economic. Low-income women often work at jobs that do not provide maternity leave. Thus, these women must return to work soon after giving birth.

Still another reason for the low rate of breast feeding in low-income women is that women need a lot of support to nurse. "People think nursing is innate, natural, and easy," says McCamman. That is not always the case, however. In some instances, getting started requires guidance and education. Without that education and support, breast feeding can be a difficult and painful experience that discourages many women. Breast feeding among all women, rich and poor, may also be discouraged by attitudes and fear of embarrassment. In fact, many open-minded people find breast feeding in public or even semipublic settings embarrassing.

Given the many benefits of breast feeding, McCamman recommends it to all mothers who can. Physicians can help by educating their patients on the benefits of breast feeding. "Doctors should present the information on breast and bottle feeding," says McCamman, "outlining the pros and cons of both methods. Then, let the woman choose. Too few doctors do that today, so women aren't making informed decisions."

(a)

(b)

▷ **FIGURE 15–14 Poison and Antidote** (*a*) Poisonous snakes like this rattler inject venom into their victims. (*b*) Venom can be milked from the snake and is used to produce antivenin, a serum containing immunoglobulins that neutralize the venom.

TABLE 15-5 Summary of Blood Types

↑ BLOOD TYPE	↑ ANTIGENS ON PLASMA MEMBRANES OF RBCs	↑ ANTIBODIES IN BLOOD	SAFE TO TRANSFUSE To	From
A	A	b*	A, AB	A, O
B	B	a	B, AB	B, O
AB	A + B	—	AB	A, B, AB, O
O	—	a + b	A, B, AB, O	

*Lowercase *b* indicates antibody to B antigen.

Vaccination Fears in the United States. Vaccines have helped lower the incidence of many infectious diseases in the United States and other Western countries. Vaccines for diphtheria, tetanus, whooping cough, polio, measles, mumps, and congenital rubella (German measles), for example, have reduced the occurrence of these often-lethal diseases in the United States by more than 99%.

Despite the successes of vaccines, publicity concerning their rare, but sometimes devastating, side effects has created something of a medical dilemma in the United States, Japan, and Great Britain. In 1976 and 1977, for example, a mass-immunization program in the United States for the swine flu, one type of influenza, resulted in the paralysis of a number of people. As a result of public concern over this and other incidents, many parents have chosen not to have their children vaccinated. Excessive media attention given to the rare but serious complications, say some critics, has harmed efforts to promote vaccination. Because of this publicity, public fear of vaccination, and lawsuits, pharmaceutical companies have become increasingly reluctant to invest the huge sums of money needed to develop and market vaccines.

Researchers also believe that another cause for the decline in vaccination stems from the success of previous immunization programs, which have greatly reduced the incidence of most infectious diseases. Parents reared in an environment free of such diseases, researchers say, are often unaware of the dangers of infectious disease. Having their children immunized seems unimportant. Public health officials are concerned that the incidence of infectious diseases such as polio and measles may increase as a consequence.

Harmful side effects from conventional vaccines are often caused by reactions to certain "nonessential" antigens on the injected microorganism. These antigens frequently play little or no role in immunity. Therefore, by eliminating these antigens from vaccines, researchers hope to develop even safer alternatives to vaccines in use today.

≈ PRACTICAL APPLICATIONS: BLOOD TRANSFUSIONS AND TISSUE TRANSPLANTATION

Although important in protecting us from microorganisms, the immune system presents something of a dilemma during blood transfusions and tissue transplants—biological interventions unwitnessed in evolution.

Blood Transfusions Require Careful Cross-Matching of Donors and Recipients

The surface of the red blood cell (RBC) membrane contains many inherited antigens (glycoproteins). One group of antigens is used to determine blood type. As you probably know, four blood types exist: A, B, AB, and O. The letters refer to one type of antigen present on the plasma membrane of RBCs of an individual.

As illustrated in Table 15–5, individuals with type A blood contain RBCs whose plasma membranes contain the A antigen. RBCs in individuals with type B blood contain type B antigens. People with AB blood have both A and B antigens, and people with type O have neither antigen.

To perform a successful blood transfusion, physicians must match donors and recipients. Individuals with type A blood, for example, can receive blood from people with the same blood type but cannot receive blood from people with type B. Individuals with type B blood can generally receive blood from people having type B blood but not from type A donors.

Cross-matching the blood is essential to prevent an immune reaction. Immune reactions during improper transfusions result from antibodies found in the bloodstream of most people. As shown in Table 15–5, each blood type carries a specific type of antibody. People with type A blood, for example, naturally contain antibodies to the B antigen. (As a result, people with type A blood cannot receive type B blood.) People with type B blood contain antibodies to the A antigen. For reasons not well understood, these antibodies appear in the blood during the first year of life.

Problems arise when incompatible blood types are mixed. For example, imagine that an individual with type A blood is accidentally given type B blood (▷ Figure 15–15) The antibodies to type B blood found in the recipient bind to the transfused RBCs (containing type B antigens), causing them to agglutinate (clump) and hemolyze (burst). Hemolysis and agglutination constitute the **transfusion reaction.**

RBC clumping restricts blood flow through capillaries, reducing oxygen and nutrient flow to cells and tissues. Massive hemolysis results in the release of large amounts of

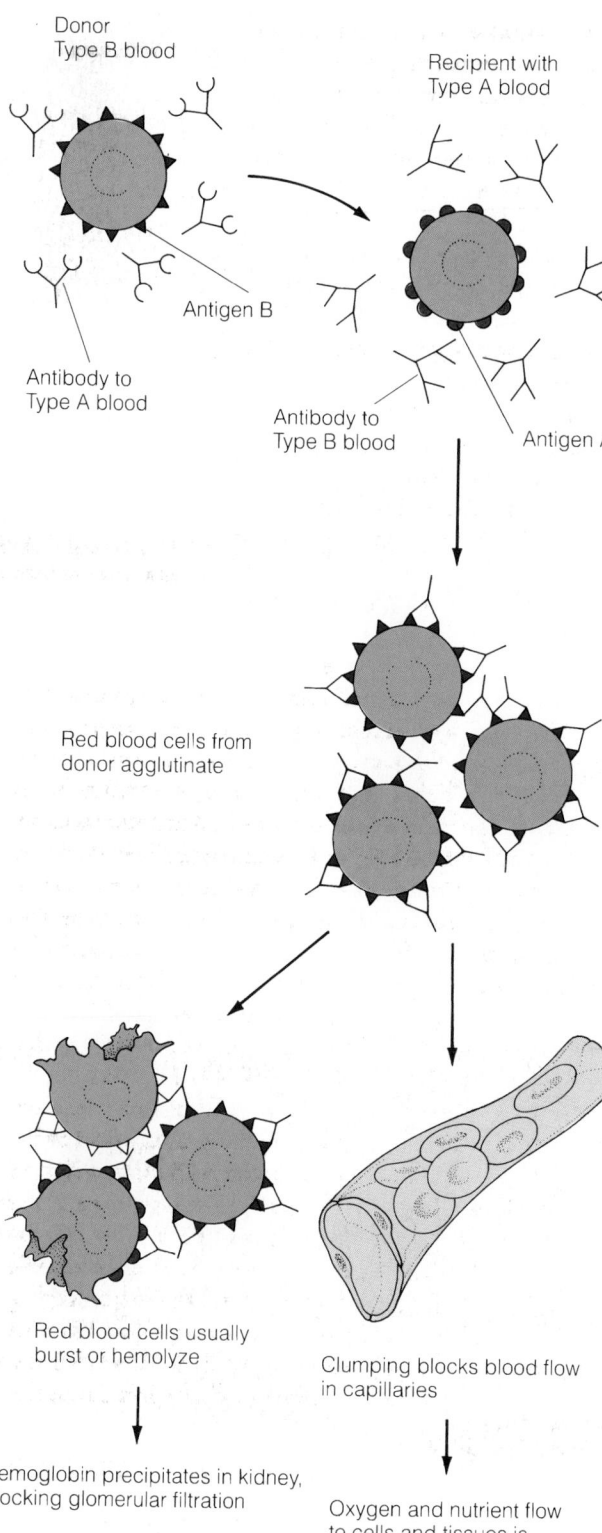

Donor
Type B blood

Recipient with
Type A blood

Antigen B

Antibody to
Type A blood

Antibody to
Type B blood

Antigen A

Red blood cells from
donor agglutinate

Red blood cells usually
burst or hemolyze

Clumping blocks blood flow
in capillaries

Hemoglobin precipitates in kidney,
blocking glomerular filtration

Oxygen and nutrient flow
to cells and tissues is
reduced

▷ **FIGURE 15–15 Transfusion Reaction** Type B blood transfused into an individual with type A blood results in a transfusion reaction, characterized by agglutination and hemolysis.

hemoglobin into the blood plasma. Hemoglobin precipitates in the kidney, blocking the tiny tubules that produce urine, which often results in acute kidney failure.

Because of the possibility of a transfusion reaction, successful transfusions require careful matching of the blood types of the donor and recipient. As Table 15–5 shows, RBCs from individuals with type O blood have neither A nor B antigens. Therefore, type O blood can be transfused into individuals with all four types: A, B, AB, and O. Type O individuals are said to be **universal donors.**

Type O blood, while free of antigens, contains antibodies to both A and B antigens. Therefore, individuals with type O blood can receive only type O blood. Any other type of blood would cause a transfusion reaction.

As shown in Table 15–5, individuals with type AB blood contain RBCs with both A and B antigens but no antibodies related to the ABO system. These people can therefore receive blood from all others and are consequently referred to as **universal recipients.** AB blood can be safely transfused only into individuals with AB blood.

The terms *universal donor* and *universal recipient* are somewhat misleading, however, because RBCs also contain other antigens that can cause transfusion reactions. The most important of these is called the Rh factor. This antigen was first identified in rhesus monkeys, hence the designation. People whose cells contain the Rh antigen, or **Rh factor,** are said to be **Rh positive.** Those without it are **Rh negative.**

Unlike the ABO system, in the Rh system antibodies are produced *only* when Rh-positive blood is transfused into the bloodstream of a person with Rh-negative blood. The first transfusion of Rh-positive blood into an Rh-negative person generally does not result in a transfusion reaction, but a second transfusion does. To reduce the likelihood of a transfusion reaction, Rh-negative people should receive only Rh-negative blood, and Rh-positive people should receive only Rh-positive blood.

The Rh factor becomes particularly important during pregnancy. Problems can arise if an Rh-negative mother has an Rh-positive baby (▷ Figure 15–16). Even though the maternal and fetal bloodstreams are separate, small amounts of fetal blood usually enter the maternal bloodstream at birth, causing the immune reaction. Rh antibodies will form in the maternal bloodstream, and the woman will be sensitized to the Rh factor.

If the woman is not treated at the time and becomes pregnant again with an Rh-positive baby, maternal antibodies to the Rh factor will cross the placenta. These antibodies cause fetal RBCs to agglutinate, then break down, eventually resulting in anemia and hypoxia (lack of oxygen to tissues). Unless the baby receives a blood transfusion (of Rh-negative blood) *before* birth and several after birth, it is likely to have brain damage and may even die.

To prevent antibody production in Rh-negative women who give birth to Rh-positive babies, physicians inject antibodies to fetal Rh-positive RBCs into the mother soon after she has given birth. (The antibody containing serum is called RhoGAM.) These antibodies bind to Rh-positive RBCs from the fetus before a woman's immune system can respond to them. This prevents a woman from being sen-

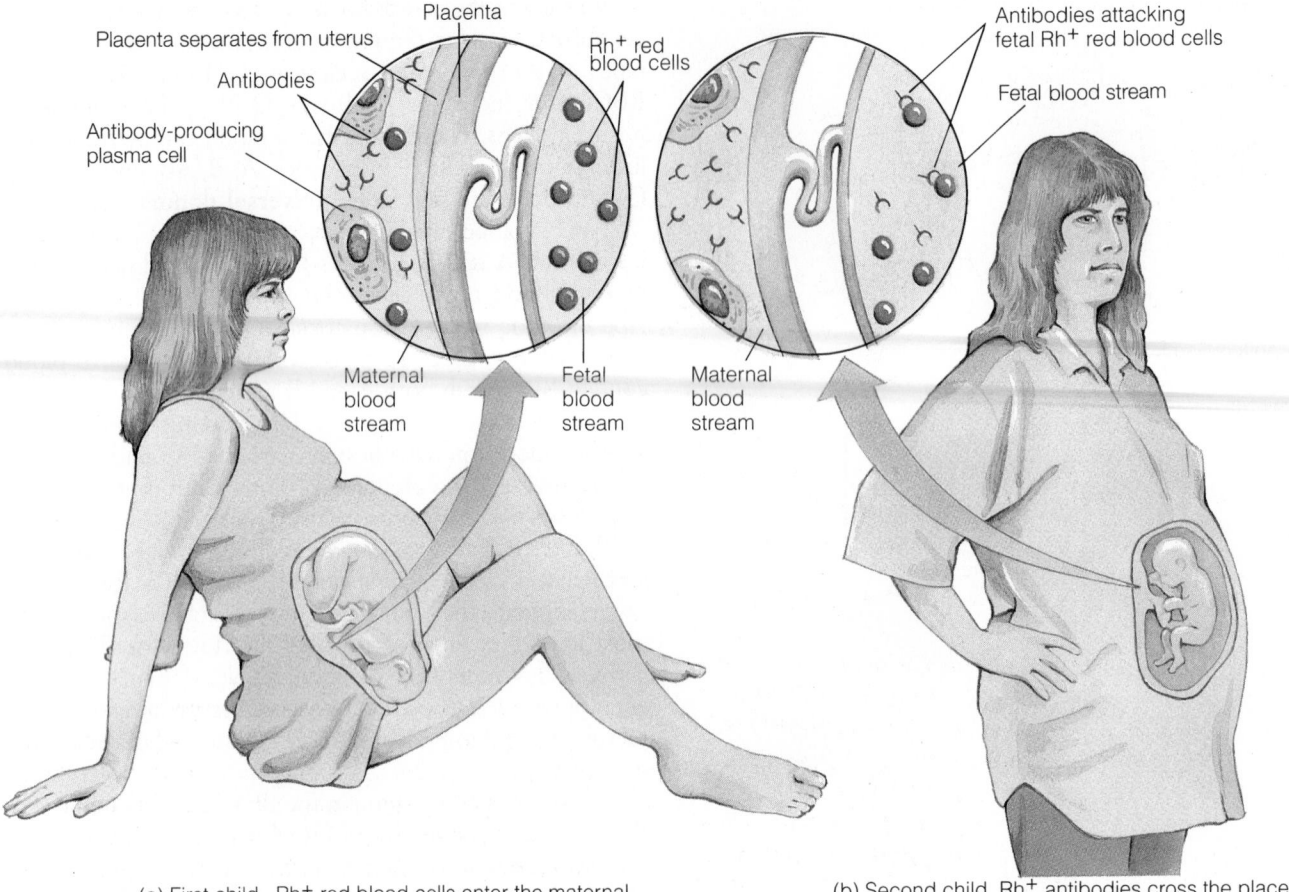

Placenta separates from uterus

Placenta

Antibodies

Antibody-producing plasma cell

Rh⁺ red blood cells

Maternal blood stream

Fetal blood stream

Antibodies attacking fetal Rh⁺ red blood cells

Fetal blood stream

Maternal blood stream

(a) First child. Rh⁺ red blood cells enter the maternal bloodstream during birth, evoking an immune reaction.

(b) Second child. Rh⁺ antibodies cross the placenta, destroying fetal red blood cells.

▷ **FIGURE 15–16 The Rh Factor and Pregnancy**
Rh-positive cells from the fetus enter the mother's blood at birth. If the mother is Rh negative, her immune system responds, producing antibodies to the Rh-positive RBCs and destroying them.

Problems arise if the mother becomes pregnant again and has another Rh-positive baby. If the mother was not treated the first time, antibodies to Rh-positive RBCs cross the placenta and destroy fetal RBCs.

sitized. In other words, RhoGAM prevents B cells from being stimulated and obviously blocks the production of memory cells to Rh factor. To be effective, however, the treatment must be given within 72 hours after the baby is born.

Tissue Transplantation Often Evokes Cell-Mediated Immunity, Which Can Be Blocked by Certain Drugs

Tissue transplantation is a much more complex matter. Only three conditions exist in which a person can receive a transplant and not reject it. One is if the tissue comes from the individual himself or herself. For burn victims, surgeons might use healthy skin from one part of the body to cover a badly damaged region. The second instance is when a tissue is transplanted between identical twins—individuals from a single fertilized ovum that split and formed two embryos. These individuals are genetically identical and therefore have identical cellular antigens.

A third instance occurs when tissue rejection is inhibited by specific drugs. For example, heart, liver, and kidney transplants are successful only when recipients are treated with immunosuppressive drugs—that is, drugs that suppress the immune system. This treatment must be continued throughout the life of the patient. Unfortunately, most immune suppressants have numerous side effects and often leave the patient vulnerable to bacterial and viral infections. In the 1980s, a new drug known as **cyclosporin** was introduced. This drug suppresses the formation of interleukin 2 by helper T cells, thus greatly reducing cell-mediated immunity without affecting B cells. Patients who receive the drug are therefore able to combat many bacterial infections with antibodies.

≈ DISEASES OF THE IMMUNE SYSTEM

The immune system, like all other body systems, can malfunction resulting in a wide range of symptoms. This section will look at two disorders: allergies and autoimmune diseases.

The Most Common Malfunctions of the Immune System Are Allergies

An **allergy** is an overreaction to some environmental substance, such as pollen or a food (▷ Figure 15–17). Antigens that stimulate allergic reactions are called **allergens.** Allergens cause the production of IgE antibodies (see Table 15–3).[5] As Figure 15–17 shows, these antibodies then bind to mast cells. **Mast cells** are found in many tissues, especially in the connective tissue surrounding blood vessels, and have large cytoplasmic vesicles containing the chemical histamine.

Allergens bind to the IgE antibodies attached to the mast cells, triggering the release of histamine from the vesicles via exocytosis, as shown in Figure 15–17. Histamine, in turn, causes nearby arterioles to dilate. Histamine released in the lungs causes the bronchioles to constrict, reducing airflow and making breathing even more difficult. This condition is called **asthma** (Chapter 14).

The allergic reaction usually occurs in specific body tissues, where it creates local symptoms that, while irritating, are not life-threatening. For example, an allergic response may occur in the eyes, causing redness and itching. It may occur in the upper respiratory tract (nose), causing stuffiness. However, the allergic response can also occur in the bloodstream, where it may cause death if not treated quickly. In certain people, penicillin or a bee venom in the bloodstream can cause the release of massive amounts of histamine and other chemicals. The sudden release of these chemicals causes extensive dilation of blood vessels in the skin and other tissues. This causes the blood pressure to fall precipitously, essentially shutting down the circulatory system. Histamine released by mast cells causes severe constriction of the bronchioles (the ducts in the lungs that open onto the alveoli), making breathing difficult. The decline in blood pressure and constriction of the bronchioles result in anaphylactic shock. Death may follow if measures are not taken to reverse the physiological nightmare. One such measure is an injection of the hormone epinephrine (commonly known as adrenalin), which rapidly reverses the constriction of the bronchioles.

Allergies are treated in three basic ways. First, patients are advised to avoid allergens—for instance, get rid of a pet dog or cat or avoid milk and milk products. Second, patients may also be given antihistamines, drugs that counteract the effects of histamine. Third, patients may also be given regular allergy shots, injections of increasing quantities of the offending allergen. In some cases, this treatment gradually makes an individual less and less sensitive to the allergen. Studies suggest that desensitization results from the production of another class of antibodies, the IgG antibodies, which bind to allergens, locking them up. This, in turn, blocks antigen from binding to the mast cells, as shown in ▷ Figure 15–18, thus preventing the release of

histamine and other chemical substances responsible for the allergic reaction.

Autoimmune Diseases Result From an Immune Attack on the Body's Own Cells

Occasionally, the immune system mounts an attack on the body's own cells. This gives rise to an **autoimmune disease.** These unusual and extremely dangerous diseases result from many causes. For example, normal body proteins may be modified by environmental pollutants, viruses, or genetic mutations so they are no longer recognizable as self.

In other cases, normal body proteins that are usually isolated from the immune system enter the bloodstream and evoke an immune response. For example, a protein called thyroglobulin is produced by the thyroid gland in the neck. Thyroglobulin is stored inside the gland and not exposed to cells of the immune system. If the gland is injured, however, thyroglobulin may enter the bloodstream. Lymphocytes encountering this essentially foreign protein may then mount an immune response to it.

Still another cause of autoimmune reaction is exposure to antigens that are nearly identical to body proteins. The bacterium that causes strep throat, for example, introduces an antigen structurally similar to one of the proteins found in the plasma membranes of the cells lining the heart valves of some individuals. The body mounts an attack on the bacterium, but antibodies may also bind to the lining of the heart valve, causing a local inflammation and scar tissue to develop, thus damaging the valve and resulting in valvular incompetence, which was discussed in Chapter 12.

≈ AIDS: FIGHTING A DEADLY VIRUS

Damion Knight, a bright, young cabinetmaker, began to lose weight and experience bouts of unexplained fever. His lymph nodes became swollen, and he began feeling weak and drowsy. Friends urged him to see a doctor. A blood test revealed that he had contracted AIDS, which is caused by a virus that attacks and weakens the immune system.

Like thousands of others, Knight died a few years after diagnosis. His doctor could not help him. Medical science has only begun to understand this disease, which could kill hundreds of thousands, perhaps millions, of people the world over in the coming decade.

AIDS was discovered in the late 1970s. Since then, one researcher has likened it to the plague that spread through Europe in the 14th and 15th centuries. In the 1300s, the plague killed one-quarter of the adult population and numerous children.

By 1985, approximately 20,000 Americans had been diagnosed with AIDS. Nearly half of them had died. By September 1990, the cases had risen to 120,000. The death toll hovered around 60,000, but all of the individuals with the virus were expected to die. In 1992, the number of Americans infected with the AIDS virus was estimated to

---

[5] Some allergies involve IgG or IgM, and some apparently do not involve antibodies at all.

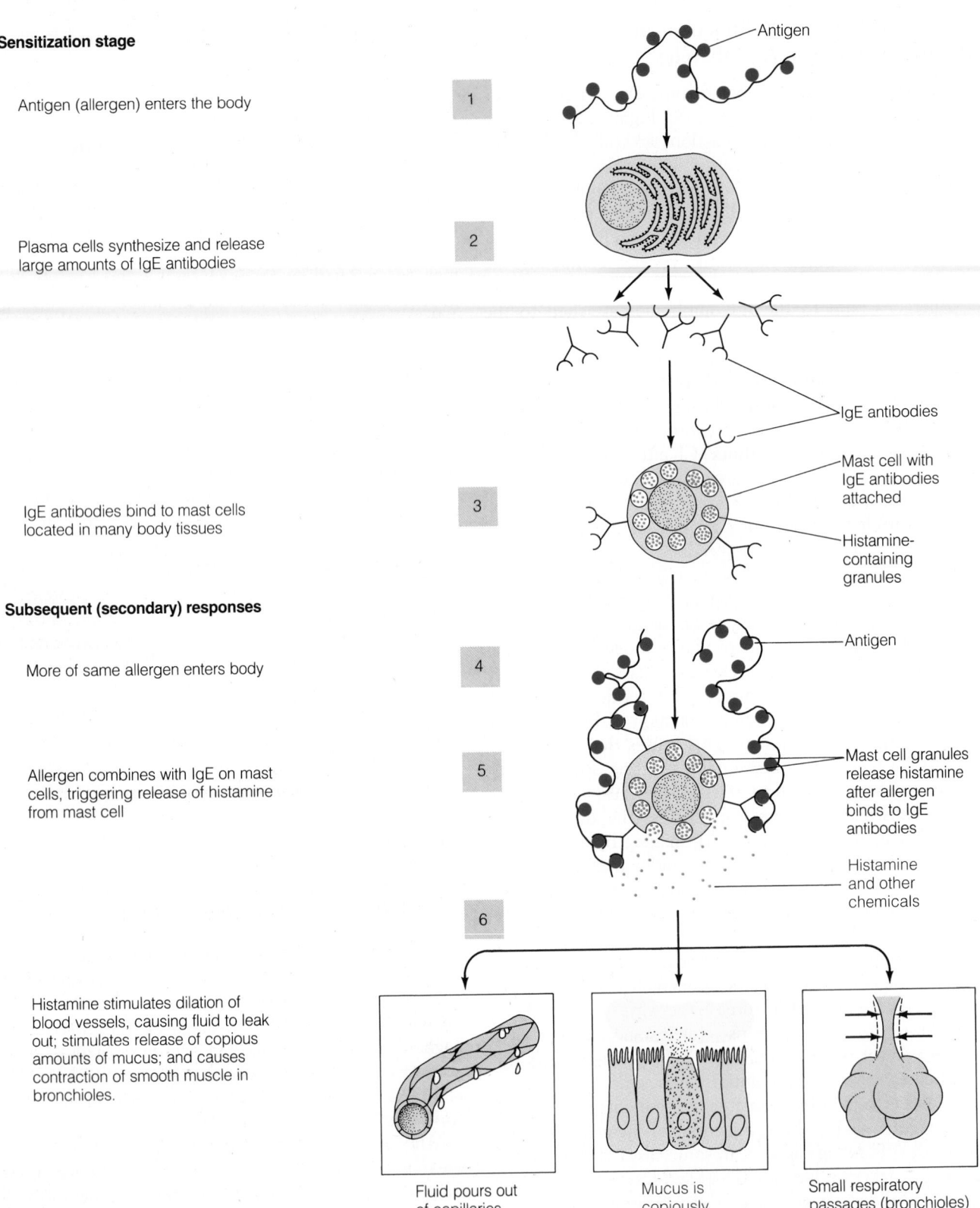

Sensitization stage

Antigen (allergen) enters the body

Plasma cells synthesize and release large amounts of IgE antibodies

IgE antibodies bind to mast cells located in many body tissues

Subsequent (secondary) responses

More of same allergen enters body

Allergen combines with IgE on mast cells, triggering release of histamine from mast cell

Histamine stimulates dilation of blood vessels, causing fluid to leak out; stimulates release of copious amounts of mucus; and causes contraction of smooth muscle in bronchioles.

1

2

3

4

5

6

Antigen

IgE antibodies

Mast cell with IgE antibodies attached

Histamine-containing granules

Antigen

Mast cell granules release histamine after allergen binds to IgE antibodies

Histamine and other chemicals

Fluid pours out of capillaries

Mucus is copiously released

Small respiratory passages (bronchioles) constrict

▷ **FIGURE 15–17 Allergic Reaction** Antigen stimulates the production of massive amounts of IgE, a type of antibody produced by plasma cells. IgE attaches to mast cells. This is the sensitization stage. When the antigen enters again, it binds to the IgE antibodies on the mast cells, triggering a massive release of histamine and other chemicals. Histamine, in turn, causes blood vessels to dilate and become leaky. This triggers the production of mucus in the respiratory tract. In some people, the chemicals released by the mast cells also cause the small air-carrying ducts in the lungs to constrict, making breathing difficult.

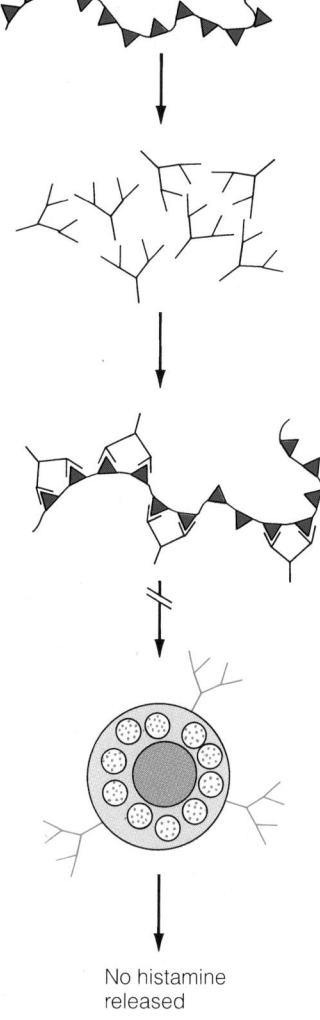

Injection of antigens

Body produces IgG antibodies

IgG antibodies bind to antigens

Binding of antigens to IgE on mast cells is reduced

No histamine released

▷ **FIGURE 15–18 How Desensitization Injections Work**

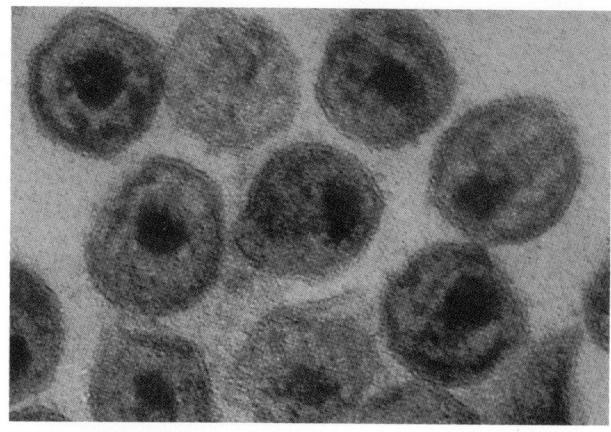

(a)

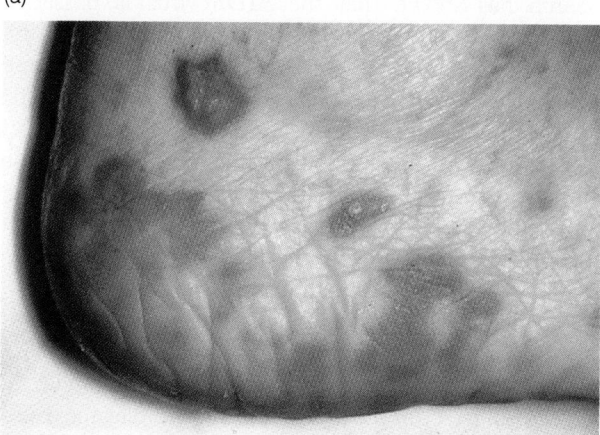

(b)

▷ **FIGURE 15–19 HIV and Kaposi's Sarcoma** (*a*) AIDS viruses. (*b*) Kaposi's sarcoma on foot.

be approximately 1 million. Without massive improvements in prevention, many more people will be infected with the virus in the coming years.

In Africa, the numbers are even more ominous. In some regions, 30% to 50% of the people test positive for the AIDS virus.

AIDS Is Caused by an RNA Virus that Infects the Helper T Cells, Weakening the Immune System

Early suspicions that the disease was caused by a virus were confirmed by researchers in France and the United States (▷ Figure 15–19a). They discovered an RNA virus—now called the **HIV virus** (human immunodeficiency virus)— that attacks helper T cells, severely impairing the immune system. AIDS patients grow progressively weaker and fall victim to other infectious agents. Many die from an otherwise rare form of pneumonia.

HIV, the Virus that Attacks the Immune System, Also Causes Cancer and Produces a Substance that May Cause Deterioration of Brain Function

HIV affects more than a person's immune system. AIDS patients, for example, may contract a rare form of skin cancer called Kaposi's sarcoma (Figure 15–19b). At first, researchers thought that Kaposi's sarcoma resulted from immunologic suppression. They hypothesized that the cancer developed because the AIDS virus had incapacitated the helper T cells. Without these cells, scientists reasoned, cancer could develop unimpeded.

Although the AIDS virus does affect the T helper cells, recent evidence suggests that Kaposi's sarcoma is caused by something else. Studies show that HIV carries a regulatory gene that is incorporated into human body cells. This gene may cause certain body cells to proliferate uncontrollably, forming a cancer. Alternatively, the gene may stimulate the production of a chemical substance that causes rapid cell growth (cancer) in neighboring cells.

AIDS patients also experience a number of neurological disorders, such as early memory loss and progressive mental deterioration. A recent laboratory study may explain how the brain is affected by this virus. AIDS viruses inside host white blood cells (helper T cells) produce a number of proteins that are incorporated into the viral capsid. One of those proteins is gp120. Researchers recently found that gp120 kills fetal brain cells in culture. Brain cells, like the immune system cells, have membrane receptors that bind to gp120. The researchers believe that gp120 may travel in the blood to the brain of some patients and kill cells, causing neurological defects.

HIV Is Transmitted In Many Ways, But Not by Casual Contact

Research has shown that the AIDS virus is passed via sexual contact, blood transfusions, and contaminated needles shared by intravenous drug abusers. Homosexual men, hemophiliacs, and drug addicts have been the chief victims in the United States. The AIDS virus is also transmitted among the heterosexual population and can even be transmitted from a mother to her baby through the placenta. Interestingly, a recent study showed that a man infected with AIDS is many times more likely to transmit the disease to a female partner than an HIV-positive female is to transmit it to an uninfected man.

Victims of the genetic disorder hemophilia, described in Chapter 13, were once at risk for AIDS. Hemophiliacs are given clotting factors from pooled human plasma. Before 1984, blood donors were not screened for AIDS. Consequently, many of the preparations were contaminated with the AIDS virus. As a result, a majority of the estimated 15,000 hemophiliacs in the United States who received clotting factors between 1975 and 1984 are thought to have HIV antibodies in their blood. Whether or not they will all develop AIDS no one knows. Evidence suggests that they will.

To prevent the spread of AIDS through blood transfusions, blood is now routinely tested for HIV. Tissues and organs for transplantation are also tested.

The AIDS virus, although lethal, does not spread as readily as the flu virus or cold viruses; individuals can protect themselves by practicing sexual abstinence before marriage, by engaging in safe sex (for example, using condoms), and by avoiding multiple sexual partners. To prevent the spread of HIV among intravenous drug users, some countries and some states distribute clean hypodermic needles to addicts.

New Study Supports Cofactor in AIDS

New research findings lend support to a controversial hypothesis: that the development of AIDS may be facilitated by a bacterium known as *Mycoplasma fermentans*. In this study, cells cultured with both *M. fermentans* and HIV die more readily than those infected with HIV alone, say a team of researchers at the Armed Forces Institute of Pathology in Washington, D.C. Human white blood cells

cultured with HIV alone died off to 20% of their original density, then recovered to 80% after two weeks. Cells infected with HIV and *M. fermentans* died off nearly completely and had grown back to 20% of their original number within two weeks.

A growing body of evidence suggests that in at least some cases, these bacteria are cofactors, independent organisms that act synergistically with the AIDS virus. Earlier studies showed that antibodies to the bacteria could block HIV infection in the test tube. Nonetheless, many virologists believe that HIV acts alone and are waiting for epidemiological evidence linking AIDS to a dual infection and more *in vitro* evidence showing how the two work together.

The Battle Against AIDS Has Been Facilitated by New Screening Tests and by Drugs that Slow Down the Development of the Disease

Health workers determine the presence of the AIDS virus by using an immunologic test, which detects antibodies to HIV in the blood. Recently, scientists announced the development of a new and more sensitive genetic test to determine the presence of the virus that could help improve the screening of blood and tissue.

Although no cure has been discovered, a drug called AZT (zidovudine) may prolong the lives of people who have tested HIV positive. AZT is costly and is also thought to be carcinogenic. New drugs are under investigation. Because of an outcry among the homosexual community, the Food and Drug Administration, which regulates all drug testing on humans, has relaxed its standards, bringing potential AIDS drugs to the public more quickly than before.

AIDS will remain a significant public health threat here and abroad for many years. Without better education and other measures to control the spread of the disease, millions of people worldwide could die from AIDS. (For a discussion of some of the social and political implications of controlling the spread of AIDS, see the Point/Counterpoint in this chapter.)

Although Some Researchers Are Optimistic About Finding a Vaccine to Stop the Virus, Not All Share Their View

The Bad News about AIDS. HIV is notorious for its ability to mutate, a feature that will make the task of developing a vaccine extremely difficult, if not impossible.[6] Even within the body, HIV mutates fairly freely. In one study, for example, researchers analyzed viruses isolated from two infected patients. Over a 16-month period, they found 9 to 17 different varieties, all thought to have been formed from the original virus.

---

[6] The AIDS virus is a retrovirus, an RNA virus whose RNA is used to produce DNA after invading a cell. This process, called reverse transcription, is fairly inaccurate and results in many mutant forms of the virus.

Making matters worse, HIV may be taking refuge in the body. In 1988, a research team announced that out of 100 homosexual men studied, 4 initially showed antibodies to the AIDS virus, but slowly lost them. This process usually occurs only in the late stages of AIDS, when the immune system is too weak to produce antibodies. These 4 men, however, had no overt symptoms of AIDS. Had the virus been conquered by the immune system, or had it simply entered a latent phase, "hiding out" as does the herpes virus?

Research suggests that HIV may take up residence in bone marrow stem cells that give rise to lymphocytes. The virus, therefore, may remain in these cells and be passed on to new white blood cells produced by cell division. Thus, once the virus is in the body, it may be there forever. As a result, completely eliminating the virus from the body may be impossible. One of the upshots is that if HIV goes into hiding, AIDS-infected blood donors may escape detection, even with the new genetic tests. AIDS-infected blood cells could unknowingly be passed to thousands of patients over the coming years.

The Good News about AIDS. Despite the discouraging news, researchers and much of the general public remain determined to find a cure for AIDS and a way to prevent it.

One encouraging development was revealed in 1988. Two California research groups announced that they had successfully transplanted parts of the human immune system into mice. Human immune cells in mice may provide scientists with a tool to study AIDS and to test potential cures for it, thus accelerating the pace of research.

On another front, researchers have developed a genetically engineered weapon that could kill cells infected with the AIDS virus, possibly curing the disease once it has developed. As noted above, cells infected with the AIDS virus produce a protein known as gp120, which is part of the capsid (▷ Figure 15–20). This protein also ends up in the plasma membrane of infected cells.

Researchers have genetically engineered a bacterium that binds to the gp120 protein. The bacterium carries with it a toxin that kills the HIV-infected cells. Preliminary studies indicate that noninfected cells are unharmed by this treatment. Although initial studies suggest that this treatment will not work, researchers hope that the technique can be improved or modified, making it possible to kill enough infected cells in AIDS patients to halt the disease. One question that must be answered before this procedure can be tried in people is whether AIDS-infected cells killed by this technique will degenerate and release active AIDS viruses that could spread to other body cells.

Researchers are also testing an HIV vaccine that could be administered as a preventive measure. The rapid rate at which the virus mutates, however, has stymied these efforts. Nevertheless, many potential vaccines are in clinical trials.

ENVIRONMENT AND HEALTH: MULTIPLE CHEMICAL SENSITIVITY

Richard Sharp was a physicist for a major aviation company in California. Today, he is confined to two stripped-down rooms equipped with special filters that remove all air contaminants.

Sharp is one of many Americans who has developed a condition known as multiple chemical sensitivity (MCS). He has become a prisoner in his own home, unable to venture forth into modern society without suffering extreme discomfort, even debilitation. MCS is thought to be caused by exposure to a number of common household and industrial chemicals, including formaldehyde, solvents, acrylic resins, mercury compounds, and pesticides. Chemicals such as formaldehyde, which is commonly found in carpeting, plywood, furniture, and many other household products, are thought to bind to naturally occurring proteins in the body. This process produces foreign substances against which the immune system mounts an attack. In other words, common household and industrial substances turn the immune system against the body.

Individuals become sensitive to low levels of chemicals over long periods. As a result, this condition is typically referred to as **hypersensitivity.** Victims also exhibit cross-reactions; that is, they react to other chemically similar substances. Massive exposures to certain chemicals may also elicit a hypersensitivity reaction.

The symptoms of MCS vary, ranging from life-threatening to mild. The most common symptoms are tension, memory loss, fatigue, sleepiness, headaches, confusion, and depression. Many victims of MCS suffer from

▷ **FIGURE 15–20 Duping the AIDS Virus** The AIDS virus has a protein, gp120, in its capsid. When it infects cells, it produces more gp120 to make new capsids. However, some gp120 ends up in the infected cell's plasma membrane, marking it. By genetically engineering a bacterium that can locate the infected cells by finding gp120, medical researchers may be able to hunt and kill infected cells, stopping the spread of the virus.

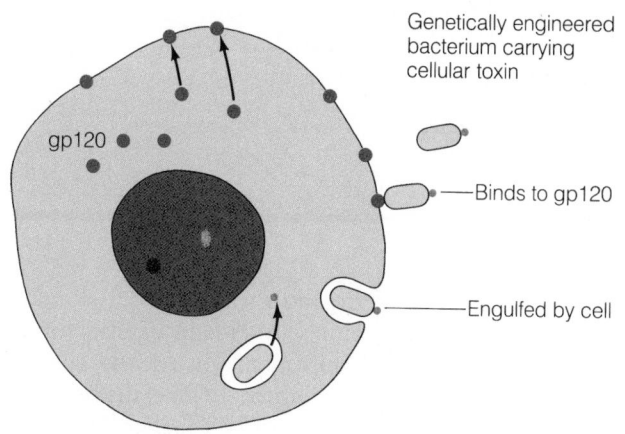

ANONYMOUS TESTING IS THE ANSWER

Earl F. Thomas

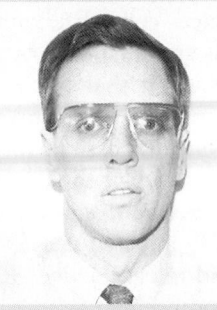

Earl F. Thomas is on the boards of directors for the Colorado AIDS Project and the People With AIDS Coalition of Colorado. Thomas was diagnosed as having AIDS in 1986.

Confidential testing for HIV (in which the names and addresses of individuals who test positive are reported to health officials) versus anonymous testing continues to be a highly controversial topic. In today's society, the word *AIDS* still breeds fear, especially the fear of discrimination. These fears, quite simply, are keeping people from being tested.

Health departments and AIDS organizations stress that persons who have reason to believe that they may have been infected with the HIV virus should be tested, for several reasons. First, the earlier a person knows he or she is HIV positive, the earlier treatment can be started to slow the progression of the disease or simply to buy time while researchers explore better treatments. Second, knowledge of one's HIV status is crucial in determining what behavioral changes need to be made.

If the testing system discourages individuals from obtaining knowledge about their HIV status, however, no one benefits. Unfortunately, name reporting tends to create an atmosphere of distrust between health officials and those who wish to be tested. These people already feel disenfranchised from society and cannot believe that the public health system is concerned about their personal rights and welfare. They believe that information concerning their HIV status goes beyond the health department to others who may have a need to know, and they fear that the information will eventually find its way to persons who have no need to know. For instance, numerous links in the information chain (nurses, laboratory workers, therapists, and secretaries) all have access to a person's medical information. Confidential medical information is not so confidential after all.

As a result of these concerns, many persons, even in high-risk groups, refuse to be tested at all.

In a survey conducted by the Educational Department of the Colorado AIDS Project in 1989, 32% of 1,112 respondents (homosexual and bisexual men) cited confidentiality concerns as their reason for not being tested. Despite the health department's insistence that fears of information leaks are poorly founded, they are very real to many individuals. More importantly, these fears prevent individuals who have engaged in high-risk behaviors from being tested.

Further information that supports anonymous testing comes from the state of Oregon. When anonymous HIV testing was offered along with confidential testing, there was a 50% increase in the demand to be tested during the first four months of the program. People from all segments of society sought out the anonymous test sites. Confidential testing sites reported *no* increase during the same period. These data strongly suggest that reporting by name is not the solution to HIV testing.

Fear of testing is justified. There is documented evidence of discrimination against people who are HIV-positive. Individuals have been denied housing and, in certain cases, have been evicted from their homes, have lost their jobs, have been denied access to public education, and have been shunned by families, friends, and co-workers.

Partner notification (contact tracing) may have a place when trying to control the spread of AIDS, but at what cost? Some feel that contact tracing is not a viable option due to the monetary cost. In six months in 1988 the state of Colorado spent $450,000 on partner notification. The result of this enormous expenditure was that 52 people were found.

The experiences of other health departments clearly show that partner notification can be carried out with reasonable success when testing is anonymous. If lists are being maintained, people shy away from testing and forfeit the possibility of early intervention treatment and partner notification information. When health officials act as contact tracers, they are perceived as police. In our society, police-state tactics will never work, and no one benefits.

Anonymous testing would reduce a serious impediment to a powerful, collective anti-AIDS effort. And most important, we could get an answer to whether people are truly avoiding testing because of reportability.

---

gastrointestinal problems as well—nausea, indigestion, and cramps. Some exhibit respiratory symptoms as well, including frequent colds, bronchitis, and shortness of breath. Skin rashes are not uncommon. Many people report allergy-like symptoms such as nasal stuffiness and sinus infections.

MCS remains one of the most mysterious of all human diseases. Victims often witness a sudden deterioration of their health. Physicians, family members, friends, and even the victims themselves are baffled by the disease. As a result, patients are often likely to be labeled "psychiatric cases." People suffering from MCS must often get rid of

NOTIFICATION WORKS

John Potterat

John Potterat is an authority on AIDS and sexually transmitted disease (STD) control. He has published numerous articles in medical journals dealing with STD and AIDS control and is currently director of the STD and AIDS programs in Colorado Springs.

The process of reporting people with the AIDS virus (HIV) by name to public health officers and in turn tracing their contacts should not be controversial. Such procedures have been standard public health practice for serious communicable diseases for nearly a century. Formal notification allows society to define the disease burden (surveillance) and to counterattack (control). You cannot control a communicable disease if you do not know who has it and who might be next to have it; moreover, you need to find those directly affected.

Notifying partners of people infected with sexually transmitted disease has been an effective control tool for 50 years. The fundamental reason that health officers are involved in this notification process is that STD patients are not good at referring their own sexual partners. Such "self-referral" fails more often than it succeeds: less than a third of STD partners are successfully referred for medical evaluation. Partner referral by HIV patients is even less successful (despite frequent assurances by patients that they "will take care of it!"). Part of this failure is due to the reluctance of HIV patients to face their partners (fear of anger or reprisal); part is due to selective notification (denial that "nice" partners can be infected); and part to failure to convince partners (partner denial). Trusting the notification process to infected people alone is a luxury that society can ill afford.

Those exposed to HIV have a right to know. Important sexual (safer practices) and reproductive (postponing pregnancy) decisions depend on knowledge of exposure and its outcome. Many persons are unaware that their partners have histories of needle exposures or of bisexuality. The duty to warn people has compelling moral, legal, and historical foundations. In free societies, notification is a straightforward, confi

dential process. Medical workers who detect HIV infection report the case by name and address to the local health officer who then discreetly contacts the patient to counsel him or her and to obtain identifying information on sexual and needle partners. People are persuaded, not coerced, into voluntary cooperation. Although counseling is "mandatory," blood testing is optional.

Even if "treatment" for partners were to consist solely of personal counseling to discourage behaviors that facilitate transmission or accelerate disease progression, partner notification would be worthwhile.

A disease control procedure should be acceptable to people. Partner notification by health officers has been well received by the affected populations. The majority (70% to 80%) of notified partners accept blood testing, and almost all who decline testing accept counseling. Although organized gay advocacy groups have generally opposed both HIV reporting by name and partner notification by health officers, when approached individually and sympathetically, gay men have generally cooperated.

Health officers are responsible for maintaining the physical security of HIV records; such records are also immune from any discovery process. They cannot be subpoenaed or released to potentially adversarial agents like insurance, police, or employer investigators. Whatever discrimination is suffered by infected people, none of it stems from disease notification to or by public health officers.

Notification initiatives are affordable, acceptable to patients, and effective in reaching high-risk people. It is well known to health officers that those at highest risk are least inclined to appear for counseling and least likely to use safer practices. While notification is not a panacea, it is one of the most useful measures for containing this tragic epidemic.

≈ SHARPENING YOUR CRITICAL THINKING SKILLS

1. Summarize Thomas's reasons for keeping AIDS testing anonymous.
2. Summarize Potterat's views in support of name reporting.
3. Do you agree or disagree with the following statement: both writers believe that their approach will provide the greatest protection to the public health, but they differ in their approach. Explain.
4. Of the two basic approaches, which do you think would be most effective in reducing the spread of AIDS?

---

all cleaning agents, pesticides, perfumes, deodorants, and other household chemicals.

At first, many physicians and scientists dismissed this disease, but a growing body of evidence shows that many common chemicals can indeed alter the immune system. The National Research Council estimates that 15% of the

U.S. population experiences some degree of chemical hypersensitivity. Studies show that 5% of the workers exposed to an agent used in the manufacture of plastics, TDI (toluene diisocyanate), develop asthmalike symptoms. TDI apparently binds to proteins in the respiratory tract, creating foreign substances that stimulate a hypersensitivity reaction. Individuals who have been hypersensitized have difficulty breathing when exposed to TDI, tobacco smoke, and air pollutants. In Japan, 15% of all cases of asthma in men can be directly attributed to industrial exposure to chemicals.

Other chemicals bind to proteins in the skin, creating foreign substances to which the immune system reacts. Formaldehyde, for example, results in a condition called **contact dermatitis,** characterized by a skin rash. T cells attack the cells of the skin, destroying them. Even low levels of formaldehyde found in newsprint dyes, some cosmetics, and photographic papers are sufficient to induce rashes.

Other chemicals apparently act by suppressing immune function, making individuals more susceptible to infectious agents. Chronic workplace exposure to benzene, for example, results in a reduction in the number of circulating lymphocytes. Benzene probably depresses lymphocyte production in the bone marrow. In rabbits, benzene exposure results in an increased susceptibility to various infectious agents.

Dioxins, PCBs, ozone, certain pesticides, and a variety of other chemicals suppress the immune response in laboratory animals and humans. However, the overall significance of immune suppression and hypersensitivity in human populations remains unknown. Nonetheless, this section illustrates the subtle but potentially far-reaching effects of toxic chemicals and underscores the importance of a healthy environment for maintaining a healthy population.

SUMMARY

THE FIRST AND SECOND LINES OF DEFENSE

1. The first line of defense against viruses, bacteria, and other infectious agents is the skin and the epithelia of the respiratory, digestive, and urinary systems. Some epithelia also produce protective chemical substances that kill microorganisms.

2. The second line of defense consists of cells and chemicals that the body produces to combat infectious agents that penetrate the epithelia.

3. One of the chief combatants in the second line of defense is the macrophage, a cell derived from the monocyte. Macrophages are found in connective tissue lying beneath the epithelia, where they phagocytize infectious agents, preventing their spread. Neutrophils and monocytes also invade infected areas from the bloodstream and destroy bacteria and viruses.

4. Another portion of the second line of defense consists of the chemicals released by damaged tissue, which stimulate arterioles in the infected tissue to dilate. The increase in blood flow raises the temperature of the wound. Heat stimulates macrophage metabolism, accelerating the rate of the destruction of infectious agents. Heat also speeds up the healing process by accelerating the metabolic rate of the cells involved in the recovery.

5. Still other chemicals increase the permeability of the capillaries, causing plasma to flow into the wound and increasing the supply of nutrients for macrophages and other cells fighting the invader.

6. The increase in blood flow, the release of chemical attractants, and the flow of plasma into the wound constitute the inflammatory response.

7. Another part of the secondary line of defense is pyrogens, chemicals released primarily by macrophages exposed to bacteria, which raise body temperature and lower iron availability, thus decreasing bacterial replication.

8. Interferons, a group of proteins released by cells infected by viruses, are also part of the second line of defense. Interferons travel to other virus-infected cells, where they inhibit viral replication.

9. The blood also contains the complement proteins, which circulate in an inactive state, becoming activated when the body is invaded by bacteria. They too are part of the second line of defense.

10. Some of the complement proteins stimulate the inflammatory response. Others embed in the plasma membrane of bacteria. There, they combine to form membrane-attack complexes, which create holes in the bacterial plasma membranes, killing these pathogens. Another complement protein binds to the invader, making it more easily phagocytized by macrophages.

THE THIRD LINE OF DEFENSE: THE IMMUNE SYSTEM

11. The immune system consists of billions of lymphocytes that circulate in the blood and lymph and take up residence in the lymphoid organs.

12. The lymphocytes recognize antigens—foreign cells and foreign molecules, mostly proteins and large-molecular-weight polysaccharides.

13. T and B cells are two types of lymphocytes produced in red bone marrow. The T cells become immunocompetent—able to respond to a particular antigen—in the thymus. B cells gain this ability in the bone marrow. During this process, the T and B cells gain specific membrane receptors that respond to specific antigens.

14. Immunocompetent B cells encounter antigens (often presented to them by macrophages) to which they are programmed to respond, then begin to divide, forming plasma cells and memory cells. Plasma cells produce huge amounts of antibodies. The memory cells enable the body to respond more quickly to future invasions by the same antigen.

15. Antibodies are small protein molecules produced by plasma cells. They bind to specific antigens, destroying them either by precipitation, agglutination, or neutralization, or by activation of the compliment system.

16. When T and B cells first encounter an antigen, they react slowly in what is called the primary response. As a result, there is often a period of illness before the pathogen is removed by the immune system. Because numerous memory cells are produced during the first assault, a reappearance of the antigen elicits a much faster and more powerful reaction,

the secondary response. Consequently, the pathogen is usually vanquished before symptoms of illness occur.

17. Resistance created by a response to an antigen is called immunity.

18. When activated by antigen, T cells multiply and differentiate, forming memory T cells, cytotoxic T cells, helper T cells, and suppressor T cells whose functions are summarized in Table 15–4.

19. A solution containing a dead or weakened virus, bacterium, or bacterial toxin that is injected into people to provide immunity is called a vaccine. Vaccines produce an active immunity, which may last for years.

20. Passive immunity can be conferred by injecting antibodies into a patient or by the transfer of antibodies from a mother to her baby through the bloodstream or milk. Passive immunity is short-lived, lasting at most only a few months.

PRACTICAL APPLICATIONS: BLOOD TRANSFUSIONS AND TISSUE TRANSPLANTATION

21. Blood transfusions require careful matching of donor and recipient blood types. The antigens on the membranes of body cells are more complex. Only cells from the same individual or an identical twin will be accepted. All others are rejected by the T cells, unless the system is suppressed with drugs, which often leave the patient vulnerable to bacterial infections.

DISEASES OF THE IMMUNE SYSTEM

22. The most common malfunctions of the immune system are allergies, extreme overreactions to some antigens.

23. Allergies are caused by IgE antibodies, produced by plasma cells. IgE antibodies bind to mast cells, which causes the cells to release histamine and other chemical substances that induce the symptoms of an allergy—production of mucus, sneezing, and itching.

24. Autoimmune diseases, another type of immune system disorder, result from an immune attack on the body's own cells. Autoimmune diseases may result when normal proteins are modified by chemicals or genetic mutations so that they are no longer recognizable as self. Another possible cause is the sudden presence of proteins that are normally isolated from the immune system. Still another cause of autoimmune reaction is the exposure to antigens that are nearly identical to body proteins.

AIDS: FIGHTING A DEADLY VIRUS

25. AIDS is a disease of the immune system caused by HIV, a virus that attacks helper T cells, severely impairing a person's immune system.

26. Patients grow progressively weaker and may develop cancer and bacterial infections because of their diminished immune response.

27. AIDS is spread through body fluids during sexual contact and blood transfusions or through needles shared by drug users.

28. Stopping the virus has proved difficult, in large part because it mutates so rapidly and appears to take refuge in body cells. Recent accomplishments in research, however, offer some promise.

ENVIRONMENT AND HEALTH: MULTIPLE CHEMICAL SENSITIVITY

29. Many individuals suffer from multiple chemical sensitivity. Chronic exposure to low levels of pollutants or short-term exposure to high levels may alter the immune system, causing a wide range of symptoms.

30. Research shows that some toxic chemicals cause immune hypersensitivity, evoking allergylike symptoms. Others stimulate autoimmune responses. Still others cause immune suppression.

EXERCISING YOUR CRITICAL THINKING SKILLS

You have been selected as a juror for a trial involving a physician who refused to perform surgery on an AIDS patient. The patient had been in an automobile accident and had suffered a ruptured spleen, causing internal bleeding. The physician refused to perform an operation that could have saved the patient's life, because she was afraid of contracting AIDS.

Consider the following facts presented by the attorneys for the plaintiffs (the AIDS patient's family). The plaintiffs argue that the physician violated her code of ethics, which obligates her to treat all patients. They also argue that she knowingly allowed a patient to die and that she should be punished by having to pay damages as compensation for the lost life. The AIDS patient had received a transfusion of HIV-contaminated blood several years earlier and had only three to six months to live. Nevertheless, these months were valuable to him and to his family.

The defense attorneys admit that the physician refused treatment and caused the premature death of her patient. They say, however, that she acted rightfully. During surgery, sharp instruments frequently pierce the protective gloves of surgeons, exposing them to the AIDS virus. Refusing to operate protected not only the physician but also her entire surgical team. That team could, over the course of years, save hundreds of lives. In refusing to operate, the physician was considering the greater good—the benefit of her services to the public. The surgeon also had a husband and two children. By protecting herself, she was also taking into account the good of her family.

How would you decide such a case? Should the physician be forced to pay damages? Why? Or was she correct in her decision? Why? What would you have done if you were the physician?

As a side note, hospitals often assemble teams of doctors and other medical personnel who are willing to treat AIDS patients to avoid problems such as the one described above.

1. A virus contains a nucleic acid core and a coat of protein called the _____ .

2. The redness and swelling around a cut or abrasion are part of the _____ _____ .

3. _____ are chemical substances, produced primarily by macrophages, that increase body temperature during a bacterial or viral infection.

4. _____ are a group of chemicals produced by virus-infected cells that impair viral replication in other virus-infected cells.

5. A(n) _____ is a large-molecular-weight carbohydrate or protein that evokes an immune response.

6. The ability of B and T cells to respond to specific foreign invaders is called _____ .

7. Antibodies provide _____ immunity, protection against bacteria and viruses occurring primarily in the blood and lymph.

8. The secondary response to an antigen occurs more rapidly than the primary response because of the formation of _____ cells.

9. Antibodies belong to a class of proteins called _____ .

10. Coating a virus or bacterial toxin with antibody is called _____ .

11. Blood cells from an incompatible source clump together, because the recipient's blood contains antibodies to the donor's RBCs. This process is called _____ .

12. T cells differentiate into a cell, the _____ _____ cell, that travels throughout the body, destroying bacteria and virus-infected cells and tumor cells directly.

13. A _____ contains dead or weakened bacteria or viruses that elicit an immune response without causing disease. It provides a form of immunity.

Answers to the Test of Terms are found in Appendix B.

1. Describe the structure of a typical virus.

2. Based on your previous studies (Chapter 3), in what ways are bacteria similar to human body cells, and in what ways are they different?

3. The human body consists of three lines of defense. Describe what they are and how they operate.

4. Describe the inflammatory response, and explain how it protects the body.

5. Define each of the following terms, and explain how they help protect the body: pyrogen, interferon, and complement.

6. The first and second lines of defense differ substantially from the third line of defense. Describe the major differences.

7. How does the immune system detect foreign substances?

8. Describe how the B cell operates. Be sure to include the following terms in your discussion: bone marrow, immunocompetence, plasma membrane receptors, primary response, plasma cell, antigen, antibody, and secondary response.

9. Describe the four mechanisms by which antibodies "destroy" antigens.

10. Describe the events that occur after a T cell encounters its antigen.

11. What is the difference between active and passive immunity?

12. A child is stung by a bee, swells up, and collapses, having great difficulty breathing. What has happened? What can be done to save the child's life?

13. What is an autoimmune disease? Explain the reasons it forms.

14. What is AIDS? What are the symptoms? What causes it?

C H A P T E R

16

The Urinary System: Ridding the Body of Wastes and Maintaining Homeostasis

≈

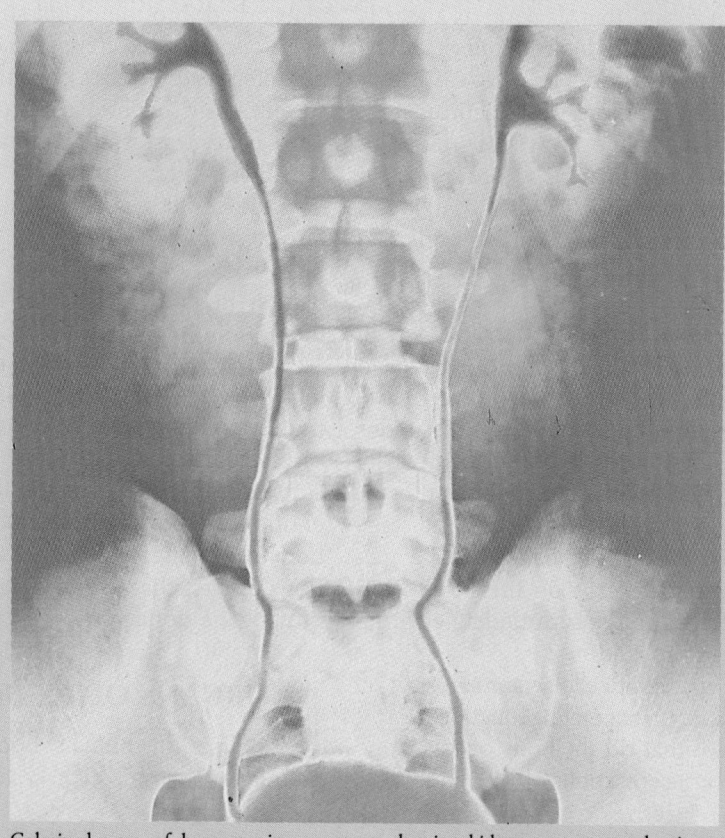

Colorized x-ray of human urinary system, showing kidneys, ureters, and urinary bladder.

eon Markowitz is lowered into a large pool of warm water in a special room in the hospital where he will spend the next few hours (▷ Figure 16–1). Physicians position a large, cylindrical device in the water in front of one of his kidneys, the organs that filter the blood and help maintain normal blood concentrations of nutrients and wastes. Over the next few hours, as Markowitz listens to tapes of Paul Simon, ultrasound waves, undetectable by the human ear, will silently smash a mass of calcium phosphate and other chemicals that has formed in his kidney and is obstructing the flow of urine and causing excruciating pain.

A decade earlier, surgeons would have had to cut an incision 15 to 20 centimeters (6 to 8 inches) long in Markowitz's side to remove the stone. He would have spent 7 to 10 days recovering in the hospital and would have had to recuperate at home eight more weeks before returning to work. With this new technique, known as ultrasound lithotripsy ("stone crushing"), patients usually return home within a day.

This chapter describes the ways in which animals rid themselves of potentially harmful wastes derived from metabolism. It focuses principally on the structure and function of the vertebrate urinary system and presents information that will be useful to you in your life. It also points out how the urinary system contributes to homeostasis and describes some common diseases, such as kidney stones, that disturb the function of this important organ system.

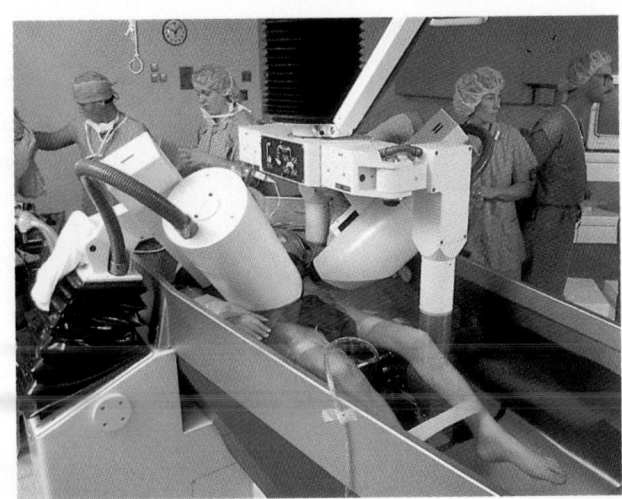

▷ **FIGURE 16–1 Lithotripsy** Ultrasound waves, undetectable to the human ear, bombard the kidney stones in this man, smashing them into sandlike particles that can be passed relatively painlessly in the urine.

≈ GETTING RID OF WASTES: EVOLUTION MEETS THE CHALLENGE

Ralph Waldo Emerson once said that as soon as there is life, there is risk. A biologist might look at things differently, noting that as soon as there is life, there is waste. Put more bluntly, all living things produce waste. Fortunately, elaborate systems of recycling have evolved in the biosphere so that the waste of one organism is a resource for another. The most obvious example is carbon dioxide, a waste product of glucose catabolism. As noted earlier in the text, carbon dioxide is one of the main raw materials of photosynthesis.

All organisms face the same challenge—how to get rid of wastes that are produced internally. For single-celled organisms, like the protists, the answer is simple: waste diffuses or is actively transported out of their bodies and is released into the medium (air, water, soil) they inhabit.

Moving up the evolutionary tree, we encounter some of the simple multicellular animals, like sponges and sea stars. Although multicellular, these organisms do not have specific waste-removing organs. The cells on their body surface eliminate wastes directly into the surrounding environment and into the extracellular fluid.

In truth, though, very few multicellular animals rely on body cells to carry out waste-management functions. Most of them have special organs that rid the body of waste and also help regulate internal concentrations of ions and water,

a function vital to homeostasis. Consider two examples: earthworms and insects.

In earthworms and other members of the group (phylum) they belong to (the annelids), fluid from the body cavity enters two small, tubular structures, the **metanephridia** (pronounced, met-uh-nef-rid-e-uh) in each segment of the body. The metanephridia collectively form a primitive kidney (▷ Figure 16–2). Fluid enters through an opening, then travels through the tubule to the **bladder,** an expanded region that stores the liquid containing wastes and other substances. Useful materials in the fluid are reabsorbed into the blood stream—that is, pass back into the blood. Wastes are eliminated through another opening that empties the waste onto the earthworm's external body surface. The nephridia rid the earthworm of water and salts; other wastes, like ammonia, diffuse across the body surface directly.

Another interesting waste-disposal system is found in terrestrial insects (▷ Figure 16–3). As illustrated, small, dead-end tubules attach to the gut. These tubules are bathed in body fluids and extract ions and water from them, thus helping get rid of excess water and wastes. The mechanism of transport is explained in the figure. Interestingly, cells that line the hindgut reabsorb useful materials, thus conserving essential ions and other substances necessary for life.

Waste disposal reaches its zenith in the vertebrates, endowed as they are with several mechanisms to rid the body of potentially harmful substances. The next two sections discuss the human organs of excretion and the anatomy of the urinary system.

≈ ORGANS OF EXCRETION

Metabolism in human body cells produces enormous amounts of waste, such as carbon dioxide, ammonia, and

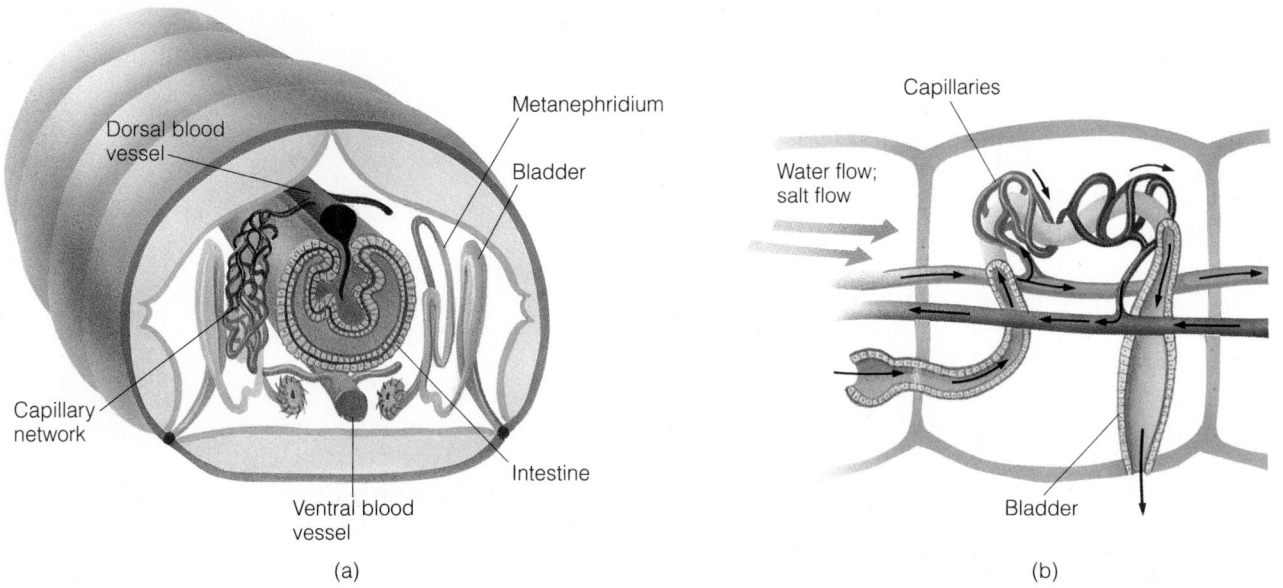

Metanephridium
Dorsal blood vessel
Bladder
Capillary network
Intestine
Ventral blood vessel
(a)
Capillaries
Water flow; salt flow
Bladder
(b)

FIGURE 16–2 Waste Removal in the Earthworm (*a*) The earthworm's body consists of many connected segments, each of which contains a pair of tubules called metanephridia. (*b*) Waste-containing fluids in the body cavity enter the tubules and are excreted through openings to the outside after useful materials have been reabsorbed into the blood stream.

FIGURE 16–3 Waste Disposal in Terrestrial Insects (*a*) Insects contain two to hundreds of tubules that connect to the gut. (*b*) Waste containing fluids enter the tubules and are expelled with the feces.

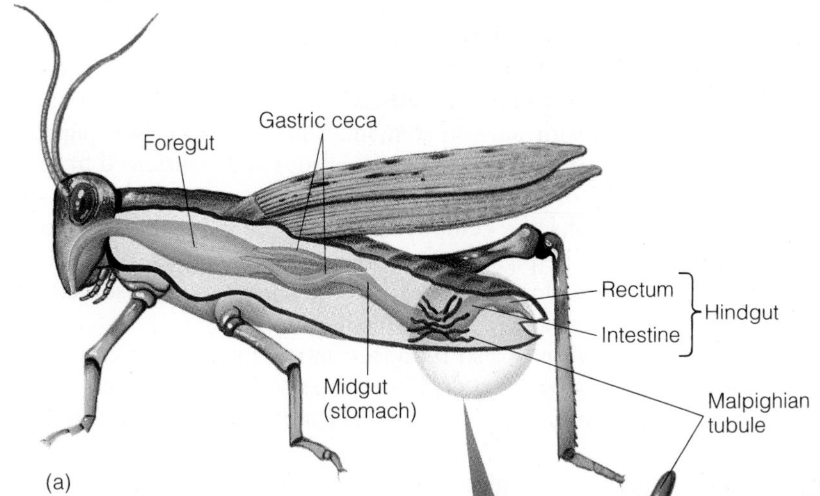

Gastric ceca
Foregut
Rectum
Intestine
Hindgut
Midgut (stomach)
Malpighian tubule
(a)

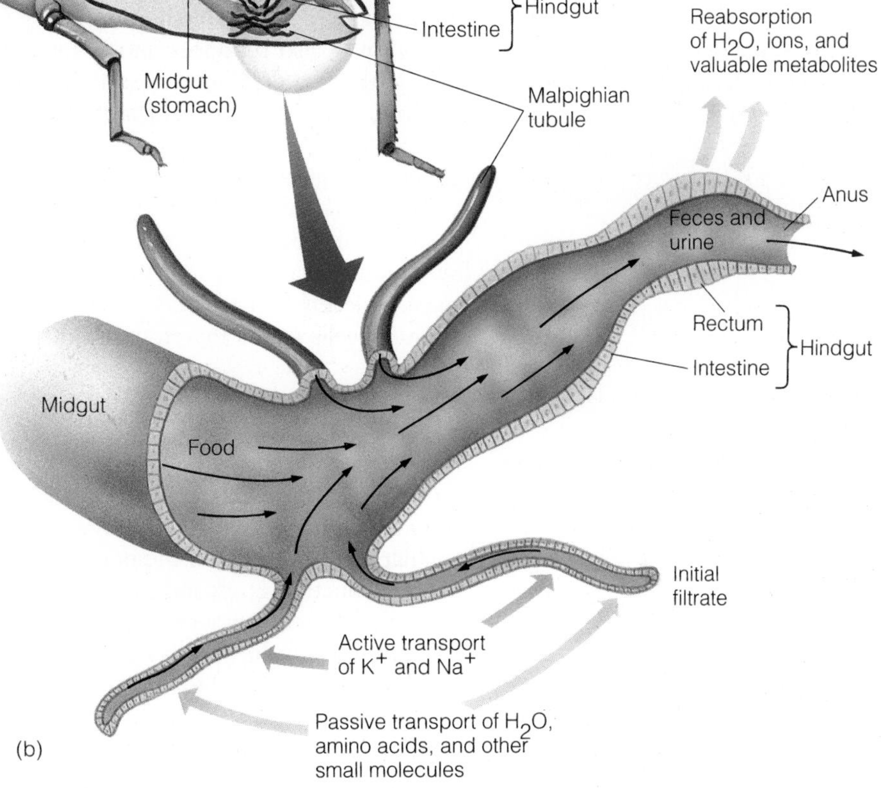

Reabsorption of H_2O, ions, and valuable metabolites
Anus
Feces and urine
Rectum
Intestine
Hindgut
Midgut
Food
Initial filtrate
Active transport of K^+ and Na^+
Passive transport of H_2O, amino acids, and other small molecules
(b)

urea. Like the toxic wastes produced by factories in an industrial society, these potentially harmful substances must go somewhere. Just as in an industrial society, however, they cannot be dumped carelessly. If these hazardous substances were discarded in one of the body's storage depots, they would surely poison body cells and kill us in a few days.

Table 16–1 lists the major metabolic wastes and other chemicals excreted from the body. Take a moment to study the list.

As you can see, the nitrogen-containing wastes like urea, uric acid, and ammonia arise from several different sources. Ammonia (NH_3), for instance, comes from the breakdown of amino acids, which occurs principally in the liver. Amino acid breakdown generally results from an excess of protein in the diet or a shortage of carbohydrate, causing the body to break down protein to acquire amino acids for energy production. The amino groups (NH_2) removed from the amino acids in a reaction called **deamination** are converted into ammonia.

Ammonia is a highly toxic chemical. A small amount of the ammonia produced by the liver is excreted in the urine, but the liver converts most ammonia to urea. The chemical structure of urea is shown in ▷ Figure 16–4.

Another by-product of metabolism in humans is **uric acid,** which the liver produces as it breaks down nucleotides. Uric acid is generally produced in very small quantities and excreted in the urine. In adults, excess production may result in the deposition of uric acid crystals in the bloodstream and in joints, where they cause considerable pain. This condition is known as **gout.** Uric acid may also

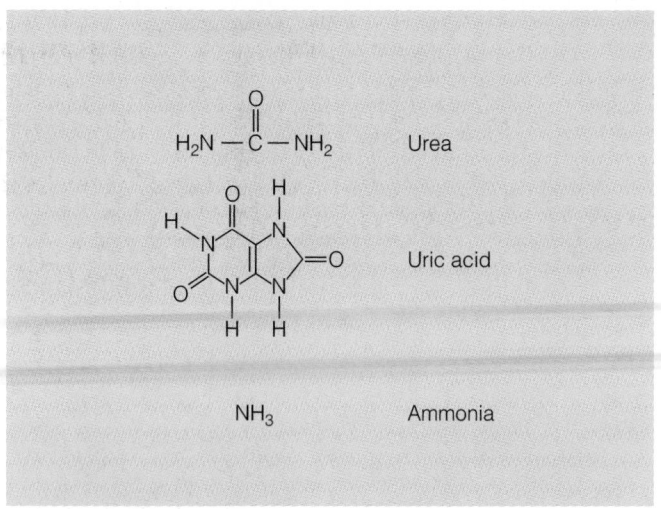

▷ **FIGURE 16–4 Structure of Urea, Uric Acid, and Ammonia** These three waste products are released by cells.

appear in the urine of some babies as orange crystals. Although alarming to a parent, their presence is generally not a sign of trouble.

The **bile pigments,** derived from the breakdown of RBCs in the liver, are another well-known waste product. Hemoglobin of the RBCs contains four protein subunits, each with an iron-containing heme group. Bile pigments are derived from the heme groups and transferred from the liver to the gallbladder, where they are stored along with bile salts, the emulsifying agents that aid in the digestion of fats. Bile pigments are released with the bile salts and are passed along the digestive tract. Some are reabsorbed, and the rest are eliminated with the feces.

The liver also produces another water-soluble pigment during the breakdown of hemoglobin. Known as **urochrome,** this yellow pigment is dissolved in the blood and passes to the kidneys, where it is excreted with the urine. Urochrome gives urine its yellowish color.

Table 16–1 also lists ions. Even though they are not end products of metabolism, ions are excreted from the body by various organs. This excretion is essential to maintain constant levels in the blood, the tissue fluid, and the cytoplasm of cells and is therefore essential to homeostasis.

≈ THE URINARY SYSTEM

Removing wastes is obviously a very important function. Not surprisingly, evolution has "provided" several avenues by which humans (and other animals) get rid of, or excrete, cellular wastes. In fact, if the number of organs involved in a body function is an indication of the importance of a process, excretion would have to rate among the most important of all. In humans, excretion of wastes occurs in the lungs, the skin, the liver, the kidneys, and even the intestines.

Of all the organs that participate in removing waste,

TABLE 16–1 Important Metabolic Wastes and Substances Excreted from the Body

CHEMICAL	SOURCE	ORGAN OF EXCRETION
Ammonia	Deamination of amino acids in liver	Kidneys
Urea	Derived from ammonia	Kidneys, skin
Uric acid	Nucleotide catabolism	Kidneys
Bile pigments	Hemoglobin breakdown in liver	Liver (into small intestine)
Carbon dioxide	Breakdown of glucose in cells	Lungs
Water	Food and water; breakdown of glucose	Kidneys, skin, and lungs
Inorganic ions*	Food and water	Kidneys and sweat glands

*Ions are not a metabolic waste product like the other substances shown in this table. Nonetheless, ions are excreted to help maintain constant levels in the body.

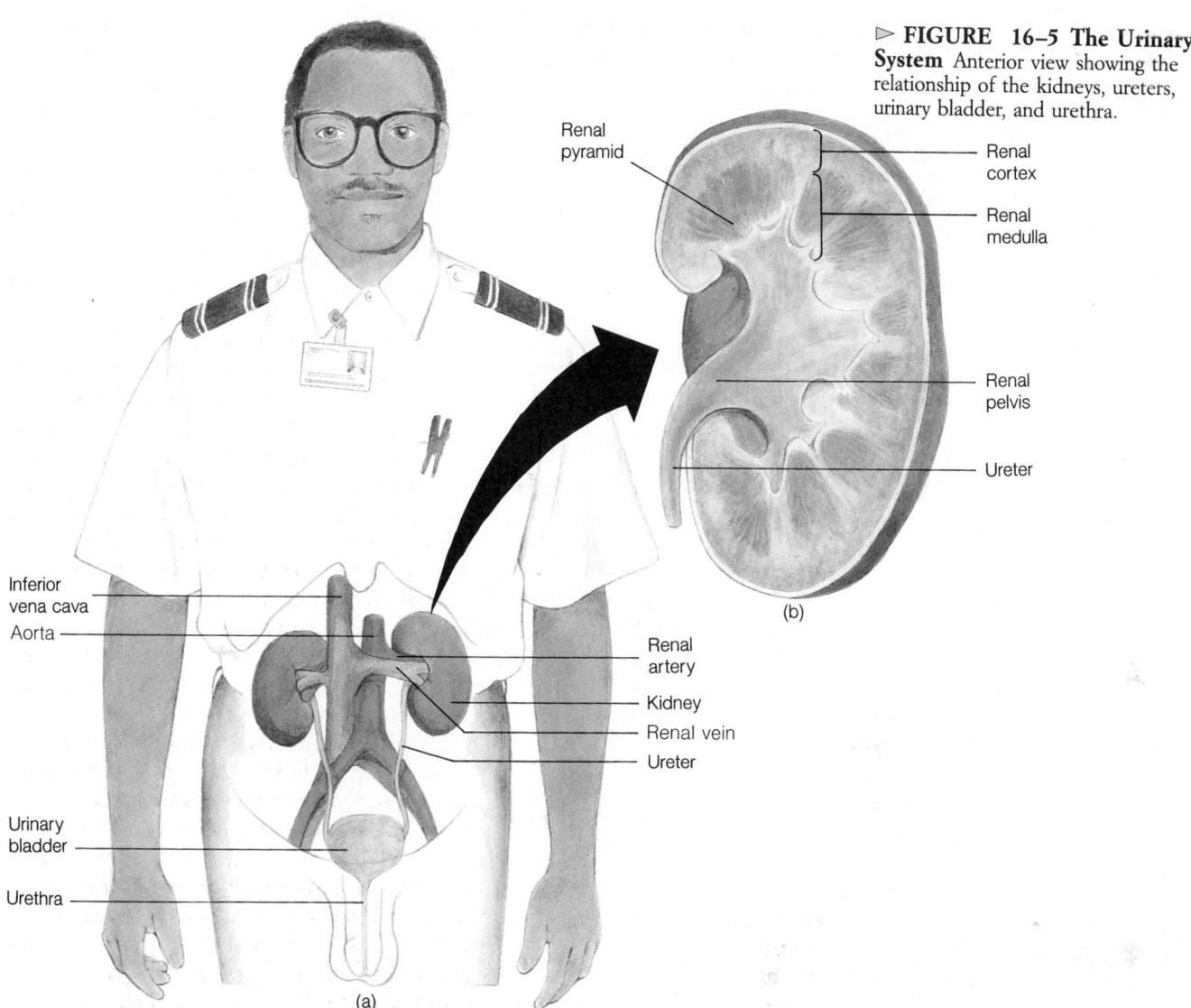

Renal pyramid

Renal cortex

Renal medulla

Renal pelvis

Ureter

(b)

Inferior vena cava

Aorta

Renal artery

Kidney

Renal vein

Ureter

Urinary bladder

Urethra

(a)

▷ **FIGURE 16–5 The Urinary System** Anterior view showing the relationship of the kidneys, ureters, urinary bladder, and urethra.

however, the kidneys rank as one of the most important, for they rid the body of the greatest variety of dissolved wastes. The kidneys also play a key role in regulating the chemical constancy of the blood.

As illustrated in ▷ Figure 16–5, humans come equipped with two kidneys, which are a part of the **urinary system.** The urinary system also includes the (1) ureters, which drain the kidneys, (2) the urinary bladder, which stores urine, and (3) the urethra, which transports the urine to the outside. The functions of these organs are described in more detail below and are summarized in Table 16–2.

The Urinary System Consists of the Kidney, Ureters, Bladder, and Urethra

The kidneys lie on either side of the vertebral column and are about the size of a person's fist. Surrounded by a layer of fat and located high in the posterior abdominal wall beneath the diaphragm, the human kidneys are oval structures, slightly indented on one side like kidney beans (Figure 16–5). Arterial blood flows into the kidneys through

the renal arteries, which enter at each indented region. The **renal arteries** are major branches of the abdominal aorta (Figure 16–5). Inside the kidney, much of the blood-borne wastes are filtered out and eliminated in the urine. **Urine** is

	TABLE 16–2 Components of the Urinary System and Their Functions	
COMPONENT	**FUNCTION**	
Kidneys	Eliminate wastes from the blood; help regulate body water concentration; help regulate blood pressure; help maintain a constant blood pH	
Ureters	Transport urine to the urinary bladder	
Urinary bladder	Stores urine; contracts to eliminate stored urine	
Urethra	Transports urine to the outside of the body	

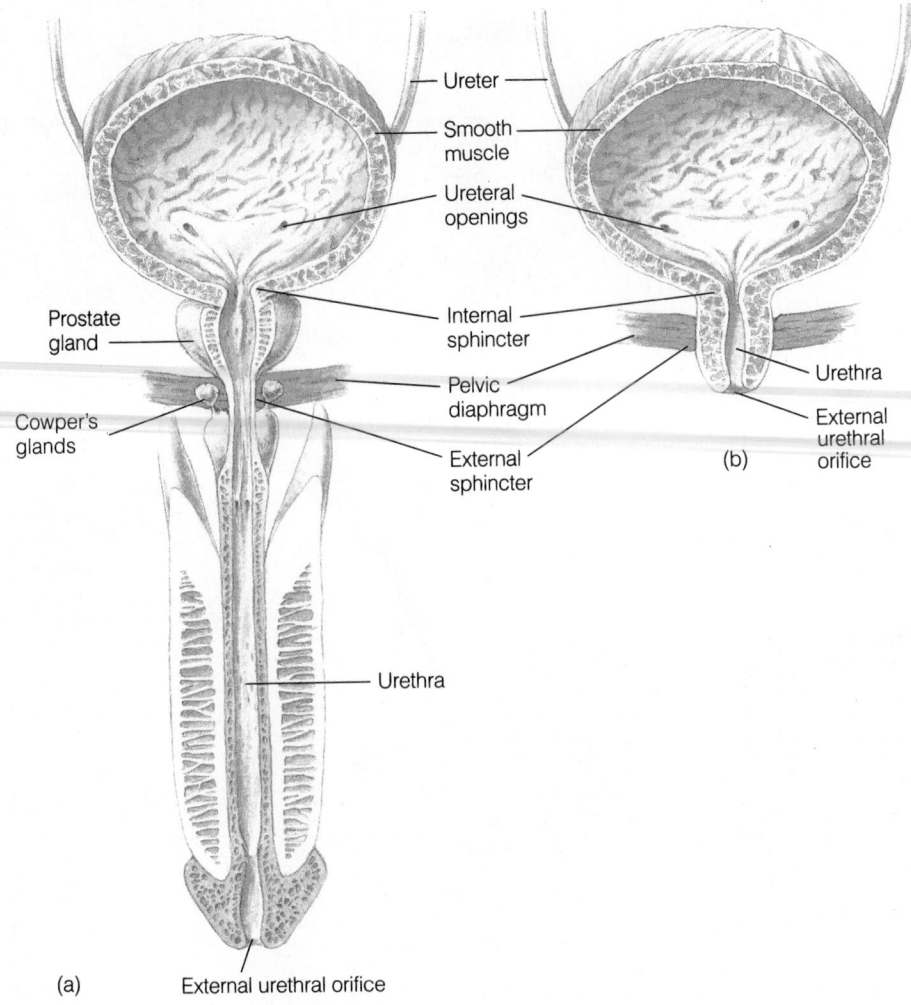

▷ **FIGURE 16–6 The Urinary Bladder and Urethra** These drawings show the differences in the urethras of men (*a*) and women (*b*). The smooth muscle at the juncture of the urinary bladder and urethra forms the internal sphincter. The pelvic diaphragm is a flat sheet of muscle covering the lower boundary of the pelvic cavity. It forms the external sphincter and is under voluntary control.

a yellowish fluid containing water, inorganic ions, nitrogenous wastes (such as urea), very small amounts of hormones, ions, and other chemical substances. After the blood has been filtered, it leaves the kidneys via the **renal veins,** which drain into the inferior vena cava, the vessel that transports venous blood to the heart.

Dissolved wastes are removed by numerous microscopic filtering units in the kidney known as **nephrons.** The nephrons produce urine, which is drained from the kidney via the ureters. The **ureters** are muscular tubes that transport urine to the urinary bladder (Figure 16–5). Waves of peristaltic contractions, involuntary smooth muscle contractions like those in the digestive tract, help propel the urine in the ureter to the urinary bladder. The **urinary bladder,** which lies in the pelvic cavity, is a temporary receptacle for urine and is located just behind the pubic bone. The bladder's distensible walls contain a relatively thick layer of smooth muscle that stretches as the bladder fills. When the bladder is full, its walls contract, forcing the urine out through the urethra.

The **urethra** is a narrow tube, measuring approximately 4 centimeters (1.5 inches) in women and 15 to 20 centimeters (6 to 8 inches) in men. The additional length in men

largely results from the fact that the urethra travels through the penis (▷ Figure 16–6). The difference in the length of the urethra between men and women has important medical implications. The shorter urethra in women, for example, makes women more susceptible to bacterial infections of the urinary bladder. Bacteria can travel up the urethra of women, invading the bladder, far more easily than they can in men.[1] Bladder infections may result in an itching or burning sensation and an increase in the frequency of urination. They may also cause blood to appear in the urine. Urinary tract infections can be treated with antibiotics. If untreated, however, infections may spread up the ureters to the kidneys, where they can seriously damage the nephrons and impair kidney function.

The Human Kidney Consists of Two Zones, an Outer Cortex and Inner Medulla

Each kidney is surrounded by a connective tissue capsule, the **renal capsule.** The internal structure of the kidney is

---

[1] Sexual intercourse increases the frequency of urinary bladder infections in many women. To help avoid the problem, physicians advise women to empty their bladders soon after intercourse.

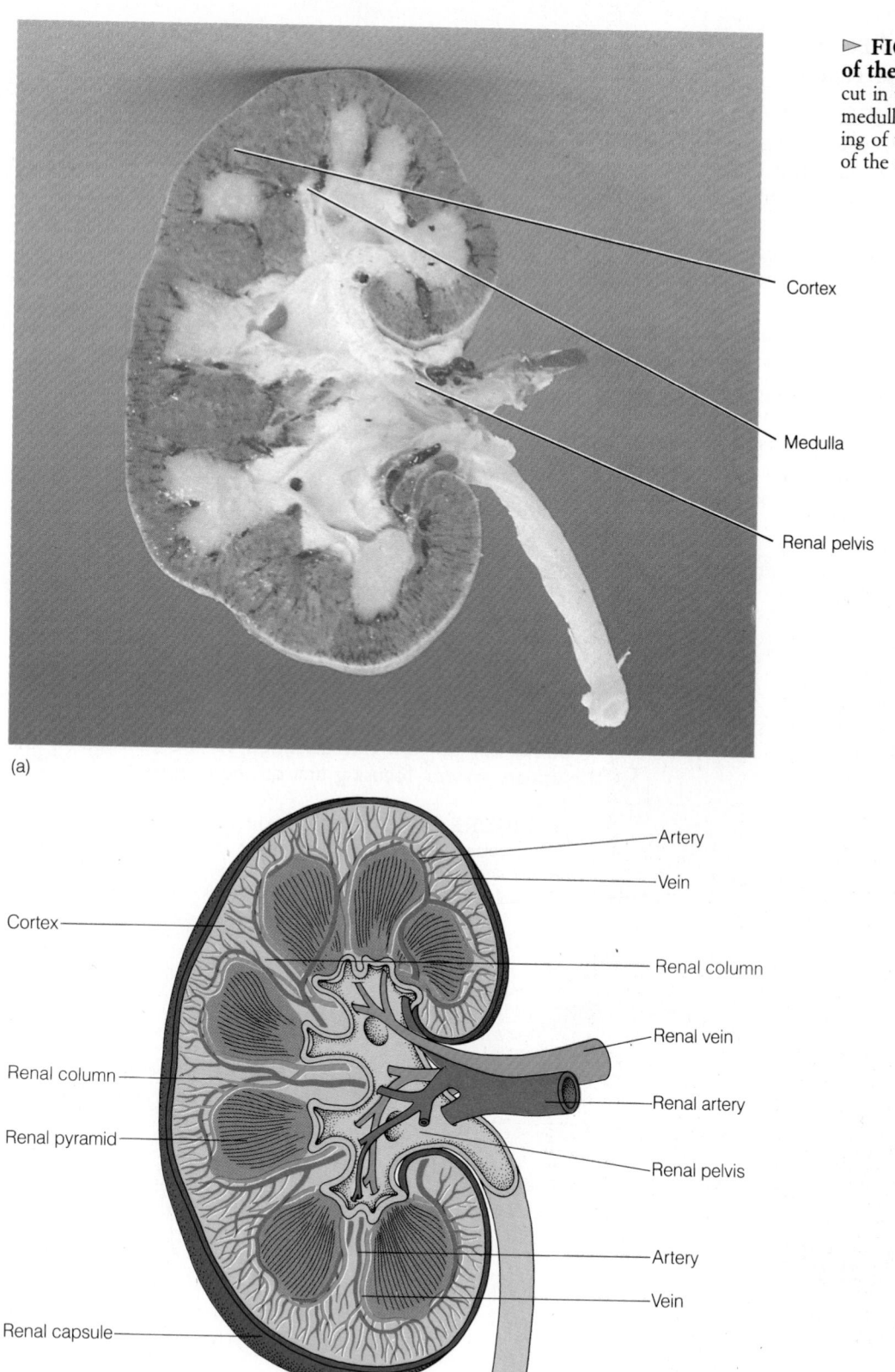

▷ **FIGURE 16–7 Cross Section of the Kidney** (*a*) Human kidney cut in two showing the cortex, medulla, and renal pelvis. (*b*) A drawing of the kidney showing the course of the arteries and veins.

Cortex

Medulla

Renal pelvis

(a)

Artery

Vein

Cortex

Renal column

Renal vein

Renal column

Renal artery

Renal pyramid

Renal pelvis

Artery

Vein

Renal capsule

Ureter

(b)

shown in ▷ Figure 16–7. As illustrated, the kidney is divided into two distinct zones. The outer zone is the **renal cortex,** containing many small filtering units, or nephrons. The inner zone, the **renal medulla,** consists of cone-shaped structures known as **renal pyramids** and intervening tissue called **renal columns.** The renal pyramids contain small ducts that transport urine from the nephrons into a central receiving chamber known as the **renal pelvis.**

Nephrons Consist of Two Parts, a Glomerulus and a Renal Tubule

Each kidney contains 1 million to 2 million nephrons, visible only through a microscope (▷ Figure 16–8a; Table 16–3). Each nephron consists of a tuft of capillaries, the **glomerulus** (Latin for "ball of yarn"), and a long, twisted tube, the **renal tubule** (Figures 16–8a and 16–8c). The renal tubule, in turn, consists of four segments: (1) Bowman's capsule, (2) the proximal convoluted tubule, (3) the loop of Henle, and (4) the distal convoluted tubule.

As illustrated in Figure 16–8c, **Bowman's capsule** is a double-walled structure that surrounds the glomerulus. The inner wall of the capsule fits closely over the glomerular capillaries and is separated from the outer wall by a small space, **Bowman's space.** To understand the relationship between the glomerulus and Bowman's capsule, imagine that your fist is a glomerulus. Then imagine that you are

holding a partially inflated balloon in your other hand. If you push your fist (glomerulus) into the balloon, the layer immediately surrounding your fist would be equivalent to the inner layer of Bowman's capsule. It is separated from the outer layer of the capsule by Bowman's space.

The outer wall of Bowman's capsule is continuous with the **proximal convoluted tubule,** a sinuous, or winding, section of the renal tubule. The tubule soon straightens, then descends and reascends, forming a thin, U-shaped segment of the renal tubule known as the **loop of Henle.** The loops of Henle of some nephrons extend into the medulla.

The loop of Henle drains into the fourth and final portion of the renal tubule, the **distal convoluted tubule,** another winding segment. Each distal convoluted tubule drains into a straight duct called a **collecting tubule.** The collecting tubules merge to form larger ducts that course through the renal pyramids and empty into the renal pelvis.

The nephrons filter large amounts of blood and produce from 1 to 3 liters of urine per day. The urinary output depends, in large part, on the intake of fluids. In general, the more fluid you drink, the more urine you produce.

≈ FUNCTION OF THE URINARY SYSTEM

With the basic anatomy of the kidney and urinary system in mind, let us turn our attention to the function of the urinary system, focusing first on the nephron.

▷ **FIGURE 16–8 The Nephron** (*a*) A microscopic view of the cortex of the kidney showing the many tubules packed together and a single glomerulus. (*b*) A cross section of the kidney showing the location of the nephrons. (*c*) A drawing of a nephron.

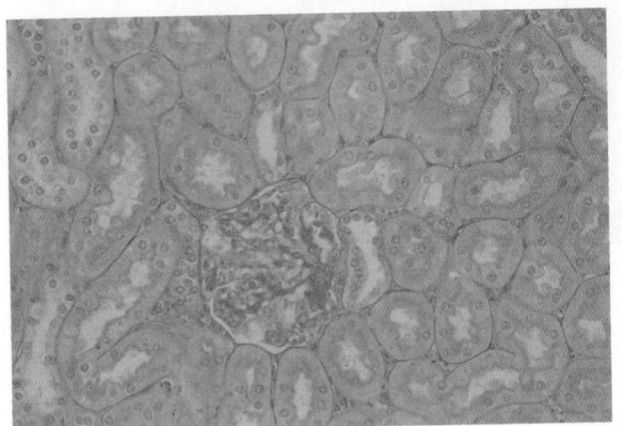

(a)

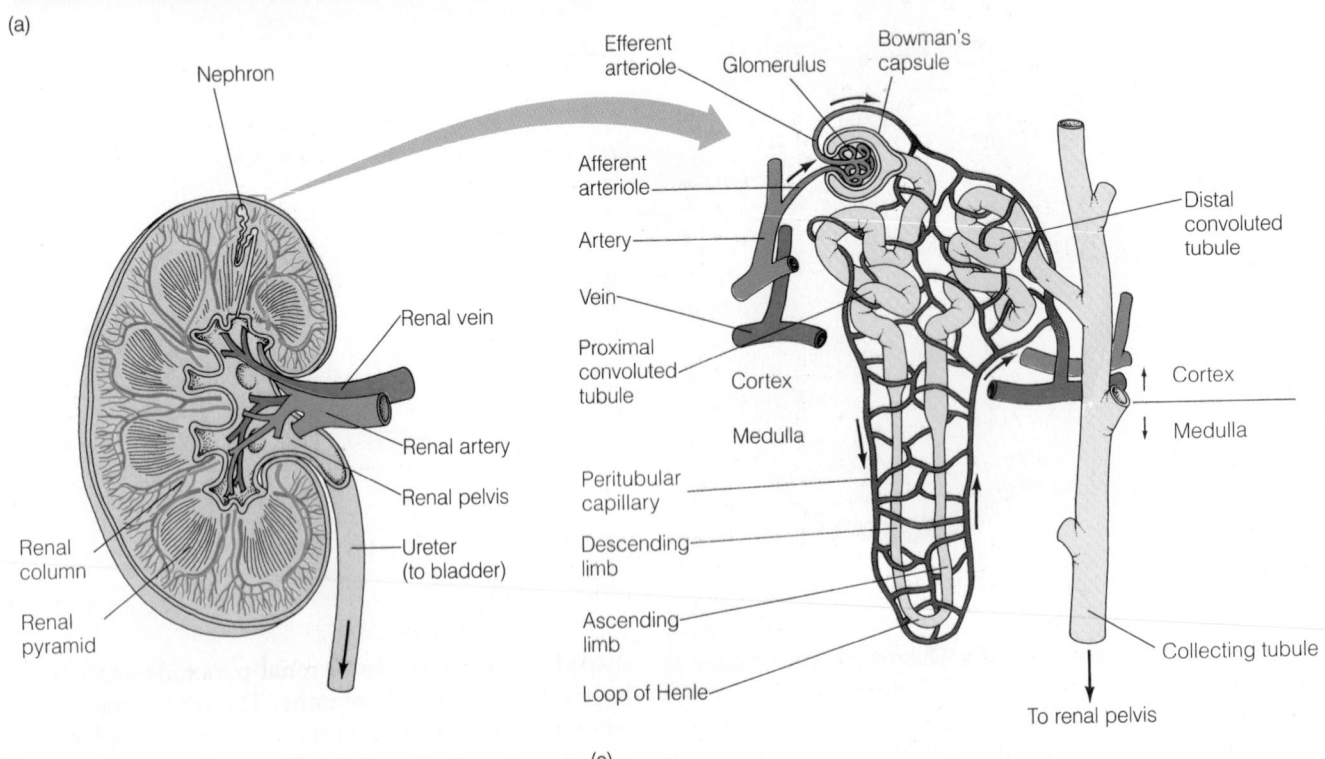

(b)

(c)

TABLE 16–3 Components of the Nephron and Their Function

COMPONENT	FUNCTION
Glomerulus	Mechanically filters the blood
Bowman's capsule	Mechanically filters the blood
Proximal convoluted tubule	Reabsorbs 75% of the water, salts, glucose, and amino acids
Loop of Henle	Participates in countercurrent exchange, which maintains the concentration gradient
Distal convoluted tubule	Site of tubular secretion of H^+, potassium, and certain drugs

Blood Filtration in Nephrons Involves Three Processes: Glomerular Filtration, Tubular Reabsorption, and Tubular Secretion

Glomerular Filtration. The first step in the purification of the blood is **glomerular filtration** ($\triangleright$ Figure 16–9). As noted earlier, the glomeruli are tufts of capillaries ($\triangleright$ Figure 16–10). Blood flows into the glomeruli through the **afferent arterioles.** As blood flows through the glomerular capillaries, water and dissolved materials are forced through the endothelium of the capillaries by blood pressure. The resulting liquid, called the **glomerular filtrate,** travels through the inner layer of Bowman's capsule and enters Bowman's space ($\triangleright$ Figure 16–11a).

Glomerular filtration is akin to the filtration that takes place in a sieve. In the kidney, the "sieve" consists of two cell layers. The innermost is the endothelium of the capillaries. Shown in the Figure 16–11a, the cells of the endothelium of the glomerular capillaries contain numerous openings, or **fenestrae** ("windows"), that permit water, ions, glucose, and even small proteins to pass through. The fenestrae, however, block larger components, such as blood cells, platelets, and large blood proteins.

The second layer of the glomerular filtration membrane is formed by the inner membrane of Bowman's capsule (Figures 16–11a and 16–11b). The inner layer of Bowman's capsule consists of highly branched cells known as **podocytes** (so named because they contain many footlike processes; *podocytes* literally means "foot cells"). The podocytes surround the glomerular capillaries (Figure 16–11c). To understand the relationship of the cells of the inner layer of Bowman's capsule to the capillaries, hold a piece of plastic tubing in your hands; a vacuum cleaner hose will do. This is the capillary. Next, wrap your hands around the tube, interlocking your fingers. Each of your hands now represents a podocyte (Figure 16–11c). The fingers resemble the branching foot processes of the cells. Looking down on your fingers, you will notice small slits between them. These openings, known as **filtration slits,**

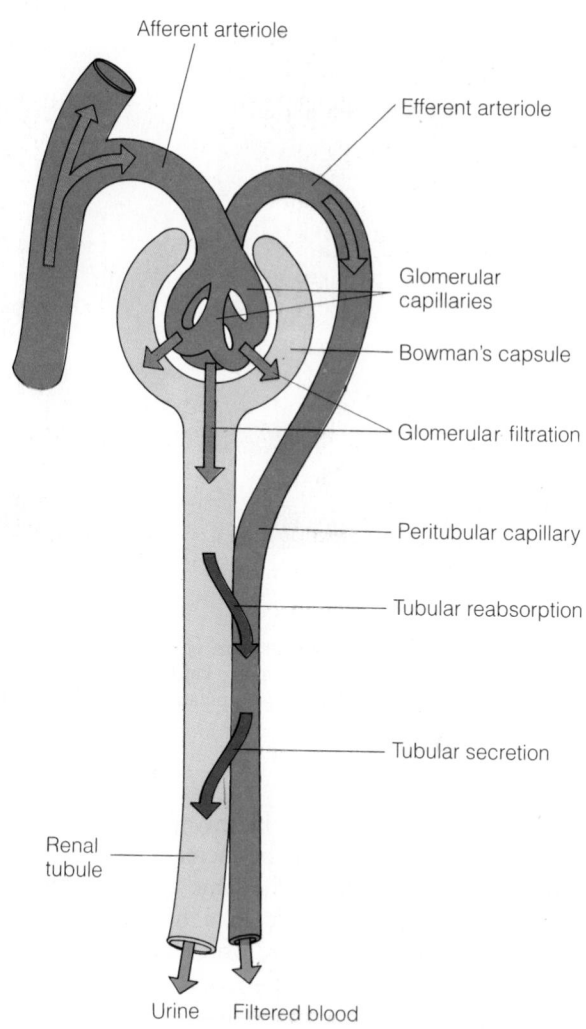

Afferent arteriole
Efferent arteriole
Glomerular capillaries
Bowman's capsule
Glomerular filtration
Peritubular capillary
Tubular reabsorption
Tubular secretion
Renal tubule
Urine   Filtered blood

$\triangleright$ **FIGURE 16–9 Physiology of the Nephron** The nephron carries out three processes: glomerular filtration, tubular reabsorption, and tubular secretion. All contribute to the filtering of the blood.

form a physical barrier that prevents the smaller blood proteins that have passed through the endothelium from entering Bowman's space (Figures 16–11b and 16–11c).

In summary, the fenestrae of the glomerular capillaries and the filtration slits allow the passage of water, ions, and many small- to medium-sized molecules but prevent the passage of blood cells, platelets, and most blood proteins. Kidney infections may damage the filtration membrane—so much so that blood cells and proteins can enter Bowman's space. In such cases, blood is excreted in the urine. Kidney infections require immediate attention to avoid permanent damage.[2]

Glomerular filtration is controlled by regulating the flow

[2]Blood in the urine is usually the result of a urinary bladder infection or a kidney stone, both of which require immediate attention as well.

Small branch of renal artery   Peritubular capillaries

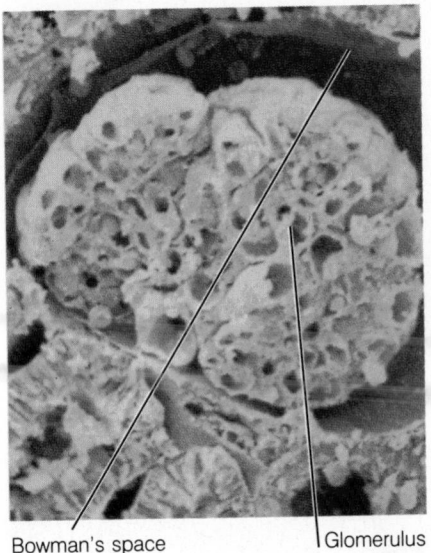

Bowman's space   Glomerulus   Glomerulus   Efferent arteriole   Afferent arteriole
(a)   (b)

▷ **FIGURE 16–10 The Glomerulus** (*a*) A scanning electron micrograph of a glomerulus and Bowman's capsule. (*b*) A scanning electron micrograph of glomerular capillaries. Part (*b*) from Richard G. Kessel and Randy H. Kardon, *Tissues and Organs: A Text-Atlas of Scanning Electron Microscopy.* Copyright © 1979 W.H. Freeman and Company. Reprinted with permission.

of blood and blood pressure in the glomerulus. Blood pressure in the glomerulus can be increased by dilation of the afferent arterioles in much the same way that opening a faucet increases water pressure in a hose fitted with a nozzle. The **efferent arterioles,** which drain blood from the glomerular capillaries, naturally raise glomerular blood pressure, because they are slightly smaller in diameter than the afferent arterioles. Slight constriction of the efferent arterioles further increases blood pressure within the glomerular capillaries in much the same way that putting a clamp on a garden hose increases pressure upstream from the point of constriction. In the capillaries, an increase in blood pressure inside the glomerular capillaries increases the rate of filtration.

Tubular Reabsorption. Each day, approximately 180 liters (45 gallons) of filtrate is formed by glomerular filtration. Despite this massive outpouring of filtrate, the kidneys produce only about 1 to 3 liters of urine each day. In other words, only about 1% of the filtrate actually leaves the kidneys as urine. What happens to the rest of the fluid filtered by the glomerulus?

Most of the fluid filtered by the glomeruli is reabsorbed, passing from the renal tubule back into the bloodstream. The movement of water, ions, and molecules from the renal tubule to the bloodstream is called **tubular reabsorption** (see Figure 16–9). During tubular reabsorption, water containing nutrients and ions passes from the renal tubule into the **peritubular capillaries,** networks of capillaries surrounding the nephrons. The peritubular capillaries are branches of the efferent arterioles.

The peritubular capillaries reabsorb water, nutrients

such as glucose, and various ions such as sodium that are filtered out of the blood in the glomerulus. Tubular reabsorption therefore helps conserve water and various dissolved materials. Table 16–4 shows that 99% of the water molecules, 99.5% of the sodium ions, and 100% of the glucose molecules filtered out of the blood during glomerular filtration are eventually reabsorbed. Without tubular reabsorption, glomerular filtration would require massive water and nutrient intake to offset the loses.

The value of tubular reabsorption lies in its selectivity—that is, its ability to selectively reabsorb valuable molecules

	TABLE 16–4 Fate of Various Substances Filtered by Kidneys	
SUBSTANCE	AVERAGE PERCENTAGE OF FILTERED SUBSTANCE REABSORBED	AVERAGE PERCENTAGE OF FILTERED SUBSTANCE EXCRETED
Water	99	1
Sodium	99.5	0.5
Glucose	100	0
Urea (a waste product)	50	50
Phenol (a waste product)	0	100

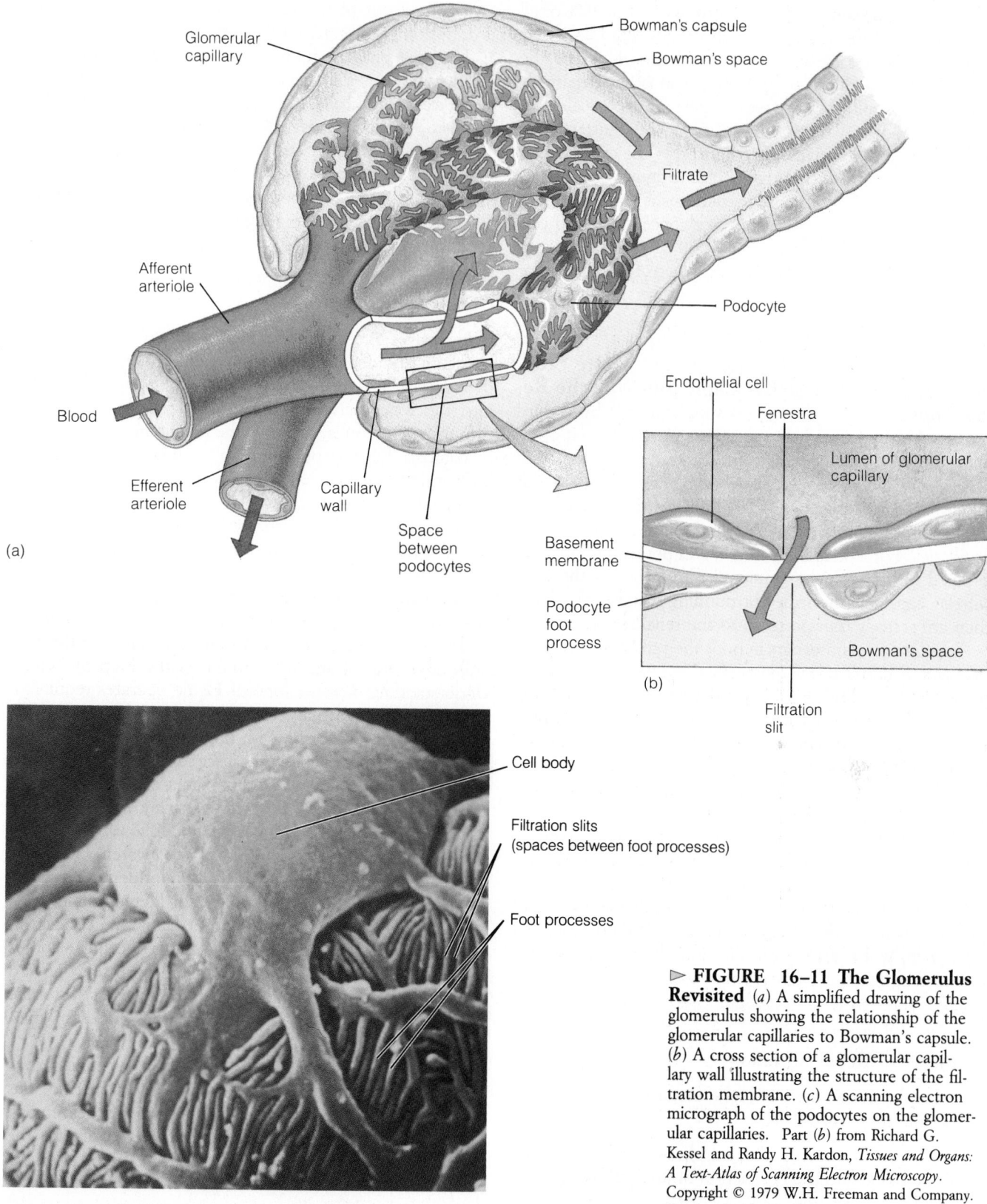

(a)

Bowman's capsule

Bowman's space

Glomerular capillary

Filtrate

Podocyte

Afferent arteriole

Blood

Efferent arteriole

Capillary wall

Space between podocytes

Endothelial cell

Fenestra

Lumen of glomerular capillary

Basement membrane

Podocyte foot process

Bowman's space

(b)

Filtration slit

Cell body

Filtration slits (spaces between foot processes)

Foot processes

(c)

▷ **FIGURE 16–11 The Glomerulus Revisited** (*a*) A simplified drawing of the glomerulus showing the relationship of the glomerular capillaries to Bowman's capsule. (*b*) A cross section of a glomerular capillary wall illustrating the structure of the filtration membrane. (*c*) A scanning electron micrograph of the podocytes on the glomerular capillaries. Part (*b*) from Richard G. Kessel and Randy H. Kardon, *Tissues and Organs: A Text-Atlas of Scanning Electron Microscopy*. Copyright © 1979 W.H. Freeman and Company. Reprinted with permission.

like water and glucose while letting wastes pass through. As a result, important chemicals are conserved, and wastes are excreted. The one exception is urea, a waste product of amino acid catabolism. As shown in Table 16–4, only 50%

of the urea that enters the nephron in the filtrate is reabsorbed, and the rest is eliminated in the urine.

Tubular reabsorption occurs in all parts of the renal tubule, but the bulk of it occurs in the proximal convoluted tubule. In the proximal convoluted tubule, about 70% of all sodium ions and water are reabsorbed. The positively charged sodium ions are actively transported out of the tubule, whereas chloride ions, urea, and other negatively charged ions follow passively, moving down an electrochemical gradient.[3] The increase in the concentration of sodium and the negatively charged ions in the extracellular fluid, in turn, causes water molecules to move outward by osmosis. As noted in Chapter 3, osmosis is the movement of water from a region of high water concentration to one of lower water concentration.

The proximal convoluted tubule is also responsible for the reabsorption of calcium ions, glucose, and vitamins from the filtrate.

Tubular Secretion Is the Transport of Waste Products From the Peritubular Capillaries Into the Renal Tubule. Waste disposal is supplemented by a third process, known as **tubular secretion.** Here is how it operates. Wastes not filtered from the blood as it passes through the glomerular capillaries remain in the blood that enters the peritubular capillaries. Some of these wastes are then transported *into* the renal tubule from the peritubular capillaries. For example, hydrogen and potassium ions that escaped filtration in the glomeruli diffuse out of the peritubular capillaries into the surrounding extracellular fluid, then are actively transported into the renal tubule.

Tubular secretion occurs in both the proximal and distal convoluted tubules, but mostly in the latter. This process helps rid the blood of wastes and also helps regulate the H^+ concentration of the blood. The latter aids in maintaining a constant pH and supplements buffers in the blood and extracellular fluid, especially bicarbonate ions.[4]

The urine leaving the nephron consists mostly of water and a variety of dissolved waste products. Blood draining from the peritubular capillaries is purified, or cleared of most wastes, and contains most of the ions, water, and nutrients that entered the kidney.

The blood draining from the peritubular capillaries empties into small veins. These veins converge to form the renal vein, which transports the filtered blood out of the kidney and into the inferior vena cava and then on to the heart.

Summary of Renal Filtration

Let me take a few moments to summarize the key points of the filtration process.

1. Glomerular filtration is the first phase in the purification of the blood and the production of urine. Blood plasma

is forced out of the glomerular capillaries through the filtration membrane into Bowman's capsule.

2. The next phase is tubular reabsorption. Valuable nutrients and ions that might otherwise be lost in the urine are reabsorbed via the renal tubule. Except for urea, waste products pass through the renal tubule untouched, eventually leaving the kidney via the ureter.

3. Some additional purification occurs via tubular secretion—the transport of unfiltered wastes from the peritubular capillaries to the renal tubule.

Countercurrent Exchange Helps Concentrate the Urine

Urine leaves the distal convoluted tubule and enters the collecting tubules. The collecting tubules descend through the medulla and converge to form larger ducts that drain into the renal pelvis. As the collecting tubules descend through the medulla, much of the remaining water escapes by osmosis, further concentrating the urine and conserving body water. What causes this outward movement of water?

As illustrated in ▷Figure 16–12, the concentration of sodium chloride (NaCl) in the extracellular fluid of the medulla increases toward the renal pelvis. As the urine flows down through the medulla, then, water moves out by osmosis through the highly permeable collecting tubules in an attempt to equalize the concentration of the two solutions.

The sodium chloride concentration gradient in the medulla is produced and maintained by the loop of Henle. Understanding how the loop of Henle operates requires a look at its permeability to various substances—notably, water and salt. We begin with the ascending limb.

As the urine moves up the ascending limb toward the distal convoluted tubule, sodium ions are actively transported out. Chloride ions follow passively, moving by diffusion (Figure 16-12).[5] Water cannot follow because the ascending limb is rather impermeable to water. The transport of sodium ions and diffusion of chloride ions is greatest in the lowest portion of the loop of Henle, as indicated in Figure 16–12 by the thick arrows. Thus, the concentration of sodium and chloride ions in the extracellular fluid is greatest at the bottom of the loop. Because the outward movement of sodium and chloride decreases as the urine moves up, the extracellular concentration of sodium chloride decreases.

Now let us turn to the descending limb. As indicated by the solid arrows in Figure 16–12, sodium chloride ions pumped out of the ascending limb enter the descending limb. These ions will soon be transported back up the ascending limb and will be pumped out as the urine ascends once again, creating a self-perpetuating closed circuit that maintains this important concentration gradient. In the medulla, an equilibrium is established in which the outward movement of sodium chloride in the ascending limb equals the inward movement of sodium chloride in the descend-

---

[3] As you recall from Chapter 3, the term *electrochemical gradient* refers to a charge and concentration difference across a membrane.

[4] To prevent the buildup of hydrogen ions inside the renal tubule, some cells of the tubule produce ammonia (NH_3). Ammonia diffuses into the renal tubule, where it combines with hydrogen ions, forming ammonium ions (NH_{4+}).

[5] Some researchers believe that the chloride ions are actively transported out and the sodium ions follow passively down the resulting electrical gradient.

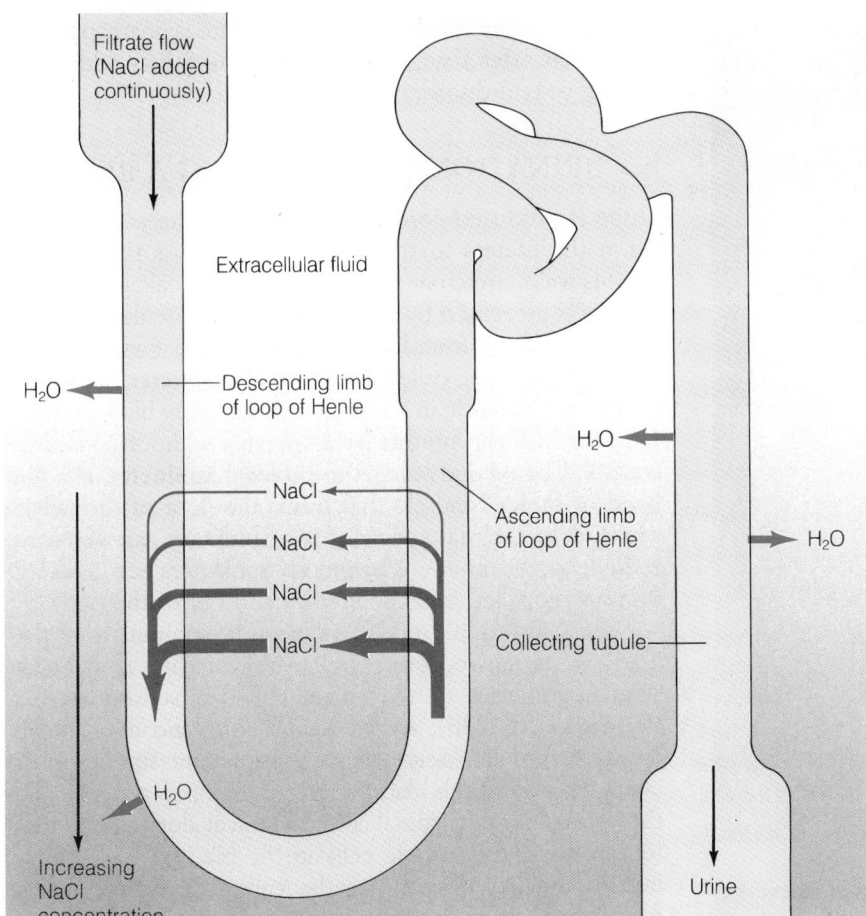

Filtrate flow
(NaCl added
continuously)

Extracellular fluid

H_2O

Descending limb
of loop of Henle

NaCl

NaCl

NaCl

NaCl

H_2O

Ascending limb
of loop of Henle

H_2O

H_2O

Collecting tubule

H_2O

Increasing
NaCl
concentration

Urine

▷ **FIGURE 16–12 The Concentration of Urine by the Nephron** A countercurrent-exchange mechanism ensures the concentration gradient in the extracellular fluid surrounding the nephron, which facilitates the movement of water out of the collecting tubule and helps conserve body water and concentrate urine. (See the text for an explanation of countercurrent-exchange mechanism.)

ing limb. The extracellular concentration therefore remains the same and is always greatest at the bottom of the loop, for the reasons described above.

This mechanism is called **countercurrent exchange.** The term *countercurrent exchange* applies to any system in which fluids, flowing in opposite directions, exchange chemical substances or heat. Arteries and veins, for example, contain blood flowing in opposite directions and often lie side by side. In the arm, warm blood traveling in the artery loses its heat to the slightly cooled venous blood returning to the heart. This mechanism helps conserve body heat and is the reason your hands are slightly colder than your torso. It is also one reason ducks can stand on ice and not be bothered by the cold.

In the kidney, countercurrent created by the descending and ascending limbs of the loops of Henle exchanges sodium and chloride ions. This exchange maintains the concentration gradient in the medulla, which, in turn, ensures the outward movement of water (by osmosis) from the descending limbs. As the urine flows through the collecting tubules and through the concentrated medulla, water also escapes via osmosis, further increasing the concentration of the urine. The water lost from the descending limb and collecting tubules is eventually reabsorbed into the bloodstream.

Practical Applications: Kidney Stones

Urine contains various dissolved wastes. Approximately 90% of the dissolved waste consists of three substances: urea, sodium ions, and chloride ions. Varying amounts of other chemical substances are also present. Increases or decreases in the level of these substances may signal underlying health problems. As a result, physicians routinely analyze the dissolved components in their patients' urine as a way of checking for underlying metabolic problems. For example, diabetes mellitus, or sugar diabetes, a defect in glucose uptake by cells in the body, results in the presence of glucose in the urine.

Excess calcium, magnesium, and uric acid in the urine, caused in many cases by inadequate fluid intake, may crystallize in the renal pelvis of the kidney, forming deposits known as **kidney stones** (▷ Figure 16–13a). These small deposits enlarge by **accretion**—the deposition of materials on the outside of the stones—causing them to grow in much the same way that a snowball enlarges as it is rolled through wet snow (Figure 16–13b).

Many kidney stones enter the ureter on their own and are passed to the bladder. As they are propelled along the ureter, however, their sharp edges can dig into its wall, stimulating pain fibers. Small stones that pass to the urinary

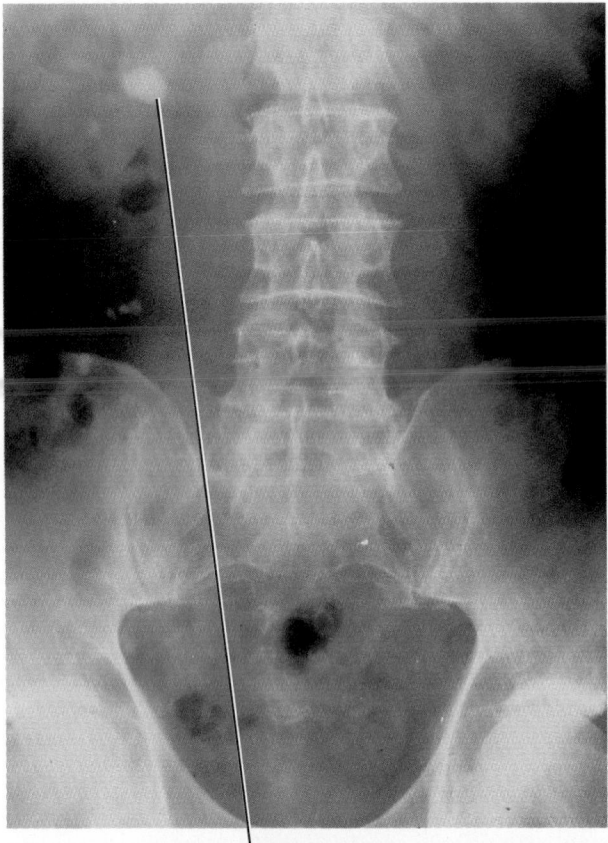

(a)   Kidney stone

(b)

▷ **FIGURE 16–13 Kidney Stones** (*a*) An X-ray of a kidney stone. (*b*) Kidney stones removed by surgery. (See nail for size.)

bladder are often excreted in the urine via the urethra. This may also be accompanied by considerable pain. Larger stones, however, often lodge in the ureters or in the renal pelvis, where they obstruct the flow of urine. Pressure increases inside the kidney, causing considerable damage to the nephrons if untreated.

For years, kidney stones were removed surgically. Today, however, the relatively new medical technique of ultrasound lithotripsy is generally used. In this technique, mentioned in the opening vignette of this chapter, physicians bombard kidneys with ultrasound waves, which shatter the stones, producing fine, sandlike grains that can be passed in

the urine without incident. The procedure is nearly painless and much safer than surgery. Table 16–5 lists additional urinary system disorders.

≈ URINATION: CONTROLLING A REFLEX

Urine is produced continuously by the kidneys and flows down the ureters to the urinary bladder. As the bladder fills, its walls stretch and become thinner. Leakage into the urethra is prevented by two sphincters—muscular "valves" not unlike those found in the stomach (see Figure 16–6). The first sphincter, called the **internal sphincter,** is formed by a smooth muscle in the neck of the urinary bladder at its junction with the urethra and operates without conscious control. The second valve, the **external sphincter,** is a flat band of skeletal muscle that forms the floor of the pelvic cavity. The external sphincter is a voluntary gateway controlled by the brain. When both sphincters are relaxed, urine is propelled into the urethra and out of the body.

When 200 to 300 milliliters of urine accumulate in the bladder, stretch receptors in the wall of the organ begin sending impulses to the spinal cord via sensory nerves (▷ Figure 16–14b). In the spinal cord, incoming nerve impulses stimulate nerve cells that supply the smooth muscle in the wall of the bladder. Nerve impulses generated in these cells leave the spinal cord and travel along nerves that terminate on the muscle cells in the bladder wall. These impulses, in turn, stimulate the cells to contract, which forces the ends of the ureters shut, preventing the backflow of urine. Contraction of the wall also forces the internal sphincter to open, letting urine escape into the urethra.

Urine does not escape, however, until the external sphincter is relaxed. The external sphincter is supplied by a continuous barrage of nerve impulses from the brain and spinal cord, which keep this muscle in a constant state of contraction. In order for the sphincter to open, these impulses must cease. As shown in Figure 16–14, in babies and very young children nerve impulses arriving in the spinal cord from the stretch fibers also inhibit the nerve cells supplying the external sphincter, permitting it to open. In babies and very young children, then, urination is a reflex; that is, there is no conscious override. Not until children grow older (2 to 3 years) can they begin to control urination.

Conscious control occurs at the external sphincter and is exerted by the brain. Thus, the brain can override the urination reflex. Here's how it works: As the bladder fills, nerve impulses from stretch receptors travel to the spinal cord; some of these impulses then travel to the brain, creating a conscious awareness of the situation. If the time and place are inappropriate, a person consciously overrides the reflex. The brain simply sends nerve impulses back down the spinal cord to the skeletal muscles of the external sphincter. These impulses keep the sphincter muscles contracted. While awaiting the signal to release this control, the bladder can expand to hold up to 800 milliliters of urine. At this stage, waiting may become quite painful.

Adults sometimes lose control over urination, resulting

TABLE16-5 Common Urinary Disorders

DISEASE	SYMPTOMS	CAUSE
Bladder infections	Especially prevalent in women; pain in lower abdomen; frequent urge to urinate; blood in urine; strong smell to urine	Nearly always bacteria
Kidney stones	Large stones lodged in the kidney often create no symptoms at all; pain occurs if stones are being passed to the bladder; pains come in waves a few minutes apart.	Deposition of calcium, phosphate, magnesium, and uric acid crystals in the kidney, possibly resulting from inadequate water intake
Kidney failure	Symptoms often occur gradually: more frequent urination, lethargy, and fatigue; should the kidney fail completely, patient may develop nausea, headaches, vomiting, diarrhea, water buildup, especially in the lungs and skin, and pain in the chest and bones.	Immune reaction to some drugs, especially antibiotics; toxic chemicals; kidney infections; sudden decreases in blood flow to the kidney—for example, resulting from trauma
Pyelonephritis	Infection of the kidney's nephrons; sudden, intense pain in the lower back immediately above the waist, high temperature, and chills	Bacterial infection

in a condition referred to as **urinary incontinence.** Urinary incontinence has several causes. In some instances, it results from traumatic injury to the spinal cord, which disrupts descending nerve fibers carrying the impulses from the brain that override the urination reflex. In such cases, the urinary reflex remains intact, and the bladder empties as soon as it reaches a certain size, much as it does in a young child.

Mild urinary incontinence is much more common. It is characterized by the escape of urine when a person sneezes or coughs and, in women, usually results from damage to the external sphincter during childbirth. Childbirth stretches the

▷ **FIGURE 16–14 Urination in Babies** (a) The bladder before and after it fills, showing how much this organ can expand to accommodate urine. (b) Stretch receptors signal the distension of the urinary bladder. Nerve impulses travel to the spinal cord. In the spinal cord, they stimulate nerve cells that send impulses back to the bladder, causing muscle contraction in the wall which forces the internal sphincter open. Nerve impulses also travel up the spinal cord to the brain signaling the need to void (not shown here). In adults the brain sends signals back to the external sphincter, causing it to relax if the time and place for urination are appropriate.

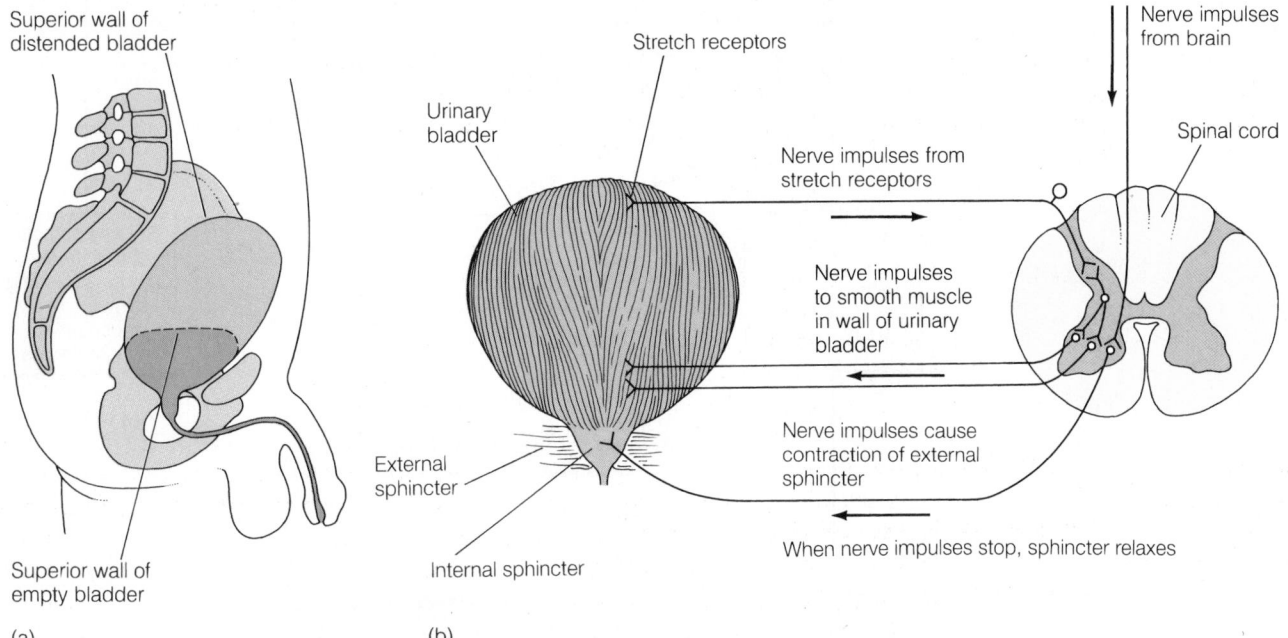

skeletal muscles that compose the external sphincter, reducing their effectiveness and making such accidents embarrassingly common. For this reason, many women undertake exercise programs to strengthen the muscles before and after childbirth. Urinary incontinence may also occur in men whose external sphincters have been injured in surgery on the prostate gland, which surrounds the neck of the urinary bladder.

≈ THE KIDNEYS AS ORGANS OF HOMEOSTASIS

The massive flow of blood through the kidneys and the high rate of glomerular filtration combine to ensure a very thorough filtering of blood borne impurities. This filtering, in turn, helps the body control the composition of the blood to maintain homeostasis.

Besides Ridding the Body of Wastes, the Kidneys Also Help Regulate the Body's Water Content

As noted earlier, much of the water filtered by the glomeruli is passively reabsorbed by the renal tubules and returned to the bloodstream. However, the rate of water reabsorption, can be increased or decreased to alter urine production. The ability to adjust water reabsorption and urine output allows

the body to rid itself of excess water when a person has consumed too much and also helps conserve body water when an individual is becoming dehydrated.

The Hormone ADH Increases the Permeability of the Distal Convoluted Tubules and Collecting Tubules, Increasing Water Reabsorption and Conserving Body Water. Water reabsorption is controlled in large part by **antidiuretic hormone (ADH)** in a negative feedback mechanism. ADH is released by an endocrine gland at the base of the brain known as the pituitary (Chapter 20).[6] ADH release is regulated by two receptors (▷ Figure 16–15). The first is a group of nerve cells in a region of the brain just above the pituitary gland known as the hypothalamus. These cells monitor the osmotic concentration of the blood. The production and release of ADH are also controlled by receptors in the heart, which detect changes in blood volume (which reflect water levels).

Consider an example. If you were to play soccer on a hot summer afternoon, you would undoubtedly perspire heavily and lose a significant amount of body water and salts. If you did not replace the water you were losing, you would become dehydrated. As a result, your blood volume

[6]ADH is manufactured by the hypothalamus and transported to the posterior lobe of the pituitary gland via modified nerve cells, called neurosecretory cells, which are described in Chapter 20.

▷ **FIGURE 16–15 ADH Secretion**
ADH secretion is under the control of the hypothalamus. When the osmotic concentration of the blood rises, receptors in the hypothalamus detect the change and trigger the release of ADH from the posterior lobe of the pituitary. Detectors in the heart respond to changes in blood volume. When it drops, they send signals to the brain, causing the release of ADH.

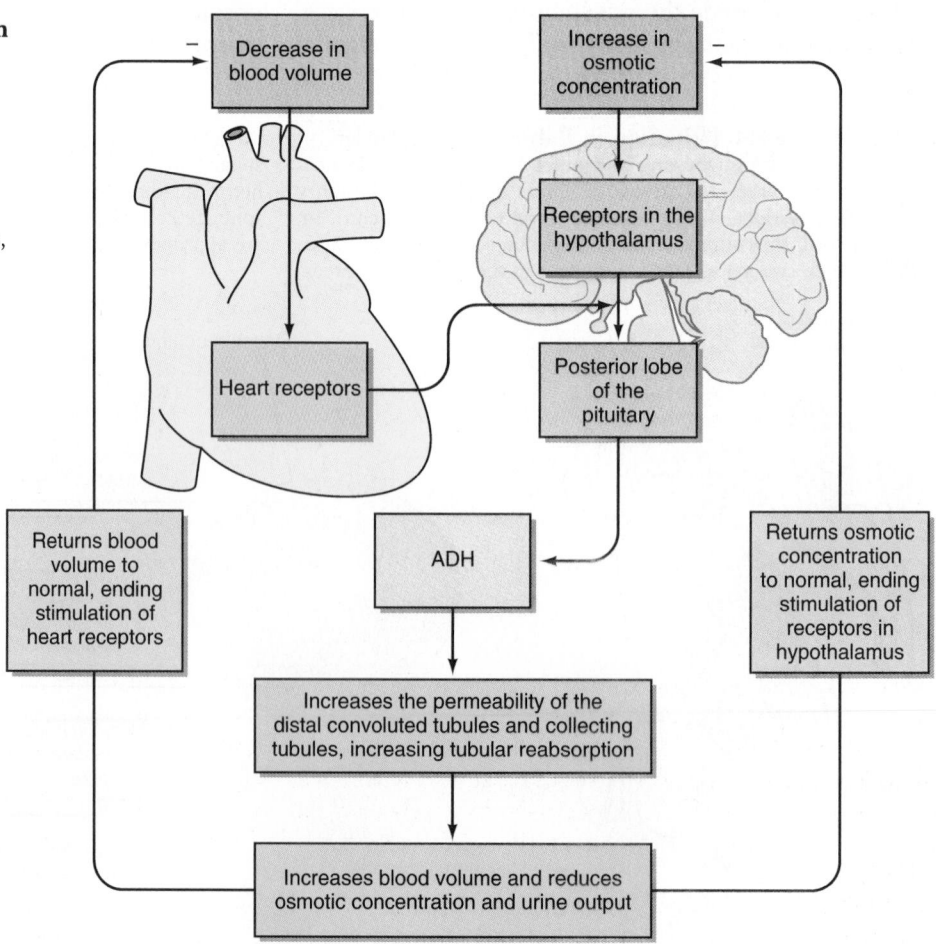

EVOLUTION'S WATER CONSERVATION CHAMPION

It is high summer in Organ Pipe Cactus National Monument in southern Arizona. The midday sun beats down. The temperature on the desert floor is nearly 120°F. Few animals stir.

Deep in its burrow, the kangaroo rat sleeps (see ▷ Figure 1). Like many other desert species, the kangaroo rat is behaviorally adapted to the desert, sleeping underground during the day to escape the fierce heat and coming out at night to feed. This behavioral adaptation is just one of many adaptations aimed at conserving water in the desert. Like other mammals, the kangaroo rat has a built-in water recycler in its nasal passages. As dry desert air is inhaled, it passes over moist tissues in the nasal cavity, picking up moisture. This hy-

drated air protects the lungs from desiccation. Evaporation of water caused by the incoming dry air cools the nasal passageway. When the moist air is later exhaled, water vapor condenses on the cool surfaces of the nasal passages, thus conserving body water.

It is important to note, however, that the countercurrent exchange of heat and moisture is not an adaptation unique to desert life. It is present in birds and lizards as well as mammals. Nevertheless, research suggests that although this evolutionary strategy is widespread, some desert rodents like the kangaroo rat are particularly good at recovering water when breathing dry air. Overall, the nasal countercurrent system of this creature reduces evaporative losses by 70% to 74%.

Desert rodents also produce relatively

dry feces, due to the efficient uptake of water by the large intestine. This adaptation conserves water and, like the nasal countercurrent system, is found in many other rodents. In the desert, it serves the kangaroo rat extremely well.

Rodents also have the ability to concentrate their urine. The urine of the laboratory white rat, for instance, is twice as concentrated as human urine. The kangaroo rat and other desert rodents produce urine about two to five times more concentrated than that of humans.

The ability to concentrate urine can be traced to the nephrons of the kidney. As a rule, the longer the loops of Henle, the greater an animal's ability to concentrate urine. The beaver, for example, which lives in aquatic environments where water conservation is not a priority, has relatively short loops of Henle. Humans have long and short loops, but kangaroo rats have uniformly long loops, giving them a remarkable ability to concentrate urine.

Collectively, the water conservation adaptations of the kangaroo rat result in such small water losses that this animal is able to get by without drinking water or consuming live plants. The kangaroo rat can, in fact, get all of the water it needs from seeds and dry plant matter. How?

As you may recall from Chapter 6, energy catabolism produces water, known as metabolic water. Metabolic water and small quantities of water in the food the kangaroo rat eats, combined with behavioral and physiological water-conservation measures, liberate the kangaroo rat from free water, making this animal ideally suited to its desert home.

▷ **FIGURE 1 Kangaroo Rat** This creature is well adapted to desert conditions and is an excellent model of water conservation.

would fall, and the osmotic concentration of your blood would rise (Figure 16–15). The decrease in blood volume obviously results from the loss of water. The increase in osmotic concentration of the blood is due to the fact that perspiration carries off water and salts but leaves behind many osmotically active chemicals, including blood proteins and glucose.

The decrease in blood volume and the rise in osmotic concentration trigger the release of ADH (Figure 16–15).

ADH circulates in the blood to the kidney and there increases the permeability of the distal convoluted tubules and the collecting tubules to water. This increases the rate of tubular reabsorption of water, which reduces urinary output and helps restore the volume and proper osmotic concentration of the blood.

Excess water intake has just the opposite effect, increasing the blood volume and decreasing its osmotic concentration. These changes reduce ADH secretion. As ADH levels in the blood fall, the permeability of the distal convoluted tubules and the collecting tubules decreases. As a result, tubular reabsorption declines, and more water is lost in the urine. This helps decrease blood volume and also helps restore the proper osmotic concentration of the blood.

Caffeine Increases Urine Output Without Affecting ADH Secretion.

Water in the fluids we drink affects urinary output through ADH secretion, but certain chemical substances in common beverages act on the kidneys directly and have a dramatic effect on urine production. Coffee and (most nonherbal) teas, for example, increase urine production because they contain caffeine.[7] Caffeine in teas, coffee, and certain soft drinks is a **diuretic**, a chemical that increases urination. Caffeine has two major effects on the kidney. First, it increases glomerular blood pressure, which, in turn, increases glomerular filtration. (As a result, more filtrate is formed.) Second, caffeine decreases the tubular reabsorption of sodium ions. As noted earlier, water follows sodium ions out of the renal tubule during tubular reabsorption. A decrease in sodium reabsorption therefore results in a decline in the amount of water leaving the renal tubule and an increase in urine output.

Ethanol Inhibits the Secretion of ADH by the Pituitary.

Alcoholic beverages also increase urination. Ethanol in alcoholic beverages is a diuretic, but works by inhibiting the secretion of ADH. This reduces water reabsorption and increases water loss, giving credence to the quip that you don't buy wine or beer, you rent it!

Diabetes Insipidus Is a Rare Disease Characterized by Excessive Drinking and Excessive Urine Output.

Severe head injuries can halt the production of ADH, leading to a disease known as **diabetes insipidus.** This condition is not to be confused with diabetes mellitus (commonly called sugar diabetes), a disorder involving the hormone insulin. **Diabetes insipidus** is characterized by frequent urination (polyuria) and excessive liquid intake (polydipsia). It results from insufficient ADH output. Patients with the disease produce up to 20 liters (5 gallons) of colorless, dilute urine per day. Diabetes insipidus gets its name from the fact that the urine is dilute and tasteless (insipid).[8] Sleeping through the night is impossible, for patients are continually awakened by thirst or the urge to urinate.

Diabetes insipidus can be treated in a variety of ways, depending on the severity of the disorder. In patients whose urine output is only slightly elevated, dietary salt restrictions and antidiuretics (drugs that reduce urine output) will work. In severe cases, patients must receive synthetic ADH. ADH can be administered by injections or

[7] Part of the increase is due to the fluid contained in these drinks.
[8] Diabetes mellitus is so named because the urine is sweet. The Latin word for "honey" is *mellifer*.

nose drops. If the pituitary damage has resulted from a head injury, treatment may be required only for a year or so, depending on the rate of recovery. If the damage is permanent, however, treatment will be required for the rest of a person's life.

The Hormone Aldosterone Also Controls Water Reabsorption in the Kidney.

ADH controls the amount of water in the body, and that, in turn, helps regulate the concentration of dissolved substances. ADH is therefore, a key component of the homeostatic mechanism that maintains the body's water and chemical balance. ADH is aided by another hormone, aldosterone. **Aldosterone** is a steroid hormone produced by the two **adrenal glands,** which sit atop the kidneys like loose-fitting stocking caps (see Figure 20–1a).

Aldosterone levels in the blood rise and fall in response to three factors: (1) blood pressure, (2) blood volume, and (3) osmotic concentration. A decrease in the blood pressure, for example, triggers the release of this hormone from the outer portion of the adrenal gland, the **adrenal cortex** (▷ Figure 16–16).

Aldosterone increases the reabsorption of sodium ions by the distal convoluted tubules and collecting tubules—that is, the passage of sodium ions from the renal tubule to the blood. Sodium enters the peritubular capillaries, thus increasing the sodium concentration of the blood. Water follows sodium, moving by osmosis into the bloodstream. The outflow of water, therefore, increases the blood volume and blood pressure and lowers the osmotic concentration.

As illustrated in Figure 16–16, the release of aldosterone is controlled by a fairly complex sequence of events. A reduction in blood pressure or a reduction in the volume of filtrate in the renal tubule causes certain cells in the kidney to produce an enzyme called **renin.** In the blood, renin cleaves a segment off a large plasma protein called **angiotensinogen.** The result is a small peptide molecule called **angiotensin I.** Angiotensin I is inactive but is converted into the active form, **angiotensin II,** by further enzymatic action. Angiotensin II stimulates aldosterone secretion.

Kidney Failure May Occur Suddenly or Gradually and Can Be Treated by Dialysis, a Mechanical Filtering of the Blood

The importance of the kidneys is most obvious when they stop working, a condition known as **renal,** or **kidney, failure.** Renal failure generally results from one of four causes: the presence of certain toxic chemicals in the blood, immune reactions to certain antibiotics, severe kidney infections, and sudden decreases in blood flow (for example, after an injury).

Renal failure may occur suddenly, over a period of a few hours or a few days, a condition referred to as **acute renal failure.** Renal function may also deteriorate slowly over many years, resulting in **chronic renal impairment** (so named because the kidneys never really quit). Chronic

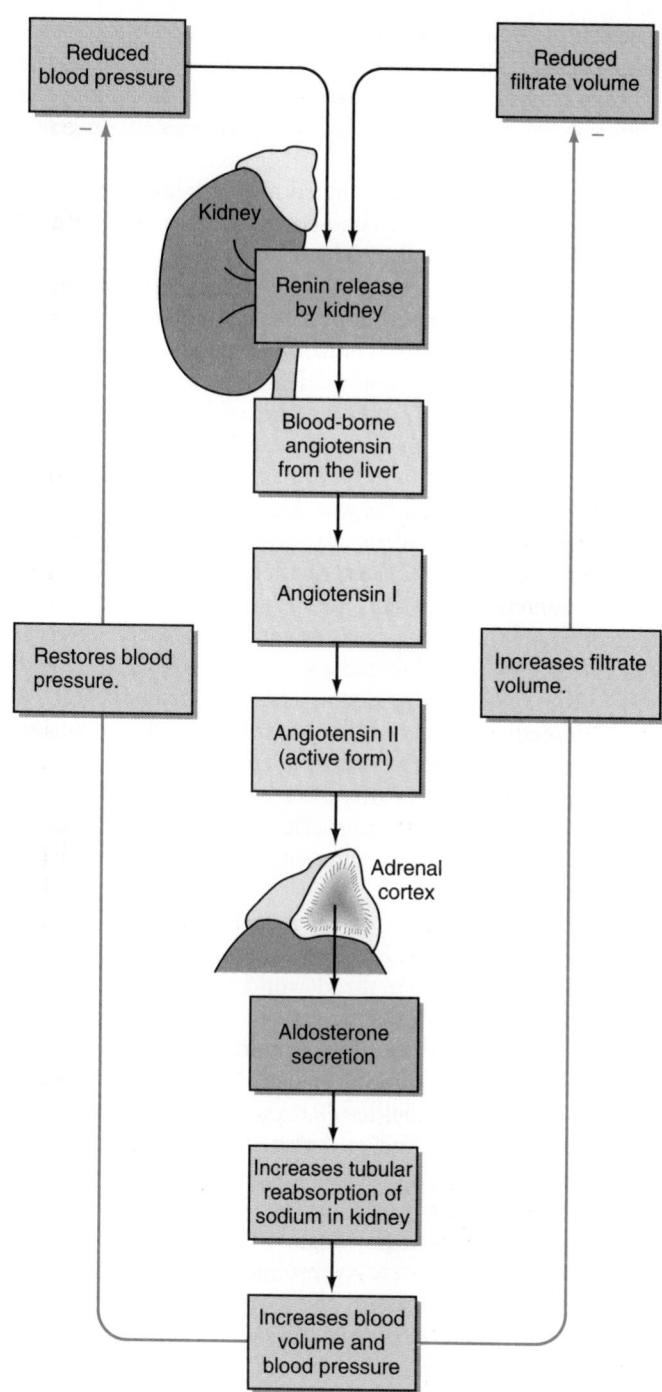

▷ FIGURE 16–16 Aldosterone Secretion Aldosterone is released by the adrenal cortex. Its release, however, is stimulated by a chain of events that begins in the kidney.

renal impairment may lead to a complete or nearly complete shutdown, known as **end-stage failure.**

Kidney failure (whether acute or end stage) is generally fatal unless treated quickly. When the kidneys stop working, water begins to accumulate in the body. Toxic wastes build up in the blood. Homeostasis is disrupted by an imbalance of chemicals that are normally regulated by the kidneys. Death generally occurs in two to three days if renal failure is untreated. Death usually results from an

increase in the concentration of potassium ions in the blood and tissue fluids. Potassium is essential for the normal function of heart muscle, but excess potassium destroys the rhythmic contraction of the heart, causing fibrillation (Chapter 12).

Treating renal failure depends on the underlying cause. If the problem is caused by an acute loss of blood, transfusions may be required. Patients whose kidneys have shut down, even temporarily, may require **renal dialysis,** a procedure in which toxic materials in the patient's blood are filtered using a machine called a **dialysis unit.** Blood is drawn out of a vein and passed through a piece of tubing that transports the blood to an osmotic filter. After filtration, the blood is pumped back into the patient's bloodstream (▷ Figure 16–17). Dialysis requires several hours and must be repeated every two or three days. Some patients have dialysis units at home and simply hook themselves up each night before they go to bed.

Complete kidney failure can be treated by kidney transplants. Transplants are generally most successful when they come from closely related family members, for reasons explained in the previous chapter.

For years, the complete or nearly complete destruction of kidney function was almost always fatal. Thanks to renal dialysis and kidney transplantation, many patients today

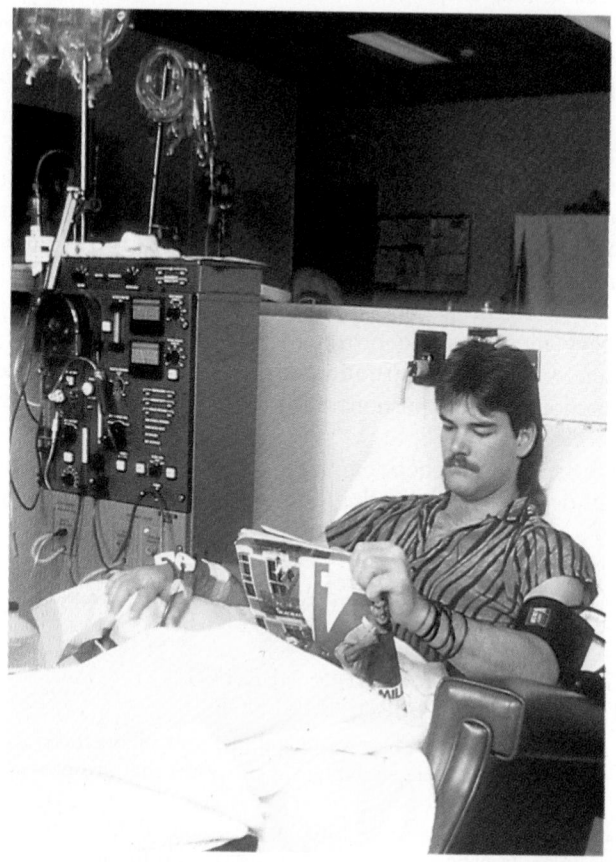

▷ FIGURE 16–17 Renal Dialysis When the kidneys fail, the blood must be filtered by a dialysis machine, which runs the blood through filters that remove the impurities.

can live normal, healthy lives. These procedures, especially transplants, are costly, however.

ENVIRONMENT AND HEALTH: MERCURY POISONING

The kidneys are elaborate filters that help maintain the proper levels of nutrients, such as glucose, amino acids, vitamins, and a number of important ions, such as potassium and chloride. The kidneys also help dispose of various endogenous (internally generated) wastes, such as urea, and eliminate numerous potentially harmful exogenous (externally produced) substances, including some drugs, food additives, pesticides, and toxic chemicals that enter the bloodstream from cigarette smoke.

Unfortunately, the kidneys can be seriously damaged by a variety of environmental contaminants. For example, most heavy metals, such as mercury and cadmium, are potent nephrotoxins–toxic chemicals that affect the nephrons. Even relatively low doses of heavy metals can damage the kidneys.

Nephrons possess several protective mechanisms to reduce the impact of heavy metals. Lysosomes inside the cells of the renal tubule, for example, bind to and engulf these toxins, helping decrease the cytoplasmic concentration of these harmful substances. As blood levels increase, however, the protective mechanisms are overwhelmed, and kidney cells begin to die.

Exposure to low levels of various heavy metals increases urinary output and increases the concentrations of amino acids and glucose in the urine—good indicators that tubular reabsorption is not working well. At higher levels, heavy metals damage the nephrons and may result in renal failure and death.

One of the portions of the nephron most sensitive to heavy metals is the proximal convoluted tubule. These metals disrupt its operation and seriously impair kidney function. Other portions of the nephron can also be affected.

One common heavy metal found in the water we drink and occasionally even the food we eat is mercury. In high concentrations in the blood, mercury produces acute renal failure. This condition results from vasoconstriction of the afferent arterioles, which reduces the flow of blood into the glomeruli. Mercury also produces numerous signs of cellular deterioration.

In the 1950s, Japan announced an outbreak of mercury poisoning in residents who had consumed fish and shellfish taken from Minimata Bay, which had been contaminated by a nearby plastics factory. Over 100 people were affected. Victims developed numbness of the limbs, lips, and tongue and lost muscular control, becoming clumsy. Most suffered from blurred vision, deafness, and mental derangement. All told, 17 people died and 23 were permanently disabled, in large part due to the inability of their kidneys to cope with the mercury and subsequent nervous system effects. The tragedy was deepened by the discovery of birth defects in 19 babies born to women who had eaten contaminated seafood. Many of their mothers showed no signs of mercury poisoning, illustrating a general principle of toxicology: the younger an organism, the more sensitive it is to harmful toxic substances.

The Minimata Bay incident was not an isolated event. A similar tragedy occurred on the Japanese island of Honshu. In Sweden in the early 1960s, mercury poisoning killed large numbers of birds that had been feeding on seeds treated with a mercury fungicide, a coating that retards mildew. Swedes who ate pheasant and other birds were also poisoned by mercury.

Fortunately, incidents of overt mercury poisoning are rare. However, mercury is one of the more common water pollutants in the industrialized world, and people are exposed to low levels from a variety of sources. For example, mercury is a by-product of the production of the plastic polyvinyl chloride (commonly called PVC plastic), which is used to manufacture children's toys, beach balls, car seats, and other products. Mercury is also emitted into waterways by a variety of chemical manufacturers and is released by coal-fired power plants and garbage incinerators that burn batteries thrown out in our trash.

Many sources of mercury production are on the rise. Without tighter controls on their emissions, it is possible that low-level mercury poisoning may become more and more common in the years to come.

SUMMARY

GETTING RID OF WASTES: MEETING THE EVOLUTIONARY CHALLENGE

1. All organisms produce waste and therefore all organisms face the same challenge: getting rid of wastes that are produced internally.
2. For single-celled organisms, waste simply diffuses or is actively transported out and is released into the medium (air, water, soil) they inhabit.
3. In some simple multicellular animals, such as sponges and sea stars, the cells on the body surface eliminate wastes directly into the surrounding environment and into the extracellular fluid.
4. Most multicellular animals have excretory organs that facilitate the removal of waste and also help regulate internal concentrations of ions and water.
5. In earthworms and other annelids, fluid from the body cavity enters two small tubular structures, the metanephridia, present in each segment of the body. They transport the waste to the earthworm's body surface.
6. In terrestrial insects dead-end small tubules attach to the gut

and are bathed in body fluids from which they extract ions and water.

7. Waste disposal reaches its zenith in the vertebrates.

ORGANS OF EXCRETION

8. Table 16–1 lists the major metabolic wastes and other molecules excreted from the body.

9. Metabolic wastes are removed by the organs of excretion, including the skin, lungs, liver, and kidneys.

THE URINARY SYSTEM

10. One of the most important organs of excretion in vertebrates is the kidney. Kidneys remove impurities from the blood but also help regulate the water levels and ionic concentrations of the blood.

11. Blood enters the kidneys in the renal arteries and is delivered to millions of nephrons.

12. The nephrons produce urine, which drains from the kidneys into the ureters, slender muscular tubes that lead to the urinary bladder. Urine is stored in the bladder, then voided through the urethra.

13. Each nephron consists of a glomerulus, a tuft of highly porous capillaries, and a renal tubule, where urine is produced.

14. The renal tubule consists of four parts: (a) Bowman's capsule, (b) the proximal convoluted tubule, (c) the loop of Henle, and (d) the distal convoluted tubule. The distal convoluted tubules of nephrons drain into collecting tubules, which converge and empty urine into the renal pelvis.

THE FUNCTION OF THE URINARY SYSTEM

15. Blood filtration is accomplished by three processes: glomerular filtration, tubular reabsorption, and tubular secretion.

16. Glomerular filtration occurs in the glomerulus, producing a liquid called the filtrate.

17. The filtrate is processed as it flows along the renal tubule. Water, ions, and nutrients are largely reabsorbed as they travel along the tubule. This process is called tubular reabsorption. Water and reabsorbed nutrients and ions pass into a network of capillaries, the peritubular capillaries, surrounding each nephron. What is left is a concentrated liquid, the urine.

18. Not all wastes are filtered from the blood in the glomerulus. Those that remain pass into the peritubular capillaries with the blood. These substances may be transported out of the peritubular capillaries into the renal tubule in a process known as tubular secretion. Hydrogen and potassium ions, for example, are secreted into the renal tubule.

19. Calcium, magnesium, and other materials can precipitate out of the urine in the renal pelvis, forming kidney stones, which can block the outflow of urine. Smaller stones may be passed along the ureters to the bladder and are often eliminated during urination. Larger stones that remain in the kidney must be removed surgically or via ultrasound lithotripsy.

URINATION: CONTROLLING A REFLEX

20. Urination is a reflex in babies and very young children. In older children and adults, the urination reflex still operates, but it is overridden by a conscious control mechanism.

THE KIDNEYS AS ORGANS OF HOMEOSTASIS

21. Each drop of blood in your body flows through the kidneys many times in a single day. This flow allows for a thorough filtering of the blood and also helps the body control the blood's chemical composition.

22. The concentration of water and dissolved substances in the blood is controlled by two hormones: ADH and aldosterone.

23. ADH is secreted by the posterior lobe of the pituitary gland. It is released when the osmotic concentration of the blood increases or when blood volume decreases.

24. ADH increases the permeability of the distal convoluted tubules and the collecting tubules to water. When ADH is present, water reabsorption increases. A lack of ADH causes diabetes insipidus, an overexcretion of urine.

25. Aldosterone is produced by the adrenal cortex. Aldosterone stimulates the reabsorption of sodium ions by the nephron. Water follows the sodium out of the renal tubule, increasing blood pressure and blood volume.

ENVIRONMENT AND HEALTH: MERCURY POISONING

26. The kidneys help regulate the levels of harmful toxins produced by the body and help eliminate toxins taken into the body from air, water, and food. But they are not immune to many potentially harmful substances. Heavy metals, for example, destroy some of the cells of the renal tubule and can restrict blood flow to the glomeruli. As a result, heavy metals can damage the kidney, impairing the function of this important homeostatic organ.

EXERCISING YOUR CRITICAL THINKING SKILLS

1. Each year, 24,000 new cases of kidney cancer are diagnosed in the United States. Ten thousand people die from the disease annually. Chemotherapy is virtually the only treatment available for patients whose cancers have spread, but it is beneficial in fewer than 10 percent of the cases. In fact, on average, patients receiving chemotherapy survive only about nine months after they have been diagnosed with the disease.

Recently, a new treatment was introduced that holds some promise for those with kidney cancer. This treatment, known as autolymphocyte therapy, or ALT for short, is marketed by a company in Newton, Massachusetts.

Costing about $22,000 per patient, ALT is designed to strengthen the immune system of patients with malignant kidney tumors. Here's how it works: Blood is withdrawn from patients with kidney cancer. Lymphocytes are extracted from the blood, then treated with monoclonal antibodies (antibodies to the kidney tumor). This process activates the lymphocytes, causing them to produce a chemical substance known as cytokine. Cytokines are natural compounds found in the body that boost the activity of lymphocytes in the immune system. The cytokines produced from this procedure are divided into six batches and stored for later use.

Once a month, for the next six months, patients donate more of their own lymphocytes. The cells are treated with cytokines, then reinjected into the patient, where they apparently go to work on the tumors.

Results of one study on the effectiveness of ALT were published in April 1990 in the British medical journal the *Lancet*. This study of 90 individuals showed that patients who had undergone the procedure survived 22 months after diagnosis, two and a half times longer than patients treated with chemotherapeutic agents.

Less than a year after the publication of this clinical trial, a treatment center opened in Boston. It offers ALT to patients who fit the profile of those who were helped by the drug in the trial.

Assume that you are considering investing in the company that markets ALT. Using your critical thinking skills and your knowledge of good scientific method, what concerns do you have, and what questions would you ask before investing in this company? Write your list on a separate page.

After preparing your list, compare your concerns and questions with mine to see how we match.

1. Concern: A clinical trial on 90 patients is extremely small. With so few patients, the reliability of the results is in question. Question: Are more clinical tests under way?
2. Question: Does the treatment have any adverse impacts?
3. Questions: Is the marketing company unbiased? Is it promoting a product that may turn out to be ineffective, opening the company to lawsuits for fraud? In other words, is it letting financial concerns outweigh the need for good scientific research and carefully controlled experiments?

Now suppose that you have kidney cancer and have the $22,000 to pay for ALT. Would you do it? Are any of the questions or concerns above still relevant? How would you go about determining whether you should try the procedure?

To be fair to all parties concerned, let me point out that approximately two-thirds of the insurance companies in the United States pay for the treatment. As a rule, most companies do not pay for experimental procedures. In other words, they must be satisfied that ALT is a useful and effective treatment before reimbursing clients. In your view, has the insurance industry decided wisely about this treatment? Why or why not?

2. After reading the following hypothetical scenario, you will be asked to make a decision. After you have made your decision, you will be asked some questions that may help you begin to clarify your values. Clarifying your values is essential to critical thinking because it helps you understand your own biases.

Here's the scenario: You are a state legislator considering legislation on prioritizing medical expenditures. Your subcommittee will make a recommendation to the legislature to adopt or reject a plan that would shift state funding from an organ transplant program to a prenatal care program. The state currently pays for organ transplants for needy families, spending over $5 million per year. A group of legislators, however, is proposing that this money be used to fund a program that would offer free checkups for pregnant women as well as advice on drug and alcohol use during pregnancy. The program would also offer information on maternal and infant nutrition to expectant mothers.

The proponents of medical care prioritization say that the money now required for one organ transplant could fund prenatal care for about a thousand mothers. By spending the money on prenatal care, the state could reach thousands of pregnant women who are too poor to see a doctor. Proponents also estimate that 20% of all newborns in your state are born addicted to cocaine, which affects mental and physical development.

Opponents of the bill point out that if needy families are denied money for organ transplants, dozens of children will die. They present the case of Jason Lowry to illustrate what will happen. Jason is 12 years old. His family lives on welfare. Jason needs a liver transplant, which will cost $100,000. Without it, the boy is certain to die. If the state chooses to fund prenatal care instead of organ transplants, dozens of children needing liver transplants will die each year.

You have a choice. Would you recommend the bill that transfers funding to prenatal care or continue funding organ transplants? Why? Make a list of reasons why you supported or opposed the bill.

Now take a moment to ponder your reasons. Was your decision based on economics? Was it based on relative benefits—that is, the benefits for a few versus the benefits for many? Was your decision based on benefits for future generations? Or were you mostly concerned with immediate effects—for example, saving a few lives now?

Imagine, if you will, that Jason was your son. How does that affect your decision? Does the issue take on a different meaning? What general observations can you make about your objectivity? Did it change as the issue came "closer to home"? To what extent do personal interests affect your decisions about other questions, such as environmental issues? Give some examples.

TEST OF TERMS

1. The kidneys are the filtering organs of the _____ system. Blood enters each of the kidneys through the _____ arteries.
2. Urine drains from the kidneys into the urinary bladder through two slender muscular tubes, the _____.
3. Urine drains from the urinary bladder through the _____ to the outside of the body.
4. The outermost region of the kidney is called the _____ .
5. Urine produced in the nephrons empties into the _____ _____ , a hollow chamber inside the kidney.
6. The nephron is the filtering unit of the kidney. It consists of two parts, the _____ , a tuft of capillaries, and the _____ _____ , a long, twisted tube.
7. The nephron's tuft of capillaries is supplied by the _____ arteriole. These capillaries are highly porous and allow much of the water and dissolved substances in the blood to enter _____ _____ , in Bowman's capsule.
8. The highly branched cells surrounding the capillaries of the glomerulus are called _____ .
9. The capillary network surrounding much of the nephron, into which water and dissolved substances are reab-

sorbed, is called the _____ _____ . The movement of materials from these capillaries into the nephron is called _____ _____ .

10. The band of smooth muscle at the neck of the urinary bladder that is under reflex control is called the _____ _____ .

11. Periodic filtering of the blood by an artificial filter is called _____ .

12. Alcohol inhibits the release of a hormone, _____ , from the posterior lobe of the pituitary. This hormone _____ the permeability of the distal convoluted tubules and collecting tubules.

13. _____ , found in coffee and other beverages, increases urine output and is known as a _____ .

14. The adrenal glands produce another hormone, _____ , which af-

fects the reabsorption of sodium ions and therefore helps control ionic balance and water levels in the body.

15. Mercury is a heavy metal that when present in high levels can destroy cells of the _____ .

Answers to the Test of Terms are found in Appendix B.

TEST OF CONCEPTS

1. Draw the various parts of the urinary system, and describe what each one does.

2. Draw a nephron, then label its parts. Explain what happens to the filtrate in each section of the nephron.

3. Trace the flow of blood into and out of the kidney. Be sure to include details of the pathway once it reaches the afferent arteriole.

4. Describe the three ways in which the kidney filters the blood.

5. A drug inhibits the uptake, or reabsorption, of water by the distal convoluted tubules and collecting tu-

bules. What effect would this drug have on urine output, urine concentration, blood pressure, blood volume, and the concentration of the blood?

6. Describe how ADH controls blood pressure and the water content of the body. Describe the hormonal and physiological changes in the body that take place when excess liquid is ingested. Do the same for dehydration.

7. How does urination differ between newborns and adults? Explain what is

meant by this statement: in older children and adults, urination is a reflex with a conscious override.

8. Aldosterone helps regulate blood pressure and water content. In what ways is this hormone different from ADH?

9. You have just finished your residency in family medicine. A patient comes to your office complaining that he drinks water all day long and spends much of the rest of the day in the bathroom urinating. What tests would you order? What diagnosis would you suspect?

The Nervous System: Integration, Coordination, and Control

≈

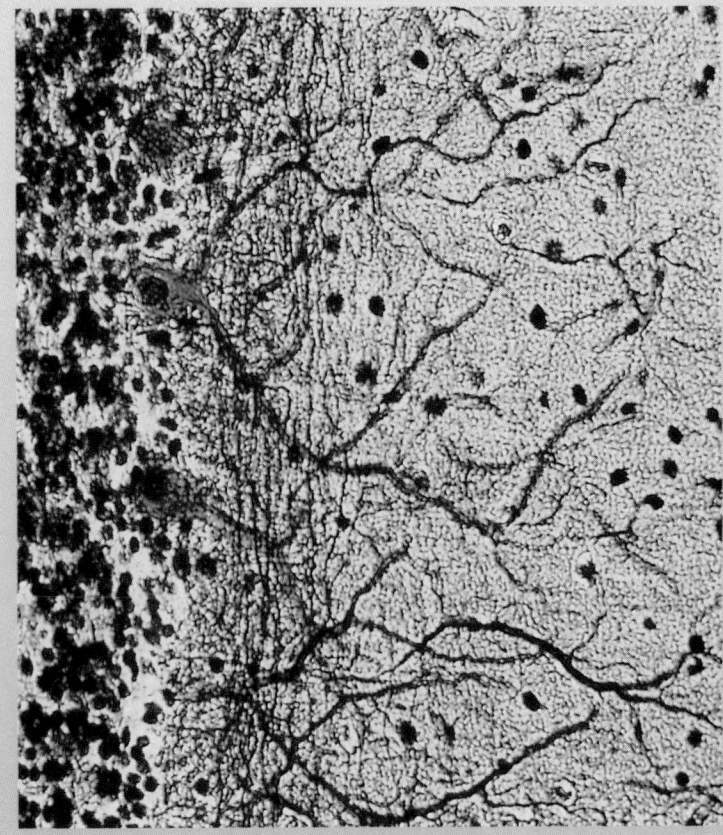

Light micrograph of neurons of the human cerebellum.

rnest Hemingway's novels and short stories won him the Nobel Prize in literature (▷ Figure 17–1). Despite success and widespread popularity, however, Hemingway was a troubled man, and he eventually committed suicide. What ultimately caused him to take his life no one can know, but some believe that he was suffering from a rare and painful nervous system disorder known as **trigeminal neuralgia.** This disease results in periodic and unexplained flashes of intense pain along the course of the **trigeminal nerve,** which supplies the face. A slight breeze or the pressure of a razor can set off this pain, which lasts a minute or more. In some patients, the pain recurs for no apparent reason every few minutes for weeks on end. Some observers believe that the pain Hemingway felt may have become unbearable. Combined with personal conflict, it may have caused the writer to take his life.

Today, physicians treat mild cases of trigeminal neuralgia with drugs. In extreme cases, however, they may elect to sever the trigeminal nerve as it leaves the brain. This procedure ends the pain, but because it cuts off the inflow of other sensory information, it leaves half of the victim's face, tongue, and oral cavity numb—a little like the feeling you get when a dentist injects novocaine into your gums.

This chapter describes the anatomy and physiology of the nervous system. You will study the nerve that may have caused Hemingway so much trouble and examine the ways in which the nervous system helps regulate homeostasis.

▷ **FIGURE 17–1 Was the Agony Too Much?** Ernest Hemingway may have suffered from trigeminal neuralgia, a painful disorder of the nervous system.

≈ OVERVIEW OF THE NERVOUS SYSTEM

The nervous system of vertebrates governs the functions of the body, exerting its control over muscles, glands, and organs. The nervous system also controls heartbeat, breathing, digestion, and urination. It helps regulate blood flow as well as the osmotic concentration of the blood.

In its governing capacity, the nervous system receives input from a large number of sources. Input from the body helps the nervous system "manage" body functions in much the same way that citizen letters and phone calls help elected officials govern society.

The human nervous system provides additional functions not seen in other animal species. For example, the brain is the site of **ideation**—the formation of ideas. It also allows us to think about and plan for the future. The brain also permits humans to reason—that is, to judge right from wrong, logical from illogical. Although some species display a rudimentary ability to reason, the function is best expressed in humankind.

Lest we forget, the nervous system also allows us to manipulate our environment for our own purposes. More than any other species alive today, humans are reshaping the planet's surface. We topple tropical forests to make room for cattle, drain and fill swamps to build homes, split atoms to generate energy, and catapult men and women into outer space. Joan McIntyre, an author and critic, once wrote, "The ability of our minds to imagine, coupled with

the ability of our hands to devise our images, brings us a power almost beyond control." Today, however, humankind has begun to realize that many of the alterations we have made in the planet actually threaten our own long-term future (Chapter 33). Our brains, then, are something of a double-edged sword. They give us a power to create but also an incredible power to destroy. In some cases, the products of our cultural evolution threaten biological evolution. Clearly, the noxious by-products of many technologies overwhelm homeostatic systems that evolved in a much cleaner world.

The Nervous System Consists of Two Main Anatomical Subdivisions, the Central and Peripheral Nervous Systems

The brain, spinal cord, and nerves make up the nervous system (▷ Figure 17–2). The brain and spinal cord constitute the **central nervous system (CNS)** and are housed in the skull and vertebral canal, respectively. Three layers of connective tissue, known as the **meninges,** surround the brain and spinal cord (▷ Figure 17–3). The outer layer consists of fibrous connective tissue and is known as the **dura mater** ("hard mother"). The middle layer is the **arachnoid layer,** so named for its spider-weblike appearance. The innermost layer of the meninges is the **pia mater** (literally, "tender mother"). The pia mater is a delicate, vascular layer that adheres closely to the brain and spinal cord. The space between the arachnoid layer and pia mater is filled with a liquid called **cerebrospinal fluid.**

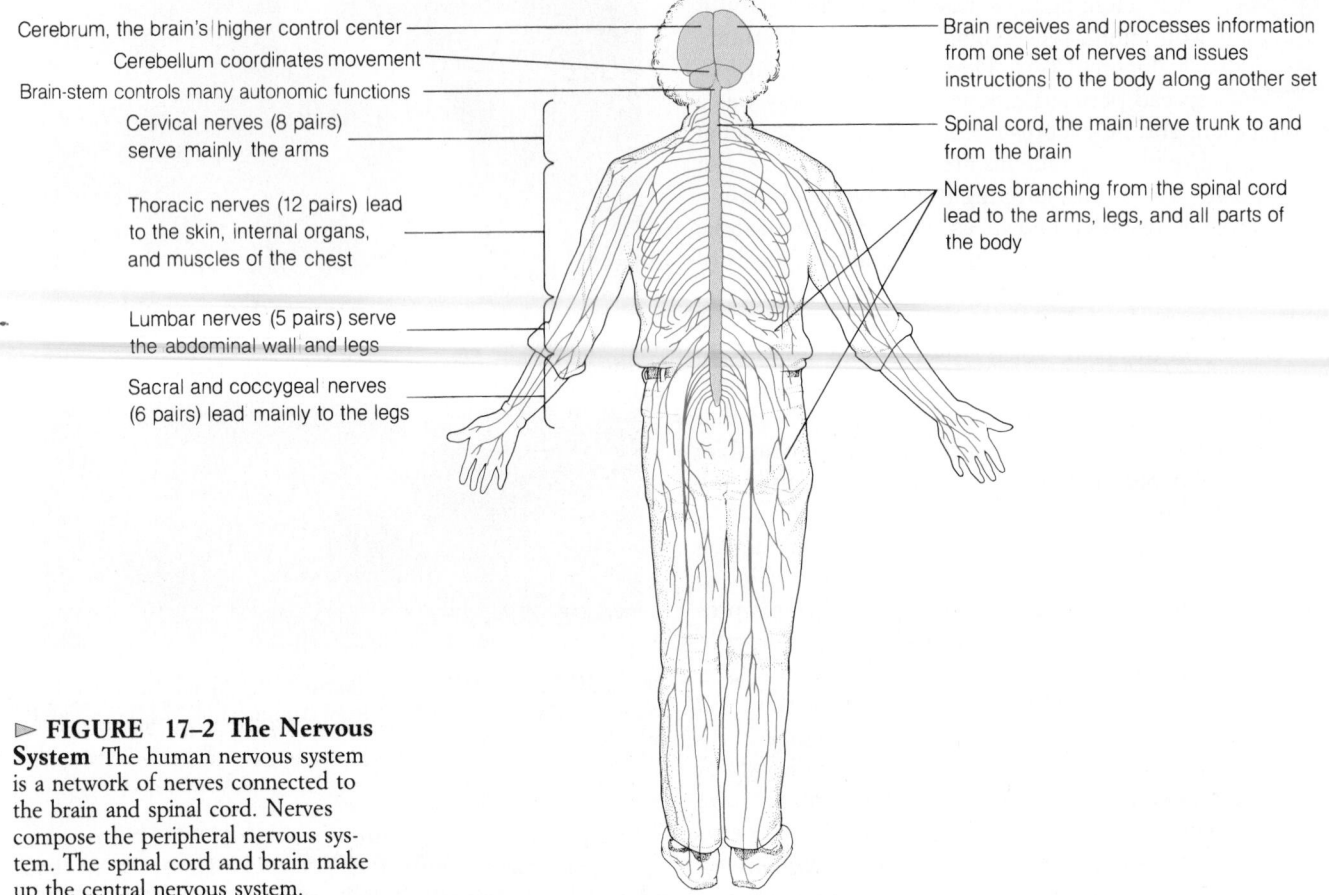

Cerebrum, the brain's higher control center

Cerebellum coordinates movement

Brain-stem controls many autonomic functions

Cervical nerves (8 pairs) serve mainly the arms

Thoracic nerves (12 pairs) lead to the skin, internal organs, and muscles of the chest

Lumbar nerves (5 pairs) serve the abdominal wall and legs

Sacral and coccygeal nerves (6 pairs) lead mainly to the legs

Brain receives and processes information from one set of nerves and issues instructions to the body along another set

Spinal cord, the main nerve trunk to and from the brain

Nerves branching from the spinal cord lead to the arms, legs, and all parts of the body

▷ **FIGURE 17–2 The Nervous System** The human nervous system is a network of nerves connected to the brain and spinal cord. Nerves compose the peripheral nervous system. The spinal cord and brain make up the central nervous system.

Nerves are bundles of nerve fibers (axons and dendrites) of nerve cells, or **neurons** (discussed below). Nerves transport messages to and from the CNS. Nerves and various receptors, which respond to a variety of internal and external stimuli, constitute the **peripheral nervous system (PNS).**

The CNS receives all sensory information from the body. Right now, for instance, your CNS is literally being bombarded with sensory impulses, which travel by nerves from receptors in your body. These receptors alert you to the room temperature, the presence of wind, traffic sounds, and the touch of the page. They transmit visual images of words and figures on the pages. This information is processed in the CNS. Some information is stored in memory. Some of it is blocked. Some incoming information is ignored, and some elicits responses. A particularly exciting section you read, for example, might accelerate your heart rate. A frightening thought might cause you to cringe.

The brain processes incoming stimuli and often responds by sending nerve impulses to muscles and glands (effectors). These impulses also travel along the nerves, but in the opposite direction. In summary, nerves carry two types of information: (1) **sensory impulses** traveling *to* the CNS from sensory receptors in the body, and (2) **motor impulses** traveling *away from* the CNS to effector muscles and glands.

Sensory information pouring into the CNS is integrated with information stored in memory. Thus, a new fact may trigger memories of previous knowledge, causing you to think about a problem in a new way. Memory also influences the way we respond to stimuli. A pet cat brushing against your leg, for example, may elicit a smile. The sensation is not startling, because your memory reminds you of the cat's presence. If you didn't own a cat, the brush of fur along your leg would probably elicit an entirely different response.

The Peripheral Nervous System Can Be Divided into Two Subdivisions, the Somatic and Autonomic Nervous Systems

As shown in ▷ Figure 17–4, the peripheral nervous system can be divided into two parts: the somatic nervous system and the autonomic nervous system. The **somatic nervous system** is that portion of the PNS that controls voluntary functions, such as muscle contractions that lead to the movement of the limbs. It also controls certain involuntary reflex actions, like the knee-jerk response. That part of the PNS that controls involuntary functions other than reflexes, such as heart rate, is known as the **autonomic nervous system (ANS)** (Figure 17–4). Breathing and di-

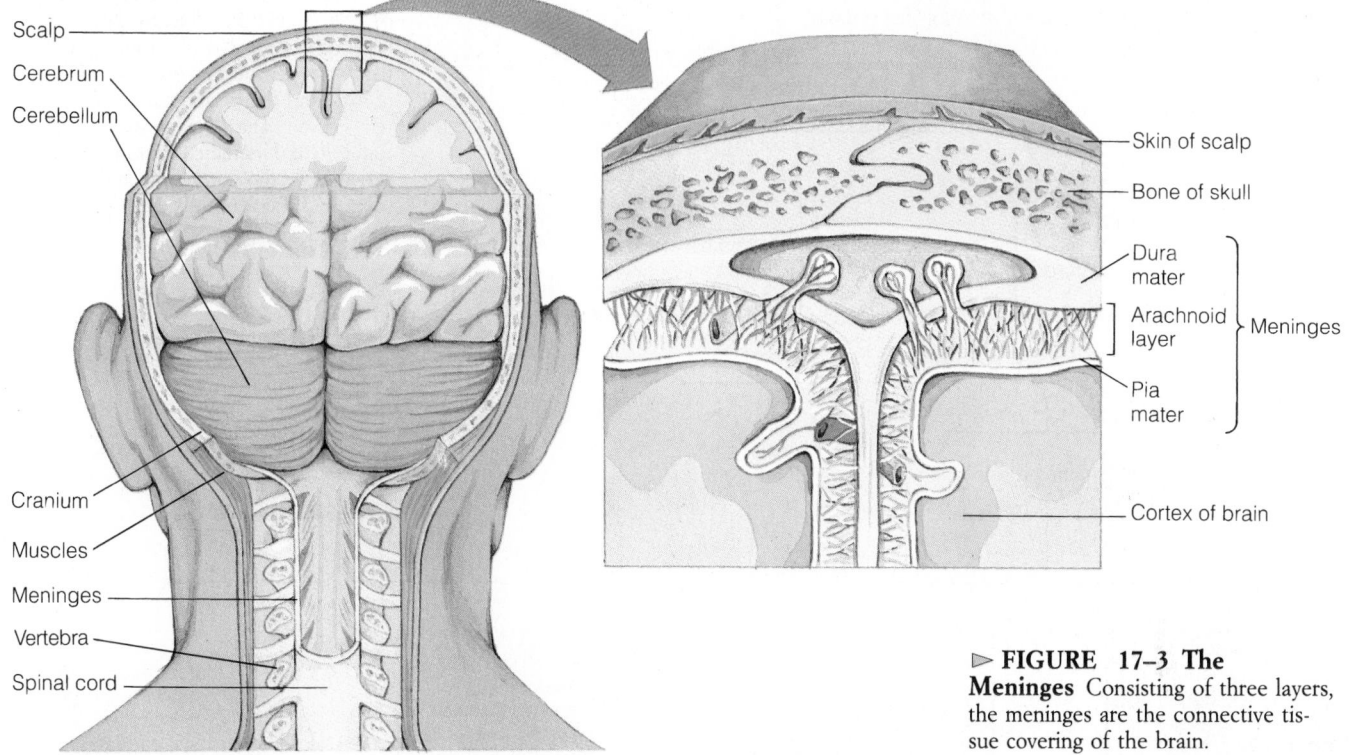

Scalp
Cerebrum
Cerebellum

Cranium
Muscles
Meninges
Vertebra
Spinal cord

Skin of scalp
Bone of skull
Dura mater
Arachnoid layer — Meninges
Pia mater
Cortex of brain

▷ **FIGURE 17–3 The Meninges** Consisting of three layers, the meninges are the connective tissue covering of the brain.

gestion are also under the control of the autonomic nervous system.[1] Many other body functions are under autonomic control and are regulated by negative feedback loops. From an evolutionary perspective, the ANS is vital to survival. Imagine, if you will, how much more difficult our lives would be if we had to consciously control our breathing and other automatic functions.

Figure 17–4 shows that each branch of the peripheral nervous system consists of two types of neurons—**sensory,** or **afferent, neurons,** which transmit information to the central nervous system, and **motor,** or **efferent, neurons,** which transmit information to the effectors.

≋ STRUCTURE AND FUNCTION OF THE NEURON

Before we look in more detail at the structure and function of the nervous system, let us look at the neuron, or nerve cell. The **neuron** is the fundamental structural unit of the nervous system. This highly specialized cell generates bioelectric impulses and transmits them from one part of the body to another—alerting us to a variety of internal and external stimuli and permitting us to respond.

All Neurons Consist of a Cell Body Containing the Nucleus

Neurons come in several shapes and sizes. Despite these differences, nerve cells share several common characteristics. All neurons, for example, consist of a more or less spherical central portion, called the cell body (Figure 17–5). The **cell body** houses the nucleus, most of the cell's cytoplasm, and numerous organelles. Metabolic activities in the cell body sustain the entire neuron, providing energy and synthesizing materials necessary for proper cell function. Two organelles of particular interest are microtubules and microfilaments. These structures form the cytoskeleton of the neuron and are responsible for the neuron's characteristic shape.

All Nerve Cells Contain Processes that Transmit Bioelectric Impulses. Neuronal processes that transmit impulses to the cell body are referred to as **dendrites.** Those that transmit impulses away from the cell body are called **axons.** Neurobiologists generally classify neurons by their anatomy—notably, the type of processes they contain. Accordingly, three distinct types of neuron are found in the human nervous system. Shown in ▷ Figure 17–6 (page 410), they are the unipolar, the bipolar, and the multipolar. The **unipolar neuron** has a single cellular process that splits into an axon and a dendrite. The **bipolar neuron** has two cellular processes, one axon and one dendrite, on opposite sides of the cell body. The **multipolar neuron** contains a single long process, the axon, and numerous short, branching dendrites.

Multipolar neurons are the most abundant and will be the focus of our discussion (see Figure 17–5).[2] In multipo-

---

[1] Note that breathing can be controlled voluntarily, but for the most part it is an involuntary action.

[2] Multipolar neurons are involved in the efferent pathways of the PNS and carry motor information to effector organs, such as glands and muscles.

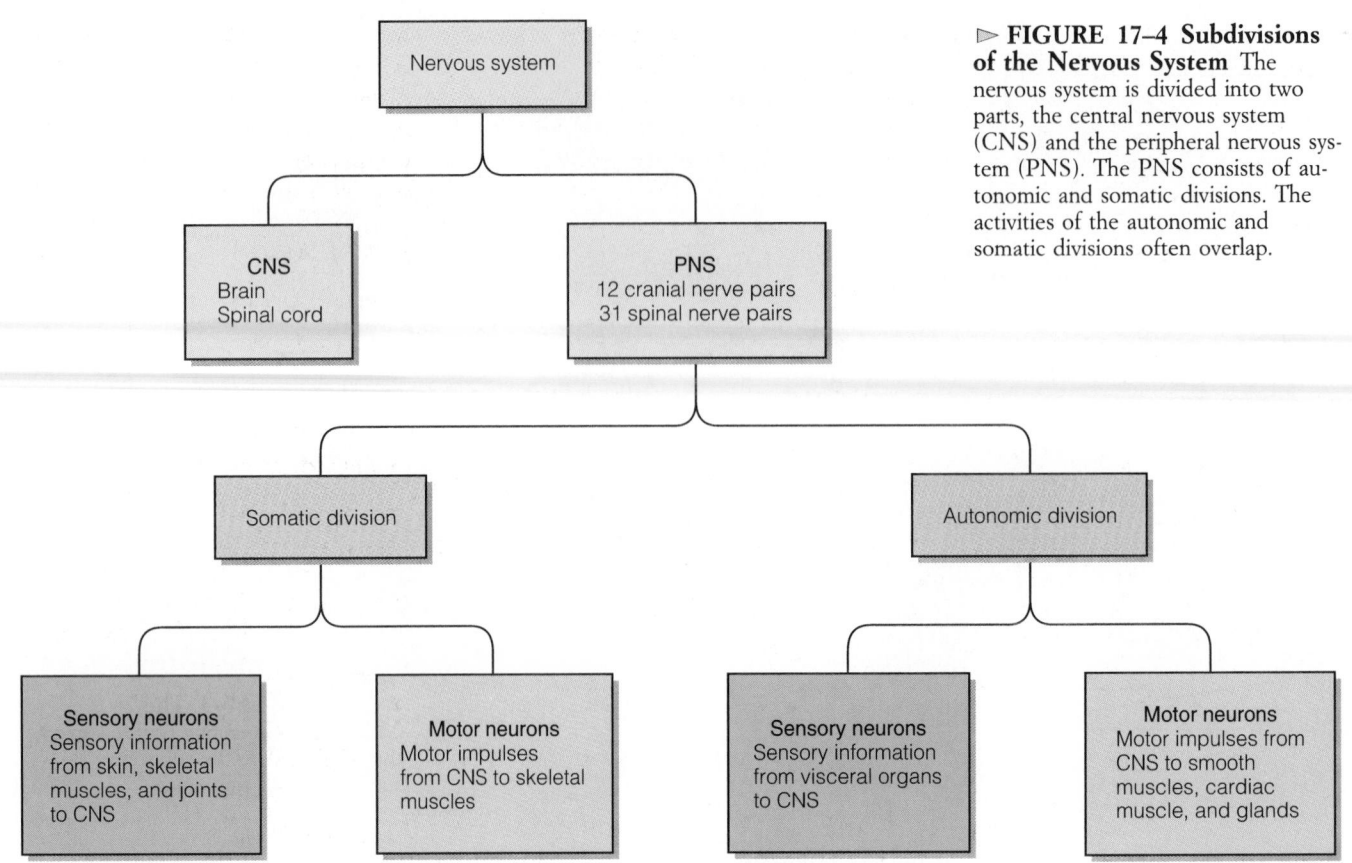

➤ FIGURE 17–4 Subdivisions of the Nervous System The nervous system is divided into two parts, the central nervous system (CNS) and the peripheral nervous system (PNS). The PNS consists of autonomic and somatic divisions. The activities of the autonomic and somatic divisions often overlap.

lar neurons, the short, branching dendrites transmit impulses to the cell body. (The arrows in Figure 17–5b indicate the direction in which an impulse travels.) The cell bodies of multipolar neurons are located in the spinal cord and brain. After reaching the cell body, impulses travel down the long, unbranched axon. Axons of multipolar neurons often leave the CNS with the nerves of the PNS or, alternatively, may connect one part of the CNS to another (thus remaining inside the CNS). Axons occasionally also give off side branches, known as **axon collaterals** (Figure 17–5b). When an axon reaches its destination, it often branches profusely, giving off many small fibers. These fibers terminate in tiny swellings known as **terminal boutons** (end buttons), or **terminal bulbs.** Nerve impulses reaching the terminal ends of axons may be transmitted to other neurons, muscle fibers, or glands.

The Myelin Sheath Increases the Rate of Transmission of Nerve Impulses and Is Found on Axons in Both the CNS and PNS. The axons of many multipolar neurons in both the central and peripheral nervous systems are coated with a protective layer called the **myelin sheath** (➤ Figure 17–7a). The myelin sheath is formed by nonconducting cells of the nervous system known as **glial cells.** In the PNS, the glial cells that form the myelin sheath are referred to as **Schwann cells,** after the early German cytologist Theodore Schwann. In the CNS, the

myelin sheath is laid down by glial cells known as **oligodendrocytes.**

During embryonic development, Schwann cells in the PNS attach to the growing axon, then begin to encircle it (Figure 17–7d). As they do, they leave behind a trail of plasma membrane, which wraps around the axon, forming many concentric layers—like an elastic bandage wrapped around your wrist (Figures 17–7b and 17–7c).

The entire myelin sheath of an axon is formed by numerous Schwann cells, which align themselves along the length of the axon. Each lays down a separate patch of myelin. Because the plasma membrane of the Schwann cell is about 80% lipid, the myelin sheath is mostly lipid (mostly triglyceride) and appears glistening white when viewed with the naked eye. As shown in Figures 17–7a and 17–7b, each segment of myelin is separated by a small unmyelinated segment known as a **node of Ranvier.** In the CNS, a single oligodendrocyte produces myelin for several axons (Figure 17–7e).

The myelin sheath permits nerve impulses to travel with great speed down axons. As illustrated in Figure 17–7a, impulses "jump" from node to node like a stone skipping along the surface of the water. (More on this process later.)

Destruction of the myelin sheath of nerve cells in the central nervous system results in a condition known as **multiple sclerosis.** The destroyed myelin is replaced by plaque that disrupts the transmission of impulses. Thought to be an autoimmune disease, multiple sclerosis can affect any part of the CNS. Early symptoms are generally mild—weakness or a tingling or numbing feeling in one part of the body. Temporary weakness may cause a person to stumble and fall. Some people report blurred vision, slurred speech, and difficulty controlling urination. In many cases, these symptoms disappear, never to return. In other cases, individuals suffer repeated attacks. Recovery after each attack is incomplete, and the patient gradually deteriorates, losing vision and becoming progressively weaker. Fortunately, many treatments are available, and only a small number of multiple sclerosis patients are crippled by the disease.

Although most axons are covered with myelin, some axons are unmyelinated. Found in both the central and peripheral nervous systems, unmyelinated axons conduct impulses much more slowly than their myelinated counterparts. The reduced rate of transmission in unmyelinated axons results from the fact that the impulse must travel along the entire membrane of the axon—more like a wave moving along the surface of a pond. As a general rule, then, the most urgent types of information are transmitted via myelinated fibers; less urgent information is transmitted via unmyelinated fibers.

Interestingly, unmyelinated axons are also associated with Schwann cells. In such instances, though, the Schwann cells merely encase the unmyelinated axons, as shown in ▷ Figure 17–8 (page 412), holding them together in a bundle.

Microtubules Inside Axons Function in Axonal Development and also Transport Materials from the Cell Body to the Axon Terminal. Inside all axons are bundles of microtubules (▷Figure 17–9, page 412). These microtubules facilitate the transport of materials

produced in the cell body to the axon terminal. They also play a key role in the development of axons. During embryonic development, for example, nerve cells start out as round cells. In the developing brain and spinal cord, these cells undergo a dramatic transformation. Axons begin to form as microtubules develop inside the cytoplasm and push outward against the plasma membrane. As the microtubules grow, the axons elongate and extend from the

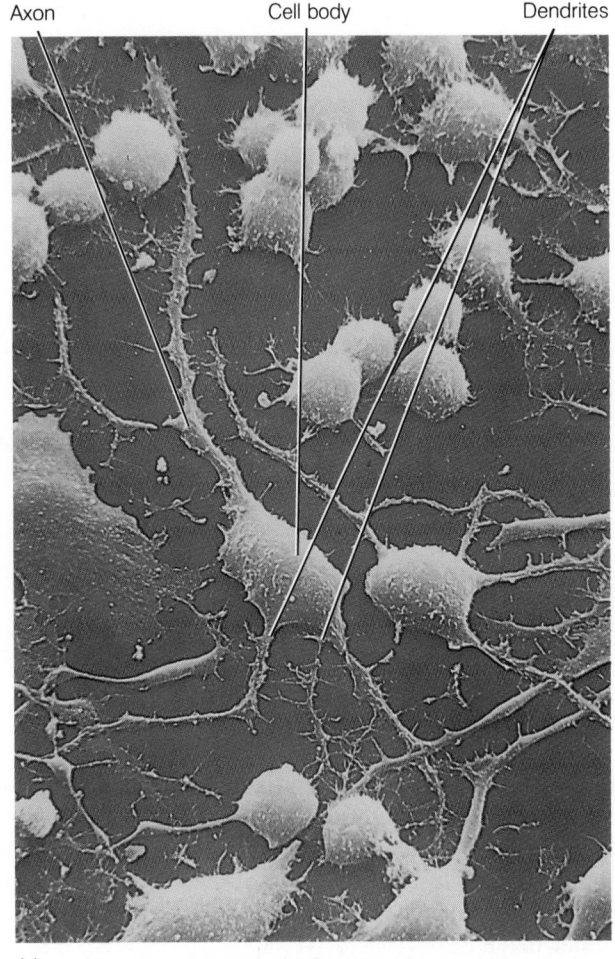

(a)

▷ **FIGURE 17–5 A Neuron** (*a*) A scanning electron micrograph of the cell body and dendrites of a multipolar neuron. The multipolar neuron resides within the central nervous system. Its multiangular cell body has several highly branched dendrites and one long axon. (*b*) Collateral branches may occur along the length of the axon. When the axon terminates, it branches many times, ending on individual muscle fibers.

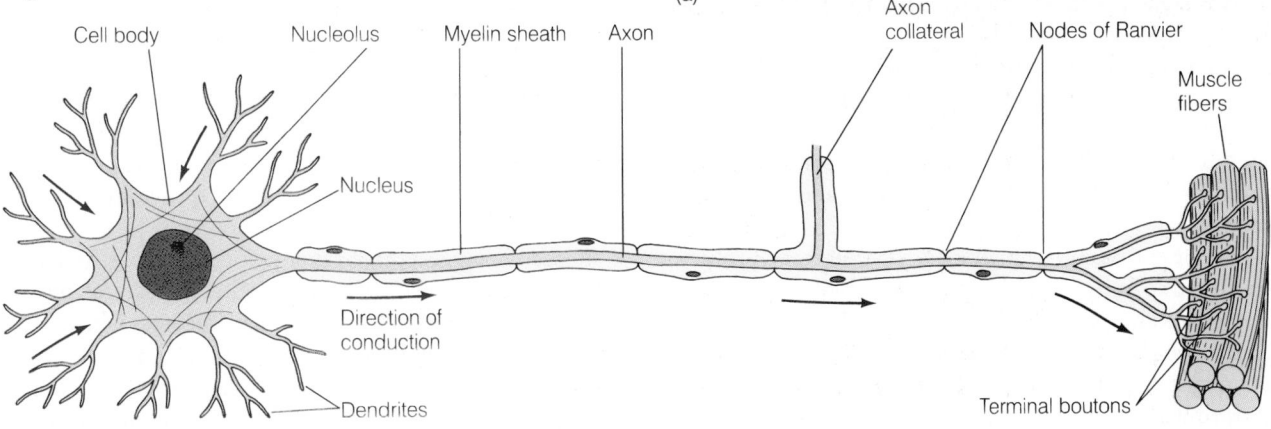

(b)

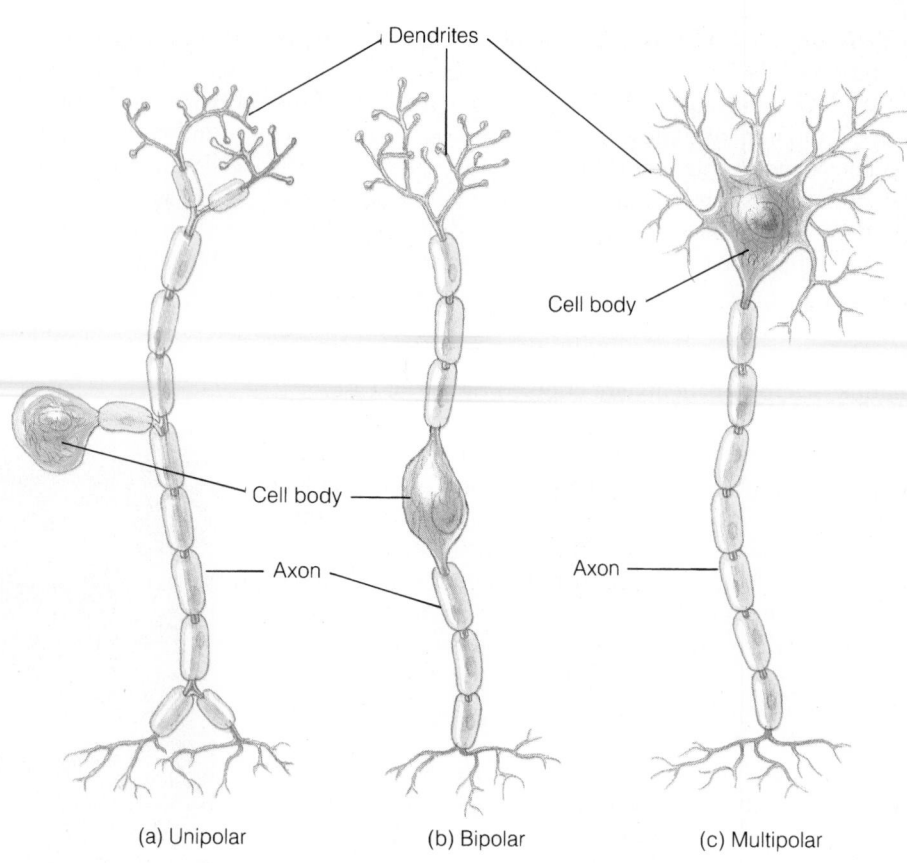

▷ **FIGURE 17–6 Three Types of Neurons** (*a*) Unipolar. (*b*) Bipolar. (*c*) Multipolar.

Dendrites

Cell body

Cell body

Axon

Axon

(a) Unipolar

(b) Bipolar

(c) Multipolar

embryonic brain and spinal cord to the muscles and glands they will supply. Some axons within the central nervous system (brain and spinal cord) extend to other regions of the CNS, thus helping connect its parts.

Neurons Are Highly Specialized Cells That Have Lost the Ability to Divide. Like a few other cells that undergo cellular differentiation during embryonic development, nerve cells lose the ability to divide. Therefore, nerve cells that die generally cannot be replaced by cell division of other nerve cells.

Fortunately, there are mechanisms to offset nerve cell damage. Consider what happens in a stroke. A **stroke** occurs when brain cells are damaged as a result of a sharp reduction in blood supply. Strokes result from one of three causes: (1) cerebral hemorrhage, a break in an artery of the brain; (2) cerebral thrombosis, a blood clot that forms in a brain artery narrowed by atherosclerosis; and (3) cerebral embolism, a blood clot from another source that lodges in an artery of the brain (▷ Figure 17–10). All three problems have the same end result: they terminate blood flow to vital brain cells, killing them.

If a person survives a stroke, undamaged neurons in the brain may take over the function of damaged brain cells, thus permitting partial neurological recovery. Recovery also occurs when cells that were injured but not killed by the

loss of blood regain their function. Recovery generally takes a long time, which explains why people who have suffered from strokes require long-term rehabilitation.

The fate of a nerve cell after injury depends on its location in the nervous system. In the central nervous system, for example, damaged axons cannot be repaired. In the peripheral nervous system, axonal regeneration is possible. Thus, a severed nerve in the arm may be able to regenerate new axons. As illustrated in ▷ Figure 17–11, a severed axon in the PNS generally degenerates from the point of injury to the muscle or gland it supplied. The segment of the axon still attached to the cell body may grow back to replace the degenerated section. During regeneration, the segment of the axon attached to the cell body elongates and expands along the hollow tunnel left in the myelin sheath by the degenerated section of axon. Eventually, the axon reestablishes connections with the muscles or glands it once supplied, providing partial or even nearly complete recovery of control.

Research suggests that regeneration of axons in the PNS is possible because Schwann cells release a neuronal growth-promoting factor. Oligodendrocytes, which are responsible for myelin formation in the CNS, do not produce a similar substance. Experiments show that Schwann cell transplants in the CNS promote axonal regeneration. Researchers hope some day to be able to isolate, purify, and

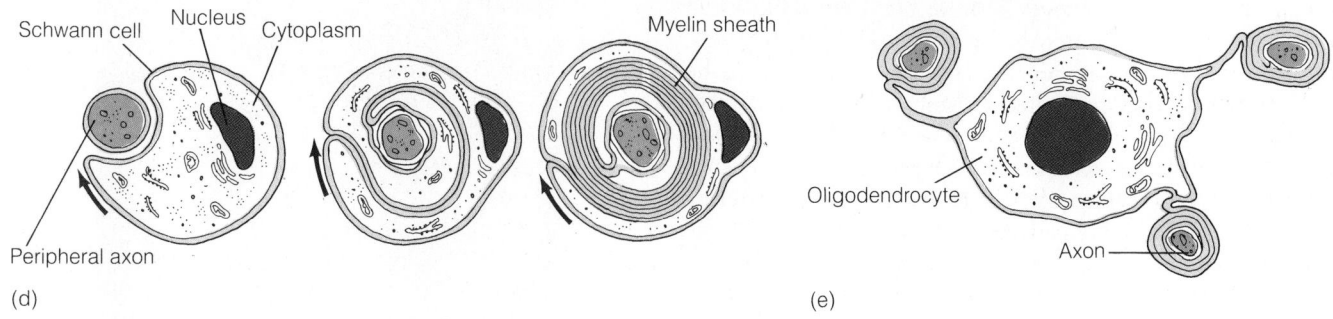

▷ **FIGURE 17–7 The Myelin Sheath and Saltatory Conduction** (*a*) The myelin sheath allows impulses to "jump" from node to node, greatly accelerating the rate of transmission. (*b*) A drawing showing the arrangement of Schwann cell membrane in the myelin sheath. (*c*) A transmission electron micrograph of an axon in cross section showing a myelin sheath. Drawings show how the myelin sheath is formed in the PNS (*d*) and CNS (*e*).

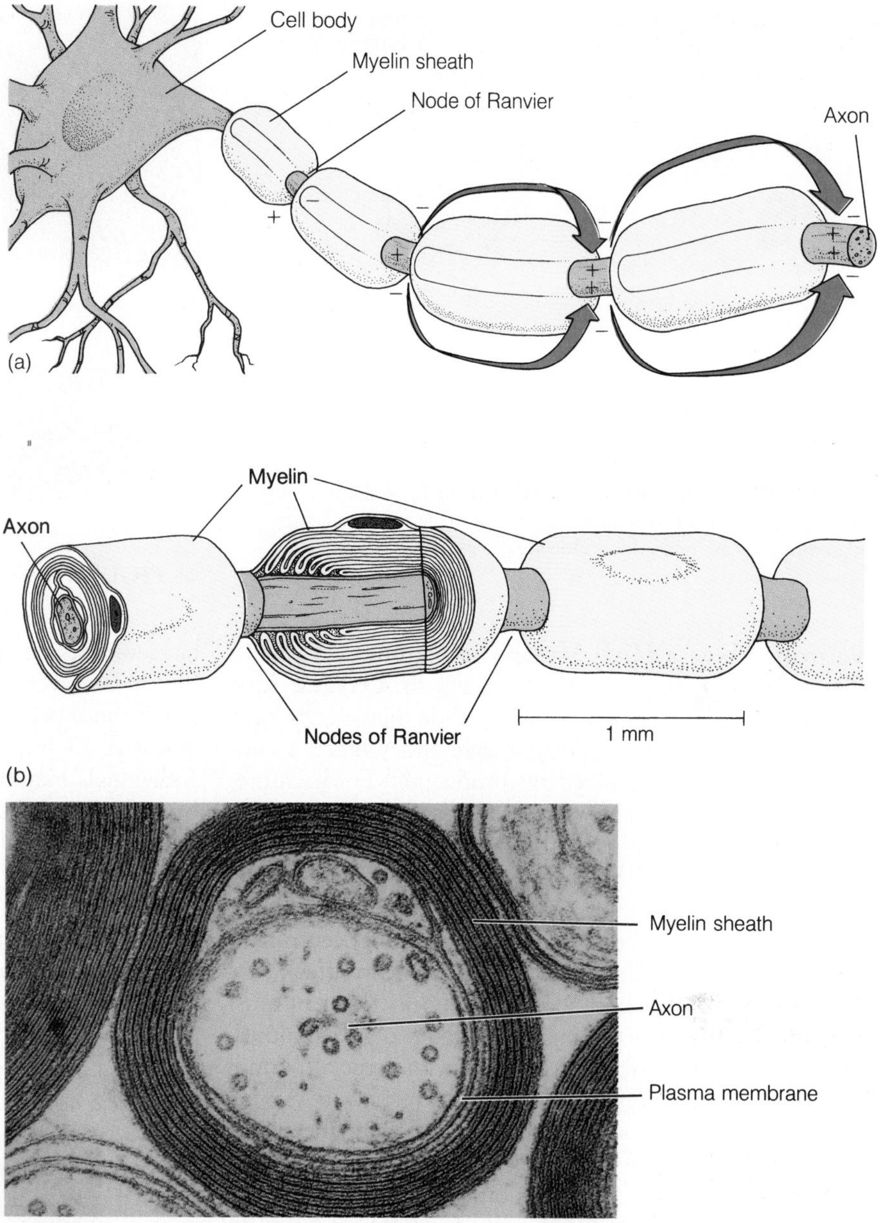

manufacture the neuronal growth-promoting factor to stimulate axonal regeneration in victims of strokes and accidents.

One final note: neurosurgeons can facilitate axonal regeneration in the PNS by microsurgery performed under a dissecting microscope more elaborate than the one you may have used in your high school biology class. Surgeons sew the severed ends of nerves together after lining up the empty myelin sheaths with the regenerating nerve fibers to facilitate axonal regrowth.

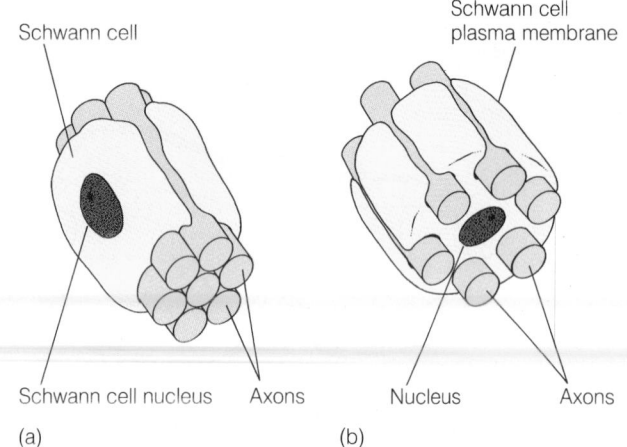

(a)   (b)

▷ **FIGURE 17–8 Unmyelinated Axons** (*a*) Schwann cells encompass groups of axons, but do not produce myelin sheaths. (*b*) Axons may be embedded individually as well.

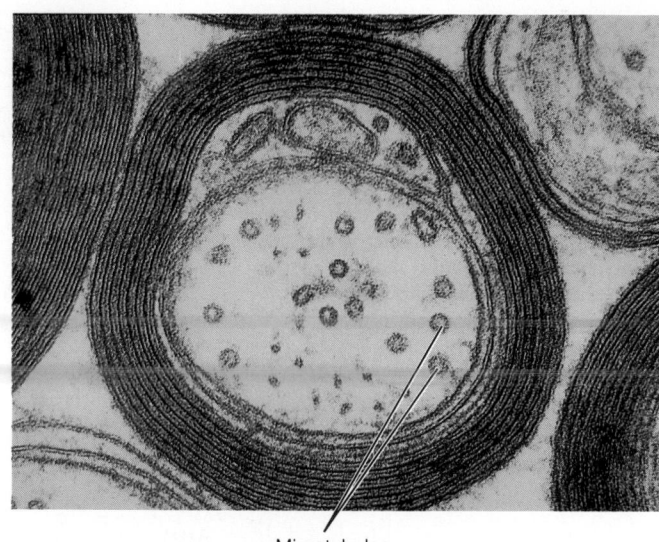

Microtubules

▷ **FIGURE 17–9 Microtubules in Axon**

Neurons Have a High Metabolic Demand, Making Them Highly Susceptible to Loss of Oxygen and Glucose.

Besides being unable to divide, nerve cells have an extraordinarily high metabolic rate and require a constant supply of oxygen for energy production. Furthermore, neurons cannot generate ATP in the absence of oxygen via fermentation (glucose breakdown to lactic acid) the way most other cells can. Thus, if the amount of oxygen flowing to the brain is drastically reduced, neurons begin to die within minutes. To prevent brain damage from occurring in someone who has collapsed with a severe heart attack, has drowned, or has suffered an electric shock, rescuers must start resuscitating the victim within four to five minutes. Although victims may be revived after this crucial period, the lack of oxygen in the brain often results in varying degrees of brain damage. Generally, the longer the deprivation, the greater the damage.

Interestingly, if a person is submerged in icy water, resuscitation may be successful if begun within an hour. In most cases, victims recover without any detectable brain damage. In one exceptional case a young boy from Fargo, North Dakota, was underwater for five hours before he was recovered and resuscitated. Much to his parents' delight, the boy not only survived the incident but also had no apparent ill effects. Why is it that people who are submerged in cold water can be revived without suffering brain damage? Cold water greatly slows metabolism in the brain, dramatically reducing oxygen demand. As a result, brain cells are preserved, and brain damage is minimized or avoided.

Nerve cells are also highly dependent on glucose for energy production. Unlike most other body cells, they cannot use fatty acids to generate energy. They also cannot store glucose as can liver and muscle cells. As a result, when blood glucose levels fall, nerve cells are first to "feel" the ill effects.

A decline in blood glucose is normally prevented by homeostatic mechanisms described in Chapter 10. In diabetics, however, blood glucose levels may fall dangerously low if too much insulin is taken or if not enough glucose is ingested. Deprived of glucose, brain cells begin to falter. Individuals may become dizzy and weak. Vision may blur. Speech may become awkward. The person may be mistaken for a drunk and may undergo seizures or become unconscious.

▷ **FIGURE 17–10 Stroke** PET (positron emission tomography) scan of brain revealing damaged region of the cerebral cortex following a stroke. This color image was generated by a computer that converts readings of the rate of emission from radioactive glucose molecules injected into the patient. Damaged (dark region) areas show the lowest glucose uptake. Highest glucose uptake and highest emissions are in red.

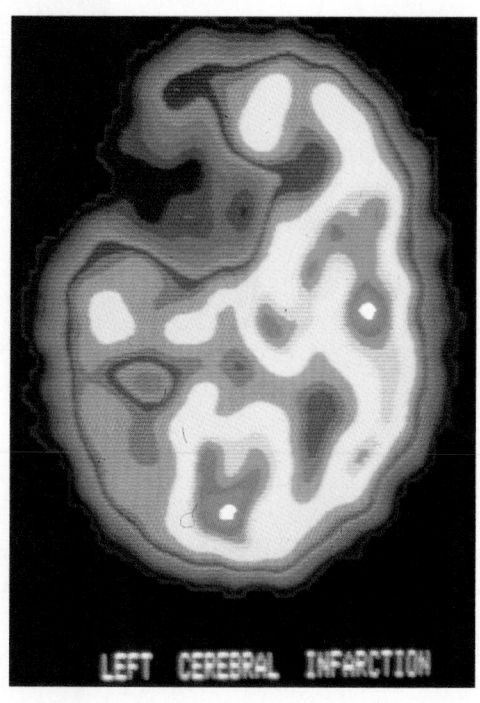

LEFT CEREBRAL INFARCTION

▷ **FIGURE 17–11 Axonal Regeneration** (*a*) A severed axon can regenerate in the peripheral nervous system. The segment from the cut to the effector organ degenerates. (*b*) The myelin sheath remains, providing a tunnel (*c*) through which the axonal stub can regrow, often reestablishing previous contacts and restoring motor function.

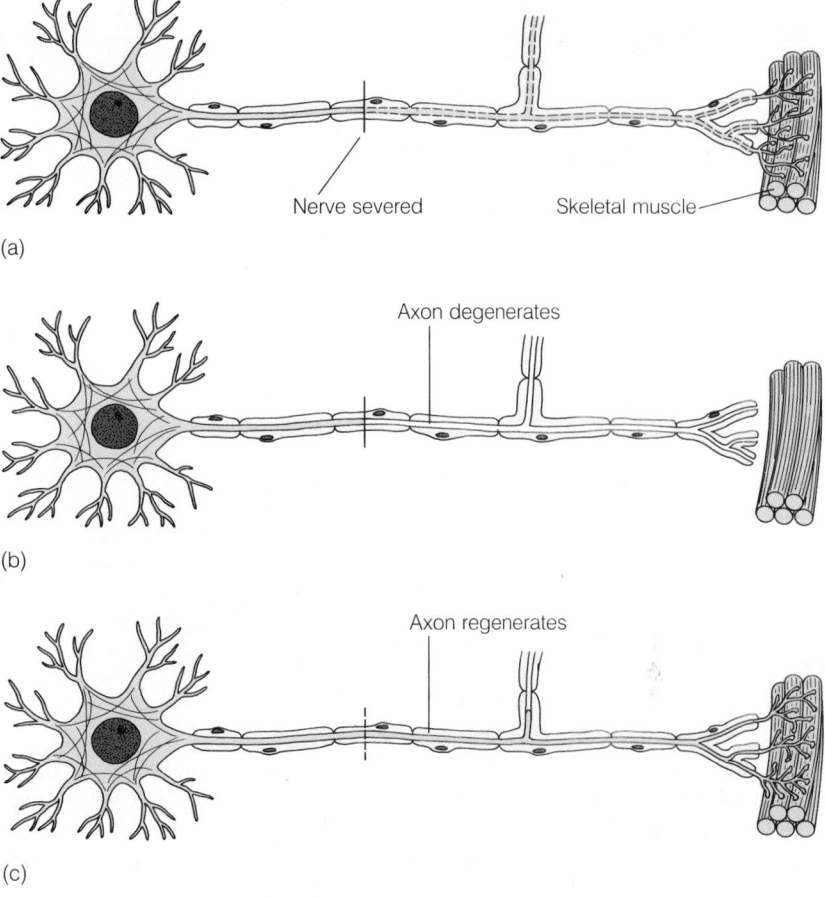

(a) Nerve severed   Skeletal muscle

(b) Axon degenerates

(c) Axon regenerates

Chronic low blood glucose (hypoglycemia) often results in severe headaches. People may become aggressive when blood sugar levels fall. In extreme cases, low blood sugar triggers convulsions and death. Other body cells are not adversely affected by a decline in blood glucose because they switch to alternative fuels, fats and proteins.

Bioelectric Impulses Traveling in Nerve Cells Result From the Flow of Ions Across the Plasma Membrane of these Cells

The evolution of the nervous system is important to the survival and propagation of animals because it keeps them attuned to changes in the environment. It also helps them make many thousands of adjustments needed to survive in an ever-changing environment.

These functions are possible because of nerve impulses transmitted through the PNS and CNS. Nerve impulses are not like the electric current that powers computers and light bulbs, which is formed by the flow of electrons. Rather, nerve impulses are small ionic changes in the membrane of the neuron that move along the plasma membrane of a nerve cell.

To understand the nerve cell impulse, or **bioelectric impulse,** so named to distinguish it from electricity, we begin by examining the plasma membrane of a neuron. If you placed a tiny electrode on the outside and inserted another on the inside of the plasma membrane of a neuron and hooked them up to a voltmeter, you would be able to measure a small voltage, much like that produced in a

battery. In the simplest terms, **voltage** is a measure of the tendency of charged particles to flow from one pole of the battery to the other. The higher the voltage, the greater the tendency for electrons to flow through a wire connected to the poles. In the nerve cell, however, electrons do not flow from one side of the membrane to another, sodium and potassium ions do.

In neurons, the potential difference, or voltage, is a measure of the force that can drive sodium ions from one side of the membrane to the other. For now, however, it is important just to remember that a small voltage exists across the plasma membrane of the neuron. It is so small, in fact, that it is measured in millivolts. A millivolt is 1/1000 of a volt.

As shown in ▷ Figure 17–12a, the potential difference in a typical neuron is about −60 millivolts. This is known as the **membrane potential,** or **resting potential,** for it is the membrane potential of a nerve cell at rest. The minus sign is added because the plasma membrane is positively charged on the outside and negatively charged on the inside, for reasons beyond the scope of this book (Figure 17–12b). It is important to note, however, that sodium ions are found in greater concentration outside the neuron, whereas potassium ions are found in greater concentration inside the cell. In fact, neurons expend a great deal of energy to maintain this concentration imbalance, which, as you will soon see, is essential to the production of nerve impulses. Cellular energy is used to power active transport

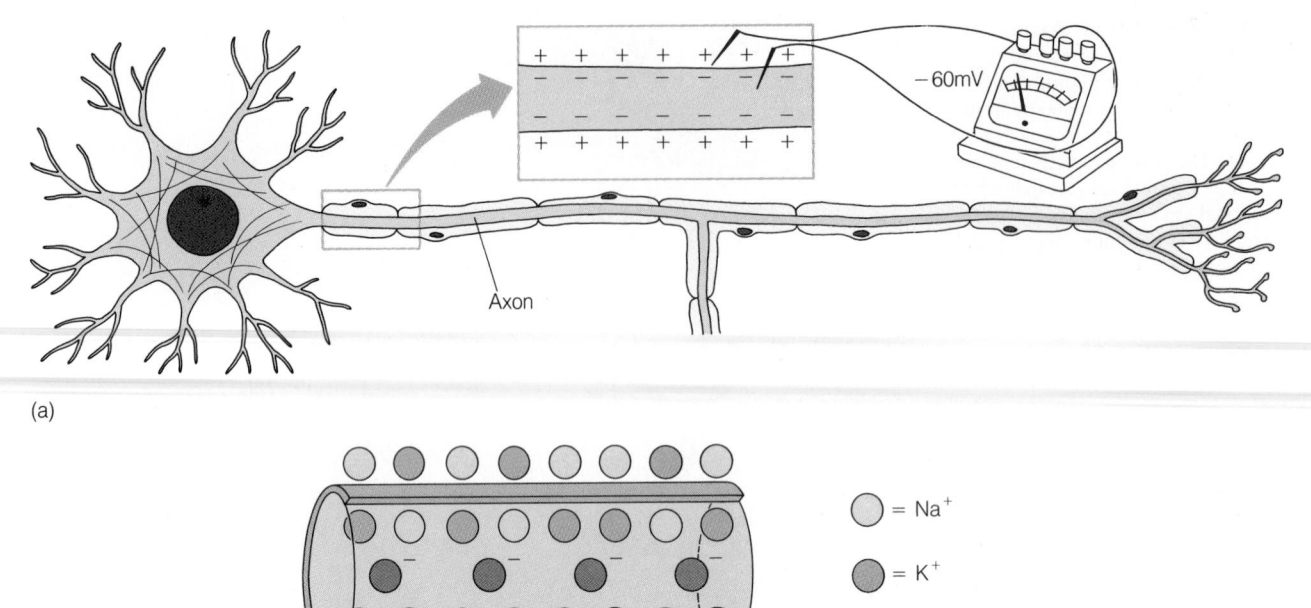

(a)

(b)

⬤ = Na⁺

⬤ = K⁺

⬤ = Organic ion

▷ **FIGURE 17–12 Resting Potential** (*a*) Electrodes placed on both sides of the plasma membrane of a neuron measure a tiny potential difference, roughly −60 millivolts (mV). This is the resting potential. (*b*) It results from a chemical disequilibrium caused by the active transport of sodium (Na⁺) ions out of the neuron and potassium (K⁺) ions into the cytoplasm and the presence of negatively charged organic ions in the cytoplasm.

pumps located in the plasma membrane of neurons. These pumps transport sodium ions that leak into the cytoplasm of the neuron back out into the surrounding fluid, helping maintain the external concentration. They also transport potassium ions that have leaked into the extracellular fluid back into the cell.

After a Nerve Cell Is Stimulated, the Membrane Undergoes Dramatic Changes in Permeability to Various Ions, Resulting in Sudden Shifts in Membrane Potential.

The plasma membrane of a nerve cell is a lot like a loaded gun. That is, it has a built-up charge. The charge consists of sodium ions concentrated on the outside of the cell. When the neuron is stimulated, the membrane undergoes a rapid change, "discharging" the load. The first change occurring in the membrane is a rapid increase in its permeability to sodium ions. Neurophysiologists believe that stimulating the nerve cell causes protein pores in the plasma membrane to open. Sodium flows into the cell through these pores.

Electrodes implanted in a nerve cell detect the sudden influx of positively charged sodium ions and register a shift in the resting potential from −60 millivolts to +40 millivolts. The change in voltage occurs at the site of stimulation and is called **depolarization.**[3] Immediately after depolarization, the membrane returns to its previous state, which is referred to as **repolarization.**

The electrical charge across the membrane is known as an **action potential** and is graphically represented in ▷ Figure 17–13a. This graph shows (1) a brief upswing, depolarization, as the voltage goes from −60 millivolts to +40 millivolts, and (2) a rapid downswing, repolarization, the return to the resting potential.

The action potential occurs so rapidly and the membrane returns to the resting state so quickly (about 3/1000 of a second) that neurons can be stimulated in rapid succession. Such brisk recovery allows us to respond swiftly and forcefully to danger and to perform rapid muscle movements. A nerve cell can also transmit many impulses in sequence, because only a small number of sodium ions are exchanged with each impulse.

As noted above, depolarization results in the rapid influx of sodium ions. During repolarization, the membrane shifts from +40 millivolts back to −60 millivolts, the resting potential. Repolarization results from two factors: a sudden decrease in the membrane's permeability to sodium ions, which stops the influx of sodium ions, and a rapid efflux of positively charged potassium ions (Figure 17–13d). The outflow of potassium ions results from concentration and electrical forces. That is to say, potassium ions flow out of the axon down a concentration gradient. Because the outside of the membrane becomes negatively charged during depolarization, an electrical gradient also comes into play.

[3] Depolarization means that the membrane loses its previous polarization.

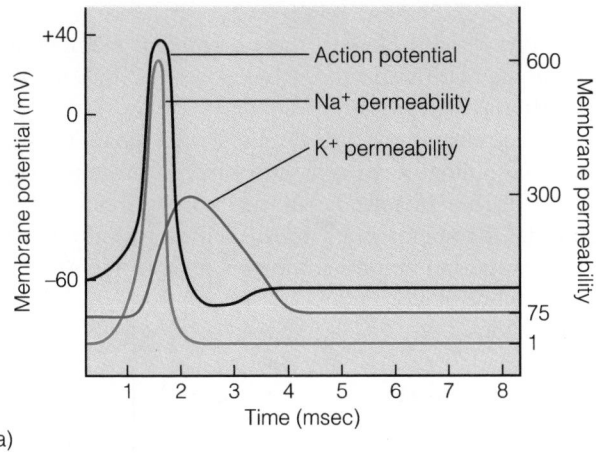

(a)

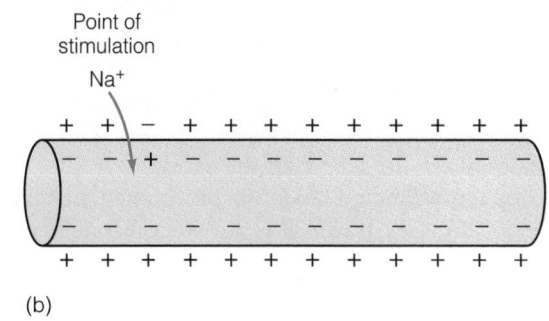

Point of stimulation

(b)

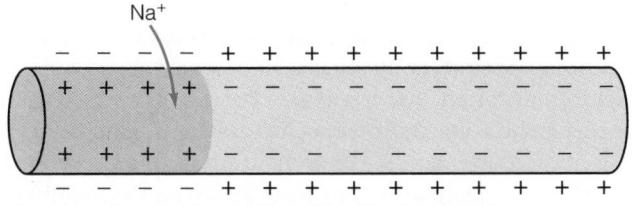

Depolarization and generation of the action potential
(c)

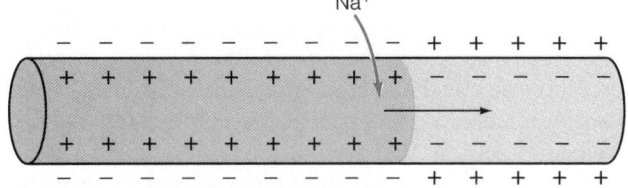

Propagation of the action potential

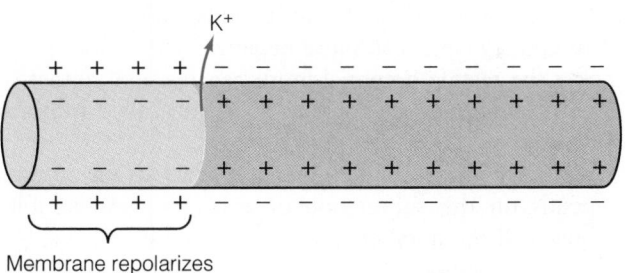

Membrane repolarizes
(d)

▷ **FIGURE 17–13 Action Potential** (*a*) Stimulating the neuron creates a bioelectric impulse, which is recorded as an action potential. The resting potential shifts from −60 millivolts to +40 millivolts. The membrane is said to be depolarized. This graph shows the shift in potential and the change in sodium (Na⁺) and potassium (K⁺) ion permeability, which is largely responsible for the action potential. (*b*) This drawing shows the influx of sodium ions and the depolarization that occur at the point of stimulation. (*c*) The impulse travels along the membrane as a wave of depolarization. (*d*) The efflux of potassium ions restores the resting potential, allowing the neuron to transmit additional impulses almost immediately.

The net outward movement of potassium ions helps reestablish the resting potential.

When a neuron is no longer being stimulated, it quickly reestablishes the proper sodium and potassium ion concentrations inside and outside the cell. It does this by pumping sodium ions that flowed into the cell during activation out of the axon and by pumping potassium ions that flowed out of the neuron during repolarization back in. The plasma membranes of nerve cells contain numerous "sodium-potassium active transport pumps" that transport sodium out of and potassium into the cell. (You may recall from Chapter 3 that active transport pumps require energy in the form of ATP.) The sodium-potassium pumps, therefore, help neurons reestablish the chemical disequilibrium so necessary for normal nerve cell function. Interestingly,

about 30% of the energy you burn at rest is used to operate this pump.[4]

The Nerve Impulse Moves Along the Membrane of the Neuron Because Depolarization in One Region Stimulates Depolarization in Neighboring Regions. Stimulating a nerve cell artificially at one point creates an inward rush of sodium ions, which dramatically shifts the resting potential. The influx of sodium ions is followed by an outward rush of potassium ions that returns the membrane potential to normal. This constitutes the bioelectric impulse. How does the bioelectric impulse travel along the nerve cell?

[4]The sodium-potassium pump is found in all cells, not just neurons.

Many studies to explore the conduction of impulses by neurons have been made using the giant unmyelinated axons of the squid. These studies show that a change in membrane permeability in the stimulated region, which results in depolarization, causes a change in the sodium permeability of neighboring regions. Thus, depolarization in one region of the plasma membrane of an axon stimulates depolarization in adjacent regions. This process continues along the length of the axon.[5]

In unmyelinated fibers in the human nervous system, nerve impulses travel like waves from one region to the next as they do in the giant squid axon. In myelinated fibers, however, the depolarization "jumps" from one node of Ranvier (the section between adjacent Schwann cells) to another. This movement is called **saltatory conduction** (from the Latin word *saltare,* meaning "to jump"). Shown in Figure 17–7a, this "jumping" of the impulse from node to node greatly increases the rate of transmission. In fact, a nerve impulse travels along an unmyelinated fiber at a rate of about 0.5 meter per second (1.5 feet per second). In a myelinated neuron, the impulse travels 400 times faster—that is, about 200 meters (650 feet) per second (or about 400 miles per hour). The difference in the rate of transmission is largely due to a difference in the total amount of axonal membrane that must be depolarized and repolarized.[6] Saltatory conduction also conserves energy, because it reduces the amount of energy needed to pump sodium and potassium ions. Like so many adaptations, it contributed significantly to the evolution of complex multicellular animals.

Nerve Impulses Travel from One Neuron to Another Across Synapses.
Nerve impulses travel from one neuron to another across a small space that separates them, as shown in ▷ Figure 17–14. The juncture of two neurons is called a synapse. A **synapse** consists of (1) a terminal bouton (or some other kind of axon terminus), (2) a gap between the adjoining neurons, called the **synaptic cleft,** and (3) the membrane of the dendrite or postsynaptic cell (Figure 17–14b). The neuron that transmits the impulse is called the **presynaptic neuron;** the one that receives the impulse is called the **postsynaptic neuron.**

When a nerve impulse reaches a terminal bouton, depolarization of the plasma membrane causes a rapid influx of calcium ions into the bouton. Calcium ions stimulate the release (by exocytosis) of a chemical substance known as a **neurotransmitter,** which is stored in small vesicles in the terminal bouton.

At least 30 chemicals serve as neurotransmitters. Produced and packaged in vesicles in the cell body of the neuron, neurotransmitters are transported down the axon along the microtubules to the terminal bouton, where they are stored until needed. When the bioelectric impulse arrives, the vesicles bind to the presynaptic membrane and release the neurotransmitter (by exocytosis) into the synaptic cleft.

Neurotransmitters diffuse across the synaptic cleft between adjoining nerve cells and bind to protein receptors in the plasma membrane of the postsynaptic (receiving) neuron. The binding of most neurotransmitters to the postsynaptic membrane stimulates a rapid increase in the permeability of the membrane of the postsynaptic cell to sodium ions.

Neurotransmitters May Excite or Inhibit the Post-Synaptic Membrane.
In the brain, a single nerve cell may have as many as 50,000 synapses. In some of these synapses, neurotransmitters stimulate the uptake of sodium ions, which slightly depolarizes the postsynaptic neuron. Synapses that depolarize the postsynaptic neuron are known as **excitatory synapses.** In other synapses, however, neurotransmitters hyperpolarize the membrane—that is, increase the voltage difference across the membrane, making it less excitable. These neurotransmitters stimulate chloride channels to open in the postsynaptic membrane. Because the chloride ion concentration outside the cell is higher than it is inside, the opening of the chloride channels results in an influx of chloride ions, making the interior of the postsynaptic cell more negative. This influx, in turn, makes the resting potential more negative and renders the neuron less excitable. Such synapses are called **inhibitory synapses.**[7]

Whether a neuron fires (generates an action potential) depends on the summation of excitatory and inhibitory impulses. If the number of excitatory impulses exceeds the inhibitory impulses, a nerve impulse will be generated. If not, the neuron will not fire. This phenomenon provides the nervous system with a way of integrating incoming information—that is, determining a response by the kinds of input it receives.

Transmission across the synapse can occur in only one direction because only terminal boutons contain neurotransmitter substances and because receptors for these substances are found only in the postsynaptic membrane.

Neurotransmitters Are Quickly Removed from the Synaptic Cleft.
Transmission across the synapse is remarkably fast, requiring only about 1/1000 of a second, or one millisecond. Synaptic transmission is also a transitory event. A short burst of neurotransmitter is released each time an impulse reaches the terminal bouton. It binds to the postsynaptic membrane, elicits a change, and is then removed from the synaptic cleft by three routes: (1) enzymatic destruction, (2) reabsorption by the terminal bouton or absorption by glial cells in the brain, and (3) diffusion away from the synapse.

---

[5] The rate of transmission actually varies depending on the diameter of the axon.
[6] In normal nerve cells, stimulation usually does not occur on the axon but, rather, on the dendrites.

[7] Inhibitory neurotransmitters may also stimulate the opening of potassium channels. Potassium ions flow out of the postsynaptic cell, making the interior more negative.

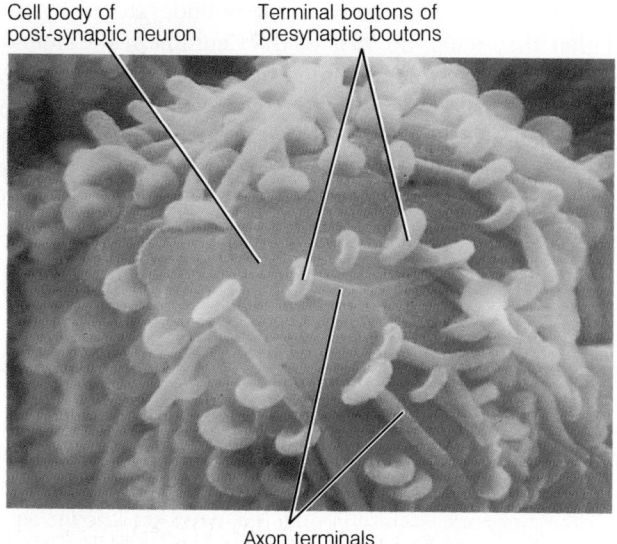

Cell body of post-synaptic neuron

Terminal boutons of presynaptic boutons

Axon terminals

(a)

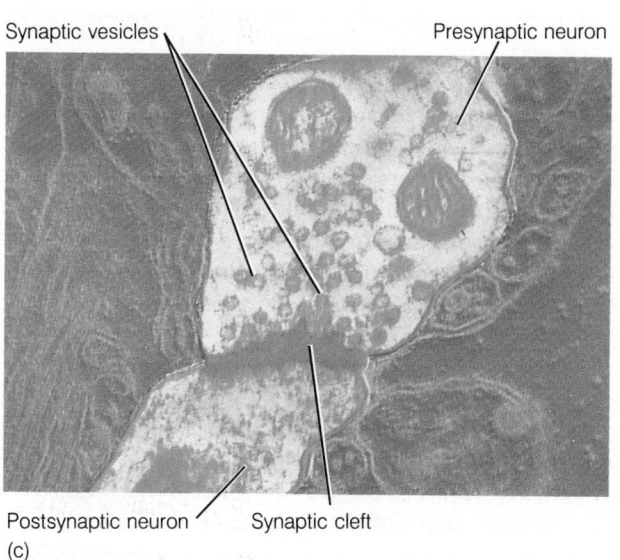

Synaptic vesicles

Presynaptic neuron

Postsynaptic neuron

Synaptic cleft

(c)

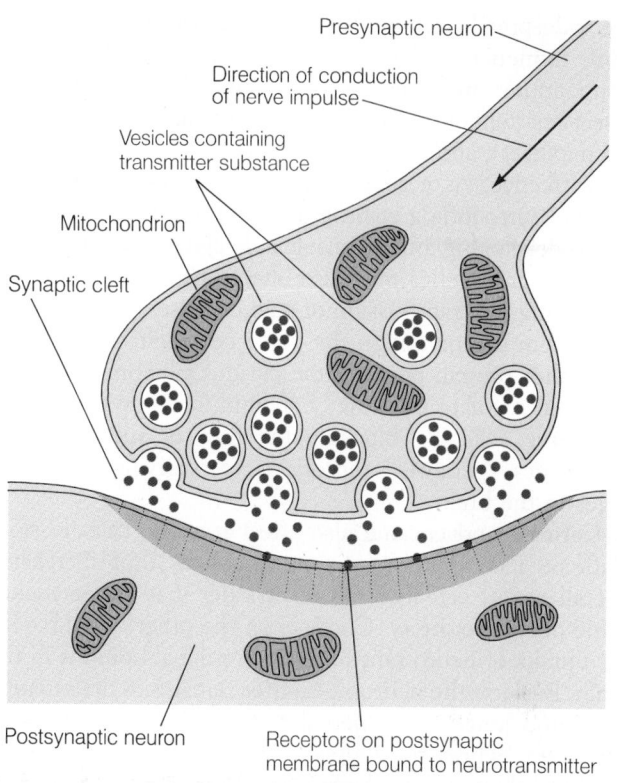

Presynaptic neuron

Direction of conduction of nerve impulse

Vesicles containing transmitter substance

Mitochondrion

Synaptic cleft

Postsynaptic neuron

Receptors on postsynaptic membrane bound to neurotransmitter

(b)

▷ **FIGURE 17–14 The Terminal Bouton and Synaptic Transmission** (*a*) A scanning electron micrograph showing the terminal boutons of an axon ending on the cell body of another neuron. (*b*) The arrival of the impulse stimulates the release of neurotransmitter held in membrane-bound vesicles in the axon terminals. Neurotransmitter diffuses across the synaptic cleft and binds to the postsynaptic membrane, where it elicits another action potential that travels down the dendrite to the cell body. (*c*) A transmission electron micrograph showing the details of the synapse.

Many Chemical Substances, Including Drugs and Insecticides, Exert their Effect by Altering Synaptic Transmission. Certain drugs and common chemicals impair the removal of a neurotransmitter from the synaptic cleft. As long as a neurotransmitter remains bound to the receptors on the postsynaptic membrane, the postsynaptic neuron remains activated. Some insecticides, for instance, act on the neurotransmitter acetylcholine (a-seat-'l-koleen). Acetylcholine is the main excitatory neurotransmitter in the neurons that innervate (supply) skeletal muscle cells. Many commonly used insecticides, however, block the action of **acetylcholinesterase,** an enzyme found in many synapses of insects and people that destroys acetylcholine after it has elicited a reaction in the muscle cell. Organophosphates, such as malathion and parathion, are two examples. Thus, insecticides that block the enzyme kill insects by disrupting nerve transmission. Insects become hyperstimulated and are so incapacitated they eventually die. Unfortunately, insecticides have the same effect on people. These nerve poisons are therefore quite harmful to farm workers and pesticide applicators, who are often exposed to high levels of the pesticides at work. In humans, blocking acetylcholinesterase greatly decreases the removal of acetylcholine from synapses, causing the postsynaptic nerve to be stimulated repeatedly. Muscles go into spasms. At low levels, these insecticides cause blurred vision, headaches, rapid pulse, and profuse sweating. At higher doses, victims may begin to writhe uncontrollably and can die in a short time.

Each year, an estimated 100,000 to 300,000 Americans (mostly farm workers) are poisoned by pesticides. By various estimates 200 to 1000 of them die. Worldwide, an estimated 500,000 people are poisoned and 5000 to 14,000 people die each year from pesticides. These problems and the widespread contamination of the environment have led some farmers to reduce pesticide use and to rely on other, nonpolluting methods of pest control. Crop rotation, insect-resistant crops, natural predators (ladybugs and praying mantises), and a variety of alternatives are practical and cost-effective. As a rule, these strategies also benefit the soil and surrounding environment and are therefore essential to developing a more sustainable system of agriculture.

Some anesthetics may also alter synaptic transmission, decreasing the transmission of pain impulses. Others, however, seem to operate on the nerve cell itself. They apparently alter protein pores in the plasma membrane of neurons that regulate the flow of sodium ions into and out of the nerve cells. By blocking the flow of sodium, these anesthetics "paralyze" sensory nerves carrying pain messages to the brain.

Caffeine and cocaine also affect synaptic transmission. Caffeine increases synaptic transmission, thus increasing overall neural activity. It is no wonder that coffee makes some people so jittery. Cocaine, on the other hand, blocks the uptake of neurotransmitters by terminal boutons in the brain. Because the neurotransmitter remains in the synaptic cleft for a longer time, neural activity is greatly increased. Increased neural transmission in the brain results in a heightened state of alertness and euphoria, commonly known as a "high," which lasts for 20 to 40 minutes. Euphoria, however, is followed by a period of depression and anxiety, which causes many people to seek another high. Excessive cocaine use can result in serious mental derangement—in particular, delusions that others are out to get the user. In this state, heavy users may become violent.

Nerve Cells Can Be Grouped into Three Functional Categories

Nerve cells can be categorized by structure or function. For our purposes, a functional classification is most useful. According to this system, nerve cells fall into three distinct groupings: (1) sensory neurons, (2) motor neurons, and (3) interneurons.

Sensory neurons carry impulses from **sensory receptors** in the body to the central nervous system. Sensory receptors come in many shapes and sizes and respond to a variety of stimuli, such as pressure, pain, heat, and movement (Chapter 18).

Motor neurons carry impulses from the brain and spinal cord to effectors, such as the muscles and glands of the body. Sensory information entering the brain and spinal cord via sensory neurons often stimulates motor neurons, creating a desired response.

In some cases, intervening neurons called **interneurons,** or **association neurons** are required to transmit impulses from the sensory neurons to motor neurons. Interneurons may also transmit impulses from sensory neurons to various parts of the CNS.

The importance of interneurons is underscored by the fact that they make up 99% of the neurons in the human central nervous system. As noted below, these neurons play an important role in coordinating complex activities. They are the neural communication network that transmits impulses from one part of the CNS to another to help bring about coordinated actions. And they are essential to many body reflexes.

≈ EVOLUTION OF THE NERVOUS SYSTEM

Like many other systems, the nervous systems of the animal kingdom exist on a continuum from the simplest to the most advanced. Those classified as primitive generally consist of a network of interconnected neurons known as a **nerve net** (▷ Figure 17–15). In these systems, sensory information is conveyed to motor neurons, which transmit impulses to effectors. Primitive nervous systems lack a central nervous system.

Advanced nervous systems are characterized by the presence of central nervous systems that consist of aggregations of nerves that process and integrate information. In these organisms, a network of nerves akin to the nerve net conveys sensory and motor information to and from the CNS. In between the two extremes on the evolutionary continuum are a multitude of organisms with nervous systems representing the evolutionary progression from the simplest to the most complex.

As one might surmise, the more advanced a nervous system is, the more sophisticated an animal's behavior will be. In organisms with primitive nervous systems, for example, behavioral responses to light and heat are usually limited to increases or decreases in the rate of movement. In those animals with more advanced nervous systems, like humans, the responses are generally more complex and may even involve reasoning.

Interestingly, the evolution of the nervous system corresponds with the evolution of complexity in body structure, as with other systems discussed in earlier chapters. That is to say, as multicellular animal life developed, the nervous system became more complex.

Perhaps one of the most significant developments in the evolution of the nervous system was the emergence of adaptations that accelerate nerve impulse conduction. In invertebrates, the presence of fibers of very large diameter permitted rapid transmission of nerve impulses vital to survival in the environment. The giant nerve axons in invertebrates, for instance, permit the sudden movement of earthworms, crayfish, and squid vital to escape from predators. In vertebrates, saltatory conduction improved the ability to respond quickly to environmental stimuli. With this overview in mind, we next take a brief look at four nervous systems along the continuum, starting with the simplest.

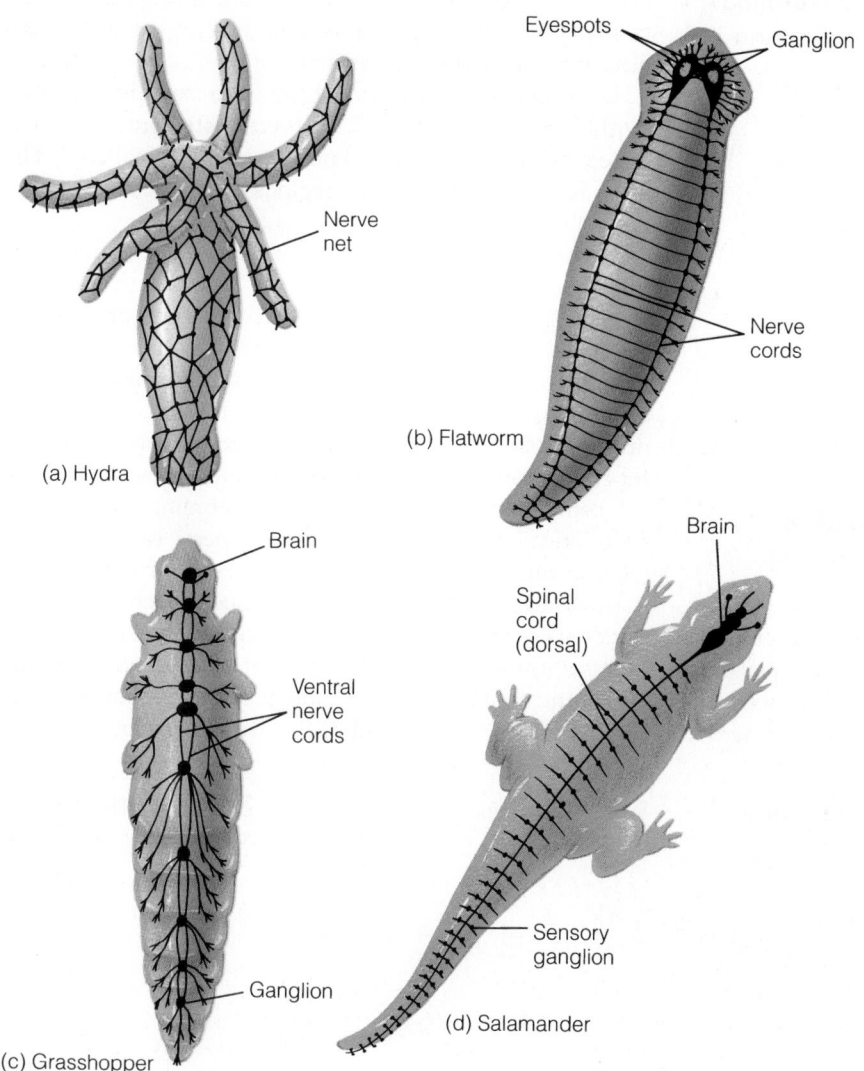

(a) Hydra

Nerve net

Eyespots

Ganglion

(b) Flatworm

Nerve cords

Brain

Ventral nerve cords

Ganglion

(c) Grasshopper

Brain

Spinal cord (dorsal)

Sensory ganglion

(d) Salamander

▷ **FIGURE 17–15 Overview of Evolution of the Nervous System** (*a*) Jellyfish and other cnidarians like this hydra contain a network of nerves that respond to stimuli and evoke responses. (*b*) In flatworms, the simple nerve net is modified to consist of two nerve cords that run the length of the body, giving off fibers to muscles. An aggregation of nerves in the head region is called the ganglion. (*c*) In arthropods, like the grasshopper, further central nervous system development is evident. (*d*) Vertebrate nervous systems have distinct brains and spinal cords that give off nerves to body structures.

The Cnidarians Contain the Simplest Nervous Systems of all Animals

Figure 17–15a shows the structure of the nervous system of a cnidarian—a group including jellyfish and hydras. As illustrated, many cnidarians contain nerve nets, loosely organized networks of nerve cells. Lacking central control, the nervous system of cnidarians is largely based on reflex pathways between epithelial receptor cells and contractile cells in the body wall. In one pathway, which is concerned with feeding, nerve cells transmit impulses from receptor cells in the tentacles to contractile cells around the mouth.

Even in the relatively simple cnidarians, some evidence of centralization occurs. In jellyfish, for instance, special clusters of nerve cells are found near sensory structures. These nerve cell clusters represent very early signs of the formation of a central nervous system and are involved in tasks such as swimming.

Some Flatworms Contain Nerve Nets, but Others Contain Specialized Aggregations of Nerve Cells

Biologists believe that more complex nervous systems evolved from the nerve nets. In the simplest flatworms, nerve nets like those described above form the entire nervous system. In more highly developed flatworms, however, sensory structures such as the paired eyespots are concentrated near the anterior (front) end of the organism. This adaptation gives a crawling animal such as planaria an advantage because the forward end is the first to encounter stimuli. Not surprisingly, the head end also contains aggregations of nerve cells forming a brainlike structure known as a **ganglion.** Ganglia (plural) help coordinate responses to the stimuli flatworms encounter. Two nerve cords also run the length of the body. Attached to the ganglia, the nerve cords give off branches that supply muscles needed for swimming and crawling.

The Nervous Systems of Annelids and Arthropods Exhibit More Advanced Forms of Centralization

Annelids (such as the earthworm) and arthropods (such as the grasshopper) have aggregations of nerve cells in their front ends, forming rudimentary brains, and single nerve cords that run the length of their bodies, giving off branches to body parts. Both animals are segmented, and each segment is served by a pair of nerves and an aggregation of nerve cells (ganglion) located in the nerve cord. Several giant axons are found in the nerve cord as well. These axons transmit nerve impulses rapidly to muscles, permitting the otherwise slow-moving worm to retract its body quickly, as anyone who has ever tried to catch an earthworm at night knows.

The Vertebrate Nervous System Is the Most Advanced of all, but Even Within Vertebrates Considerable Differences are Observed

The nervous system of vertebrates represents the culmination of a long and successful biological progression. Consisting of well-defined central and peripheral portions, the vertebrate nervous system has undergone many changes in evolution. A few general comments are appropriate before we move on to the human nervous system, the most complex of all vertebrate systems.

Located on the back side of the organism and protected by the backbone and skull, the vertebrate central nervous system consists of a well-defined brain and spinal cord. The brain is divided into distinct regions with specialized structures and functions.

Three trends are evident in vertebrate evolution. The first is an increase in the size of the brain relative to body weight. As a general rule, the more complex an organism is, the greater the ratio of brain weight to body weight. The second trend is an increase in compartmentalization. That is, although vertebrate brains have three general parts, during evolution these parts have become structurally and functionally specialized. The third trend is toward increasing complexity of a region of the forebrain. This is especially prevalent in mammals. As a general rule, the more sophisticated an animal's behavior, the better developed is its forebrain, or cerebrum, which houses such specialized functions as hearing, vision, motor control, and ideation. In humans, for instance, 80% of the total brain mass is in the cerebrum. ▷ Figure 17–16 illustrates the evolution of the forebrain (indicated by color) in several vertebrates.

▷ **FIGURE 17–16 Evolution of the Vertebrate Brain** One of the trends in the evolution of the vertebrate brain is the increase in the size of the forebrain. The forebrain, or cerebrum, houses many important sensory and motor functions. This trend is especially evident in mammals, represented here by the cat and human.

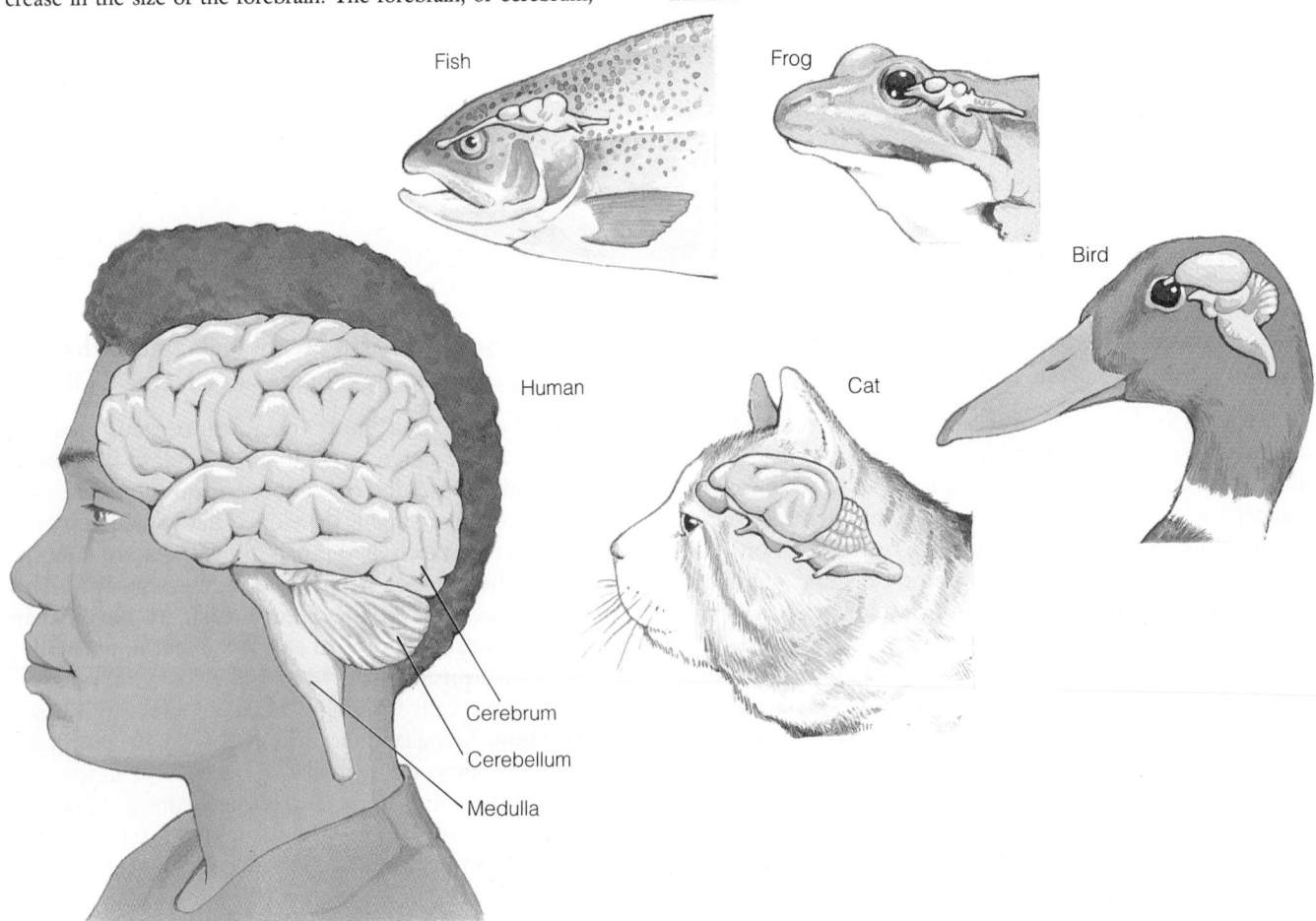

≋ THE SPINAL CORD AND NERVES

With your understanding of the neuron and this brief overview of the evolution of nervous systems, we can now turn our attention to the nervous system of humans, beginning with the spinal cord and nerves.

The Spinal Cord Is Part of the CNS, Transmits Information To and From the Brain, and also Houses Many Reflexes

The spinal cord is a long, ropelike structure about the diameter of a person's little finger. The spinal cord connects to the brain above and courses downward through the vertebral canal formed by the vertebrae (▷ Figure 17–17). The spinal cord gives off nerves along its course that innervate the skin, muscles, bones, and organs of the body. The spinal cord ends into the lower back (at about the level of the second lumbar vertebra), at which point it gives off a series of nerves called the **cauda equina** ("horse's tail"), which supply the lower sections of the body.

As shown in ▷ Figure 17–18, the central portion of the spinal cord is an H-shaped zone of gray matter. **Gray matter** consists of nerve cell bodies of interneurons and motor neurons and is so named because it appears gray to the naked eye. Surrounding the gray matter are fiber tracts consisting of axons and a much smaller number of den-

▷ **FIGURE 17–17 The Spinal Cord** The spinal cord extends from the brain to the upper lumbar region.

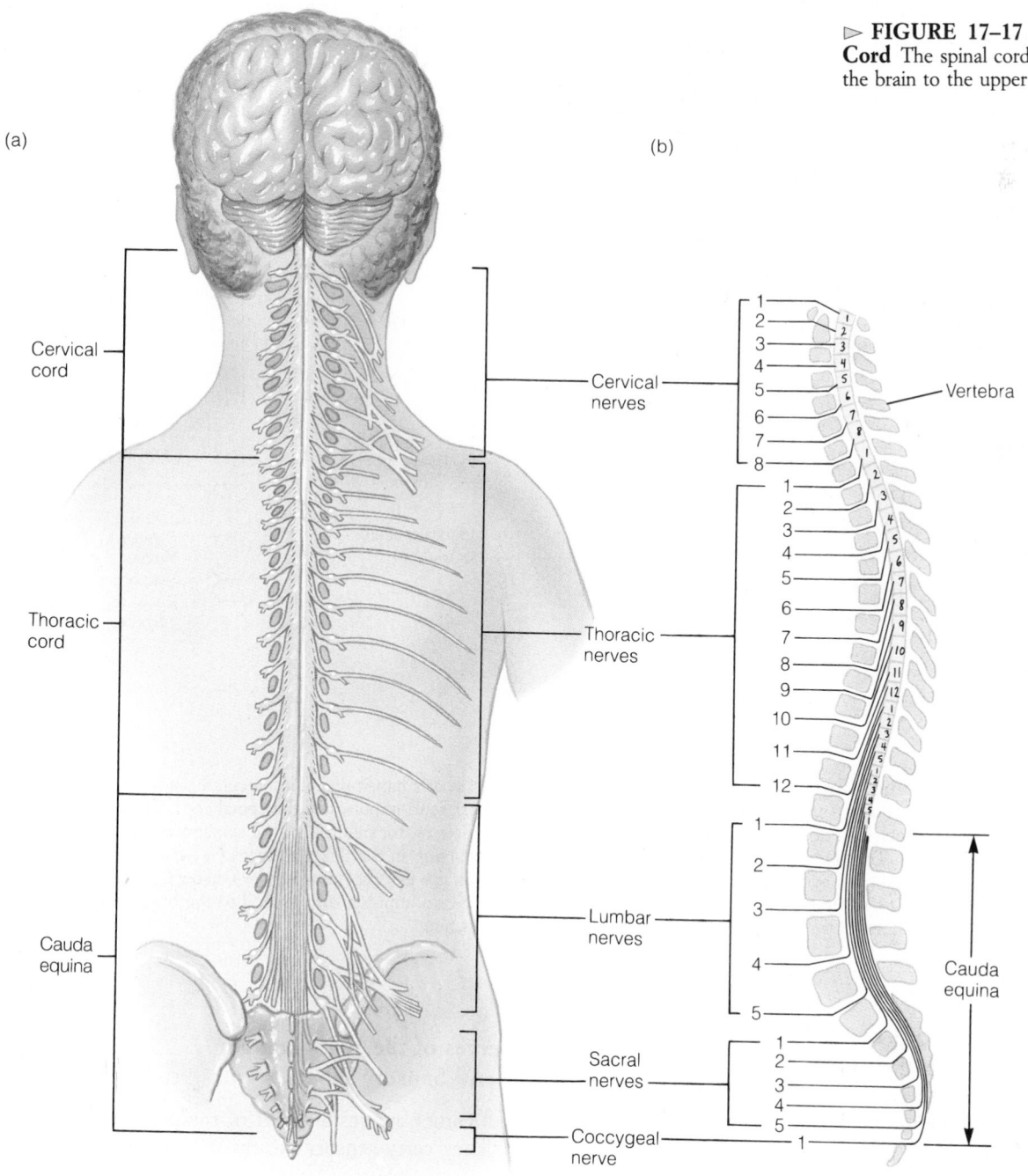

(a)

(b)

Cervical cord

Thoracic cord

Cauda equina

Cervical nerves

Thoracic nerves

Lumbar nerves

Sacral nerves

Coccygeal nerve

Vertebra

Cauda equina

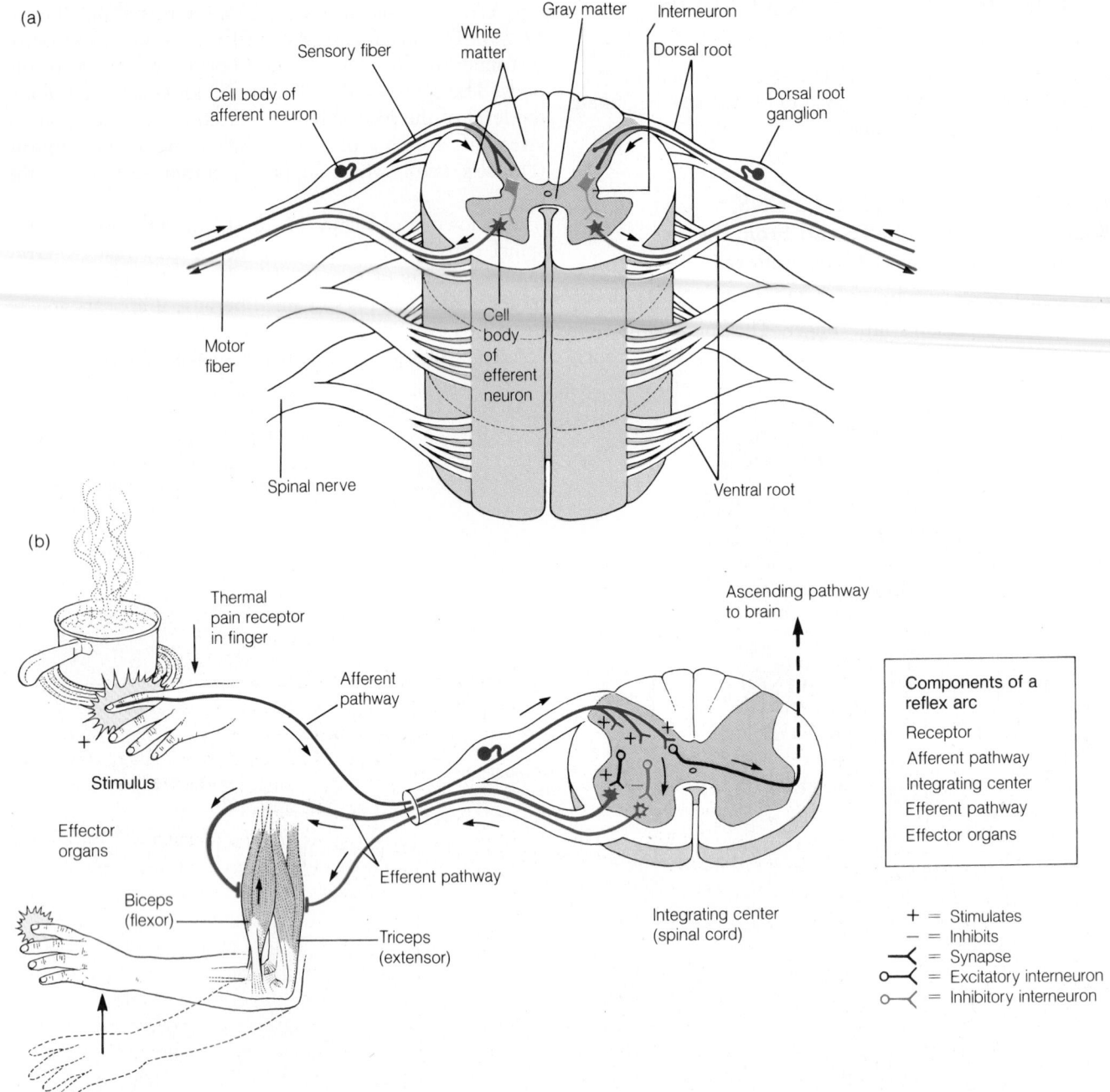

(a)

Cell body of afferent neuron

Sensory fiber

White matter

Gray matter

Interneuron

Dorsal root

Dorsal root ganglion

Motor fiber

Cell body of efferent neuron

Spinal nerve

Ventral root

(b)

Thermal pain receptor in finger

Afferent pathway

Ascending pathway to brain

Stimulus

Effector organs

Biceps (flexor)

Triceps (extensor)

Efferent pathway

Integrating center (spinal cord)

Response

Components of a reflex arc

Receptor
Afferent pathway
Integrating center
Efferent pathway
Effector organs

$+$ = Stimulates
$-$ = Inhibits
= Synapse
= Excitatory interneuron
= Inhibitory interneuron

▷ **FIGURE 17–18 The Spinal Cord and Dorsal Root Ganglia** (*a*) Spinal nerves are attached to the spinal cord by two roots, the dorsal and ventral roots. The dorsal root carries sensory information into the spinal cord. The ventral root carries motor information out of the spinal cord. The spinal nerve often contains both sensory and motor fibers. (*b*) When you accidentally touch a hot pan on the stove, you withdraw your hand before the brain even knows what's happening. This occurs because of a reflex arc. Sensory fibers send impulses to the spinal cord. The sensory impulses stimulate motor neurons in the spinal cord. This causes muscle contraction in the flexor muscles (+) and inhibits muscle contraction in the extensor muscles (−), allowing you to withdraw your hand. Nerve impulses also ascend to the brain to let it know what is happening.

drites that travel up and down the spinal cord, carrying information to and from the brain. The fiber tracts form the **white matter,** whose characteristic color comes from the myelin sheaths of the many axons coursing through this portion of the CNS.

The Nerves of the PNS Contain Motor and Sensory Fibers

As noted earlier, nerves are part of the peripheral nervous system. They carry sensory information to the spinal cord

Olfactory (I)

Optic (II)

Oculomotor (III)

Trigeminal (V)

Glosso-
pharyngeal (IX)

Vagus (X)

Spinal
accessory (XI)

Olfactory
tract

Optic tract

Trochlear (IV)

Abducens (VI)

Facial (VII)

Vestibulo-
cochlear (VIII)

Hypoglossal
(XII)

and brain and motor information out. Some nerves are strictly motor and some are strictly sensory, but many are mixed, having both motor and sensory fibers.

Nerves arising from the brain are called **cranial nerves** (▷ Figure 17–19). The cranial nerves supply structures of the head and several key body parts, such as the heart and diaphragm. The trigeminal nerves mentioned in the introduction are one of the 12 pairs of cranial nerves.

Nerves associated with the spinal cord are known as **spinal nerves** (Figure 17–18a). Each spinal nerve has two roots, the dorsal and ventral, which attach to the spinal cord. The **dorsal root** is the inlet for sensory information traveling to the spinal cord. On each dorsal root is a small aggregation of nerve cell bodies, the dorsal root ganglion. The **dorsal root ganglia** house the cell bodies of sensory neurons. As Figure 17–18a shows, the dendrites of these bipolar nerve cells conduct impulses from receptors in the body to the dorsal root ganglia; the axons, in turn, carry the impulse into the spinal cord along the dorsal root.

As noted earlier, sensory fibers entering the spinal cord often end on interneurons (Figure 17–18a). Interneurons receive input from many sensory neurons and process this information, acting like a receptionist in a busy corporate office. Interneurons transmit the impulses to nearby multipolar motor neurons, whose axons leave in the ventral root of the spinal nerve, carrying impulses to muscles and glands. This anatomical arrangement of neurons allows information to enter and leave the spinal cord quickly and forms the basis of the **reflex arc,** a neuronal pathway by which sensory impulses from receptors reach effectors without traveling to the brain (Figure 17–18b). Some reflex arcs contain interneurons, and some do not.

When a physician taps a rubber hammer on the tendon just below your kneecap (patellar tendon), he or she is testing one of your body's many reflex arcs. The tapping of the hammer on the tendon stretches it, which stimulates stretch receptors in the tendon. These receptors, in turn, generate nerve impulses that travel to the spinal cord via sensory neurons. In this reflex, each sensory neuron ends directly on a motor neuron, which supplies the muscles of the thigh. Thus, a quick tap on the tendon results in a motor impulse sent to the anterior thigh muscles (quadriceps), causing them to contract and the knee to jerk.

Reflexes are mechanisms that often protect the body from harm. Touching a hot stove, for example, elicits the withdrawal reflex. Sensory impulses stimulate the muscles of your arm to contract, so that you pull your hand away from the stove before your brain is aware of what is happening.

Babies come equipped with a number of important reflexes. Rub your finger on the cheek of a newborn, and it immediately turns its head toward your finger. This reflex helps babies find the mother's nipple. Crying is also a reflex. When a baby is hungry, thirsty, wet, or uncomfortable, it cries, a reflex sure to get attention.

The spinal cord also serves as a route by which sensory information is transmitted to the brain. Sensory impulses travel along special tracts lying outside the central H-shaped zone of the spinal cord. Because of this arrangement, sensory information that enters via the spinal nerves can also reach vital brain centers. Although the incoming sensory information may elicit a reflex, the brain is still informed of the problem, allowing for appropriate follow-up action.

The Severity of a Spinal Cord Injury Depends on the Location of the Injury and the Extent of Damage

An automobile crash, a bullet wound, or even a bad fall can damage or sever the fiber tracts of the spinal cord. Injury to the cord usually results in permanent damage, because, as noted previously, axons of the central nervous system do not usually regenerate.

The amount of damage depends on where the cord is injured and the severity of the injury. Sensory fibers traveling to the brain and motor fibers traveling from the brain to the spinal cord run in separate tracts in the white matter. If both tracts are severed—for example, by a severe vertebral fracture—all sensory and motor functions below the level of the injury are lost. As a result, muscles supplied by nerves below the injury become paralyzed and are unable to contract voluntarily. Segments of the body below the injury lose all sensation.

If the spinal cord injury occurs high in the neck (above the fifth cervical vertebra), it cuts the nerve fibers traveling to the muscles that control breathing. These muscles become paralyzed, and the person dies fairly quickly. This is how a hangman's noose kills its victim.

Damage to the cord just below the fifth cervical vertebra (the fifth vertebra in the neck) does not affect breathing, but paralyzes the legs and arms. The condition is known as **quadriplegia.** If the spinal cord injury occurs below the nerves that supply the arms, the result is **paraplegia,** paralysis of the legs.

New studies at the Craig Rehabilitation Center in Denver show that the administration of large doses of an anti-inflammatory steroid drug shortly after a head or neck injury stops swelling of the spinal cord. Swelling is thought

---

TABLE 17–1 Summary of Brain Structures and Functions

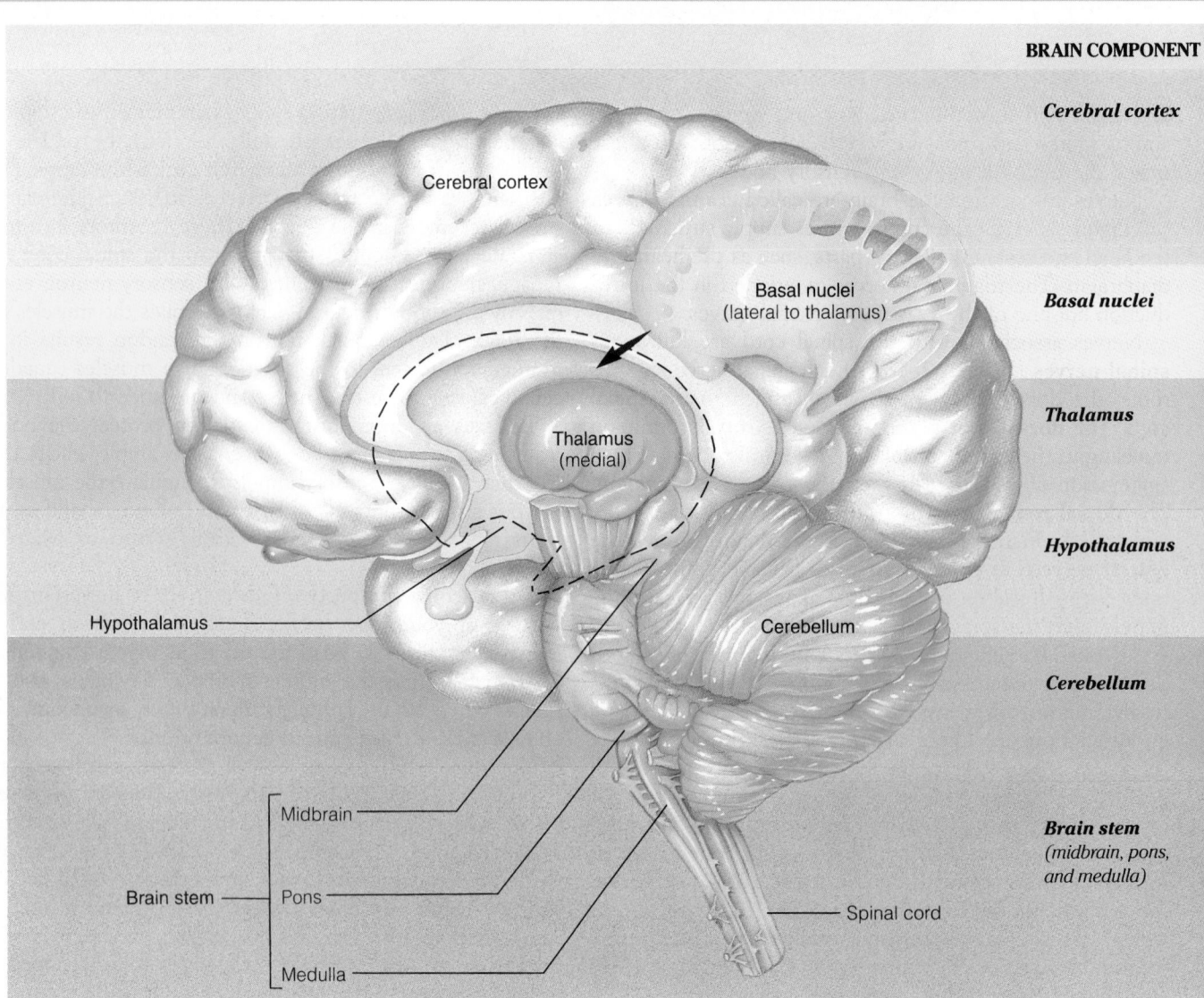

BRAIN COMPONENT

Cerebral cortex

Basal nuclei

Thalamus

Hypothalamus

Cerebellum

Brain stem
(midbrain, pons, and medulla)

to greatly increase the amount of damage resulting from an injury to the spinal cord. This treatment, therefore, may greatly reduce paralysis following an injury to the spinal cord.

Research is also under way to find ways to stimulate the regeneration of axons in the brain and spinal cord. Some early results are promising, suggesting that one day physicians may be able to offset spinal cord damage.

≋ THE BRAIN

About the size of a cantaloupe, the human brain is an extraordinary organ, a product of many millions of years of evolutionary trial and error. Responsible for producing art and music and remarkable feats of engineering and abstract reasoning, it also allows us to ponder right and wrong and remember a vast amount of information. Table 17–1

TABLE 17–1 (continued) ≋

MAJOR FUNCTIONS

1. Sensory perception
2. Voluntary control of movement
3. Language
4. Personality traits
5. Sophisticated mental events, such as thinking, memory, decision making, creativity, and self-consciousness

1. Inhibition of muscle tone
2. Coordination of slow, sustained movements
3. Suppression of useless patterns of movement

1. Relay station for all synaptic input
2. Crude awareness of sensation
3. Some degree of consciousness
4. Role in motor control

1. Regulation of many homeostatic functions, such as temperature control, thirst, urine output, and food intake
2. Important link between nervous and endocrine systems
3. Extensive involvement with emotion and basic behavioral patterns

1. Maintenance of balance
2. Enhancement of muscle tone
3. Coordination and planning of skilled voluntary muscle activity

1. Origin of majority of peripheral cranial nerves
2. Cardiovascular, respiratory, and digestive control centers
3. Regulation of muscle reflexes involved with equilibrium and posture
4. Reception and integration of all synaptic input from spinal cord; arousal and activation of cerebral cortex
5. Sleep centers

provides an overview of the structures and functions of the major components of the brain.

The Cerebral Hemispheres Function in Integration, Sensory Reception, and Motor Action

The brain is housed in the skull, a bony shell that protects it from injury. ▷ Figure 17–20a shows some of the externally visible parts of the human brain. The largest and most conspicuous is the **cerebrum,** a convoluted mass of nervous tissue that, as noted earlier, constitutes about 80% of the total brain mass.

The cerebrum is divided into two halves, the right and left **cerebral hemispheres.** Each hemisphere has a thin outer layer of gray matter, the **cerebral cortex** (Figure 17-20b). The cerebral cortex contains glial cells and the cell bodies of numerous multipolar neurons. Lying beneath the cerebral cortex is a thick central core of white matter. It contains bundles of myelinated axons that give it a white appearance. The axons transmit impulses from one part of the cerebral cortex to another, permitting the integration of the activities of various parts of the cortex. The axons of the white matter also carry impulses from the cerebral cortex to the spinal cord, permitting information to flow from the brain to motor neurons that control many of the body muscles.

As shown in Figure 17–20a, the cerebral cortex is thrown into numerous folds, called **gyri** (singular, *gyrus*). The gyri are separated by numerous valleys, known as **sulci** (singular, *sulcus*).

Deep within the white matter of the cerebral hemispheres are masses of gray matter known as the **basal ganglia.** The nerve cells of the basal ganglia inhibit motor function, helping fine-tune muscular control.

The Cerebrum Is Divided into Four Major Lobes, and Each Lobe Contains Areas that House Specific Functions. The cerebral hemispheres can be divided into four major **lobes,** shown in Figure 17–20a. They are the frontal lobe, the parietal lobe, the occipital lobe, and the temporal lobe. Within each of these lobes are areas that house specific functions, illustrating that the brain did not escape the evolutionary forces that resulted in compartmentalization and specialization (▷ Figure 17–21). The functional areas within the lobes can be broadly classified into three types: motor cortex, sensory cortex, and association cortex. The **motor cortex** stimulates muscle activity. The **sensory cortex** receives sensory stimuli, and the **association cortex** integrates information, bringing about coordinated responses.

Figure 17–21 shows the major cortical areas and lists some of their functions. As you can see, the cortex contains areas for receiving sound (primary auditory cortex) and generating speech (Broca's area), areas that receive input from the eyes (primary visual cortex), and regions that receive sensory information from the body (primary sensory cortex). It also contains regions that control muscle

▷ FIGURE 17–20 The Brain

(a) The cerebral cortex consists of lobes as shown here. The lobes, in turn, can be divided into sensory, association, and motor areas (not shown). (b) A section through the brain showing the gray and white matter of the cortex and deeper structures.

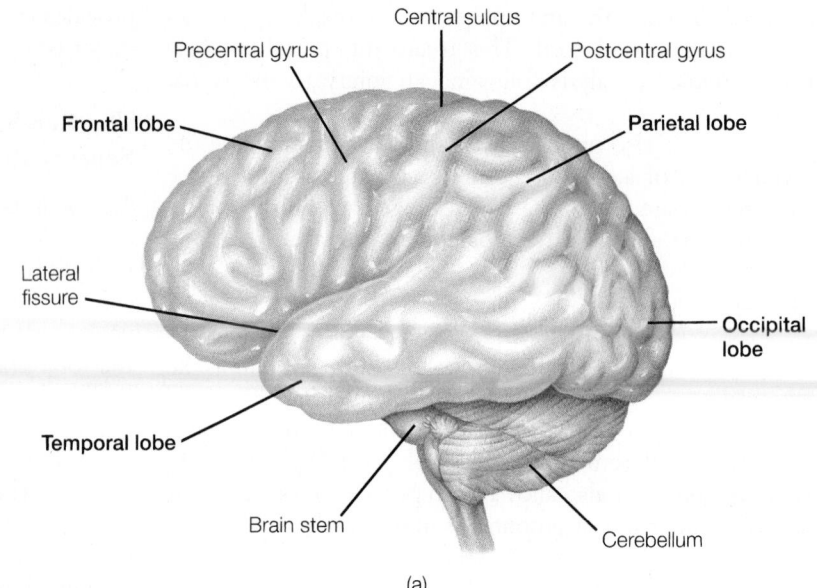

(a)

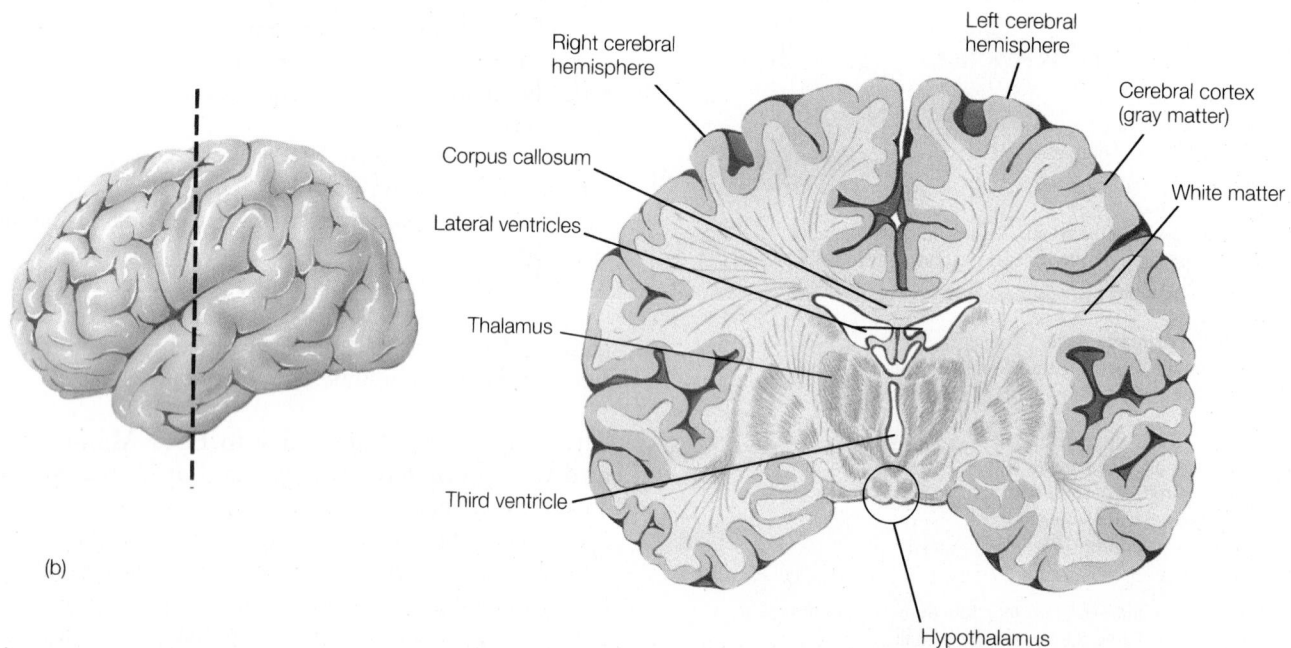

(b)

movement and areas that allow for planning (prefrontal association cortex).

The following section discusses some of the major areas of the cortex, broadening your understanding of how the brain works. We begin with the primary motor cortex.

The Primary Motor Cortex Controls Voluntary Movement. The **primary motor cortex** occupies a single gyrus (ridge) on each hemisphere, just in front of the central sulcus (Figure 17–21). The primary motor cortex controls voluntary motor activity—for example, the muscles in your hand that are turning the pages of your book.

The neurons in the primary motor cortex are arranged in a very specific order. Thus, each region of the motor cortex controls a specific body part. As illustrated in ▷ Figure 17–22a, the neurons that control the muscles of the knee are located in the uppermost region of the primary motor cortex. Hip muscle control occurs below that. Muscles of the hand are controlled by neurons located even lower.

To bring about a voluntary movement, a conscious thought stimulates the generation of nerve impulses in the primary motor cortex. The impulse then travels from the brain down the spinal cord to the motor neurons in the spinal cord. The axons of these neurons leave the spinal cord, traveling in the nerves, and terminate on the muscles of the body.

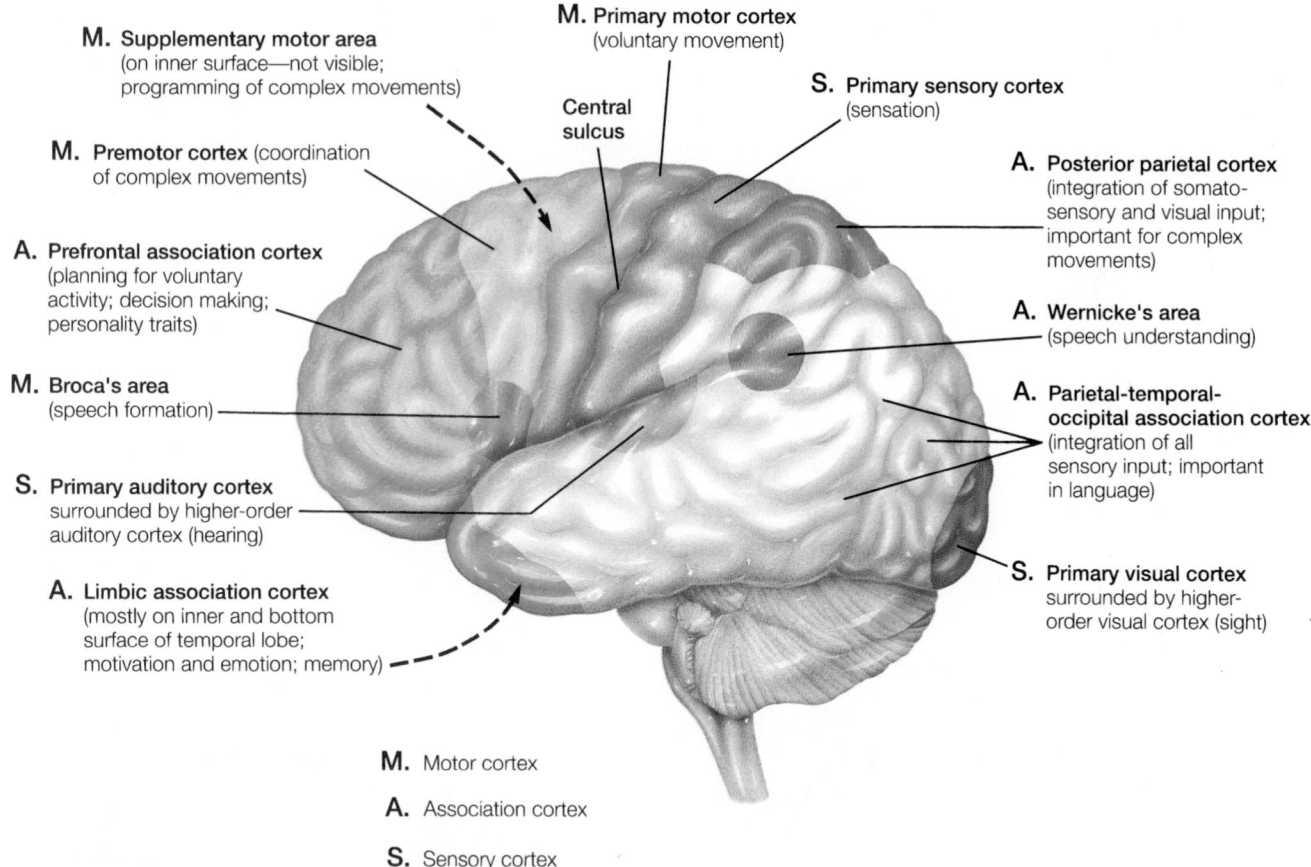

M. Supplementary motor area (on inner surface—not visible; programming of complex movements)

M. Primary motor cortex (voluntary movement)

Central sulcus

S. Primary sensory cortex (sensation)

M. Premotor cortex (coordination of complex movements)

A. Posterior parietal cortex (integration of somato-sensory and visual input; important for complex movements)

A. Prefrontal association cortex (planning for voluntary activity; decision making; personality traits)

A. Wernicke's area (speech understanding)

M. Broca's area (speech formation)

A. Parietal-temporal-occipital association cortex (integration of all sensory input; important in language)

S. Primary auditory cortex surrounded by higher-order auditory cortex (hearing)

A. Limbic association cortex (mostly on inner and bottom surface of temporal lobe; motivation and emotion; memory)

S. Primary visual cortex surrounded by higher-order visual cortex (sight)

M. Motor cortex

A. Association cortex

S. Sensory cortex

▷ **FIGURE 17–21 Functional Regions of the Cortex** The cerebral cortex has three principal functions: receiving sensory input, integrating sensory information, and generating motor responses. Special sensory areas handle vision, smell, taste, and hearing.

In front of the primary motor area is the **premotor cortex.** The premotor area is also involved in controlling muscle contraction. However, the movements that the premotor cortex controls are less voluntary—for example, typing or the fingering required to play a musical instrument.

The Primary Sensory Cortex Receives Sensory Information From the Body. Just behind the central sulcus is another long ridge of tissue running parallel to the primary motor area. Known as the **primary sensory cortex,** it is the destination of many sensory impulses traveling to the brain. As in the primary motor area, different parts of this ridge correspond to different parts of the body (Figure 17–22b). Electrical stimulation of the primary sensory area will elicit sensations that appear to be coming from specific body parts.

The Association Cortex Is the Site of Integration and Sometimes Houses Complex Intellectual Activities. The **association cortex** consists of large expanses of cerebral cortex where integration occurs. In the frontal lobe is a region of the association cortex (prefrontal asso-

ciation cortex) that houses complex intellectual activities, such as planning and ideation. This region also modifies behavior, conforming human actions with social norms. Another important association area lies posterior to the primary sensory cortex. It interprets sensory information sent to the brain and stores memories of past sensations. Other association areas interpret language in written and spoken form.

Hearing, Vision, Taste, and Smell Are also Housed in Specific Cortical Regions. Figure 17–21 also illustrates the presence of patches of sensory cortex for hearing (primary auditory cortex) and vision (primary visual cortex). Although they are not shown, separate regions also exist for taste and smell.

Unconscious Functions Are Housed in the Cerebellum, Hypothalamus, and Brain Stem

Consciousness resides in the cerebral cortex. However, many body functions occur at an unconscious level. Heartbeat, breathing, and many homeostatic functions, for ex-

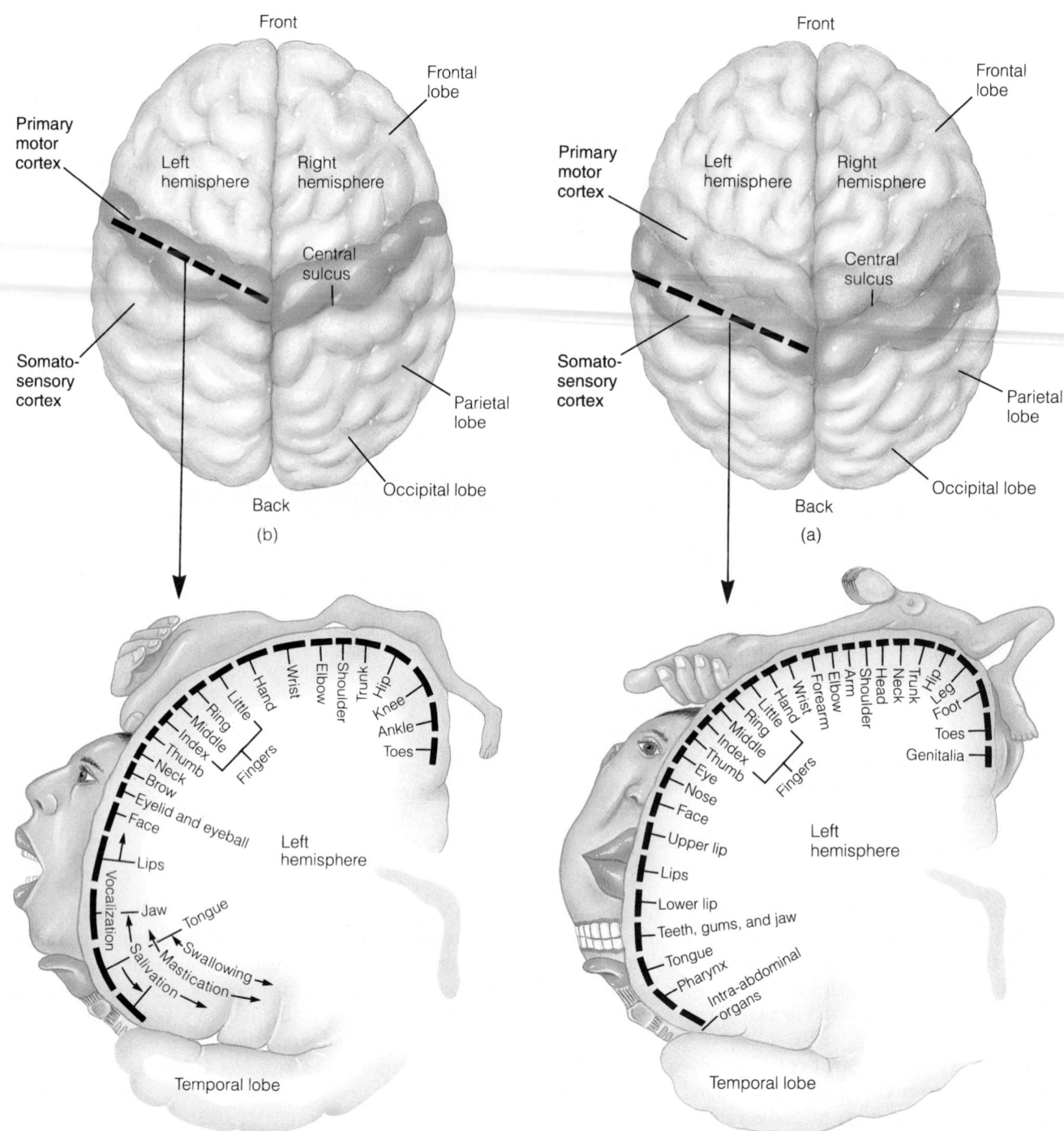

> **FIGURE 17–22 Primary Motor and Sensory Areas**
(*a*) Top view of the brain showing the primary motor cortex. Below is a map of the location of motor functions within the primary motor cortex. (*b*) Cross section through the primary sensory cortex (somatosensory cortex). Below is a map of the location of regions within the primary sensory cortex.

ample, all take place without conscious control.[8] One region of the brain that controls unconscious actions is the cerebellum. The **cerebellum** is the second largest structure

of the brain. As Figure 17–21 shows, the cerebellum sits below the cerebrum on the brain stem.

The Cerebellum Controls Muscle Synergy and Helps Maintain Posture. The cerebellum plays several key roles. One of them is synergy. Neurophysiologists define **synergy** as the coordination of skeletal muscle contrac-

---

[8] Conscious control is possible for many automatic functions, but for the most part, breathing and swallowing are controlled automatically in lower brain centers.

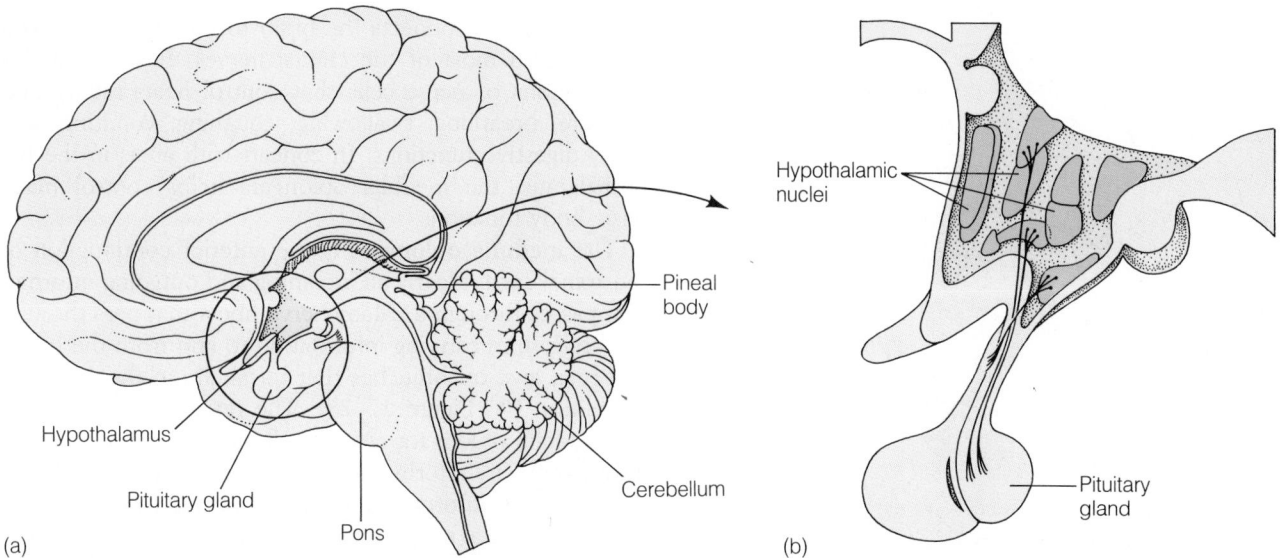

(a)   (b)

▷ **FIGURE 17–23 Cross Section of the Brain** (*a*) The hypothalamus is clearly visible in a cross section of the brain. (*b*) The hypothalamus is at the base of the brain, just above the pituitary gland. It regulates many autonomic functions and plays a particularly important role in controlling the release of hormones by the pituitary.

tion and the movement of body parts. To understand the concept, first hold your arm straight out, then bring your hand to your chest. To perform this simple action, the biceps (muscles in the front of the upper arm) contracted while the triceps (muscles in the back of the upper arm) relaxed. For smooth, coordinated movement, some muscles must contract while others relax. The cerebellum, when operating normally, ensures that opposing sets of muscles work together to bring about smooth motion, or synergy.

Damage to the cerebellum may occur during childbirth if the blood supply to the baby is interrupted—for example, if the umbilical cord accidentally wraps around the baby's neck. The lack of oxygen to the brain may cause permanent damage to the cerebellum, resulting in a loss of a synergistic control of skeletal muscles. Mild damage generally results in slight rigidity (called spasticity) and moderately jerky motions. More severe damage can result in serious impairment, with body motions becoming extremely jerky and simple tasks requiring several attempts. This condition is known as **cerebral palsy.** Cerebral palsy may also be caused by abnormal brain development, possibly due to exposure to harmful chemicals (such as alcohol) during embryonic development. Many children with this disorder have some degree of mental retardation, although some others are highly intelligent.

The cerebellum also helps maintain posture. It receives impulses from sense organs in the ear that detect body position. The cerebellum sends impulses to the muscles to maintain or correct posture.

The Thalamus Is a Relay Center. Just beneath the cerebrum is a region of the brain called the thalamus (see Figure 17–20). The **thalamus** is a relay center, like a

switchboard in a telephone system. It receives all sensory input, except for smell, then relays it to the sensory and association cortex.

The Hypothalamus Controls many Autonomic Functions Involved in Homeostasis. Beneath the thalamus is the hypothalamus. It consists of many aggregations of nerve cells, referred to as **nuclei**[9] (▷ Figure 17–23). The nuclei control a variety of autonomic functions all aimed at maintaining homeostasis. Appetite and body temperature, for example, are controlled by some of the hypothalamic nuclei, as are water balance, blood pressure, and sexual activity.

The hypothalamus is a primitive brain center. Perhaps its best known function is control of the pituitary gland, an endocrine gland that regulates many body functions through the hormones it releases (described in Chapter 20).

The Limbic System Is the Site of Instinctive Behavior and Emotion. Instincts, among other functions, reside in a complex array of structures called the **limbic system,** shown in ▷ Figure 17–24, which operates in conjuction with the hypothalamus. Instincts are among the most fundamental responses of organisms. They include the protective urge of a mother, the territorial assertions of a male, and the fight-or-flight response an animal experiences in the face of adversity.

The limbic system also plays a role in emotions—fear, anger, and so on. Electrodes in some areas of the limbic system may elicit primitive rage. In other areas, electrical stimulation elicits placidity. Stimulation of specific regions within the limbic system of humans may elicit sensations

[9]These are not to be confused with the nuclei of cells.

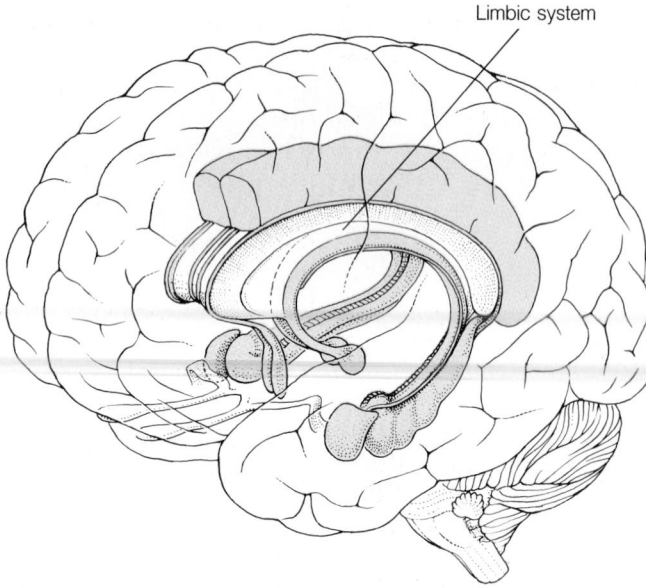

Limbic system

▷ **FIGURE 17–24 The Limbic System** This odd assortment of structures is the seat of emotions and instincts, among other functions.

patients describe as joy, pleasure, fear, or anxiety, depending on the site of stimulation.

Many Basic Body Functions Are Controlled by the Brain Stem. The **brain stem** controls additional primitive functions and consists of three parts, the medulla oblongata, the pons, and the midbrain (▷ Figure 17–25).

▷ **FIGURE 17–25 Brain Stem** The brain stem is a primitive portion of the brain and consists of the midbrain, pons, and medulla oblongata.

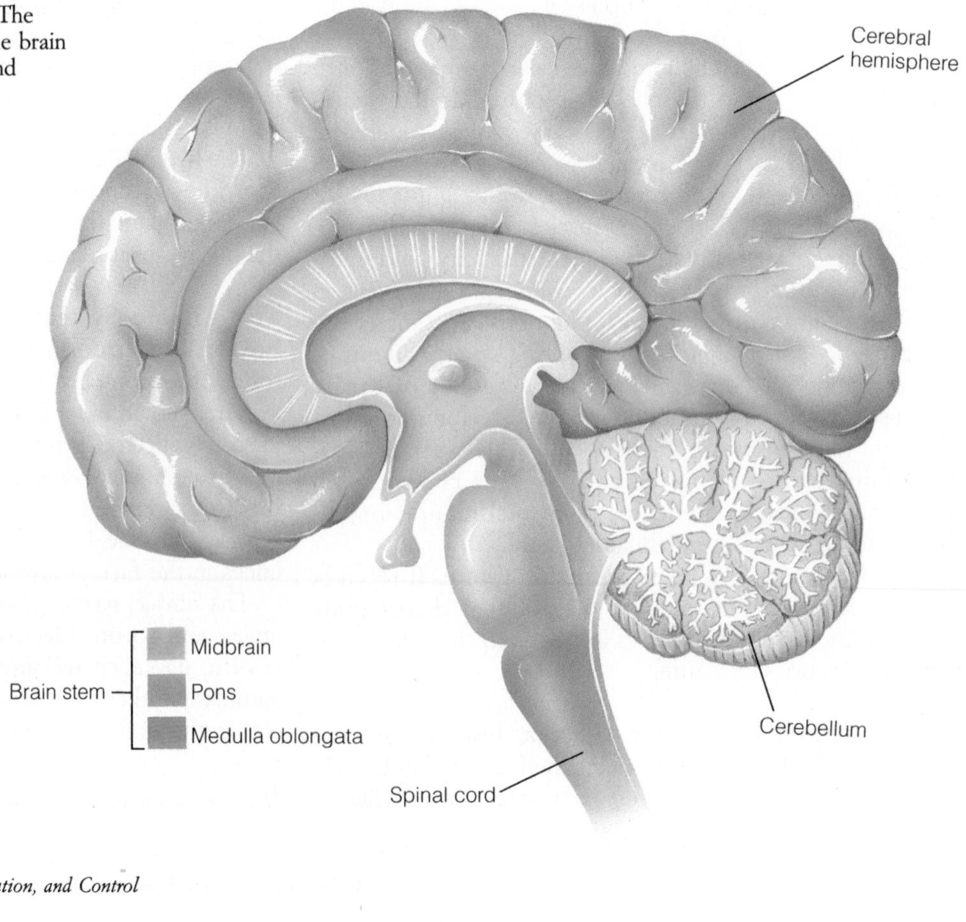

The brain stem connects the brain to the spinal cord and gives rise to most of the cranial nerves. It also contains aggregations of nerve cells that control heart rate, blood pressure, breathing, swallowing, coughing, vomiting, and many digestive functions. In concert with areas in the hypothalamus, the medulla oblongata helps control many basic body functions.

The **medulla oblongata** is the anterior continuation of the spinal cord. Nearly all incoming and outgoing information passes through it. Many nerve fibers that pass through the brain stem carrying information to and from the brain, however, give off branches that terminate in the **reticular formation** (▷ Figure 17–26). The reticular formation of the medulla oblongata receives all incoming and outgoing information, monitoring activity like a security guard at a doorway. This information is then projected to the cerebral cortex via special nerve fibers that activate cortical neurons, in much the same way that a security guard might keep the "boss" informed of the comings and goings at the door. These nerve fibers are referred to as the **reticular activating system (RAS).**

Nerve fibers of the RAS help maintain wakefulness or alertness. When we are asleep, the flow of information through the RAS is greatly reduced, and the cortex "sleeps." Sleep may be disturbed by a biting insect, which stimulates sensory nerves in the skin. These nerves transmit impulses to the spinal cord, which then transmits impulses to the brain. Impulses reaching the RAS travel to the cortex and may awaken the sleeper.

The RAS awakens us and keeps us alert, but it may also

Cerebral hemisphere

Cerebellum

Brain stem — Midbrain
Pons
Medulla oblongata

Spinal cord

prevent us from falling asleep from time to time. Pain from a bad sunburn, for example, prevents many people from falling asleep. That is because pain impulses traveling to the brain from sensors in the skin enter the RAS, generating impulses that travel to the cortex and keep the cells active despite one's fatigue.

Cerebrospinal Fluid Cushions the Central Nervous System

A watery fluid, similar to blood plasma and interstitial fluid, surrounds the brain and spinal cord, filling the arachnoid space between the inner and outer layers of the meninges and also filling internal cavities in these organs (see Figure 17–3). This fluid, called **cerebrospinal fluid (CSF),** serves as a cushion that protects the brain and spinal cord from traumatic injury. Cerebrospinal fluid also fills special cavities within the brain known as **ventricles** (▷ Figure 17–27).

Cerebrospinal fluid is produced by the highly vascularized lining of the lateral ventricles and is slowly drained out of the brain and into the bloodstream. Generally, the amount produced equals the amount removed. If the flow of CSF is blocked, however, fluid builds up in the brain, causing damage. If the circulation of CSF is blocked in a

newborn, the accumulation of fluid causes the brain and head to enlarge and produces a condition referred to as **hydrocephalus** (literally, "water on the brain") (▷ Figure 17–28). The ventricles fill with fluid, stretching the cortex and permanently damaging brain cells. If the damage is not severe, surgeons can treat hydrocephalus by implanting a plastic tube that drains the fluid from the ventricles into the thoracic cavity. This draining permits the brain to develop normally.

Because CSF bathes the central nervous system, examination of the fluid provides a means of detecting infections of the brain, spinal cord, and meninges. Samples of CSF are obtained by inserting a needle between the third and fourth lumbar vertebrae into the space between the dura mater and pia mater. This region is chosen because the spinal cord ends at the first or second lumbar vertebra. Should a needle contact one of the nerves of the cauda equina, which travel in the vertebral canal, the damage would probably be much less than if it pierced the spinal cord. CSF is then withdrawn and examined.

One dangerous condition detected by this method is **meningitis,** an infection of the meninges by certain viruses or bacteria, which, as noted in Chapter 14, usually begins in the respiratory system. Meningitis requires immediate

▷ **FIGURE 17–26 The Brain Stem and Reticular Activating System** The brain stem is composed of the pons and medulla (not labeled). The reticular formation resides in the brain stem, where it receives input from incoming and outgoing neurons. Fibers projecting from the reticular formation to the cortex constitute the reticular activating system.

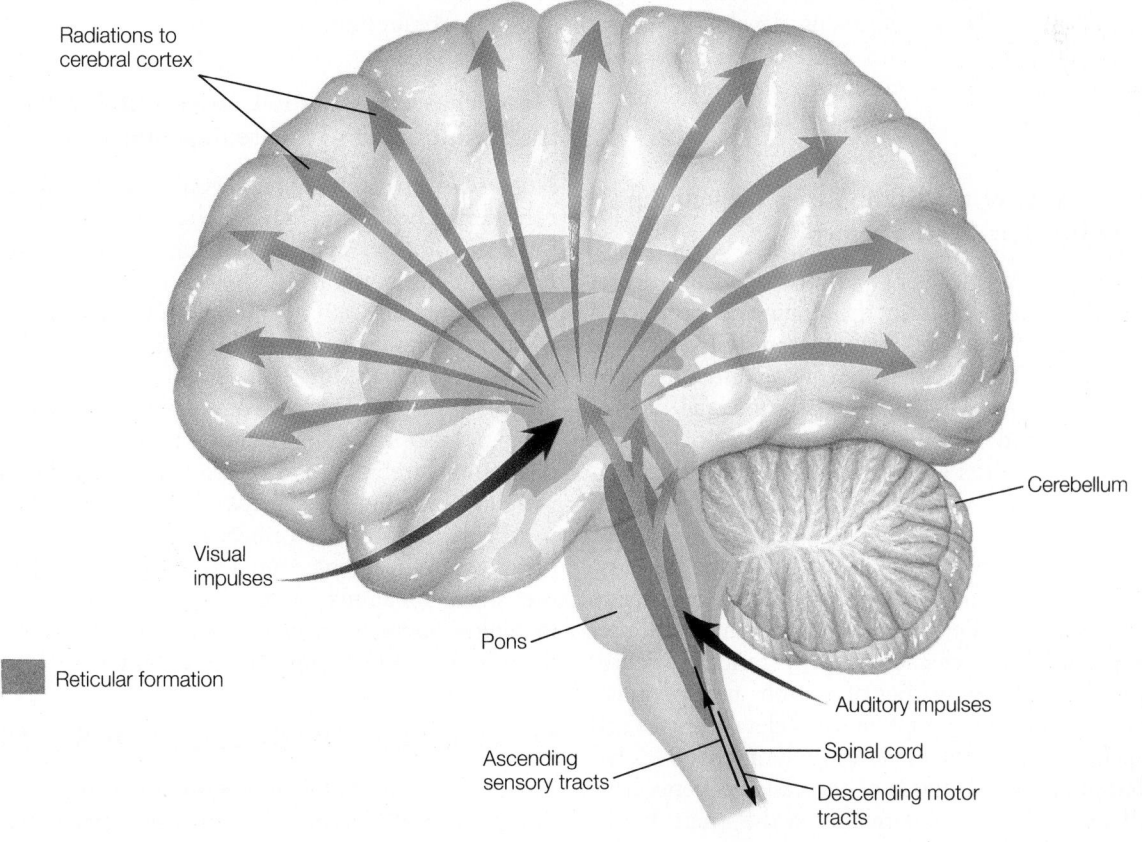

Radiations to cerebral cortex

Cerebellum

Visual impulses

Reticular formation

Pons

Auditory impulses

Spinal cord

Ascending sensory tracts

Descending motor tracts

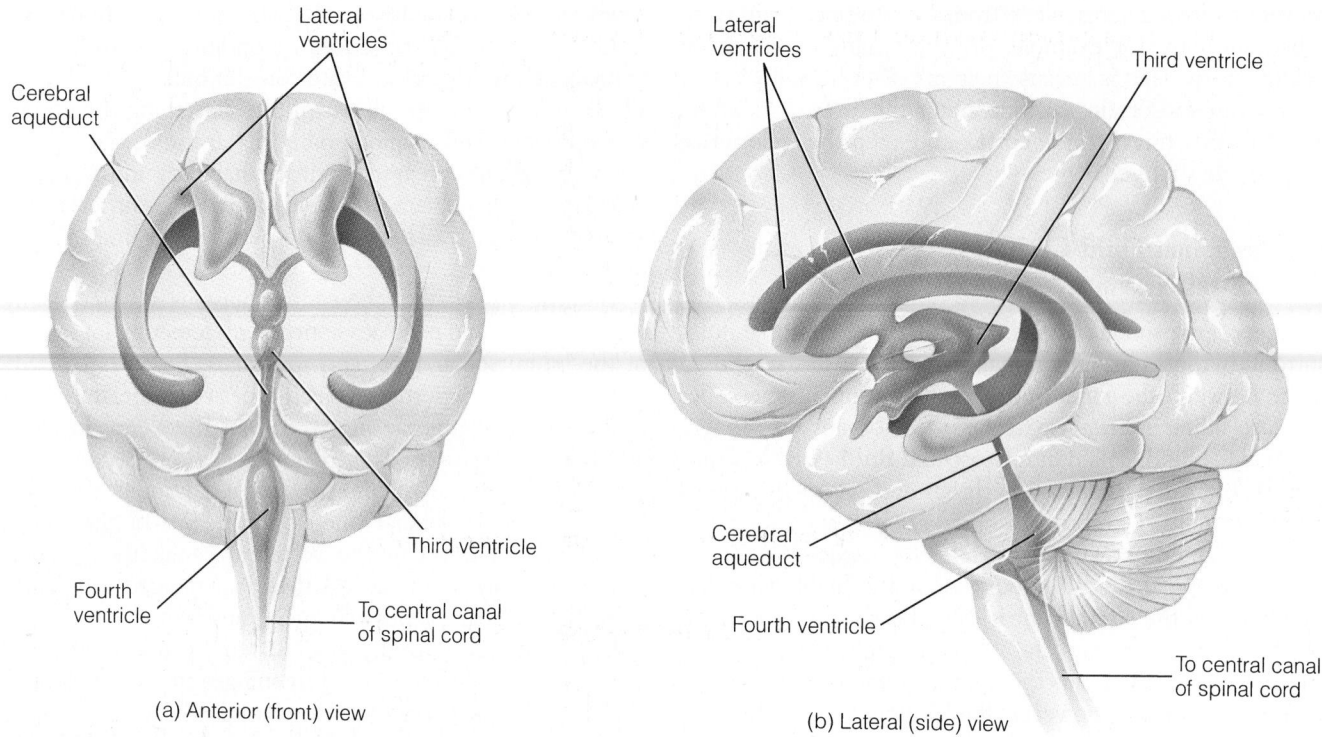

(a) Anterior (front) view

(b) Lateral (side) view

▷ **FIGURE 17–27 Ventricles**
The brain contains four internal cavities, called ventricles, which are filled with a plasmalike fluid known as cerebrospinal fluid.

attention because it can result in sudden death. In adults, meningitis is characterized by fever, headache, nausea, vomiting, a stiff neck, and an inability to tolerate bright light (photophobia). These symptoms develop over the course of a few hours. In babies and children, the symptoms are often less obvious, causing parents to overlook a potentially fatal condition until it is too late.

Electrical Activity of the Brain Varies Depending on Activity Level or Level of Sleep

Electrodes applied to different parts of the scalp detect underlying electrical activity in the brain and produce a tracing known as an **electroencephalogram (EEG).** Some sample tracings of the electrical activity at different times are shown in ▷ Figure 17–29. As illustrated, the type of brain wave recorded depends on one's level of cortical activity. The waves often appear irregular, but distinct patterns can sometimes be observed, especially during the different phases of sleep.

EEGs are used to diagnose brain dysfunction, because diseases or injuries of the cerebral cortex often give rise to altered EEG patterns. Perhaps the most common disorder is **epilepsy.** Epilepsy is characterized by periodic seizures that occur when a large number of neurons fire spontaneously, producing involuntary spasms of skeletal muscles and alterations in behavior. Intensive research on epileptics has shown that in about two-thirds of the cases, patients have no identifiable structural abnormality in the brain. In the

remaining group, epileptic seizures can usually be traced to brain damage at birth, severe head injury, or inflammation of the brain. In some instances, epileptic seizures are caused by brain tumors.

Headaches Have Many Causes But Are Rarely the Result of Life-Threatening Anomalies

No discussion of the brain would be complete without considering headaches. **Headaches** are the most common form of pain and may result from one of several causes. For example, many headaches result from tension—sustained tightening of the muscles of the head and neck when a person is nervous, stressed, or tired. Headaches also commonly result from swelling of the membranes lining the sinuses—either from infections or allergies. Eyestrain is also a common cause of headaches, as is the dilation of cerebral blood vessels, which may be associated with high blood pressure or excessive alcohol consumption.

More serious causes of headache are increased intracranial pressure, resulting from a brain tumor or bleeding into the skull. Inflammation caused by an infection of the meninges (meningitis) or the brain itself (encephalitis), although rare, requires immediate attention.

≈ THE AUTONOMIC NERVOUS SYSTEM

As noted earlier, the nervous system consists of two parts: the peripheral nervous system (the spinal and cranial

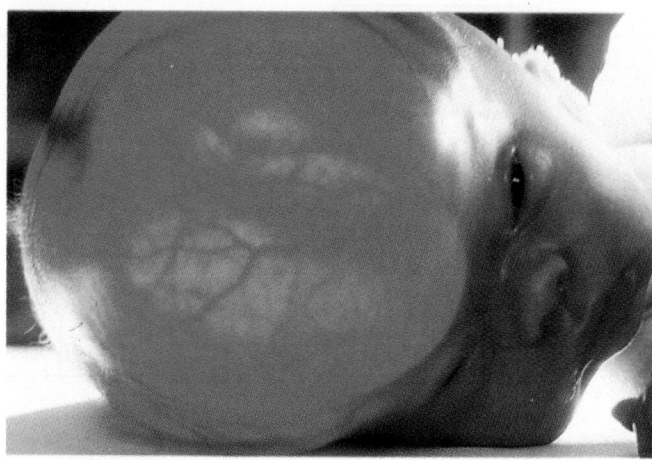

▷ **FIGURE 17–28 Hydrocephalus** This birth defect results from a blockage in the ventricles, which causes CSF to build up, thinning the cortex and causing severe brain damage.

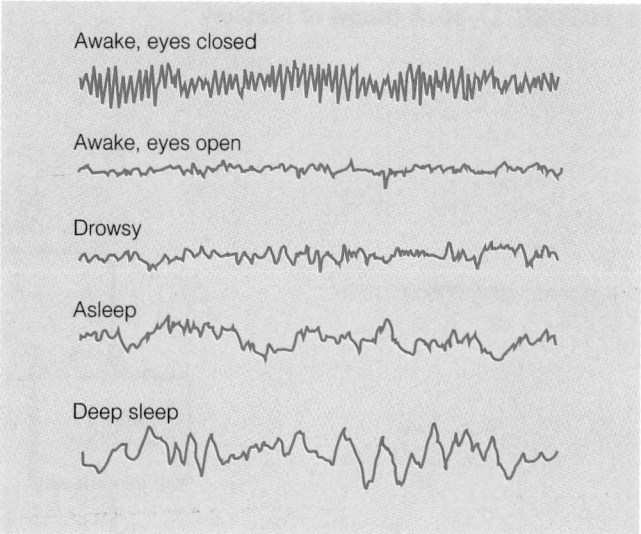

▷ **FIGURE 17–29 Electroencephalogram** Several sample tracings of brain waves during different activities and stages of sleep.

nerves) and the central nervous system (brain and spinal cord). The PNS consists of a somatic division and an autonomic division, both of which contain motor and sensory neurons (see Figure 17–4).

The **somatic division** receives sensory information from the skin, muscles, and joints, which it transmits to the CNS. Motor impulses from the CNS, in turn, are sent to skeletal muscles. In contrast, the **autonomic division** transmits sensory information from organs to the CNS, which delivers motor impulses to smooth muscle, cardiac muscle, and glands.

The autonomic nervous system (ANS) functions automatically, usually at a subconscious level. It innervates all internal organs and has two subdivisions: the sympathetic and the parasympathetic.

The **sympathetic division** of the autonomic nervous system functions in emergencies and is largely responsible for the **fight-or-flight response.** This response occurs when an individual is startled or faced with danger. A sudden scare results in sympathetic nerve impulses that bring about a whole host of responses, preparing an animal to fight or flee the scene. One change is a rapid increase in heart rate, which many of us have experienced when startled. The fight-or-flight response also results in pupil dilation and an increase in breathing rate. Blood flow to the intestines decreases, and blood flow to skeletal muscles increases, providing them with additional oxygen.

In contrast, the **parasympathetic division** of the ANS brings about internal responses associated with the relaxed state. It therefore reduces heart rate, contracts the pupils, and promotes digestion.

Most organs are supplied with both parasympathetic and sympathetic fibers. As a general rule, the parasympathetic and sympathetic fibers have antagonistic (or opposite) effects. One stimulates activity, and the other reduces activity. This provides the body with a means of fine-tuning organ function.

≈ LEARNING AND MEMORY

Few questions intrigue humans more than how the brain functions in learning and memory. **Learning** is a process in which an individual acquires knowledge and skills, usually from experience, instruction, or some combination of the two. Learning depends on one's ability to store information in the brain, a phenomenon called **memory.**

As most of us can attest, newly acquired knowledge is first stored in **short-term memory.** Short-term memory holds information for periods of seconds to hours. The phone number you just looked up or the name of a new acquaintance, for example, fall into short-term memory. If you don't do something to "fix" these memories, they will soon vanish (▷ Figure 17–30). Cramming for tests places a lot of information into short-term memory. Unfortunately, soon after the test is over, the information seems to fade into oblivion. The reason people often cannot recall details of a traumatic accident is because short-term memory has been wiped out.

Long-term memory, on the other hand, holds information for much longer periods—from days to years—and has a much greater storage capacity than short-term memory "banks."

Transferring information from short-term to long-term memory is called **consolidation.** In the *Study Skills* section at the front of the book, I described several ways to consolidate information, including repetition, mnemonics, and rhymes. All of these tools, and others, help shift information to your long-term memory.

The process of recalling information stored in either short-term or long-term memory is called **remembering.** As illustrated in Figure 17–30, short-term memory recall is generally faster than long-term memory recall. That is,

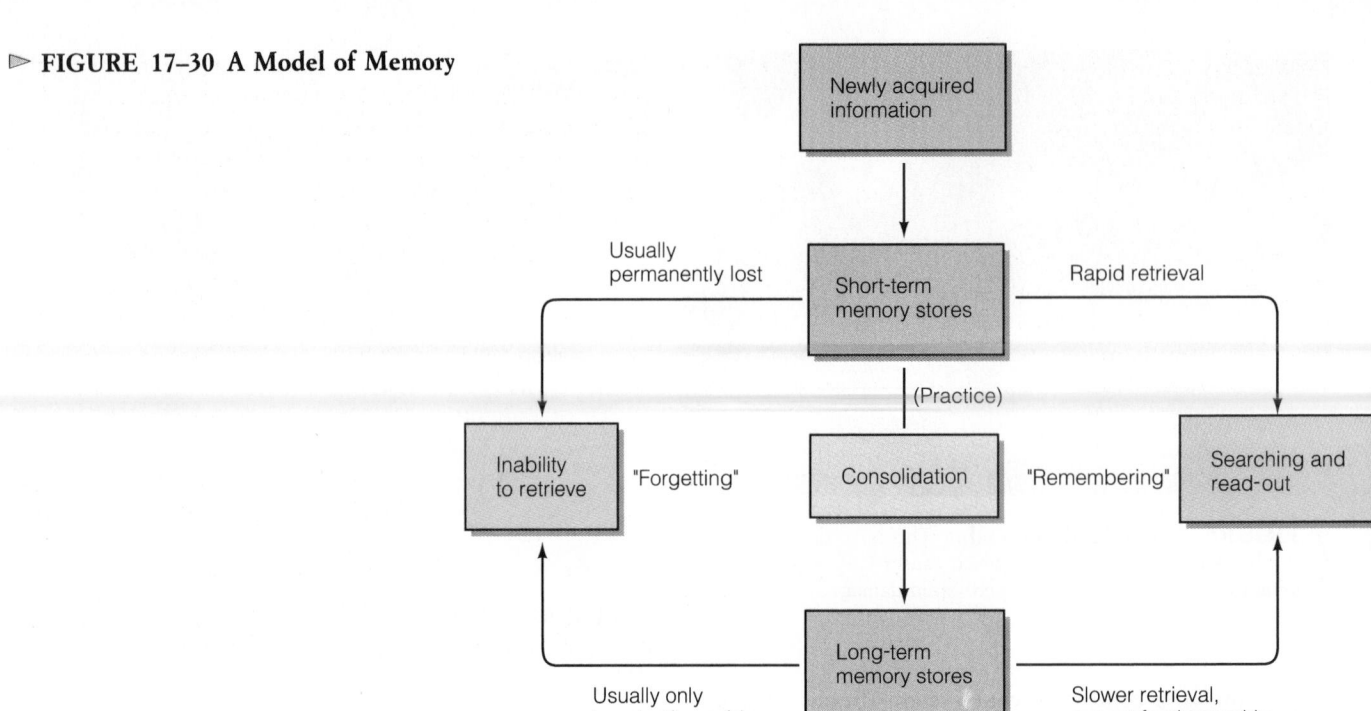

unless the long-term memory you are retrieving is one that is thoroughly ingrained (for example, your name and place of birth), it usually takes longer to recall than something in short-term memory. By the same token, information lost from short-term memory is generally gone forever. Information you cannot recall from long-term memory is often still there; it just takes awhile to remember it or some special stimulus to extract it from the cobwebs.

Memory Is Stored in Multiple Regions of the Brain

Interestingly, neurons involved in storing memories are widely distributed in the cerebral cortex (especially the temporal lobe) as well as other regions (the cerebellum and limbic system). The temporal lobes of the cerebrum and one part of the limbic system (hippocampus) appear essential for transferring short-term memories into long-term memory.

Short-Term and Long-Term Memory Appear to Involve Structural and Functional Changes of the Neurons

How are memories stored? Although no one knows how memories are stored in the brain, some evidence suggests that certain anatomical and physiological changes in the neuronal circuitry may be the basis of our memory.

Research suggests, however, that short-term and long-term memories may involve different mechanisms. Studies on slugs and snails suggest that short-term memory may involve transient modifications in the function of existing synapses. That is, synapses may become altered by experi-

ences or knowledge. Whether this mechanism holds true for humans and other animals remains to be seen.

In contrast, long-term memory involves relatively permanent structural and functional changes in the neurons of the brain. For example, studies that compared the brains of experimental animals reared in a sensory-deprived environment with those reared in a sensory-rich environment reveal marked differences in the microscopic structures of the brains of the two groups. The brains of those animals exposed to a sensory-rich environment displayed greater branching and elongation of dendrites in regions of the brain thought to serve as memory storage.

Some studies suggest that protein synthesis may be involved in long-term memory, because drugs that block protein synthesis interfere with long-term memory. What proteins are involved remains unknown.

ENVIRONMENT AND HEALTH: THE VIOLENT BRAIN

Dr. Vernon H. Mark, a neurosurgeon at the Harvard Medical School, has a special interest in the violent mind. He has studied and treated a large number of violent patients in his years of medical practice. One of the most memorable was a woman named Julie. She came to Dr. Mark at age 21, suffering from epileptic seizures and sudden, unpredictable outbursts of violence.

Julie's problems began early in life when a mumps infection spread to her brain, causing severe inflammation. At the age of 10, she began suffering from epileptic seizures. After many of her seizures, the girl would burst into fits of anger, attacking people. Depression and remorse followed her outbursts.

At age 18, Julie and her father, a physician, went to a movie together. While sitting in the theater, she felt a wave of terror overcoming her. She went to the restroom and stared at herself in the mirror, clutching a dinner knife that she carried to protect herself. Another woman in the restroom accidentally bumped into Julie, triggering her rage. Julie attacked her, plunging the knife into her chest. Her father heard the scream and raced in. He was able to administer first aid to save the woman's life.

Three years later, Julie was placed under the care of Dr. Mark. To find out where the problems were arising in her brain, he implanted a number of electrodes in her brain that would allow him to study the electrical activity of her brain during seizures. He was also able to stimulate the electrodes by remote control to study the behavioral effects of activating various parts of her brain. Stimulating most of the electrodes had little effect on her behavior. However, stimulating an electrode in a region of the limbic system called the amygdala (a-mig-da-la) had a dramatic effect on the young woman. She became unresponsive, staring vacantly in front of her, although her face turned angry. While this was occurring, the brain cells in her amygdala and hippocampus fired erratically. Dr. Mark had stimulated a full-fledged epileptic seizure. All of a sudden, Julie exploded violently.

Dr. Mark repeated the experiment later and got a similar response, confirming that he had found the focal point of Julie's seizures and rages. Over the next several months, he set out to destroy the tissue, sending radio waves along the electrode. This raised the temperature of the electrode and destroyed the cells little by little. Today, Julie's seizures and violent outbursts have ended. She lives a normal life.

This account is a remarkable testimony to the dramatic effects of brain trauma. Damage in crucial areas, no matter how slight, can upset the normal operations of the brain, resulting in erratic behavior or throwing patients into a violent rage. Dr. Richard Restak, a neurologist and author of numerous popular books on the brain, notes that alcohol and other drugs can also upset the delicate balance of the brain, creating violent outbursts. Such drugs can dismantle (at least temporarily) the brain's intricate system of checks and balances, eliciting unusual, and even violent, behavior. "It is humbling and frightening," says Restak, "to consider that our rationality is dependent on the normal function of tissue within our skulls." It is even more frightening to realize that the balance can be so easily upset. "We have the capacity, if everything is operating correctly within our brains, of composing a Bill of Rights or the Constitution," says Restak. "But in the presence of a barely measurable electrical impulse within the limbic system, our much vaunted rationality can be replaced by savage attacks and seemingly inexplicable violence."

Alcohol dampens the inhibitions that hold aggression in check in some people. Many other drugs affect levels of aggression. (For more on drugs and the brain, see Health Note 17-1). Barbiturates, for example, increase aggression. Marijuana apparently reduces aggression. Marijuana, however, is sometimes laced with PCP (phencyclidine), commonly known as angel dust. This street drug unleashes extremely violent behavior.

PCP acts through the limbic system, shutting off the inhibiting effects of the cortex on primitive rage centers. Episodes of sudden, unprovoked violence due to PCP have become common in many of our major cities, says Restak.

Perhaps one of the most shocking stories of the impacts of chemicals on the brain is that of David Garabedian, a man described by his family and friends as mild-mannered, passive, even docile. Garabedian worked for a lawn-care company in Massachusetts. On March 29, 1983, he arrived at the home of Eileen Muldoon to estimate the cost of treating her lawn with chemical insecticides. He knocked on the door but found no one at home. Garabedian decided to make an estimate and leave it for her.

When he had completed his work, however, he experienced a sudden urge to urinate. He went to the backyard and urinated near the house. Just then, Mrs. Muldoon appeared, yelling angrily at him. Garabedian became confused, apologized, and tried to explain his plight. According to his testimony, she turned away from him, refusing to listen. Garabedian tapped her on the shoulder to say he was sorry, but the woman turned on him and clawed his face with her fingernails. Garabedian exploded, grabbing the woman by the neck and strangling her. As she lay motionless on the ground, he hurled large rocks at her head, smashing her skull.

A month before the murder, Garabedian had undergone a remarkable personality change. Usually amicable, he became easily angered and abusive toward his family. He complained of tension, nervousness, impatience, and terrible nightmares. He also complained of numerous physical symptoms such as nausea, diarrhea, headaches, and frequent urination. These symptoms suggested to some that the young man was suffering from pesticide poisoning.

Dr. Peter Spencer, a toxicologist at the Albert Einstein College of Medicine in New York, testified at the trial that "David Garabedian was involuntarily intoxicated with a chemical in the lawn products that he was exposed to on a daily basis." He not only sprayed chemicals on lawns and trees but also mixed them, pouring large amounts into trucks each night in preparation for the next day's spraying.

One chemical, in particular, has been singled out as the possible culprit. It is carbaryl, a powerful inhibitor of acetylcholinesterase. Physicians believe that the carbaryl, and possibly other pesticides that Garabedian routinely handled, inhibited acetylcholinesterase in his brain. Acetylcholine then accumulated in the synaptic clefts in neurons in his limbic system, triggering uncontrollable rage.

A psychiatrist who testified at the trial said, "In my view, David did not have the capacity to control his behavior because his brain was poisoned." Despite this testimony,

THE CAUSES AND CURES OF ADDICTION

Marleen White-head lives a lie. Each morning, she kisses her husband good-bye, then drives off to work. On her way to the office, she pulls off the highway for a moment, opens a flask she keeps under the seat, and takes a drink. Throughout the day, she laces her coffee with scotch, so no one will suspect that she is drinking.

Like millions of other Americans, Whitehead's life is on the road to disaster. Her boss is aware of her drinking, and her job is at risk. What makes her, and millions of others like her who cannot get through a day without a drink or a fix from a needle, so dependent on a drug?

Many biologists think that genetics plays a key role in alcoholism. Behavioralists, however, argue that a person's upbringing and psychology create alcohol dependency. Most likely, biological and behavioral factors are both involved in alcoholism and other addictive behavior.

Let's review some of the facts. A recent study in mice showed that addictive drugs, such as alcohol and cocaine, stimulate the brain's pleasure center. The pleasure center is a region of the brain called the nucleus accumbens that, when stimulated, brings gratification.

▷ **FIGURE 1 The Pleasure Center** An electrode implanted in the pleasure center of a rat's brain can be activated when the rat presses the bar. To obtain this stimulation, the rat will forsake sex, food, and water. Addictive drugs may stimulate the pleasure center.

Stimulation of the pleasure center by drugs, researchers believe, may underlie all forms of drug addiction.

Experiments in which researchers implanted electrodes in the pleasure centers of rats showed that the rodents could be trained to press a bar to stimulate the center (▷ Figure 1). Some rats, in fact, pressed the bar hundreds of times per hour, ignoring food, water, and sex. After 15 to 20 hours of continual pressing, the rats collapsed from exhaustion. But when they awoke, they began pressing again.

Experiments with rats and mice suggest that the pleasure center is activated

Garabedian was found guilty of first-degree murder and is serving a life sentence.

Drugs, alcohol, pesticides, and industrial chemicals enter our bodies. In small amounts, they may be harmless, but in higher concentrations and in combination, they may disrupt chemical balance and cellular structure, upsetting homeostasis. Nowhere is the effect more pronounced than in the brain.

SUMMARY

OVERVIEW OF THE NERVOUS SYSTEM

1. The nervous system controls a wide range of functions in many animals and helps regulate homeostasis. The human nervous system also performs many other functions, such as ideation, planning for the future, learning, and remembering.
2. The nervous system consists of two anatomical subdivisions: the central nervous system (CNS), made up of the brain and spinal cord, and the peripheral nervous system (PNS), comprising the spinal and cranial nerves.

3. Receptors in the skin, skeletal muscles, joints, and organs, transmit sensory input to the CNS via sensory neurons. The CNS integrates all sensory input and generates appropriate responses. Motor output leaves the CNS in motor neurons.
4. The PNS has two functional divisions: the autonomic nervous system, which controls involuntary actions such as heart rate, and the somatic nervous system, which largely controls voluntary actions such as skeletal muscle contractions and provides the neural connections needed for many reflex arcs.

by cocaine and amphetamines. These drugs stimulate the production of large amounts of a neurotransmitter known as dopamine. In a recent study, researchers implanted small tubes into rats' pleasure centers and areas involved in muscle movement. The researchers administered various doses of addictive and nonaddictive drugs while the rats moved freely in their cages. The researchers then extracted fluid from the tubes to measure dopamine levels.

Drugs that are addictive to humans and rewarding to rats (amphetamine, cocaine, morphine, methadone, ethanol, and nicotine) increased dopamine concentrations in both brain areas. Levels, however, were much higher in the pleasure center. Although these results cannot be extrapolated to humans, some researchers believe that they help build a case for the theory that dopamine release in the pleasure center is a common denominator in all forms of drug addiction. Other researchers think that other neurotransmitters may also be involved in addiction.

This research helps explain the neural and biochemical basis of addiction, but what about the underlying cause? Why do some people become addicted to drugs whereas others can take them or leave them?

Research into the genetics of alcoholism has built a case for the assertion that alcoholism is largely the result of defective genes. Donald W. Goodwin of the University of Kansas Medical Center in Kansas City found that children of alcoholics had an increased risk of becoming alcoholic even when reared by adoptive (nonalcoholic) parents. This and a number of other similar studies suggest that the environment is less important than a person's genetic heritage. If one or both of your biological parents is an alcoholic, say researchers, you are much more likely to be one yourself.

Nevertheless, some researchers think that the scientific community has accepted the genetic findings too readily and uncritically. They say that the genetic theory of alcoholism may be a simplified view of the causes of the disease. Alcoholism probably results from a complex interaction of environmental factors and genetics.

New research also shows that alcoholism results from physical, personal, and social characteristics that predispose a person to drink excessively. Herbert Fingarette, a professor at the University of California at Santa Barbara who has studied alcohol addiction for years, contends that alcoholism has psychological not bi-

ological roots. Fingarette's arguments reflect the findings of many psychological studies. This research indicates that alcoholism and other addictions are more habits than diseases. Addictive behavior, the studies suggest, typically revolves around immediate gratification.

Alcohol enhances social and physical pleasure, increases sexual responsiveness and assertiveness, and reduces tension up to a point. Unfortunately, the initial physical stimulation, brought on by low doses of alcohol, can lead some people into an addictive cycle. The expectation of improved feelings drives people to drink. But higher doses of alcohol dampen arousal, sap energy, and cause hangovers. This, in turn, say psychologists, leads to a craving for alcohol's stimulating effects—that is, a craving to feel good again. The repetitive cycle of pleasure and displeasure is addiction.

The controversy over the roots of alcoholism and other addictive behaviors will undoubtedly continue for years, pitting biologists against psychologists. Although it is impossible to predict the outcome of future research, it is possible that the intermediate position will hold sway: that addiction may be explained by both genetics and psychology.

STRUCTURE AND FUNCTION OF THE NEURON

5. The fundamental unit of the nervous system is the neuron, a highly specialized cell that generates and transmits bioelectric impulses from one part of the body to another.
6. Three types of neurons are found in the body: sensory neurons, interneurons, and motor neurons.
7. All neurons have more or less spherical cell bodies. Extending from the cell body are two types of processes: dendrites, which conduct impulses to the cell body, and axons, which conduct impulses away from the cell body.
8. Many axons are covered by a layer of myelin, which increases the rate of impulse transmission. In the PNS, myelin is laid down by Schwann cells during embryonic development. In the CNS, myelin is produced by oligodendrocytes.
9. The terminal ends of axons branch profusely, forming numerous fibers that end in small knobs called terminal boutons.
10. During cellular differentiation, nerve cells lose their ability to divide. Because nerve cells cannot divide, neurons that die cannot be replaced by existing cells. Axons in the PNS may

regenerate, but axonal regeneration in the CNS is rare.
11. Nerve cells have exceptionally high metabolic rates and require a constant supply of oxygen. Because nerve cells rely exclusively on glucose for energy, decreases in blood glucose levels can have deleterious effects on the nervous system.
12. The small electrical potential across the membrane of nerve cells is known as the membrane, or resting, potential.
13. When a nerve cell is stimulated, its plasma membrane increases its permeability to sodium ions. Sodium ions rush in, causing depolarization, which spreads down the membrane.
14. Depolarization is followed by repolarization, a recovery of the resting potential stemming largely from the efflux of potassium ions. The depolarization and repolarization of the neuron's plasma membrane constitute a bioelectric impulse or action potential.
15. Nerve impulses travel along the plasma membranes of dendrites and unmyelinated axons, but in myelinated axons, impulses "jump" from node to node.
16. When a bioelectric impulse reaches the terminal bouton, it

stimulates the release of neurotransmitters contained in membrane-bound vesicles.

17. Released into synaptic clefts, neurotransmitters diffuse across the cleft, where they bind to receptors in the postsynaptic membrane. They may excite or inhibit the postsynaptic membrane. Whether a neuron fires depends on the sum of excitatory and inhibitory impulses it receives.

18. After stimulating the postsynaptic membrane, neurotransmitters may diffuse out of the synaptic cleft, be reabsorbed by the axon terminal, or be removed by enzymes.

19. Some insecticides inhibit the activity of the enzymes that deactivate neurotransmitter substance, creating a wide range of nervous system effects.

EVOLUTION OF THE NERVOUS SYSTEM

20. The nervous systems of animals exist on a continuum from the simplest to the most advanced. The simplest systems generally consist of a network of interconnected neurons, or nerve net.

21. Advanced nervous systems contain a network of nerves and a central nervous system, an aggregation of nerves where information processing and integration occur.

22. The evolution of the nervous system corresponds with the evolution of complexity in body structure.

23. Perhaps one of the most significant developments in the evolution of the nervous system was the emergence of adaptations that accelerate nerve impulse conduction—in vertebrates, saltatory conduction and in invertebrates, the presence of axons of very large diameter.

THE SPINAL CORD AND NERVES

24. The spinal cord descends from the brain through the vertebral canal to the lower back. It carries information to and from the brain, and its neurons participate in many reflexes.

25. Two types of nerves emanate from the CNS: spinal and cranial. Spinal nerves arise from the spinal cord and may be sensory, motor, or mixed. Cranial nerves attach to the brain and supply the structures of the head and several key body parts.

26. Spinal nerves are attached to the spinal cord via two roots, a dorsal root, which brings sensory information into the cord, and a ventral root, which carries motor information out. Sensory fibers entering the cord often end on interneurons, which frequently end on motor neurons in the spinal cord, forming reflex arcs. Interneurons may also send axons to the brain.

THE BRAIN

27. The brain is housed in the skull. The cerebrum with its two cerebral hemispheres is the largest part of the brain.

28. The outer layer of each hemisphere is the cortex. It contains gray matter, which houses nerve cell bodies, and underlying white matter, which contains myelinated nerve fibers that transmit nerve impulses to and from the gray matter.

29. The cerebral cortex consists of many discrete functional regions, including motor, sensory, and association areas.

30. Consciousness resides in the cerebral cortex, but a great many functions occur at the unconscious level in parts of the brain beneath the cortex.

31. The cerebellum coordinates muscle movement and controls posture. The hypothalamus regulates many homeostatic functions. The limbic system houses instincts and emotions. The brain stem, like the hypothalamus, helps regulate basic body functions.

32. A watery fluid surrounds the brain and spinal cord. Known as cerebrospinal fluid, it serves as a cushion that helps protect the brain and spinal cord from traumatic injury. CSF also fills the central canal of the spinal cord and cavities within the brain, the ventricles.

33. Electrodes applied to different parts of the scalp detect electrical activity in the brain and produce a tracing known as an electroencephalogram. The type of brain wave recorded depends on one's level of cortical activity. EEGs are used to diagnosis some brain dysfunctions.

34. Headaches are the most common form of pain and generally result from tension, swelling of the membranes lining the sinuses, eyestrain, or dilation of cerebral blood vessels.

THE AUTONOMIC NERVOUS SYSTEM

35. The autonomic nervous system (ANS) is a subdivision of the PNS and transmits sensory information from organs to the CNS and delivers motor impulses to smooth muscle, cardiac muscle, and glands.

36. The ANS innervates all internal organs and has two subdivisions: the sympathetic and the parasympathetic.

37. The sympathetic division of the ANS functions in emergencies and is responsible in large part for the fight-or-flight response, accelerating heart rate and breathing, dilating the pupils, and shunting blood to the skeletal muscles, which provides additional oxygen and nutrients needed to combat an adversary or to escape.

38. The parasympathetic division of the ANS reduces heart rate, contracts the pupils, and promotes digestion—responses associated with the relaxed state.

LEARNING AND MEMORY

39. Learning is a process in which an individual acquires knowledge and skills from experience, instruction, or some combination of the two. Learning depends on one's ability to store information in the brain—that is, memory.

40. Newly acquired knowledge is first stored in short-term memory, which retains information for periods of seconds to hours. Short-term memories may be transferred to long-term memory, which holds information for periods of days to years.

41. Memory is stored in multiple regions of the brain—in the cortex, especially the temporal lobe, as well as the cerebellum and limbic system.

42. Short-term and long-term memory appear to involve structural and functional changes of the neurons.

ENVIRONMENT AND HEALTH: THE VIOLENT BRAIN

43. Abnormal brain activity can result in bizarre behavior in humans. This is especially true if damage occurs in parts of the limbic system. Violent rage may result.

44. Damage may result from infections or harmful chemicals, such as pesticides and drugs. Alcohol can also bring on rage by eliminating normal inhibitions.

Elevated blood glucose levels in diabetics damage the kidneys and nervous system, in some instances causing renal failure and blindness. Diabetics also suffer from intestinal disorders, which researchers think may be caused by damage to the nerves that innervate the intestines.

One researcher, intent on pinpointing the cause of the intestinal problems in diabetics, decided to examine the microscopic anatomy of the paravertebral ganglia. These ganglia are clusters of nerve cell bodies along the spine that are involved in the autonomic nervous system, which as you learned in this chapter controls the function of many organs in the body, including the intestines and urinary bladder.

The researcher first examined ganglia in older diabetic rats. In his studies, he found that nerve endings (terminal boutons) in the ganglia were quite swollen, some of them 30 times their normal size. This observation led him to hypothesize that diabetes was causing the swelling, which was due to the accumulation of vesicles containing neurotransmitter.

To test this hypothesis, the researcher examined control groups composed of nondiabetic rats of the same age that had been kept under similar conditions. To his surprise, many of the nerve terminals in the nondiabetic rats' ganglia were also swollen. (This example illustrates how important a control group is to good science.)

The researcher concluded that some other factor was responsible for the swelling. From the little bit of detail given so far, can you come up with any ideas? (Read back over the material for a hint . . . it's there.)

If you said age, you are right. After thinking about the commonalities of the two groups, the researcher realized that all of his animals were old. He hypothesized that age might have been responsible for the swelling. If you were involved in the study, how would you test this new hypothesis?

If you said that you would compare the ganglia of young and old nondiabetic rats, you're right again. That's exactly what he did. When the researcher made the comparison, he found no evidence of nerve terminal swelling in the young rats.

Like other good researchers, intent on solving one problem, this one found a partial answer to another: aging. As the body ages, the autonomic nervous system can malfunction, causing a wide range of problems in elderly people, including loss of bladder control and fainting when standing too quickly.

The researcher in this example decided to see if this age-related swelling occurred in humans as well. He collected ganglia from 56 people aged 15 to 93 who had died from a variety of causes. In people under 60, there were few swollen nerve terminals, but after age 60 the number of swollen nerve terminals increased dramatically.

Although it is true that the researcher was unsuccessful in finding answers to the original problem and that much research is still needed to determine why the swollen nerve terminals exist in older animals, this case study is a good example of how using critical thinking skills and good experimental methods can lead to important results.

TEST OF TERMS

1. The brain and spinal cord are part of the _____ nervous system. The nerves belong to the _____ nervous system.

2. Heartbeat and breathing are largely controlled by the _____ nervous system.

3. Many neurons contain short, branching fibers called _____, which conduct impulses to the cell body, and a long, unbranched fiber, the _____, which carries impulses away from the cell body.

4. Many axons of the PNS have a coating, the _____, which is laid down by Schwann cells and accelerates nerve cell transmission.

5. The swellings at the ends of axons are called _____ _____. They release chemical substances called _____, which stimulate a bioelectric impulse in adjoining nerve cells.

6. The voltage measured across the plasma membrane of an inactive neuron is called the _____ _____ and is about _____ millivolts.

7. Stimulating the plasma membrane of a nerve cell results in a drastic change in the membrane's permeability to _____ _____. The change in the resting potential that occurs when the membrane is stimulated is called a(n) _____ _____.

8. When a nerve impulse reaches a terminal bouton, depolarization causes the influx of calcium ions, which stimulates the release of a neurotransmitter, which is stored in vesicles, into the _____ _____. Neurotransmitter substances bind to receptors on the _____ _____, stimulating either

the uptake of sodium or chloride ions, depending on the substance.

9. _____ is the main excitatory neurotransmitter in many synapses. Certain insecticides block _____, an enzyme that removes this transmitter from the synapse.

10. Nerves attached to the spinal cord are called _____ _____. Each of these has two roots. Motor fibers leave via the _____ root.

11. The neuron that transmits the impulse from the sensory neuron to the motor neuron in the spinal cord is called a(n) _____.

12. The portion of the cerebral cortex that controls voluntary muscle movement is the _____ _____ cortex.

13. The ridges in the cerebral cortex are called _____. The grooves between them are known as _____.

14. Sensory fibers carrying information from all over the body terminate in a part of the cerebral cortex called the _____ _____ cortex, which is located just behind the _____ _____ .

15. The _____ are three fibrous layers surrounding the brain, the _____ _____ , arachnoid layer, and _____ _____ .

16. The regions of the cortex that integrate incoming information are called the _____ cortex.

17. The _____ system has many functions. It is the site of the primitive rage response.

18. The _____ is in charge of coordinating skeletal muscle activity. Problems in this region of the brain cause spasticity.

19. Just above the pituitary and attached to it is the region of the brain called the _____ . It controls eating behavior and monitors the composition of the blood. Aggregations of nerve cell bodies in this region are called _____ .

20. The _____ _____ system is found in the medulla oblongata. It receives all incoming and outgoing information and sends impulses to the cerebral cortex, keeping it active.

21. The brain and spinal cord are bathed in a fluid, known as _____ fluid, which also fills the central canal of the spinal cord and the hollow cavities inside the brain, called the _____ .

22. Electrical activity of the neurons of the cerebral cortex can be picked up by electrodes placed on the scalp. A tracing of the electrical activity is known as an _____ , or _____ .

23. The _____ division of the autonomic nervous system increases the heart rate and breathing.

24. The _____ division of the autonomic nervous system brings about internal responses generally associated with the relaxed state.

25. Newly acquired facts are first stored in _____ - _____ memory, which holds information for periods of _____ to _____ .

Answers to the Test of Terms are found in Appendix B.

TEST OF CONCEPTS

1. The nervous system performs a great many functions. Describe them. What functions do you think are unique to humans?

2. Describe how the resting and action potentials are generated. Describe how the plasma membrane of a neuron is repolarized.

3. Draw a typical multipolar neuron, and label its parts. Show the direction in which an action potential travels.

4. Draw a typical synapse, label the parts, and explain how a nerve impulse is transmitted from one nerve cell to another.

5. Describe how each of the following substances affects synaptic transmission: caffeine, cocaine, and malathion.

6. In general, how do anesthetics function?

7. Describe the progression of the evolution of the nervous systems of animals.

8. Name and describe the various divisions of the nervous system.

9. Draw a cross section through the spinal cord showing the spinal nerves. Label the parts, and explain a reflex arc and how nerve impulses entering a spinal nerve also travel to the brain.

10. A physician can stimulate various parts of the brain and get different responses. What effects would you expect if the electrodes were placed in the premotor area? The primary motor cortex? The sensory motor cortex?

11. The brain is a delicate organ, and slight shifts in electrical activity can create bizarre behavior. Do you agree with this statement? Give examples.

The Senses

≈

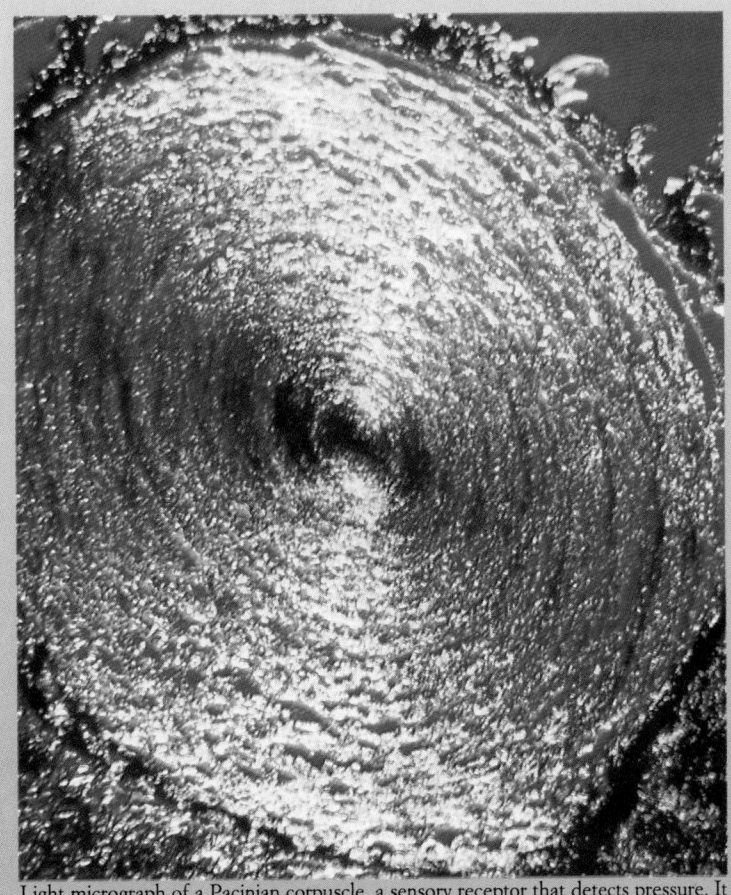

Light micrograph of a Pacinian corpuscle, a sensory receptor that detects pressure. It is found in the skin and other locations.

ebra Cartwright noticed something strange one day while she was eating dinner. For no apparent reason, the college sophomore had lost her senses of smell and taste. Doctors were puzzled at first by the finding, for Cartwright seemed to be in fine health. She was not suffering from a cold, sinus infection, or even any allergies that might have blocked her nasal passages and impaired her ability to smell and taste. When she had a blood test, though, they found the cause of her problem: low blood levels of the micronutrient zinc. Her physician prescribed a zinc supplement, and in a few days her senses of smell and taste returned.

This story illustrates the importance of a nutritionally balanced diet and the impact of a dietary deficiency on two important body functions—the senses of taste and smell.[1] It is just one of many examples presented in this book that illustrates how body functions can be altered by external factors such as stress, pollution, and diet.

This chapter examines animal senses, dividing the discussion into two broad categories: the general senses and the special senses. The chapter focuses primarily on human sensory structures. As a result, your studies will yield some information useful throughout your life and, incidentally, will explain the reason why a zinc deficiency eliminates the sense of taste. Finally, this chapter also describes sensory adaptations of other animal species that permit them to hear, to see, to smell, and to taste.

≈ THE GENERAL AND SPECIAL SENSES

Staying aware of one's environment and responding to changes in that environment require a system of surveillance not unlike that found at high-security weapons facil-

ities, where both inside and outside activities are routinely monitored to detect intruders. The surveillance system of the human body, like that of many other animals, consists of numerous receptors that detect internal and external changes. In humans, these receptors are found in the skin, internal organs, bones, joints, and muscles. They detect stimuli that give rise to the **general senses:** pain, temperature, light touch, pressure, and a sense of body and limb position (Table 18–1).

The human body is endowed with five additional senses, known as the **special senses.** They are taste, smell, vision, hearing, and balance and are generally made possible by special sensory organs (the eyes, for example). These sophisticated detectors greatly increase our ability to perceive stimuli in the environment and are one of the most remarkable developments in the evolution of complex multicellular organisms.

Receptors in the animal kingdom involved in the general and special senses fall into five functional categories: (1) mechanoreceptors, (2) chemoreceptors, (3) thermoreceptors, (4) photoreceptors, and (4) nociceptors (pain receptors). **Mechanoreceptors** are those that are activated by mechanical stimulation—for example, touch or pressure. **Chemoreceptors** are activated by chemicals in the food we eat, the air we breathe, or levels of certain chemicals (for example, hydrogen ions and oxygen) in our blood. **Thermoreceptors** are activated by heat and cold, and **photoreceptors** are sensitive to light. **Nociceptors** are stimulated by tissue damage, such as that caused by pinching, tearing, or burning. Interestingly, intense stimulation of all the other receptors also gives rise to the sensation of pain. A much rarer sense receptor, which detects minute electrical signals, is discussed in Spotlight on Evolution 18-1.

≈ THE GENERAL SENSES

Sit back in your chair for a moment, close your eyes, and concentrate on what you feel. A cold draft may be stirring

[1] Zinc deficiencies are quite rare, and dietary supplements containing zinc are generally not needed if you are eating a well-balanced diet. Furthermore, dietary supplements can be dangerous. Just a few times the recommended daily allowance of zinc can cause serious problems.

≈	TABLE 18–1 Summary of General and Special Senses		≈

SENSE	STIMULUS	RECEPTOR	
General senses	Pain	Naked nerve endings	
	Light touch	Merkel's discs; naked nerve endings around hair follicles; Meissner's corpuscles; Ruffini's corpuscles*; Krause's end-bulbs*	
	Pressure	Pacinian corpuscles	
	Temperature	Naked nerve endings	
	Proprioception	Golgi tendon organs; muscle spindles; receptors similar to Meissner's corpuscles in joints	
Special senses	Taste	Taste buds	
	Smell	Olfactory epithelium	
	Sight	Retina	
	Hearing	Organ of Corti	
	Balance	Crista ampularis in the semicircular canals; maculae in utricle and saccule	

*These may both be alternate forms of the Meissner's corpuscle and may be stimulated by light touch.

THE PLATYPUS: NATURE'S JOKE OR MARVEL OF EVOLUTION?

In 1798, scientists at the British Museum in London received an odd-looking specimen from Australia. About the size of a cat, with the bill of a duck and the tail and fur of a beaver, this web-footed creature had to be a fake, a clever joke sewn together from the parts of other animals (▷ Figure 1). Certain that their irreverent Australian colonials had foisted a joke on them, the British scientists proceeded to hunt for the sutures that held this bizarre animal together. After a painstaking search, however, they found none. Today that same specimen remains in the museum, sporting scalpel marks made by the skeptics searching for elusive stitches.

The odd-looking animal is the duck-billed platypus, an egg-laying mammal that has only recently begun to be studied, a process made difficult by the creature's lifestyle. Although it is fairly common in eastern Australia and Tasmania (an island south of Australia), the duck-billed platypus feeds underwater and only at night, spending most of its life curled up inside its underground burrow. Wary of humans, it suffers extreme stress in captivity. Too much stress and the animal dies. Accordingly, getting the animal to reproduce in captivity has been nearly impossible.

Despite its elusive nature, researchers are beginning to discover the unique adaptations that promote its survival and reproduction in the wild. One of them is a dual sensory system—one that functions on land and another that functions in water. On land, the animal depends on its eyes, ears, and nose, but when the creature dives underwater, these sense organs clamp shut. How does the platypus steer its course underwater?

In 1986, researchers found that its snout is an electronic receiver that picks up minute electrical signals. In laboratory studies, the animal responded to extremely small electrical fields as weak as those created by a shrimp flicking its tail.

In 1988, another team of researchers found minute pores on the bill's side containing nerve endings, which are the actual detectors of electrical fields. Small glands release a fluid that keeps the pores from drying out. Underwater, the animal scans back and forth with its bill, searching for prey. The bill is also extremely sensitive to touch and may help the animal feel its way around its aquatic home.

Besides these remarkable sensory adaptations, the platypus exhibits other characteristics helpful to its survival. Unlike most aquatic mammals, which propel themselves with webbed hind feet and tails, the platypus relies principally on its front legs and feet for propulsion. With its oversized front feet pulling it along, the platypus uses its tail as a rudder. Unlike the tail of a beaver, which is also used for propulsion, the platypus tail is filled with fat. Thus, the tail also serves as a fat depot for reserve energy, and at night it serves as a blanket, for when the creature curls up to sleep, it wraps its tail around its body.

▷ **FIGURE 1 The Duck-Billed Platypus**

Despite its shyness, a trapped platypus is a force to be reckoned with. Its sharp hind claws deliver a painful scratch, worsened by the release of a strong venom. Hunters who have tangled with the creature report suffering severe pain, swelling, and weeks of partial paralysis.

After laying their eggs, female platypuses remain sheltered for several months in their burrows, leaving only occasionally to feed. Taking refuge in their burrows, which they seal with mud, is thought to be an evolved behavioral strategy, which protects their offspring from predators. It does, however, pose a problem—notably, a lack of oxygen. Here too the platypus exhibits an adaptation vital to its survival. Like prairie dogs, which spend much of their time in low-oxygen burrows, the platypus has a high concentration of red blood cells.

The duck-billed platypus has confused more than British scientists over the years. The aborigines of Australia, in fact, believe that the first platypus resulted from the rape of a duck by a water rat. Early taxonomists thought that it might be a primitive mammal linking humans to reptiles. Today, the platypus is assigned to an order known as *Monotremata* which means "one hole"—referring to the single orifice used for excreting waste and reproduction. This order consists of only two other members, both spiny anteaters (one found in Australia and the other in New Guinea).

Writer Eric Hoffman notes that the platypus seems to fare well because of its adaptations. "Ultimately," he writes, "the animal that once seemed to be nature's joke could, with its extraordinary mix of specialized adaptations, prove to be nature's last laugh."

SOURCE Adapted from Eric Hoffman, "Paradoxes of the Platypus," *International Wildlife* (1990) 20(1): 18–21.

at your ankles. You may feel the pressure of the chair on your buttocks and warmth emanating from a nearby reading lamp. You may feel your cat brushing against the hairs on your arm. You may feel a slight pain from a bruise or gas pressure in your intestines from the burritos you ate for lunch. Now move your arm. Even though your eyes are closed, you can feel it moving.

The various sensations you have just experienced fall

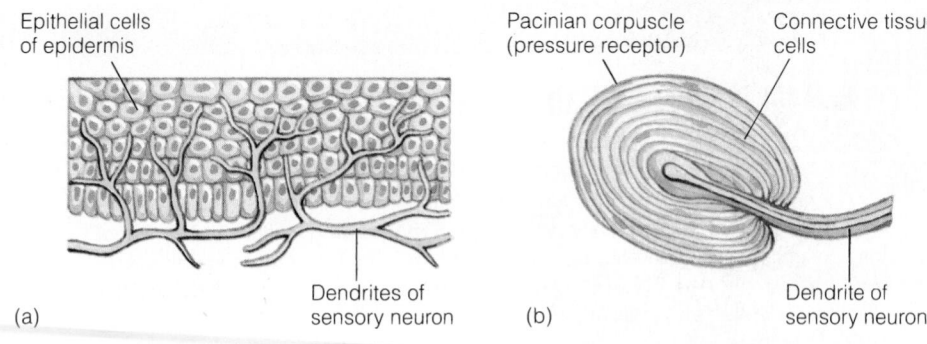

FIGURE 18–1 Receptors General sense receptors are either (*a*) naked nerve endings or (*b*) encapsulated nerve endings.

Epithelial cells of epidermis

Dendrites of sensory neuron

(a)

Pacinian corpuscle (pressure receptor)

Connective tissue cells

Dendrite of sensory neuron

(b)

into the group of general senses. Receptors for the general senses detect internal and external stimuli and relay messages to the spinal cord and brain via sensory nerves (Chapter 17). Sensory input to the central nervous system may elicit a conscious response; for example, the touch of a cat may cause you to reach down and pat your furry friend. Some stimuli will cause unconscious responses; for example, heat may cause you to perspire, or cold may cause you to shiver. Others may simply be registered in the cerebral cortex, making you aware of the stimulus. Still other stimuli may elicit no response at all.

Receptors for the general body senses come in many shapes and sizes, but they generally fit into one of two groups based on structure: naked nerve endings or encapsulated receptors, nerve endings associated with cells or groups of cells (▷ Figure 18–1). As you shall soon see, the naked nerve endings and encapsulated receptors in the body serve very specific roles. Some are thermoreceptors; others are mechanoreceptors, and so on.

Naked Nerve Endings in Body Tissues Detect Pain, Temperature, and Light Touch

Naked nerve endings are the terminal ends of the dendrites of sensory neurons. Located in the skin, bones, and internal organs and in and around joints, they are responsible for at least three sensations: pain, temperature, and light touch. Therefore, naked nerve endings may be nociceptors, thermoreceptors, or mechanoreceptors.

Pain. Two types of pain are experienced: somatic pain and visceral pain. **Somatic pain** results when receptors in the skin, joints, muscles, and tendons are stimulated. These receptors are activated when tissues are damaged and are part of a protective mechanism of considerable adaptive value in the evolution of complex multicellular organisms. They signal that something is awry and alert the brain to elicit some action.

Somatic pain receptors respond to several types of stimuli when tissue is injured. Some respond to cutting, crushing, and pinching. Others respond to temperature extremes. Still others respond to irritating chemicals released from injured tissues.

Visceral pain results from the stimulation of naked nerve endings in body organs (the viscera). Pain receptors in body organs are stimulated by distension in some and by **anoxia,** a lack of oxygen, in others. For example, intestinal pain you feel when you have a gas buildup results from the stretching of naked nerve fibers in the wall of the intestine. These nerve endings are mechanoreceptors. The pain felt

▷ **FIGURE 18–2 Referred Pain** Visceral pain is often felt on the body surface at the points indicated by the colored areas.

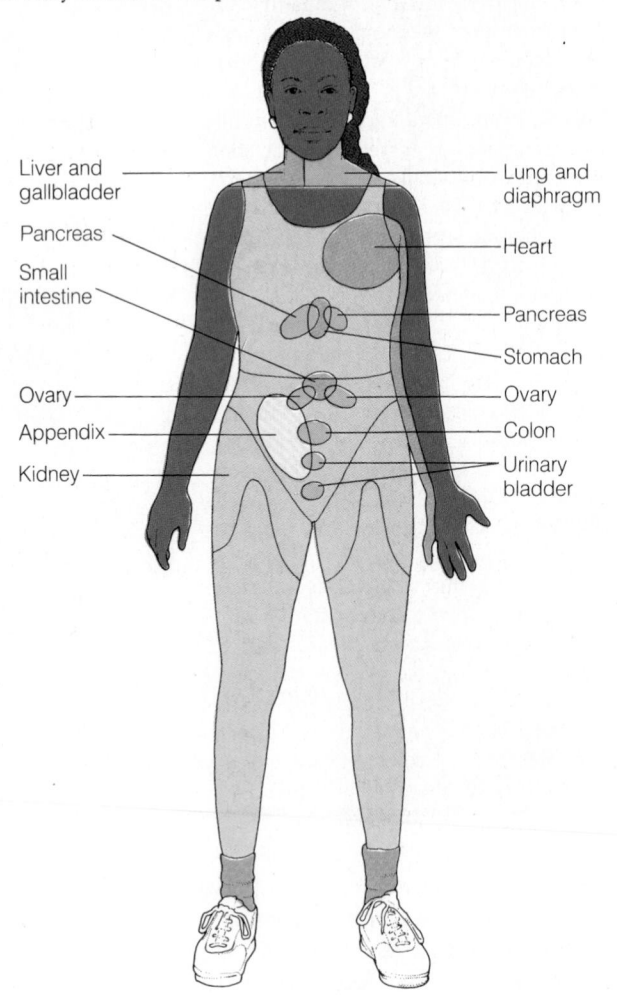

Liver and gallbladder

Pancreas

Small intestine

Ovary

Appendix

Kidney

Lung and diaphragm

Heart

Pancreas

Stomach

Ovary

Colon

Urinary bladder

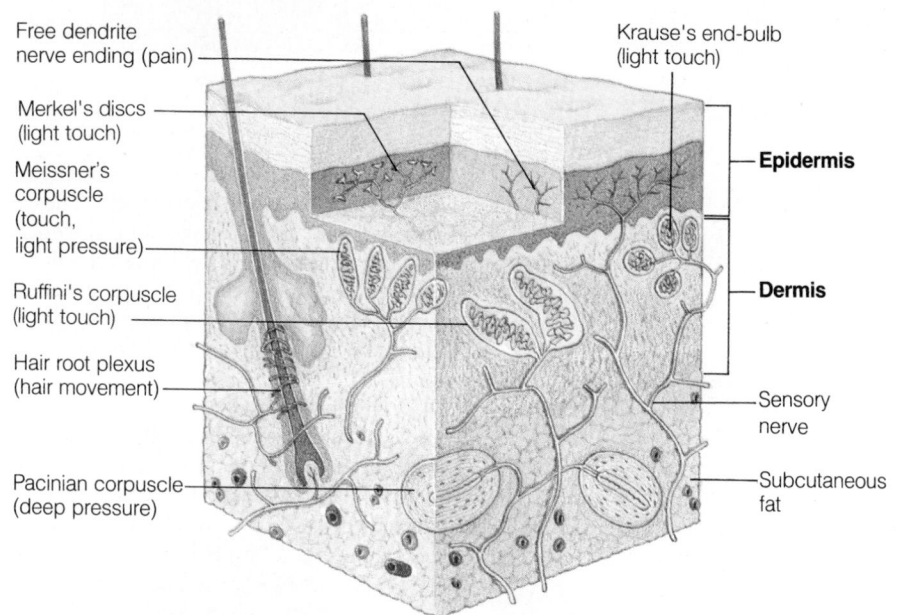

Free dendrite nerve ending (pain)

Merkel's discs (light touch)

Meissner's corpuscle (touch, light pressure)

Ruffini's corpuscle (light touch)

Hair root plexus (hair movement)

Pacinian corpuscle (deep pressure)

Krause's end-bulb (light touch)

Epidermis

Dermis

Sensory nerve

Subcutaneous fat

▷ **FIGURE 18–3 General Sense Receptors** The skin houses many of the receptors for general senses. Receptors fall into two categories: naked nerve endings and encapsulated receptors, shown here.

during a heart attack results from a lack of oxygen flowing to the heart muscle (Chapter 12).

Visceral pain and somatic pain result from quite different stimuli and are perceived very differently as well. Somatic pain is easily identified, whereas visceral pain is vague and difficult to localize. Visceral pain is generally felt on the body surface at a site some distance from its origination. For example, anginal pain, caused by a lack of oxygen to the heart muscle, appears in the chest and along the inside of the left arm (▷ Figure 18–2). Pain from the lung and diaphragm appears in the neck.

Pain that appears on the body surface away from the location of the pain is called **referred pain.** Physiologists do not know the cause of this phenomenon, but many think that it results from the fact that pain fibers from internal organs enter the spinal cord at the same location that the sensory fibers from the skin enter. The brain, they hypothesize, interprets the impulses from pain fibers supplying the organs as pain from a somatic source. (See Health Note 18–1 for some techniques to relieve pain.)

Light Touch. Light touch is perceived by two anatomically distinct mechanoreceptors. The first receptor is located at the base of the hairs in our skin. As shown in ▷ Figure 18–3, naked nerve endings (dendrites) wrap around the base of the hair follicles. When a hair is moved—for example, by a gentle touch—these nerve fibers are stimulated. The second light-touch mechanoreceptor is the **Merkel's disc.** Shown in Figure 18–3, Merkel's discs consist of small cup-shaped cells on which naked nerve endings terminate. These receptors, located in the outer layer of the epidermis of the skin, are activated by gentle pressure on the skin.

Temperature. Naked nerve endings in the skin detect heat and cold. Heat receptors respond to temperatures

from 25°C (77°F) to 45°C (113°F). If the temperature of the skin rises above this range, pain receptors are activated, creating a burning sensation. In contrast, cold receptors respond to temperatures from 10°C (50°F) to 20°C (68°F). If the temperature drops below 10°C, pain receptors respond.

Encapsulated Receptors Contain a Naked Nerve Ending Surrounded by One or More Layers of Cells that Form the Capsule

Shown in Figure 18–3, the largest encapsulated nerve ending is the **Pacinian corpuscle.** It consists of a naked nerve ending surrounded by numerous concentric cell layers. The Pacinian corpuscle (▷ Figure 18–4), resembles a small

▷ **FIGURE 18–4 Pacinian Corpuscle** This receptor, often located in the dermis of the skin, detects pressure.

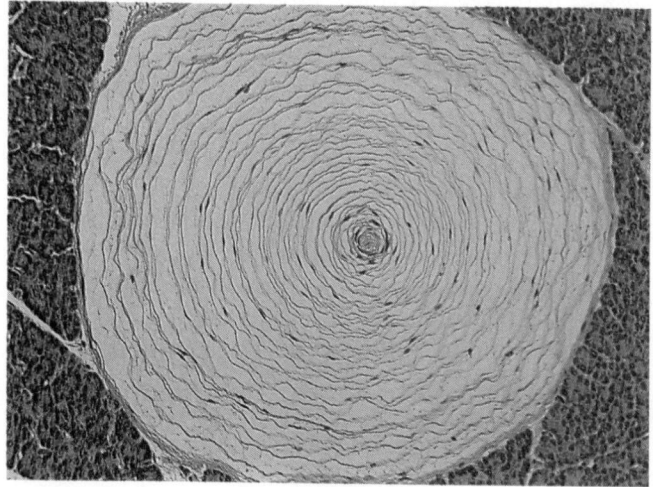

OLD AND NEW TREATMENTS FOR PAIN

Millions of people suffer from chronic or persistent pain. In the United States, for example, an estimated 70 million people are tormented by back pain. Another 20 million suffer from migraine headaches.

Despite its prevalence, pain is one of the least understood medical problems in the world. The study of pain, in fact, is so poorly funded and so widely ignored by the medical community that some have called it an orphan science. Making matters worse, physicians are often poorly trained in dealing with pain.

For many decades, pain was treated with painkillers such as morphine and codeine. Their addictive nature led researchers to look for other techniques. One of the more promising was acupuncture, a technique used by the Chinese for thousands of years. Acupuncture relies on thread-thin needles inserted in the skin near nerves and rotated very quickly (▷ Figure 1).

Neurologists are not certain how acupuncture works, but many think that it blocks pain by overloading the neuronal circuitry. They note that two types of nerve fibers transmit sensory information from the body to the central ner-

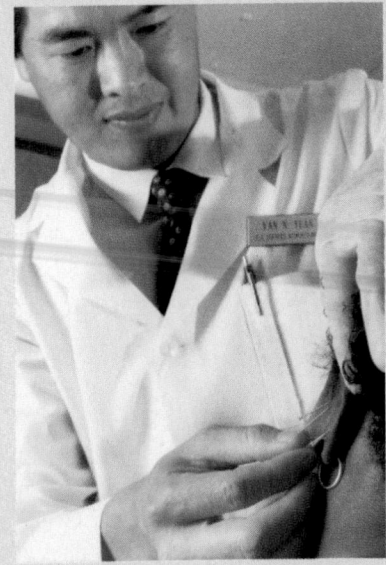

▷ **FIGURE 1 Patient Undergoing Acupuncture** Acupuncture is generally used to alleviate pain, but it has other applications as well. This patient is being treated in an experimental program to combat drug addiction.

vous system: small- and large-diameter fibers. Small-diameter fibers carry pain messages. Large-diameter fibers carry many other forms of sensory informa-

tion from receptors in the skin—for example, pressure and light touch. The dendrites of both small- and large-diameter sensory nerve cells often terminate in the same location in the spinal cord. From here they send impulses to the brain, signaling pain or some other sense.

Neurobiologists believe that acupuncture needles stimulate the large-diameter nerve fibers. This stimulation blocks nerve impulses carried by the smaller nerve fibers, thus blocking pain messages to the brain. Physicians who are trained to use drugs and surgery to solve most pain remain skeptical about the usefulness of acupuncture. Over the past 10 years, however, a small but steady stream of research has confirmed the painkilling effect of this treatment.

Joseph Helms, a physician with the American Academy of Acupuncture in Berkeley, California, performed acupuncture on 40 women with menstrual pain. Some women received real acupuncture treatment. Others received placebo treatments (shallow needle treatments that did not reach the acupuncture points). In the group of women receiving acupuncture, 10 out of 11 showed a marked decrease in pain.

---

onion pierced by a thin wire. Located in the deeper layers of the skin, in the loose connective tissue of the body, and elsewhere, Pacinian corpuscles are stimulated by pressure, such as the pressure you feel sitting in your chair.

Another common encapsulated sensory receptor is the Meissner's corpuscle. **Meissner's corpuscles** are smaller than Pacinian corpuscles. These oval receptors contain two or three spiraling dendritic ends surrounded by a thin cellular capsule (Figures 18–3 and ▷ 18–5). Like the naked nerve endings surrounding the base of the hair follicle and Merkel's discs, Meissner's corpuscles are thought to be mechanoreceptors that respond to light touch. Located just beneath the epithelium in the outermost layer of the dermis, Meissner's corpuscles are most abundant in the sensitive parts of the body, such as the lips and the tips of the fingers.

Two of the more controversial encapsulated receptors are Krause's end-bulbs and Ruffini's corpuscles (see Figure 18–3). Most scientists believe that these receptors are ac-

▷ **FIGURE 18–5 Meissner's Corpuscle** This receptor, found just beneath the epidermis, detects light touch.

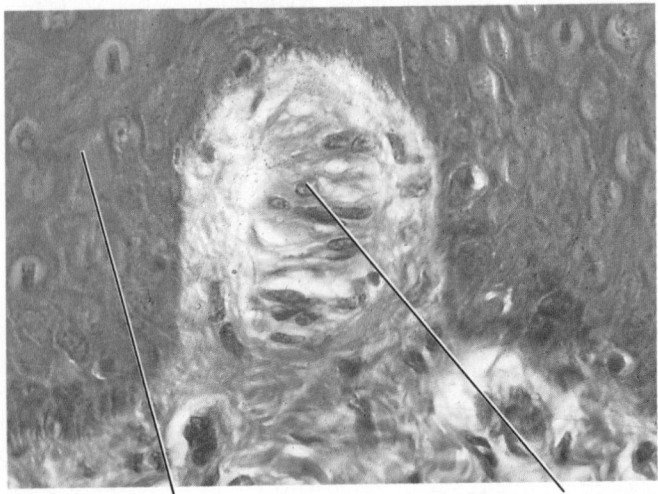

Epidermis   Meissner's corpuscle

Patients reported an approximately 50% decrease in pain. In the placebo group only 4 out of 11 reported a lessening of pain. Only 1 of 10 people given no treatment showed improvement. Acupuncture also reduced the need for painkilling drugs during treatment by over half. Remember, however, that the Critical Thinking section in Chapter 1 suggested that experiments using small numbers of subjects are themselves subject to question. Further studies are needed to confirm these results.

While acupuncture is slowly earning a respected place in the treatment of pain, researchers are also experimenting with a technique called transcutaneous electrical nerve stimulation, or TENS. Patients are fitted with electrodes that attach to the skin above the nerves that transmit pain signals to the central nervous system. A small battery supplies energy that stimulates the electrodes. When the pain begins, patients press a button on the battery pack that sends a tiny current to the electrode. The current is conducted through the skin and blocks the pain impulses.

TENS can be used to reduce pain after surgery. One study showed that this technique reduced the amount of painkillers that doctors had to administer by two-thirds and cut hospital stays by one or two days. Someday, dentists may use TENS instead of novocaine. TENS may also be used to reduce the pain of childbirth and could help treat the pain that athletes suffer.

TENS works like acupuncture, although part of its success may be psychological. In a study of 93 patients with chronic pain, researchers found that over one-third (36%) of them reported no pain or greatly reduced pain when they thought they were being stimulated but really were not. How effective was it when the battery really worked? About half of the patients reported no pain or greatly reduced pain. The slight difference suggests to some physicians that TENS may be overrated. To those suffering from chronic pain, it can be a godsend.

Severe pain can also be treated by surgery. Doctors may cut nerves or destroy small parts of the brain to get rid of chronic pain. Unfortunately, pain recurs in 9 of 10 patients who have undergone surgery, usually within a year or so. Recurring pain results from the partial regrowth of axons. Even after another operation, the pain frequently returns, often with much greater intensity. Consequently, physicians are now looking for alternative measures to eliminate chronic pain.

One promising measure is deep brain stimulation. Electrodes can be implanted in parts of the brain and stimulated to block pain impulses before they reach the sensory cortex, where pain is perceived. The electrodes are connected to a portable battery worn on the belt or implanted under the skin. When the pain begins, the patient turns on the current, blocking the pain impulses.

Research shows that deep brain stimulation is an effective blocker of even the most powerful pain stimuli. Yet it does not upset other brain functions. Unfortunately, this technique requires surgery, and implanting electrodes may cause hemorrhaging in some patients, which can result in permanent paralysis or a loss of feeling in parts of the body. Infections may also develop where the electrodes enter the skull. The 75% success rate and the reduced suffering make most patients more than willing to accept the risks.

---

tually structural variations of the Meissner's corpuscle and are therefore stimulated by light touch.

Proprioception (sense of position) is provided by special encapsulated receptors located in the joints of the body. Resembling Meissner's corpuscles, these receptors inform us of the position of our limbs and alert us to movements of the body. Proprioception is also served by the muscle spindle and the Golgi tendon organ ($\triangleright$ Figure 18–6). **Muscle spindles,** or **neuromuscular spindles,** are found in the skeletal muscles of the body. A muscle spindle consists of several modified muscle fibers with sensory nerve endings wrapped around them; a thin capsule of connective tissue surrounds the entire structure. When a muscle is extended or stretched, spindle fibers are stimulated. Nerve impulses generated by the spindle are then transmitted to the spinal cord via sensory nerves. Impulses reaching the spinal cord may ascend to the cerebral cortex, helping us remain aware of body position. The nerve impulses generated by muscle spindles (caused by stretching the muscle) may also stimulate motor neurons in the spinal cord, eliciting a reflex contraction of the muscle that is being stretched. Contraction of the muscle counteracts the stretching.

Golgi tendon organs are mechanoreceptors that are functionally similar to the muscle spindle but are located in **tendons**—the structures that connect muscles to bones. Also known as **neurotendinous organs,** Golgi tendon organs are composed of connective tissue fibers surrounded by dendrites and encased in a capsule (Figure 18–6). When a muscle contracts, the tendon stretches and stimulates the receptor. Like the muscle spindle, this receptor alerts the brain to movement and body position. Impulses from the Golgi tendon organ can also stimulate reflex contraction of muscles. The knee-jerk reflex described in the last chapter is a good example.

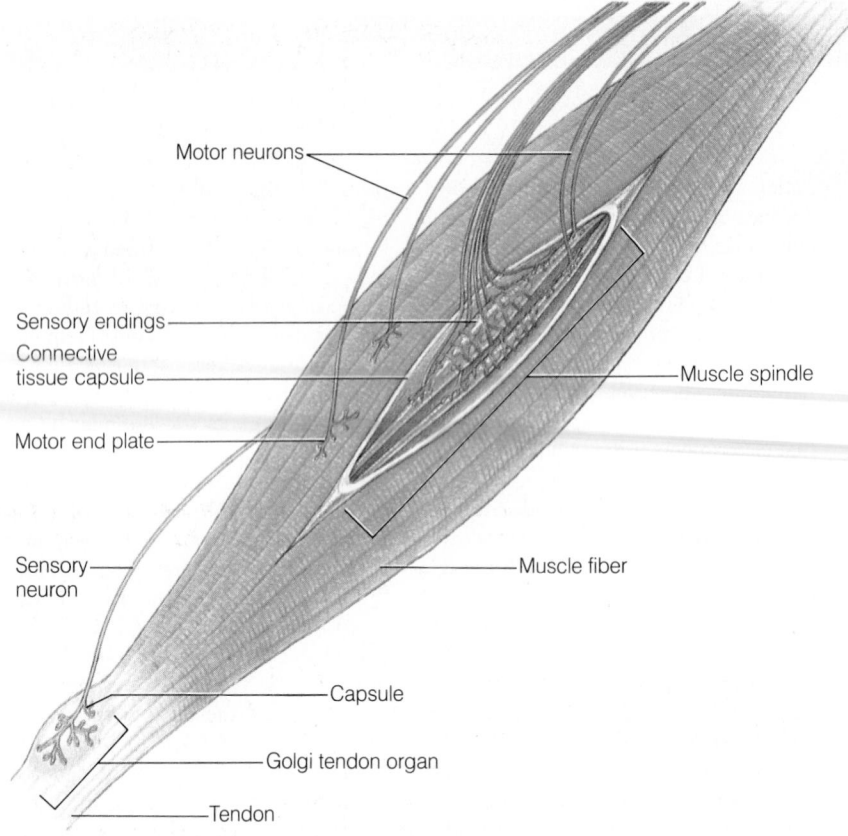

▷ **FIGURE 18–6 Stretch Receptors** The muscle spindle and Golgi tendon organ both detect muscle stretch. The dendrites of the sensory neurons from these receptors send impulses to the spinal cord. These impulses may trigger motor neurons to fire; returning signals cause the muscle to contract. Sensory signals reaching the spinal cord also ascend to the brain, alerting it to the muscle movement.

Motor neurons

Sensory endings

Connective tissue capsule

Motor end plate

Sensory neuron

Muscle spindle

Muscle fiber

Capsule

Golgi tendon organ

Tendon

Many Receptors Stop Generating Impulses After Exposure to a Stimulus For Some Length of Time

Pain, temperature, and pressure receptors are all subject to a phenomenon known as sensory adaptation.[2] **Adaptation** occurs when sensory receptors stop generating impulses even though the stimulus is still present. You have probably witnessed the phenomenon dozens of times in your lifetime. Recall, for example, the first time you wore a ring or contact lenses. At first, the sensation may have nearly driven you mad, but after a few days or perhaps a few hours, the discomfort waned, and the stimulus seemed to have disappeared. Pressure receptors that were originally alerting the brain stopped generating impulses, relieving you of what would have otherwise been unrelenting discomfort.

Interestingly, not all receptors adapt. Muscle stretch receptors and joint proprioceptors are two examples. Because the CNS must be continuously apprised of muscle length and joint position to maintain posture, adaptation of these receptors would be counterproductive.

Receptors Play an Important Role in Homeostasis

Many general sense receptors play an important role in homeostasis and no doubt evolved because of the selective

[2] Not to be confused with the adaptation occurring in evolution.

advantage they conferred on organisms. Mechanoreceptors that detect changes in blood pressure and chemoreceptors that respond to the ionic concentration of the blood are important in maintaining proper water balance and blood pressure. These receptors help regulate blood volume and blood concentration through mechanisms involving the kidney. Chemoreceptors that detect levels of carbon dioxide and hydrogen ions in the blood and cerebrospinal fluid also help regulate respiration, a function vital to normal body function. In Chapter 20, you will study additional chemoreceptors that detect levels of various hormones, nutrients, and ions in the blood and body fluids. These detectors may stimulate hormonal responses that correct potentially disruptive chemical imbalances.

≈ TASTE AND SMELL: THE CHEMICAL SENSES

The special senses are taste, smell, vision, hearing, and balance. The following sections discuss the special senses, beginning with taste.

Taste Buds Are the Receptors for Taste and Respond to Chemicals Dissolved in the Food We Eat

In humans and other mammals, the tongue contains receptors for taste. Known as the **taste buds,** these microscopic,

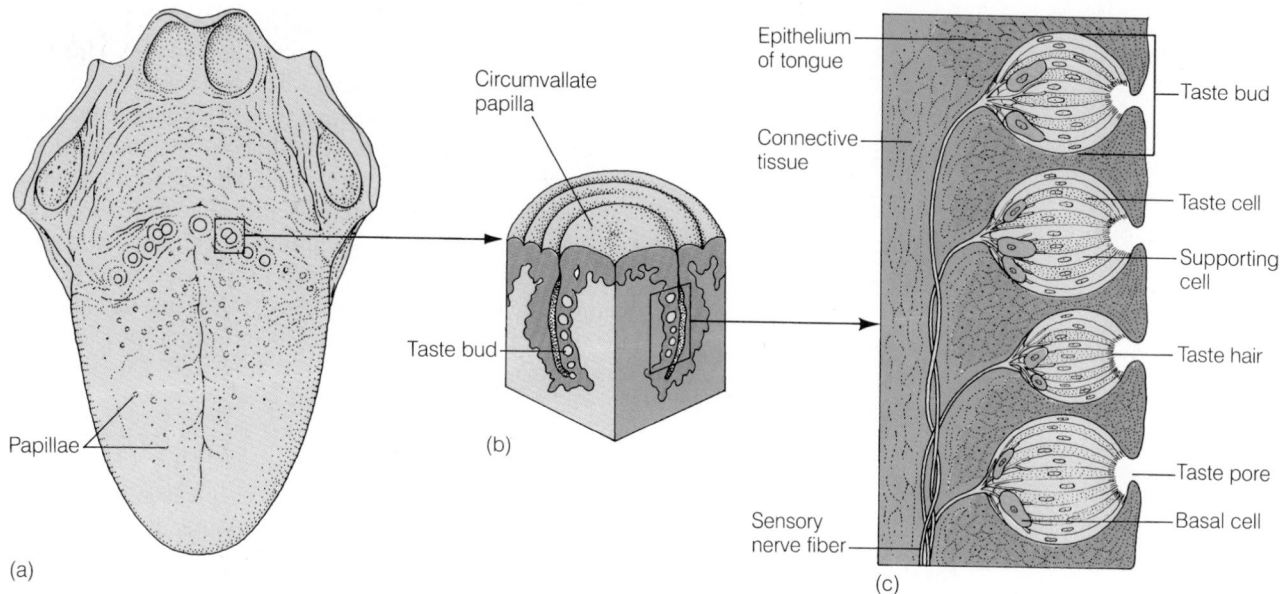

FIGURE 18–7 Taste Receptors (*a*) Taste buds are located on the upper surface of the tongue and are concentrated on the (*b*) papillae. (*c*) The structure of the taste bud.

onion-shaped structures are located in the surface epithelium of the tongue and on small protrusions, **papillae,** on the upper surface of the tongue (▷ Figure 18–7a). Taste buds are also found on the roof of the oral cavity (on the hard and soft palates), the pharynx, and the larynx, but in smaller numbers.

Taste buds are chemoreceptors and are stimulated by ions and molecules in the food we eat. These substances dissolve in the saliva and enter the **taste pores,** small openings that lead to the interior of the taste bud. As Figure 18–7c illustrates, taste buds consist of two cell types, receptor cells and supporting cells (which recent evidence suggests may be the same cell type, just in a different phase of its life cycle). The **supporting cells** lie on the outside of the taste bud and resemble the staves of a wooden barrel. The **receptor cells** lie inside, like so many pickles jammed lengthwise into a jar. The ends of the receptor cells possess large microvilli, which are known as **taste hairs.** Taste hairs project into the taste pore. The membrane of the taste hairs contains receptors that bind to food molecules dissolved in water. This stimulates the receptor cells, which then stimulate the dendrites of the sensory nerves that are wrapped around the receptor cells.[3] Impulses from the taste buds are then transmitted to the taste centers in the cerebral cortex.

In the opening paragraph of this chapter, you were introduced to an unfortunate student who lost her sense of taste because of a dietary zinc deficiency. Research has shown that zinc, which is normally found in low concentrations in the saliva, stimulates division of the cells in taste

buds. Because cells of the taste buds are lost from normal wear and tear, a zinc deficiency reduces cellular replacement, and the taste buds eventually cease operation.

The Taste Buds Respond to all Four Primary Flavors but Are Generally Preferentially Responsive To One. Humans can discriminate among thousands of taste sensations. The taste sensations are a combination of four basic flavors: sweet, sour, bitter, and salty. Sweet flavors result from sugars and some amino acids. Sour flavors result from acidic substances. Salty tastes result from metal ions (like sodium in table salt). Bitter flavors result from chemical substances belonging to a group called alkaloids, among them caffeine, but also some non-alkaloid substances, such as aspirin.

All taste buds respond, in varying degree, to all four taste sensations, but they respond preferentially to one taste. Taste buds are also distributed unevenly on the surface of the tongue. A simple experiment in which drops of various substances are placed on the tongue shows that the tip of the tongue is most sensitive to sweet flavors because it contains a higher proportion of taste buds that respond preferentially to sweet flavors (▷ Figure 18–8). The sides of the tongue are most sensitive to sour flavors, and the back of the tongue is most sensitive to bitter flavors. Salty taste is more evenly distributed, with slightly increased sensitivity on the sides of the tongue near the front.

Food contains many different flavors. What we taste, therefore, depends on the relative proportion of the four basic flavors. For example, grapefruit juice tastes sour because of the predominance of acidic substances. Adding sugar to grapefruit juice, however, gives it a sweet-sour

[3]Note: unlike the receptors in the previous section, the receptor cells of taste buds are not part of nerve cells but are independent cells.

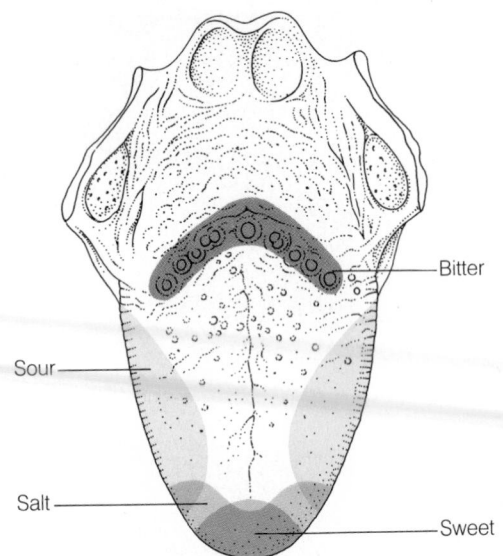

Bitter

Sour

Salt

Sweet

▷ **FIGURE 18–8 The Distribution of Taste on the Upper Surface of the Tongue**

flavor. If you add enough sugar, you can mask the sour taste almost entirely. As you will soon see, the sense of smell also plays a role in determining the taste of food.

The Olfactory Epithelium Is a Patch of Receptor Cells that Detects Odors

Smell is also a chemical sense. The receptors for smell are located in the roof of each nasal cavity in a patch of cells called the **olfactory epithelium,** or **membrane** (▷ Figure 18–9). The olfactory membrane contains receptor cells and supporting cells. **Receptor cells** are neurons whose cell bodies lie in the olfactory membrane. The dendrites of these neurons extend to the surface of the olfactory membrane, terminating in six to eight long projections, **olfactory hairs,** or **olfactory cilia** (Figure 18–9). The membranes of the olfactory hairs, like those in the receptor cells of the taste bud, contain receptors for molecules. When airborne molecules are trapped on the thin, watery layer on the surface of the cell, they bind to the receptors, activating the neurons. This causes the neurons to generate impulses that are sent to the **olfactory bulb,** a complex neural structure that contains neurons that synapse with the dendrites of the olfactory receptor cells. Axons of the neurons in the olfactory bulb cells then travel to the brain via the **olfactory nerve.** Supporting cells are interspersed among receptor cells.

Humans can distinguish tens of thousands of odors, many at very low levels. The olfactory receptors are so sensitive, in fact, that even a single molecule binding to the olfactory hairs can produce a bioelectric impulse. But the sense of smell in humans is not as sensitive as that of other animals such as dogs, wolves, and coyotes. A dog's keen sense of smell is due to the presence of an olfactory membrane about 20 times larger than that of humans.

Like taste, odor discrimination is thought to depend on combinations of primary odors. Unfortunately, neuroscientists do not agree on what the primary odors are. One system suggests seven primary odors ranging from pepperminty to floral to putrid. It is thought that molecules that produce a similar odor are similarly shaped and that specific receptor binding sites on the olfactory hairs bind to those molecules with a common shape. Various combinations of the primary odors give rise to the many odors we can perceive. Some researchers hypothesize that there may be thousands of kinds of smell receptors, providing odor discrimination.

Even Though the Olfactory Receptors are Sensitive and Able to Discriminate Many Odors, They Quickly Become Adapted. Receptors for smell adapt within a short period—about a minute. If you have ever worked on a dairy farm or lived near one, or if you have been around a baby in diapers on a long car trip, you understand (and are grateful for) olfactory adaptation.

Smell Influences the Sense of Taste, and Vice Versa. Hold a piece of hot apple pie to your nose, and take a deep breath. It smells so good you can almost taste it. In fact, you *are* tasting it. Molecules given off by the pie enter the nose, reach the mouth through the pharynx, and dissolve in the saliva, where they stimulate taste receptors. Just as odors stimulate taste receptors, food in our mouths also stimulates olfactory receptors. Molecules released by food enter the nasal cavities, dissolving in the water on the surface of the olfactory membrane and stimulating the receptor cells.

The complementary nature of taste and smell is abundantly evident when a person suffers from nasal congestion. As you have probably noticed, a viral infection in the nose often makes food seem bland. Cold sufferers complain that they cannot taste their food. This phenomenon results from the buildup of mucus in the nasal cavities, which blocks the flow of air. Mucus may also block the olfactory hairs. Therefore, food loses its "taste" when you have a stuffy nose because your sense of smell is impaired. (You can test this by holding your nose while you eat.)

The Ability to Detect Various Chemicals in the Environment—that is, Chemoreception—Is a Universal Trait of Animals and Even Protists

When it comes to finding a mate, recognizing their territory or the territory of others, or moving about in the environment, most animals depend heavily on chemoreceptors. In insects, for example, females give off small quantities of chemical substances (known as pheromones) that attract males. This adaptation greatly increases the likelihood of breeding and depends on special chemoreceptors in the males that can detect pheromones in parts-per-billion concentrations. Interestingly, many fishes have taste buds all over their body. Chemical receptors are even found in

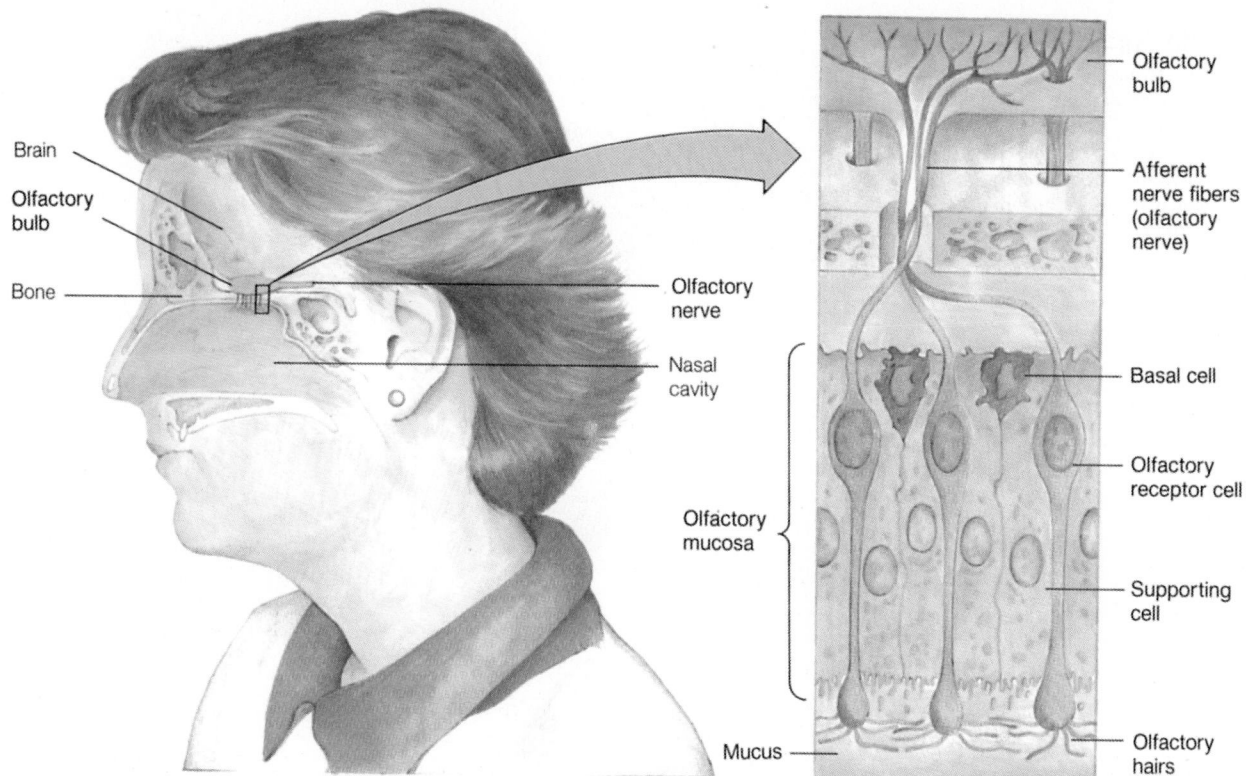

Brain
Olfactory bulb
Bone
Olfactory nerve
Nasal cavity
Olfactory mucosa
Mucus

Olfactory bulb
Afferent nerve fibers (olfactory nerve)
Basal cell
Olfactory receptor cell
Supporting cell
Olfactory hairs

▷ **FIGURE 18–9 Location and Structure of the Olfactory Epithelium** Olfactory receptors are located in the olfactory epithelium in the roof of the nasal cavity. Chemicals in the air dissolve in the watery fluid bathing the surface of the cells, then bind to receptors on the plasma membranes of the olfactory hairs. The olfactory receptors terminate in the olfactory bulb. From here, nerve fibers travel to the brain.

some of the most primitive members of the animal kingdom, such as hydra and planaria. In hydra, for example, special chemoreceptors pick up the scent of wounded prey and induce swallowing movements that cause the prey to be ingested.

Most of what is known about chemoreception in animals other than vertebrates comes from studies of insects. As shown in ▷ Figure 18–10, insects have sensory hairs on their mouthparts and feet. To taste food, all an insect has to do is land on it. The sensory hairs therefore help insects determine what is food and what is not. As a rule, the sensory hairs contain more than one receptor cell, each of which responds maximally to a different chemical. Just as in humans, the pattern of responses probably helps insects discriminate among many tastes. Besides alerting an insect to the presence of food and helping it distinguish the type of food, the sensory hairs activate or inhibit the feeding behavior.

Some sensory hairs detect airborne chemicals, such as pheromones. Flying insects, such as moths, use their olfactory sensory hairs to detect mates but also to locate food plants.

≈ THE VISUAL SENSE: THE EYE

The vertebrate eye is one of the most extraordinary products of evolution. It contains a patch of photoreceptors that permits us to perceive the remarkably diverse and colorful environment we live in.

The Vertebrate Eye Houses the Light-Sensitive Layer Known as the Retina

In humans, the eyes are roughly spherical organs located in the eye sockets, or **orbits,** cavities formed by the bones of the skull. The eye is attached to the orbit by six muscles, the **extrinsic eye muscles,** which control eye movement. Small tendons connect these muscles to the outermost layer of the eye. The location of the eye of a vertebrate—whether on the side of the head or in front—tells you a great deal about its lifestyle. If the eyes are in front, it is most likely a predator (▷ Figure 18–11a). If the eyes are on the side of the skull, the animal is most likely a prey species—that is, an animal hunted and killed by predators (Figure 18–11b). The location of the eyes on the side of the head gives a prey species a wider field of vision and increases its ability to spot a potential predator.

The Sclera and Cornea. As shown in ▷ Figure 18–12, the wall of the human eye consists of three layers (Table 18–2). The outermost is a durable fibrous layer, which consists of the **sclera,** the white of the eye, and the **cornea,** the clear part in front, which lets light into the interior of

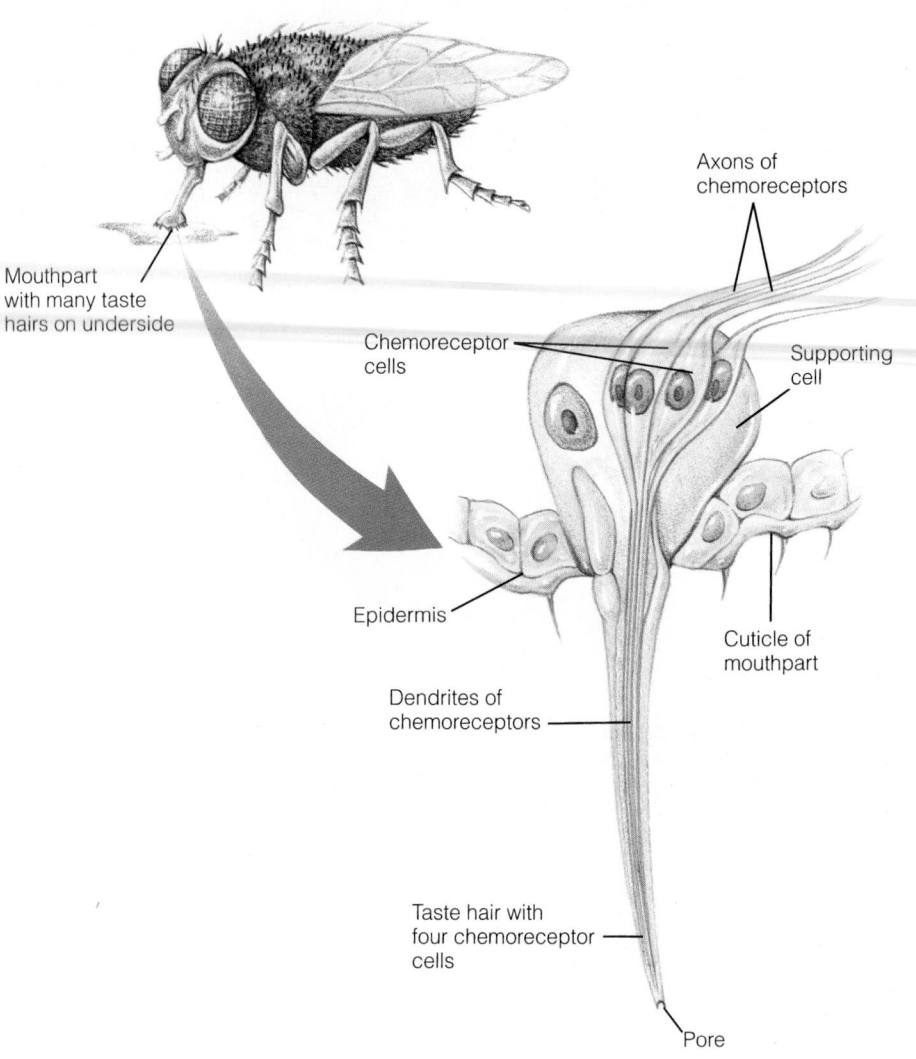

Axons of
chemoreceptors

Supporting
cell

Chemoreceptor
cells

Mouthpart
with many taste
hairs on underside

Epidermis

Cuticle of
mouthpart

Dendrites of
chemoreceptors

Taste hair with
four chemoreceptor
cells

Pore

the eye (Figure 18–12). Tendons of the extrinsic eye muscles attach to the sclera.

The Choroid, Ciliary Body, and Iris. The middle layer is a heavily pigmented and highly vascularized region. As Figure 18–12 shows, the middle layer consists of three parts: the choroid, the ciliary body, and the iris. The **choroid** is the largest portion of the middle layer. It contains a larger amount of melanin, a pigment that absorbs stray light the way the black interior of a camera does. The blood vessels of the choroid supply nutrients to the eye. Anteriorly, the choroid forms the ciliary body. The **ciliary body** contains smooth muscle fibers, which control the shape of the lens, permitting us to focus incoming light given off by objects or reflected from their surface. The third part of the middle layer is the iris. The **iris** is the colored portion of the eye visible through the cornea. Looking in a mirror, you can see a dark opening in the iris called the pupil. The **pupil** allows light to penetrate the

eye. The blackness you see through the pupil is the choroid layer and the pigmented section of the retina, discussed below. Like the ciliary body, the iris contains smooth muscle cells. The smooth muscle of the iris regulates the diameter of the pupil. Opening the pupil lets more light in, and narrowing it reduces the amount of light that can enter. The pupils open and close reflexively in response to light intensity. This reflex is an adaptation that protects the light-sensitive inner layer, the retina. The pupils also constrict when the eye focuses on a near object, a process discussed later.

The Retina. The innermost layer of the eye is the **retina.** The retina consists of an outer, pigmented layer that complements the light-absorbing function of the choroid layer, and an inner layer, the neural layer, consisting of photoreceptors and associated nerve cells. The retina is weakly attached to the choroid and can become separated from it as a result of trauma to the head. A detached retina

cells located in the outermost portion of the neural layer, adjacent to the pigmented layer. Two types of photoreceptors are present in the retina: rods and cones (Table 18–3). The **rods,** so named because of their shape, are sensitive to low light (▷ Figures 18–13b and 18–13c). Thus, the rods function on moonlit evenings, yielding grayish, somewhat vague images. The **cones,** also named because of their shape, are photoreceptors that sense colors and operate only in brighter light. They are responsible for visual acuity—sharp vision.

In order for light to stimulate the photoreceptors, it must pass through the ganglion and bipolar cell layers, two layers of neurons that lie in front of the photoreceptors in the retina (Figure 18–13). As Figure 18–13b shows, the rods and cones synapse with the bipolar neurons. These, in turn, synapse with ganglion cells. Nerve impulses from the photoreceptors travel from the bipolar neurons to the ganglion cells. The axons of the ganglion cells course along the inner surface of the retina and unite at the back of the eye to form the **optic nerve.** The optic nerve leaves at the **optic disc,** or **blind spot,** so named because it contains no photoreceptors and is therefore insensitive to light. The blood vessels that enter and leave the eye do so with the optic nerve. These blood vessels and their branches can readily be seen by an ophthalmologist by shining a light through the pupil onto the posterior wall of the eye (▷ Figure 18–14).

Rods and cones are found throughout the retina, but the cones are most abundant in a tiny region of each eye lateral to the optic disc. This spot is called the **macula lutea,**

▷ **FIGURE 18–11 Eye Location** You can tell a lot about an animal by the location of its eye. (*a*) Predators have forward-facing eyes, which give them three-dimensional vision necessary for capturing prey. (*b*) Prey species have eyes on the side of their heads for better viewing the approach of predators.

can lead to blindness if not repaired by surgery. Today, doctors of ophthalmology usually repair detached retinas with lasers.

The **photoreceptors** of the retina are modified nerve

▷ **FIGURE 18–12 Anatomy of the Human Eye**

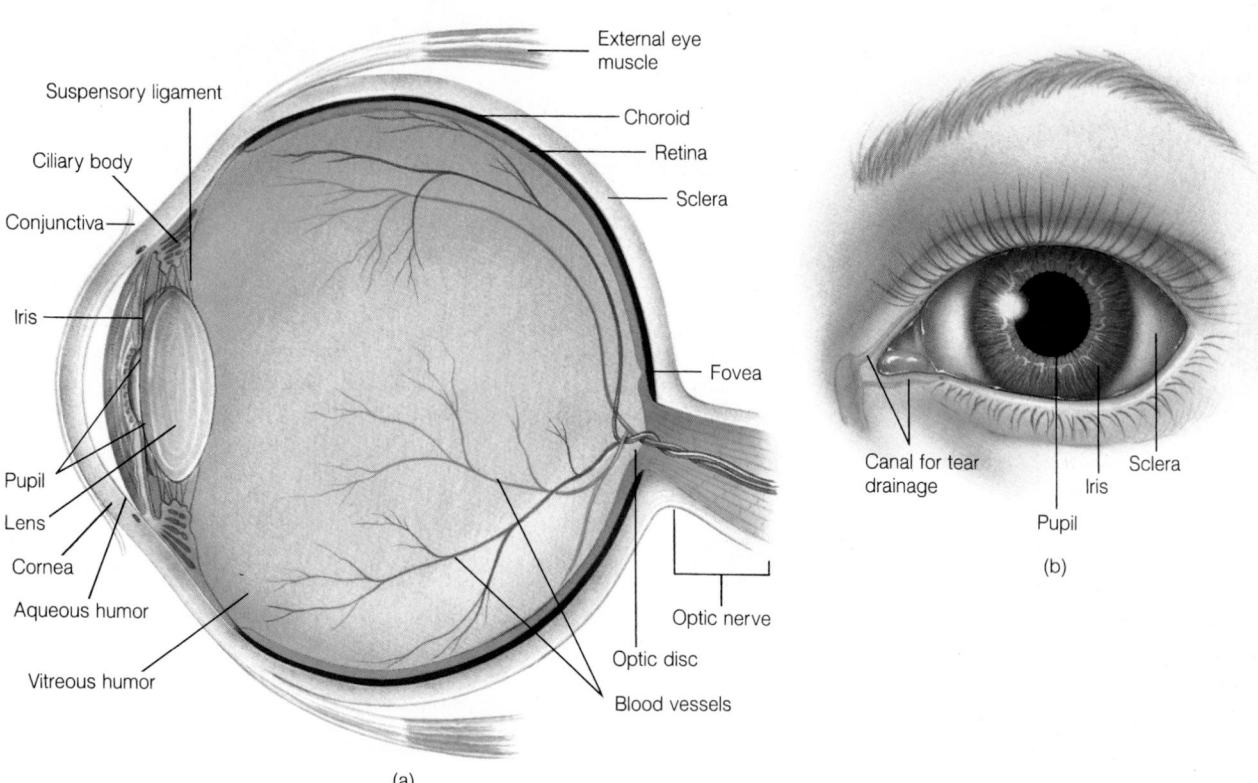

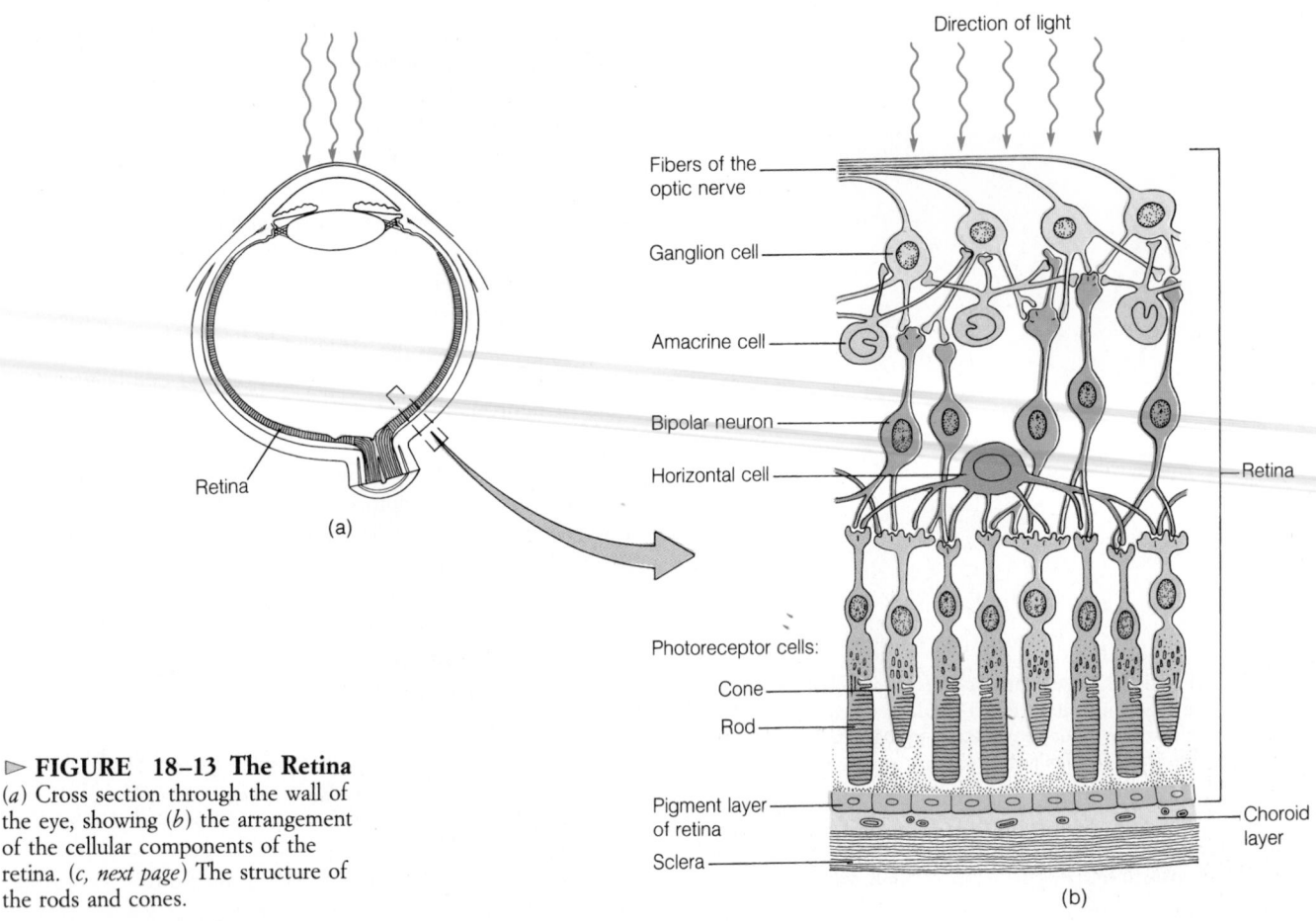

Direction of light

Fibers of the optic nerve

Ganglion cell

Amacrine cell

Bipolar neuron

Horizontal cell

Retina

Photoreceptor cells:

Cone

Rod

Pigment layer of retina

Sclera

Choroid layer

Retina

(a)

(b)

▷ **FIGURE 18–13 The Retina**
(a) Cross section through the wall of the eye, showing (b) the arrangement of the cellular components of the retina. (c, next page) The structure of the rods and cones.

≈	**TABLE 18–2 Structures and Functions of the Eye**	≈

STRUCTURE		FUNCTION
Wall		
Outer layer	Sclera	Provides insertion for extrinsic eye muscles
	Cornea	Allows light to enter; bends incoming light
Middle layer	Choroid	Absorbs stray light; provides nutrients to eye structures
	Ciliary body	Regulates lens, allowing it to focus images
	Iris	Regulates amount of light entering the eye
Inner layer	Retina	Responds to light, converting light to nerve impulses
Accessory structures and components		
	Lens	Focuses images on the retina
	Vitreous humor	Holds retina and lens in place
	Aqueous humor	Supplies nutrients to structures in contact with the anterior cavity of the eye
	Optic nerve	Transmits impulses from the retina to the brain

≈	**TABLE 18–3 Summary of Rods and Cones**		≈

PHOTORECEPTOR	DAY OR NIGHT	COLOR VISION	LOCATION
Rods	Night vision	No	Highest concentration in the periphery of the retina
Cones	Day vision	Yes	Highest concentration in the macula and fovea

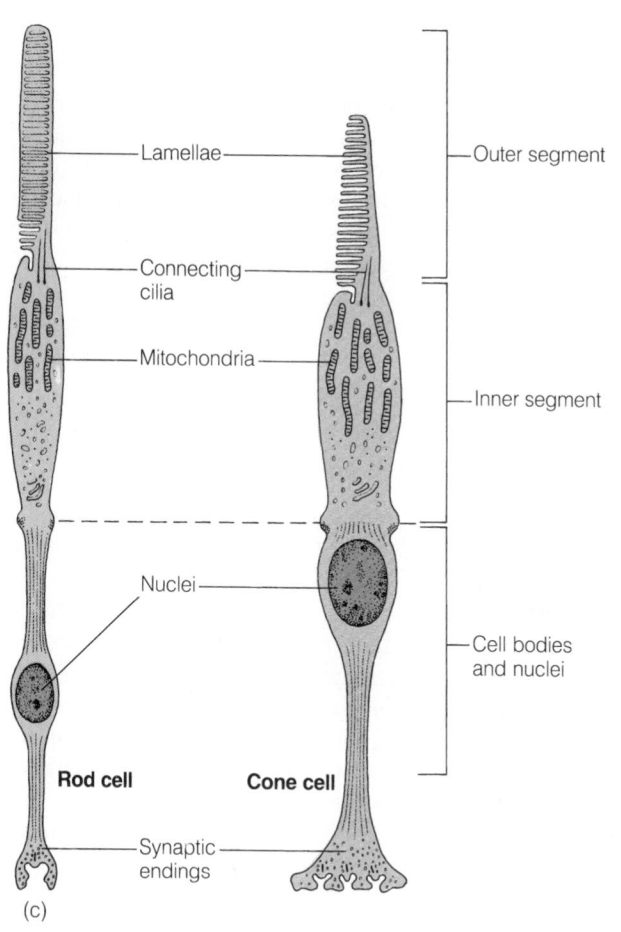

Lamellae

Outer segment

Connecting cilia

Mitochondria

Inner segment

Nuclei

Cell bodies and nuclei

Rod cell   Cone cell

Synaptic endings

(c)

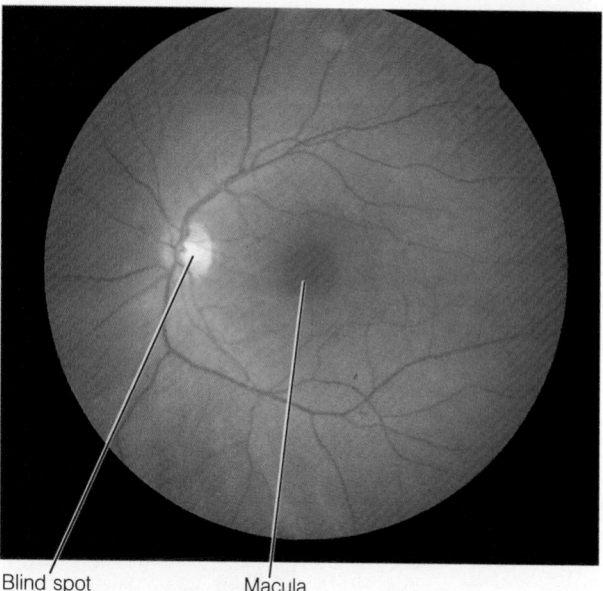

Blind spot   Macula

▷ **FIGURE 18–14 View of the Inside Back Wall of the Eye Seen through an Ophthalmoscope** The optic disk (blind spot) and macula are both indicated.

literally "yellow spot." In the center of the macula is a minute depression, about the size of the head of a pin, known as the **fovea centralis** (central depression). The fovea contains only cones. The number of cones in the retina decreases progressively from this point outward, whereas the number of rods increases. The greatest concentration of rods is found in the periphery of the retina.

Images from our visual field are cast onto the retina, and impulses are transmitted to the visual cortex of the brain's occipital lobe, where they are interpreted. When we focus on an object, the image is projected upside down onto the fovea. The sharpest vision occurs at the fovea, because it contains the highest concentration of cones and because the bipolar neurons and ganglion cells do not cover the cones in this region as they do throughout the rest of the retina. Even though the image is upside down, the brain processes the information and gives us a right-side-up image; that is, it allows us to perceive objects in their true position.

The Lens. Light is focused on the retina by the lens. The **lens** is a transparent structure that lies behind the iris (▷ Figure 18–15). This flexible, clear structure is attached to the ciliary body by thin fibers composing the **suspensory ligament.** This connection allows the smooth muscle

of the ciliary body to alter the shape of the lens, an action necessary for focusing the eye.

In older individuals, the lens may develop cloudy spots, known as **cataracts.** The loss of transparency is especially prevalent in people who have been exposed to excessive sunlight or excessive ultraviolet light at work or elsewhere. Patients with this disease complain of cloudy vision. Looking out on the world to them is a little like looking through frosted glass. Interestingly, cataract risk may increase as the Earth's ozone layer continues to be eroded by pollutants such as chlorofluorocarbons from refrigerators, air conditioning units, and other sources, as discussed in Chapter 33. As noted earlier in the book, the ozone layer blocks harmful ultraviolet light.

Recent research suggests that the color of an individual's eyes affects the risk of developing a cataract. Dark-eyed people run the highest risk of cataracts. Brown- and hazel-eyed subjects had more cataracts than did blue-, gray-, and green-eyed patients. Researchers suggest that melanin in the irises of dark-eyed people may absorb solar radiation, causing more damage to the nearby lens. To lower your risk of cataracts, many eye doctors suggest wearing sunglasses with a coating that reduces ultraviolet penetration. But beware, there is evidence that some manufacturers' claims about ultraviolet protection may be inaccurate.

Ophthalmologists once treated cataracts surgically by removing the afflicted lens. Patients were then fitted with a thick pair of glasses or a pair of contact lenses, which compensated for the missing lens. Today, however, lenses are routinely replaced by artificial plastic lenses that provide nearly normal vision.

Research has shown that cataracts may result when a

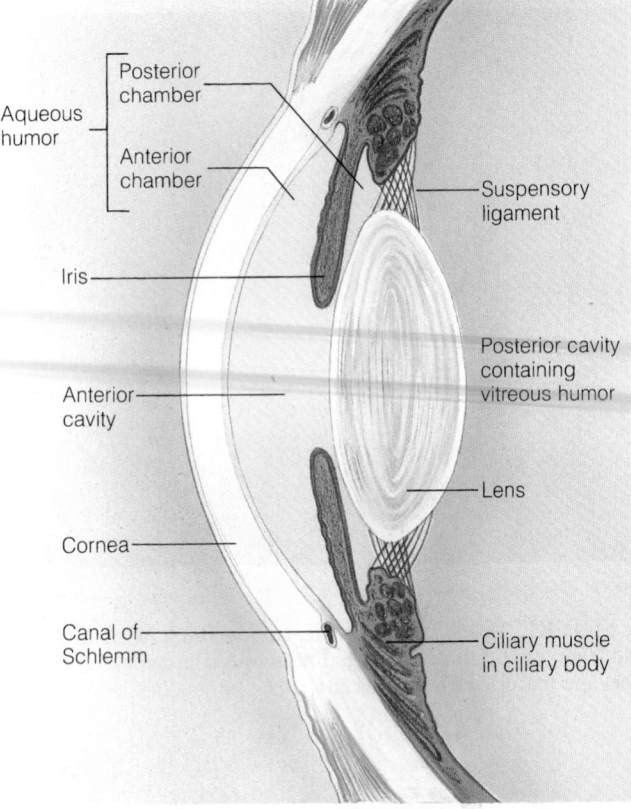

Aqueous humor

Posterior chamber

Anterior chamber

Iris

Anterior cavity

Cornea

Canal of Schlemm

Suspensory ligament

Posterior cavity containing vitreous humor

Lens

Ciliary muscle in ciliary body

▷ **FIGURE 18–15 Detailed Cross Section of the Anterior Cavity.** Arrows show the flow of aqueous humor.

lens protein called **crystallin** denatures. Researchers are hoping that new drugs may be able to reverse the process, helping restore vision without surgery.

The lens separates the interior of the eye into two cavities of unequal sizes. Everything in front of the lens is the **anterior cavity;** everything behind it is the **posterior cavity.** The posterior cavity is filled with a clear, gelatinous material, the **vitreous humor** ("glassy liquid"). Formed during embryonic development, the vitreous humor remains throughout life, holding the lens and retina in place. Being clear, the vitreous humor transmits light faithfully to the retina.

The anterior cavity is divided into two portions: the **anterior chamber** and the **posterior chamber** (Figure 18–15). The anterior chamber lies between the iris and the cornea; the posterior chamber lies between the iris and the lens. A thin liquid, chemically similar to blood plasma, fills the anterior and posterior chambers of the eye and is called the **aqueous humor** ("watery fluid").

Unlike the vitreous humor, the aqueous humor is constantly replaced. New fluid is produced by capillaries in the ciliary body. The fluid enters the posterior chamber, then flows forward into the anterior chamber, where it drains into a canal known as the **canal of Schlemm,** which is located at the junction of the sclera and cornea. From here, the plasmalike fluid flows into the bloodstream.

Aqueous humor provides nutrients to the cornea and lens and carries away cellular wastes. In normal, healthy

individuals, aqueous humor production is balanced by absorption. If the outflow is blocked, however, aqueous humor builds up inside the anterior chamber, creating internal pressure. This disease, called **glaucoma,** occurs gradually and imperceptibly. If untreated, the pressure inside the eye can damage the retina and optic nerve, causing blindness. Because the incidence of glaucoma increases after age 40, doctors recommend an annual eye examination for men and women over 40. If diagnosed early, glaucoma can be treated with eye drops that increase the rate of drainage, thus reducing intraocular pressure. In severe cases, surgery may be required to increase the outflow.

The Lens Focuses Light on the Retina

To understand how the lens operates, you have to understand a little bit about light.

Refraction. First of all, visible light travels in waves. Light waves travel at a constant rate in any given medium such as air or water. When light passes from one medium to another, however, its velocity changes. When light passes from a less dense medium, such as air, to a denser medium, such as the cornea, it slows down. Anytime light changes speed in passing from one substance to another, it bends. The bending of light is called **refraction** (▷ Figure 18–16).

Focusing the Image. Light traveling through a camera lens is bent. The lens of the camera is designed to bend light enough to focus the image on the film. The lens of the eye also bends incoming light rays, focusing images on the photoreceptors of the retina. Lying in front of the lens is the cornea; it also bends incoming light rays (▷ Figure 18–17). Although we usually think of the lens as the structure that allows us to focus, most of the bending of incoming rays takes place in the cornea. However, the cornea is a fixed structure. Like the lens on a fixed-focus camera, the cornea cannot be adjusted to focus on nearby objects. Without the adjustable lens, the eye would be unable to focus on objects close at hand.

The lens is resilient like a rubber ball. Its shape is controlled by the muscles in the ciliary body that are attached via the suspensory ligament. When the muscles of the ciliary body are relaxed, the suspensory ligament is taut, and the lens is somewhat flattened. As Figure 18–17a shows, light from distant objects comes to the eye as nearly parallel rays. The fixed refractive power of the cornea and the refractive power of the lens in its relaxed state are sufficient to bend these beams to bring them into focus on the retina.

As Figure 18–17b shows, light from nearby objects is divergent. To focus on nearby objects, the lens must become more curved. When the eye focuses on nearby objects, the ciliary smooth muscle contracts, causing the suspensory ligament to relax (▷ Figure 18–18b). The lens thickens and shortens, becoming more curved, in much the

▷ **FIGURE 18–16 Refraction** The pencil in the glass of water appears to bend. What is happening, though, is that the light rays coming to our eyes are bent as they pass through the water and glass, making the pencil appear bent.

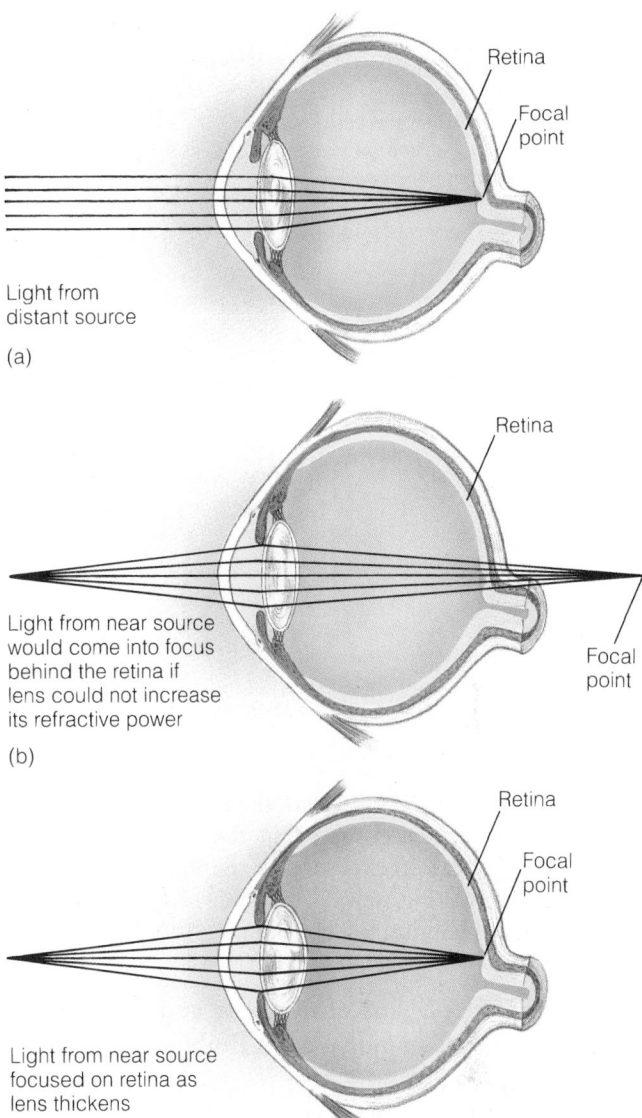

Light from distant source

(a)

Light from near source would come into focus behind the retina if lens could not increase its refractive power

(b)

Light from near source focused on retina as lens thickens

(c)

▷ **FIGURE 18–17 Refraction of Light by the Cornea and Lens** (*a*) Light rays from distant objects are parallel when they strike the eye. The refractive power of the cornea and resting lens are sufficient to bring them into focus on the retina. (*b*) Light rays from nearby objects are divergent. The cornea has a fixed refractive power and cannot change. The rays would focus behind the retina if the lens could not alter its refractive power. (*c*) To focus the image, the ciliary muscles contract. This lessens pressure on the suspensory ligaments, which allows the lens to thicken and shorten, becoming more curved and more refractive.

same way that a rubber ball flattened between your hands will return to normal when you reduce the pressure on it. The automatic adjustment in the curvature of the lens required to focus on a nearby object is called **accommodation** (Figure 18–18). Accommodation increases the refractive (bending) power of the lens.

Accommodation is enhanced by pupillary constriction. As the eyes focus on a nearby object, the pupils constrict. Pupillary constriction is a reflex that eliminates divergent rays of light that would otherwise strike the periphery of the lens. The lens would be unable to bend these rays sufficiently to bring them into focus on the retina. Thus, without pupillary constriction, images of nearby objects would be quite blurred.

Convergence. The human eyes are movable. Six muscles located outside the eye, the extrinsic eye muscles, are responsible for movement (▷ Figure 18–19). As noted earlier, these muscles attach to the orbit (the bony eye socket) and to the sclera (the white of the eye) and allow for a wide range of movements.

The eyes generally move in unison like a pair of synchronized swimmers. Synchronized movement is an evolutionary adaptation that ensures that images are focused on the foveas of both eyes at the same time. To test the synchronized movement, close one of your eyes. Place an index finger gently over the closed lid. Then hold the other hand in front of your face and move it back and forth, then up and down, following it with your opened eye. You should feel your closed eye moving in sync, even though the lid is shut.

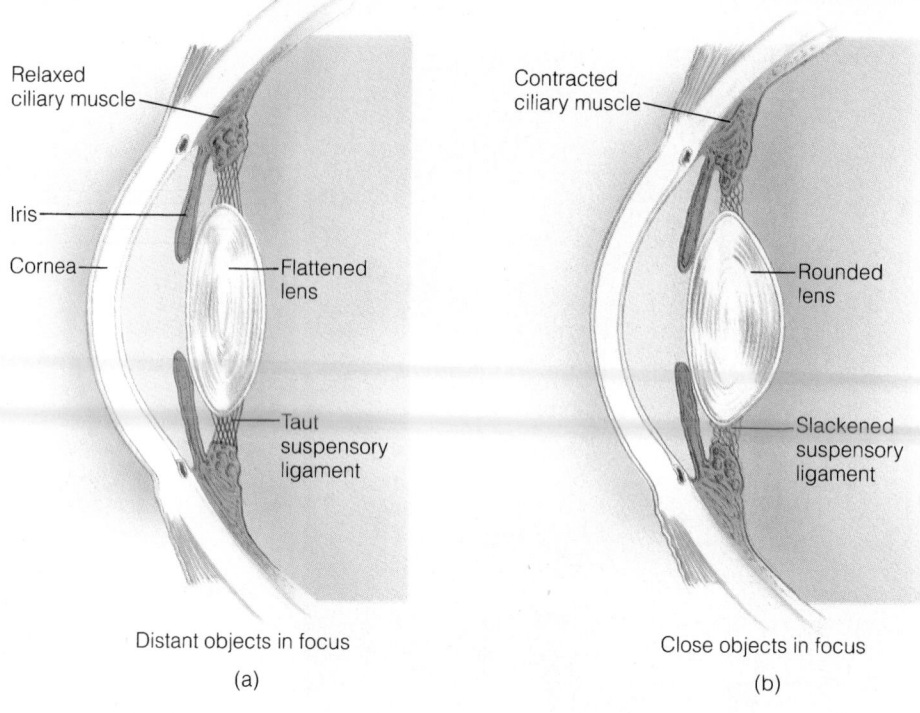

▷ FIGURE 18–18
Accommodation (*a*) The lens is flattened when the ciliary muscles are relaxed. (*b*) When the ciliary muscles contract, tension on the suspensory ligaments is reduced and the lens shortens and thickens.

Relaxed ciliary muscle

Iris

Cornea

Flattened lens

Taut suspensory ligament

Distant objects in focus
(a)

Contracted ciliary muscle

Rounded lens

Slackened suspensory ligament

Close objects in focus
(b)

When a nearby object is viewed, the eyes turn inward, or converge. Convergence ensures that the image is focused on each fovea. Convergence occurs during all near-point work—reading, writing, sewing, knitting—and puts strain on the extrinsic eye muscles, contributing to eyestrain.

Alterations in the Shape of the Lens and Eyeball Cause the Most Common Visual Problems

In the normal eye, objects farther than 6 meters (20 feet) away fall into perfect focus on the back of the relaxed eye ▷ Figure 18–20a). Many individuals, however, have imperfectly shaped eyeballs or defective lenses. These imperfections result in two visual problems: nearsightedness and farsightedness.

Myopia. Nearsightedness or **myopia,** results when the eyeball is slightly elongated (Figure 18–20b). Without corrective lenses, the parallel light rays arising from distant images fall into focus in front of the retina. In contrast, nearby images with much more divergent light rays tend to be in focus in the uncorrected eye. People with myopia, therefore, can see near objects without corrective lenses—hence the name nearsightedness. Myopia may also result when the lens is too strong—that is, too concave. This lens bends the light too much, causing the image to come into focus in front of the retina.

Myopia is quite common. Approximately one of every five Americans needs glasses to correct for it. Myopia is caused by many factors and tends to run in families; it generally appears around age 12, often worsening until a person reaches 20.

Myopia can be corrected by contact lenses or prescription glasses that cause incoming light rays to diverge (Figure 18–20b). Contact lenses fit on the surface of the cornea and bend the incoming light rays outward, compensating for the shape of the eye or a defective ocular lens.

Although useful in correcting vision, contact lenses can cause serious problems. The largest study of their use in the United States suggests that extended-wear contact lenses, those worn overnight and for up to a week at a time, carry a greater risk of serious complications than daily-wear lenses. A study of over 22,000 lens wearers showed that sight-threatening complications, such as corneal abrasion, growth of blood vessels into the cornea, corneal ulcers, and severe corneal scarring, were two to

▷ FIGURE 18–19 Extrinsic Eye Muscles These muscles move the eye in all directions. They attach to the bony orbit and the sclera.

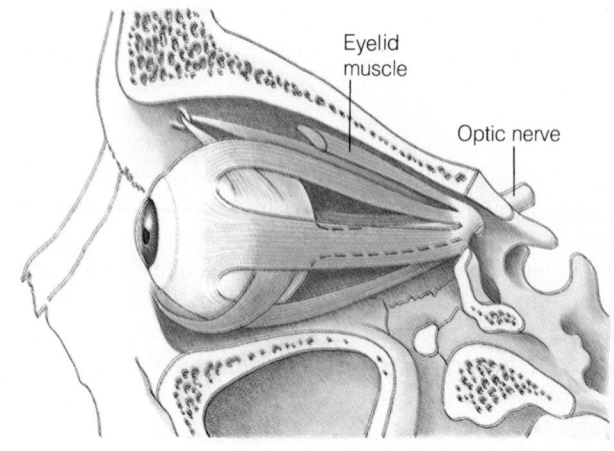

Eyelid muscle

Optic nerve

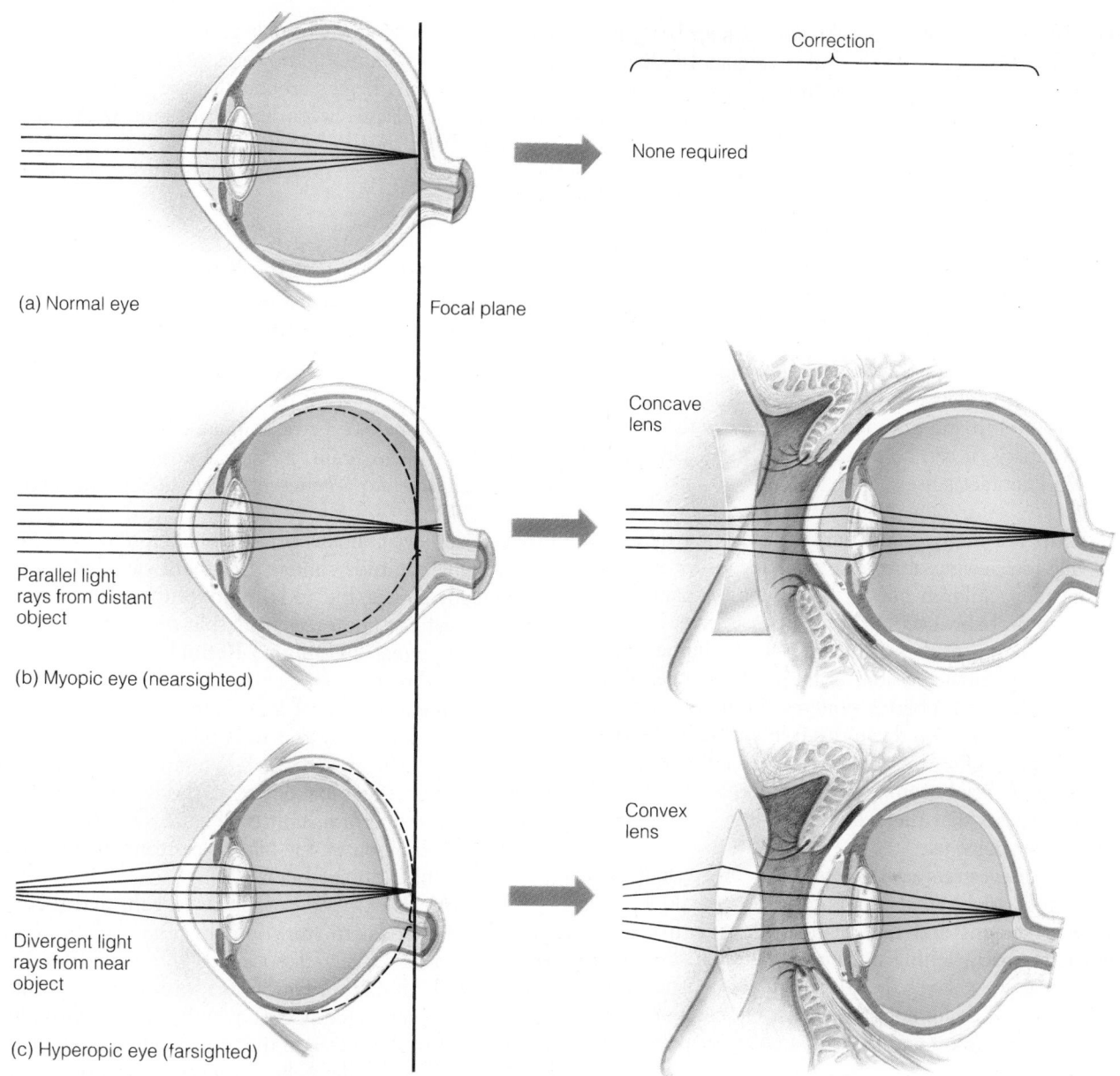

Correction

None required

(a) Normal eye

Focal plane

Concave lens

Parallel light rays from distant object

(b) Myopic eye (nearsighted)

Convex lens

Divergent light rays from near object

(c) Hyperopic eye (farsighted)

▷ **FIGURE 18–20 Common Visual Problems** (*a*) The normal eye, (*b*) the myopic (nearsighted) eye, and (*c*) the hyperopic (farsighted) eye.

four times more prevalent in users of extended-wear contacts than in those using daily-wear lenses.

Researchers found that 1 in 2000 daily-wear lens users contracted corneal ulcers; 1 in 300 suffered from other serious complications. In extended-wear users, however, 1 in 500 suffered from corneal ulcers, and 1 in 100 had other serious reactions. The increase in risk suggests one should think carefully before choosing extended-wear contacts.

Surgeons have also developed a method to decrease the refractive power of the cornea, called **radial keratotomy.** Numerous small, superficial incisions are made in the cornea, radiating from the center like the spokes of a bicycle wheel. This procedure flattens the cornea and reduces its

refractive power, causing the rays to diverge and come into focus on the back of the retina. The long-term effectiveness and safety of radial keratotomy are still under study.

Hyperopia. Hyperopia, or **farsightedness,** is the opposite of myopia. That is, it results when the eyeballs are too short or the lens is too weak (too convex). In either case, the light falls into focus behind the retina. Parallel light rays from distant objects usually fall into focus on the retina—hence the name farsightedness. But divergent rays from nearby objects cannot be focused sufficiently. Without corrective lenses, farsighted individuals see distant objects well without correction, but nearby objects are fuzzy. Glasses or contact lenses that bend the light inward help bring near objects into sharp focus on the retina (Figure 18–20c).

Hyperopia is generally present from birth and is usually diagnosed during childhood. Like myopia, it tends to run in families.

Astigmatism. In the normal eye, the cornea and lens have uniformly curved surfaces. Either of these structures may be slightly disfigured, however. The surface of the cornea, for example, may have a slightly different curvature in the vertical plane than it does in the horizontal plane. This unequal curvature is called an **astigmatism.** It creates fuzzy images because light rays are bent differently by the different parts of the lens or cornea. Astigmatism is usually present from birth and does not grow worse with age. Like other conditions, it can be corrected with specially ground glasses and contact lenses.

Eyestrain. For many years, humans have used their eyes principally for viewing objects at a distance—for example, watching their children play or wild animals roam. Near-point vision probably occupied little of their time. It should come as no surprise, then, to learn that the human eye is best suited for distance vision, as are most other vertebrate eyes. Intensive near-point vision so common in today's world strains the eyes and can result in a progressive deterioration of eyesight. Frequent readers often become more nearsighted as they become older. No one knows why, but research suggests that the eye may elongate as a result of constant near-point use.

To reduce eyestrain and deterioration of eyesight, ophthalmologists advise that computer operators look away from their screens and that readers look up from their materials regularly, letting their eyes focus on distant objects. This action relaxes the ciliary muscles, reducing eyestrain. Some ophthalmologists suggest that computer users "blink at every period and look up after every paragraph."

Presbyopia. The aging process brings with it many joys: hair loss, hearing loss, and arthritis. Aging also results in a loss in the resiliency of the lens. Thus, when the ciliary muscles contract to allow one to focus on a nearby object,

the lens responds slowly or only partially, making it difficult to focus. This condition, known as **presbyopia,** (prez-bee-ope-ee-a) usually begins around the age of 40 and can be corrected by glasses worn for near-point work such as reading.[4] With this understanding of how the eye focuses light on the retina, we now turn our attention to the retina itself.

The Rods and Cones of the Retina Contain Photosensitive Pigments that Dissociate When Struck by Light

As you have seen, the cornea and lens focus images on the retina. The image cast on the rods and cones of the retina is converted into bioelectric impulses that are transmitted to the brain. The brain, in turn, translates this flood of nerve impulses into a coherent image of our environment.

Rods. Each eye contains about 100 million rods. Sensitive to low light, the rods contain a pigment called **rhodopsin.** When light strikes the rods, rhodopsin molecules in the cytoplasm of the photoreceptors split into two component molecules, retinal and opsin. **Retinal** is a derivative of vitamin A. **Opsin** is an enzyme. In the dark, the rods release a steady stream of a neurotransmitter. The neurotransmitter released by the rods inhibits bipolar neurons from firing (▷ Figure 18–21a). When light stimulates the rods, however, the breakdown of rhodopsin inhibits the release of the neurotransmitter (Figure 18–21b). This removes the inhibition on the bipolar neurons, allowing them to send impulses to the brain.

Rhodopsin is extremely sensitive to light and therefore permits the rods to function in dim light. Because rhodopsin is so easily dissociated, the molecules break down rapidly during daylight hours, dissociating as quickly as they are regenerated. Consequently, bright light reduces the amount of rhodopsin in the rods, making them ineffective in daylight. In dim light, however, rhodopsin is regen-

---

[4]You may have seen your parents or grandparents holding a phone number at arm's length to read it. They may have argued that their eyes weren't going bad, their arms were just too short!

▷ **FIGURE 18–21 Effect of Light on Rods**

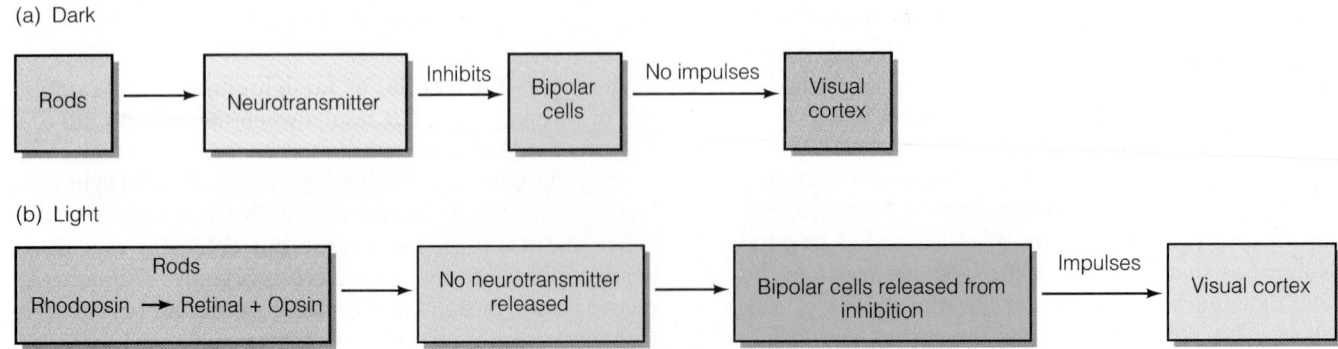

(a) Dark

(b) Light

erated, and the rods become functional. Stepping into a dark movie theater on a bright day, you have probably noticed that, at first, your vision is greatly impaired. That is because your rods, which have been bleached by bright outdoor light, require some time to recover and become fully functional. As rhodopsin molecules are regenerated and the rods begin functioning, your eyesight returns.

Rhodopsin molecules in the rods are broken down and regenerated in a continuous cycle. Although retinal and opsin are recycled to replenish rhodopsin molecules, some retinal is lost or destroyed and must be replaced. Retinal is produced in the body from vitamin A, which is found in a variety of foods, such as carrots, spinach, fortified milk, and peaches. Rhodopsin concentrations decline when vitamin A intake is reduced, decreasing the sensitivity of the rods so that they are unable to respond to dim light. Therefore, a dietary deficiency of vitamin A often leads to **night blindness,** a marked reduction in night vision that is easily corrected by eating foods rich in vitamin A.

Besides replenishing retinal supplies, vitamin A is important in maintaining the cornea. As noted in Chapter 11, a deficiency of this fat-soluble vitamin results in corneal dryness, which may, if severe, lead to ulceration. If the deficiency persists, corneal ulcers may result in permanent blindness. In the Third World, an estimated 100,000 people lose their eyesight each year because of severe vitamin A deficiency.

Cones. Each eye contains about 3 million cones. Like the rods, the cones contain photosensitive pigments. However, the cones operate under bright light and are also sensitive to different colors of light, thus making color vision possible.

Three types of cones are found in the human retina: blue cones, green cones, and red cones—so named because of their sensitivity to a particular color of light. Each type contains a unique type of pigment, which responds optimally to one particular color of light (▷ Figure 18–22). The pigments inside the cones dissociate when struck by colored light. As in the rods, the dissociation of the photopigments in the cones decreases the release of neurotransmitter, "unleashing" bipolar neurons. Nerve impulses traveling to the brain convey information the visual cortex requires to construct a visual image. To understand how color vision works, we will first look at the properties of visible light and examine why an object is a certain color.

The sun, light bulbs, and neon signs all emit visible light. White light from the sun and other sources is a blend of all of the colors of the rainbow (▷ Figure 18–23). You can test this by shining white light through a prism.

Light from the sun and other sources strikes objects in the environment. The color of an object, however, is determined by the kinds of pigments it contains. Pigments absorb some wavelengths of white light and reflect others. For example, the dye in a blue flannel shirt absorbs all of the colored light striking it except blue, which it reflects. It is the unabsorbed wavelengths that the eye detects. Put

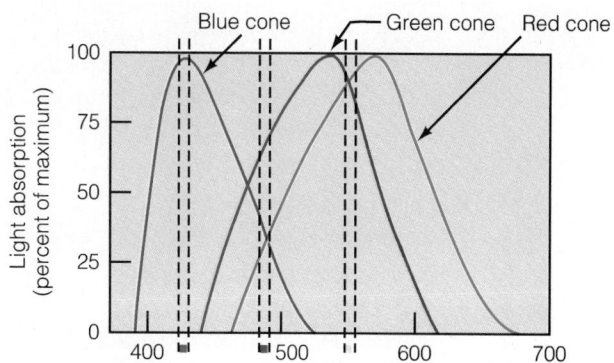

▷ **FIGURE 18–22 Sensitivity of the Three Types of Cones to Different Colors of Light** The color of light perceived is determined by the type or ratio of cones stimulated. The ratios of stimulation of the three cone types are shown for three sample colors.

another way, it is the unabsorbed wavelengths that give an object its characteristic color (▷ Figure 18–24). Thus, a leaf appears green because the pigments in the leaf absorb all of the colored light striking it except green, which they reflect.

Color vision occurs because each type of cone responds optimally to one color. Thus, a purely blue object stimulates "blue" cones (see Figure 18–22). Red light reflected from an object stimulates "red" cones. Green light stimulates "green" cones as well as some red and blue cones. How do we see so many in-between shades, such as yellow? Yellow light has a wavelength between red and green. Yellow light stimulates both red and green cones. Blue cones are not stimulated at all. The brain interprets the signal it receives from the red and green cones as yellow light. Thus, the relative proportion of different cones stimulated determines the color we perceive.

Color Blindness. About 5% of the human population suffers from color blindness. **Color blindness** is a hereditary disorder, more prevalent in men than women. Characterized by a deficiency in color perception, color blindness ranges from an inability to distinguish certain shades to a complete inability to perceive color. The most common form of this disorder is red-green color blindness. In individuals with red-green color blindness, the red or green cones may either be missing altogether or be present in reduced number. If the red cones are missing, red objects appear green. If the green cones are missing, however, green objects appear red.

Color blindness can be detected by simple tests (▷ Figure 18–25). Many color-blind people are unaware of

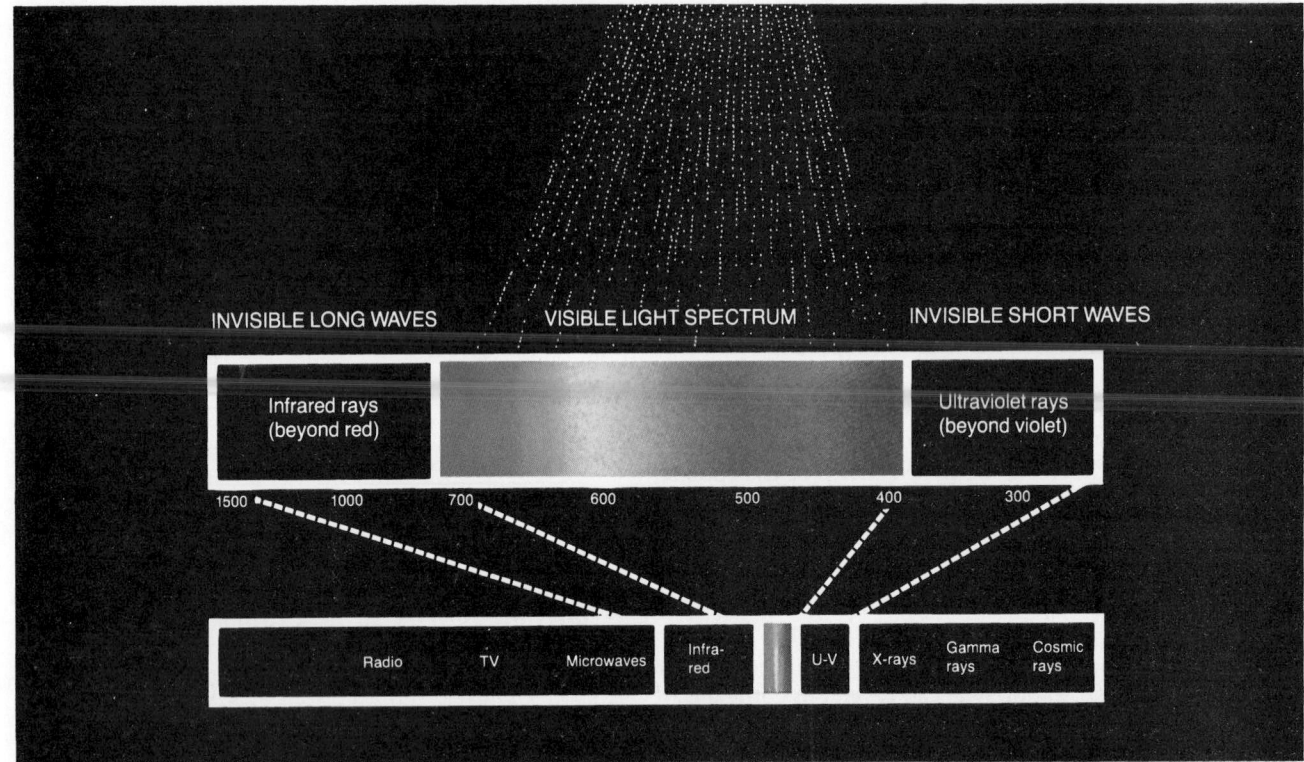

INVISIBLE LONG WAVES   VISIBLE LIGHT SPECTRUM   INVISIBLE SHORT WAVES

Infrared rays (beyond red)

Ultraviolet rays (beyond violet)

1500   1000   700   600   500   400   300

Radio   TV   Microwaves   Infra-red   U-V   X-rays   Gamma rays   Cosmic rays

▷ **FIGURE 18–23 The Electromagnetic Spectrum** The sun produces a wide range of electromagnetic radiation, a small portion of which is visible to the eye. Called visible light, it consists of a variety of colors. (Numbers on the spectrum are wavelength measured in nanometers.)

their condition or untroubled by it. They rely on a variety of visual cues, such as differences in intensity, to distinguish red and green objects. They also rely on position cues. In traffic lights, for example, the red light is always at the top of the signal; green is on the bottom. Although the colors may appear more or less the same, the position of the light helps color-blind drivers determine whether to hit the brakes or step on the gas.

▷ **FIGURE 18–24 Color** The color of an object results from the reflection of certain wavelengths of light.

Overlapping Visual Fields Give Us Depth Perception

Human eyes are located in the front of the skull, looking forward, much like those of other predatory animals. Each eye has a visual field of about 170 degrees, but the visual

▷ **FIGURE 18–25 Color Blindness Chart** People with red-green color blindness cannot detect the number 29 in this chart.

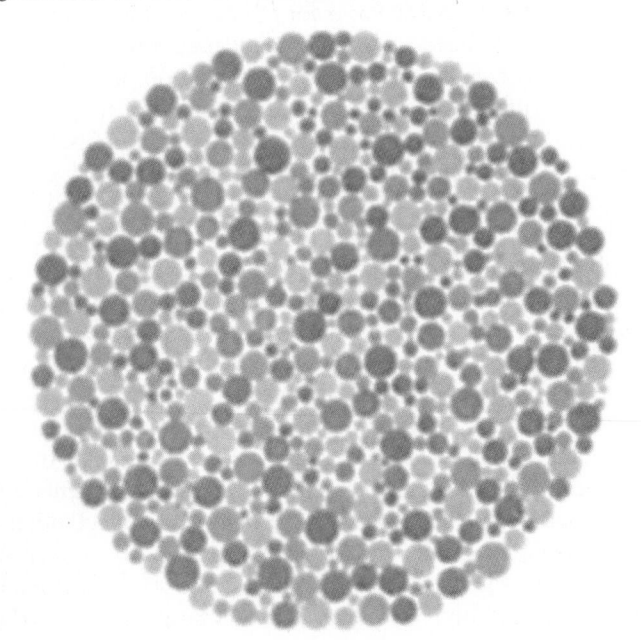

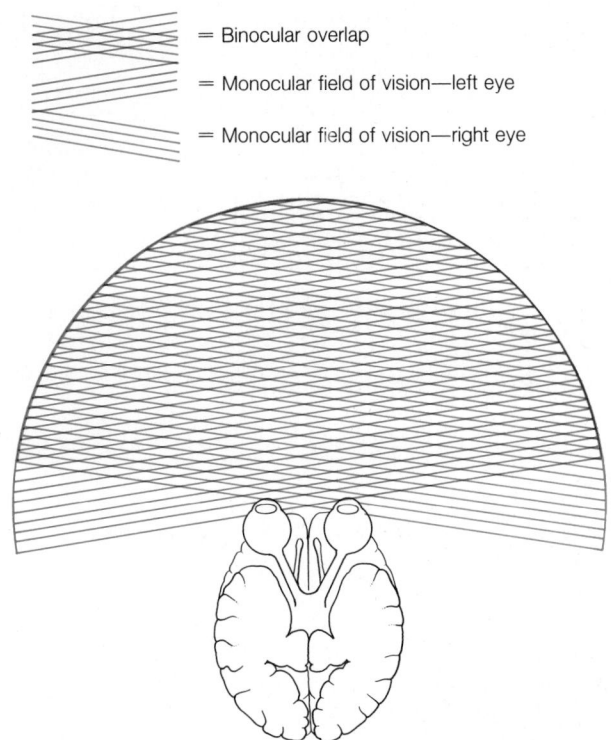

= Binocular overlap

= Monocular field of vision—left eye

= Monocular field of vision—right eye

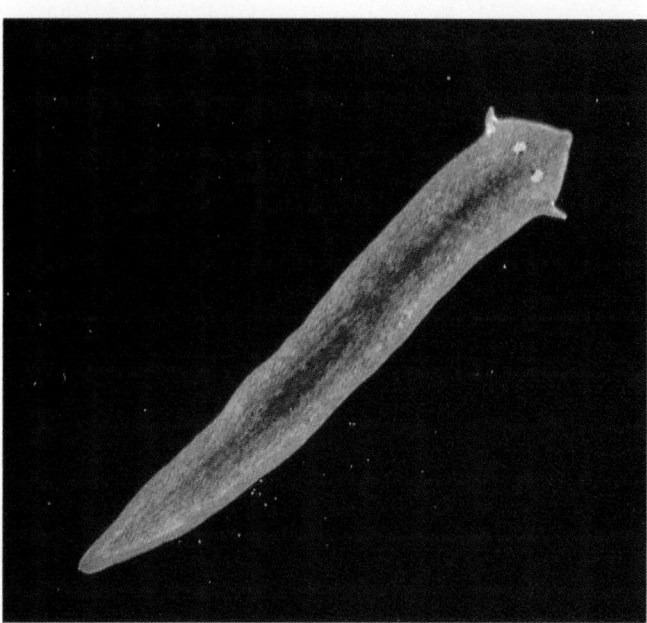

▷ **FIGURE 18–27 Planaria Eyespots** Eyespots like those found in the flatworm planaria perceive light but do not provide vision.

▷ **FIGURE 18–26 Overlapping Visual Fields** The overlap helps us perceive depth—three-dimensional relationships.

fields overlap considerably (▷ Figure 18–26). The overlapping of visual fields gives us the ability to judge the relative position of objects in our visual field; that is, it gives us **depth perception,** also known as **stereoscopic vision.** This ability is also important for predators that need to judge the location of prey with great accuracy, especially ones like the peregrine falcon, which dives at its bird prey at speeds over 150 miles per hour. A lack of depth perception in such an animal would be devastating!

Photoreceptors Appear Very Early in Animal Evolution

Specialized structures that detect light appear in some of the most primitive animals known to science. These photoreceptors range in complexity from simple cellular organelles to clusters of light-sensitive cells to complex structures like the vertebrate eye. Despite the variety, the fundamental process of light detection is quite similar. In most instances, that is, light detection depends on photochemical reactions that involve visual pigments.

A brief look at the different types of photoreceptors gives an overview of the evolutionary progression that led to the formation of the complex human eye. We begin with the eyespots of flatworms and photosynthetic protozoans (single-celled eukaryotes like *Euglena*).

Shown in ▷ Figure 18–27, **eyespots** of a freshwater planarian worm known as *Dugesia* consist of nothing more

than two cup-shaped groups of cells in the head region. These structures permit the organism to detect light, but they are not sophisticated enough to provide vision as we know it.

In the animal kingdom, primitive vision is first found in certain molluscs—a group that includes shellfish, snails, and octopi (▷ Figure 18–28). Even among molluscs, there is a wide range of eye structures, from simple eyespots to sophisticated eyes like ours with a lens, iris, and retina. The most sophisticated eyes belong to squid and octopi and are capable of forming clear images.

Most insects and crustaceans, such as crabs, have **compound eyes** consisting of many photoreceptive units (▷ Figure 18–29). Each photosynthetic unit consists of a small crystalline lens that focuses light on a photoreceptor cell beneath it. Scientists believe that each photoreceptive unit in a compound eye samples a small portion of the visual field. Information then travels to the insect brain, where it is assembled into one coherent image.

The eyes of almost all vertebrates are capable of forming extremely clear images of the world. The human eye, described earlier, is an excellent representative. Despite this commonality in structure, vertebrate eyes do contain modifications that reflect natural selection. In birds of prey (raptors), for instance, the eyes face forward to provide the depth of field vital to diving at prey, but they also contain a higher concentration of cones than those of any other vertebrate. Cones are responsible for visual acuity as well as color vision. Raptor eyes are also large in relation to head size. This is especially evident in owls and eagles, whose eyes are about the same size as the human eye. Large eye size permits a greater accumulation of light,

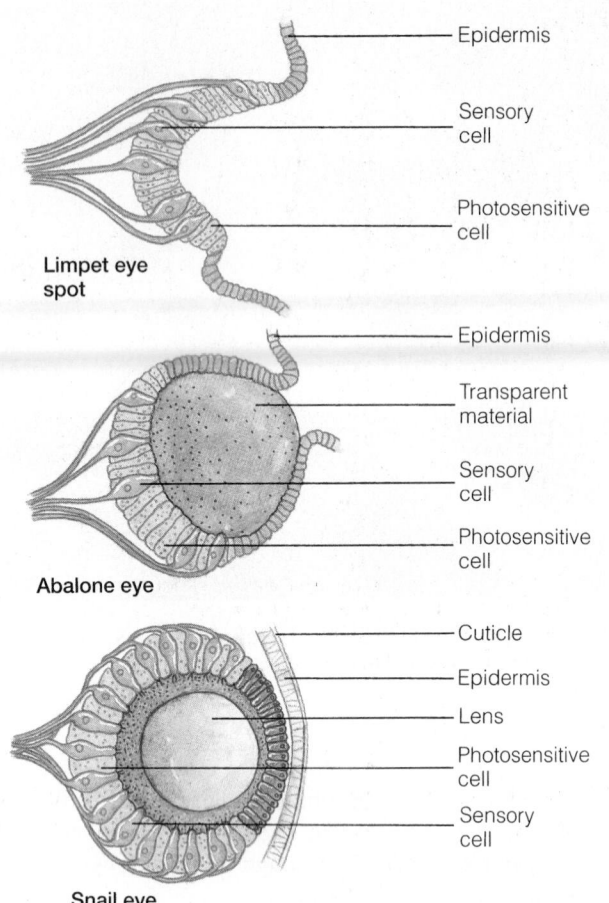

Limpet eye spot

Abalone eye

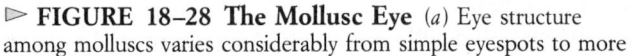

Snail eye

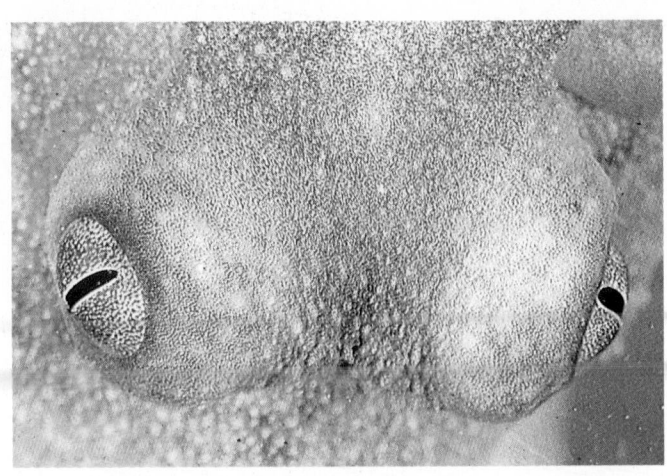

(b)

Octopus eye

▷ **FIGURE 18–28 The Mollusc Eye** (*a*) Eye structure among molluscs varies considerably from simple eyespots to more complex eyes with lenses, retinas, and other structures. (*b*) The octopus eye is similar to human eye.

necessary for seeing in dim light or at night, as in the case of the owl. The eye's large size also separates the lens from the retina, contributing to telescopic vision. (Telescopic lenses for cameras are longer than regular lenses.) Because of telescopic imaging and the higher density of cones, an eagle can see three times better than a human. An eagle can see a rabbit a mile away. You or I would have to be 500 meters, or less than a third of a mile, away to see the rabbit.

≈ HEARING AND BALANCE: THE COCHLEA AND MIDDLE EAR

The human ear is an organ of special sense. It serves two functions: it detects sound, and it detects body position, helping us maintain balance (Table 18–4).

The Ear Consists of Three Anatomically Separate Portions: the Outer, Middle, and Inner Ears

The Outer Ear. The **outer ear** consists of an irregularly shaped piece of cartilage covered by skin, the **auricle,** and the earlobe, a flap of skin that hangs down from the ear (▷ Figure 18–30a). The outer ear also consists of a short tube, the **external auditory canal,** which transmits airborne sound waves to the middle ear (Figure 18–30a). The external auditory canal is lined by skin containing modified sweat glands that produce earwax, or **cerumen.** Earwax traps foreign particles, such as bacteria, and contains a natural antibiotic substance that may help reduce ear infections.

The Middle Ear. The **middle ear** lies entirely within the temporal bone of the skull (Figure 18–30b). The eardrum, or **tympanic membrane,** separates the middle ear cavity from the external auditory canal. The tympanic membrane oscillates when struck by sound waves in much the same way that a guitar string vibrates when a note is sounded by another nearby instrument.

Inside the middle ear are three minuscule bones, the **ossicles.** Starting from the outside, they are the **malleus** (hammer), **incus** (anvil), and **stapes** (stirrup). As illustrated in Figure 18–30b, the hammer-shaped malleus abuts the tympanic membrane. When the membrane is struck by sound waves and vibrates, it causes the malleus to rock

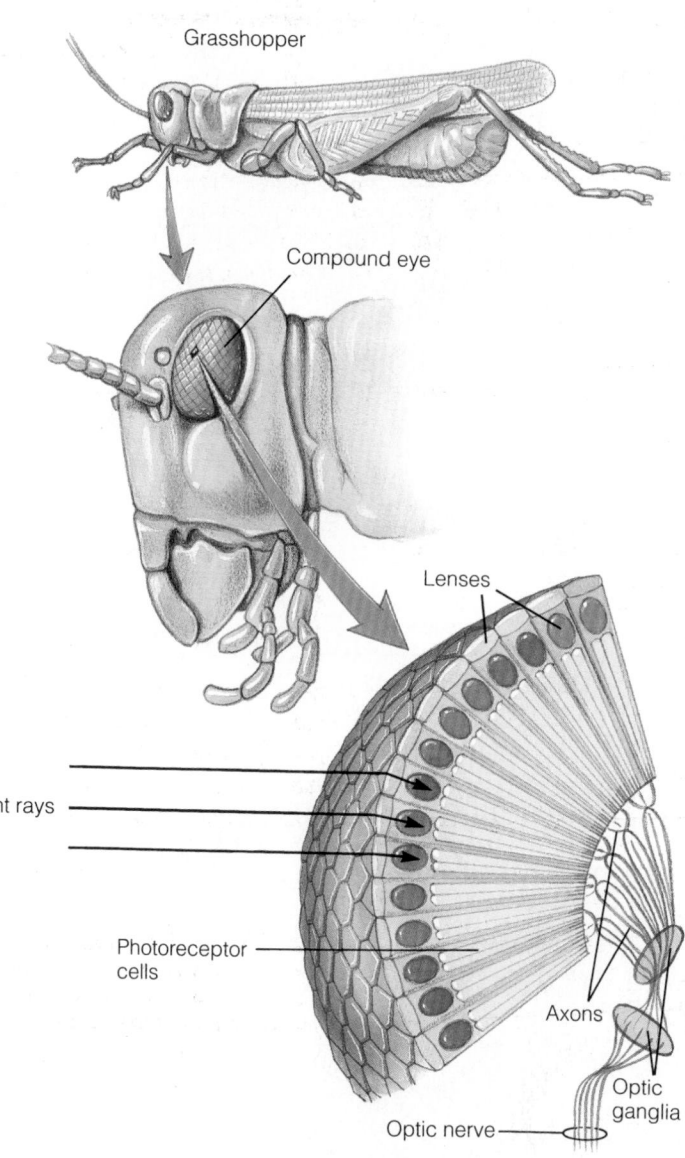

Grasshopper

Compound eye

Lenses

ght rays

Photoreceptor cells

Axons

Optic ganglia

Optic nerve

▷ **FIGURE 18–29 Compound Eye** The insect eye consists of many small units, each with a lens that focuses light on an underlying photoreceptor.

back and forth. The malleus, in turn, causes the incus to vibrate. The incus causes the stapes, the stirrup-shaped bone, to move in and out against the **oval window,** an opening to the inner ear covered with a membrane like the skin on a drum. Thus, vibrations created in the eardrum are amplified as they are transmitted to the inner ear, where the sound receptors are located.

As Figure 18–30 illustrates, the middle ear cavity opens to the pharynx via the **auditory,** or **eustachian, tube.** The eustachian tube extends downward at an angle and opens into the nasopharynx. It serves as a pressure valve. Normally, the eustachian tube is closed. Yawning and swallowing, however, cause it to open. This allows air to flow into or out of the middle ear cavity, equalizing the internal and external pressure on the eardrum, as you will notice when taking off in an airplane or ascending a mountain highway in your car.

Scuba divers and swimmers can sustain considerable damage to an eardrum if they are not careful. As a swimmer descends, pressure from the water builds, pushing the eardrum inward. To prevent the eardrum from tearing, air must be forced into the middle ear cavity. This can be done by simply holding one's nose, clamping one's mouth shut, and gently blowing. Air is forced through the eustachian tube into the middle ear cavity, equalizing the pressure. When a diver ascends, just the opposite happens: air pressure increases inside the middle ear cavity. The diver must release the pressure or else suffer a broken eardrum. This release usually occurs quite naturally, but it can be facilitated by yawning.

The Inner Ear. The inner ear occupies a much larger cavity in the temporal bone than the middle ear and contains two sensory organs, the cochlea and the vestibular apparatus. The **cochlea** is shaped like a snail shell and houses the receptors for hearing. The **vestibular apparatus** consists of two parts: the semicircular canals and the vestibule (Figure 18–30b). The **semicircular canals** are three ringlike structures set at right angles to one another. They house receptors for body position and movement. The **vestibule** is a bony chamber lying between the cochlea and

		TABLE18–4 Structures and Functions of the Ear	
PART	STRUCTURE	FUNCTION
Outer ear	Auricle Ear lobe External auditory canal	Funnels sound waves into external auditory canal Directs sound waves to the eardrum
Middle ear	Tympanic membrane, or eardrum Ossicles	Vibrates when struck by sound waves Transmit sound to the cochlea in the inner ear
Inner ear	Cochlea Semicircular canals Saccule and utricle	Converts fluid waves to nerve impulses Detect head movement Detect head movement and linear acceleration

▷ **FIGURE 18–30 Cross Section
Showing the Structure of the
Ear** (*a*) Notice the structures of the
outer, middle, and inner ears.
(*b*) Note also that the receptors for balance and sound are located in the inner ear.

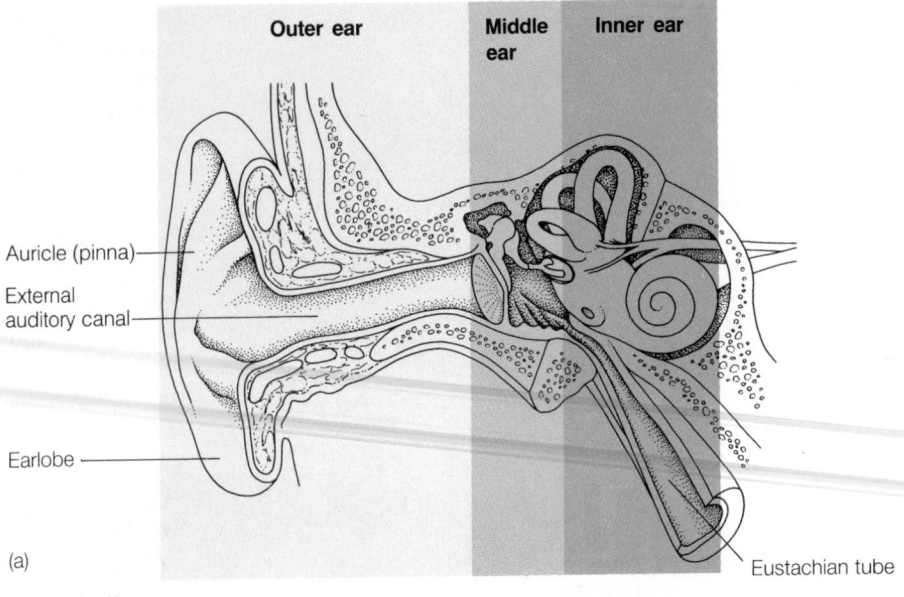

Outer ear | Middle ear | Inner ear

Auricle (pinna)

External auditory canal

Earlobe

Eustachian tube

(a)

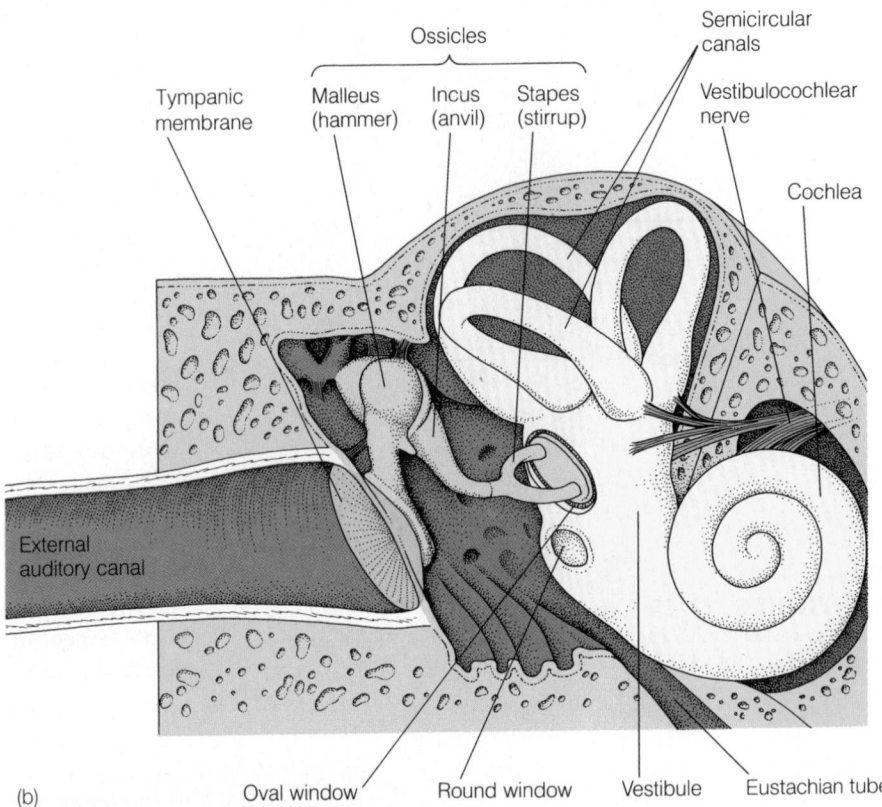

Ossicles

Semicircular canals

Tympanic membrane

Malleus (hammer)   Incus (anvil)   Stapes (stirrup)

Vestibulocochlear nerve

Cochlea

External auditory canal

(b)

Oval window   Round window   Vestibule   Eustachian tube

semicircular canals. It also houses receptors that respond to body position and movement.

Hearing Requires the Participation of Several Structures

The detection and perception of sound require the action of the eardrum, the ossicles, and the cochlea. As noted previously, sound waves enter the external auditory canal, where they strike the tympanic membrane, or eardrum. The eardrum vibrates back and forth, causing the ossicles

of the middle ear cavity to vibrate. The ossicles, in turn, transmit sound waves to the cochlea, which houses the receptor for sound.

Structure and Function of the Cochlea. The cochlea is a hollow, bony spiral. A cross section through this re-markable structure reveals three fluid-filled canals (▷ Fig-ure 18–31a). Separating the middle canal from the lower-most one is a flexible membrane called the **basilar membrane.** It supports the **organ of Corti,** the receptor organ for sound.

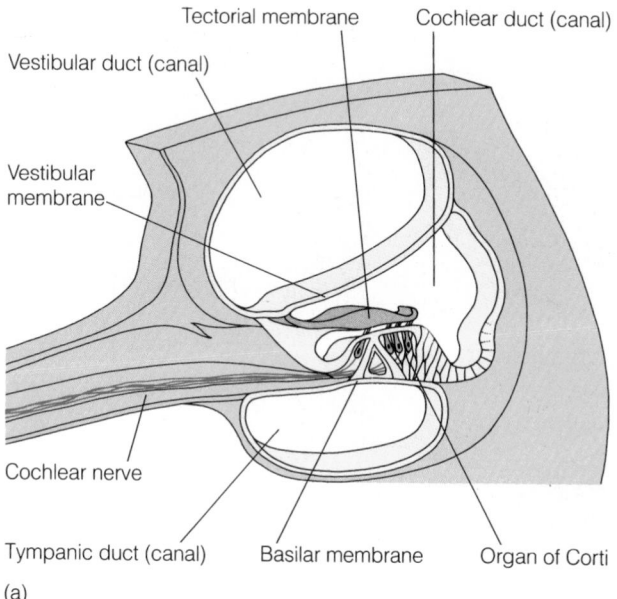

Tectorial membrane

Cochlear duct (canal)

Vestibular duct (canal)

Vestibular membrane

Cochlear nerve

Tympanic duct (canal)   Basilar membrane   Organ of Corti

(a)

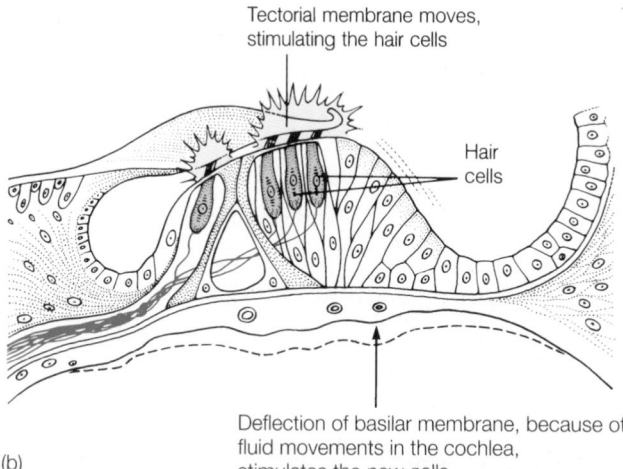

Tectorial membrane moves, stimulating the hair cells

Hair cells

Deflection of basilar membrane, because of fluid movements in the cochlea, stimulates the new cells

(b)

▷ **FIGURE 18–31 Cross Section through the Cochlea**
(*a*) Notice the three fluid-filled canals and the central position of the organ of Corti. (*b*) Hair cells of the organ of Corti are embedded in the overlying tectorial membrane. When the basilar membrane vibrates, the hair cells are stimulated.

The organ of Corti contains two rows of receptor cells, called **hair cells** (Figure 18–31b). Cellular projections from the hair cells contact the overlying **tectorial membrane.** When sound waves are transmitted from the middle ear to the inner ear, they create waves of compression and decompression in the fluid in the uppermost canal of the cochlea, the **vestibular canal.** As shown in ▷ Figure 18–32a, in which the cochlea is unwound, the stirrup transmits vibrations to the drumlike **oval window,** the opening in the bony cochlea. Fluid pressure waves are created in the vestibular canal; they then travel through the thin vestibular membrane into the middle canal, the **cochlear duct (canal).** From here, the pressure waves pass through the basilar membrane into the lowermost canal, the **tympanic**

canal. Pressure is relieved by the outward bulging of the **round window,** an opening in the bony cochlea, which, like the oval window, is spanned by a flexible membrane.

In their course through the cochlea, the pressure waves cause the basilar membrane to vibrate. This vibration stimulates the hair cells. The hair cells respond by releasing a neurotransmitter, which stimulates the dendrites that wrap around the bases of the hair cells. In the cochlea, sound waves are converted to pressure waves, then to nerve impulses. The nerve impulses leave the cochlea via the **vestibulocochlear nerve,** one of the 12 cranial nerves.

Distinguishing Pitch and Intensity. A pitch pipe helps a singer find the note to begin a song. On the pitch pipe are a range of notes from high to low. The ear can distinguish between these various pitches, or frequencies, in large part because of the structure of the basilar membrane. The basilar membrane underlying the organ of Corti is stiff and narrow at the oval window, where fluid pressure waves are first established inside the cochlea (Figure 18–32b). As the basilar membrane proceeds to the apex of the spiral, however, it becomes wider and more flexible. The change in width and stiffness results in marked differences in its ability to vibrate. The narrow, stiff end, for example, vibrates maximally when pressure waves from high-frequency sounds ("high notes") are present (Figure 18–32c). The far end of the membrane vibrates maximally with low-frequency sounds ("low notes"). In between, the membrane responds to a wide range of intermediate frequencies.

Pressure waves caused by any given sound stimulate one specific region of the organ of Corti. The hair cells stimulated in that region send impulses to the brain, which it interprets as a specific pitch. Each region of the organ of Corti sends impulses to a specific region of the auditory cortex in the temporal lobe of the brain. Thus, the auditory cortex can be mapped according to tone in much the same way the motor and sensory cortex can be mapped (Chapter 17).

The intensity of a sound, or its loudness, depends on the amplitude of the vibration in the basilar membrane. The louder the sound, the more vigorous the vibration of the eardrum. The more vigorous the vibration of the eardrum, the greater the deflection of the basilar membrane in the area of peak responsiveness. The greater the deflection of the basilar membrane, the more hair cells are stimulated.

The auditory system is remarkably sensitive, so much so that it can detect sounds that deflect the membrane only a fraction of the diameter of a hydrogen atom. It is no wonder, then, that loud noises can cause so much damage to hearing. Loud rock music or sirens, for example, cause extreme vibrations in the basilar membrane, destroying hair cells over time and causing partial deafness.

Hearing Loss. As people grow older, many lose their hearing. Hearing loss usually occurs so slowly that most people are unaware of it. In some cases, though, people

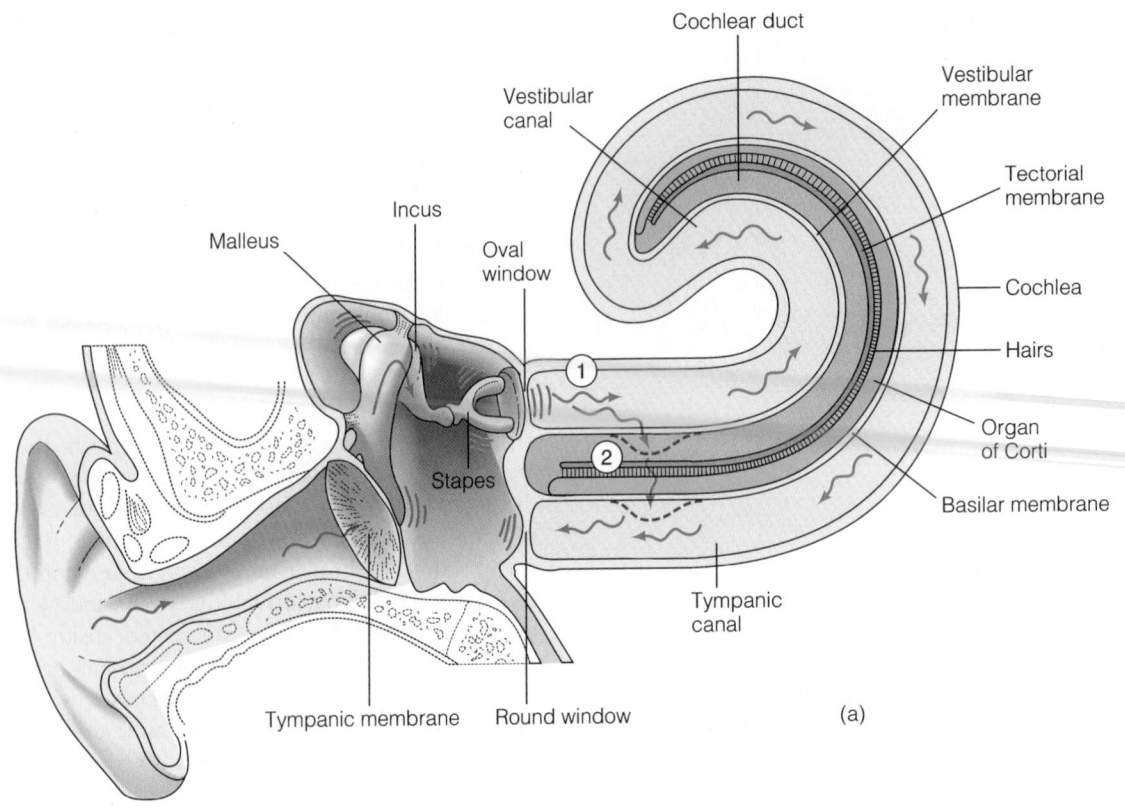

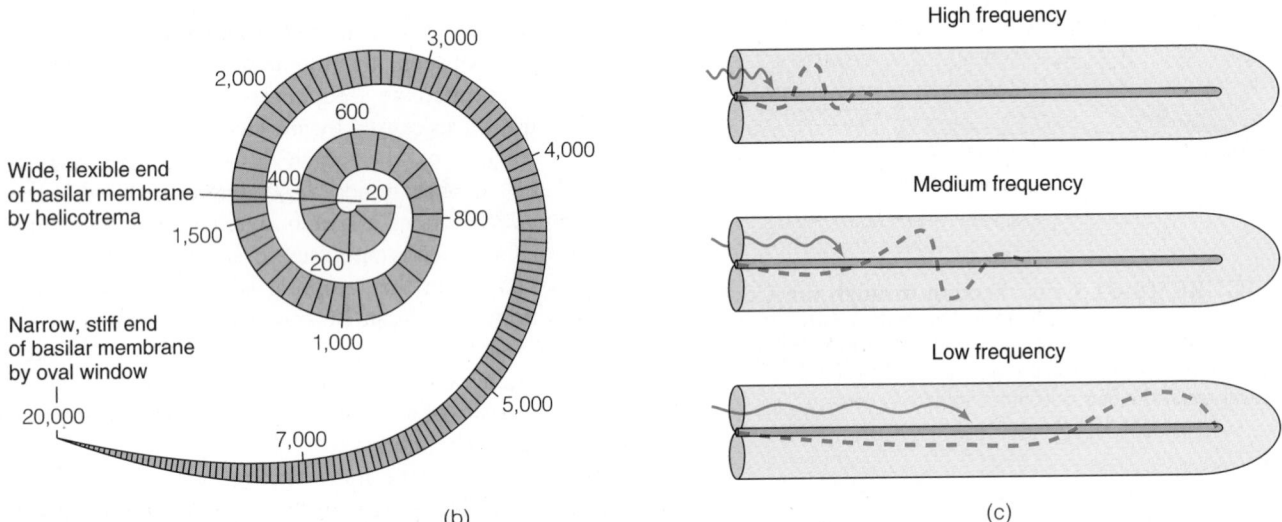

The numbers indicate the frequencies with which different regions of the basilar membrane maximally vibrate.

▷ **FIGURE 18–32 The Transmission of Sound Waves through the Cochlea** The cochlea is unwound here to simplify matters. (*a*) Vibrations are transmitted from the stirrup (stapes) to the oval window. Fluid pressure waves are established in the vestibular canal and pass to the tympanic canal, causing the basilar membrane to vibrate. (*b*) A representation of the basilar membrane, showing the points along its length where the various wavelengths of sound are perceived. Notice that the basilar membrane is narrowest at the base of the cochlea at the oval window end and widest at the apex. (*c*) High-frequency sounds set the basilar membrane near the base of the cochlea into motion. Hair cells send impulses to the brain, which interprets the signals as a high-pitch sound. Low-frequency sounds stimulate the basilar membrane where it is widest and most flexible.

lose their hearing suddenly. A loud explosion, for example, can damage the hair cells or even break the ossicles.

Hearing losses may be temporary or permanent, partial or complete. Basically, hearing loss falls into one of two categories, depending on the part of the system that is affected. The first is **conduction deafness,** which occurs when the conduction of sound waves to the inner ear is impaired. Conduction deafness may result from excessive

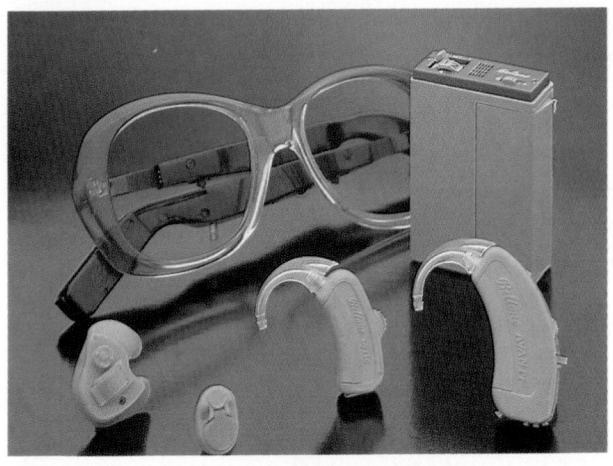

▷ **FIGURE 18–33 Hearing Aids** Worn by people with conduction deafness, hearing aids send sound impulses through the bone of the skull to the cochlea.

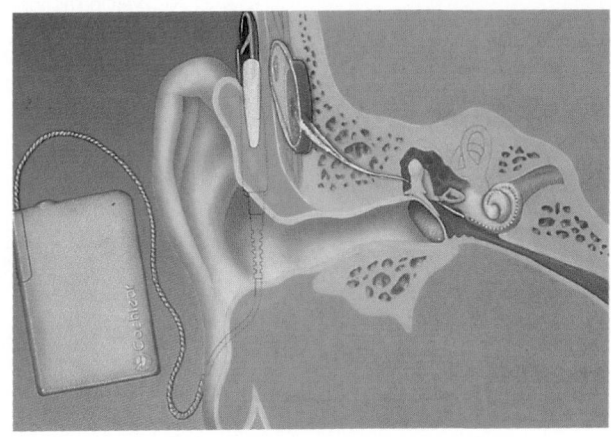

▷ **FIGURE 18–34 Cochlear Implant** The cochlear implant can correct for nerve deafness. Electrodes convey electrical impulses from a small microphone mounted in the ear to the auditory nerve.

earwax in the auditory canal or a rupture of the eardrum. Damage to the ossicles, and even a rupture of the oval window are additional causes of conduction deafness.

Conduction deafness most often results from infections in the middle ear. Bacterial infections may result in the buildup of scar tissue, causing the ossicles to fuse and lose their ability to transmit sound. Infections of the middle ear usually enter through the eustachian tube. Thus, a sore throat or a cold can easily spread to the ear, where it requires prompt treatment. Ear infections are especially common in babies and young children. If undetected, middle ear infections may slow down the development of speech. Untreated, such infections may lead to permanent deafness.

Conduction deafness is treated by hearing aids. A **hearing aid** usually fits in the ear or just behind it (▷ Figure 18–33). These devices bypass the defective sound-conduction system by transmitting sound waves through the bone of the skull to the inner ear. These cause fluid pressure waves to form in the cochlea and stimulate the basilar membrane.

The second type of hearing loss is neurological and is called **nerve,** or **sensineural, deafness.** Sensineural deafness results from physical damage to the hair cells, the vestibulocochlear nerve, or the auditory cortex. Explosions, extremely loud noises, and some antibiotics, for example, can all damage the hair cells in the organ of Corti, creating partial or even complete deafness. The auditory nerve, which conducts impulses from the organ of Corti to the cortex, may degenerate, thus ending the flow of information to the cortex. Tumors in the brain or strokes may destroy the cells of the auditory cortex.

Although the ear is quite vulnerable to loud noise and other problems, it contains a mechanism to protect itself from damage. This mechanism consists of two tiny skeletal muscles (the smallest in the body) located in the middle ear cavity. One of these muscles inserts on the malleus, and the other attaches to the stapes. As noted earlier, the malleus attaches to the eardrum, and the stapes attaches to the oval window. Loud noises stimulate a reflex contraction of the middle ear muscles, pulling them away from their membrane contacts. Whenever an individual is exposed to loud noise, this reflex reduces the conduction of the noise to the inner ear. Unfortunately, the reflex requires about 40 milliseconds, so it cannot protect the ear from explosions.

Correcting Profound Deafness. More than 2 million Americans are profoundly deaf, a condition that has until recently been considered virtually untreatable. The profoundly deaf hear nothing, not even the sound of sirens.

Children who are born deaf or are deafened before they begin to speak often fail to mature emotionally. Even reading comprehension may be impaired. Some profoundly deaf children, in fact, never advance beyond third- or fourth-grade reading levels.

Hearing aids usually cannot help individuals who are born deaf or those who suffer from nerve damage. Researchers, however, have developed a new device called a **cochlear implant,** which simulates the function of the inner ear (▷ Figure 18-34). This device picks up sound and transmits it to a receiver implanted inside the skull. The signal then travels to an electrode implanted in the vestibulocochlear nerve in the cochlea. Electrical impulses in the electrode stimulate the nerve, creating impulses that travel to the auditory cortex.

Today, hundreds of adults and children are equipped with cochlear implants. The single-electrode model, which provides the deaf with only a rudimentary hearing capacity, is quickly becoming obsolete thanks to the advent of newer multiple-electrode models. These models detect and transmit a wider range of sounds. Recipients of the new models

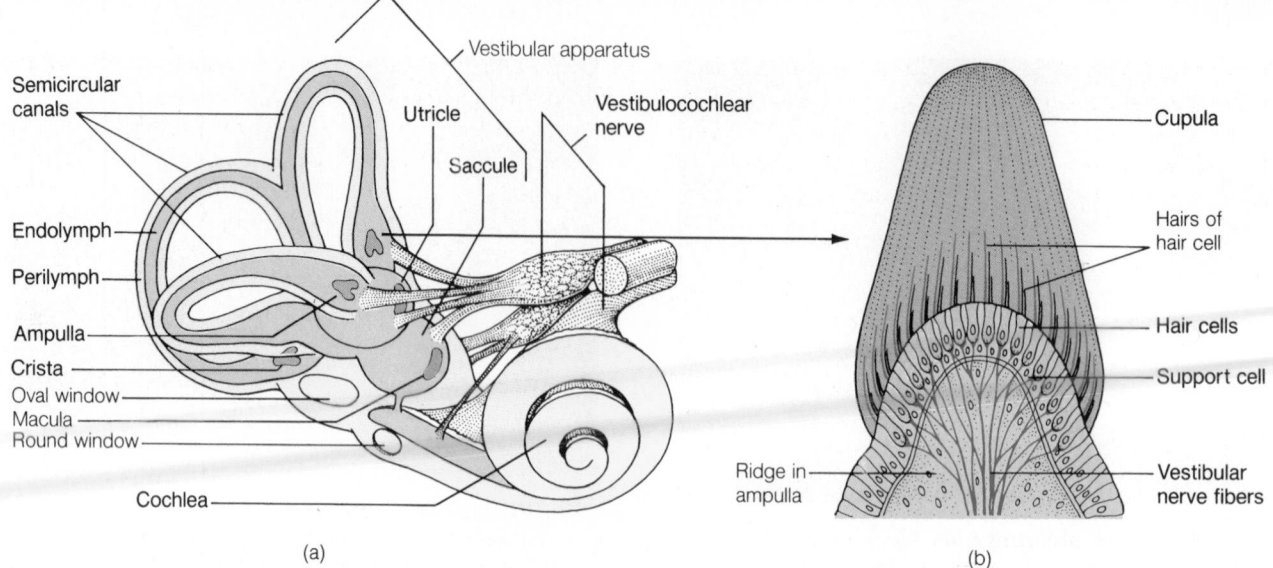

(a)   **(b)**

▷ **FIGURE 18–35 Location and Structure of the Cristae**
(*a*) This illustration shows the location of the cristae in the ampullae of the semicircular canals. The semicircular canals are filled with endolymph. (*b*) When the head spins, the endolymph is set into motion, deflecting the gelatinous cupula of the crista, thus stimulating the receptor cells.

can perceive many distinct words, not just the sound of a telephone or an automobile horn.

Although it is doubtful that "normal" hearing will ever be fully restored by such devices, people who have lived in the silent world point out that any sound is better than no sound at all. For them, the cochlear implant provides valuable outside stimuli, such as sirens and horns, the importance of which many of us take for granted. The cochlear implant also helps the deaf monitor and regulate their own voices, and it makes lip reading easier. For deaf children, a cochlear implant could mean the difference between learning to speak or a lifetime of silence.

The Vestibular Apparatus Houses Receptors that Detect Body Position and Movement

The cochlea lies alongside the **vestibular apparatus,** which as noted earlier, consists of the semicircular canals and the vestibule, a bony chamber between the cochlea and semicircular canals. The vestibular apparatus houses receptors that detect body position and movement (▷ Figure 18–35).

The Semicircular Canals. The three semicircular canals are arranged at right angles to one another. Each canal is filled with a fluid called **endolymph.** As illustrated in Figure 18–35a, the base of each semicircular canal expands to form the **ampulla.** On the inside wall of each ampulla is a small ridge of tissue, or **crista.** Each crista consists of a patch of receptor cells, each of which contains numerous microvilli and a single cilium embedded in a cap of gelatinous material, the **cupula,** as illustrated in Figure 18–35b. Dendrites of sensory nerves wrap around the base of the receptor cells, and the cupula extends into the cavity of the ampulla, where it is bathed in endolymph.

Rotation of the head causes the endolymph in the semi-circular canals to move. The movement of the fluid deflects the cupula, which stimulates the hair cells to release a neurotransmitter that excites the sensory neurons. These neurons send impulses to the brain, alerting it to the rotational movement of the head and body.

Because the semicircular canals are set in all three planes of space, movement in any direction can be detected. By alerting the brain to rotation and movement, the semicircular canals contribute to our sense of balance. They are therefore important to **dynamic balance**—helping us stay balanced when in movement.

The Utricle and Saccule. Two additional receptors play a role in balance and detection of movement, the **utricle** and **saccule** (Figure 18–35). These membranous compartments inside the vestibule provide input under two distinctly different conditions: at rest and during acceleration in a straight line. As a result, these receptors contribute to **static balance**—helping us stay balanced when we are not moving—and dynamic balance.

The utricle and saccule contain small receptor organs known as **maculae** ("spots"), which are akin to the cristae of the semicircular canals. Each macula consists of a patch of receptor cells, which is structurally similar to those found in the cristae. The cilia and microvilli of the receptor cells in the maculae are also embedded in a gelatinous cap. One notable difference with the cristae, however, is the presence of numerous small crystals of calcium carbonate, called **otoliths** (literally, "ear rocks"), which are embedded in the gelatinous material. The otoliths make the gelatin heavier than the surrounding fluid (▷ Figure 18–36a).

Although similar in structure, the maculae in the utricle and saccule are oriented in different directions. Thus, when a person is standing, the hair cells of the utricle are oriented vertically. Bending the head forward, as illustrated in Figure 18–36b, causes the otoliths and gelatinous cap to droop forward. Pulled downward by gravity, the gelatinous cap stimulates the receptor cells. Nerve impulses are transmitted to the brain, alerting it to the head movement. Moving forward—say, by walking or running—causes the gelatinous mass to slide backward, as shown in Figure

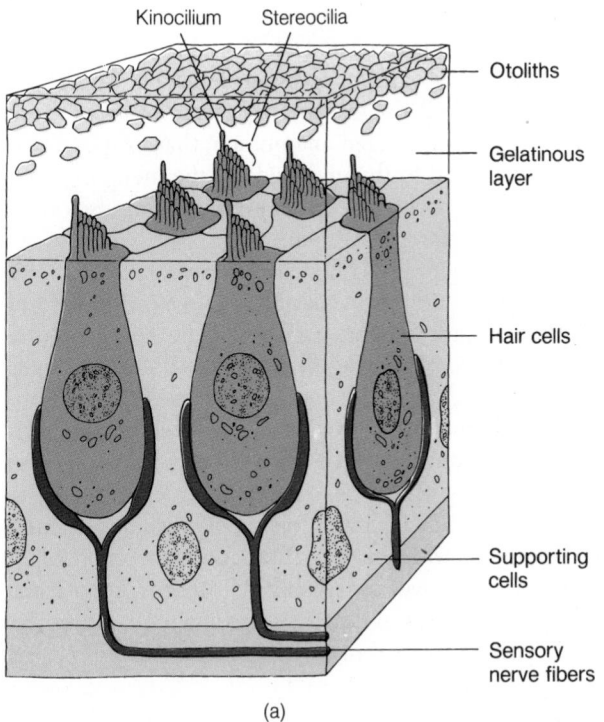

(a)

18–36c, and the utricle therefore also transmits information to the brain on linear acceleration.

In the saccule, the hair cells are oriented perpendicularly to those of the utricle. That is, they are horizontal when a person is standing or sitting. These receptor cells are therefore stimulated when the head is tilted back—for instance, when a person lies down. The saccule also responds to acceleration and deceleration but in a vertical direction—for example, when one rides on an elevator or bounces on a trampoline.

The information provided by the semicircular canals and the maculae is sent to a cluster of nerve cell bodies in the brain stem called the **vestibular nuclei.** Here all of the information the receptors generate on position and movement is integrated along with input from the eyes and from receptors in the skin, joints, and muscles, as discussed earlier in the chapter.

From the vestibular nuclei, information flows in many directions. One major pathway leads to the cerebral cortex. This path makes us conscious of our position and movement. Another path leads to the muscles of the limbs and torso. Signals to the muscles help maintain our balance and, if necessary, correct body position.

In some people, activation of the vestibular apparatus results in motion sickness, characterized by dizziness and nausea. The exact cause of motion sickness is not known.

▷ **FIGURE 18–36 The Macula** (*a*) Receptor cells in the saccule and utricle are surrounded by supporting cells. Otoliths embedded in the gelatinous cap make the layer heavier than the surrounding fluid. (*b*) Position of the macula of the utricle in an upright position and when head is tilted forward. (*c*) Deflection of the otoliths during forward motion.

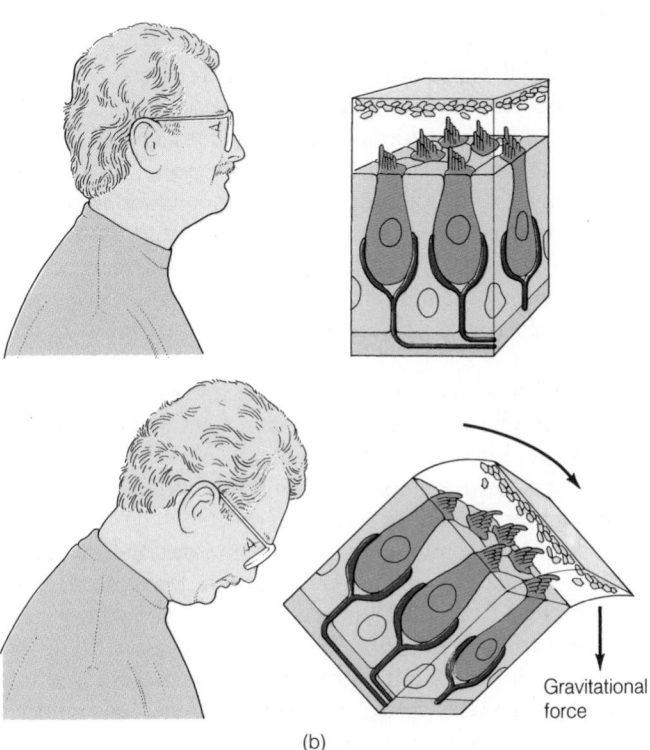

(b)

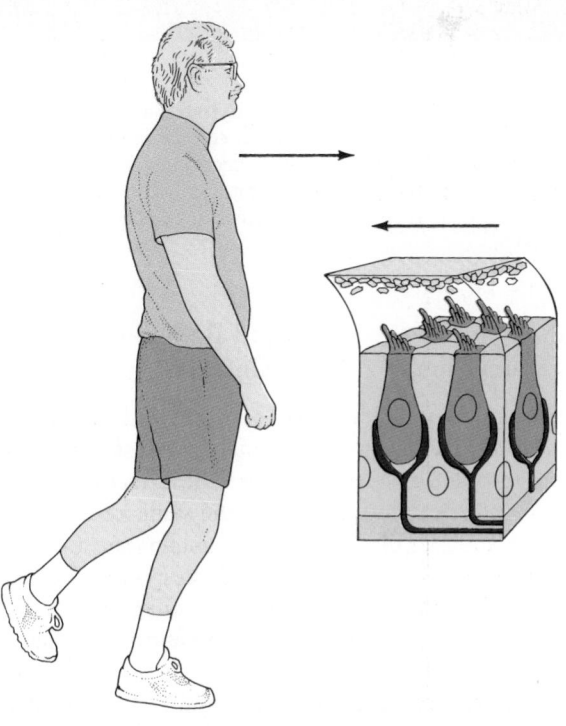

(c)

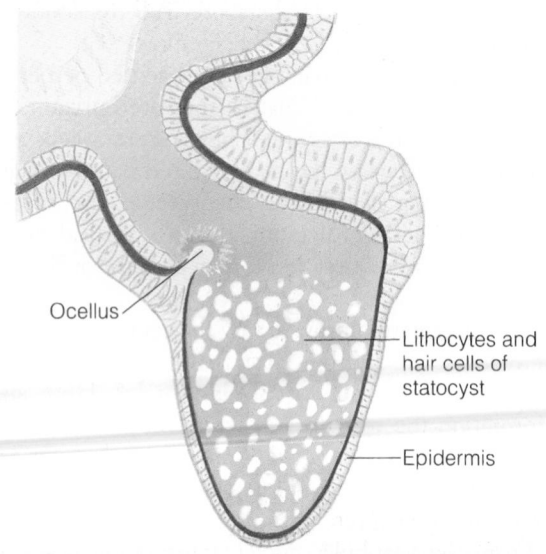

(a)

(b)

▷ **FIGURE 18–37 Statocysts in Jellyfish** Small flaps of tissue along the lower margin of the jellyfish contain sensory receptors that help the organism sense and correct body position. When the organism tilts, small crystals in the statocyst press against neighboring hair cells, alerting the nervous system of a change in body position.

The Ability to Detect Sound and Body Position Are Prevalent in the Animal Kingdom

Turn most animals upside down, and they will right themselves. Why? Because many animals possess mechanoreceptors that detect body position. These receptors contain hair cells akin to those found in the semicircular canals and utricle and saccule. Hair cells contain one or more surface processes, or hairs, that bend in response to gravity or other forces. This bending produces bioelectric signals that are sent to the brain, alerting the organism of a change in position.

A good example is the position receptor of the jellyfish, shown in ▷ Figure 18–37. As illustrated, the **statocyst**

consists of a flap of tissue containing hair cells and small crystals. When the body moves, the crystals press against the hair cells, stimulating them. They send information to the nervous system that is used to maintain body position.

Sound detection is also widespread in the animal kingdom. For example, many insects have body hairs that vibrate at the same frequency as biologically important sounds. The body hairs of certain caterpillars, for instance, vibrate in synch with the wingbeat of wasps that prey on them, thus permitting the caterpillar to take evasive action.

Some insects also have pressure-detecting organs on various parts of their bodies, such as the front legs. These organs contain a tympanic membrane that responds to vibrations. Attached to the underside of the membrane are numerous sensory cells that generate nerve impulses that, in turn, travel to the insect brain.

In many fishes, cells containing minute hair cells are located in longitudinal canals known as the **lateral lines**—that is, grooves that run along the length of the body on both sides (▷ Figure 18–38). These hair cells detect low-frequency vibrations of water flow and turbulence, allowing the fish to locate underwater obstructions and monitor the movement of other fishes swimming close by when in murky water.

Some vertebrates, including bats, dolphins, and whales, come equipped with radar. These animals emit high-frequency sound waves that echo off objects or potential predators or prey. The sound waves bounce off the object or animal and are picked up by the animals' ears.

ENVIRONMENT AND HEALTH: NOISE POLLUTION

Noise may be one of the most widespread environmental pollutants in modern industrial societies. Unknown to many, noise may be turning our nation into a country of the hearing-impaired. Traffic noise, airport noise, loud music, crowd noise, and other loud sounds so common in modern society may be slowly destroying our hearing. By the time many New York City residents reach age 20, their hearing has been so impaired that they hear only as well as a 70-year-old African Bushman who has lived a life free of city noises.

For many years, scientists attributed the decline in hearing to middle ear infections, certain antibiotics, and the natural deterioration of the hair cells. New research, however, suggests that noise is probably the principal cause of hearing loss in modern societies.

Like that of any other pollutant, the damage caused by noise is related to two factors: exposure level (how loud a noise is) and the duration of exposure (how long one is exposed to it). In general, the louder the noise, the more damaging it is. In addition, the longer you are exposed to a damaging level of noise, the more hearing loss you will suffer. Extremely loud noise can result in immediate, per-

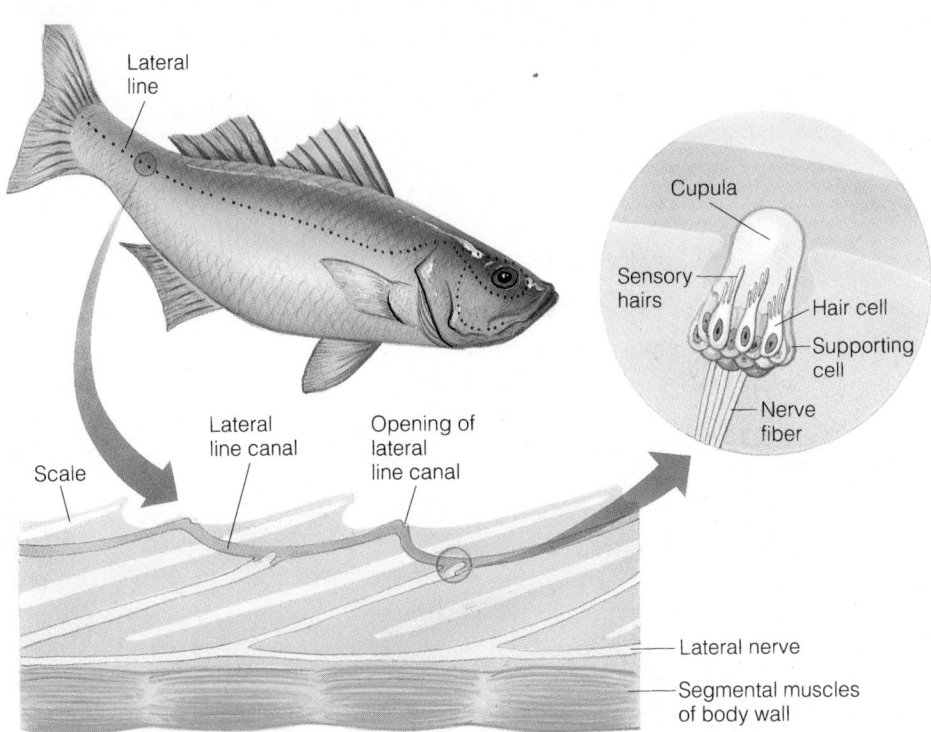

▷ **FIGURE 18–38 Lateral Line** The lateral line of fishes contains groups of sensory cells that detect pressure waves in water, alerting a fish to the presence of other fishes and obstacles.

Lateral line

Cupula

Sensory hairs

Hair cell

Supporting cell

Nerve fiber

Scale

Lateral line canal

Opening of lateral line canal

Lateral nerve

Segmental muscles of body wall

manent damage. An explosion, for instance, can rupture the eardrum or fracture the bony ossicles, resulting in conduction deafness.

The noise to which people are exposed in factories and at construction sites is sufficient to cause gradual hearing loss. A worker may notice a dulled sense of hearing after working in a noisy environment; this is called a **temporary threshold shift.** Over time, the continued assault leads to a **permanent threshold shift,** complete hearing loss in certain frequencies. In most cases, hearing loss occurs so gradually that workers do not notice it until it is too late.

The intensity of sound is measured in **decibels (db).** Table 18–5 shows the scale and lists some common sounds. Surprising new research published by the EPA shows that continuous exposure to sounds over 55 decibels can result in hearing loss. Light traffic and an air conditioner operating 6 meters (18 feet) from you are 60-decibel sounds.

Besides deafening us and cutting us off from the important sounds of modern life, noise disturbs communications, rest, and sleep. It raises our level of stress, which, in turn, shortens our lives. When hearing is impaired, communication falters, and tensions often rise. People who are losing their hearing complain that they feel inadequate in social situations.

Hearing loss is not inevitable. You can take steps to protect your hearing. Keep your stereo at a reasonable level. Avoid noisy events. Cover your ears when an ambulance or fire engine approaches. Wear ear plugs or ear guards when operating noisy equipment like vacuum cleaners, chain saws, or construction equipment. Get treatment if you develop an ear infection. Prevention is the best medicine, because once you lose your hearing, it is gone forever, and so is an important part of your life.

SUMMARY

THE GENERAL SENSES

1. The general senses are pain, light touch, pressure, temperature, and position sense. Receptors for these senses may be naked or encapsulated nerve endings.
2. Receptors fit into five functional categories: (a) mechanoreceptors, (b) thermoreceptors, (c) photoreceptors, (d) chemoreceptors, and (e) pain receptors (nocireceptors).
3. Sensory stimuli may be internal or external. Some stimuli cause reflex actions. Still others may stimulate physiological changes.

4. The naked nerve ending receptors in the body include those stimulated by pain, light touch, and temperature. The encapsulated receptors include those that detect pressure (Pacinian corpuscles), light touch (Meissner's corpuscles, Krause's end-bulbs, and Ruffini's corpuscles), and muscle extension (muscle spindles and Golgi tendon organs).
5. Many sensory receptors cease responding to prolonged stimuli, a phenomenon called adaptation. Pain, temperature, pressure, and olfactory receptors all adapt.

SOUND LEVEL, db	SOUND SOURCES	EFFECTS		
		Perceived Loudness	Damage to Hearing	Community Reaction to Outdoor Noise
180 —	Rocket engine			
170 —				
160 —				
150 —	Jet plane at takeoff	Painful	Traumatic injury	
140 —			Injurious range, irreversible damage	
130 —	Maximum recorded rock music			
120 —	Thunderclap	Uncomfortably loud		
110 —	Auto horn, 1 meter away / Riveter / Jet flying over at 300 meters		Danger zone; progressive loss of hearing	
100 —	Newspaper press			Vigorous action
90 —	Motorcycle, 8 meters away / Food blender / Diesel truck, 80 km/hr, 15 meters	Very loud	Damage begins after long exposure	Threats
80 —	Garbage disposal			
70 —	Vacuum cleaner / Ordinary conversation			Widespread complaints
60 —	Air conditioning unit, 6 meters / Light traffic noise, 30 meters	Moderately loud		Occasional complaints
50 —	Average living room			
40 —	Bedroom	Quiet		No action
30 —	Library / Soft whisper			
20 —	Broadcasting studio	Very quiet		
10 —	Rustling leaf	Barely audible		
0 —	Threshold of hearing			

TASTE AND SMELL: THE CHEMICAL SENSES

6. The special senses are served by more elaborate receptor organs, providing for taste, smell, vision, hearing, and balance.

7. Taste receptors called taste buds are located principally on the upper surface of the tongue. Taste buds contain receptor cells. Food molecules dissolve in the saliva and bind to the membranes of the microvilli of the receptor cells.

8. Taste buds respond to four flavors: salty, bitter, sweet, and sour.

9. The receptors for smell are located in the olfactory membrane in the roof of the nasal cavities. The receptor cells in the olfactory membrane respond to thousands of different molecules, which bind to membrane receptors on the olfactory hairs, stimulating nerve impulses that are transmitted to the brain via the olfactory nerve.

10. The ability to detect various chemicals in the environment is a universal trait of animals and even protists.

11. Chemoreception is used to find food and mates, recognize territories, and move about in the environment.

12. Insects have sensory hairs on their mouthparts and feet. The sensory hairs contain more than one receptor cell, each of which responds maximally to a different chemical. Some sensory hairs detect airborne chemicals, such as pheromones.

THE VISUAL SENSE: THE EYE

13. The eye is the receptor for visual stimuli and is located in the orbit, a bony socket.

14. The wall of the human eye contains three coats. The outermost coat is the fibrous layer and consists of the sclera (the white of the eye) and the cornea (the clear anterior structure that lets light shine in).

15. The middle layer consists of the choroid (the pigmented and vascularized section), the ciliary body (whose muscles control the lens), and the iris (which controls the amount of light entering the eye).

16. The innermost layer is the retina, which contains two types of photoreceptors. Rods function in dim light and provide black-and-white vision, and cones operate in bright light and provide color vision. Cones are also responsible for visual acuity. The highest concentration of cones is found in the fovea centralis.

17. Light is focused on the retina by the cornea and lens. The cornea has a fixed refractive power, but the lens can be adjusted to bend light according to need. The muscles of the ciliary body play an important role in this process. Focusing on near objects is aided by pupillary constriction.

18. The lens may become cloudy with old age or because of exposure to excess ultraviolet light. This condition, called cataracts, can be corrected by surgically removing the lens and replacing it with a plastic one.

19. The eye is divided into two cavities. The posterior cavity lies behind the lens and is filled with a gelatinous material, the vitreous humor. The anterior cavity is filled with a plasmalike fluid called the aqueous humor, which nourishes the lens and other eye structures in the vicinity. If the rate of absorption of aqueous humor decreases, however, pressure can build inside the anterior cavity, resulting in glaucoma.

20. As people age, the lens becomes less resilient and less able to focus on nearby objects, a condition called presbyopia.

21. Three additional eye problems are myopia, hyperopia, and astigmatism.

22. Myopia, or nearsightedness, results from a lens that is too strong (too concave) or an elongated eyeball. In the uncorrected eye, the image from distant objects comes in focus in front of the retina.

23. Hyperopia results from a weak lens (too convex) or a shortened eyeball. In the uncorrected eye, light rays from an image nearby would come into focus behind the retina.

24. Astigmatism is an irregularly curved lens or cornea that creates fuzzy images.

25. Three types of cones are present in the eye: red, green, and blue. Each type responds maximally to one specific color of light. Intermediate colors activate two or more types of cones.

26. Color blindness is a genetic disorder, more common in men than women. It results from a deficiency or an absence of one or more types of cones. Red-green color blindness is the most common type.

27. Structures that detect light appear in some of the most primitive animals on Earth and range in complexity from simple cellular organelles to clusters of light-sensitive cells to the vertebrate eye. Despite this variety, the fundamental process of light detection is quite similar.

28. One of the simplest photoreceptors is the eyespot of the planaria, which consists of two cup-shaped groups of cells in the head region of the organism that permit it to detect light but do not permit vision as we know it.

29. In the animal kingdom, primitive vision is first found in molluscs. Some molluscs like squid and octopi have sophisticated eyes that permit clear vision.

30. Most insects and crustaceans, such as crabs, have compound eyes that consist of many photoreceptive units, each with a small crystalline lens that focuses light on a photoreceptor cell beneath it.

HEARING AND BALANCE: THE COCHLEA AND MIDDLE EAR

31. The human ear consists of three portions: the outer, middle, and inner ears.

32. The outer ear consists of the auricle and external auditory canal, both of which direct sound to the eardrum.

33. The middle ear consists of the eardrum and three small bones, the ossicles, which transmit vibrations to the inner ear.

34. The auditory, or eustachian, tube helps equilibrate the pressure inside the middle ear cavity.

35. The inner ear contains the cochlea, which houses the organ of Corti, where the receptors for sound are located. The inner ear also houses receptors for movement and head position: the semicircular canals, utricle, and saccule.

36. The cochlea is a spiral-shaped, bony structure that contains three fluid-filled canals. Separating the middle canal from the lower one is the flexible basilar membrane, which supports the organ of Corti. Hair cells in the organ of Corti are embedded in the relatively rigid tectorial membrane.

37. Sound waves create vibrations in the eardrum and ossicles, which are transmitted to fluid in the cochlea. Pressure waves in the cochlea cause the basilar membrane to vibrate, which stimulates the hair cells.

38. Pressure waves resulting from any given sound cause one part of the membrane to vibrate maximally. The hair cells stimulated in that region send signals to the brain, which it interprets as a specific frequency.

39. Hearing loss may occur as a result of damage or blockage to the conducting system: the external auditory canal, the eardrum, and the ossicles. Damage to the hair cells, the auditory nerve, or the auditory cortex are forms of nerve or sensineural deafness.

40. The semicircular canals are three hollow rings filled with a fluid called endolymph. The receptors for head movement are located in an enlarged portion at the base of each canal, the ampulla.

41. Fluid movement inside the semicircular canals deflects the gelatinous cap (cupula) lying over the receptor cells, stimulating them and alerting the brain to head movements.

42. The semicircular canals are set in all three planes of space, so movement in any direction can be detected.

43. Two membranous sacs in the vestibule, the utricle and saccule, contain receptors called maculae that respond to linear acceleration and tilting of the head.
44. Mechanoreceptors that detect body position are common in the animal kingdom. They contain hair cells akin to those found in the vertebrates. Hair cells respond to gravity or other forces, producing bioelectric impulses that are sent to the brain, alerting the organism of a change in position.
45. Sound detection is also widespread in the animal kingdom. Many insects have body hairs that vibrate at the same frequency as biologically important sounds. Some insects also have pressure-detecting organs on various parts of their bodies, such as the front legs.

46. In many fishes, cells containing minute hair cells are located in longitudinal grooves that run along the length of the body on both sides. These receptor cells detect low-frequency vibrations of water flow and turbulence.

ENVIRONMENT AND HEALTH: NOISE POLLUTION

47. Noise damages the ears. Extremely loud noises can rupture the eardrum or break the ossicles. Less intense noises, however, generally destroy hearing gradually by damaging hair cells.
48. In most people, hearing loss occurs so gradually as to be undetected. Individuals can take steps to avoid hearing loss.

EXERCISING YOUR CRITICAL THINKING SKILLS

In an experiment to determine how chemicals affect vision, two researchers expose rats to varying levels of a toxin, chemical A. They find that at low doses, the chemical has no effect, but higher doses result in severe vision loss and blindness. The researchers immediately ask the Food and Drug Administration to ban the chemical from production for fear it might similarly affect humans. You are head of the FDA. Using your critical thinking skills, how would you go about considering the request? What factors would help you determine whether the request was valid? What studies might you want to see done?

TEST OF TERMS

1. The general senses are _____, temperature, light touch, pressure, and proprioception, or position sense.
2. Receptors fall into two broad groups: naked nerve endings and _____ _____.
3. Light touch is perceived by two sensors: naked nerve endings surrounding hair follicles and _____ _____.
4. Pressure is perceived by _____ corpuscles located in the deeper layers of the skin and around various organs.
5. The receptor situated in the superficial layer of the dermis that responds to touch is the _____ _____.
6. The _____ _____ is a stretch receptor found in skeletal muscles.
7. Stretch receptors in tendons are called _____ _____ _____.
8. _____ occurs when a sensory receptor stops sending impulses, even though the stimulus is still present.
9. Taste, hearing, and vision are three of the _____ _____.

10. The taste receptors in the oral cavity and on the dorsal (upper) surface of the tongue are called _____ _____. They are especially abundant on the _____, small protrusions on the surface of the tongue.
11. Receptors for the sense of smell are located in the _____ membrane found in the roof of each nasal cavity. Receptor cells in the membrane are modified _____ neurons.
12. The eye consists of three layers. The outermost, fibrous layer consists of the _____, the white of the eye, and the _____, the clear, anterior portion that allows light to enter.
13. The middle layer of the eye is heavily _____ and vascularized. It consists of the _____, the _____ _____, and the iris.
14. The inner layer of the eye is the light-sensitive portion of the eye and is called the _____. It contains two types of receptors, the

_____, which confer color vision, and the rods, which operate best in _____ light.
15. Nerve impulses leave the retina via the optic nerve, which is formed from the axons of the _____ cells.
16. Sharpest vision occurs when an image is cast on the _____ _____, a small depression in the retina lying lateral to the blind spot, or _____ _____.
17. The _____ is a flexible structure used to focus light coming from nearby objects on the retina. It is attached to the _____ _____ by the suspensory ligaments. It may become cloudy with age, a condition known as _____.
18. The _____ _____ is a gelatinous mass that occupies the posterior cavity of the eye.
19. _____ results from the excess buildup of _____ _____ in the anterior cavity of the eye.
20. The bending of light is called _____. It occurs anytime light waves _____ _____.

21. Eye movements are caused by the _____ _____ muscles.
22. Myopia is also called _____ . It results from a(n) _____ or a(n) _____ _____ .
23. The surgical technique that corrects for myopia is called _____ _____ .
24. An irregularly curved lens or cornea results in a condition known as _____ .
25. _____ is the pigment of the rods. It breaks down into two molecules when struck by light.
26. Color blindness is a _____-_____ trait.
27. Sound waves are directed into the _____ _____

canal to the _____ , which separates the outer ear from the middle ear.
28. Extremely loud noises may damage the _____ , the bones in the middle ear cavity, which transmit sound to the _____ _____ of the cochlea.
29. The _____ _____ leads from the middle ear cavity to the nasopharynx and serves as a pressure release valve.
30. Movement of the head is detected by sensors in the _____ of the _____ canals in the inner ear. These canals are filled with a fluid called _____ .

31. Static balance is provided in part by two receptor patches, the _____ , located in the utricle and saccule.
32. The _____ _____ _____ is the receptor organ for sound in the cochlea. It consists of three canals: the _____ canal, the cochlear duct, and the tympanic canal.
33. A rupture of the eardrum or a fusion of the bones in the middle ear results in _____ deafness.
34. A transitory loss of hearing is called a _____ _____ shift.

Answers to the Test of Terms are found in Appendix B.

TEST OF CONCEPTS

1. Define the terms *general senses* and *special senses*.
2. Using your knowledge of the senses and of other organ systems gained from previous chapters, describe the role that sensory receptors play in homeostasis. Give specific examples to illustrate your main points.
3. Make a list of both the encapsulated and the nonencapsulated general sense receptors. Note where each is located and what it does.
4. Define the term *adaptation*. What advantages does it confer? Can you think of any disadvantages?
5. Describe the receptors for taste, explaining where they are located, what they look like, and how they operate.
6. Taste buds detect four basic flavors. What are they? How do you account for the thousands of different flavors that you can detect?
7. Describe the olfactory membrane and the structure of the receptor cells. How do these cells operate? In what ways are taste receptors and olfactory receptors similar? In what ways are they different?
8. Explain the following statement: taste and smell are complementary.
9. Describe the function of chemoreception in the animal kingdom. Give some examples. Are the structures found in animals other than vertebrates like those found in humans?
10. Draw a cross section of the human eye, and label its parts.
11. Define the following terms: retina, rods, cones, fovea centralis, optic disc, ganglion cells, bipolar neurons, and optic nerve.
12. Compare and contrast rods and cones.
13. When focusing on a nearby object, your eyes go through several changes. Describe those changes and what they accomplish.
14. Define the following terms: myopia, hyperopia, presbyopia, and astigmatism.
15. You walk into a dark movie theater and find that you can barely make out the aisle. After a while,

your vision recovers. Explain both phenomena.
16. What is color blindness? What is the most common type? Explain what your world would look like if you were afflicted by red-green color blindness.
17. What is a cataract, and how is it treated?
18. Describe the disease known as glaucoma. What causes it, and how is it treated?
19. Describe the range of photoreceptor organs found in the animal kingdom. What trends do you see when progressing from invertebrates to vertebrates? Do the similarities suggest anything about evolution to you?
20. Describe the anatomy of the ear and the role of the outer, middle, and inner ears in hearing.
21. How do the semicircular canals operate? How do the utricle and saccule operate?
22. Describe the different types of deafness and how they are treated.

C H A P T E R

19

The Skeleton and Muscles

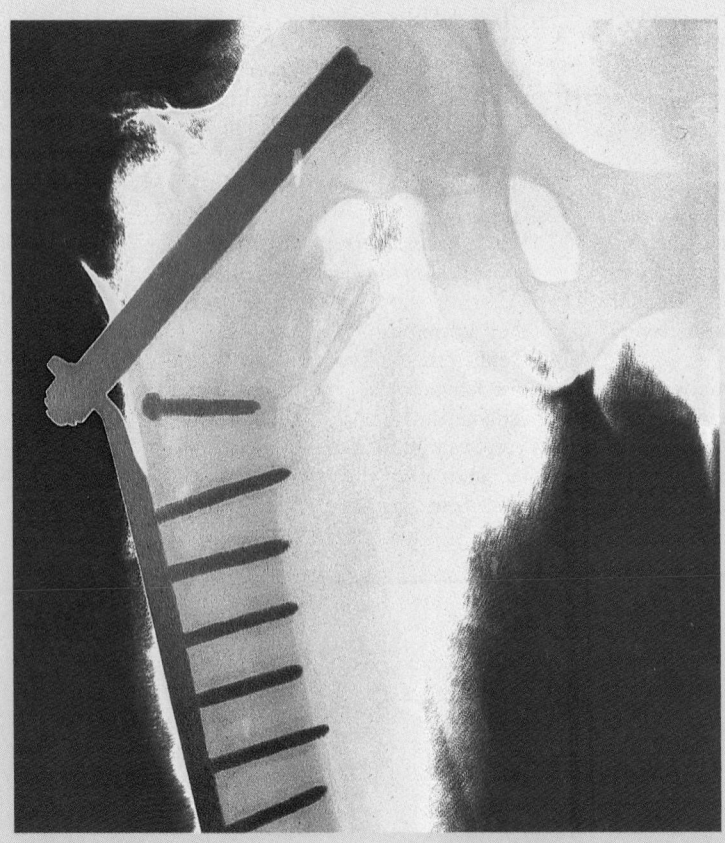

Colorized x-ray of a fractured human femur with pins and rod inserted to hold bone fragments together for repair.

A hiker on an afternoon stroll in North Carolina happens upon a white-tailed deer, munching on grass along the trail. Startled, the deer bolts away to the safety of the forest. Later, the hiker flushes a duck paddling in a quiet pond. Noisily, the duck flaps its wings, beating the water as it "runs" across the surface of the pond. When it has reached proper speed, it takes off, disappearing from sight.

These movements all depend on the skeleton and attached muscles. During evolution, these basic components have been molded by environmental forces to provide a variety of locomotive systems—the fins of fish, the wings of birds, and the legs of humans. Nonetheless, many species have a common architecture, illustrating once again that the evolutionary process is conservative; that is, it preserves what works. ▷ Figure 19–1, for example, shows the remarkable similarity of the muscles in frogs and humans.

This chapter discusses motor systems in the animal kingdom, starting with an overview of invertebrates and vertebrates. It then focuses on the human skeleton and the muscles that attach to it, the skeletal muscles. Together, the bones and muscles of vertebrates constitute the musculoskeletal system. In humans, they make up 50% to 60% of the body weight of an adult.

≈ AN OVERVIEW OF THE EVOLUTION OF MOTOR SYSTEMS

The animal kingdom is defined in part by movement—the ability to move from one place to another and the ability to move body parts to perform tasks. Movement is fundamental to animal success. It is vital to reproduction, feeding, protection, and many other daily tasks.

As in many other systems discussed in this book, animal locomotion varies considerably, from the slithering motion of planaria to the complex movements of organisms like you and me. Despite the vast differences, two common features are present in all animals. First, all animals have muscles. Second, all animals convert chemical energy in their muscles into mechanical energy (movement). The expression of these common characteristics is determined by the nature of the environment in which an animal lives. That is to say, natural selection helped shape the physical adaptations involved in movement. Animals that fly through the air, for instance, developed wings with a large surface area to push against low-density air. Paddles, fins, and flippers, much smaller than wings, developed in water-dwelling organisms in response to a denser medium that required less push.

▷ **FIGURE 19–1 Comparison of the Muscles of the Frog and Human** A glance at the muscular system and bones (not shown here) of two very different species suggests a common evolutionary ancestry.

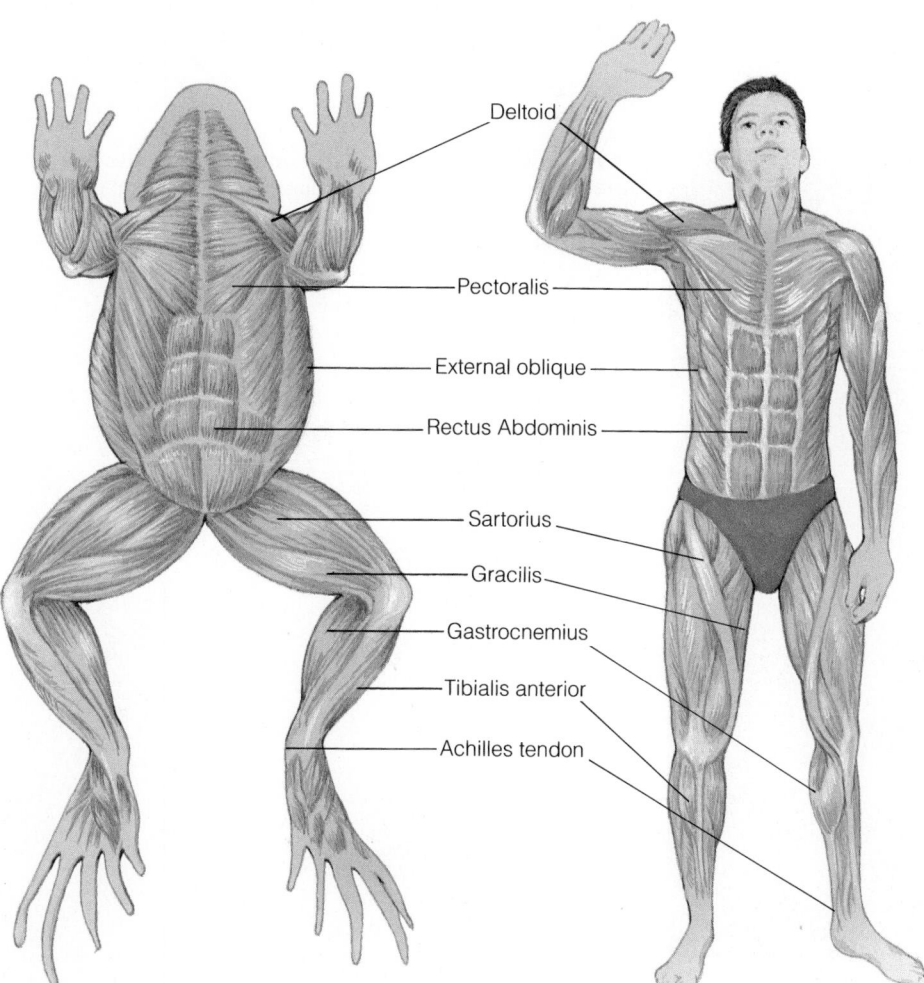

Deltoid

Pectoralis

External oblique

Rectus Abdominis

Sartorius

Gracilis

Gastrocnemius

Tibialis anterior

Achilles tendon

Invertebrates Contain Opposing Muscle Groups that Permit Primitive Movements

One of the simplest animal motor systems occurs in the sea anemone, which lives in and around coral reefs, where it attaches to the bottom (▷ Figure 19–2a). This system contains two sets of muscle fibers, circular and longitudinal. The circular muscle fibers are found in bundles that form hoops around the body axis. When they contract, the organism elongates (Figure 19–2b). The longitudinal muscle fibers occur in bundles that run the length of the animal's body (Figure 19–2c). When they contract, the organism shortens. This simple muscle system helps the anemone escape danger, as any scuba diver who has brushed past an anemone will attest.

Movement in earthworms and other annelids is also dependent on antagonistic circular and longitudinal muscles located in the body wall. When the circular muscles contract, an earthworm elongates. When the longitudinal muscles contract, the body shortens. Forward movement is aided by small bristles that protrude from the body. As the earthworm elongates, these bristles actually grip the ground, so when the body contracts, force is applied against the bristles, and the earthworm's body is pulled forward.

Interestingly, longitudinal and circular muscle fibers are found in many higher organisms, such as humans, but only in the walls of certain organs, like the small intestine, where they help propel the contents forward.

The longitudinal and circular muscles of the sea anemone and earthworm are said to be antagonistic, because they exert opposite effects. In the animal kingdom, many muscles are arranged in antagonistic groups. The muscles of your upper arm, for instance, consist of two opposing groups. The muscles in front cause the arm to flex and those in back cause it to extend. Antagonistic muscle groups, therefore, are a common characteristic of all animals.

The Presence of Exoskeletons in Invertebrates Provides a Wider Range of Movements

Anemones and annelids have no skeleton, and their movements are rather limited. Some invertebrates, like insects, crabs, and other arthropods, have hard external skeletons, or **exoskeletons,** made of chitin, proteins, and (sometimes) lipids (▷ Figure 19–3). Exoskeletons protect arthropods against prey and also reduce water loss. As you can see in Figure 19–3, the exoskeleton is not one large inflexible case but, rather, consists of several segments that connect to one another at pliable regions that act like hinges. Antagonistic muscles located internally attach to the exoskeleton and span the hinges, thus permitting movement. Exoskeletons are found in other animals as well, such as lobsters, clams, and snails.

(a)

▷ **FIGURE 19–2 Muscular Contraction in the Sea Anemone** (*a*) This sedentary animal attaches to coral reefs. (*b*) When the circular muscles contract, the organism elongates. (*c*) When the longitudinal muscles contract, it shortens.

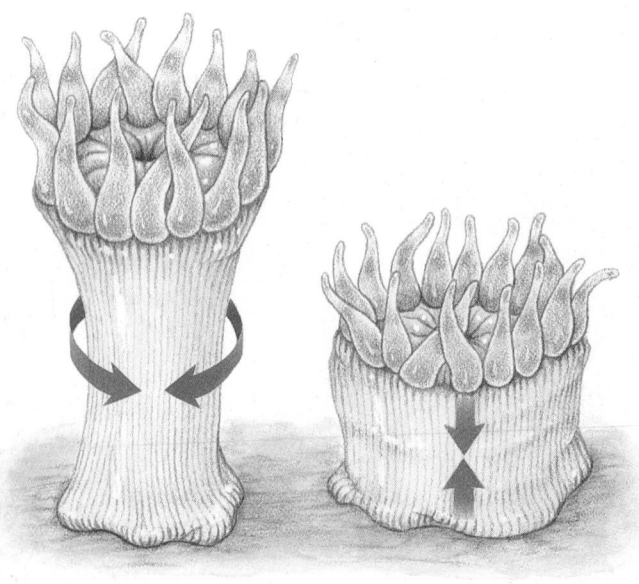

(b) (c)

▷ **FIGURE 19-3 Arthropod Exoskeleton** The hard, nonliving exoskeleton of the scarab beetle protects it from predators and reduces water loss. The exoskeleton is hinged by pliable regions. Muscles lying internally attach to the segments and are responsible for movement of the parts of the exoskeleton.

In Vertebrates, Opposing Muscle Groups Acting on Internal Skeletons Provide a Wider Range of Movements

Internal skeletons are found in a wide array of vertebrates, including birds, fish, reptiles, amphibians, and mammals. In animals such as rays and sharks, skeletons are composed entirely of cartilage. In most vertebrates, internal skeletons are made of bones, with small amounts of cartilage, as noted in Chapter 10 (▷ Figure 19–4).

The bones of the skeleton are attached by one of several types of joints, which permit a wide range of motions. Together, the muscles and bones form the musculoskeletal system. Besides aiding in movement, the musculoskeletal system is also involved in homeostasis. Bone, for instance, helps maintain constant blood calcium levels, necessary for muscle contraction. Muscles contract rhythmically when birds and mammals are cold, causing them to shiver and thus helping generate additional body heat.

≋ SKELETAL STRUCTURE AND FUNCTION

The word **skeleton** is derived from a Greek word that means "dried-up body." Many people's first impression of bone isn't much different (▷ Figure 19–5). To them, bones are dry, dead structures. **Bone,** however, is a living, metabolically active tissue. Bone tissue contains numerous cells, known as **osteocytes.** These cells are embedded in a calcified extracellular material, or matrix, which gives the bones of the body their characteristic hardness, strength, and flexibility. The extracellular material of bone consists of (1) an organic component, collagen, a protein that imparts flexibility; and (2) an inorganic component, chiefly calcium phosphate crystals that are deposited on the collagen fibers, which imparts

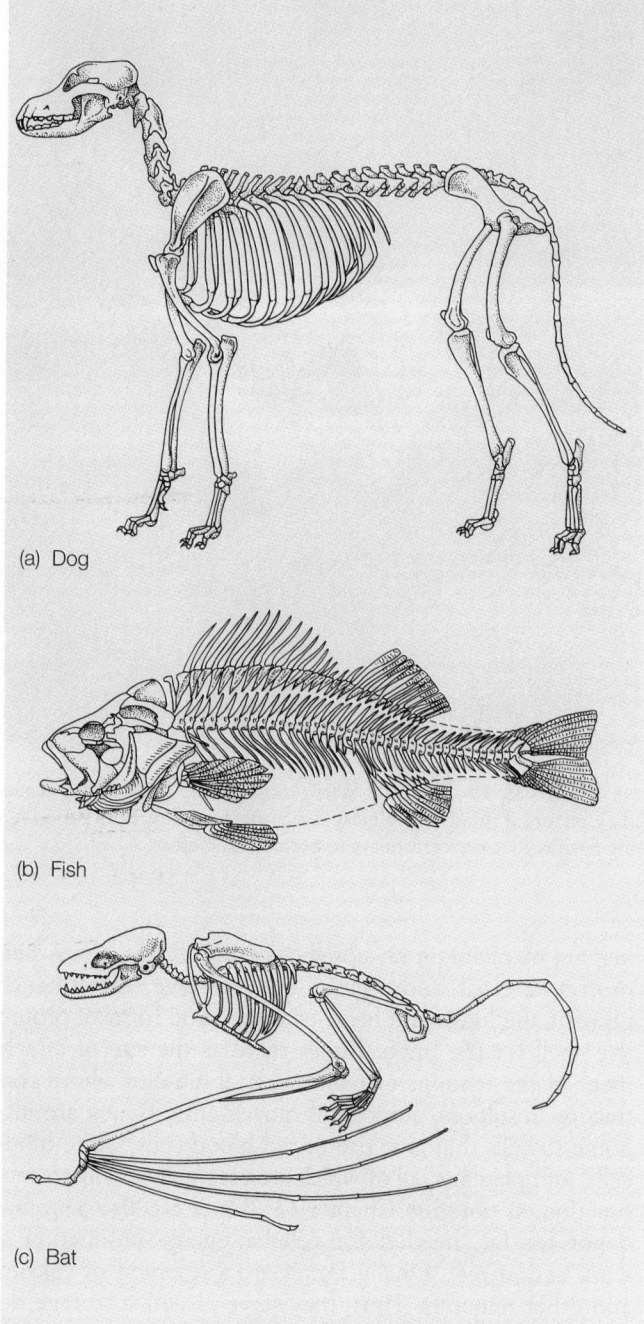

(a) Dog

(b) Fish

(c) Bat

▷ **FIGURE 19-4 Skeletal Similarities** The skeletons of the dog, fish, and bat exhibit many anatomical similarities. Each skeleton consists of a spinal column, ribs, skull, and appendages.

strength. (For more on bone, you might want to review the information presented in Chapter 10.)

Bones Serve Many Functions and Play an Important Role in Homeostasis

The human skeleton consists of 206 bones, discrete structures made of bone tissue (▷ Figure 19–6). Bones provide internal structural support, giving shape to our bodies and

► **FIGURE 19–5 Bone** When most people think about bones, they picture a dried-out, lifeless structure. In reality, bones are living tissues that perform many important functions.

helping us maintain an upright posture. Some bones help protect internal body parts. The rib cage, for instance, protects the lungs and heart, and the skull forms a protective shell for the brain. Bones serve as the site of attachment of the tendons of many skeletal muscles whose contraction results in purposeful movements. Bones are also home to cells that give rise to red blood cells, white blood cells, and platelets, all of which are essential to homeostatic function, as noted in Chapter 13. Bones are also a storage depot for fat, needed for cellular energy production at work and at rest. Finally, bones are a reservoir of calcium and other minerals. Thus, they serve as active storage depots that release and absorb calcium helping to maintain normal blood levels. Calcium is essential to muscle contraction, and disturbances in blood calcium levels can impair muscle contraction.

The Human Skeleton Consists of Two Parts

The human skeleton consists of two parts, the axial skeleton and the appendicular skeleton (Figure 19–6). The **axial skeleton** forms the long axis of the body. It consists of the skull, the vertebral column, and the rib cage. The **appendicular skeleton** consists of the bones of the arms and legs and the bones of the shoulders and pelvis, by which the upper and lower extremities are attached to the axial skeleton.

► Figure 19–7 illustrates the anatomy of the humerus, the bone found in the upper arm. As shown, the humerus, like other long bones of the body, consists of a long, narrow shaft, the **diaphysis,** and two expanded ends, the **epiphyses.** The ends of the epiphyses are coated with a thin layer of hyaline cartilage, described in Chapter 10, which reduces friction in joints. The protrusions on the bone mark the sites of muscle attachment.

All Bones Have a Hard, Dense Outer Layer that Surrounds a Less Compact Central Region

Take a moment to study the skeleton in Figure 19–6. As you examine it, you may notice that bones come in a variety of shapes and sizes. Some are long, some are short, and some are flat and irregularly shaped. Nevertheless, all bones share some characteristics. For example, all bones consist of an outer shell of dense material called **compact bone,** surrounding a spongy interior. **Spongy bone,** so named because of its spongy appearance, is less dense than compact bone and is generally concentrated in the epiphyses. As shown in Figure 19–7c, spongy bone forms a latticework.

On the outer surface of the compact bone is a layer of connective tissue, the **periosteum** ("around the bone"). The outer layer of the periosteum is composed of dense, irregularly arranged connective tissue fibers and serves as the site of attachment for many skeletal muscles (► Figure 19–8). The inner layer of the periosteum contains osteogenic (bone-forming) cells that participate in the production of new bone during remodeling or repair. The periosteum is richly supplied with blood vessels, which enter the bone at numerous locations. Blood vessels travel through small canals in the compact bone and course through the inner spongy bone, providing nutrients and oxygen and carrying off cellular wastes, such as carbon dioxide. The periosteum is also richly supplied with nerve fibers. The majority of the pain felt after a person bruises or fractures a bone results from pain fibers in the periosteum.

Spongy bone contains numerous small, adjoining cavities. They connect with the large, hollow interior of the diaphysis, together forming the **marrow cavity.** The marrow cavities in most of the bones of the fetus contain **red marrow.** Red marrow is a blood-cell factory, producing RBCs, WBCs, and platelets to replace those routinely lost each day. As an individual ages, most red marrow is slowly "retired" and becomes filled with fat, becoming **yellow marrow.** Yellow marrow begins to form during adolescence and, by adulthood, is present in all but a few bones. Red blood cell formation, however, continues in the bodies of the vertebrae, the hip bones, and a few others. Yellow marrow can be reactivated to produce blood cells under certain circumstances—for example, after an injury.

The Joints Permit Varying Degrees of Mobility

A gymnast races across the mat and leaps into space, twirling effortlessly before landing on her feet. As soon as she lands, she takes off again across the mat in a series of three

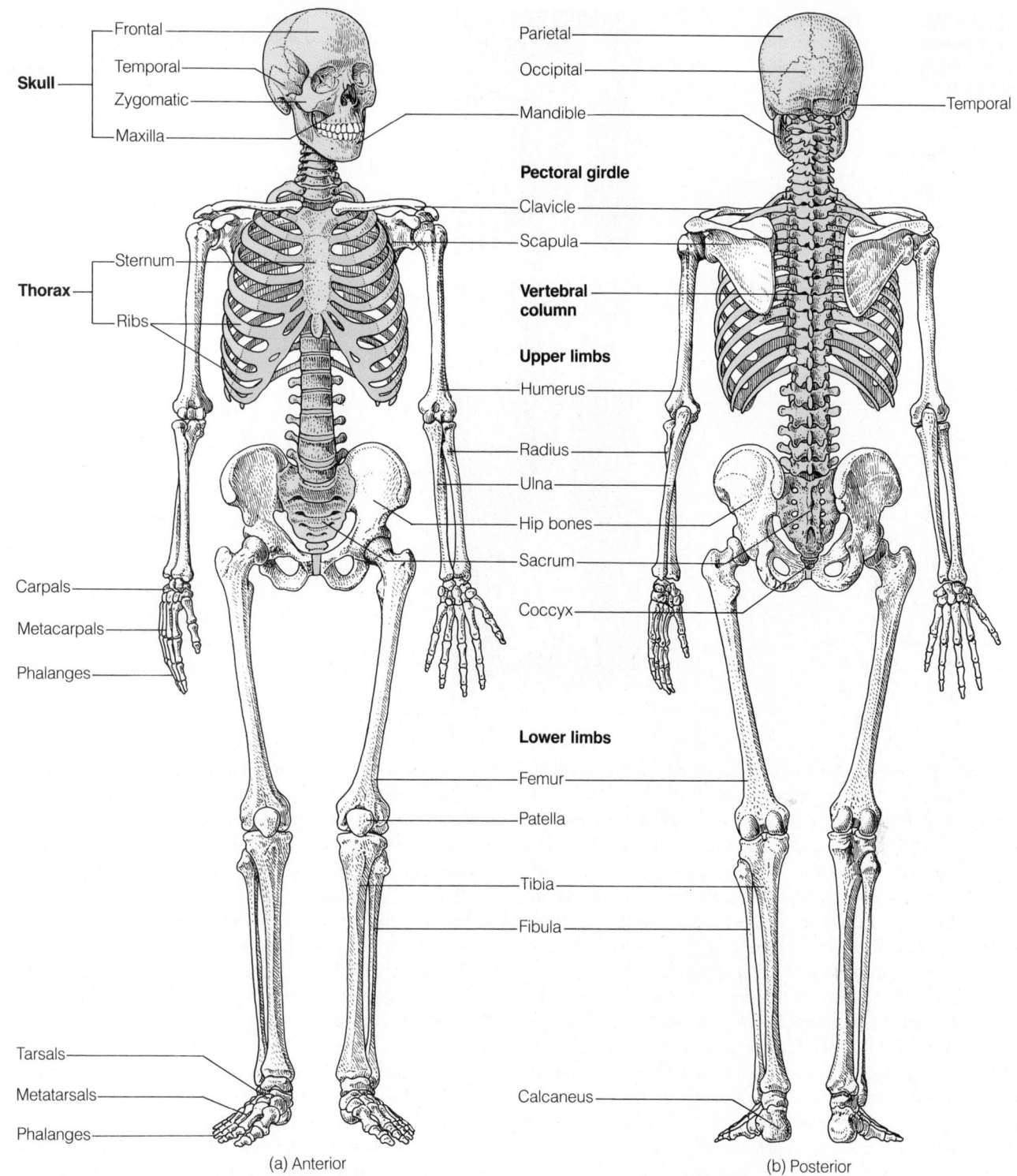

Skull
- Frontal
- Temporal
- Zygomatic
- Maxilla

Thorax
- Sternum
- Ribs

Carpals
Metacarpals
Phalanges

Tarsals
Metatarsals
Phalanges

Parietal
Occipital
Mandible

Pectoral girdle
Clavicle
Scapula

Vertebral column

Upper limbs
Humerus
Radius
Ulna
Hip bones
Sacrum
Coccyx

Lower limbs
Femur
Patella
Tibia
Fibula

Temporal

Calcaneus

(a) Anterior

(b) Posterior

▷ **FIGURE 19–6 The Human Skeleton** (*a*) Anterior view. (*b*) Posterior view. Over 200 bones of all shapes and sizes make up the skeleton. The shaded region shows the axial skeleton. The unshaded region is the appendicular skeleton.

back handsprings. These delightful movements are the result of long hours of practice and exercise and are made possible by the joints. **Joints** are the structures that connect the bones of the skeleton.

Joints can be classified by the degree of movement they permit. Those that permit no movement are called *immov-*

able joints. Those that permit some movement are *slightly movable joints,* and those that permit free movement are *freely movable joints.* Consider some examples.

The bones of the skull shown in ▷ Figure 19–9a are held together by immovable joints. As illustrated, opposing bones in the skull interdigitate (interlock). Fibrous connec-

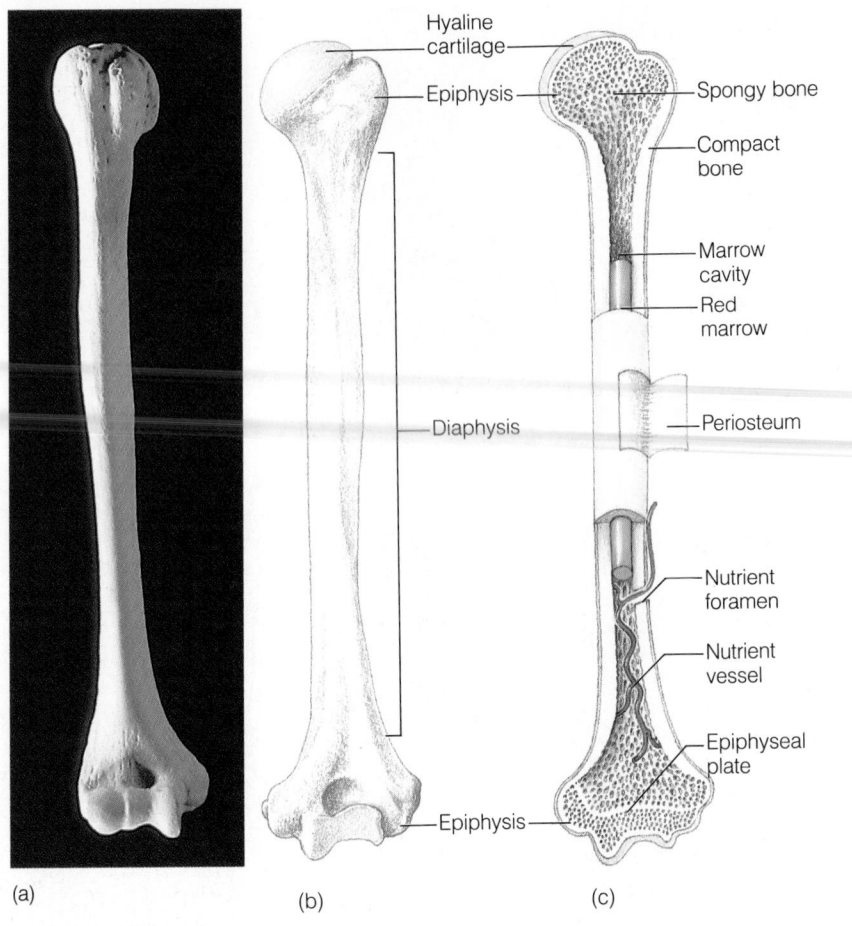

▷ **FIGURE 19–7 Anatomy of Long Bones** *(a)* Photograph and *(b)* drawing of the humerus. Notice the long shaft and dilated ends. *(c)* Longitudinal section of the humerus showing the position of the compact bone, spongy bone, and marrow.

(a) (b) (c)

tive tissue spans the small space between the interlocking bones, holding the bones together.

The pubic symphysis is the joint formed by the two pubic bones (Figure 19–9b). These bones are held in place by fibrocartilage, and the joint is basically immovable. Near the end of pregnancy, however, hormones loosen the fibrocartilage of the pubic symphysis. This allows the pelvic outlet to widen enough to permit the baby to be delivered.

The bodies of the vertebrae are united by slightly movable joints (▷ Figure 19–10). Each vertebra is separated from its nearest neighbors by an intervertebral disc. The inner portion of the disc acts as a cushion, softening the impact of walking and running, as noted in Chapter 10. The outer, fibrous portion holds the disc in place and joins one vertebra to its nearest neighbor. The joints between the vertebrae offer some degree of movement, resulting in a fair amount of flexibility. If they did not, we would be unable to bend over to tie our shoes or unable to curl up on the couch for an afternoon snooze.

The most common type of joint is the freely movable, or **synovial, joint.** The synovial joints are more complex than other types and permit varying degrees of movement. Although synovial joints differ considerably in architecture, they share several features.

The first commonality is the hyaline cartilage located on the articular (joint) surfaces of the bones (▷ Figure 19–11a). This thin cap of hyaline cartilage reduces friction

and facilitates movement. The second common feature of synovial joints is the joint capsule. The **joint capsule** is a double-layered structure that joins one bone to another in the joint (Figure 19–11a). The outer layer of the capsule consists of dense connective tissue that attaches to the periosteum of adjoining bones. Parallel bundles of dense connective tissue fibers in the outer layer of the capsule form **ligaments,** which run from bone to bone, giving additional support to the joint. As a rule, ligaments are fairly inflexible. However, some individuals have remarkably flexible ligaments and tendons. Because of this, some people can extend their thumbs well beyond the 90 degrees possible for most of us. And some can extend their fingers so much that they can touch the back of their hand. These people are said to be ''double jointed.''

The inner layer of the joint capsule is called the **synovial membrane,** and it consists of loose connective tissue with a generous supply of capillaries. The synovial membrane produces a fairly thick, slippery substance known as **synovial fluid.** It provides nutrients to the articular cartilage (the hyaline cartilage on the articular surfaces of the bone) and also acts as a lubricant, facilitating the movement of bones in joints. Normally, the synovial membrane produces only enough fluid to create a thin film on the articular cartilage. Injuries to a joint, however, may result in a dramatic increase in synovial fluid production, causing swelling and pain in joints.

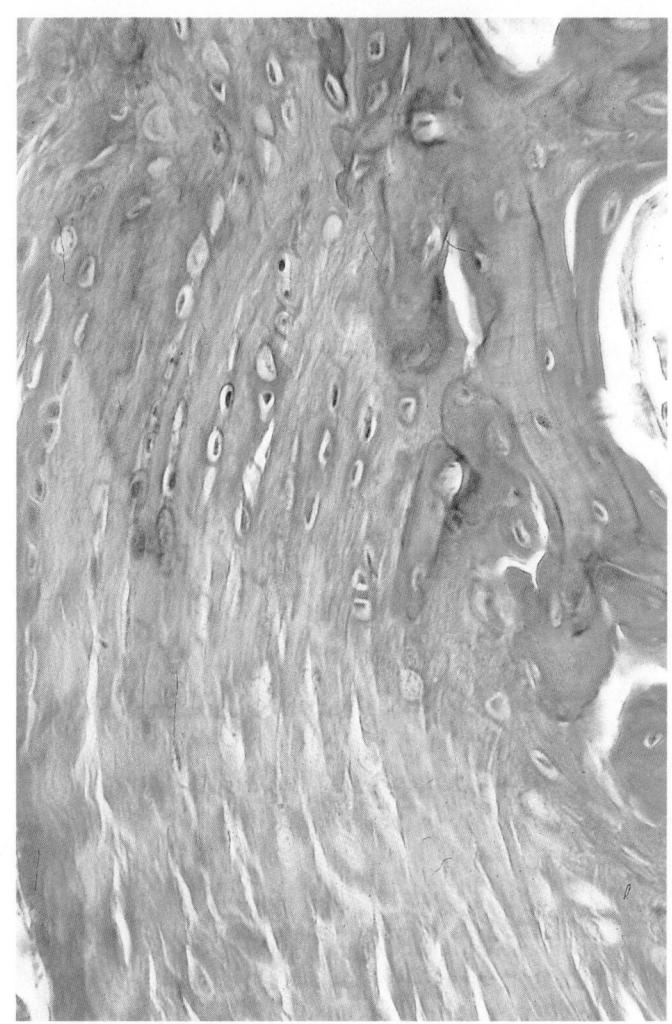

▷ **FIGURE 19-8 Tendon Attachment to Bone** The tendon attaches to the outer layer of the periosteum.

Tendons and muscles provide additional support in some joints and are commonly associated with synovial joints. In the shoulder joint, for example, the muscles of the shoulder help hold the head of the humerus in the socket (formed by the scapula). Muscles in the hip also help hold the head of the femur in place. Because muscles strengthen joints, individuals who are in poor physical shape are much more likely to suffer a dislocation on a ski trip or during exercise than someone who is in good shape. **Dislocation** is an injury in which a bone is displaced from its proper position in a joint due to a fall or some other unusual body movement. In some cases, bones will slip out of place, then back in without assistance. In others, the bone can be put back in place only by a physician or other trained health-care worker.

Synovial joints come in many shapes and sizes and are classified on the basis of structure. Two of the most common are the hinge joint and the ball-and-socket joint. The knee joint is a **hinge joint,** as are the joints in the fingers. Hinge joints open and close like hinges on a door and therefore provide for movement in only one plane. The hip and shoulder joints are **ball-and-socket joints.** They provide a wider range of motion. (Compare the movements permitted by the shoulder and hip joints to those permitted by the knee joint.)

Arthroscopy Permits Surgeons to Repair Injured Joints with a Minimum of Trauma. The acrobatics performed by modern dancers or Olympic gymnasts illustrate the wide range of movement that joints allow. The joints, however, are a biological compromise. They must allow movement, but also provide some degree of stability,

▷ **FIGURE 19-9 Two Immovable Joints** *(a)* Many of the bones of the skull are held in place by joints called sutures. The bones are linked by fibrous tissue, and the joints are immovable. *(b)* The pubic symphysis is another immovable joint. During childbirth it softens and expands to permit birth.

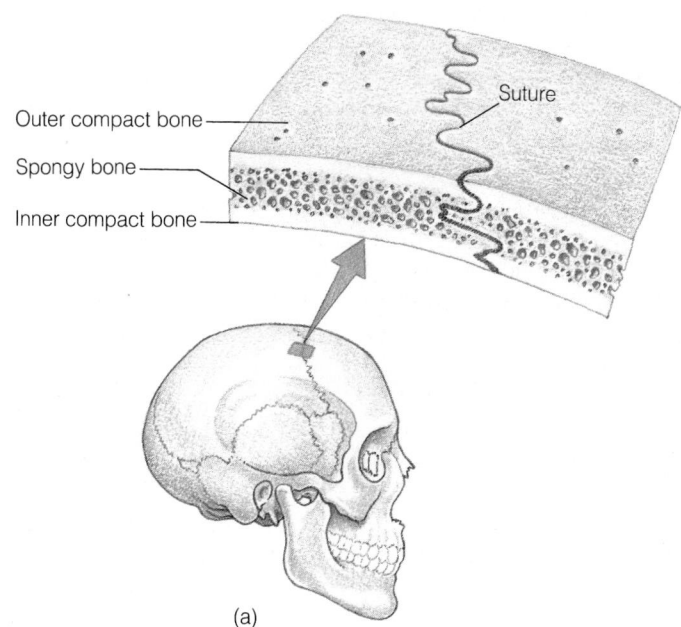

Outer compact bone

Spongy bone

Inner compact bone

Suture

(a)

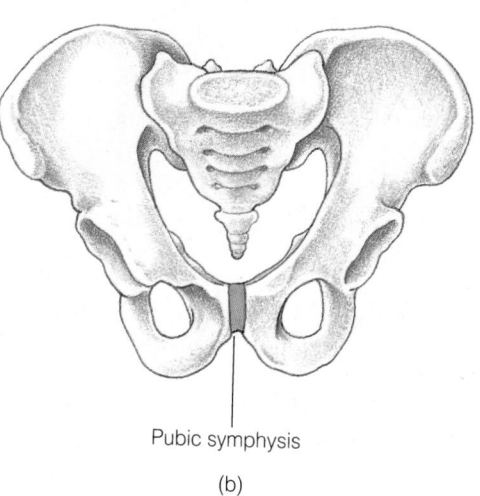

Pubic symphysis

(b)

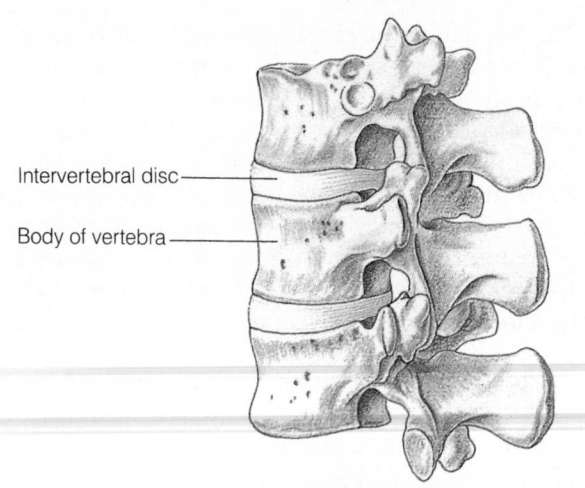

Intervertebral disc

Body of vertebra

▷ **FIGURE 19–10 A Slightly Movable Joint** The intervertebral discs allow for some movement, giving the vertebral column flexibility.

helping hold bones in place. Full mobility would compromise strength, and absolute strength would compromise mobility.

Given the need for compromise, it is not surprising that joint injuries are common among physically active people, especially athletes (Table 19–1). A hard blow to the knee of a football player, for instance, can tear the ligaments, disabling the joint. A rough fall can dislocate a skier's shoulder. Improperly lifting an object can strain the ligaments that join the vertebrae of the back, resulting in considerable pain.

Torn ligaments, tendons, and cartilage in joints heal very slowly because they are not well endowed with blood vessels.[1] For years, joint repair required major surgery that

---

[1] Because much of it is avascular, cartilage may not heal at all.

was so traumatic it put patients out of commission for several months. Today, however, new surgical techniques allow physicians to repair joints with a minimum of trauma (▷ Figure 19–12). Through small incisions in the skin over the joint, surgeons insert a device called an **arthroscope.** It allows them to view the damage inside the joint cavity and also insert special instruments to snip off damaged cartilage. Consequently, a surgeon can repair damaged cartilage without opening the joint. This reduces damage caused by large incisions and allows athletes to be back on their feet and on the playing field in a matter of weeks. New surgical techniques are also used to rebuild torn ligaments, thus returning joints to nearly their original state.

Osteoarthritis Is a Degenerative Bone Disease Caused by Wear and Tear on the Articular Cartilages.
Virtually every time you move, you use one or more of your joints. Problems in joints, therefore, are often quite noticeable. One of the most common problems is called **degenerative joint disease** or **osteoarthritis.** Although its cause is not known, degenerative joint disease may simply result from wear and tear on a joint. Over time, excess wear may cause the articular cartilage on the ends of bones to flake and crack. As the cartilage degenerates, the bones come in contact, grinding against each other during movement and causing considerable pain and discomfort. Swelling usually accompanies these changes, and swelling in a joint tends to reduce mobility.

Osteoarthritis occurs most often in the weight-bearing joints—the knee, hip, and spine—which are subject to the most wear over time. Osteoarthritis may also develop in the finger joints. The amount of swelling in victims of osteoarthritis varies considerably. Some patients experience virtually no swelling; in others, the joints become enlarged and disfigured.

Osteoarthritis is extremely common. X-ray studies of

▷ **FIGURE 19–11 A Synovial Joint** (*a*) A cross section through the hip joint (a ball-and-socket joint) showing the structures of the synovial joint. (*b*) Ligaments help support the joint.

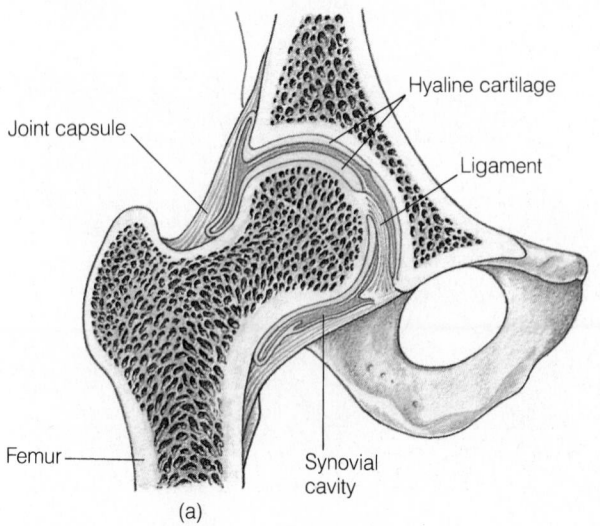

Joint capsule

Hyaline cartilage

Ligament

Femur

Synovial cavity

(a)

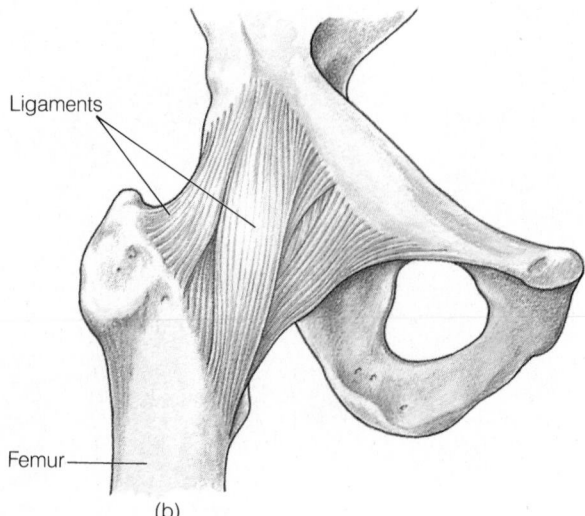

Ligaments

Femur

(b)

TABLE 19-1 Common Injuries of the Joints

INJURY	DESCRIPTION	COMMON SITE
Sprain	Partially or completely torn ligament; heals slowly; must be repaired surgically if the ligament is completely torn.	Ankle, knee, lower back, and finger joints
Dislocation	Occurs when bones are forced out of a joint; often accompanied by sprains, inflammation, and joint immobilization. Bones must be returned to normal positions.	Shoulder, knee, and finger joints
Cartilage tears	Cartilage may tear when joints are twisted or when pressure is applied to them. Torn cartilage does not repair well because of poor vascularization. It is generally removed surgically; this operation makes the joint less stable.	Cartilage in the knee is the most common

people over 40 years of age show that most people have some degree of degeneration in one or more joints. Fortunately, many people do not even notice the problem, and the disease rarely becomes a serious medical problem.

Wear and tear on joints is worsened by obesity. The extra pressure on the joints apparently wears the cartilage away more quickly. Thus, weight control can help reduce the rate of degeneration in people already suffering from the disease.[2] Painkillers, such as aspirin, and other anti-inflammatory drugs can be used to treat the symptoms (pain and swelling) that accompany osteoarthritis. Injections of steroids may help reduce inflammation, although repeated injections often damage the joint.

Rheumatoid Arthritis Is an Autoimmune Disease.
Another common disorder of the synovial joint is rheumatoid arthritis. **Rheumatoid arthritis** is the most painful and

[2]Weight control is also a preventive measure that helps people avoid the problem in the first place.

crippling form of arthritis. It is caused by an inflammation of the synovial membrane. Inflammation often spreads to the articular cartilages, causing considerable damage. If the condition persists, rheumatoid arthritis causes degeneration of the bones. The thickening of the synovial membrane and degeneration of the bone often disfigure the joints, reduce mobility, and cause considerable pain (▷ Figure 19–13). In some cases, afflicted joints may be completely immobilized. In severe cases, the bones may become dislocated, causing the joints to collapse.

Rheumatoid arthritis generally occurs in the joints of the wrist, fingers, and feet. It can also affect the hips, knees, ankles, and neck. In many cases, inflammation also occurs in the heart, lungs, and blood vessels.

Research suggests that rheumatoid arthritis results from an **autoimmune reaction**—that is, an immune response to the cells of one's own synovial membrane (Chapter 15). Rheumatoid arthritis occurs in people of all ages but most commonly appears in individuals between the ages of 20 and 40. Rheumatoid arthritis is usually a permanent condi-

▷ **FIGURE 19–12 Arthroscopic Surgery** (*a*) A physician performing arthroscopic surgery. (*b*) Inside view of the knee joint through an arthroscope.

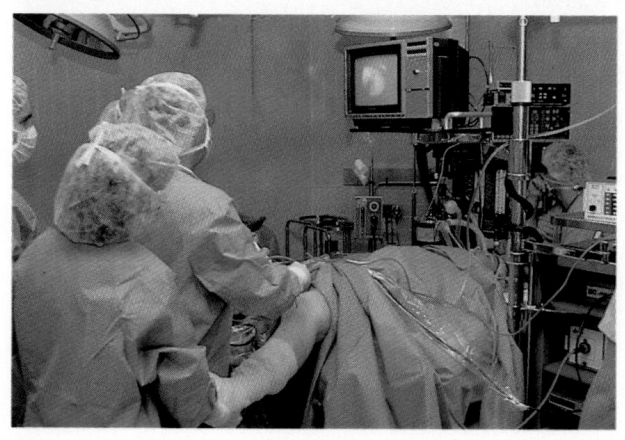

(a)

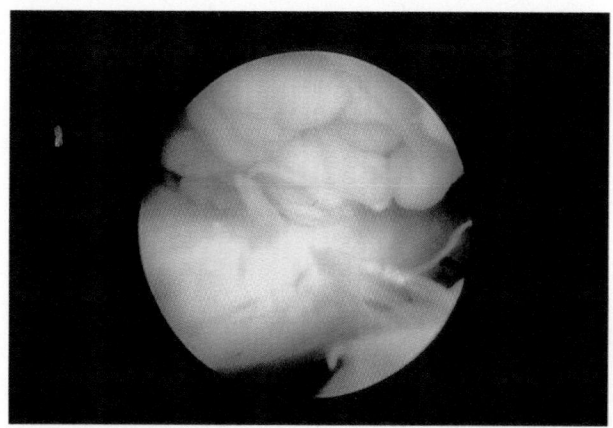

(b)

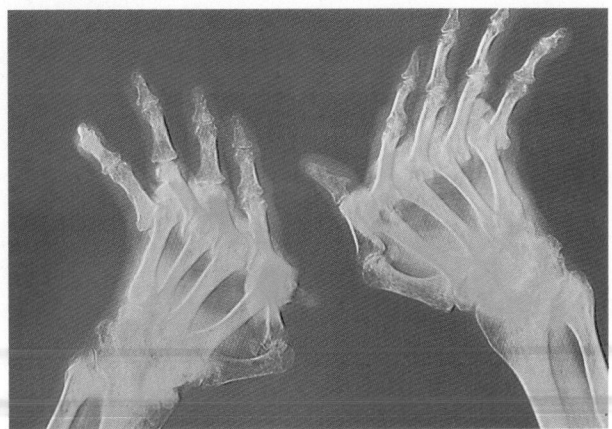

▷ **FIGURE 19–13 X-Ray of Hand Disfigured by Rheumatoid Arthritis** The image has been color-enhanced by computer.

tion, although the degree of severity varies widely. Patients suffering from it can be treated with physical therapy, pain-killers, anti-inflammatory drugs, and even surgery.

Diseased joints can also be replaced by artificial ones, or **prostheses,** restoring mobility and reducing pain. Plastic joints can be used to replace the finger joints. These prostheses greatly improve the appearance of the hands of a person with rheumatoid arthritis, eliminating the gnarled, swollen joints. Moreover, patients regain the use of previously crippled fingers. Day-to-day chores (buttoning a shirt) that often required assistance become noticeably easier. Tasks that had once been impossible because of arthritis—for example, opening screw-top jars and picking up coins—once again become possible. Severely damaged knee and hip joints are replaced with special steel or Teflon substitutes, which, if fitted properly, may last 10 to 15 years (▷ Figure 19–14). To put a new hip or knee joint in place, the surgeon first cuts away the degenerating bone. Then the prosthesis is inserted into the shaft of the bone.

▷ **FIGURE 19–14 Artificial Joints** An artificial knee joint (*a*) and hip joint (*b*).

Embryonic Development and Bone Growth

Most of the bones of the human skeleton start out as hyaline cartilage. During embryonic development, hyaline cartilage forms in the arms, legs, head, and torso where bone will eventually be (▷ Figure 19–15b). In short order, the cartilage is converted to bone. This process is known as **endochondral ossification.**

As shown in Figure 19–15a, endochondral ossification in cartilage begins in a region known as the **primary center of ossification.** Here, cartilage cells inside the template enlarge, compressing the extracellular material between them. Calcium crystals are then deposited on the extracellular material of the cartilage, and the cartilage cells soon die.

The primary center of ossification expands like a fire spreading in all directions, and a thin layer of bone is deposited around the periphery of the cartilage mass by cells of the **perichondrium,** the connective tissue layer surrounding the cartilage. As soon as bone is deposited, the perichondrium becomes the periosteum. The thin shell of bone laid down on the periphery of the cartilage will eventually become a layer of compact bone.

Blood vessels soon invade the primary center of ossification from the periosteum, carrying with them stem cells that will eventually give rise to a group of bone-forming cells called **osteoblasts.** Other stem cells carried into the primary center of ossification with the blood vessels will give rise to RBCs, WBCs, and platelets. Osteoblasts proliferate on the spicules of calcified cartilage in the interior of the embryonic bone and soon begin to secrete collagen fibers. Collagen fibers are deposited on the slightly calcified spicules, and additional calcium is then deposited on the collagen fibers. This results in the formation of a mass of spongy bone in the primary center of ossification.

Ossification centers also form in the ends (epiphyses) of the bone, but slightly later. These regions of bone formation are called the **secondary centers of ossification.** As

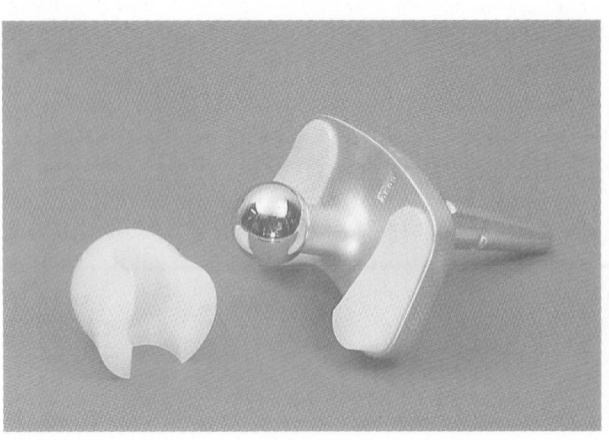

(a)

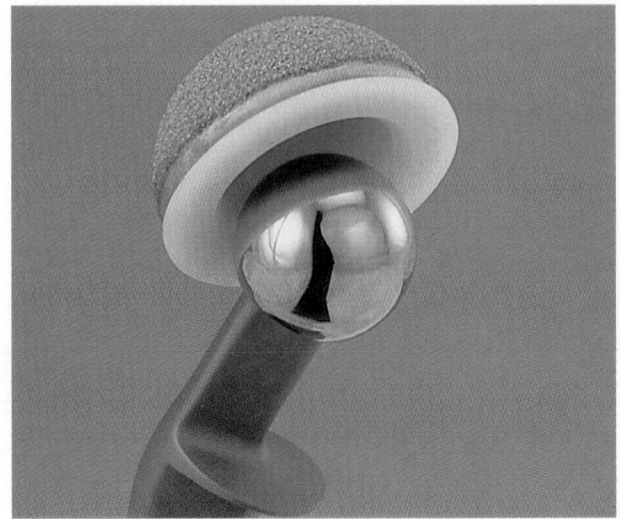

(b)

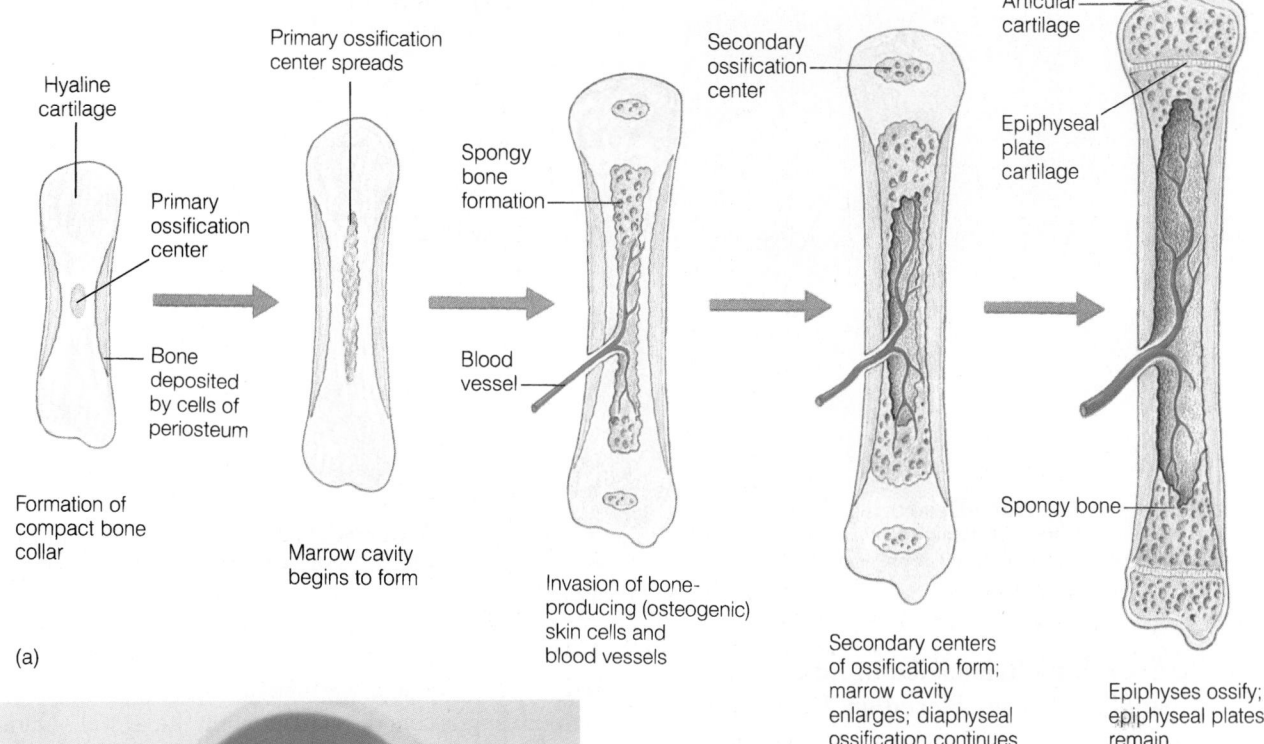

Hyaline cartilage

Primary ossification center

Bone deposited by cells of periosteum

Formation of compact bone collar

Primary ossification center spreads

Marrow cavity begins to form

Spongy bone formation

Blood vessel

Invasion of bone-producing (osteogenic) skin cells and blood vessels

Secondary ossification center

Secondary centers of ossification form; marrow cavity enlarges; diaphyseal ossification continues

Articular cartilage

Epiphyseal plate cartilage

Spongy bone

Epiphyses ossify; epiphyseal plates remain

(a)

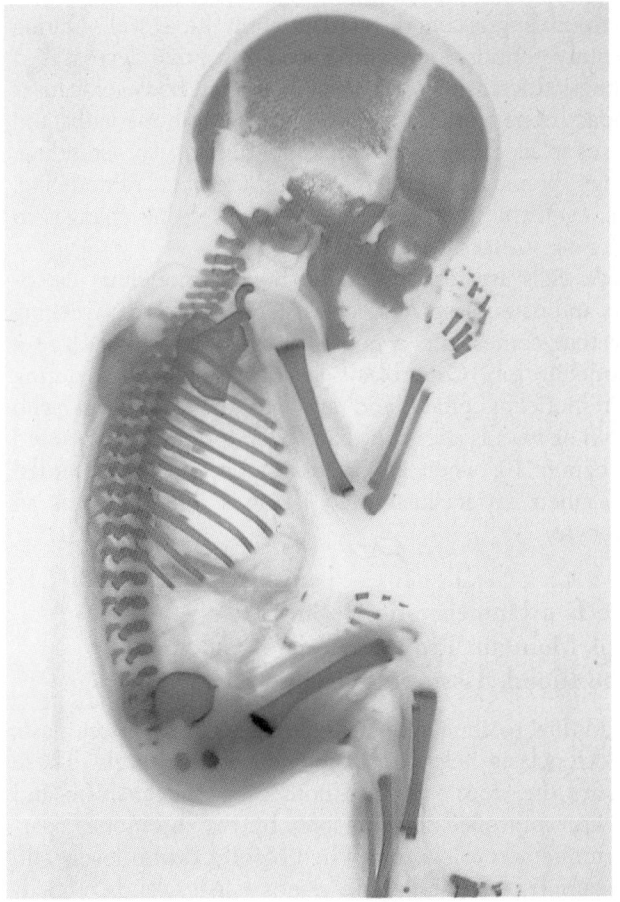

(b)

▷ **FIGURE 19–15 Endochondral Ossification**
(*a*) Stages of bone formation. (*b*) Eighteen-week-old human embryo showing bones forming by endochondral ossification.

As the bone develops, much of the calcified material in the shaft is removed by bone-digesting cells, or **osteoclasts,** soon after it is formed. Osteoclasts are multinucleated cells that, when activated, digest the extracellular material of bone (▷ Figure 19–16). They hollow out the center of the bone, forming the marrow cavity. Stem cells brought in with the blood vessels proliferate and fill the marrow. The stem cells remain in the marrow, dividing and differentiating to form RBCs, WBCs, and platelets.

When the bone is completely formed, all that remains of the cartilage is two narrow bands located between the shaft of the bone and its two ends (▷ Figure 19–17). Called the **epiphyseal plates,** these bands of cartilage contain actively dividing cells that permit bone to elongate. The process of bone elongation is beyond the scope of this book, but it basically results from the proliferation of cartilage on one side of the plate and the ossification on the other. The epiphyseal plates remain active in children and adolescents until their long bones stop growing.[3] Eventually, the increasing levels of sex steroids cause the plates to be converted to bone.

[3] Boys may continue growing until they are 20 or 21. Girls generally stop by age 17.

shown in Figure 19–15a, the primary and secondary centers of ossification spread outward, and with the aid of the periosteal bone deposition, described above, the entire cartilage mass is eventually converted to bone.

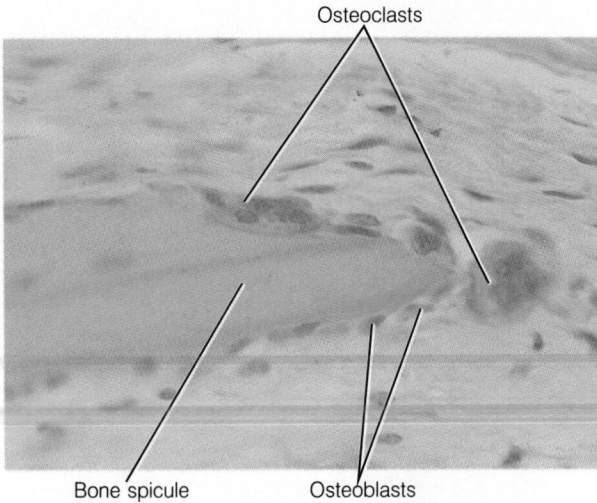

> **FIGURE 19-16 The Bone-Destroying Osteoclast**
Osteoclasts eat away at bone, releasing calcium to restore blood levels and helping remodel bone to meet changing needs.

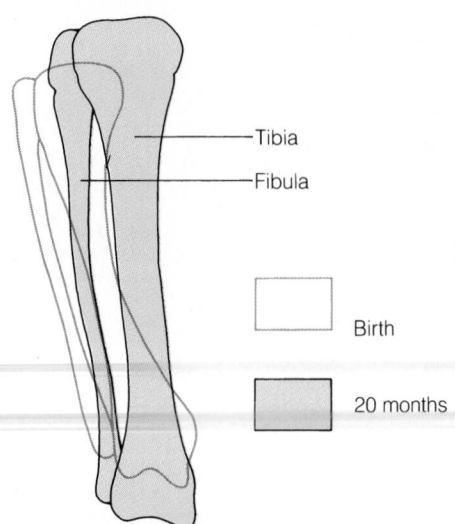

> **FIGURE 19–18 Remodeling of a Baby's Leg Bones**
Notice how dramatically the bones change to meet the changing needs of the toddler.

Bones Are Constantly Remodeled in Adults to Meet Changing Stresses Placed on Them

Bones are dynamic structures that undergo considerable remodeling in response to changes in our lives. In a newborn baby, for example, the bones of the leg (the tibia and fibula) are quite bowed. Cramped inside the mother's uterus, the bones do not grow very straight. During the first two years of life, however, as the child begins to walk and run, the leg bones generally straighten (▶ Figure 19–18). The bones are literally remodeled to meet the markedly

> **FIGURE 19–17 Epiphyseal Plate** The epiphyseal plate allows for bone elongation. Cartilage is added at the epiphyseal end while new bone is forming at the diaphyseal end, thus elongating the bone.

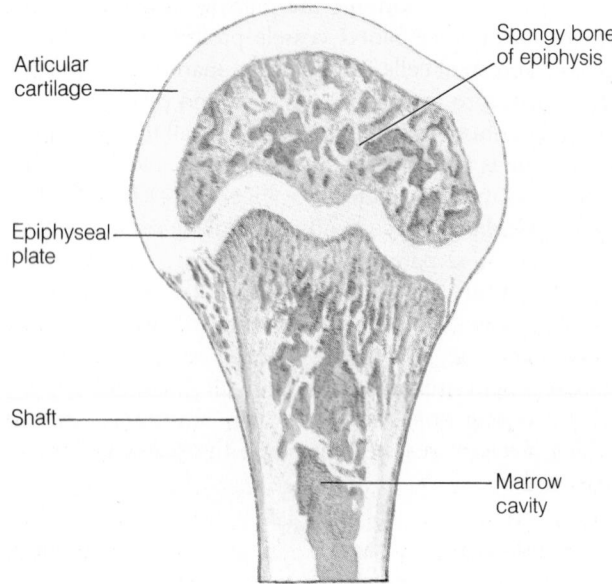

different stresses placed on them by upright posture.

Remodeling occurs throughout adult life as well. During sedentary periods, for example, compact bone decreases in thickness. Increasing one's level of activity, however, causes compact bone to thicken, thus helping the bone withstand stresses placed on it by walking, running, or standing. Spongy bone also undergoes considerable remodeling. Thus, even the internal architecture of a bone changes to meet new stresses.

Two cells are responsible for bone remodeling: osteoclasts and osteoblasts. Osteoclasts are akin to the wrecking crew that comes into a house to tear out walls before a remodeling job. Osteoblasts lay down new bone during the remodeling phase and are like the carpenters who rebuild new walls after they have been torn down. As noted in Chapter 10, when an osteoblast becomes surrounded by calcified extracellular material, it is referred to as an **osteocyte.**

Bone Is a Homeostatic Organ that Helps Maintain Proper Levels of Calcium in the Blood, Tissue Fluids, and Cells

In addition to their role in remodeling bone, osteoblasts and osteoclasts help control blood calcium levels. These cells are therefore part of a homeostatic mechanism and they are controlled in large part by two hormones: parathormone and calcitonin. When blood calcium levels fall, for example, the parathyroid glands (small glands embedded in the thyroid glands in the neck) release **parathormone (PTH).** PTH travels throughout the body in the bloodstream. When it reaches the bone, it stimulates the osteoclasts, causing them to digest bone in their vicinity. The calcium released by the activity of the osteoclasts replenishes blood calcium levels.

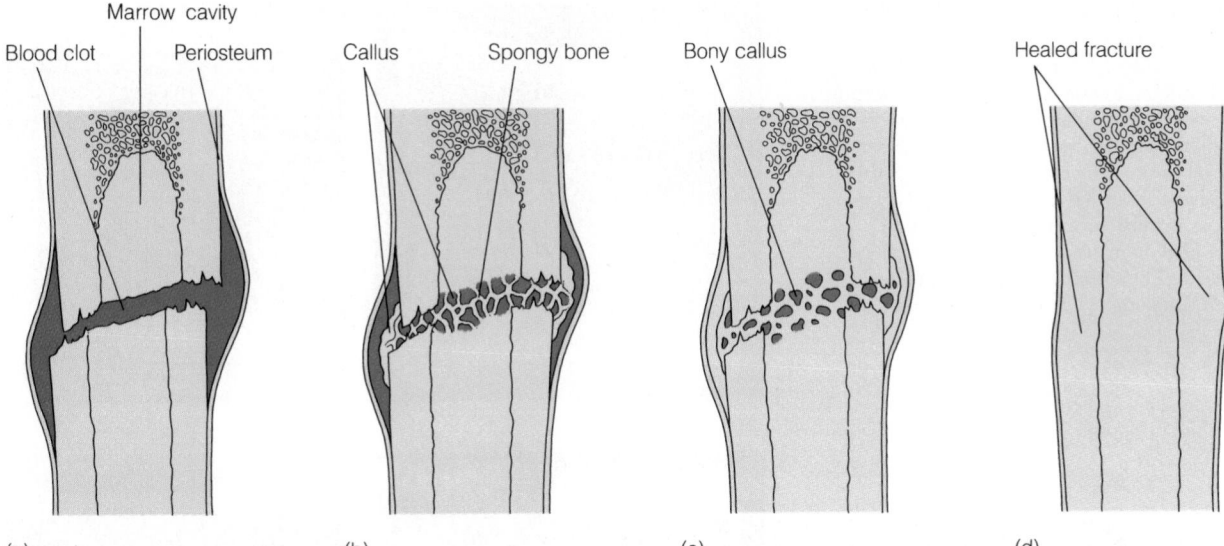

Blood clot   Marrow cavity   Periosteum   Callus   Spongy bone   Bony callus   Healed fracture

(a)   (b)   (c)   (d)

▷ **FIGURE 19–19 Fracture Repair** (*a*) A blood clot forms. (*b*) The blood clot is invaded by fibroblasts and other cells, forming the callus. (*c*) Calcium is deposited in the callus, knitting the ends together. (*d*) The fracture is repaired.

When calcium levels rise—for example, after a meal—the thyroid releases a hormone that exerts an opposite effect. This hormone, **calcitonin,** inhibits osteoclasts, stopping bone destruction. It also stimulates osteoblasts, causing them to deposit new bone. The inhibition of osteoclasts and the formation of new bone help lower blood calcium levels, returning them to normal.

Bone Fractures Are Repaired by Fibroblasts and Osteoblasts

Bone fractures can vary considerably in their severity. Some may involve hairline cracks, which mend fairly quickly. Others involve considerably more damage and take longer to repair. In order for repair to occur, the broken ends must often be aligned and immobilized by a cast. Severe fractures may require surgeons to insert steel pins to hold the bones together.

Like other connective tissues, bones are capable of self-repair, an evolutionary adaptation essential to all animals with internal skeletons. As ▷Figure 19–19 illustrates, blood from broken blood vessels in the periosteum and marrow cavity pours into the fracture, forming a clot. Within a few days, the blood clot is invaded by fibroblasts, connective tissue cells from the periosteum. Fibroblasts produce and secrete collagen fibers, thus forming a mass of cells and fibers, the **callus,** that bridges the broken ends. The callus protrudes at the surface of the bone.

The callus is next invaded by osteoblasts from the periosteum. Osteoblasts convert the callus to bone, knitting the two ends together. As Figure 19–19 shows, the callus is initially much larger than the bone itself. Excess bone, however, is gradually removed by osteoclasts, so much of the bony callus disappears.

Osteoporosis Involves a Loss of Calcium, which Results in Brittle, Easy-To-Break Bones

Helen Brockman, who is 70, gets out of bed one morning, steps on the hardwood floor, and feels a sharp pain in her back. Unknown to her, she has just fractured one of her vertebrae, the small bones that compose her spine. Like 20 million other Americans, Brockman has **osteoporosis** ("porous bone"), a condition characterized by a progressive loss of calcium from the bones of the skeleton (▷ Figure 19–20). In some individuals, calcium loss may be so severe that bones become porous and brittle—so much so that even normal activities such as getting out of bed in the morning or doing housework may cause fractures.

Osteoporosis occurs most often in postmenopausal women. Menopause occurs between 45 and 55 years of age and results from a shutdown of ovarian estrogen production. Estrogen is a reproductive hormone, but like many other hormones in the body, it performs several functions. In women, it reduces bone loss.

Osteoporosis also occurs in people who are immobilized for long periods. Hospital patients who are restricted to bed for two or three months, for example, show signs of osteoporosis. For a discussion of the way hibernating bears avoid this problem, see Spotlight on Evolution 19-1. Osteoporosis may also result from environmental factors. In the 1960s, for instance, women living along the Jintsu River in Japan developed a painful bone disease known as *itai-itai* (which literally means "ouch, ouch"). The women lived downstream from zinc and lead mines that released large amounts of the heavy metal cadmium into the river. They used river water for drinking and for irrigating rice paddies. Even though men, young women, and children were ex-

▷ **FIGURE 19–20 Osteoporosis** The loss of estrogen or prolonged immobilization weakens the bone. Bone is dissolved and becomes brittle and easily breakable. (*a*) A section of the body of a lumbar vertebra from a 29-year-old woman. (*b*) Some thinning is evident in a vertebra of a 40-year-old woman. (*c*) Bone loss is severe in an 84-year-old woman. (*d*) Bone loss is most severe in a 92-year-old woman. Osteoporosis is not inevitable and can be prevented by exercise, calcium supplements, estrogen supplements, and other measures.

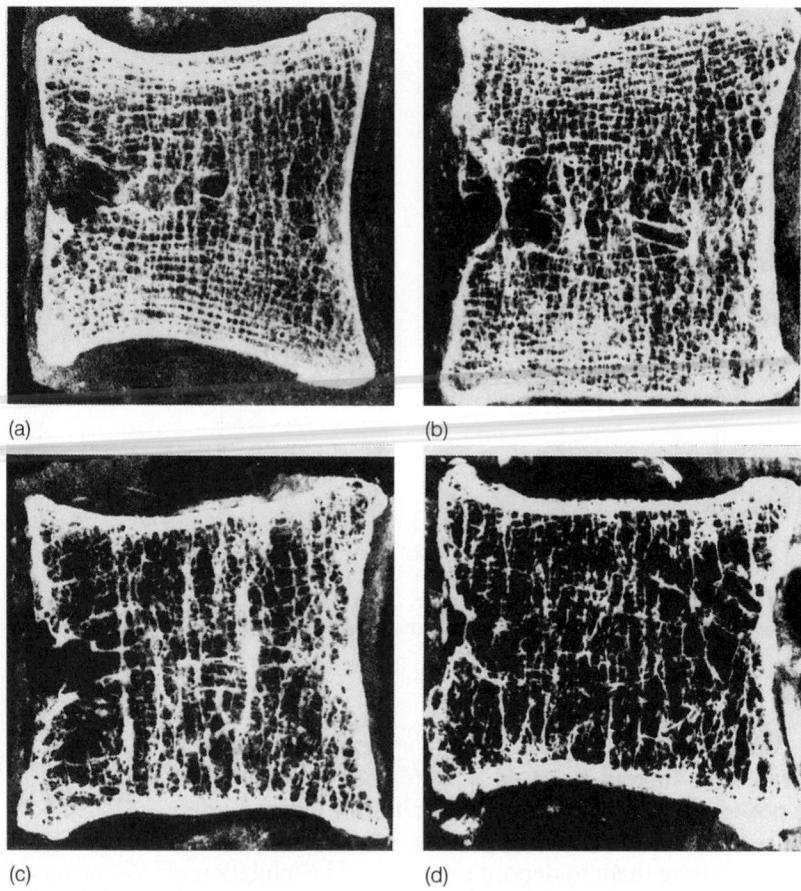

(a)

(b)

(c)

(d)

posed to cadmium, 95% of the cases occurred in post-menopausal women.

Researchers at the Argonne National Laboratory believe that cadmium may have accelerated bone loss in the post-menopausal Japanese women. To study the connection, they fed mice diets containing various levels of cadmium chloride. The group receiving the highest dose showed significant reductions in bone calcium when their ovaries had been removed.

These findings may also help explain why older women who smoke are more likely to suffer from osteoporosis. Cadmium is one of several harmful substances present in cigarette smoke. To test the connection between smoking and osteoporosis, researchers cultured fetal bone tissue in a medium containing cadmium levels similar to those found in a smoker's blood and compared the results with a control group of cells grown in a medium containing normal cadmium levels. The bone tissue exposed to higher levels of cadmium exhibited a 70% reduction in calcium content, compared with a 25% loss in the control samples, thus supporting the hypothesis that cadmium in cigarette smoking increases the incidence of osteoporosis.

The new findings provide a plausible explanation for the fact that female smokers experience more bone fractures and tooth loss than nonsmokers. Smoking, however, may also act by decreasing estrogen levels, making it doubly dangerous to a woman's health. For a discussion of ways to prevent osteoporosis, see Health Note 19–1.

≈ THE SKELETAL MUSCLES

Purposeful movement is one of the most distinctive features of animal life. As noted in the introduction to this chapter, in most vertebrates, skeletal muscles acting on bones permit movement. Most skeletal muscles are under the control of the nervous system.[4]

▷ Figure 19–21 shows some of the skeletal muscles of the body. A glance at this figure reveals that most skeletal muscles cross one or more joints. Therefore, when they contract, they produce movement. Muscles generally work in groups, rather than alone, to bring about various body movements. Groups of muscles are often arranged in such a way that one set causes one movement and another set on the opposite side of the joint causes an opposing movement. For example, the biceps is a muscle in the upper arm. When it and other members of its group contract, they cause the arm to bend, a movement called flexion. On the back side of the upper arm is another muscle, the triceps. When it contracts, the triceps causes the arm to straighten, or extend. Opposing muscles are called **antagonists.** Generally, when one muscle contracts to produce a movement, the antagonistic muscles relax. Antagonistic sets of muscles are under the control of the cerebellum.

Not all skeletal muscles are arranged so that they can

[4]Chapter 10 discussed all three types of muscle: skeletal, cardiac, and smooth.

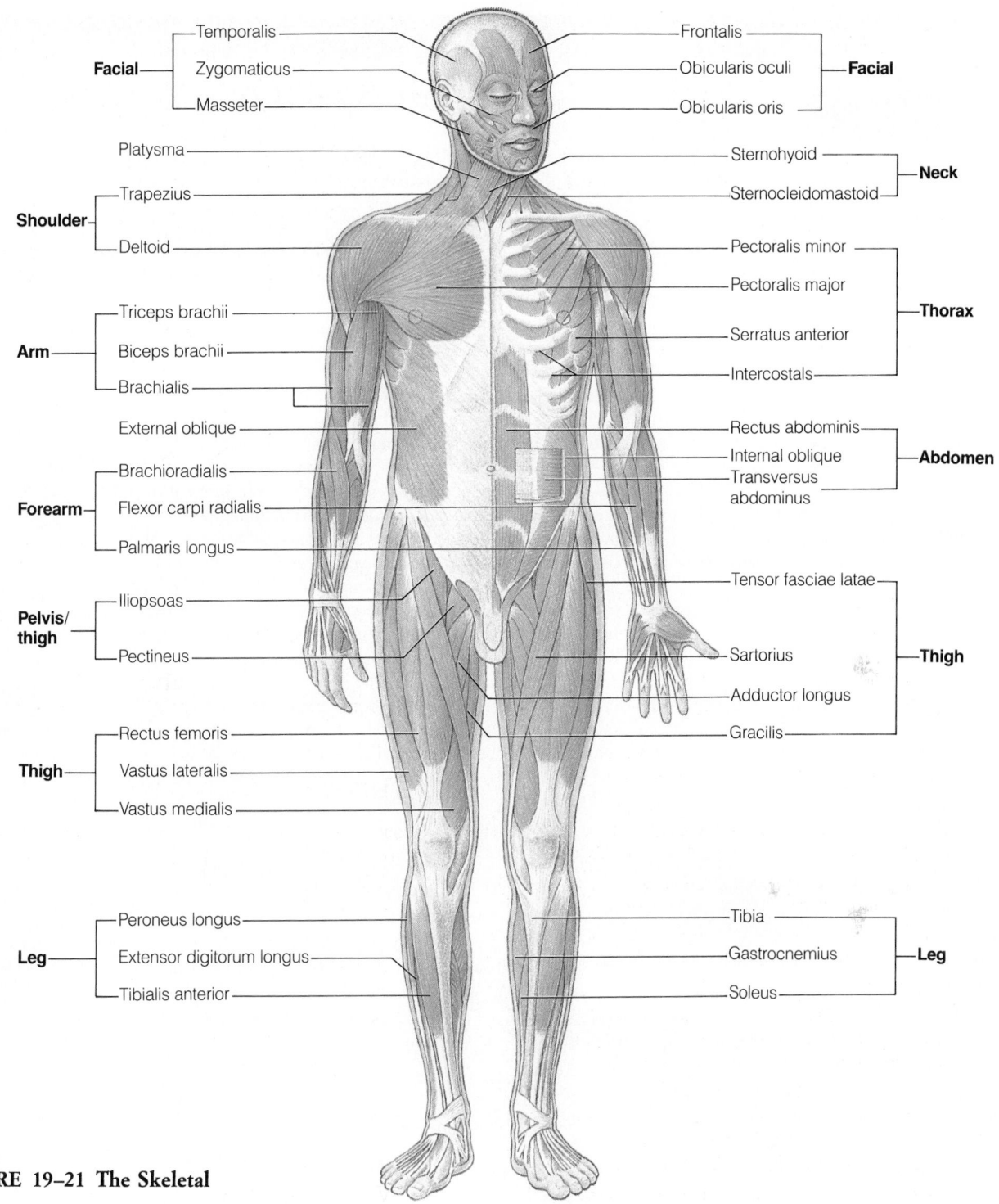

Facial ─┬─ Temporalis
            ├─ Zygomaticus
            └─ Masseter

Platysma

Shoulder ─┬─ Trapezius
              └─ Deltoid

Arm ─┬─ Triceps brachii
         ├─ Biceps brachii
         └─ Brachialis

External oblique

Forearm ─┬─ Brachioradialis
            ├─ Flexor carpi radialis
            └─ Palmaris longus

Pelvis/ thigh ─┬─ Iliopsoas
                   └─ Pectineus

Thigh ─┬─ Rectus femoris
           ├─ Vastus lateralis
           └─ Vastus medialis

Leg ─┬─ Peroneus longus
         ├─ Extensor digitorum longus
         └─ Tibialis anterior

Frontalis ─┬─ **Facial**
Obicularis oculi ─┤
Obicularis oris ─┘

Sternohyoid ─┬─ **Neck**
Sternocleidomastoid ─┘

Pectoralis minor ─┬─ **Thorax**
Pectoralis major ─┤
Serratus anterior ─┤
Intercostals ─┘

Rectus abdominis ─┬─ **Abdomen**
Internal oblique ─┤
Transversus abdominus ─┘

Tensor fasciae latae ─┬─ **Thigh**
Sartorius ─┤
Adductor longus ─┤
Gracilis ─┘

Tibia ─┬─ **Leg**
Gastrocnemius ─┤
Soleus ─┘

▷ **FIGURE 19–21 The Skeletal Muscles**

make bones move. Some muscles simply steady joints, allowing other muscles to act. These stabilizing muscles are known as **synergists.** Muscles in the face are another example of those that do not make bones move. Anchored to the bones of the skull and to the skin of the face, these muscles allow us to wrinkle our skin, open and close our eyes, and move our lips.

Muscles help us move about in the environment. Al-though few of us are aware of it, our muscles also constantly work to maintain our posture, helping us stand or sit upright despite the never-tiring pull of gravity.

The muscles of the body also produce enormous amounts of heat as a by-product of metabolism. When working, the muscles produce additional heat—so much that you can cross-country ski in freezing weather wearing only a light sweater.

HIBERNATING BEARS REVEAL SECRETS OF SKELETAL STABILITY

It is late fall in Minnesota. Black bears have begun to hole up in their dens or settle down on the forest floor, creating a nest of twigs and leaves where they will hibernate (Figure 1). During hibernation, the bears remain immobile. They do not eat or drink, urinate or defecate for up to five full months. Even more remarkably, two months into their winter hibernation, females give birth and nurse their young for three months before waking from their slumber.

Preparations for winter hibernation actually begin in the late summer, when the bears begin to gorge themselves on grubs, carrion (dead animals), berries, nuts, and insects—virtually anything they can get their paws on. Adults consume approximately 20,000 Calories a day, five times more than is required to stay alive. A layer of fat grows, reaching 5 inches in thickness.

During hibernation, the bear's heart rate and breathing drop, although its metabolism remains quite high—about 4000 Calories per day. Moreover, although the bears are inactive during hibernation, they are not asleep. They remain alert.

During hibernation, energy is supplied by fat. Very little protein is broken down. Potentially toxic nitrogenous waste in the blood is converted into protein. Therefore, a hibernating bear actually increases its lean body mass during this long fast. This adaptation prevents uremia, the potentially lethal buildup of toxic wastes in the animal's bloodstream.

Successful hibernation in the bear also depends on another as-yet poorly understood physiological adaptation—a mechanism by which the bear prevents osteoporosis during this long period of inactivity. Except for bears, all mammals kept immobile for long periods suffer from osteoporosis. Human astronauts, bedridden patients, and inactive elderly people are good examples. The bear is different. Studies show that bone mass may shift in a hibernating bear to the main pressure points where bones sup-

port the resting bear's weight, but there is no net loss of bone calcium. All in all, bone metabolism remains fairly normal.

The conservation of bone matter is probably unique to bears. Hibernating ground squirrels, for instance, lose bone mass and get rid of the excess calcium in their urine. Scientists believe that the bear's unique ability to conserve bone is probably the result of a hormonal substance, which they have set out to isolate and purify. This substance may have important implications in the prevention and treatment of osteoporosis in humans.

Scientists also believe that the bear may produce a chemical substance, dubbed hibernation induction trigger, that sets into motion a wide array of complex physiological changes associated with hibernation. This substance, already isolated and purified, is believed to be chemically related to the opiates such as morphine, which depress some nervous system functions. Injected into nonhibernating animals, it results in a dramatic decline in heart rate and many metabolic processes.

Hibernation induction trigger may prove quite useful in treating human ailments. The army, for instance, is seeking a drug that slows metabolism and reduces head swelling in trauma victims. Surgeons might be able to use it to slow metabolism and cool the body during surgery. Preliminary studies show that this hormone prevents fibrillation and thus could become an important tool in stopping heart attacks.

While scientists probe the mysteries of the bear, another winter sets in, and this remarkably adapted animal settles down for its unique and fascinating retreat, an adaptation to the rigors of winter where food is scarce and survival would be questionable.

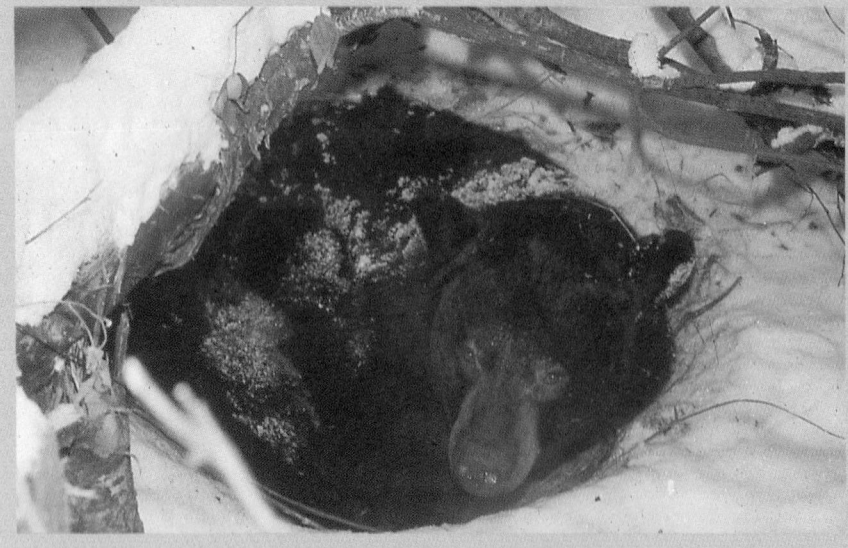

▷ **FIGURE 1 Hibernating Bear**

Skeletal Muscle Cells Are Known as Muscle Fibers and Are Both Excitable and Contractile

Skeletal muscles consist of long, unbranched cells called **muscle fibers** (▷ Figure 19–22). Muscle fibers are multi-nucleated structures formed during embryonic develop-

ment by the fusion of many smaller cells. Viewed with the light microscope, skeletal muscle fibers appear striated.

Like nerve cells, the muscle fiber is an excitable cell. A small potential difference (about -60 mv) exists across the plasma membrane of skeletal muscle fibers, as in neurons.

(a)

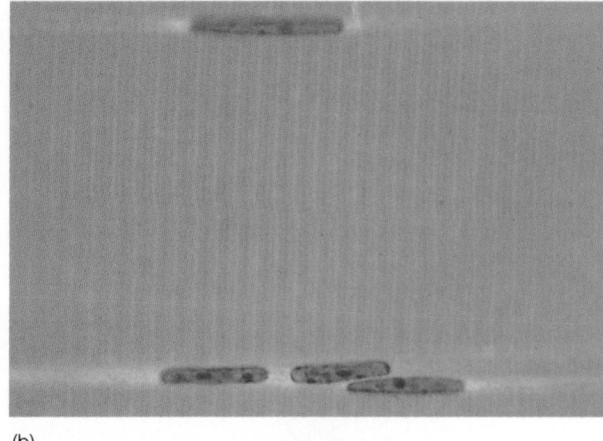
(b)

▷ **FIGURE 19–22 Light Micrographs of Skeletal Muscle** (*a*) Notice the banding pattern on these muscle fibers. (*b*) Higher magnification showing nuclei and banding pattern.

When the membrane is stimulated by a neurotransmitter from the terminal bouton of a motor neuron, a bioelectric impulse is generated. The impulse travels along the membrane of the muscle fiber in the same way a nerve impulse travels along an unmyelinated axon or dendrite.

Muscle fibers are also contractile. When stimulated, the contractile proteins inside the fibers cause the cells to shorten (explained in more detail shortly). Muscle fibers are also elastic, capable of returning to normal length after a contraction has ended.

Each muscle fiber in a skeletal muscle is surrounded by a delicate layer of connective tissue, the **endomysium** (▷ Figure 19–23). Individual fibers are joined in groups known as **fascicles.** Fascicles are also held together by a connective tissue, the **perimysium.** Numerous fascicles are bound by a connective tissue sheath that surrounds the entire muscle, the **epimysium.** This arrangement provides support and protection for muscle cells. In many muscles, the epimysium (the outermost layer) fuses at the ends of the muscle to form tendons. Tendons often attach to the periosteum of bones. Because the tendon is continuous with the epimysium and because the perimysium and endomysium are also connected to the epimysium, muscle contraction can exert a powerful force on the point of attachment.

Muscle Fibers Contain Many Small Bundles of Contractile Filaments Known as Myofibrils

Muscle cells are uniquely adapted to perform their function. Understanding how a muscle contracts, however, first requires a careful look at the muscle fiber. To begin, imagine that you could tease a single skeletal muscle fiber (cell) free from its fascicle. Under a microscope, you would find that each muscle fiber is a long cylinder wrapped in plasma membrane and containing many nuclei. Each muscle fiber is characterized by a series of dark and light bands (▷ Figure 19–24b). Inside each muscle fiber are numerous

bundles of threadlike filaments, mostly actin and myosin, the contractile proteins. Each bundle of filaments in the muscle fiber is known as a **myofibril** (Figure 19–24b); each muscle cell contains numerous myofibrils.

Myofibrils contain contractile filaments and, as illustrated in Figure 19–24c, are striated. The wide, dark bands

▷ **FIGURE 19–23 Connective Tissue Layers Investing Skeletal Muscle** Individual muscle fibers are surrounded by the endomysium. Muscle fibers are bundled together by the perimysium to form a fascicle. Fascicles are held together by the epimysium, which also forms the tendons.

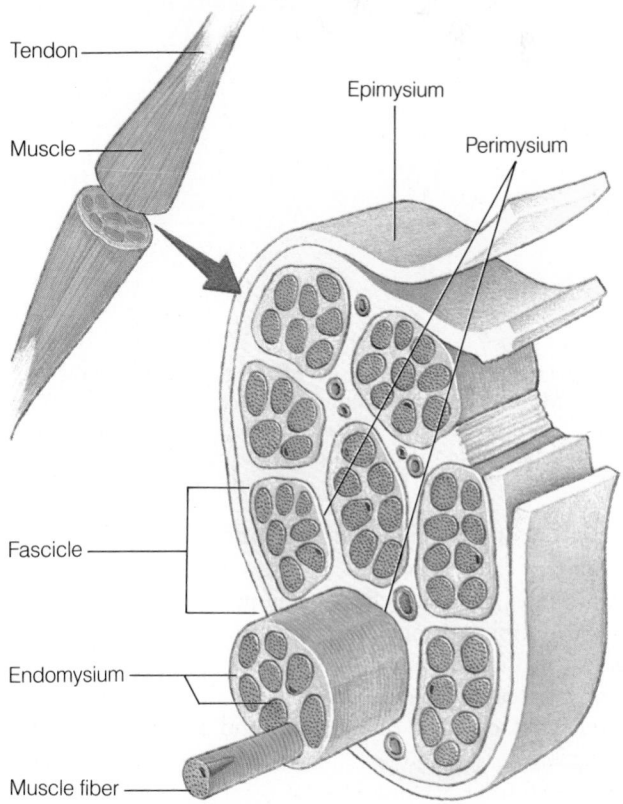

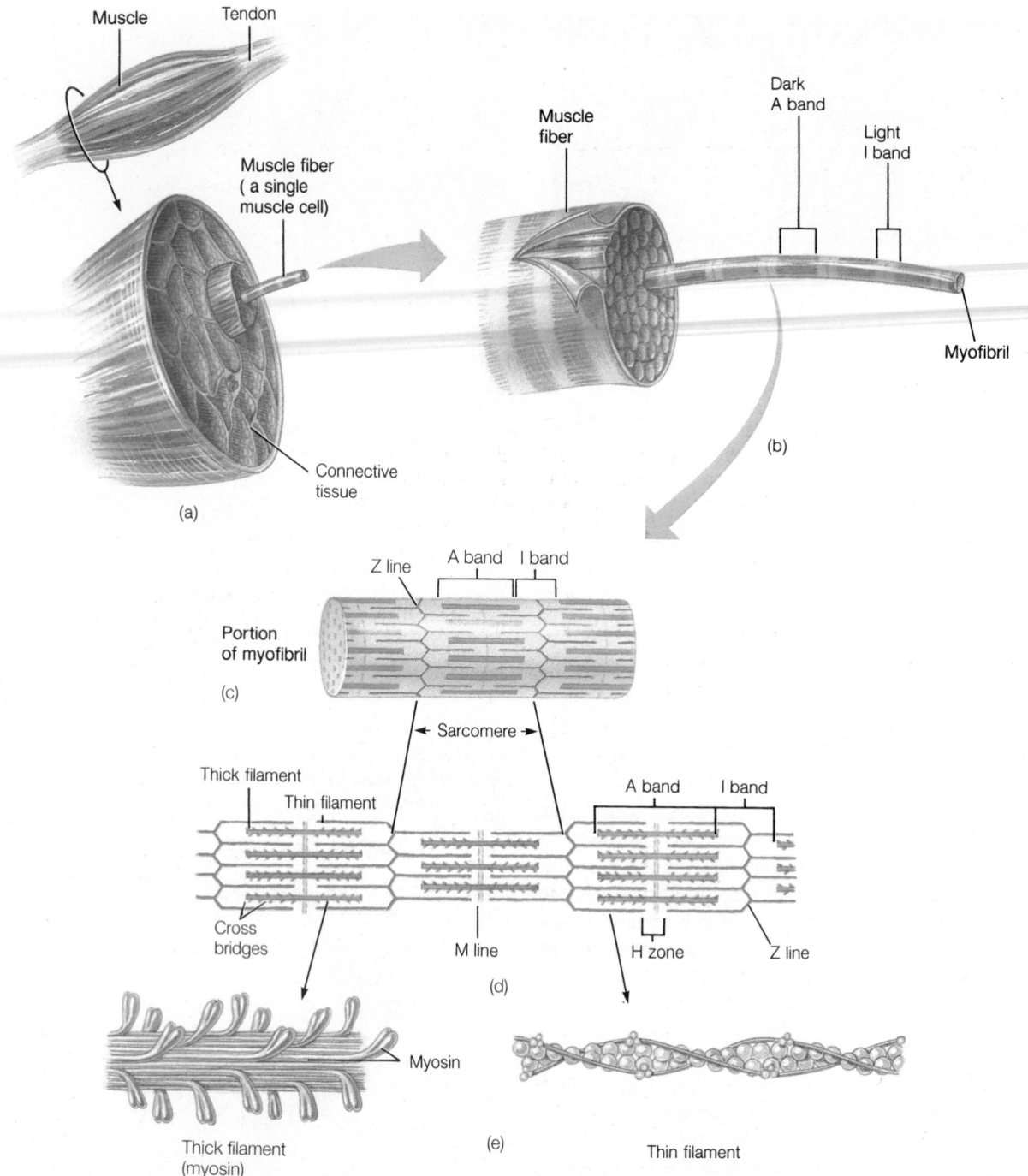

Muscle **Tendon**

Muscle fiber (a single muscle cell)

Connective tissue

(a)

Muscle fiber

Dark A band

Light I band

Myofibril

(b)

Z line   **A band**   **I band**

Portion of myofibril

(c)

← **Sarcomere** →

Thick filament

Thin filament

A band   **I band**

Cross bridges

M line

H zone

Z line

(d)

Myosin

Thick filament (myosin)

(e)

Thin filament

▷ **FIGURE 19–24 Structure of the Skeletal Muscle Fiber, Myofibril, and Sarcomere** (*a*) A single muscle fiber teased out of the muscle. (*b*) Each muscle fiber consists of many myofibrils. (*c*) Note the banded pattern of the myofibril. (*d*) Sarcomeres consist of thick and thin filaments, as shown here. (*e*) Molecular structure of the thick (myosin) and thin (actin) filaments.

of the myofibril are called **A bands;** the narrower, light bands are called **I bands.**[5] In the myofibril, the light and dark bands are arranged in a uniform pattern, which gives the myofibril and muscle fiber a striated appearance. Also shown in Figure 19–24c is a fine line that runs down the center of each I band. This jagged line looks like many letter Zs stacked on one another and is called the **Z line.**

[5] The words *dark* and *light* may help you remember which is which. Dark contains an *a*, and light contains an *i*.

The region of the myofibril between two adjacent Z lines is known as a **sarcomere** and is considered the functional unit of the muscle cell. As shown in Figure 19–24d, the sarcomere contains thick and thin filaments. The thick filaments are made of the protein myosin and lie in the middle of the sarcomere. The thin filaments are composed primarily of the protein actin and extend from the Z line toward the center of the sarcomere but do not join in the middle.

PREVENTING OSTEOPOROSIS: A PRESCRIPTION FOR HEALTHY BONES

Twenty million Americans, mostly women, suffer from a painful, debilitating disease called osteoporosis (▷ Figure 1). Caused by a gradual deterioration of the bone, osteoporosis results in nagging pain and discomfort. Bones fracture easily.

If current trends continue, one of every two American women will develop postmenopausal osteoporosis. Each year, nearly 60,000 Americans—mostly women—will die from complications resulting from the disease. Hemorrhage; fat embolisms (globules), sometimes released from the yellow marrow of broken bones; and shock are the three most common causes of death. Unfortunately, most women do not realize they have the disease until it has progressed quite far.

Recent research shows that osteoporosis begins much earlier than researchers once thought—by the time a woman reaches her mid-20s. Bone demineralization occurs very rapidly. In fact, by age 30 many women have lost one-third of their bone calcium! Between the ages of 30 and 50, many women's bones continue to deteriorate, becoming extremely brittle.

Calcium loss begins so early in American women because many women in their mid-20s avoid fatty foods such as whole milk, cheese, and ice cream to help control their weight. Although these foods are indeed fatty, they are also a major source of calcium. Milk products are also shunned by many adults who develop an intolerance to lactose (a sugar) in milk and other dairy products. Lactose intolerance results from a sharp reduction in the production and secretion of the intestinal enzyme lactase as one ages, making it difficult to digest lactose. Because of these and other factors,

▷ **FIGURE 1 Bone Deterioration**
An elderly woman suffering from osteoporosis. Notice the hunched back due to the collapse of vertebrae.

adult women often consume only about one-half of the 1000 to 1500 milligrams of calcium they need every day.

Osteoporosis may be prevented and even reversed by eating calcium-rich foods, such as spinach, milk, cheese, shrimp, oysters, and tofu (soybean curd). Calcium supplements can also help halt bone deterioration and restore calcium levels. Vitamin D supplements can also help, because vitamin D increases the absorption of calcium in the intestines. (A word of caution, however: excessive vitamin D—five times the RDA—can be toxic.)

Studies also suggest that osteoporosis can be prevented by exercise. Aerobics, jogging, walking, and tennis in conjunction with the dietary changes noted above all help prevent the disease.

Research shows that osteoporosis can

be reversed by exercise even after the disease has reached the dangerous stage. Forty-five minutes of moderate exercise (walking) three days a week, for example, greatly decreases the rate of calcium loss in older individuals. In addition, this exercise regime stimulates the rebuilding process, replacing calcium lost in previous years. Continued exercise increases bone calcium levels and decreases the rate of bone fractures and fatal complications noted earlier.

Another effective treatment for postmenopausal women is estrogen. Low doses of estrogen halt bone demineralization and promote bone formation. Because women who are given estrogen suffer an increased risk of endometrial cancer (cancer of the uterine lining), physicians often prescribe a mixed dose of estrogen and progesterone, which reduces the likelihood of this type of cancer.

Studies have also shown that high doses of fluoride and calcium stimulate bone development. Calcium fluoride treatment increases bone mass approximately 3% to 6% per year and decreases bone fractures. The average patient in one study experienced one fracture every eight months before treatment. After treatment, that figure dropped to one fracture every 4.5 years.

Unfortunately, large doses of fluoride may erode the stomach lining, causing internal bleeding. They may also stimulate abnormal bone development and cause pain and swelling in joints. To offset these problems, researchers have developed a pill that releases the fluoride gradually.

For millions of young women, early detection and sound preventive measures, including exercise, vitamin D, dietary improvements, and fluoride treatments, can prevent osteoporosis.

During Muscle Contraction, the Actin Filaments Slide Inward, Causing the Sarcomeres to Shorten

Actin and myosin filaments are surprisingly delicate yet are responsible for all muscle contraction. When a muscle contracts, each sarcomere shortens. The actin filaments slide toward the center of the sarcomere, sometimes even touch-

ing in the middle. The actin filaments are pulled inward by the myosin molecules. As ▷ Figure 19–25 shows, myosin filaments consist of numerous golf-club-shaped myosin molecules, which are arranged with their "club ends," or heads, projecting toward the actin filaments. During muscle contraction, the heads of the myosin molecules attach

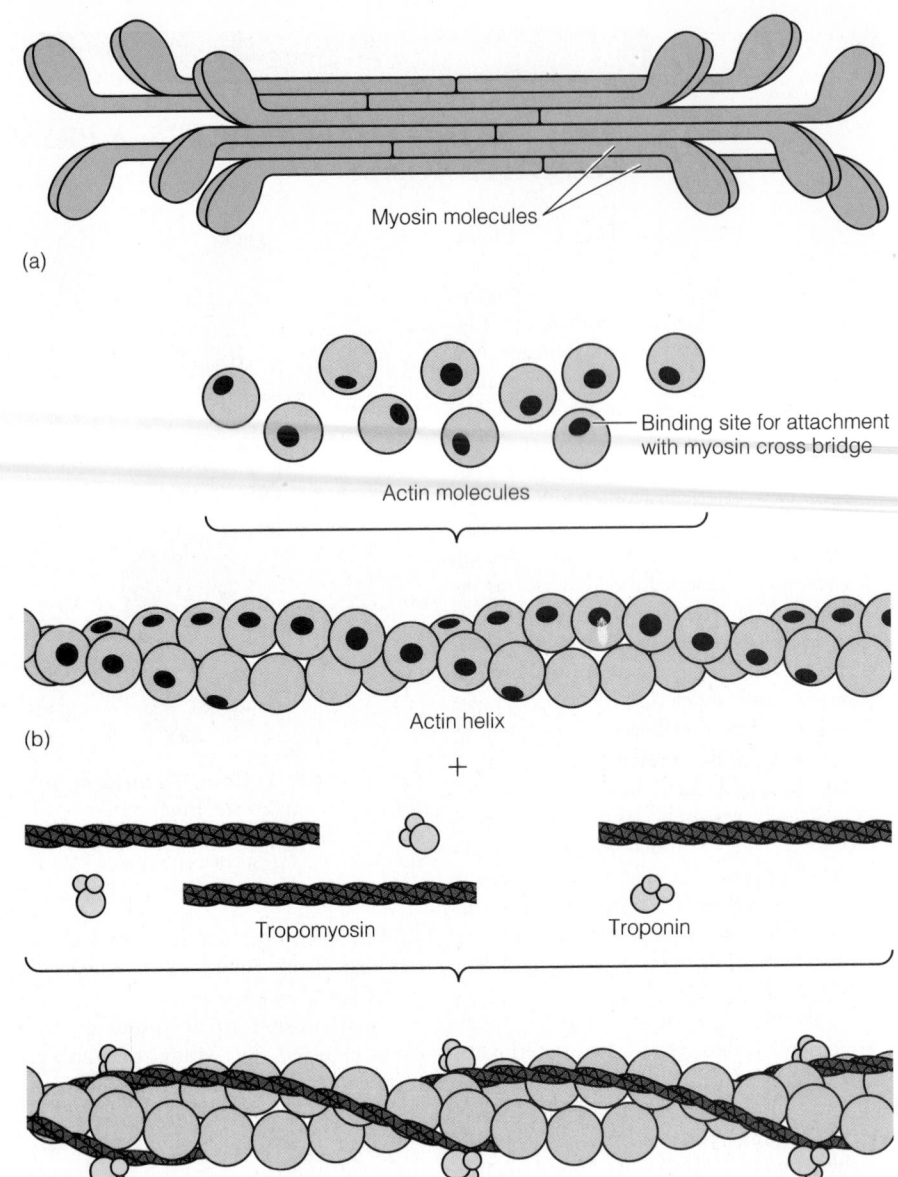

▷ **FIGURE 19–25 Structure of Myosin and Actin Filaments**
(a) Myosin molecules join to form a myosin filament. Note the presence and orientation of the heads of the myosin molecules. (b) Globular actin molecules form intertwining strands. Note the presence of the myosin binding sites. (c) Tropomyosin attaches to the binding sites, covering them. Tropomyosin molecules are held in place by the troponin.

(a)

Myosin molecules

Binding site for attachment with myosin cross bridge

Actin molecules

(b)

Actin helix

+

Tropomyosin   Troponin

(c)   Thin filament

to actin filaments, forming **cross bridges,** which tug the filaments inward, causing the sarcomere to shorten.

How do the myosin cross bridges function? To answer this question, we must first take another look at Figure 19–25. This illustration shows the molecular makeup of the actin and myosin filaments. As illustrated in Figure 19–25b, each actin filament consists of two strands of globular actin molecules joined in a double helix like two bead necklaces. Each actin molecule contains a binding site to which the club ends of the myosin molecules attach. The actin filament therefore contains many binding sites. When the heads of the myosin molecules attach to the binding sites, they undergo a change in shape that causes them to pull the actin filament toward the center of the sarcomere.

In the resting state, the binding sites on the actin molecules of the actin filament are blocked by long, stringlike

protein molecules known as **tropomyosin,** shown in Figure 19–25b. Tropomyosin prevents the heads of the myosin molecules from binding to the actin filaments when a muscle is at rest. As shown in Figures 19–25b and 19–25c, the tropomyosin molecules are held in place by another protein, **troponin.** The troponin molecules act like thumbtacks that secure the tropomyosin molecules.

Muscle contraction is stimulated when the tropomyosin is removed from the actin filaments, freeing up the binding sites on the actin filaments. In order to understand how the tropomyosin "guard" is removed, let's examine the sequence of events that occurs when a nerve impulse arrives at the muscle cell. ▷ Figure 19–26 illustrates the **neuromuscular junction,** the synapse between the terminal bouton of a motor neuron and a muscle fiber. When the nerve impulse arrives at the terminal bouton, it triggers the re-

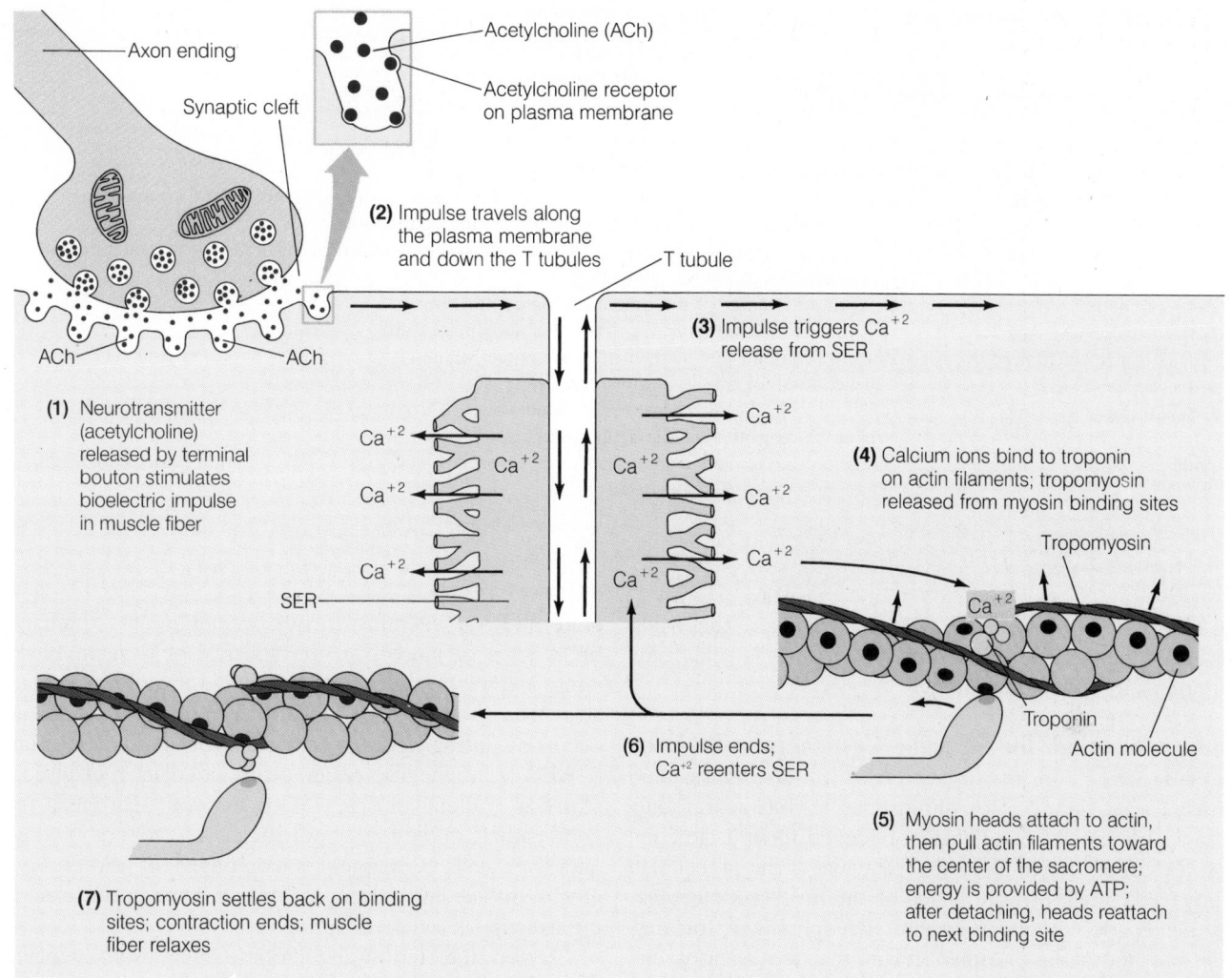

(2) Impulse travels along the plasma membrane and down the T tubules

Acetylcholine (ACh)

Acetylcholine receptor on plasma membrane

Axon ending

Synaptic cleft

ACh — ACh

T tubule

(3) Impulse triggers Ca^{+2} release from SER

(1) Neurotransmitter (acetylcholine) released by terminal bouton stimulates bioelectric impulse in muscle fiber

Ca^{+2} Ca^{+2} Ca^{+2}

Ca^{+2} Ca^{+2} Ca^{+2}

Ca^{+2} Ca^{+2} Ca^{+2}

SER

(4) Calcium ions bind to troponin on actin filaments; tropomyosin released from myosin binding sites

Tropomyosin

Ca^{+2}

Troponin

Actin molecule

(6) Impulse ends; Ca^{+2} reenters SER

(5) Myosin heads attach to actin, then pull actin filaments toward the center of the sarcomere; energy is provided by ATP; after detaching, heads reattach to next binding site

(7) Tropomyosin settles back on binding sites; contraction ends; muscle fiber relaxes

▷ **FIGURE 19–26 The Neuromuscular Junction and Muscle Contraction** Axons from motor neurons terminate on the surface of the muscle fiber, forming the neuromuscular junction.

lease of the neurotransmitter acetylcholine. Acetylcholine diffuses into the synaptic cleft, the space between the terminal bouton and the plasma membrane of the muscle fiber. It then binds to receptors in the plasma membrane.

The binding of acetylcholine to the receptors stimulates changes in the membrane permeability of the muscle fiber, resulting in membrane depolarization caused by the rapid influx of sodium ions. A wave of depolarization travels along the plasma membrane of the muscle fiber. As illustrated in Figure 19–26, the plasma membrane periodically "dips" into the muscle fiber. These deep invaginations, called **T tubules (transverse tubules),** conduct the impulse to the interior of the cell, facilitating contraction of the muscle. As it travels inward, the impulse stimulates the release of calcium ions stored inside the smooth endoplasmic reticulum (SER) of the muscle cell. In muscle fibers, the SER is called the **sarcoplasmic reticulum.** The sarcoplasmic reticulum lies close to the T tubules. Calcium ions released from the sarcoplasmic reticulum diffuse outward into the myofibril and attach to the troponin molecules,

which hold the tropomyosin molecules in place over the binding sites on the actin filaments. When calcium binds to the troponin, the troponin molecules are released, and the tropomyosin molecules slide off the binding sites on the actin filaments. This allows the heads of the myosin filaments to attach to the actin. The cross bridges thus formed contract and pull the actin filaments inward, causing the myofibrils of the sarcomere and the muscle cells to contract. Myosin cross bridges give a brief tug on the actin filaments, then detach, becoming available to bind again. This cycle repeats 50 to 100 times during each muscle contraction.

The actin filaments are pulled inward in much the same way that you would pull a boat tied to a rope toward shore. Contraction ends when the calcium ions are actively transported back into the sarcoplasmic reticulum. When calcium levels fall, tropomyosin slips back into place over the binding sites on the actin filament.

This description of the contraction of myofibrils by the inward movement of the actin filaments is called the **slid-**

TABLE 19–2 Components of Muscle Contraction

MUSCLE FIBER COMPONENT	FUNCTIONAL ROLE
Plasma membrane	Conducts impulse from terminal bouton of motor neuron
T tubule	Conducts impulse into the interior of the muscle fiber
Sarcoplasmic reticulum	Releases stored calcium, which stimulates contraction; absorbs calcium to end contraction
Tropomyosin	Blocks binding sites on actin filament, preventing contraction
Troponin	Holds tropomyosin in place on actin filament, blocking contraction; binds to calcium, releasing tropomyosin to permit contraction
Actin filaments	Slide toward center of sarcomere during contraction
Myosin filaments	Pull the actin filaments toward center of sarcomere during contraction
Heads of myosin molecules (cross bridges)	Bind to actin and pull actin filaments; contain binding site for ATP; contain myosin ATPase, which catalyzes the breakdown of ATP
Calcium ions	Released from the sarcoplasmic reticulum; bind to troponin, causing it to release tropomyosin from binding sites on the actin filaments
ATP	Binds to cross bridges of myosin filaments; broken down by ATPase in cross bridges, providing energy for muscle contraction

ing filament theory. Table 19–2 summarizes the roles played by the various components of the muscle cell. Take a moment to review them now.

Energy Needed for Muscle Contraction Is Provided by ATP.

In this discussion of the fascinating molecular events taking place during muscle contraction, one factor is missing—energy. Energy needed for muscle contraction comes from ATP, the principal form of cellular energy. ATP binds to the heads of the myosin molecules. The heads also contain an enzyme that splits ATP, forming ADP and inorganic phosphate. This reaction releases energy. When ATP is converted to ADP, the energy released is captured and stored in the myosin molecules momentarily like energy stored in a compressed spring. The energy is released when the head of the myosin molecules bind to actin, which causes the heads of the myosin molecules to change shape, drawing the actin filaments toward the center of the sarcomere.

ATP Is Regenerated by One of Several Mechanisms.

ATP is quickly regenerated by a high-energy molecule stored in muscle. Known as **creatine phosphate,** this molecule contains a high-energy bond, indicated by the squiggly line (below). Stored in muscle in high concentrations, creatine phosphate reacts with ADP as follows:

$$creatine \sim P + ADP \rightarrow ATP + creatine$$

Creatine phosphate replenishes ATP used during muscle contraction, but supplies last only 30 seconds or so. ATP is also generated by glycolysis, the citric acid cycle, and the electron transport system, as discussed in Chapter 6.

Because the electron transport process requires oxygen, vigorous muscle contraction causes oxygen supplies inside muscle cells to fall. If the circulatory system cannot replace oxygen as quickly as it is being used, the citric acid cycle and the electron transport system are shut down. To generate ATP, the cell must turn to fermentation—that is, the breakdown of glucose in the absence of oxygen. Besides being inefficient, this process results in the buildup of lactic acid. The shortage of ATP and the buildup of lactic acid occurring during vigorous exercise result in **muscle fatigue.**

Muscle fatigue also results from a depletion of glycogen stores in skeletal muscle. Glycogen, you may recall from your earlier studies, is a polysaccharide found in muscle and liver cells. Glycogen is composed of thousands of glucose molecules. Glycogen broken down in muscles during exercise is replaced during rest.

Oxygen depleted during exercise must also be replaced. This replacement occurs rather quickly. The muscle oxygen deficiency, called the **oxygen debt,** is often largely replaced right after you exercise, explaining why you keep breathing hard for a while after you stop exercising.

Individual Skeletal Muscle Fibers Contract Fully when Stimulated

Individual skeletal muscle fibers obey the **all-or-none law;** that is, when activated by an action potential, a muscle fiber contracts fully. A single contraction, followed by relaxation, is called a **twitch.** The force of that contraction is shown in ▷ Figure 19–27. You will notice that there is a brief lag after the impulse is generated in the membrane of the muscle fiber and contraction begins. Known as the **latent period,** it results from at least three factors: (1) the time required for the action potential to travel into the T tubules, (2) the time required for calcium to diffuse out of the sarcoplasmic reticulum and to bind to troponin, and (3) the time required for the filaments to begin sliding.

Peak tension in a single muscle cell occurs sometime after the impulse arrives. Relaxation requires an equal amount of time as calcium ions are pumped back into the sarcoplasmic reticulum.

Even though individual muscle fibers contract fully

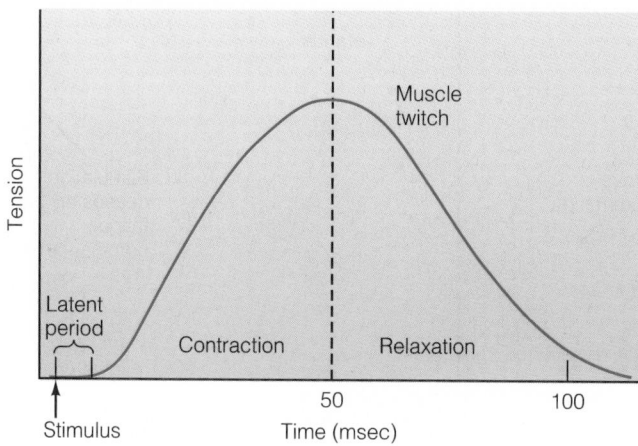

FIGURE 19–27 The Muscle Twitch A graph of muscle fiber contraction, showing the tension generated and the time required to reach maximum tension and relaxation.

when stimulated, whole skeletal muscles are capable of producing contractions of varying strength, referred to as **graded contractions.** Graded contractions result from two different processes: recruitment and wave summation.

The Strength of Muscle Contraction Can Be Increased by Stimulating (Recruiting) Additional

Muscle Fibers to Contract. To generate the force needed to move arms and legs or even eyelids requires the action of more than one muscle cell. The engagement of additional muscle fibers during muscle contraction is called **recruitment.** It is akin to what the U.S. Army might do before it goes to war.

To understand how recruitment occurs in muscle, we must take a look at the innervation of muscles by motor neurons. As shown in ▷ Figure 19–28a, the axons of motor neurons, on reaching the muscles they supply, form many branches. A single motor neuron, in fact, may innervate dozens of individual muscle fibers in a skeletal muscle. A motor neuron and the muscle fibers it supplies constitute a **motor unit.** The fewer muscle fibers in a motor unit, the finer the control. The less control, the more muscle fibers in a motor unit. Thus, the neurons that supply the strong muscles of the leg, which produce crude but strong propulsive force, may end on as many as 2000 muscle fibers. Each time an impulse is delivered to the motor unit, it stimulates all 2000 muscle fibers. In contrast, muscles involved in fine motor movements, such as the extrinsic muscles of the eye or the muscles of the hand, contain "smaller" motor units with a few dozen muscle fibers per axon, providing a greater degree of control.

How does this process relate to recruitment? To increase the force of contraction in a skeletal muscle, the

▷ **FIGURE 19–28 The Motor Unit** (*a*) Light micrograph of axon branching to terminate on many muscle fibers. (*b*) Each axon branches at its termination, supplying a few dozen to many thousand muscle fibers. A single axon and its muscle fibers constitute a motor unit.

Spinal cord

Motor neuron cell body

Nerve

Motor unit 2

Muscle

Axonal terminals

Muscle fibers

Motor unit 1

Muscle fascicle

Axon terminal

Muscle fibers

Axon

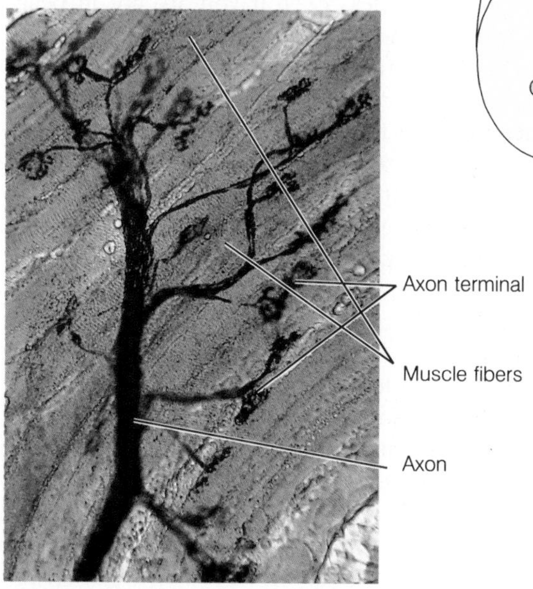

(a)

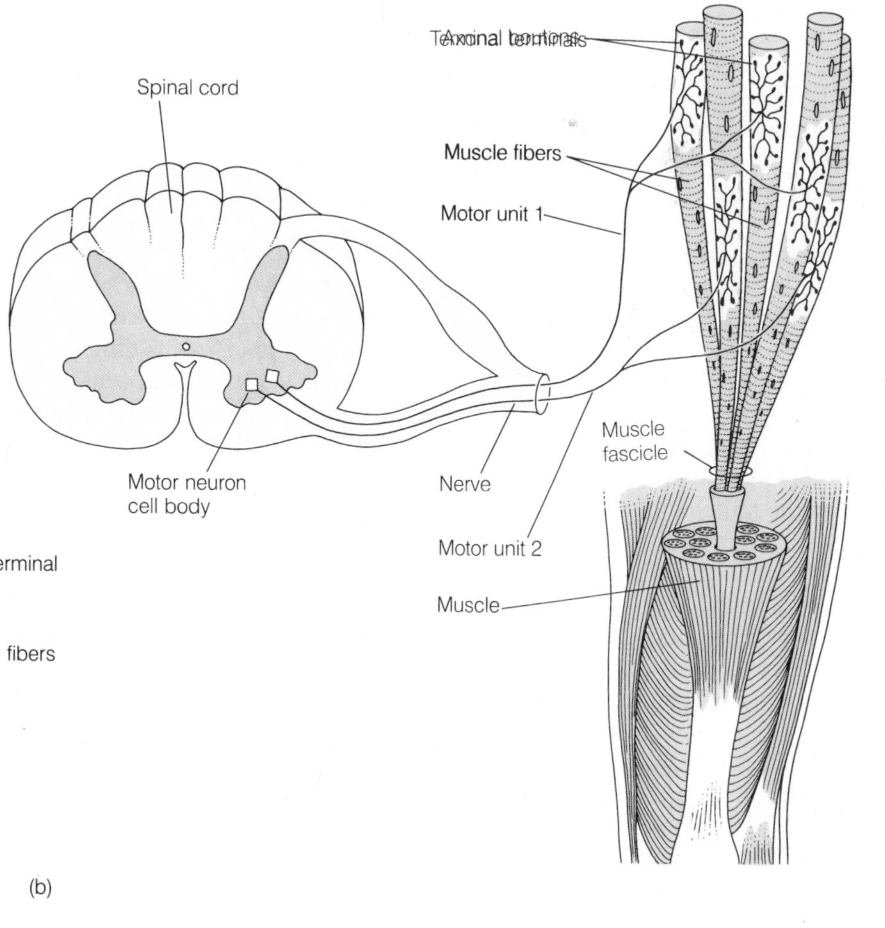

(b)

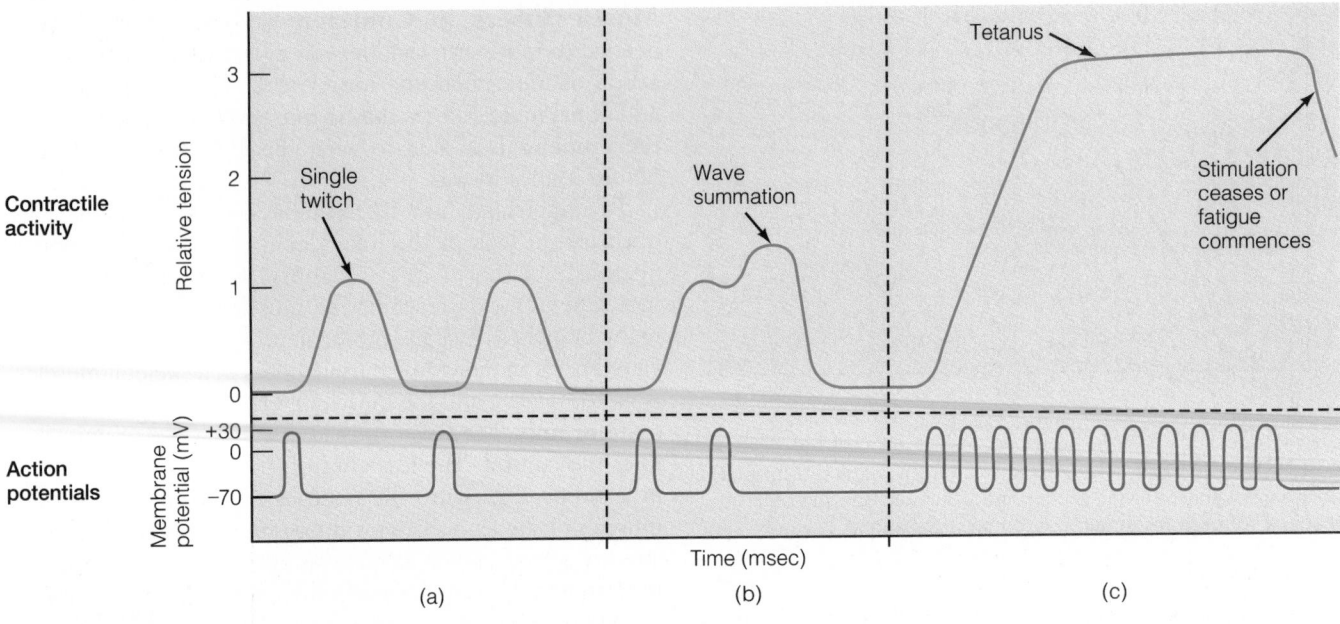

The duration of the action potentials is not drawn to scale, but is exaggerated.

▷ **FIGURE 19–29 Graph of Muscle Contraction** (*a*) The force generated by two separate stimuli (action potentials from motor neuron). (*b*) The force generated by two closely timed stimuli. (*c*) The force generated by many closely spaced stimuli.

central nervous system recruits additional motor neurons. The more motor neurons that are stimulated, the stronger the force of contraction. Thus, if you were going to lift a small piece of styrofoam, your nervous system would recruit only a tiny fraction of the muscle fibers in your arms, but if you were going to lift a 40-pound bag of dog food, your brain would recruit many more motor neurons and muscle fibers.

Additional Tension May Be Created in a Muscle Fiber by Nerve Impulses Arriving While the Muscle Fiber Is Still Contracted. Graded contractions also result from increasing the contractile force that each fiber generates. Each time a muscle fiber is stimulated, it contracts fully. ▷ Figure 19–29a shows the strength of contractions resulting from nerve impulses arriving after muscle fibers contract and relax. If, however, nerve impulses reach the muscle fiber before it has relaxed—that is, while it is still somewhat contracted—additional tension is created (Figure 19–29b). In other words, the muscle fibers contract more forcefully. In a sense, then, the second contraction "piggybacks" on the first; this process is technically referred to as **wave summation.**

If the nerve impulses arrive frequently at enough skeletal muscle fibers, a smooth, sustained contraction will occur in the muscle (Figure 19–29c). When you carry a bag of groceries in your arms, for example, your arm muscles contract to support the weight and remain contracted throughout the activity. A sustained contraction at maximal strength is called **tetanus** (not to be confused with the serious, often fatal bacterial infection of the same name).[6] Tetanic contractions eventually cause muscle fatigue. The muscle stops contracting, even though the neural stimuli may continue.

Muscle Tone Results from the Contraction of a Small Number of Muscle Fibers that Keep Muscles Slightly Tense

Touch one of your muscles. Even if you are not in peak physical condition, you will notice that the muscle is firm. This firmness is called **muscle tone.** Muscle tone is essential for maintaining posture. Without it, you would literally fall into a heap on the floor when you stood up. It also helps generate heat in warm-blooded animals.

Muscle tone results from the contraction of muscle fibers during periods of inactivity. But not all fibers contract, just enough of them to keep the muscles slightly tense.

Muscle tone is maintained, in part, by the muscle spindle, a receptor that monitors muscle stretching (Chapter 18). The spindle "alerts" the brain and spinal cord to the degree of stretching. When muscles relax, signals travel to the spinal cord, then back out motor axons, which stimulate a low level of muscle contraction, maintaining muscle tone.

[6]It is called tetanus because the toxins from the bacterial infection cause the muscle to lock.

Two Types of Muscle Fibers Are Found in Skeletal Muscle, Slow- and Fast-Twitch

Physiological studies have revealed the presence of two types of skeletal muscle fibers: fast-twitch and slow-twitch fibers.[7] **Slow-twitch muscle fibers** contract relatively slowly but have incredible endurance. Endurance athletes (for example, long-distance runners) perform for long periods without tiring because their leg muscles have a higher proportion of slow-twitch fibers than those of us who peter out quickly (▷ Figure 19–30). The flight muscles of birds, which can travel long distances without stopping, are similar.

Fast-twitch muscle fibers contract swiftly. The muscles of sprinters and other athletes whose performance depends on quick bursts of activity contain a higher proportion of fast-twitch fibers.

Slow-twitch fibers are anatomically distinct from fast-twitch fibers. Slow-twitch fibers, for example, are smaller and contain an abundance of the cytoplasmic protein myoglobin. **Myoglobin** binds to oxygen as does its counterpart in RBCs, hemoglobin, and releases oxygen when it is needed during exercise. Slow-twitch fibers also contain a slow-acting **myosin ATPase.** This enzyme forms the heads of the myosin molecules and splits ATP during muscle contraction; it is largely responsible for the slow-twitch fiber's physiological characteristics.

Fast-twitch fibers are larger than slow-twitch fibers.[8] They contain a fast-acting myosin ATPase, which permits rapid contraction.

Skeletal muscles generally contain a mixture of slow- and fast-twitch fibers, giving each muscle a wide range of performance abilities. However, a muscle that performs one type of function more often than another may have a disproportionately higher number of fibers corresponding to the type of activity it performs. The muscles of the back, for example, contain a larger number of slow-twitch fibers. These muscles operate throughout the waking hours to help maintain posture. They do not need to contract quickly but must be resistant to fatigue. In contrast, the muscles of the arm are used for many quick actions—waving, playing tennis, grasping falling objects. Fast-twitch fibers are more common in the muscles of the arm.

Genetic Differences May Account for Differences in Athletic Performance.
New research suggests that one of the reasons some people excel in certain sports whereas others do not may lie in the relative proportion of fast- and slow-twitch fibers in their skeletal muscles. As mentioned earlier, a study of the skeletal muscles of world-class long-distance runners suggests that their physical endurance is primarily due to a high proportion of slow-twitch

[7] In truth, there are three types—one kind of slow-twitch and two types of fast-twitch. For ease of discussion, only two types will be discussed in this chapter.
[8] Studies show that one type of fast-twitch fiber fatigues rather easily whereas the other type is fatigue resistant.

▷ **FIGURE 19–30 Built to Last** An abundance of slow-twitch fibers gives the Olympic long-distance runner endurance.

fibers, a trait that may be genetically determined. Biochemical studies also show that the muscle cells in endurance athletes have a higher level of ATP, both at rest and during exercise, thus providing more energy for muscle contraction. Endurance athletes start out with a larger storehouse of energy and maintain a larger supply throughout exercise.

Exercise Builds Muscles and Increases Endurance

Milo of Croton was a champion wrestler in ancient Greece who stumbled across a revolutionary way to build muscle. His method was simple. Milo secured a newborn calf, then proceeded to carry the animal around every day. Day after day, he faithfully followed this routine, so by the time the calf had become a bull, so had he. Milo's program worked, for he went on to win six Olympic championships in wrestling.

The Greek wrestler discovered a principle of muscle physiology familiar to weight lifters today: when muscles are made to work hard, they respond by becoming larger and stronger. The increase in size and strength results from an increase in the amount of contractile protein inside muscle cells.

Muscle protein is quickly made and quickly destroyed. In fact, half of the muscle you gain in a weight-lifting program is broken down two weeks after you stop exercising. In order to build muscle, then, the rate of formation must exceed the rate of destruction. If you don't keep up with your exercise program, your newly developed biceps will disappear quickly.

High-intensity exercise, such as weight lifting, builds muscle. It takes surprisingly little exercise to have an effect. Working out every other day for only a few minutes will result in noticeable changes in muscle mass. According to some sources, 18 contractions of a muscle (in three sets of six contractions each) are enough to increase muscle mass, if the contractions force your muscles to exert over 75% of their maximum capacity.

Low-intensity exercise, such as aerobics and swimming, tends to burn calories but not build muscle bulk. Aerobic exercise therefore helps increase endurance; that is, it increases one's ability to sustain muscular effort. Stamina or endurance results from numerous physiological changes. One of the most important changes occurs in the heart.

The heart responds to exercise like any other muscle— it grows stronger and even enlarges. A well-exercised heart beats more slowly but pumps more blood with each beat, as explained in Chapter 12. The net result, then, is that the heart works more efficiently, delivering more oxygen to skeletal muscles.

Increased endurance also results from improvements in the function of the respiratory system. For example, exercise increases the strength of the muscles involved in breathing. These muscles become stronger and can operate longer without tiring. Breathing during exercise becomes more efficient.

Increased endurance may also be attributed to an increase in the amount of blood in the body. An increase in blood volume, in turn, results in an increase in the number of RBCs in the body, thus increasing the amount of oxygen available to body cells. This improvement, combined with others, allows an individual to work out longer without growing tired.

When you set out on an exercise program, it is important to establish your goals first. If you are interested in increasing your endurance, you should pursue an exercise regime that works the heart and muscles at a lower intensity over longer periods. If you are after bulk, the answer lies in high-intensity exercises that result in increased muscle mass.

Many health clubs and university gyms offer programs or advice on ways to achieve your goals. And many offer a variety of machines to help you build muscles or simply tone them up. Most of the exercise machines work one particular muscle group—for example, the muscles of the upper arm or the muscles of the chest. In exercise clubs, a half-dozen or more machines, each designed for a specific muscle group, are usually placed in a line. You simply go down the line, working one set of muscles, then another, until you have exercised your entire body.

Exercise machines are popular for several reasons. First of all, they are safer than free weights—barbells and dumbbells. Progressive resistance machines eliminate the chances of your dropping a weight on your toes—or someone else's, for that matter. Moreover, it is almost impossible to strain your back if you make a mistake using one. In contrast, lifting free weights requires care and training as well as brawn.

These machines also reduce the amount of time a person needs to exercise by about half. Why? They require work when flexing and extending a joint. The biceps machine, for example, requires you to pull the weights up, then let them return slowly. In both directions, your muscles are being forced to work. The machines also allow you to lift heavier weights. The reason for this is simple: With free weights, you can only lift a weight that can safely be moved through the part of the exercise where your muscle is the weakest. Any more, and you can tear a muscle or drop a weight. One popular brand of progressive-resistance machine, the Nautilus, alters the resistance automatically as you perform an exercise. In the weakest phase of the exercise, the machine reduces the resistance to prevent damage. Throughout the rest of the exercise, however, the resistance is full.

ENVIRONMENT AND HEALTH: ATHLETES AND STEROIDS

In recent years, many Americans have become increasingly troubled by the use of illegal drugs. Even athletes have come under scrutiny—some for using cocaine and other addictive drugs and others for using anabolic steroids.

Anabolic steroids are synthetic hormones that resemble the male sex hormone, testosterone. When taken in large doses, anabolic steroids stimulate muscle formation (anabolism). They increase muscle size and strength by stimulating protein synthesis in muscle cells. High doses may also reduce inflammation that frequently results from heavy exercise, allowing athletes to work out harder and longer.

When it comes to building muscle, steroids and exercise are an unbeatable combination. Some users think that steroids may even increase aggression, which may be helpful to football players and other competitive athletes on the playing field. Despite steroids' apparent benefits, physicians are concerned about their adverse health effects.

Steroids, for example, may result in psychiatric (mental) and behavioral problems. In interviews with 41 athletes who used steroids, researchers found that one-third of them developed severe psychiatric complications. These athletes routinely took steroids in doses 10 to 100 times greater than those used in medical studies of the drugs. The athletes also reported using as many as five or six steroids simultaneously in cycles lasting 4 to 12 weeks. This practice, known as "stacking," is quite common and may be responsible for the psychiatric effects.

Athletes in the study reported episodes of severe depression during and especially after steroid use. Some reported feelings of invincibility. One man, in fact, deliberately drove into a tree at 40 miles per hour while a friend videotaped him. Some subjects reported psychotic symptoms in association with steroid use, including auditory hallucinations (hearing voices). Withdrawal from steroids results not only in depression but also in suicidal tendencies.

Making matters worse, steroids may also damage the heart and kidneys and frequently reduce testicular size in men. In women, steroids deepen the voice and may cause enlargement of the clitoris. Steroids also cause severe acne and may cause liver cancer. Unfortunately, there are no scientific studies on the long-term health effects of steroids.

Despite the fact that anabolic steroids are banned by the National Football League, the International Olympic Committee, and college athletic programs, athletes continue to use them. Most steroids used in the United States are imported illegally from Mexico and Europe. Because of a federal crackdown on the importation of steroids, some experts believe that the inflow may be slowing. Others are not so optimistic, contending that the $100-million-a-year black market will not be easily thwarted.

A recent survey of 46 public and private high schools across the United States involving over 3000 teenagers suggests that steroid use is especially prevalent in high school seniors. About 1 of every 15 senior boys reported taking anabolic steroids.

The study also suggests that the use of anabolic steroids begins early in junior high school. Two-thirds of the students surveyed said that they had started using them by age 16. Nearly half the steroid users said they took the drugs to boost athletic performance. Twenty-seven percent said their primary motive was to improve their appearance. Researchers say that adolescents who use steroids may be putting themselves at risk of stunted growth, infertility, and psychological problems.

Steroid use in the United States illustrates our dependence on quick fixes. It also illustrates the almost obsessive focus on performance and achievement in our highly competitive society, a social environment that may be endangering the health of our children and our athletes.

SUMMARY

AN OVERVIEW OF THE EVOLUTION OF MOTOR SYSTEMS

1. Animals are partly characterized by their ability to move from one place to another and to move body parts, which allows them to perform many tasks.
2. Animal locomotion varies considerably. Despite the vast differences, all animals have two common features: they possess muscles and convert chemical energy in their muscles into movement.
3. The simplest animal motor systems are found in the sea anemones and annelids, which contain two sets of antagonistic muscle fibers, circular and longitudinal.
4. The presence of exoskeletons in invertebrates, such as arthropods, crabs, snails, and clams, provides a wider range of movements.
5. In many vertebrates, opposing muscle groups acting on internal skeletons permit an even wider range of movements. Internal skeletons are found in vertebrates, including birds, fish, reptiles, amphibians, and mammals.

SKELETAL STRUCTURE AND FUNCTION

6. In vertebrates, bones provide internal support, allow for movement, help protect internal body parts, produce blood cells and platelets, store fat, and help regulate blood calcium levels.
7. Most vertebrate bones have an outer layer of compact bone and an inner layer of spongy bone. Inside the bone is the marrow cavity, filled with either fat cells (yellow marrow) or blood cells and blood-producing cells (red marrow) or with combinations of the two.
8. The joints unite bones. Three types of joints are found: immovable, slightly movable, and freely movable. The movable joints allow for flexion, extension, and other important movements.
9. Joints are often subject to injury and disease. Torn ligaments and ripped cartilage, for example, are common injuries among athletes. Wear and tear on some joints may cause the articular cartilage to crack and flake off, resulting in degenerative joint disease, or osteoarthritis. Rheumatoid arthritis results from an autoimmune reaction that produces inflammation and thickening of the synovial membrane, disfiguring and stiffening joints and causing pain.

EMBRYONIC DEVELOPMENT AND BONE GROWTH

10. Most of the bones form from hyaline cartilage.
11. Bone is constantly remodeled after birth to accommodate changing stresses. During bone remodeling, osteoclasts destroy bone. Osteoclasts are stimulated by parathormone, produced by the parathyroid glands.
12. Osteoclasts also participate in the homeostatic control of blood calcium levels. When activated, these cells free calcium from the bone, raising blood calcium levels.
13. Osteoblasts are bone-forming cells stimulated by calcitonin from the thyroid gland. Calcitonin secretion helps decrease blood calcium levels and increases the amount of calcium in bones.
14. Osteoporosis is a disease of the bone caused by progressive decalcification. The bones become brittle and easily broken. Osteoporosis is most common in postmenopausal women and results from the loss of the ovarian hormone estrogen. Osteoporosis also occurs in people who are immobilized for long periods.
15. Exercise, calcium and fluoride supplements, calcium-rich foods, vitamin D, and estrogen supplements can all help prevent or reverse this potentially fatal disease.

THE SKELETAL MUSCLES

16. Skeletal muscles are involved in body movements, help maintain our posture, and produce body heat both at rest and while we are working or exercising.

17. Skeletal muscles consist of long, unbranched, multinucleated cells known as muscle fibers, which are excitable and contractile.

18. Inside each muscle fiber are numerous myofibrils, bundles of the contractile filaments actin and myosin.

19. Contraction occurs when the heads of the myosin filaments attach to binding sites on the actin filaments, then pull the actin filaments inward.

20. Muscle contraction is stimulated by nerve impulses from motor neurons that cause the release of acetylcholine.

21. The energy for muscle contraction comes from ATP. ATP is replenished by creatine phosphate, glycolysis, and cellular respiration.

22. Individual muscle fibers obey the all-or-none law. When activated by an action potential, they contract fully. Contractions of varying strength (graded contractions) can be generated by recruitment and by wave summation.

23. Recruitment results from the engagement of many motor units. A motor unit is a motor axon and all of the muscle cells it innervates.

24. Wave summation is a piggybacking of muscle fiber contractions occurring when stimuli arrive before the fiber relaxes.

25. Muscle tone is the rigidity of resting muscle caused by low-level contraction of some muscle fibers.

26. The body contains two types of skeletal muscle fibers: slow-twitch and fast-twitch fibers. Skeletal muscles generally contain a mixture of the two types, but fast-twitch fibers are found in greatest number in muscles that perform rapid movement. Slow-twitch fibers are found in muscles such as those of the back that perform slower motions or are involved in maintaining posture.

27. Muscle mass can be increased by exercise. An increase in muscle mass results from an increase in the amount of contractile protein (actin and myosin) in muscle fibers.

28. To build mass, one must generally use more weight and do fewer repetitions.

29. Building endurance requires less weight and more repetitions. Endurance is a function of at least three factors: the condition of the heart, the condition of the muscles of the respiratory system, and the blood volume. Improvement in all three factors increases the efficiency of oxygen supply to muscles.

ENVIRONMENT AND HEALTH: ATHLETES AND STEROIDS

30. Many athletes are using synthetic anabolic steroids to improve performance and build muscle.

31. Unfortunately, massive doses of steroids have many harmful effects. They can increase aggression, cause psychiatric imbalance, such as severe depression, and result in damage to the heart and kidneys.

EXERCISING YOUR CRITICAL THINKING SKILLS

1. In a study of osteoporosis in Australian women, researchers concluded that the best prevention against bone fractures was estrogen-replacement therapy. In their study, the researchers recruited 120 nonsmokers ranging in age from 50 to 60. During the two-year study, the researchers studied the density of bone in the forearms of the women at three-month intervals. Early measurements showed very low levels of calcium—levels that indicated a high risk of fractures.

   The subjects were divided into three groups. Women in the first group received a placebo. Women in the second group received 1 gram of calcium per day, and women in the third group received daily doses of estrogen and progesterone (sex steroid hormones). In this study all women were encouraged to take two brisk 30-minute walks every day and engage in one low-impact aerobics class each week.

   The study showed that bone loss continued in the control group (exercise only) at a rate of 2.5% per year. Bone loss also continued in the group that received a daily dose of calcium, but at a slower rate (0.5% to 1.3% per year). In the group that exercised and received hormone therapy, bone density increased from 0.8% to 2.7% per year.

   Given the potential side effects of hormone therapy, the researchers recommended that only women with lowest bone densities should receive hormone therapy and that others should exercise and take calcium to reduce their risk of osteoporosis.

   Going back over the experimental design, can you pinpoint any potential flaws? If you noted that no mention was made of dietary calcium intake by the various groups, you're correct. Unfortunately, the diets of the women in this study were not monitored to determine how much calcium was being ingested through food and beverages. Second, this study focused on the bones of the forearm, not the load-bearing bones that are at greatest risk of fractures. Walking would probably influence the load-bearing bones of the legs, hips, and back, and to accurately assess the effects of the exercise regime prescribed in this study, the researchers probably should have looked at bone density in all three sites, not the arms. Given these inadequacies, do you think this study's conclusions are reliable?

2. A friend is selling an all-natural supplement guaranteed to give you more energy and help you sleep better at night. She cites a number of examples of users who felt energized and could sleep better. She also notes that the developer of the product has a Ph.D. in nutrition. She wants you to try the product, which costs $40 a bottle. Using the critical thinking rules you learned in Chapter 1, describe what you might do in this situation. How would you evaluate the evidence your friend has given supporting the use of this supplement? What additional information would you want?

1. The skull, ribs, and _____ form the _____ skeleton.

2. The _____ skeleton consists of the bones of the legs and arms and the bones of the shoulders and pelvis to which they attach.

3. The humerus and femur are examples of _____ bones. The shaft, or _____ , of this type of bone consists of a layer of _____ bone surrounding the _____ cavity.

4. Inside the epiphyses of the humerus is a network of bony spicules forming _____ bone.

5. The knee and shoulder joints are examples of _____ joints. They contain _____ fluid, which allows for easy movement of the bones. These joints are supported by four structures: the ___ _____, ligaments, muscles, and _____ .

6. Closing a joint is called _____ ; opening it is called _____ .

7. Knee surgery can be performed using a(n) _____ , a device that reduces the trauma and allows surgeons to see what they are doing.

8. Degenerative joint disease, or _____ , results from wear and tear on a joint. It occurs most often in the weight-bearing joints.

9. _____ arthritis is believed to be an autoimmune disease. It results in a thickening of the _____ membrane and degeneration of the joint.

10. An artificial joint or body part is called a(n) _____ .

11. Most bone in the body is formed during embryonic development from _____ _____ . The first location of bone formation in the diaphysis is called the _____ _____ of ossification.

12. Bone is broken down by large, multinucleated cells called _____ ; it is laid down by _____ .

13. The zone of cartilage lying between the epiphysis and diaphysis of a growing bone is called the _____ _____ .

14. Bone deposition is stimulated by the hormone _____ secreted by cells in the _____ gland. Bone dissolution is stimulated by _____ .

15. Prolonged bone degeneration, which results in brittle bones and occurs chiefly in postmenopausal women, is called _____ .

16. A skeletal muscle fiber is formed during embryonic development by the fusion of numerous embryonic muscle cells. Skeletal muscle fibers are long, unbranched cells containing many _____ . Viewed with the light microscope, they appear banded, or _____ .

17. Each muscle fiber is surrounded by a thin layer of connective tissue, the _____, which holds it in place.

18. The entire muscle is surrounded by a layer of connective tissue called _____, which forms the tendons that attach the muscle to bone.

19. Inside the muscle fiber are numerous bundles of contractile filaments. Each bundle is called a(n) _____ .

20. The _____ is the functional unit of the muscle fiber. It extends from one Z line to the next and contains a dark central band, the _____ band, and two lighter bands on either end, the _____ bands.

21. Contraction results when _____ ions are released from the _____ reticulum. These ions bind to _____ molecules, thus releasing tropomyosin from the binding sites on the _____ filament.

22. The heads of the _____ filaments pull the actin filaments inward, causing contraction.

23. The neurotransmitter _____ is released at the neuromuscular junction. It causes an action potential to be set up in the muscle membrane. The action potential penetrates the interior of the muscle fiber via the _____ _____ .

24. _____ occurs when muscle depletes its supplies of glycogen and ATP and when _____ _____ levels increase.

25. The contraction of a single muscle cell is called a muscle _____ . It is an _____-_____-_____ response to stimulation, which results in a complete contraction.

26. The strength of contraction can be increased by _____ summation, the piggybacking of contractions, and by recruiting more _____ units.

27. A world class long-distance runner would probably have an abundance of _____-_____ muscle fibers in his or her leg muscles. These fibers have a large amount of _____, which holds oxygen and releases it during activity.

28. A sprinter might have an abundance of _____ - _____ fibers. They contain a fast-acting _____ _____, which splits ATP, providing energy for muscle contraction.

29. Synthetic hormones called _____ _____ can help an athlete increase muscle mass but have many deleterious side effects.

Answers to the Test of Terms are found in Appendix B.

1. Describe the functions of bone. In what ways does bone participate in homeostasis?

2. The synovial joints move relatively freely. What structures support the

joint, helping keep the bones in place?

3. A young patient comes to your office

with swollen joints and complains about pain and stiffness in the joints. Friends have suggested that the boy has arthritis, but his parents argue that he is too young. Only old people get arthritis, they say. How would you answer them?

4. Describe the process of bone formation. Be sure to define the following terms: primary and secondary centers of ossification, osteoclasts, perichondrium, osteoblasts, and osteocytes.

5. Bone is constantly remodeled, from infancy through adulthood. Explain when bone remodeling occurs and what cells and hormones participate in the process.

6. Using what you know about bone, explain why an office worker who exercises very little is more likely to break a bone on a skiing trip than a counterpart who works out every night after work.

7. Describe how a bone heals after being fractured.

8. A 30-year-old friend of yours who smokes, exercises very little, and avoids milk products because of her diet says: "Why should I worry about osteoporosis? That's a disease of old women." Based on what you know about bone, how would you respond to her?

9. Describe the major functions of skeletal muscle.

10. Describe the detailed structure of a skeletal muscle fiber.

11. Describe the molecular events involved in muscle contraction.

12. Define the term *graded contraction* and describe how they are achieved.

13. Describe the two types of skeletal muscle fibers and explain how they differ.

14. A friend comes to you complaining that he can't seem to lose weight. He works out on barbells three times a week for an hour or so each time. His diet hasn't changed much, but with the increase in exercise, he thinks he should be losing weight, rather than staying even. Why do you think he isn't losing weight?

15. A friend is thinking about using anabolic steroids to improve his performance in gymnastics. He argues that he is young and will be taking the drug only for a year or so. He points out that other young men are using steroids and they really help build muscle and endurance. What advice would you give him?

The Endocrine System

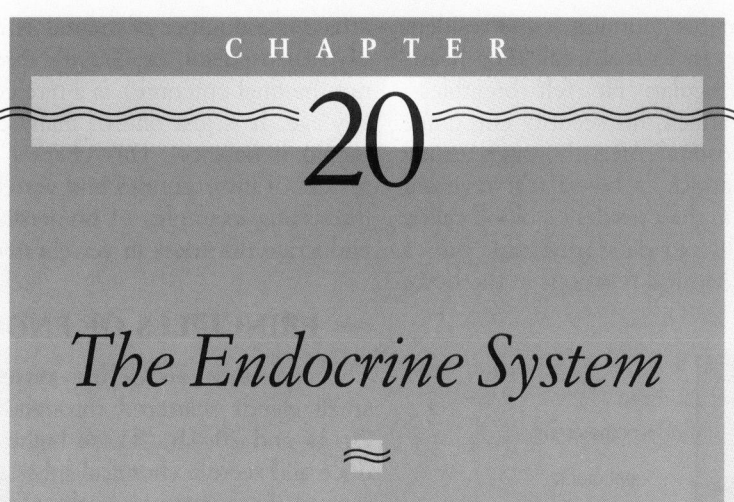

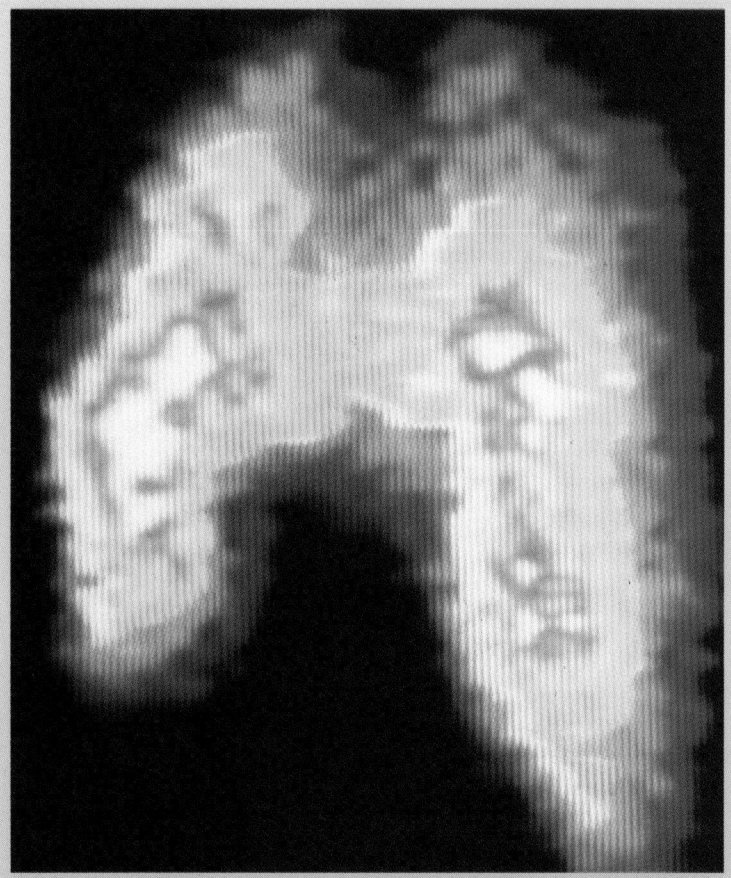

NMR scan of a normal (non-diseased) human thyroid gland. The thyroid gland is located in the neck.

One day while on his morning jog, President George Bush began to feel weak. His heartbeat became irregular. He felt breathless. Fearing a heart attack, his security entourage rushed the president to the hospital. After thorough testing showed no sign of a heart attack, a blood test revealed elevated levels of a hormone in the president's blood called thyroxine. Produced by the thyroid gland, thyroxine causes a general acceleration of all chemical reactions in the body, affecting a number of mental as well as physical processes. Hyperthyroidism, or Grave's disease (after its discoverer, not the final outcome), is a rare condition that can occur at any age. It is just one of many diseases caused by a hormonal imbalance. This chapter discusses the endocrine system of invertebrates and vertebrates and presents some interesting examples of homeostatic imbalance caused by endocrine disorders in vertebrates.

≈ PRINCIPLES OF ENDOCRINOLOGY

The vertebrate **endocrine system** consists of numerous small glands scattered throughout the body (▷ Figures 20–1a and 20–1b). These highly vascularized glands produce and secrete chemical substances known as hormones, a word that comes from the Greek *hormon,* which means "to stimulate" or "excite." A **hormone** is a chemical produced and released by cells or groups of cells that consti-

▷ **FIGURE 20–1 The Vertebrate Endocrine System**
(*a*) The endocrine system consists of a scattered group of glands that produce hormones, which help regulate growth and development, homeostasis, reproduction, energy metabolism, and behavior. (*b*) The endocrine system of the cat.

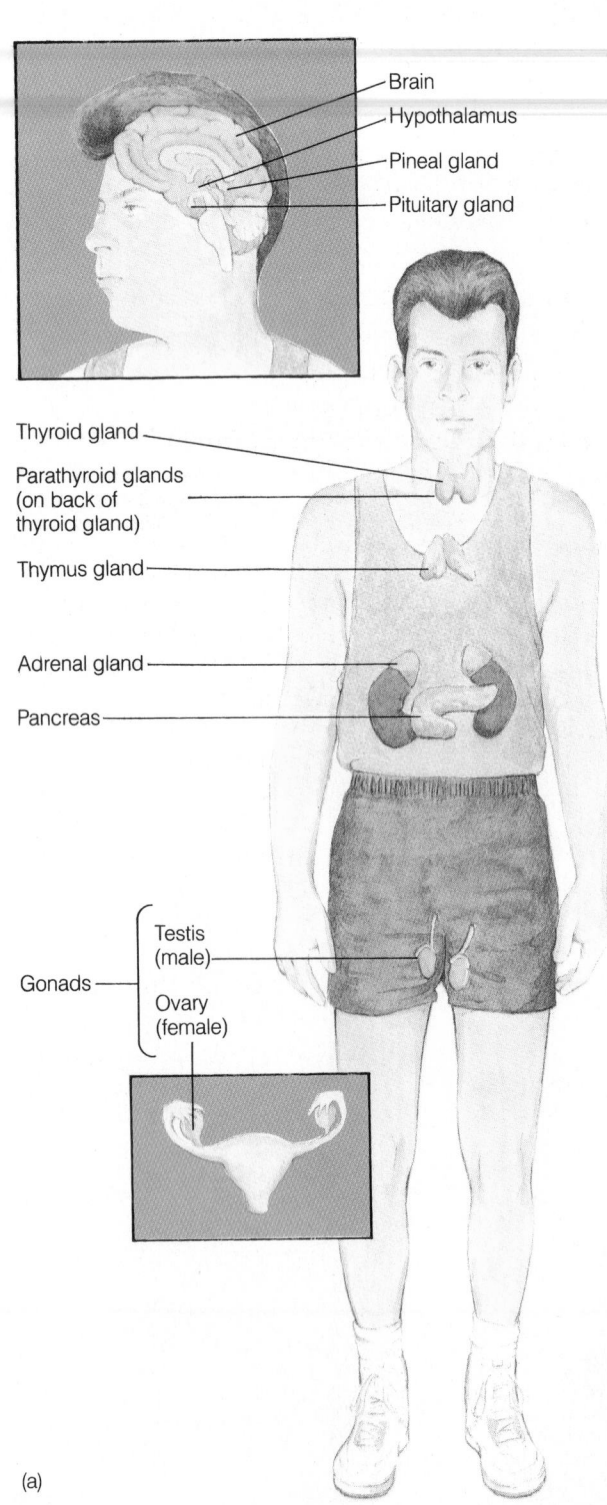

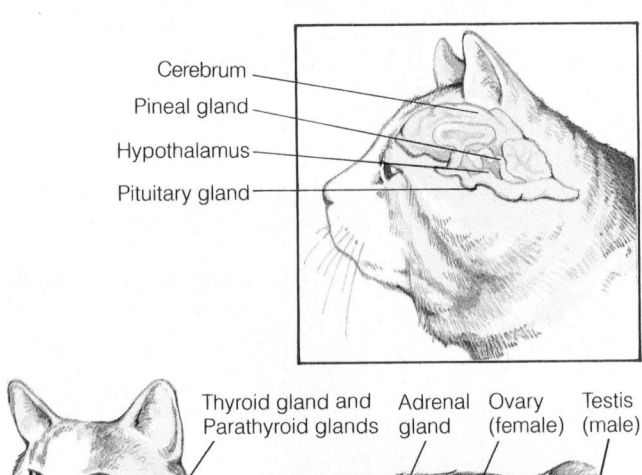

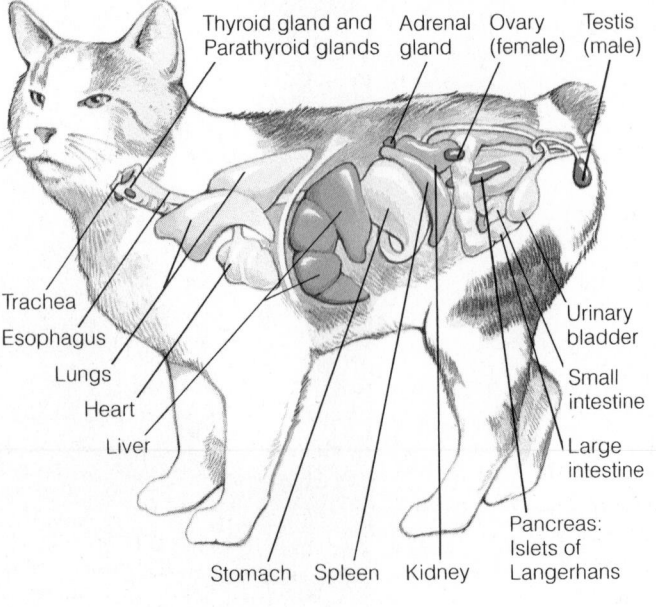

tute the **endocrine (ductless) glands.** Hormones travel in the blood to distant sites, where they elicit their response(s). The hormone insulin, for example, is produced by the pancreas and is transported in the blood to skeletal muscle and other body cells, where it stimulates glucose uptake and glycogen synthesis. The cells a hormone affects are called its **target cells.** For a discussion of some of the early research that contributed to our understanding of the endocrine system, see Scientific Discoveries 20-1.

Hormones function in five principal areas: (1) homeostasis, (2) growth and development, (3) reproduction, (4) energy production, storage, and use, and (5) behavior. At any one moment, the blood carries dozens of hormones. The cells of the body are therefore exposed to many different chemical stimuli. How does a cell keep from responding to all the signals?

Target Cells Contain Receptors for Specific Hormones

Target cells respond only to specific hormones. The "selection process" depends on protein **receptors** in the plasma membrane and in the cytoplasm of target cells. Receptors bind specifically to one type of hormone, in the same way that enzymes bind to only one substrate (Chapter 4). Thus, the cells in the body that respond to insulin contain receptors for this hormone; cells without the receptors are unaffected.

Hormones Stimulate the Synthesis and Release of Other Hormones or May Simply Activate Cellular Processes

Hormones fit into two broad categories. The first are the trophic hormones. **Trophic hormones** stimulate the production and secretion of hormones by other endocrine glands. (The word *trophic* means "to nourish.") An example is **thyroid-stimulating hormone (TSH).** Produced by the pituitary gland, TSH travels in the blood to the thyroid gland, which is located in the neck on either side of the larynx. Here, it stimulates the production and release of thyroxine, the chief function of which is to increase the metabolic rate of body cells. Thyroxine is a nontrophic hormone. **Nontrophic hormones** are those that exert their effect principally on nonendocrine tissues. That is, they *do not* stimulate the synthesis of other hormones.

Hormones can also be classified according to their chemical composition into three groups: (1) steroids, (2) proteins and polypeptides, and (3) amines.

Steroid hormones are derivatives of cholesterol. ▷ Figure 20–2 illustrates two common steroid hormones. Very small differences in the chemical structure of these molecules result in profound functional differences.

Protein and polypeptide hormones constitute the largest class of hormones. As Chapter 2 noted, proteins and polypeptides are polymers of amino acids joined by peptide bonds. Growth hormone and insulin are two examples.

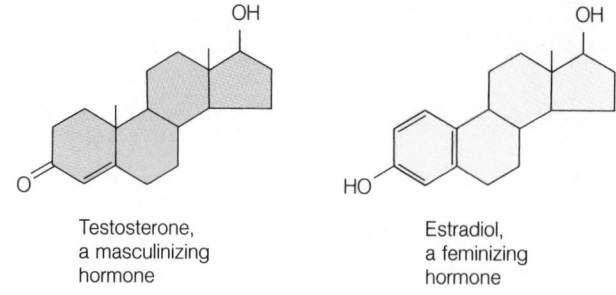

Testosterone,
a masculinizing
hormone

Estradiol,
a feminizing
hormone

▷ **FIGURE 20–2 Two Common Steroids** (*a*) Testosterone, a male sex hormone, and (*b*) estradiol, a female sex steroid.

Amine hormones are derivatives of the amino acid **tyrosine.** ▷ Figure 20–3 shows the structure of two of the four amine hormones produced in the body.

Hormone Secretion Is Often Controlled by Negative Feedback Mechanisms

Hormones help control many homeostatic mechanisms. Their production and release are generally controlled by negative feedback loops, described in detail in Chapter 10. Consider the hormone glucagon.

Glucagon is a hormone released by cells in the pancreas when glucose concentrations fall—for example, between meals. Released into the bloodstream, glucagon travels to the liver, where it stimulates the breakdown of glycogen, causing the release of glucose molecules into the blood. This helps restore blood glucose levels. When normal glucose levels are achieved, the glucagon-producing cells in the pancreas end their secretion.

Not all feedback loops are as simple as this one, however; some involve intermediary compounds. Nevertheless, all operate on the same basic principle. (Examples of other endocrine negative feedback loops are presented throughout this chapter.)

▷ **FIGURE 20–3 Representative Amine Hormones** (*a*) Thyroxine and (*b*) epinephrine (adrenalin).

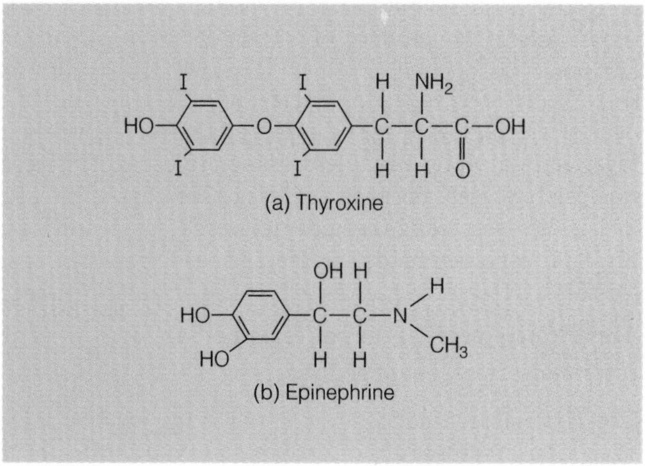

(a) Thyroxine

(b) Epinephrine

PANCREATIC FUNCTION: IS IT CONTROLLED BY NERVES OR HORMONES?

Featuring the work of Bayliss and Starling

The concept of chemical control of body functions, like many other concepts in biology, is rooted in a large number of experiments. Animal experiments in the mid-1800s, for example, showed that one could remove the testes and transplant them to another body location without the loss of secondary sex characteristics. This finding suggested that the testes produced a blood-borne substance that affected other tissues. This study and dozens of others contributed bits of information that, within 100 years, gave rise to many generalizations about hormone production, structure, and function.

Of the many experiments crucial to our understanding of endocrinology, one by two British physiologists, W. M. Bayliss and E. H. Starling, stands out. Published in 1902, this experiment set to rest a debate about the mechanism by which food entering the small intestine from the stomach stimulated the production and release of pancreatic secretions.

Several experiments before the publication of this work suggested that hydrochloric acid in chyme stimulated nerve endings in the small intestine, triggering a reflex that resulted in the release of pancreatic juices, which contain sodium bicarbonate. In these studies, researchers injected substances into the duodenum of anesthetized dogs, then studied the release of pancreatic secretions. Even though severing the nerves that supplied the small intestine did not affect the results, researchers still argued that the nervous system was involved in the control. They hypothesized that clumps of nerve cells (ganglia) and their network of interconnected fibers in the wall of the intestine participated in local reflex arcs—illustrating the influence of bias in the interpretation of scientific research.

Bayliss and Starling repeated these experiments but were careful to sever all nerve connections to the small intestine. They then tied a piece of the small intestine off and injected small amounts of hydrochloric acid into it. This treatment resulted in the production of pancreatic secretion at a rate of one drop every 20 seconds over a period of about 6 minutes. Because it was previously known that acid introduced into the bloodstream had no effect on the release of pancreatic secretions, the researchers concluded that "the effect was produced by some chemical substance finding its way into the veins of the loop of jejunum in question and being carried in the bloodstream to the pancreatic cells."

To verify their hypothesis, they tried another experiment. In this one, they scraped off the cells lining the section of the small intestine, exposed them to acid, ground them up, and then extracted the fluid, which they injected into the bloodstream. After a brief latent period, the pancreas began secreting. This experiment led to the eventual abandonment of the "nervous control hypothesis."

Bayliss and Starling referred to the mystery substance as "secretin," because it stimulated pancreatic secretion. This name remains in use today. Thankfully, most other hormones discovered since that time have far more specific names.

Besides settling the debate over the control of pancreatic function, Bayliss and Starling helped clarify our understanding of endocrinology, furnishing definitions for some of the most important terms in the field. One good example is the term *endocrine gland,* which they defined as an organ that secretes into the blood specific substances that affect some other organ or process located at a distance from the gland.

Positive feedback loops are also encountered in the endocrine system, but rarely. A positive feedback loop results when the product of a cell or organ stimulates additional production. Positive feedback loops perform some specialized functions. Ovulation, the release of the ovum from the ovary, for example, is stimulated by a positive feedback loop that is discussed in the next chapter. Fortunately, each positive feedback loop has a built-in mechanism that ends the escalating cycle, preventing the response from getting out of hand.

Most Hormones Undergo Periodic Fluctuations in Their Release

The fact that hormones are controlled by negative feedback loops does not mean that hormone concentrations in the blood are constant 24 hours a day, 365 days a year. In fact, virtually all hormones undergo periodic fluctuations in their release, causing many body functions also to fluctuate. These natural fluctuations in body function are called **biological cycles,** or **biorhythms,** and were described in Chapter 10.

Biological cycles vary in length. Some hormones, for example, are released in hourly pulses, or cycles. Others are released in daily cycles. For example, cortisol is a steroid hormone produced by the adrenal cortex in a 24-hour, or **circadian rhythm,** as shown in ▷ Figure 20–4. Cortisol increases glucose, among other functions. Its secretion increases during the night, reaching a peak just before one wakes up, then falls sharply during the day.

Other hormones are released in monthly cycles. The 28-day menstrual cycle in women, controlled by hormones,

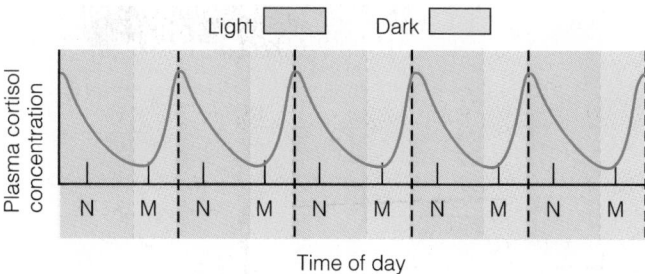

FIGURE 20–4 Cortisol Secretion Cortisol secretion follows a diurnal (daily) rhythm with highest levels occurring in the night just before waking. Adapted with permission from George A. Hedge, Howard D. Colby, and Robert L. Goodman, *Clinical Endocrine Physiology* (Philadelphia: W. B. Saunders Company, 1987), Figure 4–4, p. 80.

is a good example. During the cycle, levels of the reproductive hormones vary considerably.

Still other hormones are released in seasonal cycles. Thyroid hormone in people, for instance, is released in greater amounts during the winter months than during the summer. Thyroid hormone stimulates metabolism, and elevated levels raise body temperature. This cycle probably explains why you feel cold on the first 50-degree day in the fall but feel warm when the first 50-degree day arrives in spring.

Most hormonal cycles are controlled by the biological clocks, regions of the brain that regulate biological cycles (Chapter 10). Seasonal cycles, however, are usually controlled by environmental conditions—for example, temperature or photoperiod (day length).

The Chemical Nature of a Hormone Determines How It Is Transported in the Blood and How It Acts on Cells

Protein and polypeptide hormones are water-soluble (hydrophilic). Consequently, they dissolve in the plasma of the blood, which transports them to their target cells. In contrast, steroid hormones are lipids and do not dissolve in water. To be transported, they must be bound to much larger plasma proteins, like albumin.

Protein and Polypeptide Hormones Stimulate Changes in Target Cells by Activating a Second Messenger Inside the Target Cell. Protein molecules travel freely in the blood, but they cannot penetrate the plasma membrane of their target cells. Protein hormones must trigger internal cellular changes from outside their target cells.

A hypothesis explaining how protein and polypeptide hormones trigger intracellular changes was first presented by the late Earl Sutherland, a medical researcher at Vanderbilt University. His proposal, called the **second messenger theory,** won him a Nobel Prize in 1971. The theory states that

protein and polypeptide hormones (first messengers) stimulate intracellular changes through a second messenger, an intermediary synthesized inside the cell when the hormone binds to the plasma membrane of the target cell. The second messenger, in turn, triggers internal anatomical and physiological changes.

Since his early work, researchers have found that some amine hormones—adrenalin (epinephrine) and noradrenalin (norepinephrine) from the adrenal medulla—also bring about their effects by the second messenger mechanism.

To understand how the second messenger operates, we will study the process by steps.

1. Polypeptide hormones (and certain amines) first bind to protein receptors in the plasma membrane of the target cell (▷ Figure 20–5).
2. These receptors are linked to the enzyme **adenylate cyclase** on the inner surface of the plasma membrane.
3. Binding of the polypeptide hormone to the receptor activates adenylate cyclase.
4. The activated enzyme catalyzes the conversion of intracellular ATP to **cyclic AMP (cAMP),** a special nucleotide shown in ▷ Figure 20–6.
5. Cyclic AMP is the **second messenger.** This water-soluble compound is free to move about in the cytoplasm, where it activates another enzyme, **protein kinase.**
6. Protein kinase catalyzes the addition of phosphate groups to other enzymes. This process activates some enzymes and deactivates others. Because enzymes control chemical reactions in the cell, cyclic AMP can have profound effects on cellular function.

Cyclic AMP is not the only second messenger in body cells. In some cells, calcium ions serve as an intracellular second messenger. In other cells, cyclic GMP (guanosine monophosphate) is a second messenger. Both function independently of cAMP.

Steroid Hormones Enter the Cytoplasm of Their Target Cells and Bind to Receptors that Diffuse into the Nucleus, Activating Genes Directly. Steroid hormones stimulate intracellular change by acting directly on the genes, rather than via enzymes. The process is known as the **two-step mechanism.** Here's how it works:

1. Because of their lipid solubility, steroid hormones readily penetrate the plasma membrane of the target cell (▷ Figure 20–7).
2. Inside the cell, steroid hormones bind to protein receptors, typically found in the cytoplasm but sometimes found in the nucleus. Like their counterparts in the plasma membrane, these receptors recognize and bind to one type of steroid hormone, thus conferring specificity.
3. When a steroid binds to a cytoplasmic receptor, it causes the protein to change its shape slightly. This change activates the receptor-hormone complex. The activated receptor-hormone complex enters the nucleus.
4. The activated complex binds to the chromatin at a

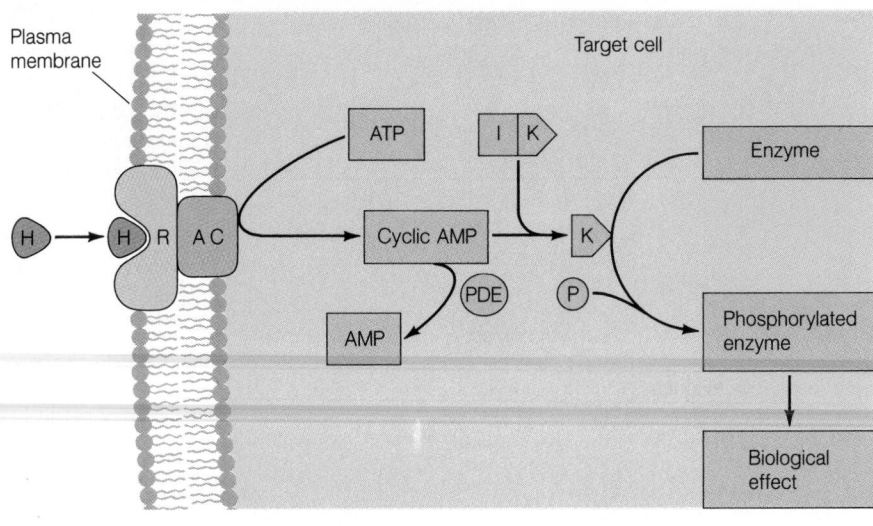

> **FIGURE 20–5 The Second Messenger Theory** Protein and polypeptide hormones bind to a membrane-bound receptor, thus stimulating the production of intracellular cyclic AMP by adenylate cyclase. Cyclic AMP activates protein kinase, which adds phosphates to intracellular enzymes, activating some and deactivating others. Phosphodiesterase (PDE) converts cyclic AMP into AMP. Adapted with permission from George A. Hedge, Howard D. Colby, and Robert L. Goodman, *Clinical Endocrine Physiology* (Philadelphia: W. B. Saunders Company, 1987), Figure 1–8, p. 18.

H = Free hydrophilic hormone
R = Surface receptor
AC = Adenylate cyclase
ATP = Adenosine triphosphate
AMP = Adenosine monophosphate

PDE = Phosphodiesterase
IK = Inactive protein kinase
K = Active protein kinase
P = Phosphate

specific site of attachment on the DNA known as an **acceptor site.**[1]

5. The binding of the receptor-steroid complex to the chromatin turns on certain genes, resulting in the transcription of DNA—that is, the production of mRNA (Chapter 9).

6. Messenger RNA, in turn, provides a template for the production of structural proteins and/or enzymes. Thus, steroid hormones affect the structure and function of target cells.

The second-messenger and two-step mechanisms will require a bit of study to master, but as you work through the details, take a moment to marvel at these remarkable systems of control. You may also want to pause to recall that they developed during evolution, a process of trial and

[1] Thyroid hormones also effect intracellular change through the two-step mechanism, but they bind directly to DNA.

error that has produced so many fascinating structures and functions.

The Endocrine System and Nervous System Are Both Control Systems, but They Exhibit Marked Differences

Now is also a good time to reflect on the similarities and differences between the body's two major control systems, the endocrine and nervous systems. Although similar in some respects—they both send signals to cells that help regulate their function and help coordinate body function—these systems differ in several key respects. First, the nervous system often elicits very rapid, short-lived responses (for example, a muscle contraction). Although there are exceptions to this rule, the endocrine system generally brings about much slower, longer lasting responses (a change in body temperature). The differences can be attributed to the

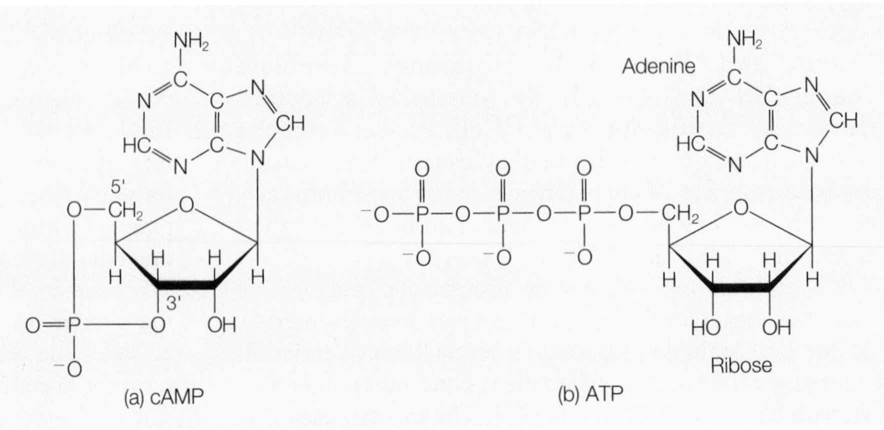

> **FIGURE 20–6 Cyclic AMP, the Second Messenger**
> (*a*) Cyclic AMP compared with (*b*) ATP.

(a) cAMP

(b) ATP

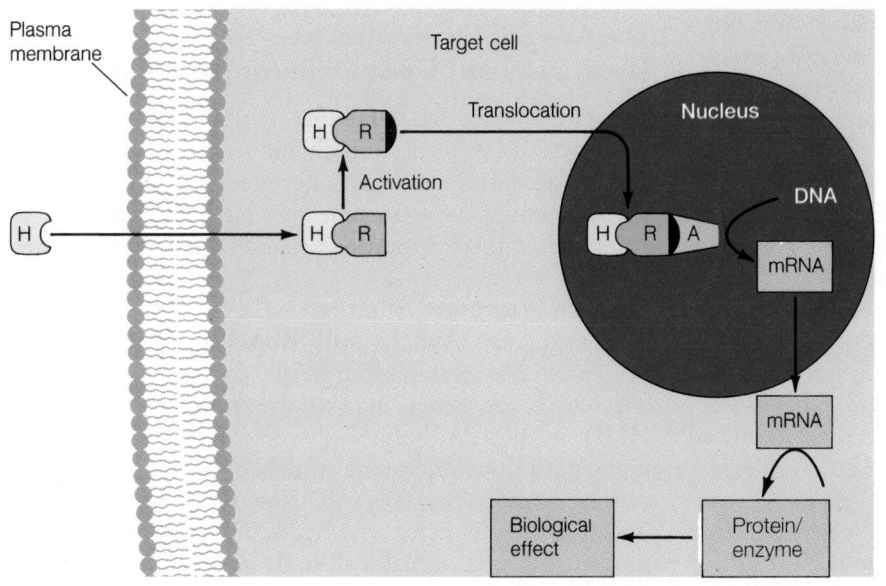

▷ **FIGURE 20–7 The Two-Step Mechanism** Steroid hormones bind to specific receptors in the cell, usually in the cytoplasm. The receptor-steroid complex then binds to the acceptor sites on the nuclear DNA, activating the genes.

Adapted with permission from George A. Hedge, Howard D. Colby, and Robert L. Goodman, *Clinical Endocrine Physiology* (Philadelphia: W. B. Saunders Company, 1987), Figure 1–9, p. 20.

H = Free steroid hormone
R = Cytoplasmic receptor
A = Nuclear acceptor site
mRNA = Messenger RNA

type of signal found in each system. In the nervous system, messages are conveyed by bioelectric impulses that travel along the nerves of the body. In the endocrine system, messages are chemical in nature and are transmitted through the bloodstream.

Despite these differences, the endocrine and nervous systems work in conjunction to ensure homeostasis. And, as you will soon see, these systems sometimes work in concert.

≈ THE PITUITARY AND HYPOTHALAMUS

Attached to the underside of the brain by a thin stalk is the **pituitary gland** (▷ Figure 20–8). About the size of a pea, the pituitary lies in a depression in the base of the skull, the **sella turcica** (literally, "Turkish saddle").

The pituitary gland is one of the most complex of all the endocrine organs. It is divided into two major parts: the anterior pituitary and the posterior pituitary. Together, they

▷ **FIGURE 20–8 The Pituitary Gland**
(*a*) A cross section of the brain showing the location of the pituitary and hypothalamus. (*b*) The structure of the pituitary gland.

(*c*) Releasing and inhibiting hormones travel via the portal system from the hypothalamus to the anterior pituitary, where they affect hormone secretion.

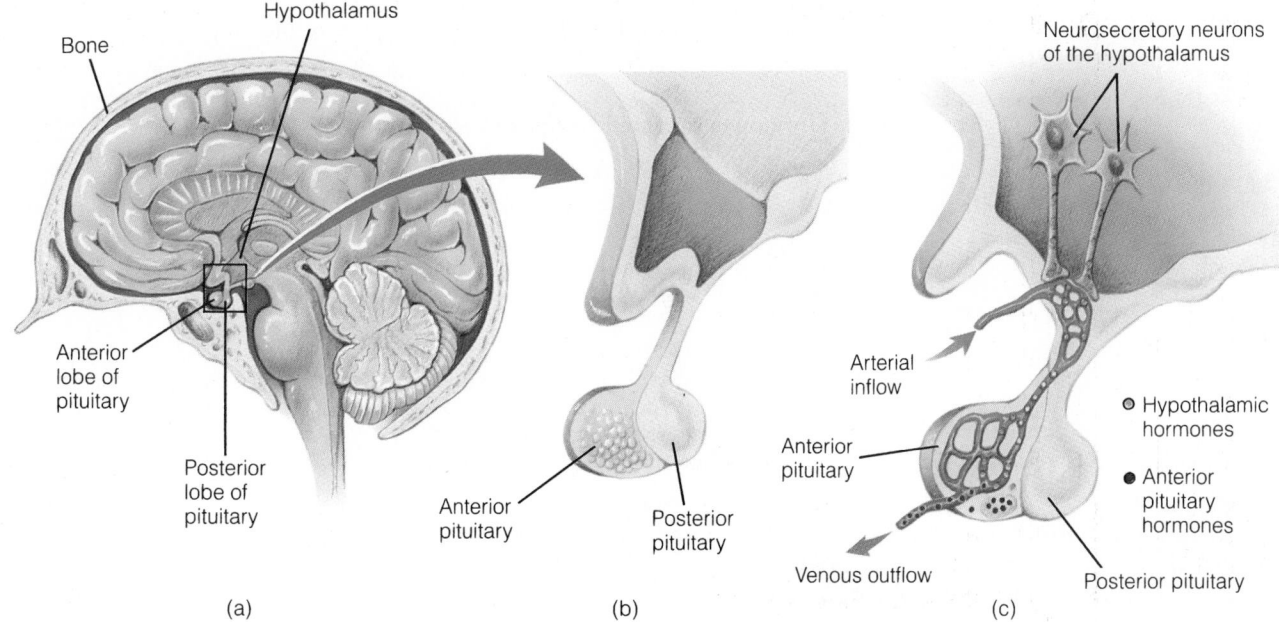

(a) (b) (c)

secrete a large number and variety of hormones, affecting a great many of the body's functions. The anterior pituitary, for example, produces seven protein and polypeptide hormones, six of which are discussed in this chapter (Table 20–1).

The release of the hormones from the anterior pituitary is controlled by a region of the brain, the **hypothalamus,** lying just above the pituitary gland. The hypothalamus contains receptors for a variety of blood-borne chemical substances. These receptors monitor blood levels of hormones, nutrients, and ions. When activated, the receptors stimulate specialized nerve cells within the hypothalamus. These nerve cells are called **neurosecretory neurons**— so named because they synthesize and secrete hormones, which act on the anterior pituitary (Figure 20–8). Some of the hypothalamic hormones stimulate the release of anterior pituitary hormones and are called **releasing hormones,** designated RH. Others inhibit the release of hormones from the anterior pituitary and are called **inhibiting hormones,** designated IH.

The releasing and inhibiting hormones travel down the axons of the neurosecretory cells, which terminate in the lower part of the hypothalamus, just above the pituitary gland. These hormones are then stored in the axon terminals of the neurosecretory cells. When released from the axon terminals, they diffuse into nearby capillaries. As shown in Figure 20–8, these capillaries drain into a series of veins in the stalk of the pituitary. The veins, in turn, empty into a capillary network in the anterior pituitary. Thus, hypothalamic hormones are transported directly to their target cells. This unusual arrangement of blood vessels in which a capillary bed drains to a vein, which drains into another capillary bed, is called a **portal system.**[2]

[2] Another portal system is associated with the liver and digestive tract.

The Anterior Pituitary Secretes Seven Hormones with Widely Different Functions

The pituitary produces a number of other hormones that in sickness and health profoundly influence our bodies. This section describes the major hormones produced by the anterior pituitary gland, some of which you may have already encountered in previous chapters or in class lectures.

Growth Hormone Stimulates Cell Growth, Primarily Targeting Muscle and Bone. We all know that people differ considerably in height and body build. These differences are largely attributable to **growth hormone (GH),** a protein hormone produced by the anterior pituitary.[3] Growth hormone stimulates growth in the body, causing cellular *hypertrophy* (enlargement) and *hyperplasia* (increase in the number of cells through division). Although it affects virtually all body cells, growth hormone acts primarily on bone and muscle. In muscle, it stimulates the uptake of amino acids and protein synthesis. In bone and cartilage, it acts through an intermediary. That is, stimulates the production of several small proteins by the liver. Known as **somatomedins,** these proteins stimulate cartilage and bone to grow. Thus, the more growth hormone produced during the growth phase of an individual, the taller and heftier he or she will be. In men, body growth is also stimulated by testosterone, an anabolic steroid produced by the testes. Testosterone stimulates bone and muscle growth, thus explaining why men are generally taller and more massive than women.

Growth hormone secretion undergoes a diurnal (daily) cycle. Like cortisol, the highest blood levels are present during sleep (▷ Figure 20–9). During the day, the level in

[3] Growth hormone levels are determined by the genes.

≋	**TABLE 20–1 Hormones Secreted by the Pituitary Gland**	≋

HORMONE	FUNCTION	
Anterior pituitary		
Growth hormone (GH)	Stimulates cell growth. Primary targets are muscle and bone, where GH stimulates amino acid uptake and protein synthesis. It also stimulates fat breakdown in the body.	
Thyroid-stimulating hormone (TSH)	Stimulates release of thyroxine and triiodothyronine	
Adrenocorticotropic hormone (ACTH)	Stimulates secretion of hormones by the adrenal cortex, especially glucocorticoids	
Gonadotropins (FSH and LH)	Stimulate gamete production and hormone production by the gonads	
Prolactin	Stimulates milk production by the breast	
Melanocyte-stimulating hormone (MSH)	Function in humans is unknown.	
Posterior pituitary		
Antidiuretic hormone (ADH)	Stimulates water reabsorption by nephrons of the kidney	
Oxytocin	Stimulates ejection of milk from breasts and uterine contractions during birth	

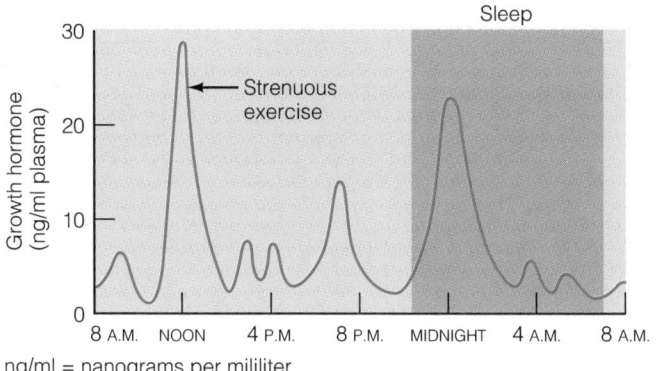

ng/ml = nanograms per mililiter

▷ **FIGURE 20–9 Growth Hormone Secretion in an Adult** Growth hormone is released during exercise and helps promote muscle growth. It is also released at night in a circadian rhythm.

the blood declines. It is no wonder that sleep is so important to a growing child. Growth hormone secretion declines gradually as we age.

Like the secretion of many other hormones of the anterior pituitary, that of growth hormone is controlled by a releasing hormone (GH-RH) *and* an inhibiting hormone (GH-IH), both produced by the hypothalamus. Levels of growth hormone in the blood participate in a negative feedback loop. Growth hormone release can also be stimulated directly through the nervous system (hypothalamus). Stress and moderate exercise, for example, stimulate the hypothalamus to release GH-RH (Figure 20–9).

Deficiencies in growth hormone can result in dramatic changes in body shape and size, depending on when the deficiency occurs. Undersecretion, or **hyposecretion,** occurring during the growth phase of a child is one cause of

stunted growth (dwarfism) (▷ Figure 20–10a). Oversecretion, or **hypersecretion,** results in **giantism** if the excess occurs during the growth phase (Figure 20–10b). If the pituitary begins producing excess growth hormone after growth is complete, the result is a relatively rare disease called **acromegaly.** Facial features become coarse, and hands and feet continue to grow throughout adulthood (▷ Figure 20–11). Growth of the vertebrae results in a hunched back. Organs also grow in response to continued secretion of growth hormone. Thus, the tongue, kidneys, and liver often become quite enlarged in patients with acromegaly.

Thyroid-Stimulating Hormone Stimulates the Thyroid Gland to Produce Thyroxine. Thyroid-stimulating hormone (TSH) is a protein hormone produced by the anterior pituitary; its release is controlled by **TSH-RH (TSH-releasing hormone)** from the hypothalamus. TSH-RH secretion is stimulated by cold and stress. Its secretion is also regulated by the level of thyroxine in the blood in a classical negative feedback loop. Receptors in the hypothalamus detect the level of circulating thyroxine. When circulating levels of the hormone are low, the hypothalamus releases TSH-RH. When the level of thyroxine increases, TSH-RH secretion declines (▷ Figure 20–12).

TSH-RH stimulates the production and release of TSH in the anterior pituitary. Thyroid-stimulating hormone, in turn, then travels in the blood to the thyroid gland, located in the neck (see Figure 20–1), where it stimulates the production and release of thyroxine and triiodothyronine.

These hormones enter the bloodstream and affect a great many cells. One of their chief functions is to stimulate the catabolism (breakdown) of glucose by body cells. Because glucose catabolism produces energy and heat, in-

▷ **FIGURE 20–10 Disorders of Growth Hormone Secretion** (*a*) Pituitary dwarves. (*b*) Pituitary giant.

(a)

(b)

(a)   (b)

(c)   (d)

▷ **FIGURE 20–11 Acromegaly** Hypersecretion of growth hormone in adults results in a gradual thickening of the bone, which is especially noticeable in the face, hands, and feet. There is no sign of the disorder at age 9 (*a*) or age 16 (*b*). Symptoms are evident at age 33 (*c*) and age 52 (*d*).

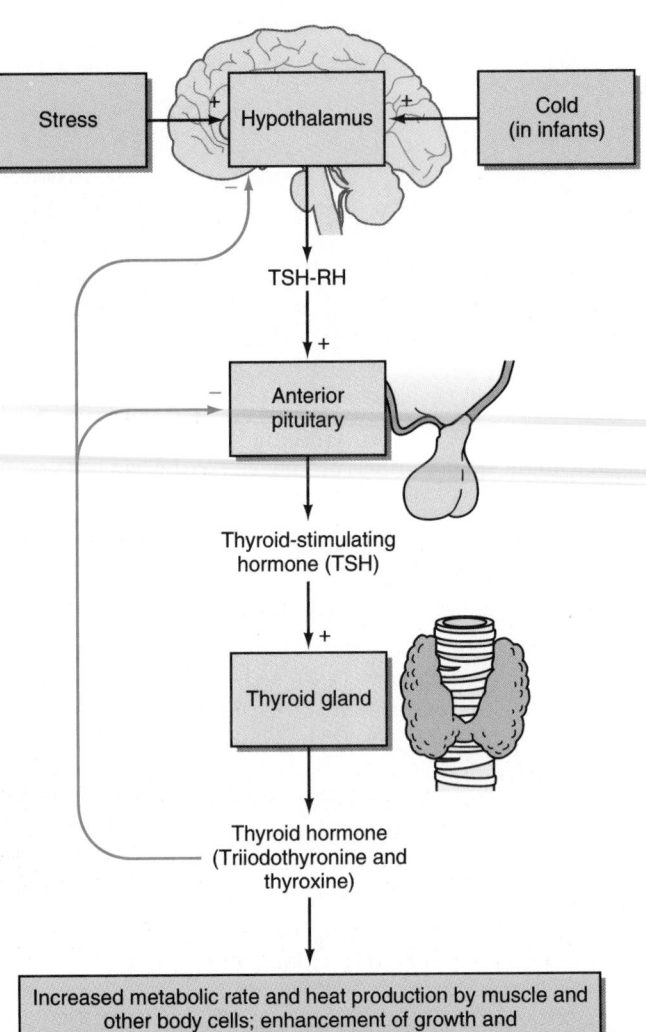

▷ **FIGURE 20–12 Negative Feedback Control of TSH Secretion** Triiodothyronine (T_3) and thyroxine regulate hypothalamic and pituitary activity. Other factors, such as stress and cold, influence the release of TSH via the hypothalamus. (A + denotes stimulation; a − denotes inhibition.)

creased levels of these hormones also raise body temperature. Interestingly, in Grave's disease, from which Bush suffers, TSH levels are normal. For reasons not well understood, the thyroid gland produces excess thyroxine without the urging of the pituitary.

Adrenocorticotropic Hormone Stimulates the Release of Hormones from the Adrenal Cortex.

A relative of yours gradually loses weight and complains of chronic fatigue and weakness. Her skin grows progressively darker, and she suffers occasional bouts of diarrhea, constipation, or mild indigestion. A blood test reveals that she is suffering from Addison's disease, which is believed to be caused by an autoimmune reaction against the **adrenal cortex,** the outer layer of the adrenal gland (see Figure 20–1). A few months after she receives steroid tablets, which replace the lost hormones, her symptoms vanish.

Like many other endocrine glands, the adrenal cortex is under control of the anterior pituitary. More specifically, it is under the control of **adrenocorticotropic hormone**

(ACTH), a polypeptide hormone released by the anterior pituitary. In response to ACTH, the adrenal cortex produces and secretes a group of steroid hormones known as the **glucocorticoids.** Glucocorticoids (like the hormone cortisol) increase blood glucose levels, thus helping maintain homeostasis (more on this later).

ACTH release is under the control of the hypothalamic releasing hormone, ACTH-RH. ACTH-RH secretion is controlled by at least three factors. The first is the level of circulating glucocorticoids, which participate in a negative feedback loop (▷ Figure 20–13). As the level of glucocorticoid decreases, ACTH-RH secretion climbs. This rise causes an increase in the release of ACTH by the pituitary.

ACTH-RH secretion is also controlled by stress, acting through the nervous system. A stress-stimulated increase in

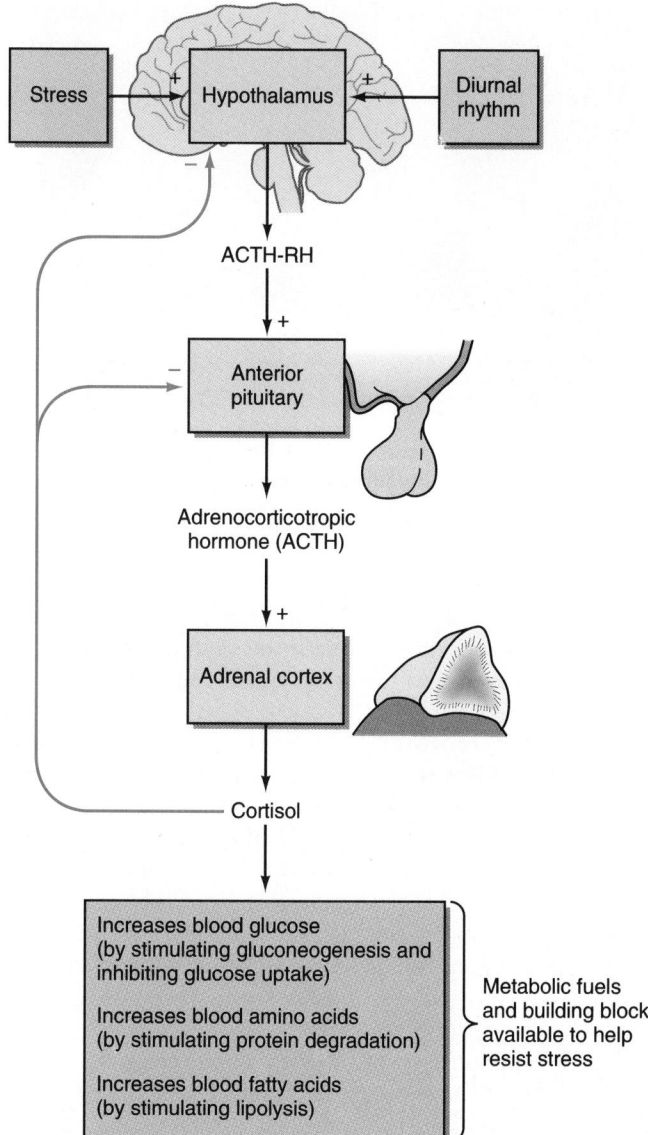

FIGURE 20–13 Feedback Control of ACTH Cortisol regulates hypothalamic and pituitary activity, but stress and the biological clock also influence the release of ACTH-RH.

ACTH-RH results in a rise in ACTH release, which, in turn, stimulates an increase in the release of glucocorticoids by the adrenal cortex. Glucocorticoids increase blood glucose levels, and this mechanism is an adaptation that ensures additional energy for cells (especially muscle) when the body is under stress.

ACTH secretion is also affected by the light-dark cycle through a biological clock in a region of the brain that monitors day length and controls ACTH-RH secretion. At night, the clock overrides the negative feedback mechanism involving ACTH-RH and glucocorticoids, which serves to keep levels fairly constant. The biological clock increases ACTH-RH secretion while you sleep. The increased production of ACTH-RH results in elevated levels of ACTH and also adrenal steroids. The highest levels of ACTH are present in the blood in the early morning, just before you awaken.

Besides producing glucocorticoids, the adrenal cortex synthesizes and releases sex steroids and aldosterone. Aldosterone was described in Chapter 16 because it plays an important role in regulating kidney function (sodium reabsorption) and blood volume. The sex steroids produced by the adrenals are released in seemingly insignificant quantities.

The Gonadotropins Are Hormones that Stimulate the Ovaries and Testes. Reproduction in both males and females is under control of the anterior pituitary. The anterior pituitary produces two hormones that affect the gonads. Known as **gonadotropins,** these hormones are discussed in detail in the next chapter. The two gonadotropins are follicle-stimulating hormone and luteinizing hormone. **Follicle-stimulating hormone (FSH)** promotes gamete formation in men and women. **Luteinizing hormone (LH)** is a trophic hormone that stimulates gonadal hormone production. In men, LH stimulates the production of testosterone, the male sex steroid, by the testes. In women, LH stimulates progesterone secretion by the ovaries. Both FSH and LH are under the control of **gonadotropin releasing hormone,** produced and secreted by the hypothalamus.

Prolactin Stimulates Milk Production in the Breasts of Women. One of the most intriguing hormones of the anterior pituitary is the protein hormone **prolactin.** Prolactin performs many functions in vertebrates. In birds, for example, it stimulates migratory behavior. In fish, it helps regulate electrolyte balance. In women, prolactin is secreted by the anterior pituitary at the end of pregnancy and stimulates milk production by the mammary glands. Suckling prolongs the release of prolactin for at least 3 to 12 months, sometimes much longer, thus maintaining milk production necessary for breast feeding.

Prolactin secretion is controlled by a **neuroendocrine reflex,** a reflex involving both the nervous and endocrine systems. As ▷ Figure 20–14 shows, suckling stimulates sensory fibers in the breast. Nerve impulses travel to the hypothalamus via sensory neurons. In the hypothalamus, these impulses stimulate the release of prolactin releasing hormone (PRH). This hormone, in turn, travels to the anterior pituitary in the bloodstream, where it stimulates the secretion of prolactin. Prolactin, in turn, stimulates milk production by the breasts.

The neuroendocrine reflex is the basis of commercial milk production. When a dairy cow gives birth, it produces milk to feed its calf. On dairy farms, however, calves are often weaned (separated from their mothers) fairly early. Left alone, the mothers would stop producing milk. Farmers prolong milk production by milking their cows, either manually or by machine (▷ Figure 20–15). Milking machines and hand milking stimulate the neuroendocrine reflex, continuing the production of milk.

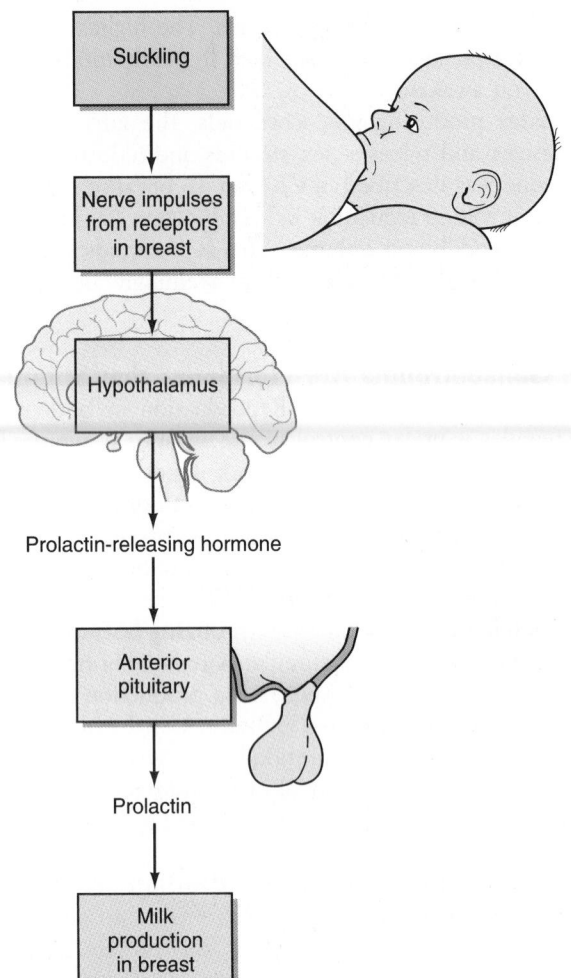

Suckling

↓

Nerve impulses from receptors in breast

↓

Hypothalamus

↓

Prolactin-releasing hormone

↓

Anterior pituitary

↓

Prolactin

↓

Milk production in breast

▷ **FIGURE 20–14 Neuroendocrine Reflex and Prolactin Secretion** Suckling stimulates prolactin release by the anterior pituitary.

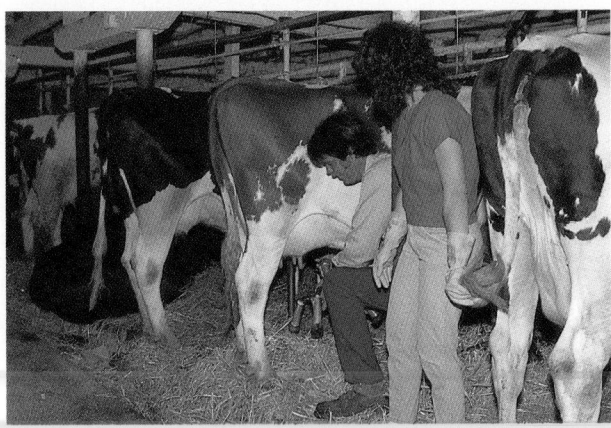

▷ **FIGURE 20–15 Stimulating a Reflex** Milking by hand or machine prolongs milk production long after a calf has been weaned from its mother.

The Posterior Pituitary Secretes Two Hormones

Have you ever wondered why alcoholic beverages make a person urinate so much or what causes milk to flow from a woman's breast? Have you ever wondered why a woman's uterus contracts when she delivers a baby? The answer to these puzzles is rather simple: hormones. More specifically, these phenomena can be attributed to two hormones from the posterior pituitary. Like the hypothalamus, the **posterior pituitary** is a neuroendocrine gland—that is, a gland made of neural tissue that produces hormones. Derived from brain tissue during embryonic development, the posterior pituitary remains connected to the brain throughout life.

The posterior pituitary consists of the axons and terminal ends of neurosecretory cells whose cell bodies are located in the hypothalamus (▷ Figure 20–16). The neurosecretory cells produce two hormones, antidiuretic hormone and oxytocin, each of which consists of nine amino acids. Oxytocin and antidiuretic hormone are synthesized in the cell bodies of the neurosecretory cells and travel down the axons of these cells into the posterior pituitary. The hormones are then stored in the axon terminals of the cells and released into the surrounding capillaries.

Antidiuretic Hormone Increases Water Absorption in the Kidney. Antidiuretic hormone(ADH) regulates water balance in humans and, as you will soon see, is noticeably affected by alcohol (see also Chapter 16). ADH travels in the bloodstream to the kidney and there increases water absorption by the distal convoluted tubules and collecting tubules (▷ Figure 20–17). As a result, water reenters the bloodstream, increasing blood volume and maintaining the normal osmotic concentration of the blood.

ADH secretion is controlled by **osmoreceptors** in the hypothalamus. These receptors monitor the concentration of dissolved substances (especially sodium ions) in the blood. When the concentration exceeds the normal level—for example, during dehydration—the osmoreceptors activate the ADH-producing cells, causing them to release ADH into the bloodstream. ADH, in turn, causes water to be reabsorbed by the kidney, diluting the blood. When the osmotic concentration of the blood approaches homeostatic levels, ADH secretion ceases (Figure 20–17).

As you might have guessed, the secretion of ADH is inhibited by ethanol in alcoholic beverages. The reduction in ADH decreases water reabsorption in the kidneys, thus increasing urine production. The increase in urinary output, in turn, can lead to dehydration—the dry mouth and intense thirst a person may experience as part of a hangover.

ADH secretion may also decline as a result of trauma. A blow to the head, for example, can damage the hypothalamus and cause a sharp decline in ADH output. In the absence of ADH, the kidneys produce several gallons of urine a day. To keep up with the water loss and to prevent dehydration, an individual must drink enormous quantities of water. This condition is known as **diabetes insipidus** (Chapter 16).

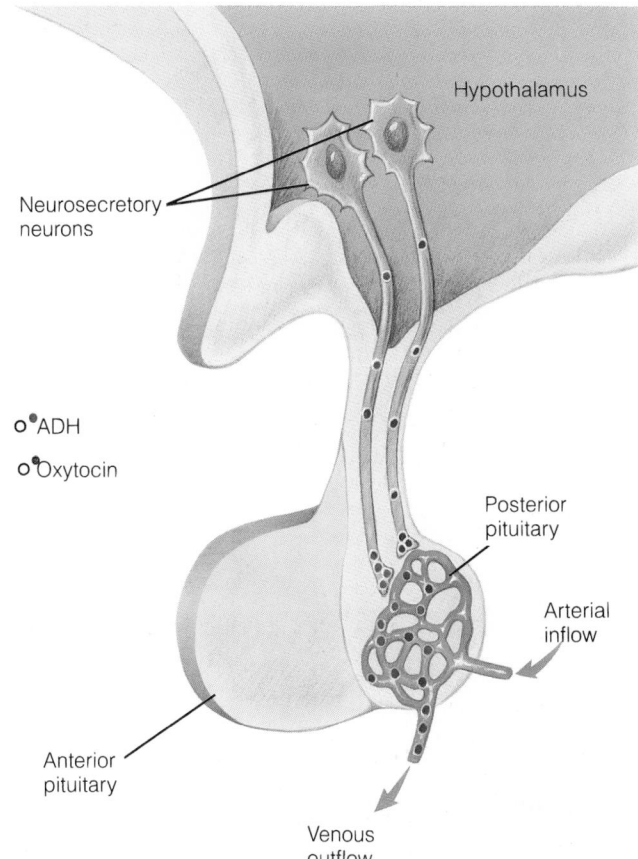

▷ **FIGURE 20–16 The Posterior Pituitary** Neurosecretory neurons that produce oxytocin and ADH originate in the hypothalamus and terminate in the posterior pituitary. Hormones are produced in the cell bodies of the neurons and are stored and released into the bloodstream in the posterior pituitary.

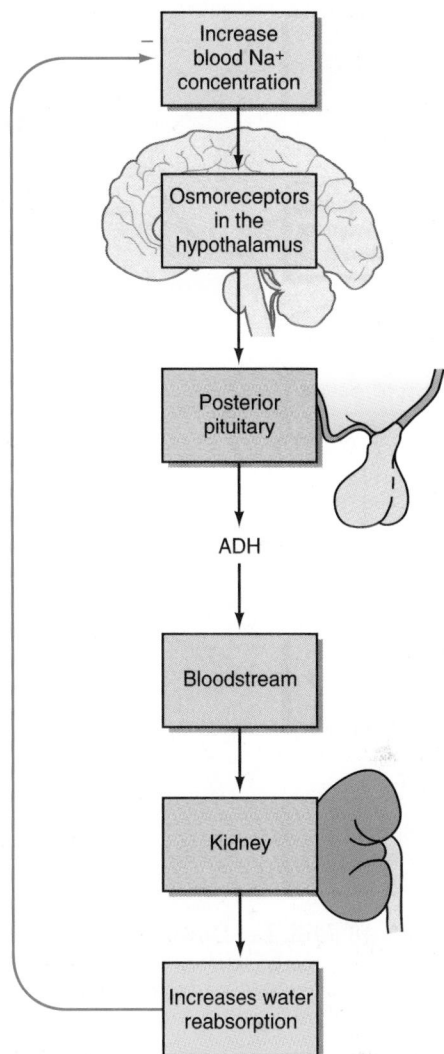

▷ **FIGURE 20–17 Role of ADH in Regulating Fluid Levels** ADH secretion is stimulated by an increase in the blood concentration of sodium caused by dehydration. ADH increases water reabsorption in the kidney, thus eliminating the stimulus for ADH secretion.

Because of its ability to increase arterial blood pressure, ADH is sometimes referred to as **vasopressin.** Under most conditions, however, it is doubtful that enough ADH is secreted to increase blood pressure. ADH secretion in levels sufficient to exert a vasopressive effect occurs when an individual suffers an extreme loss of blood—for example, in an automobile accident. When released in massive quantities, ADH constricts the walls of severed arteries, reducing blood loss. Clotting factors in the blood also help seal off damaged vessel walls. ADH also causes vasoconstriction throughout the cardiovascular system, which helps maintain blood pressure. This process is important because a large decrease in blood pressure can reduce the flow of blood to body cells and cause death.

Oxytocin Facilitates Birth and Stimulates Milk Let-down. Oxytocin from the posterior pituitary is a dual-purpose hormone. It stimulates the contraction of the smooth muscle of the uterus and stimulates contraction of the smooth-muscle-like cells around the glands in the breast.

Oxytocin release, like that of prolactin, is controlled by a neuroendocrine reflex. During childbirth, for example, the walls of the uterus and cervix stretch. Impulses from the stretch receptors in the uterus travel to a region of the hypothalamus containing the neurosecretory cells that produce oxytocin. Impulses from the stretch receptors cause the neurosecretory cells to release oxytocin into the blood of the posterior pituitary (where the neurosecretory cells terminate). Oxytocin travels in the blood to the uterus, where it stimulates smooth muscle contraction, aiding in the expulsion of the baby.

Oxytocin release is stimulated by another neuroendocrine reflex, which is activated by suckling (▷ Figure 20–18). Sensory fibers in the breast conduct bioelectric impulses to the hypothalamus, triggering the release of oxytocin. Oxytocin travels in the blood to the breast, where it stimulates **milk let-down**—the ejection of milk from the glands soon after suckling begins. Milking machines and hand milking stimulate the reflex in dairy cattle. As you may recall, prolactin from the anterior pituitary is also

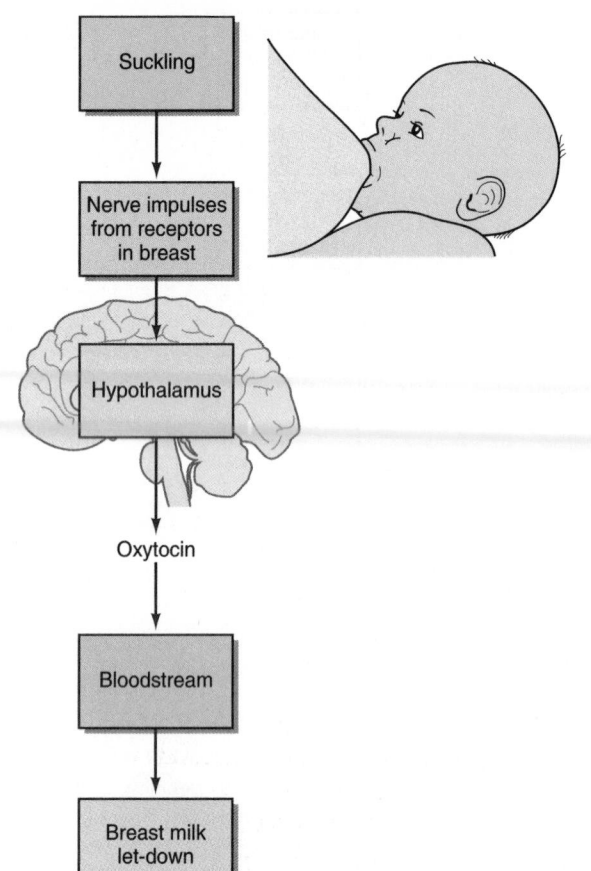

> FIGURE 20-18 Milk Let-Down Reflex Suckling stimulates oxytocin release by the posterior pituitary, causing the expulsion of milk from the glands of the breast.

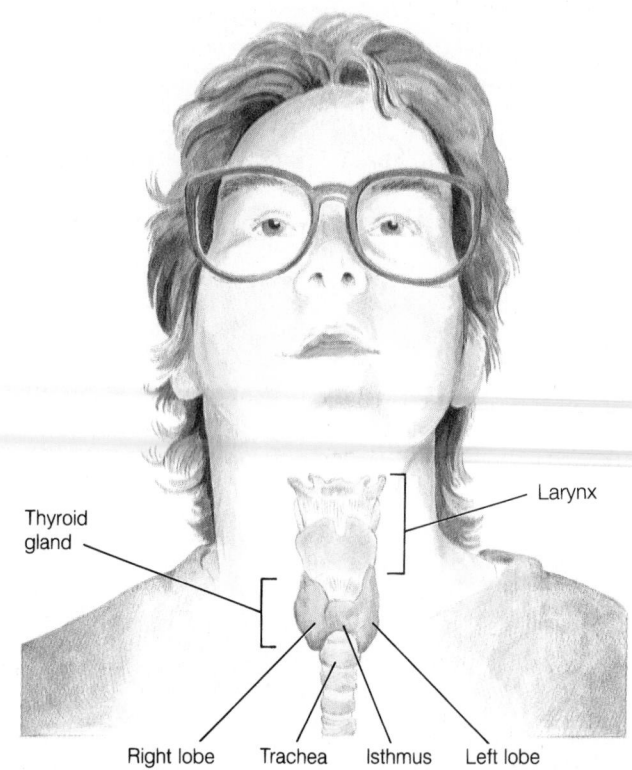

> FIGURE 20-19 Location of the Thyroid Gland

released during suckling. This hormone stimulates milk production.

≈ THE THYROID GLAND

On a 50°F day in the fall, you bundle up in a sweater but still feel cold. However, a day with the same temperature in the spring feels warm, even if you don't wear a sweater. Why the difference?

The answer lies in the thyroid gland and two of its hormones. The **thyroid gland** is a U- or H-shaped gland (it varies from one person to the next) located in the neck (> Figure 20-19). The thyroid gland produces three hormones: (1) thyroxine (tetraiodothyronine, or T_4), (2) a chemically similar compound, triiodothyronine (T_3), and (3) calcitonin. The first two are involved in controlling metabolism and heat production; the last helps regulate blood levels of calcium.

As > Figure 20-20a shows, the thyroid gland consists of large, spherical structures known as **follicles.** Each follicle consists of a central region containing a gel-like material called **thyroglobulin,** which is surrounded by a single layer of cuboidal **follicle cells.** Thyroglobulin is a glycoprotein

and is produced by the follicle cells. Thyroxine and triiodothyronine are derived from thyroglobulin. How?

Thyroglobulin molecules contain certain amino acids to which iodine ions are attached. Thyroxine and triiodothyronine are produced when two of these iodinated amino acids are covalently linked. To make T_3 and T_4, the thyroid requires large quantities of iodide, I^-. This ion must be actively transported into the cell, where it is added to certain amino acids in the thyroglobulin molecules.

Thyroglobulin is the storage form of T_3 and T_4. When the blood levels of these hormones fall, TSH is secreted by the anterior pituitary. TSH, in turn, stimulates the follicle cells to engulf a portion of the thyroglobulin reserve. Lysosomes in the cytoplasm of the follicle cells break down the ingested material, releasing T_3 and T_4. These hormones then diffuse out of the cell into the bloodstream.

Because the thyroid requires iodide, persistent dietary deficiencies of this nutrient can alter the thyroid's function, resulting in a condition known as **goiter.** Goiter is an enlargement of the thyroid gland (> Figure 20-21). A shortage of iodide in the diet results in a decline in the levels of thyroxine and triiodothyronine in the blood. The hypothalamus responds by producing more TSH. TSH causes the thyroid to increase thyroglobulin production. Without sufficient iodide, however, the thyroid gland cannot produce thyroxine and triiodothyronine. As a result, the thyroid gland continues to produce thyroglobulin. This causes the gland to hypertrophy (hy-per-truh-phee), or enlarge, sometimes forming softball-sized enlargements on the neck.

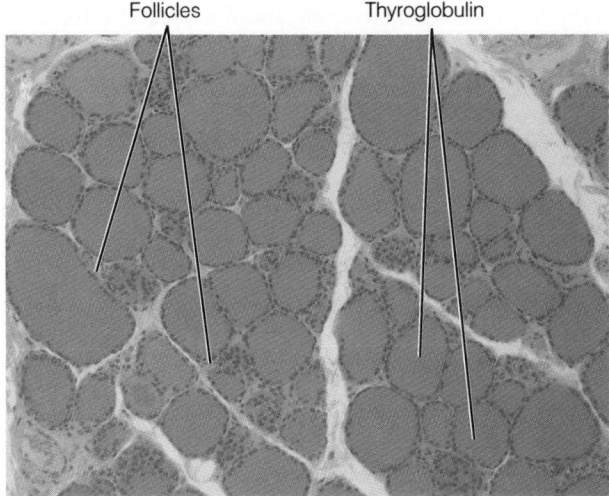

Follicles   Thyroglobulin

(a)

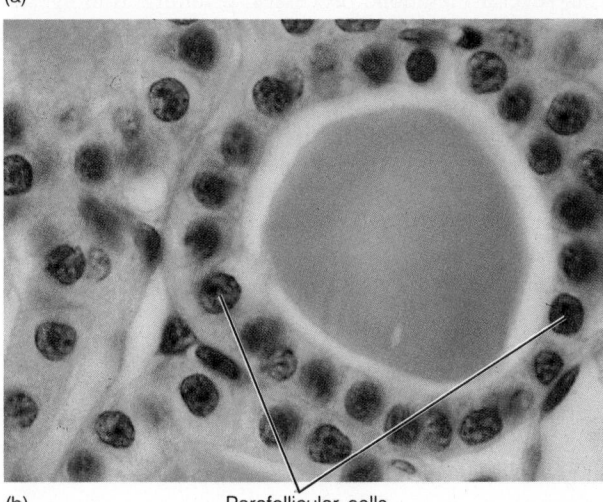

(b)   Parafollicular cells

▷ **FIGURE 20–20 Histologic Structure of the Thyroid Gland** (a) The thyroid follicles produce thyroxine and triiodothyronine. They are stored in the colloidal material called thyroglobulin. (b) Enlargement showing calcitonin-producing cells.

Goiter was once common in areas of Europe and the United States where iodine had been leached out of the soil by rain—for example, near the Great Lakes. Crops grown in iodine-poor soils failed to provide sufficient quantities. Iodine was first added to table salt, in fact, to help prevent the problem. Few people in developed countries develop goiter now because they eat iodized salt or receive sufficient quantities of iodine in their food or through mineral supplements.

Thyroxine and Triiodothyronine Accelerate the Breakdown of Glucose and Stimulate Growth and Development

As ▷ Figure 20–22 illustrates, thyroxine and triiodothyronine are virtually identical. As a result, the two hormones have nearly identical functions. Both hormones, for exam-

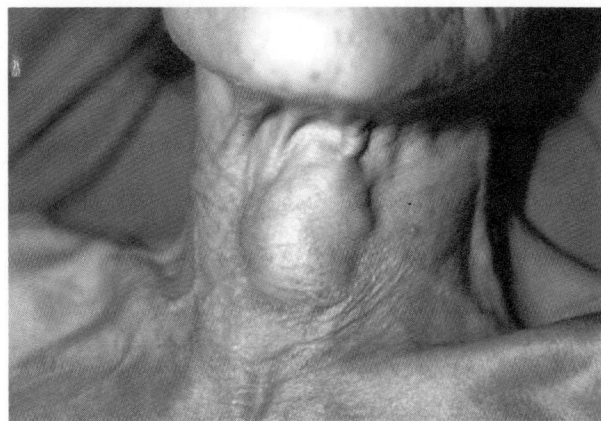

▷ **FIGURE 20–21 Goiter** An enlargement of the thyroid gland most often results from a lack of iodine in the diet.

ple, accelerate the rate of mitochondrial glucose catabolism in most body cells. Thyroid hormones also stimulate cellular growth and development. Bones and muscles are especially dependent on them during the growth phase. Even normal reproduction requires these hormones. A deficiency of T$_3$ and T$_4$, for example, delays sexual maturation in both sexes. In children, depressed thyroid output stunts mental as well as physical growth. If the deficiency is not detected and treated, the effects will be irreversible.

In adults, reduced thyroid activity, or **hypothyroidism,** is less severe and is fully reversible because growth has been completed. Hypothyroidism, however, decreases the metabolic rate, making a person feel cold much of the time. People suffering from hypothyroidism also feel tired and worn out. Even simple mental tasks become difficult. Their heart rate may slow to 50 beats per minute. Hypothyroidism is treated by pills containing artificially produced thyroid hormone.

Excess thyroid activity, **hyperthyroidism,** in adults results in elevated metabolism, excessive sweating (due to overheating), and weight loss, despite increased food intake. The increase in thyroid hormone levels results in increased mental activity, resulting in nervousness and anxiety. People suffering from hyperthyroidism often find it difficult to sleep. Their heart rate may accelerate, as in the case of President Bush, and individuals may lose their sensitivity to cold. Some people suffering from hyperthyroidism exhibit a condition called **exophthalmos,** or bulging eyes. The eyes may protrude so much that the eyelids cannot close completely. Exophthalmos may cause double vision or blurred vision.

Hyperthyroidism is treated in a number of ways. Patients may be given antithyroid medications, drugs that antagonize the effects of thyroid hormones. Surgery may be required to remove part or all of the gland if it has become cancerous. The most common treatment for hyperthyroidism, however, is radioactive iodine. As noted earlier, iodine is concentrated in the thyroid gland. Radioactive iodine accumulated in the cells of the thyroid follicle irra-

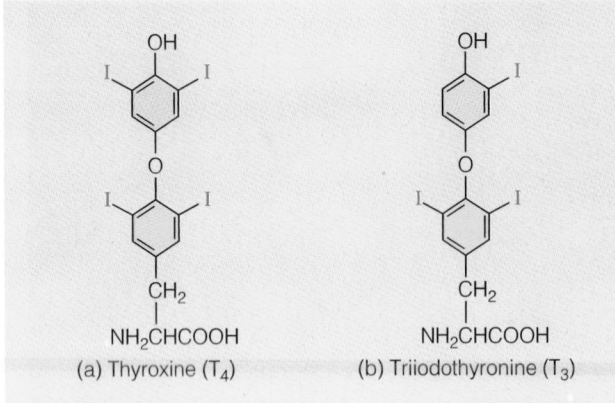

▷ **FIGURE 20–22 Structure of the Thyroid Hormones**
(*a*) Thyroxine; (*b*) triiodothyronine.

diate overactive follicle cells, damaging them and thus reducing their output of thyroid hormones. This procedure, while effective, may lead to other problems (notably cancer) later in life.

Calcitonin Decreases Blood Levels of Calcium

Large, round cells found in the perimeter of the thyroid follicles produce **calcitonin,** or **thyrocalcitonin,** a polypeptide hormone that lowers the blood calcium level (see Figure 20–20b). Calcitonin has three effects: (1) It inhibits osteoclasts, bone-resorbing cells, thus reducing the release of calcium from bone. (2) It stimulates bone-forming cells, osteoblasts, causing calcium to be deposited in bone and thus reducing blood levels. (3) It increases the excretion of calcium (and phosphate) ions by the kidneys. All three effects help lower blood calcium.

Calcitonin is involved in a simple negative feedback loop with calcium ions in the blood. When the calcium ion concentration increases, calcitonin secretion increases. As calcium concentrations fall, calcitonin secretion falls.

≈ THE PARATHYROID GLANDS

The **parathyroid glands** are four small nodules of tissue embedded in the back of the thyroid gland. These glands, once mistaken by anatomists for undeveloped thyroid tissue, are independent endocrine glands. They produce a polypeptide hormone known as **parathyroid hormone,** or **parathormone (PTH).**

As you may recall from Chapter 19, parathyroid hormone secretion is stimulated when calcium levels in the blood drop. PTH quickly reverses the decline and restores blood calcium levels by (1) increasing intestinal absorption of calcium, (2) stimulating bone destruction by osteoclasts, and (3) increasing calcium reabsorption in the kidney.

As illustrated in previous examples, hormones often act on two or more targets to bring about a desired effect. This redundancy no doubt evolved because it increased the level of homeostatic control—in the same way that redundant safety systems in a nuclear power plant provide a greater degree of control in the case of malfunction.

Calcium homeostasis is also influenced by vitamin D, which greatly increases calcium absorption in the intestine when blood levels fall. Interestingly, vitamin D also increases the responsiveness of bone to PTH. In calcium homeostasis, PTH plays a major role, supported by vitamin D.

As in other glands, the parathyroid may malfunction. **Hyperparathyroidism,** excess secretion of parathyroid hormone, is the most common condition. Hyperparathyroidism may result from a tumor of the parathyroid glands, which causes the secretion of excess PTH. Excess PTH results in a loss of calcium from the bones and teeth. It also upsets several metabolic processes, resulting in indigestion and depression. Because bones contain enormous amounts of calcium, most symptoms of hyperparathyroidism (except high blood calcium) do not appear until two to three years after the onset of the disease. Therefore, by the time the disease is discovered, kidney stones may already have formed from calcium, cholesterol, and other substances, and bones may have become more fragile and susceptible to breakage. To prevent further complications, parathyroid tumors must be removed.

≈ THE PANCREAS

A teenage girl comes to her doctor's office. She complains of frequent urination, bladder infections, fatigue, and weakness. Tests show that she is suffering from **diabetes mellitus,** a disorder of the endocrine function of the pancreas. Diabetes mellitus has several causes, but in young people it is generally caused by a lack of insulin output.

Insulin is the glucose-storage hormone; that is, it stimulates the uptake of glucose by body cells. It also stimulates the synthesis of glycogen in liver and muscle cells. Glycogen is the storage form of glucose. Supplies of glycogen in the liver increase immediately after meals, then decline in the period between feedings as the body uses these stores to supply body cells with glucose. Muscles use glycogen supplies principally during exercise.

Insulin is produced by the **pancreas,** a dual-purpose organ located in the abdominal cavity and described in Chapter 11 (▷ Figure 20–23). The head of the pancreas lies in the curve of the duodenum and its tail stretches to the left kidney. Most of the pancreas consists of tiny clumps of cells (acini) that produce digestive enzymes and sodium bicarbonate. Scattered throughout these enzyme-producing cells are small islands of endocrine cells, the **islets of Langerhans** (Figures 20–23 and ▷ 20–24). About 2 million islets are found in the human pancreas, and each islet contains about 200 cells. The islets contain four cell types, two of which will be discussed in this chapter, the alpha cell and the beta cell.

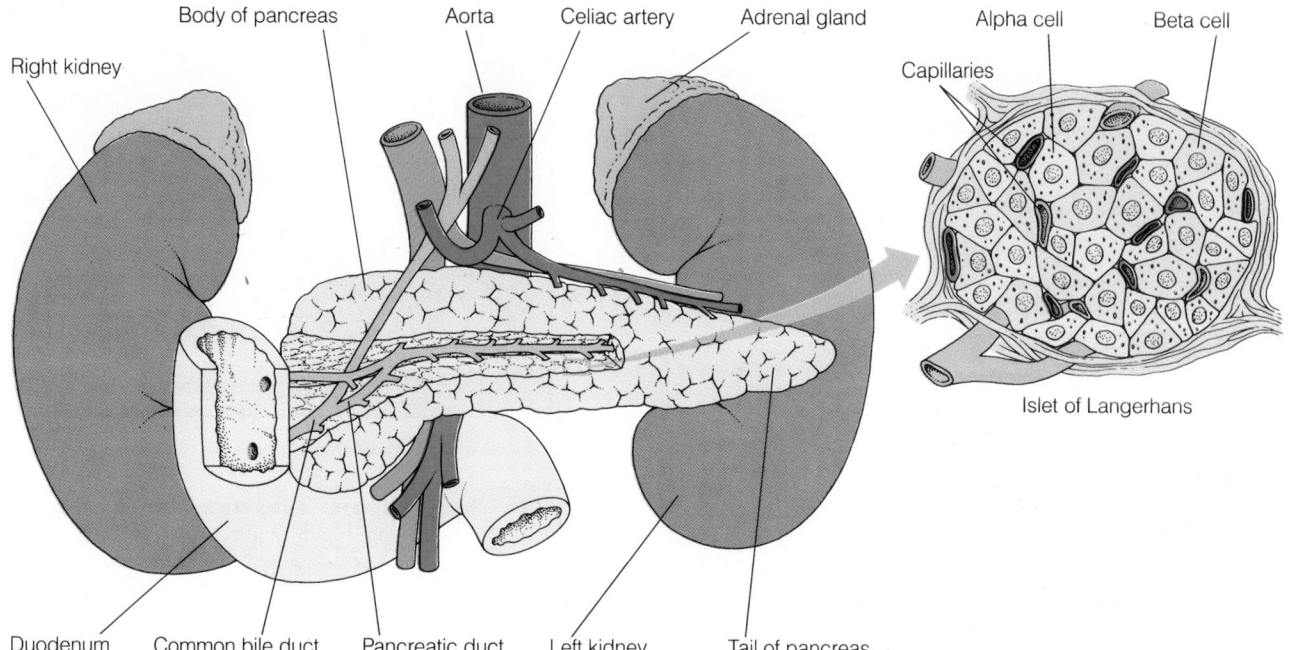

Right kidney | Body of pancreas | Aorta | Celiac artery | Adrenal gland | Capillaries | Alpha cell | Beta cell

Islet of Langerhans

Duodenum | Common bile duct | Pancreatic duct | Left kidney | Tail of pancreas

▷ **FIGURE 20–23 The Pancreas** This dual-purpose organ is located in the abdominal cavity. It produces digestive enyzmes, which it releases into the small intestine via the pancreatic duct, and two hormones, which it releases into the bloodstream.

Insulin Is a Glucose-Storage Hormone and Is Produced by the Beta Cells

The **beta cells** produce insulin. This protein hormone is released within minutes after glucose levels in the blood begin to rise. Like many other hormones, insulin plays a major role in homeostasis. It affects a number of cellular processes and a number of different cells. Its principal targets, however, are skeletal muscle cells, liver cells, and fat cells. We will examine a few of its major functions, beginning with those affecting skeletal muscles.

Skeletal muscle cells are virtually impermeable to glucose in the absence of insulin. When insulin is present, however, the transport of glucose into muscle cells increases dramatically. Glucose uptake rises because insulin stimulates facilitated transport. It also increases glycogen synthesis in skeletal muscle cells, which not only helps store glucose for later use but also lowers intracellular concentrations, accelerating diffusion. Insulin also increases the uptake of amino acids by muscle cells and stimulates protein synthesis in them, thus promoting muscle formation.

In contrast, the plasma membrane of liver cells is quite permeable to glucose, so glucose enters with great ease whether or not insulin is present. Nevertheless, insulin still increases the uptake of glucose by liver cells. It does so by stimulating the addition of phosphate groups to glucose molecules that have entered the cytoplasm. This traps glucose in the liver cell, because phosphorylated glucose cannot diffuse through the plasma membrane. Insulin also increases glycogen formation, helping store glucose for times of need (▷ Figure 20–25). Phosphorylation and glycogen synthesis decrease the cytoplasmic concentrations of glucose, thus helping maintain the concentration gradient between the blood and the cytoplasm. This helps ensure a steady influx of glucose into the liver cell.

In fat cells, insulin increases glucose uptake and also stimulates lipid synthesis, thus helping store foodstuffs for times of need.

▷ **FIGURE 20–24 The Islets of Langerhans** Scattered among the acini of the pancreas (which produce digestive enzymes) are small islands of cells, the Islets of Langerhans. They produce two hormones: glucagon, which increases blood glucose, and insulin, which lowers blood glucose.

Acini | Islet of Langerhans

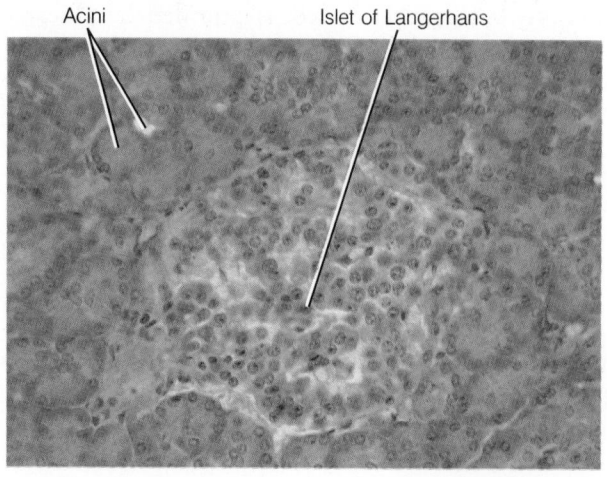

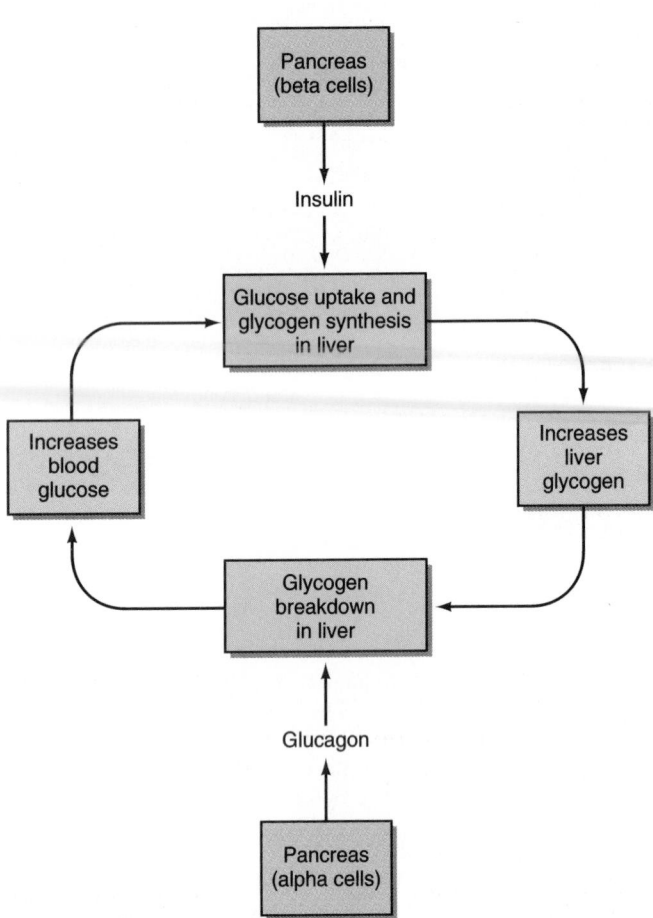

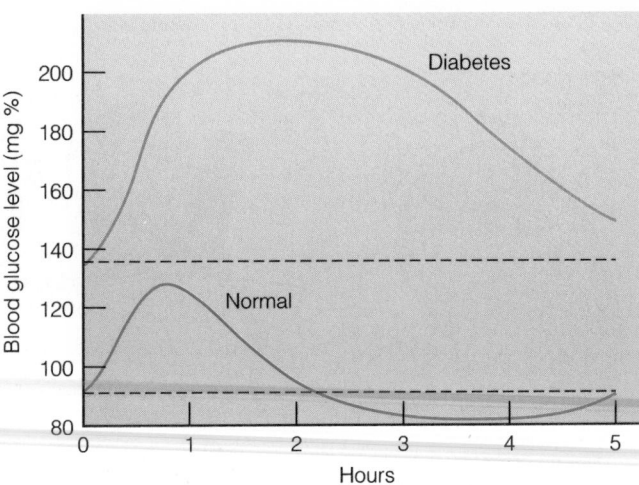

> FIGURE 20–26 **Glucose Tolerance Test** In a diabetic, blood glucose levels increase rapidly and remain high after ingestion of glucose. In a normal patient, blood glucose levels rise but are quickly reduced by insulin.

> FIGURE 20–25 **Role of Liver and Pancreas in Controlling Blood Glucose Levels** Glucagon and insulin are antagonistic hormones that regulate blood glucose levels through different mechanisms.

Glucagon Increases Blood Levels of Glucose, Thus Opposing the Actions of Insulin

A successful homeostatic mechanism often requires antagonistic—or opposing—controls. (In a car we have an accelerator and brakes to control speed.) In glucose homeostasis, the antagonistic control is provided by **glucagon.** Produced by the **alpha cells** of the Islets of Langerhans, glucagon is a polypeptide hormone that increases blood levels of glucose by stimulating the breakdown of glycogen in liver cells (Figure 20–25). This process, called **glycogenolysis,** (gly-co-je-nol-i-sis) helps maintain proper glucose levels in the blood between meals. One molecule of glucagon causes 100 million molecules of glucose to be released.

Glucagon also elevates serum glucose levels by stimulating a process known as **gluconeogenesis,** the synthesis of glucose from amino acids and glycerol molecules derived from the breakdown of triglycerides. Gluconeogenesis occurs in the liver.

Glucagon secretion, like that of insulin, is controlled principally by glucose concentrations in the blood. When

glucose levels fall, the alpha cells release the hormone into the bloodstream. Glucagon release is also stimulated by an increase in the concentration of amino acids in the blood. Because glucagon stimulates gluconeogenesis, this process ensures that excess amino acids in high-protein meals are used to produce glucose.

Diabetes Mellitus Is a Disease Resulting from an Insulin Deficiency or a Decrease in Tissue Sensitivity to Insulin

> Figure 20–26 shows a graph of blood glucose levels in normal and diabetic people. The results were obtained during a **glucose tolerance test.** During this test, physicians give their patients an oral dose of glucose, then check blood levels at regular intervals.

The bottom line on the graph in Figure 20–26 represents the blood glucose levels in a normal patient with a healthy pancreas. As illustrated, glucose levels increase slightly after the administration of glucose, but within one to two hours, glucose levels have decreased to normal, thanks to insulin secretion.

The top line illustrates the response in a diabetic patient, who produces insufficient amounts of insulin or whose cells have become unresponsive to insulin. Glucose levels rise considerably in these patients after they are given an oral dose of glucose and remain elevated for four to five hours. A similar response is seen in untreated diabetics after they eat a meal. However, blood glucose levels remain higher much longer because there is little place for the glucose to go. The absence of insulin means that virtually no glucose will enter skeletal muscle cells and that liver uptake will be greatly reduced. Glycogen synthesis in liver and muscle cells is also greatly reduced.

Glucose levels eventually decline as this nutrient is used by brain cells, red blood cells, and others. The kidneys also

TABLE 20–2 Comparison of Type I and Type II Diabetes

CHARACTERISTIC	TYPE I DIABETES	TYPE II DIABETES
Level of insulin secretion	None or almost none	May be normal or exceed normal
Typical age of onset	Childhood	Adulthood
Percent of diabetics	10% to 20%	80% to 90%
Basic defect	Destruction of beta cells	Reduced sensitivity of insulin's target cells
Associated with obesity?	No	Usually
Genetic and environmental factors important in precipitating overt disease?	Yes	Yes
Speed of development of symptoms	Rapid	Slow
Development of ketosis	Common if untreated	Rare
Treatment	Insulin injections; dietary management	Dietary control and weight reduction; occasionally oral hypoglycemic drugs

From L. Sherwood, *Human Physiology: From Cells to Systems* (St. Paul, Minn.: West, 1989).

excrete excess glucose, helping lower blood levels. Unfortunately, this process wastes a valuable nutrient and causes the kidneys to lose large quantities of water. Excess urination and constant thirst are the chief symptoms of untreated diabetes. Diabetics may have to urinate every hour, day and night.

Diabetes Is Really Two Diseases with Similar Symptoms but Different Causes. Physicians recognize two types of diabetes (Table 20-2). **Type I diabetes** occurs mainly in young people and results from a deficiency in insulin production. Type I diabetes is also called **juvenile diabetes** or **early-onset diabetes.** Type I diabetes is believed to be caused by damage to the insulin-producing cells of the pancreas. One leading theory suggests that it is caused by an autoimmune reaction. Viral infection of the pancreas may also cause Type I diabetes.

In patients with Type I diabetes, insulin production varies. In some, it is only slightly depressed; in others, it is completely suppressed. In the absence of insulin, body cells are starved for glucose. Energy must be provided by the breakdown of fats.

Type II diabetes usually occurs in people over 40 and is sometimes called **late-onset diabetes** (Table 20–2). In this disease, the beta cells produce normal or above-normal levels of insulin. However, problems arise because of a decline in the number of insulin receptors in target cells, which makes them unresponsive to insulin.

Type II diabetes is commonly associated with obesity. Heredity may also be an important contributor, as it is in Type I diabetes. Studies show that about one-third of the patients with Type II diabetes have a family history of the disease. Critical thinking rules in Chapter 1 warn us to look for other explanations. When we do, we find that it may be obesity that is inherited, rather than the tendency to develop diabetes.

Although they have different causes, both forms of diabetes exhibit many similar symptoms. Excess urination and thirst are generally the first signs of trouble. Excess glucose in the urine may result in frequent bacterial infections in the bladder. Patients often feel tired, weak, and apathetic. Weight loss and blurred vision are also common.

Although Symptoms Are Similar, Treatments for Type I and Type II Diabetes Are Quite Different. Early-onset diabetes (Type I) is treated with insulin injections and therefore is often called **insulin-dependent diabetes.** Patients receive regular injections of insulin—usually two to three times per day. Patients are also required to eat meals and snacks at regular intervals to maintain constant glucose levels in the blood and to ensure that regular insulin injections always act on approximately the same amount of blood glucose.

Insulin injections are tailored to an individual's lifestyle and body demands. To help mimic the body's natural release, medical researchers have developed a device called an **insulin pump,** which delivers predetermined amounts of insulin when needed. Because the device cannot measure blood glucose levels, it may deliver an inappropriate amount under certain circumstances. Health Note 20–1 describes the device in more detail. Medical researchers are also experimenting with ways to transplant healthy beta cells in the pancreas of a diabetic. (Chapter 3 and the Point/Counterpoint in that chapter discuss fetal cell transplantation.)

In a recent study, researchers from the University of Florida reported findings that may help prevent Type I diabetes. The researchers withdrew blood from 5000 young children who were asymptomatic—that is, showed no symptoms of early-onset diabetes. They then analyzed the blood for antibodies produced by an autoimmune reaction and followed the children's health status over time to de-

ON THE ROAD TO AN ARTIFICIAL PANCREAS

Diabetics take insulin injections two or three times a day. If they do not, they could easily go into a diabetic coma. These injections allow most victims of juvenile diabetes to live a fairly normal life. Occasionally things go awry, however. Too much insulin or too much exercise causes blood glucose levels to fall, making a diabetic feel dizzy and weak. Sometimes the balance is tipped the other way. An excess of blood sugar brings on hyperglycemia (high blood sugar) and with it depression, fatigue, irritability, and weakness.

Besides worrying over insulin doses and playing a continual balancing act with their blood sugar, most diabetics develop complications 20 to 30 years after the onset of the disease. Blindness and lethal or disabling diseases of the kidney, nervous system, and cardiovascular system are common. These complications probably result from periodic elevations in blood glucose levels, which occur despite good insulin management. High glucose levels damage blood vessels and nerves, injuring critical organs.

To prevent complications, physicians try to mimic normal insulin secretion patterns through injections three times a day just before meals. But mimicking the body's homeostatic system with regular insulin injections is crude. The pancreas normally monitors blood glucose levels minute by minute. A conscientious diabetic can measure blood glucose only three or four times a day. He or she may make adjustments for large meals or additional exercise, but such accommodations are primitive in comparison with the body's elaborate system of glucose homeostasis, a product of evolution vital to the survival and successful propagation of many multicellular animals.

Consequently, medical researchers are looking for ways to replace the daily injections. One hopeful possibility is a device called an insulin infusion pump, which delivers tiny amounts of fast-acting insulin to the body day and night via a long plastic tube and needle inserted into the skin of the thigh or abdomen (▷ Figure 1). The needle and tubing are usually replaced every three to four days. The insulin pump provides baseline insulin levels needed to maintain proper blood glucose concentrations. Worn outside the body, the pump also delivers a surge of insulin at mealtimes to offset the rise in blood sugar that accompanies a meal. To do this, one simply presses a button on the pump 30 minutes before eating. The pump delivers a preprogrammed amount of insulin. If the meal is going to be larger than anticipated, a small adjustment can be made to protect against hyperglycemia.

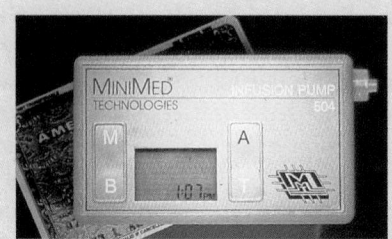

▷ **FIGURE 1 Insulin Pump** This device delivers preprogrammed doses of insulin to diabetics, mimicking the pancreas.

The newer insulin pumps capitalize on computer technology to regulate insulin flow day and night. They offer great flexibility, allowing an individual to accommodate differing levels of exercise and meals of varying size. They come equipped with a memory to store information on the exact doses given over a certain period. Using this information, physicians can fine-tune the program to an individual's needs.

The insulin pump is not a popular item among diabetics when it comes to aesthetics and comfort. When it comes to controlling blood sugar, however, the device receives high praise. For many, the freedom from daily injections outweighs the discomfort. For pregnant diabetic women, the pump may mean the difference between a normal child and no child at all, for even mild hyperglycemia can cause fetal death.

While refinements are being made to the insulin pump, researchers are also testing a biodegradable wafer that can be impregnated with insulin and inserted under the skin of a diabetic. The wafer also contains a sugar-sensitive enzyme. As blood sugar levels rise, the enzyme is activated. This action, in turn, causes a slight increase in acidity, which increases the solubility of the insulin in the wafer, allowing it to be released into the bloodstream. As a result, insulin levels respond very closely to blood glucose levels. Insulin is released only as needed in response to rising blood sugar levels.

This innovative approach could prove even more effective than the insulin pump, because it operates on a negative feedback principle as do the beta cells in the pancreas.

termine which ones developed diabetes. During the study, 12 of the subjects developed early-onset diabetes. Interestingly, blood samples from these individuals all contained an antibody that attacks a protein on the surface of the beta cells of the pancreas. The antibody was found in blood samples taken as many as seven years before the onset of diabetes.

The antibody to the surface protein of the beta cell may provide a biochemical marker, an advance notice of the disease. Eventually, researchers hope to find ways to attach chemicals to the surface proteins of the beta cells that would neutralize or destroy the antibodies that bind to them, thus preventing the disease from developing. If this research proves fruitful, regular blood tests could be used to screen children. When the antibodies are found, special treatment could be used to destroy them before they have time to destroy the islet cells. This treatment could free millions of people from a lifetime of insulin injections and the risk of blindness and other serious medical problems later in life.

Type II diabetes is often called **insulin-independent diabetes,** because insulin injections are useless given that the disease may stem from a lack of insulin receptors in target cells. In many patients, Type II diabetes can be eliminated by weight loss. In others, the disease can be controlled by diet and exercise. Physicians restrict carbohydrate intake of their patients and ask them to eat small meals at regular intervals during the day. Candy, sugar, cakes, and pies are off limits. Glucose must be supplied by complex carbohydrates, such as starches.

The treatments described above have dramatically changed the prognosis for diabetics. At one time the disease was fatal. Today, patients can live healthy, fairly normal lives. Risks are still present. Type I diabetics, for example, still suffer from diabetic comas, or unconsciousness, caused when they receive insufficient amounts of insulin—for example, if they forget their insulin injection or inject too little. Without insulin, the body cells become starved for glucose (even though blood levels are high) and begin breaking down fat. Excessive fat catabolism releases toxic chemicals (called ketones) that cause the patient to lose consciousness. Early- and late-onset diabetics may also suffer from loss of vision, nerve damage, and kidney failure 20 to 30 years after the onset of the disease, even if they are being treated. Damage to the circulatory system may cause gangrene, requiring amputation of limbs, especially the lower extremities. These serious complications result from the inevitable periodic elevations in blood glucose levels that occur over the years.

≋ THE ADRENAL GLANDS

You are standing on the banks of a raging river. Your raft is tied to a tree and floating in the calm water of an eddy above the white water. As you watch kayakers and rafters head into the rapids, your heart starts to race, and your intestines churn in excitement (and fear). Your turn is coming up.

This natural response results from the secretion of two of the many hormones produced by the **adrenal glands.** As ▷ Figure 20–27 shows, the adrenal glands perch atop the kidneys, and each consists of two zones. The central region, or **adrenal medulla,** produces the hormones that increase the heart rate and accelerate breathing when a person is excited or frightened. The outer zone, the **adrenal cortex,** produces a number of steroid hormones mentioned earlier and discussed in more detail shortly.

The Adrenal Medulla Produces Stress Hormones

The adrenal medulla produces two hormones: **adrenalin** (epinephrine) and **noradrenalin** (norepinephrine). In humans, about 80% of the adrenal medulla's output is adrenalin. Helping animals meet the stresses of life, adrenalin and noradrenalin are instrumental in the fight-

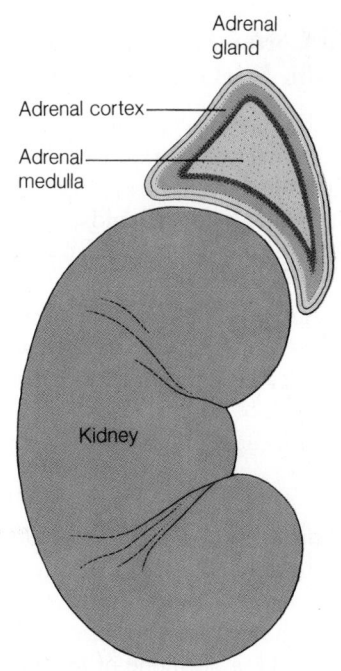

▷ **FIGURE 20–27 Adrenal Gland** The adrenal glands sit atop the kidney and consist of an outer zone of cells, the adrenal cortex, which produces a variety of steroid hormones, and an inner zone, the adrenal medulla. The adrenal medulla produces adrenalin and noradrenalin, the secretion of which is controlled by the autonomic nervous system.

or-flight response—that is, the physiological reactions that take place when an animal is threatened and facilitate its ability to fight or flee the scene (Chapter 17).

Adrenalin and noradrenalin are secreted under all kinds of stress—for example, when an angry dog leaps out at you, when a careless driver cuts in front of you in heavy traffic, or even as you wait outside a lecture hall, anticipating a final exam. Nerve impulses traveling from the brain to the adrenal medulla trigger the release of adrenalin and noradrenalin. The response is part of the autonomic nervous system (Chapter 17).

The physiological changes these hormones cause are many. For example, adrenalin and noradrenalin elevate blood glucose levels, making more energy available to body cells, particularly skeletal muscle cells. They also increase one's breathing rate, which provides additional oxygen to skeletal muscles and brain cells. These hormones cause the heart rate to accelerate as well. This action increases circulation and ensures adequate glucose and oxygen for body cells that might be called into action. Adrenalin and noradrenalin also cause the bronchioles in your lungs to dilate, permitting greater movement of air in and out of the lungs. Under the influence of these hormones, blood vessels in the intestinal tract constrict, putting digestion on temporary hold, while the blood vessels in the skeletal muscles dilate, increasing flow through them. Mental alertness increases as a result of increased blood flow and hormonal stimulation. You are ready to fight or flee.

The Adrenal Cortex Produces Three Types of Hormones with Markedly Different Roles

Surrounding the medulla is the adrenal cortex (Figure 20–27). The adrenal cortex produces three types of steroid hormones, each of which has a markedly different function. The first group, the **glucocorticoids,** affect glucose metabolism and help maintain blood glucose levels—hence the name. The second group, the **mineralocorticoids,** help regulate the ionic concentration of the blood and tissue fluids. The final group, the **sex steroids,** are identical to the hormones produced by the ovaries and testes. In healthy adults, the amount of adrenal sex steroids synthesized and released is insignificant compared to the amounts produced by the gonads.

Glucocorticoids. The prefix *gluco* reflects the fact that these steroid hormones affect carbohydrate metabolism. (Glucose, of course, is one of the many carbohydrates found in living things.) The secretion of glucocorticoid is governed by ACTH, a hormone of the anterior pituitary, as discussed earlier in the chapter. Several chemically distinct glucocorticoids are secreted, but the most important is **cortisol.** Cortisol increases blood glucose by stimulating gluconeogenesis (the synthesis of glucose from amino acids and glycerol). This action makes more glucose available for energy production in times of stress. Cortisol also stimulates the breakdown of proteins in muscle and bone, freeing amino acids that can then be chemically converted to glucose molecules in the liver.

In **pharmacological doses,** levels beyond those seen in the body, cortisol inhibits inflammation, the body's response to tissue damage. Pharmacological doses of glucocorticoids also depress the allergic reaction. Cortisol brings about its effects by inhibiting the movement of white blood cells across capillary walls, thus impeding their migration into damaged tissue. Cortisol also reduces the number of circulating lymphocytes by destroying them at their site of formation.

Because they reduce inflammation, cortisol and other glucocorticoids (particularly cortisone, a synthetic glucocorticoid-like hormone) can be used to treat inflammation resulting from diseases such as rheumatoid arthritis or physical injuries—for example, damage to a knee during a basketball game. However, the benefits must be weighed carefully against the damage that can be caused by upsetting the body's homeostatic balance (discussed below).

Mineralocorticoids. As their name implies, the mineralocorticoids are involved in electrolyte or mineral salt balance. The most important mineralocorticoid is **aldosterone** (described in Chapter 16). It is the most potent mineralocorticoid and constitutes 95% of the adrenal cortex's hormonal output.

Although mineralocorticoids regulate the level of several ions, their main function is to control sodium and potassium concentrations, and their chief target is the kidney. As noted in Chapter 16, aldosterone increases the movement of sodium ions out of the nephron and into the blood. This process is called tubular reabsorption. Aldosterone also stimulates sodium ion reabsorption in sweat glands and saliva and potassium excretion by the kidney. When sodium is shunted back into the blood in the kidney and the skin, water follows. Aldosterone therefore helps conserve body water.

As also noted in Chapter 16, aldosterone secretion is controlled by the sodium ion concentration in a negative feedback loop. As sodium levels fall, aldosterone secretion increases. As sodium levels are restored, aldosterone secretion declines. As you might expect, aldosterone secretion is also stimulated when potassium levels climb and when blood volume and blood pressure decline.

Diseases of the Adrenal Cortex. ▷ Figure 20–28 shows a patient with **Cushing's syndrome.** Cushing's syndrome generally results from pharmacological doses of cortisone, a synthetic glucocorticoid used to treat rheumatoid arthritis or asthma. In a few instances, the disease may be caused by a pituitary tumor that produces excess ACTH or a tumor of the adrenal cortex that secretes excess glucocorticoid.

Patients with Cushing's disease often suffer persistent **hyperglycemia**—high blood sugar levels—because of the presence of high levels of glucocorticoid. Bone and muscle protein may also decline sharply, because glucocorticoids stimulate the breakdown of protein. Individuals complain

▷ **FIGURE 20–28 Cushing's Syndrome** This disease results from an excess of glucocorticoid hormone, either cortisol or cortisone. It is most often caused by cortisone treatment for allergies or inflammation. The most common symptoms are a round face due to edema and excess fat deposition.

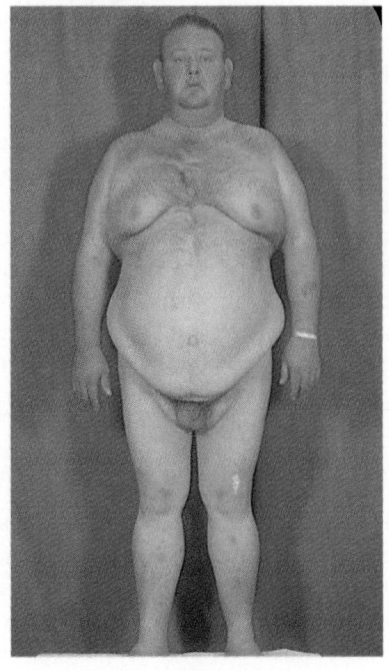

of weakness and fatigue. Loss of bone protein increases the ease with which bones fracture.

Water and salt retention are also common in Cushing's patients, resulting in tissue edema (swelling). Cushing's patients have "moon face," a rounded face resulting from edema. These symptoms occur because glucocorticoids, in high concentrations, have mineralocorticoid effects.

Because most cases of Cushing's syndrome result from steroids taken for health reasons, treatment is a simple matter. By gradually reducing the glucocorticoid dose, a physician can eliminate the symptoms. Tumors in the pituitary and the adrenal cortex can be treated with radiation or surgery.

Another disease of the adrenal cortex mentioned earlier in this chapter but worth studying more closely is **Addison's disease.** Most cases of Addison's disease are thought to be autoimmune reactions in which the cells of the adrenal cortex are recognized as foreign and destroyed by the body's own immune system.

Addison's disease is characterized by a variety of symptoms resulting from the loss of hormones from the adrenal cortex. These symptoms include loss of appetite, weight loss, fatigue, and weakness. Although insulin and glucagon are still present, the absence of cortisol upsets the body's homeostatic mechanism for controlling glucose. The body's reaction to stress is also impaired. The lack of aldosterone results in electrolyte imbalance. Because aldosterone helps maintain sodium levels and blood pressure, patients with Addison's disease have low blood sodium levels and low blood pressure. Addison's disease can be treated with steroid tablets that replace the hormones the adrenal cortex is no longer producing. Treatment allows patients to lead a fairly normal, healthy life.

THE ENDOCRINE SYSTEM OF INVERTEBRATES

Biologists have discovered hormones in arthropods and a variety of other invertebrates. Even protozoans, sponges, plants, and possibly some prokaryotes contain chemical messengers. These hormones control a wide variety of functions. Studies of the endocrinology of invertebrates suggest that vertebrate hormones evolved from those found in invertebrates but often perform quite different tasks today.

The best understood invertebrate hormones are those in insects that control metabolism, molting (periodic shedding of the exoskeleton), and reproduction. Hormones that control both fat and sugar levels have been identified in insects and, in one case, chemically characterized.

Molting occurs in many insects and crustaceans and is a process in which they shed their exoskeleton and reconstruct a new one at regular intervals to accommodate growth (▷ Figure 20–29). In insects and crustaceans, the immediate stimulus for molting is a steroid hormone known as **ecdysone** (▷ Figure 20–30). Released from a

▷ **FIGURE 20–29 Molting** Insects and other invertebrates periodically shed their exoskeletons to permit growth.

pair of glands in the bodies of insects and crustaceans, ecdysone stimulates the cells of the epidermis, causing them to secrete a new exoskeleton. The old exoskeleton is partially digested by enzymes and eventually splits apart as the new one develops underneath it.

In insects, ecdysone is controlled by a hormone from the brain appropriately named **brain hormone.** This hormone stimulates the prothoracic gland of the insect to release ecdysone. In crustaceans, ecdysone is controlled by a slightly different mechanism, an inhibitory hormone appropriately known as **molt-inhibiting hormone (MIH).** When MIH secretion stops, ecdysone is produced, and a molt occurs.

The secretion of molt-inhibiting hormone and brain hormone is controlled by a variety of internal and external factors, depending on the species. Day length, temperature, and nutritional status are the most important controls. These factors help synchronize molting with the animal's growth and development or with seasonal environmental factors such as food availability.

Hormones also control the complex life cycles of some insects and other animals. Many animals pass through two or more morphologically distinct stages in their postembryonic life (▷ Figure 20–31). Frogs, for example, develop from tadpoles. The transformation of an animal from one stage to another is called **metamorphosis.** This strategy allows an organism to exploit different habitats. In differ-

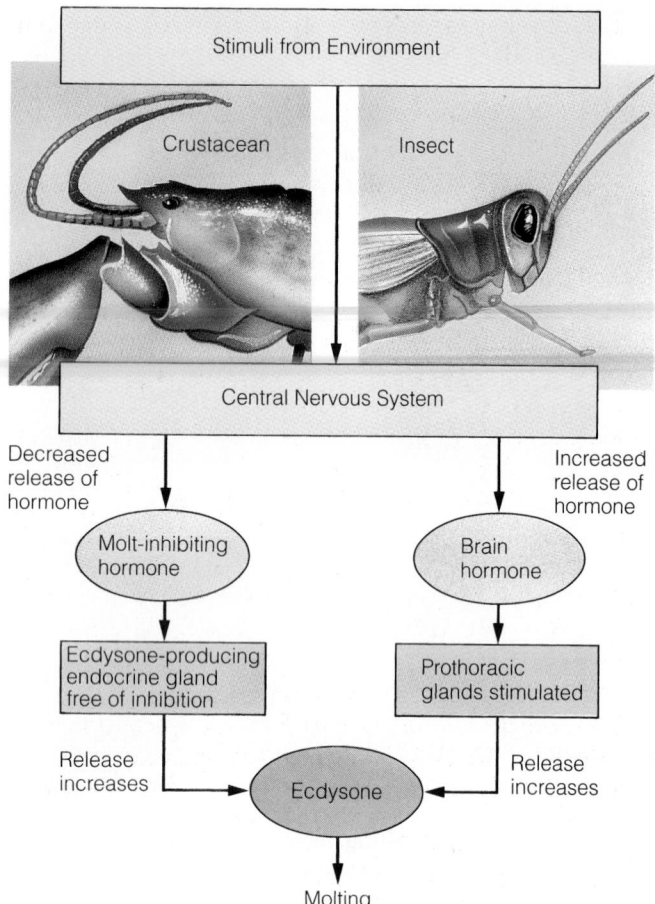

▷ FIGURE 20–30 Hormonal Control of Molting
Molting in insects and crustaceans is controlled by the hormone ecdysone. However, ecdysone is controlled by brain hormone in insects and by molt inhibiting hormone in crustaceans.

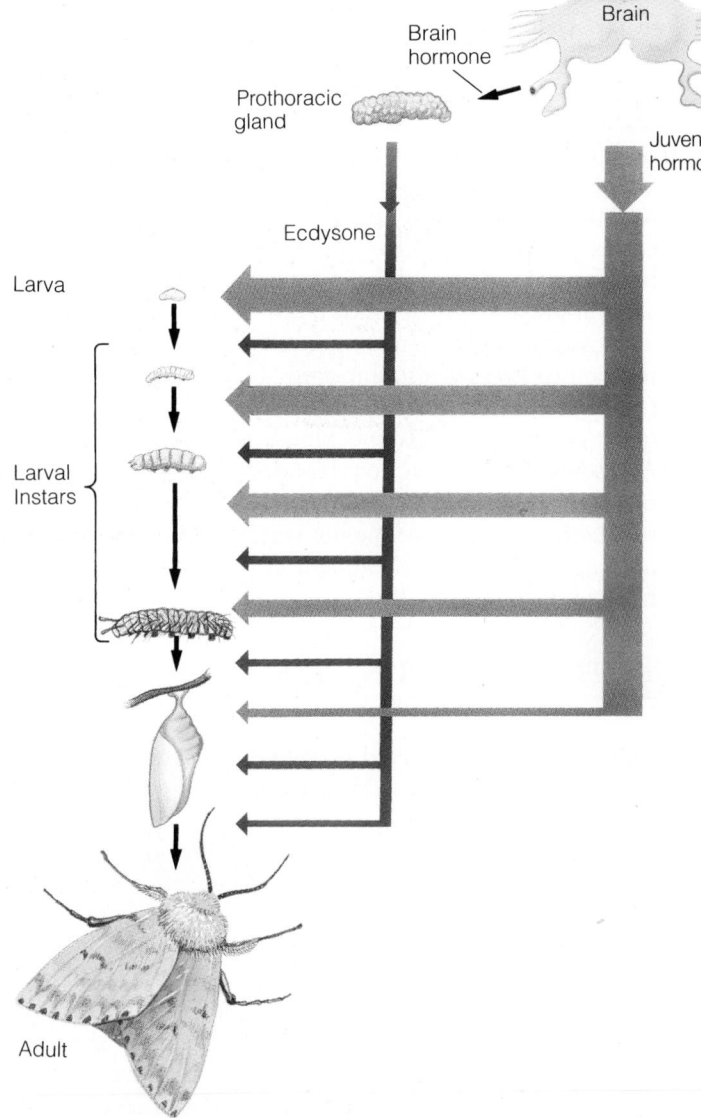

ent stages, organisms use very different modes of locomotion and feeding.

Metamorphosis is regulated by hormones. As shown in Figure 20–31, moths and butterflies develop from caterpillars (larvae). The caterpillars pass through a series of stages (larval instars), then form cocoons, or **pupae.** Inside the cocoon, radical changes occur that produce a moth or butterfly. Ecdysone and brain hormone are involved in the entire process. Another hormone, known as **juvenile hormone (JH),** is also involved. As illustrated in Figure 20–31, JH is secreted in large amounts early on. As long as JH levels are high, larva-to-larva molts occur. When metabolic changes cause JH levels to fall, though, the larva forms a cocoon in the presence of ecdysone. The ratio of juvenile hormone to ecdysone, therefore, determines the pattern of development. Interestingly, some plants have evolved chemical analogues of juvenile hormone, which when ingested by hungry larvae prevent them from pupating. This mechanism prevents the larvae from maturing and going on to produce eggs and additional generations of larvae.

▷ FIGURE 20–31 Life Cycle of the Moth Note that the moth larvae, which form from eggs, pass through several instar (larval) stages. In order to grow, the larvae must molt. Molting is controlled by the hormone ecdysone when juvenile hormone levels are high. When juvenile hormone levels decline, ecdysone which is present throughout the life cycle stimulates pupation and differentiation into an adult.

ENVIRONMENT AND HEALTH: HERBICIDES AND HORMONES

Hormones orchestrate an incredible number of body functions, creating a dynamic balance that is necessary for good health. When this balance is altered, our health suffers (Table 20–3). A major participant in homeostasis, the endocrine system, like other systems, is sensitive to outside, or environmental, factors. Stress, for example, can lead to an imbalance in adrenal hormones, resulting in high blood pressure and other complications. High blood pressure, in turn, puts strain on the heart. Hormonal balance can also be altered by toxic pollutants.

One example is dioxin. Dioxin is a contaminant found in some herbicides and in paper products. Infamous for its presumed carcinogenic properties, dioxin was found in the chemical defoliant known as Agent Orange, which was used in the Vietnam War. Oil spread on dirt roads in Times Beach, Missouri, also contained dioxin, deliberately added to the oil by a less than honest waste disposer. In 1983, after considerable public debate, the U.S. government bought up the town for $35 million, fenced it off, and moved the residents out.

New evidence suggests, however, that dioxin may be carcinogenic at high levels—those in the workplace and those associated with spills. The low levels found in Times Beach, some say, may have little effect on human cancer, and the town may have been evacuated unnecessarily.

Before you get too excited about this view, recent studies suggest that dioxin may also suppress immune functions. It may bind to plasma membrane and cytoplasmic receptors that normally bind to hormones. Some researchers are calling it an "environmental hormone." One form of dioxin, TCDD, suppresses the immune system in mice at least 100 times more effectively than corticosterone, one of the body's glucocorticoids.

TCDD causes a variety of biological responses. In some cells, it causes rapid cell growth. In others, it may alter cellular differentiation. Nonetheless, its effect on the immune system may be far more important than its impact on cancer. As for the former residents of Times Beach, they might just want to stay away a little longer until the controversy is settled.

Dioxin is not the only dangerous chemical in the environment. Studies suggest that a number of common herbicides (the thiocarbamates) may upset the thyroid's function and result in the formation of thyroid tumors. These herbicides are chemically similar to thyroid hormone and therefore block the secretion of TSH, resulting in goiter. At higher levels, they may cause thyroid cancer.

The overall impact on human health of toxic chemicals released into the environment is probably small, especially compared with the impact of tobacco smoke. However, not everyone agrees. The Point/Counterpoint in Chapter 9 debates this subject, which is an issue worth thinking about. This important debate will be around for years.

TABLE 20–3 Summary of Some Endocrine Disorders

DISEASE	CAUSE	SYMPTOMS
Gigantism	Hypersecretion of GH starting in infancy or early life	Excessive growth of long bones
Dwarfism	Hyposecretion of GH in infancy or early life	Failure to grow
Acromegaly	Hypersecretion of GH after bone growth has stopped	Facial features become coarse; hands and feet enlarge; skin and tongue thicken.
Hyperthyroidism	Overactivity of the thyroid gland	Nervousness; inability to relax; weight loss; excess body heat and sweating; palpitations of the heart
Hypothyroidism	Underactivity of the thyroid gland	Fatigue; reduced heart rate; constipation; weight gain; feel cold; dry skin
Hyperparathyroidism	Excess parathyroid hormone secretion, usually resulting from a benign tumor in the parathyroid gland	Kidney stones; indigestion; depression; loss of calcium from bones
Hypoparathyroidism	Hyposecretion of the parathyroid glands	Spasms in muscles; numbness in hands and feet; dry skin
Diabetes insipidus	Hyposecretion of ADH	Excessive drinking and urination; constipation
Diabetes mellitus	Insufficient insulin production or inability of target cells to respond to insulin	Excessive urination and thirst; poor wound healing; urinary tract infections; excess glucose in urine; fatigue and apathy
Cushing's syndrome	Hypersecretion of hormones from adrenal cortex or, more commonly, from cortisone treatment	Face and body become fatter; loss of muscle mass; weakness; fatigue; osteoporosis
Addison's disease	Gradual decrease in production of hormones from adrenal gland; most common cause is autoimmune reaction	Loss of appetite and weight; fatigue and weakness; complete adrenal failure

PRINCIPLES OF ENDOCRINOLOGY

1. The endocrine system consists of a widely dispersed set of highly vascularized ductless glands that produce hormones.
2. Hormones affect five vital aspects of an animal's life: (a) homeostasis, (b) growth and development, (c) reproduction, (d) energy production, storage, and catabolism, and (e) behavior.
3. Hormones act on specific cells. Specificity results from the presence of hormone receptors on target cells. Steroid hormone receptors are generally located in the cytoplasm, and protein and polypeptide hormone receptors are located in the plasma membrane.
4. Three types of hormones are produced in the body: (a) steroids, (b) proteins and polypeptides, and (c) amines.
5. Hormone secretion is controlled principally by negative feedback loops, many of which involve the hypothalamus.
6. Water-soluble hormones (proteins and polypeptides) act on cells via a second messenger.
7. Steroid (lipid-soluble) hormones act via the two-step mechanism.
8. The endocrine and nervous systems are similar in several respects. They both send signals that regulate cell structure and function. Both also help coordinate body functions.
9. The endocrine and nervous systems differ in several key respects. The nervous system elicits rapid, generally short-lived responses. The endocrine system elicits slower, longer lasting responses.

THE PITUITARY AND HYPOTHALAMUS

10. The pituitary is a pea-sized gland suspended from the hypothalamus by a thin stalk. It consists of two parts: the anterior pituitary and the posterior pituitary.
11. The anterior pituitary produces seven protein and polypeptide hormones. Their release is controlled by releasing and inhibiting hormones produced by the hypothalamus, which are transported to the anterior pituitary via a portal system.
12. The hypothalamic hormones are produced by neurosecretory neurons. Their release is controlled by chemical stimuli and nerve impulses.
13. The hormones of the anterior pituitary and their functions are summarized in Table 20–1.
14. The posterior pituitary is derived from brain tissue during embryonic development and consists of axons and terminal ends of neurosecretory cells whose cell bodies are in the hypothalamus. Hormones are produced in the cell bodies and are transported down the axons to the posterior pituitary, where they are stored in the axon terminals until released.
15. The posterior pituitary produces two hormones, antidiuretic hormone (ADH) and oxytocin, whose functions are also summarized in Table 20–1.

THE THYROID GLAND

16. The thyroid gland is located in the neck, on either side of the trachea near its junction with the larynx. The thyroid produces three hormones: thyroxine (T_4), triiodothyronine (T_3), and calcitonin.
17. Thyroxine and triiodothyronine accelerate the rate of glucose breakdown in most cells, increasing body heat. These hormones also stimulate cellular growth and development.

18. Thyroxine and triiodothyronine both require iodine for their synthesis. A deficiency of this element in the diet results in goiter, an enlargement of the thyroid gland.
19. Calcitonin lowers blood calcium levels by inhibiting osteoclasts, thus reducing bone destruction and the release of calcium from bone. Calcitonin also increases formation of bone by osteoblasts and increases the excretion of calcium in the kidneys.

THE PARATHYROID GLANDS

20. The parathyroid glands are located on the back of the thyroid gland and produce a polypeptide hormone called parathyroid hormone (PTH), or parathormone.
21. PTH is released when calcium levels in the blood drop. This hormone increases blood calcium levels by stimulating bone reabsorption by osteoclasts, increasing intestinal absorption, and increasing renal reabsorption of calcium.

THE PANCREAS

22. The pancreas produces two hormones, insulin and glucagon, from the islets of Langerhans.
23. Insulin is the glucose-storage hormone. It stimulates the uptake of glucose by many body cells and stimulates the synthesis of glycogen in muscle and liver cells. Insulin also increases the uptake of amino acids and stimulates protein synthesis in muscle cells, thus promoting muscle formation.
24. Glucagon is an antagonist to insulin. It raises glucose levels in the blood in the period between meals by stimulating glycogen breakdown and the synthesis of glucose from amino acids and fats (gluconeogenesis).
25. Diabetes mellitus is a disease involving insulin and blood glucose. It has two principal forms: Type I and Type II. Type I, also called early-onset diabetes, occurs early in life and may be caused by an autoimmune reaction that destroys the beta cells of the pancreas. It can be treated by insulin injections.
26. Type II, or late-onset, diabetes, results from a reduction in the number of insulin receptors on target cells. It may be caused by obesity and genetic factors and can often be treated successfully by dietary management.

THE ADRENAL MEDULLA AND ADRENAL CORTEX

27. The adrenal glands lie atop the kidneys and consist of two separate portions: the adrenal medulla, at the center, and the adrenal cortex, a surrounding band of tissue.
28. The adrenal medulla produces two hormones under stress: adrenalin and noradrenalin. These hormones stimulate heart rate and breathing, elevate blood glucose levels, constrict blood vessels in the intestine, and dilate blood vessels in the muscles.
29. The adrenal cortex produces three classes of hormones: glucocorticoids, mineralocorticoids, and sex steroids.
30. The glucocorticoids affect carbohydrate metabolism and tend to raise blood glucose levels. The principal glucocorticoid is cortisol.
31. In pharmacological doses, cortisol inhibits the immune system and allergic reactions and is used to treat allergies or inflammation caused by injury and infections. High doses, however, have many adverse impacts on the body.

32. The chief mineralocorticoid is aldosterone. It acts on the kidneys, sweat glands, and salivary glands, causing sodium and water retention and potassium excretion.

THE ENDOCRINE SYSTEM OF INVERTEBRATES

33. Hormones exist in a large number of invertebrates, where they control a wide variety of functions.
34. The best understood invertebrate hormones are those in insects that control metabolism, molting, and reproduction.
35. Many insects and crustaceans molt, shedding their exoskeleton and reconstructing a new one at regular intervals to accommodate growth.
36. In insects and crustaceans, the immediate stimulus for molting is the steroid hormone ecdysone, which stimulates the cells of the epidermis, causing them to secrete a new exoskeleton.
37. In insects, ecdysone is controlled by brain hormone; in crustaceans it is controlled by molt-inhibiting hormone. The secretion of both hormones is controlled by a variety of internal and external factors, among them day length, temperature, and nutritional status.

38. Hormones also control metamorphosis, the transformation of an animal from one stage to another. Ecdysone is vital to the transformation of larvae to adult stages, but its effects are modulated by juvenile hormone. Thus, the ratio of juvenile hormone to ecdysone determines the pattern of development.

ENVIRONMENT AND HEALTH: HERBICIDES AND HORMONES

39. Hormones orchestrate an incredible number of body functions, creating a dynamic balance necessary for good health. When this balance (homeostasis) is altered, our health suffers.
40. The endocrine system, like others, is sensitive to upset from environmental factors, including pollutants.
41. Dioxin, a contaminant in some herbicides and paper products, may exert most of its effects through the endocrine system.
42. A number of common herbicides (the thiocarbamates) upset the thyroid's function and may even cause thyroid tumors.

EXERCISING YOUR CRITICAL THINKING SKILLS

Find a copy of the *New England Journal of Medicine* in the library with an article on a discovery in endocrinology. How is the article organized? Analyze the article, using your critical thinking skills. Was the study performed on humans or laboratory animals? What were the major conclusions?

In your view, was the study performed correctly? Did the researchers include a control group? Was the control group similar to the experimental group? How many subjects were used? Should more have been used? Why? Can you tell whether the conclusions supported the data? Can you think of any alternative explanations for the data?

TEST OF TERMS

1. A _____ is a chemical produced in an endocrine gland that travels in the bloodstream to its _____ cells, where it elicits a specific response.
2. In the endocrine system, specificity is conferred by the _____ found in the cytoplasm and in the plasma membrane.
3. As a general rule, _____ hormones are those that stimulate the production and secretion of other hormones.
4. The body produces three types of hormones: _____ , proteins and polypeptides, and _____ .
5. Protein and polypeptide hormones activate cells through a _____ _____ , usually cyclic AMP. It is produced inside cells from _____ in a reaction catalyzed by a membrane-bound enzyme called _____ _____ . Cy-

clic AMP activates another enzyme, protein kinase. It in turn adds _____ to other enzymes inside the cell, turning some on and turning others off.
6. Hormones like testosterone activate cells by the _____ -_____ mechanism. Steroid hormones activate the _____ .
7. The _____ gland, located beneath the _____ , produces more hormones than any other endocrine gland.
8. The _____ _____ gland is controlled by _____ and inhibiting hormones produced by the _____ cells of the _____ .
9. _____ hormone is a protein that stimulates cellular _____ and hyperplasia. In muscle, this hor-

mone stimulates the uptake of _____ _____ and the synthesis of protein.
10. _____ is a term that describes a condition in which an endocrine gland produces less hormone than needed.
11. _____ is a hormone that stimulates the adrenal cortex to release its hormones.
12. The _____ are hormones that stimulate gamete formation and endocrine production in the gonads in both males and females.
13. Milk production in humans is stimulated by a protein hormone called _____ . Milk production begins at the end of pregnancy and can be continued by _____ .
14. Hormone secretion that is stimulated by neural impulses is called a _____ reflex.

15. The posterior pituitary is a _____ gland. It consists of nervous tissue and releases two hormones: _____ and _____.

16. The thyroid gland produces three hormones: _____, triiodothyronine, and _____.

17. The thyroid follicles contain a colloidal material called _____.

18. _____ is a condition that results from a dietary deficiency of iodine.

19. Two hormones control the level of calcium in the blood. The hormone that raises serum calcium levels is _____, and it is produced by the _____ glands. The hormone that reduces serum calcium is _____ and is produced by the _____ gland.

20. _____ is produced by the pancreas. It stimulates the uptake of _____ by muscle cells and stimulates the synthesis of _____.

21. The synthesis of glucose from amino acids and fatty acids is called _____.

22. Early-onset diabetes may result from a(n) _____ reaction.

23. Type II diabetes, or _____-_____ diabetes, occurs in older people and is often associated with _____. It is controlled by managing _____.

24. The adrenal medulla produces two hormones, _____ and _____, which stimulate heart rate and breathing.

25. Glucocorticoids are produced by the _____ _____; they increase blood glucose by stimulating _____. In high concentrations, they repress the _____ system function.

26. _____ is the principal mineralocorticoid produced by the _____ _____. It affects sodium and potassium ion concentrations.

Answers to the Test of Terms are found in Appendix B.

1. Define the following terms: endocrine system, hormone, and target cell.

2. Hormones function in five principal areas. What are they? Give some examples of each.

3. Describe the concept of specificity. How is it created in the nervous system? How is it created in the endocrine system?

4. Compare and contrast the functions of the nervous and endocrine systems.

5. The endocrine system elicits slower, longer-lasting responses than the nervous system. Do you agree or disagree? Explain your reasons.

6. Define the terms *trophic* and *nontrophic hormones,* and give several examples of each.

7. Give two examples of negative feedback loops in the endocrine system, a simple feedback mechanism and a more complex one that operates through the nervous system.

8. Define the term *neuroendocrine reflex,* and give some examples.

9. Describe the second messenger theory.

10. Describe the two-step mechanism of hormone action.

11. How are the second messenger theory and the two-step mechanism similar, and how are they different?

12. Give several biological reasons for the following observations: (a) the endocrine response tends to be delayed; (b) the endocrine response tends to be prolonged; (c) some hormones perform several different functions.

13. List the hormone(s) involved in each of the following functions: blood glucose levels, growth, milk production, milk let-down, calcium levels, and metabolic rate.

14. Describe the role of the hypothalamus in controlling anterior pituitary hormone secretion.

15. Describe the ways in which the posterior pituitary differs from the anterior pituitary.

16. Explain the hormonal reasons for each of the following conditions: acromegaly, dwarfism, and giantism. Acromegaly and giantism are both caused by the same problem. Why are these conditions so different?

17. ACTH is controlled by levels of glucocorticoid and by a biological clock. How are the controls different?

18. Describe the neuroendocrine reflex involved in prolactin secretion.

19. Offer some possible explanations for the following experimental observation: Milk production occurs late in pregnancy and is thought to be stimulated by the hormone prolactin. A nonpregnant rat is injected with prolactin but does not produce milk.

20. Where is ADH produced? Where is it released? Describe how ADH secretion is controlled. What effects does this hormone have?

21. Where is oxytocin produced? Where is it released? Describe how oxytocin secretion is controlled. What effects does this hormone have?

22. A patient comes into your office. She is thin and wasted and complains of excessive sweating and nervousness. What tests would you run?

23. A patient comes into your office. He is suffering from indigestion, depression, and bone pain. An X-ray of the bone shows some signs of osteoporosis. You think that the disorder might be the result of an endocrine problem. What test would you order?

24. How are the two basic types of diabetes mellitus different? How are they similar? How are they treated, and why are these treatments chosen?

25. Describe the physiological changes that occur under stress. What hormones are responsible for them?

26. What is gluconeogenesis? What hormones stimulate the process?

27. Cortisone depresses the allergic response. A patient comes to your office and asks that you treat her allergies with cortisone. What would you tell her?

28. Aldosterone is a mineralocorticoid. Describe its chief functions. How does it help retain body fluid? Under what conditions is aldosterone secreted?

PART

V

The Continuation of Life
Reproduction and Development

≈

Reproduction in Animals

≈

Damselflies mating. Male injects sperm into female, which fertilize the eggs.

andra Collins woke one day with a terrible pain in her abdomen, which persisted throughout the morning. Instead of calling her doctor, though, she shrugged off the pain and went shopping. While the young woman was browsing through a bookstore, the pain grew worse, she blacked out, and she was rushed to the emergency room by her husband. There, it was found that Sandra was suffering from severe internal bleeding caused by an ectopic pregnancy—that is, a fertilized ovum that had developed in the upper part of her reproductive tract, outside the uterus.

Fortunately for Sandra, physicians were able to counteract her severe blood loss with a transfusion. They then whisked her into the operating room, where they surgically removed the fetus, placenta, and surrounding tissue and repaired torn blood vessels.

Reproduction is one of the most basic body functions. And as this account shows, it doesn't always operate smoothly. Like many other body functions, reproduction is controlled by the endocrine and nervous systems. This chapter describes reproduction in the animal kingdom, with special emphasis on vertebrate reproduction as exemplified by humans. It will provide a wealth of information and insights that will help you understand Chapter 22 which deals with fertilization and development. We begin our study with a look at reproductive strategies in the animal kingdom.

≋ REPRODUCTIVE STRATEGIES: AN EVOLUTIONARY PERSPECTIVE

A friend of mine was examining a copy of a book on human biology I had written and remarked that there seemed to be an awful lot of information on reproduction. (Two chapters out of 22 were on reproduction and development.) "Isn't your coverage excessive?" he wondered out loud. Before I could answer, he retracted his question. Of course there's a lot on reproduction. It's central to the lives of all organisms!

The reproductive system of animals is not involved in the day-to-day survival of organisms, as are the digestive, nervous, and endocrine systems. It is involved with the survival of entire species. Successful reproduction, in fact, is the measure of the success of any adaptation, whether it occurs in plants, animals, or bacteria.

As you learned in the first chapter of this book, organisms reproduce in one of two ways: asexually or sexually. This section describes each of these strategies in animals and describes their importance.

Asexual Reproduction Involves only One Parent and Tends to Produce Offspring that Are Genotypically and Phenotypically Identical to the Parent

Asexual reproduction occurs when an organism produces an offspring by itself—for example, by budding or frag-

menting. Commonly encountered in invertebrates, such as sponges and flatworms, asexual reproduction produces offspring that are genetically identical to their parents and their siblings.

Budding occurs when small outgrowths, or buds, form on the body of the parent (▷ Figure 21–1). These buds often separate from the parent's body and develop into free-living forms. In some cases, however, buds remain attached and produce colonies.

Fragmentation occurs when a body part is lost due to injury. Sea stars, for example, can shed body parts when captured by predators, which permits them to escape death. The lost arm, if still attached to a section of the central disc, may form a new individual. A new arm will form to replace the one lost in the original animal's escape.

Certain free-living flatworms and annelids fragment into two or more pieces, each of which can develop into a new individual (▷ Figure 21–2). Sponges and many sea stars also fragment into numerous pieces, many of which develop into adult forms.

Like many other aspects of life, asexual reproduction has advantages and disadvantages. For example, asexual reproduction eliminates the need for males and females to find each other when mating time approaches. Energy that might otherwise have been spent finding and selecting a mate and mating is devoted to the production of offspring. Consequently, organisms that reproduce asexually often form large populations that rapidly exploit locally available resources.

The most notable disadvantage of asexual reproduction is that it produces genetically similar offspring. As men-

▷ **FIGURE 21–1 Asexual Reproduction by Budding**

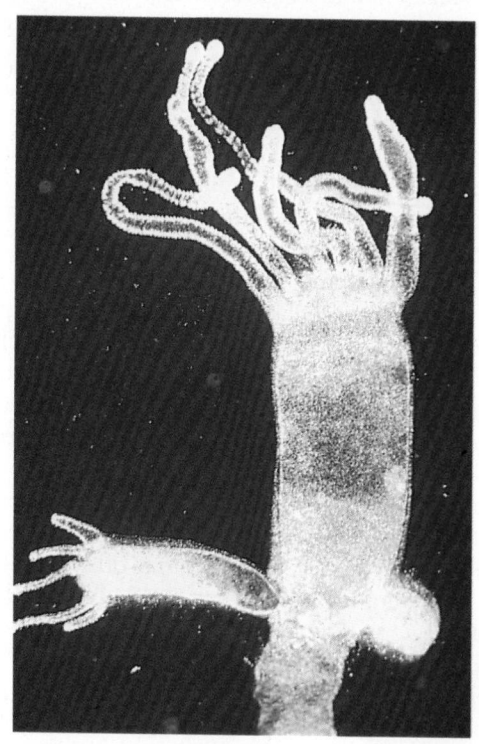

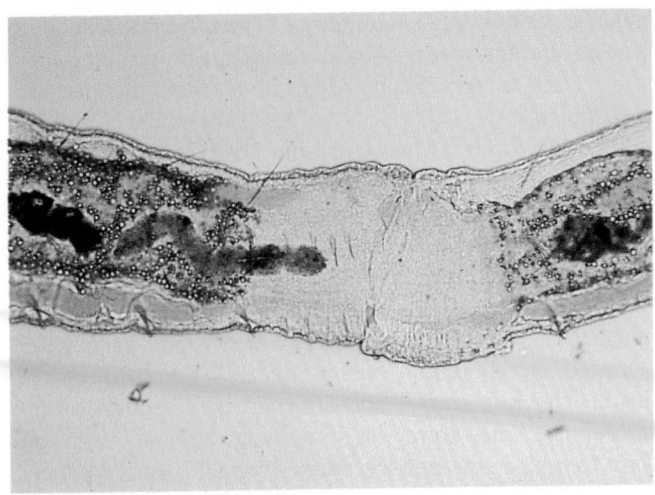

▷ **FIGURE 21–2 Asexual Reproduction by Fragmenting** This marine worm is beginning to fragment, as evidenced by the constriction shown in the center of the photo.

▷ **FIGURE 21–3 The Peccary** This animal which lives in South America, reproduces throughout the year but produces the largest number of offspring during the rainy season when there's ample water for milk production.

tioned, mutations are the only source of genetic variation. Although genetic similarity creates few problems in stable environments, it is a disadvantage when conditions change. If there are no organisms adapted to the change, an entire population can be eliminated.

Sexual Reproduction Involves Gametes that Combine to Form a New Individual

Sexual reproduction involves sex cells, or gametes, each containing half the chromosomes of the adult in which they are produced. The gametes of sexually reproducing organisms are referred to as the egg and sperm. Sperm are typically small, motile cells, and eggs are usually large, nonmotile cells. In most cases, the egg and sperm are produced by separate individuals.

Gametes combine to form a **zygote,** a cell containing a full set of chromosomes, half from the egg-producing female and half from the sperm-producing male. As a result, the offspring are genetically unlike the parents. In other words, the offspring have a different combination of genes. Thus, sexual reproduction produces considerable genetic variation in the offspring, which, in turn, provides an advantage in evolution. Unlike asexually reproducing organisms, sexually reproducing organisms are more likely to survive if environmental conditions change. That is because some members of the population will probably have the right genetic combination and subsequent adaptations that permit them to endure the change.

If you ever worked on a farm or at a zoo, you probably noticed that most sexually reproducing animals breed only once a year. Breeding is timed so that offspring are born into the world at a time when their survival is enhanced. In temperate regions, like the United States, most animals breed in the fall and give birth in the spring.

The timing of breeding in vertebrates and many invertebrates is often regulated by day length, especially in temperate environments where climate varies greatly from one season to the next. Elk and deer in North America, for example, breed in the fall as days shorten. Day length affects the level of a hormone called **melatonin,** which is produced by a small endocrine gland in the brain, the **pineal gland.** Melatonin levels affect reproductive hormones. In environments with more stable conditions such as the tropics, many animals are reproductively active all year.

Other environmental factors may also influence breeding and reproductive success, notably temperature and the availability of food and water. Peccaries, for instance, are piglike animals that live in the arid regions and forests of South America (▷ Figure 21–3). Although they breed year round, peccaries produce the largest number of young in the rainy seasons, when there is ample water for milk production. The breeding cycles of aquatic animals, such as fish and frogs, are influenced by water temperature. Amphibians, for instance, often breed when water temperatures increase.

Sexual reproduction may involve internal or external fertilization. In birds, mammals, and reptiles, for example, sperm are deposited inside the female reproductive system, where fertilization occurs. The fertilized ovum may develop inside the female, as in mammals, which give birth to live young, or it may be deposited in an egg, as in the case of birds and many reptiles (▷ Figure 21–4).

In other vertebrates, such as fish and amphibians, fertilization typically occurs after the eggs are laid, although there are some exceptions (▷ Figure 21–5). In external fertilization, organisms usually shed large numbers of gametes into the water at the same time. The fertilized eggs are often abandoned, so large numbers are necessary to ensure survival. For a unique form of external fertilization see Spotlight on Evolution 21–1.

(a)

(b)

▷ **FIGURE 21–4 Live Birth or Hatching** (*a*) In many animals, eggs are fertilized internally, and the offspring develop inside the female's reproductive system. Babies are born live. (*b*) In others, such as birds, fertilization occurs internally, but eggs are encased in a hard shell and deposited externally. Offspring hatch from the egg after a period of incubation.

▷ **FIGURE 21–5 External Fertilization** Some vertebrates, like these frogs, lay their eggs externally, where they are fertilized.

≈ HUMAN REPRODUCTION

The Male Reproductive System Produces Sperm

In mammals, the male reproductive system consists of seven basic components: (1) the two **testes,** which produce sex steroid hormones and sperm; (2) the two **epididymises,** which store sperm produced in the testes; (3) a pair of ducts, each known as a **vas deferens,** that conducts sperm from the epididymis of each testis to the urethra; (4) **sex accessory glands,** which produce secretions that make up the bulk of the ejaculate; (5) the **urethra,** which conducts sperm to the outside; (6) the **penis,** the organ of copulation, and (7) the scrotum, a sac that houses the testes (▷ Figure 21–6; Table 21–1).

The Testes Are Formed Inside the Body Cavity During Embryonic Development, then Migrate into the Scrotum. The testes are suspended in a pouch known as the **scrotum.** As Figure 21–6 shows, the scrotum is attached to the body below the attachment of the penis. Although the testes reside in the scrotum throughout a man's life, they do not originate there. During embryological development, the testes form inside the abdominal cavity near the kidneys. Toward the end of development, the testes descend into the scrotum through a small tunnel, known as the **inguinal canal,** which links the body cavity with the scrotum (▷ Figure 21–7). The testes are guided into the scrotum with the aid of a ligament that disappears soon afterward.

By the end of the eighth month of development, the testes generally complete their migration into the scrotum. In some males, however, the testes fail to descend, resulting in a condition known as **cryptorchidism.** In many of these boys, the testes descend during the first two years of life. If the testes have not descended on their own by the time a boy is 5 years old, they must be moved into the scrotum surgically to prevent permanent sterility.

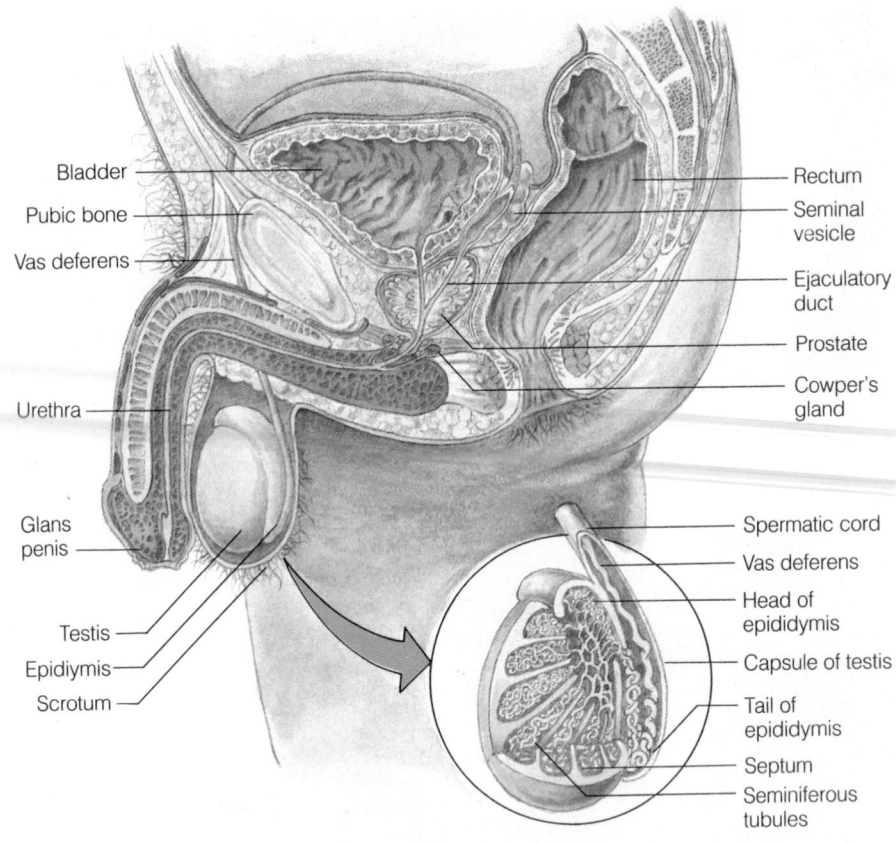

▷ **FIGURE 21–6 Anatomy of the Male Reproductive System**

Bladder

Pubic bone

Vas deferens

Urethra

Glans penis

Testis

Epidiymis

Scrotum

Rectum

Seminal vesicle

Ejaculatory duct

Prostate

Cowper's gland

Spermatic cord

Vas deferens

Head of epididymis

Capsule of testis

Tail of epididymis

Septum

Seminiferous tubules

COMPONENT	FUNCTION
TABLE 21–1 The Male Reproductive System	
Testes	Produce sperm and male sex steroids
Epididymes	Store sperm
Vasa deferentia	Conduct sperm to urethra
Sex accessory glands	Produce seminal fluid that nourishes sperm
Urethra	Conducts sperm to outside
Penis	Organ of copulation
Scrotum	Provides proper temperature for testes

▷ **FIGURE 21–7 The Inguinal Canal** The testis descends through the inguinal canal, an opening through the musculature in the lower abdominal wall. In adults, the inguinal canal provides a route for the vas deferens, blood vessels, and nerves that supply each testis.

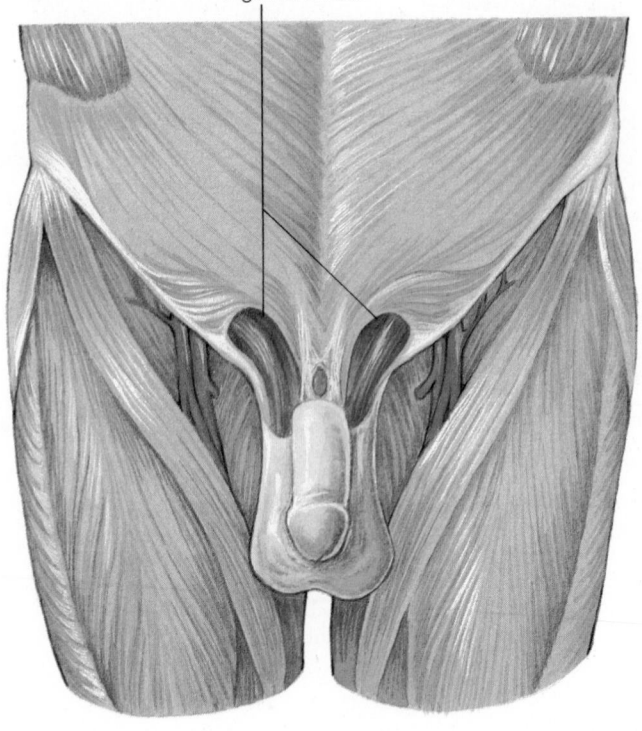

Inguinal canals

The inguinal canal through which the testes descend is a biological compromise: it is necessary for the movement of the testes into the scrotum, but remains a weak point in the lower abdomen of men. If the canal does not close properly, organs in the abdomen push aside the muscles that normally keep the canal closed. Loops of intestine may then descend into the inguinal canal, causing pain and discomfort (▷ Figure 21–8). This condition, known as an **inguinal hernia,** can be corrected by surgery in which the weakened muscles are sewn together, thus blocking the canal and preventing intestines from reentering.

The Scrotum Helps Keep the Testes Cool. The scrotum provides an environment whose temperature is

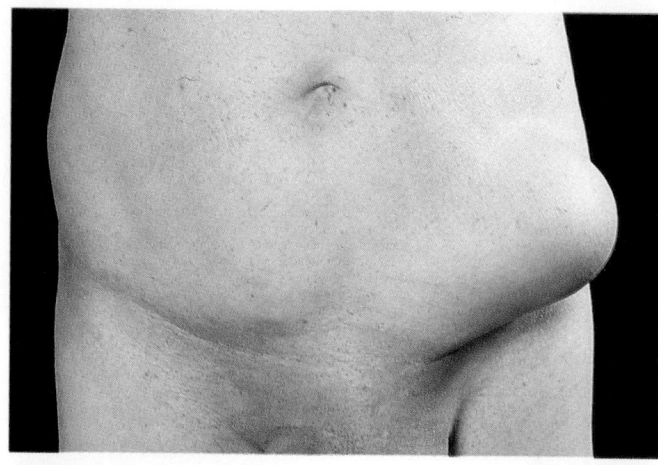

(a)

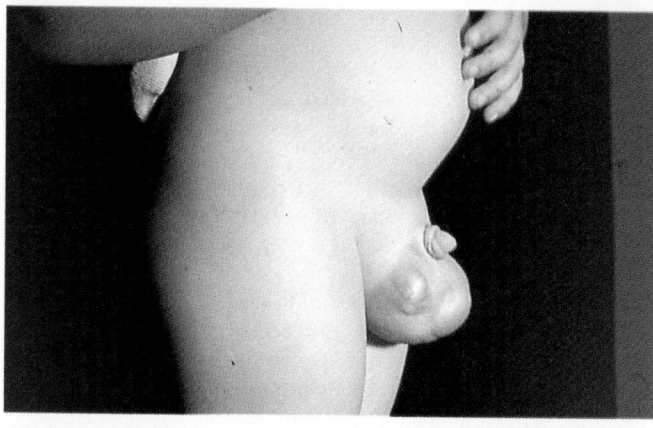

(b)

▷ **FIGURE 21–8 Inguinal Hernia** (*a*) Loops of intestine may push through the weakened musculature surrounding the inguinal canal. (*b*) In some instances, large sections of the intestine may push out through the iguinal canal.

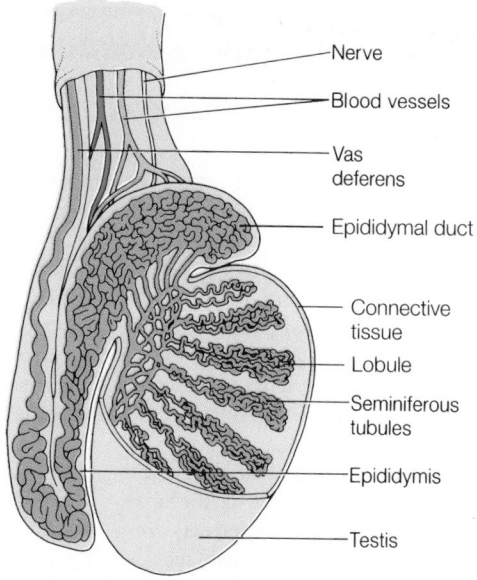

Nerve

Blood vessels

Vas deferens

Epididymal duct

Connective tissue

Lobule

Seminiferous tubules

Epididymis

Testis

▷ **FIGURE 21–9 Interior View of the Testis** The lobules and seminiferous tubules are shown.

suitable for sperm development. Inside the body cavity, the temperature is too high for sperm development, explaining why men whose testes fail to descend are usually sterile.

The influence of body heat on sperm development is illustrated by the plight of male long-distance runners, who run hundreds of miles a month. This level of exercise elevates their body temperature, and even though the testes are suspended in the scrotum, scrotal temperature may be so high that sperm formation declines. Long-distance runners may become temporarily sterile. Tight-fitting pants or shorts may have the same effect.

Sperm Are Produced in the Seminiferous Tubules and Stored in the Epididymis. As ▷ Figure 21–9 shows, each testis is surrounded by a dense layer of connective tissue. This layer is invested with numerous pain fibers, a fact to which most men will attest. The testis is divided into 200 to 400 compartments, or **lobules,** by fibrous tissue partitions that connect with the connective tissue of the outer coat. Each lobule contains one to four

highly convoluted tubes, the **seminiferous tubules,** in which sperm are formed. Stretched end to end, the seminiferous tubules of each testis would extend for about half a mile.

Sperm produced in the seminiferous tubules empty into a network of connecting tubules in the "back" of the testes. These tubules, in turn, empty into the **epididymal duct,** where sperm are stored until released during **ejaculation,** the ejection of sperm. As Figure 21–9 shows, the epididymal duct, which forms the epididymis, empties into the **vas deferens** (plural, **vasa deferentia**). These muscular ducts course upward, passing from the scrotum into the body cavity through the inguinal canals. Inside the pelvic cavity, the vasa deferentia join with the urethra. During ejaculation, the smooth muscles in the walls of the epididymal duct and vasa deferentia contract, propelling sperm to the urethra.

Semen Consists Mostly of Fluids Produced by the Sex Accessory Glands. During ejaculation, sperm are joined by fluids produced by the sex accessory glands. The **sex accessory glands** are located near the neck of the urinary bladder and include the two seminal vesicles, the prostate gland, and the two Cowper's glands, as shown in Figure 21–6. The secretions produced by the sex accessory glands empty into the urethra and the vasa deferentia and make up 99% of the volume of the **ejaculate,** or **semen.** Semen is a fluid that contains sperm and sex accessory gland secretions. The semen contains fructose, a monosaccharide that is used by sperm mitochondria to generate energy needed to help propel the sperm through the female reproductive tract. Semen also contains a buffer that helps neutralize the acidic secretions of the female reproductive tract. Prostaglandin, a chemical substance that

THE MOUTH-BREEDING FISH OF LAKE MALAWI

Angel fish are members of one of the largest families of fish in the world, known as cichlids (sick-lids). Members of this large and diverse family can be found in Central America, South America, and Africa. Although they are all interesting, one of the most intriguing is a group from eastern Africa's Lake Malawi (▷ Figure 1).

Located in the Great Rift Valley near the equator, Lake Malawi is the sixth largest in the world and reportedly contains the largest number of fish species, most of them belonging to the cichlid family. The cichlids of Lake Malawi are some of the most colorful freshwater fish known to science. Brilliant blues and yellows decorate some of these remarkable creatures. Two favorites are the Cobalt Blue and the Peacock, shown in the figure.

The diversity of Lake Malawi is often attributed to the lake's size and to the variety of geographically isolated habitats in this massive body of water. At one time, the lake was much lower, and as water levels rose, the fish spread into new habitats. In new habitats, they evolved unique coloration and unique modes of feeding, reflected largely in differences in jaw structure.

One feature that sets the African cichlid apart from most other fish is its breeding behavior. Males and females engage in a fascinating courtship unlike virtually any other fish species. Males actively pursue females, then attract their attention by holding their fins erect and vibrating wildly. If the female is ready, she approaches, aligning herself toward the male's tail. She often pecks at prominent eggs spots on the male's anal fin. Aligned head to tail, the male and female form a circle with their bodies. Then, the female drops an egg. The male and female quickly change positions, with the male situated over the egg. Hovering above it, he releases sperm. The female then picks up the fertilized egg in her mouth and deposits another one. The mating dance goes on until all of the eggs have been laid and fertilized.

Most fish produce large quantities of eggs, which are fertilized en masse by the males. After fertilization, both parents generally abandon the eggs. The lack of parental care is offset by the large number of eggs. In sharp contrast, African cichlids produce only about 5 to 30 eggs. (Larger females may lay more.) Mouth-breeding may have evolved as a means of compensation for the small number of eggs, which are held in the female's mouth throughout development. Seven to 10 days after fertilization, the eggs hatch inside the mother's mouth. The tiny fry remain in her mouth, where they live off their yolk sac and feed off algae and bacteria in the water. Ten days after hatching, the young are fully formed but still remain in the protective custody of the mother's mouth another two weeks. At this time, the young are "spit out", but they are often free to seek shelter inside their mother's mouth should danger arise. An aquarist approaching the tank will often send the swarm of fry back into the mother's mouth in a flash as if they were being vacuumed up.

During this entire development period, females of a few species nibble on food without swallowing their young, but most African cichlid females refuse to eat. During their long fast, the female's abdomen becomes concave, but she refuses food—this despite the fact she is carrying a mouthful of eggs and fry, a delicacy among fish.

The care given to the young of African cichlids may be an evolutionary adaptation to the small number of eggs and to the conditions of the lake—perhaps a reflection of heavy predation when water levels were lower. Whatever the reason, the African cichlids have evolved one of the most amazing forms of maternal behavior in the animal kingdom, a source of fascination to countless fish fanciers the world over.

▷ **FIGURE 1 African Cichlids** (*a*) Cobalt Blue and (*b*) Peacock.

causes the muscle of the uterus to contract, is also found in the ejaculate. Muscular contraction is believed to be largely responsible for the movement of sperm up the female tract.

Interestingly, lymphocytes "patrol" the testes and seminal vesicles. Some lymphocytes are expelled with the ejaculate, explaining why AIDS is transmitted in the semen of men.

The paired **seminal vesicles,** which empty into the vasa deferentia, produce the largest portion of the ejaculate.

The **prostate gland,** whose contents empty directly into the urethra, surrounds the neck of the bladder. Routine medical examinations of men over the age of 45 show that nearly all of them have enlarged prostates. Prostatic enlargement results from the formation of small nodules inside the gland. These nodules form by the condensation of prostatic secretions inside the gland (▷ Figure 21–10). Although they usually cause no trouble, in some men the nodules grow quite large, blocking the flow of urine and making urination painful. In such cases, the prostate may be reamed out by a device inserted through the penis or may be removed through surgery. The prostate is also a common site for cancer in men and therefore should be checked regularly by a physician. The **Cowper's glands,** a

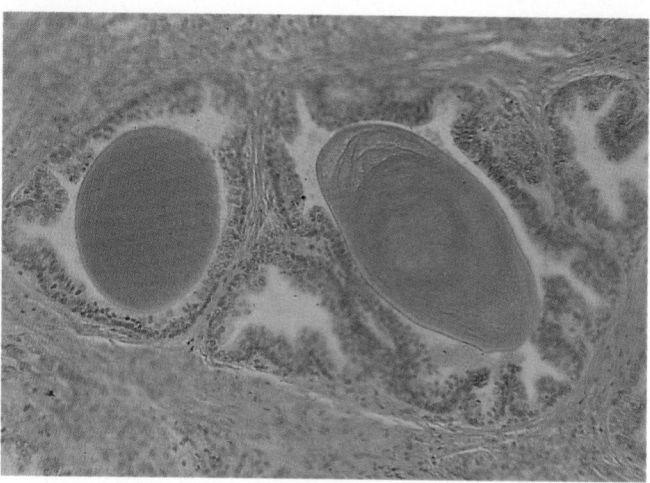

▷ **FIGURE 21–10 Prostatic Accretion**

pair of small glands located below the prostate on either side of the urethra, are the smallest of the sex accessory glands.

Sperm Are Formed from Stem Cells Known as Spermatogonia. Sperm are produced in the seminiferous tubules. ▷ Figure 21–11a shows a cross section through two seminiferous tubules. The lining of the wall of the tubule is known as the **germinal epithelium,** because its cells give rise to male germ cells, sperm. The formation

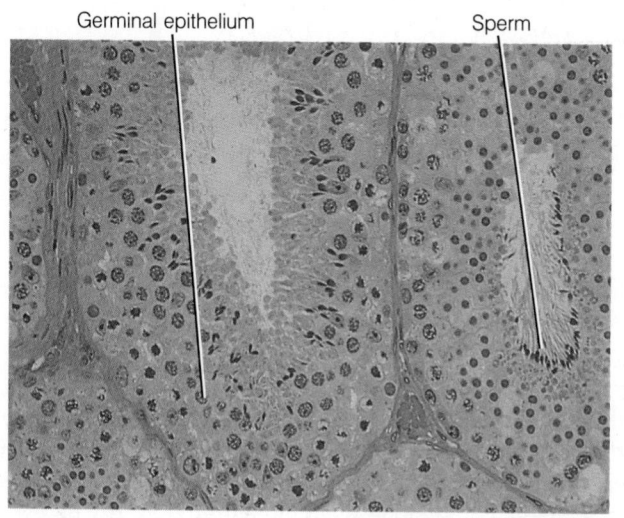

Germinal epithelium   Sperm

(a)

▷ **FIGURE 21–11 The Seminiferous Tubules** (*a*) A cross section through two seminiferous tubules showing the germinal epithelium and interstitial cells. (*b*) Details of spermatogenesis.

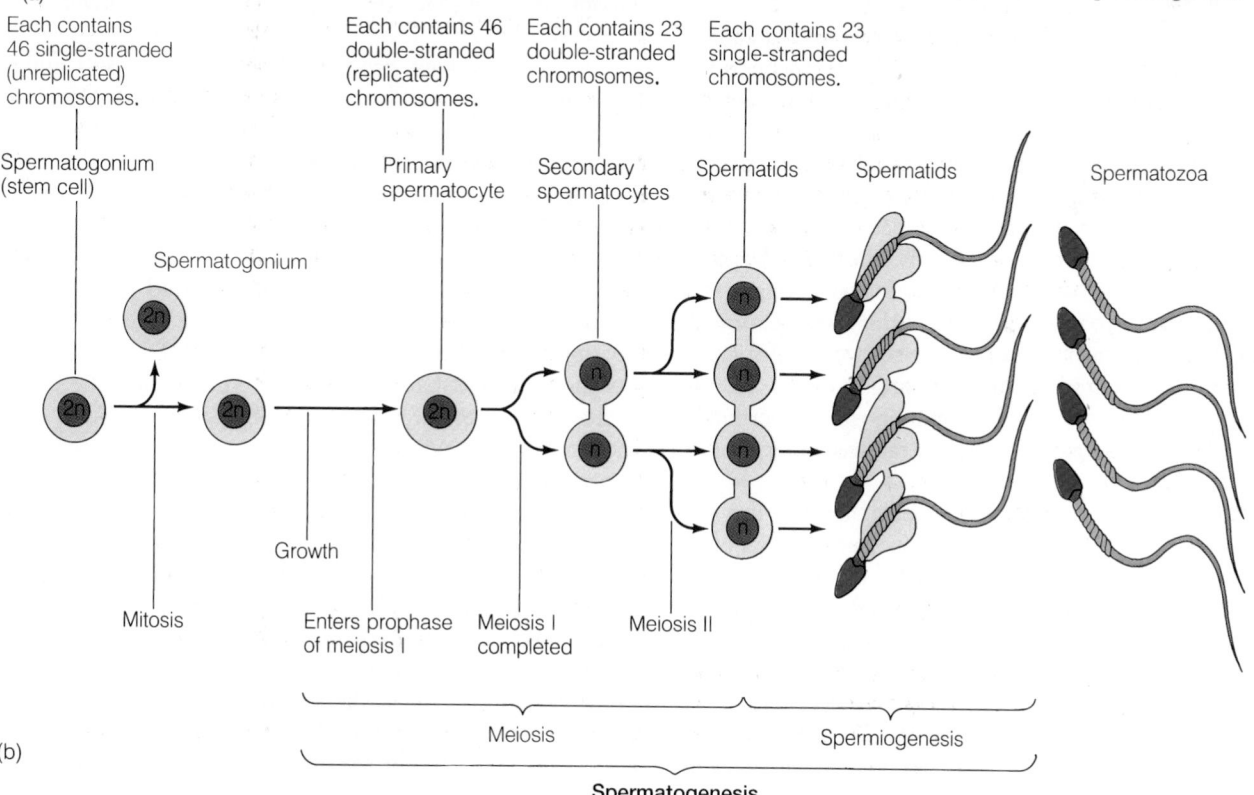

(b)

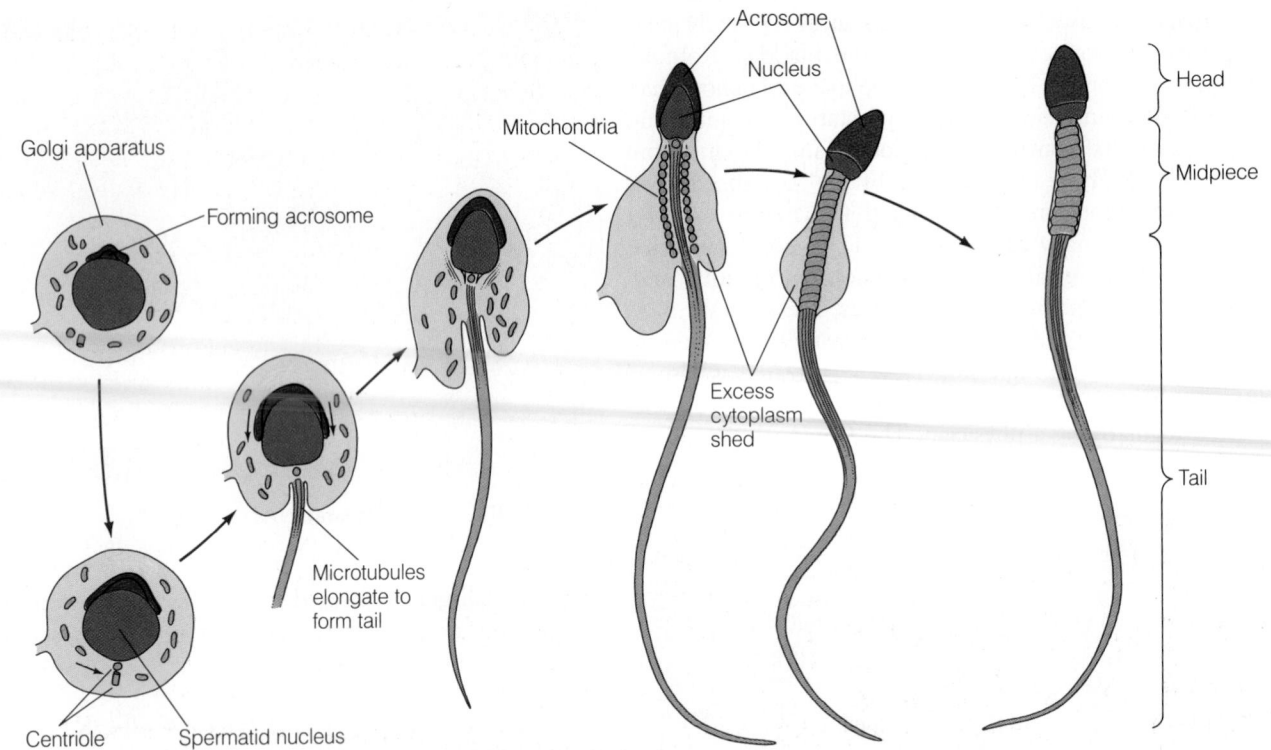

Golgi apparatus

Forming acrosome

Microtubules
elongate to
form tail

Centriole   Spermatid nucleus

Mitochondria

Acrosome

Nucleus

Excess
cytoplasm
shed

Head

Midpiece

Tail

▷ **FIGURE 21–12 Sperm Formation** Sperm form from spermatids in the germinal epithelium. Note the following changes: nuclear condensation, loss of cytoplasm, tail formation, alignment of the mitochondria, and acrosome formation.

of sperm is known as **spermatogenesis.** This phenomenon involves two processes: (1) a special type of cell division known as **meiosis** (described in Chapter 8) and (2) **spermiogenesis,** a process of cellular differentiation (Figure 21–11b).

Spermatogenesis begins with spermatogonia. Located in the periphery of the seminiferous tubule in the germinal epithelium, **spermatogonia** divide mitotically, ensuring a constant supply of sperm-producing cells. Some of the spermatogonia formed during cellular division, however, enlarge and become **primary spermatocytes.** Two meiotic divisions follow in the formation of sperm.

As you may recall from Chapter 8, the first division in meiosis is called meiosis I. During meiosis I, primary spermatocytes divide to form two **secondary spermatocytes** (Figure 21–11b). Each secondary spermatocyte contains 23 double-stranded chromosomes. During **meiosis II,** secondary spermatocytes divide, forming four spermatids. Each spermatid contains 23 single-stranded (unreplicated) chromosomes.

Spermatids soon develop into sperm. During this process, shown in ▷ Figure 21–12, the nuclear material of the spermatid condenses and most of the cytoplasm is lost, thus streamlining the cell. The sperm tail forms from the centriole, providing a means for locomotion. The mitochondria of the spermatid congregate around the first part of the tail, where they provide energy for propulsion. The Golgi apparatus enlarges and forms an enzyme-filled cap over the condensed nucleus, the head of the sperm. Called the **acrosome,** this cap will help the sperm digest its way through the coatings surrounding the ovum during fertilization (Chapter 22). The **spermatozoan,** or mature sperm, is a marvel of biological architecture. It is rid of excess cytoplasmic baggage and is streamlined for relatively swift movement.

On average, men produce 200 million to 300 million sperm every day and the average 3-milliliter ejaculate contains 240 million or more—about as many people as there are in the U.S. population. Such large numbers evolved because many sperm are eliminated as they travel through the female reproductive tract and because many sperm are required to dissolve away the ovum's outer coatings, so that one sperm cell can reach the ovum and fertilize it.

In humans, each sperm formed during meiosis contains 23 single-stranded (unreplicated) chromosomes—half the number in a normal somatic cell. Thus, when the sperm unites with an ovum (also containing 23 unreplicated chromosomes) during fertilization, they produce a zygote that contains 46 single-stranded chromosomes. One half of its chromosomes come from each parent.

Cells Lying Between the Seminiferous Tubules of the Testis, Called Interstitial Cells, Produce the Male Sex Steroid Testosterone.

The spaces between the seminiferous tubules contain clumps of large cells known as **interstitial cells** (▷ Figure 21–13). These cells produce **androgens,** steroids that exert a masculinizing effect. The most important androgen is **testosterone.** Tes-

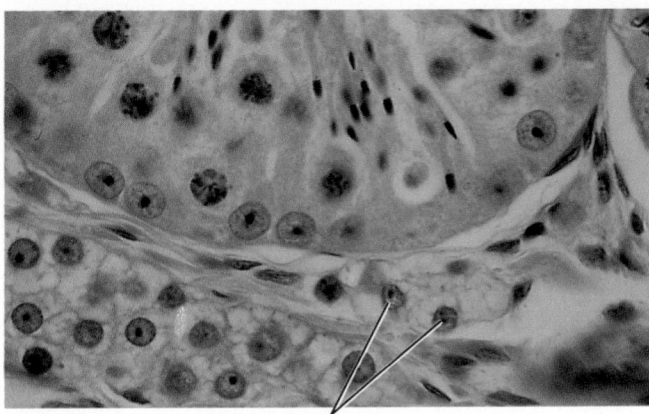

Interstitial cells

▷ **FIGURE 21–13 Interstitial Cells** Cross section of seminiferous tubules showing the interstitial cells.

tosterone diffuses out of the interstitial cells and into the seminiferous tubules, where it stimulates spermatogenesis (prophase I) and spermiogenesis, the maturation of sperm. In the absence of testosterone, sperm cell production declines, then stops, and the walls of the seminiferous tubules shrink.

▷ **FIGURE 21–14 Pattern Baldness** Some men are genetically predisposed to develop pattern baldness.

Testosterone is also transported in the bloodstream throughout the body, where it affects a variety of target cells. For example, it stimulates cellular growth in bone and muscle and accounts in part for the fact that men are generally more massive and taller than women. In addition, testosterone promotes facial hair growth and thickening of the vocal cords, typically giving men deeper voices than women. Testosterone stimulates growth of the laryngeal cartilage, producing the prominent bulge called the Adam's apple. It also stimulates cell growth in the skin, making most men's skin slightly thicker than women's.

Testosterone also affects the hair follicles on the heads of genetically predisposed men, causing pattern baldness (▷ Figure 21–14). It is not the absence of testosterone, as some believe, but the presence of testosterone and certain genes that lead to this condition.

Testosterone also stimulates the sebaceous glands of the skin. **Sebaceous glands** secrete oil (**sebum**) onto the skin, helping moisturize it (▷ Figure 21–15). During puberty (sexual maturation) in boys, testosterone levels rise dramatically, causing a marked increase in sebaceous gland activity. Dead skin cells may block the pores that normally carry the oil to the skin's surface (▷ Figure 21–16). As a result, sebum collects inside the glands. Bacteria on the skin often invade and proliferate in the small pools of oil, resulting in inflammation, pus formation, and swelling. The skin protrudes, forming an **acne pimple.**

Mild acne can be treated by washing the skin twice a day with unscented soap. Women should avoid makeup that has an oily base or use a nonoily type of foundation and wash their faces thoroughly each night. Sunlight also

▷ **FIGURE 21–15 Sebaceous Gland** The sebaceous glands are associated with hair follicles. They produce oil, which seeps onto the skin surface from the follicles.

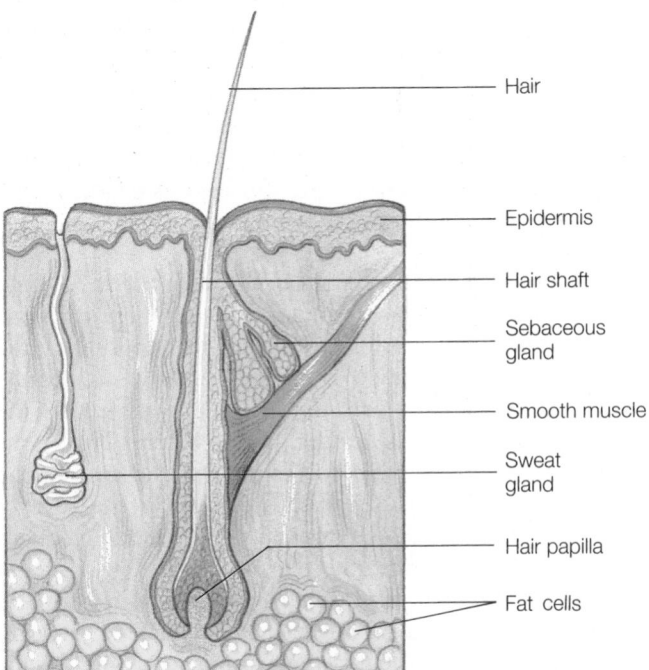

Hair

Epidermis

Hair shaft

Sebaceous gland

Smooth muscle

Sweat gland

Hair papilla

Fat cells

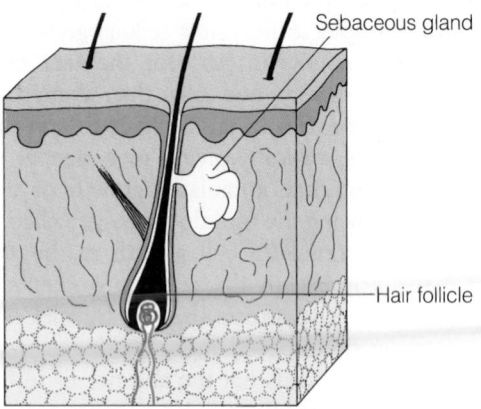

(a) Sebaceous glands associated with hair follicles secrete sebum, an oily substance that lubricates the skin and hair.

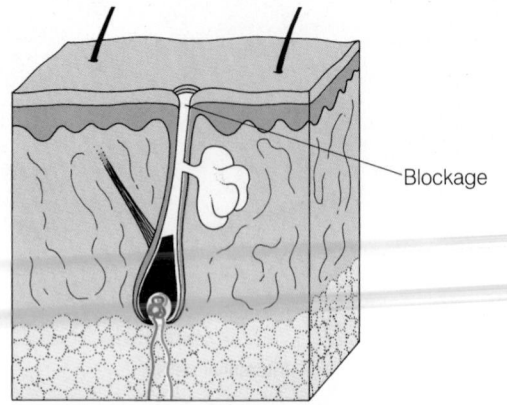

(b) A follicle may become blocked by excess sebum and dead skin cells. Unable to escape, the sebum builds up in the hair follicle.

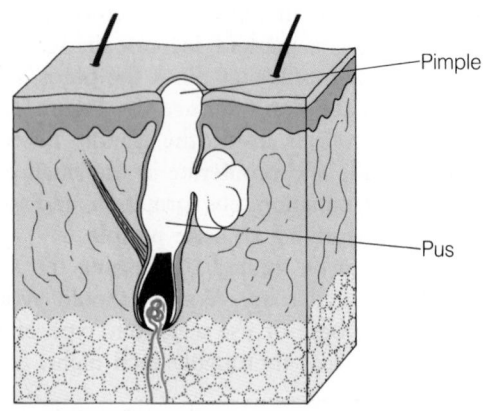

(c) Bacteria present on the skin may infect the sebum, causing inflammation; pus and swelling form an acne pimple.

▷ **FIGURE 21–16 Formation of an Acne Pimple** (*a*) Testosterone stimulates oil production in the sebaceous glands. (*b*) If the outlet is blocked, sebum builds up in the gland and (*c*) may become infected.

helps clear up acne, because it dries the oil on the skin and kills skin bacteria. Severe acne can be treated by special ointments and antibiotics.

The Penis Contains Erectile Tissue that Fills with Blood During Sexual Arousal. In order for fertilization to occur, many millions of sperm must be deposited in the female reproductive tract. The **penis** serves as the copulatory organ, which in humans deposits sperm inside the woman's vagina. As illustrated in ▷ Figure 21–17, the penis consists of a shaft of varying length and an enlarged tip, known as the **glans penis.** The glans is covered by a sheath of skin at birth, the **foreskin.** The foreskin gradually becomes separated from the glans in the first two years of life. At puberty, the inner lining of the foreskin begins to produce an oily secretion called smegma. Bacteria can grow in the protected, nutrient-rich environment created by the

foreskin, so special precautions must be made to keep the area clean.[1]

Because of potential health problems or religious reasons, parents may opt to have the foreskin removed in the first few days of their son's life. The operation, called **circumcision** (literally, "to cut around"), may help reduce penile cancer in men and may also reduce cervical cancer in the wives or sexual partners of circumcised men, as explained in the Point/Counterpoint in this chapter.

The penis becomes rigid, or erect, during sexual arousal. Erection is mediated by neurons belonging to the autonomic nervous system. The autonomic nervous system, discussed in Chapter 17, consists of two functionally and anatomically different divisions. The parasympathetic division is responsible for erection, among other functions. The sympathetic division is responsible for ejaculation.

During sexual arousal, nerve impulses in the parasympathetic division of the autonomic nervous system cause arterioles in the penis to dilate. Blood flows into a spongy **erectile tissue** in the shaft of the penis, making it harden. The growing turgidity compresses a large vein on the dorsal surface of the penis, blocking the outflow of blood and further stiffening the organ.

Coursing through the penis is the urethra, a duct that carries urine from the bladder to the outside of the body during urination. The urethra also transports semen, sperm, and secretions of the sex accessory glands during ejaculation.

Some men lose their ability to become erect or to sus-

[1] Parents must routinely clean the area in children once the foreskin becomes separated from the glans.

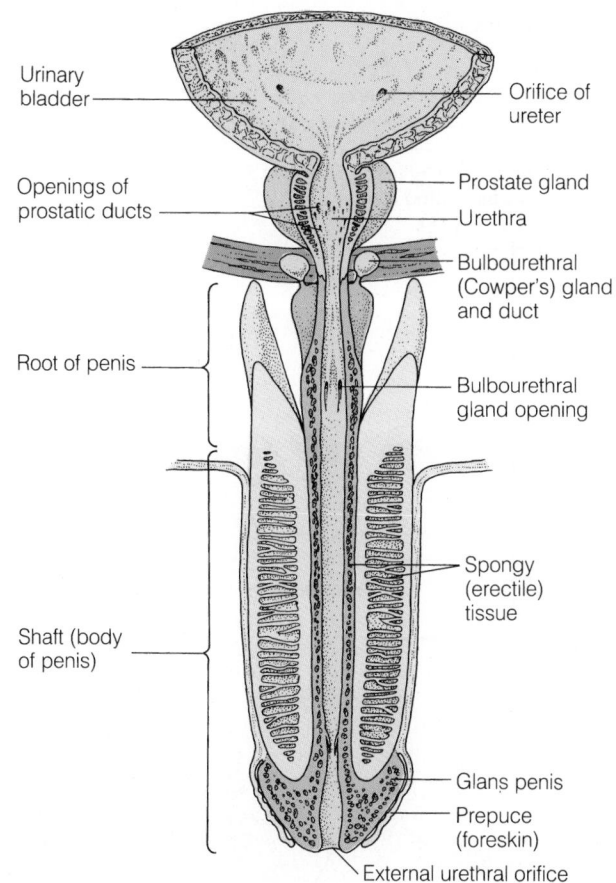

Urinary bladder

Openings of prostatic ducts

Root of penis

Shaft (body of penis)

Orifice of ureter

Prostate gland

Urethra

Bulbourethral (Cowper's) gland and duct

Bulbourethral gland opening

Spongy (erectile) tissue

Glans penis

Prepuce (foreskin)

External urethral orifice

▷ **FIGURE 21–17 Anatomy of the Penis** The penis consists principally of spongy tissue that fills with blood during sexual arousal. The urethra passes through the penis, carrying urine or semen.

tain an erection. This condition, known as **impotence,** may be caused by psychological, physical, or physiological problems. In most cases, pinpointing the exact cause is difficult. For example, marital conflict, stress, fatigue, and anxiety can all lead to impotence. If the problem is psychological, therapy is often advised. Patients with nerve damage, however, are not so lucky; permanent impotence is likely. Nerve damage may result from diabetes mellitus or from traumatic accidents. For patients with irreversible impotence, urologists can surgically insert an inflatable plastic implant in the penis. The penile implant is attached to a small, fluid-filled reservoir in the scrotum. The fluid is manually pumped into the implant upon demand, making the penis erect and permitting sexual intercourse.

Ejaculation Is a Reflex Mechanism. When sexual stimulation becomes intense, sensory nerve impulses traveling to the spinal cord activate motor neurons there. These neurons send impulses to the smooth muscle in the walls of the epididymes and vasa deferentia, causing them to contract. This action, in turn, propels sperm into the urethra. Nerve impulses emanating in the motor neurons of the spinal cord also stimulate the smooth muscle in the

walls of the sex accessory glands to contract, causing these glands to empty their secretions into the vasa deferentia and the urethra. The sperm and secretions from the sex accessory glands combine and form semen.

Semen is then propelled onward by smooth muscle contractions in the walls of the urethra, which cause the sperm to be released in spurts. The contractions in the urethra, like those in the epididymes, vasa deferentia, and sex accessory glands, are caused by nerve impulses from motor neurons involved in this spinal cord reflex.

According to the world-renowned sex researchers William Masters and Virginia Johnson, the male sexual response consists of four parts: the excitement phase, the plateau phase, the orgasm, and the resolution phase (explained in Table 21–2). Sexual arousal in the first two phases is accompanied by a tensing of the body muscles, rapid breathing, and increased blood pressure. Ejaculation occurs during the orgasm, or orgasmic phase. The "frenzy" of muscle contraction that occurs during ejaculation brings with it great pleasure.

Ejaculation is quickly followed by muscular and psychological relaxation (which explains why many men fall asleep after orgasm). This relaxation is part of the resolution phase. Soon after ejaculation, the arterioles in the penis, which were opened to let blood flow in during erection, begin to constrict. This action reduces the blood flow, and the penis becomes flaccid once again. In general, another erection is possible in younger men within 10 to 15 minutes. In older men, a repeat performance may take hours or even days.

The Male Reproductive System Is Controlled by Three Hormones, Testosterone, Luteinizing Hormone, and Follicle-Stimulating Hormone

As noted earlier, the testes produce sex steroid hormones— notably, testosterone. Besides influencing spermatogenesis, the male sex steroid hormones are responsible for **secondary sex characteristics**—that is, male physical features such as facial hair growth, greater muscle and bone development, and deeper voice.

Testosterone secretion by the interstitial cells is controlled by luteinizing hormone (LH). LH in males is also known as **interstitial cell–stimulating hormone (ICSH),** for obvious reasons. ICSH secretion, in turn, is controlled by a releasing hormone produced by the hypothalamus, known as gonadotropin-releasing hormone (GnRH).

As ▷ Figure 21–18 shows, the secretion of GnRH and ICSH is controlled by testosterone levels in the blood in a classic negative feedback loop. A decline in testosterone levels in the blood, for example, signals an increase in GnRH secretion, resulting in an increase in ICSH secretion. But when testosterone levels return to normal, GnRH and LH release subside. This feedback loop explains why athletes who use synthetic anabolic steroids (androgens) experience a decline in testicular size (Chapter 19).

The pituitary also produces the gonadotropin follicle-

TABLE 21–2 Male Responses During the Sexual Response Cycle

Excitement Phase
1. Vasocongestion (accumulation of blood) erects the penis.
2. Scrotal skin tightens.
3. Testes start to increase in size.
4. Testes and scrotum become elevated.
5. Nipples may become erect.
6. Some increase occurs in muscle tension, heart rate, and blood pressure.

Plateau Phase
1. There is a slight increase in the area of the glans penis.
2. Vasocongestion purples the glans.
3. Testes continue to elevate up into the scrotal sac until they are positioned close against the body.
4. Testes increase in size as much as 50%.
5. Cowper's glands secrete a few drops of fluid.
6. Possible sex flush and muscle tension are present, breathing is rapid, and heart rate increases (100 to 160 beats per minute).

Orgasm
1. Contractions of the vas deferens, seminal vesicles, ejaculatory duct, and prostate gland cause semen to collect in the base of the urethra.
2. Collection of semen in the base of the urethra produces feelings of ejaculatory inevitability.
3. The internal sphincter in the prostate contracts, preventing passage of urine.
4. The external sphincter in the prostate relaxes, allowing passage of semen.
5. Contractions in the urethra propel the semen out of the penis. The contractions occur four or five times at intervals of eight-tenths of a second.
6. Muscles go into spasm throughout the body, respiration increases, and blood pressure and heart rate reach a peak (about 180 beats per minute).

Resolution Phase
1. Body returns gradually to its prearoused state.
2. Male gradually loses his erection.
3. Testes and scrotum return to normal size.
4. Scrotum regains its wrinkled appearance.
5. Male enters a refractory period (unresponsive to further sexual stimulation).
6. Blood pressure, heart rate, and respiration become normal.
7. About one-third of males find their palms and soles or entire body covered with perspiration.

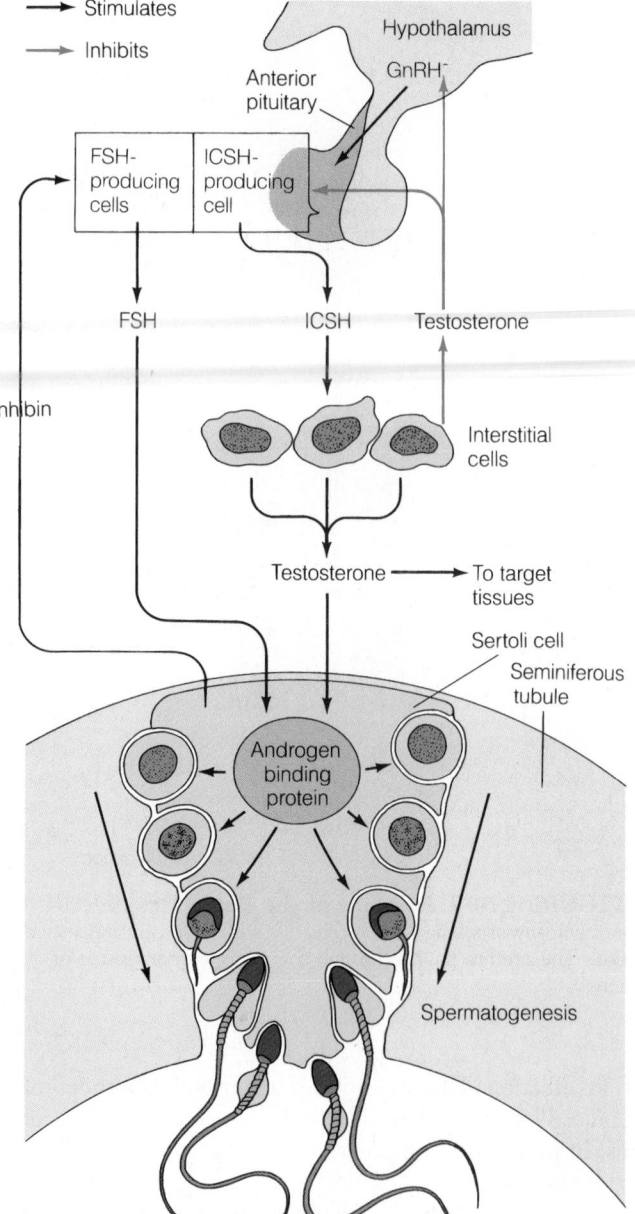

▷ **FIGURE 21–18 Hormonal Control of Testicular Function** Testosterone, FSH, and ICSH participate in a negative feedback loop. The testes also produce a substance called inhibin, which controls GnRH secretion.

stimulating hormone (FSH). As shown in Figure 21–18, FSH stimulates spermatogenesis in conjunction with testosterone. FSH does not act directly on the spermatogenic cells, however. Instead, it exerts its effects through another cell in the germinal epithelium of the seminiferous tubule, the Sertoli cell. **Sertoli cells,** shown in ▷ Figure 21–19, are large "nurse cells." The spermatogenic cells (spermatogonia, spermatocytes, and spermatids) divide and differentiate within folds in the plasma membrane of the Sertoli cell,

moving slowly to the surface of the germinal epithelium. The spermatids produced during spermatogenesis remain attached to the Sertoli cells and there differentiate into sperm.

FSH stimulates the Sertoli cells to produce a cytoplasmic receptor protein that binds to androgens, male sex steroid hormones. Called **androgen-binding protein,** this cytoplasmic receptor concentrates testosterone within the Sertoli cell. Testosterone, in turn, stimulates spermatogenesis.

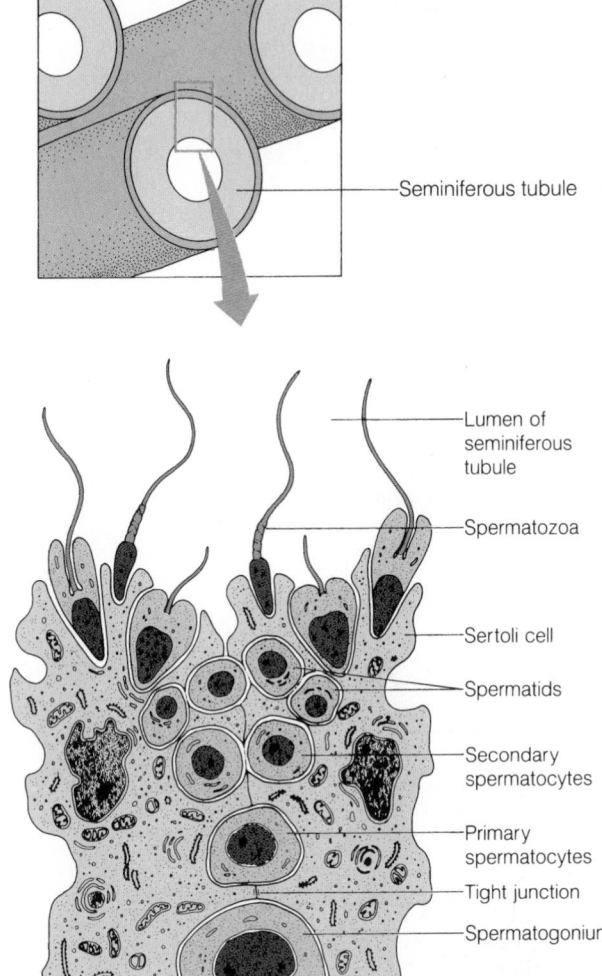

Seminiferous tubule

Lumen of
seminiferous
tubule

Spermatozoa

Sertoli cell

Spermatids

Secondary
spermatocytes

Primary
spermatocytes

Tight junction

Spermatogonium

▷ **FIGURE 21–19 Sertoli Cells** These cells encompass the spermatogenic cells as they develop in the germinal epithelium.

Like the release of LH, FSH secretion is controlled by GnRH produced by the hypothalamus. When GnRH is released from the hypothalamus, it travels to the anterior pituitary and there stimulates the release of FSH. Interestingly, the FSH-producing cells are also controlled by a peptide hormone produced by Sertoli cells (Figure 21–18). This hormone, called **inhibin,** apparently inhibits the activity of FSH-secreting cells in the anterior pituitary, thus blocking the action of GnRH. When inhibin levels are high, FSH secretion is low.

The Female Reproductive System Consists of Two Parts: the External Genitalia and the Genital, or Reproductive, Tract

▷ Figure 21–20 illustrates the female reproductive system, which consists of two parts: the external genitalia and the reproductive tract. The **female reproductive tract** consists of four structures: (1) the ovaries, (2) the uterine tubes,

(3) the uterus, and (4) the vagina. Table 21–3 offers a quick summary of the role of each structure.

The **uterus** is a pear-shaped organ about 7 centimeters (3 inches) long and about 2 centimeters (less than 1 inch) wide at its broadest point in nonpregnant women.[2] The wall of the uterus contains a thick layer of smooth muscle cells, the **myometrium.** The uterus houses and nourishes the developing embryo and fetus.

Attached to the uterus are two tubes, the **uterine tubes,** or **oviducts.** In humans, the uterine tubes are usually referred to as the **Fallopian tubes.** Ova are produced by the **ovaries,** paired, almond-shaped organs that are attached to the uterus by the ovarian ligaments. As Figure 21–20a shows, the ends of the uterine tubes are widened like a catcher's mitt and fit loosely over the ovaries. Currents created by cilia in the lining of the uterine tubes draw the ova inside and down the tubes to the uterus.

Fertilization occurs in the upper third of the uterine tubes. The fertilized ovum is then transported down the uterine tubes to the uterus. In the uterus, the embryo attaches to the lining, the **endometrium,** and embeds itself there, remaining for the duration of pregnancy.

At birth, the fetus is expelled from the uterus through the **cervix,** the lowermost portion of the uterus. As Figure 21–20 shows, the cervix protrudes into the **vagina,** a distensible, 3-inch, tubular organ that leads to the outside of the body.[3] At birth, the cervix stretches to allow the passage of the baby into the vagina. The vagina also serves as the receptacle for sperm during sexual intercourse. To reach the ovum, sperm must travel through a tiny opening and narrow canal of the cervix that leads into the lumen (cavity) of the uterus. From here, sperm move up both uterine tubes.

The **external genitalia** consist of two flaps of skin on either side of the vaginal opening (▷ Figure 21–21). The outer folds are the **labia majora.** These large folds of skin are covered with hair on the outer surface and contain numerous sebaceous glands on the inside. The inner flaps are the **labia minora.** Anteriorly, they meet to form a hood over a small knot of tissue called the **clitoris.** The clitoris

[2] The uterus is slightly larger in women who have had children and grows considerably during pregnancy to accommodate the growing embryo and fetus.

[3] The vagina is often called the birth canal.

TABLE 21–3 The Female Reproductive System	
COMPONENT	FUNCTION
Ovaries	Produce ova and female sex steroids
Uterine tubes	Transport sperm to ova; transport fertilized ova to uterus
Uterus	Nourishes and protects embryo and fetus
Vagina	Site of sperm deposition, birth canal

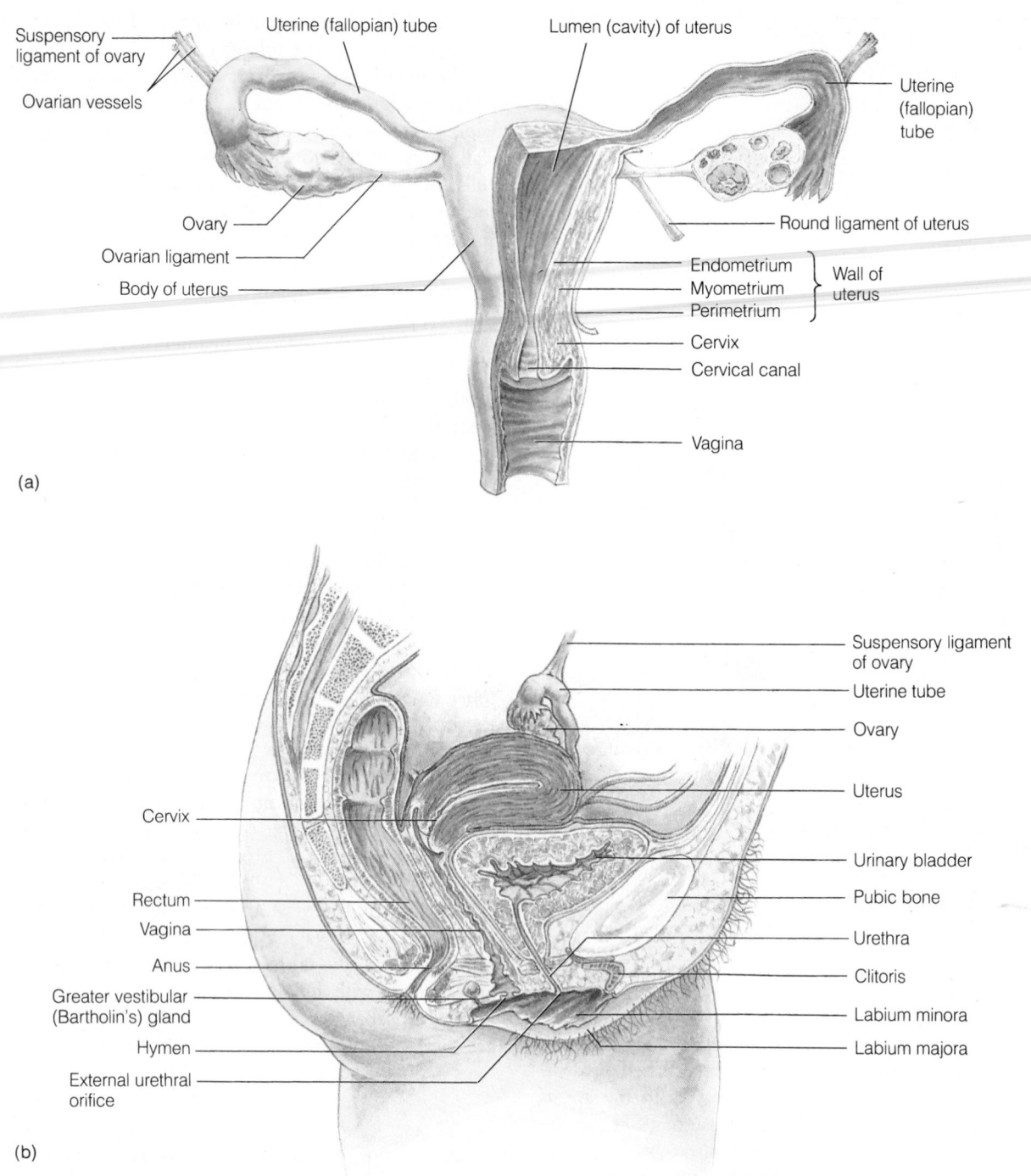

Suspensory
ligament of ovary

Ovarian vessels

Uterine (fallopian) tube

Lumen (cavity) of uterus

Uterine
(fallopian)
tube

Ovary

Ovarian ligament

Body of uterus

Round ligament of uterus

Endometrium
Myometrium
Perimetrium

Wall of
uterus

Cervix

Cervical canal

Vagina

(a)

Cervix

Rectum

Vagina

Anus

Greater vestibular
(Bartholin's) gland

Hymen

External urethral
orifice

Suspensory ligament
of ovary

Uterine tube

Ovary

Uterus

Urinary bladder

Pubic bone

Urethra

Clitoris

Labium minora

Labium majora

(b)

▷ **FIGURE 21–20 Anatomy of the Female Reproductive Tract** (*a*) Frontal view. (*b*) Midsagittal view.

consists of erectile tissue and is a highly sensitive organ involved in female sexual arousal. It is formed from the same embryonic tissue as the penis. In fact, a woman occasionally will be born with a greatly elongated clitoris.

The Ovaries Produce Ova and Release Them During Ovulation. During each menstrual cycle, one of the ovaries releases an ovum, the female gamete. The ovum oozes from the ovary and is drawn into the uterine tube.

The release of an ovum is called **ovulation.**[4] Ovulation occurs approximately once a month in women during their reproductive years—that is, from puberty (age 11–15) to menopause (age 45–55). Ovulation is temporarily halted when a woman is pregnant and may even be suppressed by emotional and physical stress.

[4]The release of the ovum is probably not an explosive event, although many women feel a sharp pain when it occurs.

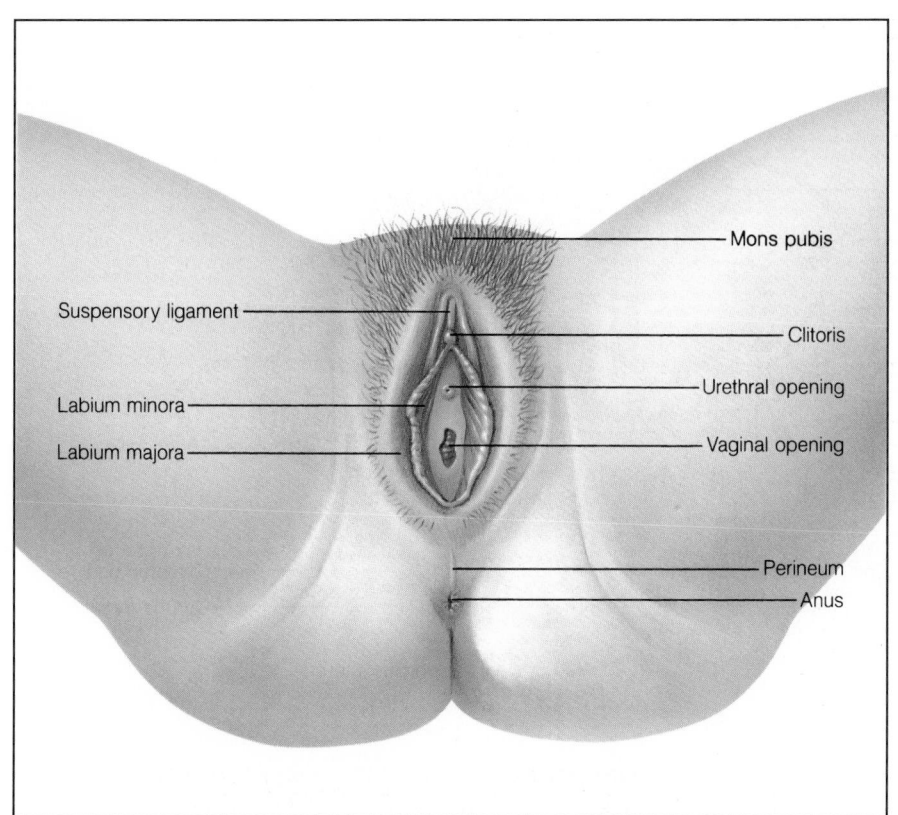

Mons pubis

Suspensory ligament

Clitoris

Urethral opening

Labium minora

Labium majora

Vaginal opening

Perineum

Anus

The structure of an ovary is shown in ▷ Figure 21–22. Several ovarian landmarks are visible. One is the germ cell. Germ cells, like those in the seminiferous tubules of the testes, undergo meiotic divisions to produce ova. This process is called **oogenesis.** The first germ cell in oogenesis is the **oogonium.** Containing 46 single-stranded (unreplicated) chromosomes, the oogonium enlarges and becomes a **primary oocyte** (▷ Figure 21–23). During this transition, the chromosomes replicate.

The formation of gametes in women also occurs by meiosis. Although meiosis is similar in oogenesis and spermatogenesis, there is one profound difference worth noting. That is, during each meiotic division in oogenesis, the nucleus divides in half, but the cytoplasm divides unequally (Figure 21–23). Therefore, the first meiotic division produces only one cell, the **secondary oocyte,** and a small package of discarded nuclear material containing 23 double-stranded chromosomes, called the **first polar body,** and a tiny amount of cytoplasm.

During the second meiotic division, the cytoplasm of the secondary oocyte also divides unequally. This "unequal division" results in the formation of an **ovum,** containing 23 single-stranded chromosomes and another "nuclear discard," the **second polar body.** The second meiotic division occurs only after a sperm penetrates the secondary oocyte. In humans and virtually all other animals, the first polar body usually does not divide.

Germ cells are housed in the ovary in special structures called follicles, shown in Figure 21–22a. A **follicle** consists of a primary oocyte surrounded by one or more layers of **follicle cells.** Follicle cells are derived from the loose connective tissue of the ovary. The most abundant of all follicles consists of a primary oocyte surrounded by an incomplete layer of flattened follicle cells. A large number of these follicles is found in the periphery of the ovary.

Each month, a dozen or so of these follicles begin to develop. During early development, the oocyte enlarges. The follicle cells divide and grow, eventually forming a complete layer around the oocyte. As follicular development proceeds, the follicle cells continue to divide, forming many layers.

The cells just outside the follicle form a layer known as the **theca folliculi** (literally, "follicular coat"). The theca soon organizes into two layers. The inner layer of cells contains many capillaries and produces a small amount of androgen (male hormone) that diffuses into the follicles, where it is converted into estrogen by the follicle cells. The outer layer is composed of tightly packed connective tissue cells.

In the largest follicles, a clear liquid begins to accumulate between the follicle cells. The fluid creates small spaces among the follicle cells, which enlarge as additional fluid is generated. Eventually, the cavities coalesce, forming one central cavity. At this point, the follicle is called an **antral follicle.**

As noted, a dozen or so follicles begin developing during each cycle. However, all but one usually degenerate. The follicle or follicles that escape degeneration, however, con-

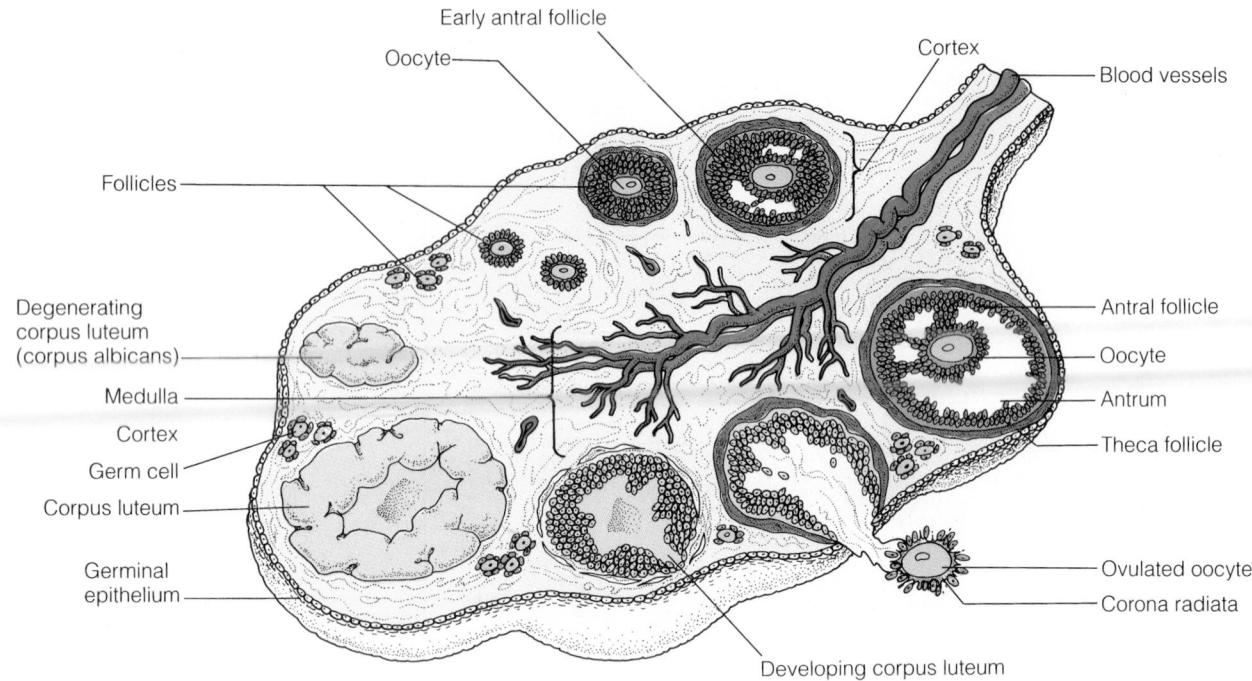

Early antral follicle

Oocyte

Cortex

Blood vessels

Follicles

Degenerating
corpus luteum
(corpus albicans)

Antral follicle

Oocyte

Antrum

Medulla

Cortex

Germ cell

Corpus luteum

Theca follicle

Germinal
epithelium

Ovulated oocyte

Corona radiata

(a)

Developing corpus luteum

▷ **FIGURE 21–22 Structure of the Ovary** (*a*) This drawing illustrates the phases of follicular development and also shows the formation and destruction of the corpus luteum (CL). Antral follicles give rise to the CL. A fully formed CL and antral follicle would not be found in the ovary at the same time. (*b*) Antral follicle.

tinue to enlarge. As fluid accumulates, the follicle bulges from the surface of the ovary like a pimple. The pressure exerted on the outside of the ovary causes the ovary's surface to stretch. Blood vessels supplying the tissue may be compressed, resulting in a region of cellular necrosis (death). This weakens the wall. Enzymes released from ovarian cells in the region are then thought to begin to digest the tissue at the weak point. Eventually, the wall of the follicle breaks down, and the oocyte is released.

Around the time of ovulation, the primary oocyte in the antral follicle completes the first meiotic division (Figure 21–23). It is then called a secondary oocyte. As illustrated in ▷ Figure 21–24, the secondary oocyte released from the ovary is surrounded by a layer of follicle cells. Immediately surrounding the oocyte is a fairly thick layer of gel-like material, called the **zona pellucida** ("clear zone"), shown in Figure 21–24. Incoming sperm must penetrate the follicle cell layer and the zona pellucida in order for fertilization to occur. Enzymes released from the acrosome of the many sperm that arrive at the site of fertilization "digest" the molecules that hold the follicle cells together, permitting sperm access to the ovum. Consequently, fertilization is a cooperative venture at first. Large numbers of sperm are needed to get through the follicle-cell barrier. Acrosomal enzymes of each sperm also digest the zona pellucida, helping each sperm penetrate this barrier. Thus, once sperm reach the zona pellucida, fertilization becomes a race. The first sperm to reach the plasma membrane of the ovum wins!

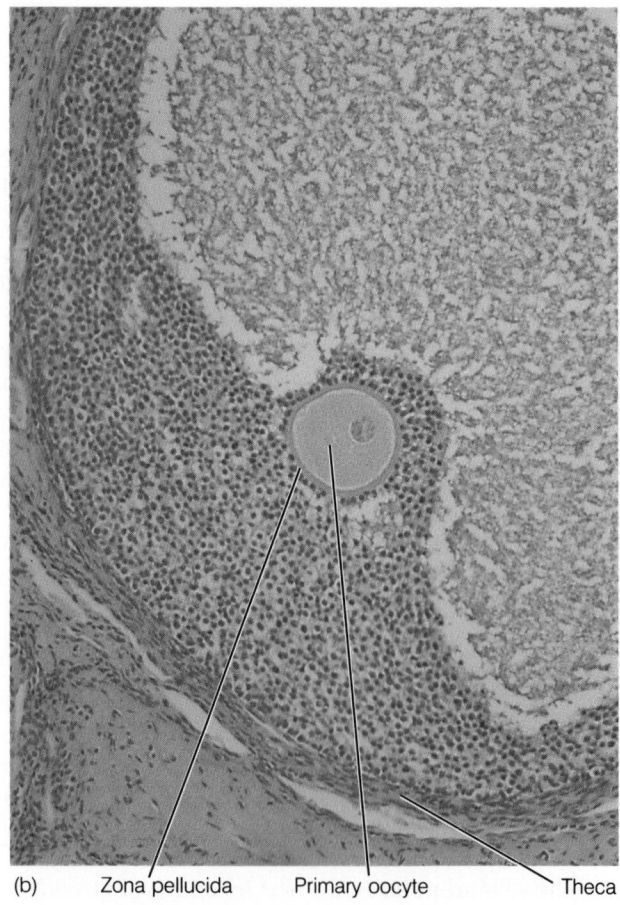

(b)   Zona pellucida   Primary oocyte   Theca

After ovulation, the ovulated follicle collapses. The remaining follicle cells enlarge and multiply. The thecal cells invade the interior of the collapsed follicle, where they proliferate and grow. The follicle and thecal cells transform

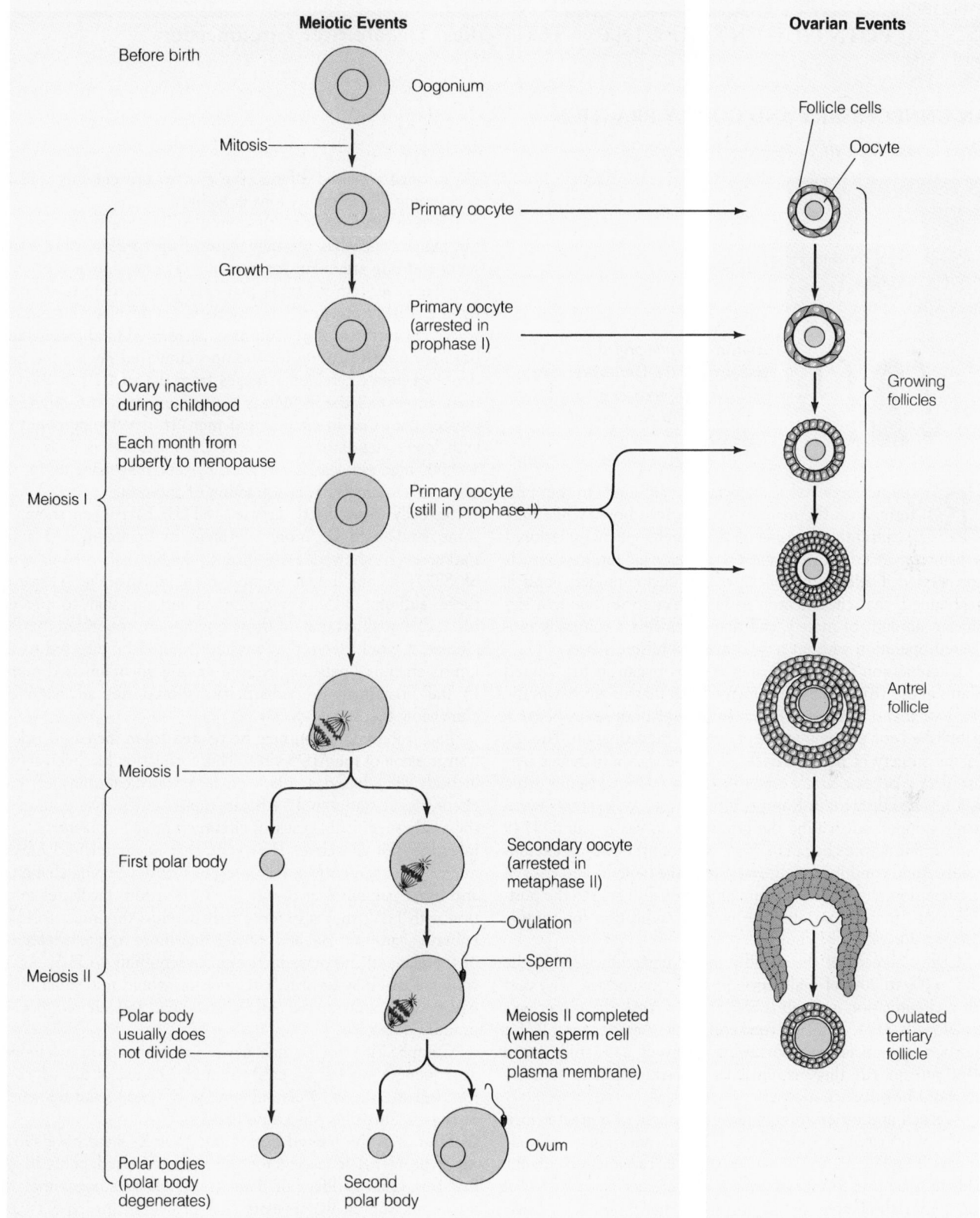

Meiotic Events

Before birth

Oogonium

Mitosis

Primary oocyte

Growth

Primary oocyte
(arrested in
prophase I)

Ovary inactive
during childhood

Each month from
puberty to menopause

Meiosis I

Primary oocyte
(still in prophase I)

Meiosis I

First polar body

Secondary oocyte
(arrested in
metaphase II)

Ovulation

Sperm

Meiosis II

Meiosis II completed
(when sperm cell
contacts
plasma membrane)

Polar body
usually does
not divide

Ovum

Polar bodies
(polar body
degenerates)

Second
polar body

Ovarian Events

Follicle cells

Oocyte

Growing
follicles

Antrel
follicle

Ovulated
tertiary
follicle

▷ **FIGURE 21–23 Oogenesis and Follicle Development**

the collapsed follicle into the **corpus luteum (CL)** ("yellow body")—so named because of the pigment it contains in cows and pigs, but not humans.

The CL produces two sex hormones, estrogen and pro-

gesterone, but the fate of the CL ultimately depends on the fate of the oocyte. If it is fertilized, the CL will remain active for several months, producing estrogen and progesterone needed for a successful pregnancy. If fertilization

AN UNNECESSARY AND COSTLY PRACTICE

Dr. Thomas Metcalf

Dr. Thomas J. Metcalf is a clinical associate professor of pediatrics at the University of Utah and practices at Willow Creek Pediatrics in Salt Lake City.

Routine neonatal circumcision continues to be performed on the majority of American boys. While data is currently emerging on the benefits of circumcision, I remain convinced that routine circumcision of the newborn is not needed if adequate hygiene of the uncircumcised penis is maintained, that circumcision costs the American public a significant amount of money, and that it remains a culturally motivated operation without a valid medical raison d'être.

Circumcision for nonreligious reasons began in the United States in the 1870s. A theory emerged that masturbation was the cause of many illnesses. Genital surgery of both sexes became established as a preventive and/or cure for masturbation. Though genital surgery of females declined, circumcision of males persisted, and became firmly established in the United States, even as it fell into disfavor in Europe. While most medical texts eventually stopped advocating the procedure, it was not until 1949 that various authors began to speak out against newborn circumcision. Amid continued controversy over the benefits and risks of circumcision, the medical community gradually came to the position that "there is no absolute medical indication for routine circumcision of the newborn."

Eighty percent of the world remains uncircumcised. In the U.S., 60% to 90% of male newborns are circumcised. The cost for a circumcision in Utah is $85; in New York the physician's fee alone is $175. Thus the total cost to the American public for newborn circumcisions is somewhere between $153 million and $360 million. Are there compelling reasons to circumcise every newborn boy at such a cost?

Wiswell and other authors have documented a greater incidence of urinary tract infections (UTIs) in uncircumcised male infants, roughly 1% to 2% versus 0.1% in circumcised infants. This may be due to more frequent and heavier colonization of the periurethral area by pathogens in uncircumcised males. However, circumcision cannot be thought of as protective against UTI—if a male infant presents with an illness and fever for which no obvious source of infection is found, the pediatrician should obtain a urine analysis to rule out the presence of a UTI, whether the child is circumcised or not. Circumcision does not do away with the need for this test. The cost to evaluate a possible UTI is currently $86. Given a UTI incidence of 1% to

2% in uncircumcised infants, the cost to prevent this UTI by performing 100 circumcisions is $8500.

There is little question that circumcision prevents cancer of the penis, which has mortality rate of up to 25%. This would seem a strong argument for neonatal circumcision as a preventive. However, other factors play a role. According to a report by the Task Force on Circumcision, in developed countries where circumcision is not routinely performed (and parents and boys are used to caring for the uncircumcised penis), the incidence of penile carcinoma ranges from 0.3 to 1.1 per 100,000 men, about half the incidence in uncircumcised U.S. men, but greater than that in circumcised men. In developing countries with lower standards of hygiene, the incidence is 3 to 6 per 100,000 men per year. Thus, good hygiene may make up for the effect of circumcision, at a fraction of the cost.

Sexually transmitted diseases (STDs) have been shown in some studies to be more prevalent in uncircumcised state. Parker et al. showed a significantly higher risk of four types of STDs among uncircumcised men in Australia. However, these authors did not recommend circumcision to prevent STDs. They stated that "if these findings are confirmed in other studies, it would seem that attention should be directed to the improvement of personal hygiene among uncircumcised men." In the final analysis, pediatricians should not advocate circumcision to prevent STDs.

Lack of circumcision may be related to an increased risk of transmission of the AIDS virus. This correlation has been shown in both clinic-based studies and in a statistical study of male circumcision status in 37 African capital cities. The authors of the study suggest that lack of circumcision is a cofactor in HIV infection, not causative. They state that uncircumcised African males "are apparently at increased risk of developing chancroid and other genital-ulcer disease," which in turn facilitates infection with HIV; they also write that perhaps the intact foreskin enhances viral survival and, finally, that more frequent infection of the glans of the penis increases susceptibility to HIV. Again, while all this may be true, circumcision would not be an effective way of solving the AIDS epidemic and should not be promoted as such.

Complications of routine newborn circumcision are indeed infrequent, and most are easily treated. The use of the xylocaine for local anesthesia, while inflicting pain, renders the remainder of the circumcision procedure painless.

Studies in the United States and New Zealand have shown more problems in caring for the uncircumcised penis in the first few years of life, but data from Europe suggest that the uncircumcised penis presents few problems for parents and boys used to dealing with it. In any event, problems that arise during care of the circumcised *or* uncircumcised penis are generally minor, requiring only one medical visit for correction.

In sum, routine newborn circumcision presents a very significant financial cost to society. Adequate hygiene of the uncircumcised penis appears to do away with the need for circumcision.

A SAFE AND BENEFICIAL PROCEDURE

Dr. Thomas Wiswell

Dr. Thomas E. Wiswell is chief of the Newborn Medicine Service at Walter Reed Army Medical Center in Washington, D.C.

Sixty percent to 90% of newborn boys (1.2 to 1.8 million) are circumcised annually in the United States. Several issues have convinced me of the benefits of the procedure: (1) the prevention of urinary tract infections (UTIs) and complications from them; (2) the prevention of penile cancer; (3) the lower incidence of sexually transmissible diseases in uncircumcised males; (4) the low risk for complications from the operation; (5) the greater incidence of penile problems among "intact" boys; and (6) recent evidence showing that circumcision protects against AIDS.

Circumcised boys are 10 to 39 times *less* likely to have UTIs during infancy. In a population of more than 400,000 children during a 10-year period, we found the higher incidence of UTIs in males compared with females to be primarily due to a lack of circumcision. From 1975 to 1984 the circumcision frequency rate decreased substantially from more than 86% to less than 71%. However, there was a concomitant significant rise in the number of male infants with UTIs, a rise solely attributable to the presence of more uncircumcised boys. We have also found uncircumcised boys aged 1 to 15 years to be 2.5 times as likely to develop UTIs compared with their circumcised counterparts. Urinary tract infections are not benign. More than 36% of 88 boys below 1 month of age with a UTI had concurrent infection in their blood stream. Furthermore, 3 of these infants had concomitant meningitis, 2 had renal failure, and 2 died. Littlewood has reported that 11% of children with a UTI during the first month of life may die. There are longer term effects of UTIs in children. Ten percent to 15% of infected infants will subsequently demonstrate kidney scarring. Of these infants, approximately 10% will develop high blood pressure, and 2% to 3% will ultimately require dialysis or kidney transplantation.

Penile cancer is the only malignancy that can be prevented categorically by a prophylactic procedure, neonatal circumcision. Of the more than 60,000 cases of penile cancer that have occurred in the United States since 1930, fewer than *10* have been in circumcised men. More than 1000 men develop penile cancer each year, and 225 to 317 annually die from it. The basic therapy for this malignancy is amputation of the penis.

Virtually all sexually transmissible diseases (STDs) have been found to occur more frequently among uncircumcised men. Fink has enumerated more than 60 references that have found STDs to occur more often among "intact" individuals. I am struck by the paucity of contrary reports. There is only one report of a venereal disease (nongonococcal urethritis) being more common in circumcised men. However, in this population more than 60% of the cases of another STD (gonorrhea) occurred in uncircumcised men.

Serious complications from routine foreskin removal are infrequent and relatively minor. We have recently examined a population of more than 100,000 circumcised boys and found complications in fewer than 2 per 1000. Two other large investigations reported complications from circumcision in 0.06% and 0.20% of circumcised boys, respectively. The majority of the complications are easily treated bleeding and minor infections. Atypical complications of the procedure (glans loss, staphylococcal scalded skin syndrome, etc.) occur infrequently and receive note due to their uniqueness. Death rarely occurs as a complication of circumcision. To date there have been a total of 3 reported deaths in the United States since 1954 that can be ascribed to neonatal circumcision. This contrasts sharply with the potentially preventable 7,500 to 11,500 deaths from penile cancer that have occurred during the same period.

Herzog and Alvarez found uncircumcised boys aged 4 months to 12 years to be more likely to have "penile problems" than were circumcised boys. Fergusson et al. similarly described uncircumcised boys as having more problems than their circumcised counterparts during the first 8 years of life. In both investigations, the "problems" largely consisted of balanitis (infection in and around the head of the penis) and phimosis (an abnormal constriction of the foreskin that prevents urine from being excreted). Finally, we found the risks from circumcision during the first month of life to be fewer than the risks from the uncircumcised state.

Reports have recently appeared suggesting that circumcision protects against AIDS. Simonsen et al. found uncircumcised men to be 9½ times as likely to become infected following exposure to the human immunodeficiency virus (HIV). A recent review in *Science* reported three corroborative studies (from Africa and the United States) that suggest a link between the lack of circumcision and AIDS virus infection.

As a pediatrician, I am a child advocate. I have pondered this issue for many years. I understand that we would have to circumcise "the many" to protect "the few." However, we have no way of identifying "the few." Neonatal circumcision is a rapid and generally safe procedure that must be performed by experienced caretakers. With the low complication rate and the many benefits of the procedure, I personally believe we should routinely circumcise newborn boys.

≈ SHARPENING YOUR CRITICAL THINKING SKILLS

1. Summarize the key points of each author, then list the data they use to support their main points.

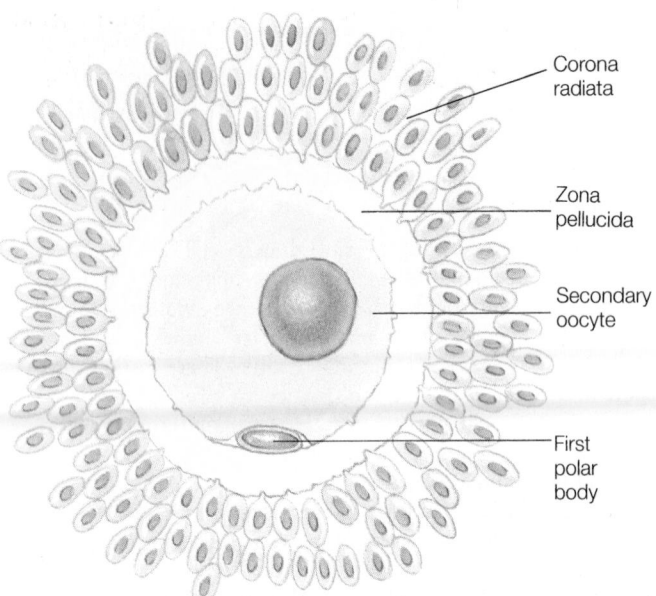

Corona
radiata

Zona
pellucida

Secondary
oocyte

First
polar
body

▷ **FIGURE 21–24 Ovulated Oocyte** Light micrograph of a recently ovulated follicle showing the corona radiata and zona pellucida.

does not occur, the CL soon disappears, and the process begins all over. (More on this process in Chapter 22.)

Hormonal Changes Taking Place in Women Produce Cyclic Variations in the Ovary and Uterus

Women of reproductive age undergo a series of interdependent hormonal, ovarian, and uterine changes each month known as the **menstrual cycle.** The length of the menstrual cycle varies from one woman to the next. In some it lasts 25 days, and in others it may last up to 35 days.[5] On average, however, the cycle repeats itself every 28 days. Ovulation usually occurs approximately at the midpoint of the 28-day cycle, or about 14 days before the onset of menstruation.

As noted above, the menstrual cycle involves three interdependent cycles (▷ Figure 21–25). The first is a hormonal cycle. The hormonal cycle, in turn, produces cyclic changes in the ovary (the ovarian cycle) and the uterus (the uterine cycle). Understanding the menstrual cycle is easiest if we begin with the hormonal and ovarian cycles.

The Hormonal and Ovarian Cycles. The first half of the menstrual cycle is known as the **follicular phase,** for during this time the follicles grow toward ovulation. The second half is called the **luteal phase,** so named because the corpus luteum forms during this time. To understand what drives these changes, we begin with the pituitary.

As shown in Figure 21–25a, FSH and LH are released

[5]The length of the menstrual cycle may also vary from month to month in the same woman.

from the anterior pituitary during the first half of the cycle, then peak in the middle, just before ovulation. FSH stimulates follicular development during the follicular phase by promoting mitosis of the follicle cells. LH stimulates estrogen production. Estrogen released from the follicle during the follicular phase also stimulates mitotic division of follicle cells, helping them grow.

As illustrated in ▷ Figure 21–26, estrogen is produced by the ovary from cholesterol. Its production begins in the cells of the theca. As illustrated, these cells convert cholesterol to androgen. This conversion requires small amounts of the hormone LH. Androgen diffuses out of the theca cells and into the follicle cells, where it is converted to estrogen. FSH is necessary for this conversion.

Estrogen secreted during the follicular phase of the menstrual cycle controls the release of both FSH and LH by inhibiting the release of GnRH from the hypothalamus in a classical negative feedback mechanism. As a result, throughout most of the follicular phase, LH and FSH levels are low and fairly constant. Just before ovulation, however, both LH and FSH secretion by the pituitary increases dramatically. These surges in LH and FSH secretion are the result of one of the body's rarest events, a positive feedback loop. Here's how it is triggered. As illustrated in Figure 21–25a, during the follicular phase of the menstrual cycle the amount of estrogen in a woman's blood creeps up fairly slowly. When estrogen reaches a certain critical level, however, both the hypothalamus and the anterior pituitary respond with a sudden outpouring of LH and FSH.

The LH surge has at least four effects: (1) it causes the primary oocyte to complete its first meiotic division, forming a secondary oocyte; (2) it stimulates the release of the enzymes that break down the ovarian wall, resulting in ovulation; (3) it stimulates estrogen production and release; and (4) it converts the collapsed follicle into a corpus luteum. The role of the preovulatory surge of FSH, if any, is not known.

During the second half of the menstrual cycle, the luteal phase, LH secretion gradually declines (Figure 21–25a). What LH is present during the luteal phase, however, stimulates the corpus luteum to produce estrogen and progesterone. As LH levels decline, though, estrogen and progesterone secretion from the CL also decline. If pregnancy does not occur, the CL stops producing hormones and degenerates (Figure 21–25b). Only a hormonal signal (discussed below) from the newly formed embryo can save it. Otherwise, the decline in levels of estrogen and progesterone permits a new cycle to begin.

The Uterine Cycle. The uterine lining, or endometrium, also undergoes cyclic changes during the menstrual cycle (Figure 21–25c). These changes result from cyclic changes in ovarian hormones, which, in turn, are controlled by changes in pituitary hormone secretion. As Figure 21–25c shows, the endometrium thickens throughout much of the cycle in preparation for pregnancy. In the absence of fertil-

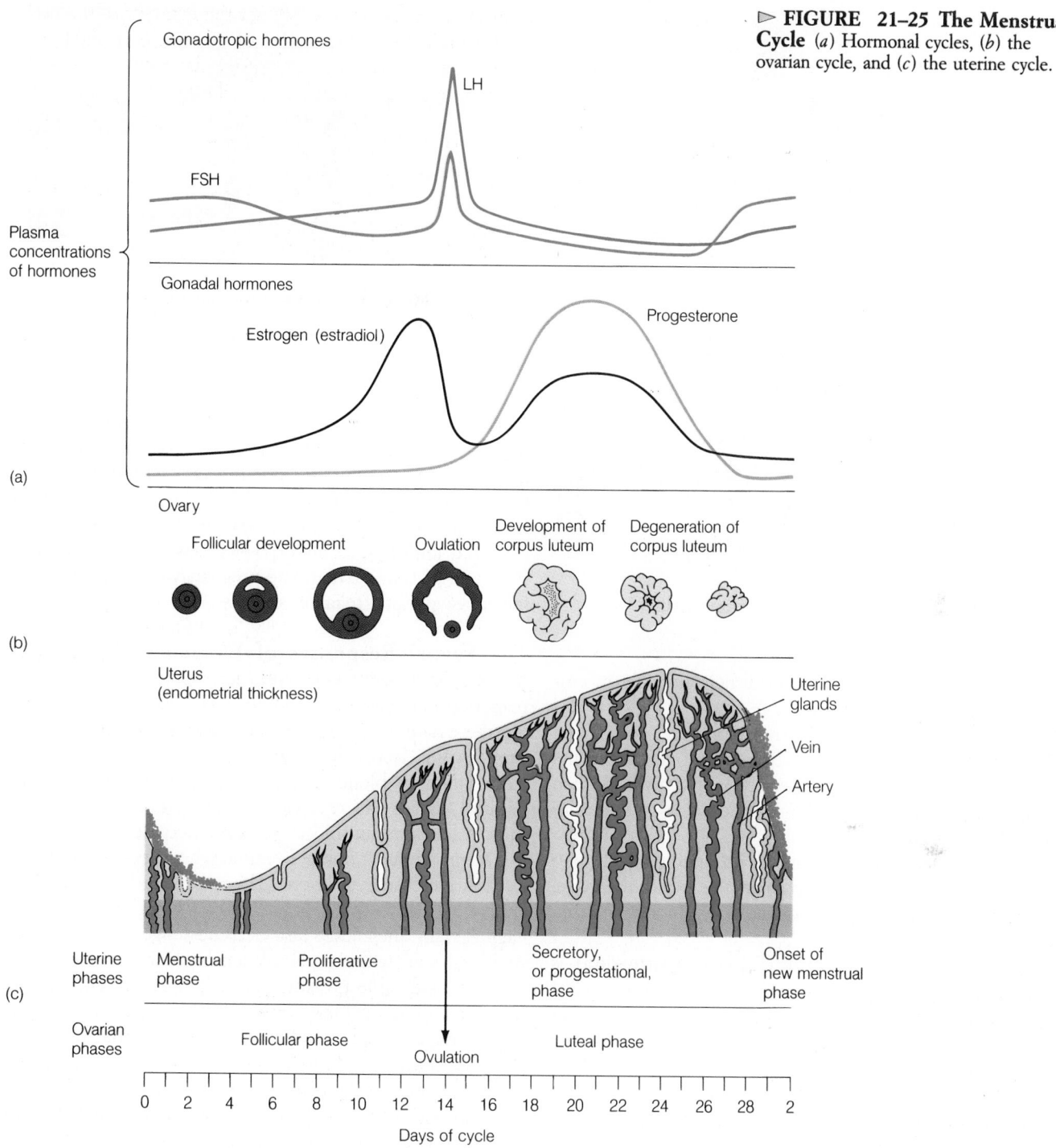

(a)

Gonadotropic hormones

LH

FSH

Plasma concentrations of hormones

Gonadal hormones

Estrogen (estradiol)

Progesterone

(b) Ovary

Follicular development

Ovulation

Development of corpus luteum

Degeneration of corpus luteum

(c) Uterus (endometrial thickness)

Uterine glands

Vein

Artery

| Uterine phases | Menstrual phase | Proliferative phase | | Secretory, or progestational, phase | Onset of new menstrual phase |

| Ovarian phases | | Follicular phase | Ovulation | Luteal phase | |

0 2 4 6 8 10 12 14 16 18 20 22 24 26 28 2

Days of cycle

ization, however, most of the thickened endometrium is shed, a process called **menstruation.**

To see how the endometrium responds to hormonal changes, we will begin on day 1 of the average 28-day menstrual cycle. Day 1 of the cycle is the first day of menstruation. During the first four or five days of the cycle, the uterine lining is shed. Tissue that formed in the previous menstrual cycle sloughs off from the lining and passes out of the uterus into the vagina along with a considerable amount of blood—on average, about 50 to 150 milliliters. The loss of blood during menstruation is the main reason

that women are more prone to develop anemia than men and why women should eat iron-rich foods or take iron supplements (Chapter 11).

As soon as the endometrium has been shed, the lining of the uterus begins to rebuild. Initial regrowth in the follicular phase is stimulated by ovarian estrogen. Estrogen stimulates the growth of glands (**uterine glands**) in the endometrium and also promotes cell division in the basal layer (deepest layer) of the endometrium—all that is left after menstruation. During the regrowth phase, also known as the **proliferative phase,** the uterine glands begin to fill

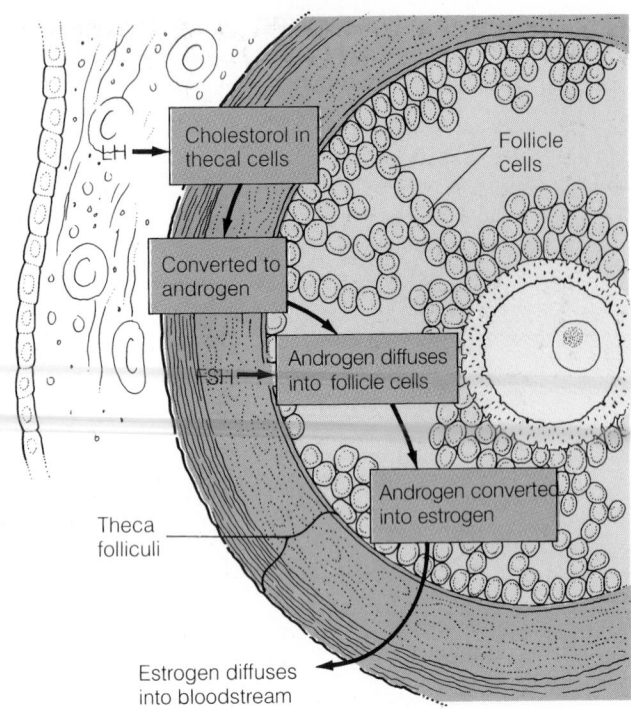

Cholesterol in
thecal cells

LH

Converted to
androgen

FSH

Androgen diffuses
into follicle cells

Androgen converted
into estrogen

Follicle
cells

Theca
folliculi

Estrogen diffuses
into bloodstream

▷ **FIGURE 21-26 Estrogen Production in the Ovary**

▷ **FIGURE 21-27 Home Pregnancy Tests**

with a nutritive secretion, which will help nourish an embryo should fertilization occur.

After ovulation the endometrium continues to thicken under the influence of both estrogen and progesterone. The uterine glands become distended with a glycogen-rich secretion. The last half of the uterine cycle is therefore called the **secretory phase.**

If fertilization does not occur, the uterine lining starts to shrink approximately four days before the end of the cycle, then begins to slough (pronounced "sluff") off, starting menstruation.

The shedding of the uterine lining (menstruation) is triggered by a decline in estrogen and progesterone concentrations in the blood. Progesterone acts as a uterine tranquilizer, inhibiting smooth muscle contraction in the myometrium. When progesterone levels fall, the uterus begins to undergo periodic contractions. These contractions propel the sloughed endometrium out of the uterus and are responsible for the cramps that many women experience during menstruation.

If fertilization occurs, the newly formed embryo produces a hormone called **human chorionic gonadotropin (HCG).** HCG is an LH-like hormone that stimulates the corpus luteum, maintaining its structure and function. When HCG is present, ovarian estrogen and progesterone continue to be secreted, and the uterine lining remains intact and suitable for embryonic attachment. A newly formed embryo arrives in the uterus and attaches to the lining approximately four days after fertilization. It then embeds in the thickened endometrium, from which it derives its nutrients.

HCG maintains the corpus luteum for approximately six months and shows up in detectable levels in a woman's blood and urine about 10 days after fertilization. Pregnancy tests available through a doctor's office or drugstore detect HCG in a woman's urine. The tests use a commercially prepared antibody to HCG, which binds to the hormone (▷ Figure 21–27). The home pregnancy tests are relatively inexpensive, fairly reliable, and fast.

The Sexual Responses of Women and Men Are Similar in Many Respects. Like that of men, the sexual response in women involves four stages (Table 21–4). During the excitement phase, cervical glands produce a secretion that lubricates the vagina. Erectile tissue beneath the labia fills with blood, and the external genitalia expand. Arousal causes the nipples to become erect. Heart rate, muscle tension, and blood pressure increase.

During the next phase, breathing, heart rate, and muscle tension increase even more. The outer third of the vagina constricts due to the engorgement of blood vessels in the vaginal wall. The clitoris also becomes engorged. The clitoris is invested with numerous sensory nerve fibers, and it yields considerable pleasure when stimulated. The nipples become more erect, and the glands near the opening of the vagina are activated, producing a lubricant that facilitates the insertion of the penis.

Sexual arousal can lead to orgasm, rhythmic contractions of the uterus and vagina. These contractions are accompanied by intense physical pleasure. Women often report feelings of warmth throughout their bodies after orgasm.

After orgasm, the body relaxes. Blood drains from the clitoris and labia. Blood pressure, heart rate, and respiration return to normal. Unlike men, women generally do not experience a refractory period (a time when orgasm is not possible) after orgasm. Thus, a woman can sometimes experience multiple orgasms during sexual intercourse.

Estrogen and Progesterone Help to Regulate the Menstrual Cycle, but also Exert Numerous other Effects. Like testosterone in boys, estrogen secretion in

TABLE 21–4 Female Responses During the Sexual Response Cycle

Excitement Phase
1. Vaginal lubrication begins.
2. Vasocongestion swells the external genitalia.
3. Labia majora flatten and retract from the vaginal opening.
4. Inner two-thirds of the vagina expands.
5. Vaginal walls thicken.
6. Breasts enlarge, and blood vessels near the surface become more prominent.
7. Nipples become erect.
8. Muscle tension, heart rate, and blood pressure increase somewhat.

Plateau Phase
1. Vasocongestion produces a narrow vaginal pathway in the outer third of the vagina.
2. Inner part of the vagina expands fully.
3. Uterus becomes elevated.
4. Clitoris withdraws beneath the clitoral hood.
5. Rosy appearance may occur on the stomach, thighs, and back.
6. Nipples become more erect.
7. Muscles tense, breathing is rapid, and heart rate increases (100 to 160 beats per minute).

Orgasm
1. Swelling of the tissues of the outer part of the vagina constricts the vaginal opening.
2. Contractions begin in the outer third of the vagina. First contractions may last 2 to 4 seconds, and later ones may last 3 to 15 seconds. They occur at intervals of 0.8 of a second.
3. Inner two-thirds of the vagina expands slightly.
4. Uterus contracts.
5. Muscles go into spasm throughout the body, respiration increases, and blood pressure and heart rate reach a peak (about 180 beats per minute).

Resolution Phase
1. Blood is released from engorged areas.
2. Rosy appearance disappears.
3. Clitoris descends to normal position.
4. Vagina, uterus, and labia gradually shrink to normal size.
5. Blood pressure, heart rate, and respiration become normal.
6. About one-third of females find their palms and soles or entire body covered with perspiration.

girls increases dramatically at puberty. As the levels of estrogen in the blood increase, the hormone begins to stimulate follicle development in the ovaries. Estrogen is also an anabolic hormone—a hormone that stimulates anabolic (synthesis) reactions. Estrogen's anabolic effects result in the pubertal growth of the external genitalia (for example, the breasts). Estrogen also stimulates growth of the internal reproductive structures: the uterus, uterine tubes, and vagina.

Estrogen promotes rapid bone growth in the early teens. Because estrogen secretion in girls usually occurs earlier

than testosterone secretion in boys, girls typically go through their growth spurt earlier. However, estrogen also stimulates the closure of the epiphyseal plates, described in Chapter 19, ending the growth spurt fairly early. Most girls, in fact, reach their full adult height by the age of 15 to 17. Boys experience their most rapid growth later in adolescence and continue growing until the age of 19 to 21.

Estrogen also stimulates the deposition of fat in women's hips, buttocks, and breasts, giving the female body its characteristic shape. In addition, estrogen stimulates the growth of ducts in the mammary glands. Progesterone works with estrogen to stimulate breast development.

Premenstrual Syndrome Is a Condition Afflicting Many Women. For reasons not yet fully understood, many women suffer premenstrual irritability, depression, fatigue, and headaches. Many complain of bloating, swelling and tenderness of the breasts, tension, and joint pain. Together these symptoms constitute a condition called **PMS,** or **premenstrual syndrome.** Premenstrual syndrome is a clinically recognizable condition characterized by one or more of these symptoms. PMS strikes 4 of every 10 women of reproductive age. All told, women report more than 150 different physical and psychological symptoms that emerge before menstruation begins.

Despite the prevalence of PMS, it may be years before medical scientists can pinpoint the cause or causes of PMS. Nevertheless, dozens of "cures," ranging from massive doses of progesterone to vitamin B-6 to L-tryptophan, have been prescribed by clinics specializing in PMS.[6] Buyers should beware, however, for very little good scientific evidence is available to indicate which, if any, of the "cures" really work. Most of the evidence consists of testimonials—individual accounts. The critical thinking skills you learned in Chapter 1 suggest that anecdotal information such as this is no substitute for controlled studies.

Work is now under way to test various treatments, but the results are not expected for several years. In the meantime, physicians recommend that women suffering from PMS see their family doctor to be certain that the symptoms are not in fact caused by some other problem.

Menopause Is the Cessation of Menstruation. The menstrual cycle continues throughout the reproductive years, but after a woman reaches 20, the ovaries gradually start to become less responsive to gonadotropins (▷ Figure 21–28). Responsiveness declines slowly. But as it declines, estrogen levels gradually fall off. Ovulation and menstruation become increasingly more erratic and eventually stop. This cessation is called **menopause.**

Menopause is attributed to a reduction in the number of ovarian follicles. At about age 45, most of the follicles

---

[6]As noted earlier in the book, L-tryptophan has been implicated in the paralysis of body muscles and death of over a dozen people. At this writing, preliminary studies suggest that the amino acid may not be at fault but, rather, that the pills taken by victims contained a contaminant that caused the ill effects.

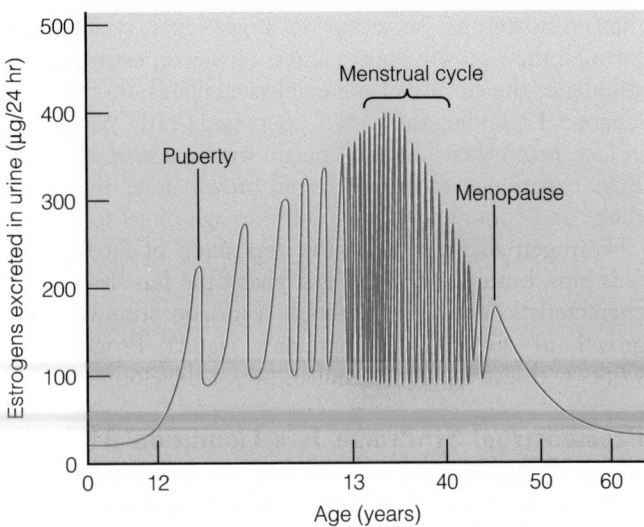

▷ **FIGURE 21–28 Ovarian Hormone Secretion** Notice that at about age 20, ovarian estrogen secretion begins a gradual decline. Adapted with permission from A. C. Guyton, *Textbook of Medical Physiology*, 7th ed. (Philadelphia: W. B. Saunders Company, 1986), Figure 81–8, p. 979.

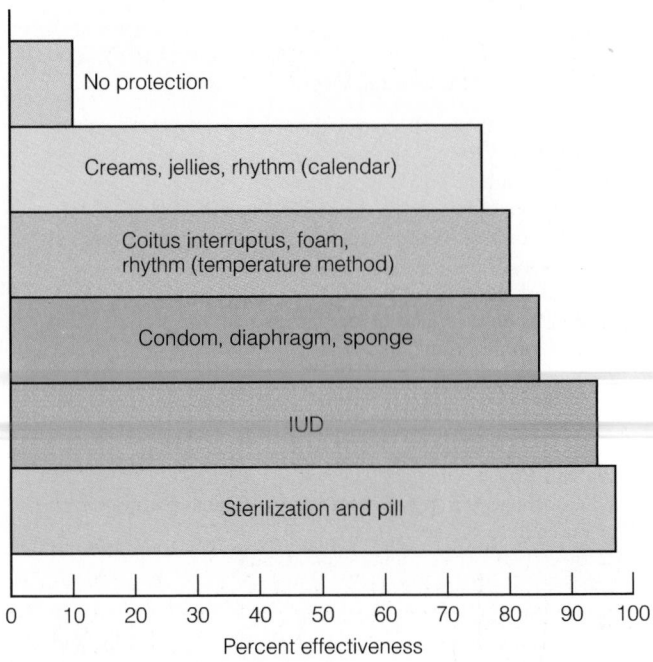

▷ **FIGURE 21–29 Effectiveness of Contraceptive Measures** Percent effectiveness is a measure of the number of women in a group of 100 who will not become pregnant in a year.

that were in the ovary at puberty have been stimulated to grow and have either degenerated or ovulated. Consequently, FSH and LH from the pituitary have no follicles to stimulate, and the production of ovarian estrogen plummets.

Ovulation and menstruation generally cease between the ages of 45 and 55. The dramatic alteration in the hormonal climate results in several important physiological changes. For example, the decline in estrogen secretion causes the breasts and reproductive organs, such as the uterus, to begin to atrophy. Vaginal secretions often decline, and in some women sexual intercourse may become painful.

The decline in estrogen levels may also result in behavioral disturbances. Many women become more irritable and suffer bouts of depression. Three quarters of all women suffer "hot flashes" and "night sweats" induced by massive vasodilation of vessels in the skin. Fortunately, these symptoms usually pass.

As noted in Chapter 19, declining estrogen levels also accelerate osteoporosis. To counter osteoporosis and other impacts of the decline in ovarian function, physicians sometimes prescribe pills containing small amounts of estrogen and progesterone, as well as a program of exercise and a diet rich in calcium and vitamin D (see Health Note 19–1).

≋ BIRTH CONTROL

Few topics in modern society create as much controversy as birth control. **Birth control** is any method or device that prevents births. Birth control measures fall into two broad categories: (1) **contraception,** ways of preventing pregnancy, and (2) **induced abortion,** the deliberate expulsion of a fetus.

Contraceptive Measures Help Prevent Pregnancy.

▷ Figure 21–29 summarizes the effectiveness of the most common means of contraception. Effectiveness is expressed as a percentage. A 95% effectiveness rating means that 95 women out of 100 using a certain method in a year will not become pregnant.

Abstinence. Not listed in the figure is a form of birth control that many of us forget to talk about, **abstinence,** refraining from sexual intercourse. This form of birth control is appropriate for many people and should not be overlooked as a strategy of reducing unwanted pregnancy and preventing the transmission of AIDS and other diseases (discussed later).

Sterilization. Except for complete abstinence, sterilization and the pill (discussed shortly) are the most effective birth control measures (Figure 21–29). In 1982, sterilization became the leading method of contraception practiced by married couples in the United States.

In women, sterilization is performed by cutting the uterine tubes. This process is called **tubal ligation** and rarely involves an overnight stay in the hospital (▷ Figure 21–30a). Surgeons usually make two small incisions in the abdomen just beneath the navel. An instrument called a **laparoscope** is inserted through each incision. The surgeon locates the uterine tubes through a special lighted viewing lens. An attachment to the laparoscope is then used to cut the uterine tubes. Surgeons then either tie off the cut ends, or cauterize them—that is, burn them using an electrical current. In some cases, surgeons clamp the uterine tubes shut with plastic or metal rings.

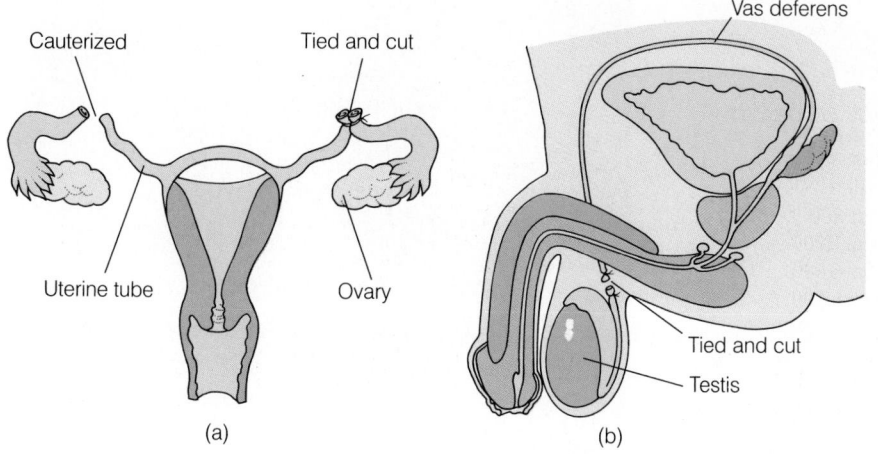

> **FIGURE 21–30 Sterilization Methods** (*a*) In a tubal ligation, the uterine tubes are cut, then tied off or cauterized. (*b*) In a vasectomy, the vasa deferentia are cut, then tied off.

Male sterilization requires a far less traumatic surgical procedure called a **vasectomy,** which can be carried out in a physician's office under local anesthesia (Figure 21–30b). To perform a vasectomy, a physician makes a small incision in the scrotum. Each vas deferens is exposed, then cut, and the free ends are tied off or cauterized.

A quick review of male anatomy will show that vasectomies only prevent the sperm from passing into the urethra during ejaculation. They do not impair sex drive and have virtually no effect on ejaculation, because 99% of the volume of the ejaculate is produced by the sex accessory glands.

Vasectomy and tubal ligation are essentially irreversible. However, special surgical methods, called **microsurgery,** can be used to reconnect the uterine tubes and the vasa deferentia. This procedure is costly and not always successful.

The Pill. The **birth control pill** is the most effective *temporary* means of birth control available (▷ Figure 21–31). Birth control pills come in several varieties, but the most common is the combined pill. It contains synthetic estrogen and progesterone, which inhibit the release of LH and FSH through the pituitary and hypothalamus. As a result, follicle development and ovulation are inhibited. A minipill containing progesterone alone is also available. Even though it is less effective than the combined pill, the minipill is more suitable for some women because it results in fewer side effects.

Birth control pills must be taken throughout the menstrual cycle. Skipping a few days may release the pituitary and hypothalamus from the inhibitory influences of estrogen and progesterone, resulting in ovulation and possible pregnancy.

Although effective, birth control pills have some adverse health effects worth noting. Even though the incidence of these adverse side effects is small, a woman considering different birth control options should carefully study them before making a decision.

One "adverse" effect is death. Table 21–5 compares the risk of death from taking birth control pills (and using other contraceptives) with a number of common risk factors. As shown, the risk of a nonsmoker dying from taking birth control pills is 1 in 63,000 in any given year, whereas the risk of dying in an auto accident is 1 in 6,000.

Death from birth control pills may result from a heart attack, stroke, or blood clot. The incidence of these life-threatening side effects is lowest in nonsmoking women under the age of 30. To reduce the risk even more, pharmaceutical companies have dramatically lowered the estrogen content of the combined pill, because estrogen is responsible for most of the adverse side effects.

Early studies showed a positive correlation between the use of birth control pills and cancers of the breast and cervix. More recent studies suggest that the new generation

> **FIGURE 21–31 The Birth Control Pill** One of the most effective means of birth control, the pill consists of a mixture of estrogen and progesterone, which is taken throughout the menstrual cycle to block ovulation. Birth control pills are packaged in numbered containers to help women keep track of them.

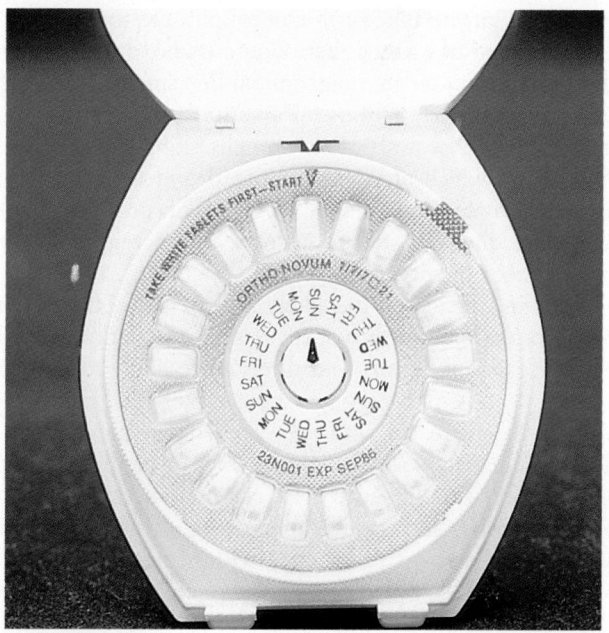

RISK	CHANCE OF DEATH IN A YEAR (UNITED STATES)
Smoking	1 in 200
Motorcycling	1 in 1,000
Automobile driving	1 in 6,000
Power boating	1 in 6,000
Rock climbing	1 in 7,500
Playing football	1 in 25,000
Canoeing	1 in 100,000
Using tampons (toxic shock syndrome)	1 in 350,000
Contracting reproductive tract infections through sexual intercourse	1 in 50,000
Preventing pregnancy:	
Oral contraception—nonsmoker	1 in 63,000
Oral contraception—smoker	1 in 16,000
Using intrauterine devices (IUDs)	1 in 100,000
Using barrier methods	None
Using natural methods	None
Undergoing sterilization:	
Laparoscopic tubal ligation	1 in 20,000
Hysterectomy	1 in 1,600
Vasectomy	None
Pregnancy:	1 in 10,000
Nonlegal abortion	1 in 3,000
Legal abortion:	
Before 9 weeks	1 in 400,000
Between 9 and 12 weeks	1 in 100,000
Between 13 and 16 weeks	1 in 25,000
After 16 weeks	1 in 10,000

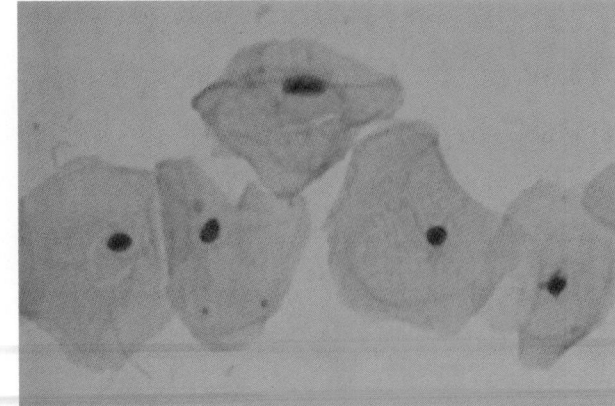

(a)

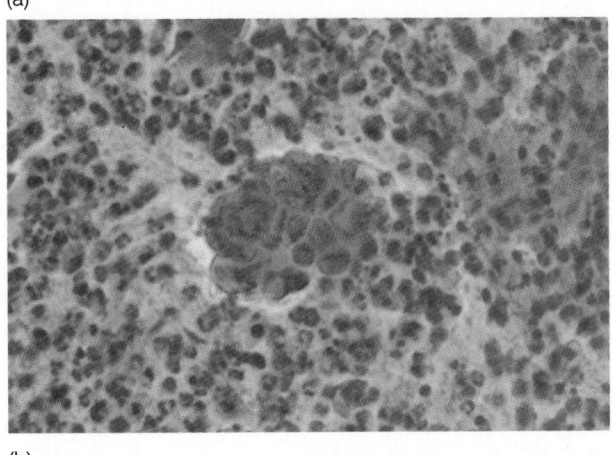

(b)

▷ **FIGURE 21–32 The Pap Smear** (*a*) A photomicrograph of a normal Pap smear showing large, flattened cells. (*b*) A photomicrograph of a cancerous smear, showing many small cancer cells.

of low-estrogen pills is less likely to cause cancer of the breast or cervix. Even with reduced estrogen levels, however, women who take birth control pills are more likely to develop cervical cancer than women who do not. Physicians, therefore, recommend annual Pap smears for women who are on the pill. During a **Pap smear,** the cervical lining is swabbed. The swab picks up cells sloughed off by the epithelium, which are later examined under a microscope for signs of cancer (▷ Figure 21–32). This procedure helps physicians diagnose cervical cancer early, increasing a woman's chances of survival.

Smoking increases the likelihood of side effects from birth control pills. If a woman is a smoker and takes the pill, for example, she is four times more likely to die from a heart attack or stroke than a nonsmoker. The risk of side effects also increases with age. To reduce the chances of developing serious side effects, women over the age of 35 who smoke should either use an alternative method of birth control or should give up smoking. Birth control pills are also not advised for women with a medical history of blood clots, high blood pressure, diabetes, uterine cancer, and cancer of the breast.

Birth control pills do have beneficial effects. First of all, they prevent pregnancy. National statistics show that one of every 10,000 women who becomes pregnant and delivers will die from complications, usually during delivery. Thus, even with the risks associated with the pill, using this mode of contraception is six times safer than pregnancy.

Birth control pills also reduce the incidence of ovarian cysts, breast lumps, anemia, rheumatoid arthritis, osteoporosis, and pelvic infection. Although birth control pills may increase the risk of cervical and breast cancer, they apparently protect a woman from cancer of the ovary and of the uterus, perhaps for life.

Intrauterine Device. The next most effective means of birth control is the **intrauterine device (IUD)** (▷ Figure 21–33). The IUD is a small plastic or metal object with a string attached to it. IUDs are inserted into the uterus by a physician, usually during menstruation, because the cervical canal is widest then and because menstrual bleeding indicates that the woman is not pregnant.

No one knows exactly how the IUD works, but there are at least two major hypotheses. Some researchers think

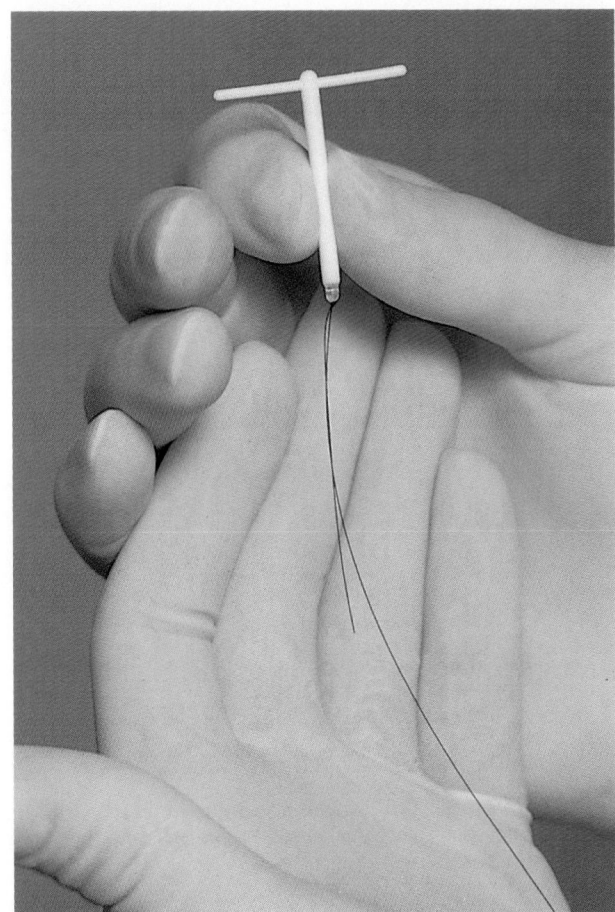

▷ **FIGURE 21–33 The IUD** IUDs come in a variety of shapes and sizes and are inserted into the uterus, where they prevent implantation. Only one type is currently legal in the United States.

that the IUD increases uterine contractions, making it difficult for the early embryo to attach and implant in the wall of the uterus. Others think that the IUD creates a local inflammatory reaction in the uterine lining, resulting in an environment inhospitable to a newly formed embryo. As a result, implantation is blocked, and the embryo dies. It is possible that both mechanisms are operating.

IUDs also have adverse effects. In some cases, the uterus expels the device, leaving a woman unprotected. Expulsion usually occurs within a month or two of insertion, so couples should check regularly during this period to be certain that the device is in place. The IUD may also cause slight pain and increase menstrual bleeding. These effects, however, are minor compared with two much rarer complications: uterine infections and perforation (a penetration of the uterine wall by the IUD). Women wearing an IUD are more likely to develop uterine infections than women practicing other forms of birth control. If not treated quickly, infections can spread to the uterine tubes, where scar tissue develops and blocks the transport of sperm and ova, causing sterility. Perforation of the uterus is a life-threatening condition requiring surgery to correct.

The Diaphragm, Condom, and Sponge. The next most effective means of birth control are the barrier methods—the diaphragm, condom, and vaginal sponge—all of which prevent the sperm from entering the uterus. The **diaphragm** is a rubber cup that fits over the end of the cervix ▷ Figure 21–34). To increase its effectiveness, a spermicidal (sperm-killing) jelly, foam, or cream should be applied to the rim and inside surface of the cup.

Diaphragms must be custom fitted by physicians. To be effective, the diaphragm must be inserted no more than two hours before sexual intercourse and must be worn for at least six hours afterward. If sexual intercourse is repeated, additional spermicidal cream or jelly should be injected into the vagina onto the diaphragm via a special applicator. If intercourse is desired after six hours, the diaphragm should be removed, washed, recoated, and then reinserted.

Smaller versions of the diaphragm, called **cervical caps,** are also available. Cervical caps fit over the end of the cervix. The cervical cap most often used is held in place by suction. When used with spermicidal jelly or cream, the caps are as effective as diaphragms.

Condoms are thin, latex rubber sheaths that are rolled onto the erect penis (▷ Figure 21–35). Sperm released during ejaculation are trapped inside (often in a small reservoir at the tip of the condom) and are therefore prevented from entering the vagina. Some condoms are prelubricated with a spermicidal chemical. Besides preventing fertilization, condoms also protect against sexually transmitted diseases, a benefit not offered by any other birth control measure except abstinence. Condoms are widely available and easy to obtain. They do not require a doctor's prescription. Because of the danger of sexually transmitted diseases, doctors in this country recommend that some women use vaginal condoms, latex devices that fit into the vagina.

Another method of birth control is the **vaginal sponge** (▷ Figure 21–36). This small absorbent sponge is impregnated with spermicidal jelly. Inserted into the vagina, the sponge is positioned over the end of the cervix. The sponge is effective immediately after placement and remains effective for 24 hours. Cervical sponges can be purchased without a doctor's prescription. Like the condom, one size fits all.

Withdrawal. One of the oldest, but least successful, means of birth control is **withdrawal,** or **coitus interruptus,** disengaging before ejaculation. This method requires tremendous willpower and frequently fails, for three reasons: because caution is often tossed to the wind in the heat of passion, because the penis is withdrawn too late, or because of preejaculatory leakage—the release of a few drops of sperm-filled semen before ejaculation. Thus, it is not an advisable method of birth control.

Spermicidal Chemicals. As mentioned earlier, spermicidal jellies, creams, and foams contain chemical agents that kill sperm but are apparently harmless to the woman

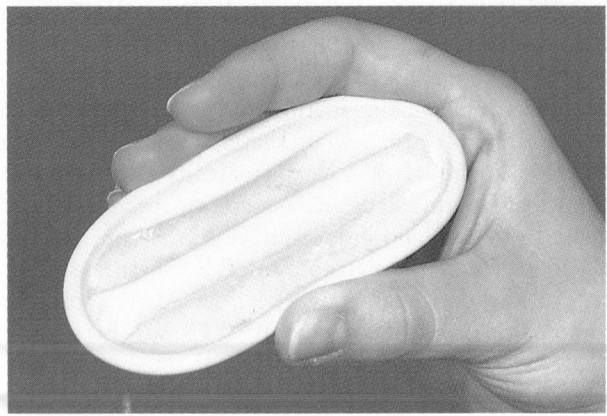

▷ **FIGURE 21–34 The Diaphragm** Worn over the cervix, the diaphragm is coated with spermicidal jelly or cream and is an effective barrier to sperm.

(▷ Figure 21–37). Spermicidal preparations are most often used in conjunction with diaphragms, condoms, and cervical caps. They can also be used alone but are only about as effective as withdrawal.

The Rhythm Method. Abstaining from sexual intercourse around the time of ovulation—the **rhythm,** or **natural, method**—can help couples reduce the likelihood of pregnancy. Because ova remain viable 12 to 24 hours after ovulation and sperm may remain alive in the female reproductive tract for up to three days, abstinence four days before and four days after the probable ovulation date should provide a margin of safety. If a couple knows the exact time of ovulation, they can time sexual intercourse to prevent pregnancy more precisely.

To practice the natural method successfully, couples must first determine when ovulation occurs (▷ Figure 21–38). Several methods are available to determine when ovulation occurs in a woman's cycle.

One approach is the **temperature method.**[7] A woman's body temperature varies throughout the menstrual cycle, as shown in ▷ Figure 21–39. In most women, body temperature rises slightly after ovulation. By taking her temperature every morning before she gets out of bed, a woman can pinpoint the day she ovulates. By keeping a temperature record over several menstrual cycles, she can deter-

---

[7] The temperature method can also be used by women who want to get pregnant, because it allows them to determine the time of ovulation.

▷ **FIGURE 21–35 The Condom** Worn over the penis during sexual intercourse, it prevents sperm from entering the vagina.

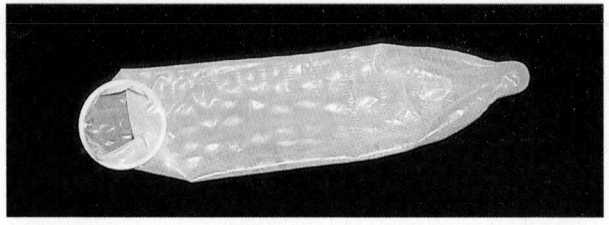

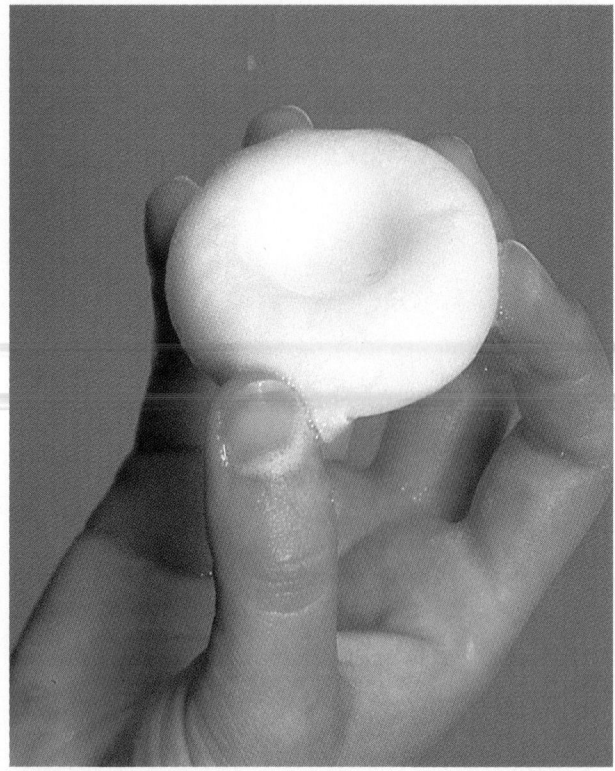

▷ **FIGURE 21–36 The Vaginal Sponge** Impregnated with spermicidal chemical, the vaginal sponge is inserted into the vagina and is effective for up to 24 hours.

mine the length of her cycle and the time of ovulation. Once the length of the cycle and the time of ovulation have been determined, days of abstinence can be determined. Erring on the safe side, some doctors recommend that couples refrain from sexual intercourse from the first day of menstruation until three days after ovulation. Translated, that means no sex for about 17 days of the 28-day cycle.

Another method used to time ovulation involves taking samples of the cervical mucus, which varies in consistency during the menstrual cycle. By testing its thickness on a

▷ **FIGURE 21–37 Spermicidal Jelly** Spermicidal preparations work alone, but they are most effective when combined with another form of protection, such as a diaphragm.

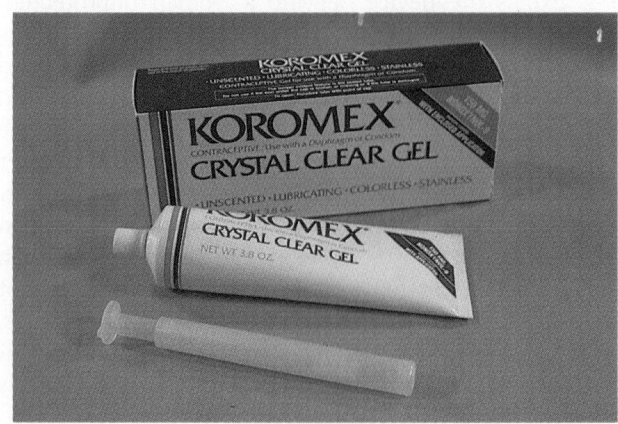

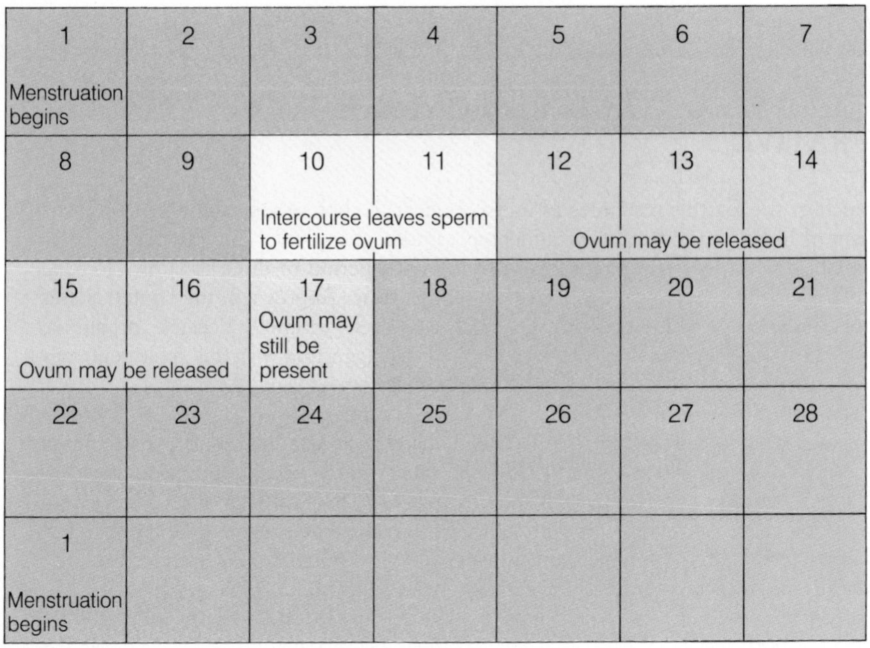

1 Menstruation begins	2	3	4	5	6	7
8	9	10 Intercourse leaves sperm to fertilize ovum	11	12 Ovum may be released	13	14
15 Ovum may be released	16	17 Ovum may still be present	18	19	20	21
22	23	24	25	26	27	28
1 Menstruation begins						

▷ FIGURE 21–38 **The Natural Method** The yellow shaded areas indicate an unsafe period for sexual intercourse, assuming ovulation occurs at the midpoint of the cycle.

daily basis, a woman can fairly accurately tell when she has ovulated.

As most people practice it, the natural method of birth control requires about eight days of abstinence during each menstrual cycle—four days before ovulation and four days after ovulation. This practice minimizes the chances of a viable sperm reaching a viable ovum. Unfortunately, some women experience the greatest sexual interest around the time of ovulation. Sexual intercourse after a period of abstinence may also advance the time of ovulation. (For a discussion of new methods of birth control, see Health Note 21–1.)

Abortion Is the Surgical Termination of Pregnancy

Some couples may elect to terminate pregnancy through **abortion.** In the United States, approximately 1 million abortions are performed every year by physicians. Although many people in our society view abortion as a legitimate means of family planning, they are quick to point out that it should not be practiced as a primary means of birth control. Contraception is less costly and less traumatic.

Abortion is not suitable or morally acceptable to many people. Pro-life advocates argue that abortion should be outlawed or severely restricted—that is, allowed only in cases of rape, incest, and threat to the life of the mother. They advise unmarried women to abstain from sexual intercourse or, if they become pregnant, to give birth and either keep the baby or give it up for adoption.

Pro-choice advocates, on the other hand, think that women should have the freedom to choose whether to terminate a pregnancy or have a child. Abortion, they say, reduces unwanted pregnancies and untold suffering among unwanted infants, especially in poor families, and gives women more options than motherhood.

In the first 12 weeks of pregnancy, abortions can be performed surgically in a doctor's office via **vacuum aspiration.** The cervix is dilated by a special instrument, and the contents of the uterus are drawn out through an aspirator tube. The operation is fairly routine and relatively painless. Usually no anesthesia is given. Women bleed for a week or so after the procedure but generally experience few complications. Between 13 and 16 weeks, a larger aspirator tube is required. The lining of the uterus may have to be scraped with a metallic instrument to ensure complete removal of the fetal tissue. After 16 weeks, abortions are more difficult and more risky. Solutions of salt, urea, or prostaglandins, which stimulate uterine contractions, can be injected into the sac of fluid surrounding the embryo to induce premature labor. The hormone oxytocin

▷ FIGURE 21–39 **Body Temperature Measurements during the Menstrual Cycle**

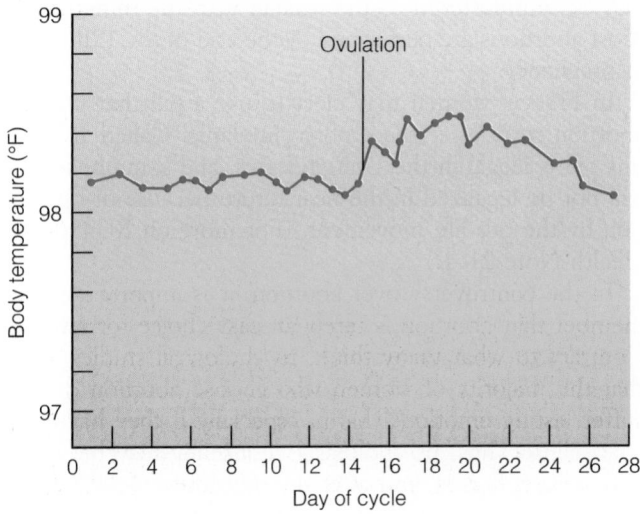

ADVANCES IN BIRTH CONTROL: RESPONDING TO A GLOBAL IMPERATIVE

Birth control is an issue of worldwide concern. In the wealthy, industrialized nations, many parents have chosen to limit their family size, for several reasons. First, many parents cannot afford more than one or two children. Raising a single child to college age could cost as much as $85,000. A recent study estimated that another $85,000 will be needed just to send a child born today to a public university. Considerably more is needed to send a student to a private university. Parents may also limit their family size to provide more personal care for their offspring. And some choose to have fewer children for environmental reasons. Consider some startling statistics: a child born in the United States today will require 16 pounds of coal, 3.6 gallons of oil, and 240 cubic feet of natural gas *each day* of his or her life.

This heavy dependence on resources results in enormous amounts of pollution and environmental damage. Every man, woman, and child, in fact, is responsible for over 1 ton of hazardous waste each year. Clearly, child rearing is an environmentally taxing activity.

Birth control is essential in the United States and other developed countries, many experts say, because our impact on the environment is so great. Each American, in fact, uses 25 to 40 times as

much of the Earth's resources as a resident of India. In terms of environmental impact, then, the United States' 248 million people are equal to 6 to 10 billion Indians, or 1.2 to 2 times the present world population.

This is not to say that birth control is unimportant in Third World nations. Each year 90 million new residents are added to the global population, or about 3 per second. Ninety percent of the new residents are born into the Third World, where hunger and starvation abound. And these countries are faced with environmental problems of epic proportions.

Family planning is making headway in Third World nations, but the task has only just begun. To help in the process, researchers are trying to develop safer, more convenient, and even more effective methods of birth control.

For example, tests are under way on the effectiveness and safety of nasal spray contraceptives for women. Research is also proceeding quickly on transdermal contraceptive patches. The small, Band-Aid-like patches are impregnated with a blend of hormones and hormone analogs, including estrogen and progesterone. The patches are worn by a woman for a week, then replaced.

In many countries outside the United States, slow-acting, injectable contraceptives are now being used. Women are

given a shot of crystalline progesterone under the skin. The crystals dissolve over a period of three months, blocking ovulation. Approval in the United States has been withheld because of animal tests suggesting that this treatment may cause some forms of cancer.

Another novel approach involves matchstick-sized capsules containing an even more potent progesterone implanted under the skin of a woman's arm; these prevent pregnancy for up to five years (▷ Figure 1). Contraceptive implants have been approved for use in 13 countries. In 1991, the Food and Drug Administration approved their use in the United States. Population experts hope that they will be widely used in Third World nations because they require virtually no effort on the part of the couple.

Researchers are also experimenting with biodegradable implants. Clinical trials are under way on a biodegradable material impregnated with progesterone that may prevent pregnancy for 18 months or more. The biodegradable material is broken down and gradually disappears.

Experimentation is continuing on a "morning-after" pill, which can be taken after sexual intercourse to prevent pregnancy. At least two morning-after pills exist. One contains a synthetic estrogen called DES (diethylstilbestrol). DES stimulates muscle contraction in the

---

may be administered to the woman with the same effect. Most abortions are performed by the end of the 12th week of pregnancy.

In France, women may elect to use a pill that induces abortion soon after the embryo implants. Called RU486, this pill is illegal in the United States, and some believe it will not be legalized in the near future because of opposition by the pro-life movement. (For more on RU486, see Health Note 21-1.)

In the controversy over abortion, it is important to remember that abortion is rarely an easy choice for anyone. Contrary to what many think, psychological studies show that the majority of women who choose abortion do not suffer lasting emotional harm, especially if they have had counseling. Thus, psychological counseling may be advisable before, during, and after the procedure.

≈ SEXUALLY TRANSMITTED DISEASES

Several years ago, a group of English professors convened to discuss some pressing questions of language. One of those questions was "What is the most melodious word in the English language?" After much deliberation, they settled on a beautiful but unlikely candidate, the word "syphilis." It rolls off the tongue with ease. By most measures, however, syphilis is hardly a thing of beauty. It is a potentially crippling or deadly disease spread through sexual contact.

Infections like syphilis that are caused by bacteria and viruses and spread from one individual to another during sexual contact are known as **sexually transmitted diseases (STDs),** or, less commonly, **venereal diseases.** Bacteria and viruses transmitted by sexual contact penetrate the

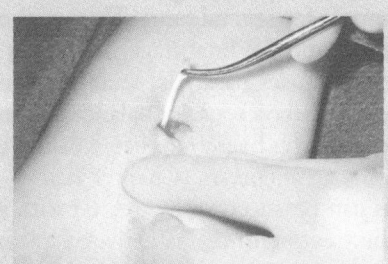

▷ **FIGURE 1 Subcutaneous Progesterone Implant** Inserted under the skin, this tiny device releases a steady stream of progesterone, blocking ovulation for months.

uterus and uterine tubes, expelling the fertilized ovum from the reproductive tract. DES is sometimes used in cases of incest or rape, but it is generally avoided because it causes vomiting and nausea.

A synthetic steroid called RU486 is also in use in France and certain other countries, but not the United States. RU486 binds to progesterone receptors in the uterus, blocking progesterone from binding. Because progesterone inhibits muscular contractions of the uterus, RU486 probably has the same effect as DES, causing an expulsion of the fertilized ovum. Vocal pro-life forces in the United States have taken a strong stand against the use of this method. At this writing, three hospitals in California

are seeking permission and funding to test the drug.

You may have noticed that when it comes to birth control, there are few options for men. Why not develop a pill for men and shift some of the contraceptive burden to them?

To be effective, a pill would have to shut down spermatogenesis. Testosterone injections would do the job, because that hormone blocks the release of pituitary FSH and LH. FSH is required for spermatogenesis, and its absence would depress sperm production. Unfortunately, testosterone injections might create aggressive behavior. Complicating matters, excess androgen in males is converted to estrogen, causing feminizing side effects. Finally, androgen treatments may depress spermatogenesis, but not enough to lower sperm count to a level where a couple would feel confident.

Another route is to selectively inhibit FSH secretion. As noted in this chapter, the seminiferous tubules produce a substance called **inhibin,** which inhibits the production of FSH by the pituitary gland. If inhibin could be produced and administered to men, it might give them a better chance to participate in birth control. It could be administered in contraceptive nasal sprays.

One thing is certain, the world's population problem will not be solved through new contraceptive technologies

alone. What is required is a change in the attitudes of millions of men and women throughout the world. Controlling family size must be a conscious decision followed by conscientious action.

In Africa, where population is doubling in some countries every 17 years, contraceptive use is a paltry 10% to 20%. Worldwide, only about half the women of reproductive age are using contraceptives. Education is needed to involve more people in a race to stem the swelling tide, which if unchecked will add 5 billion people to the world population in the next 40 years. Funds are needed to help pay for contraception and other family planning. When one condom costs more than the average person spends on medical care in a year, we can hardly expect widespread use. Many Third World countries, however, divert enormous amounts of money to pay for weapons and almost nothing to family planning. If the world population is to stabilize, if our children are to inherit a world worth living in, many experts agree, contraceptive use must increase.

Controlling population growth also requires improvements in education and job opportunities for men and women. Small-scale, sustainable economic development will give men and women options other than child bearing.

lining of the reproductive tracts of men and women and thrive in the moist, warm environment of the body. Most bacteria and viruses that cause STDs are spread by vaginal intercourse, but other forms of sexual contact, such as anal and oral sex, are also responsible for the spread of disease. AIDS, for example, can be transmitted by anal sex as well as vaginal and oral sex (Chapter 15). **Syphilis,** an STD caused by a bacterium, can be spread by oral, anal, and vaginal sex.

Although sexually transmitted diseases pass from one person to another during sexual contact, the symptoms are not confined to the reproductive tract. In fact, several STDs, including syphilis and AIDS, are primarily systemic diseases—that is, diseases that affect entire body systems.

One complicating factor in controlling sexually transmitted diseases is that some diseases, like gonorrhea, produce

no obvious symptoms in many men and women. As a result, the disease can be transmitted without a person knowing he or she is infected. In AIDS, symptoms may not appear until several years after infection. Thus, sexually active individuals who are not monogamous can spread the AIDS virus to many people before they are even aware that they were infected.

In this section, we will examine the most common STDs, except AIDS, which was discussed in Chapter 15.

Gonorrhea Is a Bacterial Infection that May Spread to Many Organs

Gonorrhea (referred to colloquially as the "clap") is caused by a bacterium that commonly infects the urethra of men and the cervical canal of women. Painful urination and a pus-like discharge from the urethra are common complaints in men. Women may experience a cloudy vaginal discharge and lower-abdominal pain. If a woman's urethra is infected, urination may be painful. Symptoms of gonorrhea usually appear about two to eight days after sexual contact.

If left untreated, gonorrhea in men can spread to the prostate gland and the epididymis. Infections in the urethra lead to the formation of scar tissue, which narrows the urethra, making urination even more difficult. In women, the bacterial infection can spread to the uterus and uterine tubes, causing the buildup of scar tissue. In the uterine tubes, scar tissue may block the passage of sperm and ova, resulting in infertility.[8] Gonorrheal infections can also spread into the abdominal cavity through the opening of the uterine tubes. If the infection enters the bloodstream in men or women, it can travel throughout the body. Fortunately, gonorrhea can be treated by antibiotics, but early diagnosis is essential to limit the damage.

Nonspecific Urethritis Is an Extremely Common Disease Caused by Several Types of Bacteria

Nonspecific urethritis (NSU) has become the most common sexually transmitted disease known to medical science and is one of several STDs whose incidence is steadily rising in the United States. Caused by any of several different bacteria, this infection is generally less threatening than gonorrhea or syphilis, although some infections can result in sterility. Approximately one-half of the reported cases of NSU are caused by a bacterium called **chlamydia.**

Many men and women often exhibit no symptoms whatsoever and therefore can spread the disease without knowing it. In men, when symptoms occur, they resemble those of gonorrhea—painful urination and a cloudy mucous discharge from the penis. In women, symptoms resemble those of a urinary tract infection. Urination becomes painful and more frequent. NSU can be treated by antibiotics, but individuals should seek treatment quickly to avoid spread of the disease and more serious complications.

Syphilis is Caused by a Bacterium and Can Be Extremely Debilitating if Left Untreated

Despite its linguistic appeal, syphilis is a serious sexually transmitted disease caused by a bacterium that penetrates the linings of the oral cavity, vagina, and penile urethra or enters through breaks in the skin. If untreated, syphilis proceeds through three stages. In stage 1, between one and eight weeks after exposure, a small, painless red sore develops, usually in the genital area. Easily visible when on the penis, these sores often go unnoticed when they occur

in the vagina or cervix of a woman. The sore heals in one to five weeks, leaving a tiny scar.

Approximately six weeks after the sore heals, individuals complain of fever, headache, and loss of appetite. Lymph nodes in the neck, groin, and armpit swell as the bacteria spread throughout the body. This is stage 2.

The symptoms disappear for several years. Then, without warning, the disease flares up. The final stage, stage 3, is an autoimmune reaction that causes paralysis, senility, or even insanity. Individuals may lose their sense of balance and may lose sensation in their legs. The bacterium can weaken the walls of the aorta, causing an aneurysm (Chapter 12).

Fortunately, syphilis can be successfully treated with antibiotics, but only if the treatment begins early. Suspicious sores in the mouth and genitals should be brought to the attention of a physician. In the late stages, antibiotics are useless. Tissue or organ damage is permanent.

Herpes Is Caused by a Virus, Is Extremely Common, and Is Essentially Incurable

Herpes is one of the most common sexually transmitted diseases. Approximately 200,000 to 300,000 people contract the disease each year. Herpes is caused by a virus, which after entering the body remains there for life. The first sign of infection is pain, tenderness, or an itchy sensation on the penis or female external genitalia, which occurs about six days after contact with someone infected by the virus. Soon afterward, painful blisters appear on the penis and female genitalia. The blisters may also form on the thighs and buttocks, in the vagina, and on the cervix.

The blisters break open and become painful ulcers that last for one to three weeks, then disappear. Because the herpes virus is a lifelong resident of the body, new outbreaks may occur from time to time, especially when an individual is under stress. Recurrent outbreaks are generally not as severe as the initial one, and in time the outbreaks generally cease altogether.

Herpes can be spread to other individuals during sexual contact, but only when the blisters are present or (as recent research suggests) just beginning to emerge. When the virus is inactive, sexual intercourse can occur without infecting a partner.

Although herpes cannot be cured, doctors may prescribe a drug called acyclovir that suppresses the virus. It reduces the incidence of outbreaks and accelerates healing of blisters.

Women who have herpes run the risk of transferring the virus to their infants at birth. Because the herpes virus can be fatal to newborns, women are often advised to deliver by cesarean section (an incision made just above the pubic bone) if the virus is active at the time of birth.

≈ INFERTILITY

A surprisingly large percentage (about one in six) of American couples cannot conceive. The inability to conceive (to

---

[8] Sexually transmitted diseases are, in fact, a leading cause of infertility in women.

become pregnant) is called **infertility.** In about 50% of the couples, infertility results from problems occurring in the woman. About 30% of the cases are due to problems in the man alone, and about 20% are due to problems in both partners.

A couple who have been actively trying for a year or more to conceive should see a physician. The physician will first check obvious problems, such as infrequent or poorly timed sex, because only intercourse around the time of ovulation will be successful. If timing is not the problem, the physician will test the man's sperm count. A low sperm count is one of the most common causes of male infertility, for reasons noted earlier.

A low sperm count may result from overwork, emotional stress, and fatigue. Excess tobacco and alcohol consumption also contribute to the problem. Tight-fitting clothes and excess exercise, which raise the scrotal temperature, also reduce the sperm count. One of the most common causes of low sperm count is an enlargement of the veins draining one or both testes, a condition called a **varicocele.** The testes are also sensitive to a wide range of chemicals and drugs, and some physicians believe that the myriad of chemicals people are exposed to in everyday life may be lowering sperm production in males (see the next section).

If infertility results from a low sperm count, a couple may choose to undergo artificial insemination, using sperm from a sperm bank. These sperm are generally acquired from anonymous donors and are stored frozen. When thawed, the sperm are reactivated, then deposited in the woman's vagina or cervix around the time she ovulates.

If sperm production and ejaculation are normal, a physician will then check the woman's reproductive tract. First comes a test of ovulation. A sample of the mucus produced by the cervix and a biopsy of the uterine lining can indicate whether or not a woman is cycling. If ovulation is not occurring, **fertility drugs** may be administered. Several kinds of drugs are available. One of the more common is HCG, which, as noted earlier, is an LH-like hormone that induces ovulation. Unfortunately, fertility drugs often result in *superovulation* (the ovulation of many fertilizable ova), leaving couples with a litter, four to six babies, instead of the one child they had hoped for. In fact, most of the multiple births you hear about on the news are the result of fertility drugs.

If tests show that ovulation is occurring normally, infertility may be caused by an obstruction in the uterine tube. A previous gonorrheal or chlamydial infection may have spread into the tubes, causing scarring and obstructing the passageway. In this case, a couple may be advised to adopt a child or to try *in vitro* fertilization. During *in vitro* fertilization, ova are surgically removed from the woman, then fertilized by the partner's sperm. The fertilized ovum can be implanted in the uterus of the woman and can grow successfully to term. This procedure is expensive and time-consuming, has a low success rate, and is not widely available. It also places heavy emotional demands on the couple.

ENVIRONMENT AND HEALTH: THE SPERM CRISIS?

In September 1979, Professor Ralph Dougherty, a chemist at Florida State University, announced findings from a study of 130 healthy, male college students. In his test group, Dougherty found extremely low sperm counts of only 20 million per milliliter of semen, compared with an expected value of 60 to 100 million per milliliter. Biochemical analyses of the testes also revealed high levels of four toxic chemicals: DDT, polychlorinated biphenyls (PCBs), pentachlorophenol, and hexachlorobenzene.

An article in the Sierra Club's magazine reporting on the results proclaimed that America was facing a "sperm crisis" caused by toxic chemicals. But not everyone agrees. Health officials, in fact, have challenged these findings. Some officials contend that the low sperm counts that Dougherty recorded may have resulted from improved counting techniques. That is to say, over the years, advances in technology have allowed scientists to count sperm more accurately. As a result, estimates of the normal sperm concentration have been markedly lowered. Dougherty maintains that such improvements, while present, are not entirely responsible for the decline.

Other health officials argue that Dougherty's findings are not representative of the American public. Floridians, they say, may be exposed to high levels of pesticides used on farms. These chemicals may be contaminating the water supplies of urban and rural residents.

Additional studies in Florida and other states show that sperm counts in American men have indeed been falling since the 1950s. Prior to 1950, the average sperm count was about 110 million per milliliter. By 1980 and 1981, sperm counts had dropped to about 60 million per millili-

TABLE 21–6 Some Agents Potentially Toxic to Male and Female Reproduction	
MALES	**FEMALES**
Natural and synthetic androgens	Natural and synthetic estrogens
Heat	Natural and synthetic progestins
Radiation	Amphetamine
Dioxin	DDT
PCBs	Parathion (insecticide)
Vinyl chloride	Carbaryl (insecticide)
Ethanol	Diethylstilbestrol
Benzene	PCBs
Diethylstilbestrol (DES)	
EDB (ethylene dibromide)	
Paraquat (herbicide)	
Carbaryl (insecticide)	
Cadmium	
Mercury	

ter. Statistical studies suggest that the decline may be related to growing pesticide use, air pollution, and other factors.

Studies in Hawaii support the belief that certain environmental chemicals may be causing a decline in sperm count. These studies show that Hawaiian men have a considerably higher sperm count than men residing in the continental United States. This observation has been attributed to a generally cleaner environment. There are, say researchers, fewer factories on the Hawaiian Islands than on the mainland. People are exposed to fewer agricultural chemicals, and frequent winds probably result in cleaner air.

Reductions in sperm count are of concern to many people because a sperm count below 20 million per milliliter is generally insufficient for fertilization. Today, low sperm counts account for a significant percentage of all infertility in U.S. couples.

Human reproduction, like other bodily processes, depends on a healthy environment. Research shows that a wide range of factors—from drugs to radiation to industrial chemicals—are toxic, or potentially toxic, to human reproduction (Table 21–6). "There has been an explosion of spermatotoxins in the environment," says Dr. Bruce Rappaport, former director of an infertility clinic in San Francisco. "The problem is environmental pollution." Today, at least 20 common industrial chemicals are known to be reproductive toxins. Ten commonly prescribed antibiotics can reduce sperm count. Even Tagamet, a drug that is used to relieve stress and treat stomach ulcers and is now the most prescribed drug in the United States, reduces sperm count by over 40%. By one estimate, at least 40 commonly used drugs depress sperm production, and thousands of other drugs and environmental pollutants have not been tested.

These facts do not necessarily mean that the United States is in a sperm crisis, but they do suggest the need for caution. Further research is needed to determine potential impacts, if any, of the many thousands of chemicals now commonly used or released into the environment. Research may prove that we need to clean up our act, or it may show that the fears are unwarranted.

SUMMARY

REPRODUCTIVE STRATEGIES

1. Organisms reproduce in one of two ways: asexually or sexually.
2. In animals, asexual reproduction occurs when an organism produces offspring by itself by budding or fragmenting. As a result, it tends to produce offspring that are genotypically and phenotypically identical to the parent.
3. Asexual reproduction permits species to proliferate quickly in the presence of adequate resources.
4. Its main disadvantage is that it results in low genetic variability in populations, a disadvantage when environmental conditions are changing.
5. Sexual reproduction involves sex cells, or gametes, each containing half the chromosomes of the adult in which they are produced.
6. Gametes combine to form a zygote whose genome is different from its parents'. Thus, sexual reproduction produces considerable genetic variation in the offspring and provides an advantage in evolution.
7. Most sexually reproducing animals breed only once a year. In temperate regions, breeding is timed so that offspring are born during a season when their survival is enhanced.
8. The timing of breeding in vertebrates and many invertebrates is regulated by day length, temperature, and the availability of food and water.
9. Sexual reproduction may involve internal or external fertilization.

HUMAN REPRODUCTION

10. The male reproductive tract consists of seven basic parts: (a) testes, (b) epididymes, (c) vasa deferentia, (d) sex accessory glands, (e) urethra, (f) penis, and (g) scrotum.
11. The testes lie in the scrotum, which provides a suitable temperature for sperm development. Each testis contains hundreds of sperm-producing seminiferous tubules. Between the seminiferous tubules are the interstitial cells, which produce testosterone.
12. Sperm produced in the seminiferous tubules are stored in the epididymis. During ejaculation, sperm pass from the epididymis to the vas deferens, then to the urethra. Secretions from the sex accessory glands are added to the sperm at this time.
13. Luteinizing hormone (LH) or interstitial cell stimulating hormone (ICSH) from the anterior pituitary stimulates the interstitial cells to produce testosterone. LH release is regulated by gonadotropin-releasing hormone from the hypothalamus.
14. Testosterone stimulates spermatogenesis, facial hair growth, thickening of the vocal cords, laryngeal cartilage growth, sebaceous gland secretion, and bone and muscle development.
15. The penis is the organ of copulation. It contains erectile tissue, which fills with blood during sexual arousal, making the penis turgid.
16. Ejaculation is under reflex control. When sexual stimulation becomes intense, sensory nerve impulses travel to the spinal cord. They stimulate motor neurons in the cord, which send impulses to the smooth muscle in the walls of the epididymes, the vasa deferentia, the sex accessory glands, and the urethra, causing ejaculation.
17. The female reproductive system consists of two basic components: the reproductive tract and the external genitalia.
18. The reproductive tract consists of (a) the uterus, (b) the two uterine tubes, (c) the two ovaries, and (d) the vagina.
19. The external genitalia consist of two flaps of skin on both sides of the vaginal opening, the labia majora and the labia minora.
20. Female germ cells are housed in follicles in the ovary. A follicle consists of a germ cell and an investing layer of follicle cells.

21. A dozen or so follicles enlarge during each menstrual cycle, but most follicles degenerate. In humans, usually only one follicle makes it to ovulation.

22. The oocyte and an investing layer of follicle cells are released during ovulation and drawn into the uterine tubes.

23. The menstrual cycle consists of a series of changes occurring in the ovaries, uterus, and endocrine system of women.

24. The first half of the menstrual cycle is called the follicular phase. It is during this period that FSH from the pituitary stimulates follicle growth and development. LH stimulates estrogen production.

25. Estrogen levels rise slowly during the follicular phase, then trigger a positive feedback mechanism that results in a preovulatory surge of FSH and LH, which triggers ovulation.

26. The ovum is expelled from the antral follicle at ovulation. The follicle then collapses and is converted into a corpus luteum (CL), which releases estrogen and progesterone.

27. In the absence of fertilization, the CL degenerates. If fertilization occurs, however, HCG from the embryo maintains the CL for approximately six months.

28. During the menstrual cycle, ovarian hormones stimulate growth of the uterine lining, which is necessary for successful implantation. If fertilization does not occur, the uterine lining is sloughed off during menstruation, which is triggered by a decline in ovarian estrogen and progesterone.

29. Like testosterone levels in boys, estrogen levels in girls increase at puberty. Estrogen promotes growth of the external genitalia, the reproductive tract, and bone. It also stimulates the deposition of fat in women's hips, buttocks, and breasts.

30. Progesterone works with estrogen to stimulate breast development. It also promotes endometrial growth and inhibits uterine contractions.

31. Many women suffer from premenstrual syndrome, characterized by irritability, depression, tension, fatigue, headaches, bloating, swelling and tenderness of the breasts, and even joint pain.

32. The menstrual cycle continues throughout the reproductive years, but after a woman reaches 20, the ovaries become progressively less responsive to gonadotropins. As a result, estrogen levels slowly decline as a woman ages. Ovulation and menstruation become erratic as a woman approaches 45.

33. Between the ages of 45 and 55, ovulation and menstruation cease. The end of reproductive function in women is known as the menopause.

34. The decline in estrogen levels results in atrophy of the reproductive organs and behavioral disturbances. Many women become irritable, suffer bouts of depression, and experience hot flashes and night sweats induced by intense vasodilation of vessels in the skin.

BIRTH CONTROL

35. Birth control refers broadly to any method or device that prevents births, and it includes two general strategies: contraception (measures that prevent pregnancy) and induced abortion (the deliberate expulsion of a fetus).

36. Figure 21–29 summarizes the effectiveness of the various birth control measures.

37. The most effective form of birth control is abstinence. Another highly effective measure is sterilization—vasectomy in men and tubal ligation in women.

38. The pill is also a highly effective means of birth control. The most common pill in use today contains a mixture of estrogen and progesterone that inhibits ovulation. In some women, however, estrogen causes adverse health effects.

39. The intrauterine device is a plastic or metal coil that is placed inside the uterus, where it prevents the fertilized ovum from implanting.

40. The diaphragm, condom, and vaginal sponge are less effective than the measures described above. The diaphragm is a rubber cap fitted over the cervix. To be fully effective, it must be coated with a spermicidal jelly, foam, or cream.

41. The condom is a thin, latex rubber sheath worn over the penis during sexual intercourse that prevents sperm from entering the vagina.

42. The vaginal sponge is a tiny, round sponge worn by the woman. It is impregnated with a spermicidal chemical.

43. One of the oldest, but least effective, methods of birth control is withdrawal, removing the penis before ejaculation. Spermicidal chemicals used alone are about as effective as withdrawal.

44. Abstaining from sexual intercourse around the time of ovulation, known as the rhythm, or natural, method, is another way to prevent pregnancy. Statistics on effectiveness show that the natural method is the least successful of all birth control measures.

45. Some couples may elect to terminate pregnancy through an abortion.

SEXUALLY TRANSMITTED DISEASES

46. Certain viruses and bacteria can be transmitted from one individual to another during sexual contact. Infections spread in this way are called sexually transmitted diseases.

47. Gonorrhea is caused by a bacterium that commonly infects the urethra in men and the cervical canal in women. Overt symptoms of the infection are frequently not present, so people can spread the disease without knowing they have it. If left untreated, gonorrhea can spread to other organs, causing considerable damage.

48. Nonspecific urethritis (NSU), the most common sexually transmitted disease, is caused by several different bacteria, but most commonly by chlamydia. It is less threatening than gonorrhea or syphilis. Many men and women show no symptoms of NSU and therefore can spread the disease without knowing it. Symptoms, when they occur, resemble those of gonorrhea.

49. Syphilis is a serious sexually transmitted disease caused by a bacterium that penetrates the linings of the oral cavity, vagina, and penile urethra. If untreated, syphilis proceeds through three stages. It can be treated with antibiotics during the first two stages, but in stage 3, when damage to the brain and blood vessels is evident, treatment is ineffective.

50. Herpes is also a very common sexually transmitted disease. It is caused by a virus. Once the virus enters the body, it remains for life. Blisters form on the genitals and sometimes on the thighs and buttocks. The blisters break open and become painful ulcers. At this stage, an individual is highly infectious. New outbreaks of the virus may occur from time to time, especially when an individual is under stress.

INFERTILITY

51. The inability to conceive is called infertility. Infertility may result from a variety of problems in men and women: poorly timed sex, low sperm count, failure to ovulate, or obstruction in the uterine tubes.

52. A variety of drugs and chemical pollutants affect sperm de-velopment and may be causing a decline in the sperm count of American men.

EXERCISING YOUR CRITICAL THINKING SKILLS

1. Devise an experiment or set of experiments to test the hypothesis that the United States is in the midst of a sperm crisis, a decline in male sperm production caused by chemicals in the environment. What type of evidence would support or refute your hypothesis? How can you devise your experiment to avoid bias?
2. You are a journalist for a major urban newspaper. You receive a press release from the local medical school announcing that one of the researchers has discovered that a chemical found in a common household cleaning agent reduces fertility in rats and mice. Large doses were given to both males and females before conception. The results showed a statistically significant decline in the litter size and several abnormalities. The researcher suggests that the chemical should be banned from use in homes. Using your critical thinking skills, what questions would you ask before writing your article? What other information would you seek out?

TEST OF TERMS

1. The testes lie in the _____ , a sac that provides a suitable temperature for the development of sperm. Sperm are produced inside the _____ tubules.
2. Sperm are stored in the _____ and delivered to the urethra during ejaculation via the _____ _____, two muscular ducts.
3. The ejaculate consists of secretions from three glands, known as the _____ _____ glands.
4. Spermatogenic cells are found in the _____ epithelium. Spermatogenesis begins with _____ , which divide mitotically. Some of these cells, however, enlarge to form the _____ _____ , which undergo the first meiotic division, forming two _____ _____ .
5. The second meiotic division produces four _____ , each containing _____ single-stranded (unreplicated) chromosomes in humans.
6. The nucleus of a mature sperm is capped by an enzyme-containing structure called the _____ , which helps the sperm penetrate the barriers around the ovum.
7. Male sex steroid hormones are produced by the _____ cells in the testes. One of the chief steroids is _____. Its secretion is controlled by a pituitary hormone called _____ _____ .
8. The glans penis is covered by a flap of skin at birth called the _____ . Surgical removal of this part of the penis is called _____ .
9. The penis contains _____ tissue, which fills with blood during sexual arousal.
10. During _____ , which is under reflex control, muscle contractions in the reproductive tract cause sperm to be propelled to the outside. As the sperm pass along the tract, they are mixed with secretions from certain glands along the way. Sperm and the secretions of these glands constitute the _____ .
11. The _____ is a pear-shaped organ that houses and nourishes the developing embryo. Ova are produced by the ovaries and picked up by the _____ _____ .
12. Sperm are deposited in the _____ and make their way through the _____ canal.
13. The _____ _____ are two flaps of skin covered with hair that are part of the external genitalia in women.
14. The _____ in females is formed from the same embryological tissue that forms the penis in males.
15. Release of an ovum from the ovary is called _____ . It occurs at the midpoint of the _____ cycle.
16. A(n) _____ follicle consists of many layers of follicle cells and a large central cavity. The primary oocyte is surrounded by a gel-like layer called the _____ _____ .
17. During oogenesis, the primary oocyte divides during the _____ meiotic division, producing a secondary oocyte and the first _____ body containing [give a number] _____ -stranded chromosomes in humans.
18. The _____ _____ is a structure in the ovary produced from a follicle that releases its ovum. It produces two steroid hormones, _____ and _____ .
19. The first half of the menstrual cycle is called the _____ phase. LH released from the pituitary at this time stimulates _____ production by the large follicles in the ovary, while the pituitary hormone _____ stimulates follicle growth.
20. The lining of the uterus, called the _____ , thickens during the first half of the menstrual cycle. It continues to grow throughout the cycle, but in the absence of fertilization, it is sloughed off. The loss of blood and tissue from the lining is called _____ .
21. In women, depression, irritability, and swelling and tenderness of the breast are symptoms of _____ _____ .
22. Ovarian function ceases in women sometime after age 45. This is called the _____ . The decline in the secretion of _____

from the ovaries at this time results in behavioral changes and signs of atrophy in the breasts and uterus.

23. Male sterilization, a surgical procedure in which the vasa deferentia are cut, is called _____ . In women, sterilization is achieved by severing the _____ _____ , an operation called a(n) _____ _____ .

24. The combined birth control pill contains two synthetic femalesteroid hormones, _____ and _____ .

25. Cervical cancer can be diagnosed by a _____ _____ , a swab of the cervical lining.

26. A(n) _____ _____ is a plastic coil inserted in the uterus that prevents pregnancy.

27. A(n) _____ is a rubber cup that fits over the cervix as it protrudes into the vagina. It is coated with _____ jelly or foam.

28. A thin, latex rubber device worn over the penis during sexual intercourse to reduce disease and the chance of pregnancy is called a _____ .

29. Timing sexual intercourse to avoid the deposition of sperm in the vagina near the time of ovulation constitutes the _____ method and is one of the least effective contraceptive techniques.

30. A woman can determine when she ovulates by taking daily _____ measurements.

31. Gonorrhea and syphilis are _____

_____ _____ caused by bacteria.

32. _____ _____ is the most common sexually transmitted disease known to medical science and is one of several STDs whose incidence is steadily rising in the United States.

33. _____ is a viral disease transmitted during sexual intercourse and results in small blisters in the genital region, thighs, and buttocks.

Answers to the Test of Terms are found in Appendix B.

TEST OF CONCEPTS

1. Organisms reproduce in one of two ways. What are they? How are they different? What are the advantages and disadvantages of each?

2. You are a fertility specialist. A young woman arrives in your office complaining that she has been trying to get pregnant for two years but to no avail. Describe how you would go about determining whether the problem was with her, her husband, or both of them.

3. Describe the anatomy of the male reproductive system. Where are sperm produced? Where are they stored? What structures produce the semen?

4. Describe the process of spermatogenesis, noting the cell types and the number of chromosomes in each type.

5. Why is the first meiotic division called a reduction division?

6. List the hormones that control testicular function. Where are they produced, and what effects do they have on the testes?

7. You are a family doctor. A man comes to your office complaining of impotence. What are the possible causes? How would you go about testing for the causes?

8. Trace the pathway for a sperm from the seminiferous tubule to the site of fertilization.

9. Describe the process of ovulation and its hormonal control.

10. What is the corpus luteum? How does it form? What does it produce? Why does it degenerate at the end of the menstrual cycle if fertilization does not occur?

11. What is menstruation? What triggers its onset?

12. Describe the effects of estrogen and progesterone on the reproductive tract and the body.

13. A woman comes to your office. She is 47 years old and complains of irritability and depression. She asks for the name of a reliable psychiatrist who could help her. She says that she wakes up in the middle of the night in a sweat. Would you give her the name of a psychiatrist? Why or why not? If not, what would you do?

14. Describe each of the following birth control measures, explaining what they are and how they work: the pill, IUD, diaphragm, cervical cap, condom, spermicidal jelly, and natural method.

15. Describe ways to prevent the spread of sexually transmitted diseases.

CHAPTER

22

Animal Development and Aging

≈

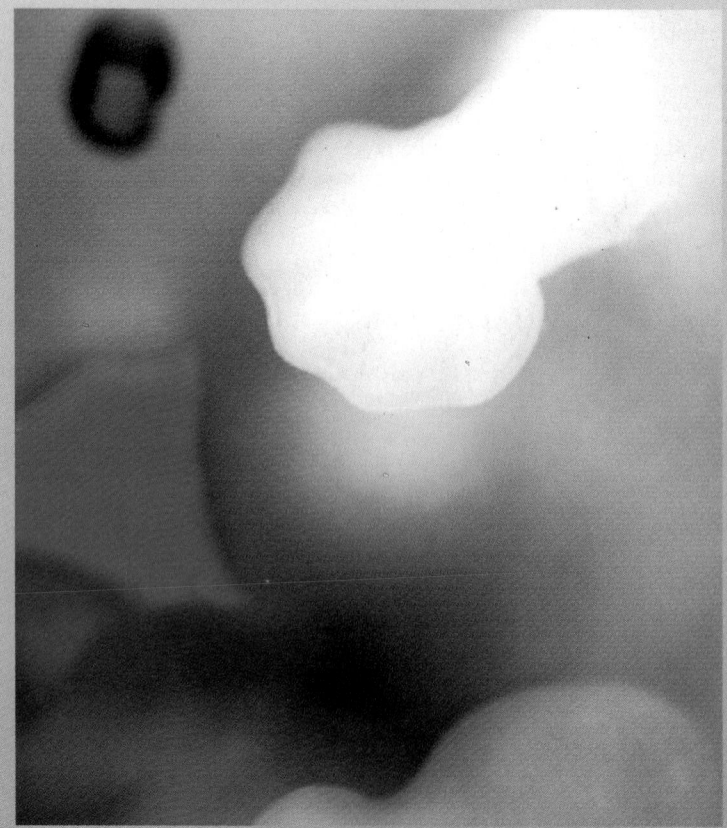

Hand of a human embryo at 40 days. The webbing connecting the fingers is removed by enzymes from lysosomes at a genetically determined stage of development.

Reproduction is one of the most varied of all biological functions. Like many other organ systems, the reproductive system has taken many twists and turns along the course of evolution, producing a fascinating array of ways to achieve the same end: procreation, or producing offspring. As you saw in the last chapter, reproductive strategies fall into one of two categories, sexual and asexual, but the variation in each major group is incredible. Among those animal species that reproduce sexually, for example, many fertilize externally. In many fishes, for instance, females lay the eggs on the bottom of a body of water or the walls of a rocky cave, and males deposit sperm on them. The parents then often abandon the fertilized eggs and move on to other activities. In other complex animal species, internal fertilization is the rule. Males deposit sperm inside the female reproductive system, where they unite with eggs. In some species, like turtles and lizards, the eggs are expelled (usually buried), then abandoned. But in others animals, the internally fertilized eggs develop internally, protecting the offspring from the harsh realities of predation, adverse weather, disease, and other factors to which their biological cousins are exposed. Internal development occurs in some species of fishes, like sharks, and in some reptiles, such as snakes. It is also found in virtually all mammals. Internal development required adaptations that permit the nourishment of offspring within the female tract for long periods.

This chapter focuses on fertilization and development, primarily on human development. It also presents a brief discussion of human growth after birth and of aging and death.

≈ AN OVERVIEW OF ANIMAL DEVELOPMENT

In sexually reproducing animals, new life begins when the sperm and egg unite, forming a diploid zygote. This process is referred to as **fertilization** (▷Figure 22–1). As you learned in the last chapter, the sperm and egg are produced in special reproductive organs (gonads) via meiosis.

In vertebrates and invertebrates, sperm contact the plasma membrane of eggs and are engulfed (phagocytized) by them. The nuclei of the egg and sperm cell then fuse to form a diploid zygote.

Fertilization is followed by mitotic division that converts the zygote into a two-celled structure. These cells continue to divide, producing a multicelled structure, known as a **morula** (from the Latin word *morus,* meaning "mulberry") (▷ Figure 22–2). The successive division of embryonic cells, or **cleavage,** continues for a while. In many species, such as frogs and sea urchins, a hollow cavity forms in the morula (Figure 22–2). The resulting structure is called a **blastula.**

Cleavage of the embryonic cells soon begins to slow, and the cells begin to migrate about in distinct patterns, dramatically changing the appearance of the embryo. As illustrated in Figure 22–2, some cells of the embryo invagi-

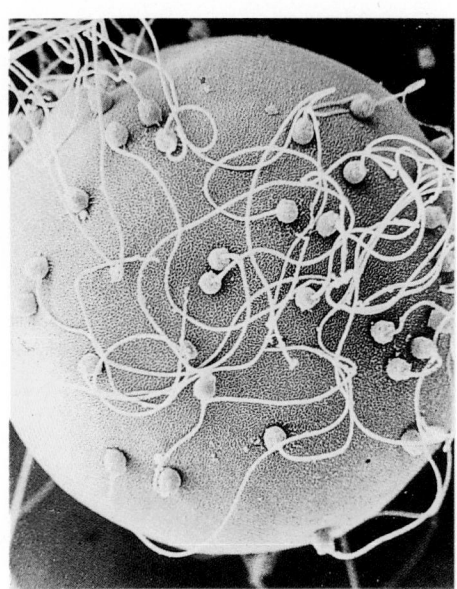

▷ **FIGURE 22–1 Fertilization** Although many sperm gather around this clam egg, only one will enter. Fertilized sperm are phagocytized by the egg.

nate to form a tube that will later become the gastrointestinal tract. This stage of development is referred to as **gastrulation,** and the embryo is called a **gastrula.** During gastrulation, three distinct layers of cells form in the gastrula: an internal layer, the **endoderm;** a middle layer, the **mesoderm;** and an outer layer, the **ectoderm.**

Following this remarkable redistribution and reorganization of cells, the three cell layers differentiate further, forming organs. The endoderm gives rise to the lining of the gut and various organs associated with the gut. The mesoderm gives rise to bone, muscle, cartilage, connective tissue, and blood. And the ectoderm gives rise to the skin and nervous system. The formation of organs is referred to as **organogenesis.**

Organogenesis is a complicated process beyond the scope of this book. What is important to remember is that once the organs form, they usually continue to grow until an individual reaches the adult stage. In some animals, such as nematodes (wormlike creatures that are often parasites), all organs grow at the same rate until adulthood is reached. In others, such as birds, reptiles, and mammals, they grow at different rates. Thus, changes in body proportions also occur in postembryonic growth and development (▷ Figure 22–3).

All animal embryos proceed pretty much the same through the morula, blastula, and gastrula stages, but there are differences in embryonic development. In chickens, for example, the egg contains so much yolk that cell division concentrates at one pole (▷ Figure 22–4). The embryonic cells therefore form a small, disklike structure atop a mass of yolk that differentiates into three layers. These layers, in turn, give rise to the organs.

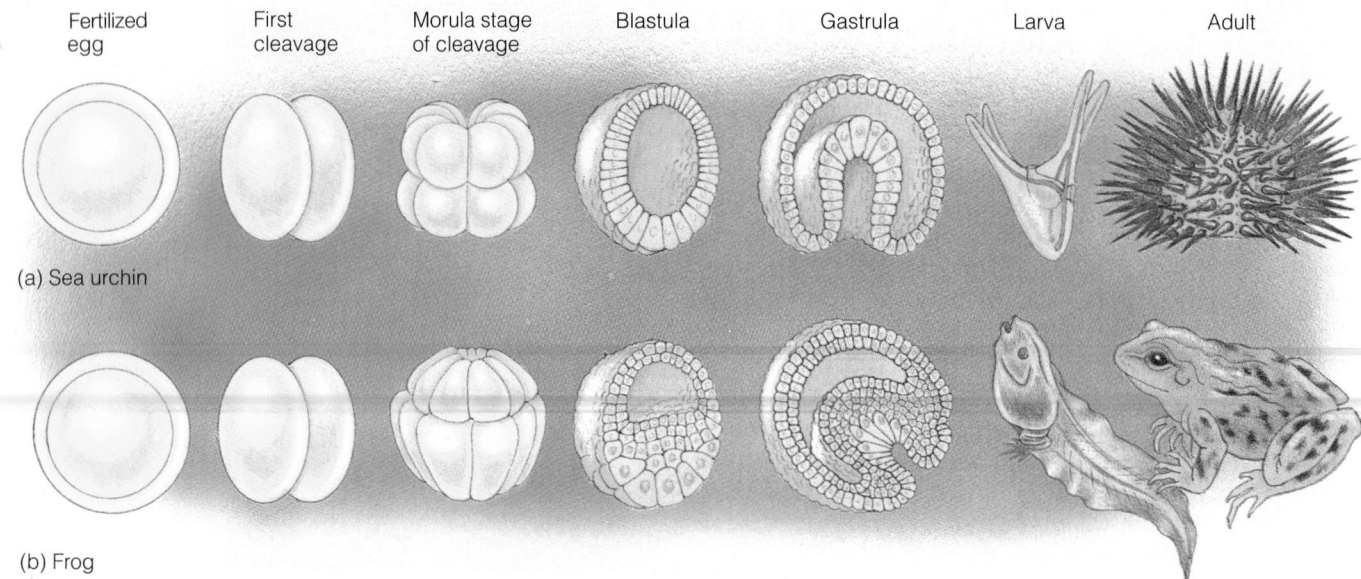

| Fertilized egg | First cleavage | Morula stage of cleavage | Blastula | Gastrula | Larva | Adult |

(a) Sea urchin

(b) Frog

▷ **FIGURE 22–2 Early Embryonic Development** Embryonic development proceeds through several key phases as shown here in the (*a*) sea urchin and (*b*) frog.

≈ HUMAN REPRODUCTION AND DEVELOPMENT

With this overview of animal development, we turn our attention to fertilization and development in humans.

Human Development Begins with Fertilization

In humans, the sperm and the ovum unite in the upper third of the uterine tube (▷ Figure 22–5). As you learned in the last chapter, oocytes are released from the ovary during ovulation and are drawn into the uterine tube in part by the rhythmic beating of cilia inside the tube, which create an inward-flowing current. This process is aided by fingerlike projections of the upper end of the uterine tube that, like massaging fingers, contract rhythmically and help sweep oocytes inside.

In humans, sperm are deposited in the vagina and quickly make their way up the reproductive tract (▷ Figure 22–6). Within a few minutes of ejaculation, they enter the cervical canal; 30 minutes later, they arrive at the junction of the uterus and uterine tube. Sperm then travel to the upper portion of the uterine tube, where they may encounter an ovum. Along the way, many millions of sperm die, killed by acidic secretions of the vagina and cervix. Studies based on laboratory animals suggest that only 1 in 3 sperm makes it through the cervical canal and 1 in 1000 makes it through the uterus to the uterine tubes (Figure 22–6).

Although sperm are motile, the principal driving force through the female reproductive tract is muscular contractions in the walls of the uterus and uterine tubes. Some evidence suggests that these contractions are stimulated by prostaglandins, hormonelike substances in the semen.

▷ **FIGURE 22–3 Changes in Body Proportions** In many animal species, like humans, after organ systems have formed, further growth and development involve changes in body proportions.

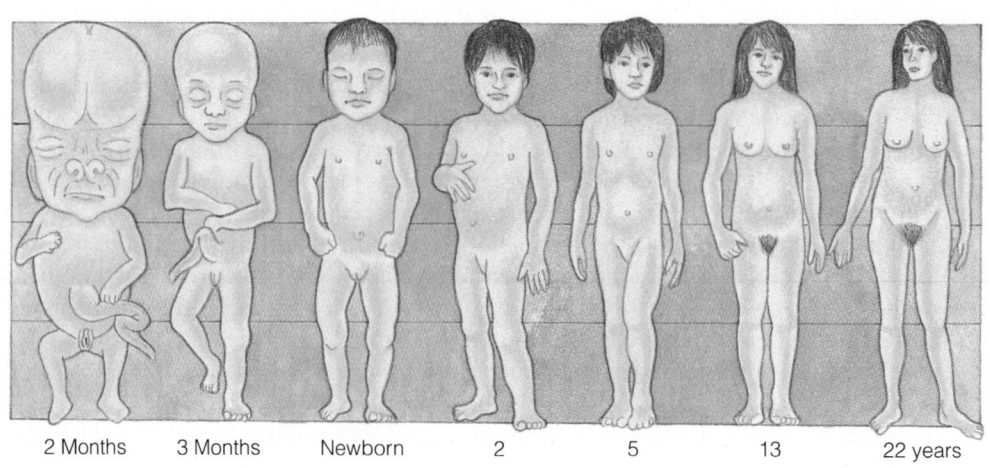

| 2 Months | 3 Months | Newborn | 2 | 5 | 13 | 22 years |

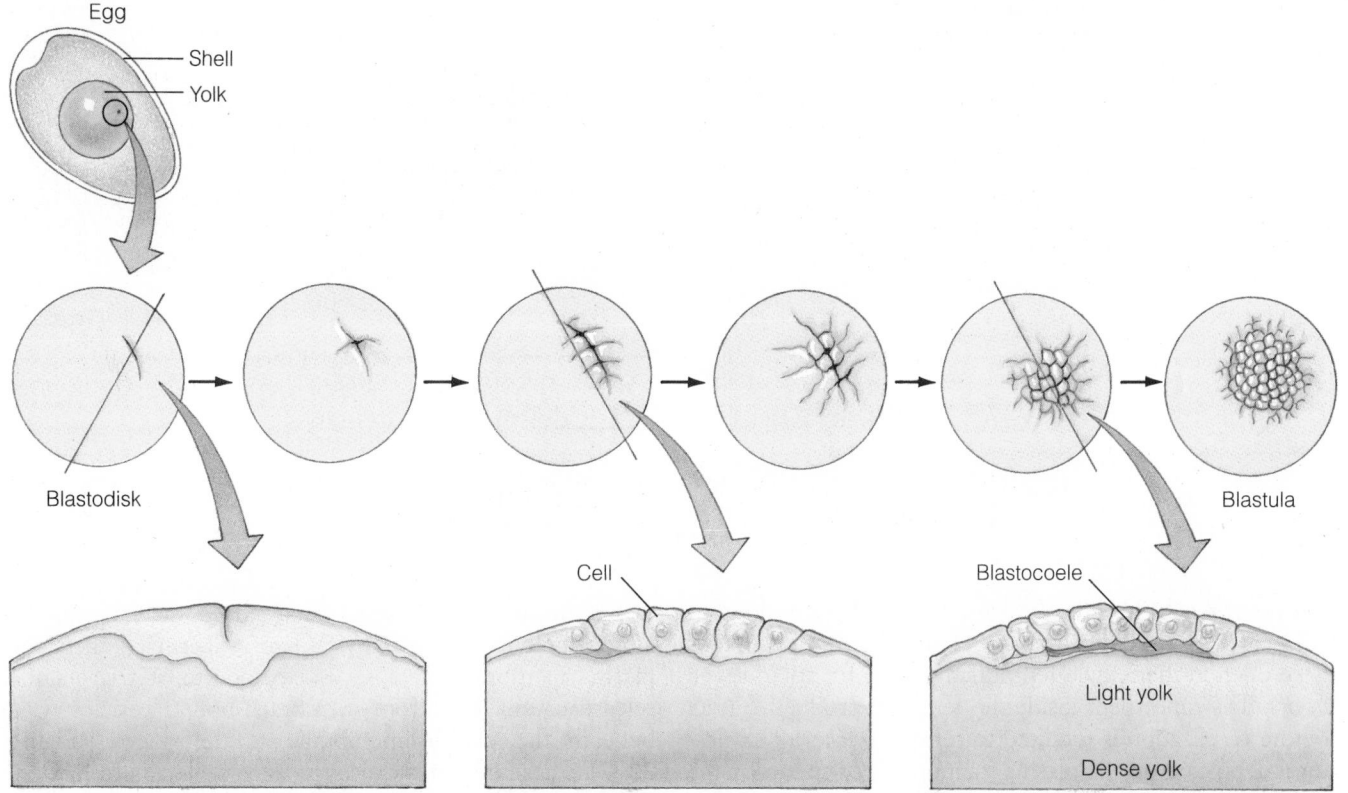

▷ **FIGURE 22–4 Early Embryonic Development in Chickens** Because of the large mass of yolk that nourishes the embryos of birds, cell division in early embryonic development is concentrated at one pole. The embryo becomes a flattened disk.

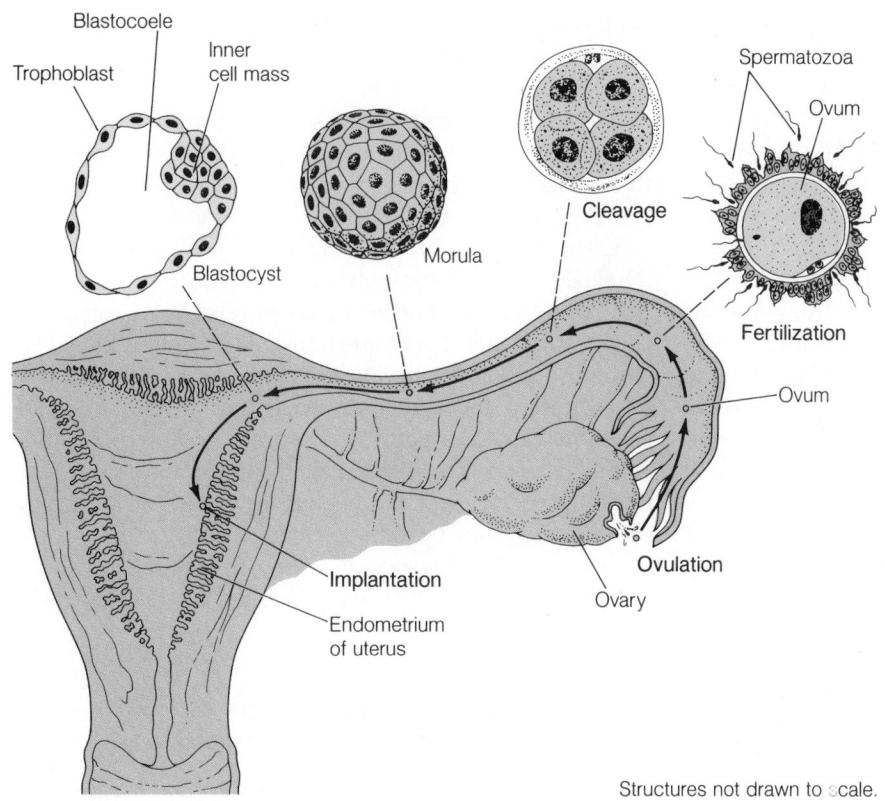

▷ **FIGURE 22–5 Fertilization and Early Embryonic Development** In humans, fertilization usually occurs in the upper third of the uterine tube. In the next three days, the zygote becomes a morula, and the morula develops into a blastocyst. The blastocyst then enters the uterus, where it will implant.

Structures not drawn to scale.

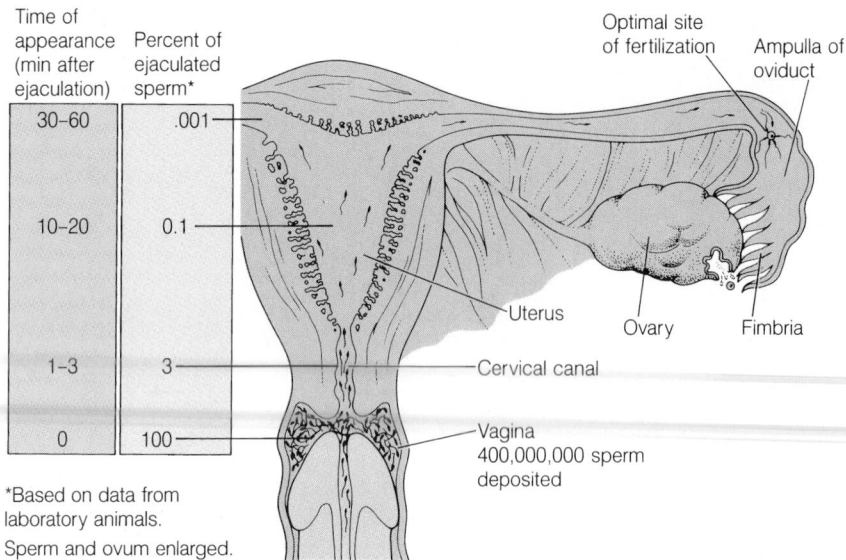

▷ **FIGURE 22–6 Sperm Transport in the Female Reproductive System** Sperm move rapidly up the female reproductive tract of humans and other vertebrates principally as a result of contractions in the muscular walls of the uterus and uterine tubes. Notice the rapid decline in sperm number along the way.

Time of appearance (min after ejaculation)	Percent of ejaculated sperm*
30–60	.001
10–20	0.1
1–3	3
0	100

*Based on data from laboratory animals.
Sperm and ovum enlarged.

Sperm reach the site of fertilization an hour or so after ejaculation, but they cannot fertilize an ovum until they have been in the female reproductive tract for six to seven hours. The time spent inside the female reproductive tract before fertilization is required to remove a layer of cholesterol deposited on the plasma membranes of sperm by the secretions of the sex accessory glands. The cholesterol coat helps stabilize the sperm plasma membranes and protects sperm as they move through the female reproductive tract. Removal of the coat renders the sperm's membranes fragile and disruptible, a prerequisite for fertilization.

After this process, the plasma membrane over the head of the sperm fuses with the outer membrane of the acrosome, a caplike structure filled with digestive enzymes (▷ Figure 22–7). Tiny openings develop at the points of fusion, allowing acrosomal enzymes to leak out.

As Figure 22–7 shows, the follicle cells around the ovum become elongated and radiate outward like the spokes of a wheel. Acrosomal enzymes of the sperm that swarm around the oocyte like so many bees around a hive dissolve the extracellular material that binds the cells together. After passing through this layer, sperm must digest their way through the zona pellucida, a gel-like layer surrounding the oocyte, with the aid of acrosomal enzymes (Figure 22–7).

Sperm cells that traverse the zona pellucida enter the space between the plasma membrane of the oocyte and the zona. As a rule, the first sperm cell to come in contact with the plasma membrane of the oocyte will fertilize it; all other sperm are excluded.

Sperm are excluded by two known mechanisms. The first is called the **fast block to polyspermy** ("many sperm"). The fast block to polyspermy occurs when the sperm cell contacts the plasma membrane of the oocyte, triggering membrane depolarization—which, as you may recall from Chapter 16, is a change in the potential difference across the plasma membrane resulting from the influx of sodium ions. For reasons not well understood, depolarization blocks other sperm from fusing with the oocyte.

Sperm are also excluded by a slower mechanism, the **slow block to polyspermy.** When a sperm contacts the plasma membrane of a secondary oocyte, it triggers the release of enzymes from membrane-bound vesicles lying beneath the plasma membrane of the oocyte (Figure 22–7). These enzymes cause the zona pellucida to harden, which blocks other sperm from reaching the oocyte. These secretions may also cause the "extra sperm" that have attached to the plasma membrane of the oocyte to detach.

These two mechanisms are important evolutionary developments insuring that one and only one sperm penetrates an egg. Without them, fertilized eggs might quickly overload with extra nuclear material, a condition that would most likely impair subsequent cell divisions and result in embryonic death.

Sperm contact with the plasma membrane of the oocyte also triggers the second meiotic division, thus converting the secondary oocyte into an ovum. Once the sperm nucleus enters, the ovum is called a zygote.

Once inside the oocyte, the sperm cell nucleus with its 23 single-stranded (unreplicated) chromosomes swells. At this stage, the nuclei of the ovum and sperm are referred to as the male and female **pronuclei** (▷ Figure 22–8). The chromosomes in the pronuclei replicate as the pronuclei move toward the center of the ovum. A mitotic spindle assembles in the zygote. After chromosome replication is complete, the chromosomes of the male and female pronuclei condense, and the nuclear envelopes of the pronuclei disintegrate. The spindle fibers attach to the chromosomes, and the chromosomes line up on the equatorial plate. The zygote is now ready for the first mitotic division (Figure 22–8).

Pre-Embryonic Development Begins at Fertilization and Ends at Implantation

Human development is divided into three stages: pre-

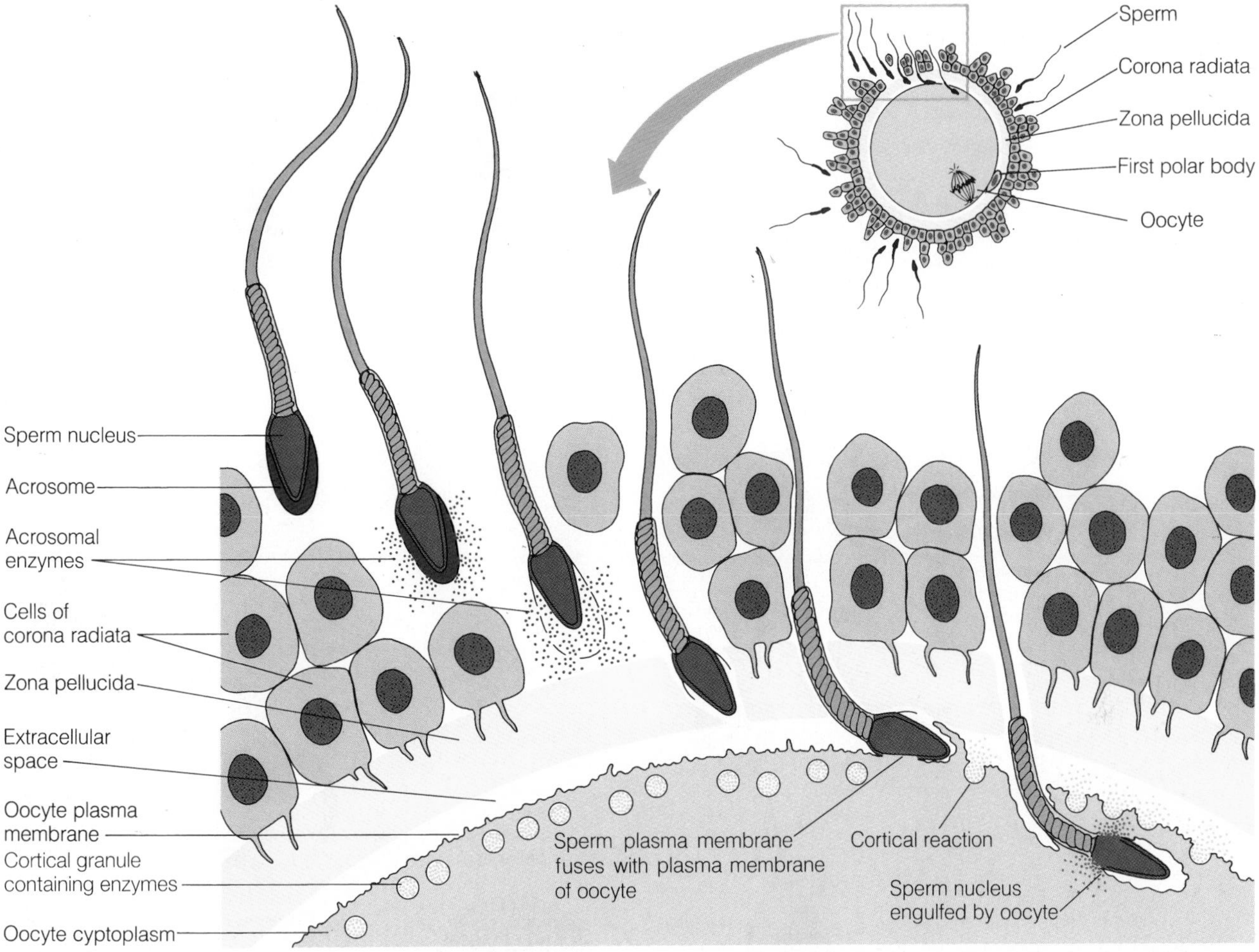
Sperm
Corona radiata
Zona pellucida
First polar body
Oocyte
Sperm nucleus
Acrosome
Acrosomal enzymes
Cells of corona radiata
Zona pellucida
Extracellular space
Oocyte plasma membrane
Cortical granule containing enzymes
Oocyte cytoplasm
Sperm plasma membrane fuses with plasma membrane of oocyte
Cortical reaction
Sperm nucleus engulfed by oocyte

▻ **FIGURE 22–7 Fertilization and Cortical Reaction** The plasma membrane of the sperm and the outer membrane of the acrosome fuse and the membranes break down, releasing enzymes that allow the sperm to penetrate the corona radiata. Sperm digest their way through the zona pellucida via enzymes associated with the inner acrosomal membrane. Sperm are engulfed by the oocyte plasma membrane. Cortical granules are released when the sperm cell contacts the membrane. These granules cause other sperm in contact with the membrane to detach.

embryonic, embryonic, and fetal. **Pre-embryonic development** includes all the changes that occur from fertilization to the time an embryo implants in the uterine wall. During this phase, the zygote undergoes rapid cellular division and is converted into a morula, not much bigger than the fertilized ovum (▻ Figure 22–9). The morula is nourished by secretions produced by the epithelium of the uterine tubes. Approximately three to four days after ovulation, the morula enters the uterus.

Fluid soon begins to accumulate in the morula, converting it into a **blastocyst,** a hollow sphere of cells slightly larger than the morula.[1] As Figure 22–9 shows, the blastocyst consists of two parts: a clump of cells, the **inner cell mass,** which will become the embryo, and a ring of flattened cells called the **trophoblast** (meaning "to nourish the blastocyst"). The trophoblast gives rise to the embryonic portion of the **placenta,** an organ that supplies nutrients to and removes wastes from the growing embryo.

The blastocyst remains unattached in the uterine lumen for two to three days. During this period, it is nourished by secretions of uterine glands.

If the Embryo Is to Survive, It Must Embed in the Wall of the Uterus. The blastocyst attaches to the uterine lining and digests its way into the endometrium. This process, called **implantation,** begins six to seven days after fertilization. Interestingly, in some species implantation can be delayed for long periods. For a discussion of this phenomenon, see Spotlight on Evolution 22-1.

Most embryos implant high on the back wall of the uterus. The cells of the trophoblast first contact the endometrium, then adhere to it, but only if the uterine lining is healthy and properly primed by estrogen and progesterone (▻ Figure 22–10a). If the endometrium is not ready or is "unhealthy"—for example, because of the presence of an IUD or a endometrial infection—the blastocyst fails to implant. Blastocysts may also fail to implant if their cells

[1] The term blastocyst comes from the Greek *blastos* meaning germ and *kystis* meaning cyst.

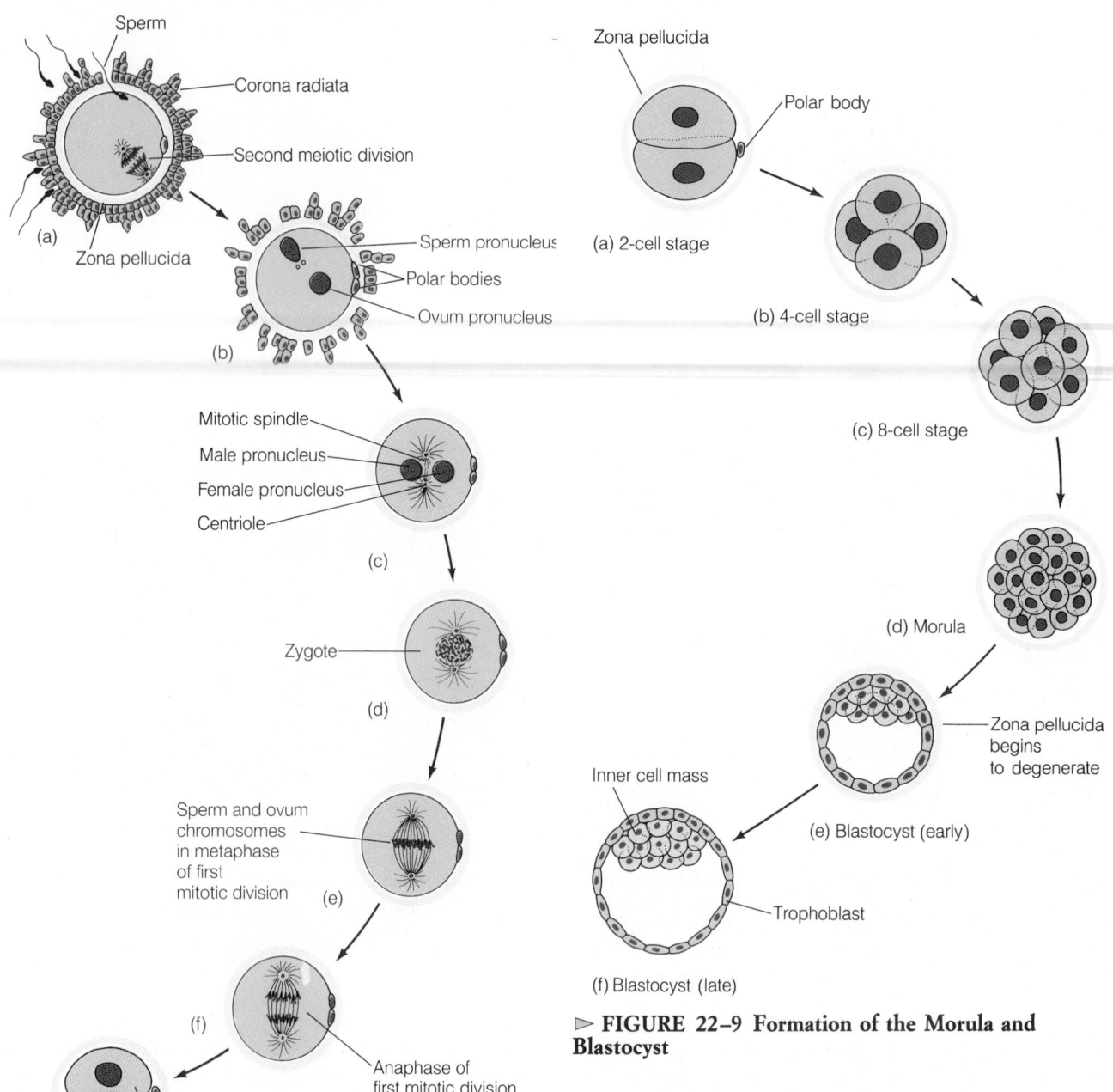

FIGURE 22–8 The Zygote Prepares for Division
(a) The sperm contacts the plasma membrane of the oocyte. Second meiotic division takes place. (b) Sperm and oocyte pronuclei form. (c) The pronuclei migrate toward the center of the cell. Chromosomes condense, and a mitotic spindle forms. (d) Chromosomes condense further, and nuclear membrane breaks down. (e) Metaphase plate is formed. (f) Anaphase of the first mitotic division. (g) Two daughter cells form.

FIGURE 22–9 Formation of the Morula and Blastocyst

phy). Enzymes released by the cells of the trophoblast digest a small hole in the endometrium, and the blastocyst "bores" its way into the uterine lining (Figure 22–10b). As it "eats" its way into the layer of enlarged endometrial cells, the blastocyst feeds on nutrients released from the endometrial cells it destroys. This action helps sustain the blastocyst before the placenta forms. By day 14, the uterine endometrium grows over the blastocyst, walling it off from the uterine lumen (cavity).

Endometrial cells respond to the invasion of the blastocyst by producing prostaglandins.[2] Prostaglandins increase the development of uterine blood vessels and, therefore, help ensure an ample supply of blood and nutrients for the blastocyst.

contain certain genetic mutations. If it is unable to bind, the blastocyst perishes and is either reabsorbed (phagocytized by the cells of the endometrium) or shed during menstruation.

In cases where implantation occurs, the endometrium responds to the embryo by cellular enlargement (hypertro-

[2]You may recall from earlier discussions that prostaglandins have many functions in the body.

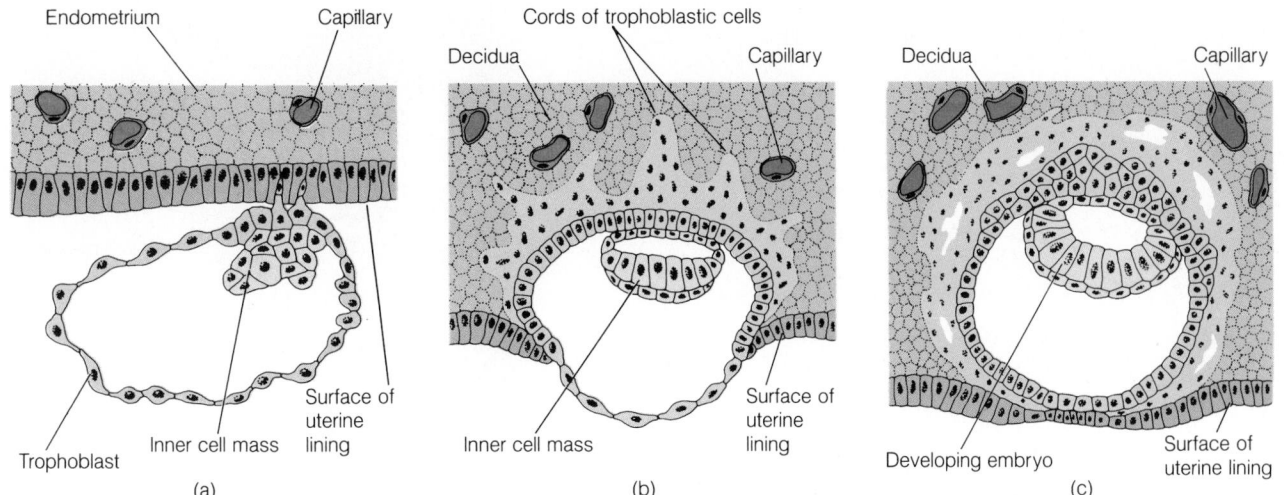

Endometrium   Capillary   Cords of trophoblastic cells

                                            Decidua   Capillary   Decidua   Capillary

                                            Inner cell mass   Developing embryo
Trophoblast   Inner cell mass   Surface of
                            Surface of   Inner cell mass   Surface of   uterine lining
                            uterine   uterine
                            lining   lining
        (a)   (b)   (c)

▷ **FIGURE 22–10 Implantation** (*a*) The blastocyst fuses with the endometrial lining. Endometrial cells proliferate, forming the decidua. (*b*) The blastocyst digests its way into the endometrium. Cords of trophoblastic cells invade, digesting mater-nal tissue and providing nutrients for the developing blastocyst. (*c*) The blastocyst soon becomes completely embedded in the endometrium.

The Placenta Forms from Embryonic and Maternal Tissue. *Placenta* is the Latin word for "cake". Shaped somewhat like a cake, the placenta forms from maternal and embryonic tissue. Figures 22–10 and ▷ 22–11 illustrate this complex process. As shown in Figure 22–11a, soon after implantation begins, the cells of the outer layer of the trophoblast begin to proliferate and invade the endometrium. Cavities form among the cells of the outer layer of the trophoblast as it digests its way into the uterine lining (Figure 22–11b). During its inward march, this layer of cells severs the walls of maternal blood capillaries. Blood pours out of the capillaries and fills the cavities (Figure 22–11c).

The inner layer of the trophoblast invades sometime later, forming fingerlike projections called placental villi (Figure 22–11c). The **placental villi** carry blood vessels from the embryo, which absorb nutrients from the pools of maternal blood in the outer layer of the trophoblast. Because the blood vessels of the placental villi connect to the developing embryo, they provide a route for nutrients to flow from the maternal blood to the embryo.

The placental villi grow and divide, increasing the total surface area for diffusion of nutrients and wastes. As they grow, they continue to invade the maternal tissue. At the same time, the blood-filled cavities enlarge, forming even bigger pools of maternal blood. Villi projecting into the blood-filled cavities absorb nutrients but also release embryonic wastes, such as carbon dioxide and urea, into the maternal blood. Because the walls of the villi are thin, wastes and nutrients can diffuse between the embryonic and maternal blood with ease. Some villi span the blood-filled cavities and anchor the embryonic portion of the placenta to the maternal tissue.

Besides providing nutrients and getting rid of embryonic wastes, the placenta produces a variety of hormones needed to maintain pregnancy. The placenta, therefore, is a respiratory, nutritive, excretory, and endocrine organ.

Placental Hormones Are Essential to Reproduction. This section discusses three placental hormones: human chorionic gonadotropin (HCG), estrogen, and progesterone (Table 22-1).

▷ Figure 22–12 shows blood levels of these three hormones during pregnancy. As illustrated, HCG levels peak in the second month of pregnancy, then drop off by the end of the third month. Levels of estrogen and progesterone, which are produced chiefly by the placenta during pregnancy, rise dramatically throughout gestation (the period between fertilization and childbirth).

HCG is an LH-like hormone produced by the embryo early in pregnancy (Chapter 21). As noted in the previous chapter, HCG prevents the corpus luteum (CL) in the ovary from degenerating when an embryo is present. It also stimulates estrogen and progesterone production from the CL, which is essential to maintaining pregnancy early in the gestational period. HCG stimulates estrogen and progesterone production by the CL for only about 10 weeks. When HCG levels decline, the CL degenerates. Estrogen and progesterone production needed to maintain pregnancy is taken over by the placenta.

High Levels of Hormones During Pregnancy May Cause Morning Sickness. Many women experience nausea (morning sickness) during the first two to three months of pregnancy. Although it often occurs in the morning hours, in some women "morning sickness" may last all day. The exact cause is not known, but some researchers believe that HCG may stimulate the brain directly, creating nausea. Other researchers blame high levels of estrogen and progesterone during pregnancy are responsible.

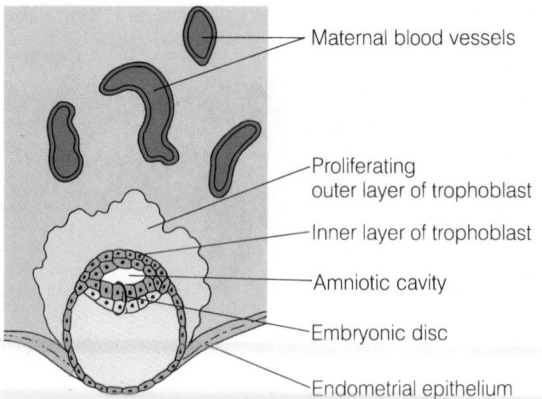

(a) 7½-day implanting blastocyst

- Maternal blood vessels
- Proliferating outer layer of trophoblast
- Inner layer of trophoblast
- Amniotic cavity
- Embryonic disc
- Endometrial epithelium

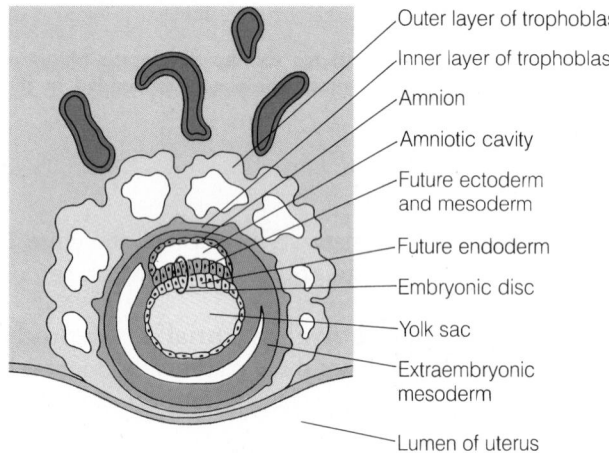

(b) 9-day implanted blastocyst

- Outer layer of trophoblast
- Inner layer of trophoblast
- Amnion
- Amniotic cavity
- Future ectoderm and mesoderm
- Future endoderm
- Embryonic disc
- Yolk sac
- Extraembryonic mesoderm
- Lumen of uterus

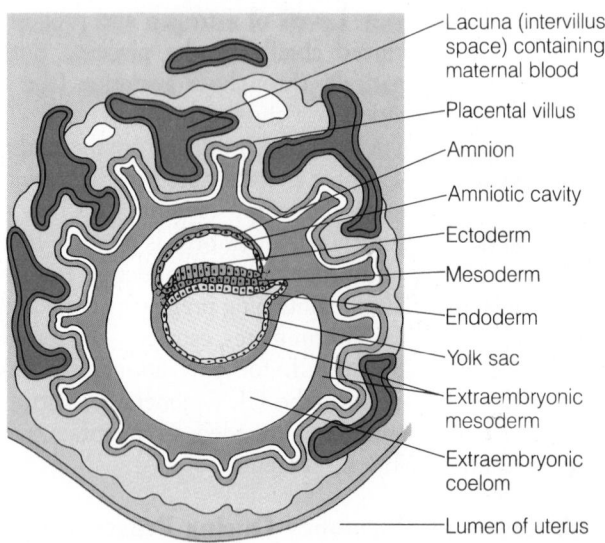

(c) 16-day embryo

- Lacuna (intervillus space) containing maternal blood
- Placental villus
- Amnion
- Amniotic cavity
- Ectoderm
- Mesoderm
- Endoderm
- Yolk sac
- Extraembryonic mesoderm
- Extraembryonic coelom
- Lumen of uterus

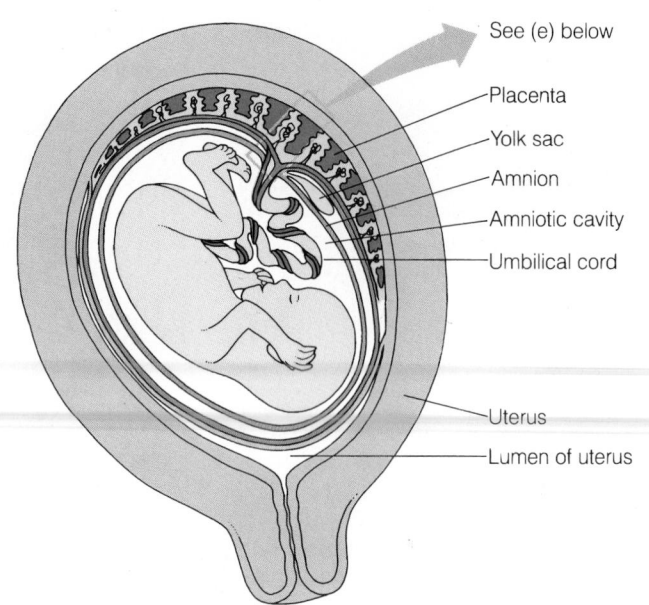

(d) 13-week fetus

- See (e) below
- Placenta
- Yolk sac
- Amnion
- Amniotic cavity
- Umbilical cord
- Uterus
- Lumen of uterus

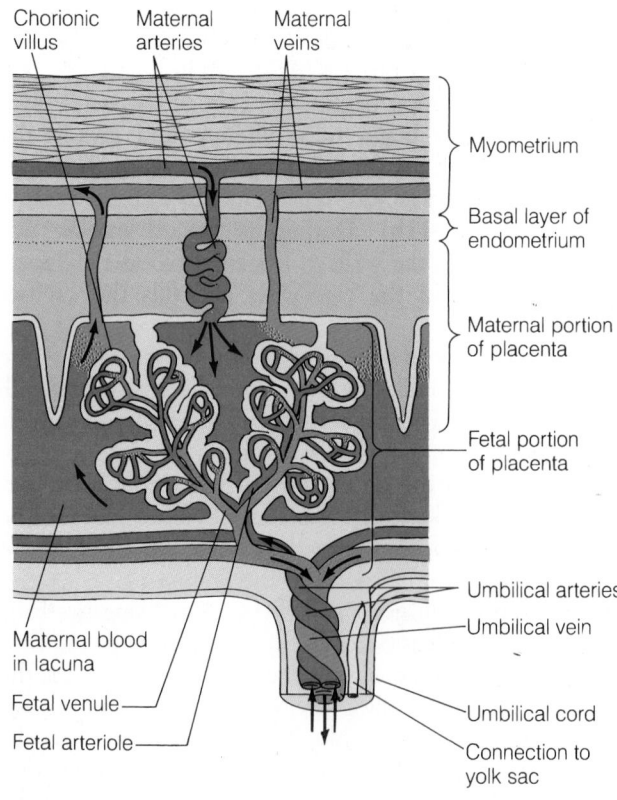

(e) Placental structure

- Chorionic villus
- Maternal arteries
- Maternal veins
- Myometrium
- Basal layer of endometrium
- Maternal portion of placenta
- Fetal portion of placenta
- Maternal blood in lacuna
- Fetal venule
- Fetal arteriole
- Umbilical arteries
- Umbilical vein
- Umbilical cord
- Connection to yolk sac

▷ **FIGURE 22–11 Placental Formation** (*a*) Invasion of the maternal tissue by the outer layer of the trophoblast. (*b*) Invasion continues. Cavities form. Note the presence of extraembryonic mesoderm from which blood vessels and blood cells will form. (*c*) The inner layer of the trophoblast and blood vessels invade, forming placental villi. (*d*) Fully formed placenta showing rich vascular supply. (*e*) Enlarged view showing the relationship of maternal and fetal blood vessels in the placenta. Note that the fetal blood vessels are in the placental villus, which is bathed in a pool of maternal blood.

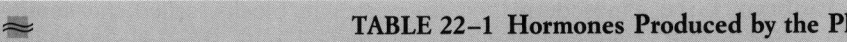 **TABLE 22–1 Hormones Produced by the Placenta**

HORMONE	FUNCTION
Human chorionic gonadotropin (HCG)	Maintains corpus luteum of pregnancy
	Stimulates secretion of testosterone by developing testes in XY embryos
Estrogen (also secreted by corpus luteum of pregnancy)	Stimulates growth of myometrium, increasing uterine strength for parturition (childbirth)
	Helps prepare mammary glands for lactation
Progesterone (also secreted by corpus luteum of pregnancy)	Suppresses uterine contractions to provide quiet environment for fetus
	Promotes formation of cervical mucus plug to prevent uterine contamination
	Helps prepare mammary glands for lactation
Human chorionic somatomammotropin	Helps prepare mammary glands for lactation
	Believed to reduce maternal utilization of glucose so that greater quantities of glucose can be shunted to the fetus
Relaxin (also secreted by corpus luteum of pregnancy)	Softens cervix in preparation for cervical dilation at parturition
	Loosens connective tissue between pelvic bones in preparation for parturition

Estrogen and Progesterone, First From the Ovary and then from the Placenta, Serve a Number of Functions Essential for Pregnancy. Together, estrogen and progesterone stimulate the growth of the uterine endometrium, which is essential to successful pregnancy. By itself, estrogen stimulates growth of the smooth muscle

▷ **FIGURE 22–12 Blood Levels of Placental Hormones**
Human chorionic gonadotropin levels peak in the second month of pregnancy, then drop off by the end of the third month. Levels of estrogen and progesterone, produced chiefly by the placenta, continue to rise.

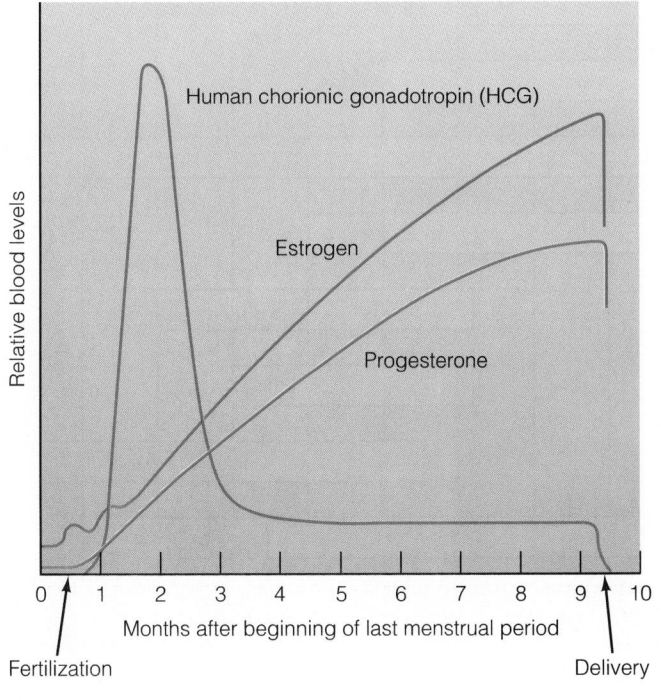

cells of the myometrium, allowing the uterus to expand to many times its original size during pregnancy, thus accommodating the growing baby but also providing additional propulsive force needed to expel the child at birth.

Progesterone helps calm the uterine musculature during pregnancy, preventing premature expulsion of the embryo and fetus. Progesterone also stimulates the production of mucus by the cervix and the formation of a plug of mucus that prevents bacteria from entering the uterus and infecting the growing embryo.

The Amnion Forms from the Inner Cell Mass. As the placenta begins to form, the inner cell mass (ICM) of the blastocyst undergoes some remarkable changes of its own. As Figure 22–11 shows, early in development a layer of cells separates from the ICM to form the **amnion**. A small cavity, the **amniotic cavity,** forms between the ICM and the amnion. The amniotic cavity fills with a watery fluid known as **amniotic fluid,** which forms a protective cushion surrounding the baby during development.[3]

Embryonic Development Begins with Implantation and Ends When the Organs Are More or Less Formed

After the amnion forms, the cells of the inner cell mass differentiate, forming three distinct germ cell layers, the ectoderm, mesoderm, and endoderm, which are known as the **primary germ layers.** The formation of the primary germ layers marks the start of **embryonic development.**

[3] A portion of the amniotic fluid is produced by the fetus. The rest apparently comes from the amniotic membranes.

The Primary Germ Layers of the Embryo Give Rise to the Organs of the Body. As noted earlier, during embryonic development the primary germ layers differentiate a begin to form organs. This process is called **organogenesis.** As shown in ▷ Figure 22–13, organogenesis begins about the third week of pregnancy.

One of the first events of organogenesis is the formation of the central nervous system (the spinal cord and brain), which is produced from ectoderm (Table 22–2). Early in embryonic development, the ectoderm invaginates (folds inward) and forms an indentation, the **neural groove,** which runs the length of the dorsal (back) surface of the embryo (▷ Figure 22–14b). The neural groove deepens and eventually closes off, forming the **neural tube** (Figure 22–14d). The walls of the neural tube thicken, eventually forming the spinal cord. Anteriorly, the neural tube expands to form the brain.

The nerves that attach to the spinal cord (spinal nerves) and brain (cranial nerves) develop from groups of ectodermal cells (the **neural crest**) that lie on either side of the neural tube throughout most of its length (Figures 22–14c and 22–14d). These cells sprout processes that extend into the body and attach to organs, muscle, bone, and skin, among others.

The middle germ layer, the mesoderm, gives rise to muscle, cartilage, bone, and other structures. Much of the mesoderm first aggregates in blocks, called the **somites,** situated alongside the neural tube (Figures 22–14d and ▷ 22–15). The somites form the vertebrae (the backbone) and the muscles of the neck and trunk. Mesoderm lateral to the somites becomes the dermis of the skin, connective tissue, and the bones and muscles of the limbs.

The endoderm, the "lowermost" germ layer of the inner cell mass, forms a large pouch under the embryo called the **yolk sac** (Figures 22–14a and ▷ 22–16). In birds, reptiles, and amphibians, the yolk sac nourishes the growing embryo, but in humans, nourishment comes from the placenta. The human yolk sac gives rise to blood cells and primitive germ cells, called **primordial germ cells.** Interestingly, these cells migrate from the wall of the yolk sac via amoeboid motion to the developing testes and ovaries, where they become spermatogonia or oogonia, depending on the sex of the embryo. Finally, as shown in Figure 22–16c, the uppermost part of the yolk sac becomes the lining of the intestinal tract.

Fetal Development Begins After the Organs Have Formed

As shown in Figure 22–13, organogenesis is well on its way by the end of the eighth week of development. All of the organ systems have begun to develop, and some such as

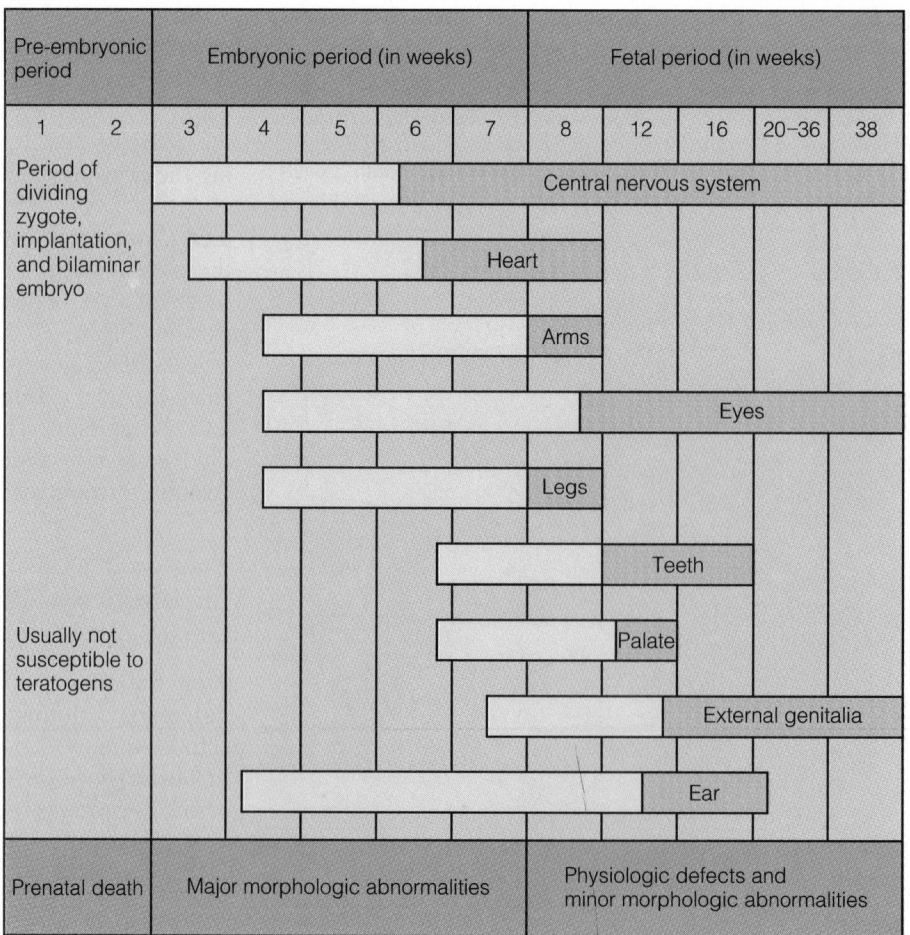

▷ **FIGURE 22–13**
Organogenesis Human development is divided into three periods, or stages: pre-embryonic, embryonic, and fetal. Organogenesis occurs during the embryonic stage. Each bar indicates when an organ system develops. The yellow shaded area indicates the periods most sensitive to teratogenic agents (agents that can cause birth defects).

Ectoderm	Mesoderm	Endoderm
Epidermis	Dermis	Lining of the digestive system
Hair, nails, sweat glands	All muscles of the body	Lining of the respiratory system
Brain and spinal cord	Cartilage	Urethra and urinary bladder
Cranial and spinal nerves	Bone	Gallbladder
Retina, lens, and cornea of eye	Blood	Liver and pancreas
Inner ear	All other connective tissue	Thyroid gland
Epithelium of nose, mouth, and anus	Blood vessels	Parathyroid gland
Enamel of teeth	Reproductive organs	Thymus
	Kidneys	

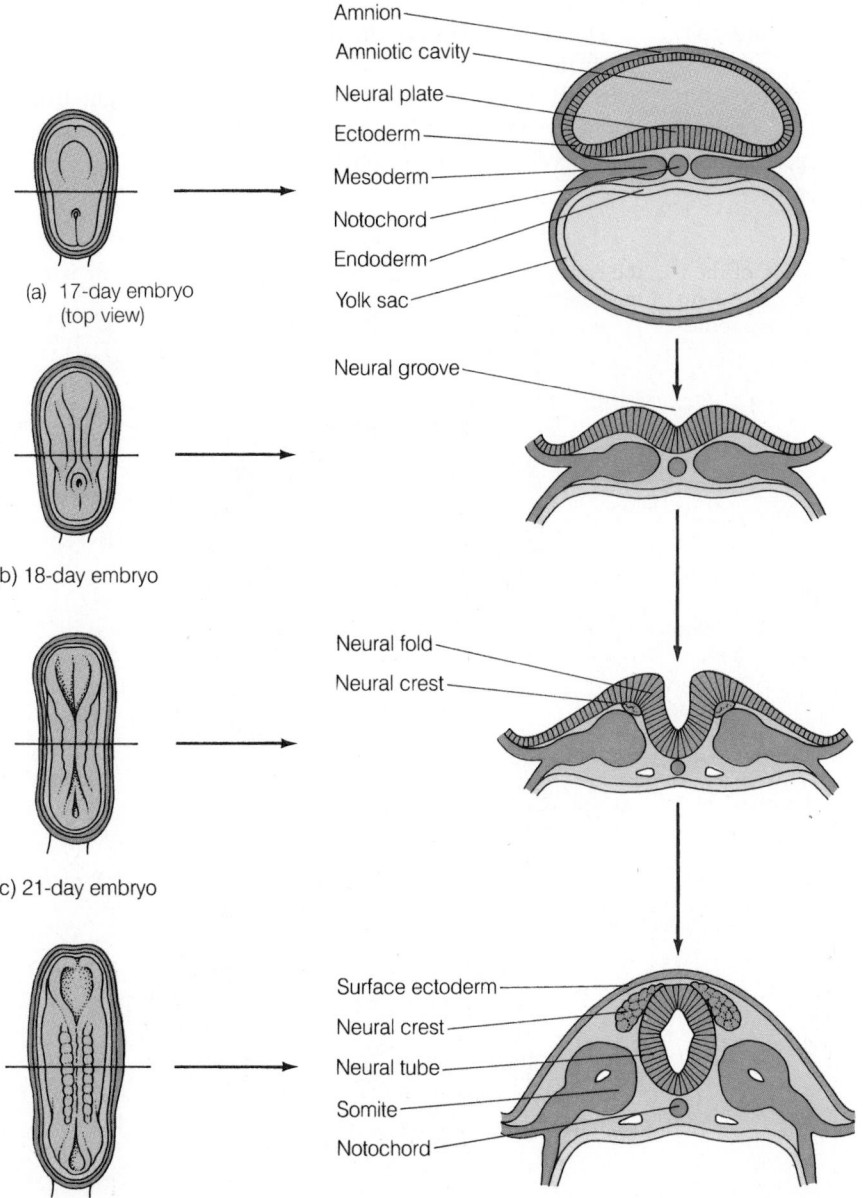

Amnion
Amniotic cavity
Neural plate
Ectoderm
Mesoderm
Notochord
Endoderm
Yolk sac

(a) 17-day embryo (top view)

Neural groove

(b) 18-day embryo

Neural fold
Neural crest

(c) 21-day embryo

Surface ectoderm
Neural crest
Neural tube
Somite
Notochord

(d) 23-day embryo

▷ **FIGURE 22–14 Formation of the Spinal Cord** (*a*) A top view and cross section of a 17-day human embryo showing the relationship between the three embryonic tissues and the amnion and yolk sac. (*b*) The neural groove begins to form from ectoderm. (*c*) The neural groove deepens. (*d*) The neural tube forms. Note the presence of the neural crest, ectodermal cells that give rise to nerves.

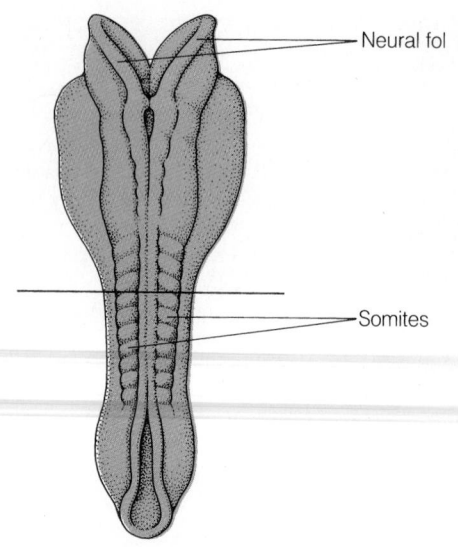

(a)

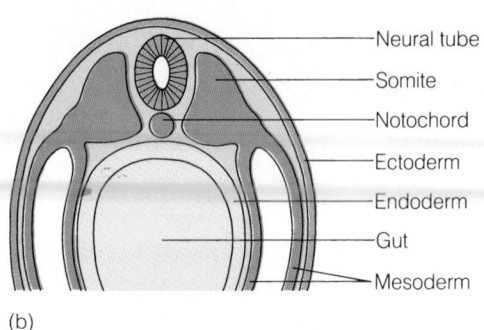

▷ **FIGURE 22–15 Somites** (*a*) Blocks of mesoderm lying lateral to the neural tube in the embryo give rise to the vertebrae of the spinal column and muscles of the back. (*b*) A cross section through the embryo showing the relationship between the somites and the neural tube.

Neural fol

Somites

Neural tube
Somite
Notochord
Ectoderm
Endoderm
Gut
Mesoderm

(b)

the circulatory system (blood vessels and heart) are fully operational.

Fetal development begins in the eighth week and ends at birth. It involves two basic processes: (1) continued organ development and growth and (2) changes in body proportions—for example, elongation of the limbs.

The fetus grows rapidly during the fetal period, increasing from about 2.5 centimeters (1 inch) to 35 to 50 centimeters (14 to 21 inches) and increasing in weight from 1 gram to 3000 to 4000 grams. The fetus also undergoes considerable change in physical appearance.

▷ **FIGURE 22–16 The Yolk Sac** (*a*) A cross section of the embryo. The yolk sac forms from embryonic endoderm. (*b*) A longitudinal section showing how the upper end of the yolk sac forms the embryonic gut, (*c*) which will become the lining of the intestinal tract.

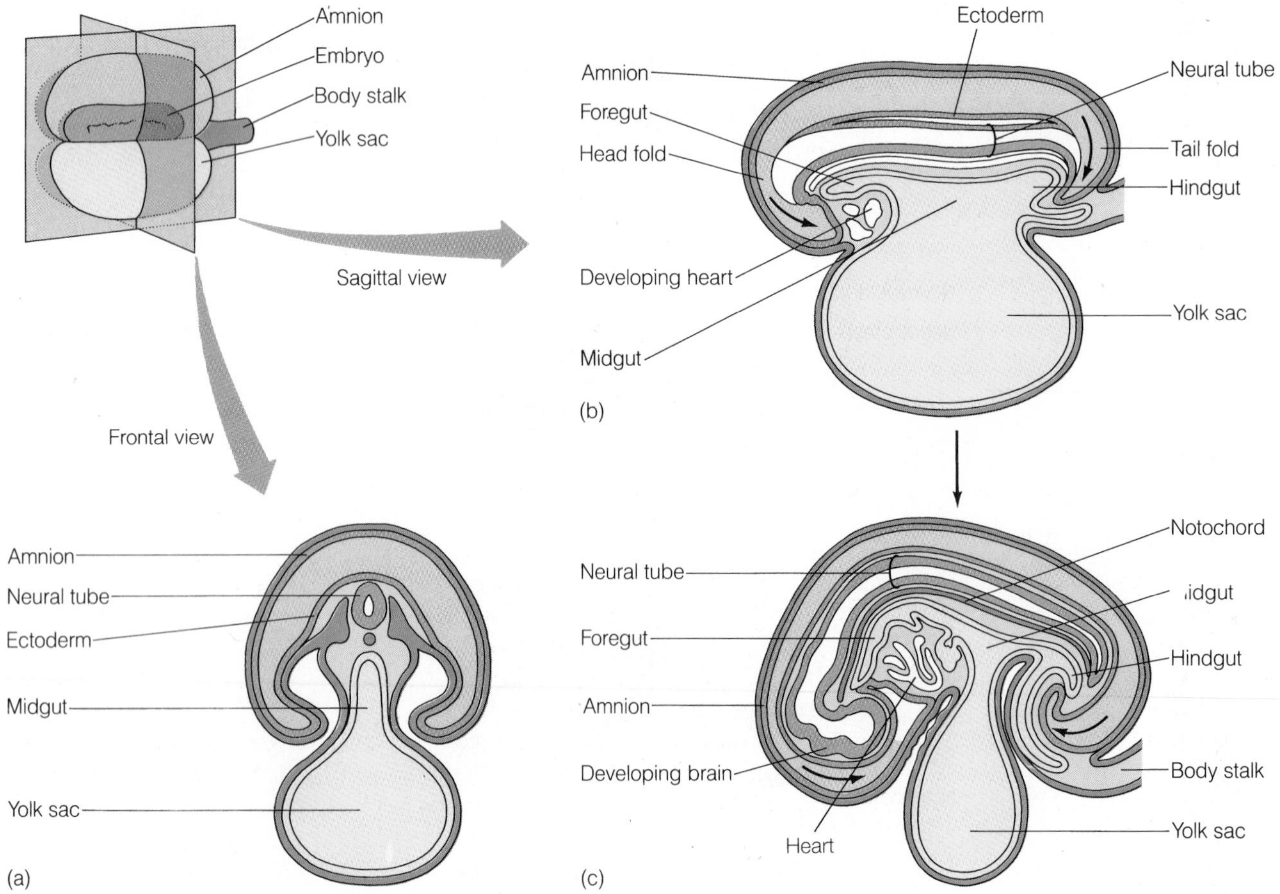

Amnion
Embryo
Body stalk
Yolk sac

Sagittal view

Frontal view

Amnion
Neural tube
Ectoderm
Midgut
Yolk sac

(a)

Ectoderm
Amnion
Foregut
Head fold
Neural tube
Tail fold
Hindgut
Developing heart
Yolk sac
Midgut

(b)

Neural tube
Foregut
Amnion
Developing brain
Heart
Notochord
idgut
Hindgut
Body stalk
Yolk sac

(c)

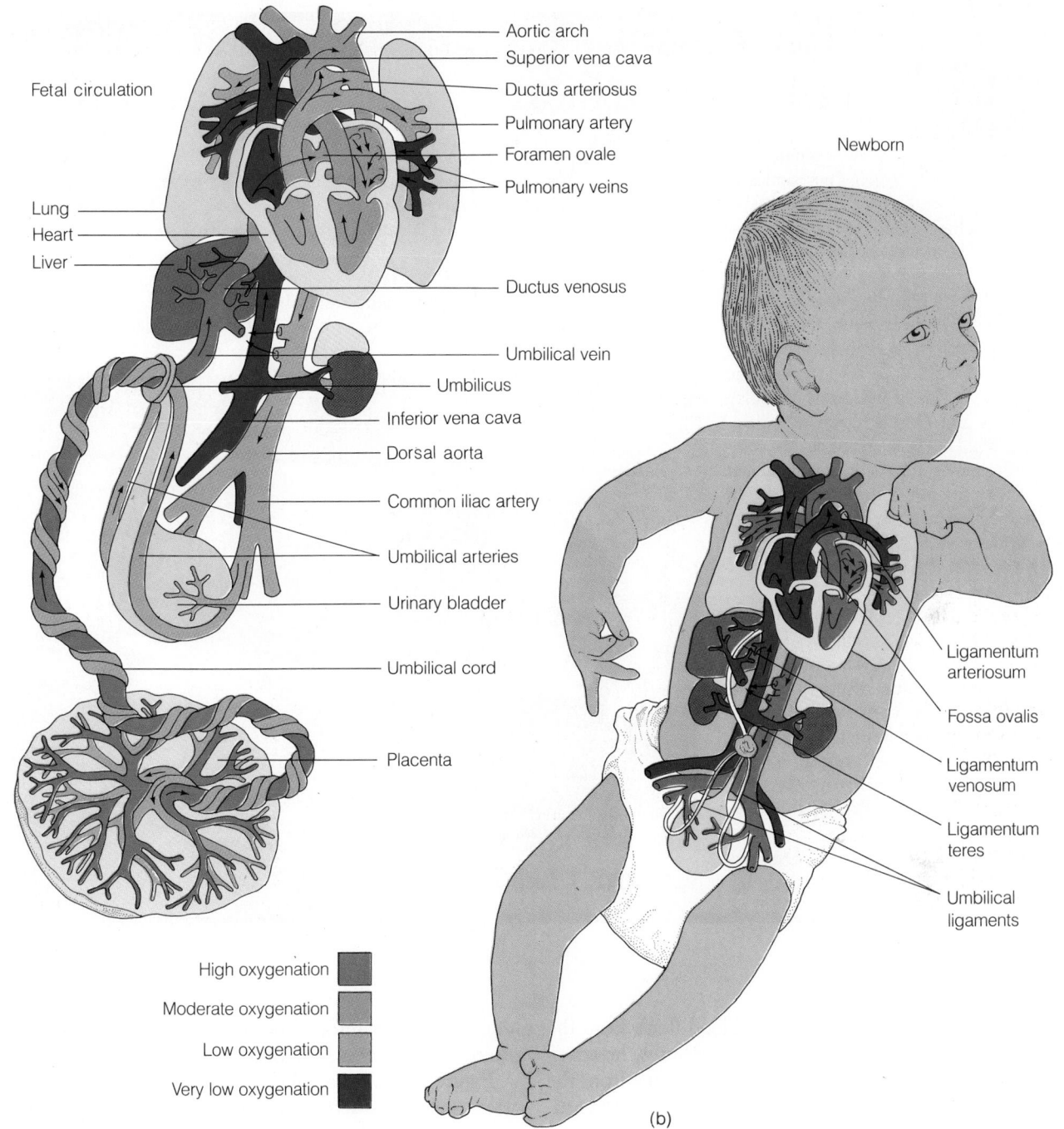

Fetal circulation

Aortic arch
Superior vena cava
Ductus arteriosus
Pulmonary artery
Foramen ovale
Pulmonary veins

Lung
Heart
Liver

Newborn

Ductus venosus

Umbilical vein

Umbilicus

Inferior vena cava

Dorsal aorta

Common iliac artery

Umbilical arteries

Urinary bladder

Umbilical cord

Placenta

Ligamentum arteriosum

Fossa ovalis

Ligamentum venosum

Ligamentum teres

Umbilical ligaments

High oxygenation
Moderate oxygenation
Low oxygenation
Very low oxygenation

(b)

▷ **FIGURE 22–17 Fetal Circulation** (*a*) Before birth. (*b*) After birth. The umbilical arteries and umbilical vein in the fetus shrivel and disappear.

The Fetal Circulatory System Largely Bypasses the Liver and the Lungs, But These Bypasses Disappear at Birth. Space limitations prevent a discussion of the development of each organ system. Instead we will focus our attention on the circulatory system. As shown in ▷ Figure 22–17a, fetal circulation is similar to adult circulation with the exception of several bypasses (adaptations) that divert most of the blood around various organs that are not functional in the fetus. In the fetus, blood travels to and from the placenta in the umbilical cord. The cord

contains two **umbilical arteries** and a single **umbilical vein.** The umbilical vein carries oxygen- and nutrient-rich blood from the placenta to the fetus. Some of this blood flows into and through the fetal liver, as shown in Figure 22–17a. From the liver, the blood passes into the inferior vena cava and on to the heart. Because the liver is not yet functional, most of the blood bypasses the organ via a shunt, the **ductus venosus,** that connects the umbilical vein to the inferior vena cava.

Blood from the inferior vena cava flows into the right

DELAYED IMPLANTATION: EVOLUTION'S UNSOLVED MYSTERY

A woman and her husband who have been trying to have a baby receive a call from their doctor telling them that the latest test was positive. Joyously, they spend much of that evening talking about all the things they want to do for their child. The next week, though, the husband goes to work only to find he will be laid off in a few weeks. The company in his small town is going bankrupt and closing down all operations. With unemployment in his community running at 10%, it is unlikely that he will find a job with decent pay. His wife, who owns a restaurant that caters to workers at her husband's company, realizes that her earnings will probably drop.

Imagine, if you will, that the couple could simply put the pregnancy on hold for five or six months while they found ways to make ends meet. And, once they had gotten their lives back on track, they could restart the pregnancy. Sound like science fiction? It is. Humans can't delay pregnancy, any more than we can change the Earth's orbit around the sun.

In contrast, a number of other animals have a remarkable, but poorly un-

derstood, capacity to put pregnancy on hold. In fact, current estimates suggest that as many as 100 species experience delayed implantation, holding blastocysts in suspended animation in the uterine lumen for periods of a few days to more than a year. Among them are the black bears, western spotted skunks,

weasels, otters, bats, armadillos, and kangaroos (▷ Figure 1).

Consider the fisher, a weasel-like creature that lives in the northeastern United States and much of Canada. In Maine, most female fishers give birth in mid-March. Soon after delivery, the female mates, but the blastocyst remains

▷ **FIGURE 1 Pregnancy on Hold**

atrium of the heart. In an adult, the blood flows from the right atrium into the right ventricle. In the fetus, however, blood follows two paths. Part of the blood flows from the right atrium into the right ventricle. From the right ventricle, it is pumped to the developing lungs via the pulmonary artery, supplying the growing but nonfunctional lung tissue with oxygen and nutrients and picking up waste products of cellular metabolism. The rest of the blood in the right atrium is shunted through a hole in the atrial wall, known as the **foramen ovale.** Blood entering the left atrium passes into the left ventricle, then is delivered to the head and the rest of the body through the aorta and its branches. The foramen ovale essentially diverts much of the blood from the fetal lungs, which like the liver are not yet functional.

The foramen ovale closes at birth in most infants, establishing the adult circulation pattern. In some children, however, the hole remains open and must be corrected surgically. Babies born with an open foramen ovale usually appear bluish, because much of their blood bypasses the

lungs and is therefore low in oxygen. These infants are often referred to as blue babies.

Figure 22–17a also illustrates a third shunt, the ductus arteriosus. The **ductus arteriosus** lies between the pulmonary artery and the aorta. Like the foramen ovale, it helps divert blood away from the lungs. The lungs will need this blood when the baby is born and starts breathing, but during fetal development, the lungs' requirement for blood is considerably smaller.

Blood pumped to the body through the aorta and its branches returns to the heart via the inferior vena cava. Fetal blood, however, also returns to the placenta, where it can pick up nutrients and rid itself of wastes. (Remember, the placenta is serving as the fetus's lungs and digestive system.) A return route, shown in Figure 22–17a, is provided by the **umbilical arteries,** branches of the large arteries of the legs (the iliac arteries). The umbilical arteries carry blood to the placenta, where it is cleansed of much of its waste and recharged with nutrients.

in the uterine cavity for 9 to 10 months before implanting. That is, it does not implant until the following winter. This example raises an important question. Why delay implantation? Why not breed in the fall or early winter and avoid delayed implantation altogether?

Frankly, researchers are not sure. Bill Berg of the Minnesota Department of Natural Resources says that predators like the fisher have to be swift. Carrying blastocysts around instead of developing fetuses makes them lighter and better able to survive through the fall and early winter. Put in evolutionary terms, delayed implantation may have increased the survival and reproductive success of the mother during this critical period. Another wildlife biologist remarked that "delayed implantation allows animals to tailor their reproductive cycle to their yearly food and weather cycles." Again, in evolutionary terms, delayed implantation may be an adaptation to such food and climate cycles.

The answer may lie in some ecological factor or in a development that took place hundreds of thousands of years ago. Here is one purely hypothetical possibility: Perhaps at one time fishers that bred after giving birth in March did

indeed give birth again in the fall. Offspring would have had to survive a harsh winter and were probably selected against. The females may also have been selected against. That is, natural selection may have weeded out those that were pregnant (and slower moving) during the summer and fall. These females may not have been able to get enough to eat or may have fallen prey to predators. Continuing with this hypothesis: Perhaps by chance, some females evolved the ability to delay implantation. Remaining sleeker through the spring, summer, and fall, these individuals had a selective advantage over their pregnant counterparts. They therefore passed their genes on more successfully, so that today all fishers are capable of delayed implantation.

Black bears also exhibit delayed implantation. Unlike fishers, bears mate in the summer, long after giving birth. Nonetheless, black bears still delay implantation until the beginning of the winter.

In black bears, delayed implantation may reduce stress on females as they prepare for winter hibernation; that is, it may help ensure that females receive enough food to get through the winter

themselves. It may also increase the chances of a successful pregnancy, for if a female is not adequately nourished, she will often abort.

The advantage of delayed implantation to black bear cubs is not evident at first glance, because it puts birth in the middle of winter while females are denning. However, consider what would happen if a female black bear gave birth in the fall before she hibernated. In this case, cubs would have very little time to build up body stores of fat to make it through the long winter hibernation. Few would probably succeed and grow to reproductive age. As it is, cubs are born in the dead of winter, often in a protected den where they feed off their mother's milk while she is hibernating. When spring comes, the cubs start eating on their own. By the following fall, they are able to store up enough fat to hibernate.

As in the case of the fisher, these hypotheses, which describe the advantage of delayed implantation to males and females, fail to explain why males and females simply don't wait until the fall to breed. For now, we'll have to wait for an answer, or several answers, as biologists probe this fascinating phenomenon.

At birth, the fetal circulation pattern is replaced by the adult pattern (Figure 22–17b). The umbilical cord is tied off and cut, preventing blood flow to and from the placenta. Inside the fetus, the umbilical arteries and umbilical veins shrivel, becoming ligamentous structures. The foramen ovale closes, so that all the blood entering the right atrium now travels to the right ventricle and the lungs. The ductus venosus and the ductus arteriosus close down and become ligamentous structures, which disappear altogether as a child ages.

Most Ectopic Pregnancies Occur In the Uterine Tubes

An **ectopic pregnancy** occurs when a fertilized ovum develops outside the uterus, usually in the uterine tube.[4]

Thus, most ectopic pregnancies are also called **tubal pregnancies.**

In a tubal pregnancy, the zygote implants in the lining of the uterine tube. However, placental development damages the uterine tubes, causing internal bleeding and severe abdominal pain. Because the uterine tube cannot sustain the embryo, a tubal pregnancy cannot generally proceed to term. Surgery is required to remove the embryo.

Interestingly, about 1 in every 200 pregnancies is ectopic. Usually diagnosed within the first two months, tubal pregnancies occur as a result of congenital defects in the uterine tubes or scar tissue from a previous infection, which impedes the transport of the blastocyst to the uterus.

Embryonic and Fetal Development Can Be Altered by Outside Influences, Sometimes Resulting in Birth Defects and Miscarriages

By several estimates, 31% of all fertilizations end in a

[4] Ectopic pregnancies can also occur in the abdominal cavity.

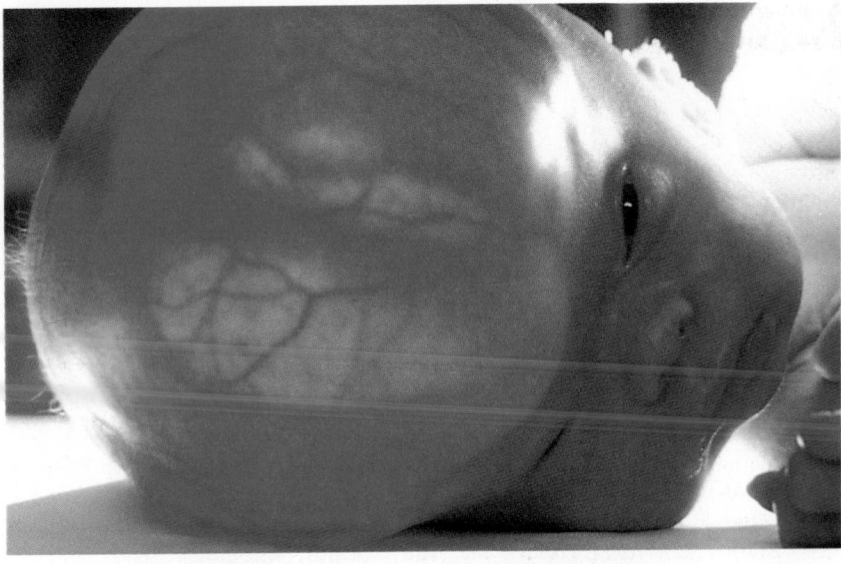
(a)

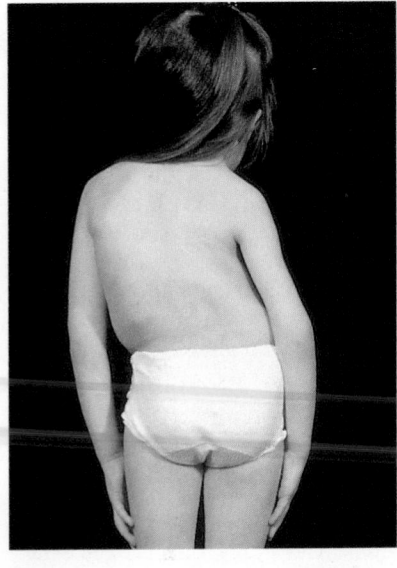
(b)

▷ **FIGURE 22–18 Common Birth Defects** (*a*) Hydrocephalus, or "water on the brain," is caused by an enlargement of the brain's ventricles. (*b*) Scoliosis, a lateral curvature of the spine.

miscarriage—that is, a spontaneous abortion. Two of every three of these miscarriages occur before a woman is even aware that she is pregnant. Why such a high rate? Biologists hypothesize that early miscarriage is nature's way of "discarding" defective embryos. In other words, it is an evolutionary mechanism that helps ensure a healthier population.

Because nature generally eliminates defects, you might expect that most children born into the world would be free of defects. Unfortunately, that is not the case. Humans experience a surprisingly high rate of **birth defects,** physical or physiological abnormalities present in newborns. By various estimates, 10% to 12% of all newborns have some kind of birth defect, ranging from minor biochemical or physiological problems, which are not even noticed at birth, to gross physical defects (▷ Figure 22–18).[5]

The study of birth defects is called **teratology.** The word teratology comes from the Greek *teratos,* meaning "monster" and reflects some of the more gruesome or disfiguring birth defects. Some experts, however, point out that the term is an unfair characterization of all birth defects. Many defects are minor, and people with them are clearly not monsters.

Although there is much to be learned about the causes of birth defects, scientists believe that most arise from chemical, biological, and physical agents, collectively known as **teratogens.** Table 22–3 lists known and suspected teratogens in humans. For example, the virus that produces rubella (German measles) is a known teratogen. So are alcohol, radiation, and megadoses of vitamins A and D.

The effect of a teratogenic agent on the developing embryo is related to three factors: (1) the time of exposure, (2) the nature of the agent, and (3) the dose. Consider

[5] About 2% of all births show gross malformations.

time first. Because the organ systems develop at different times, the timing of exposure determines which systems are affected by a given teratogen. Organ systems are usually most sensitive to potentially harmful agents early in their development, as indicated by the yellow bars in Figure 22–13. The central nervous system, for example, begins to develop during the third week of **gestation,** or pregnancy. Because most women do not know they are pregnant for three to four weeks after fertilization, the central nervous system is at risk. In contrast, the teeth, palate, and genitalia

TABLE 22–3 Known and Suspected Human Teratogens	
KNOWN AGENTS	POSSIBLE OR SUSPECTED AGENTS
Progesterone	Aspirin
Thalidomide	Certain antibiotics
Rubella (German measles)	Insulin
Alcohol	Antitubercular drugs
Irradiation	Antihistamines
	Barbiturates
	Iron
	Tobacco
	Antacids
	Excess vitamins A and D
	Certain antitumor drugs
	Certain insecticides
	Certain fungicides
	Certain herbicides
	Dioxin
	Cortisone
	Lead

do not begin to form until about the sixth or seventh week of pregnancy. Exposure to teratogens during the seventh week therefore might affect the teeth, palate, and genitalia but might have little effect on the central nervous system.[6]

The nature of the teratogen also influences the outcome of exposure. Some teratogens are broad-acting agents; that is, they affect a variety of developing systems. Others are narrow-spectrum agents, affecting only one system. Ethanol in alcoholic beverages, for example, is a broad-acting teratogen. Consequently, children who are born to alcoholic mothers—*or even women who have consumed one or two drinks early in pregnancy*—may have a variety of birth defects, including facial, heart, and skeletal defects. Behavioral problems and learning disabilities are also common in children of alcoholic mothers. These symptoms are part of a condition referred to as **fetal alcohol syndrome.** In contrast, other teratogenic agents are more selective, targeting only one system. Methyl mercury found in fish and shellfish, for instance, damages the central nervous system, creating defects in the brain and spinal cord, but has little effect on other systems.

Finally, consider the dose. Like most toxins, teratogens generally follow a dose-response relationship: the greater the dose, the greater the effect.

One of the best-known teratogens is not a chemical agent but, rather, a biological agent, the virus that causes German measles. Rubella is less common and less contagious than ordinary measles. If a pregnant woman contracts the disease during the first three months of pregnancy, however, she has a one in three chance of giving birth to a baby with a serious birth defect. Deafness, cataracts, heart defects, and mental retardation are common defects. Moreover, 15% of all babies with congenital rubella syndrome die within one year of birth.

Many birth defects can be avoided. German measles vaccine can be given to young girls and may protect them throughout their childbearing years. Vaccinating boys so they do not spread the virus also helps reduce the risk to society. As with many other diseases, prevention is the best policy. During the first eight weeks of pregnancy, women should avoid alcohol and other potential teratogens to protect the developing organ systems. As noted earlier, though, most women do not know that they are pregnant until two to four weeks after conception. Thus, a woman who is trying to get pregnant should carefully control what she eats and drinks.

Once the period of high sensitivity has passed, good nutrition and a healthy environment remain essential. That is because fetal development is also influenced by a number of physical and chemical agents in our homes and places of work. Toxins can stunt fetal growth and, in higher quantities, can even kill a fetus, resulting in a stillbirth. Proper nutrition is essential because nutrient, vitamin, and mineral deficiencies can retard growth.

[6] Note that the central nervous system is by no means fully formed at this time. Some teratogens, such as the rubella virus, have effects through the first three months of gestation.

Women Undergo Many Reversible Physiological and Physical Changes During Pregnancy

During pregnancy a woman's body undergoes dramatic (largely reversible) changes. The uterus, for example, grows to about 20 times its original weight (excluding the weight of the fetus and placenta). About the size of a pear or a woman's fist at the onset of pregnancy, the uterus (with its contents) expands to fill most of the pelvic and abdominal cavities ($\triangleright$ Figure 22–19). The enlarged uterus pushes up on the diaphragm and rib cage, sometimes making breathing difficult. It also presses against the abdominal organs, such as the stomach. This crowding may cause acid from the stomach to percolate up into the esophagus, creating heartburn. The uterus also pushes out on the abdominal wall, which accentuates the normal curvature of the lower back and strains the ligaments of the spinal column, often resulting in back pain in the last month or two of pregnancy. In addition, the uterus compresses the urinary bladder, resulting in more frequent urination.

During pregnancy, the uterus and vagina become engorged with blood. In some women, the rise in vaginal vascularity increases its sensitivity to touch, heightening sexual pleasure. The breasts also enlarge during pregnancy as a result of growth in the glands and ducts. These changes help prepare the breasts for milk production, or **lactation** (described below).

A woman's blood volume increases during pregnancy by about 30%, which is partly responsible for the 20 to 30 pounds most women gain through the gestational period. Additionally, weight gain results from the growth of the fetus and placenta, buildup of amniotic fluid, and enlargement of the uterus and breasts (Table 22–4).

During pregnancy, a woman is eating, breathing, and excreting (urinating) for two living beings. Therefore, the presence of an embryo and fetus results in three additional changes during pregnancy: (1) an increase in appetite, (2) an increase in the rate of breathing, and (3) an increase in urinary output. However, the humorist Dave Barry points out in his book *Babies and Other Hazards of Sex* that one of the two beings, early on, is only the size of a golf ball. In other words, just because she is eating for two, a woman doesn't have to double her food intake.

The embryo and fetus are akin to an internal parasite, acquiring nutrients from its host, sometimes at the host's expense. If the fetus needs calcium, for example, it draws on maternal supplies even if that means a thinning of her bones. If it needs iron, it generally gets what it needs, even if the mother becomes anemic in the process. This is not to imply that a fetus can always acquire sufficient nutrients from its mother. If a woman's diet is deficient and body reserves are small or nonexistent, fetal deficiencies may occur. If a mother's diet contains inadequate protein, for example, fetal growth may be stunted. Accordingly, women are advised to eat carefully during pregnancy, not only for their own nutrition but also for the nourishment of their offspring.

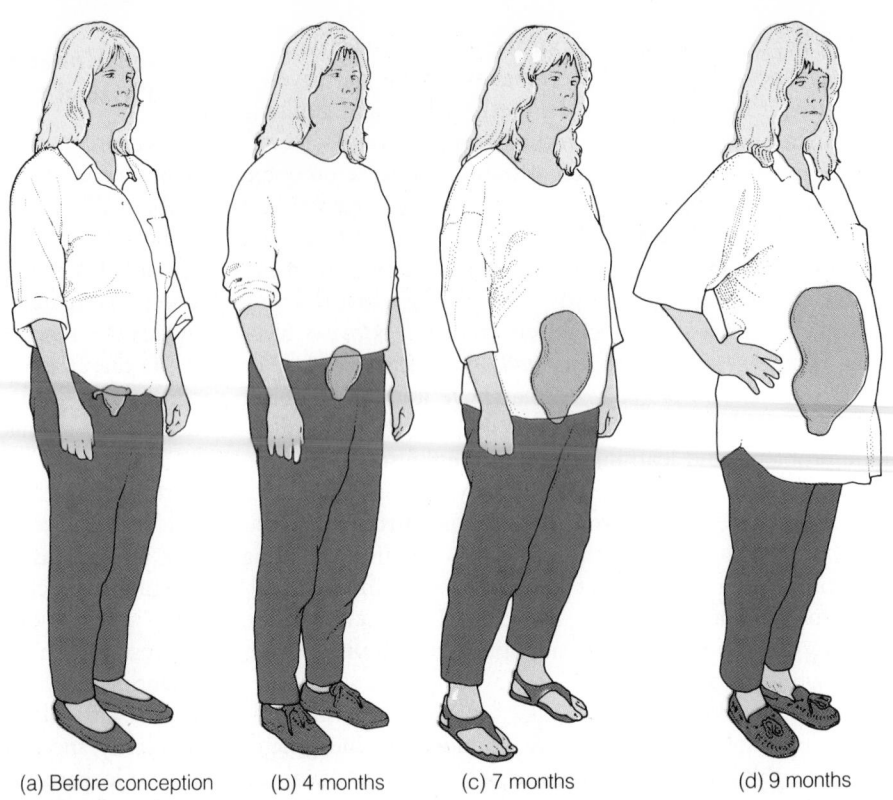

(a) Before conception   (b) 4 months   (c) 7 months   (d) 9 months

After the baby is delivered, the uterus shrinks, and the blood volume returns to normal. The breasts remain engorged and begin producing milk, but if the baby is not breast-fed, the breasts will return to normal in a matter of weeks. Many women are able to return to their prepregnancy weight.

≋ CHILDBIRTH AND LACTATION

For most expectant parents, the birth of a child is one of the most exciting events of their lives. Childbirth, or labor, begins with mild uterine contractions, which increase in intensity and frequency until the baby is born.

Uterine Muscle Contractions Are Stimulated by a Change in Hormonal Levels

During the last few weeks of pregnancy, estrogen levels in the blood of a woman increase dramatically. Research sug-

≋	TABLE 22–4 Average Weight Gain during Pregnancy	≋
SOURCE OF WEIGHT GAIN		POUNDS
Total		24
Fetus		7–8
Amniotic fluid, placenta, and fetal membranes		4–5
Uterus		2
Blood		6–7
Breasts		2

gests that this rise in estrogen levels stimulates the production of oxytocin receptors in the smooth muscle cells of the uterus. As noted in Chapter 20, oxytocin is a hormone released by the posterior pituitary.

Oxytocin stimulates uterine muscle contractions. Thus, the increase in the number of receptors in the smooth muscle cells renders them increasingly responsive to the small amounts of oxytocin released from the posterior pituitary at the end of pregnancy.

Researchers believe that the high levels of estrogen at the end of pregnancy may block the "calming" influence of placental progesterone. Consequently, the uterus begins to contract at irregular intervals. Irregular contractions usually begin a month or two before childbirth and are known as false labor, or **Braxton-Hicks contractions.** As the due date approaches, false labor contractions often occur with greater frequency, causing many anxious couples to race to the hospital only to be told to go home and wait for a few weeks for the real thing.

True labor contractions are more intense and occur more frequently than those of false labor. These contractions at the beginning of true labor are thought to be triggered by the fetus itself. The fetal pituitary gland releases small quantities of oxytocin right before birth. Fetal oxytocin travels across the placenta, entering the mother's blood. In the mother, it stimulates uterine contractions in the sensitized musculature. Fetal oxytocin also stimulates the release of placental prostaglandins that act on smooth muscle. Prostaglandins, acting in concert with fetal oxytocin, stimulate more powerful and more frequent uterine contractions.

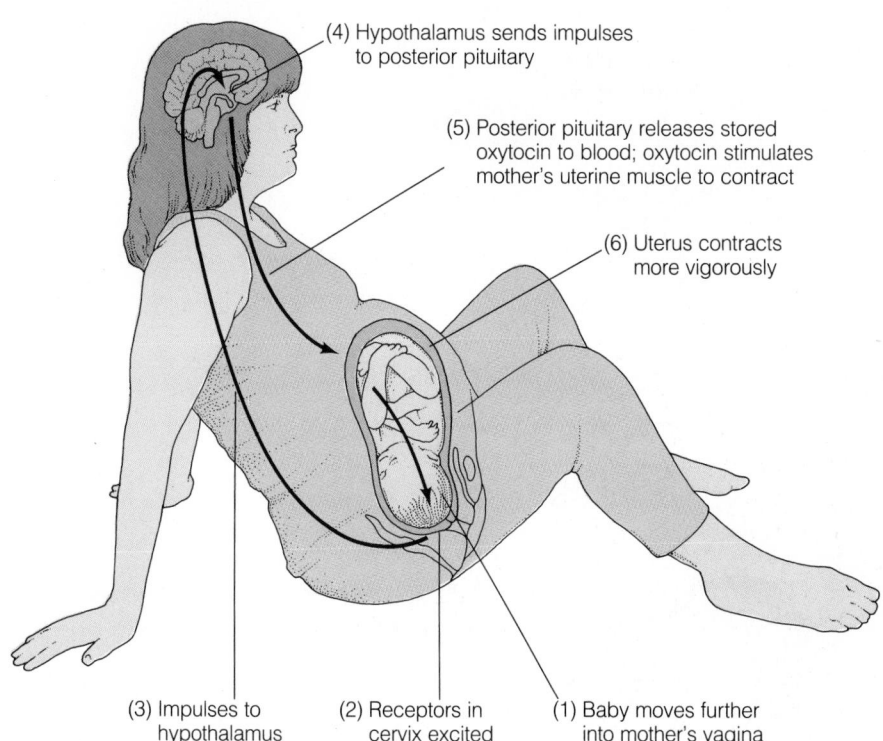

(4) Hypothalamus sends impulses to posterior pituitary

(5) Posterior pituitary releases stored oxytocin to blood; oxytocin stimulates mother's uterine muscle to contract

(6) Uterus contracts more vigorously

▷ **FIGURE 22–20 Oxytocin Positive Feedback Mechanism in Birth** The mechanism continues to cycle until interrupted by the birth of the baby.

(3) Impulses to hypothalamus

(2) Receptors in cervix excited

(1) Baby moves further into mother's vagina

Emotional and physical stress in the mother may also play a role in triggering childbirth. Uterine contractions and discomfort create stress, which is believed to trigger the release of maternal oxytocin. Maternal oxytocin, in turn, augments muscle contractions that are already underway (▷ Figure 22–20). Thus, a positive feedback mechanism is triggered. As uterine muscle contraction increases, maternal oxytocin release increases. This stimulates even stronger contractions, resulting in additional oxytocin release, a cycle that continues until the baby is delivered.

Childbirth also requires the hormone **relaxin,** produced by the ovary and the placenta. Released near the end of gestation, relaxin has two effects. First, it softens the fibrocartilage uniting the pubic bones. This allows the pelvic cavity to widen and thus greatly facilitates childbirth. Relaxin also softens the cervix, allowing it to expand to allow the baby to pass. Without this remarkable adaptation, delivery would be a problematic event indeed!

Childbirth Occurs in Three Stages

▷Figure 22-21 illustrates the three stages of childbirth.

During the Dilation Stage, the Baby Is Pushed Against the Cervix, which Causes Cervical Dilation.
Stage 1, the **dilation stage,** gets its name from the dilation of the cervix. This phase begins when uterine contractions start and generally lasts 6 to 12 hours, but sometimes much longer. At the beginning of stage 1, uterine contractions typically last only 30 seconds and may come every half an hour or so. As time passes, however, uterine contractions become more frequent and powerful. Uterine contractions generally rupture the amnion early in

the dilation phase, causing the release of the amniotic fluid, an event commonly referred to as "breaking the water."

Uterine contractions push the infant's head against the relaxin-softened cervix, causing it to stretch and become thin (Figure 22–21b). By the end of stage 1, the cervix has dilated to about 10 centimeters (4 inches), approximately the diameter of a baby's head.

The baby is pushed downward by the uterine contractions and descends into the pelvic cavity. When the head is "locked" in the pelvis, the baby is said to be **engaged.**

During the Expulsion Stage, the Baby Is Pushed Out of the Uterus, Through the Cervix and Vagina.
Stage 2, the **expulsion stage,** begins after the cervix is dilated to 10 centimeters and the baby is engaged (Figure 22–21c). At this time, uterine contractions usually occur every 2 or 3 minutes and last 1 to 1.5 minutes each. For most women having their first child, 50 to 60 minutes are required for delivery. If a woman is delivering her second child, only 20 to 30 minutes of uterine contraction are required to push the infant's head through the cervix and the vagina. The expulsion stage ends when the child is pushed through the vagina into the waiting hands of a doctor or midwife.

To facilitate the delivery, physicians or midwives often make an incision to widen the vaginal opening. This procedure, called an **episiotomy,** is performed when the baby's head enters the vagina in stage 2 of labor. The incision enlarges the vaginal orifice, prevents tearing, and allows the infant to pass quickly. The incision is stitched up immediately after the baby is born.

Once the baby's head emerges from the vagina, the rest of the body slips out almost instantly. However, the baby

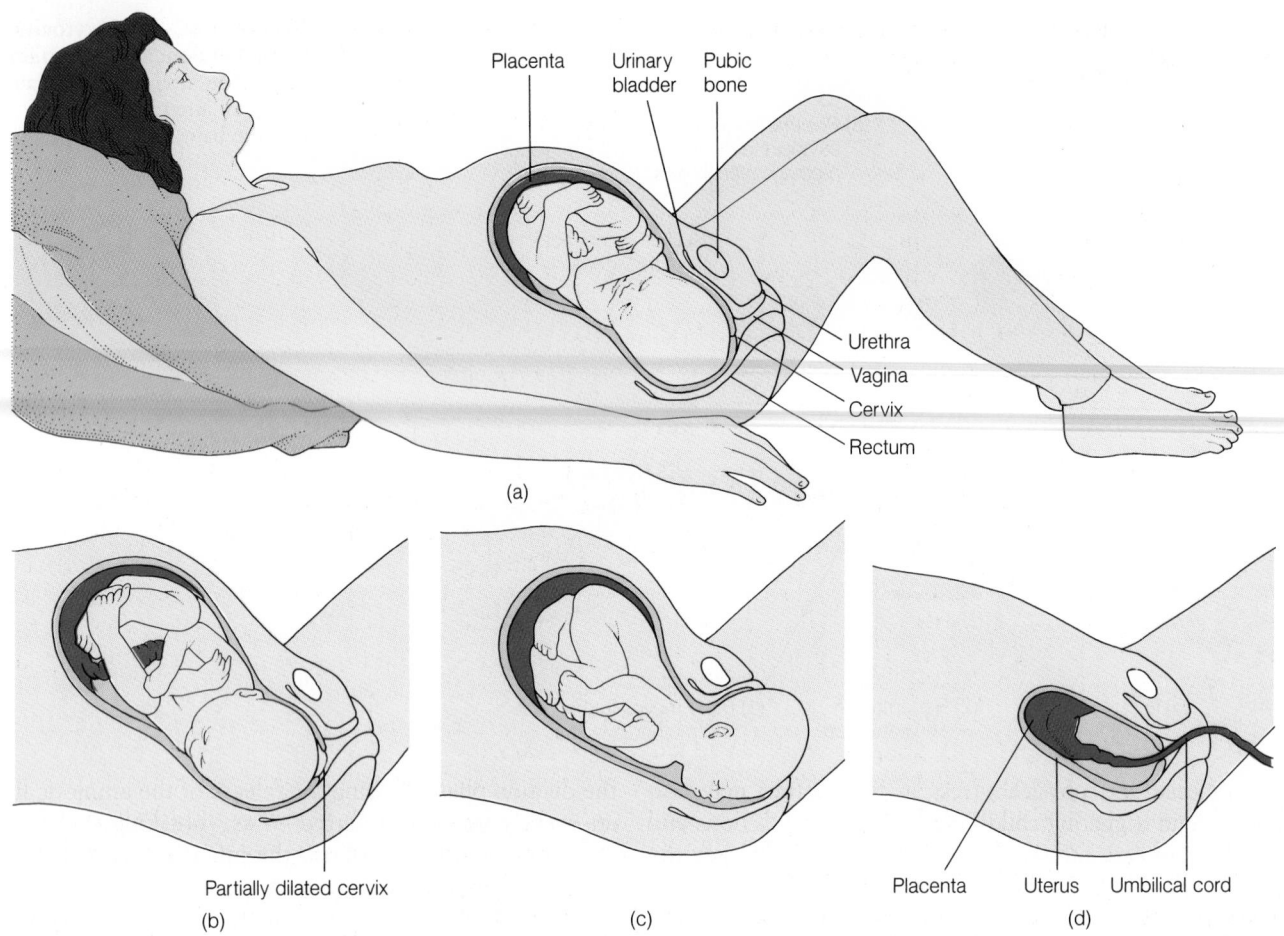

Placenta   Urinary   Pubic
            bladder   bone

Urethra
Vagina
Cervix
Rectum

(a)

Partially dilated cervix

(b)   (c)

Placenta   Uterus   Umbilical cord

(d)

▷ **FIGURE 22–21 Stages of Labor** (*a*) Position of the fetus near birth. (*b*) Dilation stage. Uterine contractions push the fetal head lower in the uterus and cause the relaxin-softened cervix to dilate. (*c*) Expulsion stage. Fetus is expelled through the cervix and vagina. (*d*) Placental stage. Placenta is delivered.

remains attached to the placenta via the umbilical cord, which continues to deliver blood to the baby for a minute or so. Consequently, many health-care workers wait a minute or two to allow the blood remaining in the placenta to be pumped into the newborn before they tie off and cut the umbilical cord.

Most babies (95%) are delivered head first with their noses pointed toward the mother's tailbone (Figure 22–21). Occasionally, however, babies may be oriented in other positions. This makes delivery more difficult, time-consuming, and hazardous for the mother and the baby. The most common alternative delivery is the **breech birth,** in which the baby is expelled rear-end first. Breech births require more time and may cause extreme fatigue in the mother and brain damage in the baby. The umbilical cord sometimes wraps around the baby's neck, cutting off its supply of blood. To avoid complications, breech babies are often physically turned by physicians before birth by applying pressure to the woman's abdomen. If a fetus cannot be turned, the baby is usually delivered by a **cesarean section,** a horizontal incision through the abdomen just at the pubic hair line. Cesarean sections may be performed for other

reasons as well—for example, if labor is prolonged or if the mother has an active infection caused by herpes or some other sexually transmitted disease that might be transferred to her child as it passed through the vagina.

The Placental Stage Is Marked by the Delivery of the Placenta. The final stage of delivery is the **placental stage** (Figure 22–21d). The placenta, sometimes called the afterbirth, remains attached to the uterine wall for a short while, then is expelled by uterine contractions, usually within 15 minutes of childbirth. After the placenta is expelled, the uterine blood vessels clamp shut, preventing hemorrhage, although the mother continues to lose some blood for 3 to 6 weeks after delivery.

The uterus gradually returns to its normal size after delivery. Uterine involution (shrinkage) results from the rapid decline in estrogen and progesterone and is accelerated by oxytocin released by the posterior pituitary when a woman breastfeeds her baby. Complete involution in women who breast-feed usually occurs within four weeks of pregnancy. In women who do not, the process usually takes six weeks.

The Pain Associated with Childbirth Can Be Relieved by Drugs and by Special Birthing Methods that Seek to Reduce Tension

The level of pain a woman feels during childbirth varies greatly and is partly governed by her level of fear and tension. The more tense a woman is, generally, the more pain she feels. For this reason, many hospitals provide comfortable birthing rooms and relaxation training to expectant mothers. Drugs can also be given to reduce tension and pain, but they must be administered well before the birth of the baby to be effective. If given just before delivery, the drug offers little relief to the mother and may retard the baby's breathing.

Painkilling drugs can be injected into the wall of the vagina with a syringe. This procedure, known as a **pudendal block,** is performed if an episiotomy is likely or if forceps are going to be used to facilitate birth. Many physicians routinely offer **epidural anesthesia.** An anesthetic agent that temporarily deadens the sensory nerves in the vagina and elsewhere is injected into the bony canal that houses the spinal cord, just outside the dura mater. The drug blocks pain in the nerves that supply the lower body, stopping all sensations. This anesthetic may also block motor nerve impulses to the muscles of the abdomen, rendering a woman unable to push the baby out. In such cases, forceps may be required to facilitate the delivery, but forceps can damage a newborn and are generally avoided whenever possible.

Many couples and health-care workers believe that drugs may be harmful to mothers and their babies. This belief and a desire to do things "naturally" have spawned the **natural childbirth** movement. Natural childbirth means different things to different people. In general, it refers to drug-free deliveries. Couples receive training in relaxation and special breathing techniques.

The most popular natural childbirth method today is known as the Lamaze technique. Parents who are going to use this method attend childbirth classes in which they learn special breathing techniques. Shallow breathing, for example, is used when uterine contractions begin; it keeps the diaphragm from pressing down on abdominal organs, reducing tension and pain. It also ensures an adequate supply of oxygen to the fetus. Other breathing techniques are also learned.

The second most popular method is the Bradley method. The Bradley method teaches women how to relax during uterine contractions, thus reducing pain and the need for drugs. No special breathing is required. The woman helps push, as in the Lamaze method, but only near the end of childbirth.

Milk Production, or Lactation, Is Controlled By the Hormone Prolactin

A baby emerges into a novel environment at birth. No longer connected to the placental lifeline, the newborn must now breathe to get its oxygen and dispose of carbon dioxide, and it must also find a new way of acquiring food. For most of human evolution, newborns have been fed milk from their mothers' breasts.

The breasts of a nonpregnant woman consist primarily of fat and connective tissue interspersed with milk-producing glandular tissue. The glands are drained by ducts that lead to the nipple (▷ Figure 22–22). As noted earlier, during pregnancy the ducts and glands proliferate under the influence of placental and ovarian estrogen and progesterone. Milk production is induced by prolactin and a placental hormone, HCS (described below).

Prolactin secretion in the mother begins during the fifth week of pregnancy and increases throughout gestation, peaking at birth (▷ Figure 22–23). The placenta also secretes a mildly lactogenic (milk-producing) hormone, **human chorionic somatomammotropin (HCS).** Despite the presence of these lactogenic hormones, milk production does not begin until two or three days after birth. Why? Because high levels of estrogen and progesterone throughout pregnancy, which are necessary for breast development, inhibit the action of the prolactin and HCS. When estrogen and progesterone levels decline after birth, HCS and prolactin can exert their effect.

Although the breasts do not produce milk for two or three days, they immediately begin producing small quantities of colostrum. **Colostrum** (co-loss-trum) is a fluid rich

▷ **FIGURE 22–22 The Lactating Breast** The glandular units enlarge considerably under the influence of progesterone and prolactin. Milk is expelled by contraction of musclelike cells surrounding the glandular units. Ducts drain the milk to the nipple.

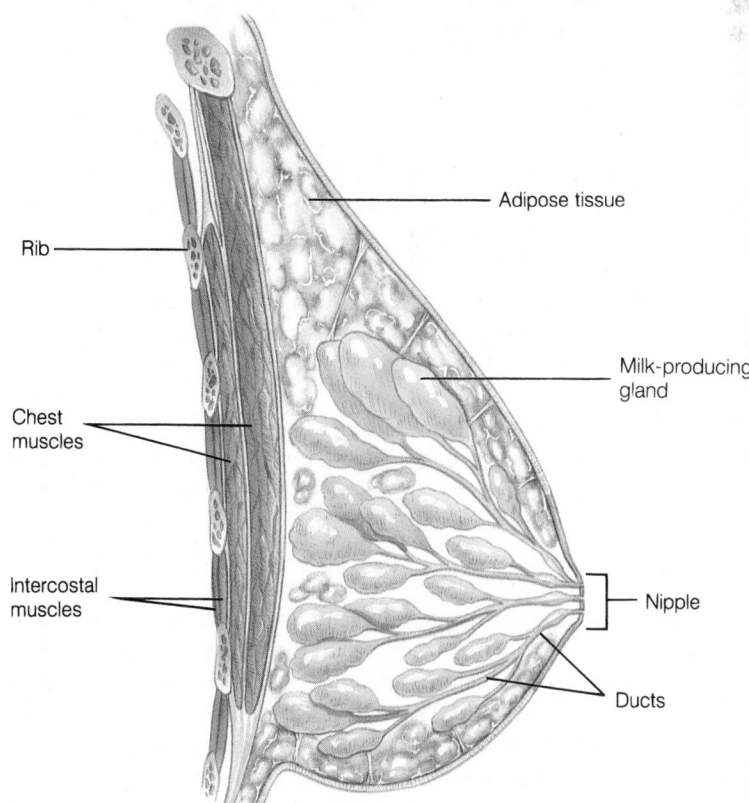

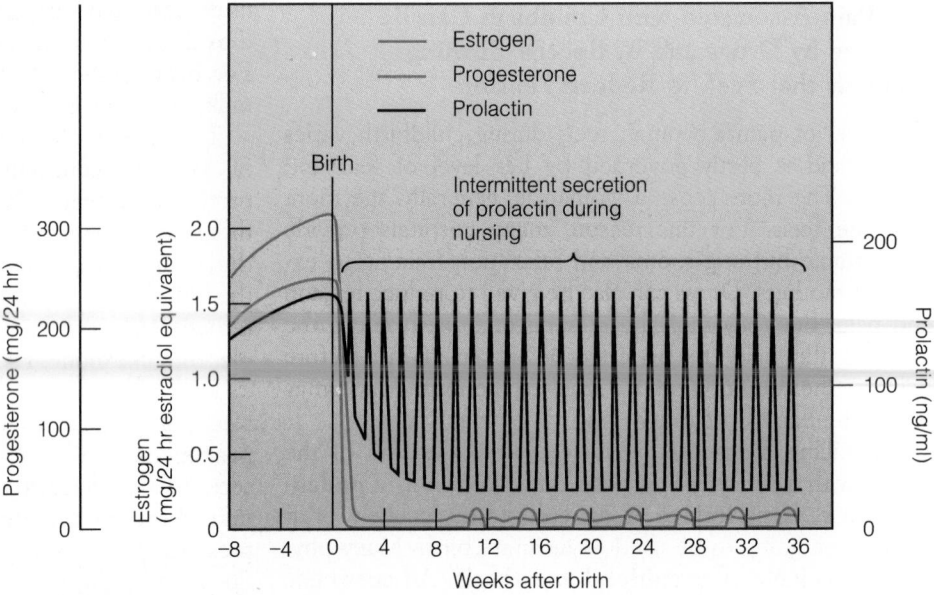

in protein and lactose, but lacking fat. It contains antibodies that protect the infant from bacteria. A newborn subsists on colostrum for the first few days. (Health Note 15–1 describes the benefits of colostrum and breast feeding in general.)

Milk production begins in earnest when estrogen and progesterone levels plummet (Figure 22–23). Thus, within two to three days, the breasts start to produce large quantities of milk. Milk production is facilitated by growth hormone, cortisol, and parathyroid hormone.

Prolactin Secretion Is Maintained After Childbirth by Suckling. Besides stimulating oxytocin release, suckling causes a surge in prolactin secretion. Prolactin levels remain elevated for approximately one hour after each feeding. As a result, each surge stimulates milk production needed for the next feeding. Milk production continues as long as the baby suckles. If nursing is interrupted for three or four days, however, the breasts stop producing milk. In most women, milk production begins to decline by the seventh to ninth month of lactation, a time at which her baby is beginning to feed on semisolid food and breast-feeds less frequently.

Prolactin Secretion Is Controlled By a Neuroendocrine Reflex. Suckling generates nerve impulses that travel to the hypothalamus, causing the release of **prolactin releasing hormone (PRH)**. PRH, in turn, stimulates prolactin secretion. When suckling ends, PRH secretion stops.

In order for milk to reach the nipple, it must be actively propelled from the glandular units and through the ducts. This is the function of oxytocin, the release of which is also controlled by a neuroendocrine reflex, described in Chapter 20. Neuroendocrine reflexes are important adaptations that provide services through hormones on demand. Such systems help conserve energy and nutrients and contribute to the evolutionary success of multicellular animals.

⪢ INFANCY, CHILDHOOD, ADOLESCENCE, AND ADULTHOOD

Describing the changes that occur from birth to adulthood in one section of one chapter is a little like trying to define the universe in a paragraph. Nevertheless, there are some important milestones along this exciting journey worth noting. Much of this information will be relevant if you have, or are planning to have, children.

Infancy Lasts From Birth to the End of the Second Year of Life

At birth, the infant emerges from its warm, relatively secure uterine home—where its needs were catered to automatically—to the home environment, where it must play a much more active role to satisfy its needs. The human newborn survives in its new environment largely because of parental care and because it comes equipped with several built-in reflexes. One of the most important is the **suckling reflex.** Within minutes of birth, the baby begins to suckle the breast (or anything else put in its mouth).

Babies also come equipped with a **rooting reflex,** which helps them find the nipple. Pressure from a nipple (or any stimulus) touching the cheek causes infants to turn their head toward the stimulus. Interestingly, crying is also a reflex in the newborn. Babies cry when they are hungry, hurt, or uncomfortable. The crying reflex alerts parents that something is wrong. Usually, a simple remedy will cure the crying and restore quiet—at least for a while.

During infancy, the child undergoes a dramatic physical and mental transformation. Although children develop in predictable stages (▷ Figure 22–24), the timetable varies from one child to the next. One infant, for example, may learn to crawl at 5 months but another may not crawl until its 7th or 10th month. This variability is natural and to be expected.

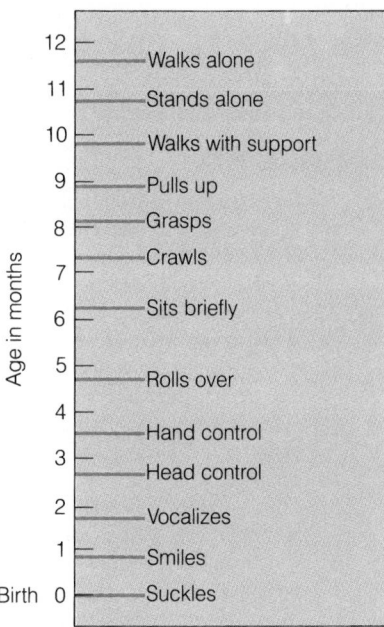

Age in months	
12	Walks alone
11	Stands alone
10	Walks with support
9	Pulls up
8	Grasps
7	Crawls
6	Sits briefly
5	
4	Rolls over
3	Hand control
2	Head control
1	Vocalizes
Birth 0	Smiles / Suckles

▷ **FIGURE 22–24 Motor Development in the Year after Birth**

Within the first year, the child's body weight triples. Even though growth continues at a fairly fast pace, however, the rate of growth begins to decline. Rapid physical growth is replaced by a growth in sensory, motor, and intellectual capacity.

At birth, the child is fairly immobile but capable of moving its arms and feet in joy, protest, or play. By the 5th month, the baby learns to roll over, a maneuver that provides surprising mobility. Next, the infant learns to crawl, often slithering along the floor like a snake. During the 7th month, the baby begins to crawl on its hands and knees. By the 11th or 12th month, the child stands upright. From an upright position, the child quickly masters the art of walking. A parent's life is forever changed. Throughout this period, the brain is changing as myelin is laid down on axons.

During infancy, the child also becomes more aware of its environment and more and more interactive with parents and peers. Child development involves important emotional and intellectual achievements. One of the first emotional milestones of normal human development is **bonding,** the establishment of an intimate relationship between the infant and one (or preferably both) of its parents or caregivers. Some psychologists think that bonding begins within an hour or so of birth. The initial contact, they say, is crucial to developing the bond. Others disagree. They think that bonding may require a period of several months.

What is important is for children to develop a close relationship with caregivers and to learn that they can be comforted by another human and that they can trust another person explicitly. Successful bonding or attachment is important for subsequent development and may lead to emotional security later in life.

As the brain develops, infants become more curious. They begin to remember things and show rudimentary signs of reasoning. With these changes come improvements in motor control.

Research suggests that intelligence, like motor skills, improves with use. Children are eager explorers, ready to tear apart drawers or closets to explore the objects of the adult world. Parents are encouraged to foster their children's curious nature, to let them explore, and to provide opportunities for a rich and varied experience. A child's innate curiosity, then, is a seed of intellectual development that a parent can nurture. Parents are also advised to provide ample stimuli for their children—toys, books, and personal interaction. Parents should read to their baby even before the infant can talk, should point out and name objects, and should talk to the child whenever possible.

Childhood Lasts From Infancy to Puberty, the Beginning of Adolescence

Growth continues throughout childhood, but as in infancy, the rate of growth continues to decline. During childhood, children display increased motor abilities. Language skills improve enormously. Curiosity abounds, and mental capacity increases remarkably. In early childhood (even during infancy), children start to assert their own will. They may resist or even disobey parents. Asserting one's will and developing an identity are important and natural phases of growth, and parents should be careful not to quash a child's developing will. That task requires the judicious use of discipline, and, of course, lots of love.

Adolescence Lasts From Puberty, or Sexual Maturation, to Adulthood

Puberty, or sexual maturation, marks the onset of **adolescence,** often a rather tumultuous period of growth and development that lasts until adulthood (age 18 to 20). For girls, puberty begins about age 10 or 11. For boys, it begins at age 12 or 13. As in all developmental processes, individual variation is to be expected.

Sexual maturation, discussed briefly in Chapter 21, results from the production of sex steroid hormones in both boys and girls. Research suggests that the pituitary, hypothalamus, and gonads begin functioning long before puberty, but the hypothalamus is extremely sensitive to circulating levels of sex steroid hormones. These hormones inhibit gonadotropin release through a negative feedback mechanism. Thus, the hypothalamus's high degree of sensitivity to sex steroid hormones inhibits gonadotropin release. As an individual ages, however, the hypothalamus becomes less sensitive to the negative feedback influence of these hormones. GnRH release increases, and the pituitary output of gonadotropins climbs. This rise, in turn, increases gonadal hormone production, inducing puberty.

Tables 22–5 and 22–6 summarize some of the impor-

TABLE 22–5 Physical Development in Adolescent Girls

PHYSICAL CHANGE	AVERAGE AGE WHEN CHANGE BEGINS	AVERAGE AGE WHEN NOTICEABLE CHANGE USUALLY STOPS	REMARKS
Increase in rate of growth	10 to 11	15 to 16	If conspicuous growth fails to begin by 15, consult your physician.
Breast development	10 to 11	13 to 14	Noticeable development of breasts (one of which may begin to "bud" before the other) is usually the first sign of puberty. If changes do not occur by 16, there may be cause for concern.
Emergence of body hair	Pubic hair: 10 to 11 Underarm hair: 12 to 13	13 to 14 15 to 16	Age at first appearance of body hair is extremely variable. Pubic hair usually darkens and thickens as puberty progresses.
Development of sweat glands	12 to 13	15 to 16	The apocrine sweat glands are responsible for increased underarm sweating, which causes a type of body odor not present in children.
Menstruation	11 to 14	45 to 55	Menstruation often begins with extremely irregular periods, but by 17 a regular cycle (3 to 7 days every 28 days) usually becomes evident. If menstruation begins before 10 or has not begun by 17, consult your physician.

SOURCE: From *The American Medical Association Family Medical Guide.* Copyright © 1982 by The American Medical Association. Reprinted by permission of Random House, Inc.

tant physical changes taking place during puberty. The physical changes induced by ovarian and testicular steroids during puberty were discussed in Chapter 21. Effects on bone were discussed in Chapter 19.

Adolescents also undergo a tremendous change in mental capacity. They become capable of dealing with abstract concepts, and their thinking skills often improve considerably. Rudiments of critical thinking are apparent. A teenager may, in fact, use the scientific method to test hypotheses.

Adolescence is a time of emerging identity. That identity may contrast sharply with a parent's. It may also run counter to a parent's desire for the child, leading to conflict. Accepting the emerging personality and maintaining a healthy relationship require that parents learn to tolerate, better yet appreciate, differences. Love, support, and understanding are as important now as at any time in human development.

Adulthood Begins When Adolescence Ends

In adults, physical growth stops, but for many people social and psychological development continues throughout adult life. Not everyone continues to grow during adulthood, however. Many people seem to congeal, remaining more or less the same throughout their adult life. They develop fixed patterns of behavior and live accordingly—often to the frustration of those around them.

In their 20s, many adults begin to show signs of aging. With age, wrinkles may begin to appear, weight may increase, and gray hairs may develop.

≈ AGING AND DEATH

Aging is part of the life process, but it remains one of the great mysteries of biology. **Aging** is a progressive deterioration of the body's structure and function. One function of extreme importance is homeostasis. As homeostatic mechanisms falter, the body becomes less resilient. Healing may take longer. Individuals may become more susceptible to disease.

The most notable changes occurring with age are physical changes—wrinkling, loss of hair, and stooping. Less obvious is the gradual deterioration of the function of body organs. As one ages, vision and hearing both decline. Muscular strength also ebbs as a result of a decrease in the number of myofibrils (bundles of contractile filaments) in the skeletal muscles. Bones tend to thin, and joints often show signs of wear. Aging is also accompanied by a gradual reduction in cardiac output and pulmonary function. The decrease in pulmonary function results in a reduction in oxygen absorption and the amount of air that can be inhaled and exhaled. The number of functional nephrons in the kidney declines, as does renal function. The immune system also becomes less able to respond to antigens. The

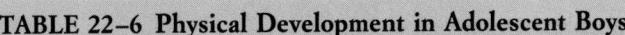

PHYSICAL CHANGE	AVERAGE AGE WHEN CHANGE BEGINS	AVERAGE AGE WHEN NOTICEABLE CHANGE USUALLY STOPS	REMARKS
Increase in rate of growth	12 to 13	17 to 18	If conspicuous growth fails to begin by 15, you should consult your physician.
Enlargement of genitals	Testicles and scrotum: 11 to 12	16 to 17	As the testicles grow, the skin of the scrotum darkens. The penis usually lengthens before it broadens. Ability to ejaculate seminal fluid usually begins about a year after the penis starts to lengthen.
	Penis: 12 to 13	15 to 16	
Emergence of body hair	Pubic hair: 11 to 12	15 to 16	Development of hair is extremely variable and largely dependent on genetic inheritance. The spread of hair up the abdomen and onto the chest usually continues into adulthood.
	Underarm hair: 13 to 15	16 to 18	
Development of sweat glands	13 to 15	17 to 18	See remarks in the accompanying table for girls.
Voice change	Enlargement of the larynx, or voice box, begins at 13 to 14, and the voice deepens at 14 to 15	16 to 17	Growth of the larynx may make the adam's apple more prominent. The voice may change rapidly or gradually. If childhood voice persists after 16, consult your physician.

SOURCE: From *The American Medical Association Family Medical Guide.* Copyright © 1982 by The American Medical Association. Reprinted by permission of Random House, Inc.

nervous system does not escape the aging process. Memory deteriorates, and reaction time decreases. Although these changes are part of the aging process, not all of them begin at the same time.

Aging May Be Brought About By a Decline in Cell Numbers and a Decline in the Function of Cells

The decline in homeostatic systems and general body function results from at least two factors: a decrease in the number of cells in the organs and a decline in the function of existing cells. What causes the decline in cell number?

Cell number is determined by the balance between cell division and cell death. Laboratory experiments show that body cells grown in culture divide a certain number of times, then stop. The number of divisions a cell undergoes is directly related to the age of the organism from which the cell is taken: the older the organism, the fewer divisions are possible.

Laboratory studies such as these have also uncovered a correlation between the number of divisions in culture and the life span of the species from which the cells were taken. In other words, cells from species with long life spans undergo more divisions than cells from species with short life spans. These data and others suggest that the end of cell division and, therefore, aging, is genetically programmed.

If cells are programmed to divide a given number of times, why don't all humans live to be the same age? One answer is that people differ genetically. Genetic differences

may account for the difference in life span. In addition, people live markedly different lifestyles and eat very different diets. Research shows that the better one lives, the longer the life span (see Health Note 22–1). Diet, exercise, and stress management are keys to a healthy life.

Scientists are beginning to discover why proper diet and exercise help increase the life span. Laboratory studies, for instance, show that by manipulating the chemical environment of cells, one can produce more cell divisions than normal. Vitamin E in large quantities, for example, increases the number of cell divisions in tissue cultures; that is, it increases the cell's life span. Whether vitamin E will extend a person's life is not known.

To understand aging, scientists are looking at the cell itself to see what causes its function to deteriorate. Studies show that the decline in cell function results from problems arising in the DNA, RNA, and cell proteins. Cells are exposed to many potentially harmful factors in the course of a lifetime. Natural and anthropogenic (produced by human activities) radiation and other potentially harmful agents, such as chemicals in the home and work environments, may damage cells and impair their function. As the damage accumulates, body function gradually deteriorates.

Although the decline in cell number and the loss of cell function are thought to be the two leading causes of aging, some researchers think that aging is largely the result of a gradual decline in immune system function caused by a reduction in cell number and a loss of cell function. Although the immune system does falter, this decline surely cannot explain the other signs of aging.

CAN WE REVERSE THE PROCESS OF AGING?

Suppose a drug were invented that would help you live to be 100. Would you take it? If you are like most people, you probably would, as long as your additional years would be healthy and fairly trouble free.

Because most of us would like to beat this aging thing that nature imposes on us, we secretly hope for a medical breakthrough that would prolong our lives. Unfortunately, the search for a "cure" for old age has been fraught with frustration. Over the years, numerous treatments have been tried and have failed. As pointed out in this chapter, although the average U.S. life span has increased, we really are not living much longer than our grandparents.

There is some encouraging news, however. Scientists recently discovered a protein called **stomatin,** which they isolated from fibroblasts that had stopped dividing in tissue culture. Scientists hope that stomatin will stop other cells from dividing as well. If this turns out to be true, researchers may have discovered an important clue in the aging puzzle, the signal that ends cell division. Moreover, they may have found a way to prolong human life.

Their reasoning goes as follows: If this protein signals the end of cellular division, there must be a gene that controls its production. If the stomatin gene can be located, researchers may be able to find a way to inactivate the gene. A drug or genetically engineered chemical, for example, could be injected in people to block the gene. The tissues of these people, researchers think, might continue to regenerate beyond the genetically programmed life span.

Efforts to reverse the aging process have met with some success in another arena as well—slowing down the aging of skin. Because age-related changes in the skin mirror those occurring in other aging tissues, scientists have long studied the skin to expand their understanding of the aging process in general. Their studies have shown that skin aging results from two processes: intrinsic, chronological aging, which may be genetically programmed, and extrinsic aging, resulting from accumulated environmental damage from sunlight and other factors.

Many skin scientists doubt whether anything can be done about intrinsic aging. Their studies suggest that after a skin cell has lived out its lifetime, its plasma membrane receptors become insensitive to growth factors that stimulate DNA replication and cell division.

New research suggests, however, that extrinsic aging, especially that induced by sunlight, is preventable and reversible. Unknown to many, sunlight damages epithelial cells of the epidermis and the fibroblasts in the underlying dermis. When exposed to sunlight, molecules in

▷ **FIGURE 1 Beware: Skin Damage**

side various skin cells become electrically excited, and the energy they absorb from the sun may drive reactions with other molecules in the cell. These reactions, in turn, may lead to the dissociation of chemical bonds, which may injure important molecules essential for cellular function. Plasma membranes, in fact, can be damaged within a few minutes of exposure to sunlight (▷ Figure 1). Sunlight energy can also be absorbed by oxygen in the tissue, resulting in the formation of oxygen free radicals, highly destructive chemicals that are primarily responsible for extrinsic aging.

In 1988, medical researchers found that Retin-A, a derivative of vitamin A used to treat severe cases of acne, reduces wrinkling of the skin induced by sunlight. A study of Retin-A that followed patients for 22 months or more after treatment showed that the drug reduces wrinkles, age spots, and roughness for at least 22 months as long as it is applied regularly.

Researchers do not know how the drug works, but studies suggest that it does indeed have several beneficial effects. For example, Retin-A stimulates the growth of new blood vessels in the dermis, which may nurture the regeneration of damaged skin cells. It also detoxifies oxygen free radicals in tissues and can inhibit the destruction of collagen fibers in the dermis. Some researchers think that it may regulate genes that play a role in cell growth and differentiation.

Despite the discovery of stomatin and Retin-A, there is as yet no evidence that medical scientists can prevent or even retard the overall rate of aging. The best way to live a long and healthy life is to live well: learn ways to reduce stress, eat well, exercise regularly, take alcohol in moderation or not at all, and avoid harmful practices such as smoking.

Aging Is Often Associated With Certain Diseases, But the Likelihood of Contracting a Disease of Old Age Depends in Large Part on Lifestyle

Many diseases are associated with old age. Osteoporosis, arthritis, atherosclerosis, and cancer are examples. Although these diseases are more prevalent in older people, they are emphatically not the inevitable consequence of aging. That is, they are not unavoidable signs of aging like gray hair. As previous chapters have shown, the likelihood of developing these diseases can be greatly reduced by exercise, diet, and other lifestyle adjustments. In other words, people can age in a healthy manner by living healthy lives.

A Careful Analysis of the Data on Longevity Shows that Modern Medicine Is Not Letting Us Live Much Longer

"Thanks to improvements in medicine, people are living longer." You have heard this statement dozens of times in one form or another. The assertion is that advances in medicine are actually increasing how long you and I will live. This statement is so often repeated that most of us believe it implicitly. But is it true?

Medical scientists measure longevity by a statistic called **life expectancy at birth.** Life expectancy at birth is the number of years, on average, a person lives after he or she is born. As we all know, life expectancy at birth has increased dramatically in the last century. In 1900, for example, on average, white American females lived only 50 years. Today, life expectancy is 81 years. For males, a similar trend is observed. In 1900, for example, the life expectancy of a white American male was 47 years; today, it is 74 years.

Most people take these statistics to mean that men and women are actually living to a much older age. In reality, something very different is happening: the increase in the life expectancy at birth is largely the result of declining infant mortality. That is, thanks to improvements in medicine, more children are living through the first year of life. In the early 1900s, 100 of every 1000 babies died during the first year of life. Today, that number has fallen to 12. This reduction has greatly affected average life expectancy.

This is not to say that all of the gain in life expectancy results from a decline in infant mortality. Medical advances have increased life expectancy after infancy, but these changes are small in comparison to those brought about by lowering infant mortality. In fact, about 85% of the increase in life span in the last century is the result of decreased infant mortality.

Death Is the Final Act of Living

Aging results in a deterioration of function that eventually leads to death. But death can also result from traumatic injury to the body—for example, severe damage to the brain or a sudden loss of blood—and many other causes.

When I began to research this topic for the book, I thought the task would be easy. After all, what can you say about death? It is when a person is no longer alive, isn't it? What I found was that death is difficult to define precisely. Although we can all agree that death is the cessation of life, trouble begins when people try to define it in cases requiring life-support systems. In 25 states, a person is considered dead when his or her heart stops beating and breathing ceases. In the remaining 25 states, however, death is defined more liberally as an irreversible loss of brain function. A woman in a coma, for example,

often shows little or no brain activity other than that needed to keep the heart beating and the lungs working. Is this woman alive? In some states, she is; in others, she is not. What is your opinion on this issue? If you were lying in a coma with little or no brain function, what would you want your family to do?

As you can well imagine, maintaining a person on life support is extremely costly and can be financially and emotionally crippling to a family and to society, if, for example the patient is receiving federal benefits through Medicare or Medicaid. The money spent on maintaining a life could arguably be used to save dozens of lives through preventive measures. These issues lead to another biomedical dilemma, the controversy over euthanasia, a word derived from the Greek meaning "easy death." **Euthanasia,** an act or method of causing death painlessly, may be either passive or active. Passive euthanasia consists of deliberate actions and decisions to withhold treatment that might prolong life. Active euthanasia consists of actions and decisions that actively shorten a person's life—for example, injecting a lethal substance to terminate a patient's life.

Opinions on euthanasia vary widely, and the controversy will, no doubt, be debated for many years. (For a debate on physician-assisted suicide, see the Point/Counterpoint in this chapter.) As a final note, individuals who want to save their relatives emotional and legal turmoil can sign a living will. This is a legal document stipulating conditions under which doctors should allow a person to die.

ENVIRONMENT AND HEALTH: VIDEO DISPLAY TERMINALS

Computer monitors, or video display terminals (VDTs), produce numerous types of radiation with frequencies ranging from X-rays to radio waves. Protective shields prevent most of this radiation from escaping. It is what escapes the shield, however, that has some health officials concerned.

Most scientists agree that small amounts of middle- and high-frequency radiation escaping from the protective shielding are not a threat to health. Lower-frequency radiation (radio frequencies) is another story. Laboratory experiments show that electromagnetic fields generated by extremely low-frequency radiation can alter fetal development in chickens, rabbits, and swine.

In addition, researchers in Oakland, California, recently published results of a medical study on the incidence of miscarriage in nearly 1600 women. The study showed that women who sit in front of a computer VDT for more than 20 hours a week during the first three months of pregnancy are nearly twice as likely to miscarry as women in similar jobs not using computers.

What causes these miscarriages? Researchers suspect that they result from radiation emitted from the VDTs. However, other factors may also be involved. For example,

THE RIGHT TO A PHYSICIAN WHO WILL NOT KILL

Rita Marker

Rita L. Marker is Director of the International Anti-Euthanasia Task Force and the author of Deadly Compassion.

S hould physician administered euthanasia—the direct and intentional killing of a patient by a physician—be legalized? Common sense has always said NO. And laws have wisely banned euthanasia.

Now, however, an effort is underway to change these laws. Those who seek to legalize euthanasia have framed their arguments in terms of personal rights and freedom. They've clearly recognized the importance of words in molding public opinion: carefully crafted verbal engineering is being employed to transform the appalling crime of mercy killing into an appealing matter of patient choice.

Proposals appearing in state after state have such benign titles as "Death With Dignity Act," and the deceptively soothing term "aid-in-dying" is often substituted for "euthanasia."

As efforts of euthanasia activists increase, careful examination of what is at stake has become increasingly important. This examination should include answers to key questions.

Who Will Be the Recipients of New "Rights" If Euthanasia Is Legalized?

The law would benefit doctors, not patients. Competent adults already have the legal right to refuse medical treatment, as well as the right to make their wishes known about such interventions for the future. Additionally, neither attempting nor carrying out the very personal and tragic decision to commit suicide is a criminal offense. Bluntly put, people can refuse medical care, or can kill themselves without physician intervention.

Doctors, however, are prevented by law from killing their patients. Passage of the Death With Dignity Act would give doctors both the right to directly and intentionally kill patients and the promise of immunity. Legalization of euthanasia would transform an action which is previously considered homicide into a "medical service."

What Types of Safeguards Are Contained in the Death With Dignity Act?

Initial efforts to pass the Death With Dignity Act have failed. In the Fall of 1991, it was turned down by Washington voters. A subsequent unsuccessful effort placed it on California's ballot, where its advocates claimed that the measure contained ade-

quate safeguards. However, the so-called safeguards had as little substance as the Emperor's new clothes. For example:

- A signed request for euthanasia, made years before the diagnosis of a terminal condition, could have been put into effect without any additional witnessed documentation.
- No requirement existed that the family be notified before a loved one was killed.
- No waiting period or counseling about care and symptom alleviation was required.
- The mental health or competency of a patient who was to be killed could not have been assessed unless the patient approved a psychological evaluation.
- Physicians could have offered euthanasia as an "option" and encouraged their patients to "choose" it.
- Only the most general information (such as the number of patients killed) would need to have been reported, thus preventing any serious investigation of abuses.

What Has Happened Where Euthanasia Has Been Practiced?

Euthanasia is technically illegal in Holland, but is nonetheless widely practiced as a result of court decisions which set forth guidelines under which it may be administered. Even with these guidelines, which are far tighter than the "safeguards" proposed in the Death With Dignity Act, a 1991 Dutch government study released horrifying information. The study found that in tiny Holland—a country with only one-half the population of the State of California—Dutch physicians deliberately and intentionally end the lives of more than 11,000 people each year by lethal overdoses or injections. And, the study found that more than half of those killed had *not* requested euthanasia.

What Will Happen If Euthanasia Is Legalized?

While Death With Dignity proposals pose a threat to everyone, they are particularly dangerous for the many people who lack health insurance. Euthanasia as a medical option may present a choice for the rich. For the poor, however, it may become the only affordable "medical treatment."

Expansion of the Death With Dignity Act would also be a problem. Indeed, expansion is already being discussed among euthanasia leaders. For example, in *Final Exit,* Derek Humphry wrote that when the Death With Dignity Act becomes law everywhere, it will provide a means of handling the dilemma of "terminal old age." He has also called for expansion of euthanasia to include those who are physically and mentally disabled.

Those who think that euthanasia, once unleashed, can be controlled are making a deadly mistake. Euthanasia is not a means of dying with dignity, nor is it a matter of liberty. It does not legislate compassion. It only legalizes killing.

THE RIGHT TO CHOOSE TO DIE

Derek Humphry

Derek Humphry was the principal founder of the Hemlock Society and its executive director from 1980 to 1992. He is the author of four books on euthanasia, including the best-seller Final Exit (1991).

If we are truly free people and if our bodies belong to ourselves and not to others, then we have the right to choose when and how to die. Death comes to us all in the end, although modern medicine can often help us improve and extend our life spans. Thus, given today's high technology and ruinous cost of medicine, it is wise if we all give advance thought to the manner of our dying. Such decisions should be transferred to paper (a living will and durable power of attorney for health care), because 46 states now legally recognize one's wishes in respect to the withdrawal of life-support equipment.

Advance-declaration documents, of course, deal only with the legal and ethical problems of medical equipment and treatments. Less than half of dying people are connected to such equipment. For many more patients, technology does nothing for their terminal illness, so—in effect—there is no "plug to pull." Therefore, the cutting edge of the right-to-die debate in the 1990s centers on assisted death and voluntary euthanasia.

Hemlock Society supporters feel not only that hopelessly sick people should have an unfettered right to accelerate their end to save them pain, distress, and indignity but also that willing doctors should be able to help them die. (To clarify some terms: *self-deliverance* (suicide) is ending your own life to be free of suffering; *assisted suicide* is helping another die; *euthanasia* is the direct ending of another's life by request. Currently, suicide in any form, for any reason, is not illegal, but assisted suicide and euthanasia are technically crimes.)

Most people at the close of life do not wish to suffer, desiring a quick and painless demise. The sophisticated pain-management drugs now available, when properly administered, control some 90% of terminal pain. But they do nothing to alleviate indignities, psychic pain, and loss of quality of life associated with some debilitating terminal illnesses.

A complaint that any hastening of the end interferes with God's authority can be answered for many through a belief that *their* God is tolerant and charitable and would not wish to see them suffer. Other people have no faith in God. Pious people differ with this view, of course, and I respect that.

Numerous opinion polls in the United States and other Western countries indicate that two-thirds of people want the right to have a doctor lawfully assist them to die. People in the states of Washington, Oregon, and California are engaged in political action to achieve this law reform.

The proposed law in these West Coast states is called the Death With Dignity Act, and the broad outline of its purpose is as follows:

1. The patient wanting physician-assisted dying would have to be a mature adult suffering from a terminal illness likely to cause death within about six months.
2. People with emotional or mental illness (especially depression) could not get help to die under this law.
3. The request would have to be in writing, and the signature would have to be witnessed by two independent persons.
4. The physician could decline to help the patient die on grounds of conscience but would then cease to be the treating physician. The patient could then seek a physician who was willing.
5. The family would have to be informed and its views, if any, taken into account. But the family could neither promote nor veto the patient's request to die.
6. The physician would have to be satisfied that the patient was fully aware of his or her condition, had been informed of all possible alternatives, and was competent to make this request.
7. If the physician was unsure of the patient's mental state, a mental-health professional could be called to make an evaluation.
8. The time and manner of the assisted dying would have to be negotiated between patient and physician, with the patient's wishes paramount.
9. At any time the patient could orally or in writing revoke the request for assisted dying.
10. Any person who pressured another person to get assistance in dying, who forged such a request, or who ignored a revocation would be subject to prosecution.
11. After helping the patient die, the physician would have to report the action in confidence to a state health agency.

The right-to-die movement feels that these conditions, plus others in the Death With Dignity Act too numerous to describe here, are an intelligent and humane approach to euthanasia with the necessary safeguards against abuse.

Dying on one's own terms is not only an idea whose time has come but also the ultimate civil and personal liberty.

≋ SHARPENING YOUR CRITICAL THINKING SKILLS

1. Summarize the key points of each author and their supporting arguments.
2. What is the basis for their fundamental disagreement?
3. Which viewpoint corresponds to yours? Why?
4. How did your values and experiences affect your opinion in this matter?

the stress of using a computer may be the cause. Only more research will tell. Preferring to err on the conservative side, some scientists advise pregnant women to minimize their exposure to radiation from computer monitors.

Should the link between radiation from VDTs and miscarriage be substantiated by further research, it would once again illustrate the importance of a healthy environment to overall human health and reproduction.

SUMMARY

AN OVERVIEW OF ANIMAL DEVELOPMENT

1. The sperm and egg of sexually reproducing animals unite to form a zygote, a process called fertilization.
2. In vertebrates and invertebrates, sperm contact the plasma membrane of eggs and are phagocytized by them. The egg and sperm cell nuclei then fuse to form a diploid zygote.
3. Fertilization is followed by mitotic division, or cleavage. This process produces a multicelled structure, the morula.
4. In many species, such as frogs and sea urchins, a hollow cavity forms in the morula. The resulting structure is a blastula.
5. When cells of the blastula invaginate, gastrulation begins. Three distinct layers of cells form in the gastrula: the endoderm, mesoderm, and ectoderm. Each layer gives rise to specific organs during organogenesis.
6. All animal embryos go through much the same process of development, although there are some notable differences. In chickens and other birds, embryonic cells form a small, disklike structure atop a mass of yolk. It differentiates into three layers that give rise to the organs.

HUMAN REPRODUCTION AND DEVELOPMENT

7. In humans, fertilization usually occurs in the upper third of the uterine tube.
8. Sperm deposited in the vagina reach the site of fertilization with the aid of muscular contractions in the walls of the uterus and uterine tube.
9. Sperm bore through the zona pellucida and contact the plasma membrane. The first one to contact the membrane fertilizes the oocyte. Further sperm penetration is blocked.
10. Sperm are engulfed by the oocyte. The chromosomes of the sperm and oocyte duplicate and merge in the center of the cell, where mitosis begins.
11. Human development during gestation is divided into three stages: pre-embryonic, embryonic, and fetal.
12. Pre-embryonic development begins at fertilization and ends at implantation. The zygote undergoes rapid cellular division and forms a morula.
13. Next, a cavity forms in the morula, converting it to a blastocyst. The blastocyst consists of a clump of cells, the inner cell mass, which becomes the embryo, and the trophoblast, which gives rise to the embryonic portion of the placenta.
14. The morula arrives in the uterus three to four days after fertilization and is converted into a blastocyst, which implants in the uterine wall two to three days later.
15. Soon after implantation begins, the trophoblast differentiates into two layers. The outer layer invades the endometrium. Cavities form in this layer and fill with blood.
16. Fingerlike projections of the inner layer carrying fetal blood vessels then invade the trophoblast, forming the placental villi. The villi are bathed in maternal blood and provide a means of acquiring oxygen and nutrients from the mother and disposing of embryonic wastes.
17. While the placenta is forming, a layer of cells from the inner cell mass of the blastocyst separates from it and forms the

amnion. The amnion fills with fluid and enlarges during embryonic and fetal development, eventually surrounding the entire embryo and fetus. The amniotic fluid protects the embryo and fetus during development.
18. After the amnion forms, the cells of the inner cell mass differentiate into the three germ cell layers: ectoderm, mesoderm, and endoderm. The formation of the three primary germ layers marks the beginning of embryonic development.
19. The organs develop from the three basic tissues during organogenesis. Table 22–2 lists the organs and tissues formed from each of the layers.
20. Fetal development begins eight weeks after fertilization. Because most of the organ systems have developed or are under development, fetal development is primarily a period of growth.
21. The placenta produces several hormones that play an important part in reproduction. Human chorionic gonadotropin maintains the corpus luteum during pregnancy. Progesterone and estrogen stimulate uterine growth and the development of the glands and ducts of the breast.
22. About 10% to 12% of all newborns enter the world with some form of birth defect. Birth defects arise from chemical, biological, and physical agents known as teratogens.
23. The effect of teratogenic agents is related to the time of exposure, the nature of the agent, and the dose. A defect is most likely to arise if a woman is exposed to a teratogen during the embryonic period when the organs are forming.
24. Many physical and chemical agents are toxic to the human fetus and when present in sufficient quantities can kill a fetus or retard its growth.
25. During pregnancy a woman's body undergoes incredible change. The uterus and breasts enlarge considerably, blood volume increases, respiration rate climbs, and urination increases.

CHILDBIRTH AND LACTATION

26. Labor consists of intense and frequent uterine contractions that are believed to be caused by the release of small amounts of fetal oxytocin prior to birth. Fetal oxytocin stimulates the release of prostaglandins by the placenta. Oxytocin and prostaglandins stimulate contractions in the sensitized uterine musculature.
27. Emotional and physical stress in the mother may trigger maternal oxytocin release, augmenting muscle contractions.
28. As uterine contractions increase, they cause more maternal oxytocin to be released, which stimulates stronger contractions and more oxytocin release, a positive feedback that continues until the baby is born.
29. Labor consists of three stages, the dilation, the expulsion, and the placental.
30. The breasts of a nonpregnant woman consist primarily of fat and connective tissue interspersed with milk-producing glandular tissue and ducts.

31. During pregnancy, the glands and ducts proliferate under the influence of placental and ovarian estrogen and progesterone.
32. Milk production is induced by maternal prolactin and human chorionic somatomammotropin but does not begin until two to three days after birth.
33. Before milk production begins, the breasts produce small quantities of a protein-rich fluid, colostrum. A newborn can subsist on colostrum for the first few days and derives antibodies from it that help protect it from bacteria.
34. Suckling causes a surge in prolactin secretion. Each surge stimulates milk production needed for the next feeding.

INFANCY, CHILDHOOD, ADOLESCENCE, AND ADULTHOOD

35. The newborn survives because of parental care and because it comes equipped with several built-in reflexes, including the suckling reflex, rooting reflex, and crying reflex.
36. During infancy, the child transforms physically and mentally. Children develop in predictable stages, although the timetable varies considerably.
37. During infancy, sensory, motor, and intellectual capacity increase dramatically. Children develop emotionally as well. One of the first emotional milestones of normal human development is bonding, the establishment of an intimate relationship between the infant and its caregivers. Successful bonding is important for later development and may facilitate emotional security later in life.

38. Research suggests that intelligence, like motor skills, improves with use. Parents are encouraged to foster their children's curious nature, to let them explore, and to provide opportunities for a rich and varied experience.
39. Childhood lasts from infancy to puberty. During childhood, children undergo dramatic improvements in motor skills and in the use and comprehension of language.
40. Puberty, or sexual maturation, marks the onset of adolescence, which lasts until adulthood (age 18 to 20). Sexual maturation results from the production of sex steroids.
41. Adolescence is also a time of emerging identity. Conflict between teenagers and their parents is common during this period, but it can be reduced if parents learn to accept the adolescent's personality and maintain a supportive relationship.
42. Adulthood begins when adolescence ends. Physical growth stops, but, for many people, social and psychological development continues.

AGING AND DEATH

43. Aging is the progressive deterioration of the body's homeostatic abilities and the gradual deterioration of the function of body organs.
44. These changes result from at least two factors: a decrease in the number of cells in the organs and a decline in the function of existing cells.
45. Death results from aging, traumatic injury, or infectious disease.

EXERCISING YOUR CRITICAL THINKING SKILLS

The frequency of Down syndrome babies increases with maternal age. Because 95% of Down syndrome babies can be attributed to maternal chromosomal nondisjunction, it has long been thought to be the result of the age of the oocytes in a woman's ovaries. New research suggests, however, that the reason older women have a higher percentage of Down syndrome babies may be that they are more likely to carry one to term than younger women.

In an international study conducted by 19 scientists, researchers found that the extra chromosome 21 that is responsible for this genetic disorder is most often incorporated during the first meiotic division. Much to their surprise, however, the frequency of nondisjunction in older and younger women was nearly equal.

If the frequency of nondisjunction is the same in mothers of all ages, one possible reason why older mothers have a higher incidence of Down syndrome babies is that their bodies fail to recognize an abnormal embryo as well as those of younger mothers. As a result, they are more likely to carry a Down syndrome baby to term.

Further research is needed to determine whether this hypothesis is valid. Can you think of any way to test it?

After thinking about it for a while, consider this suggestion: one direct way of testing the hypothesis, of course, would be to compare the ages of women who spontaneously abort Down syndrome fetuses.

TEST OF TERMS

1. In humans, the ovum and sperm unite to form a(n) _____ during fertilization, which takes place in the upper third of the _____ _____ .
2. Residence in the female reproductive tract for a period of six to seven hours renders sperm membranes fragile, allowing the sperm to release enzymes from the _____ .
3. When a sperm cell contacts the plasma membrane of the oocyte, it triggers an immediate change in the membrane that prevents other sperm from entering. This is called the

_____ block to_____ .
4. During pre-embryonic development, the cells first multiply to form a solid ball, the _____ . In humans, a cavity soon forms in this structure, converting it to a blastocyst. The blastocyst consists of a clump of cells, the

_____ _____

_____ , which gives rise to the embryo, and a ring of flattened cells, the _____ , which will give rise to the embryonic portion of the placenta.

5. The blastocyst attaches to the uterine lining, and digests its way into it in a process called _____ , which begins _____ to _____ days after fertilization.

6. The placenta forms from embryonic and maternal tissue. Blood vessels invade the forming placenta with the inner layer of the trophoblast. Fingerlike projections form and are bathed by maternal blood. These structures are called _____ .

7. Early in embryonic development, the _____ forms from the cells of the blastocyst that will eventually give rise to the embryo proper. It eventually forms a complete fluid-filled sac that surrounds the fetus.

8. The organs form during embryonic development; this process is called _____ .

9. Ectoderm forms the neural tube, which later develops into the _____ and _____ _____ .

10. Blocks of mesoderm along the neural tube, called _____ , form the muscles of the neck and trunk as well as the _____ .

11. The lining of the intestines comes from embryonic _____ .

12. The umbilical _____ carries blood rich in oxygen and nutrients from the placenta to the fetus. The _____ _____ shunts blood from this vessel to the _____ _____ _____ , thus bypassing the liner.

13. The hole in the interatrial septum shunts blood from the _____ to the _____ and is called the _____ _____ .

14. The placental hormone _____ maintains the corpus luteum and stimulates production of _____ and _____ by the CL.

15. The study of birth defects is called _____ .

16. _____ , a hormone produced by the placenta and ovaries of pregnant women, loosens the connective tissue in the pubic symphysis and softens the cervix in preparation for childbirth.

17. Stage 1 of childbirth is also known as the _____ phase. During this phase, the baby descends into the pelvic cavity, and the _____ dilates.

18. Childbirth in which the baby is born buttocks first is called a(n) _____ birth.

19. Muscular contractions in the wall of the uterus are stimulated by the hormone _____ , secreted by the _____ _____ gland.

20. The final stage of childbirth is the _____ stage.

21. An anesthetic injected into the vagina during childbirth is called a(n) _____ block.

22. Lactation, the production of milk from the breasts, is stimulated by the maternal hormone _____ , secreted from the anterior pituitary gland.

23. Expulsion of the milk by the glands in the breast is induced by the hormone _____ , whose release is stimulated by suckling.

24. The protein-rich secretion produced by the breasts during the first few days after birth is called _____ .

25. A baby turns its head toward a nipple that brushes against its cheek. This is called the _____ reflex.

26. One of the first emotional milestones of normal human development is _____ , the establishment of an intimate relationship between the infant and one (or preferably both) of its parents or caregivers.

27. Sexual maturation occurs during _____ .

28. _____ _____ at birth is the average life span of an individual.

Answers to the Test of Terms are found in Appendix B.

TEST OF CONCEPTS

1. Give an overview of the process of development in animals. Describe each stage and what is happening at it.

2. The formation of the gastrula results in the formation of endoderm, mesoderm, and ectoderm and the rearrangement of cells. Why is this process so important to the future development of the embryo?

3. Describe the process of fertilization in humans in detail. Use drawings to elaborate your points, and label all drawings.

4. Define the terms *inner cell mass* and *trophoblast*.

5. Where does the human embryo acquire nutrients before the placenta forms?

6. Describe the formation of the placenta in humans.

7. What are the major functions of the human placenta?

8. Describe the flow of blood from the placenta to the fetus and back. What is the role of the various shunts?

9. List and describe the function of the five placental hormones.

10. Describe the changes that occur in pregnant women. Describe the factors that contribute to maternal weight gain during pregnancy.

11. Discuss how the nature of a teratogenic agent and the time of exposure affect teratogenesis, the production of birth defects.

12. What factors trigger labor?

13. What factors are responsible for cervical dilation during labor?

14. Design a humane experiment to test the two major hypotheses about bonding presented in this chapter. Describe how you would go about the experiment and the special considerations that would be necessary to eliminate bias. Could such an experiment be objective? Could such an experiment be ethical?

15. Describe some of the reflexes of the newborn and their importance to the infant's survival.

16. Define the term *aging,* and describe some of the hypotheses that attempt to explain it.

17. Thanks to modern medicine, men and women are living longer, says a friend. Using your understanding of average life expectancy data and the reason for an increase in life expectancy in the last century, explain why this statement is not quite true. 22

CHAPTER 23

Evolution: Five Billion Years of Change

Clown fish hides in the sea anemone, another member of the animal kingdom.

cientists believe that the atoms that make up both the living world and the nonliving world (rocks, for example) once existed in outer space as a giant cloud of cosmic dust and gas (▷ Figure 23–1). About 4.5 billion years ago, this huge cloud began to condense. No one knows what triggered the condensation, but some scientists think that a nearby exploding star (supernova) may have been the catalyst. As the cloud condensed, scientists believe, dust particles in its center compacted more rapidly. This region would give rise to the sun. When the mass of the forming sun reached a critical density, heat and pressure caused hydrogen and helium in it to unite, forming larger atoms. This process, known as **fusion,** releases large amounts of energy in the form of heat, light, X-rays, gamma rays, radio waves, and other radiation.

Cosmic dust lying outside the forming sun also condensed, creating the planets (Greek for "wanderers"). The third planet from the sun was the Earth, the only planet in our solar system that supports life.

When it originated, the Earth was a solid mass of rock and ice. Scientists believe that radioactive decay, intense solar heat, and other sources of heat caused the Earth to melt, turning it into a mass of molten material. As the Earth grew hotter, water and the lighter elements, such as hydrogen, escaped into the atmosphere.

In the millennia that followed, the Earth gradually began to cool, and a thick, rocky crust formed. Today, the crust encircles a molten core, a reminder of the planet's fiery past (▷ Figure 23–2). As the Earth cooled, though, water in the atmosphere began to rain down from the skies, creating lakes and oceans. Today, the oceans cover nearly 70% of the planet. Scientists believe that it is in these oceans, or on their shores, that life began.

This chapter traces the emergence of life from the Earth's beginnings and describes the process of *evolution,* which scientists believe is responsible for the rich and varied life-forms that make this planet their home.

≈ THE EVOLUTION OF LIFE: AN OVERVIEW

In the broadest sense, **evolution** is a process in which existing life-forms change. As you will soon see, evolution can also produce entirely new organisms. The evolution of life on Earth can be divided into three phases: (1) chemical evolution, (2) cellular evolution, and (3) the

▷ **FIGURE 23–2 The Earth in Cross Section** (*a*) The molten core is a remnant of the Earth's early history. (*b*) Lava pouring out of the Earth is evidence of the planet's molten core.

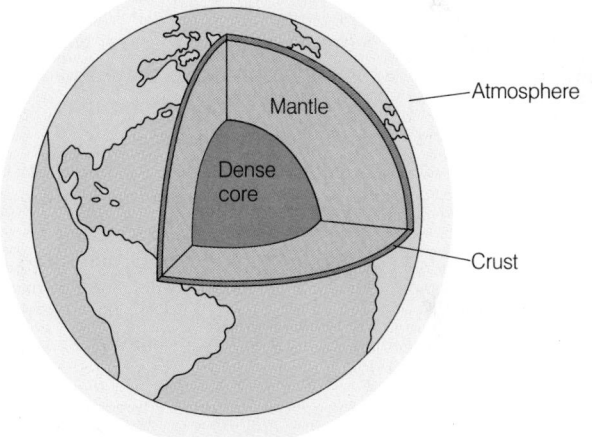

(a)

(b)

▷ **FIGURE 23–1 Cosmic Clouds** Outer space is riddled with enormous clouds of dust and gas from which stars and planets are formed.

evolution of multicellular organisms. This section outlines key events during each phase, providing an overview of the history of life on Earth and an introduction to the evolutionary process.

Chemical Evolution Led to the Formation of Organic Molecules from Inorganic Molecules in the Earth's Primitive Atmosphere

When they first formed, the seas were lifeless bodies of water. The landmasses, which today are richly carpeted with grasses, trees, and other plants, were barren rock. How could life have formed from such an inauspicious start?

In 1924, a Russian scientist, A. I. Oparin, suggested one hypothesis (▷ Figure 23–3). He conjectured that inorganic molecules in the Earth's early atmosphere had reacted to form the very first organic molecules. These molecules, in turn, combined to form polymers. Polymers then combined to form the very first primitive cells.

The formation of organic molecules from the Earth's primitive atmosphere that Oparin hypothesized is now referred to as **chemical evolution.** Chemical evolution began about 4 billion years ago—or 500 to 600 million years after the Earth formed. At that time, scientists believe, the Earth's atmosphere contained a mixture of water vapor, methane, ammonia, and hydrogen. It may also have included lesser quantities of nitrogen, carbon monoxide, and hydrogen sulfide but was devoid of oxygen.

The theory of chemical evolution holds that as the Earth cooled, atmospheric methane, ammonia, and hydrogen dis-

▷ **FIGURE 23–3 Oparin's Hypothesis** Oparin hypothesized that the organic molecules necessary for life formed from the Earth's primitive atmosphere. Today, this idea is called the theory of chemical evolution and is supported by a considerable body of evidence.

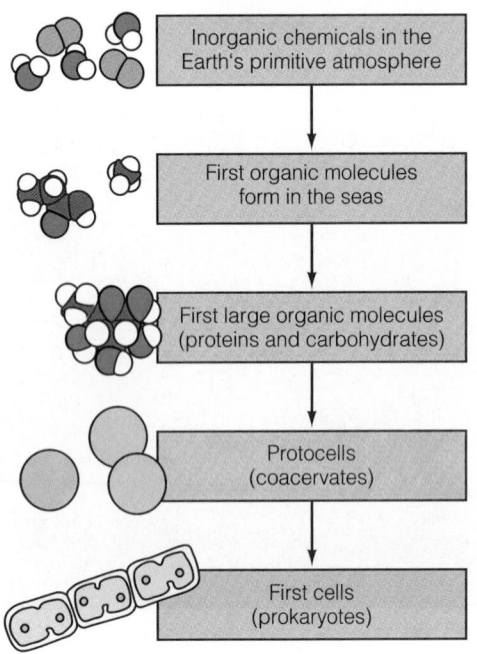

solved in rainwater. In the shallow waters of the newly formed seas, sunlight, heat from volcanoes, or lightning energized these molecules, causing them to react with one another in much the same way that molecules in a heated test tube react. These reactions produced a variety of simple organic molecules, including monosaccharides (simple sugars) and amino acids.

According to the theory of chemical evolution, these organic molecules, in turn, began to react, forming small polymers. As a result, primitive proteins and perhaps even rudimentary RNA or DNA molecules may have formed in the shallow waters of the seas.

Little by little, and over tens of thousands of years, all of the organic molecules (monomers and polymers) necessary for life began to emerge. These molecules, in turn, combined to form aggregates, or **protocells** (literally, "first cells"), the precursors of cells, discussed in detail shortly.

The Theory of Chemical Evolution Is Supported by Some Rather Convincing Experiments. Although chemical evolution may sound like science fiction, a considerable body of research now supports this theory. The first supporting evidence came from an ingenious American graduate student, Stanley Miller. While studying for his Ph.D. in chemistry at the University of Chicago in the early 1950s, Miller devised an apparatus to test Oparin's hypothesis (▷ Figure 23–4). To his closed, sterilized, glass apparatus, Miller added three gases thought to have existed in the Earth's primitive atmosphere: methane (CH_4), ammonia (NH_3), and hydrogen (H_2). A sparking device that simulated lightning provided energy. Boiling water created steam that circulated through the device, carrying with it the reactant gases and the products of the reactions occurring inside.

After several days, the water in the apparatus turned brown. A chemical analysis of the brown, soupy liquid revealed the presence of several biologically important organic compounds, including amino acids, urea, and lactic acid (Table 23–1). In only a few days, Miller had created exciting proof that organic molecules could indeed form abiotically (in the absence of life) from chemicals thought to exist in the Earth's primitive atmosphere.

Although Miller's experiment did not *prove* Oparin's hypothesis, it created a groundswell of excitement and a flurry of additional research. Experiments to confirm his results and expand our understanding of abiotic synthesis were performed by several other scientists.

These researchers found that a variety of organic molecules, including the building blocks of DNA and RNA (purines and pyrimidines), could be created abiotically. Together, this research strongly supports Oparin's hypothesis that the small molecules present in the Earth's early atmosphere reacted to form the building blocks of life.

But what about the next step? Could these molecules combine to form polymers? Researchers eagerly sought answers to this intriguing question, and to their delight, soon found that the organic building blocks produced

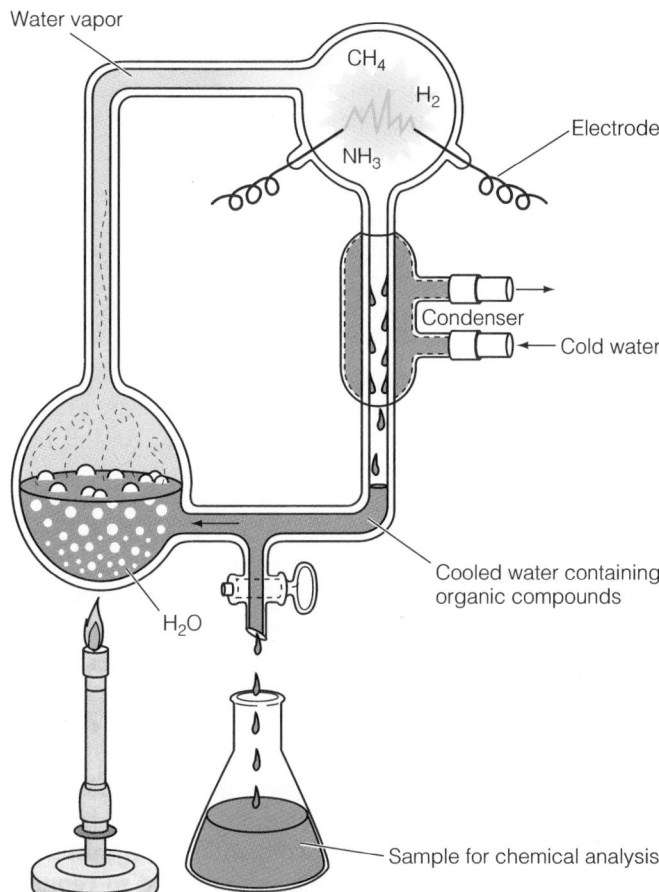

Water vapor

CH₄

H₂

NH₃

Electrode

Condenser

Cold water

H₂O

Cooled water containing organic compounds

Sample for chemical analysis

▷ **FIGURE 23–4 Miller's Apparatus** This device showed that organic molecules could be produced from the chemical components of the Earth's early atmosphere.

abiotically in the laboratory could indeed assemble into larger polymers in the absence of living organisms. One notable experiment was performed by Sidney Fox of the University of Miami. Fox found that when he heated amino acids in air, these molecules joined to form small polymers, which he called **proteinoids** (meaning "like proteins"). Biologists hypothesize that dilute solutions of amino acids that formed during chemical evolution may have been splashed onto hot rocks or lava, where the amino acids polymerized. Polymerization may also have occurred on clay particles, which tend to concentrate

≈	**TABLE 23–1 Chemical Products Produced Abiotically in Miller's Experiment**	≈
Glycine	Formic acid	
Alanine	Lactic acid	
Aspartic acid	Urea	
Butyric acid	Succinic acid	
Glutamic acid	Aldehyde	
Acetic acid	Hydrogen cyanide	

amino acids that are dissolved in water. Binding to the charged surface of the clay particles, the amino acids may have united via peptide bonds (Chapter 2).

Nucleic acids and polymers of carbohydrates have also been produced in similar experiments, thus confirming the second step in Oparin's hypothesis. Scientists next turned their attention to the third and final step, perhaps the most difficult to imagine of all, the formation of living cells from polymers.

The Theory of Cellular Evolution Suggests that the Precursors of Cells Were Aggregations of Polymers

The origin of the very first cells remains an enigma, but experimental evidence suggests a way in which these early cells may have formed from the mixture of organic molecules in the shallow coastal waters. Noteworthy again are the experiments of Fox. He immersed the proteinoids he had produced abiotically (described above) in a boiling salt solution, then cooled the solution. This procedure resulted in the formation of tiny globules of protein, which he called **microspheres** (▷ Figure 23–5). Much to everyone's surprise, they look a lot like bacteria. They are delimited by a membrane that resembles the plasma membrane of modern cells, and they also grow and reproduce. In fact, when a microsphere reaches a critical size, tiny protrusions form and break off, forming new ones. This process resembles budding, a form of asexual reproduction described in Chapter 1 and Chapter 21.

Researchers have also found that mixing dilute solutions of several polymers produces small globules similar to microspheres. Called **coacervates,** these microscopic globules of variable size and composition selectively incorporate molecules from their environment, in much the same way that cells selectively transport materials across their plasma membrane.[1] Coacervates also grow and divide.

[1] Coacervates differ chemically from Fox's microspheres, because they contain a mixture of molecules.

▷ **FIGURE 23–5 Microspheres** Sidney Fox showed that his abiotically produced proteinoids formed small spherical structures that resemble cells. Could they have acquired genetic material and enzymes to become the first living cells?

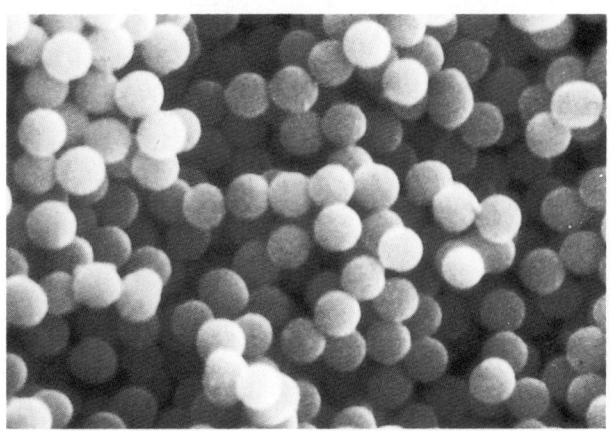

Some researchers hypothesize that coacervates and microspheres were the precursors of the Earth's first true cells. Some of the proteins in the precursor cells, or protocells, may have had enzymatic properties. These primitive enzymes may have allowed the protocells to synthesize some of their own molecules and catabolize others to generate energy. Researchers hypothesize that small molecules of DNA that formed during chemical evolution may have been incorporated into protocells, providing a primitive mechanism of heredity. Research also suggests that RNA present at the time may even have acted as enzymes.

The Very First Cells Were Probably Heterotrophic Fermenters.

The evolution of cells at this point is subject to great speculation. Many biologists believe that protocells gave rise to the first true cells. The latter probably contained primitive enzymes, rudimentary genes, and a selectively permeable membrane. However, the first real cells probably received nourishment by absorbing organic molecules, such as glucose, from their environment. These molecules were then broken down via anaerobic glycolysis to produce energy. (Remember: there is no oxygen in the atmosphere yet.) Thus, the first true cells were probably **heterotrophic fermenters.** ("Heterotrophic" means that they nourished themselves by eating others; fermentation is the anaerobic breakdown of glucose.) The emergence of these cells marks the beginning of the kingdom Monera (▷ Figure 23–6).

Many biologists believe that the early heterotrophic monerans may have given rise to early autotrophs—cells (or organisms) that can synthesize their own food. Plants and algae are common modern-day examples. These photosynthetic organisms use carbon dioxide, water, and solar energy to synthesize organic food molecules. In many modern-day autotrophs, water acts as a source of electrons for photosynthesis.

The very first autotrophs to evolve, however, were probably not photosynthetic but, rather, single-celled **chemosynthetic organisms**—organisms that produce their own organic molecules using energy from inorganic molecules in the environment. Some chemosynthetic bacteria can be found today.[2]

The Evolution of Photosynthesis Was a Pivotal Event in the Life History of the Earth.

Over many millions of years, photosynthesis evolved. The emergence of photosynthesis depended on the evolution of metabolic pathways that produce the light-absorbing pigment chlorophyll. The depletion of organic molecules in the shallow waters of the seas may have been the evolutionary driving force behind the emergence of chlorophyll. As the primordial soup became depleted, organisms that could capture sunlight and make their own organic foodstuffs—that is, photosynthetic organisms, or autotrophs—may have had an advantage over heterotrophic organisms that were dependent on food from the sea. In other words, organisms capable of photosynthesis were more likely to flourish, reproducing and passing on their genes more successfully than existing heterotrophs that fed on organic molecules produced abiotically. The autotrophs eventually predominated.

Photosynthesis probably evolved in two phases, although concrete evidence on this point is sorely lacking. Scientists believe that the first photosynthetic organisms to evolve contained chlorophyll, which they used to capture sunlight energy. Trapped solar energy was used to produce organic molecules. However, scientists believe that only the metabolic machinery of photosystem I was present early on and that the early photosynthesizers were therefore unable to produce oxygen. Electrons needed to replenish those stripped from photosystems came from inorganic and organic molecules present in the environment.

Next came the evolution of photosystem II. It allowed autotrophs to tap the Earth's abundant supplies of water as a source of electrons. As you may recall from your study of photosynthesis, water is broken down into oxygen and hydrogen ions. Because the early atmosphere was devoid of free oxygen, the evolution of photosystem II may have caused the Earth's first global pollution disaster. As populations of aerobic organisms expanded, oxygen levels in the water and atmosphere increased. Biologists believe that many of the anaerobic organisms living at the time were unable to cope with the oxygen and perished. Others retreated to oxygen-free environments, such as the mud or sediment beneath lakes and oceans, where many of their descendants (the anaerobic bacteria) remain today. Still others evolved mechanisms that allowed them to live in the oxygen-rich atmosphere. These organisms included aerobic monerans (aerobic bacteria) and the eukaryotes (organisms with true nuclei).

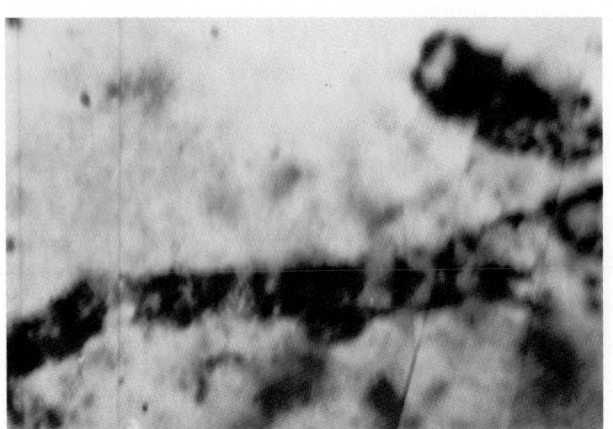

▷ **FIGURE 23–6 The First Prokaryote** A composite photograph of the first prokaryote fossil, discovered in western Australia approximately 30 years ago and believed to be about 3.5 billion years old.

[2] On the seafloor, chemosynthetic bacteria live on hydrogen sulfide or natural gas.

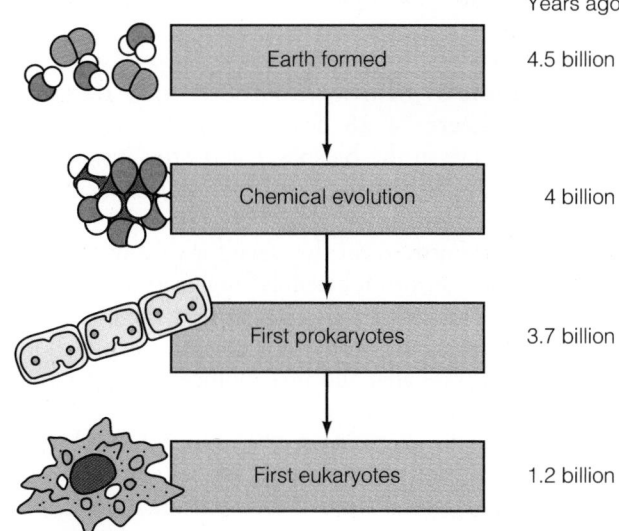

	Years ago
Earth formed	4.5 billion
Chemical evolution	4 billion
First prokaryotes	3.7 billion
First eukaryotes	1.2 billion

▷ **FIGURE 23–7 Summary of the Evolution of the Earth and Life**

Eukaryotes Probably Arose by Endosymbiotic Evolution.

The theory of chemical evolution describes how molecules necessary for life emerged. The theory of cellular evolution explains how cells arose from the organic molecules. The theory of **endosymbiotic evolution** explains the formation of eukaryotic cells. The very first eukaryotes were members of the kingdom Protista. (For a timetable of these events, see ▷ Figure 23–7.)

The theory of endosymbiotic evolution states that eukaryotes arose from preexisting cells approximately 1.2 billion years ago. In this section, we will consider the origin of mitochondria, chloroplasts, and nuclei.

According to the theory, mitochondria of the eukaryotes came into existence when a host cell (perhaps a heterotrophic fermenter) phagocytized a smaller, oxygen-respiring bacterium. The bacterium set up residence inside the host cell, using oxygen that entered its host and thus helped reduce internal levels, which was of great benefit to its host. As atmospheric oxygen levels increased, such unions may have conferred a selective advantage on this new biological composite.

Chloroplasts are believed to have come about in a similar fashion, notably when a host cell phagocytized a photosynthetic bacterium. The photosynthetic bacterium provided nutrients to the host cell and has remained inside ever since.

The theory of endosymbiotic evolution also holds that as time went on, the relationship between host cells and their internal partners (the predecessors of mitochondria and chloroplasts) became hereditary. Thus, when host cells divided, they passed the bacteria they contained to their daughter cells.

The theory of endosymbiotic evolution clearly offers a plausible explanation for the emergence of mitochondria, chloroplasts, and even flagella—cellular organelles found in eukaryotic cells. But how did the nucleus form? The nucleus of the eukaryote, its most characteristic feature, is believed to have evolved from infoldings of the plasma membrane that came to surround the DNA. Evidence of such infoldings is found in some species of bacteria. In these species, the plasma membrane folds inward and envelops the DNA, forming packages called **mesosomes** (described in more detail in the next chapter). Although mesosomes remain attached to the plasma membrane, it is thought that similar structures may have formed during cellular evolution and lost their connection, giving rise to the true nucleus.

Evidence for endosymbiotic evolution is circumstantial, or indirect. The first line of evidence can be classified as "proof by resemblance."

Comparisons of mitochondria and bacteria (free-living organisms), for example, show several striking similarities. First, mitochondria contain circular DNA, similar to that found in all bacteria. Second, mitochondria contain ribosomes structurally similar to those found in bacteria. Third, they contain enzymes needed to produce some of their own proteins. Fourth, mitochondria are capable of dividing. These similarities suggest that mitochondria may have been free-living, bacterialike organisms at one time.

Continuing along this same line, mitochondria also contain two membranes. Biologists believe that the outer membrane may have been formed when the bacteria that gave rise to this organelle were first phagocytized by the host cells during endosymbiotic evolution many millions of years ago. The outer membrane, say biologists, may have been derived from the membrane of the host cell; the inner membrane may be the phagocytized bacterium's own plasma membrane.

The theory of endosymbiotic evolution is also supported by the fact that endosymbiosis (one organism living inside another) is a rather common occurrence in biology. Termites, for example, contain a microorganism (a flagellate) in their gut that digests the wood they eat. More to the point, the flagellate living in the gut of the termite contains its own internal symbiont—a bacterium that lives inside. Dozens of other examples exist, and many of the internal partners are passed from one generation to another during reproduction. The important point is that endosymbiosis is not an anomaly among living things but, rather, a seemingly common strategy for survival.

Similar evidence in support of this phenomenal theory can be cited for chloroplasts. Together, the anatomical evidence and the common occurrence of endosymbiosis do not *prove* that endosymbiotic evolution occurred. They do, however, strongly suggest that it is a plausible explanation for the emergence of eukaryotic cells.

The first eukaryotic cells, with their distinct organelles and membrane-bound nuclei, were the free-living, single-celled organisms that compose the kingdom Protista. Common members of the kingdom include such familiar examples as the amoeba and paramecium.

The previous discussion is not meant to suggest that

scientists know all there is to know about chemical and cellular evolution. The truth is, scientists know very little about the origin of life and the emergence of prokaryotic and eukaryotic cells. They have many hypotheses, and the subject is one of the most controversial and open-ended in biology.

The Evolution of Multicellular Organisms Occurred Rather Rapidly Compared with Previous Stages

The evolution of eukaryotes from prokaryotes took an enormous amount of time, but it marked a major turning point in the history of life on Earth, for it opened the door to the evolution of multicellular eukaryotic organisms—plants, animals, and fungi. ▷ Figure 23–8 shows the evolutionary relationship of these three kingdoms to their moneran and protist predecessors. You may want to take a moment to study the figure before you read the next material.

The Evolution of Multicellular Life also Began in the Sea.
Chapters 24–28 describe the diverse array of life-forms that live in the biosphere and offer insights into their evolution. As a result, the following material will highlight some of the key evolutionary events in the emergence of life on Earth to set the stage for your understanding of the evolution of multicellular organisms.

Because the first eukaryotic cells arose in the oceans and remained there for hundreds of millions of years, it should come as no surprise that the first multicellular plants and animals also evolved in the seas.

The first aquatic plants were the algae, of which there are three types: brown, red, and green. Scientists believe that each form evolved from a different ancestor. Brown

▷ **FIGURE 23–8 The Five Kingdoms**

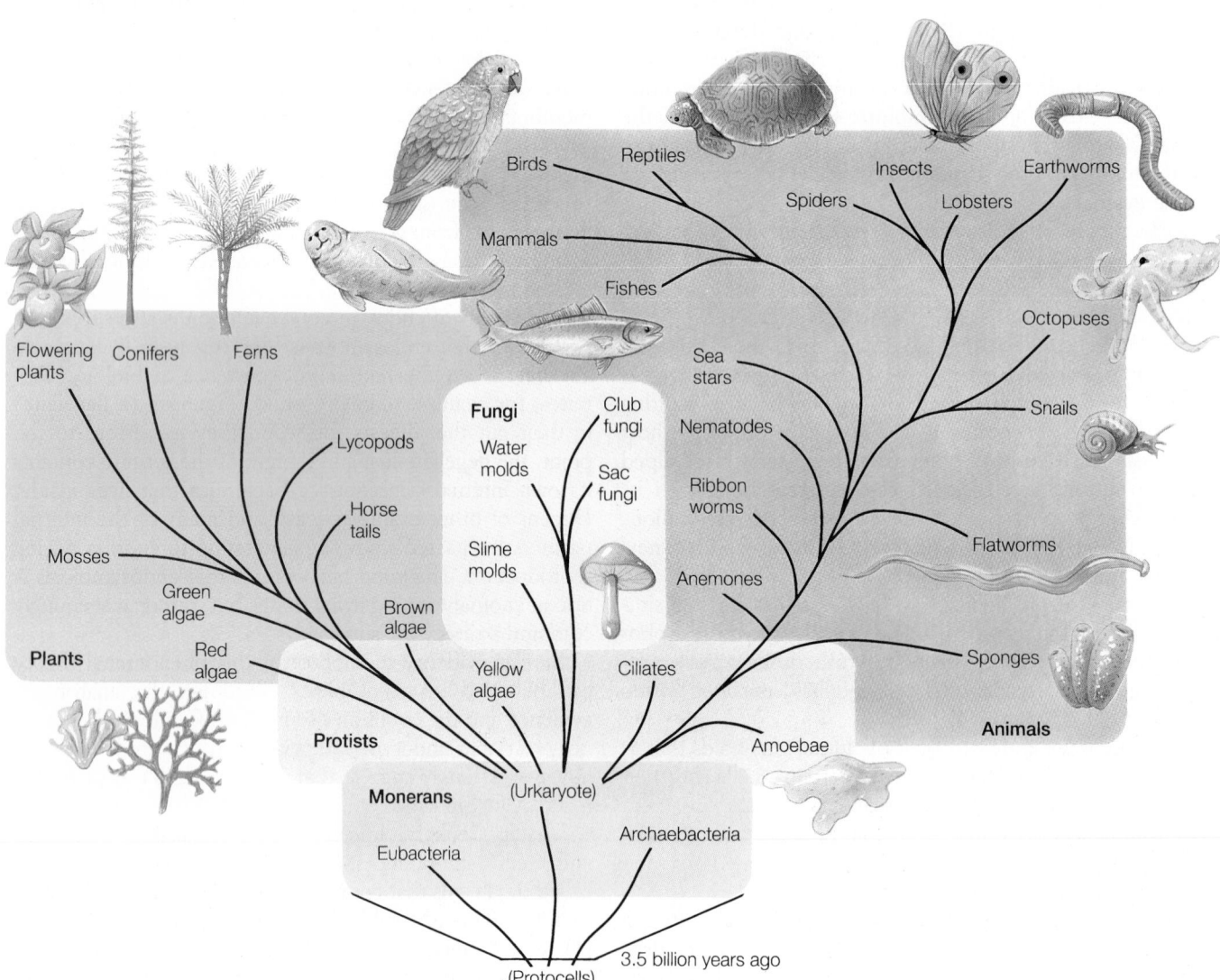

(a)

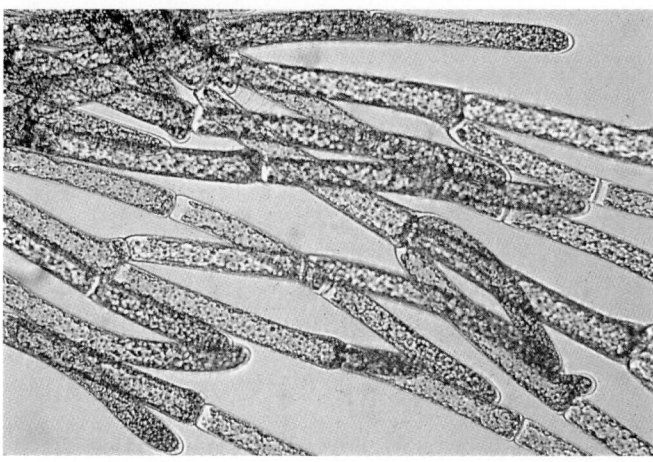

(b)

▷ **FIGURE 23–9 Algae** (*a*) This aquarium plant is really a multicellular alga displaying a remarkable degree of specialization. (*b*) Algal colony.

▷ **FIGURE 23–10 Freshwater Descendant** Sea lettuce.

and red algae remained in the water. Green algae also remained in the water, but are believed to have given rise to all land plants. Because of this, the following discussion will be limited to them.

The first green algae were probably free-floating photo-synthetic protists (eukaryotes with chloroplasts). These autotrophic organisms absorbed sunlight and carbon dioxide and released oxygen into the waters. Single-celled photosynthetic protists evolved the ability to form colonies, consisting of aggregations of cells, or filamentous forms, consisting of long strings of cells (▷ Figure 23–9). Others evolved into multicellular organisms, more complex forms whose cells are arranged to give a tissuelike appearance. An example is the sea lettuce, shown in ▷ Figure 23–10. Freshwater descendants include many common aquarium plants like the one shown in Figure 23–9b.

A variety of multicellular animals also arose in the sea around the same time. Studies of the fossil record suggest that a number of groups arose independently from early protists (single-celled eukaryotes). One of the first multi-

cellular animals may have been the sponge (▷ Figure 23–11). Sponges are sedentary (immobile) invertebrates (animals without backbones) that attach to rocks and other hard surfaces, usually in shallow water. Sponges consist of many cells arranged around a hollow interior. Water enters them through numerous tiny openings. Food particles are strained from the water, which is then expelled through one or more larger pores. Sponges exhibit a modest degree of specialization, or division of labor.

Another early multicellular invertebrate was a soft-bodied organism that resembled modern-day jellyfish. Vaselike in their shape, jellyfish and their relatives have a single opening that serves as both mouth to ingest food and anus to expel wastes from food digested in the body cavity. Soon, more efficient digestive systems evolved, which provided one opening for food to enter and a sepa-

▷ **FIGURE 23–11 The Sponge** Probably one of the first multicellular animals, the sponge consists of a central cavity into which it draws nutrients. The sponge exhibits a modest degree of specialization.

(a–1)

(a–2)

(b)

▷ **FIGURE 23–12 Early Sea Animals** (*a*) Trilobites were one of the most common life-forms in the ocean 600 to 500 million years ago. (*b*) Also present in large numbers were the predatory ammonites.

rate one for wastes to be released. This form of digestion is found in virtually all multicellular animals today.

Jellyfish were only one of many species to evolve. Molluscs (shellfish) and arthropods (crabs and lobsters) were among this extraordinary profusion of life. One notable arthropod was the armored trilobite (▷ Figure 23–12a). Trilobites crawled along and burrowed into the sand and mud of the bottom of the ocean. They were preyed on by other animals like the ammonite (Figure 23–12b).

Plants and Animals on Land First Arose from the Seas. Between 400 and 600 million years ago, many animals evolved in the oceans, some of which would eventually move onto land, giving rise to an equally impressive array of terrestrial life-forms. Scientists believe that the invasion of the Earth's barren landmasses was prevented for a long time in part by ultraviolet radiation from the sun. Ultraviolet radiation is lethal to plants and other unprotected organisms. The evolution of life on land may have been made possible in part by the emergence of the **ozone layer,** a thin layer of air containing a slightly higher concentration of **ozone molecules (O_3)** 20 to 30 miles above the Earth's surface (▷ Figure 23–13).

The ozone layer probably formed from oxygen released by photosynthesis. As noted earlier, oxygen first came from photosynthetic monerans, then from autotrophic protists, which evolved later. The more photosynthetic organisms were present, the greater the oxygen release. In the upper

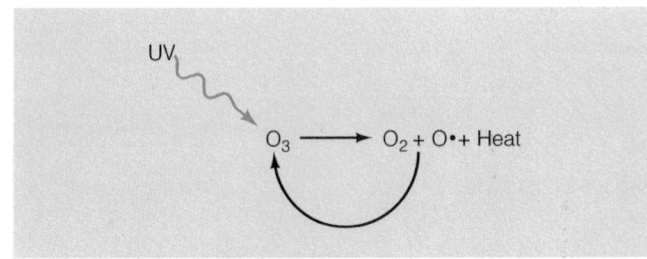

▷ FIGURE 23–13 Action of Ozone Molecules in the Ozone Layer When ultraviolet light strikes an ozone molecule, it causes the molecule to dissociate. The energy absorbed by the molecule is given off as heat, and the ozone molecule re-forms.

atmosphere, oxygen molecules react to form ozone molecules. Over time, the concentration of ozone in the upper atmosphere increased, thus creating the ozone layer. Ozone molecules absorb ultraviolet light, reducing the amount of harmful radiation that strikes the Earth by about 99%.

The colonization of the continents by plants occurred approximately 400 to 500 million years ago. As noted above, the first land plants probably evolved from green algae. Scientists believe that those algal species living in the intertidal zone (the region between high and low tides) that could survive periods of dryness persisted and eventually evolved into the first terrestrial species, the bryophytes (▷ Figure 23–14). The **bryophytes** (bry-o-fites) are mosses and other low-lying plants, all of which live in moist environments.

Moist environments are essential to bryophyte survival and reproduction for at least two reasons. First, these plants do not have sophisticated means of transporting water. Because they lack structures to transport food and water, bryophytes are called nonvascular plants. Second, bryophytes require a watery environment for reproduction. They have swimming sperm, which travel to the eggs in an aqueous environment. As a result, the bryophytes must live in swamps and marshes or areas of abundant rainfall where the ground is covered with water at least some part of the year.

The bryophytes and other nonvascular plants were followed by **vascular plants,** an enormous group containing specialized tissues to transport water from the soil to leaves and stems. Evolutionary biologists believe that the lineage giving rise to the vascular plants probably arose soon after plants colonized the land (▷ Figure 23–15).

The presence of vascular tissues gave these newcomers a decided advantage over bryophytes. It allowed them to grow much larger and more importantly, to colonize a much more diverse set of habitats.

The earliest vascular plants, such as the ferns, usually live in moist environments, which they require for fertilization. They persisted for over 200 million years and were largely replaced by more advanced vascular plants, which were capable of withstanding and reproducing in drier conditions. As you will see in Chapter 25, two groups of advanced vascular plants are the gymnosperms (the evergreens) and angiosperms (the flowering plants).

The advanced vascular plants succeeded in part because they evolved a protective coating (a layer of wax called the cuticle) on leaves and stems that reduces evaporative loss. Unlike the early vascular plants, discussed above, the advanced vascular plants also produce seeds, which gave them a distinct selective advantage, as discussed in Chapter 25.

Soon after the terrestrial plants began to evolve, the first animals invaded from the seas. Evolutionary biologists be

▷ FIGURE 23–14 The First Land Plants Bryophytes similar to (*a*) liverworts and (*b*) mosses were some of the earliest plants to inhabit land. These short-stalked plants live in moist environments.

(a)

(b)

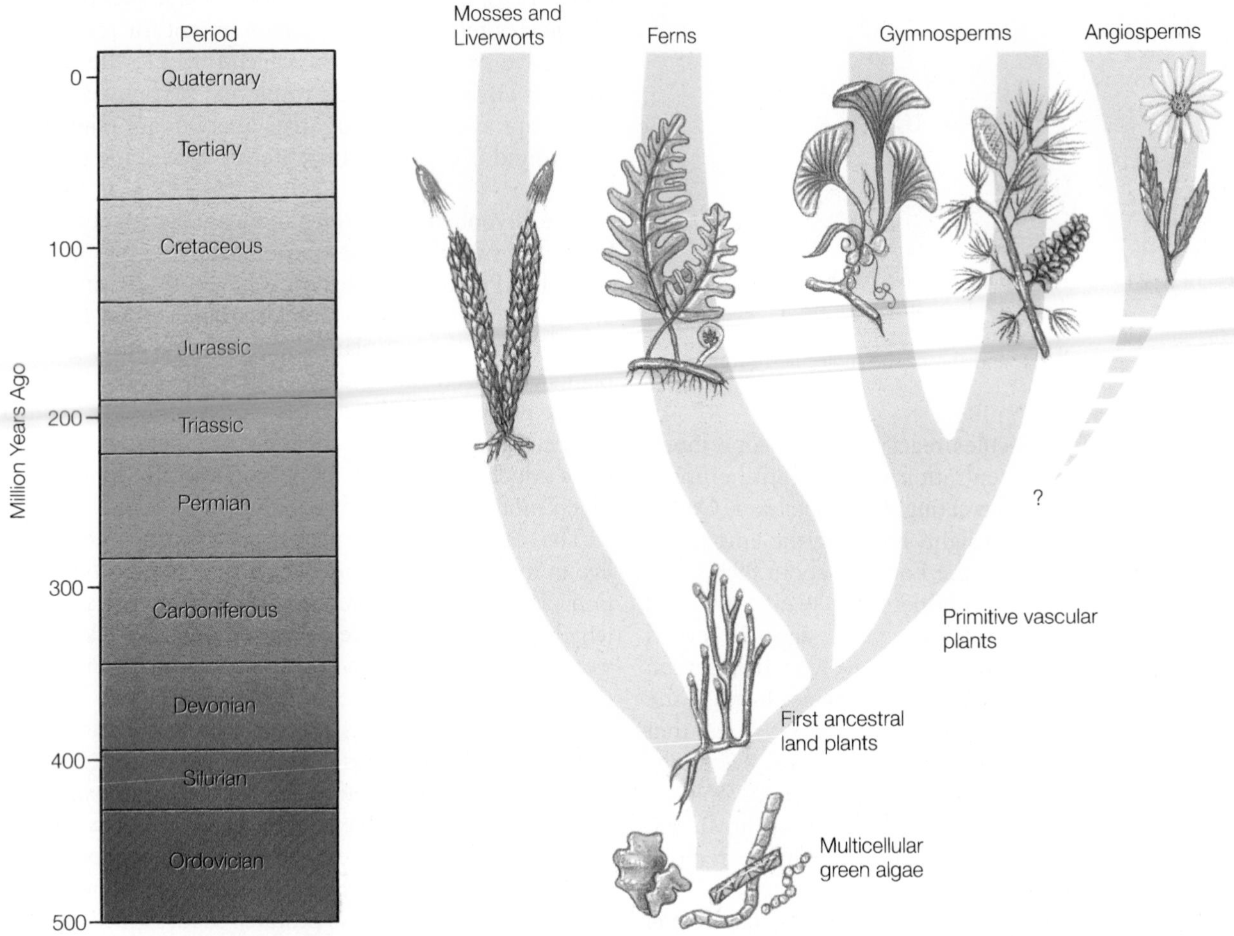

Period

Million Years Ago	
0	Quaternary
	Tertiary
100	Cretaceous
	Jurassic
200	Triassic
	Permian
300	Carboniferous
	Devonian
400	Silurian
	Ordovician
500	

Mosses and Liverworts

Ferns

Gymnosperms

Angiosperms

?

Primitive vascular plants

First ancestral land plants

Multicellular green algae

▷ **FIGURE 23–15 Plant Lineage** This diagram shows the approximate time of emergence of various plants from the first an-cestral land plants. Scientists divide the Earth's history into distinct periods shown on the left.

lieve that the previous emergence of plants provided the first land animals an abundant supply of food, necessary for survival.

The first animals to invade the land were probably the **arthropods,** invertebrates with segmented bodies and jointed legs. Two well-known examples are insects and crustaceans. Biologists believe that the very first terrestrial arthropods were the ancestors of the scorpion (▷ Figure 23–16).

Arthropods have hard outer skeletons that reduce evaporation and provide support. This feature, already acquired in the ocean, made the transition to terrestrial life that much easier. Organisms with characteristics that make them suitable for other environments are said to be **preadapted** to the new environment.

The arthropods had the land and its plants to themselves for millions of years. But in the seas and lakes of the newly forming world, fishes began to emerge. One group, the lobe-fins, is believed to have evolved in fresh water. Colonizing shallow ponds and streams, lobe-fins had lungs that allowed them to breathe out of water. They could also crawl out of the water and "walk" on dry land, using their strong, fleshy fins. This adaptation enabled them to travel

to new bodies of water when a pond dried up or ran out of food.

Evolutionary biologists believe that the lobe-fins gave rise to the amphibians. Modern amphibians include organ-

▷ **FIGURE 23–16 The First Land Animals** Arthropods, perhaps the early ancestors of the scorpion, were the first to invade the land.

(a)

(b)

▷ **FIGURE 23–17 The Dinosaurs** (*a*) Probably one of the most successful species ever to inhabit the land, the dinosaurs lived on Earth for about 100 million years. Why they disappeared remains a mystery. (*b*) Alligators and crocodiles are modern descendants of the dinosaurs.

▷ **FIGURE 23–18 Flying Reptiles** This pterosaur flew in prehistoric skies.

isms such as frogs, toads, and salamanders, which are dependent on both land and water for their subsistence. Most amphibians must return to water to lay their eggs. The earliest amphibians resembled the lobe-fins in several key respects: they could walk on land and could breathe air. Although the earlier amphibians shared many traits with lobe-fin fishes, they also possessed some unique adaptations that helped them survive and prosper on land—for example, a musculoskeletal system better suited to terrestrial locomotion and a skin that permitted gas exchange.

Reptiles (represented today by snakes, lizards, and turtles) evolved from amphibians many millions of years after the first amphibians emerged. The very first reptiles (called cotylosaurs) gave rise to a group known as the thecodonts, which, in turn, gave rise to the dinosaurs and many modern reptiles (▷ Figure 23–17). Thanks to their thick, scaly skin and eggs covered by a thick, rubbery shell, reptiles survived nicely on land.

As most schoolchildren will attest, the dinosaurs were magnificent creatures that lived for about 125 million years. Some consider them one of the most successful of all

animals that have inhabited the planet. Ranging in size from catlike creatures to the 80-foot-long ultrasaurus, dinosaurs were both herbivores (plant eaters) and carnivores (meat eaters). Some researchers believe that large dinosaurs were warm-blooded—that is, capable of maintaining a constant body temperature by metabolism of food molecules. However, most evidence suggests that, like other reptiles, they maintained their temperature by absorbing heat from their environment. Massive dinosaurs probably retained much of this heat overnight.

The cotylosaurs and their descendants were terrestrial animals, but many reptilian forms reinvaded the water, competing with fishes and amphibians. Flying reptilian forms also evolved from the early reptiles and are known as pterosaurs (tear-uh-sores) (▷ Figure 23–18). Some pterosaurs were as small as chickens, but others had wing spans up to 50 feet!

Approximately 65 million years ago, the dinosaurs vanished from the face of the Earth. What caused these beasts to disappear? No one knows for sure. By one count, 86 hypotheses have been proposed to explain their extinction. Some think that diseases may have wiped out the dinosaurs. Others contend that climatic change may have gradually driven them to extinction. Most likely, there are several or even many reasons for the demise of these magnificent creatures. (For a discussion of extinction, see the Point/Counterpoint in this chapter.)

In 1980, Luis Alvarez, Walter Alvarez, and several other scientists hypothesized that a comet or an asteroid might have struck the Earth, creating a huge cloud of dust that cooled the atmosphere. The cloud may have persisted for months, creating conditions too cold for dinosaurs to survive. Studies of sedimentary rock clearly reveal a layer of

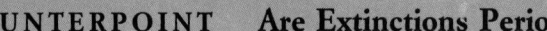

THE NEMESIS THEORY FOR A 26-MILLION-YEAR CYCLE

Richard Muller

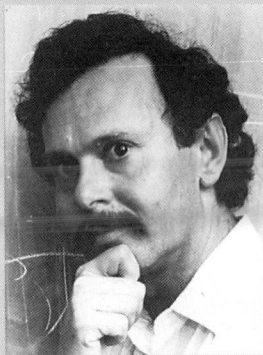

Dr. Richard A. Muller, Professor of Physics at the University of California at Berkeley, and Faculty Senior Scientist at the Lawrence Berkeley Laboratory, is the recipient of the Alan T. Waterman Award, the highest award bestowed by the American National Science Foundation. He is the author of Nemesis: The Death Star.

H ave the mass extinctions on Earth occurred on a regular 26-million-year time schedule? This apparently absurd idea, presented in 1984 by paleontologists David Raup and Jack Sepkoski, has attracted a great deal of interest. If it is true, our picture of evolution must be dramatically changed.

Suppose that the periodicity is real; then how could we explain it? The most carefully studied model that works is the Nemesis theory, which postulates a companion star to the sun that orbits close every 26 million years, triggering a storm of comets, several of which hit the Earth. The ecological disaster that would follow from their impacts may account for the extinctions.

But the Nemesis theory is worth considering only if the paleontologic data truly show periodicity. Do they? This issue is now being hotly debated by scientists and statisticians. Over

the years there have been many false claims of periodicity in geologic phenomena, in everything from earthquakes to volcanic eruptions. This is probably because the mathematics of periodic phenomena contains many potential traps. If you reject claims of periodicity in all geologic phenomena, you will be right 95% of the time. However, you may also miss a potentially great discovery. Periodicity in the Earth's climate is now widely accepted, and attributed to periodic changes in the Earth's orbit around the sun.

Are the mass extinctions periodic? There are ways to analyze the data that show no significant effect, but that does not mean that there is no periodicity. If you flip a coin 100 times and detect 52 heads and 48 tails, you may think you are following random statistics, until someone points out that the first 52 tosses were heads, and the last 48 tails. Then you realize that the tosses were not random at all.

Thus we should beware of claims that some analysis suggest that extinctions were random. In fact, for a true periodicity, there will be many analyses that show nothing. For example, an analysis that depends on a perfect periodicity might not discover a quasi-periodic signal, one in which the cycles vary slightly from period to period. This is the kind of periodicity expected from the Nemesis model, since the period of Nemesis is expected to undergo slight perturbations from other passing stars.

▷Figure 1 shows the original data published by Raup and Sepkoski, but I have altered their plot to take into account the uncertainty in the age scale. Each family extinction is represented by a normal curve, with the width of the curve equal to the uncertainty in the date of the extinction, and the area of the curve proportional to the intensity of the extinction. Periodicity in the plot is evident to my eye, and a mathematical analysis that verifies my intuition, demonstrating that the probability of such apparent periodicity appearing by chance is less than 1 in 1000.

But scientists have fooled themselves many times with such arguments, and it is important to try to look at the data in new ways. Such analysis continues. Sepkoski has recently shown that the same 26-million-year period appears when the data are analyzed in terms of fossil genera (rather than fossil families). Ultimately, it will be new data that will convince most scientists that the periodicity is, or isn't, there.

In the meantime, I am placing my bet by searching for Nemesis. If we find a star orbiting the sun with a 26-million-year period, it will provide convincing proof that the theory was real. And Nemesis could be the first astronomical object whose discovery will be due to a careful study of fossils.

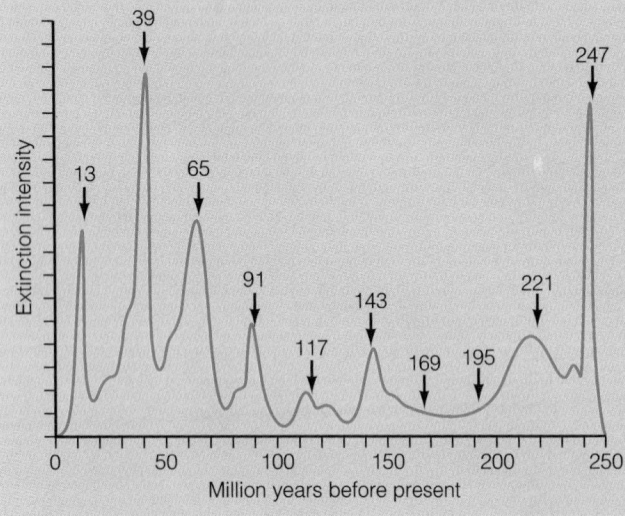

▷ **FIGURE 1 Evidence of Periodicity?** The data of the paleontologists Raup and Sepkoski show great mass extinctions occurring every 26 million years, as replotted by the author. This plot inspired the Nemesis theory.

DELAYED RECOVERY EXPLAINS THE SPACING OF MAJOR EXTINCTIONS

Steven M. Stanley

Steven M. Stanley is a professor of paleobiology at Johns Hopkins University. He is the author of several books, including Extinction *and* Earth and Life Through Time.

I n 1984, David M. Raup and J. John Sepkoski suggested that major global extinctions of life on Earth during the past 250 million years had been periodic, occurring every 26 million years. A truly periodic pattern would have two implications. First, this pattern would imply that a single agent caused all major extinctions. Second, this agent of extinction would probably be of extraterrestrial origin. We would not expect earthly causes, such as volcanic activity or lowering of sea level, to follow a regular rhythm. On the other hand, many astronomical phenomena are periodic because they entail cyclical motion of extraterrestrial bodies.

We now have strong evidence that a comet or meteorite caused one of the great extinctions of the geologic past: the one that swept away the dinosaurs and many other forms of Mesozoic life. The evidence is found in the very youngest Mesozoic stratum in many regions of the world. It includes the presence of (1) a relatively high concentration of the element iridium, which is very rare on Earth but more common in meteorites; (2) shocked mineral grains of a type not known to have formed on Earth except from the enormous pressures generated when a large extraterrestrial object has landed; and (3) tiny, almost spherical grains of glassy rock of a type known to have formed when extraterrestrial impacts have melted rock and sent drops of liquid rock high into the atmosphere.

There are, however, two problems with the idea that evenly spaced episodes of extraterrestrial bombardment caused all major extinctions. One is that the extinctions actually appear not to have been evenly spaced. The second is that we have evidence that some of these great crises had earthly causes.

The major extinction of Late Eocene time illustrates both of these points. Since Raup and Sepkoski compiled their data, experts have revised the date of this event, which wiped out groups of mammals on the land and other forms of life in the ocean. The Eocene ended about 34 million years ago, not 38 million years ago, as previously believed. The shift of this date expands the interval between the dinosaur extinction and this one to 31 or 32 million years, and it shrinks the next interval to 22 or 23 million years. In other words, the discrepancy between the lengths of these adjacent intervals is at least 8 million years.

In addition, we have abundant evidence that climatic change throughout the world caused the Late Eocene extinctions. Prior to the crisis, the world had been extremely warm by modern standards. A tropical jungle had covered southeastern England, for example, and palm trees had lived as far north as Alaska. While the extinction was occurring, these plants disappeared from high latitudes, and species adapted to cooler climates took their place. In fact, this climatic transition has never been reversed: regions far from the equator have never been so warm again. We even have a geological explanation for the great climatic change at the end of the Eocene Epoch. Continental movements altered current patterns in the oceans, causing polar regions to cool down. Antarctica became isolated over the South Pole at this time, and the ice cap that still covers most of Antarctica began to form at this time.

Even though the great extinctions have not been distributed evenly through geologic time, we do know that they have not occurred at random intervals. In fact, they appear to be "spaced out." Never have two occurred in close succession. I suggest that this spacing resulted from the presence of a recovery interval after each crisis. The environment remained harsh for a time, and only hardy species that were adapted to the harsh conditions remained or came into being during this period. After more favorable conditions returned, "fragile" species evolved and ecosystems once again vulnerable to mass extinction. The fossil record offers evidence that a recovery interval of this kind followed many mass extinctions. For example, most global extinctions severely damaged tropical reef communities, and it normally took several million years for these and neighboring tropical communities to flourish once again. Only after such a recovery could another severe crisis have struck again in the tropics.

≋ SHARPENING YOUR CRITICAL THINKING SKILLS

1. Summarize the key points of both authors.
2. Can you decide which is presenting a more plausible hypothesis?
3. Do you see any errors in reasoning in either essay, or any assumptions that may be affecting the authors' views?

dust that settled on the Earth, suggesting a period of intense meteoric activity or a single asteroid collision of immense magnitude. Recent evidence shows, however, that the dinosaurs were dying out long before this time and that some species persisted well after this period. Thus, even though the extinction of the dinosaurs was abrupt in geologic time, it was not as abrupt as those supporting the asteroid theory might believe.

Dinosaurs perished from Earth, but many reptiles remained, among them alligators, crocodiles, snakes, lizards, and turtles. As shown in ▷ Figure 23–19, birds are believed to have evolved from reptiles about 225 million years ago at a time when the dinosaurs were beginning to diversify. The fossil bones of what may have been the first bird, called **protoavis** (literally, "first bird"), were recently discovered. Protoavis is believed to have evolved sometime after the flying reptiles emerged and from a distinctly different lineage. Protoavis had hollow bones and other features that were vital to flight. It may have been a transitional form between reptiles and birds.

Perhaps the best studied of the early birds is the archaeopteryx (ar-key-op-ter-ix). Shown in ▷ Figure 23–20, this bird appeared about 75 million years after protoavis. About the size of a crow, archaeopteryx had wings and feathers very much like those of modern birds but had a lizardlike tail, suggesting its evolutionary link to reptiles.

Studies of the fossil record suggest that mammals probably also evolved from reptiles, as illustrated in Figure 23–19. Many mammal-like reptiles were present about 230 million years ago, around the time of the emergence of dinosaurs, and many became extinct. The first truly recognizable mammals appear in rock about 180 million years

▷ **FIGURE 23–19 Evolutionary Relationships of the Early Reptiles and Their Descendents.** Note that ancient reptiles gave rise to dinosaurs, birds, and, of course, modern reptiles.

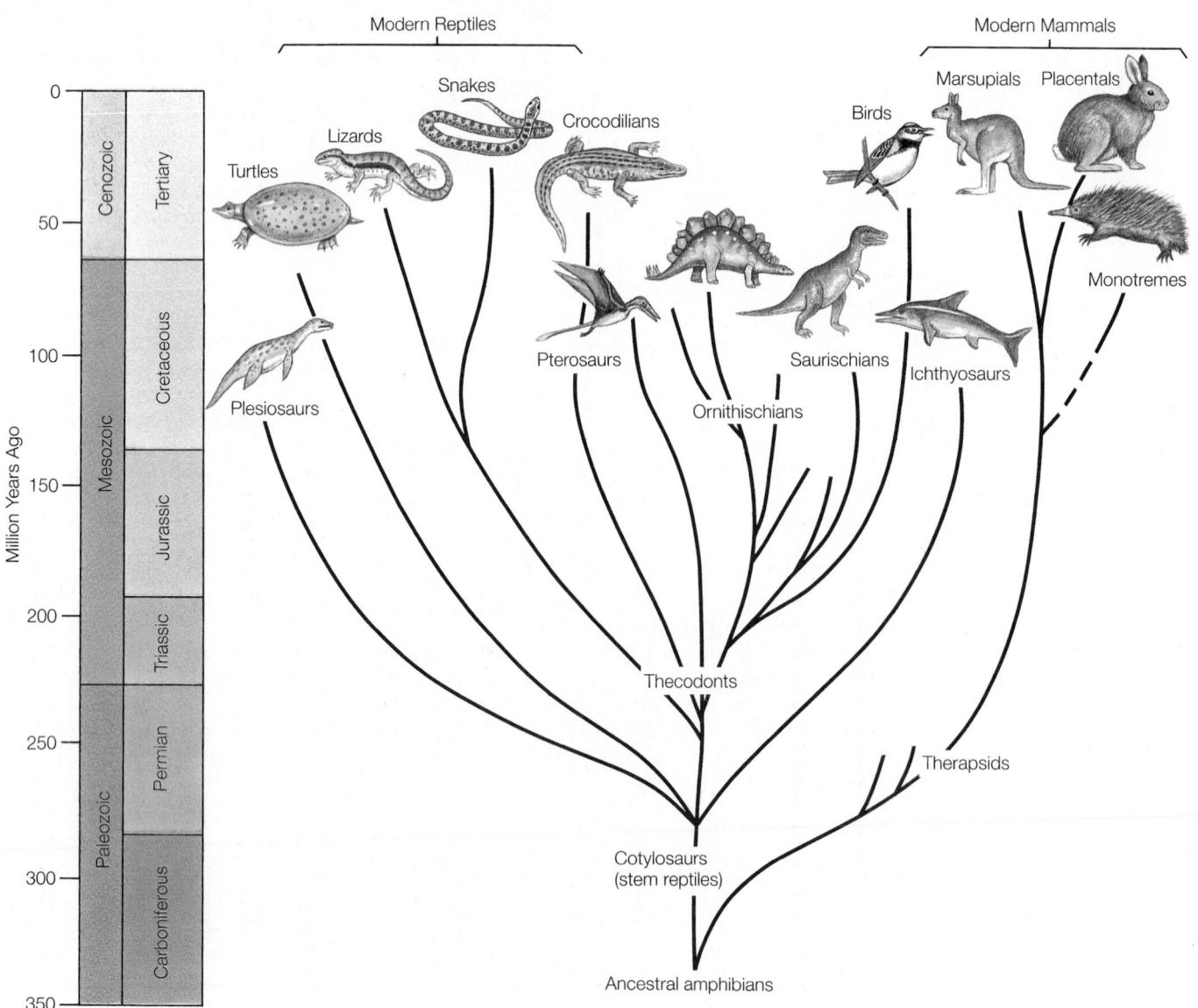

> **FIGURE 23–20**
Archaeopteryx About the size of a crow, the archaeopteryx had wings and feathers very much like those of modern birds but had teeth and a lizardlike tail.

old. Living among the dinosaurs for millions of years, the earliest mammals were rather small creatures that inhabited trees and were active principally at night.

Mammals are characterized by mammary glands and hair. As they evolved, they developed increasingly sophisticated means of regulating body temperature. Body size increased greatly, and the limbs elongated and diversified, a feature that provides many styles of locomotion—from flight (bats) to swimming (beavers, whales) to running (cheetahs, deer). Brain size also increased during evolution, and today mammals have the largest brains (relative to body size) of all organisms on Earth.

The primates, which include the apes and humans, are a group of mammals that evolved from a tree-dwelling mammal that lived among the dinosaurs. As you will see in Chapter 28, the early primates persisted long after the dinosaurs perished and gave rise to human beings.

≈ HOW EVOLUTION WORKS

The theory of evolution explains why life forms change. It also explains how life became so diverse. This section describes the mechanisms behind evolution.

Human knowledge of evolution comes from many sources. One of the most important sources is fossils. **Fossils** consist of the remains (wood from trees, bones and eggs from animals) or impressions of organisms that lived on Earth in times past.[3] Over the past 100 years, thousands of bones of dinosaurs and other animals have been excavated from sites all over the globe. How were these bones preserved?

The most common explanation is that animals that died in regions of rapid sediment deposition (shallow lakes, mud flats, and swamps) decayed over time, leaving behind

bones that were buried in sediment. In these sites, the minerals of the bones were often replaced by minerals in the water that seeped through them. (A similar process is responsible for the formation of petrified wood.) Some creatures, like the dinosaurs, also left footprints in the mud that hardened into stone. These footprints are part of the vast body of fossil evidence of earlier life forms (▷ Figure 23–21a). Imprints of leaves also make up part of the fossil record (Figure 23–21b). Some ancient creatures (frogs, insects, and flowers) were preserved in amber, resins released by ancient trees (Figure 23–21c). Blocks of amber, like plastic blocks containing biological specimens, provide a window to our past. Gas bubbles in amber have even been analyzed to determine the composition of the atmosphere in earlier times.

Genetic Variation Is the Raw Material of Evolution

For years, biologists believed that evolution was a slow and gradual process during which organisms changed, sometimes even forming entirely new species or new groups (for example, when reptiles gave rise to birds). As you shall soon see, this view has changed dramatically in recent years. For our purposes, a species is a group of organisms whose members are anatomically and physiologically similar to one another. The members of a species interbreed successfully; that is, they produce viable, reproductively functional offspring.

During evolution, the changes that occur in species take place because of changes in the genetic composition of entire populations of organisms. Biologists refer to the genes of a population as a **gene pool.** Genetic changes in a population come from five processes: (1) random mutations, (2) crossing over, or genetic recombination, (3) the independent assortment of genes during meiosis, (4) new genetic combinations produced during sexual reproduction (that is, the combination of genetically different gametes), and (5) gene flow. Of these sources of genetic variation, the only one that creates *new* forms of the genes is mutations. The rest only shuffle existing genes, creating new combinations that may lead to favorable adaptations. This section will discuss each process.

Mutations in the genes of organisms occur naturally and randomly. Many of these errors are corrected by the cells, but some remain.

Mutations occur in the body cells (somatic cells) and germ cells. In the evolution of multicellular organisms, however, somatic mutations are meaningless, for they cannot be passed to future generations. It is germ cell mutations that are of importance.

Like somatic cell mutations, germ cell mutations can be harmful, even lethal. In fact, many human embryos that abort spontaneously are the result of harmful germ cell mutations in the sperm, the ovum, or both (Chapter 22).[4]

---

[3] The word *fossil* comes from a Latin word meaning something "dug up".

[4] A mutation in a somatic cell in a morula or blastocyst could also lead to embryonic or fetal death and miscarriage.

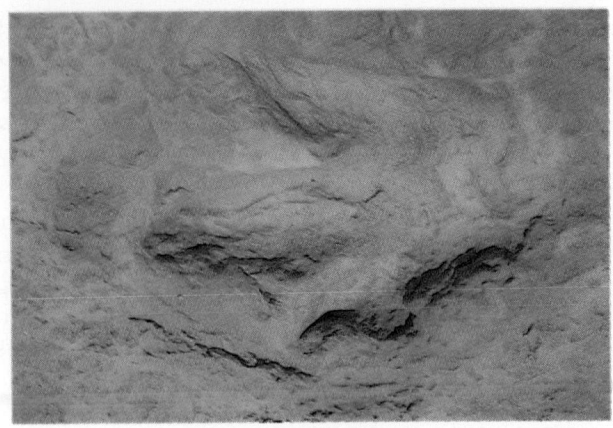

(a)

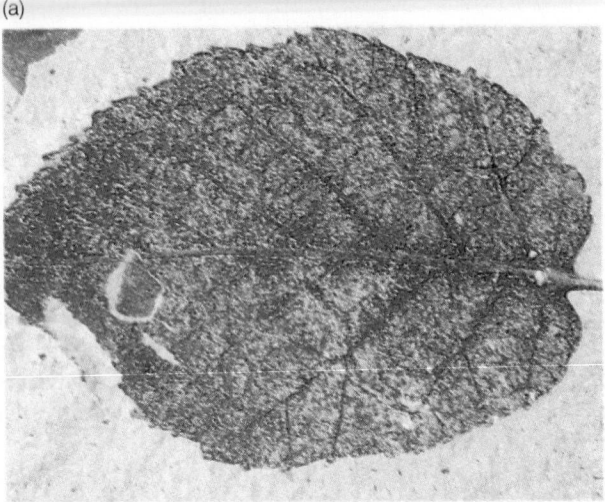

(b)

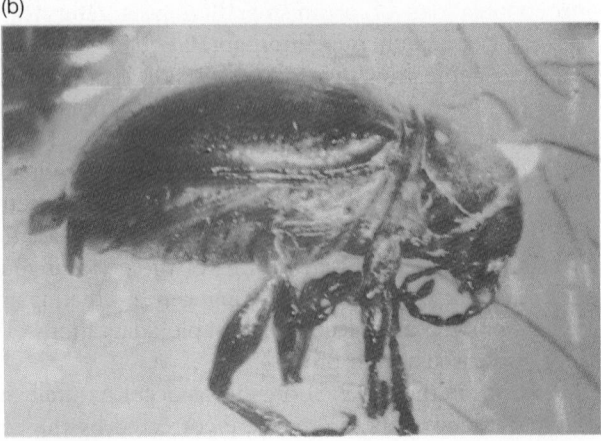

(c)

▷ **FIGURE 23–21 Fossils** (*a*) Dinosaur tracks made in a Texas streambed about 120 million years ago. (*b*) Imprint of a leaf. (*c*) Insect embedded in amber about 40 million years ago.

tions. For example, a random mutation in the germ cell of a fish may make its offspring more efficient in their digestion of food. This efficiency, in turn, increases the likelihood that their offspring will survive and reproduce. On the other hand, members of the same population that do not share this trait are less likely to survive and reproduce and will very probably produce fewer offspring. Over time, the better adapted fish will leave a larger number of offspring. These fish are also more likely to survive and reproduce than others. As a result, the gene pool changes over time, and future populations contain proportionately more offspring of the fish with the beneficial mutation. On a genetic level, the allele that gave these organisms an advantage increases in frequency in the population. As a result, beneficial mutations are often said to be the raw material of evolution. This is true for sexually as well as asexually reproducing organisms (bacteria).

As noted earlier, mutations are a source of new alleles and thus a major source of phenotypic **variation** in a population—that is, differences in anatomical, physiological, and even behavioral characteristics.[5] In sexually reproducing organisms, variation also results from recombination. **Recombination** is the formation of new combinations of genes during meiosis via crossing over (Chapter 8). During meiosis, homologous chromosomes pair up and often exchange segments in a process called crossing over (Chapter 9). Crossing over results in new genetic combinations and, therefore, variation in a population. If the new combination of genes produces advantageous adaptations, they will increase in frequency in the population.

The independent assortment of chromosomes during meiosis may be even more important in producing new genetic combinations that result in phenotypic variation. Independent assortment, described in Chapter 8, simply means that maternal and paternal chromosomes in the forming germ cells segregate independently of one another during meiosis, creating new genetic combinations.

Still another obvious source of variation is sexual reproduction. When male and female gametes combine they produce new genetic combinations, some of which may provide advantageous adaptations.

The final source of genetic variation is called gene flow. **Gene flow** is the introduction of new alleles to a population by individuals from another population of the same species.[6] This process occurs when members of previously isolated populations mate. A male bird capable of flying long distances while using less energy than other members of the population, for example, might cross a mountain range and mate with a female from a previously isolated population. Her offspring would then carry the gene that permits more efficient flight. Variation has been introduced into the population.

---

[5] For bacteria and other organisms that reproduce asexually (by dividing or budding), mutation is the only source of variation.
[6] The new genes are most likely to have arisen from a mutation.

Other germ cell mutations, however, are beneficial to the offspring. That is, they produce characteristics that give the offspring an advantage over other members of the population. Genetically based characteristics that increase an organism's chances of passing on its genes are called **adapta-**

Natural Selection Is a Process by which Organisms Become Better Adapted to their Environment

A sociologist, Andrew Schmookler, once wrote that evolution employs no author, only an extremely patient editor. That is to say, evolution is not directed. There is no author. Instead, mutations arise spontaneously. Recombination, gene flow, mutations, and new combinations from the independent assortment of genes and from sexual reproduction produce genotypic and phenotypic variants. If beneficial, the genetically based characteristics (adaptations) that arise from these sources tend to persist. The species changes over time, becoming better suited to its environment. In some cases, a whole new species may evolve. The gene pool shifts because of evolution's patient "editor," a process called **natural selection.**

Charles Darwin, who with Alfred Wallace originated the idea of natural selection, describes it as a process in which slight variations, if useful, are preserved (▷ Figure 23–22). Thus, natural selection is often a process by which organisms become better adapted to their environment.

Two principal factors influence a population: **biotic factors,** or other living organisms, and **abiotic factors,** the physical and chemical environment (temperature, rainfall, and so on). Abiotic and biotic factors are evolution's editors. They influence survival by "selecting" the fittest—those best able to reproduce and pass on their genes. If these conditions change, those organisms best adapted to the new conditions tend to remain and pass on their genes to subsequent generations. This process causes a shift in the frequency of certain genes in the gene pool of a population.

▷ **FIGURE 23–22 Evolutionary Voyage** (*a*) Darwin as a young man proposed the Theory of Evolution by Natural Selection, helping solve one of the key puzzles of biology: how species evolve. (*b*) From 1831 to 1836, Darwin made a famous voyage on the *Beagle*, collecting and cataloging thousands of diverse species.

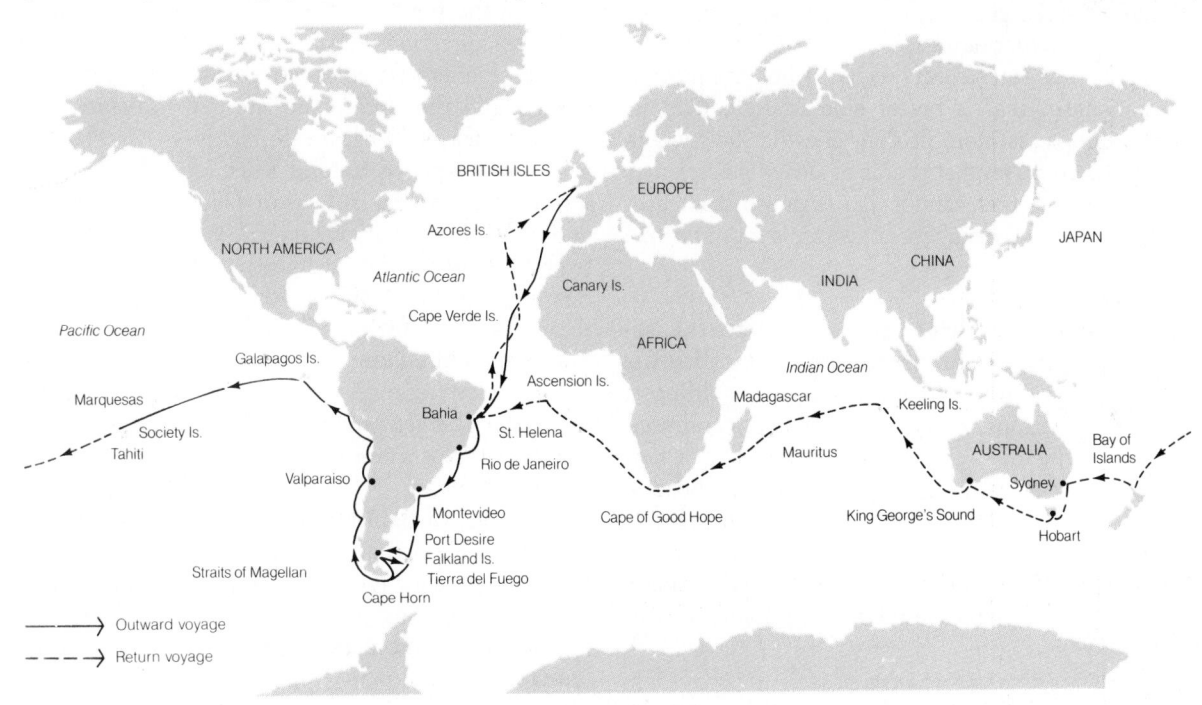

Darwin was a 19-century British naturalist. In his time, evolution was widely discussed among naturalists and other scientists, but the mechanism by which it occurred remained a mystery. Darwin dedicated many years to the search for an answer, traveling around the world by ship and cataloging species (Figure 23–22b).

A careful scientist, Darwin spent about 20 years of continued research after coming up with the idea of natural selection in looking for flaws in his own reasoning. In 1858, much to his surprise, Darwin received a paper from Wallace, a respected naturalist who had proposed the same concept. Darwin sent Wallace's paper to some of his colleagues and suggested that they publish it. Fortunately, Darwin's colleagues, who were aware of his own theory, prevailed on him to write a paper of his own. In 1858, Wallace's and Darwin's papers were both presented to the Linnaean Society. In 1859, Darwin published his now-famous book, *On the Origin of Species by Means of Natural Selection*.

As in the case of Mendel, Darwin's ideas took many years to be understood and accepted. Not until the 1940s, about 60 years after his death, did natural selection become widely known and appreciated. Today, it is one of the central tenets of modern biology.

Natural Selection Ensures that the Fittest Organisms in a Population Survive and Reproduce.
Darwin used the phrase "survival of the fittest" to describe how natural selection worked.[7] Survival of the fittest is commonly interpreted as "survival of the strongest." To a biologist, however, **fitness** is a measure of reproductive success and, therefore, of the genetic influence an individual has on future generations. By definition, then, the fittest individuals leave the largest number of descendants. Their influence on the gene pool will be greater than that of their less-fit contemporaries.

Fitness does not necessarily result from strength or speed. An organism that is better able to hide than its cohorts, for example, is more fit than one that falls victim to predators. An organism that is able to digest grasses may be more fit than one that cannot. An organism that uses water and energy with the highest efficiency is more fit than its inefficient cohorts.

Biologists Recognize Three Types of Natural Selection, Depending on Environmental Conditions.
Natural selection occurs under three widely different environmental conditions. In the first instance, conditions are changing rapidly, or organisms are exposed to a new environment with conditions very different from those in which they have evolved. Under these circumstances, extreme phenotypes best adapted to the new or changing conditions persist. This phenomenon is called **directional selection** and can be represented by a simple series of

[7]Herbert Spencer is thought to have coined the phrase.

graphs, shown in ▷ Figure 23–23a. The bell-shaped curve in the top panel represents a characteristic in a hypothetical population—say, for example, the ability of a population of field mice to withstand cold. This graph shows a wide range of ability. The shaded area indicates the mice that are best able to withstand cold. If the climate of an area suddenly shifts, becoming colder, or the mice are moved to a colder climatic zone, the ones best suited to cold will be more likely to survive and reproduce. The others will die. As a result, the frequency of the gene that permitted them to survive and reproduce in a colder climate increases in the population. This phenomenon is called directional selection, because the agents of natural selection unknowingly "direct" the course of evolution.

The second variation on natural selection is **stabilizing selection.** It occurs when an environment is unchanging and when a species is already well adapted to that particular environment. In this case, naturally occurring variation in the population will produce phenotypes that are not necessarily any better off than the average. New variants will be eliminated; natural selection will favor the "average" individual.

Stabilizing selection is illustrated in Figure 23–23b. As shown, stabilizing selection favors the phenotypes close to the average. By eliminating the fringes, this type of natural selection increases the frequency of alleles responsible for the fittest phenotype.

To help you understand this phenomenon, consider an example: wildflowers that grow on the forest floor in New England. The wildflowers emerge in early spring, after the last hard frost and before the leaves come out on the trees. After the trees "leaf out," the forest floor is covered with shadow, too dark to support flowers. As a result, most wildflowers sprout, flower, and set seed in the brief period between the last frost and full leaf development. Variants in either direction may occur, but they will be selected against. Plants that sprout too early will be eliminated by frost. Plants that sprout too late may not survive to flower when the leaves block the sunlight.

A third type of natural selection is called disruptive. **Disruptive natural selection** favors phenotypes at both ends of the phenotypic range (Figure 23–23c). Alternatively, environmental change such as the appearance of a new predator or a shift in food supply may select against the average phenotype. The result is a split in the population. A large predator, for example, may be unable to catch smaller prey because they are too fast or can hide more effectively. Larger prey may be aggressive and too risky to bother with. As a result, the medium-sized prey fall victim to their predator.

All forms of natural selection cause a shift in gene frequency in a population—favoring the fittest individuals. Interestingly, they tend to decrease variation within a population, countering increases in variation from gene flow, mutation, recombination, independent assortment, and sexual reproduction.

 (a) Directional selection

 (b) Stabilizing selection

(c) Disruptive selection

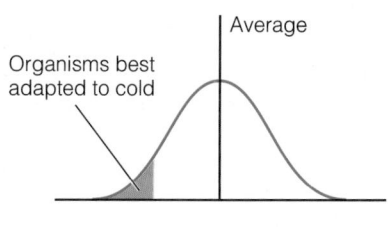

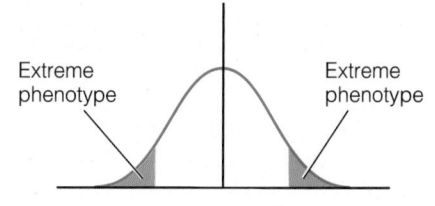

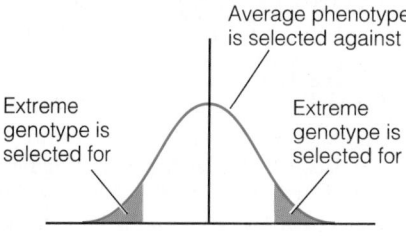

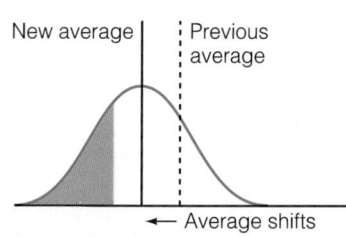

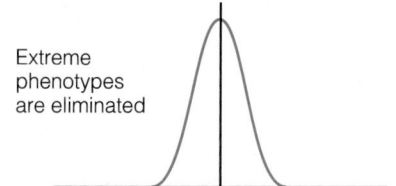

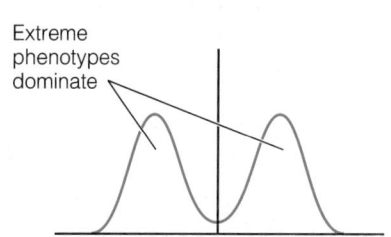

Proportion of cold-adapted individuals in population increases

Average remains the same, but number of "average" individuals increases

▷ **FIGURE 23–23 Three Major Types of Natural Selection** (*a*) In directional selection, environmental factors select the fittest individuals. This causes a shift in the gene pool. Over time, the population shifts so that these individuals and their offspring represent a larger proportion of the population. (*b*) In sta-

bilizing selection, environmental factors eliminate fringes, preserving the average phenotype. (*c*) In disruptive selection, environmental factors favor the fringes and eliminate the average phenotype.

Allele Frequencies Can Shift Because of Chance Events

Imagine a population of mountain goats in which 3 members—out of, say, thirty—possess a remarkable ability to climb extremely steep, rocky slopes. One day those three animals climb a ridge to reach some food. The rest of their herd stays on lower ground. That afternoon a lightning bolt strikes the ridge, killing all 3 goats.

This chance event causes a shift in the frequency of alleles in the population, essentially wiping out the gene that improved the three goats' climbing ability. A shift in allele frequencies like this that is due to a chance occurrence is referred to as **genetic drift.** This process is most rapid when a population is small, as in the example above.

If in this example an airplane crashed on lower ground and killed the 27 less-fit animals, leaving only the 3 exceptional climbers, the allele frequency would also shift—this time in favor of improved climbing ability. Over time, a new population would emerge. This extreme case of genetic drift is also known as the **founder effect,** because a small number of individuals establish a new and genetically distinct population.

The founder effect is extremely important in the colonization of islands and other isolated habitats. It is also important when a few individuals of a species are introduced into new habitat.

Genetic drift also takes place in populations that are severely reduced through natural or human causes. The

extreme reduction in the number of individuals in a population is called the **bottleneck effect.** In the late 1800s, for instance, the northern elephant seal was nearly hunted to extinction. The 20 or so remaining seals formed the basis of the new population, which today numbers around 30,000. This much larger population has almost no genetic variation compared with other seal populations that have not gone through similar bottlenecks.

Cheetahs are also genetically fairly homogeneous, so much so that they can receive skin grafts from other members of the population without graft rejection, a rare phenomenon indeed. Some scientists think that cheetahs may have gone through a bottleneck.

Genetic uniformity resulting from these cases of genetic drift make populations highly susceptible, because they do not contain the wide genetic base vital to evolutionary change in the event that environmental conditions change. This is one reason why it is prudent not to let wild populations of any species be diminished.

The Hardy-Weinberg Equation Provides a Means of Measuring Genetic Change

It is important to remember that individuals do not evolve, populations do. As the previous discussions have shown, populations evolve as a result of changes in the frequencies of genes in their gene pool through natural selection.

In the early 1900s, two scientists working independently

A PRIMER ON POPULATION GENETICS

The Hardy-Weinberg equation helps predict the frequency of alleles in a population. It was derived by two scientists, G. H. Hardy and W. Weinberg, working independently. This formula allows one to estimate the frequency of alleles in populations that fit certain (rather formidable) criteria. Those criteria include the following. First, the population is large. Second, no mutations are occurring in the population. Third, the population is isolated from other populations of the same species, eliminating gene flow. Fourth, mating within the population is purely random. And fifth, there is no natural selection under way. These criteria are a rather tall order for a population, and thus the equation that these scientists derived (shown below) only provides a means of measuring gene frequency in an ideal state or in departures from that state.

To understand how the equation works, we begin with a hypothetical population carrying a gene with two alleles, A and a. In the Hardy-Weinberg equation, the frequency of the A allele (dominant) is indicated by the letter p, and the frequency of the a allele (recessive) is indicated by the letter q. Frequency is a measure of occurrence—that is, how often an allele appears—and is indicated by decimals rather than percentages. A heterozygous individual (Aa) therefore has one A and one a allele. In this example, p is equal to .5, and q is equal to .5. (That is to say, one-half (50%) of the genes are A, and 50% are a.) The sum of p and q is always 1.

Now suppose that you cross a heterozygous individual (Aa) with another heterozygote (Aa). The result is shown in ▷ Figure 1. Using the Punnett square, we find the frequency of the genotypes produced in a random combination of gametes. In the four offspring produced by this cross, there are 4 A and 4 a alleles. As a result, $p = .5$, and $q = .5$. Next, plug these values into the Hardy-Weinberg equation:

$$p^2 + 2pq + q^2 = 1$$

In this equation, p^2 represents the frequency of homozygous-dominant individuals (AA). Thus, $p^2 = .5 \times .5$, or .25. The frequency of homozygous dominants is .25, or one in four. The second factor in the equation, $2pq$, represents the frequency of heterozygous individuals (Aa). Thus, $2pq = 2 \times .5 \times .5 = .5$. In other words, 50% of the offspring are heterozygous (Aa). Finally, q^2 is equal to the frequency of homozygous individuals (aa). Thus, $q^2 = .5 \times .5 = .25$.

In this example, we knew the frequency of the dominant and recessive alleles in a population. In most cases, however, we know the frequency of only one genotype. With this information, we can mathematically determine the frequency of all the alleles (assuming equilibrium conditions). Consider a simple example, dimples.

Dimples are present in homozygous-recessive individuals. In this example, suppose that 1% of the members of a population under study has dimples. One percent converted to a decimal is .01. To calculate the number of individuals that is homozygous dominant and heterozygous, first remember that in the Hardy-Weinberg equation q^2 is equal to

the number of homozygous-recessive individuals. Therefore, in this case $q^2 = .01$, and $q = 0.1$.[1] Now that you know the value of q, you must find the value of p. Recall that the sum of p and q is always equal to one, when two alleles are involved. If $q = 0.1$, then $p = 0.9$. Next, recall from the previous discussion that p^2 equals the number of homozygous-dominant individuals in a population. Accordingly, $.9 \times .9 = .81$. This is to say, 81% of the population is homozygous-dominant. The remainder (19%) is heterozygous.

As noted earlier, the Hardy-Weinberg equation holds only for populations in which the gene pool stays constant, a rare event indeed. Nonetheless, this technique permits one to approximate the frequency of alleles and is useful for that reason.

As noted in the text, the Hardy-Weinberg equation is an important tool for measuring genetic change in populations. That is, the degree to which deviations occur from a hypothetical reference point provides an approximate measure of the rate of evolutionary (genetic) change occurring in a population.

▷ **FIGURE 1 Test Cross Illustrating Use of the Hardy-Weinberg Equation**

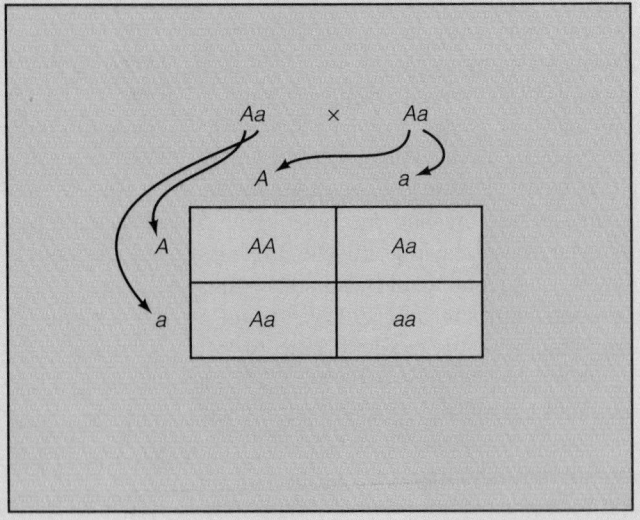

---

[1]To solve this equation, you simply take the square root of both sides.

derived an equation that can be used as a baseline against which changes in gene or allele frequencies can be measured. It is called the Hardy-Weinberg equation and is explained in the accompanying box.

≋ THE EVOLUTION OF NEW SPECIES

Evolution is a process by which species change over time, becoming better adapted to their environment. As noted earlier, evolution can also produce new species; in fact, scientists believe that all of the species alive today have evolved from others. A good example is the evolution of amphibians from the lobe-fin fish. Another is the evolution of birds from reptiles.

The evolution of a new species is called **speciation.** New species arise in large part as a result of a phenomenon called **geographic isolation**. Geographic isolation occurs when members of a population of organisms are physically separated by some barrier—a mountain range, a lake, an ocean, a river, or simply a great distance. In their separate

habitats, the members of the subpopulations are exposed to different environmental influences. Over time, two distinct species may form, each adapted to its environment. If a population is separated into more than two subgroups, many new species may arise.

One of the best known examples of this phenomenon occurred on the Galapagos Islands in the Pacific Ocean off the coast of Ecuador. On these islands, over a dozen species of finches (sometimes called Darwin's finches) evolved from a single population of seed-eating birds that arrived from South America sometime in the past 500,000 years (▷ Figure 23–24). As the ancestral population settled on different islands, the birds began to evolve in different ways, primarily because of differences in food sources. Major changes took place in body size and the shape of the bill. The differences in beak size and shape correspond to different ways of obtaining food.

Darwin's finches illustrate how new species develop from a common ancestor. This phenomenon is called **divergent evolution.** The term refers to the fact that species

▷ **FIGURE 23–24 Darwin's Finches** All but one species (the Cocos Island finch) are believed to have evolved from a single ancestral flock that arrived on the Galapagos Islands within the past 500,000 years.

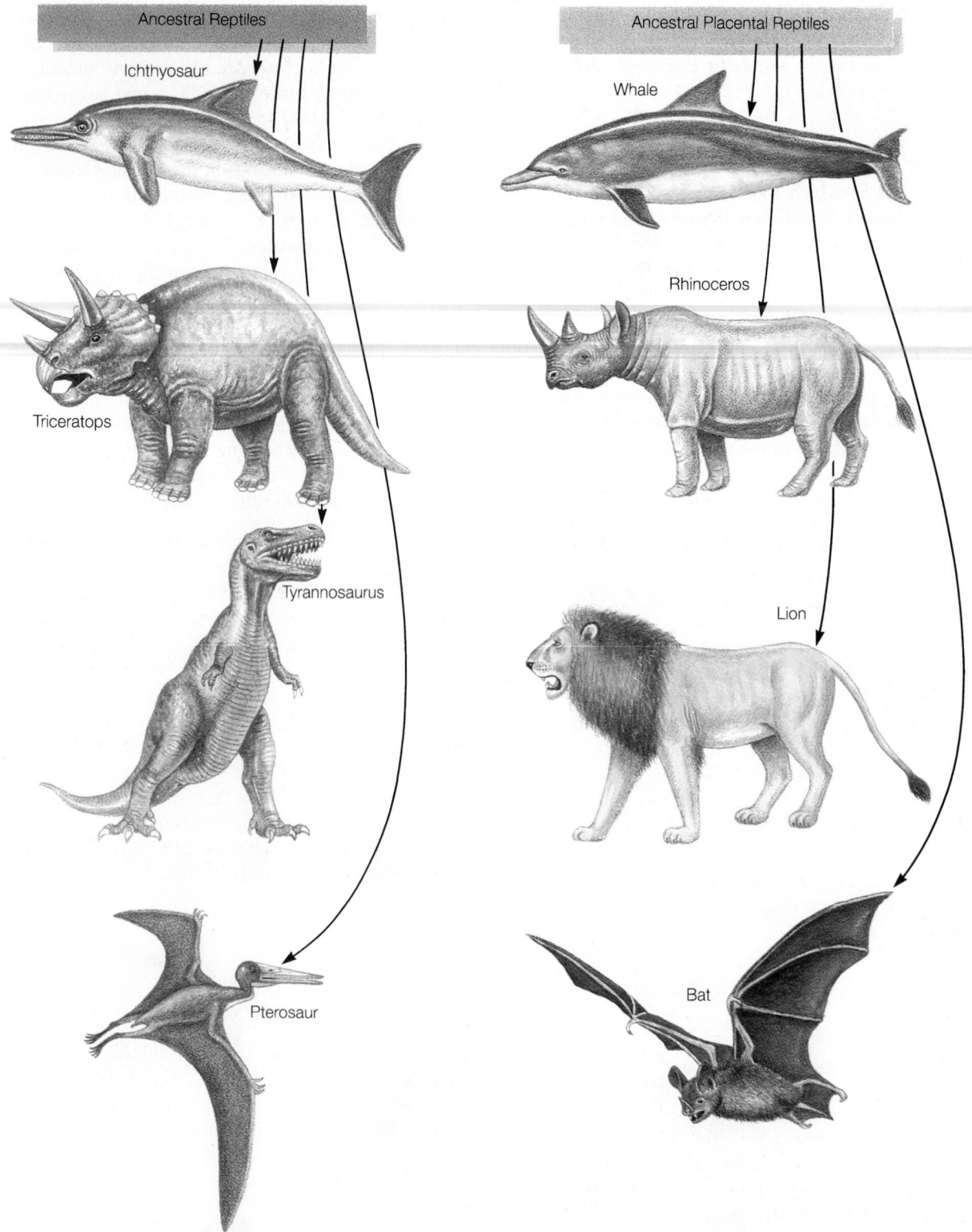

▷ **FIGURE 23–25 Divergent Evolution** From common ancestors, new species emerge, each adapted to its environment.

diverge genotypically and phenotypically in response to differing environmental conditions. ▷ Figure 23–25 shows divergent evolution among ancestral placental animals which gave rise to a wide assortment of species, some adapted to life in the water, and others adapted to life on land or in the air. The term **adaptive radiation** is used to describe this process of rapid evolution of numerous species from a common ancestry, each of which is specially adapted to its environment.

Certain genetic lineages may produce a number of different species over tens of millions of years. Others produce numerous species over a relatively short period.

During geographic isolation, organisms from the same population may lose the ability to interbreed. Biologists then say that the members of the original population have become **reproductively isolated** from one another. According to some, the first step in isolation in animal populations is the emergence of behavioral differences among the two groups. Courtship patterns, for example, may change so that females from one population no longer respond to males of the other. New species emerge when geographical isolation results in reproductive isolation.

Table 23-2 lists several reproductive isolating mechanisms. As you can see, those mechanisms fall into one of two broad categories: (1) those factors that prevent mating, such as different mating times, and (2) those that prevent the production of viable offspring and/or reproduction—for example, the inability of sperm to reach ova or sterility in offspring.

Figure 23-25 also shows that unrelated species may adapt to similar environments in similar ways. In this example, mammals and reptiles both evolved flying forms (in fact, 25% of all mammal species are bats). In addition, mammals and reptiles both gave rise to carnivores, as represented in the figure by the tyrannosaurus and the lion. Evolutionary biologists call this tendency for organisms to develop the same types of adaptations in response to similar environmental conditions **convergent evolution.**

As the preceding material illustrates, when geographic isolation leads to reproductive isolation, new species are formed. When the new species emerge in geographically separate regions, the process is called **allopatric speciation** (*allopatric* is derived from the Greek for "other" and "fatherland"). New species may also form without geographical isolation. This process, discussed shortly, is common in plants and is known as **sympatric speciation** ("same fatherland").

Species Often Evolve in Concert with Each Other

So far, the discussion of evolution has focused principally on abiotic factors as agents of natural selection. Abiotic factors, such as climate, are only part of the picture. Organisms themselves are agents of natural selection. Predatory pressure, for example, may act as an agent of natural selection in the evolution of prey populations. Owls hunt mice at night. Changes in the owls' ability to locate mice can have a profound effect on their prey population. If the eyesight of a population of owls improves, for instance, only the quickest mice may survive and reproduce, changing the mouse population over time.

Because biological organisms interact, two populations like the owl and mouse can evolve in concert. Owls can influence the evolution of mouse populations, and mouse populations, in turn, can influence the evolution of the owl.

Changes in the ability of the mice to hide, for example, might eliminate all but the most skilled owl hunters. Over

TABLE 23-2 Reasons for Reproductive Isolation

Factors that prevent mating
Mating occurs at different times.
Different cues (songs and coloration) develop.
Different courtship behaviors develop.
Genital incompatibility develops.
Factors that prevent production of viable offspring and/or reproduction
Sperm cannot reach ova.
Hybrid offspring die in utero or shortly after birth.
Hybrid offspring survive but are sterile.
Hybrid offspring survive but have lower fitness.

time, then, the owl population would improve. As the owl population became better able to hunt and catch swift mice, the mouse population would change.

When members of two interacting species affect each other's gene pool, the species evolve in concert. This process is called **coevolution.** In the example above, coevolution is like an arms race between predator and prey. Each improvement in a predator's ability to catch its prey is followed by an improvement in the prey's ability to avoid or resist attack.

Coevolution occurs in relationships other than predator-prey interaction. Defense mechanisms in plants, for instance, may result in coevolutionary changes in insects that eat them. A toxic or noxious-tasting chemical, for example, may evolve in some plant species, persisting because it helps ward off hungry insects. In time, some insects may evolve enzymes capable of detoxifying the harmful chemical.

Plants Can Form New Species Rapidly Via Polyploidy

For years, biologists believed that new species formed rather slowly from a parent stock split apart for one of a variety of reasons. That is not always the case. Studies show that the genetic split needed to produce two new species from a common gene pool can occur quite rapidly in plants.

New plant species emerge by **polyploidy,** the accumulation of one or more additional sets of chromosomes. That is to say, fertile offspring are produced with three or more of each type of chromosome.

Polyploidy occurs as a result of the complete nondisjunction of chromosomes during meiosis. This in turn produces diploid gametes. If a diploid gamete combines with a haploid gamete, the offspring has three sets of chromosomes. Other mechanisms are also possible.

Fertile offspring may be produced by this measure. Because they are unable to be fertilized by gametes from the parental stock, these new plants are new species. Polyploidy is one mechanism of sympatric speciation, defined earlier. Artificially induced polyploidy has provided human society with a number of economically important crops, including cotton, wheat, sugar cane, and coffee.

≋ A MODERN VERSION OF EVOLUTIONARY THEORY

Darwin's theory of evolution by natural selection can be summarized in three principles. First, natural variations exist in all species. Second, which members of the species survive to reproduce or reproduce at a greater rate is determined by inherited (genetically determined) variations (adaptations). Third, natural selection "determines" which organisms survive and reproduce.

For many years, biologists thought that evolutionary changes occurred gradually over many millions of years, a process called **gradualism.** If gradualism does indeed occur, evolutionary biologists now reason, the fossil record should contain many intermediate forms of plants and animals. In most cases, it doesn't.

As a rule, new species in the fossil record appear rather abruptly, persist for several million years, then vanish as abruptly as they arrived on the scene. Based on these and other data, Stephen Jay Gould of Harvard University and Niles Eldredge of the American Museum of Natural History have proposed an alternative hypothesis stating that evolution occurs in spurts. Their hypothesis is called **punctuated equilibrium.** According to them, evolution consists of long periods of relatively little change (equilibrium) interspersed (punctuated) with brief periods of relatively rapid change, although these periods are still many thousands of years long. During these periods of rapid change, some species become extinct, and new species arise. Proponents of punctuated equilibrium believe that species undergo most of their morphological change when they first diverge from their parent species. Thus, a species will appear rather suddenly, then will undergo little or no change for long periods.

≋ EVIDENCE SUPPORTING EVOLUTION

Evolution by natural selection is one of the central theories of modern biology. Although biologists may argue over some of the details, such as gradualism versus punctuated equilibrium, they agree about the basic tenets of the theory presented in the previous material. From the outside, this bickering may appear to be evidence that the theory of evolution is on shaky ground. Nothing could be further from the truth. What is disputed is the time scale and other details, not the fact that evolution has occurred. The next section examines some of the evidence in support of the theory of evolution. For a discussion of some of the consequences of evolution, see Spotlight on Evolution 23-1.

The Fossil Record Yields Some of the Best Supporting Evidence for the Existence of Evolution

The fossil remains of species no longer in existence have intrigued humans for 200 years and lend support to the theory of evolution. Because fossils and the rocks they come from can be dated using radioactive techniques, scientists can determine when different species lived. Studying when a species lived and tracking the types of life forms present at different times during the Earth's history permits scientists to piece together an evolutionary history of the planet. These studies have shown that the oldest rock contains relatively simple single-celled organisms and that successively younger rock holds fossils representing increasingly more complex life forms.

To date, scientists have discovered fossils belonging to about 250,000 species, most of them from rock formed within the last 600 million years. In this still-growing record, some lineages are nearly complete. The best records come from prehistoric environments like shallow seas where sedimentation rates (and thus preservation) were high. Many other lineages are incomplete and may never be fully understood. This is particularly true of animals that lived on dry land, as opposed to swamps. Land animals die and are eaten by scavengers. Their bones may have been devoured or scattered far and wide by hungry animals. Without sedimentation to bury them, the bones eventually crumbled into oblivion. Even in environments with rapid sedimentation, not all species left their mark. Soft-bodied creatures, without bones, outer skeletons, or shells, perished without a trace.

▷ Figure 23–19 shows a much simplified diagram of the evolution of mammals, birds, and reptiles. This schematic is called an **evolutionary tree** and is based principally on the study of fossils and the anatomy of modern species. It tracks the progression of life from simple to complex over a period of nearly 350 million years. Evolutionary trees also show how various species are related to one another.

Because evolution appears to involve sudden bursts of activity, in which one species gives off numerous others, evolutionary trees often look more like bushes, with lots of branches. A good example is represented in the evolutionary origin of the modern horse (▷ Figure 23–26). This figure traces the history of the horse over the last 60 million years. As illustrated, horses evolved from dog-sized, woodland browsers (*Eohippus*) through a series of increasingly larger, plains-dwelling grazers. Structural similarities in the fossil remains of the horse's ancestors suggest that this progression is indeed real.

You will note in Figure 23–26 that the evolution of horses was not a linear progression. A number of adaptive radiations (new species arising from a common ancestor) occurred along the way, as indicated by the circles in the evolutionary scheme. All of these species became extinct, except, of course, the modern horse, which is doing very well.

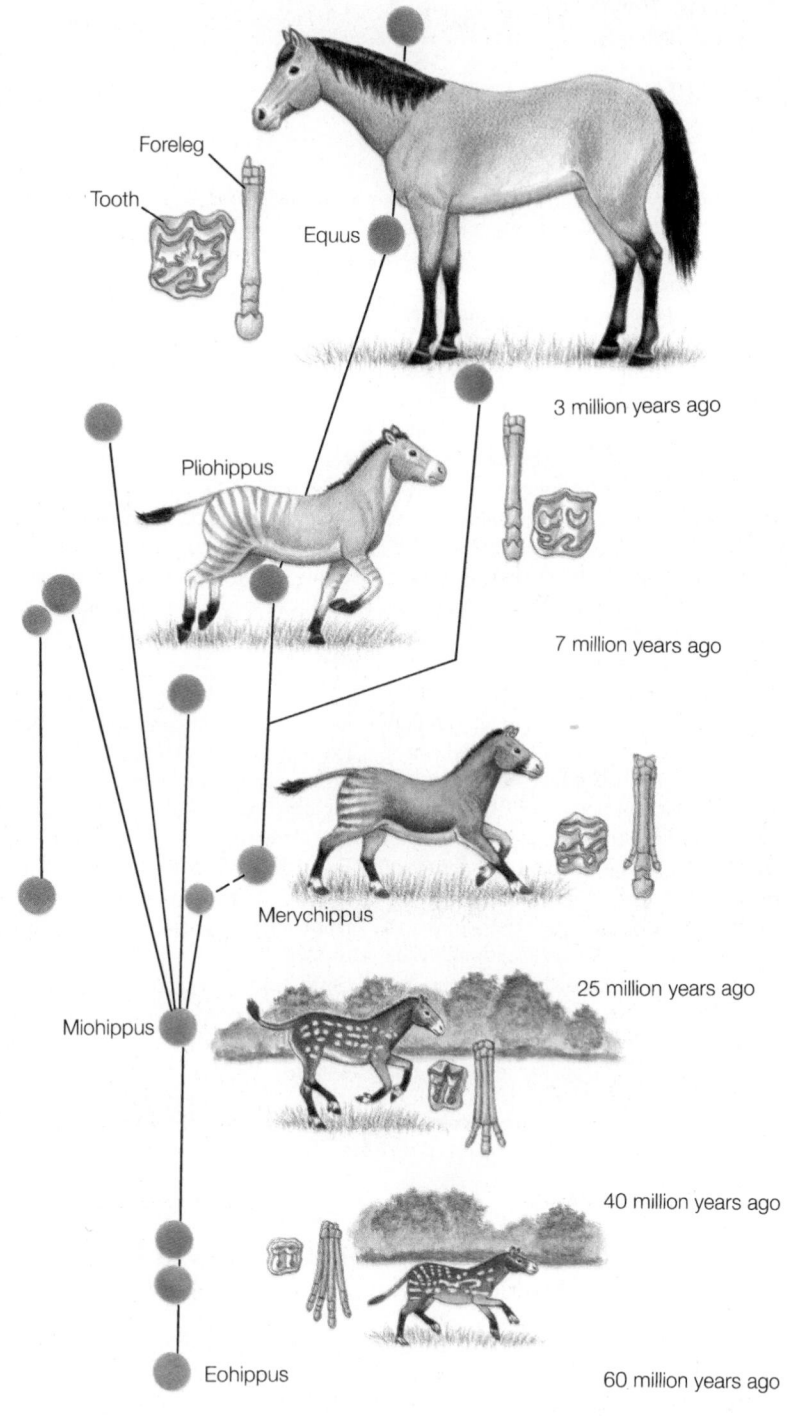

Recent

Pleistocene

Pliocene

Miocene

Oligocene

Eocene

Foreleg

Tooth

Equus

3 million years ago

Pliohippus

7 million years ago

Merychippus

25 million years ago

Miohippus

40 million years ago

Eohippus

60 million years ago

The fossil record also suggests how new species or new groups of organisms arose. As noted earlier, biologists believe that during evolution a primitive fish gave rise to the amphibians. Amphibians, they believe, gave rise to the reptiles. Birds and mammals evolved from reptiles. This progression is supported by two lines of evidence from the fossil record. The first is structural evidence; that is, species thought to have given rise to new groups bear a remarkable resemblance to their descendants. Thus, primitive fishes (lobe-fins) bear a remarkable resemblance to the fossil remains of the early amphibians (▷ Figure 23–27). Similarly, the early reptiles share many features in common with their amphibian ancestors. The second line of evidence is chronological. That is, the amphibians, reptiles, mammals, and birds appear in the fossil record in this order. No one has ever found a mammal or a bird that predated the amphibians. Thus, the structural and chronological evidence can be interpreted to support evolutionary theory.

FIRE SIRENS AND FIRE HYDRANTS: WHY WOLFDOGS MAKE LOUSY PETS

We've all heard the dogs in the neighborhood howl at fire engine sirens and reflected on the purpose and origins of this seemingly odd behavior. Rooted in the dog's ancient past, howling is an evolutionary remnant of a behavior derived from the wolf, from which modern dogs evolved through artificial selection—human breeding programs that seek to enhance certain traits. This process began with hunters and gatherers. Artificial selection has brought us a wide assortment of creatures, many of which bear very little resemblance to the wolf (▷ Figure 1).

Although the look of the wolf is gone from most breeds, the howl remains, an evolutionary legacy of the wolf's past. It is doubtful that humankind will ever breed the howl out of the dog. Nor are we likely to succeed in breeding out other behaviors that were vital to the wolf's survival in the wild. One of them is scent marking—the deposition of urine containing scent. Wolves do it to mark the boundaries of their territories, setting up invisible fences delimiting territories that packs of wolves lay claim to. In evolutionary terms, scent marking is a low-energy way of reducing competition

for food among wolf packs. Like their ancestors, domestic dogs, especially males, leave their scent on rocks, trees, car tires, and fire hydrants.

Another troubling evolutionary carryover is the wounded-prey response, especially evident in crosses between wolves and dogs, commonly known as wolfdogs. Biologists who study wolves believe that when confronted with a wounded, screaming prey, the wolves will automatically go for the neck. In the wild, this reflex ensures a quick kill. But this reflex may also be elicited in captivity by a crying child. When it is triggered, the reflex causes a seemingly friendly wolfdog pet to lunge for the throat of a frightened child, often killing it. Within a few seconds, the animal may wander off to find its master, whom it greets with a friendly face lick. Contrary to what many may think, this animal isn't a cold-blooded killer, it is a "victim" of its own instincts, a victim of evolution and of people.

"Wolfdogs make lousy pets," advises Lori Schmidt of the International Wolf Center in Minnesota. If you're thinking about buying one, she says, don't. Her advice is important, but often ignored. In the past three years, at least six chil-

dren have been killed by wolfdogs in the United States. One animal shelter in Fort Walton Beach, Florida, for example, advertised a wolfdog that it had received as "the pet of the week." But within a few hours of arriving at its new home, the animal had attacked and killed a neighbor's child.

Many other children have been severely mauled by hybrids. One of the most gruesome attacks occurred in New Jersey in 1990 when a wolfdog severed the right arm of a 16-month-old boy. "The hybrids are wild and will always be wild," says Lisa Barrington, a dog trainer who has found wolfdogs impossible to train.

Unfortunately, many people like the idea of owning a wolfdog. To many, this exotic pet promises to give them distinction among their peers. To some, owning a hybrid makes them macho. Others are attracted to the animal's wildness. For still others, owning a wolfdog seems like the perfect way to keep strangers from invading their homes. In truth, though, many wolfdogs are often quite timid and will run and hide when a stranger enters.

No matter what the reason, owning a wolfdog is not advisable. Even the

▷ **FIGURE 23–27 Fossil of Early Amphibian** This early amphibian resembles the lobe-fin fishes in many respects, suggesting an evolutionary relationship.

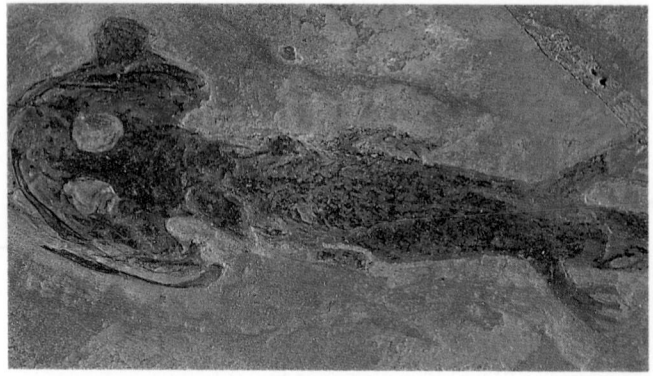

Common Anatomical Features in Very Different Species also Support the Theory of Evolution

Today's organisms are adapted to a wide range of conditions and display a dazzling diversity of appearance. Despite the diversity of life, many organisms exhibit similar anatomical features, suggesting a common evolutionary ancestry. Structures thought to have arisen from common ancestors are known as **homologous structures.** ▷ Figure 23–28, for example, illustrates the bones of several functionally different structures in birds and mammals—wings, flippers, arms, and legs. As shown, the bones in the wing of a bird and bat, the flipper of a whale, the arm of a human being, and the leg of a horse and cat are remarkably similar, even though these appendages perform quite different functions. Had each of these creatures evolved separately, it is unlikely that the bones would be so similar. Similarities

staunchest supporters of the hybrid agree that special precautions must be taken to own one—among them, double fences and locked gates. Some breeders even refuse to sell to families with children.

▷ **FIGURE Evolutionary Cousins?** (*a*) The wolf is the ancestor of (*b*) dogs which often show little resemblance to their predecessor.

Humans may have been able to domesticate the wolf, but it probably took thousands of years to breed out the wildness. Wolves are pack animals, and wolfdogs retain that hereditary preference. Unfortunately, many people buy a single wolfdog and chain it up in the backyard or leave their "pet" alone in the house while they're off at work. This kind of life is a jail sentence for a pack animal, and it is not unusual for wolfdog owners to return to find their home torn to shreds by the family pet, who in boredom has chewed the sofa and spread the stuffing all over the living room.

Caged wolfdogs are almost always a depressing sight. The animals often pace back and forth in their pens, a clear sign of boredom. Isolated from social contact, they live out their lives as you and I would in solitary confinement. "These animals are trapped between two worlds," says Randall Lockwood of the Humane Society of the United States. "They can't live in the wild, or as successful companion animals." Despite many people's fascination with the wolf and our love for the domestic dog, the wolfdog is a creature best avoided.

such as these correspond to what you would expect if birds and mammals evolved from a common ancestor.

The presence of homologous structures among life's diverse forms represents another piece of evidence in support of evolution. In contrast, **analogous structures** are those that perform similar functions but have quite different evolutionary origins. The wing of an insect and the wing of a bat are good examples. Table 23–3 compares homologous and analogous structures.

Creationists, who believe that life on Earth was created by God, argue that the presence of common anatomical features among diverse organisms and the presence of analogous structures lend support to their ideas. The common criticism from scientists is that creationism is not a science. It cannot be subjected to experimental method. It is, instead, based on faith. Nonetheless, creationists still argue that creation may be a valid explanation for the emergence and evolution of life on Earth.

Vestigial Structures Support Evolutionary Theory. The preservation of homologous structures during evolution also explains the presence of **vestigial structures,** which serve no apparent purpose in an organism. An example is the pelvic bones of whales. The pelvis generally attaches an organism's legs to the axial skeleton, but because whales have no legs, the pelvis is nonfunctional. Then why are pelvic bones found in whales? Evolutionary biologists believe that whales evolved from terrestrial mammals that had four legs and an obvious need for pelvic bones. During evolution, whales retained the pelvic bones.

Vestigial structures are "evolutionary baggage." They support the notion of evolution by modification. Had the

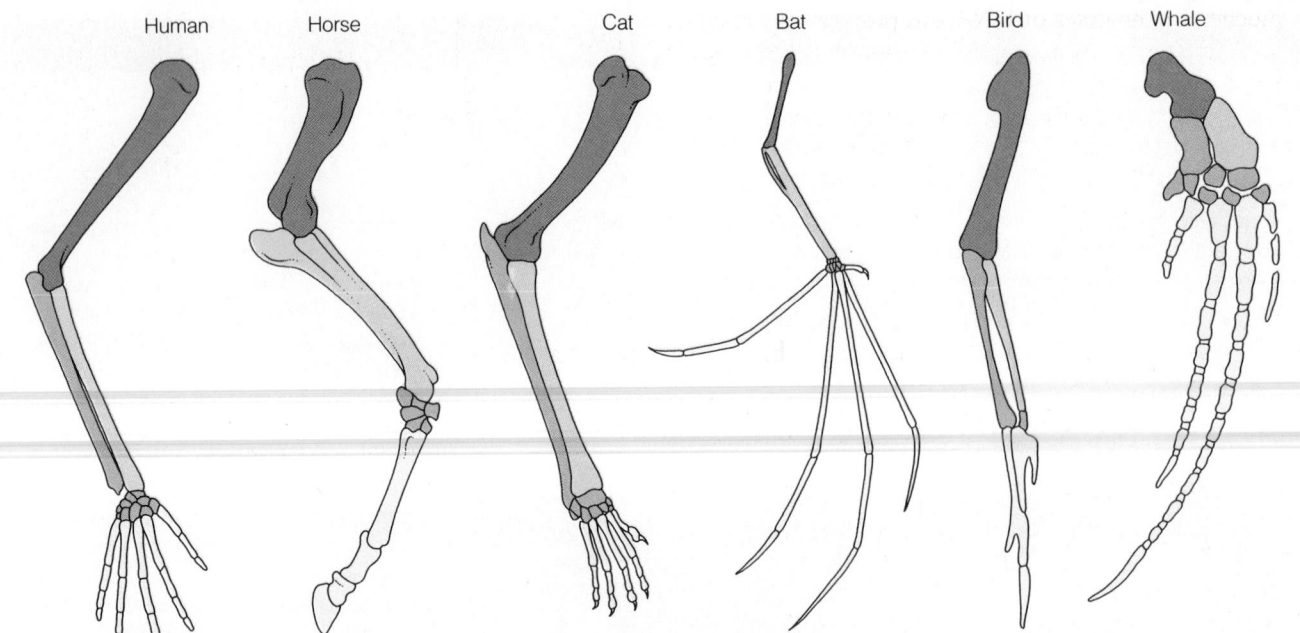

Human Horse Cat Bat Bird Whale

▷ **FIGURE 23-28 Homologous Structures among Vertebrates** The presence of homologous structures in vertebrates, and other groups, supports the theory of evolution.

whale evolved separately, it is doubtful that a pelvis would have been included as standard equipment.

Humans contain many vestigial structures. Hair on the body, muscles that move the ears, the appendix, and the pointed canine teeth are examples. The presence of these structures should come as no surprise, because we share so many genes with our ancestors.

Comparative Anatomy Also Permits Scientists to Classify Organisms. Studies of the anatomy of different species have long been used to determine "relatedness" among organisms, based on the belief that the more similar the internal structures of two species were, the more closely related they must be. Structural and functional similarities among species thus also allow scientists to classify organisms into groups, as illustrated in the next four chapters.

The Common Biochemical Makeup of Organisms Also Supports the Theory of Evolution

Organisms as distantly related as roses and rhinos are made of the same basic biochemicals—ATP, DNA, RNA, and protein. All living organisms store genetic information in DNA, and all use ATP to collect, store, and transport energy. In addition, many biochemical pathways in organisms, such as those involved in producing energy, bear a remarkable similarity. Are these coincidental, or are they the result of evolution? Probably the latter.

To many biologists, the common biochemistry of the Earth's diverse organisms lends further support to the notion of a common ancestry and thus to the theory of

TABLE 23–3 Comparison of Homologous and Analogous Structures

STRUCTURE	EVOLUTIONARY ORIGIN	FUNCTION
Homologous	Same	Often different
Analogous	Different	Same

evolution. (To creationists, it supports the notion of a designer using similar blueprints.) Interestingly, although many species have similar enzymes and other proteins, differences do exist. Hemoglobin, for example, is found in vertebrate red blood cells, but differences in the composition of hemoglobin are common among vertebrates. These differences are useful tools. By comparing the amino acid sequence of certain proteins (such as hemoglobin) from different species, evolutionary biologists can determine how closely related they are. The more differences, the more distant the evolutionary relationship. Similar studies of the sequence of nucleotides in the DNA of different species also help evolutionary biologists determine the degree of relatedness. Chimpanzees and humans, for instance, share about 90% of the same genes, suggesting that they are indeed very closely related.

Analysis of the amino acid sequence of proteins and the composition of the DNA have allowed evolutionary biologists to create their own evolutionary trees. Not surprisingly, the evolutionary trees they have developed correspond well to those created by comparative anatomists.

Biochemical analyses of DNA and protein have recently been expanded to extinct organisms (museum specimens) and even fossils. DNA extracted from the cells of extinct species can be cloned via genetic engineering. This process provides sufficient amounts of DNA to determine its sequence and make comparisons with other organisms, both living and dead, thus helping evolutionary biologists study the relationships among many more species.

Another Line of Evidence that Supports the Theory of Evolution Comes from Comparative Embryology

Studies show that the embryos of many different groups of organisms develop similarly. Thus, the embryos of chickens, turtles, mice, fishes, and humans—all vertebrates—bear a remarkable resemblance during their early stages of development. For example, each of them has a tail and gill slits, even though only adult fishes and amphibian larvae have gills.

One plausible explanation for the similarity of vertebrate embryos is that all of these species contain the genes that control the development of tails and gills and that these genes were passed on from a common ancestor. In humans, however, these genes are active only during embryological development. The structures they produce either become inconspicuous, as in the case of the tail (all that is visible is the tailbone), or become other structures, as in the case of the gill slits.

Experimental Evidence also Supports the Theory of Evolution by Natural Selection

A long list of experiments supports the theory of evolution by natural selection. One classic study is presented in Scientific Discoveries 23–1. It shows how genetic differences in wild populations of fruit flies can be attributed to differing environmental conditions (natural selection) and offers support for this hypothesis from an exhaustive series of field and laboratory studies.

Another elegant study supporting the theory of evolution by natural selection was performed by Stevan Arnold. Arnold examined two separate populations of garter snakes in California. One population lived in moist coastal habitat where it subsisted, in part, on banana slugs, shown in ▷ Figure 23–29. The other population lived inland and fed primarily on tadpoles and tiny fish.

When kept in captivity, the inland population rarely ate slugs. Even newborns who were offered banana slugs would rarely touch them. Newborn coastal garter snakes, true to form, almost always devoured this tasty offering.

The difference in food preference, Arnold contends, probably stems from feeding preferences present at birth. More important, these differences were probably genetically controlled. To test this hypothesis, Arnold interbred the two types of snakes and tested the offsprings' taste preference once again. He found that the offspring showed an intermediate response. The ''hybrid'' young tended to

▷ **FIGURE 23–29 Banana Slug**

eat slugs more frequently than members of the inland population but less often than members of the coastal population.

These results support the hypothesis that food preference is genetically based. Arnold suggested, in fact, that in inland populations the genetic preference for eating slugs had, over time, been selected against. The most important reason was that banana slugs do not live inland. This hypothesis would hold true only if the inland population had been derived from the coastal population. If it had been the other way around, an alternative explanation might be forwarded—notably, that the taste preference for banana slugs found in the coastal population over time might have been selected for. The relative abundance of this food type might have favored those garter snakes that fed on them.

ENVIRONMENT AND HEALTH: THE PESTICIDE TREADMILL

After World War II, chemical pesticides became the cornerstone of modern agriculture. Chemical pesticides kill insects and other organisms that destroy crops. To combat weeds and animal pests worldwide, farmers, homeowners, and others spray 2.5 million metric tons of pesticides on the land each year. These chemicals cause a number of problems. First, many broad-spectrum pesticides kill beneficial insects that control pests naturally. Spider mites were once only a minor crop pest in California, but the heavy use of chemical pesticides to control them has inadvertently killed the mites' natural predators, species that were more sensitive to spraying than the mites. As a result, spider mites today cause twice as much damage in California as all other insect pests combined.

A second problem of great concern is **genetic resistance.** Genetic resistance results from genetic variation—the pres-

NATURAL SELECTION IN THE WILD

Featuring the work of Dobzhansky

In the early to mid-1900s, critics of the theory of evolution argued that the lack of examples of natural selection in biology invalidated the idea. Proponents of the theory, however, noted that natural selection was such a slow process that one could not expect to find evidence of it within the span of a human lifetime. Fortunately, both groups were wrong. Research has shown otherwise and has helped silence some of the critics of evolutionary theory.

By studying organisms with a short generation time, such as fruit flies and bacteria, scientists have found a number of compelling examples of natural selection. In addition, they have been able to quantify the rate of change and the forces responsible for evolutionary change.

One noteworthy contribution to this endeavor was the work of Theodosius Dobzhansky, which was published in 1947. Dobzhansky studied fruit flies from six different states: Texas, New Mexico, Colorado, Arizona, Nevada, and California. He examined variations in chromosome 3 of the fruit fly. Chromosome 3 has at least 15 different gene arrangements, and the population of fruit flies in any one location can be described in terms of the relative frequencies of different gene arrangements (▷ Figure 1). The blue bars in the figure represent the standard gene arrangement (ST). As you can see, the frequency of the standard arrangement of genes drops as one moves east. Another gene arrangement, called the arrowhead arrangement (AR), is indicated by the red bars. Its frequency increases as one heads east to New Mexico, then drops. The third gene arrangement, called the Pikes Peak arrangement (PP), is indicated by yellow bars. It is prevalent in Texas.

The distance covered in this experiment was roughly 1200 miles. Dobzhansky, however, also found remarkable variation in the frequency of gene arrangements between populations that live much closer together but at different altitudes. He looked at fruit flies from deserts (800 feet), dry, pinon pine forests (4000 feet), and higher, ponderosa pine forests (4500 feet). The standard gene arrangement, for in

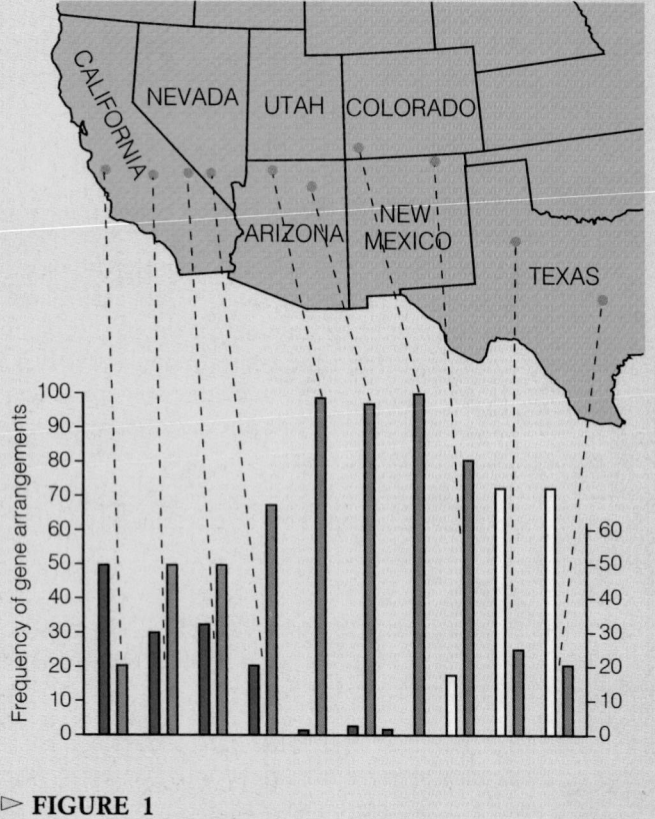

▷ **FIGURE 1**

ence of a mutation that makes a small percentage (about 5%) of any insect population resistant to chemical pesticides. Therefore, when farmers spray their fields with chemical pesticides, they kill only the nonresistant insects. Surviving the onslaught, the small subpopulation of genetically resistant insects breeds and produces a new population that is resistant to spraying. To kill it, farmers must apply additional chemical pesticide or switch to another. Once again, though, they find that a small segment of the population is genetically resistant to the higher dose or the new chemical preparation. That group survives the spraying—is selected for—and produces an even more resistant population, thus continuing an ever-escalating cycle often referred to as the "pesticide treadmill."

The pesticide treadmill is an excellent example of artificial selection at work. Some scientists warn farmers that they can never win in the battle against pests. Thanks to genetic variation in a population, no matter what pesticide

stance, was most prevalent at low altitudes and decreased as one ascended.

Dobzhansky concluded that the differences in the frequency of gene arrangements in fruit flies taken from different elevations and different regions of the United States were adaptive and were produced by natural selection. This hypothesis is supported by data on seasonal variation in the frequency of gene arrangements on chromosome 3. Studies of one population throughout the year showed that the frequency of certain genes varied markedly from January to December as climatic conditions changed. These cycles occurred quite regularly over a six-year period. ▷ Figure 2 illustrates the change in gene frequency of the arrowhead and standard arrangements.

The changes, asserts Dobzhansky, can be accounted for by natural selection. For example, the carriers of ST leave more surviving offspring during the summer months but are less successful in April, May and June. Another gene combination under study is known as the Chiricahua arrangement (CH). Flies with this arrangement seem to do best during the summer when the ST-carrying flies are doing their worst.

With these results in mind, Dobzhansky decided to test the hypothesis that temperature acted as an agent of natural selection by performing experiments in the laboratory under controlled conditions. In each of approximately 30 different experiments, Dobzhansky exposed flies with known gene arrangements to different temperatures and light conditions, then followed their reproductive success over long periods. For example, in one experiment using CH- and ST-carrying flies, he found that the percentage of the ST-carrying flies increased rapidly at 25°C, whereas the CH-carrying flies decreased. This view supported the hypothesis that warm weather favored reproduction and survival of ST-carrying flies and selected against the CH arrangement, as noted above.

Being a thorough researcher, however, Dobzhansky pushed onward in an effort to determine the stage in the life cycle of the fruit fly at which selection takes place. Through an elaborate set of experiments, beyond the scope of this book, he found that natural selection was operating sometime between the egg and the adult stage–most likely at the larval stage.

Dobzhansky's experiments are a monument to thorough, painstaking research. They are considered an outstanding contribution to the modern synthesis of genetics and evolution. Moreover, they are part of a decisive set of experiments that has supported Darwin's theory of evolution by natural selection.

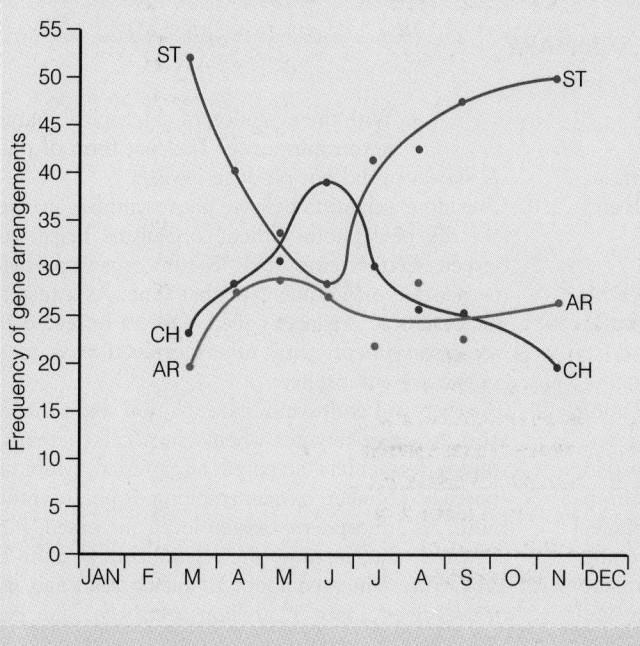

▷ **FIGURE 2**

they use, or how much they use, there will always be a resistant strain. Ever more powerful and ever more frequent applications at higher doses will be necessary just to stay even. As an example, a few decades ago, farmers in Central America applied pesticides about 8 times a year; today, 30 to 40 applications are the norm on any given field. Increased spraying pollutes the environment and kills fishes, birds, and other beneficial animals.

Unfortunately, farmers and chemical pesticide manufac-turers fail to take into account the existence of genetic variation and the selection process at work in the fields. As a result, over 450 species of insects are resistant to one or more chemical pesticides. Twenty of the worst pests are now resistant to all types of insecticide. Even weeds develop resistance to herbicides.

Knowledge of this phenomenon suggests alternative strategies. One alternative is **crop rotation.** This practice successfully reduces damage by maintaining pest populations at manageable levels. Practiced years ago on many farms but now largely abandoned, crop rotation requires farmers to alternate the crops they plant in a given field. One year corn might be grown in the field, the next year, beans, and the third year, alfalfa. This practice not only helps maintain soil nutrient levels but also effectively cuts down on pests. How does it work?

Many insect pests are specialists, preferring one crop over another. A field of corn, for example, provides an abundant supply of food for corn borers. At the end of the season, corn borers lay their eggs in the soil or in organic material on the surface. If the same crop is planted the next year, the newly hatched corn borers will have an abundant food supply. The insect population will increase quickly, causing considerable damage. If a different crop is planted the second year, however, the corn borer population will very likely decline. Continued rotation helps hold down pests year after year without costly and potentially harmful pesticide applications.

Farmers can also plant several crops in the same field, a practice called **heteroculture.** A field planted with corn and peanuts, for example, can reduce corn borers by as much as 80%. Heteroculture works because different insects attack different plants and because it also limits the amount of food available to any one kind of insect.

Chemical resistance is a rather common occurrence in the modern world. As noted above, weeds may also become resistant to herbicides; even microorganisms develop resistance to antibiotics, which is one reason many physicians prescribe antibiotics judiciously to patients. They fear, and rightly so, that the more antibiotics we use in society, the more likely it is that a resistant strain will emerge. Antibiotics are used in livestock feed to protect closely housed animals and to stimulate growth; Health Note 23–1 discusses the problems that can arise from this common practice.

In sum, heavy pesticide use only accelerates the pesticide treadmill. A little knowledge of genetics and natural selection can help farmers find alternative ways of controlling pests, methods that reduce overall pesticide use and help protect their own groundwater from contamination.

SUMMARY

THE EVOLUTION OF LIFE: AN OVERVIEW

1. Scientists believe that all atoms and molecules in our solar system came from the same source, an enormous cloud of cosmic dust and gas that gave rise to the Earth and the sun.
2. The Earth formed about 4.5 billion years ago.
3. The evolution of life probably began in the sea and can be divided into three phases: chemical evolution, cellular evolution, and the evolution of multicellular organisms.
4. Chemical evolution, scientists hypothesize, began about 4 billion years ago. At that time, the Earth's primitive atmosphere contained a mixture of water vapor and several gases. As rains drenched the Earth, the oceans formed. Dissolved in the waters of early seas were many inorganic molecules (from the atmosphere) that combined to form small organic molecules. Energy needed to drive the reaction may have come from sunlight or other sources. Over time, the small organic molecules combined to form polymers—small proteins and nucleic acids.
5. The theory of chemical evolution holds that organic polymers combined to form aggregates called coacervates and microspheres, the precursors of the cell. These aggregates exhibit structural and functional characteristics of living organisms.
6. Cellular evolution is the evolutionary development of cells from cell precursors.
7. The first true cells probably contained primitive enzymes, rudimentary genes, and selectively permeable membranes. These cells probably derived nourishment from organic molecules they absorbed from their environment and are thought to have been heterotrophic fermenters.
8. Autotrophs, organisms capable of synthesizing their own food, arose next. The earliest autotrophs were probably chemosynthetic bacteria that acquired energy from chemicals in the environment.
9. With the evolution of chlorophyll, photosynthetic autotrophs arose. The first form of photosynthesis, however, did not produce oxygen.
10. Over time, scientists believe, photosynthesis evolved further, and the new photosynthetic organisms began to produce oxygen. Oxygen was toxic to the organisms living in the oxygen-free environment of that time. As a result, many of the anaerobic organisms died. Others retreated to oxygen-free environments. Still others evolved ways to survive an oxygen-rich atmosphere.
11. The theory of endosymbiotic evolution states that eukaryotes arose from prokaryotes when a host cell (perhaps a heterotrophic fermenter) acquired an internal symbiotic partner (possibly a smaller, oxygen-respiring or photosynthetic bacterium). As atmospheric oxygen levels increased, these unions persisted, and, in time, the relationship became hereditary.
12. Prokaryotes emerged about 3.5 billion years ago, and eukaryotes evolved about 1.2 billion years ago. The evolution of eukaryotes opened the door for the evolution of multicellular organisms.
13. A variety of multicellular plants and animals evolved from single-celled eukaryotes in the oceans. As the ozone layer developed, life on land became possible.
14. The first species to invade the land were plants that thrive in moist environments. Soon after the land plants invaded, animals came ashore. The very first of these animals were probably scorpionlike creatures, which were preadapted to life on land.

HOW EVOLUTION WORKS

15. Evolution has produced a great diversity of organisms.
16. Evolution takes place because of genetic variation and natural selection.
17. Genetic variation in a species arises from mutations, gene

CONTROVERSY OVER ANTIBIOTICS IN MEAT

Livestock growers began adding antibiotics to cattle and pig feed over 30 years ago to protect animals confined to pens from disease as they were being fattened for market ($\triangleright$ Figure 1). Antibiotics were first given to help control disease, which can run rampant under crowded conditions. However, farmers soon found that the drugs had an unanticipated beneficial effect: for reasons still not fully understood, antibiotics accelerated the rate of body growth. That meant farmers could turn a higher profit.

Not surprisingly, today 70% of all cattle and 90% of all veal calves and pigs are reared on feed laced with penicillin or tetracycline. Nearly half of the antibiotics sold in the United States, in fact, are used for livestock feed.

The addition of antibiotics to feed has been sharply criticized by microbiologists and health officials who fear that widespread use could promote the evolution of superstrains of bacteria that are immune to antibiotics. In an editorial in the New England Journal of Medicine, a Tufts University microbiologist, Stuart Levy, noted that "every animal . . . taking an antibiotic . . . becomes a factory producing resistant strains" of bacteria. Resistant bacteria, in turn, could transfer their resistance to other bacteria, creating highly lethal strains that could infect humans. This is called the crossover effect.

Scientists are also concerned that some resistant bacteria in cattle, such as Salmonella, could be transmitted directly to people in meat or milk. The effects could be grave.

Despite these concerns, efforts to reduce the use of antibiotics in feed have been soundly defeated. In 1978, for instance, the Food and Drug Administration, which controls food additives such as antibiotics, proposed cutting back on penicillin and tetracycline use in feeds. Livestock producers, feed producers, and the multimillion-dollar drug industry, however, fought vigorously. They argued that microbiologists' concerns had not been proven.

New evidence in recent years, however, confirms the medical suspicions. Dr. Thomas O'Brien of the Harvard University School of Medicine published a study in the New England Journal of Medicine in 1982, showing that bacteria that commonly infect humans and other animals share genetic information quite freely through the exchange of plasmids, tiny segments of DNA separate from the bacteria's chromosomes. O'Brien argued that drug resistance could be transferred easily from bacterium to bacterium.

Another study, by researchers at the Centers for Disease Control, published in Science in 1984, confirmed the suspicion that antibiotic-resistant bacteria—artificially selected by adding antibiotics to cattle feed—could be transferred directly from meat to humans. This research showed that some of the outbreaks of illness caused by antibiotic-resistant Salmonella in the previous decade could be traced to meat from animals that had been fed antibiotic-treated grains. The research also showed that 20% to 30% of the Salmonella outbreaks involved antibiotic-resistant strains. About 4.2% of the people contracting the antibiotic-resistant bacteria died, compared with only 0.2% of the victims of the normal bacteria.

Proponents of antibiotics believe that the link between antibiotics and human disease is still weak and that further research is needed. Even if these findings are substantiated by further research, proponents believe, the benefits of using antibiotics outweigh the potential health effects. Banning antibiotics or cutting back on their use, they say, could have enormous economic impacts that must be weighed against sickness and loss of life. But health experts hope that the United States, like Europe, which strictly limited the use of antibiotics in animal feed in the early 1970s, will find the political will to end this activity.

$\triangleright$ **FIGURE 1 Feed Lot Cattle** Antibiotics reduce infections in cattle in tight quarters, and also accelerate growth. Microbiologists, however, worry that they may be stimulating the evolution of super strains of bacteria.

flow, recombination, independent assortment, and new combinations resulting from sexual reproduction. Genetic variation results in phenotypic variation—the presence of many different phenotypes. Some phenotypes confer a selective advantage on certain offspring, giving them a better chance of surviving and reproducing and passing the genes on to future generations.

18. Beneficial traits are preserved in a population by natural selection. Abiotic and biotic factors are the agents of natural selection.

19. It is important to remember that natural selection affects the survival and reproduction of individuals in a population but that individuals do not evolve, populations do.

THE EVOLUTION OF NEW SPECIES

20. Natural selection results in a shift in the gene pool of a population and produces organisms that are better adapted to their environment. Dramatic changes in the gene pool may lead to the evolution of entirely new species.
21. Geographical isolation is one of the most common mechanisms by which new species evolve. It occurs when two subpopulations of the same species become separated. Subject to different environmental conditions, the populations may evolve independently. Over time, they may become reproductively isolated—that is, unable to interbreed. When this occurs, two different species are said to have formed.
22. Geographical isolation results in divergent evolution—the emergence of two or more species from a common ancestor. Another term that describes this phenomenon is adaptive radiation—the evolution of numerous related species, each of which is specially adapted to its environment.
23. Species often evolve in concert with each other, a phenomenon called coevolution. Predatory pressure, for example, may act as an agent of natural selection in the evolution of prey populations. Changes in the prey population may also affect the evolution of the predator population.

A MODERN VERSION OF EVOLUTIONARY THEORY

24. Many evolutionary biologists believe that evolution occurs in spurts. According to the theory of punctuated equilibrium, long periods of relatively little change (equilibrium) are punctuated by briefer periods of relatively rapid change, many thousands of years long.

EVIDENCE SUPPORTING EVOLUTION

25. The scientific knowledge in support of evolution is rich and varied. The fossil record, anatomical similarities in groups of organisms, the occurrence of vestigial structures, the common biochemical makeup of organisms, similar embryological development among many groups of organisms, and experimental evidence all support the theory.

ENVIRONMENT AND HEALTH: THE PESTICIDE TREADMILL

26. Artificial selection can be witnessed today in modern agriculture. Chemical pesticides used to control insects select resistant organisms.
27. Farmers who spray their fields to kill insects leave behind a genetically resistant subpopulation that breeds and repopulates farm fields. A second application at a higher dose or an application of another pesticide leaves behind another subset that often becomes a further pest, forcing farmers to try higher doses or still another pesticide. This escalation in the war against pests is called the pesticide treadmill.
28. Chemical resistance is a rather common occurrence in the modern world. Weeds can become resistant to herbicides, and even microorganisms develop resistance to antibiotics.

EXERCISING YOUR CRITICAL THINKING SKILLS

For more than a century, biology teachers have been relating a fascinating story of mimicry among butterflies that involves two colorful varieties, the monarch and viceroy (▷Figure 23–30). The monarch larvae, the story goes, feed on milkweed and ingest a heart toxin (cardiac glycoside) produced by the plant, which remains in the bodies of the butterflies through adulthood.[8] Some birds that prey on the monarch die, but most simply become ill and from that point on avoid these colorful but poisonous insects.

According to common belief, birds also avoid similarly colored viceroy butterflies. Long thought to be edible, viceroys evolved from the tasty but drab-looking admiral butterfly. Over time, we thought, viceroys evolved a coloration similar to the monarch butterfly, and this coloration (mimicry) persisted because it protected the mimic from predation.

Two researchers at the University of Florida in Gainesville, however, have recently challenged this classic example. They conducted an avian taste test, feeding the abdomens of seven different butterfly species, including viceroys and monarchs, to red-winged blackbirds. In their experiments, they found that viceroys were just as unappetizing as monarch butterflies.

This finding throws into question the belief that viceroys are palatable Batesian mimics—organisms that evolved in such a way

as to acquire another's protective system. Henry Bates was a 19th century British naturalist who studied the phenomenon.

Interestingly, the viceroy butterfly feeds on nontoxic willows, leading the Florida researchers to hypothesize that the viceroy actually produces its own chemical defense, rather than absorbing poison from toxic plants such as milkweed. This hypothesis, if it turns out to be true, dispels a long-held belief that all butterflies depend on plant poisons for the protective chemicals they harbor.

This new research suggests that instead of being Batesian mimics, exploiting another's protective mechanism, the viceroy is actually a Mullerian mimic (after zoologist Fritz Muller who first described the phenomenon). This type of mimicry evolves among two equally distasteful butterfly species, which gain greater protection from predators by evolving nearly identical coloration. Mullerian mimicry is therefore mutually beneficial to both species. A bird that eats either a viceroy or a monarch would very probably leave both species of butterfly alone.

This example illustrates a failure of critical thinking—notably, the failure of scientists to question underlying assumptions in the formulation of theories. One key assumption here was that the viceroy was actually a palatable species. This conclusion was based on knowledge that the viceroy evolved from the rather tasty admiral butterfly. If it came from a palatable species, scien-

---

[8]Cardiac glycosides may have evolved to protect milkweed from herbivores.

▷ **FIGURE 23–30 Monarch and Viceroy Butterflies**
New studies suggest that long-standing beliefs about the evolution of the Viceroy butterfly may be wrong.

tists reasoned, it must be edible. A test of this assumption may have led to the downfall of a classic biological example.

A second assumption that may have led scientists astray has to do with the evolution of color and wing pattern. Because the viceroy's ancestors are dark-colored, scientists believed that the viceroy probably evolved to mimic the monarch. If the two species turn out to be Mullerian mimics, it may be that they evolved simultaneously, benefiting from similar wing and color patterns.

TEST OF TERMS

1. The process in which organic chemicals formed from chemicals in the atmosphere is called _____ evolution.

2. Organic molecules combined to form polymers, and they, in turn, gave rise to the first primitive cells, _____.

3. Stanley Miller demonstrated the plausibility of Oparin's hypothesis, showing that organic molecules could be produced _____ in conditions similar to those thought to have existed early in the Earth's history.

4. Microspheres and _____ are small, spherical structures exhibiting some characteristics of life and containing organic polymers.

5. The first true cells may have absorbed glucose and broken it down by anaerobic glycolysis and are, therefore, called _____ _____ .

6. The first autotrophic organisms were probably _____ bacteria that acquired electrons needed to synthesize their food from chemicals, such as hydrogen sulfide, in the environment.

7. The evolution of chemical mechanisms that allowed cells to synthesize _____ made possible the evolution of photosynthesis.

8. An organism that is better adapted to its environment than its cohorts is said to have a(n) _____ _____ over others.

9. The evolution of photosynthesis was a turning point in evolutionary history. One of its chief products, _____ , changed the course of evolutionary history and may have been a major stimulus for the evolution of _____ .

10. The evolution of eukaryotic cells containing cellular organelles, such as the mitochondrion and chloroplast, is explained by the theory of _____ evolution.

11. Life first evolved in the seas, but with the development of the _____ _____ , plants were able to invade the land.

12. Animals invaded the land sometime after plants. The first to invade were probably the _____ , organisms with jointed legs and segmented bodies.

13. A(n) _____ is a group of organisms that are structurally and functionally similar and producevi-able, reproductively competent offspring.

14. Mutations are one source of genetic _____ in a population— that is responsible for differences in structural, functional, and behavioral characteristics among the members of that population.

15. A genetically based characteristic that increases an organism's chances of passing on its genes is a(n)_____ .

16. _____ _____ is a process in which slight variations, if useful, are preserved.

17. _____ is a measure of reproductive success and of the genetic influence an individual has on future generations.

18. The evolution of a new species, _____ , occurs when _____ isolation leads to _____ isolation.

19. The process in which one species gives rise to several others that occupy different environments is called _____ _____ .

20. According to the theory of _____ _____ , long periods of relatively little change are broken up by briefer periods of relatively rapid change, still lasting many thousands of years.

1. In 1924, a Russian scientist, A. I. Oparin, suggested that life arose from nonliving matter. Explain his hypothesis, and discuss the research supporting it.
2. Describe how the first cells may have arisen during evolution. What critical requirements must have been met for life to begin?
3. If life formed abiotically in the shallow waters of the sea, would you expect the process to be occurring now? Why or why not?
4. Describe the theory of endosymbiotic evolution and the evidence supporting it.
5. The development of photosynthesis and the emergence of eukaryotes were pivotal events in evolution. Why?
6. Numerous plants and animals had evolved in the sea before life emerged on land. Why?
7. A critic of evolution says, "Life-forms are too diverse to have come from a common ancestor." How would you respond?
8. What is meant by the phrase "the conservative nature of evolution"?
9. Over the course of the Earth's history, mountain ranges rise and are gradually worn down. Explain what might happen to a species population that was geographically isolated by the emergence of a mountain range.
10. How do random mutations in germ cells contribute to the evolutionary process?
11. Define the following terms: adaptation, variation, natural selection, biotic factors, abiotic factors, selective advantage, and fitness.
12. Discuss the following statement: natural selection is nature's editor.
13. Discuss the evidence supporting the theory of evolution.
14. How do crop rotation and heteroculture keep pest populations in check?

The Worlds of Microorganisms and Fungi

Slime mold fruiting.

On a hike near her Massachusetts home, Marge Simpson was bitten by a tick. The next morning, she noticed a red spot developing on her leg. Over the next few days, the spot got larger. When she went to her doctor, he found she had contracted Lyme disease, a bacterial infection transmitted by ticks. Although readily treatable in its early stages with antibiotics, Lyme disease can lead to serious health consequences, including stiff joints and loss of mobility, if untreated. It is just one of many diseases caused by microorganisms, one of the main topics of this chapter.

This chapter examines 3 kingdoms: Monera (bacteria), Protista (protozoans and algae), and Fungi (mushrooms, molds, and yeasts). It also examines viruses, biological agents whose evolutionary agent is unknown. We begin our journey with viruses.

▷ **FIGURE 24–1 The Culprit?** The green monkey from Africa is believed to be the original source of the HIV, the virus responsible for AIDS.

🎓 VIRUSES: CELLULAR PIRATES

Viruses often confound students of biology. Although clearly agents of disease, viruses are not technically considered living organisms. Why? The answer is quite simple: viruses, unlike living organisms, cannot reproduce on their own and do not respond to stimuli. Viruses are not cells, because they lack cytoplasm and organelles. They have no metabolism either. So what are they?

Viruses Are Biological Agents That Function Inside Other Cells

Viruses are cellular parasites that invade a wide variety of cells, from simple bacteria to plant cells to almost every known animal. After invading a cell, viruses commandeer the cell's metabolic machinery, using it to produce additional viruses. The new viruses are then released from the host and may go on to infect other cells, creating a cascading effect that continues until the immune system puts a stop to the virus.

Some viruses can cross species boundaries. The rabies virus, for example, can be transmitted from dogs to humans. The AIDS virus is thought to have been passed to humans from the African green monkey, either through contact or through the consumption of viral-contaminated meat (▷ Figure 24–1). The AIDS virus is transmitted from person to person by sexual contact, blood transfusions, shared needles, and very rarely by accidental exchange of blood (for example, involving medical workers).

Even though viruses do cross species boundaries, such examples are the exception rather than the rule. Most viruses have a specific host range; that is, they infect very specific hosts. For example, some viruses infect only plants, and then only certain types of plants; other viruses attack only birds. Viruses also show specificity within an organism. The viruses that cause the common cold, for example, attack the cells lining the respiratory tract, and the AIDS virus attacks certain cells (T helper cells) of the immune system (Chapter 15).

Viruses Consist of a Nucleic Acid Core Surrounded by a Layer of Protein, the Capsid

Viruses are tiny packets of nucleic acid, either DNA or RNA, bounded by a protein coat (▷ Figure 24–2a). The protein coat is called the **capsid** and consists of 100 to 3000 globular proteins known as **capsomeres.**

The capsomeres protect the virus when it is outside the host. The proteins of the capsid also serve other functions. Some, for instance, bind to specific receptor proteins in the plasma membrane of the host cells they infect, a step essential for uptake. Viruses will bind only to cells with proper receptors. Other capsomeres are enzymes that digest holes in the plasma membranes or cell walls of host cells, permitting viruses to enter.

Some viruses are endowed with an additional protective coat, the envelope, which lies outside the capsid (Figure 24–2b). The **envelope** consists of a layer of lipid and protein forming a membrane that resembles the plasma membrane of cells (Chapter 3).

Viruses Come in Many Shapes and Sizes. Viruses fit into four distinct groups: globular, polyhedral, cylindrical, and irregular. The simplest of all viruses is the *globular virus* (▷ Figure 24–3a). *Polyhedral* (many-sided) *viruses* resemble multifaceted diamonds and are the most common form (Figure 24–3b). *Cylindrical viruses* consist of a nucleic acid core surrounded by capsomeres arranged to form long, hollow cylinders (Figure 24–3c). The last group are the *irregular* (odd-shaped) *viruses.* One of the most intriguing of these viruses is the bacteriophage T4 virus shown in Figure 24–3d. The bacteriophage T4 virus invades bacterial cells—hence the name "bacteria eater." Resembling a lunar landing module from a science fiction story, the T4 virus consists of a head containing the nucleic acid core, a tail, and tail filaments.

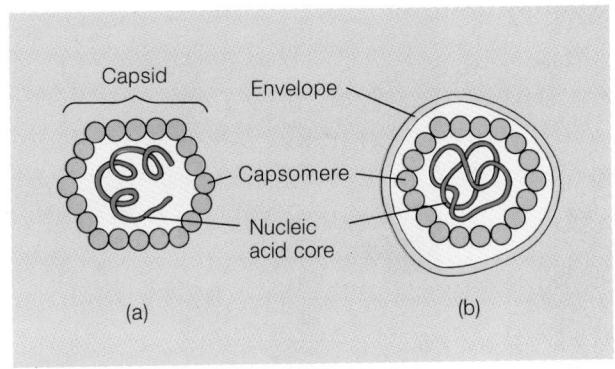

> **FIGURE 24–2 General Structure of Viruses** (*a*) Naked virus with no envelope. (*b*) Enveloped virus.

Viruses Enter Cells Via Two Routes. Some viruses bind to the cell walls or plasma membranes of host cells, digest holes in them, then "inject" their nucleic acid. The bacteriophage T4 virus, for example, attaches to the surface of a bacterial cell with its tail down. Enzymes in the tail section digest a hole in the plasma membrane of the host cell, and the virus then injects its nucleic acid into the bacterium. Inside the cell, the nucleic acid replicates, forming exact copies required to create more viruses.

Other viruses enter cells via phagocytosis. In order to be phagocytized, these viruses must first bind to protein receptors on the plasma membrane of the host cell (▷ Figure 24–4, page 652). Inside the cell, these viruses are encased in a vacuole consisting of the host cell's plasma membrane. In order to become active, the viral DNA must escape from the vacuole and enter the cytoplasm. Once inside the cytoplasm, the nucleic acid duplicates and directs the synthesis of new capsomeres.

The energy and chemical building blocks needed to synthesize additional viral nucleic acid and capsomeres come from the cell's reserves. Therefore, when a cell is commandeered by viruses, it stops producing the materials it needs to survive and makes new viruses. The new viruses are assembled from the capsomeres and viral nucleic acid synthesized in the cytoplasm of the host cell.

New viruses may be released a few at a time via a process virologists refer to as **budding.** As Figure 24–4b illustrates, budding adds the envelope found in some viruses. Viruses may also be released en masse when the host cell dies and breaks down.

A cell infected by a virus can produce as many as 80,000 to 120,000 new viruses. Viruses released by a host cell enter the tissue fluid surrounding the cell, then are transported in the blood or lymph throughout the body to new sites.

Unfortunately, the spread of viruses within the body cannot be stopped by antibiotics, because antibiotics are specific for bacteria (Chapter 9). Viruses, however, can be destroyed by the immune system, as described in Chapter 15. Even though the immune system normally brings viral infections under control, some viruses remain in the body in hiding. The herpes simplex virus is a case in point. The herpes simplex virus consists of several strains (▷ Figure 24–5, page 653). One strain, herpes simplex type II, produces genital lesions and afflicts an estimated 20 million American men and women. Soon after an individual is infected by the virus, blisters appear on the genitals, thighs, and buttocks (Figure 24–5b). These blisters soon rupture, leaving painful ulcers in the skin. Although the ulcers heal within one to three weeks, the virus is not completely destroyed. During the infection, the herpes simplex II virus enters a small clump of nerve cells alongside the sacrum, the wedge-shaped bone just above the tailbone. Here the virus remains, periodically flaring up under certain conditions. Stress, menstruation, sex, and even exposure to sunlight can cause the virus to reemerge. (For more on the herpes simplex virus, see Chapter 21.)

Viruses Are Often Transmitted Externally from Host to Host

Viruses travel by many routes. Some may be carried in water; others may be carried on fine dust particles in the air or in tiny moisture droplets released when an infected person sneezes. Viruses may also be transmitted via bodily contact—either by a friendly handshake or by sexual contact. Viruses may also be transmitted on inanimate objects.

The "common cold" can be caused by about 200 different viruses, spread chiefly by hand contact. Physicians therefore recommend washing your hands frequently during the cold season, especially if you have been around people suffering from colds. They also suggest keeping your hands away from your eyes, nose, and mouth after you've been around someone with a cold. Avoiding airports and other crowded buildings can also help reduce the likelihood of contracting a cold during the "cold season."

Modern transportation, urban living, and our high-mobility lifestyles team up to spread viruses from one person to another with remarkable speed. So fast is the spread of viruses, in fact, that a new strain of the influenza virus spreads to half the world's population within two years of its emergence.

DNA Viruses Produce New Viral Components Inside a Host Cell. Inside a host cell, viral DNA produces complementary strands of DNA using the host cell's enzymes (DNA polymerase) and nucleotides (▷ Figure 24–6a, page 653). The complementary DNA strands produced in the cell synthesize viral messenger RNA, using the host cell's RNA polymerase and RNA nucleotides. Viral messenger RNA, in turn, serves as a template for the production of protein capsomeres. Transfer RNAs and amino acids needed to make viral protein are supplied by the host cell.

Viruses have a debilitating effect on cells because they take over, or commandeer, the metabolic machinery. Viral DNA replication, RNA synthesis, and protein synthesis

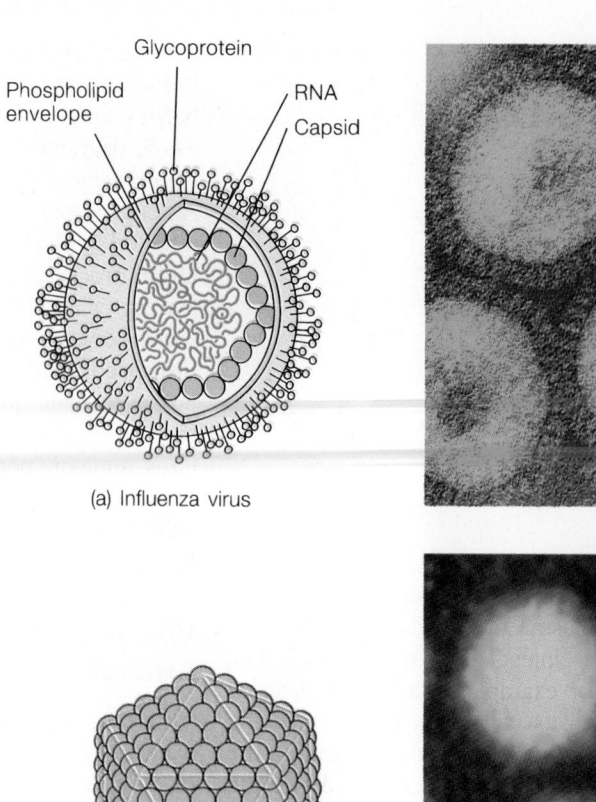

▷ **FIGURE 24–3 Viral Architecture** Viruses have varied structures. Each virus consists of a nucleic acid core surrounded by a capsid made of proteins. Some, like the influenza virus, have an additional coating, an envelope made chiefly of phospholipid molecules. This illustration shows the four main viral shapes: (*a*) globular, (*b*) polyhedral, (*c*) cylindrical, and (*d*) odd-shaped.

(a) Influenza virus

(b) Adenovirus

occur at the expense of a cell's normal metabolic functions. The cell has to forgo activities required to survive. Deprived of vital nutrients, viral-infected cells often die.

RNA Viruses also Replicate Inside Host Cells but Require Special Enzymes to Do So. In eukaryotic cells, RNA is not a self-replicating molecule. All new RNA is synthesized on a template of DNA. The RNA from most viruses, however, is capable of self-replication.

RNA-containing viruses replicate by one of two mechanisms, depending on the type of virus. In most RNA viruses, the single strand of RNA actually serves as a template for the production of a complementary strand of RNA, in much the same way that DNA viruses produce complementary DNA strands in DNA viruses and eukaryotic cells (Figure 24–6b). The complementary strands of RNA then serve as templates for the production of additional viral RNA (original strands). RNA replication from an RNA template requires a special enzyme called **RNA replicase**.

The second group of RNA viruses, known as the **retroviruses**, uses its RNA as a template for the production of complementary strands of DNA (Figure 24–6c). To perform this unusual feat, these viruses bring with them a special enzyme called **reverse transcriptase** (the reverse of transcription). This enzyme catalyzes the synthesis of DNA on the virus's own RNA. DNA produced in the host cell is then used to synthesize complementary strands of RNA—the original viral RNA. The viral RNA, in turn, is used to make new viruses and also serves as a template for the production of viral capsomeres and enzymes.

Viral Infections Can Cause a Variety of Health Problems

Viruses cause many human diseases. As noted above, approximately 200 different viruses are responsible for the common cold, a condition characterized by a runny nose, sneezing, sore throat, and coughing. Many other viruses produce influenza (the flu). Influenza is characterized by chills, fever, sneezing, headache, muscular aches, and sore throat.

If you're suffering from a viral cold or flu, your physician will probably simply let the virus "run its course," counting on your immune system to battle the infection. You may receive an antihistamine to reduce mucus buildup and a cough suppressant. Aspirin may be prescribed to relieve your achy joints. But that's about all the relief you will

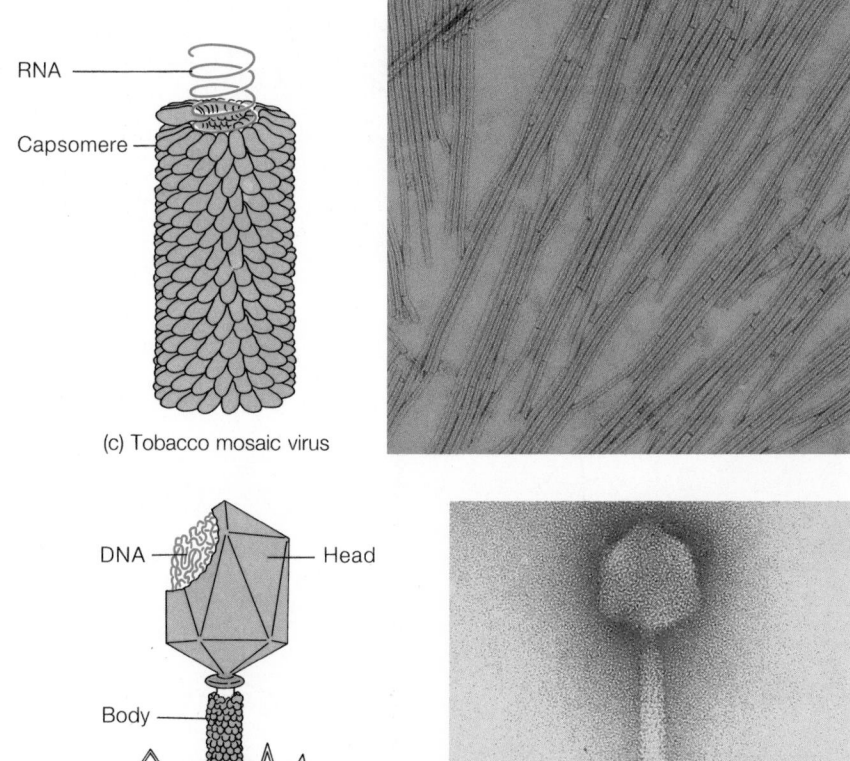

(c) Tobacco mosaic virus

RNA

Capsomere

DNA — Head

Body

Tail filament

(d) Bacteriophage T4

get. (Some recent research suggests that aspirin may reduce viral replication, helping combat the viral infections directly.)

Another common viral disease is viral pneumonia, an infection of the lungs. Viral pneumonia, although treatable, can be quite debilitating and often requires hospitalization. Characterized by rapid deterioration of the patient, viral pneumonia can cause death. Pneumonia may also be caused by bacteria, with equally devastating consequences. Jim Hensen, creator of the Muppets, died in 1990 from a severe case of bacterial pneumonia.

Still other viruses cause cancer. As noted in Chapter 7, these viruses may insert cancer-causing genes in the host cell genome, causing uncontrollable cellular replication. For a discussion of ways that a virus may someday be used to treat diabetes, see Health Note 24-1.

Viral Diseases Can Be Prevented with Vaccines, but Treatment of the Disease Once it has Developed Has Proved Difficult

Although vaccines can be used to prevent viral infections, as Chapter 15 explains, once a viral infection has begun, traditional wisdom holds that a cure is possible only if the immune system mounts a successful attack. Frustrated by this problem, researchers have been looking for ways to cure viral diseases once they have started. (The search for a cure for AIDS is a good example.)

Consider the sometimes deadly hepatitis B virus. In the United States, approximately 1 million people carry the virus without showing symptoms, and more than 300,000 actual cases of acute hepatitis B occur every year. This virus causes cirrhosis of the liver and liver cancer.

In 1982, a vaccine for hepatitis B became available to stop the virus. Nonetheless, the incidence of viral hepatitis has continued to rise. Why? Unfortunately, few people receive the vaccine, and even if they do, the vaccine is not fully effective. Researchers, however, have devised a new treatment that may actually stop viral hepatitis once it has developed. This treatment combines two earlier approaches and involves two steps.

In the first phase, physicians administer a steroid that suppresses the immune system. After a brief treatment, patients are taken off the drug. This causes their immune systems to rebound, resulting in enhanced immune activity. In the second phase of the treatment, physicians administer a chemical called alpha interferon, a small protein naturally produced by white blood cells. As you may recall from

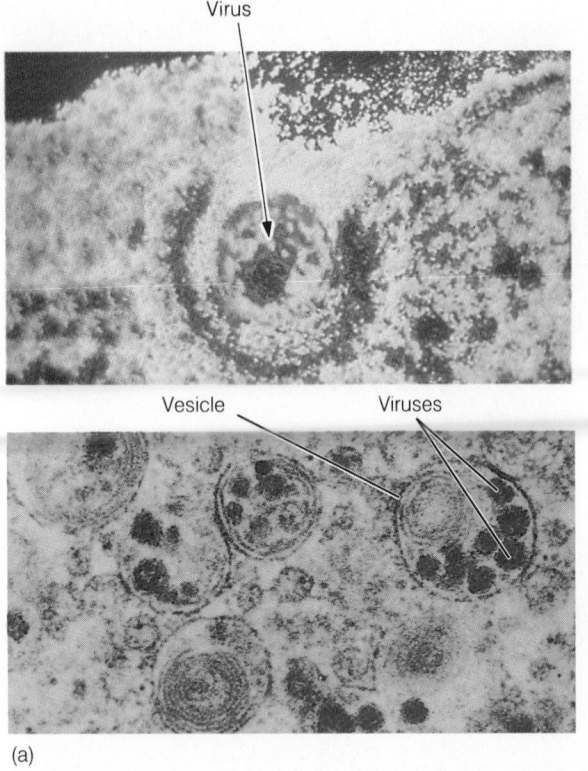

Virus

Vesicle   Viruses

(a)

Viruses budding from plasma membrane

Virus-with envelope

Viruses

Viruses being phagocytized by cell

New viruses assemble in cell

Nucleic acid and viral protein formed

Nucleus

Viral DNA replicates inside cell

Viral DNA "escapes" from phagasome

(b)

▷ **FIGURE 24–4 Viral Infection** (*a*) Viruses enter a cell either by phagocytosis or by fusing with the plasma membrane and injecting their nucleic acid into the cell. Viruses that are phagocytized fuse with the membrane of the phagosome and inject their nucleic acid into the cytoplasm. (*b*) Once inside the cell, the viral nucleic acid replicates and codes for the proteins needed for the capsid. Viruses reassemble inside the cell and are released either via exocytosis or when the cell dies and its plasma membrane disintegrates.

Chapter 15, interferons halt the spread of viruses from cell to cell.

Researchers from Washington University in St. Louis gave patients an immune-suppressing steroid for six weeks followed by regular injections of alpha interferon. The results were remarkable: half of the patients showed no replicating forms of the hepatitis virus. Treatment with interferon alone was successful in only about 30% of patients.[1] Further work such as this may help open the doors for cures for previously untreatable viral diseases.

At the beginning of the chapter, you learned that viruses are not considered to be living organisms because they fail to meet some of the criteria that distinguish what is living from what is dead. In fact, viruses are said to lie on the threshold of life, straddling the line between living and nonliving. Important as they are, no one knows for sure how viruses arose. One school of thought suggests that they arose independently of the monerans, as a separate line in evolution that never went very far. The other school of thought suggests that viruses may have evolved from the monerans. As such, the viruses would represent a degenerate offshoot of the bacterial line. This we do know: viruses evolve by the same basic mechanisms that operate in the living world.

≈ THE KINGDOM MONERA: THE BACTERIAL WORLD

Bacteria are single-celled organisms that are larger than many viruses but smaller than eukaryotic cells. Equipped with all of the metabolic machinery necessary for survival and reproduction, bacteria are truly living organisms and belong to the kingdom Monera (Chapter 23).

Bacteria Are Relatively Simple Cells Containing only Ribosomes, Circular DNA, and Cytoplasm and Are Surrounded by a Protective Cell Wall

A typical bacterium is delimited by a plasma membrane, which, in turn, is surrounded by a thick, protective cell wall (chemically different from the plant cell wall) (▷ Figure 24–7). The cell wall protects bacteria against osmotic rupture caused by swelling and yields several distinct shapes. The three most common shapes—rod-shaped, spherical, and spiral—are shown in ▷ Figure 24–8.

Bacteria contain a single circular molecule of DNA. Unlike the DNA found in eukaryotic cells, bacterial DNA is not surrounded by a nuclear envelope. Bacteria also

---

[1] This study showed encouraging results only for patients who had been infected for a short time. These individuals responded better than those who had been infected for a longer period.

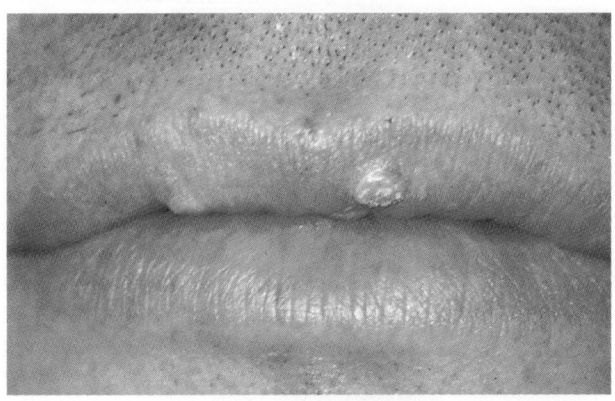

(a)

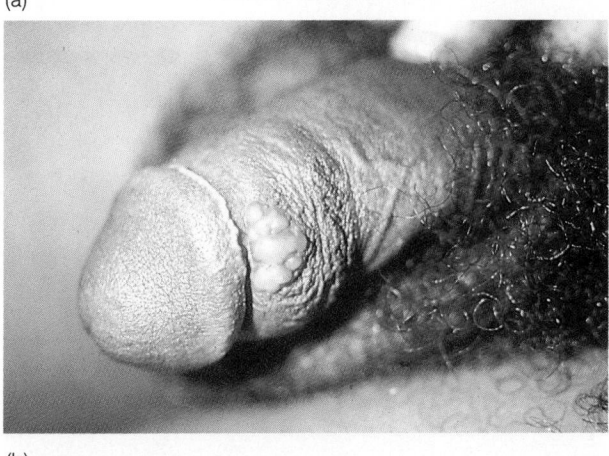

(b)

▷ **FIGURE 24–5 Herpes Simplex Virus** This virus comes in several forms. Two of the most common cause (*a*) cold sores and (*b*) genital lesions.

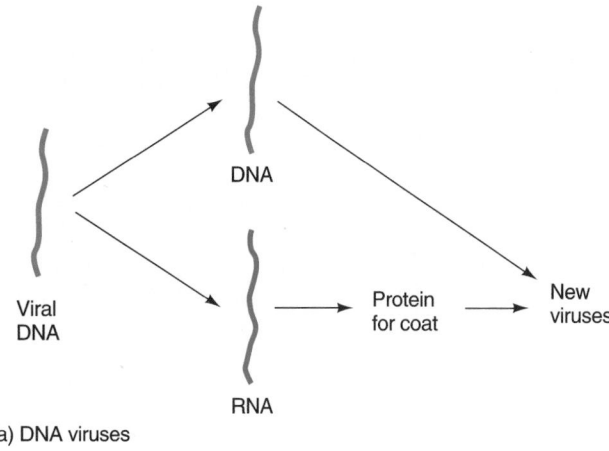

(a) DNA viruses

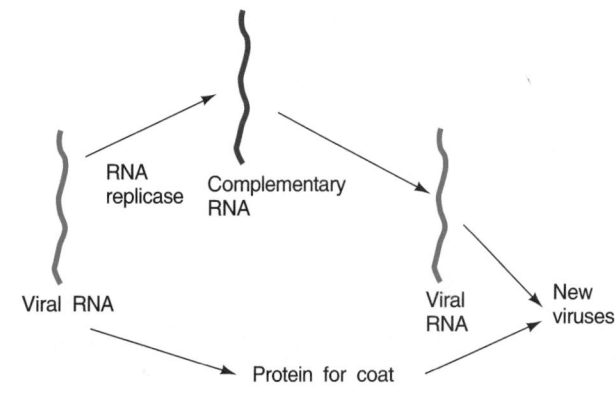

(b) Most RNA viruses

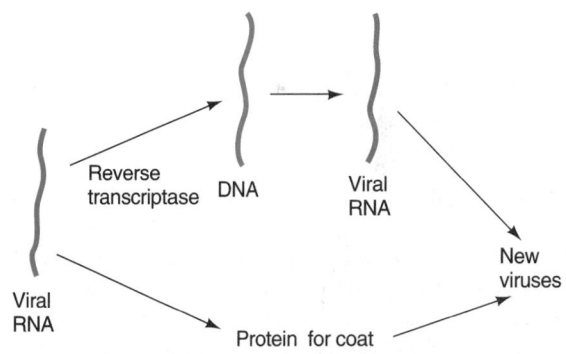

(b) Retroviruses

▷ **FIGURE 24–6 Details of Viral Replication** (*a*) DNA viruses produce additional DNA, needed to make more viruses. The DNA is also transcribed, making RNA needed to make protein for the capsid. (*b*) Inside host cells, most RNA viruses reproduce by forming complementary strands of RNA from the RNA template. This requires a special enzyme called RNA replicase. (*c*) Retroviruses carry with them an enzyme, reverse transcriptase, used to produce complementary DNA from the RNA. The DNA, in turn, makes viral RNA, allowing the virus to replicate itself.

contain smaller circular "strands" called **plasmids.** These contain a small number of genes not essential to bacterial function. Bacteria lack cellular organelles, except ribosomes (Figure 24-7). Some degree of intracellular organization like that seen in eukaryotes may be provided by a convoluted, whorled structure known as the **mesosome** (▷ Figure 24–9). It is thought that transport proteins involved in cellular respiration may be attached to this structure, providing some degree of internal organization. In photosynthetic bacteria, the mesosome may act as a simple type of chloroplast. Some bacteria have flagella, similar only in function to those found in eukaryotic cells (Chapter 3).[2] Flagella allow bacteria to travel through water in search of food or favorable living conditions.

Bacteria Occupy Many Different Habitats

Bacteria are the most abundant organisms on Earth. Billions of bacteria, belonging to many different species, inhabit a single handful of soil. Bacteria also inhabit a wide

[2] Morphologically, bacterial flagella are very different from eukaryotic flagella.

range of habitats—from the tops of mountains to the depths of the oceans. Some bacteria live in the scalding water of hot springs; others can survive in the hot thermal vents at the ocean bottom, where water pours out at even higher temperatures (350°C, or 660°F).

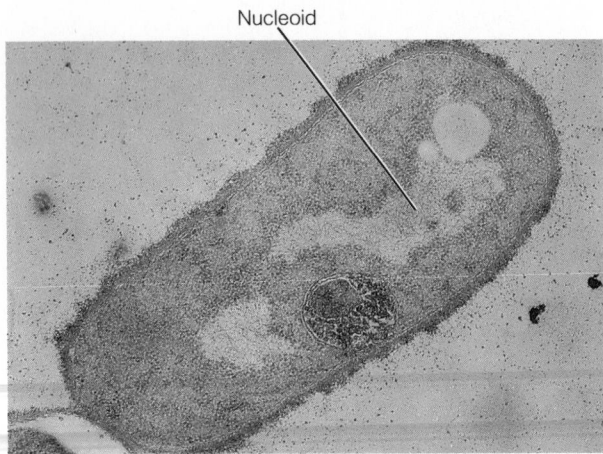
Nucleoid

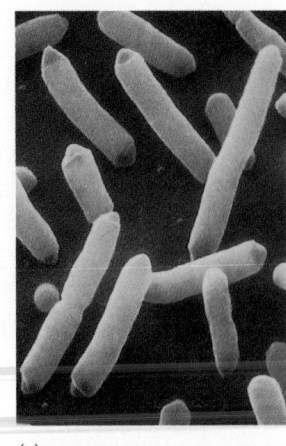
(a)

▷ **FIGURE 24–7 General Structure of a Bacterium**
Bacteria consist of a circular strand of naked DNA in the nucleoid.
The cell membrane is encased by a thick cell wall lying just outside the plasma membrane. The only organelle present is the ribosome, needed for protein synthesis.

Scientists have recently found that several species of
bacteria live in deep aquifers (water-containing zones) over
330 meters (1000 feet) below the Earth's surface. These
bacteria produce carbon dioxide, which dissolves in the
water of the aquifer and forms carbonic acid. This acid
helps eat away the limestone rock of aquifers, enlarging the
pores and enabling water to flow more freely. It also increases the amount of water an aquifer can hold. Fortunately, the bacteria found in these aquifers do not adversely
affect the quality of drinking water, nor are they harmful to
health. Bacteria also inhabit the human body. The bacterium *E. coli,* first described in Chapter 11, for example,
lives within the large intestine feeding off undigested
foods. Another bacterium lives in the mouth, forming troublesome plaque, or tartar.

Bacteria Generally Reproduce
Asexually by Binary Fission

Bacterial replication is a rather simple affair. When a bacterium grows to a certain size, its cytoplasm divides, as does its
replicated DNA (▷ Figure 24–10). Called **binary fission,**
this process generally produces two identical offspring. Unlike eukaryotes, bacteria do not have a distinct cell cycle.
They divide more or less continuously as long as an adequate
supply of nutrients is available. Under ideal conditions, a
bacterium can divide every 20 minutes. New strains arise
from random mutations caused by chemicals in the environment and by ultraviolet light and other radiation.

Bacteria also exchange genetic material in a kind of
quasi-sexual reproduction called *conjugation* (▷ Figure
24–11).[3] During conjugation, thin cytoplasmic strands

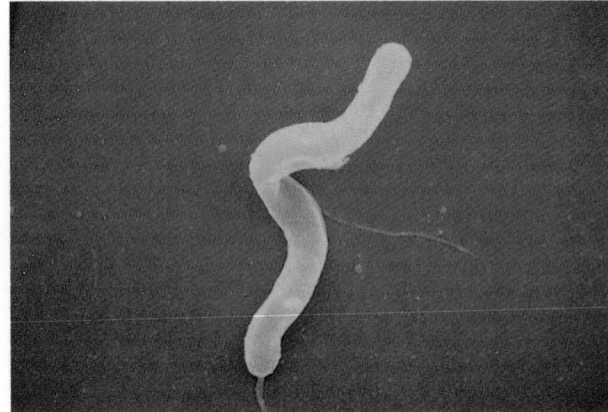
(b)

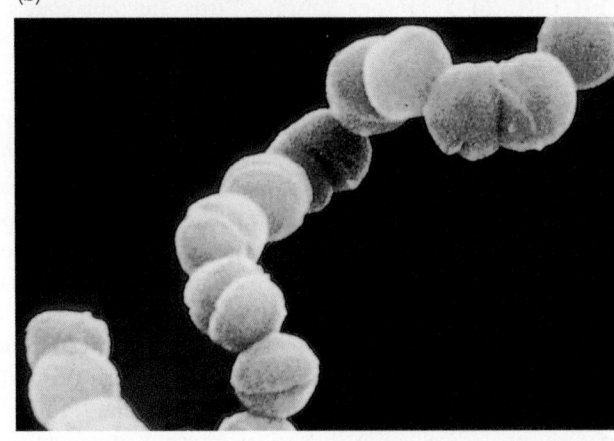
(c)

▷ **FIGURE 24–8 Bacterial Architecture** Bacteria come in
many shapes and sizes. The three predominant types are shown
here. (*a*) Rod-shaped, (*b*) spiral and (*c*) spherical.

form and temporarily connect two "mating" bacteria. Plasmids and genetic material from the circular DNA can be
transferred from one bacterium to another through these
thin connections. The transferred DNA is incorporated
into the genome of the receiving bacterium, creating a new
genetic combination. New, more hardy strains may arise
from conjugation. In fact, genes responsible for antibiotic
resistance may be transferred in this manner.

[3] Sexual reproduction is generally defined as the combination of germ cells
produced by meiosis, so conjugation cannot be considered a true form of
sexual reproduction.

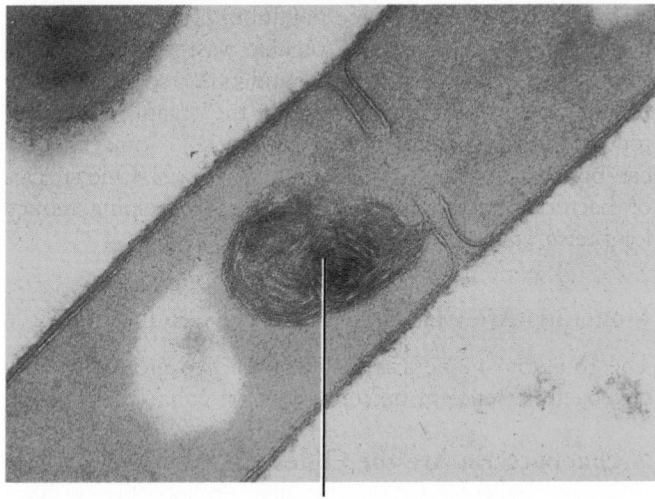

Mesosome

▷ **FIGURE 24–9 The Mesosome** Although bacteria lack cellular organelles, except ribosomes, the mesosomes shown here maintain some degree of intracellular organization akin to that seen in eukaryotes. In photosynthetic bacteria, the mesosome may act as a simple type of chloroplast.

Conjugation is a rather costly occurrence, because the donor bacterium, deprived of some of its genetic material, may die. The persistence of this event may be explained by the evolutionary advantages it confers; that is, it serves as an important source of genetic variation in bacteria.

Many species of bacteria exhibit another evolutionary adaptation vital to bacterial success—the production of resistant **endospores,** resting cells. Endospores are produced when environmental conditions become unfavorable. These resistant structures house the circular strand of DNA and a tiny amount of cytoplasm—just enough to start active metabolism when conditions become favorable for survival. A thick cell wall encases the inert spore.

Spores can withstand extremely harsh conditions that would kill normal bacterial cells. Some, for example, can survive boiling for an hour; others have been found in the intestines of mummies over 2000 years old. Still others can withstand harsh chemical treatment.

Spores may be transported by wind or water. When they encounter favorable growing conditions, they develop rapidly. Water absorbed by the spore in such an environment causes the wall to break, and the spore develops into an active bacterium.

Bacterial Metabolism and Feeding Strategies Vary Widely

Another evolutionary adaptation that helps account for the success of bacteria is their wide range of metabolic strategies for harvesting energy from their environment to power internal events. This, in turn, results in their presence in many different environments. Some bacteria live in oxygen-free environments—for example, in the deep layers of mud in a stream bed or lake bottom. These bacteria are known as

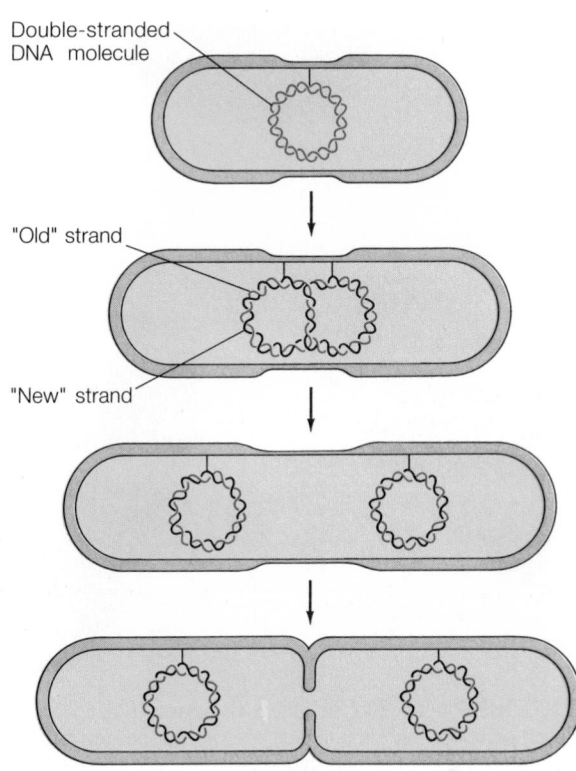

Double-stranded DNA molecule

"Old" strand

"New" strand

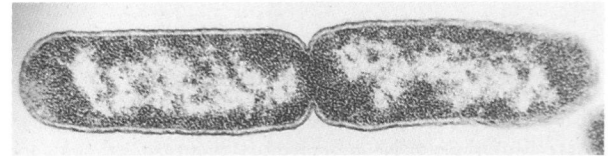

▷ **FIGURE 24–10 Binary Fission** Bacteria reproduce by a splitting of the cells. The circular DNA molecule reproduces prior to fission, as shown here; then the cell splits, forming two new bacteria.

obligate anaerobes (meaning they are obligated to live under these conditions). Obligate anaerobes perish in the presence of oxygen. Botulism (a kind of food poisoning) and tetanus are caused by two such bacteria. The bacteria responsible for these diseases are both spore formers. Their spores can withstand the presence of oxygen.

Other bacteria do not require oxygen but can function in its presence. These are called **facultative anaerobes.** (Facultative means they are capable of living under varying conditions.) Most bacteria, however, are **aerobic** organisms—that is, organisms that require a constant supply of oxygen to maintain cellular metabolism.

Bacteria acquire nutrients in a variety of ways. Some are autotrophic, or capable of making their own food via photosynthesis or chemosynthesis (Chapter 23). Most bacteria, however, are heterotrophic; that is, they eat others. Heterotrophic bacteria are also classified as **saprophytes** because they release enzymes that digest large food molecules externally. Small food molecules produced in the process may then be absorbed by the bacterium.

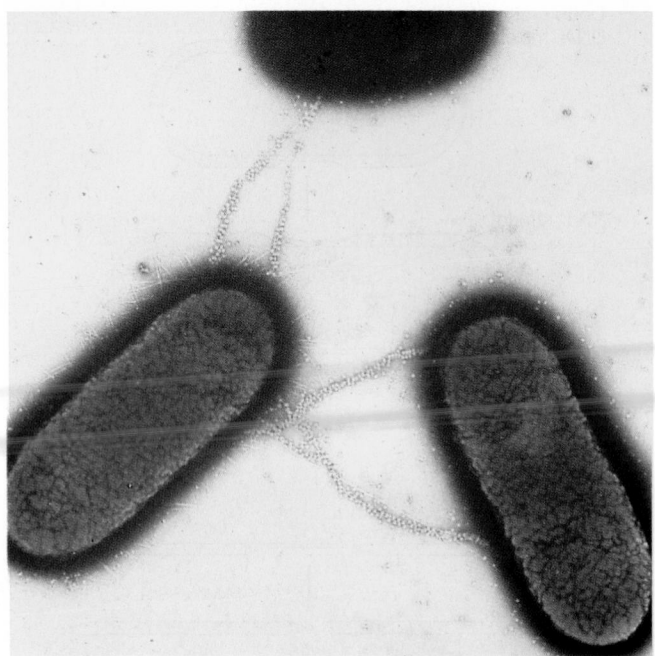

▷ **FIGURE 24–11 Bacterial Conjugation** Bacteria can "reproduce" sexually. Genetic material can be transferred from one bacterium to another through thin cytoplasmic connections. The new genetic information is inserted into the genome of the recipient.

Some Bacteria Cause Diseases, but Many Provide Useful Services

Like viruses, bacteria can be harmful to other organisms. Most pathogenic (disease-causing) bacteria release some kind of chemical toxin or enzyme that injures cells or disrupts their function. Bacteria cause many common diseases, such as strep throat, some forms of pneumonia, food poisoning, and urinary tract infections.

Not all bacteria are harmful. Many, in fact, are useful. Bacteria in the stomach of cows and other hoofed animals, for example, allow these animals to digest cellulose. Without them, cows would be unable to digest cellulose, because they lack the proper enzyme. In humans, intestinal bacteria produce several vitamins that are absorbed by the large intestine (Chapter 11).

Many bacteria are essential to the continuation of life on Earth. Certain species in the soil, for example, break down animal wastes and the remains of plants and animals (Chapter 31). The decomposition of these organic materials releases nutrients, which replenish the soil. These nutrients are incorporated into plants. Because animals feed on the plants, soil bacteria help maintain the cycle necessary to the perpetuation of the web of life.

Other soil bacteria form symbiotic relationships with the roots of certain plants. These bacteria absorb nitrogen gas from the air in the soil and convert it into a chemical form that plants can use to make amino acids and nucleic acids. It is from plants that humans and most other animals ultimately derive their nutrients. (Chapter 31 describes the role of these bacteria in more detail.)

Bacteria also help reduce pollution. Certain bacteria in water, for example, feed on organic wastes from sewage-treatment plants, farms, lawns, and other sources. These naturally occurring bacteria degrade the organic chemicals, reducing levels of pollution (Chapter 31). Some bacteria can be used to clean up other human messes. One species of bacterium, for example, degrades oil, helping reduce long-term environmental damage from a spill.

Monerans Are Classified Into Two Groups

The Kingdom monera is divided into two groups, the archaebacteria (ancient bacteria) and the eubacteria.

Archaebacteria Are the Oldest Bacteria. Archaebacteria are believed to be the first prokaryotes to have evolved during cellular evolution. They live in the most extreme environments—for example, in water with high salt concentrations or in extremely hot water. Such extreme environments may be a reflection of the conditions on the Earth when bacteria first formed.

The archaebacteria differ markedly from the eubacteria. Their plasma membranes, for instance, contain a very different kind of lipid. Their cell walls also differ from those of the eubacteria. Even their ribosomal RNA is different. These differences suggest to some biologists that the archaebacteria should be placed in a separate kingdom of their own.

There Are Many Different Kinds of Eubacteria. Eubacteria are a large and diverse group of monerans, a discussion of which is beyond the scope of this book. A few examples will help to familiarize you with them, however. Gram-positive eubacteria, (so named because they appear purple when stained by a dye developed by Christian Gram, a Danish physician) have thick cell walls and are responsible for a number of diseases, including strep throat and botulism. Spirochetes are gram-positive, spiral-shaped eubacteria that move about with the aid of flagella. One well-known spirochete, *Treponema pallidum,* infects humans, causing syphilis. Another spirochete causes the debilitating disorder called Lyme disease, transmitted by ticks.

The Cyanobacteria Are Photosynthetic Monerans. The photosynthetic eubacteria are known as the **cyanobacteria.** Formerly known as the blue-green algae, cyanobacteria are represented by approximately 200 different species. Cyanobacteria were once called blue-green algae because they contain blue and green pigments and were thought to be algae. Upon closer examination, it became clear that they more closely resembled bacteria.[4]

Cyanobacteria are single rod-shaped or spherical cells

---

[4]In addition to chlorophyll a, cyanobacteria usually contain accessory photosynthetic pigments, and therefore various cyanobacteria appear black, brown, yellow, or red.

VIRUSES MAY HELP PREVENT DIABETES

In 1988, a California researcher reported that injecting a certain virus into mice prone to developing diabetes seems to prevent the disease. Michael Oldstone of the Research Institute of Scripps Clinic in La Jolla (pronounced "la hoya") found that a virus, known as LCMV, stops the destruction of insulin-producing cells in mice by the mice's own immune system.

The mice used in these studies become diabetic when their own pancreatic cells are destroyed via an autoimmune reaction thought to be a cause for early-onset diabetes in humans (Chapter 15). By 6 months of age, the mice develop life-threatening diabetes.

When diabetes-prone mice were injected at birth with the LCMV, however, not one mouse developed the disease within the next nine months. If the injection was delayed until the sixth week of life, the virus still reduced the incidence of diabetes to about 6%; 95% of the untreated mice in the control group developed the disease.

Evidence suggests that this virus attacks and incapacitates the helper T lymphocytes. Researchers believe that the virus either destroys or alters the function of these cells, which are involved in the autoimmune reaction.

Researchers believe that the LCMV could be used to prevent many cases of diabetes in humans. Because the LCMV causes chronic (long-term) infection in mice, however, Oldstone does not advocate injecting whole viruses as a treatment for diabetes. He hopes that further research may uncover a component of the virus that gives protection without causing infection.

Ironically, researchers have thought for many years that certain viruses cause early-onset diabetes. Now, it appears that viruses may also have a role in preventing the disease.

---

that occur in small clusters or long filamentous strands (▷ Figure 24–12a). In these chains, a rudimentary division of labor often occurs. A few cells may "specialize," becoming able to capture and fix atmospheric nitrogen, while the rest of the cells photosynthesize. Some cells act as spores.

In cyanobacteria, photosynthesis resembles that occurring in plants (Chapter 6). Cyanobacteria contain photosystems I and II, release oxygen, and rely on water as a source of electrons. Chlorophyll is housed in special plasma membrane invaginations in the cell (Figure 24–12b).

Cyanobacteria live primarily in fresh water and in soil and on moist surfaces. Some species dwell in rather harsh conditions—for example, hot springs. Other cyanobacteria form a symbiotic relationship with certain fungi (described later), forming lichens, organisms capable of living on bare rock.

≋ THE KINGDOM PROTISTA

Chapter 23 noted that during cellular evolution, the first nucleated cells, the eukaryotes, formed about 1.2 billion years ago. Early eukaryotes belong to the kingdom Protista, which today contains over 30,000 recognizable species. During the course of evolution, the protists gave rise

▷ **FIGURE 24–12 Cyanobacterial Filament** (*a*) Some bacteria form filamentous strands with rudimentary cellular differentiation. Several bacterial cells in the strand shown here are specialized to fix nitrogen. (*b*) Chlorophyll is embedded in the membranes within the cytoplasm of a cyanobacterium.

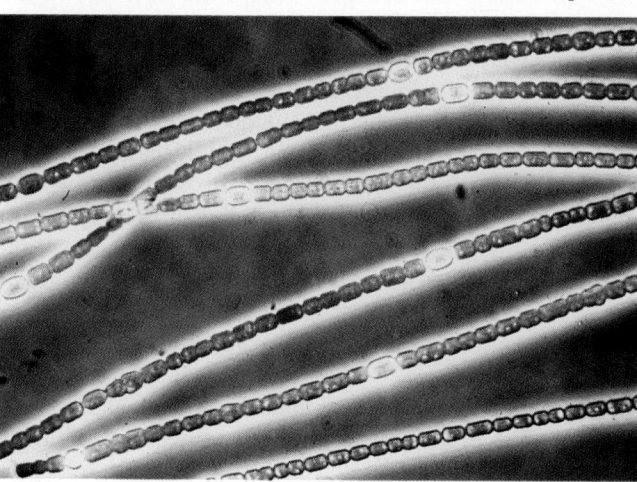

(a)

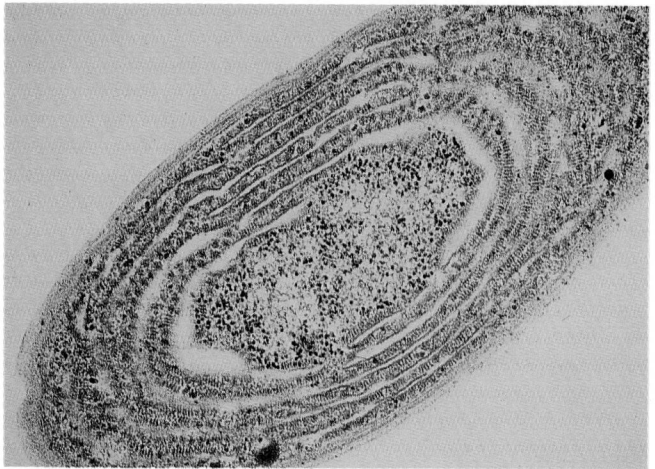

(b)

to three groups of multicellular organisms, the plants, fungi, and animals.

Trying to define the protists is difficult, because they are so diverse. Protists are best described as a hodgepodge of organisms. This section explores the kingdom Protista, looking at two major groups—plantlike and animal-like protists. Even here, the distinction is not as clear as one might like. Many protists, in fact, could fit into either class. For example, the single-celled *Euglena* is classified as a plantlike protist (▷ Figure 24–13), for it contains chloroplasts and is capable of photosynthesis. However, these organisms contain eyespots that detect light and are also capable of swimming about with the aid of flagella, both animal characteristics.

Reproduction among protists generally occurs asexually. However, many protists are capable of reproducing sexually. This section examines the lives of plantlike and animal-like protists, beginning with plantlike organisms, the unicellular algae.

▷ **FIGURE 24–14 Diatoms** These unicellular algae are encased in often-ornate, glasslike shells and are the most abundant form of algae found in the ocean.

Unicellular Algae Are Widely Distributed in Fresh Water and Salt Water

The **plantlike protists** are microscopic, photosynthetic organisms that capture sunlight and produce food for many aquatic organisms (Chapter 31). These organisms are part of a larger group known as **phytoplankton** (literally, "floating plants"). Phytoplankton include many different photosynthetic organisms, most of them single-celled but not all of them protists. Phytoplankton are a vital food source in freshwater and marine environments and are an important source of atmospheric oxygen. Marine phytoplankton are perhaps the most important source of oxygen, because they are responsible for about 70% of all photosynthesis occurring on the planet. This section describes three types (divisions) of plantlike protists: diatoms, dinoflagellates, and euglenoids.[5]

Diatoms Are Photosynthetic Organisms Encased in Protective Coverings. Perhaps the best known and ecologically most important of all unicellular algae are the diatoms. **Diatoms** are free-living organisms found in fresh and salt water (▷ Figure 24–14). These exquisite organisms are distinguished by their unique cell walls. Each cell wall consists of two halves, one slightly larger than the other and fitted together like the top and bottom of a pillbox. The outer portion of the cell wall is composed of silica, a hard, glassy material that is often fashioned into exquisite designs. The rigid wall is riddled with tiny holes, often arranged in elaborate patterns. These holes are pores that allow nutrients to enter the organism and wastes to escape.

Diatoms are the most abundant form of algae in the ocean and therefore are an essential source of food in the

▷ **FIGURE 24–13 Euglena** This plant-like protist actually has characteristics of both animals (flagellum, lack of cell wall) and plants (chloroplasts).

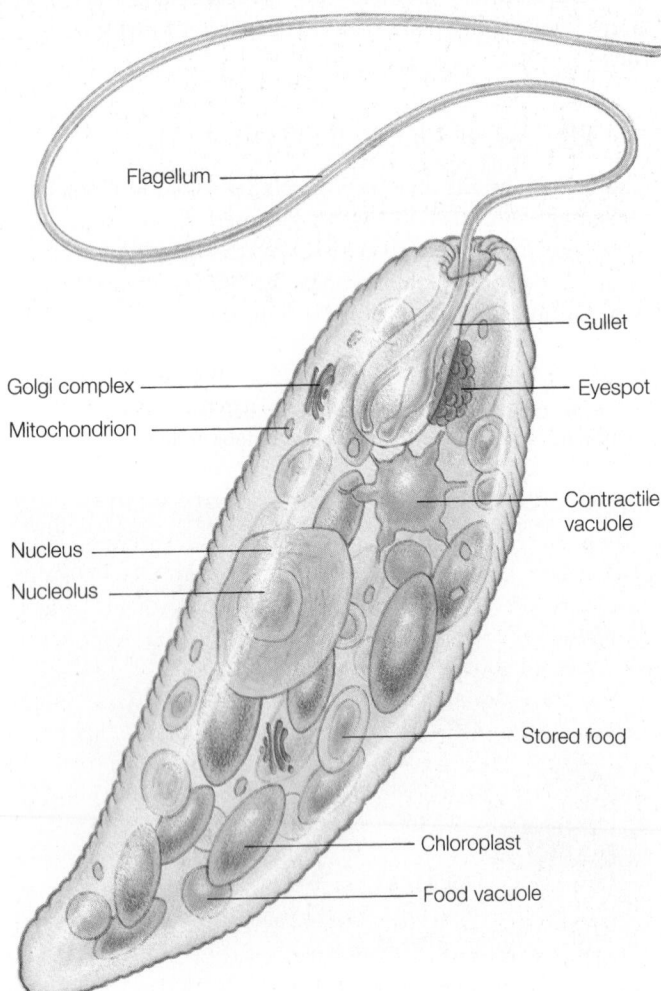

Flagellum

Gullet

Golgi complex

Eyespot

Mitochondrion

Contractile vacuole

Nucleus

Nucleolus

Stored food

Chloroplast

Food vacuole

[5] Botanists use the term *division* instead of *phylum*.

seas. When diatoms die, their glassine "shells" settle to the bottom. Accumulating on the ocean floor for centuries, the shells may produce thick deposits sometimes hundreds of meters deep. The slightly abrasive material is used in a variety of products, including some toothpastes, silver polish, swimming pool filters, and soundproofing materials.

Biologist Norman Wessels has noted that diatom reproduction is as interesting as the organism is beautiful. Like other single-celled organisms, diatoms reproduce asexually by cell division. During this process, the two halves of the diatom shell detach, and each produces another, slightly smaller box inside itself. When these diatoms divide, they produce an even smaller shell until the cell becomes so small that a meiotic division is stimulated. The minute cells make two gametes equipped with flagella. The gametes swim away in search of another. When they find another gamete, the haploid cells unite, forming a diploid diatom that produces a new cell wall of the original size.

Dinoflagellates Are Photosynthetic Protists Propelled by Two Flagella.

Dinoflagellates live in fresh and salt water and possess two flagella (▷ Figure 24–15a). One of the flagella causes these organisms to spin like tops. In fact, in Greek the word *dinos* means "rotation." The other flagellum propels them forward. Dinoflagellates, therefore, are rotating, flagellated organisms. Dinoflagellates also contain light-absorbing chlorophyll pigments in distinct chloroplasts, which they use to trap sunlight for photosynthesis. Many dinoflagellates have cell walls consisting of numerous smaller plates of cellulose (Figure 24–15b).

Dinoflagellates are most abundant in salt water, and they provide an important source of food for marine life. They also form many symbiotic relationships. Coral are members of the animal kingdom and live in a symbiotic relationship with certain dinoflagellates (Chapter 31). The photosynthetic dinoflagellates provide nutrients for the coral, and the coral provide protection. When deprived of their symbiotic partners, coral grow much slower, and some may even die. Recent studies show that the symbiotic relationship is being disturbed worldwide. Scientists believe that global warming, a gradual rise in the Earth's temperature caused by pollution and other factors, may be causing a rise in the temperature of the seas. This warming, in turn, may be the reason for the demise of the dinoflagellates so essential for the survival of a coral reef. If the waters continue to warm, some scientists believe, many coral reefs could stop growing and die, leaving only the skeletal remains of these once magnificent living creatures.

Although virtually all dinoflagellates contain chlorophyll, it is often masked by red pigments (carotenoids), which capture light and participate in photosynthesis. In warm ocean waters that are rich with nutrients (often from sewage-treatment plants or farm fields), one troublesome species of dinoflagellates often flourishes. So abundant is this organism that it turns the water red and creates a condition known as **red tide** (▷ Figure 24–16). The dinoflagellate responsible for red tide produces a nerve toxin,

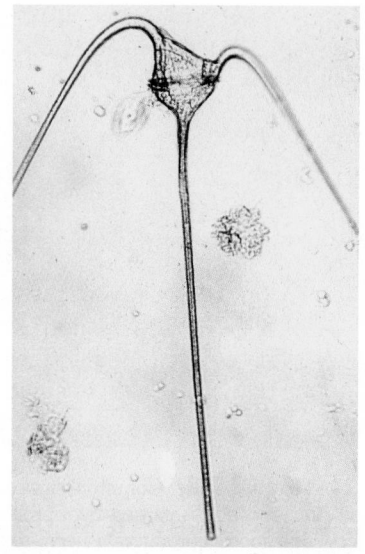

(a)

(b)

▷ **FIGURE 24–15 Dinoflagellate** (*a*) Light micrograph of a dinoflagellate showing the flagella. (*b*) Electron micrograph illustrating the cellulose plates.

which in high concentrations kills fish, dolphins, whales, and other vertebrates. Shellfish may feed on the dinoflagellate, unaffected by the poison. However, shellfish concentrate the poison in their tissues. Humans eating contaminated shellfish may become sick and die.

Recently, scientists have measured an increase in the incidence of red tide throughout the world. They warn that this phenomenon is a symptom of the growing pollution of our planet and could have devastating effects on marine life if left unchecked.

If you have ever been on a boat on the ocean at night, you may have witnessed one of the most exciting light shows known to humankind. Water disturbed by a boat propeller or breaking waves gives off a brilliant blue-green light. This light is emitted from certain species of di-

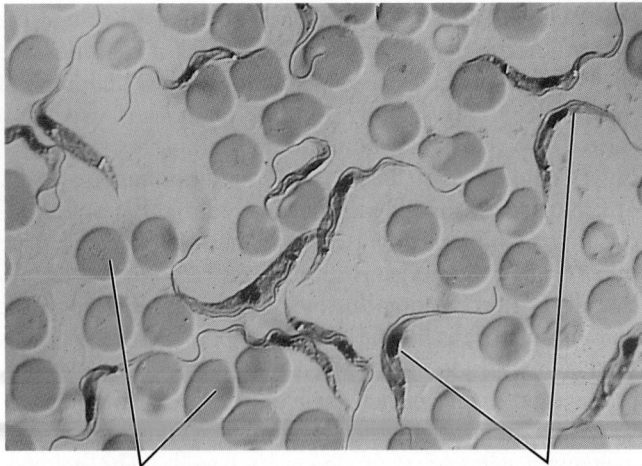

(a) Red blood cells   Trypanosome parasites

(b)

▷ **FIGURE 24–16 Red Tide** Common in nutrient-rich (often polluted) ocean water, red tide is caused by a proliferation of dinoflagellates. At least one species produces a nerve toxin that is lethal to fish and to humans who feed on shellfish that have concentrated the poison in their tissues.

▷ **FIGURE 24–17 Zooflagellates** (*a*) The trypanosome is a parasitic zooflagellate that is responsible for malaria. (*b*) This organism is *Giardia*, a zooflagellate that causes diarrhea and cramps.

noflagellates and bacteria. The phenomenon is called bioluminescence.

Euglenoids Are Freshwater Organisms with Characteristics of Animals and Plants. The euglenoids are truly puzzling creatures. The early botanists observed their ability to photosynthesize and called them animal-like plants. Early zoologists noted their mobility and classified them as plantlike animals. Today, modern biologists group them with the photosynthetic protists. Like dinoflagellates, the **euglenoids** are single-celled protists with characteristics of both plants and animals. They receive their name from the best-known of their kind, the genus *Euglena* (see Figure 24–13). Flexible and motile like animal-like protists, most euglenoids contain chloroplasts and propel themselves through fresh water with the aid of a flagellum. Unlike diatoms and dinoflagellates, the euglenoids lack rigid outer coatings. Although photosynthesis is the chief source of food for most euglenoids, when these creatures are deprived of light, they lose their chloroplasts and change their feeding strategy, absorbing nutrients from their surroundings. Some species are heterotrophic—capable of phagocytizing food particles and digesting them intracellularly.

Euglena contains a light-sensitive organelle, a photoreceptor called the *eyespot*. Located at the base of the flagellum, the eyespot allows the organism to detect the direction and intensity of light. Euglenoids reproduce asexually by cell division. Each mitotic division is followed by cytokinesis, which results in identical offspring.

The Protozoans Are Animal-Like Protists

The animal-like protists, or **protozoans** (literally, "first animals"), are unicellular eukaryotic organisms that survive by

eating bacteria and other protists. This section describes four major groups (phyla).

The Zooflagellates Are Protozoans that Move about Via Flagella. Flagellated protozoans are often called **zooflagellates** to distinguish them from flagellated plantlike (photosynthetic) protists (▷ Figure 24–17). Zooflagellates live in water and in moist soils. These organisms contain one or sometimes several flagella, by which they propel themselves through their watery environment. Flagella are multipurpose organelles, also serving as sense organs. Some organisms even use their flagella to snare prey.

Many zooflagellates are free-living organisms; others form symbiotic relationships, living within other organisms in a mutually beneficial relationship. One of the best studied symbiotic relationships exists between the zooflagellate *Trichonympha collaris* and a wood-eating termite. *Trichonympha* lives inside the gut of the termite, where it digests cellulose, a feat that the termite is incapable of performing alone.

Some zooflagellates are parasitic. Perhaps the best known example is the zooflagellate of the genus *Trypanosoma* (Figure 24–17a). Trypanosomes can dwell in the blood of people, causing a potentially fatal disease known as African sleeping sickness. The tiny parasites are transmitted from one infected person to another by the blood-sucking tsetse fly.

One increasingly common parasite is the zooflagellate *Giardia* (Figure 24–17b). *Giardia* is carried by humans, sheep, raccoons, beavers, and muskrats. It reproduces in the body and produces cysts that are released with the feces. Feces contaminate streams where campers and backpackers get their water and may even contaminate community water supplies, causing the disease to spread to local residents. *Giardia* cysts develop into adult forms in the intestine of the host. An infection results in recurrent bouts of diarrhea, dehydration, nausea, vomiting, and cramps. Untreated, a victim may be sick for several months and may lose a considerable amount of weight, but the disease is rarely fatal and readily treatable with a drug known as Flagyl.

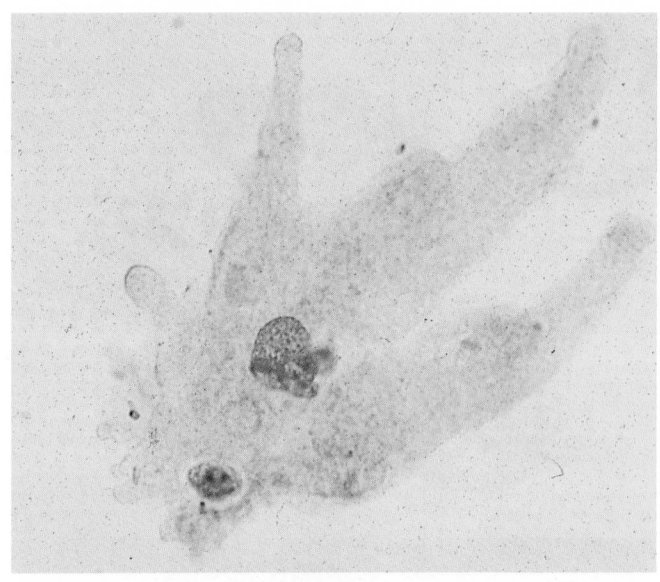

▷ **FIGURE 24–18 Amoeboid Protozoans** Amoebae propel themselves by cytoplasmic extensions called pseudopodia. They engulf their prey via phagocytosis.

The Amoeboids Are Protists that Propel Themselves by Amoeboid Motion.
▷ Figure 24–18 shows a typical member of the amoeboid protists. As noted in Chapter 3, during amoeboid movement, small pseudopodia extend from the body of amoebae, attaching to a solid surface. Cytoplasm then flows into the pseudopodia, filling them and causing the organism to grow at the "front end" (the direction of movement) and shrink at the back end.

Often seen as small masses of cytoplasm without definite shape, amoebae contain nuclei and several other organelles (Figure 24–18). One prominent organelle is the **contractile vacuole,** which collects excess water from the cytoplasm. When full, the vacuoles contract, expelling the water through a temporary opening in the plasma membrane.

Amoebae are heterotrophs that phagocytize bacteria and other small prey. As explained in Chapter 3, phagocytized material is packaged in a **food vacuole,** a membrane-bound vacuole in which the food is enzymatically digested.

Although most amoebae lack definite form, many species have distinct shells made of silica, calcium carbonate, or other materials (▷ Figure 24–19). Silica-shelled amoebae are called **radiolarians,** and amoebae with calcium carbonate shells are called **foraminiferans.** Holes in the shells permit the extrusion of pseudopodia, which are used to capture food.

When radiolarians and foraminiferans die, their shells sink to the bottom of the ocean. Over time, they may form extremely thick layers. In fact, the limestone cliffs of Dover were once a seafloor deposit formed by countless foraminiferans. Limestone deposits such as these often occur near deposits of phytoplankton, which, when trapped in sediment, form oil over many millions of years.

One of the best known amoebae is a parasitic species found in warm climates—for example, in Mexico. This organism, *Entamoeba histolytica,* causes amoebic dysentery in humans, an intestinal disorder characterized by diarrhea, cramps, and nausea. *Entamoeba* multiplies inside the cells lining the intestinal tract. If this disease is not treated, the intestinal lining can be seriously damaged, and death may occur due to severe dehydration.

Ciliates Are Protozoans that Move Through Their Aqueous World with the Aid of Cilia.
The **ciliates** are the most structurally complex of all protozoans. They get their name from the many cilia projecting from their surface, which propel these organisms through the fresh and salt water where they live. Cilia may cover the entire organism or may be concentrated in certain regions.

Perhaps the best known representative of the ciliates are the members of the genus known as **Paramecium** (▷ Figure 24–20). These organisms feed by sweeping food into their gullets, where it is then phagocytized. Food vacuoles form inside the organisms and serve as a site of digestion. Soluble nutrients pass out of the vacuoles into the cytoplasm, and undigested waste is eventually eliminated by exocytosis at the anal pore.

Some ciliates acquire food by spearing it with explosive, dartlike projections called **trichocysts.** Trichocysts are housed in oval capsules in the outer cytoplasm. When stimulated, the trichocysts explode like a harpoon gun, releasing a threadlike structure that impales the prey and delivers a fatal dose of toxin. The prey is then swept into the oral groove via cilia. Ciliates also contain contractile vacuoles that rid the organism of excess water.

Biologists have recently discovered that one ciliate species can change into a parasite in the presence of predatory mosquito larvae. Research suggests that the mosquito larvae secrete a water-soluble substance that causes these ciliates, which normally feed on bacteria and other micro-

(a)

(b)

▷ **FIGURE 24–19 Shelled Amoeboids** (*a*) Silica-shelled radiolarians and (*b*) carbonate-shelled foraminiferans.

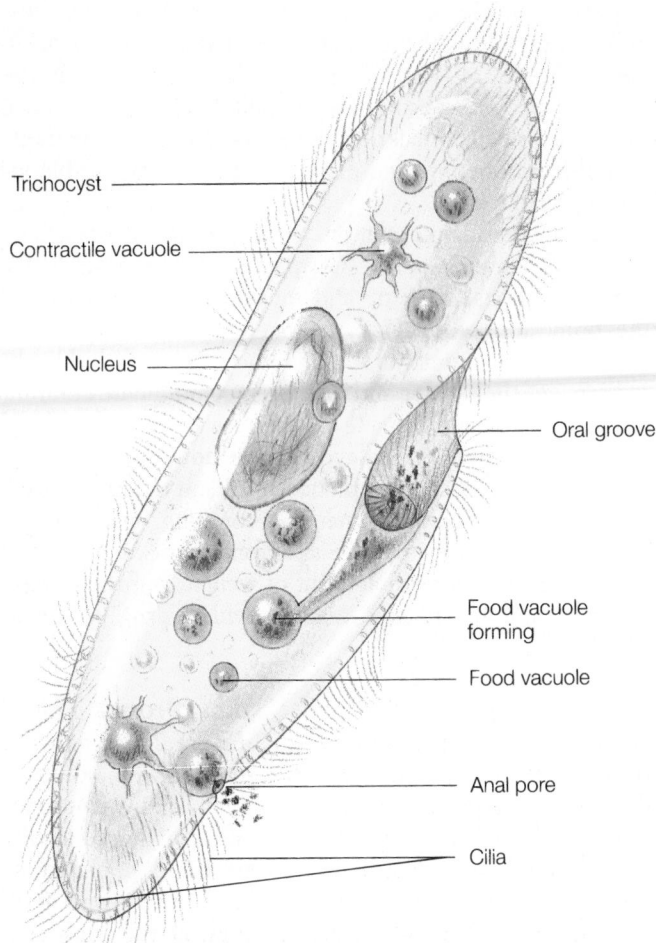

Trichocyst

Contractile vacuole

Nucleus

Oral groove

Food vacuole forming

Food vacuole

Anal pore

Cilia

▷ **FIGURE 24–20 Paramecium** The paramecium is a common ciliate and illustrates many important ciliate organelles.

organisms, to undergo the rapid transformation to parasites. The parasitic ciliates infect the larvae and kill their would-be predators. In parasitic form, this ciliate may prove useful as a biological control against mosquitoes, replacing potentially dangerous chemical insecticides.

Another interesting "behavior" among the ciliates is their ability to change shape when in the presence of substances released by predators. One ciliate shaped like a lemon transforms into a flattened disk much larger and more difficult to consume. This adaptation, like others, is testimony to the process of evolution!

The Sporozoans Are Parasitic Protozoans that Live Inside the Bodies of their Hosts. **Sporozoans** are nonmotile, parasitic protists, so named because of their ability to produce infectious, sporelike structures that can be passed from an infected animal to a noninfected animal through food, water, and other means (▷ Figure 24–21). Sporozoans infect a wide range of organisms, including

birds, mammals, fishes, and even some plants. The adult forms of these protists are nonmotile, but immature forms can move about via flagella and pseudopodia.

Perhaps the best known of all the sporozoans is an organism called *Plasmodium vivax*. This organism causes one form of malaria (Chapter 13). Transmitted by infected **Anopheles** mosquitoes, this protozoan first infects the liver, where it replicates. It then reenters the blood, infecting red blood cells, in which it also replicates. Infected red blood cells eventually burst, releasing the organisms and toxic substances they produce, which cause the recurrent chills and fever of malaria. Noninfected mosquitoes may become infected when they draw blood from a person afflicted with this unicellular parasite, thus ensuring the spread of the disease. Efforts to control the disease, which occurs principally in tropical regions, have relied principally on the insecticide DDT, used to control the mosquitoes. Initial spraying greatly reduced the incidence of malaria in India and other countries, but in many places mosquitoes are now resistant to the insecticide, and malaria is once again on the rise.

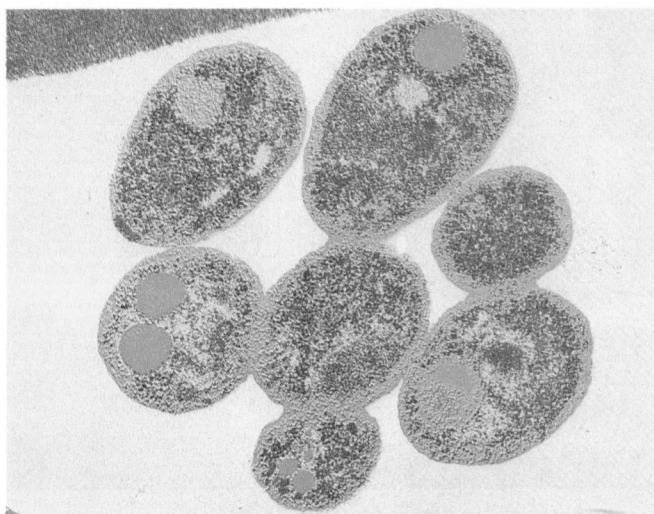

▷ FIGURE 24–21 Malaria-Causing Sporozoan This sporozoan causes one form of malaria. It is transmitted from one person to the next by mosquitos. Noninfected mosquitos become infected when they bite a person carrying the unicellular organism.

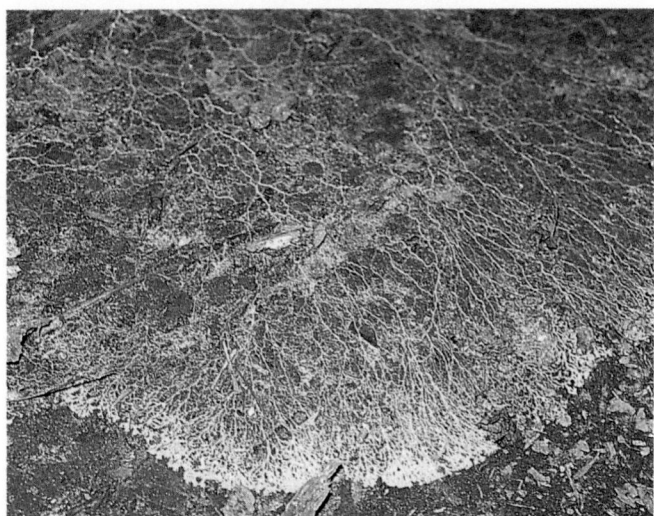

▷ FIGURE 24–22 Acellular Slime Mold

The use of DDT to control malaria illustrates the connection between environment and ecosystem health extraordinarily well. In Borneo, DDT was sprayed in and around thatched-roof huts to control mosquitoes, successfully reducing the incidence of malaria. However, the DDT also poisoned many other insects. These insects were eaten by lizards that lived in the thatched roofs. The lizards were quite beneficial, for they held thatch-eating insects not sprayed with DDT in check. When the lizards disappeared, the thatch roofs began to crumble. To make matters worse, household cats ate the dead lizards and died. Rat populations, once controlled by cats, skyrocketed, creating another problem for villagers. Biological control agents, like the parasitic ciliate mentioned earlier, can be used to help control mosquitoes without all of the chain-reaction effects of DDT.

The Funguslike Protists Digest Foodstuffs Externally and Include the Slime Molds

The kingdom Protista is truly one of the most fascinating of all kingdoms, displaying a wide array of adaptations to life on Earth. One adaptation, also seen in the bacteria, is saprophytic existence—living by digesting foodstuffs externally. This feature is found in the **funguslike protists,** the slime molds. (The term *slime mold* comes from their glistening, moldlike appearance.) Slime molds crawl about by amoeboid motion. Two types of slime mold are found: cellular and acellular (or plasmodial).

Cellular slime molds are single-celled organisms that live in the soil, engulfing food particles as they move about. When food is scarce, cellular slime molds release a chemical substance that attracts others. The individual cells then come together, forming a massive aggregation. Moving

along the surface of the soil en masse, the slime mold acts like a single multicellular organism.

The individual cells in a cellular slime mold remain distinct, and many differentiate, taking on specific functions that benefit the whole. Some cells, for example, form a reproductive structure that generates spores that are released into the air. These spores form new individuals.

The **acellular slime molds,** or true slime molds, consist of a mass of cytoplasm encased in a single membrane. The highly branched, fan-shaped, or solid masses they form consist of thousands of nuclei but no distinct cells, hence the name. Acellular slime molds often spread out over large areas—occasionally covering several square meters or more (▷ Figure 24–22).

Acellular slime molds may be thought of as giant, multinucleated cells that move about phagocytizing bacteria and protists found in decaying leaves, wood, and other organic matter. They also ingest organic molecules produced by the action of enzymes they release. Often colorful, slime molds produce spores that grow into entirely new aggregates.

When environmental conditions change and an acellular slime mold enters an area low on food, it then produces stalk-bearing spore capsules. The spores, produced by meiotic division, are haploid cells that are released from the structure and dispersed by wind and rain. When they encounter conditions conducive to growth, the spores germinate, producing haploid cells that fuse to form diploid cells, which give rise to a new slime mold.

≈ THE KINGDOM FUNGI

Fungi are probably the least appreciated of all the Earth's organisms. Anyone who has walked in a forest and seen a mushroom or has opened a container of food left in the refrigerator too long is already acquainted with the kingdom Fungi.

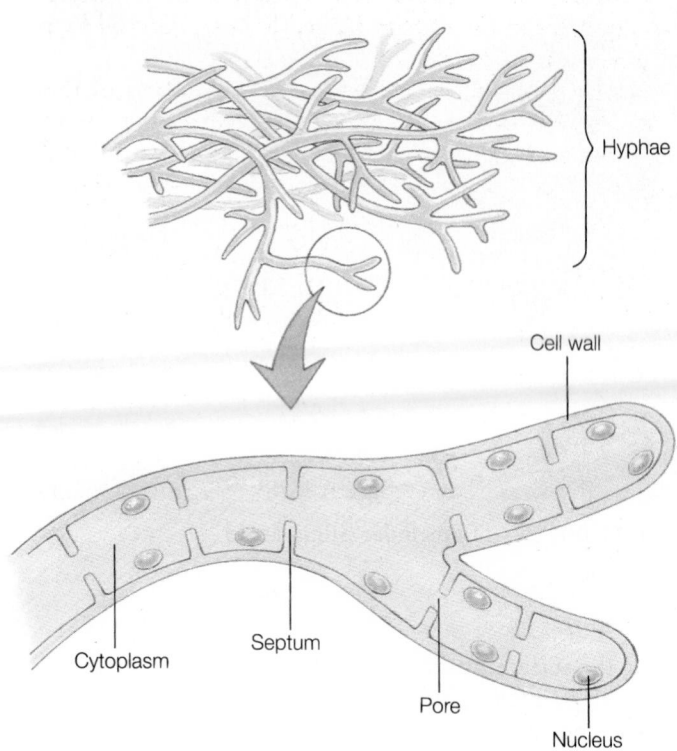

▷ FIGURE 24–23 Hyphae and Mycelia The bodies of most true fungi consist of filamentous strands called hyphae, which unite into masses known as mycelia.

Fungi Have Distinct Cell Walls and Include Single-Celled Yeasts and the Multicellular Fungi

The kingdom Fungi contains a striking variety of multicellular organisms, including puffballs, toadstools, mushrooms, and bread molds. It also contains unicellular organisms, the yeasts.

Multicellular fungi consist of microscopic filaments, or strands, called **hyphae** (▷ Figure 24–23). In some fungi, the hyphae consist of single, elongated cells containing many nuclei; in others, the hyphae are subdivided by partitions, forming many individual cells. Pores in partitions facilitate the flow of cytoplasm, allowing nutrients to flow from one cell to another. Hyphae grow together, forming an interwoven mass called a **mycelium** (Figure 24–23).

Most fungal cells are enclosed by a rigid cell wall composed of a material called chitin. (You will recall that plant cell walls are composed mostly of cellulose.) **Chitin** is a durable polysaccharide also found in the exoskeletons (external skeletons) of arthropods (insects, crustaceans, and spiders).

Like many bacteria, fungi are saprophytic; that is, they release enzymes that break down dead organisms, then absorb food molecules. Leaves and rotting tree trunks, for example, are common sources of food in forests. Fungi that grow in the refrigerator feed on a variety of organic food materials. Consequently, most fungi are essential for decomposition.

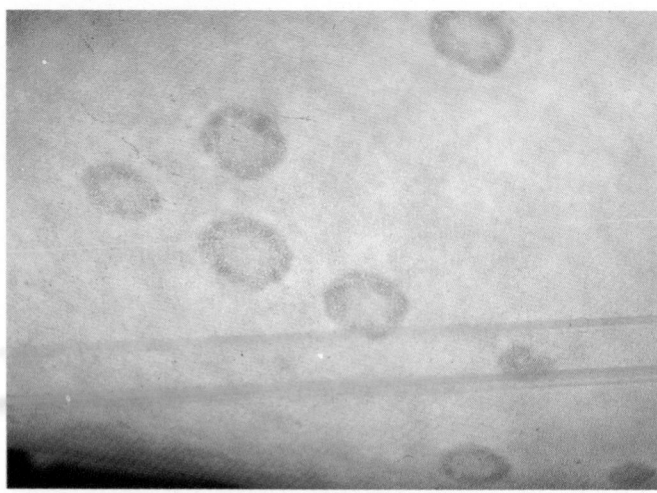

▷ FIGURE 24–24 Ringworm This skin disease is caused by a parasitic fungus—that is, a fungus that feeds on living tissue. It is not caused by a worm.

One group of fungi is parasitic—that is, it feeds on living food sources. Several species of parasitic fungi, for example, attack humans, causing such problems as ringworm, athlete's foot, and vaginal infections. The ringworm fungus lives on skin cells and causes scaly, round, itchy patches, usually on the head, trunk, or feet (▷ Figure 24–24). The fungus that causes athlete's foot grows on the skin between the toes, causing itching and pain. Vaginal yeasts feed on cells and secretions present in a woman's vagina.

Fungi have intriguing life cycles, and many can reproduce both sexually and asexually. In the simplest form of asexual reproduction, a mycelium fragments, forming many pieces. Each piece may form a new individual. Most fungi, however, produce spores that develop into a new fungus under proper conditions. Spores are often dispersed by the wind and may be produced either sexually or asexually, as noted in later sections.

The following sections describe three common groups of fungi: zygote, sac, and club.

Zygote Fungi Include Black Bread Molds. The **zygote fungi** consist of about 600 species, including one of the best known, black bread molds ▷ Figure 24–25). Zygote fungi are so named because the hyphae actually unite, or fuse, as illustrated in Figure 24–25b. When the haploid cells of the hyphae unite, they form diploid zygotes known as **zygospores.** These spores are rather resistant structures that are dispersed through the air. Zygospores may remain dormant for some time until conditions for growth are met. When conditions are right, the zygospore undergoes meiosis, producing filamentous hyphae containing haploid cells again.

Hyphae from single spores may fuse with those of other strains of the same species, thus providing the opportunity for sexual reproduction. Asexual reproduction occurs when hyphae of the zygote fungi grow upright and form spherical bodies called **sporangia** (Figure 24–25b). Inside these bodies, thousands of haploid spores are formed. Spores are released and transported by air currents to new sites, where, if conditions are suitable, they develop into new hyphae. This is a form of asexual reproduction.

Sac Fungi Include Common Molds that Grow on Food, Single-Celled Yeast, and Others.

Sac fungi are a large group that also reproduce sexually and asexually. They get their name from the saclike cases in which spores are produced during sexual reproduction. Sac fungi include many common molds that attack food, including fruit (▷ Figure 24–26a). The sac fungi also include delicacies such as truffles and morels (Figure 24–26b). Dutch

elm disease, which has annihilated nearly all of the American elm trees in the eastern United States, is caused by a member of this group.

Many yeasts (single-celled fungi) are members of this group as well. All yeasts are unicellular organisms and most often reproduce asexually (by budding). As you may recall from Chapter 6, under anaerobic conditions, yeasts break

(a)

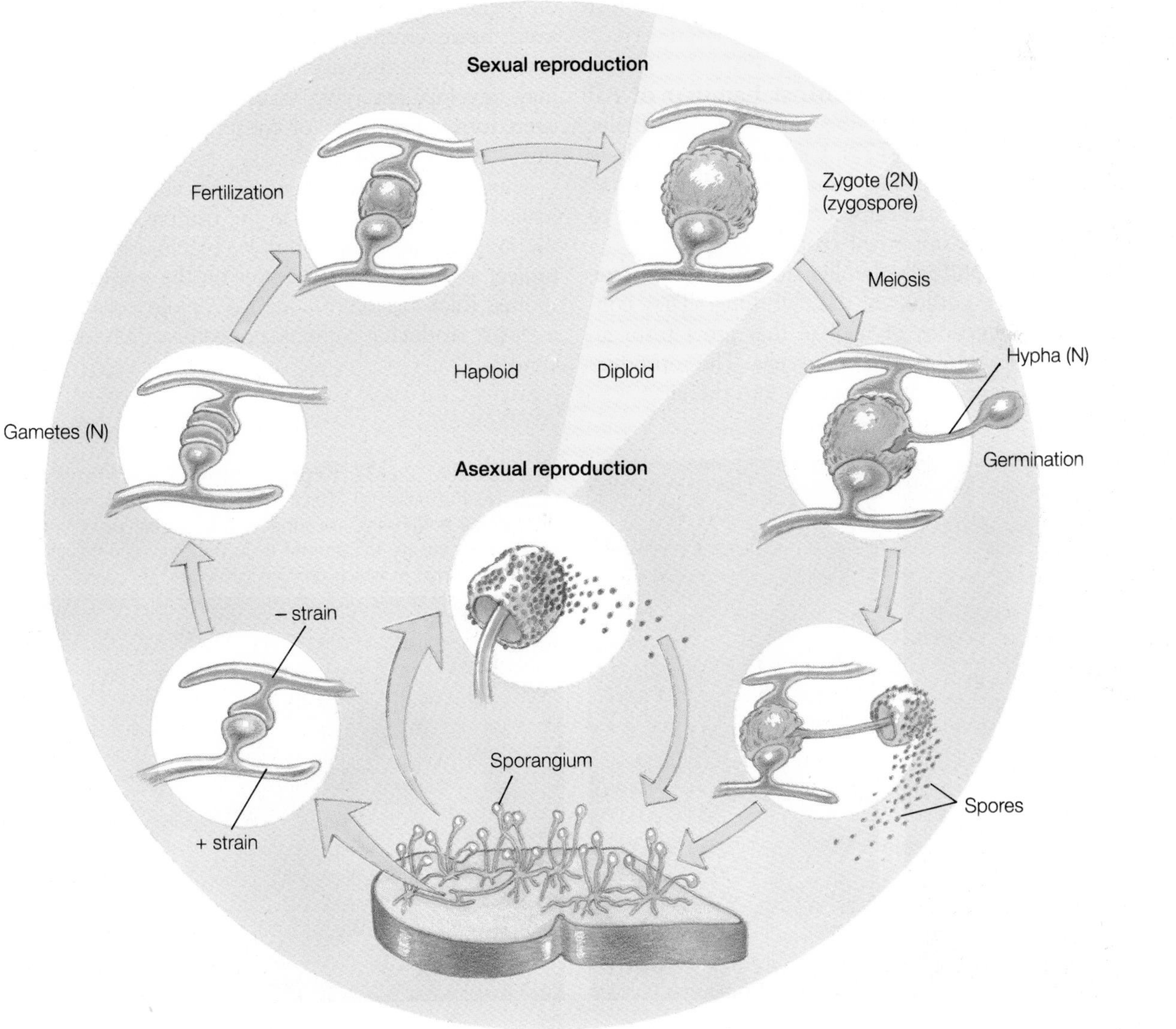

▷ **FIGURE 24–25 Zygote Fungus and Its Reproduction**
(*a*) Black bread mold. (*b*) The two modes of reproduction in the black bread mold fungus. In sexual reproduction, hyphae from two strains (indicated by + and −) fuse and form two haploid (N) gametes that fuse to form a diploid zygote (2N). The zygote or zygospore may remain dormant for several months. Meiosis occurs in the zygote, producing haploid spores that develop into new mycelia.

Sexual reproduction

Fertilization

Zygote (2N)
(zygospore)

Meiosis

Haploid   Diploid

Hypha (N)

Gametes (N)

Germination

Asexual reproduction

− strain

+ strain

Sporangium

Spores

(b)

> **FIGURE 24-26**
> **Representative Sac Fungi**
> (a) Penicilium fungus grow-
> ing on an orange. (b) The mo-
> rel, a fungus that is a delicacy.

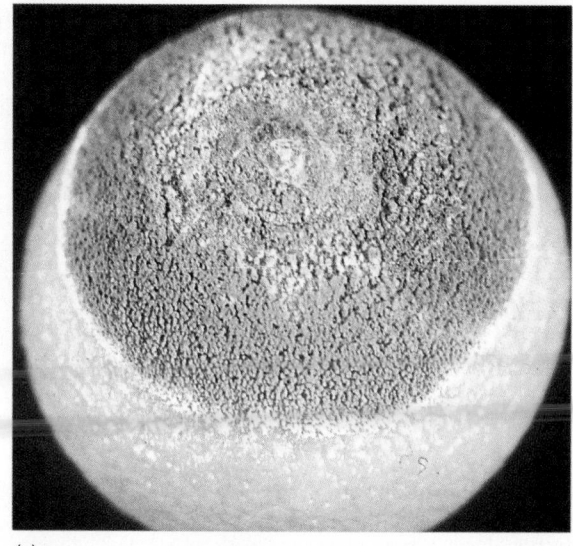

(a)

(b)

down glucose, forming carbon dioxide and alcohol, and are used to produce alcoholic beverages. Yeast is also used to make bread. Added to the dough, yeast breaks down glucose, forming bubbles of carbon dioxide that cause a loaf of bread to rise.

Club Fungi Are Probably the Most Familiar of All Fungi.
The **club fungi** include the mushrooms, puffballs, and shelf fungi (▷ Figure 24–27). Club fungi also include many species that infect plants. Commonly called "rusts" or "smuts," these fungi can cause considerable damage to wheat, corn, and other important crops.

Mushrooms and puffballs, familiar to almost everyone, are the reproductive portions of the club fungi. They consist of dense aggregations of hyphae that grow from an extensive underground network of mycelia. The reproductive structures release billions of spores into the air.

When spores fall on a suitable location, they may begin growing. The spores first give rise to an underground network of mycelia. The mycelia often grow outward, feeding on organic material in the soil. As the mycelia move outward, however, those in the center die as their food supply is depleted. As the mycelia move peripherally, mushrooms may develop, emerging from the ground and sometimes even forming a circle of mushrooms known as a "fairy ring" (▷ Figure 24–28).

Club fungi get their name from club-shaped structures (called *basidia*) that form in the fruiting body (in this case, the mushroom), as shown in ▷ Figure 24–29. The basidia project from gill-like structures on the underside of mushrooms. Each basidium contains diploid cells that undergo meiosis, producing haploid spores that are released into the wind.

> **FIGURE 24–27 Representative Club Fungi** (a) Shelf fungi. (b) Puffballs. (c) Mushrooms. The structures seen here are the fruiting bodies and are supported by an extensive network of mycelia beneath the ground or in the tree. The mycelia digest organic material in which they are growing.

(a)

(b)

(c)

Fungi Enter into Many Intriguing Symbiotic Relationships

Fungi frequently enter into symbiotic relationships with other organisms. One of the best known examples is the **lichen** (▷ Figure 24–30). Lichens consist of two organisms living in a symbiotic relationship that is presumably beneficial to both.[6] They are found on rock, trees, and other substrates. The most visible of the two symbionts in this composite organism is the fungus, usually a sac fungus. Living within the fungus is a unicellular alga or cyanobacterium (a photosynthetic bacterium). In this unique partnership, the photosynthetic algae produce organic food materials for themselves and their fungal host. The fungus "provides" water, minerals, and a secure place to live. Some biologists argue that the association is not mutually beneficial to both organisms. They say that the fungus provides little or nothing to the algae and that algae can live on their own without the fungi. As such, the fungus acts as a parasite—taking from the algae. There are many environments, however, in which algae could not survive

▷ **FIGURE 24–28 Fairy Ring** Mushrooms sprouting from an underground network of mycelia growing outward from a central point where a spore landed and germinated.

alone, among them bare rock. In these environments, the fungus does seem to provide services, like protection from sunlight and drying out.

The lichen is a composite organism that can inhabit harsh environments. That is to say, it is an evolutionary development that has extended the range of living organisms into environments that would otherwise be inhospitable. Interestingly, lichens are the first organisms to colonize

[6]A symbiotic relationship beneficial to both organisms is referred to as a mutualistic relationship.

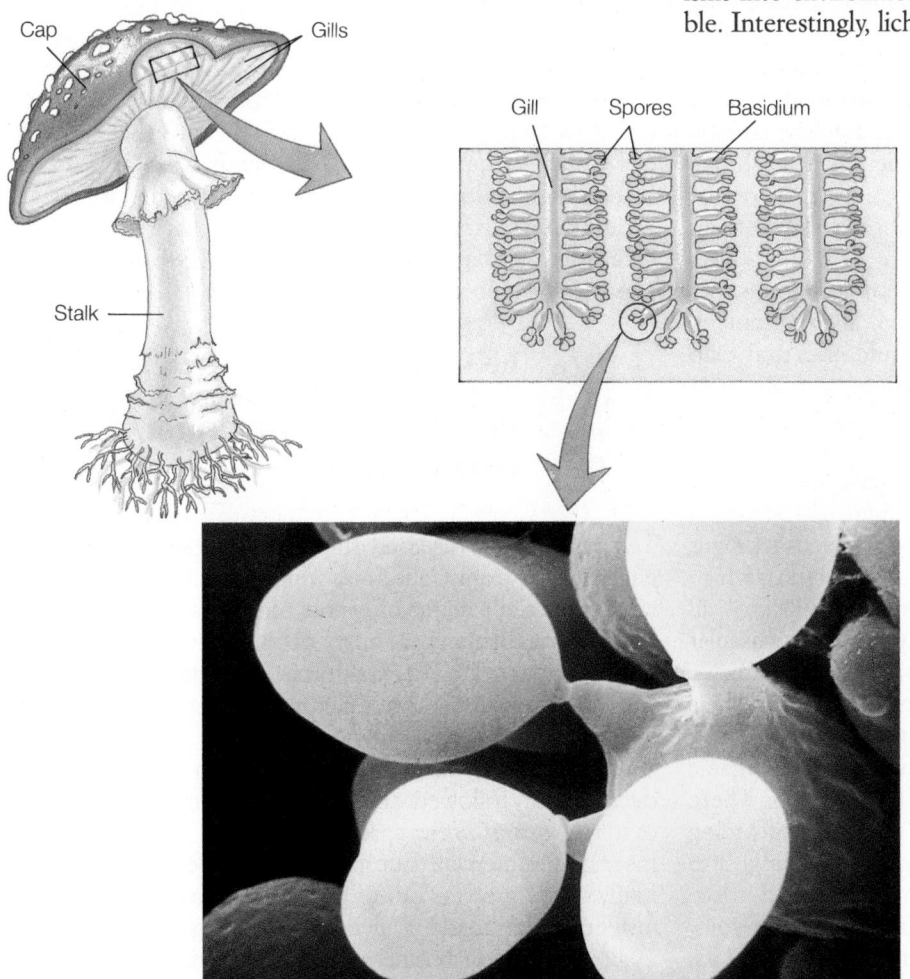

Cap

Gills

Stalk

Gill   Spores   Basidium

▷ **FIGURE 24–29 Club Fungi** These fungi are so named because of the presence of clublike structures that produce spores.

▷ **FIGURE 24–30 Lichens** These composite organisms are actually symbiotic partnerships between fungi and unicellular algae or cyanobacteria. They grow on rocks or bark and live off moisture from rain. The photosynthetic partner synthesizes organic food molecules used by it and the fungus.

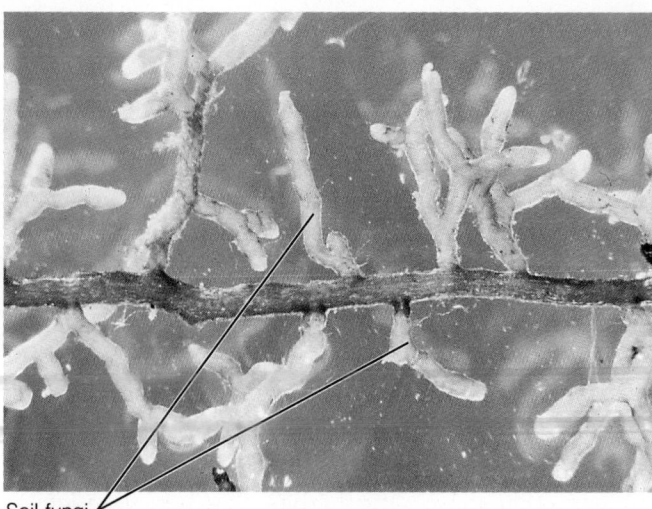

Soil fungi

▷ **FIGURE 24–31 Mycorrhizal Fungi** These soil fungi aid plants in acquiring water and nutrients from the soil. The fungi reap a benefit as well, acquiring nutrients made by the plant's photosynthetic cells.

newly formed volcanic islands or rocks exposed by melting glaciers (Chapter 31). Here, they secrete weak acids that erode the surface of rock, beginning the long, slow process of soil formation. Thus, lichens are essential to the formation of new biotic communities. As testimony to their hardy nature, lichens can also be found in regions as diverse as the frigid Arctic and the hot, arid desert. To date, 25,000 different species of lichens have been identified; each one consists of a unique combination of symbiotic partners. Hardy as they may be, lichens are extremely sensitive to air pollution.

Another noteworthy symbiotic relationship exists between fungi (mostly club and zygote fungi) and the roots of many plants. Over 5000 species of zygote, sac, and club fungi form intimate symbiotic relationships with the root cells of plants. The thin, branching filaments of the mycelium of these fungi often fuse to the root cells. These tiny extensions of the root greatly increase the area for absorption of water and nutrients (▷ Figure 24–31).

The mutually beneficial relationship between a fungus and a root is called a **mycorrhiza** (my-kuh-rize-uh), which literally means "fungus root." Approximately 90% of all trees and most other vascular plants (plants with vascular tissue to transport water and nutrients) have mycorrhizae.

As a rule, plants that participate in such relationships grow larger and grow more quickly than those that do not, especially in poor soils. Why? The fungi digest organic materials in the soil, some of which they absorb. These nutrients can then pass directly into the root, providing nourishment. As noted above, the mycorrhizal fungi also absorb water and minerals dissolved in water found in the soil. These fungi are especially advantageous in arid soils, where soil moisture is low, and in tropical rain forests, where the soils are generally poor.

This relationship also benefits the mycorrhizal fungi. Sugars produced by the leaves of plants are transported to the root system, where they nourish root cells and the symbiotic fungi.

No one knows how the kingdom Fungi evolved. Some scientists believe that it formed from unicellular eukaryotes (protists) lacking chloroplasts. Some believe that the various groups of fungi arose independently from prokaryotes (Monerans). Whatever their lineage, fungi are important organisms vital to the process of decay that returns nutrients to the air and soil, a process essential to the continuation of the web of life.

ENVIRONMENT AND HEALTH: BAD ENVIRONMENTAL PRACTICES MAKE FOR BAD ECONOMICS

Ten years ago a diseased shipment of a species of European oysters from California, which were released into the wild, triggered a series of events that devastated the oyster industry of Brittany, a huge peninsula jutting from the French coastline. The California imports were infected with a protozoan parasite called *Bonami ostrea,* which translates into "good friend of the oyster."

Unfortunately, the parasite is far from a friend. In the decade that followed the introduction of the European oysters, *Bonami ostrea* spread throughout the region, decimating commercial European oyster beds and putting many oyster growers out of business. In fact, 50,000 oyster farmers once made a living in the region; today, only 10,000 remain in business, eking out a meager existence, growing another strain from Japan. In 1978, Brittany harvested 4,000 metric tons of European oysters, but by 1987, the harvest was down to 2,000 tons, and most of these were Japanese oysters. Mature European oysters have become an expensive delicacy. Today, European oysters account for just 10% of Brittany's oyster harvest.

Although the Japanese oyster is apparently immune to *Bonami*, optimal breeding conditions for it are found only in waters south of Brittany where many oyster growers have moved, further affecting the once-lucrative oyster business in the north.

All of this could have been avoided had oyster growers heeded the advice of Henri Grizel, an internationally known mollusc pathologist. Early in the infestation, Grizel warned Brittany's oyster growers not to transfer their oysters from bay to bay, a practice common in the region. The Bretons often move their oysters from one location to another as many as five times before selling them, a practice that speeds up growth and improves the flavor.

Unconvinced of the severity of the situation, the growers refused to listen and insisted on moving the diseased oysters to areas free of the disease, thus facilitating the spread of the protozoan and destroying their own industry.

Help may be on the way. Researchers are now trying to reestablish European oysters in the cold waters of Brittany by reseeding oyster beds with genetically resistant three- and four-year-old oysters that have survived the dreaded protozoan. Apparently immune to *Bonami*, these oysters are being transferred to regions still free of the disease. In time, they could become the basis of a thriving oyster business. Grizel is also breeding oysters that resist the disease in hopes of creating a new generation to harvest in the years to come.

This example illustrates the environmental and economic impacts of introducing a disease organism. Unfortunately, it is only one of many such examples of the devastating effects of species introduction. It should be a lesson to future generations, but few will heed it. Today, officials are proposing to introduce sport fishes into Lake Malawi in Africa. Biologists fear that introductions will devastate the hundreds of unique species that inhabit the lake.

SUMMARY

VIRUSES: CELLULAR PIRATES

1. A virus consists of a nucleic acid core bound by a protein coat, the capsid. Viruses invade cells and take over their metabolic machinery, using it to produce new viral components and often killing the host cell in the process.

2. Viruses infect a wide variety of organisms—from bacteria to humans—but generally have a rather specific host range.

3. Viruses produced in a host cell leave either one at a time via budding or en masse, an exodus caused by the breakdown of the plasma membrane of the host.

THE KINGDOM MONERA: THE BACTERIAL WORLD

4. Monerans inhabit many different environments. They are divided into two groups, the archaebacteria (ancient bacteria) and the eubacteria (true bacteria).

5. Archaebacteria generally live in the most extreme environments and are believed to be the first prokaryotes to have evolved during cellular evolution.

6. The eubacteria are a large and diverse group, containing many different genera.

7. Bacteria contain cytoplasm enveloped by a plasma membrane. Outside the plasma membrane is a thick cell wall.

8. Bacteria usually reproduce by binary fission and acquire nutrients in a variety of ways. Some bacteria are photosynthetic; many are saprophytes, releasing enzymes to digest food materials externally; and a few are chemosynthetic.

9. Bacteria perform many useful functions, such as decomposition and nitrogen fixation, but some bacteria cause human illness and death.

10. The cyanobacteria are photosynthetic eubacteria. They contain photosystems I and II, release oxygen, and rely on water as a source of electrons.

THE KINGDOM PROTISTA

11. The protists are unicellular eukaryotic organisms that generally reproduce asexually by cell division. They consist of three major groups: plantlike, animal-like, and funguslike protists.

12. Plantlike protists are photosynthetic organisms generally known as unicellular algae. Three main types are discussed in this chapter: diatoms, dinoflagellates, and euglenoids.

13. Diatoms are found in fresh and salt water and are encased in elaborate "shells," the outer portions of which are made of silica.

14. Dinoflagellates have two flagella used for propulsion and have rigid cell walls consisting of numerous smaller plates of cellulose. Dinoflagellates are most abundant in salt water.

15. Euglenoids get their name from *Euglena*. Although they contain chloroplasts, euglenoids propel themselves through fresh water with the aid of flagella. They lack rigid outer coatings, unlike most other unicellular algae.

16. Animal-like protists are often called protozoans. All protozoans are unicellular eukaryotic organisms that feed by engulfing their prey. Four phyla are discussed in the chapter: zooflagellates, ciliates, amoeboids, and sporozoans.

17. Zooflagellates move about via flagella. They live in water and in moist soils.

18. Many zooflagellates are free-living organisms; others form symbiotic relationships, living within other organisms in a mutually beneficial relationship. Others are parasitic.

19. The amoeboids are animal-like protists that propel themselves by amoeboid motion. One of the best known is a parasitic species found in warm climates, *Entamoeba histolytica*, which causes amoebic dysentery.

20. Ciliates are the most structurally complex of all protozoans and move through their watery world with the aid of numerous cilia. Perhaps the best known representative of the ciliates are members of the genus *Paramecium*.

21. The sporozoans are parasitic protozoans that live inside the bodies, and sometimes the cells, of their hosts. Sporozoans

produce infectious spores, resistant structures that are passed from an infected animal to a noninfected animal through food, water, and other means.

THE KINGDOM FUNGI

22. Fungi include a wide variety of multicellular organisms, including puffballs, toadstools, mushrooms, and bread molds. The kingdom also contains many unicellular organisms, the yeasts.

23. All multicellular fungi consist of microscopic filaments, hyphae, which often grow together, forming an interwoven mass called a mycelium.

24. Most fungal cells are encased in a rigid cell wall made of chitin. Most fungi are saprophytes. Some fungi are parasitic and feed on living food sources—for example, human skin.

25. Most fungi produce spores, specialized cells that are dispersed by the wind and can develop into new organisms under proper conditions.

26. Three common groups of true fungi were discussed: zygote, sac, and club. They have cell walls, have filamentous bodies, and feed by absorption.

27. The zygote fungi (for example, the black bread molds) are so named because the hyphae fuse, forming diploid structures known as zygospores. These zygospores remain dormant until conditions for growth are met, at which time they undergo meiosis, producing haploid cells again. These cells grow and elongate, forming new hyphae. Hyphae may fuse with those of other strains.

28. Hyphae may also grow upright and form spherical bodies, sporangia, that produce haploid spores that develop into new hyphae. This is asexual reproduction.

29. Sac fungi include common molds that grow on food, single-celled yeasts, and delicacies such as truffles and morels. The sac fungi get their name from the saclike cases in which spores are produced during sexual reproduction.

30. Yeasts are single-celled sac fungi. Most yeasts reproduce asexually by budding.

31. Club fungi include mushrooms, puffballs, and many species that infect plants. The latter, commonly called "rusts" or "smuts," can cause considerable damage to crops.

32. Mushrooms and puffballs are the reproductive portions of the club fungi and consist of dense aggregations of hyphae that grow from an extensive underground network of mycelia. The reproductive structures release billions of spores into the air.

33. Club fungi acquire their name from club-shaped structures (called basidia) that often form in the fruiting body (the visible structure, like the mushroom). Each basidium contains a diploid nucleus that undergoes meiosis, producing haploid spores that are released into the wind.

34. Fungi enter into many symbiotic relationships. For example, hyphae of many soil-dwelling fungi fuse with the roots of many plants and serve as root hairs, increasing the absorption of water and nutrients. This mutually beneficial relationship is called a mycorrhiza (fungus root).

ENVIRONMENT AND HEALTH: BAD ENVIRONMENTAL PRACTICES MAKE FOR BAD ECONOMICS

35. Ten years ago, a protozoan parasite was accidentally introduced into France in a shipment of European oysters from California. The protozoan spread throughout coastal waters of Brittany, killing off European oysters and putting many oyster growers out of business.

36. This economically devastating event could have been avoided had growers not transferred their oysters from bay to bay, a practice that facilitated the spread of the organism.

EXERCISING YOUR CRITICAL THINKING SKILLS

In the United States, zoos are participating in the reintroduction of more than 50 endangered species. They hope that reintroducing members of these species into their native habitats will save populations from extinction. Increasing the numbers of individuals in a population increases the genetic diversity of the population. Maximum genetic diversity is essential for survival because it increases the vigor of the species and allows populations to evolve in response to changes in environmental conditions.

Unfortunately, the zoos where the animals are kept before reintroduction present risks to the animals. The animals are exposed to many foreign parasites and pathogens, which spread easily because of the proximity of hundreds of species and the presence of insects and rodents, which transmit these organisms from one animal to another. In new hosts, previously nonthreatening parasites and disease organisms can decimate populations.

Shortly before the scheduled release of golden lion tamarins, a type of primate, that were being held at the National Zoological Park, it was discovered that 1 of the 11 animals had antibodies to the callitrichid hepatitis virus, commonly known as CHV. Nearly a dozen U.S. zoos have recently reported outbreaks of CHV in

their primate populations. The virus is indigenous to Old World primates, and it apparently spread to the tamarins, New World primates, through the baby mice they were fed.

It is impossible to predict the consequences of introducing CHV into the wilds of Brazil. Because CHV can be transmitted from one animal to another through rodents and is not found in Brazil, it is possible that the virus might spread to other primates and devastate other species. It is also possible that the virus might die out with no effect.

If the decision was yours, would you authorize the release of the disease-free tamarins? Why or why not?

As it turned out, the pathologists who found CHV in the tamarin decided against release. Although animals are quarantined and tested for viruses before they are released, it is not possible to detect all potential viruses. Sophisticated equipment is sometimes necessary, and viruses evolve quickly, making detection even more difficult. The risks are unknown, and the outcomes are uncertain.

In response to these problems, some countries have instituted strict import regulations that essentially prohibit the reintroduction of endangered species. However, researchers believe that reintroduction programs are critical for the survival of endangered species. Supporters point out that other human activities, from agriculture to tourism, present a much greater threat of introducing new viruses.

The problem in this issue and in others is that all the answers are not known. It is critical that the big picture be considered. The reintroduction programs affect not only the target species but also other species and the environment. Do you see any other solutions to this dilemma?

Some argue that the best solution would be on-site conservation efforts—that is, steps to protect species in their native habitat. The animals would be managed and would breed in their natural environments, minimizing the risk of exposure to new viruses. Unfortunately, these programs are costly.

TEST OF TERMS

1. A virus contains a nucleic acid core and a coat of protein called the _____.

2. The enzyme _____ _____ allows retroviruses to synthesize DNA from their own RNA.

3. Viruses are released from a cell one at a time in a process virologists call _____.

4. RNA replication from an RNA template requires a special enzyme called _____ _____.

5. Viral disease can be prevented by _____.

6. Bacteria belong to the kingdom _____.

7. Bacteria divide by _____ _____.

8. Bacteria may "reproduce" sexually via a process called _____ .

9. When environmental conditions become unfavorable, some bacteria produce _____ , resistant structures that house the circular chromosome and a tiny amount of cytoplasm and are encased in a thick cell wall.

10. Some bacteria live in oxygen-free environments and die if exposed to oxygen; these are called _____ anaerobes.

11. Other bacteria do not require oxygen but can survive in its presence. These are called _____ anaerobes.

12. Most bacteria are _____ , organisms that absorb food molecules produced by the enzymatic digestion of large food molecules externally.

13. The _____ are believed to be the oldest of all bacteria.

14. The most abundant photosynthetic bacteria are the _____ , which were once called blue-green algae.

15. Single-celled organisms like diatoms and euglenoids belong to the kingdom _____ .

16. Plantlike protists, or unicellular _____ , are members of a large group known as phytoplankton.

17. _____ are plantlike protists found in fresh and salt water that have unique cell walls. The outer portion of the cell wall is composed of silica, a glasslike material that is often fashioned into exquisite designs.

18. _____ are protists with two flagella, the beating of which causes them to spin like tops.

19. The _____ are single-celled organisms with characteristics of both plants and animals. They get their name from the best-known of their kind, *Euglena.*

20. The animal-like protists are called _____ .

21. _____ are animal-like protists that live in water and in moist soils and contain one, sometimes several, flagella.

22. The _____ _____ of amoeboid protists collects excess water from the cytoplasm. When full, the vacuoles contract, voiding the water through a temporary opening in the plasma membrane.

23. The _____ are the most structurally complex of all animal-like protists. They get their name from the many cilia projecting from their surface. These organisms often acquire food by spearing them with explosive dartlike projections called _____ .

24. _____ are nonmotile, animal-like protists so named because of their ability to produce infectious spores.

25. Multicellular fungi consist of microscopic filaments, or strands, called _____ . These strands grow together, forming an interwoven mass called a(n) _____ .

26. Most fungal cells are enclosed by a rigid cell wall composed of a material called _____ .

27. _____ _____ have distinct cell walls and include single-celled yeasts and the multicellular fungi.

28. _____ _____ are fungi that move about by amoeboid motion, engulfing food from the soil.

29. Fungi frequently enter into symbiotic relationships. One of the best known examples is the _____ , which consists of a fungus and a unicellular alga or cyanobacterium.

30. Some fungi fuse to the root cells of plants and root hairs. This mutually beneficial relationship between a fungus and a root is called a(n) _____ .

Answers to the Test of Terms are found in Appendix B.

1. Describe the structure of a typical virus, how it gets into cells, and how it reproduces.
2. Why are viruses not considered to be alive?
3. In what ways are bacteria similar to human body cells, and in what ways are they different?

4. Describe some of the environmental benefits of bacteria.
5. How do bacteria reproduce, and how do they feed themselves?
6. In what ways are cyanobacteria different from other photosynthetic bacteria?

7. What are the two major types of protists? Describe each one, and give some examples.
8. What are trichocysts, and which protozoans have them?
9. What role do fungi play?
10. Why are fungi called saprophytes?
11. How are the true fungi and slime molds different?

The Plant Kingdom

≈

The decidous trees represent only a small portion of planet's dazzling array of flowering plants.

You wake up in the morning to the sound of your alarm, buzzing obnoxiously. You tramp to the kitchen for a bowl of cereal and a glass of orange juice. You read the morning newspaper, take a deep breath, and head out the door for class. What do all these activities have in common?

They all depend on plants. Prehistoric plants made the coal that is used to generate electricity to run your clock radio. The cereal came from plants, and so, too, did the orange juice. The newspaper was made out of wood pulp from trees, and even the oxygen you breathed came courtesy of plants.

Virtually every organism in the world depends on plants for a host of free services. This chapter will acquaint you with the many and diverse forms of plants that make the planet their home. It will describe a little about their evolution and outline some of their modes of reproduction and life cycles. This information will help you understand the basic nature of plants. The knowledge you gain will help you understand the differences and similarities between plants and other living organisms and will help you appreciate the importance of plants to the Web of Life.

≈ AN OVERVIEW OF THE PLANT KINGDOM

The plant kingdom comprises an amazingly vast assemblage of species, almost as diverse in appearance and life requirements as those found in the animal world. From microscopic algae in our rivers and oceans to towering redwoods in a California forest, plants exhibit an enormous range of size and form. From species that thrive on glacial ice or bare rock to those requiring the heat and humidity of an equatorial climate, plants inhabit almost every place on Earth where light is available in sufficient amounts to support their photosynthetic activities.

Plants Are Autotrophic Organisms with Cell Walls and Chloroplasts

Plants are multicellular organisms characterized by the presence of a cell wall and chloroplasts, two structures missing from the cells of animals. Another essential characteristic of plants is their ability to synthesize their own food. As was noted in Chapter 5, plants and other photosynthetic organisms capture solar energy and use it to synthesize carbohydrates from carbon dioxide and water, which they obtain from their environment. All animals, including humans, depend on plants for their food and oxygen, both products of photosynthesis. As was noted in Chapter 23, scientists believe that our oxygen-rich atmosphere developed as a result of the photosynthetic activities of early plants and that the evolution of higher forms of life would have been impossible in the absence of atmospheric oxygen.

The Plant Kingdom Consists of Many Diverse Groups

The plant kingdom includes a wide range of organisms, among them the common flowering plants that grow along our highways and our gardens; many nonflowering species, such as evergreens; ferns and their relatives, the mosses and liverworts, which live in moist environments, and many algae and seaweeds that make their home in water. As noted previously, these diverse groups have several features in common, among them, chloroplasts and the rigid cell walls that surround each cell.

Angiosperms, commonly called the flowering plants, are clearly the dominant form of vegetation around human habitats. We grow them for the lovely flowers and delicious fruits and vegetables they produce. Angiosperms are the most recently evolved and most widely distributed members of the plant kingdom. These plants produce seeds that are contained within specialized structures known as **fruits.** The seeds are dispersed by a number of agents, such as wind, water and animals, resulting in their extensive distribution within the biosphere.

The **gymnosperms** (pines, spruces, firs, redwoods), which produce seeds in cones, make up the second major group of land plants. Non-seed-producing plants, which constitute another segment of the terrestrial flora, are the ferns, horsetails, and club mosses. Also included in this group are the whisk ferns, mosses, liverworts, and hornworts. The seaweeds and other algae, which are almost exclusively aquatic and much less conspicuous, make up the remaining segment of the plant kingdom.

Plant classification systems have changed over the years. Certain systems group all algae, regardless of their complexity, in the kingdom Protista. In this text, however, only the unicellular algae (diatoms, dinoflagellates, and euglenoids) are considered protists (Chapter 24), and the multicellular algae are included in the plant kingdom.

Plants Evolved in the Seas and then Spread to the Land

The first aquatic plants probably evolved from plantlike protists. Land plants, in turn, probably evolved from green algae living in the intertidal zone. This conclusion is based on studies of fossilized algae in ancient rocks and other fossilized plants. The establishment of plants on land is a relatively recent event on the geologic time scale—a mere 400 million years out of 4.5 billion years of life on Earth. The transition from an aquatic to a terrestrial habitat must have taken place very gradually, with the first land plants getting a foothold at the edge of the water.

The successful establishment of plants on land required several adaptations. These include coverings that protect plants from dehydration and exposure to ultraviolet rays and specialized tissues to transport food, water, and minerals to different parts of the plant as well as to support upright growth. The algae that gave rise to the land plants lacked all of these features.

The ancestors of the first land plants to evolve, for example, the mosses, today straddle the land and water. These and other similar species still have swimming sperm, like aquatic plants, and require water for fertilization. Only the most recently evolved groups of plants, the gymnosperms and angiosperms, have mechanisms to bring about fertilization in the absence of a watery environment.

Evolutionary changes help plants adapt and survive in a changing environment. As in animals, the genetic variation resulting from crossing over during meiosis, mutations, and long-term reproductive isolation is vital to the evolution of new plant species.

Although evolution, in general, has taken place in an aquatic-to-terrestrial direction, some flowering plants, such as the water lily, water hyacinth, and duckweed, have successfully reverted to an aquatic habitat. A similar phenomenon can be found among the marine mammals—notably, the whales—which are believed to have evolved from terrestrial species.

≈ PLANT REPRODUCTION AND LIFE CYCLES

Reproduction is a characteristic of all living things and is vital to the success of all species. Reproduction among plants can be either sexual or asexual. Remarkably, even the most highly evolved plants have retained the ability to reproduce both ways. This ability stands in sharp contrast to higher forms of animals, which reproduce only sexually.

As you may recall from Chapter 21, asexual reproduction results in progeny genetically identical to the parent, whereas sexual reproduction produces offspring that are genetically different from the parents.

Asexual Reproduction in Plants Occurs in Several Different Ways

In its simplest form, asexual reproduction may involve nothing more than a cellular division. In species of multicellular algae, which form long, filamentous strands, it is not unusual for wave action in the oceans or lakes to disrupt the strands. The cells of the resulting fragments divide to form new filaments, identical to the ones from which they broke off.

Asexual reproduction is highly evolved among certain groups of algae. These algae produce reproductive structures known as **spores,** which are either free-swimming or free-floating. Each spore has the potential to develop into a new algal filament. Because a single algal filament can generate numerous spores, population size can increase rapidly within a short period when conditions are optimal.

Asexual reproduction can be as simple as starting a plant from a cutting. This process, called **vegetative propagation,** can be used to grow many house plants such as coleus, impatiens, and African violets. Many varieties of plants are mass produced from pieces of tissue grown in culture, a form of asexual reproduction.

Successful mass production of plants to obtain desired characteristics such as shape, color, flavor, or texture can be achieved only through asexual reproduction. This is the reason why crops such as the potato, for example, are planted as "seed potatoes"—pieces of tuber that contain at least one bud, or "eye". Each bud sprouts to form a new plant genetically identical to the parent. The seeds produced as a result of sexual reproduction are genetically different from the plant that produces them and give rise to offspring that exhibit a wide range of variability.

Sexual Reproduction in Plants Involves the Fusion of Gametes to Form Diploid Organisms

How sexual reproduction evolved among plant ancestors is not known, but it gives certain advantages to plants that utilize it. As noted earlier, sexual reproduction introduces genetic variation into the population. This genetic diversity is essential for natural selection to operate. Thus, sexual reproduction facilitates evolution.

As you learned in Chapter 21, asexual reproduction involves two important processes, fertilization and meiosis. During fertilization, a diploid zygote is produced by the fusion of two haploid gametes, usually a male sperm and a female egg. Meiosis is generally involved in gamete formation. It reduces the number of chromosomes by half and therefore maintains a constant chromosome number from one generation to the next. It also helps ensure genetic variation in the progeny. Without meiosis, the number of chromosomes would double with every successive generation.

Unlike asexual spores, gametes are usually incapable of developing directly into new organisms. Two gametes of the opposite kind have to fuse, and only the resulting zygote can produce a new organism.

Among many algae, sexual reproduction takes place only toward the end of the growing season, with the resulting zygotes developing a protective covering and remaining dormant until the following spring. Interestingly, most green algae are haploid organisms. Therefore, before a new generation of algae can develop from a diploid zygote, the number of chromosomes has to be reduced by half. The diploid zygotes of some algae, however, develop directly into diploid organisms, which give rise to haploid gametes by meiosis. (More on this process later.)

The diploid zygotes of land plants always develop into multicellular embryos, which in turn give rise to the new diploid plants. The embryos of angiosperms and gymnosperms develop inside seeds, which give these evolutionarily advanced plants a distinct survival advantage over early forms. Seeds contain a supply of stored food to aid the development of embryos into seedlings, but they also provide a protective covering, enabling embryos to survive in a dormant state for long periods when environmental conditions are not suitable. Various adaptations for seed dispersal also ensure a wider range of geographic distribution for the progeny of these plants than is possible for species with unprotected embryos.

Several Variations in Sexual Reproduction Are Evident in the Plant Kingdom

The most primitive method of sexual reproduction, which is common among certain algae, is known as **isogamy.** This process is characterized by the fusion of gametes that are morphologically similar (▷ Figure 25–1a). Such algae cannot be distinguished as male and female.

The majority of plant species exhibit a form of sexual reproduction known as **oogamy,** in which a small sperm cell fuses with a much larger egg, resulting in formation of the zygote (Figure 25–1b). The egg cells of all plants are nonmotile, but sperm cells of most plant species, except for angiosperms and some gymnosperms, are motile.

In plants, male and female reproductive organs can be present on separate plants or on the same plant. Maple, mulberry, willow, and ash trees are common unisexual trees. Corn, bean, pea, snapdragon, rose, pine, oak, walnut, and birch have both male and female reproductive organs.

The Life Cycle of All Land Plants and Many Algae Exhibits Two Distinct Generations

Many plants exhibit a phenomenon called **alternation of generations.** That is, during part of their life cycle they exist as a haploid generation that produces gametes, and during another part they exist as a diploid generation that produces spores. The haploid generation is referred to as the **gametophyte,** and the diploid generation is referred to as the **sporophyte.**

▷ **FIGURE 25–1 Sexual Reproduction in Plants** (a) Isogamy and (b) oogamy occur in plants. Both result in the formation of a zygote. Gametes are always haploid, and the zygote is always diploid. Morphologically similar gametes are often motile.

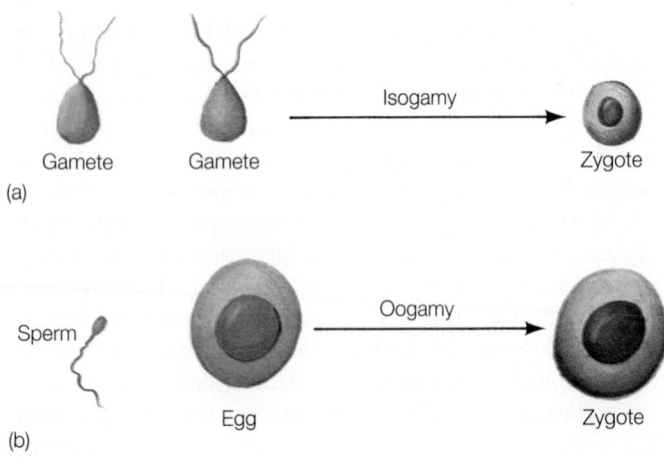

Gamete   Gamete   Isogamy   Zygote

(a)

Sperm   Oogamy

(b)   Egg   Zygote

Fertilization initiates the diploid sporophyte generation, and meiosis initiates the haploid gametophyte generation. The alternation between these two generations is an important characteristic of plant life cycles. In the descriptions of the different life cycles later in this chapter, you will find that in ferns the two generations are completely independent. In mosses, gymnosperms, and angiosperms, however, one generation is dependent on the other. The dominant generation in a life cycle is structurally complex and lives longer.

Although the life cycles of animals do not have a distinct alternation of generations, fertilization and meiosis are still essential for the completion of their life cycles. In short, fertilization and meiosis are common processes for all organisms that reproduce sexually.

Alternation of generations will become more familiar to you as we study the different plant groups, starting with algae, in the following sections of this chapter.

≈ THE MULTICELLULAR ALGAE

The term **alga** (plural, *algae*) is rather inexact and is commonly used to denote a heterogeneous group of aquatic plants. Although most algae are microscopic, some such as the seaweeds, which grow attached to the intertidal rocks along coastlines, are multicellular. The green scum that accumulates on the surface of ponds and lakes is a mass of algae.

Algae are distributed worldwide in both marine and freshwater habitats, with the majority found in oceans. Some of the algae, however, grow in damp terrestrial environments, and a few also grow in a mutually beneficial association with other organisms.

Algae consist of three major groups: green, red, and brown. The classification of algae into these three taxonomic groups is based on the type of photosynthetic pigments and stored food reserves present in each group. Differences in the amount and types of pigments contribute to their characteristic color. These pigments absorb sunlight filtered through the seawater. (For a discussion of a plant pigment of importance to medicine, see Health Note 25–1.

Scientists do not know whether these three groups of algae evolved from a common ancestor or from several different ancestral lines. As you learned earlier, the green algae probably gave rise to the land plants. Brown and red algae, however, represent independent evolutionary lines that did not give rise to new groups of plants.

The Algae Exhibit Considerable Morphological Diversity

Most algae consist of a single cell or a chain of cells attached end to end to form filaments. Some species form colonies in which a few to hundreds of cells stay together, behaving as one organism. Members of certain colonial forms are held together by protoplasmic connections. Only

PLANT PIGMENT MAY PROTECT PEOPLE FROM CARDIOVASCULAR DISEASE

Carotenoids are plant pigments. Ranging from yellow to red, these pigments are found in many fruits and vegetables. To date, over 500 chemically different carotenoids have been discovered. One of the most common is beta carotene.

Beta carotene is responsible for the deep orange color of carrots, mangoes, papayas, and apricots (see ▷ Figure 1). Beta carotene is also found in green leafy vegetables, where its color is masked by the more abundant pigment, chlorophyll.

In the body, beta carotene is converted into vitamin A. Several years ago, the endocrinologist Daniel Steinberg tested the effects of beta carotene in rabbits that were genetically prone to atherosclerosis, a disease briefly discussed in Chapter 2. Steinberg's studies showed that the beta carotene reduced the risk of atherosclerosis by half.

In 1990, a study of the effects of beta carotene on 333 men with signs of atherosclerotic plaque buildup in the arteries of the heart, such as chest pain,

showed similar results. Men taking 50 milligrams of beta carotene every other day for six years suffered half as many strokes and heart attacks as men taking placebos (pills containing sugar).

Researchers do not know how beta carotene reduces the risk of developing atherosclerosis, but some believe that it inhibits the formation of a particularly damaging kind of protein-cholesterol complex, an oxidized low-density

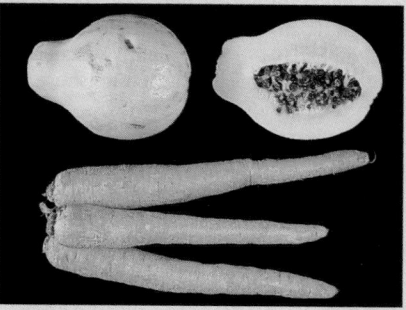

▷ **FIGURE 1 Beta Carotene** Beta carotene is responsible for the color of these fruits.

lipoprotein (LDL) cholesterol. When oxidized, LDL cholesterol is believed to damage cells that line blood vessels. This damage leads to a buildup of cholesterol plaque in the wall of the vessel, eventually blocking blood flow. Beta carotene is a powerful antioxidant and thus may reduce the early damage that leads to plaque formation.

Cholesterol research on humans, although impressive and backed up by animal studies, must be carried out on larger numbers of people. Studies are under way to test the effects of beta carotene on healthy individuals. Researchers want to know if it will keep healthy men and women from developing atherosclerosis in the first place.

The 50-milligram dose used in the 1990 study appears to be completely safe, but researchers worry that individuals suffering from atherosclerosis will view beta carotene as a quick fix. As a result, many may forgo changing important risk factors, such as cigarette smoking or high-fat diets, which also improve one's chances of avoiding stroke and heart attacks.

---

a few groups of algae have cells arranged to give a tissue-like appearance. The filamentous and tissuelike forms are either free-floating or attached to intertidal rocks and other substrates, and the colonial forms are either free-floating or motile.

Algae Are Vital Producers in Aquatic Food Chains

Algae synthesize their own food and along with the photosynthetic protists are the primary producers in virtually all aquatic food chains. These organisms' contribution to the total global photosynthetic output is very significant. All aquatic organisms that do not synthesize their own food depend on algae, directly or indirectly, for their energy. Algae are also a source of oxygen for aquatic organisms. A few species of seaweeds provide food for humans. Pollution of our lakes, rivers, and oceans by toxic chemicals that harm the normal growth of algae can have grave repercussions for fish populations. In recent years, oil spills at sea have adversely affected the marine algae that float near the surface of the ocean in large numbers. Some scientists fear

that death of these microscopic plants could lead to the collapse of entire aquatic food chains.

Three Types of Life Cycles Are Known Among Algae, Based on the Chromosome Number of the Free-Living Forms

▷ Figure 25–2 illustrates the three basic life cycles of algae. This section will introduce them, and more detailed examples will be presented in the next sections.

Most green algae have a haploid, free-living phase (Figure 25–2a). These organisms are called gametophytes because they produce haploid gametes. The gametes combine to form diploid zygotes, which constitute the 2n, or diploid, phase of the algal life cycle. Interestingly, though, the zygotes of these algae do not develop into full-fledged organisms as in animals. Instead, they undergo meiosis themselves, producing haploid spores that develop into free-living haploid algae (gametophytes).

The second life cycle consists of two free-living forms, the diploid sporophyte and the haploid gametophyte (Fig-

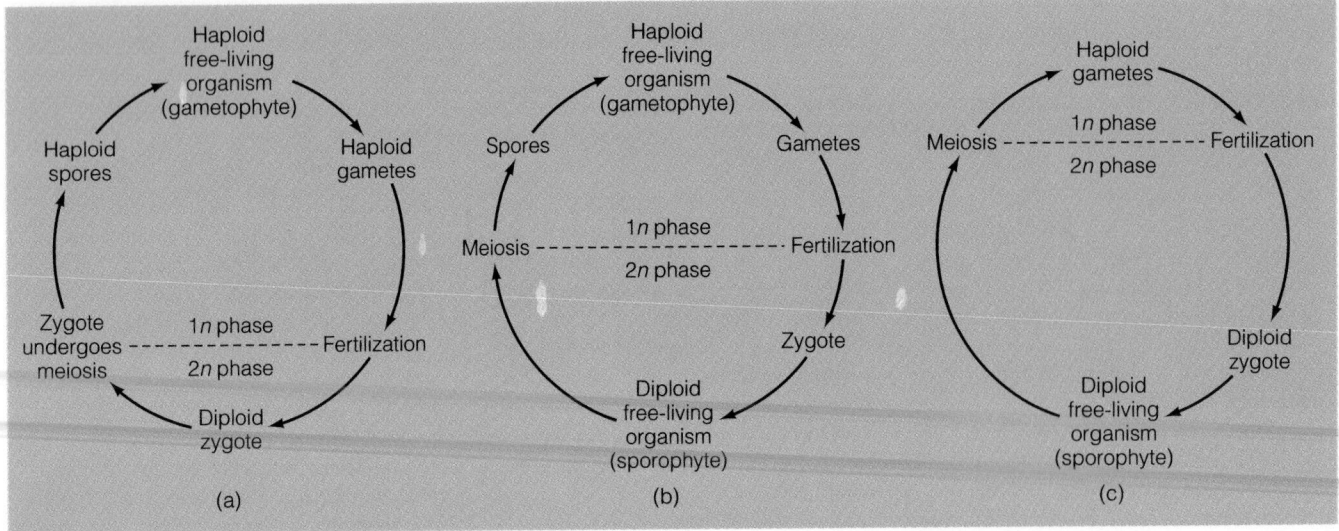

FIGURE 25–2 Algal Life Cycles (*a*) Most green algae such as *Spirogyra* and *Oedogonium* have a life cycle in which the free-living phase is haploid. (*b*) *Ulva*, another green alga, is an example that has a life cycle in which both haploid and diploid phases are free-living. (*c*) This life cycle is typical of the brown alga *Fucus* in which the free-living phase is diploid. Due to the consid-erable variation in life cycles among members of each algal division, it is not possible to make broad generalizations about life cycles of any single algal division. A group of land plants known as the early vascular plants (ferns and the like) has a life cycle similar to the one shown in (*b*).

ure 25–2b). Both forms are conspicuous, but they have very different roles to play. The diploid sporophyte, for example, gives rise to spores that form during meiosis. The haploid spores develop into free-living gametophytes. The gametophytes give rise to gametes that combine to form diploid sporophytes. This type of reproduction is found in some multicellular algae.

The third life cycle is very similar to sexual reproduction in vertebrates. It consists of two forms, a free-living diploid organism, the sporophyte, and haploid gametes. The mature sporophyte gives rise to gametes via meiosis. These gametes combine to form a diploid zygote, which then develops into free-living sporophyte (Figure 25–2c). The third type is found in some of the more structurally advanced algae.

The Green Algae Exhibit Many Different Forms of Reproduction

The green algae are members of the division **Chlorophyta.** A division is a taxonomic unit equivalent to the phylum in the animal kingdom. The grass-green color of these algae is due to the presence of photosynthetic pigments identical to those in land plants. The green algae also store starch as their reserve food, as land plants do, suggesting an evolutionary relationship to the latter. Although green algae are found in all aquatic habitats, the great majority of them are freshwater forms, such as *Spirogyra* and *Oedogonium*. In these species, many cells are attached end to end to form filaments. Both unbranched and branched filaments are common among the green algae.

Different types of sexual reproduction are present in the members of Chlorophyta. However, the majority of green algae, such as *Spirogyra* and *Oedogonium,* rely on the first strategy outlined earlier. That is, they have a haploid free-living gametophyte, with the zygote being the only diploid structure in their life cycle. As shown in ▷ Figure 25–3, the gametophyte of *Oedogonium* produces haploid eggs and sperm. A sperm that is released into the water swims to an egg. After a period of dormancy, the zygote undergoes meiosis and produces spores that develop into a new generation of free-living haploid gametophytes. Individual swimming spores produced during asexual reproduction may also form new filaments on their own.

Ulva, another marine green alga, commonly called sea lettuce because of its similarity to common leaf lettuce, illustrates the second type of life cycle. As illustrated in ▷ Figure 25–4a, free-living gametophyte (haploid) and free-living sporophyte (diploid) generations are present in its life cycle. In this case, the haploid gametophyte generation gives rise to haploid gametes. They combine to form a zygote that gives rise to a free-living diploid sporophyte. It, in turn, produces haploid spores via meiosis that develop into haploid gametophytes.

The marine alga *Acetabularia* illustrates the third type of life cycle. As shown in Figure 25–4b, the free-living form (mature sporophyte) is diploid. This structure produces haploid gametes via meiosis. The gametes combine to form a diploid zygote, which develops into a mature sporophyte.

Brown Algae are Highly Evolved Forms Displaying Considerable Differentiation

All brown algae are in the division **Phaeophyta** and are the most common species of algae along the northeastern coast of the United States. In addition to the green pigment

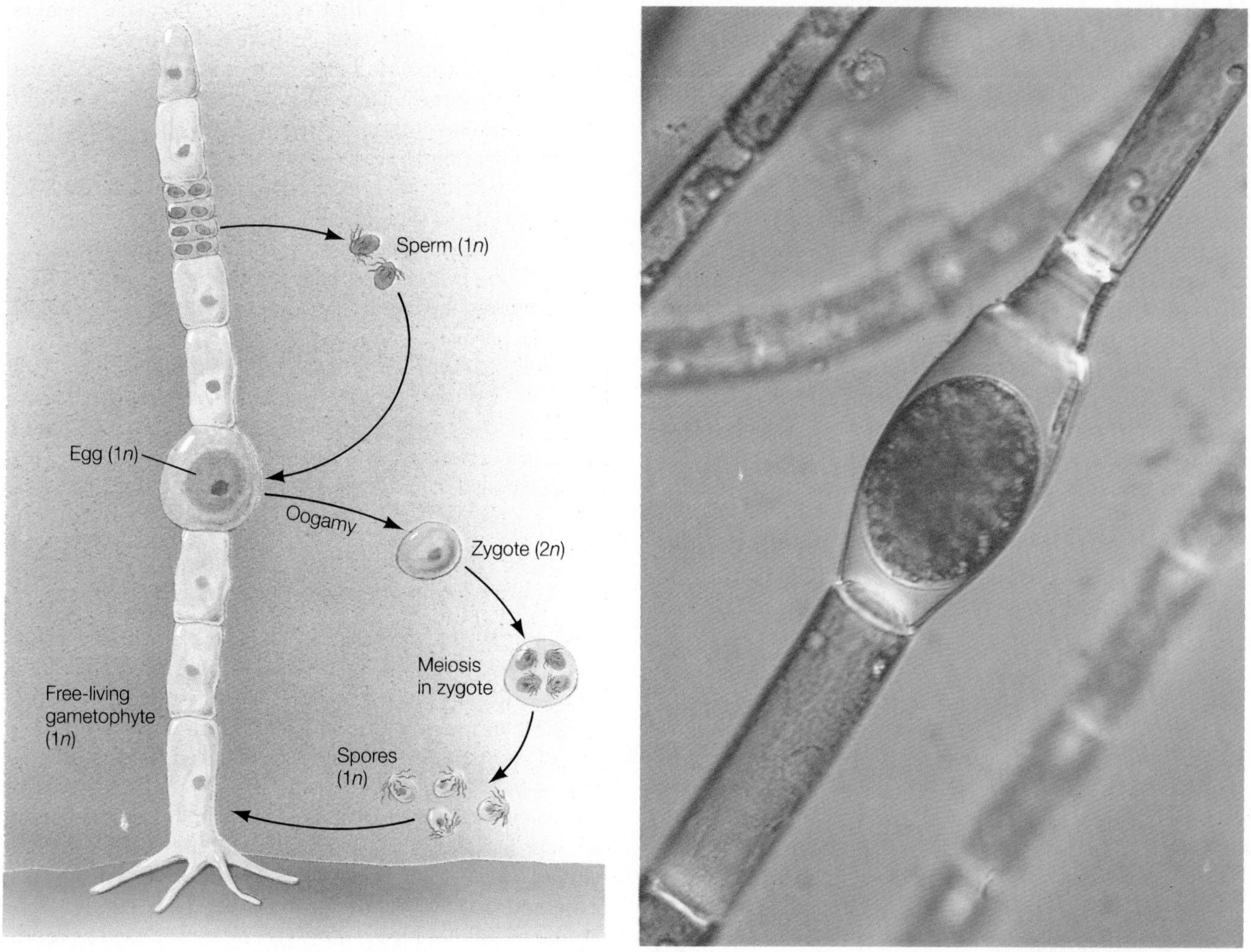

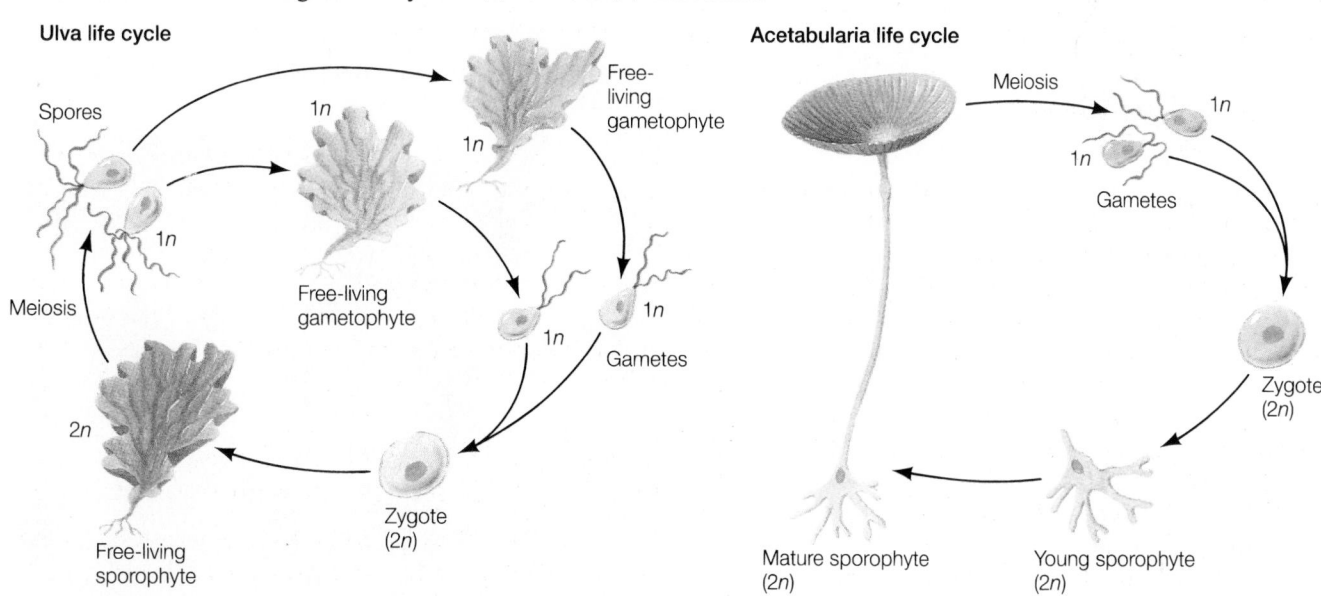

(a)

(b)

▷ **FIGURE 25–3 Oedogonium** (*a*) The life cycle of this green alga. (*b*) Photomicrograph of *Oedogonium* filaments showing sperm produced in short cylindrical cells and eggs in large cells.

▷ **FIGURE 25–4 Two Algal Life Cycles.** (*a*) *Ulva* and (*b*) *Acetabularia*.

chlorophyll, the brown algae contain large amounts of two yellow photosynthetic pigments, which are responsible for their color. Brown algae are almost exclusively marine and usually grow in the intertidal zone in colder waters. Most species—such as *Laminaria,* or kelp (▷Figure 25–5), *Fucus,* or rockweed (▷Figure 25–6), and *Macrocystis,* or Pacific kelp—grow attached to the bottom or to intertidal rocks. The intertidal species remain submerged during high tide but are exposed when the tide is low. A gelatinous covering keeps them from dehydrating when they are exposed to the air. A few species, such as *Sargassum,* are free-floating. An extensive area of the Atlantic Ocean near the West Indies is known as the Sargasso Sea because of the high density of *Sargassum* in that area.

Brown algae are highly evolved forms, displaying considerable tissue differentiation. Air bladders, commonly found by beach goers on shore along the U.S. West Coast, serve as flotation devices. Another region, the holdfast, anchors the alga to a solid substrate. A short stalk connects to the holdfast and an expanded leaflike blade. The Pacific forms of kelp are the largest algae known, attaining lengths of 150 feet or more and forming extensive stands of underwater "forests" covering large areas of the ocean.

In the life cycle of rockweed *(Fucus),* the conspicuous free-living stage is diploid, and only the gametes represent the haploid phase. (This is the third life cycle shown in Figure 25–2.) In contrast, kelp show a distinct alternation of generations between a large sporophyte generation and a microscopic filamentous gametophyte generation.

Brown algae contain considerable amounts of algin, a gelatinous secretion present on the surface of the plants. Algin is used as a stabilizer and moisturizer in ice cream, mayonnaise, frostings, and other products. Several species of kelp are also valuable as a source of iodine.

Red Algae Are Predominantly Filamentous and Live Almost Exclusively in Warm Marine Waters

Members of the red algae, classified in the division **Rhodophyta,** are mostly found in warm, marine habitats. Red algae usually grow attached to a substrate and seldom survive when free-floating. The characteristic color of these algae is due to the predominance of a red photosynthetic pigment.

Most species of red algae are highly branched filaments; some are flat and expanded, and a few are unicellular and colonial. Sexual reproduction is highly evolved in this group, and oogamy is the general rule. The red alga *Porphyra,* shown in ▷ Figure 25–7, is commonly used as food, especially in parts of the Orient. The food stabilizer carrageenin is obtained from certain species of red algae. The culture medium agar, which is used to grow bacteria and fungi in the laboratory, is obtained from the red alga *Gelidium.* Agar becomes a liquid when heated but turns solid

▷ **FIGURE 25–5 Brown Alga** *Laminaria* (kelp) showing rootlike holdfast, stalk, and expanded leaflike blade.

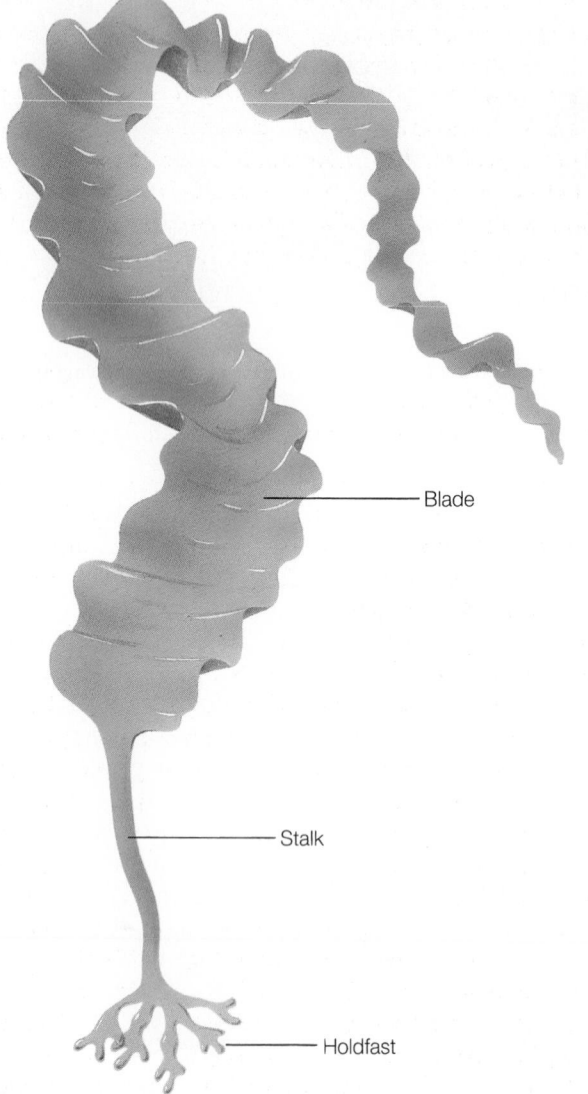

Blade

Stalk

Holdfast

▷ **FIGURE 25–6 *Fucus* (Rockweed)** This is a species of brown algae, which tend to live in cool marine waters.

▷ **FIGURE 25–7 Red Alga** Photomicrograph of *Porphyra*, a species of red algae that lives in warm marine waters.

▷ **FIGURE 25–8 Moss** The sporophytes, represented by the stalk and capsule, are attached to the gametophytes, which have small, leaflike expansions.

when cooled to room temperature, even after being heated to a very high temperature. This special property allows agar to be sterilized by heat. The agar medium that is used for making a throat culture in a pathology laboratory has nutrients added to it. The nutrients provide the energy, and the agar serves as a substrate for the growth of the bacteria.

≈ THE FIRST LAND PLANTS: THE NONVASCULAR PLANTS

The majority of land plants are structurally more complex than the algae, with plant bodies composed of tissues organized into roots, stems, and leaves. Land plants further differ from algae in having a protective outer covering, the **epidermis,** and an upright pattern of growth. The evolution of such structural features made it possible for the descendants of aquatic plants to establish themselves on land. The development of the zygote into a multicellular embryo, retained within the female reproductive structure, is common in land plants but rarely seen in algae. In all land plants, except bryophytes, the diploid sporophyte generation, rather than the haploid gametophyte generation, is the dominant phase in the life cycle.

The land plants are divided into two broad categories, **nonvascular plants** and **vascular plants,** based on the absence or presence of tissues for the transport of food and water, which are discussed in Chapter 26. All nonvascular land plants are classified under one division, Bryophyta. Besides lacking vascular tissues, nonvascular plants, such as the mosses and liverworts, also lack roots, stems, and leaves. The vascular plants are classified under several divisions.

Bryophytes Are Adapted to Live in Moist Environments

The division **Bryophyta** includes such plants as mosses

(▷ Figure 25–8), liverworts, and hornworts. Bryophytes are relatively small plants, because the absence of vascular tissues severely restricts water transport to distant tissues. Most bryophytes require very moist habitats, with some species, such as *Sphagnum,* living exclusively in boggy areas. The sperm of bryophytes are motile and must swim to the eggs in order to fertilize them. As a result, a watery environment is needed for successful propagation. As a rule, bryophytes are of no special economic value, except for one species, *Sphagnum,* which is a source of peat moss that is commonly used as a soil conditioner in gardens and flower beds.

Mosses Reproduce Sexually, and the Life Cycle Consists of Two Anatomically Distinct Generations

The haploid gametophyte generation is the dominant phase in the moss life cycle. The gametophyte consists of a short, upright plant with a stemlike stalk and leaflike outgrowths. The plant is anchored to a substrate by hairlike structures that also function to absorb water and minerals.

Gametes are produced in two anatomically distinct structures that form at the tips of mature gametophytes. Motile sperm are produced in club-shaped **antheridia** (singular, *antheridium*), shown in Figure 25–9a. The eggs form in flask-shaped structures known as **archegonia** (singular, *archegonium*), illustrated in Figure 25–9b. Sperm swim into the archegonium to reach the egg, and after fertilization the zygote remains in the archegonium, where it develops into a multicellular embryo. The embryo then develops into a sporophyte. The sporophyte forms a stalk and capsule that grows into the gametophyte tissue (Figure 25–8).

The diploid stalk and capsule—that is, the sporophyte—remain attached to the gametophyte. In fact, the sporophyte cannot survive on its own if removed from the gametophyte. When the sporophyte matures, haploid

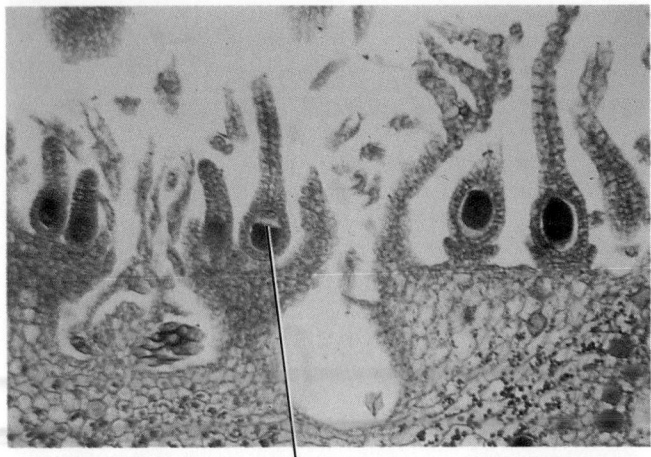

(a)   Archegonium

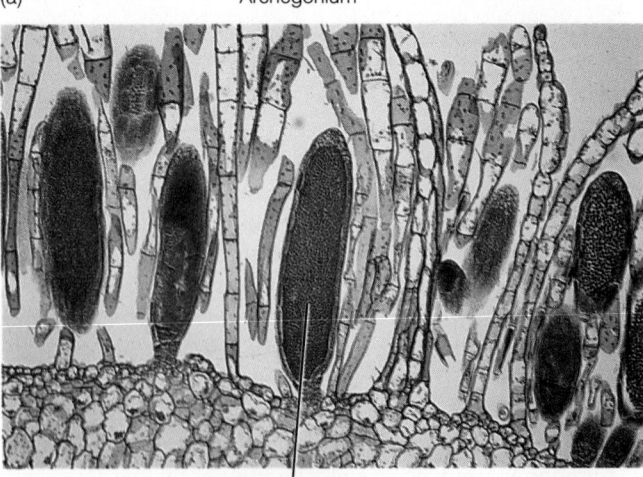

(b)   Antheridium

▷ **FIGURE 25–9 Moss Gamete Production** (*a*) Photomicrograph of tip of a female moss plant showing flask-shaped archegonia. (*b*) A male plant showing club-shaped antheridia. Each archegonium contains one egg, and each antheridium contains many motile sperm.

spores are produced within the capsule by meiosis and are dispersed by wind. On moist soil, each haploid spore germinates into a branched filamentous structure known as the **protonema,** shown in ▷ Figure 25–10. Within a few days many new gametophyte plants emerge from small buds that develop on the protonema (▷ Figure 25–11), thus completing the life cycle, which is summarized in ▷ Figure 25–12. Other bryophytes, such as the liverworts and hornworts, have life cycles very similar to that of a moss.

Liverworts and Hornworts Are Less Common Bryophytes

Liverworts got their name from the resemblance of gametophytes produced by certain members of this group to the human liver. For example, one common species, known as *Marchantia,* consists of several expanded, liver-shaped lobes joined to form the gametophyte plant (▷ Figure 25–13). The antheridia and archegonia of such plants are grouped

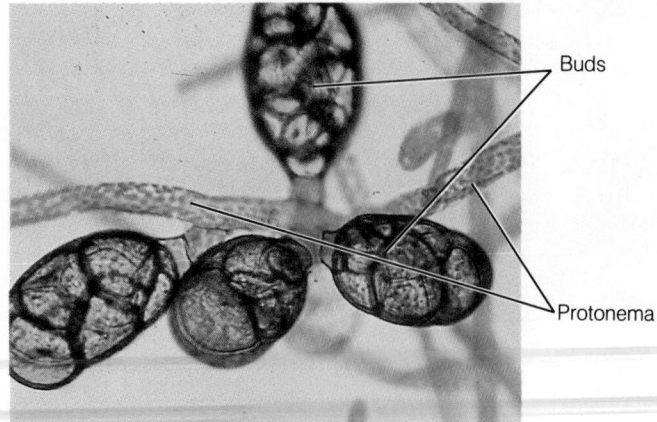

▷ **FIGURE 25–10 Moss Protonema** A moss spore develops into a protonema, a highly branched, filamentous structure that produces many buds.

together separately on stalked, upright branches. The gametophytes of hornworts resemble those of *Marchantia.*

≈ THE VASCULAR PLANTS

All vascular plants contain two types of vascular tissue, **xylem** for the transport of water and minerals and **phloem** for the transport of food. The evolution of specialized conductive tissues gave vascular plants a decided selective advantage over bryophytes, allowing them to grow much larger and to inhabit a much more varied range of habitats.

Among the vascular plants, the dominant phase in the life cycle is the sporophyte, which consists of stems, leaves, and roots. Some of these sporophytes have the potential to live for hundreds of years, as in the case of angiosperm and gymnosperm trees.

The gametophyte, on the other hand, is microscopic and short-lived, composed of a few cells and without vascular tissues. Because of their small size, these gametophytes are not easily recognized as plants.

▷ **FIGURE 25–11 Moss Gametophyte** The gametophyte plant develops from a bud on the protonema.

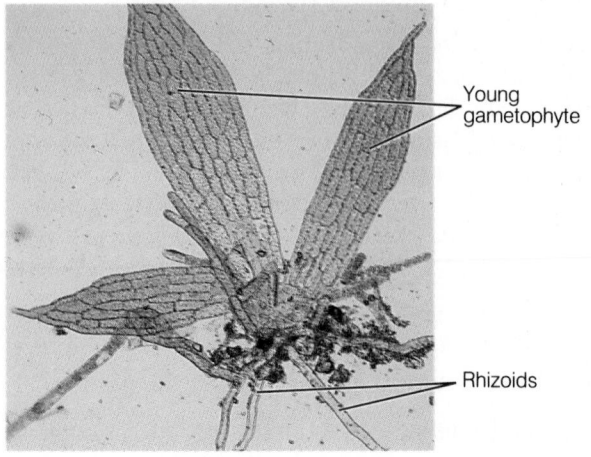

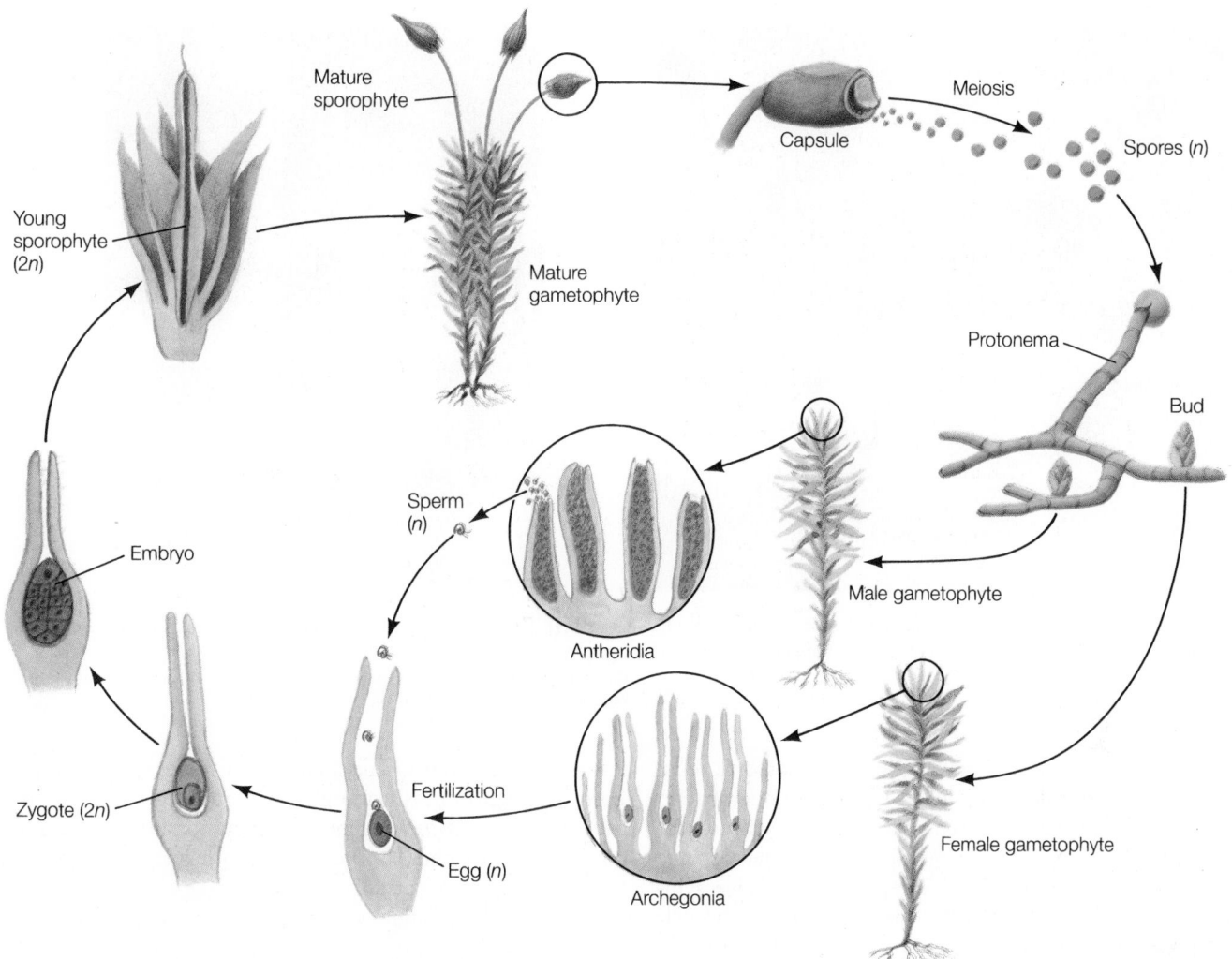

Labels on figure:
- Mature sporophyte
- Capsule
- Meiosis
- Spores (n)
- Young sporophyte (2n)
- Mature gametophyte
- Protonema
- Bud
- Sperm (n)
- Antheridia
- Male gametophyte
- Embryo
- Zygote (2n)
- Fertilization
- Egg (n)
- Archegonia
- Female gametophyte

▶ **FIGURE 25–12 Moss Life Cycle** Germination of a spore produces the protonema, the first stage in the gametophyte generation. Eventually, buds form on the protonema and produce the male and female gametophytes. Sperm produced in antheridia of the male gametophyte swim through water to reach an egg in the archegonium of the female gametophyte. After fertilization, the zygote develops into an embryo, beginning the sporophyte generation. The mature sporophyte consists of a stalk and a capsule in which spores are produced by meiosis. Spores released from the capsule continue the cycle.

Based on the presence or absence of certain specialized structures produced during their life cycle, the vascular plants are commonly divided into two major groups, early vascular plants and advanced vascular plants.

The Early Vascular Plants Are Generally Restricted to Moist Environments

The **early vascular plants** are characterized by several common features: the absence of seeds, the need for water to achieve fertilization, and the presence of a highly developed underground stem, known as a **rhizome.** When the aboveground portions of the early vascular plants die during the winter months, the rhizome survives, giving rise to new growth in the spring.

All early vascular plants produce morphologically distinct sporophyte and gametophyte plants, which remain independent of each other. The similarity exhibited in the life cycles of these plants is indicative of the close relationship between the different groups.

The early vascular plants are classified under four divisions: Psilophyta (whisk ferns), Microphyllophyta (club mosses), Arthrophyta (horsetails), and Pterophyta (true ferns). These divisions constitute a relatively minor proportion of the vascular plants. Their members are found in moist, shady habitats—for example, along river banks and in wooded areas.

Whisk Ferns. *Psilotum,* one of two surviving genera (singular, *genus*) of whisk ferns, is the most primitive of all the vascular plants (▷ Figure 25–14). Its evolutionary significance is its close resemblance to an extinct vascular plant known from fossil records. The sporophyte generation of *Psilotum* consists of a green, characteristically branched stem with minute, scalelike protuberances and tri-lobed, spore-producing structures known as sporangia. In the ab-

▷ **FIGURE 25–13 Liverwort** *Marchantia*, with a liver-shaped, lobed plant body. Liverworts are one of several types of bryophytes—nonvascular plants.

▷ **FIGURE 25–15 Club Moss** This species, *Lycopodium*, has upright branches and an underground rhizome (not shown). The conelike structures at the tip are clusters of sporangia. It is an early vascular plant.

sence of leaves, the stem itself carries out photosynthesis. Fine, rootlike extensions produced from the underground rhizome absorb water and minerals. Haploid spores produced in the sporangia develop into almost microscopic gametophyte plants, which are independent and live symbiotically with soil fungi. Other groups of early vascular plants, such as the club mosses, horsetails, and true ferns, have similar life cycles.

Club Mosses and Horsetails. The division **Microphyllophyta** includes a number of early vascular plants commonly called club mosses. The sporophytes of these plants seldom grow taller than a foot and have many small, scalelike

leaves on their stems. The stems of one species known as *Lycopodium* grow mostly upright (▷ Figure 25–15), whereas the stems of *Selaginella* generally grow horizontally with a few upright branches. Spore-producing leaves are grouped together in conelike arrangements at the tips of the upright branches, and sporangia are produced on them. The club mosses also have well-developed underground rhizomes

▷ **FIGURE 25–16 Horsetail** *Equisetum*, another early vascular plant, has upright branches that are produced from the rhizome, with leaves clustered around the joints of the stem and terminal cone. In this species, the fertile shoot lacks chlorophyll and appears quite different from the vegetative shoot (right).

▷ **FIGURE 25–14 Whisk Fern** This plant (*Psilotum*) is composed entirely of stem tissue. The aboveground portion, the stem, shows a typical branching pattern. This is an early vascular plant.

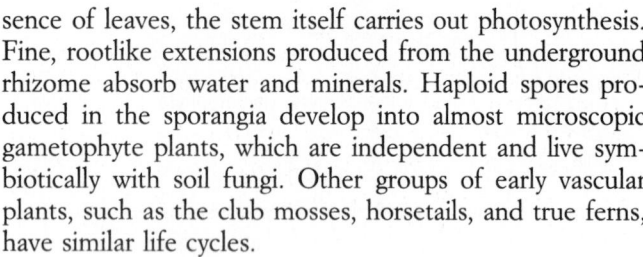

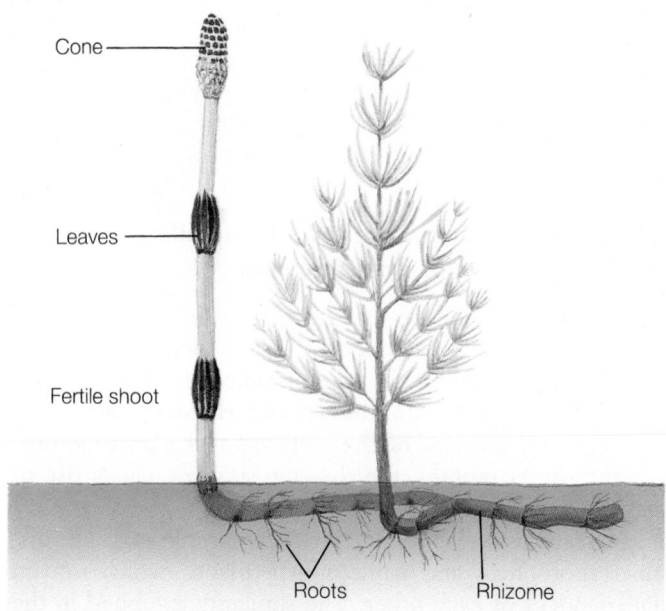

Cone

Leaves

Fertile shoot

Roots   Rhizome

> **FIGURE 25–17 True Fern** The fern sporophyte shown here consists mostly of leaves, commonly called fronds. Ferns are early vascular plants.

with true roots for absorption. Gametophytes developed from haploid spores are very small, but independent.

Horsetails *(Equisetum),* which may grow to a height of several feet, have segmented stems with small, scalelike leaves arranged where the stem segments are joined (▷ Figure 25–16). The sporangia of horsetails are also arranged in conelike structures at the end of the stems. Because of the high concentration of silica in the cell walls of horsetails, these plants were used by early American settlers for scouring pots and pans.

Ferns. The true **ferns** comprise a heterogeneous group of plants with a wide distribution, including a few aquatic

species. Typically, ferns grow in moist, shady places. What most people call a fern is the sporophyte plant and consists of a well-developed rhizome from which roots and leaves are produced. The aboveground portion of a fern plant is predominantly leaves, commonly called fronds (▷ Figure 25–17). The unopened young leaves resemble the head of a violin, hence their common name of "fiddlehead." Some tropical ferns contain an unbranched, upright stem that can grow 20 feet or more, at the end of which is a crown of leaves.

Stalked sporangia are produced in groups on the undersurface of leaves of ferns in a variety of patterns that vary among species (▷ Figure 25–18). Meiosis takes place in specialized cells within the sporangia, and the resulting haploid spores spring from the sporangia when they break open. A spore landing on moist soil develops into a small, independent, heart-shaped gametophyte plant known as the **prothallus** (▷ Figures 25–19 and ▷ 25–20). Egg- and sperm-producing archegonia and antheridia develop on the lower surface of the prothallus.

A watery environment is essential for the motile sperm to reach the egg in the archegonium. After fertilization, the zygote remains in the archegonium and develops into an embryo and then into the new sporophyte plant to complete the life cycle (Figure 25–19).

Virtually all living members of the early vascular plants are relatively small and restricted in their distribution. However, they are well adapted to the moist habitats where they are found. The gametophytes of these plants are short-lived. Water from periodic rains prevents these nonvascular gametophytes from drying up and allows them to complete growth and reproduction within a very short period.

> **FIGURE 25–18 Fern Sporangia.** Sporangia of ferns are produced in various patterns on the lower surface of fern leaves.

The circular structures seen on the leaves are groups of sporangia. Three different patterns are shown here.

(a)

(b)

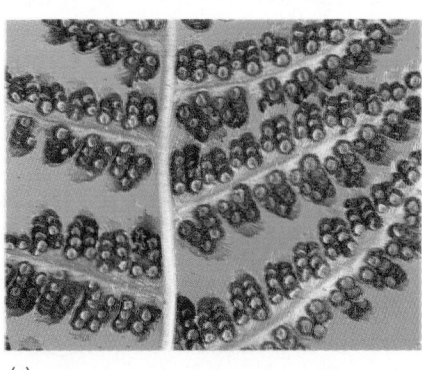

(c)

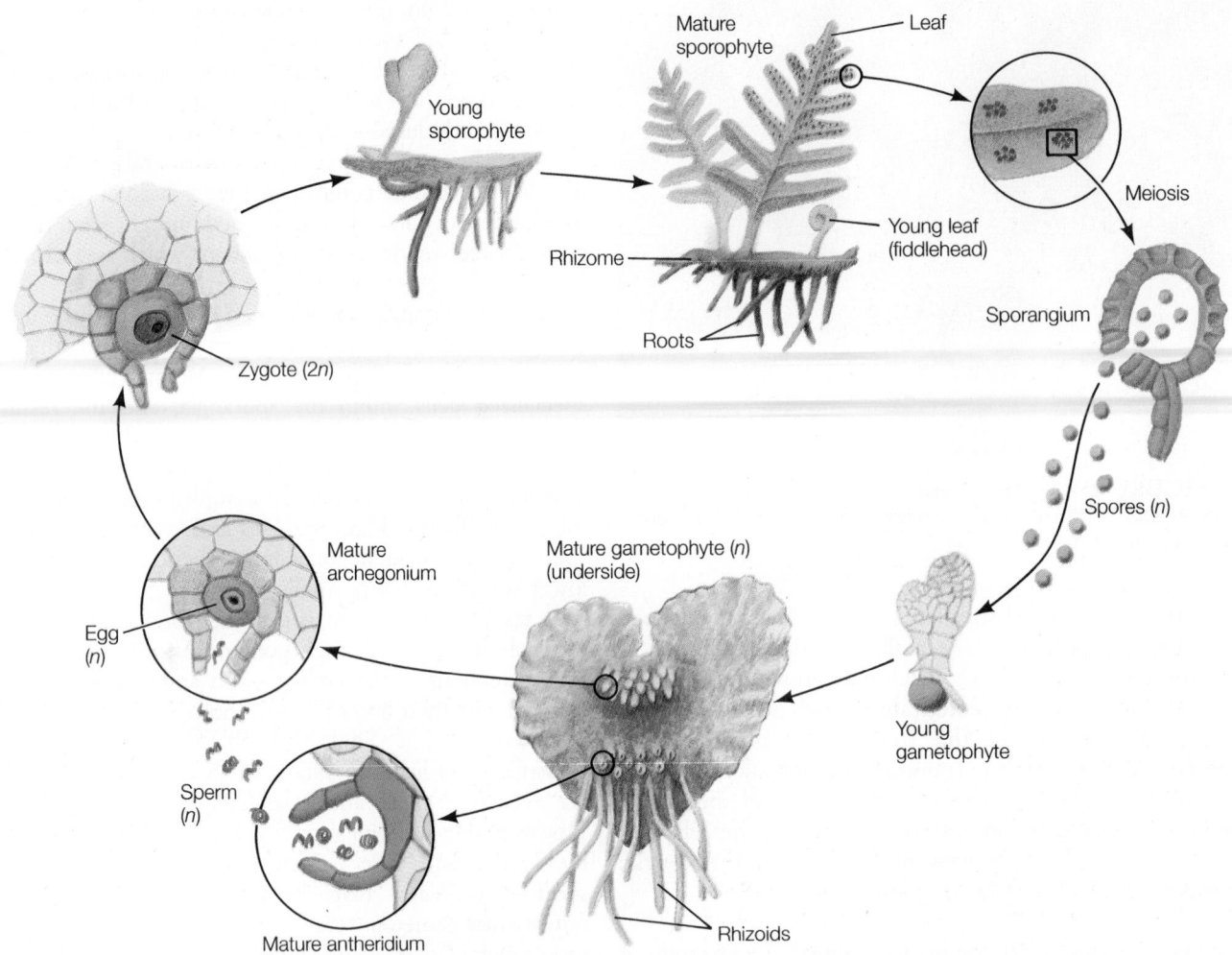

Mature sporophyte
Leaf
Young sporophyte
Young leaf (fiddlehead)
Rhizome
Roots
Meiosis
Sporangium
Zygote (2n)
Spores (n)
Mature archegonium
Mature gametophyte (n) (underside)
Egg (n)
Young gametophyte
Sperm (n)
Mature antheridium
Rhizoids

▷ **FIGURE 25–19 Fern Life Cycle** Fronds of the mature sporophyte (upper right) produce sporangia. Haploid spores develop in the sporangia by meiosis. Germination of the spore gives rise to a gametophyte plant called the prothallus on which the antheridia and archegonia form. Sperm produced in the antheridia swim to the egg in an archegonium. The resulting zygote becomes the embryo, which gives rise to the young sporophyte and, finally, the mature sporophyte.

According to fossil records from the Carboniferous Period (300 million years ago), members of the early vascular plants were the major component of the flora at that time, and they grew as large trees in swampy areas, unlike their descendants. Evidence attesting to the ecological dominance of early vascular plants during that period can be seen in the vast deposits of coal found in many parts of the world that were once covered by tropical marshes. Present-day coal deposits are believed to be the fossilized remains of the extinct members of the early vascular plants. Coal is formed when undecomposed plant material is flooded and buried under sediment.

The Success of Advanced Vascular Plants Can Be Attributed to Several Evolutionary Adaptations

The forests of the world are dominated by angiosperms and gymnosperms, **advanced vascular plants.** Over a period of 250 million years, the advanced vascular plants gradually replaced the once-dominant early vascular plants.

One of the most significant characteristics of gymnosperms and angiosperms that distinguishes them from the early vascular plants is seed production. The evolution of seeds has played an important role in the successful colonization of the Earth by gymnosperms and angiosperms. Two evolutionary trends that permitted seed production were (1) the development of separate male and female gametophytes and (2) the development of an ovule, which becomes the seed after fertilization.

The success of the advanced vascular plants was also due to several evolutionary adaptations that made it possible for gymnosperms and angiosperms to survive under much drier conditions than their predecessors. Three important adaptations were (1) the retention of gametophytes within the sporophyte, (2) the transfer of sperm to the egg through a tubelike structure, and (3) the development of the embryo-containing seed, which remains dormant until conditions are favorable for the establishment of a new plant.

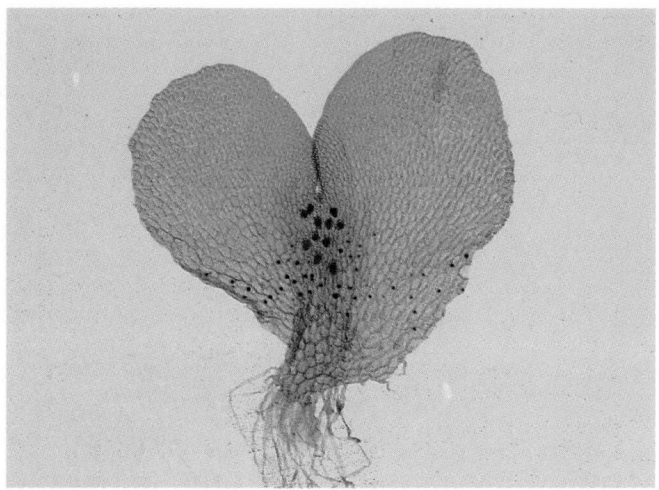

▷ **FIGURE 25–20 Fern Gametophyte** The fern gameto-phyte, called a prothallus, is a free-living plant anchored to the sub-strate by rootlike structures. Although the antheridia are produced in a scattered fashion on the prothalus, the archegonia are clus-tered at the base of the cleft.

▷ **FIGURE 25–21 Cycad** A tropical gymnosperm with an un-branched stem and a crown of leaves. This is one of many advanced vascular plants.

Gymnosperms. The gymnosperms, such as the pines, spruces, and firs, are commonly called evergreens. Most gymnosperms retain their needle-shaped or scalelike leaves during winter, unlike the deciduous plants, which shed all their leaves in the fall.

The gymnosperms, in general, are large, woody trees that produce their seeds in cones or conelike structures. Based on the oldest known fossils, the gymnosperms are believed to have evolved approximately 350 million years ago. The fossilized ancestors of the present-day gymno-sperms resemble tree ferns and were mistakenly classified as ferns until recently.

Gymnosperms include several species that grow to phe-nomenal heights. The world records for the tallest (369 feet) and largest (102 feet in diameter at the base) living things are held by redwood trees growing along the coastal regions of northern California. The world age record of 4600 years for a living organism is held by a bristlecone pine growing in Inyo National Forest in California. Eco-nomically, gymnosperms are among the most commercially valuable plants. They are extensively harvested for lumber and paper and thousands of other products.

Interestingly, gymnosperms vary considerably in their appearance. Cycads, which are common in southern Flor-ida, are tropical plants that resemble palm trees (▷ Figure 25–21). Ginkgo is a large, highly branched tree native to mainland China. It has characteristic fan-shaped leaves and is widely grown in the United States as an ornamental tree (▷ Figure 25–22). The seeds produced on the female tree have a fleshy covering that causes an unpleasant odor when mature. *Gnetum,* shown in ▷ Figure 25–23, is a woody vine native to tropical forests, and *ephedra,* shown in ▷ Figure 25–24, is a shrub restricted to arid regions, including those of the southwestern United States. *Wel-*

witschia, native to the deserts of southwestern Africa, is the most unusual looking of all gymnosperms, with a conical stem, mostly underground, and two large leaves (▷ Figure 25–25). Conifers, the most common gymnosperms, are widely distributed in the temperate regions of the world. The life cycle of all gymnosperms is very similar to that of a pine, which is a familiar conifer.

Pine *(Pinus)* is a large tree with needle-shaped leaves. The pine tree is the sporophyte plant and is the dominant phase in the life cycle (▷ Figure 25–26). The short-lived gametophytes are produced on the cones during reproduc-tion. Two kinds of cone are found on different parts of a pine tree, **pollen cones,** which occur in groups, and **ovu-**

▷ **FIGURE 25–22 Ginkgo** A branch of a female tree. Note the characteristic fan-shaped leaves and the seeds with fleshy covering, often mistaken for fruits. The foul smell associated with female ginkgo trees is produced by the seed coat.

▷ **FIGURE 25–23 Gnetum** This woody gymnosperm is a tropical vine with leaves.

▷ **FIGURE 25–24 Ephedra** This gymnosperm is a shrub with scale leaves.

late cones, which appear singly or in groups of two or three ▷ Figure 25–27a and 25–27b). The pollen cones are much smaller than the ovulate cones and are short-lived. Both types of cone are composed of many scales. The terms *male cone* and *female cone* are commonly, but incorrectly, used to describe the pollen cone and ovulate cone, respectively.

Haploid pollen grains are produced by meiosis within sporangia of the pollen cones. The pollen grains, which represent the male gametophytes of the pine, have winglike expansions and are dispersed by air currents to the ovulate cones (▷ Figure 25–28). After the transfer of pollen to the ovulate cone, each pollen grain produces two nonmotile sperm.

The scales of the ovulate cone contain two ovules, each of which contains a cell that produces four haploid spores via meiosis (see Figure 25–26). One of these spores develops into the female gametophyte generation. The female gametophyte produces two to six egg-containing archegonia.

Interestingly, fertilization in pines usually takes place approximately 12 to 15 months after **pollination,** the transfer of pollen to the ovulate cone. The sperm in the pollen grain are transported to the egg by a tube produced by the pollen grain. One of the sperm fuses with the egg, and the resulting zygote develops into an embryo. Hormonal changes brought about by fertilization transform the ovule into a seed, which will remain on the ovulate cone until dispersed by wind. The embryo in the seed develops into a new sporophyte plant when the seed germinates. As you will soon see, the angiosperms share many of these life cycle features with the gymnosperms.

Angiosperms. Angiosperms are the best known and the most abundant of all plant groups, with a wide distribution in all known habitats. Morphologically, they are the most diverse group of plants, ranging in size from the tiny duckweeds to the gigantic oaks. The most characteristic feature of all angiosperms is the production of flowers. As a result, the angiosperms are commonly called the flowering plants and are classified under one division, **Anthophyta.**

Angiosperms first appeared in the fossil record approximately 160 million years ago, giving them the distinction of being the most recently evolved group of plants. Within a relatively short evolutionary period, though, the angiosperms have acquired considerable structural diversity and remarkable adaptation for survival over a wide range of environmental conditions.

As in gymnosperms, the sporophyte is the dominant generation in the life cycle of angiosperms, with the short-lived gametophyte highly reduced and dependent on the sporophyte. Unlike gymnosperms, however, angiosperms develop seeds inside a fruit. The evolution of the fruit has given the angiosperms an added advantage in the dispersal of their seeds, as you will soon see.

▷ **FIGURE 25–25 Welwitschia** This unusual looking gymnosperm has two large leaves that split longitudinally as the plant grows older, giving it the appearance of having many leaves. The conical stem is mostly underground.

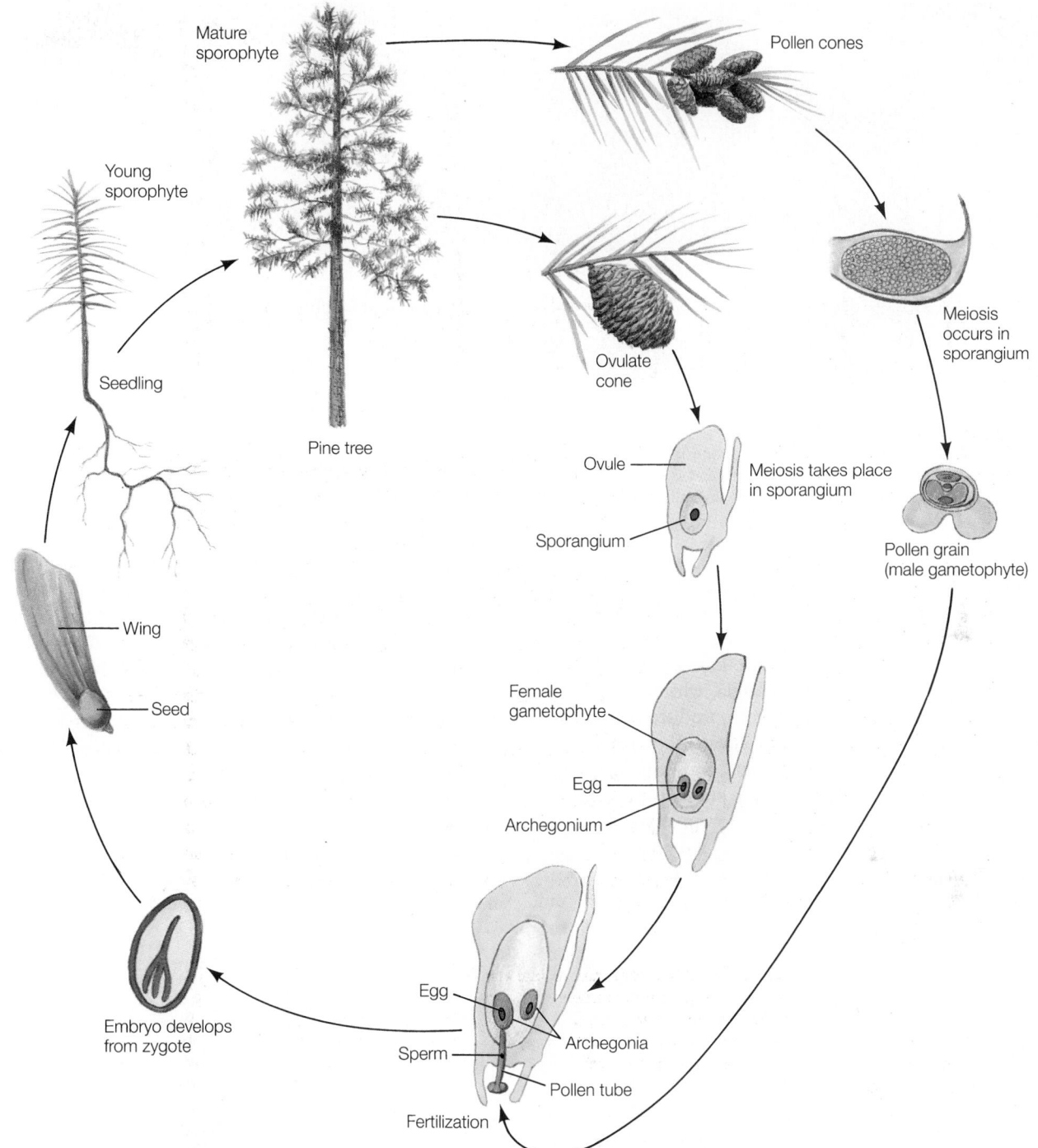

Mature sporophyte

Young sporophyte

Seedling

Pine tree

Pollen cones

Meiosis occurs in sporangium

Ovulate cone

Ovule

Meiosis takes place in sporangium

Sporangium

Pollen grain (male gametophyte)

Wing

Seed

Female gametophyte

Egg

Archegonium

Embryo develops from zygote

Egg

Sperm

Archegonia

Pollen tube

Fertilization

▷ **FIGURE 25–26 Pine Life Cycle** Haploid pollen grain (male gametophyte) produced in a sporangium of a pollen cone is transferred by air current to an ovule on the ovulate cone. Ovules also contain sporangia. After pollination, one of the haploid spores produced in the sporangium of the ovule develops into the female gametophyte, which in turn gives rise to archegonia, each with an egg. During fertilization, nonmotile sperm produced by the pollen grain are transported to an archegonium via the pollen tube. After fertilization, the zygote develops into an embryo enclosed within the seed. A new sporophyte plant is initiated when the seed germinates, completing the life cycle.

Angiosperms, without doubt, are the most important source of food for humans and many other animals. In addition to food, we obtain a variety of other products from flowering plants, such as wood to make furniture and buildings, natural fibers for garments and other uses, spices used in cooking, and medicinal drugs.

The rainbow of color that spreads across the land as winter turns to spring is due to the countless violets, bluebells, daffodils, tulips, and a myriad of other blossoms that are the reproductive organs of angiosperms. A close examination of a flower will show that it has four parts: sepals, petals, stamens, and pistil (▷ Figure 25–29). The **sepals,**

(a)

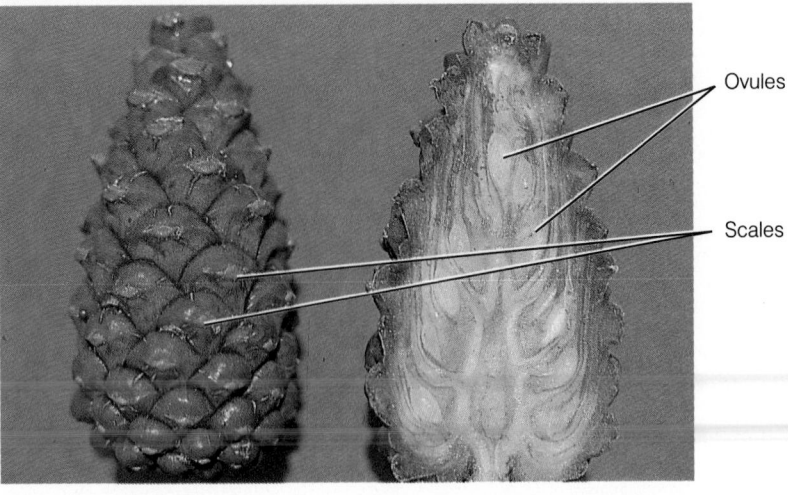

Ovules

Scales

(b)

▷ **FIGURE 25–27 Cones** (*a*) Pollen cones of a pine tree are produced in clusters during early spring and are short-lived. (*b*) Ovulate cones of pine are much larger than the pollen cones and remain on the tree for two or more years. On the left is a complete cone, and on the right is a cut-open view with ovules attached to scales.

the green, leaflike portion of the flower, and **petals** do not have a direct role in reproduction, except to attract pollinating agents. The **stamen** is a filamentous structure that ends with a small, bulbous structure known as the **anther.** The anther consists of four sporangia, which contain specialized cells that undergo meiosis to produce haploid pollen grains, the male gametophytes. A variety of agents, such as wind, insects, birds, other animals, and rain, help transfer pollen from an anther to the **stigma,** the terminal portion of the **pistil.** The pistil of a flower, shown in Figure 25–29, consists of the terminal stigma, the basal ovary, and a tube that connects them, known as the **style.** The **ovary** contains one or more ovules, each with a sporangium. Meiosis takes place in a specialized cell within the sporangium, giving rise to haploid spores. One of the spores

develops into the female gametophyte, containing an egg and several other cells.

After landing on the stigma, each pollen grain produces two nonmotile sperm. A tube produced by the pollen grain carries the sperm to the ovule. The pollen tube literally digests its way down the style to the ovary of the flower (▷ Figure 25–30). The pollen tube is a highly evolved

▷ **FIGURE 25–29 A Typical Flower** The anthers are located on the ends of the stamens, and the ovary is located at the base of the pistil. Many pollen grains are produced within an anther, and one or more ovules are produced within an ovary.

▷ **FIGURE 25–28 Pine Pollen** Pollen grains of pine with winglike appendages that aid in wind-borne dispersal.

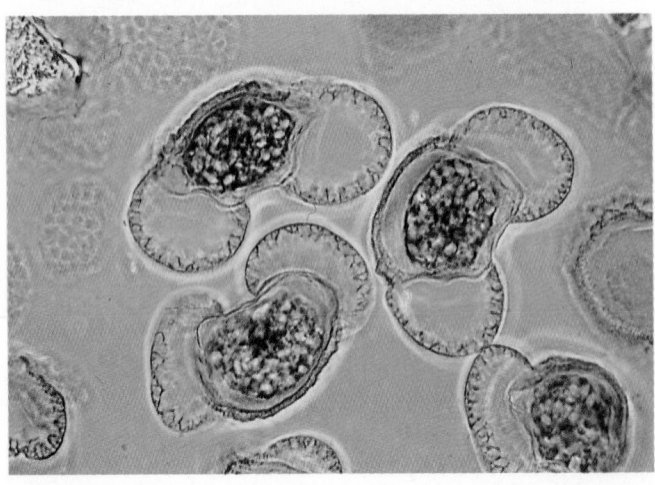

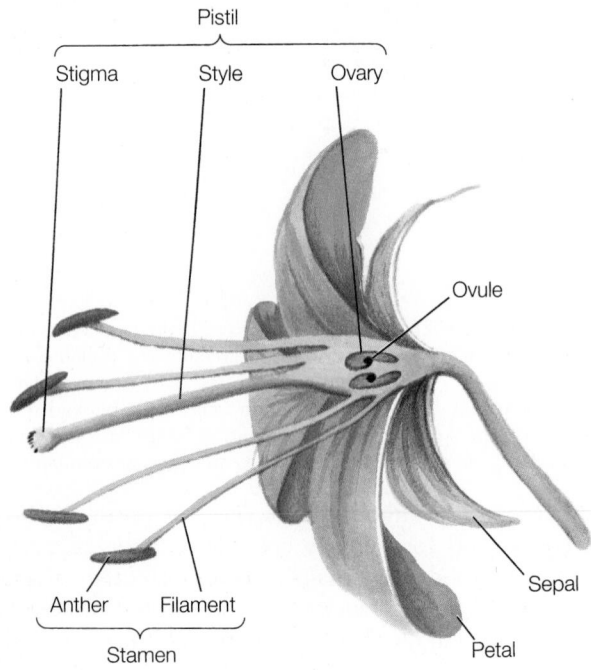

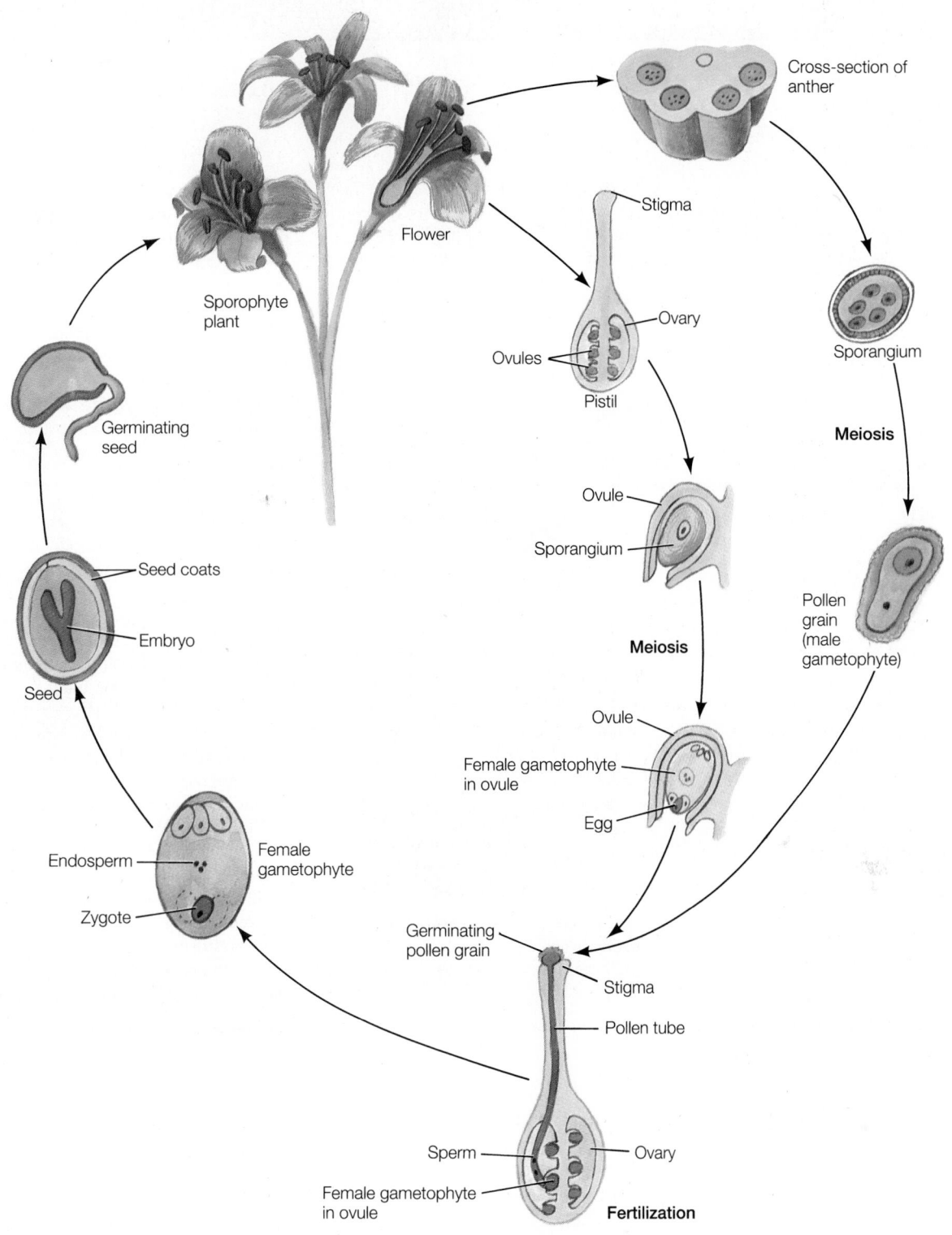

Labels within the figure:

- Cross-section of anther
- Flower
- Stigma
- Ovary
- Ovules
- Pistil
- Sporangium
- **Meiosis**
- Sporophyte plant
- Germinating seed
- Ovule
- Sporangium
- **Meiosis**
- Pollen grain (male gametophyte)
- Seed coats
- Embryo
- Seed
- Ovule
- Female gametophyte in ovule
- Egg
- Endosperm
- Female gametophyte
- Zygote
- Germinating pollen grain
- Stigma
- Pollen tube
- Sperm
- Ovary
- Female gametophyte in ovule
- **Fertilization**

▷ **FIGURE 25–30 Angiosperm Life Cycle** During reproduction of an angiosperm plant, haploid pollen grains (male gametophytes) produced in a sporangium of the anther are transferred
to the stigma of a flower. Each pollen grain develops a tube that
transports the two nonmotile sperm to an ovule within the ovary.
One of the haploid spores produced in the sporangium of the
ovule develops into the female gametophyte containing an egg and
several other cells. One sperm fuses with the egg, forming a zygote, and the other sperm fuses with two other cells in the middle
of the female gametophyte, resulting in the formation of the endosperm. The embryo that develops from the zygote is part of the
seed enclosed within the fruit derived from the ovary. Germination of the seed into a new sporophyte plant completes the life
cycle.

WILDFLOWERS AND HIGH-ALTITUDE ADAPTATIONS

In the Rockies, Sierra Nevadas, and Cascade Mountains, two miles or more above sea level, wildflowers grow in remarkable diversity, gracing the landscape with a display of unparalleled beauty. Like the animals that make these high-mountain regions their homes, alpine wildflowers have evolved a variety of adaptations to the harsh climatic conditions they face.

With winds gusting up to 200 miles per hour, temperatures below freezing, subsoil that remains permanently frozen, and high doses of ultraviolet light, the tundra represents the limit of Earth's habitability. The growing season on the tundra is only one and a half months long, and most of the precipitation comes as snow, which is promptly swept off the tundra by fierce winds.

To survive here, wildflowers have evolved short stems and small leaves, keeping the plants close to the ground for protection from the wind and for warmth. Buds form near the plants' bases and often take years to develop. Much of the energy is devoted to creating an extensive root system. Despite the fact that alpine wildflowers appear stunted, the blossoms have not shrunk in proportion. This fact under scores the evolutionary importance of the flower as an attractant for pollinating insects.

The alpine sunflower survives thanks to many of the above-mentioned adaptations. For many years, the sunflower grows almost imperceptibly, all the while storing nutrients in its root system. Eventually, a bud forms on the plant, which takes several seasons to mature. Then, the flower appears. Four inches in diameter, the flower seems incongruous on the relatively small stem (▷ Figure 1). The flower lasts only a few days, producing seeds that will fall to the ground to start a new plant or feed a hungry bird or rodent.

Wildflowers have also evolved some interesting ways to protect themselves from the cold, the intense sunlight, and evaporation. One adaptation is hairs

▷ **FIGURE 1 The Alpine Sun Flower** Like other tundra species, the sunflower produces a large flower on a compact stem.

that appear on the leaves of many species. These hairs diffuse the light, reduce water loss, and trap heat.

Some species contain red pigments in their stems and leaves, which appear in the spring and fall. These pigments absorb solar energy and release heat to the cells of the plant on cold days.

Interestingly, alpine plants are often subject to stresses similar to those encountered in the desert. Many species, for example, have evolved fleshy internal tissues that absorb water and thick, tough skins like those of cacti that reduce evaporation.

Because seed production is a risky affair, many alpine wildflowers have also evolved an alternative mode of reproduction—vegetative propagation, which depends on rhizomes or small bulbs. Rhizomes grow horizontally from a parent plant and periodically send down roots. Stalks develop at these points and grow upward. The new plant feeds off the parent plant until it is established. Some plants also produce small bulbs on their stalks, which frequently sprout roots before the bulb falls to the ground, thus giving it a head start.

Alpine wildflowers have adapted remarkably to their cold, inhospitable domain, but though they endure the harsh conditions of the tundra, these plants are remarkably fragile. Off-road vehicles, livestock, hikers, and campers can easily damage and kill them. Even moderate use results in long-term damage.

---

mechanism for the transfer of sperm to the egg and eliminates the dependence on water for fertilization.

After entering the female gametophyte, the pollen tube breaks and releases the two sperm. One of the sperm fertilizes the egg, located at one end of the female gametophyte, and produces a diploid zygote. The other sperm fuses with two cells in the middle of the ovule to produce the endosperm, which will serve as a source of food for the embryo that develops from the zygote. Because two separate fusion events take place, this process is called **double fertilization,** a phenomenon considered to be unique to the angiosperms.

After fertilization, the ovules develop into seeds with the embryos inside. The remainder of the ovary develops into a fruit that houses the seeds. A new plant is produced when a seed germinates, thus completing the life cycle. For a discussion of wild flower adaptations to high-altitude life, see Spotlight on Evolution 25–1.

≈ EVOLUTIONARY TRENDS

As discussed previously, the land plants are believed to have evolved from a filamentous green algal ancestor with a gametophyte-dominant life cycle. One of the evolutionary trends that we see when comparing the life cycles of plants is the increasing prominence of the sporophyte generation (▷Figure 25–31). The more advanced plants also exhibit greater morphological differentiation, as witnessed by the

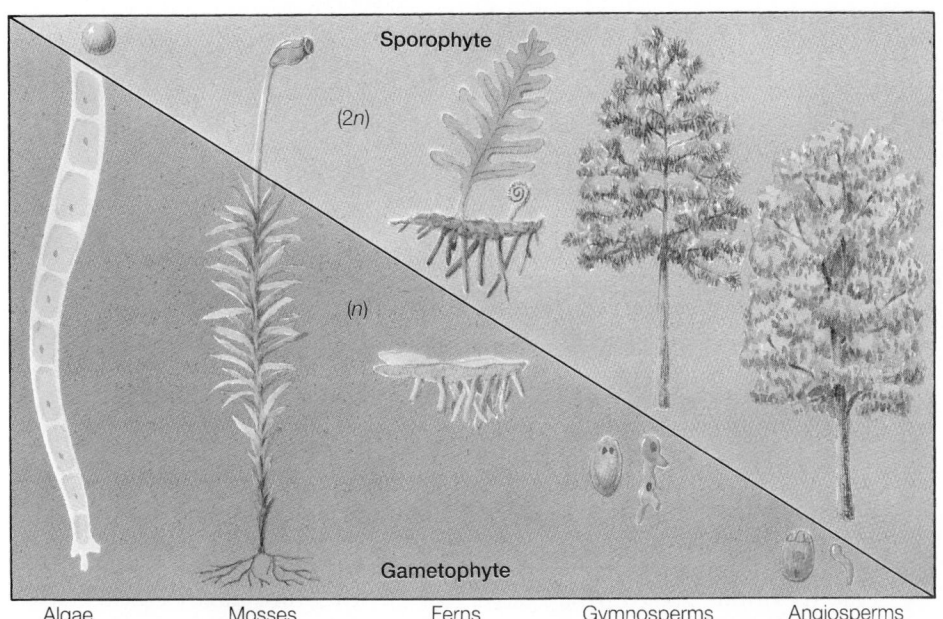

Sporophyte

(2n)

(n)

Gametophyte

Algae
(Ulothrix)   Mosses   Ferns   Gymnosperms   Angiosperms

presence of stems, leaves, and roots. Plant organs are also composed of specialized tissues. Specialization provides a more efficient way of carrying out essential functions, such as transport of water and food.

The production of flowers and fruits represents the culmination of the evolutionary development of sexual reproduction in plants. This ability has given the flowering plants a decisive advantage over other plant groups, resulting in their dominance and wide distribution on Earth today.

〰 PLANTS AND ANIMALS: A COMPARISON

As the preceding discussion indicates, the life cycle of all land plants and some algae is characterized by distinct haploid and diploid phases, a phenomenon known as the alternation of generations. In all land plants, meiosis results in the production of haploid spores (not gametes), which then develop into gametophyte structures. These haploid plants, in turn, give rise to egg and sperm, which fuse and develop into the diploid sporophyte plant.

By contrast, meiosis in animals results in the direct production of egg and sperm, which are the only haploid structures in their life cycle. Therefore, animals lack a clear-cut alternation of generations.

Another significant difference between plants and animals is that plants have the ability to synthesize their own food. More obvious, however, is the ability of animals to move from one place to another, whereas plants are generally anchored. For a discussion of one type of movement found in plants, see Scientific Discoveries 25–1.

At the cellular level, plant cells are surrounded by a cellulose cell wall, a structure that is absent from animal cells. The cell wall gives rigidity and support to the plant body.

The presence of a large, centrally located vacuole is another characteristic of plant cells. In the absence of an excretory system, plants use the vacuoles to store metabolic byproducts that might otherwise interfere with normal cellular functions.

In addition, plant cells contain plastids, such as chloroplasts. Plastids are responsible for the synthesis and storage of food. A large portion of a plant's body, unlike an animal's body, is composed of dead cells that transport water and provide structural support.

Finally, plants and animals show a fundamental difference in development. In animals, after the embryo develops into an adult organism, further cell growth or development is very limited, resulting in a finite life span for all animals. However, most vascular plants have specialized tissues at the tips of roots and stems that perpetually remain in an embryonic state. These embryonic tissues continue to produce new growth year after year and have the potential to keep a plant alive for an indefinite period. Several of the bristlecone pines are known to be between 4000 and 5000 years old, a feat no animal can match!

ENVIRONMENT AND HEALTH: LICHENS AND AIR POLLUTION

Lichens are one of the oldest and slowest growing organisms on Earth. A greenish, hand-size lichen on a rock in Canada's Arctic tundra began to grow when Mesopotamian civilization was at its height. When the temple in Jerusalem was built by Solomon it was dime-size. It grew to the size of a dollar coin by the time of Christ. Today, it is four times that size and continues to grow at the average lichen

THE OTHER SIDE OF DARWIN
Featuring the work of Charles and Francis Darwin

"There is no single figure who has influenced the direction of modern biological thought more than the distinguished Charles Darwin," write Mordecai Gabriel and Seymour Fogel in *Great Experiments in Biology*. Although renowned for his discoveries about evolution, Darwin is virtually unrecognized for his experimental abilities and his contribution to our understanding of plant physiology. Little known to many, it was he and his son Francis who laid the groundwork for the discovery of plant hormones in the 20th century.

The Darwins' studies showed that plant growth depends on the production of some active material by the growing tips of plants. This revolutionary concept arose from a lengthy set of experiments carried out on a number of plant species.

First, the Darwins confirmed the hypothesis that young plants bend toward light (a phenomenon called phototropism), by exposing small potted plants to a lateral light source. In these experiments, they found that the top of the plant bent toward the light first and that the bending gradually extended down to the base. But did the lower sections of plants bend because of the overhanging weight of the bent upper section?

To test their belief that the upper part of the shoots regulated the bending of the lower parts when exposed to lateral light, Charles and Francis Darwin performed many experiments. In one experiment, they placed small, tinfoil caps over the plants and exposed them to lateral light. No bending occurred. In one particularly compelling experiment, they covered the tips with little caps that were impermeable to light. Controls were covered with transparent caps. When exposed to lateral light, the groups reacted quite differently. Those with transparent caps bent toward the light; those with blackened caps failed to bow or bowed ever so slightly.

From these and other carefully conducted experiments, the Darwins concluded that the exclusion of light from the upper part of the shoot prevented the lower part from bending. Thus, in normal plants, some "influence" was being transmitted from the upper to the lower part, causing it to bend.

In 1910, 30 years after the Darwins' work, a Danish researcher, Peter Boysen-Jensen, performed an experiment suggesting that the "influence" the Darwins had speculated on was a substance that was transmitted downward from the tip. Boysen-Jensen separated the tips of the growing shoots from the lower sections by inserting a thin plate of mica, which blocked the diffusion of chemical substances, then exposed the young plants to light. As expected, the plants did not bend. In other plants, he inserted thin blocks of gelatin, which do not block the flow of chemicals. These plants bent toward the light, confirming the hypothesis that the influence postulated by Charles and Francis Darwin was capable of diffusing through a gelatin barrier. This experiment, though simple, shows how researchers can often arrive at important insights without complex equipment.

In 1926, F. W. Went, a young plant physiologist from Holland, performed another study that further clarified the matter. Went removed the tip of young shoots and placed them on blocks of agar, a gelatinous material derived from one kind of algae, hoping that the chemical substance would diffuse into the agar. If so, the agar block would then be able to substitute for the tip. After exposing the tips embedded in agar blocks to light, he placed the blocks on "decapatitated" shoots of plants that had been grown in the dark. Sure enough, the blocks caused the shoots to bend. Went coined the term *auxin* for this substance that still remains unidentified.

Since that time, the chemical nature of this hormone has been determined, and a variety of other hormones have also been discovered, capping a search begun by one of biology's most capable and hardworking scientists, Charles Darwin.

growth rate of about 2 inches in diameter every thousand years.

Lichens grow on rock, wood, or soil and are amazingly hardy and resilient. They can endure a wide variety of climates and are sometimes found where few other life forms exist. Lichens are also extremely drought-tolerant. They are capable of maintaining dormancy for decades, perhaps even centuries, waiting for rain to fall. There are 2,500 species of lichens in the Arctic, but only 900 species of more complex vascular plants. In Chile's Atacama Desert, lichens are found on rocks. The climate there changes from freezing cold at night to boiling hot during the day, and fog may be the only source of moisture for many years.

In 1867, the botanist Simon Schwendener discovered that lichens are made up of two organisms, an algae and a fungus. Later, it was discovered that this association is symbiotic—the fungus stores water and provides protection for the delicate green algae, and the photosynthetic algae provides food for itself and the fungus.

Humans have found many uses for lichens throughout the centuries. They were used to embalm the dead in Egypt. In California, the Achomawi Indians used the lichen *Letharia* to make poisoned arrows. In Bolivia, lichen poultices are used to treat festering wounds and inflammations. Today, broad spectrum antibiotics produced from this lichen are used to treat inflammation and serious skin infections. In chemistry laboratories, acidity is gauged with litmus made from lichen. One type of lichen, called rock tripe, is known as the "food of starving explorers," and the manna lichen of the Middle East is called "bread from Heaven" by the Kurds.

Lichens' most important role in nature is as a plant pioneer. Lichens are the first colonizers in rocky areas once covered by glaciers. In this inhospitable location, the lichens slowly break down the rocks they are attached to. After a long period of time, the debris from the lichens and the rocks turns into soil that nourishes other plants.

Lichens obtain nourishment from the air, rain, and snow, and this makes them extremely sensitive to air pollution. In the 1800s, scientists began to notice that lichens were adversely affected by pollution when these organisms began to die off in European cities. Thirty-two species of lichens were collected in the Jardin de Luxembourg in Paris in the 1860's. By 1896, not a single species survived.

In 1901, 21 lichen species were found in Lille, France, but only two survived in 1951. In the town of Mendlesham in Suffolk, England, 129 species were recorded between 1912 and 1921. In 1973, only 67 remained. Lichens were absorbing heavy metals and pollution, and wherever pollution increased, the lichens died.

Regional air-quality and the spread of pollution from industrial emission sources can be determined by mapping the presence or absence of lichens, or by chemically analyzing the lichens in an area. Many lichen species disappeared from Kvarntorp, Sweden when a factory opened in 1942, only to begin to reappear in 1966 after the factory closed. An air-quality map of the British Isles was based on a lichen survey undertaken by 15,000 British school children in 1971.

The "caribou moss," or "reindeer moss," is one of the world's most important lichens. It is found in vast areas of tundra and the taiga forest. The one million wild caribou of North America and the 2.5 million semi-domesticated reindeer of Eurasia eat up to 12 pounds of this lichen daily. A Soviet study showed that, during winter, up to 95% of reindeer nourishment comes from "reindeer moss."

Caribou and reindeer herds are essential to the hunting and herding peoples of the north. In Sweden, the Saami (Lapps) eat an average of 8 to 10 reindeer per person per year. In Siberia, the economy and culture of many traditional herding groups are based on the reindeer. Caribou products are essential to the diets of many Inuit groups.

During the 1950's and 1960's when above-ground nuclear testing was common, lichens absorbed radioactivity from the atmosphere. Lapps who ate mostly reindeer meat consumed 30 times as much strontium-90 and cesium-137 as the Finnish people who lived further south. The milk of Inuit mothers in Canada was found to be contaminated by the same radioactive materials. With the advent of underground nuclear testing, this threat ceased, but on April 28, 1986, the accident at Chernobyl again put a cloud of radioactivity into the atmosphere. The radioactive cloud spread across Scandinavia, and the contaminants, primarily cesium-137, were absorbed by lichens. Eight months after the accident, reindeer in Norway and Sweden were too radioactive for human consumption and were destroyed and buried, or the meat was dried and sold to fox and mink farms. The level of contamination was astounding. In Norway, marketable reindeer meat is allowed up to 6,000 becquerel per kilogram (bq/kg), and in Sweden up to 1,500 bq/kl is allowed. (A becquerel is a unit of radioactive measurement, representing one nuclear disintegration per second.) After Chernobyl, radiation measurements had increased in some parts of Norway and Sweden to up to 70,000 bq/kg, and went as high as 137,000 bq/kg in some areas.

For thousands of years, lichens have been able to survive under the most difficult conditions that nature has to offer. Today, lichens face new threats and their survival is not certain, especially in cities and other polluted areas. And, though lichens may be able to withstand radioactive fallout, the contamination will be passed on to other organisms that may not be as fortunate.

SUMMARY

AN OVERVIEW OF THE PLANT KINGDOM

1. The plant kingdom is composed of multicellular organisms that are either aquatic or terrestrial and are capable of synthesizing their food.

PLANT REPRODUCTION AND LIFE CYCLES

2. Reproduction among plants can be either sexual or asexual. Remarkably, even the most highly evolved plants have retained the ability to reproduce both ways.
3. In its simplest form, asexual reproduction may involve nothing more than cellular division.
4. Asexual reproduction is highly evolved among certain groups of algae, which produce spores having the potential to develop into new algal filaments.
5. Asexual reproduction can be as simple as starting a plant from a cutting, a process known as vegetative propagation.
6. Sexual reproduction in plants involves the fusion of haploid gametes to form diploid organisms.

7. Several variations in sexual reproduction are evident in the plant kingdom. The most primitive method, common among certain algae, is known as isogamy, a process characterized by the fusion of gametes that are morphologically similar.
8. The majority of plant species exhibit a form of sexual reproduction known as oogamy, in which a small sperm cell fuses with a much larger egg, resulting in the formation of a zygote.
9. Male and female reproductive organs can be present on separate plants or on the same plant.
10. The life cycle of all land plants and many algae exhibits two distinct generations, a phenomenon known as alternation of generations. That is, during part of its life cycle a plant exists as a haploid generation (gametophyte) and during the other part it exists as a diploid generation (sporophyte).

THE ALGAE

11. Algae are primarily aquatic and show considerable diversity in

structure and reproduction.

12. Classification of algae into different divisions is based on the types of photosynthetic pigment they contain.

13. Although an asexual mode of reproduction is common among most algae, sexual reproduction by isogamy and oogamy is also present.

14. Although green algae are of universal distribution, most of the brown algae are restricted to cool coastal waters, and the red algae to warmer open seas.

15. Algae contribute significantly as primary producers in aquatic food chains.

THE FIRST LAND PLANTS (NONVASCULAR PLANTS)

16. Land plants are believed to have evolved from the green algae.

17. Development of an epidermis and rootlike structures for the absorption of water and minerals were early evolutionary features that made it possible for the land plants' ancestors to successfully colonize the land.

18. The development of vascular tissues and highly differentiated plant bodies composed of roots, stems, and leaves was significant in fully establishing plants on land.

19. Land plants are divided into nonvascular and vascular plants.

20. Nonvascular bryophytes, such as the mosses, liverworts, and hornworts, all require watery environments to complete their life cycle.

21. Nonvascular plants are usually small and are characterized by a dominant gametophyte stage and a sporophyte that is dependent on the gametophyte.

22. Bryophytes are dispersed by means of haploid spores, a fact that limits their distribution.

THE VASCULAR PLANTS

23. Vascular plants, by contrast, have a dominant sporophyte phase in which specialized water- and food-conducting tissues are present. Vascular plants are divided into early and advanced groups.

24. The predominance of the early vascular plants during past geological periods is evidenced by the vast coal deposits found in many parts of the world.

25. Today, however, the early vascular plants represent only a small component of the world's flora.

26. The most recently evolved vascular plants, gymnosperms and angiosperms, make up the vast majority of land plants today.

27. The gymnosperms and angiosperms produce embryos inside seeds, an adaptation that has promoted their survival and dispersal.

28. Separate male and female gametophytes are produced by both gymnosperms and angiosperms.

29. The pollen grains that represent the male gametophytes are produced in the pollen cones of gymnosperms and in the anthers of the angiosperm flower. In both gymnosperms and angiosperms, nonmotile sperm develop inside pollen grains.

30. Eggs are produced inside ovules on female cones in gymnosperms and inside the ovaries of flowers in angiosperms.

31. Sperm are transported to the egg within the ovule via pollen tubes, thus eliminating the need for water to complete the life cycle of gymnosperms and angiosperms.

32. After fertilization, the ovules of gymnosperms and angiosperms become the seeds.

EVOLUTIONARY TRENDS

33. At least three evolutionary trends are apparent when life cycles of different plant groups are compared: (a) an increasing importance of the sporophyte generation and a corresponding reduction of the gametophyte generation, (b) the gradual development of vascular tissues in the sporophyte plants, and (c) the production of flowers for reproduction.

PLANTS AND ANIMALS: A COMPARISON

34. The major differences between plants and animals are the ability of plants to synthesize their own food, the alternation of generations, and the presence of cell walls, plastids, and vacuoles, all of which are lacking in animals.

EXERCISING YOUR CRITICAL THINKING SKILLS

Star thistle is an attractive plant with blue-green leaves and yellow flowers. Its needle-sharp spikes, however, deter people and cattle from entering the fields where it grows. Horses that graze on star thistle develop "chewing disease," which destroys brain cells and leads to starvation as the facial muscles become paralyzed.

Star thistle is not native to the United States. Its seeds probably arrived in the late 1800s with alfalfa seeds brought from Mediterranean Europe. There were no natural parasites, pathogens, or predators of star thistle in its new habitat to keep its spread under control. Today, it covers large areas of the Northwest, including 8 million acres in California.

In the late 1950s, researchers began looking for a natural predator of the star thistle. The search was focused around the Mediterranean, the plant's native habitat. There the researchers could not find a field fully covered by star thistle because it was controlled by its natural predators. Dozens of insects found on the plant were collected for study. Years of tests followed in which the insects were given the choice of starving or of surviving by eating important American crop plants, ornamentals, and na-

tive plants that closely resemble the star thistle. If the insects showed any inclination to eat the plants, they were considered too risky for release and were disqualified.

One species, a weevil known as *Bangasternis orientalis,* ate nothing but the yellow star thistle and the purple star thistle and was deemed safe for use in the United States. In 1985, the weevil was released in California, Oregon, and Washington. The weevil is well established now and can be found in fields of star thistle in those states.

Biological control is also being used in the management of other weed species, including hydrilla, water lettuce, water hyacinth, Eurasian water milfoil, leafy spurge, field bindweed, and three species of knapweed. In an article in *Science,* Jack Coulson, head of the Agricultural Research Service's Biological Control Documentation Center in Maryland notes that annual savings from our successful biocontrol programs for both native and introduced species pests, based primarily on the costs of pesticides no longer required, total over $155 million per year.

Biological control will probably increase in the future as the use of herbicides and pesticides continues to decrease due to environmental regulations and ineffectiveness in controlling pests. Though the introduction of native pests to control introduced plant species is generally considered to be a safe practice, some have questioned the sensibility of introducing any new species, regardless of the results of testing programs. The pests may transmit diseases to vulnerable species.

Can you think of any way of controlling weeds that has not been considered in this discussion? You may want to read some books on gardening to get some ideas. If you did, you probably found out that invasive plants (weeds) usually populate areas that have been degraded by human use, particularly overgrazing of livestock. Star thistle is a good example. What does this tell you?

One method of preventing the plant's spread to other areas may be to improve the land management practices in areas adjacent to star thistle infestation. By maintaining the condition of the native species at a high quality, farmers and others can slow, or even stop, the spread of thistles.

The critical thinking lesson is to seek additional information and look at the big picture. This example clearly shows that by addressing the root causes of the problem (weed invasion), not simply treating the symptoms (the weeds themselves), one can find significant solutions to serious problems.

TEST OF TERMS

1. A rigid structure that is present around the cells of plants, but missing from those of animals, is the _____ _____ .
2. _____ , or flowering plants, are the dominant form of terrestrial plants.
3. Pines, firs, and spruces are members of a group called the _____ .
4. Terrestrial plants are believed to have evolved from _____ _____ living in the intertidal zone.
5. _____ _____ is the term used to describe one form of asexual reproduction in flowering plants in which new plants are produced from cuttings.
6. Sexual reproduction in plants involves two processes, _____ , which reduces chromosome number, and _____ , the union of the gametes.
7. The multicellular structure that develops directly from a zygote of flowering plants is the _____ .
8. _____ occurs when two morphologically similar gametes fuse.
9. The gametophyte generation of land plants and many algae is _____ and produces _____ .
10. The sporophyte generation of land plants and many algae is _____ and produces _____ .
11. Fertilization initiates the 2n _____ generation; meiosis initiates the 1n _____ generation.
12. The green algae are members of the division _____ .
13. _____ algae are almost exclusively marine and live in relatively cold water.
14. Land plants are divided into two broad categories, _____ and _____ plants. The most primitive are the _____ plants.
15. During the moss life cycle, the spore develops into a filamentous structure called the _____ , from which new gametophyte plants emerge.
16. An underground stem that is characteristic of all early vascular plants is the _____ .
17. The tissue responsible for the transport of water in vascular plants is the _____ . The tissue responsible for the transport of food in vascular plants is the _____ .
18. The embryos of angiosperms and gymnosperms are found within structures called _____ .
19. The specialized reproductive structures of a pine tree are known as _____ .
20. The specialized reproductive structures of all angiosperms are known as _____ .
21. Pollen grains are produced in structures of a flower that are known as the _____ .
22. Ovules are produced inside the structure of a flower known as the _____ .
23. The seeds of angiosperms are always produced inside a(n) _____ .
24. The process that gives rise to both a zygote and an endosperm during the life cycle of a flowering plant is called _____ _____ .
25. The transfer of pollen from the pollen cone to the ovulate cone is called _____ .

TEST OF CONCEPTS

1. Discuss the trends in plant evolution with respect to the relative importance of gametophyte and sporophyte generations.
2. Discuss the significance of asexual reproduction in plants.
3. Discuss the significance of sexual reproduction to plants.
4. Describe the types of sexual reproduction in terms of size and mobility of gametes.
5. Compare and contrast the life cycles of mosses and ferns.
6. What evolutionary advantages do you see in a method of reproduction involving seed and fruit production?

26

The Structure and Function of Plants

≈

Blackberry leaves on a frosty fall morning

In the jungles of Borneo, a man wanders in search of medicinal plants for his tribe. Thousands of miles away in the concrete and steel jungle of New York, a woman wanders the isles of a drugstore in search of medicine for a headache. After a short period, the tribesman finds a vine, removes some bark, and carries it home in a small leather pouch. His urban counterpart finds her medicine, neatly packaged, pays for it, drops it in her purse, then heads for the subway.

What do these seemingly disconnected acts have in common? Obviously, the common denominator is medicine. In the case of the tribal witch doctor, the medicine comes from the finely ground bark of a vine in his rain forest home. His counterpart in New York City finds her medicine on the shelf of an air conditioned drug store. Although it was made by a pharmaceutical company, it, too, owes its origin to the tropical rain forest. In fact, half of all prescription and nonprescription medicines sold in the United States originally came from a plant, many of them living in the tropics.

This chapter will look more closely at plants, focusing on the flowering plants because they have evolved special features that give them a decided advantage for life on land. Casual observation of a plant gives the impression that it has a relatively simple structure. Like the higher animals, though, flowering plants are complex organisms. In fact, a microscopic examination of the different plant organs such as stems, leaves, and roots shows that these parts are made up of groups of cells arranged in distinct patterns. Just as in animals, groups of cells with specialized functions are called tissues. The cells within a tissue often show characteristic shapes, which are commonly used to identify tissue types.

In this chapter, we will examine the basic structure and function of plant organs and briefly look at the different cell types and tissues that compose them. We will also study plant growth and look at factors that influence growth and other vital functions.

As you learned in the last chapter, the end result of sexual reproduction in flowering plants is the production of seeds, which are housed within fruit. Seeds give rise to new plants. In this chapter, we begin our discussion with the nature of fruits and seeds.

≈ FRUITS

Among angiosperms, the entire reproductive process takes place within flowers and culminates in the formation of a fruit that contains one or more seeds. The transformation of the ovary of a flower into a fruit is triggered by hormonal changes within these structures initiated by fertilization. (See Chapter 25 for a summary of the flower parts.)

A **fruit,** by definition, is a structure derived from the ovary of a flower. As ▷ Figure 26–1 shows, many types of fruit are formed.

Certain fruits—for example, apples—form from other parts of the flower as well as the ovary. The core of an

▷ **FIGURE 26–1 Fruit Types** Flowering plants produce many different fruits, among them grapes, oranges, cucumbers, peaches, soybeans, corn, and walnuts.

apple, which contains the seeds, is derived from the ovary. However, the edible portion is derived from the base of the flower petals (▷ Figure 26–2a). The red, fleshy portion of strawberries is formed from the upper end of the flower stalk, as shown in Figure 26–2b.

Vegetables, such as green beans, pea pods, and cucumbers, are actually fruits. So are nuts. **Nuts** are fruits in which the entire fruit wall becomes hard. Fruits produced by grasses are unique, because the seed and the fruit fuse and remain inseparable. Such fruits are called **grains.**

▷ **FIGURE 26–2 Accessory Fruits** Accessory fruits contain one or more floral parts in addition to the ovary. (*a*) The base of the petals of an apple blossom fuse to the ovary and develop into the edible portion. (*b*) The edible portion of a strawberry is the swollen end of the flower stalk.

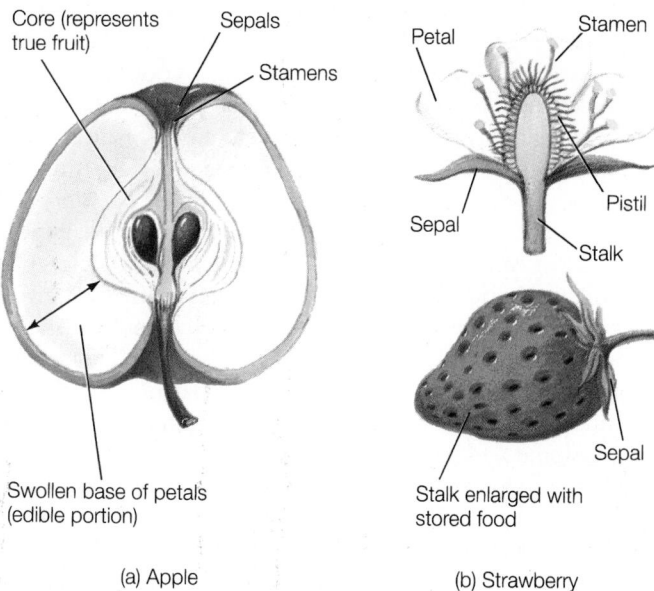

Core (represents true fruit) Sepals
Stamens
Swollen base of petals (edible portion)

(a) Apple

Petal Stamen
Pistil
Sepal Stalk
Stalk enlarged with stored food
Sepal

(b) Strawberry

Fruits of maples, ashes, and elms have winglike expansions for dispersal by wind.

The majority of angiosperms produce seeds and fruits only after the eggs have been fertilized. However, some flowering plants produce fruits without the fertilization of eggs. In such cases, the fruits are characteristically seedless. Some familiar examples are seedless grapes, oranges, and melons. In most bananas, however, seeds are not formed even after fertilization. Seedless varieties of plants are produced commercially from cuttings of stems and roots or by cloning, both forms of asexual reproduction. Cloning is carried out by growing a new plant in an artificial culture medium from an excised piece of plant tissue. The culture medium contains nutrients and plant hormones necessary for the development of an undifferentiated piece of tissue into a complete plant. This method of reproduction results in genetically identical progeny, ensuring the desired traits that were present in the parent from which the clones are made.

The production of seeds inside fruits is a highly efficient method for seed dispersal, as fruits are often eaten by mammals and birds and dispersed in their droppings. Dispersal by water is seen commonly among coconut fruits, which can drift across hundreds of miles of ocean before washing ashore and sprouting. Dispersal of seeds and fruits by birds, wind, and water currents is the primary mechanism that initiates plant life on barren islands that are formed by volcanic action in the oceans, a subject described in more detail in Chapter 31.

≈ SEEDS

Seeds are derived from the ovules and are the end result of sexual reproduction among the gymnosperms and angiosperms, as mentioned in the last chapter. The embryos within seeds give rise to a new generation of plants.

Most seeds remain dormant because of the relatively low quantity of water they contain. Water accounts for about 30% of the fresh weight of most seeds. In contrast, the water content of actively growing parts of plants is as high as 95%. Hence, keeping seeds in a cool, dry place increases their longevity. When given the proper conditions for germination, such as water, oxygen, and a proper temperature range, the embryo in the seed develops into a seedling (▷ Figure 26–3).

Seeds are an important resource for plant breeders interested in improving the yield and disease resistance of commercial varieties of plants. Unfortunately, many native plants are being wiped out as the human population expands into fields and forests. To protect native plants from around the world, the United States Department of Agriculture has undertaken an ambitious program in Fort Collins, Colorado, to preserve seeds of economically important species. Seeds from these plants are gathered in the wild and shipped to Colorado, where they are stored in liquid nitrogen at 196°C below zero. Scientists believe that seeds will remain viable indefinitely at this temperature and hope

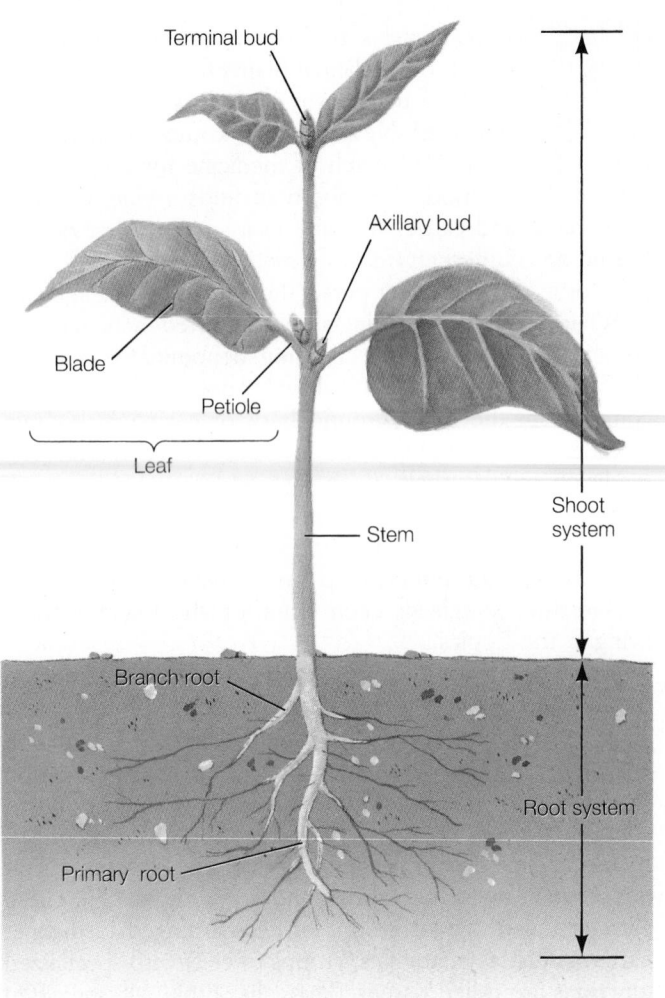

▷ **FIGURE 26–3 Seedling** Future branches are produced by the axillary buds.

that they will serve as a gene pool for plant-breeding projects in the future.

Seeds Contain Plant Embryos and Often Contain Food Required for Early Growth

The embryo housed inside a seed consists of three parts, (1) an **epicotyl,** which gives rise to the shoot system; (2) a **hypocotyl,** which produces the primary root of the seedling; and (3) one or more leaflike structures called **cotyledons,** which are commonly, but not exclusively, used for the storage of food needed for the germination of an embryo (▷ Figure 26–4). Based on the number of cotyledons in a seed, angiosperms are divided into two broad groups, **monocots,** containing one cotyledon, and **dicots,** containing two (Figure 26–4). Gymnosperms, by contrast, usually have more than two cotyledons in their seeds.

In cereal grains (wheat and barley, for example) and some other monocots and dicots, the food is stored in a nonembryonic structure of the seed called the **endosperm.**

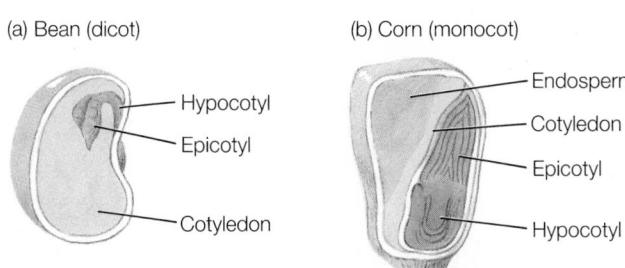

(a) Bean (dicot)

- Hypocotyl
- Epicotyl
- Cotyledon

(b) Corn (monocot)

- Endosperm
- Cotyledon
- Epicotyl
- Hypocotyl

▷ **FIGURE 26–4 Typical monocot and dicot seeds**
(*a*) Corn is a monoct seed, with one cotyledon. The embryonic portion consists of an epicotyl and a hypocotyl. In the seed of corn, food is stored in a nonembryonic structure called the endosperm. (*b*) The bean is a dicot seed because it has two cotyledons. The epicotyl develops into the shoot system, and the hypocotyl develops into the root system. Food is stored in the cotyledon.

During germination of such seeds, the cotyledons release enzymes that digest the food in the endosperm, making it available to the developing embryo.

From time immemorial, human beings have been dependent on the food stored in seeds for their own nourishment. Today, in fact, the seeds of just three plant species—rice, wheat, and corn—provide over half of the total Calories consumed by humans. In addition, the seeds of beans and peas are the main source of proteins for those in countries where meat is rare or avoided because of religious beliefs, as among the orthodox Hindus in India.

Our early ancestors probably spent much of their time gathering seeds and other edible plant products. It is interesting to note that modern agriculture had its beginnings in our ancestors' discovery 10,000 years ago that planting wild seeds resulted in a more reliable source of food than that available through random gathering. Today, a small fraction of our population grows most of the food we eat.

Seed Germination Results in the Formation of a Primary Root and a Shoot System

As noted previously, water, oxygen, and proper temperature are essential for the germination of seeds. The first structure to emerge from a germinating seed is the **primary root,** from which many branch roots are produced. Soon after the emergence of the root, parts of the shoot system emerge, pushing above the soil level and establishing a seedling (see Figure 26–3). In the garden bean and many other dicots, the entire seed is pulled aboveground by the elongating embryo after the primary root is formed (▷ Figure 26–5a). The cotyledons that are dragged aboveground serve as photosynthetic organs until the first pair of leaves develops. After this, the cotyledons wither and fall off. In corn and many other monocots, the seed remains underground (Figure 26–5b).

As a seedling develops, the embryonic tissues differentiate to produce a variety of additional tissues that can carry out all functional needs of a mature plant.

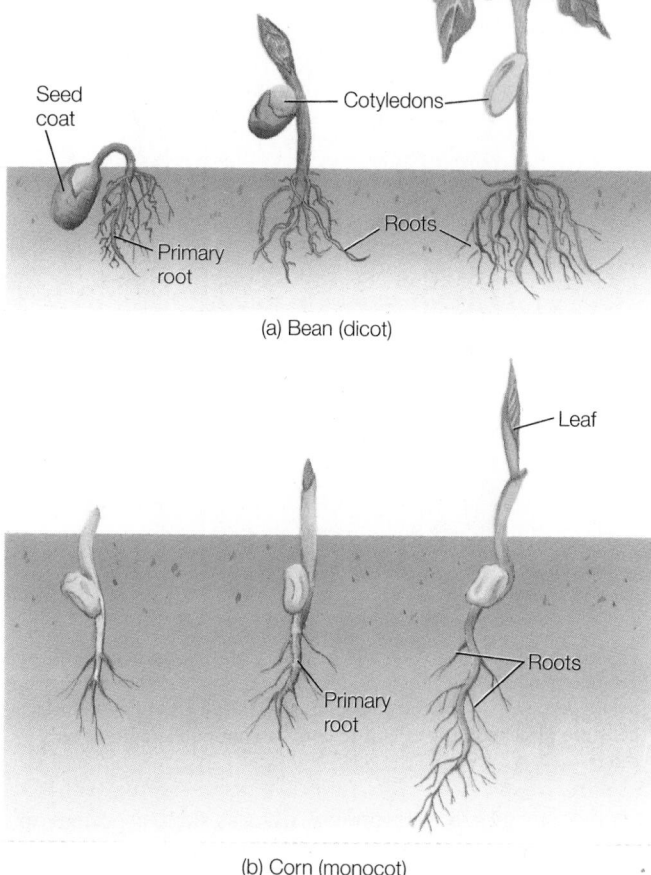

(a) Bean (dicot)

- Seed coat
- Cotyledons
- Roots
- Primary root

(b) Corn (monocot)

- Leaf
- Roots
- Primary root

▷ **FIGURE 26–5 Germinating Seeds of Bean and Corn**
(*a*) In a bean plant, the entire seed is brought aboveground during the germination. (*b*) The corn seed remains underground during germination.

≈ PLANT TISSUES

Like all other multicellular organisms, plants are composed of cells arranged into different tissue types. As noted in Chapter 10, a tissue is a group of cells that carries out a specific function. Tissues allow for a division of labor and thus increase the functional ability of an organism. In plants, some tissues have only one cell type, which may be living or dead. Other tissues have more than one type of cell, some types living and other types dead. As you will soon see, even dead cells in a plant are able to carry out certain functions, such as transporting water or providing strength.

Plant tissues are divided into four groups based on their function: meristematic, dermal, vascular, and ground. **Meristematic tissues** are responsible for the production of all other tissues and are located either at the tips of the stem and the root or along the entire length of the stem and root. The terminally located meristems are known as **primary meristems,** and the laterally located ones are the **secondary meristems.** Primary meristems give rise to primary tissues, such as the parenchyma and sclerenchyma, which are discussed in the next section. The secondary

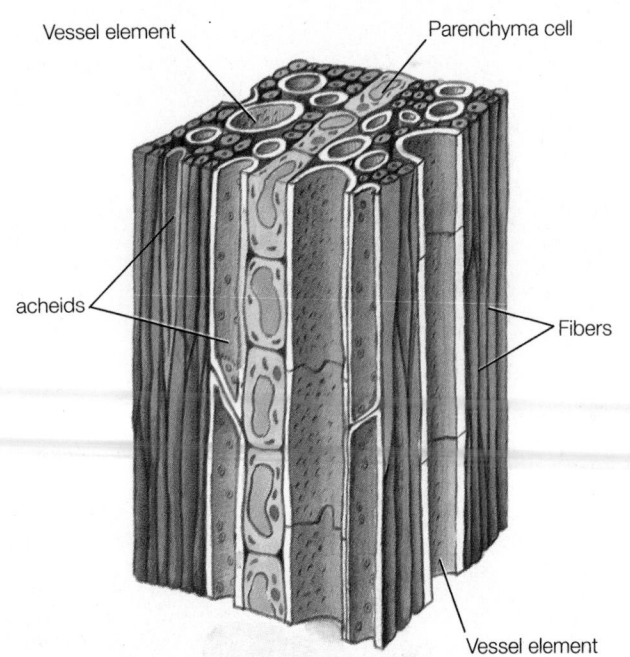

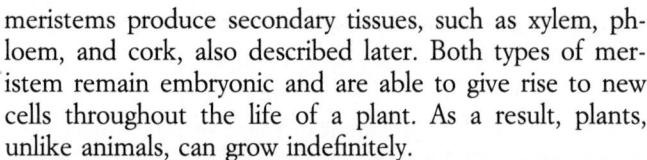

▷ **FIGURE 26–6 Xylem** Cross section of wood showing the arrangement of cells. Vessel elements and tracheids are the cells in xylem responsible for the transport of water in the plant. These cells are attached end to end to form a continuous system of ducts.

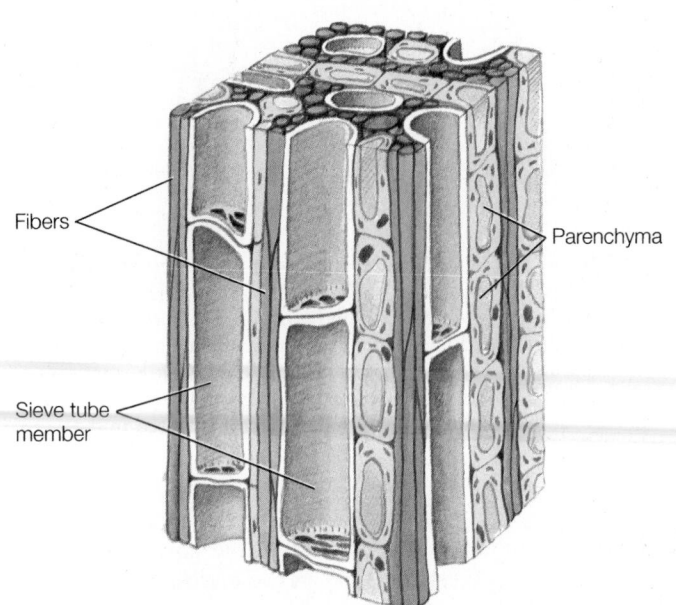

▷ **FIGURE 26–7 Phloem** Cross section of the inner bark showing the arrangement of cells. Sieve tube members are the cells in phloem through which food is transported. They are attached end to end to form a continuous duct system.

meristems produce secondary tissues, such as xylem, phloem, and cork, also described later. Both types of meristem remain embryonic and are able to give rise to new cells throughout the life of a plant. As a result, plants, unlike animals, can grow indefinitely.

The **dermal tissues** are the plant's outermost, protective tissues. Two types are encountered, epidermis and cork. A layer of epidermis, for example, covers the surface of all leaves and the young stems and roots. This cellular layer prevents the loss of water and the entry of pathogens. Cork replaces the epidermis in the older parts of a plant. This is particularly apparent in woody plants, where cork can accumulate in thick layers.

Vascular tissues are responsible for the transport of materials within a plant. Two types of vascular tissue are present, as noted in the last chapter: xylem, which transports water and minerals, and phloem which transports food. Several cell types are found in xylem. Two of them, known as **vessel elements** and **tracheids,** are mainly responsible for the transport of water in plants (▷ Figure 26–6). Both vessel elements and tracheids are tubelike cells that form a pipe system when attached end to end. Because they are thick-walled dead cells, they also provide support.

Of the different cell types present in phloem, only the **sieve tube member** has a direct role in transporting food. Similar to the water-transporting cells, sieve tube members form a system of tubules within the plant (▷ Figure 26–7). Although living, a mature sieve tube member does not have a nucleus. See Scientific Discoveries That Changed Our World 26–1 for a discussion of the discovery of circulation in plants.

Ground tissues include a diverse group that serve an equally diverse number of functions, such as food synthesis, food and water storage, and physical support. The most common ground tissue, known as **parenchyma,** is present in all parts of a plant and is the location of food production and storage. Plants such as the cacti growing in arid regions also store water in their parenchymal tissues. Other ground tissues, mainly one known as **sclerenchyma,** provide strength and support to different parts of the plant body. Sclerenchyma tissue is usually composed of dead fiber cells, similar to the ones used to make yarn and rope.

Plant organs, such as stems, leaves, and roots, are composed of the different tissues, arranged in specific patterns to optimize their function.

≈ STEMS

The shoot system, mentioned earlier, is the aboveground portion of a plant and is composed of stems and leaves. At the very end of all stems are the **terminal buds,** or, more commonly, **buds.** These structures contain meristematic tissue (primary meristems) that is protected by a series of unopened leaves (▷ Figure 26–8). The unopened leaves give rise to the future leaves. The primary meristematic tissues beneath them give rise to the shoot system and still more leaves. Future stems and leaves arise via cell divisions and the subsequent elongation of the divided cells.

As shown in Figure 26–3, **axillary buds** are present at the junction of the stem and leaves. These buds give rise to new branches and their leaves. Certain axillary buds also give rise to flowers.

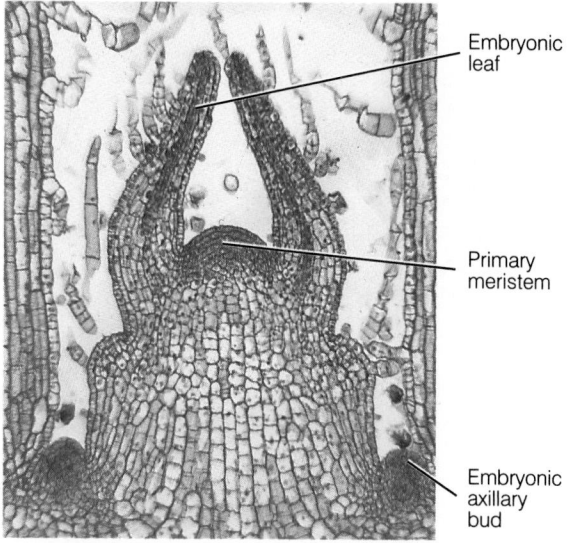

> **FIGURE 26–8 Structure of the Bud** Photomicrograph of a longitudinal section through a terminal bud showing the primary meristems and the embryonic leaves that protect them. These meristems are responsible for the elongation of the stem.

Lengthwise growth of the stem and branches takes place in segments. Consequently, annual growth can be observed for a period of several years along the length of a branch. This is possible because the terminal bud always leaves an encircling scar on the stem. These scars indicate the position of the terminal bud during the previous years. The distance between the terminal bud and the first scar indicates the amount of growth during the most recent period. By counting the number of scars, the age of a stem or branch can be determined. The distance between two adjacent scars varies depending on conditions, particularly temperature and rainfall, during the growing season.

The primary meristematic cells in the terminal bud are replenished by cell division during growth. Because terminal buds remain at the end of all stems and branches, these structures can continue to grow indefinitely.

Herbaceous Stems Differ Markedly in Internal Structure in Monocots and Dicots

Herbaceous plants, such as corn, soybeans, sunflowers, and alfalfa, have stems that are usually small and short-lived, in contrast to those of the woody plants.[1] The stem of a dicot herbaceous plant, such as the sunflower, is made up largely of parenchyma tissue and a ring of vascular bundles, as shown in ▷ Figure 26–9a. In a cross-sectional view, each vascular bundle consists of xylem, which is located toward the center of the stem, and phloem, with a cap of **fiber cells** toward the outside. A layer of vascular cambium may be present between the xylem and the phloem. **Vascular cambium** is a secondary meristem that gives rise to new xylem and phloem. In sharp contrast, the stems of monocot herbaceous plants, such as corn, contain randomly scattered vascular bundles (▷ Figure 26–10).

The outermost tissue in most herbaceous stems is the epidermis, and other tissues next to the epidermis provide support.

In some plants, the entire stem or portions of the stem are located underground and are modified for food storage. Examples include the iris, potato, and onion. The modified stems include rhizomes, tubers, and bulbs (▷ Figure 26–11). The aboveground parts of these plants die in late fall, and new shoot systems are produced from buds on the underground stems.

---

[1] A herbaceous plant is one that is composed mainly of primary tissues.

> **FIGURE 26–9 Dicot Herbaceous Stem** (*a*) Cross section of the stem showing the arrangement of vascular bundles in a ring. (*b*) Enlarged view of a single vascular bundle. The bundles are arranged with the xylem toward the center of the stem and phloem toward the outside. A meristematic tissue called the vascular cambium separates the xylem and phloem.

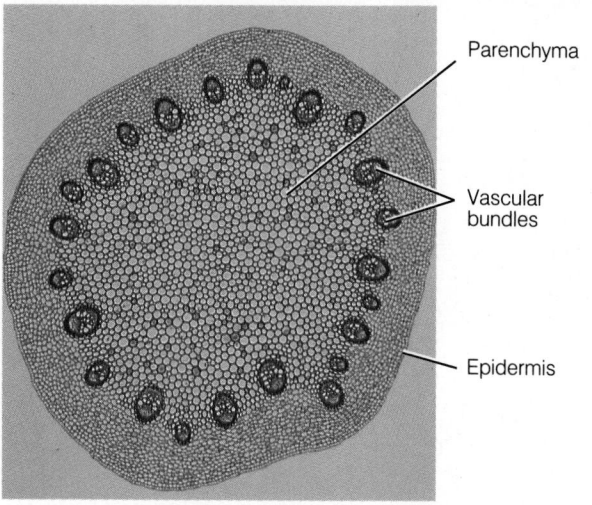

(a)

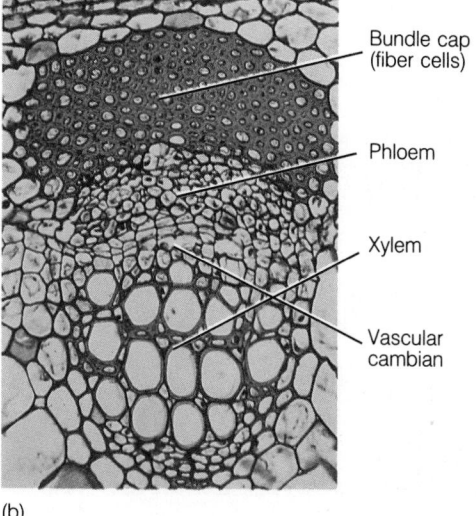

(b)

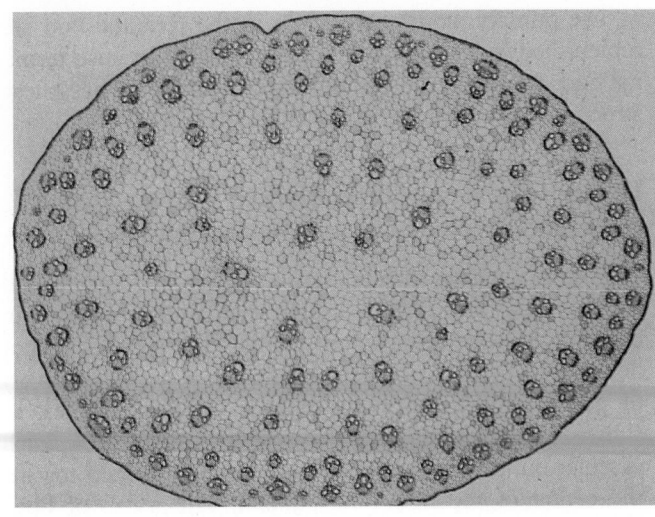

(a)

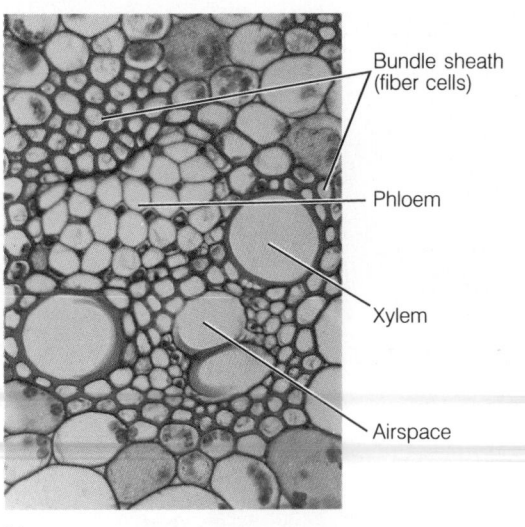

Bundle sheath
(fiber cells)

Phloem

Xylem

Airspace

(b)

▷ **FIGURE 26–10 Monocot Herbaceous Stem** (*a*) Cross section of the entire stem showing the random arrangement of vascular bundles. (*b*) Enlarged view of a single vascular bundle. As in a dicot bundle, the xylem is toward the center of the stem, and the phloem is toward the outside. There is no vascular cambium in a monocot, and the fiber cells go all around the bundle.

Woody Stems Consist of Bark and Wood

All woody plants are either dicot angiosperms or gymnosperms, and the arrangement of tissues in them is very similar, with only minor differences. As shown in ▷ Figure 26–12, the stems of woody plants consist of the **bark** and **wood.** The outer part of the bark is made up of cork for protection, and the inner bark consists of phloem for food transport. Although cork is composed of dead cells, the inner bark (phloem) consists of living cells, which transport food to different parts of the tree. To clear forests for farmland, the colonists of North America often

removed a ring of bark, which stopped the flow of food materials. By so doing, they deprived the roots of food, and the tree died.

Wood consists of xylem and is the tissue through which water and minerals are transported from the roots to the leaves. Unlike the xylem and phloem of herbaceous stems, shown in Figure 26–9b and 26–10b, those of a woody plant are arranged in continuous rings. The xylem and phloem are separated by a single layer of a secondary meristem known as the **vascular cambium** (Figure 26–12). The vascular cambium in a woody plant produces new

▷ **FIGURE 26–11 Stem Modifications**
(*a*) Rhizomes are horizontally growing, underground stems. (*b*) Tubers are terminal portions of rhizomes, as in the potato. (*c*) Bulbs are storage sites for molecules made in the leaves.

(a)

(b)

(c)

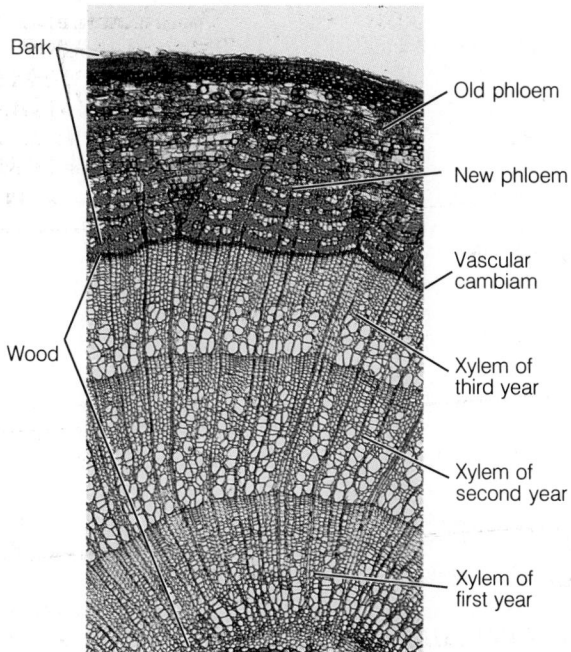

► **FIGURE 26–12 Woody Stem** Cross section of woody stem showing the arrangement of tissues. Xylem makes up the wood, and phloem and cork (not shown) make up the bark. The vascular cambium separates the bark from the wood and gives rise to new phloem and xylem. Note the growth rings in the wood.

xylem and phloem during the growing season, thus adding to the girth of the tree. The newest xylem and phloem of a tree therefore always lie next to each other, separated by the vascular cambium.

Another secondary meristem, known as the **cork cambium,** is interposed between the cork and the phloem in the bark. This tissue gives rise to new cork. The vascular cambium and cork cambium are responsible for the increase in diameter of a woody plant.

The cork that is commonly used as stoppers in bottles comes from species of oak that grow in several Mediterranean countries, especially Portugal. These trees produce unusually thick cork layers, some of which can be removed periodically without hurting the trees.

The xylem produced during each year remains as discrete concentric rings known as **growth rings** (Figures 26–12 and ► 26–13). Counting the growth rings on a core sample taken from the base of the trunk can help one determine the age of the tree without destroying it.

The patterns seen on lumber are also produced by the growth rings. Pieces of lumber cut away from the center of

► **FIGURE 26–13 Growth Pyramid** (*a*) The growth rings within the trunk of a tree can be visualized as a series of cones, each representing one year's growth, with the outermost being the newest. Only the base of a 5-year-old tree will have all five growth rings. This diagram also shows that tree trunks are conical and not cylindrical in shape. (*b*) A cross section from a 30-year-old tree showing growth rings.

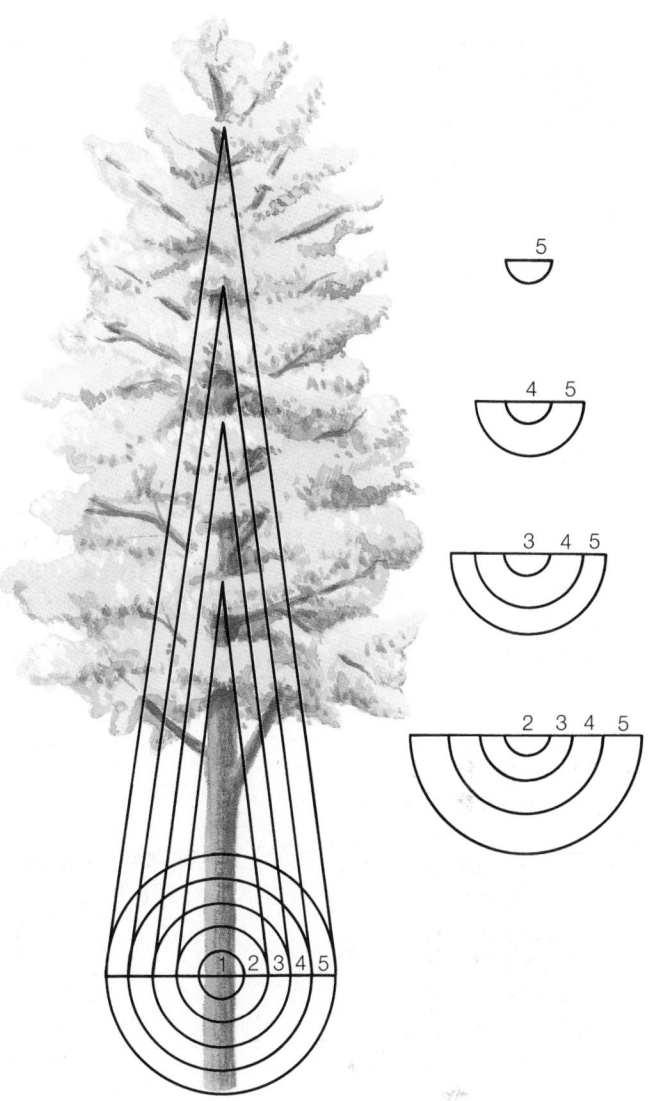

(a)

(b)

THE CIRCULATION OF SAP IN PLANTS
Featuring the work of Hales

Stephen Hales, an 18th-century English scientist, dabbled in many areas. His earliest research investigated the circulatory system of animals and addressed many questions that had been left unanswered since William Harvey's pioneering work 60 years earlier. In 1724, however, after successfully studying blood pressure, blood flow, and other aspects of the cardiovascular system, Hales turned his attention to plants. He began a series of experiments that would establish the main outlines of plant physiology. He studied exchanges between plants and their environment and the movement of sap. He showed how water drawn into the roots of a plant is transported to the leaves, where it evaporates (transpiration).

But Hales was a prominent citizen as well. He was a trustee of the Colony of Georgia and a regular member of various commissions set up to look into public health matters, such as conditions aboard ships of the Royal Navy and alleged wonder cures. One challenge that occupied much of his time for a while was developing ways of delivering fresh air into confined spaces, such as hospitals, prisons, and the living quarters of ships. In fact, many ventilating devices in current use today owe their origin to Hales. It is his work on plants that is the focus of this disucssion.

Like other scientists, Hales stood on the shoulders of giants. One of them was Robert Hooke, an early microscopist who had made careful descriptions of the cellular structure of various plant parts. Early microscopic studies by Hooke and others revealed the presence of branching systems of tubes that ran from the roots to the stems and branches of plants and then to the leaves. This discovery suggested the presence of a circulatory system in plants.

It was up to Hales to demonstate the validity of this hypothesis through scientific experimentation. One of his first experiments was designed to determine if the movement of water from roots to leaves was powered by water pressure in the roots or by some "drawing process" in the leaves. In order to answer this question, Hales cut an apple branch and immersed it in water. He then measured the drop in water level in the vessel. Without roots, the tree limb imbibed considerable amounts of water, suggesting to him the "great power of perspiration," that is, the power of evaporation from the leaves to draw water up the limb.

To confirm his hypothesis, Hales tried to determine if water did indeed evaporate from the leaves. He enclosed them in an airtight vessel, then collected the liquid that evaporated. To his satisfac-tion, most of the "perspired" liquid was water.

Next, Hales turned his attention to the movement of fluids within the plant. Some scientists hypothesized that plants had circlatory systems like those of animals. Watery sap from the roots, they said, moves upward in the inner part of the stems of plants and down in the outer part in a closed system.

To test this hypothesis, Hales took another apple branch and immersed it in water. By carefully cutting away sections of the bark and wood, he observed water ascending, but not descending. He thus concluded that water flowed upward in the plant but was not recirculated to the roots.

Today, we know that water moves in the xylem in a unidirectional flow from roots to leaves. The outer tubules of the phloem discovered by Hales's predecessors carry a carbohydrate-rich sap, as described in this chapter.

Hales performed many ingenious experiments and contributed greatly to our understanding of plant physiology. To his credit, over the next century his successors added very little by comparison to our understanding of plant physiology.

---

a tree trunk tend to show more patterns than those cut closer to the center.

As a tree grows older, the cells in the inner rings of xylem tissue become clogged and eventually lose their ability to transport water. These blocked cells are darker than the younger cells toward the outside and make up the heartwood. The lighter colored functional xylem is often called the sapwood (▷ Figure 26–14).

≋ LEAVES

The most conspicuous structures on the stem are the **leaves,** the food-synthesizing organs of most plants. A typical leaf, such as that of a maple or oak, has a flat, expanded **blade** at the end of a **petiole,** a stalk that is attached to the stem at a position called the **node** (Figures 26–3 and ▷ 26–15a). In plants such as roses and horse chestnut, hickory, and locust trees, the leaf blade is divided into a number of leaflets (Figure 26–15b and 26–15c).

The leaves on a plant are arranged in such a way as to maximize their exposure to sunlight, which is needed for photosynthesis. The leaves of many plants also have the ability to track the sun, altering their orientation during the day as the Earth rotates.

All leaves contain veins, which consist of the vascular tissues, xylem and phloem. As in animals, the vascular system of a plant is continuous, carrying water and minerals to the leaf and transporting food synthesized in the leaf to other parts of the plant.

A leaf is covered on both sides by the epidermis, which is coated with a waxy layer called the **cuticle.** The cuticle prevents the loss of water. As you learned in Chapter 5,

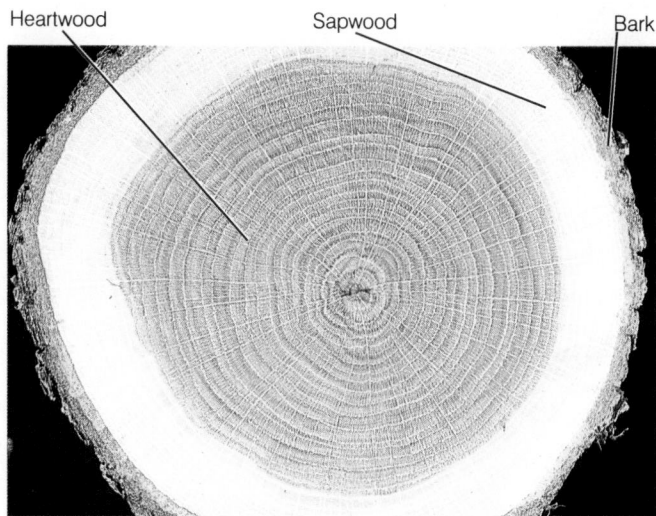

Heartwood   Sapwood   Bark

▷ **FIGURE 26–14 Heartwood and Sapwood** Cross section from the trunk of a tree showing the dark-colored, nonconducting heartwood and the light-colored, functional sapwood. The ratio of heartwood to sapwood increases as the tree grows older.

numerous small openings, **stomata,** are found in the epidermis, usually on the lower surface in dicots and on both surfaces in monocots. The stomata are regulated by **guard cells,** specialized epidermal cells that can change shape depending on their water content (▷ Figure 26–16). When the cells are full and internal pressure in the guard cells is high, the stomata remain open. When water levels fall, the cells wilt, and the stomata close. Gases such as oxygen and carbon dioxide, needed for respiration and photosynthesis, respectively, also diffuse in and out through these openings.

Leaves Contain Parenchymal Cells that Carry Out Photosynthesis

Between the two surfaces of the leaf are the parenchyma and the vascular tissues (▷ Figure 26–17a). The cells of the parenchyma (mesophyll cells) in the leaf contain large numbers of chloroplasts, and the primary function of the parenchyma cells in the leaf is energy and food production. A vascular bundle in a leaf is a section through a vein. Vascular bundles are located midway between the upper and lower epidermis. Only leaves of dicot plants have the columnlike arrangement of cells in the upper half of the leaf, as shown in Figure 26–17a. The loosely arranged, spongy parenchyma cells in the lower half allow easy movement of gases throughout the leaf interior.

Leaves Have Evolved Many Forms to Meet Different Environmental Conditions

Even a casual observer can recognize a great variation in leaf size and shape among different species of plants. In some plants, leaves are reduced in size to minimize water loss. In others, like cacti, they are completely modified to form spines, which also minimize water loss and protect the plants from being eaten. (For a discussion of other modes of protection, see Spotlight on Evolution 26–1.) As noted earlier, the stems of cactus and certain other species of succulents store large amounts of water and are green. Such stems also carry out photosynthesis, a process normally restricted to leaves. Another unusual leaf modification is seen in plants such as the Venus flytrap, sundew, and pitcher plant. These species, which grow in marshy areas where nitrogen supplies are inadequate, have specially modified leaves that trap and digest insects, making up for the lack of nitrogen in the soil.

▷ **FIGURE 26–15 Leaf Types** The leaf blade of the basswood (lower center) is undivided and is called a simple leaf. The blade of compound leaves is divided into leaflets as in (left to right) black locust, horse chestnut, and ash.

▷ **FIGURE 26–16 Guard Cells** Pairs of guard cells with unevenly thickened inner walls regulate the opening and closing of the stoma.

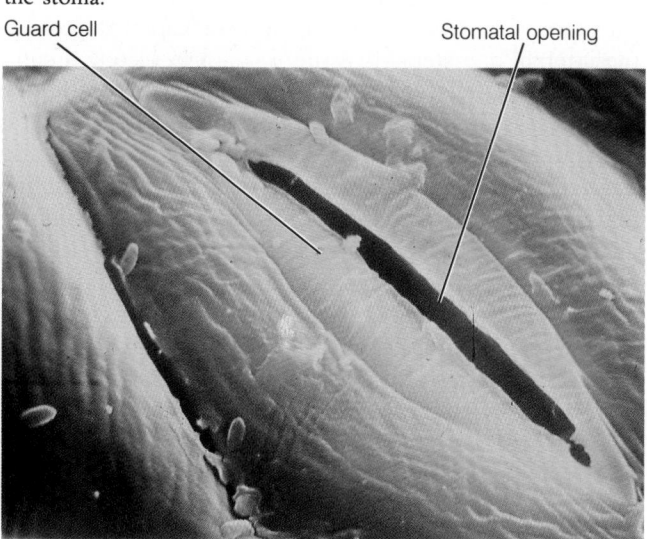

Guard cell   Stomatal opening

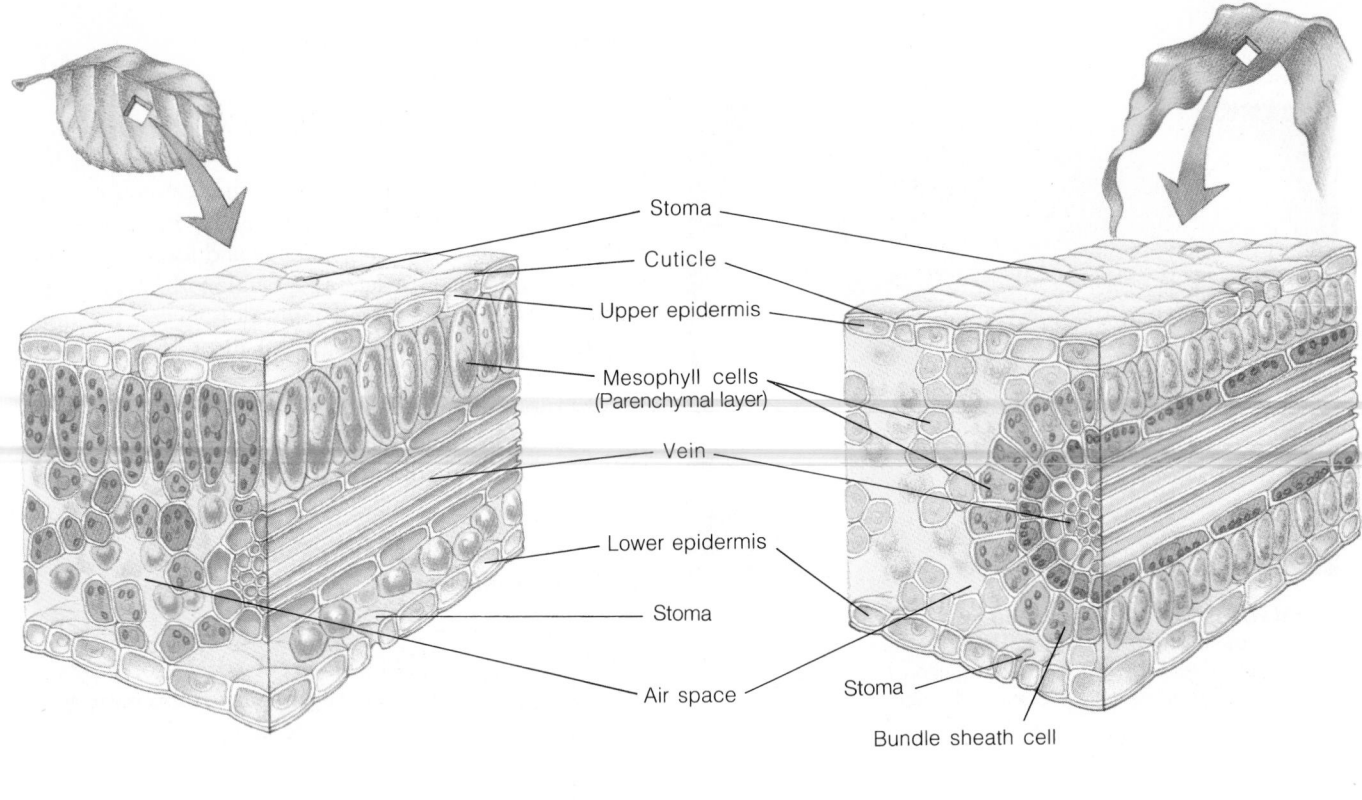

Stoma
Cuticle
Upper epidermis
Mesophyll cells
(Parenchymal layer)
Vein
Lower epidermis
Stoma
Air space
Stoma
Bundle sheath cell

(a) Dicot leaf

(b) Monocot leaf

▷ **FIGURE 26–17 Leaf Cross Sections** (*a*) A dicot leaf with stomata and guard cells on the lower epidermis. (*b*) A mono- cot leaf. Note that the column-shaped parenchyma layer (mesophyll cells) is absent in monocot leaves.

≈ THE ROOT SYSTEM

As a rule, the underground parts of a plant make up the **root system,** which functions to anchor the plant, absorb water and minerals from the soil, and store food. Although not very apparent, the root system of a plant is as extensive as its shoot system, sometimes more so. The largest number of roots is usually in the upper 4 feet of soil, where most of the minerals are present.

The terminal portions of all roots are similar in structure. Most root tips have a thimblelike **root cap** that protects the internal meristematic tissue. Most also secrete a slimy substance that allows the root to slide through the soil as it elongates. A short distance from the tip is a region that is characterized by a large number of **root hairs.** Root hairs are extensions of epidermal cells that absorb water and minerals. Root hairs are easily broken off when a plant is pulled up from the soil and thus are not commonly seen on uprooted plants. Another external structure found in the root system is the **branch root.** Branch roots are found on more mature sections of the root system.

Between the tip and the root hair zone lies the primary meristem. Elongation of roots, as in stems, results from cell divisions in the meristems and the subsequent enlargement of the new cells. Lying outside the region of the primary meristem where the cells elongate is an area where cells develop into specialized tissues.

The arrangement of tissues in the mature part of the dicot root can be seen in ▷ Figure 26–18. As illustrated, the xylem and phloem are in the center of a root, surrounded by parenchyma and an outer epidermis. A ring of dead and living cells that forms a layer known as the **endodermis** separates the parenchyma from the central core.

Food is usually stored in the parenchymatous layer, as in carrots. Water and minerals moving through this layer reach the xylem through the living cells of the endodermis. Branch roots arise from pericycle present immediately inside the endodermis (▷ Figure 26–19).

The major difference in root anatomy between the dicot and the monocot, which is shown in ▷ Figure 26–20, is in the location of the vascular tissues. In dicot roots, the xylem occupies a central position, but in monocots it is present in isolated groups around the centrally located parenchyma.

Roots of woody plants have vascular cambium and cork cambium, and become woody. These roots also increase in diameter by the yearly accumulation of xylem and cork, as does the stem. Consequently, growth rings similar to those seen in woody stems can be seen in woody roots.

Many plant species are characterized by aboveground roots adapted to perform functions not typically associated with roots. Some notable examples include banyan trees,

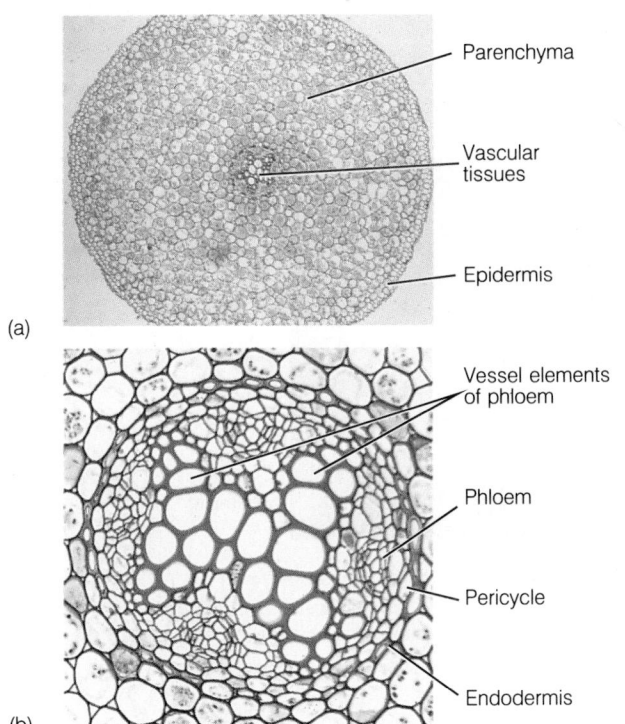

(a)

(b)

▷ **FIGURE 26–18 Cross-Sectional Views of Dicot Root**
(*a*) This photomicrograph shows an outer cortex consisting of parenchyma and central core of vascular tissues. (*b*) Enlarged view of the central core of the same root showing the star-shaped arrangement of centrally located xylem, isolated groups of phloem, a prominent endodermis, and the pericycle.

found in tropical areas, where large, pillarlike roots grow downward to the ground from heavy, horizontal branches and serve primarily as props that prevent the branches from breaking. Various species of ivy produce climbing roots along their stems that attach the weak-stemmed plants to a wall (▷ Figure 26–21a). Certain tree species, such as crab apple and cherry, send up "suckers" from their roots, which give rise to new trees. Roots can also grow from the cut end of a stem or leaf, making it possible to clone many kinds of plants from stem or leaf cuttings. Two species of parasitic plants, mistletoe and dodder, send their roots directly into the vascular system of the plants they parasitize, obtaining food and water from their hosts (Figure 26–21b).

≈ INTERNAL TRANSPORT

As in animals, many substances are transported within plants. However, the vascular system in plants, unlike that in animals, operates without a pump. Concentration gradients and energy from sunlight are primarily responsible for transporting water and food within plants. For example, molecules that can diffuse freely through plasma membranes, such as ions, move along concentration gradients without any energy expenditure. Molecules moving against a concentration gradient require an expenditure of energy.

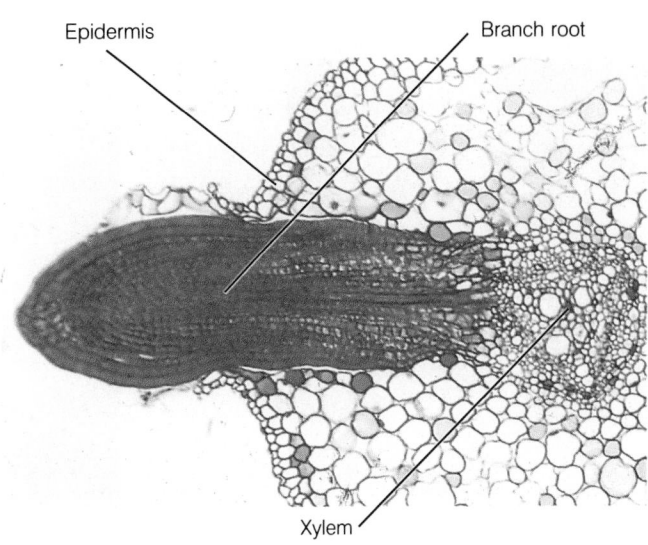

▷ **FIGURE 26–19 Branch Root** Cross section of a willow root showing the origin of a branch root from the pericycle, which allows for the flow of water and minerals from the soil through the root to the vascular system.

These processes in plants are similar to those discussed in Chapter 3.

The pressure that develops inside a plant cell due to the uptake of water is known as **turgor pressure.** Loss of water reduces the turgor pressure and causes the plant to wilt.

Water Is Transported Through the Bodies of Plants by Two Basic Mechanisms

Water enters plants from the soil through their roots and is transported upward to their parts through the xylem. As you have seen, both vessel elements and tracheids of the

▷ **FIGURE 26–20 Cross Section through Monocot Root**
Note the location of the xylem and phloem, and compare with the dicot root shown in Figure 26–18.

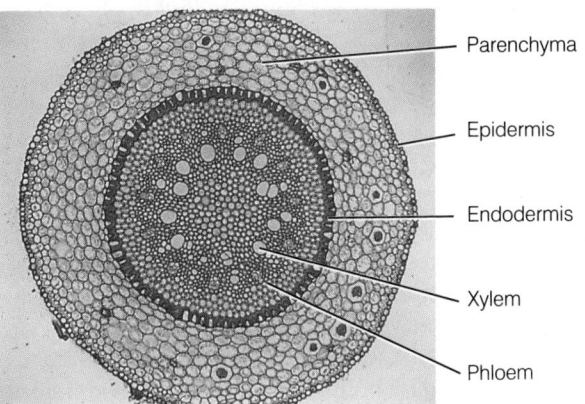

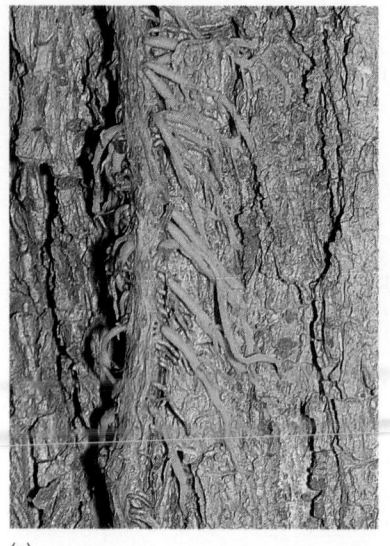

(a)

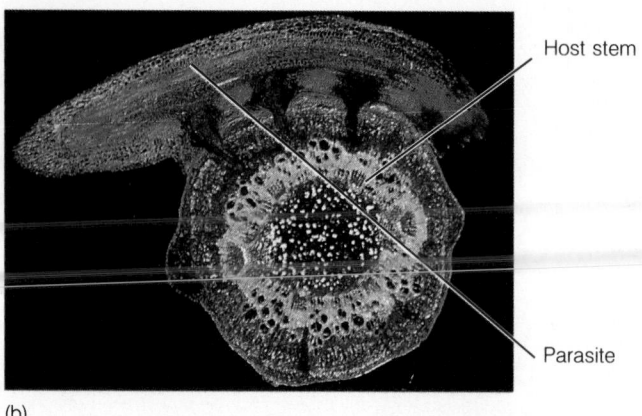

Host stem

Parasite

(b)

▷ **FIGURE 26–21 Modified Roots** (*a*) Climbing roots produced from the nodal regions of a stem of Virginia Creeper vine. Such roots help a weak-stemmed plant attach itself to a substrate for support. (*b*) The haustorium is a root modification that allows a parasitic plant to absorb food and water from its host. This photomicrograph shows a section of a haustorium attached to the stem of a host plant.

xylem are attached end to end to form a continuous pipe system from roots to leaves. Although there is some lateral movement of water within plants, the bulk of the water moves upward.

The evaporation of water from leaf surfaces, called **transpiration,** is responsible for the upward movement of water from a plant's roots. According to the **transpiration-cohesion-tension theory,** transpiration generates a suction in the vessel elements and tracheids similar to that created inside a straw when a person draws liquid into it. Similarly, transpiration pulls columns of water all the way from the roots to the leaves through the vessel elements and tracheids. The strong cohesion between water molecules, resulting from hydrogen bonding, and the high tensile strength of the narrow columns of water in the vessel elements and tracheids keep these columns intact. Water from the soil moves passively into the roots as this process continues. The thick secondary walls of the vessel elements and tracheids prevent them from collapsing under tension.

Plants use only a small fraction of the water that is taken up from the soil. Most of the water is lost to the atmosphere through stomata in the leaf epidermis. As mentioned earlier, the two guard cells that form each stoma regulate transpiration. The walls of the guard cells are thicker along the inside than the other parts (▷ Figure 26–22). When water moves into the guard cells by osmosis, the increasing turgor pressure causes the cells to bulge outward, opening the stoma. When there is excessive transpiration, water moves out of the guard cells, allowing them to return to their relaxed state and closing the opening, which prevents excessive loss of water.

Transpiration also benefits the plant by cooling its leaves and facilitating the distribution of mineral ions to different parts. In addition, transpiration plays a significant role in the hydrological cycle, moving water from the soil to the air where it may fall as rain or snow. As a result, plants can have a modifying influence on our weather and on the composition of our landscapes. This has become painfully apparent in recent years in areas of the world where excessive deforestation has caused desertification of once forested areas, a topic detailed in Chapter 33.

Water can also move into plants in the absence of transpiration, a phenomenon that occurs quite commonly in herbaceous plants when the air is saturated with moisture. In these plants, guard cells generally close at night, which ends transpiration. However, because the concentration of water is

▷ **FIGURE 26–22 A Stoma in Action** (*a*) When water diffuses out of the guard cells, they resume a more elliptical shape, closing the gap between the paired cells and preventing excessive loss of water. (*b*) As water moves into the guard cells, they acquire a kidneylike shape, causing the stoma to open.

(a) Stoma open (turgor high)　　　(b) Stoma closed (turgor low)

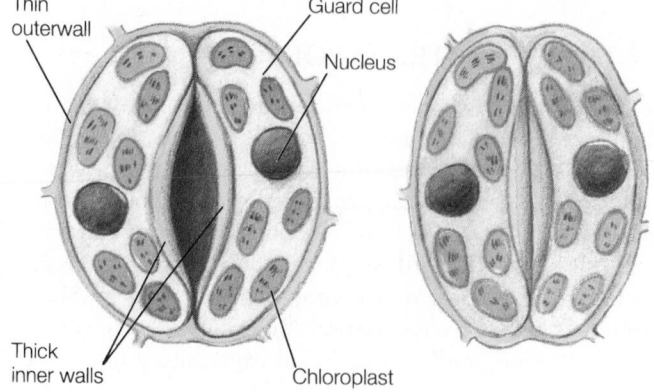

Thin outerwall

Guard cell

Nucleus

Thick inner walls

Chloroplast

THE CHEMICAL WEAPONRY OF PLANTS

Don't be fooled for a minute by the helpless and vulnerable appearance of a plant. It's a guise. In truth, many plants are tough combatants in life and have evolved a variety of measures that protect themselves from hungry animal predators, especially insects and herbivores.

One useful mechanism is the growth and regeneration strategy—that is, the ability to regenerate parts consumed by other organisms. Many plants have evolved spines that deter others. Nonetheless, the main form of protection comes from an arsenal of protective chemicals—toxins that taste bad or are outright poisons. Both tend to deter predation. To date, scientists have extracted and identified over 10,000 defensive chemicals made by plants, including morphine, phenol, caffeine nicotine, and tannin.

Many of the chemicals extracted from plants have already proved useful to humans in battling crop insects or diseases such as cancer. Organic gardeners, for example, use a naturally occurring chemical extracted from certain plants. This substance, called rotenone, protects many vegetables, such as cabbage and cucumber, from hungry caterpillars, beetles, and other insects.

By various estimates, half of all over-the-counter and prescription drugs come from plants. Worldwide, these medicines contribute well over $40 billion a year to the human economy.

One striking example of a plant that packs a punch is a grass called *Ajugo remota*, found in Africa. Witch doctors use the leaves to treat a variety of disorders, among them malaria and dysentery. Extremely resistant to locusts, the plant also produces chemical substances virtually identical to insect hormones that promote molting, the periodic loss of an insect's exoskeleton that is required for growth. More potent than the regular hormones, this naturally occurring chemical causes rapid molting. In laboratory studies, several common insect pests that were fed the plant were unable to shed their exoskeleton com-

pletely when they molted, and they were eventually accumulated as many as three unshed head capsules. The capsules covered the insects' mouths and prevented them from eating. Death followed in quick order.

Natural selection has also provided many other plants with a chemical that protects against predators. For example, oak leaves produce a group of chemicals called tannins, which are stored in the central vacuoles of the leaf cells. When the leaves are consumed by caterpillars, the vacuoles break down and release their tannins. The tannins bind to plant proteins and make them indigestible. This slows the growth of caterpillars, decreasing chances that they will survive and develop into reproductive adults.

Some plants, such as black cherry, produce cyanide, a chemical lethal to most insects. A group of plants known as the legumes (including peas, beans, locust trees, and vetch) produce toxic alkaloids, among them nicotine, strychnine, opium, and peyote. These substances are poisonous to heart muscle, nervous tissue, and other cells. The lovely orchid produces calcium oxalate, which gives it a bad taste and causes illnesses among its would-be predators (▷Figure 1).

As you might expect, however, through the process of coevolution many insects have evolved mechanisms to detoxify or avoid poisons in plants. The leaf-mining beetles, for example, burrow into the interior of oak leaves and eat the cells in the inner layers which do not contain tannins. The cat erpillar of the monarch butterfly accumulates a toxin it acquires from the milkweed plants it feeds on, and stores them in its body. Subsequently, birds that feed on adult forms become sick and avoid further predation on this butterfly.

Far from being passive, plants also respond to attack by manufacturing defensive chemicals. When a tomato leaf is chewed by an insect, for instance, the cells release the plant's wound hormone. Circulated throughout the plant, this hormone stimulates the production of a group of chemicals called "antinutrients" that inhibit digestive enzymes in the intestines of insects. This, in turn, makes it difficult for an insect to digest its food. Antinutrient production continues as long as the plant is under attack and increases if the attack is intensified.

By identifying the genes that produce protective substances and introducing them into crop plants, scientists may be able to transfer the ability to make protective chemicals into important food sources. This could reduce or even eliminate the need for hazardous chemical pesticides.

Naturally occurring toxins could also be produced commercially for direct application on plants. Scientists hope that these toxins might provide protection while decaying more rapidly than synthetic insecticides, thus producing fewer environmental impacts.

Clearly, plants produce a wealth of chemicals important to human use. Unfortunately, many useful species are being destroyed by the rampant deforestation in the tropics. Protecting the tropical forests is therefore vital to preserving evolution's legacy of protective chemicals.

▷ **FIGURE 1 Don't Be Fooled by Its Beauty** The orchid contains a bad-tasting chemical, calcium oxalate, which protects it from hungry predators.

usually higher in the soil than it is inside the root cells, water moves in by osmosis. The increased pressure within the root cells forces the water up the plant. This water comes out through open veins along the leaf tips or margins. The wetness one experiences when walking across a lawn or through a field of corn or tomatoes in the morning is often caused by this phenomenon.

Food Molecules Are Transported from Source to Sink

As you know very well by now, plants, unlike animals, can synthesize their own food through photosynthesis. The food that they synthesize is in the form of sugar phosphate (Chapter 5). Using this basic building block, plants synthesize a variety of additional carbohydrates, such as glucose, fructose, sucrose, and starch.

Food is usually transported in plants as sucrose, common table sugar, which is readily soluble in water. However, food is stored as insoluble starch inside plant cells. Unlike the unidirectional movement of water from roots to leaves, transport of food takes place in both directions.

As mentioned earlier, food is transported through living cells, the sieve tube members, which are part of the phloem tissue. Sieve tube members are tubular cells and are attached end to end, forming a continuous pipe system from leaves to roots. Food always moves from source to sink. The source is the site where food is synthesized or readily available, and the sink is where the food is used or stored.

In the growing season, food synthesized in the upper leaves of a plant is usually transported upward to the buds. Food synthesized in the lower leaves generally moves downward for use in roots and for storage. However, during early spring, before the leaves open, most of the food moves upward from the areas where it is stored to parts of the plant where new growth is occurring.

Food transport is explained by the **pressure-flow hypothesis,** which states that food molecules move from regions of high pressure to regions of low pressure. When newly synthesized sucrose dissolves in water within the leaf cells (source), water moves into the cells by osmosis. The increased cell content raises the turgor pressure in these cells. Because phloem is a closed system, increased pressure at one end of the system forces the food molecules to parts of the system in which the pressure is lower (sink). Thus, during the growing season, food molecules generally move from the leaves, where food is synthesized, to parts of the plant where food is used or stored. Because food is usually stored in the form of insoluble starch, there is no change in pressure in the cells in which starch is formed and stored. In the early spring, however, starch in the roots is converted to sugar, and a reverse process takes place. Water flows into the root cells, and the increased pressure in them forces food molecules outward to areas where the pressure is low. A good example of this phenomenon is the sugary sap removed from sugar maples in the early spring, which is used to make maple syrup. The sap is concentrated by boiling.

Minerals Are Actively Transported Into Root Cells and Passively Carried Through the Xylem with Water

Plants use a variety of minerals, as do animals. Various deficiency symptoms, such as yellowing and mottling of leaves and stunted growth, are common when plants are deprived of essential minerals. Yellowing of grass is a common occurrence in a lawn that is deficient in nitrogen and other nutrients.

The minerals that are required for normal plant growth are present in the soil as positively or negatively charged ions. Plants take up these mineral ions from the soil and use them in a variety of processes. Animals obtain many of the minerals they require by eating plants.

Some minerals, such as nitrogen, potassium, and phosphorus, are required in larger quantities than others, such as copper, iron, and zinc, which are needed only in trace amounts. Intensive farming methods practiced in the United States and other developed countries, in which crops are grown year after year in the same field, may deplete the soil of many minerals. To avoid depletion, farmers often add fertilizer to their fields. A balanced fertilizer contains all essential minerals in the proper concentrations needed for normal plant growth. Organic farmers often use natural fertilizers—for example, manure. This practice is often better for the soil, because it replaces minerals and valuable organic matter, which acts as a sponge, holding water for crops. Organic matter also feeds bacteria and other microorganisms that are essential for proper soil nutrient recycling, as discussed in Chapter 33.

Unlike water, minerals in the soil do not move freely into root cells, primarily because their concentrations are generally higher inside the root cells than they are in the soil. In order to enter the root cells, the ions must move against concentration gradients. This process requires specialized active transport proteins that are found in the plasma membrane of the root cells (Chapter 3). Different ions are transported at varying rates, and once the ions are inside the plant, they are transported to the different parts with water through the xylem.

≈ PLANT GROWTH AND RELATED PHENOMENA

Plant growth is an exceedingly complex process involving many unresolved mechanisms. Because of its significance, plant growth has received considerable attention and research funds in recent years.

The growth of a plant is controlled by its genetic makeup, acquired over millions of years of evolution. But several other factors, both internal and external, also influence plant growth and development.

Numerous Hormones Control Plant Growth and Other Vital Functions

Hormones either enhance or inhibit growth, depending on the hormone and its concentration. In general, hormones are made in the very young parts of a plant. As in animals, hormones can influence growth in areas away from where they are synthesized. Too, only small amounts of hormones are needed to bring about an irreversible growth change.

Several classes of plant hormones exist. The proper concentration and interaction of these hormones are responsible for normal growth. One group of hormones, the **auxins,** are responsible for cell enlargement. Cell division in plants is controlled by a hormone known as **cytokinin.** The auxins and cytokinin together regulate the transformation of undifferentiated tissues into stems and roots. Another well-studied hormone is **gibberellin,** best known for its ability to induce stem elongation. **Abscisic acid** is the hormone believed responsible for the induction of leaf abscission, the separation of the leaf from the stem in the fall. Leaf abscission results from the formation of a layer of cork tissue at a point where leaves are attached to the stem. This tissue prevents the transport of food and water in and out of the leaves, resulting in their death and abscission (severance) in the autumn. Abscisic acid is also known to have an important role in triggering stomatal closing when excessive transpiratory water loss occurs. **Ethylene,** another plant hormone, induces the ripening of fruits.

Tropism Is a Plant's Response to a Unidirectional External Stimulus

Plants respond to external stimuli. This phenomenon, called **tropism,** may be elicited by any of a number of stimuli, such as light, gravity, and touch. Plants usually respond by bending toward or away from a stimulus, the former being a positive response and the latter a negative one. The bending of a plant toward light (described in Scientific Discoveries 25–1) results from the differential elongation of cells in the stem; that is, the cells on one side of the stem become longer than those on the other side. The stem always bends toward the side where the cells are smaller. As a rule, only young, growing regions of stems and roots can respond to tropic stimuli.

Phototropism Is the Growth Response of a Plant to Light From One Direction.
▷ Figure 26–23 shows a familiar form of tropism, the bending of stems toward light, or **phototropism,** often witnessed in houseplants on a windowsill. Most people avoid this problem by rotating the pot every few days. Plants usually bend toward the bright light, which ensures a more adequate rate of photosynthesis.

Phototropism is brought about by an auxin, which is normally distributed evenly at the growing tips of stems. When plants are exposed to light from one direction, however, the auxin is transported away from the lighted side of

▷ **FIGURE 26–23 Phototropism** Bean seedlings bend toward light from one direction.

the plant. The increase in auxin concentration in the shaded side induces the cells there to elongate. This results in the bending of the stem toward light. As a sidelight, several widely-used herbicides are auxin-like chemicals that stimulate weeds to grow so quickly that they die. For a discussion of the use of herbicides and genetically-engineered herbicide resistance in crop species, see the Point/Counterpoint in this chapter.

Geotropism Is a Response to Gravity. When a plant falls over on its side, the stem often bends upward, and the roots bend downward (▷ Figure 26–24). In addition, the

▷ **FIGURE 26–24 Geotropism** The stem of a coleus plant placed on its side bends upward, showing a negative response to gravity.

THE BENEFITS OF GENETICALLY ENGINEERED HERBICIDE RESISTANCE

Charles J. Arntzen

Dr. Charles J. Arntzen is the Director of the International Plant Biology Program in the Institute of Biosciences and Technology in the Texas Medical Center in Houston, Texas. Dr. Arntzen was elected to the U.S. National Academy of Sciences in 1983 as a result of his pioneering research in photosynthesis and plant molecular biology. He is also Chairman of the National Biotechnology Policy Board of the National Institutes of Health, and is a member of the Editorial Board of Science.

One of the principal outcomes of plant genetic engineering has been a rapid expansion in our understanding of plant cell biology, genetics, and molecular controls over the structure and function of plants. This understanding is now helping to solve practical problems such as developing crops which are insect or disease resistant, thereby decreasing needs for chemical pesticides. Another approach is to create herbicide resistant plants, which will allow farmers more choices in weed control, and to switch herbicide uses to newer, rapidly biodegradable compounds.

Herbicides are chemicals sprayed on crops to control weeds. Their effect is selective—that is, they kill weeds without significantly harming crops. All herbicides currently available to farmers began with an evaluation of the sensitivity of weeds and crops to experimental chemicals. This evaluation is accomplished by applying potentially useful chemicals to test samples of weed and crop seedlings. Compounds that kill weeds but not crops are then subjected to animal toxicology studies. Those deemed acceptable are developed for farmers' use.

The success of herbicides is based on the fact that crop species contain enzymes that convert chemical compounds, including herbicides, to inactive forms. These enzymes tend to be crop specific. For instance, the enzymes that render soysbeans immune to 30 or more commercial herbicides are different from the enzymes that make corn resistant to other herbicides. This can be a benefit when corn or soybeans are grown in rotation. In such instances, both the unwanted "volunteer" corn and weed seeds that germinate in a soybean field can be controlled by the soybean herbicide.

Interestingly, certain weeds have the same enzymes as the crop species they invade. This is often true when the weeds and crops are somewhat closely related, such as wild oats in a field of wheat or barley. Thus, they are resistant to the same herbicides. Consequently, the farmer has few or no choices of chemical weed control since herbicides that would kill the weeds would also kill the crop.

Through genetic engineering, researchers have devised ways to make crops resistant to new classes of herbicides. That is, they found ways to give crops such as soybeans additional enzymes to augment their natural herbicide-resistance. By transferring genes that code for enzymes conferring herbicide resistance, scientists can give farmers a broader set of options when selecting herbicides. If successful, these broader options will solve problems where there currently is no weed control, or where the herbicides available are prohibitively expensive or environmentally damaging.

This line of research may also provide ways to give the farmer a simpler means to control weeds through the use of a single, more effective herbicide selected on the basis of several traits, including herbicide resistance, reduced cost, and, ideally, greater safety in the environment.

This explanation of the benefits of genetic research aimed at making crops more resistant to herbicides may not satisfy the reader who is asking, "Why don't they simply stop using all herbicides?" The main answer is economics for the farmer. Weed control is essential for all crops. Left unchecked, weed populations can reduce crop yields to near zero. Prior to the 1950s, weeds were controlled strictly by mechanical means—that is, cultivation or hand pulling. This was a labor intensive and costly process. By the 1980s, more than 95% of all major row crops in the United States were produced using herbicides. Farmers based this decision (to apply herbicides rather than using a cultivator or a hoe) on costs. An application of herbicides at the time of planting, costing as little as $10 per acre, is many times cheaper than mechanical cultivation. Weed control through herbicides is a factor in the low food prices we enjoy in the United States. It also saves on energy use, and when properly used, does not harm the environment.

With the continuing development of herbicide-resistant crops as part of an integrated weed control strategy, crop production cost can be held at low levels, while newer, but as yet unavailable, methods for weed control will be discovered.

THE PERILS OF GENETICALLY ENGINEERED HERBICIDE RESISTANCE

Margaret Mellon

Margaret Mellon has been the Director of the Biotechnology Policy Center at the National Wildlife Federation since its founding in 1987. The Policy Center comments on proposed rules concerning releases of engineered organisms, testifies before Congress, and organizes grassroots campaigns. Dr. Mellon lectures widely on biotechnology issues and has appeared frequently on television and radio talk shows. Among her recent publications is a contribution to The Genetic Revolution, *Bernard Davis ed., published in 1991 by the Johns Hopkins University Press.*

Environmentalists were active in the first wave of debate about genetic engineering, and their interest has increased since that time. The current wave of *public* interest in genetic engineering is based on the prospect of commercial production of a broad range of genetically engineered organisms—for example, bacteria and viruses. These organisms could have direct and adverse impacts on our environment and our health.

Public concern over genetic engineering is now also focused on indirect environmental consequences, the best example of which is herbicide-resistant plants. The development of major crops that are tolerant of chemical herbicides seems likely to lead to increased or prolonged use of these dangerous agricultural chemicals.

Herbicide-resistant crops are being developed by large, transnational chemical companies. Worldwide, at least 28 enterprises have launched more than 65 research programs to develop herbicide-resistant crops. Fifteen major crops are involved, including cotton, corn, soybeans, wheat, and potatoes. Engineered crops being made resistant to a company's own herbicides will surely increase that company's market share in chemicals. Thus, industry analysts expect herbicide-resistant plants to be big business, mainly for pesticide and herbicide producers.

Herbicide-resistant products run directly counter to the promise of a reduced dependence on agricultural chemicals put forward by biotechnology advocates. Rather than weaning agriculture from chemicals, these crops will be shackled to herbicide use for the foreseeable future. Herbicides represent an estimated 65% of the chemical pesticides used in agriculture. By continuing to promote the use of herbicides, the biotechnology industry greatly restricts its potential for reducing overall chemical use.

While herbicide-resistant crops offer no hope of environmental benefits in agriculture, there are alternatives to these products that *do* promise substantial reduction in overall pesticide use. These alternatives fall under the rubric of sustainable agriculture.

Sustainable agriculture employs a variety of agricultural practices such as crop rotation, intercropping, and till fields (that produce long ridges of soil on which plants are grown). These practices control pests without the application of synthetic pesticides and fertilizers. Used by knowledgeable farm managers, these techniques work to make farms more profitable and environmentally sound. Growing different crops in successive growing seasons, for example, dramatically reduces pests by sequentially removing the hosts on which the pests depend. With fewer weeds or insects to contend with, the farmers reduce the need for costly chemical pesticides. Once thought impractical, sustainable agriculture is rapidly gaining support in national policy forums, including the National Academy of Sciences.

The environmental advantage of such an approach is clear. Crop rotations that reduce pests reduce the need for pesticides now and into the future. Thus, developing sustainable practices should be the highest priority for pesticides now and into the future. Thus, developing sustainable practices should be the highest priority for agriculture as it moves toward the twenty-first century.

Although a minor part of the sustainable agriculture picture up to now, biotechnology could help by developing new crop varieties usable in sustainable systems—for example, faster germinating, cold-tolerant crop varieties could enhance low-input sustainable systems by enabling more effective weed control. Engineered products such as these would fulfill the promise of biotechnology, and would benefit farmers, the environment, and the rural economy alike.

≈ SHARPENING YOUR CRITICAL THINKING SKILLS

1. State the main thesis of each author in your own words.
2. List the data or arguments used to support each hypothesis and the data used by the authors to refute the other's point of view, if any.
3. Can you determine if there are any flaws of reasoning in either essay?
4. Which viewpoint do you agree with? Why?

shoot system of a germinating seedling always grows upward and the root system grows downward, irrespective of the direction in which the seed is planted in the ground. These phenomena are examples of **geotropism,** a process controlled by hormones and highly beneficial to plants.

Geotropism is also stimulated by an auxin. For example, when stems and roots are horizontal, the auxin is transported downward, resulting in increased concentration on the lower surface of these organs. In the stem, the increased auxin concentration induces cell enlargement on the lower surface, which causes the stem to bend upward. In the roots, however, the increased concentration on the lower surface of a root makes the cells smaller than on the upper surface, resulting in the downward bending of the root.

Note that in both phototropism and geotropism, the respective stimuli, light and gravitational force, come from one direction only. Plants do not bend if exposed to light evenly from all directions.

▷ **FIGURE 26–25 Action of Gibberellin** Spraying gibberellic acid on California poppies (right) causes them to grow much larger than untreated plants (left).

Changing Water Pressure Inside Plants Is Responsible for Some Movements

Mail-order catalogs commonly advertise "prayer plants" as exotic items. The praying plants are usually species of *Maranta.* These plants, like many others, including the shamrock (a species of oxalis), fold their leaves in the evening, an adaptation that conserves water. The folded leaves of *Maranta* have the appearance of hands folded during prayer.

This phenomenon, known as **sleep movement,** is brought about by reversible changes in turgor pressure within a group of specialized cells at the base of the leaf stalk (petiole). Reversible plant responses influenced by turgor pressure are also responsible for the sudden closing of leaves of the sensitive touch-me-not and the insectivorous Venus flytrap.

A Lack of the Hormone Gibberellin Is Responsible for Dwarf Varieties of Plants

Several common varieties of plants, such as pea, come in dwarf varieties. In addition, in plants such as cabbage and lettuce the distance between successive leaves on the stem is very short. Studies have shown that dwarf varieties of plants and leaf patterns on cabbages and lettuce result from naturally low levels of gibberellin, the hormone responsible for stem elongation. In fact, by spraying dwarf peas and normal cabbage with a low concentration of gibberellic acid, one can produce pea plants of normal height and cabbage plants that grow 6 to 10 feet tall (▷ Figure 26–25)!

The Apical Bud of a Plant Inhibits Growth of Axillary Buds

Have you ever wondered why plants tend to become bushier after being trimmed? The answer is that pruning removes the terminal buds and allows the development of the adjacent axillary buds, resulting in more branches and a fuller appearance. Studies have shown that the apical bud exerts an inhibitory influence on the growth of the axillary buds. This phenomenon, called **apical dominance,** is hormonally controlled. Normally, auxin produced in the terminal bud moves down the stem to the axillary buds. The closer an

▷ **FIGURE 26–26 Apical Dominance** (*a*) Plant with intact apical bud. (*b*) Removal of the apical bud induces the development of the axillary buds into branches. An application of an auxin after the removal of the apical bud prevents the development of the axillary buds into branches.

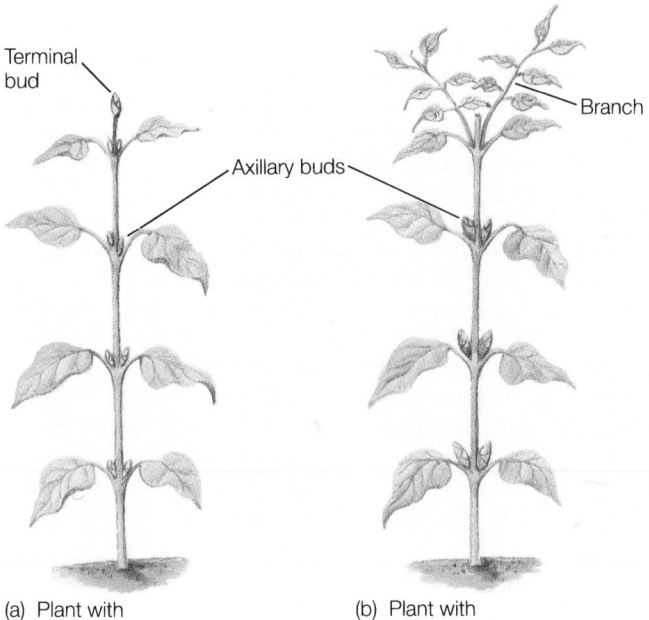

(a) Plant with intact terminal bud

(b) Plant with terminal bud removed

axillary bud is to the terminal bud, the higher the concentration of auxins. Elevated auxin levels in the axillary buds exert an inhibitory effect that prevents the axillary buds from developing into branches. Thus, when the terminal bud is removed, axillary buds are released from inhibition and tend to produce more branches (▷ Figure 26–26).

Flower Induction Is Brought About by a Plant Pigment and May Also Be Hormonally Controlled

Flowering, culminating in the production of seeds and fruits, is an important phase in the life cycle of all angiosperm plants. You may have noticed that many plants, such as the maple, cherry, crab apple, tulip, and daffodil, produce flowers only in early spring, whereas others, such as chrysanthemums, bloom only in the fall. Still others, among them roses and impatiens, bloom from spring until a hard frost kills them.

In many plants, the timing of flower production (flower induction) is controlled by the relative length of day and night, a phenomenon known as **photoperiodism.** As you learned in Chapter 21, reproduction in many animals is also influenced by the light-dark cycle.

Based on photoperiod requirements, plants fall into three groups: short-day plants, long-day plants, and day-neutral plants. In general, **short-day plants** (strawberry, chrysanthemum, tulip, crocus, dandelion, ragweed, and many trees) flower during early spring and/or late summer when the nights are relatively long and the days are short. **Long-day plants** (potato, onion, wheat, mint, and lettuce) flower during late spring and early summer when the nights are short and the days are long. Day-neutral plants, such as rose, impatiens, snapdragon, and tomato, do not have a photoperiod requirement for flowering.

Studies have shown that uninterrupted night periods of specific duration are critical for initiating flower production. A break in the night period, in fact, will inhibit flowering of short-day plants and promote flowering of long-day plants (▷ Figure 26–27).

Studies have also shown that a plant pigment called **phytochrome** is responsible for flowering and a number of other processes. Phytochrome apparently detects the

▷ **FIGURE 26–27 Photoperiodism** On the top is a short-day plant, and on the bottom is a long-day plant. Yellow areas represent day and blue areas represent night. Note that the short-day plant (top 1) flowers only when the days are shorter than 14 hours and the long-day plant (bottom 2) flowers when the days are longer than 14 hours. A short interruption of the night with light (shown in the far right drawings) has opposite effect on the short-day and long-day plants. The short-day plant (on the top) does not flower even though the days are shorter than 14 hours, while the long-day plant (on the bottom) flowers even when the days are shorter than 14 hours.

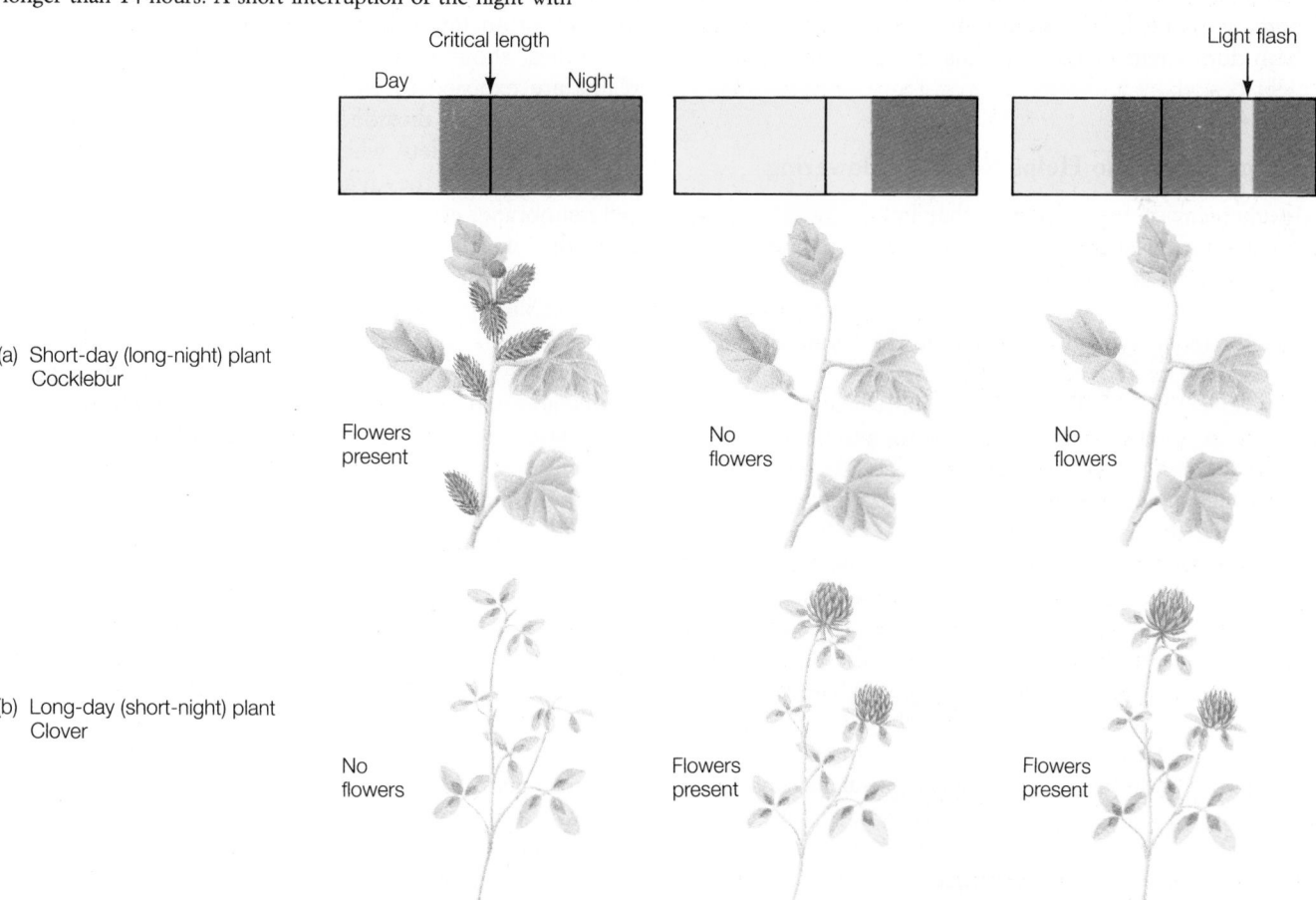

Critical length

Day | Night

Light flash

(a) Short-day (long-night) plant
Cocklebur

Flowers present | No flowers | No flowers

(b) Long-day (short-night) plant
Clover

No flowers | Flowers present | Flowers present

length of the day-night cycle. The plant organ responsible for day-length detection is the leaf.

Phytochrome exists in two chemical forms, phytochrome red (Pr) and phytochrome far-red (Pfr). During the day, most of the Pr is converted into Pfr, the active form—that is, the form that induces plant responses. During the night, Pfr is converted back into Pr. Thus, the length of the day determines how much Pfr is produced. Any light introduced during the night would convert Pr back to Pfr and reset the chemical clock.

This mechanism has been used in the past to explain the induction of flowering, especially in short-day plants. Recent experimental data strongly suggest, however, that although Pr-Pfr conversion is a factor, it cannot fully explain the flowering process.

Each plant species appears to have a critical photoperiod requirement with regard to flower production. Depending on species, plants have to be exposed to the proper photoperiod for a few to many days to initiate flower bud development. The formation of the flower buds themselves, however, is believed to be controlled by an as-yet-unidentified hormone.

Knowledge of photoperiodism has made it possible for horticulturists to induce flowering of specific flowers out of season. Easter lily and poinsettia are two common plants that are induced to bloom in time for Easter and Christmas, respectively. Long-day plants can be forced to bloom during the winter months by using lights in a greenhouse. Similarly, short-day plants are made to bloom by covering them during part of the day, thus artificially increasing the night period.

Temperature Also Helps Regulate Flowering

Another environmental factor that influences flowering is temperature. Plants such as winter wheat, apples, pears, and perennial grasses that grow in temperate regions have to be exposed to a temperature near freezing for several days for them to produce flowers in the spring. Artificially exposing plants to cold, a process called **vernalization,** overcomes this requirement. In winter wheat, for example, this requirement is met by exposing the seeds to cold before planting them in the spring. The artificial exposure to cold allows these plants to produce flowers and set seed during the same year.

A cold requirement to produce flowers is the reason why many of the fruit trees common in the temperate countries are not found in the tropics. Interestingly, the cold requirement of Christmas cactus and several of the other plants—like poinsettias— can be met by exposing them to a short photoperiod instead. However, it should be noted that several of the plants with a cold requirement also require the proper photoperiod after their exposure to cold. Practical problems in exposing large plants to either cold or a short photoperiod limit the use of vernalization.

Plants are truly intriguing organisms. They brighten up our lives. They provide food, shelter, and clothing. They furnish us with oxygen and many medicines. They control erosion and participate in the global hydrological cycle. Last but not least, their processes, many of them discussed in this chapter, provide endless fascination to those who take the time to study them.

ENVIRONMENT AND HEALTH: AIR POLLUTION AND TREES

Air pollution affects many species, and scientists think it will have a growing influence in the years to come, especially on plants. The effect that air pollution has on plants is not well understood, and scientists are still looking for ways to discern damage. One obvious way is by looking at the leaves, which frequently show overt signs of physical damage when exposed to air pollution. However, the presence of visible damage in many cases is not always associated with a reduction in plant growth or any underlying physiological damage.

Some scientists have proposed that air pollution damage can be assessed by measuring the level of ethanol production in leaves. Many tree species have the ability to produce ethanol under anaerobic conditions. Scientific studies suggest that ethanol is produced in anaerobic root tissue when the cytoplasm becomes acidic due to the production of organic acids. Ethanol apparently stabilizes the cytoplasmic pH, protecting cells from damage.

Studies show that ethanol production occurs when plants are exposed to elevated levels of nitrogen dioxide (NO_2) and sulfur dioxide (SO_2), two common pollutants from modern society which are converted to nitric and sulfuric acid when they react with water. These gases cross cell membranes and enter the cytoplasm. Scientists hypothesize that if a large enough concentration of these acid-precursors enter the cytoplasm, a plant's cells would become acidified, stimulating ethanol production.

To test this idea, scientists studied trees in the Ohio River Valley, where approximately 90 coal-fired power plants are currently operating. Trees in both rural and urban areas were studied from a variety of habitats. Leaves from the trees in the study area were collected and sealed in vials. After several hours, the gases in the vials were analyzed to detect the presence of ethanol.

In eastern cottonwood trees, ethanol production occurred in 79 of the 271 leaf samples, or 29%. Ethanol production was also observed in trial samples from yellow-poplar, American sycamore, and sweetgum. Ethanol production was detected more often in urban areas, but also in trees growing in Paradise, Kentucky which is a rural site downwind from a large coal-fired power plant.

Researchers also sampled eastern cottonwood, black cherry, black locust, hackberry, and boxelder and found that 21% of the trees from these species sampled, except hackberry, produced ethanol. Thirty-five percent of the hackberry trees sampled produced ethanol. Because ethanol was produced by different species at the same site, at the same time, scientists hypothesize that ethanol production was caused by the same factor(s).

To determine what that factor might be, scientists examined many possibilities besides pollution levels, among them flooding, fungal infections, temperature, and drought. The only consistent relationship found was the relationship between the concentration of NO_2 levels in the atmosphere and ethanol production.

The researchers concluded that leaves of trees in the Ohio River Valley were undergoing unusual metabolic changes manifested by transient ethanol production, which implies that significant metabolic disturbances were occurring. They conclude that these changes were regional, and were related to urbanization and industrialization as well as warm, hazy weather, and elevated NO_2 levels in some areas.

Photosynthesis in leaves is highly sensitive to the pH of the cytoplasm. The optimum pH for photosynthesis is 7.5 to 8.5. Below a pH of 7.0, photosynthetic activity is reduced or completely stopped. Thus, ethanol production, which indicates a lowered rate of energy catabolism, may also be an indicator of reduced carbon fixation. Other pH-dependent processes in plant cells may also be affected.

Air pollution may affect trees. Studies show that large forested regions are suffering extreme die-off, possibly linked to air pollution. Trees in the mountains east of Los Angeles, trees of the northeastern United States, and trees of the Black forest of Germany are three notable examples. More research is needed to understand the mechanisms involved, but ultimately protecting and maintaining the productivity of the Earth's vegetation will require efforts to clean up our air.

SUMMARY

FRUITS

1. Fruits are structures derived either from ovaries alone or from ovaries and other flower parts, such as the petals.
2. Fruits include nuts and vegetables.

SEEDS

3. Seeds, which are the transformed ovules, contain plant embryos that remain dormant until conditions are proper for germination.
4. Flowering plants are divided into two groups, monocots and dicots, based on the number of cotyledons present in their seeds.
5. As a seed germinates, the upper end of the embryo gives rise to the shoot system, and the lower end produces the root system.

PLANT TISSUES

6. The plant body of most vascular plants is differentiated into stems, leaves, and roots, each composed of specialized tissues that help carry out specific functions.
7. Meristematic tissue that is responsible for elongation of the stems and roots is present at the tips of these structures.
8. The two dermal tissues in plants are the epidermis, in the case of herbaceous plants, and cork, in woody plants.
9. Vascular tissues, xylem and phloem, transport water and food, respectively. All other functions are carried out by ground tissues such as parenchyma and sclerenchyma.
10. The vascular cambium and cork cambium are forms of secondary meristems that are responsible for growth in diameter.

STEMS

11. Vascular bundles in the stems of herbaceous plants are arranged in a ring, whereas those of a monocot stem are randomly arranged.
12. For the most part, herbaceous plants are composed of primary tissues—that is, tissues derived from primary meristems. Woody plants have large amounts of secondary tissues.

13. Bark and wood are major components of all woody plants.
14. Wood, which is composed of xylem, persists throughout the life of a tree and consists of growth rings that indicate the age of the tree.

LEAVES

15. A typical leaf consists of a blade at the end of a petiole that is attached to a stem at the node.
16. Leaves contain veins that consist of vascular tissues.
17. A leaf is covered on both surfaces by an epidermis containing many stomata, openings that regulate transpiration and gas exchange.
18. The centrally located veins of a leaf are surrounded by parenchyma cells containing many chloroplasts, which carry out photosynthesis, the leaf's primary function.

THE ROOT SYSTEM

19. Water and minerals are absorbed through the root hairs, extensions of the root epidermis.
20. The cells of the endodermis regulate the movement of water into the xylem.
21. The xylem in a dicot root is centrally located. In the monocot, the xylem lies in a ring surrounding the parenchyma.
22. Branch roots are produced from the pericycle, which is located just inside the endodermis.

INTERNAL TRANSPORT

23. Transpirational pull is considered to be the force that transports water from the soil to the leaves through the cells of the xylem.
24. The uptake of minerals by the roots requires transport proteins and an expenditure of energy.
25. Food is transported in the form of sucrose bidirectionally through the phloem.
26. Movement of food is based on concentration differences

between an area where food is available (source) and an area where it is used or stored (sink).

PLANT GROWTH AND RELATED PHENOMENA

27. Virtually all aspects of plant growth are regulated by the concentration and interaction of hormones.
28. The different plant hormones are the auxins, cytokinin, gibberellin, abscisic acid, and ethylene.
29. Phototropic and geotropic bending of stems and roots are responses to unidirectional stimuli that are regulated by an auxin.
30. The inhibitory influence of the apical bud on the development of the axillary buds, known as apical dominance, is controlled by an auxin.
31. Induction of flowering in short-day plants and long-day plants is controlled by the duration of the night period, which is detected by the pigment phytochrome.

EXERCISING YOUR CRITICAL THINKING SKILLS

Corn is a staple for 200 million people, including nearly half of the world's chronically malnourished people. Because corn is such an important source of calories and protein throughout the world, researchers have developed a new strain of corn called quality-protein maize (QPM).

QPM has about the same amount of protein as common maize but has twice the usable amount because its protein is more complete; that is, it supplies more of the essential amino acids. Normal maize has about 40% of the biological value of milk protein. QPM approaches that of milk, a common standard of nutritional excellence.

In countries where maize is a staple, QPM could double the efficiency of subsistence farming. A study by the National Research Council reports that QPM could allow families to combat malnutrition without outside help. QPM could prove helpful in Mexico, Central America, and Africa, where hunger and starvation are common.

Using your critical thinking skills, examine this optimistic finding. What questions come to mind about the conclusion that QPM could help reduce malnutrition? You may want to refer to an environmental science book (see the Suggested Readings in Chapter 33) to find out more on world hunger—that is, to get a view of the big picture.

Prepare a list of questions on this issue, then consult the list below. How does this list compare with your list?

1. Will QPM be affordable to peasants in the Third World?
2. Will it be more susceptible to insect pests than currently used native strains?
3. Will it require costly pesticides?
4. Will it require additional irrigation?
5. Will it deplete the soil?
6. How have efforts to improve crop yield worked in the past?
7. How does QPM compare economically with other measures?
8. What are other solutions to the problem?
9. Should QPM be supplemented by other strategies?

TEST OF TERMS

1. The shoot system of a seedling is derived from the upper portion of the embryonic axis, which is called the _____.
2. The part of an embryo that gives rise to the root system is called the _____.
3. In some seeds, the food is stored in a nonembryonic structure called the _____.
4. A plant that produces seeds with one cotyldon is called a(n)_____.
5. Primary tissues are derived from the _____ _____ _____.
6. The structure at the very tip of the shoot system is known as the _____ _____.
7. A structure on the stem that gives rise to a branch is the _____ _____.
8. The major tissue in plants that is responsible for the synthesis and storage of food is the _____.
9. The tissue through which water is transported in a plant is known as the _____.
10. Food is transported through a tissue called the _____.
11. The tissue present on the outermost part of a leaf blade is the_____.
12. The outermost tissue on woody stems and roots is the _____.
13. A common synonym for the xylem of a woody plant is _____.
14. The secondary meristem responsible for producing growth rings in a woody stem is the _____.
15. Cork cambium is an example of a(n) _____ _____.
16. The tissue that makes up the growth rings in a woody stem is the _____.
17. The discolored central core of a tree trunk is commonly known as _____.
18. A leaf consists of two parts, an expanded _____ and a(n) _____ that attaches it to the stem.
19. In some leaves, the blade is divided into several _____.
20. The small openings in the leaf epidermis are regulated by specialized cells known as _____.
21. The protective covering at the growing tip of root is called the _____ _____.
22. Water and mineral absorption is carried out by extensions of epidermal

cells called _____ _____ .

23. Food is stored in _____ cells in the root.

24. The evaporation of water from leaf surfaces that is responsible for the upward movement of water is known as _____ .

25. The plant hormone that controls the bending of stems toward light is a(n) _____ .

26. The plant hormone that regulates the downward bending of roots is a(n) _____ .

27. The plant hormone responsible for the ripening of fruits is _____ .

28. The growth response of a stem to unidirectional light is known as _____ .

29. The growth response of stems and roots to gravity is known as _____ .

30. The inhibitory influence of the apical bud on the growth of the axillary buds is called _____ _____ .

31. The length of the dark period is detected by a plant pigment known as _____ .

32. _____ -day plants produce flowers during early spring and/or late summer.

33. _____ -day plants produce flowers during late spring and early summer.

34. Day- _____ plants do not have a day-length requirement for flower production.

35. _____ is the artificial exposure of seeds or seedlings to cold to overcome the cold requirement for flowering.

Answers to the Test of Terms are found in Appendix B.

TEST OF CONCEPTS

1. What are the four major classes of tissues in plants? What is the basis for their classification into these groups? Give an example of each.
2. Describe how a stem increases in length.
3. Distinguish between a herbaceous stem and a woody stem.
4. Explain the importance of bark to a tree.
5. How is a root morphologically and anatomically adapted for its role (a) as an absorptive organ and (b) as a storage organ?
6. Distinguish between a monocot plant and a dicot plant. List as many differences as possible.
7. Explain how water can reach the top of tall trees.
8. Select three plant hormones, and describe a special role each plays in the life of a plant.
9. How is flower production regulated in a short-day plant?

The Animal Kingdom

≈

Protected by numerous spines, these sea urchins move across the ocean floor in search of food.

People living in the tropical rain forest or the desert of southern Africa know and can name hundreds of kinds of plants and animals in their immediate surroundings. They know where these animals live, what they do, and many uses for them as food or medicine. It was not until the 18th century, however, that the Swedish naturalist Carl von Linné (who signed himself "Carolus Linnaeus," because like other learned people of his day he wrote in Latin) actually formalized this vast folk knowledge into a system of scientific nomenclature that is still in use today.

The 19th century was an age of extensive biological exploration, especially in temperate, more populated regions of the globe. Naturalists like Charles Darwin and Alfred Wallace journeyed to the tropics and returned deeply moved by the diversity and exuberance of tropical life. Armed with their data, they proposed a revolutionary new theory about the mechanism underlying evolution. However, the 19th century left most of the tropical biota undescribed and unknown.

By the late 20th century, about 1.7 million species of organisms had been named, and most experts agreed that the job Linnaeus had started was about half finished. In the 1970s, an insect biologist from the Smithsonian Institution, Terry Erwin, traveled to Panama to study the insects in an unexplored ecosystem: the canopy of the tropical rain forest. What he discovered changed his thinking and ours, too. Erwin was struck by the huge number of species he captured, only a fraction of which had been described by scientists. He was also impressed by the fact that many species lived in very specific places; for example, a species might be limited to leaves of a certain size on branches at a certain height in trees of certain species. Because he was studying only a few trees in a tract of lowland rain forest that might contain a thousand species of trees, he estimated that the number of insect species in the world might be on the order of 20 million to 30 million. When the other animal species are added to the list, some scientists think, as many as 50 million species may presently share the planet with us.

Like every thoughtful scientist, Erwin began to put his observations and his insights into the context of here and now. Approximately 25% to 40% of the tropical rain forest has been destroyed or severely disturbed in this century. Every year, an area of rain forest roughly the size of the state of Washington is destroyed. At present rates of destruction, most of the world's tropical rain forest could be gone within your lifetime. Erwin estimated that the next generation might witness the extinction of about half of all species alive today, and most of what we destroy will never have been cataloged and studied by scientists.

As is so often the case in science, a single question leads not only to answers but also to many additional questions. In this case, Erwin's answers were biological, but his lingering question was cultural. What will we humans, armed with our growing but still woefully incomplete knowledge of Earth's magnificent biota, do? "Do we have the resolve," he asked, "to rise above profit and greed?"

This chapter will not answer that question, but it will give you a view of the immense and stunningly diverse array of animal species that make the Earth their home. It will provide insights into animal evolution and perhaps some impetus for us to learn to tread more carefully in the web of life.

≋ WHAT IS AN ANIMAL?

To define animals, it is often helpful to contrast them with the other kingdoms of living things. First, the cells of all animals have membrane-bound nuclei, which contain the genetic material. That makes them eukaryotic (unlike the monerans). Second, animals are multicellular eukaryotes, as are fungi and plants. This feature sets them apart from protists. Third, unlike plants, animals (and fungi) are heterotrophic. In other words, they depend on autotrophs (plants, protists, or monerans) to produce their food. Finally, in contrast to fungi, animals lack cell walls.

In summation, **animals** are eukaryotic, multicellular heterotrophs that lack cell walls. This simple list of characteristics is common to all members of the extraordinarily diverse kingdom.

In nearly all ecosystems, the most conspicuous organisms are animals and plants. The most obvious contrast between animals and plants is that most animals are mobile. Indeed, this mobility lends the kingdom its name: animals are "animated," (literally, "spirited"); that is, they are quickly and visibly responsive to environmental stimuli. Animals move around, which is vital to their heterotrophic lifestyle. In fact, most animals must actively seek food, because their food is dispersed and rooted or secretive (▷ Figure 27–1). See Spotlight on Evolution 27–1 for an example of one of the most mobile of all animals.

▷ **FIGURE 27–1 Animated Eater** Animals are heterotrophs, and most seek food actively. Here, a cougar pursues a hare.

THE MOUNTAIN GOAT: UNDISPUTED KING OF THE MOUNTAIN

A biologist looking at mountain goats high on a perilous ridge in Montana was captivated one day by an animal working its way along a narrow ledge that fell off steeply into a deep canyon. Walking along the tiny shelf, the animal came to an impasse. With no room to turn around, and apparently not enough room to back up, the goat braced its front hooves, then carefully walked its rear legs up the sheer rock wall and over its head in a slow-motion reverse cartwheel. In a few moments, the animal was back on all fours, facing in the opposite direction as if nothing had happened.

What the Montana scientist saw, says writer Gary Turbak, was simply "business as usual for the mountain goat, a supremely sure-footed creature that thrives in a narrow, high-country niche where few animals other than birds dare even visit. Uniquely adapted in every respect to its rugged and inhospitable habitat, the goat is the unquestioned "king of the mountain."

Mountain goats live above timberline, often 3000 meters (10,000 feet) above sea level in what is best described as a cold and perilous environment frequented by few animal species. "Theirs is a land of wind, ice, and snow, of pinnacles, ledges and landslides," writes Turbak. Their habitat is so dangerous

that females frequently position themselves below their young to protect them from falls.

Now living exclusively in North America, the mountain goat has short legs that provide a low center of gravity and thus the great stability it needs to survive. Three inches of cashmerelike wool and long, hollow, outer guard hairs, which provide additional insulation, protect the animal from fierce winds and temperatures that frequently plummet well below 0° F. Its white coat provides additional protection, blending the animal with a habitat that is covered with snow much of the year.

Another one of the keys to mountain goat survival is the animal's hooves. Unlike the hard, concave hoof of deer and other ungulates, the mountain goat's hooves contain rubbery, convex pads. Pressed against rock, these pads grip tightly, holding the goat in rather precarious positions on rock walls that make observers nervous. In addition, the two toes on the front feet splay outward, acting as brakes when the animal moves downhill, giving additional stability.

"Just when you imagine that the goat you're watching must be clinging for dear life to a shred of ledge," says Douglas Chadwick, a Montana wildlife biologist, "it calmly lifts a rear foot and begins scratching an ear with it."

Mountain goats seem to be born with an ability to climb. Within a day or two after birth, a newborn can go most places its mother travels.

The goat's agility is legendary. In fact, one biologist perched high in an observation tower was surprised one day to come face to face with a goat that had climbed the steep stairs to see who had invaded its territory.

Like the tundra where it lives, the mountain goat is a paradox. In many ways, the species is a paragon of toughness, enduring winter hardships that few other species could tolerate. At the same time, it is a fragile species easily disturbed by human intervention. Seismic testing and frequent helicopter traffic associated with oil and gas exploration upset goat reproduction in one herd in Montana, resulting in a 50% decline in the population in a four-year period. In Canada, timber harvesting in nearby forests has nearly exterminated some goat populations. The problem is that although goats have adapted perfectly to the rigors of their habitat, they are unequipped to confront humanity's intrusion. The good news is that despite human intrusion, goats continue to thrive in the remaining habitat.

SOURCE: Adapted from Gary Turbak, "Life on the Edge," *National Wildlife* (1991) 29(5): 40–45.

≈ THE ORIGINS OF ANIMALS

The evidence is indirect, but animals—like plants—probably originated from flagellated protists 600 to 800 million years ago.

As already noted, the diversity of animals is immense. Animals inhabit every conceivable habitat, from the deepest, darkest oceanic trenches to lofty mountain peaks, and from humid tropical forests to polar deserts. Animals range in size from mites and roundworms that are a fraction of a millimeter long to blue whales 35 meters (110 feet) long and weighing 150 tons.

This array of animal forms faces a common set of environmental challenges. All animals must gather resources (food and energy) and dispose of wastes. They need protection and physical support (especially if they are

large and live on land). They need a means of locomotion, some means of sensation, and some form of communication (to permit coordination among individual cells and among individual organisms). They also need some mode of reproduction.

The variety of responses to these challenges distinguishes the broad groups (phyla) of animals. In other words, each phylum represents a distinct way of being an animal—a eukaryotic, multicellular, usually mobile heterotroph. Scientists distinguish about two dozen phyla of animals, nearly all of which are represented today. This chapter will discuss the major phyla listed in Table 27–1.

Doubtless, the actual time and place of the emergence of animals from their flagellated protist ancestry will never be known. Indeed, animals may have arisen on more than

TABLE 27–1 Comparison of the Major Animal Phyla

PHYLUM	DIGESTION	CIRCULATION	NITROGENOUS WASTES	GASEOUS EXCHANGE	NERVOUS INTEGRATION	SENSATION	MOVEMENT	SUPPORT
Porifera (sponges)	Intra-cellular	None	Diffusion	Diffusion	None	Irritability	Passive	Water
Cnidaria (jellyfish)	Extra-cellular and intra-cellular	None	Diffusion	Diffusion	Radial nerve net	Minimal; cnido-cysts	Contractile fibrils	Water, mesoglea
Platy-helminthes (flatworms)	Intra-cellular and extra-cellular	None	Proto-nephridia	Diffusion	Nerve cords, "brain"	Eyespot, chemo-sensation	Mesodermal muscles	Minimal hydrostatic skeleton
Annelida (segmented worms)	Extra-cellular	Closed	Nephridia	Diffusion, gills	Ventral nerve cord, brain	Complex	Muscles	Hydrostatic skeleton, chitin
Mollusca (molluscs)	Extra-cellular	Mostly open	Nephridia	Gills, land lungs	Variable, advanced	Variable, complex	Muscles	Exoskeleton or endo-skeleton, shell
Arthropoda (arthropods)	Extra-cellular	Open	Variable, Malpighian tubules	Spiracles, booklungs, gills	Ventral nerve cord, advanced brain	Various, advanced, complex	Muscles	Exoskeleton, chitin
Echinodermata (echinoderms)	Extra-cellular	Various, open, coelomic	Diffusion	Diffusion via tube feet, gills, respiratory tree, stomach	Radial nerves, central ganglion	Various	Muscles	Endoskeleton, calcareous
Chordata[a] (Chordates)	Extra-cellular	Closed	Kidney	Gills, lungs, diffusion	Dorsal, hollow nerve cord, brain	Various	Muscles	Endoskeleton, bone

[a]Using vertebrate chordates as example

one occasion. After all, "animalness" was a "promising" evolutionary development.

Two of the trends you will identify in this survey of the animal kingdom are a progressive increase in size and the development of multicellularity. Large size probably conferred immediate advantages. The larger an organism, the more resources it can control and the less likely it is to be eaten. However, because a single nucleus can control only a limited amount of cytoplasm and because the plasma membrane of a very large cell would be inadequate as an exchange surface, large size in the animal kingdom was achieved through multicellularity—groups of small cells rather than a single large cell.

Of course, a multicellular organism faces challenges that single-celled organisms do not. If it gets large enough,

some cells will be separated from supplies of food, oxygen, and other resources. Multicellular organisms, in order to function smoothly, required some form of internal circulation.

The Sponges Are Multicellular Animals that Exhibit a Modest Amount of Specialization

There is no fossil record of the earliest animals. These organisms were probably small, soft-bodied organisms that were unlikely to be preserved as fossils. However, we are fortunate to have a kind of model of earliest animal life, the **sponges** (▷ Figure 27–2).

Sponges are rather simple animals that live mostly in salt water. They exemplify an early stage of animal evolution

(a)   (b)

▷ **FIGURE 27–2 The Sponges** (*a*) Glass sponge. (*b*) Bath sponge.

somewhat more complex than a simple aggregation of one-celled organisms, but they fall short of the "tissue grade" of organization found in other animals. Scientists do not believe that sponges gave rise to more complex phyla. But they do think that the lineage leading to more complex animals went through a spongelike stage. That is because the hallmark of the "sponge grade" of organization is cellular specialization, and such division of labor certainly was the first step toward getting multicellular organisms organized.

Sponge colonies are immobile, and most sponges are asymmetrical (Figure 27–2). The colony imports resources rather than moving around to gather them. The colony is perforated by numerous holes (hence the name of the phylum, **Porifera,** "pore-bearer"). Water moves into the colony through the numerous incurrent pores and leaves through the excurrent opening, directed by the beating of the flagellae of specialized **collar cells** (Figure 12–2). Water flowing through the interior of the sponge contains oxygen and bits of organic matter that are ingested by the cells of the sponge.

Sponges are covered with an outer **epidermis.** The cells of the sponge are supported by a variety of secretions. Some sponges, in fact, have a "skeleton" of spicules of lime (calcium carbonate); the glass sponges have a supportive framework composed of silica. The bath sponges contain a flexible fibrous protein, known as spongin, that creates a supportive framework.[1]

Sponges readily repair or replace lost parts, a process called **regeneration.** By straining an individual sponge colony gently through a fine mesh, one can separate the component cells. Within a few hours, the individual cells reorganize into several new sponges.

[1] The sponge in your bath is probably a synthetic substitute made from plant fibers, because commercial harvest has depleted natural stocks of bath sponges.

Sponges reproduce asexually, a process that takes several forms. For example, some sponges form small buds that break off and form new sponge colonies. Several kinds of freshwater sponges, adapted to seasonal changes in water level, form **gemmules,** internally produced buds that contain food stores within a drought-resistant cover. When the parent sponge dies, the gemmule remains, protected and ready to "germinate" and grow into a new colony the next season.

Sponges also reproduce sexually via sperm and eggs. A fertilized egg develops into a **larva,** a mobile developmental stage that swims away to form a new colony. The larval stage allows dispersal.

Sponges exhibit the barest rudiments of cellular specialization: collar cells for internal circulation, epithelial cells for protection, and sperm and eggs for sexual reproduction. Cellular secretions provide support. What sponges lack is coordination—a nervous and endocrine system—and organization of specialized cells into distinct tissues and organs. Because sponges are asymmetrical, they have no particular orientation to the environment.

The Cnidarians Are Characterized by Distinct Tissues and Primitive Nervous Systems

Coordination and symmetry are first seen in the jellyfishes and their kin, the **cnidarians** (ny-dair-ee-uns). The cnidarians belong to a phylum that contains such familiar and beautiful marine animals as corals and sea anemones (▷ Figure 27–3).

Cnidarians are characterized by distinct tissues, of which there are two general types, an **ectoderm** ("outer skin") and an **endoderm** ("inner skin"). Between those tissue layers is a gelatinous **mesoglea,** the "jelly" of jellyfishes (Figure 27–3c).

Hydra, a freshwater relative of the jellyfish, is a good example of this phylum (Figure 27–3c). Like other cnidarians, hydras exhibit **radial symmetry;** that is, body parts are arranged around a central axis, like the spokes of a bicycle wheel. **Tentacles** located around the mouth draw water (and food materials in the water) into a **gastrovascular cavity.** Digestion in the hydras and other cnidarians may be extracellular, as in most higher animals, with cellular secretions flowing to the food in the gastrovascular cavity, or intracellular, taking place within the cell, as in sponges and phagocytic protists. Cells in the ectoderm and endoderm contain contractile proteins that move the tentacles and the body wall. A network of nerve cells runs through the body, connecting **sensory cells** and **contractile fibrils.** Although there is no brain, the **nerve net** provides for simple reflex responses. These reflexes are vital to all active animals, ourselves included.

Cnidarians are equipped with specialized cells called **cnidocysts** (ny-doe-sists) (Figure 27–3c) that aid in acquiring food and defending themselves. The cnidocyst consists of a thread coiled inside a fluid-filled capsule with a trigger mechanism. The trigger responds to the taste of prey.

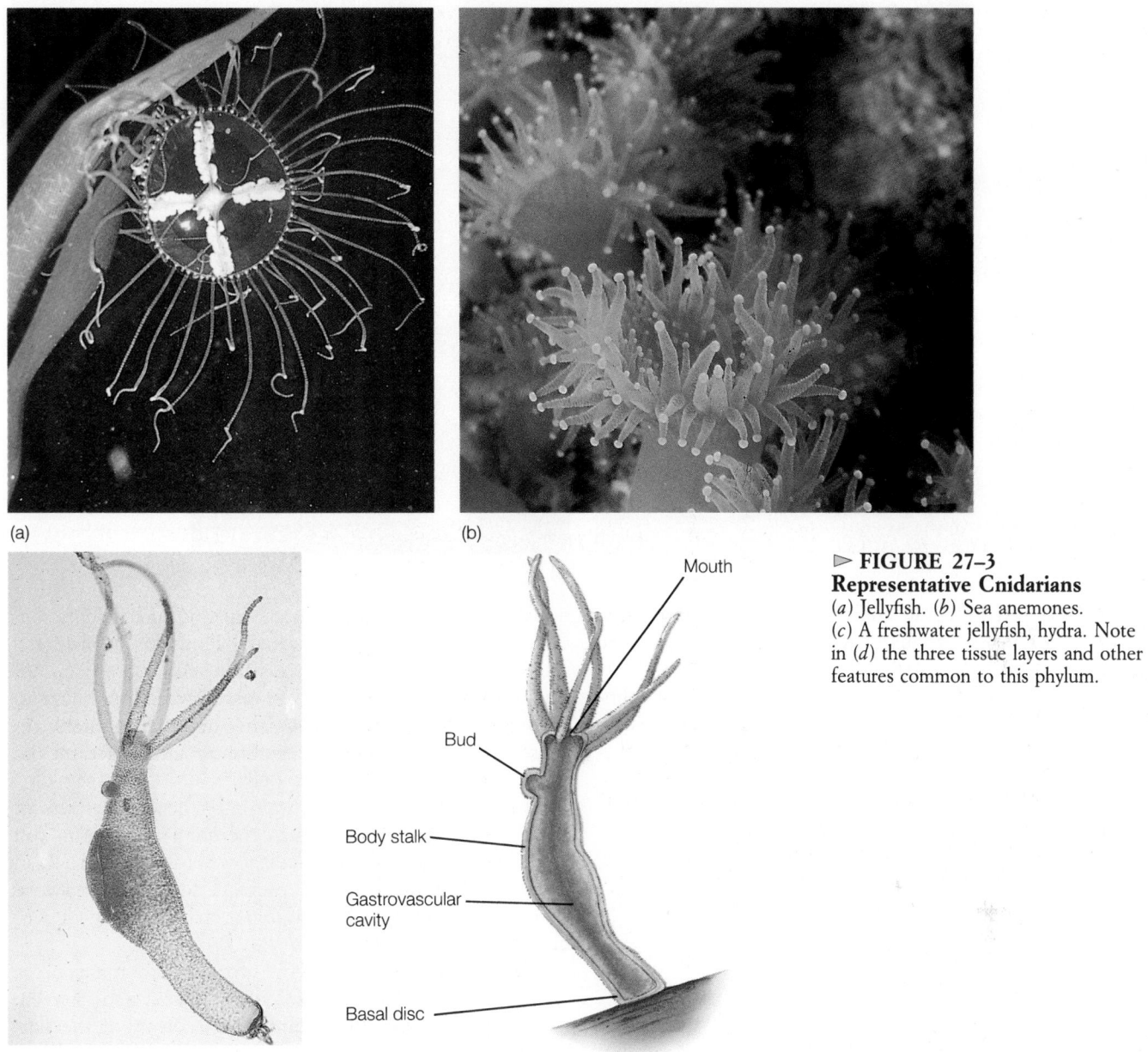

Mouth

Bud

Body stalk

Gastrovascular cavity

Basal disc

▷ **FIGURE 27–3**
Representative Cnidarians
(*a*) Jellyfish. (*b*) Sea anemones.
(*c*) A freshwater jellyfish, hydra. Note in (*d*) the three tissue layers and other features common to this phylum.

(a)

(b)

(c)

(d)

When stimulated, the cnidocyst releases the thread, which harpoons or sticks to its prey.

Reproduction in hydras occurs by budding (a form of asexual reproduction) throughout most of the growing season. During times of environmental stress, however, gonads form in the ectoderm and produce sperm and eggs. A new generation is produced sexually.

Many cnidarians (such as sea anemones, corals, and sea fans) are **sessile**—that is, attached to the seafloor or some other surface. Sexual reproduction produces free-swimming larvae that permits dispersal that allows offspring to escape competition with the parents. It also permits them to escape an exhausted environment and find a promising new one.

Many species of cnidarians are beautiful animals; others are simple and unimpressive. All are important from an evolutionary standpoint because they mark the beginning of distinct tissues and nervous coordination.

Some cnidarians are also important geologically. Consider the corals. No other group of animals (other than ourselves, of course) has been of geologic influence on as vast a scale. The reef-building corals have built shorelines in tropical seas around the world (▷ Figure 27–4). Australia's Great Barrier Reef is 1900 kilometers (1200 miles) long and up to 300 kilometers (190 miles) wide.

Corals are related to sea anemones. Most individuals are tiny, but they sometimes join to form huge colonies. Individual coral animals produce walls or platforms of lime. Like other cnidarians, corals have tentacles that conduct food particles into the gastrovascular cavity.

Many species of coral are host to photosynthetic protists that live in their endoderm. These symbionts produce car-

(a)

(b)

bohydrates that the corals consume. They also apparently assist in removing carbonate from seawater to form the coral's limestone "skeleton."

Coral reefs are among the most productive biotic communities on Earth, and they are extraordinary nurseries for biodiversity (Figure 27–4). They flourish only in warm water, with temperatures above about 22°C (about 72°F). Hence, corals are limited to tropical waters, such as the Pacific and Indian oceans and the Caribbean Sea. Because their symbiotic, photosynthetic, algal colleagues need light, corals live only in clear, shallow waters.

The Flatworms Are Characterized by Bilateral Symmetry, Cephalization, and Three Distinct Tissue Layers

Over half of the known animal phyla are organisms that the casual observer would call "worms," a term that describes their general shape. These apparently similar phyla actually differ considerably in structure, function, reproduction, development, and life history. Interestingly, zoologists suspect that the wormlike body form has evolved repeatedly. "Wormness" is an effective body form that provides directionality—that is, a distinct head region and tail region.

As you learned earlier, the sponge is asymmetric; it has no particular orientation (▷ Figure 27–5). The jellyfish exhibits radial symmetry, the same mode of organization as a wheel. A worm, by contrast, exhibits **bilateral symmetry.** In bilaterally symmetrical organisms, body parts are arranged as mirror images on either side of a medial plane (Figure 27–5). Bilateral symmetry occurs in nearly all animals from flatworms on. It is even the body form of human beings. Many plants also exhibit it.

Bilateral symmetry offers an important advantage: direc-

tional orientation. Round objects have no such orientation. (Where is the front of a baseball?) However, a bilaterally symmetrical organism can have a front end and a rear end. That means that the ends of the organism can specialize to perform different functions. The head encounters resources and danger and is therefore a logical place for the sense organs and the mouth.

The concentration of functions into a head end is known as **cephalization.** As you study the animal kingdom, you will see that increasing cephalization is a major trend in evolution. It is of obvious importance to humans because we are the most cephalized of organisms, extremely dependent upon our heads for survival.

The basic body plan of the worms (and many more complex forms of animals) was established in the **flatworms (phylum Platyhelminthes).** Many of the flatworms (tapeworms, flukes) are internal parasites, living within and taking nourishment from host organisms (▷ Figure 27–6); other forms, such as the free-living planarians, also occur and exhibit the important features of the group.

The body of planarians has three distinct tissue layers, an inner lining (endoderm), an outer covering (ectoderm), and a middle cellular layer known as the **mesoderm.** The digestive system of these organisms is highly branched, delivering food near each of the cells. Planarians have a single opening that serves as a port of entry for food and an exit for wastes. With food and waste using the same portal, there is no specialization of function along the digestive tract.

Flatworms have no circulatory system. Gaseous exchange occurs by diffusion through the body wall, which restricts their size (although their flatness does permit them to achieve a modest size while still allowing the body cells to be within diffusion distance of the exterior).

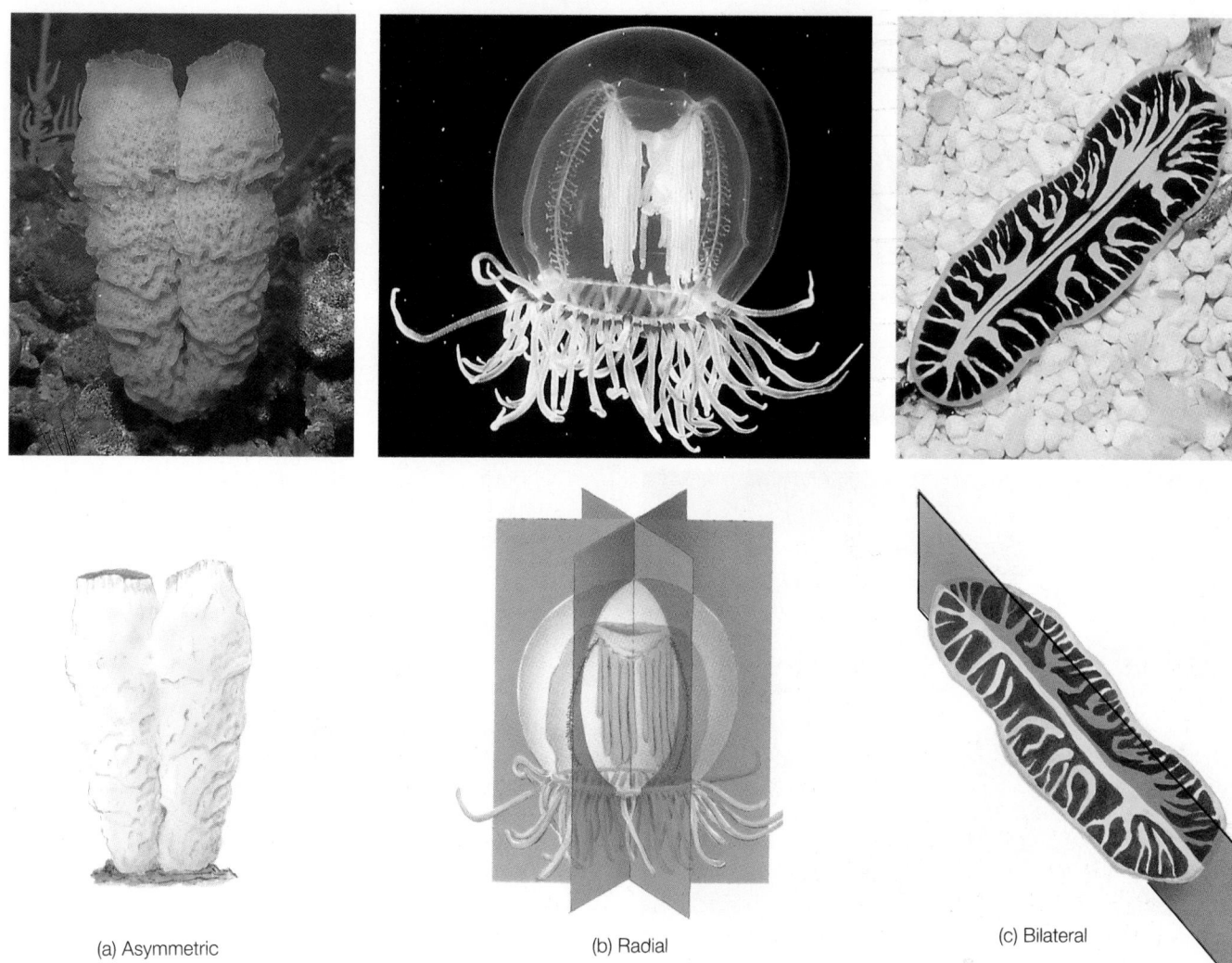

(a) Asymmetric

(b) Radial

(c) Bilateral

▷ **FIGURE 27–5 Three Forms of Symmetry** (*a*) The sponge is asymmetric. (*b*) The jellyfish exhibits radial symmetry. (*c*) The earthworm is characterized by bilateral symmetry, a feature found in many higher animals.

Most flatworms are **hermaphrodites**—organisms in which a single individual has both female and male gonads and therefore can produce both eggs and sperm. Reproduction occurs through cross-fertilization, the exchange of sperm from one organism to another.

Flatworms are not a particularly successful group. Only about 15,000 species are known. The body plan is something of an evolutionary dead end. Flatworms cannot get very large, and they do not dominate the ecosystems in which they live. However, building on the evolutionary "inventions" of the flatworms—notably, bilateral symmetry, cephalization, and three tissue layers—other groups succeeded on a grand scale.

Three Important Evolutionary Developments Contributed Significantly to the Diversification of the Animal Kingdom

Somewhere after the flatworm stage of evolution, some unknown (but no doubt wormlike) organisms evolved three extraordinary features: a circulatory system, a one-way digestive tract, and the **coelom**—that is, a body cavity lined with mesoderm. Then the evolving "family tree" branched, giving rise to many new organisms characterized by these important features.

To understand how animals came to be the conspicuous and dominant organisms they are today, we need to explore each of these innovations, then trace two main evolutionary branches, each of which led to great success in modern ecosystems.

The Evolution of Circulatory Systems Permitted the Evolution of Large Bodies. The circulatory system carries gases, nutrients, and waste products to and from the cells. In most organisms, circulating fluids also carry information in the form of hormones. As you learned in Chapter 12, some circulatory systems are closed (▷ Figure 27–7). These systems contain a pump (or pumps) and a series of

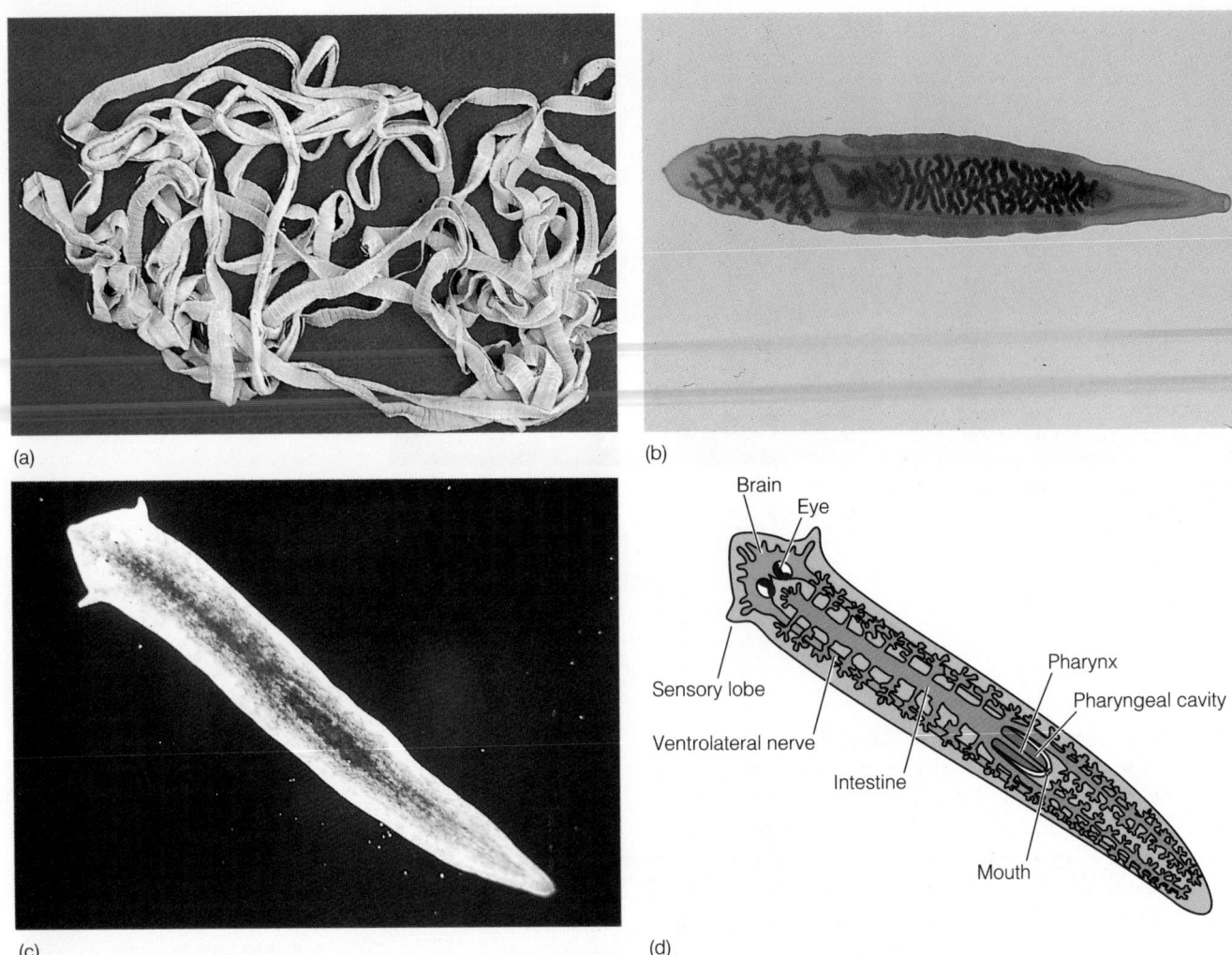

(a)

(b)

Brain
Eye
Sensory lobe
Ventrolateral nerve
Intestine
Pharynx
Pharyngeal cavity
Mouth

(c)

(d)

▷ **FIGURE 27–6 A Gallery of Flatworms** (*a*) Tapeworm. (*b*) Liver fluke. (*c*) Free-living planarian.
(*d*) Notice the pharynx, eyespots, branched gut, and other characteristic features of the planarian.

vessels that transport the blood throughout the body. Other circulatory systems are open, with blood flowing out of the vessels into the body cavity itself. Regardless of their form, all circulatory systems bathe cells in a resource-rich fluid.

The alternative to internal transport of fluid is simple diffusion. Because diffusion beyond a few layers of cells is so slow, however, inner cells of a large animal relying on this form of transport would either suffocate or starve to death. Animals without some sort of internal circulatory system are doomed to remain small, or very flat. Put another way, circulation allowed the evolution of larger and more active forms.

The Evolution of the One-Way Digestive System Also Permitted the Evolution of Larger Bodies. As you may recall from Chapter 11, higher animals have a one-way digestive system with two openings, a mouth and an anus (▷ Figure 27–8). The front end of the tract meets fresh, resource-rich environments. The posterior end of the tract leaves behind resource-diminished (or even toxic) waste. The evolution of one-way digestive systems represents a major step beyond the flatworm's digestive system with its single opening, serving as both mouth and anus, because it allows for specialization along the length of the digestive tract. This, specialization, in turn, permits an efficient, stepwise digestion and absorption of food. Efficient processing permitted the evolution of larger size.

The Coelom Is the Body Cavity, a Space Between the Body Wall and the Internal Organs. Coeloms of various groups differ in structure and embryologic origin, but they function similarly. Animals with a coelom are built as a tube-within-a-tube (▷ Figure 27–9). The inner tube—the gut—is separated from the outer tube—the muscles and coverings of the body wall—by the coelom. Therefore, the gut can move independently of the rest of the body. Furthermore, structural change (evolutionary or developmental) can occur in the inner tube without direct impacts on the outer tube (and vice versa).

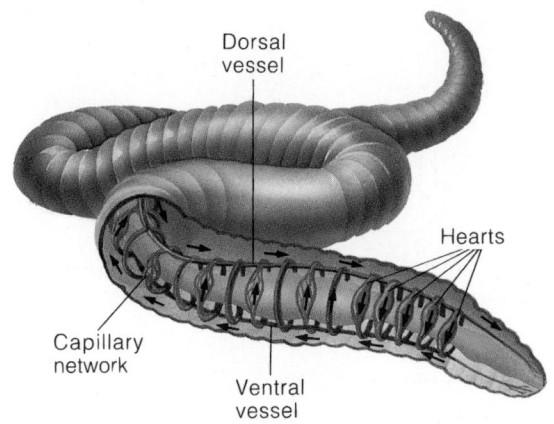

FIGURE 27–7 Closed Circulatory System Closed circulatory systems, such as the one found in an earthworm, consists of a pump (or pumps) and a series of blood vessels that conduct blood throughout the body. Nutrients and wastes must diffuse in and out of the blood vessels.

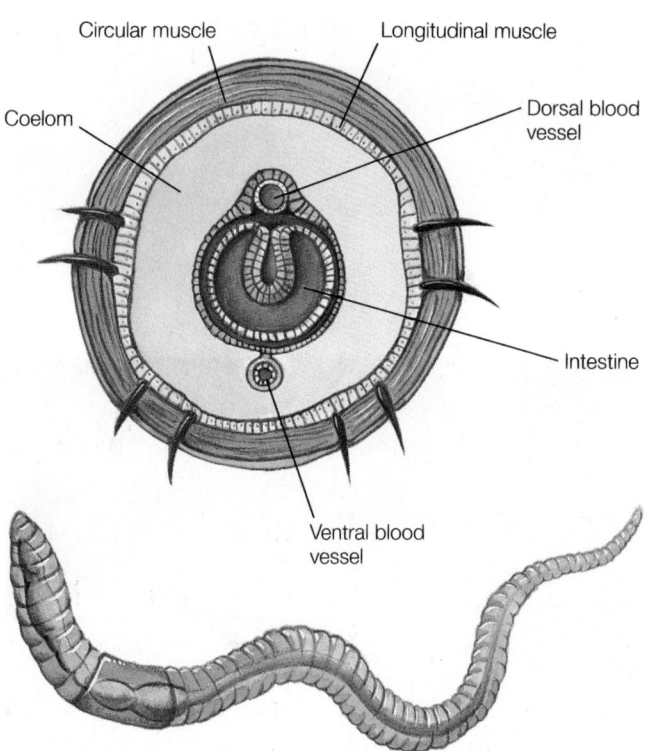

FIGURE 27–9 Coelom Animals with a coelom consist of a tube (digestive system) within a tube (body wall).

The coelom has evolved several times, so details of embryologic development vary, providing strong clues to evolutionary relationships. Embryologic development of the coelom is one of a number of differences between the two broad branches of the evolutionary tree of animal life. So distinctive are these two branches that they are treated as two different superphyla. The **annelid superphylum** includes the segmented worms, the arthropods, and the molluscs. The **echinoderm superphylum** includes the starfishes and their immediate kin and also the chordates, which includes all vertebrates.

The coelom, circulatory systems, and a one-way digestive tract are features of the soft anatomy, not readily preserved in fossil form. Consequently, the origin of these three characteristics is unknown.

≈ THE ANNELID SUPERPHYLUM

The Annelid Worms Are Characterized by Segmented Bodies

Annelid worms include the familiar earthworm and leech and thousands of beautiful marine forms. This phylum includes nearly 10,000 species, most of which live in the oceans (▷ Figure 27–10). All annelids have closed circulatory systems and a complete digestive tract with a distinct mouth and anus. The annelids are well cephalized, with a

FIGURE 27–8 The One-Way Digestive System Higher animals have one-way digestive systems with a mouth for gathering food and an anus for disposing of wastes.

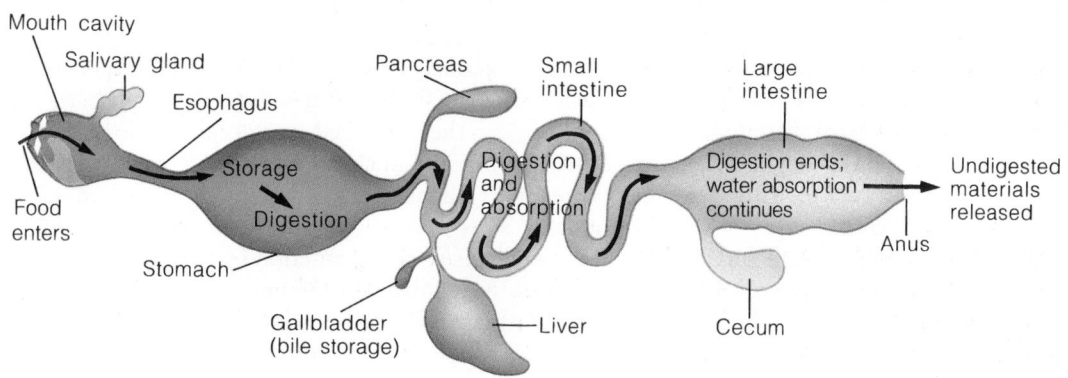

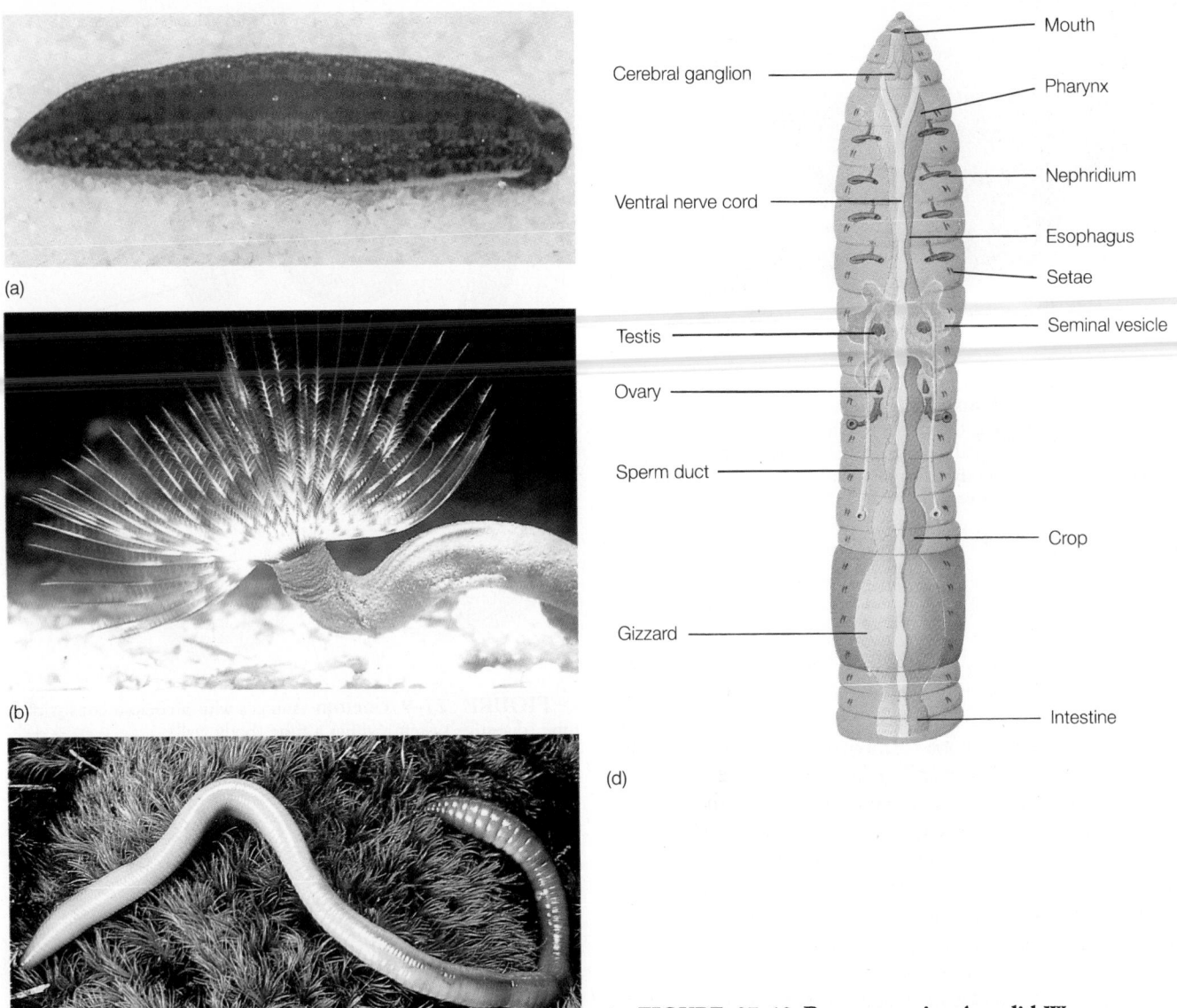

Mouth
Cerebral ganglion
Pharynx
Nephridium
Ventral nerve cord
Esophagus
Setae
Testis
Seminal vesicle
Ovary
Sperm duct
Crop
Gizzard
Intestine

(a)
(b)
(c)
(d)

▷ **FIGURE 27–10 Representative Annelid Worms**
(*a*) Leech. (*b*) Tubeworm. (*c*) Earthworm. (*d*) The major features common to many annelids.

nervous system and sensory receptors that detect touch, light, and chemicals.

The most distinguishing feature of the annelids is **segmentation,** the serial repetition of body parts. Each segment of an earthworm has muscles, nerves, and paired excretory structures. Five of the forward segments each contains a pair of hearts.

Segmentation represents a major evolutionary advance in the animal kingdom, for it provides considerable evolutionary leeway. Why? Segmented organisms essentially come equipped with spare parts. Evolutionary change can occur in some segments while others carry on their former activities.

Segmentation has evolved at least twice in the higher animals: in annelids and their descendants, the arthropods, and in our phylum, the Chordata (think about your ribs and your backbone). The common ancestor of annelids and chordates, however, was probably not segmented.

Arthropods Are an Extremely Diverse Group United by One Common Feature, Jointed Appendages

The **arthropods** belong to a phylum characterized by jointed appendages (legs and feelers, for example). The name *arthropod,* in fact, means "jointed foot." This phylum, which contains almost a million named species, is extraordinarily diverse. It includes the 6-legged insects, the 8-legged spiders, the 10-legged crustaceans, and the many-legged centipedes and millipedes. Many arthropods inhabit freshwater and marine ecosystems, but the vast majority live on land.

If we use diversity as the criterion of success, the arthropods have to be considered *the* most successful group of organisms to have evolved on Earth.

The jointed legs that are found in the species of this diverse group exhibit a variety of forms and functions. This

▷ **FIGURE 27–11 An abundance of Joints** The appendages (legs, feelers, pincers) of the Atlantic lobster, a crustacean, exemplify the evolutionary opportunity presented by segmentation.

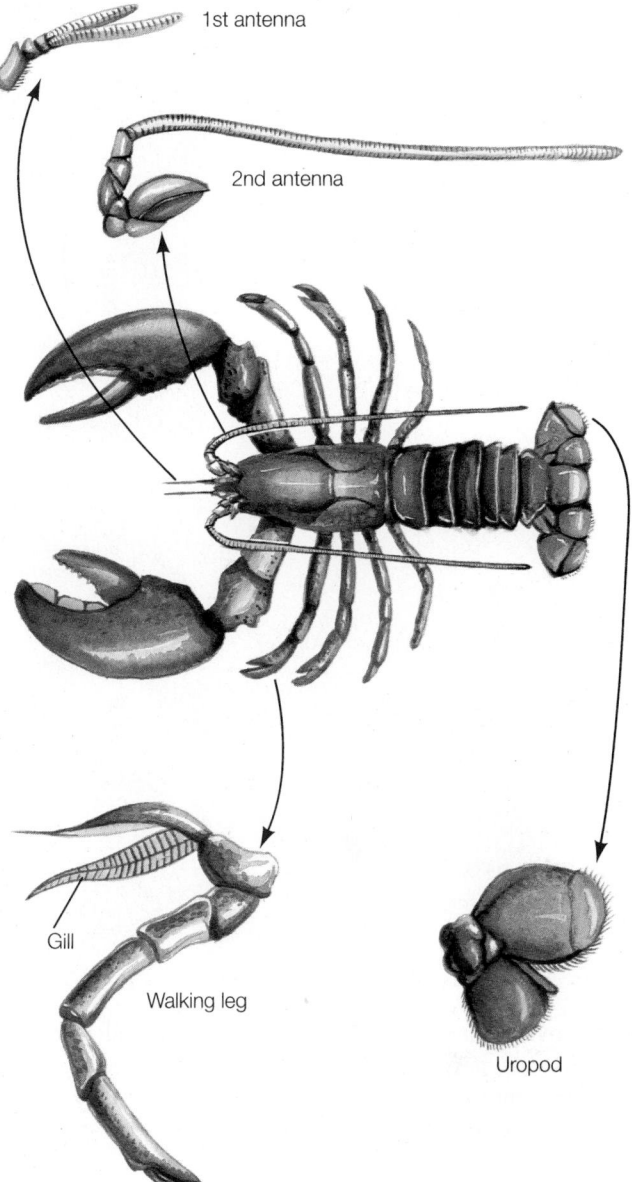

great diversity is possible because limbs of arthropods are segmented. Consider the appendages of a lobster (▷ Figure 27–11). In the lobster, jointed appendages detect stimuli, handle and chew food, mate, ward off enemies, and swim. Even so, there are plenty left over for just walking.

Another factor in the success of the arthropods is **chitin,** a complex, waterproof, chemical-resistant molecule made of both protein and carbohydrate (Chapter 2). Chitin forms the **exoskeleton** of arthropods, protecting them from desiccation and enemies and providing the point of attachment for muscles, as noted in Chapter 19. Despite its advantages, an external skeleton has drawbacks. It is relatively heavy and makes growth difficult. In fact, to accommodate internal growth, arthropods must shed their "armor," a feat that temporarily exposes a layer of soft, fresh chitin that will soon harden. Often, the most vulnerable times in the life of an arthropod are those when an old skeleton is being shed and the new one has yet to harden.

Arthropods are familiar, conspicuous animals, and most people have encountered representatives from the major classes in the phylum: crustaceans, arachnids, insects, and centipedes and millipedes.

Crustaceans. The **class Crustacea** includes about 20,000 species, among them crabs, shrimps, crayfish, and lobsters. Although generally appreciated for their contribution to the human diet, crustaceans also comprise most of the zooplankton and thus are critical components of freshwater and saltwater food chains. Crustaceans are thought to be derived from the trilobites, fossils of Cambrian times, described in Chapter 23.

Arachnids. Mites, ticks, scorpions, and spiders are **arachnids** (▷ Figure 27–12). About 60,000 species of arachnids have been named, but the actual number of species may be 10 times greater. Why? Most mites are extremely small and thus escape notice, even though they live in abundance in such mundane habitats as household dust and garden soil.

Insects. Insects are by far the most successful of arthropods. Indeed, the **class Insecta** is the most diverse group of organisms on the Earth. Nearly a million species of insects have been described to date. By various estimates, somewhere between 3 and 30 times that number may yet be discovered.

One of the keys to the success of insects was their invasion of the land and subsequent exploitation of the expanding terrestrial flora (plantlife) about 350 million years ago. Once they had established themselves on land, the next great "achievement" of insects was the evolution of flight. Flight allowed the exploitation of still another habitat, the air, and facilitated dispersal and escape from predators and stressful environments.

With the evolution of flowering plants about 100 million years ago, insects achieved even greater success and diversity. The most diverse insect orders today are the beetles

▷ **FIGURE 27–12 Representative Arachnids** (*a*) Mite. (*b*) Tick. (*c*) Scorpion. (*d*) Spider.

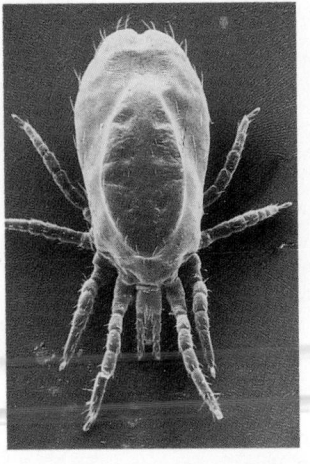

(a)

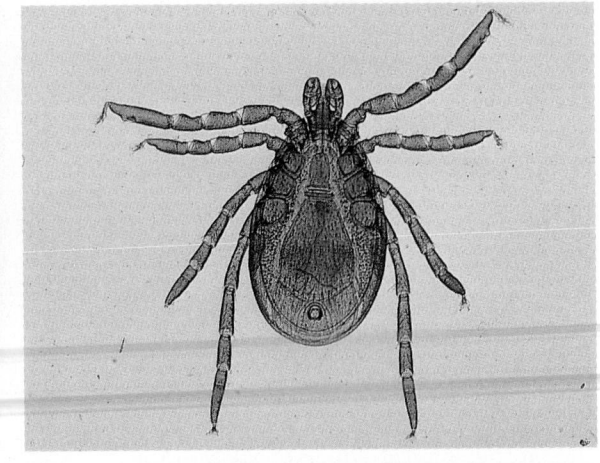

(b)

(c)

(d)

(with over 300,000 named species); the social insects—bees, wasps, and ants (with over 100,000 species); butterflies and moths (over 100,000 species); and flies (nearly 100,000 species). Representatives of each order are shown in ▷ Figure 27–13.

The evolution and diversification of insects and higher plants represent a kind of biological positive feedback system. That is, the more species of plants, the more species of insects there can be, because angiosperms (flowering plants) and higher insects are involved in many **symbiotic associations,** such as predator-prey relationships, host-specific parasitism, and mutualistic interactions in pollination systems which are described in Chapter 30. The increasing diversity of angiosperms has fostered a wide variety of insects, and this diversification of insects has promoted a wider diversity of flowering plants. In many cases, insects that are specialized to feed on particular plants have specialized predatory or parasitic insects that feed on them. Insects are abundant in virtually every habitat on Earth, except the ocean, an environment in which flowering plants (the insects' coevolutionary "colleagues" in generating biotic diversity) are also scarce.

Insects' complex life cycle is another secret to their success. Adults and larvae of beetles and other insects that undergo metamorphosis have markedly different diets

and therefore do not compete with each other for food (▷ Figure 27–14). For example, caterpillars (larval forms of moths and butterflies) eat leaves, whereas the adult form, the moths and butterflies, sip nectar. Maggots eat carrion; their adult form, houseflies, eat bacteria. The different insect forms manifest during the life cycle perform very different functions. Larvae, for example, are efficient feeders but are fairly restricted in habitat. Flying adults are considerably more mobile and can readily find mates, disperse, and migrate. Between larval and adult stages there is frequently a nonfeeding stage, the *pupa*. The pupa is very active undergoing wholesale tissue destruction and reassembly, transforming the larva into an adult organism.

Insects are fascinating organisms. Some forms such as butterflies and beetles, are extremely beautiful. Others, among them the pollinators, scavengers, and predators who feed on insect pests, have great economic value. Of course, some insects carry disease to people and their domestic animals and plants. Some are external parasites on people, livestock, and plants. Others, such as mosquitoes, bedbugs, and biting flies, feed on body fluids. Ants, bees, and wasps inflict painful bites or stings. As any farmer will tell you, insects can be major competitors with humans for the food produced on farmland. Modern agriculture, which promotes

▷ **FIGURE 27–13 Insects** The most diverse orders of insects are (*a*) beetles, (*b*) social insects, (*c*) butterflies and moths, and (*d*) flies, all groups that have evolved in communal interactions with flowering plants.

(c)

(d)

▷ **FIGURE 27–13 Insects** The most diverse orders of insects are (*a*) beetles, (*b*) social insects, (*c*) butterflies and moths, and (*d*) flies, all groups that have evolved in communal interactions with flowering plants.

huge fields of single-crop species, is particularly vulnerable, and under these conditions insects may take a third of the crop before harvest and another 5% to 15% of the food when it is in storage. By practicing heteroculture, that is, planting several crop species in a single field, farmers can greatly reduce insect population and damage.

Centipedes and Millipedes. The familiar terrestrial arthropods—centipedes and millipedes—are frequently confused, but they represent distinctly different classes with markedly different characteristics and lifestyles. Centipedes, for instance, have only one pair of appendages per segment, whereas millipedes have two pairs.

▷ **FIGURE 27–14 Insect Life Cycle** Beetles and many other insects exhibit complete metamorphosis, with separate (*a*) egg, (*b*) larva, (*c*) pupa, and (*d*) adult stages.

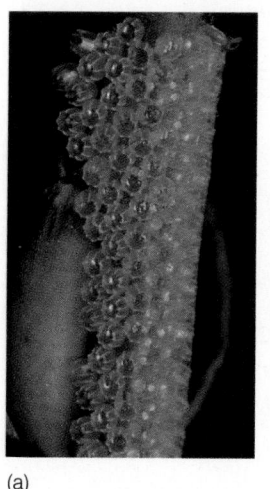
(a)

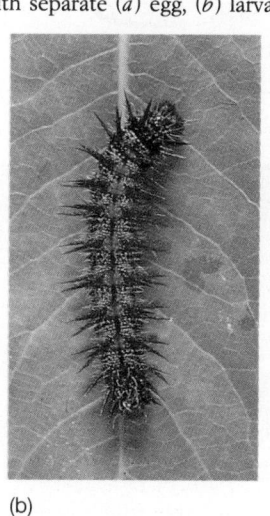
(b)

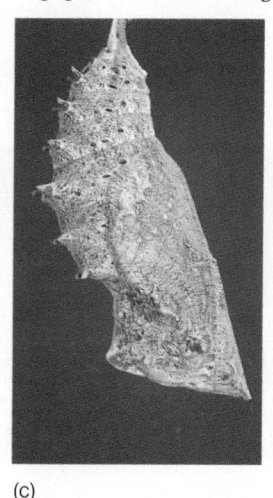
(c)

(d)

Although the names *centipede,* (hundred feet) and *milli-pede* (thousand feet) are numerically inaccurate, they do serve to emphasize the fact that both groups have lots of jointed legs. The number of legs ranges from about 10 to nearly 200 in both groups.

Although centipedes and millipedes are segmented, the segments are not very differentiated. The appendages of the first segment of the predatory centipedes have become piercing fangs that deliver a toxin to stun prey. Millipedes are scavengers.

Molluscs Are a Diverse Phylum Whose Members Share Four Features

The **molluscs** compose the second largest animal phylum, containing over 100,000 species. In some ways this may seem a peculiar group to have been so successful. As you may recall, most higher animals are bilaterally symmetrical; at least superficially, adult clams and snails are not. Most advanced animals show strong cephalization; some molluscs do not. Most higher animals are segmented; most molluscs are not.

Molluscs include such divergent members as snails, squids, abalone, and octopi (▷ Figure 27–15), but all exhibit four common characteristics: (1) a **mantle,** a membranous structure that surrounds the internal organs and secretes calcium carbonate to build shells; (2) a **radula,** a filelike feeding structure in the mouth, roughly comparable to a tongue with teeth; (3) a **foot,** used for locomotion; and (4) a **shell,** which provides protection.

Each of these structures has been modified during evolution to meet specific challenges. The most advanced molluscs, the squids and octopi, exhibit a radically evolved molluscan body plan. Unlike those of other molluscs, the

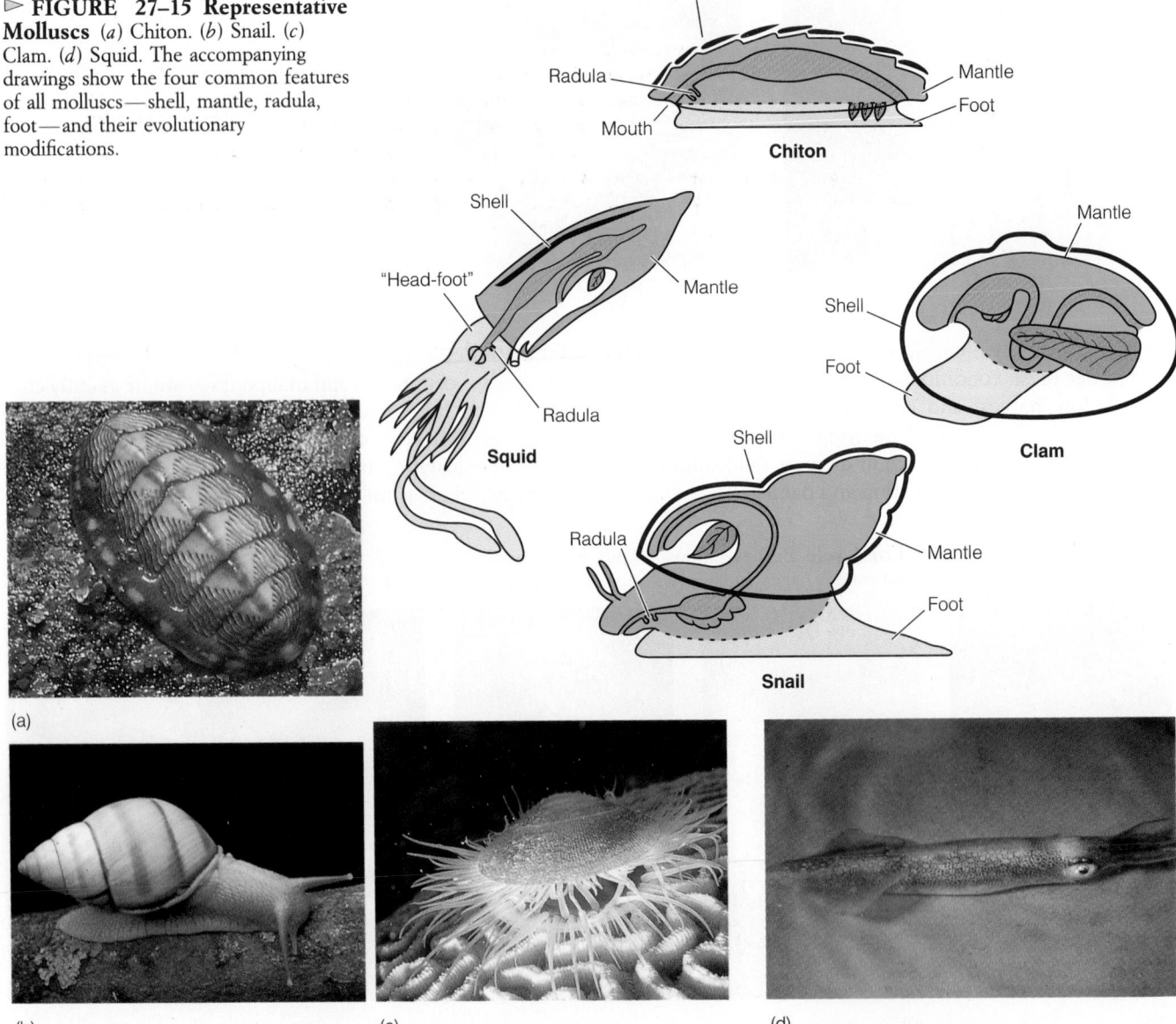

▷ **FIGURE 27–15 Representative Molluscs** (*a*) Chiton. (*b*) Snail. (*c*) Clam. (*d*) Squid. The accompanying drawings show the four common features of all molluscs—shell, mantle, radula, foot—and their evolutionary modifications.

shells of squids and octopi are internal. The mantle has evolved into a water jet propulsion system. The foot has evolved into a series of grasping tentacles. The advanced nervous system has image-forming eyes.

The molluscs include the largest of all invertebrates, the giant squids, which may be up to 15 meters (50 feet) in length and may weigh more than two tons. Most molluscs live in marine or freshwater habitats, but there are many kinds of land snails.

≋ THE ECHINODERM SUPERPHYLUM

The annelid superphylum includes the vast majority of animal species alive today. The echinoderm superphylum includes just two phyla, **Echinodermata** (starfishes, sea urchins, and others) and **Chordata** (birds, reptiles, amphibians, mammals, and others). Although these groups may seem unrelated, evidence showing their kinship is strong.

The Echinoderms Are Radially Symmetric Organisms that Gave Rise to the Chordates, the Phylum To Which We Belong

The most conspicuous feature of **echinoderms**—the starfishes, sea urchins, and sea cucumbers shown in ▷ Figure 27–16—is their spiny skin, which is the basis for their formal name (*echinos,* meaning "spiny," and *derm,* meaning "skin"). The echinoderm's skin provides protection from prey.

Adult echinoderms are radially symmetric, and there is no evidence of a head, both features of some of the earliest animal forms. Unlike the early animals, however, echinoderms have a complete digestive system with a distinct mouth and anus.

Echinoderms have internal skeletons composed of shell-like plates. They also have a distinctive **water vascular system,** a circuit of calcareous pipes filled with seawater.

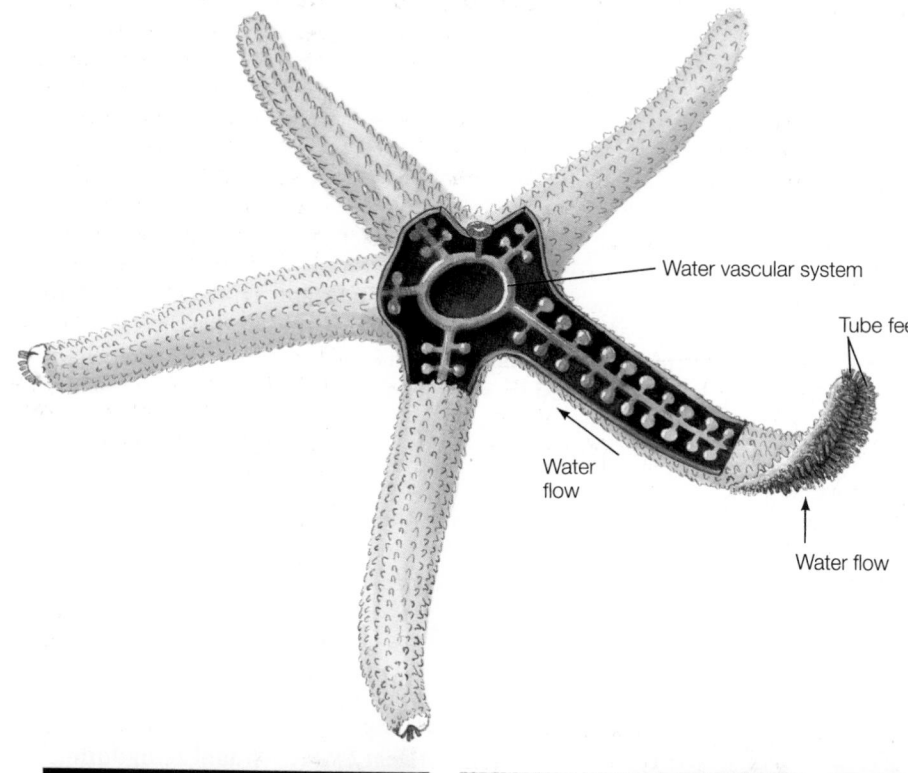

▷ **FIGURE 27–16 Representative Echinoderms** (*a*) Starfish drawing showing the water vascular system, tube feet, and other features. (*b*) Starfish. (*c*) Brittlestar. (*d*) Sea cucumbers.

Water vascular system

Tube feet

Water flow

Water flow

(b)

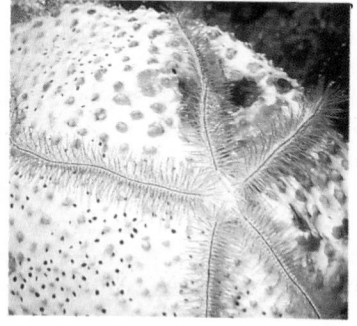

(c)

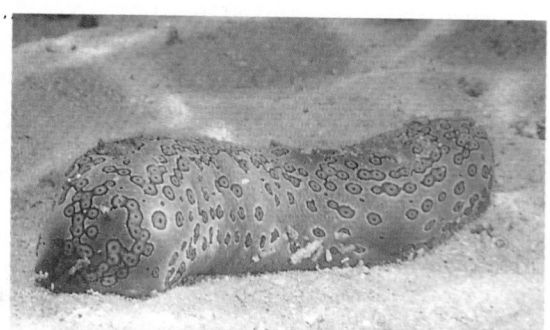

(d)

(a)

(b)

(c)

(d)

▷ **FIGURE 27–17 Larvae Tell Stories** (*a*) Larva, with (*b*) adult sea urchin, an echinoderm. (*c*) Larva, with (*d*) adult acorn worm, a chordate. Note the similarity of the two larvae.

This system helps them move about and prey on other animals. As illustrated in Figure 27–16c, when water is withdrawn from the small tube feet, a partial vacuum is formed, and the tube foot acts as a "suction cup" that can grasp the rocks of the ocean floor or the shell of a clam.

Echinoderms are neither a diverse group, with just over 6000 species, nor are they important ecologically. In fact, they are the only major animal phylum that never exploited the land or made its way into fresh water. One species, the crown-of-thorns starfish, has become a serious pest, destroying coral over large expanses of Australia's Great Barrier Reef. This potential catastrophe is the result of human actions—notably, widespread predation on the starfish's natural enemies.

To many biologists, the most impressive features of echinoderms are their larvae and their evolutionary descendants (which include ourselves). As noted earlier, a larva is an active stage between embryo and adult. Larvae are important to sedentary species because it is a dispersal or feeding stage that does not compete with adult forms. Echinoderm larvae—tiny, translucent, bilaterally symmetric bags of cells propelled by fringes of cilia (shown in ▷ Figure 27–17a)—also provide a clue to the origin of the chordates.

The echinoderm larva looks very much like the larva of an acorn worm, as shown in Figure 27–17b, and the acorn worm is one type of chordate, the phylum to which we and

other vertebrates belong. Evolutionary biologists believe that in some ancestral echinoderm species metamorphosis changed, with sexual maturation and reproduction taking place in the free-living larval form, rather than in the adult. This slight change may have opened a new branch of animal evolution, a complex and beautiful branch of which we humans are one twig.

A number of other features are common to echinoderms and chordates—among them, certain details of embryonic development (including the origin of the mouth and the anus) and the development of the coelom. Although there is no fossil record of chordate origins, because they were soft-bodied forms at first, biologists have constructed an interesting hypothesis to explain their roots.

The Invertebrate Chordates Are a Link Between the Invertebrates and the Rest of the Animal Kingdom

All chordates have three features in common (in part or all of their life cycle): (1) a hollow dorsal **nerve cord** that runs the length of the body, (2) a supporting, cartilaginous **notochord,** and (3) **gill slits** in the pharyngeal (neck) region.

Interestingly, some fairly simple animals—such as acorn worms, sea squirts, and the lancelet—have these characteristics but do not have a vertebral column (backbone). Consequently, these organisms are classified as **invertebrate chordates** (▷ Figure 27–18). These animals are links be-

(a)

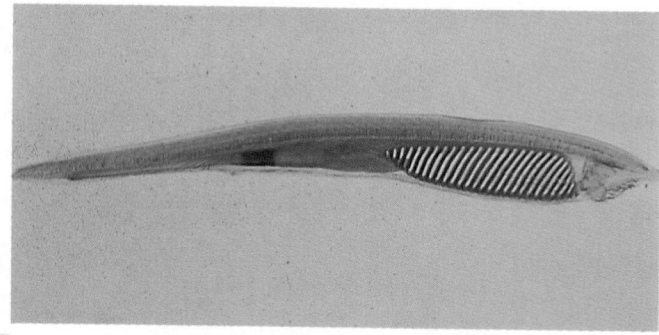

(b)

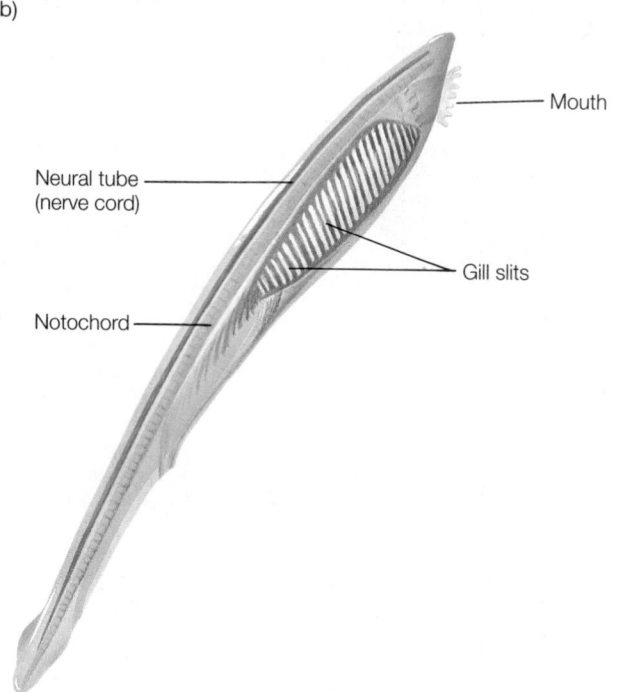

Mouth

Neural tube (nerve cord)

Gill slits

Notochord

▷ **FIGURE 27–18 Representative Invertebrate Chordates** (*a*) Sea squirt. (*b*) Lancelet. The accompanying drawing (*c*) shows notochord, neural tube, and pharyngeal gill slits.

tween the invertebrates and the rest of the animal kingdom.

The invertebrate chordates are filter-feeders, as are echinoderm larvae. In these organisms, cilia conduct water into their mouths. Food particles are removed by cells in the digestive tract, and the water is returned to the sea through the pharyngeal gill slits.

The First Vertebrates, the Jawless Fishes, Evolved in the Seas

The earliest known vertebrates are **jawless fishes (class Agnatha),** a group represented today by the lamprey and the hagfish (▷ Figure 27–19). Lampreys have larvae that resemble the invertebrate chordate called the lancelet (see Figure 27–18b). Jawless fishes first appear in the fossil record approximately 450 million years ago. These slow-moving bottom-feeders were covered with tilelike plates of bone, as shown in Figure 27–19, which no doubt protected them against giant water scorpions, the dominant predators of the seas at that time. The bones of the jawless fish are a complex phosphate mineral, one of the most durable (hence fossilizable) materials ever evolved.

The jawless fishes of today are not dominant life-forms in the ecosystems they occupy. Because they lack fins, their locomotion is not as efficient as it might be. The absence of jaws greatly limits their feeding options.

The Evolution of Jaws and Paired Fins Marked a Turning Point in Vertebrate History

A group of primitive fishes, the acanthodians, were the first to evolve jaws and paired fins. These fishes gave rise to the two groups of more advanced, modern fishes: the sharks (or cartilaginous fishes) and the bony fishes.

The presence of jaws makes feeding much more efficient and opens up a wide range of possibilities. But evolution is not conscious engineering; that is, it cannot foresee favorable adaptations. Rather, evolution is more like tinkering. New structures are made from old parts. If they provide a selective advantage, they tend to persist.

Jaws evolved from pharyngeal gill slits, described in Chapter 23, and are one evolutionary product of segmentation. Evolutionary biologists believe that jaws arose from the bony gill bars located between the pharyngeal gill slits. At about the same time, paired fins evolved, but the first fins were merely bony supports that formed in fleshy flaps of skin already present as stabilizing "outriggers."

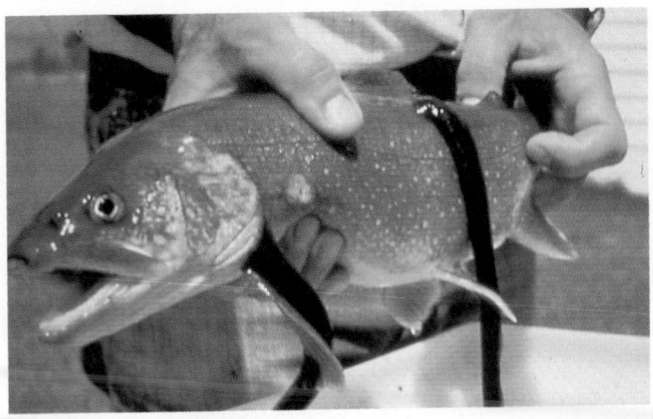

(a)

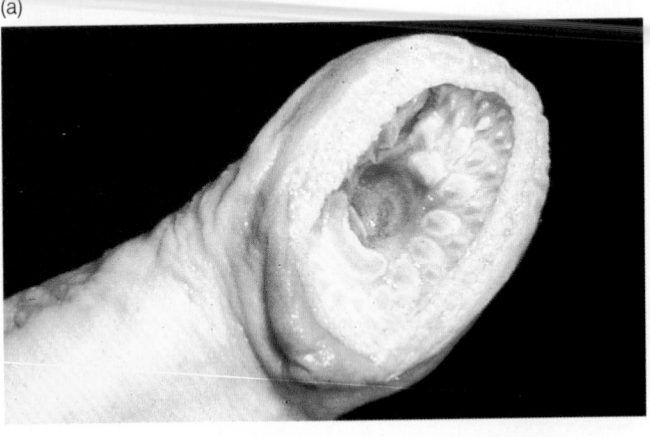

(b)

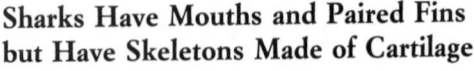

▷ **FIGURE 27–19 Representative Jawless Fishes** (*a*) The Great Lakes lamprey is a highly specialized external parasite on bony fishes. (*b*) A closeup on the mouth of a lamprey.

(a)

(b)

▷ **FIGURE 27–20 Representative Cartilaginous Fishes** (*a*) The great white shark is a carnivore, known to attack people on occasion. (*b*) The huge whale shark feeds on microscopic plankton.

Sharks Have Mouths and Paired Fins but Have Skeletons Made of Cartilage

Although sharks are impressive organisms, they are an evolutionary dead end; that is, they gave rise to no other animal group. Today, there are only about 550 living shark species, nearly all saltwater forms. Sharks belong to a class called **Chondrichthyes,** or the **cartilaginous fishes,** because they have internal skeletons made out of cartilage. Sharks have jaws (with lots of enamel-covered teeth) and paired fins (▷ Figure 27–20).

The Bony Fishes had Gills, but a Few Evolved Lungs, an Adaptation that Was Important to the Evolution of Amphibians

If sharks are an evolutionary dead end, the bony fishes (**class Osteichthyes**) are an "evolutionary thoroughfare" (▷ Figure 27–21). The bony fishes have more species— about 25,000—than all other vertebrates combined.

It is important to note, however, that bony fishes did not originate bone; jawless fishes did. However, bony fishes have more elaborate bone development. Every bone in your body (except the knee cap) is traceable to a homolo-

gous bone in the fish body. Bony fishes probably emerged 400 million years ago, well before the sharks.

Lungs. Among vertebrates, there are two organs for gaseous exchange, gills and lungs. **Gills** are adapted to life in the water. As described in Chapter 14, water flows over the gills, whose vessels are filled with oxygen-poor blood. Because concentrations in the water are greater than those in the blood, oxygen diffuses inward. Carbon dioxide, of course, moves in the opposite direction.

Lungs are internal exchange surfaces adapted to dry environments. They provide a moist surface across which gases are exchanged directly with the air. The climate of the Earth at the time fishes evolved lungs was somewhat like that of the present, characterized by wide climatic fluctuations and seasonal drought. During the wet season, streams and ponds were full, and fishes could extract oxygen from water with their gills. During dry spells, however, gills would not work. In response to the challenge of drought, primitive bony fishes evolved an internal surface for gaseous exchange, a lung, that permitted them to acquire oxygen directly from the air (Figure 27–21a).

The lung evolved into a **swim bladder,** an organ that

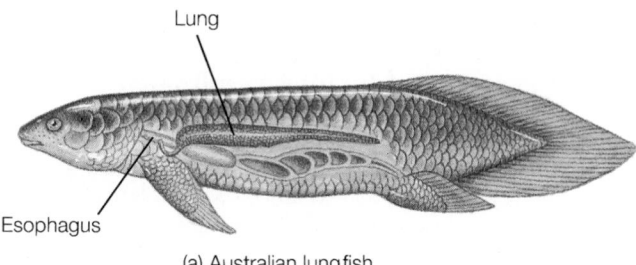

Lung

Esophagus

(a) Australian lungfish

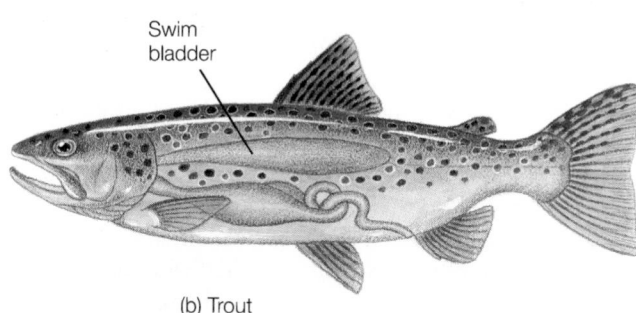

Swim
bladder

(b) Trout

▷ **FIGURE 27–21 Representative Bony Fishes** (*a*) Australian lungfish, showing the location of the lung. (*b*) Trout, showing the location of the swim bladder.

regulates buoyancy (Figure 27–21b). By controlling the amount of gas in the swim bladder, a fish can control its buoyancy in accordance with the depth of the water. The majority of the fishes today have swim bladders, but a few species like the Australian lungfish have retained the primitive lungs and are therefore capable of surviving periodic drought, moving to a new pond or puddle when the one they are occupying dries up.

Fins. Most of the early bony fishes had graceful fins adapted to efficient swimming. These species gave rise to the predominant groups of modern fishes. Others had fins with single, large bony supports and smaller, secondary supports. This group of fishes was never very successful in its aqueous home, but it eventually evolved into land vertebrates.

Natural selection favored the evolution of lungs and stumpy, leglike fins, because these adaptations helped the fishes survive periodic drought, allowing them to move from pond to pond. Because they could survive such calamities, these species persisted. Land limbs may also have allowed these fishes, the ancestors of amphibians, to drag themselves out on shore to bask in the sun. Such behavioral heat regulation could increase the rate at which they grew. Scientists speculate that the ancestors of amphibians may have come ashore to breed, spawning in quiet side channels away from hungry predatory fishes.

Biologists speculate that the invasion of the land offered many additional evolutionary advantages. On land was an expanding terrestrial flora and an expanding arthropod fauna, and there were no other vertebrates to exploit these

(a)

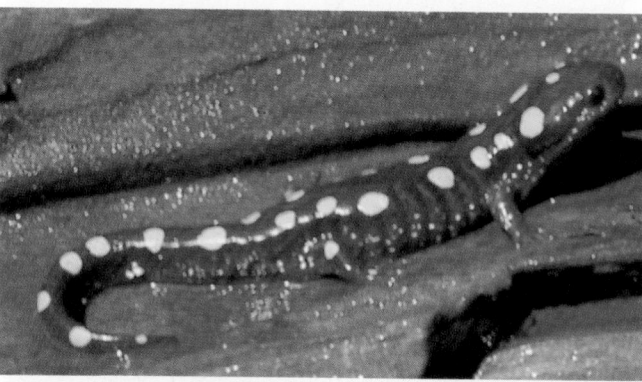

(b)

(c)

▷ **FIGURE 27–22 Representative Amphibians** (*a*) Marsupial toad. (*b*) Salamander. (*c*) Caecilian.

food resources. Natural selection undoubtedly favored individuals that could exploit the land's expanding resource base.

The Amphibians Inhabit Water and Land

Approximately 2500 living species belong to the **class Amphibia,** which includes mostly frogs and toads but also some salamanders and a few limbless, wormlike organisms known as caecilians (▷ Figure 27–22). The word *amphibia,*

(b)

▷ **FIGURE 27–23 Amphibian Eggs** (*a*) A female frog with her newly laid eggs. (*b*) Amphibian eggs must be deposited in water for successful development.

literally translated, means "double life," which reflects the fact that most species of this class spend part of their lives on land and part in the water.

Amphibians are a transitional group, whose lives straddle water and land and whose locomotion straddles that of swimming fishes and efficient quadrupedal (four-footed) animals on land. Even those amphibians that exhibit efficient terrestrial locomotion must return to the water to reproduce. Despite 350 million years of existence, amphibians have never quite escaped the water.

Amphibians reproduce sexually. Females produce unshelled, fishlike eggs (▷ Figure 27–23), and fertilization occurs in the external environment. The amphibian embryo that develops from the fertilized egg acquires oxygen and gets rid of wastes directly with the water that surrounds it. Water also protects the eggs and embryos from desiccation and from mechanical and thermal shock. Water is a great place for eggs to develop as long as it is permanent, still, and free of predators. It is not always an ideal environment. Eggs are subject to desiccation (as ponds dry up), predation, or being swept downstream by currents.

One group of amphibians evolved terrestrial reproduction, moving into a whole new adaptive zone and beginning a major new branch on the evolutionary tree, the reptiles.

Reptiles Are a Diverse and Successful Group Whose Evolution Depended in Part on the Emergence of Internal Fertilization

Reptiles (class Reptilia) differ from amphibians in several respects, but the basic distinction is that they lay eggs with shells. The reptilian egg solved all of the problems faced by amphibian eggs simultaneously because it is, in effect, an embryo's private pond.

Eggs are produced by the ovary of the female, and shells are laid down after fertilization occurs. As the embryo develops, it forms several important membranes. The **yolk sac,** for example, contains abundant blood vessels and surrounds the nutritious yolk of the egg (▷ Figure 27–24). Another membrane, the **allantois,** serves as an embryonic bladder, storing nitrogen-rich wastes. The entire embryo is then surrounded by a fluid-filled sac, the **amnion,** which prevents the embryo from desiccating. Finally, a tough, outer **chorion** is formed directly beneath the shell membrane. This membrane is richly supplied with blood vessels and provides a surface for gaseous exchange.

Internal Fertilization. The complex land egg of the reptile is a major evolutionary advance that protects the parents' genetic investment. However, the shelled egg requires internal fertilization, an adaptation that occurred in some early amphibians that then gave rise to the reptiles.

Why did internal fertilization evolve? Many species of amphibians were present in the Age of the Amphibians, and most of them probably moved to ponds to reproduce, just as their descendants do today. In the pond, some eggs might be fertilized by alien sperm and hence be wasted on sterile hybrids. Evolutionary biologists believe that natural selection would favor patterns of mating that brought the sexes of the same species sufficiently close that the risk of

▷ **FIGURE 27–24 The Reptilian Land Egg** Successful colonization of the land by reptiles depended in part on the shelled egg, which provides an aqueous environment for the growing embryo and a ready supply of food.

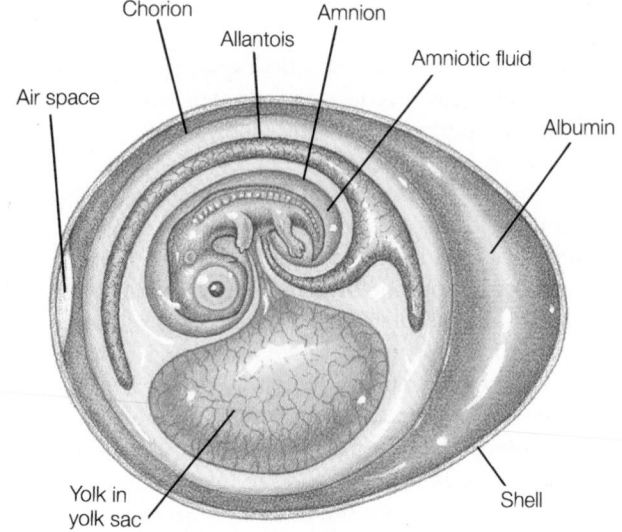

contamination by another species' sperm was minimized. The most secure strategy would involve actual insertion of male gametes into the female body. Amphibians that evolved that strategy were probably the ancestors of the reptiles.

The land egg was a monumental evolutionary advance. It allowed reptiles to exploit the dry land, to diversify, and to dominate ecosystems for 100 million years. No other vertebrate group has been as successful as the reptiles for as long a time.

Reptilian Diversification. A simplified family tree of reptiles hints at their diversity (▷ Figure 27–25). As shown, ancestral reptiles (or stem reptiles) arose from the amphibians and subsequently gave rise to turtles, fishlike ichthyosaurs, plesiosaurs, and the mammal-like therapsids. Somewhat later, a line leading to lizards and snakes branched off from the stem reptiles. The stem reptiles gave rise to

another group called the thecodont ("socket-toothed") archosaurs ("ruling reptiles"). "Ruling" is a good description, because this group included flying pterosaurs, stegosaurs, and numerous dinosaurs. Two terminal twigs of the archosaur tree remain today, the crocodiles and their kin, and the birds.

Because some of the archosaurs were enormous, they have stimulated the popular imagination ever since their discovery. In recent decades, the thinking about dinosaurs has changed dramatically. Scientists once envisioned them as dim-witted and sluggish, but today increasing evidence suggests that some dinosaurs were quite active, perhaps even warm-blooded. Some may have cared for their young and formed cooperative groups for hunting and defense.

Of course, many aspects of the biology of archosaurs must remain unknown. Perhaps the biggest question is, "Why did they become extinct?" This subject was discussed in Chapter 23.

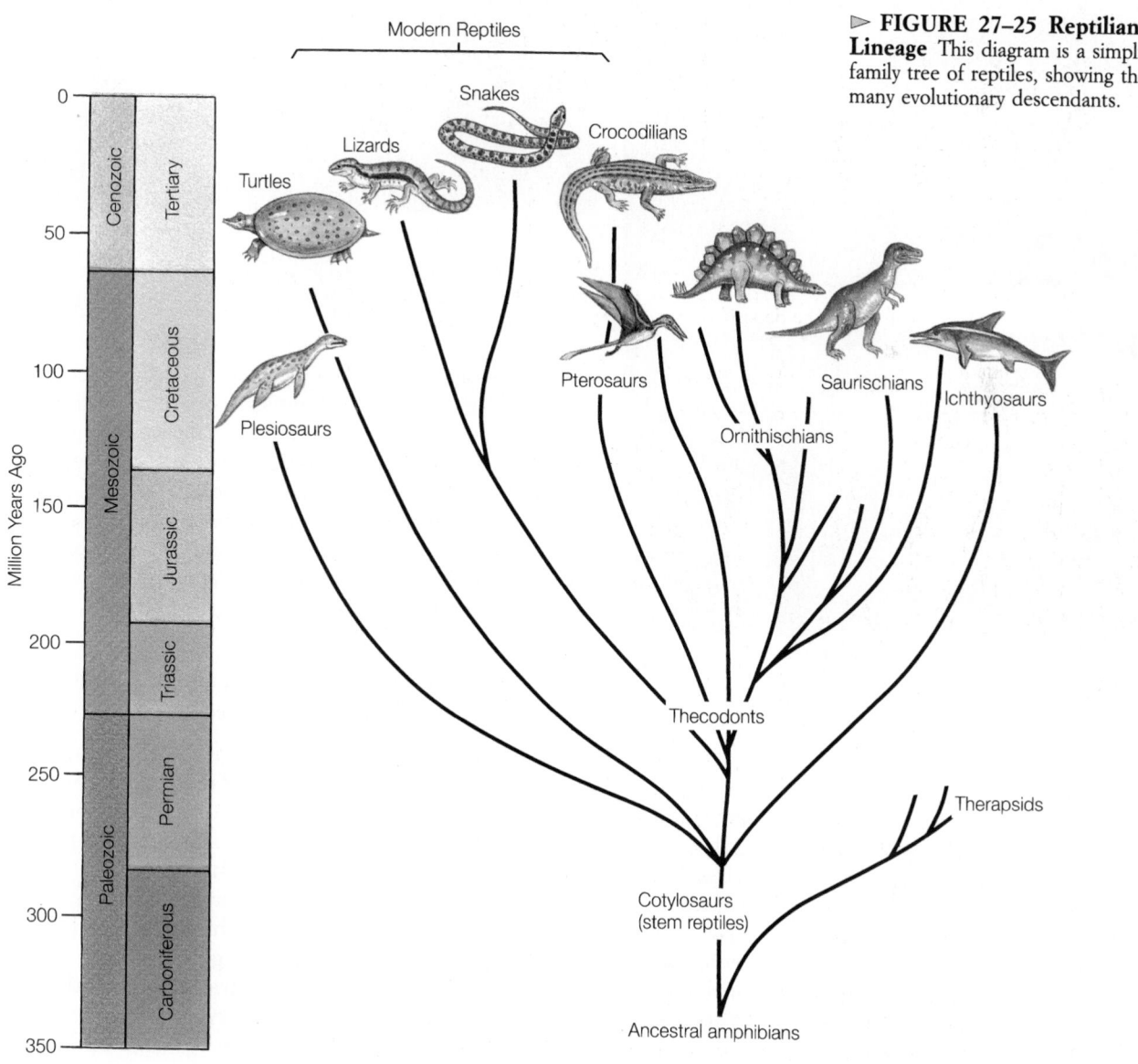

▷ **FIGURE 27–25 Reptilian Lineage** This diagram is a simplified family tree of reptiles, showing their many evolutionary descendants.

(a)

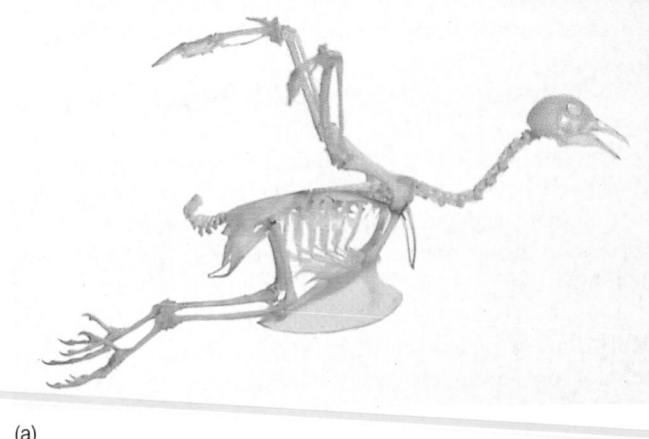

(a)

(b)

(b)

> **FIGURE 27-27 A Bird Is a Toothless Dinosaur?**
Stripped of their feathers, birds betray their ancestry. (*a*) Pigeon skeleton. (*b*) Dinosaur.

(c)

> **FIGURE 27-26 Representative Reptiles** (*a*) Garter snake giving birth to live young. (*b*) Chameleon. (*c*) Alligator.

Approximately 6000 species of reptiles are known today, including 250 species of turtles and a dozen kinds of crocodiles and alligators (▷ Figure 27-26). The lizards and snakes are also found in abundance in certain regions.

Modern reptiles are most abundant and diverse in tropical and subtropical areas. Thus, we can expect new species to be reported as equatorial rain forests are explored, and we can expect many species to become extinct as these forests are destroyed. As one moves out of the tropics toward the poles, fewer and fewer reptile species are encountered.

A common peculiarity of reptiles in the more challenging environments is **viviparity:** production of live young, as in the garter snake (Figure 27-26a). Successful movement into cold or even seasonally cold environments and reproduction in these habitats demands control of body temperature. One offshoot of the dinosaurs solved the problem of temperature regulation with feathers and endothermy (internal heat production); we call this group birds.

Birds Evolved From Reptiles and Are Characterized by Feathers and Warm-Bloodedness

Birds (class Aves) are among the most familiar animals in our surroundings, and they are important members of most biological communities on Earth. (For a discussion of a prominent threat to birds, see Health Note 27-1.) Most birds are active by day, colorful, and noisy, just as we primates are. People of most cultures appreciate birds. Stripped of their feathers, however, birds seem to be just fancy little toothless dinosaurs (▷ Figure 27-27).

The fundamental distinction between birds and their reptilian forebears is feathers, which evolved from reptilian scales. (You can still see reptilelike scales on a bird's legs

PESTICIDES' DEADLY LEGACY

Americans and many other people the world over love vast green expanses, and will go to great lengths to produce the perfect lawn or golf course fairway. Unfortunately, birds have often fallen victim to our zeal for the perfect grassy patch. In the United States, more than 67 million pounds of chemicals are used each year to control fungus, insects, and weeds on our lawns.

One pesticide, diazinon, was commonly used on sod farms and golf courses throughout the United States to control insects. In 1984, three fairways were treated with diazinon at a Hempstead, New York golf course. In the two days following treatment, 700 Atlantic brant geese, 7% of the state's population, died of acute diazinon poisoning.

Diazinon was also used to treat nine fairways at a golf course in Bellingham, Washington. After application, the area was irrigated, as recommended by the pesticide manufacturer, to decrease pesticide concentrate on the turf. Nevertheless, 85 American wigeon (a kind of duck) perished after grazing on one of the treated fairways on the day of application.

The EPA received 60 reports of bird kills from 18 states in 1984. Twenty-three bird species were affected and 20 of the incidents occurred on golf courses. In 1986, the EPA banned the use of diazinon on golf courses and sod farms. The ban was based largely on potential hazards to waterfowl, although human illness had also been reported. Since the bird kills appeared to be associated with large areas of diazinon use, the pesticide is still available for use on home lawns.

Birds have also been dramatically affected by the use of granular carbofuran, which was developed in 1970 by the FMC Corporation of Philadelphia. Carbofuran is the name of the poison in Furadan 15G. Farmers throughout the United States use Furadan 15G to eradicate nematodes and insects from corn, rice, and other crops. While agricultural carbofuran apparently poses no threat to humans, a songbird can die after ingesting a single granule. Many isolated incidents of bird kills associated with its use have occurred. Twenty-three states have reported carbofuran-poisoned birds since 1972.

By 1989, EPA records showed that 2 million birds were dying of carbofuran poisoning each year. In 1990, more than 200 dead birds were poisoned in eastern Virginia near the Rappahannock River. The dead birds included sparrows, eastern bluebirds, red-winged blackbirds, blue jays, goldfinches, and starlings. FMC Corporation acknowledged a problem but faulted the farmers for misusing carbofuran to bait predators, or for mishandling it by spilling it.

In 1991, FMC issued new instructions for the use of carbofuran. Concerned parties in Virginia decided to monitor the results of that year's pesticide application on 900 acres of treated farmland. After application, 62 bird carcasses, 10 sick birds, and 47 "feather spots" (indicating a dead bird eaten by a scavenger) were found. The monitors probably found only half of the carcasses on the 900 acres. Considering this, and the hundreds of thousands of acres of farmland across the state that had been treated with Furadan that spring, it is possible that tens of thousands of birds are dying of pesticide poisoning every year.

The EPA recommended banning Furadan by canceling its registration in the early 1980s. But no action was ever taken. According to John Bascietto, a biologist formerly with the EPA, "If a chemical is not going to cause a problem for humans, it generally won't be regulated for wildlife concerns." In 1991, after the monitoring study was complete, concerned citizens in Virginia successfully persuaded the Pesticide Control Board to issue an emergency ban on Furadan use. The ban on both the sale and use of Furadan in Virginia became effective June 1, 1991. Wildlife supporters hailed this ban as being the first time a chemical had been banned solely because of its potential for harming wildlife.

Furadan has since been banned nationwide by the EPA. By September 1994, only 2,500 pounds will be allowed to be sold. This compares to recent sales of up to 10 million pounds per year. Though the ban will likely have a positive impact on birds, it is not faultless. Export of Furadan will continue and will probably increase to make up for shrinking domestic sales. So, migratory birds will still be threatened.

---

and feet.) Feathers insulate a bird's body. Although insulation may have been their initial function, feathers are also vital to flight, which permits birds to access a wide range of resources. Insulative feathers also permit a high and constant body temperature independent of external environmental temperature (warm-bloodedness, or endothermy). Feathers basically allow a tropical internal environment to be carried to temperate and polar zones.

Ornithologists (scientists who study birds) have identified about 8600 species of birds alive today. These remarkable animals exhibit many adaptations to the environments they inhabit. Most birds can fly. On some islands, however, where predators do not exist, some birds have lost the ability to fly. The wings of the penguin are more like flippers that allow the bird to "fly" through water. The feet of birds are uniquely adapted to different habitats. (Contrast waders with swimmers with perching birds, for example.) Beaks have also diversified in response to variety of diets. (Contrast hunting hawks, fishing cormorants, nectar-feeding hummingbirds, and seed-eating finches.)

Birds arose from reptiles about 190 million years ago, somewhat later than the mammals. Today, there are about twice as many species of birds as there are mammals. Nonetheless, an unbiased observer would have to admit that there is more similarity in form among an ostrich, a warbler, and a penguin—all birds—than among mammals such as a blue whale, a rhinoceros, and a bat.

Although the variety of birds is great at the species level, the class itself exhibits a rather limited diversity of body plans, probably because flying evolved early and birds were forever constrained by the requirements of successful flight. The mammals are quite a different story.

Mammals, Which Also Evolved from Reptiles, Are Characterized by Body Hair for Insulation and by Mammary Glands for Milk Production

Mammals (class Mammalia) arose from reptiles 240 million years ago. However, living mammals differ from living reptiles in several obvious ways. These differences evolved over tens of millions of years. Because many of the characteristics that distinguish mammals do not fossilize well, scientists do not know exactly when reptilelike mammals arose from mammal-like reptiles. Most paleontologists use one readily fossilizable criterion to define a mammal, the bones of the jaw and ear.

Jawbones Become Ear Bones. The lower jaw of reptiles is composed of three bones that articulate with a fourth bone on the skull (▷ Figure 27–28). In mammals, the jaw is a much simpler affair, consisting of a single bone, the **mandible,** which is hinged directly to the skull. This radical simplification of the jaw, say evolutionary biologists, probably left the remaining reptilian jawbones available for novel uses. The spare jaw parts evolved into **ear ossicles,** tiny bones in the middle ear that transmit vibrations from the eardrum to auditory organs of the inner ear (Chapter 18).

The reptilian jaw has three bones, and the ear just one; the mammalian ear has three bones, the jaw just one. Although the mammalian jaw is less mobile than that of the reptile, it has much greater mechanical strength. The increased strength is important to full utilization of mammalian teeth.

Teeth. Mammalian teeth have three distinctive features. They are (1) specialized from front to back, (2) set in sockets, and (3) occur in two sets, the **deciduous teeth** and the **permanent teeth.** The deciduous teeth (baby teeth) are lost as the permanent teeth develop underneath. In contrast, the more primitive vertebrates (including most living reptiles) have teeth that are about the same size and shape, are fastened to the jaw bones, and can usually be replaced as needed throughout life (▷ Figure 27–29).

Primates (including humans) exhibit a fairly primitive dentition and have four kinds of teeth: (1) **incisors** (for nipping), (2) **canines** (for grasping prey), (3) premolars (often called bicuspids in humans), and (4) molars, the latter two for crushing, shearing, or grinding food. Molars are present only in the permanent teeth.

This specialization of teeth permits a more efficient processing of food. Because food is ground more completely before swallowing, digestion in the intestine is more rapid, providing fuel and building blocks for active animals.

Hair. A hair is a cylinder of a protein known as **keratin,** extruded from a hair follicle in the epidermis. The original function of hair was probably insulation, which is essential to endothermy. However, hair has been modified to serve a variety of other functions, some of which are shown in ▷ Figure 27–30.

Attached to each hair is a tiny muscle that raises an animal's hairs, trapping an insulating layer of warm air in the coat. In humans (minimal body hair), these muscles are vestigial, and their contraction raises goose bumps.

Glands. Each hair has a sebaceous gland, which secretes an oily material called sebum that lubricates and waterproofs the hair. Sweat glands produce a solution of water and salts that evaporates from skin and hair to cool the body. Dogs lack sweat glands and cool themselves by panting. Hair glands of many mammals produce musky odors that function in sexual attraction or individual or species recognition.

▷ **FIGURE 27–28 Reptiles Versus Mammals I** (*a*) The reptilian jaw has three bones, the ear just one. (*b*) The mammalian jaw has one bone, and the ear has three.

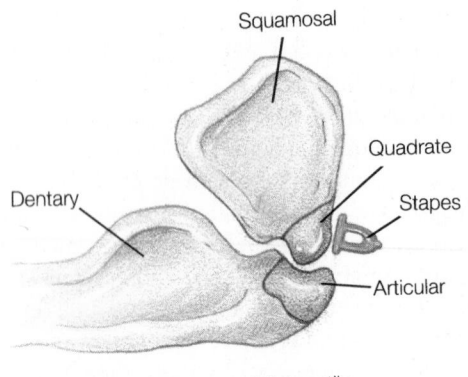

(a) Advanced mammal-like reptile

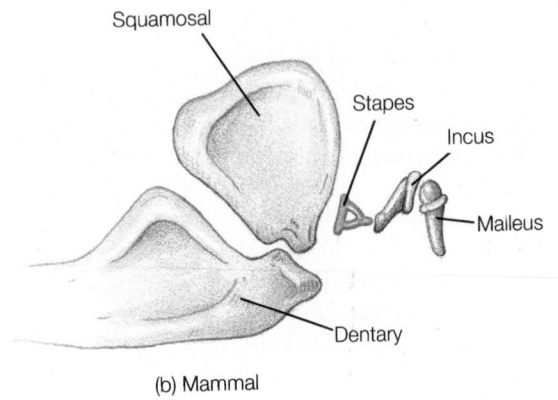

(b) Mammal

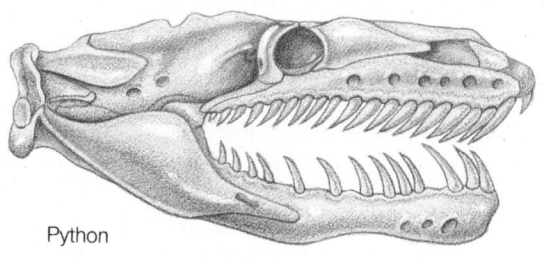

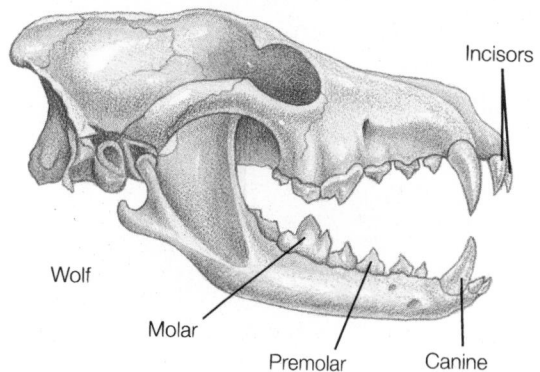

▷ **FIGURE 27–29 Reptiles Versus Mammals II** Reptilian teeth are all fairly similar. Mammalian teeth, as seen in the opossum, are of four kinds: (*a*) incisors, (*b*) canines, (*c*) premolars, and (*d*) molars.

Mammary Glands. All mammals contain milk-producing **mammary glands,** which are believed to represent evolutionary modifications of sebaceous and sweat glands (▷ Figure 27–31). The class Mammalia derives its name from the mammary glands, which have been fundamental to the success of mammalian species. As you learned in Chapter 22, milk (a mixture of water, carbohydrates, fats, proteins, and salts) produced by the mammary glands provides newborns with food. Furthermore, nursing makes young mammals a captive audience and forces a period of interaction with the mother (and often with siblings) that allows training and socialization (Figure 27–31).

Heart. Like the birds, mammals have an efficient, four-chambered (double-pump) heart, with separate circuits for oxygen-poor and oxygen-rich blood. This system allows efficient delivery of oxygen (and removal of carbon dioxide), permitting high rates of activity. Studies of comparative anatomy and embryology tell us that the four-chambered hearts of mammals and birds developed independently; the common ancestor of birds and mammals was a stem reptile that probably had a three-chambered heart.

Endothermy. The presence of hair and a structurally advanced heart are essential to endothermy. The body temperature of both birds and mammals is fairly high, closer to the upper limit of environmental temperatures than to the lower limit. Scientists believe that this phenomenon is a result of the fact that heating is easier for animals than is cooling.

▷ **FIGURE 27–30 Hair** The basic function of hair is insulation. During evolution, hair has been modified for (*a*) sensation, (*b*) camouflage, (*c*) warning, and other functions.

(a)

(b)

(c)

▷ **FIGURE 27–31 Milk-Producing Mammary Glands** A distinctive feature of mammals, the mammary glands evolved from sebaceous and sweat glands. Milk nourishes the young, and nursing makes the young a captive audience.

Animal cooling systems are usually based on evaporation, and water loss is costly for terrestrial animals.

Internal Development. Except for the duck-billed platypus and spiny anteater of Australia, mammals do not lay eggs. Rather, embryos develop within the mother and are then born at different stages of development. In mammals other than marsupials (the pouched mammals of the Americas and Australia), an intimate relationship develops between mother and embryos across the placenta, a structure built of maternal and embryonic tissues that supplies food and nutrients to the embryo and removes its waste, as discussed in Chapter 22.

The mammalian embryo is formed from membranes that owe their origin to reptiles. As you may recall from Chapter 22, the amnion retains its reptilian function as a fluid-filled, shock-absorbing bag. The chorion, not mentioned in Chapter 22, is present in mammals and provides the surface for gaseous exchange. However, the reptilian allantois and yolk sac are no longer needed. In the reptile, the yolk sac stores food; embryonic mammals derive nutrients from the mother's bloodstream. The reptilian allantois is a receptacle for embryonic wastes; embryonic mammals pass wastes to the maternal circulation. Thus, these structures are "surplus," but they have not simply been discarded by "evolutionary progress"—they are used for different purposes. The yolk sac and the allantois contain blood vessels that lead to and from the embryo. In mammals, these blood vessels are part of the umbilical cord, which connects the mammalian embryo to the placenta.

The membranes of the mammalian yolk sac and allantois lend mechanical strength to the umbilicus.

In addition to providing an exchange surface, the placenta helps (in ways that are only beginning to be understood) prevent the mother's rejection of the embryo, which is, genetically speaking, only half like its mother and hence should be rejected as an alien protein.

Brain. An advanced brain is another element in the success of the mammals. The mammalian brain differs dramatically from the reptilian brain, but it differs in degree, not in kind. All parts of the mammalian brain, in fact, were present in ancestral reptiles and even in fish. In the evolution of mammals, the parts have evolved further, increasing in size and complexity. A good example is the cerebral cortex, which is involved in learning (Figure 16–16). Mammals are more highly dependent on learning than are other kinds of animals (Chapter 29). They often lack instinctive solutions to the challenges they face and instead must piece together solutions based on experience and other learned information.

The Diversity of Mammals. To date, approximately 4400 species of living mammals have been identified, considerably fewer than there are species of birds, reptiles, or fishes, let alone several classes of invertebrates (insects, snails, and so forth). The mammals are therefore impressive not for the number of species but for the number of different body plans they exhibit.

The common ancestor of both placental mammals and marsupials was a rat- to opossum-sized, quadrupedal, omnivorous, terrestrial animal. From those beginnings, numerous and diverse mammalian forms evolved, species as different as bats, whales, and elephants (▷ Figure 27–32). Different in outward appearance, the mammals also differ considerably in size, from tiny shrews weighing 3 grams (the weight of a penny) to 150-ton whales (the mass of three full railway boxcars). Shrews are half as long as your little finger; blue whales are half again as long as a tennis court.

Mammals show immense dental diversity, which reflects adaptive radiation. Herbivores, such as rodents and rabbits, that feed on plants have no canine teeth but have simplified incisors and large molars with increased surface for grinding their food. The molars of horses and meadow mice grow continuously because they are worn down by a diet of fairly abrasive grasses (▷ Figure 27–33). The incisors of beaver grow continuously and are self-sharpened to offset the wear caused by gnawing. The front teeth of vampire bats are as sharp as the physician's lance (and serve the same function—to draw blood). Tusks of elephants, walruses, narwhals, and warthogs are teeth turned to tasks beyond feeding. Some whales and anteaters have no teeth at all. Toothed whales, by contrast, have proliferated teeth; some have over 200 of them.

Another arena for adaptive radiation is locomotion. The basic mammalian stock was quadrupedal. Each foot had

(a)

(b)

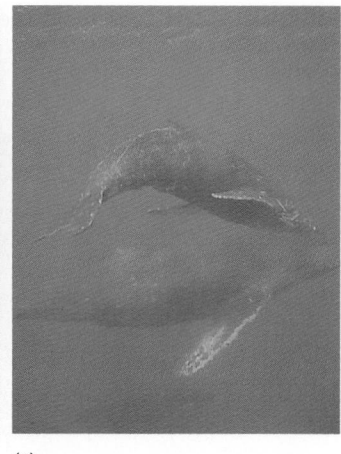
(c)

▷ **FIGURE 27–32 Representative Mammals** (*a*) African elephant. (*b*) Big brown bat. (*c*) Humpback whales.

five toes, a pattern traceable back through reptiles to ancestral amphibians. Humans retain this primitive arrangement, although it has changed in detail, of course, to accommodate our unique bipedal gait. In other mammalian lineages, the basic plan has changed in many ways, as is evident in the limbs and locomotion of squirrels, bats, giraffes, and manatees (Figure 27–34).

≈ TRENDS IN ANIMAL EVOLUTION

A quick recap of trends in animal evolution reveals four basic changes. As you review each one, consider the extent to which we humans exemplify these trends:

1. One of the most obvious changes in animal evolution was an increase in size and the emergence of multicellularity. Multicellularity led to cellular specialization and a subsequent division of labor that eventually resulted in the evolution of distinct tissues and organ systems. Humans are obviously large and multicellular with an extraordinary degree of specialization among our cells.

2. The evolution of the animal kingdom also reveals a

▷ **FIGURE 27–33 Dental Diversity in Some Mammals** (*a*) Horse. (*b*) Beaver.
l(*c*) Common vampire bat. (*d*) Narwhal.

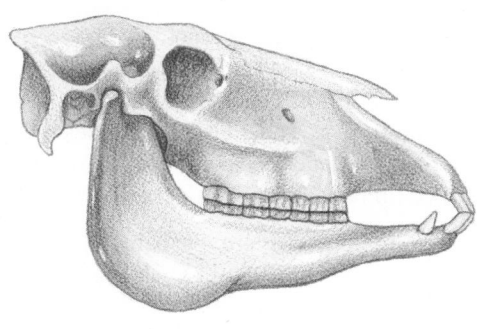
a. Horse

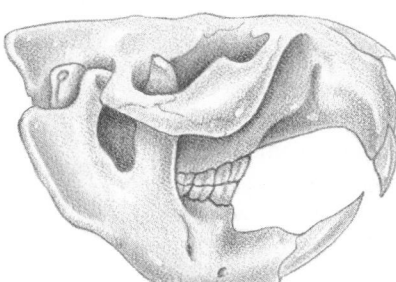
b. Beaver

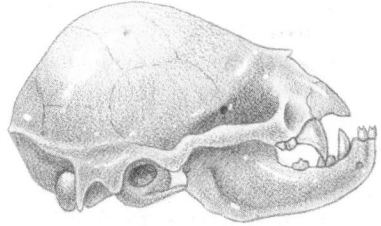
c. Common vampire bat

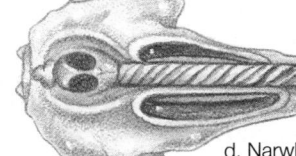
d. Narwhal (top view with bone dissected away)

▷ **FIGURE 27–34 Some of the Diversity of Mammalian Hind Limbs.** (*a*) Squirrel. (*b*) Cheetah. (*c*) Giraffe. (*d*) Manatee.

(a)

(c)

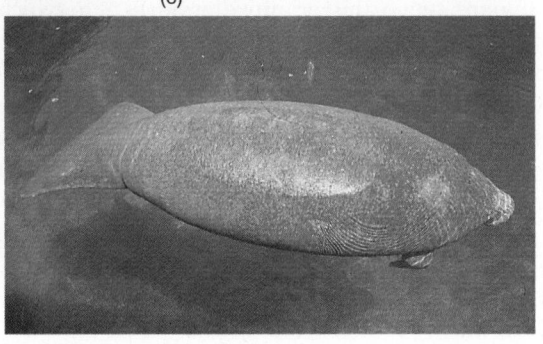

(b)

(d)

trend from asymmetry to radial symmetry to bilateral symmetry. Bilateral symmetry allows directional orientation in the environment and allows and encourages cephalization, the concentration of functions (sensation, information processing, food getting) in that end of the organism that confronts stimuli first, the head. Humans are the most cephalized of living things, the most completely dependent on their head for survival.

3. Our survey of animal evolution showed that successful new adaptations arose early in the history of many ancestral groups, before the lineages had become specialized. Many new classes emerged from these forms. Amphibians, for example, arose from primitive bony fishes, not from more evolutionarily advanced forms, such as angelfishes or seahorses. Reptiles emerged from early amphibians, not from specialized amphibians such as frogs. Mammals arose from primitive reptiles, not from snakes.

4. The evolutionary story shows that the usual trend has been from generalist to specialist—that is, from organisms capable of living in a wider range of conditions to those more specially adapted to specific habitats. Species fit into environments, but environments change. Populations of organisms may change apace. However,

specialists may not evolve quickly enough in the face of environmental change. This failure to adapt leads to extinction.

The trend in evolution is generalization followed by specialization, which is sometimes followed by extinction. Where do we humans fit on that continuum?

In the history of life, humans are probably the first species ever to have evolved the ability to ask this important question. Also, we are apparently the only animal with the ability to choose the course of our own evolution and literally shape the future of the biosphere, the living skin of the Earth, a lovely planet we share with millions of species billions of years in the making.

ENVIRONMENT AND HEALTH: WHAT GOES AROUND COMES AROUND

More than 70,000 chemicals are in commercial use in the United States. Relatively few members of this legion of chemical substances have been tested for toxic effects or are subject to regulation, and many of them eventually end up in streams, rivers, and lakes. Even chemicals that were banned in the 1970s, such as the PCBs and DDT, continue to trickle into bodies of water, leaching from the soil. Such chemicals may also be present in bottom sediments, from which they are slowly released into the water.

In Lake Michigan, 9 out of 10 lake trout are tainted with PCBs and DDT. One-quarter of these fishes contain levels too high for human consumption. Forty percent of the chinook salmon taken from the lake exceed public health standards for safe levels of PCBs, and nearly every salmon tested contains mercury, dieldrin, DDT, and other toxins in addition to PCBs.[2] Those fishing on Lake Michigan have been cautioned against eating contaminated fish. Pregnant women, women who intend to have children, and children under the age of 15 have been advised not to eat chinook salmon, walleye, yellow perch, brown trout, and lake trout from Lake Michigan.

Fishes are clearly an avenue by which our pollutants come back to haunt us. But they are not passive transmitters of toxins from the environment to human consumers. Industrial chemicals may injure fishes, causing tumors and other defects, or even kill them outright. One of the most widespread effects is liver cancer, which occurs most often in freshwater and marine species that feed on the bottom. Studies show that all of the sauger (a type of fish) in Michigan's Torch Lake have liver cancer. Ninety percent of the tom cod in New York's Hudson River and 60% of the brown bullheads in the Black River near Cleveland also have cancerous livers.

The Smithsonian Institution maintains a fish tumor registry in its Department of Invertebrate Zoology. In 1964, the first unusually high incidence of tumors was reported in Deep Creek Lake in Michigan. That outbreak was found to have been caused by a naturally occurring virus. Today, however, 35 to 50 known tumor clusters (places where fishes suffer from cancer) have been identified, most of which are located near industrial areas, a fact that suggests a causal relationship.

In 1987, Paul R. Bowser and his colleagues compared the occurrence of liver lesions in brown bullheads from two sites on the Hudson River in New York, one contaminated

[2] Dieldrin and DDT are pesticides; PCBs are a class of chemicals used as electrical insulators and additives to plastics.

with potentially toxic chemicals, the other relatively free of them. Fish movement between the sites was prevented by a waterfall and a dam. Fish taken from the upstream site were the control group used for comparison. Fish from the downstream site, near Fort Edward, New York, which was formerly a PCB discharge area, were the "experimental group."

Sediment in the downstream site contained elevated levels of PCBs, lead, chromium, and cadmium. Although industrial and municipal discharges had made their way into the upstream site, concentrations were considerably lower than those in the downstream section of the river. Researchers collected 24 fish from the contaminated site and 14 from the control site. The livers were removed and preserved. Specimens were coded so that the collection site was not known by the examiner (to eliminate bias.) The livers were then examined under the microscope to detect any abnormalities.

Researchers found that the incidence of three liver abnormalities was significantly higher in fishes from the contaminated site than in those from the upstream section. Because the liver functions to concentrate and remove toxic materials from the blood and may alter them chemically so they can be excreted, the higher incidence of liver lesions in fish from the downstream site suggests a causal relationship.

Like many other experiments, this one is not conclusive. It does not prove that the higher incidence of liver lesions was the result of chemical contamination. However, experiments in which rats are fed PCBs also produce similar liver abnormalities. Studies also suggest that heavy metals such as cadmium, lead, and chromium cause the breakdown of red blood cells in the liver, resulting in the development of lesions.

Fishes from the less contaminated site had abnormalities, but a lower number. Because this area was not free from pollution, abnormalities found in the fishes there may have been caused by nearby industrial sources. The causative factor is not known, and it is possible that an unknown factor or factors could be responsible for the liver damage. Nevertheless, a growing body of evidence shows that polluted water is not good for us. It can contaminate our food supply and kill and poison our fish and wildlife.

SUMMARY

WHAT IS AN ANIMAL?

1. Animals are eukaryotic, multicellular heterotrophs that lack cell walls; most are mobile.

THE ORIGINS OF ANIMALS

2. Animals originated from flagellated protists 600 to 800 million years ago.
3. Scientists distinguish two dozen phyla of animals, nearly all of which are represented today. Each phylum represents a distinct strategy for survival.

4. There is no fossil record of the earliest animals, but scientists believe that these organisms were probably small, soft-bodied organisms, somewhat like modern-day sponges.
5. Sponges exemplify an early stage of animal evolution, being somewhat more complex than a simple aggregation of one-celled organisms and falling short of organisms with distinct tissues.
6. Scientists do not believe that sponges gave rise to more complex phyla.

7. Jellyfishes and their relatives are radially symmetrical and have a tissue grade of organization, with an outer ectoderm, an inner endoderm, and a gelatinous mesoglea between.

8. Over half of known animal phyla are worms of one kind or another. Worms are bilaterally symmetric and cephalized.

9. The basic body plan of familiar animals was established by the flatworms, which have three distinct tissue layers: endoderm, mesoderm, and ectoderm.

10. After the flatworm stage, the circulatory system, the complete, one-way digestive tract, and the coelom evolved.

11. The circulatory system carries resources and waste products to and from the cells. Organisms with a circulatory system can be larger than those that depend on simple diffusion.

12. The evolution of the one-way digestive system with a distinct mouth and anus allowed specialization of segments of the tract, permitting food to be digested in a stepwise, orderly way.

13. The coelom is a space between the body wall and the internal organs. Animals with a coelom are built as a tube-within-a-tube. Therefore, the digestive tract and associated organs can move independently of the rest of the body.

14. In Precambrian time, a fundamental branching in the animal family tree occurred, leading to two great superphyla of higher animals, the annelid superphylum and the echinoderm superphylum.

THE ANNELID SUPERPHYLUM

15. The annelid superphylum contains three major phyla: annelids, arthropods, and molluscs.

16. Annelid worms are segmented. Segmentation represents a major evolutionary advance in the animal kingdom, for segmented organisms essentially come equipped with spare parts. Evolutionary change can occur in some segments while others carry on previous activities.

17. Another phylum of segmented animals is the arthropods, which include crustaceans, arachnids, insects, and centipedes and millipedes. Arthropods are distinguished by jointed appendages and chitinous exoskeletons.

18. Among arthropods, the insects are the most successful. They diversified on land, exploiting the expanding terrestrial flora. Advanced insects achieved flight, which allows dispersal, escape from predators, and migration in response to seasonal environmental change.

19. Molluscs, the second largest animal phylum, have four anatomic characteristics in common: the mantle, shell, radula, and foot. Each of these structures has been modified during evolution to meet specific challenges.

THE ECHINODERM SUPERPHYLUM

20. The echinoderm superphylum includes two phyla, echinoderms and chordates.

21. Adult echinoderms have spiny skin and a water vascular system with tube feet. Echinoderms are radially symmetric and lack obvious cephalization.

22. The echinoderm larva looks very much like the larva of an acorn worm, one type of chordate, the phylum to which we and other vertebrates belong.

23. Evolutionary biologists believe that in some ancestral echinoderm species metamorphosis changed, so that sexual reproduction took place in the free-living larval form rather than in the adult. This slight change may have opened the door to the evolution of the chordates.

24. All chordates have, in all or part of their life cycle, (a) a hollow, dorsal nerve cord, (b) a supporting, cartilaginous notochord, and (c) pharyngeal gill slits.

25. Invertebrate chordates—sea squirts, acorn worms, and the lancelet—are filter-feeders, as are echinoderm larvae. Invertebrate chordates may be a link between echinoderms and chordates.

26. The earliest vertebrates were jawless fishes, the first animals to have bone.

27. Acanthodians are the fishes believed to be the first to have evolved jaws and paired fins; they gave rise to the two groups of modern fishes, the sharks and the bony fishes.

28. Sharks are evolutionary dead ends; they gave rise to no other group of vertebrates, and their modern diversity is low.

29. Bony fishes are the most diverse group of vertebrates, with more species today than all other classes combined.

30. Vertebrates have two kinds of organs for gaseous exchange, gills (suited to life in water) and lungs (adapted to life on land).

31. Lungs first evolved in fishes and were retained in some species. In others, the lungs became swim bladders.

32. Fishes with lungs that could walk on land may have given rise to the amphibians.

33. Amphibians are a transitional group, whose lives straddle water and land. Their locomotion is somewhere between that of swimming fishes and efficient quadrupedal land animals.

34. Amphibians today have not escaped the need to return to water to reproduce. They produce unshelled eggs; fertilization is usually external.

35. Amphibians gave rise to reptiles, which lay shelled eggs.

36. The shelled land egg requires internal fertilization, which probably evolved first in reptiles' amphibian ancestors.

37. The land egg helped reptiles exploit the dry land, diversify, and dominate ecosystems for 100 million years. No other vertebrate group has been as successful as the reptiles for as long a time.

38. Birds arose from reptiles. Feathers evolved from reptilian scales and served as insulation conducive to endothermy; later, some feathers formed a flight surface.

39. The diversity of birds is great at the species level, but the class is limited in its variety of body plans, perhaps because the requirements of flight constrained divergence.

40. Mammals, which evolved from reptiles, are characterized by numerous features, among them (a) body hair for insulation, (b) mammary glands for milk production, (c) a single bone in the lower jaw and three ossicles in the middle ear, (d) endothermy, and (e) a four-chambered heart with separate circuits for oxygen-poor and oxygen-rich blood.

41. In most mammals, the embryo develops to some degree inside the mother's reproductive system. In all mammals other than marsupials, a placenta is formed of embryonic tissues and maternal tissues.

42. The mammalian brain differs from that of other vertebrates in the degree of development of the cerebrum. Mammals rely less on instinctive behavior and more on learning than do other vertebrates.

TRENDS IN ANIMAL EVOLUTION

43. Several trends are evident in animal evolution: (a) increasing size and multicellularity, (b) the evolution of radial symmetry and then bilateral symmetry with cephalization, and (c) specialization of organisms.

An endangered species has a population so small that it is in danger of becoming extinct without human intervention. In 1973, the U.S. Congress passed the Endangered Species Act, which requires the U.S. Fish and Wildlife Service to list species that are endangered or threatened. The act also makes it unlawful to kill, injure, or disturb an endangered species and prohibits trade in endangered species or products derived from them. In addition, it prohibits federal financial support of projects that would injure or disturb such species or their habitats and strictly forbids projects on federal land that would have a similar effect.

To date, a thousand such species have been listed. Listing or unlisting species is strictly a biological decision, not an economic one, but some people would like to see the listing take economic considerations into account. You will be asked to make an assessment of this proposal based on some additional information about endangered species—notably, the black-footed ferret.

The black-footed ferret is a medium-sized member of the weasel family that once lived throughout the Great Plains and the intermountain West, wherever prairie dogs, its principal prey animal, were abundant. However, because prairie dogs can become agricultural pests they have often been poisoned or shot. An unintended consequence of prairie dog control was the decimation of populations of black-footed ferrets, a species that apparently was never particularly abundant anywhere in its range.

In the early 1970s, the only known population of black-footed ferrets was under study in South Dakota. The population disappeared, and some thought that the species had become extinct. However, a new population of the animals was located sometime later in Wyoming. The Wyoming population was studied inten

sively for a number of years, but when an outbreak of distemper threatened the population, all individuals still living in the wild were brought into captivity for treatment. Wildlife biologists hoped that reproduction in the captive population would eventually allow restoration of the animals in the wild.

Extinction is the fate of all species. Most species that ever lived are now extinct. Extinction results from failure to adapt to changing environmental conditions. Moreover, highly specialized species such as the black-footed ferret are susceptible to extinction after only minor environmental change. Some people believe that by saving the ferret we are interfering in the natural process of extinction of a species that may have already been on its way out. Do you agree? Why or why not?

Ferrets are secretive, nocturnal animals that live in the semiarid West. Even if they were abundant on their native range, not one American in a hundred thousand would ever see one. Should we be concerned about a species that lives only in the places where most people are unlikely ever to go, and even if they do, will probably never see a ferret?

The black-footed ferret lives in areas underlain by vast resources of coal as well as natural gas and oil, resources essential to any program of "energy independence" based on fossil fuels. Should an obscure animal like the black-footed ferret be allowed to slow or halt "progress"?

Outline a debate about the following proposition, "Be it resolved that the Endangered Species Act of 1973 should be revised to consider the economic impact of saving endangered species." Use the black-footed ferret as an example. Remember, a debate includes arguments pro and con, and a logical conclusion.

TEST OF TERMS

1. Animals are _____ , which means that they consume other organisms, and _____ , which means that they contain nucleated cells.
2. An organism without orderly organization is called _____ .
3. An organism organized like a wheel exhibits _____ symmetry. An organism with parts arranged as mirror images on either side of a medial plane exhibits _____ symmetry.
4. A group of cells (and intervening secretions) similar in origin, structure, and function is a _____ .
5. The outer layer of animals is the _____ ; the inner layer is the _____ .
6. In jellyfishes, between the inner and outer tissue layers lies the gelatinous _____ . The interior of a jellyfish is a hollow _____ cavity.

7. The _____ _____ is a network of nerve cells linking the cnidarian body.
8. Stinging cells of cnidarians, the _____ , are used to capture and kill prey and for protection.
9. Many cnidarians are affixed to the seafloor; such a sedentary lifestyle necessitates a mobile developmental stage, a _____ .
10. Flatworms exhibit a concentration of sensory and feeding structures in a head end, a phenomenon known as _____ .
11. Most individual flatworms have gonads of both sexes and are therefore _____ .
12. The _____ is a space between the body's outer wall and the inner digestive tube.
13. A _____ digestive tract has a mouth and an _____ and

allows specialization along its length.
14. _____ is the serial repetition of body parts found in annelids and other animals.
15. Arthropods have an external skeleton, or _____ , built of _____ .
16. Diversification of insects can be seen as a _____ feedback system in which diversity leads to greater diversity.
17. Many kinds of insects undergo _____ in their life cycle that involves an egg, a feeding _____ , a superficially inanimate _____ , and a breeding adult form.
18. Molluscs have four structures in common, a _____ , which secretes the _____ , a _____ (variously modified for locomotion), and a _____ , a rasplike feeding device.
19. Adult echinoderms have a _____

_____ system of calcareous pipes that produces the suction pressure to operate the _____ _____, which function in locomotion and food acquisition.

20. All chordates have at some stage in their life cycle a hollow, dorsal _____ _____, a cartilaginous, supporting _____, and pharyngeal _____ _____ that allow filter-feeding and gaseous exchange.

21. Chordates that lack a vertebral column are _____ chordates.

22. The earliest vertebrates were bottom-dwelling filter-feeders. They had a skeleton of _____ .

23. Sharks have a skeleton of _____, but they evolved from bony ancestors.

24. _____ are external surfaces for gaseous exchange, usually functioning in water. _____ , by contrast, are internal structures that maintain a moist surface for gaseous exchange; they allow breathing on dry land.

25. Amphibians are mostly four-footed, or _____ , animals.

26. The hallmark of reptiles is the _____ egg. Reptilian eggs have an abundant food supply, the _____ , which is contained in the _____ _____ and nourishes the developing embryo.

27. In reptile embryos, waste is deposited in the _____ . The _____ is a fluid-filled bag that prevents drying and shock, and the _____ is a membrane for gaseous exchange.

28. Before the land egg was possible, _____ _____ had to evolve, so that the embryo could be packaged in an impenetrable shell.

29. The feathers of birds provide insulation conducive to _____ , or warm-bloodedness.

30. Mammals are distinguished from reptiles by the fact that the lower jaw, or _____ , consists of a single bone. In mammals there are three _____ in the middle ear.

31. Hair is made of a protein called _____ , which is extruded from a hair _____ .

32. The milk-producing glands of mammals are called _____ .

33. Internal development of most mammal embryos is supported by a structure composed of fetal and maternal tissues, the _____ . Blood vessels in the _____ _____ connect the developing young to the placenta.

Answers to the Test of Terms are found in Appendix B.

TEST OF CONCEPTS

1. Knowing what you know about the coelom, please explain why it is almost impossible to stop one's "stomach" from "growling."

2. Explain how diversification of higher insects (beetles, social insects, butterflies and moths, flies) and diversification of flowering plants operate in a positive feedback fashion.

3. What does it mean to say that sharks are an evolutionary dead end?

4. There are about two dozen known animal phyla. List the animal phyla that have representatives living on dry land.

5. Several groups of animals (sponges, corals, barnacles, many parasites, and sea squirts, for example) are sedentary at some stage of their life cycle. What advantages does such an existence have? What problems does it raise?

6. Gills and lungs are both organs for gaseous exchange. What do they have in common? How do they differ?

7. Why should viviparity in reptiles increase with increased latitude or elevation?

8. Active flight has evolved in four groups of animals. Name them. What is the adaptive value of flight? What are the problems or constraints involved in flight?

9. Why do you think there are about twice as many species of birds as there are species of mammals?

10. Mammals generally seem to rely less on instinctive behavior than do other animals and to be more dependent, instead, on learned behavior. Can you think of any advantages and disadvantages of the latter?

11. The point has been made frequently that evolution is more like "tinkering" than it is like engineering design. From your knowledge of animal evolution, provide at least four examples to illustrate that point.

Tracing Our Roots: The Story of Human Evolution

We've come a long way since our early days as hunters and gatherers, but we've got a long way to go if we're going to live sustainably on the planet.

The humorist Will Cuppy once quipped that "all modern men are descended from a wormlike creature, but it shows more in some people." In reality, humans belong to a group (order) called primates. Primates are believed to have evolved not from worms but from a mammalian insectivore—an insect-eating mammal—that probably resembled the modern-day tree shrew found in Southeast Asia (▷ Figure 28–1).

This chapter examines the evolutionary development of the primates, beginning with these insectivores, but focuses principally on the evolution of human beings. It highlights some of the important evolutionary advances taking place in primate evolution and ends with a brief discussion of human cultural evolution, laying a foundation for an understanding of the environmental problems facing humankind.

▷ **FIGURE 28–1 Look Familiar?** An organism resembling this tree shrew is believed to have been the early ancestor of the primates.

≈ EARLY PRIMATE EVOLUTION

Primates consist of two suborders: (1) the prosimians (premonkeys) and (2) the anthropoids. The prosimians were the first primates to inhabit the Earth and today consist of tree shrews, lemurs, and tarsiers (Figures 28–1 and ▷ 28–2). During the course of evolution, the prosimians gave rise to anthropoids, a group that includes monkeys, apes, and humans. Today, most primates—prosimians and anthropoids alike—live in tropical or subtropical regions in forests, grasslands, and mixed woodlands, consisting of both grassland and forest. Most primate species alive today are tree-dwellers. An important exception to these generalities is *Homo sapiens*, which ranges widely over the Earth, inhabiting virtually every available biome.

Primates are characterized by several features. These include grasping hands, which permit them to pick up objects, among other things, and forward-directed eyes, which permit varying degrees of stereoscopic (three-dimensional) vision. Primates also have the largest brains of all mammals in proportion to their body size.

The Fossil Evidence of Primate Evolution Is Limited, Making It Difficult to Determine the Exact Progression of Early Human Ancestors

Unlike the dinosaurs, early primates did not live in habitats conducive to fossil preservation. Whereas the mud in swamps or along rivers preserved many a giant dinosaur, the forests and grasslands in which primates lived were not an environment where bones of a dead animal would be readily covered with sediment and preserved. Another reason for the scarcity of early primate fossils is that these creatures were rather small. The bones of animals that died and were eaten by scavengers were no doubt scattered across the landscape.

Because the primate fossil record is incomplete, the origin of primates remains uncertain. Sweeping conclusions about primate evolution have been made on rather skimpy evidence. The scheme presented here represents a synthe-

sis of current thinking on the evolution of primates. New evidence could radically change the present interpretation of the fossil record.

The Earliest Primates Probably Evolved From Tree Dwelling-Mammals that Lived During the Age of the Dinosaurs

The primates are one of the oldest living orders of placental animals. As noted above, evolutionary biologists believe that the very first primates were the prosimians, which probably evolved from small, tree-dwelling insectivores similar to tree shrews. The fossil remains of these animals can be found in rock approximately 80 million years old. Although the insectivorous tree-dwelling mammals also gave rise to many other (nonprimate) mammals, including ground-dwelling shrews, water shrews, moles, and bats, we will focus our attention here on the primate lineage, beginning with the prosimians.

Although no one knows for sure, the earliest prosimians probably emerged about 65 million years ago. As a group,

▷ **FIGURE 28–2 Prosimians** The prosimians (premonkeys) were probably the first primates. The earliest ones probably resembled modern-day (*a*) tarsiers and (*b*) lemurs.

(a)　　　　　　　　　　　　　(b)

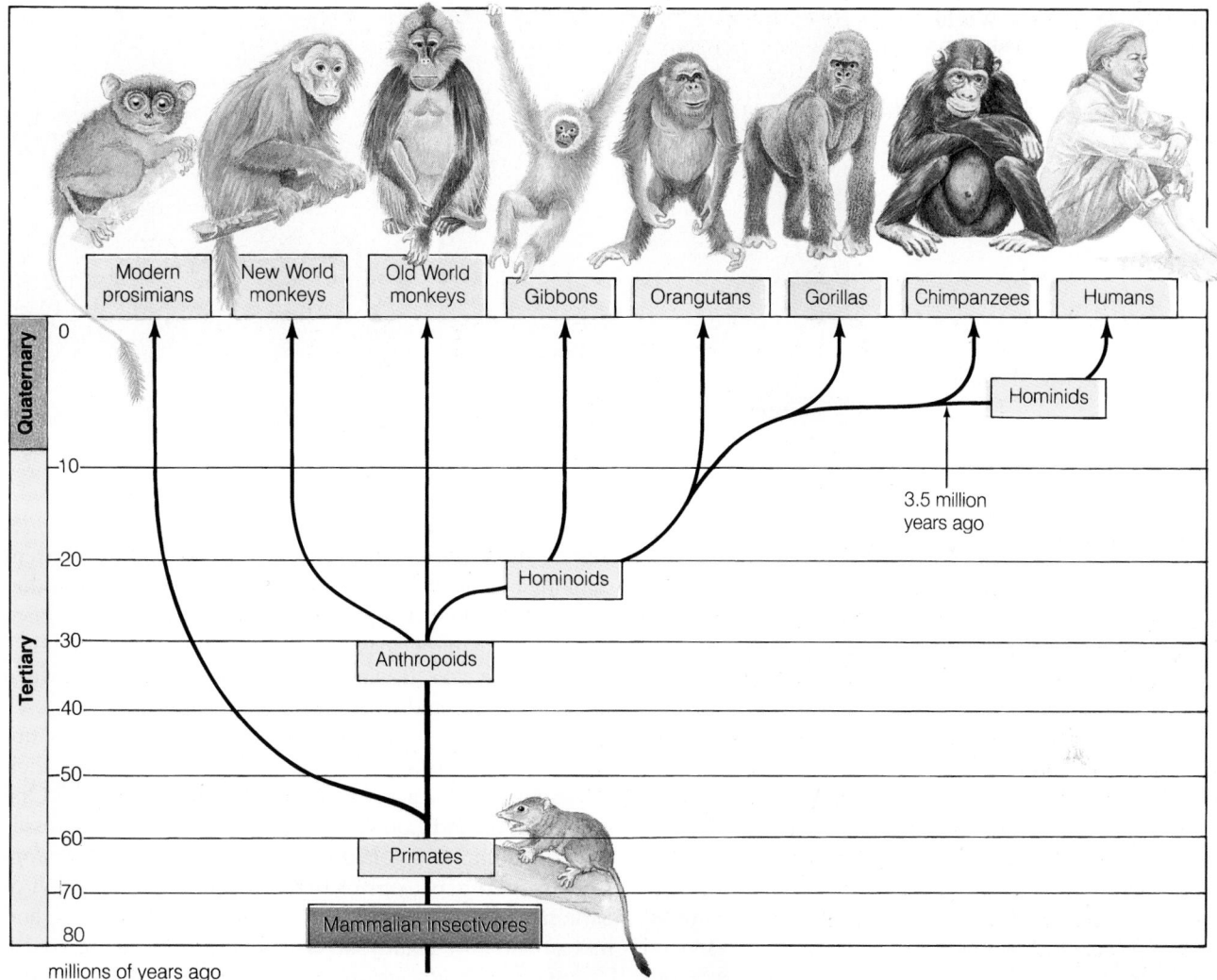

Modern prosimians | New World monkeys | Old World monkeys | Gibbons | Orangutans | Gorillas | Chimpanzees | Humans

Hominids

3.5 million years ago

Hominoids

Anthropoids

Primates

Mammalian insectivores

millions of years ago

> **FIGURE 28–3 The Evolution of Primates**

the early prosimians became quite abundant and geographically widespread. Fossils show that they lived for tens of millions of years. Approximately 55 million years ago, they gave rise to the modern-day prosimians: tree shrews, lemurs, and tarsiers (▷ Figure 28–3).

The Fossil Evidence Suggests that One of the Early Prosimian Lines Gave Rise to the Anthropoids

Sometime within the next 15 million years, the anthropoids appeared. The very first anthropoids were monkeylike primates believed to have evolved from prosimian stock in Africa and possibly earlier in Asia. **Anthropoids** today include three major groups: (1) the New World monkeys, (2) the Old World monkeys, and (3) the **hominoids**— the apes, gibbons, and humans (Figure 28–3).

The New World monkeys, which inhabit Central and South America, include the familiar squirrel and spider monkeys (▷ Figure 28–4). All New World monkeys live in the trees. The Old World monkeys occupy the tropical regions on the continents of Africa and Asia. Some familiar examples are the baboon and proboscis monkey (▷ Figure

28–5). They include ground-dwelling as well as arboreal species. Most monkeys in both groups live in bands and are active during the day.

The relationship between Old World and New World monkeys is not clearly delineated. Most evolutionary biologists believe that Old World monkeys gave rise to the New World monkeys. First emerging in Africa and Asia, the Old World monkeys may have migrated across North America into Central and South America or may have arrived on rafts of logs or floating debris.

The hominoid line, which today consists of apes, gibbons, and humans, diverged from the Old World monkeys over 20 million years ago. The very first hominoids were apelike creatures. Two distinct apes were present at this time and are considered potential ancestors of humans and modern apes.

The first family, the **dryopithecines** (dry-o-pith-a-seens), originated about 20 million years ago in Africa and spread to Eurasia about 14 million years ago. Unfortunately, no complete skulls or skeletons of the dryopithecines have been found, and our knowledge of these animals is based on fragments of skulls, jaws, and limb bones. This evidence

(a)   (b)

suggests that dryopithecines had relatively small brains, apelike teeth and jaws, and apelike faces (▷ Figure 28–6). The structure of their limbs suggests to scientists that dryopithecines walked about on four legs, much like chimpanzees and gorillas, but spent most of their time in the trees.

The second possible ancestor of apes and humans lived slightly later—from about 17 million years to 7 million years ago. Called **ramapithecines,** these creatures were similar to the dryopithecines in many respects.

Dryopithecines and ramapithecines may have been related or may have evolved separately. Unfortunately, fossil evidence is not complete enough to establish the relationship between them or to determine their exact relationship to apes and humans. For now, all we can do is content ourselves with knowing that these creatures existed and hypothesize that they represent a link between early primates and more modern primates, including gorillas, chimpanzees, and humans.

≈ EVOLUTION OF THE AUSTRALOPITHECINES

Archaeological evidence suggests that the first humanlike primates, or **hominids,** belonged to the genus *Australopithecus* ("southern apeman") (▷ Figure 28–7). Current thinking holds that the australopithecines arose from either ramapithecines or dryopithecines. Unfortunately, fossil evidence of ramapithecines and dryopithecines ends about 7 million years ago, leaving a 3.5-million-year gap between *Australopithecus* and its supposed ancestor(s).

The oldest known australopithecine skeleton was unearthed by Donald Johanson, Yves Coppens, and their co-workers in Africa and is about 3.5 million years old. One specimen, named Lucy, is one of the most complete fossils of early hominids (▷ Figure 28–8). Known as ***Australopithecus afarensis*** (*afarensis* means from the Afar region of Ethiopia), these hominids stood only about 3 feet high and had a brain only slightly larger than an ape's (Figure 28–8). The skull of *A. afarensis* has many apelike features, including massive brow ridges, a low forehead, and a forward-jutting jaw. The shape of its pelvis suggests that *A. afarensis* walked erect.

About 3 million years ago, *A. afarensis* vanished and was replaced by another species, *A. africanus,* which lived for about 1.5 million years (Figure 28–7). *A. africanus* was slightly taller than its predecessor and had a slightly larger brain.

▷ FIGURE 28–5 Old World Monkeys (a) Olive Baboon and (b) Rhesus monkey.

(a)   (b)

▷ **FIGURE 28–6 Dryopithecus** A possible forerunner of modern apes and humans.

Studies of the fossil record suggest that about 2.3 million years ago another species of *Australopithecus* emerged, *A. robustus*—so named because it was heavier and taller and had a larger brain than *A. africanus* (Figure 28–7). About 2.2 million years ago, the fourth species, *A. boisei*, appeared. It was even more robust than *A. robustus.*

Members of the genus *Australopithecus* had many com-

mon features. For example, all of them were **bipedal;** that is, they walked on two legs. They all had brains larger than chimpanzees but considerably smaller than modern humans and ranged from 1 meter to 1.7 meters in height. The differences between the four species are relatively minor and are mostly a matter of degree. As the previous discussion mentioned, the brain got larger, and height increased as the genus evolved. This enlargement of the brain is believed to be the result of natural selection. (For an opposing, now defunct, theory, see Scientific Discoveries 28–1.)

▷ Figure 28–9 indicates when the various species of *Australopithecus* lived. As illustrated, anthropologists believe that all three "younger" species coexisted, at least for a while, during the time span ranging from 1 million to 3 million years ago. Over 1 million years ago, however, Australopithecines disappeared. Why they vanished no one knows.

⇌ EVOLUTION OF THE GENUS *HOMO*

The exact origin of the genus to which modern humans belong is in dispute. Most paleontologists believe that the

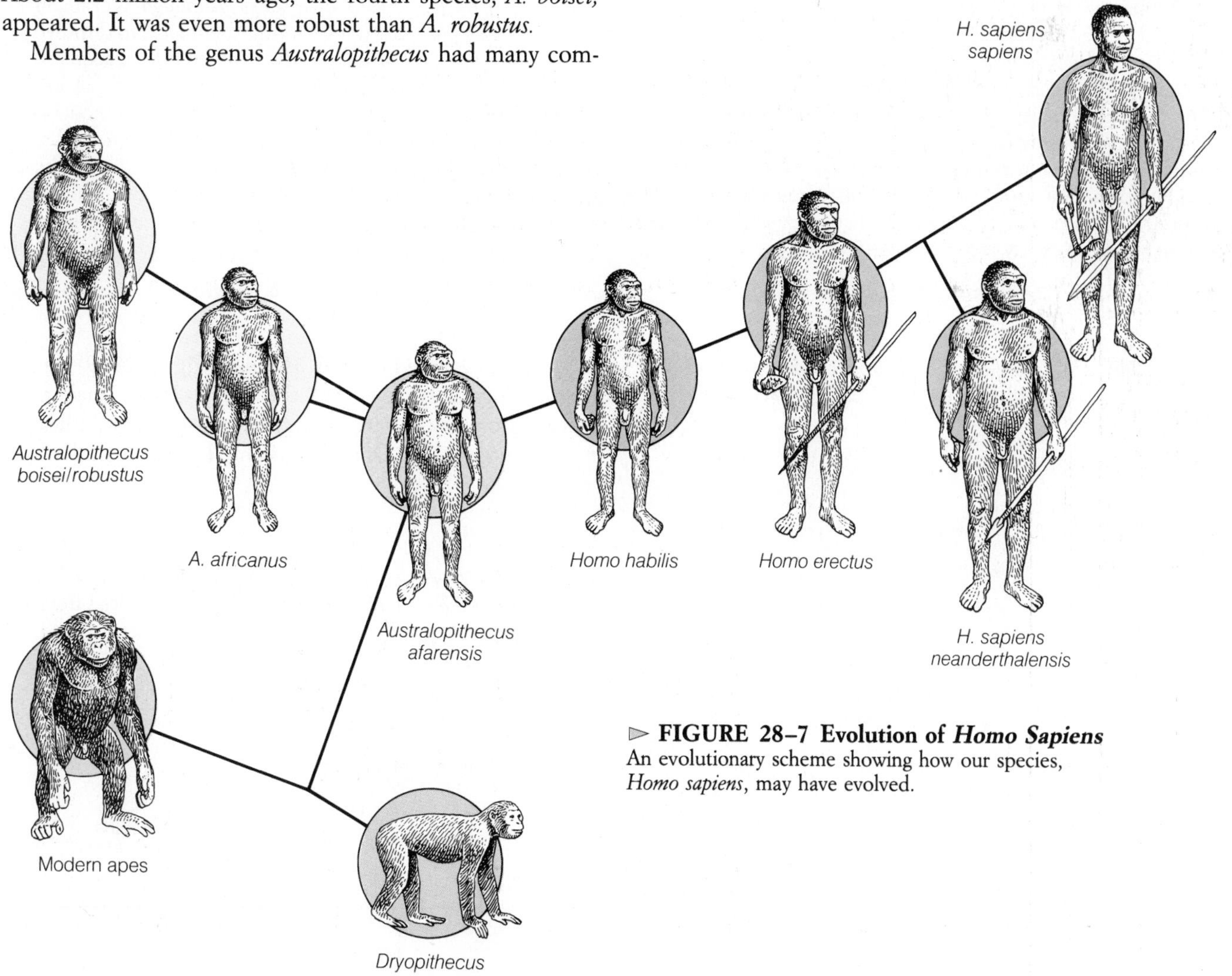

H. sapiens sapiens

Australopithecus boisei/robustus

A. africanus

Australopithecus afarensis

Homo habilis

Homo erectus

H. sapiens neanderthalensis

Modern apes

Dryopithecus

▷ **FIGURE 28–7 Evolution of *Homo Sapiens***
An evolutionary scheme showing how our species, *Homo sapiens*, may have evolved.

DEBUNKING THE NOTION OF INHERITANCE OF ACQUIRED CHARACTERISTICS

Featuring the work of Lamarck, Weismann, Castle, and Phillips

The 18th-century French naturalist Jean-Baptiste Lamarck argued that life had been created long before his time in a relatively simple state. Over time, he said, life forms gradually improved as a result of an innate drive for perfection. He further hypothesized that this inherent drive was centered in nerve cells and that these cells released a "fluida" (chemical substance) that traveled to different body parts needing improvement.

Consider Lamarck's explanation of the evolution of the giraffe. According to him, the giraffe of years past was a short-necked animal. The elongation of the giraffe neck, over time, resulted from stretching of its neck—a feat required to feed on leaves that were out of reach of other animals. The act of stretching, said Lamarck, directed fluida to the neck. This stimulus made the neck grow longer. What is more, the slightly stretched neck that an adult acquired was then transmitted to its offspring. The offspring, in turn, stretched their necks to reach food, causing fur-

ther elongation. Generation after generation of giraffes, each stretching to find food, said Lamarck, led to the modern giraffe.

Lamarck proposed an evolutionary theory based on the "use and disuse" of organs. It stated that an individual acquired traits during its lifetime and that such traits were in some way incorporated into the hereditary material and passed to the next generation, thus explaining how a species could change over time. Conversely, disuse of organs led to their disappearance.

The inheritance of acquired characteristics was doubted by many of Lamarck's contemporaries, in large part because he failed to support his hypothesis with observations or experiments.

His concept is based on what is viewed today as a fundamental fallacy; that is, the inheritance of acquired characteristics supposes that all of the organs of one's parents produce the hereditary factors that form corresponding parts in the offspring. Thus, any change that occurred in an organ (an acquired charac-

teristic) before one transmitted gene material to offspring would result in th production of a hereditary factor tha would reflect the altered organ. If you were lifting weights and had developed extensive musculature before the conception of your child, your child would also grow up to be brawny.

After the discovery of fertilization, it was assumed that the organs transmitted parcels of hereditary information to the gametes. In this way, acquired traits were passed on to the offspring.

August Weismann, a late-19th-century German scientist, attempted to bring some sense to the debate in a series of extraordinary essays, based largely on the work of others and also on his own ideas. Although Weismann did not perform experiments to disprove the notion of the inheritance of acquired characteristics, his work remains an important contribution to modern thought. For example, he argued that transmission of traits from one generation to the next depended on the sex cells. The sex cells, he said, are capable of reproducing

▷ **FIGURE 28–8 Skeleton of *Australopithecus afarensis*** This skeleton, estimated to be about 3.5 million years old, is popularly known as Lucy.

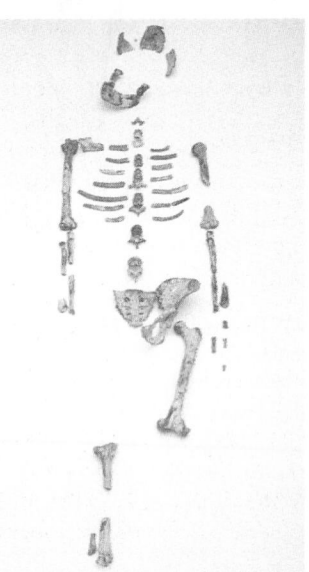

genus *Homo* evolved from *A. afarensis* (Figure 28–9). Some anthropologists believe that the genus *Homo* may have evolved from an as-yet-unidentified ancestor, as indicated in ▷ Figure 28–10. Only time will tell if this hypothesis is correct.

▷ **FIGURE 28–9 Australopithecine Evolution** Conventional thinking holds that the genus *Homo* evolved from *Australopithecus afarensis*.

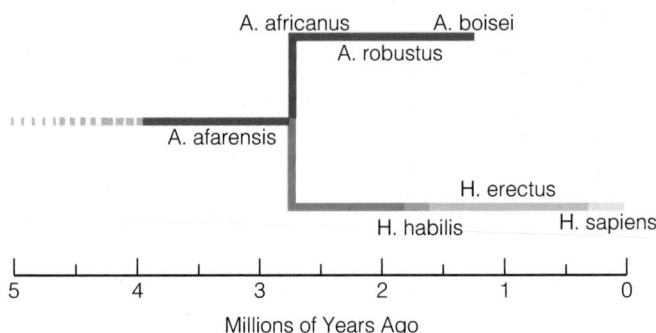

all of the "peculiarities of the parent body in the new individual."

Furthermore, Weismann asserted that the germ cell was not an extract of the whole body—that is, that the cells of an organism did not dispatch small hereditary particles to germ cells, from which the latter derived their power of heredity. Rather, he contended that heredity was brought about by the transference from one generation to another of a substance with a definite molecular constitution. He called it "germ-plasm" and agreed with others that it was probably found in the nucleus of germ cells. Today, we know it as DNA.

Although Weismann himself propagated a few erroneous ideas, his notion that there was a chemical substance that contained the hereditary information responsible for the faithful transmission of traits profoundly influenced the thinking of many biologists and marked a turning point in our understanding of genetics. It also helped debunk Larmarck's notion of the inheritance of acquired characteristics and his errone-

ous view of evolution driven by such a mechanism.

One of the most rigorous tests of Weismann's hypothesis that germ plasma was the basis for heredity came about 20 years later in experiments by W. E. Castle and John C. Phillips. Castle and Phillips transplanted ovaries from black guinea pigs (homozygous dominant) into albino guinea pigs (homozygous recessive) who had had their ovaries removed. Later, the albino females were bred to albino males. Previous studies had shown that matings between albino males and females resulted in 100% albino offspring.

If acquired characteristics were inherited, Castle and Phillips argued, the offspring of these matings might be white, having acquired their coloration from the host. When bred to an albino male, however, the females produced only black offspring. The young, they said, "are such as might have been produced by the black guinea pig herself, had she been allowed to grow to maturity and been mated with the albino male used in the experiment."

Lamarck did have many good ideas—for instance, that species changed over time and that the environment was a factor in this change. He also contributed to the science of biology in other ways. For example, he was the first scientist to distinguish between animals with backbones (vertebrates) and those without backbones (invertebrates). He went on to classify many invertebrates into the categories of arachnids, crustaceans, and echinoderms and also wrote a text on invertebrate systems and a seven-volume treatise on the natural history of invertebrates. But history generally knows Lamarck for his erroneous concepts of inheritance and evolution. In fact, despite the general criticism of Lamarck's views, the 20th-century Russian agronomist Trofim D. Lysenko adopted a Lamarckian viewpoint and, with Stalin's backing, established it as Soviet doctrine in genetics research and teaching. The result was a significant setback in genetic research in the Soviet Union during the Stalinist era.

The First Truly Humanlike Creatures Were Called *Homo Habilis*

The earliest member of the genus *Homo* is a 1.8-million-year-old skull and partial skeleton found in Tanzania in 1960 by Mary Leakey. She and her husband, Louis, called

▷ **FIGURE 28–10 Alternative Evolutionary Tree** Some scientists believe that the genus *Homo* may have evolved from an as-yet-unidentified primate, not *A. afarensis* as indicated in the previous figure.

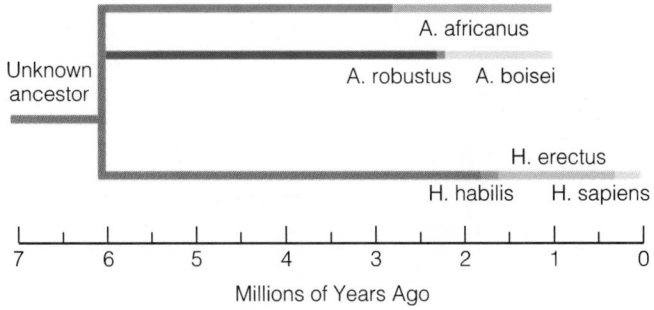

this species *Homo habilis* ("skillful man"). Archaeological evidence indicates that *Homo habilis* apparently made tools and butchered large animals, but its tools were crude fractured rock. Although they do not appear to be deliberately fashioned, these primitive tools have been found in large number in association with broken and cut animal bones at excavations in Tanzania, Kenya, and Ethiopia.

As in other species, the exact origin of *Homo habilis* is uncertain. Some paleontologists think that it evolved from *A. afarensis;* others contend that it arose from the slightly more modern *A. africanus.*

Debate over its exact origin aside, what was *Homo habilis* like? The skull of *H. habilis* is similar to the skull of *Australopithecus,* suggesting an evolutionary relationship. However, the brain of *Homo habilis* is 50% larger than its presumed australopithecine relatives. The skeleton of *H. habilis* displays many apelike features, especially relatively long arms and a small body. Because of this, some paleontologists contend that skeletons classified as belonging to *Homo habilis* are really members of the genus *Australopithecus.*

Homo habilis Gave Rise to *Homo erectus*

Two hundred thousand years after the emergence of *Homo habilis,* the first unmistakable member of the genus *Homo,* known as **Homo erectus** ("upright man"), arose (see Figure 28–7). Skeletons of *Homo erectus* appear in geological formations 300,000 to 1.6 million years old in the Old World. The most complete fossilized human skeleton to have been found is that of a 12-year old boy (▷ Figure 28–11).

Unlike *Australopithecus* and *H. habilis,* which remained in Africa, *H. erectus* first appeared in Africa then spread to Europe and Asia (India, China, and Indonesia). *H. erectus* was the first to leave the warmth and abundance of the tropics and subtropics and take up residence in the temperate zone, characterized by warm summers but cold winters. Consequently, anthropologists believe that *Homo erectus* may have been better adapted to deal with seasonal changes and climate extremes than *Homo habilis.*

Homo erectus stood about 5 feet tall, used fire, and made more sophisticated tools and weapons than its predecessor, *Homo habilis.* With a brain slightly smaller than ours, *Homo erectus* is believed to be the direct ancestor of **Homo sapiens,** the self-proclaimed "thinking man." For a debate on the origins of modern humans, see the Point/Counterpoint in this chapter.

Modern Humans Belong to Homo sapiens

Homo sapiens emerged about 300,000 years ago. *Homo sapiens* consists of two subspecies, **Homo sapiens neanderthalensis,** the Neanderthals, and **Homo sapiens sapiens,** modern humans (see Figure 28–7).

Widely distributed in Europe and Asia, the Neanderthals lived in caves and camps. Their name comes from the Neander Valley in Germany, where the first specimens were discovered.

Archaeological evidence shows that the Neanderthals gathered fruits, berries, grains, and roots and hunted animals with weapons. They cooked some of their food on fires. Neanderthals stood erect and walked upright. Archaeological evidence indicates that they lived in small clans and buried their dead in elaborate rituals.

The skeleton of the Neanderthals resembles that of humans but differs in several ways. One important difference is the size of the skull. Interestingly, the skulls of Neanderthals were, on average, actually larger than those of modern humans. Their skeletons were also more massive and heavily muscled than those of modern humans, and they had rather short lower limbs, much like Eskimos and other cold-adapted people.

Many people think of the Neanderthals as dim-witted and brutish, shuffling along bent over like an ape. This view is based on an interpretation of a skeleton found in 1908 in France. At the time of the discovery, however, the researchers failed to recognize that the skeleton under study was bent over because the Neanderthal had suffered from arthritis of the hip and had had a disease of its vertebrae. It

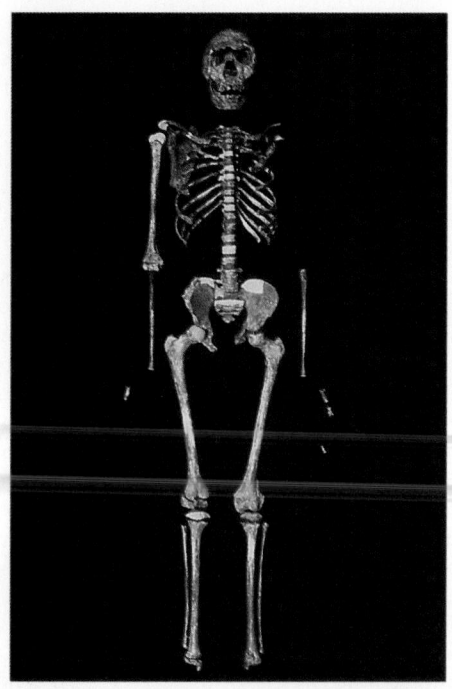

▷ **FIGURE 28–11 Skeleton of *Homo Erectus*** This, the most complete fossilized skeleton of *Homo erectus* ever found, belonged to a boy who lived about 1.6 million years ago.

is interesting, though, how this hasty conclusion from one observation spawned such a long-standing myth.

Some anthropologists wonder whether Neanderthals, despite many humanlike behaviors, should really be considered members of *Homo sapiens.* Differences in physical appearance between them are quite striking. With broad faces, large projecting brow ridges, and heavily built bodies, they should perhaps be placed in a species all their own (▷ Figure 28–12). Nonetheless, if one compares skeletons of australopithecines, early species of the genus *Homo,* and modern *Homo sapiens,* the Neanderthal does seem to fit well into the progression between *Homo erectus* and *Homo sapiens sapiens* (▷ Figure 28–13).

▷ **FIGURE 28–12 Neanderthal** Notice the projecting brow ridges of this reconstruction of a Neanderthal man.

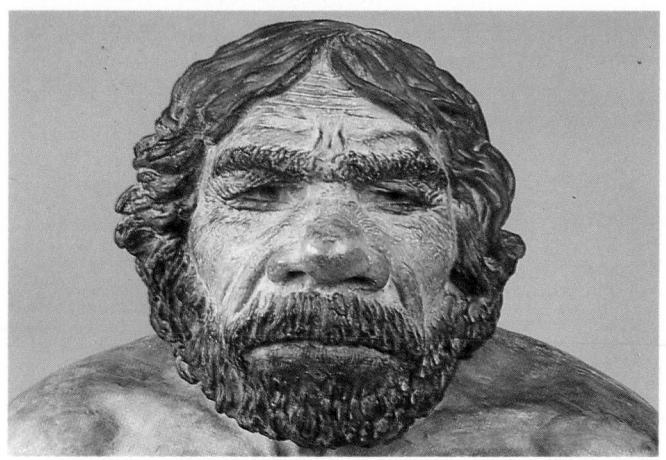

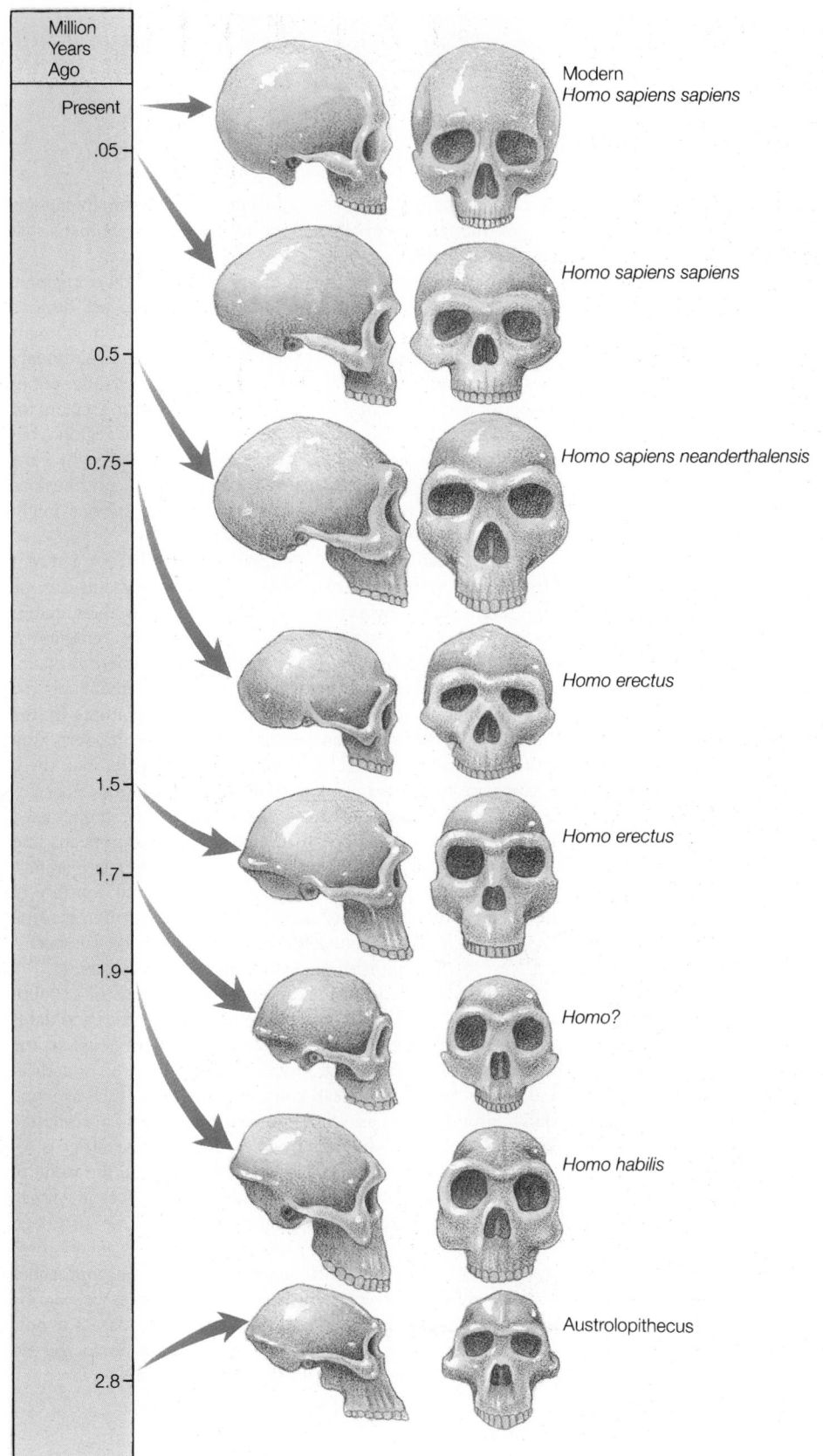

▷ **FIGURE 28–13 Evolutionary Progression in Skull Development** Notice the changes that occurred over time in the shape of the human skeleton, starting with the apelike *Australopithecus* at the bottom. Note that the dates in the left column indicate age of skulls in drawing not the evolutionary emergence of each species.

Million Years Ago

Present
.05
0.5
0.75

1.5
1.7

1.9

2.8

Modern
Homo sapiens sapiens

Homo sapiens sapiens

Homo sapiens neanderthalensis

Homo erectus

Homo erectus

Homo?

Homo habilis

Austrolopithecus

FOSSIL EVIDENCE SUPPORTS THE MULTIREGIONAL EVOLUTION MODEL

Fred H. Smith

Fred H. Smith is a paleoanthropologist with specific interests in the paleobiology of European Neandertals and the broader issue of the origins of modern people. He is Professor and Chairman of the Department of Anthropology at Northern Illinois University.

S cientists interested in paleobiology use information from several sources to reconstruct the evolutionary history of organisms. Anatomical and genetic structures of living creatures, for example, help scientists determine how related two species are. Such information yields important insights into evolutionary history. However, as valuable as this information is, the *most* direct evidence of evolutionary history is to be found in the fossil record. Paleobiologists study fossils to verify the evolutionary relationships suggested by studies of living creatures.

The degree to which scientists can piece together evolutionary history from the fossil record depends on how adequate the record is for the organisms under study. Those interested in human evolution are relatively fortunate, because the human fossil record over the last 4 million years is an excellent one. This does not mean that there are no gaps in the record nor any disagreements about interpreting these fossils. Nonetheless we can learn a great deal about human biological history from the fossils, and any attempted reconstruction of that history must be consistent with the fossil evidence.

The emergence of modern humans (people fundamentally like us) is a particularly fascinating phenomenon on which we have considerable fossil evidence. Before approximately 100,000 years ago, archaic humans (such as the well-known Neandertals of Europe and western Asia) existed throughout all but the harshest areas of the Old World (Africa, Asia, and Europe). By 30,000 years ago, these people were gone, and modern human populations ubiquitous. How did this occur?

Recently, a model has emerged that is based largely on interpretations of the genetic variability (particularly in mitochondrial DNA) in people today. It argues that modern humans evolved from archaic humans in Africa at around 100,000 years ago. This new species then spread into Eurasia and totally replaced the human populations living there. According to this African origins model, archaic European and Asian (Neandertal) contributed essentially *nothing* to the modern humans that succeeded them.

An alternative model, the multiregional evolution or regional continuity model, argues that modern human are not derived entirely from any single population. Rather, modern humans emerged in various regions of the Old World from evolutionary lines extending back through archaic humans to regional populations of *Homo erectus*. This does not constitute a claim for independent origins of modern people in different regions, because the various regional lines were always connected by gene flow. Neither does it deny that important genetic contributions to the emergence of modern humans in some regions might have been introduced from other regions.

If all modern people are completely derived from a recent African source, the earliest modern people of Asia and Europe would not exhibit any anatomical features found in their archaic human predecessors in that region that are not also found in the earliest modern Africans. However, there are numerous examples from Eurasia that contradict this requirement of the African origin model. In China, for example, features of the cheek area of the skull can be traced back from modern Chinese to archaic Chinese to Chinese *Homo erectus*, but these features are not present in the skulls of early modern Africans. In Europe, a distinctive feature of the rear of the skull, called the occipital bun, is common in early modern Europeans and European Neandertals but is not present in either late archaic or the earliest modern Africans. Such findings run contrary to the predictions of the African origins model and indicate that archaic Eurasians do contribute to their local modern successors, just as the multiregional evolution model predicts.

Furthermore, there is no evidence of a spread of modern people out of Africa at the requisite time. The best candidates for the earliest appearance of anatomically modern people come from west Asia (Israel), not Africa. Neither is there any indication of the spread of African cultural complexes (such as stone tools) into Eurasia, as would be expected with a complete replacement of local archaics by migrating modern Africans.

Supporters of multiregional evolution recognize the value of genetic data but deny that these data prove a recent African ancestry for all living humans. Many geneticists have also been critical of the African origins interpretation of the genetic data as well. Clearly, we need further refinements in our approaches to both genetic and paleontological data before we can be certain about the details of modern human origins. As it now stands, however, the bulk of available paleoanthropological evidence supports the multiregional evolution model.

GENETIC EVIDENCE SUPPORTS THE AFRICAN ORIGINS MODEL

Rebecca L. Cann

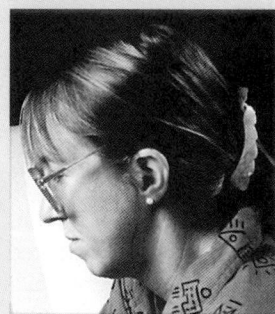

Dr. Rebecca L. Cann is associate professor of genetics and molecular biology at the John A. Burns School of Medicine of the University of Hawaii at Manoa.

Geneticists would like to understand the origins of modern people. Since humans, like all living species, carry a record of their evolutionary past in every gene of their body, geneticists have been working with the DNA sequences of genes to reconstruct the genes of our ancestors. A concept called *absolute time* is central to this approach. According to molecular archaeology, absolute time is proportional to the rate at which a gene mutates and recombines. This rate of mutation and recombination becomes a kind of "molecular clock" that can be used to estimate when an individual first carrying the ancestral sequence lived. For example, if you count the number of mutations that separate any two existing DNA sequences in living people, then determine the amount of time required to make this number of mutations, you now have a measure with which you can get at the time when their last common ancestor carrying that sequence must have lived.

Molecular biologists estimate that there are 100,000 genes in the chromosomes of the human cells. However, in the cytoplasm of the cell, there are only 37 separate genes in the mitochondria, and these mitochondrial genes are especially useful for tracing ancestry and looking at historical changes in populations. This is because they are inherited, with no recombination, *only* from the female parent, and they mutate so fast that almost every nonmaternally related individual is unique in a population.

Research using mitochondrial DNA (mtDNA) has led to the hypothesis that all living people trace their maternal ancestry to a population, originating in Africa, estimated to have existed about 200,000 years ago. This hypothesis does not state that all people alive today stem from one woman, as misrepresented in the popular press by the use of the biblical term "Eve." The hypothesis instead suggests that while archaic human populations are known to have existed in Europe, Africa, China, and Indonesia, only the African population persists in an unbroken maternal genetic line, thereby being alone in its ability to spread out and replace those earlier populations. Using this hypothe-

sis, any similarities among ancient (700,000 year-old) and modern Chinese populations are presumed to be due to morphological convergences—that is, to a convergence of form and structure—and not due to inheritance.

We don't know yet how an African source population could apparently spread and eventually displace residents, but some geneticists as well as archaeologists have speculated that the replacement may have been tied to increased communication skills. Such skills could have given ths group an advantage over other existing populations. Archaeologists see some evidence that a major behavioral shift in the way all human populations used available resources occurred about 40,000 years ago, and they believe that some of these changes may have first started in Africa with the early production of blade tools.

Some paleoanthropologists disagree with the African origins theory, and instead favor the idea that each archaic population evolved into modern populations (regional continuity). This debate is an active focus of research in both the molecular and paleontological communities. Issues in question focus on structural characteristics of particular fossils, whether dates attributed to them are accurate, and whether a single fossil or small number of fossils can be representative of an entire population. Geneticists promoting regional continuity doubt that the African origins model of speciation for humans is workable. Similar models have not worked for any other vertebrate species; they break down because the great distances involved are a hindrance to maintaining the genetic flow required to keep the population united.

Molecular anthropologists promoting the molecular clock have taken the stance that while fossils and stone tools can help provide some perspectives on these questions, paleontologists must make too many subjective judgements for their models to be reliable. Molecular anthropologists recognize that they too make assumptions about rates of change that could be in error. However, they believe that DNA sequences are objective, quantitative indicators of evolutionary change that are free of the biases that have hampered previous generations of biologists in their understanding of human history. The roots of the human tree will be debated for many years, but quantitative evidence from nuclear genes and mtDNA sequences gives the best indication that Africa was our recent home.

≈ SHARPENING YOUR CRITICAL THINKING SKILLS

1. State the hypotheses of both authors in your own words.
2. List the data used to support each hypothesis and data used by the authors to refute the other's point of view.
3. Can you determine if there are any flaws of reasoning in either essay?

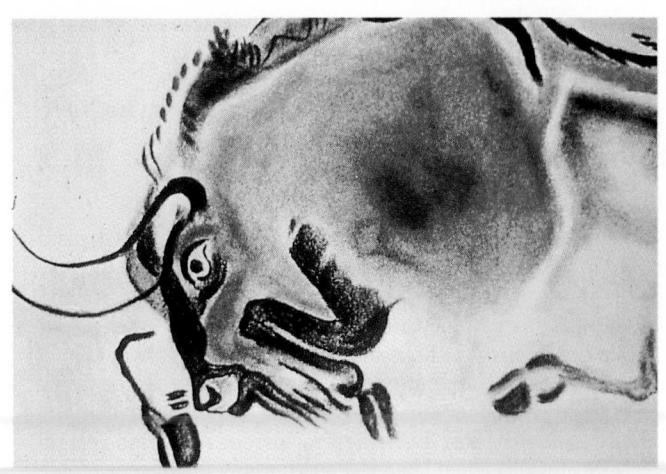

▷ **FIGURE 28–14 Cro-Magnon Art**

Neanderthals disappeared approximately 40,000 years ago for reasons still not understood. Some archaeologists believe that they were replaced by the **Cro-Magnons,** the earliest known members of *Homo sapiens sapiens.* Cro-Magnons arose in Africa and invaded Europe and northern Asia, wiping out the Neanderthals or possibly interbreeding with them.

Archaeological evidence shows that Cro-Magnons used sophisticated tools and weapons, including the bow and arrow, and were highly skilled nomadic hunters, following great herds of animals during their seasonal migrations. They may have had a well-developed language.

Cro-Magnons lived in caves and rock shelters, in groups of 50 to 75 people, and are best known for the elaborate art work that adorned the walls of their caves (▷ Figure 28–14).

Approximately 10,000 years ago, modern humans emerged. But the changes were not great. In fact, over the past 40,000 years, human evolution has produced little noticeable change in the physical appearance of humans. A Cro-Magnon on the streets of Los Angeles, in fact, would probably go unnoticed. Those 40,000 years have not been without change, however, for during this period, *Homo sapiens* has developed a rich and varied culture, complex language, and extraordinarily sophisticated tools.

Human Races Result from Variations Caused by Geographic Separation

As noted in Chapter 23, species that become subdivided into isolated populations undergo changes in response to their environment. If the changes are profound, the subpopulations often lose the ability to interbreed and produce new species.

Over the course of time, the human population has wandered far and wide on the planet and has fragmented into distinct subpopulations, living in very different environments. This fragmentation has resulted in the formation of a number of phenotypically distinct subpopulations, or **races** (▷ Figure 28–15). It is important to remember, however, that races are not distinct species, just regional variants of the one species *H. sapiens.*

Most of us are familiar with regional characteristics. We know, for instance, that Mexicans tend to have brown eyes, dark skin, and black hair and that Scandinavians tend to have blue eyes and blond hair. Many differences found in the races are undoubtedly adaptations to differing environmental conditions. The darker skin that evolved in tropical and subtropical populations, for instance, may be an adaptation that protects people from harmful ultraviolet light.

How many races are there? Some scientists recognize three main races: Mongoloids, Negroids, and Caucasians. Others see these three and two others: American Indians and Australian aborigines (Figure 28–15d and 28–15e). Still others recognize 30 different races. What all this tells us is that the races are arbitrary categories. As time goes on, the distinction between various races may decrease. Why? As people move about the planet and interbreed, races will mix, blurring the lines between the world's peoples.

▷ **FIGURE 28–15 Human Races** Humans, no matter what they look like, belong to one species, *Homo sapiens.* The various races are phenotypic variants of that species. Many scientists recognize five distinct races: (*a*) Mongoloid, (*b*) Negroid, (*c*) Caucasian, (*d*) American Indian, and (*e*) Australian aborigine.

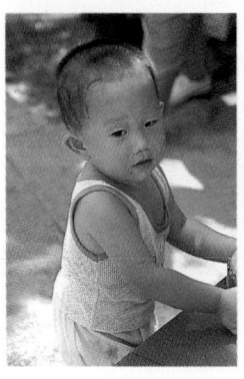

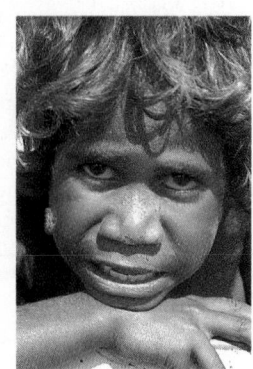

(a)　　　　(b)　　　　(c)　　　　(d)　　　　(e)

TRENDS IN PRIMATE EVOLUTION

The evolutionary path of primates leading to humans is marked by a number of significant changes. One important change was the evolution of bipedal (two-legged) locomotion, which freed the front legs for other useful tasks. In other words, bipedal locomotion may have permitted or facilitated the evolution of highly mobile front limbs.

In order to understand the importance of this development, we begin by examining the four-legged ancestors of primates, the earliest prosimians. In these quadrupeds, the limbs function primarily as a means of propulsion. In most quadrupeds, the limbs tend to move back and forth but cannot be rotated or spread very far from the body. In sharp contrast, the limbs of monkeys, apes, and humans are highly mobile. In tree-dwelling primates, this evolutionary advance permitted free movement among the branches. In humans, it permitted the use of tools and, of course, weapons. For more on the evolutionary advantages of upright posture, see Spotlight on Evolution 28–1.

Another important change taking place during the evolution of primates has been the increased mobility of the digits (fingers). This mobility permitted monkeys to grasp branches and fruit and permitted apes to manipulate objects, even primitive tools. Chimpanzees, for instance, use sticks to dig termites out of the ground. In humans, dexterity permits intricate crafts, the playing of musical instruments, and many other fine motor skills.

Yet another important advance, no doubt related to the arboreal life of the early primates, was the development of eyes on the front of the head. This placement permits stereoscopic (three-dimensional) vision, essential to the quick pace of life in the trees. Swinging from branch to branch, often jumping long distances, monkeys would be nearly helpless without three-dimensional vision.

Another important development throughout the evolution of primates to humans has been an increase in brain size. Larger brains were probably selected for because they gave organisms an advantage over others. Precise hand-eye coordination afforded by a larger brain no doubt facilitated life within the trees. And a larger brain has been a boon to humans. A large, highly capable brain and extraordinary manipulative abilities have permitted humans to devise elaborately powerful tools to reshape the environment, sometimes with untoward consequences.

The next section describes some of the technological and cultural changes in human society largely made possible by the evolution of larger brains and skillful hands.

ENVIRONMENT AND HEALTH: HUMAN CULTURAL EVOLUTION

Human culture has evolved through three phases: hunting and gathering, agricultural, and industrial. For 99% of human evolutionary history, though, humans survived by gath-

▷ **FIGURE 28–16 Hunters and Gatherers** Some anthropologists believe that the early hunters and gatherers acquired most of their food by gathering nuts, fruits, berries, roots, and seeds. Hunting may have provided a supplementary food source.

ering grains, fruits, nuts, berries, and roots and hunting animals (▷ Figure 28–16). Recent archaeological evidence suggests that our earliest ancestors were also scavengers, eating animals that died from natural causes. Modern ways of life and agricultural practices are really newcomers in human evolution.

Based on studies of modern-day hunters and gatherers, whom some anthropologists call our "living ancestors," early hunter-gatherers were probably extraordinarily knowledgeable about the environment. For example, they knew hundreds of edible and medicinal plants and were adept at locating insects, grubs, and other foodstuffs. Some studies also suggest that the lives of many hunters and gatherers, especially in warmer climates, were not as difficult as we often imagine. In many locales, they did not live in constant danger of starvation and did not spend a great deal of time finding food. Their lives were often leisurely and healthy.

Today, remaining hunters and gatherers live in relative harmony with nature, taking what they need and generally causing little ecological disruption, especially compared with modern humans. Ancient hunting-and-gathering societies may have also lived in relative harmony. Their lack of environmental damage can be attributed to at least three facts of life: (1) many were nomadic, wandering in search of food and a favorable climate, (2) their numbers were small, and (3) their tools were primitive.

This is not to say that all hunters and gatherers were benign. Some North American Indians, in fact, set fire to the prairie to drive buffalo off cliffs, starting massive fires and killing large numbers of animals. In addition, the extinction of several North American species parallels the migration of ancient hunters and gatherers across the continent.

New archaeological research suggests that many groups of hunters and gatherers grew their own food and raised animals to feed their people. Some may even have engaged in trade with other groups. Over time, more and more of

THE EVOLUTION OF UPRIGHT POSTURE IN HUMANS

The first truly humanlike apes belonged to a group known as the hominids, the first member of which was *Australopithecus afarensis*. This short, humanlike primate lived about 3.5 million years ago on the grasslands of Africa. Judging from the structure of its pelvis, *A. afarensis* probably walked upright on two feet; its primate ancestors, however, walked on all fours.

The morphological changes required to make the shift to bipedalism are well known. They include anatomical alterations of the legs and feet, pelvis, and vertebral column. More specifically, they include the acquisition of relatively straight, downward-projecting femurs, changes in the pelvis, and development of an inward curve of the lumbar vertebrae, all shown in ▷ Figure 1.

Of all of the unique features that define hominids, bipedalism is one of the oldest. In fact, it probably came well before the evolution of the large brain. In his book *The Human Species,* John Relethford argues that the anatomical alterations required for bipedalism probably required little genetic change.

But why was bipedalism selected for in the first place? Four-legged animals, as a rule, can run faster and maintain their balance more easily than two-legged animals. As in many areas of science, a number of hypotheses have been offered to explain why bipedalism evolved. Today, most scientists believe that it offered a selective advantage over other forms of movement in three situations: avoiding predators, reproducing, and acquiring food.

When hominids left the forests and moved to the grasslands (the savanna), they were no doubt preyed on by four-legged mammals, such as lions, which are able to run faster than humans over short distances. From this standpoint, bipedalism would be a deficit. However, some biologists believe that bipedalism also permitted early hominids to better spot predators from a distance. Because the human nose is no asset whereas vision is relatively sharp, bipedalism permitted the earliest humans to see predators long before they were a danger, then escape from potential harm.

Bipedalism also frees the hands, allowing adults to carry their offspring, a trait that may have enhanced reproductive success. Besides freeing the hands to carry babies, bipedalism may have allowed for the simultaneous care of more than one offspring. That is, when the hands are free to carry food and babies, more than one child can be cared for at a time. The result may have been greater reproductive success, which, in turn, led to the further selection of this trait.

Another major selective advantage of bipedalism is in the acquisition of food. Anthropologists believe that by freeing up the hands, bipedalism would have allowed our early ancestors to carry more food, an advantage in grasslands, where food was widely distributed.

Bipedalism is a rather efficient stance. Relatively little muscular exertion is required to maintain an upright posture when standing or walking. One study, in fact, showed that ambulatory bipedal humans require far less energy than knuckle-walking chimpanzees and normal quadrupeds of the same size. According to one of the most popular hypotheses, this efficiency permits hominids to walk for long distances in search of food with very little energy expenditure.

"In a changing environment," writes Relethford, "such as that found at the end of the Miocene, food resources would be scattered. The ability to move long distances in search of food would be an advantageous trait, and the shift to a hominid form of bipedalism would provide this ability."

Although bipedalism has its disadvantages, it no doubt evolved because it offered more advantages.

---

our early ancestors began to cultivate food crops, slowly giving rise to a new form of life, the agricultural society.

Agricultural societies emerged between 10,000 and 6,000 years ago. In the moist rain forests of Southeast Asia, farmers cleared small jungle plots to raise their crops. They grew a variety of vegetables but also raised pigs and other domesticated animals for additional food.

Seed crops originated in a wider region extending from China west to India and eastern Africa. In these regions, farmers cleared forests to plant crops, and with the advent of the plow, they began to till the rich grassland soils.

The plow allowed for larger fields and higher grain production. Farming, which had generally supplied the immediate needs of a farm family, began to change. Now a farmer could produce food for many families. With this development arose towns and cities. People who were no longer needed on the farm congregated in cities and towns, where they began trades, making clothing, pots, tools, and weapons.

Several important changes in the human-environment interaction were also evident as agriculture grew. First, humans began to drastically modify the natural environment. Poor soil management, overgrazing, and heavy timber cutting, however, destroyed large regions. The rich Tigris-Euphrates river valley, often referred to as the cradle of civilization, is now part of Iran and Iraq. Once lush and productive, much of this landscape is now parched and unproductive because of human intervention. Land abuse from poor farm management continues today in many parts of the world, threatening long-term food production.

The second major change during the shift to agriculture came with an upsurge in commerce. Commerce demands natural resources—metals, energy, and stone. These materials came from the outlying countryside. The towns and

▷ **FIGURE 1 Anatomical Changes Responsible for Bipedalism** (*a*) The chimp's skeleton is adapted for knuckle-walking. (*b*) It can stand upright, but the pelvis and thigh bones make it difficult to attain erect posture. (*c*) In order to stand erect, the chimp would lean forward in an unbalanced stance not seen in chimps. (*d*) Changes in the anatomy of the backbone, hips, and thighs permit upright posture in humans.

cities, therefore, drew heavily on the surrounding land, often causing considerable environmental damage.

The third change, possibly the most important of all, was a weakening of the link between humans and nature. As noted above, many hunter-gatherers took little from the land and lived within the bounds of nature. Agriculturalists, however, attempted to harness the forces of nature. Humans may have begun to see themselves as separate from the environment, as its masters (▷ Figure 28–17). This change in attitude followed humans into the next phase of cultural evolution, the industrial society. With the advent of trade and commerce in agricultural societies, humans began to regard the natural world more and more as a source of wealth.

The industrial society is a recent occurrence in human history. In fact, if the Earth's history were condensed into a one-year-long movie, the Industrial Revolution would

▷ **FIGURE 28–17 The Control of Nature** Many forms of agriculture attempt to control or dominate the forces of nature, reshaping landscapes to human liking.

occur in the last half-second of the film. The **Industrial Revolution,** the advent of mechanized production, began in England in the 1700s and in the United States in the 1800s. Although machines took over much manual production, they required enormous amounts of energy and produced enormous amounts of air and water pollution (▷ Figure 28–18).

Mechanization swept the farms, too, in the industrial era. As a result, still fewer people were needed to raise food. Cities grew. Pollution increased. Streams once rich in fish turned putrid with the stench of human and factory wastes. The countryside was often pillaged to provide energy and materials for factories. Pollution and species extinction were two of the principal environmental problems that arose during the Industrial Revolution.

Changes occurring during the agricultural and industrial revolutions planted the seeds of a dramatic increase in human population. Between 1850 and 1992, human population increased from 1 billion to over 5.4 billion. The rise in population occurred for many reasons. The most important were (1) modern medicines, which lowered infant mortality, (2) improved sanitation, which stopped the spread of disease in the increasingly crowded urban environment, and (3) greater food production.

Humans' relationship with nature grew even more strained. Philosophers, economists, and some theologians argued that people must seek power over nature. Control became the byword. Survival, as many saw it, required complete domination of nature. Today, efforts at domination continue, but they have spawned some rather serious

▷ **FIGURE 28–18 A Product of Cultural Evolution**

global environmental problems. Some think they are a warning sign that we have pushed too far.

Current problems, such as global warming and ozone depletion, threaten to alter global climate and create widespread species extinction. It is no exaggeration to say that the survival of humans and the other species that share this planet with us are in danger. Our health and survival on the planet depend on reestablishing a balance with nature. Many people see the challenge today as finding ways to live and conduct business on the planet that are sustainable. Our future depends on finding ways to meet our own legitimate needs without bankrupting the Earth and foreclosing on future generations and the millions of species that share the planet with us.

SUMMARY

EARLY PRIMATE EVOLUTION

1. Humans belong to the order called primates, which includes two suborders: the prosimians (premonkeys) and the anthropoids (monkeys, apes, and humans).
2. Today, most primates live in tropic and subtropic forests and are well adapted to an arboreal way of life. The main exception is humans, which inhabit a wide range of habitats.
3. Primates are characterized by grasping hands, forward-looking eyes, and large brains (in proportion to body size).
4. Based on fossil evidence, it appears that the primates evolved from a mammalian insectivore that resembled the modern-day tree shrew and lived about 80 million years ago.
5. The first primates to evolve were the tree-dwelling prosimians. Modern-day prosimians include the lemurs and tarsiers.
6. The prosimians gave rise to the New World and Old World monkeys. The hominoid line, which today consists of apes, gibbons, and humans, diverged from the Old World monkeys over 20 million years ago.
7. The very first hominoids were apelike creatures. Two distinct apes present at that time are considered potential ancestors of humans and modern apes.
8. The first family, the dryopithecines, originated about 20 million years ago in Africa and spread to Eurasia about 14 million years ago. The scant fossil remains of these creatures

suggest that they walked about on four legs, much like chimpanzees and gorillas, but spent most of their time in the trees.
9. The second possible ancestor of apes and humans lived slightly later—from about 17 million years ago to 7 million years ago—and are called ramapithecines.

EVOLUTION OF THE AUSTRALOPITHECINES

10. The first truly human primate belonged to the genus *Australopithecus* ("southern apeman") and may have evolved from either the dryopithecines or ramapithecines.
11. The oldest known australopithecine skeleton was unearthed in Africa and is believed to be about 3.5 million years old. It belongs to a group called *Australopithecus afarensis.*
12. *A. afarensis* stood only about 3 feet high and had a brain only slightly larger than an ape's, but it probably walked erect.
13. About 3 million years ago, *A. afarensis* was replaced by another species, *A. africanus,* which was slightly taller than its predecessor and had a slightly larger brain.
14. About 2.3 million years ago *A. robustus* emerged. It was taller and heavier and had a larger brain than its predecessors.
15. About 2.2 million years ago, the fourth species, *A. boisei,* appeared. It was even more robust than *A. robustus.*

EVOLUTION OF THE GENUS *HOMO*

16. Many paleontologists believe that *Australopithecus afarensis* also gave rise to the genus *Homo*, the ancestors of modern humans, *Homo sapiens*. Others believe that the genus *Homo* may have evolved from an as-yet-unidentified ancestor.

17. The earliest discovered member of the genus *Homo* is a 1.8-million-year-old skull and partial skeleton found in Tanzania. It is called *Homo habilis.*

18. Two hundred thousand years after the emergence of *Homo habilis, Homo erectus* arose. Skeletons of *Homo erectus* appear in geological formations 300,000 to 1.6 million years old in the Old World.

19. Unlike *Australopithecus* and *H. habilis,* which remained in Africa, *H. erectus* moved from Africa to Europe and Asia.

20. *Homo erectus* stood about 5 feet tall, used fire, and made more sophisticated tools and weapons than its predecessors. With a brain slightly smaller than ours, *Homo erectus* is believed to be the direct ancestor of *Homo sapiens.*

21. *Homo sapiens* emerged about 300,000 years ago and consists of two subspecies: *Homo sapiens neanderthalensis,* the Neanderthals, and *Homo sapiens sapiens.*

22. The Neanderthals lived in caves and camps in Europe and Asia. Approximately 40,000 years ago, they disappeared. Some archaeologists believe that they were replaced by modern humans, the Cro-Magnons, the earliest known members of *Homo sapiens sapiens.*

23. The Cro-Magnons arose in Africa and invaded Europe and northern Asia, perhaps wiping out the Neanderthals or possibly interbreeding with them.

24. Over the course of time, the human population has wandered far and wide on the planet, inhabiting a wide range of climatic zones. This has resulted in the formation of a number of phenotypically distinct subpopulations, or races. Many differences found in the races are thought to be adaptations to differing environmental conditions.

TRENDS IN PRIMATE EVOLUTION

25. The evolutionary history of primates leading to humans is marked by important changes, including (1) the evolution of bipedal locomotion, (2) an increase in brain size, (3) modifications of the hands, which improves dexterity, and (4) the emergence of stereoscopic vision.

ENVIRONMENT AND HEALTH: THE IMPLICATIONS OF HUMAN CULTURAL EVOLUTION

26. Over the past 40,000 years, human evolution has provided little noticeable change in physical appearance but tremendous cultural change.

27. Human culture has evolved through three phases: hunting and gathering, agriculture, and industry. During that transition, humans have sought ways to control the environment, sometimes with disastrous consequences.

28. Today, our attempts to dominate nature continue. But current global environmental problems are warning signs that we may have pushed too far, that our society is out of balance with the natural world, threatening our own existence.

EXERCISING YOUR CRITICAL THINKING SKILLS

Some critics make a blanket statement that genetic engineering in humans allows scientists to tinker with our evolutionary process and therefore should be avoided. Using your knowledge of human evolution, genetics, and genetic engineering and your critical thinking skills, how would you analyze this statement?

Hint: You might want to start by examining what genetic engineering in humans attempts to do. Then determine if that could affect our evolution as a species. Is there any possibility that it will help or hinder our evolution? How would evolutionary changes be brought about by genetic engineering?

TEST OF TERMS

1. The _____ were the first primates to evolve. They developed from small, insectivorous mammals that probably resembled modern-day tree shrews.

2. The _____ include three major groups: New World monkeys, Old World monkeys, and the hominoids, which include _____, gibbons, and _____ .

3. Australopithecines and the genus *Homo* are thought to have arisen from one of two primitive apes, _____ or _____ .

4. Louis and Mary Leakey discovered what may be the first member of the genus *Homo*, which they called *Homo* _____.

5. _____ _____ is believed to be the direct ancestor of *Homo sapiens,* modern humans. It arose in _____ and spread to Europe and _____ .

6. Neanderthals were members of _____ _____ . They lived in small nomadic groups

7. and had brains that were _____ than those of modern humans.

8. Neanderthals mysteriously disappeared and were replaced by the _____, which arose in _____ .

9. A _____ is a phenotypically distinct subpopulation of a species.

Answers to the Test of Terms are located in Appendix B.

1. Why is the primate fossil record incomplete? What problems does this gap create in tracing the evolutionary history of primates?

2. Draw a flow diagram indicating the progression of primate evolution, beginning with the mammalian insectivores and ending with humans. Briefly describe each organism, and note how it differed from the early prosimians.

3. When you have finished question 2, describe the major anatomical changes that have taken place over the course of primate evolution.

4. Describe the major changes that took place in human cultural evolution. How have they affected the health of the planet and the future of humankind?

P A R T

VII

Interactions

*Behavior, Ecology, and
the Environment*

≈

An Introduction to Animal Behavior

≈

In marshlands around North America, the spring air is filled with the sounds of red-winged blackbirds preparing for the breeding season. In this annual event, males defend small tracts of marshland that they will share with their mates and their offspring (▷ Figure 29–1). They advertise their readiness to defend this territory with a precise behavioral display. One component of territorial behavior is their song, which to other males indicates a willingness to engage in combat if necessary. Males perch on the cattails to show off the bright red shoulder patches on their wings, a display that also seems to convey their intention to ward off intruders.

Females are attracted to successful males who hold territories with adequate food and cover. Some extremely successful males, in fact, may attract 10 or more females to their territories; others may attract only 1. Some males fail to establish territories and do not mate at all. The key to male reproductive success is having exclusive access to prime marshland habitat containing an abundance of cattails and insects. The key to maintaining its domain, though, is a bird's behavior.

What is behavior? **Behavior** is an individual organism's response to environmental stimuli. The "environmental" in this definition includes other individuals of the same species—for example, a bird's offspring, neighbors, or potential mates. It also includes members of another species, such as a potential predator or a potential meal. In addition, it may refer to some aspect of the organism's physical environment, including resources such as water, shelter, and light or hazards like wind, cold, and rain.

Behavior is such an important feature of an organism's

biology that two chapters are devoted to the subject. This chapter focuses first on general definitions, then turns to various kinds of behavior between individuals and their physical environment and between individuals of the same species. In other words, this chapter concerns itself with behavior within populations, technically referred to as **intraspecific behavior.** Chapter 30 deals with the behavioral interactions of members of one species with those of other species. In other words, it deals with behavior within communities of organisms, with symbiosis, in the broad sense of the term—in short, with **interspecific behavior.**

≈ RUDIMENTS OF BEHAVIOR

Next to an organism's structure, color, and general appearance, behavior is one of its most readily observable features. In fact, behavior may even help us identify an animal. For example, if a bird that is perched on a telephone pole is drumming on it, it must be a woodpecker.

Because behavior is so obvious and familiar, we often take it for granted and forget that it is among the most complex, and least understood, aspects of the biology of organisms.

Innate Behaviors Are Apparently Genetically Programmed and Ready To Be Put To Use When Needed

Some behavior is **instinctive,** or **innate**—in other words, "inborn." Ducklings, for example, run to water no matter where they are raised. They swim and dive for food without any instruction. Chicks peck at food, drink liquids, and shake their bodies dry soon after hatching without any coaching from their parents. These behaviors are automatic and are apparently genetically programmed. That is, they are fixed responses to stimuli, independent of previous experience. Thus, two key features of all innate behaviors are that they are in place and operational when organisms require them.

Taxes and Tropisms Are Innate Mechanisms Found in Less Complex Organisms. One of the simplest of all innate animal behaviors is the **taxis** (plural, **taxes**), the orientation of an animal in relation to some stimulus. Planarians (flatworms), for example, orient themselves toward light (phototaxis). Leaf-cutting army ants join a column of fellow workers by aligning themselves to small quantities of chemical (chemotaxis) substances released by the leader, as shown in ▷ Figure 29–2.

In sum, taxes are orientational movements of whole organisms. The ability of ants and flatworms to orient themselves quickly and accurately in response to environmental cues improves the likelihood of their finding food, shelter, and other resources needed for survival and reproduction.

As you may recall from Chapter 26, plants exhibit similar responses, known as **tropisms.** Phototropism, for exam-

▷ **FIGURE 29–1 Red-winged Blackbird on a Cattail Marsh.** The first male to arrive at the site makes his presence known by singing and displaying his red shoulder patches to other males in the vicinity.

▷ **FIGURE 29–2 Taxis** Leaf-cutting army ants align themselves along an invisible chemical trail released by the leader.

▷ **FIGURE 29–3 Venus Flytrap** The cagelike leaves of this carnivorous plant can close in seconds, trapping an unsuspecting insect.

ple, is the bending of a plant toward light. Tropisms usually involve movement of a plant by the growth of cells rather than by an actual change in the position of the plant. Like taxes, tropisms have obvious adaptive value.

Numerous Reflexes Are Displayed by Many Animals and Provide Obvious Advantages. Reflexes are another form of innate behavior. Human babies, for example, grasp a finger tightly when it is placed in their hand. Baby chicks raise their heads and open their mouths wide whenever their nests are shaken. These reflexive behaviors occur without forethought or prior learning.

As explained in Chapter 17, reflexes in vertebrates involve receptors, afferent nerves, association neurons, efferent nerves, and effectors. Reflexes are not under voluntary control. Indeed, they occur without the brain ever being involved, although the brain may be "informed" that an action has occurred, after the fact.

Beyond simple taxes, tropisms, and reflexes, animal behavior gets rather complicated.

Plants Exhibit Distinct Behaviors that Are Generally Slow Responses

When we think of behavior, we usually think of animals. Animals are "animated," after all. Most of them move around and exhibit a wide variety of behaviors: feeding, escape, aggression, courtship, play—the list is endless. As the previous discussion noted, plants also display distinct behaviors, but most plant behaviors are fairly slow responses.

Most plants stay in one place; to use the colloquial idiom, they simply "vegetate." This kind of behavior is, of course, quite appropriate for organisms that are rooted to their environment, which serves as a source of water and minerals. For such species, there is no need to move about in search of an energy supply, because their energy flows to them as sunlight.

Much plant behavior is slow because, as noted earlier, it

is based on growth. Roots grow down, in response to gravity; leaves grow upward, toward light.

Some plant behavior occurs more rapidly and is based on changes in turgor pressure within cells. A morning glory, for example, produces spiraled tendrils at the tips of its growing shoots. The shoot waves slowly back and forth through the air. When the tendril contacts a solid object—a tree, a post, or the gardener's trellis—it twists around it, providing an attachment that lends support to the growing plant.

Another even more rapid behavior is found in the Venus flytrap. The cagelike leaves of the plant, shown in ▷ Figure 29–3, can close within seconds on hapless insects that trip an internal trigger that responds to touch. Another example is the sensitive plant that responds instantly to the touch by folding its leaves.

Two important generalizations are worth noting at this point. First, behavior does not require multicellularity. Microbes behave. Protists such as amoebae and paramecia behave, moving quickly to orient with respect to positive or negative stimuli—chemicals, light, or temperature. Photosynthetic protists like *Euglena* move even more quickly, propelled by whiplike flagella. Second, all organisms exhibit some form of behavior. How they respond, however, depends in large part on their complexity and their requirements.

Although the remainder of this chapter is—with due apology—devoted mostly to the behavior of animals, as a general science of behavior develops, it will surely include more on the study of nonanimals as well. At present, though, the science of behavior is still emerging. The main reason is that behavior is one of the most difficult aspects of biology to study. That may come as a surprise because, as noted earlier, behavior is one of the most obvious features of an organism. In short, it is easy to observe behavior but often difficult to interpret its meaning. Why does a dog wag its tail? Is it happy? Why does a red-winged blackbird let out its metallic cry? Is it telling potential intruders to stay away?

≈ THE ROOTS OF CLASSICAL ETHOLOGY

Over the past century, most scientists who studied animal behavior have come from one of two academic disciplines, psychology and zoology. For much of the century, these scientists have pursued independent lines of inquiry and have generated two distinctly different schools of thought regarding behavior, with different methods of study and even different subjects. The two schools are **ethology** (animal behavior) and comparative psychology. The early **ethologists** were largely European zoologists who conducted detailed, descriptive studies of the instinctive behavior of animals, especially birds, fishes, and insects, under natural conditions. Early **comparative psychologists,** by contrast, were largely North Americans who conducted experimental studies of learned behavior, especially on mice, rats, and college freshmen, under controlled laboratory conditions. Unfortunately, about all that these scientists had in common was an interest in behavior. For many years, the only real contact that ethologists and comparative psychologists had with each other was to argue about which was *the* right way to understand behavior.

More recently, however, scientists from both camps have come to realize that both modes of study have their merits and that scientists in both groups are arriving at interesting answers to different questions. As the validity of each discipline is acknowledged, the term *ethology* is coming to encompass the entire field of animal behavior. Such integration is viewed by many as a positive development, as long as one continues to appreciate the distinct contributions of classical ethology and comparative psychology.

Three Scientists Contributed Mightily to Our Understanding of Animal Behavior

Classical ethology grew out of work by three pioneering researchers, who together were awarded the Nobel Prize in Physiology and Medicine in 1973. This illustrious group included an Austrian, Konrad Lorenz; a Dutchman who worked mostly in England, Nikolaas Tinbergen; and a German, Karl von Frisch (▷ Figure 29–4). Lorenz's most famous work focused on the behavior of geese, Tinbergen's studies focused on gulls, and von Frisch's work centered on honeybees. With contributions from co-workers, these three zoologists provided many insights into animal behavior.

Classical Ethologists Were Interested in the "Anatomy" of Behavior

Classical ethologists analyzed sequences of behavior (through slow-motion photography, for example) and observed that complex behaviors seemed to be composed of simpler parts. The parts, they noted, can be assembled in various ways to accomplish different needs. For example, a bird's courtship ritual may combine bits of behavior that are also used in feeding, nest building, and aggressive territorial defense. In other words, those pieces of behavior can be assembled to answer a particular biological need. According to this view, a behavior is analogous to any other complex organ, such as a hand or a foot. The organ is made up of simpler materials (muscle, bone, connective tissue,

▷ **FIGURE 29–4 Pioneers of Classical Ethology** These three men shared the Nobel Prize in Physiology and Medicine in 1973: (*a*) Konrad Lorenz, (*b*) Niko Tinbergen, and (*c*) Karl von Frisch.

(a)   (b)   (c)

and nervous tissue), which are also used to build organs elsewhere in the body. In the hand, the pieces are assembled in particular configurations to meet particular needs. In the feet, those components are assembled to produce a structurally similar arrangement, but with markedly different function.

Fixed Action Patterns Are Innate Components of Many Other Behaviors

Because their interest is in the component parts of complex behaviors, ethologists first build an **ethogram,** a list, or "menu," of the component parts of a species' behavior. More specifically, an ethogram consists of stereotyped pieces of behavior, coordinated sequences of neuromuscular activity. These stereotyped behaviors are known as **fixed action patterns (FAPs).** Examples of FAPs are the ear movements of a dog, the butting motion of a goat, the tail flash of a deer, or the neck movements of a strutting rooster or a cock pheasant (▷ Figure 29–5).

You can probably think of a few FAPs yourself. If you cannot, carefully watch any animal—a dog, a squirrel, or a pigeon—for a while, and you will begin to pick out repeated components of behavior. These components are about the same in all normal members of a species. They tend to develop even in isolated individuals, without their having been learned. Consequently, fixed action patterns are innate.

If you watch your chosen organism long enough, you will note that a given fixed action pattern has a consistent environmental antecedent, or **stimulus.** The male robin, for example, attacks other male robins, not female robins and not sparrows. Squirrels handle, carry, and bury nuts, not tiny seeds or pieces of trash. A gull broods eggs, not egg-sized pebbles. Specific antecedents of particular FAPs are called **sign stimuli (SS),** or **releasers.**

Sign stimuli tend to be quite specific. If you present a

live male robin with a beautifully stuffed male robin with its breast painted blue, it will not perform the FAPs typical of territorial defense. If, by contrast, you present that same robin with a poorly shaped ball of red feathers, it will probably attack it. (Does this suggest to you why male and female birds are so often colored differently?) As another example, a male stickleback fish fights in response to a red patch on the belly of a rival. If you paint over the patch of its rival—or paint a fake one on its back—the male will not attack.

Some wonderful experiments help further our understanding of sign stimuli. For example, Tinbergen modified the eggs of the ringed plover, intensifying the contrast between the dark splotches and the pale background (▷ Figure 29–6). The retouched eggs elicited stronger brooding behavior than normal eggs. He also found that the larger the egg, the stronger the response, even when artificial eggs became too large to sit on! Such sign stimuli are called **supernormal releasers,** because they elicit excessive performance of fixed action patterns.

Environmental sign stimuli elicit a predictable and fixed, or stereotyped, behavioral response in individuals. However, the response can change over time. Toads, for example, capture insects on their sticky tongues. A naive toad snaps (FAP) at anything (sign stimulus) that flies within range that will fit inside its mouth, until it captures a bee. Thereafter, it snaps at anything that moves except bees. In this instance, the range of suitable sign stimuli has been narrowed by experience.

A young goose innately follows (FAP) anything larger than itself that moves (sign stimulus). At some critical period in the gosling's development, however, the range of

▷ **FIGURE 29–6 Supernormal Releaser** Eggs of the ringed plover retouched to intensify the contrast between the dark splotches and the pale background elicit a stronger brooding behavior than normal eggs.

▷ **FIGURE 29–5 Fixed Action Pattern** The neck movements of a male pheasant are an example of a fixed action pattern, a stereotyped behavior that may be used as a part of more complex behavioral displays.

sign stimuli that elicit the FAP narrows, and the young bird follows only one large, moving object, which under normal circumstances is its mother. Under experimental circumstances, Lorenz found that one could artificially narrow the range of sign stimuli so that the gosling exclusively followed a person or a roller skate on the end of a string.

In goslings and other birds, such as ducks, this process in which the range of sign stimuli narrows occurs in a fairly short time, known as the **critical period.** The attachment to its mother, or the artificial substitute, is called **imprinting.** Imprinting is vital to the survival of goslings and ducklings. Inappropriately imprinted animals grow up with behavioral problems. For example, if a male gosling imprints on a roller skate or an ethologist, as a sexually mature gander the bird may try to court and attempt to mate with a skate or an ethologist. (For more on imprinting, see Scientific Discoveries 29–1.)

The fact that the relationship between an FAP and sign stimulus can change suggests that the individual is capable of changing with experience; that is, the individual has the capacity to learn. This ability has led ethologists to propose the presence of **innate releasing mechanisms (IRMs),** distinct neural units in the nervous systems of animals that control responses to specific sign stimuli. Because the relationship between a fixed action pattern and a sign stimulus can change over time and because the range of relevant sign stimuli can be narrowed, ethologists proposed something changeable in the individual that mediated between the FAP and the SS (▷ Figure 29–7). That something is the IRM. Although ethologists cannot see it or localize it (yet), they assume that it exists because of its observable effects.

In summary, classical ethologists have primarily focused their interest on instinctive behavior, behavior that is somehow "built in." They have tried to understand the structure of behavior—that is, its component parts. Lorenz captured the essence of the ethological view of behavior in the phrase "parliament of instincts." This delightful characterization of behavior reflects the fact that behaviors are complexes of instincts. At any moment, an individual's observable behavior results from animated "debate" in the "parliament." Despite their fascination with instinct, ethologists recognize that behavior changes as a result of experience; that is, they appreciate the importance of learning.

≈ COMPARATIVE PSYCHOLOGY

As noted earlier, comparative psychologists tended historically to focus on learned behavior, which they studied under controlled laboratory conditions. Hence, they studied animals, including humans, that depend heavily on learning in their daily lives. The studies of comparative psychologists often focus on mammals, such as laboratory mice, rats, dogs, and cats, and also human beings, although far simpler animals and even plants have been studied as well. In these studies, human behavior is often compared with the behavior of other animals.

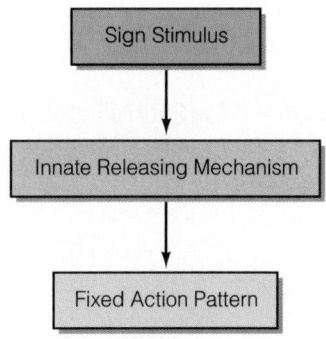

▷ **FIGURE 29–7 The Constructs of Classical Ethology** This drawing shows the relationship of the three principle constructs of classical ethology: the fixed action pattern, the innate releasing mechanism, and the sign stimulus.

Although comparative psychology has been tremendously productive, its understanding is still a long way from complete. Nonetheless, many insights gained from this discipline are both fascinating and instructive.

Learning Leads to Changes in Behavior

As noted above, many animals respond innately to stimuli but seem to store information about the various experiences they have had. This stored information may lead to changes in behavior. This process is known as **learning.** Learning requires that animals recall the information they have stored at appropriate times and therefore allows an animal to modify its responses to deal specifically with current situations. Learning seems to require more complex neural circuitry than innate behavior.

Apparently, not all learning is alike. Although the internal, neural mechanisms involved in learning remain mostly unknown, learning has been classified by outward circumstances.

Conditioning. A type of learning familiar to almost everyone is known as conditioning, a response studied by the famous Russian physiologist Ivan P. Pavlov. (For more on Pavlov's work, see Scientific Discoveries 29–2.) In his research on digestive secretions, Pavlov noted that dogs salivated when they smelled food. That is, the presence of a stimulus, food, led to a specific response, salivation. Pavlov then sounded a bell just before feeding the dogs and found that they soon learned to associate the sound of the bell with feeding. Eventually, the canines began to salivate at the sound of the bell, even in the absence of food.

Pavlov called this response a **conditioned reflex.** Today, the process is called **conditioning,** or **associative learning,** because an animal has learned to make a connection (association) between a new stimulus (bell) and a familiar one (food). Put another way, in associative learning a previous stimulus-response relationship is transferred to a novel stimulus—in this case, the bell.

IMPRINTING: SETTLING THE RIVALRY
Featuring the work of Lorenz

Why is it that experiences early in life often have lasting effects? According to one psychologist, Dennis Coon, part of the answer "lies in the existence of critical periods for acquiring specific behaviors." A critical period, he says, is a time of increased sensitivity to environmental influences. Studies suggest that experiences during critical periods may dramatically influence one's future development. This influence often persists throughout one's life.

Studies of animal behavior suggest that certain events must occur during a critical period for an animal to develop normally. When I was a teenager in rural western New York, for instance, I raised fancy pigeons. One of my prized pairs laid eggs, then incubated them, but after 21 days when the egg did not hatch, the parents abandoned the nest. Thinking the egg was sterile or the chick dead, I decided to crack it open. Much to my surprise, the chick was alive! Still, its parents wanted little to do with the youngster, so I fed the pigeon by hand. Pretty soon, the two of us became the best of friends. Charley, who turned out to be a female, would ride on my shoulder and on the handle bar of my bike as I peddled down the road to visit my friend. She would even fly from the pigeon coop to the roof of the house and then onto my shoulder as I opened the door to go in. Clearly, Charley had imprinted on me. As far as she could tell, I was her parent.

Had her parents provided food and care, no doubt, Charley would have imprinted on them. I didn't mind at the time, though, I quite enjoyed having so close a companion from a different species.

Imprinting is a very natural event in the lives of birds and other animals, but unless offspring imprint on a member of the same species, their behavior is bound to be unusual. Had my pet pigeon not died while I was away at camp, I suspect that she might have been a little confused come breeding time. Her imprinting on me might have precluded her from mating with her own kind.

The first recorded observation of imprinting is credited to a Viennese naturalist, Oskar Heinroth. He noted that young geese imprinted on humans and regarded the human as their parent. One of Heinroth's devoted followers was Konrad Lorenz, who died in 1989 (▷ Figure 1). Although Lorenz is often credited with the discovery of imprinting, his contribution was really the systematic exploration of the conditions under which imprinting occurs. One of his most noteworthy findings is his demonstration that imprinting occurs during very definite periods in a bird's life. After imprinting has occurred, he showed, it is not easily undone.

Of Lorenz's many experiments, we will focus our attention on just one. This one was designed to determine whether all instinctive behaviors (instincts) resulted from one imprinting event. Put another way, does each of these behaviors require imprinting at separate stages of a bird's development.

Under study was a bird called a jackdaw (also known as the European starling). One bird Lorenz studied had been reared in complete isolation from others of its kind. During the period when the jackdaw would normally have started flying with others of its kind, Lorenz introduced it to a flock of hooded crows. The bird joined the crows as one of the flock. Soon thereafter, the bird was placed in the jackdaw colony, and it was living among its kind during the time when Lorenz suspected that imprinting for sexual reproduction took place. So what happened when the bird became sexually mature?

During most of the year, the bird cavorted with the hooded crows. During mating season, however, it began to live a split existence. Although it hung around and bred with other jackdaws, each day when the time came to fly, it took off to join the hooded crows. These results strongly suggested that each major behavior was imprinted separately.

Imprinting obviously serves to attach a young animal to its mother, but, as this example shows, it also guides the selection of mates of the same species once a bird reaches sexual maturity. Lorenz was to find this out rather graphically one day. Snoozing on the lawn one day, he was awakened by one of his birds, which was—picture this—attempting to stuff worms into his mouth. Although this is strange behavior for most humans, it is, I'm told, a normal part of the mating ritual of adult jackdaws. When the baffled scientist refused the gift, his would-be mate stuffed the worm in his ear.

In conclusion, Lorenz did not discover imprinting. He merely settled a rivalry between two competing hypotheses about the timing of imprinting of various behaviors within one species. For his contribution, he was awarded the Nobel Prize.

▷ **FIGURE 1 Lorenz and his Subjects** Konrad Lorenz, renowned for his studies of imprinting, is shown with the geese he studied.

CLASSICAL CONDITIONING: DOES THE NAME PAVLOV RING A BELL?

Featuring the work of Pavlov

Scientific discovery is not always painstaking and deliberate. In fact, a scientist hot on the trail of one question may by chance make an observation that leads to other important discoveries.

One such example is the discovery of classical conditioning by the Russian physiologist Ivan Pavlov. Pavlov was not studying animal behavior, but rather digestion. In studies of salivation, he placed meat powder or some other tasty morsel on a dog's tongue. After a while, he noticed that dogs started salivating before the food was delivered. Later, the dogs began to salivate at the sight of the scientist, a phenomenon that may have caused him a little concern.

Pavlov reasoned that some kind of learning had taken place and he called this learning **conditioning.** Because of its importance in the history of psychology, it is now called classical conditioning.

Pavlov set about to test his hypothesis with a now-famous set of experiments. To begin, he rang a bell. The dog did not respond, because most dogs don't associate bells with much of anything. Each time he rang the bell, however, Pavlov gave the dog some meat powder, which, of course, caused the animal to salivate. He repeated this procedure many times. After a while, the bell, which had previously caused no response, began to produce the same response as meat powder.

A dog's response to meat powder is an unconditioned response. That is, it is unlearned. Salivation in response to the bell is a **conditioned,** or **learned, response.**

Conditioned responses explain one of the most common mistakes dog owners make: punishing a dog when it fails to come when called. This practice associates the owner's call with fear and withdrawal. The next time, the dog will be even more reluctant to come.

Conditioned responses also help explain some human emotions. That is to say, some responses result from conditioning. A painful experience at a dentist's office, for example, may explain why your heart pounds and your palms sweat before you go in for your next visit.

Fears that persist even when no real danger exists (phobias) may also be the result of conditioning. Fear of water, animals, or fire, for instance, may be linked to a previous experience when a person was frightened or hurt by the stimulus.

Operant Conditioning. Another form of associative learning is known as **operant conditioning.** In this process, a reinforcing stimulus, a punishment or reward, is given to an animal after it performs a particular behavior by chance. For example, a chicken in a cage with a pecking bar of a particular color may be rewarded with some food every time it pecks the bar. Over time, the subject begins to associate the reward with the action and therefore begins to perform the action routinely. Circus trainers reward "tame" bears, lions, and seals with food if they display a particular behavior. Such conditioning can be used to encourage highly unnatural behaviors (▷ Figure 29–8).

Latent Learning. Many animals spend a good deal of time exploring their surroundings with no immediate reward. Later, however, it becomes obvious that learning has taken place but has been **latent**—that is, stored for later use. A mouse scurries about its home range poking its nose under every rock and log. Its movements seem to be random. Or at least they seem to be random until the mouse is faced with a weasel or an owl. Then the mouse heads straight for cover, putting to survival use the **latent learning** that it has stored up as a mental map of its surroundings.

Imitation. Anyone who has ever watched young children knows that they learn a great deal by imitating their parents.

Imitation is another form of learning. Many young birds, for instance, learn how to sing from their parents or other adults. Birdsong is an important form of communication, announcing territory or initiating courtship. To do its job, though, a

▷ **FIGURE 29–8 Conditioning and Animal Training**
Conditioning can be used to encourage highly unnatural behaviors.

song must be sung correctly. If, during the critical period for such learning, the bird is caged with a bird of a different species, the youngster will sing inappropriately.

An exciting example of imitation occurred in a tribe of Japanese macaques (a kind of monkey) that were fed wheat that had been thrown on a beach. At first, the monkeys painstakingly picked the seeds out of the sand. By chance, however, one monkey discovered that by throwing wheat and sand into the water and scooping up the more buoyant wheat, it could avoid the tedious process of separating its food from the inedible sand. The process spread rapidly by imitation.

Play. Play is a form of behavior with no obvious immediate product. Play often involves imitation. More importantly, it seems frequently to involve the practice of skills that may eventually be useful in responding to stimuli in the "real world." You have surely seen a cat stalk a ball of yarn or a dog shake an old sock into submission. These behaviors may help develop species-typical hunting tactics. You may also have seen "play mounting" in young monkeys or dogs. The playful behavior of otters is well known. It takes real discipline not to dismiss playful behavior as *just* play, which is peripheral to the needs of an animal. Because play requires energy, however, we had best assume (until demonstrated otherwise) that it is important and adaptive. It may be practice (working on the cerebellum) or it may lead to socialization and establish an individual's status, or it may just be a tension-release valve, clearing the way for other behavior (like a quick game of Frisbee as a study break).

Habituation. Animals can learn *not* to respond to stimuli as well. This phenomenon is called **habituation.** Listen carefully to the noises around you—the refrigerator or outside traffic. All are sounds that you were ignoring but that are clearly audible if you focus on them. Habituation, described in Chapter 18, is a highly adaptive type of learning, without which humans and other animals would probably quickly suffer from sensory overload. Occasionally, individuals do indeed become overloaded with stimuli. In fact, it was through such pathology (and subsequent experimentation) that habituation was localized in a region of the brain called the reticular formation, the so-called "window of consciousness."

Sensitization. Roughly the opposite of habituation is **sensitization.** Sensitization occurs when an individual learns to pay heightened attention to stimuli. Fishes, for example, sense movements in water through the lateral line organ, described in Chapter 18. A fish in an aquarium is usually unresponsive to minor motion in the water. Place the aquarium on a shaker, and the fish will habituate and become unresponsive to quite violent vibrations. However, that same fish, exposed just briefly to even a small piranha, can become quite sensitive to tiny movements in its tank.

Animals Often Migrate Long Distances, Using Environmental Cues to Find Their Way

Two of the most remarkable animal behaviors are **migration,** the movement of an animal from one region to another, often over long distances, and **homing,** returning to a familiar (home) location. These behaviors seem so extraordinary in large part because we humans do not exhibit them. After hatching on the beaches of Central America, one species of sea turtle migrates to the waters around Ascension Island in the South Atlantic, where it matures. Seven years later, when the turtles reach sexual maturity, they migrate back to their natal beach 2000 miles away to breed. The latter migration is a form of homing behavior.

Many species of birds and bats migrate thousands of miles when the season changes to mate in preferred areas or to find food. When the season changes again, they return home. Homing behavior has been particularly well documented in pigeons. Homing pigeons can be transported hundreds of miles from their home and, after being released, return to the loft where they are kept, even flying alone.

Migration and homing routes seem to be learned, but the ability to navigate is at least partially innate. Pigeons, for example, seem to orient by the sun during the day and by the stars at night, and they may also use the Earth's magnetic field to guide them.

Salmon spend their first three years of life swimming downstream to the ocean, where they complete their maturation. When sexually mature, salmon migrate back to fresh water to spawn, following a chemical scent unique to the stream in which they hatched. This scent, imprinted during their earliest experiences, leads them to the spot where they hatched (▷ Figure 29–9).

The ability of migratory birds to find their way home may also be based on another kind of imprinting. Stephen Emlen, for example, showed that indigo buntings imprint on the first night sky they see. Indigo buntings breed in the northeastern United States but winter in Central America. When only a few months old, the young buntings take off on their southern migration, traveling alone. To see how they successfully make their journey, Emlen reared buntings in a planetarium where he could control the apparent position of the stars. Through careful study, Emlen found that the birds used the position of the stars as a guide in their nocturnal flight, orienting to the Big Dipper, the North Star, and other constellations. When Emlen reared the buntings under atypical sky conditions, they flew in random directions.

Innate Behaviors May Be Influenced by Internal Conditions

You study hard because you are motivated to do well. You know that if you get straight A's, you are bound to be accepted to a prestigious medical school or will be a prime

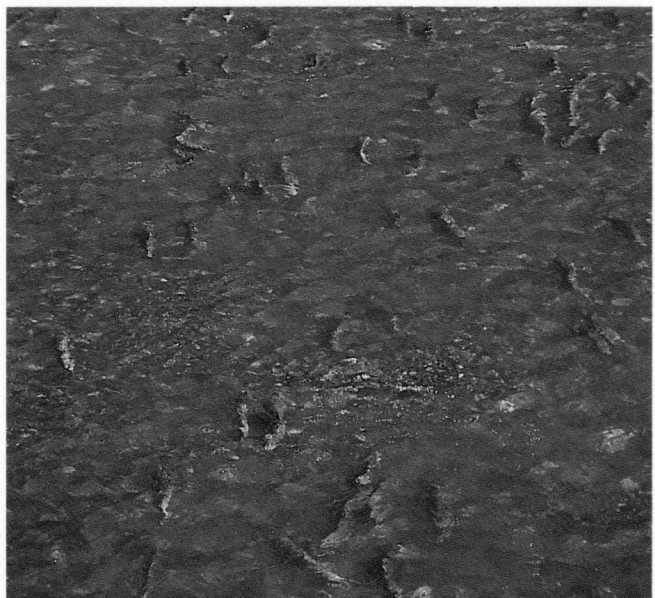

▷ FIGURE 29–9 Chemical Imprinting Salmon return to the fresh water where they hatched, following chemical cues recognized from their earliest experiences.

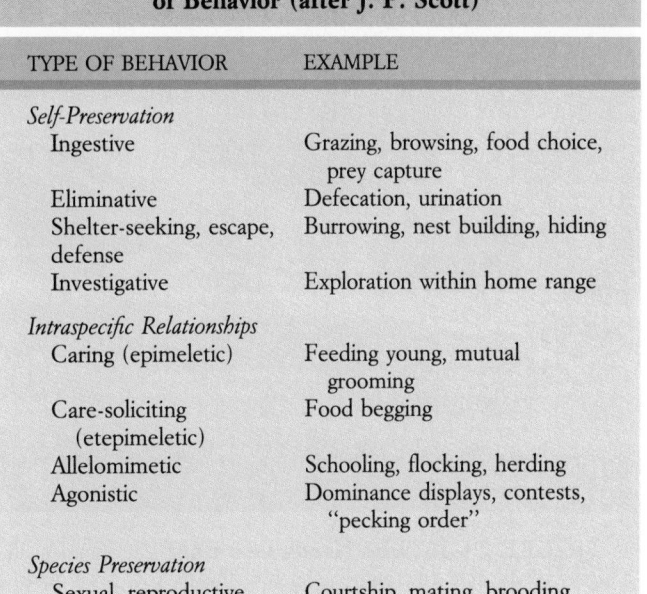

TABLE 29–1 A Functional Classification of Behavior (after J. P. Scott)

TYPE OF BEHAVIOR	EXAMPLE
Self-Preservation	
Ingestive	Grazing, browsing, food choice, prey capture
Eliminative	Defecation, urination
Shelter-seeking, escape, defense	Burrowing, nest building, hiding
Investigative	Exploration within home range
Intraspecific Relationships	
Caring (epimeletic)	Feeding young, mutual grooming
Care-soliciting (etepimeletic)	Food begging
Allelomimetic	Schooling, flocking, herding
Agonistic	Dominance displays, contests, "pecking order"
Species Preservation	
Sexual, reproductive	Courtship, mating, brooding

candidate for graduate study at your favorite university. Your diligent study behavior in this case is motivated by some conscious desire.

Innate behaviors can also be motivated by internal conditions. A hungry animal, for instance, seeks food. Hunger is said to be a motivating factor. After the animal acquires food, it no longer hunts, because the motivation is gone.

The internal state of an organism, its **motivation,** or drive, evokes certain behaviors. In female canaries, for example, a change in day length affects the maturation and development of the gonads, which in turn steps up its production of the hormone estrogen. In the canary and other species, estrogen stimulates interest in males; that is, it is a motivating factor. In the presence of a courting male, the ovaries develop more rapidly, and more estrogen is produced sooner. This leads to increased interest in the male and reciprocal courtship display, which further enhances estrogen production, which then stimulates nest-building behavior. Nest-building stimulates further gonadal maturation. This, in turn, leads to development of the oviducts and the brood patch, and stimulates the production of the hormones that elicit egg laying. In this example, an external stimulus evokes an internal change, which alters the animal's behavior.

≈ THE FUNCTIONS OF BEHAVIOR

Table 29–1 classifies numerous animal behaviors. This categorization does not portray the underlying mechanisms of behavior but, rather, what various behavior accomplish—in other words, their outcomes or functions. One thing that you will see as you study this list is that the various types of

behavior are not mutually exclusive. Investigative behavior, for example, may through latent learning facilitate escape or food-searching behavior.

Behavior, beyond the simplest forms such as taxes, is far too complicated a matter to have yielded such an elegant and simplified classification. Nonetheless, the table helps us focus on the functions or accomplishments of various forms of behavior.

In Most Animal Species, Learning Is Rather Rare

As you examine Table 29–1, you will see that some of the examples represent learned behavior (food choice, for instance) and some clearly represent instinct (like the schooling of fish). Others involve combinations of the two. Mice, for instance, have an innate tendency to explore their home range, and in so doing they learn for future episodes of ingestive (feeding) or escape behavior.

As you study the following material, you will also note that learning plays a relatively greater role in behaviors that function to preserve individuals—for example, feeding behavior. Learning what food tastes good and learning the best escape routes are self-preservation behaviors. Instinct plays a greater role in behaviors that function to preserve genes and hence species—for example, courtship behavior. This is one example where stereotyped innate behavior is quite appropriate.

The relative importance of learning and instinct in organisms depends in part on the environmental conditions in which they live. Consider food-getting behavior. If an organism lives in a highly predictable environment, there is no need for learned food choice. A caterpillar hatched on a

▷ **FIGURE 29–10 Who Needs Learning?** An organism in a highly predictable environment, such as this hickory horn devil caterpillar on a hickory tree, has no particular need for learned food choice.

▷ **FIGURE 29–11 Adaptive Behavior** Black-headed gulls, like other species of shore birds, remove empty eggshells from the nest soon after the young are hatched, a behavior that may protect their young from sharp-eyed predators.

food plant is a good example (▷ Figure 29–10). For organisms such as this, it is enough to distinguish food from nonfood, innately. However, an organism in a less predictable environment, with resources scattered in patches, needs to be programmed to profit from experience—that is, to learn. The blue jay learns to avoid monarch butterflies by aversive conditioning, coupling a distasteful experience with a particular motor action. If it could not learn, it would very probably become ill or die in a complex resource environment that included poisonous butterflies.

Learning has been demonstrated in organisms as humble as flatworms, but most animals (especially invertebrates) do not depend on learning to any appreciable degree. Among the vertebrates, fishes, amphibians, reptiles, and birds tend to be far less dependent on learning than do mammals. Even among mammals, there are differences. Primates seem to rely on learning for a greater spectrum of their behavioral repertoire than do the closely related insectivores and bats. Humans are the present-day culmination of this general, phylogenetic trend. Even on our worst days, we humans are the best learners ever to have evolved, and we surely are more dependent for our survival on learning than is any other organism.

Behavior Is Influenced by Genes and Is Subject to Natural Selection

Whether a behavior is innate or learned, genes are at the root of all behaviors. An organism's genes, for example, are responsible for the establishment of circuits in the nervous system that control innate behaviors. Genes also clearly underlie the ability to learn. Because genes underlie behavior, directly or indirectly, behavior is subject to natural selec-

tion, a fact that has been verified repeatedly by experiment.

Some 50 years ago, Robert Tryon noticed that some rats were adept at locating food in a complex maze, whereas others took much longer. He selected the "smart" rats and bred them to one another. He also cross-bred slow learners. He then ran offspring of fast learners through the maze. Consistent with the hypothesis that learning was under genetic influence, they learned to find the food even more rapidly than their parents. Moreover, the performance of the slow learners was poorer than that of their parents. This experiment illustrates that learning ability is genetically based and can be artificially selected. Thus, natural selection of genetically based behavior probably also occurs in the wild.

The adaptive value of innate behavior has also been verified experimentally. Black-headed gulls, for example, remove eggshells from the nest soon after the young are hatched (▷ Figure 29–11). This behavior is innate, because the birds do it whether or not they have seen it done. Naturalists have wondered why this housekeeping task occurs almost universally among shore birds. In a classic study, Tinbergen found that predators were attracted to otherwise well-camouflaged nests by the white inner surface of broken eggshells. He also found that newly hatched chicks of parents that removed the broken eggshells were less vulnerable to predation. Thus, gulls that removed shells were more successful in rearing offspring than those that did not.

So far, our discussion has been mostly about animal behavior important for the survival of individuals. Egg-laying behavior is quite another matter, however, for it functions to preserve the species. Egg-laying behavior is one of a vast range of behaviors that involves groups of individuals, behaviors that we consider—in the broadest sense of the word—social.

≈ SOCIAL BEHAVIOR

What is social behavior? **Social behavior** involves interactions between members of the same species. As shown in Table 29–1, several of the functional kinds of behavior involve such interactions: sexual and reproductive behavior, schooling, flocking, caring, soliciting care, and dominance. Such behaviors have evolved repeatedly in the history of living organisms. This section focuses on these behaviors, describing their evolution and other important aspects.

Social Groupings Evolved Because They Offer More Benefits than Costs

Many different species exhibit a characteristic called **sociality,** a tendency to form interactive groupings. Knowing what we do about evolution, sociality at first glance seems paradoxical. Similar organisms tend to compete for limited resources. The evolution of sociality is the evolution of tolerance for close competitors, which comes at a certain cost. In order for sociality to endure, it must have been counterbalanced by greater benefits. What might they be? The answer lies, in part, in the fact that social groups are often cooperative arrangements that provide greater protection than solitary existence does.

Defense. One of the most obvious values of sociality is defense against predators. A herd of 100 gazelles has 200 eyes; an individual gazelle has only 2. Even in the hungriest foraging herd, a few animals will be on the alert for potential predators (▷ Figure 29–12). Flocks of birds and schools of fishes benefit similarly.

Flocking and herding behavior are far more common in open country (grasslands, for example) than in forested habitats or rocky terrain where long-range visibility is minimal. In open country, the herd or flock provides protection that the landscape does not. Many species of fishes exhibit pronounced schooling, with hundreds or even thousands of individuals compounding sensory awareness. Such large groups may also provide deception, appearing to be a very large (dangerous or at least unmanageable) "superfish." Flocks, herds, or schools may scatter at a given signal, startling a would-be predator. Also, flocks or schools groups may contribute to predator confusion over which individual to have for lunch.

Modifying the Physical Environment. Social behavior has adaptive value because it permits modification of the physical environment. Geese, for example, fly in a V formation, presumably because it reduces friction and thus makes flying easier. Schooling may have evolved in fish, in part, for similar reasons.

Honeybee colonies maintain a nest temperature about 35°C (95°F). In the heat of summer, the bees fan the air,

▷ **FIGURE 29–12 One Advantage of Sociality** In even the hungriest foraging herd, a few animals are always alert to potential predators.

which cools it. In cooler weather, they huddle in a compact ball, their heat keeping the nest warm (▷ Figure 29–13). The hive temperature does not drop below about 17°C (62°F), even in subzero weather.

Foraging Efficiency. Social groups may also increase foraging efficiency. Shared foraging may free time and energy for other tasks. Cooperative hyenas, for example, can subdue larger prey.

Reproductive Functions. Some species require flocks or herds for successful reproduction—for example, mating swarms of mayflies and other insects. Others animals come together for mating as well. Each spring, for instance, male sage grouse congregate at a specific dancing, or "booming," ground. Cocks inflate their chests, tilt their bodies back, hold their feathers erect, strut around, and "boom."

▷ **FIGURE 29–13 Temperature Regulation in the Hive** In cold weather, honeybees huddle in a compact ball, which slows the flow of metabolic heat from the core. The bees take turns in the outermost insulative layers.

▷ **FIGURE 29–14 Courtship among Grebes** Male and female Western grebes perform mutual courtship displays that reach a height early in the breeding season.

(A boom is the noise the bird makes.) Through these displays, the males sort themselves out in an orderly manner before the females appear. When the females arrive, they are attracted to the strutting males.

Courtship behaviors such as these, which take place during mating time, are observed in many animal species and serve a variety of functions. They are performed mostly by males to entice females to mate, and they may have evolved to allow males to advertise their relative rank in flocks and herds. Courtship may be reciprocal. Male and female Western grebes, for instance, perform mutual courtship displays that reach a crescendo early in the breeding season (▷ Figure 29–14).

Courtship behavior also stimulates physiological changes necessary for successful reproduction. The bowing and cooing of male ring doves, for example, stimulate hormonal changes in females, which lead to a rapid increase in the size of their reproductive organs. A week after the onset of courtship, ovulation begins.

In some species, courtship behavior serves not only to attract females but also to direct them toward a mating site. Male three-spined stickleback fish, for instance, swim in a zigzag pattern that leads the female to the nest, where she lays her eggs.

Some scientists believe that courtship may have evolved initially to prevent mating with members of other species, an event that is futile from a reproductive standpoint. Behavioral differences between species allow accurate identification and hence ensure that sexual activity does not waste time and energy and that gametes are not expended in the production of sterile hybrids.

Protection of Young. Sociality may increase the chances of survival among offspring. Female green sea turtles, for instance, dig flask-shaped nests in beach sand in which they deposit about 100 eggs. The young all hatch at once, then clamber out to the surface. Studies show that a single young cannot reach the surface by itself, but a group as small as four can. So why lay 100 eggs? Studies suggest that there is safety in numbers. After they hatch, the young are preyed on by gulls and raccoons as they scramble toward the water. In the chaos of 400 flailing feet, a few hatchlings from every nest manage to escape to the safety of the sea.

Communication Is Involved in Many Forms of Social Behavior

Cooperative behavior in social groups is greatly enhanced by **communication.** Communication is a signal from one individual that alters the probability of behavior of another individual. Communication signals may be visual, chemical, or tactile (involving touch). Notice that in defining communication, we do not say that the signal causes a behavior in another individual; it merely "alters the probability" that a behavior will occur. That is to say, the receiving individual may or may not respond to a signal from the sender, but the likelihood of its responding has been changed by the signal.

Communication requires organisms with two types of organs, those that send or transmit some signal and those that receive them. Individuals typically have both types of organs, and communication is therefore a reciprocal event. Organs that send signals change the environment in the vicinity of the receiving individual. For example, they may change the quality or quantity of reflected light, the chemical composition of the air or water, or the character of mechanical vibrations in soil, air, or water. The receiver is an organ that detects these changes.

Visual Communication. Visual communication is universally found in organisms that are active during the day. It is also prevalent in organisms that inhabit open country where senders and receivers are not separated by physical barriers or where they are in communication at close range. Organisms that live in dark habitats depend very little on visual communication. A good example is the cave-dwelling salamander, which is blind. In some cases, denizens of the darkness such as fireflies and some deep-sea fishes actually emit light. Visual communication in these species is extremely important.

Auditory Communication. Unlike reflected light, which moves in straight lines, sound is propagated as waves in some medium (air or water) that move in all directions from the source. Therefore, communication through sound can take place in complex environments (for example, in dense forests) and in the dark. The howl of a coyote and the chirp of the cricket are two examples of the latter.

Not all sound occurs within the range of human hearing. Bats, shrews, and some mice, for instance, produce sounds of very high frequencies (ultrasound) that are involved in communication. Elephants produce very-low-frequency infrasound that can carry "silently" (from our biased viewpoint) for miles.

Chemical Communication. Chemical communication is especially well developed in terrestrial species. Ants, for instance, lay odor trails to guide their colleagues to food supplies. Some insects release chemicals known as **pheromones,** "external hormones" that attract potential mates, who find the releaser by following the chemical concentration gradient. Males of many species of mammals locate sexually receptive females by the odor they emit when in a state of sexual readiness. Chemical cues move more slowly than light or sound and are also more persistent in the environment.

Tactile Communication. Communication by touch is highly developed in many social species, among them the higher primates. Tactile communication is frequently involved in courtship, and it often occurs between a mother and her young. In many cases, tactile communication supplements olfactory communication. A good example is the nuzzling behavior that is so common among mammals. In primates, relatives are more likely to groom each other than are nonrelatives (▷ Figure 29–15). This contact may encourage cooperation in a broad range of social settings.

These are just a few of the means by which animals communicate. If they seem familiar to you, it is because they are the modes we rely on most of the time to send signals to one another.

Interestingly, other species communicate by means not available to us. For example, some water bugs communicate via vibrations of the water's surface. Several groups of fishes, including electric eels, respond to electrical voltage gradients they produce. Because we personally lack sensors to detect the signals transmitted from one member to another, these forms of communication may be very difficult to study. Only with the use of special equipment can we study these fascinating means of communication.

Communication in a Social Group Serves Many Functions that Enhance the Group's Survival

Communication has many functions or outcomes. It allows organisms to monitor the position of others and hence stay in contact or remain at a respectful distance. It also allows individuals to recognize one another and determine relative social status, promoting appropriate interaction. The communication of an animal's status in a social group and its intent is extremely important in a general type of behavior referred to as **agonistic.** Agonistic behavior includes dominance displays and contested encounters (two males competing for a potential mate). Communication may solicit aid (food, grooming) or announce an animal's intention to provide aid.

Communication is also important because it helps protect a social group from outside harm. Alarm signals, for

▷ **FIGURE 29–15 Grooming Behavior** In primates, such as these baboons, relatives are more likely to groom each other than are nonrelatives. Grooming helps knit the groups together.

instance, are common in social groups, as are signals of distress. Often a signal conveys both meanings, encouraging retreat in some individuals and a protective response in others.

Communication may initiate group activities; for example, it may signal the time to assemble to hunt. Chimpanzees howl to announce the discovery of a good fruit tree. Some rain forest birds, such as parrots, send loud signals to their flock as well. Communication may consist of simple signals, or it may be considerably more complex, verging on language (a term generally reserved for humans until recent years). The "language" of honeybees, which was elucidated by von Frisch, is a case in point. Information on food sources is shared among hive members by elaborate "dances" performed by those bees returning from successful food-gathering flights (▷ Figure 29–16). These dances communicate the location of the food source. A round dance tells a bee's mates that food is near the hive. A waggle tells them that the source is farther away. Several other visual cues given during the dance inform them of the exact location of the food source. Honeybees rely on a combination of chemical, visual, and tactile stimuli that permits cooperative foraging and enhances the survival of the colony.

This chapter stresses communication in the social context within a species, but communication also takes place between individuals of different species, as you will see in Chapter 30.

Agonistic Behavior Is Vital to the Establishment of Dominance Hierarchies and Territoriality

Agonistic behavior involves some form of contest between individuals of a species. The term comes from the Greek word *agon,* which denotes the conflict between characters in classical Greek drama. Agonistic behavior is a peculiar sort

▷ **FIGURE 29–16 The Honeybee's Dance Speaks Volumes** Von Frisch elucidated the hidden meaning behind the elaborate dance of the honeybee, showing that humans are not the only species with language.

of social behavior. On the surface, it does not appear to involve cooperation. But in a broad sense of cooperation—operating together for a common end—agonism is indeed cooperative. That is, agonism operating in an individual's self-interest may have the net effect of helping the species as well. In some species, for example, fathers and mothers aggressively protect their young from danger. Such behavior is not overtly self-protection, because offspring are not quite "self," but genetically they are the nearest thing to self. Protecting them helps protect the future of the species.

Status. Agonistic behavior is used to establish and maintain an animal's status in a social group. Chickens, for example, establish a linear **pecking order**—that is, a graded series from most dominant to most subordinate (▷ Figure 29–17). In a flock of mixed sexes, hens and cocks form parallel, separate pecking orders. Pecking order gets its name from the fact that it is maintained by pecking. The lead bird can (and occasionally, just as a reminder, does) peck every other bird. The second in line can peck with impunity every bird other than the lead bird. The last bird cannot peck any bird without being summarily counterpecked.

This linear array of dominance sets an order of precedence for feeding, roosting, and even mating. Suppose, for example, that you had six hens in a pen into which you tossed three kernels of corn. Would they divide each kernel in half? Not likely. Would they squabble over it? Only occasionally. Typically, if two hens are equidistant from a single bit of food (a resource in short supply), you can predict on the basis of observations of previous behavior which hen will get the corn. It will almost always be the one higher in the pecking order.

In a pecking order, one contest generally serves to establish a structure. This system eliminates the need for many separate contests, resulting in a considerable saving of energy. In flocks of chickens where pecking orders are experimentally disturbed, growth is stunted, and egg production falls off substantially.

The linear pecking order is simpler than most systems of dominance. In many species, dominance is **hierarchical.** That is, there are several levels of status in the society, with one or more individuals at each level. You are probably familiar with hierarchies in human society. In the army, for instance, there is a distinct hierarchy, shown in descending order of importance and power: general > colonel > major > captain > lieutenant > sergeant > corporal

▷ **FIGURE 29–17 The Pecking Order** Can you pick out the lowest ranking chicken in this photo?

> private. This is a chain of authority, command, and dominance. It is maintained by symbols (pins with stars, birds, and stripes) and gestures (salutes and stance).

Hierarchies of dominance are also found in nonhuman societies. These are also usually maintained by systems of gestures. In several species of ground-dwelling, Old World primates, the troops consist of several adults of each sex and their offspring. The age-sex groups are arranged hierarchically, with adult males mostly dominant to females, subadult males, and the young. Within each age-sex level, there is a tendency for other dominance relationships to be established. Established by combat, they are maintained by threat postures and gestures, like staring and "yawns" that bare the canine teeth.

Hierarchies such as this one may establish precedence in mating. In autumn, for example, bull elk joust for dominance with their massive antlers (▷ Figure 29–18), and their superiority gives them mating rights with a harem of cows. Male mountain sheep spend time early in the breeding season in ritual head butting, sorting into a "mating lineup" that determines access to receptive females. Such behavior is common among hoofed mammals.

It is important to note that such combat is not aggression for aggression's sake. It has a reproductive and evolutionary function. Note, too, that the aggression does not usually injure but simply makes a point. The point is made by a ritual, such as antler wrestling, rather than by mortal combat. In **rituals,** one behavior comes to represent another. Ritualization is typically found in species where individuals are equipped to do each other bodily harm.

Subordinate individuals in many hierarchies exhibit **appeasement behavior,** such as greeting by lowering the face and neck. This behavior is common in social carnivores such as wolves and hyenas (▷ Figure 29–19). Subordinate behavior frequently mimics juvenile patterns and reduces friction in a pack.

Territoriality. Agonistic behavior may help maintain optimal spacing. At any particular time, individuals of a population exhibit one of three general types of spacing, or dispersion, patterns: random, contagious (clumped), or uniform (hyperdispersed) (▷ Figure 29–20).

In **random dispersion,** there is generally no interaction among individuals. This pattern is commonly found in species like the grizzly bear, which as a rule remains solitary throughout the year except for a brief time when mating occurs.

In **contagious distributions,** caring, cooperative, and flocking behaviors prevail. A group may consist of a mother and her offspring or a nuclear family (a mated pair and its offspring). In some cases, it may consist of a flock that forages together or a group of bachelors that migrate and feed together. Whatever its composition, the group offers several key advantages to the individuals that compose it.

In **uniform distributions** in which large numbers live together, individuals may simply avoid or actively repel each another. Agonism is therefore frequently prevalent in

▷ **FIGURE 29–18 Autumn Jousting** In autumn, bull elk compete for dominance and breeding priority.

species characterized by uniform distributions. The observed spacing often results from **territorial behavior,** active measures that establish and defend a territory.

What is a territory? In its daily routine, an individual moves through an area or a volume of habitat, called its **home range.** Every mobile organism has a home range. In many species, a part of the home range is defended against others of the same species. The defended area is known as its **territory.**

Territories are marked and defended in various ways. Dogs, for example, mark their territories with urine. Songbirds sing to announce their territory. Another male of the same species that ventures into the territory is actively chased out. Male birds are often brightly colored and contrast sharply with their dull-colored mates. Interestingly, it is only the bright coloration of males that elicits agonistic behavior from holders of territory. Once the breeding season is over, males in many bird species lose their bright plumage, territories break down, and agonism wanes.

▷ **FIGURE 29–19 Appeasement Behavior** Common in canids, like the gray wolf, appeasement behavior is displayed by subordinate individuals in the dominance hierarchy.

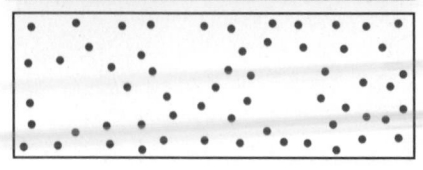

(a) Random dispersal

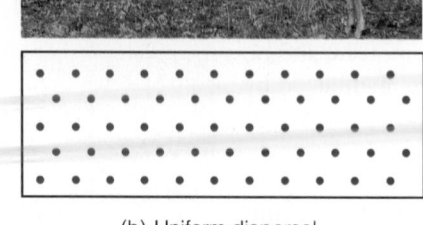

(b) Uniform dispersal

(c) Contagious (clumped) dispersal

▷ **FIGURE 29–20 Dispersal Patterns** Three kinds of relative spacing, or dispersion: (*a*) random, (*b*), contagious ("clumped"), and (*c*) uniform ("hyperdispersed"). Can you guess what functional types of behavior listed in Table 29–1 are likely to be responsible for each of the patterns observed?

Males and females and their young become part of a flock for the rest of the year.

Territories are usually established around food, although some species defend other resources, such as nesting materials or escape routes. Territoriality tends to reduce crowding. Surplus individuals are forced to inhabit the margins of suitable habitat, where subsistence is less certain.

It is not hard to imagine how territorial behavior evolves. Individuals who efficiently defend adequate resources survive to reproduce. To the extent that territorial behavior is genetic, it is selected for strongly through the increased fitness of those who are good at it.

Many Behaviors Are Altruistic, or Overtly Beneficial to Other Members of the Same Species

Dominance and territoriality may seem like selfish forms of behavior, and in a genetic sense they are. Natural selection favors behaviors that take care of genes and get them into the next generation. However, contrary to what we might expect, selfish behavior is not the only kind of behavior observed in groups. In many group settings, overtly selfish behaviors seem to be limited or nonexistent.

Scientists have discovered in nature many instances of behaviors that appear to be **altruistic**—that is, beneficial to others at the expense of the organism performing the act. Behaviors that appear to be altruistic are common in species living in social groups (such as baboon troops) that are formed from extended families.

The evolution of altruism and cooperation in nonhuman species is more difficult to understand than the evolution of agonism and competition. As you may recall from Chapter 23, evolutionary fitness is measured by an organism's success in passing on its genes. If an individual hurts or kills itself doing favors for others that increase the others' fitness, the altruistic individual's relative fitness is clearly lowered. "Do-gooders" (and their "do-gooder genes") are doomed to extinction. Despite this fact, we see in social organisms countless examples of acts that appear to be altruistic. For example, if you or some other dangerous-looking organism gets too near the nest of a killdeer, the female will flutter away, calling and dragging a wing as if it were broken (▷ Figure 29–21). This behavior diverts a would-be predator from the nest, but it may be more dangerous to the mother than simply flying away.

Explanations for the evolution of altruistic behavior such as this are several. Before we look at them, first bear in mind that natural selection operates on individuals. Darwin was correct in believing that arguments about the evolution of traits for "the good of the species" would not do. Because selection operates on the level of the individual, "selfishness" will always prevail.

In the case of the killdeer, the female clearly endangers herself by drawing a potential predator away from her nest. By so doing, however, she saves a whole nest of young who carry her genes. This act thus increases the probability that her genes will be passed on to future generations and that future mothers will act similarly. This example of apparent altruism is really nothing more than a genetically based characteristic that promotes the survival of the genes that created it.

Behavioral biologists use the term **kin selection** to refer to altruistic behavior that increases the likelihood that an individual's genes will be successfully transmitted to future generations. The killdeer is a perfect example of *direct kin selection,* aid given by adults to their children. *Indirect kin selection* occurs when a relative and others with the same genotype act to protect an offspring in its social group. The helper's genes, of course, may be endangered by an act, but because the offspring contains those same genes, the

▷ **FIGURE 29–21 Altruistic Behavior?** The "broken-wing" display of the killdeer draws potential predators from the nest.

helper is contributing to the transmission of advantageous genes.

Some apparently selfless behavior is the result of **reciprocal altruism.** In other words, one individual acts kindly toward another because it was the recipient of a similar act at an earlier time. If what one individual does for another results in reciprocation, then a seemingly selfless act may not be selfless at all. It is merely a shrewd investment, a form of insurance. (You scratch my back, I'll scratch yours.)

In monkeys and apes, mutual give-and-take plays an important role in establishing social cliques. Individuals, particularly juveniles, form strong social "in groups" that remain together in a cooperative arrangement providing protection and support when needed.

Sociobiology Is a Relatively New Field of Science that Attempts to Understand Behavior in Biological and Evolutionary Terms

In 1975, the Harvard biologist E. O. Wilson, an authority on ants and other social insects, published a monumental book entitled *Sociobiology: The New Synthesis.* The book outlines a new branch of biology, the general biology of social systems. Wilson ambitiously merges several distinct disciplines into one line of inquiry, among them population biology, ethology, comparative psychology, sociology, social psychology, anthropology, and even history. "When the same parameters and quantitative theory are used to analyze both termite colonies and troops of rhesus macaques," Wilson writes, "we will have a unified science of sociobiology."

Wilson's bold and ambitious idea has sparked the most spirited (even acrimonious) biological debate since Darwin's *The Origin of Species.* Some social scientists do not like zoologists meddling in their academic business. Some biologists resent the broadening of their already difficult field to include the problems of human social behavior, the most complex problems imaginable. Some thinkers reject

the seeming genetic determinism in some sociobiological thought. But no thoughtful biologist—or sociologist either—thinks Wilson's book is unimportant. It makes anyone who can think and inquire and probe do just that. That is important to us all.

Sociobiology seeks a unified view of the biology of social behavior. Because humans are clearly social organisms, we are within the legitimate scope of such a discipline. Wilson urges us to "consider man[kind] in the free spirit of natural history, as though we were zoologists from another planet completing a catalog of social species on Earth."

That is a big order, but an important one. Investigation must proceed with the knowledge that life on Earth is a product of history, the "evolutionary play in the ecological theater," according to Wilson. In the most recent couple of million years, the drama has been complicated. Biological evolution selected a species with the capacity for a new kind of evolution, **cultural evolution.** Cultural evolution is based on the nongenetic (symbolic) transfer of information from one generation to the next. Two evolutionary plays—biological and cultural—are now showing in the same theater. The plays interact and sometimes look alike, but they are not the same. If this reality is kept in mind and questions of sociobiology are explored with tough-minded fairness, the potential implications for human welfare are great.

 ## ENVIRONMENT AND HEALTH: THE INNER CITY SYNDROME

Cities are important centers of commerce, industry, and culture. These contributions notwithstanding, many cities of the world are noisy, polluted, and overcrowded. They are centers of poverty, crime, and untold human misery.

Crowding in urban centers has been implicated in a variety of behavioral disorders and social problems. Many social psychologists assert that social instability, divorce, mental illness, and drug and alcohol abuse are due, at least in part, to stress from overcrowding. Prenatal death and rising crime rates may also be attributable to overcrowding. The complex of behavioral aberrations caused by crowding is known as the inner city syndrome.

Can crowding influence the way we act toward one another? Psychologist John Calhoun studied behavioral changes in rats that were confined to a specially built room and allowed to breed freely. He found that as the population increased, the rats displayed increased violence and aggression as well as abnormal sexual behavior. Adults cannibalized their young, and mothers paid little attention to their offspring.

Hans Selye, a physiologist, performed similar experiments in which he observed hormonal imbalances in rats induced by stress. These disruptions led to a variety of physical symptoms not uncommon in people: ulcers, hypertension, kidney disease, hardening of the arteries, and increased susceptibility to other diseases. Selye's studies, like those of Calhoun, underscore the influence of our environment on not just our physical health but our mental well-being as well.

In 1990, about 45% of the world's people lived in cities; by 2000, researchers estimate that over half of the world's people will reside in urban areas. Without steps to make our cities considerably more sustainable, crowding, and the many problems that result from it, will probably worsen, as will the fate of those who live in them.

SUMMARY

RUDIMENTS OF BEHAVIOR

1. Behavior is an individual organism's response to stimuli.
2. Behaviors that are genetically programmed and ready to be put to use when needed are called innate behaviors.
3. The simplest of animal innate behaviors are taxes, behaviors that orient an animal toward or away from a particular stimulus.
4. Plants exhibit tropisms, movements that occur because of directional growth rather than by actual change in body position.
5. Reflexes are innate behaviors produced without forethought. They are beyond the individual's voluntary control.

THE ROOTS OF CLASSICAL ETHOLOGY

6. Ethologists are scientists who study animal behavior. Early ethologists were European zoologists who conducted detailed, descriptive studies of the instinctive behavior of animals under natural conditions.
7. Classical ethology grew out of work by three towering pioneers, Konrad Lorenz, Nikolaas Tinbergen, and Karl von Frisch.
8. Classical ethologists were interested in the "anatomy" of behavior, having observed that complex behaviors seemed to be constructed of simpler parts.
9. Instinctive behavior was explained in terms of three basic constructs: (a) fixed action patterns (stereotyped responses), (b) sign stimuli (environmental antecedents of particular FAPs), and (3) innate releasing mechanisms (neural circuits that brought about responses).
10. Imprinting at a particular critical period in development narrows the range of sign stimuli that elicits an FAP.

COMPARATIVE PSYCHOLOGY

11. The early comparative psychologists were North Americans who conducted experimental studies of learned behavior in mammals.
12. A learned behavior is one that results from previous experience. Learned behaviors include conditioning, latent learning, imitation, play, habituation, sensitization, and a variety of other forms.
13. Migration and homing are remarkable behaviors that may be partly learned and partly innate.
14. The internal state of an organism, its motivation or drive, evokes certain innate behaviors.

THE FUNCTIONS OF BEHAVIOR

15. Behavior has a variety of functions or outcomes, among them seeking shelter, escape, and defense.

16. As a rule, learning plays a relatively greater role in behaviors that function to preserve individuals—for example, feeding behavior. Instinct plays a greater role in behaviors that function to preserve genes and hence species—for example, courtship behavior.
17. Learning is also more effective for species whose environments are unpredictable, whereas instinct is more efficient where environments are more constant.
18. Humans appear to be more dependent on learning than any other species.
19. Behavior, whether innate or learned, is influenced by genes and is therefore subject to natural selection.

SOCIAL BEHAVIOR

20. Many species exhibit sociality, a tendency to form interactive groupings.
21. Within these groups, numerous social behaviors are seen—among them, flocking, caring, soliciting care, and dominance.
22. Social behavior offers many advantages, including improved defense, the ability to modify physical environments to improve survival and reproduction, and an increase in foraging efficiency.
23. Communication is involved in many social behaviors.
24. Communication may be visual, auditory, tactile, or chemical. It may also involve modes that are poorly developed or absent in humans.
25. Language is complex, versatile, dynamic communication. It is not limited to humans but is found in varying degrees in other animals.
26. Agonistic behavior involves some form of contest between individuals of a species. It is used to establish and maintain an animal's status in a social group, and it is vital to establishing and maintaining territories.
27. Agonistic behavior is ritualized.
28. Altruistic behavior is also observed in some species. Apparent altruism can usually be explained in evolutionary terms as a means of promoting the transmission of favorable genes to future generations.
29. Sociobiology is a branch of biology that has emerged over the past two decades, comprising the general biology of social systems. If the questions of sociobiology are explored with tough-minded fairness, its potential is great for insights important to human welfare.

Some sociologists believe that inner city syndrome, which is characterized by social instability, crime, divorce, mental illness, and drug and alcohol abuse, is due, at least in part, to stress caused by crowded urban conditions. The Environment and Health section of this chapter cites evidence from animal studies that support a causal relationship between socially deviant behavior and overcrowding.

Some observers argue that other factors are involved in the inner city syndrome, such as poverty, poor nutrition, and inadequate housing. Are there any connections between these problems and overcrowding? Do these problems increase stress levels? Do the animal studies of Calhoun provide any insight into the debate over cause and effect?

TEST OF TERMS

1. Some behavior is instinctive, or _____—literally, "inborn."
2. A _____ orients an animal toward or away from a particular stimulus.
3. Plants exhibit _____, which involve movement by growth rather than by actual change of position.
4. _____ are innate animal behaviors that do not involve the higher centers of the brain.
5. _____ _____ _____ are stereotyped behaviors elicited by _____ _____, or releasers, specific environmental antecedents.
6. _____ is a kind of learning that involves a narrowing of the range of sign stimuli that will elicit a particular fixed action pattern at some _____ _____ in the individual's development.
7. _____ is a change in behavior as a result of experience.
8. _____ _____ involves transfer of a previous stimulus-response relationship to a novel stimulus.
9. In _____ _____, a reinforcing stimulus, a punishment or reward, is given to an animal after it performs a particular behavior by chance.
10. _____ _____ may involve production of a "mental map" of an organism's surroundings.
11. _____ occurs when an individual copies the behavior of another individual, often at a critical period in development.
12. _____ is a type of learning in which an individual learns to ignore unimportant background stimuli. _____, by contrast, is a type of learning in which an individual learns to pay heightened attention to stimuli.
13. The internal state of an organism, that is, its _____ or "drive," partly determines its response in a certain stimulus.
14. _____ is a term that refers to aggressive behavior within a species. Often it is _____, which avoids bodily harm and saves energy.
15. _____ is a process in which the probability of behavior in one individual is affected by a signal from another.

16. _____ is a versatile and dynamic form of communication.
17. A _____ _____ is a linear arrangement of organisms in a herd or flock from the most dominant to the most subordinate.
18. In a dominance _____, there are several levels of status in the society, with one or more individuals at each level.
19. The area through which an individual moves in its daily routine is its _____ _____.
20. A part of that area defended against others of the species is called a _____.
21. _____ behavior is that in which individuals endanger or harm themselves to benefit others.
22. _____ _____ describes the process in which an individual helps its relatives and their young.
23. _____ _____ occurs when a behavior takes place with the expectation that it might be repaid.
24. _____ is the general biology of social systems.

Answers to the Test of Terms are found in Appendix B.

TEST OF CONCEPTS

1. In your view, what is the extent or importance of instinctive behavior in humans? How would one determine that a particular trait was innate? What difference would it make to know that a particular behavior was innate? (As a starting point, you might think about your reaction to being stared at.)
2. Craft a definition of *language* that will encompass audible human speech, American Sign, and the dances of bees.
3. Bees use language, and chimpanzees use tools. What does this say about the distinctiveness of human beings?
4. All behavior has some genetic component. (An ability to learn is clearly inherited.) A species' pool of genetic information is shaped over time by natural selection. For over 99% of our tenure on Earth, we human beings were preliterate. For over 99.9% of our tenure on Earth, we were preindustrial. Would this have any relevance to human behavior?

C H A P T E R

30

Symbiosis and Community Interactions

≈

This bobcat has caught its prey. Such interaction is part of the Web of Life.

The brightly colored monarch butterfly is one of the more conspicuous insects in North America. Its black-lined, orange wings are a common sight in the spring and summer ▷ Figure 30–1).

When the monarch butterfly hatches from its egg, it begins independent life as a caterpillar feeding on the milkweed plant. Milkweeds contain a potent toxin in their tissues, known as a glycoside, that protects them from hungry herbivores. Concentrated in the leaves, this poison is a secondary chemical; that is, it has no primary physiological function in the plant. Its sole function is defense. This chemical is an emetic: it causes caterpillars to vomit. If they consume too large a quantity of milkweed leaves, they will die.

One notable exception is the caterpillar of the monarch butterfly. Over time, this species has evolved the ability to tolerate the poison in its tissues. Rather than digesting or detoxifying the glycoside—that is, rather than breaking it down—the caterpillar stores it in specialized structures called fat bodies. The chemical, stored intact, is then passed on to the adult butterfly.

The relationship between monarchs and milkweeds is an example of **coevolution,** described in Chapter 23. Coevolution is a process in which two or more species act as mutually important selective forces on each other. In this instance, evolutionary changes in the milkweed led to an evolutionary response in the monarch over time. The relationship is reciprocal. In other words, as resistant insects evolved over time, the milkweeds became even more toxic. Adaptation takes place continuously in both species. Coevolutionary responses are part of the continuous process of change that occurs in ecosystems. Plants adapt to being eaten by evolving more potent chemical defenses. Animals adapt by improving their counterstrategies, evolving ways to store toxic substances before they can activate nervous system responses that bring on negative reactions. (For a discussion of plant poisons, see Spotlight on Evolution 26–1.)

As adults, monarch butterflies benefit from the presence of the glycoside stored in their tissues, because the chemical is poisonous to one of their predators, the blue jay. The butterfly profits from an evolutionary development in the milkweed. It becomes poisonous without investing energy and biochemical reserves in making its own noxious chemicals.

Blue jays that try to eat monarch butterflies are immediately affected by the stored toxin and regurgitate their prey. They learn to associate the bad experience with the characteristic orange and black markings of the monarch. Over time, they begin to avoid monarchs and turn to other food sources. Thus, in a sense, the entire biotic community—plant, herbivore, and carnivore—is evolving together.

A **biotic community** is the living part of an ecosystem, the myriad plants, animals, fungi, and microbes in an environment. In the previous chapter, we discussed some ways in which individual organisms cope with the challenges of survival. In this chapter, we will examine some of the interactions and evolutionary trade-offs that allow the coexistence and survival of species in a biotic community.

In particular, we will explore ways in which behavior affects communities. In the last chapter, you saw that species adapt to their biotic and physical environments. You will now see that adaptation does not occur in isolation. Evolutionary changes in one species have consequences for other members of the biological community.

≈ SYMBIOSIS

Because no two species have the same "genetic program," no two species tolerate the same environmental conditions or have the exact same resource needs. Nonetheless, some species are more alike than others. Species with similar

▷ **FIGURE 30–1 Monarch Butterfly and Caterpillar on Milkweed** (a) The milkweed produces a toxin that causes most insects to become sick. The monarch's caterpillar, however, consumes the poison without becoming ill and stores it in its body. (b) The poison is passed to the adult and offers the species a remarkable degree of protection.

(b)

requirements and similar geographic distribution tend to occur together in communities. Let me hasten to add, however, that species in a community do not just occur together, they interact, directly or indirectly. Interactions with other species are unavoidable and take many forms. Some are strong and readily observable; others may be quite subtle.

A biotic interaction within a community is known as **symbiosis** (*sym* = "together" and *biosis* = "life"). Symbioses are the connections that structure biotic communities into integral wholes. They are the strands in the web of life. Symbiotic species also affect each other's evolutionary prospects, through the process of coevolution.

What sorts of interactions between organisms are possible? Consider the simple case of two hypothetical species, A and B. Let us symbolize a beneficial effect with "+," a detrimental effect with "−," and no effect with "0." The possible interactions between species A and B are shown (and named) in Table 30–1. With this scheme as a guide, we can explore the varieties of symbiosis present in biotic communities.

≈ MUTUALISM

Mutualism is a symbiotic relationship in which both species "benefit." *Benefit* is a tricky word in scientific language, because it implies "goodness," and what is good (or bad) is for us a matter of judgment. Value judgment is something that scientists very much try to avoid. However, value judgments can be avoided if one defines *benefit* in terms of some external, measurable standard, such as increased population size or rate of energy flow. A relationship is mutualistic if it is beneficial by one of these criteria for both of the species involved. (In some older literature, the word *mutualism* is used as a synonym for *symbiosis,* leaving no general heading for all the other modes of interaction explored in this chapter.)

Examples of mutualism are many. They were collected by the early evolutionists as examples of the evolutionary process (before it was generally understood that the biosphere as a whole and every particular bit of it exemplify evolutionary process). Even earlier, the Natural Theologians of the 18th and early 19th centuries cataloged examples of mutualism as evidence of the intricacy and cleverness of divine design. Mutualism still has a lot of popular appeal, perhaps because the cooperative ethic (you scratch my back and I'll scratch yours) makes a lot more ecological and social sense than the competitive ethic of the athletic field or the rental car business.

The bull-horn acacia is a shrub that belongs to the pea family and is found in Central America ('▷ Figure 30–2). At the base of the thorns is a pair of small glands, known as Beltian bodies, named for Thomas Belt, a 19th-century naturalist who explored Nicaragua. These glands produce a sweet carbohydrate that attracts ants. The ants live inside the thorns and feed on the sugary secretions of the Beltian bodies. In return, they do what tropical ants do best: they attack and bite. The ants attack and bite insects that alight on the tree, they chew on hikers who chance along, and they bite plants that grow too close. In short, they protect the plant from other forms of symbiosis, such as competition and predation.

Many other mutualistic relationships have evolved. Some are more easily recognizable than others. A lichen, for example, is a mutualistic association between a fungus and a photosynthetic bacterium or alga. Lichens, described in Chapter 24, grow on impoverished substrates, such as bare rock or tree limbs. These composites of two species prosper together in a mutualistic relationship.

Mycorrhizae (from the Greek *myco,* "fungus," and *rhizo,* "root") are another example of mutualism discussed in Chapter 24. As you may recall, mycorrhizae are thin, branching filaments of certain fungi that often fuse to a plant's root cells, acting as extensions of the root that greatly increase the area for absorption of water and nutrients, thus benefiting the plant. The fungi derive nutrients, mainly carbohydrates, from the root cells. Mycorrhizae are

▷ **FIGURE 30–2 Mutualism** The bull-horn acacia and ant are part of a mutualistic relationship. The acacia produces a sugary secretion that feeds the ant, and the ant protects the acacia from predators.

TABLE 30–1 A Symbolic Classification of Symbioses[a]		
FORM OF SYMBIOSIS	SPECIES A	SPECIES B
Mutualism	+	+
Predation and parasitism	+	−
Commensalism	+	0
Competition	−	−
Amensalism	−	0
Neutralism	0	0

[a] + = benefit, − = detriment, 0 = no effect.

OF COLUMBINES AND MOTHS

As noted in this chapter, in many symbiotic relationships one species becomes an important selective force in the evolution of another. Rapid locomotion among grazing animals, for example, may have stimulated the evolution of rapid locomotion among some predators, and vice versa.

Other predators evolved hunting strategies that rely more on stealth or on cooperative social behavior. As hunters became more capable, for example, the hunted undoubtedly became faster, more adept at avoiding their predators, or more dependent on cooperative social behavior—for example, forming flocks or herds which provide many advantages when it comes to avoiding predators. Thus, predator and prey constantly evolve in the presence of other species with which they interact. In a word, interactive species are coevolving.

The term *coevolution* was coined to describe the evolution of mimicry and mutualistic interactions, especially pollination systems. The term is now used more broadly, to encompass the evolutionary development of all symbiotic species and relationships. When one considers the broader picture, it becomes clear that coevolution has been a powerful force in the evolution of the diversity of our world.

One intricate example of coevolution is the columbine and the sphinx moth, which depends on the plant for survival (▷ Figure 1). Columbines are beautiful flowers, found in the wild and, increasingly, in gardens. Wild Colorado columbines are among the showiest of the species. Columbines are pollinated by

▷ **FIGURE 1 Evolutionary Colleagues** Colorado columbines are pollinated by sphinx moths, which visit the flowers to collect nectar from the spurs, shown on the left.

certain species of sphinx moths, which visit the flowers to collect a sweet nectar. This nutrient-rich secretion is produced in long spurs (which are modified petals) shown in Figure 1. As the moth hovers above the nectar, which it laps up with its long tongue, its head becomes covered with pollen. The pollen is then carried to the next flower it visits. In this mutualistic relationship, the plant provides food to the moth, and the moth, in return, carries pollen to other columbines, ensuring cross-fertilization.

This mutualistic symbiosis is no doubt the product of coevolution. How can we tell? The evidence is circumstantial but still rather compelling. Suppose a moth with a longer tongue evolved. This moth could acquire its nectar without getting dusted with pollen. In such cases, one would expect to find columbines with longer spurs, so that the larger moth still had to come in close enough to bump its head.

Remarkably, this is precisely what ecologists have discovered. Where long-tongued moths are present, the flowers have longer spurs.

Coevolution has led to complex patterns of geographic variation in mutualists. It continues to diversify and enrich our environments and thus our lives.

---

particularly beneficial in forests where trees must acquire the nutrients they need from poor soil. Any help they can get in obtaining nutrients more efficiently greatly improves their chances of survival. Some plant species, such as orchids, require mycorrhizal fungi to survive. Without the fungus, germination cannot occur.

Mutualism probably evolved early in the history of life on Earth. In fact, it may be responsible for the evolution of eukaryotic organisms. The endosymbiotic hypothesis of the origin of eukaryotic cells, described in Chapter 23, suggests that chloroplasts and mitochondria evolved from independent moneran ancestors. Moreover, the presence of producer (chloroplast) and consumer (mitochondria) organelles in the cells of photosynthetic organisms may represent mutualism.

Mutualism is common in pollination. Flowering plants produce energy-rich nectar, which is consumed by insects, birds, bats, or other animals. These organisms inadvertently pick up a little pollen on their bodies, which they unknowingly carry with them to the next flower. In so doing, they help pollinate a wide variety of plant species. Spotlight on Evolution 30–1 describes an interesting example of mutualism and coevolution.

≈ COMPETITION

Competition occurs when two or more individuals (or populations of individuals) vie for a common resource. Charles Darwin noted in *The Origin of Species* that the "struggle for existence" would be even more intense if organisms were more closely related, because of overlapping resource needs. Some of the most intense competition occurs among members of the same species, which have identical resource needs. However, competition also occurs between different species.

For competition to occur, a resource must be limited; that is, it must be in short supply relative to the needs of the species. As shown in Table 30–1, in competitive interactions both species A and species B are adversely affected. Competition is generally seen as a "lose-lose" situation. The farmer competes with grasshoppers for wheat. If there were no grasshoppers, there would be more wheat for the farmer. If the farmer did not harvest the wheat, on the other hand, there would be more for the grasshoppers. The rancher's cattle compete with jackrabbits. Thirty black-tailed jackrabbits are reported to have the energy impact on rangeland of one sheep; 150 jackrabbits are equivalent in impact to a cow (and vice versa).

Exploitation Is a Type of Competition in which Two or More Organisms Vie for the Same Resource

Competition consists of two basic types: exploitation and interference. In **exploitation,** both competitors have access to a resource, and they actually compete for it. The "winner" is the individual or population that can appropriate the most resource for its use. Exploitation is therefore a kind of ecological "scramble."

The classic studies of the Russian ecologist G. J. Gause demonstrate exploitation. He grew populations of two species of the ciliated protist *Paramecium* in laboratory cultures with a supply of yeast cells for food. ▷ Figures 30–3a and 30–3b indicate the growth curve of each species living separately. Gause then placed the two species in the same culture (Figure 30–3c). After a brief initial lead, *Paramecium caudatum* began to decline, then was wiped out entirely. It was outcompeted by *P. aurelia* for the resources contained in its environment.

Interference Is a Type of Competition in which One Organism Blocks Another from a Resource

Interference competition occurs when one organism prevents another from gaining access to a resource. Within a given species, this type of competition often takes the form of territorial behavior or some other form of "aggressive" interaction, as described in Chapter 29. Interference competition is common among many plant species, especially in habitats where nutrients are relatively scarce. In desert and grassland habitats, for instance, many plants release chemicals into the soil that prevent the growth or reduce the germination rates of seeds of potential competitors.

In the shrublands along California's coast, many species of sage produce toxic chemicals called terpenes in their leaves (▷ Figure 30–4). Terpenes are washed into the soil by the rain where they inhibit germination and growth of nearby plants. Consequently, a bare patch of soil often exists around these plants, ensuring an exclusive supply of nutrients, moisture, and light for them.

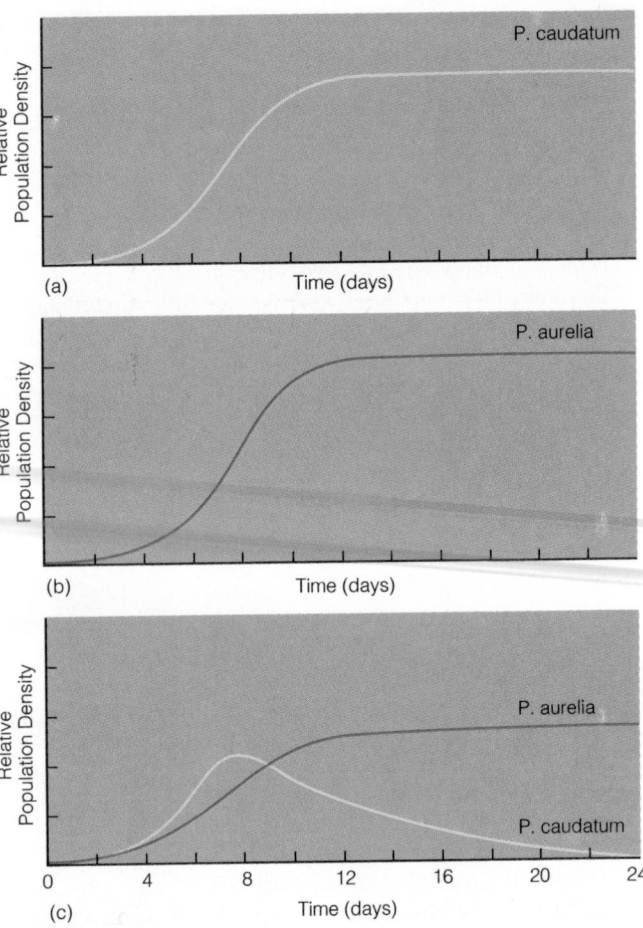

▷ **FIGURE 30–3 Exploitive Competition** (*a*) This graph shows the growth of *Paramecium aurelia* in culture by themselves. (*b*) The growth of *Paramecium caudatum* in culture by themselves. (*c*) The growth of *Paramecium caudatum* and *P. aurelia* together.

Living space is another resource for which animal species frequently compete. In the intertidal zone of the Pacific Coast, for example, two species of barnacles occur on the same rocks, submerged part of the day and exposed to the air at other times, depending on the tides. The species of larger barnacles prefers rocks that are rarely exposed to the air; the species of smaller barnacles lives on rocks more often exposed. The two species' habitats overlap in an intermediate zone, but the smaller one is crowded out by the larger species. When the larger barnacles are removed from the lower zone, however, the smaller species colonize the deeper waters, suggesting that the larger barnacles block the smaller ones from living in the deeper waters.

Some competitive symbioses are quite complex and may involve both interference and exploitive competition. Red squirrels, for example, eat seeds of conifer cones and are active by day. Squirrels are highly territorial and compete with one another (interference). During the night, however, other seed-eating rodents, such as deer mice and red-backed voles, burrow into cone caches (places where the red squirrels hide their cones) and pilfer the seeds. The

▷ **FIGURE 30–4 Interference Competition in Plants**
Species of sage in coastal shrub communities in California produce
terpenes, which inhibit the germination and growth of nearby
plants, ensuring the sage exclusive and adequate resource supply.

nocturnal rodents also feed on fungi that grow on the
cones and on the other organic debris in the squirrel's
larder. Red squirrels and mice therefore compete for food.
Deer mice also compete among themselves for seeds and
fungi, as do the red-backed voles. These interactions are
forms of exploitive competition. The fungus, which also
consumes seeds, is also in exploitive competition with the
mice and squirrels.

In forests, the trees compete with one another. The
more dense the trees, the less light there is per individual.
The less light per individual, the less photosynthesis and
the less energy for growth and reproduction. The density of
coniferous trees influences the amount of seed production.
The amount of seed production, in turn, influences the
population density of the various seed predators.

This example illustrates both interference and exploitive
competition and also provides examples of both intra- and
interspecific competition. It is by no means the most
complex system that can be cited, but it is sufficiently
complex to illustrate how competition helps shape biotic
communities.

Competition Results in Evolutionary Change and Distinct Ecological Responses

Over time, organisms that compete for resources may
evolve adaptations that avoid competition. If two species
require the same resources, they will spend considerable
energy competing directly for the resource, or they will
both exploit the resource to the detriment of the other. In
other words, neither population will grow to its full potential. Obviously, natural selection would favor any adaptation that moved one species out of competition with another and into a lifestyle based on other resources. We see
this tendency clearly in the phenomenon of **character
displacement.**

Character Displacement. Natural selection favors individuals whose resource demands differ from those of the
coexisting species. Generally speaking, such individuals
spend less energy on resource acquisition and hence have
more energy for reproduction.

Morphological divergence, or **character displacement,**
is the process in which species evolve different resource
demands. It generally occurs between closely related, anatomically similar species. Character displacement can be
seen in species whose ranges overlap. When this overlapping occurs, populations are more divergent. When the
ranges of the species do not overlap, the organisms are
quite similar in appearance.

This phenomenon occurs in Eurasian nuthatches. The
rock nuthatch is a little bird that ranges from Yugoslavia
eastward to Iran. The Persian nuthatch ranges from Turke-
stan westward to Armenia (▷ Figure 30–5). In Iran, where
the two species occur together, they are quite distinctive,
differing markedly in bill length and facial pattern. But in
places where their ranges do not overlap—that is, where
they do not occur together—only the most experienced
student of nuthatches can distinguish individuals of the two
species.

Many other examples of character displacement are
available. From such examples we learn something about
the effect of symbiosis (competition) on the evolution of
diversity.

Habitat Segregation. Character displacement is an evolutionary product of competition. An ecological product of
competition may be **competitive exclusion,** where one
population outcompetes another and causes it to become
extinct, as discussed above in Gause's study of paramecia.

Competitive exclusion is fairly easy to demonstrate in
the laboratory. In two-species systems consisting of protists, beetles, or fruit flies, one species eventually outcompetes the other and drives it to local extinction. Such
situations are rarely observed, however, because competitive exclusion is a transitory process. We are much more
likely to see the product of competitive exclusion—**habitat
segregation,** or the separation of geographic ranges—than
the actual process of exclusion.

Habitat segregation is readily seen in the field. Several
kinds of white-footed mice (genus *Peromyscus*) live on the
Colorado Plateau (▷ Figure 30–6). One species, pinyon
mice, lives in woodlands of junipers and pinyon; another,
deer mice, lives in more open habitat, such as grasslands. If
pinyon mice are removed from their habitat, though, deer
mice tend to move into it. The usual ecological segregation
of these two species suggests a history of competition,
although competition (or any other interaction) is mostly
avoided today by differential habitat selection.

It may come as a surprise, but competition is often more
intense among plants than among animals. Plants are obviously rooted in place and cannot move to avoid interac-

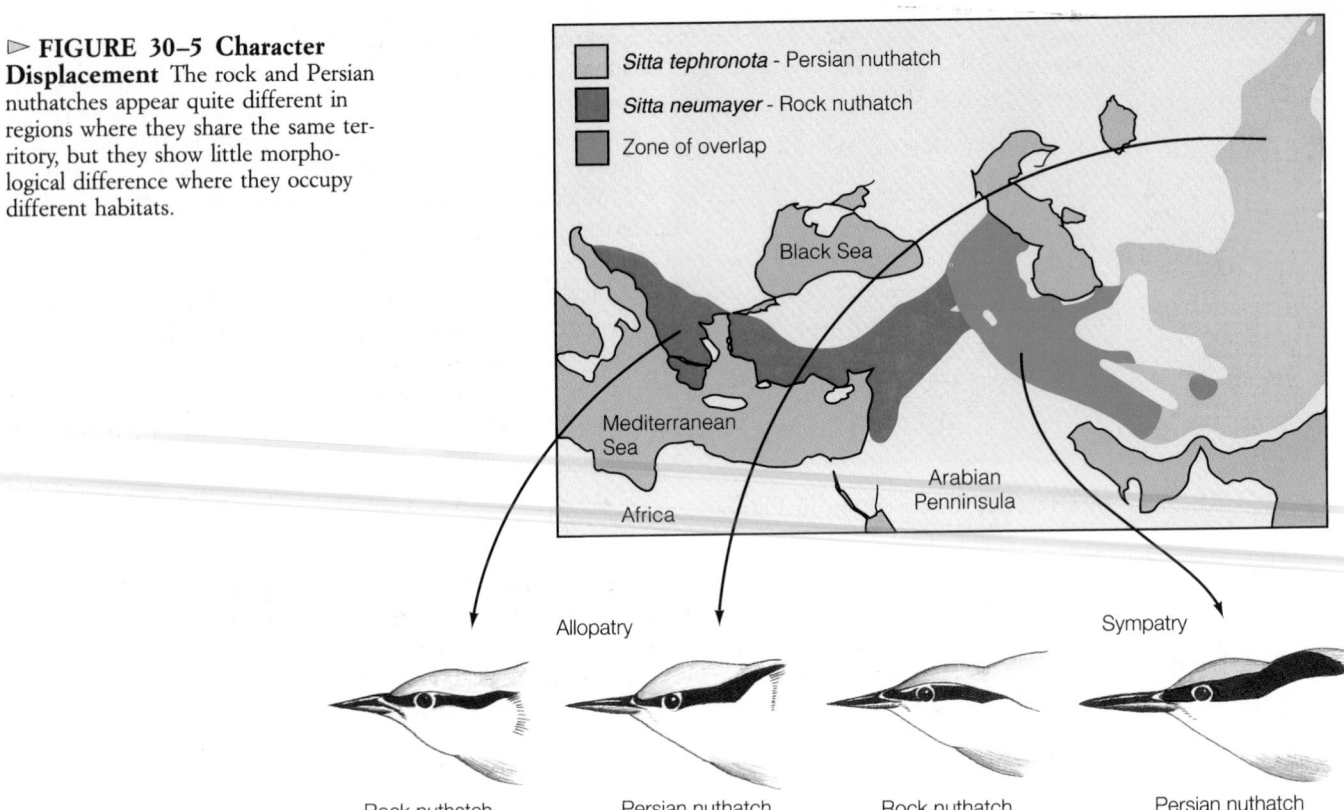

▷ **FIGURE 30–5 Character Displacement** The rock and Persian nuthatches appear quite different in regions where they share the same territory, but they show little morphological difference where they occupy different habitats.

Sitta tephronota - Persian nuthatch
Sitta neumayer - Rock nuthatch
Zone of overlap

Black Sea

Mediterranean Sea

Africa

Arabian Penninsula

Allopatry

Rock nuthatch   Persian nuthatch

Sympatry

Rock nuthatch   Persian nuthatch

tion. Thus, plants need to be spaced a certain distance apart to allow each individual to acquire minimal amounts of minerals, water, light, and the other resources essential for survival. In nature, direct competition is usually avoided by species partitioning the resource, either by growing their leaves to different heights in the canopy or by growing their roots into different depths in the soil.

Farmers have long been concerned about competition between weeds and crops. **Weeds** are plants that are fast-growing, early colonizers of severe habitats, such as a newly plowed field. Weed species tend to outcompete crop plants early in the growing season when resources are abundant. Although crop plants can coexist with weeds, their growth is slowed considerably as the weed population grows. Crop yield is also significantly lower.

The Ecological Niche

Studies of the tangled web of symbiotic relationships in a biotic community show that all species evolved particular roles. This species-specific role is termed its **ecological niche**. The niche is the "ecological space" occupied by a species population, the sum of its resource needs. If a species' habitat is its residence, its niche is its "profession" or "lifestyle," in all its myriad aspects.

The role of niches in community organization is described in a classic field study of warblers by the late Robert MacArthur. He found that birds avoided direct competition by parceling out the habitat and feeding in different ways within their own subdivision, thereby minimizing overlap. For example, five different species of warblers coexist in the coniferous forests of northeastern United States, each feeding in different locations in the trees (▷ Figure 30–7). Although these five species live in the same spruce trees and all eat budworms, they forage in different parts of the tree, rarely coming into direct contact with one another. Small differences in feeding behavior and slightly different reproductive and nesting habits also help minimize overlap. Their ecological niches are sufficiently different to allow coexistence. The concept of the niche is developed more fully in Chapter 31.

≈ PREDATION AND PARASITISM

Predation and **parasitism** are variants on a common theme, forms of interaction that are a benefit to one kind of organism and a detriment to another. That is, they represent what is commonly known as a "win-lose" situation. The two terms are generally used in reference to feeding relationships.

(a)

(b)

(c)

▷ **FIGURE 30–6 Habitat Segregation** (*a*) Pinyon mice live in (*b*) woodlands of juniper and pinyon on the Colorado Plateau. (*c*) Deer mice live in more open habitat, such as grasslands. If pinyon mice are removed from their habitat, deer mice move in. The segregation of these two species suggests a history of competition.

▷ **FIGURE 30–7 Niche Segregation** Five species of warblers coexist in the coniferous forests of the northeastern United States, feeding at different strata to reduce competition.

Cape May warbler

Blackburnian warbler

Black-throated green warbler

Bay-breasted warbler

Myrtle warbler

Some authors distinguish them by noting that if the individual feeding on the other is larger than the food species, the interaction is predation. If the one feeding on the other is smaller than its food, the relationship is parasitism. This distinction misses an important point. Predators kill their prey; parasites normally do not.

An example will help clarify the distinctions between these two relationships. The cheetah is a predator that chases down and kills small antelope (▷ Figure 30–8). Inside the average antelope lives a variety of parasites: muscle worms, lung worms, liver flukes, and bloodworms; outside live fleas and mites. The cheetah serves as host to a parasitic menagerie of its own. These parasites normally occur at populations low enough that they do not adversely influence the health of the host.

In this example, a healthy antelope, alert and fast enough to escape the cheetah, is a wonderful provider for its many parasites. If the parasites adversely affect the health of the antelope host, however, it may be slowed or weakened. Easy prey for the cheetah, the antelope will be killed, and the parasites may end up being eaten along with the antelope.

An analogy may help further clarify the distinction between predation and parasitism. Predation is similar to an armed robbery; parasitism is akin to embezzlement. Robbery is a one-time event; the robber cleans out the safe. The wise embezzler—in distinct contrast—is patient, skimming just enough to meet his or her needs without crippling the "host" enterprise.

Of course, this is only an analogy; do not take it literally. In science, one must avoid value judgments about predators and parasites. Robbery and embezzlement are both crimes. Neither predation nor parasitism is a crime. Rather, both are merely ways of making a living in ecosystems. Virtually all consumer organisms (heterotrophs) are either predators or parasites. Many species (including us humans) are both. Even scavengers, organisms such as the hyena, that feed on dead organisms, prey on others.

Predators are often viewed with contempt. Whether this disdain is cultural (based on such lore as *The Three Little Pigs*) or genetic (based on our species' long evolutionary history as potential prey), most of us tend to cheer for the rabbit and not for the hawk. Yet the hawk is only feeding itself and its young as it is programmed genetically to do.

Although parasites are generally much less harmful than predators, they can grow in excess and cause serious troubles for the host. When numbers exceed the host's ability to support them, a parasitic disease results. These diseases can lead to crippling deformity or even death.

Animals must feed on other species to survive. Hence, animals have evolved a variety of behaviors that allow them to find and eat their food. Some animals are **specialists,** eating only limited types of food; others are **generalists,** eating a broader variety of items.[1] One of the most specialized animals is the koala of Australia, which feeds only on leaves of some species of eucalyptus trees (▷ Figure 30–9). The giant panda of China is another highly specialized animal, which feeds only on bamboo. Vampire bats eat only the blood of vertebrates and among the vampires, there is further specialization, with one species feeding on mammals, and two species feeding mostly on birds. Many herbivores are generalists in one sense, eating many different plant species, but in another sense they are specialists, feeding only on particular parts of plants. Pocket gophers specialize on roots and tubers, rabbits and deer are browsers (eating twigs and branches), elk are mostly grazers (eating grasses), and beavers eat mostly bark. Some herbivores change their diet with the season. For example, porcupines eat mostly the inner bark of trees in winter but eat green leaves and other softer plant tissues during the growing season.

Insectivorous bats are among the most fascinating of predators. As noted in Chapter 18, these bats emit ultrasonic pulses that allow them to locate flying insects. Bats fly toward their prey until they are very close. They then flick at the insect with the tip of a wing to knock it off balance, scoop it up in the flap of skin between their hind legs, and fling the morsel into their mouth, sometimes neatly clipping the wings off of the prey as it passes into the mouth. Bats have been trained to locate and capture mealworms thrown by researchers. The bats avoid similar inanimate objects thrown in the air, even if the objects give back similar echoes. Bats must, therefore, be able to make very subtle distinctions in ways that still are not quite understood.

▷ **FIGURE 30–8 Predator and Parasites** The cheetah is a predator on small antelope. Inside the antelope is a variety of parasites.

[1] The terms *specialist* and *generalist* also refer to other aspects of their niche, including habitat.

(a)

(b)

▷ **FIGURE 30–9 Two Specialized Mammalian Herbivores** (*a*) The koala feeds only on leaves of some species of eucalyptus. (*b*) The giant panda is highly specialized to feed on bamboo.

Prey Species Avoid Predation through Active and Passive Means

In a biotic community, the predatory behavior of one species elicits self-preservation behavior on the part of potential prey. Predator-prey relationships are, therefore, a perpetual arena of evolutionary interactions. Prey species have evolved two basic means by which to avoid predation: passive avoidance and active defense (including prudent retreat). Many species display combinations of the two.

Passive Predator Avoidance. Perhaps the most effective way in which prey have adapted to the possibility of predation is by becoming difficult to find. **Camouflage** is one form of passive avoidance. Camouflage is a color or color pattern that makes a species blend with its surroundings (▷ Figure 30–10).

The effectiveness of an animal's color or color pattern depends on the predator's sensory mechanisms and lifestyle. In many ways, then, predators drive the evolution of camouflage. Most birds, for example, are active during the day and have acute color vision. Animals on which birds prey are often color-matched to their background. Many grasshoppers are grass- or soil-colored and are therefore passively concealed from birds.

Animals whose principal predators are color-blind generally display cryptic (concealing) color patterns. Cottontail rabbits, for example, are active at dawn and dusk. Their dappled, brownish coats conceal them in their brushy habitats. Movement will reveal them, but not their color.

Numerous other examples of passive concealment exist in the animal kingdom. A few examples will illustrate the variety. Animals living in snow-covered habitats often have white fur that makes them difficult to see. Insects that live on leaves often blend into the foliage. Stick caterpillars resemble twigs and when motionless are almost impossible to detect. Some species of insects look like bird droppings on leaves.

Passive concealment is also widespread in plants. Many desert plants, for example, especially succulents, resemble rocks on the desert floor.

Structural modes of concealment are often complemented by specific behavioral patterns. A twig-shaped caterpillar, for example, must orient properly to really look like a leaf (▷ Figure 30–11).

Vertebrate predators tend to be highly intelligent and not easily fooled. Predators, such as these, become selective agents for prey, in a continual game of coevolutionary tag.

Concealment works on both sides of the predator-prey interaction. That is to say, predators are often as well camouflaged as prey. The lines of a tiger, for example, camouflage the animal in tall grass. Ermines (a type of weasel) and snowy owls, with white coloration, disappear against a snow-white horizon, especially at a distance.

▷ **FIGURE 30–10 Camouflage** Camouflage renders an organism difficult to locate because the organism blends in with its immediate surroundings. Could you find a moth meal in this environment?

▷ **FIGURE 30–11 Behavior Complements Structure**

▷ **FIGURE 30–13 Batesian Mimicry** Some nonpoisonous flies, like the one shown here, resemble the poisonous yellowjackets. This evolutionary adaptation provides an enormous benefit to the flies.

Many potential prey are brightly colored and extremely obvious in their surroundings. What advantage could such a blatant display of colors provide? Biologists believe that the bright coloration of these species is a means of warning potential predators. That is, it is another passive means of avoiding predation. Brightly colored coats, or **warning coloration,** must, of course, be backed up in some way with an active antipredator defense, often a toxin (▷ Figure 30–12). More discussion on this topic follows soon.

A number of prey species have evolved color patterns similar to those of species that possess effective active defense mechanisms against predators. Prey species, for instance, may evolve bright color patterns similar to those found in animals that contain toxins. This strategy, in which one species evolves a warning coloration and benefits from its outward similarity, is referred to as *Batesian mimicry,* after Henry Walter Bates, who first described this evolutionary phenomenon in the 1860s. A Batesian mimic is a

▷ **FIGURE 30–12 Predators, Beware** Warning coloration, such as that shown here in the brightly colored poison-arrow frog, often indicates the presence of a toxin.

species that produces some form of warning but does not back it up with overt defense. Batesian mimics benefit from the fact that predators avoid them because they resemble animals that predators associate with a previous bad experience. Some nonpoisonous flies, for example, resemble poisonous yellowjackets (▷ Figure 30–13). The flies have a distinctive yellow and black band on their back, but they produce no toxin like that of the yellowjacket. Although their coloration is a kind of "false advertisement," it is enough to protect them from experienced predators.

The classic example of a Batesian mimic is the viceroy butterfly. Naturalists once believed that the viceroy evolved to resemble the poisonous monarch butterfly, which protected it from birds. However, as noted in the section on Exercising Your Critical Thinking Skills in Chapter 23, this "classic" example may be an example of a hasty conclusion.

Mimicry as a defense has obvious benefits for a species. Mimicking other species in the environment, is fine, however, only as long as potential predators remain fooled. The strategy will fail if there are too many individuals of the mimicking species or if there are too few of the "teaching" model (the poisonous counterpart). It will also fail if there is an increase in the discriminatory ability of predators. Of course, the continuing evolutionary give and take between predator and prey may lead over time to ever-more-elaborate means of deception and to predators that are ever more difficult to confuse. That is the nature of coevolution.

Mullerian mimics, named after Fritz Muller, who was the first to record the phenomenon, occurs when different species, each toxic to some extent, evolve similar color patterns. Mullerian mimicry is seen among poisonous coral snakes (▷ Figure 30–14). By resembling one another, they take advantage of a potential predator's previous negative experiences with any of the species involved. The harmless milk snake may be a Batesian mimic of the coral snakes (Figure 30–14b).

(a)

(b)

(c)

▷ **FIGURE 30–14 Mullerian and Batesian Mimicry** (*a*) Coral snakes, shown in (*a*) and (*b*), are highly poisonous, and exhibit a prominent warning coloration. (*c*) The milk snake may be a Batesian mimic of the coral snakes.

Distinctions between these two kinds of mimicry are not absolute, and distinguishing one kind of mimicry from another may be difficult.

Active Defense. As mentioned above, animals or plants that contain poisons advertise their bad taste with bright colors or striking color patterns. Experienced predators can easily recognize such species and avoid them. Warning a predator is a defense against being eaten but also against

▷ **FIGURE 30-15 Active Defense** A male baboon displaying his longs fangs and growling may frighten away even the hungriest leopard.

being injured. Rattlesnakes, for instance, do not rattle as predators; they rattle as prey. They rattle to warn other animals that might step on them of their presence. The toxin they contain is offensive, not defensive. For a venomous snake to waste energy striking an animal that it cannot swallow makes no sense. Rattlesnakes are silent when they hunt.

Defensive mechanisms are sometimes more effective in one situation than another. Skunks, for example, have anal scent glands that emit a powerfully pungent odor that warns others of their presence. Their flash of white fur, moreover, makes skunks recognizable even at night. The skunk's distinctive scent and color pattern are a powerful defensive combination backed up by the animal's spray. The skunk's protective mechanisms work only with predators that are capable of learning—that is, capable of changing their behavior as a result of experience.

Unfortunately for the skunk, its defensive arsenal is not effective against all enemies. Skunks are frequent highway casualties, far out of proportion to their actual population size. When confronted with a potential predator, the skunk stands its ground, releasing its spray, thus teaching or reminding a would-be predator. Although this is a fine strategy for defense against wily coyotes, it is not at all adaptive when faced with a speeding dump truck.

Active physical defense is another strategy exhibited by some prey. Being large and frightening can be effective; even the mere appearance of being fierce can deter a predator. A group of male baboons displaying their long fangs and growling may startle even the hungriest leopard (▷ Figure 30–15). Some moths have patterns on their wings that look like the eyes of much larger animals; these patterns may fool or frighten would-be predators.

When an animal cannot defend itself any longer, escape may be the only option. Animals have evolved elaborate means to escape predators. Cornered animals may confuse predators at the last moment by producing an unexpected noise. Some beetles spray chemicals at predators, distracting them long enough to make a getaway. The cryptic grass-green grasshopper may flash brilliant red and orange underwings as it takes flight, startling a predator for the brief moment needed for escape.

Distraction is seldom sufficient; prey must be capable of fleeing danger. Some adaptations that permit animals to escape are quite remarkable. The tails of kangaroo rats are not mere decoration. (Given the nature of the evolutionary process, few things in "real life" are.) Pursued by an owl, the rat moves in graceful bounds across the desert, using the tail as a balancing organ. Further, during a prodigious leap, the tail can be used as a rudder, allowing a sudden, midflight change in course. Finally, the tail is tipped with a conspicuous black flag. That is an obvious target for the owl, but it is expendable. Occasionally, rats escape by sacrificing the tip of the tail, but they are otherwise intact, ready to feed (and breed) another day.

The pronghorn is the fastest North American mammal, capable of speeds over 60 miles per hour (▷ Figure 30–16). Pronghorn live and have lived in the Great Plains and the Great Basin. They are herbivores and prey to other species—notably, coyotes and, at one time, gray wolves. Their great speed, however, must have to do with predator avoidance, because their own "prey"—sagebrush and other vegetation—is surely not going anywhere fast. Interestingly, none of the animal's predators can approach sustained speeds of 60 miles per hour. How does seeming excess evolve? One hypothesis is that this speed evolved under conditions different from those found today. The fossil record shows that only a few thousand years ago, the

pronghorn's environment included not only native horses and camels but a 60-mph carnivore: a North American cheetah, which is now extinct.

Populations of Predators and their Prey Fluctuate Together, Sometimes in Close Unison and at Other Times in a Delayed Manner

Predators can influence the size of their prey populations. As prey populations grow, predator populations may expand thanks to the surplus of prey. However, as prey populations decline, predators—especially those that feed on one or a few prey species—may reduce the population size of prey to low levels before the predators face food shortages. This lag effect can produce large swings in populations of predators and prey.

In some cases, predator-prey relationships are stable over time. The cyclamen mite, for instance, invades strawberry fields soon after planting. These mites feast on the crop throughout the growing season and can become significant pests, especially during the second year.

A predatory carnivorous mite species, however, also colonizes the strawberry fields, feeding on the cyclamen mites. As the prey population grows, the predatory mite population increases. The population of the herbivorous cyclamen mite is eventually kept in check by the predator. The two mite species tend to coexist in strawberry fields in approximately equal numbers, and their population size remains fairly constant. In fields where predatory mites are absent, however, the cyclamen mite population is about 20 times greater than it is in fields where predators are present.

The synchronous population cycles observed in some predator-prey interactions, such as those between these mites, result largely from the ability of both predator and prey to take advantage of a food resource as it becomes available. Both species have the potential for rapid population growth. This ability leads to a close match between the growth rate of the prey population and the eventual expansion of the predator population (Figure 30–17a). The high reproductive potential of the predator allows it to benefit quickly from an increase in its food supply, so the two populations are closely linked.

Sometimes the population growth rate of predator populations is slower than that of their prey, because predators are often larger and have longer life spans. This produces predator-prey cycles that are not closely synchronized. Rather, growth of the predator population lags behind that of the prey (Figure 30–17b).

≈ PARASITISM

As noted earlier, the distinguishing characteristic of parasites is that they feed on other organisms, but they do not usually kill them. By definition, **parasitism** is a "win-lose" situation (see Table 30–1). However, the host's "loss" is

▷ **FIGURE 30–16 Speed, an Evolutionary Legacy** The pronghorn is the fastest of North American mammals, seemingly faster than it needs to be for escape on today's Great Plains. One hypothesis is that high speed evolved in the presence of the North American cheetah.

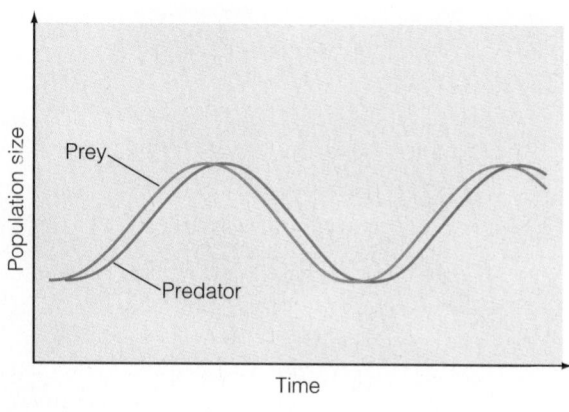

(a)

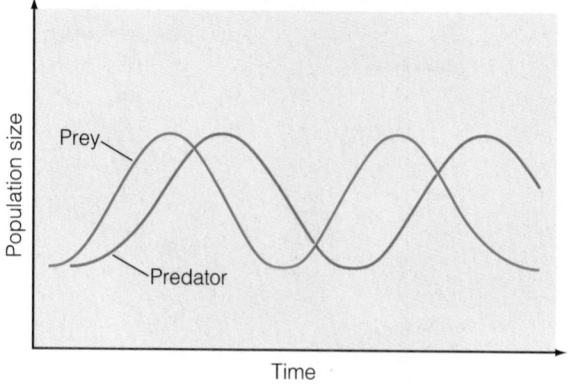

(b)

▷ **FIGURE 30–17 Population Cycles** (*a*) Synchronous cycle showing that populations increase and decrease nearly simultaneously as explained in the text. (*b*) Asynchronous cycle showing that predator population size lags behind prey population.

generally minimal, because parasites have evolved strategies to keep the host alive and well.

Unknown to many, parasitic organisms are found in each of the five kingdoms. Why is parasitism so prevalent? Thinking about parasitism forces us to look beyond our prejudices and our own adaptations and to think about what the evolutionary process really is and is not. Strange as it may seem to many human beings, evolution is not a process of getting "bigger and better." It is not a process of change that leads toward perfection, which is impossible in the continually changing environments of the biosphere. Rather, evolution is a process driven by differential reproduction. Success is measured as individual ability to transmit genes into the gene pool of succeeding generations. To do that may not mean getting bigger, faster, or fancier. It may mean getting simpler.

Parasites exemplify this alternative road to evolutionary success. They depend on other organisms for resources, defense, movement, and many other necessities of life while they concentrate on reproduction.

Many bacteria, protists, and fungi are parasites. In addition, viruses—which are below the level of life in the definition of most biologists—may also be considered parasites (Chapter 24). Indeed, being essentially nothing but genetic material—the very essence of the reproductive and evolutionary processes—viruses may well be the quintessential parasite. As you will recall from Chapter 24, viruses reproduce inside the cells of other organisms, and in some cases they take over the metabolic machinery of their hosts. Outside the host, they are dormant. Most parasites, however, do not take over the biological function of their hosts in the same way.

Microbial parasites include the familiar agents of human disease, the "staph" and "strep" bacteria, amoebae that cause dysentery, and fungi that cause athlete's foot. These agents can cause infection as the host's immune system reacts to their presence. Nonetheless, many microbial parasites can live at very low levels in our bodies (and the bodies of other organisms) without causing any obvious symptoms, because their numbers remain very low. A change in the environment (in this case, the body of the host) can lead to a population explosion of the parasite and thus to obvious disease.

Although plants are mostly photosynthetic, a few species have become parasites, depending on other plants to one degree or another. The Indian paintbrush is photosynthetic and produces its own food, but it also parasitizes the roots of neighboring plants from which it derives some of its water. Dwarf mistletoe is a flowering plant with no photosynthetic pigments (▷ Figure 30–18). It germinates on the branches of conifers, such as the ponderosa pine, and its roots penetrate the bark to remove resources from the host.

Among animals, parasitism has arisen in a number of phyla, but the most familiar parasites are in three phyla: the flatworms, roundworms, and arthropods. Flukes and tapeworms are examples of parasitic flatworms. Many kinds of flatworm parasites are well-known agents of human disease. In healthy hosts, moderate populations of tapeworms and flukes can persist unnoticed for years. In hosts debilitated by malnutrition or other disease, however, the parasites may cause death.

One well-known flatworm parasite is the blood fluke, which causes a debilitating disease known as schistosomiasis that presently afflicts hundreds of millions of people in the Third World. Female flukes inside adult humans release eggs that evoke a series of immune responses that can lead to serious damage to the liver, kidneys, heart, and urinary bladder.

Many flatworms have complex life cycles involving hosts

▷ **FIGURE 30–18 Dwarf Mistletoe** This flowering plant is a parasite that grows on the ponderosa pine.

▷ **FIGURE 30–19 Plant Parasite** Aphids are familiar ectoparasites of plants. They suck the juices out of their prey.

of different species at different stages. In some blood flukes, eggs pass from humans to fresh water in feces. In the river, the eggs hatch and form a free-swimming stage that enters snails, their intermediate host. Leaving the snail as another distinctive stage, the flukes then penetrate the skin of humans who are swimming or bathing in the water. The greater the population density and the more concentrated the use of the stream, the greater the incidence of infection will be.

Many nematodes (roundworms) parasitize plants or animals. Among nematode parasites of humans is the filaria worm, which lives in organs such as lymph nodes. In severe cases, the worms block lymphatic drainage and cause the grotesque and debilitating swelling known as **elephantiasis** (Figure 11–23). This disease is spread by certain biting flies. Obviously, such an infestation can incapacitate the host.

Parasites that habitually debilitate or kill the host have reached an evolutionary dead end. The evolution of measures that control population numbers (perhaps some kind of interference competition) or the strategies that permit the parasite to infect other hosts are necessary for its continuation.

We have considered parasites that live inside their hosts, which are known as **endoparasites.** Other parasites live on the surface of the host and are, therefore, called **ectoparasites.** Many ectoparasites are arthropods. A good example is the mosquito, which lives on body fluids from humans and other animals, taking energy and other food resources without killing the host. Similarly, the ticks and fleas on your dog and the lice on your canary are ectoparasites. Many kinds of mites are ectoparasitic; the larval mites called chiggers are a familiar example if you live in a warm, humid environment. Aphids and scale insects are familiar ectoparasites of plants (▷ Figure 30–19).

Just as not all insects are parasites, neither are all mites. Indeed, most mites are nonparasitic, free-living organisms. Some of the free-living forms, however, may be found on larger organisms, where they eat bacteria or bits of dead skin or other waste materials.

Some insects are referred to as **parasitoids.** Unlike true parasites, parasitoids kill their hosts. Many wasps are parasitoids, laying their eggs on or in larval or pupal stages of their hosts. The eggs hatch and devour their hosts. Generation times of parasitoids are generally very short, and their reproductive output is enormous. Because they kill the host, parasitoids are really predators. Unlike the most familiar predators, they occur in numbers much higher than those of their hosts, and they seem to have a much greater ecological impact on host populations than typical predators have on prey.

Parasitoids are increasingly used in agriculture as biological agents to control insect pests. They can decrease populations of target insects faster than predators or true parasites because they are usually much more specialized to their hosts.

The term *parasitism* is also used in connection with the nesting behavior of some species of birds. Brown-headed cowbirds (▷ Figure 30–20), for instance, are referred to as **nest parasites.** Cowbirds lay their eggs in the nests of several other species. The unsuspecting surrogate parents incubate the eggs, then feed the young cowbirds. In some cases, the cowbird nestlings outcompete the parents' real offspring.

Although the young cowbirds do not feed directly on their host, this relationship is considered a form of parasitism because the nestlings take resources from their host without killing them. Like other parasitic behavior, to persist it must be finely tuned to the resource base—in this case, populations of nesting host species. Overexploitation could lead to the demise of both host and parasite species. Scientists have observed overexploitation of some species in some regions, especially where nesting birds (hosts) are already under pressure from other environmental changes,

▷ **FIGURE 30–20 Nest Parasite** The brown-headed cowbird lays its eggs in the nests of a number of other species of songbirds. In this nest, the two larger eggs belong to a brown-headed cowbird. The three smaller eggs belong to the nest holder, a pair of vesper sparrows.

such as pesticides or destruction of nesting habitats by human development.

While we are expanding our concept of parasitism, consider the relationship between a mammalian mother and her offspring. The developing embryo is a dependent individual, genetically distinct from its mother. It derives nourishment from the mother through the placenta. This relationship is akin to that between many parasites and their hosts. After birth, the mammalian newborn becomes an ectoparasite, living off resources—in the form of milk—provided by the mother. This is a remarkable case of parasitism, seldom called by that name.

Although this relationship may be a form of parasitism, it makes extraordinary sense from an evolutionary standpoint, for it represents an interaction in which one individual provides protection and care for her genetic investment in the next generation.

≋ COMMENSALISM

A relationship in which one species benefits and the other is unaffected is a form of **commensalism.** Although it is easy to define, finding a good example is a little tricky, because actually demonstrating that one organism has no influence on another is difficult, if not impossible. An influence may be too small to be detectable or too small to matter, but it may still be there.

Large, hoofed animals often provide a temporary home for birds and insects that ride on their backs (▷ Figure 30–21a). The larger animal usually tolerates the presence of the smaller animals but apparently gets nothing in return, an example of a commensal relationship. On closer examination, though, one can find that the birds may feed on biting insects, which is of obvious benefit to both.

Many marine organisms attach themselves to mobile animals for transportation and protection. Limpets and barnacles, for example, adhere to larger organisms such as

(a)

(b)

(c)

▷ **FIGURE 30–21 Commensalism** (*a*) Cattle egrets follow elephants, cattle, and other ungulates around as they graze, often hitching a ride on an animal's back. The egrets eat insects that the animals frighten. The grazers may not benefit at all from this relationship. (*b*) Red crab covered by commensal kelp scallops. The scallops receive a ready supply of food scraps from their host. (*c*) Orchids are epiphytes that derive several benefits from their life among the trees.

sponges and fishes, gaining some protection (Figure 30–21b). Commensal relationships are common in tropical reef fishes, some of which live in close relationship with much larger fishes, or other marine organisms, living off scraps of food rejected by their hosts.

Commensal relationships are found in the plant kingdom as well. Orchids, for instance, are known as **epiphytes** (*epi* = "upon," *phyt* = "plant"). Found in tropical rain forests, orchids grow on the branches or bark of trees (Figure 30–21c). In most instances, epiphytes seem to have no effect on the host; that is, the survival, growth, and reproduction of the host are not affected. The orchid, however, benefits by the relationship, growing high in the forest canopy where it receives more light than it would on the ground. It has, in effect, the advantages of being a tree with little or none of the expense. Spanish moss (which is neither Spanish nor moss but, rather, a New World relative of the pineapple) is an epiphyte of the wet forests of the South—Alabama, South Carolina, and Georgia, for instance.

≈ AMENSALISM

As Table 30–1 shows, the next relationship, known as **amensalism,** is a "−/0" relationship. That is, it has a negative impact on one organism and no impact on the other. A good example of amensalism is the relationship between the mold *Penicillium* and bacteria. The mold secretes a substance (penicillin) that destroys the cell walls of bacteria, but the mold seems to get nothing in return.

An African proverb says, "When the elephants mate, the grass trembles." Elephants thrashing through the bush have a detrimental effect on numerous other species with no particular gain to themselves. This is an amensal relationship. So, too, are many human-environment interactions. Often, we thoughtlessly perform "just for fun" activities that have little or no real benefit to us biologically or culturally yet have untoward effects on other species. Driving an off-road vehicle in the California desert, for example, exemplifies amensalism (▷ Figure 30–22).

≈ NEUTRALISM

Neutralism is the final type of "symbiosis" shown in Table 30–1. Neutralism is a kind of biological independence, a relationship in which neither organism interacts with the other. Consequently, one could argue that this relationship is not really symbiosis—life together—but life apart.

Examples of neutralism are hard to come by and impossible to prove. Surely, things would be much simpler if we ignored neutralism altogether. But including this concept draws our attention to a very important point about biotic communities. One cannot assert positively that there is no relation whatsoever between two organisms. The symbiotic interaction between a fruit-eating bat and an earthworm, for example, may be highly indirect, but it probably is there. Careful study may reveal a subtle, indirect relationship.

ENVIRONMENT AND HEALTH: ACID RAIN, PLANTS, AND ANIMALS

Acid deposition, rain and snow containing acids from pollutants in the atmosphere, is a product of the modern fossil-fuel based economy. Two gases, sulfur and nitrogen dioxides, released from such sources as automobiles and power plants, are converted into sulfuric and nitric acid in the atmosphere. These acids eventually deposit onto land and bodies of water with sometimes devastating effects. A study of the effects of acid shows the connection between two prominent kingdoms, plants and animals, and shows how pollution can interrupt one of the most fundamental community interactions (animal predation on plants) on Earth.

Cadmium is a toxic metal that accumulates in certain plants living in regions exposed to acid rain and snow. In a study of deer and moose in Sweden, the concentrations of cadmium in body tissues was found to be age-dependent, showing this toxic metal accumulates in the plant-eating animals also. To date, no wildlife deaths have been directly attributed to high concentrations of metals in body tissues. However, sublethal concentrations have appeared to affect animal behavior and performance. Some scientists think that it may be only a matter of time before lethal concentrations are encountered.

Acid rain may also affect land mammals by reducing habitat or food resources. For example, the large caribou herds in northern Canada are dependent on lichens. Lichens are pH-sensitive, and lichen communities may be damaged by acid rain. If so, the caribou herds may suffer population declines.

Habitat resources are also affected by acid rain. Forest

▷ **FIGURE 30–22 Amensalism** The actions of some species harm others but apparently offer no biological advantage to the perpetrator. Off-road vehicles damage public lands in the Mojave Desert.

die-back due to acidic conditions may have profound impacts on wildlife populations. However, ecosystems are extremely complicated, and predicting the loss of wildlife as a result of habitat changes is extremely difficult.

Essential nutritional elements may be removed from food sources by acid rain and snow. Selenium is a trace element that is necessary for the normal health of all vertebrates. Plants absorb selenium from the soil, but sulfuric acid in acid rain lowers the pH, which makes selenium less soluble and decreases the plant's uptake. Consequently, herbivores which acquire selenium from plants, may suffer from deficiencies. A number of diseases may result. One disease is known as white muscle disease that is well known in livestock. The incidence of such diseases is increasing worldwide. Widespread effects on wildlife populations in areas with acid rainfall and low selenium concentration in plants is possible.

The effects of acids from rain and snow on terrestrial mammals are indirect, through their food supply—their prey. As you will learn in later chapters, alterations in the base of the food chain, the plants, can have severe repercussions on the web of life.

SUMMARY

SYMBIOSIS

1. Species with similar requirements and geographic distribution tend to occur together in communities and to interact, directly or indirectly, in many ways.
2. A biotic interaction within a community is generally referred to as symbiosis.
3. Symbiotic species often affect each other's evolutionary development through coevolution.

MUTUALISM

4. Mutualism is a symbiotic relationship in which both species benefit, as measured by increases in population size or rate of energy flow.
5. Mutualistic relationships are numerous and include pollination systems, lichens, and mycorrhizal associations.

COMPETITION

6. Competition occurs when two or more individuals (or populations of individuals) vie for a limited resource.
7. In exploitation competition, both competitors have access to a resource and compete directly for it.
8. Interference competition occurs when one organism prevents another from gaining access to a resource. Within a given species, this activity often takes the form of territorial behavior.
9. Competition often leads to evolutionary changes that reduce competition, such as character displacement. It may also lead to habitat segregation.
10. Competition in plants may be more intense than that in animals. Because plants cannot move, some species avoid competition by releasing chemicals into the soil that prevent the growth or reduce the germination rates of seeds of potential competitors.
11. A species' ecological niche is its role in a community, or the sum of its resource demands—that is, the "ecological space" it occupies.

PREDATION AND PARASITISM

12. Predation and parasitism are fundamentally similar in that they benefit one species and harm another, creating a "win-lose" situation. Predators kill their prey; parasites normally do not.
13. Some heterotrophs are specialists, eating only limited types of food; others are generalists, eating a broader variety of items.

14. In the evolutionary history of a biotic community, the predatory behavior of one species tends to result in the evolution of self-preserving behavior in prey.
15. An effective protection against predation is to avoid the predator. Avoidance can involve activities such as burrowing or structural adaptations such as camouflage.
16. Warning coloration is usually backed up by an active defense.
17. Batesian mimicry occurs when a species (the mimic) without an active defense against predators evolves characteristics similar to another species (the model) that has an effective defense.
18. Mullerian mimicry occurs when several different species, all effectively defended to some extent, evolve a resemblance to one another.
19. Active defense may include toxins, strong scents, frightening or startling displays, standing one's ground, or fleeing.
20. Predators can affect the size of their prey populations; in some systems, prey populations may decline considerably because predator populations change more slowly than they do. In others, the predator and prey populations cycle more or less in synchrony.

COMMENSALISM

21. In a commensal relationship, one species benefits and a second is not affected, as in the relationship between epiphytic orchids and rain forest trees.

AMENSALISM

22. Amensalism is a relationship in which one species is influenced negatively while another is not affected.

NEUTRALISM

23. Neutralism is biological independence. In the complexity of biotic communities, neutralism is impossible to demonstrate. Influences may be minute, but in living systems, minute influences can make a difference.

ENVIRONMENT AND HEALTH
ACID RAIN, PLANTS, AND ANIMALS

24. Acid deposition, rain and snow containing acids from pollutants in the atmosphere, is a product of the modern fossil-fuel based economy.
25. Two gases, sulfur and nitrogen dioxides, released from such sources as automobiles and power plants, are converted into sulfuric and nitric acid in the atmosphere.

26. These acids deposit onto land and bodies of water with sometimes devastating effects on plants and animals.
27. Acids may mobilize toxic metals in the soil, which end up in plants and the animals that eat them. Sublethal concentrations may affect animal behavior and performance.
28. Acid rain may also affect land mammals by reducing habitat or food resources, such as lichen, which is vital to the large caribou herds in northern Canada.
29. Habitat resources are also affected by acid rain. Forest dieback due to acidic conditions may have profound impacts on wildlife populations.
30. Acid rain may also make some essential nutrition unavailable to plants, which affects those species that feed on the plants.

EXERCISING YOUR CRITICAL THINKING SKILLS

Many human actions are said to be amensal; that is, they negatively affect another species but have no noticeable biological and cultural impact on the perpetrator. One example is given in the text, off-road-vehicle use. Do you agree with the characterization of this act as amensal? Why or why not?

What criteria did you use to make your analysis? What critical thinking rules did you need to apply? When you looked at the issue a little closer, did you find that the chapter had oversimplified the matter?

TEST OF TERMS

1. The interaction of two organisms within a community is known as _____ .
2. _____ is the process in which two or more species become mutually important _____ forces in their evolution.
3. _____ is a symbiotic relationship in which both species "benefit."
4. A(n) _____ is a composite organism that is a mutualistic association between a(n) _____ and a photosynthetic bacterium or alga.
5. _____ occurs when two or more individuals (or populations of individuals) vie for a limited resource.
6. In _____ , both competitors have access to a resource, and they actually contest for it. By contrast, _____ competition occurs when one organism blocks another from getting a resource through such mechanisms as territorial behavior.
7. _____ _____ is evidenced by a pattern of minimal phenotypic difference when ranges of two closely related species do not overlap and greater phenotypic difference when their ranges do overlap.
8. _____ _____ of species suggests a history of competition that eliminates one species from shared habitat.
9. _____ are fast-growing, early colonizers of severe habitats.
10. _____ and _____ are relationships that benefit one kind of organism at the expense of another.
11. Some animals are known as dietary _____ , eating only limited kinds of food; others are known as _____ , eating a broader variety of foods.
12. _____ coloration, as seen in skunks, needs to be backed up with an active antipredator defense.
13. _____ mimics are prey species that have evolved coloration or color patterns similar to other organisms in the community that have an effective defense against predators.
14. _____ mimics are two or more species that are all toxic to some extent and have evolved similar coloration.
15. Schistosomiasis is caused by the blood fluke, a type of parasite known as a(n)_____ . The disease _____ is caused by a roundworm that blocks lymphatic drainage.
16. _____ , such as flukes, tapeworms, and many microbial parasites, live within the host; _____ , such as fleas, lice, and ticks, live on the exterior surface of the host.
17. Many wasps are _____ , laying their eggs on or in larval or pupal stages of their hosts and eventually killing them.
18. Brown-headed cowbirds are _____ parasites, laying their eggs in the nests of other species.
19. In a(n) _____ relationship, one species benefits and a second is not affected.
20. Orchids are _____ , growing as _____ on the branches or bark of tropical trees.

Answers to the Test of Terms are found in Appendix B.

1. In describing symbiosis, why is it valuable to insist on a biological definition of "benefit" that includes increased fitness and energy flow?

2. List the different kinds of symbiosis. Define each one, and then give an example.

3. Define the term *coevolution.* How does coevolution contribute to the survival of species?

4. In what ways does the biological definition of *weed* correspond with your understanding of the word? Can you think of some animal species that might be classified as weeds? Why?

5. Define the terms *exploitive competition* and *interference competition.* How are they different? How are they similar?

6. Define the term *character displacement,* and give an example.

7. Define the term *habitat segregation,* and give an example. How is it similar to character displacement, and how is it different?

8. Describe the ecological niche of some organism you know well, such as a houseplant, your goldfish, or your roommate. Because the ecological niche is a characteristic of a species, not of an individual, how likely is it that your description accurately portrays the species you selected? How large a sample of individuals would it take to make you comfortable with your description of the niche of the species you chose?

9. Coevolution is a strong positive force for generating biotic diversity. Given the fact that phenomena like mutualism and host-specific parasitism exist, the more species there are, the more there can be. Close symbiosis also presents risks that may reduce biotic diversity. Can you think of some?

10. Define Batesian and Mullerian mimicry. In what ways are they similar? In what ways are they different?

11. What must be the relative population sizes of a Batesian mimic and its model for the defense to work effectively?

12. Broadly speaking, the HIV—the virus that causes AIDS—is a parasite. Would you consider it a very poorly adapted parasite? Why? Will the reasons it is poorly adapted lead to its extinction? Why or why not?

13. Some vegetarians justify their potentially very healthy eating habits by asserting their unwillingness to kill other organisms. In other words, they elect life as plant and animal parasites rather than as animal predators. What does this choice mean? List foods that such confirmed plant parasites would eat and those that they would avoid. (A visit to a cafeteria or supermarket might stimulate your thinking.)

14. Explain how bird species that live in the same forest might avoid direct competition for food.

15. Why is a mammalian embryo or fetus a kind of endoparasite? Why is an infant an ectoparasite? How is the characterization of the human fetus and newborn as parasites misleading or dangerous?

Principles of Ecology: Understanding The Economy of Nature

≈

Satellite image of Europe and northern Africa.

ost of us live our lives seemingly apart from nature. We make our homes in cities and towns, surround ourselves with concrete and steel, and inadvertently drown out the sound of birds with our noise. The closest many of us get to nature is a romp with the family dog on the grass in our backyard.

Raymond Dasmann, a world-renowned ecologist, wrote that despite what many of us may think, a human apart from nature is an abstraction. No such being exists. Human life depends heavily on the environment. The clothes we wear, our morning coffee, and even the breakfast cereal we eat are all products of nature. So is the oxygen we breathe. In fact, one-half of the oxygen in the atmosphere is replenished annually by plants and algae. Without them, humans and other animal species could not survive. Trees, grasses, and other plants provide other free services. For example, plant life protects the watersheds near our homes, preventing flooding and erosion. Swamps help purify the water we drink. Birds control insect populations, and predators control rodent populations. Clearly, nature "serves" us well. Thus, although we may have isolated ourselves from nature, we have not emancipated ourselves. The ties that bind us cannot be broken.

This chapter examines **ecology,** the study of living organisms and the web of relationships that binds all of us together in nature. Ecology takes as its domain the entire living world. Ecologists study how organisms interact with one another and how they interact with the abiotic, or nonliving, components of the environment. Throughout this chapter, I will point out our connections to the living world and ways we can lessen our impact on living systems.

≈ PRINCIPLES OF ECOLOGY: ECOSYSTEM STRUCTURE

Ecology, like all disciplines in science, is a body of knowledge and a process of inquiry that seeks to understand the mysteries of nature. You will find that a great many of the lessons learned in the study of ecology can be applied to human society.

Ecology, however, probably ranks as one of the most misused words in the English language. Banners proclaim, "Save Our Ecology." Speakers argue that "our ecology is in danger," and others talk about the "ecological movement." These common uses of the word *ecology* are incorrect. Why? Ecology is a scientific field of inquiry. It is not synonymous with the word *environment.* It does not mean the web of interactions in the environment. We can save our ecology department and ecology textbooks, but we cannot save our ecology. Our ecology is not in danger, our environment is. You cannot join the ecology movement, but you would be a welcome addition to the environmental movement.

The Biosphere Is the Zone of Life

The science of ecology, unlike many other branches of scientific endeavor, often focuses on systems. The largest biological system is the **biosphere.** The biosphere can be thought of as the thin skin of life on the planet. As shown in ▷ Figure 31–1, the biosphere forms at the intersection of air, water, and land. All organisms consist of components derived from all three contributing spheres. The carbon atoms in protein come from carbon dioxide in the atmosphere, captured, of course, by plants. The minerals in your bones come from the soil in which plants grow. Water comes from streams and lakes.

The biosphere extends from the bottom of the ocean to the tops of the highest mountains. Although that may seem like a long way, it's not—at least in comparison with the size of the Earth. In fact, if the Earth were the size of an apple, the biosphere would be about the thickness of its skin. Although life exists throughout the biosphere, it is rare at the extremes, where conditions for survival are marginal.

The biosphere is a **closed system,** much like a sealed terrarium. By definition, a closed system receives no materials from the outside.[1] The only outside contribution is sunlight, which is vital to the health and well-being of virtually all life.

As noted previously in the book, sunlight powers almost all life on the planet. Even the energy released by the combustion of coal, oil, and natural gas, which we use to light our homes and run our factories, owes its origin to the sunlight that fell on the Earth several hundred million years ago.

Because the Earth is a closed system, all materials necessary for life must be recycled over and over. The carbon dioxide you exhale, for instance, may be used by a rice plant during photosynthesis next week in Indonesia. Those carbon dioxide molecules will be incorporated in carbohydrate produced by the plant and stored in the seed. Consumed by an Indonesian boy, the carbohydrate will be broken back down during cellular respiration, and the carbon dioxide molecules will be released once again into the atmosphere to continue its never-ending cycle. Without this and dozens of other recycling processes, all life on the planet would grind to a halt. Part of the reason for protecting the environment is to protect components of the global recycling systems on which we all depend.

The Biosphere Can Be Divided into Distinct Regions Called Biomes and Aquatic Life Zones

Viewed from outer space, the Earth resembles a giant jigsaw puzzle, consisting of large landmasses and vast expanses of ocean (▷ Figure 31–2). The landmasses, or continents, can be divided into large subregions or **biomes** (▷ Figure 31–3). A biome is a region characterized by a distinct climate and a characteristic assemblage of plants and animals adapted to it.

Chapter 32 describes the biomes in detail. This section

---

[1] Cosmic dust settles on the Earth, but virtually all of the materials necessary for life come from the Earth itself.

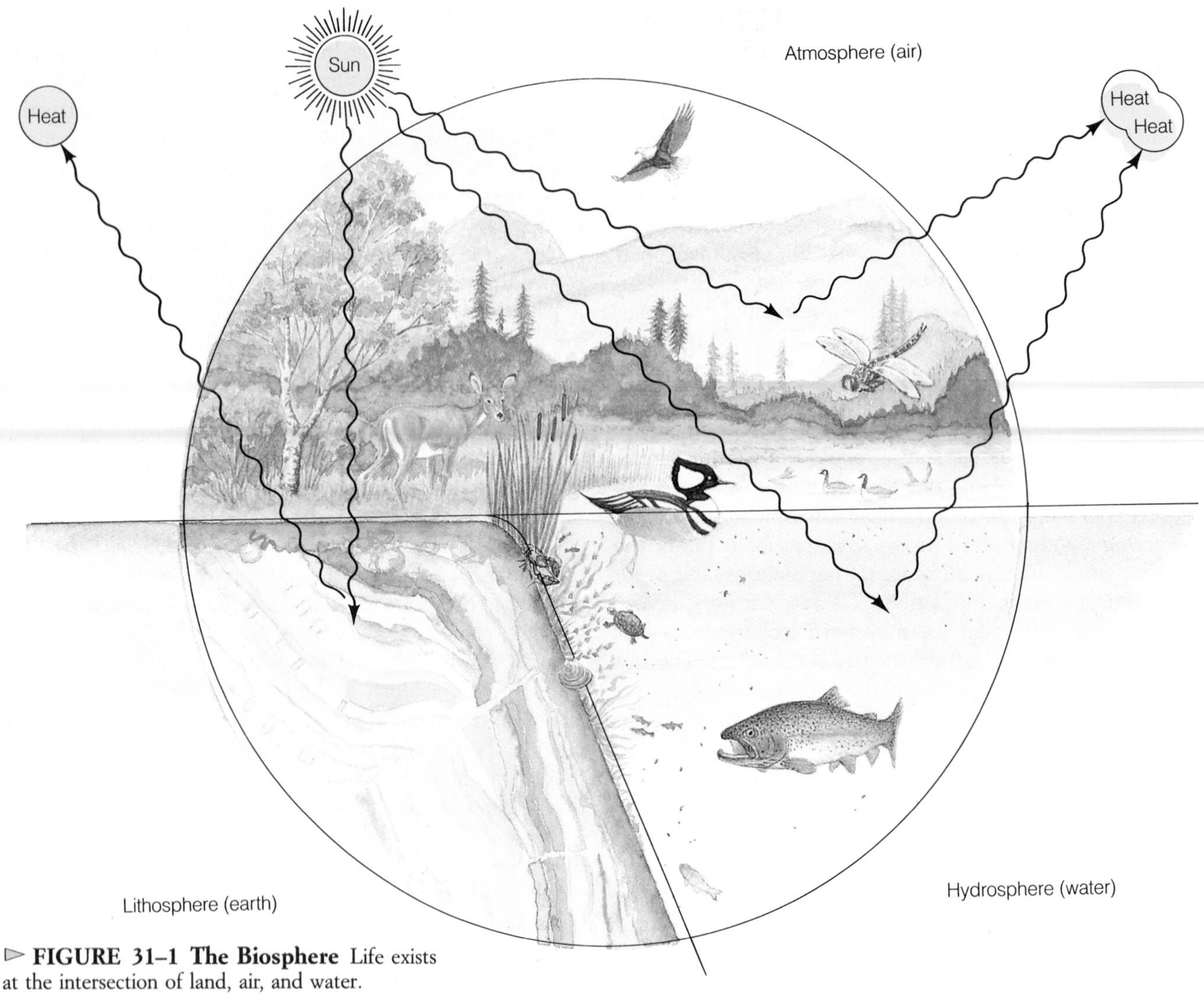

Heat

Sun

Atmosphere (air)

Heat
Heat

Lithosphere (earth)

Hydrosphere (water)

▷ **FIGURE 31–1 The Biosphere** Life exists at the intersection of land, air, and water.

▷ **FIGURE 31–2 The Earth from Outer Space**

gives an overview, making some important points about some of the major biomes of North America. As illustrated in Figure 31–3, the North American continent contains seven major biomes, five of which are discussed here.

Starting in the frozen north is the **tundra,** a region of long, cold winters and rather short growing seasons (▷ Figure 31–4a). The rolling terrain of the tundra supports grasses, mosses, lichens, wolves, musk oxen, and other animals adapted to the bitter cold. Trees cannot grow on the tundra because of the short growing season and because the subsoil (permafrost) remains frozen all year round, preventing the deep root growth necessary for trees.

Immediately south of the tundra lies the **taiga** (tie-ga), the northern coniferous, or boreal, forest. The taiga's milder climate and longer growing season result in a greater diversity and abundance of plant and animal life than exists on the tundra. Evergreen trees, bears, wolverines, and moose are characteristic species (Figure 31–4b).

East of the Mississippi River lies the **temperate deciduous forest biome,** characterized by an even warmer climate and more abundant rainfall (Figure 31–4c). Broad-

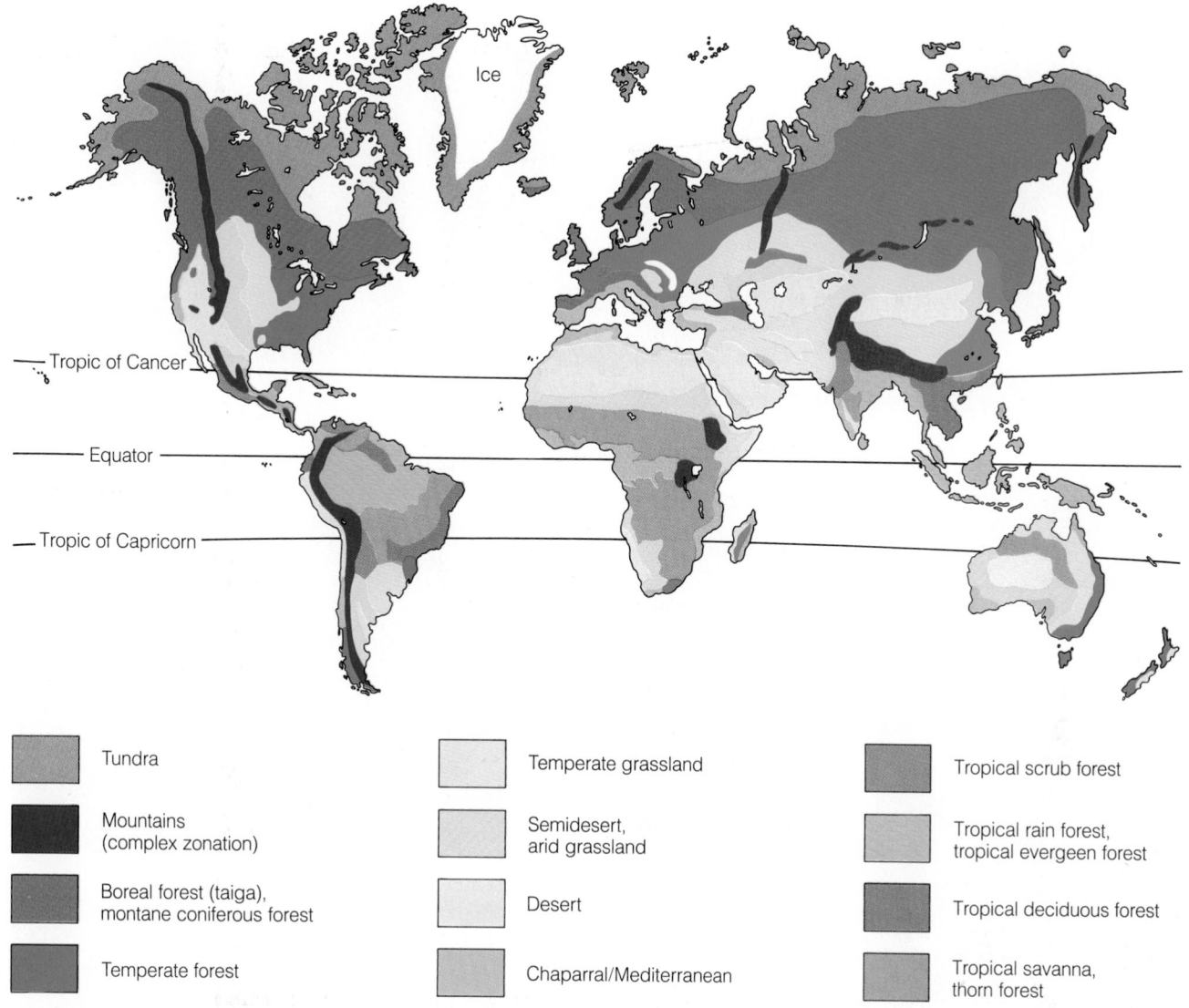

Ice

Tropic of Cancer

Equator

Tropic of Capricorn

	Tundra		Temperate grassland		Tropical scrub forest
	Mountains (complex zonation)		Semidesert, arid grassland		Tropical rain forest, tropical evergeen forest
	Boreal forest (taiga), montane coniferous forest		Desert		Tropical deciduous forest
	Temperate forest		Chaparral/Mediterranean		Tropical savanna, thorn forest

▷ **FIGURE 31–3 The Biomes**

leafed trees make their home in this biome. Opossums, black bears, squirrels, and foxes are characteristic animal species.

West of the Mississippi lies the **grassland biome** (Figure 31–4d). Inadequate rainfall and periodic drought prevent trees from growing on the grasslands, except near rivers, streams, and human habitation. Over the years, deep-rooted grasses have evolved on the plains. These grasses tolerate drought and can withstand grazing. Coyotes, hawks, and voles are characteristic animal species.

In the Southwest, where even less rain falls, is the **desert biome** (Figure 31–4e). Contrary to what many people think, the desert often contains a rich assortment of plants and animals uniquely adapted to the aridity and heat. Cacti, mesquite trees, rattlesnakes, and a variety of lizards all make their home in this seemingly inhospitable environment. The scorching hot deserts of Saudi Arabia receive less moisture than the desert around Tucson and therefore contain far fewer plants and animals.

The oceans can also be divided into subregions, known as **aquatic life zones.** Aquatic life zones are the aquatic equivalent of biomes. Like their land-based counterparts, each of these regions has a distinct environment and characteristic plant and animal life adapted to conditions of the zone. Four major aquatic life zones exist: coral reefs, estuaries (the mouths of rivers where fresh and salt water mix), the deep ocean, and the continental shelf. (These zones are discussed in detail in the next chapter.)

As you will see in Chapter 32, humans inhabit all biomes on Earth in testimony to our remarkable adaptability. Within these regions, humans benefit in numerous ways from the microbes, plants, animals, soil, water, and air that compose the biome. We tend to think of biomes, though, in terms of their natural resources (coal, timber, oil). Although vital to our economic health, these resources are but a fraction of the services we receive from the biome. Trees, for instance, provide oxygen and may remove some pollutants. Grasses and trees protect the soil and reduce flooding. Soil bacteria recycle nutrients on farms. Too, many other species also depend on these same services.

▷ **FIGURE 31–4 North American Biomes** (*a*) Tundra; (*b*) taiga; (*c*) temperate deciduous forest; (*d*) grassland; (*e*) desert.

Ecosystems Consist of Organisms and Their Environment

The biosphere is a global **ecological system,** or **ecosystem.** An ecosystem consists of organisms and their environment. Innumerable interactions are possible within an ecosystem. For the sake of convenience, ecologists often limit their study to small ecological systems—for example, a pond, a rotting log, or a grassy meadow. Even in these small ecosystems, the number of organisms present and the number of possible interactions can be astounding.

Abiotic Components. Reduced to a minimum, ecosys-

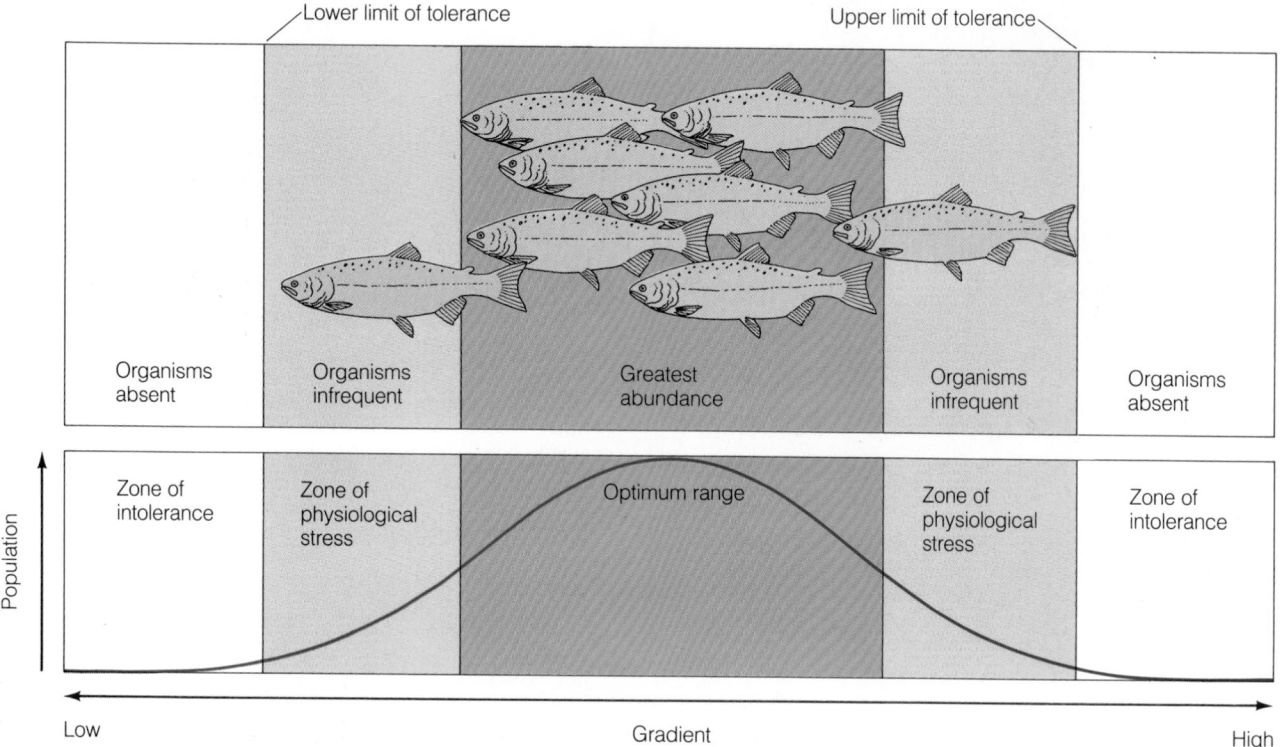

FIGURE 31–5 Range of Tolerance Organisms tolerate a range of conditions but thrive within an optimum range.

tems consist of two basic components: abiotic and biotic. The **abiotic components** are the physical and chemical factors necessary for life, including sunlight, precipitation, temperature, and nutrients. The **biotic components** are the organisms that live in an ecosystem.

The abiotic conditions within an ecosystem often vary. To live in most ecosystems, then, organisms must be able to survive a range of conditions. The range of conditions to which an organism is adapted is called its **range of tolerance.** As ▷ Figure 31–5 shows, organisms do best in the optimum range. Outside of that are the **zones of physiological stress,** where survival and reproduction are possible but not optimal. Outside of these zones are the **zones of intolerance,** where life for that organism is not possible.

Humans often alter the physical and chemical components of their environment. If conditions change drastically, some species may perish. Dams, for example, create lakes in streambeds, changing the water temperature, water flow, and other abiotic factors. As a result, many fishes that live in the stream perish. Native razorback suckers and other species that once thrived in the Colorado River, for example, are now endangered species (in danger of extinction) because of the large dams constructed along the river (▷ Figure 31–6). These dams release extremely cold water from the bottom of huge reservoirs, changing the temperature of the river and wiping out most of the native fish. The Point/Counterpoint in this chapter gives two opposing views on human-caused extinction.

Although species are sensitive to all of the abiotic factors in their environment, one factor often turns out to be more important than others in regulating growth. This factor, which ultimately regulates the growth of a population, is called a **limiting factor.**

In freshwater lakes and rivers, for example, dissolved phosphate is a limiting factor. Phosphate is needed by plants and algae for growth, but phosphate concentrations are naturally low. As a result, plant and algal growth is held in check. When phosphate is added to a body of water, however, plants and algae proliferate. Algae often form dense surface mats, blocking sunlight (▷ Figure 31–7). As a result, plants rooted on the bottom may die, and oxygen levels in the deeper waters may decline, killing fishes and other aquatic organisms.

Phosphates are added to lakes and rivers from several sources. One of the major sources is the sewage-treatment plant. The phosphates released by these plants come from detergents and human waste. People can help lessen the problem by using low- or no-phosphate detergents.

On land, precipitation tends to be the limiting factor. At any given temperature, the more moisture that falls, the richer the plant and animal life.

Biotic Components. Within biomes and aquatic life zones, organisms of a given species often occupy specific regions. A group of organisms of the same species occupying a specific region constitutes a **population.** The members of a population may remain together throughout all or much of the year. Bighorn sheep, for example, remain in herds throughout the year. Other organisms, such as the grizzly bear, are solitary for the most part, keeping to

(a)

(b)

▷ **FIGURE 31–6 Altering Native Fish Life Along the Mighty Colorado** (*a*) The Glen Canyon Dam in Arizona not only inundated a canyon some believe to have been as magnificent as the Grand Canyon but also changed the water temperature in the Colorado River below the dam. Extremely cold water flows out of the bottom of the reservoir into the river and has nearly eliminated several fish species, including the (*b*) razorback sucker.

themselves except for mating. In any given ecosystem, several populations exist together and form a **community,** a network of plants, animals, and microorganisms. All organisms in a community, even humans, are part of an interdependent web of life, the members of which interact in many ways.

Competition May Occur Between Species Occupying the Same Habitat if Their Niches Overlap

If asked to give a brief description of yourself, you would probably begin by describing the place where you live. You would then probably discuss the work you do, the friends you have, and other important relationships that describe your place in society. A biologist would do much the same when describing an organism. He or she would start with a

▷ **FIGURE 31–7 Algal Bloom** This pond is choked with algae due to the abundance of plant nutrients from human sources.

description of the place an organism lives—that is, its **habitat.** Next, he or she would describe how the organism "fits" into the ecosystem, its **ecological niche,** or simply **niche.** An organism's niche includes its habitat and all of its relationships with its environment. This niche includes what an organism eats, what eats it, its range of tolerance for various environmental factors, and other important elements.

Organisms in a community occupy the same habitat, but most of them have quite different niches. This phenomenon minimizes competition for resources and is a condition favored by natural selection. The fact that organisms occupy separate niches provides for a wider use of an ecosystem's resources, especially food.

Niches do overlap somewhat. For example, two species may feed on some of the same foods. The more two species' niches overlap, however, the more the organisms compete with each other. Just as in the human economy, competition in the natural world can be a good thing. In the biological world, competition is an agent of natural selection that may lead to the evolution of new adaptations. When niches overlap considerably, competition becomes intense and one species usually suffers. If two species occupy identical niches, competition will eliminate one of them. As a result, two species cannot occupy the same niche for long. This rule is called the **competitive exclusion principle.**

The concept of the niche is very important to us. For example, successful control of an insect pest is best achieved through an understanding of the species' niche. An analysis of an insect's niche might show that one species of bird or a particular insect feeds on the pest. By simply encouraging these beneficial species—say, by providing trees for the birds or food preferred by the pest-eating

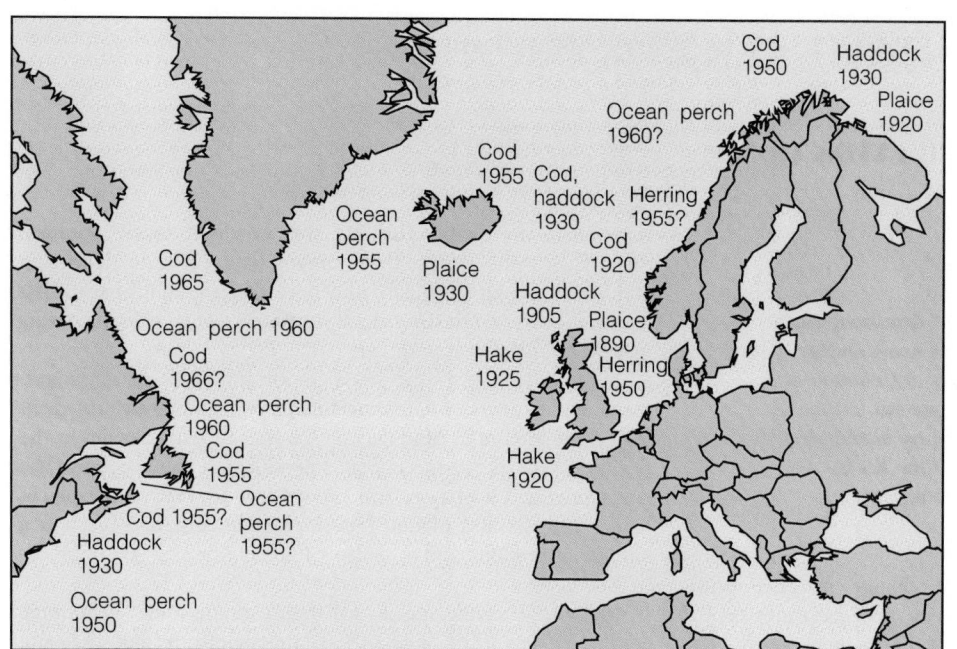

▷ **FIGURE 31–8 Depletion of North Atlantic Fisheries** The dates on the map indicate the approximate time when various commercial fisheries were depleted by overfishing.

insect—farmers can hold pest populations down and save themselves enormous sums of money on chemical pesticides. Folks interested in saving species have also learned that to protect a plant or animal, you have to protect the ecosystem it lives in. These efforts help protect the niche of the species in question.

Humans Have Become a Major Competitive Force in Nature

Humans compete with many other species for food and living space. However, humans possess a marked advantage provided by our technology and enormous population size. As our population grows, as our demand for food and resources climbs, and as our technological power increases, more and more of the habitats of wild animals and plants will be lost, and habitat loss is the number one cause of extinction. For a discussion of extinction, see the Point/Counterpoint in this chapter.

Already, commercial overfishing has depleted dozens of the world's fisheries (▷ Figure 31–8). Overfishing has had a ripple effect on other species, reducing populations of seals and other fish-eating animals. On every continent and every nation, humans are outcompeting the species that share this planet with us. According to estimates by various experts, 40 to 100 species become extinct every day, largely because of tropical deforestation. Unless we do something, hundreds of thousands of species will become extinct in the next decade. Many ecologists and environmentalists warn that none of us will be immune to the impacts of such widespread biological impoverishment. Cutting down the rain forests, for example, could alter global climate, making the Earth hotter. We could feel the effect in higher utility bills to cool our homes, hotter summers, higher food costs,

and higher taxes. If conditions became bad enough, massive food shortages could occur. Although all this human intrusion is rather depressing, there is hope. We can find ways of coexisting with nature. We have to.

≈ ECOSYSTEM FUNCTION

Life on land and in the Earth's waters is possible principally because of the existence of one group, the **producers.** These organisms include the algae and plants, which, as you learned earlier, absorb sunlight and use its energy to synthesize organic foodstuffs from atmospheric carbon dioxide and water via photosynthesis.[2] These organic molecules are used by the producers themselves but also provide nourishment for all the other organisms in the web of life. As a result, producers literally form the foundation of the living world.

Another large group of other organisms is the **consumers.** Ecologists place consumers into four general categories, depending on the type of food they eat. Some, such as deer, elk, and cattle, feed directly on plants and are called **herbivores.** Others, such as wolves, feed on herbivores and other animals and are known as **carnivores.** Humans and a great many other animal species subsist on a mixed diet of plants and animals and are known as **omnivores.** The final group feeds on animal waste or the remains of plants and animals. These organisms are called **detritivores,** or **decomposers.**[3] This important group includes many bacteria and fungi, and insects.

---

[2] Some producers are chemosynthetic organisms—that is, organisms capable of using methane and hydrogen sulfide as a source of energy to produce organic molecules.

[3] Remember that detritus is waste material.

HUMANS ARE ACCELERATING EXTINCTION

David M. Armstrong

David M. Armstrong teaches science for non-scientists at the University of Colorado at Boulder and has written several books on mammals and ecology of the Rocky Mountain region.

Evolution is the process of change in gene pools through time. When one gene pool becomes reproductively independent of another, a new species has formed. Such speciation generates species; extinction takes them away. Simply put, extinction is a failure to adapt to change, the termination of a gene pool and the end of an evolutionary line.

Extinction is a natural process. Most species that have ever lived are now extinct. The 3 to 30 million species on Earth today are no more than 1 to 10 percent of the species that have evolved since life began about 3.5 billion years ago. Given these facts, why are thoughtful people concerned about endangered species? After all, history makes it clear that—given enough time—all species will become extinct.

The basis for concern is that today the natural process of extinction proceeds at an unnatural rate. Let us estimate how much human activity has accelerated rates of extinction. The lifespan of species seems to average from 1 to 10 million years. Assume (to be conservative) that the average longevity of species of higher vertebrates is 1 million years. In round numbers (to make calculations easy) there are 10,000 species of birds and mammals. So, on average, one species ought to go extinct each century. However, between 1600 and 1980, at least 36 species of mammals and 94 species of birds became extinct. That is about 0.29 species per year, 29 times the natural rate.

What does it mean to increase a rate by 29 times? The speed limit is 55 miles per hour. Exceed the speed limit by 29-fold, and you are moving 1,595 miles per hour, over twice the speed of sound. The difference between natural rates of extinction and present, human-influenced rates is analogous to the difference between a casual drive and Mach 2! Is that a problem? You decide: concern is a moral construct, not a scientific one.

Several human activities have contributed—mostly inadvertently—to accelerating rates of extinction. The dodo and the passenger pigeon were extinguished by overhunting. Wolves and grizzly bears were exterminated over much of their ranges as threats to livestock. The black-footed ferret was driven to the verge of extinction because prairie dogs, its staple food, were poisoned as an agricultural pest. The smallpox virus was exter-

minated in the "wild" (but survives in a half-dozen laboratories). This is the closest that humans have come to deliberate elimination of an organism, and note that thoughtful scientists with the power to destroy smallpox chose not to do so, electing instead to manage it with care.

Habitat change is the most important cause of endangerment and extinction. Clearing forests for agriculture has decimated the lemurs of Madagascar. Chemical pesticides led to the decline of the peregrine falcon. Introducing exotic species (like goats on the Galapagos and mongooses in Hawaii) displaces native animals and plants. Developing the Amazon Basin is a habitat alteration, and a cause of extinction, on an unprecedented scale.

Many urge saving species for their esthetic value. Whooping cranes *are* beautiful, and part of the beauty is that they are products of a marvelous evolutionary process. Most concern about accelerated extinction, however, stresses economic value. A tiny fraction of Earth's seed plants are used commercially. Perhaps an obscure plant like jojoba will be a source of oil more reliable than Saudi Arabia. Wild grasses have furnished genes that improved disease-resistance in wheat. Numerous wild animals (like musk ox, kudu, and whales) could contribute protein to the human diet. Wild species may have medical value; penicillin, after all, was once merely an obscure mold on citrus fruit. Some sensitive species are useful monitors of environmental quality, and the presence of healthy populations of many species may promote greater stability or resilience of ecosystems. Naturalist Aldo Leopold noted that we humans have a way of "tinkering" with the ecosphere, seeing how it works. And, since we can work it better, we ought to remember the first rule of tinkering: never throw away any of the parts.

"Extinction is forever," and extinction impoverishes both Earth's ecosystems and the potential richness of human life. Borrowing again from Aldo Leopold, I believe we should be concerned about unnaturally rapid extinction because such concern is part of a "right relationship" between people and the landscapes that nurture and inspire them.

Biologist Sir Julian Huxley noted that "we humans find ourselves, for better or worse, business agents for the cosmic process of evolution." We hold power over the future of the biosphere, the power to destroy or preserve. German philosopher George Hegel noted that freedom (including the power to destroy species) implies responsibility (to preserve them). I agree. The question of human-accelerated extinction boils down to a simple ethical question, "does posterity matter?" Some of us have ethics that are human-centered. We ask simply, "Do my children deserve a life as rich, with as much opportunity, as mine?" If they do, then we have the responsibility to choose restraint. (Perhaps my life has been rich enough without the dodo, but I am reluctant to make that judgment for future generations.)

EXTINCTION IS THE COURSE OF NATURE

Norman D. Levine

Norman D. Levine is a professor emeritus at the College of Veterinary Medicine and Agricultural Experiment Station, University of Illinois at Urbana. His research interests cover parasitology, protozoology, and human ecology. This article, ©1989, is provided by the American Institute of Biological Sciences.

E volution is the formation of new species from preexisting ones by a process of adaptation to the environment. Evolution began long ago and is still going on. During evolution those species better adapted to the environment replaced the less well adapted. It is this process, repeated year after year for millenia, that has produced the present mixture of wild species. Perhaps 95% of the species that once existed no longer exist.

Human activities have eliminated many wild species. The dodo is gone, and so is the passenger pigeon. The whooping crane, the California condor, and many other species are on the way out. The bison is still with us because it is protected, and small herds are raised in semicaptivity. The Pacific salmon remains because we provide fish ladders around our dams so it can reach its breeding places. The mountain goat survives because it lives in inaccessible places. But some thousands of other animal species, to say nothing of plants, are extinct, or soon will be.

Some nature lovers weep at this passing, and collect money to save species. They make lists of animals and plants that are in danger of extinction and sponsor legislation to save them.

I don't. What the species preservers are trying to do is to stop the clock. It can not and should not be done.

Extinction is an inevitable fact of evolution, and it is needed for progress. New species continually arise, and they are better adapted to their environment than those that have died out.

Extinction comes from failure to adapt to a changing environment. The passenger pigeon did not disappear because of hunting alone, but because its food trees were destroyed by land clearing and farming. The prairie chicken cannot find enough of the proper food and nesting places in the cultivated fields that once were prairie.

And you cannot necessarily introduce a new species, even by breeding it in tremendous numbers and putting it out into the wild. Thousands of pheasants were bred and set out year after year in southern Illinois, but in the spring of each year there were none left. Another bird, the capercaillie, is a fine,

large game bird in Scandinavia, but every attempt to introduce it into the United States has failed. An introduced species cannot survive unless it is preadapted to its new environment.

A few introduced species are preadapted and some make spectacular gains. The United States has received the English sparrow, the starling, and the house mouse from Europe, and also the gypsy moth, the European corn borer, the Mediterranean fruit fly, and the Japanese beetle. The United States gave Europe the gray squirrel and the muskrat, among others. The rabbit took over in Australia, at least for a time.

The rabbit and the squirrel were successful on new continents because their requirements are not as narrow as those of other species that failed. Today, adjustment to human-made environments may be just as difficult as adjustment to new continents. The rabbit and the squirrel have succeeded in adjusting to the backyard habitat but most wild animals have disappeared.

Human-made environments are artificial. People replace mixed grasses, shrubs, and trees with rows of clean-cultivated corn, soybeans, wheat, oats, or alfalfa. Variety has turned into uniform monotony, and the number of species of small vertebrates and invertebrates that can find the proper food to survive has become markedly reduced. But some species have multiplied in these environments and have assumed economic importance; the European corn borer in this country is an example.

Would it improve Earth if even half of the species that have died out were to return? A few starving, shipwrecked sailors might be better off if the dodo were to return, but I would not be. The smallpox virus has been eliminated, except for a few strains in medical laboratories. Should it be brought back? Should we bring panthers back into the eastern states? Think of all the horses that the automobile and tractors have replaced, and of all the streets and roads that have been paved and the wild animals and plants killed as a consequence. Before people arrived in America about 10,000 years ago, the animal-plant situation was quite different. What should we do? Should we all commit suicide?

Evolution exists, and it goes on continually. People are here because of it, but people may be replaced someday. It is neither possible nor desirable to stop it, and that is what we are trying to do when we try to preserve species on their way out. It can be done, I think, but should we do it to them all? Or to just a few, as we are doing now?

≋ SHARPENING YOUR CRITICAL THINKING SKILLS

1. Summarize the key points of both authors.
2. Do you see any flaws in the reasoning of either author?
3. Which viewpoint do you adhere to? Why?

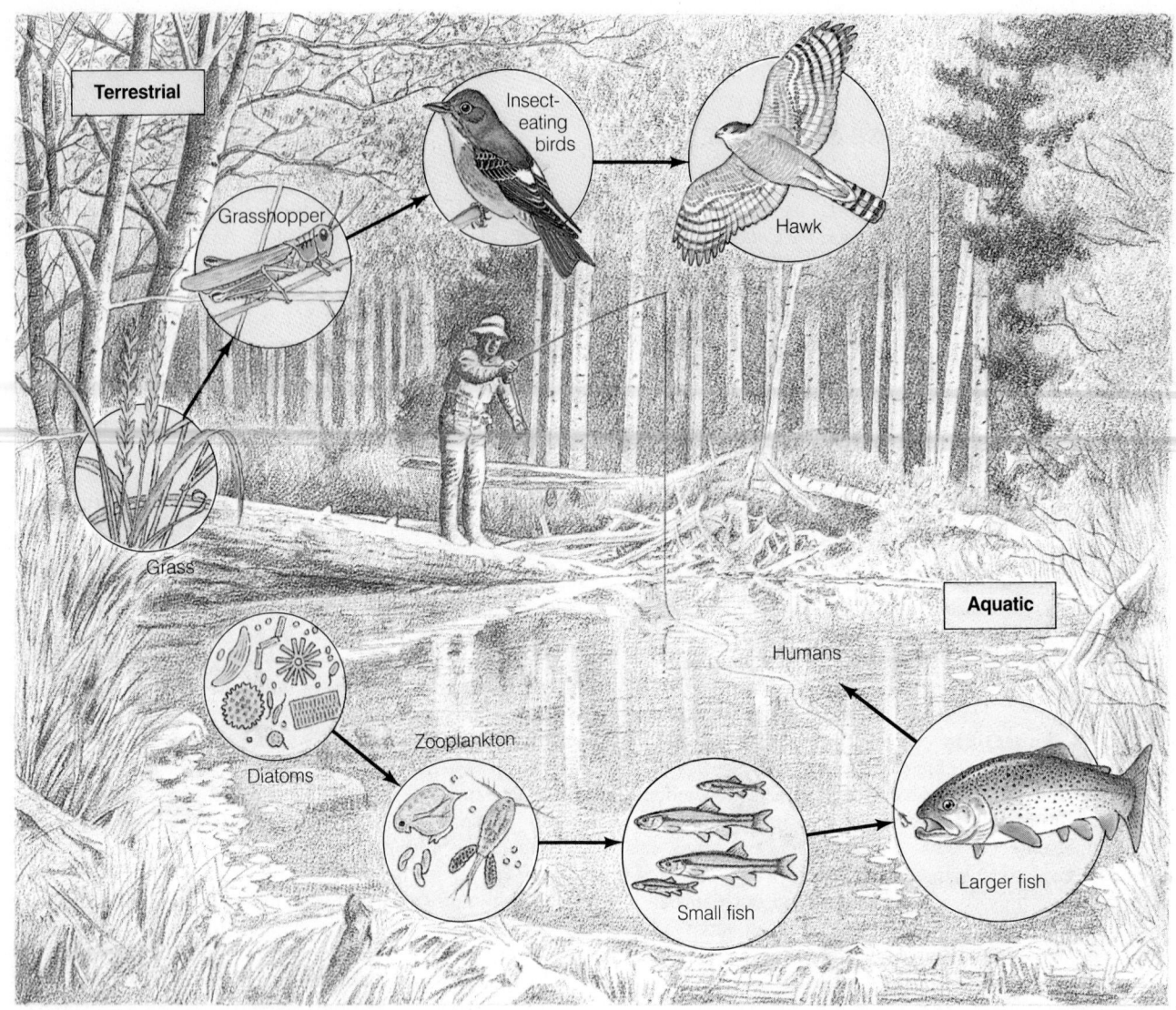

▷ **FIGURE 31–9 Simplified Food Chains** Two grazer food chains are shown, a terrestrial one and an aquatic one.

Food and Energy Flow Through Food Chains that Are Generally Part of Much Larger Food Webs in Ecosystems

Biological communities consist of numerous food chains. A **food chain** is a series of organisms, each one feeding on the organism preceding it (▷ Figure 31–9). All organisms in the community are members of one or more food chains.

Biologists recognize two general types of food chains: grazer and decomposer. **Grazer food chains** begin with plants and algae. These organisms are consumed by herbivores, or **grazers.** Herbivores, in turn, may be eaten by carnivores. **Decomposer food chains** begin with dead material—either animal wastes (feces) or the remains of plants and animals (▷ Figure 31–10a). In ecosystems, decomposer and grazer food chains are tightly linked (Figure 31–10b). Thus, waste from the grazer food chain enters the decomposer food chain. Nutrients liberated by the

decomposer food chain enter the soil and water and are reincorporated into plants at the base of the grazer food chain. The success of the living world depends on the link between these two food chains.

Food chains exist only on the pages of textbooks; in a community of living organisms, all food chains are part of a much more complex network of feeding interactions known as **food webs** (▷ Figure 31–11). Food webs present a complete picture of the feeding relationships in any given ecosystem.

As with so many other topics in ecology, an understanding of food chains and food webs is essential to living successfully on the planet. For instance, efforts to protect the ozone layer are important to protect us from cancer but also to protect phytoplankton, which form the base of aquatic food chains. If ozone levels continue to fall, ultraviolet radiation could kill phytoplankton and cause a collapse of many aquatic food chains. Scientists in Antarctica have recorded a dramatic decline in two populations of

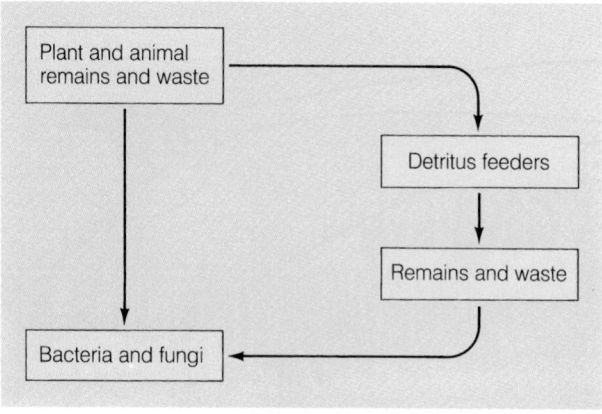

Decomposer food chain

(a)

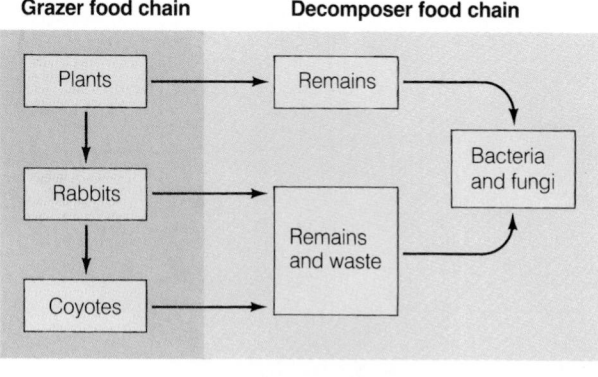

Grazer food chain | **Decomposer food chain**

(b)

▷ **FIGURE 31–10 Food Chains** (*a*) A decomposer food chain. (*b*) A grazer food chain and a decomposer food chain, showing the connection between the two.

penguins and think the reason might be a decline in krill, the shrimp-like creatures on which the birds feed. The decline in krill populations may be due to increased ultraviolet radiation acting on phytoplankton populations. Studies are under way to test this hypothesis. If it does prove correct, and ozone depletion continues, commercially valuable fish species could also decline.

Food Chains Provide Avenues for the Flow of Energy and the Cycling of Nutrients Through the Environment

Energy and nutrients both flow through food webs, but in very different ways. We begin with energy. As noted in previous chapters, solar energy is captured by plants and algae and used to produce organic food molecules. Energy from the sun is stored in the covalent bonds of these molecules. In the food chain, organic molecules pass from plants to animals, where they are broken down in mitochondria, which release stored solar energy. This energy is used to power numerous cellular activities.

During cellular respiration, much of the energy stored in organic food molecules is simply lost as heat. Heat escap-

ing from plants and animals is radiated into the atmosphere and then into outer space. It cannot be recaptured and reused by plants or animals. Because all solar energy is eventually converted to heat, energy is said to flow unidirectionally through food chains and food webs. Put another way, energy cannot be recycled.

In sharp contrast, nutrients flow cyclically; that is, they are recycled over and over. Nutrients in the soil, air, and water are first incorporated into plants and algae, then passed from plants to animals in various food chains. Nutrients in the food chain eventually reenter the environment by one of three paths: (1) the excretion of wastes, (2) the decomposition of dead organisms and (3) the decomposition of waste.

Every time you exhale, for example, you release carbon dioxide that reenters the atmosphere for reuse. Thus, through the act of breathing you play an important role in the global recycling system that makes life possible.

Nutrients also reenter the environment through the decomposition of dead organisms. When a plant or animal dies, it decomposes. Decomposition is caused by bacteria and fungi. As you may recall from Chapter 24, these organisms release enzymes that decompose organic matter. Although these microorganisms absorb many nutrients released during this process, some nutrients escape, entering the soil and water for reuse. And, of course, when a bacterium dies, it too breaks down, releasing nutrients into its environment.

Finally, many nutrients reenter the environment through animal waste. The feces of a rhinoceros, for example, are broken down by bacteria, which liberate carbon dioxide, nitrogen, and numerous minerals.

One way or another, all nutrients eventually make their way back to the environment for recycling. Each new generation of organisms, therefore, relies on the recycling of material in the biosphere. Even the atoms in your body have been recycled since the beginning of life on Earth. Who knows—some of those atoms may have been in the very first cell that lived in the shallow seas.

The Organisms of a Food Chain Exist on Different Trophic Levels

Ecologists classify the organisms in a food chain according to their position, or **trophic level** (literally, "feeding" level). The producers are the base of the grazer food chain and are, therefore, members of the first trophic level. The grazers are members of the second trophic level. Carnivores that feed on grazers are members of the third trophic level, and so on.

Most terrestrial food chains are limited to three or four trophic levels. Longer terrestrial food chains are rare. The reason for this is that food chains generally do not have a large enough producer base to support many levels of consumers. Put another way, in the longer food chains, less food is available for the top-level consumers. Why?

Plants absorb only a small portion of the sunlight that

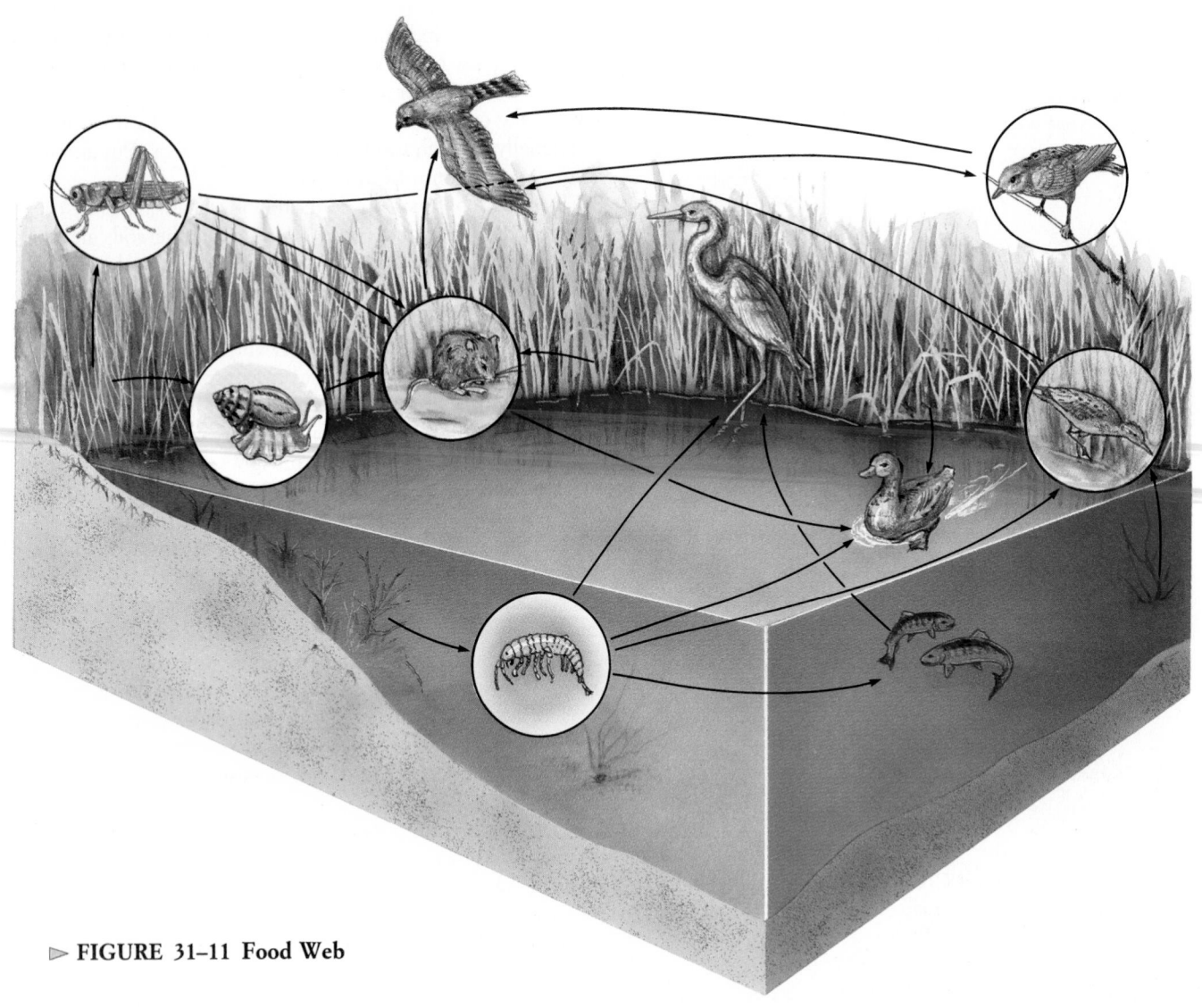

▷ **FIGURE 31–11 Food Web**

strikes the Earth (only 1% to 2%), which they use to produce organic matter, or biomass. Technically, **biomass** is the dry weight of living material in an ecosystem. The biomass at the first trophic level is the raw material for the second trophic level. The biomass at the second trophic level is the raw material for the third trophic level, and so on.

As shown at the top of ▷ Figure 31–12, not all of the biomass produced by plants is converted to grazer biomass. At least three reasons account for the incomplete transfer of biomass from one trophic level to the next. These are illustrated at the bottom of the figure. First, some of the plant material, such as the roots, is not eaten. Second, not all of the material that the grazers eat is digested. Third, some of the digested material is broken down to produce energy and heat and therefore cannot be used to build biomass in the grazers (Figure 31–12). As a rule, only 5% to 20% of the biomass at any one trophic level can be passed to the next. (The amount varies depending on the organisms involved in the food chain.)

When plotted on graph paper, the biomass at the various trophic levels forms a pyramid, the **biomass pyramid**

(▷ Figure 31–13). Because biomass contains energy (stored in the covalent bonds), the biomass pyramid can be converted into a graph of the chemical energy in the various trophic levels. This graph is called an **energy pyramid.**

In most food chains, the number of organisms also decreases with each trophic level, forming a **pyramid of numbers.** Knowledge of ecological pyramids will help you understand why people in many Third World countries generally subsist on a diet of grains (corn, rice, or wheat) rather than meat. To understand what I mean, take a look at ▷ Figure 31–14, which presents energy pyramids for two food chains. In the grain → human food chain on the right, 20,000 kilocalories of grain can feed 10 people for a day. If that grain is fed to a steer and the beef is fed to humans, however, only 1 person can subsist on the 20,000 kilocalories, as shown on the left. Why? In the grain → steer → human food chain, the 20,000 kilocalories fed to the cow produces only 2000 kilocalories of food, barely enough to feed 1 person for a day (assuming a 10% transfer of biomass). Thus, the shorter the food chain, the more food is available to top-level consumers.

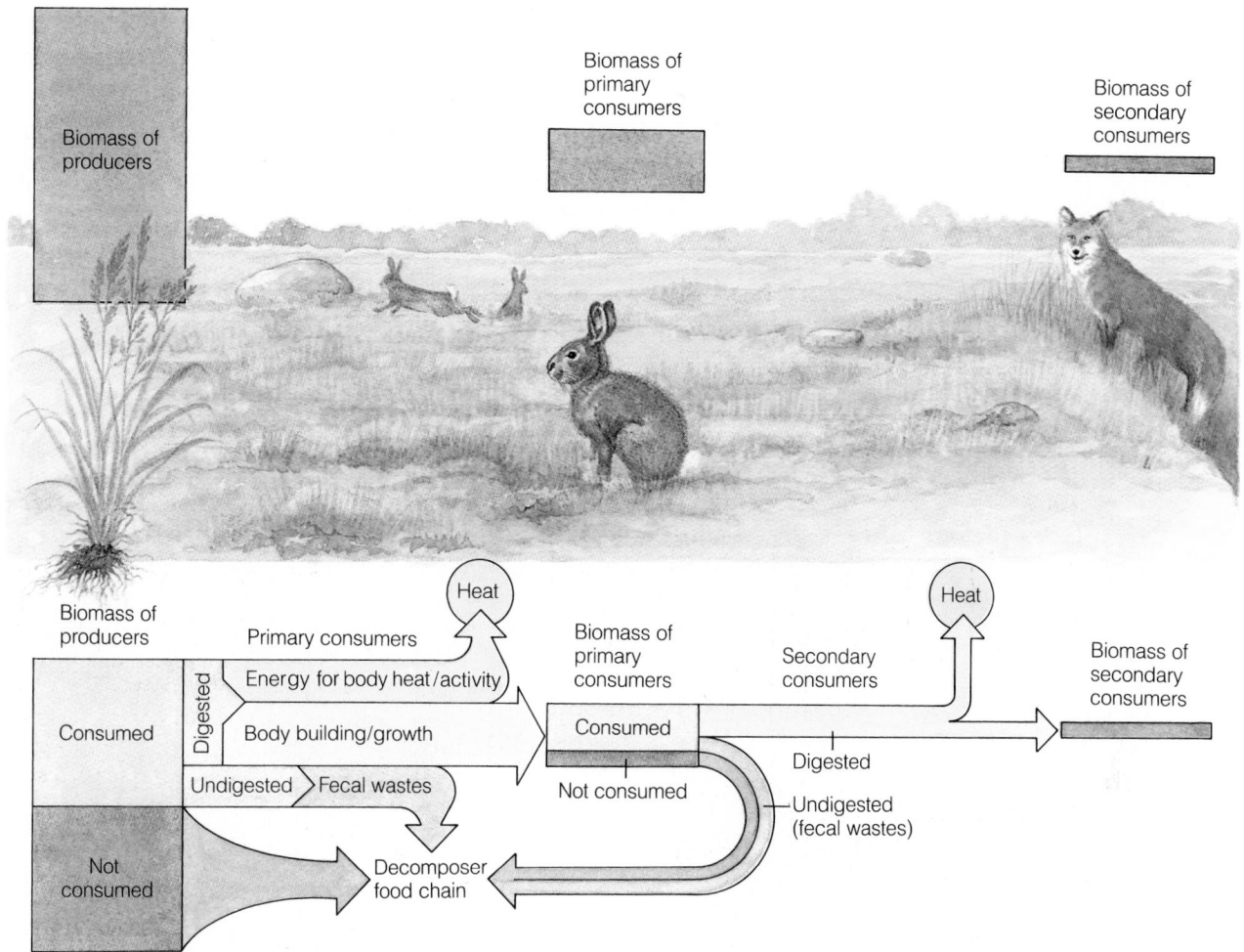

▷ **FIGURE 31–12 Flow of Energy and Biomass through a Food Chain** Not all biomass from one trophic level ends up in the next level, for reasons shown in the diagram and discussed in the text. Note that all energy is eventually lost as heat.

This simple rule has profound implications for the human race. The human population increases by about 90 million people a year. Feeding these people poses an enormous challenge. How can new residents be fed most efficiently?

The most efficient food source will be crops such as corn, rice, and wheat that are fed directly to people. It is obviously far less efficient to feed corn and other grains to cattle and other livestock, which are then slaughtered for human consumption. Vegetarianism, say proponents, is not only good for your health but better for the environment, because it requires less grain production than a nonvegetarian diet. People, however, require an adequate intake of protein, as noted in Chapter 11. Protein can be supplied by fish, meat, and a mixture of legumes. (For more on feeding the world's hungry, see Chapter 33.)

Nutrient Cycles Consist of Two Phases, the Environmental and the Organismic

Nature has its own economy: a system of exchange, competition, and cooperation. The economy of nature is driven by the sun and is dependent on the circular flow of many nutrients through food webs. The term *nutrients* is used here to refer to all ions and molecules used by living organisms. As noted earlier, nutrients flow from the environment through food webs, then are released back into the environment. This circular flow constitutes a **nutrient cycle,** also known as a **biogeochemical cycle.**

Nutrient cycles can be divided broadly into two phases: the environmental and organismic (▷ Figure 31–15, page 890). In the **environmental phase,** a nutrient exists in the air, water, or soil or sometimes in two or more of them simultaneously. In the **organismic phase,** nutrients are found in the biota—the plants, animals, and microorganisms.

Dozens of global nutrient cycles operate continuously to ensure the availability of chemicals vital to all living things, present and future. Unfortunately, however, a great many human activities disrupt important nutrient cycles. These activities can profoundly influence the survival of a species. This section looks at two of the most important nutrient cycles, the carbon and nitrogen cycles, and the ways they

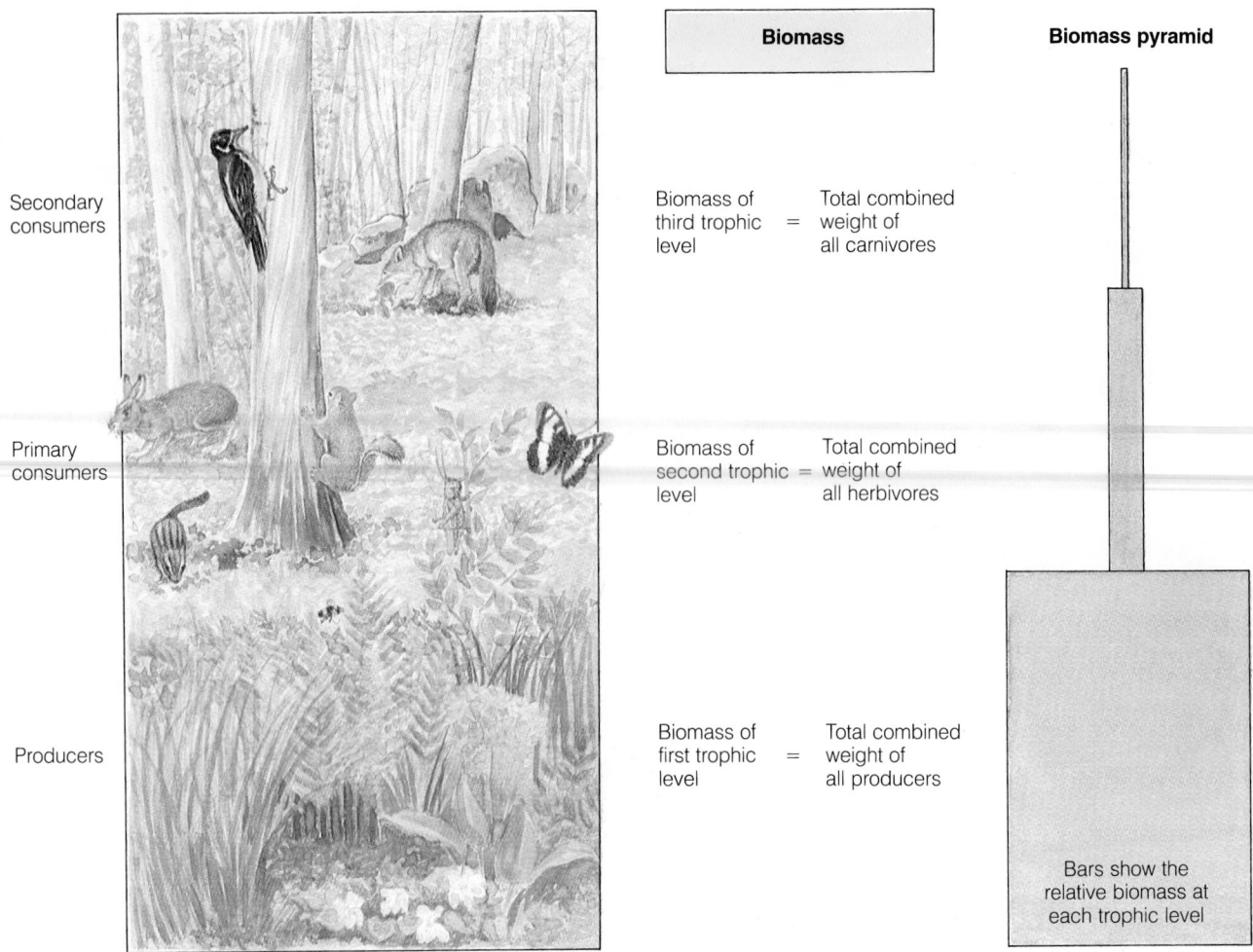

Secondary
consumers

Primary
consumers

Producers

Biomass of
third trophic = Total combined
level   weight of
               all carnivores

Biomass of
second trophic = Total combined
level   weight of
               all herbivores

Biomass of
first trophic = Total combined
level   weight of
               all producers

Biomass pyramid

Bars show the
relative biomass at
each trophic level

▷ **FIGURE 31–13 Biomass Pyramid** In most food chains, biomass decreases from one trophic level to the next higher one.

are being altered. (Solutions to the problems are described in Chapter 33.) Although these cycles are most often discussed in ecology chapters, they are only two of several dozen nutrient cycles essential to life.

The Carbon Cycle. The carbon cycle is illustrated in ▷ Figure 31–16 in a simplified form. We begin with free carbon dioxide. In the environmental phase of the cycle, carbon dioxide resides in two reservoirs, or sinks: the atmosphere and the surface waters (oceans, lakes, and rivers). Figure 31–16 deals primarily with the terrestrial component of the carbon cycle. As illustrated, atmospheric carbon dioxide is absorbed by plants and photosynthetic protists, thus entering the organismic phase of the cycle. Autotrophs convert carbon dioxide into organic food materials, which travel along the food chain from one trophic level to the next. Carbon dioxide reenters the environmental phase via cellular energy production (cellular respiration) of the organisms in the grazer and decomposer food chains.

For tens of thousands of years, our ancestors lived in harmony with nature. With the advent of the Industrial

Revolution, however, human beings began to interfere with natural processes on a large scale. One of the victims of our technological development has been the global carbon cycle. The widespread combustion of fossil fuels (which releases carbon dioxide) and rampant deforestation (which reduces carbon dioxide uptake) have seriously overwhelmed the cycle.

For many years before the Industrial Revolution, global carbon dioxide production equaled carbon dioxide absorption by plants and algae. Today, approximately 6 billion tons of carbon dioxide is added to the atmosphere each year. Three-quarters of the increase results from the combustion of fossil fuels like the gasoline in our cars; the remaining quarter stems from deforestation. How does deforestation add carbon dioxide to the atmosphere? Trees absorb enormous amounts of carbon dioxide for photosynthesis. As forests are cleared, then, the amount of atmospheric carbon dioxide they absorb declines. Making matters worse, many forests are burned after cutting, further adding to the carbon dioxide levels in the atmosphere.

In the past 100 years, global atmospheric carbon dioxide levels have increased 25%. In the atmosphere, carbon

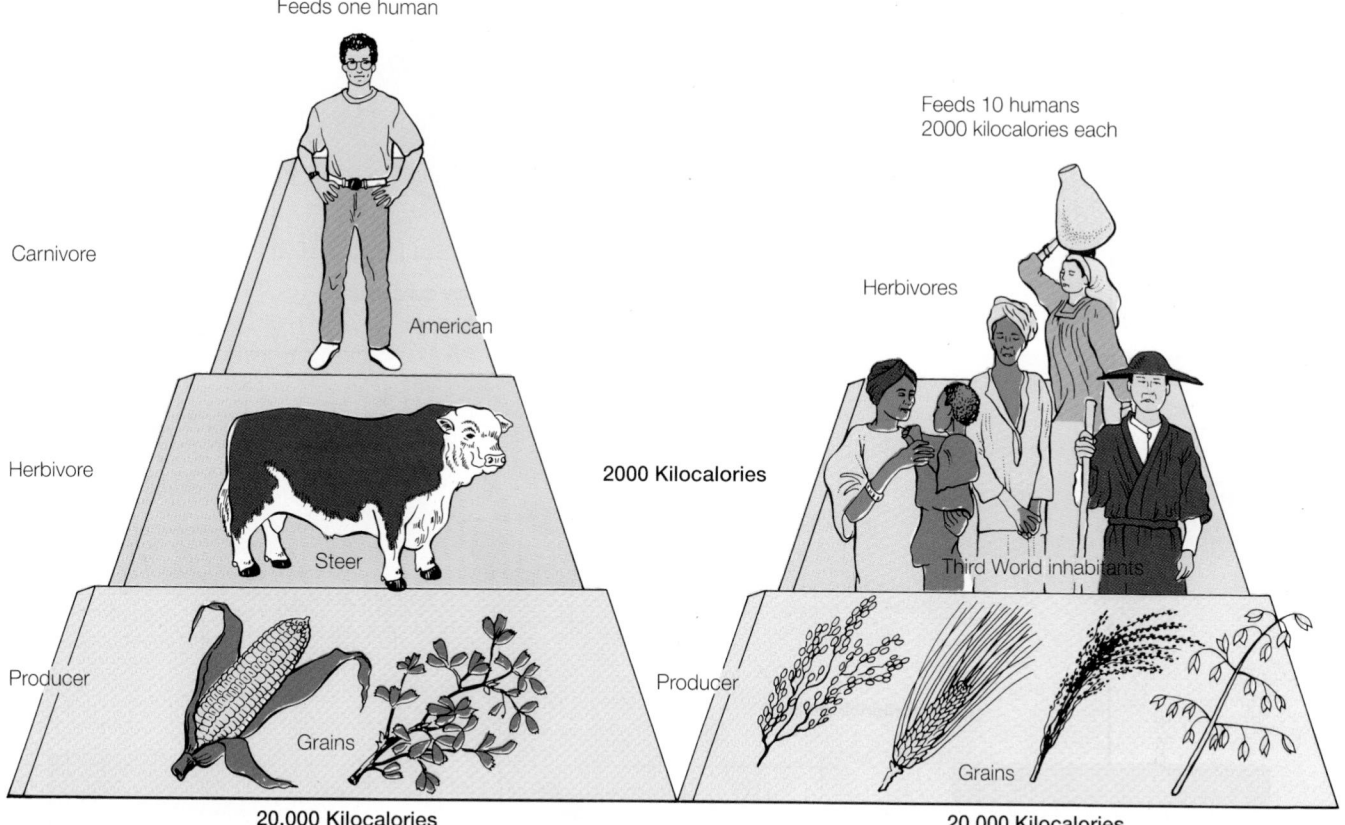

Feeds one human

Carnivore

American

Herbivore

2000 Kilocalories

Steer

Producer

Grains

20,000 Kilocalories

Feeds 10 humans
2000 kilocalories each

Herbivores

Third World inhabitants

Producer

Grains

20,000 Kilocalories

▷ **FIGURE 31–14 Energy Pyramids in Two Food Chains** (*a*) The typical American diet. Note that the 20,000 kilocalories of corn fed to cows feeds only one American for a day.

(*b*) In the shorter food chain in the Third World nations, the same 20,000 kilocalories will feed 10 people directly.

dioxide traps heat escaping from Earth and reradiates it to the Earth's surface. As carbon dioxide levels increase, global temperature may rise dramatically. Such a rise could have devastating effects on global climate. It could shift rainfall patterns, destroy agricultural production in many regions, and wipe out thousands of species. A rising global temperature might cause glaciers and the polar ice caps to melt, raising the sea level and flooding many low-lying coastal regions. Fortunately, there are many cost-effective strategies for reducing our dependency on fossil fuel. The most notable are energy efficiency and renewable fuels—solar energy and wind, for example. (The topic of global warming is detailed in Chapter 33.)

The Nitrogen Cycle. Nitrogen is an element that is essential to many important biological molecules, including amino acids, DNA, and RNA. The Earth's atmosphere contains enormous amounts of it, but atmospheric nitrogen is in the form of nitrogen gas (N_2), which is unusable to all but a few organisms. As a result, atmospheric nitrogen must first be converted to a usable form—either nitrate or ammonia.

The conversion of nitrogen to ammonia is known as

nitrogen fixation. How does it occur? As ▷ Figure 31–17 (page 832) shows, the roots of leguminous plants (peas, beans, clover, alfalfa, vetch, and others) contain small swellings called **root nodules.** Inside the nodules are symbiotic bacteria that convert atmospheric nitrogen to ammonia. Ammonia is also produced by cyanobacteria bacteria that live in the soil. Once ammonia is produced, other soil bacteria convert it to nitrite and then to nitrate. Nitrates are incorporated by plants and used to make amino acids and nucleic acids. All consumers ultimately receive the nitrogen they require from plants.

Nitrate in soil also comes indirectly from the decay of animal waste and the remains of plants and animals. As shown on the right side of Figure 31–17, this process returns ammonia to the soil for reuse. Ammonia is converted to nitrite, then to nitrate and reused. Some nitrate, however, may be converted to nitrite and then to nitrous oxide (N_2O) by denitrifying bacteria, as illustrated in Figure 31–17. Nitrous oxide is converted to nitrogen and released into the atmosphere.

Humans alter the nitrogen cycle in at least four ways: (1) by applying excess nitrogen-containing fertilizer on farmland, much of which ends up in waterways, (2) by

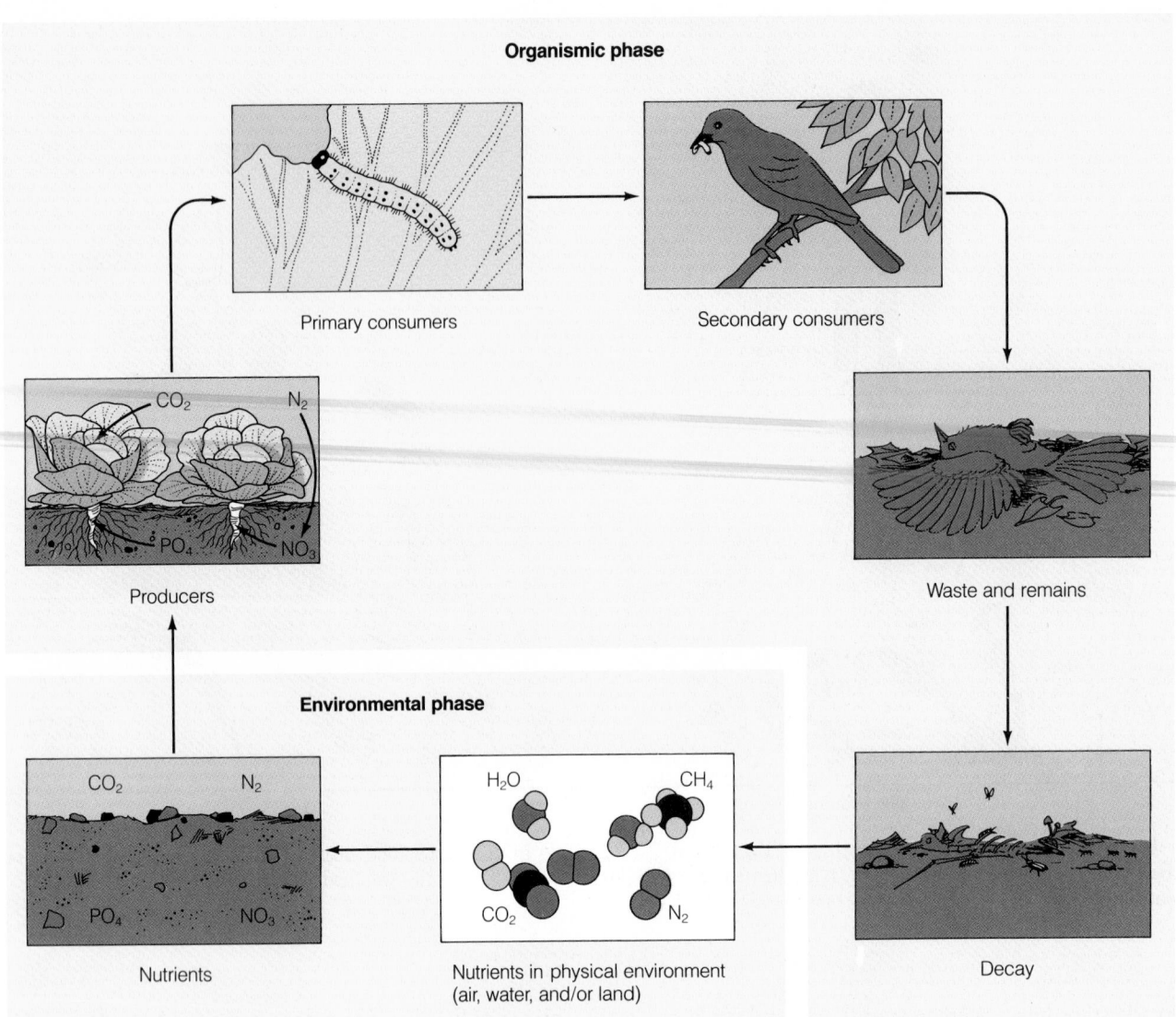

Organismic phase

Primary consumers

Secondary consumers

Producers

Waste and remains

Environmental phase

CO_2 N_2

PO_4 NO_3

Nutrients

H_2O CH_4

CO_2 N_2

Nutrients in physical environment
(air, water, and/or land)

Decay

▷ **FIGURE 31–15 General Structure of a Nutrient Cycle** Nutrients exist in the organismic and environmental phase and cycle back and forth between them.

disposing of nitrogen-rich municipal sewage in waterways, (3) by raising cattle in feedlots adjacent to waterways, and (4) by burning fossil fuels, which release a class of chemicals known as nitrogen oxides into the atmosphere. The first three activities increase the concentration of nitrogen in the soil or water, upsetting the ecological balance. Nitrogen oxides released into the atmosphere by power plants, automobiles, and other sources are converted to nitric acid, which falls with rain or snow. Besides changing the pH of soil and aquatic ecosystems, nitric acid also adds nitrogen to surface waters and may be responsible for 25% of the nitrogen pollution in some coastal waters in the United States.

Nitrogen, like phosphate, is a plant nutrient. It stimulates the growth of aquatic plants and causes rivers and lakes to become congested with dense mats of vegetation, making them unnavigable. Sunlight penetration to deeper levels is also impaired by the growth of plants, causing

oxygen levels in deeper waters to decline. In the autumn, when aquatic plants die and decay, oxygen levels may fall further, killing aquatic life.

≋ ECOSYSTEM HOMEOSTASIS

This book has pointed out that human health depends on homeostasis at two levels. First, our health requires internal balance. A breakdown of internal homeostasis can lead to disease and death. Second, human health is dependent on the environment we live in. Thus, healthy physical, chemical, social, and psychological environments are crucial to preserving our own health. Of course, the health of the environment depends on the well-being of homeostatic mechanisms operating in the biosphere to maintain ecosystem balance. **Ecosystem balance** is a kind of dynamic equilibrium. In balanced ecosystems, for example, populations grow and decline in natural cycles (thus, they are

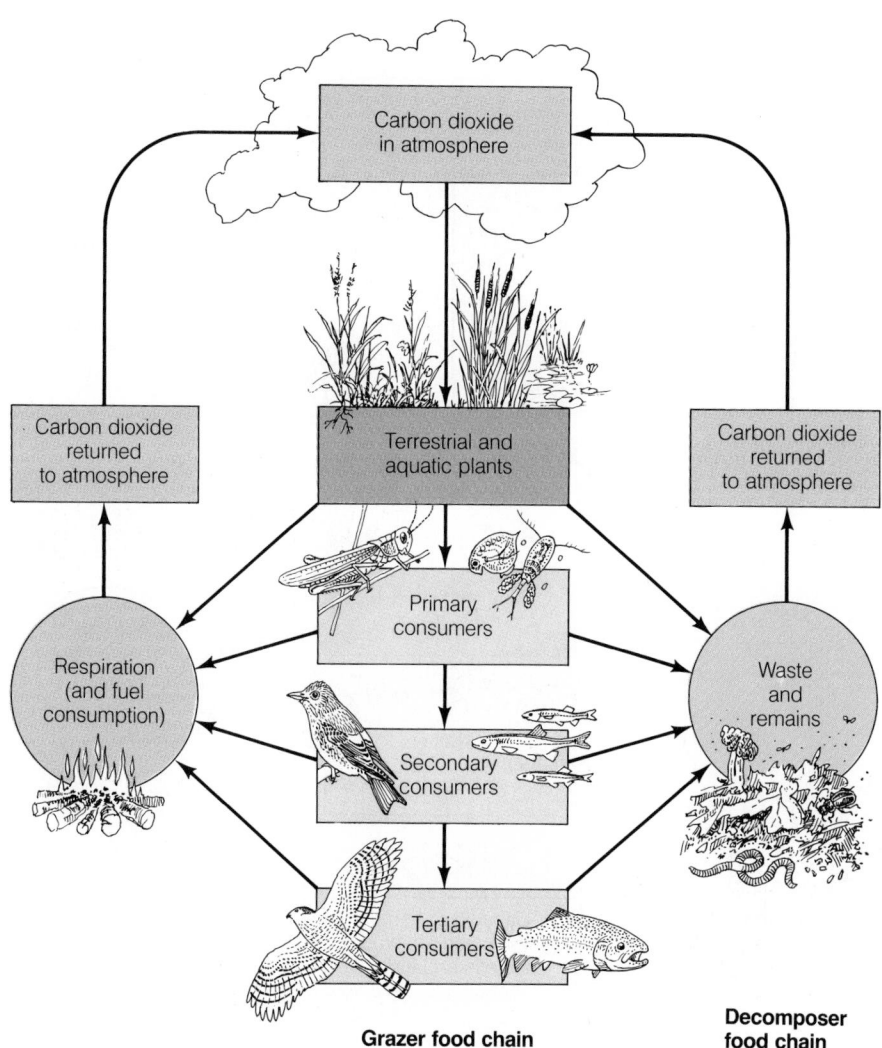

▷ **FIGURE 31–16 The Carbon Cycle**

Carbon dioxide in atmosphere

Carbon dioxide returned to atmosphere

Terrestrial and aquatic plants

Carbon dioxide returned to atmosphere

Primary consumers

Respiration (and fuel consumption)

Waste and remains

Secondary consumers

Tertiary consumers

Grazer food chain

Decomposer food chain

---

dynamic). But from year to year, they remain more or less the same (therefore, they are in a state of equilibrium).

Ecosystem Balance Is the Result of Many Interacting Factors

Ecosystem balance results from opposing forces that act on individual populations within an ecological system (▷ Figure 31–18). The first set of forces is those that cause populations to grow. I call them **growth factors.** Favorable weather, ample food supplies, and a high reproductive rate are three of many factors that cause populations to increase. The second set of factors includes those that cause populations to decline, the **reduction factors.** Adverse weather, lack of food, disease, and predation, for instance, can cause populations to decline. Reduction factors collectively constitute **environmental resistance**— so named because they resist population growth.

Numerous growth and reduction factors operate simultaneously in ecosystems. As Figure 31–18 shows, each set of factors has two components: abiotic and biotic. To understand how they work, consider a simplified example.

In wet years, grasses and other plants of the Midwest grow well. Mice and other rodents thrive on the plains because of the abundance of food, and their populations increase. This increase, in turn, results in larger populations of hawks and owls, which feed on rodents. The more food that is available, the more owl and hawk young that will survive.

In this simplified chain of events, favorable weather conditions shift ecosystem balance. Conditions are partly restored, however, by the increase in the hawk and owl populations. Increased predatory pressure from these birds will tend to decrease the mouse population. A decline in the mouse population may also be caused, in part, by a reduction in food caused by the increased number of mice. A decline in the number of mice, in turn, causes a decline in predatory birds and allows vegetation to recover.

Balance can also be restored by abiotic conditions. A harsh winter, for example, might cause the mouse population to decline. Because the number of mice decreases, the owl and hawk populations will fall.

This example illustrates some mechanisms by which slight ecosystem imbalances are corrected. Owls and hawks

CHAPTER 31   *Principles of Ecology: Understanding The Economy of Nature*   831

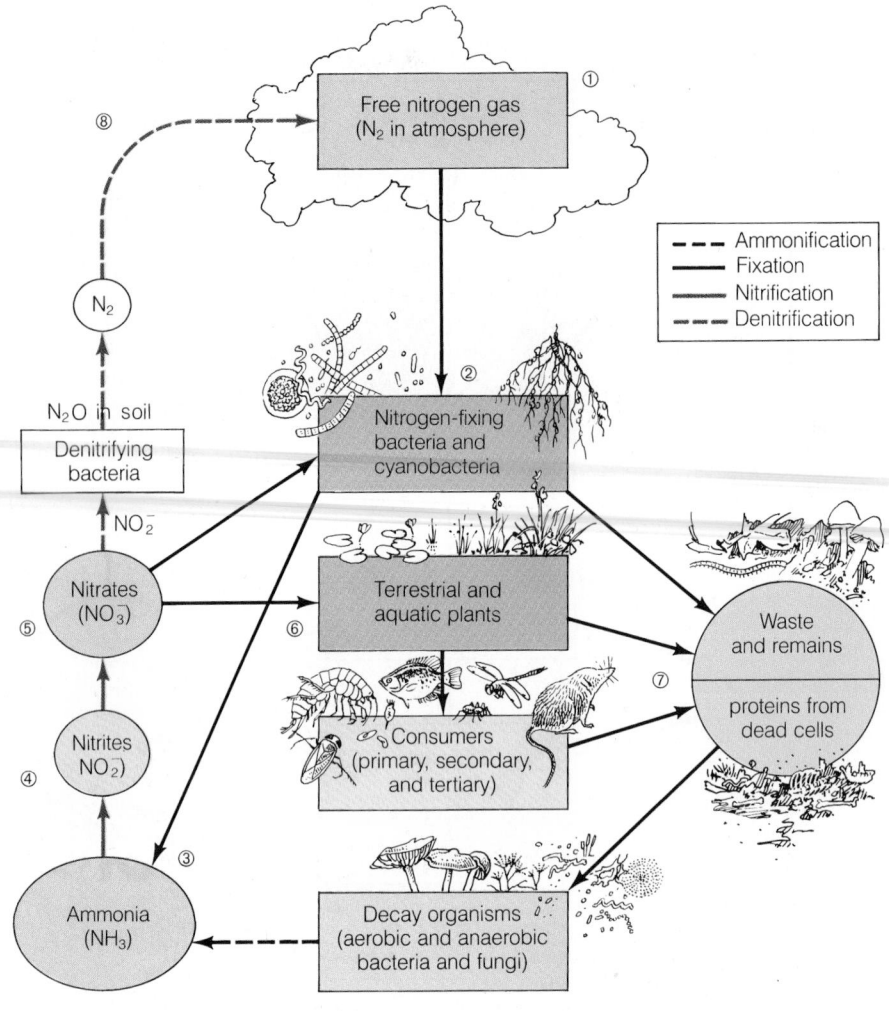

> ▷ **FIGURE 31–17 The Nitrogen Cycle** Nitrogen in the atmosphere is converted to NH₃ (ammonia) by bacteria and cyanobacteria in soil. Ammonia is converted to nitrates and taken up by plants. To follow the flow of nitrogen, follow the circled numbers.

are biotic reduction factors that help maintain ecosystem stability. In contrast, adverse winter weather is an abiotic reduction factor that helps offset the rise in the rodent population and even helps reduce the size of the hawk and owl populations.

Changes in abiotic and biotic conditions occur with great regularity in ecosystems. Minor fluctuations are of little consequence, however, as the simplified example shows. Thus, when minor changes occur, ecosystems appear to remain fairly resilient—capable of bouncing back.

As the previous discussion shows, ecosystems house an elaborate set of checks and balances that help preserve the integrity of the whole. These checks and balances also tend to minimize human impact. Sewage dumped into a river, for example, adds organic nutrients which are consumed by naturally occurring aquatic bacteria whose population is normally low. As a result, the number of bacteria increases. Because bacteria consume oxygen in the process of cellular respiration, the level of dissolved oxygen in the stream often drops. This decline, in turn, kills fishes and other organisms. As the organic pollution moves down the stream and is consumed by bacteria, however, oxygen levels in the water generally return to normal—that is, if

further spills do not occur. Fish populations below the pollution zone may be unaffected. This is an example of resilience.

Human populations are also subject to the forces that act on other populations. The impact of environmental resistance—for example, adverse weather—is generally not as noticeable in the more well-to-do nations of the world as it is in the poor countries, where large numbers frequently die in natural catastrophes such as typhoons. One reason the influence of the reduction factors, such as disease and drought, is not visible in wealthy countries can be summed up in one word: technology. Technology frees us from some of the constraints felt by other species. Nonetheless, they do not free us entirely. The epidemic of AIDS and the periodic outbreaks of influenza are just two reminders that we are still subject to the laws of nature.

Some observers assert that technology is a double-edged sword. Although some forms may increase our chances of survival, the same ones or others may threaten our overall well-being. For more on this important point, you may want to refer to the Exercising Your Critical Thinking section at the end of the chapter when you have finished reading the following material.

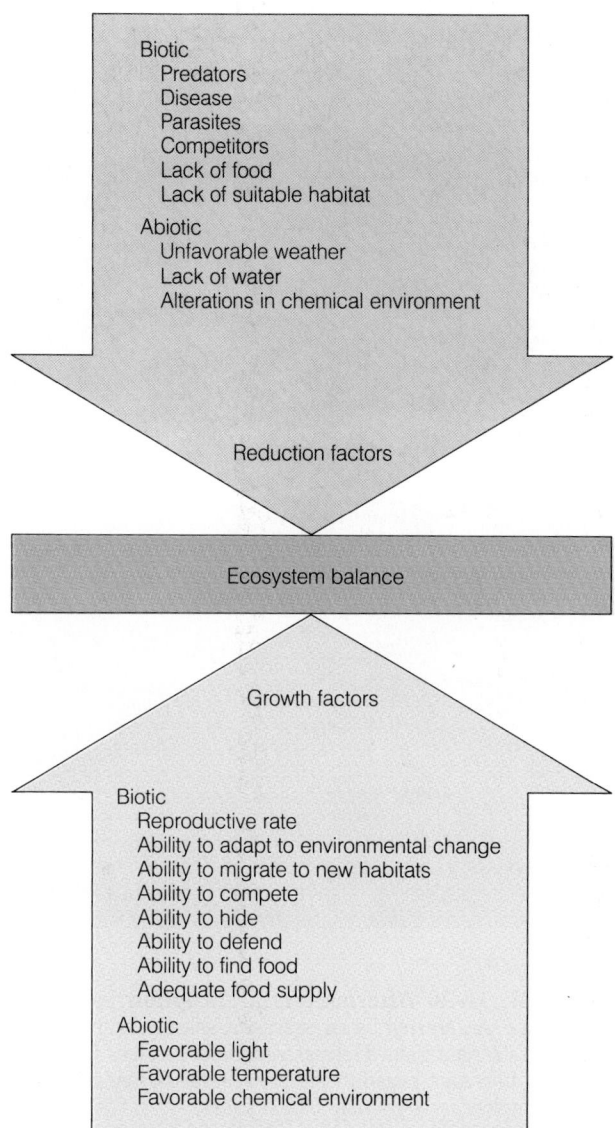

Biotic
   Predators
   Disease
   Parasites
   Competitors
   Lack of food
   Lack of suitable habitat
Abiotic
   Unfavorable weather
   Lack of water
   Alterations in chemical environment

Reduction factors

Ecosystem balance

Growth factors

Biotic
   Reproductive rate
   Ability to adapt to environmental change
   Ability to migrate to new habitats
   Ability to compete
   Ability to hide
   Ability to defend
   Ability to find food
   Adequate food supply
Abiotic
   Favorable light
   Favorable temperature
   Favorable chemical environment

▷ **FIGURE 31–18 Ecosystem Balance** Ecosystem balance is a complex equilibrium brought about by biotic and abiotic growth and reduction factors. From Daniel D. Chiras, *Environmental Science*, 3rd ed. Copyright © 1991, the Benjamin/Cummings Publishing Company, Redwood City, Calif. Reprinted by permission.

Ecologists Group the Components of Environmental Resistance (Reduction Factors) into Two Broad Categories: Density-Independent Factors and Density-Dependent Factors

Ecosystem balance is a complex issue, but it is basically the net result of the interplay of factors that increase and decrease the size of constituent populations. **Density-independent factors** are those that limit the growth of populations irrespective of their size or their density (the number of organisms per acre). Droughts, heat waves, cold spells, tornadoes, floods, and storms are good examples. An unexpected freeze in the usually warm tropics, for example, will kill most native butterflies, no matter what the population density.

Density-dependent factors are those factors whose effect is influenced by the density of a population. The greater the density, the greater the impact these factors have. Ecologists recognize at least four major density-dependent factors: competition, predation, parasitism, and disease.

Let's take predation as an example. As noted in the last chapter, **predation** is the hunting and killing of other animals, **prey.** Those organisms that hunt and kill for food are known as **predators.** As a rule, as the population density of a prey species increases, the percentage of organisms killed by predators also increases. One reason is that as the population density of prey increases, predators find it easier to locate prey. Furthermore, increased competition in the dense prey population may result in a larger number of weakened organisms. They become easy targets for predators. Prey may also be forced into less suitable habitat and may become weakened or diseased and, therefore, more likely to be captured and killed by a predator. Ecologists have also found that when the density of a prey population increases, predators that eat a variety of prey tend to shift their attention to the most concentrated ones.

Parasitism, also described in the last chapter, is another density-dependent factor. Parasites, such as tapeworms, feed on the bodies of other living organisms. As a rule, though, they do so without killing their hosts. From an evolutionary standpoint, killing one's host is suicidal for a parasite species.

Parasites spread from organism to organism in a population; thus, the higher the population density, the more readily parasites spread. In Colorado in the late 1970s, for example, bighorn sheep offspring died in record numbers because of a parasite called the lungworm (▷ Figure 31–19). Lungworms weaken sheep, making them more susceptible to pneumonia and other diseases. Under normal (uncrowded) conditions, the lungworms are found in sheep but cause little harm. Lungworm eggs are deposited in the sheep's feces and are often ingested by adults as they graze. In populations that move freely through a large territory, however, the number of eggs ingested each year is small, and lungworm infestations are inconsequential. But in populations that are pressed for space, females pick up large numbers of eggs as they graze. In Colorado, human activity limited the range of the bighorn sheep, causing the population density on the remaining habitat to increase. In the late 1970s, record numbers of parasites were found in the state's bighorns. In pregnant females, the parasitic worms pass through the placenta and infect their fetuses. Newborns, too weak to withstand the parasite, perished in record numbers. Fortunately, wildlife biologists discovered the problem and have since then captured, treated, and relocated many sheep to suitable habitat in the Rockies where human encroachment is minimal.

The incidence of infectious diseases like the flu also increases with rising population density. In World War I, for example, large numbers of American soldiers, crowded into barracks and trenches, contracted a fatal influenza

(a)

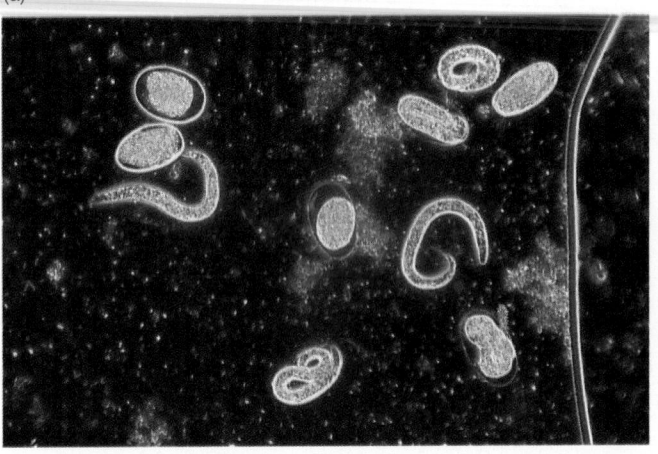

(b)

▷ **FIGURE 31–19 Host and Parasite** (*a*) Colorado's state animal, the bighorn sheep, was threatened by overcrowding, which increased the spread of (*b*) a parasitic lungworm, which passes through the placenta to the fetus. Newborns cannot survive long when infested with lungworm.

(a)

(b)

▷ **FIGURE 31–20 Aftermath of an Eruption** (*a*) This barren landscape was created by an enormous volcanic eruption at Washington's Mount Saint Helens. Thousands of acres were destroyed. (*b*) Life soon began to take root in the barren volcanic ash, but complete recovery may take decades.

virus. Because organisms that cause infectious diseases are usually transmitted by contact, through food, and by animal vectors (insects), as the population density rises, so does the spread of disease.

Succession Is the Progressive Development of Biological Communities, Either on Barren Ground or in Damaged Ecosystems

Changes in both the density-independent and the density-dependent reduction factors are commonplace in nature. The ability of ecosystems to recover from minor changes gives them a measure of stability. Not all changes can be quickly reversed, and ecosystem balance can be drastically upset by natural forces, such as volcanoes, or by human forces (▷ Figure 31–20). In such instances, ecosystems may recover, but the recovery is quite slow. A deciduous forest that is cleared, then planted with crops, for example, will return to forest on its own if it is abandoned, but full recovery generally requires 70 years. This process of resto-

ration, described more fully shortly, is one example of secondary succession and is a means by which natural systems are restored following severe disruption.

Succession, in general, is a process of sequential change in which one community is replaced by another until a mature, or climax, ecosystem is formed. A mature ecosystem is one that has reached a state of long-term dynamic balance. It is characterized by a stable level of species diversity (explained below). Two types of succession exist: primary and secondary.

Primary succession occurs where no biotic community previously existed—for example, when deep-sea volcanoes form islands or when glaciers retreat, exposing barren rock. On volcanic islands in the tropics, a rich paradise can form on the barren rock, but it takes tens of thousands of years. Seeds for plants may be carried to the islands by waves or may be dropped by birds. Over time, the plants take root, then spread to cover the entire island. Birds may settle on the island as well. In the ensuing years, new species may evolve from the old.

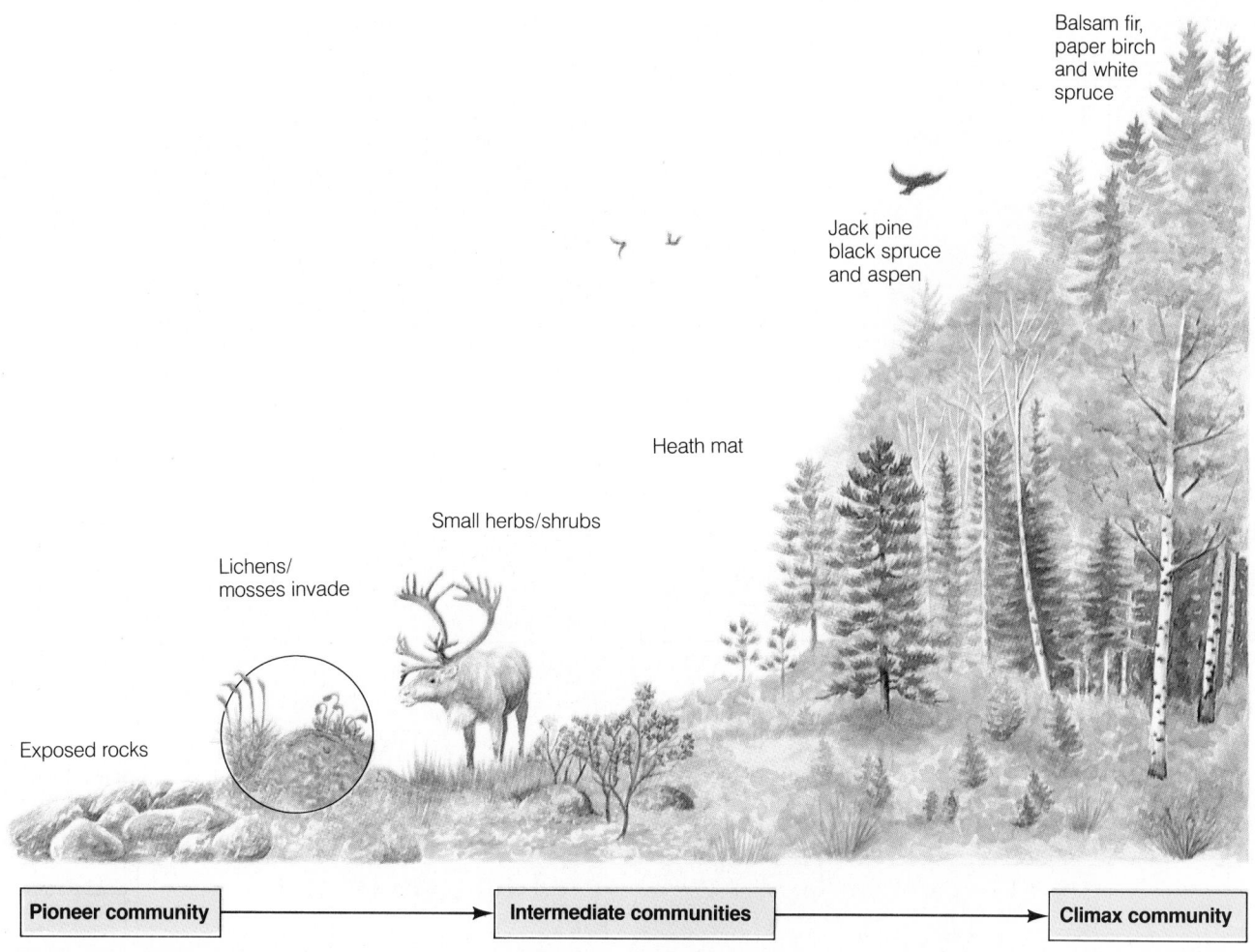

Balsam fir, paper birch and white spruce

Jack pine black spruce and aspen

Heath mat

Small herbs/shrubs

Lichens/ mosses invade

Exposed rocks

| Pioneer community | Intermediate communities | Climax community |

▷ **FIGURE 31–21 Primary Succession** As rock is exposed by a retreating glacier, biological communities develop in a process that may take hundreds or thousands of years.

The process of primary succession on rock exposed by the retreat of glaciers is shown in ▷ Figure 31–21. You may want to take a few minutes to study this process.

Secondary succession occurs where a biotic community had previously existed but was destroyed by natural forces or human actions. Secondary succession is nature's means of restoring ecosystems and reestablishing balance.

▷ Figure 31–22 illustrates secondary succession on an abandoned eastern farm field. As shown in this figure, crabgrass first invades the abandoned field. Crabgrass is well suited to life in open, sunny fields. Unlike many other plants, it can tolerate hot, sunny locations because it is a C4 plant. As noted in Chapter 5, C4 plants are capable of fixing atmospheric carbon dioxide when other plants have closed down their stomata to conserve water.

In the next two years, grasses and other herbaceous plants take root from seeds in the soil. Seeds may also be blown in from neighboring fields or carried in by animals. Over the next three decades, pine seedlings take root and begin to grow. As they get larger, sun-loving pine trees produce shade—so much, in fact, that shade-tolerant hardwood trees begin to grow (▷ Figure 31–23). Hardwoods eventually grow so large that they "shade out" many of the

pine trees. In ecological terms, they outcompete the pines. Seventy years after this process began, a mature, or climax, forest is formed, consisting mostly of hardwoods and an occasional pine. A climax forest is in a state of dynamic equilibrium. It will remain more or less the same if it is not disturbed by fire, timber cutting, or some catastrophic event.

Secondary succession in the abandoned farm field illustrates some key principles of succession. One of those is that colonizers, such as crabgrass, and transitional species, such as pines, thrive in their new habitat because environmental resistance is low and the conditions necessary for growth (growth factors) are favorable. These same species, however, gradually alter the environment, creating conditions suitable for the growth of other species. The new species, in turn, alter environmental conditions so much that they eliminate the species that came before them. As noted in the previous example, pines invade the open sunny fields but begin to produce shade, which is suitable for hardwood seedlings but not for pine seedlings. Hardwood seedlings grow and produce so much shade that most conifers are eliminated.

Secondary succession occurs much more rapidly than

Abandoned farmland		10–30 years Established pine forest
0–1 years Crabgrass colonizes first		
1–3 years Tall grass/ herbaceous plants		30–70 years Hardwoods invade
3–10 years Pines invade		70+ years Hardwood climax forest Succession complete

▷ **FIGURE 31–22 Secondary Succession** On an abandoned eastern farm (top left), nature begins the process of restoring a biotic community.

primary succession because soil is already present. In primary succession, soil must be formed from rock, gravel, or sand, which takes 100 to 1000 years.

One final note: damaged ecosystems are not always repaired by secondary succession. In Vietnam, for example, thousands of acres of tropical forests were destroyed by Agent Orange, an herbicide sprayed on forests from planes and helicopters by U.S. forces and their allies to reduce the likelihood of ambush. Broadleaf trees are quite susceptible to Agent Orange and died quickly. The forests, however, were invaded by a hardy brush that some scientists think may prevent mature forests from reestablishing.

As another example, clear-cutting large tracts of tropical rain forests may also prevent or severely retard secondary succession (▷ Figure 31–24). Clear-cutting exposes the land to intense sunlight, baking the soil in some areas to a bricklike consistency and preventing plants from taking root. Clear-cutting can also result in serious soil erosion, making full recovery difficult, if not impossible.

On the island of Krakatoa, a volcano destroyed once-rich tropical forests a century ago. Today, the forest has

▷ **FIGURE 31–23 Evolution of a Forest** This growing forest was once a farm field. Notice the hardwood trees growing among the well-developed pine trees.

> **FIGURE 31–24 Tropical Deforestation** Large tracts of tropical rain forest are clear-cut to provide timber and to make way for ranches and farms, which often fail because tropical soils are poor in nutrients. Abandoned clear-cuts may never recover because of severe soil erosion and other factors.

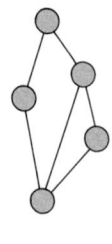
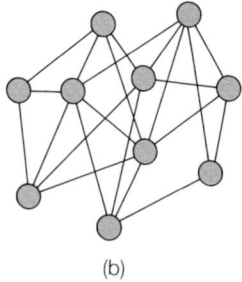

(a)　　　　　　　　　(b)

> **FIGURE 31–25 Diversity and Stability** (*a*) In a simplified ecosystem, the loss of one species generally has a more profound impact on the food web than the loss of one species in (*b*) a more complex ecosystem.

regrown. To the untrained eye it appears quite normal. But ecologists have found that only 80 tree species inhabit the forest, compared with an estimated 800 to 1000 tree species in equivalent forests. Thousands of years may be required to fully restore the forest.

Ecosystem Stability May Be Related to Species Diversity

Ecosystem stability results from the interplay of growth and reduction factors, which are part of nature's homeostatic mechanisms. Some ecologists believe that ecosystem stability may also result from species diversity. Generally speaking, **species diversity** is a measure of the number of different species in a given community or ecosystem. The more species there are, the more diverse an ecosystem is said to be.

Ecologists who believe that species diversity leads to ecosystem stability support this contention with the following facts. First, extremely stable ecosystems, such as tropical rain forests, are characterized by a remarkably high species diversity. By various estimates, tropical rain forests cover about 10% of the Earth's land surface but are home to about 60% of the world's species. In contrast, some ecosystems, such as the tundra, contain populations that tend to oscillate widely. Ecologists sometimes attribute this instability to their low species diversity. A second fact that supports the notion that species diversity is related to ecosystem stability is that intentionally simplifying an ecosystem (removing species) tends to make it unstable. Farm fields are a good example. Large fields containing one crop are known as monocultures. Monocultures are much more vulnerable to pests and disease than heterocultures—that is, fields containing several different crops. Both monoculture and heteroculture crops are more vulnerable than nature's heteroculture grasslands, which contain many more species of plants and animals.

To understand why scientists think diversity leads to stability, consider the simple models shown in ▷ Figure 31–25. As illustrated, in the simplified ecosystem, there are few species and few connections. Thus, the loss of one species—for example, a producer—in a simplified ecosystem would be far more noticeable than the loss of a producer in the much more diverse one where several producers are present. The waters off Antarctica, for example, are surprisingly rich, but compared with other life zones, Antarctica is a rather simple ecosystem. The aquatic food chain in this area is largely dependent on phytoplankton. Some biologists are concerned that the dramatic decline in ozone in the skies over Antarctica could damage phytoplankton. Loss of phytoplankton, they say, would eliminate krill, a major source of food for many aquatic species. As noted earlier, there is some evidence to suggest that this scenario may be happening.

Not all ecologists agree with the preceding ideas. Some say that diverse ecosystems such as tropical rain forests are stable because of their uniform climate, not because of their species diversity. In other words, biological stability may be the product of a constant climate. Diversity may be an outcome of the climate, which provides ample food for a great variety of species. By the same token, instability in the tundra could be a result of the volatile climate, not of the low level of diversity.

Despite differences of opinion, some generalizations can be drawn from the evidence. The most important is that intentionally reducing species diversity makes ecosystems unstable. That is important advice for farmers and foresters, who tend to convert diverse ecosystems into much simpler ones. By creating artificially simplified ecosystems, they guarantee the need for costly and sometimes environmentally harmful ways of management. By maintaining diversity, managers reduce the cost and incidental damage they create.

Scientists debate the concept of ecosystem stability with vigor these days, because some recent research has cast doubt on the concept of a naturally maintained equilibrium. Some researchers believe that populations and ecosystems rarely, if ever, return to equilibrium once disturbed.

The debate makes one thing certain: a return to normal equilibrium depends in large part on the nature and severity of the disturbance. Evidence suggests that small perturbations like changes in rainfall or short-term drought are resisted or adequately compensated for. More severe alterations, such as deforestation of the tropics, may render a system unable to recover.

Unfortunately, the topic is so complex and so easily misconstrued that more confusion than clarity seems to be coming out of the debate. What can be said at this time is that our understanding of ecosystem homeostasis is a long way from complete. As the critical thinking rules outlined in Chapter 1 suggest, sometimes a little uncertainty is necessary.

 ## ENVIRONMENT AND HEALTH: THE CUMULATIVE IMPACT

Humans impact the environment in two principal ways: (1) by altering the abiotic components and (2) by altering the biotic components. Let's first consider some examples of human tampering with abiotic factors—chemical and physical components.

Chemical pollutants from factories, power plants, lawns, and even gardens enter streams and lakes. Here they alter the chemical environment, often killing fish and other aquatic organisms. Hot water is another pollutant. Released from power plants and factories into lakes and streams, it alters the physical environment of these bodies of water, killing aquatic life. Changing the physical and chemical environment has an impact on plants, nonhuman animals, and microorganisms. It can also affect humans.

Human populations also directly impact the biotic components of the environment. Overfishing, for example, eliminates valuable members of aquatic food chains. In the Pacific Northwest, for example, overfishing reduces food for seals and other sea mammals. Another important direct impact occurs when foreign species are introduced into an ecosystem. In their new home, these species often outcompete native species of plants and animals. Two examples are shown in ▷ Figure 31–26. The first is a plant called kudzu. Intentionally introduced to the South to help control soil erosion, kudzu literally takes over fields, growing over trees and covering abandoned homes. The second is the water hyacinth, a plant accidentally introduced into Florida waterways from South America. The water hyacinth grows rapidly, choking off streams and making navigation virtually impossible.

Eliminating species can have equally disruptive effects. The elimination of wolves, mountain lions, and bears in the Kaibab plateau on the north rim of the Grand Canyon in the early 1900s, for instance, may have resulted in an explosion in the local deer population from 4000 to 100,000 by 1924. The dramatic increase was followed by widespread environmental damage caused by overgrazing

(a)

(b)

▷ **FIGURE 31–26 Foreign Competitors Introduced to the United States** (*a*) Kudzu grows wild in the south, covering fields and homes. (*b*) Water hyacinth, introduced from South America, proliferates in southern waters, choking out native plants and making navigation difficult.

and an equally dramatic crash in the deer population. Approximately 60,000 deer died in the following winters.

It is important to point out that all organisms have an impact on their environment. That is to say, all living things alter the abiotic and biotic components of the ecosystem in which they live. But the impact of human populations is different from that of other organisms, for several reasons. First, our advanced technological development has given us incredible power to alter the face of the Earth. Second, the human population has reached unprecedented numbers. Today, over 5.4 billion people live on the planet. The cumulative impacts of all these individual actions are enormous and may be endangering the biosphere. Chapter 33 examines the evidence behind this assertion and describes ways to reduce human impact and build a sustainable future. The key to success, many experts believe, is restoring global ecosystem balance and building a relationship with nature that does not endanger the homeostatic mechanisms that maintain the planet's health and the health and well-being of the millions of species that make the Earth their home.

PRINCIPLES OF ECOLOGY: ECOSYSTEM STRUCTURE

1. Most of us see our lives as apart from nature. But in truth we depend heavily on our environment for food, oxygen, clothing, materials, and much else.

2. Ecology is the study of ecosystems. It examines the relationship of organisms to their environment and the many interactions between the abiotic and biotic components of ecosystems.

3. The living "skin" of the planet is called the biosphere. It extends from the bottom of the oceans to the top of the highest mountains.

4. The biosphere is a closed system in which materials are recycled over and over. The only outside contribution is sunlight, which powers all biological processes.

5. The Earth's surface is divided into large biological regions, or biomes, each with a characteristic climate and characteristic plant and animal life.

6. The oceans can also be divided into biological regions, known as aquatic life zones.

7. An ecosystem consists of a community of organisms, its environment, and all of the interactions between them.

8. Organisms are adapted to a range of conditions in the ecosystem in which they live. This is called their range of tolerance.

9. Human activities can alter the abiotic and biotic conditions in an ecosystem, causing considerable harm.

10. In any ecosystem, one factor tends to limit growth and is therefore called a limiting factor.

11. A group of organisms of the same species living in a specific region constitutes a population. Two or more populations occupying that region form a community.

12. The physical space a species occupies is called its habitat. A species' niche includes its habitat, its position in the food chain, its range of tolerance, and so on.

ECOSYSTEM FUNCTION

13. Virtually all life on Earth depends on the producers, organisms that synthesize organic materials from sunlight, carbon dioxide, and water. The major producers are the plants, photosynthetic protists, and photosynthetic bacteria.

14. Organisms dependent on producers and other organisms for food are called consumers. Four types of consumers are present: herbivores, carnivores, omnivores, and detritivores.

15. All organisms are part of food chains. A food chain represents a feeding relationship in an ecosystem. Food chains that begin with plants consumed by grazers (herbivores) are known as grazer food chains. Those that begin with animal waste or the remains of plants, animals, and microorganisms are known as decomposer food chains.

16. Food chains are simplified components of larger networks, called food webs, that represent a truer picture of the feeding relationships in an ecosystem.

17. Food chains and food webs are a conduit for the one-way flow of energy through an ecosystem and provide avenues for the cycling of minerals and other nutrients through the ecosystem.

18. The position of an organism in a food chain is called its trophic level. In a grazer food chain, plants (producers) are on the first trophic level; herbivores are on the second; carnivores are on the third.

19. Nutrients necessary for life are involved in nutrient cycles. Nutrients in the environment enter the organismic phase through producers. The nutrients are then shunted through the food chain (organismic phase of the cycle) and eventually reenter the environment (environmental phase), where they are generally available for reuse.

20. Humans can interrupt nutrient cycles, locally and globally. The nitrogen cycle, for example, is flooded with nitrogen from human sources (fertilizer, sewage, and animal waste) in towns, cities, and rural communities.

21. The global carbon cycle is flooded with carbon dioxide from the worldwide consumption of fossil fuels. Deforestation also increases atmospheric carbon dioxide.

ECOSYSTEM HOMEOSTASIS

22. Human health is dependent on homeostasis at two levels: within the environment and within ourselves. Healthy physical, chemical, social, and psychological environments are crucial to a healthy body.

23. The environment contains numerous homeostatic mechanisms. Ecosystem balance is the net result of the interplay of biotic and abiotic growth and reduction factors.

24. Environmental resistance is the sum of the reduction factors. These factors fall into two broad categories: density-independent factors and density-dependent factors.

25. Density-independent factors are those events that limit the growth of populations irrespective of their size or density (number of organisms per acre), such as drought, heat waves, cold spells, and storms.

26. Density-dependent factors are those conditions or events whose influence in controlling population size increases or decreases depending on the density of the population: competition, predation, parasitism, and disease.

27. Ecosystems can recover from minor disturbances rather easily, but more significant changes may require more time. In some cases, damage may be so severe that recovery is impossible.

28. The term *succession* refers to a series of changes in an ecosystem in which one community replaces another until a mature, or climax, ecosystem is produced. Primary succession occurs where no community previously existed. Secondary succession occurs where a community was destroyed by natural or human events.

29. Ecosystem stability may also result from high species diversity or from favorable abiotic factors, such as climate.

ENVIRONMENT AND HEALTH: THE CUMULATIVE IMPACT

30. Humans change the environment by altering the abiotic and the biotic components.

31. All organisms have an impact on their environment. The impact of human populations, however, is different from that of other organisms because of our technological prowess and because our population has reached unprecedented numbers. Even small acts are now beginning to add up and may be putting many other species and our own population at risk.

EXERCISING YOUR CRITICAL THINKING SKILLS

1. The following statements express the view of some citizens in this country: Technology can solve all problems. We don't need to worry about global warming, overpopulation, and other environmental problems, because scientists will find technical answers to address them.

   This view is often labeled as pure technological optimism. Do you agree or disagree with the viewpoint expressed above? Why or why not? Give specific examples that support your case. Is bias entering into your opinion? In other words, do facts support your contention?

2. You work for your state's Department of Natural Resources and are asked to comment on the proposed introduction of a deer from a remote part of the Soviet Union. The rationale behind the introduction is that it would enhance hunting in the state and would improve the economy of several rural areas. How would you evaluate the proposal? On what ecological criteria would you rely to determine whether it was a good idea?

TEST OF TERMS

1. _____ is the study of ecosystems, which include biotic and abiotic components and the numerous interactions that exist between them.
2. The region of the planet that supports life is called the _____ .
3. A(n) _____ is a region characterized by a distinct climate and distinct plants and animals.
4. An ecosystem consists of two basic components; _____ and _____ .
5. The spectrum of conditions under which an organism can survive is called its _____ _____ _____ .
6. One factor is often more important than others in regulating growth in an ecosystem. It is therefore called a(n) _____ _____ .
7. A group of like organisms occupying a specific region constitutes a(n) _____ . In any given ecosystem, several populations exist together and form a(n) _____ .
8. An ecological _____ includes the habitat of an organism and its place in the ecosystem.
9. Plants and algae are also called _____ , because they synthesize organic molecules necessary for all other organisms, which are collectively referred to as _____ .
10. Animals that eat only plants are called grazers, or _____ ; animals that eat plants and animals are called _____ ; and organisms that consume waste and the remains of animals and plants are _____ .
11. Two types of food chains exist, _____ and _____ .
12. Nutrients are recycled in ecosystems, but all sunlight energy is eventually dissipated into space as _____ .
13. The position of an organism in the food chain is called its _____ level.
14. _____ is the dry weight of the living material in an ecosystem.
15. Each nutrient cycle consists of two basic parts: the _____ phase and the _____ phase.
16. The conversion of nitrogen gas to nitrate and ammonia by root nodules and soil bacteria is called _____ _____ .
17. Ecosystem balance, or stability, results from the interplay of two sets of factors: _____ factors, which tend to increase the size of populations, and _____ factors, which tend to decrease the size of populations.
18. Competition and predation are two _____ -dependent components of environmental resistance.
19. _____ is a process of sequential change in which one community is replaced by another until a mature, or climax, ecosystem is formed. When this process occurs on a newly formed island, it is called _____ _____ .

Answers to the Test of Terms are found in Appendix B.

TEST OF CONCEPTS

1. Humans are a part of nature. Do you agree or disagree with this statement? Support your answer.
2. Define the term *ecology* and give examples of its proper and improper use.
3. The Earth is a closed system. What are the implications of this statement?
4. Define the following terms: biosphere, biome, aquatic life zone, and ecosystem.
5. Define the term *range of tolerance*. Using your knowledge of ecology, give some examples of ways in which humans alter the abiotic and biotic conditions of certain organisms, and describe the potential consequences of such actions.
6. Describe ways in which humans alter their own range of tolerance for various environmental factors.
7. What is a limiting factor? Give some examples.
8. Define the following terms: habitat, niche, producer, consumer, trophic level, food chain, and food web.
9. Explain why the biomass at one trophic level is lower than the biomass at the next lower trophic level.
10. Outline the flow of carbon dioxide through the carbon cycle, and de-

scribe ways in which humans adversely influence the carbon cycle.

11. Using what you have learned about ecology, describe why it is important to protect natural ecosystems and other species.

12. With the knowledge you have gained, explain why it is beneficial to set aside habitat to protect an endangered species.

13. Looking back over the principles you have learned in this chapter, write a set of guidelines for human society that would help us live sustainably on the Earth.

The Biomes

≈

Bridger Mountains, Montana. In order to have a clean, healthy environment and a strong economy, we must change our ways, patterning our society after nature.

T he Earth formed approximately 4.5 billion years ago. At that time, the planet was a barren mass of rock and ice. Over billions of years, however, life evolved on the planet. Today, the Earth's surface is carpeted with a rich and diverse array of life forms existing in a complex web of life, the biosphere. But as anyone who has traveled across the North American continent knows, the biosphere is divided into distinct regions, which ecologists call **biomes.** In other words, the biosphere is a mosaic of distinct habitats. Biomes differ from one another in climate—that is, rainfall, sunlight, temperature, and other abiotic factors. These discrepancies lead to marked differences in the types and abundance of species that live in the Earth's many biomes. Put another way, the differing abiotic conditions of the biomes profoundly affect the distribution of life forms.

This chapter describes some of the major biomes and outlines some threats to them. It also offers suggestions for protecting the world's rich biological legacy and points out how important protection is for the survival of the many other species that share this planet with us. Finally, like other chapters, this one stresses the importance of a healthy biosphere to the immediate health and long-term future of humankind.

≈ THE TERRESTRIAL BIOMES

We begin our journey through the Earth's diverse biomes in the north and work our way south, passing from the frozen Arctic to the hot, humid jungles of the equatorial region. On this journey, you will learn a little about the climate of several key biomes and the unique biological adaptations that permit plants and animals to thrive within them.

Remember as you study each biome that adaptations arise as a result of mutations and other events (Chapter 23). Favorable or beneficial adaptations persist through natural selection. Many abiotic factors are agents of natural selection and thus shape life within the biomes.

As you will see, each biome is characterized by the dominant form of vegetation. But that does not mean that vegetation is identical throughout the biome. Within any given biome, regional differences in climate (and other factors like soil) can alter the composition and abundance of species. Also bear in mind that the boundaries between biomes are not always clear. Most biomes merge with their nearest neighbors because geographic changes in climate are usually gradual.

The study of the biomes underscores an important lesson, presented in the previous chapter: life can adapt to a wide range of conditions. In the frozen tundra, for example, some organisms survive in air temperatures as low as $-70°C$ ($-90°F$). In contrast, some desert species can tolerate $50°C$ ($120°F$).

The Tundra Is the Northernmost Biome

Stretching across the northernmost portions of North America, Europe, and Asia is the Arctic **tundra.** The Arctic tundra is the northern limit to plant growth.[1] Covering about 10% of the land mass of the Earth, the tundra lies between a region of perpetual ice and snow to the north and a band of coniferous forests to the south (▷ Figure 32–1). The tundra is a treeless tract, characterized by grasses, shrubs, and matlike vegetation (mosses and lichens) adapted to the harsh climate.

The Arctic tundra receives very little precipitation (less than 25 centimeters, or 10 inches, a year), and most precipitation occurs in the summer. During the long, cold winters, the mean average temperature remains well below zero. Because Arctic summers are short and winters are so cold, the deeper layers of soil remain frozen throughout the year. Permanently frozen soils are called the **permafrost.**

The permafrost and the harsh winters of the Arctic tundra prevent deep-rooted plants, such as trees, from growing there. In the brief Arctic summer, however, the long days and relatively warm temperatures permit the superficial layers of soil to melt. Because evaporation is low and melted snow and ice cannot percolate downward into the soil in the summer, the tundra becomes a soggy landscape, dotted with shallow ponds, lakes, and bogs (▷ Figure 32–2). The water-saturated soil also limits the types of plants that can grow there.

During the summer months, the tundra comes alive with insects (mosquitoes and black flies) and birds that migrate north to nest on the rolling plains. The birds feed on the swarms of insects, raise their young, then migrate south with their offspring, often in great flocks.

Despite the harsh conditions, a variety of animals live year round on the tundra. Ptarmigan, musk oxen, and Arctic hares, for example, are adapted to the extreme cold (▷ Figure 32–3). Animals survive by living in burrows or by having thick layers of insulation or a large body size (which retains heat well). Many animals, like the birds mentioned above, are migratory. (For a discussion of adaptations that permit some animals to survive in the tundra, see Spotlight on Evolution 6–1.)

The tundra is a fragile environment, easily damaged by human actions. The short growing season means there is little time for vegetation to recover from damage caused by mining, oil and gas development, and other human activities. Tire tracks on the tundra that destroy vegetation take several decades to repair. Damage caused by spills of oil or hazardous waste may take far longer.

Nowhere is the threat to the tundra more worrisome than in northern Alaska. Alaska's northern coastline is approximately 1100 miles long, and all of it except a 115-mile stretch is open to oil and gas exploration and development. Today, however, many oil companies want the remaining 115 miles opened—even though it is part of the Arctic National Wildlife Refuge (ANWR), one of the last great wilderness areas on the planet and home to dozens of

[1] Note that the word *Arctic* is pronounced ark-tik, not art-ik.

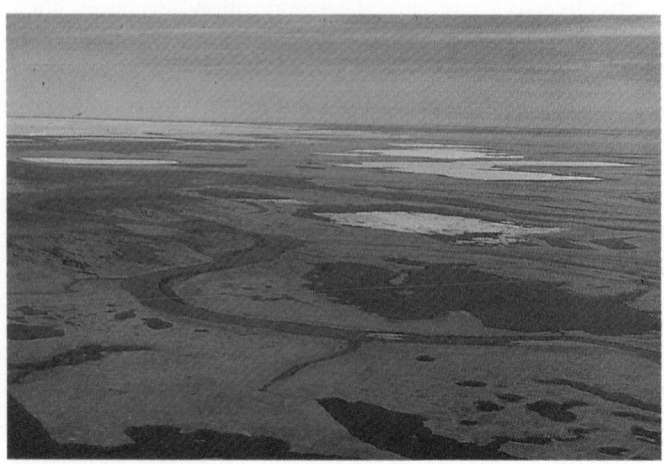

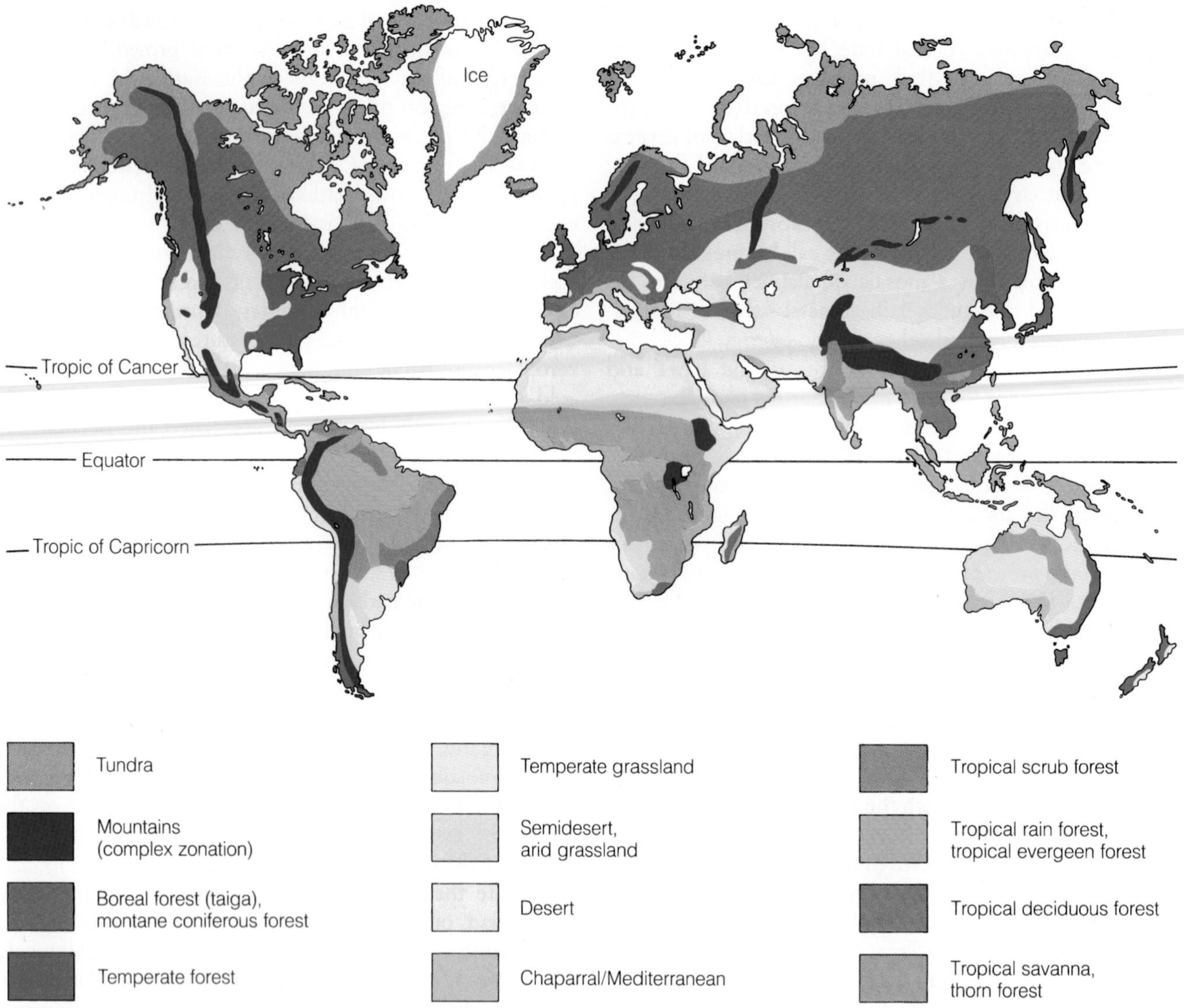
Ice
Tropic of Cancer
Equator
Tropic of Capricorn
Tundra
Mountains
(complex zonation)
Boreal forest (taiga),
montane coniferous forest
Temperate forest
Temperate grassland
Semidesert,
arid grassland
Desert
Chaparral/Mediterranean
Tropical scrub forest
Tropical rain forest,
tropical evergeen forest
Tropical deciduous forest
Tropical savanna,
thorn forest

▻ **FIGURE 32–1 The Earth's Biomes**

species. Should the oil companies be allowed to explore for oil in this part of the ANWR?

If oil is discovered in the ANWR, the region would be turned into a network of roads, airports, power plants, oil platforms, waste ponds, buildings, and gravel pits (▻ Figure 32–4). Nonetheless, proponents of oil development in the ANWR believe that wildlife and oil development can peacefully coexist. Critics scoff at the idea and point to estimates by wildlife biologists suggesting that oil development will result in a 20% to 40% decline in the large caribou herds that spend their summers in the region. Biologists also predict that oil development will wipe out half of the existing musk oxen population. Populations of grizzly bears, polar bears, wolverines, and other animals are likely to suffer enormously.

Oil development in the region would also result in air pollution, water pollution, and noise. Oil spills and hazardous waste disposal on the tundra are common in nearby

▻ **FIGURE 32–2 The Tundra** In the summer, this biome becomes dotted with lakes, ponds, and bogs. Water collects because evaporation is low and water percolation into the ground is prevented by the permafrost.

(a)

(b)

▷ **FIGURE 32–3 Tundra Animals** (*a*) Ptarmigan (tar-mi-gun) and (*b*) musk oxen.

Prudhoe Bay. In fact, the U.S. Department of Interior has recorded over 17,000 oil spills in the Arctic since 1973. Oil spilled on the tundra kills plants.

The Taiga Is a Band of Coniferous Trees Spreading Across the Northern Continents Just Below the Tundra

A wide band of cone-bearing, or coniferous, trees (pine, fir, and spruce) extends across Canada, part of Europe, and Asia and constitutes the **taiga,** or **northern coniferous forest biome** (see Figure 32–1). The taiga, which a student of mine once defined as a "carnivorous forest" on his test, is characterized by a longer growing season than its northern neighbor, the tundra.[2] It also receives far more precipitation.

In the summer, the subsoil of the taiga thaws, permitting

[2] I gave him full credit for humor.

▷ **FIGURE 32–4 Arctic Nightmares** Oil development on Alaska's North Slope causes significant damage to the delicate tundra.

deep-rooted plants, such as trees, to live. But the summer growing season is still short in comparison with that in the southern biomes (grassland and temperate deciduous forest).

The rather cold, snowy winters and limited growing season of the taiga have resulted in numerous adaptations. Few organisms illustrate these adaptations as well as the conifer. **Conifers** are trees that remain green throughout the year. They have narrow, pointed leaves called needles. Needles contain photosynthetic cells and are retained throughout the year, allowing conifers to continue to photosynthesize (albeit slowly) throughout the winter. The presence of needles throughout the year also permits the plant to make the most of the growing season. Unlike the deciduous hardwoods of the south, which lose their leaves in the winter and must develop new ones each growing season, conifers can take immediate advantage of the sunlight, moisture, and warmer weather of spring.

Retaining needles may help conifers survive in the taiga's short growing season, but it also creates a problem, for the needles tend to capture snow, which builds up on the tree. Instead of breaking under the weight of snow, though, the branches of conifers are adapted to bend and shed snow, preventing it from accumulating.

Another adaptation of great importance to conifers is the waxy coating found on their needles which greatly reduces evaporation. Because water transport up the tree from the ground is restricted during the winter, the waxy coating prevents the needles from drying out and helps the tree survive.

As anyone who has flown east-west over Canada will attest, the dense coniferous forests of the taiga extend for miles on end. These forests however, are interspersed with meadows, lakes, ponds, and rivers, and together they support a variety of species. Bears, moose, beavers, and deer are common inhabitants of the taiga. Many smaller mammals also make their home in the taiga, including foxes, marten, wolverines, and snowshoe hares. Because of the relatively cold winter and relatively short growing season, however, species diversity in the taiga is fairly low compared with that in biomes to the south.

Because of Its Remoteness and Relatively Harsh Conditions, the Taiga Still Supports many Large Populations of Wild Animals. Wolves and grizzly bears are two important species in the taiga. At one time, the grizzly population extended well into the United States—from the Mississippi River to California and from Canada to Arizona. As many as 1.5 million grizzly bears may have lived in the lower 48 states in the 1800s. However, predator control programs, ranching, farming, hunting, and other human activities have virtually exterminated the grizzly from its former U.S. range; today, only about 600 to 800 grizzlies remain in the lower 48, mostly in Montana and Wyoming. Wolf populations were also pretty well exterminated by 1935; today, only remnant populations remain in a few isolated spots (Isle Royale, Michigan, northern Minnesota, and northern Montana).

In the northern limits of their range, the taiga, grizzlies and wolves still thrive. Because of the relatively harsh winters and remoteness of the taiga, human activities have been somewhat limited. As humans move northward in search of timber and other resources, however, many fear that the survival of these animals will be threatened.

Heavy Timber Cutting in Parts of the Taiga Is Destroying Large Tracts of Forest and the Species that Live in Them. Along the west coast of North America, the forests of the taiga are bathed in moisture from the Pacific Ocean. Because the ocean also moderates temperature, this region enjoys a much longer growing season. These two factors combined result in forests that contain some of the world's largest trees. Heavy pressure from timber companies and consumers, however, is now destroying many of these magnificent trees and the species that make these forests their home ($\triangleright$ Figure 32–5).

Efforts are under way to reduce the overcutting, but in many areas of Canada and the United States, powerful logging companies are resisting. Nonetheless, progress toward reducing timber cutting is being made. The U.S. Congress, for example, recently reduced cutting in the Tongass National Forest of Alaska and set aside a large tract of the forest for permanent protection. The U.S. government also set aside 7 million acres of old-growth forest in the Pacific Northwest to protect this nearly extinct ecosystem and the species that live there. You can help reduce the nation's demand for wood by recycling cardboard, paper shopping bags, newspaper, and other paper products. You can also participate by purchasing recycled paper products (copy paper, toilet paper, paper towels, and greeting cards). When it comes time to build a house, build a smaller one to help reduce the growing pressure on the world's forests.

The Temperate Deciduous Forest Biome Occurs in Regions Characterized by Abundant Rainfall and Long Growing Seasons

The **temperate deciduous forest biome** is located in the eastern United States, Europe, and northeast China. In the United States, this biome, which occupies the eastern half of the country, is home to about half of the human population. Characterized by abundant precipitation (Table 32–1) and a long growing season (five to six months), this biome supports a wide variety of plants and animals.

The dominant plants of the temperate deciduous forest biome are **deciduous trees**—the broad-leaved trees that shed their leaves each year in the fall. Maple, oak, black cherry, and beech trees are good examples ($\triangleright$ Figure 32–6). The loss of leaves is believed to be an adaptation that greatly reduces evaporation at a time when the supply of liquid water is limited. (It is interesting to note how differently hardwoods and conifers respond to water shortages and cold weather.)

In the spring, new leaves develop from buds. In the relatively brief period after the ground has thawed and before the leaves have fully developed, numerous species

$\triangleright$ **FIGURE 32–5 Clear-Cut in Canada** This forest has been clean cut. It once contained trees 500 to 700 years old. Old-growth forests such as these contain a number of unique species that cannot survive in younger forests.

TABLE 32-1 Precipitation in Major Biomes	
BIOME	ANNUAL PRECIPITATION (INCHES)
Tundra	Under 10
Taiga	15 to 40
Temperate deciduous	30 to 60
Grassland	10 to 30
Desert	Under 10
Tropical rain forest	60 to 160

▷ **FIGURE 32–6 Temperate Deciduous Forest** Broadleaf trees in this biome lose their leaves each winter. Years of leaf litter have produced rich soils that human society has long used for farming.

of wildflowers sprout on the sunny forest floor. But the shade provided by deciduous trees greatly limits plant growth on the forest floor throughout most of the rest of the growing season.

Deciduous trees can be thought of as nutrient pumps. Nutrients are drawn up through the roots and incorporated into the leaves. As winter approaches, the leaves detach and fall to the ground. There, they decay, releasing their nutrients into the soil. As a result, the soils of the temperate deciduous forest biome are generally rich and productive. The fertile soil and abundant plant life of the forest support a rich and varied population of insects, microorganisms, birds, reptiles, amphibians, and mammals (▷ Figure

32–7). Common mammals include the racoon, white-tailed deer, red fox, and black bear.

The Temperate Deciduous Forest Biome Has Been Heavily Altered by Human Activities. Ernest Hemingway once wrote that "a continent ages quickly once we come." In fact, few biomes have been so heavily altered by human activities as the temperate deciduous forest. Destruction of the forests began when colonists first settled the continent.

Early settlers cleared the forests for farmland, taking advantage of thousands of years of soil production. They also cut trees to make room for their homes, towns, orchards, and roads. The colonists and those that followed, in fact, have so dramatically changed the landscape that only about 10% of the land east of the Mississippi is still forested, and only about 0.1% of the original forests remain. Much of the existing forest is second-, third-, and fourth-growth—that is, the second, third, or fourth generations of trees to grow in an area since humans arrived and cut down the trees that were present 200 years ago.

Unfortunately, many early settlers failed to practice soil conservation on their farms. Heavy rains often washed the soil away. Making matters worse, farmers usually did little to replenish soil nutrients. In a cycle that would repeat itself many times, farmers depleted the land, destroying soil that had taken centuries to form, then moved westward.

The Grassland Biome Occurs in Regions of Intermediate Precipitation

Grasslands exist in temperate and tropical regions in areas that receive intermediate levels of precipitation–that is, less precipitation than forested regions, but more than deserts (see Table 32–1). As shown in Figure 32–1, grasslands can be found in North America, South America, Africa, Europe, Asia, and Australia.

▷ **FIGURE 32–7 Animals of the Temperate Deciduous Forest** (*a*) Black bear, (*b*) white-tailed deer, and (*c*) painted turtle.

(a)　　　　　　　(b)　　　　　　　(c)

▷ **FIGURE 32–8 Grassland** This biome does not receive enough moisture to support trees. The only trees found in grasslands, therefore, were planted by humans or live along streams.

▷ **FIGURE 32–9 Farming the Grassland** Huge farms now occupy most of North America's grasslands. Very little native grassland remains.

In North America, grasslands form a continuous, wedge-shaped zone extending from the Gulf of Mexico northward through Canada, reaching the taiga (see Figure 32–1). All grasslands, however, bear a remarkable similarity (▷ Figure 32–8). Most are on flat or slightly rolling terrain. Carpeted in thick grasses, the soils are probably the richest in the world. The deep, nutrient-rich soil has resulted from thousands of years of plant growth and decay.

In the United States, the grasslands begin just east of the Rocky Mountains.[3] As you will soon see, moisture from air masses sweeping across the country from the Pacific Northwest is largely captured by the Cascade Mountain range of Washington and Oregon and its eastern cousins, the Rocky Mountains, which extend from Canada to Mexico. As the air moves eastward from the Rockies, however, precipitation gradually increases. The meager amounts of moisture that fall on the eastern plains of Colorado, Wyoming, and Montana support a segment of the grassland biome known as the short-grass prairie. As rain and snowfall increase, the short-grass prairie gives way to what was the tall-grass prairie of the Dakotas, Nebraska, and Kansas, which has been lost to human development, mostly farming, and exists today in only small patches.

Grasslands are usually devoid of trees because of the lack of sufficient annual precipitation and periodic drought. Thus, the few trees that can be found in the Great Plains of the United States are those that have been planted around homes and farms or those (cottonwoods) that live along streams and rivers.

Grasses, however, require less water than trees and are well adapted to periodic drought. Many species of grass, for example, send their roots deep into the Earth, drawing on reliable moisture supplies far below the surface. Grasses are also well adapted to periodic fires caused by lightning.

[3] A relatively small patch of grassland can also be found in the Great Basin, between the Cascade Mountains on the west and the Rockies on the east.

Fires burn the plant above the ground but do not harm the roots, from which new life can spring.

Several Native Species of the North American Grasslands Have Been Exterminated by Human Settlement. The grassland biome of North America once supported an enormous population of bison—perhaps as many as 60 million animals. Roaming the prairies, bison herds were so large that it would take men traveling on horseback three to four days to pass through a single herd.

The grassland biome of North America also supported grizzly bears and elk. As humans settled the grasslands and began to farm and raise cattle, grizzlies were exterminated, and the elk herds fled to the protection of the Rocky Mountains, where they remain today. Many species, including deer, pronghorn, badgers, and coyotes, still persist on the grasslands of North America, sharing habitat with cattle, horses, sheep, and people.

Grasslands Are Vital to World Food Production, But Short-Sighted Land Management Threatens many Regions. In a contest for the most drastic human alteration of a biome, the grasslands would have to rank among the top candidates. Like the temperate deciduous forests, grasslands have been radically altered by human activities. Farming has probably had the most significant impact of all human actions (▷ Figure 32–9). Today, many of the world's grasslands have been plowed under and planted. In the United States, in fact, very little of the tall-grass prairie still exists. Most of it has been plowed to plant corn, wheat, soybeans, and other crops. The short-grass prairie, being drier and less productive, has fared much better. Except where irrigation water has allowed farmers to cultivate the land, much of the short-grass prairie still exists. This land, however, is often grazed by cattle and, in some places, overgrazed.

Poor agricultural practices on farms and ranches

▷ **FIGURE 32–10 Shelterbelts** Trees planted alongside farm fields reduce wind erosion, helping protect soil. Shelterbelts also capture snow, increasing soil moisture, and provide habitat for many species.

▷ **FIGURE 32–11 Sagebrush** This woody plant (right of fence) invades overgrazed rangeland, making it useless for grazing.

throughout the world have resulted in widespread soil erosion and desertification in the grassland biome (Chapter 33). The most dramatic evidence of farmland abuse came in the 1930s in the western and midwestern United States. An extended drought combined with fence-row-to-fence-row planting spawned one of the most significant environmental disasters of human history, the dust bowl. Millions of tons of topsoil were lost from US farms in huge dust storms.

The dust bowl stimulated a rash of conservation efforts to help protect farmland. Perhaps the most significant step was the planting of long rows of trees alongside fields to reduce wind erosion (▷ Figure 32–10). Today, farmers sometimes leave wheat stubble and residues from crops on the ground over the winter to protect soils. Unfortunately, many of the improvements in soil management made in the immediate post-dust-bowl era are being lost. Many farmers are once again planting fence row to fence row in an effort to increase their output and make up for low grain prices. It is not surprising that gargantuan dust storms are becoming more and more common in some areas, such as Texas and California. Mismanagement of farmland occurs in many other countries as well and now threatens the long-term future of agriculture. (For more on the subject, see Chapter 33.)

Overgrazing Greatly Alters the Short-Grass Prairies.
Although the short-grass prairie is less often farmed, over the years ranchers have allowed their cattle to overgraze many areas in the private and public domain. Continual overgrazing eliminates many hardy grasses, and in semiarid grasslands of North America (Colorado, for example) it creates dry soil conditions suitable for the growth of weedy species such as sagebrush. Because cattle will not eat sagebrush, the woody plant thrives in overgrazed fields, eventually taking over. Throughout the West, once productive pastures have become overgrown with sagebrush

(▷ Figure 32–11). As you travel in eastern Wyoming and Colorado, look for fields of sagebrush; chances are they are the result of overgrazing. With a little care, such an ecological travesty could have been avoided.

The Desert Biome Is Characterized by Dry, Hot Conditions but Abounds with Life Forms Adapted to the Harsh Climate

Deserts exist throughout the world, and some cover vast regions. The Sahara, for example, stretches across northern Africa and is about the size of the United States. In North America, deserts exist primarily on the downwind side of mountain ranges. As mentioned earlier, mountains rob air masses passing over them of moisture. As moist, relatively warm air reaches a mountain range, it is thrust upward and cooled. Cooling causes moisture in air masses to condense, forming larger droplets, which fall from the sky as rain (▷ Figure 32–12). In the winter, ice crystals in clouds grow larger, then fall from the sky as snow. On the downwind side of mountains, the air mass falls. The air expands and warms. Because warm air holds more moisture than cold air, precipitation in the region immediately downwind from a mountain range is greatly reduced, creating semiarid or arid (desert) conditions. This phenomenon is called the **rainshadow effect.**

Deserts, although dry, are not devoid of precipitation (see Table 32–1). But rain often comes in violent downpours, causing flash flooding and severe erosion. Flash floods, in fact, have played a key role in sculpting the magnificent canyons of the desert Southwest (▷ Figure 32–13).[4] Many species of desert wildflowers are well adapted to the infrequent but intense spring rains. These

---

[4]The dramatic topography also results from the retreat of the oceans that once covered much of the North American continent. As waters retreated, they scoured the landscape, leaving magnificent geological formations.

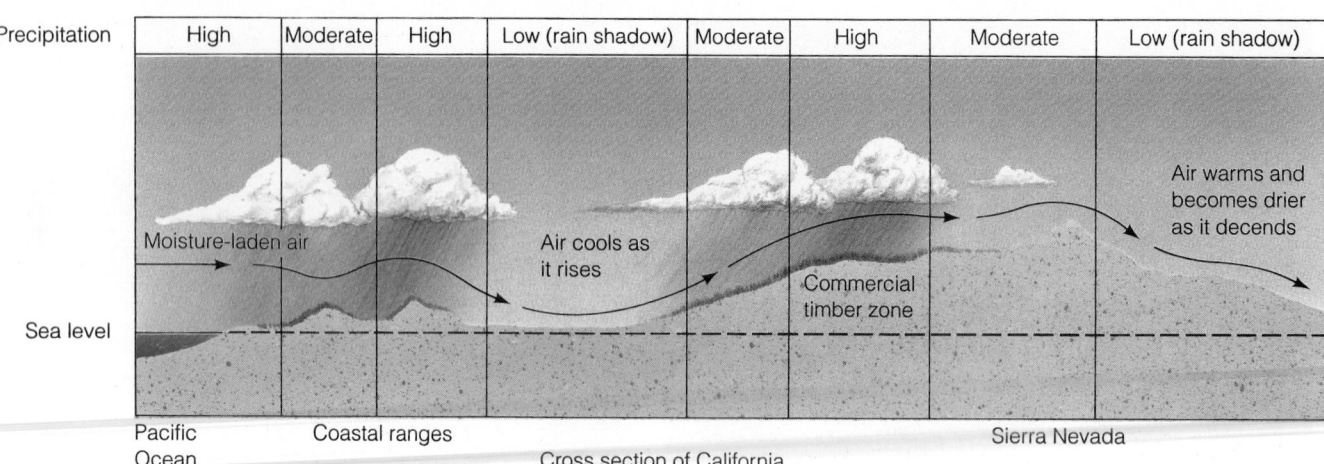

Precipitation	High	Moderate	High	Low (rain shadow)	Moderate	High	Moderate	Low (rain shadow)

Moisture-laden air

Air cools as it rises

Commercial timber zone

Air warms and becomes drier as it decends

Sea level

Pacific Ocean

Coastal ranges

Cross section of California

Sierra Nevada

▷ **FIGURE 32–12 The Rain Shadow Effect** Mountains rob air masses of moisture, often creating deserts downwind.

species have an accelerated life cycle, in which they grow and produce flowers within a few days of a thunderstorm, turning the desert into a colorful garden almost overnight (▷ Figure 32–14).

Plants that live in the desert must also be adapted to tolerate a wide range of temperatures. Daily temperatures in the desert may reach 50°C (120°F) during the day but drop to near freezing at night. The temperature in the desert fluctuates so much in large part because the air is dry; the absence of moisture in the atmosphere reduces heat retention. The baking sun warms up the desert floor during the day, but the heat quickly escapes at night.

Winter in the desert often brings freezing temperatures and snow for several days and sometimes weeks on end. Desert plants must also be able to tolerate such cold.

Plants in the desert are adapted to low soil moisture,

because water evaporates quickly from the soil. Many desert plants, such as cacti, have shallow root systems that extend laterally from the plant. Running just below the surface, these roots absorb rain and melted snow, then transport the water to the main body of the plant, where it is stored. Other desert plants, like the mesquite tree, have deep tap roots, which extend downward sometimes over 30 to 60 meters (100 to 200 feet) to moist soil. Tap roots anchor the plant and also provide a relatively reliable source of water.

Water absorbed by the roots of many desert plants is stored in succulent, water-retaining tissues, giving the plant an ample supply on which to draw during the rainless months. In most desert plants, water supplies are protected by the presence of thick outer layers and waxy coats that reduce moisture loss.

Cacti and other plants are often widely spaced on the desert floor. This spacing reduces competition for water, ensuring an adequate supply for plants. But how do plants

▷ **FIGURE 32–13 Rainfall Sculpts the Desert** Rainfall carves the loose sand and rock, creating magnificent formations in the desert Southwest.

▷ **FIGURE 32–14 Desert Garden** Sudden spring rains cause desert flowers to bloom.

(a)

(b)

▷ **FIGURE 32–15 The Cactus** (*a*) Spines protect cacti from hungry animals and also provide shade. (*b*) Some cacti have white hairs that block out the sun.

space themselves? Some plants release growth-inhibiting chemicals into the soil; these chemicals deter competitors from taking root, thus creating a region around each plant that is relatively free from competition (Chapter 30).

The thorns of cacti are yet another adaptation that reduces water loss. Thorns protect cacti from being eaten, but they also give some degree of protection from the sun

by providing shade and reflecting some sunlight from the plant (▷ Figure 32–15a). White, fluffy hairs found on some species of cacti also reflect sunlight and provide shade (Figure 32–15b).

Many insects and other animals also make the desert biome their home. Like the plants, the animals are well adapted to desert conditions. The thick scales of snakes and lizards, for example, minimize water loss, permitting these creatures to thrive in the dry, hot conditions (▷Figures 32-16a and 32-16b). Snakes and lizards also excrete a highly concentrated urine that reduces water loss. Lizards

(a)

▷ **FIGURE 32–16 Animals of the Desert** (*a*) Gila monster, (*b*) desert tortoise, and (*c*) ringtail.

(b)

(c)

and other species, such as mice, also survive by avoiding daytime heat—resting in caves or burrows and coming out only at night. The ever-delightful ringtail, for example, sleeps all day and comes out at night to find its food (Figure 32–16c).

Because water is a rare commodity in the desert, many species acquire the moisture they need from cellular respiration. As noted in Chapter 6, metabolic water is produced by the electron transport system, the final stage of cellular respiration. In some species, metabolic water provides nearly all of the water needed to survive. The kangaroo rat, as you may recall, receives all of the moisture it needs from cellular respiration and from plants and insects. Although cellular respiration does not produce large amounts of water, it can be enough if species possess adaptations that help them conserve body water. The kangaroo rat, for example, excretes a highly concentrated urine, a physiological adaptation to the hot, dry conditions of the desert. (For more on water conservation in desert species, see Spotlight on Evolution 16–1.)

Overall, Human Culture Is Expanding Desert Conditions.

Partly to escape cold winters, many people have made the desert their home. Large cities have sprung up in many of the world's deserts. To supply inhabitants, food is often shipped in from farms hundreds of miles away. Water is often pumped from deep aquifers or is transported in extensive pipelines. Phoenix, Arizona, for example, receives much of its water from the Colorado River several hundred miles to the northwest through a gigantic (and costly) canal and pipeline.

The continuing expansion of cities and growing water demand have created serious problems, however. Southeast of Phoenix, over 300 square kilometers (120 square miles) of land has subsided (sunk) more than 2 meters (6 feet) because of intensive groundwater overdrafting. Cracks in the Earth's surface have developed in subsided areas. Some of the largest cracks are 3 meters (10 feet) wide and 3 meters (10 feet) deep and run for 300 meters, or about the length of a football field.

Subsidence and cracks in the Earth's surface can interrupt pipelines and can destroy homes and highways. Groundwater overdrafting can eliminate the supply of water for native plant species.

Each year, millions of acres of new desert form on semiarid grasslands. Research suggests that deserts are expanding principally because of human intervention. As noted earlier, for instance, heavy grazing destroys grasses in semiarid regions. The loss of vegetation may reduce rainfall, creating desertlike conditions.

Stopping desertification will require dramatic improvements in land management, especially grazing. The effectiveness of proper grazing management was dramatically illustrated in the early 1970s when a scientist who was examining satellite photographs of northern Africa discovered a huge green patch in the middle of the drought-stricken Sahel, a region that borders the southern Sahara.

▷ **FIGURE 32–17 Tropical Rain Forest** This biome contains hundreds of thousands—probably millions—of species of microorganisms, plants, insects, birds, amphibians, reptiles, and mammals.

When he went to investigate, he found that the oasis was a 100,000-hectare (250,000-acre) ranch. Unlike the surrounding territory, which had been overgrazed by cattle, this ranch had been properly managed by its owner. He had divided his land into five sections, each of which was grazed once every five years, infrequently enough to protect the grasses and the long-term productivity of the land. (For more on desertification, see Chapter 33.)

The Tropical Rain Forest Is One of the Most Diverse Biomes on Earth

Heading south to the equator, we find one of the most endangered biomes on Earth, the **tropical rain forest** (▷ Figure 32–17). Once covering a region approximately the size of the United States, only about half of the forest remains.

Tropical rain forests exist near the equator in South and Central America, Africa, and Asia (see Figure 32–1). By far the most complex and diverse of all the Earth's biomes, tropical rain forests support a wealth of plants, animals, and microorganisms. A small tropical island, for example, may have as many butterfly species (500 to 600) as the entire United States. Two hundred and fifty different tree species exist in a single hectare (2.5 acres) of forest; in the temperate deciduous forest biome of the United States, a similar tract may have 20 to 30 species.

The tropics actually contain a variety of different ecosystems, of which the rain forest is the largest and best known. With 200 to 400 centimeters (60 to 160 inches) of rain falling per year, trees of the tropical rain forest grow to heights of 60 meters (200 feet) or more. The trees are rather shallow-rooted, which is believed to be an adapta-

▷ **FIGURE 32–18 The Buttress** This majestic but shallow-rooted tree in a tropical rain forest is supported by buttresses that flare out from its trunk at the base of the tree.

▷ **FIGURE 32–19 Eroded Tropical Rain Forest Soil** This land was once covered with trees. It was cleared for grazing cattle and now, after severe rains, has turned into a barren landscape. Recovery, if possible, may take 500 years or more.

tion to the abundance of rain. Deeper roots are unnecessary because water is so abundant at the surface. Many trees in tropical rain forests have evolved wide bases (buttresses) that support their massive weight and prevent them from toppling over (▷ Figure 32–18).

The tops of the tallest trees in a tropical rain forest form a dense canopy that blocks out much of the incoming sunlight. Shorter trees form a lower canopy that intercepts most of the remaining light. As a result, only about 1% of the incoming sunlight reaches the ground. Because so little light strikes the forest floor, ground-level vegetation is sparse. Only those species adapted to low light (for example, African violets, philodendrons, and ferns) can survive in such conditions.

The lack of ground-level vegetation relegates most life to the forest canopy, the treetops. Dwelling in the tops of trees are monkeys, birds, and countless species of insects. In fact, the treetops of the tropical rain forest are probably the most heavily and diversely populated ecosystems in the world. In a recent study, one scientist found as many as 20,000 insect species living in a single tree.

Although Biologically Rich, Most Tropical Rain Forests Have the Poorest Soil Known to Science. The tropical rain forest is a paradox to the untrained observer. Although it is the richest and most diverse biome on Earth, most of its soils are thin and nutrient-poor. A section of the forest growing on poor soil that is cleared to make room for crops or cattle ranches generally produces for four or five years before its nutrients are depleted. Large-scale farming and ranching on many rain forest soils are therefore generally doomed to fail.

An understanding of basic ecology explains the paradox of the tropical rain forest. The reason for the poor soil lies in the decomposer food chain. Dead trees and leaves and other biomass (dead animals) that fall to the ground are consumed in part by the numerous insects. Because the tropics are warm and moist throughout the year, material that remains is rapidly decomposed by bacteria, fungi, and other organisms. Nutrients released into the soil by bacterial decay are quickly absorbed by the roots of trees located just beneath the ground's surface. In essence, then, tropical soils are so poor because life is so rich. Virtually all of the nutrients are incorporated into and locked up in plants and animals.

Many rain forest soils are also useless for farming and ranching because they contain large amounts of iron. These soils are called **lateritic soils** (*later* is Latin for "brick"). When cleared and exposed to sunlight, lateritic soils bake as hard as bricks. The impenetrable crust that develops one to two years after clearing makes the soils virtually impossible to cultivate.

Cleared Forests Are Subject to Erosion and Are Exceedingly Slow to Recover. Abundant rainfall in the tropical rain forests also makes clearing for agriculture a chancy proposition. Heavy rainfall in the wet season erodes soils, rendering the new farmland useless and choking nearby streams and rivers with sediment (▷ Figure 32–19).

The loss of the thin topsoil also makes recovery from damage more difficult and protracted. Some forestry scientists believe that it could take large forest clearings 500 years or more, if they can recover at all.

Tropical Deforestation Continues Unabated and Threatens Global Climate. "Tropical forests are the lungs of the planet," says the actress Meryl Streep in a PBS special **Race to Save the Planet**.[5] They "breathe in" vast

[5] As noted in previous chapters, marine algae also produce enormous amounts of oxygen.

amounts of carbon dioxide and release oxygen vital to animals (Chapters 5 and 6). By absorbing carbon dioxide, tropical rain forests help reduce global warming, a problem discussed in Chapters 4, 5, and 33. Tropical rain forests also help regulate planetary climate. One computer simulation showed that widespread destruction of rain forests in Africa, for example, could dramatically reduce rainfall as far north as Europe. The impact on agriculture would be substantial.

Despite worldwide concern for the fate of tropical rain forests, timber cutting continues. A 1980 study by the United Nations suggested that about 11 million hectares (27 million acres) of tropical rain forest were being cleared each year. A more recent study of tropical deforestation, also by the United Nations, suggests that deforestation is far greater—about 17 million hectares (42 million acres) per year. That is equivalent to an area about the size of the state of Washington. At this rate, the remaining tropical forests will be gone in 100 years and with them, tens of thousands, perhaps millions, of species. Computer studies suggest that the widespread destruction of tropical forests could upset global rainfall patterns and could affect global climate by reducing the amount of carbon dioxide removed from the atmosphere each year, as explained in the next chapter.

As the previous material suggests, we destroy the tropical rain forests at our own peril. Chapter 33 suggests some ways to end the destruction and save these ecologically vital regions of the world.

Altitude Affects Climate and Therefore Also Affects the Distribution and Abundance of Life

As the previous sections illustrate, all living things are a product of their environment. That is, they evolve in certain ways as a result of natural selection. Natural selection weeds out those individuals ill-adapted to the abiotic and biotic factors in their environment.

On land, the two principal determinants of species composition are temperature and precipitation. The relationship between these climatic factors and species composition is dramatically illustrated in mountains (▷ Figure 32–20). Traveling from the base of a mountain to its top, one progresses through very distinct life zones, known as **altitudinal biomes.** Like the biomes you have been studying, these zones are determined by temperature and rainfall.

Altitudinal biomes more or less mirror the latitudinal biomes described earlier. Starting at the top of the mountain, for example, one encounters a cold, treeless region similar to the Arctic tundra. Known as the **alpine tundra,** this area is characterized by a short growing season and long, cold winters (▷ Figure 32–21). But unlike the Arctic tundra, the alpine tundra receives lots of moisture. Most of the precipitation comes in the form of snow, but most blows away, ending up in the next lower level.

Climbing downward, one encounters a band of conifers, akin to the taiga. In the Rocky Mountains, this zone borders on grassland. On the west slope of the Cascade

▷ **FIGURE 32–20 Altitudinal Biomes**

▷ **FIGURE 32–21 Alpine Tundra** This biome is, in many ways, like the Arctic tundra, except that it receives far more moisture. Because the growing season is short, plants remain small.

Mountains, however, the taigalike region borders deciduous forests.

≈ FRESHWATER LIFE ZONES

Aquatic systems are also divided into distinct regions. Instead of calling them biomes, however, biologists generally refer to these regions as **aquatic life zones.** This chapter examines the two basic types: (1) freshwater and (2) saltwater.

As noted above, precipitation and temperature are the chief determinants of the distribution and abundance of life on land. In aquatic life zones, however, water is found in abundance, and temperature is relatively constant. Thus, the abundance and diversity of life forms are determined principally by two other factors—energy and nutrients. As in terrestrial biomes, energy ultimately comes from sunlight. Nutrients chiefly come from the land.

In both freshwater and saltwater life zones, many food chains begin with a group of organisms collectively known as phytoplankton. **Phytoplankton** are microscopic, free-floating organisms, mostly algae and diatoms (▷ Figure 32–22). These photosynthetic organisms capture solar energy, using it to produce carbohydrates from carbon dioxide that is dissolved in the water.

Phytoplankton are consumed by microscopic **zooplankton,** single-celled protozoans and multicellular crustaceans. Zooplankton form the second trophic level of many aquatic food chains (▷ Figure 32–23). Zooplankton, in turn, are consumed by small fish, which are a food source for larger fish and a variety of other organisms.

Lakes Are Divided into Four Ecologically Distinct Zones

Freshwater life zones consist of lakes, ponds, rivers, and streams. We'll begin by looking at ponds and lakes.

Ponds are relatively shallow, relatively small bodies of water. Because they are shallow, sunlight often penetrates to the bottom, providing plenty of energy for all aquatic life.

Lakes are deeper than ponds and generally contain four distinct zones. The first is the littoral zone, from the Latin word *litoralis,* meaning "seashore" or "coast" (▷ Figure 32–24). The **littoral zone** consists of the shallow waters at the margin of a lake where rooted vegetation often grows. Some rooted vegetation—for example, cattails—extends above the water's surface; other plants—such as water lilies—have leaves and flowers that float on the water's surface. Still others are submerged, rarely breaking the surface. Phytoplankton are also common in the shallow littoral zone. Energy for phytoplankton and rooted vegetation comes from sunlight.

As Figure 32–24 shows, the **limnetic zone** is a region we commonly call "open water." Extending downward to

▷ **FIGURE 32–22 Phytoplankton** Unicellular algae like these are called phytoplankton and form the basis of many aquatic food chains.

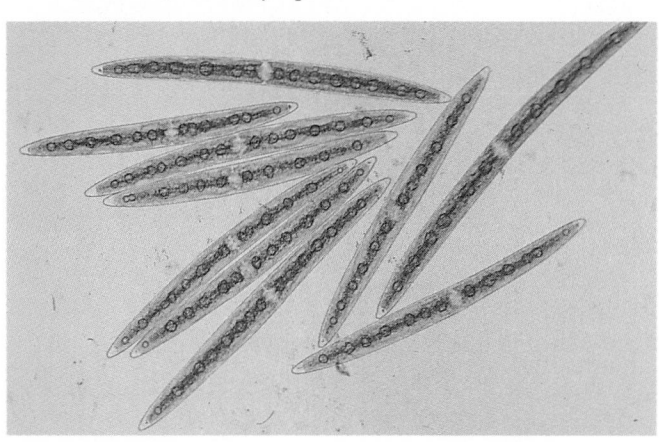

(a)

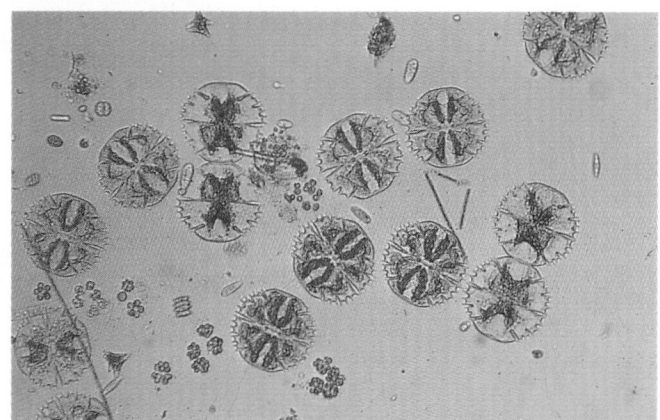

(b)

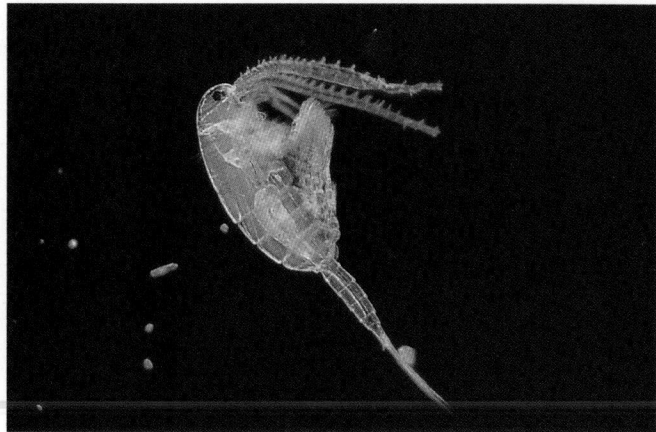

▷ **FIGURE 32–23 Zooplankton** These organisms feed on algae and other phytoplankton and are consumed by small fishes.

the point at which light no longer penetrates, the limnetic zone is the main photosynthetic body of a lake.

In many lakes, the limnetic zone supports abundant phytoplankton—so many, in fact, that phytoplankton biomass may exceed the biomass of rooted vegetation along the shoreline. Therefore, the limnetic zone is a biological factory, producing food that supports most aquatic life forms. It is also a major source of oxygen, required by zooplankton, many bacteria, and animals, such as fishes.

The **profundal zone** lies beneath the limnetic zone in deeper lakes and extends to the bottom (Figure 32–24). Because little or no sunlight penetrates the profundal zone, conditions are not favorable for plant and algal growth. Without sunlight and plants, oxygen levels remain fairly

low. Fishes can survive in the region, but they rely on food produced in the limnetic and littoral zones.

The bottom of a lake is called the **benthic zone.** It is home to organisms that tolerate cool temperatures and low oxygen levels, such as snails, clams, crayfish, various aquatic worms, and insect larvae like those of the mayfly, dragonfly, and damselfly (Figure 32–24). The larvae emerge during the spring and become free-flying insects. Mayflies, for example, generally breed within a day of emergence, then die. Many of them are eaten by fish and are therefore an important source of food in freshwater aquatic ecosystems.

Deeper Lakes Contain Three Distinct Thermal Layers During the Summer Months. To understand a

▷ **FIGURE 32–24 Ecological Zones of a Lake**

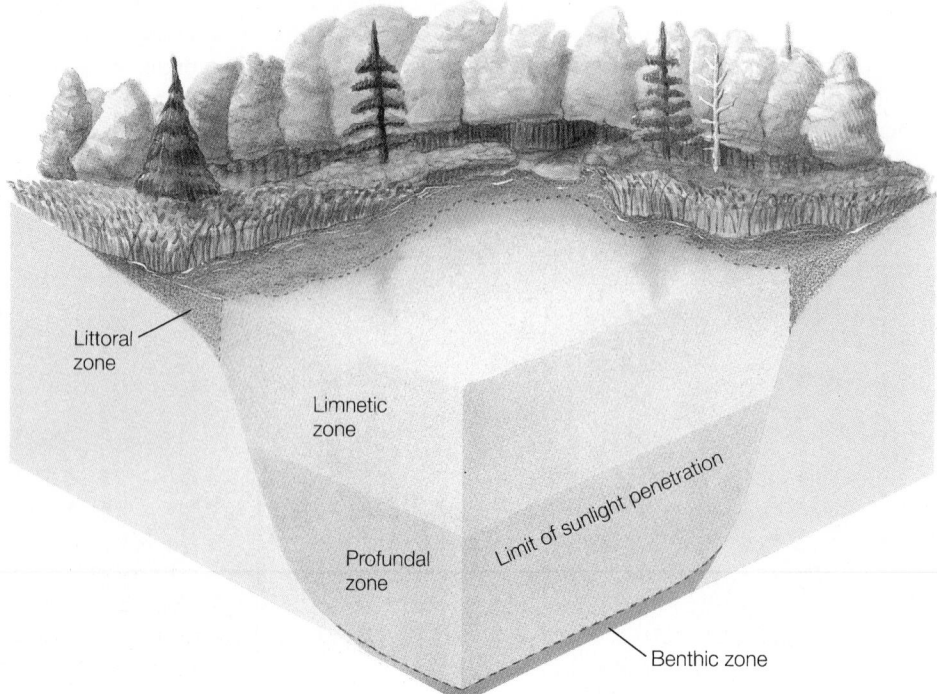

Littoral zone

Limnetic zone

Profundal zone

Limit of sunlight penetration

Benthic zone

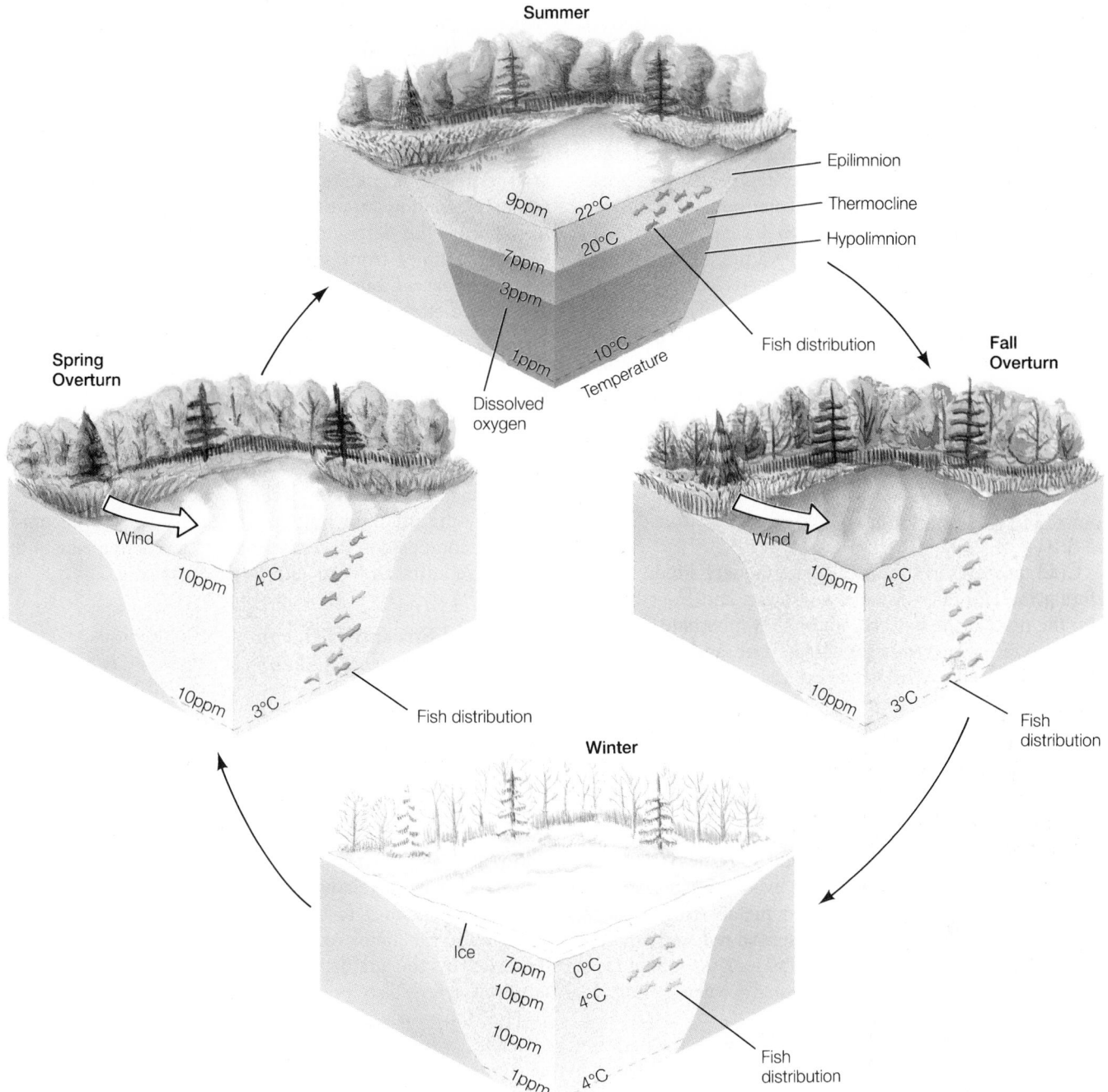

▷ FIGURE 32–25 Thermal Stratification of Lakes
(a) During the summer months, deep lakes become thermally stratified. (b) During the fall, the surface waters cool, and winds cause the lake to destratify. (c) During the winter, ice forms, but the lake is more or less thermally uniform. (d) During the spring, ice melts, and the water mixes. As the days grow warmer, the lake becomes thermally stratified once again.

lake, you must also understand thermal stratification—that is, thermal layering. During the summer in temperate climates (such as North America), the water of most lakes forms three distinct thermal layers (▷ Figure 32–25).

The warm surface water of a lake is called the **epilimnion** (Figure 32–25a). Just beneath the epilimnion is a region of abrupt temperature change, the **thermocline.** Swimmers can experience the thermocline by diving toward the bottom of a lake. (The epilimnion and thermocline correspond to the limnetic zone.) Below the thermocline is a region of fairly uniform temperature, the **hypolimnion.** In the summer, the hypolimnion contains dense, cold water. (The hypolimnion corresponds to the profundal zone.)

As Figure 32–25a shows, the three thermal layers in a summer lake also differ in oxygen levels. The oxygen concentration is highest in the epilimnion, where light penetration and photosynthesis are greatest, and falls rather rapidly from the surface to the bottom as photosynthesis declines. The lowest levels are found along the bottom.

Twice a Year, Lake Water "Turns Over," Thus Mixing Surface and Deeper Waters.
Thermal stratification is not static. During the fall, for example, the surface waters gradually cool, and the water temperature of a lake eventually becomes fairly uniform from top to bottom (Figure 32–25b). Thermal stratification disappears. In many lakes, winds churn the water, causing a thorough mixing of surface and deep water. As a result, oxygen levels also become fairly uniform from top to bottom. This mixing of surface and bottom waters is known as the **fall overturn.**

In late fall and early winter, the surface waters of a lake cool even more. Ice may form on the lake, often covering the entire surface. Because ice is less dense than liquid water, it floats, leaving an ice-free zone in deeper lakes where fish can survive. At this time, water oxygen levels are fairly uniform from top to bottom, although levels are lowest in the deepest waters and at the bottom (Figure 32–25c).

Cold temperatures drastically alter aquatic life in a lake. During the late fall, for example, turtles and frogs burrow into the mud at the bottom, where they hibernate throughout the cold winter months. Heart rate and metabolism slow dramatically. When the water warms in the spring, they reemerge.

In contrast, fishes remain active and continue to feed throughout the winter. They can survive in lakes even when a layer of ice forms over its surface, cutting the lake off from oxygen. How?

Healthy lakes often have a sizable reserve of oxygen produced during the summer months. Fishes and other organisms use up that supply, but even though fishes remain active, cold water lowers their metabolism, reducing oxygen demand. Algae are also present in reduced number. They continue to photosynthesize as long as sunlight can penetrate the ice, producing additional oxygen to help replenish water levels.

In the spring, the ice melts, and the water temperature is uniform once again. Winds agitate the water, causing a mixing, known as the **spring overturn** (Figure 32–25d). As the days get longer and warmer, however, the surface waters begin to warm, and the lake becomes thermally stratified.

Lakes Are Vulnerable to Pollution.
Lakes and the organisms they support are extremely vulnerable to pollution. Toxic wastes from factories, for instance, can kill fishes outright. Heavy metals and other chemicals released by factories may cause cancer in fishes or impair reproduction.

With over 5.4 billion people living on the planet, it is no surprise that one of the most common pollutants in lakes is human sewage. Rich in organic matter, sewage dumped into lakes is normally broken down by aerobic (oxygen-requiring) bacteria. If large quantities of organic matter enter a lake, bacterial decomposition of organic wastes may cause oxygen levels in the water to fall. Because oxygen levels in the deeper waters are already quite low in the summer, many fishes and benthic (bottom-dwelling) organisms may perish.

Two other pollutants of concern are nitrate and phosphate. Nitrate and phosphate are present in the effluent of sewage-treatment plants. Nitrate entering water bodies comes from human urine and feces; phosphate comes largely from the detergents used to clean our clothes. Nitrate also comes from artificial fertilizer often applied by farmers in excess. Surplus nitrate washes into nearby lakes, ponds, and streams during heavy rains.

Both nitrates and phosphates are plant nutrients and stimulate the growth of algae and aquatic plants. Rapid growth of algae may result in the formation of thick algal mats (called algal blooms). When nitrate and phosphate levels are high, rooted vegetation along the shoreline also proliferates, making it difficult to navigate near the shore.

In the fall, as daylight declines, plants and algae produced by the influx of nutrients begin to die and decompose. Decomposition decreases oxygen levels in a lake and may prove lethal to fishes and other organisms.

Rivers and Streams Differ in Many Respects from Lakes

Rivers and streams are complex ecosystems. As with lakes, no two streams are alike. In many areas, streams begin in mountains or hilly terrain, collecting water that falls to the Earth as rain or snow. The region drained by a stream is called a **watershed.** Small streams join to form rivers and rivers flow downhill to the sea.

Streams and rivers are generally well oxygenated because they have a relatively large surface area (relative to their water volume) to absorb oxygen from the air. The current also facilitates oxygenation. Current velocity is determined by the gradient—that is, the steepness of the terrain. Current also influences dissolved oxygen levels.

The fast-moving currents of mountain streams, for example, produce waves and rapids as the water collides with rocks or drops over ledges. This agitation greatly increases oxygenation. For these and other reasons, photosynthesis is a less important source of oxygen in rivers and streams than it is in ponds and lakes. Certain fish are adapted to different currents. Trout, for example, generally inhabit cold, oxygen-rich mountain streams where the current is quite rapid. Black bass live in rivers, where the waters are warmer, less oxygenated, and slower.

Unlike many lakes and ponds and terrestrial biomes, streams are rather open ecosystems; that is, they receive a great many nutrients from bordering ecosystems. In some streams, much of the biologically available energy actually comes from nearby terrestrial vegetation—for example, leaves that fall into the stream. Animal feces, insects, stems, nuts, and other biomass may also be washed into streams during heavy rainstorms. All of this material produced on land "feeds" the aquatic food web. Many primary consumers in streams, therefore, are detritivores—organisms that feed on waste or remains of plants and animals. Streams do have their own producers, mostly algae and rooted vegetation, but in some cases these organisms play only a minor role in providing food.

Like Lakes and Ponds, Streams Are Extremely Vulnerable to Pollution.

Despite the mixing and replenishment of oxygen in streams and rivers, these bodies of water are quite vulnerable to pollutants. In fact, toxic wastes can destroy fish populations. The Rhine River of Europe, for example, once supported 150 species of fishes. Today, only about 15 species remain. Why?

The Rhine travels through seven countries and is probably the most heavily used river in the world. Hundreds of factories, towns, farms, and sewage-treatment plants are located along its banks, each releasing a variety of pollutants. In streams, plant nutrients from farms and sewage cause algal blooms and stimulate the growth of rooted vegetation, clogging smaller rivers and streams. Sediment washing from construction sites, poorly built roads, clear-cuts, farm fields, and other places muddies the waters, reducing photosynthesis and oxygen levels. Sediment settles on the streambed, covering gravel where fish spawn. If fish eggs are present, sediment may smother and kill them, thus reducing population size. Sediment buildup also causes streams to flood more often, damaging farms and homes.

To Protect a Stream or River, One Must Protect Its Watershed.

Protecting rivers and streams requires active measures to protect the land around them, their watersheds. These measures are aimed at reducing all forms of pollution from two sources: point and nonpoint.

Point sources are distinct, easily identifiable sources—for example, factories and sewage-treatment plants whose wastes empty directly into streams and rivers. Controlling pollution at point sources is relatively straightforward: all one needs to do is add a pollution-control device that separates out the harmful materials. Unfortunately, pollutants captured from the waste stream must go somewhere and usually end up in landfills. Landfills have their own set of problems. One of the most significant is that many landfills leak wastes into groundwater. Groundwater sometimes empties into rivers and lakes.

Recognizing that pollution must always go somewhere, many companies have taken a new strategy—waste minimization. **Waste minimization,** or **waste reduction,** means not producing waste in the first place or drastically reducing its production. Waste minimization is a pollution-prevention strategy. Companies are finding that simple, often inexpensive changes in their manufacturing processes will eliminate the need for toxic chemicals or greatly reduce their output. Where there is no waste, there is no need for waste treatment. Waste minimization, therefore, can save companies enormous amounts of money in waste disposal costs and liability for harm done by pollution.

Nonpoint sources are those that do not empty via distinct, easily identifiable routes. Farm fields, suburban lawns, city streets, and parking lots are good examples. Toxins and other pollutants deposited on the land often make their way to rivers, lakes, and streams. For example, rain washes excess fertilizer or pesticides sprayed on lawns into city sewers or nearby streams. Oil from streets and parking lots is also washed away in storms, ending up in waterways. Sediment from poorly designed and executed clear-cuts washes into streams.

Nonpoint sources are more difficult to control because they are diffuse and not generally amenable to pollution control devices. Nonpoint sources must be controlled by the actions of thousands of individuals and businesses within a watershed. Homeowners can help cut water pollution by reducing the amount of fertilizer and pesticides they apply to their lawns or by eliminating these chemicals altogether. Timber companies can avoid clear-cutting near rivers to reduce erosion and sedimentation. Farmers can practice better soil conservation.

Streams and Rivers Can Recover from Pollution, But Human Society Often Overwhelms their Natural Recuperative Capacity.

Water flowing in streams and rivers flushes out pollutants, cleansing them. The cleansing effect, however, also results from bacteria and other microorganisms, which break down many organic pollutants. As polluted water flows downstream from a sewage-treatment plant, for instance, microorganisms degrade organic wastes, purifying the water.

Water flow can also dilute waste to harmless levels. Therefore, although pollution may poison a portion of a stream or river, downstream life may go on unaffected.

Unfortunately, the cleansing ability of streams and rivers is often overwhelmed. Numerous sources of pollution along the course of a river and extremely large producers of waste, for example, can simply overpower a stream's recuperative capacity.

≈ SALTWATER LIFE ZONES

The oceans cover over 70% of the Earth's surface. Like freshwater systems, the ocean can be divided into ecologically distinct life zones. This section discusses some of the most important ones.

In the ocean, like all other bodies of water, the distribution and abundance of life are dependent on many factors, but the most important are energy and nutrients. We begin by looking at three coastal life zones: estuaries, seashores, and coral reefs.

The Coastal Life Zone

Estuaries Are Nutrient-Rich Zones at the Mouths of Rivers. **Estuaries** are places where fresh water mixes with salt water. The combination is called **brackish water.** As a rule, estuaries are rich life zones, for two reasons:

▷ **FIGURE 32–26 Coastal Wetland** Aerial view of estuary and mangrove swamp (dark green).

▷ **FIGURE 32–27 Tidal Pool** Numerous species live along the rocky shorelines of the world's coasts.

First, streams and rivers transport many nutrients from the land. Second, incoming tides carry nutrients into estuaries from the ocean. Incoming tides also tend to prevent the escape of nutrients delivered by the river. These nutrients support abundant plant and algal growth and sizable populations of fish and molluscs (clams, oysters, and scallops).

Most estuaries are located near **coastal wetlands**—salt marshes, mangrove swamps, and mud flats (▷ Figure 32–26). Together, estuaries, coastal wetlands, and mud flats form the **estuarine zone,** one of the most beleaguered places on the Earth. Environmentalists and scientists point out that the destruction of the estuarine zone is suicidal, because two-thirds of the world's commercially valuable fishes and molluscs depend on the estuarine zone at some point of their life cycle. For example, many marine fishes spawn in the estuarine zone. Their eggs are laid on the bottom and are extremely susceptible to the deposition of sediment. The larvae that hatch from the eggs feed off phytoplankton and zooplankton that flourish in the nutrient-rich waters of the estuary. The estuarine zone also provides food, shelter, and breeding grounds for millions of waterfowl and fur-bearing animals, such as muskrats. Protecting the estuaries and coastal wetlands—even restoring those that have been damaged by human activities—is therefore vital to humans and a host of other species.

Organisms that live in estuarine zones must be able to tolerate dramatic changes in salt levels. When tides go out, fresh water from rivers bathes the zone. Salt concentrations in estuarine waters fall. When the tide returns, however, the salt concentrations increase. Over a 24-hour period, salt concentrations may change by a factor of 10.

The Shoreline Is Home to Many Organisms Adapted to Rising and Falling Tides and Turbulent Water.
The Earth's coastlines are generally rocky or sandy regions that support a variety of organisms. Rocky shorelines are home to seaweed (algae), sea urchins, barnacles, and a variety of sea stars (▷ Figure 32–27). Abundant

sunlight and nutrients account for the diversity of life in this zone. Many of these organisms anchor themselves to the rocks and therefore are able to withstand turbulence created by the pounding of waves. Oysters and mussels are anchored in place by filaments (called byssal threads). Limpets, snaillike creatures that eat algae on rock, have a cone-shaped shell that fits tightly to a rock as they feed. Barnacles secrete an adhesive that literally glues them in place. Organisms that do not attach themselves, such as sea urchins, remain in rocky crevices, largely protected from water currents. Various multicellular algae (rock weed, for example) anchor themselves to rocks as well.

The coastline is also home to a variety of shorebirds that feed on insects, crustaceans, and other organisms. In addition, the coastline serves as a nesting site for turtles and is a home to breeding colonies of seals and walruses. Protecting the coastlines from offshore oil spills and human encroachment is therefore extremely important.

Coral Reefs Are often Located Near Shore.
Coral reefs are biologically rich life zones found in relatively warm and shallow waters in the tropics or nearby regions (subtropics). A coral reef consists of calcium carbonate or limestone produced by calcarous red and green algae and by colonies of organisms called stony corals (▷ Figure 32–28). Stony corals are organisms (cnidarians) with stiff, calcium carbonate exoskeletons that persist long after their death, creating a base on which other corals grow.

Coral reefs take thousands of years to form and become home to a great variety of organisms. Various forms of multicellular algae, for example, live on and around coral reefs. The coral reef's rich and colorful fish life is nothing short of astounding (▷ Figure 32–29a). Living off phytoplankton and fishes, inhabitants of the reef find protection within the crevices formed in the reef, which also shelter moray eels and many sedentary filter feeders—organisms that feed themselves by filtering microscopic food particles from the seawater (Figure 32–29b). The filter feeders in-

▷ **FIGURE 32–28 Coral Reef** Coral reefs are akin to tropical rain forests. They support a great variety of colorful species.

clude sponges and fan worms. Larger fishes also frequent the coral reefs looking for food. The barracuda, fierce-looking but generally harmless to humans, often lurks in the waters, feeding on other fishes.

All Coastal Life Zones Are Threatened by Human Activities. Coastal life zones are essential to life in the sea. Because humans depend so heavily on the sea for food, protecting coastal life zones is also vital to long-term human welfare. Unfortunately, the coastal life zones are in danger. Health Note 32–1 describes the tragic decimation of real populations off the coast of Europe, caused by toxic water pollution.

As noted earlier, perhaps the most endangered coastal life zone is the estuarine zone. Dams built on rivers to control flooding, to supply drinking or irrigation water, or to provide recreational activities, for example, block the natural flow of water. Reservoirs capture sediment that would otherwise travel to the estuary. Sediment contains

▷ **FIGURE 32–29 Coral Inhabitants** (*a*) Coral reefs house some of the most spectacular fishes known to humankind. (*b*) Brightly colored anemones filter food from the water.

(a)

many nutrients vital to estuarine phytoplankton. Dams, therefore, reduce the nutrient input.

The impact of dams is aptly illustrated in Egypt. In the early 1960s, Egypt constructed the huge Aswan Dam on the Nile River to provide irrigation water for farms. However, the dam also nearly eliminated the flow of nutrients into the river's estuary, causing a collapse in the sardine fishery in the Mediterranean Sea. In fact, the commercial sardine catch in the Mediterranean dropped from 18,000 tons per year before the dam was built to about 500 tons per year today.

Coastal wetlands have been drained and filled with dirt to build homes, highways, recreational facilities, and factories. These activities destroy habitat for fish and other species. In the United States, over 40% of the coastal wetlands have been destroyed. In California, an estimated 90% have been destroyed.

Like estuaries, coral reefs are also under siege. Ships running aground on the coral reefs of Florida, for example, are causing considerable damage to these fragile structures. Divers are taking their toll as well. Careless divers can break delicate coral with their swim fins. Each year, thousands of divers visit the coral reefs of Florida. Over the years, many of the most popular reefs have been destroyed.

Sediment is a particularly troublesome pollutant for coral reefs. It clouds seawater, reducing photosynthesis in a group of protists (microscopic organisms called dinoflagellates) that live in a mutualistic relationship with many spe-

(b)

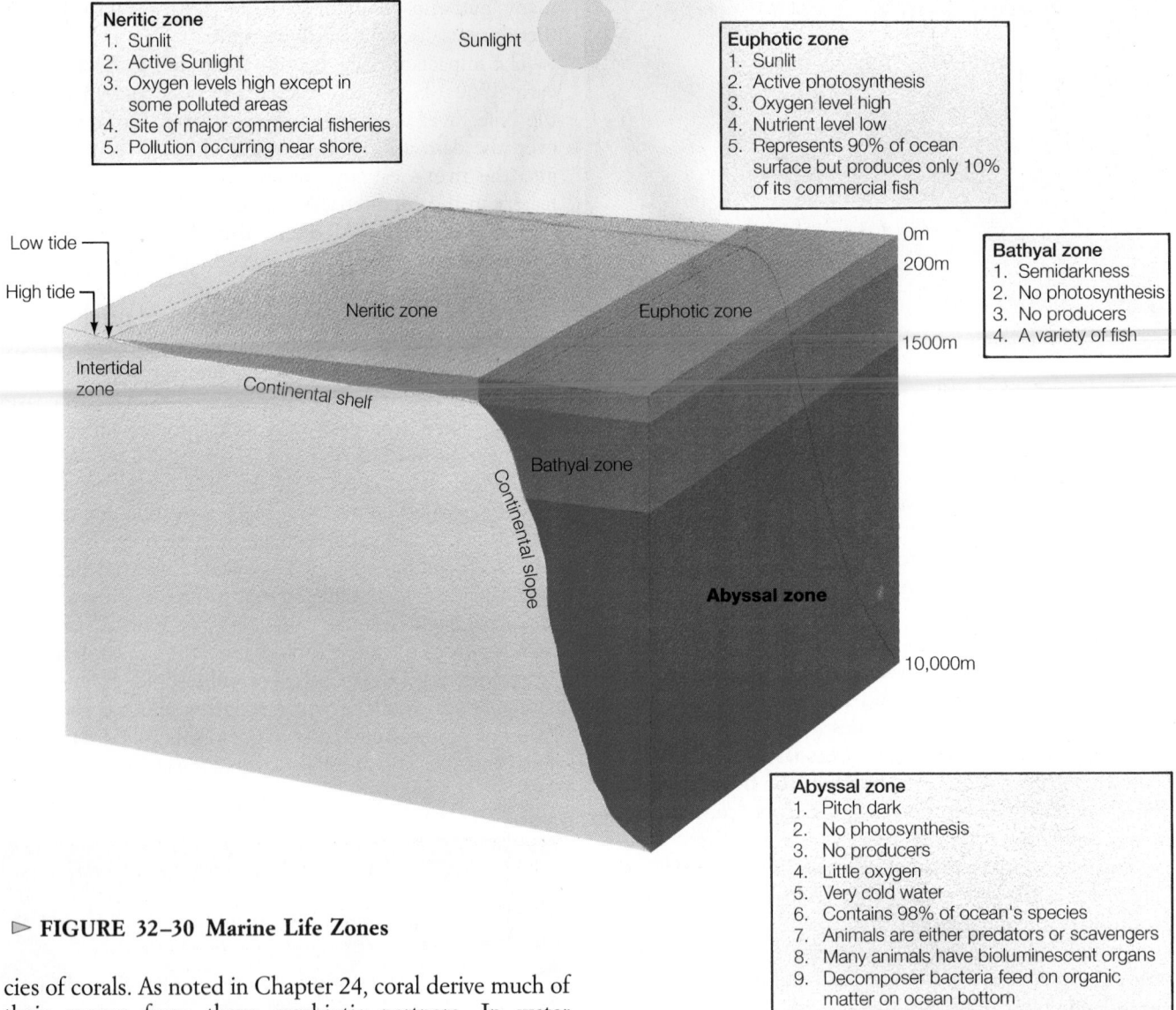

Sunlight

Neritic zone
1. Sunlit
2. Active Sunlight
3. Oxygen levels high except in some polluted areas
4. Site of major commercial fisheries
5. Pollution occurring near shore.

Euphotic zone
1. Sunlit
2. Active photosynthesis
3. Oxygen level high
4. Nutrient level low
5. Represents 90% of ocean surface but produces only 10% of its commercial fish

Bathyal zone
1. Semidarkness
2. No photosynthesis
3. No producers
4. A variety of fish

Low tide
High tide

Intertidal zone
Continental shelf
Neritic zone
Euphotic zone

0m
200m
1500m

Continental slope

Bathyal zone

Abyssal zone

10,000m

Abyssal zone
1. Pitch dark
2. No photosynthesis
3. No producers
4. Little oxygen
5. Very cold water
6. Contains 98% of ocean's species
7. Animals are either predators or scavengers
8. Many animals have bioluminescent organs
9. Decomposer bacteria feed on organic matter on ocean bottom
10. High water pressure
11. High in nutrients on ocean floor

▷ **FIGURE 32–30 Marine Life Zones**

cies of corals. As noted in Chapter 24, coral derive much of their energy from these symbiotic partners. In water clouded by sediment, the amount of food available to coral is reduced, and reefs grow more slowly. Particularly heavy sedimentation can even bury a reef, choking the life out of it. Several large reefs near Honolulu, in fact, have been destroyed by sediment washed from construction sites, highways, and farms. Housing development on the Florida Keys is also increasing sedimentation, which threatens nearby coral reefs.

The Marine Ecosystem Consists of Four Ecologically Distinct Life Zones

▷ Figure 32–30 shows a cross section of the ocean floor. As illustrated, the ocean floor slopes downward and away from land masses, then drops more steeply. The gradually sloping region is called the **continental shelf.** The more steeply falling region is the **continental slope,** and the bottom of the deep ocean is called the **abyssal plain.**

As illustrated in Figure 32–30, the ocean is divided into four ecologically distinct life zones: the neritic, euphotic,

bathyal, and abyssal. The **neritic zone** is equivalent to the littoral zone of lakes and overlays the continental shelf. The continental shelf varies in width, ranging from 10 to 200 miles. As illustrated, the neritic zone is relatively shallow and therefore receives abundant sunshine. Its waters are relatively warm and well oxygenated.

Nutrients in the neritic zone are supplied from two sources: stream discharge and upwelling. **Upwelling,** illustrated in ▷ Figure 32–31, occurs when currents drive nutrient-rich water from the floor of the ocean up the continental slope. These nutrients, which had been deposited from shallower water above, support abundant phytoplankton, zooplankton, and fishes.

In the neritic zone, sunlight normally penetrates to the ocean bottom. Ample sunlight supports large populations of algae and rooted plants, which, in turn, support a great many other species. In fact, most commercial fishing oper-

SEALS AND POLLUTION

The study of ecology reminds us that all organisms are a connected part of the web of life through the media of air and water, which allows nutrients to flow from one system to another. This connectedness also causes problems. The seas, for example, receive extraordinary amounts of pollution produced by human activities on land. Pollution enters the seas through rainfall and through polluted rivers that empty into the world's oceans.

Numerous studies show that levels of certain pollutants in the world's oceans are rising. One group of especially troublesome chemicals is the organochlorine compounds, organic compounds containing chlorine atoms, among them pesticides such as DDT and a group of chemicals called PCBs. PCBs were once widely used as insulating fluid in electrical insulators, and as a plasticizer, an agent that helps preserve plastic. The cumulative world production of PCBs was estimated in 1989 at 200 million tons, 57% of which is still in use. Scientists estimate that 16% to 30% of these PCBs have reached the environment.

Some of the PCBs in the world's ocean also come from ever-present plastics. Many plastics contain large quantities of PCBs that are released over time. Tens of thousands of tons of plastic are dumped from inattentive anglers or sailors. Plastics have many effects on marine animals. Sea turtles, for instance, may eat plastic which accumulates in their stomach or intestines, blocking digestion and eventually killing the animals.

The effect of PCBs is indirectly exerted by poisoning the food chain. The most severely affected organisms are marine predators, seals and other organisms that feed at the top of the food chain. These animals tend to accumulate organochlorines in fatty tissues. They acquire their pollutants from the fish they eat, which in turn have accumulated pollutants from their food sources.

Because organochlorines accumulate in fatty tissues of fish-eating marine predators, reaching levels many thousands of times higher than concentrations in sea water, scientists have directed a considerable amount of attention to the study of the impacts of these compounds. Their studies show that organochlorines may cause reproductive failure and immune system suppression. Studies suggest that the presence of PCBs and DDT in the tissues of seals may be responsible for reproductive failure in southern Wadden Sea harbour seals, and for obstructions in the uterus in Baltic Sea grey and ringed seals and the grey seals in Liverpool Bay, United Kingdom. Experimental evidence suggests that suppression of the immune system in harbour seals is the result of consuming fish contaminated with PCBs.

During the late 1980s, seals throughout the waters of Europe died in great number. For example, approximately 60% of the harbour seals in the Wadden Sea colony perished. Studies showed that the epidemic was caused by the phocine distemper virus (PDV), a new virus to science. However, a group of biologists, veterinarians, and toxicologists based in London concluded that water pollution could not be excluded as a contributing factor in the seal deaths, especially since the seals were highly contaminated with pesticides and related compounds. Some scientists hypothesized that immune suppression made the seals more susceptible to the virus.

Although the epidemic has abated, several countries that experienced massive seal die-offs are concerned about a repeat occurrence. With some seal colonies at an all-time low, these countries know that environmental stresses that could impact seal populations must be avoided or at least minimized to give the colonies a chance to rebound. The countries around the Wadden Sea, for example, agreed to protect the harbour seals by minimizing disturbance and pollution of critical seal habitat.

This example illustrates how our actions affect the health, and even the survival, of other species. It suggests that a healthy environment is not desirable simply because it helps people, but because it is essential for all species that are a part of the web of life.

---

ations in the ocean concentrate their operations in the neritic zone.

The **euphotic zone** is the oceanic equivalent of the limnetic zone of lakes.[6] It is the open water region that extends to the lower limits of sunlight penetration, or about 200 meters (650 feet) below the ocean's surface. Abundant light supports numerous species of phytoplankton, including diatoms, dinoflagellates, and others. Phytoplankton support a variety of zooplankton, mostly minute crustaceans.

Phytoplankton also produce plenty of oxygen. Given the high levels of oxygen and ample supply of sunlight, one might think that the euphotic zone would be extremely productive. Unfortunately, there is one factor missing: nutrients. The euphotic zone is not very productive because the waters are nutrient-poor. Thus, even though the euphotic zone covers 90% of the ocean's surface, it produces only about 10% of the commercial fish taken each year.

Beneath the euphotic zone is a region of semidarkness, the **bathyal zone.** The bathyal zone is too dark to support photosynthesis. It is therefore characterized by a lack of photosynthetic organisms and low oxygen levels. Despite these factors, the bathyal zone is home to a variety of fish

---

[6] Euphotic means "well-lighted." Many books simply refer to this area as the photic zone.

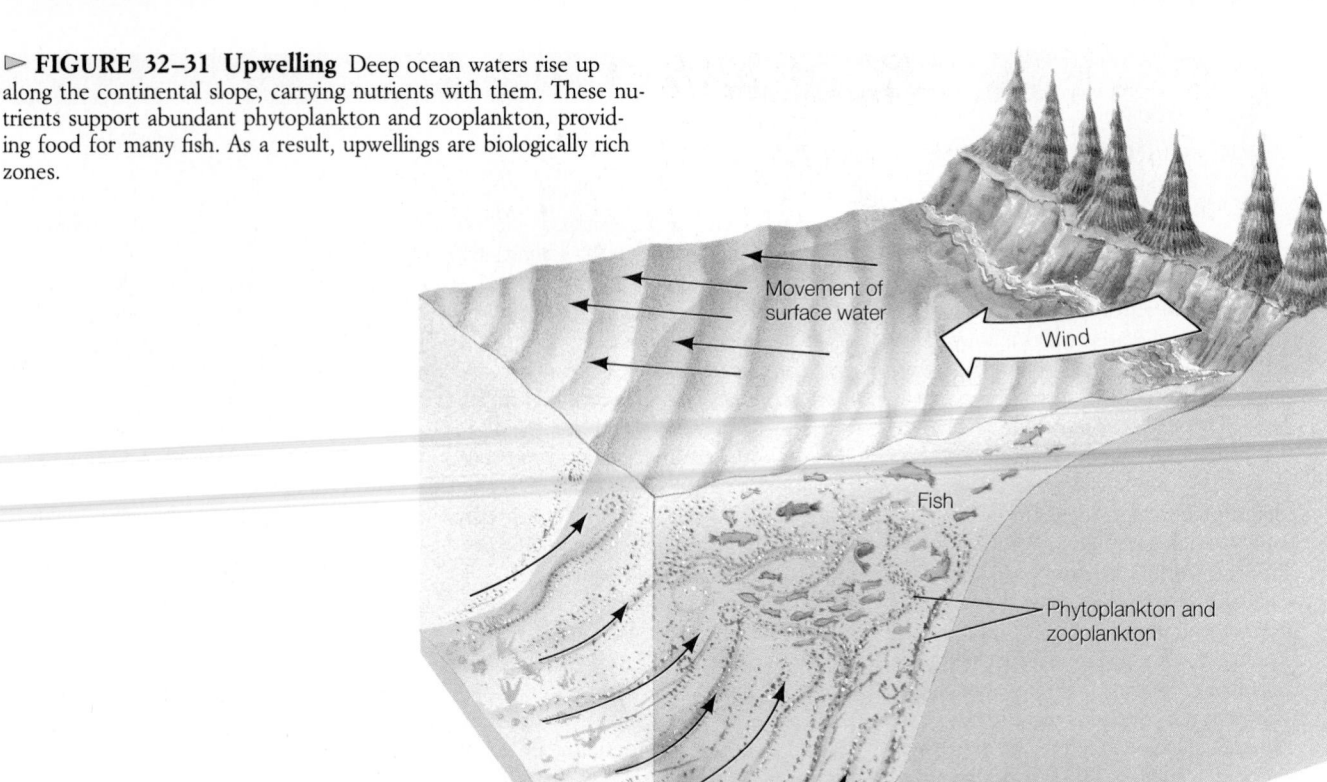

▷ **FIGURE 32–31 Upwelling** Deep ocean waters rise up along the continental slope, carrying nutrients with them. These nutrients support abundant phytoplankton and zooplankton, providing food for many fish. As a result, upwellings are biologically rich zones.

and other organisms, such as shrimp and squid, which feed on organisms "raining" down from above.

Beneath the bathyal zone is the **abyssal zone.** The abyssal zone is a region of complete darkness. It contains no photosynthetic organisms and is characterized by low oxygen levels. Animals that live there are either predators or scavengers. To live in this zone, an animal must be adapted to extremely cold water, high water pressure, low oxygen, and complete darkness.

Sediment in the abyssal zone is often rich in nutrients. And much to the surprise of many, the deep ocean floor is populated by a variety of strange-looking creatures that make use of these nutrients. Some species of fishes have evolved luminescent organs that shine in the dark, presumably helping them attract food and mates.

 **ENVIRONMENT AND HEALTH: WETLANDS, WILDLIFE, AND WATER TREATMENT**

The United States government and local municipalities have spent billions of dollars building sewage plants to treat human waste. Most plants put human sewage through two treatment phases. During primary treatment, screens first filter out solid objects (cigarette butts and the like). The remaining waste, rich in organic matter, is in a liquid form that flows into a grit chamber, where the sediment settles out (▷ Figure 32–32). The remaining waste, still high in organic matter, then travels to a settling tank that permits the dissolved solids (organic matter) to settle out.

In a facility with secondary treatment, the organic sludge is treated biologically. In some cases, air is pumped into the sludge (Figure 32–32). Bacteria break down the organic material, thus reducing the amount of sludge that must be disposed of. In other plants, liquid sludge is sent to a trickling filter system (▷ Figure 32–33). The sludge drips out of long pipes that rotate slowly over a bed of stone, populated by bacteria, protozoans, fungi, snails, and worms. These organisms feed on the sludge and on one another, greatly reducing the organic content of the sludge. The liquid remaining is then chlorinated to kill bacteria and viruses and other potentially harmful organisms and is usually discharged into streams or lakes or nearby oceans.

Primary and secondary treatment removes about 90% of the organic wastes but only 30% of the phosphates and 50% of the nitrates. As noted earlier, sewage-treatment plants are therefore significant sources of water pollution. Organic material from them stimulates bacterial growth in aquatic ecosystems, reducing oxygen levels, which, in turn, affects fishes and other aquatic organisms. Nitrates and phosphates stimulate algal blooms, sometimes covering

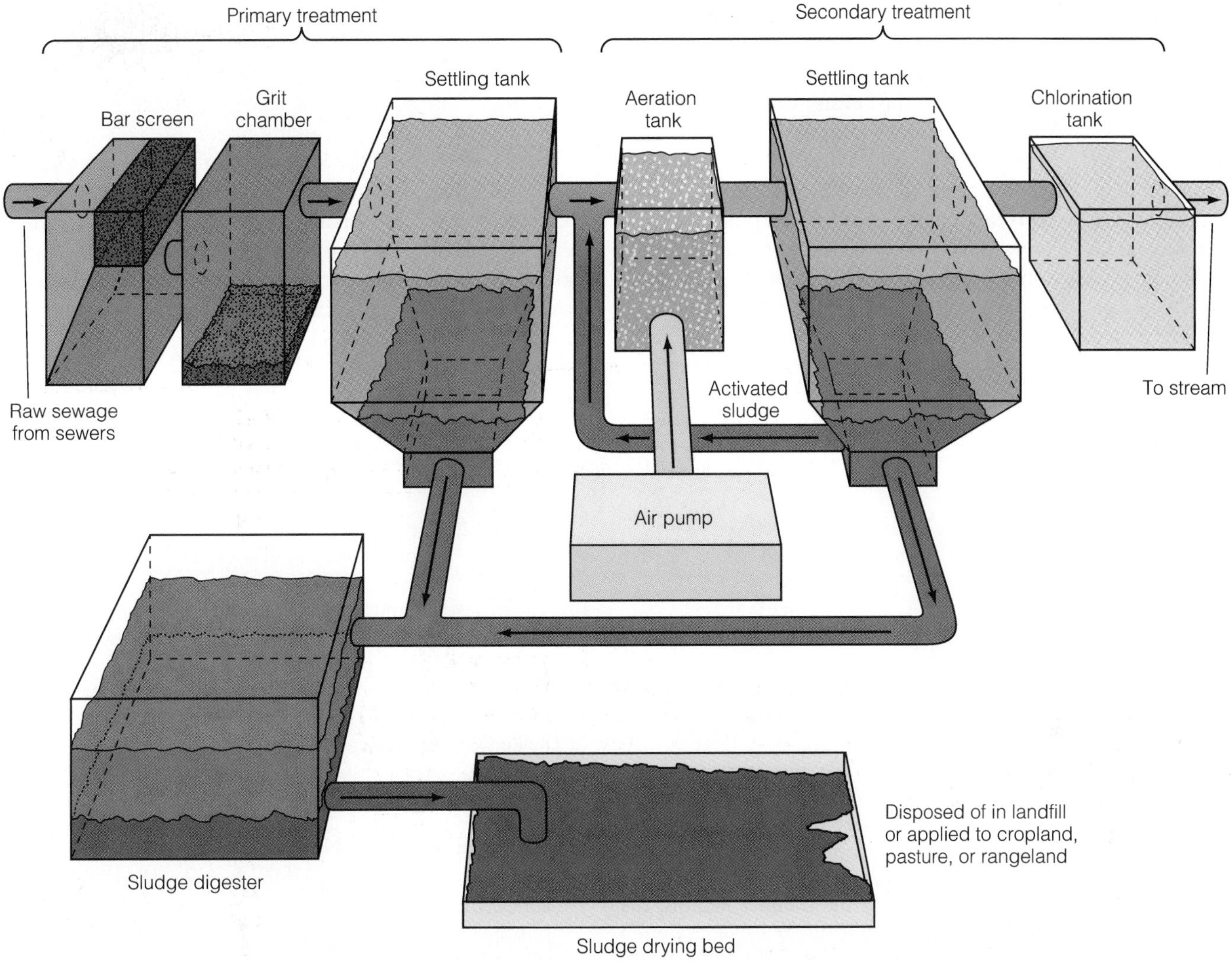

Primary treatment | Secondary treatment

Bar screen
Grit chamber
Settling tank
Aeration tank
Settling tank
Chlorination tank

Raw sewage from sewers

Activated sludge

To stream

Air pump

Sludge digester

Sludge drying bed

Disposed of in landfill or applied to cropland, pasture, or rangeland

▷ **FIGURE 32–32 Sewage Treatment Plant** Primary treatment removes suspended and dissolved solids, producing sludge that must be dumped in a landfill. Secondary treatment allows bacteria and other decomposers to degrade organic wastes, greatly reducing organic matter in human sewage.

lakes and rivers with a dense mat of algae. Algae can block off sunlight to deeper waters, reducing photosynthesis and also oxygen levels. When winter comes, the algae rot, further decreasing oxygen levels.

Given the inadequacies of sewage-treatment plants, some scientists are looking for alternatives that might do a better job at less cost. One possibility is the artificial wetland—a pond or marsh planted with typical wetland vegetation. In Arcata, California, for example, the city's sewage now pours into a series of interconnected artificial marshes. Solid materials settle out and decay on the bottom, releasing carbon dioxide. Nitrates and phosphates are incorporated into wetland vegetation. By the time the wa

▷ **FIGURE 32–33 Trickling Filter** Found in many secondary-waste-treatment facilities, these devices drip liquid sludge over a bed containing bacteria and other organisms. They live off the organic material, removing it from solution, and help purify the water in the process.

ter leaves the system, it is cleaner than effluent from sewage-treatment plants. Such a system is far cheaper, because it basically lets nature do the work.

Working with nature is the basis behind a similar scheme devised by a biologist and his wife, John and Nancy Jack Todd. In 1986, they built a solar aquatic waste-treatment plant in Vermont. A few years later, they built another facility to treat septage, solid waste pumped from septic tanks. The facility consists of 21 giant aquariums, containing a variety of species, that surround a small, artificial marsh. The system takes in sewage with many toxic contaminants, including lead, mercury, and copper, and emits pure water, virtually free of contaminants. Of 14 carcinogenic pollutants in the waste, 13 are 100% removed. The 14th (toluene) is 99.9% removed. The pure water produced by the plant is now returned to the groundwater, replenishing the town's drinking water.

This example, and the artificial wetlands of Arcata, California, illustrate how we can create a healthier environment, not just for humans but for all species. Artificial wetlands provide habitat for many species and produce water potentially much cleaner than that from sewage-treatment plants. This example shows that working with nature, we can help protect nature.

SUMMARY

1. The biosphere is a mosaic of habitats, or biomes, each characterized by a distinct climate and assemblage of organisms.

THE TERRESTRIAL BIOMES

2. Each biome contains a dominant form of vegetation, but significant variations occur within any given biome because of regional differences in soil, temperature, rainfall, and other factors.

3. The Arctic tundra is a treeless zone that constitutes the northern limit of plant growth. It supports grasses, shrubs, mosses and lichens, and a number of species of animals adapted to the long, cold winters.

4. The subsoil of the tundra remains permanently frozen and is known as the permafrost. During the summer months, the superficial layers of soil thaw, and the tundra becomes dotted with ponds and bogs. Abundant insects support large populations of migratory birds.

5. Although conditions are harsh, the tundra is a rather fragile environment, easily damaged and requiring many decades for natural recuperation.

6. The taiga, or northern coniferous forest biome, consists of a band of conifers that extends across the northern continents below the tundra. Conifers are well adapted to the cold, snowy winters.

7. Generally warmer and wetter than the tundra, the taiga supports a wide variety of species. Timber cutting and other human activity now threaten many regions of this biome.

8. The temperate deciduous forest biome, located in the eastern United States and elsewhere, receives abundant precipitation and supports large stands of deciduous trees.

9. The long growing season, abundant rainfall, and rich soils make this biome excellent for crops, so much so that most of the original forests have been cleared.

10. Temperate grassland biomes occur in regions of intermediate precipitation. These regions do not support trees but support grasses that are well adapted to periodic drought. Their rich soils have made them prime targets for agricultural development.

11. The desert biome is characterized by minimal precipitation. Many deserts occur downwind from mountain ranges, which "rob" air masses of moisture. Plants and animals that live in the desert are well adapted to the dry, hot conditions.

12. The desert biome is expanding, in large part because of overgrazing and poor agricultural practices.

13. The richest and most diverse terrestrial biome on Earth is the tropical rain forest, which is blessed with abundant precipitation and warm temperatures year round.

14. Tropical rain forests are vital to the ecological health of the planet, but they are being progressively cut down often to make way for unsustainable human practices.

15. Mountains display distinct life zones (altitudinal biomes) that result from variations in climate due to elevation.

FRESHWATER LIFE ZONES

16. Freshwater and saltwater biomes are called aquatic life zones. In aquatic life zones, solar energy and nutrients are principal determinants of the diversity and abundance of life.

17. In aquatic life zones, many grazing food chains begin with phytoplankton, which are consumed by zooplankton.

18. Lakes are divided into four ecologically distinct zones, (a) the littoral zone, the shallow waters along the shoreline; (b) the limnetic zone, the open water of a lake extending to the limit of sunlight penetration; (c) the profundal zone, the deep waters of the lake where photosynthesis does not occur; and (d) the benthic zone, the lake's bottom.

19. Deeper lakes thermally stratify during the summer, with the warmest layer (the epilimnion) on top and the coldest water (the hypolimnion) on the bottom. Between these two layers is an intermediate layer, the thermocline, characterized by a rapid change in water temperature. Oxygen levels are highest in the upper layer.

20. During the fall, the surface waters cool, and the lake's water becomes agitated by winds, causing a mixing of water. The lake loses its thermal stratification, and oxygen levels become more or less uniform.

21. Ice forms on the lake during the winter, but fishes survive thanks to greatly reduced oxygen demands, oxygen surpluses generated during the summer months, and continued photosynthesis by some aquatic plants and algae.

22. During the spring, the surface water warms again, and the lake becomes thermally uniform. Wind agitates the water again, causing a complete mixing.

23. Streams and rivers are generally well oxygenated, but oxygen levels vary considerably. Faster moving streams have higher

levels and are home to species that require more oxygen.

24. All freshwater ecosystems are vulnerable to pollutants, which enter from discrete (point) or nondiscrete (nonpoint) sources. Lakes, streams, and rivers are capable of recovering from many pollutants, but human activities often overwhelm their recuperative capacity.

SALTWATER LIFE ZONES

25. The oceans are also divided into distinct zones. The coastal zones include estuaries, shorelines, and coral reefs.
26. Estuaries are nutrient-rich zones at the mouths of rivers. Estuaries and coastal wetlands constitute the estuarine zone and are home to many species of fish and molluscs essential to human society.
27. Shorelines are rocky or sandy coastal regions. Sandy shorelines contain relatively few species, but rocky shorelines house a great many species that anchor themselves to rocks or take refuge in crevices.

28. Coral reefs are biologically rich zones whose organisms live on, in, or around giant deposits of coral.
29. All coastal life zones are threatened by human activities.
30. The ocean is divided into four ecologically distinct life zones: the neritic, euphotic, bathyal, and abyssal. The characteristics of each zone are shown in Figure 32–30.

ENVIRONMENT AND HEALTH: WETLANDS, WILDLIFE, AND WATER TREATMENT

31. Primary and secondary treatment removes about 90% of the organic wastes but only 30% of the phosphates and 50% of the nitrates. Sewage treatment plants, therefore, are a significant source of water pollution.
32. Given the inadequacies of sewage treatment plants, some scientists are looking for alternatives that might do a better job at less cost. One possibility is the artificial wetland—a pond or marsh planted with typical wetland vegetation.

EXERCISING YOUR CRITICAL THINKING SKILLS

A recent article in *Science News* reports on the studies of a British scientist who examined sonar measurements of ice thickness in the Arctic Sea. He studied the measurements made by two British submarines, which took similar routes across the Arctic, but years apart. The first was in 1976; the second was in 1987. During the first trip, the average ice thickness over a region the size of Nevada was 5.3 meters. In the second trip, the average thickness was 4.5 meters.

To some scientists, these data suggest that the Arctic ice is melting, a sign that global warming may be occurring. What questions would you ask to determine the validity of this conclusion? Write them down.

After thinking about it, see how your questions compare with

some questions that scientists have asked. Could the differences in ice thickness have been the result of normal variation. (Yes, possibly.) Were the measurements of the two submarines taken during the same time of year? (No.) Did the submarines follow the same path? (Actually, they followed slightly different routes.) Could winds have altered ice buildup between the two periods? (Some research suggests that wind patterns were quite different, and winds do alter ice buildup.) Have other studies shown thinning? (Yes. A study of ice measurements by two submarines on a virtually identical route across the Canada basin during 1958 and 1970 shows that the mean ice cover was thinner in 1970 by 0.69 meter.)

TEST OF TERMS

1. The _____ _____ is the northern limit of plant growth and supports grasses, lichens, shrubs, and mosses.
2. Permanently frozen subsoil is called the _____ .
3. The band of coniferous trees stretching across the North American continent is part of the biome known as the _____ .
4. The _____ _____ _____ biome is characterized by broadleaf trees that shed their leaves each fall.
5. Trees that shed their leaves are called _____ trees.
6. The _____ biome con-

sists of grass-covered plains or rolling hills, usually lacks trees except along rivers, and receives intermediate precipitation.
7. Moisture in air masses is often captured by mountain ranges, creating desert on the _____ side.
8. By far the most diverse of all terrestrial biomes is the _____ _____ _____ .
9. Traveling from the base of a mountain to its top, one moves through distinct life zones, or _____ .
10. On the tops of many high mountains

you will encounter a cold, treeless region similar to the Arctic tundra and known as the _____ _____ .
11. Aquatic systems are divided into distinct regions known as _____ _____ _____ .
12. In aquatic systems, the abundance and diversity of life forms result from two factors, _____ and _____ .
13. _____ are microscopic, free-floating organisms that capture sunlight energy, using it to produce carbohydrates from carbon dioxide that is dissolved in the water. These organisms are consumed by

microscopic _____ , single-celled organisms (crustaceans and protozoans), which form the second trophic level of many aquatic food chains.

14. The _____ zone of a lake consists of the shallow waters at the margin of the lake where rooted vegetation can grow.

15. The _____ zone of a lake is a region commonly called the "open water." It extends downward to the point at which light no longer penetrates and is, therefore, the main photosynthetic zone of a lake.

16. The deepest water of a lake, into which very little light penetrates, is called the _____ zone. The bottom of a lake is called the _____ zone.

17. In the summer, lakes thermally stratify. The warmest water always occurs along the surface and forms the _____. The coldest water lies deeper and constitutes the _____. Between them is a zone of rapid temperature change, the _____ .

18. During the fall, the water in a lake becomes thermally uniform, and winds cause the water to mix. This phenomenon is known as the _____ _____ .

19. The region drained by a river is called its _____ .

20. A factory represents a _____ source of pollution, whereas an agricultural area represents a _____ source.

21. The estuarine zone consists of _____ and _____ and is one of the richest coastal life zones.

22. _____ _____ are biologically rich life zones found in relatively warm and shallow waters in the tropics or nearby regions (subtropics). They consist of calcium carbonate or limestone produced by calcarous red and green algae and by colonies of organisms called stony corals.

23. The ocean floor gradually drops off from the coastline, forming the _____ _____ .

24. The _____ zone consists of the open waters of the ocean and extends to the depth at which sunlight no longer penetrates.

25. The ocean region of semidarkness, characterized by the absence of photosynthesis, constitutes the _____ _____ .

Answers to the Test of Terms are found in Appendix B.

TEST OF CONCEPTS

1. What is a biome? Why is one biome so different from another? Give some examples.

2. Animals are generally well adapted to the biome in which they live. Give some examples of plants and animals and their adaptations.

3. Explain why the tundra is one of the most fragile biomes.

4. Explain why there are regional differences in vegetation and animal life within a biome.

5. Which biome do you live in? Describe it. How has it been altered?

6. List the pros and cons of oil development in the ANWR. Which side do you take on the issue? Why?

7. Explain why the tropical rain forest is the richest terrestrial biome yet has the poorest soil.

8. Why do many deserts form downwind from mountain ranges?

9. Explain why the desert biome is spreading.

10. Explain why clearing large sections of tropical rain forest is often an invitation to disaster.

11. What are the primary determinants of species composition and abundance in terrestrial and aquatic habitats?

12. Describe each of the four zones of a lake (littoral, limnetic, profundal, and benthic) in terms of light availability, nutrient levels, oxygen concentration, and life forms.

13. Describe thermal stratification in deep lakes. Why does it occur during the summer and disappear throughout most of the rest of the year?

14. Why is current so important in determining the abundance and type of life in a river?

15. Explain the differences between nonpoint and point sources of water pollution. Why is controlling pollution from these sources so different?

16. Explain how streams and rivers recover from pollution and why natural recovery methods are so often overwhelmed.

17. What is an upwelling? How is it similar to an estuary? How is it different?

18. Describe ways in which human activities affect coral reefs and the estuarine zone.

Environmental Issues: Population, Pollution and Resources

Floating barriers help restrain the monstrous oil spill of the Exxon Valdez in Alaska's Prince William Sound in 1989.

Nature has its own economy, a complex system of exchange, competition, and cooperation. In many ways, nature's economy is like our economy. For example, the economy of nature contains producers and consumers and avenues of exchange between them, food chains and food webs. But lest we forget, the human economy is like a small gear driven by the economy of nature. All wealth, all food and materials come from the soil, water, air, and biota (living organisms). Unfortunately, the small wheel of the human economy is moving faster and becoming more powerful. In the process, the human economy is beginning to destroy the larger economy that drives it. After 200 years of industrial growth and significant gains in human welfare in many countries, the human economy has come to threaten the workings of nature and the health of the planet. Not just human existence is in danger, but the existence of many other life forms as well.

The threat to the environment is commonly referred to as the "environmental crisis." To many people, the phrase "environmental crisis" may seem like an exaggeration. To them, such a proclamation seems unwarranted. But a close examination of trends in population growth, pollution, and resource use and depletion yields a disturbingly different view. Many experts who have studied the trends believe that the problems are severe enough to warrant the term "crisis." Jay Hair, a zoologist and president of the National Wildlife Federation, believes that humanity has at best 10 or 15 years to come to grips with the menacing trends. If we do not, our future 100 years from now will be far from optimum.

This chapter looks at the major environmental trends, outlining a host of problems, the true dimensions of which most people are unaware or only partly aware. It focuses on three principal areas—population, pollution, and resource use—and describes several of the most pressing environmental problems under each category. A close examination of these trends, although at times depressing, is essential at this point in human history. Unless we understand the full dimensions of the environmental crisis, we cannot mount a response adequate to the task.

This chapter also discusses a variety of solutions for individuals, corporations, and governments and concludes with a discussion of ways by which human society can steer onto a sustainable course that is needed to ensure long-term ecological stability on the planet—a condition essential for a healthy human existence.

≈ OVERSHOOTING THE EARTH'S CARRYING CAPACITY

The environmental problems you hear about on the news and read about in the paper occur in both the rich, industrialized nations and the poor, less-developed nations. Although the problems vary from one nation to another, they all have a common root. *In a phrase, environmental problems result from people living beyond the means of the environment.*

▷ **FIGURE 33–1 Food Shortages** Overpopulation, political turmoil, and mismanagement of farmland are chief causes of scenes such as this one. Children die by the thousands every day in similar camps.

In the language of ecologists, human society is exceeding the Earth's carrying capacity.

Carrying capacity is the number of organisms an ecosystem can support indefinitely—that is, the number it can sustain. Carrying capacity for all organisms, including humans, is determined by at least three factors: (1) food production, (2) resource supply, and (3) the environment's ability to assimilate pollution. Let's look at each one.

In Many Places the Human Population Is Exceeding Food Production

As the global human population grows, many nations are finding it more and more difficult to meet the demands for food. Starvation abounds in the Third World, and millions of people perish each year from malnutrition and starvation or from diseases worsened by hunger (▷ Figure 33–1). These conditions provide strong evidence for the view that some human populations are living beyond the carrying capacity.

Many Resources Are in Short Supply and Will Be Depleted in the Near Future

Human populations also require a great many other resources, such as fuel, fiber, and building materials. Those resources that are finite, such as oil, natural gas, and minerals, are known as **nonrenewable resources.** Resources that replenish themselves via natural biological and geological processes are called **renewable resources.** Wind, hydropower, trees, and fishes are examples.

Today, some important nonrenewable resources like oil are on the decline. As you will soon see, some resources are just about used up. Nonrenewable resources like tropical forests are also being depleted faster than they can be replenished. These trends, illustrated more fully in the chapter, suggest that in meeting human needs, we may be reducing the Earth's carrying capacity.

Pollution from Human Activities Exceeds the Environment's Assimilative Capacity

The final determinant of carrying capacity is the capability of the environment to assimilate and degrade pollutants. In natural ecosystems, wastes are usually diluted to harmless levels and are recycled in nutrient cycles, thus putting them to good use. Animal feces, for instance, are broken down by soil bacteria. These bacteria release carbon dioxide, nitrogen, and other nutrients into the environment for reuse. In nature, nutrient cycles ensure a steady supply of resources for all generations. Unfortunately, many human activities interrupt these vital cycles. Too much animal waste in an aquatic environment, for example, may shift the concentration of nitrogen and phosphorus, starting a series of changes that results in rapid growth of algae and aquatic plants; algae and plants decay in the fall, consuming oxygen in the process and killing fish.

Human populations produce enormous amounts of wastes that overwhelm natural systems. In the United States, according to a study by the Conservation Foundation, sewage, garbage, and hazardous wastes from factories, pollution from automobiles and homes, and wastes from agriculture and mining amount to 50,000 pounds per year for every man, woman, and child. The U.S. population represents only 6% of the world population but produces about 25% of the world's pollution.

The environment can dilute some of this waste to harmless levels. It can break down other wastes, rendering them harmless, but in many locations, the environment's assimilative capacity is being severely overtaxed. In lakes, rivers, and oceans, pollution is taking a toll on fish and wildlife. Discarded plastic fishnets, six-pack yokes, and other plastic garbage annually kill an estimated 100,000 marine mammals, such as seals. In U.S. waters, tighter regulations have reduced the outflow of noxious pollutants from factories and sewage treatment plants, helping reduce water pollution. Despite exhaustive efforts and billions of dollars, many lakes and rivers have shown little improvement over the past decade. Why? Many of these bodies of water are contaminated by pollutants that are being washed from urban lawns, golf courses, parking lots, and farm fields, or from the sky by rain and snow. These pollutants have offset many of the gains in pollution control of our factories and

▷ **FIGURE 33–2 Air Pollution** Studies show that air pollutants in some locales can increase respiratory disease, such as emphysema and lung cancer. Air pollution is aesthetically unappealing, and levels that do not affect human health can cause considerable environmental damage.

sewage treatment plants. Growth in population and economic activity also contribute to the mounting pollution burden. Even though efforts have been made to reduce the amount of pollution our society produces, continued economic expansion and growing numbers of people often offset these gains.

The stress on the assimilative capacity of the atmosphere is evident around most cities with populations over 50,000. These pockets of air and water pollution pose a threat to human health and a threat to a great many species that make this planet their home (▷ Figure 33–2). Today, over 110 million Americans (nearly half of our population) live in cities whose air is deemed harmful to their health.

One of the "advances" of the modern industrial society was the advent of synthetic organic chemicals, such as the insecticide DDT and plastics. Many of these products cannot be broken down by natural processes or are degraded very slowly because naturally occurring bacteria are not equipped with the enzymes needed to decompose these new molecules. As a result, DDT, PCBs, and a number of other toxic organic chemicals often persist in the environment for decades, accumulating in the tissues of organisms, impairing reproduction, and threatening their hosts' future.

Clearly, the environmental crisis is multifaceted. In all instances, the problem is basically the same: humans are exceeding the carrying capacity. Correcting that imbalance is one way of alleviating the problems.

≈ OVERPOPULATION: PROBLEMS AND SOLUTIONS

Overpopulation occurs when populations exceed the carrying capacity of the environment. The world population is 5.4 billion and is increasing at a rate of 1.8% per year. Although the growth rate may seem small, if the current rate continues, world population could reach 10 billion people by the year 2030—that is, in less than 40 years.

Overpopulation Is a Problem in Virtually All Countries, Rich and Poor

The annual **growth rate** of the world population is calculated by the following formula:

$$\text{growth rate} = \text{birth rate} - \text{death rate}$$

The world birth rate is 28 per 1000. In other words, 28 children are born for every 1000 people living on the Earth each year. The global death rate is 10 per 1000. To calculate the growth rate, you simply subtract the death rate from the birth rate. The global growth rate is therefore equal to 28/1000 − 10/1000, or 18 per 1000. This means that 18 new residents are added to the planet each year for every 1000 people alive. To convert the growth rate (18/1000) to a percentage, simply multiply by 100. Thus, the growth rate is equal to 18/1000 × 100, or 1.8%.

Growth rates are important measures of population dynamics, but they can be deceiving to the uninitiated. For this reason, **demographers,** scientists who study populations, often convert growth rates into **doubling time,** the time it takes a population to double. Doubling time is calculated by the following equation:

$$\text{doubling time} = 70/\text{growth rate (in percent)}$$
$$= 70/1.8$$
$$= 39 \text{ years}$$

The figure *70* in this equation is a demographic constant. The global growth rate is an average of all countries and thus masks regional growth differences. For example, the U.S. growth rate is about 0.7%. In other words, 7 new residents are added to the population each year for every 1000 people alive.[1] If this growth rate continues, our present population of 248 million people will double in 100 years. In contrast, in Africa the growth rate is 2.9%, yielding an alarming doubling time of 24 years!

---

[1] Population growth in any country is the result not only of the difference between birth rate and death rate but also of the net migration—that is, the number of people entering or leaving the country. In the United States, about 40% of our annual growth results from legal and illegal immigration.

 TABLE 33–1 Growth Rate and Doubling Time

REGION	GROWTH RATE (%)	DOUBLING TIME (YEARS)
World	1.8	39
Developed countries	0.6	117
Less-developed countries	2.1	33
Africa	2.9	24
Asia	1.9	38
North America	0.7	100
Latin America	2.1	33
Europe	0.3	230
Soviet Union	1.0	68
Oceania	1.2	56

Today, the most rapid growth is occurring on three continents: Africa, Asia, and Latin America (Table 33–1). The slowest growth is occurring in Europe, where the population is, on average, rising at a rate of 0.3% per year. At this rate, Europe's population will double in about 230 years.

Most people view overpopulation as a problem of the Third World. Actually, overpopulation is a problem in all countries. It is as serious in the United States as it is in Bangladesh.

The reason is the fairly high standard of living in the rich, industrialized countries. Citizens of these nations have an enormous impact on the environment. One baby born in the United States will use 25 to 40 times as many resources as a baby in India. The 248 million Americans alive today cause as much damage as 6 billion to 10 billion people in the Third World.

The Human Population Is Growing Exponentially

The human population has not always been so large, nor has it always grown so rapidly. As ▷ Figure 33–3a shows, it was not until the last 200 years that global human population began to skyrocket, and then only because of better sanitation, improvements in medicine, and advances in technology.

As you may have noticed, the graph of human population growth in Figure 33–3a is J-shaped. This graph is also called an exponential curve.

What is exponential growth? To begin, consider an example. Suppose your parents invested $1000 in a savings account at 10% interest the day you were born. Suppose also that the interest your money earned was applied to the balance, so that it too earned interest. If this were the case, your bank account would be said to be growing exponentially. **Exponential growth** occurs anytime a value, such as population size, grows by a fixed percentage if the interest (increase) is applied to the base amount.

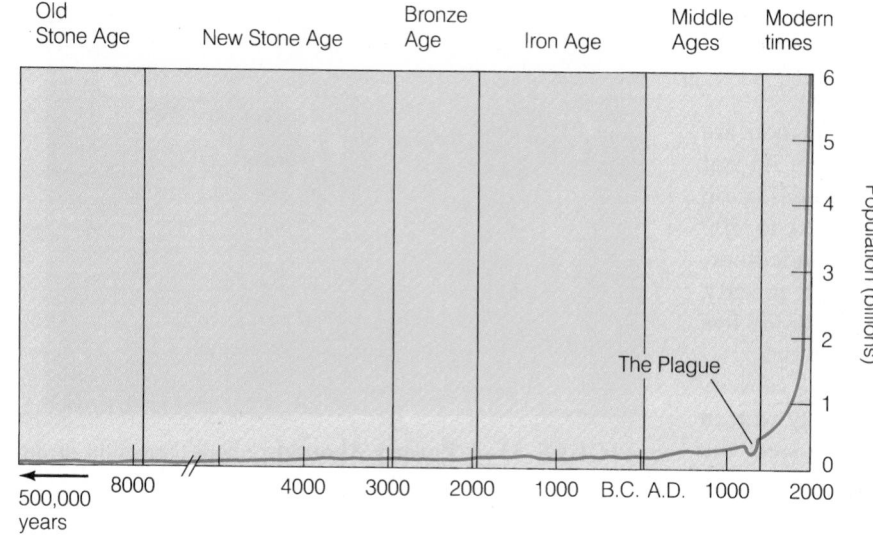

500,000 years   8000   4000   3000   2000   1000   B.C. A.D.   1000   2000

The Plague

Population (billions)

(a)

▷ **FIGURE 33–3 Exponential Growth of the World Population** (*a*) World Population. (*b*) Exponential growth of a bank account starting with $1000 at 10% interest.

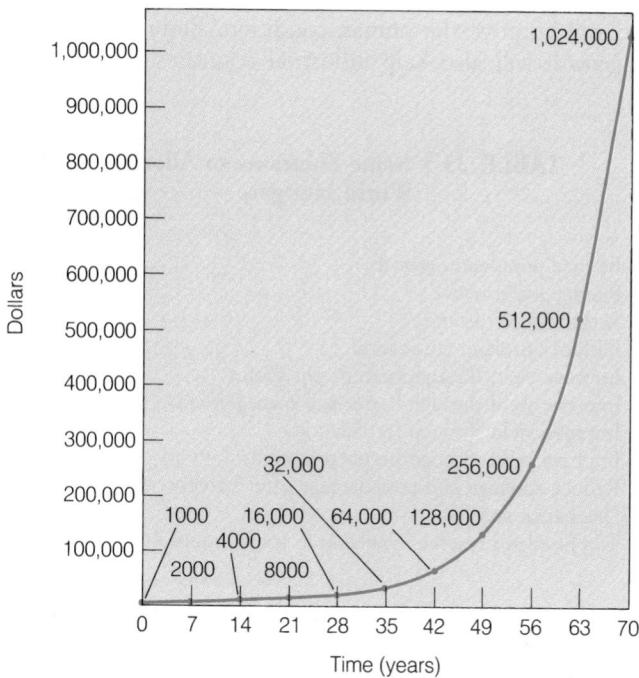

(b)

of growth is constant throughout the entire period. In this example, for instance, it took 49 years for your account to grow from $1000 to $128,000. In the next 21 years, however, your account grew by $900,000. Why?

Although your account doubled every seven years, it was not until the base amount was very large that the doublings amounted to much. Thus, once the base amount reached a certain level, each doubling yielded what appeared to be incredible gains—even though the account grew at a steady 10% per year.

The human population has taken over 3 million years to reach its current size of 5.4 billion, but because of exponential growth, it could increase by another 5.4 billion in the next 40 years, if current growth continues (Table 33–2).[2] Pollution and resource demand will also grow exponentially as the human population skyrockets, and therein lies the concern of many people. Even though supplies of many resources seem large, exponential growth could deplete them in a very short time. Just to show you how insidious exponential growth is, consider a resource with a billion-year life span at the current rate of consumption. If the rate of demand increases 5% per year, that billion-year resource will last only about 500 years.

Exponential growth is deceptive. In this example, the growth in your account seems small at first. At 10% interest, the $1000 will double in 7 years, yielding $2000 (Figure 33–3b). It will double again in another 7 years, yielding $4000. The next doubling, occurring when you are 21 years old, will give you $8000. By age 42, the account will have grown to $64,000. By the time you were 49, your account would hold $128,000. At age 56, you'd have over $250,000. If you waited 7 more years, the account would grow to over $500,000. In 7 more years, at 70 years of age, you'd be a millionaire.

What is deceptive about exponential growth is that it begins slowly, then seems to go wild, even though the rate

[2]It is important to point out that the human population has not been growing at 1.8% per year for the 3 million years.

≈ TABLE33–2 Global Population Growth ≈		
POPULATION SIZE	YEAR	TIME REQUIRED TO DOUBLE
1 billion	1850	All of human history
2 billion	1930	80 years
4 billion	1975	45 years
8 billion (projected)	2017	42 years

The Human Population Problem Can Be Summed Up in a Phrase: Too Many People, Reproducing Too Quickly

In many countries, large segments of the population are living in extreme poverty. Many of these people do not have enough to eat. People live in makeshift homes or on the streets (▷ Figure 33–4). An estimated 700 to 800 million people are malnourished or severely undernourished. About 2 billion people are on the edge of poverty, with barely enough to eat. Thus, nearly three of every five people on the planet are living in horrible conditions.

Because humans are living beyond the carrying capacity of the environment now, further growth will only worsen problems. Unfortunately, little is being done to stem the swelling tide of humanity. Each year, in fact, 90 million new residents are added to the world population. Approximately 9 of every 10 of these people are born in the Third World, where hunger and poverty have become a way of life (▷ Figure 33–5). In African nations, such as Kenya, where the human population is expected to double in 17 years, efforts aimed at keeping up the substandard food supplies will put enormous strains on the economy and the environment. Making improvements in diet to reduce starvation and persistent hunger seems impossible. In this struggle, many believe that nature will restore a more equitable balance. Millions of people will die unless something is done to reduce population growth and increase food supply.

Solving World Hunger and Living Sustainably on the Planet Require Many Actions

There is no easy answer to world hunger. Solving such a complex issue and furthermore learning to live sustainably on the Earth require a multifaceted response (Table 33–3).

Slowing the Rate of Growth. One of the most important steps is to reduce population growth. Most world

▷ **FIGURE 33–4 Poverty Abounds** About three-fifths of the world's people live in poverty. Nearly one of every five people on this planet does not have enough to eat. Many live in makeshift shelters like these in Rio de Janeiro.

leaders agree that greatly reducing the rate of increase will help Third World countries produce the food they need and will improve the human condition. Reducing the rate of growth will also help industrial countries reduce their

TABLE 33-3 Some Solutions to Alleviate World Hunger
Reduce population growth.
Reduce soil erosion.
Reduce desertification.
Reduce farmland conversion.
Improve yield through better crop strains.
Improve yield through better soil management.
Improve yield through fertilization.
Improve yield through better pest control.
Reduce spoilage and pest damage after harvest.
Use native animals for meat production.
Tap farmland reserves available in some countries.

▷ **FIGURE 33–5 Growth of Human Population** The graph shows the relative size of the populations in the less-developed regions and the developed regions and projected growth. Ninety percent of the growth is occurring in the Third World. Today, about one of every five people in the world lives in an industrialized country. By 2100, if current growth rates continue, the ratio will be about 1 to 10.

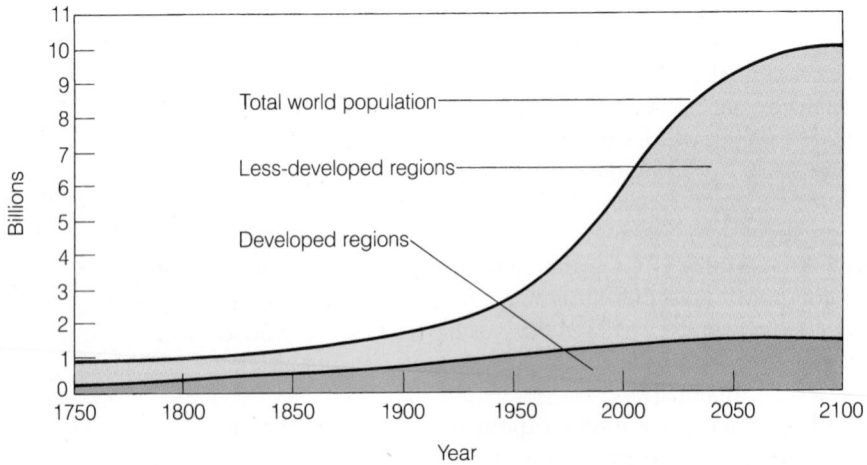

Total world population

Less-developed regions

Developed regions

Billions

Year

> **FIGURE 33-6 Habitat Destruction** Wetlands like these in Southern Florida are a valuable natural resource but have been destroyed by home building and other activities.

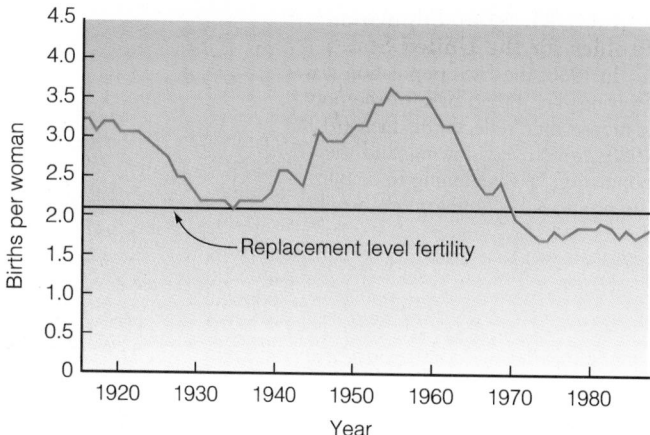

> **FIGURE 33-7 U.S. Total Fertility Rate** Since 1972, the total fertility rate in the United States has been below replacement-level fertility. Yet overall population growth continues because of immigration and because of the large number of women having children.

demand for resources and reduce pollution, habitat destruction, and a host of other environmental problems (▷ Figure 33–6).

Reducing Numbers through Attrition. Judging from the state of the environment, today's population of 5.4 billion people is well beyond the carrying capacity. Slowing the growth of human population, some experts believe, will not be sufficient to build a sustainable human presence. Over the next 50 years, it may be necessary to reduce the size of the human population. This will permit human society to live within the Earth's carrying capacity. The principal means of reducing population size is through attrition—reducing the birth rate so that it falls below the death rate.

Reducing the Total Fertility Rate. Reducing the population size of a country, indeed the world, is not as difficult as many might believe, or as undesirable. Already, a dozen or so countries in Europe have reached a stage of extremely slow growth, no growth, or even "negative" growth (shrinkage), among them Hungary, Denmark, Italy, and Austria. The decline in population growth is achieved,

in large part, because couples are having smaller families.

The number of children a woman has in her lifetime is called the **total fertility rate.** For a population to remain stable in countries such as ours, the total fertility rate must be maintained at 2.1.[3] This means that each woman of reproductive age has, on average, 2.1 children. This level of fertility is called the **replacement-level fertility.** It is the number of children that will replace a couple when they die. A replacement-level fertility rate of 2.1 means that every 10 couples in a population must have 21 children to maintain a steady population size. The additional child accounts for typical mortality and for childless couples.

When the total fertility rate is below the replacement-level fertility, a population will decline, but only if there is no net immigration—that is, there are no newcomers from other countries. In the United States, the total fertility rate has been below replacement-level fertility since 1972 (▷ Figure 33–7). Despite this important development, the U.S. population continues to grow. Why?

One of the reasons for growth is the steady inflow of illegal and legal immigrants—about a million people each year. Growth also occurs because of the age structure of our population. ▷ Figure 33–8 is a **population histogram,** or **population profile,** of the United States in 1960 and 1988. It shows the number or percentage of males and females in each age group. As you can see, the profile for 1960 is bottom-heavy. There are more people in the lower age groups than in the upper ones. Many of the people in the lower age groups are products of the baby boom era; that is, they were born soon after World War II.

After the war, America's prospects looked bright. Judging from the rise in total fertility to well over 3, people must have been extremely optimistic and happy to be done with the brutal war. The large postwar families caused the U.S. population to swell (▷ Figure 33–9).

[3] Assuming there is no immigration.

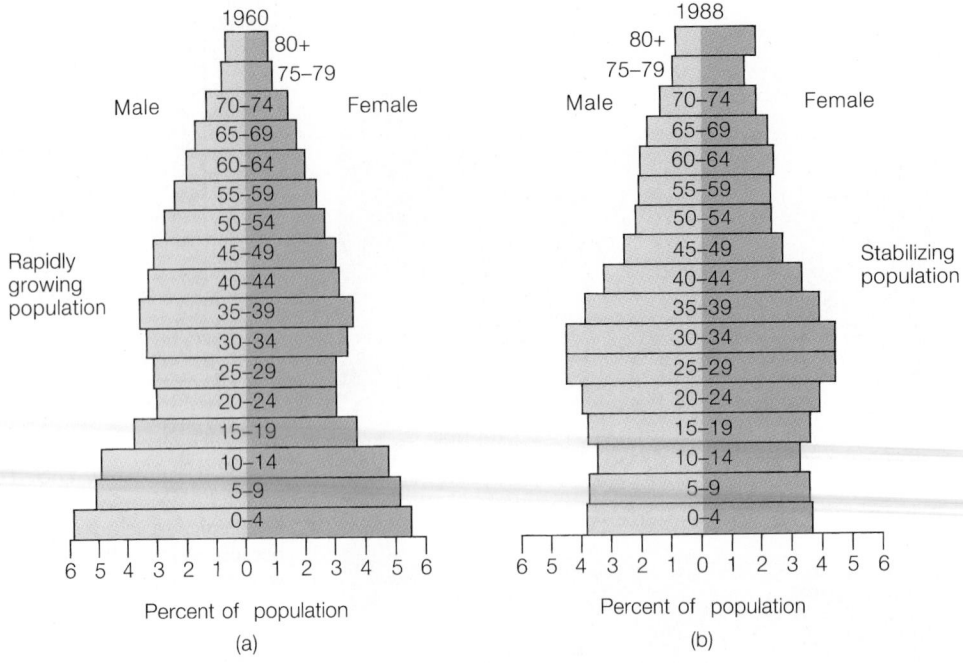

FIGURE 33-8 Population

▷ **FIGURE 33-8 Population Profiles for the United States** (*a*) In 1960, the U.S. population was growing. The broad base of the population profile reflects this fact. (*b*) In 1988, growth was slowing, and the population was beginning to stabilize the profile is becoming more boxlike

Today, the baby boomers are having children of their own, and even though couples are having far fewer children than their parents did, the reproductive age group is larger today than it was after the war. Growth in the U.S. population today, therefore, occurs not because women are having more children, but because more women are having children.

If immigration quotas were not increased and if the total fertility rate remained the same, the U.S. population profile should become more boxlike, or stationary (▷ Figure 33–10). At this point, the population would cease growing, reaching a steady state called **zero population growth.** However, Congress passed legislation in 1991 that will greatly increase the number of legal immigrants and could accelerate population growth in the United States.

Worldwide, the population profile today looks a lot like the U.S. population profile did shortly after the war—the histogram is triangular, or expansive. Today, 33% of the world's people are under the age of 15. Soon, they will be entering the reproductive age group and will start having families. Unless these children restrict their own reproduction, they will cause massive population growth in the next four decades.

≈ RESOURCE DEPLETION: ERODING THE PROSPECTS OF ALL ORGANISMS

The human population depends on a variety of resources for its survival and well-being—among them, forests, soil, water, minerals, and oil. Today, however, many of these resources are in danger. According to one estimate, 5 billion trees are cut down every day, and the worldwide rate of timber harvest now exceeds regrowth. Productive soils are being eroded by careless farming practices, and once-productive farmland is turning to desert at an alarming rate. Water for drinking and irrigation is being withdrawn

▷ **FIGURE 33-9 The Baby Boom Effect** This figure follows the babies born from 1955 to 1959 through the year 2010. Population profiles provide a means of projecting trends in populations.

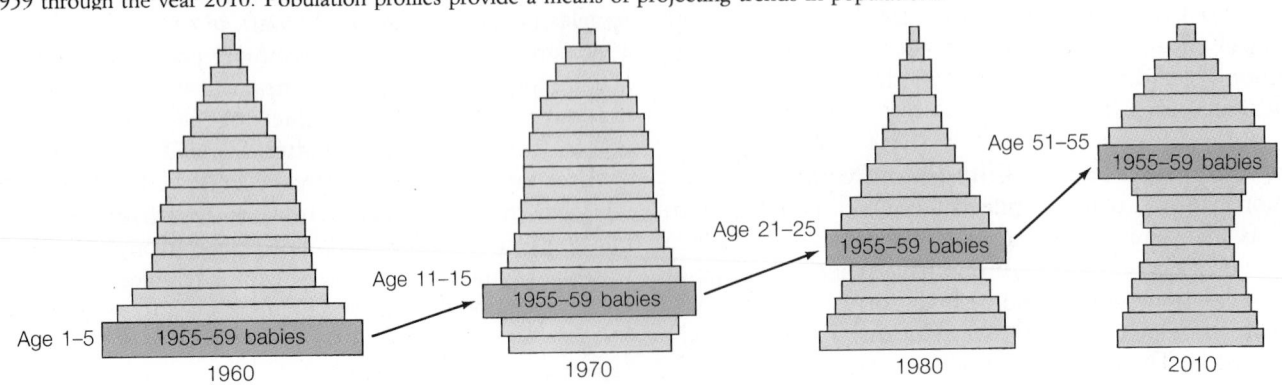

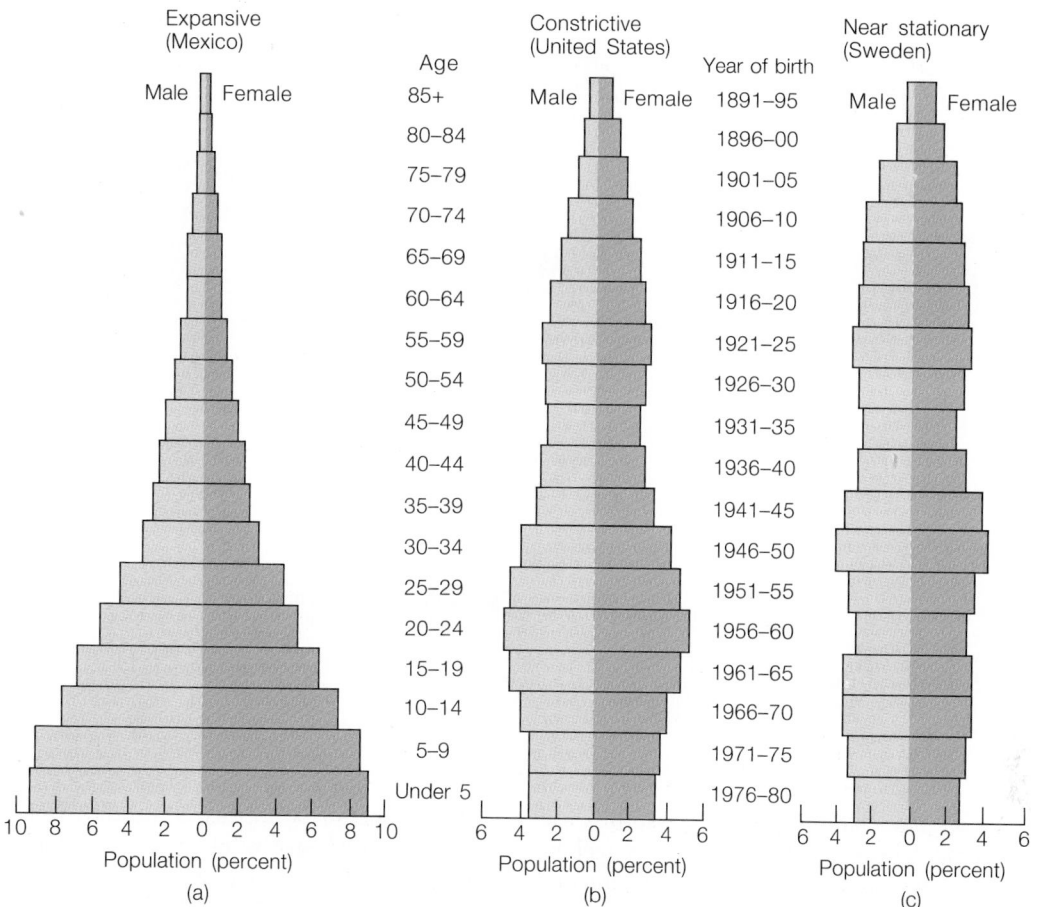

Expansive (Mexico)
Male | Female

Constrictive (United States)
Male | Female

Near stationary (Sweden)
Male | Female

Age	Year of birth
85+	1891–95
80–84	1896–00
75–79	1901–05
70–74	1906–10
65–69	1911–15
60–64	1916–20
55–59	1921–25
50–54	1926–30
45–49	1931–35
40–44	1936–40
35–39	1941–45
30–34	1946–50
25–29	1951–55
20–24	1956–60
15–19	1961–65
10–14	1966–70
5–9	1971–75
Under 5	1976–80

(a) Population (percent) 10 8 6 4 2 0 2 4 6 8 10

(b) Population (percent) 6 4 2 0 2 4 6

(c) Population (percent) 6 4 2 0 2 4 6

▷ **FIGURE 33–10 Three Possible Population Profiles**
(*a*) Mexico's population profile is typical of a rapidly growing population. (*b*) The population profile of the United States reflects its much slower growth. (*c*) Sweden's population profile is nearly stationary.

from wells faster than it can be replaced, creating regional shortages in many parts of the world that are bound to worsen in the near future. Minerals are being extracted in ever-increasing quantities, and many vital mineral supplies will last no more than four decades. This section describes the problems and discusses some of the solutions.

Humanity Is Destroying the World's Forests, but Broad Replanting Efforts Could Reduce or Even Stop Deforestation

At one time, tropical rain forests covered a region about the size of the United States (▷ Figure 33–11). Today, about half of those forests are gone. Because of an exponential increase in population size and timber harvests to provide a host of products from teak for yachts to disposable chopsticks, most of the remaining tropical rain forests could be destroyed by the year 2000. Along with them, thousands of species, perhaps as many as a million, would vanish. Making matters worse, in the tropics, only 1 tree is replanted for every 10 trees that are cut down. In tropical Africa, the ratio is 1 to 29.

The decline of the world's forests is not limited to the tropics. In the United States, for example, 45% of the forested land present when the eastern seaboard was first colonized has been converted to other uses. In the Pacific Northwest, 90% to 95% of the old-growth forests, consisting of trees from 250 to 1000 years old, have been cut (▷ Figure 33–12). India's forests decline by 1.5 million hectares (3.7 million acres) per year. In 1989, logging in British Columbia was estimated to be 30% over the sustainable yield. Softwood harvest on the West Coast of the United States during the 1980s exceeded sustainable yield by 25% on privately-owned land and 61% on National Forests.

Since 1920, however, the overall rate of timber harvest in the United States has more or less equaled the rate of regeneration. That is due, in large part, to the reforestation of abandoned farmland in the East and more stringent policies in national forests. The rolling hills of Connecticut, for example, once cleared for farming, now support healthy young forests. Expected increases in the demand for timber and wood products, however, could shift the balance once again.

Because global deforestation exceeds reforestation and because population and demand for resources continue to

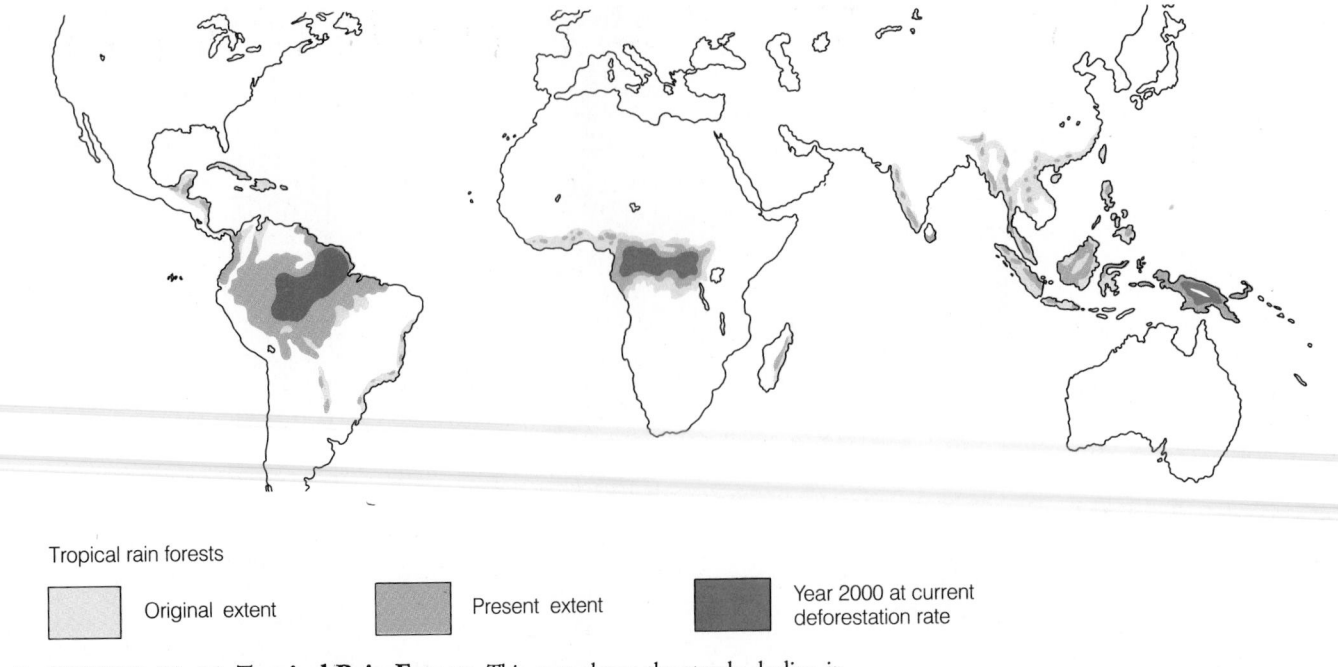

Tropical rain forests

[ ] Original extent   [ ] Present extent   [ ] Year 2000 at current deforestation rate

▷ **FIGURE 33–11 Tropical Rain Forests** This map shows the steady decline in tropical forests and projected remaining trees in the year 2000.

climb, many scientists have little hope for the world's forests. In your lifetime, numerous tropical forests will be destroyed. Consequently, many environmentalists are urging sharp cutbacks in current demand and other strategies that will ease the pressure on forests. Smaller homes, paper recycling, widespread tree planting, and better forest management could all help avert a shortage of timber in the coming years and protect remaining forests. Increasing the rate of paper recycling by 30% in the United States alone would save an estimated 350 million trees per year.

▷ **FIGURE 33–12 Old Growth Forest** (a) These ancient trees in Washington are protected in Olympic National Park. (b) But 90–95% of the old-growth stands, home to the endangered spotted owl and many other species, have been cut, and despite recent protection large stands of old-growth forest on private land and on state and federal land is slated for harvest.

(a)   (b)

Saving forests is not just a matter of ensuring a steady supply of wood and wood products. Forest conservation also protects the habitat of wild species, helps purify air, and reduces carbon dioxide buildup in the atmosphere. It protects watersheds, reduces soil erosion, and protects recreational opportunities as well. Tropical forests are a source of new medicines and potentially new food plants that could help feed the world's people.

Destroying tropical forests also disrupts native cultures, forcing people who have lived off the land for centuries to move to towns and cities. Protecting forests is a way of protecting these people and their cultures. New efforts are now under way to save the forests by developing markets for sustainable products, such as nuts and fruits. Several recent studies suggest that such sustainable harvesting of the forest is far more profitable than timber harvesting and cattle ranching. (For a discussion of efforts by one U.S. company to help save the tropical rain forests, see Health Note 33–1.)

Soil Erosion, Like Deforestation, Is Also a Worldwide Phenomenon

Although soil erosion occurs naturally, it is often greatly accelerated by human activities, such as farming, construction, and mining (▷ Figure 33–13). In the past 100 years, about one-third of the fertile topsoil on American farmland has been eroded away by water and wind, largely because of poor land management (▷ Figure 33–14). Each year, an estimated 1 billion tons of topsoil are eroded by wind and water from U.S. farmland. Over half of America's once-rich farmland is eroding faster than it can be replaced.

PLANTING TREES TO OFFSET POLLUTION

In the tropics, an estimated 3 million square miles of land has been cleared of magnificent forests. Norman Myers, a wildlife biologist, notes that about 2 million square miles of now-treeless land could be replanted, helping eliminate the massive output of carbon dioxide from the burning of coal, oil, natural gas, and gasoline. Replanting half this amount would cost an estimated $100 billion, a sizable sum of money until one calculates the cost of global warming. Building levees to protect the East Coast of the United States from the projected rise in sea level from global warming is estimated at $75 to $100 billion. Modifying U.S. irrigation systems and hydropower facilities would add another $100 billion to the global warming price tag. Planting 1 million square miles of trees would require 1/10th of the annual expenditure on weapons by the world's nations.

Unfortunately, replanting efforts have fallen far behind deforestation. As noted in this chapter, in the Third World only 1 tree is replanted for every 10 trees cut down.

On October 11, 1988, however, a public-minded utility based in Arlington, Virginia, announced plans to build a small coal-fired power plant in Connecticut. Its chief executive officer, a former U.S. Energy Department official, contacted the World Resources Institute for advice on ways to offset the 15.5 million tons of carbon dioxide that would be emitted by the facility over its 40-year life span. The institute found a suitable project in Guatemala proposed by CARE, the international relief and development agency.

With volunteers from a number of

▷ **FIGURE 1**

private groups, both Guatemalan and American, and the assistance of local residents, the utility hopes to plant 52 million trees in Guatemala, covering a region of 400 square miles (▷ Figure 1). Calculations show that the trees will absorb the excess carbon dioxide produced by the power plant at a cost to the company of about $2 million. Additional support, amounting to about $13.5 million in cash and tree-planting services, will come from CARE, the U.S. Agency for International Development, the Peace Corps, and the Guatemalan government.

Tree-planting operations such as these will not prevent global warming, but like so many issues that require a lot of different actions, they can help make a dent in a serious problem. Not only do they remove carbon dioxide pollution, but they also help restore forest land that protects watersheds. Although planted forests will not display the same species diversity, they will provide additional habitat for some species. If properly managed, they could provide a sustainable source of income. All in all, it's better for the health of the planet than leaving the land barren.

▷ **FIGURE 33–13 Soil Erosion** Millions of acres of farmland are destroyed each year because of poor land management that leads to severe soil erosion.

▷ **FIGURE 33–14 American Topsoil—Then and Now**

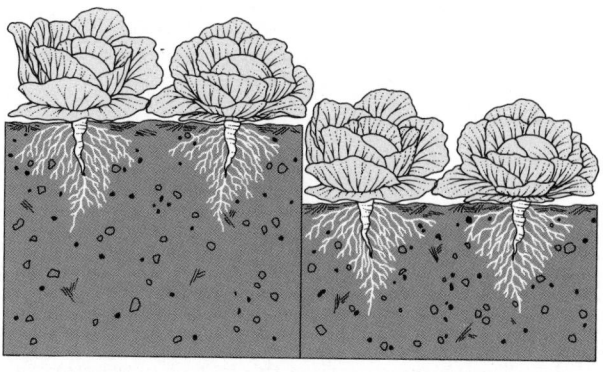

1780's   1980's

9 inches of topsoil   6 inches of topsoil

Average thickness

▷ **FIGURE 33–15 Farmland Conversion** Cities are often surrounded by excellent farmland—soils that drain well and are flat and highly productive. Many of the features that make soils suitable for farmland also make them suitable for building. This scene, unfortunately, is becoming all too common as urban areas expand.

▷ **FIGURE 33–16 Desertification** Poor land management and overgrazing are turning valuable farmland and rangeland into desert, further compounding global food production problems. The rangeland behind the fence has been properly grazed, ensuring its productivity. The land on the right has been overgrazed and ruined as grazing land.

Worldwide, about 25 billion tons of soil is eroded each year. Annual erosion rates are 18 to 100 times greater than rates at which soil is reformed. (The average renewal time is 500 years, but it ranges from 200 to 1000 years.) Erosion of soil by wind and water robs enough topsoil each year to fill a train that would encircle the Earth at the equator 150 times.

The annual loss of topsoil causes farmers to retire millions of acres of once-productive farmland and, in many cases, to cut down forests or plow up grassland to replace what has been lost. Approximately 25% of the tropical rain forests cut down each year are leveled to replace farmland destroyed by human activities.

Making matters worse, millions of acres of farmland are lost to urban and suburban sprawl, highway construction, and other human activities (▷ Figure 33–15). In the United States, 7000 acres of actual or potential farmland, pasture, and rangeland are lost every day. That's equivalent to 2.5 million acres a year, or a strip 0.6 mile wide extending from New York City to San Francisco. Overgrazing and poor land management are destroying millions of acres of farmland as well. Worldwide, an area the size of Ohio becomes desert each year, principally because of overgrazing by livestock and poor land management practices (▷ Figure 33–16).

These figures paint a rather grim picture for the long-term future of food production here and abroad, especially when viewed against the inevitable increase in human population. But trend need not be destiny. The loss of topsoil can be stopped, and soils can be replenished. However, such efforts will require a reduction in population growth and worldwide conservation measures. See Table 33–3 for additional measures to reduce the loss of productive farmland and rangeland. For these actions to be effective, society must begin work soon.

Many Areas of the World Are Facing Water Shortages or Will Soon Face Them as Population and Demand Increase

▷ Figure 33–17 is a map of areas in the United States that will face a water shortage in the near future. Regional water shortages result because too many people are drawing on limited water supplies. Long-term prospects here and abroad appear bleak. Between 1975 and 2000, irrigated agriculture worldwide is expected to double to meet the rising demand for food. Industry's demand for water is expected to increase 20-fold. By the end of the century, water demand is expected to exceed supply in at least 30 countries.

▷ Figure 33–18 shows the location of an enormous **aquifer,** a porous underground zone that contains water. Known as the **Ogallala aquifer,** this zone supplies irrigation and drinking water for farms in Nebraska, Kansas, Colorado, Oklahoma, and Texas. Unfortunately, a severe **groundwater overdraft,** withdrawals that exceed natural replenishment, may put an end to irrigated farming in these states. Texas alone is expected to lose half of its irrigated farmland by 2000.

Reducing the mining of groundwater is essential to meeting future needs. That will require strict conservation efforts, such as lining irrigation ditches with concrete or using pipes rather than open ditches to transport water to fields, to reduce evaporation. More efficient sprinklers, computerized systems that monitor soil moisture so that farmers know exactly how much irrigation is required, and other measures can also help.

Many Essential Minerals Will Be Depleted in the Next Four Decades

Metals, such as steel and aluminum, are produced from the Earth's mineral deposits. More than 100 minerals, worth

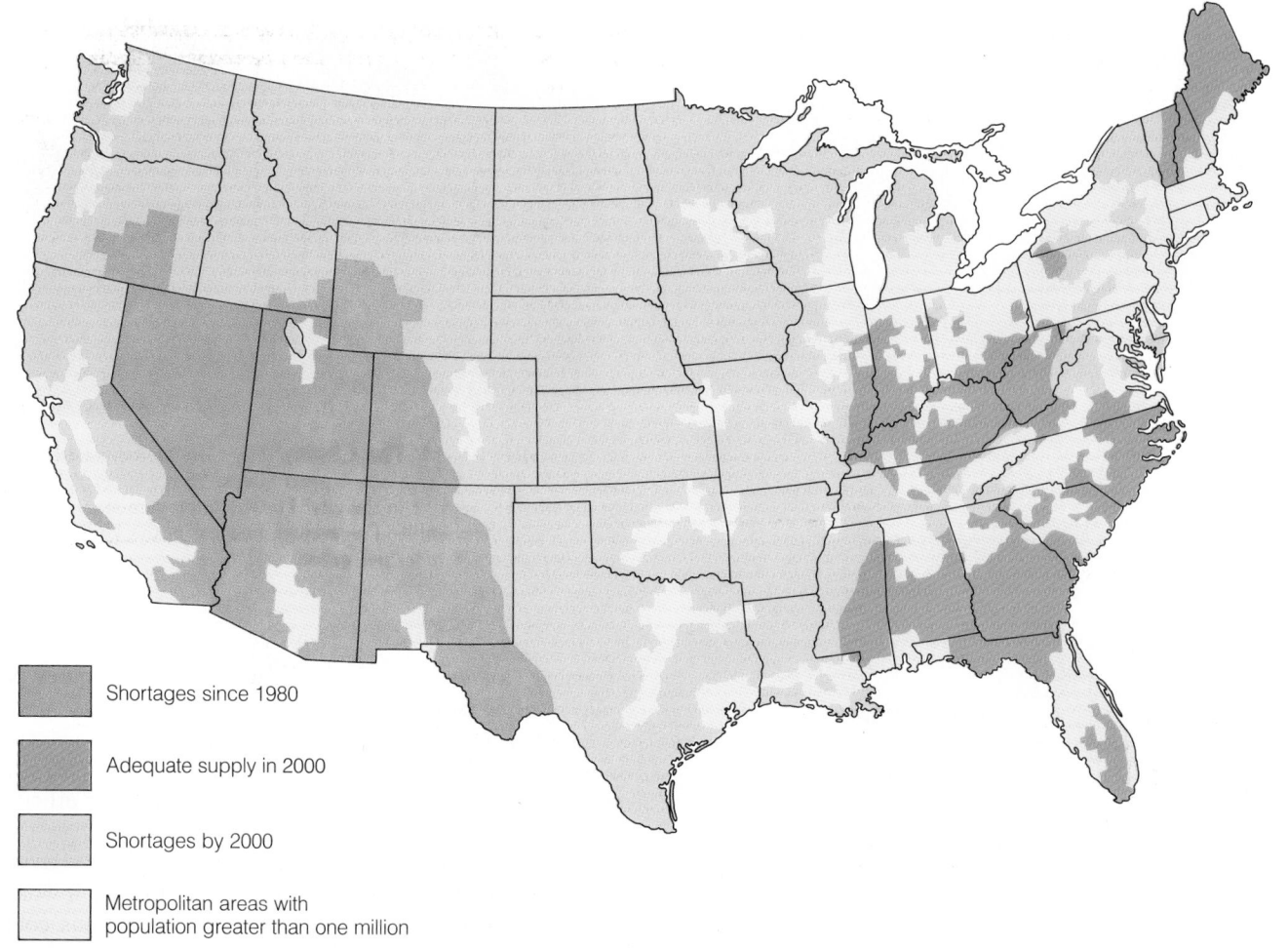

FIGURE 33–17 Water-Short Regions in the United States

Shortages since 1980

Adequate supply in 2000

Shortages by 2000

Metropolitan areas with population greater than one million

billions of dollars to the global economy, are traded on the world market each year. Several dozen of these minerals are so important to modern society that if they were suddenly no longer available or were unavailable at a reasonable price, the U.S. economy, like those of other industrialized countries, would come to a standstill.

At least 18 economically important minerals will fall into short supply in the next 40 years, even if countries expand their recycling programs. Silver, mercury, lead, sulfur, tin, tungsten, and zinc are all candidates. Even if new discoveries and new technologies make it possible to extract five times the currently known reserves, this group will be 80% depleted by or before 2040.

Oil Supplies Are Limited, and Most Students Alive Today Will See the End of Oil in Their Lifetimes

Oil is the lifeblood of modern society. In the United States, for example, oil supplies 43% of our annual energy demand. But the supply of oil is finite. The proven global reserves of oil (the amount known to exist and to be economically recoverable) are about 900 billion barrels. Although that may sound like a lot of oil, it will last only

about 40 years *at the current rate of consumption*. The undiscovered global reserves (oil thought to exist and to be economically recoverable) amount to 525 billion barrels. That's enough oil for another 25 years at the current rate of consumption.

Unfortunately, except for two brief periods, global energy use has risen 5% per year since 1860—a doubling of demand every 14 years for over 100 years. Should this rate of increase continue, all the oil will be gone by 2018. Unless replacements are found, and soon, a new oil crisis could cripple the world economy.

Some petroleum geologists believe that global oil supplies are possibly 3000 billion barrels greater than the figure noted above. But a 5% annual increase in use would deplete even this additional reserve, if it indeed exists, by the year 2038.

The outlook for oil in the United States is even grimmer than the global prospects. By various estimates, about 100 billion barrels remain in U.S. territories, enough to last only 17 years at the current rate of consumption.

The United States has become increasingly dependent on foreign oil in recent years, and that dependence is bound to grow worse, given the relatively small national oil

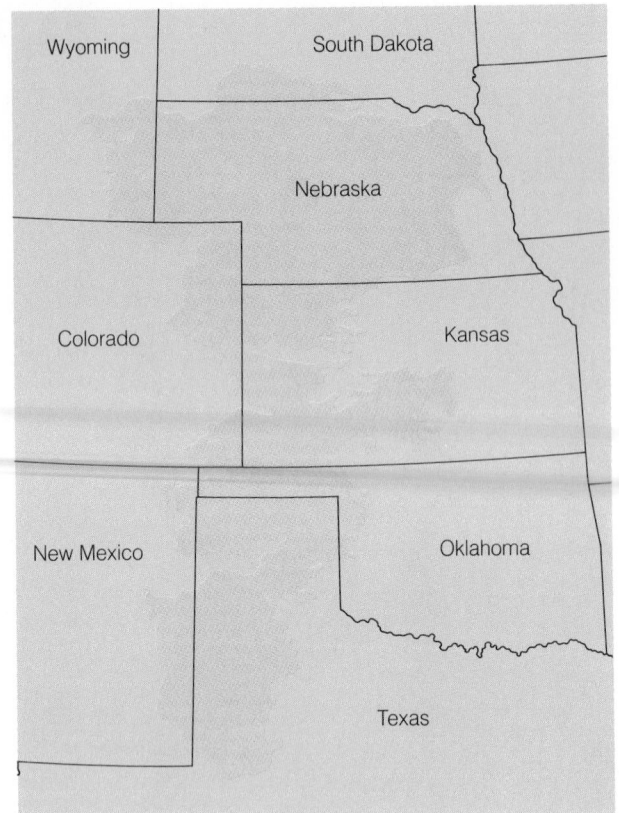

▷ **FIGURE 33–18 The Ogallala Aquifer** Map of the Ogallala aquifer. The water of the Ogallala is replenished very slowly and is currently being withdrawn much faster than it can be replaced. Farmers' wells are running dry, and eventually many irrigated farms will go out of business.

▷ **FIGURE 33–19 The Champ** The Geo Metro gets the best gas mileage of any car in the United States: 58 miles per gallon on the highway and 53 in the city. Further improvements in gasoline mileage are possible. The average new car in the United States gets only about 28 miles per gallon.

supplies. Unless a clean, economical substitute is found, and soon, our country could face a severe economic crisis.

Clearly, time is running out for oil. You will probably see the end of oil within your lifetime. So what do we do?

The first step in meeting future demand is to make current energy use much more efficient. Efficiency will no doubt occur as fuel supplies rise and citizens shift to more efficient modes of transportation. Waiting until that time, say many experts, wastes precious fuel and unnecessarily pollutes the planet.

Enormous supplies of energy exist today—in our waste. The efficient use of oil and oil products, such as gasoline, diesel fuel, and home heating oil, can help us stretch current oil supplies considerably while actually improving the economy. In fact, many of the leading economic powers, like Japan and West Germany, are far more energy-efficient than the United States.

Energy-efficient investments like storm doors, storm windows, and added insulation coupled with measures to cut air inflow (infiltration) are often inexpensive and can reduce home energy consumption by 30% to 50%. More efficient cars, such as the Geo and Honda CVCC, already use fuel twice as efficiently as the average new car (▷ Figure 33–19). Even higher mileage is possible. The

Japanese, for example, have a car that gets 98 miles per gallon on the highway and seats four people!

Another important step in increasing energy efficiency is mass transit—buses and trains to transport commuters to and from work. Mass transit is four times more efficient than the automobile.

Energy efficiency is the cheapest and most cost-effective means of meeting future demand. Amory Lovins, a world-renowned energy expert, calculates that Americans could reduce oil demand by 80% without changing lifestyles by using energy-efficient technologies currently on the market. Lovins's calculations also show that U.S. electrical demand could be cut by 75% by using energy-efficient light bulbs, motors, and other technologies *that are currently available* (▷ Figure 33–20). These measures could save us hundreds of billions of dollars a year.

Alternative fuels can also help meet future demand. Bear in mind that the immediate need is for an alternative source of energy to replace oil, which is used chiefly in transportation and home heating.

The more traditional—and, incidentally, the more environmentally damaging—approach to meeting demand is the development of other fossil fuel resources. One of those is oil shale, a rock that when heated gives off an oily substance that can be refined to make gasoline. Despite what many would have you believe, oil shale deposits are rather small in relation to global oil demand, and producing oil from shale is very costly, inefficient, and dirty (▷ Figure 33–21). Oil shale development would cause considerable air pollution in the West, where the deposits are located, and would destroy large tracts of winter habitat vital to deer, elk, and other wildlife.

A more acceptable option from an environmental standpoint is ethanol. Produced from corn, wheat, and other crops, ethanol could be used to power automobiles and trucks. Grown on special fuel farms, ethanol could virtually

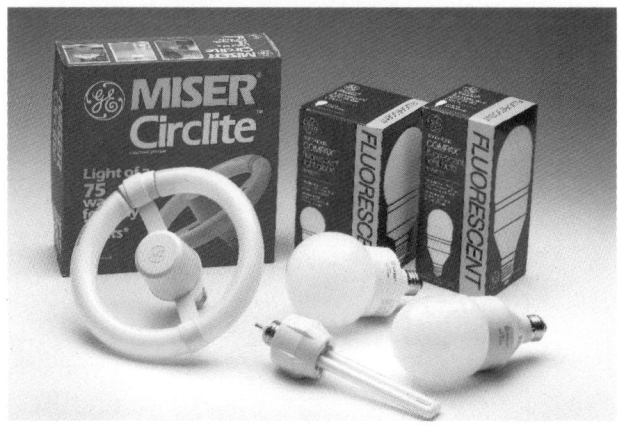

▷ **FIGURE 33–20 Energy-Efficient** Many new light bulbs use only 25% of the energy a standard light bulb requires. Althought they cost more, they last as long as 10 standard bulbs and save $20 to $40 in electricity over their lifetime.

▷ **FIGURE 33–21 Oil Shale** This sedimentary rock contains an organic material called kerogen. When heated, oil shale releases its kerogen, forming shale oil. Shale oil can be refined to produce gasoline, kerosene, jet fuel, and other products, much like crude oil. Shale oil production, however, is expensive, limited by supply, and environmentally harmful.

replace gasoline, but a shift to ethanol in the United States would not be without its costs. One of those might be a reduction in food output. Brazil is a leader in ethanol use. Ninety-eight percent of the new cars sold in Brazil are equipped to use it.

Renewable energy could provide enormous amounts of fuel in the years to come. Unknown to many, the renewable energy supply is enormous and, in some cases, can be tapped quickly and inexpensively. According to one estimate, nonrenewable energy reserves (coal, oil, natural gas, and so on) would provide the equivalent of 8.8 trillion barrels of oil. Renewable energy could provide 10 times that amount of energy every year!

≈ POLLUTION: FOULING OUR NEST

Like death and taxes, waste is an inescapable fact of life. All organisms produce it. Humans, however, are by far the most prolific generators of waste on the planet. Today, waste from human society is overwhelming many nutrient cycles, poisoning other species (and ourselves), and altering planetary homeostasis. This section recaps four of the most serious waste problems.

Global Warming Results from the Release of Carbon Dioxide and Other Greenhouse Gases

Carbon dioxide is produced during cellular respiration and the combustion of all organic materials—most importantly, fossil fuels. In some respects, atmospheric carbon dioxide is like a prescription drug: it is beneficial at low levels but potentially harmful at higher concentrations. In normal concentrations in the atmosphere, carbon dioxide has a warming effect on the planet. Acting much like the glass in a greenhouse, it traps heat escaping from the Earth and radiates it back to the surface. Carbon dioxide is, therefore, also known as a **greenhouse gas.** A little bit of carbon

dioxide is essential to life on Earth; too much, however, may lead to overheating.

Over the past 100 years, global carbon dioxide levels have increased about 25%, principally as a result of industrialization powered by the combustion of fossil fuels and deforestation. Today, atmospheric carbon dioxide levels continue to rise at a rate of about 0.5% per year.

Several other pollutants also contribute to global warming. Methane, chlorofluorocarbons, nitrous oxide, and even water vapor all radiate heat back to the Earth, causing the atmosphere to heat up. Methane is released from livestock and the manure they produce; humankind's nearly 1 billion cattle annually release about 73 million metric tons of this gas (providing another good reason to reduce or eliminate one's consumption of beef). Methane production has increased 435% in the last century. Incidentally, termites, which thrive in the deforested tropics, also excrete methane. Methane output from all sources is increasing at a rate of about 1% per year and is 20 times more effective at trapping heat than carbon dioxide.

Chlorofluorocarbons, or CFCs, are known for their effect on the ozone layer and are released from spray cans, refrigerators, air conditioners, and freezers.[4] One class of CFCs is used to clean circuit boards used in computers and other electronic equipment. In the lower atmosphere, CFCs also trap heat, contributing to the greenhouse effect.

Because of the dramatic increase in the release of greenhouse gases, many atmospheric scientists believe that global temperature may rise dramatically in the coming decades, causing an extraordinary shift in climate. The graph in ▷ Figure 33–22 shows average global temperature over the past century and suggests that we may indeed be witnessing a gradual warming. But some scientists think that the rise may result from normal climatic variation. (For

---

[4]CFCs used in spray cans were banned in the United States and several other countries in the late 1970s.

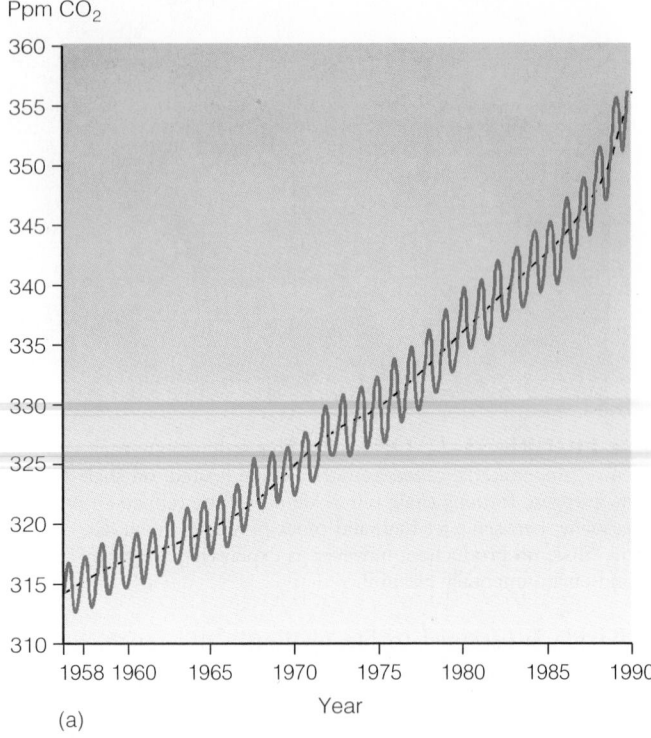

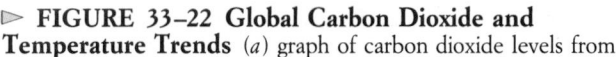

(a)
Year

(b)
Year

▷ **FIGURE 33–22 Global Carbon Dioxide and Temperature Trends** (*a*) graph of carbon dioxide levels from before 1958 to 1986 showing steady rise, and (*b*) graph of global temperature since 1880 with two possible scenarios.

more on this subject, see the accompanying Point/ Counterpoint on pages 886–887.)

The Impacts of Global Warming Could Be Ecologically Devastating. According to computer models based on the present rate of increase in greenhouse gases, global temperatures could be 2° to 5°C hotter within 40 years. The models suggest that global rainfall patterns will shift dramatically as a result of warming (▷ Figure 33–23). The models suggest that the Midwest and much of the western United States will be drier and hotter than they are today. If that happens, many midwestern farmers will be driven out of business. As rainfall declines in this agriculturally productive region, farming may intensify in the northern states. Overall agricultural productivity in the United States may fall, because northern soils are not as rich as those in the Midwest. Food shortages caused by the decline in U.S. agriculture in the Midwest and rising food prices could affect the economy in profound ways. Computer climate models predict that the southern United States and Pacific Coast may be wetter but hotter. The models suggest that, in Dallas, the number of days over 90°C may increase from 17 today to 78 by 2030. In Washington, D.C., the number of days over 90°C may increase from 36 to 87.

A rise in global temperature is expected to melt glaciers and the polar ice caps (▷ Figure 33–24). Warmer temperatures would also expand the volume of the seas. Together, melting ice and expanding oceans would result in an increase in sea level. In the past 50 years, sea level has risen

10 to 12 centimeters (4 to 6 inches). By 2050, the computer models suggest, sea level will rise 50 to 100 centimeters (2 to 3 feet) because of global warming. What impact would this have?

In the United States, approximately half of the population lives within 50 miles of the ocean, and many large cities like Miami are located only a few feet above sea level (▷ Figure 33–25). Rising sea level may flood many low-lying regions. Storms could cause more damage than they do now because high water would be able to move farther inland. Expensive dikes and levees would be needed to protect cities such as Miami and New Orleans. Other coastal cities would have to be rebuilt on higher ground, but relocation would be expensive.

A rising sea level would be particularly hard on Bangladesh and other Asian countries with extensive lowland rice paddies. Approximately 17% of Bangladesh would be reclaimed by the sea if sea level rose 2 to 3 feet.

Rising temperatures would also have an impact on many species. Some might adapt or find new habitat; many plants, however, would be wiped out by a relatively rapid increase in temperature, for temperature would probably increase far faster than they could adapt. As a result, forests would dry out and die off. The incidence of forest fires may increase, further adding to global warming.

These predictions may sound dire—and they are. But as you will see in the Point/Counterpoint, a small number of scientists (represented here by Fred Singer) believe that we don't know enough about the atmosphere and the impact

▷ **FIGURE 33–23 Projected Rainfall Patterns Resulting from Global Warming**

▷ **FIGURE 33–24 Mount Rainier Glaciers** This enormous glacier and others could melt as the Earth gets warmer, raising sea level.

of greenhouse gases to know for certain whether global warming will actually occur and, if it does, how bad it will be. The reluctance of the U.S. government to take action on global warming is largely attributed to this uncertainty.

▷ **FIGURE 33–25 Miami Underwater?** Rising sea level caused by global warming could flood many coastal cities the world over.

GLOBAL WARMING IS REAL

Stephen H. Schneider

Stephen H. Schneider is head of the Interdisciplinary Climate Systems Section at the National Center for Atmospheric Research. His research interests include climatic change, global warming, and modeling of human impacts on climate.

Observations have already established beyond doubt that atmospheric constituents, such as water vapor, clouds, carbon dioxide (CO_2), methane, (CH_4), nitrous oxide (N_2O), and chlorofluorocarbons (CFCs), trap heat escaping from the Earth's surface, causing the greenhouse effect. Likewise, it is virtually certain that an unprecedented 25% increase in CO_2 and 100% increase in CH_4 over the past 150 years have resulted from increased use of fossil fuels and expanded deforestation. It is also well accepted that the buildup of these gases has trapped 2 extra watts of radiative energy averaged over every square meter of Earth. What then is the basis of the debate over global warming?

First of all, translating 2 watts per square meter of heating into X degrees of temperature rise requires calculations that are based on not-yet-verified assumptions about how clouds, soils, forests, ice, and oceans will respond to this heating. These factors could change in ways that feed back on the heating, either reducing it, as global warming critics like to point out, or enhancing it, as most present climate models project. Such models predict that the past hundred years should have experienced some 1°C of warming from the extra 2 watts per square meter of heating, provided all other factors were constant, a dubious prospect. Already, half a dozen national and international assessment bodies have suggested that if the CO_2, CH_4, and CFC trends continue, global warming of some 1.5° to 4.5°C can be expected in the next century. Already, 0.5° ± 0.2°C of global warming has been observed from 1890 to 1990.

Some critics note that there is not a perfect match between decade-to-decade fluctuations in climate and the buildup of greenhouse gases. Unfortunately, such critics often fail to mention that no knowledgeable scientist would ever expect such

agreement, because fluctuations of several 10ths of a degree Celsius per decade occur naturally. Thus, it is illogical to look for a decade-by-decade match. Only long-term, global trends can verify that a warming trend due to the buildup of greenhouse gases has been detected. A high level of certainty will take 10 to 20 years more to establish, not a few years as some critics of immediate changes argue. Moreover, waiting for such scientific certainty is not cost-free, because the Earth could then be forced to adapt to a greater amount and rate of climate change than if we acted now.

Over the past 10 years, half a dozen government-sponsored assessments have agreed that there is a better than 50% chance that current trends in population growth, fossil-fuel use, and land utilization practices will cause climatic changes of 2°C or more in the next century. Moreover, the rate of projected human-induced changes is some 10 times greater than the long-term rate of natural, global climate changes.

Critics suggest that three to five more years of testing is needed before taking action. Although this sounds prudent, such testing will not provide definitive answers on climatic change or its implications for ecosystems, forestry, agriculture, water supplies, human health, sea level, and severe storms.

I'm not a planetary gambler. I'd prefer to slow down the rate of buildup of greenhouse gases rather than gamble that things may work out all right in the end. Study after study has shown that the best way to start to reduce the buildup rate of greenhouse gases is to make cost-effective improvements, such as controlling population growth, increasing the efficiency of energy use, and virtually eliminating the production of CFCs. The Yale economist William Nordhaus, a critic of severe cuts in CO_2 emissions, has argued nevertheless that modest cuts in CO_2 emissions would, at present, yield economic benefits in excess of the costs. And his calculations did not even include the free extras: less acid rain, less air pollution, a lower balance-of-payments deficit from importing foreign oil, and lower long-term energy costs of manufactured goods.

To me, reducing CO_2 and other measures is a kind of climate change "insurance" that pays other dividends. We need to eliminate the billions of dollars of government subsidies to inefficient current fossil-fuel uses and deforestation practices and move toward lower real costs and an environmentally more stable society. Political rhetoric about uncertainty only commits the future to greater risks.

Solving Global Warming Requires Massive Action, and Soon. While scientists and politicians debate global warming, the Earth appears to be getting hotter. Eight of the hottest years in the past 100 years, in fact, have occurred since 1980. Can anything be done to stave off global warming?

The answer is yes, but changes will have to be dramatic and swift. Sharp reductions in fossil fuel consumption through energy efficiency, conservation, and the use of alternative sources of fuel will be required. Global reforestation, most experts agree, is a must. But for each family of four in the United States, 6 acres of fast-growing trees

TOO EARLY TO TELL

S. Fred Singer

S. Fred Singer, professor of environmental sciences at the University of Virginia, has served as deputy assistant administrator of the Environmental Protection Agency and as the first director of the U.S. weather satellite program in the Department of Commerce. An atmospheric and space physicist, he predicted the increase of atmospheric methane due to human activities.

Greenhouse warming (GW) has emerged as the issue of the 1990s. Wide acceptance of the Montreal Protocol, which reduces the manufacture of chlorofluorocarbons, considered a threat to the stratospheric ozone layer, has encouraged environmental activists to call for similar controls on carbon dioxide. They have expressed disappointment with the White House for not supporting immediate action on CO_2. Should the United States assume "leadership" in this campaign, or would it be more prudent to first assure through scientific research that the problem is both real and urgent?

The scientific base for GW includes some facts, lots of uncertainty, and just plain ignorance. What is needed are more observations, better theories, and more extensive calculations. There is consensus about an increase in greenhouse gases (CO_2, CFCs, methane, nitrous oxide, ozone) in the Earth's atmosphere. There is some uncertainty about their rate of generation and their rate of removal. There is major uncertainty and disagreement about whether this increase has caused a change in the climate during the last hundred years; many observations do not fit the theory. There is also major disagreement in the scientific community about predicted changes from further increases in greenhouse gases; the models used to calculate future climate are not yet refined enough to simulate nature. As a consequence, we cannot be sure whether the next century will bring a warming that is negligible or significant. Finally, even if there is a warming and associated climate changes, it is debatable whether the consequences will be good or bad; likely, we will get some of each.

Has the observed increase of greenhouse gases in the last decades had an effect on climate? The data are ambiguous to say the least. Advocates of immediate action profess to see a global warming of about 0.5°C since 1880 and point to record temperatures experienced in the 1980s. Others tend to be more cautious; they call attention to the fact that the strongest increase occurred *before* the major rise in greenhouse gas concentration; it was followed by a quarter-century decrease, between 1940 and 1965. Since then, temperatures have begun to climb again. Some researchers consider the warming observed before 1940 to be a recovery from the "Little Ice Age" that prevailed from 1600 to about 1850.

We can sum up our conclusions in a simple message: *the scientific base for greenhouse warming is too uncertain to justify drastic action at this time.* There is little risk in delaying policy responses to this century-old problem, because there is every expectation that scientific understanding will be substantially improved within a few years. Instead of panicky and premature actions, we will then be able to apply specific remedies as necessary. That is not to say that steps cannot be taken now; indeed, many kinds of energy conservation and efficiency increases make economic sense even *without* the threat of greenhouse warming.

Drastic, precipitous, and, especially, unilateral steps to delay the putative greenhouse impacts can cost jobs and prosperity without being effective. The Yale economist William Nordhaus, one of the few who has been trying to deal quantitatively with the economics of the greenhouse effect, has pointed out that "those who argue for strong measures to slow greenhouse warming have reached their conclusion without any discernible analysis of the costs and benefits." It would be prudent to complete the ongoing and recently expanded research so that we will know what we are doing before we act. "Look before you leap" may still be good advice.

≈ SHARPENING YOUR CRITICAL THINKING SKILLS

1. Summarize the major points made by each author. Are there areas where they are looking at nearly the same data but reaching different conclusions? If so, how is this possible?
2. Using your critical thinking skills, analyze each essay. What flaws do you see in the logic, if any?

would have to be planted to offset the carbon dioxide the family will produce during its life. Additional strategies are shown in Table 33–4.

In addition to the strategies in Table 33–4, individuals can also help. On average, each gallon of gasoline you consume in your automobile produces 5 pounds of carbon dioxide. Every hour you watch television results in the production of 0.64 pound. Your frost-free refrigerator accounts for nearly 13 pounds of carbon dioxide a day (Table 33–5). The lesson is that individuals are partly responsible for global warming. We can help reduce the problem by using energy much more efficiently: using mass transit,

TABLE 33–4 Measures to Reduce Global Warming

Reduce the rate of population growth.
Switch from coal-fired and oil-fired power plants to natural gas, which produces much less carbon dioxide per unit of electricity.
Implement the technologies that burn coal more efficiently.
Expand cogeneration—processes that trap waste heat and put it to good use.
Boost automobile efficiency.
Expand mass transit.
Develop alternative liquid fuels for the transportation sector.
Improve the efficiency of industry.
Make new and existing homes more energy-efficient through insulation, weather stripping, storm doors, and storm windows.
Build many new homes that use solar energy for space heating.
Reduce global deforestation.
Begin a massive global reforestation effort.
Phase out all CFCs and other CFC damaging chemicals soon.
Reduce consumption of unneccesary items.
Expand recycling efforts.

 ## TABLE 33–5 Carbon Dioxide Production from Common Activities

ELECTRICAL APPLIANCES	POUNDS OF CARBON DIOXIDE ADDED TO ATMOSPHERE*
Color television	0.64 per hour
Steam iron	0.85 per hour
Vacuum cleaner	1.70 per hour
Air conditioner, room	4.00 per hour
Toaster oven	12.80 per hour
Ceiling fan	4.00 per day
Refrigerator, frost-free	12.80 per day
Waterbed heater with thermostat	24.00 per day 12.80 per day
Clothes dryer	10.00 per load
Dishwasher	2.60 per load
Toaster	0.12 per use
Microwave oven	0.25 per 5-minute use
Coffeemaker	0.50 per brew

*At room temperature and sea level, every pound of carbon dioxide occupies 8.75 cubic feet, about half the size of a refrigerator.

SOURCE: Reprinted from the February/March 1990 issue of *National Wildlife*. Copyright © 1990 by the National Wildlife Federation.

recycling, insulating our homes, and a great many other strategies (Table 33–6).

Large Portions of the World Are Threatened by Acid Deposition

In the Adirondack Mountains of New York, hundreds of lakes are dying. Fishes have vanished as the lakes turn

TABLE 33–6 Individual Actions That Can Reduce Global Warming

Automobile energy savings
  Buy energy-efficient vehicles.
  Reduce unnecessary driving.
  Car-pool, take mass transit, walk, or bike to work.
  Combine trips.
  Keep your car tuned and your tires inflated to the proper level.
  Drive at or below the speed limit.
Home energy savings
  Increase your attic insulation to R30 or R38.
  Caulk and weather-strip your house.
  Add storm windows and insulated curtains.
  Install an automatic thermostat.
  Turn the thermostat down a few degrees in winter, and wear warmer clothing.
  Replace furnace filters when needed.
  Lower water heater setting to 120 to 130° F.
  Insulate water heater and pipes, install a water heater insulation blanket, and repair or replace all leaky faucets.
  Take shorter showers.
  Use cold water as much as possible.
  Avoid unnecessary appliances.
  Buy energy-efficient appliances.
  Use low-energy light bulbs.
Reducing waste and resource consumption
  Recycle at home and at work.
  Avoid products with excessive packaging.
  Reuse shopping bags.
  Refuse bags for single items.
  Use a diaper service instead of disposable diapers.
  Reduce consumption of throwaways.
  Donate used items to Goodwill, Disabled American Veterans, the Salvation Army, or other charities.
  Buy durable items.
  Give environmentally sensitive gifts.

acidic (▷ Figure 33–26). A similar phenomenon is occurring in southeastern Canada and in Sweden and Norway, where dying lakes number in the thousands.

The acids falling from the skies are produced from two atmospheric pollutants, sulfur dioxide and nitrogen dioxide, arising chiefly from the combustion of fossil fuels: coal, oil, and natural gas. In the atmosphere, these gases, or **acid precursors,** combine with water and oxygen to form sulfuric and nitric acids, which are responsible for **acid deposition.** Acid deposition may be wet or dry. Rain and snow wash acids from the sky and constitute wet deposition. Fog and clouds also carry heavy loads of acids that are deposited on trees. This is another form of wet deposition. **Dry deposition** occurs when particulates in the atmosphere also attach to airborne acids, then fall to the Earth.

In the United States and Europe, acid deposition has been growing worse for three decades, largely as a result of increased fossil-fuel combustion. The map in ▷ Figure

▷ **FIGURE 33–26 Victims of Acid** Acid deposition kills fishes and other aquatic species. These fishes were confined in a cage in a stream polluted by acid rain.

33–27 shows that the region of the United States affected by acid deposition is growing larger and that the level of acidity, measured by pH, is increasing. Today, acid deposition is common downwind from virtually any major population center. It comes from power plants, motorized vehicles, factories, and even our own homes.

Acid deposition changes the pH of lakes and streams, killing fishes and other aquatic organisms. Acids on land dissolve toxic minerals, such as aluminum, from the soil and wash them into bodies of water. Aluminum causes the gills of fishes to clog with mucus, suffocating them. Extensive acidification of lakes and rivers is putting resort owners out of business in the Northeast, upper Midwest, and southern Canada.

▷ Figure 33–28 shows areas most susceptible to acid deposition. These regions are generally mountainous and contain soils with little capacity to neutralize acids. Consequently, acids that fall on the land quickly wash to nearby lakes and streams.

▷ **FIGURE 33–27 Extent and Strength of Acid Rain in the United States in 1955 and 1988** Remember, as pH goes down, acidity goes up.

1955

1988

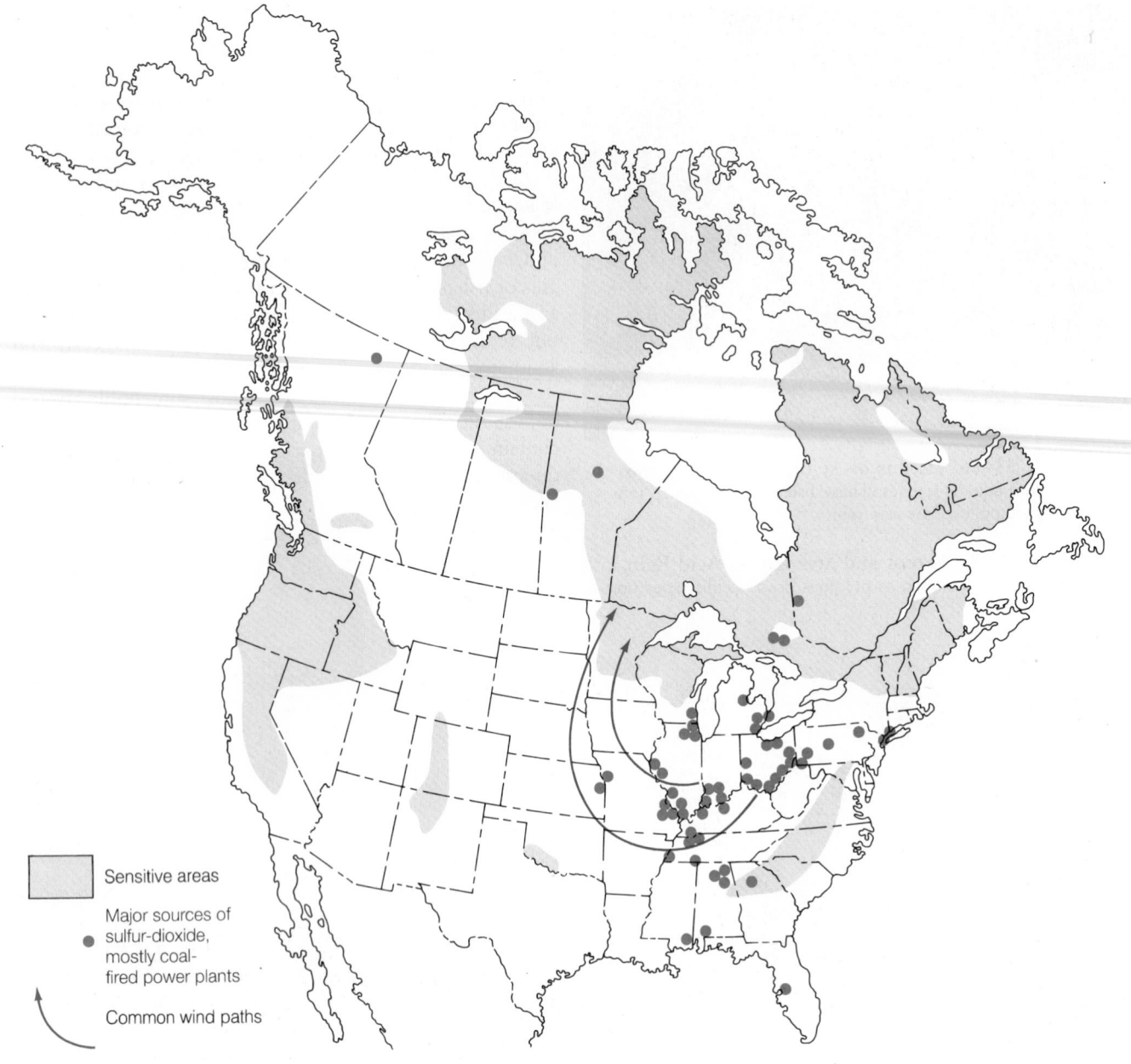

Sensitive areas

Major sources of sulfur-dioxide, mostly coal-fired power plants

Common wind paths

▷ **FIGURE 33–28 Acid-Sensitive Regions in North America and Major Sources of Acid Precursors**

Acids can also damage crops and trees. The damage may be direct or indirect. Direct damage results when acids damage growing buds or leaves. Indirect damage occurs when acids alter soil chemistry, inhibiting soil bacteria necessary for nutrient recycling. Acids may also leach important minerals from the soil, resulting in slower plant growth.

Some scientists believe that massive forest diebacks occurring throughout the world may be the result of acidic rainfall and acid fog that often blankets forests (▷ Figure 33–29). Finally, acid deposition also damages buildings, statues, and other structures. The estimated cost in the United States is about $5 billion per year.

Reducing the deposition of acids will require a dramatic reduction in the release of sulfur dioxide and nitrogen dioxide from power plants, factories, and automobiles. One way of reducing sulfur dioxide is the smokestack scrubber, a device that traps sulfur dioxide gas escaping from power plants, removing up to 95% of this pollutant. Installing and operating a scrubber is rather expensive, and some utilities have objected to this strategy. Using low-sulfur coal can also help, but eastern coal producers find this strategy objectionable, because their coal is especially high in sulfur. Because of these and other objections, progress in reducing acid deposition has been painfully slow. A far better strategy is energy efficiency. By using less fossil-fuel energy, we cut the combustion of polluting fuels. This action reduces both sulfur dioxide and nitrogen diox-

▶ **FIGURE 33–29 Forest Die-off** (*a*) Ghostly remains of trees killed by acid deposition. (*b*) Brown patches are trees killed by acid deposition in New England.

ide release and also reduces carbon dioxide, a major contributor to global warming. Energy efficiency therefore has multiple environmental benefits. Add the economic benefits, and it becomes an even more attractive strategy for environmental protection.

The Ozone Layer Is Endangered by Human Pollutants

Encircling the Earth, 20 to 30 miles above its surface, is a stratum of the atmosphere, the **ozone layer,** containing a slightly elevated level of ozone gas. As noted in Chapter 23, the ozone layer shields the Earth from potentially harmful ultraviolet radiation coming from the sun. During evolution, in fact, the generation of the ozone layer probably "allowed" the colonization of land by plants and animals.

Today, however, the ozone layer is being gradually destroyed by chemicals released into the atmosphere by modern society. The most potent destroyer of ozone is a class of chemicals known as the chlorofluorocarbons (CFCs).[5] Used as spray can propellants (not in the United States),

[5] Nitric oxide from high-flying jet airplanes, such as the supersonic transport, can also destroy the ozone layer.

refrigerants, blowing agents for plastic foam, and cleansing agents, CFCs are highly stable molecules. Scientists, in fact, selected them for use in spray cans in large part because of their lack of chemical reactivity. In the early 1970s, however, a U.S. scientist discovered that CFCs were broken down by sunlight. Another scientist showed that CFCs released from human activities gradually drift into the upper atmosphere. Here, the scientists hypothesized, the breakdown products react with and destroy ozone molecules.

This hypothesis sparked intense research and many dire predictions. Some scientists predicted massive declines in the ozone layer. One study, in fact, suggested a 16% decline in the ozone layer by the year 2000. What worried the scientists, and a great many others, was a projection that each 1% decline in ozone would increase human skin cancer by at least 2% (some think it may be 4%). A 16% ozone decline would therefore increase the skin cancer rate by 32%. Even though skin cancer is not, as a rule, highly lethal, this dramatic increase would result in tremendous escalation in medical costs and would kill 4000 to 12,000 people each year in the United States alone. Worldwide, the death toll would be staggering. Ozone depletion could be harmful to a great many other species, especially plants that are damaged by ultraviolet light.

Other scientists projected smaller declines. Still others thought that the fears were totally unfounded. Estimates of ozone depletion came and went over the years, but scientists were hard pressed to detect any measurable decline, in large part because ozone levels fluctuate naturally from year to year. In 1988, however, a panel of atmospheric scientists reviewing 20 years of satellite data on ozone levels concluded that the ozone layer was indeed on the decline. What they found was that satellite measurements of ozone automatically logged into their computer had shown a consistent decline. It hadn't shown up because the computer had been programmed to delete low levels on the belief that such readings represented instrument errors. When scientists re-examined the data, they found that over North America ozone levels had fallen 1.7% to 3% (▶ Figure 33–30). Much larger declines were evident at the poles. In fact, scientists found that each year a giant hole in the ozone layer about the size of the United States formed over Antarctica. Ozone levels in the hole declined by as much as 50%. In 1992, the ozone hole reached a record high—almost three times the size of the United States. Studies strongly suggested that the main reason for the decline in the ozone layer was the accumulation of chlorofluorocarbons.

In a show of solidarity on September 1988, 23 nations met in Montreal to sign an agreement to cut back on the production of many ozone-destroying chemicals. By 1999, the treaty would achieve a 50% reduction in CFC manufacture and release. Some critics objected, saying that a 50% cut, while important, would still result in a 10% decline in the ozone layer. Clearly, greater reductions were needed. Because of these concerns and the cooperation of the CFC industry, the Montreal treaty was soon revised. The new plan

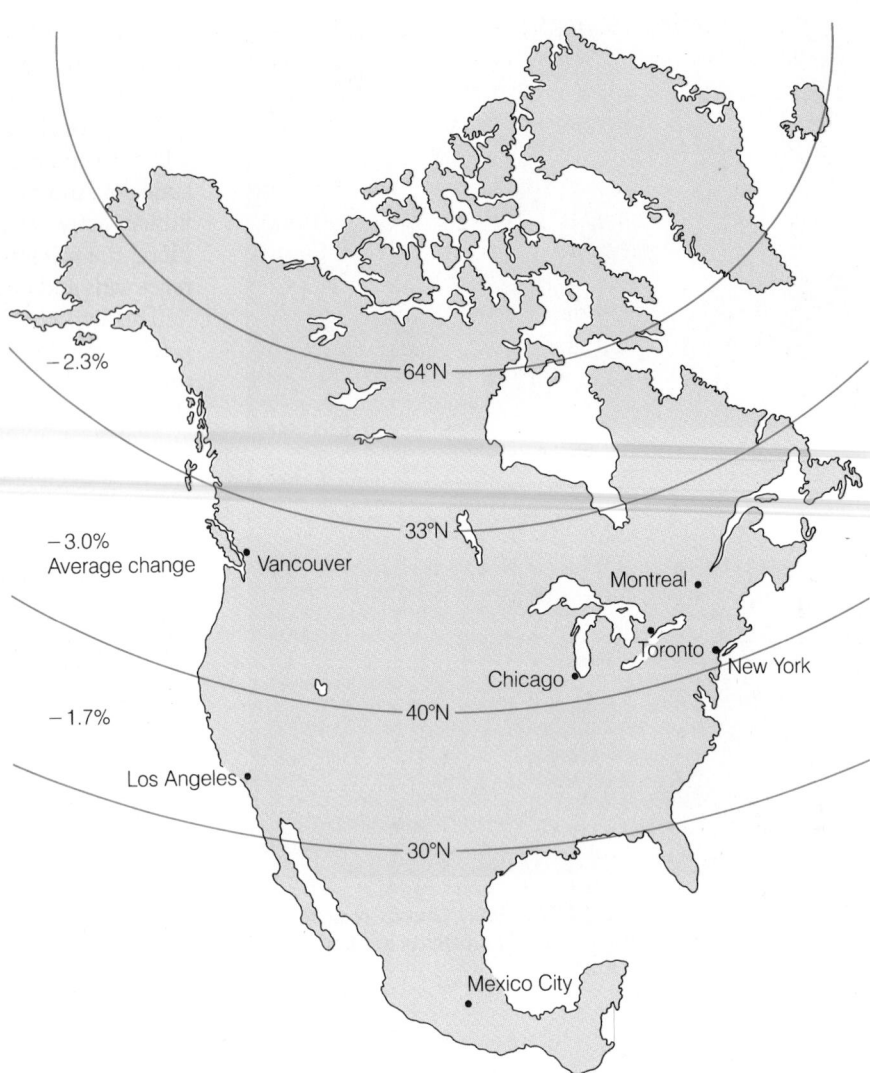

> **FIGURE 33–30 Decline in the Ozone Layer**

signed in London in 1990 called for a complete phase-out of many CFCs by 1999. After recent findings that suggest an even greater loss than previously suspected, this achievement may not be enough. Consequently, negotiations on an earlier phase-out are under way.

ENVIRONMENT AND HEALTH: BUILDING A SUSTAINABLE FUTURE

The trends and problems discussed in the previous sections add up to a crisis of epic proportions. Solving it will require tougher laws and regulations, reductions in population growth, more efficient technologies, and changes in human behavior. Many people, according to opinion polls, believe that broader societal changes are required. One suggestion is that we build a **sustainable society,** which will live within the carrying capacity of the environment. Living within the limits of nature implies achieving a population size and a way of life that do not exceed the planet's ability to supply food and other resources and to handle wastes. It is a society that manages its affairs in a way that protects the Earth's homeostatic mechanisms.

For much of human history, our ancestors lived in a sustainable relationship with nature. They took little, principally what they needed to survive. They did not waste raw materials. They relied on renewable plant and animal resources for food and clothing. They were participants in the Earth's elaborate recycling mechanisms, the nutrient cycles discussed in Chapter 31. What our ancestors threw away was recycled and made available for reuse. Sustainability also arose from the fact that for much of our early history, human populations were small, being held in check by disease and other mechanisms. In most cases, any damage our ancestors caused was restored by natural processes.

Interestingly, the sustainability of early human populations depended on the very same factors responsible for the sustainability of ecosystems. Ecosystems remain stable and can persist thousands of years, barring major disaster or human intervention, because of six principles: conservation, recycling, renewable resources, restoration, population control, and adaptability.

In contrast, the modern human economy is built on

waste, much of it resulting from the disposable products that flood the markets in developed countries. Modern society recycles only a fraction of the materials it uses. In fact, in the United States, despite the recent upsurge in recycling, we still recycle only about 13% of municipal garbage. By various estimates, 80% to 90% of our municipal garbage could be recycled or composted. American society, like many others, also relies heavily on nonrenewable resources, such as oil, natural gas, and minerals. We tap only a small fraction of our renewable resource potential. Acid deposition and global warming, two of the major pollution problems we now face, result from pollutants released from fossil fuel combustion. In contrast to most "natural" populations, human society has, for years, overcome population control and has pushed natural systems, among them ocean fisheries and many tropical rain forests, beyond their recuperative abilities. Finally, human society seems largely unwilling to change its ways to prevent global environmental disasters.

Building a sustainable society will require a shift to a pattern seen in nature. It will require that we become more efficient in our use of resources, seriously engage in recycling, shift to renewable resources wherever possible, restore damage, and stabilize, if not reduce, world population.

Building a sustainable society will also depend on a profound change in attitudes. The prevalent frontier notions of unlimited resources and human dominance over nature, say proponents, must be replaced with a new ethic, a **sustainable-Earth ethic.** This new ethic is based on three tenets: (1) the world has a limited supply of resources, which must be shared with all living things; (2) humans are a part of nature and subject to its rules; and (3) nature is not something to conquer but, rather, a force with which we must learn to cooperate. New attitudes will not come easily, but they are needed to strike a balance with nature.

Building a sustainable society will require a lifetime of commitment on the part of businesses, governments, and individuals. Given the course of modern society, it means taking a 180-degree turn. By comparison, the moon landing will seem like a weekend fix-it project.

Some changes required to create an enduring human presence may come as a result of legislative actions—new laws and regulations that force us to rely more heavily on conservation, recycling, and renewable resources and to restore damaged ecosystems and reach a stable, sustainable population size. Technological innovations will also help us become a sustainable society. For example, a new line of compact fluorescent light bulbs is now available. These bulbs fit into a regular light socket and use one-fourth as much energy as a regular incandescent bulb (see Figure 33–20). Photovoltaics, thin silicon wafers that produce electricity from sunlight, could also help us make the transition to solar energy. Water-sensing computers can help farmers prevent overwatering. Improvements in automobile efficiency can help us cut back on fossil-fuel consumption and clean up the air. But new technologies are not the only, or even the most important, answer. Many existing technologies simply need to be installed. Insulation, weather stripping, flow restrictors for shower heads and faucets, and efficient appliances already on the market can make tremendous inroads into waste.

Many proponents also believe that we will eventually need to reduce consumption and alter our lifestyles. Taking shorter showers, shaving without the water running, turning off lights, and hundreds of other small actions on the part of individuals, when combined with similar actions by millions of other people, can result in significant cuts in resource demand.

The days of profligate resource use are quickly coming to an end. Ending the waste and pollution—and soon—is essential if we are to protect the environment and ourselves. The balance of nature is, after all, the balance that sustains us. We risk upsetting it at our own peril. Protecting the planet is the ultimate form of self care.

SUMMARY

1. The economy of nature and the human economy are at odds. Today, that "conflict" manifests itself in a global environmental crisis characterized by overpopulation, resource depletion, and pollution.

OVERSHOOTING THE EARTH'S CARRYING CAPACITY

2. Although environmental problems vary from one nation to another, they are all the result of human populations exceeding the carrying capacity of the environment.
3. Carrying capacity is the number of organisms an ecosystem can support indefinitely. It is determined by food and resource supplies and by the capacity of the environment to assimilate or destroy waste products of organisms.

OVERPOPULATION: PROBLEMS AND SOLUTIONS

4. Overpopulation occurs anytime a population overshoots its carrying capacity. It is manifest in shortages of food, lack of other resources, or excessive pollution—sometimes all three.
5. The world population is about 5.4 billion and is growing at a rate of 1.8% per year, a doubling time of 40 years.
6. The most rapid growth in human population is occurring in the Third World, in parts of Africa, Asia, and Latin America. Resource depletion and food shortages are the most common problems in these regions.
7. Many experts believe that the industrialized countries are also overpopulated. Judging from the quality of our air, water, and soils, we are clearly exceeding the Earth's carrying capacity.
8. Because of our excessive resource demand, in fact, an average American has 25 to 40 times as much impact on the environment as a resident of a Third World country.

9. The human population is growing exponentially. Exponential growth occurs anytime a value, such as population size, grows by a fixed percentage with the "interest" (increase) being applied to the base amount.

10. Exponential growth of population, resource demand, and pollution are at the heart of all environmental problems.

11. The human population problem can be summed up in six words: too many people, reproducing too quickly. The size of the population and the rapid rate of growth result in resource shortages, excessive pollution, and poverty.

12. A decline in total fertility rate, the number of children a woman will have over her lifetime, to replacement-level fertility, the rate at which parents replace themselves, is required to reduce the rate of population growth.

RESOURCE DEPLETION: ERODING THE PROSPECTS OF ALL ORGANISMS

13. The human population requires a variety of renewable and nonrenewable resources for survival, and many of these resources are being mismanaged. Some are even in danger of depletion.

14. Worldwide, forests are being cut faster than they can regenerate. Forests are important sources of wood and wood products, but they also provide wildlife habitat, protect watersheds, regenerate oxygen, and remove pollutants, such as carbon dioxide.

15. To prevent the destruction of the world's forests, tree planting, paper recycling, and other strategies are needed.

16. Worldwide, soils are being eroded from rangeland and farmland at an unsustainable rate, and millions of acres of farmland are being destroyed by human encroachment. The destruction of productive soils threatens the long-term prospects for food production. Soil conservation and population control measures can help ensure an adequate supply of soil, but serious efforts must begin soon.

17. Because of regional overpopulation, many areas suffer water shortages. The rise in population and the rise in demand are likely to cause further shortages over the coming decades. Reducing the growth of the human population and strict water conservation are needed.

18. At least 18 economically important metals could fall into short supply in the next 20 years. The more efficient use of metals, recycling, and a reduction in demand are needed to help offset inevitable shortages.

19. Oil supplies, like mineral supplies, are also finite. At best, the world has 65 years of oil remaining at the current rate of consumption. A rise in consumption would cut the oil supply sharply, resulting in shortages within the next 10 years.

20. By reducing population growth, using oil much more efficiently, and seeking clean, renewable alternative fuels, modern society can help make a smooth transition to a sustainable society.

POLLUTION: FOULING OUR NEST

21. All organisms produce waste, but humans are by far the most prolific generators of waste on the planet. Today, our waste is overwhelming nutrient cycles, poisoning other species (and ourselves), and destroying the planet's homeostatic mechanisms.

22. One of the most serious threats from pollution comes from carbon dioxide. Global atmospheric concentrations of carbon dioxide have increased 25% in the past century, in large part due to the combustion of fossil fuels and deforestation.

23. Carbon dioxide is a greenhouse gas, trapping heat in the Earth's atmosphere. A little carbon dioxide is important, but too much will increase the planet's surface temperature, altering its climate, shifting rainfall patterns and agricultural zones, flooding low-lying regions, and destroying many species that cannot adapt to the sudden change in temperature.

24. Chlorofluorocarbons and methane are also greenhouse gases that are on the increase.

25. Many scientists believe that global warming has begun. To slow it down or stop it, they recommend massive reforestation projects and dramatic improvements in efficiency of fossil-fuel combustion. Alternative fuels and reductions in population growth can also help.

26. Sulfur dioxide and nitrogen dioxide are two gaseous pollutants released from power plants, factories, automobiles, and other sources. In the atmosphere, they are converted to sulfuric and nitric acid, respectively.

27. Acids fall from the sky in wet and dry deposition. Acids alter the pH of lakes and streams, killing fish and other organisms. They also leach from soils toxic metals that kill fishes. They destroy trees and crops and deface buildings, costing society billions of dollars a year.

28. Pollution-control devices called scrubbers, which are now in use on some power plants, can help remove sulfur dioxide from smokestack gases, but they do not reduce nitrogen dioxide emissions. Stopping acid rain requires a multifaceted approach, but conservation, use of cleaner coal, and pollution control devices are the three prominent strategies.

29. The ozone layer encircles the Earth, trapping ultraviolet light. It is being destroyed by chlorofluorocarbons, or CFCs, and other pollutants.

30. Fears over the decline in ozone led to an international agreement to reduce CFC production by 100% by the year 1999.

ENVIRONMENT AND HEALTH: BUILDING A SUSTAINABLE FUTURE

31. Solving our problems will require tougher laws and tighter regulations, but it will also require a profound change in the way we live and conduct business.

32. A sustainable society is built on five operating principles: conservation, recycling, renewable resources, restoration, and population control.

33. It is based on respect for nature and a willingness to cooperate with natural processes.

One of the most dramatic changes on the planet in the last two decades has been the steady march of the Sahara, the largest desert on Earth. Studies published in the 1970s and 1980s documented an impressive southward expansion of the desert at a rate of 5 kilometers (3 miles) per year and attributed the problem to drought, overgrazing, and agricultural land abuse in semiarid lands bordering the vast Sahara. This projection, however, was based on measurements of the southward spread in a few isolated locations, assuming that they represented the entire desert.

More recent satellite observations of vegetation over the entire continent of Africa, however, show that the Sahara advances and retreats like a tide, largely in response to rainfall. In the period from 1980 to 1984, for example, the desert's southern boundary moved 240 kilometers south. Between 1984 and 1985, the southward migration reversed itself by 110 kilometers. In 1987, the boundary between desert and semiarid lands shifted northward again by 55 kilometers and in 1988 it shifted northward by 100 kilometers. In 1989 and 1990, however, the desert boundary shifted 77 kilometers southward.

Although the southern border of the desert in 1990 was 130 kilometers farther south than in 1980, some researchers believe that the shift does not reflect a long-term trend but, rather, differences in year-to-year rainfall. Clearly, continued tracking is necessary.

Many critics of the theory of global warming will find this report encouraging and use it to argue that desertification caused by climatic shift is not occurring. Critically analyze the conclusion of the study. Can you arrive at an alternative explanation?

Note: One alternative interpretation is that although the Sahara may ebb and flow, it may still be marching southward overall as a result of drought, overgrazing, and other poor land management practices. Although heavier rainfalls may result in vegetative recovery, it is possible that the recovery is temporary. Furthermore, if stress continues, the recuperative ability of the ecosystem may be overwhelmed.

TEST OF TERMS

1. The _____ _____ is the number of organisms an ecosystem can support on a sustainable basis.
2. Wind, sunlight, and trees are examples of _____ resources.
3. _____ is a condition occurring when populations exceed the carrying capacity of the environment.
4. To calculate the _____ _____ of the human population, divide 70 by the annual _____ _____ .
5. _____ growth occurs anytime a value, such as population size, grows by a fixed percentage with the "interest" (increase) being applied to the base amount.
6. The total number of children a woman is expected to have during her lifetime is the _____ _____ rate. The number of children a couple must have to replace themselves is called _____ fertility.
7. In a population, _____ _____ growth occurs when the birth rate and death rate are equal and there is no net migration.
8. A porous underground layer that is saturated with water is called a(n) _____ .
9. Carbon dioxide, methane, and chlorofluorocarbons are all considered _____ gases because they contribute to global warming.
10. Sulfur dioxide and nitrogen dioxide are considered to be _____ _____ because they are converted in the atmosphere to sulfuric and nitric acid, respectively.
11. Rain and snow are a form of acid deposition called _____ deposition.

Answers to the Test of Terms are found in Appendix B.

TEST OF CONCEPTS

1. In what ways are the economy of nature and the human economy working in opposite directions? In your estimation, is this a serious problem, and, if so, how can it be reduced or eliminated?
2. Human populations in both the industrialized and nonindustrialized countries are overshooting the Earth's carrying capacity. Do you agree or disagree with this statement? Be sure to describe what is meant by overshooting the carrying capacity.
3. The global human population is growing at a rate of 1.8% per year, so there's nothing to worry about. Do you agree or disagree with this statement? Support your position.
4. Why is overpopulation as much a problem in the United States as it is in Bangladesh?
5. The U.S. population fell below replacement-level fertility in 1972, yet it continues to increase. Why?
6. Describe key trends in the use of forests, soils, water, minerals, and oil suggesting that humanity is on an unsustainable course. List and discuss

solutions to each of the problems.

7. Given trends in natural resource use, many experts believe that global population growth must stop. Do you agree or disagree? Why? Is your position supported by scientific fact or based more on a general feeling?

8. Describe the cause and impacts of each of the following: global warm-ing, acid deposition, stratospheric ozone depletion, and hazardous wastes.

9. Make a list of solutions for each of the problems you identified in question 8. How could conservation (efficiency), recycling, renewable resources, and population control factor into the solutions?

10. Explain what is meant by the following statement: A sustainable society is based on a design from nature.

11. Outline the operational and ethical principles of a sustainable society. Using your critical thinking skills, determine how they differ from the principles of modern society.

A P P E N D I X   A

Periodic Table of Elements

≈

	Group IA	Group IIA						Group VIII			Group IB	Group IIB	Group IIIB	Group IVB	Group VB	Group VIB	Group VIIB	Group 0
Period 1	1 **H** 1.008																	2 **He** 4.003
Period 2	3 **Li** 6.941	4 **Be** 9.012											5 **B** 10.81	6 **C** 12.01	7 **N** 14.01	8 **O** 16.00	9 **F** 19.00	10 **Ne** 20.18
Period 3	11 **Na** 22.99	12 **Mg** 24.31	Group IIIA	Group IVA	Group VA	Group VIA	Group VIIA						13 **Al** 26.98	14 **Si** 28.09	15 **P** 30.97	16 **S** 32.06	17 **Cl** 35.45	18 **Ar** 39.95
Period 4	19 **K** 39.10	20 **Ca** 40.08	21 **Sc** 44.96	22 **Ti** 47.90	23 **V** 50.94	24 **Cr** 52.00	25 **Mn** 54.94	26 **Fe** 55.85	27 **Co** 58.93	28 **Ni** 58.70	29 **Cu** 63.55	30 **Zn** 65.38	31 **Ga** 69.72	32 **Ge** 72.59	33 **As** 74.92	34 **Se** 78.96	35 **Br** 79.90	36 **Kr** 83.80
Period 5	37 **Rb** 85.47	38 **Sr** 87.62	39 **Y** 88.91	40 **Zr** 91.22	41 **Nb** 92.91	42 **Mo** 95.94	43 **Tc** (98)	44 **Ru** 101.1	45 **Rh** 102.9	46 **Pd** 106.4	47 **Ag** 107.9	48 **Cd** 112.4	49 **In** 114.8	50 **Sn** 118.7	51 **Sb** 121.8	52 **Te** 127.6	53 **I** 126.9	54 **Xe** 131.3
Period 6	55 **Cs** 132.9	56 **Ba** 137.3	57 **La** 138.9	72 **Hf** 178.5	73 **Ta** 180.9	74 **W** 183.9	75 **Re** 186.2	76 **Os** 190.2	77 **Ir** 192.2	78 **Pt** 195.1	79 **Au** 197.0	80 **Hg** 200.6	81 **Tl** 204.4	82 **Pb** 207.2	83 **Bi** 209.0	84 **Po** (209)	85 **At** (210)	86 **Rn** (222)
Period 7	87 **Fr** (223)	88 **Ra** (226.0)	89 **Ac** (227)	104 **Unq**	105 **Unp**	106 **Unh**	107 **Uns**		109 **Une**									

Lanthanides (rare earth metals)

58 **Ce** 140.1	59 **Pr** 140.9	60 **Nd** 144.2	61 **Pm** (145)	62 **Sm** 150.4	63 **Eu** 152.0	64 **Gd** 157.3	65 **Tb** 158.9	66 **Dy** 162.5	67 **Ho** 164.9	68 **Er** 167.3	69 **Tm** 168.9	70 **Yb** 173.0	71 **Lu** 175.0

Actinides

90 **Th** 232.0	91 **Pa** (231)	92 **U** 238.0	93 **Np** (244)	94 **Pu** (242)	95 **Am** (243)	96 **Cm** (247)	97 **Bk** (247)	98 **Cf** (251)	99 **Es** (252)	100 **Fm** (257)	101 **Md** (258)	102 **No** (259)	103 **Lr** (260)

KEY:

16 — Atomic number
S — Symbol of element
32.06 — Atomic mass

■ Metals
□ Nonmetals
■ Metalloids
□ Noble gases

Answers to the Test of Terms

CHAPTER 1

1. biosphere; 2. Homeostasis; 3. insulin; 4. Asexual; 5. metabolism; 6. Irritability; 7. natural selection; 8. scientific method; 9. Hypotheses; 10. theory.

CHAPTER 2

1. matter; 2. Elements; 3. electrons, neutrons; 4. nucleus, positive; 5. atomic number; 6. neutrons; 7. Radiation; 8. ion; 9. covalent; 10. hydrogen; 11. carbon, hydrogen, covalent; 12. glycogen; 13. hydroxide, hydrogen; 14. acid, base; 15. neutral, greater; 16. buffer; 17. monosaccharides, polysaccharides; 18. fatty acid, glycerol; 19. phosphoglyceride; 20. Cholesterol; 21. peptide; 22. Enzymes; 23. primary, tertiary; 24. nucleotides; 25. energy, inorganic phosphate.

CHAPTER 3

1. bacteria; 2. nucleus, DNA; 3. cytoskeleton, enzymes; 4. metabolic pathway; 5. bimolecular (double), phosphoglyceride, integral; 6. semipermeable or selectively permeable; 7. glycoprotein, RER, Golgi complex; 8. diffusion; 9. carrier; 10. phagocytosis; 11. osmotic pressure; 12. isotonic, hypotonic, swell; 13. envelope, pores; 14. chromatin, chromosomes; 15. Ribosomal RNA, protein; 16. mitochondrion; 17. chloroplast, plastids; 18. RER, messenger RNA; 19. Golgi complex; 20. lysosome, food vacuole; 21. flagellum, nine, 9 + 2; 22. cilia.

CHAPTER 4

1. catabolic, anabolic; 2. release, absorb, coupled; 3. reactants; 4. equilibrium; 5. catalysts, active site; 6. specificity; 7. allosteric; 8. anabolic, endergonic, or reduction; 9. reduction; 10. work; 11. conservation, energy, created, converted; 12. lower, activation.

CHAPTER 5

1. carbon dioxide, chloroplasts; 2. light-dependent, light-independent; 3. chlorophyll; 4. wavelengths, visible; 5. photosystems; 6. reaction, electrons; 7. electron transport system or photoelectron transport system; 8. oxygen, ATP, NADPH, light-independent; 9. water; 10. ATP, electron transport system, hydrogen, protein, ATP; 11. carbon dioxide, cyclic photophosphorylation; 12. Calvin-Benson, or C_3; 13. Hatch-Slack, or C_4; 14. stomata.

CHAPTER 6

1. Cellular respiration; 2. pyruvate, or pyruvic acid; two; 3. cytoplasm; 4. acetyl CoA, or acetyl Coenzyme A; 5. matrix, or inner compartment; oxaloacetate; 6. NAD, $FADH_2$; 7. electron transport system, chemiosmosis; 8. fats, or fatty acids; amino acids, or proteins; 9. phosphofructokinase, ATP, citric acid; 10. Fermentation; pyruvate; lactate, or lactic acid; 11. two; 12. 38.

CHAPTER 7

1. cell division, interphase; 2. S; 3. gene; 4. mitosis, cytokinesis; 5. haploid; 6. contact inhibition; 7. chromatin; 8. centromere; 9. karyotype; 10. mitotic spindle; 11. metaphase; 12. microfilamentous network; 13. cell plate; 14. metastasis, secondary; 15. transformation; 16. carcinogen; 17. ozone, CFCs.

CHAPTER 8

1. Gregor Mendel; 2. hereditary, segregation; 3. recessive; 4. allele; 5. heterozygous; 6. phenotype; 7. monohybrid; 8. independent assortment; 9. XX, autosomes; 10. autosomal recessive, carrier; 11. autosomal dominant, homozygous, heterozygous; 12. incomplete dominance; 13. multiple; 14. Codominant; 15. polygenic; 16. linked; 17. sex-linked; 18. recessive, X; 19. sex-influenced; 20. nondisjunction; 21. trisomy 21; 22. nondisjunction; 23. polyploidy; 24. deletion, translocation; 25. chorionic villus

CHAPTER 9

1. hydrogen, double helix; 2. nucleotide; 3. thymine, guanine; 4. transcription, translation; 5. codon; 6. tRNA; 7. Complementary base; 8. structural, operon; 9. inducible; 10. heterochromatin; 11. Enhancers; 12. introns, exons; 13. oncogenes; 14. proto-oncogenes.

CHAPTER 10

1. ectoderm, mesoderm, endoderm; 2. extracellular material; 3. muscle, nervous, epithelial, connective; 4. exocrine; 5. epidermis, epithelial; 6. connective; 7. fibroblast; 8. hyaline cartilage; 9. fibrocartilage; 10. central, canaliculi, osteocytes; 11. compact; 12. plasma; 13. red blood cells, platelets; 14. smooth, actin, myosin; 15. neuron; 16. organ system; 17. set point; 18. reflexes;

19. integration; 20. effectors; 21. hormone, paracrine; 22. circadian; 23. suprachiasmatic; 24. jet lag.

CHAPTER 11

1. macronutrients; 2. carbohydrates (glucose) and fat (triglycerides); 3. fat; 4. essential, complete; 5. fiber, colon; 6. Vitamins, water-soluble, water-insoluble; 7. trace minerals; 8. amylase; 9. taste buds; 10. pharynx; 11. peristalsis; 12. gastroesophageal; 13. chyme; 14. pepsin, pepsinogen; 15. pyloric; 16. liver, bile salts; 17. sodium bicarbonate, digestive enzymes; 18. circular folds, villi, microvilli; 19. lymphatic (lacteals); 20. cholecystokinin (CCK).

CHAPTER 12

1. pulmonary, systemic; 2. right atrium; 3. aorta, elastic; 4. semilunar; 5. atrioventricular; 6. sinoatrial, right atrium; 7. atrioventricular bundle; 8. tunica intima, tunica media, tunica adventitia; 9. Elastic fibers; 10. systolic; 11. capillaries, venules; 12. Varicose veins; 13. edema; 14. valves; 15. lymphatic capillaries, interstitial (tissue); 16. Lymph nodes.

CHAPTER 13

1. plasma; 2. immunoglobulins or antibodies; 3. RBC, hemoglobin; 4. red bone; 5. erythropoietin, RBCs; 6. porphyrin; 7. anemia; 8. neutrophils, monocytes; 9. lymphocyte; 10. Leukemia; 11. platelet; 12. Fibrinogen; 13. Plasmin; 14. Hemophilia; 15. Carbon monoxide.

CHAPTER 14

1. Emphysema; 2. pharynx; 3. trachea, bronchi; 4. bronchioles; 5. mucous; 6. surfactant; 7. dust or alveolar macrophage; 8. vocal cords; 9. olfactory epithelium or membrane; 10. hemoglobin, bicarbonate ions; 11. brain, breathing center; 12. inhalation, diaphragm, intercostal muscles; 13. exhalation; 14. chronic bronchitis; 15. tidal, expiratory reserve.

CHAPTER 15

1. capsid; 2. inflammatory response; 3. Pyrogens; 4. Interferons; 5. antigen; 6. immunocompetence; 7. active; 8. memory; 9. immunoglobulins; 10. neutralization; 11. agglutination; 12. cytotoxic T; 13. vaccine.

CHAPTER 16

1. urinary, renal; 2. ureters; 3. urethra; 4. cortex; 5. renal pelvis; 6. glomerulus, renal tubule; 7. afferent, Bowman's space; 8. podocytes; 9. peritubular capillaries, tubular secretion; 10. internal sphincter; 11. dialysis; 12. ADH, increases; 13. Caffeine, diuretic; 14. aldosterone; 15. nephron.

CHAPTER 17

1. central, peripheral; 2. autonomic; 3. dendrites, axon; 4. myelin sheath; 5. terminal boutons, neurotransmitters; 6. resting (membrane) potential, -60; 7. sodium ions, action potential; 8. synaptic cleft, postsynaptic membrane; 9. Acetylcholine, acetylcholinesterase; 10. spinal nerves, ventral; 11. interneuron; 12. primary motor; 13. gyri, sulci; 14. primary sensory, central sulcus; 15. meninges, dura mater, pia mater; 16. association; 17. limbic; 18. cerebellum; 19. hypothalamus, nuclei; 20. reticular activating; 21. cerebrospinal, ventricles; 22. electroencephalogram, EEG; 23. sympathetic; 24. parasympathetic; 25. short-term, seconds, hours.

CHAPTER 18

1. pain; 2. encapsulated receptors; 3. Merkel's disks; 4. Pacinian; 5. Meissner's corpuscle; 6. muscle spindle; 7. Golgi tendon organs; 8. Adaptation; 9. special senses; 10. taste buds, papillae; 11. olfactory, bipolar; 12. sclera, cornea; 13. pigmented, choroid, ciliary body; 14. retina, cones, night; 15. ganglion; 16. fovea centralis, optic disk; 17. lens, ciliary body, cataracts; 18. vitreous humor; 19. Glaucoma, aqueous humor; 20. refraction, change velocity; 21. extrinsic eye; 22. nearsightedness, elongated eyeball, strong lens; 23. radial keratotomy; 24. astigmatism; 25. rhodopsin; 26. sex-linked; 27. external auditory, eardrum; 28. ossicles, oval window; 29. eustachian tube or auditory tube; 30. ampullae, semicircular, endolymph; 31. maculae; 32. organ of Corti, vestibular; 33. conduction; 34. temporary threshold.

CHAPTER 19

1. vertebrae, axial; 2. appendicular; 3. long, diaphysis, compact, marrow; 4. spongy; 5. synovial, synovial, joint capsule, tendons; 6. flexion, extension; 7. arthroscope; 8. osteoarthritis; 9. Rheumatoid, synovial; 10. prosthesis; 11. hyaline cartilage, primary center; 12. osteoclasts, osteoblasts; 13. epiphyseal plate; 14. calcitonin, thyroid, parathormone or PTH; 15. osteoporosis; 16. nuclei, striated; 17. endomysium; 18. epimysium; 19. myofibril; 20. sarcomere, A, I; 21. calcium, sarcoplasmic, troponin, actin; 22. myosin; 23. acetylcholine, T tubules; 24. Muscle fatigue, lactic acid; 25. twitch, all-or-none; 26. wave, motor; 27. slow-twitch, myoglobin; 28. fast-twitch, myosin ATPase; 29. anabolic steroids.

CHAPTER 20

1. hormone, target; 2. receptors; 3. trophic; 4. steroids, amines; 5. second messenger, ATP, adenylate cyclase, phosphates; 6. two-step, genes or DNA; 7. pituitary, hypothalamus; 8. anterior pituitary, releasing, neurosecretory, hypothalamus; 9. Growth, hypertrophy, amino acids; 10. Hyposecretion; 11. ACTH; 12. gonadotropins; 13. prolactin, suckling; 14. neuroendocrine; 15. neuroendocrine, ADH, oxytocin; 16. thyroxine, calcitonin; 17. thyroglobulin; 18. goiter; 19. parathormone or PTH, parathyroid, calcitonin, thyroid; 20. Insulin, glucose, glycogen; 21. gluconeogenesis; 22. autoimmune; 23. insulin-dependent, obesity, diet; 24. adrenalin or epinephrine, noradrenalin or norepinephrine; 25. adrenal cortex, gluconeogenesis, immune; 26. Aldosterone, adrenal cortex.

CHAPTER 21

1. scrotum, seminiferous; 2. epididymes, vasa deferentia; 3. sex accessory; 4. germinal, spermatogonia, primary spermatocytes, secondary spermatocytes; 5. spermatids, 23; 6. acrosome; 7. interstitial, testosterone, luteinizing hormone or LH; 8. foreskin, circumcision; 9. erectile; 10. ejaculation, semen; 11. uterus, uterine tubes; 12. vagina, cervical; 13. labia majora; 14. clitoris; 15. ovulation, menstrual; 16. tertiary or antral, zona pellucida; 17. first, polar, 23, double; 18. corpus luteum or CL, progesterone, estrogen; 19. follicular, estrogen, FSH; 20. endometrium, menstruation; 21. premenstrual syndrome; 22. menopause, estrogen; 23. vasectomy, uterine tubes, tubal ligation; 24. estrogen, progesterone; 25. Pap smear; 26. intrauterine device or IUD; 27. diaphragm, spermicidal; 28. condom; 29. rhythm or natural; 30. temperature; 31. sexually transmitted diseases; 32. Nonspecific urethritis; 33. Herpes.

CHAPTER 22

1. zygote, uterine tube or Fallopian tube; 2. acrosome; 3. fast, polyspermy; 4. morula, inner cell mass, trophoblast; 5. implantation, 5, 7; 6. placental villi; 7. amnion; 8. organogenesis; 9. brain, spinal cord; 10. somites, spine or vertebrae; 11. endoderm; 12. vein, ductus arteriosus, inferior vena cava; 13. right, left, foramen ovale; 14. human chorionic gonadotropin, estrogen, progesterone; 15. teratology; 16. Relaxin; 17. dilation, cervix; 18. breach; 19. oxytocin, posterior pituitary; 20. placental; 21. pudendal; 22. prolactin; 23. oxytocin; 24. colostrum; 25. rooting; 26. bonding; 27. puberty; 28. Life expectancy.

CHAPTER 23

1. chemical; 2. protocells; 3. abiotically; 4. coacervates; 5. hetertrophic fermenters; 6. chemosynthetic; 7. chlorophyll; 8. selective advantage; 9. oxygen, eukaryotes; 10. endosymbiotic; 11. ozone layer; 12. arthropods; 13. species; 14. variation; 15. adaptation; 16. natural selection; 17. fitness; 18. speciation, geographic, reproductive; 19. divergent evolution or adaptive radiation; 20. punctuated equilibrium.

CHAPTER 24

1. capsid; 2. reverse transcriptase; 3. budding; 4. RNA replicase; 5. vaccines; 6. Monera; 7. binary fission; 8. conjugation; 9. spores; 10. obligate; 11. facultative; 12. saprophytes; 13. archaebacteria; 14. cyanobacteria; 15. Protista; 16. algae; 17. Diatoms; 18. Dinoflagellates; 19. euglenoids; 20. protozoans; 21. Zooflagellates; 22. contractile vacuole; 23. ciliates; trichocysts; 24. Sporozoans; 25. hyphae, mycelium; 26. chitin; 27. True fungi; 28. Slime molds; 29. lichen; 30. mycorrhiza.

CHAPTER 25

1. cell wall; 2. angiosperms; 3. gymnosperms; 4. green algae; 5. vegetative propagation; 6. meiosis, fertilization; 7. embryo; 8. oogamy; 9. haploid, gametes; 10. diploid, spores; 11. sporophyte, gametophyte; 12. chlorophyte; 13. brown; 14. avascular, vascular, avascular; 15. protonema; 16. rhizome; 17. xylem, phloem; 18. seeds; 19. cones; 20. flowers; 21. anthers; 22. ovary; 23. fruit; 24. double fertilization; 25. pollination.

CHAPTER 26

1. epicotyl; 2. hypocotyl; 3. endosperm; 4. monocot; 5. primary meristem; 6. terminal bud; 7. axillary bud; 8. parenchyma; 9. xylem; 10. phloem; 11. epidermis; 12. bark; 13. sapwood; 14. vascular cambium; 15. secondary meristem; 16. xylem; 17. heartwood; 18. blade, petiole; 19. leaflets; 20. guard cells; 21. root cap; 22. root hairs; 23. parenchymal; 24. transpiration; 25. auxin; 26. auxin; 27. ethylene; 28. phototropism; 29. geotropism; 30. apical dominance; 31. phytochrome; 32. Short; 33. Long; 34. neutral; 35. Vernalization.

CHAPTER 27

1. heterotrophic, eukaryotic; 2. asymmetric; 3. radial, bilateral; 4. tissue; 5. ectoderm, endoderm; 6. mesoglea, gastrovascular; 7. nerve net; 8. cnidocysts; 9. larva; 10. cephalization; 11. hermaphroditic or hermaphrodites; 12. coelom; 13. complete, anus; 14. Segmentation; 15. exoskeleton, chitin; 16. positive; 17. metamorphosis; larva, pupa; 18. mantle, shell, foot, radula; 19. water vascular, tube feet; 20. nerve cord, notochord, gill slits; 21. invertebrate; 22. bone; 23. cartilage; 24. Gills, Lungs; 25. quadrupedal; 26. shelled, yolk, yolk sac; 27. allantois, amnion, chorion; 28. internal fertilization; 29. endothermy; 30. mandible, ossicles; 31. keratin, follicle; 32. mammary glands; 33. placenta, umbilical cord.

CHAPTER 28

1. prosimians; 2. anthropoids, apes, humans; 3. dryopithecus, ramapithecus; 4. *habilis*; 5. *Homo erectus,* Africa, Asia; 6. *Homo sapiens*; larger; 7. Cro-Magnons, Africa; 8. race.

CHAPTER 29

1. innate; 2. taxis; 3. tropisms; 4. Reflexes; 5. Fixed action patterns, sign stimuli; 6. Imprinting, critical period; 7. Learning; 8. Classical conditioning; 9. operant conditioning; 10. Latent learning; 11. Imitation; 12. Habituation, Sensitization; 13. motivation; 14. Agonism, ritualized; 15. Communication; 16. Language; 17. pecking order; 18. hierarchy; 19. home range; 20. territory; 21. Altruistic; 22. Kin selection; 23. Reciprocal altruism; 24. Sociobiology.

CHAPTER 30

1. symbiosis; 2. Coevolution, selective; 3. Mutualism; 4. lichen, fungus; 5. Competition; 6. exploitation, Interference; 7. Character displacement; 8. Ecological segregation; 9. Weeds; 10. Predation, parasitism; 11. specialists, generalists; 12. Warning; 13. Batesian; 14. Mullerian; 15. flatworm, Elephantiasis; 16. Endoparasites, ectoparasites; 17. parasitoids; 18. nest; 19. commensal; 20. epiphytes, commensals.

CHAPTER 31

1. Ecology; 2. biosphere; 3. biome; 4. biotic, abiotic; 5. range of tolerance; 6. limiting factor; 7. population, community; 8. niche; 9. producers, consumers; 10. herbivores, omnivores, detritivores; 11. grazer, decomposer; 12. heat; 13. trophic; 14. Biomass; 15. organismic, environmental; 16. nitrogen fixation; 17. growth, reduction; 18. density; 19. Succession, primary succession.

CHAPTER 32

1. Arctic tundra; 2. permafrost; 3. taiga; 4. temperate deciduous forest; 5. deciduous; 6. temperate grassland; 7. downwind; 8. tropical rain forest; 9. altitudinal biomes; 10. alpine tundra; 11. aqautic life zones; 12. energy, nutrients; 13. Phytoplankton, zooplankton; 14. littoral; 15. limnetic; 16. profundal, benthic; 17. epilimnion, hypolimnion, thermocline; 18. fall overturn; 19. watershed; 20. point, nonpoint; 21. estuaries, coastal wetlands; 22. Coral reefs; 23. continental shelf; 24. euphotic; 25. bathyal zone.

CHAPTER 33

1. carrying capacity; 2. renewable; 3. Overpopulation; 4. doubling time, growth rate; 5. Exponential; 6. total fertility, replacement level; 7. zero population; 8. aquifer; 9. greenhouse; 10. acid precursors; 11. wet.

The Metric System

≈

In the United States, the metric system is a system of measurement used principally by scientists. In our day-to-day lives, though, most Americans use the English system of measurement—miles, inches, feet, pounds, tons, and so on.

Many countries like New Zealand use the metric system for weights and other measures. If you travel abroad, road signs will list distance in kilometers rather than miles. Other linear measurements will be given in meters and centimeters instead of yards, feet, and inches. The weight of objects will be expressed in kilograms or grams instead of pounds and ounces. Liquids will be measured in liters rather than quarts or gallons.

It can be confusing if you don't know how to convert from one system of weights and measures to another. The lists below show some of the most common units you will encounter in biology and other sciences and compare them to their English equivalents.

≈	**Most Common English Units and the Corresponding Metric Units**	≈
	ENGLISH UNIT	METRIC UNIT
Weight	tons	metric tons
	pounds	kilograms
	ounces	grams
Length	miles	kilometers
	yards	meters
	inches	centimeters
Square Measure	acres	hectares
	square miles	square kilometers
Volume	quarts and gallons	liters
	fluid ounces	milliliters

—See the next page for more Metric System tables

Converting English Units to Metric Units

	ENGLISH UNIT		METRIC UNIT
Weight	1 ton (2000 pounds)	=	0.9 metric tons
	1 pound	=	0.454 kilograms
	1 ounce	=	28.35 grams
Length	*1 mile	=	1.6 kilometers
	1 yard	=	0.9 meters
	*1 inch	=	2.54 centimeters
Square Measure	1 acre	=	0.4 hectares
	1 square mile	=	2.59 square kilometers
Volume	1 quart	=	0.95 liters
	*1 gallon	=	3.78 liters
	1 fluid ounce	=	29.58 milliliters

*Most useful conversions to know.

Converting Metric Units to English Units

	METRIC UNIT		ENGLISH UNIT
Weight	*1 metric ton	=	2204 pounds
	1 metric ton	=	1.1 tons
	*1 kilogram	=	2.2 pounds
	1 gram	=	28.35 ounces
Length	*1 kilometer	=	.6 miles
	1 meter	=	1.1 yards
	1 centimeter	=	.39 inches
Square Measure	*1 hectare	=	2.47 acres
	1 square kilometer	=	0.386 square miles
Volume	1 liter	=	1.057 quarts
	1 liter	=	.26 gallons
	1 milliliter	=	0.0338 fluid ounces

*Most useful conversions to know

Glossary

≈

Abiotic factors Physical and chemical components of an organism's environment.

Abyssal zone Deepest waters of the ocean. Characterized by complete darkness.

Accommodation Change in the shape of the lens caused by contraction or relaxation of the smooth muscle of the ciliary body. Through accommodation, the lens adjusts the degree to which incoming light rays are bent, permitting objects to be focused on the retina.

Acetylcholine Neurotransmitter substance in the central and peripheral nervous systems of humans.

Acetylcholinesterase Enzyme that destroys the neurotransmitter acetylcholine in the synaptic cleft.

Achondroplasia Genetic disease that results from an autosomal dominant gene. Individuals with the disease have short legs and arms, but a relatively normal body size.

Acid deposition Deposition of sulfuric and nitric acids in the atmosphere onto the Earth's surface. Damages buildings, lakes, streams, crops, and forests. Acids come from sulfur dioxide and nitrogen dioxide produced during the combustion of fossil fuels.

Acid precursors Sulfur dioxide and nitrogen dioxide gases that combine with water and oxygen to form sulfuric and nitric acids in the atmosphere.

Acrosome Enzyme-filled cap over the head of a sperm. Helps the sperm dissolve its way through the corona radiata and zona pellucida.

Actin microfilaments Contractile filaments made of protein and found in cells as part of the cytoskeleton. Especially abundant in the microfilamentous network beneath the plasma membrane and in muscle cells.

Action potential Recording of electrical change in membrane potential when a neuron is stimulated.

Activation energy The energy needed to force the electron clouds of reactants together before the formation of products; also, the energy needed to break internal chemical bonds.

Active immunity Immune resistance gained when an antigen is introduced into the body either naturally or through vaccination.

Active site The region of an enzyme molecule that binds substrates and performs the catalytic function of the enzyme.

Active transport Movement of molecules across membranes using protein molecules and energy supplied by ATP. Moves molecules and ions from regions of low to high concentration.

Adaptation Genetically based characteristic that increases an organism's chances of passing on its genes.

Adaptive radiation Process in which one species gives rise to many others that occupy different environments. Also known as divergent evolution.

Adenosine diphosphate (ADP) Precursor to *ATP*; consists of adenine, ribose, and two phosphates; *see also* Adenosine triphosphate.

Adenosine triphosphate (ATP) A molecule composed of ribose sugar, adenine, and three phosphate groups. The last two phosphate groups are attached by "high-energy bonds" that require considerable energy to form but release that energy again when broken. ATP serves as the major energy carrier in cells.

Adenylate cyclase Enzyme bound to the inner surface of the plasma membrane. Linked to plasma membrane hormone receptors. Responsible for the conversion of ATP to cyclic AMP (a second messenger).

Adipose tissue Type of loose connective tissue containing numerous fat cells. Important storage area for lipids.

Adolescence Period of human life from puberty until adulthood. Characterized by sexual maturity.

Adrenal cortex Outer portion of the adrenal gland. Produces a variety of steroid hormones, including cortisol and aldosterone.

Adrenal gland Endocrine gland located on top of the kidney. Consists of two parts: adrenal cortex and medulla, each with separate functions.

Adrenal medulla Inner portion of the adrenal glands. Produces epinephrine (adrenalin) and norepinephrine (noradrenalin).

Adrenalin (epinephrine) Hormone secreted under stress. Contributes to the fight-or-flight response by increasing heart rate, shunting blood to muscles, increasing blood glucose levels, and other functions.

Adrenocorticotropic hormone (ACTH) Polypeptide hormone produced by the anterior pituitary. Stimulates the cells of the adrenal cortex, causing them to synthesize and release their hormones, especially glucocorticoids.

Aerobic exercise Exercise, such as swimming, that does not deplete muscle oxygen. Excellent for strengthening the heart and for losing weight.

Age-structure diagram (population histogram) Graphical representation of the number or percentage of males and females in various age groups in a population.

Agglutination Clumping of antigens that occurs when antibodies bind to several antigens.

Aging Inevitable and progressive deterioration of the body's function, especially its homeostatic mechanisms.

AIDS Acquired immune deficiency syndrome. Fatal disease caused by the HIV virus, which attacks T helper cells, greatly reducing the body's ability to fight infection.

Albinism Genetic disease resulting in a lack of pigment in the eyes or the eyes, skin, and hair. An autosomal recessive trait.

Aldosterone Steroid released by the adrenal cortex in response to a decrease in blood pressure, blood volume, and osmotic concentration. Acts principally on the kidney.

Alga (algae, plural) Heterogeneous group of aquatic plants consisting of three major groups: green, red, and brown. Important producer essential to aquatic food chains.

Alleles Alternative form of a gene.

Allergen Antigen that stimulates an allergic response.

Allergy Extreme overreaction to some antigens, such as pollen or foods. Characterized by sneezing, mucus production, and itchy eyes.

Allosteric inhibition Enzyme regulation in which an inhibitor molecule binds to an enzyme at a site away from the active site, changing the shape or charge of the active site so that it can no longer bind substrate molecules.

Allosteric site Region of an enzyme where products of metabolic pathways bind, changing the shape of the active site. In some enzymes this prevents substrates from binding to the active site; in others, it allows them to bind. Thus, allosteric sites can either turn on or turn off enzymes.

Alpine tundra A relatively cold, treeless region similar to the Arctic tundra but found on the tops of mountains.

Alternation of generations Phenomenon exhibited by all land plants and many algae where the plant alternately exists as a haploid generation (produces gametophytes) and as a diploid generation that produces spores (sporophytes).

Altitudinal biome Distinct region on a mountain, characterized by an assemblage of organisms resulting from the climatic conditions at an altitutde.

Alveoli Tiny, thin-walled sacs in the lung where oxygen and carbon dioxide are exchanged between the blood and the air.

Ameboid motion Cellular locomotion common in single-celled organisms and some cells in the human body. The cells send out slender cytoplasmic projections that attach to the substrate "ahead" of the cell. The cytoplasm flows into the projections, or pseudopodia, advancing the organism.

Amenalism Relationship between two organisms in which one organism is negatively impacted while the other is unaffected.

Amniocentesis Procedure whereby physicians extract cells and fluid from the amnion surrounding the fetus. The cells are examined for genetic defects, and the fluid is studied biochemically.

Amnion Layer of cells that separates from the inner cell mass of the embryo and eventually forms a complete sac around the fetus.

Amniotic fluid Liquid in the amniotic cavity surrounding the embryo and fetus during development. Helps protect the fetus.

Amoeboids Group of protozoans that move by way of amoeboid motion.

Amphibians Transitional class of vertebrates that spend part of their lives on land and part in the water. Sexual reproduction generally occurs through fertilization of fish-like eggs in the external environment.

Ampulla Enlarged area of each semicircular canal that houses receptor cells for movement.

Amylase Enzyme in saliva that helps break down starch molecules.

Anabolic reaction A reaction in which substances are formed; for example, the synthesis of glucose during photosynthesis is an anabolic reaction.

Anabolic steroids Synthetic androgen hormones that promote muscle development.

Anaerobe An organism that does not require oxygen to survive. Some anaerobes are actually killed by oxygen.

Analogous structures Anatomical structures that function similarly but differ in structure; for example, the wing of a bird and the wing of an insect.

Anaphase Phase of mitosis during which the chromatids of each chromosome begin to uncouple and are pulled in opposite directions with the aid of the mitotic apparatus.

Androgen-binding protein Cytoplasmic receptor protein that binds to and concentrates testosterone within the Sertoli cell. Production of ABP is stimulated by FSH.

Androgens Male sex steroids such as testosterone produced principally by the testes.

Anemia Condition characterized by an insufficient number of red blood cells in the blood or insufficient hemoglobin. Often caused by insufficient iron intake.

Aneuploidy Describes a genetic condition in which there is an abnormal number of chromosomes.

Aneurysm Ballooning of the arterial wall caused by a degeneration of the tunica media.

Angiosperms Recently evolved and widely distributed land plants that produce seeds contained within specialized structures known as fruits. Commonly called flowering plants.

Anoxia Lack of oxygen.

Antagonistic Refers to hormones or muscles that exert opposite effects.

Anterior pituitary Major portion of the pituitary gland, which is controlled by hypothalamic hormones. Produces seven protein and polypeptide hormones.

Anthropoids Monkeys, the great apes, and humans.

Antibodies Proteins produced by immune system cells that destroy or inactivate antigens, including pollen, bacteria, yeast, and viruses.

Anticodon loop Part of the transfer RNA molecule that bears three bases that bind to the three bases of the codon on messenger RNA.

Anticodon Sequence of three bases found on the transfer RNA molecule. Aligns with the codon on messenger RNA and helps control the sequence of amino acids inserted into the growing protein.

Antidiuretic hormone (ADH) Hormone released by the posterior pituitary. Increases the permeability of the distal convoluted tubule and collecting tubules, increasing water reabsorption.

Antigens Any substance that is detected

as foreign by an organism and elicits an immune response. Most antigens are proteins and large molecular weight carbohydrates.

Anvil (incus) One of three bones of the middle ear that helps transmit sound waves to the receptor for sound in the inner ear.

Aorta Largest artery in the body; carries the oxygenated blood away from the heart and delivers it to the rest of the body through many branches.

Appendicular skeleton Bones of the arms, legs, shoulders, and pelvis. Contrast with axial skeleton.

Aquatic life zones Ecologically distinct regions in fresh water and salt water.

Aqueous humor Liquid in the anterior and posterior chambers of the eye.

Aquifer Porous underground zone containing water.

Arachnids Class of arthropods that includes mites, ticks, scorpions, and spiders.

Arctic tundra Massive biome north of taiga characterized by low precipitation and cold temperatures.

Arteries Vessels that transport blood away from the heart.

Arteriole Smallest of all arteries; usually drains into capillaries.

Arthropods Phylum of invertebrate animals with jointed appendages and chitinous exoskeletons such as insects, arachnids, and crustaceans. Many arthropods inhabit freshwater and marine ecosystems, but the vast majority live on land.

Arthroscope Device used to examine internal joint injuries.

Asexual reproduction Reproductive strategy common in single-celled organisms, such as the amoeba. Reproduction occurs by cell division.

Association cortex Area of the brain where integration occurs.

Association neurons Nerve cells that receive input from many sensory neurons and help process them, ultimately carrying impulses to nearby multipolar neurons.

Aster Array of microtubules found in the cell in association with the spindle fibers during cell division.

Asthma Respiratory disease resulting from an allergic response. Allergens cause histamine to be released in the lungs. Histamine causes the air-carrying ducts (bronchioles) to constrict, cutting down airflow and making breathing difficult.

Astigmatism Unequal curvature of the cornea (sometimes the lens) that distorts vision.

Atom Smallest particles of matter that can be achieved by ordinary chemical means, consisting of protons, neutrons, and electrons.

Atomic mass units Unit used to measure atomic weight. One atomic mass unit is 1/12 the weight of a carbon atom.

Atomic weight Average mass of the atoms of a given element, measured in atomic mass units.

Atrioventricular bundle Tract of modified cardiac muscle fibers that conduct the pacemaker's impulse into the ventricular muscle tissue.

Atrioventricular node (AV node) Knot of tissue located in the right ventricle. Picks up the electrical signal arriving from the atria and transmits it down the atrioventricular bundle.

Atrioventricular valves Valves between the atria and ventricles.

Auditory (eustachian) tube Collapsible tube that joins the nasopharynx and middle ear cavities and helps equalize pressure in the middle ear.

Auricle (or pinna) Skin-covered cartilage portion of the outer ear.

Autocrines Chemical substances produced by cells, which affect the function of the cells producing them.

Autoimmune reaction Immune response directed at one's own cells.

Autonomic nervous system That part of the nervous system not under voluntary control.

Autosomal dominant trait Trait that is carried on the autosomes and is expressed in heterozygotes and homozygote dominants.

Autosomal recessive trait Trait that is carried on the autosomes and is expressed only when both recessive genes are present.

Autosomes All human chromosomes except the sex chromosomes.

Autotrophs Organisms such as plants that, unlike animals, are able to synthesize their own food.

Auxins Group of plant hormones responsible for cell enlargement. Plays a key role in phototropism. Together with cytokinin, auxin regulates the transformation of undifferentiated tissues into stems and roots.

Axial skeleton The skull, vertebral column, and rib cage. Contrast with appendicular skeleton.

Axon Long, unbranched process attached to the nerve cell body of a neuron. Transports bioelectric impulses away from the cell body.

Basal body Organelle located at the base of the cilium and flagellum. Consists of nine sets of microtubules arranged in a circle. Each "set" contains three microtubules.

Basilar membrane Membrane that supports the organ of Corti in the cochlea.

Batesian mimicry Strategy in which one species evolves a warning coloration similar to a species with active defense mechanisms such that the imitating species benefits from the outward similarity.

Bathyal zone Region of semidarkness in salt waters where photosynthesis does not occur; located below the euphotic zone but above the abyssal zone.

Behavior An individual's response to environmental stimuli such as other individuals of the same species, members of other species, or some aspect of the physical environment.

Benign tumor Abnormal cellular proliferation. Unlike a malignant tumor, the cells in a benign tumor stop growing after a while, and the tumor remains localized.

Benthic zone Bottom layer of a lake; supports life forms that can survive in water with very low oxygen levels.

Beta cells Insulin-producing cells of the islets of Langerhans in the pancreas.

Bile Fluid produced by the liver and stored and concentrated in the gallbladder.

Bile salts Steroids produced by the liver, stored in the gallbladder, and released into the small intestine where they emulsify fats, a step necessary for enzyme digestion.

Binary fission Bacterial cellular division.

Bioelectric impulse Nerve impulse resulting from the influx of sodium ions along the plasma membrane of a neuron.

Biomass The dry weight of living material in an ecosystem.

Biomass pyramid Diagram of the amount

of biomass at each trophic level in an ecosystem or, more commonly, a food chain.

Biome Terrestrial region characterized by a distinct climate and a characteristic plant and animal life.

Biorhythms (biological cycles) Naturally fluctuating physiological process.

Biosphere Region on Earth that supports life. Exists at the junction of the atmosphere, lithosphere, and hydrosphere.

Biotic factor Biological components of ecosystems.

Bipedal Refers to the ability to walk on two legs.

Birds Class of vertebrates characterized by feathers and endothermy (warmbloodedness). Most, but not all, of these species retain the ability to fly.

Birth control Any method or device that prevents conception and birth.

Birth control pill Pill generally taken to inhibit ovulation. The most commonly used pills contain both synthetic estrogen and progesterone.

Birth defects Physical or physiological defects in newborns. Caused by a variety of biological, chemical, and physical factors.

Blastocyst Hollow sphere of cells formed from the morula. Consists of the inner cell mass and the trophoblast.

Blood Specialized form of connective tissue. Consists of white blood cells, platelets, red blood cells, and plasma.

Blood clot Mass of fibrin containing platelets, red blood cells, and other cells. Forms in walls of damaged blood vessels, halting the efflux of blood.

B-lymphocytes Type of lymphocyte that transforms into a plasma cell when exposed to antigens.

Bonding Process in which an infant establishes an intimate relationship with one or both of its parents or caretakers. Essential for emotional health.

Bone (organ) Structure comprised of bone tissue. Provides internal support, protects organs, and helps maintain blood calcium levels.

Bone (tissue) Tissue consisting of a calcified extracellular material with numerous cells (osteocytes) embedded in it.

Bony fishes Class of vertebrates with more elaborate bone development than jawless fishes. Contains more species than all other vertebrates combined.

Bowman's capsule Cup-shaped end of the nephron that participates in glomerular filtration.

Bowman's space Cavity between the inner and outer layer of Bowman's capsule.

Brain stem Part of the brain that consists of the medulla and pons. Houses structures, such as the breathing control center and reticular activating systems, that control many basic body functions.

Braxton-Hicks contractions Contractions that begin a month or two before childbirth. Also known as false labor.

Breathing center Aggregation of nerve cells in the brain stem that controls breathing.

Breech birth Delivery of a baby feet first.

Bronchi Ducts that convey air from the trachea to the bronchioles and alveoli.

Bronchioles Smallest ducts in the lungs. Their walls are largely made of smooth muscle that contracts and relaxes, regulating the flow of air into the lung.

Bryophytes Small plants such as mosses and figworts that lack vascular tissue, which severely restricts water transport to distant tissues. Most bryophytes require moist habitats to live and reproduce.

Bulimia Eating disorder characterized by recurrent binge eating followed by vomiting.

Bundle sheath cells Cells surrounding veins in leaves. They participate in the C_4 cycle by breaking down malic acid, producing carbon dioxide that can be used in the C_3 cycle.

C_3 cycle The cyclic series of reactions whereby carbon dioxide is fixed into carbohydrates, using energy from the light-dependent reactions. Also called the Calvin-Benson cycle.

C_4 cycle The series of reactions in certain plants that fixes carbon dioxide into organic acids for later use in the C_3 cycle of photosynthesis.

Calcitonin (thyrocalcitonin) Polypeptide hormone produced by the thyroid gland that inhibits osteoclasts and stimulates osteoblasts to produce bone, thus lowering blood calcium levels.

Calvin-Benson cycle See C_3 cycle.

Canal of Schlemm Network of channels located at the junction of the sclera and cornea that drains aqueous humor from the anterior chamber of the eye.

Canaliculi Tiny canals in compact bone that provide a route for nutrients and wastes to flow to and from the osteocytes.

Cancer Disease characterized by the uncontrollable replication of cells.

Capacitation Process in which the outer protective coat of the sperm is dissolved away in the female reproductive tract. Makes the membrane fragile and disruptible, allowing the acrosome to break down and release its enzymes.

Capillaries Tiny vessels in body tissues whose walls are composed of a flattened layer of cells that allow water and other molecules to flow freely into and out of the tissue fluid.

Capillary bed Branching network of capillaries supplied by arterioles and drained by venules.

Capsid Protein coat of a virus.

Capsomere Globular proteins that make up the capsid of viruses.

Carbohydrate Organic compound consisting of carbon, hydrogen, and oxygen. A structural component of plant cells, it is used principally as a source of energy in animal cells.

Carbon fixation The initial steps in the C_3 cycle, in which carbon dioxide reacts with ribulose bisphosphate to form two stable molecules of 3-phosphoglyceric acid, or PGA.

Carbonic anhydrase Enzyme found in red blood cells that catalyzes the conversion of carbon dioxide to carbonic acid.

Carcinogens Cancer-causing agents.

Cardiac muscle Type of muscle found in the walls of the heart; it is striated and involuntary.

Carnivores Meat eaters; organisms that feed on grazers and other animals.

Carotenoid A family of pigments, usually yellow, orange, or red, found in chloroplasts of plants and serving as accessory light-gathering molecules in thylakoid photosystems.

Carrier proteins Class of proteins that transport smaller molecules and ions across the plasma membrane of the cell. Are involved in facilitated diffusion.

Carriers Individuals who carry a gene for a particular trait that can be passed on to their children, but who do not express the trait.

Carrying capacity The number of organisms an ecosystem can support on a sustainable basis.

Cartilage Type of specialized connective tissue. Found in joints on the articular surfaces of bones and other locations.

Catabolic reaction A reaction in which molecules are broken down; for example, the breakdown of glucose is a catabolic reaction.

Catalyst Class of compounds that speed up chemical reactions. Although they take an active role in the process, they are left unchanged by the reaction. Thus, they can be used over and over again. Catalysts lower the activation energy of a reaction. *See also* Enzyme.

Cataracts Disease of the eye resulting in cloudy spots on the lens (and sometimes the cornea) that cause cloudy vision.

Cell body Part of the nerve cell that contains the nucleus and other cellular organelles; the center of chemical synthesis.

Cell culture Glass bottle or shallow dish containing nutrient medium and designed to permit cells to grow in the laboratory under controlled conditions.

Cell cycle Repeating series of events in the lives of many cells. Consists of two principal parts: interphase and cellular division.

Cell division Process by which the nucleus and the cytoplasm of a cell are split, creating two daughter cells. Consists of mitosis and cytokinesis.

Cellular respiration The complete breakdown of glucose in the cell, producing carbon dioxide and water. Comprised of four separate but interconnected parts: glycolysis, the intermediate reaction, the citric acid cycle, and the electron transport system.

Central nervous system The brain and spinal chord.

Centriole Organelle consisting of a ring of microtubules, arranged in nine sets of three. Structurally identical to basal bodies, but associated with the spindle apparatus. Gives rise to the basal body in ciliated cells.

Centromere Region on each chromatid that joins with the centromere of its sister chromatid.

Cerebellum Structure of the brain that lies blow the cerebral cortex. It has many important functions including synergy.

Cerebral cortex Outer layer of each cerebral hemisphere, consisting of many multipolar neurons and nerve cell fibers.

Cerebral hemisphere Convoluted mass of nervous tissue located above the deeper structures, such as the hypothalamus and limbic system. Home of consciousness, memory, and sensory perception; originates much conscious motor activity.

Cervical cap Birth control device consisting of a small cup that fits over the tip of the cervix.

Cervix Lowermost portion of the uterus; it protrudes into the vagina.

Cesarean section Delivery of a baby via an incision through the abdominal and uterine walls.

Chemical equilibrium The point at which the "forward" reaction from reactants to products proceeds at the same rate as the "backward" reaction from products to reactants, so that there is no net change in chemical composition.

Chemical evolution Formation of organic molecules from inorganic molecules early in the history of the Earth.

Chemiosmosis Production of ATP in the chloroplast's and the *mitochondrion's* electron transport systems. H^+ flowing through pores in ATPase molecules drives the endergonic synthesis of ATP.

Chemosynthetic organisms Cells that lack chlorophyll and acquire electrons from inorganic molecules.

Childhood Period of human life that lasts from infancy to puberty.

Chitin Complex, waterproof, chemical-resistant molecule made of both protein and carbohydrate. Forms the rigid cell wall of fungi and the exoskeleton of insects and other arthropods.

Chlamydia Bacterium that causes nonspecific urethritis, a type of sexually transmitted disease.

Chlorofluorocarbons (CFCs) Chemical substances often used as spray can propellants (outside the United States and several other countries) and refrigerants that drift to the upper stratosphere and dissociate. Chlorine released by CFCs reacts with ozone, thus eroding the ozone layer.

Chlorophyll A green pigment that acts as the primary light-trapping molecule for photosynthesis.

Cholecystokinin (CCK) Hormone produced by cells of the duodenum when chyme is present. Causes the gallbladder to contract, releasing bile.

Chordae tendineae Tendinous chords that anchor the atrioventricular valves to the inner walls of the ventricles.

Chordates Phylum that has three features in common: a hollow dorsal nerve cord, a supporting, cartilaginous notochord, and gill slits in the neck region.

Chorionic villus biopsy Medical procedure to detect genetic defects; involves removing a small portion of the villi and then examining it for chromosomal abnormalities.

Choroid Middle layer of the eye that absorbs stray light and supplies nutrients to the eye.

Chromatid Strand of the chromosome consisting of DNA and protein.

Chromatin Long, threadlike fibers containing DNA and protein in the nucleus.

Chromosome-to-pole fibers Microtubules of the spindle that extend from the centriole to the chromosome where they attach. They play a crucial role in separating the double-stranded chromosomes during mitosis.

Chronic bronchitis Persistent irritation of the bronchi, which causes a mucus buildup, coughing, and difficulty breathing.

Chyme Liquified food in the stomach.

Ciliary body Portion of the middle layer of the eye that is located near the lens. Contains smooth muscle that constricts, thus helping control the shape of the lens and permitting the eye to focus.

Ciliates The most structurally complex of all animal-like protists. They get their name from the many cilia projecting from their surface.

Circadian rhythm Biorhythm that occurs on a daily cycle.

Circulatory system Organ system consisting of the heart, blood vessels, and blood.

Circumcision Operation to remove the foreskin of the penis. Generally performed on newborns.

Cisterna Channels of the endoplasmic reticulum and Golgi.

Citric acid Six-carbon compound produced in the first reaction of the citric acid or Krebs cycle. Formed when oxaloacetate reacts with acetyl Coenzyme A.

Citric acid cycle A cyclic series of reactions in which the pyruvates produced by glycolysis are broken down to CO_2, accompanied by the formation of *ATP* and electron carriers. Occurs in the matrix of mitochondria.

Class Subgroup of a phylum.

Clitoris Small knot of tissue located where the labia minora meet. Consists of erectile tissue.

Cloning Technique of genetic engineering whereby many copies of a gene are produced.

Closed system System that receives no materials from the outside.

Cnidarians Phylum characterized by distinct tissues and primitive nervous systems, coordination, and symmetry. Includes jellyfish, corals, and sea anemones. Reproduction is asexual.

Coacervates Microscopic globules that selectively incorporate molecules from their environment. Coacervates existing on Earth over 3 billion years ago may have given rise to the earliest cells.

Cochlea Sensory organ of the inner ear that houses the receptor for hearing.

Codominant Refers to two equally expressed alleles.

Codon Three adjacent bases in the messenger RNA that code for a single amino acid.

Coelom Body cavity (a space between the body wall and internal organs) lined by mesoderm.

Coevolution Process in which two or more species act as selective forces on each other, resulting in anatomical, behavioral, and functional changes in each other.

Collecting tubules Tubules in the kidney into which nephrons drain. They converge and drain into the renal pelvis.

Color blindness Condition that occurs in individuals who have a deficiency of certain cones. The most common form involves difficulty in distinguishing between red and green.

Colostrum Protein-rich product of the breast, produced for two to three days immediately after delivery.

Commensal Relationship in which one species benefits while the other is unaffected.

Community All of the plants, animals, and microorganisms in an ecosystem.

Compact bone Dense bony tissue in the outer portion of all bones.

Competition Struggle by two or more individuals for the same limited resource.

Complement Group of blood proteins that circulate in the blood in an inactive state until the body is invaded by bacteria; then they help destroy the bacteria.

Complementary base pairing Unalterable coupling of purine adenine to pyrimidine thymine, and purine guanine to pyrimidine cytosine. Responsible for the accurate transmission of genetic information from parent to offspring.

Conditioning Type of learning in which an animal learns to make a connection (association) between a new stimulus and a familiar one—that is, a previous stimulus-response relationship is transferred to a novel stimulus.

Condom Birth control device, consisting of a thin latex rubber sheath or other material that is rolled onto the erect penis. Prevents sperm from entering the vagina and helps prevent the spread of sexually transmitted diseases.

Conduction deafness Loss of hearing that occurs when the conduction of sound waves to the inner ear is impaired. May be caused by ruptured eardrum or damage to ossicles.

Cone Type of photoreceptor that operates in bright light; is responsible for color vision.

Conjugation Process in which bacteria exchange genetic material; considered a primitive form of sexual reproduction.

Connective tissue One of the primary tissues. It contains cells and varying amounts of extracellular material and holds cells together, forming tissues and organs.

Connective tissue proper Name referring to loose and dense connective tissue; supports and joins various body structures.

Consumers Organisms that eat plants and algae (producers) and other consumers.

Contact inhibition Cessation of growth that results when two or more cells contact each other. A feature of normal cells but absent in cancer cells.

Continental shelf Gradually sloping ocean bottom next to continents.

Contraceptive Any measure that helps prevent fertilization and pregnancy.

Contractile vacuole Vacuole in amoeboid protists that collects excess water from the cytoplasm and, when full, contracts, voiding the water through a temporary opening in the plasma membrane.

Convergence Inward turning of the eyes to focus on a nearby object.

Coral reef Biologically rich life zones, found in relatively warm and shallow waters in the tropics or nearby regions consisting of calcium carbonate or limestone produced by calcareous red and green algae and by colonies of organisms called stony corals.

Corepressor Molecule that binds to a repressor protein, allowing it to bind to the operator site, which, in turn, shuts down the structural genes by blocking RNA polymerase.

Cornea Clear part of the wall of the eye continuous with the sclera; allows light into the interior of the eye.

Cork cambium Secondary meristem interposed between the cork and the phloem in the bark that gives rise to new cork, part of the bark.

Coronary bypass surgery Surgical technique used to reestablish blood flow to the heart muscle by grafting a vein to shunt blood around a clogged coronary artery.

Corpus luteum (CL) Structure formed from the ovulated follicle in the ovary; produces estrogen and progesterone.

Cortical granules Secretory vesicles lying beneath the plasma membrane of the oocyte that are released when a sperm cell contacts the oocyte membrane. They block additional sperm from fertilizing the ovum.

Cortisol Glucocorticoid hormone that increases blood glucose by stimulating gluconeogenesis. Also stimulates protein breakdown in muscle and bone.

Cotyledon Part of the embryo that is commonly, but not exclusively, used for the storage of food needed in the germination of an embryo.

Coupled reactions A pair of reactions, one exergonic and one endergonic, that are linked so that the energy produced by the exergonic reaction provides the energy to drive the endergonic reaction.

Cowper's gland Smallest of the sex accessory glands; empties into the urethra.

Cranial nerves Nerves arising from the brain and brain stem.

Creatine phosphate High-energy molecule in muscle.

Cristae Folds formed by the inner membrane of the mitochondrion.

Cro-Magnons Earliest known members of *Homo sapiens sapiens.*

Cross bridges Part of the myosin molecule that attaches to and pulls actin molecules inward causing the sarcomere to shorten.

Crossing over Exchange of chromatin by homologous chromosomes during prophase I. Results in considerably more genetic variation in gametes and offspring.

Crustaceans Class of arthropods that comprise most of the zooplankton as well as crabs, shrimp, and lobster.

Crystallin Protein inside the lens that may denature, causing cataracts.

Culture The ideas, customs, skills, and arts of a given people in a given time that can change or evolve over time.

Cushing's disease Disease that results from pharmacologic doses of cortisone usually administered for rheumatoid arthritis or allergies.

Cyanobacteria The most abundant of the photosynthetic bacteria; once called blue-green algae.

Cyclic AMP Nucleotide derived from ATP. Its synthesis is stimulated when protein and polypeptide hormones bind to the plasma membrane of cells. In the cytoplasm, it activates protein kinase, which, in turn, activates other enzymes.

Cyclic photophosphorylation Production of ATP in plants and photosynthetic protists when $NADP^+$ levels are low; instead of being passed to $NADP^+$, electrons are transferred to the electron transport system, producing ATP. Chief means of ATP production in photosynthetic bacteria.

Cyclosporine Drug used to suppress graft rejection.

Cystic fibrosis Autosomal recessive disease that leads to problems in sweat glands, mucus glands, and the pancreas. Pancreas may become blocked, thus reducing the flow of digestive enzymes to the small intestine. Mucus buildup in the lungs makes breathing difficult.

Cytokinesis Cytoplasmic division brought about by the contraction of a microfilamentous network lying beneath the plasma membrane at the midline. Usually begins when the cell is in late anaphase or early telophase.

Cytoplasm Material occupying the cytoplasmic compartment of a cell. Consists of a semifluid substance, the cytosol, containing many dissolved substances, and formed elements, the organelles.

Cytoskeleton A network of protein tubules in the cytoplasmic compartment of a cell. Attaches to many organelles and enzyme molecules and thus helps organize cellular activities, increasing efficiency.

Cytotoxic cells Type of T cell (T-lymphocyte) that attacks and kills virus-infected cells, parasites, fungi, and tumor cells.

Daughter cells Cells produced during cell division.

Decibel Unit used to measure the intensity of sound.

Deciduous Referring to trees that shed their leaves during the fall.

Decomposer food chain Series of organisms that feed on organic wastes and the dead remains of other organisms.

Defibrillation Procedure to stop fibrillation (erratic electrical activity) of the heart.

Deletion Loss of a piece of a chromosome.

Demographer Scientist who studies populations.

Dendrite Short, highly branched fiber that carries impulses to the nerve cell body.

Dense connective tissue Type of connective tissue that consists primarily of densely packed fibers, such as those found in ligaments and tendons.

Deoxyribonuclease Pancreatic enzyme that breaks RNA and DNA into shorter chains.

Depth perception Ability to judge the relative position of objects in our visual field.

Dermis Layer of dense irregular connective tissue that binds the epidermis to underlying structures.

Desert Biome characterized by low rainfall and a hot climate. Contains organisms well adapted to these conditions.

Detrivores Organisms that feed on animal waste or the remains of plants and animals.

Diabetes insipidus Condition caused by

lack of ADH. Main symptoms are polydipsia (excessive drinking) and polyuria (excessive urination).

Diabetes mellitus Insulin disorder either resulting from insufficient insulin production or decreased sensitivity of target cells to insulin. Results in elevated blood glucose levels unless treated.

Dialysis Procedure used to treat patients whose kidneys have failed. Blood is removed from the body and pumped through an artificial filter that removes impurities.

Diaphragm (birth control) Birth control devise consisting of a rubber cup that fits over the end of the cervix. Used in conjunction with spermicidal jelly or cream.

Diaphragm (muscle) Dome-shaped muscle that separates the abdominal and thoracic cavities.

Diaphysis Shaft of the long bones. Consists of an outer layer of compact bone and an inner marrow cavity.

Diastolic pressure The pressure at the moment the heart relaxes. The lower of the two blood pressure readings.

Diatom Plantlike protist found in fresh and salt water. It has a unique cell wall composed of silica, a glassy material that is often fashioned into exquisite designs.

Dicots Angiosperms that contain two cotyledons.

Differentiation Structural and functional divergence from the common cell line. Occurs during embryonic development.

Dihybrid cross Procedure where one plant is bred with another to study two traits.

Dinoflagellate Photosynthetic protists with two flagella, the beating of which cause them to spin like tops.

Diplopia Double vision. May occur when the eyes fail to move synchronously.

Distal convoluted tubule Section of the nephron that connects the loop of Henle to the collecting tubule. Site of tubular reabsorption.

Divergent evolution Process in which organisms evolve in different directions due to exposure to different environmental influences.

Diverticulitis Expansion of the large intestine due to obstruction.

DNA polymerase Enzyme that helps align the nucleotides and join the phosphates and sugar molecules in a newly forming DNA strand.

Dominant Adjective used in genetics to refer to an allele that is always expressed in heterozygotes. Designated by a capital letter.

Dorsal root (of a spinal nerve) Inlet for sensory nerve fibers to the spinal cord.

Double helix Describes the helical structure formed by two polynucleotide chains making up the DNA molecule.

Doubling time Time it takes a population to double.

Down syndrome Genetic disorder caused by an additional chromosome 21 that results in distinctive facial characteristics and mental retardation. Also known as Trisomy 21.

Dryopithecus Genus of apelike creatures that is thought to have given rise to the gibbons, gorillas, orangutans, and chimpanzees.

Ductus arteriosus Shunt that lies between the pulmonary artery and the aorta, helping divert blood from the lungs.

Ductus venosus Shunt that connects directly from the umbilical vein to the inferior vena cava.

Duodenum First portion of the small intestine; site where most food digestion and absorption takes places.

Dust cell Cell found in and around the alveoli; phagocytizes particulate matter that has entered the lung.

E. coli Common bacterium that lives in the large intestine of humans and digests leftover glucose and other materials from food. Used in much genetic research.

Echinoderms Phylum of radially symmetric marine animals with a spiny skin, a complete digestive system, and an internal skeleton composed of shell-like plates.

Ecological niche An organism's habitat and all of the relationships that exist between that organism and its environment.

Ecological system (ecosystem) System consisting of organisms and their environment and all of the interactions that exist between these components.

Ecology Study of living organisms and the web of relationships that binds them together in the economy of nature. The study of ecosystems.

Ecosystem balance Dynamic equilibrium in ecosystems. Maintained by the interplay of growth and reduction factors.

Ectoderm One of the three types of cells that emerges in human embryonic development. Gives rise to the skin and associated structures, including the eyes.

Ectoparasites Parasites that live on the surface of the host.

Edema Swelling resulting from the buildup of fluid in the tissues.

Effector General term for any organ or gland that is controlled by the nervous system.

Ejaculation Ejection of semen from the male reproductive tract.

Elastic arteries Arteries that contain numerous elastic fibers interspersed among the smooth muscle cells of the tunica media.

Elastic cartilage Type of cartilage containing many elastic fibers found in regions where support and flexibility are required.

Electron Highly energetic particle carrying a negative charge that orbits the nucleus of an atom.

Electron carrier A molecule that can reversibly gain and lose electrons. Electron carriers generally accept high-energy electrons produced during an exergonic reaction and donate the electrons to acceptor molecules that use the energy to drive endergonic reactions.

Electron cloud A region surrounding the nucleus of an atom where electrons orbit.

Electron transport system Series of protein molecules in the inner membrane of the mitochondrion that pass electrons from the citric acid cycle from one to another, eventually donating them to oxygen. During their journey along this chain of proteins, the electrons lose energy, which is used to make ATP. *See also* Chemiosmosis.

Elements Purest form of matter; substances that cannot be separated into different substances by chemical means.

Emphysema Progressive, debilitating disease that destroys tiny air sacs in the lung (alveoli). The fastest growing cause of death in the United States.

Endergonic reaction A chemical reaction that requires an input of energy to proceed.

Endocrine glands Glands of internal secretion. They produce hormones that are secreted into the bloodstream.

Endocrine system Numerous, small, hormone-producing glands scattered throughout the body.

Endocytosis Process by which cells engulf solid particles, bacteria, viruses, and even other cells.

Endoderm One of the three types of cells that emerges during embryonic development. Gives rise to the intestinal tract and associated glands.

Endolymph Fluid inside the semicircular canals that deflects the cupula, signaling rotational movement of the head and body.

Endometrium Uterine endothelium or lining.

Endoparasites Parasites that live within their host.

Endoplasmic reticulum Branched network of channels found throughout the cytoplasm of many cells. Formed from flattened sheets of membrane derived from the nuclear membrane.

Endosymbiotic evolution Theory that accounts for the development of the first eukaryotes. Says that free-living bacteria-like organisms were engulfed by other cells and became internal symbionts. Internal symbionts later became the organelles of eukaryotes.

Endothelium Single-celled lining of blood vessels.

End-product inhibition The inhibition of an enzyme by a product of the chemical reaction it catalyzes or by the product of a series of chemical reactions of which the enzyme is a part; may result from binding of the end product to the allosteric site or active site of the enzyme.

Energy The capacity to do work.

Energy carrier A molecule that stores energy in "high-energy" chemical bonds and releases the energy again to drive coupled endergonic reactions. ATP is the most common energy carrier in cells; *NAD* and *FAD* are others.

Energy pyramid Diagram of the amount of energy at various trophic levels in a food chain or ecosystem.

Enhancer Segment of DNA that increases the activity of nearby genes several hundred times.

Envelope Protective membrane of some viruses; lies outside the capsid.

Enzyme A protein catalyst that speeds up the rate of specific biological reactions.

Epicotyl Part of the embryo that gives rise to the shoot system.

Epidermis Outermost layer of the skin that protects underlying tissues from drying out and from bacteria and viruses.

Epididymal duct Duct within the epididymis; site where sperm are stored until ejaculation.

Epididymis Storage site of sperm. Located on the testis, it consists of a long, tortuous duct.

Epiglottis Flap of tissue that closes off the trachea during swallowing.

Epilimnion The warm-water layer of a lake located along the surface.

Epiphyseal plate Band of cartilage cells between the shaft of the bone and the epiphysis. Allows for bone growth.

Epiphysis Expanded end of the long bones.

Epiphytes Plants that obtain their moisture and food from the air and rain rather than from the plants on which they grow.

Episiotomy Surgical incision that runs from the vaginal opening toward the rectum. Enlarges the vaginal opening, easing childbirth.

Epithelium One of the primary tissues. Forms linings and external coatings of organs.

Erectile tissue Spongy tissue of the penis that fills with blood during sexual excitement, making the penis turgid.

Erythropoietin Hormone produced by the kidney when oxygen levels decline. Stimulates red blood cell production in the bone marrow.

Esophagus Muscular tube that transports food to the stomach.

Essential amino acid One of nine amino acids that must be provided in the human diet.

Estuarine zone Estuary and coastal wetland. One of the richest coastal life zones.

Ethology Study of animal behavior. Encompasses both the comparative psychology approach of experimental studies of learned behavior and the classical approach of careful observation of instinctive behavior.

Euchromatin Metabolically active chromatin.

Euphotic zone The open waters of the ocean, extending to the depth at which sunlight no longer penetrates.

Euglenoids Protists with characteristics of both plants and animals. They get their name from the best-known of their kind, *Euglena*.

Evolution Process that leads to structural and functional changes in species, making them better able to survive in their environment; also leads to the formation of new species. Results from natural genetic variation and environmental conditions that select for organisms best suited to their environment.

Exhalation Expulsion of air from the lungs.

Exergonic reaction A chemical reaction that releases energy.

Exocrine gland Gland of external secretion; empties its contents into ducts.

Exocytosis Process by which cells release materials stored in secretory vesicles. The reverse of endocytosis.

Exon Expressed segment of DNA.

Experiment Test performed to prove or disprove a hypothesis.

Exponential growth Type of growth that occurs when a value grows by a fixed percentage and the increase is applied to the base amount.

Extension Movement of a body part (limbs, fingers, and toes) that opens a joint.

External auditory canal Channel that directs sound waves to the eardrum.

External genitalia External portion of the female reproductive system consisting of the clitoris, labia minor, and labia majora.

External sphincter of the bladder Voluntary muscular valve that controls urine release under conscious control. Formed by a flat band of muscle that forms the floor of the pelvic cavity.

Extrinsic eye muscles Six muscles located outside the eye that are responsible for eye movement.

Facilitated diffusion Process in which carrier proteins shuttle molecules across plasma membranes. The molecules move in response to concentration gradients.

Feces Semisolid material containing undigested food, bacteria, ions, and water; produced in the large intestine.

Feedback mechanism A mechanism in which the product of one process regulates the rate of the process, either turning it on or shutting it off.

Fenestrae Minute openings in capillary walls that help permit movement of molecules to and from the capillary.

Fermentation Process occurring in eukaryotic cells in the absence of oxygen, during which pyruvic acid is converted to lactic acid. Also occurs in prokaryotes.

Fertilization Union of sperm and ovum.

Fiber Any of the indigestible polysaccharides in fruits, vegetables, and grains.

Fibrillation Cardiac muscle spasms occurring during heart attacks due to a loss of synchronized electrical signals.

Fibrin Fibrous protein produced from fibrinogen, a soluble plasma protein. Helps form blood clots.

Fibrinogen Protein in plasma that forms fibrin.

Fibroblast Connective tissue cell, found in loose and dense connective tissues that produces collagen, elastic fibers, and a gelatinous extracellular material; responsible for repairing damage created by cuts or tears to connective tissue.

Fibrocartilage Type of cartilage whose extracellular matrix consists of numerous bundles of collagen fibers. Principally found in the intervertebral disks.

Fimbriae Fingerlike projections of the end of the oviduct that sweep the oocyte into the oviduct.

First law of thermodynamics Law stating that energy can be converted from one form to another but is never created or destroyed.

First polar body Cast-off nuclear material produced during the first meiotic division during oogenesis.

Fitness Measure of reproductive success of an organism and, therefore, the genetic influence an individual has on future generations.

Fixed action patterns (FAPs) Stereotyped behavior or response to specific stimuli.

Flagellum Long, whiplike extension of the plasma membrane of certain protozoans and sperm cells in humans. Used for motility.

Flatworms Phylum characterized by bilateral symmetry, cephalization, and three distinct tissue layers. Most flatworms are hermaphrodites (contain both male and female sex organs), but reproduction occurs through cross-fertilization.

Flavine adenine dinucleotide (FAD) A molecule that carries high-energy electrons from one chemical reaction to another; found in the citric acid cycle.

Flexion Movement of a limb, finger, or toe that involves closing a joint.

Follicle (ovary) Structure found in the ovary. Each follicle contains an oocyte and one or more layers of follicle cells that are derived from the loose connective tissue of the ovary surrounding the follicle.

Follicle (thyroid) Structure found in the thyroid gland. Consists of an outer layer of cuboidal cells surrounding thyroglobulin.

Follicle-stimulating hormone (FSH) Gonadotropic hormone that promotes gamete formation in both men and women.

Food chain Series of organisms in an ecosystem in which each organism feeds on the organism preceding it.

Food vacuole Membrane-bound vacuole in a cell containing material engulfed by the cell.

Food web All of the connected food chains in an ecosystem.

Foramen ovale Hole in the interatrial septum of the embryonic heart that diverts blood from the right atrium to the left atrium, reducing the flow of blood to the pulmonary arteries and lungs.

Foreskin Sheath of skin that covers the glans penis.

Fossil Remains or imprints of organisms that lived on Earth many years ago, usually embedded in rocks or sediment.

Fovea centralis Tiny spot in the center of the macula of the eye that contains only cones. Objects are focused onto the fovea for sharp vision.

Fruit Structure derived from the ovary of a flower. Some types of fruit contain seeds and some do not. Some fruits, called grains, fuse with their seeds and remain inseparable.

Fusion Joining of two atoms, which releases large amounts of energy.

Gallbladder Sac on the underside of the liver that stores and concentrates bile.

Gametoyphyte Haploid generation of a plant exhibiting alternation of generations. Produces gametes.

Gastrin Stomach hormone that stimulates HCl production and release by the gastric glands.

Gastroesophageal sphincter Ring of muscle located in the lower esophagus that opens when food arrives, allowing food to pass into the stomach, and then closes to keep food and stomach acid from percolating upward.

Gene Segment of the DNA that controls cell structure and function.

Gene flow Introduction of new genes into a population when new individuals join the population.

Gene pool All the genes of all of the members of a population or species.

Genera Plural of genus.

Genome Genes of an organism.

Genotype Genetic makeup of an organism.

Genus Subgroup of a family of organisms.

Geographic isolation Physical separation of a population by some barrier. Sometimes results in reproductive isolation and the formation of new species.

Geotropism Growth response of a plant or seed to gravity—that is, plant stems bend upward and plant roots bend downward.

Germinal epithelium Germ cells in the wall of the seminiferous tubule that give rise to sperm.

Gestation The period of pregnancy.

Glans penis Slightly enlarged tip of the penis.

Glaucoma Disease of the eye caused by pressure resulting from a buildup of aqueous humor in the anterior chamber.

Glomerular filtration Movement of materials out of the glomeruli into Bowman's capsule in the kidney.

Glomerulus Tuft of capillaries that make up part of the nephron; site of glomerular filtration.

Glucagon Hormone released by the pancreas that stimulates the breakdown of glycogen in liver and muscle and the release of glucose molecules, thus increasing blood levels of glucose.

Glucocorticoids Group of steroid hormones produced by the adrenal cortex that stimulate gluconeogenesis.

Gluconeogenesis Synthesis of glucose from fatty acids and amino acids. Takes place in the liver where amino acids and fatty acids are stored.

Glycogenolysis Breakdown of glycogen, releasing glucose.

Glycolysis Metabolic pathway in the cytoplasm of the cell, during which glucose is split in half, forming two molecules of pyruvic acid. The energy released during the reaction is used to generate two molecules of ATP.

Glycoproteins Proteins that have carbohydrate attached to them.

Goiter Condition in which the thyroid gland enlarges due to lack of dietary iodide.

Golgi complex Organelle consisting of a series of flattened membranes that form channels. It sorts and chemically modifies molecules and repackages its proteins into secretory vesicles.

Golgi tendon organs Special receptors found in tendons that respond to stretch. Also known as neurotendinous organs.

Gonadotropin General term for FSH and LH, which are produced by the anterior pituitary and target male and female gonads.

Gonadotropin-releasing hormone (GnRH) Hormone produced by the hypothalamus that controls the release of FSH (ICSH in males) and LH.

Gonorrhea Sexually transmitted disease caused by a bacterium.

Granum A stack of thylakoid disks found in chloroplasts.

Gray matter Gray, outermost region of the cerebral cortex.

Grazer Herbivorous organism.

Grazer food chain Food chain beginning with plants and grazers (herbivores).

Greenhouse gas Gas, such as carbon dioxide and chlorofluorocarbons, that traps heat escaping from the Earth and radiates it back to the surface.

Growth factor Any biotic or abiotic factor that causes a population to grow.

Growth hormone A protein hormone produced by the anterior pituitary that stimulates cellular growth in the body, causing cellular hypertrophy and hyperplasia. Its major targets are bone and muscle.

Growth rate (of a population) Determined by subtracting the death rate from the birth rate.

Gymnosperms Vascular land trees such as conifers that produce seeds in cones or conelike structures. Also includes cycads and other plants.

Habitat Place in which an organism lives.

Habituation Condition where sensory receptors stop generating impulses, even though a stimulus is still present.

Hammer (malleus) One of three bones of the middle ear. Abuts the tympanic membrane and helps transmit sound from the eardrum to the inner ear.

Helper cell Type of T-lymphocyte that stimulates the proliferation of T and B cells when antigen is present.

Heme group Subunit of the hemoglobin molecule. Consists of a porphyrin ring and a central iron ion to which oxygen binds.

Hemoglobin Protein molecules inside red blood cells; binds to oxygen.

Hemophilia Disease caused by a gene defect occurring on the Y chromosome. Results in absence of certain blood-clotting factors.

Herbicide Any chemical applied to crops to control weeds.

Herbivore Any organism that feeds directly on plants. Also known as a grazer.

Herpes One of the most common sexually transmitted diseases; caused by a virus.

Heterochromatin Inactive chromatin that is slightly coiled or compacted in the interphase nucleus.

Heterotrophic fermenter Evolutionarily probably one of the first cells. Absorbed glucose from the environment and broke it down by anaerobic glycolysis.

Heterotrophs Organisms such as animals that, unlike plants, are unable to synthesize their own food. Consume plants and other organisms.

Heterozygous Adjective describing a genetic condition in which an individual contains one dominant and one recessive gene in a gene pair.

High-density lipoproteins (HDLs) Complexes of lipid and protein that transport cholesterol to the liver for destruction.

Histamine Potent vasodilator released by certain cells in the body during allergic reactions.

Histone Globular protein thought to play a role in regulating the genes.

Homeostasis A condition of stability or equilibrium within any biological or social system. Achieved through a variety of automatic mechanisms that compensate for internal and external changes.

Hominid First humanlike creatures.

Hominoids Subgroup of anthropoids.

Homo sapiens neanderthalensis The Neanderthals. Subspecies of *Homo sapiens.*

Homo sapiens sapiens Species of modern humans that emerged about 400,000 years ago.

Homologous structures Structures thought to have arisen from a common origin.

Homozygous Adjective describing a genetic condition marked by the presence of two identical alleles for a given gene.

Hormone Chemical substance produced in one part of the body that travels to another where it elicits a response.

Human chorionic gonadotropin (HCG) Hormone produced by the embryo that stimulates the corpus luteum to produce estrogen.

Humoral immunity Immune reaction that protects the body primarily against viruses and bacteria in the body fluids via antibodies produced by plasma cells.

Hyperglycemia High blood glucose levels.

Hyperopia (farsightedness) Condition that occurs when the eyeball is too short or the lens is too weak, resulting in poor focus on nearby objects.

Hypertension High blood pressure.

Hypertonic Adjective describing a solution with a higher solute concentration than the cell's cytoplasm, causing the cell to shrivel.

Hyphae Microscopic filaments or strands of which most fungi are composed.

Hypocotyl Part of the embryo that gives rise to root system.

Hypolimnion The coldest water of a lake, which lies below the thermocline.

Hypothalamus Structure in the brain located beneath the thalamus. It consists of many aggregations of nerve cells and controls a variety of autonomic functions aimed at maintaining homeostasis.

Hypothesis Tentative and testable explanation for a phenomenon or observation.

Hypotonic Adjective describing a solution with a solute concentration lower than the cell's cytoplasm, resulting in a swelling of the cell.

I gene Gene that controls blood type through the synthesis of glycoproteins on the plasma membrane of the red blood cell.

Immune system Diffuse system consisting of trillions of cells that circulate in the blood and lymph and take up residence in the lymphoid organs, such as the spleen, thymus, lymph nodes, and tonsils, as well as other body tissues. Helps protect the body against foreign cells, such as bacteria and viruses, and protects against cancer cells.

Immunity Term referring to the resistance of the body to infectious disease.

Immunocompetence Process in which lymphocytes mature and become capable of responding to specific antigens.

Immunoglobulins Antibodies.

Implantation Process in which the blastocyst embeds in the uterine lining.

Impotency Inability of a male to achieve an erection.

Imprinting Attachment of young birds to their mothers or artificial substitutes during a critical period.

Innate behaviors Genetically programmed behaviors—literally in-born behavior—ready to be put to use when needed.

Instinctive behaviors Genetically programmed responses to stimuli that are fixed, automatic, and independent of previous experience.

In vitro Term referring to any procedure carried out in a test tube or petri dish, such as *in vitro* fertilization.

Incomplete dominance Partial dominance. Occurs when an allele exerts only partial dominance over another allele, resulting in an intermediate trait.

Incontinence Inability to control urination.

Induced abortion Deliberate expulsion of a fetus or embryo.

Inducer Chemical substance that activates inducible genes.

Inducible operon Set of genes that remains inactive until needed. Activated by inducers.

Infectious mononucleosis White blood cell disorder caused by a virus. Characterized by a rapid increase of monocytes and lymphocytes.

Inferior vena cava Large vein that empties deoxygenated blood from the body below the heart into the right atrium of the heart.

Infertility Inability to conceive; can be due to problems in either the male or the female.

Inflammatory response Response to tissue damage including an increase in blood flow, the release of chemical attractants, which draw monocytes to the scene, and an increase in the flow of plasma into a wound.

Inhalation Process of air being drawn into the lungs.

Inhibin Substance produced by the seminiferous tubules that inhibits the production of FSH by the anterior pituitary.

Inhibiting hormone Hormone from the hypothalamus that inhibits the release of hormones from the anterior pituitary.

Initiator codon Codon found on a messenger RNA strand that marks where protein synthesis begins.

Inner cell mass Cells of the blastocyst that become the embryo and amnion.

Insects Most successful class of arthropods with nearly a million species named to date. A complex life cycle, a symbiotic relationship with angiosperms, an ability to fly, and seasonal migration have contributed to their phenomenal success.

Insulin Hormone that stimulates the uptake of glucose by body cells, especially muscle and liver cells. Stimulates the synthesis of glycogen in liver and muscle cells.

Insulin-dependent diabetes Type of diabetes that can only be treated with injections of insulin. May be caused by an autoimmune reaction. Also known as early-onset diabetes.

Insulin-independent diabetes Type of diabetes that often occurs in obese people. In most patients, it can be controlled by diet. Also known as late-onset diabetes.

Integral protein Large protein molecules in the lipid bilayer of the plasma membrane.

Integration Process of making sense of various nervous inputs so that a meaningful response can be achieved.

Intercostal muscles Short, powerful muscles that lie between the ribs. Involved in inspiration and active exhalation.

Interferon Protein released from cells infected by viruses that stops the replication of viruses in other cells.

Interleukin 2 Chemical released by helper cells that activates T and B cells, stimulating cell division.

Internal sphincter (of the bladder) Involuntary muscular valve that relaxes reflexively, releasing urine. Formed by a smooth muscle in the neck of the bladder at the junction of the bladder and the urethra.

Internodes Segments of the axon between nodes of Ranvier.

Interphase Period of cellular activity occurring between cell divisions. Synthesis and growth occur in preparation for cell division.

Interstitial cells Cells located in the loose connective tissue between the seminiferous tubules of the testes. Produce testosterone.

Interstitial cell stimulating hormone (ICSH) Luteinizing hormone in males. Regulates testosterone secretion.

Interstitial fluid Fluid surrounding cells in body tissues. Provides a path through which nutrients, gases, and wastes can travel between the capillary and the cells.

Intervertebral disks Shock-absorbing material between the bones of the spine.

Intrauterine device (IUD) Birth control device that consists of a small plastic or metal object with a string attached that is inserted into the uterus through the cervix. Prevents implantation.

Intron Segment of DNA that is not expressed. Lies between exons (expressed segments).

Invertebrates Animals without a spinal column.

Ion Atom that has gained or lost one or more electrons. May be either positively or negatively charged.

Ionic bond Weak bond that forms between oppositely charged ions.

Iris Colored segment of the middle layer of the eye visible through the cornea.

Irritability Ability to perceive and respond to stimuli.

Islets of Langerhans Group of endocrine cells found in the pancreas that produce insulin and glucagon.

Isotonic Having the same solute concentration as a cell or body fluid.

Isotope Alternative form of an atom; differs from other atoms in the number of neutrons in the nucleus.

Jawless fishes Earliest known vertebrates. Slow-moving, bottom-feeders. Not dominant life forms in the ecosystems they occupy.

Joint capsule Connective tissue that connects to the opposing bones of a joint and forms the synovial cavity. The inner layer of the joint capsule produces synovial fluid.

Kidney Organ that rids the body of wastes and plays a key role in regulating the chemical constancy of blood.

Kin selection Altruistic behavior that increases the likelihood that an individual's genes will be transmitted to future generations.

Klinefelter syndrome Genetic disorder that results from an XXY genotype.

Labia majora Outer folds of skin of the external genitalia in women.

Labia minora Inner folds of the external genitalia in women.

Labor The process or period of childbirth.

Lactation Milk production in the breasts.

Laparoscope Instrument used to examine internal organs through small openings made in the skin and underlying muscle.

Larynx Rigid but hollow cartilaginous structure that houses the vocal cords and participates in swallowing.

Learning Process in which stored information can lead to changes in innate behavior.

Lens Transparent structure that lies behind the iris and in front of the vitreous humor. Focuses light on the retina.

Leukemia Cancer of white blood cells.

Leukocytosis An increase in the concentration of white blood cells, which often occurs during a bacterial or viral infection.

Lichen Association resulting from the symbiotic relationship between a fungus and a unicellular alga or cyanobacterium.

Life expectancy Average length of time a person will live.

Ligament Connective tissue structure that runs from bone to bone, located alongside and sometimes inside the joint. Offers support for joints.

Light-independent reactions The second state of photosynthesis, in which the energy obtained by the light-dependent reactions is used to fix carbon dioxide into carbohydrates; occurs in the stroma of chloroplasts.

Limbic system Array of structures in the brain that work in concert with centers of the hypothalamus. Site of instincts and emotions.

Limiting factor One factor that is most important in regulating growth in an ecosystem.

Limnetic zone The region of a lake commonly called the "open water"; extends downward to the point at which light no longer penetrates and is the main photosynthetic body of a lake.

Lipase Enzyme that removes some of the fatty acids from the glycerol molecule, forming a monoglyceride. Produced by the salivary glands and the pancreas.

Lipid Commonly known as fats. Water-insoluble organic molecules that provide energy to body cells, help insulate the body from heat loss, and serve as precursors in the synthesis of certain hormones. A principal component of the plasma membrane.

Liposuction Technique used to remove subcutaneous fat.

Littoral zone The shallow waters at the margin of a lake where rooted vegetation can grow.

Liver Organ located in the abdominal cavity that performs many functions essential to homeostasis. It stores glucose and fats, synthesizes some key blood proteins, stores iron and certain vitamins, detoxifies certain chemicals, and plays an important role in digestion by producing bile.

Long bones Bones of the skeleton that form parts of the extremities.

Loose connective tissue Type of connective tissue that serves primarily as a packing material. Contains many cells among a loose network of collagen and elastic fibers, especially cells that help protect the body from foreign organisms.

Low-density lipoproteins (LDLs) Complexes of protein and lipid that transport cholesterol, depositing it in body tissues.

Lungs Two large saclike organs in the thoracic cavity where the blood and air exchange carbon dioxide and oxygen.

Luteinizing hormone (LH) Hormone that stimulates gonadal hormone production. In men, LH stimulates the production of testosterone, the male sex steroid. In women, LH stimulates estrogen secretion.

Lymph Fluid contained in the lymphatic vessels. Similar to tissue fluid, but also contains white blood cells and may contain large amounts of fat.

Lymph node Small nodular organ interspersed along the course of the lymphatic vessels. Serves as a filter for lymph.

Lymphatic system Network of vessels that drains extracellular fluid from body tissues and returns it to the circulatory system.

Lymphocyte Type of white blood cell. *See also* B-lymphocyte and T-lymphocyte.

Lymphoid organs Organs, such as the spleen and thymus, that belong to the lymphatic system.

Lymphokine Chemical released by suppressor T cells that inhibits the division of B and T cells.

Lysosome Membrane-bound organelle that contains enzymes. Responsible for the breakdown of material that enters the cell by endocytosis. Also destroys aged or malfunctioning cellular organelles.

Lysozyme Enzyme produced in saliva that dissolves the cell wall of bacteria, killing them.

Macronutrients Nutrients required in relatively large amounts by organisms. Includes water, proteins, carbohydrates, and lipids.

Macrophage Phagocytic cell derived from monocytes that resides in loose connective tissues and helps guard tissues against bacterial and viral invasion.

Macula lutea Region of the retina located lateral to the optic disc where cones are most abundant.

Maculae Receptor organs in the saccule and utricle that play a role in position sense.

Malignant tumor Structure resulting from uncontrollable cellular growth. Cells often spread to other parts of the body.

Mammals Class of vertebrates with a single jaw bone, specialized teeth, hair, mammary glands, a four-chambered heart, and an advanced brain. Reproduction is the most complex of animals.

Marfan's syndrome Autosomal dominant genetic disorder that affects the skeletal system, the eye, and the cardiovascular system.

Marrow cavity Cavity inside a bone containing either red or yellow marrow.

Mast cell Cell found in many tissues, especially in the connective tissue surrounding blood vessels. Contains large granules containing histamine.

Matrix Extracellular material found in cartilage. Also the material in the inner compartment of the mitochondrion.

Matter Anything that has mass and occupies space.

Medulla Term referring to the central portion of some organs; for example, the adrenal medulla.

Megakaryocyte Large cell found in bone marrow that produces platelets.

Meiosis Type of cell division that occurs in the gonads during the formation of gametes. Requires two cellular divisions (meiosis I and meiosis II). In humans, it reduces the chromosome number from 46 to 23.

Meiosis I First meiotic division.

Meiosis II Second meiotic division.

Meissner's corpuscle Encapsulated sensory receptor thought to respond to light touch.

Membranous epithelium Refers to any sheet of epithelium that forms a continuous lining on organs.

Memory cells T or B cells produced after antigen exposure. They form a reserve force that responds rapidly to antigen during subsequent exposure.

Menopause End of the reproductive function (ovulation) in women. Usually occurs between the ages of 45 and 55.

Menstrual cycle Recurring series of events in the reproductive functions of women. Characterized by dramatic changes in ovarian and pituitary hormone levels and changes in the uterine lining that prepare the uterus for implantation.

Menstruation Process in which the endometrium is sloughed off, resulting in bleeding. Occurs approximately once every month.

Meristematic tissues Tissues responsible for the production of all other tissues. Located either at the tips of the stem (primary meristems) or along the entire length of the stem and root (secondary meristems). Meristems remain embryonic and give rise to new cells throughout the life of the plant.

Merkel disk Light touch receptor. Consists of dendrites that end on cells in the epidermis.

Mesoderm One of the three types of cells that emerge in human embryonic development. Lies in the middle of the forming embryo and forms muscle, bone, and cartilage.

Messenger RNA (mRNA) Type of RNA that carries genetic information needed to synthesize proteins to the cytoplasm of a cell.

Metabolic pathway Series of linked chemical reactions in which the product of one reaction becomes the reactant in another reaction.

Metabolic water Water produced during cellular respiration by the addition of protons (hydrogen ions) and electrons to oxygen.

Metabolism The chemical reactions of the body, including all catabolic and anabolic reactions.

Metaphase Stage of cellular division in which chromosomes line up in the center of the cell.

Metarterioles Arterioles that serve as circulatory short cuts, connecting arterioles with venules in a capillary bed. Also known as thoroughfare channels.

Metastasis Spread of cancerous cells throughout the body, through the lymph vessels and circulatory system or directly through tissue fluid.

Microfilament Solid fiber consisting of contractile proteins that is found in cells in a dense network under the plasma membrane. Forms part of the cytoskeleton.

Micronutrients Nutrients required in small quantities. They include two broad groups, vitamins and minerals.

Microspheres Small globules consisting of protein that may have been precursors of the first cells. Also known as proteinoids.

Microsurgery Type of surgery performed under dissecting microscopes. Used to reconnect axons, blood vessels, and other small structures.

Microtubules Hollow protein tubules in the cytoplasm of cells that form part of the cytoskeleton. Also form spindles.

Microvilli Tiny projections of the plasma membranes of certain epithelial cells that increase the surface area for absorption.

Middle ear Portion of the ear located within a bony cavity in the temporal bone of the skull. Houses the ossicles.

Migration Movement of an animal from one region to another often over long distances.

Mineralocorticoids Group of steroid hormones produced by the adrenal cortex. Involved in electrolyte or mineral salt balance.

Mitochondrion Membrane-bound organelle where the bulk of cellular energy production occurs in eukaryotic cells. Houses the citric acid cycle and electron transport system.

Mitosis Term referring specifically to the division of a cell's nucleus. Consists of four stages: prophase, metaphase, anaphase, and telophase.

Mitotic spindle Array of microtubules constructed in the cytoplasm during prophase. Microtubules of the mitotic spindle connect to the chromosomes and help draw them apart during mitosis.

Mollusks Phylum that exhibits four common characteristics: a mantle, a radula, a foot, and a shell.

Monerans Kingdom containing prokaryotic organisms, bacteria.

Monocots Angiosperms that contain one cotyledon.

Monocyte White blood cell that phagocytizes bacteria and viruses in body tissues.

Monohybrid cross Procedure in which one plant is bred with another to study the inheritance of a single trait.

Monosomy Genetic condition caused by a missing chromosome.

Morning sickness Nausea that often occurs in the first two to three months of pregnancy.

Morula Solid ball of cells produced from the zygote by numerous cellular divisions.

Motor unit Muscle fibers supplied by a single axon and its branches.

Mucus Thick, slimy material produced by the lining of the respiratory tract and parts of the digestive tract. Moistens and protects them.

Mullerian mimicry Phenomenon of several species, each with toxic characteristics, evolving similar color patterns.

Multipolar neuron Motor neuron found in the central nervous system. Contains

a prominent, multiangular cell body and several dendrites.

Muscle fiber Long, unbranched, multinucleated cell found in skeletal muscle.

Muscle spindles Stretch receptors found in skeletal muscle. Also known as neuromuscular spindles.

Muscle tone Inherent firmness of muscle, resulting from contraction of muscle fibers during periods of inactivity.

Muscular artery Any one of the main branches of the aorta. Tunica media consists primarily of smooth muscle cells.

Mutation Technically, a change in the DNA caused by chemical and physical agents. Also refers to a wide range of chromosomal defects.

Mutualism Symbiotic relationship in which both species benefit.

Mycelia Aggregations of hyphae.

Mycorrhiza Fungus root; a symbiotic relationship between certain soil-dwelling fungi and root cells of many vascular plants; acts as root hairs.

Myelin sheath Layer of fatty material coating the axons of many neurons in the central and peripheral nervous systems. Formed by Schwann cells.

Myofibril Bundle of contractile myofilaments in skeletal muscle cells.

Myoglobin Cytoplasmic protein in muscle cells that binds to oxygen.

Myometrium Uterine smooth muscle.

Myopia (nearsightedness) Visual condition that results when the eyeball is slightly elongated or the lens is too strong. In the uncorrected eye, light rays from distant images come into focus in front of the retina.

Myosin Protein filament found in many cells in the microfilamentous network. Also found in muscle cells.

Myosin ATPase Enzyme found in the myosin cross bridges that splits ATP during muscle contraction.

Naked nerve ending Unmodified dendritic ending of the sensory neurons. Responsible for at least three sensations: pain, temperature, and light touch.

Natural childbirth Childbirth without the use of drugs.

Natural selection Evolutionary process in which environmental abiotic and biotic factors "weed" out the less fit—those organisms not as well adapted to the environment as their counterparts.

Nephron Filtering unit in the kidney. Consists of a glomerulus and renal tubule.

Neritic zone Shallow, biologically rich coastal life zone overlying the continental shelf.

Nerve Bundle of nerve fibers. May consist of axons, dendrites, or both. Carries information to and from the central nervous system.

Nerve deafness Loss of hearing resulting from nerve or brain damage.

Nervous tissue One of the primary tissues. Found in the nervous system and consists of two types of cells: conducting cells (neurons) and supportive cells.

Neural groove Ectodermal groove that forms early in embryonic development and runs the length of the embryo, later forming the neural tube.

Neural tube Tube of ectoderm that arises from the neural groove and will become the spinal cord.

Neuroendocrine reflex A reflex involving the endocrine and nervous systems.

Neuron Highly specialized cell that generates and transmits bioelectric impulses from one part of the body to another.

Neurosecretory neurons Specialized nerve cells of the hypothalamus and posterior pituitary that produce and secrete hormones.

Neurotransmitter Chemical substance released from the terminal ends (terminal boutons) of axons when a bioelectric impulse arrives. May stimulate or inhibit the next neuron.

Neutron Uncharged particle in the nucleus of the atom.

Neutrophil Type of white blood cell that phagocytizes bacteria and cellular debris.

Nicotine adenine dinucleotide (NAD) A molecule that carries high-energy electrons from one chemical reaction to another.

Nicotinamide adenine dinucleotide (NAD) Electron acceptor molecule that shuttles energetic electrons from glycolysis, the transition reaction, and the citric acid cycle to the electron transport system.

Nitrogen fixation Process in which bacteria and a few other organisms convert atmospheric nitrogen to nitrate or ammonia, forms usable by plants.

Node of Ranvier Small gap in the myelin sheath of an axon; located between segments formed by Schwann cells. Responsible for saltatory conduction.

Noncyclic photophosphorylation Production of ATP using photosystem I and II and the electron transport system.

Nondisjunction Failure of a chromosome pair or chromatids of a double-stranded chromosome to separate during mitosis or meiosis.

Nonpoint source (of pollution) A source that does not release pollutants via an easily identifiable route. *See also* Point source.

Nonspecific urethritis (NSU) One of the most common sexually transmitted diseases. Caused by several different bacteria.

Nonvascular plants Land plants that not only lack vascular tissue, but also lack roots, stems, and leaves. All nonvascular plants are bryophytes.

Noradrenaline (norepinephrine) Hormone produced by adrenal medulla and secreted under stress. Contributes to the fight-or-flight response.

Nuclear envelope Double membrane delimiting the nucleus.

Nuclear pores Minute openings in the nuclear envelope that allow materials to pass to and from the nucleus.

Nucleoli Temporary structures in the nuclei of cells during interphase. Regions of the DNA that are active in the production of RNA.

Nucleus (atom) Dense, center region of the atom that contains neutrons and protons.

Nucleus (cell) Cellular organelle that contains the genetic information that controls the structure and function of the cell.

Nutrient cycle Circular flow of nutrients from the environment through the various food chains back into the environment.

Olfactory membrane Receptor for smell; found in the roof of the nasal cavity.

Olfactory nerve Nerve that transmits impulses from the olfactory membrane to the brain.

Omnivores Organisms that feed on both plants and animals.

Ontogeny Development of an organism starting with fertilization.

Oogenesis Production of ova.

Oogonium Germ cell in ovary that contains 46 double-stranded chromosomes. Forms primary oocytes.

Operant conditioning Type of associative learning in which the repeated use of a reinforcing stimulus (punishment or reward) elicits a desired behavior.

Operator site Region of the DNA molecule adjacent to the structural genes that acts as a switch to turn the operon on or off.

Operon Functional unit of the DNA of bacteria. Consists of structural and regulatory genes.

Optic disk Site in the retina where the optic nerve exits. Also known as the blind spot.

Optic nerve Nerve that carries impulses from the retina to the brain.

Order Taxonomic term that refers to a subgroup of a class.

Organ Discrete structure that carries out specialized functions.

Organ of Corti Receptor for sound; located in the inner ear within the cochlea.

Organ system Group of organs that participate in a common function.

Organogenesis Organ formation during embryonic development.

Osmosis Diffusion of water across a selectively permeable membrane.

Osmotic pressure Force that drives water across a selectively permeable membrane. Created by differences in solute concentrations.

Ossicles Three small bones inside the middle ear that transmit vibrations created by sound waves to the organ of Corti.

Osteoarthritis Degenerative joint disease caused by wear and tear that impairs movement of joints.

Osteoblast Bone-forming cell; secretes collagen.

Osteoclast Cell that digests the extracellular material of bone. Stimulated by the parathyroid hormone.

Osteocyte Bone cell derived from osteoblasts that has been surrounded by calcified extracellular material.

Osteoporosis Degenerative disease result-

**ing in the deterioration of bone. Due to in-activity in men and women and loss of ovarian hormones in postmenopausal women.

Outer ear External portion of the ear.

Oval window Membrane-covered opening in the cochlea where vibrations are transmitted from the stirrup to the fluid within the cochlea.

Ovary Enlarged, rounded basal portion of the flower pistil containing one or more ovules, each with a sporangium. Produces the female gametophyte containing an egg and several other cells.

Overpopulation Condition in which a species has exceeded the carrying capacity of the environment.

Ovulation Release of the oocyte from ovary. Stimulated by hormones from the anterior pituitary.

Ovum Germ cell containing 23 single-stranded chromosomes. Produced during the second meiotic division.

Oxaloacetate Four-carbon compound of the citric acid cycle. It is involved in the very first reaction of the cycle and is regenerated during the cycle.

Oxidation The loss of hydrogens or electrons from a substance.

Oxytocin Hormone from the posterior pituitary hormone. Stimulates contraction of the smooth muscle of the uterus and smooth-muscle-like cells surrounding the glandular units of the breast.

Ozone O$_3$ Molecule produced in the stratosphere (upper layer of the atmosphere) from molecular oxygen. Helps screen out incoming ultraviolet light. *See also* Ozone layer.

Ozone layer Region of the atmosphere located approximately 12 to 30 miles above the Earth's surface where ozone molecules are produced. Helps protect the Earth from ultraviolet light.

Pacinian corpuscle Large encapsulated nerve ending that is located in the deeper layers of the skin and near body organs. Responds to pressure.

Pancreas Organ found in the abdominal cavity under the stomach, nestled in a loop formed by the first portion of the small intestine. Produces enzymes needed to digest foodstuffs in the small intestine and hormones that regulate blood glucose levels.

Pap smear Procedure in which cells are

retrieved from the cervical canal to be examined for the presence of cancer.

Papillae Small protrusions on the upper surface of the tongue. Some papillae contain taste buds.

Paracrines Chemicals released by cells that elicit a response in nearby regions.

Parasitism Process in which an individual feeds (usually without killing) on another larger individual—the host.

Parasympathetic division (of the autonomic nervous system) Portion of the autonomic nervous system responsible for a variety of involuntary functions.

Parathyroid glands Endocrine glands located on the posterior surface of the thyroid gland in the neck. Produce parathyroid hormone.

Parathyroid hormone (PTH) Hormone that helps regulate blood calcium levels. Stimulates osteoclasts to digest bone, thus raising blood calcium levels. Also known as parathormone.

Parturition Childbirth.

Passive immunity Temporary protection from antigen (bacteria and others) produced by the injection of immuno-globulins.

Penis Male organ of copulation.

Pepsin Enzyme released by the gastric glands of the stomach. Breaks down proteins into large peptide fragments.

Pepsinogen Inactive form of pepsin.

Perforin Chemical released by cytotoxic cells that destroys bacteria. Binds to plasma membrane of target cells, forming pores that make the target cells leak and die.

Perichondrium Connective tissue layer surrounding most types of cartilage. Contains blood vessels that supply nutrients to cartilage cells.

Periodic table of elements Table that lists elements by ascending atomic number. Also lists other vital statistics of each element.

Peripheral nervous system Portion of the nervous system consisting of the cranial and spinal nerves and receptors.

Peristalsis Involuntary contractions of the smooth muscles in the wall of the esophagus, stomach, and intestines, which propel food along the digestive tract.

Peritubular capillaries Capillaries that surround nephrons. They pick up water,

nutrients, and ions from the renal tubule, thus helping maintain the osmotic concentration of the blood.

Permafrost Permanently frozen subsoil in the Arctic tundra.

Permanent threshold shift Permanent hearing loss caused by repeated exposure to noise. Results from damage to hair cells of the organ of Corti.

Pharynx Chamber that connects the oral cavity with the esophagus.

Phenotype Outward appearance of an organism.

Phloem Type of vascular tissue in gymnosperms and angiosperms that transports food through stem, petioles, and leaves.

Phonation Production of sound.

Photoelectron transport system Protein carriers that transport electrons from photosystem II to photosystem I and produce ATP from the energy released by the electrons.

Photophosphorylation Production of ATP by the electron transport system.

Phosphoglyceraldehyde A three-carbon monosaccharide produced during glycolysis by the splitting of the glucose molecules; also produced in the light-independent reactions of photosynthesis.

Phosphofructokinase Enzyme in the glycolytic pathway that catalyzes the conversion of glucose-6-phosphate to fructose-6-phosphate; regulated by ATP and citric acid.

Photoreceptors Modified nerve cells that respond to light. Located in the retina of humans and other animals.

Photosynthesis The series of chemical reactions in which the energy of light is used to synthesize high-energy organic molecules, usually carbohydrates, from low-energy inorganic molecules, usually carbon dioxide and water.

Photosystem In thylakoid membranes, a light-harvesting complex and its associated electron transport system.

Phototropism Growth response of a plant to light coming from one direction.

Phylogeny Evolutionary development of a species.

Phylum (phyla, plural) Major grouping of organisms. Animal kingdom contains about two dozen primary phyla, nearly all of which are represented today.

Phytoplankton Microscopic, free-floating organisms, principally algae, which capture sunlight energy, using it to produce carbohydrates from carbon dioxide dissolved in the water.

Pineal gland Small gland located in the brain that secretes a hormone thought to help control the biological clock.

Pistil Portion of flower that consists of terminal stigma, basal ovary, and style of a flower that connects the stigma and ovary.

Pituitary gland Small pea-sized gland located beneath the brain in the sella turcica. It produces numerous hormones and consists of two main subdivisions: anterior and posterior pituitary.

Placenta Organ produced from maternal and embryonic tissue. Supplies nutrients to the growing embryo and fetus and removes fetal wastes.

Plasma Extracellular fluid of blood. Comprises about 55% of the blood.

Plasma cell Cell produced from B-lymphocytes (B cells); synthesizes and releases antibodies.

Plasma membrane Outer layer of the cell. Consists of lipid and protein and controls the movement of materials into and out of the cell.

Plasmids Small circular strands of DNA found in bacterial cytoplasm separate from the main DNA.

Plasmin Enzyme in the blood that helps dissolve blood clots.

Plasminogen Inactive form of plasmin.

Plasmodium Single-celled parasite responsible for malaria.

Platelet Cell fragment produced from megakaryocytes in the red bone marrow. Plays a key role in blood clotting.

Podocyte Type of cell forming the inner lining of Bowman's capsule of the nephron. Part of the filtration mechanism in the glomerulus.

Point source (of pollution) A discrete, easily identifiable source, usually releasing pollutants directly into waterways via pipes.

Polar body Discarded nuclear material produced during meiosis I and meiosis II of oogenesis.

Pole-to-pole fibers Type of microtubule found in the spindle. Extend from one centriole to·the other.

Pollination Transfer of pollen grains (male gametophytes) to the ovulate cone in gymnosperms and to the ovaries of flowers in angiosperms.

Polygenic inheritance Transmission of traits that are controlled by more than one gene.

Polyploidy Term referring to a genetic disorder caused by an abnormal number of chromosomes. Includes tetraploidy and triploidy.

Polyribosome Also known as polysome. Organelle formed by several ribosomes attached to a single messenger RNA. Synthesizes proteins used inside the cell.

Population Group of like organisms occupying a specific region.

Porphyrin ring Part of the chlorophyll molecule that absorbs sunlight.

Portal system Arrangement of blood vessels in which a capillary bed drains to a vein, which drains to another capillary bed.

Posterior chamber Posterior portion of the anterior cavity of the eye.

Posterior pituitary Neuroendocrine gland that consists of neural tissue and releases two hormones, oxytocin and antidiuretic hormone.

Precapillary sphincters Tiny rings of smooth muscle that surround the capillaries arising from the metarterioles.

Predation Process in which an individual kills and feeds on another smaller individual.

Premature birth Birth of a baby before 37 weeks of gestation.

Premenstrual syndrome (PMS) Condition that occurs in some women in the days before menstruation normally begins. Characterized by a variety of symptoms such as irritability, depression, fatigue, headaches, bloating, swelling, and tenderness of breasts, joint pain, and tension.

Premotor area Region of the brain in front of the primary motor area. Controls muscle contraction and other less voluntary actions (playing a musical instrument).

Presbyopia Visual impairment caused by aging. Lens becomes stiffer, making it more difficult to focus on nearby objects.

Primary center of ossification Region in the interior of a cartilage mass that first becomes bone.

Primary electron acceptor Protein molecule in the assembly that accepts the energized electron from chlorphyll a in the reaction center.

Primary follicle Structure in the ovary consisting of a primary oocyte and a complete single layer of cuboidal follicle cells.

Primary motor area Ridge of tissue in front of a central groove (the central sulcus). Controls voluntary motor activity.

Primary oocyte Germ cell produced from oogonium in the ovary. Undergoes the first meiotic division.

Primary response Immune response elicited when an antigen first enters the body.

Primary sensory area Region of the brain located just behind the central sulcus. The point of destination for many sensory impulses traveling from the body into the spinal cord and up to the brain.

Primary spermatocyte Cell produced from spermatogonium in the seminiferous tubule. Will undergo first meiotic division.

Primary succession Process of sequential change in which one community is replaced by another. Occurs where no biotic community has existed before.

Primary tissue One of major tissue types, including epithelial, connective, muscle, and nervous tissue.

Primary tumor Cancerous growth that gives rise to cells that spread to other regions of the body.

Primates An order of the kingdom Animalia. Includes prosimians (premonkeys), monkeys, apes, and humans.

Primordial follicle Structure in the ovary that consists of a primary oocyte surrounded by a layer of flattened follicle cells. Gives rise to the primary follicle.

Primordial germ cells Cells that originate in the wall of the yolk sac and eventually become either spermatogonia or oogonia.

Principle of independent assortment Mendel's second law. Hereditary factors are segregated independently during gamete formation. Occurs only when genes are on different chromosomes.

Principle of segregation Mendel's first law, which states that hereditary factors separate during gamete formation.

Producers Generally refers to organisms that can synthesize their own foodstuffs. Major producers are the algae and plants that absorb sunlight and use its energy to synthesize organic foodstuffs from water and carbon dioxide.

Profundal zone The deepest water of a lake, into which very little light penetrates.

Prolactin Protein hormone under control of the hypothalamus. In humans, it is responsible for milk production by the glandular units of the breast.

Promoter Region of the operon between the regulator gene and operator site. Binds to RNA polymerase.

Pronuclei Name of the ovum and sperm cell nuclei shortly after fertilization occurs. Each contains 23 chromosomes.

Prophase First phase of mitosis during which chromosomes condense, the nuclear membrane disappears, and the spindle forms.

Proprioception Sense of body and limb position.

Prosimians Premonkeys; tarsiers and lemurs.

Prostaglandins Group of chemical substances that have a variety of functions. Act on nearby cells.

Prostate gland Sex accessory gland that is located near the neck of the bladder and empties into the urethra. Produces fluid that is added to the sperm during ejaculation.

Proteinoids Spherical structures composed of small amino acid chains formed when amino acids are heated in air. May have been an early precursor of the first cells.

Protists Unicellular eukaryotic organisms such as protozoans and amoebae. Some are autotrophic and some are heterotrophic.

Proton Subatomic particle found in the nucleus of the atom. Each proton carries a positive charge.

Proto-oncogenes Genes in cells that, when mutated, lead to cancerous growth.

Protozoans Animal-like protists, including zooflagellates, ameoboids, ciliates, and sporozoans.

Puberty Period of sexual maturation in humans.

Pulmonary circuit (or circulation) Short circulatory loop that supplies blood to the lungs and transports it back to the heart.

Pulmonary veins Veins that carry oxygenated blood from the lungs to the left atrium.

Punctuated equilibrium Hypothesis explaining how evolutionary change occurs. States that long periods of relatively little change are broken up by briefer periods of relatively rapid evolution.

Pupil Opening in the iris that allows light to penetrate deeper into the eye.

Purine Type of nitrogenous base found in DNA nucleotides. Consists of two fused rings.

Purkinje fiber Modified cardiac muscle fiber that conducts bioelectric impulses to individual heart muscle cells.

Pus Liquid emanating from a wound. Contains plasma, many dead neutrophils, dead cells, and bacteria.

Pyloric sphincter Ring of smooth muscle cells in the lower portion of the stomach where it joins the duodenum. Serves as a gate valve. Opens periodically after a meal, releasing spurts of chyme (liquified, partially digested food) into the small intestine.

Pyramid of numbers Diagram of the number of organisms at various trophic levels in a food chain or ecosystem.

Pyrimidine One of two types of nitrogen base found in DNA nucleotides. Consists of one ring.

Pyrogen Chemical released primarily from macrophages that have been exposed to bacteria and other foreign substances. Responsible for fever.

Radial keratotomy Procedure to correct nearsightedness. Numerous, small superficial incisions are made in the cornea, flattening it and reducing its refractive power.

Radioactivity Tiny bursts of energy or particles emitted from the nucleus of some unstable atoms. Results from excess neutrons in the nuclei of some atoms.

Radionuclide Radioactive isotope of an atom.

Range of tolerance Range of conditions in which an organism is adapted.

Reaction center In the light-harvesting complex of a photosystem, containing chlorophyll a, a molecule to which light energy is transferred by the other pigment molecules.

Receptor Any structure that responds to internal or external changes. Three types of receptors are found in the body: encapsulated, nonencapsulated (naked nerve endings), and specialized (e.g., the retina and semicircular canals).

Recessive Term describing an allele of a gene that is expressed when the dominant factor is missing.

Recombinant DNA technology Procedure in which scientists take segments of DNA from an organism and combine them with DNA from other organisms.

Recombination Process of crossing over during meiosis, resulting in new genetic combinations.

Red blood cells (RBCs) Enucleated cells in blood that transport oxygen in the bloodstream.

Red bone marrow Tissue found in the marrow cavity of bones. Site of blood cell and platelet production.

Reduction The addition of hydrogens or electrons to a molecule.

Reduction factor Any of the factors that cause populations to decline.

Reflex Automatic response to a stimulus. Mediated by the nervous system.

Refraction Bending of light.

Regulator gene Gene that codes for the synthesis of repressor protein in an operon.

Relaxin Hormone produced by the corpus luteum and the placenta. It is released near the end of pregnancy and softens the cervix and the fibrocartilage uniting the pubic bones, thus facilitating birth.

Releasers *See* sign stimuli.

Releasing hormone Any of a group of hormones that stimulates the release of other hormones by the anterior pituitary.

Renal pelvis Hollow chamber inside the kidney. Receives urine from the collecting tubules and empties into the ureter.

Renal tubule That portion of the nephron where urine is produced.

Renewable resources Resources that replenish themselves via natural biological and geological processes, such as wind, hydropower, trees, fish, and wildlife.

Replacement-level fertility Number of children that will replace a couple when they die.

Repressible operon Operon whose genes remain active unless turned off. Found in bacteria and may be present in eukaryotes as well.

Repressor protein Protein produced by a regulator gene. Binds to a region of the DNA molecule (the operator site) adjacent to the structural genes. Blocks RNA polymerase from transcribing structural genes.

Reproductive isolation Condition in which two groups of similar organisms derived from the same parent stock lose the ability to interbreed. Often due to geographic isolation.

Reptiles Class of vertebrates in which fertilization occurs internally. Includes snakes, lizards, and turtles.

Respiratory distress syndrome Disease of premature babies that results from an insufficient amount of surfactant in the infant's lungs, causing alveoli to collapse. Also known as hyaline membrane disease.

Resting potential Minute voltage differential across the membrane of neurons. Also known as the membrane potential.

Restriction endonuclease Enzyme used in recombinant DNA technology. Cuts off segments of the DNA molecule for cloning and splicing.

Reticular activating system (RAS) Region of the medulla that receives nerve impulses from neurons transmitting information to and from the brain. Impulses are transmitted to the cortex, alerting it.

Retina Innermost, light-sensitive layer of the eye. Consists of an outer pigmented layer and an inner layer of nerve cells and photoreceptors (rods and cones).

Retrovirus Special type of RNA virus that carries an enzyme enabling it to produce complementary strands of DNA on the RNA template.

Reverse transcriptase Enzyme that allows the production of DNA from strands of viral RNA.

Rheumatoid arthritis Type of arthritis in which the synovial membrane of the joint becomes inflamed and thickens. Results in pain and stiffness in joints. Thought to be an autoimmune disease.

Rhodopsin (visual purple) Pigment contained in the rods.

Rhythm method Birth control method in which a couple abstains from sexual intercourse around the time of ovulation. Also known as the natural method.

Ribosomal RNA (rRNA) RNA produced at the nucleolus. Combines with protein to form the ribosome.

Ribosome Cellular organelle consisting of two subunits, each made of protein and ribosomal RNA. Plays an important part in protein synthesis.

RNA polymerase Enzyme that helps align and join the nucleotides in a replicating RNA molecule.

RNA replicase Enzyme produced by RNA viruses in host cells that allows them to produce complementary strands of RNA from an RNA template.

Rod Type of photoreceptor in the eye. Provides for vision in dim light.

Root nodule Swelling in the roots of certain plants (legumes) containing nitrogen-fixing bacteria.

Rough endoplasmic reticulum (RER) Ribosome-coated endoplasmic reticulum. Produces lysosomal enzymes and proteins for use outside the cell.

Saccule Membranous sac located inside the vestibule. Contains a receptor for movement and body position.

Salivary gland Any of several exocrine glands situated around the oral cavity. Produces saliva.

Saltatory conduction Conduction of a bioelectric impulse down a myelinated neuron from node to node.

Saprophyte Organism that releases enzymes that digest food materials externally. Smaller food molecules generated in the process are absorbed by the organism. Includes most fungi and nonphotosynthetic bacteria.

Sarcomere Functional unit of the muscle cell. Consists of the myofilaments, actin, and myosin.

Sarcoplasmic reticulum Term given to the smooth endoplasmic reticulum of a skeletal muscle fiber. Stores and releases calcium ions essential for muscle contractions.

Schwann cell Type of neuroglial cell or supportive cell in the nervous system. Responsible for the formation of the myelin sheath.

Science Body of knowledge on the workings of the world and a method of accumulating knowledge. *See also* Scientific method.

Scientific method Deliberate, systematic process of discovery. Begins with observation and measurement. From observations, hypotheses are generated and tested. This leads to more observation and measurement that supports or refutes the original hypothesis.

Sclera Outermost layer of the eye.

Scrotum Skin-covered sac containing the testes.

Sebum Oil excreted by sebaceous glands onto the surface of the skin.

Second law of thermodynamics Law stating that no energy conversion is ever 100% efficient.

Second messenger mechanism Describes how protein hormones and others effect intracellular change by binding to a receptor, activating adenylate cyclase, which leads to the production of cyclic AMP. Cyclic AMP, the second messenger, then activates a cytoplasmic enzyme, protein kinase, which activates or inactivates other enzymes.

Secondary center of ossification Region of bone formation that occurs in the ends (epiphyses) of bones.

Secondary response Generally, a powerful, swift immune system response occurring the second time an antigen enters the body. Much faster than the primary response.

Secondary sex characteristics Distinguishing features of men and women resulting from the sex steroids. In men, includes facial hair growth and deeper voices. In women, includes breast development and fatty deposits in the hips and other regions.

Secondary succession Process of sequential change in which one community is replaced by another. It occurs where a biotic community previously existed, but was destroyed by natural forces or human actions.

Secondary tumor Cancerous growth formed by cells arising from a primary tumor.

Secretin Hormone produced by the cells of the duodenum. Stimulates the pancreas to release sodium bicarbonate.

Secretory vesicles Membrane-bound vesicles containing protein (hormones or enzymes) produced by the endoplasmic reticulum and packaged by the Golgi complex of some cells. They fuse with the membrane, releasing their contents by exocytosis.

Selective permeability Control of what moves across the plasma membrane of a cell.

Sensitization Type of associative learning in which an individual learns to pay heightened attention to stimuli.

Semen Fluid containing sperm and secretions of the secondary sex glands.

Semicircular canal Sensory organ of the inner ear. Houses the receptors that detect body position and movement.

Semilunar valve Type of valve lying between the ventricles and the arteries that conduct blood away from the heart.

Seminal vesicles Sex accessory glands that empty into the vas deferens. Produce the largest portion of ejaculate.

Seminiferous tubule Sperm-producing tubule in the testis.

Sertoli cell Cell in the germinal epithelium of the seminiferous tubule. Houses spermatogenic cells as they develop.

Sex accessory gland One of several glands that produce secretions that are added to sperm during ejaculation.

Sex chromosomes X and Y chromosomes that help determine the sex of an individual.

Sex-linked trait Trait produced by a gene carried on a sex chromosome.

Sex steroid Steroid hormones produced principally by the ovaries (in women) and testes (in men). Help regulate secretion of gonadotropins and determine secondary sex characteristics.

Sexually transmitted diseases (venereal diseases) Infections that are transmitted by sexual contact.

Sickle-cell anemia Genetic disease common in African Americans that results in abnormal hemoglobin in red blood cells, causing cells to become sickle shaped when exposed to low oxygen levels. Sickling causes cells to block capillaries, restricting blood flow to tissues.

Sign stimuli Specific environmental antecedents (stimuli) to particular fixed action patterns.

Sinoatrial node The heart's pacemaker. Located in the wall of the right atrium, it sends timed impulses to the heart muscle, thus synchronizing muscle contractions.

Skeletal muscle Muscle that is generally attached to the skeleton and causes body parts to move.

Skeleton Internal support of humans and other animals. Consists of bones joined together at joints.

Sliding filament mechanism Sliding of actin filaments toward the center of a sarcomere, causing muscle contraction.

Slime molds Fungi that move about by ameboid motion, engulfing food from the soil.

Smooth endoplasmic reticulum (SER) Endoplasmic reticulum without ribosomes. Produces phosphoglycerides used to make the plasma membrane. Performs a variety of different functions in different cells.

Smooth muscle Involuntary muscle that lacks striations. Found around circulatory system vessels and in the walls of such organs as the stomach, uterus, and intestines.

Somite Block of mesoderm that gives rise to the vertebrae, muscles of the neck, and trunk.

Special sense Vision, hearing, taste, smell, and balance.

Speciation Formation of new species resulting from geographic isolation.

Species Group of organisms that is structurally and functionally similar. When members of the group breed, they produce viable, reproductively competent offspring. Also a subgroup of a genus.

Specificity The property of an enzyme allowing it to catalyze only one or a few chemical reactions.

Spermatogenesis Formation of sperm in the seminiferous tubules.

Spermatogonia Sperm-producing cells in the periphery of the germinal epithelium of the seminiferous tubules.

Spermatozoan Sperm.

Spinal nerve Nerve that arises from the spinal cord.

Sponges Phylum of simple, immobile animals that live in colonies, mostly in saltwater. Demonstrate somewhat more complexity than one-celled organisms but less than organisms with distinct tissues.

Spongy bone Type of bony tissue inside most bones. Consists of an irregular network of bone spicules.

Spores (bacterial) Resistant structures produced when environmental conditions become unfavorable. They house the circular chromosome and a tiny amount of cytoplasm and are encased in a thick cell wall.

Sporophyte Diploid generation of a plant exhibiting alternation of generations that produces asexual spores.

Sporozoans Nonmotile, animal-like protists so named because of their ability to produce infectious spores.

Stamen Filamentous structure of a flower that ends with a small bulbous structure known as the anther that produces the male gamete.

Sterilization Procedure to render a man or woman sterile or infertile. In men, the method is generally a vasectomy; in women, it is usually tubal ligation.

Stoma (plural, *stomata*) Adjustable openings in plant leaves. Most gas exchange between leaves and the air occurs through the stoma.

Stirrup (stapes) One of three bones of the middle ear that conducts vibrations from the eardrum to the inner ear.

Stroma The semifluid medium of chloroplasts, in which the membranous grana are embedded; the site of the light-independent reactions.

Structural gene Any gene of an operon that codes for the production of enzymes and other proteins.

Subatomic particles Electrons, protons, and neutrons. Particles that can be separated from an atom by physical means.

Substrate Molecule that fits into the active site of an enzyme.

Succession Process of sequential change in which one community is replaced by another until a mature or climax ecosystem is formed. *See also* Primary succession and Secondary succession.

Sulcus Indented region or groove in the cerebral cortex between ridges.

Suppressor cell Cell of the immune system that shuts down the immune reaction as the antigen begins to disappear.

Suprachiasmatic nucleus Clump of nerve cells in the hypothalamus. Thought to play a major role in coordinating several key functions and several other control centers. Sometimes called the "master clock."

Surfactant Detergent-like substance produced by the lungs. Dissolves in the thin watery lining of the alveoli; helps reduce surface tension, keeping the alveoli from collapsing.

Suspensory ligament Zonular fibers that connect the lens to the ciliary body.

Sustainable-Earth ethic Ethic based on

three tenets: (1) the world has a limited supply of resources that must be shared with all living things, (2) humans are a part of nature and subject to its rules, and (3) nature is not something to conquer, but rather a force we must learn to cooperate with.

Sustainable society Society that lives within the carrying capacity of the environment.

Symbiosis Refers to a number of different types of biotic interaction within a community, including mutualism and commensalism.

Sympathetic division Division of the autonomic nervous system that is responsible for many functions, especially those involved in the fight-or-flight response.

Synapse Juncture of two neurons.

Synaptic cleft Gap between an axon and the dendrite or effector (e.g., gland or muscle) it supplies.

Synergy Coordination of the workings of antagonistic muscle groups.

Synovial fluid Lubricating liquid inside joint cavities. Produced by the synovial membrane.

Synovial membrane Inner layer of the joint capsule.

Syphilis Potentially serious, sexually transmitted disease caused by a bacterium.

Systemic circulation System of blood vessels that transports blood to and from the body and heart, excluding the lungs.

Systolic pressure Peak pressure at the moment the ventricles contract. The higher of the two numbers in a blood pressure reading.

Taiga The northern coniferous forests biome.

Taste bud Receptor for taste principally found in the surface epithelium and certain papillae of the tongue.

Taxis (taxes, plural) Orientation of an animal in relation to some stimulus. An innate mechanism found in less complex organisms.

T cell *See* T-lymphocyte.

Telophase Final stage of mitosis in which the nuclear envelope reforms from vesicles and the chromosomes uncoil.

Temperate deciduous forest Biome that in the United States lies east of the Missis-sippi River and is characterized by broad-leafed trees.

Temporary threshold shift Temporary loss of hearing after being exposed to a noisy environment.

Tendons Connective tissue structures that generally attach muscles to bones.

Teratogen Chemical, biological, or physical agent that causes birth defects.

Teratology Study of birth defects.

Terminal boutons Small swellings on the terminal fibers of axons. They lie close to the membranes of the dendrites of other axons or the membranes of the effectors, and transfer bioelectric impulses from one cell to another.

Terminator codon Codon found on each strand of messenger RNA that marks where protein synthesis should end.

Testes Male gonads. They produce sex steroids and sperm.

Testosterone Male sex hormone that stimulates sperm formation and is responsible for secondary sex characteristics, such as facial hair growth and muscle growth.

Tetraploidy Condition in which an individual is endowed with two complete sets of chromosomes. Instead of having 46 chromosomes, he or she has 92.

Theories Principles of science—the broader generalizations about the world and its components. Theories are supported by considerable scientific research.

Thermocline A layer of water between the epilimnion and hypolimnion; characterized by rapid temperature change.

Thoracic duct Duct carrying lymph to the circulatory system. Empties into the large veins at the base of the neck.

Thoroughfare channel Vessel that connects the arterioles with the venules, thus allowing blood to bypass a capillary bed. Also known as a metarteriole.

Thylakoid A disk-shaped, membranous sac found in chloroplasts, the membranes of which contain the photosystems and ATP-synthesizing enzymes used in the light-dependent reactions of photosynthesis.

Thyroid gland U- or H-shaped gland located in the neck on either side of the trachea just below the larynx. Produces three hormones: thyroxin, triiodothyronine, and calcitonin.

Thyroid-stimulating hormone (TSH) Hormone produced by the pituitary gland. Stimulates production and release of thyroxine and triiodothyronine by the thyroid gland.

Thyroxin Hormone produced by the thyroid gland that accelerates the rate of mitochondrial glucose catabolism in most body cells and also stimulates cellular growth and development.

Tissue Component of the body from which organs are made. Consists of cells and extracellular material (fluid, fibers, and so on).

T-lymphocyte Type of lymphocyte responsible for cell-mediated immunity. Attacks foreign cells, virus-infected cells, and cancer cells directly. Also known as T cell.

Total fertility rate Number of children a woman is expected to have during her lifetime.

Trachea Duct that leads from the pharynx to the lungs.

Transcription RNA production on a DNA template.

Transfer RNA (tRNA) Small RNA molecules that bind to amino acids in the cytoplasm and deliver them to specific sites on the messenger RNA.

Transformation Conversion of a normal cell to a cancerous one.

Transition reaction Part of cellular respiration in which one carbon is cleaved from pyruvic acid, forming a two-carbon compound, which reacts with Coenzyme A. The resulting chemical enters the citric acid cycle.

Translation Synthesis of protein on a messenger RNA template.

Translocation Process in which a segment of a chromosome breaks off but reattaches to another site on the same chromosome or another one.

Transpiration Process by which water evaporates from leaf surfaces. Creates a tension (reduced pressure) in the xylem that pulls columns of water from the roots to the stems.

Trichocysts Harpoonlike projections found in ciliates and used to capture prey.

Triiodothyronine Hormone produced by the thyroid gland. Nearly identical in function to thyroxin.

Triploidy Genetic disorder in which cells have 69 chromosomes instead of 46.

Trisomy Genetic condition characterized by the presence of one extra chromosome.

Trophic hormones Hormones that stimulate the production and secretion of other hormones. Also known as tropic hormones.

Trophic level Feeding level in a food chain.

Tropism Plant response to stimuli, such as light, gravity, and touch, in which it bends toward (positive response) or away from (negative response) the stimulus.

Trophoblast Outer ring of cells of the blastocyst that form the embryonic portion of the placenta.

True fungi Organisms with distinct cell walls, including single-celled yeasts and the multicellular fungi but not slime molds.

TSH-releasing hormone (TSH-RH) Hormone secreted by the posterior lobe of the pituitary gland. Stimulates thyroxin secretion by the thyroid gland.

T tubules (transverse tubules) Invaginations of the plasma membrane of skeletal muscle fibers that conduct an impulse to the interior of the cell.

Tubal ligation Sterilization procedure in women. Uterine tubes are cut, preventing sperm and ova from uniting.

Tubular reabsorption Process in which nutrients are transported out of the nephron into the peritubular capillaries.

Tubular secretion Process in which wastes are transported from the peritubular capillaries into the nephron.

Tumor Mass of cells derived from a single cell that has begun to divide. In malignant tumors, the cells divide uncontrollably and often release clusters of cells or single cells that spread in the blood and lymphatic systems to other parts of the body. Benign tumors grow to a certain size, then stop.

Tundra Northernmost biome with long, cold winters and a short growing season.

Turner syndrome Genetic disorder in which an offspring contains 22 pairs of autosomes and a single, unmatched X chromosome. Phenotypically female.

Twitch Single muscle fiber contraction.

Tympanic membrane (eardrum) Membrane between the external auditory canal and middle ear that oscillates when struck by sound waves.

Type I diabetes Form of diabetes that occurs mainly in young people and results from an insufficient amount of insulin production and release. Brought on by damage to insulin-producing cells of the pancreas. Also known as early-onset diabetes.

Type II diabetes Form of diabetes that occurs chiefly in older individuals (around age of 40) and results from a loss of tissue responsiveness to insulin. Also known as late-onset diabetes.

Type II alveolar cell Cell found in the lining of the alveoli. Produces surfactant.

Umbilical artery One of two arteries in the umbilical cord that carries blood from the embryo to the placenta.

Umbilical vein Vein in the umbilical cord that carries blood from the placenta to the fetus.

Ureter Hollow, muscular tube that transports urine by peristaltic contractions from the kidney to the urinary bladder.

Urethra Narrow tube that transports urine from the urinary bladder to the outside of the body. In males, it also conducts sperm and semen to the outside.

Urinary bladder Hollow, distensible organ with muscular walls that stores urine. Drained by the urethra.

Urine Fluid containing various wastes that is produced in the kidney and excreted out of the urinary bladder.

Uterus Organ that houses and nourishes the developing embryo and fetus.

Utricle Membranous sac containing a receptor for body position and movement. Located inside the vestibule of the inner ear.

Vaccine Preparation containing dead or weakened bacteria and viruses that, when injected in the body, elicits an immune response. *See also* Active immunity.

Vagina Tubular organ that serves as a receptacle for sperm and provides a route for delivery of the baby at birth.

Vagus nerve Nerve that terminates in the stomach wall and stimulates HCl production by cells in the gastric glands.

Variation Genetically based differences in physical or functional characteristics within a population.

Varicose vein Vein whose wall balloons out because the flow of blood downstream is obstructed.

Vascular plants Land plants having a specialized conducting tissue (xylem and phloem) for the transport of food and water.

Vascular cambium Tissue in woody plants that produces new xylem and phloem.

Vascular tissues Responsible for the transport of materials within a plant, including water, minerals, and food. Includes xylem and phleom.

Vas deferens Duct that carries sperm from the testis to the urethra. Contracts during ejaculation.

Vasectomy Contraceptive procedure in men in which the vas deferens is cut and the free ends sealed to prevent sperm from entering the urethra during ejaculation.

Vasopressin Also known as anti-diuretic hormone, which in high concentrations increases blood pressure.

Vena cava One of two large veins that empty into the right atrium of the heart.

Vein Type of blood vessel that carries blood to the heart.

Venule Smallest of all veins. Empties into capillary networks.

Vertebrates Animals with a spinal column (backbone).

Villi Fingerlike projections of the lining of the small intestine that increase the surface area for absorption.

Virus Nonliving entity consisting of a nucleic acid—either DNA or RNA—core surrounded by a protein coat, the capsid. Viruses are cellular parasites, invading cells and taking over their metabolic machinery to reproduce.

Visible light Electromagnetic radiation visible to humans and other animals.

Vitamin Any of a diverse group of organic compounds. Essential to many metabolic reactions.

Vitreous humor Gelatinous material found in the posterior cavity of the eye.

Vocal cords Elastic ligaments inside the larynx that vibrate as air is expelled from the lungs, generating sound.

Warning coloration Passive means of avoiding predation by warning potential predators, usually backed up with some active defense such as a toxin.

Watershed A region drained by a river.

White blood cells (WBCs) Cells of the blood formed in the bone marrow. Principally involved in fighting infection.

White matter The portion of the brain

and spinal cord that appears white to the naked eye. Consists primarily of white, myelinated nerve fibers.

Xylem Vascular tissue that transports water and minerals in vascular plants from roots to leaves or needles. Comprises the bulk of wood in a tree.

Yellow marrow Inactive marrow of bones in adults containing fat. Formed from red marrow.

Yolk sac Embryonic pouch formed from endoderm. Site of early formation of red blood cells and germ cells.

Zero population growth Condition in which a population stops growing.

Zona pellucida Band of material surrounding the oocyte.

Zonular fibers Thin fibers that attach the lens to the ciliary body.

Zooflagellates Animal-like protists that live in water and in moist soils and contain one or sometimes several flagella.

Zooplankton Crustaceans and protozoans in aquatic ecosystems that feed on phytoplankton.

Zygote Cell produced by a sperm and ovum during fertilization. Contains 46 chromosomes.

Index

and herbicides, 533
inhibiting, 516
nontrophic, 511
pancreatic, 5
polypeptide, 511
releasing, 516
role of, in controlling heart, 291–292
role of, in reflexes, 246
secretion of, 511–512
steroid, 511
and target cells, 511, 513–514
trophic, 511
thyroid-stimulating, 511
Hornworts, 681, 682
Horsetails, 684f–685
Human chorionic gonadotropin (HCG), 560, 583
Human chorionic somatomammotropin (HCS), 597
Human cultural evolution, 767f–770f
Human development
adolescence, 599–600
adulthood, 600
aging, 600–603
and birth defects, 591–593, 592f
and body proportions, 578f
childbirth, 594–597, 595f, 596f
childhood, 599
death, 603
ectopic pregnancy, 591
fertilization in, 578f, 579f, 580, 581f
fetal development, 587, 588f, 588–591
implantation, 585–586f
infancy, 598–599f
lactation, 597f–598
pre-embryonic development, 580–583f, 581f, 582f, 584f, 585t
pregnancy, 593–594f
Human fetal tissues, in treating human disease, 60f, 62–63
Human Genome Project, 181
Human immunodeficiency virus (HIV). See HIV virus
Humans
as competitive force, 821f
evolution of upright posture in, 768, 769f
mendelian genetics in, 173
autosomal-dominant traits, 175–177f, 176f
autosomal-recessive traits, 173–175f, 174f
roots of modern, 764–765
taxonomic classification, 13t
unique characteristics of, 13
Humerus, anatomy of, 482, 484f
Hummingbird, 244f
Humoral immunity, 358t, 360–362
Humphry, Derek, 605f
Hungry browsers, 108
Hunting and gathering, 767–768
Huntington's disease, 174t
Huxley, Sir Julian, 822
Hyaline cartilage, 232–233f, 327, 332f
Hyaline membrane disease, 332–333
Hydra, 11f, 726, 727f
reproduction in, 727
Hydrilla, 696
Hydrocephalus, 431, 433f, 433f
Hydrochloric acid (HCl), 270, 353

Hydrogen, 27, 37
isotopes of, 29
Hydrogen bonds, 36f
Hydrolysis, 256f
Hydroxide ions, 37
Hyman, Flo, 176, 177f
Hypercholesterolemia, 260
Hyperextendable thumb, 174t
Hyperglycemia, 530
Hyperopia, 459
Hyperparathyroidism, 524
Hyperplasia, diurnal cycle of, 516–517f
Hypersensitivity, 375
Hypertension, 298
and cadmium, 306
causes and cures of, 302–303
Hyperthyroidism, 510, 523–524
Hypertonic, 74
Hypertrophy, 522
diurnal cycle of, 516–517f
Hyperventilation, 345
Hyphae, 664f
Hypocotyl, 700
Hypolimnion, 857
Hypophosphatemia, 182, 184f
Hyposecretion, 517
Hypothalamus, 354, 429, 516
Hypothermia, adaptive, 244
Hypothesis, 15
Hypothyroidism, 523
Hypotonic, 74

I

I bands, 496f
Ichthycsaurs, 743
Ideation, 405
IgE, 371
IgG, 371
Imitation, 781–782
Immune reaction, 357
Immune system, 356
active immunity, 364–366, 368
antibodies in, 360–361f
B cells in, 358f, 359f
as challenge in tissue/organ transplantation, 70–71
diseases of, 370–371
first line of defense in, 352–353
lymphocytes in, 356, 357
macrophages in, 361–362, 364f
passive immunity, 364–366, 368
polysaccharides in, 357
primary response in, 360f
proteins in, 357
secondary response in, 360f
second line of defense in, 353–356
T cells in, 357–358f, 362–364, 365t
Immunity, 360
and AIDS, 371, 373–75f
and blood transfusions, 368–370f, 369f
from breast milk, 366f, 367
cellular, 317
comparison of humoral and cell-mediated, 358t
and multiple clinical sensitivity, 375–378
and tissue transplantation, 370
Immunocompetence, 358f

Immunoglobulins, 360–361f, 366
and functions of, 362t
Implantation, 583, 585–586f
delayed, 590, 590f–591
Imprinting, 779, 780
chemical, 783f
Incisors, 746
Incomplete dominance, 177f–178
Incomplete proteins, 259
Incus, 464
Independent assortment, 172
Indian paintbrush, 807
Inducible operons, 208, 209f
Industrial revolution, 770
Industrial society, 769–770f
Infancy, 598–599f
Infectious mononucleosis, 317–318
Inferior vena cavae, 289
Infertility, 570–571
Inflammatory response, 353, 354f
Inguinal canal, 541–542f
Inhalation, 340–341t, 341t
Inheritance. See Genetics; Heredity
Inhibin, 551
Inhibitory synapses, 416
Innate behaviors, 775–776, 782–783
Innate releasing mechanisms, 779
Inner cell mass, 581, 585
Inner city syndrome, 791–792
Inorganic molecules, 36
Insectivorous bats, 802f–803
Insects, 733–735f
adaptations for feeding, 267f
Inspiration, 340–341t
Inspiratory reserve volume, 341
Insulin
actions of, 5f
production of human, 216, 239, 511, 524, 525
Insulin-independent diabetes, 529
Insulin pump, 527, 528f
Integral proteins, 69
Intensity, distinguishing pitch and, 467, 468f
Intercostal muscles, 341
Interference competition, 798–799f
Interferons, 355
functions of, 355t
Interleukin 1, 362, 364f
Interleukin 2, 363
Internal anal sphincter, 277
Internal constancy, homeostasis as state of, 4
Internal sphincter, 394
Interphase, 140–141ft, 210–211
Interspecific behavior, 775
Interstitial cells, 546–548, 547f
Interstitial cell-stimulating hormone (ICSH), 549
Interstitial fluid, 303
Intervertebral disks, 233f
Intestinal epithelium, 275
Intrapulmonary pressure, 341
Intraspecific behavior, 775
Intrauterine device, 564–565f
Introns, 211
Invertebrate chordates, 738–739f
Invertebrates
endocrine system of, 531–532
in exoskeletons, 480, 481f
opposing muscle groups in, 480f

Credits

≈

Photo Credits

Part I Opener, p. 1: © Comstock, Incorporated
Chapter 1 Opener, p. 2: © George Lepp/Comstock
Fig. 1-1, p. 3: © J. Serrao/Visuals Unlimited
Fig. 1-2, p. 4: © Jan L. Wassink/Visuals Unlimited
Fig. 1-5a, p. 8: © Milton H. Tierney, Jr./Visuals Unlimited
Fig. 1-5b, p. 8: © Walt Anderson/Visuals Unlimited
Fig. 1-6a, p. 9: © Walt Anderson/Visuals Unlimited
Fig. 1-6b, p. 9: © A Kerstitch/Visuals Unlimited
Fig. 1-6c, p. 9: © Patrice Ceisel/Visuals Unlimited
Fig. 1-6d, p. 9: © A. Kerstitch/Visuals Unlimited
Fig. 1-6e, p. 9: © John Gerlach/Visuals Unlimited
Fig. 1-6f, p. 9: © D. Cavagnaro/Visuals Unlimited
Fig. 1-8a, p. 11: © Carolina Biological/Visuals Unlimited
Fig. 1-8b, p. 11: © Visuals Unlimited
SE 1-1, p. 12: © Kjell B. Sandved/Visuals Unlimited
Fig. 1-9a, p. 14: © Grant Heilman/Grant Heilman Photography
Fig. 1-9b, p. 14: © Grant Heilman/Grant Heilman Photography
Fig. 1-11a, p. 19: © William Banaszewski/Visuals Unlimited
Fig. 1-11b, p. 19: © John D. Cunningham/Visuals Unlimited

Chapter 2 Opener, p. 25: © Michael W. Davidson
Fig. 2-1, p. 26: © SIU/Visuals Unlimited
Fig. 2-3, p. 27: © Richard Treptow/Visuals Unlimited
Fig. 2-5, p. 32: © PEGCO/Visuals Unlimited
Fig. 2-13, p. 37: © Walt Anderson/Visuals Unlimited
Fig. 2-15, p. 38: © Bill Beatty/Visuals Unlimited
Fig. 2-19, p. 41: © John D. Cunningham/Visuals Unlimited
Fig. 2-21a, p. 43: © Carolina Biological/Visuals Unlimited
Fig. 2-21b, p. 43: © Veronika Burmeister/Visuals Unlimited
Fig. 2-22, p. 43: © Nicholas Desciose/Photo Researchers
Fig. 2-23, p. 44: © John D. Cunningham/Visuals Unlimited
Fig. 2-24a, p. 44: © Carolina Biological/Visuals Unlimited
Fig. 2-24b, p. 44: © William Ober/Visuals Unlimited
Fig. 2-31a, p. 49: © Stanley Flegler/Visuals Unlimited
Fig. 2-31b, p. 49: © Stanley Fleger/Visuals Unlimited

Part II Opener, p. 57: © Cibisco/Visuals Unlimited
Chapter 3 Opener, p. 58: © Michael Gabridge/Visuals Unlimited
Fig. 3-1a, p. 59: © Ralph J. Slepecky/Visuals Unlimited
Fig. 3-1b, p. 59: © K.G. Murti/Visuals Unlimited
Fig. 3-3a, p. 61: © R. Calentine/Visuals Unlimited
Fig. 3-3b, p. 61: © Bruce Iverson/Visuals Unlimited

Fig. 3-3c, p. 61: © M. Abbey/Visuals Unlimited
Fig. 3-4a, p. 65: © W. Ormerod/Visuals Unlimited
Fig. 3-4b, p. 65: © Michael Dykstra/Visuals Unlimited
Fig. 3-4c, p. 65: © Michael Dykstra/Visuals Unlimited
Fig. 3-5, p. 65: © Science Source/Photo Researchers, Inc.
Fig. 3-7a, p. 69: © M. Schliwa/Visuals Unlimited
Fig. 3-12, p. 74: From *Introduction to The Fine Structure of Plant Cells* by Myron C. Ledbetter and Keith R. Porter. © 1970 by Springer-Verlag used by permisson.
Fig. 3-13a, p. 75: © K.G. Murti/Visuals Unlimited
Fig. 3-13b, p. 75: © Lester Bergman & Associates, Inc.
Fig. 3-14, p. 76: © K.G. Murti/Visuals Unlimited
Fig. 3-15a, p. 77: © G. Musil/Visuals Unlimited
Fig. 3-16a, p. 78: © George B. Chapmen & Priscilla Devadeoss/Visuals Unlimited
Fig. 3-17b, p. 79: © K.G. Murti/Visuals Unlimited
Fig. 3-17c, p. 79: © David M. Phillips/Visuals Unlimited
Fig. 3-19b, p. 80: © David M. Phillips/Visuals Unlimited
Fig 3-20a, p. 81: © David M. Phillips/Visuals Unlimited
Fig. 3-21, p. 81: © Biophoto Associates/Science Source/Photo Researchers, Inc.
Fig. 3-22a, p. 82: © David M. Phillips/Visuals Unlimited
Fig. 3-22b, p. 82: © David M. Phillips/Visuals Unlimited
Fig 3-23a, p. 83: © David M. Phillips/Visuals Unliited
Fig. 3-23b, p. 83: © David M. Phillips/Visuals Unliited
Fig. 3-23c, p. 83: © David M. Phillips/Visuals Unlimited
Fig. 3-23d, p. 83: © David M. Phillips/Visuals Unlimited
Fig. 3-25, p. 84: © Bruce Iverson/Visuals Unlimited

Chapter 4 Opener, p. 88: © M. Eichelberger/Visuals Unlimited
Fig. 4-2, p. 91: © AP/Wide World Photos
HN4-1, p. 92: © Custom Medical Stock Photo
Fig 4-6, p. 94: © A. Gurmankin/Visuals Unlimited

Chapter 5 Opener, p. 100: © Cibisco/Visuals Unlimited
Fig. 5-1, p. 101: © Stanley L. Flegler/Visuals Unlimited
Fig. 5-3a, p. 103: © M. Powell/Visuals Unlimited
Fig. 5-6, p. 104: © Steve McCutcheon/Visuals Unlimited
Fig. 5-7, p. 105: © Science VU/Visuals Unlimited
Fig. 5-9, p. 105: © William J. Weber/Visuals Unlimited
SE5-1a, p. 108: © John D. Cunningham/Visuals Unlimited
SE5-1b, p. 108: © John D. Cunningham/Visuals Unlimited

Fig. 5-18a, p. 113: © John D. Cunningham/Visuals Unlimited
Fig. 5-18b, p. 113: © Robert E. Lyons/Visuals Unlimited

Chapter 6 Opener, p. 117: © Don Fawcett/Visuals Unlimited
SE6-1, p. 125: © Kjell B. San/Visuals Unlimited
Fig. 6-14, p. 132: © Stephen J. Lang/Visuals Unlimited
Fig. 6-15, p. 133: © Ray Coppinger, Hampshire College Dog Project

Part Opener III, p. 138: © K.G. Murti/Visuals Unlimited
Chapter 7 Opener, p. 139: © John D. Cunningham/Visuals Unlimited
Fig. 7-1, p. 140: © Stan W. Elems/Visuals Unlimited
Fig. 7-4, p. 144: © Carolina Biological/Visuals Unlimited
Fig. 7-5, p. 145: © David M. Phillips/Visuals Unlimited
SD 7-1a, p. 146: © Bettman Archive
SD 7-1b, p. 147: © Science UV/NIHLBL/Visuals Unlimited
Fig. 7-6b, p. 148: ©Science VU/Visuals Unlimited
Fig. 7-7a, p. 149: © SIU/Visuals Unlimited
Fig. 7-8a, p. 150: © Michael Abbey/Visuals Unlimited
Fig. 7-8b, p. 150: © John D. Cunningham/Visuals Unlimited
Fig. 7-8c, p. 150: © John D. Cunningham/Visuals Unlimited
Fig. 7-8d, p. 150: © John D. Cunningham/Visuals Unlimited
Fig. 7-8e, p. 150: © John D. Cunningham/Visuals Unlimited
Fig. 7-11b, p. 152: © David M. Phillips/Visuals Unlimited
Fig. 7-12a, b, d, e, f, p. 153: © John Cunningham/Visuals Unlimited
Fig. 7-12c, p. 153: Calentine/Visuals Unlimited
Fig. 7-14, p. 154: © Cytographics, Inc./Visuals Unlimited
Fig. 7-15a, p. 155: © Linda H. Hopson/Visuals Unlimited
Fig. 7-15b, p. 155: © Jack M. Bostrack/Visuals Unlimited
HN 7-1, p. 157: © Cabisco/Visuals Unlimited

Chapter 8 Opener, p. 163: © CNRE, Science Photo Library/Photo Researchers
Fig. 8-1, p. 164: © Science VU/Visuals Unlimited
Fig. 8-7a, p. 170: © Walt Anderson/Visuals Unlimited
Fig 8-7b, p. 170: © Malcolm Gutter/Visuals Unlimited
Fig. 8-11, p. 174: © Joe McDonald/Visuals Unlimited
Fig. 8-12a, p. 175: © John D. Cunningham/Visuals Unlimited
Fig. 8-12b, p. 175: © David V. Schidlow/Visuals Unlimited
Fig. 8-13, p. 175: © Jeffrey Reed/Medichrome
Fig. 8-14a, p. 176: © Bruce Berg/Visuals Unlimited
Fig. 8-14b, p. 176: © Bruce Berg/Visuals Unlimited
Fig. 8-15, p. 176: © Dr. Ira Rosenthal, Dept. of Pediatrics, University of Illinois at Chicago
Fig. 8-16, p. 177: © AP/Wide World Photos
Fig. 8-19a, p. 179: © Joe McDonald/Visuals Unlimited
Fig. 8-19b, p. 179: © Tim Hauf/Visuals Unlimited
Fig. 8-19c, p. 179: © M. Long/Visuals Unlimited
Fig. 8-19d, p. 179: © Marek Litman/Visuals Unlimited
Fig. 8-19e, p. 179: © Mark D. Cunningham/Visuals Unlimited
Fig. 8-20, all p. 180: © Ralph Somes/Visuals Unlimited
Fig. 8-22a, p. 182: © John Cabisco/Visuals Unlimited
Fig. 8-24, p. 184: © Biophote Associates/Science Source/Photo Researchers
Fig. 8-26, p. 185: © Ron Spomer/Visuals Unlimited
Fig. 8-28a, p. 187: © Dr. Ira Rosenthal, Dept. of Pediatircs, University of Illinois at Chicago
Fig. 8-28b, p. 187: © M. Coleman/Visuals Unlimited
Fig. 8-30a-b, p. 188: © Martin M. Rotker
Fig. 8-31a-b, p. 188: © Dr. Ira Rosenthal, Dept. of Pediatrics, University of Illinois at Chicago
Fig. 8-33, p. 191: © SC Reuman/Visuals Unlimited

Part IV Opener, p. 225: © CNRI, Science Photo Library/Photo Researchers
Chapter 9 Opener, p. 196: © Lawrence Livermore Laboratory/Visuals Unlimited

Fig. SD9-1, p. 207: © AP/Wide World Photos
Fig. 9-12a, p. 208: © Ralph Slepecky/Visuals Unlimited
Fig. 9-12b, p. 208: © K.G. Murti/Visuals Unlimited
Fig. 9-18, p. 220: © Dana Richter/Visuals Unlimited
Fig. 9-19, p. 220: Martha Cooper/Peter Arnold, Inc.
Fig. 9-20, p. 221: Courtesy of Impact/Gentle Earth Ltd.

Chapter 10 Opener, p. 226: © NIBSC, Science Photo Library/Photo Researchers
Fig. 10-7a, p. 231: © John D. Cunningham/Visuals Unlimited
Fig. 10-7b, p. 231: © Stan Elems/Visuals Unlimited
Fig. 10-8, p. 232: © J. Boliver/Custom Medical Stock Photos
Fig. 10-9a, p. 233: © Fred Hossler/Visuals Unlimited
Fig. 10-9b, p. 233: © John D. Cunningham/Visuals Unlimited
Fig. 10-9c, p. 233: © Carolina Biological Supply Co./Visuals Unlimited
Fig. 10-11a, p. 235: © Lester Bergman & Associates, Inc.
Fig. 10-11b, p. 235: © John D. Cunningham/Visuals Unlimited
Fig. 10-11c, p. 235: © John D. Cunningham/Visuals Unlimited
Fig. 10-13a, p. 238: © John D. Cunningham/Visuals Unlimited
Fig. 10-13b, p. 238: © John D. Cunningham/Visuals Unlimited
Fig. 10-13c, p. 238: © Fred Hossler/Visuals Unlimited
Fig. 10-14, p. 238: © John D. Cunningham/Visuals Unlimited
SE10-1, p. 244: © Joe McDonald/Visuals Unlimited
Fig. 10-25, p. 250: © Reuters/Bettmann
Fig. 10-26, p. 250: © C.C. Duncan/Medical Images, Inc.

Chapter 11 Opener, p. 254: © Cibisco/Visuals Unlimited
Fig. 11-1, p. 255: United States Department of Agriculture
Fig. 11-4, p. 258: © L. V. Bergmann & Associates, Inc.
Fig. 11-5b, p. 258: © Bernd Wittich/Visuals Unlimited
Fig. 11-6, p. 259: © AP/Wide World Photo
HN11-1a, p. 261: © Bernard Wittich/Visuals Unlimited
HN11-1b, p. 261: © Bernd Wittich/Visuals Unlimited
Fig 11-8, p. 264: © Tony Cubillas-BPEI/Visuals Unlimited
Fig. 11-9, p. 264: © M. Abbey/Visuals Unlimited
Fig. 11-10b, p. 265: © James R. McCullagh/Visuals Unlimited
SE11-1, p. 267: © Stanley L. Flegler/Visuals Unlimited
Fig. 11-15b, p. 271: © Triarch/Visuals Unlimited
Fig. 11-19, p. 273: © L.V. Bergman & Associates, Inc.
Fig. 11-22a, p. 276: From Edward J. Reith and Michael H. Ross, *Atlas of Descriptive Histology*, 3/e, 1977 by Harper & Row. Used by permission of J.B. Lippincott Company
Fig. 11-22b, p. 276: © John D. Cunningham/Visuals Unlimited
Fig. 11-22c, p. 276: © John D. Cunningham/Visuals Unlimited
Fig. 11-22d, p. 276: © Michael Webb/Visuals Unlimited
Fig. 11-28, p. 282: © Adrian Caston/Visuals Unlimited

Chapter 12 Opener, p. 285: © Carolina Biological Supply/Visuals Unlimited
Fig. 12-7c, p. 291: © Science Photo Library/Photo Researchers
Fig. 12-9a, p. 293: © Custom Medical Stock Photos
Fig. 12-11, p. 295: © T. Kuwabara-D.W. Fawcett/Visuals Unlimited
Fig. 12-12, p. 295: © Carolina Biological/Visuals Unlimited
SD 12-1, p. 296: © Bettmann Archive
Fig. 12-14a, p. 298: © Triarch/Visuals Unlimited
Fig. 12-14b, p. 298: © Carolina Biological/Visuals Unlimited
Fig. 12-14c, p. 298: © William Ober/Visuals Unlimited
Fig. 12-16a, p. 300: © John D. Cunningham/Visuals Unlimited
Fig. 12-16b, p. 300: © R. Bollrnder/Visuals Unlimited
Fig. 12-16c, p. 300: © Don W. Fawcett/Visuals Unlimited
Fig. 12-16d, p. 300: © Dan Friend-Don Fawcett/Visuals Unlimited
Fig. 12-17b, p. 301: © Fawcett-Vehara-Suyama, Science Source/Photo Researchers
HN12-1, p. 303: © CNRI, Science Photo Library/Photo Researchers
Fig. 12-19b, p. 304: © Keith/Custom Medical Stock Photos

Fig. 12-23, p. 306: © Science UV-Fred Marsik/Visuals Unlimited
Fig. 12-24, p. 306: © John Schlden/Visuals Unlimited

Chapter 13 Opener, p. 310: © David M. Phillips/Visuals Unlimited
Fig. 13-2a, p. 313: © David M. Phillips/Visuals Unlimited
Fig. 13-2b, p. 313: © Don W. Fawcett/Visuals Unlimited
Fig. 13-3c, p. 313: © Stanley Flegler/Visuals Unlimited
Table 13-2 all, p. 314: © John D. Cunningham/Visuals Unlimited
Fig. 13-3, p. 315: © Stanley Flegler/Visuals Unlimited
HN 13-1, p. 316: © John D. Cunningham/Visuals Unlimited
Fig. 13-6, p. 317: © R. Calentine/Visuals Unlimited
Fig. 13-7, p. 318: © G. Prance/Visuals Unlimited
Fig. 13-8a, p. 318: © T. Gula/Visuals Unlimited
Fig. 13-8b, p. 318: © Peter K. Ziminski/Visuals Unlimited
Fig. 13-11, p. 320: © David M. Phillips/Visuals Unlimited
Fig. 13-12, p. 321: © Peter K. Ziminski/Visuals Unlimited
Fig. 13-13, p. 321: © Daniel D. Chiras

Chapter 14 Opener, p. 324: © Michael W. Gabridge/Visuals Unlimited
Fig. 14-5, p. 330: From *Tissues and Organs: A Text Atlas of Scanning Electron Microscopy* by Richard G. Kessel and Randy H. Kardon. © 1979 by W. H. Freeman and Company. Reprinted with permission.
Fig. 14-7a-b, p. 332: © John D. Cunningham/Visuals Unlimited
Fig. 14-8b, p. 332: © David M. Phillips/Visuals Unlimited
Fig. 14-8c, p. 332: © Cabisco/Visuals Unlimited
Fig. 14-9a-b, p. 333: © David M. Phillips/Visuals Unlimited
Fig. 14-9c, p. 333: © Fawcett-Gehr/Photo Researchers
Fig. 14-10b, p. 334: © D.W. Fawcett/Visuals Unlimited
Fig. 14-10c, p. 334: © D.W. Fawcett/Visuals Unlimited
Fig. 14-11a, p. 332: © R. Calentine/Visuals Unlimited
Fig. 14-11b, p. 335: © Dick Thomas/Visuals Unlimited
Fig 14-12, p. 335: © Science VU-NPS/Visuals Unlimited
Fig. 14-13b, p. 336: © Custom Medical Stock
Fig. 14-19b1-b2, p. 343: © SIU/Visuals Unlimited
Fig. 14-20, p. 344: SIU/Visuals Unlimited
HN14-1a, p. 347: © O. Auerbach/Visuals Unlimited
HN14-1b, p. 347: © F. Sloop-William Ober/Visuals Unlimited

Chapter 15 Opener, p. 351: © David M. Phillips/Visuals Unlimited
HN15-1, p. 366: © John D. Cunningham/Visuals Unlimited
Fig. 15-14a, p. 367: © Tom J. Urich/Visuals Unlimited
Fig. 15-14b, p. 367: © Leonard Lee rue III/Visuals Unlimited
Fig. 15-19a, p. 373: © David M. Phillips/Visuals Unlimited
Fig. 15-19b, p. 373: © Science VU-AFIF/Visuals Unlimited

Chapter 16 Opener, p. 381: © CNRI, Science Photo Laboratory/Photo Researchers
Fig 16-1, p. 382: © SIU/Visuals Unlimited
Fig. 16-7, p. 387: © Carolina Biological/Visuals Unlimited
Fig. 16-8, p. 388: © Fred Hossler/Visuals Unlimited
Fig. 16-10a, p. 390: © M. Webb/Visuals Unlimited
Fig. 16-10b, p. 390: From *Tissues and Organs: A Text Atlas of Scanning Electron Microscopy* by Richard G. Kessel and Randy H. Kardon. © 1979 by W. H. Freeman and Company. Reprinted with permission.
Fig. 16-11c, p. 391: © F. Spinelli/Visuals Unlimited
Fig. 16-13a, p. 394: © NMSB/Custom Medical Stock
Fig 16-13b, p. 394: © Runk/Schoenberger/Grant Heilman Photography
SE16-1, p. 397: © Tom McHugh/Photo Researchers
Fig. 16-17, p. 400: © SIU/Visuals Unlimited

Chapter 17 Opener, p. 404: © Cabisco/Visuals Unlimited

Fig. 17-1, p. 405: © The Bettmann Archive
Fig. 17-5a, p. 409: © David M. Phillips/Visuals Unlimited
Fig. 17-9, p. 412: © C. Raines/Visuals Unlimited
Fig. 17-10, p. 412: © Howard Sochurek/Medical Image, Inc.
Fig. 17-14a, p. 417: © Science VU/E.R. Lewis, T.E. Everhart, and Y.Y. Zeevi, University of California/Visuals Unlimited
Fig. 17-14c, p. 417: ©T. Reese-D.W. Fawcett/Visuals Unlimited
Fig. 17-28, p. 433: © SIU/Visuals Unlimited

Chapter 18 Opener, p. 441: © Cabisco/Visuals Unlimited
SE 18-1, p. 443: © Tom McHugh/Photo Researchers, Inc.
Fig. 18-4,, p. 445: © Cabisco/Visuals Unlimited
HN 18-1, p. 446: © AP/Wide World Photos
Fig. 18-5, p. 446: © Biophoto Associates/Photo Researchers, Inc.
Fig. 18-14, p. 455: © A.L. Blum/Visuals Unlimited
Fig. 18-16, p. 457: © Bill Beatty/Visuals Unlimited
Fig. 18-24, p. 462: © Daniel D. Chiras
Fig. 18-27, p. 463: © T.E. Adams/Visuals Unlimited
Fig. 18-28b, p. 464: © Kjell B. Sandved/Visuals Unlimited
Fig. 18-33, p. 469: © Beltone Electronics
Fig. 18-37, p. 472: © Dave B. Fleetham/Visuals Unlimited

Chapter 19 Opener, p. 478: © Science Photo Library/Photo Researchers
Fig. 19-2a, p. 480: © Marty Snyderman/Visuals Unlimited
Fig. 19-3, p. 481: © Thomas Gula/Visuals Unlimited
Fig. 19-5, p. 482: © Joe McDonald/Visuals Unlimited/Visuals Unlimited
Fig. 19-7a, p. 484: © Jame Stevenson/Science Photo Library/Photo Researchers
Fig. 19-8, p. 485: © Calentine/Visuals Unlimited
Fig. 19-9, p. 485: © Mathta Cooper/Peter Arnold, inc.
Fig. 19-12a-b, p. 487: © SIU/Visuals Unlimited
Fig. 19-13, p. 488: © CNRI/Science Photo Library/Photo Researchers
Fig. 19-14a-b, p. 488: SIU/Visuals Unlimited
Fig. 19-15b, p. 489: © Science VU/Visuals Unlimited
Fig. 19-16, p. 490: © John D. Cunningham/Visuals Unlimited
Fig. 19-20, p. 492: Reprinted with permission from *Calcified Tissue Research*, 1967
SE19-1, p. 494: © Stephen J. Lang/Visuals Unlimited
Fig. 19-22a, p. 495: © John D. Cunningham/Visuals Unlimited
Fig. 19-22b, p. 495: © R. Calentine/Visuals Unlimited
HN-1 p. 497: © Lester Bergman & Associates
Fig. 19-28a, p. 501: © John D. Cunningham/Visuals Unlimited
Fig. 19-30, p. 503: © Bruce Berg/Visuals Unlimited

Chapter 20 Opener, p. 509: © VU/SIU/Visuals Unlimited
Fig. 20-10a-b, p. 517: © AP/Wide World Photos
Fig. 20-11a-d, p. 518: Reprinted with permission, *American Journal of Medicine*, 20(1956)
Fig. 20-15, p. 520: © William Banaszewski/Visuals Unlimited
Fig. 20-20a, p. 523: © R. Calentine/Visuals Unlimited
Fig. 20-20b, p. 523: © David M. Phillips/Visuals Unlimited
Fig. 20-21, p. 523: © Ken Greer/Visuals Unlimited
Fig. 20-24, p. 525: © John D. Cunningham/Visuals Unlimited
HN20-1, p. 528: © NASA/Science Photo Library/Photo Researchers
Fig. 20-28, p. 530: © John Serrao/Visuals Unlimited
Fig. 20-29, p. 531: © John Serrao/Visuals Unlimited

Part Opener V Opener, p. 537: © Tom McHugh/Photo Researchers
Chapter 21 Opener, p. 538: © John Gerlach/Visuals Unlimited
Fig. 21-1, p. 539: © Cabisco/Visuals Unlimited
Fig. 2-2, p. 540: © R. Calentine/Visuals Unlimited
Fig. 21-3, p. 540: © William Grenfell/Visuals Unlimited
Fig. 21-4a, p. 541: © William J. Weber/Visuals Unlimited
Fig. 21-4b, p. 541: © S. Maslowski/Visuals Unlimited

Fig. 26-14, p. 707: © Carolina Biological Supply Co./Visuals Unlimited
Fig. 26-15, p. 707: © R. Calentine/Visuals Unlimited
Fig. 26-16, p. 707: © C.G. VanDyke/Visuals Unlimited
Fig. 26-18a, p. 709: © S. Elmes/Visuals Unlimited
Fig. 26-18b, p. 709: © John D. Cunningham/Visuals Unlimited
Fig. 26-19, p. 709: © Triarch/Visuals Unlimited
Fig. 26-20, p. 709: © Gustav Verderberg/Visuals Unlimited
Fig. 26-21a, p. 710: © Bill Beatty/Visuals Unlimited
Fig. 26-21b, p. 710: © Science VU-Polaroid/Visuals Unlimited
SE 26-1, p. 711: © Daniel D. Chiras
Fig. 26-23, p. 713: © David Newman/Visuals Unlimited
Fig. 26-24, p. 713: © Joel Arrington/Visuals Unlimited
Fig. 26-25, p. 716: © Robert E. Lyons/Visuals Unlimited

Chapter 27 Opener, p. 722: © Daniel W. Gotshall/Visuals Unlimited
Fig. 27-1, p. 723: © Leonard Rue III/Visuals Unlimited
Fig. 27-2a, p. 726: © Don W. Fawcett/Visuals Unlimited
Fig. 27-2b, p. 726: © Rudolf Arndt/Visuals Unlimited
Fig. 27-3a, p. 727: © Gary R. Robinson/Visuals Unlimited
Fig. 27-3b, p. 727: © Daniel W. Gotshall/Visuals Unlimited
Fig. 27-3c, p. 727: © John D. Cunningham/Visuals Unlimited
Fig. 27-4a, p. 728: © Dave B. Fleetham/Visuals Unlimited
Fig. 27-4b, p. 728: © Edward Hodgson/Visuals Unlimited
Fig. 27-5a, p. 729: © R. Wallace/Visuals Unlimited
Fig. 27-5b, p. 729: © James R. McCullash/Visuals Unlimited
Fig. 27-5c, p. 729: © A. Kerstitch/Visuals Unlimited
Fig. 27-6a-b, p. 730: © R. Calentine/Visuals Unlimited
Fig. 27-6c, p. 730: © T.E. Adams/Visuals Unlimited
Fig. 29-10a, p. 732: © Cabisco/Visuals Unlimited
Fig. 27-10b, p. 732: © Glenn M. Oliver/Visuals Unlimited
Fig. 27-10c, p. 732: © Cabisco/Visuals Unlimited
Fig. 27-11a, p. 733: © Alan Desbonnet/Visuals Unlimited
Fig. 27-12a, p. 734: © Science VU/Visuals Unlimited
Fig. 27-12b, p. 734: © Cabisco/Visuals Unlimited
Fig. 27-12c-d, p. 734: © John D. Cunningham/Visuals Unlimited
Fig. 27-13a, p. 735: © John D. Cunningham/Visuals Unlimited
Fig. 27-13b, p. 735: © P. Starborn/Visuals Unlimited
Fig. 27-13c, p. 735: © Bill Beatty/Visuals Unlimited
Fig. 27-13d, p. 735: © Nada Pecnik/Visuals Unlimited
Fig. 27-14a-c, p. 735: © John D. Cunningham/Visuals Unlimited
Fig. 27-14d, p. 735: © A. Kerstitch/Visuals Unlimited
Fig. 27-15a, p. 736: © John D. Cunningham/Visuals Unlimited
Fig. 27-15b, p. 736: © William J. Weber/Visuals Unlimited
Fig. 27-15c, p. 736: © W. Ober/Visuals Unlimited
Fig. 27-15d, p. 736: © Wm. C. Jorgensen/Visuals Unlimited
Fig. 27-16b, p. 737: © James R. McCullagh/Visuals Unlimited
Fig. 27-16c, p. 737: © W. Ober/Visuals Unlimited
Fig. 27-16d, p. 737: © Marty Snyderman/Visuals Unlimited
Fig. 27-17a, p. 738: © Cabisco/Visuals Unlimited
Fig. 27-17b, p. 738: © Daniel Gotshall/Visuals Unlimited
Fig. 27-17c, p. 738: © Animals, Animals, Earth Scenes
Fig. 27-17d, p. 738: © Carolina Biological Supply Co./Visuals Unlimited
Fig. 27-18a, p. 739: © Marty Snyderman/Visuals Unlimited
Fig. 27-18b, p. 739: © John D. Cunningham/Visuals Unlimited
Fig. 27-19a, p. 740: © Science VU/Visuals Unlimited
Fig. 27-19b, p. 740: © S. Maslowski/Visuals Unlimited
Fig. 27-20a, p. 740: © Marty Snyderman/Visuals Unlimited
Fig. 27-22a, p. 741: © Nathan W. Cohen/Visuals Unlimited
Fig. 27-22b, p. 741: © G. Twiest/Visuals Unlimited
Fig. 27-22c, p. 741: © Peter Arnold, Inc.
Fig. 27-23a, p. 742: © Dale Jackson/Visuals Unlimited
Fig. 27-23b, p. 742: © Glenn M. Oliver/Visuals Unlimited
Fig. 27-24, p. 742: © Rudolf Arndt/Visuals Unlimited
Fig. 27-26a, p. 744: © Jim Merli/Visuals Unlimited
Fig. 27-26b, p. 744: © Joe McDonald/Visuals Unlimited
Fig. 27-26c, p. 744: © Paul Gier/Visuals Unlimited
Fig. 27-27a, p. 744: © Carolina Biological Supply Co./Visuals Unlimited
Fig. 27-27b, p. 744: © A.D. Copley/Visuals Unlimited

Fig. 27-30a, p. 747: © William J. Weber/Visuals Unlimited
Fig. 27-30b, p. 747: © Kjell B. Sandved/Visuals Unlimited
Fig. 27-30c, p. 747: © William J. Weber/Visuals Unlimited
Fig. 27-31, p. 748: © Don W. Fawcett/Visuals Unlimited
Fig. 27-32a, p. 749: © John D. Cunningham/Visuals Unlimited
Fig. 27-32b, p. 749: © Stephen Dalton/Photo Researchers
Fig. 27-32c, p. 749: © Dave B. Fleetham/Visuals Unlimited
Fig. 27-34a, p. 750: © Tom J. Urich/Visuals Unlimited
Fig. 27-34b, p. 750: © Kjell B. Sandved/Visuals Unlimited
Fig. 27-34c, p. 750: © Leonard Rue III/Visuals Unlimited

Chapter 28 Opener, p. 755: © Fred Espenak, NASA/Science Photo Library/Photo Researchers
Fig. 28-1, p. 756: © Milton H. Tierney, Jr./Visuals Unlimited
Fig. 28-2a, p. 756: © San Diego Zoo/R. D. Schmidt
Fig. 28-2b, p. 756: © Walt Anderson/Visuals Unlimited
Fig. 28-4a, p. 758: © Thomas C. Boyden
Fig. 28-4b, p. 758: © Charles Rushing/Visuals Unlimited
Fig. 28-5a, p. 758: © Walt Anderson/Visuals Unlimited
Fig. 28-5b, p. 758: © John D. Cunningham/Visuals Unlimited
Fig. 28-8, p. 760: © Photo Researchers
Fig. 28-11, p. 762: © Science VU-National Museum of Kenya/Visuals Unlimited
Fig. 28-12, p. 762: © Cabisco/Visuals Unlimited
Fig. 28-14, p. 766: © Science VU/Visuals Unlimited
Fig. 28-15a, p. 766: © Heni Sommer/Visuals Unlimited
Fig. 28-15b, p. 766: © Emily Strong/Visuals Unlimited
Fig. 28-15c, p. 766: © Daniel D. Chiras
Fig. 28-15d, p. 766: © Charles Sykes/Visuals Unlimited
Fig. 28-15e, p. 766: © J. Cancalosi/Peter Arnold, Inc.
Fig. 28-16, p. 767: © G. Prance/Visuals Unlimited
Fig. 28-17, p. 769: © D. Cavagnaro/Visuals Unlimited

Part VII Opener, p. 773: © Stephen Sharnoff/Visuals Unlimited
Chapter 29 Opener, p. 774: © G. Dimijian/Photo Researchers
Fig. 29-1, p. 775: © Barbara Gerlach/Visuals Unlimited
Fig. 29-2, p. 776: © Cabisco/Visuals Unlimited
Fig. 29-3, p. 76: © Cabisco/Visuals Unlimited
Fig. 29-4a–c, p. 777: © UPI/Bettmann
Fig. 29-5, p. 778: © Steve McCutchen/Visuals Unlimited
SD 29-1a, p. 780: © UPI/Bettmann
Fig. 29-8, p. 781: © John D. Cunningham/Visuals Unlimited
Fig. 29-9, p. 783: © Steve McCutcheon/Visuals Unlimited
Fig. 29-10, p. 784: A. Kerstitch/Visuals Unlimited
Fig. 29-11, p. 784: No credit
Fig. 29-12, p. 785: Jan L. Wassink/Visuals Unlimited
Fig. 29-13, p. 785: © John D. Cunningham/Visuals Unlimited
Fig. 29-14, p. 786: © Don Enger/Animals, Animals, Earth Scenes
Fig. 29-15, p. 787: © Don W. Fawcett/Visuals Unlimited
Fig. 29-16, p. 788: © W. Ormerob/Visuals Unlimited
Fig. 29-17, p. 788: © Bill Beatty/Visuals Unlimited
Fig. 29-18, p. 789: © Michael S. Quinton/Visuals Unlimited
Fig. 29-19, p. 789: © McAlonan/Visuals Unlimited
Fig. 29-20a, p. 790: © Steve McMutcheon/Visuals Unlimited
Fig. 29-20b, p. 790: © Bill Kamin/Visuals Unlimited
Fig. 29-20c, p. 790: © Dick Poe/Visuals Unlimited
Fig. 29-21, p. 791: © Tom J. Ulrich/Visuals Unlimited

Chapter 30 Opener, p. 794: © Alan Carey/Photo Researchers
Fig 30-1a, p. 795: © Walt Anderson/Visuals Unlimited
Fig. 30-1b, p. 795: © S. Maslowski/Visuals Unlimited
Fig. 30-2, p. 796: © John L. Pontier/Animals, Animals, Earth Scenes
SE 30-1, p. 797: © Daniel D. Chiras
Fig. 30-4, p. 799: © Frank T. Awbrey/Visuals Unlimited
Fig. 30-6a, p. 801: © Michael Dick/Animals, Animals, Earth Scenes
Fig. 30-6b, p. 801: D. Newman/Visuals Unlimited

Fig. 30-6c, p. 801: © William J. Weber/Visuals Unlimited
Fig. 30-8, p. 802: © David L. Pearson/Visuals Unlimited
Fig. 30-9a, p. 803: © Bruce Berg/Visuals Unlimited
Fig. 30-9b, p. 803: © Kjell B. Sandved/Visuals Unlimited
Fig. 30-10, p. 803: © Kjell B. Sandved/Visuals Unlimited
Fig. 30-11, p. 804: © J. Alcock/Visuals Unlimited
Fig. 30-12, p. 804: © A. Kerstitch/Visuals Unlimited
Fig. 30-13, p. 804: © J. Alcock/Visuals Unlimited
Fig. 30-14a, p. 805: © John D. Cunningham/Visuals Unlimited
Fig. 30-14b, p. 805: © William J. Weber/Visuals Unlimited
Fig. 30-14c, p. 805: © Nathan W. Cohen/Visuals Unlimited
Fig. 30-15, p. 805: © Lenard Rue III/Visuals Unlimted
Fig. 30-16, p. 806: © William J. Weber/Visuals Unlimited
Fig. 30-18, p. 808: © Van A Truan/Visuals Unlimited
Fig. 30-19, p. 808: © Richard Walters/Visuals Unlimited
Fig. 30-20, p. 809: © Pat Armstrong/Visuals Unlimited
Fig. 30-21a, p. 809: © Walter Anderson /Visuals Unlimited
Fig. 30-21b, p. 809: © Marty Snyderman/Visuals Unlimited
Fig. 30-21c, p. 809: © R.F. Ashley/Visuals Unlimited
Fig. 30-22, p. 810: © Photo Researchers, Inc.

Chapter 31 Opener, p. 814: © European Space Agency/Science Photo
 Library/Photo Researchers
Fig. 31-2, p. 816: Science VU/NASA/Visuals Unlimited
Fig. 31-4a, p. 818: © Steve McCutcheon/Visuals Unlimited
Fig. 31-4b, p. 818: © Albert J. Copley/Visuals Unlimited
Fig. 31-4c, p. 818: © William J. Weber/Visuals Unlimited
Fig. 31-4d, p. 818: © Ron Spomer/Visuals Unlimited
Fig. 31-4e, p. 818: © John Cunningham/Visuals Unlimited
Fig. 31-6a, p. 820: © Frank M. Hanna/Visuals Unlimited
Fig. 31-6b, p. 820: © John H. Rinne/Visuals Unlimited
Fig. 31-7, p. 820: © Richard Thom/Visuals Unlimited
Fig. 31-19a, p. 834: © Tom J. Urich/Visuals Unlimited
Fig. 31-19b, p. 834: © Science VU-VMBMSU/Visuals Unlimited
Fig. 31-20a, p. 834: © Photo Vulcain/Explorer/Photo Researchers, Inc.
Fig. 31-20b, p. 834: © Peter K. Ziminski/Visuals Unlimited
Fig. 31-23, p. 836: © John D. Cunningham/Visuals Unlimited
Fig. 31-24, p. 837: © J. Tork/Visuals Unlimited
Fig. 31-26a, p. 838: © Richard Ashley/Visuals Unlimited
Fig. 31-26b, p. 838: © G. Prance/Visuals Unlimited

Chapter 32 Opener, p. 842: Daniel D. Chiras
Fig. 32-2, p. 844: © Will Troyer/Visuals Unlimited
Fig. 32-3a, p. 845: © Jan L. Wassink/Visuals Unlimited
Fig. 32-3b, p. 845: © Tom J. Ulrich/Visuals Unlimited
Fig. 32-4, p. 845: © S. McCutcheon/Visuals Unlimited
Fig. 32-5, p. 846: © Stephen & Sylvia Sharnoff/Visuals Unlimited

Fig. 32-6, p. 847: © Bill Beatty/Visuals Unlimited
Fig. 32-7a, p. 847: © J. Serrao/Visuals Unlimited
Fig. 32-7b, p. 847: © Tom Edwards/Visuals Unlimited
Fig. 32-7c, p. 847: © Joe McDonald/Visuals Unlimited
Fig. 32-8, p. 848: © Ron Spomer/Visuals Unlimited
Fig. 32-9, p. 848: © C.P. Hickman/Visuals Unlimited
Fig. 32-10, p. 849: © Science VU/Visuals Unlimited
Fig. 32-11, p. 849: © John D. Cunningham/Visuals Unlimited
Fig. 32-13, p. 850: © Doug Sokell/Visuals Unlimited
Fig. 32-14, p. 850: © Doug Sokell/Visuals Unlimited
Fig. 32-15a-b, p. 851: © John D. Cunningham/Visuals Unlimited
Fig. 32-16a, p. 851: © A Kerstitch/Visuals Unlimited
Fig. 32-16b, p. 851: © John Gerlach/Visuals Unlimited
Fig. 32-16c, p. 851: © Bill Kamin/Visuals Unlimited
Fig. 32-17, p. 852: © Nada Pecnik/Visuals Unlimited
Fig. 32-18, p. 853: © Thomas C. Boyden
Fig. 32-19, p. 853: © Walt Anderson/Visuals Unlimited
Fig. 32-21, p. 855: © David Matherly/Visuals Unlimited
Fig. 32-22a, p. 855: © Cibisco/Visuals Unlimited
Fig. 32-22b, p. 855: © T.E. Adams/Visuals Unlimited
Fig. 32-23a, p. 856: © John D. Cunningham/Visuals Unlimited
Fig. 32-23b, p. 856: © Tom Adams/Visuals Unlimited
Fig. 32-26, p. 860: © Joel Arrington/Visuals Unlimited
Fig. 32-27, p. 860: © Daniel W. Gotshall/Visuals Unlimited
Fig. 32-28, p. 861: © Dave B. Fleetham/Visuals Unlimited
Fig. 32-29a, p. 861: © Robert F. Myers/Visuals Unlimited
Fig 32-33, p. 865: © Martin Bond/Science Photo Library/Photo Researchers

Chapter 33 Opener, p. 869: © B Nation/Sygma
Fig. 33-1, p. 870: © AP/Wide World Photos
Fig. 33-2, p. 871: © Jerom Wyckoff/Visuals Unlimited
Fig. 33-4, p. 874: © Sylvan Wittwer/Visuals Unlimited
Fig. 33-6, p. 875: © Frank M. Hanna/Visuals Unlimited
Fig. 33-12a, p. 878: © Daniel D. Chiras
Fig. 33-12b, p. 878: © William Grenfell/Visulas Unlimited
HN 33-1, p. 879: Courtesy of AES
Fig. 33-13, p. 879: © William Banaszewski/Visuals Unlimited
Fig. 33-15, p. 880: © Science VU, SCS/Visuals Unlimited
Fig. 33-16, p. 880: © Dennis Paulson/Visuals Unlimited
Fig. 33-19, p. 882: Courtesy Chevrolet Division/General Motors
Fig. 33-20, p. 883: Courtesy General Electric
Fig. 33-21, p. 883: © Science VU, API/Visuals Unlimited
Fig. 33-24, p. 885: © Martin G. Miller/Visuals Unlimited
Fig. 33-25, p. 885: © Max & Bea Hunn/Visuals Unlimited
Fig. 33-26, p. 889: © David S. Addison/Visuals Unlimited
Fig. 33-29a, p. 891: © Joe McDonald/Visuals Unlimited
Fig. 33-29b, p. 891: © Tim Perkins/Visuals Unlimited

Art Credits

Chapter 1
Fig 1-3, page 5: Publications Services
Fig 1-4, page 6: Cyndie Wooley
Fig 1-7, page 10: Carlyn Iverson
Fig 1-10, page 14: layout by Georg Klatt and computer conversion by Pub-
 lications Services
SD Fig 1, page 17: J/B Woolsey Associates

Chapter 2
Fig 2-2, page 27: layout by George Klatt and computer conversion by Pub-
 lications Services
Fig 2-4, page 28: Georg Klatt

Fig 2-6, page 33: J/B Woolsey Associates
Fig 2-7, page 33: Georg Klatt
Fig 2-8, page 34: Georg Klatt
Fig 2-9, page 34: Georg Klatt
Fig 2-10, page 35: Publications Services
Fig 2-11, page 35: layout by Georg Klatt and computer conversion by Pub-
 lications Services
Fig 2-12, page 36: Georg Klatt
Fig 2-14, page 38: Publications Services
Fig 2-16, page 39: layout by Georg Klatt and computer converson by Pub-
 lications Services
Fig 2-17, page 40: layout by Georg Klatt and computer conversion by Pub-
 lications Services

Biological Roots

One of the major challenges of a biology course is mastering new terminology. As noted in the Study Skills section of this book, one of the best ways of learning and remembering technical terms is to first learn their component parts—that is, their roots. But studying common Latin and Greek roots for biological terms before you begin your study can also help. Spend a few minutes early in your course studying these common roots. They will not only help you understand new terms, but also make them easier to remember.

a-, an- [Gk. *an-*, not, without, lacking]: anaerobic, abiotic, anemia

ad- [L. *ad-*, toward, to]: adrenalin

amphi- [Gk. *amphi-*, two, both, both sides of]: amphibian

ana- [Gk. *ana-*, up, up against]: anaphase, anabolic, anatomy

andro- [Gk. *andros*, an old man]: androgen

anti- [Gk. *anti-*, against, opposite, opposed to]: antibiotic, antibody, antigen, antidiuretic hormone

arthro- [Gk. *arthron*, a joint]: arthropod, arthritis

auto- [Gk. *auto-*, self, same]: autoimmune, autotroph

bi-, bin- [L. *bis*, twice; *bini*, two-by-two]: binary fission, binocular vision, bicarbonate

bio- [Gk. *bios*, life]: biology, biomass, biome, biosphere, biotic

blasto-, -blast [Gk. *blastos*, sprout; pertains to embryo]: blastula, trophoblast, osteoblast

broncho- [Gk. *bronchos*, windpipe]: bronchus, bronchi, bronchiole, bronchitis

carb- [L. *carbo*, coal]: carbon, carbohydrate

carcino- [Gk. *karkin*, a crab, cancer]: carcinogen

cardio- [Gk. *kardia*, heart]: cardiac, myocardium, electrocardiogram

cat- [Gk. *kata*, down, downward]: catabolic

chloro- [Gk. *chloros*, green]: chlorophyll, chloroplast, chlorine

chromo- [Gk. *chroma*, color]: chromosome, ??? chromatin

???, -coel [Gk. *koilos*, hollow, cavity]: coelom

com-, con-, col-, co- [L. *cum*, with, together]: coenzyme, covalent

cranio- [Gk. *kranios*, L. *cranium*, skull]: cranial, cranium

cuti- [L. *cutis*, skin]: cutaneous, cuticle

cyto-, -cyte [Gk. *kytos*, vessel or container; now, "cell"]: cytoplasm, cytokinesis, erythrocyte, leucocyte

de- [L. *de-*, away, off, removal, separation]: deciduous, decomposer, dehydration

derm-, dermato- [Gk. *derma*, skin]: dermis, epidermis, ectoderm, endoderm, mesoderm

di- [Gk. *dis*, twice, two, double]: disaccharide, dioxide

dia- [Gk. through, passing through, thorough, thoroughly]: diabetes, dialysis, diaphragm

diplo- [Gk. *diploos*, two-fold]: diploid

eco- [Gk. *oikos*, house, home]: ecology, ecosystem, economy

ecto- [Gk. *ektos*, outside]: ectoderm

endo- [Gk. *endon*, within]: endoderm, endometrium

epi- [Gk. *epi*, on, upon, over]: epidermis, epididymis, epiglottis, epithelium

equi- [L. *aequus*, equal]: equilibrium

eu- [Gk. *eus*, good; *eu*, well, true]: eukaryote

ex-, exo-, ec-, e- [Gk., L. out, out of, from, beyond]: emission, ejaculation, excretion, exergonic, exhale, exocytosis, exoskeleton

extra- [L. outside of, beyond]: extracellular, extraembryonic

-fer [L. *ferre*, to bear]: fertile, fertilization, conifer

gam-, gameto- [Gk. gamos, marriage; now usually in reference to gametes (sex cells)]: gamete

gastro- [Gk. *gaster*, stomach]: gastric, gastrin, gastrovascular cavity

gen- [Gk. *gen*, born, produced by; Gk. *genos*, race, kind; L. *genus*, *generare*, to beget]: polygenic genotype, geneology, glycogen, pyrogen, heterogenous

gluco-, glyco- [Gk. *glykys*, sweet; now pertaining to sugar]: glucose, glycogen, glycolysis, glycoprotein

hemo-, hemato-, -hemia, -emia [Gk. *haima*, blood]: hematology, hemoglobin, hemophilia

hepato- [Gk. *hepar*, *hepat-*, liver]: hepatitis, hepatic portal system

hetero- [Gk. *heteros*, other, different]: heterogeneous, heterozygote

histo- [Gk. *histos*, web of a loom, tissue; pertains to biological tissues]: histology, histamine, antihistamine

homo-, homeo- [Gk. *homos*, same; Gk. *homios*, similar]: homeostasis, homogeneous, homologous, homozygote